MEDICAL RADIOLOGY
Radiation Oncology

Editors:
L. W. Brady, Philadelphia
H.-P. Heilmann, Hamburg
M. Molls, Munich

S. H. Levitt · J. A. Purdy · C. A. Perez
S. Vijayakumar (Eds.)

Technical Basis of Radiation Therapy

Practical Clinical Applications

4th Revised Edition

With Contributions by

R. R. Allison · R. J. Amdur · J. D. Bradley · D. J. Brenner · T. A. Buchholz · H. R. Cardenes
W. H. Choi · L. C. Cho · H. Choy · F. Claus · T. F. DeLaney · K. E. Dusenbery · R. E. Drzymala
J. F. Fowler · D. P. Garwood · B. J. Gerbi · S. Gorty · L. L. Gunderson · M. G. Haddock
E. J. Hall · D. C. Harmon · F. J. Hornicek · H. Hricak · E. Klein · C. K. K. Lee · S. H. Levitt
Z. Li · D. A. Low · W. Lu · J. A. Martenson Jr. · K. P. McMullen · M. D. McNeese
W. M. Mendenhall · G. S. Merrick · J. M. Michalski · G. C. Morton · S. B. Motwani
M. T. Munley · S. Mutic · S. Nag · S. Narayan · S. Nilsson · J. R. Palta · M. Pearse · C. A. Perez
J. A. Purdy · K. M. Rich · A. E. Rosenberg · M. Rotman · A. R. Schulsinger · E. G. Shaw
J. R. Simpson · V. W. Stieber · E. A. Strom · B. Tinnel · P. Tripuraneni · S. Vijayakumar
I. Zoberi · R. D. Zwicker

Series Editor's Foreword by
L. W. Brady · H.-P. Heilmann · M. Molls

With 507 Figures in 799 Separate Illustrations, 175 in Color and 146 Tables

 Springer

Seymour H. Levitt, MD
Professor, Department of Therapeutic Radiation
Oncology MMC 436
University of Minnesota Hospital
420 Delaware St SE
Minneapolis, MN 55455
USA

and

Foreign Adjunct Professor
Karolinska Institute Stockholm, Sweden

James A. Purdy, PhD
Professor and Vice Chairman
Department of Radiation Oncology
Chief, Physics Section
UC Davis Medical Center
4501 X Street, Suite G140
Sacramento, CA 95817
USA

Carlos A. Perez, MD
Professor, Department of Radiation Oncology
Washington University Medical Center
4511 Forest Park Boulevard, Suite 200
St. Louis, MO 63108
USA

Srinivasan Vijayakumar, MD
Chair, Department of Radiation Oncology
4501 X Street, Suite G126
Sacramento, CA 95817
USA

Medical Radiology · Diagnostic Imaging and Radiation Oncology
Series Editors: A. L. Baert · L. W. Brady · H.-P. Heilmann · M. Molls · K. Sartor

Continuation of Handbuch der medizinischen Radiologie
Encyclopedia of Medical Radiology

Library of Congress Control Number: 2005920466

ISBN-10   3-540-21338-4 Springer Berlin Heidelberg New York
ISBN-13   978-3-540-21338-3 Springer Berlin Heidelberg New York

Springer is part of Springer Science+Business Media

http//www.springer.com
© Springer-Verlag Berlin Heidelberg 2006
Printed in Germany

Medical Editor: Dr. Ute Heilmann, Heidelberg
Desk Editor: Ursula N. Davis, Heidelberg
Production Editor: Kurt Teichmann, Mauer
Cover-Design and Typesetting: Verlagsservice Teichmann, Mauer

Printed on acid-free paper – 21/3111 – 5 4 3 2 1   SPIN: 11962083

To my wife and children,
whose support and
encouragement make it all happen.

Seymour H. Levitt

Foreword

This fourth revised edition of "Technical Basis of Radiation Therapy: Practical Clinical Applications", edited by S. H. Levitt, J. A. Purdy, C. A. Perez, and S. Vijaykumar, continues this publication's outstanding excellence in the definition of the technical advances for radiation therapy. The previous three editions were milestones in the definition of new technologies and how they would be applied in clinical practice. The present volume presents significant and important concepts of treatment planning, not only with regard to appropriate treatment plans but also how various auxiliary technologies can be used to achieve the best clinical outcome with the minimum complication.

The first portion of the book deals not only with the advances in imaging technology and their role in defining more precisely the extent of the tumor but also how these imaging techniques can be used in devising treatment plans. Clearly, three-dimensional treatment planning and conformal therapy along with intensity-modulated radiation therapy, stereotactic radiosurgery and radiotherapy, new technologies in brachytherapy as well as advances in cyberknife, tomotherapy and image-guided radiation therapy can lead to better outcomes. Consonant with these considerations is the impact on second malignancies following radiation therapy and the clinical applications for chemoradiation.

In the second part of the book, the practical clinical applications are defined precisely for essentially all major tumor sites. Each tumor site is dealt with in depth, and the authors show how the new techniques can improve the potential outcome in terms of management. In 2005, 60% of all malignant tumors were cancers of the lung, breast, prostate, and colorectum. Particular attention is directed toward these tumor sites and how the new methods can improve the overall outcome.

There is significant emphasis on the utilization of various brachytherapy techniques and how they may be integrated to produce better outcomes at each tumor site.

The excellence of the text is obvious, the data submitted important. This volume makes a major contribution to patient management.

LUTHER W. BRADY
HANS-PETER HEILMANN
MICHAEL MOLLS

Preface

This is the 4th edition of a book which was originally initiated as a supplement to a post-graduate refresher course in radiation oncology held at the University of Minnesota. This program was instituted in 1970 and ended 25 years later. The idea of the course and the book that followed was to acquaint radiation oncologists with the concepts, policies and treatment methods in the cutting-edge radiation oncology departments.

The inspiration for this program came from Dr. Gilbert Fletcher and from Dr. Norah Tapley. Indeed, the first edition of this book was entitled "Levitt and Tapley's". Unfortunately these wonderful colleagues are no longer with us and their friendship and advice and contributions are much missed.

This edition has been completely rewritten. There are several new chapters, and all of the original chapters have been updated. We, the editors and authors, feel that this edition truly reflects the best approach to the technical basis of radiation oncology at this time.

Obviously our discipline is moving forward at break-neck speed, and it is essential for the practitioners of radiation oncology to have as many tools as possible to help them be at the cutting edge of our practice. We believe this edition achieves just that.

All of the chapter authors are experts in their areas, and have many high demands on their time. We are grateful for their efforts..

Minneaopolis	SEYMOUR H. LEVITT
Sacramento	JAMES A. PURDY
St. Louis	CARLOS A. PEREZ
Sacramento	SRINIVASAN VIJAYAKUMAR

Contents

Part I: Basic Concepts in Treatment Planning............................... 1

1   Practical Time-Dose Evaluations, or How to Stop Worrying and Learn to
    Love Linear Quadratics
    JACK F. FOWLER 3

2   Second Malignancies Following Radiotherapy
    ERIC J. HALL 33

3   Clinical Principles and Applications of Chemoirradiation
    DAN P. GARWOOD, L. CHINSOO CHO, and HAK CHOY 39

4   Imaging in Radiation Therapy. Recent Advances and Their Role in Radiotherapy
    FILIP CLAUS and HEDVIG HRICAK 57

5   Physics of Treatment Planning in Radiation Oncology
    JAMES A. PURDY, SRINIVASAN VIJAYAKUMAR, CARLOS A. PEREZ,
    and SEYMOUR H. LEVITT 69

6   The Simulation Process in the Determination and Definition
    of the Treatment Volume and Treatment Planning
    SASA MUTIC, JAMES A. PURDY, JEFF M. MICHALSKI, and CARLOS A. PEREZ 107

7   Clinical Applications of High-Energy Electrons
    BRUCE J. GERBI.. 135

8   Treatment Aids for External Beam Radiotherapy
    ERIC KLEIN, SASA MUTIC, and JAMES A. PURDY...................... 167

9   Three-Dimensional Treatment Planning and Conformal Therapy
    JAMES A. PURDY, JEFF M. MICHALSKI, JEFFREY BRADLEY,
    SRINIVASAN VIJAYAKUMAR, CARLOS A. PEREZ, and SEYMOUR H. LEVITT........ 179

10 Intensity-Modulated Radiation Therapy
    DANIEL A. LOW, WEI LU, JAMES A. PURDY, CARLOS A. PEREZ,
    and SEYMOUR H. LEVITT 203

11 Stereotactic Radiosurgery and Radiotherapy
    JOSEPH R. SIMPSON, ROBERT E. DRZYMALA, and KEITH M. RICH 233

12 Physics and Clinical Aspects of Brachytherapy
    ZUOFENG LI.. 255

13 Radiobiology of Low- and High-Dose-Rate Brachytherapy
    ERIC J. HALL and DAVID J. BRENNER 291

14 Clinical Applications of Low Dose Rate and Medium Dose Rate Brachytherapy
    CARLOS A. PEREZ, ROBERT D. ZWICKER, and ZUOFENG LI 309

15 Clinical Applications of High-Dose-Rate Brachytherapy
    SUBIR NAG 379

16 Quality Assurance in Radiation Oncology
    JAMES A. PURDY, ERIC KLEIN, SRINIVASAN VIJAYAKUMAR, CARLOS A. PEREZ,
    and SEYMOUR H. LEVITT 395

Part II: Practical Clinical Applications 423

17 Central Nervous System Tumors
    VOLKER W. STIEBER, KEVIN P. MCMULLEN, MICHAEL T. MUNLEY,
    and EDWARD G. SHAW 425

18 Head and Neck Cancer
    WILLIAM M. MENDENHALL, ROBERT J. AMDUR, and JATINDER R. PALTA 453

19 Breast Cancer
    SABIN B. MOTWANI, ERIC A. STROM, MARSHA D. MCNEESE,
    and THOMAS A. BUCHHOLZ 485

20 Carcinoma of the Esophagus
    JEFFREY D. BRADLEY and SASA MUTIC 511

21 Carcinoma of the Lung
    SAMIR NARAYAN, SRINIVASAN VIJAYAKUMAR 525

22 Cancers of the Colon, Rectum, and Anus
    JAMES A. MARTENSON JR., MICHAEL G. HADDOCK, and LEONARD L. GUNDERSON . . 545

23 Bladder Cancer - Technical Basis of Radiation Therapy
    ALAN R. SCHULSINGER, RON R. ALLISON, WALTER H. CHOI,
    and MARVIN ROTMAN 561

24 Radiation Therapy for Cervical Cancer
    KATHRYN E. DUSENBERY and BRUCE J. GERBI 579

25 Technical Aspects of Radiation Therapy in Endometrial Carcinoma
    HIGINIA R. CARDENES and BRENT TINNEL 599

26 Vulva
    CARLOS A. PEREZ and IMRAN ZOBERI 631

27 Carcinoma of the Vagina
    HIGINIA R. CARDENES 657

28 Prostate
    Jeff M. Michalski, Gregory S. Merrick, and Sten Nilsson 687

29 Testicular Cancer
    Maria Pearse and Gerard C. Morton 739

30 Extremity Soft Tissue Sarcoma in Adults
    Thomas F. DeLaney, David C. Harmon, Andrew E. Rosenberg,
    and Francis J. Hornicek 755

31 Total Body Irradiation Conditioning Regimens in Stem Cell Transplantation
    Kathryn E. Dusenbery and Bruce J. Gerbi 785

32 Radiotherapy for Hodgkin's Disease
    Chung K. K. Lee 805

33 Techniques of Intravascular Brachytherapy
    Sri Gorty and Prabhakar Tripuraneni 837

Subject Index 843

List of Contributors 857

Part I:
Basic Concepts in Treatment Planning

1 Practical Time-Dose Evaluations, or How to Stop Worrying and Learn to Love Linear Quadratics

Jack F. Fowler

CONTENTS

Glossary *3*

1.1 Introduction *6*
1.2 The Simplest Modeling *7*
1.2.1 The Seven Steps to LQ Heaven – Brief Summary *7*
1.3 The Seven Steps to LQ Heaven – The Details *8*
1.3.1 Development of the Simple LQ Formula E = nd (1+d/α/β) *9*
1.3.2 Biologically Effective Dose *10*
1.3.3 Relative Effectiveness *11*
1.3.4 Overall Treatment Time *12*
1.3.5 Acute Mucosal Tolerance *13*
1.3.6 To Convert from BED to NTD or EQD$_{2\,\text{Gy}}$ *14*
1.3.7 One Example *14*
1.3.8 What is the Standard of Precision of These Estimates of BED or NTD? Gamma Slopes *15*
1.4 Rejoining Point for Those Who Skipped: How to Evaluate a New Schedule – Brief Summary *15*
1.5 Now Let Us Study Some of the Best-Known Schedules for Head-And-Neck Tumor Radiotherapy *16*
1.5.1 Standard Fractionation *16*
1.5.2 Hyperfractionation *17*
1.5.3 Radiation Therapy Oncology Group Four-Arm Fractionation Trial (RTOG 90-03) *17*
1.5.4 Head-and-Neck Schedules That Were Initially "Too Hot" in Table 1.2 *18*
1.5.5 Shortening the Wang 2-Fraction-a-Day Schedule Using BED to Adjust Individual Doses *19*
1.5.6 General Considerations of Head-and-Neck Radiotherapy *19*
1.5.7 A Theoretical Calculation of "Close to Optimum" Head & Neck Schedules: 3 Weeks at Five Fractions per Week *20*
1.5.8 Conclusions Re Head-and-Neck Schedules *21*
1.5.9 Concurrent Chemotherapy *22*
1.6 Hypofractionation for Prostate Tumors *22*
1.7 Summary *23*
1.8 Appendix: Is This a Mistaken Dose Prescription? *24*
1.8.1 For Equal "Late Complications" *25*
1.8.2 For Equal Tumor Effect *25*
1.8.3 Acute Mucosal Effects *26*
1.9 Line-by-Line Worked Examples: Details of Calculations of the Schedules Discussed in This Appendix *27*
References *29*

This chapter is written mainly for those who say "I don't understand this α/β business – I can't be bothered with Linear Quadratic and that sort of stuff." Well, it might seem boring – depending on your personality – but it is easy, and it makes so many things in radiation therapy wonderfully and delightfully clear. Experienced readers can turn straight to Section 1.4.

J. F. Fowler, DSc, PhD
Emeritus Professor of Human Oncology and Medical Physics, Medical School of University of Wisconsin, Madison, Wisconsin, USA; Former Director of the Gray Laboratory, Northwood, London, UK
Present address:
150 Lambeth Road, London, SE1 7DF, UK

Glossary

α, alpha	Intrinsic radiosensitivity. Log$_e$ of the number of cells sterilized non-repairably per gray of dose of ionizing radiation.
β, beta	Repair capacity. Log$_e$ of the number of cells sterilized in a repairable way per gray squared.
α/β, alpha/beta ratio	the ratio of "intrinsic radiosensitivity" to "repair capability" of a specified tissue. This ratio is large (>8 Gy) for rapidly proliferating tissues and most tumors. It is small (<6 Gy) for slowly proliferating tissues, including late normal-tissue complications. This difference is vital for the success of radiotherapy. When beta (β) is large, both mis-repair and good-repair are high. It is the mis-repair that causes the cell survival curve to bend downward.

Accelerated fractionation	fractionated schedules with shorter overall times than the conventional 7 (or 6) weeks.
BED	Biologically effective dose, proportional to log cell kill and therefore more conceptually useful as a measure of biological damage than physical dose, the effects of which vary with fraction size and dose rate. Formally, "the radiation dose equivalent to an infinite number of infinitely small fractions or a very low dose-rate". Corresponds to the intrinsic radiosensitivity (α) of the target cells when all repairable radiation damage (β) has been given time to be repaired. In linear quadratic modeling, BED = total dose × relative effectiveness (RE), where RE = $(1 + d/\alpha/\beta)$, with d = dose per fraction, α = intrinsic radiosensitivity, and β = repair capacity of target cells.
bNED	Biochemically no evidence of disease. No progressive increase of prostate specific antigen (PSA) level in patients treated for prostate cancer.
CI	Confidence interval (usually ±95%).
CTV	Clinical tumor volume. The volume into which malignant cells are estimated to have spread at the time of treatment, larger than the gross tumor volume (GTV) by at least several millimeters, depending on site, stage, and location. See also GTV and planning treatment volume (PTV).
Δt	Time interval between fractions, recommended to be not less than 6 h.
EBR	External beam radiation.
EGFR	Epithelial growth factor receptor, one of the main intracellular biochemical pathways controlling rate of cell proliferation.
EQD	Biologically equivalent total dose, usually in 2-Gy dose fractions. The total dose of a schedule using, for example, 2 Gy per fraction that gives the same log cell kill as the schedule in question. If so, should be designated by the subscript EQD2 Gy.
EUD	Equivalent uniform dose. A construct from the DVH of a non-uniformly irradiated volume of tissue or tumor that estimates the surviving proportion of cells for each volume element (voxel), sums them, and calculates that dose which, if given as a uniform dose to the same volume, would give the same total cell survival as the given non-uniform dose. Local fraction size is taken into account by assuming an α/β ratio for the tissue concerned.
Gamma, γ-50, γ-37	Slope of a graph of probability, usually tumor control probability (TCP), versus total fractionated dose (NTD or EQD), as percentage absolute increase of probability per 1% increase in dose. The steepest part of the curve is at 50% for logistic-type curves and at 37% for Poisson-type curves. Tumor TCP is usually between a gamma-50 (or -37) of 1.0 and 2.5. The difference between γ-50 and γ-37 is rarely clinically significant.
Gy, gray	The international unit of radiation dose: one joule per kilogram of matter. Commonly used radiotherapy doses are approximately 2 Gy on each of 5 days a week.
Gy10, Gy3, Gy1.5	Biologically effective dose (BED), with the subscript representing the value of that tissue's α/β ratio = 10 Gy for early radiation effects, 3 Gy for late radiation effects and 1.5 Gy for prostate tumors. The subscript confirms that this is a BED, proportional to log cell kill, and not a real physical dose.
GTV	Gross tumor volume. The best estimate of tumor volume visualized by radiological, computed tomography (CT) scan, magnetic resonance, ultrasound imaging, or positron emission tomography.
HDR	High dose rate. When the dose fraction is delivered in less than five or ten minutes; that is, much shorter than any half-time of repair of radiation damage.
Hyperfractionation	More (and smaller) dose fractions than 1.8 Gy or 2 Gy.
Hypofractionation	Fewer (and larger) dose fractions than 1.8 Gy or 2 Gy.
Isoeffect	Equal effect.

LC	Local control (of tumors).
LDR	Low dose rate. Officially (ICRU), less than 2 Gy/h; but this is deceptive because any dose rate greater than 0.5 Gy/h will give an increased biological effect compared with the traditional 0.42 Gy/h (1000 cGy per day). For example at 2 Gy/h, the biological effects will be similar to daily fractions of 3.3 Gy and 2.8 Gy on late complications and on tumors respectively.
Linear effect	Directly proportional to dose.
Ln $\log_e$	Natural logarithm, to base e. One $\log_{10}$ is equal to 2.303 $\log_e$.
$\log_{10}$	Common logarithm, to base 10. "Ten logs of cell kill" are 23.03 $\log_e$ of cell kill.
LQ	Linear quadratic formula: $\log_e$ cells killed $= \alpha \times$ dose $+ \beta \times$ dose-squared.
Logistic curve	A symmetrical sigmoid or S-shaped graph relating the statistically probable incidence of "events", including complications, or tumors controlled, at a specified time after treatment, to total dose (NTD). This curve is steepest at the probability of 50%.
LRC	Loco-regional tumor control. LC would be local control.
NTCP	Normal tissue complication probability.
NTD	Normalized total dose of any schedule. The total dose of a schedule using 2 Gy per fraction that gives the same log cell kill as the schedule in question. The NTD will be very different for late effects (with $\alpha/\beta = 3$ Gy and no overall treatment time factor) than for tumor effect (with $\alpha/\beta = 10$ Gy and an appropriate time factor).
Poisson curve	A near-sigmoid graph of probability of occurrence of "events", such as tumor control at X years, versus total dose or NTD. Based on random chance of successes among a population of tumors or patients, the probability of curve $P = \exp(-n)$, where an average of n cells survive per tumor after the schedule, but 0 cells must survive to achieve 100% cure. If an average of 1 cell survives per tumor, $P = 37\%$. If 2 cells survive, $P = 14\%$. If 0.1 cells survive on average, $P = 90\%$. This curve is steepest at the probability of 37%.
PTV	Planning treatment volume – larger than CTV to allow for set-up and treatment-planning errors.
PSA	Prostate-specific antigen: can be measured in a blood specimen as a measure of activity of the prostate gland. Often taken as a measure of activity of prostate cancer.
Quadratic	Effect proportional to dose squared, for example from two particle tracks passing through a target.
QED	Quod Erat Demonstrandum – Latin for "That's what we wanted to show!"
RE	Relative effectiveness. We multiply total dose by RE to obtain BED. $RE = (1 + d/[\alpha/\beta])$ where d is the dose per fraction.
RTOG	Radiation Therapy Oncology Group, USA.
SF	Surviving fraction after irradiation, usually of cells.
Tpot	Potential doubling time of cells in a population; before allowing for the cell loss factor. Tpot is the reciprocal of cell birth rate. It can only be measured in a tissue before any treatment is given to disturb its turnover time.
Tp	Cell doubling time in a tissue during radiotherapy; probably somewhat faster than Tpot. Determined from gross tumor (or other tissue) results when overall time is altered.
Tk	Kick-off or onset time: the apparent starting time of rapid compensatory repopulation in tumor or tissue after the start of treatment, when it is assumed that there are just two rates of cell proliferation during radiotherapy: zero from start to Tk, then constant doubling each Tp days until end of treatment at T days. Accelerating repopulation is discussed in Section 1.5.6.
TCP	Tumor control probability.

1.1
Introduction

It is well known that the simplest description of radiation dose, the total dose, is not adequate because its effect varies with size of dose per session (the dose fraction) and with dose rate. If we double the dose per fraction from 2 Gy to 4 Gy (keeping total dose constant), the effect is 20% greater for tumors but 100% greater for late complications. Further, if a given physical dose is spread evenly over 24 h instead of 2 min, its effect is reduced by 20% for most tumors, but to about half for late complications. We need a way of expressing radiation "dose" in some quantitative way that is more proportional to the observed biological effect. This is the object of calculating a biologically effective dose (BED), and an equivalent dose in 2 Gy fractions (EQD or NTD), so that a 20% increase or decrease of BED or NTD or $EQD_{2\,Gy}$ will lead to a reasonable approximation of a 20% increase or decrease of the expected biological effect. The interesting point is that the same change in physical dose is likely to alter the incidence of late complications to double or half of its effect on tumors. So how can we deal with that?

The basic truth in radiotherapy is that any change in the schedule of dose delivery has a different effect on tumors from its effect on late complications, unless both dose per fraction and dose rate are kept constant. These differences provide some of the remarkable advantages of radiation therapy, and also some puzzles until they are explained. BED can take these differences into account, and preferably explain them.

In the 25 years since the linear quadratic (LQ) formula has been used for the evaluation of radiotherapy schedules, it has proved remarkably reliable. It is now the main and generally accepted method of rationalizing the improved time–dose-fractionation schedules that have been developed to replace, in some body sites, the standard "2 Gy given five times a week for 6 or 7 weeks" schedules. It was first of all useful in identifying the important difference in the effect of dose-fraction size between rapidly proliferating tissues (most tumors) and slowly proliferating tissues (most late complications). This explained the blindly used, but not always wrong, predominance of multi-small-fraction schedules, such as 1.8 or 2 Gy five times a week for 6 or 7 weeks. As explained below, the theoretically ideal overall time would be close to the time at which rapid repopulation in the tumor kicks off, designated Tk days after starting treatment.

In the early years of the development of the LQ formulation, there was no overall treatment-time factor (DOUGLAS and FOWLER 1976; BARENDSEN 1982; WITHERS et al. 1983). This was added later (TRAVIS and TUCKER 1987; VAN DE GEIJN 1989; FOWLER 1989), based on LQ-aided analyses of animal and clinical data (DENEKAMP 1973; TURESSON and NOTTER 1984a,b; THAMES and HENDRY 1987). Since then, the strong effect of repopulation of tumor cells during radiotherapy has been well substantiated so that a repopulation term has been added for tumors (FOWLER 1978, 1989; WITHERS et al. 1988; FOWLER and LINDSTROM 1992; HENDRY et al. 1996). More recently, a different set of parameters has been described to predict acute mucosal reactions in human patients (FOWLER et al. 2003c).

Although the accuracy and even the nature of the LQ factors has been queried a few times, for example whether the parameters are unique or distributed (KING and MAYO 2000; BRENNER and HALL 1999, 2000; KING and FOWLER 2002; DASU et al. 2003; MOISEENKO 2004), the LQ formulation has remained solidly useful and has aided in the design of clinical trials that have changed the practice of radiotherapy (THAMES et al. 1983). Examples include the design of hyperfractionated (more and smaller fractions) and accelerated fractionation (shorter) trials, the avoidance of gaps in radiation treatment, the development of high dose-rate brachytherapy, and a better understanding of when to use or avoid hypofractionation (fewer and larger fractions). The recent growth of stereotactic body radiotherapy is a subset of the latter category (FOWLER et al. 2004).

One of the most interesting series of modeling investigations concerns oral and laryngeal cancers in which the overall times were deliberately shortened until the acute reactions became too severe, in several well known schedules in different countries. Each schedule was then moderated in some way until it became tolerable. The modeling then showed that not only the acute mucosal reactions fell into a narrow band of BED, but the modeled tumor responses were then all close to 11 $\log_{10}$ of predicted tumor cell kill for a variety of different time–dose schedules. This story will be told in this chapter.

Although the numerical results of modeling depend to some extent on the values assumed for the parameters, ratios of parameters such as α/β ratios and time–dose trade-offs (grays per day) are often known sufficiently well for reasonable variations to lead to no clinically significant differences in predicted total dose or BED or NTD. In the modeling described below, we take care to limit the assumed values of parameters to a small library of values selected from experience, avoiding "elegant variation". Then, results that are useful and self-consistent are obtained.

Unlike some other attempts at modeling, the number of initially viable cells per mm^3 is not essential in LQ modeling for time–dose evaluations, because it is largely cancelled out against radiosensitivity α in the standard BED formulation. This is also true of equivalent uniform dose (EUD; NIEMIERKO 1997). Both BED and EUD enjoy similar stability for this reason. The most essential biological factor in LQ formulation is the α/β ratio of the tissue concerned, which appears in the BED with weighting equal to dose per fraction. The other significant factors are the "kick-off time" of rapid tumor repopulation Tk and the doubling time of repopulating tumor cells Tp, which together with the α value all appear in the repopulation term which is usually, but not always, a small proportion of BED.

A glossary is attached at the beginning of this chapter of terms that are best for readers to know when reading about these topics.

1.2
The Simplest Modeling

Years ago, mathematical models were regarded with suspicion, or with derision as playthings for children – but not any more. Modeling has become an important scientific tool in the design and evaluation of topics from global warming to engineering design of aircraft and the pharmacological development of drugs, replacing expensive experimentation in many cases (Prof. Gordon Steel 1990, personal communication). With the aid of computers, mathematical modeling using optimization and new imaging are continuing to revolutionize treatment planning in radiotherapy, as other chapters in this book will show.

"There are good reasons for believing that the primary effects of radiation on biological tissues are cell damage and cell depopulation in renewing populations" (THAMES and HENDRY 1987). This is still true whether the concern is damage to normal tissues or the elimination of every malignant cell in tumors. However, certain indirect biological end points, such as radiation sickness or extent of late fibrosis, do not appear to depend only on numbers of cells sterilized, although it does not mean that

they are not a strong function of cells sterilized. The phantom of immunological response keeps rearing its head, with little practical effect. The additive effect of chemotherapy seems to apply most effectively when it is used concomitantly, and up to now amounts to some 10% of the total cell kill compared with radiotherapy. The strength of radiation as a treatment strategy is that it can and does reach wherever the physical plan puts it, with increasing efficiency of positional accuracy.

The object of the present modeling is to find similar biological effects to the radiotherapy treatments with which we are familiar, from different scheduling of time and dose. That is the "isoeffect modeling" that has much history, passing through the "cube root law" of the 1930s, the "Strandqvist slope" of the 1940s, and the "Nominal Standard Dose" (NSD) or "Time-dose Factor" (TDF) of Dr Frank Ellis in the 1960s and 1970s – and then to evaluate methods of doing better. We shall start by writing down some ideas for explaining the non-linear action of ionizing radiation in damaging biological cells.

Radiation sterilizes cells, meaning that they do not die immediately, but at the time of the next cell division, or a few divisions later. An important factor is the repair occurring in the cells between irradiation and their next cell division. This repair in cells, i.e., recovery in tissues, depends on the turnover rate of renewing tissues – a day or so for rapidly proliferating tissues and most tumors, but many months for the organs that normally proliferate slowly. That is the important difference in the tissues that LQ modeling helped to bring out, with the help of the famous or infamous ratio α over β.

1.2.1
The Seven Steps to LQ Heaven – Brief Summary

First, before explaining them in detail, we list here the "Seven (algebraic) Steps to LQ Heaven", so that readers who wish to do so can skip several pages and turn to Section 1.4. That's where the story gets exciting.

Alpha is the intrinsic radiosensitivity of the cells, defined as how many logs (to the exponential base "e") are killed (sterilized) per gray, in a "non-repairable" way. Beta is the repairable portion of the radiation damage, requiring 6 h or more for complete repair. It can be regarded as the result of two charged-particle tracks passing through a sensitive target in the cell nucleus in less than 6 h, so this term has to be multiplied by d squared. E is the log$_e$ sum of the non-repairable α term and the partly repairable β term. So for n fractions of d Gy dose each:

$$E = n(\alpha d + \beta d^2) \qquad [1.1]$$

$$E = nd(\alpha + \beta d) \qquad [1.2]$$

$$E/\alpha = nd(1 + d\beta/\alpha) \qquad [1.3]$$

Dividing through by α to express α/β as a ratio.

$$\frac{E}{\alpha} = nd(1 + \frac{d}{\alpha/\beta}) \underline{E = nd(1+\underline{d})} \qquad [1.4]$$

With no repopulation considered; as for most types of late complication.

This is the limited definition of BED. It applies to late effects. $BED = E/\alpha = $ total dose $\times RE$ where $RE = (1 + d/[\alpha/\beta])$, and it is very useful.

Next, we subtract the log cell kill due to repopulation of any cells during radiotherapy, after the "kick-off" or onset time Tk, where T is overall time and Tp is the average cell-number doubling time (in days) between Tk and T.

$$E = nd(\alpha + \beta d) - (T - Tk) \times \text{ rate of repopulation per day} \qquad [1.5]$$

$$E = nd(\alpha + \beta d) - \log_e 2(T - Tk)/Tp \qquad [1.6]$$
$$(\log_e 2 = 0.693)$$

To transform the total log cell kill E into the total BED requires the same division throughout by α that we carried out in step 3 (Eq. 1.3) above:

$$BED = \frac{E}{\alpha} = nd(1 + \frac{d}{\alpha/\beta}) - \frac{0,693}{\alpha Tp}(T - Tk) \qquad [1.7]$$

Final step = LQ heaven!

BED can be expressed as Gy_3 (or Gy_2) for late complications, or as Gy_{10} (or Gy_x) for tumor or early normal-tissue reactions, the subscript referring to the α/β value used in its calculation. Gy_3 and Gy_{10} values must not be mixed, as USA and Canadian or Hong Kong dollars cannot. However, several segments of a schedule can have their Gy_{10} values added together and, separately, their Gy_3 values added, for a comparison of total BEDs amounting to a "therapeutic ratio" of Gy_{10}/Gy_3 – representing tumor cell damage divided by late normal-tissue damage. This notation should always be used, or confusion quickly sets in. It both reminds us that this is a BED, not the real physical dose; and confirms which α/β ratio was used, thus

for which tissue each BED or $EQD_{2\,Gy}$ or NTD is calculated. We should state each time "late" or "tumor" or "early" BED or NTD (FOWLER 1989).

A further measure of radiation damage from these formulae is:

Log_e cell kill $= E = BED \times \alpha$, so that in the "common log to base 10 scale, $\log_{10}$ cell kill $= \log_e$ cell kill/2.303.

To convert from BED to EQD 2 Gy or NTD (total equivalent dose in 2 Gy fractions):
For late complications, divide Gy_3 by 1.667. For tumor or early effects, divide Gy_{10} by 1.2.

The explanation comes from the identity BED1=BED2, where "1" is for "d Gy per fraction" and "2" is for "2 Gy per fraction", so for identical BEDs, we have:

Total dose $1 \times RE1 = $ total dose $2 \times (RE$ for 2 Gy and the same $\alpha/\beta)$.
$NTD \times (1 + 2/\alpha/\beta) = BED$. Then solve for NTD!

1.3
The Seven Steps to LQ Heaven – The Details

This is the section that experienced modelers might wish to skip by several pages, and go to Sections 1.4 and 1.5, where comparisons of actual schedules are tabulated and discussed.

The mathematics of the LQ formula are very simple and are taught in courses before the end of the high school syllabus at age 15 or 16 years. Solving an algebraic quadratic equation and manipulating the conversion of an exponential to a logarithmic form are the only steps requiring an effort of memory and are only necessary if you get into more calculations than you will normally need. If you don't deal with this sort of arithmetic every week, you normally need no more than the four simplest keys (+, –, ×, /) on your hand calculator, or even just the back of an envelope. The "Seven Steps to LQ Heaven" are intended to ensure that you, the reader, are never puzzled by such simple calculations again. Having followed, and understood, the seven steps that follow, you should be able to use them with confidence for any comparisons of radiotherapy schedules you wish to make in terms of BED (FOWLER 1989, 1992).

BED is proportional to log cell kill for cells of the specified α/β ratio, and so is a strong function of biological effect. BED is itself a ratio (E/α), the two parameters of which are not individually important. Mathematically, E/α is simply a linkage term showing equivalence between two schedules – their "iso-effect" in the time-dose scale – as well as providing a useful number, the BED. Its numerical value is a convenient number, representing a dose without the repairable component, that is avoiding dose per fraction or dose rate, until relative effectiveness (RE) is built in. We could talk about $\alpha = 0.35 \log_e$ per Gy and $E = 10 \log_{10}$ of cell kill, or equally $\alpha = 0.035 \log_e$ per Gy and $E = $ one $\log_{10}$ of cell kill, and the BED would be the same because it is E/α.

1.3.1
Development of the Simple LQ Formula
$E = nd\,(1 + d/\alpha/\beta)$

Cell depopulation is the main effect of radiation, both for eliminating tumors and for damaging normal tissues. In addition, some genes are activated, which can be relevant in those cells that survive irradiation, and some apoptosis (cell death independent of mitosis) may be caused. Although apoptosis does not appear to be the major effect of radiation, when it happens it adds to the effect. The promising strategy of damaging tumors by depleting their blood supply, with pharmacological or enzyme-pathway aid (Fuks et al. 1994), can be regarded as considering the neovasculature (that is, rapidly proliferating endothelial cells (Hobson and Denekamp 1984) as legitimate oncogenic targets, as well as the directly malignant tumor clonogens (Hahnfeldt et al. 1999). Radiation can reach them all.

If a dose of radiation D sterilizes a proportion of cells in a given tumor or normal organ so that the number of viable cells is reduced from the initial No to Ns: the surviving proportion is Ns/No, which is designated S.

This process is represented as $S = e^{-D/Do} = \exp(-D/Do)$, in its simplest form, where Do is the average dose that would sterilize one cell. This shows that the surviving proportion of cells is reduced exponentially with radiation dose. That is, each successive equal increment of dose reduces the surviving cells by the same proportion, not by the same number. The proportion of surviving cells would then decrease from 1 to 0.5, to 0.25, to 0.125, etc.; if higher doses per fraction were used, from 0.1 to 0.01 to 0.001.

Another way of writing this is: $\log_e S = -D/Do$, which plots out as a graph of $\log_e S$ vertically versus dose horizontally to give a straight line of slope

minus D/Do. This would be called a "single-hit" curve. The dose Do would reduce survival by one $\log_e$ which is to 37% survival. You can see that $Do = 1/\alpha$ in the LQ formula, that is $1/$(the dose to reduce cell survival by one $\log_e$).

Plotting graphically the proportion of surviving mammalian cells against single dose of radiation gives a curve, not a straight line. This curve starts from zero dose with a definite non-zero slope, gradually bending downward into a "shoulder" at less than 1 gray of dose, and continuing to bend downward until at least 10 or 20 grays. At higher doses, the cell survival curve appears to become nearly straight again but of course steeper than at the origin (Gilbert et al. 1980). This is the well-known "cell survival curve" where the logarithm of the surviving proportion of cells is plotted downward, against radiation dose on the horizontal axis (Fig 1.1). It is the same curve as the negative logarithm of the fraction of cells "killed" – actually "sterilized" so that they die later, after the next cell division or a few divisions.

This is the plot of $\log_e$ proportion of cells surviving: $S = -\alpha d - \beta d^2$.

Therefore also: $\log_e$ proportion of cells sterilized (killed) $= +\alpha d + \beta d^2$.

It is this logarithm of the number of cells sterilized that can be divided into one part proportional to dose αd; and another part proportional to dose-

SIMPLY LQ

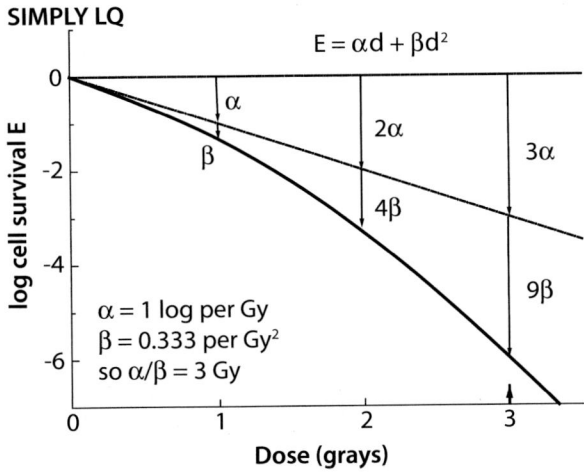

Fig. 1.1. The simple cell survival curve for linear quadratic cell kill versus radiation dose, for a single dose of radiation delivered within a few minutes. The alpha component increases as shown linearly with dose. The beta component is added to this in a curving pattern, increasing with the square of the dose. This example is numerically correct for the α/β ratio of 3 Gy

squared βd^2 (that is "quadratically", where two sub-lesions combine, each produced in number proportional to dose). The logarithm (proportion) of lethal events caused by a dose d is then:

$$E = \alpha d + \beta d^2$$

The linear component is found to be not repairable beyond a few milliseconds after the irradiation, but this does not mean it cannot be altered, by oxygen if present at the time of irradiation as one major example. However, the dose-squared damage gradually fades over a few hours. It is repaired by several processes within the biological cells, mostly within the DNA, so that cell survival recovers toward the straight initial α slope. Cells and tissues are said to "recover" and biochemical lesions in DNA are said to be "repaired".

Let us call this cell-number damage E, the logarithm of the number of cells sterilized by a dose d in grays (Gy). Then, we just write the first equation as:

$$E = \alpha d + \beta d^2$$

where α and β are the coefficients of the linear component and the dose-squared component, respectively. This is the first step on the ladder to explaining the LQ formulation, commonly taught as the "Seven Steps to Heaven". If you remember this little starting formula you will have no trouble at all with the next three steps. "LQ Heaven" is reached when

you understand the LQ steps well enough not to forget them within the next few days.

The next step is the simple one of adding several fractions of daily doses, each of d grays, to obtain the total dose if n fractions (daily doses) are given. Figure 1.2 illustrates the sequence of equal fractions, giving a total curve that is made up of a sequence of small shoulders, in toto an exactly linear locus, of slope depending on the value of Eq. 1.1 at each dose d:

$$E = n(\alpha d + \beta d^2) \qquad [1.1]$$

The dose per fraction d can be taken outside the parentheses, nd being of course the total dose:

$$E = nd(\alpha + \beta d) \qquad [1.2]$$

We usually know the ratio of the two coefficients, α/β, for given cells and tissues much better than we know their individual values, so the next step is to express E in terms of this ratio. It is most usefully done by dividing both sides of Eq. 1.3 by α (Barendsen 1982). If instead we divided by β, then the resulting BED would be in terms of dose squared, which would be awkward.

$$E/\alpha = nd(1 + d\beta/\alpha) \qquad [1.3]$$

which is also of course identical to:

$$\frac{E}{\alpha} = nd\left(1 + \frac{d}{\alpha/\beta}\right)\underline{E} = nd(1+\underline{d}) \qquad [1.4]$$

Because α is defined as a number (log number of cells sterilized) "per gray", the term E/α has the dimensions of dose. It is in fact the BED that we are seeking to calculate, so Eq. 1.4 gives it to us. We are halfway up the "Steps to Heaven" and this equation enables us to do many useful things in predicting biological damage (Fowler 1989; also Chapters 12 & 13 in Steel 2002).

1.3.2
Biologically Effective Dose

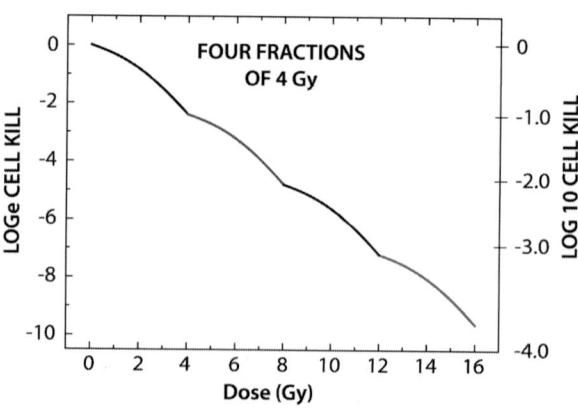

Fig. 1.2. The survival curve for four equal radiation doses given sequentially, with sufficient time – at least 6 h – between them to allow complete repair of the beta component of radiation damage. Since the shape of each is then a repetition of the previous dose, the track of the result after each dose fraction is a straight line when plotted as log cell kill against dose as shown

The basic concept of BED was defined by Barendsen (1982), who at first called it extrapolated tolerance dose (ETD), meaning that dose which, if given as an infinite number of infinitely small fractions (along the initial slope of the cell survival curve), or at a very low dose rate (so that all the quadratic damage has been repaired), would cause the same log cell kill as the schedule under consideration, thinking of the maximum dose that a normal tissue would tolerate. Since it was obvious that this conceptual extrapolation to very small dose per fraction could be applied to any level of damage, not just to the maximum tolerated level or only to normal tissues, it was soon renamed extrapolated response dose (ERD) and later to the more general BED (FOWLER 1989). It is illustrated graphically in Figure 1.3. Because BED is defined in relation to the initial slope, i.e., the linear component of damage, it is E/α. Because α has the dimensions of 1/dose, E/α has the dimensions of dose, as we require. We are talking here about an averaged value of α during the weeks of radiotherapy. Different values of α could be applied to different segments of a schedule but there is not yet evidence to justify this.

Since the definition of BED is the ratio E/α, the individual values of E and α are irrelevant for estimating relative total doses. The ratio E/α is mathematically a link function, signifying biological equivalence between two schedules having equal effect. The individual values of E or α have no necessary biological significance in LQ formulation. Values of α are particularly vulnerable to variation of tumor size, stage of tumor, and accuracy of dose, but provided that the ratio between E and α (and $\sqrt{\beta}$) does not vary – between one prospective population in a clinical trial and another – there are no effects on ratios of doses between schedules, which are what we want to compare. This is the important reason why BED, like EUD (NIEMIERKO 1997), is robust. They are relatively stable if parameters are varied by modest amounts.

These biological ratios between log cell kill and cellular radiosensitivity have in fact been determined in certain biological experiments concerning skin and intestinal clones in mice (WITHERS 1967, 1971), so these tissues can be put on to a more exact basis concerning number of cells per millimeter of mucosal surface. However, the use of a number to designate log cell kill in tumors depends on a specific assumption of how many viable malignant cells were present per mm^3 of tumor volume, which is usually unknown. The number has wildly varied between ten thousand million (10^{10}) and one-tenth per tumor among various authors for various tumors (the latter being curable with 90% probability). Because of the real variability between tumors and patients, a value for α [derived for example from a gamma-50 slope of dose versus tumor control probability (TCP)] cannot be regarded as a reliable guide to cell numbers, as some writers have assumed and argued about. Heterogeneity decreases the extracted value of α from populations. What is interesting is that, as more detailed and precise descriptions for stage of tumors are developed, as they are by various international agreements, the recorded gamma-50 slopes from clinical data become steeper (or α values higher), illustrating less heterogeneity (HANKS et al. 2000; REGNAN et al. 2004). The log$_{10}$ cell kill values quoted in the tables below are all assuming α=0.35 log$_e$ per Gy.

1.3.3
Relative Effectiveness

The BED thus comprises the total dose "nd" multiplied by a parenthesis $(1 + d/\alpha/\beta)$. This parenthetic term is called the RE, and is one of the most useful concepts of the LQ formulation. RE is what determines the strength of any schedule when multiplied by total physical dose. BED is simply total dose multiplied by RE:

$$BED = nd\,(1 + d/\alpha/\beta) = nd \times RE \qquad [1.4]$$

or

$$BED = \text{total dose} \times RE$$

"Effect" (log cell kill) defined re Initial Slope, instead of Single Dose (BARENDSEN 1982)

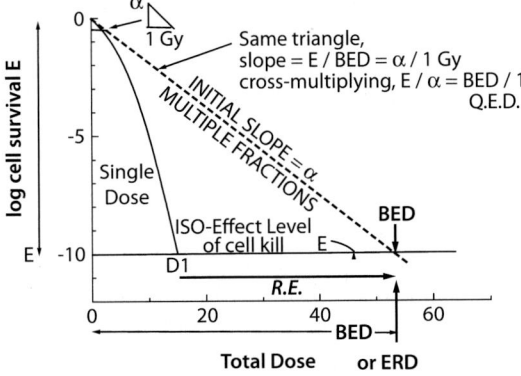

Fig. 1.3. Graphical illustration of the concept of biologically effective dose. The *dashed line* represents the log cell kill of 20 (or more) equal fractions of radiation. BED is the dose that would cause the same log cell kill E as the schedule with n×d fractions, if it could be delivered in an infinite number of infinitely small fractions, or at very low dose rate. All the beta component would then have been repaired, and the straight line representing the log cell kill versus total dose would follow just the initial slope alpha as shown, its slope determined by the value of alpha only. The ratio of BED to the real physical dose is of course the RE, relative effectiveness=BED/total dose

When dose rate varies, the term containing β includes, instead of just d, the dose rate and the recovery rate of tissues as a ratio, as discussed in detail elsewhere (BARENDSEN 1982; THAMES and HENDRY 1987; FOWLER 1989).

The concern of RE is obviously dose-per-fraction size, in relation to the α/β ratio for different tissues. RE is larger for larger dose per fraction and for smaller ratios of α/β. Dose per fraction is therefore one of the three major factors in LQ formulation, the co-equal second being the α/β ratio. The third is overall time, which we deal with in the remaining three "steps to LQ heaven" below. The major biological importance of RE depends on the fact that α/β is large for rapidly proliferating tissues such as most tumors and small for slowly proliferating tissues such as late complications. This is attributed to the fact that slowly proliferating tissues, with long cell cycle times, have more time to carry out repair of radiation damage than do short cell cycles. The exact molecular causes of these differences in ratio of radiosensitivity to repair are not yet known but inhibition of epidermal growth factor receptors (EGRFs) is likely to be involved (FOWLER 2001). It is these differences that enable conventional radiotherapy to succeed, often using a large number of small fractions such as 30 or $35F \times 2$ Gy. We shall discuss the differences in α/β in the next section.

RE is of course high when d is high and α/β is low, for example as in hypofractionation (fewer and larger fraction sizes than normal), which could therefore cause excessive damage in slowly responding, late-reacting tissues, i.e., in late complications. In contrast, RE is naturally low for hyperfractionation (more and smaller fractions) and for low dose-rates (lower than 100 cGy/h; ICRU can be criticized for defining low dose rates as "up to 200 cGy/h", because their BEDs differ significantly from those for 50 or 100 cGy/h). It is the difference in doses per fraction above or below 2 Gy that has biased radiotherapy in the direction of many small fractions. Total doses at 2 Gy per fraction should be above 60 Gy for cure of all except a few radiosensitive types of tumor; and preferably above 70 to 90 Gy $NTD_{2\,Gy}$ for many tumors of stage II or III sizes. Equations 1.4 and 1.4a enable a great many aspects of radiotherapy to be compared simply and quickly.

A question often asked is "what new total dose should be used if dose per fraction is changed from d_1 to d_2?" This is easily answered by taking the inverse ratio of the two values of RE. If the two total doses are D_1 and D_2, of which D_1 is known and D_2 is the value sought, both of the BEDs have to be equal, by definition:

$$BED = \text{total dose} \times RE = D_1 \times RE_1 = D_2 \times RE_2;$$

So that
$$\frac{D_2}{D_1} = \frac{RE_2}{RE_1} = \frac{(1 + d_1/\alpha/\beta)}{(1 + d_2/\alpha/\beta)} \qquad [1.4b]$$

This is an example of a comparison of schedules where the only biological factor is the ratio α/β, treated as if it were a single parameter.

1.3.4
Overall Treatment Time

The major effects of radiation are diminished by any proliferation occurring in the cell populations during the weeks of radiotherapy, obviously by replacement of some of the cells that were sterilized by irradiation. In slowly proliferating tissues, naturally those in whom the reactions appear late, this replacement by repopulation is negligible (as in nerve tissue) (STEWART 1986). In rapidly proliferating tissues, however, including most tumors, it can counteract up to one-third of the cell-killing effect of the radiotherapy, although there is a delay of some days before it begins in tumors; it is more rapid in normal mucosa, at 7 days in oral mucosa.

This delay in tumor repopulation is because the tumor needs time to shrink due to the treatment, so as to bring more of the surviving cells into range of the generally poor blood supplies in the tumor. In solid tumors above a few millimeters in diameter, the daily production of cells is much faster than the volume growth rate would suggest. In most carcinomas, there is a cell loss factor of 70–95% due to nutritional failure and, to a lesser extent, due to apoptosis. This causes the volume doubling times of tumors to be as slow as months, although the number of cells born would cause the population to double in only a few days if no such cell loss occurred. Therefore, the clinically observed doubling time of a tumor is no guide to the cell repopulation rate inside the tumor. This is one of the curious facts in cancer that has only been apparent from kinetic population studies in the 1970s, using originally radioisotope labels and recently immunological fluorescent labels with flow cytometry (BEGG and STEEL 2002). Most human carcinomas have volume doubling times varying from 1 month to 3 months, but their cell population birth rates yield cell-number doubling times of 2–10 days, with 3–5 days commonly found. The cell doubling times are called Tpot (potential doubling times, meaning cell birth rate "in the absence of nutritional or apoptotic cell loss"). Prostate tumors are exceptionally slowly proliferating with median cell doubling times of 42 days (range 15 days to >70 days).

So, if a number S of cells survives a certain phase of treatment, this number will increase at the rate of $Se^{+1/Tp}$ per day and $Se^{+t/Tp}$ in t days. This means the number of cells increases by a constant proportion per day. This rate is conventionally described as the time in days required to double the number of cells, Tp, so the population increases by $\log_e 2/Tp$ per day. Therefore, the $\log_e$ number of cells killed *decreases* by $\log_e 2/Tp$ per day, meaning its change is $-0.693/Tp$ per day and $-0.693\, t/Tp$ in t days.

It is very important that we do not forget the delay in start of repopulation mentioned above, after a "kick-off" or "onset" time designated Tk days. Allowing for this, the time available for repopulation is of course t = T – Tk days, where T is the overall time in days. It is important to note that the treatment begins at day 0, not day 1. The $\log_e$ amount of repopulation is then $0.693(T-Tk)/Tp$, and it has to be subtracted from the $\log_e$ cell kill E that we calculated above in Eq. 1.3. Thus, the total amount of log cell kill, allowing for repopulation, is of course:

E = nd $(\alpha + \beta d)$ – (T – Tk) × rate of repopulation per day   [1.5]

E = nd $(\alpha + \beta d)$ – $\log_e 2(T - Tk)/Tp$ [1.6]

where T = overall treatment time (in days, the first day being 0 not 1), Tp = cell population doubling time during treatment, Tk = starting time of repopulation, and Tp = average cell-number birth rate during the irradiation. Note that Tp may be different from Tpot, which can only be measured before any treatment is given; Tp is possibly somewhat faster. This simple modeling, assuming zero repopulation up to Tk days and a constant doubling time Tp after Tk, is certainly an approximation to a smooth curve of accelerating Tp, but several more complex models to add different times at which faster Tp is assumed have not succeeded in giving better correlations with clinical outcome (van Dyk et al. 1989; Denham and Kron 2001), although Denham and Kron (2001) made a persuasive case for some type of accelerating curve in tumors. Fenwick has recently proposed an ingenious Delayed Feedback (from the level of cell depletion reached) algorithm which deals elegantly with this concept (Fenwick 2006).

To transform E into the total BED requires the same division throughout by α that we carried out in step three (Eq. 1.3), and this is the final equation, the seventh step:

$$\text{BED} = \frac{E}{\alpha} = nd\left(1 + \frac{d}{\alpha/\beta}\right) - \frac{0{,}693}{\alpha TP}(T - Tk) \qquad [1.7]$$

Putting it into words:

$$\text{BED} = \frac{E}{\alpha} = \text{Total Dose} \times \text{Relative Effectiveness}$$

minus

$$\frac{(\log_e 2) \times (\text{Days availible for repop})}{\text{alpha} \times \text{Cell Doubling Time}} \qquad [1.7a]$$

It will be noticed that this LQ formula now involves three more parameters, so we might think that LQ has "lost its innocence" when repopulation is included. A value must be chosen for radiosensitivity α and the start of rapid repopulation in tumors Tk, and for the doubling time Tp of viable cells thereafter. There are, however, some clinical data that enable ratios of these parameters to be estimated. (There is a practical snag in that it is easy to forget the α in the last term! But, if you remember that this repopulation term is usually about 0.5–0.8 Gy_{10} per day for rapidly proliferating tissues including most tumors, you can easily make a rough check on the subtraction from the simpler BED of step four (Eq. 1.4). This term is somewhat greater if doses per fraction are larger than 2 Gy.)

In radiotherapy practice, late-responding normal tissues generally have a normal turnover time of many months, and these are reflected first of all in their kick-off time of acceleration, Tk, after starting radiotherapy (Stewart and van der Kogel 2002), and also obviously in the value of Tp after compensatory repopulation has started (Hendry et al. 1996; Fowler and Chappell 2000), which may be a little shorter than Tpot measured, as it can only be, before any treatment. It is this repopulation aspect of radiobiological modeling that has brought most radiotherapy departments to pro-actively avoid gaps in treatment, and in many to choose shorter schedules (Hendry et al. 1996). To a radiobiology modeler, it is surprising that this took so long to realize (Fowler 1978; Withers et al. 1988; Fowler and Lindstrom 1982; Hendry et al. 1996). The huge difference between gross tumor volume doubling time and the cell birth doubling time, Tpot, is due to the high cell loss factor (70–90%) in carcinomas. The kick-off time for tumor repopulation, Tk, could be between 21 days and 32 days in human head and neck tumors (Brenner 1993; Roberts and Hendry 1999).

1.3.5
Acute Mucosal Tolerance

A particular application of Eq. 1.7 above has been a long-term development of a collection of data about head and neck radiotherapy schedules that are judged to be clinically just tolerable from the acute reaction point of view. A comprehensive review of published head and neck schedules with consequential recommendations for a "tolerance zone of early BED" was recently published by the present author and colleagues (FOWLER et al. 2003c). The best matching with these not particularly small treatment fields, for "nearly intolerable" acute mucosal reactions in many types of fractionation regime, was found to be at BED = 59–63 Gy$_{10}$ if the following specific parameters in Eq. 1.7 were used: $\alpha/\beta = 10$ Gy; $\alpha = 0.35$ Gy per log$_e$; Tk (starting time of repopulation) = 7 days for oral mucosa (DÖRR et al. 2002); and Tp = 2.5 days (average mucosal cell doubling time during irradiation). There is provisional information that the range of BED may be somewhat higher in rectal mucosa, for example 64–69 Gy$_{10}$ together with a mucosal area limitation down to a few centimeters squared for doses of 78–80 Gy NTD (HUANG et al. 2002; VARGAS et al. 2005), but further confirmation is required. A practical conclusion from our acute mucosal review (FOWLER et al. 2003c) was that if a new schedule causes too strong acute reactions, these can be avoided easily by extending the overall time by a few days. Acute BED for mucosa is normally reduced at a rate of about 0.8 Gy$_{10}$ per day (with 2 Gy fractions), after the time Tk = 7 days from starting daily radiotherapy (DÖRR et al. 2002).

1.3.6
To Convert from BED to NTD or EQD$_{2\,Gy}$

We have talked in terms of BED in Gy$_3$ and Gy$_{10}$ mainly, and familiarity with these terms and numbers is recommended. However, those who do not use them regularly are often more comfortable with their conversion into the biological equivalent of total dose in 2 Gy fractions. This conversion of BEDs into EQD$_{2\,Gy}$ or NTD (normalized total dose/normalized to 2-Gy fractions; MACIEJEWSKI et al. 1986; JOINER et al. 2002) is a further mathematical step, which experienced modelers may not wish to do, but it is very simple, as described below. It does, and this is important, give different EQD and NTD values for late complications from the EQD and NTD for tumors in the same schedule, so that subscripts for BEDs in Gy$_3$ or Gy$_{10}$, etc. must be kept rigorously through every step of every calculation. Further, the translation from "tumor BED" or "late complications BED" into "tumor NTD" or "late complications NTD" must be made when the results are summa-

rized. Such designation of Gy$_3$ or Gy$_{10}$, and "Late" or "Tumor" or "Acute" is necessary every time, or readers (and you the writer!) will quickly become confused.

We can see, from the obvious equality of: BED of any schedule 1, called BED1, required to be equal in radiobiological effect to BED2, shows that: total dose D1 × "the RE for its dose per fraction" must be equal to NTD$_{2\,Gy}$ × RE for 2 Gy fractions, with the appropriate α/β ratio so we obtain the formula:

$$NTD_{2\,Gy} \times RE2 = BED1; \text{ therefore}$$
$$NTD_{2\,Gy} = BED1 / RE2$$
$$= BED1 / (RE \text{ for 2 Gy fractions}). \text{ QED}$$

Therefore, an NTD$_{2\,Gy}$ or EQD$_{2\,Gy}$ can be derived from any BED by simply dividing that BED by the RE for 2 Gy fraction size and the same α/β ratio as used to calculate the BED (Table 1.1).

1.3.7
One Example

First let us look at the simple example of a 60 Gy dose given with different fraction sizes, starting with 2 Gy × 30F in 6 weeks. The "late complications BED" is usually calculated first, assuming $\alpha/\beta = 3$ Gy, and from Eq. 1.4 it is:

$$BED = nd (1 + d / (\alpha/\beta))$$
$$60 (1 + 2/3) = 60 \times 1.667 = \mathbf{100\ Gy3}$$

The "tumor response BED" is calculated assuming $\alpha/\beta = 10$ Gy and is therefore:

$$60 (1 + 2/10) = 60 \times 1.2 = \mathbf{72\ Gy10}$$

Both the late BED of 100 Gy$_3$ and the tumor BED of 72 Gy$_{10}$ correspond to an NTD to 2 Gy fractions or EQD$_{2\,Gy}$ of 60 Gy, of course.

Table 1.1. Conversion of BED to NTD or EQD$_{2\,Gy}$

Type of tissue	α/β for BED	Divide the BED by:
Late complications	3 Gy	$(1 + 2/3) = 1.667$
Early and tumor effects	10 Gy	$(1 + 2/10) = 1.2$
Exceptions would be:		
Late CNS with $\alpha/\beta = 2$ Gy	2 Gy	$(1 + 2/2) = 2.0$
Prostate tumors	1.5 Gy	$(1 + 2/1.5) = 2.333$
$\alpha/\beta = 1.5$ Gy		

For a stronger schedule of 70 Gy = 2 Gy×35F the corresponding "late" (complications) BED and the early (tumor) BED would be 70×1.667 = **117 Gy3** and 70×1.2 = **84 Gy10**, respectively. Both the late BED of 117 Gy3 and the tumor BED of 84 Gy10 correspond of course to the NTD or $EQD_{2\,Gy}$ of 70 Gy. Those four values of BED form useful reference points when considering other schedules. In particular, the late complications BED of 117 Gy3 should not be exceeded except in special circumstances, such as research or IMRT-related "new" schedules. The NTD can be obtained by dividing the BED by 1.667 for "late complications" or by 1.2 for the "early or tumor effects", respectively, as described in Table 1.1 above.

Other examples have been worked out in detail in the Appendix (Section 1.8), starting with 60 Gy in 4 Gy fractions.

1.3.8
What is the Standard of Precision of These Estimates of BED or NTD? Gamma Slopes

How much margin of tolerance can we allow before being concerned about an overdose to normal tissues, above the arbitrary but currently practical limit of 117 Gy_3? And how much should we be concerned about tumor BEDs below the 70 Gy NTD or $EQD_{2\,Gy}$, that is below a BED of 84 Gy_{10}? These questions require whole papers to answer them properly, involving the slopes of dose–response curves. These slopes are often called gamma-50 or gamma-37, a term derived from photographic film density, meaning a change in absolute percentage response for a 1% change in total dose or NTD (and BED if fraction size is not changed). However, a common sense answer is that about a 7% change in BED or total dose at 2 Gy fraction size would probably be large enough to be clinically noticeable, and a 3% change would probably not, until very many hundreds of patients had been treated in the same way. The degree of anxiety about an over- or under-dose would also depend on the actual circumstance of what risk is most important for the individual patient.

Gamma-50 or gamma-37 is used to define the steepest part of a dose–response curve, which is at 50% probability for logistic modeling, but at 37% for Poisson probability modeling. The difference is rarely important in comparison with the precision of clinical results. For many types of tumor gamma-50 or -37 slopes found are about 1.5 to 2 times the percentage change in total dose (and BED if dose per fraction is unchanged). This is in the middle range of the sigmoid response curve between about 20% and 70% for local control (LC). So tumor response often corresponds to a gamma slope of 1.5 or 2. For doses above or below that middle range, the slopes will of course be smaller. For certain types of tumor that have been well classified in detail so that narrow risk categories are available, the change of response with BED may be greater than twice the percentage (HANKS et al. 2000; REGNAN et al. 2004). In general a 10% change in tumor or $NTD_{2\,Gy}$ or EQD means a change in LC of approximately 20%. WADSLEY and BENTZEN (2004) showed that there was a highly statistical relationship (P=0.00017) between loco-regional control (LRC) and later overall survival in head-and-neck cancer, 5-year overall survival averaging two-thirds of the 2-year LRC.

For normal tissues, the dose–response curves are often steeper than for tumors, because normal tissues are less heterogeneous in physiological properties than tumors are. Gamma-50 slopes of three or four in the steepest range of late complications are found, although slopes are shallower at the lower incidences of a few percent that are common for serious complications; the "toe" of the curve (and at the top, should we see any result close to 90% or 100%).

1.4
Rejoining Point for Those Who Skipped: How to Evaluate a New Schedule – Brief Summary

In case some of these details might have slipped your mind, we now summarize the three main weapons in our armamentarium for assessing any proposed schedule, before irradiating any patient. For example, we may wish to add an introductory phase-I dose-escalation study of effects in normal tissues before deciding on a particular dose level in a long clinical trial. A calculation of BEDs or NTDs would enable any arm to be evaluated theoretically in relation to known schedules. Such theoretical comparisons do not replace phase-I clinical trials, but they can put them into critical perspective and save time-wasting steps with dose intervals that may be too wide or too narrow.

- Calculate late complications BED, $\alpha/\beta = 3$ Gy, with no time factor [Eq.1. 4], keeping below 117 Gy_3.
- Estimate tumor BED including repopulation, as Gy10 [Eq. 1.7]. For head-and-neck or lung tumors assume:

Tk = 21d, Tp = 3d, α = 0.35 log$_e$/Gy, and α/β = 10 Gy, until further data allow the parameters to be reestimated; or as *log$_{10}$ cell kill = E = α × BED/2.303;* and compare with known schedules.

- Calculate acute mucosal BED Gy$_{10}$; assume Tk = 7d, Tp = 2.5d, α/β = 10 Gy, a = 0.35 [Eq. 1.7]; keeping below 59–63 Gy$_{10}$, for oral and pharyngeal mucosa, until further evidence is obtained (FOWLER et al. 2003c). Possibly higher than 63 Gy$_{10}$ for late rectal mucosa, but with strict volume limits.

cluded that an absolute benefit on LRC of 7%, from 46% to 53%, at 5 years was obtained, depending on type of schedule. Many of those trials and some more recent ones are included in the present analysis.

Table 1.2 lists some of the strongest published radiotherapy schedules that have been used to treat patients with oral and pharyngeal cancer. The first five were successful schedules, with one possible exception, but the bottom three in the table gave too many complications, so that they were later revised to become clinically acceptable. We can learn from these three pairs of later-modified schedules.

1.5
Now Let Us Study Some of the Best-Known Schedules for Head-And-Neck Tumor Radiotherapy

Take a deep breath and read each line as slowly as you like! Make sure a glass or cup of your favorite drink is at your elbow. Stop skipping for once!

BAUMANN et al. (2002) have written an excellent general review of "Altered Fractionation", which includes both hyperfractionation (smaller doses per fraction and higher total doses) and accelerated fractionation (shorter overall times and lower total doses). BOURHIS et al. (2004) reviewed the results of 15 randomized trials between 1970 and 1998. They con-

1.5.1
Standard Fractionation

The first line in Table 1.2 shows a standard treatment of 2 Gy×35F = 70 Gy in 7 weeks (SF). This schedule also gives the standard top level of "late BED" at 116.7 Gy$_{10}$. The acute mucosal BED of 53.1 Gy$_{10}$ for this schedule is comfortably below our recently derived "tolerance zone" of 59–63 Gy$_{10}$ for oral and pharyngeal irradiations (FOWLER et al. 2003c) and so acute mucosal reactions appear tolerable for the tissue volumes usual in head and neck radiotherapy, including some coning down, in accord with known clinical experience for 7-week schedules.

Table 1.2. Modeling landmark head-and-neck schedules. Parameters for tumor log$_{10}$ cell kill: α/β = 10 Gy; α = 0.35 ln/Gy; Tk = 21 days; Tp = 3 days; of which α is inversely proportional both to log cell kill and to BED (= E/α) in this formulation; so its actual value is not a sensitive factor. For acute mucosal BED: α/β = 10 Gy; α = 0.35 ln/Gy; Tk = 7 days; Tp = 2.5 days

Schedule (reference)	Dose/fraction × number of fractions	Total dose (Gy)	Overall time (days)	Tumor log$_{10}$ cell kill	Late effect BED (Gy$_3$)	Acute mucosal BED (Gy$_{10}$)
1. Standard 7 weeks	2 Gy×35F	70	46	10.26	116.7	53.1
2. EORTC HFX 7 weeks (HORIOT et al. 1992)	1.15×70F	81.5	46	11.13	113	58.9
3. RTOG HFX 7 weeks (FU et al. 2000)	1.2 Gy×68F	81.6	45	11.48	114	61.3
4. Concomitant boost 6 weeks (KNEE et al. 1985)	1.8 Gy×30F + 1.5 Gy×12F	72	39	11.03	113	59.1
5. Wang split course 6 weeks (WANG 1988)	1.6 Gy×42F	67.2	39	10.04	103	52.6
6. 4–5 weeks (LEBORGNE et al. 2000)	1.6 Gy×42F– ×–44F	64– –70.4	28– –33	11.15 11.21	103 108	61.3 61.1
7. GORTEC 1 3 weeks (BOURHIS et al. 2000)	2 Gy×32F	64	21	11.47	107	64.9 ??
8. CAIR 1 5 weeks (MACIEJEWSKI et al. 1996)	2 Gy×35F	70	34	11.46	117+?	62.6 ?
9. HARDE 1 5 weeks 1.4×20F+1.6×10F+ (McGINN et al. 1993)	1.2 Gy×20F + 2 Gy×4F	76	33	12.01	112	66.2 ???

? "Probably too high to be clinically tolerable"

1.5.2
Hyperfractionation

The second row of Table 1.2 shows the EORTC hyperfractionation clinical trial of 1.15 Gy twice a day for 70 F = 81.5 Gy in 7 weeks, which was the first randomized clinical trial to show an advantage for a nonstandard fractionation schedule, with 325 patients in two arms (HORIOT et al. 1992). The control arm was 70 Gy in 7 weeks (Row 1). The increase of 5y LC at 3 years was from 40% to 59% with good significance at $P=0.02$. The difference in log cell kill with our parameters was 0.8 $\log_{10}$, so if the gamma-50 slope was 2.0 for these tumors, a difference in LC of about 16% might have been expected. The observed difference of 19% was not statistically significantly different from this expected gain. The late complications were not significantly different in the two arms. Although they were slightly higher in the hyperfractionated arm, which was not expected, the low incidence of late effects did not allow the difference to be significant. The interval Δt between two fractions per day is emphasized as being critical in any schedule. The acute mucosal reactions in Gy_{10} were strong at 58.9 Gy10, just below the lower end of our arbitrary "tolerance zone" (FOWLER et al. 2003c), but tolerable. There is less repairable damage to worry about if α/β is 10 Gy than if it is 3 Gy, of course, in the RE bracket. It is also less for smaller doses per fraction.

1.5.3
Radiation Therapy Oncology Group Four-Arm Fractionation Trial (RTOG 90-03)

The Radiation Therapy Oncology Group (RTOG) was then encouraged to set up its four-arm randomized clinical trial 90-03, which ran from 1990 to 2000 and accumulated close to 280 patients in each arm (FU et al. 2000). Table 1.2 shows that it consisted of:

Row 1: the standard 2 Gy×35F in 7 weeks, five treatments a week (designated SF).
Row 3: a hyperfractionation arm [of 68F×1.2 Gy twice a day (b.i.d.) = 81.6 Gy in 7 weeks (HFX)].
Row 4: the concomitant boost at 72 Gy in 6 weeks, with 1.8 Gy given daily for 6 weeks plus 1.5 Gy at two fractions a day in the final 12 treatment days, designated AF-C.
Row 5: the CC Wang split-course accelerated schedule of 1.6 Gy 2 fractions a day×42F = 67.2 Gy in 6 weeks, designated AF-S. This had a 2-week planned gap in it.

All these schedules are summarized in the first five lines of Table 1.2, including their "late BED" in Gy_3, tumor cell kill in log10, and "acute mucosal BED" in Gy_{10}. Two of these schedules in RTOG 90-03 yielded 54% 2y LC, and two yielded 46%, the difference being significant (FU et al. 2000). Both of the 54% arms gave more than an estimated 11 $\log_{10}$ cell kill, using our stated parameters. Both of the 46% arms gave barely over 10 $\log_{10}$ of cell kill, as shown in Table 1.2. These results give a guide to the calibration of our currently used radiobiological model parameters for tumor BED. A difference of one $\log_{10}$ represents a difference of about 10% in log cell kill, from 10 $\log_{10}$ to 11 $\log_{10}$ and, therefore, up to about 20% in predicted change of percentage LC for tumors whose gamma-50 is about two. Non-uniformity between tumor sizes and stages could flatten this dose–response gamma-50 from 2 to about 1, as appears to have happened in this trial. About 30% of the patients were T4 or N2A, whereas those in the EORTC trial previously mentioned were T2-T3 and N_0, so the difference in gamma-50 is easy to explain.

How do the parameters for complications in normal tissues hold up for consistency? Only one of the "late" BEDs in the table and none of these first five "acute mucosal" BEDs were above the limits we have described above (117 Gy_3 for "late" and 59–63 Gy_{10} for "acute mucosal"). However the acute Gy_{10} for concomitant boost is in the modeled "maximum mucosal tolerance zone" at its lower edge, which does not contradict its acute reactions reported as the highest of the four arms from the RTOG trial. Concomitant boost is an ingenious method of shortening a previous 8-week schedule so that the overall time was reduced to 6 weeks (KNEE et al. 1985). The coned-down boost dose was inserted as a second dose of 1.5 Gy on the last 12 treatment days of the basic 6 weeks of 1.8 Gy fractions (with $\Delta t=6$ h), so that two fractions a day were required in only the last 2.5 weeks.

The finally listed arm of RTOG 90-03 was the C.C. Wang schedule using 1.6 Gy fractions twice a day, with a planned break of 2 weeks after 32 Gy in 2 weeks (row 5, WANG 1988). This offered the lowest acute mucosal BED, the lowest "late effect" BED of 103 Gy_3 and, not surprisingly, the lowest predicted tumor log cell kill, equal to that of the standard 70 Gy in the 7-week schedule. The originator, Dr. C.C. Wang, often added one or two fractions if reactions in individual patients justified doing so, but this was not done in RTOG 90-03. However, Drs. J. and F. Leborgne in Montevideo have explored that elective variability in 471 head and neck radiotherapy patients. They compared the acute reactions caused by one or two more fractions of 1.6 Gy, together with one or two days shorter or longer gap in the planned split after 32 Gy, so as to cancel out some of the increase in reaction level due to the increased dose. In this way, they have reduced the 2-week split so that the overall time has been shortened to just over 4 weeks, still with acceptable mucosal reactions as described in the next section (LEBORGNE et al. 2000). The 7-year LC was increased from 46% to 59% by a median reduction of 13 days overall time. Row 6 summarizes the best of their schedules.

It is interesting that most of the first six schedules listed in Table 1.2 predict $\log_{10}$ cell kill values of 11 or slightly above, except for the two schedules tested in RTOG 90-03, which came in with significantly lower LC by about 8%, and which were predicted by this modeling as being close to 10 $\log_{10}$ instead of 11 logs. The remaining three rows show tumor cell kill values well above 11 logs10, but they also show acute mucosal BEDs above the middle of our recommended maximum acute mucosal tolerance zone. All three schedules were judged too strong for continued use, and were modified to reduce their BED.

1.5.4
Head-and-Neck Schedules That Were Initially "Too Hot" in Table 1.2

Turning now to Table 1.3, three of the same schedules as above are listed: GORTEC (BOURHIS et al. 2000), CAIR (Continuous Accelerated Irradiation; MACIEJEWSKI et al. 1996), and HARDE (High-Dose Accelerated Dose-Per-Fraction Escalation; MCGINN et al. 1993), rows 1 to 6, together with the moderated versions. These regimes were first set at an apparently just too high total dose or BED, were found to give too many complications (rows 1, 3, and 5), and were subsequently reduced in dose per fraction and in extended overall time by 4 days in the case of CAIR, to obtain clinically acceptable acute reaction levels (rows 2, 4, and 6). The final column shows that this was achieved. The modeling results found for log cell kill of the ultimately acceptable schedules are then remarkably similar, all in the range 10.9–11.1 $\log_{10}$ cell kill, and also similar to the highest acceptable log cell kill values given in Table 1.2. So also are the results for acute mucosal BEDs as they should be (58–61 Gy_{10}). Indeed it was these types of data that enabled our "tolerance zone" to be investigated and defined originally (FOWLER, HARARI, LEBORGNE, LEBORGNE 2003c).

Table 1.3. Head-and-neck schedules moderated to avoid too-severe, acute normal tissue reactions. Parameters for tumor $\log_{10}$ cell kill: $\alpha/\beta = 10$ Gy; $\alpha = 0.35$ ln/Gy; Tk = 21 days; Tp = 3 days: of which α is inversely proportional both to log cell kill and to BED (= E/a) in this formulation; so its actual value is not a sensitive factor. For acute mucosal BED: $\alpha/\beta = 10$ Gy; $\alpha = 0.35$ ln/Gy; Tk = 7 days; Tp = 2.5 days

Schedule (reference)	Dose/fraction × number of fractions	Total dose (Gy)	Overall time (days)	Tumor $\log_{10}$ cell kill	Late effect BED (Gy_3)	Acute mucosal BED (Gy_{10})
1. GORTEC 1: 2F/day (BOURHIS et al. 2000)	2 Gy×32F	64	21	11.47	107	64.9 ??
2. GORTEC 2: 2F/day (J. BOURHIS 2002, personal communication)	1.75×36F	63	23	11.05	100	61.4
3. CAIR 1: 7F/week (MACIEJEWSKI et al. 1996)	2 Gy×35F	70	34	11.46	117+?	62.6 ?
4. CAIR 2: 7F/week (SKLADOWSKI et al. 2000)	1.8 Gy×39F	70.2	38	10.9	112	58.3
5. HARDE 1: 2F/day	1.2 Gy×20F+1.4×20F +1.6×10F+2 Gy×4F	76	33	12.01	112	66.2 ??
6. HARDE 2: 2F/day (P. Harari, personal communication)	1.2 Gy×36F+ 1.5 Gy×20F	73.2	37	11.0	106	59.1
7. Wang Split 2F/day (WANG 1988)	1.6 Gy×42F	67.2	39	10.04	103	52.6
8. 2F/day (LEBORGNE et al. 2000)	1.6 Gy×40F	64	25	10.88	98.1	60.0
"	1.6 Gy×40F	64	28	10.59	98.1	57.6
"	1.6 Gy×41F	65.6	29	10.76	100.6	59.5
"	1.6 Gy×42F	67.2	29	11.05	103	60.5
"	1.6 Gy×43F	68.8	29	11.33	105.5	62.4
"	1.6 Gy×43F	68.8	30	11.23	105.5	61.6
"	1.6 Gy×44F	70.4	30	11.51	107.9	63.4 ??
"	1.6 Gy×44F	70.4	33	11.21	107.9	61.1

? "Probably too high to be clinically tolerable

1.5.5
Shortening the Wang 2-Fraction-a-Day Schedule Using BED to Adjust Individual Doses

One other set of data added to and confirmed our choice of a "maximum tolerance zone" (FOWLER et al. 2003c) for radiotherapy-only acute mucosal tolerance in oral and pharyngeal irradiation. These data are summarized in Table 1.3, row 8, by a single-institute set of protocols (LEBORGNE et al. 2000) using the 1.6 Gy b.i.d. regime with, initially, a planned 2-week gap based on C.C. Wang's schedule (WANG 1988) as in RTOG 90-03 (row 7). This protocol was gradually shortened from 6 weeks to 4 weeks, testing one less or one more fraction against one more or one less day of the gap, depending on the theoretical acute BED Gy_{10} and the reaction in the patients. In the end, an "optimum estimated tumor log cell kill" can be estimated, as a maximum value before acute mucosal BEDs rises above 63 Gy_{10} and the late BED rises too close to 117 Gy_3.

This appears, from both the tables above, to be a log_{10} cell kill value of 11.1 to 11.2 log_{10}, which is also 25–26 log_e. This was for the nominal value of tumor BED of 72–74.5 Gy_{10}, and the acute mucosal range of 59–63 Gy10 (FOWLER et al. 2003c) in many types of schedule, assuming $\alpha = 0.35$ ln/Gy. It appears difficult to find any fractionation schedule that provides more tumor cell kill than 10.9–11.3 log_{10} without causing too many late or early complications in head and neck irradiations, more than 117 late Gy_3, and more than 63 acute mucosal Gy_{10}. Until more years of IMRT or stereotactic body radiotherapy or concurrent chemotherapy have been tested, this appears to be the current optimum tumor cell kill obtainable for head-and-neck radiotherapy at the time of writing, as identified by LQ modeling. The BED value of 72–74.5 Gy_{10} is, however, a less parameter-dependent figure than log cell kill (because log cell kill requires the assumption of a value for α, which is less certain than the ratio α/β). This BED of 72–74.5 Gy_{10} corresponds to 60–62 Gy NTD or EQD in 2 Gy fractions, for head-and-neck tumors, if it could be given in an overall time less than Tk.

1.5.6
General Considerations of Head-and-Neck Radiotherapy

There are three other head and neck schedules worth discussing briefly.

The first ingenious head and neck radiotherapy schedule is that designed at Umeå, Sweden, by ZACKRISSON et al. (1994). This regime delivered a 1.1 Gy fraction in the morning and a 2 Gy fraction in the afternoon. The idea was that if a slow component of recovery of radiation damage was present, the smaller dose fraction would have less damage to repair in the 5-h or 6-h interval, while the larger fraction would have a longer interval overnight to repair the greater repairable damage (defined as greater by the $d/\alpha/\beta$ term in the RE). This 5-week schedule's specifications are:

Total dose = 68 Gy in 5 weeks (at the longer end of likely Tk values)

Late BED=106.7 Gy_3; late $EQD_{2\,Gy}$ = 64 Gy NTD (well below the nominal late tolerance of 70 Gy NTD)

Tumor BED = 73.4 Gy_{10}; log_{10} cell kill = 11.20 (among the highest values of any H&N schedule; Table 1.3)

Acute mucosal BED = 61.4 Gy10 (in the middle of the acute mucosal tolerance zone for H&N, as for concomitant boost)

These specifications can be seen to be good, when compared with the moderated schedules from Table 1.3, rows 2, 4, and 6. Long-term tumor results remain to be reported, but this schedule has been in use satisfactorily for ten years now.

The second interesting schedule is the Trans-Tasmanian Radiation Oncology Group (TROG) randomized phase-III trial 91.01 (POULSEN et al. 1999, 2001). Advanced tumors of stages III and IV were treated. Two fractions a day of 1.8 Gy were given 5 days a week for 33 fractions in 23 days. The total dose was 59.4 Gy, giving an estimated 10.45 log_{10} of tumor cell kill on our scale, which is slightly above the 70 Gy control tumor estimate of 10.26 Gy_{10}. This is a difference of 2% in BED, predicting a 4–5% increase in LRC in the accelerated arm at the dose used. With a total of only 170 patients entered in each of two arms, however, a difference of 15% in LRC would be necessary for significance at the $P=0.05$ level and 90% power (12% difference in LRC for 80% power). It is therefore no surprise that the long-term tumor results are not significantly different from those of the control arm; they could not be with those patient numbers if our modeling was anywhere near correct, which it appears to be. There was a small but non-significant percentage in favor of the accelerated arm, as predicted by our modeling. The 5-year LRC was increased by 5% (from 47% to 52%), and the disease-free and disease-specific survivals both increased by 6% (POULSEN, DENHAM, PETERS et al. 2001).

Fewer late complications were seen, as expected from the lower late BED (95 Gy_3), except for the late mucosal effects, attributed to the "consequential late effects" of strong acute reactions, previously thought to occur after failure of acute reactions to heal up. Acute reactions started sooner, became more intense, then healed more quickly, as expected, and healing was complete in both arms. Because in the accelerated arm the acute reactions did heal up but still showed the same incidence of late mucosal effects as in the other arm, the description of "consequential" should be altered slightly to "after sufficiently severe early reactions, whether healed up or not".

DENHAM and KRON (2001) carried out some good mathematical modeling of tumor control, testing several descriptions of gradually accelerating proliferation rates after a range of Tk values. They modestly claimed not to find an ideal model, but they were convincing in showing that some biomathematical model incorporating an acceleration of repopulation during the radiotherapy would give better matching than the present two-rate model of "zero up to Tk and thereafter a constant Tp". This approach should be developed further (FENWICK 2006). The modeling of tumor response and of acute mucosal response must, of course, be done separately. The acute reactions were reported as 1157 naso-gastric feeding days for the accelerated arm versus 1154 days for the control arm, and the Australian dollar costs were $11,750 and $11,587, respectively.

The similarity in tumor response in the two arms is due to the relatively small number of patients, and does not contradict our assumptions of Tk = 21 days and Tp = 3 days. The late complications BED was a "safe" 95.0 Gy_3, (provided $\Delta t = 6$ h); so, although small numbers might prevent a significant difference between the two arms, it would be expected that extended follow-up might show an advantage for the accelerated arm with its lower total dose and BED. The shorter overall time is a potential advantage for patients, and for resources if the dollar cost is no greater. This is a treatment at least as good as the standard 70 Gy, but in just over 3 weeks, which has to be balanced against the 7-week duration of a once-daily treatment.

The third noteworthy schedule is the highly practical Danish development of six fractions a week, to shorten overall time from 7 weeks to 6 weeks. The total doses in the two randomized arms were 33F×2 Gy=66 Gy in either 6 weeks or 7 weeks. Two fractions a day are given on just one of the five working days each week ($\Delta t = 6$ h) or on a Saturday. The late BED is a warm but safe 110 Gy_3, equally for the 6-week or the 7-week arm (both giving 66 Gy), and no significant difference was seen in the incidence

of late reactions at 5 years. The acute mucosal BED increased from a very safe 49.9 Gy_{10} to 53.9 Gy_{10}. The frequency of confluent mucositis was increased from 33% in the 7-week arm to 53% in the 6-week arm, but all were healed within 3 months of the start of treatment (OVERGAARD et al. 2003). The calculated tumor cell kill, allowing for tumor cell repopulation at a doubling time of Tp=3 days after Tk=21 days, increased from a modest 9.5 $\log_{10}$ to a respectable 10.2 $\log_{10}$ on our scale. Disease-specific survival at 5 years was reported to increase by 7% absolute, and LRC by 12% absolute (OVERGAARD et al. 2003). Both of those clinical gains conform well to the modeled increase of BED by 7.5%, since LRC always shows a higher percentage gain than overall survival (WADSLEY and BENTZEN 2004). To be over-precise, the modeled BED gain of 7.5% had predicted an approximate 15% gain in LC by this 1-week shortening; an acceptable agreement with the observed 12%. It is interesting that our modeling from retrospective head and neck radiotherapy in 1992 predicted 14% per week, averaged for a dozen data sets before clinical trials came available (FOWLER and LINDSTROM 1992) – a good mark for modeling, especially for retrospective data! The DAHANCA curve for clinical results of overall survival had a gamma-50 of unity, and the slope for LRC had a gamma-50 of 1.7 instead of 2.

It only remains to emphasize that whatever we can do with one fraction a day, we can do somewhat better in relation to late BED with two fractions a day, as long as tumor α/β is as high as about 10 Gy, late effects are as low as 2–4 Gy, and intervals of more than 6 h are used. Modeling calculations can show how much better any proposed protocol might be. Treatments 7 days a week should be avoided because of a risk of unrepaired damage at 24 h being compounded if there is no gap longer than 24 h (FOWLER 2002; MACIEJEWSKI et al. 1996; SKLADOWSKI et al. 2000).

1.5.7
A Theoretical Calculation of "Close to Optimum" Head & Neck Schedules: 3 Weeks at Five Fractions per Week

Table 1.4 shows a set of schedules calculated using the same LQ parameters, keeping all the Gy_3 and Gy_{10} values below the theoretical maximum values described above, and also yielding close to 11 $\log_{10}$ for tumor cell kill, for an overall time of 3 weeks (Monday–Friday). These were calculated with the assumption that the starting time of rapid tumor repopulation was Tk=21 days. The first choice was one rather than two fractions a day, simply because that is a current USA favorite; although schedules delivering two fractions a day of size 1.6–1.8 Gy have by no means exhausted their potential – depending on what an optimum T_k for these tumors turns out to be, presumably between 2 weeks and 5 weeks. The first three rows show the progression of acute mucosal BED and could be considered for a mini-trial of escalating fraction size, with the third row possibly too high, as shown by the black typeface.

The fourth and fifth rows show the actual clinical schedule that Manchester, UK, has used for many years, using 16 days starting on a Wednesday (say) and ending on a Wednesday, 1 day beyond 3 weeks later (SLEVIN et al. 1992). The similarity is obvious, but the fine-tuning to an extra 3 days and slightly lower dose per fraction in the actual Manchester schedule brings up the tumor log cell kill and brings down the acute mucosal Gy_{10} compared with the 15-fraction options. This treatment is used with comparatively small fields, especially the 54 Gy schedule.

1.5.8
Conclusions Re Head-and-Neck Schedules

A "best choice" for head-and-neck radiotherapy will depend on values of T_k and T_p still to be measured for individual patients. When can we determine the starting time of rapid repopulation T_k? Will the pressure of resources urge us to the possible, but more sensitively balanced, shorter treatment times? We still await the ideal situation when inividual patients will have their tumours' T_k and T_p measured so that they may be designated suitable candidates for accelerated or conventional schedules.

There are three schedules using 6-week overall times that are predicted to provide better tumor cell kill than the conventional 70 Gy in 7 weeks that pro-

vides 10.26 log10. One is concomitant boost (KNEE et al. 1985) with 11.0 logs, another is HARDE II (P. Harari 2003, personal communication) also with 11.0 logs. The third is the 7-days-a-week schedule CAIR II (53) which delivers 10.9 logs in 38 days without any days off at weekends.

A difference in overall time of 7 days at fixed total dose and fraction size should make a difference in tumor BED, NTD, and log cell kill predicted to be about 5%, meaning approximately 10% in LRC and 7.5% in 5-year overall survival (WADSLEY and BENTZEN 2004). This is consistent with the result from RTOG 90-03 which showed a gain of 8% for concomitant boost at 6 weeks versus the conventional 70 Gy in 7 weeks.

The DAHANCA schedule of the somewhat lower dose of 2 Gy×33 F=66 Gy in 6 weeks comes into this analysis at nearly the same log cell kill as 70 Gy in 7 weeks (QVERGAARD et al. 1997), that is at 10.23 versus 10.26. To do better than the standard 70 Gy in 7 weeks, as two of the schedules in the RTOG 90-03 trial were reported to do, DAHANCA would have to utilize its shortest possible overall time of 37 days instead of 39 days; and/or T_p would have to average less than 3 days or T_k be less than 21 days. However DAHANCA's own control was 66 Gy in 7 weeks, and the 6-week arm was reported to do better than its own control by 10% LRC (OVERGAARD et al. 2003), in good agreement with the present predictions.

Whether ultimate clinical results will reflect the expected difference of about 7–10% in LRC and 5–7% in survival between the several "stronger" schedules which give 11 logs, on the one hand, and DAHANCA at 6F/wk or the conventional 70 Gy in 7 weeks (10.2 logs or 10.3 logs), on the other, remains to be seen. Much would depend on agreement among dose calibrations and dose specifications in different countries. It is interesting to watch for continuing results.

Table 1.4. Three theoretical and two practical head-and-neck 3-week schedules

Schedule	Dose per fraction × number	Total dose (Gy)	Overall time (days)	Tumor log10 cell kill	Late BED (Gy3)	Acute mucosal BED (Gy10)
Hypothetical examples:						
1) 3-week	3.3 Gy×15F	49.5	18	10.01	104.0	57.1
2) 3-week	3.4 Gy×15F	51.0	18	10.39	108.8	59.6
3) 3-week	3.5 Gy×15F	52.5	18	10.77	113.8	62.2 ?
Clinically used: (SLEVIN et al. 1992) Starting and ending on same day of week						
4) Manchester	3.281 Gy×16F	52.5	21	10.60	110.0	58.0
5) Manchester	3.375 Gy×16F	54.0	21	10.98	114.8	61.1

? "Possibly too high to be clinically tolerable"

Several good schedules are using 5 weeks overall time. One is the Umeå schedule of 1.1 Gy+2 Gy per day giving 11.2 logs (ZACKRISSON et al. 1994), another is one of the Wang shortenings at Montevideo (44F×1.6 Gy; LEBORGNE et al. 2000; but not in as short as 30 days) which also gives 11.2 logs in 33 days on our scale.

A few schedules use shorter overall times of 3 weeks or 4 weeks, together with reduced total doses so that their predicted late complications should also be less. These include Gortec II (BAUMANN et al. 2002; BOURHIS et al. 2000, 2004) which delivers 11.1 $\log_{10}$ in 24 days, Manchester's traditional 16 fractions (SLEVIN et al. 1992) which gives 11.0 $\log_{10}$ in 21 days at its 54 Gy level.

Since the present modeling assumed Tk=21 days for this chapter, no theoretical advantage is shown for times shorter than 3 weeks. We still do not know whether an average value of Tk for head-and-neck tumors is longer or shorter than 21 days. We still need to narrow the range of likely values for Tk. Fuller analyses which present tumor BED and $\log_{10}$ cell kill as a function of Tk and Tp can be seen in graphical form (FOWLER 1989) and are interesting in case Tp turns out to be even shorter than 3 days, as suggested from RTOG 90-003.

Finally, if repopulation can be slowed down by a particular choice of concomitant chemotherapy (e.g., anti-EGFR agents such as cetuximab or erlotinib; HARARI 2004), then the shortest schedules would have less biological advantage for tumor cell kill but could be useful for resource economy and for sparing late effects.

The schedules that had to be moderated as described in section 1.5.4 above were reduced to bring acute mucosal reactions down rather than late BEDs, and the same appears to be occurring with the new schedule described by SANGUINETTI et al (2004). This delivers 60 F×1.3 Gy at two fractions a day, but in 43 days instead of the maximum possible time of 39 days. Its late BED is 112 Gy_3, which appears unexceptional.

All of the commonly accepted schedules in Table 1.3 deliver no greater late BEDs than 111 Gy_3. The five or six top-scoring tumor-cell-killing schedules [that deliver 11–11.2 $\log_{10}$ in the present modeling, including concomitant boost (KNEE et al 1985) and SANGUINETTI et al (2004)] all consist of two-fraction-day schedules. Their overall times range from 23 days to 43 days, with a modal value of 5 weeks. Within this group, the shorter schedules provide the lowest late complication BEDs – 100–106 Gy_3 for up to 30 days overall time versus 107–113 Gy_3 for the

longer schedules up to 6 weeks. Is this difference in late BEDs an incidental bonus of the lower total doses that must be used to keep acute mucosal reactions tolerable in the shorter overall times? Or is it a necessary limitation because of hitherto unexpected incomplete repair in late effects when two fractions a day are used (FOWLER 2002)? These are questions that remain to be settled before true optimum schedules for head and neck cancer can be determined. In that determination, it can be remembered that lower doses-per-fraction are associated with smaller problems of incomplete repair.

1.5.9
Concurrent Chemotherapy

The addition of chemotherapy to radiotherapy appears to require concurrent administration if higher LC is to be obtained (PIGNON et al. 2000). An overview of 63 clinical trials covering 10,741 patients reported an average survival benefit at 2 years and 5 years of 4%.

The equivalent in terms of extra radiation dose in concomitant chemoradiotherapy is reported to be two or three extra 2 Gy fractions, by which the total dose has to be reduced to keep the acute reactions tolerable. This represents a BED of 4.8–7.2 Gy_{10}, which is 5–7% of the total log cell kill by radiotherapy. A survival benefit averaging 4% or 5% is rather less than expected from this estimate. Whether this additional tumor effect is obtained with no more than an equivalent extra effect in normal-tissue acute (or any) reactions remains to be determined by more detailed clinical observations.

1.6
Hypofractionation for Prostate Tumors

Table 1.4 gives examples of hypofractionation (with fewer and larger fractions), a strategy which has been used at modest levels of larger dose per fraction for many years in some countries, with results reported to be "almost as good" as the more biologically, but less economically, efficient hyperfractionation. These results should be examined more critically as a function not only of schedule but of targeted type of tumor.

Hypofractionation can indeed be biologically advantageous when tumors repopulate very rapidly, as in head-and-neck and lung tumors, although individual measurements of the repopulation would be a major advance, and further determinations of Tk are still needed, as explained above. Perhaps some kinds of chemotherapy work by slowing down repopulation, making the shorter, hotter, schedules unnecessary as a biological advantage, although they could still be useful for resource economy and for limiting total dose and hence late injury.

There is another situation in which the biology of the tumors makes hypofractionation the best clinical rationale available, as well as a convenient one. That is in the exceptional situation of tumors proliferating so slowly that their α/β ratios are significantly lower than those of late-responding normal tissue complications. This has been reported for malignant melanomas (BENTZEN et al. 1989), although these are notoriously variable in response. The outstanding example is prostate tumors, with a median Tpot of 42 days (range 15 days to >70 days) compared with Tpot = 4–10 days in most other types of tumor, especially carcinomas (HAUSTERMANS and FOWLER 2000). The α/β ratio for prostate tumors was reported as 1.5 Gy (±0.8 Gy 95% confidence interval) in 1999 (BRENNER and HALL 1999, 2000; BRENNER 2000) and 1.5±0.3 Gy in 2001 when updated with more patients and centers (FOWLER et al. 2001). Some other estimates have however suggested α/β to be about 3 Gy, similar to the commonly accepted α/β ratio for late rectal complications (WANG et al. 2003; KAL and VAN GELLEKOM 2003). If that were true, hypofractionation would be possible and safe, but would give no therapeutic advantage over conventional treatment (2 Gy fractions up to 76 Gy or 78 Gy in 1.8 Gy or 2 Gy fractions). That estimate of prostate tumor α/β of about 3 Gy, however, assumed that prostate tumor cells repopulate as rapidly as head-and-neck tumors. We did not agree with this assumption (FOWLER et al. 2003a) and still do not (DASU and FOWLER 2005). Much has been written about the topic of low α/β ratios in prostate tumors (BRENNER and HALL 2000; KING and FOWLER 2002; BRENNER 2000, 2003; BRENNER et al. 2002; AMER et al. 2003; FOWLER et al. 2003b). Only recently have clinical results of good tumor control begun to mature at the 5-year follow-up required to be entirely convincing (LIVSEY et al. 2003; LUKKA et al. 2003; CHAPPELL and FOWLER 2004). They continue to support the low value of α/β about 1.5 Gy. At the same time, evidence is also accumulating that the α/β ratio for late complications in rectum might be greater than the usually assumed value of 3 Gy, perhaps about 5 Gy (BRENNER 2004). Clinical results will resolve this issue within a few years, but the use of hypofractionation for prostate cancer appears to be a unique opportunity to obtain both better tumor control and economy of resources.

1.7
Summary

LQ analysis of radiotherapy schedules can give reliable comparisons of the biological effects of various schedules, provided that consistency is maintained in the choice of biological parameters. The assumption is not that these are necessarily absolutely correct, but that they are constant for similar groups of patients treated by the different schedules.

We have described the risk of late complications by calculating a "late BED or $NTD_{2\,Gy}$ or $EQD_{2\,Gy}$", assuming that $\alpha/\beta = 3$ Gy generally, with the exception of CNS (brain and spinal cord) or kidney – for which $\alpha/\beta = 2$ Gy is usually used. No overall time factor is normally assumed for late complications, that is zero repopulation. A maximum late BED of 117 Gy_3 can seldom be exceeded.

We have also described the various manifestations of damage to tumors as estimated by a "tumor and early BED or $NTD_{2\,Gy}$ or $EQD_{2\,Gy}$". A subtraction is necessary for tumor BED because of repopulation during treatment for any overall time longer then the "kick-off time" of repopulation Tk days, which might be 21–32 days or even shorter in rapidly growing tumors, especially head-and-neck and lung tumors. For them, a repopulation cell doubling time of 3 days is assumed, and 5 days for other carcinomas. In addition, an estimate of the log cell kill can be obtained by multiplying the BED in Gy_{10} by the assumed value of α to obtain the log_e cell kill, and dividing this by 2.303 to convert to log_{10}. Further, the BED or log cell kill can be converted to an estimate of TCP by knowing clinical results from published tumor response versus dose data, or more generically assuming a reasonable gamma slope (of say 2% TCP per 1% increase of BED) and knowing a good estimate for the D_{50} (total dose that yields 50% TCP at a stated follow-up time).

The third factor that can now be estimated is the acute mucosal response. This is a relatively new proposal to ensure that acute mucosal response should be tolerable enough to allow a schedule to be completed without an unplanned gap. Using selected parameters of $\alpha/\beta = 10$ Gy, $\alpha = 0.35 \, log_e$ per Gy, Tk = 7 days and Tp = 2.5 days, a "tolerance zone" of 59–63 Gy_{10} for oral and pharyngeal radiotherapy is recommended, with no special volume constraint, although tolerance dose would be expected to be larger with much smaller volumes of high dose. There is preliminary information that when the high-dose mucosal wall area is limited to a few square centimeters, as in rectal wall irradiation for prostate cancer radiotherapy, the limiting value of Gy_{10} on this scale might be larger, in the region of 65 Gy_{10} or above, but more clinical results are needed to clarify this point.

When therapeutic ratios are evaluated using the ratio of tumor BED divided by late-complications BED, the tumor BED must always be calculated allowing for repopulation appropriately.

We have also explained how any BED can be converted to an equivalent total dose in 2 Gy fractions, that is NTD or $EQD_{2\,Gy}$, by dividing a "late" BED in Gy_3 by 1.667 and an "early or tumor" BED in Gy_{10} by 1.2, depending on the α/β of the tissue under consideration being 3 Gy or 10 Gy, respectively. We have emphasized that the use of these descriptors and suffixes is important to maintain clarity both during calculations and in presentations.

Finally, we emphasize again that BED when used with a consistent and strictly limited library of parameters, as here, is a robust and remarkably consistent way of comparing, or ranking, the biological effects of different radiotherapy schedules. It can be particularly useful in setting up and critiquing proposed clinical trials.

1.8
Appendix:
Is This a Mistaken Dose Prescription?

Imagine that a radiation oncologist decides to give 60 Gy in 4 Gy fractions instead of 2 Gy fractions – that is 4 Gy×15F=60 Gy – perhaps under the impression that a 3-week treatment might give better tumor control than 60 Gy in 6 weeks (which it might), but under the illusion that 15 fractions of 4 Gy, to the same total dose, might be the way to do it safely (which it would not be – and how can we tell quantitatively?).

The late BED would be $60(1+4/3) = 60 \times 2.333 = 140\ Gy_3$ and the tumor BED would be $60(1+4/10) = 84\ Gy_{10}$. Neither of these BEDs corresponds to an NTD or $EQD_{2\,Gy}$ of 60 Gy in 2 Gy fractions, because of course of the different fraction size. With 4 Gy fractions they are instead 84 Gy NTD instead of 60 Gy for "late" but only 70 Gy NTD instead of 60 Gy for "tumor" tissues; in $EQD_{2\,Gy}$ or NTD in 2 Gy fractions. This difference illustrates the way that therapeutic ratio becomes worse (for most types of tumor except prostate Ca) when larger doses per fraction are used. The late BED is then 20% above the normally safe upper limit of 117 Gy_3, so it is far from safe and this schedule should not be used. This criterion alone is sufficient to exclude this schedule from practical use.

Further, the tumor BED is only raised to the same BED of 84 Gy_{10} as would be given by a 70 Gy NTD schedule (with 2 Gy fractions). The increase in late effects would be obviously dangerous, without giving a proportionately high gain in tumor BED. It demonstrates the loss of therapeutic gain that occurs for *most tumors* when hypofractionation (larger and fewer fractions) is used. There are certain exceptional circumstances in which hypofractionation could give an increased therapeutic gain – especially in prostate cancer – and these are discussed in Section 1.6 of this chapter.

Since the excessive "late complications" BED or NTD is sufficient to exclude 60 Gy in 4 Gy fractions from being used on any patient, it will be no surprise to find that the "acute mucosal" BED is also too high, being 75.3 Gy10, well above the recommended "tolerance zone" of 59–63 Gy10 (FOWLER et al 2003c), if the 15 fractions were given in the minimum time of 18 days. However, it might or might not surprise you that if the 15F×4 Gy was spread over the original 6 weeks (39 days), the acute mucosal BED would be acceptable at 58.7 Gy10. This could be deceptive, if this same overall time had been chosen, so that the first patients tested safely for acute reactions, but would suffer later intolerable reactions because of the high "late" BED. This did happen in the 1950s, for some breast treatments (SAMBROOK 1974). At only 5 weeks overall time, however, the acute mucosal BED would be 64.2 Gy10, possibly too high to be accepted in the first one or two patients seen by an observant oncologist – or possibly not.

If 4 Gy fractions were required for some good reason, such as making sure the overall time is shortened by using fewer fractions, because the tumor is known to be very rapidly proliferating, then a reduced total dose must be used. This should be carefully calculated and LQ is ideally suited to do this. But should it be matched to equal late complications or to equal tumor effect? The consequences will be different, but how much difference would there be? Would it matter? If the difference appears to be substantial, a phase-I clinical trial could be set up, based on the LQ calculations to select a safe starting level. The length of follow-up would have to be long enough to detect the type of reaction at risk.

The simplest way to find a new total dose for 4 Gy fractions would be to use formula [Eq. 1.4d] in Section 1.3d above, which is $D_2 = D_1 \times RE_1/RE_2$. Here we experience the divergence in results for "equal tumor effects" contrasted with "equal late complications". There are two calculations that have to be made. They are described briefly here and given in line-by-line detail at the end of this Appendix. Please try to take in the numbers in as you read them – don't skip!

1.8.1
For Equal "Late Complications"

$D_2 = 60 \times 1.667/2.333 = 60 \times 0.714 = 42.86$ Gy in 4 Gy fractions, requiring 10.7 fractions instead of 15 fractions. (Fractional numbers of fractions should not be approximated with large fractions; just give the last fraction as a smaller dose. Reducing all of the fractions leads to larger changes of BED.) The new "late" BED would be $42.85 \times 2.333 = 99.97$ Gy$_3$, as required, and the schedule would be just as safe for late reactions as the original 60 Gy in 2 Gy fractions. No correction for overall treatment time is normally assumed for "late effects" in such calculations. However, certain late complications may have an element of treatment-time correction, especially after high doses (DÖRR and HENDRY 2001).

The tumor effect would be reduced, however, being only $42.85 \times 1.4 = 59.4$ Gy$_{10}$ instead of the 72 Gy$_{10}$ from the original NTD of 2 Gy×30F = 60 Gy. (This BED of 59.4 Gy$_{10}$ would be "worth" an NTD of $59.4/1.2 = 49.5$ Gy in 2 Gy fractions.) This is a serious loss of 17.5% in BED and NTD, meaning a possible loss of 35% in TCP, assuming a gamma-50 of 2 for the tumor – until we figure in the effect of repopulation in the tumor, below.

1.8.2
For Equal Tumor Effect

However, $D_2 = 60 \times 1.2/1.4 = 60 \times 0.857 = 51.42$ Gy real dose, requiring 12.85 fractions of 4 Gy. The new "equal tumor effect" BED would be $51.42 \times 1.4 = 71.96$ Gy$_{10}$, a close approximation to the original 72 Gy$_{10}$. However, the corresponding "late complications BED" would be $51.42 \times 2.333 = 120$ Gy$_3$. This is 2.8% above the theoretical upper limit of 116.7 Gy$_3$ that we discussed above. Whether it might be risked will depend on the details of this patient, especially his or her general medical condition and the volume irradiated.

Discussions might take place about a compromise schedule, involving say 12 instead of 12.85 fractions of 4 Gy, which would give a "late BED" of $48 \times 2.33 = 111.8$ Gy$_3$, comfortably below the recommended maximum Gy$_3$ of 117 Gy$_3$ mentioned above.

The calculation so far has ignored tumor repopulation. If we now take it into account, it changes our perspective on the whole strategy, and we should consider this. There is no change in the BED in Gy$_3$ for late complications, of course. However, we are likely to find a decreased BED for tumors in the original 60 Gy in 6 weeks, which was less effective in tumor cell kill than we might have thought. However, this decrease in tumor cell kill by repopulation in 6 weeks could be avoided if fewer (say 12) fractions of 4 Gy were given in 2.5 weeks instead. To evaluate this possibility, let us complete the calculations for tumor BED, using the 7th LQ formula allowing for cell proliferation, first for the original 2 Gy×30F = 60 Gy in 6 weeks:

$$BED = [\text{total dose} \times RE] \text{ minus } [\ln 2(T-Tk)/\alpha Tp] \quad [1.7]$$

Let us assume $\alpha/\beta = 10$ Gy, $\alpha = 0.35$ ln per Gy, T = 39 days, Tk = 21 days (reported range 21–35 days) and Tp = 3 days (as in head-and-neck or lung tumors; FOWLER and CHAPPELL 2000).

For the original 60 Gy in 39 days, the proliferation subtraction is $0.693/(0.35 \times 3)$ Gy$_{10}$ per day = $1.98/3 = 0.66$ Gy$_{10}$ per day of repopulation, close to the value originally reported by WITHERS et al. in 1988 and confirmed by others since then (HENDRY et al. 1996). The duration of repopulation here is $39-21 = 18$ days. Therefore, instead of a tumor BED of 72 Gy$_{10}$, the original 60 Gy in 6 weeks including repopulation would have given a tumor BED of $72-(0.66 \times 18)$ Gy$_{10} = 72-11.9$ Gy$_{10} = 60.1$ Gy$_{10}$ (this is an NTD of 50 Gy).

Please note that this BED, diminished as it is by repopulation in the tumor, is purely coincidentally equivalent to the physical dose of 60 Gy in 6 weeks at 2 Gy fractions. In fact this schedule lost the equivalent of 10 Gy (in 2 Gy fractions), that is 17% of the whole dose, because of repopulation in the tumor. In contrast, in the short hypofractionated schedule of 2.5 weeks, the subtraction of BED for tumor repopulation is zero, so the calculated tumor BED for 4 Gy×12F will be $48 \times 1.4 = 67.2$ Gy$_{10}$ instead of 60.1 Gy$_{10}$. Thus, a gain is demonstrated in tumor BED of 67.2/60.1, which is 12% using the shorter 12 fraction schedule; together, however, with the 12% increase in "late complication BED" mentioned above. The "late" BED was increased from 100 Gy_3 to 112 Gy_3 by the change from 2 Gy×30F to 4 Gy×12F. This pair of 12% values are again a coincidence. Although confusing, they are convenient.

The percentage gains in tumor or "late" BED are similar here – they are not always so in altered fractionation. Which gain might be more important clinically has to be tested in a clinical trial, essentially randomized and stratified. In this case, the gamma-50 slope for late complications is probably greater than that for tumor control, so a general change to hypofractionation for most types of tumor would probably come out badly. What makes this an uncertain prediction, more true of past experience than of 3D or IMRT today, is that if the physics were such that smaller volumes of normal tissue could be irradiated to the prescribed dose, then the gamma-50 slope of the "late complications" would be shallower because the percentage would be lower, so the gain in tumor control identified above might not be counteracted by a larger increase in late complications. This is likely to be tested by IMRT in the near future.

If, instead, the theoretically "equal late Gy_3" had been chosen to be maintained, with 10.7 fractions of 4 Gy, then the tumor BED would have been 59.9 Gy_{10}, which is also coincidentally a good match to the 2 Gy fraction equivalent dose of the original 60 Gy schedule, which was reduced from a tumor BED of 72 Gy_{10} to 60.1 Gy_{10} by repopulation in the 6-week overall time. (The 12 Gy_{10} loss in BED is exactly equal to an NTD of 10 Gy in 2 Gy fractions). We have demonstrated by these calculations that the standard and one of the hypofractionated schedules could give a well-matched BED, one for "equal late BED or NTD" and the other for "nearly equal tumor BED or NTD" types of tissue. Such coincidences do not always occur, and careful calculations have to be made to obtain any new total dose to give the specific "equal effect" that you require. The traditional choice has to be made between equal late complications and poorer tumor results, on one hand, or equal tumor effects and greater risk of late complications on the other. You can see the care that has to be taken to avoid getting confused by these numbers, and it is worth while.

1.8.3
Acute Mucosal Effects

We have now explored in detail the main effects of changing from a standard to a nominally "biologically equivalent" shorter regime, and the delicate effects of changes in dose per fraction together with total dose

have been illustrated. There is one more biological effect that should be looked at in case it is not equivalent in the schedules, and that is the acute mucosal reaction, which might appear before the end of a standard schedule and should heal within 3 months. Although acute reactions are transient, they must not become too severe or the patients might not complete the regime without a gap in the radiotherapy, which has been demonstrated to be significantly bad for tumor cure (HENDRY et al. 1996).

There has recently been a large review of head-and-neck data with consequential recommendations for a "tolerance zone of early BED" by the author and colleagues (FOWLER et al. 2003c). The best matching to "borderline intolerable" acute mucosal reactions (in a fairly large volume of tissue) in many types of fractionation regime was found to be at BED = 59–63 Gy_{10} if the following specific parameters in Eq. 1.7 were used: $\alpha/\beta = 10$ Gy; $\alpha = 0.35$ Gy per $\log_e$; Tk (starting time of repopulation) = 7 days for oral mucosa; and Tp = 2.5 days (average mucosal cell doubling time during irradiation). There is provisional information that this range of BED may be somewhat higher in rectal mucosa, for example 64–69 Gy_{10} together with a mucosal area limitation down to 2–5 cm^2 for doses of 78–80 Gy NTD (HUANG et al. 2002; VARGAS et al. 2005), but further confirmation is required. A practical conclusion from this review (FOWLER et al. 2003c) was that if a new schedule causes too strong acute reactions, these can be avoided easily by extending the overall time by a few days. Acute BED for mucosa is normally reduced at a rate of about 0.8 Gy_{10} per day (with 2 Gy fractions), after the time Tk=7 days from starting daily radiotherapy (DÖRR et al. 2002).

Now let us look at our standard long and the two short schedules above:

After 2 Gy×30F/6 weeks (39 days), the acute mucosal BED = 46.7 Gy_{10}
After 4 Gy×10.7F/2.2 weeks (14 days) = 54.4 Gy_{10}
After 4 Gy×12F/2.4 weeks (15 days) = 60.9 Gy_{10}

All of these are therefore predicted to be "safe" with respect to acute reactions. The 2 Gy×30F is comfortably underdosed; and every radiation oncologist knows that it is safe in head-and-neck treatments, although it will only cure very small tumors. The 4 Gy×10.7F is, however, significantly below the tolerance range of 59–63 Gy_{10}, so should be safe except for concomitant skin diseases. The 4 Gy×12F is in the middle of the tolerance range, so its safety could be compromised by large fields or concomitant skin disease.

1.9
Line-by-Line Worked Examples: Details of Calculations of the Schedules Discussed in This Appendix

Linear-quadratic analyses, using the parameters described above, offer substantial progress in comparing radiotherapy schedules quantitatively, but first let me remind you more precisely of the tools, in case any of the relevant details have slipped your mind.

These parameters have given rather reliable comparisons, even if they become modified as further clinical data accumulate. New data are indeed expected as a result of the rapid progress in planning and data collection with IMRT (intensity modulated radiation therapy) – including helical tomotherapy – and advanced imaging.

The examples in this Appendix provide working frameworks on "how to calculate BED, etc, in steps"

for the comparisons of schedules described above – "Is this a mistaken dose prescription?" The starting point was "If 60 Gy in 2 Gy fractions is safe and fairly effective, what would be the effect of giving 60 Gy in fewer 4 Gy fractions?" We had quickly excluded the same total dose, i.e., 15F×4 Gy, as giving too many late complications and, if given in the minimum 3 weeks, too many acute mucosal reactions as well. However, carefully reduced total doses could be given by either of two strategies, (a) equal late complications but less tumor cell kill or (b) equal tumor effects but a higher risk of late complications. These are the predictions. Note that the first mathematical step is always to calculate relative effectiveness (RE).

To compare (a) 2 Gy×30F in 6 weeks with (b) 4 Gy×10.7F and (c) 4 Gy×12F in 2.5 weeks, for the late (L), tumor (T) and early normal-tissue (E) effect calculations, as in Section 1.8.

(aL) Calculate Late Effects BED and EQD$_{2\,Gy}$ or NTD

Use $\alpha/\beta=3$ Gy (unless CNS then use $\alpha/\beta=2$ Gy)	
No overall time factor is normally used for late effects	
Schedule: 2 Gy×30F	= 60 Gy in 6 weeks (39 days)
Total dose (TD)	= 60 Gy
Dose per fraction	= 2 Gy
RE = $(1+d/\alpha/\beta)$	= 1+2/3 = 1.667
BED = TD×RE	= 60×1.667 = 100 Gy$_3$
ED$_{2\,Gy=}$ "late" NTD=BED/1.667	= 100/1.667 = 60 Gy in 2 Gy fractions

(aT) Calculate Tumor Effects BED and EQD$_{2\,Gy}$ or NTD

Use $\alpha/\beta=10$ Gy (unless prostate Ca then use $\alpha/\beta=1.5$ Gy)	
Schedule: 2 Gy×30F	= 60 Gy in 6 weeks (39 days)
Total dose (TD)	= 60 Gy
Dose per fraction	= 2 Gy
RE=$(1+d/\alpha/\beta)$	= 1+2/19 = 1.2
BED=TD×RE	= 60×1.2 = 72 Gy$_3$ not allowing for repopulation
ED$_{2\,Gy=}$ "tumor" NTD=BED/1.2	= 72/1.2 = 60 Gy in 2 Gy fractions not allowing for repopulation

(aT) Now allow for repopulation of tumor cells if T>Tk. (Assume α=0.35 ln/Gy; T=overall time; Tk=21 days or other; Tp=3 days or other.)

Overall time T days	= 39 days
Kick-off time Tk assumed	= 21 days
Cell doubling time assumed Tp	= 3 days
Gy$_{10}$/day lost from BED 0.693/(αTp)	= 0.693/(0.35×3) = 1.98/3 = 0.66 Gy$_{10}$/day
Duration of repopulation (T–Tk)	= 39–21d = 18d
Loss of tumor BED due to repopulation	= 18×0.66 = 11.88 Gy$_{10}$ allowing for repopulation
Net (final) "tumor" BED	= 72–11.9 = 60.1 Gy$_{10}$ allowing for repopulation
Net EQD$_{2\,Gy}$ = tumor NTD = BED/1.2	= 60.1/1,2 = 50.1 Gy in 2 Gy fractions, including repopulation

(aE) Finally, let us calculate the acute mucosal BED in Gy_{10}:

assuming $\alpha/\beta = 10$ Gy; $\alpha = 0.35$ ln/Gy; $Tk = 7$ days; $Tp = 2.5$ days as described by FOWLER et al. (2003c): $BED = 60 \times 1.2 = 72$ $Gy_{10} - 0.693(39-7)/(0.35 \times 2.85) = 72 - 22.18/0.998 = 72 - 21.1 = 49.8$ Gy_{10}. This is well below the "tolerance zone" of $59-63$ Gy_{10} described by FOWLER et al. (2003c), therefore "safe".

We now have the BED values for the standard 60 Gy in 2 Gy fractions for the 6-weeks schedule, which is our baseline.

Now we do the same summaries for the two versions of 4 Gy schedules, so in two columns:

Calculate Late Effects BED and $EQD_{2\,Gy}$ or NTD

Use $\alpha/\beta = 3$ Gy (unless CNS then use $\alpha/\beta = 2$ Gy)

No overall time factor is normally used for late effects

(bL) Late complications constant; tumor effect less	(cL) Tumor effect more nearly constant; complications greater
Schedule: 4 Gy $\times$ 10.7F = 42.8 Gy/2.1 weeks (14 days)	4 Gy$\times$12F = 48 Gy/2.4 weeks (15 days)
Total dose (TD) = 42.8y	TD=48 Gy
Dose per fraction = 4 Gy	Dose per fraction = 4 Gy
RE = $(1+d/\alpha/\beta) = 1 + 4/3 = 2.333$	RE = $(1+d/\alpha/\beta) = 1+4/3 = 2.333$
BED = TD$\times$RE = $42.8 \times 2.333 = 99.85$ Gy_3	BED = TD$\times$RE = $48 \times 2.333 = 111.98$ Gy_3
$ED_{2\,Gy}$ = NTD = BED/1.667 = 99.85/1.667 = 59.91 Gy late NTD	$ED_{2\,Gy}$ = NTD = BED/1.667 = 99.85/1.667=67.2 Gy late NTD

Calculate Tumor Effects BED and $EQD_{2\,Gy}$ or NTD

Use $\alpha/\beta = 10$ Gy (unless prostate Ca then use $\alpha/\beta = 1.5$ Gy)

(bT) Schedule: 4 Gy $\times$ 10.7F = 42.8 Gy/14 days	(cT) Schedule: 4 Gy$\times$12F = 48 Gy/15 days
Total dose (TD) = 42.8 Gy	$\rightarrow$ 48 Gy
Dose per fraction = 4 Gy	$\rightarrow$ 4 Gy
RE = $(1+d/\alpha/\beta) = 1+2/10 = 1.4$	$\rightarrow$ 1.4
BED = TD$\times$RE = $42.8 \times 1.4 = 59.9$ Gy_{10}	$48 \times 1.4 \rightarrow 67.2$ Gy_{10}
$EQD_{2\,Gy}$ = NTD = BED/1.2 = 59.9/1.2 = 49.1 Gy tumor NTD	$67.2/1.2 \rightarrow 56$ tumor Gy NTD

Because the overall time is now shorter than the Tk of 21 days, no repopulation has to be subtracted from these tumor BEDs or NTDs (FOWLER 1989, 1992).

Now estimate the acute mucosal BED in Gy_{10} to compare with the expected maximum tolerance range of $59-63$ Gy_{10}

(bE) 4 Gy$\times$10.7F = 42.8 Gy:	(cE) 4 Gy $\times$ 12F = 48 Gy:
Acute mucosal Gy10 = 54.4 Gy10	Acute Gy10 = 60.9 Gy_{10}
– Well below 59 Gy10, therefore safe	– Probably safe, but exercise caution

Finally, suppose we had gone up to the theoretical 4 Gy$\times$12.85F mentioned above to get a tumor $EQD_{2\,Gy}$ or NTD to match the original tumor response? To cut the story short, the "late" BED would have been 119.5 Gy3 which is ("only") a 2.5% overdose for late complications. If the full 13 fractions of 4 Gy had been given, the increase in dose would have been 1.2% above 4 Gy$\times$12.85F, but the increase in BED would have been 1.5%, illustrating the "double trouble" that arises when dose per fraction and total dose are both increased, making the late BED 4% higher than the nominal limit 116.7 Gy_3.

The acute mucosal BED would have been 64.8 Gy_{10} – just above the theoretical upper limit of 63 Gy_{10}. With this marginal overdose, it does not disagree with the "late effects" BED, which we also marginally overdosed by matching the tumor effect. You'll see that the balance for therapeutic benefit is quite delicate, but can be checked reliably by appropriate modeling as described here.

References

Amer AM, Mott J, MacKay RI et al (2003) Prediction of the benefits from dose-escalated hypofractionated intensity-modulated radiotherapy for prostate cancer. Int J Radiat Oncol Biol Phys 56:199–207

Barendsen GW (1982) Dose fractionation, dose rate, and isoeffect relationships for normal tissue responses. Int J Radiat Oncol Biol Phys 8:1981–1997

Baumann M, Saunders MI, Joiner MC (2002) Modified fractionation. In: Steel GG (ed) Basic clinical radiobiology, 3rd edn, chap 14. Arnold, London, pp 147–157

Begg AC, Steel GG (2002) Cell proliferation and the growth rate of tumours. In: Steel GG (ed) Basic clinical radiobiology, 3rd edn, chap 2. Arnold, London, pp 8–22

Bentzen SM, Overgaard J, Thames HD et al (1989) Clinical radiobiology of malignant melanoma. Radiother Oncol 16:168–187

Bourhis J, de Crevoisier R, Abdulkarim B et al (2000) A randomized study of very accelerated radiotherapy with and without amifostine in head and neck squamous cell carcinoma. Int J Radiat Oncol Biol Phys 46:1105–1108

Bourhis J, Etessami A, Pignon JP et al (2004) Altered fractionated radiotherapy in the management of head and neck carcinomas: advantages and limitations. Curr Opin Oncol 16:215–219

Brenner DJ (1993) Accelerated repopulation during radiotherapy. Quantitative evidence for delayed onset. Radiat Oncol Invest 1:167–172

Brenner JD (2000) Towards optimal external-beam fractionation for prostate cancer (editorial). Int J Radiat Oncol Biol Phys 48:315–316

Brenner JD (2003) Hypofractionation for prostate cancer therapy. What are the issues? (Editorial.) Int J Radiat Oncol Biol Phys 57:912–914

Brenner DJ (2004) Fractionation and late rectal injury (editorial). Int J Radiat Oncol Biol Phys 60:1013–1015

Brenner DJ, Hall EJ (1999) Fractionation and protraction for radiotherapy of prostate cancer. Int J Oncol Biol Phys 43:1095–1101

Brenner DJ, Hall EJ (2000) In response to Drs King and Mayo. Low α/β ratios for prostate appear to be independent of modeling details. Int J Radiat Oncol Biol Phys 47:538–539

Brenner DJ, Martinez AA, Edmundson GK et al (2002) Direct evidence that prostate tumors show high sensitivity to fractionation (low α/β ratio) comparable to late-responding normal tissue. Int J Radiat Oncol Biol Phys 52:6–13

Carlson DJ, Stewart RD, Li X, et al (2004) Comparison of in vitro and in vivo α/β ratios for prostate cancer. Phys Med Biol 49:4477–4491

Chappell RJ, Fowler JF (2004) New data on the value of alpha/beta : evidence mounts that it is low. Int J Radiat Oncol Biol Phys 60:1002–1003

Dasu A, Fowler JF (2005) Comments on the "Comparisons of in vitro and in vivo α/β ratios for prostate cancer". Phys Med Biol (in press)

Dasu A, Toma-Dasu I, Fowler JF (2003) Should single or distributed parameters be used to explain the steepness of tumor control probability curves? Phys Med Biol 48:387–397

Denekamp J (1973) Changes in the rate of repopulation during multifraction irradiation of mouse skin. Br J Radiol 46:381–387

Denham JW, Kron T (2001) Extinction of the weakest. Int J Radiat Oncol Biol Phys 51:807–819

Dische S, Saunders M, Barrett A et al (1997) A randomised multicentre trial of CHART versus conventional radiotherapy. Radiother Oncol 44:123–36

Douglas BG, Fowler JF (1976) The effect of multiple small doses of×rays on skin reactions in the mouse and a basic interpretation. Radiat Res 66:401–426

Dörr W, Hendry JH (2001) Consequential late effects in normal tissues. Radiother Oncol 61:223–231

Dörr W, Hamilton CS, Boyd T et al (2002) Radiation-induced changes in cellularity and proliferation in human oral mucosa. Int J Radiat Oncol Biol Phys 52:911–917

Fenwick JD (2006) Delay differential equations and the dose-time dependence of early radiotherapy reactions I: model development. Med Phys (in press).

Fowler JF (1978) Int J Radiat Oncol Biol Phys. Final comments on Rome Symposium on the biological bases of tumor radioresistance. Int J Radiat Oncol Biol Phys 8:115–116

Fowler JF (1989) Review article: the Linear-Quadratic formula and progress in fractionated radiotherapy: a review. Br J Radiol 62:679–694

Fowler JF (1992a) Brief summary of radiobiological principles in fractionated radiotherapy. Semin Radiat Oncol 2:16–21

Fowler JF (1992b) Intercomparisons of new and old schedules in fractionated radiotherapy. Semin Radiat Oncol 2:67–72

Fowler JF (2001) Biological factors influencing optimum fractionation in radiation therapy. Acta Oncol 40:712–717

Fowler JF (2002) Repair between dose fractions: a simpler method of analyzing and reporting apparently biexponential repair. Radiat Res 158:141–151

Fowler JF, Chappell RJ (2000) Non-small-cell lung tumors repopulate rapidly during radiation therapy (letter). Int J Radiat Oncol Biol Phys 46:516–517

Fowler JF, Lindstrom MJ (1992) Loss of local control with prolongation in radiotherapy. Int J Radiat Oncol Biol Phys 23:457–467

Fowler JF, Chappell RJ, Ritter MA (2001) Is α/β for prostate tumors really low? Int J Radiat Oncol Biol Phys 50:1021–1031

Fowler JF, Ritter MA, Fenwick JD, Chappell RJ (2003a) How low is the α/β ratio for prostate cancer? In regard to Wang et al 2003. IJROBP 55:194–203

Fowler JF, Ritter MA, Chappell RJ, Brenner JD (2003b) What hypofractionated protocols should be tested for prostate cancer? Int J Radiat Oncol Biol Phys 56:1093–1104

Fowler JF, Harari PM, Leborgne F, Leborgne JH (2003c) Acute radiation reactions in oral and pharyngeal mucosa: tolerable levels in altered fractionation schedules. Radiother Oncol 69:161–168

Fowler JF, Tome WA, Fenwick JD, Mehta MP (2004) A challenge to traditional radiation oncology. Int J Radiat Oncol Biol Phys 60:1241–1256

Fu KK, Pajak TF, Trotti A et al (2000) A Radiation Therapy Oncology Group (RTOG) phase III randomized study to compare hyperfractionation and two variants of accelerated fractionation to standard fractionation radiotherapy for head and neck squamous cell carcinomas; first report of RTOG 90-03. Int J Radiat Oncol Biol Phys 48:7–16

Fuks Z, Persaud RS, Alfieri A et al (1994) Basic fibroblast growth factor protects endothelial cells against radiation-induced programmed cell death in vitro and in vivo. Cancer Res 54:2582–2590

Gilbert CW, Hendry JH, Major D (1980) The approximation in the formulation for survival S = exp – (αd + βd^2). Int J Radiat Biol 37:469–471

Hahnfeldt P, Panigrahy D, Folkman J. Hlatky L (1999) Tumor development under angiogenic signaling: a dynamical theory of tumor growth, treatment response, and postvascular dormancy. Cancer Res 59:4770–4775

Hanks GE, Hanlon AL, Pinover WH et al (2000) Dose selection for prostate cancer patients based on dose comparison and dose response studies. Int J Radiat Oncol Biol Phys 46:823–832

Harari PM (2004) Epidermal growth factor receptor inhibition strategies in oncology. Endocr Relat Cancer 11:689–708

Haustermans K, Fowler JF (2000) A comment on proliferation rates in human prostate cancer (letter). Int J Radiat Oncol Biol Phys 48:303

Hendry JH, Bentzen SM, Dale RG et al (1996) A modelled comparison of the effects of using different ways to compensate for missed treatment days in radiotherapy. Clin Oncol (Roy Coll Radiol) 8:297–307

Hobson B Denekamp J (1984) Endothelial proliferation in tumours and normal tissues: continuous labelling studies. Br J Cancer 49:405–413

Horiot JC, Le Fur R, N'Guyen T et al (1992) Hyperfractionation versus conventional fractionation in oropharyngeal carcinoma: final analysis of a randomized trial of the EORTC cooperative group of radiotherapy. Radiother Oncol 25:231–241

Huang EH, Pollack A, Levy L et al (2002) Late rectal toxicitry: dose-volume effects of conformal radiotherapy fot prostate cancer. Int J Radiat Oncol Biol Phys 54:1314–1321

Joiner MC, Bentzen SM Baumann M (2002) Time-dose relationships: the linear-quadratic approach and the model in clinical practice. In: Steel GG (ed) Basic clinical radiobiology, 3rd edn, chap 12 and 13. Arnold, London, pp 120–146

Kal HB, van Gellekom MP (2003) How low is the α/β ratio for prostate cancer? Int J Radiat Oncol Biol Phys 57:1116–1121

King CR, Mayo CS (2000) Is the prostate alpha/beta ratio of 1.5 from Brenner and Hall a modeling artifact? Int J Radiat Oncol Biol Phys 47:536–538

King CR, Fowler JF (2002) Yes the alpha/beta ratio for prostate cancer is low or "methinks the lady doth protest too much…." about a low α/β ratio, that is. Int J Radiat Oncol Biol Phys 54:626–627

Knee R, Fields RS, Peters LJ (1985) Concomitant boost radiotherapy for advanced squamous cell carcinoma of the head and neck. Radiother Oncol 4:1–7

Leborgne F, Zubizaretta E, Fowler JF et al (2000) Improved results with accelerated hyperfractionated radiotherapy of advanced head and neck cancer. Int J Cancer (Radiat Oncol Invest) 90: 80–91

Livsey JE, Cowan RA, Wylie JP et al (2003) Hypofractionated conformal radiotherapy in carcinoma of the prostate: five-year outcome analysis. Int J Radiat Oncol Biol Phys 57:1254–1259

Lukka H, Hayter C, Warde P et al (2003) A randomized trial comparing two fractionation schedules for patients with localized prostate cancer (abstract 26). Radiother Oncol 59 [Suppl]:S7

Maciejewski B, Taylor JMG, Withers HR (1986) Alpha/beta value and the importance of size of dose per fraction for late complications of the supraglottic larynx. Radiother Oncol 7:323–326

Maciejewski B, Skladowski K, Pilecki B et al (1996) Randomized clinical trial on accelerated seven days per week fractionation in radiotherapy for head and neck cancer. Preliminary report on acute toxicity. Radiother Oncol 40:137–145

McGinn CJ, Harari PM, Fowler JF et al (1993) Intensification in curative head and neck cancer radiation therapy: linear quadratic analysis and preliminary assessment of clinical results. Int J Radiat Oncol Biol Phys 27:363–369

Moiseenko V (2004) Effect of heterogeneity in radiosensitivity on LQ based isoeffect formalism for low alpha/beta cancers. Acta Oncol 43:499–502

Niemierko A (1997) Reporting and analyzing dose distributions: a concept of equivalent uniform dose. Med Phys 24:103–110

Overgaard J, Hansen HS, Lena S et al (2003) Five compared with six fractions per week of conventional radiotherapy of squamous-cell carcinoma of head and neck: DAHANCA 6&7 randomized controlled trial. Lancet 362:933–940

Poulsen M, Denham J, Spry et al (1999) Acute toxicity and cost analysis of a phase III randomized trial of accelerated and conventional radiotherapy for squamous carcinoma of the head and neck: a Trans-Tasmanian radiation Oncology group study. Australas Radiol 43:487–494

Poulsen M, Denham J, Peters L et al (2001) A randomized trial of accelerated and conventional radiotherapy for stage iii and IV squamous carcinoma of the head and neck: a Trans-tasman radiation Oncology group Study (TROG 91.01). Radiother Oncol 60:113–122

Pignon JP, Bourhis J, Domenge C et al (2000) Chemotherapy added to locoregional treatment for head and neck squamous-cell carcinoma; three meta-analysis of updated individual data. Lancet 355:949–955

Regnan R, Rosenzweig HE, Yorke E et al (2004) Improved local control with higher doses of radiation in large-volume stage III non-small-cell lung cancer. Int J Radiat Oncol Biol Phys 60:741–747

Roberts SA, Hendry JH (1999) Time factors in larunx tumor radiotherapy: lag times and intertumor heterogeneity in clinical datasets from four centers. Int J Radiat Oncol Biol Phys 45:1247–1257

Sanguinetti G, Sosa M, Endres E, et al (2004) Hyperfractionated IMRT (HF-IMRT) alone for locally advanced oropharyngeal carcinoma: a phase 1 study. Radiother Oncol 73(Suppl 1):S300

Sambrook DK (1974) Limited surgical treatment of carcinoma of the breast: radiotherapy experiences. Proc R Soc Med 67:476

Skladowski K, Maciejewski B, Golen M et al (2000) Randomized clinical trial on 7-day-continuous accelerated irradiation (CAIR) of head and neck cancer – report on 3-year tumour control and normal tissue toxicity. Radiother Oncol 55:101–110

Slevin NJ, Hendry JH, Roberts SA et al (1992) The effect of increasing the treatment time beyond three weeks on the control of T2 and T3 laryngeal cancer using radiotherapy. Radiother Oncol 42:215–220

Steel GG (ed) (2002) Basic clinical radiobiology, 3rd edn. Arnold, London

Stewart FA (1986) Mechanisms of bladder damage and repair

after treatment with radiation and cytostatic drugs. Br J Cancer 53 [Suppl VII]:280–291

Stewart FA, van der Kogel A (2002) Proliferative and cellular organization of normal tissues. In: Steel GG (ed) Basic clinical radiobiology, 3rd edn, chap 3. Arnold, London, pp 23–29

Thames HD, Hendry JH (1987) Fractionation in radiotherapy. Taylor and Francis, London

Thames HD, Peters LJ, Withers HR, Fletcher GH (1983) Accelerated fractionation vs hyperfractionation: rationales for several treatments per day. Int J Radiat Oncol Biol Phys 9:127–138

Travis EL, Tucker SL (1987) Isoeffect models and fractionated radiation therapy. Int J Radiat Oncol Biol Phys 13:283–287

Turesson I, Notter G (1984a) The influence of fraction size in radiotherapy on the late normal tissue reaction I. Comparison of effects of daily and once-a-week fractionation on human skin. Int J Radiat Oncol Biol Phys 10:593–598

Turesson I, Notter G (1984b) The influence of the overall treatment time in radiotherapy on the acute reaction: comparison of the effects of daily and twice-a-week fractionation on human skin. Int J Radiat Oncol Biol Phys 10:607–619

Van de Geijn J (1989) Incorporating the time factor into the linear-quadratic model (letter). Br J Radiol 62:296–2988

Van Dyk J, Mah k, Keane T (1989) Radiation-induced lung damage; dose-time fractionation considerations. Radiother Oncol 14:55–69

Vargas C, Kestin LL, Martinez AA et al (2003) Dose-volume analysis of predictors fpr chronic rectal toxicity following treatment of prostate cancer with high-dose conformal radiotherapy (ASTRO abstract # 2093). Int J Radiat Oncol Biol Phys 57(2S):S398–S399

Wadsley JC, Bentzen SM (2004) Investigation of relationship between change in locoregional control and change in overall survival in randomized controlled trials of modified radiotherapy in head-and-neck cancer. Int J Radiat Oncol Biol Phys 60:1405–1409

Wang CC (1988) Local control of oropharyngeal carcinoma after two accelerated hyperfractionation radiation therapy schemes. Int J Radiat Oncol Biol Phys 14:1143–1146

Wang JZ, Guerrero M, Li AX (2003) How low is the alpha/beta ratio for prostate cancer? Int J Radiat Oncol Biol Phys 55:194–203

Withers HR (1967) Recovery and repopulation in vivo by mouse skin epithelial cells during fractionated irradiation. Radiat Res 32:227–239

Withers HR (1971) Regeneration of intestinal mucosa after irradiation. Cancer 28:75–81

Withers RH, Thames HD, Peters LJ (1983) A new isoeffect curve for change in dose per fraction. Radiother Oncol 1:187–191

Withers HR, Taylor JMG, Maciejewski B (1988) The hazard of accelerated tumor clonogen repopulation during radiotherapy. Acta Oncol 27:131–146

Yan D, Kestin LL, Krauss D, Lockman DL, Brabbins DS, Martinez AA (2005) Phase II dose escalation study of image-guided adaptive radiotherapy for prostate cancer: use of dose-volume constraints to achieve rectal isotoxicity. Int J Radiat Oncol Biol Phys 63:141–149

Zackrisson B, Franzen L, Henriksson R, Littbrand B (1994) Tolerance to accelerated fractionation in the head and neck region. Acta Oncol 33:391–396

2 Second Malignancies Following Radiotherapy

ERIC J. HALL

CONTENTS

2.1   Introduction   33
2.1.1   Continued Lifestyle   33
2.1.2   Genetic Susceptibility   33
2.1.3   Treatment-Related Malignancies   34
2.1.3.1 Radiation Therapy for Prostate Cancer   34
2.1.3.2 Radiation Therapy for Carcinoma of the Cervix   35
2.1.3.3 Second Cancers Among Long-Term Survivors
        of Hodgkin's Disease   36
2.2   The Impact of IMRT on the Incidence
        of Radiation-Induced Second Cancers   36
2.3   Protons Compared with γ-Rays   37
2.4   The Bottom Line   37
        References   38

2.1
Introduction

The use of radiation has such an established place in the practice of medicine, both for the diagnosis of multiple ailments and for the therapy of cancer, that it would be difficult to imagine modern medicine without X-rays. Each year worldwide, 2 billion diagnostic X-ray procedures are performed, while 5.5 million patients receive radiotherapy. With so many individuals exposed to an agent that is a known and proven human carcinogen, it is prudent to ask whether there is a price tag.

It has been estimated that between 6% and 13% of all patients presenting at major cancer centers in the U.S. do so with a second malignancy. Second cancers arise from: (a) continued lifestyle, (b) genetic susceptibility, or (c) they are treatment related.

2.1.1
Continued Lifestyle

In those instances where cancer is a direct result of the excesses of a particular lifestyle, then continuance of that lifestyle is likely to result in multiple cancers. For example, an individual with head and neck cancer due to excessive use of alcohol and tobacco is then at a high risk of developing another malignancy (a second or even a third) in an adjacent tissue. For example, a cancer of the tongue may be followed by a carcinoma of the buccal mucosa, or an esophageal cancer.

2.1.2
Genetic Susceptibility

When the treatment of cancer was extended to include a study of its causes, it soon became apparent that some patients, albeit a small minority, had strong family histories of cancer, often in Mendelian patterns, suggesting inherited susceptibility (LI 1996; LI and STOVALL 1998). The diversity of site-specific familial cancers suggested the existence of multiple predisposing genes. Examples include the RB1 gene for retinoblastoma; the WT1 gene for Wilms' tumor; germline p53 mutations in families with the Li-Fraumeni syndrome; the NF1 and NF2 genes for neuroblastomatosis, types 1 and 2; the VHL gene for renal cancer and other tumors associated with Von Hippel-Lindau disease; the APC gene for adenomatous polyposis coli; the BRCA1 gene for hereditary breast and ovarian cancer; and the mismatch repair genes for colon and other common cancers. While individuals with a known genetic disorder may have an alarmingly high risk for second and even third malignancies, they account for a relatively small fraction of human cancers.

E. J. HALL, D.Phil., D.Sc, FACR, FRCR
Higgins Professor of Radiation Biophysics, Center for Radiological Research, Columbia University, 630, West 168th St., New York, NY 10032, USA

2.1.3
Treatment-Related Malignancies

In this chapter, we direct attention to radiation-induced second malignancies. There are many single-institution studies in the literature involving radiotherapy for a variety of sites that concluded that there was no increase in second malignancies, although a more accurate assessment would have been that the statistical power of the studies was too limited to detect a relatively small increased incidence of second malignancies induced by the treatment (Movas et al. 1998).

Most radiation oncologists who see a limited number of patients with any given type of tumor do not see second malignancies as a serious problem. There are well-known exceptions, such as the significant incidence of breast cancer in young women receiving radiotherapy for Hodgkin's lymphoma (Bhatis et al. 1996; Nyandoto et al. 1998; Travis et al. 1996), where the effect is too large to be missed. However, in most instances, it is difficult to get a reliable estimate for the incidence of second cancers following radiotherapy because, in the first place, most patients do not live long enough for second cancers to develop and, in the second place, a truly appropriate control group is not available. The two principal exceptions to the lack of a control group are carcinoma of the cervix in women and carcinoma of the prostate in men; in both of these examples, patient survival is good, and surgery and radiotherapy are alternative choices, thus the patients surgically treated constitute the ideal control.

2.1.3.1
Radiation Therapy for Prostate Cancer

In the year 2000, through a collaborative project with the Radiation Epidemiological Branch of the National Cancer Institute, we completed the largest ever study of second malignancies in patients treated for prostate cancer. Data regarding the rate of incidence from the Surveillance, Epidemiology, and End Results (SEER) Program cancer registry (1973–1993) (Brenner et al. 2000) were used to compare directly second malignancy risks in 51,584 men with prostate carcinoma who received radiotherapy (3549 of whom developed second malignancies) with 70,539 men who underwent surgery without radiotherapy (5055 of whom developed second malignancies). Data were stratified by latency periods, age at diagnosis, and site of the second malignancy.

Radiotherapy for prostate carcinoma was associated with a small, statistically significant increase in the risk of solid tumors relative to treatment with surgery. Among patients who survived 5 years or more, the increased relative risk reached 15%, and was 34% for patients surviving 10 years or more (Fig. 2.1). The pattern of excess second malignancies among men treated with radiotherapy was consistent with radiobiological principles in terms of site, dose, and latency. In absolute terms, 1 in 70 patients who receive radiotherapy for prostate cancer will develop a second malignancy if they survive for 10 years following treatment.

A closer look at this study of prostate cancer patients reveals some interesting biological insights.

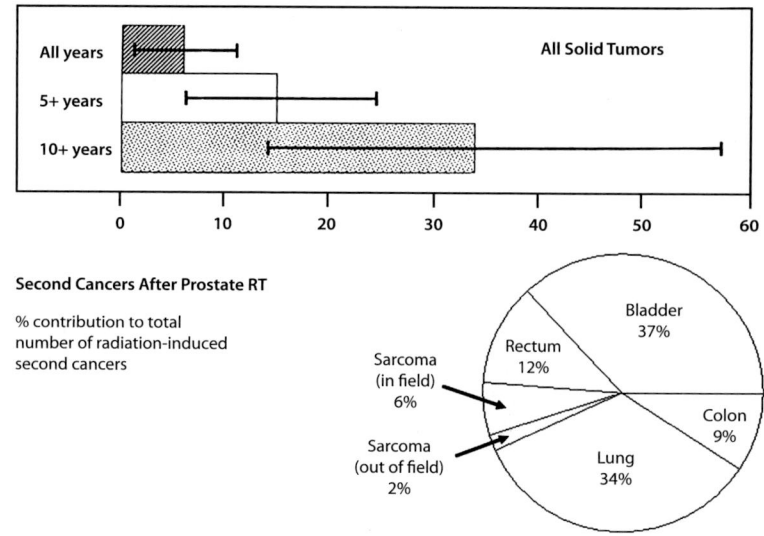

Fig. 2.1. The upper panel shows the percentage increase in relative risk for all solid tumors as a function of time after radiotherapy for prostate cancer. The *error bars* represent 95% confidence limits. "*All years*" refer to all years post-treatment; the standard error is smaller in this case because of the larger number of patients; most did not survive to 10 years. The *lower panel* shows the distribution of the principal radiation-induced cancers, namely bladder, lung, rectum and colon. There are also a smaller number of sarcomas that appear in heavily irradiated areas. (Data from Brenner et al. 2000. Figure courtesy of Dr. David Brenner)

Analyzing the solid tumors site by site, there were significant radiation-associated increases in bladder carcinoma, rectal carcinoma, and lung carcinoma, as well as sarcomas in or near the treatment field. The distribution of second cancers is also shown in Figure 2.1. It is interesting to note that the increase in relative risk for carcinoma of the lung, which was exposed to a relatively low dose (about 0.5 Gy), is of the same order as that for carcinomas of the bladder, rectum, and colon, all of which were subject to much higher doses (typically more than 5 Gy).

Although the larger number of radiation-associated malignancies clearly are carcinomas, as in the Japanese A-bomb survivors, the largest increase in relative risk is for in-field sarcomas, where it reaches over 200% at 10 years. This is a category of malignancy not observed in excess in the A-bomb survivors. In this, as in the majority of other studies, radiation-induced sarcomas occur only in heavily irradiated sites, close to the treatment volume. These observations most likely reflect a different mechanism for the induction of sarcomas compared with carcinomas. Carcinomas arise in tissues where, even in the adult, cells are turning over and/or are under hormonal control. By contrast, the target cells for sarcoma typically are dormant cells and large doses are needed to produce sufficient tissue damage to stimulate cellular proliferation. The sarcoma data in prostate patients appear to follow this pattern, with significant radiation-associated risks being observed for sites in and close to the treatment volume but not for more distant sites, which received lower doses.

The most probable reason that so few sarcomas were observed in the prostate patients is that most lived for such a short time after radiation therapy. A comparison with animal data is enlightening. A study at the National Institute of Health in the US involved irradiating Beagle dogs with large single doses in order to determine the tolerance of various organs in preparation for a program of Intraoperative Radiation Therapy (IORT) (JOHNSTONE et al. 1996). An unexpected observation was that 25% of the dogs that received 25 Gy or more developed an in-field sarcoma with a latency of 3.6 years. This was an incidental observation and not the purpose of the study. Dr. A. van der Kogel has irradiated a large number of rats with the primary purpose to study radiation myelopathy. It was again an incidental observation that 50% of the animals who received 50 Gy developed a sarcoma, while 20% of those exposed to 20 Gy developed a sarcoma (A. van der Kogel, personal communication). Two decades ago, Herman Suit studied the incidence of radiation-induced sarcoma in

defined flora and specific pathogen free mice, which had a life expectancy of 900–1000 days (SUIT et al. 1978). He showed that 50% of the animals developed a sarcoma by 480 days after a dose of 6.5–7.5 Gy, and 85% of the animals developed a sarcoma by 800 days. In comparing the animal data with the human experience, the latency periods must be thought of relative to the life span of the animals, i.e., the animals were observed for a much longer period post-irradiation relative to their life than were the radiotherapy patients, as illustrated in Figure 2.2. The conclusion is that the incidence of sarcomas in heavily exposed tissues approaches 100% if a sufficiently long period is available for study following radiation.

2.1.3.2
Radiation Therapy for Carcinoma of the Cervix

In this large study, BOICE et al. (1988) studied the risk of second malignancies in a wide range of organs and tissues as a consequence of the treatment by radiation of carcinoma of the uterine cervix. This huge international study was a "tour de force". The paper had 42 authors from 38 institutions representing both sides of the Atlantic. Such a collaboration allowed the accumulation of 150,000 patients to be studied. This study is strengthened enormously by the fact that an ideal control group is available for comparison. This malignancy is equally well treated by radiation or surgery. The results can be grouped under the following headings.

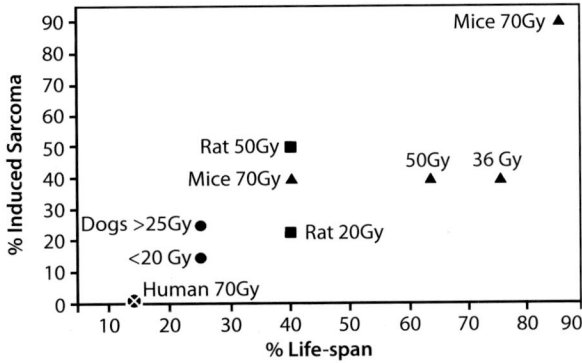

Development of Sarcomas with Time After Irradiation

Fig. 2.2. Percentage radiation-induced sarcomas as a function of time after irradiation, expressed as a percentage of normal life-span, for humans, dogs, rats, and mice. The number of sarcomas is also dependent on the radiation dose, but, in particular, it increases with time. The fact that radiation-induced sarcomas are rare in radiotherapy patients reflects the fact that most patients do not live for a large fraction of their life-span after treatment

i) Very high doses, on the order of several hundred Gy, were found to increase the risk of cancer of the bladder, vagina, possibly bone, uterine corpus, cecum and non-Hodgkin's lymphoma. The risk ratios vary from a high of 4.0 for the bladder to a low of 1.3 for bone. For all female genital cancers combined, a steep dose-response was observed, with a fivefold excess at doses of more than 150 Gy.

ii) Doses of several Gy increased the risk of stomach cancer and leukemia.

iii) Perhaps surprisingly, radiation was found not to increase the overall risk of cancers of the small intestine, colon, ovary, vulva, connective tissue, breast, Hodgkin's disease, multiple myeloma or chronic lymphocytic leukemia.

The overall conclusion of this study was that excess cancers were certainly associated with radiotherapy, as opposed to surgery, and that the risks were highest among long-term survivors and concentrated among women irradiated at relatively young ages.

2.1.3.3
Second Cancers Among Long-Term Survivors of Hodgkin's Disease

The biggest study of this kind, published by TRAVIS, CURTIS and BOICE (1996) evaluated 3869 women in population-based registries participating in the National Cancer Institute's Surveillance Epidemiology and End Results (SEER) Program. All of these women received radiotherapy as an initial treatment for Hodgkin's disease. Breast cancer developed in a total of 55 patients, which represents a ratio of observed to expected cases of 2.24. However, the risk of breast cancer was 60.57% in women treated before the age 16 years with most tumors appearing 10 or more years later. The risk of breast cancer decreased with increasing age at the time of therapy and was only slightly elevated in women who were 30 years old or older when treated.

2.2
The Impact of IMRT on the Incidence of Radiation-Induced Second Cancers

The move from three-dimensional conformed radiotherapy (3D-CRT) to intensity-modulated radiation therapy (IMRT) involves more treatment fields. The dose-volume histograms (Fig. 2.3) show that, as a consequence, a larger volume of normal tissue is exposed to lower doses in the case of IMRT compared with 3D-CRT. In addition, the number of monitor units is increased by a factor of 2–3, increasing the total body exposure due to leakage radiation from the accelerator head. Both factors will tend to increase the risk of second cancers. Before an estimate can be made of the consequences of these two factors, we must arrive at a dose-response relationship for radiation-induced cancer. For single whole-body exposures, the relationship between mortality from solid tumors among the atomic bomb survivors is consistent with linearity up to about 2.5. Sv. There is considerable uncertainty concerning the shape of the dose-response relationship for higher doses in the context of radiotherapy, where limited volumes of tissue receive doses of 20, 30, or even 70 Gy, while a much larger volume receives a lower dose because it is exposed to only some of the treatment fields.

Several possibilities can be entertained. First, it might be expected that the risk of inducing cancer would fall off sharply at higher doses due to cell killing, on the grounds that dead cells cannot give rise to a malignancy. However, none of the dose-response curves for radiation-induced cancer in humans have this shape. It must be regarded, therefore, as an extreme possibility. The other extreme possibility, suggested by the data from some human studies, is that the risk of solid tumors shows a leveling off at 4–8 Gy with no decline thereafter. An intermediate case is represented by women who have been treated with radiation for cervical cancer and have an increased risk of developing leukemia, but the dose-response relationship is complex: the risk increases with doses up to about 4 Gy and decreases slowly at higher doses.

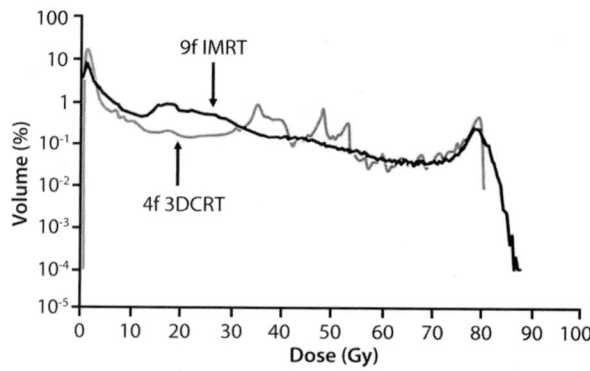

Dose-volume histogram for 4f 3DCRT and 9f IMRT

Fig. 2.3. Dose volume histograms for two typical treatment plans for prostate cancer; a four-field conformal plan and a nine-field plan using intensity modulation (IMRT)

Figure 2.4 shows data for excess relative risk over a wide range of doses for three types of human cancers. The low-dose data came from the A-bomb survivors, and the high-dose data came from radiotherapy patients. It is quite evident that excess relative risk is not a linear function of dose, but rather it tends to plateau after rising steeply with dose up to about 5 Gy. These data imply that there is comparatively little change in relative risk from 5 Gy to 50 Gy, so that in this range it is the volume of normal tissue exposed that dominates the magnitude of the risk.

A simple way to compare 3D-CRT and IMRT is to assume, as a first approximation, that the cancer risk associated with irradiating part of the trunk is directly proportional to the volume irradiated. By a comparison of dose volume histograms for 3D-CRT and IMRT, it was estimated that IMRT might increase the risk of radiation-induced carcinomas by perhaps 0.5% (HALL and WUU 2003).

Delivery of a specified dose to the isocentre from a modulated field, delivered by either dynamic IMRT or the step and shoot method of IMRT, will, in general, require the accelerator to be energized for longer (hence more monitor units are needed) compared with delivering the same dose from an unmodulated field. Some years ago, we made measurements of scattered and leakage radiation using an anthropomorphic "Randoman" phantom (HALL et al. 1995). We used ionization chambers to measure the dose to a breast while a four-field technique was used to deliver a dose of 70 Gy to the cervix. Using a 6-MV

LINAC, the breast dose was 0.25 Gy, while, with a 20-MV LINAC, the dose consisted of 0.5 Gy of X-rays plus a photoneutron component of about 1 cGy. We need only consider the data for the 6-MV LINAC, since higher energies are not usually used for IMRT. The breast dose of 0.25 Gy translates into a risk of radiation-induced cancer of about 0.5%, using a risk estimate of 2%/Sv. The total extra cancer risk posed by IMRT is the sum of that due to the extra volume of normal tissue exposed (0.5%) and the total body dose due to extra leakage resulting from a doubling of the number of monitor units (0.5%); in other words, the change to IMRT results in about a doubling of the incidence of second cancers observed, compared with more conventional radiation therapy.

2.3
Protons Compared with γ-Rays

Protons offer the possibility of reducing the volume of normal tissue involved, which one might expect to reduce the risk of second malignancies. However, there is an inherent problem with the present generation of proton therapy installations. The beam extracted from a synchrotron is a small pencil beam and, in order to produce fields large enough for treatment, the beam is directed onto a scattering foil. This process inevitably produces neutrons that give the patient a larger total body equivalent dose than that from a conventional photon LINAC. A more sophisticated technique is to "scan" the pencil beam to provide a treatment field of the required size and shape. This avoids the production of neutrons and leads to a leakage dose that is much less even than that associated with conventional LINACs. Scanning beams are available on a few facilities in Europe but, to date, not on facilities in the US. In summary, for facilities where passive modulation is used (i.e., scattering foils), the total body neutron dose will more than negate the gains from the proton dose distribution. The use of a scanning beam greatly reduces the production of neutrons and, in this situation, the full potential advantage of protons can be realized.

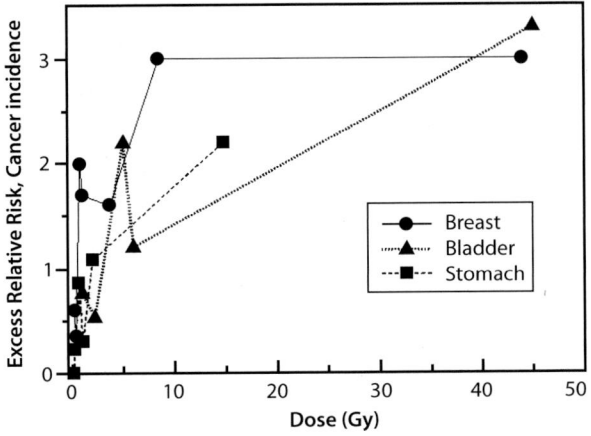

Dose Response for Carcinogenesis At High Radiation Doses

Fig. 2.4. Excess relative risk as a function of dose for three types of radiation-induced human solid cancers. The low-dose data came from the A-bomb survivors, while the high-dose data refer to radiotherapy patients. (Data compiled by Dr. Elaine Ron)

2.4
The Bottom Line

In Western countries, rather more than half of all cancer patients receive radiotherapy at some stage in the management of their disease. Because of the latent period between exposures to radiation and the appearance of a radiation-induced cancer, studies show that the incidence of second malignancies following radiotherapy increases with time after treatment. In patients that survive 10 years, about 1.5% will develop a radiation-induced second cancer. This percentage is likely to be approximately doubled by new sophisticated techniques, such as IMRT, which deliver a higher curative dose to the primary cancer, but result in more radiation to adjacent organs and to the whole body. This may not be too serious in older adults, but needs to be considered in children who are about ten times as sensitive to radiation carcinogenesis as adults (ICRP 1991). In this case, a doubling of the second cancer incidence would not be acceptable.

Second cancers become an increasing problem as treatment techniques improve, since patients must survive the first cancer in order to develop a second. It also becomes more of a problem as younger patients become candidates for radiotherapy. Protons may alleviate the problem, but only if scanning beams are available.

We felt that, in a book on the technological basis of radiation therapy, it was extremely relevant and important to include the subject of second malignancy that potentially may be induced by this modality. An effort must be made, when techniques for irradiation of patients are designed, to minimize as much as possible unnecessary irradiation to normal tissues around the target volumes or at distant sites.

References

Bhatia S, Robison LL, Oberlin O et al (1996) Breast cancer and other second neoplasms after childhood Hodgkin's Disease. N Engl J Med 334:745–751

Boice JD Jr, Engholm G, Kleinerman R, Blettner M, Stovall M, Lisco H, Moloney WC, Austin DF, Bosch A, Cookfair DL, Krementz ET, Latourette HB, Merrill JA, Peters LJ, Schulz MD, Storm HH, Björkholm E, Pettersson F, Bell CMJ, Coleman MP, Fraser P, Neal FE, Prior P, Choi NW, Hislop TG, Koch M, Kreiger N, Robb D, Robson D, Thomson DH, Lochmüller H, von Fournier D, Frischkorn R, Kjørstad KE, Rimpela A, Pejovic MH, Kirn VP, Stankusova H, Berrino F, Sigurdsson K, Hutchison GB, MacMahon B (1988) Radiation dose and second cancer risk in patients treated for cancer of the cervix. Radiat Res 116:3–55

Brenner DJ, Curtis RE, Hall EJ et al (2000) Second malignancies in prostate carcinoma patients after radiotherapy compared with surgery. Cancer 88:398–406

Hall EJ, Martin SG, Amols H et al (1995) Photoneutrons from medical linear accelerators - radiobiological measurements and risk estimates. Int J Radiat Onc Biol Phys 33:225–230

Hall EJ, Wuu CS (2003) Radiation-induced second cancers: the impact of 3D-CRT and IMRT. Int J Radiat Onc Biol Phys 56:83–88

International Commission on Radiological Protection: Recommendations (1991) Report no 60. Pergamon, New York

Johnstone PAS, Laskin WB, DeLuca AM et al (1996) Tumors in dogs exposed to experimental intraoperative radiotherapy. Int J Radiat Onc Biol Phys 34:853–857

Li FP (1996) Hereditary cancer susceptibility. Cancer 78:553–557

Li FP, Stovall EL (1998) Long term survivors of cancer. Cancer Epidemiol Biomakers Prevent 7:269–270

Movas B, Hanlon AL, Pinover W et al (1998) Is there an increased risk of second primaries following prostate irradiation? Int J Radiat Oncol Biol Phys 41:251–255

Nyandoto P, Muhonen T, Joensuu H (1998) Second cancers among long-term survivors from Hodgkin's Disease. Int J Radiat Oncol Biol Phys 42:373–378

Suit HD, Sedlacek R, Fagundes L et al (1978) Time distributions of recurrences of immunogenic and nonimmunogenic tumors following local irradiation. Radiat Res 3:251–266

Travis LV, Curtis RE, Boice JD Jr (1996) Late effects of treatment for childhood Hodgkin's disease. N Engl J Med 335:352–353

3 Clinical Principles and Applications of Chemoirradiation

DAN P. GARWOOD, L. CHINSOO CHO, and HAK CHOY

CONTENTS

3.1 Clinical Principles 39
3.2 Specific Chemotherapeutic Agents 43
3.3 Clinical Use of Chemoirradiation in Specific Sites 44
3.3.1 Small Cell Lung Carcinoma 44
3.3.2 Non-Small Cell Lung Cancer 44
3.3.3 Head and Neck 45
3.4 Uterine Cervix 47
3.5 Urinary Bladder 47
3.6 Anus 48
3.7 Esophagus 48
3.8 Rectum 49
3.8.1 Central Nervous System 51
3.9 Conclusion 51
    References 51

For more than a century, ionizing radiation has been used as an effective local agent against neoplastic disease. Increasing sophistication in computer-assisted treatment planning and delivery has brought vast strides in the accuracy and distribution of radiation dose; even so, local control of malignant tumors often still remains elusive. In parallel to the modern development of radiotherapy, the use of systemic chemotherapeutic agents has blossomed since it was first possible to produce significant clinical impact on hematological malignancies using nitrogen mustard around the time of World War II (GILMAN 1963). Since that time, an increasingly broad array of chemotherapeutic agents have entered the clinical armamentarium. It is only natural that researchers have sought to combine the two approaches, in hopes of building upon the strengths of each and achieving more than can be reached by either modality alone. Chemoirradiation is the science of applying combined modality therapy – local irradiation with the concurrent use of systemic

D. P. GARWOOD, MD; L. C. CHO, MD; H. CHOY, MD
Department of Radiation Oncology, UT Southwestern Medical Center at Dallas, 5801 Forest Park, Dallas, TX 75390-9183, USA

drugs – in the treatment of cancer. From the initial use of fluorinated pyrimidines (e.g., 5-fluorouracil; 5-FU) in the late 1950s (HEIDELBERGER et al. 1957) to the recent application of vascular endothelial growth factor (VEGF) (WILLETT et al. 2004) and epidermal growth factor receptor (EGFR) inhibition (BONNER et al. 2004), chemoradiation holds great promise for the future.

This chapter will first examine the biological basis for combining chemotherapy with ionizing radiation and then look briefly at some of the specific agents involved. Lastly, a limited tour will be taken through some of the clinical experience in specific sites of malignancy.

3.1 Clinical Principles

Both radiation and chemotherapeutic drugs are cytotoxic to tumor and normal tissue cells, a lack of specificity which is a major limitation in their use when applied either as individual treatments or in combination. Toxicity is often accentuated when the two agents are combined and when they affect the same tissue. The goals of combining chemotherapeutic drugs with radiotherapy are to increase patient survival by improving local-regional tumor control, decreasing or eliminating distant metastases, or both, while preserving organ and tissue integrity and function. Combined modality treatment can further improve positive therapeutic outcome of individual treatments through a number of specific strategies, which STEEL and PECKHAM (1979) classified into four groups: "spatial cooperation", independent toxicity, enhancement of tumor response, and protection of normal tissues.

1. "Spatial cooperation" was the initial rationale for combining chemotherapy with radiotherapy, where the action of radiation and chemotherapeutic drugs is directed toward different anatomical sites. Localized tumors would be the domain of

radiotherapy, and chemotherapeutic drugs would likely be more effective in eliminating disseminated micrometastases than in eradicating larger primary tumors. Thus, the cooperation between radiation and chemotherapy is achieved through the independent action of two agents. "Spatial cooperation" is the basis for adjuvant chemoradiotherapy, where radiation is given first to control the primary tumor, and chemotherapy is given at a later time to cope with micrometastases. The concept of "spatial cooperation" is also applied in the treatment of hematological malignancies that have spread to "sanctuary" sites, such as the brain, poorly accessible to chemotherapeutic agents.

2. Independent toxicity is another strategy to increase the ratio of efficacy to toxicity of chemoradiotherapy. Combinations of radiation and drugs would be better tolerated if the drugs were selected such that there was no overlap of toxicities to specific cell types and tissues from each modality, or minimally so. Careful drug selection based on knowledge of mechanisms of drug toxicity, mode of action, and drug pharmacokinetics may minimize normal tissue damage while retaining antitumor efficacy when combined with radiotherapy.

3. Another strategy in chemoradiotherapy is exploitation of the ability of chemotherapeutic agents to enhance tumor radiation response. The enhancement denotes the existence of some type of interaction between drugs and radiation at the molecular level, resulting in an antitumor effect greater than would be expected on the basis of additive actions. The enhancement must be selective or preferential to tumors compared with critical normal tissues in order to achieve therapeutic gain. The ability of chemotherapeutic agents to enhance tumor radiation response by counteracting determinants associated with tumor radioresistance is a major rationale for concurrent radiotherapy.

4. In addition, protection of normal tissues in order to deliver higher doses of radiation to the tumor is important. This can be achieved through technical improvements in radiation delivery or administration of chemical or biological agents that exert selective or preferential protection of normal tissues against the damage by radiation or drugs.

Any drug considered for use in combination with radiotherapy needs to undergo preclinical evaluation for its interaction with radiation both in in vitro cell culture systems and in vivo, with the aim of assessing antitumor activity and normal tissue toxicity. The interaction between two agents is more easily defined and quantified in vitro because complete cell survival curves are readily obtained. Cell survival is determined after treatment with a drug or radiation alone, given at different doses, or after treatment with both agents, in which case the cells are exposed to the drug before, during, or after irradiation.

In vitro testing is often followed by in vivo exploration of drug–radiation interactions, which allows assessment of the combined treatment on both tumors and normal tissues. The efficacy of the treatment is determined by the extent of tumor growth delay or the rate of tumor cure. In normal tissues, the effect of chemotherapeutic drugs on radiation response of acutely and late responding tissues can be assessed using a variety of available assays. Some of these assays are clonogenic, such as the jejunal crypt assay, where the endpoint depends directly on the reproductive integrity of individual cells. More frequently, however, dose–response relationships for normal tissues are based on functional endpoints (such as breathing rate in lung damage and paralysis in spinal cord damage).

Radiation induces many different lesions in the DNA molecule, which is the critical target for radiation damage. DNA double-strand breaks (DSBs), and chromosome aberrations that occur as a consequence, are generally considered to be the principal damage that results in cell death (RADFORD 1986). Any agent that makes DNA more susceptible to radiation damage may enhance cell killing. Certain drugs, such as halogenated pyrimidines, incorporate into DNA and make it more susceptible to radiation damage (KINSELLA et al. 1987).

Both sublethal (ELKIND and SUTTON 1959) and potentially lethal (LITTLE et al. 1973) damage inflicted by radiation can be repaired. While sublethal damage repair (SLDR) denotes the increase in cell survival when radiation dose is split into two fractions of radiation separated by a time interval, potentially lethal damage repair (PLDR) designates the increase in cell survival as the result of post-irradiation environmental conditions. SLDR is rapid, with a half time of about 1 h, and is complete within 4–6 h after irradiation. This time between two radiation fractions allows radiation-induced DSB in DNA to rejoin and repair. SLDR is expressed as the restitution of the shoulder on the cell survival curve for the second dose. PLDR occurs when environmental conditions prevent cells from dividing for several hours, such as keeping in vitro growing cells in plateau phase after irradiation. Preventing cells from division allows completion of the repair of DNA lesions that would have been lethal had DNA undergone replication within several hours after irradiation. PLDR is considered to be a major determinant responsible for radioresistance in some tumor types, such as melanomas.

Many chemotherapeutic agents used in chemoradiotherapy interact with cellular repair mechanisms and inhibit repair; hence, they may enhance cell or tissue response to radiation. The above-mentioned halogenated pyrimidines enhance cell radiosensitivity not only through increasing initial radiation damage but also by inhibiting cellular repair (KINSELLA et al. 1987; WANG et al. 1994). Nucleoside analogs, such as gemcitabine, are a class of chemotherapeutic agents potent in inhibiting the repair of radiation-induced DNA and chromosome damage (PLUNKETT et al. 1995; LAWRENCE et al. 1997; GREGOIRE et al. 1999). They have been shown to strongly enhance tumor radiation response in preclinical studies and are at present being extensively investigated for such activity in cancer patients (GREGOIRE et al. 1999; MILAS et al. 1999a).

Both chemotherapeutic agents and radiation are more effective against proliferating than nonproliferating cells. Their cytotoxic action further depends on the position of cells in the cell cycle. Cell-cycle dependency in response to radiation was first reported almost 30 years ago (TERASIMA and TOLMACH 1963). TERASIMA and TOLMACH (1963) reported that radiosensitivity of cell response to radiation widely varied depending on which phase of the cell cycle the cells were in at the time of irradiation, and that cells in the G_2 and M cell-cycle phases were about three times more sensitive than cells in S phase.

The influence of cell cycle on cell response to cytotoxic agents can be therapeutically exploited in chemoradiotherapy using cell-cycle redistribution strategies. For example, some chemotherapeutic drugs, such as taxanes, can block transition of cells through mitosis with the result that cells accumulate in the radiosensitive G_2 and M phases of the cell cycle. Radiation delivered at the time of significant accumulation of cells in both the G_2 and M phases results in an enhanced radiation response of cells in vitro (TISHLER et al. 1992; CHOY et al. 1993) and of tumors in vivo (MILAS et al. 1995, 1999b). However, this cell-cycle mechanism of taxane-induced enhancement of tumor radiation response is dominant only in tumors that are resistant to paclitaxel

or docetaxel as a single treatment. Although tumor growth in taxane-resistant tumors is not substantially affected by the drug, tumors do exhibit significant transient accumulation of cells in mitosis 6–12 h after the treatment (MILAS et al. 1999b).

Elimination of the radioresistant S phase cells by the chemotherapeutic agents may be another cell-cycle redistribution strategy in chemoradiotherapy. Nucleoside analogs, such as fludarabine or gemcitabine, are good examples of the agents that become incorporated into S phase cells and eliminate them by inducing apoptosis (GREGOIRE et al. 1999; MILAS et al. 1999a). In addition to purging S phase cells, the analogs induce the surviving cells to undergo parasynchronous movement to accumulate in G_2 and M phases of the cell cycle between 1 day and 2 days after drug administration, a time when the highest enhancement of tumor radiation response was observed (MILAS et al. 1999a). It should be noted that tumors with a high cell growth fraction are likely to respond better to the cell-cycle redistribution strategy in chemoradiotherapy than tumors with a low cell growth fraction.

Solid malignant tumors are generally characterized by defective vascularization, both in the number of blood vessels and vessel function. Because of this, blood supply to tumor cells is inadequate, cells lack oxygen and nutrients, and multiple tumor microregions become hypoxic, acidic, and eventually necrotic. Hypoxic cells are 2.5–3 times more resistant to radiation than well-oxygenated cells. Combining chemotherapeutic agents with radiotherapy can reduce or eliminate hypoxia or its negative influence on tumor radiation response. Most chemotherapeutic drugs preferentially kill proliferating cells, which are primarily found in well-oxygenated regions of the tumor. Since these regions are located at a close proximity to blood vessels, they are easily accessible to chemotherapeutic agents. Destruction of tumor cells in these areas will lead to an increased oxygen supply to hypoxic regions and hence re-oxygenate hypoxic tumor cells. It was recently shown that tumor reoxygenation is a major mechanism underlying the enhancement of tumor radiation response induced by taxanes in tumors sensitive to these drugs (MILAS et al. 1995).

The constant balance between cell production and cell loss maintains the integrity of normal tissues. When this balance is perturbed by cytotoxic action of chemotherapeutic drugs or radiation, the integrity of tissues is reestablished by an increased rate of cell production. The cell loss after each fraction of radiation during radiotherapy induces compensatory cell regeneration (repopulation), the extent of which determines tissue tolerance to radiotherapy. Chemotherapeutic drugs, due to their cytotoxic or cytostatic activity, can reduce the rate of proliferation when given concurrently with radiotherapy and, hence, increase the effectiveness of the treatment. Caution must be taken to select drugs that preferentially affect rapidly proliferating cells and preferentially localize in malignant tumors. However, the main limitation of concurrent chemo- and radiotherapy is the enhanced toxicity of rapidly dividing normal tissues, because most available chemotherapeutic agents show poor tumor selectivity. Moreover, accelerated repopulation induced by chemotherapeutic drugs may have a negative influence on the outcome of tumor response to radiation when drugs are used in induction or neoadjuvant chemotherapy protocols. Using this strategy, chemotherapy precedes radiotherapy. Treatment outcomes following induction chemotherapy followed by radiotherapy have not been overly encouraging in terms of both local tumor control and patient survival, even if a large proportion of tumors initially responded with total or partial clinical regression by the time of radiotherapy implementation. Some experimental evidence suggests that the drug-induced accelerated cell repopulation can actually make the tumor more difficult to control with radiation (STEPHENS and STEEL 1980; MILAS et al. 1994).

Most clinical chemoradiotherapy regimens evolved empirically: drugs known to be active against a tumor type were combined with radiation, and the doses of both agents and their administration schedules were selected for safety. Increasingly, however, information from preclinical studies is being considered in planning optimal timing of drug administration in relation to radiotherapy. Depending on the principal aim of the therapy, drugs are administered before (induction or neoadjuvant chemotherapy), during (concurrent or concomitant chemotherapy), or after (adjuvant chemotherapy) the course of radiotherapy. The advantages and disadvantages of each approach are summarized in Table 3.1.

Induction (neo-adjuvant) chemotherapy is aimed at both the disseminated disease and the primary tumor. It is initiated soon after tumor diagnosis to cope with metastatic foci, while these still contain a small number of tumor cells. With regard to the primary tumor, induction chemotherapy may reduce the number of clonogenic cells and cause the reoxygenation of the surviving hypoxic cells, both of which render tumors more controllable by radiation. Additionally, chemotherapy-induced tumor shrinkage may allow the use of smaller radiation fields, in which case less normal tissue is exposed and damaged by radiation. This treatment approach is often used in therapy of solid tumors in children and of lymphomas. Induction chemotherapy precedes radiotherapy for a few weeks to a few months, which improves tolerability of the combined treatment.

Induction chemotherapy has resulted in therapeutic improvement in a number of clinical trials when compared with radiotherapy; but, in general, the therapeutic benefits are below expectations. A number of factors could account for this, including accelerated proliferation of tumor cell clonogens

Table 3.1. Advantages and disadvantages of different chemoradiation sequencing strategies

Strategy	Advantages	Disadvantages
Sequential chemoradiation	• Least toxic • Maximize systemic therapy • Smaller radiation fields if induction shrinks tumor	• Increased treatment time • Lack of local synergy
Concurrent chemoradiation	• Shorter treatment time • Radiation enhancement	• Compromised systemic therapy • Increased toxicity • No cytoreduction of tumor
Concurrent chemoradiation and posterior chemotherapy	• Maximize systemic therapy • Radiation enhancement • Both local and distant therapy delivered up front	• Increased toxicity • Increased treatment time • Difficult to complete chemotherapy after chemoradiation
Induction chemotherapy and concurrent chemoradiation	• Maximize systemic therapy • Radiation enhancement	• Increased toxicity • Increased treatment time • Difficult to complete chemoradiation after induction therapy

and selection or induction of drug-resistant cells that are cross-resistant to radiation. The preclinical findings provide solid evidence for the existence of accelerated repopulation in tumors treated with chemotherapeutic agents. However, although development of drug resistance is a significant problem in chemotherapy, the evidence that cells that acquire drug resistance are also resistant to radiation is not convincing.

The treatment approach in which chemotherapeutic agents are given during a course of radiotherapy is referred to as concurrent chemotherapy. This form of treatment is intended to cope with both disseminated lesions and the primary tumor, but it takes the advantage of drug–radiation interactions to maximize tumor radiation response. The drug scheduling in relation to individual radiation fractions is highly important, and the selection of optimal timing of drug administration must be based on mechanisms of tumor radioenhancement by a given drug, the drug's normal tissue toxicity, and conditions under which the highest enhancement is achieved. The data from preclinical studies can greatly contribute to the selection of the most optimal schedules. For example, it has been demonstrated that murine tumors sensitive to taxanes show enhanced radiation response, but the best effect is achieved if drug treatment precedes radiation by 1–3 days (MILAS et al. 1995). A major mechanism for tumor radioenhancement was reoxygenation of hypoxic cells. Based on this preclinical information, one would anticipate that in clinical protocols such tumors would best respond to a bolus of a taxane given once or twice weekly during radiotherapy. In contrast, tumors resistant to taxanes on their own would call for daily administration of a taxane, since they show accumulation of radiosensitive G_2 and M cells 6–12 h after drug administration. If the objective is to counteract rapid repopulation of tumor cell clonogens induced by radiation, then administration of cell-cycle-specific chemotherapeutic agents during the second half of radiotherapy, when accelerated repopulation is more expressed, might be more effective. Optimal scheduling is essential in concurrent chemotherapy, not only to maximize tumor radiation response but also to minimize increases in toxicity to critical normal tissues. At present, the enhancement in normal tissue complications remains the major limitation of concurrently combining chemotherapy with radiotherapy. Nevertheless, concurrent chemo- and radiotherapy has provided better clinical results in terms of both local tumor control and patient sur-

vival than have other modes of chemoradiotherapy combinations (MUNRO 1995; MORRIS et al. 1999; EIFEL et al. 2004).

Adjuvant chemotherapy designates a treatment modality in which chemotherapeutic drugs are given some time after completion of surgery or radiotherapy. The primary objective is to eradicate disseminated disease; however, the control of the primary tumor may also be improved by the ability of drugs to deal with tumor cells that have survived radiation.

Technical improvements in radiotherapy, such as three-dimensional treatment planning and conformational or intensity-modulated radiotherapy, are other approaches likely to minimize the toxicity, consequently enhancing the effectiveness, of chemoradiation (HOLLOWAY et al. 2004). Either the use of radioprotective compounds or the implementation of technical advances may enable administration of higher doses of radiation, chemotherapeutic drugs, or both, which may result in superior treatment outcome.

3.2
Specific Chemotherapeutic Agents

Many classes of drugs have been used with radiotherapy. Anti-metabolites such as 5-FU were among the earliest radiosensitizers (BAGSHAW 1961; BUCHHOLZ et al. 1995). A newer nucleoside analog, gemcitabine, is a potent sensitizer (SCALLIET et al. 1998). Platinum-based drugs have been widely used, including cisplatin (DEWIT 1987), carboplatin (O'HARA et al. 1986; BEGG et al. 1987), and more recently oxaliplatin (FREYER et al. 2001). The taxanes paclitaxel and docetaxel inhibit the mitotic spindle by promoting microtubule assembly and inhibiting disaggregation (ROWINSKY 1997); this leads to cellular arrest in the G_2M phase of the cell cycle, a point of increased radiosensitivity (HALL 1994). The camptothecins, such as irinotecan and topotecan, target DNA topoisomerase I (HSIANG et al. 1985; HSIANG and LIU 1988). Other agents have included mitomycin C91, which targets hypoxic cells which are relatively radioresistant (BRISTOW and HILL 1998).

Newer strategies of combining systemic therapy with irradiation involve exploiting targets in the signal transduction pathways of cells, such as EGFR, or angiogenic factors supporting the growth of tumor vasculature (MASON et al. 2001). EGFR is involved in tumor growth and response to cytotoxic agents, including ionizing radiation, and expression of the receptor in a cancer is often associated with an aggressive neoplasm that is resistant to chemotherapy (MENDELSOHN and FAN 1997; SCHMIDT-ULLRICH et al. 2000). The formation of blood vessels necessary for tumor growth is dependent on angiogenic factors such as VEGF, and inhibitors have been shown to improve the efficacy of irradiation (TEICHER et al. 1995; MAUCERI et al. 1998).

There are numerous other new, targeted agents that hold the promise of improving outcomes from therapy not discussed above. They will aim to specifically block the action of numerous specific targets, including: cyclin-dependent kinase, mitogen-activated protein kinase, farnesyl transferases, mitogen-activated protein kinase, PI 3'-kinase, matrix metalloproteinases, and Bcl-2 (RABEN et al. 2004).

3.3
Clinical Use of Chemoirradiation in Specific Sites

The combination of chemotherapy and ionizing radiation has increased steadily, with the goal of increasing local control and sometimes overall survival, as well as allowing organ preservation. Some illustrative examples follow.

3.3.1
Small Cell Lung Carcinoma

While the natural history of this malignancy involves early seeding of distant metastases, patients diagnosed with localized or limited stage tumors are potentially curable. Chemotherapy remains the mainstay of therapy, although up to 80% of patients treated solely in this manner suffer a relapse, some of whom have no noted distant disease (WARDE and PAYNE 1992). Meta-analyses that examined the value of thoracic radiation in the limited stage of this disease performed by both WARDE and PAYNE (1992) and PIGNON et al. (1992) demonstrated an improvement in 2 to 3-year survivals of 5%. Warde's analysis showed that thoracic in early limited disease radiation improved local control by 25%. FRIED et al. (2004), in a meta-analysis of randomized trials, found that with early administration of thoracic irradiation there was a significant increase in 2-year compared with late

administration when radiation therapy was combined with cisplatin. The use of cisplatin and etoposide in concurrent therapy is generally accepted, as these drugs lack many of the toxicities caused when the cyclophosphamide–doxorubicin–vincristine (CAV) regimen is delivered with radiation. A recently published randomized trial has examined the benefit of delivering hyperfractionated radiation with the first cycle of chemotherapy and found that the 2-year and 5-year survivals with the hyperfractionated regimen were 47% and 26% compared with only 41% and 21% with the once daily regimen (TURRISI et al. 1999). On further analysis, it was found that not only was the local control improved in the group that received hyperfractionated radiation but there was also a decreased incidence of simultaneous local and distant failure. This suggests that improved local control may lead to improved survival, even with a malignancy that tends to disseminate systemically. The principle of spatial cooperation enters the treatment realm in this disease wherein the value of prophylactic cranial radiation is associated with a decrease in the incidence of brain metastases. This decreased incidence of brain metastases is associated with an increased survival in patients with limited stage disease who have had a complete response to therapy (AUPERIN et al. 1999; KOTALIK et al. 2001). As such, prophylactic cranial radiation forms a piece of the standard part of treatment of limited stage small cell lung cancer.

3.3.2
Non-Small Cell Lung Cancer

There are multiple trials that have compared the delivery of standard radiation with chemoradiation in non-small cell lung cancer (NSCLC) (LE CHEVALIER et al. 1991; SCHAAKE-KONING et al. 1992; DILLMAN et al. 1996; SAUSE et al. 2000). The median survival improvement from 9.6 months with radiation alone to 13.7 months with sequential chemoradiation in the original Dillman study is illustrative of the benefit of the addition of chemotherapy (DILLMAN et al. 1996). A meta-analysis confirmed that cisplatin-based chemotherapy in combination with radiation does improve outcome relative to radiation alone in cases of NSCLC at the expense of increased toxicity (PRITCHARD and ANTHONY 1996).

Two more recent trials have moved the paradigm one step further in that they both found a benefit to delivering radiation and chemotherapy concurrently. Furuse et al. presented the results of their Japanese trial, in which MVP (mitomycin C, vindesine and cisplatin) with 56 Gy were given in both sequential and concurrent fashions (FURUSE et al. 1999). There was an improved median survival of 16.5 versus 13.3 months with the concurrent therapy. The concurrent therapy was also well tolerated as the radiation was delivered in a split-course fashion with comparable rates of esophagitis in both treatment arms. The Radiation Therapy Oncology Group (RTOG) 9410 randomized phase-III experience has confirmed the benefit of concurrent therapy (CURRAN et al. 2000). This trial used cisplatin-based therapy delivered in one of three fashions: (1) sequentially with vinblastine followed by radiotherapy, (2) concurrently with vinblastine and once daily radiotherapy, and (3) concurrently with etoposide and twice daily hyperfractionated radiotherapy. Arm 2 of the trial, during which chemotherapy was delivered concurrently with once daily radiation, revealed a statistical benefit over the sequential administration of chemotherapy and radiation with a median survival of 17 months versus 14.6 months (P=0.038).

CAKIR and EGAHAN (2004) reported on another randomized trial in which 176 patients with stage-III NSCLC were allocated to radiation therapy alone (64 Gy, 2-Gy fractions) or combined with 20 mg/m^2 cisplatin given 1 h before irradiation on days 1–5 of the 2nd and 6th treatment weeks. The 3-year survival was 10% with combined chemoirradiation and 0% in the group treated with radiation therapy alone (P=0.00001). Many of the current ongoing trials are incorporating new chemotherapies and dose schedules into the treatment of NSCLC. Concurrent weekly paclitaxel and carboplatin, a treatment which tries to maximize the enhancing effects of both drugs, is based largely on the phase I–II experience of investigators (CHOY et al. 1998a,b). Efforts at including other newer agents including gemcitabine, vinorelbine, irinotecan, and docetaxel are ongoing (PENLAND and SOCINSKI 2004; ROWELL and O'ROURKE 2004). Strategies seeking to maximize radiation dose with the delivery of high-dose conformal therapy remain investigational, although preliminary results are encouraging (BRADLEY et al. 2005). No trials suggest a substantial benefit to post-operative therapy (DAUTZENBERG et al. 1999; KELLER et al. 2000). Some reports have described a lower incidence and severity of pneumonitis and esophagitis when amifostine is given before administration of chemoradiation in patients with NSCLC (KOMAKI et al. 2004).

3.3.3
Head and Neck

Traditional management of locally advanced squamous cell cancers of the head and neck has involved a combination of surgery and radiation in most cases. Despite aggressive therapy with significant morbidity, treatment often yields poor long-term survivals when there is unresectable disease with 5-year survivals in the range of 30%.

Alternative fractionation schemes that exploit the differential ability of cells to repair radiation-induced damage, thus allowing for delivery of a higher tumor dose, or those that attempt to deliver therapy in a shorter overall treatment time to combat accelerated repopulation of tumors have become more popular therapies due to some trials suggesting a benefit (HORIOT et al. 1992; FU et al. 2000).

Approximately 70 randomized trials have been performed to examine the contribution of combined chemoradiation on local control and overall survival. Many studies have been small, with inadequate power to detect a significant benefit to the addition to chemotherapy in a heterogeneous population of tumors. As such, several meta-analyses have been undertaken to assess a larger patient population and to help determine the absolute benefits of the addition of chemotherapy (MUNRO 1995; EL-SAYED and NELSON 1996; BOURHIS and PIGNON 1999; PIGNON et al. 2000).

The MACH-NC study, which evaluated 63 trials and a total of 10,741 patients, is the largest of the meta-analyses (PIGNON et al. 2000). Individual data, rather than literature-based data, with the inclusion of updated data and unpublished trials were assessed including individual data updated in 66% of the trials as far out as a median follow-up of 6.8 years. Subcategorization into locoregional treatment with and without concomitant chemotherapy, induction/adjuvant chemotherapy, and laryngeal preservation with induction chemotherapy rather than definitive treatment for laryngeal and hypopharyngeal tumors were reported. No benefit was detected for neoadjuvant or adjuvant chemotherapy, while a trend toward a statistically significant benefit (4%) was reported for concurrent or alternating chemoradiotherapy (P=0.23).

At present, the literature does justify the use of neoadjuvant chemotherapy in the limited setting of advanced laryngeal or hypopharyngeal primaries with the dual goals of organ preservation and the treatment of micrometastatic disease. Induction chemotherapy has been considered appropriate in this setting as it does improve laryngeal preservation while not compromising overall survival. The landmark Veteran's Affairs Laryngeal Study randomized patients into two treatment arms: (1) induction with cisplatin/5-FU for three cycles followed by radiation or (2) laryngectomy followed by radiation (ANONYMOUS 1991). There were 332 patients entered into the study. An evaluation occurred after two cycles of chemotherapy. Patients with a partial response received a third cycle of chemotherapy followed by radiotherapy. Those patients without an initial response to induction chemotherapy received a laryngectomy followed by radiation therapy. Patients with residual disease following the completion of radiotherapy underwent surgical resection. With a median follow-up of 33 months, an estimated 2-year survival of 68% in both groups failed to demonstrate a difference in overall survival (P=0.9846), while a majority of patients (64%) were able to preserve function of the larynx. Recurrences differed between the two groups with increased local-regional control (P=0.0005) and decreased metastases (P=0.016) in the induction chemotherapy group. Given that there was no compromise of overall survival, induction therapy is felt to be feasible in the setting of laryngeal carcinoma in order to allow organ preservation without compromise of overall survival. These results have been confirmed by the European Organization for the Research and Treatment of Cancer (EORTC) (LEFEBVRE et al. 1996).

The updated results of the intergroup trial R91-11 add to the picture (FORASTIERE et al. 2003). A total of 547 patients with stage III and stage IV potentially resectable carcinoma of the larynx were randomized to receive one of three treatments. In arm A, induction with cisplatin 100 mg/m^2 and continuous infusion of 5-FU 1000 mg/m^2 per day for three cycles was used followed by 70 Gy of radiation in responding patients. In arm B, concurrent cisplatin at 100 mg/m^2 was used on days 1, 22, and 43 along with 70 Gy, while in arm C patients received radiation alone. The rate of loco-regional tumor control was 78% with concurrent chemoirradiation, 61% with induction by cisplatin followed by radiation therapy, and 56% with radiation therapy alone. The proportion of patients with larynx preservation was 88%, 75%, and 70%, respectively. At 2 years, the laryngectomy-free survival rates for the treatments were: A 58%, B 66%, C 52%. No significant difference in laryngectomy-free survival or overall survival was found when comparing arms B or C to the control arm A. The time to laryngectomy was significantly better for arm B than arm A (P=0.0094). While this study was not powered to compare arms B and C, a secondary analysis shows that concurrent therapy yields a superior laryngectomy-free survival (P=0.02). This confirms that concurrent chemotherapy and radiation is the preferred therapy for this population when organ preservation is desired.

Al-Sarraf and colleagues have performed a large randomized, prospective, phase-III intergroup trial of 185 patients with locally advanced nasopharyngeal cancer randomized to radiation therapy alone or concomitant chemoradiation therapy (AL-SARRAF et al. 1998). All patients received 35–39 fractions of daily radiotherapy and were randomized to receive concomitant cisplatin (100 mg/m^2 on days 1, 22, and 43) followed by three cycles of adjuvant cisplatin (80 mg/m^2, day 1) and continuous infusion 5-FU (1000 mg/m^2, days 1–4) every 28 days. The superiority of combined treatment was seen in the concomitant chemoradiation therapy arm with a 3-year progression-free survival of 69% versus 24% (P<0.001) and a 3-year overall survival of 78% versus 47% (P=0.005). Hence, the recommended standard of care in treating patients with more advanced nasopharyngeal carcinoma has become concomitant chemoradiotherapy.

The contribution of the newer generation of chemotherapies including the taxanes and gemcitabine continues to be investigated. The early results of a phase-I trial that incorporated C225 – an antibody directed against the extracellular domain of the EGFR – are encouraging, with 9 of 13 patients (69%) who had received greater than 50 mg/m^2 of C225 along with chemoradiation achieving disease stabilization (BASELGA et al. 2000). Equally interesting are reports like those from GLASER et al. (2001) suggesting that locoregional control benefits when patients are treated with neoadjuvant radiation (50 Gy) and chemotherapy (5-FU and mitomycin C) with human recombinant erythropoietin prior to radical tumor resection. Results of a retrospective comparison of anemic patients (Hb<14.5 g/dl pretreatment) treated similarly with or without erythropoietin reveal that the pathological complete response rate increased from 17% to 61% (P<0.001) with the addition of the drug. This has led to a significantly improved rate of local control (P<0.001) and 2-year survival. The ongoing RTOG phase-III trial is examining this further.

A phase-III study of the EGFR-blocking antibody cetuximab in locoregionally advanced squamous cell carcinoma of the oropharynx, hypopharynx, or larynx has been reported by BONNER et al. (2004), with patients randomized to radiation alone or to radiation plus weekly cetuximab. This trial showed that the addition of cetuximab provided a statistically significant prolongation in overall survival, while showing only a minimal increase in the toxicity expected with radiotherapy.

3.4
Uterine Cervix

While the exact indications and the regimen of choice remains controversial, there is convincing evidence from recent studies that concurrent chemotherapy can improve outcome in patients requiring radiation for locally advanced cervix cancer. While several studies with debatable results have been conducted using hydroxyurea and 5-FU, the weight of the data suggests that a cisplatin-containing concurrent regimen is now the treatment of choice for many patients. There is no evidence to support the use of neoadjuvant or adjuvant chemotherapy at present (THOMAS 1999).

A GOG trial in which patients with locally advanced disease who had negative para-aortic nodes at lymphadenectomy were randomized to receive concurrent hydroxyurea plus radiation therapy versus concurrent cisplatin and FU plus radiation therapy has been completed and has shown a benefit for the cisplatin-containing arm (ROSE et al. 1999). In a second trial, two more aggressive cisplatin-containing chemotherapy regimens showed benefit over a hydroxyurea and radiation regimen (KEYS et al. 1999). Both cisplatin-containing treatments yielded dramatic, highly significant improvements in local disease control and survival.

Over the same time period the RTOG designed a trial comparing a combination of cisplatin, 5-FU and pelvic irradiation with extended-field irradiation alone in RTOG 90-01 (MORRIS et al. 1999; EIFEL et al. 2004). Patients were required to have negative para-aortic lymph nodes based on a lymphangiogram or retroperitoneal lymph node dissection. The radiation alone arm was based on a previous study that found a survival benefit when prophylactic para-aortic irradiation was added to standard pelvic irradiation. This trial was published early, after an interim analysis revealed a highly significant improvement in overall survival, disease-free survival, local disease control, and rate of freedom from distant metastases in the combined modality arm. A later update (EIFEL et al. 2004) confirmed the initial observations.

Two further trials (PEARCEY et al. 2000; PETERS et al. 2000) have demonstrated a survival benefit when cisplatin is added to radiotherapy in the setting of earlier stage disease followed by an extrafascial hysterectomy or when cisplatin is added to pelvic radiation in those patients who have already undergone a radical hysterectomy.

Of the recent trials looking at the addition of cisplatin-based chemotherapy, only the NCI Canada trial has failed to demonstrate a survival benefit to the addition of concurrent chemotherapy (PEARCEY et al. 2000). However, the authors of this study maintain that the optimization of the radiation as it was delivered in their trial may account for the lack of benefit. The issues of how to best integrate newer chemotherapies such as the taxanes, which have considerable radiation sensitization properties, targeted biological agents, and agents that may optimize the oxygenation status of the tumors are still under investigation. The differences and the possible explanations for them were carefully analyzed by LEHMAN and THOMAS (2001).

3.5
Urinary Bladder

The natural history of muscle-invasive bladder cancer is much more aggressive than superficial disease, with a 5-year survival of only 50%. The strategy of using radiation, chemotherapy or a maximal transurethral resection of a bladder tumor in isolation to achieve lasting pelvic control pales in comparison with the modern radical cystectomy. This surgical therapy yields local control in better than 90% of all cases. Issues related to overall survival benefits as well as quality of life endpoints have led to the pursuit of a combined modality strategy that incorporates all elements of therapy in an attempt to preserve organ function (COPPIN et al. 1996; KUCZYK et al. 2003).

A randomized trial to show a benefit to the addition of concurrent chemotherapy in the definitive treatment of bladder cancer is that of the NCI Canada, which showed a significant improvement in local control ($P=0.036$) and suggested a survival difference (47% versus 33%, $P=0.34$), with addition of concurrent cisplatin to local therapy (KAUFMAN et al. 1993). This study was, unfortunately, small and not adequately powered to show a survival benefit.

In 1993, investigators at the Massachusetts General Hospital published the results of a single-arm institutional study, which has become the model for several subsequent trials (Dunst et al. 1994); 53 patients underwent maximal TURBT, followed by two cycles of CMV, then 40 Gy with two cycles of concurrent single-agent cisplatin. At this point, they underwent endoscopic re-evaluation, and if they had an incomplete response to therapy they underwent a cystectomy, if medically feasible, while complete responders were consolidated with an additional 24.8 Gy and an additional cycle of cisplatin. A total of 42 patients completed therapy, and there was no chemotherapy-related mortality. Radical cystectomy was required in a total of 15 patients, including 3 that had a salvage surgery. After 48 months of follow-up, 53% of the patients were alive and 42% had no evidence of disease. An updated report on 106 patients found that 34% of patients ultimately required a salvage cystectomy, with 49% of patients alive, and 43% alive with their native bladders intact (Dunst et al. 1994).

Series from several other centers or groups (Housset et al. 1993; Tester et al. 1993; Shipley et al. 2003) have tested bladder conservation strategies with reasonable results and far more patients failing with distant disease than local-only failures. While there is not likely to be a definitive randomized trial comparing a bladder conservation approach to radical cystectomy, it would appear that combined modality therapy is not an unreasonable option for selected patients (George et al. 2004). Ongoing studies are looking at integrating the taxanes and other biological agents into therapy to improve outcome.

3.6
Anus

In the case of tumors originating in this location, the early experience of Nigro et al. (1974) suggested that there may not be a need to perform surgery as part of the initial therapy of this cancer, reserving the abdominoperineal resection for local recurrence. Three patients treated with pre-operative radiation, 5-FU and mitomycin C were found to have had a complete pathological response at the time of their surgery. This work has been expanded upon by Cummings et al. (1991) in their series of patients who were treated by various concurrent regimens over time and by several large intergroup studies (Flam et al. 1996; UKCCCR 1996; Bartelink et al. 1997) that have demonstrated the

value of concurrent chemotherapy and radiation in this disease.

The UKCCCR trial showed that the combined modality arm improved 3-year actuarial local tumor control from 29% to 61% and was significant. This trial did not show a survival benefit to the addition of chemotherapy, and the combined arm had more early grade-4 toxicity. Similar benefit in terms of local control was also seen in the EORTC trial with the addition of 5-FU and mitomycin C (5-year colostomy-free survival was 72% versus 40%). However, once again, a survival benefit was not seen. The Intergroup trial, which randomized patients to treatment with or without mitomycin C, confirmed its benefit to therapy with a higher complete response rate (92% versus 85%) and a significantly lower colostomy rate (9% versus 22%). There was, however, no significant survival benefit. The RTOG 98-11 is ongoing, examining the benefit of two cycles of induction using 5-FU and cisplatin.

The EORTC 22861 trial confirmed that radiation chemotherapy combination is the standard treatment for locally advanced anal cancer. The EORTC phase-II study no. 22953 tests the feasibility of reducing the gap between sequences to 2 weeks, to deliver mitomycin C in each radiation sequence and to administer 5-FU continuously. The initial dose was 36 Gy/4 weeks, mitomycin C (10 mg/m^2 on day 1), and 5-FU (200 mg/m^2 on days 1–26).

The second sequence consisted of 23.4 Gy/17 days combined with the same doses of chemotherapy. The 3-year results for the above trials are 68% and 88% local tumor control, 72% and 81% colostomy-free survival, and 70% and 81% survival, respectively (Bosset et al. 2003). Similar results were reported in 305 patients with stage T1–T4 anal cancer treated with external pelvic irradiation and brachytherapy (Deniand-Alexandre, 2003). Further, Chauveire et al. (2003) noted that in 67 patients with anal cancer (24 with T3–4 tumors) chemoirradiation was administered to only 55% of the patients with T3-4 lesions, mostly because they were deemed too old, emphasizing the importance of patient selection.

3.7
Esophagus

While the ideal approach to the management of locally advanced disease is controversial, the evidence from several randomized trials shows that chemoradiation is associated with an improved survival when compared with radiation alone (AL SARRAF et al. 1997; SMITH et al. 1998). The Intergroup trial RTOG 85-01 has had a profound influence of patterns of practice (SMITH et al. 1998). In this phase-III study, patients were randomized to treatments consisting of 5-FU (1000 mg/m^2 per day for 96 h), cisplatin (75 mg/m^2, day 1) and 50 Gy in 25 fractions of daily irradiation starting on the first day of chemotherapy or 64 Gy of daily radiation in 2-Gy fractions. This chemotherapy was administered every 4 weeks during radiation and every 3 weeks after its completion. Concurrent therapy was associated with significant benefits in terms of 5-year overall survival (26% versus 0%, $P<0.0001$) as well as decreased local failure (45% versus 68%, $P=0.0123$). Concurrent therapy as delivered in this trial is associated with significant toxicity, i.e., 20% grade-4 toxicity and one treatment-related death. Based on these studies, concurrent chemotherapy and radiation has become the standard of care for the non-surgical management of esophageal cancer, with radiation alone being reserved for those patients unable to tolerate the addition of chemotherapy. Intensifying the radiation to deliver a dose of 64.8 Gy was more recently tested in a phase-III intergroup study (INT 0123/RTOG 94-05) with the intent of further improving local control and, potentially, survival. Preliminary results revealed no significant difference with the intensified radiation in the median survival (12.9 months versus 17.6 months), the 2-year survival (29% versus 38%) or the locoregional failure rates (59% versus 52%). After the first interim analysis, the trial was ended (MINSKY et al. 2000). Efforts to improve primary chemoradiation through the incorporation of novel radiosensitizing chemotherapies continue.

The realm of treating resectable esophageal cancer is far less clear. While several randomized studies have been completed comparing neoadjuvant chemoradiation followed by esophagectomy to surgery alone, they either are underpowered, have used unconventional fractionation schemes, have used split course radiation, or may have had unbalanced treatment arms (WALSH et al. 1996; Bosset et al. 1997; URBA et al. 2001). What may have been the more definitive trial (the CALGB 9781) was closed due to poor accrual. Given that the evidence from these randomized trials is plagued by design-related issues and conflicting results, the literature does not clearly support the use of pre-operative chemoradiation outside of a clinical trial. Efforts to incorporate new agents including the taxanes, UFT, and irinotecan are ongoing.

3.8
Rectum

The location of the rectum within the confines of the bony pelvis and its intimate relationships with adjacent organs make resection of these tumors with wide radial margins difficult unless a total mesorectal excision is undertaken. Tumors originating in the rectum are often associated with a higher risk of local failure than extrapelvic colon cancers on a stage-by-stage basis. Neoadjuvant therapies are often given in an effort to improve the respectability of tumors and to hopefully increase sphincter preservation rates. The value of adjuvant therapy with chemoradiation to improve local control as well as overall survival is well recognized.

The GITSG has performed a four-arm randomized trial looking at the benefit of adjuvant therapies with patients allocated to surgery alone, post-operative pelvic radiation, post-operative chemotherapy, or both post-operative chemotherapy and radiation in patients with B2 or C disease (GASTROINTESTINAL TUMOR STUDY GROUP 1985). Local recurrence was reduced in the chemoradiation arm from 25% to 11%, and overall survival was improved from 14% to 56%. Confirmation of these results was found with the NSABP-R01 (FISHER et al. 1988) and, as such, the National Institutes of Health issued a clinical announcement recommending adjuvant 5-FU-based chemotherapy and concurrent radiation therapy for patients with stage B2, B3, or C rectal cancer (NIH CONSENSUS CONFERENCE 1990). Although these initial trials included semustine, several randomized studies examined the value of adding this drug to therapy and their results have led to the dropping of semustine from adjuvant treatment, as it was not associated with a significant benefit (GASTROINTESTINAL TUMOR STUDY GROUP 1992; O'CONNELL et al. 1994; WOLMARK et al. 2000).

An Intergroup trial (INT-0114) randomized patients to pelvic irradiation plus 6 months of bolus 5-FU versus bolus 5-FU and levamisole, leucovorin, or both (TEPPER et al. 1997). A preliminary analysis revealed no significant difference in disease-free survival or overall survival (78–80%) among the four treatment arms and, as expected, toxicity was greatest with the three-drug regimen. Efforts to further refine the delivery of 5-FU-based therapy are ongoing and include the use of a prolonged venous infusion, which is associated with less myelosuppression but more diarrhea (MEHTA et al. 2003). The current intergroup trial examines the benefit of this regimen relative to bolus infusion of the drug with leucovorin and levamisole.

The value of pre-operative therapy is well recognized, and includes sphincter preservation and less bowel-related toxicity, at the possible expense of over treatment and increased wound-healing difficulties. A German trial randomized patients staged by means of endoscopic ultrasound to pre-operative or post-operative radiation with concurrent 5-FU by prolonged venous infusion for two cycles followed by maintenance 5-FU (SAUER et al. 2000). An interim analysis of toxicity reported fewer cases of grade-3+ diarrhea with pre-operative chemoradiation, and no difference in post-surgical complications. In a subsequent report, SAUER et al. (2004) updated results of the trial, in which 421 patients were randomly assigned to receive pre-operative and 402 post-operative chemoradiotherapy. The 5-year local relapse rate was 6% and 13%, respectively ($P=0.006$), and overall survival was 76% and 74% ($P=0.8$). Grade-3 to grade-4 acute toxicity occurred in 27% of the patients in the pre-operative group compared with 40% with post-operative chemoradiotherapy ($P=0.001$). The incidence of long-term sequelae was 14% and 24%, respectively ($P=0.01$). This trial documented improved local tumor control and a reduction in treatment morbidity with pre-operative chemoradiotherapy, although there was no impact on overall survival. WILLETT et al. (2004) have shown that the concurrent administration of the VEGF-specific antibody bevacizumab with 5-FU and radiation to the pelvis showed antivascular effects in human rectal cancer.

There have been several reported publications regarding pre-operative chemoirradiation in rectal cancer. BOSSET et al. (2004) reported on the toxicity of the EORTC 22921 protocol, a four-arm study that compared pre-operative chemoirradiation with pre-operative irradiation alone or post-operative chemotherapy with no further therapy in T3-T4 M0 resectable rectal cancer. In the various groups, 6 patients died pre-operatively and 8 within 30 days after surgery. The addition of 5-FU and Leucovorin to pre-operative irradiation slightly increased acute toxicity, without affecting compliance to the radiation protocol.

At MD Anderson Cancer Center, CRANE et al. (2003) reported on 403 patients with clinical stage T3-T4 rectal cancer treated with pre-operative radiation therapy alone or combined with concurrent chemotherapy (continuous infusion of 5-FU). The use of concurrent chemotherapy and irradiation improved tumor response and down staging (62%

versus 42%) with radiation alone ($P=0.001$). BONNEN et al. (2004), from the same institution, published the results of a study of 431 patients with clinical stage-T3 rectal cancer treated with pre-operative chemoradiation followed by surgical resection. Similar pre-operative treatment was followed by total surgical excision in 405 patients. With 46 months mean follow-up, two intrapelvic recurrences were observed in 26 patients treated with local excision (6% at 5 years), compared with 8% pelvic recurrence in patients treated with mesorectal-excision and 6% in a subgroup of patients with complete clinical response to pre-operative chemoradiation treated with mesorectal excision. Actuarial overall 5-year survival was 86% in the local excision group compared with 81% in the mesorectal-excision patients, and 85% in the patients with complete clinical response to chemoradiation followed by mesorectal excision by abdominal perineal resection or lower anterior resection.

Similar results were reported by BUJKO et al. (2004) in a randomized trial comparing short-term radiation therapy with conventional fractionation chemoradiation. In the study in which 316 patients were randomized, the sphincter preservation rate was 61% in the short-term pre-operative irradiation group and 58% in the radiochemotherapy arm ($P=0.57$).

Some studies have addressed the efficacy of post-operative radiation therapy alone or combined with chemotherapy. CAFIERO et al. (2003) reported on a study of 218 patients randomized to be treated with 50 Gy/2-Gy fractions administered post-operatively or a treatment consisting of 5-FU plus levamisole and radiation therapy to the same dose schedule as above. There was no significant difference with regard to disease-free survival or overall survival in the two arms. The local-regional recurrence rate was about 9%. The chemoirradiation regimen was associated with higher toxicity, which impaired patient compliance to chemotherapy.

A number of new chemotherapeutic agents are being used in the treatment of patients with colorectal cancer, usually in combination with pelvic radiation therapy. Among the newer drugs, oxaliplatin-based drugs or irinotecan and capecitabine have been used, frequently combined with 5-FU (MINSKY 2004a, b).

ZHU and WILLETT (2005) recently reviewed the gains that have been achieved with the integration with radiation therapy, chemotherapy, and surgery in the management of patients with localized rectal cancer.

3.8.1
Central Nervous System

High-grade gliomas such as glioblastoma multiforme are aggressive malignancies, with very poor control and survival rates. Therapy has traditionally been based on surgical extirpation to the maximal extent feasible, followed by limited field radiotherapy. A variety of chemotherapeutic regimens, mostly based around a nitrosourea, have been employed either concomitantly with radiation or sequentially, but the role of chemotherapy has remained controversial (LONARDI et al. 2005). However, a recent, phase-III randomized EORTC study compared a regimen of irradiation with concurrent and adjuvant temozolomide to radiotherapy alone (STUPP et al. 2005). This study showed a clinically meaningful and statistically significant survival benefit for the chemoirradiation arm, with the median survival being 14.6 months with radiotherapy plus temozolomide versus 12.1 months for radiotherapy alone. The 2-year survival in the combined modality arm was 26.5%, compared with only 10.4% for the radiotherapy alone arm. Toxicity with the combined regimen was modest.

3.9
Conclusion

The preceding discussion has been a brief overview of the varied ways in which chemotherapy and radiotherapy can be combined in treating patients with cancer. The practice of chemoirradiation is truly multidisciplinary and, as knowledge in the field expands, should play an increasing role in the control of neoplasms. As molecular biology brings us new targeted therapies, it is hoped that existing treatment regimens can be improved, in terms of both survival and ameliorating acute and late toxicity of therapy.

References

Al Sarraf M, Martz K, Herskovic A et al (1997) Progress report of combined chemoradiotherapy versus radiotherapy alone in patients with esophageal cancer: an intergroup study. J Clin Oncol 15:277–284

Al-Sarraf M, LeBlanc M, Giri PG et al (1998) Chemoradiotherapy versus radiotherapy in patients with advanced nasopharyngeal cancer: phase III randomized intergroup study 0099. J Clin Oncol 16:1310–1317

Anonymous (1991) Induction chemotherapy plus radiation compared with surgery plus radiation in patients with advanced laryngeal cancer. The Department of Veterans Affairs Laryngeal Cancer Study Group. N Engl J Med 324:1685–1690

Auperin A, Arriagada R, Pignon JP et al (1999) Prophylactic cranial irradiation for patients with small-cell lung cancer in complete remission. Prophylactic Cranial Irradiation Overview Collaborative Group. N Engl J Med 341:476–484

Bagshaw MA (1961) Possible role of potentiators in radiation therapy. Am J Roentgenol 85:822–833

Bartelink H, Roelofsen F, Eschwege F et al (1997) Concomitant radiotherapy and chemotherapy is superior to radiotherapy alone in the treatment of locally advanced anal cancer: results of a phase III randomized trial of the European Organization for Research and Treatment of Cancer Radiotherapy and Gastrointestinal Cooperative Groups. J Clin Oncol 15:2040–2049

Baselga J, Pfister D, Cooper MR et al (2000) Phase I studies of anti-epidermal growth factor receptor chimeric antibody C225 alone and in combination with cisplatin. J Clin Oncol 18:904–914

Begg AC, van der Kolk PJ, Emondt J et al (1987) Radiosensitization in vitro by cis-diammine (1,1-cyclobutanedicarboxylato) platinum(II) (carboplatin, JM8) and ethylenediammine-malonatoplatinum(II) (JM40). Radiother Oncol 9:157–165

Bonnen M, Crane C, Vauthey JN et al (2004) Long-term results using local excision after preoperative chemoradiation among selected T3 rectal cancer patients. Int J Radiat Oncol Biol Phys 60:1098–1105

Bonner J, Harari P, Giralt N et al (2004) Cetuximab prolongs survival in patients with locoregionally advanced squamous cell carcinoma of head and neck: a phase III study of high dose radiation therapy with or without cetuximab. J Clin Oncol Annu Meeting Proc Abstract no. 5507

Bosset JF, Gignoux M, Triboulet JP et al (1997) Chemoradiotherapy followed by surgery compared with surgery alone in squamous-cell cancer of the esophagus. N Engl J Med 337:161–167

Bosset JF, Roelofsen F, Morgan DA, Budach V, Coucke P, Jager JJ, Van der Steen-Banasik E, Triviere N, Stuben G, Puyraveau M, Mercier M; European Organization for Research and Treatment of Cancer. Radiotherapy and Gastrointestinal Cooperative Groups (2003) Shortened irradiation scheme, continuous infusion of 5-fluorouracil and fractionation of mitomycin C in locally advanced anal carcinomas. Results of a phase II study of the European Organization for Research and Treatment of Cancer. Radiotherapy and Gastrointestinal Cooperative Groups. Eur J Cancer 39:45–51

Bosset JF, Calais G, Daban, A et al (2004) Preoperative chemoradiotherapy versus preoperative radiotherapy in rectal cancer patients: assessment of acute toxicity and treatment compliance. Report of the 22921 randomised trial conducted by the EORTC Radiotherapy Group. Eur J Cancer 40:219–224

Bourhis J, Pignon JP (1999) Meta-analyses in head and neck squamous cell carcinoma. What is the role of chemotherapy? Hematol Oncol Clin North Am 13:769–775, vii

Bradley J, Graham MV, Winter K, Purdy JA, Komaki R, Roa WH, Ryu JK, Bosch W, Emami B (2005) Toxicity and outcome results of RTOG 9311: a phase I–II dose-escalation study using three-dimensional conformal radiotherapy in patients with inoperable non-small-cell lung carcinoma. Int J Radiat Oncol Biol Phys 61:318–328

Bristow RG, Hill RP (1998) Molecular and cellular basis of radiotherapy. In: Tannock IF, Hill RP (eds) The basic science of oncology, 3rd edn. McGraw-Hill, Montreal, pp 295–321

Buchholz DJ, Lepek KJ, Rich TA et al (1995) 5-Fluorouracil-radiation interactions in human colon adenocarcinoma cells. Int J Radiat Oncol Biol Phys 32:1053–1058

Bujko K, Nowacki MP, Nasierowska-Guttmejer A et al (2004) Sphincter preservation following preoperative radiotherapy for rectal cancer: report of a randomized trial comparing short-term radiotherapy versus conventionally fractionated radiochemotherapy. Radiother Oncol 72:15–24

Cafiero F, Gipponi M, Lionetto R et al (2003) Randomised clinical trial of adjuvant postoperative RT vs. sequential postoperative RT plus 5-FU and levemisole in patients with stage II–III respectable rectal cancer: a final report. J Surg Oncol 83:140–146

Cakir S, Egehan I (2004) A randomised clinical trial of radiotherapy plus cisplatin versus radiotherapy alone in stage III non-small cell lung cancer. Lung Cancer 43:309–316

Choy H, Akerley W, Safran H et al (1998a) Multiinstitutional phase II trial of paclitaxel, carboplatin, and concurrent radiation therapy for locally advanced non-small-cell lung cancer. J Clin Oncol 16:3316–3322

Choy H, Safran H, Akerley W et al (1998b) Phase II trial of weekly paclitaxel and concurrent radiation therapy for locally advanced non-small cell lung cancer. Clin Cancer Res 4:1931–1936

Choy H et al (1993) Investigation of taxol as potential radiation sensitizer. Cancer 71:3774–3778

Coppin CML, Gospodarowicz M, James K et al (1996) Improved local control of invasive bladder cancer by concurrent cisplatin and preoperative or definitive radiation. J Clin Oncol 14:2901–2907

Crane CH, Skibber JM, Birnbaum EH et al (2003) The addition of continuous infusion 5-FU to preoperative radiation therapy increases tumor response, leading to increased sphincter preservation in locally advances rectal cancer. Int J Radiat Oncol Biol Phys 57:84–89

Cummings BJ, Keane TJ, O'Sullivan B et al (1991) Epidermoid anal cancer: treatment by radiation alone or by radiation and 5-fluorouracil with and without mitomycin C. Int J Radiat Oncol Biol Phys 21:115–125

Curran W, Scott C, Langer C et al (2000) Phase III comparison of sequential vs concurrent chemo-radiation for pts with unresected stage III non small cell lung cancer (NSCLC): report of Radiation Therapy Oncology Group (RTOG) 9410 (abstract 303). Lung Cancer 29[Suppl 1]:93

Dautzenberg B, Arriagada R, Chammard AB et al (1999) A controlled study of postoperative radiotherapy for patients with completely resected nonsmall cell lung carcinoma. Groupe d'Etude et de Traitement des Cancers Bronchiques. Cancer 86:265–273

Dewit L (1987) Combined treatment of radiation and cis-diamminedichloroplatinum (II): a review of experimental and clinical data. Int J Radiat Oncol Biol Phys 13:403–426

Dillman RO, Herndon J, Seagren SL et al (1996) Improved survival in stage III non-small-cell lung cancer: seven-year follow-up of Cancer and Leukemia Group B (CALGB) 8433 trial. J Natl Cancer Inst 88:1210–1215

Dunst J, Sauer R, Schrott KM et al (1994) An organ sparing treatment of advanced bladder carcinoma: a 10-year experience. Int J Radiat Oncol Biol Phys 30:261–266

Eifel PJ, Winter K, Morris M, Levenback C, Grigsby PW, Cooper J, Rotman M, Gershenson D, Mutch DG (2004) Pelvic irradiation with concurrent chemotherapy versus pelvic and para-aortic irradiation for high-risk cervical cancer: an update of radiation therapy oncology group trial (RTOG) 90-01. J Clin Oncol 22:872–880

Elkind MM, Sutton HF (1959) X-ray damage and recovery in mammalian cells in culture. Nature 184:1293

El-Sayed S, Nelson N (1996) Adjuvant and adjunctive chemotherapy in the management of squamous cell carcinoma of the head and neck region. A meta-analysis of prospective and randomized trials. J Clin Oncol 14:838–847

Fisher B, Wolmark N, Rockette H et al (1988) Postoperative adjuvant chemotherapy or radiation therapy for rectal cancer: results from NSABP protocol R-01. J Natl Cancer Inst 80:21–29

Flam M, John M, Pajak TF et al (1996) Role of mitomycin in combination with fluorouracil and radiotherapy, and of salvage chemoradiation in the definitive nonsurgical treatment of epidermoidcarcinoma of the anal canal: results of a phase III randomized intergroup study. J Clin Oncol 14:2527–2539

Forastiere AA, Goepfert H, Maor M, Pajak TF, Weber R, Morrison W, Glisson B, Trotti A, Ridge JA, Chao C, Peters G, Lee DJ, Leaf A, Ensley J, Cooper J (2003) Concurrent chemotherapy and radiotherapy for organ preservation in advanced laryngeal cancer. N Engl J Med 349:2091–2098

Freyer G, Bossard N, Romestaing P et al (2001) Addition of oxaliplatin to continuous fluorouracil, l-folinic acid, and concomitant radiotherapy in rectal cancer: the Lyon R 97-03 phase I trial. J Clin Oncol 19:2433–2438

Fried DB, Morris DE, Poole C et al (2004) A metaanalysis of the timing of chest irradiation in combined modality treatment of limited-stage small cell lung cancer. J Clin Oncol 22:4837–4845

Fu KK, Pajak TF, Trotti A et al (2000) A Radiation Therapy Oncology Group (RTOG) phase III randomized study to compare hyperfractionation and two variants of accelerated fractionation to standard fractionation radiotherapy for head and neck squamous cell carcinomas: first report of RTOG 9003. Int J Radiat Oncol Biol Phys 48:7–16

Furuse K, Fukuoka M, Kawahara M et al (1999) Phase III study of concurrent versus sequential thoracic radiotherapy in combination with mitomycin, vindesine, and cisplatin in unresectable stage III non-small-cell lung cancer. J Clin Oncol 17:2692–2699

Gastrointestinal Tumor Study Group (1985) Prolongation of the disease-free interval in surgically treated rectal carcinoma. N Engl J Med 312:1465–1472

Gastrointestinal Tumor Study Group (1992) Radiation therapy and fluorouracil with or without semustine for the treatment of patients with surgical adjuvant adenocarcinoma of the rectum. J Clin Oncol 10:549–557

George L, Bladou F, Bardou VJ, Gravis G, Tallet A, Alzieu C, Serment G, Salem N (2004) Clinical outcome in patients with locally advanced bladder carcinoma treated with conservative multimodality therapy. Urology 64:488–493

Gilman A (1963) The initial clinical trial of nitrogen mustard. Am J Surg 105:574–578

Glaser C, Millesi W, Kornek G et al (2001) Impact of hemoglobin level and the use of recombinant erythropoietin on efficacy of preoperative chemoradiation therapy for squamous cell carcinoma of the oral cavity and oropharynx. Int J Radiat Oncol Biol Phys 50:705–715

Gregoire V, Hittelman WN, Rosier JF, Milas L (1999) Chemoradiotherapy: radiosensitizing nucleoside analogues. Oncol Rep 6:949–957

Hall EJ (1994) Radiobiology for the radiobiologist, 4th edn. Lippincott, Philadelphia

Heidelberger C, Chaudhari N, Dannenberg P et al (1957) Fluorinated pyrimidines: new class of tumor-inhibiting compounds. Nature 179:663

Holloway CL, Robinson D, Murray B, Amanie J, Butts C, Smylie M, Chu K, McEwan AJ, Halperin R, Roa WH (2004) Results of a phase I study to dose escalate using intensity modulated radiotherapy guided by combined PET/CT imaging with induction chemotherapy for patients with non-small cell lung cancer. Radiother Oncol 73:285–287

Horiot JC, Le Fur R, N'Guyen T et al (1992) Hyperfractionation versus conventional fractionation in oropharyngeal carcinoma: final analysis of a randomized trial of the EORTC cooperative group of radiotherapy. Radiother Oncol 25:231–241

Housset M, Maulard C, Chretien YC et al (1993) Combined radiation and chemotherapy for invasive transitional cell carcinoma of the bladder: a prospective study. J Clin Oncol 11:2150–2157

Hsiang YH, Liu LF (1988) Identification of mammalian DNA topoisomerase I as an intracellular target of the anticancer drug camptothecin. Cancer Res 48:1722–1726

Hsiang YH, Hertzberg R, Hecht S et al (1985) Camptothecin induces protein-linked DNA breaks via mammalian DNA topoisomerase I. J Biol Chem 260:14873–14878

Kaufman DS, Shipley WU, Griffin PP et al (1993) Selective bladder preservation by combination treatment of invasive bladder cancer. N Engl J Med 329:1377–1382

Keller SM, Adak S, Wagner H et al (2000) A randomized trial of postoperative adjuvant therapy in patients with completely resected stage II or IIIA non-small-cell lung cancer. Eastern Cooperative Oncology Group. N Engl J Med 343:1217–1222

Keys HM, Bundy BN, Stehman FB et al (1999) Cisplatin, radiation, and adjuvant hysterectomy compared with radiation and adjuvant hysterectomy for bulky stage IB cervical carcinoma. N Engl J Med 340:1154–1161

Kinsella TJ, Dobson PP, Mitchell JB, Fornace AJ Jr. (1987) Enhancement of X ray induced DNA damage by pre-treatment with halogenated pyrimidine analogs. Int J Radiat Oncol Biol Phys 13:733–739

Komaki R, Lee JS, Milas L, Lee HK, Fossella FV, Herbst RS, Allen PK, Liao Z, Stevens CW, Lu C, Zinner RG, Papadimitrakopoulou VA, Kies MS, Blumenschein GR Jr, Pisters KM, Glisson BS, Kurie J, Kaplan B, Garza VP, Mooring D, Tucker SL, Cox JD (2004) Effects of amifostine on acute toxicity from concurrent chemotherapy and radiotherapy for inoperable non-small-cell lung cancer: report of a randomized comparative trial. Int J Radiat Oncol Biol Phys 58:1369–1377

Kotalik J, Yu E, Markman BR et al (2001) Practice guideline on prophylactic cranial irradiation in small-cell lung cancer. Int J Radiat Oncol Biol Phys 50:309–316

Kuczyk M, Turkeri L, Hammerer P, Ravery V, European Society for Oncological Urology (2003) Is there a role for bladder preserving strategies in the treatment of muscle-invasive bladder cancer? Eur Urol 44:57–64

Lawrence TS, Eisbruch A, Shewach DS (1997) Gemcitabine-mediated radiosensitization. Semin Oncol 24:S-7,24–S7,28

Le Chevalier T, Arriagada R, Quoix E et al (1991) Radiotherapy alone versus combined chemotherapy and radiotherapy in nonresectable non-small-cell lung cancer: first analysis of a randomized trial in 353 patients. J Natl Cancer Inst 83:417–423

Lefebvre JL, Chevalier D, Luboinski B et al (1996) Larynx preservation in pyriform sinus cancer: preliminary results of a European Organization for Research and Treatment of Cancer phase III trial. EORTC Head and Neck Cancer Cooperative Group. J Natl Cancer Inst 88:890–899

Lehman M, Thomas G (2001) Is concurrent chemotherapy and radiotherapy the new standard of care for locally advanced cervical cancer? Int J Gynecol Cancer 11:87–99

Little JB, Hahn GM, Frindel E, Tubiana M (1973) Repair of potentially lethal radiation damage in vitro and in vivo. Radiology 106:689

Lonardi S, Tosoni A, Brandes A (2005) Adjuvant chemotherapy in the treatment of high grade gliomas. Cancer Treat Rev 31:79–89

Mason KA, Komaki R, Cox JD, Milas L (2001) Biology-based combined-modality radiotherapy: workshop report. Int J Radiat Oncol Biol Phys 50:1079–1089

Mauceri HJ, Hanna NN, Beckett MA et al (1998) Combined effects of angiostatin and ionizing radiation in antitumour therapy. Nature 394:287–291

Mehta VK, Cho C, Ford JM, Jambalos C, Poen J, Koong A, Lin A, Bastidas JA, Young H, Dunphy EP, Fisher G (2003) Phase II trial of preoperative 3D conformal radiotherapy, protracted venous infusion 5-fluorouracil, and weekly CPT-11, followed by surgery for ultrasound-staged T3 rectal cancer. Int J Radiat Oncol Biol Phys 55:132–137

Mendelsohn J, Fan Z (1997) Epidermal growth factor receptor family and chemosensitization. J Natl Cancer Inst 89:341–343

Milas L, Nakayama T, Hunter N et al (1994) Dynamics of tumor cell clonogen repopulation in a murine sarcoma treated with cyclophosphamide. Radiother Oncol 30:247–253

Milas L, Hunter NR, Mason KA et al (1995) Role of reoxygenation in induction of enhancement of tumor radioresponse by paclitaxel. Cancer Res 55:3564–3568

Milas L, Fujii T, Hunter N et al (1999a) Enhancement of tumor radioresponse in vivo by gemcitabine. Cancer Res 59:107–114

Milas L, Milas MM, Mason KA (1999b) Combination of taxanes with radiation: preclinical studies. Semin Radiat Oncol 9:12–26

Minsky BD (2004a) Combined-modality therapy of rectal cancer with irinocetan-based regimens. Oncology (Huntingt) 18:49–55

Minsky BD (2004b) Combined-modality therapy of rectal cancer with oxaliplatin-based regimens. Clin Colorectal Cancer 4[Suppl 1]:S29–S36

Minsky BD, Berkey B, Kelsen D et al (2000) Preliminary results of Intergroup INT 0123 Randomized Trial of Combined Modality Therapy (CMT) for esophageal cancer: standard vs. high dose radiation therapy. Proc Am Soc Clin Oncol 19:239a

Morris M, Eifel PJ, Lu J et al (1999) Pelvic radiation with concurrent chemotherapy compared with pelvic and para-aortic radiation for high-risk cervical cancer. N Engl J Med 340:1137–1143

Munro AJ (1995) An overview of randomised controlled trials of adjuvant chemotherapy in head and neck cancer. Br J Cancer 71:83–91

Nigro ND, Vaitkevicius VK, Considine B Jr (1974) Combined therapy for cancer of the anal canal: a preliminary report. Dis Colon Rectum 17:354–356

NIH Consensus Conference (1990) Adjuvant therapy for patients with colon and rectal cancer. JAMA 264:1444–1450

O'Connell MJ, Martenson JA, Wieand HS et al (1994) Improving adjuvant therapy for rectal cancer by combining protracted-infusion fluorouracil with radiation therapy after curative surgery. N Engl J Med 331:502–507

O'Hara JA, Double EB, Richmond RC (1986) Enhancement of radiation-induced cell kill by platinum complexes (carboplatin and iproplatin) in V79 cells. Int J Radiat Oncol Biol Phys 12:1419–1422

Pearcey RG, Brundage MD, Drouin P et al (2000) A clinical trial comparing concurrent cisplatin and radiation therapy versus radiation alone for locally advanced squamous cell carcinoma of the cervix carried out by the National Cancer Institute of Canada Clinical Trials Group. Proc Am Soc Clin Oncol 19:378a

Penland SK, Socinski MA.(2004) Management of unresectable stage III non-small cell lung cancer: the role of combined chemoradiation. Semin Radiat Oncol 14:326–334

Peters WA 3rd, Liu PY, Barrett RJ 2nd et al (2000) Concurrent chemotherapy and pelvic radiation therapy compared with pelvic radiation therapy alone as adjuvant therapy after radical surgery in high-risk early-stage cancer of the cervix. J Clin Oncol 18:1606–1613

Pignon JP, Arriagada R, Ihde DC et al (1992) A meta-analysis of thoracic radiotherapy for small-cell lung cancer. N Engl J Med 327(23):1618–1624

Pignon JP, Bourhis J, Domenge C et al (2000) Chemotherapy added to locoregional treatment for head and neck squamous-cell carcinoma: three meta-analyses of updated individual data. MACH-NC Collaborative Group. Meta-analysis of chemotherapy on head and neck cancer. Lancet 355:949–955

Plunkett W, Huang P, Xu YZ et al (1995) Gemcitabine: metabolism, mechanisms of action, and self-potentiation. Semin Oncol 22[4 Suppl 11]:3–10

Pritchard RS, Anthony SP (1996) Chemotherapy plus radiotherapy compared with radiotherapy alone in the treatment of locally advanced, unresectable, non-small-cell lung cancer. A meta-analysis. Ann Intern Med 125:723–729

Raben D, Helfrich B, Bunn PA Jr. (2004) Targeted therapies for non-small-cell lung cancer: biology, rationale, and preclinical results from a radiation oncology perspective. Int J Radiat Oncol Biol Phys 59[2 Suppl]:27–38

Radford IR (1986) Evidence for a general relationship between the induced level of DNA double-strand breakage and cell-killing after X-irradiation of mammalian cells. Int J Radiat Biol 49:611–620

Rose PG, Bundy BN, Watkins J et al (1999) Concurrent cisplatin-based chemotherapy and radiotherapy for locally advanced cervical cancer. N Engl J Med 340:1144–1153

Rowell NP, O'Rourke NP (2004) Concurrent chemoradiotherapy in non-small cell lung cancer. Cochrane Database Syst Rev 18:CD002140

Rowinsky EK (1997) The development and clinical utility of the taxane class of antimicrotubule chemotherapy agents. Annu Rev Med 48:353–374

Sauer R, Fietkau R, Martus P et al (2000) Adjuvant and neoadjuvant radiochemotherapy or advanced rectal cancer – first results of the German multicenter phase-III-trial (abstract). Int J Radiat Oncol Biol Phys 48[3 Suppl]:119a

Sauer R, Fietkau C, Wittekind C et al (2004) Pre-operative versus post-operative chemoradiotherapy for rectal cancer. N Engl J Med 351:1731–1740

Sause W, Kolesar P, Taylor S et al (2000) Final results of phase III trial in regionally advanced unresectable non-small cell lung cancer: Radiation Therapy Oncology Group, Eastern Cooperative Oncology Group, and Southwest Oncology Group. Chest 117:358–364

Scalliet P, Goor C, Galdermans D et al (1998) Gemzar (Gemcitabine) with thoracic radiotherapy – a phase II pilot study in chemonaive patients with advanced nonsmall cell lung cancer (NSCLC) (abstract). Proc Am Soc Clin Oncol 17:1923

Schaake-Koning C, Van den Bogart W, Dalesio O et al (1992) Effects of concomitant cisplatin and radiotherapy on inoperable non-small cell lung cancer. N Engl J Med 326:524–530

Schmidt-Ullrich RK, Dent P, Grant S et al (2000) Signal transduction and cellular radiation responses. Radiat Res 153:245–257

Shipley WU, Kaufman DS, Tester WJ, Pilepich MV, Sandler HM, Radiation Therapy Oncology Group (2003) Overview of bladder cancer trials in the Radiation Therapy Oncology Group. Cancer 97[8 Suppl]:2115–2119

Smith TJ, Ryan LM, Douglass HO Jr. et al (1998) Combined chemoradiotherapy vs. radiotherapy alone for early stage squamous cell carcinoma of the esophagus: a study of the Eastern Cooperative Oncology Group. Int J Radiat Oncol Biol Phys 42:269–276

Steel GG, Peckham MJ (1979) Exploitable mechanisms in combined radiotherapy-chemotherapy: the concept of additivity. Int J Radiat Oncol Biol Phys 5:85–91

Stephens T, Steel GG (1980) Regeneration of tumors after cytotoxic treatment. In: Meyn R, Withers HR (eds) Radiation biology in cancer research. Raven, New York, pp 385–295

Stupp R, Mason W, van den Bent M et al (2005) Radiotherapy plus concomitant and adjuvant temozolomide for glioblastoma. N Engl J Med 352:987–996

Teicher BA, Holden SA, Dupuis NP et al (1995) Potentiation of cytotoxic therapies by TNP-470 and minocycline in mice bearing EMT-6 mammary carcinoma. Breast Cancer Res Treat 36:227–236

Tepper JE, O'Connell MJ, Petroni GR et al (1997) Adjuvant postoperative fluorouracil-modulated chemotherapy combined with pelvic radiation therapy for rectal cancer: initial results of Intergroup 0114. J Clin Oncol 15:2030–2039

Terasima T, Tolmach LJ (1963) Variations in survival responses of HeLa cells to X-irradiation during the division cycle. Biophysics J 3:11–33

Tester W, Porter A, Asbell S et al (1993) Combined modality program with possible organ preservation for invasive bladder carcinoma: results of the RTOG protocol 85-12. Int J Radiat Oncol Biol Phys 25:783–790

Thomas GM (1999) Improved treatment for cervical cancer-concurrent chemotherapy and radiotherapy. N Engl J Med 340:1198–1200

Tishler RB, Geard CR, Hall EJ, Schiff PB (1992) Taxol sensitizes human astrocytoma cells to radiation. Cancer Res 52:3495–3497

Turrisi AT 3rd, Kim K, Blum R et al (1999) Twice-daily compared with once-daily thoracic radiotherapy in limited small-cell lung cancer treated concurrently with cisplatin and etoposide. N Engl J Med 340:265–271

UKCCCR (1996) Epidermoid anal cancer: results from the UKCCCR randomised trial of radiotherapy alone versus radiotherapy, 5-fluorouracil, and mitomycin. UKCCCR Anal Cancer Trial Working Party. UK Coordinating Committee on Cancer Research. Lancet 348:1049–1054

Urba SG, Orringer MB, Turrisi A et al (2001) Randomized trial of preoperative chemoradiation versus surgery alone in patients with locoregional esophageal carcinoma. J Clin Oncol 19:305–313

Walsh TN, Noonan N, Hollywood D et al (1996) A comparison of multimodal therapy and surgery for esophageal adenocarcinoma. N Engl J Med 335:462–467

Wang Y, Pantelias GE, Iliakis G (1994) Mechanism of radiosensitization by halogenated pyrimidines: the contribution of excess DNA and chromosome damage in BrdU radiosensitization may be minimal in plateau-phase cells. Int J Radiat Biol 66:133–142

Warde P, Payne D (1992) Does thoracic irradiation improve survival and local control in limited-stage small-cell carcinoma of the lung? A meta-analysis. J Clin Oncol 10:890–895

Willett C, Boucher Y, di Tomaso et al (2004) Direct evidence that the VEGF-specific antibody bevacizumab has antivascular effects in human rectal cancer. Nat Med 10:145

Wolmark N, Wieand HS, Hyams DM et al (2000) Randomized trial of postoperative adjuvant chemotherapy with or without radiotherapy for carcinoma of the rectum: National Surgical Adjuvant Breast and Bowel Project Protocol R-02. J Natl Cancer Inst 92:388–396

Zhu AX, Willett CG (2005) Combined modality treatment for rectal cancer. Semin Oncol 32:103–112

4 Imaging in Radiation Therapy

Recent Advances and Their Role in Radiotherapy

Filip Claus and Hedvig Hricak

CONTENTS

4.1 Introduction 57
4.2 Imaging Modalities 58
4.2.1 Computed Tomography 58
4.2.2 Magnetic Resonance Imaging 59
4.2.3 Positron Emission Tomography 64
    Addendum 66
    References 66

4.1 Introduction

In their landmark article of 1993, Black and Welch declared, "Over the past two decades a vast new armamentarium of diagnostic techniques has revolutionized the practice of medicine" (Black and Welch 1993). Another decade later, their statement is even more pertinent. However, in what they called "The Cycle of Increased Intervention," the authors warned of a possible misperception of disease prevalence and therapeutic effectiveness based on improved imaging technology, and the inherent risk of promoting a cycle of increasing medical intervention. Lead time bias (Ad1), length time bias (Ad2) and stage migration (Will Rogers phenomenon; Ad3) can distort the natural history of disease and its response to medical intervention, as has been well-illustrated in the past by screening programs for lung and breast cancer. This cycle distracts the physician from the fundamental question: how should patients with this newly detectable subclinical disease be treated? In other words,

can early imaging discriminate between indolent and aggressive disease? Some cancers progress very slowly and may not affect an individual during his or her lifetime, as is demonstrated by the high prevalence of certain tumor types in autopsy series (Sakr et al. 1994). Today, distinguishing between aggressive and indolent disease remains one of the biggest challenges in oncological imaging.

Fortunately, imaging has also made substantial progress in analyzing tumor heterogeneity and metabolism during the last decade; its role is no longer limited to detection, pretreatment staging and surveillance after curative treatment, but has expanded to include tailoring, and monitoring the response to, therapy. Conventional site-specific treatment regimens, based mainly on tumor stage, are being replaced by individualized and targeted treatment approaches based on individual tumor characteristics. Current technology for delivering 3D conformally shaped external beam radiation therapy (3D-CRT), and in particular intensity modulated radiation therapy (IMRT), may have exceeded our ability to localize tumors and normal tissues using conventional imaging techniques. With IMRT, it is possible to deliver different dose prescriptions to multiple target volumes with steep dose gradients between tumor and normal tissues. Dose escalation studies, driven by the hypothesis that dose non-uniformity within the planning target volume may lead to an increase in local control, have boosted the interest in more precise morphological and biological delineation of target volume. Increasingly important is the role of imaging in providing noninvasive, objective measures of tumor response to therapy in order to validate the biological target volume concept. This multidimensional conformal radiotherapy advocates the full incorporation of molecular medicine into the radiation planning process (Ling et al. 2000; Chapman et al. 2003; Brahme 2004).

Figure 4.1 summarizes the different roles of imaging as applied to radiotherapy. In this chapter, we will discuss some recent advances in computed tomography (CT), magnetic resonance imaging (MRI) and

F. Claus, MD, PhD
Memorial Sloan-Kettering Cancer Center, 1275 York Avenue, New York, NY 10021, USA
H. Hricak, MD, PhD
Chairman, Department of Radiology, Carroll and Milton Petrie Chair, Professor of Radiology, Weill Medical College, Cornell University, Memorial Sloan-Kettering Cancer Center, 1275 York Avenue, New York, NY 10021, USA

Pre-treatment imaging

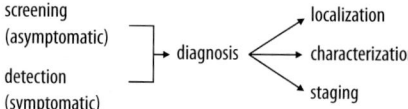

Imaging for treatment planning

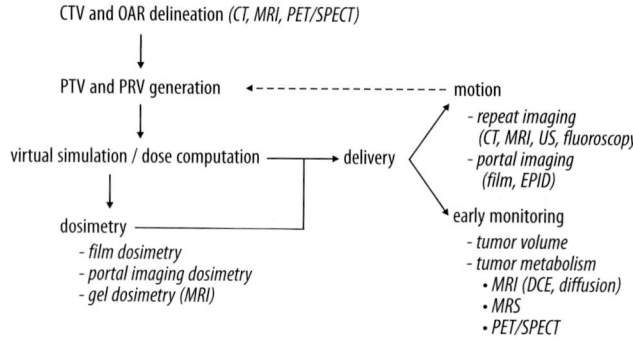

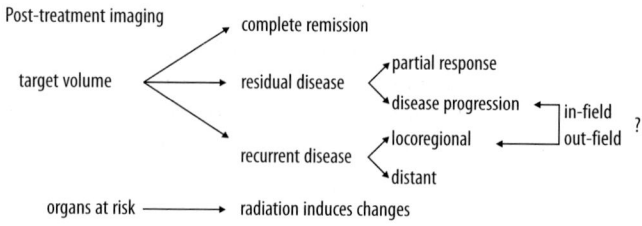

Fig. 4.1. Overview of the role of imaging in radiotherapy. *CTV* clinical target volume, *OAR* organ at risk, *PTV* planning target volume, *PRV* planning risk volume, *CT* computed tomography, *MRI* magnetic resonance imaging, *US* ultrasound, *EPID* electronic portal imaging device, *DCE* dynamic contrast-enhanced, *MRS* magnetic resonance spectroscopy, *PET* positron emission tomography, *SPECT* single photon emission computed tomography

positron emission tomography (PET), with special emphasis on their application for radiation treatment planning and early treatment monitoring.

4.2
Imaging Modalities

4.2.1
Computed Tomography

CT remains the engine for radiation treatment planning; it is the modality of choice for the delineation of target volumes and organs at risk, for virtual simulation and dose computation. Two recent advances in imaging technology have reinforced the role of CT: (1) the advent of multi-slice helical CT scanners and (2) the combined use of PET and CT.

Setup inaccuracy and internal motion, inter- and intrafraction, limit the ability to reduce margins in

radiation treatment planning. Adding margins to account for respiratory motion increases the volume of healthy tissues exposed to high doses. In efforts to reduce these margins, plans with sharper dose gradients, for example IMRT plans, are particularly susceptible to inadequate coverage of the target volume. In the case of intrathoracic (lung, mediastinum) and intra-abdominal (liver, pancreas) treatments, intrafraction motion can be significant. Four-dimensional radiotherapy refers to the addition of time to the 3D treatment process (KEALL 2004). In 4D-CT imaging, a sequence of CT image sets are acquired over consecutive segments of a breathing cycle. This allows encoding of the tumor and organ motion information in the 4D image set. The advent of multi-slice helical CT scanners, combining high-quality images with a very short acquisition time (in a single breath-hold), has greatly facilitated the use of CT for this purpose. Another advantage of acquiring a planning CT scan at one phase of the respiratory cycle is the reduction of volume and shape

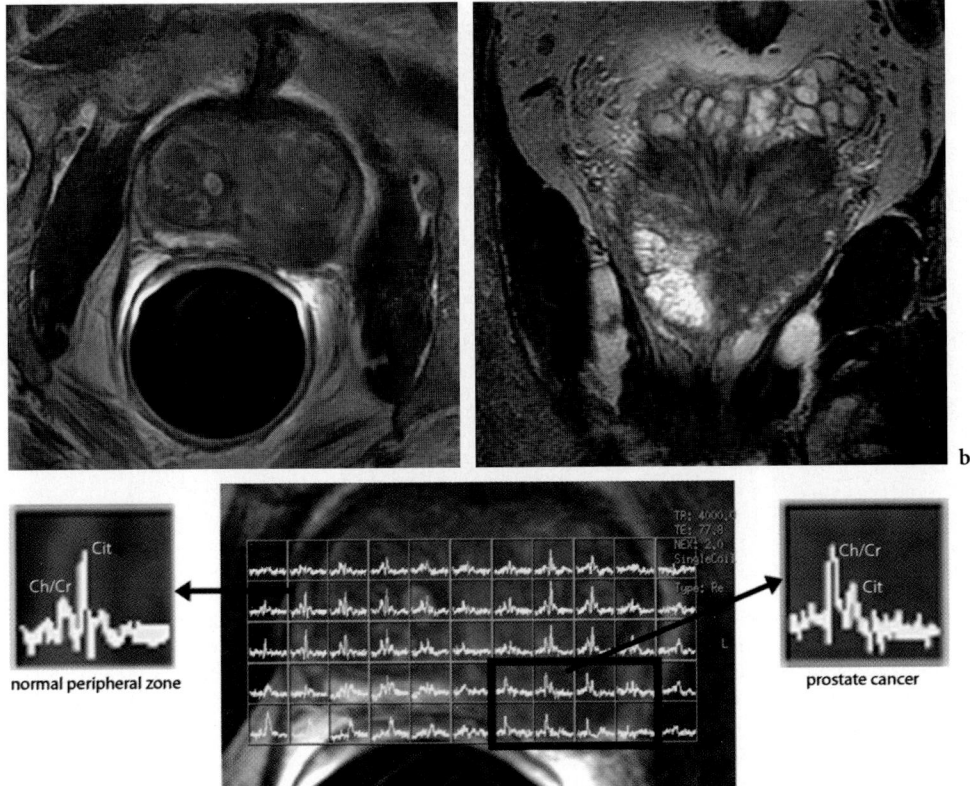

Fig. 4.2a–c. A 70-year-old male with biopsy-proven prostate adenocarcinoma (PSA 5.65 ng/ml, Gleason 8); axial (a) and coronal (b) magnetic resonance T2-weighted images show low signal intensity in the left peripheral zone extending from the base to the apex. The obliteration of the left recto-prostatic angle, visible on axial imaging, is highly suggestive of extracapsular extension of the tumor. The spectroscopic findings (c) for the axial slice in panel *a* show elevated choline and reduced citrate peaks in the voxels corresponding to the abnormal low-signal-intensity lesion in the left posterior peripheral zone (the voxels outlined in *black*). *Ch/Cr* choline/creatine, *Cit* citrate

artifacts due to motion. Repeat CT scans can also be used to determine the reproducibility of organ position, helpful in assessing the efficacy of interventions to reduce organ motion (DAWSON et al. 2001; REMOUCHAMPS et al. 2003).

PET/CT combination scanners have become increasingly popular for radiotherapy treatment imaging and planning (GREGOIRE 2004). (For a more detailed discussion, see subsection 4.2.3.) Similarly, the utility of gated PET acquisition for lung cancer, in particular the combination of 4D-CT with 4D-PET, promises to be an exciting development in radiotherapy treatment planning for lung cancer (NEHMEH et al. 2004).

4.2.2
Magnetic Resonance Imaging

MRI has become the modality of choice in a variety of diagnostic applications, in particular head and neck and pelvic disease sites. Due to its high soft tissue contrast and ability to provide information about tumor physiology and microenvironment, MRI and magnetic resonance spectroscopy (MRS) have the potential to contribute to the choice, planning and monitoring of treatment.

A good example of the utility of MRI and MRS is in the management of prostate cancer. The inclusion of MRI/MRS findings in clinical nomograms improves prediction of cancer extent, thereby improving patient selection for local therapy (COAKLEY et al. 2003; POULAKIS et al. 2004; WANG et al. 2004). MRS has also shown great promise in discriminating indolent from aggressive disease. The improved assessment of cancer extent and aggressiveness may result in a better stratification of patients in clinical trials, the possibility of monitoring the progress of patients who select watchful waiting or other minimally aggressive cancer management options, and the guidance and assessment of emerging local prostate cancer therapies (PICKETT et al. 2004).

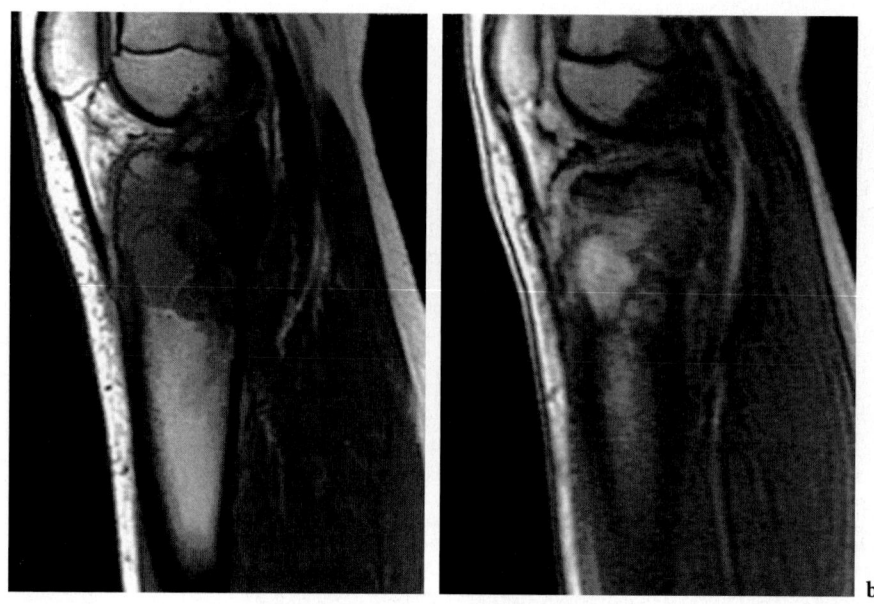

a   b

Fig. 4.3a,b. Dynamic contrast-enhanced magnetic resonance image (MRI) in a 15-year-old male with tibial osteosarcoma. *Left*: sagittal T1-weighted MRI (TR 650 ms/TE 9 ms) obtained after chemotherapy shows heterogeneous tumor in proximal tibial epiphysis and metaphysic. *Right*: dynamic gadolinium-enhanced sagittal MRI (TR 7.9/TE 1.7) shows early, intense enhancement in portion of tumor, compatible with viable tumor. Quantitative assessment of enhancement (calculated from the weighted average of the slope histogram mean values) yielded an estimated necrotic fraction of 40%. Courtesy of Dr. David Panicek, Department of Radiology, Memorial Sloan Kettering Cancer Center, New York

Because of its ability to show tumor location, extent and volume, MRI can be of tremendous help in 3D conformal or IMRT treatment planning, both of which require knowledge of tumor location, volume and extent. Several retrospective analyses and the early results of clinical trials indicate that an increased radiation dose is associated with reduced rates of biochemical failure and may therefore increase local control rates and decrease the risk of distant metastasis and the overall mortality rate (LEIBEL et al. 2002; POLLACK et al. 2004). This observation is important for management of intermediate- and high-risk prostate cancer patients; however, increased radiation doses may be associated with a risk of treatment morbidity. Both 3D-CRT and IMRT offer the possibility of combining very high radiation doses in areas of high tumor-cell density within the prostate gland without significantly increasing the risk of normal tissue damage. Further implementation of these technological advances has increased the interest in imaging techniques that are able to map tumor volume or localize more aggressive regions within the tumor, such as highly proliferating or hypoxic foci. Such imaging-optimized dose delivery has shifted radiotherapy treatment planning toward a more individualized treatment approach. MRI is more accurate than

CT in defining the prostate volume, with the highest non-agreement in the apex and posterior parts of the gland and the seminal vesicles (ROACH et al. 1996). CT-derived prostate volumes are larger than MRI-derived volumes (RASCH et al. 1999). MRI is also more accurate for the anatomic delineation of surrounding organs, such as the rectal wall, the sigmoid, the urethra and the penile bulb, which are at risk for radiation-induced tissue damage. The dose delivered to the rectal wall and bulb of the penis is significantly reduced with treatment plans based on the MRI-delineated prostate anatomy compared with treatment plans based on the CT-delineated prostate anatomy, decreasing the risk of rectal and urological complications (STEENBAKKERS et al. 2003). A more precise delineation of organs at risk also allows more accurate assessment of dose toxicity (BUYYOUNOUSKI et al. 2004). Metabolic mapping of the prostate gland by MRSI offers the possibility of tumor-targeted radiotherapy (i.e., intraprostatic dose escalation), applicable by both external radiation treatment and brachytherapy (DIBIASE et al. 2002; MIZOWAKI et al. 2002).

The value of MRS for treating cancer is not limited to prostate cancer. For other sites, such as brain, breast or liver, MRS is used to investigate tumor metabolism in order to individualize therapeutic

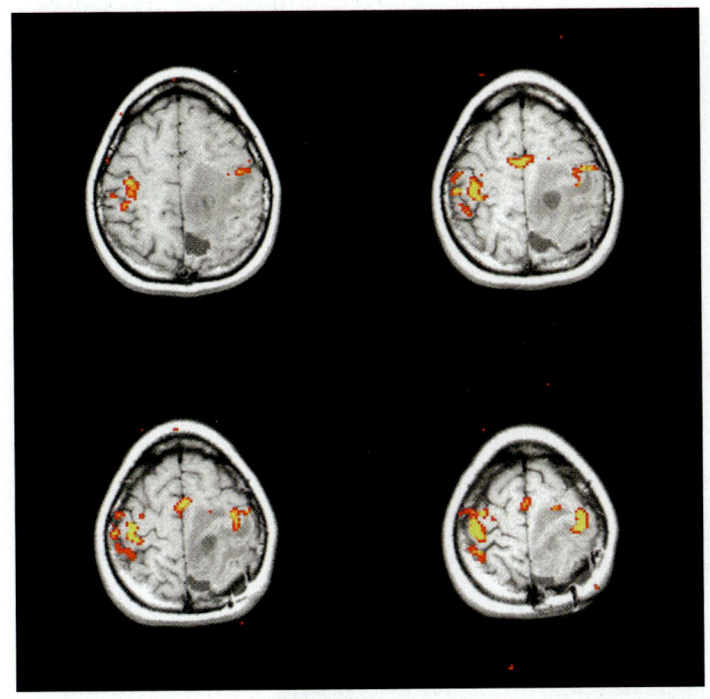

Fig. 4.4. Axial T1-weighted images with co-registered functional magnetic resonance image (fMRI) data (in *red, orange* and *yellow*) from a bilateral finger tapping paradigm. The fMRI data depict the primary motor cortices bilaterally as well as the supplementary motor areas (SMAs). The relationship of the tumor to the ipsilateral motor cortex is clearly seen. Courtesy of Dr. Andrei Holodny, Department of Radiology, Memorial Sloan-Kettering Cancer Center, New York

planning and assess treatment response (ZAKIAN et al. 2001; PIRZKALL et al. 2004). In vivo applications of phosphor MRS assess phospholipid anabolites and catabolites. A common characteristic in cancer cells and solid tumors is the elevation of phosphocholine- and total choline-containing compounds. Phospholipid precursors and catabolites are biochemical indicators of tumor progression and response to therapy. Arias-Mendoza et al. have shown that the normalized sum of phosphoethanolamine and phosphocholine measured before treatment successfully predicts long-term response to treatment and time to treatment failure in non-Hodgkin's lymphoma (ARIAS-MENDOZA et al. 2004).

Angiogenesis (the sprouting of new capillaries from existing blood vessels) and vasculogenesis (the de novo generation of blood vessels) are the two primary methods of vascular expansion that compensate for an increased nutrient demand during tumor proliferation. A number of distinguishing features characteristic of malignant vasculature can be studied by means of MRI. Examples include the spatial heterogeneity and chaotic architecture of the tumor, the replacement of normal arterioles, veins and capillaries by poorly formed leaky vessels, the high permeability to macromolecules of the latter vessels, intermittent or unstable blood flow, and the heterogeneity of vascular density and areas of spontaneous hemorrhage. By tracking the contrast uptake dynamically in malignant tissue, MRI is able to yield information on the angiogenic status of proliferating tissues. Dynamic contrast-enhanced MRI (DCE-MRI) is able to distinguish malignant from benign and normal tissues by exploiting differences in contrast agent behavior in their respective microcirculations. Several groups have reported on the value of DCE-MRI in the detection and characterization of primary and recurrent prostate cancer (BARENTSZ et al. 1999; ROUVIERE et al. 2004). Other applications include the differentiation between benign and malignant breast lesions and the estimation of the necrotic fraction of sarcomas during therapy (PADHANI 2002; DYKE et al. 2003).

MR contrast media can be distinguished in terms of contrast enhancement, paramagnetic agents generating positive contrast enhancement and super-paramagnetic agents generating negative contrast enhancement. The most widely used contrast agents are gadolinium (Gd)-based relaxation agents, paramagnetic extracellular low-molecular-weight metal-lochelates. Gd is not directly seen on an MR image but facilitates the relaxation of tissue hydrogen protons (high T1 relaxivity). The Gd ion is chelated to a ligand such as diethylenetriamine pentaacetic acid (DTPA) or tetraazacyclododecane tetraacetic acid (DOTA). These compounds distribute in the intravascular and interstitial space and allow the evaluation of physiological parameters, such as perfusion, and show a differential uptake in various tissues. The enhancement is not really specific to the type of tissue but is helpful in the classification of pathol-

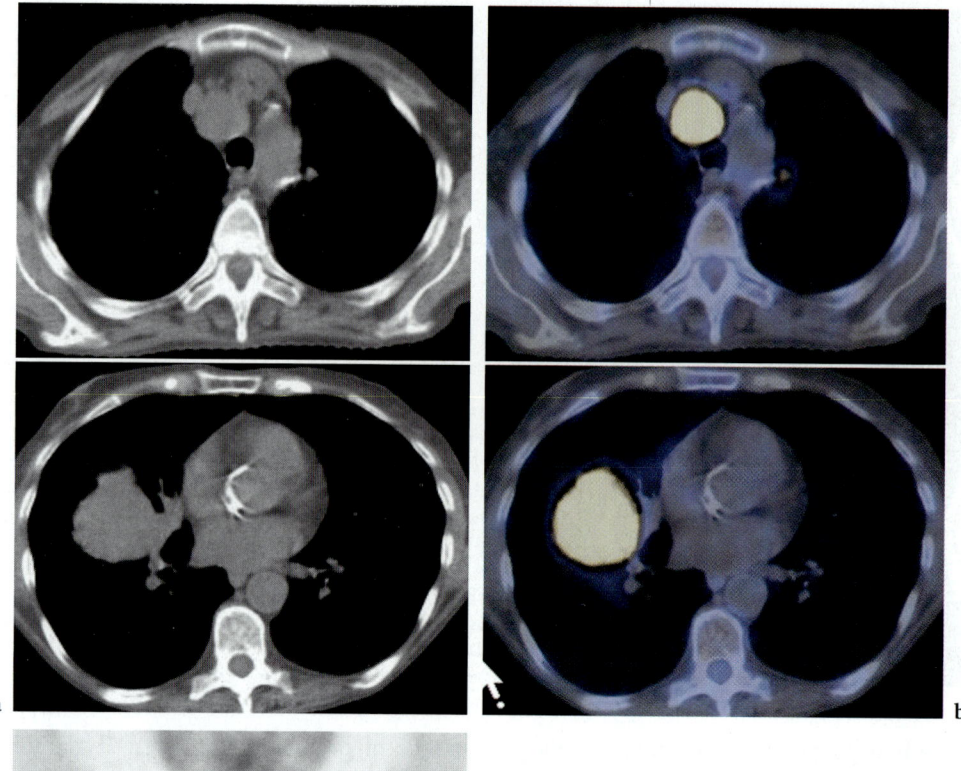

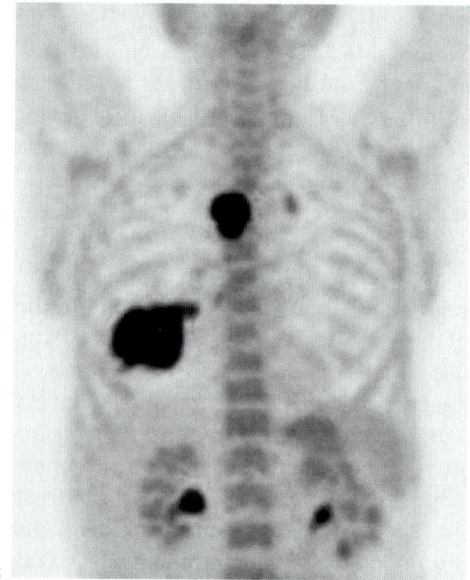

Fig. 4.5a–c. Initial staging in a 73-year-old female with lung carcinoma. **a** axial computed tomography (CT) slices at the upper and lower thoracic levels show an enlarged lymph node in the upper mediastinum and the primary tumor in the right lower lung field. **b** superimposed positron emission tomography (PET)-CT scan of the slices shown in **a**; 18F-deoxyglucose (FDG) uptake visible in the primary tumor and in the lymph node in the upper mediastinum. **c** FDG-PET scan in coronal plane showing FDG uptake visible in the primary tumor in the right lower lung field and in the lymph node in the upper mediastinum. Courtesy of Dr. Chaitanya Divgi and Dr. Neeta Pandit-Taskar, Department of Radiology, Memorial Sloan-Kettering Cancer Center, New York

ogy. Analysis can be performed dynamically (DCE-MRI) or statically (at a particular time point after administration).

Highly specific molecular paramagnetic contrast agents, targeting selected molecular epitopes, are of great interest (Artemov 2003; Weinmann et al. 2003). So far, their use is mainly investigated in preclinical models. Because of the rather low expression of antigens or receptors and the inherent limited sensitivity of MRI, compared with isotope imag-

ing, high doses of these 'molecular imaging agents' are required. The rather low contrast-to-noise ratio (which means a lack of sensitivity) can be increased by conjugating Gd chelates to carriers such as particles, dendrimers or other constructs. By delivering larger payloads of paramagnetic agents, sufficient binding can be achieved to obtain diagnostic contrast to noise levels (Morawski et al. 2004).

Ultrasmall superparamagnetic iron oxides (USPIOs) are very promising in staging lymph node

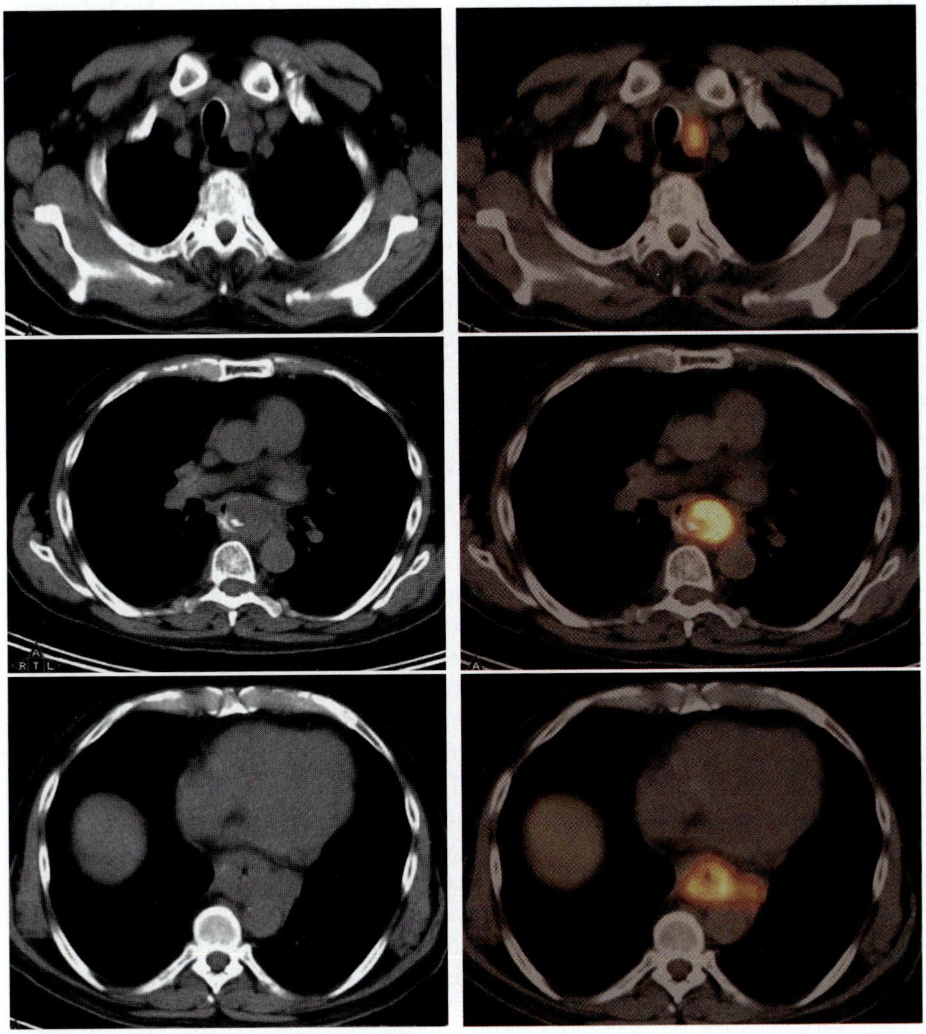

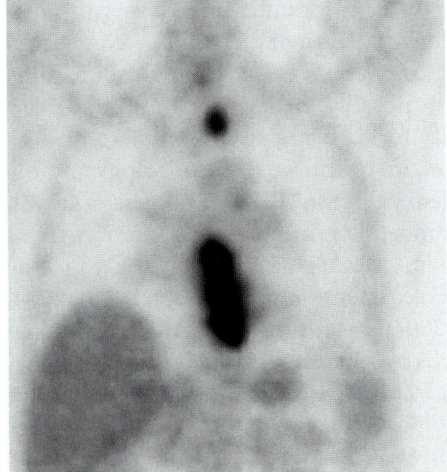

Fig. 4.6a–c. Initial staging in a 54-year-old male with esophageal carcinoma. **a** axial computed tomography (CT) slices at the upper and lower thoracic levels show an enlarged lymph node in the upper mediastinum and a thickening of the distal esophageal wall. **b** superimposed positron emission tomography (PET)-CT scan of the slices shown in the left column; 18F-deoxyglucose (FDG) uptake visible in the distal esophagus and in the lymph node in the upper mediastinum. **c** FDG-PET scan in coronal plane showing FDG uptake visible in the distal esophagus and in the lymph node in the upper mediastinum. Courtesy of Dr. Chaitanya Divgi and Dr. Neeta Pandit-Taskar, Department of Radiology, Memorial Sloan-Kettering Cancer Center, New York

disease, and clinical data for these agents have been reported (ANZAI 2004; BULTE and KRAITCHMAN 2004). Endolymphatic uptake of these lymphotropic nanoparticles results in a darkening of the lymphatic vessels and lymph nodes on the MR image, typically observed 24–36 h after IV administration (HARISINGHANI et al. 2003). It is well accepted that determination of the presence of tumor in normal or only mildly enlarged lymph nodes is impossible and results in staging inaccuracies. In prostate cancer, nodal staging is not only an important prognostic parameter, but also has a direct influence on the choice of treatment. Since the specificity and negative predictive value of nodal staging according to size are relatively low, a more accurate image-based method of distinguishing malignant from nonmalignant lymph nodes is needed. Harisinghani and colleagues demonstrated that 71% of malignant nodes detected using MRI with lymphotropic superparamagnetic nanoparticles were smaller than the threshold size (10 mm) used to identify nodal disease on conventional imaging (HARISINGHANI et al. 2003). The use of iron particles has proven to be very valuable in identifying metastatic infiltration in nodes measuring 5–10 mm, important for selecting patients for extended lymphadenectomy or for delineating radiotherapy fields. For nodes measuring less than 5 mm, detection of metastatic disease is still troublesome.

Functional MRI (fMRI), based on the different magnetic properties of oxy- and deoxyhemoglobin, is used for a variety of functional and physiological studies of the brain. In radiation treatment of brain tumors, fMRI is useful to map the function of critical structures surrounding the tumor in order to minimize radiation-induced damage of these structures (LIU et al. 2000). Diffusion-weighted magnetic resonance imaging (DWMRI) enables noninvasive characterization of biological tissues based on their water diffusion properties. Possible applications of DWMRI include the pretreatment prediction of response to radiation therapy in patients with brain tumors, the characterization of malignant breast lesions (for example cystic lesions) and the monitoring of response to therapy (for example the differentiation between viable tumor tissue and necrosis) (MARDOR et al. 2004).

Similarly to fast cine CT, high speed MRI is a suitable modality for determining the magnitude of organ motion and deformation occurring over a short time period. Examples include the use of dynamic MRI to assess the movement of prostate tumors, hepatobiliary tumors, pancreatic lesions and kidneys (MAH et al. 2002; BUSSELS et al. 2003).

4.2.3
Positron Emission Tomography

The rapid expansion in oncological applications has brought PET and molecular imaging to the clinic. Hybrid PET/CT scanners are widely installed and form the basis for merging anatomical information with functional molecular information to further advance cancer management (GAMBHIR 2002). Anatomic information and electron density maps for radiotherapy dose calculations are combined with more accurate staging and quantitative data for prognosis and therapeutic monitoring. During a combined PET/CT exam, the patient stays immobilized in the same position, which facilitates the co-registration of the data sets. Apart from patient motion during scanning, the misalignment error can be limited to values in the range of the PET spatial resolution. Due to its unique tracer specificity, only trace amounts of nuclear medicine agents are necessary for PET imaging. Although PET is rarely the modality of choice for the diagnosis of disease, it has an important role in cancer staging, diagnosis of recurrent disease and evaluation of response to therapy. Despite its inferior resolution compared with CT and MRI, PET proves to be very valuable after surgery or radiation therapy, when the normal anatomy may be distorted such that standard anatomical imaging methods may be extremely difficult to interpret. An example is the use of PET to distinguish scars from residual tumor or tumor recurrence. The development of radiolabeled markers associated with tumor prognosis and prediction of radioresistance and/or outcome are very promising. Prediction of treatment response, considered one of the major challenges in oncological imaging, would make it possible to select the optimal treatment modality for a given patient population.

PET is now in routine use in oncology mainly due to 18F-deoxyglucose (FDG). FDG has taken on increasing importance in staging, radiation treatment planning, therapeutic monitoring and the prediction of treatment outcome (ALAVI et al. 2004; ALLAL et al. 2004; DAISNE et al. 2004). FDG is taken up into the cells by glucose transporters and is subsequently trapped because of the lack of phosphorylation. Uptake of FDG corresponds to regions of higher metabolic rate (glycolysis pathway). Although FDG correlates strongly with cancer, uptake can also be explained by other causes, such as infection. The greatest advantage of imaging with FDG is probably its ability to improve staging by detecting nodal and other metastases. Functional imaging with FDG has demonstrated improved staging for many disease sites, such as non-small cell lung cancer, esophageal cancer, colorectal cancer, melanoma and lymphomas (PIETERMAN et al. 2000; KALFF et al. 2001; GAMBHIR 2002). Lardinois et al. have shown that combining PET with CT in one exam provides better results than visual correlation of PET and CT images arranged side by side or of PET alone (LARDINOIS et al. 2003). Integrated PET-CT was shown to be more accurate diagnostically in the staging of non-small cell lung cancer than PET alone, CT alone or visual correlation of PET and CT. Tumor staging and nodal staging were both significantly more accurate with integrated PET-CT, as was metastasis staging in some patients. The problem with CT and MR for staging of non-small cell lung carcinoma is that it is based on size rather than functional criteria, which results in relatively poor accuracy in all the reviewed literature. On PET imaging, nodules with metabolic activity of a standardized uptake value greater than 2.5 are generally malignant. Nevertheless, false-positive FDG uptake can occur in lung nodules (tuberculosis, other granulomata), and false-negative results can occur if the nodules are very small (less than 6 mm). Also for the detection of colorectal cancer recurrences, PET imaging is particularly important because resection of limited recurrence by metastatectomy significantly improves prognosis (DELBEKE and MARTIN 2004).

A very exciting use of nuclear imaging is the use of radiolabeled markers to map the hypoxic microenvironment of cancerous lesions. Tumor hypoxia presents a severe problem for radiation therapy because radiosensitivity rapidly decreases when the O2 partial pressure in a tumor is less than 25 mmHg. Hypoxia is known to be a significant prognostic variable, and it is likely that the use of tracers to monitor hypoxia before, during and after treatment will continue to increase. Multiple clinical trials are ongoing, the majority using nitromidazoles such as 18F-MISO, 123I-AZA, 123I-AZGP or EF5. Chao et al. reported on the use of 60CU-ATSM, a marker relying on the reduction of a chelated metal, to investigate its use for hypoxia imaging-guided IMRT (CHAO et al. 2001). Although hypoxia imaging looks very promising, confirmation using needle oximetry or an equivalent gold standard is still needed. For most radiopharmaceuticals, a low tumor-to-background ratio has been reported, and the difficulty of determining a cutoff value to separate normoxic from hypoxic tumors limits the clinical usefulness of these markers. New tracers with higher tumor-to-background ratios are required.

Cell proliferation is one of the hallmarks of malignant transformation. It is characterized by cells undergoing uncontrolled DNA replication. Because growth requires energy, FDG has often been used to assess tumor proliferation. The use of nucleosides and amino acids is a more specific approach for visualizing tumor proliferation (VAN DE WIELE et al. 2003). Examples of radiolabeled nucleoside analogues to target DNA replication are 11C-thymidine, 131I-deoxyuridine and 18F-fluorothymidine. The latter might be superior to FDG, given its lower uptake in inflammatory tissues and lower uptake in normal brain tissue. The rapid catabolism of thymidine in vivo and the short half-life of 11C are major disadvantages for clinical imaging.

Apoptosis (programmed cell death) can be used to predict tumor response. This could ultimately contribute to an individually tailored treatment. In vivo imaging of apoptosis is feasible using radiolabeled Annexin, 99mTc-radiolabeled annexin V, an endogenous human protein with a high affinity for membrane-bound lipid phosphatidylserine, which becomes exposed at the outer leaflet of the plasma membrane bilayer at an early stage of the apoptotic process. Accumulation of the tracer demonstrates complete or partial response after treatment, whereas absence of significant uptake increase early after the start of treatment can predict stable or progressive disease (HAAS et al. 2004; KARTACHOVA et al. 2004).

Challenges in PET imaging remain the cost of combined scanners, the acquisition time, organ motion and gated image acquisition (for example for thoracic lesions). Other limitations include the need for specialized radiopharmaceutical in-house expertise and the close proximity of a cyclotron. Unlike morphological imaging modalities, such as CT and MR, tumor edges on PET are not well defined, a possible source of interobserver variability in gross tumor volume delineation. The definition of subclinical tumor response remains a challenge for modern diagnostic imaging modalities. At present, tumor volume is still the standard accepted by most oncologists. New criteria, based on cellular metabolic indices, will need to be identified and agreed upon after careful clinical follow-up imaging studies. PET may play an important role in this context. Lastly, clinicians should be aware that more accurate staging by means of PET imaging, as in the case of non-small cell lung cancer, might improve outcome for patients regardless of any improvement in treatment–the so called Will Rogers phenomenon (FEINSTEIN et al. 1985).

Addendum

(Ad1)
Lead time bias: early diagnosis falsely appears to prolong survival.
(Ad2)
Length bias: screening over represents less aggressive disease.
(Ad3)
Will Rogers (1879–1935): "When the Okies left Oklahoma and moved to California, they raised the average intelligence level in both states." The Will Rogers phenomenon refers to improvements in diagnostic imaging shifting patients with clinically silent metastatic lesions into more advanced stages. This shift can improve the stage-specific survival of both stages, without overall improvement for the entire cohort.

References

Alavi A, Lakhani P, Mavi A, et al. (2004) PET: a revolution in medical imaging. Radiol Clin North Am 42:983–1001

Allal AS, Slosman DO, Kebdani T, et al. (2004) Prediction of outcome in head-and-neck cancer patients using the standardized uptake value of 2-(18F)fluoro-2-deoxy-D-glucose. Int J Radiat Oncol Biol Phys 59:1295–1300

Anzai Y (2004) Superparamagnetic iron oxide nanoparticles: nodal metastases and beyond. Top Magn Reson Imaging 15:103–111

Arias-Mendoza F, Smith MR, Brown TR (2004) Predicting treatment response in non-Hodgkin's lymphoma from the pretreatment tumor content of phosphoethanolamine plus phosphocholine. Acad Radiol 11:368–276

Artemov D (2003) Molecular magnetic resonance imaging with targeted contrast agents. J Cell Biochem 90:518–524

Barentsz JO, Engelbrecht M, Jager GJ, et al. (1999) Fast dynamic gadolinium-enhanced MR imaging of urinary bladder and prostate cancer. J Magn Reson Imaging 10:295–304

Black WC, Welch HG (1993) Advances in diagnostic imaging and overestimations of disease prevalence and the benefits of therapy. N Engl J Med 328:1237–1243

Brahme A (2004) Recent advances in light ion radiation therapy. Int J Radiat Oncol Biol Phys 58:603–616

Bulte JWM, Kraitchman DL (2004) Iron oxide MR contrast agents for molecular and cellular imaging. NMR Biomed 17:484–499

Bussels B, Goethals L, Feron M, et al. (2003) Respiration-induced movement of the upper abdominal organs: a pitfall for the three-dimensional conformal radiation treatment of pancreatic cancer. Radiother Oncol 68:69–74

Buyyounouski MK, Horwitz EM, Uzzo RG, et al. (2004) The radiation doses to erectile tissues defined with magnetic resonance imaging after intensity-modulated radiation therapy or iodine-125 brachytherapy. Int J Radiat Oncol Biol Phys 59:1383–1391

Chao KS, Bosch WR, Mutic S, et al. (2001) A novel approach to overcome hypoxic tumor resistance: Cu-ATSM-guided intensity-modulated radiation therapy. Int J Radiat Oncol Biol Phys 49:1171–1182

Chapman JD, Bradley JD, Eary JF, et al. (2003) Molecular (functional) imaging for radiotherapy applications: an RTOG symposium. Int J Radiat Oncol Biol Phys 55:294–301

Coakley FV, Qayyum A, Kurhanewicz J (2003) Magnetic resonance imaging and spectroscopic imaging of prostate cancer. J Urol 170:S69–S75

Daisne JF, Duprez T, Weynand B, et al. (2004) Tumor volume in pharyngolaryngeal squamous cell carcinoma: comparison at CT, MR imaging, and FDG PET and validation with surgical specimen. Radiology 233:93–100

Dawson LA, Brock KK, Kazanjian S et al. (2001) The reproducibility of organ position using active breathing control (ABC) during liver radiotherapy. Int J Radiat Oncol Biol Phys 51:1410–1421

Delbeke D, Martin WH (2004) PET and PET-CT for evaluation of colorectal carcinoma. Semin Nucl Med 34:209–223

DiBiase SJ, Hosseinzadeh K, Gullapalli RP, et al. (2002) Magnetic resonance spectroscopic imaging-guided brachytherapy for localized prostate cancer. Int J Radiat Oncol Biol Phys 52:429–438

Dyke JP, Panicek DM, Healey JH, et al. (2003) Osteogenic and Ewing sarcomas: estimation of necrotic fraction during induction chemotherapy with dynamic contrast-enhanced MR imaging. Radiology 228:271–278

Feinstein AR, Sosin DM, Wells CK (1985) The Will Rogers phenomenon. Stage migration and new diagnostic techniques as a source of misleading statistics for survival in cancer. N Engl J Med 312:1604–1608

Gambhir SS (2002) Molecular imaging of cancer with positron emission tomography. Nat Rev Cancer 2:683–693

Gregoire V (2004) Is there any future in radiotherapy planning without the use of PET: unraveling the myth. Radiother Oncol 73:261–263

Haas RL, de Jong D, Valdes Olmos RA, et al. (2004) In vivo imaging of radiation-induced apoptosis in follicular lymphoma patients. Int J Radiat Oncol Biol Phys 59:782–787

Harisinghani MG, Barentsz J, Hahn PF, et al. (2003) Noninvasive detection of clinically occult lymph-node metastases in prostate cancer. N Engl J Med 348:2491–2499

Kalff V, Hicks RJ, MacManus MP, et al. (2001) Clinical impact of (18)F fluorodeoxyglucose positron emission tomography in patients with non-small-cell lung cancer: a prospective study. J Clin Oncol 19:111–118

Kartachova M, Haas RL, Valdes Olmos RA, et al. (2004) In vivo imaging of apoptosis by (99m)Tc-Annexin V scintigraphy: visual analysis in relation to treatment response. Radiother Oncol 72:333–339

Keall P (2004) 4-Dimensional computed tomography imaging and treatment planning. Semin Radiat Oncol 14:81–90

Lardinois D, Weder W, Hany TF, et al. (2003) Staging of non-small-cell lung cancer with integrated positron-emission tomography and computed tomography. N Engl J Med 348:2500–2507

Leibel SA, Fuks Z, Zelefsky MJ, et al. (2002) Intensity-modulated radiotherapy. Cancer J 8:164–176

Ling CC, Humm J, Larson S, Amols H, et al. (2000) Towards multidimensional radiotherapy (MD-CRT): biological imaging and biological conformality. Int J Radiat Oncol Biol Phys 47:551–560

Liu WC, Schulder M, Narra V et al. (2000) Functional magnetic resonance imaging aided radiation treatment planning. Med Phys 27:1563–1572

Mah D, Freedman G, Milestone B, et al. (2002) Measurement of intrafractional prostate motion using magnetic resonance imaging. Int J Radiat Oncol Biol Phys 54:568–575

Mardor Y, Roth Y, Ochershvilli A, et al. (2004) Pretreatment prediction of brain tumors' response to radiation therapy using high b-value diffusion-weighted MRI. Neoplasia 6:136–142

Mizowaki T, Cohen GN, Fung AY, et al. (2002) Towards integrating functional imaging in the treatment of prostate cancer with radiation: the registration of the MR spectroscopy imaging to ultrasound/CT images and its implementation in treatment planning. Int J Radiat Oncol Biol Phys 54:1558–1564

Morawski AM, Winter PM, Crowder KC, et al. (2004) Targeted nanoparticles for quantitative imaging of sparse molecular epitopes with MRI. Magn Reson Med 51:480–486

Nehmeh SA, Erdi YE, Pan T, et al. (2004) Quantitation of respiratory motion during 4D PET/CT acquisition. Med Phys 31:1333–1338

Padhani AR (2002) Dynamic contrast-enhanced MRI in clinical oncology: current status and future directions. J Magn Reson Imaging 16:407–422

Pieterman RM, van Putten JW, Meuzelaar JJ, et al. (2000) Preoperative staging of non-small-cell lung cancer with positron-emission tomography. N Engl J Med 343:254–261

Pickett B, Ten Haken RK, Kurhanewicz J, et al. (2004) Time to metabolic atrophy after permanent prostate seed implantation based on magnetic resonance spectroscopic imaging. Int J Radiat Oncol Biol Phys 59:665–673

Pirzkall A, Li X, Oh J, et al. (2004) 3D MRSI for resected high-grade gliomas before RT: tumor extent according to metabolic activity in relation to MRI. Int J Radiat Oncol Biol Phys 59:126–137

Pollack A, Hanlon AL, Horwitz EM, et al. (2004) Prostate cancer radiotherapy dose response: an update of the fox chase experience. J Urol 171:1132–1136

Poulakis V, Witzsch U, de Vries R, et al. (2004) Preoperative neural network using combined magnetic resonance imaging variables, prostate-specific antigen, and Gleason score for predicting prostate cancer biochemical recurrence after radical prostatectomy. Urology 64:1165–1170

Rasch C, Barillot I, Remeijer P, et al. (1999) Definition of the prostate in CT and MRI: a multi-observer study. Int J Radiat Oncol Biol Phys 43:57–66

Remouchamps VM, Letts N, Yan D, et al. (2003) Three-dimensional evaluation of intra- and interfraction immobilization of lung and chest wall using active breathing control: A reproducibility study with breast cancer patients. International Int J Radiat Oncol Biol Phys 57:968–978

Roach M 3rd, Faillace-Akazawa P, Malfatti C, et al. (1996) Prostate volumes defined by magnetic resonance imaging and computerized tomographic scans for three-dimensional conformal radiotherapy. Int J Radiat Oncol Biol Phys 35:1011–1018

Rouviere O, Valette O, Grivolat S, et al. (2004) Recurrent prostate cancer after external beam radiotherapy: value of contrast-enhanced dynamic MRI in localizing intraprostatic tumor-correlation with biopsy findings. Urology 63:922–927

Sakr WA, Grignon DJ, Crissman JD, et al. (1994) High grade prostatic intraepithelial neoplasia (HGPIN) and prostatic adenocarcinoma between the ages of 20-69: an autopsy study of 249 cases. In Vivo 8:439–443

Steenbakkers RJ, Deurloo KE, Nowak PJ, et al. (2003) Reduction of dose delivered to the rectum and bulb of the penis using MRI delineation for radiotherapy of the prostate. Int J Radiat Oncol Biol Phys 57:1269–1279

Van de Wiele C, Lahorte C, Oyen W, et al. (2003) Nuclear medicine imaging to predict response to radiotherapy: a review. Int J Radiat Oncol Biol Phys 55:5–15

Wang L, Mullerad M, Chen HN, et al. (2004) Prostate cancer: incremental value of endorectal MR imaging findings for prediction of extracapsular extension. Radiology 232:133–139

Weinmann HJ, Ebert W, Misselwitz B, et al. (2003) Tissue-specific MR contrast agents. Eur J Radiol 46:33–44

Zakian KL, Koutcher JA, Ballon D, et al. (2001) Developments in nuclear magnetic resonance imaging and spectroscopy: application to radiation oncology. Semin Radiat Oncol 11:3–15

5 Physics of Treatment Planning in Radiation Oncology

James A. Purdy, Srinivasan Vijayakumar, Carlos A. Perez, and Seymour H. Levitt

CONTENTS

5.1 Introduction *69*
5.2 Dosimetry Parameters *70*
5.3 Monitor Unit and Dose Calculation Methods *75*
5.3.1 Monitor Unit Calculation for Fixed Fields *75*
5.3.2 Monitor Unit Calculation for Rotation Therapy *76*
5.3.3 Monitor Unit Calculations for Irregular Fields *76*
5.3.4 Monitor Unit Calculations for Asymmetric X-ray Collimators *77*
5.3.5 Monitor Unit Calculations for Multileaf Collimator System *78*
5.4 Dose Calculation Algorithms and Correction Factors *78*
5.4.1 Ratio of Tissue–Air Ratio or Tissue–Phantom Ratio Method *78*
5.4.2 Effective Source–Skin Ratio Method *79*
5.4.3 Isodose Shift Method *79*
5.5 Correction for Tissue Inhomogeneities *79*
5.5.1 Ratio of Tissue–Air Ratio Method *80*
5.5.2 Isodose Shift Method *80*
5.5.3 Power Law TAR Method *80*
5.6 Clinical Photon-Beam Dosimetry *81*
5.6.1 Single-Field Isodose Charts *81*
5.6.2 Depth Dose Build-up Region *82*
5.6.3 Depth Dose Exit Dose Region *83*
5.6.4 Interface Dosimetry *83*
5.6.4.1 Air Cavities *84*
5.6.4.2 Lung Interfaces *84*
5.6.4.3 Bone Interfaces *85*
5.6.4.4 Prostheses (Steel and Silicon) *85*
5.6.5 Wedge Filter Dosimetry *85*
5.7 Treatment Planning: Combination of Treatment Fields *86*
5.7.1 Parallel-Opposed Fields *86*
5.7.2 Multiple-Beam Arrangements *88*
5.7.3 Rotational Therapy *89*
5.8 Field Shaping *91*
5.8.1 Low Melting Alloy Blocks *91*
5.8.2 Multileaf Collimation *92*
5.8.3 Asymmetric Collimator Jaws *94*
5.9 Compensating Filters *95*
5.10 Bolus *96*
5.11 Treatment-Planning Process *96*
5.11.1 Target Volume Localization *96*
5.11.2 Patient Positioning, Registration, and Immobilization *97*
5.11.3 Beam Direction and Shape *98*
5.11.4 Patient Contour and Anatomic Measurements *99*
5.11.5 Dose and Target Specification *99*
5.12 Separation of Adjacent X-Ray Fields *100*
5.12.1 Field Junctions *100*
5.12.2 Orthogonal Field Junctions *101*
References *103*

J. A. PURDY, PhD
Professor and Vice Chairman, Department of Radiation Oncology, Chief, Physics Section, University of California Davis Medical Center, 4501 X Street, Suite G140, Sacramento, CA 95817, USA

S. VIJAYAKUMAR, MD
Chair, Department of Radiation Oncology, University of California Davis Medical Center, 4501 X Street, Suite G126, Sacramento, CA 95817, USA

C.A. PEREZ, MD
Professor, Department of Radiation Oncology, Mallinckrodt Institute of Radiology, Washington University School of Medicine, St. Louis, MO 63110, USA

S.H. LEVITT, MD
Department of Therapeutic Radiation Oncology, University of Minnesota, Minneapolis, MN 55455, USA

5.1 Introduction

The radiation oncologist, when planning the treatment of a patient with cancer, is faced with the difficult problem of prescribing a treatment regimen with a radiation dose that is large enough to potentially cure or control the disease, but does not cause serious normal tissue complications. This task is a difficult one because tumor control and normal tissue effect responses are typically steep functions of radiation dose, i.e., a small change in the dose delivered ($\pm$5%) can result in a dramatic change in the local response of the tissue ($\pm$20%; FISCHER and MOULDER 1975; HERRING and COMPTON 1971; HERRING 1975; STEWART and JACKSON 1975). Moreover, the prescribed curative doses are often, by necessity, very close to the doses tolerated by the normal tissues; thus, for optimum treatment, the radiation dose must be planned and delivered with a high degree of accuracy.

One can readily compute the dose distribution resulting from radiation beams of photons, electrons, or mixtures of these impinging on a regularly shaped, flat-surface, homogeneous unit density phantom; however, the patient presents a much more complicated situation because of irregularly shaped topography and many tissues of varying densities and atomic composition (called heterogeneities). In addition, beam modifiers, such as wedges and compensating filters or bolus, are sometimes inserted into the radiation beam to achieve the desired dose distribution.

This chapter reviews the physics and dosimetric parameters, and the basic physics concepts, used in planning the cancer patient's radiation therapy.

5.2
Dosimetry Parameters

Several dosimetric parameters have been defined for use in calculating the patient's dose distribution and the treatment machine settings to deliver the prescribed dose. These parameters are only briefly reviewed here, but more details are given by KHAN (2003).

Percentage depth dose (PDD) is defined as the ratio, expressed as a percentage, of the absorbed dose on the central axis at depth d to the absorbed dose at the reference point d_{max} (Fig. 5.1):

$$PDD(d, d_o, S, f, E) = \frac{D_d}{D_{d_o}} \times 100$$

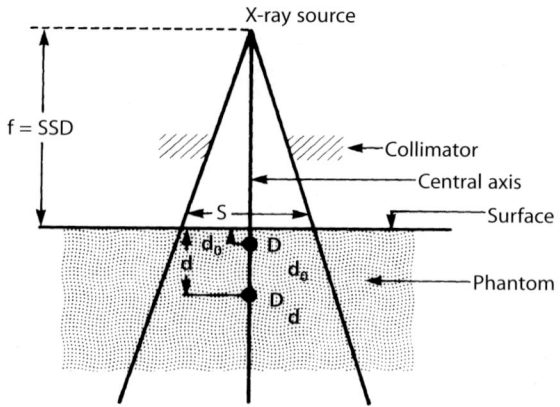

Fig. 5.1. Definition of percentage depth dose where d is any depth and d_o is the reference depth, usually d_{max}

The functional symbols have been inserted in the foregoing equation to make it clear that the PDD is affected by a number of parameters, including d, d_{max}, field dimension S, source-to-surface distance f, and radiation beam energy (or quality) E. S refers to the side length of a square beam at a specified reference depth. Nonsquare beams may be designated by their equivalent square. Field shape and added beam collimation also can affect the central axis depth dose distribution.

The central-axis PDD expresses the penetrability of a radiation beam. Table 5.1 summarizes beam characteristics for X-ray and γ-ray beams typically used in radiation therapy and lists the depth at which the dose is maximum (100%) and the 10-cm

Table 5.1. Beam characteristics for photon-beam energies of interest in radiation therapy. SSD source-to-skin distances, HVL half-value layer

200 kVp, 2.0 mm Cu HVL, SSD=50 cm	Cobalt-60, SSD=80 cm	4-mV X-ray, SSD=80 cm	6-mV X-ray, SSD=100 cm	15-18-mV X-ray, SSD=100 cm
Depth of maximum dose=surface	Depth of maximum dose=0.5 cm	Depth of maximum dose=1.0-1.2 cm	Depth of maximum dose=1.5 cm	Depth of maximum dose=3.0-3.5 cm
Rapid fall-off with depth due to (a) low energy and (b) short SSD	Increased penetration (10 cm %DD=55%)	Penetration slightly greater than cobalt (10 cm %DD=61%)	Slightly more penetration than 60Co and 4 mV (10 cm %DD=67%)	Much greater penetration (10 cm %DD=80%)
Sharp beam edge due to small focal spot	Beam edge not as well defined; penumbra due to source size	Penumbra smaller	Small penumbra	Small penumbra
Significant dose outside beam boundaries due to Compton scattered radiation at low energies	Dose outside beam low since most scattering is in forward direction	"Horns" (beam intensity off-axis) due to flattening filter design can be significant (14%)	"Horns" (beam intensity off-axis) due to flattening filter design reduced (9%)	"Horns" (beam intensity off-axis) due to flattening filter design reduced (5%)
	Isodose curvature increases as the field size increases			Exit dose often higher than entrance dose

depth PDD value. Representative PDD curves are shown in Fig. 5.2 for conventional source-to-skin distances (SSDs). As a rule of thumb, an 18-mV, 6-mV, and ^{60}Co photon beam loses approximately 2, 3.5, and 4.5% per cm, respectively, beyond the depth of maximum dose, d_{max} (values are for a 10×10-cm field, 100-cm SSD). There is no agreement as to what is the single optimal X-ray beam energy; instead, institutional bias or radiation oncologist training typically influences its selection, and it is generally treatment site specific. Most modern medical linear accelerators (linacs) are multimodality, and provide a range of photon and electron beam energies ranging from 4 to 25 mV with 6 mV and 15- or 18-mV X-ray beams the most common.

The tissue–air ratio (TAR) is defined as the ratio of the absorbed dose D_d at a given point in the phantom

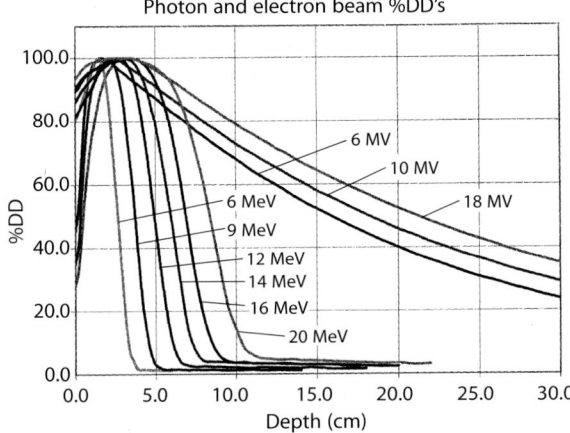

Photon and electron beam %DD's

Fig. 5.2. Examples of typical photon beam and electron beam central axis percent depth dose curves for a 10×10-cm field for selected energies

by the absorbed dose in free space, D_{fs}, that would be measured at the same point but in the absence of the phantom, if all other conditions of the irradiation (e.g., collimator, distance from the source) are equal (Fig. 5.3). The TAR is expressed as:

$$TAR(d,S_d,E) = \frac{D_d}{D_{fs}},$$

where d is depth, E is radiation beam energy, and S_d is the beam dimension measured at depth d. The TAR depends on depth, field size, and beam quality, but is, for all practical purposes, independent of the distance from the source.

The TAR at the depth of maximum dose is called the peakscatter factor. It is perhaps better known as the backscatter factor, but because of the finite depth d_o, this tends to be misleading. Figure 5.4 shows the peakscatter factors for various field sizes and beam qualities.

The concepts of tissue–phantom ratio (TPR) and tissue–maximum ratio (TMR) were proposed for high-energy radiation as alternatives to TAR in response to arguments raised against the use of in-air measurement for a photon beam with a maximum energy greater than 3 MeV (HOLT et al. 1970; KARZMARK et al. 1965). As originally defined, TPR is given by the ratio of two doses:

$$TPR(d,d_r,S_d,E) = \frac{D_d}{D_{d_r}},$$

where D_{d_r} is the dose at a specified point on the central axis in a phantom with a fixed reference depth, d_r, of tissue-equivalent material overlying the point; D_d is the dose in phantom at the same spatial point as before but with an arbitrary depth, d, of overlying material; and S_d is the beam width at the level of measurement (Fig. 5.5). In each instance, underlying material is sufficient to provide for full backscatter. The TPR is intended to be analogous to the *TAR* but has an advantage because the reference dose, D_{d_r}, is directly measurable over the entire range of X-rays and γ-rays in use, eliminating problems in obtaining a value for the dose in free space when the depth for electronic build-up is great.

The original TMR definition is similar to the definition of TPR, except that the reference depth, d_r, is specifically defined as the depth of maximum dose, d_{max}; however, the d_{max} for megavoltage X-ray

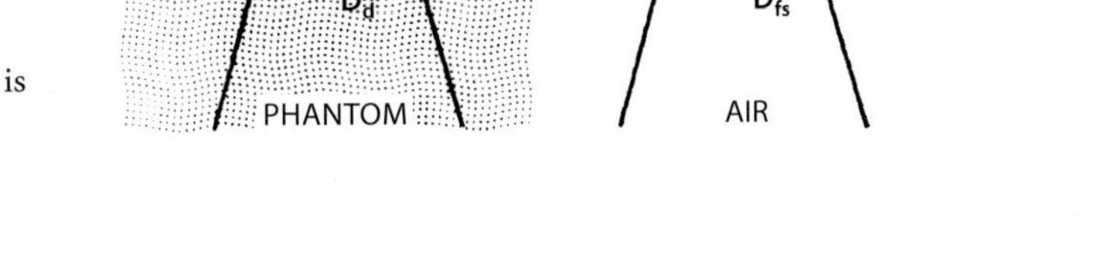

Fig. 5.3. Definition of tissue–air ratio, where d is the thickness of overlying material

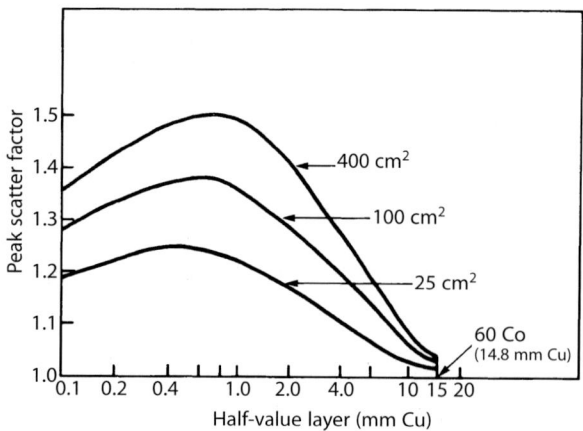

Fig. 5.4. Variation of peakscatter factor with beam quality (half-value layer). (From JOHNS and CUNNINGHAM 1983)

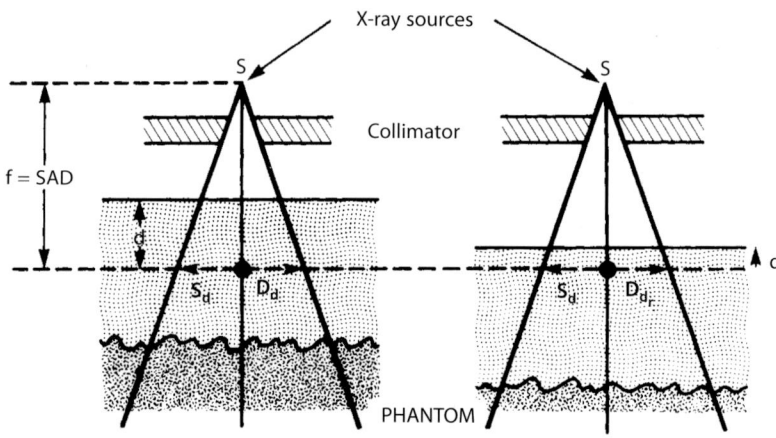

Fig. 5.5. Definition of tissue-phantom ratio and tissue-maximum ratio, where d is the thickness of overlying material and d_r is the reference thickness

beams varies significantly with field size and also depends on SSD; thus, the definition of TMR creates a measurement inefficiency because a variable d_r is required. hA modification by KHAN and co-workers (1980) redefines the TMR so that the reference depth, d_r, must be equal to or greater than the largest d_{max}. This definition has now become the accepted practice.

The scatter-air ratio (SAR) can be thought of as the scatter component of the TAR (CUNNINGHAM 1972). It is defined as follows:

$$SAR(d,S_d,E) = TAR(d,S_d,E) = TAR(d,0,E).$$

SAR is the difference between the TAR for a field of finite area and the TAR for a zero-area field size. The zero-area TAR is a mathematical abstraction obtained by extrapolation of the TAR values measured for finite field sizes.

Similarly, the scatter-maximum ratio (SMR), the scatter component of the TMR, is defined as follows:

$$SMR(d,S_d,E) = TMR(d,S_d,E) \cdot \frac{S_p(S_d,E)}{S_p(0,E)} - TMR(d,0,E),$$

where S_p is defined as the phantom scatter correction factor, which accounts for changes in scatter radiation originating in the phantom at the reference depth as the field size is changed.

The output factor (denoted by $S_{c,p}$) for a given field size is defined as the ratio of the dose rate at d_{max} for the field size in question to that for the reference field size (usually 10×10 cm), again measured at its d_{max}. The output factor varies with field size (Fig. 5.6) as a result of two distinct phenomena. As the collimator jaws are opened, the primary dose, D_p at d_{max} on the central ray per motor unit (MU), increases due to larger number of primary X-ray photons scattered out of the flattening filter. In addition, the scatter dose, $D_s(d_{max},r)$, at the measurement point per unit D_p increases as the scattering volume irradiated by primary photons increases with increasing collimated field size. Note that these two components

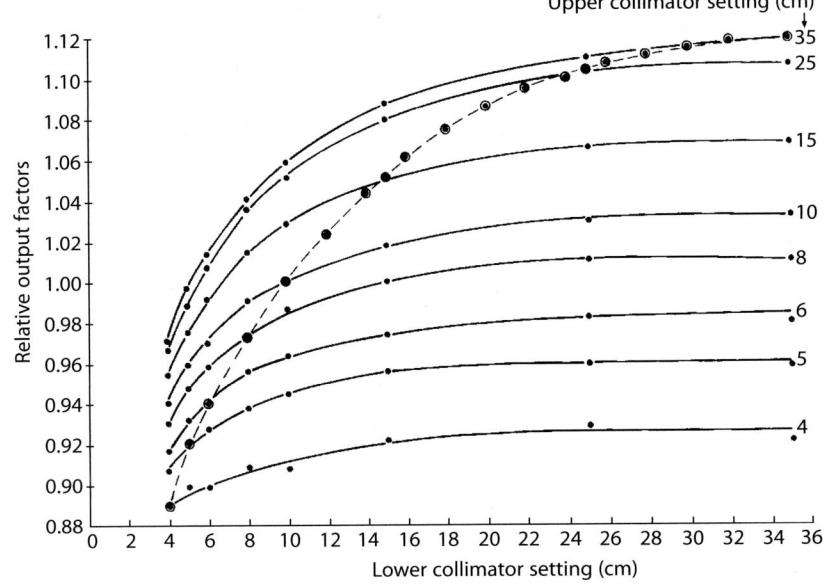

Fig. 5.6. Example of output factor as a function of lower and upper collimator settings for a medical linear accelerator 18-mV X-ray beam

can vary independently of one another if nonstandard treatment distances or extensive secondary blocking is used.

KHAN and associates (1980) described a method for separating the total output factor, $S_{c,p}$, into two components, as given by the following formula:

$$S_{c,p}(r) = S_c(r_c) \cdot S_p(r).$$

The two components are the collimator scatter factor, $S_c(r_c)$, which is a function only of the collimator opening, r_c, projected to isocenter, and the phantom scatter factor, $S_p(r)$, which is a function only of the cross-sectional area, or effective field size, r, that is irradiated at the treatment distance. In practice, the total and collimator scatter factors are both measured and the phantom scatter factor calculated using the relationship above. $S_c(r_c)$ is measured in air using an ion chamber fitted with an equilibrium-thickness build-up cap and given by the ratio of the reading for the given collimator opening to the reading for a reference field (typically 10×10 cm) collimator opening. The overall output factor is measured in-phantom using the standard treatment distance and is given by the reading relative to that for a 10 × 10-cm field size.

By carefully extrapolating this measured ratio to zero-field size, the zero-field size phantom scatter factor, $S_p(o)$, is obtained. If a small ion chamber is positioned axially in the beam, it is possible to measure $S_{c,p}$ for field sizes as small as 1 × 1 cm. Because of the loss of lateral secondary electron equilibrium

encountered near the edges of high-energy photon beams, S_p deviates significantly from unity. Such an extrapolation is needed to calculate SMR values from broad-beam TMR data. Consistent separation of primary and scatter dose components significantly improves the accuracy of dose predictions near block edges and under blocks, overcoming many of the dose-modeling problems presented by use of extensive customized blocking. In addition, the extrapolation procedure implicit in this formalism leads to irregular field calculations that more accurately model the dose falloff near beam edges due to lateral electron disequilibrium.

An isodose curve represents points of equal dose. A set of these curves, normally given in 10% increments normalized to the dose at d_{max}, can be plotted on a chart (i.e., isodose chart) to give a visual representation of the dose distribution in a single plane (Fig. 5.7). Beam parameters, such as source size, flattening filter, field size, and SSD, play important roles in the shape of the isodose curve.

A dose profile is a representation of the dose in an irradiated volume as a function of spatial position along a single line. Dose profiles are particularly well suited to the description of field flatness and penumbra. The data are typically given as ratios of doses normalized to the dose at the central axis of the field (Fig. 5.8). The profiles, also called off-axis factors or off-center ratios, may be measured in-air (i.e., with only a build-up cap) or in a phantom at selected depths. The in-air off-axis factor gives only the variation in primary beam intensity; the in-

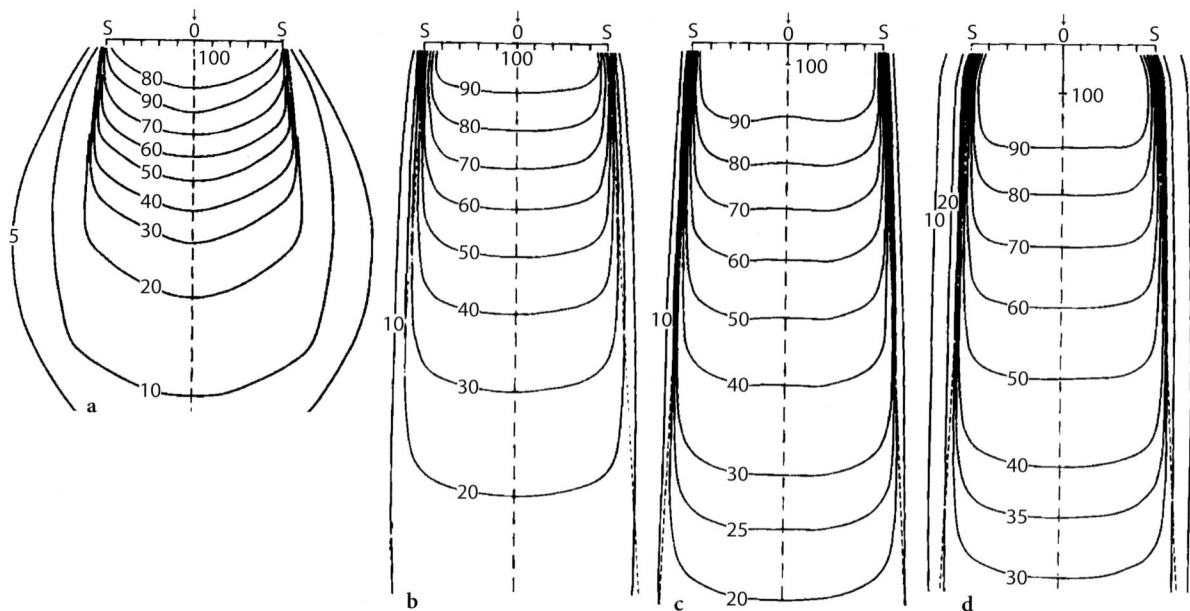

Fig. 5.7a–d. Isodose distributions for different-quality radiations. **a** 200 kVp, SSD = 50 cm, HVL = 1 mm Cu, field size = 10×10 cm. **b** ^{60}Co, SSD = 80 cm, field size = 10×10 cm. **c** 4-mV X-rays, SSD = 100 cm, field size = 10×10 cm. **d** 10-mV X-rays, SSD = 100 cm, field size = 10×10 cm. (From KHAN 1994)

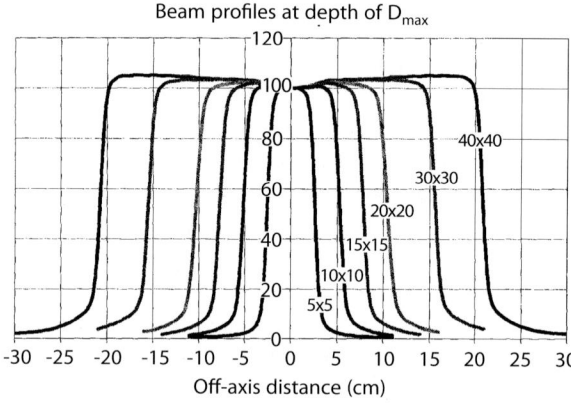

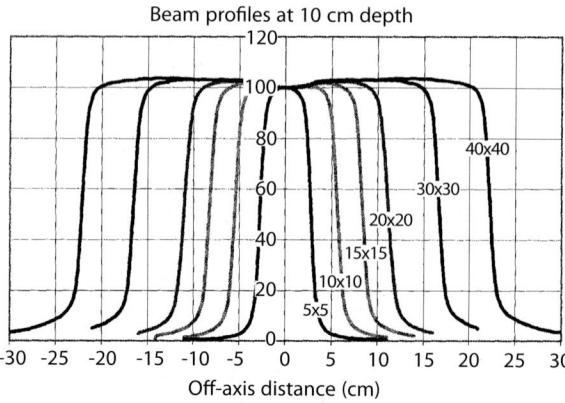

Fig. 5.8. Example of dose profiles for an 18-mV linear accelerator X-ray beam measured at depths of 3 and 10 cm

phantom off-center ratio shows the added effect of phantom scatter.

Wedge filters, first introduced by ELLIS and MILLER (1944), are generally constructed of brass, steel, or lead, and when placed in the beam they progressively decrease intensity across the field, going from the thin edge to the thick edge of the filter, causing the isodose distribution to have a planned asymmetry (Fig. 5.9).

The wedge angle is defined as the angle the isodose curve subtends with a line perpendicular to the central axis at a specific depth and for a specified field size. Current practice is to use a depth of 10 cm. Past definitions were based on the 50th percentile isodose curve and, more recently, the 80th percentile isodose curve. The wedge angle is a function of field size and depth.

The wedge factor is defined as the ratio of the dose measured in a tissue-equivalent phantom at the depth of maximum build-up on the central axis with the wedge in place to the dose at the same point with the wedge removed.

Wedge isodose curves are generally normalized to the dose at d_{max} of the unwedged beam, resulting in percentiles greater than 100% under the thin portion of the wedge; however, this normalization is not always used. The normalization and the use of the wedge factor in calculating machine settings should be clearly understood before being used clinically.

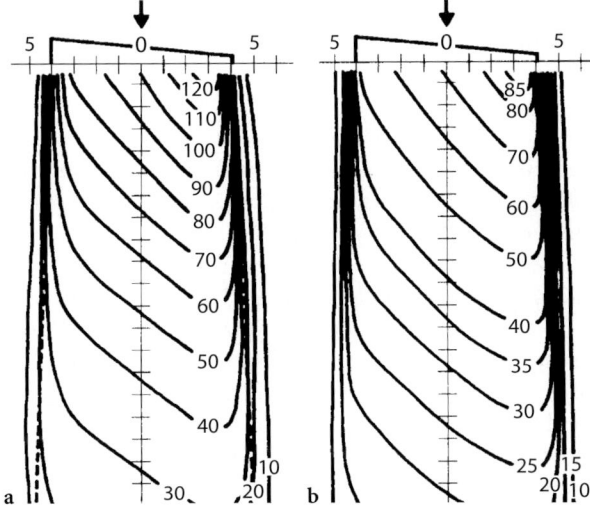

Fig. 5.9a,b. Isodose curves for a wedge filter. **a** Normalized to D_{max}. **b** Normalized to D_{max} without the wedge. ^{60}Co, wedge angle=45°, field size=8×10 cm, SSD=80 cm. (From KHAN 1994)

and d_{cal}, respectively. Usually, but by no means universally, it is assumed that

$$SCD = SAD \text{ (source–axis distance)} + d_{max}$$
$$r_{cal} = 10 \times 10 \text{ cm}$$
$$d_{cal} = d_{max}.$$

Normal incidence and open-beam geometry (i.e., absence of trays or any beam modifying filters) are assumed. The above setup parameters are described as fixed-SSD calibration geometry. Treatment machines are also calibrated isocentrically with the point of MU specification located at distance SAD rather than at distance $SAD + d_{max}$ as described above. For isocentric calibration, $SCD = SAD$.

If the machine is a linear accelerator, it is calibrated by adjusting the sensitivity of its internal monitor transmission chamber so that for the reference geometry

$$D(SCD, r_{cal}, d_{cal}) / MU = 1\,Gy/1\,MU.$$

Several reports on monitor unit calculations are now available (DUTREIX et al. 1997; GIBBONS 2000; GEORG et al. 2001).

5.3.1
Monitor Unit Calculation for Fixed Fields

When the patient is to be treated isocentrically, the point of dose prescription is located at the isocenter regardless of the target depth. The MU needed to deliver a prescribed tumor dose to isocenter (TD_{iso}) for a depth d of overlying tissue on the central ray is given by

$$MU = \frac{TD_{iso}}{TMR(d, r_d) \cdot S_c(r_c) \cdot S_p(r_d) \cdot TF \cdot WF \cdot \left(\dfrac{SCD}{SAD}\right)^2},$$

where TF and WF denote the tray and wedge factors, respectively. They are defined as the ratio of the central ray dose with the tray or wedge filter in place relative to the dose in the open beam geometry. The collimated field size is denoted by r_c and is usually described as the square field size equivalent to the rectangular collimator opening projected to isocenter. The effective field size is denoted by r_d and is always specified to the isocenter distance, SAD. The inverse-square law factor accounts for the difference in distances from the source-to-point of dose prescription relative to the point of MU specification.

5.3
Monitor Unit and Dose Calculation Methods

Monitor unit calculations relate the dose at any point on the central ray of the treatment beam, regardless of depth, treatment distance, secondary blocking configuration, or collimator opening selected, to the calibrated output of the treatment machine (described in units commonly described as "dose per monitor unit"). This is accomplished by using the various dosimetric quantities described in the preceding section to relate the dose corresponding to an arbitrary set of treatment parameters to a single standard treatment setup where the output of the machine is specified in terms of Gy/MU. The reference distance, field size, and depth of output specification are denoted by the symbols SCD, r_{cal},

Beam hardening occurs when a wedge is inserted into the radiation beam, causing the PDD to increase at depth. Differences in PDD of nearly 7% have been reported for a 4-mV X-ray, 60° wedge field, compared with the open field at a depth of 12 cm, and as much as 3% difference between the 60° wedge field and the open field for a 25-mV X-ray beam (SEWCHAND et al. 1978; ABRATH and PURDY 1980).

When isocentric calibration is used, this factor is unity. Note that collimator-defined field size is used for lookup of S_c, whereas effective field size projected to isocenter is used for lookup of TMR and S_p. By separately accounting for the effect of collimator opening on the primary dose component and the influence of cross-sectional area of tissue irradiated, most of the difficulties in accurately delivering a dose in the presence of extensive blocking are overcome.

When a fixed distance between the target and entry skin surface, SSD, is used to treat the patient, a dose-calculation formalism based on PDD is used rather than one based on isocentric dose ratios. The MU needed to deliver a prescribed tumor dose to depth d *(TDd)* on the central axis is given by

$$MU = \frac{TD_d \cdot 100}{PDD(SSD,d,r) \cdot S_c(r_c) \cdot S_p(r) \cdot TF \cdot WF \cdot \left(\frac{SCD}{SSD + d_{max}}\right)^2}$$

The field size (or its equivalent square) on the skin surface at central axis is denoted by r and is used for lookup of both PDD and S_p. The collimated field size r_c at the isocenter must be used for lookup of S_c. Note that when an extended treatment distance is used, the collimated field size at isocenter differs significantly from that at the projected skin surface of the patient. Also, when this dose calculation formalism for highly extended treatment distances is used, such as encountered in administering whole-body irradiation, care must be taken to verify the validity of inverse square law at these distances. It is recommended that such setups always be verified by ion chamber measurement. Because of the large scatter contribution to effective primary dose originating from the flattening filter and other components in the treatment head, the virtual source of radiation may be as much as 2 cm proximal to the target of the accelerator.

All MU calculation formalisms require some means of estimating the square field size, r, that is equivalent, in terms of scattering characteristics, to an arbitrary rectangular field of width a and length b. Such an equivalence is of great practical importance because it reduces the dimensionality of table lookups by one. In addition, those formalisms that distinguish between overall and effective field size require some means of estimating the square or rectangular field size that is equivalent to an arbitrary irregular field. Perhaps the most widely used rectangular equivalency principal is the "A/P" rule. It states that a square and a rect-

angle are equivalent if they have the same area/perimeter ratio, i.e.,

$$r = \frac{2(a \times b)}{(a+b)}$$

This relationship leads to remarkably accurate *PDDs, scatter factors,* and isocentric dose ratios for all but the most elongated fields.

Another widely used approach to reducing rectangular estimates of effective field size to square field sizes is the equivalent square table published in the "British Journal of Radiology," Supplement 25 (BJR 1996). The problem of estimating the effective field size equivalent to a clinical irregular field is most accurately handled by irregular field calculations (CUNDIFF et al. 1973).

5.3.2
Monitor Unit Calculation for Rotation Therapy

The MU calculations for rotation therapy can be computed using the following equation:

$$MU = \frac{ID}{TAR_{avg} \cdot \dot{D}_{fs} \left(\frac{SCD}{SAD}\right)^2}$$

and the MU per degree setting is given by

$$MU/\deg = \frac{monitor\ unit\ setting}{degrees\ of\ rotation},$$

where the symbols have the previous meaning and TAR_{avg} is an average TAR (averaged over radii at selected angular intervals, e.g., 20°).

5.3.3
Monitor Unit Calculations for Irregular Fields

For large, irregularly shaped fields and at points off the central axis, it is necessary to take account of the off-axis change in intensity (relative to the central axis) of the beam, the variation of the SSD within the field of treatment, the influence of the primary collimator on the output factor, and the scatter contribution to the dose. Changes in the beam quality as a function of position in the radiation field also should be considered (HANSON and BERKLEY 1980; HANSON et al. 1980).

The general method used for irregular-field calculations consists of summation of the primary and scatter irradiation at each point of interest, with allowance for the off-axis change in intensity (off-axis factor) and SSD (Cundiff et al. 1973). The MUs required to deliver a specified tumor dose at an arbitrary point in an irregular field (Fig. 5.10) can be calculated as follows:

$$MU = \frac{TD}{[TAR(d,0) + \overline{SAR}(d)] \cdot \dot{D}_{fs}(SSD + d_{max}, r_c) \cdot TF \cdot OAF \cdot \left(\frac{SSD + d_{max}}{SSD + g + d}\right)^2},$$

where

$TAR(d,0)$ = zero-field size TAR at depth d

$SAR(d)$ = average SAR for point in question at depth d determined using the Clarkson technique

$\dot{D}_{fs}$ = Gy/MU in a small mass of tissue, in air, on the central axis at normal $SSD+d_{max}$ for the collimated field size

SSD = nominal SSD for treatment constraints

d_{max} = depth of dose maximum

TF = blocking tray attenuation factor

G = vertical distance between skin surface over point in question and nominal SSD (beam vertical)

D = vertical depth, skin surface to point in question

OAF = in-air off-axis factor

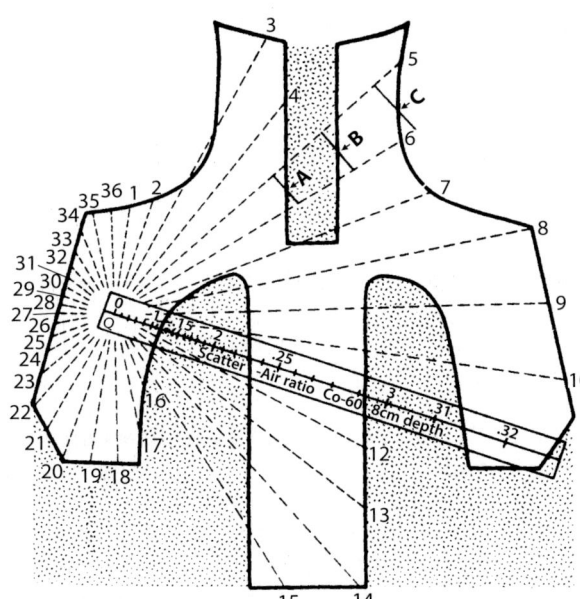

Fig. 5.10. Outline of mantle field illustrates method of determining scatter-to-air ratio, used for irregular-field dose calculations. (From Cundiff et al. 1973)

Several modifications to the original method have been suggested. These include using the expanded field size at a depth for the SAR calculation; determining the off-axis factor using the distance from the central axis to the slant projection of the point of calculation to the SSD plane along a ray from the source; and determining the zero-area TAR using the slant depth along a ray going from the source to the point of calculation. It is generally accepted that the off-axis factor should be multiplied by the sum of the zero-area TAR and the SAR, as originally proposed.

Beam quality is a function of position in the field for beams generated by linear accelerators (Hanson and Berkley 1980; Hanson et al. 1980). The TAR_0 may be expressed as a function of position in the beam so that changes in beam quality can be incorporated into calculations, and it can be related to the half-value layer (HVL) of water by the following equation:

$$TAR(d,0,r) = e^{\left[\frac{-0.693(d - d_{max})}{HVL(r)}\right]}$$

where d is the depth of the point of reference, d_{max} is the depth of maximum dose, r is the radial distance from the central axis of the beam to the point of calculation, e is the base of the natural logarithm, and $HVL(r)$ is the beam quality expressed as the HVL measured in water.

One final point about irregular-field or off-axis dose calculations concerns the off-axis factor. In some computerized treatment planning systems, the calculation of dose to points off the central axis is based on the assumption that the off-axis factor can be represented by a separable function given by

$$OAF(x, y) = OAF(x, 0) \cdot OAF(0, y)$$

where x and y are the symmetry axes perpendicular to the beam axis and the functions $OAF(x,0)$ and $OAF(0,y)$ are equal for a square open field. For some accelerator-generated beams, this assumption is invalid because measured values differ from those predicted by the above equation by as much as 20% (Lam and Lam 1983).

5.3.4
Monitor Unit Calculations for Asymmetric X-ray Collimators

Asymmetric X-ray collimators (also referred to as independent jaws) allow independent motion of an individual jaw and may be available for one jaw pair or both pairs. Because MU calculations and treatment planning methods generally rely on symmetric jaw data, the dosimetric effects for asymmetric jaws must be understood before being implemented into the clinic. Several investigators have examined the effects of asymmetric jaws on PDD, collimator scatter, and isodose distributions; the reader is referred to other sources for more details (PALTA et al. 1988; SLESSINGER et al. 1993). In general, the only change to MU calculations for asymmetric fields is the need to incorporate an off-axis factor OAR(x) to account for off-axis beam intensity changes. The PDD is only minimally affected, but isodose curve shape can be altered and must be investigated for the particular treatment unit.

Monitor unit calculations for asymmetric jaws are only slightly more complex than for symmetric jaws (GIBBONS 2000). Typically, one simply applies an off-axis ratio (OAR) or off-center ratio (OCR) correction factor that depends only on the distance from the machine's central axis to the center of the independently collimated open field (SLESSINGER et al. 1993; PALTA et al. 1996). A more complex system that accounts for independent jaw settings, field size, and depth also has been described (CHUI et al. 1986; ROSENBERG et al. 1995). Calculations for asymmetric wedge fields follow similar procedures by simply incorporating a wedge OAR or OCR (ROSENBERG et al. 1995; KHAN 1993).

5.3.5
Monitor Unit Calculations for Multileaf Collimator System

Multileaf collimators (MLCs) are rapidly being deployed in clinics around the world as a replacement for alloy field shaping. Several investigators have examined the effects of MLC on PDD, collimator scatter, and isodose distributions (KLEIN et al. 1995). The effects due to field area shaped by the MLC on PDD and beam output parameters are similar to those resulting from Cerrobend field shaping; thus, dose/MU calculation methods as discussed previously simply use the equivalent area as defined by the MLC. The collimator scatter factor is determined using the X-ray collimator jaw settings, with an off-axis factor applied for asymmetric jaw settings; however, PALTA and colleagues (1996) found that the MLC field shape was a determining factor in selecting the appropriate output factor for their system. Their results

showed that MLC dosimetry is clearly dependent on machine-dosing differences, and so because treatment machine MLC designs are still changing, each institution is advised to carefully study the impact of the MLC on their basic MU calculation procedure before use in the clinic.

5.4
Dose Calculation Algorithms and Correction Factors

Current dose calculation algorithms can be broadly classified into correction-based and kernel-based models (MACKIE et al. 1996). Correction-based models correct the dose distribution in a homogeneous water phantom for the presence of beam modifiers, contour corrections (or air gaps), and tissue heterogeneities encountered in treatment planning of real patients. The homogeneous dose is obtained from broad field measurements obtained in a water phantom. Kernel-based models (also called convolution methods) directly compute the dose in a phantom or patient. The convolution methods take into account lateral transport of radiation, beam energy, geometry, beam modifiers, patient contour, and electron density distribution. The reader is referred to review articles for more details regarding the rigorous mathematical formalism of these type dose calculation algorithms (MACKIE et al. 1996; AHNESJÖ and ASPRADAKIS 1999).

Because basic dose distribution data are obtained for idealized geometries (e.g., flat surface, unit density media), corrections are needed to determine the dose distribution in actual patients. Typically, the dose data calculated for idealized geometry are multiplied by a correction factor to obtain the revised dose distribution. The methods commonly used to correct for the air gap are discussed briefly.

5.4.1
Ratio of Tissue–Air Ratio or Tissue–Phantom Ratio Method

In the ratio of TAR (or ratio of TPR) method (RTAR), the surface (along a ray line) directly above point A is unaltered, so the primary dose distribution at this point is unchanged (Fig. 5.11). For relatively small changes in surface topography, the scatter component is essentially unchanged; thus,

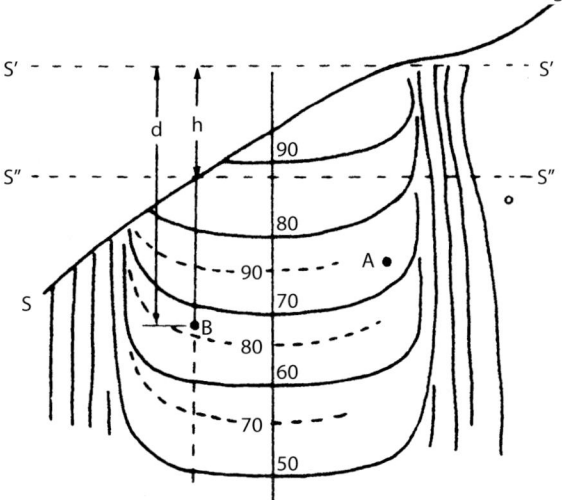

Fig. 5.11 Tissue–air ratio and effective SSD methods for the correction of isodose curves under a sloping surface. (From ICRU 1976)

the dose at point A can be considered as unaltered by patient shape; however, for point B, where there are considerable variations in the patient's topography, both the primary and scatter components of the radiation beam are altered. The correction factor *(CF)* can be determined using two TARs or TPRs as follows:

$$CF = \frac{T(d-h,s_d)}{T(d,s_d)},$$

where h = air gap.

5.4.2
Effective Source–Skin Ratio Method

In the effective SSD method, the isodose chart to be used is placed on the contour representation, positioning the central axis at the distance for which the curve was measured (Fig. 5.11). It is then shifted down along the ray line for the length of the air gap, h. The PDD value at point B is read and modified by an inverse-square calculation to account for the effective change in the peak dose. The CF can be expressed as follows:

$$CF = \frac{P(d-h,d_o,S,f,E)}{P(d,d_o,S,f,E)} \cdot \left(\frac{f+d_o}{f+h+d_o} \right)^2$$

5.4.3
Isodose Shift Method

Manual construction of an entire dose distribution for an actual patient with the previous methods would be time-consuming. The isodose shift method (Fig. 5.12), although simplistic, is efficient and gives satisfactory results in most cases. In this method, the isodose chart is moved down along a diverging ray by a fractional amount of the air gap, h. The intersection of the isodose lines with this ray are read off directly. For ^{60}Co radiation, a shift of two-thirds h is used, and for 18-mV X-rays, a shift of one-half h is used.

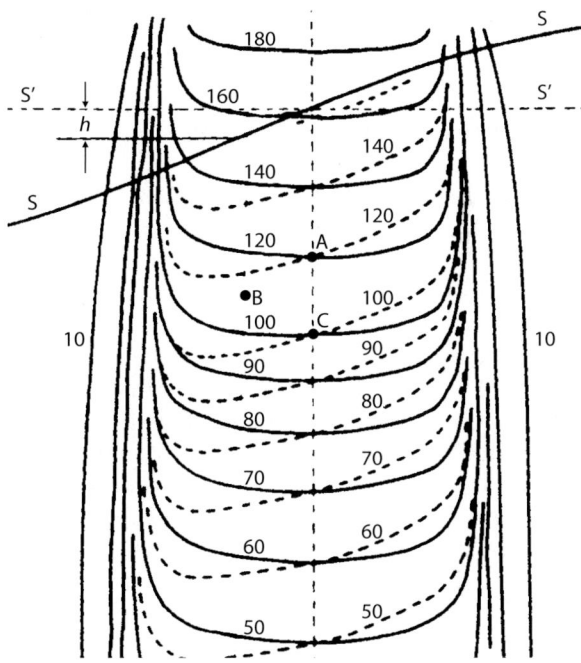

Fig. 5.12 Isodose shift method of correcting isodose curves under a sloping surface. (From ICRU 1976)

5.5
Correction for Tissue Inhomogeneities

Most of the correction-based models used for clinical dose calculations for radiation therapy treatment planning still rely mainly on "one-dimensional" effective pathlength (EPL) approaches, some of which were developed before X-ray CT. These models consider the effect of patient structure only along the ray joining the point of computation and the source of radiation and are limited in accuracy under some circumstances. It should be well understood that in correcting water-based dose calculations for tissue inhomogeneities, obtaining correct anatomical information is as important, if not more, than the type of dose algorithms used.

Patient tissues may differ from water by composition or density, which alters the dose distribution. Four inhomogeneities are usually encountered in treatment planning dose calculations: air cavities; lung; fat; and bone. To correct fully for these inhomogeneities, it is necessary to know their size, shape, and position, and to specify their electron density and atomic number. Manual correction methods for tissue inhomogeneities closely resemble the three methods discussed for correcting for patient contour changes and are described below.

5.5.1
Ratio of Tissue–Air Ratio Method

The ratio of TAR (RTAR) method of correction for inhomogeneities is given by

$$CF = \frac{T(d_{eff}, S_d)}{T(d, S_d)},$$

where the numerator is the TAR for the equivalent water thickness, d_{eff}, and the denominator is the TAR for the actual thickness, d, of tissue between the point of calculation and the surface along a ray passing through the point. S_d is the dimension of the beam cross section at the depth of calculation. The RTAR method accounts for the field size and depth of calculation. It does not account for the position of the point of calculation with respect to the heterogeneity. It also does not take into account the shape of the inhomogeneity, but instead assumes that it extends the full width of the beam and has a constant thickness (i.e., a slab-type geometry).

5.5.2
Isodose Shift Method

Isodose lines are shifted by an amount equal to a constant times the thickness of the inhomogeneity as measured along a line parallel to the central axis and passing through the calculation point. Values for the shift constant empirically determined for ^{60}Co and 4-mV X-rays are –0.6 for air cavities, –0.4 for lung, +0.5 for hard bone, and +0.25 for spongy bone. The isodose curves are shifted away from the surface for lung and air cavities and toward the surface for bone.

5.5.3
Power Law TAR Method

The power law TAR method was proposed by BATHO (1964) and generalized by YOUNG and GAYLORD (1970). This method, sometimes called the Batho method, attempts to account for the nature of the inhomogeneity and its position relative to the point of calculation; however, it does not account for the extent or shape of the inhomogeneity. The correction factor for the point P is given by

$$CF = \left(\frac{T(d_2, S_d)}{T(d_1, S_d)}\right)^{\rho_2 - 1},$$

where d_1 and d_2 refer to the distances from point P to the near and far side of the water-equivalent material, respectively, S_d is the beam dimension at the depth of P, and ρ_2 is the relative electron density of the inhomogeneity with respect to water.

SONTAG and CUNNINGHAM (1978) derived a more general form of this correction factor, which can be applied to a case in which the effective atomic number of the inhomogeneity is different from that of water and the point of interest lies within the inhomogeneity. The correction factor in this situation is given by

$$CF = \frac{T(d_2, S_d)^{(\rho_b - 1)}}{T(d_1, S_d)^{(\rho_b - \rho_a)}} \cdot \frac{(\mu_{en}/\rho)_a}{(\mu_{en}/\rho)_b},$$

where ρ_a is the density of the material in which point P lies at a depth d below the surface and ρ_b is the density of an overlying material of thickness $(d_2 - d_1)$; $(\mu_{en}/\rho)_a$ and $(\mu_{en}/\rho)_b$ are the mass energy absorption coefficients for the medium a and b.

At an interface between two materials of different composition, there is a loss of equilibrium as a result of changes in the electron fluence, i.e., the number of electrons generated from the photon interactions; therefore, the dose over a distance comparable with the range of the electrons is perturbed. For ^{60}Co radiation, the alteration in the dose distribution occurs in only the few millimeters surrounding the interface; however, for high-energy photon beams, the region extends for several centimeters.

Table 5.2, modified from WONG and PURDY (1990), summarizes the results of comparisons of measured and calculated results for several of the methods of dose calculations commonly used for radiation therapy dose calculations. The comparisons are for a single Co-60 beam directly incident on a semi-infinite heterogeneous phantom geometry.

Table 5.2. Summary of the comparisons of calculated correction factors and measurements for different methods of tissue inhomogeneity corrections. The differences are given as (calculations - measurements)/measurements × 100%. The table lists the mean differences and one standard deviation for n number of points of comparison. *RTAR* ratio of tissue–air ratio (Modified from WONG and PURDY 1990)

Phantoms	Methods				
	RTAR	BATHO	ETAR	Convolution	Superposition
Phantom 1 (n=14)	6.46±2.96	−0.21±1.54	0.18±2.26	1.77±0.73	−0.28±2.04
Phantom 2 (n=13)	8.92±4.97	−0.17±1.71	0.45±0.83	0.64±0.63	0.42±1.95
Phantom 3 (n=14)	−0.18±5.21	3.00±4.54	−0.01±1.97	3.74±3.40	−0.09±2.96
Phantom 4 (n=14)	1.88±7.72	2.52±4.14	1.23±2.96	0.98±2.58	−0.09±3.83

Phantom 1: Slab phantom composed of polystyrene/cork/polystyrene/polystyrene
Phantom 2: Slab phantom composed of polystyrene/polystyrene/cork/polystyrene
Phantom 3: Slab phantom composed of polystyrene/cork/aluminum/polystyrene
Phantom 4: Slab phantom composed of polystyrene/aluminum/cork/polystyrene

The importance of verifying dose distributions generated by a treatment planning system should not be understated. Modern systems are extremely complicated and require well-defined dose calculation verification techniques. Traditional 2D methods using depth doses and selected profiles are difficult to apply in the 3D situation. It is the task of the clinical physicist to be aware where these calculations may fail and to take proper precautions, such as verifying with measurements, in those situations where the errors will lead to undesirable clinical outcome.

5.6
Clinical Photon-Beam Dosimetry

Photon-beam PDD for a specific depth increases with increasing energy, SSD, and field size. The depth of the 50th percentile increases from approximately 14 cm for 4-mV X-rays to nearly 23 cm for 25-mV X-rays. The depth of maximum dose varies from about 1 cm for 4-mV X-rays to over 3.5 cm for 25-mV X-rays; however, the depth of maximum dose position is not unambiguously defined by the energy of the X-ray beam, but depends on the field size and on the treatment-head design of the particular machine. This shift in dose maximum is principally the result of electron scattering from the X-ray collimator. The specification of X-ray PDD in terms of the maximum electron energy impinging on the X-ray target is not sufficient to characterize the X-ray beam. Various methods of specifying beam quality are now in use, including the tissue phantom ratio at 20 cm depth to that at 10 cm depth (TPR_{10}^{20}), the PDD at 10 cm depth, and the depth of the 80% dose level.

In general, the dose to the surface and in the build-up region for megavoltage photon beams increases with an increase in field size, with the introduction of a blocking tray into the beam, and with a decrease in the distance between skin and blocking tray. The surface dose tends to decrease with increasing X-ray beam energy.

The exit dose is frequently calculated by multiplying the maximum dose by the central axis PDD value corresponding to the patient's thickness. Sometimes the thickness value is reduced by the depth of maximum dose. However, this technique overlooks the fact that there is insufficient material beyond the exit surface to provide the total scatter dose in many situations; thus, the actual dose received by the tissues at or near the exit surface is less than that calculated using this method. It has been shown that if there is no backscatter material, the dose to the skin at the exit surface (≈ 4 mg/cm^2 depth) is 15–20% less for ^{60}Co γ-rays and about 10% less for 25-mV X-rays than the dose with full backscatter (GAGNON and HORTON 1979). The skin dose increases sharply as the thickness of material is increased beyond the exit surface, until full backscatter conditions are obtained with about 0.5 g/cm^2 of added material.

5.6.1
Single-Field Isodose Charts

Isodose charts, such as those shown in Figure 5.7, provide much more information about the radiation beam characteristics than do central-axis PDD data alone; however, even isodose charts are limited in that they represent the dose distribution in only one plane (typically the one containing the beam's central axis) and are usually available only for square or rectangular fields.

Isodose charts are generally measured in a water phantom with the radiation beam directed perpendicular to the phantom's flat surface. When the dose distribution is calculated for a patient, the isodose curves must then be corrected for the effects of irregular surface topography, oblique incidence, and inhomogeneities encountered in the path of the beam, using the methods discussed previously.

Isodose curves show the relative uniformity of the beams across the field at various depths, and also provide a graphical depiction of the width of the beam's penumbra region. Cobalt-60 teletherapy units exhibit a relatively large penumbra, and their isodose distributions are more rounded than those from linac X-ray beams. This is due to the relatively large source size (typically 1–2 cm in diameter vs only a few millimeters for linacs). Linac beam penumbra width does increase slightly as a function of energy and if unfocused MLC leaves are used, but is still much less than that for ^{60}Co units. In addition to the smaller penumbra, linac X-ray isodose distributions have relatively flat isodose curves at depth; however, at shallow depths, particularly at d_{max}, linac X-ray beams typically exhibit an increase in beam intensity away from the central axis. This beam characteristic is referred to as the dose profile "horns" and is dependent on flattening filter design. In general, each treatment unit has unique radiation beam characteristics, and thus, isodose distributions must be measured, or at least verified, for each specific unit.

Another important point to understand is how the radiation field size is defined. The radiation field size dimensions refer to the distance perpendicular to the beam's direction of incidence that corresponds to the 50% isodose at the beam's edge. It is defined at the skin surface for SSD treatments, and at the axis depth for source-to-axis distances (SADs) for isocentric treatments. The linac's light field is typically set using this definition (radiation-light field agreement tolerance is typically ±2 mm).

5.6.2
Depth Dose Build-up Region

When a photon beam strikes the tissue surface, electrons are set in motion, causing the dose to increase with depth until the maximum dose is achieved at depth d_{max}. As the energy of the photon beam increases, the depth of the build-up region is increased. The subcutaneous tissue-sparing effects

of higher energy X-rays, combined with their great penetrability, make them well suited for treating deep lesions. For specific X-ray energy, the magnitude of the skin dose generally increases with increasing field size, and with the insertion of plastic blocking trays in the beam (Fig. 5.13). The blocking trays should be at least 20 cm above the skin surface because skin doses are significantly increased for lesser distances. Copper, lead, or lead glass filters beneath plastic trays can be used to remove the undesired lower energy electrons that contribute to skin dose, but this is rarely done routinely in the clinic (PURDY 1986; RUSTGI and RODGERS 1985).

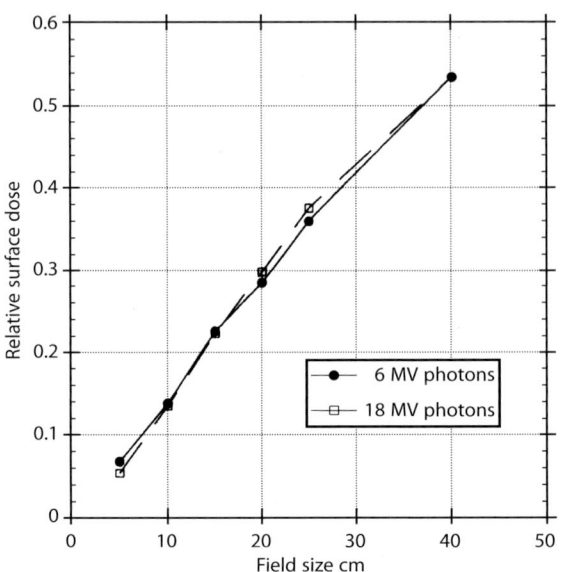

Fig. 5.13. Relative surface dose vs field size with blocking tray in place for 6-mV and 18-mV photons. (From KLEIN and PURDY 1993)

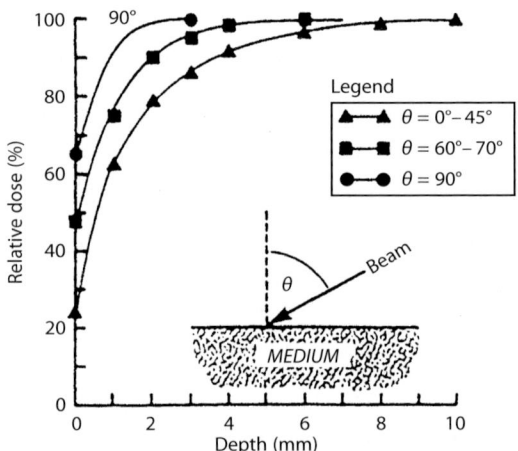

Fig. 5.14. The variation of surface dose and depth of maximum dose as a function of the angle of incidence of the X-ray beam with the surface (4 mV, 10×10 cm)

As the angle of the incident radiation beam becomes more oblique, the surface dose increases, and d_{max} moves toward the surface (Fig. 5.14). This is due to more secondary electrons being ejected along the oblique path of the beam (GAGNON 1979; GERBI et al. 1987; SVENSSON et al. 1977).

5.6.3
Depth Dose Exit Dose Region

The skin and superficial tissue on the side of the patient from which the beam exits receive a reduced dose if there is insufficient backscatter material present. The amount of dose reduction is a function of X-ray beam energy, field size, and the thickness of tissue that the beam has penetrated reaching the exit surface. For a 6-mV beam, PURDY (1986) measured a 15% reduction in dose with little dependency on field size. This work was repeated for 18-mV beams by KLEIN and PURDY (1993), who reported a 11% reduction in exit dose. Generally, the addition of a thickness of tissue-equivalent material on the exit side equivalent in thickness to about two-thirds of the d_{max} depth is sufficient to provide full dose to the build-down region on the exit side. Figure 5.15 shows the effects of various backscattering media when placed directly behind the exit surface.

5.6.4
Interface Dosimetry

The dose distribution within the patient in transition zones (interfaces of different media) depends on radiation field size (scatter influence), distance between interfaces (e.g., air cavities), differences between physical densities and atomic number of the interfacing media, and the size and shape of the different media. Near the edge of the lungs and air cavities, the reduction in dose can be larger than 15% (KORNELSEN and YOUNG 1982). For inhomogeneities with density larger than water, there will be an increase in dose locally due to the generation of more electrons; however, most dense inhomogeneities have atomic numbers higher than that of water, so that the resulting dose perturbation is further compounded by the perturbation of the multiple coulomb scattering of the electrons. Near the interface between a bony structure and water-like tissue, large hot and cold dose spots can be present.

Measurements are generally done with parallel-plate ionization chambers. Corrections should be

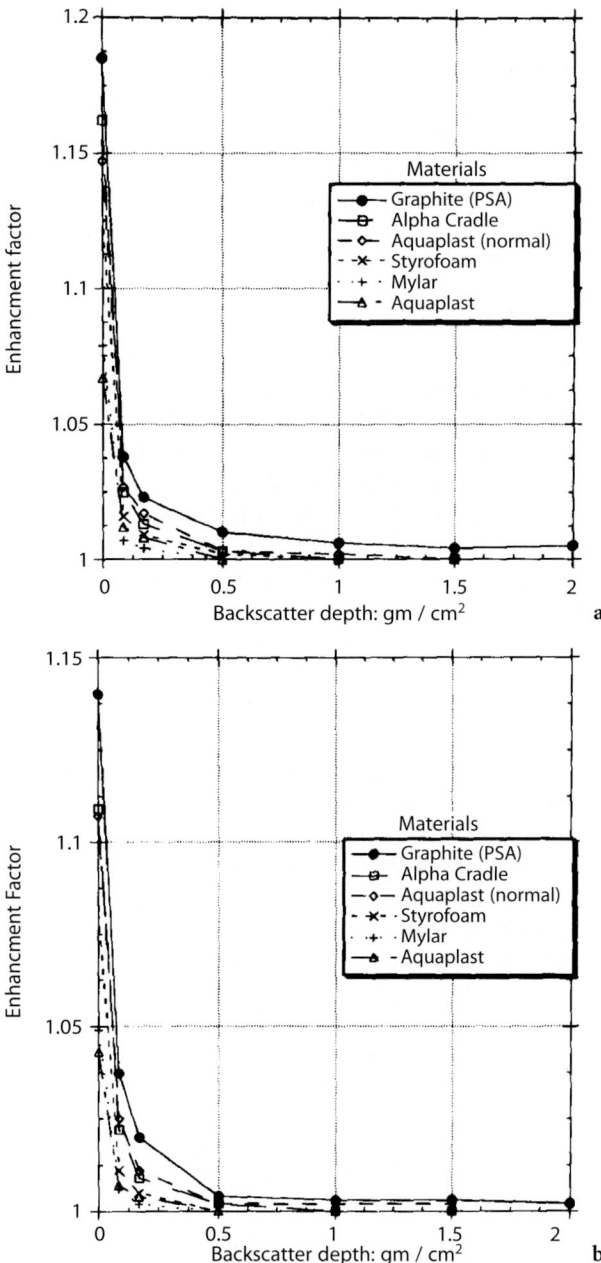

Fig. 5.15a,b. Enhancement of exit dose for **a** 6-mV and **b** 18-mV photons for a 15 × 15-cm field at 100 cm SAD vs backscatter depth for various backscattering materials. (From KLEIN and PURDY 1993)

used to account for plate separation, energy (ionization ratio), and guard width (GERBI and KHAN 1990). Thermoluminescent dosimeters (TLDs) and film also have been used for transition zone measurements, but the problems associated with thickness and atomic number (respectively) and the associated QA needed make measurements with these dosimeters more laborious, and the results typically have a greater uncertainty. Several benchmark measurements have been reported for various geometries simulating clinical situations such as air cavities (larynx), lung (mediastinum), bone (femur), and prostheses (steel for hip and silicon for breast); these are discussed below.

5.6.4.1
Air Cavities

EPP and colleagues (1958) performed measurements with a parallel-plate ionization chamber for ^{60}Co beams that showed significant losses of ionization on the central axis after traversing air cavities of varying dimensions. The losses, which were due to lack of forward scattered electrons, were approximately 12% for a typical larynx-like air cavity, but recovered within 5 mm in the new build-up region. KOSKINEN and SPRING (1973) confirmed these measurements with ultra-thin (20 mm) LiF-Teflon dosimeters and reported similar responses in the proximal region of the air cavity due to lack of backscatter. NILLSON and SCHNELL (1976) used even thinner LiF disks (10 mm) and reported data for 6- and 42-mV photons, with the higher-energy beam showing fewer effects. EPP and colleagues (1977) reported a 14.5% loss at the distal interface for 10-mV photons with a build-up curve that reached a plateau within 20 mm behind the interface. BEACH and associates (1987) measured losses at the distal interfaces with an extrapolation chamber and recommended minimum field sizes

to be used in irradiation of the larynx to balance losses due to forward scatter. KLEIN and colleagues (1993) measured distributions about air cavities for 4- and 15-mV photons with a parallel-plate chamber in both the distal and proximal regions. They combined the distributions in a parallel-opposed fashion and observed a 10% loss at the interfaces for an air cavity of $2\times2\times20$ cm for 4×4-cm parallel-opposed fields for either energy. They also observed losses at the lateral interfaces perpendicular to the beam on the order of 5% for the 4-mV beam. Finally, OSTWALD et al. (1996) used TLD for discrete measurements in simulated larynx geometries.

5.6.4.2
Lung Interfaces

Although the problem of reestablishing equilibrium for lung interfaces is not as severe as with air cavities, a transition zone region at the lung–tissue interface still exists over the range of typical clinical photon beam energies. RICE and colleagues (1988) measured responses within various simulated lung media using a parallel-plate chamber and phantom constructed of simulated lung material (average lung material density, $\rho = 0.31$ g/cm^3). They measured correction factors (CFs) with a 10-cm layer of lung material vs water and observed minor differences at the interface compared with regions beyond the lung and a small dependence on field size (7% for 4 mV; Fig. 5.16a). A considerable build-up curve was

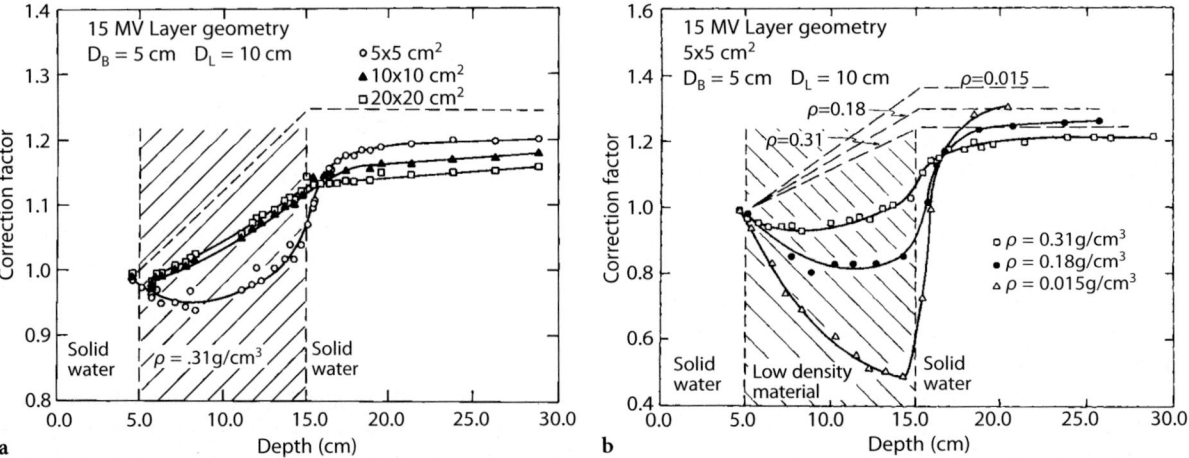

Fig. 5.16. a Dose correction factors as a function of depth for a transition zone geometry that simulates a lung–tissue interface for three different field sizes and a lung thickness of 10 cm for 15-mV X-rays. The modification to the primary dose only on the central axis (shown by the *dashed curve*) is independent of field size. (From RICE et al. 1988). **b** Dose correction factors as a function of depth for a transition zone geometry that simulates a lung–tissue interface for three different densities, a 5×5-cm field, and a lung thickness of 10 cm for 15-mV X-rays. The modification to the primary dose only on the central axis is shown by the *dashed curve*. (From RICE et al. 1988)

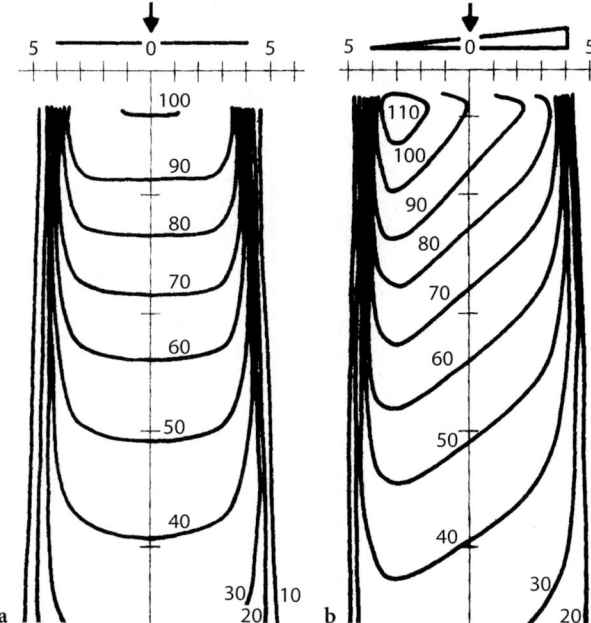

Fig. 5.17a,b. Isodose distributions for a 6-mV X-ray beam with an 8×8-cm field size. **a** Open field. **b** Field with a 45° wedge. (From KHAN 1994)

observed (10% change in CF) for a 5×5-cm field for the 15-mV beam, which began in the distal region of the lung and reached a plateau beyond the lung (Fig. 5.17b). KLEIN et al. (1997) measured the effects of nonequilibrium for tissue equivalent volumes within lung media and found significant underdosage, especially for small volumes and high-energy beams.

5.6.4.3
Bone Interfaces

DAS and colleagues measured dose perturbation factors (DPFs) proximal and distal for simulated bone–tissue interface regions using a parallel plate chamber for both 6- and 24-mV X-ray beams (DAS and KHAN 1989; DAS et al. 1990). They reported DPFs of 1.1 for the 6-mV beam and 1.07 for the 24-mV beam at the proximal interface. A 7% enhancement (build-down) was measured for the 24-mV beam at the distal interface, whereas the 6-mV beam exhibited a new build-up region distally with a DPF of 0.95 at the interface. KLEIN and co-workers (1988) made similar measurements for 4- and 15-mV photons with similar results, except that the 15-mV beam exhibited no enhancement. These build-up or build-down regions dissipated within a few millimeters in

the tissue-like media. The perturbations were independent of thickness and lateral extent of the bone or radiation field size.

5.6.4.4
Prostheses (Steel and Silicon)

DAS et al. (1990) measured forward dose perturbation factors (FDPFs) following a 10.5-mm-thick stainless-steel layer simulating a hip prosthesis geometry. They measured an enhancement of 19% for 24-mV photons, but only 3% for 6-mV photons. They also measured backscatter dose perturbation factors (BDPFs) for various energies for many high-Z materials, including steel. They reported an enhancement of 30% for steel due to backscattered electrons independent of energy, field size, or lateral extent of the steel. These interface effects dissipated within a few millimeters in polystyrene. Other reports dealing with dosimetry perturbations due to metal objects are included in the references (SIBATA et al. 1990; THATCHER 1984).

KLEIN and KUSKE (1993) reported on interface perturbations with silicon breast prostheses. Such prostheses have a density similar to breast tissue but have a different atomic number. They observed a 6% enhancement at the proximal interface and a 9% loss at the distal interface.

5.6.5
Wedge Filter Dosimetry

When a wedge filter is inserted into the beam, the dose distribution is angled at some specified depth to some desired angle relative to the incident beam direction over the entire transverse dimension of the radiation beam (Fig. 5.17). For cobalt units, the depth of the 50% isodose is usually selected for specification of the wedge angle, whereas for higher energy linacs, higher percentile isodose curves, such as the 80% curve, or the isodose curves at a specific depth (10 cm), are used to define the wedge angle.

Cobalt unit wedges are typically designed for specific field sizes (nonuniversal wedges) to keep the dose rate of the unit within a useful clinical range. Linacs are typically equipped with multiple wedges (universal wedges) that may be used with an allowed range of field sizes. Some linacs (Elekta) feature a single wedge, referred to as a motorized wedge, located in the treatment head, and the desired wedged dose distribution is obtained by the proper combination of wedged and unwedged treatment.

Although wedges can be designed for any desired wedge angle, 15, 30, 45, and 60° wedges are most common. The wedged isodose curves can be normalized in two ways. In some older systems, the wedge dose distributions have the wedge factor (i.e., the ratio of the measured central axis dose rate with and without the wedge in place) incorporated into the wedged isodose distribution. More commonly, the wedge isodose curves are normalized to 100% at d_{max}, and a separate wedge factor is used to calculate the actual treatment monitor units or time. McCullough et al. (1988) noted that wedge factors measured at d_{max} are generally accurate to within 2% for depths up to 10 cm, but at greater depths can be inaccurate to 5% or more. The inclusion (or noninclusion) of the wedge factor is an extremely important point to understand, as serious error in dose delivered to the patient can occur if used improperly.

Sewchand et al. (1978) and Abrath and Purdy (1980) pointed out that beam hardening results when a wedge is inserted into the radiation beam. The percent depth dose (PDD), therefore, can be considerably increased at depth. Differences reported were nearly 7% for 4-mV 60° wedge field PDD from the open-field PDDs at 12-cm depth and 3% difference in depth dose values between the wedge field and the open field for a 60° wedge using 25-mV X-rays.

When the patient's treatment is planned, wedged fields are commonly arranged such that the angle between the beams, the hinge angle ϕ, is related to the wedge angle θ by the following relationship (Fig. 5.18):

$$\theta = 90\ degrees - \phi/2$$

For example, as shown in Figure 5.19, 45° wedge fields orthogonal to one another yield a uniform dose distribution.

Modern computer-controlled medical linacs now have software features that allow the user to create a wedge-shaped dose distribution by moving one collimator jaw across the field in conjunction with adjustment of the dose rate over the course of the daily single field treatment (see Fig. 5.20). In the case of Varian linacs, this feature is referred to as enhanced dynamic wedge (EDW; Leavitt et al. 1990). This technology provides superior dose distributions and eliminates the above-mentioned beam-hardening problem seen in physical wedges.

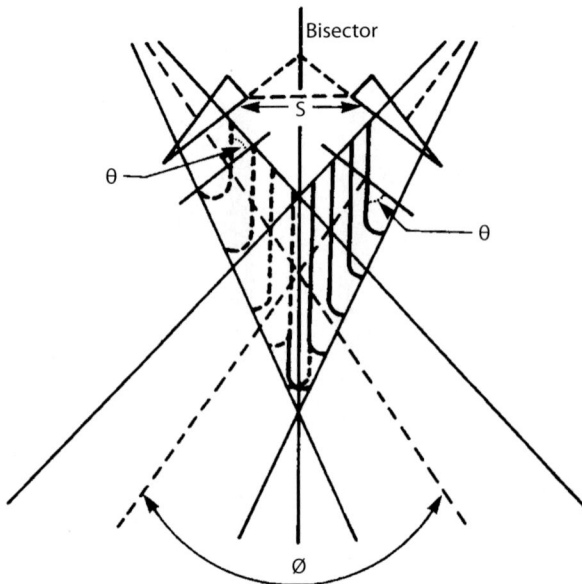

Fig. 5.18. Parameters of the wedge beams: θ is the wedge angle, ϕ is the hinge angle, and S is separation. Isodose curves for each wedge field are parallel to the bisector. (From Khan 1994)

This feature can deliver a greater number of wedge angles (10, 15, 20, 25, 30, 45, and 60°) and over larger field sizes, including asymmetric field sizes (30 cm in the wedge direction, with 20 cm toward the wedge "heel" and 10 cm toward the wedge "toe").

5.7
Treatment Planning: Combination of Treatment Fields

5.7.1
Parallel-Opposed Fields

When only two unmodified X-ray beams are used in conventional radiation therapy (i.e., non-IMRT), they usually are parallel-opposed beams (i.e., directed toward each other from opposite sides of the anatomic site with the central axes coinciding). Figure 5.21 presents the normalized relative axis dose profiles from parallel-opposed photon beams for a 10×10-cm field at an SSD of 100 cm and for patient diameters of 15–30 cm in 5-cm increments. The weight of a beam denotes a numerical value assigned to the beam at some normalization point. For SSD beams, the weight specifies the relative dose assigned to the beam at d_{max}, and for isocentric beams, at isocenter.

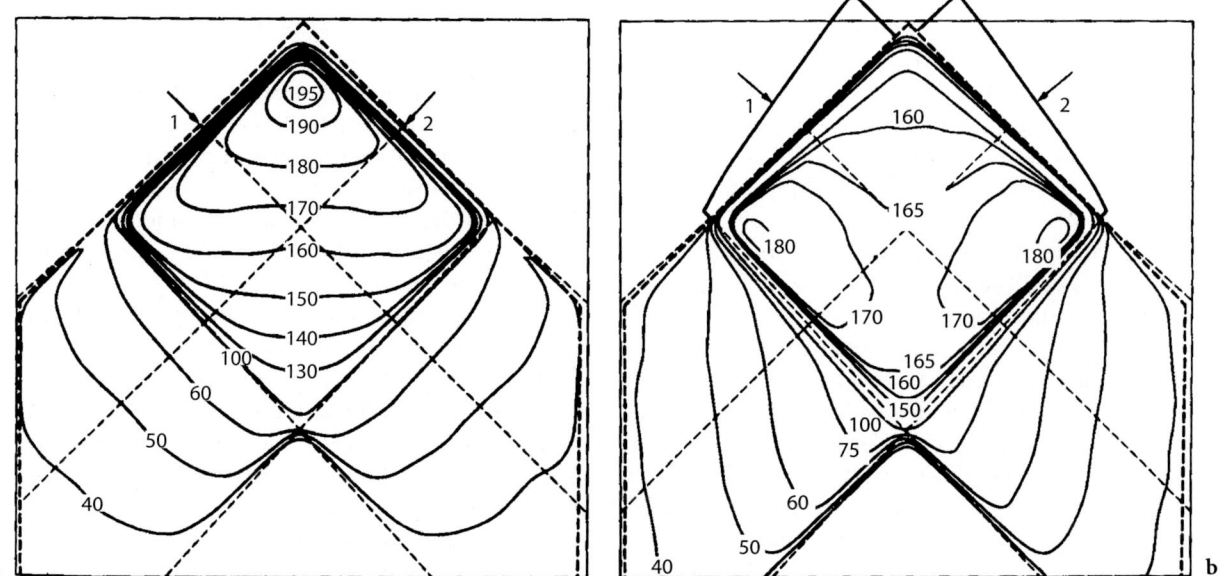

Fig. 5.19a,b. Isodose distribution for two angled beams. **a** Without wedges. **b** With wedges: 4-mV; field size = 10 × 10 cm; SSD = 100 cm; wedge angle = 45°. (From KHAN 1994)

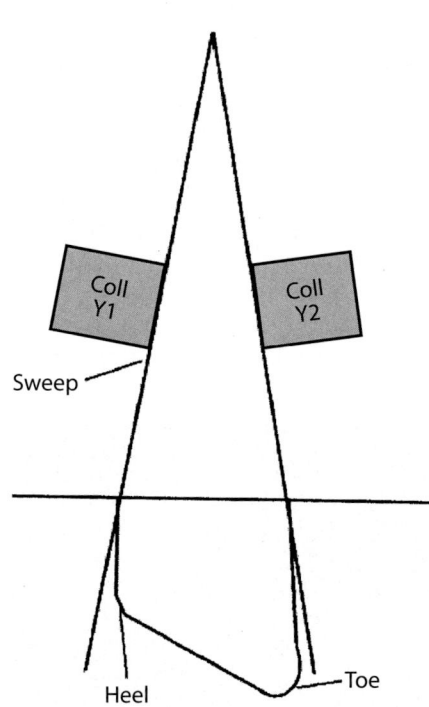

Fig. 5.20. Principles of Varian's Enhanced Dynamic Wedge. Modern computer-controlled medical linacs now have software features that allow the user to create a wedge-shaped dose distribution by moving one collimator jaw (Y1) across the field (Y2 is held fixed) in conjunction with adjustment of the dose rate over the course of the daily single field treatment.

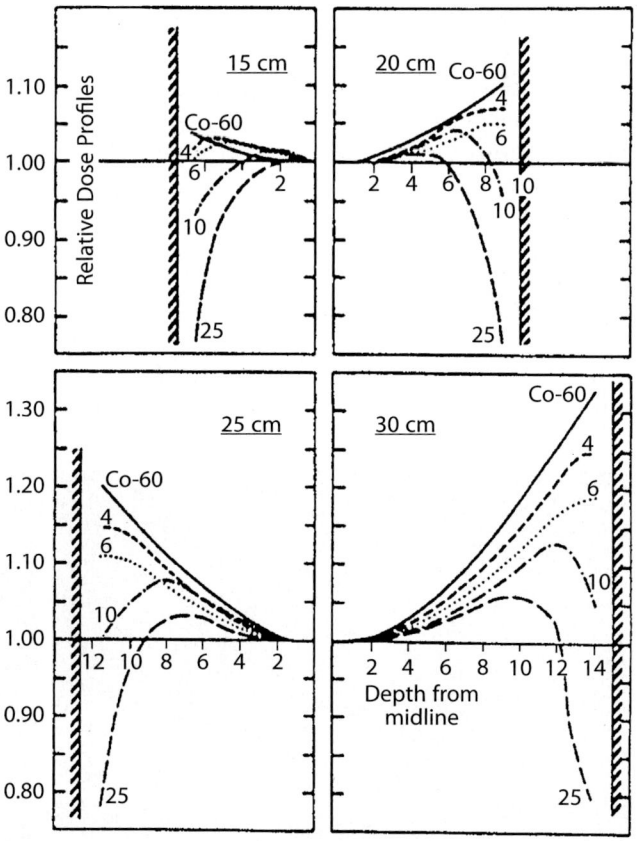

Fig. 5.21. Relative central-axis dose profiles as a function of X-ray energy (^{60}Co or 4, 6, 10, and 18 mV) and patient thickness (15, 20, 25, and 30 cm). The parallel-opposed beams are equally weighted, and the profiles are normalized to unity at midline. Because of symmetry, only half of each profile is shown.

The maximum patient diameter easily treated with parallel-opposed beams for a mid-plane tumor requiring 50 Gy or less with low-energy megavoltage beams is about 18 cm. For "thicker" patients, higher X-ray energies produce improved dose profiles with less dose variation along the central axis without resorting to more complex multibeam arrangements.

For some treatment sites, the underdosing achieved near the skin surface with very high-energy, parallel-opposed X-ray beams is a highly advantageous feature, but in others it may be desirable to achieve a higher dose nearer to the skin. With very high energy X-ray beams traversing small anatomic thicknesses, the exit dose can exceed the entry dose, and the exact dose distribution in the regions beneath the entry and exit surfaces from parallel-opposed high-energy X-ray beams must be carefully evaluated to consider properly the contribution from both entrance and exit components.

Unequal beam weightings are advantageous if the target volume is not midline. The greater the unequal weighting, the greater will be the shift of the higher-dose region toward one surface and away from midline. Although in some anatomic sites unequal weighting may be advantageous, special attention must be directed to the anatomic structures in the high-dose volume.

5.7.2
Multiple-Beam Arrangements

Figure 5.22 shows three commonly used coaxial three-field beam arrangements. A direct anterior field with two anterior oblique fields can be used to generate a high-dose region where the three fields overlap, whereas a low-dose region exists beyond this intersection point. For example, if this arrangement is used for treating the mediastinum, the spinal cord might be included in the anterior beam but spared by the anterior oblique beams. Moving the anterior oblique fields laterally to form a parallel-opposed pair yields a rectangular isodose region with a more uniform dose gradient; however, the magnitude of the dose gradient is determined by the relative weighting of the beams and the thickness of tissue traversed. An anterior field with two symmetrically placed posterior oblique beams yields elongated isodose curves. The degree of elongation is determined by the relative thickness of tissue each beam traverses to the point of intersection and by the relative weights of the beams. Three-field arrangements are often useful for treating tumors lateral to the midline of a patient.

Three-field nonaxial (noncoplanar) arrangements are readily achieved with linacs by rotating the table and gantry. A common technique for treating pituitary tumors uses two lateral fields and a vertex field with the beam entering through the top of the head. Astrocytomas often are treated with parallel-opposed lateral fields and a frontal field entering through the forehead. A 90° couch rotation is used with the gantry rotated laterally for the vertex or frontal fields. The lateral fields are also rotated via collimator to ensure that the "heel" of the wedges are in the plane of the vertex/frontal field trajectory.

Four-field techniques are typically used in sites such as the abdomen or the pelvis. In most instances the arrangements consist of pairs of par-

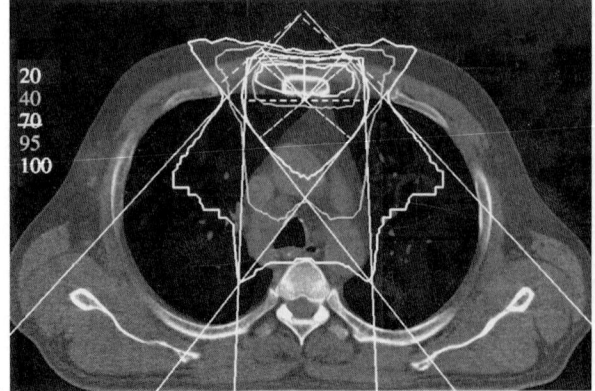
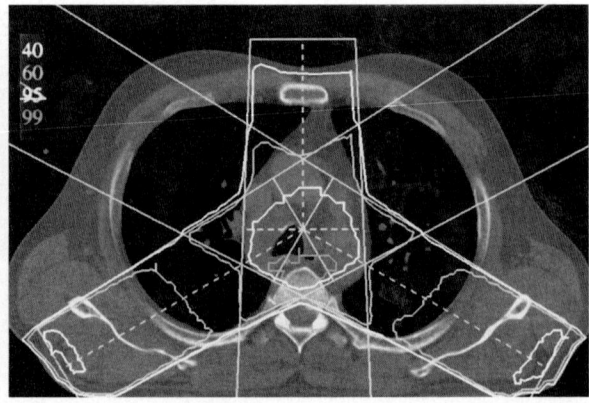

a   b

Fig. 5.22a,b. Dose distribution for two different beam arrangements using 6-mV X-ray beams, 8×10-cm, 100-cm SSD. Isodose curves have been renormalized to show the 100% line almost encompassing the target volume. **a** Anterior field with two anterior oblique fields at 40° off the midline, all equally weighted. **b** Anterior field with a weight of 0.8 with two equally weighted (1.0) posterior oblique fields separated by 120°.

allel-opposed fields, with a common intersecting point, which yield a "box-like" isodose distribution. Figure 5.23 compares the dose distributions achieved with a four-field box-like technique for 6- and 18-mV X-ray beams. The central dose distribution is similar for all beam energies, but the greater penetrability of the higher energy beams yields a lower dose to the region outside the box. Variations in the dose gradient are achieved by differential weighting of each pair of beams. Figure 5.24 shows other possible four-beam arrangements. Angulations of the beams yield a diamond-shaped dose distribution. A butterfly-shaped distribution is achieved if each pair of beams has a point of intersection lying on a common line but separated by a few centimeters.

Treatments involving more than four gantry angles, historically required with orthovoltage X-ray units to treat deep, midline lesions, have rarely been used with modern 18-mV megavoltage therapy

units; however, with the advent of three-dimensional conformal radiation therapy (3D CRT) and intensity-modulated radiation therapy (IMRT; see Chaps. 9 and 10) there has been an increase in multifield treatments such as the 3D CRT six-field technique commonly used for the treatment of prostate carcinoma.

5.7.3
Rotational Therapy

Rotational (or arc) therapy techniques, in which the treatment is delivered while the gantry (and thus the radiation beam) rotates around the patient, can be thought of as an infinite extension of the multiple-field techniques already described. This technique is most useful when applied to small, symmetrically shaped, deep-seated tumors, and is usually limited to field sizes less than about 10 cm in width

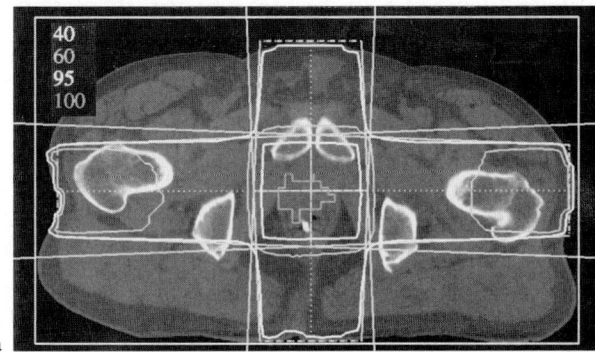
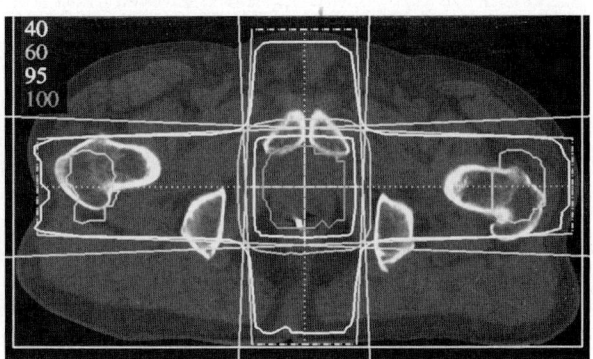

Fig. 5.23a,b. Dose distribution for four-field "box" technique with equal beam weightings: **a** 6-mV X-ray beams; **b** 18-mV X-ray beams. Note the improved dose distribution with the higher energy beam technique (more uniform dose in the target region and lower doses near the femoral head region of the lateral fields) as a result of the increased percent depth for 18-mV X-rays.

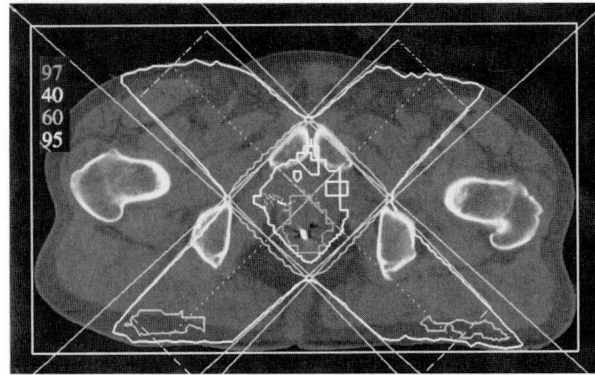
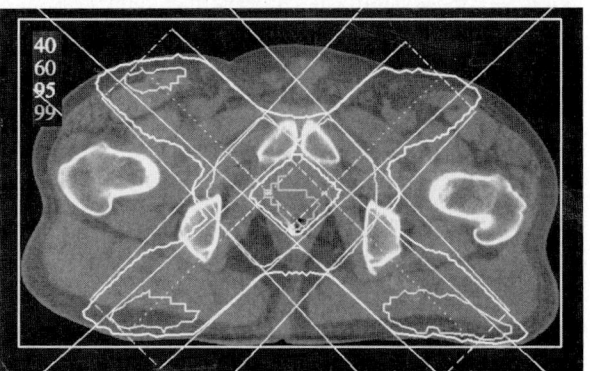

Fig. 5.24a,b. Dose distribution for four-field oblique beam technique for 6-mV X-rays with equal beam weightings. **a** With common isocenter resulting in a diamond-shaped dose distribution. **b** Each beam pair intersecting at two different points on a common line resulting in a butterfly-shaped isodose distribution.

for the treatment of centrally located lesions (i.e., have approximately an equal amount of tissue in all directions around the lesion).

Dose distributions generated by rotational techniques are not very sensitive to the energy of the photon beam. Figure 5.25 illustrates this fact, showing the dose distribution achieved using a 6-mV X-ray beam, and also the distribution using an 18-mV X-ray beam. There is a little less elongation in the direction of the shorter dimension of the patient's anatomy for the 18-mV beam and the dose distribution in the periphery is slightly lower.

In arc therapy techniques, one or more sectors of a 360° rotation are skipped to reduce the dose to critical normal structures. When a sector is skipped, the high-dose region is shifted away from the skipped region; therefore, the isocenter must be moved toward the skipped sector. This technique is referred to as past-pointing. Examples are shown in Figure 5.26.

The prostate, bladder, cervix, and pituitary are clinical sites that have been treated, either initially or for boost doses, with rotation or arc therapy techniques. Although the dose distributions achieved by rotation or arc therapy yield high target volume doses, these techniques normally result in a greater volume of normal tissue being irradiated (albeit at low doses) than fixed, multiple-field techniques. Moreover, the dose gradient at the edge of the target volume is never as sharp with a rotational technique as that achieved with a multiple-field technique.

In the past decade, a form of IMRT called tomotherapy, which literally means "slice therapy," has been developed and utilizes an arc therapy approach (CAROL 1994; GRANT 1996). Also, another IMRT rotational technique called intensity-modulated arc therapy (IMAT) has been developed and is in use at the University of Maryland (YU et al. 2002). Details on both of these arc-based IMRT techniques are given in Chapter 10.

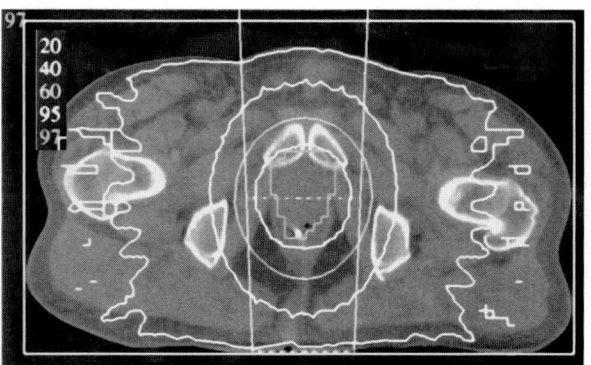 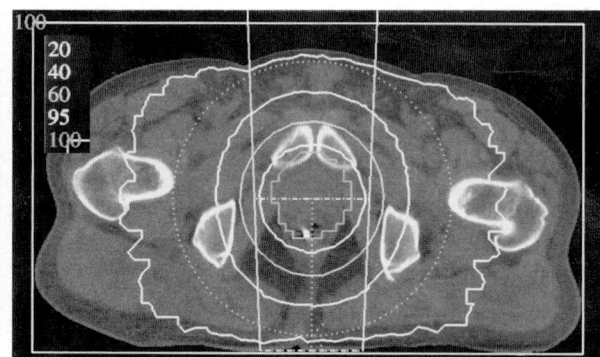

a   b

Fig. 5.25a,b. Dose distribution for 360° rotational therapy technique: **a** 6-mV X-ray beams; **b** 18-mV X-ray beams. Note that there is little difference in the dose distribution when using a higher-energy beam as a result of the offsetting effects of increased percent depth vs higher exit dose.

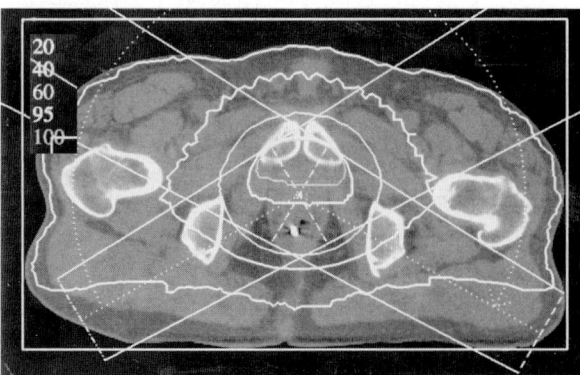 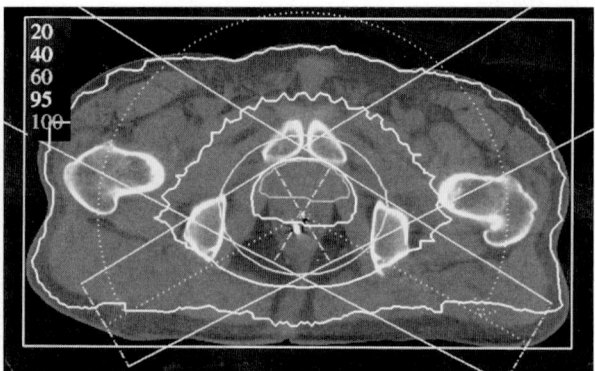

a   b

Fig. 5.26a,b. Dose distribution for arc therapy technique for 6-mV X-rays: **a** 240° arc. Note that when a sector of the full 360° rotation is skipped, the high-dose isodose curves are shifted away from the skipped sector. **b** A 240° arc, but patient is positioned so that isocenter is 2 cm lower toward the skipped sector (a technique called "past-pointing"). Note that high-dose isodose curves now encompass the target volume.

5.8
Field Shaping

A major constraint in the treatment of cancer using radiation is the limitation in the dose that can be delivered to the tumor because of the dose tolerance to the tissue (critical organs) surrounding or near to the target volume. Shielding normal tissue and critical organs has allowed the radiation oncologist to increase the dose to the tumor volume while maintaining the dose to critical organs below some tolerance level. The frequently used tolerance doses for these organs are not absolute and larger doses are sometimes given to fractional volumes of these organs (EMAMI et al. 1991). Shielding is usually accomplished using low melting point alloy blocks or MLCs, in which the beam aperture (field shape) is customized for individual patients.

5.8.1
Low Melting Alloy Blocks

The Lipowitz metal (Cerrobend) shielding block system was introduced by POWERS et al. (1973). Lipowitz metal consists of 13.3% tin, 50.0% bismuth, 26.7% lead, and 10.0% cadmium. The physical density at 20°C is 9.4 g/cm^3 as compared with 11.3 g/cm^3 for lead. The block fabrication procedure is illustrated in Figure 5.27. More details using this form of field shaping are given by LEAVITT and GIBBS (1992).

Computer-controlled adaptations of the hot-wire cutting technique have evolved as an adjunct to 3D treatment planning, in which the treatment field shape is defined based on beam's-eye-view displays. The shaped field coordinates are transferred directly to the computer-controlled blockcutting system, thereby eliminating potential errors in manual tracing, magnification, or image reversal. The other

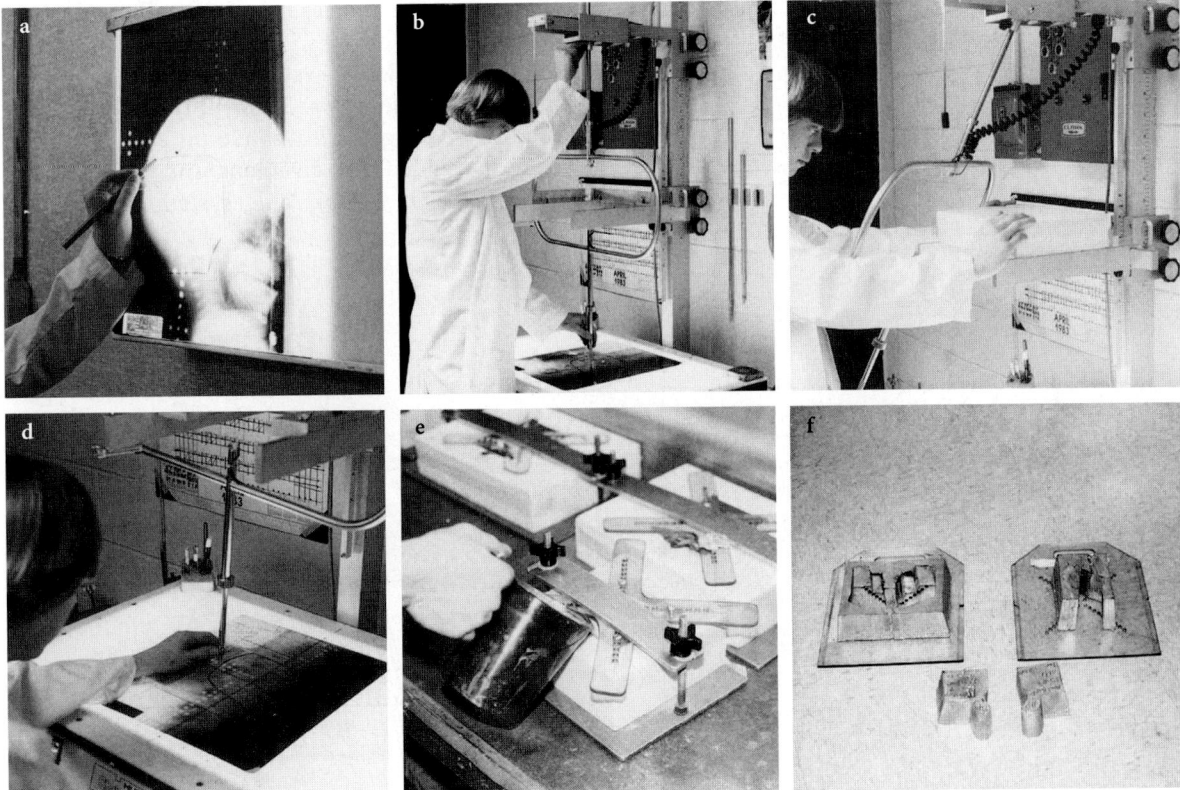

Fig. 5.27a–f. Composite photographs illustrate the low melting alloy shielding block design and fabrication process. **a** Physician defining the treatment volume on the X-ray simulator radiograph. **b** Physics technician adjusting the source–skin distance and skin–film distance of a hot-wire cutter to emulate simulator geometry. **c** Proper-thickness foam block aligned to the central axis of the cutter. **d** Foam mold cut with hot-wire cutter. **e** Foam pieces aligned and held in place using a special clamping device. Molten alloy is poured into the mold and allowed to harden. **f** Examples of typical shielding blocks cast using this system. (From PURDY 1983)

steps in the block-forming and verification process remain similar to the manual procedure.

Doses to critical organs may be limited by using either a full thickness block, usually five HVLs (3.125% transmission) or six HVLs (1.562% transmission), or a partial transmission shield, such as a single HVL (50% transmission) of shielding material. The actual dose delivered under the shielded area is generally greater than these stated transmission levels because of scatter radiation beneath the blocks from adjacent unshielded portions of the field. The scatter component of the dose increases with depth as more radiation scatters into the shielded volume beneath the block; thus, the dose to the blocked area is a function of block material, thickness (and width), field size, and energy. Figure 5.28 shows the attenuation of Lipowitz metal of X-rays produced at 2, 4, 10, and 18 MeV and ^{60}Co gamma rays (HUEN et al. 1979). Alloy blocks made from the standard thickness (7.6 cm) of foam molds reduce the primary beam intensity to 5% of its unattenuated value. Increasing the block thickness any greater is generally not worthwhile as it makes the block heavier, whereas the scatter radiation contributes an equal or greater share of the dose under the blocks.

5.8.2
Multileaf Collimation

Multileaf collimation, first introduced in Japan in the 1960s, has now gained widespread acceptance and has replaced alloy blocking as the standard of practice for field shaping in modern radiation therapy clinics (TAKAHASHI 1965). The different manufacturers' MLC systems vary with respect to field-size coverage, leaf design, and MLC location. The leaves are typically carried on two opposed carriages that transport the leaves in unison (Fig. 5.29). The leaves have individual controls that are computer assigned and positioned. Initially, most commercial systems were designed to serve as a block replacement but now provide for IMRT delivery.

Elekta (previously Philips Medical Systems) first introduced its MLC system in the late 1980s (HOUNSELL et al. 1992). Their MLC replaces one set of the photon collimator jaws, and therefore, the maximum field size can open to a full 40×40 cm. The MLC system is augmented by parallel diaphragms, which increase the leaf's attenuation by an additional two HVLs.

The Varian MLC system is placed below the photon collimator jaws. Their initial version consisted of 52 tungsten leaves (26 on each side) that are round ended, nondivergent, and 5.65 cm thick (GALVIN et al. 1993b). This system was quickly

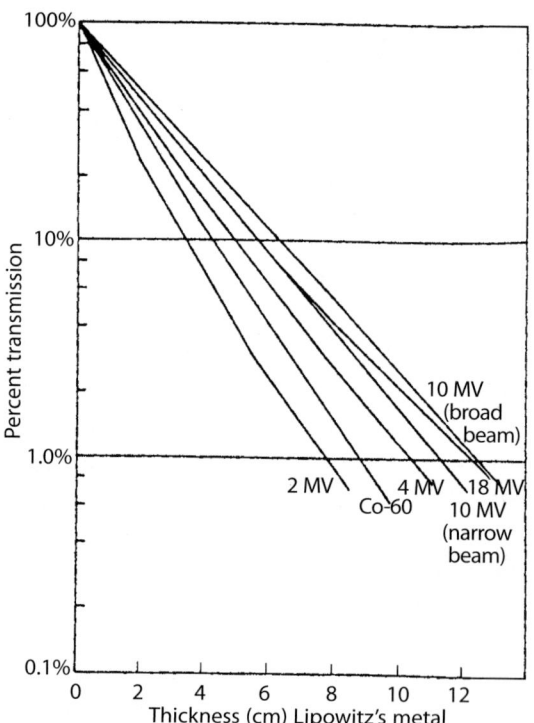

Fig. 5.28. Attenuation in Lipowitz's metal of X-rays produced at 2, 4, 10, and 18 mV and gamma rays from cobalt-60. (From HUEN et al. 1979)

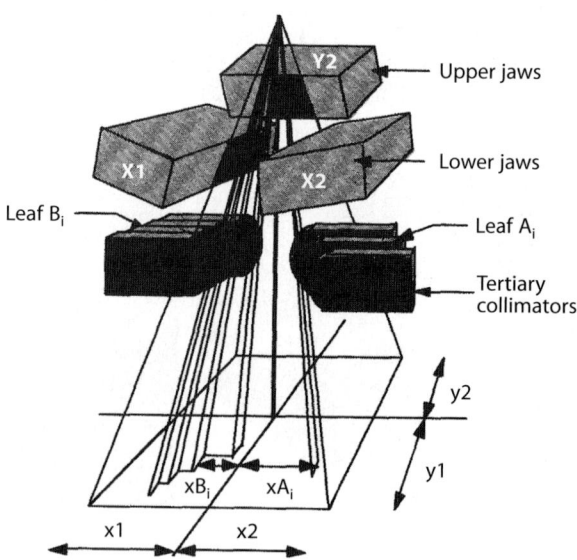

Fig. 5.29. Geometry of multileaf collimator for Varian linacs. The x-direction is the field width across each leaf pair, and the y-direction is the field length. (Courtesy of Varian Associates, Palo Alto, Calif.).

replaced with an 80-leaf system (40 on each side), increasing the maximum field size to 40×40 cm. The latest Varian MLC is a 120-leaf MLC system, in which the middle 20 cm consists of 0.5 cm wide leaves, whereas the outer 20 cm leaves still project to 1.0-cm widths. The narrower leaves in the middle provide smoother shaping. The leaves project to 16.0-cm in length at isocenter, and the leaf span range (maximum–minimum positions on the same carriage) is limited to 14.5 cm. The leaves move perpendicular to the beam's central axis. The distance from the X-ray target to the bottom of the leaves (on central axis) is 54.0 cm. The leaves fan away from central axis so that their sides are divergent with the beam's fan lines. The leaves are interdigitated by a tongue-and-groove design. Siemens also introduced a MLC system in which the lower collimating jaws are replaced with a double-focused leaf system (DAS et al. 1998). GALVIN et al. (1993a) and KLEIN et al. (1995) reported leaf-transmission values of 1.5–2.0% for a Varian 6-mV beam, and 1.5–5% for an 18-mV beam. Transmission through the screw attachment plane was 2.5%. These values are lower than those found for alloy blocks (3.5%), but higher than those for collimator jaw transmission (<1.0%). Transmission through abutted (closed) leaf pairs was as high as 28% for 18-mV photons on the central axis. The abutment transmission decreased as a function of off-axis distance to as low as 12%.

Figure 5.30 shows a comparison of MLC and alloy blocks regarding penumbra. The discrete steps of the MLC systems introduce undulations in the isodose lines. This effect causes an apparent increase in penumbra with wave patterns after the undulations. Some investigators describe this apparent penumbra increase as an "effective" penumbra accounting for the maximum and minimums of the undulations. Single, focused MLC systems have a slightly larger penumbra than do alloy shields and have an even larger difference in comparison with collimator jaws. BOYER et al. (1992) found the penumbra (80 to 20%) generated by leaf ends to be wider than those generated by upper collimator jaws by 1.0–1.5 mm, and 1.0–2.5 mm compared with the lower jaws, depending on energy and field size. POWLIS et al. (1993) compared MLC and alloy field shaping and found few differences. LoSASSO and KUTCHER (1995) found similar results and concluded that geometric accuracy is even improved with MLC.

The penumbras measured for the leaf sides are comparable with those found for upper jaws, due to their divergent nature. The penumbra increase and stair-stepping effect are most prominent at d_{max}. The effects diminish at depth due to the influence of scattered electrons and photons as the scatter-to-primary ratio increases with depth. Adding an opposed beam leads to further smoothing of the undulations and penumbra differences become less significant. For multibeam arrangements, the differences in dose distribution between MLC and alloy shields are negligible.

Two methods for designing the optimal MLC configurations to fit the treatment plan's field apertures have evolved: (a) configuring the MLC using a digitized film image using a dedicated MLC workstation (with or without automated optimization); and (b) configuring the MLC using treatment-planning system software. The main limitation in optimizing the MLC leaf settings to conform to the shaped field is the discrete leaf steps. Most field shapes require only minor adjustment of collimator angle to achieve minimal discrepancy between the desired and resultant field shape. The criteria for optimizing the MLC leaf settings are governed by placing the most leaf ends tangent to the field and also maintaining the same internal area as originally prescribed. The MLC shaping systems typically provide an option to place the leaf ends entirely outside the field (exterior), entirely within the field (interior), or crossing the field at mid-leaf (leaf-center insertion). The last is the most widely used criterion because the desired field area is more closely maintained; however, this choice leads to regions in which some treatment areas are shielded and some normal tissues are irradiated. ZHU et al. (1992) reported on a variable insertion technique in which leaves are placed only far enough into the field to cause the 50% isodose con-

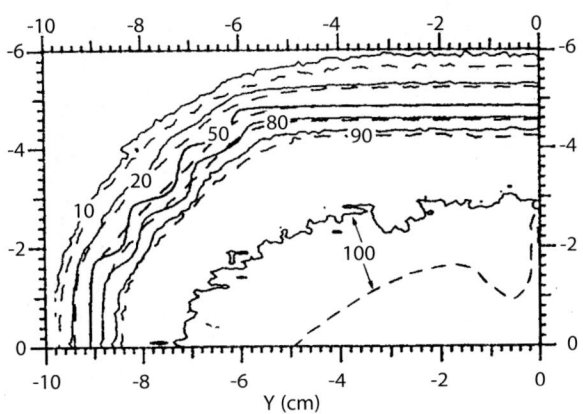

Fig. 5.30. Comparison of beam's eye view isodose curves at 10-cm depth for MLC (*solid line*) and Cerrobend-shaped (*dashed line*) beam apertures for 18-mV photons. (From KLEIN et al. 1995)

tour to undulate outside and up to the desired contour. LoSasso et al. (1993) reported on a method in which each leaf is inserted such that the treatment area covered by the leaf equals the normal tissue area that is not spared. Brahme (1988) also demonstrates optimal choices for choosing a collimator angle to optimize leaf direction, depending on whether the field shape is convex or concave. Du et al. (1995) reported on a method that defines optimal leaf positioning in combination with optimal collimator angulation. Typically, the optimal direction for the leaf motion is along the narrower axis. For a simple ellipse the optimal leaf direction is parallel to the short axis.

Klein et al. (1995) studied the effects of tissue heterogeneities on penumbra and resultant field definition and found lung to increase penumbra (especially for 18-mV photons) and bone to decrease penumbra for both alloy blocks and MLC. When multibeam arrangements were used, the summed doses consistently showed a superior dose distribution for the MLC fields, despite the stair-stepping effects, as opposed to alloy blocks.

As indicated previously, because MLC systems are still evolving, a careful evaluation of the effect of MLC on monitor unit calculations must be performed before clinical use. Extensive testing over the clinical range of field sizes and shapes should be undertaken before the MLC system is used clinically.

prudent to label the four settings on the simulator collimator and even the block trays. This is especially useful when collimator rotation is used alone or in conjunction with some of the newer technology devices such as MLC and enhanced dynamic wedge (EDW), which are oriented in a particular direction along an independent jaw set (e.g., X_1 or X_2 MLC, and Y_1 or Y_2 EDW).

Depth-dose characteristics for asymmetric fields are similar to those of symmetric fields as long as the degree of asymmetry is not too extreme; however, there are noticeable differences at the field edges between alloy-shaped fields and collimator-shaped fields, because the block aperture and collimators are at different distances from the patient and there are differences in scatter at the beam edges. Most treatment planning systems do not rigorously account for asymmetric fields. In most cases, simulating the field shapes as alloy blocks approximates them.

Clinical sites where asymmetric jaws are typically used include breast (Fig. 5.32), head and neck, craniospinal, and prostate. In addition, the use of asymmetric jaws as beam splitters, for field reductions, and with MLC is helpful for most sites. Rosenow et al. (1990), and later Marshall (1993), described the use of a single set of asymmetric jaws to match supraclavicular and tangential fields in the longitudinal plane for breast irradiation. Klein et al. (1994) described the use of dual asymmetric jaws

5.8.3
Asymmetric Collimator Jaws

Field shaping and abutted field radiation techniques have been made even more versatile with the asymmetric jaw feature found on modern-day linacs. This feature allows each set of jaws to open and close independently of each other (Fig. 5.31). The collimator jaw provides greater attenuation than the alloy shield, thus providing an advantage (which is readily apparent on portal films) in reducing dose to blocked regions. In addition, there is a practical advantage in reducing the size of the block alloy needed to create the beam aperture, thus reducing the physical effort needed to lift heavy blocks, along with providing some cost savings, since less of the alloy is needed.

However, the four independent jaw settings (Y_1, Y_2, X_1, X_2) can lead to some confusion in the clinic, so it is necessary that the patient's treatment record clearly denote the independent jaw settings. It is also

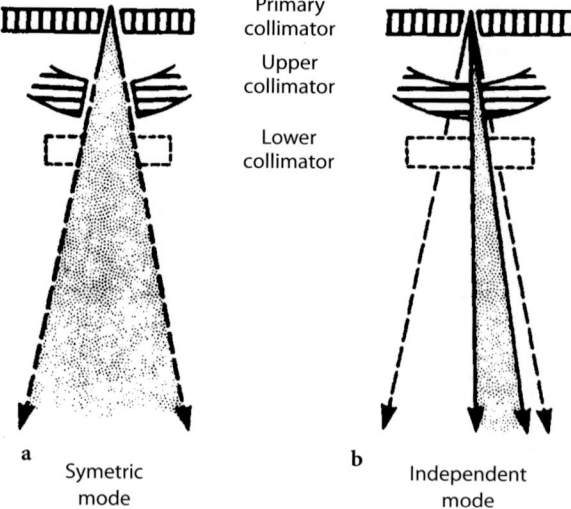

Fig. 5.31a,b. Independent or asymmetric collimators. **a** Conventional symmetric pairs of collimators. **b** Asymmetric collimators in which collimator jaws are allowed to move independently of each other

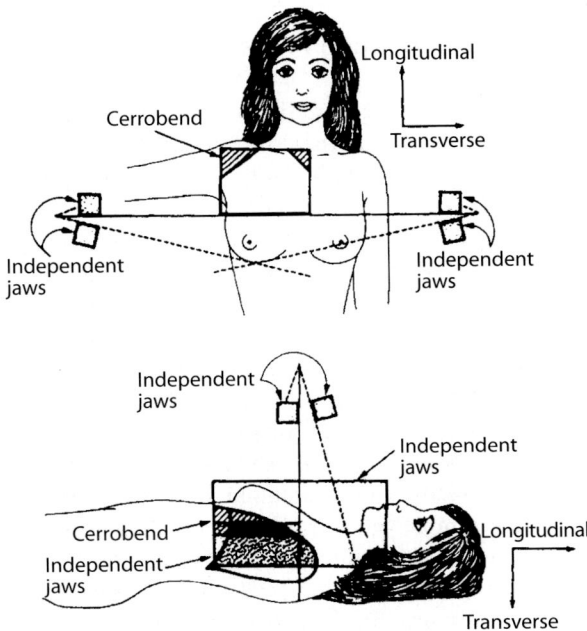

Fig. 5.32. Treatment technique for breast cancer using independent collimators. (From KLEIN et al. 1994)

to create nondivergence along the chest wall for the tangential beams. This technique allows a single setup point for all of the treatment fields, including the posterior axillary field. The Y-jaws can beam split the caudal and cephalic regions for the supraclavicular and tangential beams, respectively, and the X-jaws are used to shield the ipsilateral lung and contralateral breast; hence, a common match plane with one common isocenter can be used for all portals, including a posterior axilla beam, eliminating the need to move the patient between portals, thus reducing overall patient setup time by almost a factor of two. In addition, the increased attenuation by the jaws reduces the dose to the contralateral breast and lung (Foo et al. 1993). A technique for matching lateral head and neck fields and the supraclavicular field using independent jaws was described by SOHN et al. (1995).

5.9
Compensating Filters

The compensating filter, introduced by ELLIS and colleagues (1959), counteracts the effects caused by variations in patient surface curvature while still preserving the desirable skin-sparing feature of

megavoltage photon beams. This is accomplished by placing the custom-designed compensating filter in the beam, sufficiently "upstream" from the patient's surface.

A compensating filter system can be separated into several distinct subsystems, including a method to measure the missing tissue deficit, a means to demagnify patient topography, a method for constructing the compensating filter, a method of aligning and holding the filter in the beam, and a means of quality control inherent to it. Figure 5.33 shows a typical situation in which the principles of tissue compensation are illustrated. An air gap or tissue deficit, x, exists over point P. All points at depth d are to receive the same dose as the point P. These points all lie in a plane, each at an effective depth (d-x); thus, one must determine the tissue deficit, x, and the distance, y, from some reference point (usually the central axis) for all points within the irradiated area. This surface topography must then be demagnified back to the level at which the compensating filter will be placed in the radiation beam.

The appropriate filter thickness depends primarily on the material used. It also depends on the field size, X-ray beam energy, depth of target volume, and distance of compensator from the topographic deficit. Compensators made of material that is nearly tissue equivalent generally must have a thickness less than the missing tissue deficit to adequately correct for the lack of scatter produced as a result of the missing tissue; hence, compensators are designed to compensate to a specified depth, for a given geometry and beam energy. Over-compensation (less dose) usually occurs above and under-compensa-

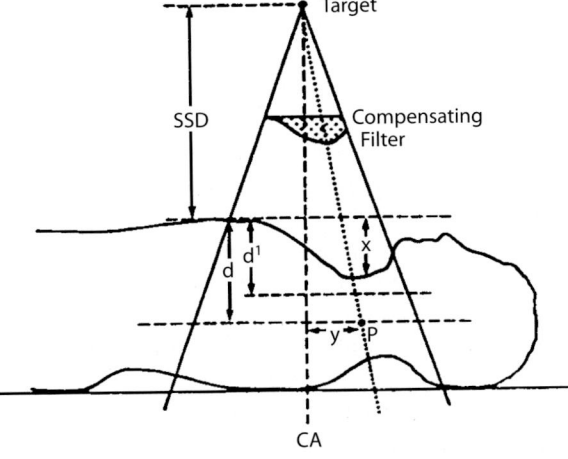

Fig. 5.33. Typical geometry used in the design of a compensator filter to account for patient's irregularly shaped surface. *SSD* source-to-skin distances

tion (more dose) below the specified compensator depth.

Several different compensator systems are in clinical use. They vary in complexity and technological sophistication from the simple Ellis technique using an array of stacked aluminum and brass blocks to the latest imaged-based planning systems. Early methods required the patient's presence for long periods of time (typically 45 min) for acquisition of topographic surface data and actual fabrication of the compensator shape. New approaches separate these two operations and require the patient's presence for only a very brief time. The new methodologies use computerized milling units to construct the actual filter or filter mold. These systems require limited human interaction during the construction process and provide registration guides that improve filter alignment. The reader is referred to review papers by Boyer (1982) and (Reinstein 1992) for more details regarding the different compensator techniques.

5.10
Bolus

Tissue-equivalent material placed directly on the patient's skin surface to reduce the skin sparing of megavoltage photon beams is referred to as bolus. A tissue-equivalent bolus should have electron density, physical density, and atomic number similar to that of tissue or water and be pliable so that it conforms to the skin-surface contour. Inexpensive, nearly tissue-equivalent materials used as a bolus in radiation therapy include slabs of paraffin wax, rice bags filled with soda, and gauze coated with petrolatum. These forms of bolus are not as tissue equivalent as some of the newer synthetic-based substances, such as Super-Flab (Moyer et al. 1983), Super-Stuff (Binder and Karcher 1977), Elastomeric Polymer (Moyer et al. 1983), and Elasto-Gel (Chang et al. 1989).

Thin slabs of bolus that follow the surface contour increase the dose to the skin beneath the bolus with a maximum reduction when the bolus thickness is approximately equal to d_{max} depth for the photon beam. In addition, adding bolus to fill a tissue deficit may smooth an irregular surface. A bolus also can be shaped to alter the dose distribution, but normally missing tissue compensators or wedges are used to alter the dose distribution for megavoltage photon beams in order to retain skin sparing.

5.11
Treatment-Planning Process

Treatment planning in radiation therapy is synonymous with specifying the dose prescription, including the dose fractionation schedule, and designing the beam arrangement to achieve the desired radiation dose distribution for a planning target volume (PTV) and any critical normal tissues. The radiation beams selected for treatment and their arrangement clearly depend on the location and shape, in all three dimensions, of the PTV, which consists of the gross tumor volume (GTV) and clinical target volume (CTV) with adequate margins for positional uncertainties (ICRU 1993). The selection of a particular treatment technique is dictated by the armamentarium available to the radiation oncologist in the treatment facility, which presently includes X-ray simulators, CT simulators, 2D and 3D treatment-planning systems, megavoltage treatment machines, improved methods of beam modification, improved methods of patient immobilization and repositioning, and other advanced technologies previously unavailable. As indicated previously, this chapter addresses only the 2D treatment-planning process.

The planning process begins with a clear designation of the treatment site and the critical radiation-sensitive organs in or near the tumor. Treatment-planning simulation includes appropriate positioning of the patient on the conventional simulator, as well as selection and preparation of patient positioning and immobilization devices, selection and simulation of radiation field shapes, and beam entry and exit points to encompass the target volume, preparation of organ shielding blocks, measurements of patient topography and anatomic thicknesses, decisions about the need for beam modifiers, such as wedges and compensating filters, and any considerations for dose calculation heterogeneity corrections. The dose prescription and isodose computations to determine the suitability of the treatment plan are the final steps of this process. In many clinics, the patient is returned to the conventional simulator for verification of the selected treatment method, using all treatment aids, which shows any deficiencies in the proposed treatment plan and allows modifications before radiation therapy is started.

5.11.1
Target Volume Localization

The conventional X-ray simulator, in combination with modern imaging studies such as CT, positron emission tomography (PET), and magnetic resonance imaging (MRI), complement physical examinations to provide the radiation oncologist with an effective tool to specify the target to be treated and provides an efficient system to generate the template (radiography film) for designing the shielding block. A special report by the British Institute of Radiology concluded that two simulators can support about five therapy units and allow the therapy units to be fully used for patient treatment (BOMFORD et al. 1981).

Through traditional radiographic (or fluoroscopic) methods using radiopaque rulers placed or projected on the patient's surface, dimensions of treatment areas on films can be determined to cover projected target volumes. Textbooks are available that provide site-specific examples of simulation procedures (BENTEL 1996; MIZER et al. 1986; WASHINGTON and LEAVER 1986).

Multiple off-axis CT images enhance the use of the convention simulator for treatment planning purposes, in which the target area and critical organs can be defined in each plane. Note that the patient's position should be identical to the position to be used during therapy for the planning CT scan.

5.11.2
Patient Positioning, Registration, and Immobilization

Ensuring accurate daily positioning of the patient in the treatment position and reduction of patient movement during treatment is essential to deliver the prescribed dose and achieve the planned dose distribution. Several studies in the literature quantify this effect, including the early study by HAUS and MARKS (1973) who, in an analysis of anatomic and geometric precision achieved in the treatment of mantle and pelvic portals, observed that about one-third of the localization errors were caused by patient movement. The reproducibility achievable in the daily positioning of a patient for treatment depends on several factors other than the anatomic site under treatment, including the patient's age, general health, and weight. In general, obese patients and small children are the most difficult to position.

The fields to be treated are typically delineated in the simulation process using either visible skin markings or marks on the skin visible only under an ultraviolet light. In some instances, external tat-

toos are applied. These markings are used in positioning a patient on the treatment machine, using the machine's field localization light and distance indicator and laser alignment lights mounted in the treatment room that project transverse, coronal, and sagittal light lines (or dots) on the patient's skin surface.

Numerous patient restraint and repositioning devices have been designed and used in treating specific anatomic sites. For example, the disposable foam plastic head holder provides stability for the head when the patient is in the supine position. If the patient is treated in the prone position, a face-down stabilizer can be used. This device has a foam rubber lining covered by disposable paper with an opening provided for the patient's eyes, nose, and mouth. It allows comfort and stability as well as air access for the patient during treatment in the prone position.

A vacuum-form body immobilization system is commercially available. This system consists of a vacuum pump and an outer rubber bag filled with plastic minispheres. The rubber bag containing the minispheres is positioned to support the patient's treatment position. A vacuum is then applied, causing the minispheres to come together to form a firm solid support molded to the patient's shape. The bite block (Fig. 5.34) is another device used as an aid in patient repositioning in the treatment of head and neck cancer. With this device, the patient, in the treatment position, bites into a specially prepared dental impression material layered on a fork that is attached to a supporting device. When the material hardens, the impression of the teeth is recorded. The bite-block fork is connected to a support arm, which is attached to the treatment couch, and may be used either with or without scales for registration.

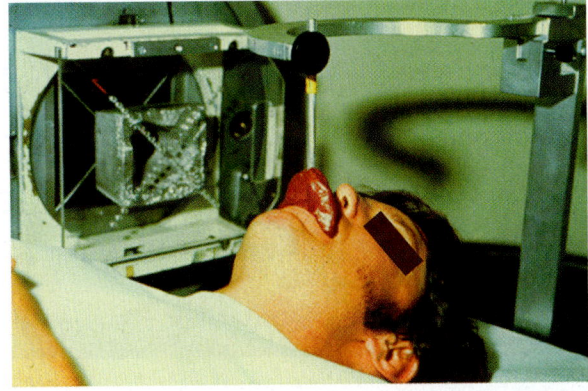

Fig. 5.34. Example of a bite-block registration and immobilization system used in treatment of head and neck cancer

The traditional plaster casting technique is still used in some clinics but has not gained widespread use in the United States. Transparent form-fitting plastic shells that are fabricated using a special vacuum device are also used extensively in Great Britain and Canada, but again have not gained acceptance in the United States. Both methods are described in detail by WATKINS (1981). Thermal plastic molds are now widely used in the United States (Fig. 5.35). A plastic sheet is placed in warm water, draped over the site, and hardens on cooling (GERBER et al. 1982). The use of thermal plastic masks allows treatments with few skin marks made on the patient because most of the reference lines can be placed on the mask. Treatments can be given through the mask; however, there is some loss of skin sparing. When skin sparing is critical, the mask may be cut out to match the treatment portal, although some of the structural rigidity is lost.

Custom molds constructed from polyurethane formed to patient contours have now gained widespread use as aids in immobilization and repositioning (Fig. 5.36). The constituent chemicals for the polyurethane foam are mixed in liquid form and are allowed to expand and harden around the patient while the patient is in the treatment position. These molds are used for treatment of Hodgkin's disease with the mantle irradiation technique, in thorax and prostate cancer patients, and for extremity repositioning/immobilization.

In the past, we have used either a bite block system or a thermal plastic facemask system to immobilize our head and neck cancer patients. Our experience has shown that patients immobilized with the bite block system typically require a larger number of

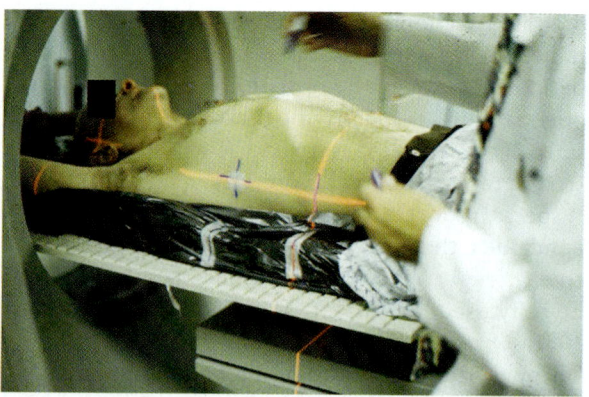

Fig. 5.36. Examples of foam mold registration and immobilization systems used in treatment of the thorax. Mold is registered to table and the patient is registered to the mold by fiducial markings.

adjustments than when more effective systems, such as, the thermal facemask, are used. Also, patients prefer the facemask because most of the reference marks are on the mask rather than on the skin. Radiation therapists have greater confidence in the accuracy of the treatment with the facemask; however, the final assessment of accuracy and reproducibility of the daily treatment is obtained by radiographic imaging of the area treated because there is the possibility of patient movement within the mask, especially if significant tumor shrinkage or weight loss has taken place.

5.11.3
Beam Direction and Shape

The selection of radiation beams, their entry and exit points on the patient, and their shapes in the planes perpendicular and parallel to their incident direction is the next crucial step in treatment simulation.

Beam direction on conventional simulators and therapy units is visualized by an optical system that projects a light field and cross hair onto the skin surface to denote the entry point of the beam, whereas the distance to the skin surface is indicated with an optical distance indicator. A reticule with a radiopaque grid that projects 1 cm apart at 100 cm, with every fifth centimeter delineated, is standard and is used for simulation localization radiographs for direct field size measurements. It is now generally used for portal films on therapy units as well. Laser back pointers are now commonly used to denote the beam exit point.

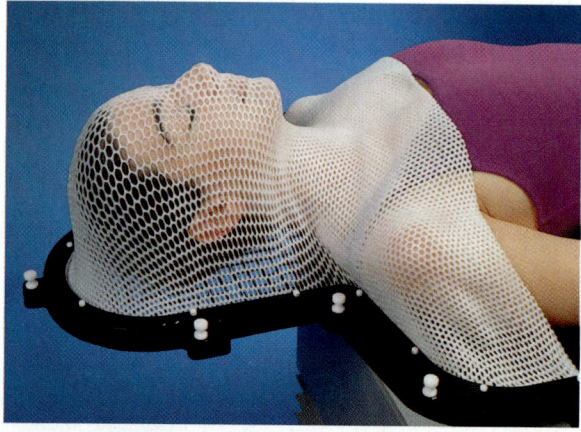

Fig. 5.35. Example of a thermal plastic mask registration and immobilization system used in treatment of head and neck cancer. (Courtesy of MED-TEC, Inc., Orange City, Iowa)

5.11.4
Patient Contour and Anatomic Measurements

The patient should be in the treatment position during measurements of the patient's contour and anatomic thickness, because the patient's position alters the thickness of some anatomic regions. Flexible plastic-coated lead (solder) wires or other devices, combined with anatomic thickness measurements made with calipers, are common methods of measuring the patient's topography. Normally, the shape registration device is placed over the anatomic region and transferred to graph paper onto which the shape is traced. The anatomic thicknesses measured are recorded on a patient planning diagram and the entry and exit points of the beams, as well as their field edges, are marked for isodose computations. Outlines of the tumor and the target volume, as well as internal organs of interest, are added to the diagram from information obtained from orthogonal films, CT scans, other imaging modalities, or anatomic atlases if other means are not available.

5.11.5
Dose and Target Specification

The International Commission on Radiation Units and Measurements (ICRU) has addressed the issue of consistent volume and dose specification in radiation therapy in ICRU report 29 in 1978 and ICRU report 50 in 1993 (ICRU 1978, 1993). The ICRU report 50 definitions are discussed in detail in Chap. 9. Briefly, the target volume is separated into three distinct boundaries: (a) visible tumor; (b) a region to account for uncertainties in microscopic tumor spread; and (c) a region to account for positional uncertainties. These boundaries create three volumes. The GTV is the gross extent of the malignant growth as determined by palpation or imaging studies. We have used the terms "$GTV_{primary}$" and "GTV_{nodal}" to distinguish between primary disease and other areas of macroscopic tumor involvement such as involved lymph nodes.

The CTV is the tissue volume that contains the GTV and/or subclinical microscopic malignant disease. In specifying the CTV, the physician must consider microextensions of the disease near the GTV and the natural avenues of spread for the particular disease and site, including lymph node, perivascular, and perineural extensions. The GTV and CTV are an anatomic–clinical concept that must be defined before choosing a treatment modality and technique.

The PTV is defined by specifying the margins that must be added around the CTV to compensate for the effects of organ, tumor, and patient movements and inaccuracies in beam and patient setup. The PTV is a static, geometric concept used for treatment planning and specification of dose. Its size and shape depend primarily on the GTV and CTV, the effects caused by internal motions of organs and the tumor, as well as the treatment technique (beam orientation and patient fixation) used. The PTV can be considered an envelope in which the tumor and any microscopic extensions reside. The GTV and CTV can move within this envelope, but not through it.

Another important point is the added block margin needed around the PTV. Physicians (and physicists) have confused this margin in the past by thinking of the PTV boundary as the beam edge, which it is not. Once the PTV is defined, an additional margin (block aperture), on the order of 7 mm in most cases, must be added to allow for penumbra and beam arrangements.

The prescription of radiation treatment includes the designation of the pertinent volumes along with the prescription of dose and fractionation. For dose reporting, ICRU 50 defines a series of doses including the minimum, maximum, and mean dose. In addition, an ICRU reference dose is defined at what is called the ICRU reference point. The ICRU reference point is chosen based on the following criteria: It must be clinically relevant and be defined in an unambiguous way. It must be located where the dose can be accurately determined, and it cannot be located in a region where there are steep dose gradients. In general, this point should be in the central part of the PTV. In cases where the beams intersect at a given point, it is recommended that the intersection point be chosen as the ICRU reference point. Note that the prescription can be specified by the radiation oncologist using any one of the above-named doses.

A common method of dose prescription specifies the dose at some depth in the patient (e.g., target depth, mid-plane depth, or depth determined by a common intersection point of multiple fields). A proper dose prescription states not only the total absorbed dose, daily total dose, and fractionation schedule, but also where and how the dose is prescribed, so that ambiguities are avoided. Generally, it is desirable to have a uniform dose within the target volume, with the ratio of maximum dose to minimal dose not exceeding 1.10, a dose variation within the target volume of 10% or less. A clear statement of the dose to critical organs is also required.

Dose prescriptions must be unambiguous. Misunderstanding about whether a prescribed tumor dose is a point dose at some convenient point within the target volume or is the dose to an isodose curve (e.g., 95%) encompassing the target volume can lead to either undertreatment of tumor at the edges of the target volume or overdosing of normal tissues within the target volume, with subsequent complications.

By carefully considering the dose distribution in terms of coverage of the target volume and the dose to adjacent critical structures, the radiation oncologist approves a beam arrangement, which should be both practical and cost-effective.

5.12
Separation of Adjacent X-Ray Fields

5.12.1
Field Junctions

The numerous methods of matching adjacent X-ray fields have been reviewed by HOPFAN et al. (1977), and DEA (1985) reviewed radiographic methods of confirming gaps. The geometries of adjacent fields, either a single pair or parallel-opposed pairs, are illustrated in Figure 5.37. A commonly used method matches adjacent radiation fields at depth d and is illustrated in Figure 5.38a. The separation between adjacent field edges necessary to produce junction doses similar to central-axis doses follows from the similar triangles formed by the half-field length and SSD in each field. The field edge is defined by the dose at the edge that is 50% of the dose at d_{max}. For two contiguous fields of lengths L_1 and L_2, the separation, S, of these two fields at the skin surface can be calculated using the following equation:

$$S = {\textstyle\frac{1}{2}}L_1\left(\frac{d}{SSD}\right) + {\textstyle\frac{1}{2}}L_2\left(\frac{d}{SSD}\right).$$

A slight modification of this equation is needed when sloping surfaces are involved, as shown in Figure 5.38B (KEYS and GRIGSBY 1990). Typically, the skin gap location is moved frequently to reduce the hot and cold spots that arise with this technique. Figure 5.39 illustrates the dose distribution for three different field separations (JOHNSON and KHAN 1994).

Beam divergence may be eliminated by using a "beam splitter," created using a five- or six-HVL block over 50% of the treatment field. The central

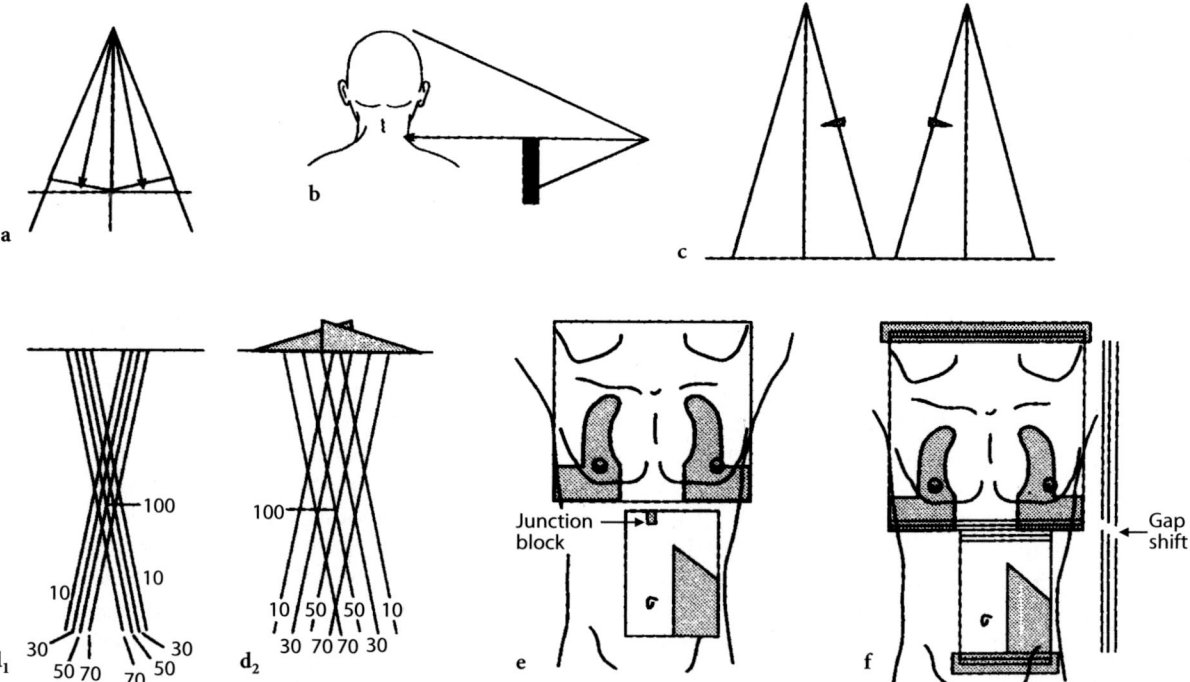

Fig. 5.37a–f Different techniques for matching adjacent fields. **a** Beam's central rays are angled slightly away from one another so that the diverging beams are parallel. **b** Half-beam block to eliminate divergence. **c** Penumbra generators (small wedges) to increase width of penumbra as illustrated in d_1 and d_2. **e** Junction block over spinal cord. **f** Moving gap technique. (From BENTEL 1996)

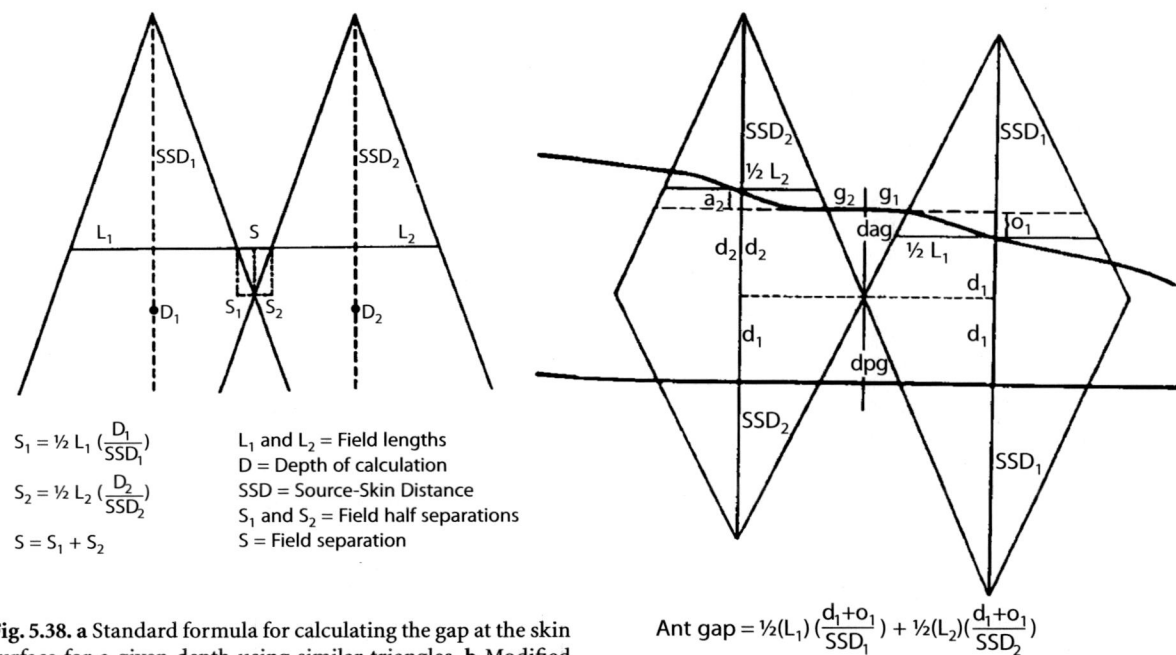

$$S_1 = \tfrac{1}{2} L_1 \left(\frac{D_1}{SSD_1}\right)$$

$$S_2 = \tfrac{1}{2} L_2 \left(\frac{D_2}{SSD_2}\right)$$

$$S = S_1 + S_2$$

L_1 and L_2 = Field lengths
D = Depth of calculation
SSD = Source-Skin Distance
S_1 and S_2 = Field half separations
S = Field separation

a

Fig. 5.38. a Standard formula for calculating the gap at the skin surface for a given depth using similar triangles. b Modified formula for calculating the gap for matching four fields on a sloping surface. (From KEYS and GRIGSBY 1990)

$$\text{Ant gap} = \tfrac{1}{2}(L_1)\left(\frac{d_1+o_1}{SSD_1}\right) + \tfrac{1}{2}(L_2)\left(\frac{d_1+o_1}{SSD_2}\right)$$

$$\text{Post gap} = \tfrac{1}{2}(L_1)\left(\frac{d_1}{SSD_1}\right) + \tfrac{1}{2}(L_2)\left(\frac{d_1}{SSD_1}\right)$$

b

axes of the adjacent fields, where there is no divergence, are then matched. As previously discussed, this is a useful method on linacs with the independent jaw feature. Match-line wedges or penumbra generators that generate a broad penumbra for linac beams have been reported but have not found widespread use (FRAASS et al. 1983). Here the intent is to broaden the narrow penumbra of the linacs so that it is not so difficult to match the 50% isodose levels. The resulting dose distributions are similar to those obtained with a moving gap technique.

There are numerous reports of edge-matching techniques based on the mathematical relationships between adjacent beams and the allowed angles of the gantry, collimators, and couch. CHRISTOPHERSON and co-workers (1984) developed a useful nomograph for field matching for treating cancer of the breast.

5.12.2
Orthogonal Field Junctions

Figure 5.40 illustrates the geometry of matching abutting orthogonal photon beams. Such techniques are necessary, particularly in the head and neck region where the spinal cord can be in an area of beam overlap, in the treatment of medulloblastoma with multiple spinal portals and lateral

brain portals, as well as in multiple-field treatments of the breast (VAN DYK et al. 1977; SIDDON et al. 1981). A common method of avoiding overlap is to use a half-block, as previously discussed, so that abutting anterior and lateral field edges are perpendicular to the gantry axis (KARZMARK et al. 1980). In addition, a notch in the posterior corner of the lateral oral cavity portal is commonly used to ensure overlap avoidance of the spinal cord when midline cord blocks cannot be used on anteroposterior portals irradiating the lower neck and matched to the oral cavity portals. Other techniques rotate the couch about a vertical axis to compensate for the divergence of the lateral field (SIDDON et al. 1981). The angle of rotation is given by

$$\tan\theta^{-1} = \left(\frac{\tfrac{1}{2}\,field\ width}{SAD}\right).$$

Another technique is to leave a gap, S, on the anterior neck surface between the posterior field of length L and lateral field edges (GILLIN and KLINE 1980; WILLIAMSON 1979). S can be calculated using the formula below where d is the depth of the spine beneath the posterior field:

$$S = \frac{1}{2}(L)\left(\frac{d}{SAD}\right).$$

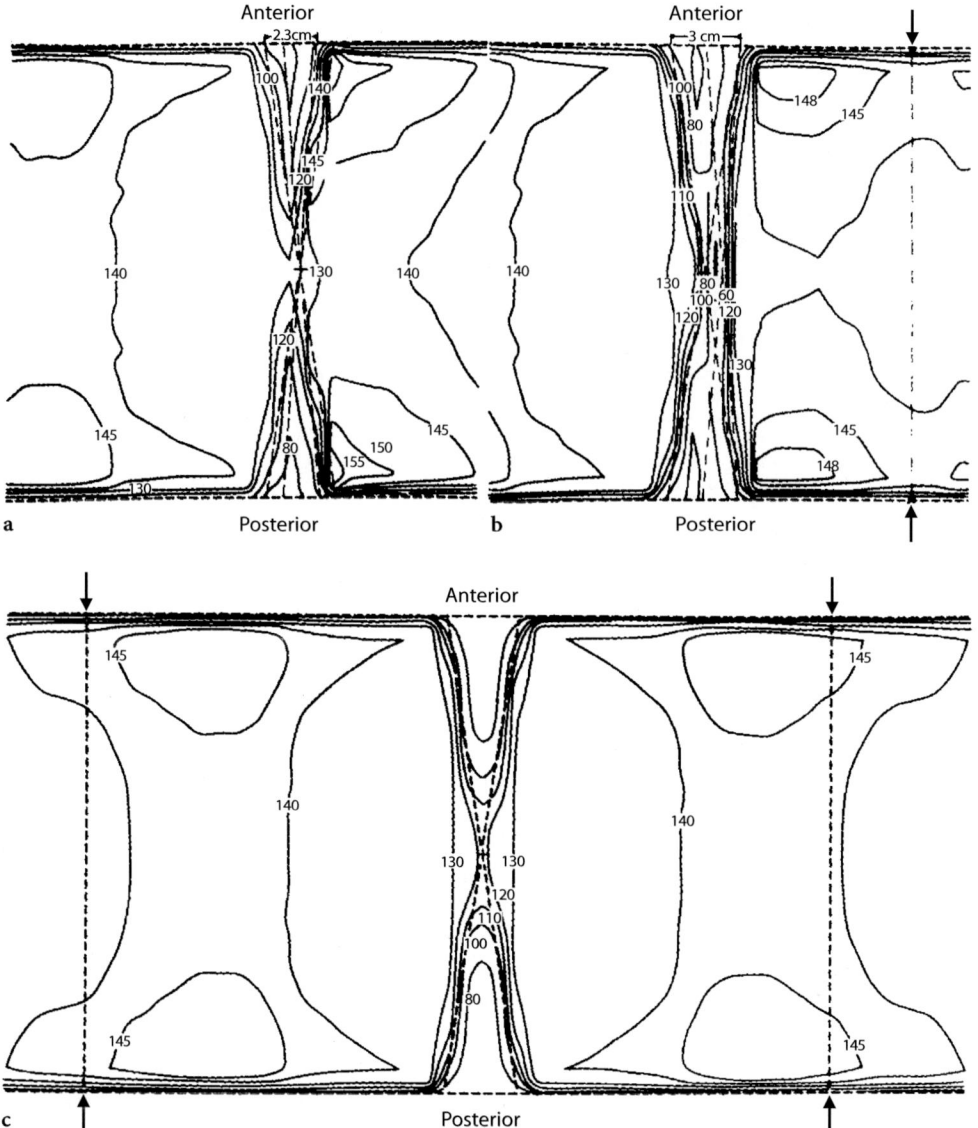

Fig. 5.39a–c. Dose distribution for geometric separation of fields with all four beams intersecting at midpoint. Adjacent field sizes: 30×30 and 15×15 cm; SSD = 100 cm; AP thickness = 20 cm; 4-mV X-ray beams. **a** Field separation at surface is 2.3 cm. A three-field overlap exists in this case because the fields have different sizes but the same SSD. **b** The adjacent field separation increased to eliminate three-field overlap on the surface. **c** Field separation adjusted to 2.7 cm to eliminate three-field overlap at the cord at 15-cm depth from anterior. (From KHAN 1994)

Craniospinal irradiation is well established as a standard method of treating suprasellar dysgerminoma, pineal tumors, medulloblastomas, and other tumors involving the central nervous system. Uniform treatment of the entire craniospinal target volume is possible using separate parallel-opposed lateral cranial portals rotated so that their inferior borders match with the superior border of the spinal portal, which is treated with either one or two fields, depending on the length of the spine to be treated.

LIM (1985, 1986) excellently describes the dosimetry of optional methods of treating medulloblastoma with diagrams. Two junctional moves are made at one-third and two-thirds of the total dose. The spinal field central axis is shifted away from the brain by 0.5 cm and the field size length reduced by 0.5 cm with corresponding increases in the length of the cranial field, so that a match exists between the inferior border of the brain portal and the superior border of the spine portal. To achieve the match, the

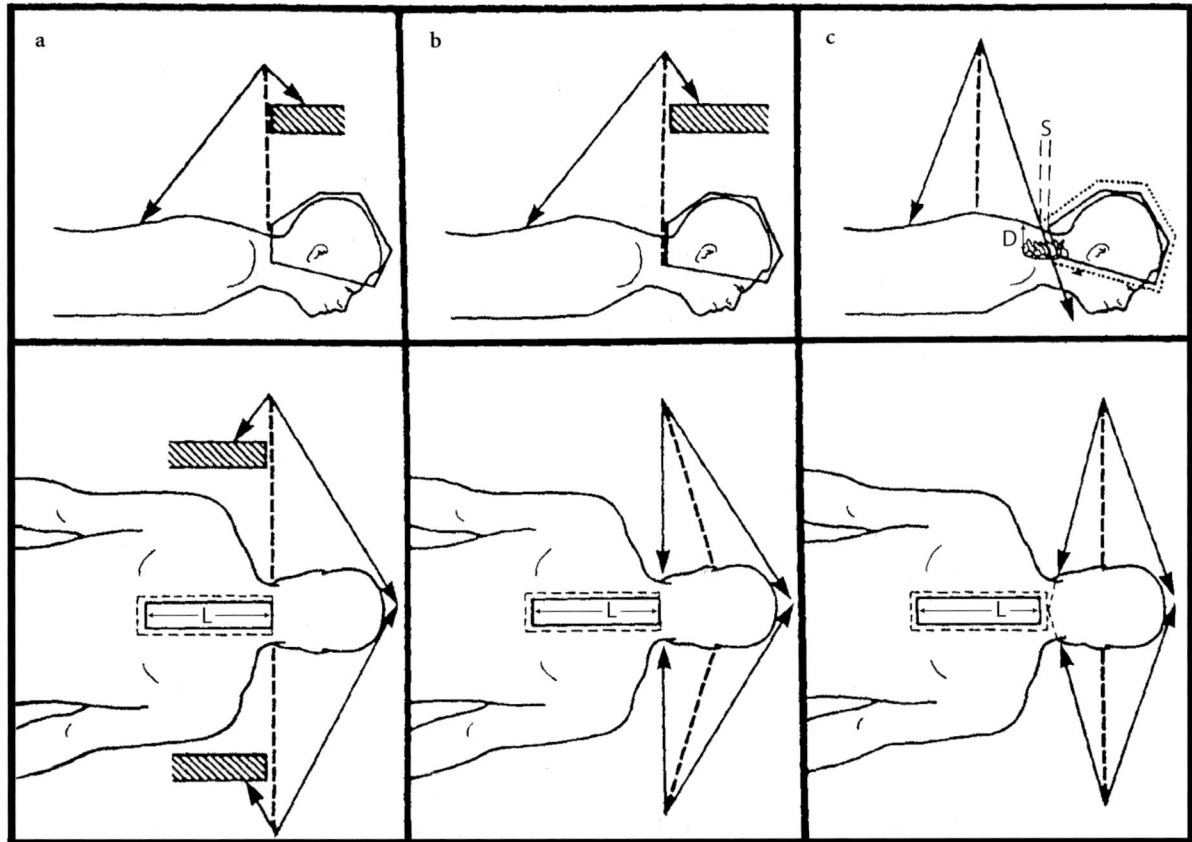

Fig. 5.40a–c. Some solutions for the problem of overlap for orthogonal fields. **a** A beam splitter, a shield that blocks half of the field, is used on the lateral and posterior fields and on the spinal cord portal to match the nondivergent edges of the beams. **b** The divergence in the lateral beams may also be removed by angling the lateral beams so that their caudal edges match. Because most therapy units cannot be angled like this, the couch is rotated through small angles in opposite directions to achieve the same effect. **c** A gap technique allows the posterior and lateral field to be matched at depth using a gap S on the skin surface. The *dashed lines* indicate projected field edges at depth D, where the orthogonal fields meet. (From WILLIAMSON 1979)

whole-brain portals are rotated by an angle given by the following relationship:

$$\tan\theta^{-1} = \left| \frac{\frac{1}{2}\, spinal\ field\ length}{SAD} \right|.$$

To eliminate the divergence between the cranial portal and spinal portal, the table is rotated through a floor angle

$$\tan\alpha^{-1} = \left| \frac{\frac{1}{2}\, cranial\ field\ length}{SAD} \right|.$$

Potential for the occurrence of radiation myelopathy resulting from the potentially excessive dose from misaligned overlapping fields is always a concern when central nervous system tumors are treated.

References

Abrath FG, Purdy JA (1980) Wedge design and dosimetry for 25-mV x rays. Radiology 136:757–762

Ahnesjö A, Aspradakis MM (1999) Dose calculations for external photon beams in radiotherapy. Phys Med Biol 44: R99–R155

Batho HF (1964) Lung corrections in cobalt 60 beam therapy. J Can Assoc Radiol 15:79–83

Beach JL, Mendiondo MS, Mendiondo OA (1987) A comparison of air-cavity inhomogeneity effects for cobalt-60, 6- and 10-mV X-ray beams. Med Phys 14:140

Bentel GC (ed) (1996) Radiation therapy planning, 2nd edn. McGraw-Hill, New York

Binder W, Karcher KH (1977) "Super-stuff" als bolus in der strahlentherapie. Strahlentherapie 153:754

BJR (1996) Central axis depth dose data for use in radiotherapy. Br J Radiol (Suppl 25)

Bomford CK et al. (1981) Treatment simulators. Br J Radiol 16 (Suppl):1–31

Boyer AL et al. (1992) Clinical implementation of a multileaf collimator. Med Phys 19:1255–1261

Brahme A (1988) Optimization of stationary and moving beam radiation therapy techniques. Radiother Oncol 12:129–140

Carol MP (1994) Integrated 3D conformal multivane intensity modulation delivery system for radiotherapy. In: Proc 11th International Conference on the Use of Computers in Radiation Therapy. Medical Physics Publishing, Madison, Wisconsin

Chang F, Benson K, Share F (1989) The study of elasto-gel pads used as surface bolus material in high energy photon and electron therapy (abstract). Med Phys 16:449

Christopherson D, Courlas GJ, Jette D (1984) Field matching in radiotherapy. Med Phys 3:369

Chui C, Mohan R, Fontanela D (1986) Dose computation for asymmetric fields defined by independent jaws. Med Phys 15:92

Cundiff JH et al. (1973) A method for the calculation of dose in the radiation treatment of Hodgkin's disease. Am J Roentgenol 117:30–44

Cunningham JR (1972) Scatter-air ratios. Phys Med Biol 17:42–51

Das IJ, Khan FM (1989) Backscatter dose perturbation at high atomic number interfaces in megavoltage photon beams. Med Phys 16:367–375

Das IJ et al. (1990) Validity of transition-zone dosimetry at high atomic number interfaces in megavoltage photon beams. Med Phys 17:10–16

Das IJ et al. (1998) Beam characteristics of a retrofitted double-focused multileaf collimator. Med Phys 25:1676–1684

Dea D (1985) Dosimetric problems with adjacent fields: verification of gap size. Am Assoc Med Dosim J 10:37

Du MN et al. (1995) A multi-leaf collimator prescription preparation system for conventional radiotherapy. Int J Radiat Oncol Biol Phys 32:513–520

Dutreix A et al. (1997) Monitor unit calculation for high energy photon beams. Garant, Leuven, Apeldoorn

Ellis F, Miller H (1944) The use of wedge filters in deep X-ray therapy. Br J Radiol 17:90

Ellis F, Hall EJ, Oliver R (1959) A compensator for variations in tissue thickness for high energy beams. Br J Radiol 32:421–422

Emami B et al. (1991) Tolerance of normal tissue to therapeutic irradiation. Int J Radiat Oncol Biol Phys 21:109–122

Epp ER, Lougheed MN, McKay JW (1958) Ionization build-up in upper respiratory air passages during teletherapy units with cobalt-60 irradiation. Br J Radiol 31:361

Epp ER, Boyer AL, Doppke KP (1977) Underdosing of lesions resulting from lack of electronic equilibrium in upper respiratory air cavities irradiated by 10 mV X-ray beams. Int J Radiat Oncol Biol Phys 2:613

Fischer JJ, Moulder JE (1975) The steepness of the dose-response curve in radiation therapy. Radiology 117:179–184

Foo ML et al. (1993) Doses to radiation sensitive organs and structures located outside the radiotherapeutic target volume for four treatment situations. Int J Radiat Oncol Biol Phys 27:403

Fraass BA et al. (1983) Clinical use of a match line wedge for adjacent megavoltage radiation field matching. Int J Radiat Oncol Biol Phys 9:209

Gagnon WF (1979) Measurement of surface dose. Int J Radiat Oncol Biol Phys 5:449–450

Gagnon WF, Horton JL (1979) Physical factors affecting absorbed dose to the skin from cobalt-60 gamma rays and 25 mV X-rays. Med Phys 6:285

Galvin JM, Smith AR, Lally B (1993a) Characterization of a multileaf collimator system. Int J Radiat Oncol Biol Phys 25:181–192

Galvin JM, Xuan-Gen C, Smith RM (1993b) Combining multileaf fields to modulate fluence distributions. Int J Radiat Oncol Biol Phys 27:697–705

Georg D, Huekelom S, Venselaar J (2001) Formalisms for mu calculations, estro booklet 3 versus NCS report 12. Radiother Oncol 60:319–328

Gerber RL, Marks JE, Purdy JA (1982) The use of thermal plastics for immobilization of patients during radiotherapy. Int J Radiat Oncol Biol Phys 8:1461

Gerbi BJ, Khan FM (1990) Measurement of dose in the buildup region using fixed-separation plane-parallel ionization chambers. Med Phys 17:17–26

Gerbi BJ, Meigooni A, Khan FM (1987) Dose buildup for obliquely incident photon beams. Med Phys 14:393

Gibbons JP (ed) (2000) Monitor unit calculations for external photon and electron beams. Advanced Medical Publishing, Madison, Wisconsin

Gillin MT, Kline RW (1980) Field separation between lateral and anterior fields on a 6-mV linear accelerator. Int J Radiat Oncol Biol Phys 6:233

Grant W III (1996) Experience with intensity modulated beam delivery. In: Palta J, Mackie TR (eds) Teletherapy: present and future. Advanced Medical Publishing, College Park, Maryland, pp 793–804

Hanson WF, Berkley LW (1980) Off-axis beam quality change in linear accelerator X-ray beams. Med Phys 7:145–146

Hanson WF, Berkley LW, Peterson M (1980) Calculative technique to correct for the change in linear accelerator beam energy at off-axis points. Med Phys 7:147

Haus AG, Marks JE (1973) Detection and evaluation of localization errors in patient radiation therapy. Invest Radiol 8:384

Herring DF (1975) The consequences of dose response curves for tumor control and normal tissue injury on the precision necessary in patient management. Laryngos 85:119–125

Herring DF, Compton DMJ (1971) The degree of precision required in radiation dose delivered in cancer radiotherapy. In: Glicksman AJ, Cohen M, Cunningham JR (eds) Computers in radiotherapy. Br J Radiol: Spec Rep series no. 5

Holt JD, Laughlin JS, Moroney JP (1970) The extension of the concept of tissue-air (tar) to high energy X-ray beams. Radiology 96:437

Hopfan S et al. (1977) Clinical complications arising from overlapping of adjacent fields: physical and technical considerations. Int J Radiat Oncol Biol Phys 2:801

Hounsell AR et al. (1992) Computer-assisted generation of multileaf collimator settings for conformation therapy. Br J Radiol 65:321–326

Huen A, Findley DO, Skov DD (1979) Attenuation in Lipowitz's metal of X-rays produced at 2, 4, 10, and 18 mV and gamma rays from cobalt-60. Med Phys 6:147

ICRU (1976) Determination of absorbed dose in a patient irradiated by beams of X or gamma rays in radiotherapy procedures. ICRU report 24, Washington, DC

ICRU (1978) Dose specification for reporting external beam therapy with photons and electrons. I.C.o.R.U.a. Measurements. International Commission on Radiation Units and Measurements, report no. 29, Bethesda, Maryland

ICRU (1993) Prescribing, recording, and reporting photon beam therapy. International Commission on Radiation Units and Measurements, Bethesda, Maryland

Johns HE, Cunningham JR (1983) The physics of radiology, 4th edn. Charles C. Thomas, Springfield, Illinois)

Johnson JM, Khan FM (1994) Dosimetric effects of abutting extended source to surface distance electron fields with photon fields in the treatment of head and neck cancers. Int J Radiat Oncol Biol Phys 28:741

Karzmark CJ, Deubert A, Loevinger R (1965) Tissue-phantom ratios: an aid to treatment planning. Br J Radiol 38:158–159

Karzmark CJ et al. (1980) Overlap at the cord in abutting orthogonal fields: a perceptual anomaly. Int J Radiat Oncol Biol Phys 6:1366

Keys R, Grigsby PW (1990) Gapping fields on sloping surfaces. Int J Radiat Oncol Biol Phys 18:1183

Khan F (1993) Dosimetry of wedged fields with asymmetric collimation. Med Phys 20:1447

Khan FM (1994) The physics of radiation therapy, 3rd edn. Williams and Wilkins, Baltimore

Khan FM (2003) The physics of radiation therapy, 3rd edn. Lippincott Williams and Wilkins, Philadelphia

Khan FM et al. (1980) Revision of tissue-maximum ratio and scatter-maximum ratio concepts for cobalt 60 and higher energy X-ray beams. Med Phys 7:230–237

Klein EE, Kuske RR (1993) Changes in photon dosimetry due to breast prosthesis. Int J Radiat Oncol Biol Phys 25:541–549

Klein EE, Purdy JA (1993) Entrance and exit dose regions for clinac-2100c. Int J Radiat Oncol Biol Phys 27:429–435

Klein EE et al. (1988) Influence of aluminum and bone on dose distributions for photon beams (abstract). Med Phys 15:122

Klein EE et al. (1993) The influence of air cavities on interface doses for photon beams (abstract). Int J Radiat Oncol Biol Phys 27:419

Klein EE et al. (1994) A mono-isocentric technique for breast and regional nodal therapy using dual asymmetric jaws. Int J Radiat Oncol Biol Phys 28:753–760

Klein EE et al. (1995) Clinical implementation of a commercial multileaf collimator: dosimetry, networking, simulation, and quality assurance. Int J Radiat Oncol Biol Phys 33:1195–1208

Klein EE et al. (1997) A volumetric study of measurements and calculations of lung density corrections for 6 and 18 mV photons. Int J Radiat Oncol Biol Phys 37:1163–1170

Kornelsen RO, Young MEJ (1982) Changes in the dose-profile of a 10 mV X-ray beam within and beyond low density material. Med Phys 9:114–116

Koskinen MO, Spring E (1973) Build-up and build-down measurements with thin LiF-Teflon dosimeters with special reference to radiotherapy of carcinoma of the larynx. Strahlentherapie 145:565

Lam WC, Lam KS (1983) Errors in off-axis treatment planning for a 4 mev machine. Med Phys 10:480

Leavitt DD, Gibbs FA Jr (1992) Field shaping. In: Purdy JA (ed) Advances in radiation oncology physics: dosimetry, treatment planning, and brachytherapy. American Institute of Physics, New York, pp 500–523

Leavitt DD et al. (1990) Dynamic wedge field techniques through computer-controlled collimator motion and dose delivery. Med Phys 17:87–91

Lim MLF (1985) A study of four methods of junction change in the treatment of medulloblastoma. Am Assoc Med Dosim J 10:17

Lim MLF (1986) Evolution of medulloblastoma treatment techniques. Am Assoc Med Dosim J 11:25

LoSasso T, Kutcher GJ (1995) Multi-leaf collimation vs Cerrobend blocks: analysis of geometric accuracy. Int J Radiat Oncol Biol Phys 32:499–506

LoSasso T, Chui CS, Kutcher GJ (1993) The use of multileaf collimator for conformal radiotherapy of carcinomas of the prostate and nasopharynx. Int J Radiat Oncol Biol Phys 25:161

Mackie TR et al. (1996) Photon beam dose computations. In: Palta J, Mackie TR (eds) Teletherapy: present and future. Advanced Medical Publishing, College Park, Maryland, pp 103–136

Marshall M (1993) Three-field isocentric breast irradiation using asymmetric jaws and a tilt board. Radiother Oncol 28:228–232

McCullough EC, Gortney J, Blackwell CR (1988) A depth dependence determination of the wedge transmission factor for 4–10 mV photon beams. Med Phys 15:621

Mizer S, Scheller RR, Deye JA (1986) Radiation therapy simulation workbook. Pergamon Press, New York

Moyer RF et al. (1983) A surface bolus material for high energy photon and electron therapy. Radiology 146:531

Nillson B, S Schnell (1976) Build-up effects at air cavities measured with thin thermoluminescent dosimeters. Acta Radiol Ther Phys Biol 15:427–432

Ostwald PM, Kron T, Hamilton CS (1996) Assessment of mucosal underdosing in larynx irradiation. Int J Radiat Oncol Biol Phys 36:181–187

Palta JR, Yeung DK, Frouhar V (1996) Dosimetric considerations for a multileaf collimator system. Med Phys 23:1219–1224

Palta JR, Ayyangar KM, Suntharalingam N (1998) Dosimetric characteristics of a 6 mV photon beam from a linear accelerator with asymmetric collimator jaws. Int J Radiat Oncol Biol Phys 14:383–387

Powers WE et al. (1973) A new system of field shaping for external-beam radiation therapy. Radiology 108:407–411

Powlis WD et al. (1986) Initiation of multileaf collimator conformal radiation therapy. Int J Radiat Oncol Biol Phys 25:171–179

Purdy JA (1983) Secondary field shaping. In: Wright AE, Boyer AL (eds) Advances in radiation therapy treatment planning. American Institute of Physics, New York

Purdy JA (1986) Buildup/surface dose and exit dose measurements for 6-mV linear accelerator. Med Phys 13:259

Reinstein LE (1992) New approaches to tissue compensation in radiation oncology. In: Purdy JA (ed) Advances in radiation oncology physics: dosimetry, treatment planning, and brachytherapy. American Institute of Physics, New York, pp 535–572

Rice RK, Mijnheer BJ, Chin LM (1988) Benchmark measurements for lung dose corrections for X-ray beams. Int J Radiat Oncol Biol Phys 15:399–409

Rosenberg I, Chu JC, Saxena V (1995) Calculation of monitor units for a linear accelerator with asymmetric jaws. Med Phys 22:55–61

Rosenow UF, Valentine ES, Davis LW (1990) A technique for treating local breast cancer using a single set-up point and asymmetric collimation. Int J Radiat Oncol Biol Phys 19:183–188

Rustgi SN, Rodgers JE (1985) Improvement in the buildup characteristics of a 10-mV photon beam with electron filters. Phys Med Biol 30:587

Sewchand W, Khan FM, Williamson J (1978) Variations in depth-dose data between open and wedge fields for 4-mV X-rays. Radiology 127:789–792

Sibata CH et al. (1990) Influence of hip prostheses on high energy photon dose distribution. Int J Radiat Oncol Biol Phys 18:455–461

Siddon RL, Tonnesen GL, Svensson GK (1981) Three-field techniques for breast treatment using a rotatable half-beam block. Int J Radiat Oncol Biol Phys 7:1473

Slessinger ED et al. (1993) Independent collimator dosimetry for a dual photon energy linear accelerator. Int J Radiat Oncol Biol Phys 27:681–687

Sohn JW, Suh JH, Pohar S (1995) A method for delivering accurate and uniform radiation dosages to the head and neck with asymmetric collimators and a single isocenter. Int J Radiat Oncol Biol Phys 32:809–814

Sontag MR, Cunningham JR (1978) The equivalent tissue–air ratio needed for making absorbed dose calculations in a heterogeneous medium. Radiology 129:787–794

Stewart J, Jackson A (1975) The steepness of the dose response curve for both tumor cure and normal tissue injury. Laryngoscope 85:1107–1111

Svensson GK et al. (1977) Superficial doses in treatment of breast and tangential fields using 4-mV X-rays. Int J Radiat Oncol Biol Phys 2:705

Takahashi S (1965) Conformation radiotherapy. Rotation techniques as applied to radiography and radiotherapy of cancer. Acta Radiol (Suppl) 242:1–142

Thatcher M (1984) Perturbation of cobalt 60 radiation doses by metal objects implanted during oral and maxillofacial surgery. J Oral Maxillofac Surg 42:108–110

Van Dyk J et al. (1977) Medulloblastoma: treatment technique and radiation dosimetry. Int J Radiat Oncol Biol Phys 2:993

Washington CM, Leaver DT (eds) (1996) Principles and practice of radiation therapy: introduction to radiation therapy, vol 1. Mosby, St. Louis

Watkins DMB (1981) Radiation therapy mold technology. Pergamon Press Toronto

Williamson TJ (1979) A technique for matching orthogonal megavoltage fields. Int J Radiat Oncol Biol Phys 5:111

Wong JW, Purdy JA (1990) On methods of inhomogeneity corrections for photon transport. Med Phys 17:807–814

Young MEJ, Gaylord JD (1970) Experimental tests of corrections for tissue inhomogeneities in radiotherapy. Br J Radiol 43:349–355

Yu CW et al. (2002) Clinical implementation of intensity-modulated arc therapy. Int J Radiat Oncol Biol Phys 53:453–463

Zhu Y, Boyer AL, Desorby GE (1992) Dose distributions of X-ray fields as shaped with multileaf collimators. Phys Med Biol 37:163–173

6 The Simulation Process in the Determination and Definition of the Treatment Volume and Treatment Planning

Sasa Mutic, James A. Purdy, Jeff M. Michalski, and Carlos A. Perez

CONTENTS

6.1   Introduction   107
6.2   Technology Overview   109
6.2.1   Conventional Simulator   110
6.2.1.1   Imaging Chain   110
6.2.1.2   Simulation Software   110
6.2.2   CT Simulator   111
6.2.2.1   Large-Bore CT   111
6.2.2.2   Multislice CT   112
6.2.2.3   CT-Simulator Tabletop   113
6.2.3.4   Patient-Marking Lasers   114
6.2.3   MR Simulator   114
6.2.4   PET/CT Simulator   116
6.2.4.1   Stand-Alone PET   116
6.2.4.2   Combined PET/CT   116
6.2.5   Virtual Simulation Software   117
6.3   Multimodality Imaging   118
6.3.1   Detection   118
6.3.2   Staging   118
6.3.3   Target Definition and Altered Dose
          Distributions   118
6.3.4   Evaluation of Response to Therapy and
          Follow-up   119
6.4   Patient Positioning and Immobilization   119
6.5   Simulation Process   124
6.5.1   Conventional Simulation   124
6.5.1.1   Patient Positioning and Immobilization   125
6.5.1.2   Verification of Patient Position Using
          Fluoroscopic Imaging   125
6.5.1.3   Determination of the Isocenter Location   125
6.5.1.4   Beam-Placement Design   126
6.5.1.5   Transfer of Simulation Information for
          Treatment Planning and Treatment   126
6.5.2   CT Simulation   126
6.5.2.1   Scan and Patient Positioning   126
6.5.2.2   Image Transfer and Registration   129
6.5.2.3   Target and Normal Structure Delineation   130
6.5.2.4   Treatment Techniques   131
6.6   Discussion and Conclusion   131
          References   132

S. Mutic, MS; J. M. Michalski, MD; C. A. Perez, MD
Department of Radiation Oncology, Washington University
School of Medicine, St. Louis, MO 63110, USA
James A. Purdy, PhD
Department of Radiation Oncology, University of California
Davis Medical Center, 4501 X Street, Sacramento, CA 95817,
USA

6.1 Introduction

Radiation therapy is a continually evolving medical specialty, especially considering the technology used for treatment planning, treatment, and delivery verification. During the past two decades, the field has evolved from treatment planning based primarily on planar radiographs to planning based on volumetric study sets. Currently, the vast majority of radiotherapy treatment plans are based on volumetric study sets. This significant increase in use of volumetric imaging in radiotherapy is due to the overwhelming acceptance of conformal radiation therapy as a standard of care for many treatment sites. The four primary imaging modalities employed in modern radiation therapy treatment planning process include computed tomography (CT), magnetic resonance imaging (MR), positron emission tomography (PET), and ultrasound (US). Planar X-ray radiography remains an important component of treatment planning and treatment verification process. Due to the current importance of treatment planning based on volumetric images, the primary emphasis of this chapter is CT simulation and incorporation of other imaging modalities in the CT simulation process. The use of conventional simulator in the age of CT simulation is addressed accordingly.

The radiation therapy simulator has been an integral component of the treatment-planning process for over 30 years. Conventional simulators are a combination of diagnostic X-ray machine and certain components of a radiation therapy linear accelerator. The conventional simulator (Fig. 6.1) consists of a diagnostic X-ray unit and fluoroscopic imaging system. The treatment table and the gantry are designed to mimic functions of a linear accelerator. The gantry head is designed to accommodate different beam modification devices (blocks, wedges, compensating filters), similar to a linear accelerator. The images are transmission radiographs with field collimator setting outlined by

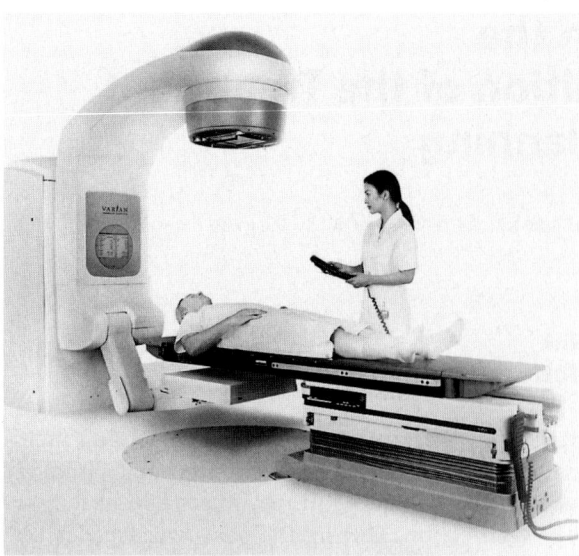

Fig. 6.1 Modern version of a conventional simulator. (Courtesy of Varian Medical Systems, Palo Alto, Calif.)

delineator wires. Using primarily bony landmarks, the physician outlines areas to receive therapeutic radiation doses. A shortcoming of a conventional simulation process is that very little anatomy, other than bony anatomy, is available for design of treatment portals. Shortly after the introduction of clinical CT scanners in early 1970s, it was realized that this imaging modality has much to offer in a radiation oncology setting. The CT images provide information not only about target volumes but about critical structures as well. Using CT images for radiation therapy treatment planning has enabled us to improve dose delivery to target volumes while reducing dose to critical organs. The CT images also provide density information for heterogeneity-based dose calculations.

A major weakness of CT imaging is a relatively limited soft tissue contrast. This limitation can be overcome by using CT images in conjunction with MR studies for treatment planning. The PET images can be used to add physiological information. Ultrasound has also been useful for imaging in brachytherapy. Multimodality-imaging-based treatment planning and target and normal structure delineation offer an opportunity to better define the anatomic extent of target volumes and to define their biological properties.

TATCHER (1977) proposed treatment simulation with CT scanners. This short article described the feasibility of a CT simulator and indicated potential economical benefits. GOITEIN and ABRAMS (1983)

and GOITEIN et al. (1983) further described multidimensional treatment planning based on CT images. They described a "beam's-eye-view" (BEV) function which "provides the user with an accurate reproduction of anatomic features from the viewpoint of a treatment source." They also described how "projection through the CT data from any desired origin provides an alignment film simulation which can be used to confirm accuracy of treatment, as well as help establish anatomic relationships relative to the margins of a treatment field." In reality, this was a description of the major characteristics of a system that we know today as a CT simulator or virtual simulator. An alignment film created from a divergent projection through the CT study data is commonly known as a digitally reconstructed radiograph (DRR). Additionally, use of DRRs in radiation therapy has been developed (SHEROUSE et al. 1990a).

SHEROUSE et al. (1987, 1990b) described a CT-image-based virtual simulation process which they referred to as a "software analog to conventional simulation." They described software tools and addressed technical issues that affect the present CT-simulation process. They pointed out the need for fast computers and specialized software, but also the need for improved patient immobilization and setup reproducibility.

The radiation oncology community eagerly embraced the concept of virtual simulation and in early 1990s commercial packages became available. These systems consisted of a diagnostic CT scanner, external laser positioning system, and a virtual simulation software workstation. One of the early commercial CT simulation packages is shown in Fig. 6.2. (AcQSim Oncodiagnostic Simulation/Localization System, Philips Medical Systems).

The CT simulators have matured to a point where they are one of the cornerstones of modern radiation oncology facilities. The present systems incorporate specially designed large-bore CT scanners, multislice CT scanners, high-quality laser positioning systems, and sophisticated virtual simulation packages. Many systems incorporate dose calculation capabilities and treatment-plan analysis and evaluation tools.

Additional virtual simulation software features and functions along with increased efficiency and flexibility have enabled CT simulators to replace conventional simulators in many facilities. This trend seems to be further fueled by the increased demand for imaging studies for 3D and IMRT treatment planning where conventional simulators are of limited value.

Fig. 6.2 A CT-simulator room layout. (Courtesy of Philips Medical Systems, Cleveland, Ohio)

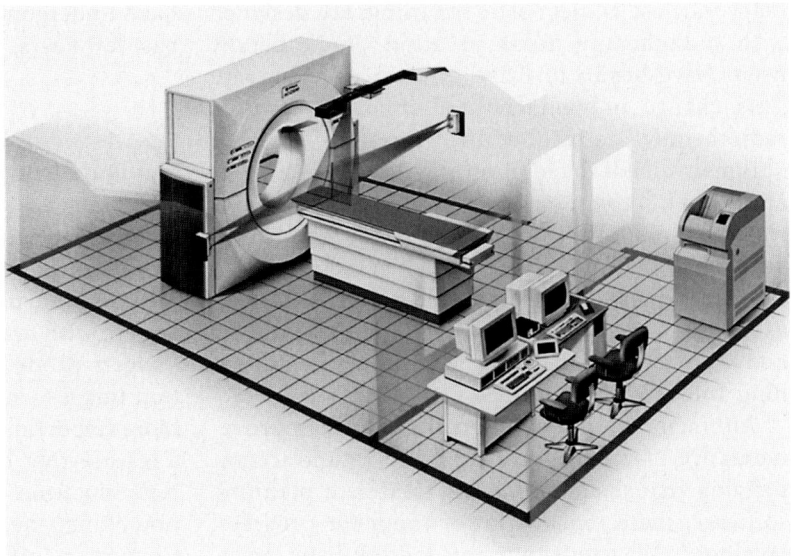

Figure 6.3 shows the place of CT simulation in the treatment-planning process. The implementation of simulation and treatment-planning process varies greatly between radiation oncology departments. This diversity is in part driven by significant technical differences between simulation and treatment-planning systems offered by different manufacturers. It would be impractical to address all possible variations in implementation of these processes. This chapter discusses the most common points and describes general differences between some popular approaches.

6.2
Technology Overview

One of the major recent changes in radiotherapy imaging is the approach of imaging equipment manufacturers towards radiation oncology and its unique imaging needs. Conventional simulators were always designed specifically for radiotherapy purposes and provided tools for accurate and efficient simulation and treatment planning. The scanners used for volumetric imaging were, on the other hand, historically designed with diagnostic radiology needs in mind, with little or no concern for radiation therapy needs. A compounding factor to this problem is that scanner characteristics which are extremely important in radiotherapy are often not a significant concern or not needed in diagnostic radiology. The scanners used in radiotherapy should have flat table tops, larger openings to accommodate immobilization devices and patients in conventional treatment positions, and software tools which can improve patient positioning and target delineation (MUTIC et al. 2003). Typically, diagnostic scanners were modified to meet the radiotherapy needs; flat tabletops and external patient positioning lasers were added to these scanners. This process worked well but had many limitations for simulation and treatment planning of certain tumor sites. The CT simulation process was also not as efficient as possible due to technological limitations. To rectify this, in the past few years, several CT scanners have been introduced with features designed specifically for radiotherapy. Even PET/CT scanners, whose pri-

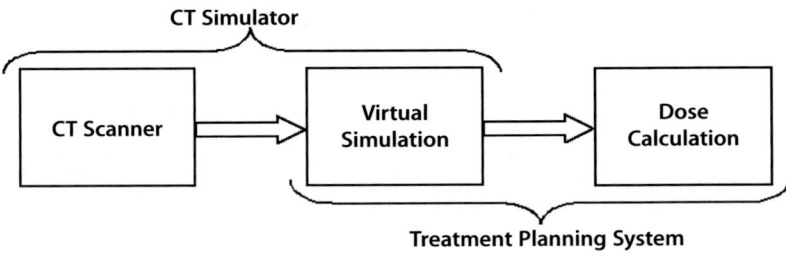

Fig. 6.3 Place of CT simulation in radiotherapy treatment-planning process. (From MUTIC et al. 2003)

mary purpose is diagnostic scanning, are designed with radiotherapy needs in mind. Radiotherapy needs have begun to drive product development. This change in manufacturer approach towards radiotherapy has resulted in a variety of imaging options available to radiation oncology departments and it has improved our ability to image patients for treatment-planning purposes. This development in volumetric scanning capabilities has inevitably led to improvements in conventional simulator design as well, as this technology has to be able to compete and keep up with other advances in treatment planning and delivery techniques.

Although there have been numerous improvements in CT technology for radiotherapy and accompanying virtual simulation and treatment-planning software, there remains lot of room for progress. Dynamic CT acquisition for radiotherapy treatment planning needs significant development and is really just in its infancy. Virtual simulation software (contouring, isocenter and beam placement, and port definition capabilities) needs to reside directly on CT-scanner control consoles to improve simulation efficiency. The CT acquisition protocols for radiotherapy need further improvements to capitalize on the fact that radiation dose from a CT scan is not a significant concern for radiotherapy patients, and that significant increases in CT acquisition techniques are acceptable in order to improve soft tissue contrast and image quality. Correspondingly, virtual simulation and treatment planning software needs to be able to accommodate increased image-acquisition capabilities of multislice CT scanners, as described later in this chapter. The vast majority of the modern treatment-planning systems are practically limited to 200–300 images per patient study. Multislice CT scanners can produce several hundred to several thousand images per scan which can be used to improve tumor definition, understanding of tumor and normal structure breathing motion, and verification of patient treatments accuracy.

6.2.1
Conventional Simulator

The conventional simulator (Fig. 6.1), consists of a fluoroscopic imaging chain (X-ray tube, filters, collimation, image intensifier, video camera, etc.; BUSHBERG et al. 2002), generator, patient support assembly (treatment table), laser patient positioning / marking system, and simulation and connectivity software. The imaging chain and simulator software

have undergone several improvements during the past few years.

6.2.1.1
Imaging Chain

One of the major changes in the imaging-chain design for conventional simulator was the replacement of the image intensifier and video camera system with amorphous silicon detectors. The new imagers produce high spatial and contrast resolution images which approach film quality (Fig. 6.4). More importantly, these images are distortion-free, a feature that is important for accurate geometric representation of patient anatomy. The introduction of high-quality digital imagers in conventional simulation further facilitates the concept of filmless radiation oncology departments.

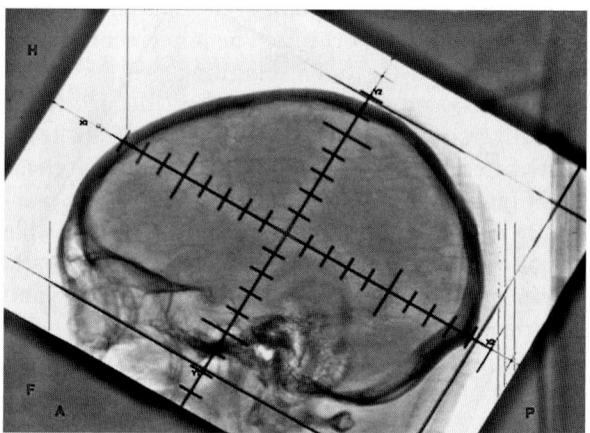

Fig. 6.4 Digital image of a head from a modern conventional simulator equipped with an amorphous silicon imager. (Courtesy of Varian Medical Systems, Palo Alto, Calif.)

6.2.1.2
Simulation Software

The conventional simulation software has also undergone many improvements. Modern simulators have Digital Image Communications in Medicine (DICOM) standard import capabilities (NEMA 1998). Treatment field parameters can be imported directly from the treatment-planning computer. The software can then automatically set the simulator parameters according to the treatment plan. This facilitates efficient and accurate verification of patient treatment setup on the conventional simulator. These simulators also have DICOM export capabilities which enable transfer of treatment setup parameters directly to a record and verify system or to a treatment-planning computer. The ability to import and capture digital images enables conventional simulators to have tools for automatic correlation of treatment planning and verification fields.

Vendors also offer solutions for some shortcomings of older conventional simulators. For example, older simulators were not equipped with tools to verify portal shapes created with multileaf collimators (MLCs). Newer simulators have features which can project MLC shapes directly on the patient's skin or on the portal films (Fig. 6.5).

A possibility for future of conventional simulation imaging is cone-beam CT. Since newer simulators are equipped with digital imagers, it is possible that cone-beam CT technology, which is available on linear accelerators, can also be implemented on conventional simulators. This will significantly improve imaging capabilities and usefulness of these devices. Figure 6.6 shows a cone beam CT image from a conventional simulator.

While it is often mentioned that conventional simulators can be completely replaced with CT simulators, new features and usefulness of conventional simulators are slowing down this process. Conventional simulator continues to be an important component of radiotherapy process, even though its use for treatment planning of many tumor sites has been significantly reduced.

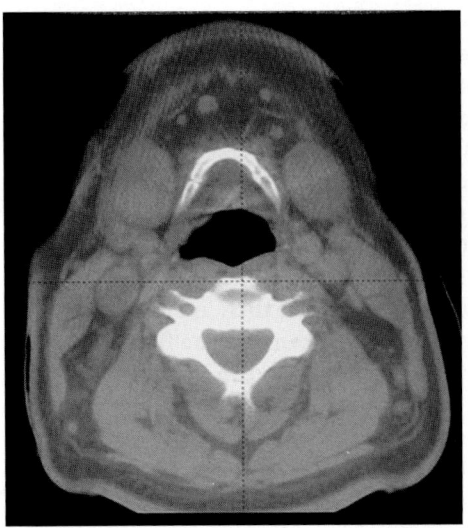

Fig. 6.6 Cone-beam CT image of a head acquired on a conventional simulator. (Courtesy of Varian Medical Systems, Palo Alto, Calif.)

6.2.2
CT Simulator

The CT simulator consists of a CT scanner, laser patient positioning / marking system, virtual simulation / 3D treatment planning software, and different hardcopy output devices. The CT scanner is used to acquire volumetric CT scan of a patient which represents the virtual patient and the simulation software creates virtual functions of a conventional simulator. The three most significant changes in CT-simulation technology, in recent years, have been the introduction of a larger gantry bore opening (large-bore CT; GARCIA-RAMIREZ et al. 2002), multislice image acquisition (multislice CT; KLINGENBECK et al. 1999), and addition of CT-simulation software directly on the CT scanner control console. These innovations improve efficiency and accuracy of CT-simulation process. They also improve patient experience by allowing patients to be positioned in more comfortable positions and reducing the simulation procedure time.

6.2.2.1
Large-Bore CT

Large-bore CT scanners were specifically designed with radiation therapy needs in mind. One of the requirements in treatment of several cancer sites (breast, lung, vulva, etc.) is for extremities to be

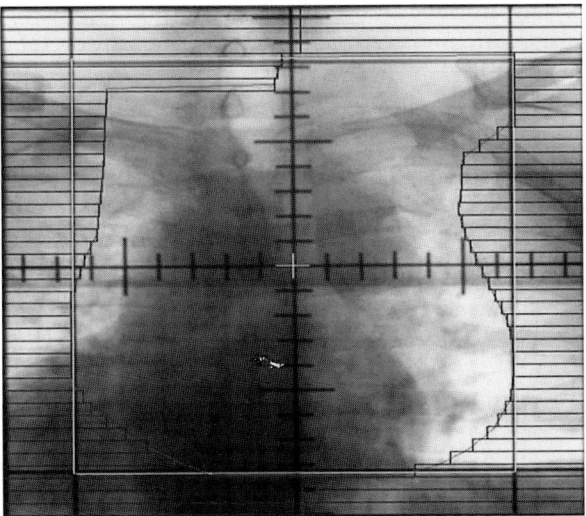

Fig. 6.5 Digital image of a chest from an amorphous silicon imager with multileaf collimator shape projected on the image. (Courtesy of Varian Medical Systems, Palo Alto, Calif.)

positioned away from the torso. When acquiring a CT scan with a patient in such treatment position, extremities often cannot fit through a conventional 70-cm-diameter scanner bore opening. In such situations, patient positioning needs to be modified to acquire the scan. This can result in less than optimal treatment position (patient may be less comfortable and therefore the daily setup reproducibility may be compromised). Large immobilization devices (slant board, body molds) are also difficult to fit through a conventional diameter scanner. The first large-bore CT-simulator was introduced in 2000, and several additional models with enlarged bore opening have been introduced since then.

Large-bore scanners also have increased scan field of view (SFOV). The SFOV determines the largest dimension of an object that can be fully included in the CT image and it is typically 48–50 cm on most conventional 70-cm bore opening scanners. For treatment-planning purposes it is necessary to have the full extent of the patient's skin on the CT image. Lateral patient separation can often be larger than 48–50 cm and the skin is then not visible on CT images. Increased SFOV available on large bore scanners solves this problem. There are, however, differences in implementation of extended SFOV and validity of quantitative CT values (quantitative CT) at larger image sizes. The CT numbers for some scanners are accurate only for smaller SFOVs and the values towards the periphery of large SFOV images are not reliable. This can be a concern for heterogeneity based dose calculations as inaccurate CT numbers can lead to dose calculation errors. The impact of CT number accuracy for increased SFOV images on dose calculation accuracy should be evaluated during scanner commissioning.

6.2.2.2
Multislice CT

In 1992 Elscint (Haifa, Israel) introduced a scanner which had a dual row of detectors and could acquire two images (slices) simultaneously. Since then, multislice CT has gained widespread acceptance, and scanners which can acquire 4, 8, 10, 16, 32, 40, 64, etc. (typically with sub-second rotation times) are now available from all major vendors. The basic premise behind the multislice CT technology is that multiple rows of detectors are used to create several images for one rotation of the X-ray tube around the patient. The detector design and arrangement varies among the vendors. Figure 6.7 shows an example of implementation for a 16-slice scanner available from a major vendor.

Although the scanner is considered a 16-slice scanner, there are 24 rows of detectors or detector elements. The center 16 have 0.75-mm collimated width at the isocenter and the outer four on either side have 1.5-mm collimated width at the isocenter. The total length coverage at the isocenter is then 24 mm. The thinnest nominal slice thickness that the scanner can produce is slightly larger than 0.75 mm, but for practical purposes it can be considered here as 0.75 mm. With proper collimation (16×0.75) on the X-ray tube side, signal from the center 16 detector elements can be used to acquire 16 0.75-mm-thick images at a time. If the collimation is increased to 16×1.5, so the X-ray beam includes the outer eight detectors, 16 1.5-mm-thick images can be acquired. In this situation, signals from the adjoining pairs of 0.75-mm detectors are combined to create 1.5-mm-thick images. Similarly, larger slice thicknesses can be created by combining signal from multiple detector elements. The primary advantage of multislice scanners is the ability to acquire image studies many times faster than single-slice scanners.

One of the obstacles for radiation therapy scanning with single-slice scanners is the limited tube heat loading capability. Often, fewer images are taken, slice thickness is increased, mAs is decreased, or scan pitch is increased to reduce the amount of heat produced during the scan and to allow for the entire scan to be acquired in a single acquisition. Due to the longer length of imaged volume per tube rotation (multiple slices acquired simultaneously), the tube heat loading for a particular patient volume is lower for multislice than for single-slice

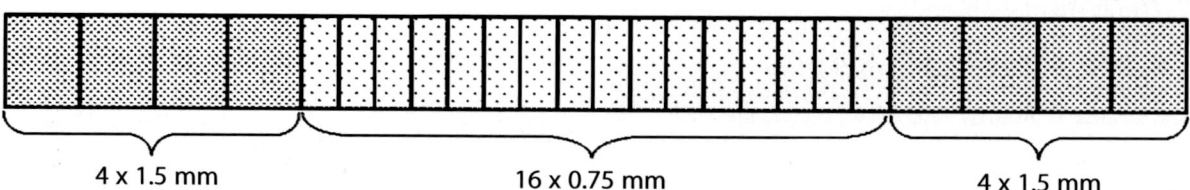

Fig. 6.7 A detector array for a 16-slice CT scanner

4 x 1.5 mm   16 x 0.75 mm   4 x 1.5 mm

scanners, and multislice scanners are generally not associated with tube head-loading concerns. Faster acquisition times and decreased tube loading of multislice scanners (which will allow longer volumes to be scanned in a single acquisition) can provide an advantage over single-slice systems for treatment-planning purposes. Multislice technology can be especially beneficial for imaging of the thorax where breathing artifacts can be minimized with faster scanning. Multislice technology also facilitates dynamic CT scanning, often referred to as 4D or 5D CT (Low et al. 2003). This application of multislice CT in radiation therapy has yet to be fully explored.

Multislice scanners are also capable of acquiring thinner slices which can result in better quality DRRs and more accurate target delineation (better spatial resolution; Fig. 6.8). Studies with thinner slices result in an increased number of images to process. Target volumes and critical structures have to be delineated on an increased number of images and treatment-planning systems have to handle larger amounts of data. Currently, this can result in increased time and labor required for treatment planning. Software vendors are creating tools which will allow easier manipulation of larger study sets, but that will likely take several years to implement. In the meantime, the numbers of CT images that are acquired for a treatment plan needs to be balanced between resolution requirements and the ability to process larger number of images.

6.2.2.3
CT-Simulator Tabletop

This section and discussion about simulator tabletops applies equally to all simulators used in radiation therapy (conventional, MRI, CT, and PET) and treatment machines. Tabletops used for patient support in radiation therapy during imaging or treatment should facilitate easy, efficient, reproducible, and accurate patient positioning. It is not only important that a tabletop improve patient positioning on a single device (i.e., treatment machine), but the repositioning of a patient from one imaging or treatment device to another also has to be considered. A great improvement in this process would be if all tabletops involved in patient simulation and treatment had a common design. They do not necessarily have to be identical, but they should have same dimensions (primarily width), flex and sag under patient weight, and

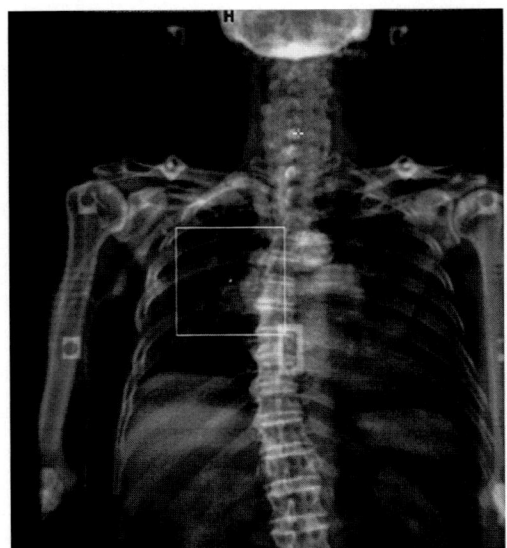

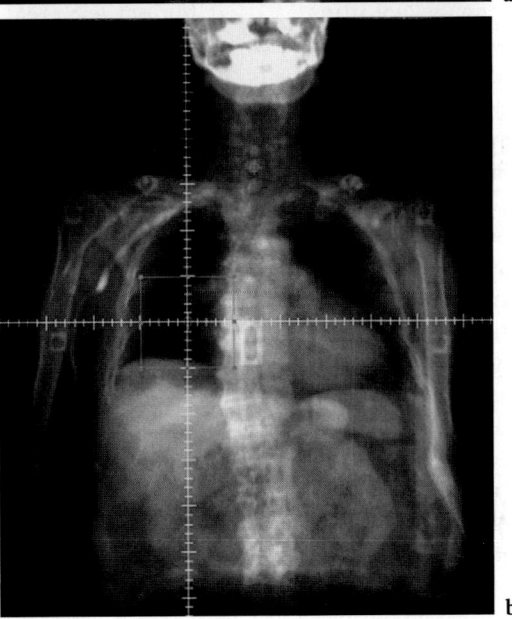

Fig. 6.8 a A 0.8-mm- and **b** a 3-mm-slice-thickness CT digitally reconstructed radiograph. Image **a** contains much more detail than **b**

they should allow registration (indexing) of patient immobilization devices to the tabletop. Figure 6.9 demonstrates this concept. The CT simulator tabletop has the same width as the linear accelerator used for patient treatment and both allow registration of patient immobilization system to the treatment couch. Ability to register the immobilization device and the patient to a treatment table is extremely important and improves immobilization, setup reproducibility, accuracy, and efficiency. The patient is always positioned in the same place on

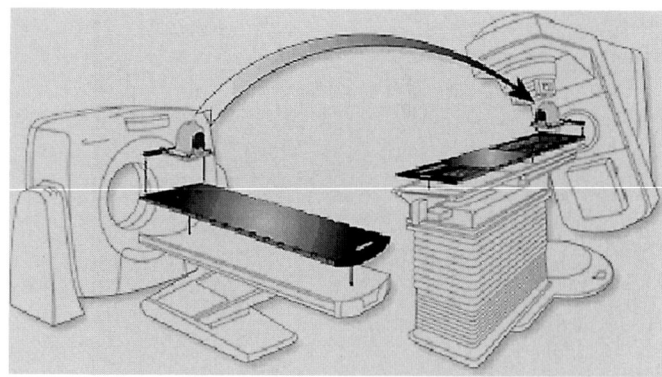

Fig. 6.9 Similarity in design of simulator and treatment machine tabletops allows efficient and accurate reproducibility of patient positioning. (Courtesy of MED-TEC, Inc., Orange City, Iowa)

the treatment machine and patient daily setup can be facilitated using the treatment-couch positions. Actually, if the patient is registered to the treatment couch, couch coordinates used for patient treatment can become a part of parameters that are set and tracked in the record and verify system. The tolerance for the couch parameters can be set according to type of treatment that the patient is receiving. For example, for conformal radiotherapy treatments the coordinates should allow minimal deviations (comparable to margins used for target delineation) in daily couch positioning. The therapist can then first place the treatment couch to the coordinates set in the record and verify system and then evaluate patient positioning. If the patient is well immobilized, minimal adjustments should be needed in patient setup.

6.2.3.4
Patient-Marking Lasers

A laser system is necessary to provide reference marks on patient skin or on the immobilization device. Figure 6.2 shows a laser system for a CT simulator.

Wall lasers
Wall lasers are vertical and horizontal, mounted to the side of the gantry. These lasers can be fixed or movable.

Sagittal lasers
The sagittal laser is a ceiling- or wall-mounted single laser, preferably movable. Scanner couch can move up/down and in/out but cannot move left/right; therefore, the sagittal laser should move left/right to allow marking away from patient midline.

Scanner lasers
Internally mounted, vertical and horizontal lasers on either side of the gantry and an overhead sagittal laser.

Lasers should be spatially stable over time and allow positional adjustment. Properly aligned simulator lasers greatly improve accuracy of patient treatments. Misaligned simulator lasers can introduce systematic errors in patient treatment; therefore, simulator laser alignment should be checked daily and the alignment tolerance should be within 2 mm (MUTIC et al. 2003).

6.2.3
MR Simulator

The MR images for radiotherapy treatment planning are usually acquired in diagnostic radiology, and very few radiation oncology departments have a dedicated MR scanner. Furthermore, the vast majority of MR studies in radiotherapy are currently limited to brain imaging. Magnetic resonance imaging has a superior soft tissue contrast compared with CT imaging, and there are several benefits that MR can offer for target delineation based on this advantage. There have been several reports describing use of MR scanners for imaging and treatment simulation in radiotherapy (POTTER et al. 1992; OKAMOTO et al. 1997; BEAVIS et al. 1998; SCHUBERT et al. 1999; MAH et al. 2002). Some of these reports have suggested that MR studies can be used alone for radiotherapy treatment planning. Indeed, if spatial distortions (the geometry of imaged objects is not always reproduced correctly), which is the largest concern with MR imaging, can be removed or minimized, MR studies can be used as the primary imaging modality for several treatment sites. Superior soft tissue contrast provided by MR can also be an advantage for treatment planning of certain extracranial tumor sites such as prostate (LEE et al. 2003; CHEN et al. 2004).

Conventional MR scanners are not well suited for extracranial imaging for treatment planning. The main difficulty is placement of patient in treatment position with immobilization device in the scanner. Small diameter and long length of conventional MR scanner openings severely limits patient positioning options for imaging. Open MR scanners, however, do not have this problem and patients can be scanned in conventional treatment positions. At least one manufacturer offers an open MR scanner which has been modified to serve as a radiotherapy simulator (Fig. 6.10). The scanner table is equipped with a flat top and external patient alignment lasers. The geometry of the scanner is similar to the CT simulator shown in Figure 6.2. Another manufacturer offers a 70-cm-diameter gantry opening conventional MRI scanner. The depth of the scanner opening is 125 cm. The dimensions of this scanner are very similar to a conventional CT scanner and in fact the scanner could be mistaken for a CT scanner. The ergonomics of this scanner are also well suited for radiotherapy simulation. One of the major problems with MR imaging for radiotherapy treatment planning are geometric distortions in acquired images. The MR scanners are often equipped with correction algorithms which minimize geometrical distortions. These corrections do not affect the entire image and only the center portion of the image (center 20–35 cm diameter) is adequately correct (within 2 mm); therefore, the representation of patient's skin and peripheral anatomy for larger body sections may be inaccurate. The effect of these inaccuracies must be evaluated if dose distributions and monitor units will be calculated directly on MR images.

Virtually all treatment-planning systems allow import of MR images and image registration with CT study. Some treatment-planning systems also allow design of treatment portals and display of isodose distributions on MR images directly. If the treatment-planning system can calculate doses directly on MR images, and if it was determined that geometric distortions are not significant, then there may be no need for CT images and MR study may be the only image set used for treatment planning. There should be a way to create images from the MR study which are equivalent to simulation radiographs for comparison with port films from the treatment machine. Another potential problem with MR images is that they do not contain information which can be related to electron density of imaged tissues for heterogeneity based corrections. This is not a significant problem, as bulk density corrections can be applied in the majority of treatment-planning systems. Due to availability of CT images in modern radiation oncology departments, it may be easiest if a CT study set is always acquired to complement the MR data and facilitate easier and more accurate heterogeneity-based dose calculations.

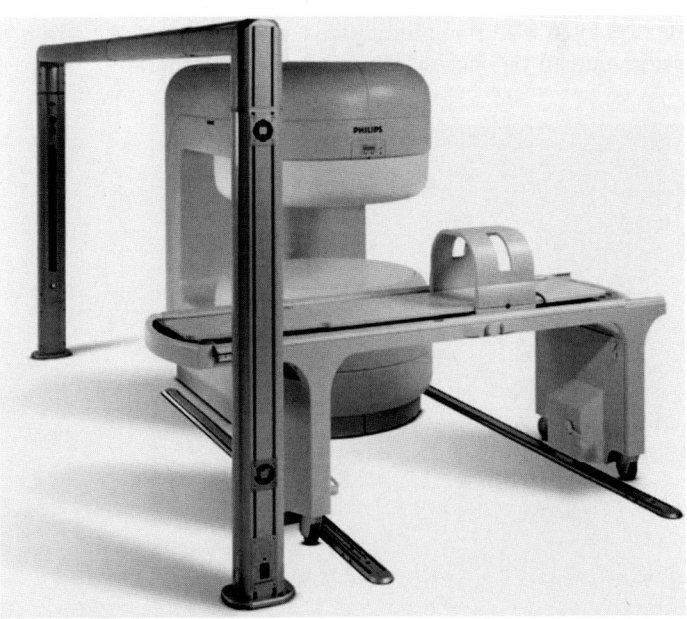

Fig. 6.10 An MR simulator. (Courtesy of Philips Medical Systems, Cleveland, Ohio)

6.2.4
PET/CT Simulator

The PET images for radiotherapy planning can come from a stand-alone PET scanner or a combined PET/CT unit. Combined PET/CT scanners are being installed in radiation oncology departments and are used for PET scanning, but also these machines can be used for CT scanning only without PET acquisition. Due to this purpose, these scanners can be classified as CT simulators, although the PET/CT simulator term may be more appropriate. Combined PET/CT scanners offer several advantages for radiotherapy imaging and are generally preferred over stand-alone units.

6.2.4.1
Stand-Alone PET

One of the major limitations of stand-alone PET scanners is a relatively small gantry bore opening, typically 55–60 cm. These scanners were designed to optimize image quality and not necessarily to accommodate radiotherapy patients in treatment positions with immobilization devices. This design feature of stand-alone PET scanners can severely limit the size of immobilization devices and patient position that is used for treatment. The immobilization devices used with a stand-alone PET scanner are limited in size and cannot be wider than approximately 50 cm. This is a major limitation for scanning patients with lung cancer where the patients need to have their arms positioned above the head. Also, larger patients may not be able to be scanned in an immobilization device. Even with these limitations, stand-alone PET scanners can be successfully used for radiotherapy

imaging and very good registrations (within 3 mm) can be achieved for the majority of patients. A stand-alone PET scanner has been used for radiotherapy scanning for several years at the Mallinckrodt Institute of Radiology in St. Louis (Missouri). Successful registrations have been achieved for head and neck, thorax, abdomen, and pelvis scans.

6.2.4.2
Combined PET/CT

The first combined PET/CT prototype was introduced in 1998 at the University of Pittsburgh (BEYER 2000). Since then, all major manufacturers have produced several commercial models. The key description of PET/CT scanners is that a PET and a CT scanner are "combined" in the same housing (Fig. 6.11), meaning that there are two gantries (PET and CT) combined in one housing sharing a common couch. Image reconstruction and scanner operation is increasingly performed from one control console.

Combined PET/CT scanner design varies among different vendors with respect to PET detectors, image quality and resolution, speed, image field of view; number of slices for the CT part, scanner couch design, gantry bore opening, and other considerations. All of the commercially available scanners have a 70-cm gantry opening for the CT portion, although large-bore CT scanners will likely become part of PET/CT scanners in the future. The PET gantry opening ranges in diameter from 60 to 70 cm, meaning that some of the commercial scanners have a non-uniform gantry opening as the patient travels from the CT portion of the scanner to the PET side. More importantly, the scanners with the smaller

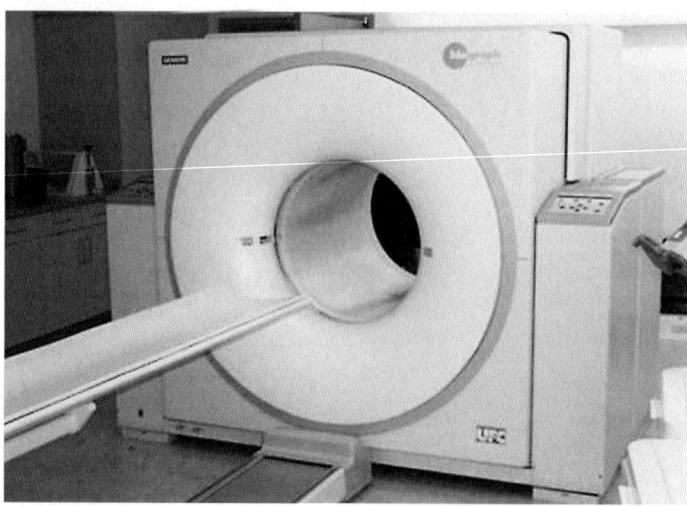

Fig. 6.11 A combined PET/CT scanner

gantry opening on the PET side will pose the same difficulties for radiotherapy scanning as stand-alone PET scanners. Again, the size of patient immobilization devices and patient scan/treatment position will have to be adapted to the size of the gantry opening.

The combined PET/CT technology offers two major benefits for radiotherapy planning. Firstly, because the images are acquired on the same scanner, providing that the patient does not move between the two studies, the patient anatomy will have the same coordinates in both studies. These images have been registered using hardware registration rather than software registration. The second benefit of the combined PET/CT units is that CT images are used to measure attenuation correction factors (ACFs) for the PET emission data, obviating the need for a time-consuming PET transmission scan (BAILEY 2003; BAILEY et al. 2003). The use of CT images to generate PET ACFs reduces the scan time up to 40% and also provides essentially noiseless ACFs compared with those from standard PET transmission measurements (TOWNSEND et al. 2004). Shorter scan times can benefit radiotherapy patients who are scanned in treatment position which often can be uncomfortable and difficult to tolerate for prolonged periods of time. One of the concerns with ACFs generated from CT images is mismatch or misalignment between CT and PET images. The PET images are acquired during many cycles of free breathing and CT images are acquired as a snapshot in time at full inspiration, partial inspiration, or some form of shallow breathing. The breathing motion will cause mismatch in anatomy between PET and CT images in the base of lung and through the diaphragm region. This mismatch can result in artifacts in these areas which may influence diagnosis and radiotherapy target definition in this region. There are various gating methods that can be used during image acquisition to minimize the motion component and essentially acquire *true*, motionless, images of patient anatomy. Gated or 4D CT (with time being the fourth dimension) can be used to generate more reliable ACFs and also for radiotherapy treatment planning where gated delivery methods are being used.

Contrast-enhanced CT images can cause inaccurate ACFs due to artificially increased attenuation through anatomy which contains contrast material. The most obvious way to avoid this problem is to acquire a routine CT with contrast and another non-contrast CT. There is also an option to use software tools to correct for these artifacts. For radiotherapy scanning, it is preferred to acquire two separate scans. The attenuation correction CT can be a whole-body, low-dose scan with greater slice thickness if desired.

The second CT would be a treatment-planning scan with thin slices for better resolution and DRR quality. This scan is acquired only through the volume of interest, thus limiting the number of images and memory requirements to manipulate these images in the treatment-planning computer. This second scan can then be contrast enhanced if desired.

6.2.5
Virtual Simulation Software

As with all software programs, user-friendly, fast, and well-functioning virtual (CT) simulation software with useful features and tools will be a determining factor for success of a virtual simulation program. Commercially available programs far surpass in-house written software and are the most efficient approach to virtual simulation. Several features are very important when considering virtual simulation / 3D treatment-planning software:

1. Contouring and localization of structures: contouring and localization of structures is often mentioned as one of the most time-consuming tasks in the treatment planning process. The virtual simulation software should allow fast user-friendly contouring process with help of semi-automatic or automatic contouring tools. An array of editing tools (erase, rotate, translate, stretch, undo) should be available. An ability to add margins in three dimensions and to automatically draw treatment portals around target volumes should be available. An underlining emphasis should be functionality and efficiency.

2. Image processing and display: virtual simulation workstation must be capable of processing large volumetric sets of images and displaying them in different views as quickly as possible (near realtime image manipulation and display is desired). The quality of reconstructed images is just as important as the quality of the original study set. The reconstructed images (DRRs and multiplanar reconstruction) are used for target volume definition and treatment verification, and have a direct impact on accuracy of patient treatments.

3. Simulator geometry: a prerequisite of virtual simulation software is the ability to mimic functions of a conventional simulator and of a medical linear accelerator. The software has to be able to show gantry, couch, collimator, jaw motion, SSD changes, beam divergence, etc. The software should facilitate design of treatment portals with blocks and multileaf collimators.

6.3
Multimodality Imaging

Imaging is involved in all steps of patient management, disease detection, staging, treatment modality selection (intramodality and intermodality), target volume definitions, treatment planning, and outcome estimation and patient follow-up. An overall goal of imaging in radiotherapy is to accurately delineate and biologically characterize an individual tumor, select an appropriate course of therapy, and predict the response at the earliest possible time. The requirement to biologically characterize an individual tumor means that an imaging modality must be capable of imagining not only the gross anatomy but also recording information about physiology, metabolism, and the molecular makeup of a tumor; therefore, the image information used in radiotherapy can be classified as anatomical and/or biological. The four primary imaging modalities used in radiation therapy are CT, MRI, ultrasound (US), and nuclear medicine imaging.

No single imaging modality provides all the necessary information for treatment planning and patient management for several cancer sites, but multiple imaging modalities can be used to complement each other and improve disease detection, staging, therapy selection, target design, outcome prognosis, and follow-up. Figure 6.12 shows the information content possibilities of the imaging modalities used in radiation therapy. The maximum benefits may be realized if anatomical and biological imaging modalities complement each other.

6.3.1
Detection

Imaging of disease with CT or MRI (non-functional) is based on anatomic or physiological changes that are a late manifestation of molecular changes that underlie the disease. By detecting changes in the

molecular and biochemical process, biological imaging (PET or functional MRI) can demonstrate disease before it becomes anatomically detectable. Changes in tumor detection capabilities can lead to modification in radiation therapy target volumes and dose prescriptions.

6.3.2
Staging

Positron emission tomography has improved patient staging in several treatment sites (DIZENDORF 2003). Better knowledge of the true extent of the patient's disease can significantly alter patient management. For some patients, who would otherwise undergo curative radiotherapy, PET may demonstrate distal disease or alter the extent of local disease and indicate that a palliative course of therapy is more appropriate. These patients would not only be spared the side effects of futile curative treatment, but the overall health care costs could also be lowered due to PET findings.

In addition to more accurate staging, PET may also be able to provide information about individual tumor biology (phenotype). This would allow further stratification of patients within the same clinical stage. So rather than basing therapy selection for an individual patient on the stage alone, which is statistically appropriate for a large group of patients, biological properties of an individual tumor can then be used for therapy selection. The tumor phenotype information may affect intermodality and intramodality patient management depending on suspected radiation or chemotherapy sensitivity of an individual tumor. If we know more about biological properties of an individual tumor, it may be possible to incorporate biological response models in the therapy selection process to maximize the therapeutic ratio.

6.3.3
Target Definition and Altered Dose Distributions

The true extent of the disease may extend beyond anatomically defined volumes, and biological imaging with PET has already been shown to be valuable for defining the extent of target volumes. Furthermore, PET can be used to differentiate areas of biological importance within the boundaries of target volumes. LING et al. (2000) have described a concept

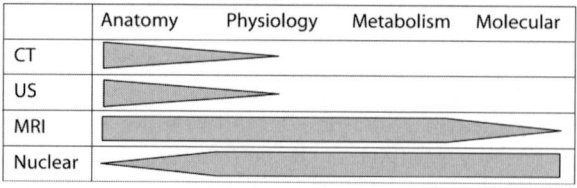

Fig. 6.12 Information content of current imaging modalities in radiotherapy

of biological target volumes (BTVs). In addition to recommendations for target volume definitions proposed by the International Commission on Radiation Units and Measurements (ICRU) reports 50 and 62 (ICRU 1993, 1999), portions of target volumes would be identified as having increased growth activity or radioresistance. Identification of these volumes would be performed with biological imaging and these volumes would be labeled as BTVs. Biological target volumes would then have a special consideration during the treatment-planning process and would be subject to dose escalation.

For example, CHAO et al. (2001) have shown how PET imaging based hypoxia measurement technique with Cu(II)-diacetyl-bis(N^4-methylthiosemicarbazone) (Cu-ATSM) tracer can be used to identify a BTV for head and neck tumors. Experiments have shown that increased Cu-ATSM uptake can be used to identify hypoxic tissues, which are also associated with increased radioresistance. The proposed treatment technique is based on the idea that Cu-ATSM can be used to identify the hypoxic BTV and IMRT delivery can be used to deliver escalated doses to overcome the radioresistance of the BTV.

6.3.4
Evaluation of Response to Therapy and Follow-up

Currently, tumor control and effectiveness of radiotherapy is evaluated in the weeks and months following the completion of treatments. The evaluation, similar to detection and diagnosis, relies largely on anatomical changes, which take time to manifest. If the planned approach of radiotherapy is not effective and the patient has a persistent disease or new growth, it is too late to make any modifications, as the therapy has already been completed. Additionally, by the time it is determined that a local tumor control has not been achieved, it may be too late to initiate a second line of therapy. Biological imaging may be used to detect response to therapy on a molecular level and allow evaluation of therapy effectiveness sooner after completion of treatments (YOUNG et al. 1999). Ideally, biological imaging may be used shortly after initiation of treatments to image tumor changes. This approach has had limited success thus far, but research in this area is active and it eventually may be possible to evaluate tumor response after initiation of therapy.

We are just beginning to exploit benefits of multimodality imaging in management of radiotherapy patients. With time, use of several image types will be a commonplace for treatment planning of many cancer sites. This has already, to an extent, taken place for treatment planning of central nervous system tumors where CT images are complemented with MRI studies for a significant fraction of patients.

One concern with utilization of novel imaging data for treatment planning and management of radiation therapy patients is that the information contained in the images may be misinterpreted or may be incorrect resulting in inappropriate patient treatments. It is imperative for radiation oncologists to understand potential pitfalls and shortcomings of individual imaging modalities, and also to realize that the best results can be achieved if newer imaging techniques are used to supplement existing staging and tumor delineation processes. This is especially true if biological or functional information is used for target delineation where possibility of false positive or negative findings exists. The correlation of biological or functional signals with anatomic abnormalities detected by CT or MRI can provide an important validation in the target delineation and patient management process.

6.4
Patient Positioning and Immobilization

The success of conformal radiation therapy process begins with proper setup and immobilization. One of the primary rules for positioning patients for simulation and radiotherapy treatments is that the patient should be as comfortable as possible. Patients who are uncomfortable typically have poor treatment setup reproducibility. An uncomfortable position that a patient was able to tolerate during simulation may be impossible to successfully reproduce for treatment. Immobilization devices tremendously improve reproducibility and rigidity of the setup. Another important consideration in patient positioning is that immobilization devices custom made for individual patients can significantly improve intra and inter fraction immobilization and setup reproducibility. Standard immobilization devices often do not provide an adequate fit for all patients, they work for many, but not for all, patients. This is well accepted and understood for body molds, as described in Chapter 7, and also for thermoplastic masks for immobilization of head and neck region. One very important point that is often overlooked is design of head supports (head cups) for treatment of head and neck region. There is a tendency to use a standard head cup and a custom thermoplastic mask. An analogy could be made that the head cup is a foundation for a house and thermoplastic mask is framing for the walls. If the foundation is not appropriately constructed, the entire structure will be unstable; therefore, for conformal treatments of head and neck region improved setup reproducibility can be achieved with custom head supports made from body mold material as shown in Figure 6.13.

Similar approach and forethought can be applied to other treatment sites and respective immobilization devices. Figure 6.14 shows a body mold that is used at the Washington University School of Medicine for treatment of patients with breast cancer. This device was designed to facilitate CT-based treatment planning and to improve patient reproducibility from the simulator to the treatment machine. In the inside of the body mold are Styrofoam wedges which elevate the patient and provide adequate positioning for the breast or chest wall. The handle improves patient comfort by providing a solid grip point for the arm, and the non-skid material prevents patients from sliding in the body mold. The ear mold, which is made from dental wax, improves the head position reproducibility. The device also registers to the treatment table so the couch coordinates are tracked during patient treatment in the record and verify system. Other authors have proposed an even more elaborate positioning device for the breast, such as the prone breast board which has an opening for the breast to hang freely beneath the patient.

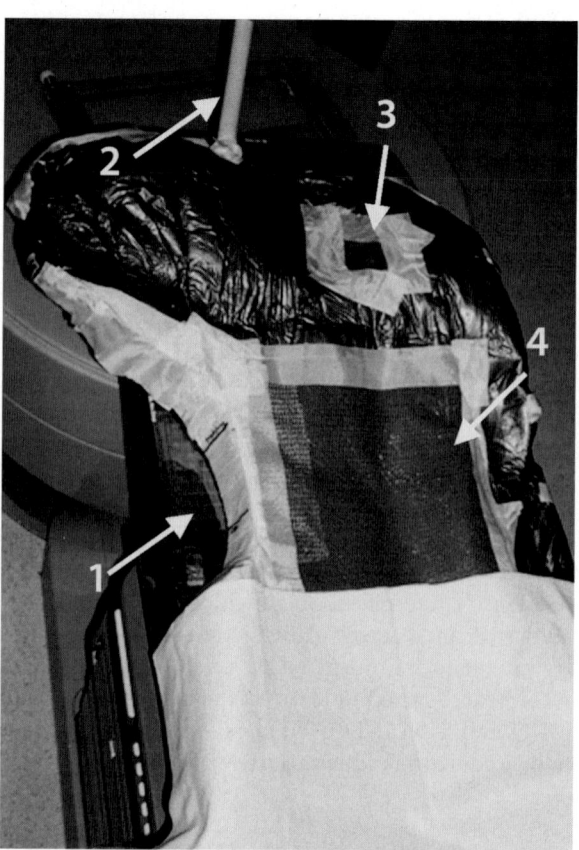

Fig. 6.14 Breast treatment immobilization device. *1* Portion of the mold removed on the ipsilateral side; *2* arm grip; *3* ear mold made from dental wax; *4* non-skid surface

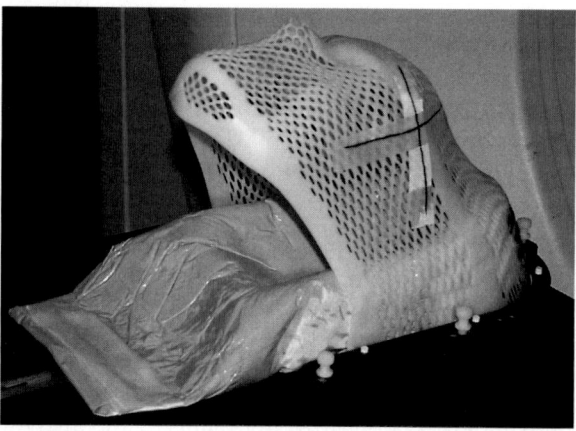

Fig. 6.13 Head and neck immobilization device with custom headrest

The third important point about patient positioning is that patients should be aligned straight on the simulator table, and patients should not be rotated or slanted. It is much easier to reproduce a straight patient position than a rotated one. Simulation and treatment policies should include positioning and immobilization specifications for individual treatment sites. Patient setup design should consider location of critical structures and target volumes, patient overall health and flexibility, possible implants and anatomic anomalies, and available immobilization devices. Immobilization devices should not produce artifacts on CT images. Table 6.1 shows a treatment-site-specific set of instructions CT simulation instructions. The instructions are designed to be used with a single-slice CT scanner and non-uniform CT slice thickness and spacing is used to minimize heat production on the X-ray tube while obtaining excellent image quality. Table 6.2 shows the CT scan parameters that accompany instructions in Table 6.1.

Table 6.1. An example of treatment site specific CT simulation instructions. Initial reference refers to the initial set of positioning marks placed on the patient. In most instances; these marks will be the same as the treatment isocenter. The CT scan protocol parameters are provided in Table 6.2. *SSD* source-to-skin distance, *SAD* source–axis distance. *SCV* supraclavicular, supraclavicle, *SIM* simulation, *CAX* central axis, *SI* sacroiliac, *IC* iliac crest, *BIT* bottom of ischial tuberosity, *AC* alpha cradle

Site	Patient position	Immobilization	Setup	Protocol	Slice (mm)	Index (mm)	Scan limits	Contrast	Special instructions
Head and neck IMRT	Supine, head straight	Thermoplastic mask; custom head mold; IMRT registration device, no arm stretcher; no pad; initial reference midline and midplane per doctor's instructions	Wire initial reference	Head 5×5	5	5	Top of orbits/top of skull	None	Triangle sponge under knees; no dentures; arms on abdomen; wire areas of interest; tongue blade or mouth piece on doctor's request
				Head 3×3	3	3	Top of shoulders/top of orbits		
				Head 5×5	5	5	3 cm below clavicles/top of shoulders		
Brain IMRT	Supine, head straight	Initial reference per doctor	Initial reference: midline and midbrain or per doctor's instructions	Head 3×3	3	3	Base of skull/top of skull	None	Have MR images ready for simulation, if necessary
				Head 5×5	5	5	Chin/base of skull		
Head and neck conventional	Supine, head straight	Thermoplastic mask; clear head rest; arm stretcher, (clear mouth piece or dental wax for dentures per doctor's instructions), no pad	Initial reference just above shoulders at 3 cm depth; wire initial reference	Head 5×5	5	5	Mid-orbits/top of skull	100 ml Optiray 300	Head position per doctor's instructions: neutral/clear „C"; extended clear „D"; hyper-extended clear „D"; chin tucked/clear „F" reversed; tongue blade on doctor's request
				Head 3×3	3	3	Top of shoulders/mid-orbits		
				Onc med thorax 5×5	5	5	3 cm below clavicle/top of shoulders		
Orbits	Supine, head straight	Thermoplastic mask; clear head rest; bolus may be used. Bolus made from thick thermoplastic material	Initial reference: vertical-lateral canthus; long-middle orbit; sagittal midline (bilateral); center of orbit (ipsilateral)	Head 5×5	5	5	3 cm above orbits/top of skull	None	Per doctor's instructions wire lateral edge of canthus; -cut mask over orbits out; tape with silk tape; have patient look straight up during scan; head neutral, Clr. „F", if this tucks head clear „C"; start scan at bottom
				Head 3×3	3	3	3 cm below/3 cm above orbits		
				Head 5×5	5	5	Bottom of chin/3 cm below orbits		

Table 6.1. (Continued)

Site	Patient position	Immobilization	Setup	Protocol	Slice (mm)	Index (mm)	Scan limits	Contrast	Special instructions
Parotid	Supine, slightly oblique, „chicken wing"	Thermoplastic mask, clear head rest, triangle sponge to support oblique position	Initial reference: at angle of mandible or in center of MD field, affected side parallel to floor	Head 5×5	5	5	Top of orbits/top of skull	None	Remove dentures; pad can be used; wire areas of interest
				Head 3×3	3	3	Top of thorax/top of orbits		
				Onc med thorax 5×5	5	5	3 cm below clavicle/top of thorax		
Brain	Supine, head straight, arms on abdomen	Thermoplastic mask, clear head rest, pad on table, triangular sponge under knees	Initial reference: midline and midbrain or per doctor's instructions	Head 3×3	3	3	Base of skull/top of skull	100 ml Optiray 300	Neutral/Clear „F"/chin tucked or neck flexed/clear „F" reversed; mask reinforced; contrast per doctor's instructions; have MR images ready for simulation, if necessary
				Head 5×5	5	5	Bottom of chin/base of skull		
Breast/SCV/SSD/SAD	Supine. Bent elbow to clear scanner opening, if necessary	Body mold registered to table with arm holder and dental wax ear mold	Doctor to mark upper, lower, medial, and lateral borders; place *thick* wires on marks	Onc med/large thorax 5×5	5	5	1–2 cm above upper border	None	See breast simulation procedures for determining isocenter
				Onc med/large thorax 5×5	3	3	Through SCV/breast field		
				Onc med/large thorax 5×5	5	5	1–2 cm below lower border		
Lung 3D/CT-SIM	Supine; chin extended; arms above head, folded, may rest on 5- or 7-cm sponge	3D: body mold registered to table; CT-SIM: none, unless necessary per doctor's instructions; no pad; triangle sponge under knees	Initial reference at carina and midplane or per doctor's instructions on simulation request form	Onc med/large thorax 5×5	5	5	Chin-to-lung apex	125 ml Optiray 320 and/or two tablespoons of Esophocat	CAX drawn on patient's anterior and lateral surfaces. Per doctor's instructions the CAX can be placed mid-depth and midplane at highest level of thorax; contrast given just before the scan; start scanning after half contrast in
				Onc med/large thorax 3×3	3	3	Through target and CAX, most of lung		
				Onc med/large thorax 5×5	5	5	Through rest of the lung		
				Onc med/large thorax 8×8	8	8	Top of kidneys or per MD request		
Esophagus	Same as lung	Same as lung	Same as lung	Same as lung	Same as lung	Same as lung	Same as lung	None or Esophocat	Same as lung

Site	Position	Immobilization	Reference	Field			Field extent	Contrast	Notes
Abdomen	Supine; head on 5- or 7-cm sponge, as comfortable	Body mold. Arms above head, hands on 7 sponge.	Initial reference at T12/L1 interspace, midplane and midline, or per doctor's instructions	Onc med/large body 5×5	5	5	Above diaphragm to above initial reference	450 ml; Readi-Cat 2, 1 h prior to scan	Readi-Cat 2 given to visualize small bowel on the scan
				Onc med/large body 3×3	3	3	Through initial reference		
				Onc med/large body 5×5	5	5	To IC or through pelvis per doctor's instructions		
Standard pelvis	Supine; arms on upper abdomen or chest	Feet banded with rubber band; none otherwise, unless requested by doctor; head on 5- to 7-cm sponge	Initial reference midline and midplane of area requested by doctor	Onc med/large body 5×5	5	5	Top of L3-above upper port edge	450 ml; Readi-Cat 2	Standard pelvis: 16.5-20, inferior border/bottom of ischial tuberosity; 16.5×16.5, inferior border/bottom of obturator foramen
				Onc med/large body 3×3	3	3	Above/below port edge		
				Onc med/large body 5×5	5	5	Below port edge/ peritoneum		
Pelvis and peri-aortic	Supine; arms above head	Same as above	Same as above	Onc med/large body 5×5	5	5	Diaphragm/L3/L4 interspace	450 ml; Readi-Cat 2	Upper port edge: T12-L1; lower port edge: bottom of obturator foramen; place center between upper and lower border
				Onc med/large body 3×3	3	3	L3/L4 interspace/ ischial tuberosity		
				Onc med/large body 5×5	5	5	Ischial tuberosity peritoneum		
3D/IMRT prostate	Supine; arms on upper chest	Body mold, feet banded with rubber band, AC close on lateral thighs, lines on body mold, registration; head on 5- to 7-cm sponge (or as comfortable)	Initial reference midline and at level of the prostate	Onc med/large body 5×5	5	5	Physician will specify upper margin; if not, go to L3/L4	10–15 cc Conray 30	Rectal markers; option 1: flexible Foley; option 2: rectal marker with metal balls; urethrogram by doctor; after body mold is finished, have patient get up and then reposition the patient
				Onc med/large body 5×5	5	5	Mid-SI joints/top of iliac crest /JMM		
				Onc med/large body 3×3	3	3	BIT/mid-SI joints		
				Onc med/large body 5×5	5	5	Below urethral stricture/BIT		
Brachy prostate	Supine; arms folded on chest	None; no pad; head on 7-cm sponge; triangle sponge under knees	Do not need initial reference	Onc med/large prostate implant	3	3	Bottom of SI joints/ bottom of ischial tuberosity	None	Patient to drink a glass of water 30 min prior to scan

Table 6.2. An example set of CT scan parameters for a single-slice scanner. *FOV* field of view

Protocol	kV	mA	Time (s)	Display FOV (cm)	Pitch	Thickness (cm)	Spacing (cm)	Pilot length
Head 5×5	130	300	1	48	1.3	5	5	450
Head 3×3	130	300	1	48	1.3	3	3	450
Onc child brain	130	280	1	35	1.3	3	3	450
Onc neck	130	280	1	48	1.3	3	3	450
Onc medium thorax 8×8	130	230	1	48	1.5	8	8	650
Onc medium thorax 5×5	130	230	1	48	1.5	5	5	650
Onc medium thorax 3×3	130	230	1	48	1.7	3	3	650
Onc large thorax 8×8	130	250	1	55	1.5	8	8	650
Onc large thorax 5×5	130	250	1	55	1.5	5	5	650
Onc large thorax 3×3	130	250	1	55	1.7	3	3	650
Child thorax 5×5	130	150	1	35	1.5	5	5	450
Child thorax 3×3	130	150	1	35	1.7	3	3	450
Onc medium body 5×5	130	250	1	48	1.5	5	5	650
Onc medium body 3×3	130	250	1	48	2	3	3	650
Onc large body 5×5	130	300	1	55	1.5	5	5	650
Onc large body 3×3	130	300	1	55	2	3	3	650
Onc child body 5×5	130	180	1	35	1.5	5	5	512
Onc child body 3×3	130	180	1	35	2	3	3	512
Onc medium prostate implant	130	250	1	30	1.5	3	5	650
Onc large prostate implant	130	300	1	30	1.5	3	3	650
Onc extremity	130	200	1	48	2	5	5	650

6.5
Simulation Process

Like the other areas of radiotherapy treatment planning and treatment, simulation requires a team approach involving physicians, physicists, dosimetrists, therapists, nurses, etc. The team needs to understand individual components of the process and their specific technical requirements. A well-informed and knowledgeable personnel is needed to fully exploit benefits of modern treatment simulation equipment. Furthermore, treatment-site-specific written procedures can significantly improve efficiency, consistency, and accuracy of simulations. Table 6.1 shows an example of such procedures. Written procedures are also helpful for training of new staff and performing simulations for less frequent treatment procedures. A well-designed and simple simulation process greatly increases treatment planning efficiency and improves patient setup reproducibility between the simulator and treatment machine. This section describes conventional and CT simulation processes. The CT simulation is given larger emphasis as this has become the primary source of treatment planning information for large number of radiation oncology departments.

6.5.1
Conventional Simulation

As described previously, the dependence on conventional simulators has decreased over the past several years as conformal radiation therapy has become the standard of care for several treatment sites, and CT simulators have evolved to overcome most of their initial limitations (gantry opening size, long scan times, connectivity, etc.). In this new era, many radiation oncology departments have determined that they can replace conventional simulators with CT simulators, and there are many departments that no longer have a conventional simulator. Other institutions have reduced the number of conventional simulators and/or number of conventional simulations. For example, during the 1990s, the Department of Radiation Oncology at the Washington University School of Medicine operated with three conventional simulators and one CT simulator. In 2000, the department replaced two of the conventional simulators for another CT simulator for a total of one conventional simulator and two CT simulators, while treating the same number of patients. Even though the conventional simulator use has been reduced, certain procedures can be performed much easier and more efficiently on a conventional simulator than with a CT simulation. For example, palliative treatments for bone metastasis, whole pelvis irradiation for gynecological tumors, pelvis irradiation for rectal cancer, certain extremity treatments, brachytherapy treatment planning, and some other treatment procedures are very simple to simulate using conventional technology. One of the advantages of conventional simulators is that there are virtually no limitation on available patient positions and on size and shape of immobilization devices that are used for simulation. If a patient cannot lie down, the sitting position in a special treatment chair can be accommodated with a conventional simulator where a CT scanner would not be an option. For some treatments of hands and arms it may be desirable for the patient to stand next to the treatment table; again, this type of simulation can only be performed with conventional simulator. With better imaging capabilities, cone-beam CT, and better connectivity with the treatment-planning system and treatment machine, conventional simulators are actually becoming more valuable than in the past.

Conventional simulation process consists of the following:

- Patient positioning and immobilization
- Verification of patient positioning using fluoroscopic imaging
- Determination of isocenter location
- Beam placement design
- Marking of patient and immobilization devices based on isocenter coordinates
- Acquisition of X-ray films
- Outlining of treatment portals on the X-ray films
- Transferring or acquiring of patient setup data for the record and verify system
- Transferring of simulation data to dosimetry for treatment planning and monitor unit calculation
- Preparation of documentation for treatment
- Performance of necessary verifications and treatment-plan checks

6.5.1.1
Patient Positioning and Immobilization

The goals for patient positioning and simulation described earlier should be followed for conventional simulation. The flexibility in size of immobilization devices greatly simplifies patient positioning for treatment. If the patient's conventional simulation will be followed with a CT scan, then the limitations of the CT scanner should be considered in patient positioning and immobilization.

6.5.1.2
Verification of Patient Position Using Fluoroscopic Imaging

Prior to construction of the immobilization device, the patient should be aligned to lay straight on the treatment table. This means that the patient's head, vertebra, and possibly extremities should be parallel with the longitudinal axis of the treatment table. The patient should also lay flat on the table with no rotation. It is much easier to reproduce a straight patient position on the treatment table than if the patient is rotated. The verification of patient position is performed under fluoroscopic guidance. After it has been verified that the patient is straight, an initial set of skin marks should be placed on the patient so this position can be reproduced throughout the simulation. Patient alignment should be monitored throughout the simulation procedure.

6.5.1.3
Determination of the Isocenter Location

Treatment isocenter is typically placed based on physician instructions. For the majority of standard treatments, the isocenter placement should be predetermined and outlined in treatment and simulation policies. It is desirable to place the isocenter on a stable location on the patient where the skin or patient anatomy does not move significantly. If the treatment isocenter must be placed in a position where the overlaying external anatomy does move, then treatment setup point should be used. Treatment setup point is a set of marks which are placed on a stable position on patient's anatomy (like sternum). For treatment, the patient is first aligned to the setup point and then shifted to the isocenter location using the shifts which were determined during the simulation. Other considerations for placement of isocenter include limitations/capabilities of treatment machines and desired dose distributions. These considerations are beyond the scope of this chapter.

6.5.1.4
Beam-Placement Design

Beams should be placed according to the treatment policies. Fluoroscopic capabilities of the conventional simulator are used for this purpose. Other chapters in this book outline common treatment techniques. Outlining of treatment portals based on simulation X-ray films is also better addressed in other chapters in this book.

6.5.1.5
Transfer of Simulation Information for Treatment Planning and Treatment

The final step in simulation process is transfer of patient setup data to dosimetry for calculation of monitor units and possibly for some simple treatment planning. The setup information is also transferred to the linear accelerator. Depending of the connectivity of the conventionally simulator and its ability to acquire patient setup information electronically, some or all of the patient setup data can be exported from the simulator electronically. Integrity of captured and exported electronic data should be verified through periodic quality assurance process.

6.5.2
CT Simulation

The CT-simulation process consists of the following steps:
- Patient positioning and immobilization
- Patient marking
- CT scanning
- Transfer to virtual simulation workstation
- Localization of initial coordinate system
- Localization of targets and placement of isocenter
- Marking of patient and immobilization devices based on isocenter coordinates
- Contouring of critical structures and target volumes
- Beam placement design, design of treatment portals

- Transfer of data to treatment-planning system for dose calculation
- Prepare documentation for treatment
- Perform necessary verifications and treatment plan checks

Again, this process and its implementation vary from institution to institution. The system design is dependent on available resources (equipment and personnel), patient workload, physical layout and location of different components, and proximity of team members. Communication channels need to be well established to avoid errors and unnecessary re-simulations. A simulation request form can be used to communicate simulation specifics between the physician and other team members (Fig. 6.15). The following is a general description of major steps in the CT simulation process.

6.5.2.1
Scan and Patient Positioning

The CT simulation scan is similar to conventional diagnostic scans; however, there are several differences. Patient positioning and immobilization are very important. Scan parameters and long scan volumes with large number of slices often push scanners to their technical performance limits. The CT-simulator staff must be aware of scanner imaging performance capabilities and limitations and also geometrical accuracy limitations. Imaging capabilities should be exploited to achieve high image quality and geometrical limitations need to be considered when positioning and marking patients.

Patient Positioning and Immobilization
General patient positioning and immobilization considerations are as described earlier in this chapter. Larger bore scanners typically can afford more comfortable patient positions and larger immobilization devices and offer a definitive advantage over conventional CT scanners. Pilot (scout) images are a very efficient tool for evaluation of patient positioning prior to the actual CT scan. After patient initial immobilization, a preliminary pilot scan should be imaged to assure that the patient positioning is straight. Immobilization devices should not produce artifacts on CT images.

Scan Protocol
The CT scan parameters should be designed to optimize both axial and DRR image quality and rapidly

BARNES-JEWISH HOSPITAL - DEPARTMENT OF RADIATION ONCOLOGY
SIMULATION INSTRUCTION WORKSHEET

Patient Name: _____Therapy #:_____ DOB:_____ M.D.: _____

Anatomical Site (be specific): _____ SSN:_____ Previous TX: ☐Y ☐N

Drop off Date/Time_____ Nurse contact Date/Time _____ Weight _____lbs

Simulation procedure: ☐Convent. Sim ☐CT Sim ☐3D ☐Tomotherapy ☐SMLC ☐DMLC ☐Brachy

Patient Status: ☐In-Patient Room #: _____   ☐Out-Patient Protocol: ☐Y ☐N #_____

Sim Date/Time: _____ Treat. Start Date/Time: _____ Machine: _____ ☐Permit Signed

M.D. to indicate scan volume (CT sim), approximate port edges, and the isocenter location.

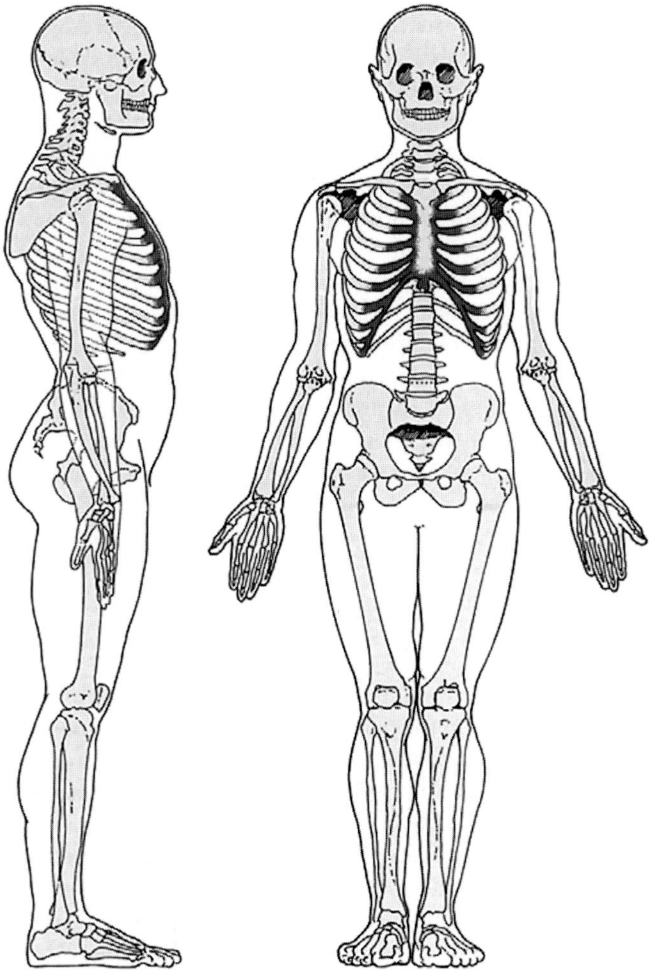

Preview: ☐Y ☐N

Energy: [☐6 ☐10 ☐18] MV ☐e⁻

Number of Fractions:_____

Port Description:
☐AP   ☐PA   ☐RL   ☐LL
☐RAO ☐LAO ☐RPO ☐LPO
☐TANGENTS   ☐Electrons
Other: _____

Pt. Position:
☐Supine ☐Prone ☐ Decubitus

Head:
☐Neutral ☐Extended ☐Flexed
Arm:_____
Other:_____

Treatment Aids/Devices: ☐None
☐Alpha Cradle ☐I.M.
☐Block/MLC☐Full Bladder
☐Tongue Blade
☐Bowel Compression Device
☐Bolus Type:_____
☐Slant Board ☐Wedges/Filters

Landmark Wire:
☐Nodes ☐Canthus ☐Anus
☐Scars Other: _____

Contrast:
IV:_____
Barium:_____
Urethrogram:_____
Rectal probe:_____
Precautions:_____
Patient on Metformin: ☐Y ☐N
Date of Last Blood Test:_____

Special instructions:

Attending Physician Signature: _____ Date:_____

Fig. 6.15 A sample simulation request form

acquire images to minimize patient motion (CURRY et al. 1990; CONWAY and ROBINSON 1997; COIA et al. 1995; BAHNER et al. 1999; McGEE et al. 1995; YANG et al. 2000). The parameters influencing axial and DRR image quality include kVp, mAs, slice thickness, slice spacing, spiral pitch (KALENDER and POLACIN 1991; KALENDER et al. 1994), data acquisition, reconstruction algorithms, scanned volume, total scan time, field of view (FOV), and size of image reconstruction matrix. Modern scanners come with preset protocols. Often, these include "oncology" protocols, which take the needs of virtual simulation process into consideration. Preset protocols should be evaluated for adequacy and modified according to treatment-planning needs. Physicians, dosimetrists, therapists, and physicists should be involved in protocol parameter selection. This is a very important component of CT simulation implementation process. The quality of images from the same scanner can vary significantly and the information contained in these images may be inadequate if care is not taken to properly select scan acquisition parameters. Suboptimal scan protocols can cause significant inefficiencies and potential errors in treatment planning. The best protocol selection can only be implemented with thorough understanding of properties of individual scan parameters and reconstruction algorithms. Increased mAs, decreased slice thickness and spacing, and decreased spiral pitch improve axial image and DRR quality to varying degrees. For spiral scanning with single slice scanners, all these factors (except slice spacing) significantly affect tube loading during a scan acquisition and limit the length of scanned volume before tube heat storage capacity has reached its limit. For adequate DRRs, long volumes often need to be scanned. If an X-ray tube reaches heat limit, the scan will be interrupted for the tube to cool. The cooling time can take several minutes allowing patients to move and degrade the spatial accuracy of data. Multislice scanners are generally not affected with tube-heating issues, and minimal compromises are required in protocol selection. Table 6.2 shows an example of scan protocols for a single-slice scanner used in radiation oncology (this is for illustration purposes only).

Scan Limits

Scan limits should be specified by the physician and should encompass area at least 5 cm away from the anticipated treatment volumes. Slice thickness and spacing do not have to be constant throughout the entire scanned volume. Areas of interest can be scanned with narrow (3 mm) thickness and spacing, whereas large slices (5 mm) can be used for scanning surrounding volumes. This will maintain good DRR quality while minimizing tube heat. An anatomical drawing can help when designing scan limits.

Contrast

For several treatment sites contrast can be used to help differentiate between tumors and surrounding healthy tissue. Contrast is not always useful and should be used carefully. Care should be taken to identify any contraindications. For heterogeneity-based calculations, contrast can cause dose distribution errors due to artificial CT numbers and corresponding tissue densities. For implementation of contrast in radiotherapy scanning, especially if the scanner is located in the radiation oncology department, a diagnostic radiologist should be consulted.

Special Considerations and Instructions

Each treatment site has unique considerations; these should be specified in CT scan procedures (see Table 6.1). Special considerations include: individual physician preferences; wiring of surgical scars for identification on CT images; scanning of patients with pacemakers and other implants; scanning of pediatric patients; scanning of patients under anesthesia, etc. A communication chain and responsibilities should be established for new problems and scans of patients with special needs.

Reference Marks

During the CT scan a set of reference marks must be placed on the patient so the patient can be positioned on the treatment machine. When and in relationship to which anatomical landmarks the reference marks are placed can be done in two different ways:

1. No shift method: for this method the patient is scanned and, while the patient is still on the CT scanner couch, images are transferred to the virtual simulation workstation. The physician contours the target volume and the software calculates the coordinates for the center of the contoured volume. During this time, the patient should remain still on the couch in the treatment position. The calculated coordinates are transferred to the CT scanner, the couch and the movable lasers are placed at that position, and the patient is marked. On the first day of treatment, the patient will be positioned using these marks on the treatment machine.

This method requires the physician to be present during the CT scan and the patient scan procedure is longer; however, the marks marked during the CT scan can be used for patient positioning without any shifts. This method can be greatly simplified if the virtual software is located directly on the CT scanner control console, obviating the need to transfer the CT study set to another computer. If the software is located on the scanner, the physician can start contouring the preliminary set of target contours directly on the scanner console as soon as images are reconstructed. This minimizes the time between the CT scan and the time when the patient is marked. This is preferred as the patient must remain still in the treatment position on the scanner couch while the physician contours target volumes. If the patient moves between the scan and the time when alignment marks are placed on the skin, the marks will not correspond to contoured tumor volumes resulting in potentially significant treatment errors. Contouring on the CT scanner ensures the shortest possible time between the CT scan and placement of alignment marks. Also, the scanner software is aware of absolute scanner couch coordinates relative to target volumes contoured by the physician. If the contouring is performed on the scanner console, absolute couch coordinates can be used for patient positioning for placement of alignment marks. If independent software package is used for contouring, the patient marking usually involves relative shifts to some initial set of reference marks. Relative shifts can be inaccurate and also can result is significant errors if shifts are applied in wrong direction or magnitude.

2. Shift method: this method does not require physicians to be present for the CT scan. Prior to the scan procedure, based on the diagnostic workup (CT, MRI, PET, palpation, etc.), the physician instructs CT simulator therapists where to place reference marks on the patient. For example, "place reference marks at the level of carina, 4 cm left from patient midline and midplane." The intention is to place these initial marks as close to the final isocenter as possible. Prior to the CT scan, the reference marks are marked on the patient and then radiation-opaque markers are placed over the skin marks. The radiation-opaque markers allow the reference marks to be visible on the CT study. The markers can be constructed from thin solder wire, aluminum wire, commercial markers, etc. After the CT scan, the patient can go home and images are transferred to the virtual simulation workstation. Later, the physician contours target volumes and determines the treatment isocenter

coordinates. Shifts (distances in three directions) between the reference marks drawn on the CT scanner and the treatment isocenter are then calculated. On the first day of treatment, the patient is first positioned to the initial reference marks and then *shifted* to the treatment isocenter using the calculated shifts. Initial reference marks are then removed and the treatment isocenter is marked on the patient.

This method is commonly used when a dedicated radiation oncology CT scanner is not available. With proper planning (from diagnostic work-up), the initial marks can be placed very close to the center of target volume. With asymmetric jaws, the initial reference may be used as the isocenter eliminating need for shifts.

6.5.2.2
Image Transfer and Registration

The CT study set is almost always the primary data on which the isodoses are computed and displayed, due to its high spatial resolution and fidelity. The exceptions are some stereotactic radiosurgery and brachytherapy applications where MRI or US, respectively, are used as the primary studies. When properly calibrated and free of image artifacts, CT images can provide electron-density information for heterogeneity-based dose calculations. As previously described, CT images do have shortcomings, and other imaging modalities can offer unique information about tumor volumes. If other imaging modalities are used in the treatment-planning process, they are typically considered secondary data sets and must be spatially registered to the CT study to accurately aid in tumor volume delineation. One of the important functions of a CT study set is defi-

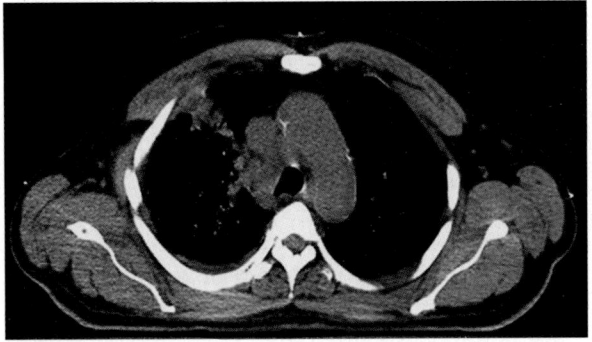

Fig. 6.16 A CT image with radiation opaque markers on patient's anterior and left and right sides

nition of the patient/treatment coordinate system. Usually the orientation of the coordinate system (X, Y, and Z) is predetermined by the treatment-planning software; however, the origin of the coordinate system is defined, in most instances, by the location of the reference marks which were placed on the patient during the simulation scan as described in the previous section. If the patient was marked during the simulation using the "no shift" method, there will not necessarily be any visible landmarks on the patient's scan which correlate with the location of the origin of the coordinate system. In this situation, the accuracy of the treatment relies on the ability to accurately transfer coordinates from the simulation software to the treatment-planning software. This feature should be thoroughly tested during commissioning for all scanners and for different patient orientations on the scanner (supine, prone, head first, feet first, etc.). Errors in transfer can result in significant treatment errors. If the patient was marked using the "shift" method, a set of radiation-opaque markers which are placed on top of the skin marks can be seen on CT images, and the location of the coordinate system can be defined using these marks. Figure 6.16 shows a CT image with such marks.

Transfer of image studies and registration of multimodality images is a several-step process requiring multifunction software capable of image set transfer, storage, coordinate transformation, and voxel interpolation. These features enable image study registration (transforming images to a common reference frame and resampling to a common pixel grid) and fusion (the display of a combination of pixel intensities from registered image studies). Registered and "fused" image studies can then be used for radiotherapy treatment planning.

Images and other treatment planning data are transferred between modern systems using DICOM standard. DICOM is a standard for representing and exchanging medical imaging data. This standard has greatly simplified image exchange between scanners and software manufactured by various vendors. Some limitations still exist but are being eliminated gradually.

There are several methods for image registration:

1. Surface-based registration requires contouring of the same structure (internal or external) on the two data sets, and the studies are then registered by aligning the contours. This method is well suited for CT–CT or CT–MR registration. Edges of organs on PET images are poorly defined, and surface-based registration typically cannot be used for image registration.

2. Image-based registration involves displaying CT data set in background, in grayscale, and superimposing the other study set image in color-wash or in grayscale on top of the CT study. The two studies are typically simultaneously viewed in transverse, sagittal, and coronal orientations. The studies are then registered by manipulating the secondary-study-set images in three displayed planes.

3. Point-based registration is based on identifying a set of at least three corresponding points in both data sets and performing image transforms to align these points. Point-based registration works well for CT and MRI as there are numerous anatomical points which are identifiable on both studies. Point-based registration with PET, using anatomic points, is virtually impossible due to poor image resolution; however, point-based registration of PET images with external fiducial markers is extremely useful (MUTIC et al. 2001).

As a part of initial implementation of multimodality-imaging-based radiotherapy treatment planning, tests should be performed to verify that transferred images have correct geometry (e.g., pixel size, spatial fidelity, slice thickness, and spacing), orientation (e.g., prone/supine, head–foot orientation, and left–right orientation), scan text information, and grayscale values (MUTIC et al. 2001; LAVELY et al. 2004). For routine treatment planning, images should always be inspected for any distortions, misalignments, and artifacts. This should be a part of a routine quality-assurance program (MUTIC et al. 2003).

6.5.2.3
Target and Normal Structure Delineation

Virtual simulation process typically consists of contouring target and normal structures, computation of the isocenter, manipulation of treatment machine motions for placement of the beams, design of treatment portals, generation of DRRs, and related treatment setup information. This process is dependent largely on the virtual simulation software capabilities. There are well-designed methods for simulating specific treatment sites. Several publications describe these methods in detail (BUTKER et al. 1999; VAN DYK and TAYLOR 1999; VAN DYK and MAH 2000; COIA et al. 1995).

Target and normal structurede lineation requirements for conformal radiotherapy have been addressed by the ICRU (1993, 1999) and are described throughout this book for various treatment sites. This is often the most time-consuming portion of the virtual simulation process and care should be taken to simplify this task as much as possible. Well-designed contouring software package is a prerequisite and should be one of the main concerns when selecting virtual simulation software. Another important component is to predefine which structures are to be contoured for individual treatment sites. This will obviate unnecessary work. It is also important to predefine the anatomical extent of each structure to be contoured. Uniform extent of critical structures makes outcome and complication analysis more meaningful.

6.5.2.4
Treatment Techniques

Treatment techniques for individual cancer sites are described throughout this book. Implementation of these treatment techniques also requires simulation procedures. Simulation techniques for these individual sites depend greatly on the available simulation, treatment planning, and treatment technology and staff expertise and understanding of this technology; therefore, the actual simulation processes can vary significantly among radiation oncology facilities and description of these processes is beyond the scope of this chapter. The goal is to implement treatment and simulation techniques which best serve the individual institution's patients based on the available equipment and staff. When developing simulation processes, it is important to understand capabilities of local resources. Simulation techniques will inevitably continually evolve as these resources change.

6.6
Discussion and Conclusion

As radiotherapy treatment planning and delivery technology and techniques change, so does the treatment simulation. The most significant change in the recent past has been the wide adoption of CT simulation to support conformal radiotherapy and 3D treatment planning. The CT simulation has gone from a concept practiced at few academic centers to several available sophisticated commercial systems located in hundreds of radiation oncology departments around the world. The concept has been embraced by the radiation community as a whole. The acceptance of virtual simulation comes from improved outcomes and increased efficiency associated with conformal radiation therapy. Imaged-based treatment planning is necessary to properly treat a multitude of cancers and CT simulation is a key component in this process. Due to demand for CT images, CT scanners are commonly found in radiation oncology departments. In our radiation oncology department, we have two CT simulators and a large number of our patients are also scanned using MR and PET scanners located in the diagnostic radiology department. As CT technology and computer power continue to improve, so will the simulation process, and it may no longer be based on CT alone. The CT/PET combined units are commercially available and could prove to be very useful for radiation oncology needs. Several authors have described MR simulators where the MR scanner has taken the place of the CT scanner. It is difficult to predict what will happen over the next 10 years, but it is safe to say that image-based treatment planning will continue to evolve.

One great opportunity for an overall improvement of radiation oncology is better understanding of tumors through biological imaging. Biological imaging has been shown to better characterize the extent of disease than anatomical imaging and also to better characterize individual tumor properties. Enhanced understanding of individual tumors can improve selection of the most appropriate therapy and better definition of target volumes. Improved target volumes can utilize the full potential of IMRT delivery. Biological imaging can also allow evaluation of tumor response and possibly modifications in the therapy plan if the initial therapy is deemed not effective.

Future developments in the radiotherapy treatment-planning simulation process will involve integration of biological imaging. It is likely that this process will be similar to the way that CT scanning was implemented in radiotherapy. Where the imaging equipment is initially located in diagnostic radiology facilities, and as the demand increases, the imaging is gradually moved directly to radiation oncology.

References

Bahner ML, Debus J, Zabel A, Levegrun S, Van Kaick G (1999) Digitally reconstructed radiographs from abdominal CT scans as a new tool for radiotherapy planning. Invest Radiol 34:643–647

Bailey DL (2003) Data acquisition and performance characterization in PET. In: Valk PE, Bailey DL, Townsend DW, Maisey MN (eds) Positron emission tomography: basic science and clinical practice. Springer, Berlin Heidelberg New York, pp 69–90

Bailey DL, Karp JS, Surti S (2003) Physics and instrumentation in PET. In: Valk PE, Bailey DL, Townsend DW, Maisey MN (eds) Positron emission tomography: basic science and clinical practice. Springer, Berlin Heidelberg New York, pp 41–67

Beavis AW, Gibbs P, Dealey RA et al (1998) Radiotherapy treatment planning of brain tumors using MRI alone. Br J Radiol 71:544–548

Beyer T, Townsend DW, Brun T, Kinahan PE, Charron M, Roddy R et al (2000) A combined PET/CT scanner for clinical oncology. J Nucl Med 41:1369–1379

Bushberg JT, Seibert JA, Leidholdt EM, Boone JM (eds) (2002) Fluoroscopy. In: The essential physics of medical imaging, 2nd edn. Lippincott Williams and Wilkins, Philadelphia, pp 231–254

Butker EK, Helton DJ, Keller JW, Crenshaw T, Fox TH, Elder ES et al (1999) Practical implementation of CT-simulation: the Emory experience. In: Purdy JA, Starkschall G (eds) A practical guide to 3-D planning and conformal radiation therapy. Advanced Medical Publishing, Middleton, Wisconsin, pp 58–59

Chao KSC, Bosch WR, Mutic S, Lewis JS, Dehdashti F, Mintun MA, Dempsey JF, Perez CA, Purdy JA, Welch MJ (2001) A novel approach to overcome hypoxic tumor resistance Cu-ATSM-guided intensity-modulated radiation therapy. Int J Radiat Oncol Biol Phys 49:1171–1182

Chen L, Price R, Wang L, Li J, Qin L, McNeeley S, Ma C, Freedman G, Pollack A (2004) MRI-Based treatment planning for radiotherapy: dosimetric verification for prostate IMRT. Int J Radiat Oncol Biol Phys 60:636–647

Coia LR, Schultheiss TE, Hanks G (eds) (1995) A practical guide to CT simulation. Advanced Medical Publishing, Middleton, Wisconsin

Conway J, Robinson MH (1997) CT virtual simulation. Br J Radiol 70:S106–S118

Curry TS, Dowdey JE, Murry RC (1990) Computed tomography. Chriestensen's physics of diagnostic radiology, 4th edn. Lea and Febiger, Malvern, Pennsylvania, pp 289–322

Dizendorf EV, Baumert BG, Schulthess GK von, Lutolf UM, Steinert HC (2003) Impact of whole-body 18F-FDG PET on staging and managing patients for radiation therapy. J Nucl Med 44:24–29

Garcia-Ramirez JL, Mutic S, Dempsey JF, Low DA, Purdy JA (2002) Performance evaluation of an 85 cm bore X-ray computed tomography scanner designed for radiation oncology and comparison with current diagnostic CT scanners. Int J Radiat Oncol Biol Phys 52:1123–1131

Goitein M, Abrams M (1983) Multi-dimensional treatment planning: I. Delineation of anatomy. Int J Radiat Oncol Biol Phys 9:777–787

Goitein M, Abrams M, Rowell D, Pollari H, Wiles J (1983) Multi-dimensional treatment planning: II. Beam's eye-view, back projection, and projection through CT sections. Int J Radiat Oncol Biol Phys 9:789–797

International Commission of Radiation Units and Measurements (1993) ICRU report no. 50. Prescribing, recording, and reporting photon beam therapy. International Commission of Radiation Units and Measurements, Bethesda, Maryland

International Commission of Radiation Units and Measurements (1999) ICRU report no. 62. Prescribing, recording and reporting photon beam therapy (Supplement to ICRU report 50). International Commission of Radiation Units and Measurements, Bethesda, Maryland

Kalender WA, Polacin A (1991) Physical performance characteristics of spiral CT scanning. Med Phys 18:910–915

Kalender WA, Polacin A, Suss C (1994) A comparison of conventional and spiral CT: an experimental study on the detection of spherical lesions J Comput Assist Tomogr 18:167–176

Klingenbeck-Regen K, Schaller S, Flohr T, Ohnesorge B, Kopp AF, Baum U (1999) Subsecond multi-slice computed tomography: basics and applications. Eur J Radiol 31:110–124

Lavely WC, Scarfone C, Cevikalp H, Rui L, Byrne DW, Cemelak AJ, Dawant B, Price RR, Hallahan DE, Fitzpatrick JM (2004) Phantom validation of coregistration of PET and CT for image-guided radiotherapy. Med Phys 31:1083–1092

Lee YK, Bollet M, Charles-Edwards G et al (2003) Radiotherapy treatment planning of prostate cancer using magnetic resonance imaging alone. Radiother Oncol 2:203–216

Ling CC, Humm J, Larson S, Amols H, Fuks Z, Leibel S, Koutcher JA (2000) Towards multidimensional radiotherapy (MD-CRT): biological imaging and biological conformality. Int J Radiat Oncol Biol Phys 47:551–560

Low DA, Nystrom M, Kalinin E, Parikh P, Dempsey JF, Bradley JD, Mutic S, Wahab SH, Islam T, Christensen G, Politte DG, Whiting BR (2003) A method for the reconstruction of 4-dimensional synchronized CT scans acquired during free breathing. Med Phys 30:1254–1263

Mah D, Steckner M, Palacio E et al (2002) Characteristics and quality assurance of dedicated open 0.23 T MRI for radiation therapy simulation. Med Phys 29:2541–2547

McGee KP, Das IJ, Sims C (1995) Evaluation of digitally reconstructed radiographs (DRRs) used for clinical radiotherapy: a phantom study. Med Phys 22:1815–1827

Mutic S, Dempsey JF, Bosch WR, Low DA, Drzymala RE, Chao KSC, Goddu SM, Cutler PD, Purdy JA (2001) Multimodality image registration quality assurance for conformal three-dimensional treatment planning. Int J Radiat Oncol Biol Phys 51:255–260

Mutic S, Palta JR, Butker E, Das IJ, Huq MS, Loo LD, Salter BJ, McCollough CH, Van Dyk J (2003) Quality assurance for CT simulators and the CT simulation process: report of the AAPM Radiation Therapy Committee Task Group no. 66. Med Phys 30:2762–2792

National Electrical Manufacturers Association (NEMA) (1998) DICOM PS 3 (set). Digital Imaging Communictaions in Medicine (DICOM)

Okamoto Y, Kodama A, Kono M (1997) Development and clinical application of MR simulation system for radiotherapy planning: with reference to intracranial and head and neck regions. Nippon Igaku Hoshasen Gakkai Zasshi 57:203–210

Potter R, Heil B, Schneider L et al (1992) Sagittal and coronal planes from MRI for treatment planning in tumors of brain, head and neck: MRI assisted simulation. Radiother Oncol 23:127–130

Schubert K, Wenz F, Krempien R, Schramm O, Sroka-Perez G, Schraul P, Wannenmacher M (1999) Possibilities of an open magnetic resonance scanner integration in therapy simulation and three-dimensional radiotherapy planning. Strahlenther Onkol 175:225–331

Sherouse G, Mosher KL, Novins K, Rosenman EL, Chaney EL (1987) Virtual simulation: concept and implementation. In: Bruinvis IAD, van der Giessen PH, van Kleffens HJ, Wittkamper FW (eds) Ninth International Conference on the Use of Computers in Radiation Therapy. North-Holland Publishing, The Netherlands, pp 433–436

Sherouse GW, Novins K, Chaney EL (1990a) Computation of digitally reconstructed radiographs for use in radiotherapy treatment design. Int J Radiat Oncol Biol Phys 18:651–658

Sherouse GW, Bourland JD, Reynolds K (1990b) Virtual simulation in the clinical setting: some practical considerations. Int J Radiat Oncol Biol Phys 19:1059–1065

Tatcher M (1977) Treatment simulators and computer assisted tomography. Br J Radiol 50:294

Townsend DW, Carney JP, Yap JT, Hall NC (2004) PET/CT today and tomorrow. J Nucl Med 45 (Suppl 1):4S–14S

Van Dyk J, Mah K (2000) Simulation and imaging for radiation therapy planning. In: Williams JR, Thwaites TI (ed) Radiotherapy physics in practice, 2nd edn. Oxford University Press, Oxford, pp 118–149

Van Dyk J, Taylor JS (1999) CT-simulators. In: Van Dyk J (ed) The modern technology for radiation oncology: a compendium for medical physicist and radiation oncologists. Medical Physics Publishing, Madison, Wisconsin, pp 131–168

Yang C, Guiney M, Hughes P, Leung S, Liew KH, Matar J et al (2000) Use of digitally reconstructed radiographs in radiotherapy treatment planning and verification. Australas Radiol 44:439–443

Young H, Baum R, Cremerius U, Herholz K, Hoekstra O, Lammertsma AA, Pruim J, Price P (1999) Measurement of clinical and subclinical tumour response using [18F]-fluorodeoxyglucose and positron emission tomography: review and 1999 EORTC recommendations. European Organization for Research and Treatment of Cancer (EORTC) PET Study Group. Eur J Cancer 13:1773–1782

7 Clinical Applications of High-Energy Electrons

Bruce J. Gerbi

CONTENTS

7.1 Introduction 135
7.2 Historical Perspective 135
7.3 Electron Interactions 136
7.4 Central Axis Percentage Depth-Dose
    Distributions 137
7.4.1 Central Axis Percentage Depth–Dose
    Dependence on Beam Energy 138
7.4.2 Central Axis Percentage Depth–Dose
    Dependence on Field Size and SSD 138
7.4.3 Flatness and Symmetry –
    Off-Axis Characteristics 139
7.5 Isodose Curves 140
7.5.1 Change in Isodose Curves Versus SSD 141
7.5.2 Change in Isodose Curves Versus Angle of
    Beam Incidence 142
7.5.3 Irregular Surfaces 142
7.6 Effect of Inhomogeneities on
    Electron Distributions 143
7.6.1 Lungs 144
7.6.2 Bones 144
7.6.3 Air Cavities 145
7.7 Clinical Applications of Electron Beams 146
7.7.1 Target Definition 146
7.7.2 Therapeutic Range – Selection of Beam Energy 147
7.7.3 Dose Prescription – ICRU 71 147
7.7.3.1 ICRU 71 Recommendations –
    Intraoperative Radiation Therapy 147
7.7.3.2 ICRU 71 Recommendations –
    Total Skin Irradiation 148
7.7.4 Field Shaping and Collimation 148
7.7.5 Internal Shielding 149
7.7.6 Bolus 151
7.7.6.1 Custom, Compensating Bolus 151
7.7.6.2 Field Abutment 152
7.7.6.3 Electron–Electron Field Matching:
    Sloping, Curved Surfaces 155
7.7.6.4 Electron–Photon Field Matching 156
7.7.7 Intracavitary Irradiation 157
7.7.7.1 Intraoral and Transvaginal Irradiation 157
7.7.7.2 Intraoperative Radiation Therapy 158
7.8 Special Electron Techniques 159
7.8.1 Electron Arc Irradiation 159
7.8.2 Craniospinal Irradiation 160
7.8.3 Total Skin Electron Therapy 162
    References 164

7.1 Introduction

The basic physics of electron beams has been discussed in several books and in several excellent chapters of standard radiation therapy textbooks (Khan 2003; Hogstrom 2004; Strydom et al. 2003). As with many of these previous works, the purpose of this chapter is to discuss the role of electron beams in radiation therapy, describe their physical characteristics, and describe how this information is relevant to clinical practice in radiation therapy. In addition, several techniques using electron beams that have been found to be useful in clinical settings will also be presented.

7.2 Historical Perspective

In modern radiation therapy departments, high-energy electrons are a useful and expected modality. Although, they have been available for many years, it was not until the 1970s when linear accelerators became widely available that electrons moved into the mainstream in radiation therapy. Similar to advances in photon beam treatments in radiation therapy, there have been several key advances in the 1970s that improved dramatically the ability to deliver optimized electron beam treatments. These developments were: (1) computed tomography (CT) scanners that paved the way for CT-based treatment planning, (2) improvements in electron treatment-planning algorithms (electron pencil beam

B. J. Gerbi, PhD, Associate Professor
Therapeutic Radiology – Radiation Oncology, University of Minnesota, Mayo Mail Code 494, 420 Delaware St SE, Minneapolis, MN 55455, USA. gerbi001@umn.edu

algorithms) to calculate accurately and display dose deposition using CT data, and (3) improvement in linac designs resulting in improved depth dose, off-axis uniformity, and the physical characteristics of electron beams. Central to these latter improvements were the development of dual scattering foil systems and improvements in electron beam applicators (HOGSTROM 2004).

Modern linear accelerators are capable of producing several electron beam energies in addition to two or more photon energies. Electron beams in the range of 6–20 MeV are most clinically useful with the intermediate beam energies being the most commonly used. The energy designation of a clinical beam is described using the most probable energy at the surface of the phantom at the standard treatment or source–skin distance (SSD). This most probable energy is the energy possessed by the majority of the incident electrons and is represented on the linear accelerator console by the closest integer value to the actual electron beam energy. Figure 7.1 shows the nature of the electron energy spectrum as a function of location in the beam. Before the accelerator beam exit window, the electrons are fairly monoenergetic in nature. The spectrum of electron energies in the beam is spread more widely as it hits the surface of the phantom with the spectrum centered on the most probable energy. As the beam penetrates into the patient, the width of the energy spectrum increases while the average energy of the beam decreases with increasing depth in the medium. The

rate of this decrease is about 2 MeV/cm in unit density material such as muscle.

7.3
Electron Interactions

Two basic properties of electrons are that they possess a negative charge (we are not considering positrons at this time) and that they are low in mass having approximately 1/2000 of the mass of a proton or neutron. Being charged particles, they are directly ionizing, meaning that they interact directly with the material on which they are incident. They are attracted to charges of opposite sign and are repelled by charges of like sign as they travel through a medium. These forces of attraction or repulsion are called Coulomb force interactions and lead directly to ionizations and excitations of the absorbing material. Due to their relatively low mass, their direction of travel can be changed easily during these interactions. When electrons pass through a medium, their mean energy decreases with depth and they scatter to the side of their original path. Thus, the locations to which electrons scatter dictates where bonds are broken and ultimately where dose is deposited.

Interactions undergone by electrons can be either elastic in which no kinetic energy is lost or inelastic in which some portion of the kinetic energy is changed into another form of energy. Elastic colli-

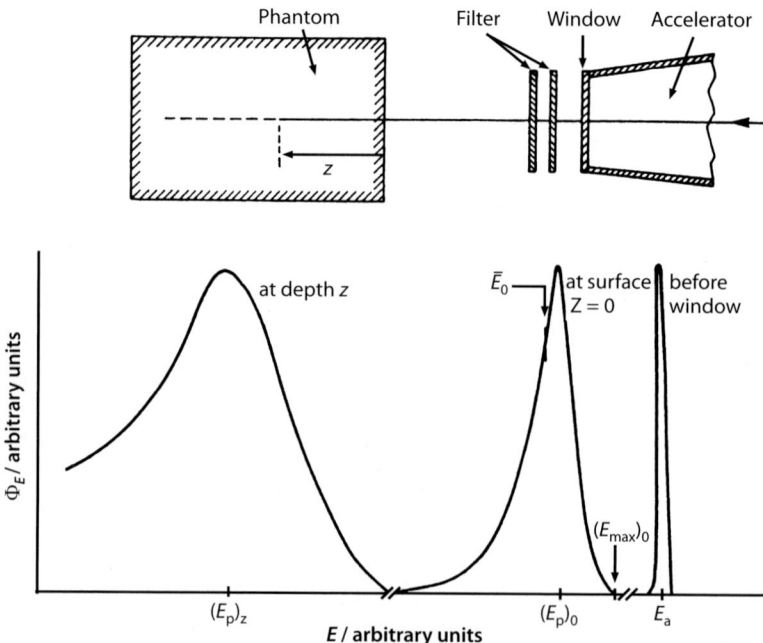

Fig. 7.1. Representation of the energy of electrons in a linear accelerator beam as a function of the location in the beam. $E_{p'_o}$ represents the most probable energy at a depth of zero, the surface; $\bar{E}_0$ is the average mean energy at the surface; and $E_{p'_z}$ is the most probable energy at the depth, z. [Reprinted with permission from BRAHME and SVENSSON (1976). Specification of electron beam quality from the central-axis depth absorbed-dose distribution.]

sions occur with either atomic electrons or atomic nuclei, resulting in elastic scattering. These interactions are characterized by a change in direction of the incident electron with no loss of energy. In the collision process between the incident electron and atomic electrons, it is possible for the ejected electron to acquire enough kinetic energy to cause additional ionizations of its own. These electrons are called secondary electrons or delta rays and they can go on to produce additional ionizations and excitations. Inelastic collisions can occur with electrons resulting in ionizations and excitations of atoms or inelastic collisions of nuclei that result in the production of Bremsstrahlung X-rays (radiative losses). The energy of these X-rays can be as high as the maximum accelerating potential of the beam and contribute dose to tissues deep within the body. The magnitude of this photon contamination component can range from about 1% to as high as 5% for 6-MeV and 20-MeV electron beams, respectively.

The typical energy loss in tissue for a therapeutic electron beam, averaged over its entire range, is about 2 MeV/cm in water. The rate of energy loss for collisional interactions depends on the energy of the electrons and on the electron density of the medium. The rate of energy loss per gram/cm^2 is greater for low atomic number (low Z) materials than for high Z materials. This is because high Z materials have fewer electrons per gram than low Z materials. In addition, the electrons in high Z materials are more tightly bound and are thus not available for these types of interactions. Keeping the above types of interactions of electrons in mind will help to explain many of the clinical situations presented later. The most important of these is scattering at edges of tissue or lead shielding materials.

7.4
Central Axis Percentage Depth-Dose Distributions

The shape of the central axis depth–dose curve of electron beams depends on many factors, most notably the beam energy, field size, source surface distance, collimation, depth of penetration, and angle of beam incidence. A typical central axis percentage depth–dose curve for high-energy electrons is shown in Figure 7.2. This figure represents some of the quantities used to describe electron beams. D_s represents dose at the surface and is defined at a depth of 0.05 cm. D_m is the dose maximum at depth

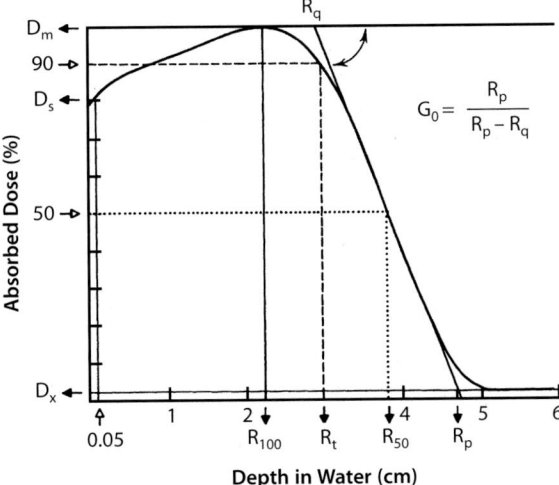

Fig. 7.2. Central-axis depth–dose curve for an electron beam with parameters indicated that can be used to characterize electron beams. The therapeutic range is at the 90% depth–dose level and all other parameters are discussed in ICRU (1984)

while D_x is the dose due to X-ray contamination. R_{100} is the depth of the maximum dose while R_{90} is the depth of the 90% depth dose. R_{50} is the depth of the 50% depth–dose curve. R_p is the practical range, which is close to the most probable energy of the electron beam at the surface, E_0, divided by two in centimeters. Thus, a 20-MeV electron beam would have a practical range of 10 cm in tissue. R_p is determined by taking the depth of intersection of the straight line descending portion of the central axis percentage depth–dose curve with a line drawn representing the photon contamination. G_0, as shown in the figure, gives what has been termed the reduced dose gradient. This is a measure of how quickly the dose decreases beyond the therapeutic range. This factor depends on the quantity R_q, which is the depth at the tangent of the central axis percentage dose curve as the inflection point meets the level of dose maximum, D_{max}, and its relationship to the practical range, R_p. As shown in the figure, $G_0 = R_p/(R_p - R_q)$.

The portion of the central axis percentage depth–dose curve at depths deeper than the practical range, R_p, represents the photon contamination present in the electron beam. Photon contamination results from electron interactions with the exit window of the linear accelerator, the scattering foils, beam ion chambers, the collimator jaws, the cones and intervening air which produce Bremsstrahlung X-rays. Additional X-rays are also produced within the patient, although this is a small source of photon generation. Photon contamination varies with both beam energy and type of linear accelerator. The amount of photon contamination is low for low-energy beams usually less than approximately 1% and increases to approximately 5% at 20 MeV. Linear accelerators employing scanning electron beams do not use scattering foils to spread the beam and thus produce the least amount of photon contamination. In most cases, the magnitude of the photon contamination is very acceptable for patient treatment. It does become a consideration when performing total skin electron irradiation where multiple electron fields and patient positions are required. Consequently, techniques designed to treat the total skin using high-energy electrons must strive to keep the magnitude of the photon contamination as low as possible and ideally less than 1%. Thus, the total body dose can be held as low as possible, since treating the skin surface is the primary objective of this particular treatment regimen.

7.4.1
Central Axis Percentage Depth–Dose Dependence on Beam Energy

Figure 7.3 shows the change in the central axis percentage depth–dose curves as a function of beam energy. With modern linear accelerators, the percentage depth-dose at the surface for 6-MeV electron beams is approximately 70–75%. The surface dose increases with increasing beam energy to about 95% for 20-MeV beams. Tables 7.1 and 7.2 show the dose at the surface and at 0.5- and 1.0-cm depths for a Varian Clinac 2300CD for a 10×10-cm^2 cone at 100 cm SSD. This characteristic of clinical electron beams where low-energy electrons have a lower surface dose than high-energy electrons exists because low-energy electrons scatter into wider angles than higher energy electrons. Thus, in comparison with the amount of scatter exhibited at the surface, at the depth of D_{max} lower energy electrons scatter more than higher energy electrons. Knowledge of the dose at the surface is important clinically because the target to be treated using electrons often includes the skin, and adequate dosage must be ensured in these areas.

Both the depth of D_{90} and the R_p increase with increasing beam energy. Additionally, the central axis percentage depth doses for lower energy electron beams decrease more rapidly beyond the depth of D_{90} than do higher energy beams. As a result, when using higher energy electron beams, more distance

must be placed between the target to be treated and sensitive structures than when using lower energy beams.

7.4.2
Central Axis Percentage Depth–Dose Dependence on Field Size and SSD

There is a significant change in the central axis percentage depth dose for electron beams with change in field size when the field size decreases to less than the practical range for that electron beam energy (MEYER et al. 1984). There is little change in the percentage depth–dose curve for field sizes greater than the practical range. The reason for this field size dependence is loss of side scatter equilibrium with decreasing field size. Figure 7.4 shows

Table 7.1. The percentage depth dose at the surface and in superficial regions of high-energy electron beams for a Varian 2300CD for a 10×10-cm^2 cone at 100 cm source–skin distance (SSD). The data were taken using an Attix plane-parallel chamber

Depth (cm)	Electron percentage depth dose					
	6 MeV	9 MeV	12 MeV	15 MeV	18 MeV	22 MeV
0	70.8%	76.5%	82.0%	86.6%	88.4%	89.1%
0.5	82.5%	84.7%	89.5%	93.7%	96.0%	97.0%
1.0	94.0%	90.0%	92.6%	96.4%	98.7%	98.9%

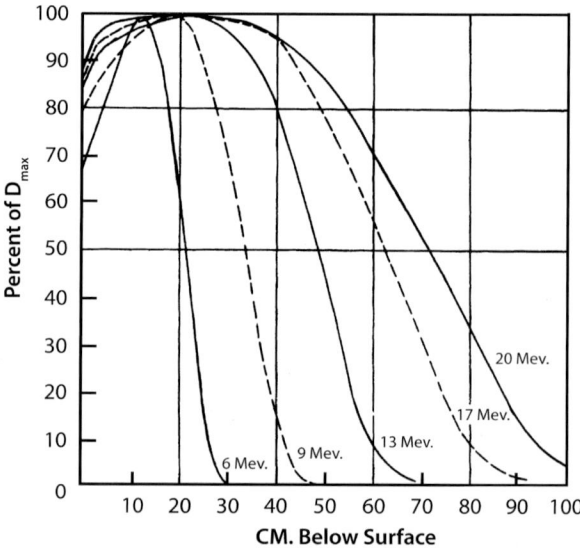

Fig. 7.3. Central axis percentage depth dose as a function of beam energy from 4 MeV to 20 MeV

Table 7.2. Depths at which D_{max} and D_{90} occur for various electron beam energies

	6 MeV	9 MeV	12 MeV	15 MeV	18 MeV	22 MeV
D_{max} depth (cm)	1.4	2.2	2.9	2.9	2.9	2.2
D_{90} depth (cm)	1.8	2.8	3.9	4.8	5.4	5.8

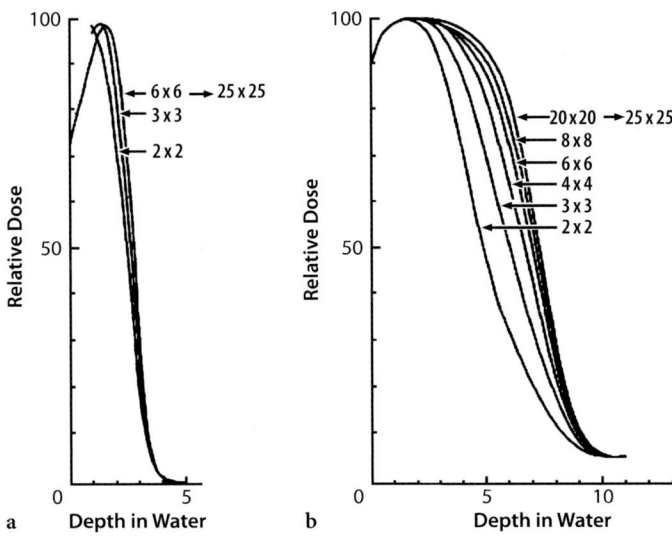

Fig. 7.4a,b. Change in percentage depth dose versus field size for a 7-MeV (**a**) and 18-MeV (**b**) electron beam for 2×2 to 25×25-cm², 100 cm SSD field sizes produced from a Siemens Mevatron 80 accelerator. [Reprinted with permission from MEYER et al. (1984)]

the change in percentage depth dose with change in field size for 7-MeV and 18-MeV electron beams for field sizes ranging from 2×2 to 25×25 cm². As shown, there is greater change in the percentage depth dose for higher beam energies when the field size is decreased below the practical range. For all beam energies where this situation exists, there is a shift in both the depth of dose maximum and the D_{90} dose towards the surface with decreasing field size. The surface dose increases with increasing amounts of field restriction for lower energy electrons while there is no change in the surface dose for the higher energy beams with field restriction. The practical range, R_p, is unchanged by field restriction since the overall beam energy is not affected by a change in field size.

The central axis percentage depth–dose curve for high-energy electron beams is only slightly dependent on SSD. Figure 7.5 shows no discernable difference in the central axis data for 9 MeV electrons at 100 cm versus 115 cm SSD. For the 20-MeV beam, the 80–95 isodose lines are only a few millimeters deeper at 115 cm SSD than their depth at 100 cm SSD. Isodose curves lower than 80% are negligibly more penetrating at extended SSDs since inverse square effects are not that great for electrons over their short depth of penetration.

7.4.3
Flatness and Symmetry – Off-Axis Characteristics

A typical dose profile is shown in Figure 7.6. A dose profile represents a plot of the beam intensity as a function of distance off axis. The variation in the dose distribution in the direction perpendicular to the central axis can be described by the off-axis ratio. The off-axis ratio is defined as the ratio of dose at a point away from the central axis divided by the dose at the central axis of the beam at the same depth.

Specifications for flatness are given by the IEC (International Electrotechnical Commission) at the depth of maximum dose. The flatness specification consists of two requirements: (1) the distance between the 90% dose levels and the geometric beam edge should not exceed 10 mm along the major axis and 20 mm along the diagonals of the beam and (2) the maximum value of the absorbed dose anywhere within the region bounded by the 90% isodose contour should not exceed 1.05 times the absorbed dose on the axis of the beam at the same depth. The specification for symmetry according to the IEC at the depth of D_{max} for high-energy electron beams is that the crossbeam profile should not differ by more than 2% for any pair of symmetric points with respect to the central axis value. The AAPM has also made rec-

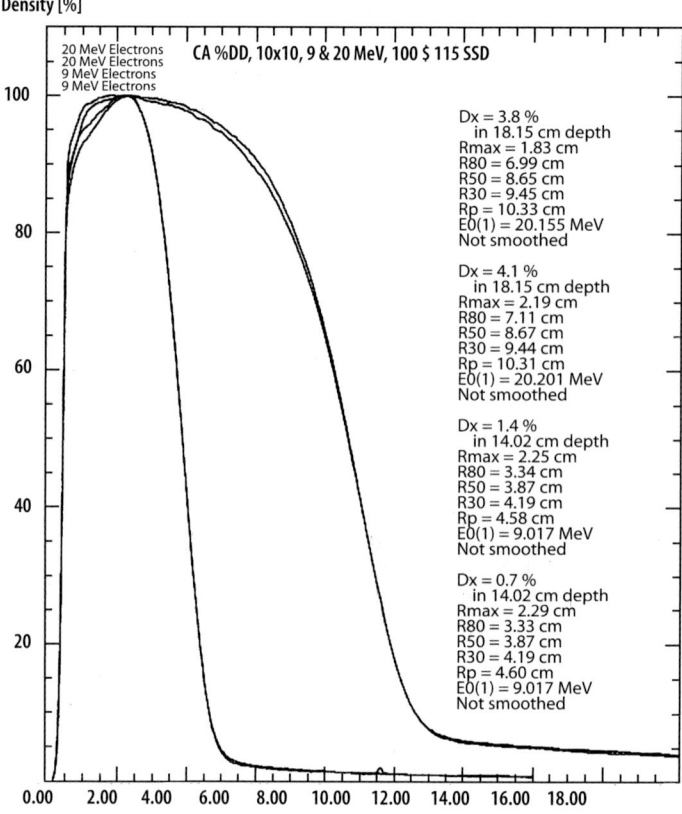

Density [%]

20 MeV Electrons
20 MeV Electrons
9 MeV Electrons
9 MeV Electrons

CA %DD, 10x10, 9 & 20 MeV, 100 $ 115 SSD

Dx = 3.8 %
  in 18.15 cm depth
Rmax = 1.83 cm
R80 = 6.99 cm
R50 = 8.65 cm
R30 = 9.45 cm
Rp = 10.33 cm
E0(1) = 20.155 MeV
Not smoothed

Dx = 4.1 %
  in 18.15 cm depth
Rmax = 2.19 cm
R80 = 7.11 cm
R50 = 8.67 cm
R30 = 9.44 cm
Rp = 10.31 cm
E0(1) = 20.201 MeV
Not smoothed

Dx = 1.4 %
  in 14.02 cm depth
Rmax = 2.25 cm
R80 = 3.34 cm
R50 = 3.87 cm
R30 = 4.19 cm
Rp = 4.58 cm
E0(1) = 9.017 MeV
Not smoothed

Dx = 0.7 %
  in 14.02 cm depth
Rmax = 2.29 cm
R80 = 3.33 cm
R50 = 3.87 cm
R30 = 4.19 cm
Rp = 4.60 cm
E0(1) = 9.017 MeV
Not smoothed

Fairview – UMC Radiation Therapy

Depth [cm]

Fig. 7.5. Comparison of central axis depth–dose curves for 9 MeV and 20 MeV electrons using a 10×10-cm^2 cone at 100 cm and 115 cm SSD. Data are taken on a Varian 2100C linac using Kodak XV2 film in a solid water phantom and scanned using a Wellhofer scanner running WP700 software

ommendations (TG-25; AMERICAN ASSOCIATION OF PHYSICISTS IN MEDICINE 1991) concerning field flatness and symmetry. They specify that the flatness be specified in a plane perpendicular to the central axis at the depth of the 95% isodose line beyond the depth of dose maximum. The variation in the beam at this depth should not vary by more than ±5% versus the value at the central axis and optimally should be within ±3% at a point 2 cm within the geometric field edge (the 50% isodose width) for fields greater than or equal to 10×10 cm^2. Knowledge of the beam flatness is important from a clinical standpoint in determining an adequate field size to treat a particular region, particularly in setting an adequate margin around the target.

7.5
Isodose Curves

Isodose curves are generated by connecting points of equal dose. The curves are usually drawn at intervals of absorbed dose separated by 10% and expressed as a percentage of the dose at the reference point compared with that at the D_{max} point on the central

axis of the beam. The shape of individual electron isodose curves varies with beam energy, field size, beam collimation, SSD, and the level of the isodose curve. Typical isodose curves are shown in Fig. 7.7 for both a 6×6-cm^2 and 15×15-cm^2 12-MeV, 100 cm SSD electron beam. The 6×6-cm^2 field is overlaid on the 15×15-cm^2 field to illustrate that the shape of

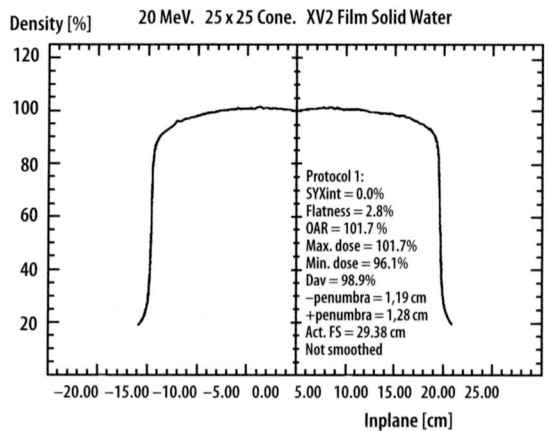

Density [%]   20 MeV. 25 x 25 Cone. XV2 Film Solid Water

Protocol 1:
SYXint = 0.0%
Flatness = 2.8%
OAR = 101.7 %
Max. dose = 101.7%
Min. dose = 96.1%
Dav = 98.9%
–penumbra = 1,19 cm
+penumbra = 1,28 cm
Act. FS = 29.38 cm
Not smoothed

Inplane [cm]

Fig. 7.6. Electron dose profile showing both field flatness and symmetry. Data are for a 20-MeV, Varian 2100C, 25×25-cm^2 cone in place. Data are taken using Kodak XV2 film in a solid water phantom and scanned using a Wellhofer scanner running WP700 software

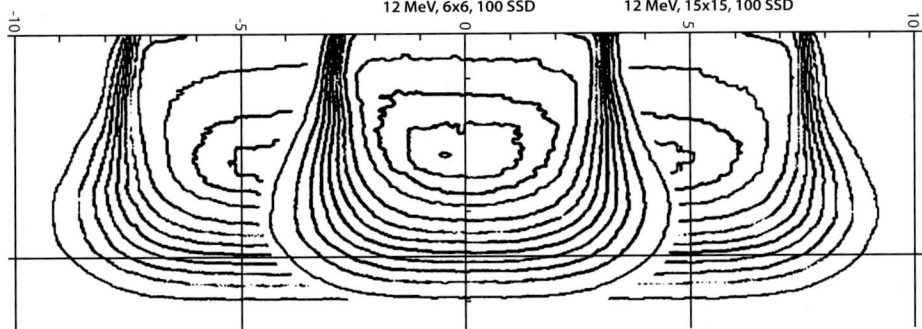

Fig. 7.7. Isodose curves for a 6×6-cm² field overlaid on a 15×15-cm² field for a 12-MeV, 100-cm SSD electron beam from a Varian 2100C Clinac. The central axis percentage depth–dose values are the same for the 6×6-cm² and 15×15-cm² fields as is the shape of the isodose curves in the penumbral region. Isodose values represented are the 98, 95 and 90 to 10 by steps of 10 levels. Data are taken using Kodak XV2 film in solid water phantom, scanned using a Wellhofer WP700 system

the isodose curves in the penumbra region, the distance from the 80% to the 20% isodose curve, changes very little as a function of field size. An investigation of these isodose curves shows that the 50% isodose curve penetrates almost straight down perpendicular to the surface, assuming a flat patient surface. The edge of both the 90% and 80% isodose curves moves in toward the central axis, while that in the isodose curves less than 50% spreads out from the 50% isodose line. Keeping these beam characteristics in mind is important to ensure adequate field size coverage of the specific target region by the 90% isodose curves. Usually a 1-cm margin around the desired treatment area provides an adequate margin for electron beams at standard treatment distances. When electron fields are abutted with other electron or photon fields, the widening of the penumbra exhibited by the lower value isodose curves influences the dose in any overlap regions or in fields adjacent to the electron field.

The overall shape of isodose curves can be influenced by many factors. These include patient curvature, inhomogeneous materials such as air, bone, lung, and high Z materials, and the effects of extended SSDs and field-shaping devices. These field-shaping devices include cerrobend inserts, lead shields or cutouts on the patient's skin surface, and internal shields designed to protect underlying sensitive or normal tissue.

7.5.1
Change in Isodose Curves Versus SSD

Figure 7.8 shows the change in the shape of isodose curves with changing SSD. The data is for a 10×10-cm² cone at 100 cm SSD and 115 cm SSD, creating an

11.5×11.5-cm² field at 115 cm SSD, the reason that the right side of the figure is wider than the left side. It can be seen at both 9 MeV and 20 MeV that the shapes of the isodose curves change dramatically

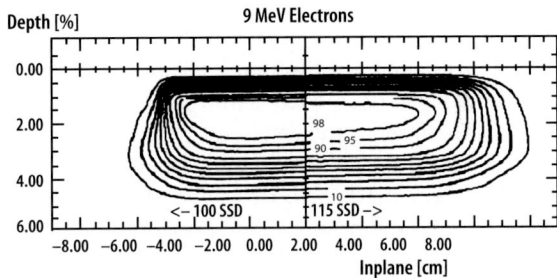

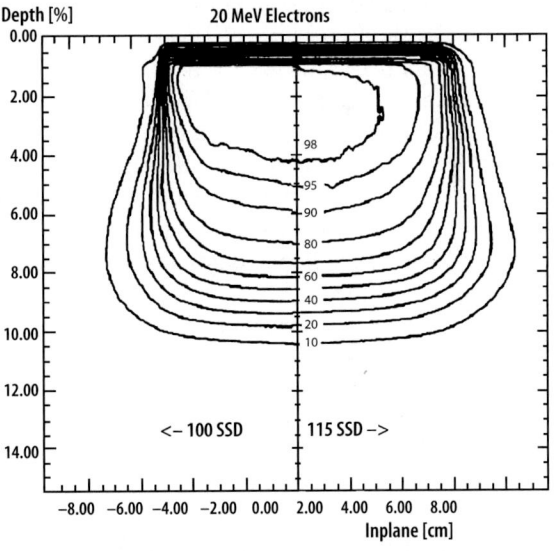

Fig. 7.8. Isodose curves for 9-MeV (*top*) and 20-MeV (*bottom*) electron beams, 10×10 cm² at both 100 cm and 115 cm SSD. Isodose values represented are the 98, 95 and 90 to 10 by steps of 10 isodose levels. Data are taken using Kodak XV2 film in solid water phantom, scanned using a Wellhofer WP700 system

at extended distance. Near the surface, the isodose
lines less than 50% extend much farther outside the
field edge than at 100 cm SSD. Isodose lines greater
than 50% also are changed markedly from what is
exhibited at the standard 100 cm SSD. In general, the
90% field width is decreased at extended SSDs while
more dose is contributed outside of the field edge at
extended SSDs. This change in the shape of the isodose curves is important to keep in mind so that the
target is adequately covered at extended distances and
for field abutment at extended SSDs. As was shown
in Figure 7.5, there is very little change in the central
axis percentage depth dose with increasing SSD.

7.5.2
Change in Isodose Curves Versus Angle of Beam Incidence

Many times in clinical situations, electrons are incident on the patient's surface at oblique angles. This is
most notable when treating the chest wall, the scalp, or
extremities of the body. For obliquely incident beams
whose angle of incidence is greater than 30°, there is
a significant change in the shape of the central axis
percentage depth dose. With reference to Figure 7.9,
as the angle of beam incidence increases, the depth
of the dose maximum decreases. At slightly increased
angles of beam incidence, the slope of the central
axis percentage depth–dose curve and the practical
range remain unchanged. As the angle of incidence
increases beyond 60°, the shape of the central axis
percentage depth–dose curve changes significantly,
and the D_{max} increases dramatically when compared
with the dose at d_{max} for normally incident beams
(EKSTRAND and DIXON 1982; KHAN et al. 1985; KHAN
2003). Table 7.3 gives a numerical description of the
change in the central axis dose distribution as a function of oblique beam incidence for different electron
beam energies as a function of depth for a 20×20-cm²
field (KHAN et al. 1985).

Clinical examples where sloped or curved surfaces are encountered include chest wall treatments,
treatment of the limbs, and treatments of the scalp.
Figure 7.10 shows the isodose distributions for a
beam of electrons incident vertically on a curved
surface, such as the chest wall. The isodose curves
follow roughly the curvature of the contour but there
is significant change in the beam coverage due to
the attenuation of the beam, loss of scatter, inverse
square decrease due to distance, and obliquity effects.
The magnitude of these changes also depends on the
radius of curvature of the surface to be treated.

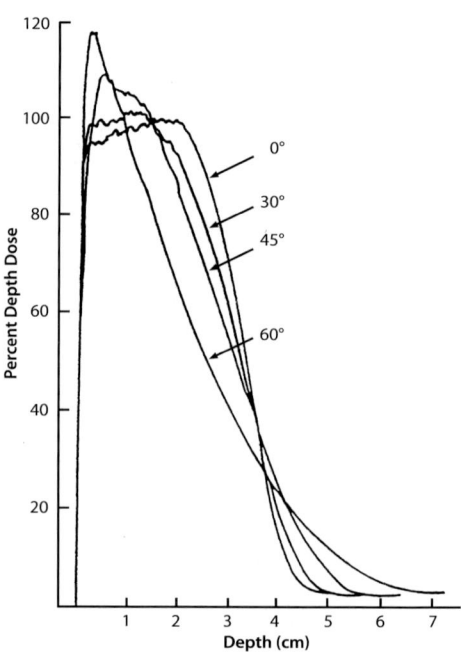

Fig. 7.9. Change in dose versus depth for 9-MeV electrons incident on water. The curves are normalized to the 100% point
for the 0° incident beam. [Reprinted with permission from
EKSTRAND and DIXON (1982)]

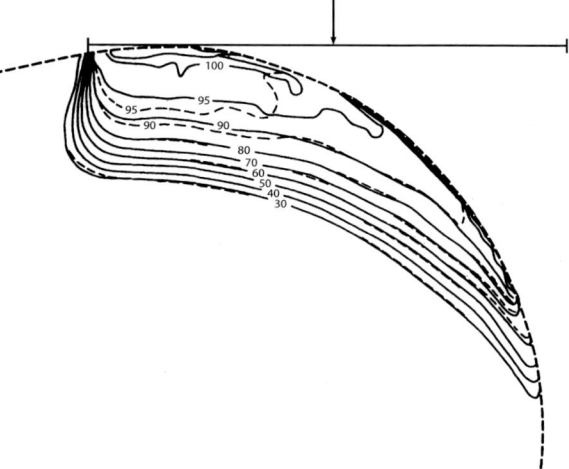

Fig. 7.10. Measured (*solid lines*) and calculated (*dashed lines*)
isodose distribution for an electron beam incident on a cylindrical polystyrene phantom. [Reprinted with permission from
KHAN (1984)]

7.5.3
Irregular Surfaces

Irregular skin surfaces involving abrupt changes
in the patient surface are encountered primarily
during the treatment of the nose, eye, ear and ear

Table 7.3. Obliquity factors for electron beams. These factors show the relative change in the beam intensity at angle θ compared with that at normal beam incidence. [Reprinted with permission from KHAN et al. (1985)]

(a) $\theta = 30°$

Z^b/R_ρ	E_0(MeV)					
	22	18	15	12	9	6
0.0	1.00	0.98	0.98	1.00	0.94	1.01
0.1	1.00	1.00	1.00	1.00	1.00	1.08
0.2	1.00	1.00	1.01	1.02	1.05	1.11
0.3	1.01	1.00	1.02	1.03	1.05	1.06
0.4	1.01	1.01	1.02	1.00	1.00	0.96
0.5	1.00	1.00	0.98	0.96	0.92	0.86
0.6	0.95	0.94	0.92	0.90	0.86	0.79
0.7	0.92	0.90	0.87	0.86	0.86	0.83
0.8	0.93	0.85	0.82	0.90	1.00	0.96
0.9	1.09	1.00	1.20	1.11	1.44	1.00
1.0	1.42	1.54	1.50	1.50	1.30	1.00

(b) $\theta = 45°$

0.0	1.03	1.02	1.03	1.05	0.98	1.14
0.1	1.03	1.04	1.04	1.06	1.10	1.14
0.2	1.05	1.06	1.07	1.11	1.12	1.12
0.3	1.06	1.07	1.09	1.09	1.05	1.07
0.4	1.04	1.04	1.04	1.01	0.93	0.92
0.5	1.00	0.99	0.92	0.92	0.80	0.77
0.6	0.93	0.90	0.86	0.82	0.70	0.69
0.7	0.84	0.84	0.82	0.77	0.70	0.76
0.8	0.87	0.83	0.85	0.86	0.83	1.10
0.9	1.30	1.00	1.43	1.20	1.40	1.46
1.0	2.17	2.31	2.19	2.50	2.00	2.14

(c) $\theta = 60°$

0.0	1.06	1.06	1.10	1.14	1.14	1.30
0.1	1.10	1.12	1.17	1.20	1.23	1.21
0.2	1.12	1.14	1.15	1.16	1.17	1.08
0.3	1.07	1.07	1.07	1.02	0.98	0.90
0.4	1.00	0.96	0.93	0.86	0.79	0.70
0.5	0.87	0.84	0.79	0.74	0.67	0.56
0.6	0.75	0.74	0.69	0.63	0.58	0.51
0.7	0.70	0.68	0.67	0.62	0.57	0.56
0.8	0.75	0.71	0.67	0.74	0.77	0.87
0.9	1.21	1.00	1.29	1.14	1.60	1.40
1.0	2.31	2.46	2.75	3.0	3.2	2.45

canal, and in the groin area. Surgical excisions can also create treatment areas with abrupt changes in the surface of the body. For areas of the body where sharp surface gradients exists (nose, ear), there will be a loss of side scatter equilibrium resulting in a hot spot beneath the distal edge of the step and a cold spot beneath the proximal surface. This is illustrated in Figure 7.11 for an experimental situation. Figure 7.12 shows a clinical example treating a region around the nose and the high-dose areas to the side of the nose (CHOBE et al. 1988). This figure shows that the magnitude of the high-dose region can be increased as much as 20% with a low-dose region directly adjacent.

In clinical practice, the potential for the creation of these types of high-dose regions has to be monitored constantly so that this effect can be minimized by the use of bolus material when possible to smooth out the surface irregularity.

7.6
Effect of Inhomogeneities on Electron Distributions

Commissioning data for electron beams is performed using a flat, homogeneous, water phantom. This idealized data set provides a good starting point for clinical dosimetry for electron beams. However, the presence of inhomogeneous material in the body is a fact of life in real world situations. The inhomogeneities of greatest concern are air cavities (primarily in the head and neck region), lung, and bone. Electron depth–dose distributions in a medium are dependent on electron density (electrons/cm^3). Since the number of electrons per gram is the same for all materials (except hydrogen which has about twice the number of electrons per gram), the physical density (g/cm^3) of the material determines the depth of penetration of the beam. Therefore, the depth of penetration of the beam of electrons can be determined by scaling the depth of penetration in water by the physical density of the inhomogeneous medium. This can be represented by the following equations:

$$z_{water} = z_{med}\, \rho_{med} \tag{1}$$

where z is the depth in the indicated material, water or medium, and ρ_{med} is the physical density of the medium or inhomogeneity in this example. For beams passing through lung material of density 0.25 g/cm^3, the depth of penetration of the beam in the lung would be:

$$z_{lung} = z_{water}\, /\rho_{lung}. \tag{2}$$

Thus, a beam that would penetrate 1 cm of normal, unit density material such as water would penetrate to a 4-cm depth in lung having a density of 0.25 g/cm^3. This is a quick rule of thumb that can be applied in the clinic to determine the amount of penetration into materials, the density of which differs from that of normal tissue. The actual situation is more complicated due to scattering of the electron beam and interface effects.

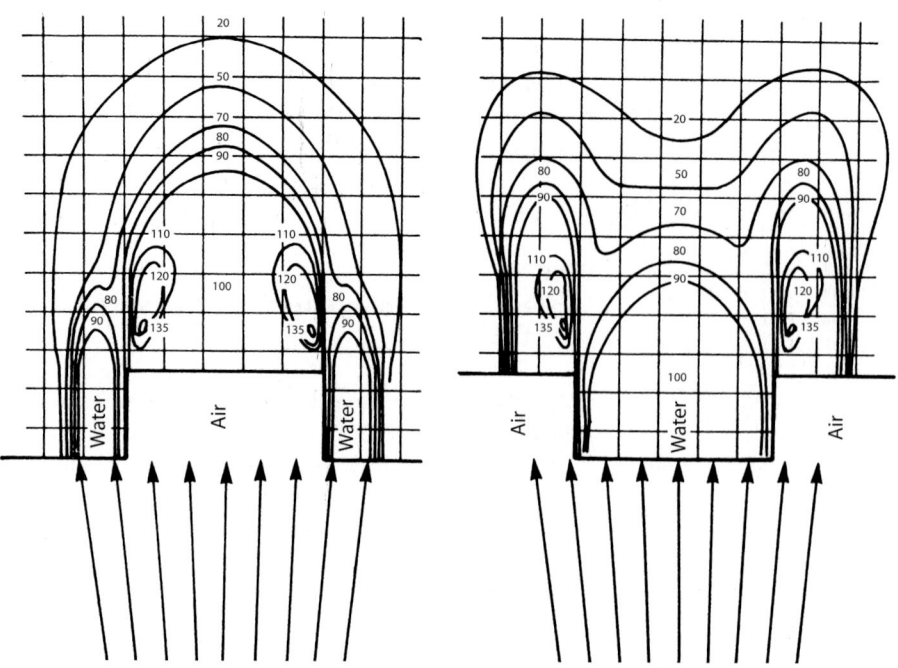

Fig. 7.11. High-dose regions produced by an electron beam incident on a water phantom have sharp, step-like irregularities in the surface contour. [Reprinted with permission from DUTREIX (1970)]

7.6.1
Lungs

Figure 7.13 illustrates the increased penetration of electron beams into lung tissue. Figure 7.13a shows a 12-MeV beam incident on the chest wall of a patient without taking the density of the lung into account (KHAN 2003). Figure 7.13b shows the dramatic increase in dose to the lung when this inhomogeneity is taken into account in the calculation. For simplicity, the effects of the ribs have not been considered in this illustration. As mentioned above, the energy of the beam is selected to place the 80% isodose line at the lung chest wall interface rather than using the 90% isodose line. Insuring that this 80% isodose line lies at the interface often requires the use of a bolus or compensating bolus plus a judicious selection of the best electron energy to employ.

7.6.2
Bones

Inhomogeneities in the form of bones are often present within the electron treatment field. Bone density can range from 1.0 g/cm³ to 1.10 g/cm³ for the spongy bone of the sternum to 1.5 g/cm³ to 1.8 g/cm³ for hard bones such as those of the mandible, skull, and other bones that provide structural support for

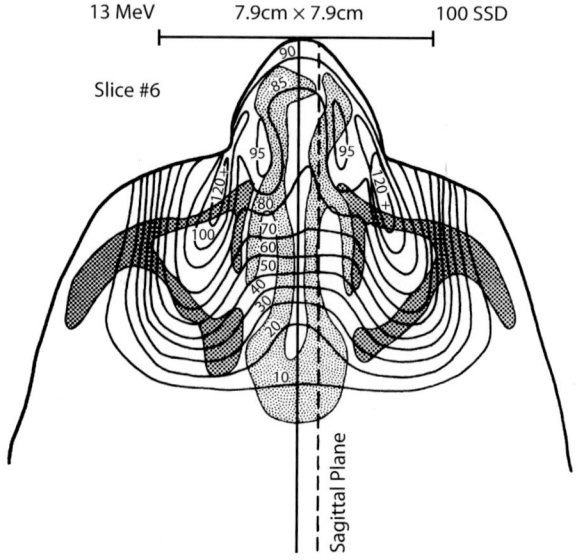

Fig. 7.12. Effect of the abrupt change in contour by the nose and the effect on the final isodose distribution. The plan is done assuming unit density material and not taking into account the air cavities particular to this region of the body [Reprinted with permission from CHOBE et al. (1988)]

the body. Additionally, the density of bone is not uniform throughout their cross section. Figure 7.14 illustrates the effect of hard bone on electron isodose curves (HOGSTROM and FIELDS 1983). Beneath the bone, the electron isodoses are shifted toward the surface due to extra attenuation or the shielding

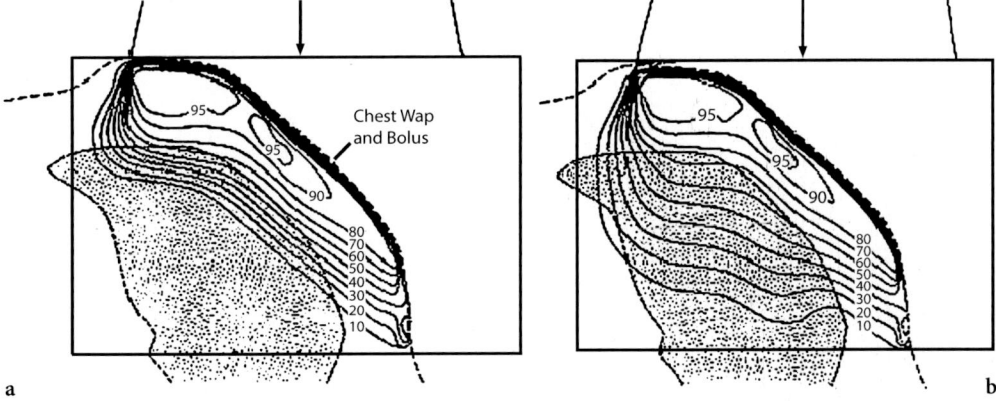

Fig. 7.13a,b. The effect of lung corrections on the overall dose distribution without correction for lung (**a**) and with bulk correction for lung using a density of 0.25 g/cm³ for lung tissue (**b**). A bolus was used to maximize the dose on the surface. 10-MeV, 100-cm SSD (68 cm SSD effective) electrons were used. [Reprinted with permission from KHAN (2003)]

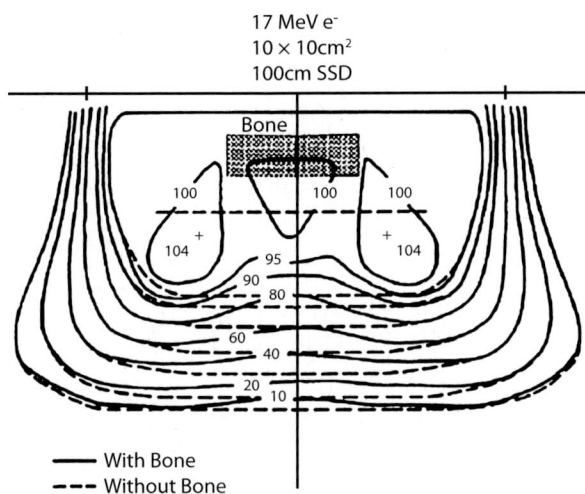

Fig. 7.14. The effect of hard bone on electron isodose curves. [Reprinted with permission from HOGSTROM (2004)]

7.6.3
Air Cavities

Because of the low physical density of air (0.0013 g/cm³), electrons pass easily through this medium. In addition, complicated scatter situations are created at the interface between air cavities and other tissues. Figure 7.15 shows the dose distribution using an anterior electron field to treat the nasal region using high-energy electrons. The first illustration (Fig. 7.15a) shows the dose distribution neglecting corrections for inhomogeneities. Figure 7.15b shows the dramatic difference in the dose distribution when air cavities and bone are taken into account. Very high doses penetrating into the brain and other underlying tissues can easily be seen from this diagram. If this increased dose is not considered, large doses to these underlying structures can result. It is common practice when treating this region to put a bolus into the nose to help counteract this increased penetration. Figure 7.15c shows the improvement in dose distribution by the use of both internal and external bolus placement. However, the bolus can be difficult to place so that it fills completely the internal cavity especially if immobilization masks are used. Thus, protection of underlying structures from increased penetration through these air cavities might be assumed by the placement of this bolus, when, in actual fact, little protection is actually provided. Figure 7.15d–f shows an example of the actual field and bolus placement along with clinical results obtained at 2 years post-treatment (MCNEESE 1989).

Another significant effect is the reduction in the dose to the tissues directly adjacent to the air cavi-

effect of the bone. In addition, the dose outside or at the edge of the bone–tissue interface is increased by about 5%, while that at the edge of the interface but beneath it is decreased, by approximately the same amount. This is due to the loss of side scatter equilibrium. It should be noted that actual bone–tissue interfaces are more rounded than those depicted in the figure, which would lead to lesser deviations than those indicated. However, some investigators (BOONE et al. 1967, 1969) showed, using thermoluminescent dosimeter capsules, that at 9 MeV the intercostal doses in dogs were increased by an appreciable amount. This effect is even more dramatic when using 6 MeV electrons.

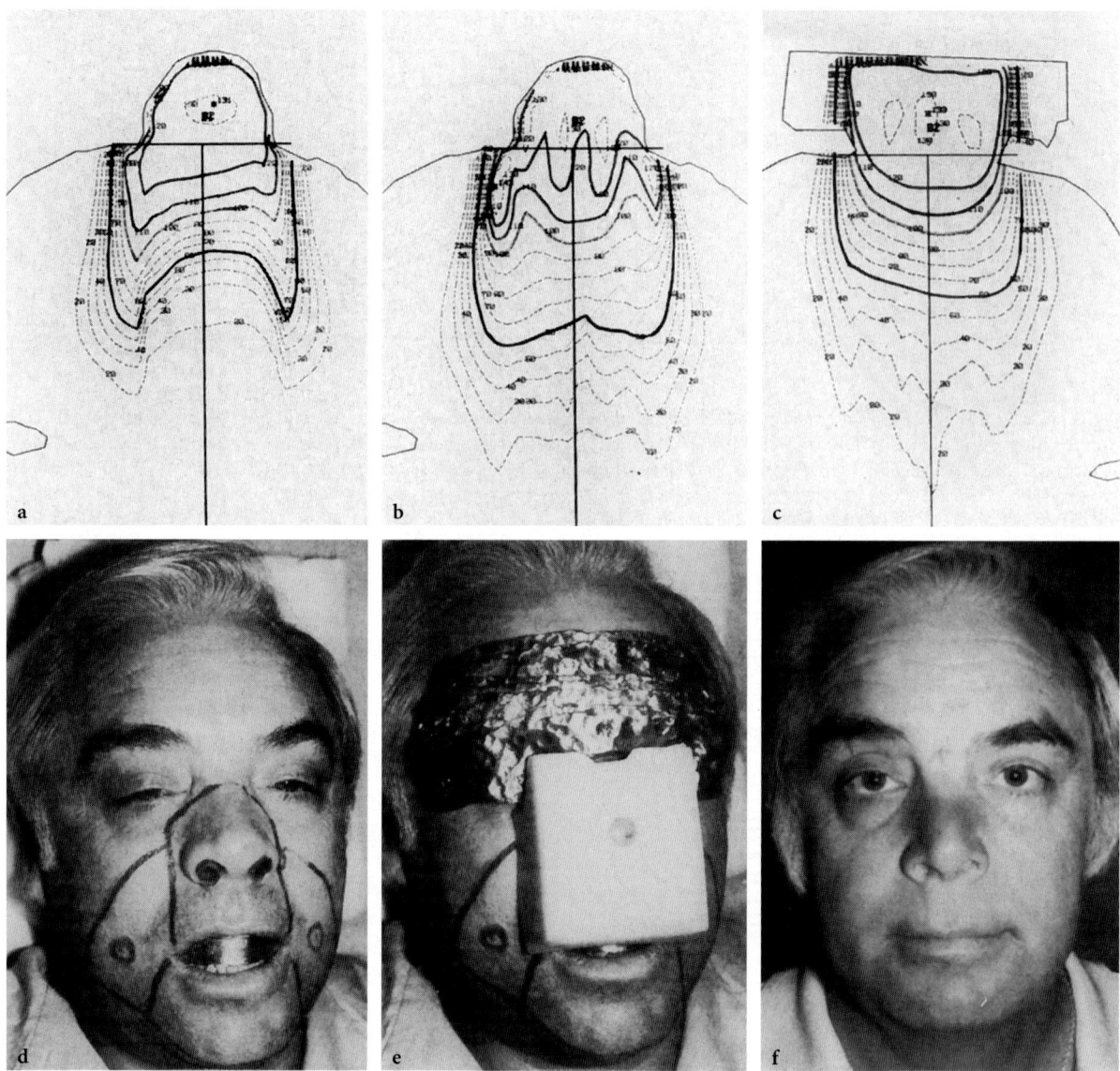

Fig. 7.15a–f. a Dosimetry without heterogeneity correction gives false impression of isodose distribution. **b** Dosimetry with heterogeneity correction shows more accurate isodose distribution. **c** Improved isodose distribution with use of internal and external bolus. **d** Actual treatment fields, with bolus placed in nostrils and intraoral stent in place. [Reprinted with permission from McNEESE (1989)]. **e** Completion of treatment setup with external wax bolus and lead eyeshield in place. **f** 2 years after completion of therapy. [Reprinted with permission from McNEESE (1989); CHOBE et al. (1988)]

ties. Electrons scatter preferentially into low-density regions (air cavities), and electrons scattering from the adjacent tissue into air are not replaced because the air is not able to scatter an equal number of electrons back to the higher density tissues. This can lead to clinical under dosing of the tissues in this area by approximately 10% which could result in the loss of control of the primary target site.

7.7
Clinical Applications of Electron Beams

7.7.1
Target Definition

As with photon beam treatments, the first step in the initiation of electron therapy is to determine accurately the target to be treated. All available diagnostic, operative, and medical information should be consulted to determine the extent and the final planning target volume (PTV) with appropriate margins to be treated before simulation and placement of the electron fields is initiated. Treatment-planning CT scans may be helpful and even required to make this determination along with the determination of the optimum beam placement and energy selection.

7.7.2
Therapeutic Range – Selection of Beam Energy

The depth of the 90% isodose level (D_{90}) is a very common therapeutic depth for electron beam therapy. The electron energy for treatment should be selected such that the depth of the 90% isodose line covers the distal or deepest portion of the region to be treated in addition to an approximate 5-mm additional depth beyond the treatment region. This depth of R_{90} can be approximated by dividing the energy of the electron beam in MeV by four ($E_0/4$) in centimeters of water. The 80% isodose level (D_{80}) is also frequently used as a treatment parameter for defining the therapeutic treatment range. It is most commonly used for chest wall treatments where the D_{80} would be placed at the lung-chest wall interface. The 80% point is chosen in this region since it adequately covers the chest wall without depositing an excessive amount of radiation in the underlying lung and heart tissue. The depth of D_{80} is approximately equal to $E_0/3$ in centimeters of water.

7.7.3
Dose Prescription – ICRU 71

In 2004, the International Commission on Radiation Units and Measurements (ICRU) published Report 71 detailing new recommendations for "Prescribing, Recording, and Reporting Electron Beam Therapy" (GAHBAUER et al. 2004). They recommended the same general approach for dose prescriptions for electron treatments as those taken for photons as specified in the previous reports ICRU 50 and 62 for photon beams (INTERNATIONAL COMMISSION ON RADIATION UNITS AND MEASUREMENTS 1993, 1999). For treatment-planning purposes and to maintain consistency with the previous ICRU reports dealing with photon beam treatments, the concepts of gross tumor volume (GTV), clinical target volume (CTV), planning target volume (PTV), treated volume, organs at risk (OAR), and planning organ at risk

volume (PRV) are defined as in previous reports and are to be used.

They indicated that the treatment should be specified completely, including time–dose characteristics and making no adjustments in the relative biological effectiveness differences between photons and electrons. Specifically, they recommended the selection of a reference point for reporting electron doses which is referred to as the "ICRU reference point". This point should always be selected at the center (or central part) of the PTV and should be clearly indicated. In general, the beam energy is selected so that the maximum of the depth–dose curve on the beam axis is located at the center of the PTV. If the peak dose does not fall in the center of the PTV, then the ICRU reference point for reporting should be selected at the center of the PTV and the maximum dose should also be reported. For reference electron irradiation conditions, they recommend that the following dose values be reported (GAHBAUER et al. 2004):

- the maximum absorbed dose to water
- the location of and dose value at the ICRU reference point if not located at the level of the peak-absorbed dose
- the maximum and minimum doses in the PTV, and dose(s) to OARs derived from dose distributions and/or dose–volume histograms

For small and irregularly shaped beams, the peak absorbed dose to water for reference conditions should be reported. It is also recommended that when corrections for oblique incidence and inhomogeneous material are applied, the application of these corrections should be reported.

7.7.3.1
ICRU 71 Recommendations – Intraoperative Radiation Therapy

ICRU Report 71 also provides direction for the special electron beam techniques of intraoperative radiation therapy (IORT) and total skin irradiation (TSI). In IORT, high-energy electrons are used to deliver a large, single-fraction dose to a well-defined anatomical area after surgical intervention. The CTV is defined as accurately as possible by both the surgeon and the radiation oncologist during the procedure.

All devices specific to IORT need to be recorded such as the IORT applicator system including type, shape, bevel angle, and size of the applicator. The

ICRU reference point for reporting is always selected in the center or central part of the PTV and, when possible, at the level of the maximum dose on the beam axis.

The ICRU recommended that the following dose values be reported for IORT (ICRU 2004):

- the peak absorbed dose to water, in reference conditions, for each individual beam (if the beam axis is perpendicular to the tissue surface)
- for oblique beam axis, the maximum absorbed dose in water on the "clinical axis" (i.e., the axis perpendicular to the surface of the tissues, at the point of intersection of the central axis of the beam with the tissue surface)
- the location of and dose value at the ICRU reference point (if different from above)
- the best estimate of the maximum and minimum doses to the PTV. Usually the irradiation conditions (electron energy, field size, etc.) are selected so that at least 90% of the dose at the ICRU reference point is expected to be delivered to the entire PTV

7.7.3.2
ICRU 71 Recommendations – Total Skin Irradiation

For total skin irradiation (TSI), the aim is to irradiate the total skin surface as homogeneously as possible. For patients with superficial disease, TSI can be delivered with one electron energy. In other clinical situations, the thickness of the skin disease may vary with stage, pathology, and location on the body surface. For such cases, several CTVs need to be identified and different beam penetrations have to be used. For each anatomical site, an ICRU reference point for reporting at or near the center of the PTVs/CTVs has to be selected. The reference point may be at the level of the peak dose if it is located in the central part of the PTV. In addition, an ICRU reference point, clinically relevant and located within the PTV, can be selected for the whole PTV.

For TSI treatments, reporting of the following dose values are recommended:

- the peak absorbed dose in water for each individual electron beam
- the location of and dose value at the ICRU reference point for each anatomical area (the ICRU reference point may or may not be at the level of the peak dose)
- the best estimate of maximum and minimum dose to each anatomical area

- the location and absorbed dose at the ICRU point for the whole PTV, and best estimate of the maximum and minimum doses for the whole PTV
- any other dose value considered as clinically significant

7.7.4
Field Shaping and Collimation

Cerrobend or lead cutouts are used to restrict the cone-generated electron fields to the desired area to be treated. Cerrobend is the material of choice for these field restriction devices due to the ease by which they can be constructed. After the field has been defined on the surface of the patient, it is extremely easy to outline on a clear piece of acrylic the insert to be constructed that will be inserted into the electron cone in exactly the orientation in which it will be used for treatment. This design process can be done in either the simulator or in the linac treatment room. Alternatively, a lead cut out can be manufactured from commercially available sheets of lead. They are more difficult and time consuming to construct and require that a store of lead sheets be kept in house. Lead cutouts are placed directly on the patient's surface to define the treatment area and offer the advantage of producing a field with sharper edges than what can be accomplished with cerrobend inserts. Lead cutouts should be considered for small fields, low electron energies, when critical areas lie directly adjacent to sensitive structures, to sharpen the field edge either at standard or extended treatment distances, or when electron arc treatments are employed. Figure 7.16 shows the difference in the isodose curves for a 6-MeV, 3×3-cm^2 electron field when a cerrobend insert is placed 10 cm above the skin surface to define the field as opposed to lead blocking placed directly on the skin surface HOGSTROM 1991).

The thickness of lead required to stop the primary electrons has been investigated by GIARRATANO et al. (1975), who concluded that a lead thickness in millimeters equal to the most probable electron energy at the surface (in MeV) divided by two is adequate to provide shielding. An extra millimeter of lead can be added to provide an additional margin of safety. Thus, a 20-MeV electron beam would require a lead shield thickness of 10 mm of lead plus 1 mm for safety for a total thickness of 11 mm. Sheet lead is commercially available in 1/16 inch increments, so the conversion needs to be made between millimeters and inches.

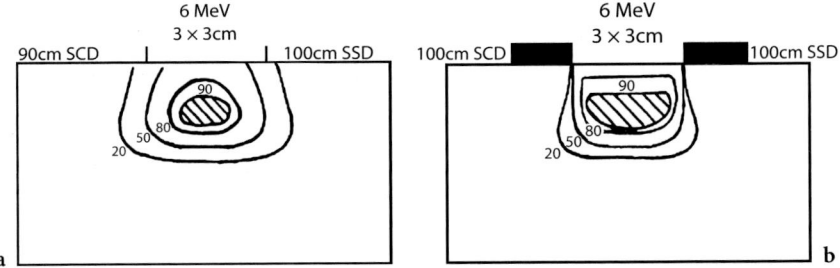

Fig. 7.16a,b. The effect of collimating with a field defining device at 10 cm from the skin surface (**a**) as opposed to defining the field at the skin surface (**b**) for a 6-MeV, 3×3-cm² electron field at 100 cm SSD. [Reprinted with permission from HOGSTROM (1991)]

Required thicknesses of cerrobend for electron shielding have been studied by PURDY et al. (1980). Since low melting point alloy has approximately 82% of the density of lead (9.3 g/cm³ versus 11.34 g/cm³, respectively), 20% additional thickness needs to be employed for adequate shielding. In the example above for lead shielding for 20 MeV electrons, the required thickness of cerrobend would be 13 mm. It is interesting to note that the addition of high-Z material for electron shielding increases slightly the amount of photon contamination in the beam treating the patient and thus the deep dose (PURDY et al. 1980; ZHU et al. 2001).

When lead cutouts are placed directly on the skin surface, electrons scattered from the edges of the lead into the treated field become a clinical consideration. Figure 7.17 shows the isodose distribution at the edge of a lead shield placed at the water surface.

Shown in Fig. 7.17a are the angles α and β, which represent the angles of maximum and minimum dose changes, respectively. Figure 7.17b shows how α and β are affected by the energy of the incident electron beam. In general, both α and β decrease with increasing beam energy since electron scatter becomes more forward directed as the beam energy increases (POHLIT and MANEGOLD 1976).

7.7.5
Internal Shielding

In some instances, internal shields need to be used to protect underlying sensitive structures. This is most commonly seen when using fields to treat the lip, buccal mucosa, and eyelid lesions. Internal shields are also commonly used for intraoperative radio-

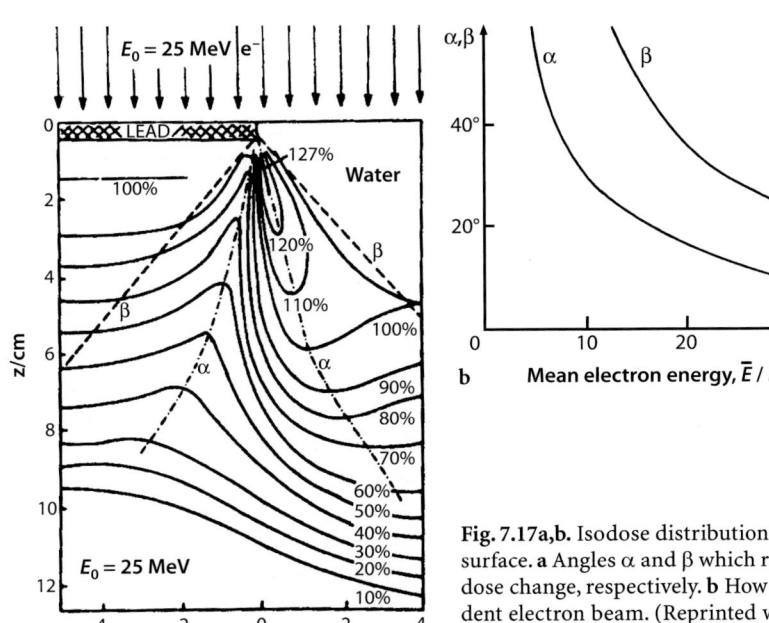

Fig. 7.17a,b. Isodose distribution at the edge of a lead shield placed at the water surface. **a** Angles α and β which represent the angle of maximum and minimum dose change, respectively. **b** How α and β are affected by the energy of the incident electron beam. (Reprinted with permission from POHLIT and MANEGOLD (1976)]

therapy treatments. Important considerations when designing internal shields are to ensure an adequate shield thickness at the depth of shield placement, to ensure that the electrons backscattered from the lead surface do not dangerously increase dose at the interface, and to consider the high-dose edge effects mentioned above. Lead is the most common material used for the production of internal shields because of its availability and ease of use. The required thickness of the shield depends on the energy of the electron beam at the location of the internal shield, the fact that electrons decrease in energy by 2 MeV/cm in muscle, and that 1 mm of lead is required as shielding for every 2 MeV of electron energy (plus 1 mm for safety, as mentioned above). Thus, if 9 MeV of electrons are used to treat the buccal mucosa of thickness 1 cm, a shield placed beneath the cheek to protect the oral cavity would have to be 4.5 mm thick. This is because the electrons would decrease to 7 MeV after penetrating 1 cm of tissue, and that 3.5+1 = 4.5 mm of lead would be required to shield 7 MeV electrons. For final shield design, the backscatter of electrons from the lead surface has to be taken into account. As shown in Fig. 7.18, the amount of electron backscatter that would be produced by placement of a shield made solely of lead would produce an increase in dose of approximately 50% at the lead–tissue interface (KLEVENHAGEN et al. 1982; KLEVENHAGEN 1985). Reduction of the increased dose due to backscatter from the lead surface can be accomplished by the addition of some low-Z material – dental acrylic, bolus material, or, in some cases, aluminum. We find it convenient to apply a coating of dental acrylic to the surface of the lead followed by layers of dental boxing wax to the surface of the lead facing the beam. The dental acrylic seals the lead, which is itself toxic, and makes cleaning and sterilization much easier to accomplish between treatments. Using two half-value layers of wax is usually sufficient to reduce the amount of backscatter to an acceptable level. At 6 MeV, one half-value layer is approximately 3.5 mm of unit density material (LAMBERT and KLEVENHAGEN 1982). Thus, the final intraoral shield would consist of 4.5 mm lead, a coating of dental acrylic, plus about 7 mm of wax.

Internal shields placed under the eyelid to protect the underlying eye are particularly challenging to design and manufacture because of the limited space for shield placement and the required thicknesses of materials to provide adequate protection. In some instances, an effective shield cannot be designed that will allow the use of the desired electron energy while making the shield sufficiency

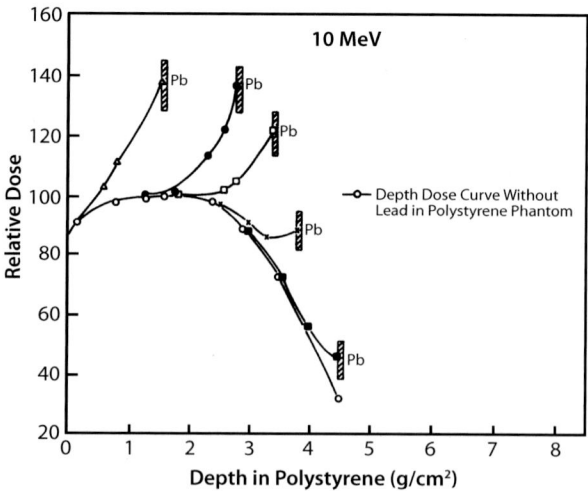

Fig. 7.18. Increase in the dose upstream from a lead shield thickness = 1.7 mm) at various depths in polystyrene. [Reprinted with permission from KHAN et al. (1976)]

thick and of the proper combination of materials. By using tungsten (Z = 74, density = 17.3 g/cm³) instead of lead (Z = 82, density = 11.34 g/cm³), acceptably thin eye shields can be made to fit under the eyelid that are capable of shielding 9 MeV electrons (SHIU et al. 1996). An added benefit is that tungsten is of a lower Z value than lead which also leads to a slightly lower amount of backscatter from the surface of the tungsten shield. Commercially available eye shields (Radiation Products Design, Inc., Albertville, MN) made from tungsten and aluminum have also been shown to provide excellent protection for the underlying eye (WEAVER et al. 1998).

Pencil eye shields can be utilized to protect the lens of the eye when the eyelid is involved and needs to be irradiated using electrons. A cerrobend shield 1.3 cm in diameter and at least 1 cm thick was shown to be adequate in protecting the lens when 6 MeV or 9 MeV of electrons were used for the treatment (RUSTGI 1986). Protection of the lens was optimal when the shield was places 1 cm or closer to the surface of the eye. In our clinic, we use a 1.2-cm-diameter, 2-cm-thick cerrobend shield coated with acrylic to reduce electron scatter from the side of the shield into the eye. An acrylic support rod is attached perpendicular to the side of the cylindrical shield so that a small ring stand can be used for daily shield placement. This arrangement allows for very easy placement of the shield to less than the desired 1-cm distance from the surface of the eye. Figure 7.19 shows that the 2-cm-thick shield is effective in reducing even 12 MeV electrons to an acceptable level of protection for the lens.

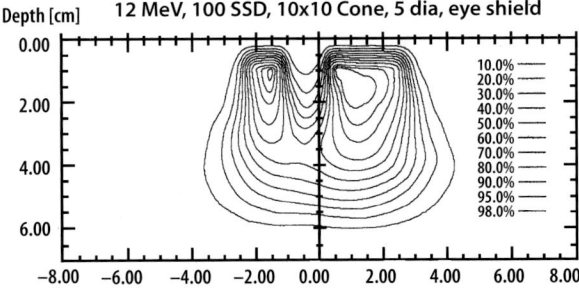

Fig. 7.19. Isodose distribution produced using a Varian 2100C, 12-MeV, 5-cm diameter circular electron field incident on a 2-cm thick, 1.2-cm diameter cylindrical cerrobend pencil eye shield in contact with a solid water phantom. The data is taken using Kodak XV2 film and scanned using a Wellhofer scanning densitometer system running WP700 software

7.7.6
Bolus

A bolus is used for several reasons in electron beam treatments: to increase the dose on the skin surface, to replace missing tissue due to surface irregularities, and as compensating material to shape the coverage of the radiation to conform as closely as possible to the target volume while sparing normal tissue. For modern linear accelerators, the surface dose can be quite low and, to treat to the 90% point, a bolus would be required to cover adequately the superficial regions of the body.

The ideal electron bolus material would be equivalent to tissue in both stopping power and scattering power. Additionally, it should be flexible and moldable to best conform to the variations in surface topology of the patient. This is an important requirement since small air gaps between the bolus and skin surface can promote in-scattering of electrons into the lower density air spaces resulting in local high-dose regions. Figure 7.20 shows the effect of an acrylic plate placed at a distance of 5 cm from a flat surface (HOGSTROM 1991). The reduction in dose to the patient and decreased field coverage from what would be desired is quite dramatic and would lead to a significant deviation in the delivered dose and insufficient treatment of the target area.

Several commonly available materials can be used as bolus material. These are paraffin wax, polystyrene, acrylic (PMMA), Super Stuff, Superflab, and Super-flex. Additionally, solid sheets of thermo-plastics (3 mm thickness per sheet) can be used. When hot, the material can be held in contact with the skin surface so that it conforms almost perfectly to the underlying contours. Additional layers of bolus can be added to this initial layer to produce the final desired thickness. An additional benefit of thermoplastic material is that it is transparent when hot. The transparency of the material makes it easy to transfer any marks made on the skin during setup to the surface of the bolus so as to aid in the accurate placement of the bolus.

7.7.6.1
Custom, Compensating Bolus

A custom compensating bolus can be designed for complex situations to eliminate or decrease the

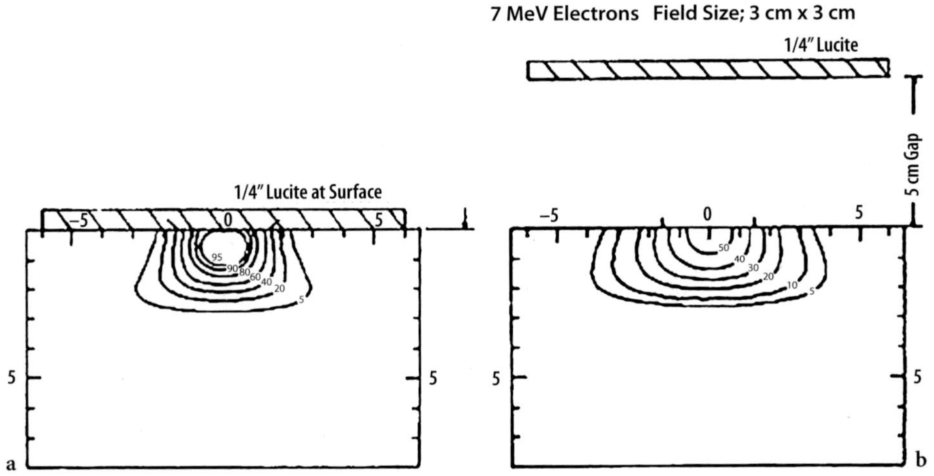

7 MeV Electrons   Field Size; 3 cm x 3 cm

Fig. 7.20a,b. The effect of location of an acrylic plate on the shape and magnitude of the dose distribution. An acrylic plate placed on the surface of a phantom (**a**) preserves both the shape of the original isodose curves and the magnitude of the delivered dose. An acrylic plate placed at a distance of 5 cm from the surface (**b**) produces not only a significant change in the shape of the isodose curves but also a large decrease in the overall delivered dose to the skin surface and at depth. [Reprinted with permission from HOGSTROM (1991)]

effect of tissue heterogeneities, irregular patient surface structure, distance or curvature effects, or other parameters that would affect the production of an optimal dose distribution. A custom bolus can be designed by hand using individual CT scans of the region to be treated (ARCHAMBEAU et al. 1981). The electron energy is selected by choosing the electron energy that would penetrate to cover the deepest extent of the target to be treated. A bolus is added to the top of the CT scans to create an equal depth of penetration along fan lines on all scans to cover the target (Fig. 7.21). The proposed bolus is added to each individual CT scan and a computerized treatment plan is completed to ascertain the accuracy of the bolus design. The technique is repeated until a final acceptable bolus design and plan is achieved. A grid describing the bolus location on the patient and the thickness of the bolus at those locations is produced and the compensator is made to these specifications (Fig. 7.22). The grid also serves as an alignment tool so that the compensating bolus can be placed accurately on the patient from one treatment to the next. This hand technique has several limitations: it is iterative in nature, time consuming, and one-dimensional which neglects multiple Coulomb scattering, and it does not account for the full three-dimensional nature of the problem (Low et al. 1992; ANTOLAK et al. 1992).

Computer-based techniques (Low et al. 1992) have been devised to address these difficulties and limitations. The distribution in Figure 7.23 gives an idea of the capabilities of these approaches in covering the target volume and limiting the high-dose regions (PERKINS et al. 2001; KUDCHADKER et al. 2002). The computerized bolus technique has been paired with electron intensity modulation to produce theoretical dose distributions for challenging target locations, although currently there is no easy way to deliver the treatment (KUDCHADKER et al. 2003).

7.7.6.2
Field Abutment

Field abutment is employed in most instances either to cover a larger treatment area or region or to change electron energy in a particular region to more adequately cover the target at depth. Alternatively, electron fields are abutted to photon fields so that a superior dose distribution can be achieved. In any case, optimal coverage of the target area is the goal of treatment and complicated field arrangements may be required to treat the patient properly.

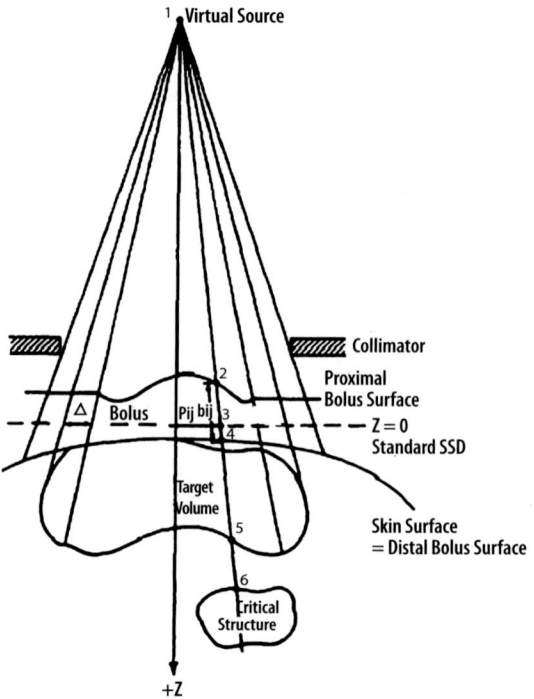

Fig. 7.21. Schematic representation of the patient contour, target volume, and compensating bolus designed to optimize the coverage of the target while minimizing the dose to the underlying critical structure [Reprinted with permission from Low et al. (1992)]

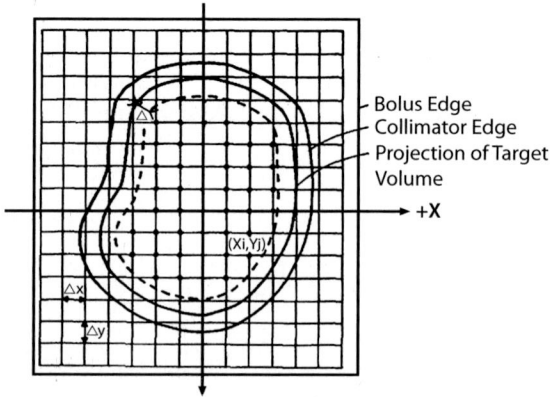

Fig. 7.22. Compensating bolus diagram for construction and placement of the bolus for patient treatment [Reprinted with permission from Low et al. (1992)]

Since most targets where electrons are used involve the surface of the patient or are very close to the patient's surface, no gaps in the dose coverage at the surface can be allowed. With field abutment, hot spots beneath the surface will be created which may be acceptable depending on the size, the magnitude of the high-dose region, and the location in the

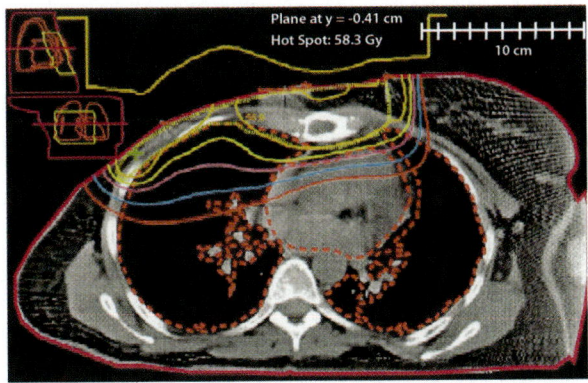

Fig. 7.23. Isodose distribution (Gy) using the custom 3D electron bolus technique. A dose of 50 Gy was prescribed to 100% of the given dose using 16 MeV electrons. The bolus was designed to deliver 90% of the given dose to the target volume. The plan shows the outline of the bolus above the skin surface (*yellow*), along with dose minimization to the ipsilateral lung and underlying cardiac tissues. [Reprinted with permission from PERKINS et al. (2001)]

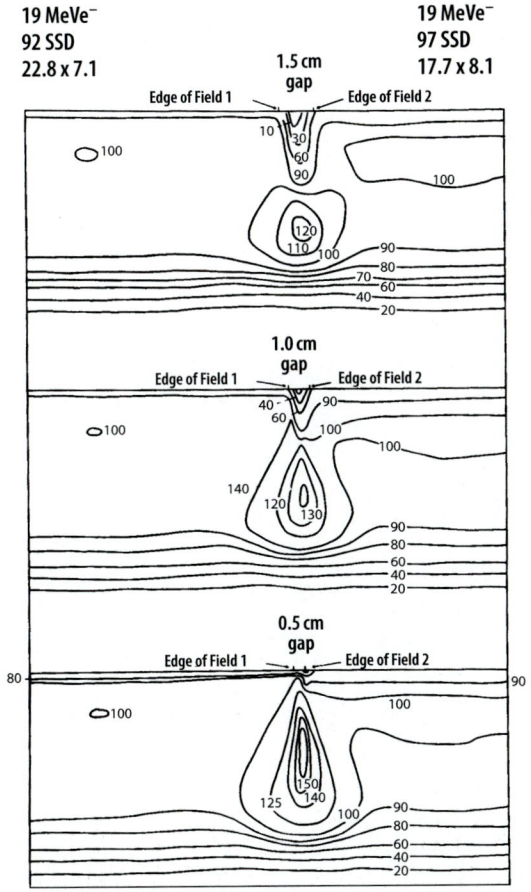

Fig. 7.24. Isodose curves showing the dose at the junction location for matching electron beams of the same energy incident on a flat surface as a function of different gap widths: 0.5-cm gap on bottom, 1.0-cm gap middle figure, 1.5-cm gap top figure. (Reprinted with permission from ALMOND (1976)]

patient of the high-dose region. Figure 7.24 shows the resultant isodose distributions when gaps of 0.5, 1.0, and 1.5 cm are placed between two electron fields of the same beam energy with parallel central axes incident on a flat surface. With a 0.5-cm gap, high-dose regions of 140–150% result. As the gap is increased, the magnitude of the high-dose region decreases to a more acceptable level, but low-dose regions that may be clinically important start to become evident near the surface. Figure 7.25 shows the effect of field matching with different electron beam energies incident on a flat surface as a function of different gap widths on the skin surface (0, 0.5, and 1.0-cm gaps). The magnitude of dose in the overlap region is not as severe as when identical beam energies are employed but the high-dose regions can still lead to significant treatment consequences.

Figure 7.26 shows a comparison of several abutting beam configurations. As shown in 7.26a, the extent and magnitude of the high-dose region can be minimized by angling the central axis of each beam away from each other so that a common beam edge is formed. Figure 7.26b represents overlap that can occur when the central axis of the beams are parallel. Figure 7.26c shows converging beam central axes that result in the greatest amount of overlap with the highest doses and largest high-dose regions.

Figure 7.27 shows a complicated collection of treatment fields used to cover the postmastectomy chest wall and supraclavicular region. Using this approach, the chest wall is usually treated with 6–9 MeV electrons with a 0.5- to 1.0-cm bolus to ensure an adequate skin dose. Higher beam energies can be used if a thicker chest wall is encountered but care must be taken to ensure that the underlying lung or heart does not receive too high a dose. The internal mammary chain is treated with 12–16 MeV electrons, depending on the thickness of the chest and the depth of the target at this location. The supraclavicular area can be treated with either electrons or photons depending on the anatomical characteristics of the patient and the depth of treatment for the supraclavicular nodes.

Figure 7.28 shows a clinical example of abutting electron fields in chest wall treatment (HOGSTROM 2004). The dose homogeneity is acceptable at the border of the internal mammary chain and medial chest wall fields because central axes are parallel and field widths are small. However, the dose homogeneity is unacceptable at the border of the medial and lateral chest wall fields because the central axes are converging. Figure 7.28 shows the smoothing effect of moving the junction by 1 cm twice during the treat-

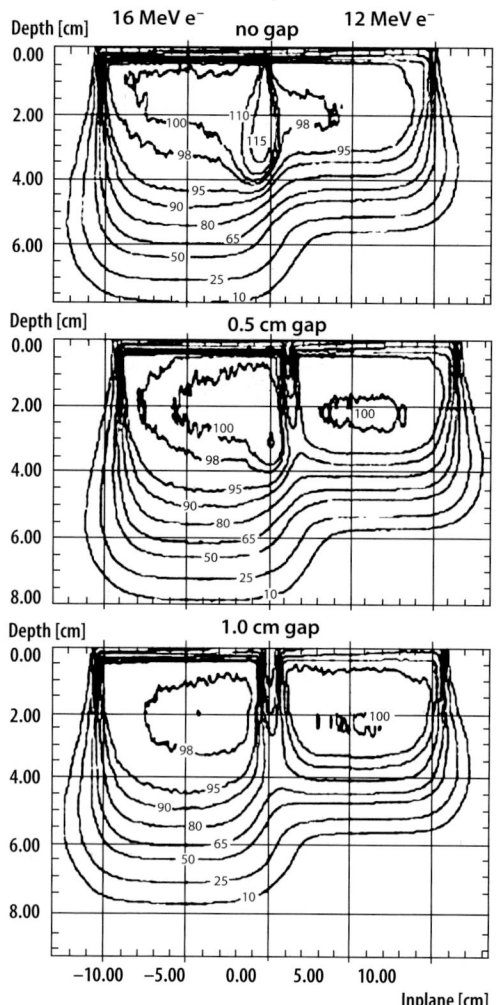

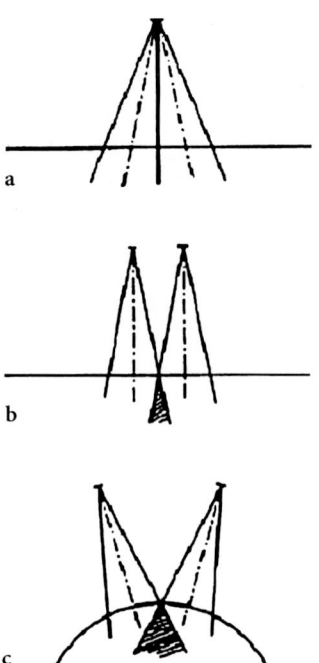

Fig. 7.26a–c. A collection of abutment geometries. a Diverging central axes form a common edge with the least amount of beam overlap. The central axes are angled away from each other by the angle $\theta = tan^{-1}$ (0.5 w/SSD) where w is the common field width. b Parallel central axes lead to overlapping fields and high-dose area beneath the patient surface. c Converging beam axes resulting in the highest dose regions and the greatest amount of overlap. [Reprinted with permission from ICRU Report No. 35: (1984)]

Fig. 7.25. Electron field matching isodose curves for different electron beam energies for a Varian 2100C, 10×10 cone, 100 cm SSD, 16 MeV electrons on the left and 12 MeV fields on the right. No gap on the skin surface (*top*), 0.5-cm gap on the surface (*middle*), and 1.0-cm gap on the surface (*bottom*). Data taken using Kodak XV-2 film in a solid water cassette and scanned using a Wellhofer isodensitomer and WP700 software

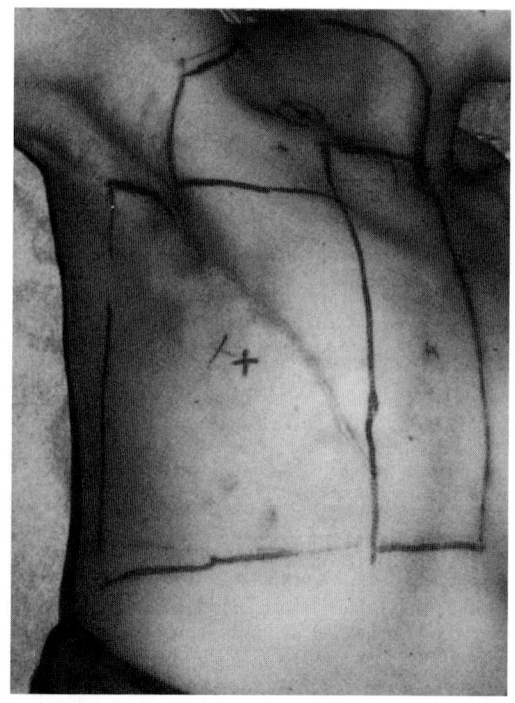

Fig. 7.27. Treatment field arrangement for postmastectomy patient. The chest wall is usually treated with 6–9 MeV electrons with a 0.5- to 1-cm bolus to ensure an adequate skin dose. The internal mammary chain is treated with 12–16 MeV electrons depending on the thickness of the chest at this location. The supraclavicular area can be treated with either electrons or photons depending on the anatomical characteristics of the patient and the depth of treatment for the supraclavicular nodes. [Reprinted with permission from LEVITT and TAPLEY (1992)]

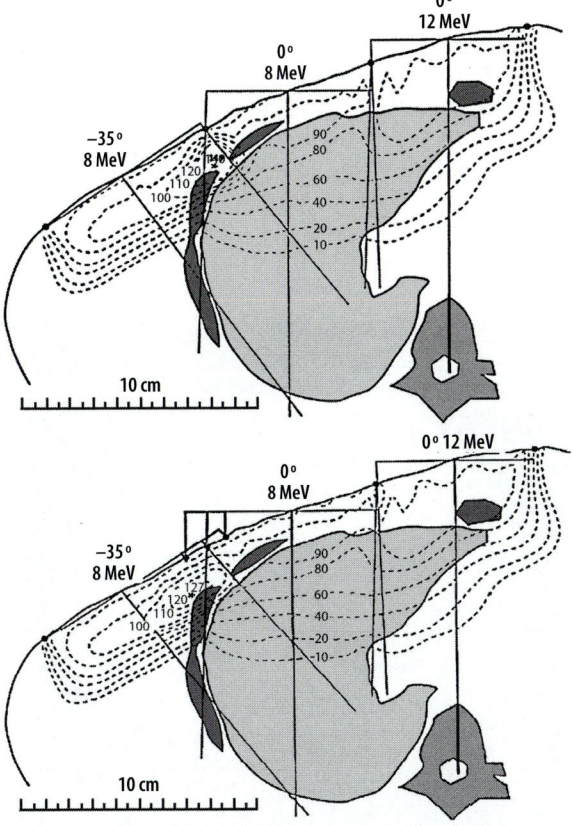

Fig. 7.28. Clinical examples of abutting electron fields in chest wall treatment. Dose homogeneity is acceptable at the border of the internal mammary chain and medial chest wall fields because central axes are parallel and field widths are small. Dose homogeneity is unacceptable at the border of the medial and lateral chest wall fields because central axes are converging. *Bottom figure*: Dose homogeneity is improved in this region by moving the match line twice during treatment by 1 cm. [Reprinted with permission from HOGSTROM (2004)]

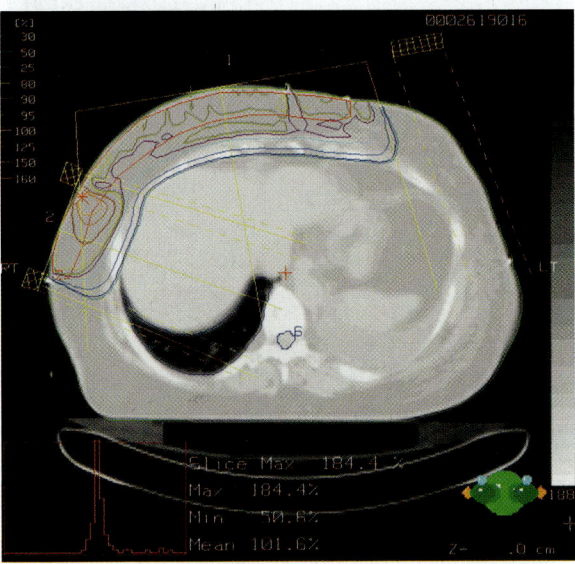

ment. A 50% high-dose region can be reduced to +27% by moving the junction in this manner. A shift in the junction of 1 cm is adequate in most instances but the absolute amount depends on the size of the overlap region and the total dose. It is recommended that the amount of the shift and the number of junction changes be made to ensure that the high-dose region does not exceed the prescription dose by more than 15–20% in any region, if possible.

7.7.6.3
Electron–Electron Field Matching: Sloping, Curved Surfaces

Matching electron fields on a curved surface, such as those that exist in clinical situations, only tends to exaggerate the magnitude and size of the overlap region. This occurs since each of the electron fields is usually positioned perpendicular to the skin surface (Fig. 7.29). The dose in the overlap region increases with decreasing radius of curvature of the external body contour. Situations such as those illustrated in Figure 7.29 should be closely monitored in the clinic, and the location of the junction should be repositioned with sufficient frequency to limit the risk of a complication.

7.7.6.3.1
Total Limb Irradiation

Treatment of the entire periphery of body extremities can be accomplished using electron fields spaced uniformly around the limb. The advantage of using electrons over treating simply with parallel opposed photon fields is that the central uninvolved regions of the limb can be spared from unnecessary radiation. The disadvantages of the technique lie in a more complicated field arrangement, additional dosimetry to verify the accuracy of the technique, and more treatment time on the linear accelerator per day to deliver the treatment. Figure 7.30 illustrates a technique described in the literature (WOODEN et al. 1996) using six equally spaced 5-MeV electron beams to treat a 9-cm diameter cylinder. Each beam is wide enough to cover the entire width of the limb and provide grazing radiation to all surfaces. The

Fig. 7.29. Representation of the magnitude of the high-dose regions when two electron fields abut at the skin surface. Both beams are perpendicular to the skin surface. Represented are 12-MeV electron fields using Varian CadPlan beam model

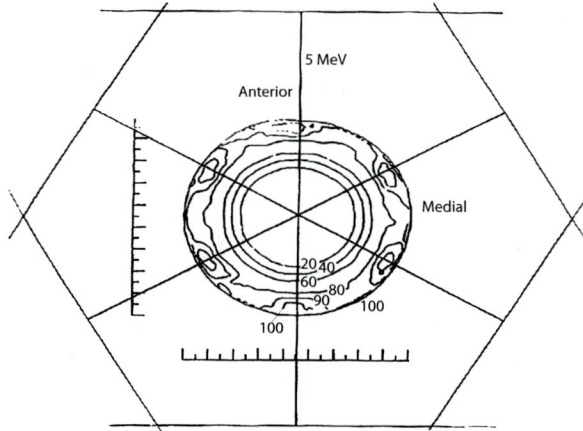

Fig. 7.30. Resultant dose distribution obtained for total limb irradiation using six equally spaced 17-cm-wide, 5-MeV electron fields on a 9-cm diameter phantom. [Reprinted with permission from WOODEN et al. (1996)]

resultant distribution shows that a surface dose of 90% of the average maximum is delivered compared with 70% for a single electron beam, and that the depth of the 90% isodose decreases to a depth of 8–10 mm versus 15 mm for a single beam.

7.7.6.4
Electron–Photon Field Matching

Often in clinical situations, a combination of electron and photon fields yields a dose distribution superior to that obtained using electron fields exclusively. Adjacent photon and electron fields are commonly used in treatment of the head and neck region and again, for postmastectomy chest wall situations. It is extremely challenging to treat extensive regions of the chest wall using electron fields alone. Figures 7.31 and 7.32 illustrate possible combinations of photon and electron fields to treat large regions of the chest wall. The use of combined photon and electron fields with moving junctions allows for the treatment of a large amount of chest wall with acceptable high-dose regions. This approach could be extended to treat the entire periphery of the chest wall if necessary.

Treatments of the head and neck region commonly use abutted photon and electron fields where right/left lateral 6-MV fields are used to treat the anterior neck, and electrons are used to treat the posterior neck nodes. A high-dose region is created at the junction location in the photon field and a corresponding low-dose region is created in

the electron field. Since the posterior neck electron fields are often treated at an extended distance of 110 cm SSD instead of the nominal 100 cm, changes in the electron field brought about by this increased distance cause the high-dose region in the photon fields to be larger while the coverage in the electron fields is decreased over what is achieved at the usual

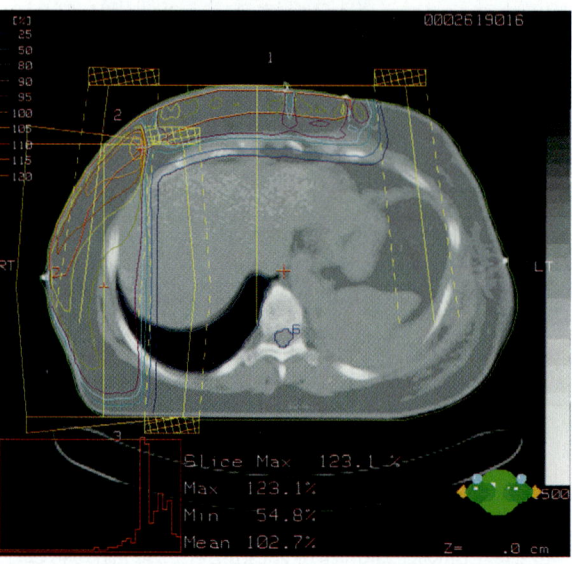

Fig. 7.31. Combination of tangential 6 MV photons with 12 MeV electrons to cover a large portion of the chest wall. Varian CadPlan beam models

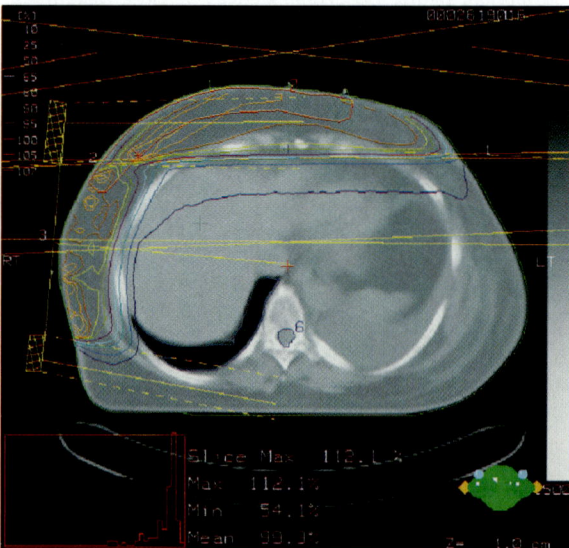

Fig. 7.32. An alternate combination of tangential 6 MV photons matched with 12 MeV electrons to treat a large portion of the chest wall. Varian CadPlan beam models

100-cm electron treatment distance. This effect is shown in Figure 7.33.

Total scalp treatments present a particularly challenging situation where the entire periphery of the scalp needs to be treated while sparing underlying brain tissue. Electron beams alone have been used to treat this area but involve very extensive field matching, special lead shielding, and junctioning techniques (TAPLEY 1976; ABLE et al. 1991). A simpler and dosimetrically superior technique using right/left lateral 6-MV photon fields to treat the rind of skin of the scalp while avoiding the brain plus matched electron fields to treat the lateral surface of the scalp was developed at the University of California, San Francisco (AKAZAWA 1989) and later modified by TUNG et al. (1993). Figure 7.34 shows the photon and electron field arrangement used for the technique. In the technique by TUNG et al., the outer edge of the electron field overlaps the inner edge of the photon field by 3 mm to account for the divergence of the contralateral 6-MV photon field whose central axis is placed approximately in the middle of the brain. The border of the photon field is placed initially 0.5 cm interior to the inner table of the skull. Midway through treatment, the junction between the fields is shifted by 1 cm toward the central axis to improve the dose homogeneity. To ensure that the dose to the scalp is maximized, a 6-mm wax bolus is used for both the photon and electron treatments.

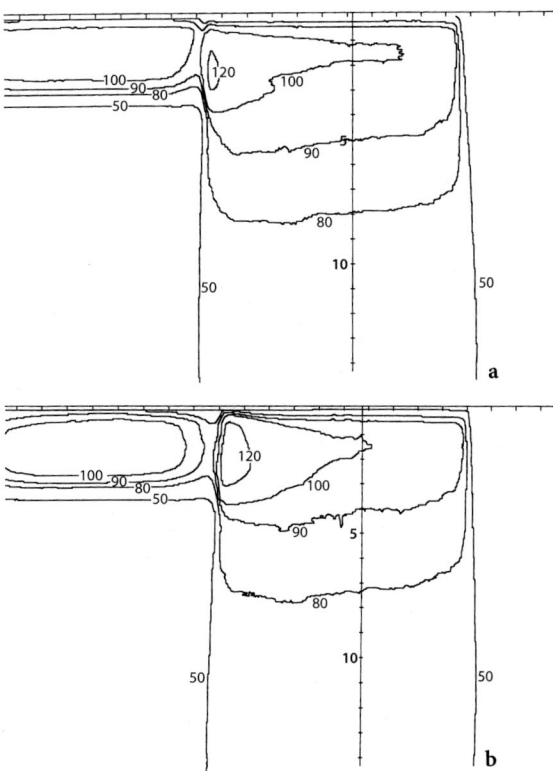

Fig. 7.33a,b. Composite isodose distribution created by abutting photon and electron fields. 9-MeV electron beam; field size=10×10 cm², 6-MV photon beam, SSD=100 cm. **a** Electron beam at standard SSD of 100 cm. **b** Electron beam at extended SSD of 120 cm. [Reprinted with permission from JOHNSON and KHAN (1994)]

7.7.7
Intracavitary Irradiation

7.7.7.1
Intraoral and Transvaginal Irradiation

Intracavitary radiation is performed for treatment of intraoral or transvaginal areas of the body. Additionally, IORT can be considered an intracavitary electron technique. Intracavitary electron irradiation is primarily used as a boost for particular sites offering the advantage of delivering a high dose to a specific and well-defined area and allowing the sparing of closely adjacent normal or sensitive tissue. In the case of intraoperative radiotherapy, the sensitive structures are exposed and then physically moved from the beam before delivery of the radiation. The use of intraoral cones for boost treatments has been described in the literature indicating a benefit for the treatment of oral lesions presenting in the floor of the mouth, tongue, soft palate, and retromolar

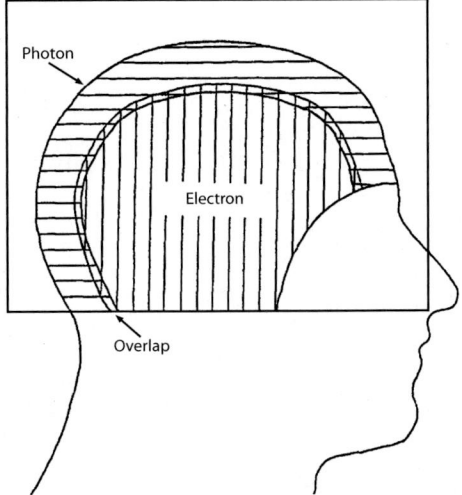

Fig. 7.34. Combination of right–left lateral photon fields with abutting electron fields to treat the entire scalp region. The overlap between the photon and electron fields is approximately 3 mm. Both the photon and electron fields have a common central axis placed approximately in the center of the brain. [Reprinted with permission from TUNG et al. (1993)]

trigone. Even though not a randomized prospective study, the data presented showed that intraoral cone electron beam boost technique was superior to interstitial implant and kilovoltage intraoral treatments for boosting early carcinoma of the tongue (WANG 1989, 1991).

For all intracavitary irradiation, specially designed treatment cones are required. In addition, an adapter to attach the cone to the linear accelerator has to be available which should incorporate a good system by which to visualize the area being treated. Figure 7.35 shows a commercially available intraoral/intravaginal cone system (Radiation Products Design, Inc., Albertville, MN). Cones are available with internal diameters from 1.9 cm to 9.5 cm and bevel angles from no angle to 60° angulation in steps of 15°. In clinical practice, the cone size and bevel angulation are chosen to provide adequate field coverage with the best contact between the end of the cone and the treated surface.

Special dosimetry for the electron cones has to be performed before initiating an intracavitary electron treatment program. The size of the opening of the adjustable linac jaws has a large impact on both the output and on the flatness of the electron field emanating from the end of the cone. A jaw size that yields a uniform field for all available cones is required to be set as the default size when the cone adapter is inserted into the treatment head. For each cone to be used clinically, isodose curves in the direction of the beam and perpendicular to the beam have to be obtained. These data are required to determine accurately the depth of penetration of the beam, the depth of the 90% isodose line, and to ensure that the treatment field is large enough to cover the target with a sufficient margin. Figure 7.36 shows 6-, 12-, and 16-MeV electron isodose curves for a 5-cm diameter cone designed for use in IORT. For comparison, isodose curves in the plane of the beam for a 5-cm diameter cone with a 0° and 22.5° bevel are shown. The depth of penetration for the angled beam is significantly different from the normally incident beam showing less penetration perpendicular to the skin surface (NYERICK et al. 1991).

7.7.7.2
Intraoperative Radiation Therapy

IORT is a very involved technique that requires a great deal of time and effort to create a dedicated program for large volumes of patients. The specifics of this program development are presented in great detail in AAPM Task Group Report 48 (PALTA et al. 1995). Either a dedicated linear accelerator room that can meet the requirements of operating room (OR) sterile conditions or new mobile electron linacs (ELLIS et al. 2000) that can be transported to a shielded OR need to be used. Because of the need to have the patient under anesthesia, efficient field positioning and convenient dosimetry need to be

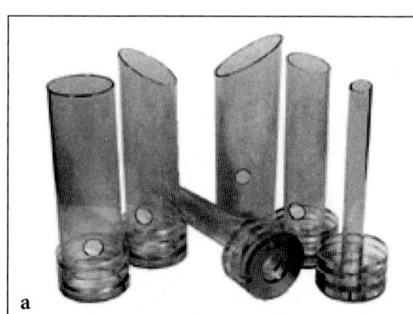
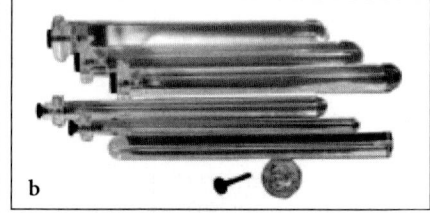
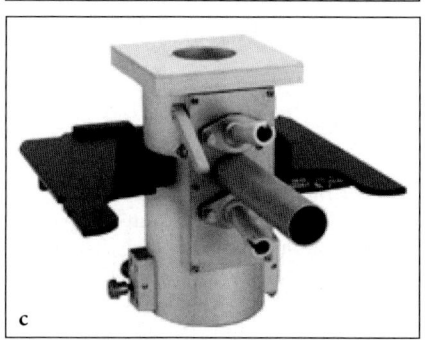

Fig. 7.35a–d. Intraoral/intravaginal cone system showing an assortment of cone sizes and bevel angles (**a**), vaginal obturators to aid in insertion of the cone (**b**), the docking assembly showing the periscopic visualization system (mirror, light pen holders, periscope) (**c**), and cone mated to the periscopic attachment system (**d**). (With permission of Radiation Products Design, Inc., Albertville, MN)

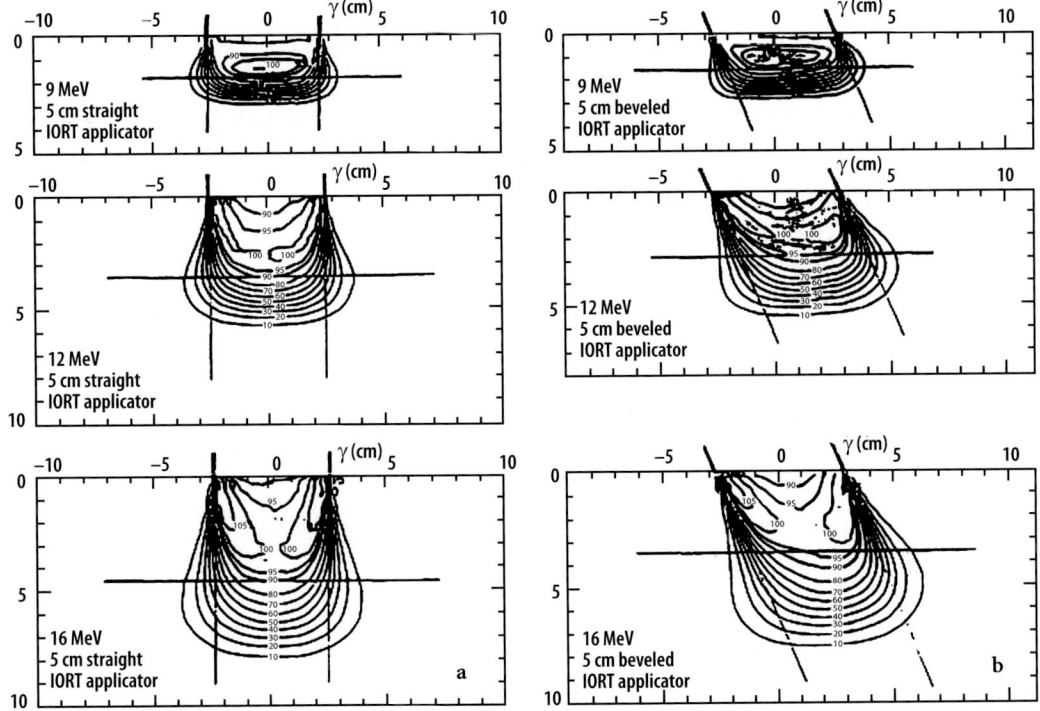

Fig. 7.36a,b. Isodose curves for a 5-cm inner diameter IORT cone for 6-, 12-, and 16-MeV electron beams angled perpendicular to the surface of the phantom (**a**) and at an angle of 22.5° (**b**) [Reprinted with permission from NYERICK et al. (1991)]

available at the time of the operation. Thus, isodose tables and graphs for each applicator and energy combination should be readily available in the OR, as well as output factors for each possible clinical condition. A clinical team needs to be assembled consisting of surgeons, radiation oncologists, physicists, therapists, and engineering and support staff. Each of these members must be thoroughly familiar with their role and how their role affects other members of the team. If a dedicated unit is not available in the OR suite, which is the usual situation, then the route of travel from the OR to the treatment unit must be chosen to minimize distance while ensuring maintenance of sterility, security, and patient confidentiality.

7.8
Special Electron Techniques

7.8.1
Electron Arc Irradiation

Electron arc therapy is employed in the treatment of postmastectomy chest wall and of other areas of the body such as the ribs and limbs. This technique can

provide an excellent dose distribution in these areas. It is most effective when a constant radius of curvature is present for the patient to be treated since the isocenter needs to be placed at a constant distance from the surface of the patient. Electron arc therapy is seldom employed at the University of Minnesota due to the difficulty of fabricating the necessary tertiary shielding, the amount of time to perform dosimetric verification of the technique, and competing techniques for covering large areas of the chest wall or limbs that are much easier to accomplish. In addition, the depth–dose distribution is such that the dose to the skin surface is significantly lower for electron arc treatments than for a single stationary electron beam. This is due to the "velocity effect" where a deeper point is exposed to the beam longer than a shallower point resulting in a higher dose at depth than for a static field. Because of this, a bolus is often required so that an adequate dose is delivered to the skin surface. Finally, treatment-planning computers lack the ability to perform electron arc calculations and the distribution has to be approximated using several stationary beams placed around the arc track or by direct measurement. An excellent and detailed description of all aspects of electron arc therapy is given in the AAPM 1990 Summer School Proceedings (LEAVITT et al. 1992).

For the electron arc technique, three levels of beam definition are required: the opening of the adjustable X-ray collimators, a secondary cerrobend insert placed below the adjustable collimators, and tertiary shielding placed on the patient's surface to define sharply the area of treatment. The adjustable X-ray collimator setting is usually determined by the linac manufacturer and it automatically selected when electron arc mode is selected. The secondary cerrobend insert is placed at a distance from the patient's surface so that it does not collide with the patient during the rotation of the gantry during arc treatment. Any field width can be used on the secondary cerrobend insert but geometric field widths of 4–8 cm at the isocenter are most suitable for clinical situations. It must be kept in mind that the smaller the field at isocenter, the lower the overall dose rate resulting in higher photon contamination. The radius of curvature usually decreases superiorly from the curvature at the isocenter. This decreased radius of curvature leads to a higher dose in this region. The width of the defined field using the secondary collimator can be tailored to make the dose more uniform throughout the treatment region taking these changes in radius of curvature into account.

Since the secondary cerrobend insert is far from the patient's skin surface, the dose falloff at the treatment field borders is gradual relative to ordinary static electron fields due to air scatter. To re-establish a sharper dose falloff at the ends of the treatment arc, the electron arc is extended approximately 15° beyond each end of the treatment arc. A tertiary shield is fabricated to not only sharpen the edges of the treatment area but also protect the uninvolved regions that would be irradiated by the extended arc. Figure 7.37 shows an example of an electron arc distribution both with and without a lead tertiary shield in place (KHAN et al. 1977). Typically, the shield is made from sheets of lead of the requisite thickness. Due to the large area to be covered, the thickness of the shield, and the extra shielding needed beyond the edge of the field, the shield is not only time consuming to produce but also very heavy to use.

Calibration of the dose rate for electron arc therapy can be done either by integrating stationary beam profiles or by direct measurement. Direct measurement using an ionization chamber in a cylindrical phantom of polystyrene (acrylic, or solid water) provides a direct means by which to determine the dose rate. Holes are drilled at d_{max} in the phantom to accommodate the chamber and standard correc-

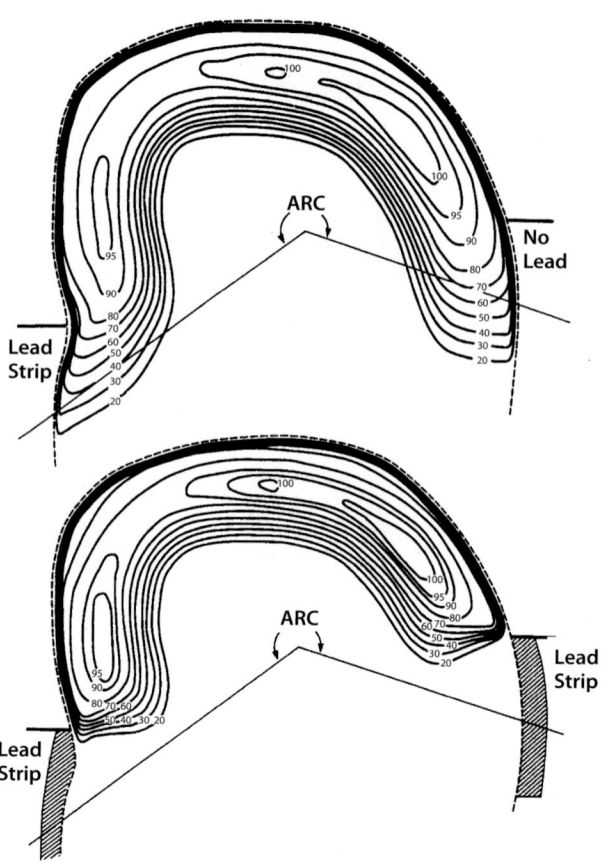

Fig. 7.37. Isodose distributions showing the difference in the sharpness at the ends of the electron arc when lead is in place. Data is taken in Rando phantom using 10 MeV electrons, an arc length of 236° and average radius of curvature of 10 cm. [Reprinted with permission from KHAN et al. (1977)]

tions are applied to convert the integrated charge into dose (KHAN 2003). Thermoluminescent dosimeters can be placed on the surface of the phantom to quantify the surface dose for the treatment. For dosimetry, the depth of the isocenter has to be the same as for the treatment, even though the radius of curvature of the phantom need only be approximately that of the patient (KHAN 1982).

7.8.2
Craniospinal Irradiation

Craniospinal irradiation is used to manage brain tumors that seed along the entire length of the cerebral spinal fluid. Medulloblastoma, malignant ependymoma, germinoma, and infratentorial glioblastoma are all candidates for this irradiation approach. Commonly employed techniques treat the patient in a prone position and use right/left lateral photon beams to treat the brain in addition to a posteriorly directed photon beam to treat the spinal cord. Replacement of the posterior photon field with a high-energy electron field can reduce greatly the exit dose to the upper thorax region, especially the heart, and the lower digestive tract. This is especially important for pediatric patients and results in reductions of both acute and late complications. Key challenges in the use of this technique involve matching of the right/left lateral photon fields with the posterior electron spine fields, selection of the proper electron energy to cover adequately the spinal canal all along its length, and the production of a posterior electron field of adequate length to treat the entire involved region. Techniques have been published addressing these concerns with consequent solutions to the above-stated problems (MAOR et al. 1985, 1986; ROBACK et al. 1997). The first two techniques using high-energy electrons employ conventional treatment distances for the posterior spine field of 110 cm and 115 cm SSD and use two adjacent electron fields if one field is not adequate. The technique of ROBACK et al. (1997) for production of a larger posterior electron field uses an extended SSD of approximately 120 cm. Special considerations in the application of this technique hinge on the change in the isodose distributions exhibited at the extended SSD. Figure 7.39 and 7.40 show the difference in the sharpness of the electron field both with and without tertiary collimation at the depths of D_{max} and 4 cm, respectively. This approach can be extended to larger treatment distances (140 cm SSD) providing that adequate dosimetry is performed at this distance.

Figure 7.38 shows the basic field arrangement for the M.D. Anderson technique (MAOR et al. 1985). The lateral photon fields are rotated through an angle θ to match the divergence of the posterior electron field. The central axis of the photon beams is placed as close to the junction region as possible to eliminate divergence in the superior–inferior direction. The superior field edge of electron field "e_1" is not moved during the treatment but the inferior border of the photon fields is shifted 9 mm to feather the junction location (positions y_1, y_2, and y_3). To achieve the most uniform dose per fraction in the region of the junction, one-third of the photon treatments are delivered with the inferior border of the two photon fields coincident with the electron field edge. The next one-third of the photon treatments are delivered with the edge of one photon field

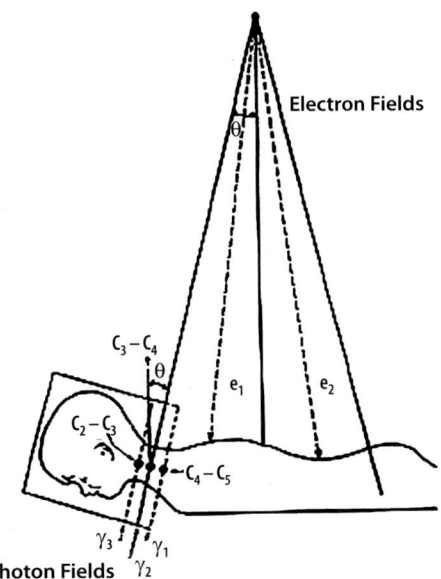

Fig. 7.38. Craniospinal field arrangement showing the prone patient treatment position and the arrangement of the right/left lateral photon fields and the posterior electron fields. Two electron fields are shown in the diagram but some patients are small enough such that one posterior electron field covers adequately the entire spine. The lateral photon fields are rotated through an angle θ to match the divergence of the posterior electron field. The superior field edge of electron field "e_1" is not moved during the treatment but the inferior border of the photon fields is shifted 9 mm to feather the junction location (positions y_1, y_2, and y_3). The central axis of the photon beams is placed as close to the junction region as possible to eliminate divergence in the superior–inferior direction. [Reprinted with permission from MAOR et al. (1985)]

moved 9 mm superior to the electron field edge and the edge of the second photon field moved 9 mm inferiorly to the electron field edge. The final one-third of the photon treatments are delivered with the edges of the photon fields reversed from their previous position. Specifically, the photon field that was shifted 9 mm superior to the electron field edge is now positioned 9 mm inferior to the electron field edge while the photon field edge that was 9 mm inferior to the electron field edge is now 9 mm superior to the electron field edge.

The overall length of the cord to be treated often exceeds the field size that can be covered using a 25×25 cm² cone at either 110 cm or 115 cm SSD. A small increase in overall field size can be accomplished by rotating the collimator 45° to produce a field size of approximately 30–35 cm in length. If the entire length of the cord cannot be covered in one electron field, then a second posterior field must be abutted to the inferior border of the first electron

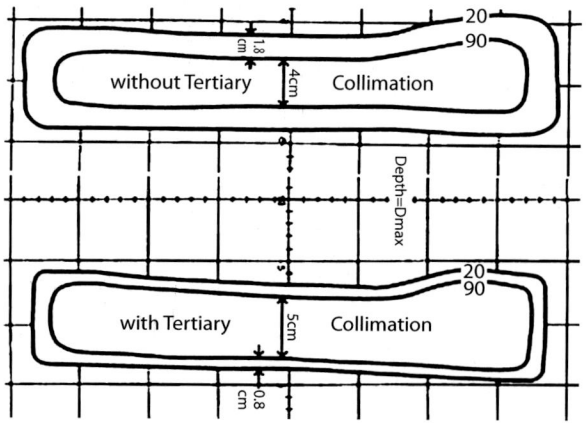

Fig. 7.39. Isodose curves demonstrating therapeutic and penumbra widths at the depth of dose maximum (d_{max}=2.5 cm). *Top illustrations* are for the field without tertiary collimation, while the *bottom* is with tertiary collimation. Data are for 16 MeV electrons at 120 cm SSD using Kodak XV2 film. [Reprinted with permission from Roback et al. (1997)]

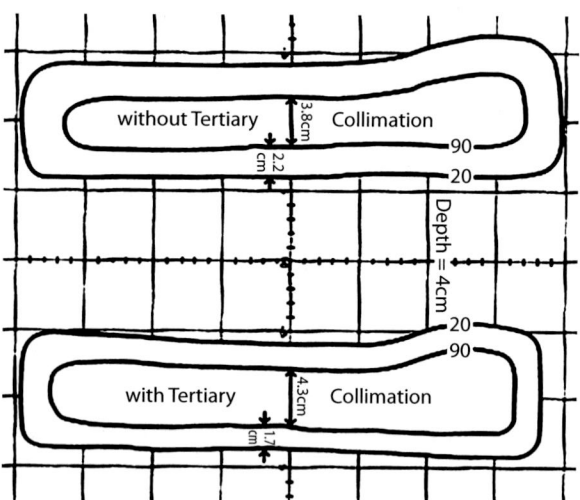

Fig. 7.40. Isodose curves demonstrating therapeutic and penumbra widths at the approximate deepest depth of a child's spinal cord (4 cm). *Top illustrations* are for the field without tertiary collimation, while the *bottom* is with tertiary collimation. Data are for 16 MeV electrons at 120 cm SSD using Kodak XV2 film. [Reprinted with permission from Roback et al. (1997)]

field. The addition of this second field requires that the couch be rotated 90° and that the angle of the two electron fields be rotated by an angle θ (Figure 7.38) to account for the divergence of each of these electron fields and to produce a common field edge (Maor et al. 1985).

A simulation of the patient is done to establish the treatment position and properly place the photon and electron fields and to provide documentation

for subsequent patient treatment. A lateral radiograph is taken to define the depth of the cord along its entire length and to show the changes in the SSD along the length of the cord. A computerized treatment plan of this sagittal plane can be done easily using this information.

The electron energy is selected so that the 90% isodose surface covers the target to be treated. The energy should be selected such that the 90% isodose should exceed the maximum depth of the cord by 7 mm – 4 mm to account for the increased absorption of bone and 3 mm for a margin of error to ensure coverage of the target. If the depth of the spinal cord or the SSD to the patient skin surface varies significantly, then a bolus can be added to the spinal cord to conform the 90% isodose surface to the anterior border of the cord. With modern 3D-treatment-planning computers, the overall plan can be calculated before treatment is begun.

7.8.3
Total Skin Electron Therapy

Total skin electron treatments are employed in the management of mycosis fungoides (Duvic et al. 2003). Numerous techniques for treating the entire skin surface using electron beams have been devised and each has its particular advantage. For all techniques, the objective is to deliver as uniform a dose to the entire skin surface as possible. This is quite a challenging goal considering the various surfaces and individual variations that may be encountered. Report 23 of the AAPM (American Association of Physicists in Medicine 1987) goes into great detail on the various techniques, the dosimetry, and the proper steps required to initiate successfully a total skin electron treatment program. At the University of Minnesota, many different total skin electron techniques have been developed (Sewchand et al. 1979; Gerbi et al. 1989). Our preferred technique is the modified Stanford technique, while others have been devised for patients unable to stand for the entire course of therapy. We use 9 MeV for our treatments and the high-dose rate total skin electron insert for our linac which automatically sets the standard linac jaws to 36×36 cm² and allows the unit to operate at between 800 and 900 monitor units per minute.

The first requirement for total skin electron treatments is a uniform electron field large enough to cover the entire patient in a standing position from head to foot and in the right to left direction. This is accomplished by treating the patient at an extended distance (410 cm), angling the beams superiorly and inferiorly (θ=±16.7°), and using a large sheet of plastic (3/8-inch thickness acrylic at 20 cm from the patient surface) to scatter the beam (Figure 7.41). This beam angulation not only produces a large treatment field but also limits the amount of photon contamination that is directed onto the patient. This is because the lateral scattering of the electrons at the patient surface extends beyond the edges of the diverging photon field. In addition, the acrylic plate decreases the energy of the beam from 9 MeV at the exit window to about 6 MeV at the surface of the patient. The scatter produced by the acrylic sheet aids in providing a more uniform dose around the periphery of the patient. The beam angulation of ±16.7° is particular to this accelerator, the treatment distance and the thickness and distance of the scatterer from the patient. The optimum angulations have to be determined for each specific set of treatment parameters.

Several different patient positions need to be used to ensure that the entire surface of the body is covered uniformly. This is accomplished using the six different patient positions indicated in Figure 7.42. Each of the positions is rotated at a 60° interval from the other. Only three of the six fields are treated per day to help expedite the treatment on a per day basis (Fig. 7.43). Thermoluminescent dosimeters are placed at multiple locations on the third and fourth treatment days to measure the dose uniformity at those locations. Published material (WEAVER et al. 1995; ANTOLAK et al. 1998) gives an indication of the amount of variation that should be expected for various measurement locations. The usual areas of low dose – the perineum, under breast tissue, the

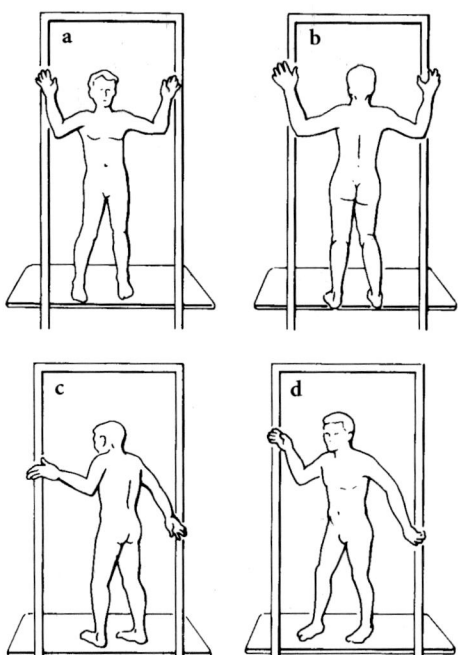

Fig. 7.42a–d. Four of the six standard treatment positions for the modified Stanford technique for total skin electron beam treatments. The anterior (a), left posterior oblique (c), and right posterior oblique (similar to C but for the right posterior) are treated on day 1, while the posterior (b), left anterior oblique (d), and the right anterior oblique (similar to d but for the right anterior) are treated on day 2

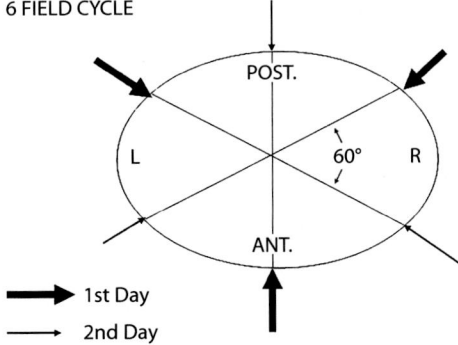

Fig. 7.43. The six field cycle for total skin electron treatments for the positions indicated in Fig. 7.42

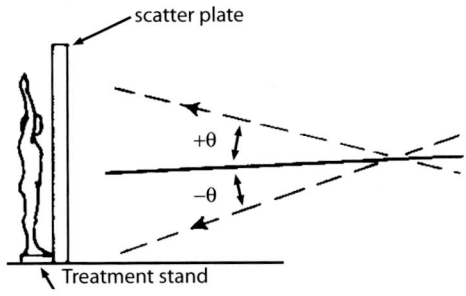

Fig. 7.41. Diagram illustrating total skin electron treatment position. Beam angulation provides a larger superior to inferior treatment field, along with extended treatment distance. The scatter plate (3/8-inch-thick acrylic) is placed approximately 20 cm from the patient surface while the treatment stand is designed to place the average-height patient at about the middle of the overall treatment field

top of the head – need to be boosted with separate electron fields to make up for the dose deficit experienced from the normal total skin treatment. In contrast, high-dose areas such as the finger tips, toes, and tops of the feet receive too much dose and need to be shielded for a large portion of the treatment. The fingers and toes receive a higher dose since they are irradiated by more than three of the six electron fields and their lack of thickness for the fingers.

Eye shields are very often used (the tungsten/aluminum shields described above) for the entire treatment to shield the eyes while treating the overlying eyelid. The eye is first anesthetized, a non-prescription contact lens is inserted to protect the cornea from the shield, and then the tungsten/aluminum eye shield is placed under the lid. Once the eye shield is in place, extreme caution and constant monitoring of the patient must be conducted to ensure that they do not lose their balance and fall from the treatment stand.

Total skin electron treatments involve a substantial amount of time and effort on the part of the department. Commissioning of the technique requires numerous hours from the physics staff while actual treatments require more than 30 minutes of linac time for the six field treatment to be completed. Once the boost fields are added, an hour of linac time can be consumed in the treatment of one patient. However, total skin electron beam therapy has been shown to be highly effective in the treatment of early-stage mycosis fungoides without adjuvant therapy. In addition, the management of relapses with local radiotherapy or using second total skin electron treatment is an effective means of treatment for this disease (HOPPE 2003; YSEBAERT et al. 2004).

References

Able CM, Mills MD, McNeese MD, Hogstrom KR (1991) Evaluation of a total scalp electron irradiation technique. Int J Radiat Oncol Biol Phys 21:1063–1072

Akazawa C (1989) Treatment of the scalp using photon and electron beams. Med Dosim. 14(2):129–131

Almond PR (1976) Radiation physics of electron beams. In: Tapley N (ed) Clinical application of the electron beam. John Wiley & Sons, New York

American Association of Physicists in Medicine (1987) American Association of Physicists in Medicine: Report 23. Total skin electron therapy: technique and dosimetry. American Institute of Physics, New York

American Association of Physicists in Medicine (1991) American Association of Physicists in Medicine: Task Group 25 Report. Clinical electron beam dosimetry. Med Phys 18:73–109

Antolak JA, Scrimger JW, Mah E (1992) Optimization of a cord shielding technique for electrons. Australas Phys Eng Sci Med 15:91–94

Antolak JA, Cundiff JH, Ha CS (1998) Utilization of thermoluminescent dosimetry in total skin electron beam radiotherapy of mycosis fungoides. Int J Radiat Oncol Biol Phys. 40:101–108

Archambeau JO, Forell B, Doria R, et al. (1981) Use of variable thickness bolus to control electron beam penetration in chest wall irradiation. Int J Radiat Oncol Biol Phys 7:835–842

Boone ML, Jardine JH, Wright AE, Tapley ND (1967) High-energy electron dose perturbations in regions of tissue heterogeneity. I. In vivo dosimetry. Radiology 88:1136–1145

Boone ML, Almond PR, Wright AE (1969) High-energy electron dose perturbations in regions of tissue heterogeneity. Ann N Y Acad Sci 161:214–232

Brahme A, Svensson H (1976) Specification of electron beam quality from the central-axis depth absorbed-dose distribution. Med Phys 3:95–102

Chobe R, McNeese M, Weber R, Fletcher GH (1988) Radiation therapy for carcinoma of the nasal vestibule. Otolaryngol Head Neck Surg 98:67–71

Dutreix J (1970) Dosimetry. In: Gil G, Gayarre G (eds) Symposium on high-energy electrons. General Directorate of Health, Madrid

Duvic M, Apisarnthanarax N, Cohen DS, Smith TL, Ha CS, Kurzrock R (2003) Analysis of long-term outcomes of combined modality therapy for cutaneous T-cell lymphoma. J Am Acad Dermatol. 49:35–49

Ekstrand KE, Dixon RL (1982) The problem of obliquely incident beams in electron-beam treatment planning. Med Phys 9:276–278

Ellis RJ, Nag S, Kinsella TJ (2000) Alternative techniques of intraoperative radiotherapy. Eur J Surg Oncol Nov 26 Suppl A:S25–S27

Gahbauer R, Landberg T, Chavaudra J, Dobbs J, et al. (2004) Prescribing, recording, and reporting electron beam therapy. J ICRU vol.4

Gerbi BJ, Khan FM, Deibel FC, Kim TH (1989) Total skin electron arc irradiation using a reclined patient position. Int J Radiat Oncol Biol Phys. 17(2):397–404

Giarratano JC, Duerkes RJ, Almond PR (1975) Lead shielding thickness for dose reduction of 7- to 28MeV electrons. Med Phys 2:336–337

Hogstrom KR (1991) Clinical electron beam dosimetry: basic dosimetry date. In: Purdy JA (ed) Advances in radiation oncology physics: dosimetry, treatment planning, and brachytherapy. American Institute of Physics, Inc., Woodbury, pp 390–429

Hogstrom KR (2004) Electron beam therapy: dosimetry, planning, and techniques. In: CA Perez, Brady LW, Halperin EC, Schmidt-Ullrich RK (eds) Principles and practice of radiation oncology. Lippincott Williams & Wilkins, Philadelphia

Hogstrom KR, Fields RS (1983) Use of CT in electron beam treatment planning: current and future development. In: Ling CC, Rogers CC, Morton RJ (eds) Computed tomography in radiation therapy. Raven, NY

Hoppe RT (2003) Mycosis fungoides: radiation therapy (review). Dermatol Ther 16:347–354

International Commission on Radiation Units and Measurements (1984) ICRU Report No. 35: radiation dosimetry: electron beams with energies between 1 and 50 MeV. International Commission on Radiation Units and Measurements, Washington, D.C.

International Commission on Radiation Units and Measurements (1993) ICRU Report 50: prescribing, recording, and reporting photon beam therapy. International Commission on Radiation Units and Measurements, Washington, D.C.

International Commission on Radiation Units and Measurements (1999) ICRU Report 62: prescribing, recording and reporting photon beam therapy (supplement to ICRU

Report 50). International Commission on Radiation Units and Measurements, Washington, D.C.

Johnson JM, Khan FM (1994) Dosimetric effects of abutting extended source to surface distance electron fields with photon fields in the treatment of head and neck cancers. Int J Radiat Oncol Biol Phys 28:741–747

Khan FM (1982) Calibration and treatment planning of electron beam arc therapy. In: Paliwal B (ed) Proceedings of the symposium on electron dosimetry and arc therapy. AAPM. American Institute of Physics, New York, p 249

Khan FM (1984) The physics of radiation therapy. Williams & Wilkins, Baltimore

Khan FM (2003) The physics of radiation therapy, 3rd edn. Williams & Wilkins, Baltimore

Khan FM, Moore VC, Levitt SH (1976) Field shaping in electron beam therapy. Br J Radiol 49:883

Khan FM, Fullerton GD, Lee JM, Moore VC, Levitt SH (1977) Physical aspects of electron-beam arc therapy. Radiology 124:497–500

Khan FM, Deibel FC, Soleimani-Meigooni A (1985) Obliquity correction for electron beams. Med Phys 12:749

Klevenhagen SC (1985) Physics of electron beam therapy. Adam Hilger, Ltd., Bristol

Klevenhagen SC, Lambert GD, Arbabi A (1982) Backscattering in electron beam therapy for energies between 3 and 35 MeV. Phys Med Biol 27:363–373

Kudchadker RJ, Hogstrom KR, Garden AS, McNeese MD, Boyd RA, Antolak JA (2002) Electron conformal radiotherapy using bolus and intensity modulation. Int J Radiat Oncol Biol Phys 53:1023–1037

Kudchadker RJ, Antolak JA, Morrison WH, Wong PF, Hogstrom KR (2003) Utilization of custom electron bolus in head and neck radiotherapy. J Appl Clin Med Phys A4:321–333

Lambert GD, Klevenhagen SC (1982) Penetration of backscattered electrons in polystyrene for energies between 1 and 25 MeV. Phys Med Biol 27:721–725

Leavitt DD, Stewart JR, Moeller JH, Earley L (1992) Electron beam arc therapy. In: Purdy JA (ed) Medical Physics Monograph 19, Advances in Radiation Oncology Physics: Dosimetry, Treatment Planning, and Brachytherapy. American Institute of Physics, Inc. Woodbury, NY, p. 430ff

Levitt SH and Tapley N duV (1992) Technological basis of radiation therapy: practical clinical applications. Levitt SH, Khan FM, Potish RA (eds) 2nd edn. Lea & Febiger, Philadelphia

Low DA, Starkschall G, Bujnowski SW, Wang LL, Hogstrom KR (1992) Electron bolus design for radiotherapy treatment planning: Bolus design algorithms. Med Phys 19:115–124

Maor MH, Fields RS, Hogstrom KR, van Eys J (1985) Improving the therapeutic ratio of craniospinal irradiation in medulloblastoma. Int J Radiat Oncol Biol Phys 11(4):687–697

Maor MH, Hogstrom KR, Fields RS, et al. (1986) Newer approaches to cerebrospinal irradiation in pediatric brain tumors. In: Brooks BF (ed) Malignant tumors of childhood. The University of Texas Press, Austin, pp 245–254

McNeese MD (1989) Cancer Bulletin 41:88

Meyer JA, Palta JR, Hogstrom KR (1984) Demonstration of relatively new electron dosimetry measurement techniques on the Mevatron 80. Med Phys 11:670–677

Nyerick CE, Ochran TG, Boyer AL, Hogstrom KR (1991) Dosimetry characteristics of metallic cones for intraoperative radiotherapy. Int J Radiat Oncol Biol Phys 21:501–510

Palta JR, Biggs PJ. Hazle JD. Huq MS, Dahl RA. Ochran TG-. Soen

J. Dobelbower RR Jr. McCullough EC (1995) Intraoperative electron beam radiation therapy: technique, dosimetry, and dose specification: report of task force 48 of the Radiation Therapy Committee, American Association of Physicists in Medicine. Int J Radiat Oncol Biol Phys 33:725–746

Perkins GH, McNeese MD, Antolak JA, Buchholz TA, Strom EA, Hogstrom KR (2001) A custom three-dimensional electron bolus technique for optimization of postmastectomy irradiation. Int J Radiat Oncol Biol Phys 51:1142–1151

Pohlit W, Manegold KH (1976) Electron-beam dose distribution in inhomogeneous media. In: Kramer S, Suntharalingam N, Zinnenger GF, (eds) High energy photons and electrons. John Wiley & Sons, New York, pp 243

Purdy JA, Choi MC, Feldman A (1980) Lipowitz metal shielding thickness for dose reduction of 6-20 MeV electrons. Med Phys 7(3):251–253

Roback DM, Johnson JM, Khan FM, Engeler GP, McGuire WA (1997) The use of tertiary collimation for spinal irradiation with extended SSD electron fields. Int J Radiat Oncol Biol Phys 37(5):1187–1192

Rustgi SN (1986) Dose distribution under external eye shields for high energy electrons. Int J Radiat Oncol Biol Phys 12(1):141–144

Sewchand W, Khan FM, Williamson J (1979) Total-body superficial electron-beam therapy using a multiple-field pendulum-arc technique. Radiology 130:493–498

Shiu AS, Tung SS, Gastorf RJ, Hogstrom KR, Morrison WH, Peters LJ (1996) Dosimetric evaluation of lead and tungsten eye shields in electron beam treatment. Int J Radiat Oncol Biol Phys 35:599–604

Strydom W, Parker W, Olivares M (2003) Electron beams: physical and clinical aspects. In: Podgorsak EB (ed) Review of radiation oncology physics: a handbook for teachers and students. International Atomic Energy Agency, Vienna

Tapley N duV (1976) Clinical applications of the electron beam. John Wiley & Sons, New York

Tung SS, Shiu AS, Starkschall G, Morrison WH, Hogstrom KR (1993) Dosimetric evaluation of total scalp irradiation using a lateral electron–photon technique. Int J Radiat Oncol Biol Phys 27:153–160

Wang CC (1989) Radiotherapeutic management and results of T1N0, T2N0 carcinoma of the oral tongue: evaluation of boost techniques. Int J Radiat Oncol Biol Phys 17:287–291

Wang CC (1991) Intraoral cone for carcinoma of the oral cavity. In: Vaeth JM, Meyer JL (eds) Frontiers of radiation therapy and oncology, vol. 25: the role of high energy electrons in the treatment of cancer. Karger AG, Basel. pp 128–131

Weaver RD, Gerbi BJ, Dusenbery KE (1995) Evaluation of dose variation during total skin electron irradiation using thermoluminescent dosimeters. Int J Radiat Oncol Biol Phys 33:475–478

Weaver RD, Gerbi BJ, Dusenbery KE (1998) Evaluation of eye shields made of tungsten and aluminum in high-energy electron beams. Int J Radiat Oncol Biol Phys 41:233–237

Wooden KK, Hogstrom KR, Blum P, Gastorf RJ, Cox JD (1996) Whole-limb irradiation of the lower calf using a six-field electron technique. Med Dosim 21:211–218

Ysebaert L, Truc G, Dalac S, Lambert D, Petrella T, Barillot I, Naudy S, Horiot JC, Maingon P (2004) Ultimate results of radiation therapy for T1-T2 mycosis fungoides (including reirradiation). Int J Radiat Oncol Biol Phys 58:1128–1134

Zhu TC, Das IJ, Bjärngard BE (2001) Characteristics of bremsstrahlung in electron beams. Med Phys 8:352

8 Treatment Aids for External Beam Radiotherapy

Eric Klein, Sasa Mutic, and James A. Purdy

CONTENTS

8.1 Introduction *167*
8.2 Patient Registration Systems *167*
8.3 Immobilization *168*
8.3.1 Head and Neck Immobilization *168*
8.3.2 Body Immobilization *169*
8.4 Collimation *170*
8.4.1 Low Temperature Melting Alloy Blocks *172*
8.5 Compensation Systems *172*
8.5.1 Physical Wedges *173*
8.5.2 Dynamic Wedging *173*
8.5.3 Compensating Systems *173*
8.5.4 Bolus *173*
8.5.5 External and Internal Shields *175*
8.5.6 Treatment Devices for Special Procedures *175*
8.5.6.1 Total Body Irradiation *175*
8.5.6.2 Total Skin Electron Therapy *176*
References *177*

8.1 Introduction

Patient positioning for radiotherapy is one of the most important components of the entire planning and treatment process. According to the International Commission on Radiation Units and Measurements (ICRU) Report 24 (ICRU, 1993), the planning volume must be irradiated uniformly and accurately within 5% for adequate tumor control. With the advent of conformal therapy and more recently advancement of intensely modulated radiation therapy (IMRT) and image-guided radiation therapy (IGRT), doses are being escalated to control tumors often near critical organs. With increased accuracy of treatment planning systems to plan higher doses and quality assurance to ensure that the mapped doses are accurate, patient positioning becomes even more crucial. The purpose of this chapter is to review devices and processes that allow for excellent accuracy and reproducibility for both intra- and inter-treatment stability. We will first review methods of patient registration, moving from simulation devices such as computed tomography (CT) scanners to treatment. We will also review beam shaping and modifying devices such as wedges and compensating filters. We will also examine external immobilization systems and devices for external and internal shielding. Finally, we will finish with methods for treatment aids for special procedures such as total body photon irradiation.

8.2 Patient Registration Systems

It is vital to have accurate methods for relocating patients from CT simulator or conventional simulators to the treatment machine. Although the treatment planning system is key to migrating the patient information between simulation and treatment, it is imperative that there be a well-understood reference system and method of shifting to locate the isocenter, as prescribed by treatment planning. Most modern-day immobilization systems, and for that matter treatment tables, work in concert so that the patient is aligned within a confined system for scanning and simulation, and more importantly these devices allow positioning of the patient in the same location on the treatment table for subsequent treatments. The key is to have the immobilization systems lock onto the treatment table by an interface device. Once the immobilization device has been locked ("registered") into an ideal position, the patient then lies in the immobilization device – whether it is a body cast or a mask system – and also becomes registered to the treatment table (Fig. 8.1).

E. KLEIN, MS; S. MUTIC, MS
Department of Radiation Oncology, Washington University School of Medicine, St. Louis, Missouri 63110, USA
J. A. PURDY, PhD
Professor and Vice Chairman, Department of Radiation Oncology, Chief, Physics Section, UC Davis Medical Center, 4501 X Street, Suite G140, Sacramento, CA 95817, USA

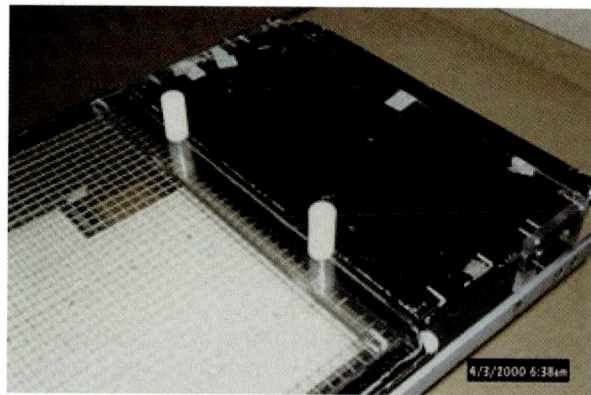
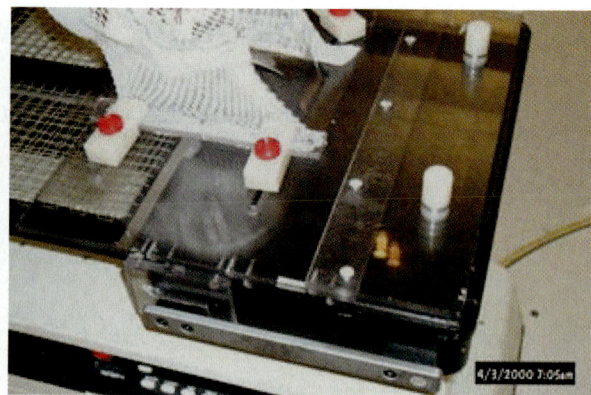

Fig. 8.1. a Photograph of platform that functions to register the immobilization device to the treatment table. The device locks onto the table in a fixed position and has acrylic posts that join the body cast to the platform. **b** Photograph of interface platform as appropriated to connect a head and neck immobilization system. **c** Photograph of acrylic posts ready to join the body cast that has been formed around the posts before scanning

The end result is that there is now a set of coordinates from the simulation that correlates to a patient positioning reference point (Fig. 8.2) and later to the treatment isocenter. During the treatment planning process, there are instructions on how to move from the reference point to the isocenter, if the reference point itself is not the isocenter. Care must be taken to understand the directionality of the movements to move from reference point to isocenter as the coordinate systems of different treatment planning and delivery devices are not necessarily uniform (FRAASS et al. 1998).

8.3
Immobilization

Immobilization systems ensure that patient position is reproducible and that the patient does not move during the treatment. It has been shown that localization errors can be caused by patient motion (HAUS and MARKS 1973). For this reason, immobilization systems are designed with specific sites in mind.

8.3.1
Head and Neck Immobilization

Most head and neck immobilization is performed using thermoplastic masks (KLEIN and PURDY 1993). These masks start as plastic sheets, which after placing in warm water become pliable and are placed over the patient's face and head. In turn they are molded to the face surface while simultaneously adhering to locking devices peripheral to the head that are used daily to lock the mask in place (Fig. 8.3). The advantages of the thermoplastic masks are that they avoid having to demarcate the patient's face with reference points or setup marks. The potential disadvantage is a potential increase in surface dose. However, depending on the size of the field and location, the area of treatment can be cut out from the mask. As previously described, these mask systems can be configured so they are placed daily with an interface plate that is placed on the treatment table at exactly the same location. Mask systems must be evaluated as they have been known to shrink or stretch to a degree that effects reproducibility. Another device used for head and neck immobilization is a bite block. The bite block

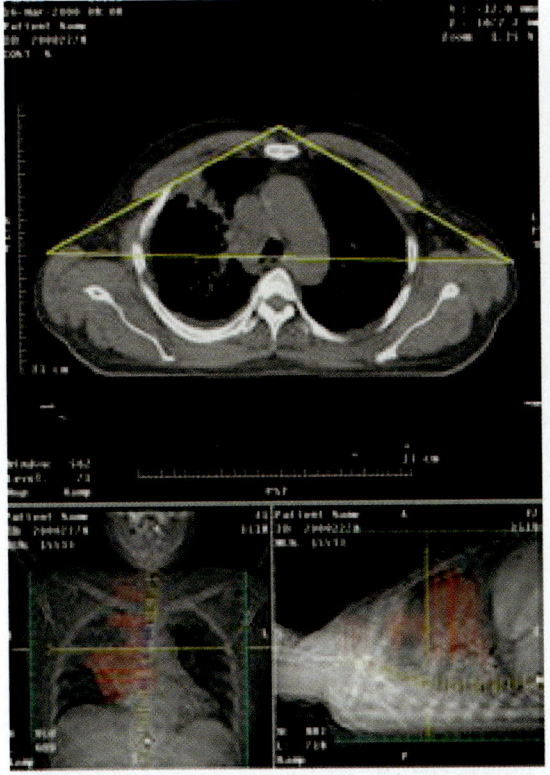

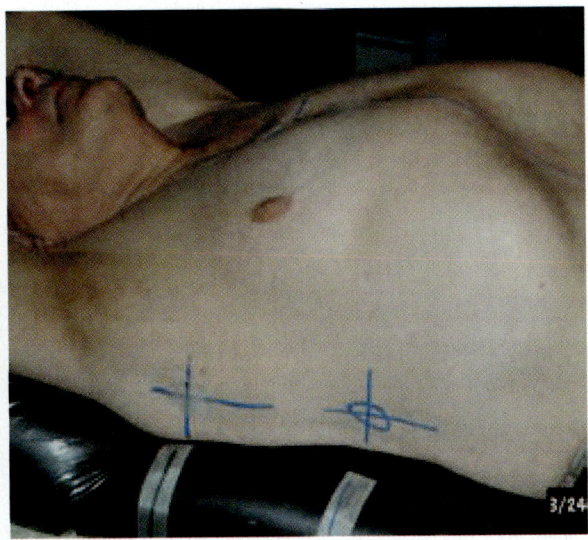

Fig. 8.2. a Screen captures of multiple planes indicating fiducial reference points for a 3-point setup. b Patient with external fiducial marks lying in a body cast that has corresponding demarcations

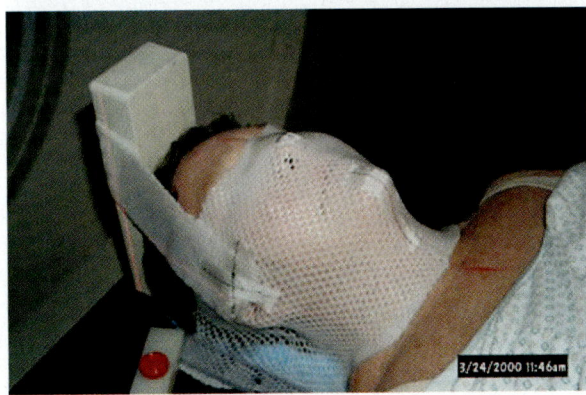

Fig. 8.3. Patient within head and neck immobilization system. There is a solid foam piece placed behind the patient's head to strengthen support

consists of a mold that is configured for the patient to latch their teeth onto. The advantage of bite block is that there is no buildup material on the patient's surface. However, not all patients may be able to tolerate a device in their mouth for a long period of time for setup and treatment, especially after a few weeks of treatments and radiation effects in the mouth. Finally, there is an option for treating patients prone in a prone pillow (Fig. 8.4), which is

a foam device that can be placed on the treatment table in exactly the same place everyday. These foam devices have openings for the patient's eyes, nose, and mouth. They are quite comfortable and allow air access. Unfortunately, they are not as rigid as the mask or bite block systems. One key to these systems is that an ideal head position must to be chosen with future thought on how treatment planning might be affected. Therefore, neutral positions or extended chin positions must be considered before the patient's mask or bite block is made. For head and neck immobilization systems custom materials may be used to aid immobilization. The backing material behind the patient's head could be custom polyethylene foam if desired. It could also be standard head and neck supports that shape to the head with an option of arching in order to create a neutral or extended head position.

8.3.2
Body Immobilization

Body immobilization systems, which can be used for treating the upper or lower body, come in two main forms. The most common form is the custom poly-

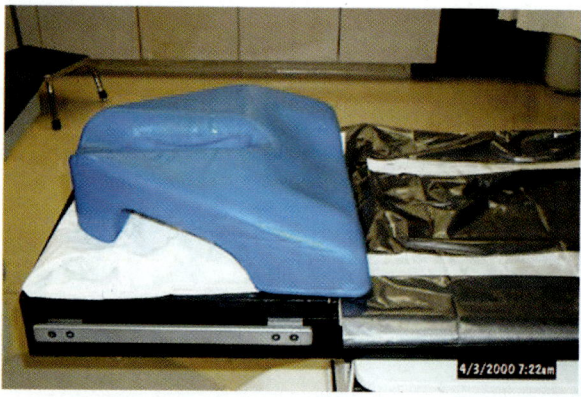

Fig. 8.4. Photograph of prone pillow device used for immobilization

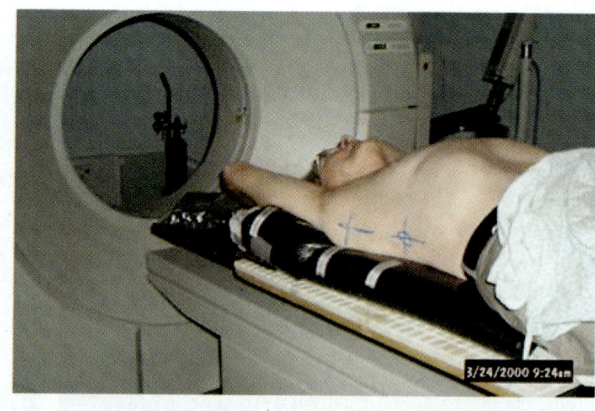

Fig. 8.5. Patient within registered body cast on computed tomography (CT) scanner table. Patient's fiducial marks were made by laser projections from lasers within CT scanner room

ethylene foam with the patient lying either supine or prone (BENTEL et al. 1997). Chemicals are mixed in liquid form, filled into a bag surrounding the patient, and as the chemicals react cooling takes place. The result is that the polyethylene foam forms around the patient rising up along the sides and in between legs and arms according to how the patient is positioned (Fig. 8.5). These custom foams are registered to the treatment table or simulation table with an interface device to allow for the patient registration. The other body mold type is the vacuum form body immobilization system. In this scenario, the patient lies amongst a bag filled with loose plastic mini-spheres. A vacuum is applied that removes the air allowing the remaining mini-spheres to collapse with the result of a taut bag custom fitted for the patient. Though this system is environmentally cleaner to work with, one disadvantage is that it does not allow for removal of sections of the body mould to allow visualization of beam projection on patient skin for some treatment sites (i.e., breast).

In the case of the foam or vacuum body forms, Styrofoam wedges can be placed to help angle the patient, such as the case for breast cancer where the chest wall is desired to be parallel with the treatment table. Other devices for immobilization include angle boards, such as breast boards, that can be used to elevate the patient and articulate arms as desired for treatment. Finally, there may also be the utility for compression devices for patients lying prone (Fig. 8.6) where the small bowel can be pushed outside of the treatment area. Another alternative to allow small bowel out of the treatment field is to allow the pannus to fall into an opening built within a foam cradle (DAS et al. 1997) or Styrofoam block configuration.

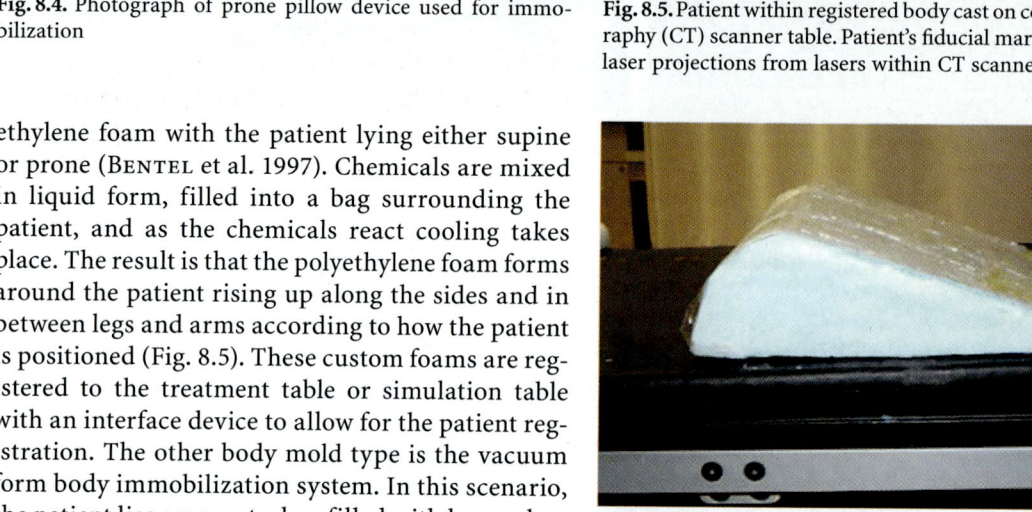

Fig. 8.6. Styrofoam wedge used for prone patients treated for gastrointestinal (GI) cancers. The wedge is designed to "push" bowel superiorly

One additional option for patient position and immobilization is for upright treatments in a treatment chair. Investigators (KLEIN et al. 1995b) have demonstrated advantages of upright treatments for Hodgkin's disease (mantle irradiation) in order to shrink the mediastinum and increase the width of lung blocking. Commercial treatment chairs are available to facilitate not only mantle irradiation, but also that of thorax and head & neck (Fig. 8.7).

8.4
Collimation

There are three main collimation systems available in radiotherapy. The first and most obvious is the collimating jaws in the treatment head. Historically, these jaws moved in two independent planes and

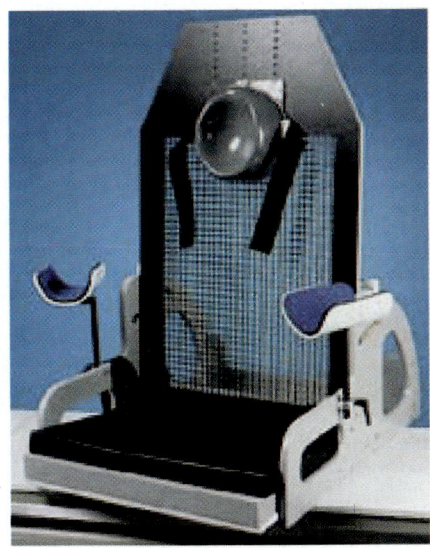

Fig. 8.7. Treatment-chair photograph, provided by Med-Tec, Inc.; facilitates upright treatments

had little to do with tertiary collimation. However, modern accelerators have integrated multileaf collimation (MLC) within the gantry heads, which in some cases have replaced jaws systems. Therefore, independent collimating jaws (SLESSINGER et al. 1993) and MLC must be considered as complete collimation systems. One system for jaws and multi-collimation is the tertiary multi-collimator that is installed below the independent collimating jaws. In this system, the collimating jaws move in either the longitudinal plane or the transverse plane in the non-rotated position. An example of this is the Varian collimating jaw system that places the Y-jaws longitudinally and the X-jaws transversely (KLEIN et al. 1995a). The Y-jaws, which are located closer to the beam target, are smaller and have a greater range where a jaw can move pass the central axis by as much as 10 cm. The X-jaws, which are below the Y-jaws and closer to the patient, have a little less flexibility and can move 2 cm past the isocenter. Again, all four jaws have the ability to move independently. In the upper jaw replacement system, whereby the multi-collimator is the uppermost collimating system close to the target, there are backup diaphragms that work in concert with the multileaf collimator. This system is found in the Elekta machine, where the leaves are placed closer to the target and backup diaphragms are located below the multi-collimating jaws (JORDAN and WILLIAMS 1994). Finally, in the lower jaw replacement system, such as the Siemens machine, there are independent jaws that move along the longitudinal plane and the

lower jaws are replaced by a MLC system that moves along the X (or transverse) plane (DAS et al. 1998). Typically, the collimating jaws found in the Varian System and the Siemens System, which are full collimating jaws, are considered to have transmissions on the order of six half-value layers (HVL) as they are made of tungsten. As the collimating jaws in all systems can move independently, they also have the ability to beam split the field if the jaws are placed along the central axis plane. This is important for matching adjacent treatment fields. The MLC systems found within these three different types of accelerators have similarities and also unique attributes. In the upper jaw replacement system as found in Elekta, the leaves move perpendicular to the central axis and are curve-ended. They have the ability to create a custom field shape up to 40×40 cm, although the irregularity of the field is somewhat limited as the leaves do not interdigitate. In the Varian tertiary system that falls below the collimating jaws (Fig. 8.8), there are a few varieties. The most popular system is the 120 leaf MLC system, for which the central 20-cm area of the leaves is configured with 5-mm leaf widths projected at the isocenter plane. The regions outside of the central 20 cm are collimated with 1-cm wide leaves. The width of the leaves is important as it allows the field shapes either to be smooth, as in the case of the 5-mm leaves, or to have larger stepping, as in the case of the 1-cm leaves. The Varian leaves move perpendicular to the central axis and therefore do not purely follow beam divergent as in the case of collimating jaws. Varian also supplies tertiary collimation systems that are 80-leaf systems where the leaf projection width is 1 cm for all leaves. The maximum field width that can be configured for a shape treatment field for Varian System is 29 cm wide due

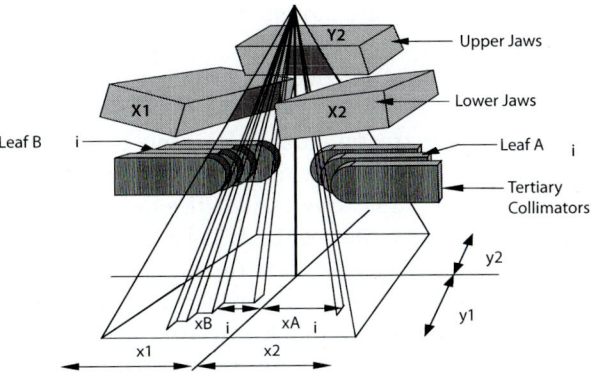

Fig. 8.8. Graphic of tertiary multileaf collimation (MLC) system in configuration with independent jaws. Leaf-span projections are noted by X and Y dimensions, with limits noted

to the fact that they are limited on these tertiary carriage rails, limiting again the full range of the leaves themselves. Finally, the Siemens system is a lower jaw replacement and has full 40-cm capability. These leaves also move in a pendulum fashion similar to the collimating jaws and therefore do follow beam divergence. They do however project to a 1-cm width across for each leaf. Beyond this, there are also additional tertiary multi-collimation systems such as those provided by BrainLab that have very fine resolutions as small as 3 mm and are idealized for stereotactic radiosurgery or radiotherapy (XIA et al. 1999). Most MLC systems made of tungsten possess 5.5 HVL of absorption.

Fig. 8.9. Photograph displaying molten alloy being poured into the form of shaped foam for purposes of casting a shaped field block

8.4.1
Low Temperature Melting Alloy Blocks

Use of Cerrobend shielding block, otherwise known as Lipowitz's metal, was first introduced in the 1960s (POWERS et al. 1973). This metal consists of bismuth, lead, tin, and cadmium. It possesses a density on the order of 9.4 g/cm^3. As it is a molten metal alloy, it is stored in solid or liquid form and can be poured into custom-crafted Styrofoam molds, which are cut out to allow molding of the eventual shaped alloy block (Fig. 8.9). The Styrofoam is cut to follow beam divergence so that the resulting collimating block will also follow beam divergence. Not only can these blocks be used to shape the outskirts of the field for a particular irregular shaped field, but also can be used to create internal shields, which is something that other collimating systems (jaws and multileaf collimators) cannot do. These blocks are typically constructed to have heights of 7.5 cm and a transmission of 5 HVL. Once the blocks have been cast and cooled, they are mounted onto trays known as blocking or shadow trays that sit below the accelerator head in a particular slotted tray location. This has the disadvantage of removing some clearance from the treatment head in terms of potential collision with patient or table.

There are some obvious differences between the two systems. Most notably is that the position of the collimating jaws with the MLC is appropriated by computer control. Meanwhile, the alloy block system on the block tray requires manual placement of the blocks onto the trays. For each treatment field, the therapist must enter the room to place the correct block. With the promotion of dose escalation requiring more treatment fields to

be used for patients, the low melting alloy block can reduce efficiency. In addition, the construction of the blocks including its ergonomic and environmental problems makes this system less desirable. However, one should not dismiss the fact that the Cerrobend blocks can be cut exactly to the shape of the field, where the MLC systems are limited according to the step size. This is further limited by the fact that collimator rotation may orientate the leaf direction in a non-desirable direction.

8.5
Compensation Systems

Photon beams leave the treatment machine homogeneously. There may be reason to change the fluence of the beam before it reaches the patient. The two main reasons for compensation are missing tissue and beam intersections. In the case of missing tissue, if the desire is to deliver a homogeneous plane of dose at depth, an irregular surface poses a dilemma that can be easily compensated with wedges or compensating filters. If the missing tissue is restricted to one plane, the wedges work ideally. However, if it involves missing tissue in two planes, then it is desired to have customized compensators. Often treatment planning leads to the use of wedges. Depending on the treatment technique, wedges counteract high dose regions. due to beams intersecting. We will now discuss the various systems.

8.5.1
Physical Wedges

Physical wedges come in the form of either tertiary wedges (SEWCHAND et al. 1978) or wedges built-in within the treatment head. In the case of tertiary wedges, such as the Varian System, the wedges are configured for four different angles (15°, 30°, 45°, 60°) and are limited in the wedge plane to field sizes of either 20 cm or 15 cm – although in the non-wedge direction the field limit is 40 cm. The Varian physical wedges may be placed in any four directions, thereby placing the thick compensation known as the 'heel' or thin compensation known as the 'toe' along the desired plane. The other type of physical wedge is an internal wedge such as in the case of the Elekta or the Siemens System. These are limited to one plane of heel to toe direction. In the case of the Elekta System, the maximum field size in the wedge direction is a 30-cm symmetric field (PETTI and SIDDON 1985). Although these wedges are size limited they are more efficient as they do not require daily placement by therapist.

8.5.2
Dynamic Wedging

There are two versions of dynamic wedges, one being the Varian enhanced dynamic wedge (KLEIN et al. 1998), the other the Siemens virtual wedge. The Varian enhanced dynamic wedge allows nominal wedge angles of 10, 15, 20, 25, 30, 45, or 60°. The field size is 30 cm, although asymmetric width is 20 cm toward the heel and 10 cm toward the toe with 40 cm in the non-wedge plane. These wedges are derived using a collimating jaw that moves while the beam is on, thereby shrinking the field during treatment (Fig. 8.10). This desired gradient of fluence results in a wedge-shaped distribution. The advantage of using this virtual type of wedging rather than physical wedging is that the beam energy spectrum is minimally affected, and that the therapists are not manually placing the wedges daily. The downfall is that the wedging direction is restricted to one plane in direction along the moving jaw.

8.5.3
Compensating Systems

Physical compensators, which are configured depending on missing tissue, and/or can be designed by a treatment planning system, have the ability to compensate along two planes and not necessarily in a fixed gradient (ELLIS et al. 1959). For example, a concavity or defect may be uniquely compensated for. This is quite desirable as in the case of breast cancer where compensation is often desired in the anterior/posterior and also in the superior/inferior aspects surrounding the breast tissue (Fig. 8.11).

8.5.4
Bolus

The desire to pull doses close to the surface for either photon or electron beams is facilitated using

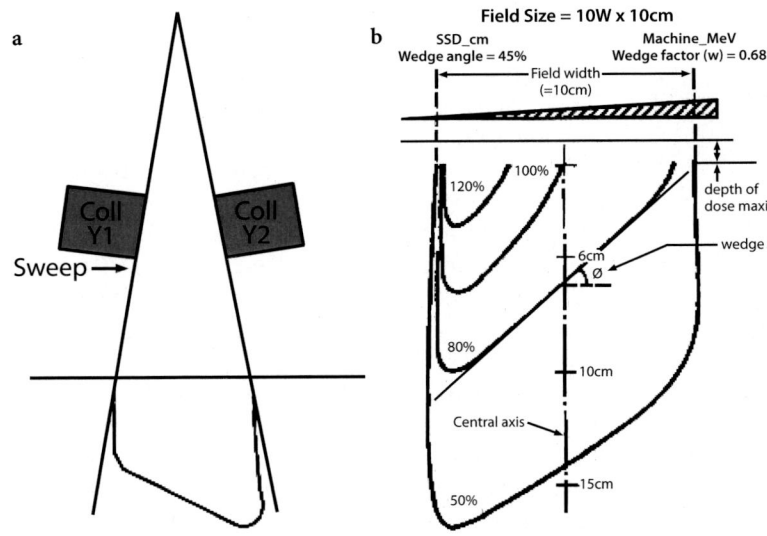

Fig. 8.10. a Diagram depicting independent jaw motion in order to produce enhanced dynamic wedge field on Varian treatment machines. **b** Depiction of wedge-shaped isodoses in reference to wedge delivery, where the wedge angle is defined by the isodose tilt at a 5-cm depth

a

b

Fig. 8.11. a Photograph of "Ellis" type-two dimensional filter constructed of brass and aluminum filters according to a map of missing tissue. **b** Photograph of a smooth milled two dimensional filter constructed of Cerrobend metal alloy according to a map of missing tissue

tissue equivalent materials known as bolus. The most common bolus is 'layer', which is used over the entire portal (Fig. 8.12). These bolus materials are commercially available and have attributes, such as physical and electron density, very close to that of tissue. They are available in different sheet sizes and thicknesses. The thickness used depends on the dosimetry. For example, use of a high energy photon field, where dose needs to increase to underlying tissues, a 1-cm bolus might be used. A 5-mm-thick bolus may be used for tangential breast irradiation for inflammatory breast disease with the use of 6-MV photons. These sheets can not only be cut into layers that encompass the entire treatment area of a

Fig. 8.12. Photograph of commercially available, tissue equivalent layer bolus

patient's surface, but can also be cut into narrower and custom shapes. A custom bolus may be used for a particular patient in order to build up the dose in a certain location. It may be used to fill in cavities or particular region of interest where the dose is desired to be pulled toward the surface. For tissue sparing, a custom bolus may be used in a portion of the field. This is very typical, for example with breast irradiation where a bolus is used to enhance the dose to the scar region. A bolus can also be used for electron beams to pull up the surface dose, as there is minimal (but not negligible) skin sparing for electron beams. However, more importantly, for electron beams, is the desire to use a bolus to pull up dose so that underlying tissue can be further spared.

Bolus material does not necessarily have to be purchased commercially, but can be in-house materials, such as paraffin wax. Use of custom wax is ideal for filling in cavities, such as nasal cavities or ear canals. However, wax can also be used to even surfaces, such as surgical cavities. This is especially used for en-face electron beams where cavities and surface irregularities can distort the isodose distribution significantly. Finally, there is also the potential for using non-tissue equivalent materials as a bolus in order to reduce the amount of physical thickness of the bolus material. Materials such as gypsum have been shown to be fully effective without introducing very high atomic number scattering that would unnecessarily enhance surface dose.

8.5.5
External and Internal Shields

There are many devices that have been developed either in-house or commercially for reduction of dose by either peripheral or internal means. Two of the more common external shields used are for treatment of pregnant patients and for breast irradiation. If it is determined that a pregnant patient must be treated, there are shields that have been designed that situate over the patient's stomach, ideally where the fetus is located at a given time. These shields are most often portable with wheels and can be positioned in place and ideally can be used independent of gantry angle. The AAPM Task Group 36 Report describes some of these external shields. One of the most important aspects of such external shields is the patient's safety and ideally these devices should be made as solid, uniform pieces (STOVALL et al. 1995; MUTIC and KLEIN 2000). For breast irradiation, a portable shield can be wheeled into place to spare the contralateral breast. Again, these devices are designed to be safe and portable enough that gantry rotation will not be the cause of a potential collision.

Internal shields are used most often for electron beam irradiation. One of the most common is an eye shield placed often on top of, or potentially underneath the lid of the eye if the lid of the eye needs (WEAVER et al. 1998) to be included in the treatment field (Fig. 8.13). These are typically made of lead with gold plating in order to reduce the scattering of the high atomic number lead as the gold works as a mediator to absorb scatter. Other internal shields are often placed in the oral cavity to again stop electron beams from penetrating beyond the desired depth in order to spare underlying tissue. For example, treatment of the lip is often complemented by having an internal lead shield placed below the lip to spare the gums. This lead shield is again coated with a low density material such as wax in order to absorb scatter that comes off the high atomic number lead.

Finally, an important device for external shielding is needed to absorb scatter, both external and internal, to the testes (FRAASS et al. 1985). There are custom-made commercial devices that are meant to clamp around the testes in order to reduce scatter to them from both head leakage and internal scatter (Fig. 8.14). Other devices that are implantable as a treatment device include gold markers. These gold markers, which are commercially available, are meant to be viewed during online portal imaging.

8.5.6
Treatment Devices for Special Procedures

8.5.6.1
Total Body Irradiation

Treatment devices for total body irradiation depend on the technique used (AAPM 1986). Total body irradiation patients can be treated in a fetal position, lying on a stretcher, or standing up with support devices. Other techniques include lying flat on a stretcher underneath the radiation beam while the stretcher moves through the treatment beam to spread out the dose over the entire body. Therefore, we will address treatment aids depending on the technique. For treatment of the patient lying on a couch in a fetal position using lateral total body irradiation beams, the stretcher itself must be able to

Fig. 8.13. Photograph of protective eye shield composed of lead with a gold covering

Fig. 8.14. Testicular shields that function to surround testes during photon beam irradiation of nearby treatment sites

be raised to place the patient centered in the direct beam from the rotated gantry. The stretcher must be flexible enough to elevate the patient's back and also allow the patient to bring up their knees toward their chest to compact the patient into the field, which can be problematic in a small room where there may be only 3 m from the source to the patient, thereby only allowing 120×120 cm² field size. With this lateral technique, there is also a need to use compensation to equalize the dose to the patient. Typically, compensators are used to compensate for missing tissue above the shoulders. There is also the option to use custom filters for the lower limbs. These filters are placed on the block tray at the gantry head location. For all techniques, there is often a need for scattering screen in front of the patient to help build up surface dosing without building up the skin dose itself. These are usually rolling devices that are placed in front of the patient, positioned at least 10 cm from the patient to prevent build up of the surface dose itself.

The other technique was the patient standing upright to be treated AP/PA with the gantry pointed in a lateral position toward the patient. As the patient is standing up for the entire treatment, they must be supported. The example shown here from the University of Minnesota (Fig. 8.15) is a technique whereby a bicycle seat is used for the patient to sit onto and also hand holders placed in strategic positions to allow the patients to hold onto during the

treatments, which typically take a total of 20 min. This AP/PA technique facilitates use of lung blocks that are often desired due to the lung transmission for this technique. These blocks are often placed close the patient on a screen that works both as a scatter and holder of the blocks.

8.5.6.2
Total Skin Electron Therapy

Total skin electron therapy is used for treatment of mucosis fungoides and Kaposi's sarcoma. As the goal is to treat the skin surface and immediate underlying tissue, electrons are often used. In order to cover the entire body, an extended treatment distance is utilized. As the patients are treated in multiple positions, and altering arm and leg configurations, there must be direct methods for the patients to be in these proper positions for treatment. The classic use of the Stanford technique calls for six different treatment positions with the patient again standing upright. Therefore, arm positioning and leg placement mats are often used to achieve the proper placement of arms and legs. In addition, a scattering screen is used to build-up surface doses for this electron beam therapy.

Finally, a fairly new technique for extracranial stereotactic body irradiation is a self-contained immobilization and positioning system. Using this technique, patients are scanned and treated within a self-contained frame system that has (1) both fiducial coordinates to localize the tumor to bring the ideal center of the tumor to the isocenter, (2) immobilization within the frame and body cast system, and (3) compression devices to minimize respiratory motion (Fig. 8.16).

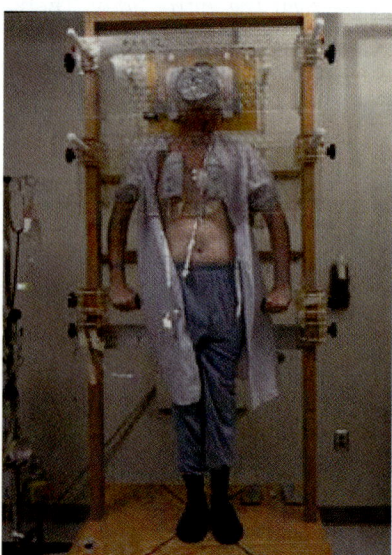

Fig. 8.15. Photograph, courtesy of University of Minnesota, showing patient treated upright for total body irradiation using AP/PA fields. Patient is supported by hand rests and the lungs are shielded using custom-shaped Cerrobend blocks

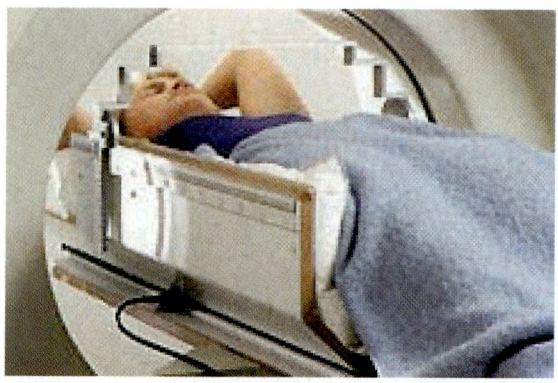

Fig. 8.16. Patient within stereotactic body frame system. Peripheral scales are used to appropriate coordinates configured during planning to be used for treatment

References

AAPM (1986) Report of Task Group 29. AAPM Report 17. The physical aspects of total and half body photon irradiation

Bentel GC et al. (1997) Comparison of two repositioning devices used during radiation therapy for Hodgkin's disease. Int J Radiat Oncol Biol Phys 38:791–795

Das IJ et al. (1997) Efficacy of a belly board device with CT-simulation in reducing small bowel volume within pelvic irradiation fields. Int J Radiat Oncol Biol Phys 39:67–76

Das IJ et al. (1998) Beam characteristics of a retrofitted double-focused multileaf collimator. Med Phys 25:1676–1684

Ellis F, Hall EJ, Oliver R (1959) A compensator for variations in tissue thickness for high energy beams. Br J Radiol 32:421–422

Fraass BA et al. (1985) Peripheral dose to the testes: the design and clinical use of a practical and effective gonadal shield. Int J Radiat Oncol Biol Phys 11:609–616

Fraass B et al. (1998) AAPM Radiation Therapy Committee Task Group 53: quality assurance for clinical radiotherapy treatment planning. Med Phys 25:1773–1829

Haus AG, Marks JE (1973) Detection and evaluation of localization errors in patient radiation therapy. Invest Radiol 8:384

ICRU (1993) Prescribing, recording, and reporting photon beam therapy. International Commission on Radiation Units and Measurements: Bethesda, MD

Jordan TJ, Williams PC (1994) The design and performance characteristics of a multileaf collimator. Phys Med Biol 39:231–251

Klein EE, Purdy JA (1993) Entrance and exit dose regions for Clinac-2100C. Int J Radiat Oncol Biol Phys 27:429–435

Klein EE et al. (1995a) Clinical implementation of a commercial multileaf collimator: dosimetry, networking, simulation, and quality assurance. Int J Radiat Oncol Biol Phys 33:1195–1208

Klein EE, Wasserman TH, Ermer B (1995b) Clinical introduction of a commercial treatment chair to facilitate thorax irradiation. Med Dosimetry 20:171–176

Klein EE et al. (1998) Mutliple machine implementation of enhanced dynamic wedge. Int J Radiat Oncol Biol Phys 40:977–985

Mutic S, Klein EE (2000) A reduction in the AAPM TG-36 reported peripheral dose distributions with tertiary multileaf collimation. Int J Radiat Oncol Biol Phys 44:947–953

Petti PL, Siddon RL (1985) Effective wedge angles with a universal wedge. Phys Med Biol 30:985–991

Powers WE et al. (1973) A new system of field shaping for external-beam radiation therapy. Radiology 108:407–411

Sewchand W, Khan FM, Williamson J (1978) Variations in depth-dose data between open and wedge fields for 4-MV X rays. Radiology 127:789–792

Slessinger ED et al. (1993) Independent collimator dosimetry for a dual photon energy linear accelerator. Int J Radiat Oncol Biol Phys 27:681–687

Stovall M et al. (1995) Fetal dose from radiotherapy with photon beams: report of AAPM Radiation Therapy Committee Task Group No. 36. Med Phys 22:63–82

Weaver RD, Gerbi BJ, Dusenbery KE (1998) Evaluation of eye shields made of tungsten and aluminum in high-energy electron beams. Int J Radiat Oncol Biol Phys 41:233–237

Xia P et al. (1999) Physical characteristics of a miniature multileaf collimator. Med Phys 26:65–70

9 Three-Dimensional Treatment Planning and Conformal Therapy

James A. Purdy, Jeff M. Michalski, Jeffrey Bradley, Srinivasan Vijayakumar, Carlos A. Perez, and Seymour H. Levitt

CONTENTS

9.1 Introduction 179
9.2 Three-Dimensional Radiation Therapy Treatment Planning 180
9.2.1 Patient Treatment Position/Immobilization and Planning CT Scan 181
9.2.2 Tumor, Target Volume and Critical Structure Delineation, and Dose Prescription 182
9.2.3 Designing Beam Arrangement and Field Apertures 184
9.2.4 Dose Calculation 186
9.2.5 Plan Evaluation and Improvement 186
9.2.6 Plan Implementation and Treatment Verification 187
9.3 Volume and Dose Specification for 3D CRT 188
9.3.1 Definition of Target Volumes 188
9.3.2 Definition of Organs at Risk 189
9.3.3 Dose Reporting and Dose Prescription 190
9.3.4 Using the GTV, CTV, and PTV Concepts 190
9.4 Integration of Multimodality Image Data for 3D Planning 193
9.5 Three-Dimensional Dose Calculation Algorithms 193
9.6 Dose-Volume Histograms 194
9.6.1 Differential Dose-Volume Histograms 194
9.6.2 Cumulative Dose-Volume Histograms 195
9.6.3 Dose-Volume Statistics 196
9.6.4 Plan Evaluation Using Dose-Volume Histograms and Dose-Volume Statistics 196
9.7 Biological Models 197
9.7.1 Normal Tissue Complication Probability 197
9.7.2 Tumor Control Probability 198
9.7.3 Equivalent Uniform Dose 198
9.8 Management of Three-Dimensional Treatment Planning Data 198
9.9 Summary and Conclusion 200
References 200

J. A. Purdy, PhD; S. Vijayakumar, MD
Department of Radiation Oncology, University of California Davis Medical Center, Sacramento, CA 95817, USA
J. M. Michalski, MD; J. Bradley, MD; C. A. Perez, MD
Department of Radiation Oncology, Mallinckrodt Institute of Radiology, Washington University School of Medicine, St. Louis, MO 63110, USA
S. H. Levitt, MD
Department of Therapeutic Radiation Oncology, University of Minnesota, Minneapolis, MN 55455, USA

9.1 Introduction

Several technological developments have combined to move radiation oncology into what is now referred to as the three-dimensional (3D) radiation therapy or 3D conformal radiation therapy (3D CRT) era (Purdy 1996a). Modern anatomic imaging technologies, such as X-ray computed tomography (CT) and magnetic resonance imaging (MRI), provide a fully 3D model of the cancer patient's anatomy, which is often complemented with functional imaging, such as positron emission tomography (PET) or magnetic resonance spectroscopy (MRS), and now allows the radiation oncologist to more accurately identify tumor volumes and their relationship with other critical normal organs. Powerful X-ray CT-simulation and three-dimensional treatment-planning systems (3D TPS) have now been commercially available for over a decade and have replaced the conventional radiation therapy X-ray simulator and two-dimensional (2D) dose-planning process in modern-day radiotherapy clinics (Meyer and Purdy 1996; Purdy and Starkschall 1999). In addition, modern day medical linear accelerators now come equipped with sophisticated computer-controlled multileaf collimator systems (MLCs) that provide beam aperture and/or beam-intensity modulation capabilities that allow precise shaping of the patient's dose distributions.

The 3D CRT plans generally use an increased number of radiation beams that are shaped using beam's-eye-view (BEV) planning to conform to the target volume (Goitein et al. 1983; McShan et al. 1979; Reinstein et al. 1978). To improve the conformality of the dose distribution, conventional beam modifiers (e.g., wedges and/or compensating filters) are sometimes used. Figure 9.1 shows comparison isodose clouds on the room-view display for a 2D anteroposterior/posteroanterior technique vs a 3D four-field technique utilizing wedges for a lung cancer patient. Note that the high-dose region is more conformal using the 3D technique. A more advanced form of conformal therapy, called inten-

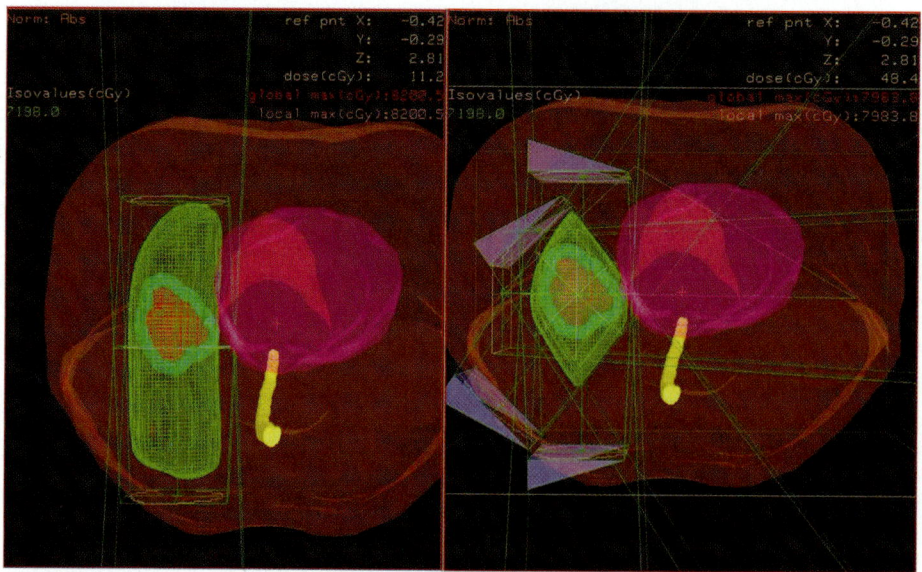

Fig. 9.1 Simple anteroposterior/posteroanterior opposed-field technique for a patient with stage-I non-small cell lung cancer (left panel). On the right is a 3D conformal four-field technique. *Red* GTV, *light blue* PTV, *green* 7198 cGy isodose cloud

sity-modulated radiation therapy (IMRT), has recently evolved and is reviewed in detail in Chapter 10 (IMRT 2001). In this chapter, the physics and clinical aspects of 3D treatment planning and conformal therapy are discussed.

Three-dimensional treatment planning is not just an add-on to the 2D radiation treatment-planning process. Instead, it represents a radical change in practice, particularly for the radiation oncologist. The 2D treatment-planning approach emphasizes the use of a conventional X-ray simulator utilizing bony landmarks visualized on planar radiographs for designing beam portals for standardized beam arrangement techniques. In contrast, 3D treatment planning emphasizes a volumetric image-based virtual simulation approach for defining tumor and critical structure volumes for the individual patient (PURDY and STARKSCHALL 1999); hence, it should be understood that the 3D planning process puts increased demands on the radiation oncologist to specify target volume(s) and critical structure(s) with far greater accuracy than before. Moreover, this technology also places increased demands on the radiation oncology physicist to insure adequate quality assurance measures are in place to accommodate the 3D CRT process, e.g., the need for increased precision in tumor imaging, patient set-up reproducibility, organ motion assessment, and treatment-delivery verification.

It is important to further clarify the use of the terms 2D and 3D as descriptors for the planning pro-

cess. Planning the cancer patient's treatment is (and always has been) a 3D problem (at least with regard to the spatial distribution of dose), and when the authors refer to 2D planning, we are referring to the process and tools used. Also, 3D treatment planning does not require the use of "noncoplanar" beams, a common misconception. The reader will be able to appreciate the 3D planning approach much more fully if they view it as a treatment-planning process, rather than viewing it as a particular beam configuration, or considering it simply as the purchase of a new planning system.

9.2
Three-Dimensional Radiation Therapy Treatment Planning

Three-dimensional treatment planning typically involves a series of procedures summarized in Table 9.1; these include establishing the patient's treatment position, constructing a patient repositioning immobilization device when needed, obtaining a volumetric image data set of the patient in treatment position, contouring target volume(s) and critical normal organs using the volumetric planning image data set, determining beam orientation and designing beam apertures, computing a 3D dose distribution according to the dose prescription, evaluating the treatment plan, and if needed, modifying the plan

Table 9.1 The 3D treatment planning and delivery process. *TPS* treatment-planning system. *DRR* digitally reconstructed radiographs

Patient treatment position, immobilization, and planning CT scan
Position patient in proposed treatment position
Fabricate immobilization devices
Place radiopaque markers and mark repositioning lines on patient and immobilization devices
Obtain topograms to check patient alignment
Perform volumetric CT scan of patient in treatment position
Transfer CT images to 3D TPS or virtual simulation workstation
Tumor, target volume and critical structure delineation, and dose prescription
Evaluate plan and modify, if necessary
Plan implementation and treatment verification
Evaluate plan and modify until plan is found to be acceptable
Physician approves treatment plan
Calculate treatment machine monitor unit setting
Transfer patient's plan to patient's chart (electronic medical record) and record and verify system
Physicist checks plan and transfer of data to record and verify system
Verify patient position and isocenter placement on treatment machine using orthogonal DRRs against orthogonal port films
Check field shapes by comparing treatment field DRRs with treatment beam port films
Check first-day treatment with diode measurement, and record and verify
Periodic verification checks during treatment (orthogonal DRRs/films or DRRs/port films, record and verify)

(e.g., beam orientations, apertures, beam weights, etc.) until an acceptable plan is approved by the radiation oncologist. The approved plan must then be implemented on the treatment machine and the patient's treatment verified using appropriate quality-assurance procedures. All of these tasks make up the 3DCRT process and are discussed herein.

9.2.1
Patient Treatment Position/Immobilization and Planning CT Scan

In the initial part of the 3DCRT process (pre-planning), the proposed treatment position of the patient is determined, and the immobilization device to be used during treatment is fabricated. In should be clearly understood that repositioning patients and accounting for internal organ movement for fractionated radiation therapy, in order to accurately reproduce the planned dose distribution, remains a difficult technical aspect of the 3DCRT process. Errors may occur if patients are inadequately immobilized, with resultant treatment fields inaccurately aligned from treatment to treatment (interfraction). In addition, patients and/or their tumor volume may also move during treatment (intrafraction) because of either inadequate immobilization or physiological activity. Accounting for all of the uncertainties in the 3DCRT planning and delivery process remains a challenge for radiation oncology and continued research and development is needed.

Determining the treatment position of the patient and construction of the immobilization device can

be performed on a conventional radiation therapy simulator, but more preferably is now done in a dedicated radiation therapy CT-simulator facility (Fig. 9.2). A radiation therapy CT simulator consists of a diagnostic-quality CT scanner, laser patient positioning / marking system, virtual simulation 3D treatment-planning software, and various hardcopy output devices (PEREZ et al. 1994; MUTIC et al. 2003). The CT scanner is used to acquire a volumetric planning CT scan of a patient in treatment position, which is then used to create a virtual patient model for use with the virtual simulation software that mimics the functions of a conventional radiation therapy simulator. The CT simulation is a complicated team process involving the radiation oncologist, medical physicist, medical dosimetrist, and radiation therapist. All team members should

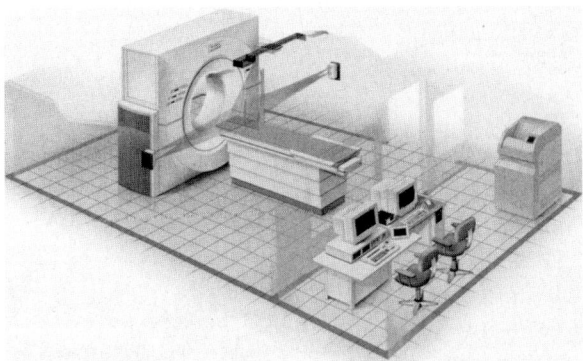

Fig. 9.2 Typical CT simulation suite shows the scanner, flat tabletop, orthogonal laser system, virtual simulation workstation, and hardcopy output device. (Courtesy of Philips, Inc.)

be knowledgeable about the CT-simulator features and overall virtual simulation process.

The CT scan must be performed with the patient in the treatment position, as determined in the pre-planning step. The CT topograms should be generated first and reviewed prior to acquiring the planning scan to insure that patient alignment is correct; adjustments are made if needed. Radiopaque markers are typically placed on the patient's skin and the immobilization device to serve as fiducial marks to assist in any coordinate transformation needed as a result of 3D planning and eventual plan implementation. An example of a typical immobilization repositioning system used for patients undergoing radiation therapy for head and neck cancer is shown in Figure 9.3. Other aids in radiation treatment planning, such as the use of intravenous contrast to help delineate target volumes, need to be considered during simulation. Figure 9.4 shows chest CT images of a patient with a right lower lobe lung cancer involving hilar adenopathy. The use of intravenous contrast clarifies the hilar mass in this example.

Planning CT scan protocols are tumor-site dependent and typically range from 2- to 8-mm slice thicknesses and 50–200 slices. In general, a 3-mm slice thickness provides adequate quality digitally reconstructed radiographs (DRR). The planning CT data set provides an accurate geometric model of the patient as well as the electron density information needed for the calculation of the 3D dose distribution that takes into account tissue heterogeneities.

As shown in Figure 9.5, some complex treatment techniques, such as tangential breast irradiation, are difficult or nearly impossible to set up because of older CT scanner bore-size limitations. A large-bore

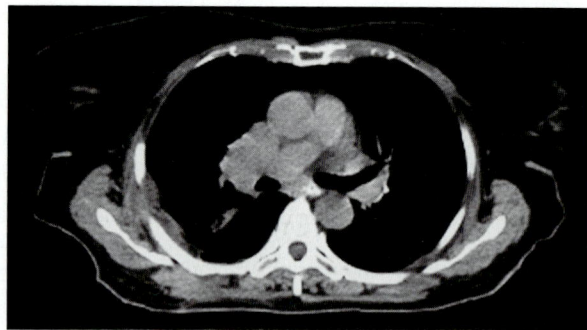

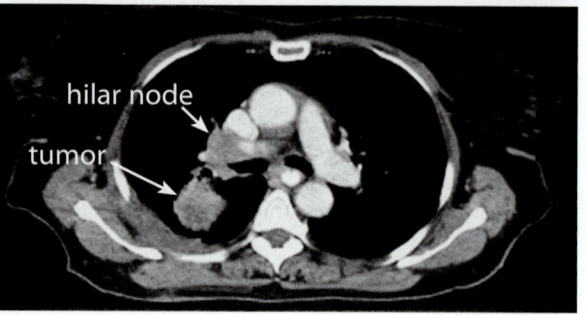

Fig. 9.4. a Intravenous non-ionic contrast is helpful in distinguishing hilar nodes from the pulmonary artery. A CT scan of the chest without using intravenous contrast is shown. Note that the right hilar mass cannot be distinguished from the pulmonary artery. **b** Same patient with intravenous contrast. Note that the hilar mass is easily clarified using contrast

(85-cm) CT scanner developed by Marconi (now part of Philips, Inc.) designed specifically for radiation oncology applications has largely solved this problem (GARCIA-RAMIREZ et al. 2002).

The planning CT data set is typically transferred to a 3DTPS or virtual simulation computer workstation via a computer network. Data-transfer issues are discussed in more detail in a later section.

9.2.2
Tumor, Target Volume and Critical Structure Delineation, and Dose Prescription

Delineation of tumor/target volume and organs-at-risk contours using the volumetric CT data set is typically performed by the radiation oncologist and the medical dosimetrist, working as a team. The CT data are displayed and contours are drawn manually using a computer mouse on a slice-by-slice basis. Some organs at risk with distinct boundaries (e.g., skin, lung) can be contoured automatically, with only minor editing required; others (e.g., brachial plexus, optic chiasm) require the "hands-on" effort of the radiation oncologist. When modern 3DTPS

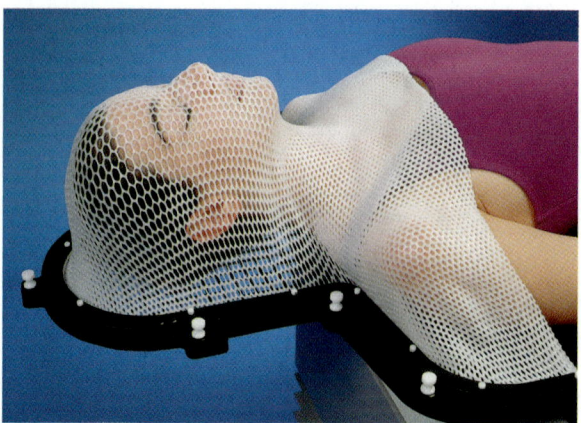

Fig. 9.3 Example of immobilization repositioning system used for patients undergoing radiation therapy for head and neck cancer. (Courtesy of MEDTEC, Inc.)

image segmentation software is used (Fig. 9.6), contouring generally takes 1–2 h depending on disease site; however, for some complex sites, such as head and neck cancer, where many organs at risk and complex tumor/target volumes are often the norm, this task can take up to 4 h.

These contours are used to generate realistic solid-shaded surface graphic representations of structures. Such 3D displays of the volumes are powerful planning tools but can be confusing when large numbers of structures overlap; color, transparency, and interactive manipulation of the image help to clarify the planning display.

One of the most important factors that has undoubtedly contributed to the success of the current 3D treatment-planning process is the standardiza-

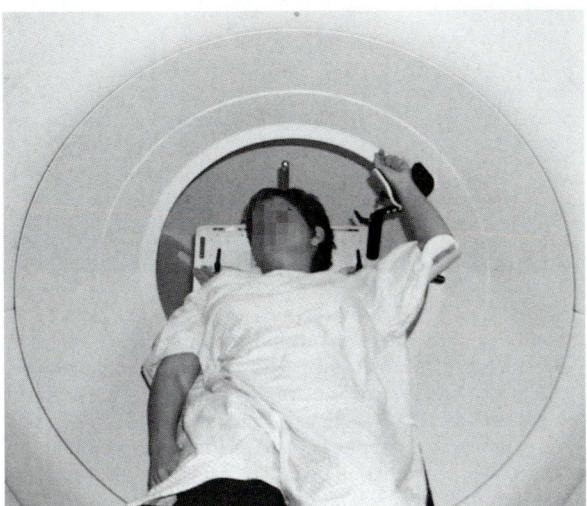 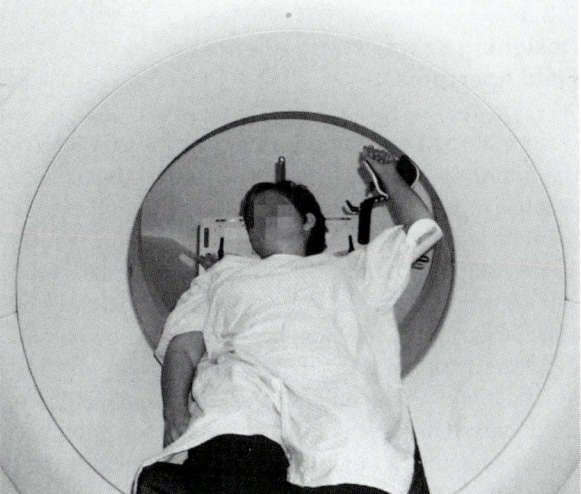

Fig. 9.5 Complex treatment techniques, such as tangential breast irradiation, can be problematic to setup on diagnostic CT scanners or older CT simulators (70-cm bore) because of bore-size limitations. The modern large-bore (85-cm) CT scanner developed by Marconi (now part of Philips) designed specifically for radiation oncology applications has solved this problem.

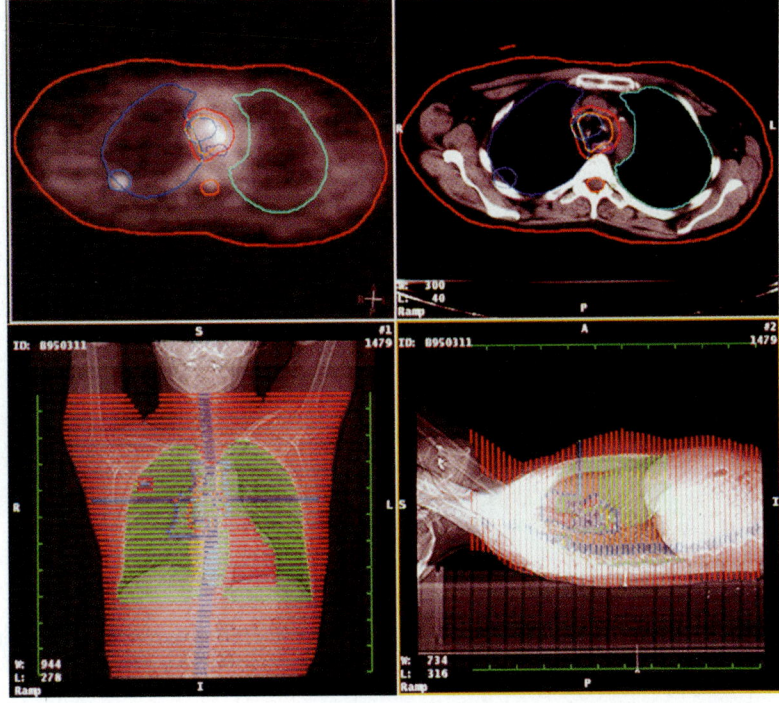

Fig. 9.6 Three-dimensional treatment-planning-system (TPS) image segmentation software provides effective tools for radiation oncologists and treatment planners to delineate critical structures, tumor, and target volumes for 3D planning. The CT data are displayed and contours are drawn by the treatment planner/radiation oncologist around the tumor, target, and normal tissues on a slice-by-slice basis, as seen in the *upper right panel*. At the same time, planar images from both anteroposterior and lateral projections are displayed in the *bottom right and left panels. Upper left panel* shows PET scan data with overlying contours after image registration with the CT data

tion of nomenclature and methodology for defining the volume of known tumor, suspected microscopic spread, and marginal volumes necessary to account for setup variations and organ and patient motion published in the International Commission on Radiation Units and Measurements (ICRU) reports 50 and 62 (ICRU 1993a, 1999). Details on the use of this methodology are discussed in a later section.

9.2.3
Designing Beam Arrangement and Field Apertures

Design of the beam arrangement is the next step in the planning process. For 3D planning, the 3D TPS must have the capability to simulate each of the treatment machine motion functions, including gantry angle, collimator length, width and angle, MLC leaf settings, couch latitude, longitude, height, and angle. This ability to orient beams in 3D allows one to develop treatment plans that use noncoplanar beams; however, when noncoplanar beam arrangements are used, care must be taken to avoid the selection of a gantry and couch angles that results in table/gantry collisions or other treatment-room restrictions.

An essential feature in a 3D TPS is the BEV display (Fig. 9.7), in which the observer's viewing point is at the source of radiation looking out along the axis of the radiation beam (GOITEIN et al. 1983; MCSHAN et al. 1979; REINSTEIN et al. 1978). This type of display of the patient model is analogous to the simulator radiograph or port film. The BEV display allows the planner to easily view the critical structure volumes and the target volume so that shielding blocks or MLC apertures can be drawn using a computer mouse.

Another powerful display feature in a 3D TPS is the room's eye view (REV; Fig. 9.8), in which the planner can simulate any arbitrary viewing location within the treatment room (PURDY et al. 1987, 1993). The REV display complements the BEV in the beam design phase of treatment planning, particularly in positioning of beam isocenter depth and in visualizing all, or selected, beams, to better appreciate the beam arrangement geometry. The REV display is even more valuable in the plan evaluation phase in which "dose clouds" are used to evaluate where hot or cold spots occur in the dose distribution, as shown in Figure 9.9.

Another powerful 3D TPS display feature is digitally reconstructed radiographs (DRRs; Fig. 9.10; GOITEIN et al. 1983; SIDDON 1985; SHEROUSE et al. 1990). The DRRs provide planar reference images

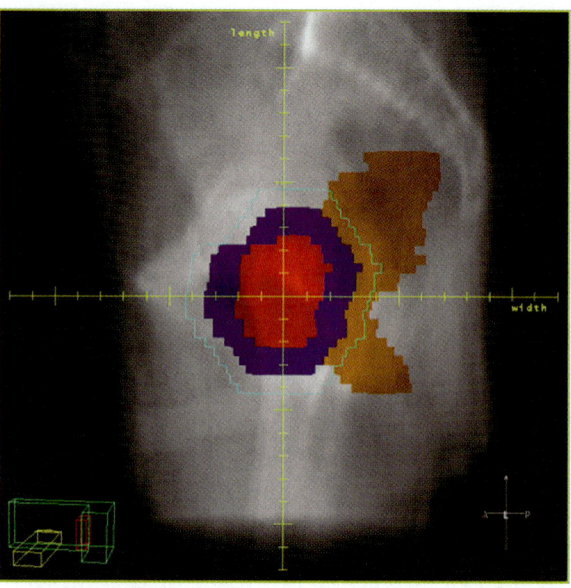

Fig. 9.7 A 3D-TPS beam's-eye-view (BEV) display is useful in identifying the best gantry, collimator, and couch angles at which to irradiate target and avoid irradiating adjacent normal structures. Critical structures and target volumes are outlined on patient's serial CT sections. Contours are seen in perspective, as though observer's eye is at radiation source looking out along axis of the radiation beam. Outline of multileaf collimator aperture or beam-shaping block can be displayed.

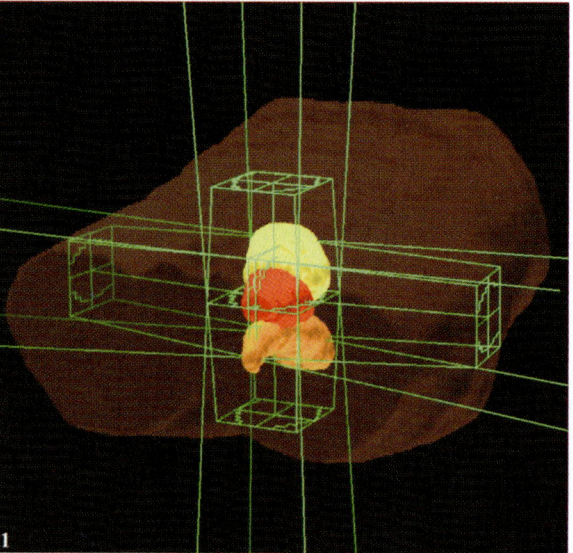

Fig. 9.8 A 3D-TPS room's-eye-view (REV) display shows multiple beam arrangement, external skin surface, prostate planning target volume, rectum, and bladder. The REV display helps the treatment planner to better appreciate the overall treatment technique geometry and placement of the isocenter. (From PURDY et al. 1993)

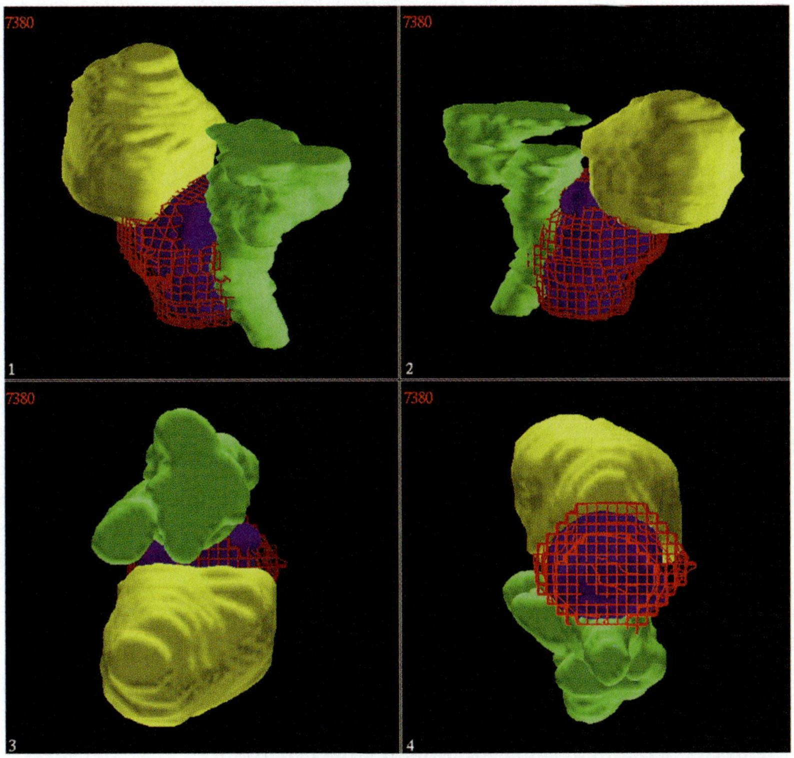

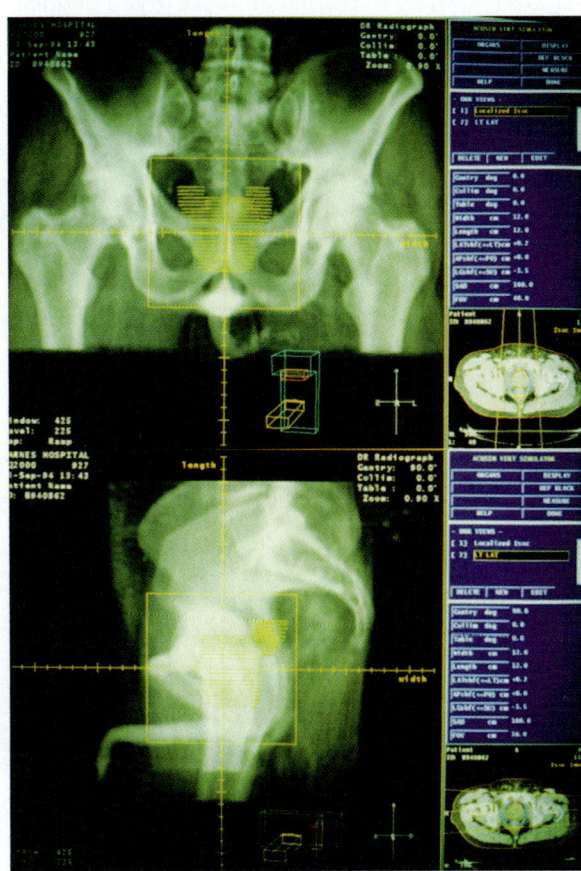

Fig. 9.9 A 3D-TPS REV 3D isodose surface display with real-time interactivity is a valuable tool for evaluation of 3D dose distributions in terms of adequate coverage of target volumes and sparing of critical structures. The REV display enables radiation oncologists to view target volume or normal tissue volume with superimposed isodose surfaces or "dose clouds" from any arbitrary viewing angle. Shown is a two-panel REV display of the 73.8-Gy isodose volume, the prostate PTV, bladder, and rectum of a prostate cancer patient treated with a six-field technique. The locations of the PTV region not covered by the specified dose level are easily discernible using the REV display. (From PURDY 1998)

Fig. 9.10 A 3D-TPS digitally reconstructed radiograph (DRR) display. Orthogonal-setup DRRs of a prostate cancer patient are displayed with surface prostate target volume contours and collimator positions. *Icon at lower right* depicts gantry and treatment couch position. *Icon at far lower right* depicts patient orientation. (From PURDY 1997)

that can be used for transferring the 3D treatment plan to the clinical setting; thus, their role is similar to that of conventional simulation radiographs used for treatment portal design and for verification of treatment delivery by comparison with port films or electronic portal images obtained on a treatment machine. The DRR, however, provides additional value by allowing the target volume and critical structure contours to be clearly shown on the computed image. Modern 3D planning systems provide for relatively fast generation of DRRs that can be archived and viewed on image-viewing workstations or printed on film using laser printers and are stored in the patient's film jacket.

The digitally composite radiograph (DCR; Fig. 9.11) is a type of DRR that allows different ranges of CT numbers that relate to a certain tissue type to be selectively suppressed or enhanced in the image. This is analogous to a transmission radiograph through a virtual patient where certain tissue types have been removed, leaving only the organs of interest to be displayed. The DCRs are very useful when designing treatment portals as they can allow for better visualization of organs of interest. Also, portions of an overlaying organ of interest can be removed entirely to further enhance organ delineation.

9.2.4
Dose Calculation

After the initial beam geometry is designed, the 3D dose distribution is calculated. Dose calculation methods have traditionally been based on parameterizing radiation beams measured in water phantoms under standard conditions and applying correction factors to the beam representations for the nonuniform surface contour of the patient or the obliquity of the beam, tissue heterogeneities, and beam modifiers such as blocks, wedges, and compensator; however, state-of-the-art 3D TPS now use more advanced models, such as the convolution/superposition algorithm, which compute the dose more from first principles (Purdy 1992; Mackie et al. 1996). Details on dose-calculation algorithms are discussed in a later section.

9.2.5
Plan Evaluation and Improvement

The 3D CRT plan evaluation/improvement process involves an iterative, interactive approach. Typically, the initial beam arrangement is selected based primarily on clinical experience using BEV

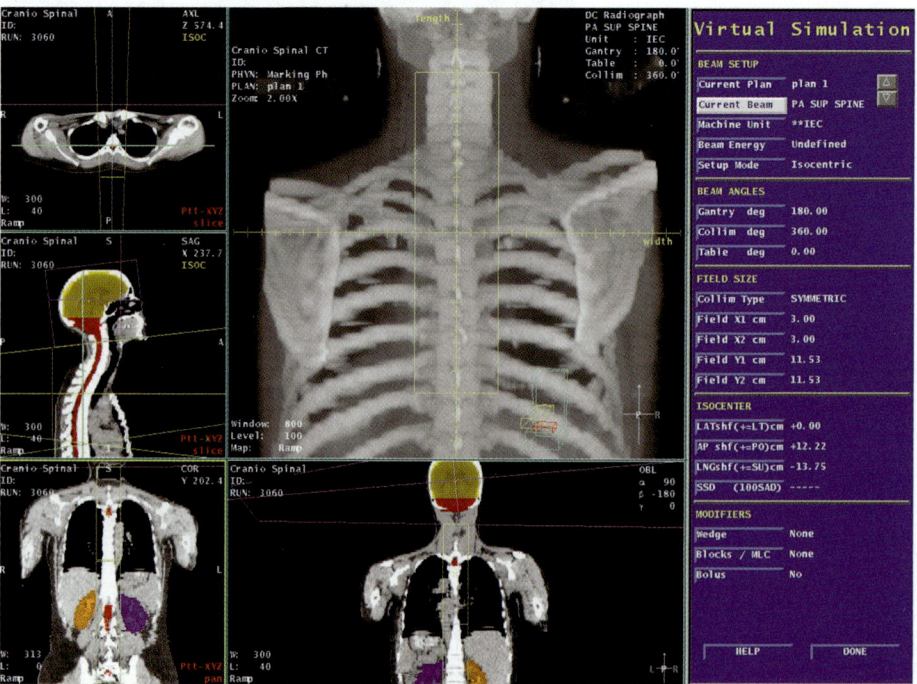

Fig. 9.11 A 3D-TPS digital composite radiograph (DCR). The CT numbers are grouped into ranges corresponding to bone, fat, and muscle, and are modified by a weighting factor and re-displayed to provide greater enhancement of the specified tissue range compared with normal DRR. (From Purdy 1999)

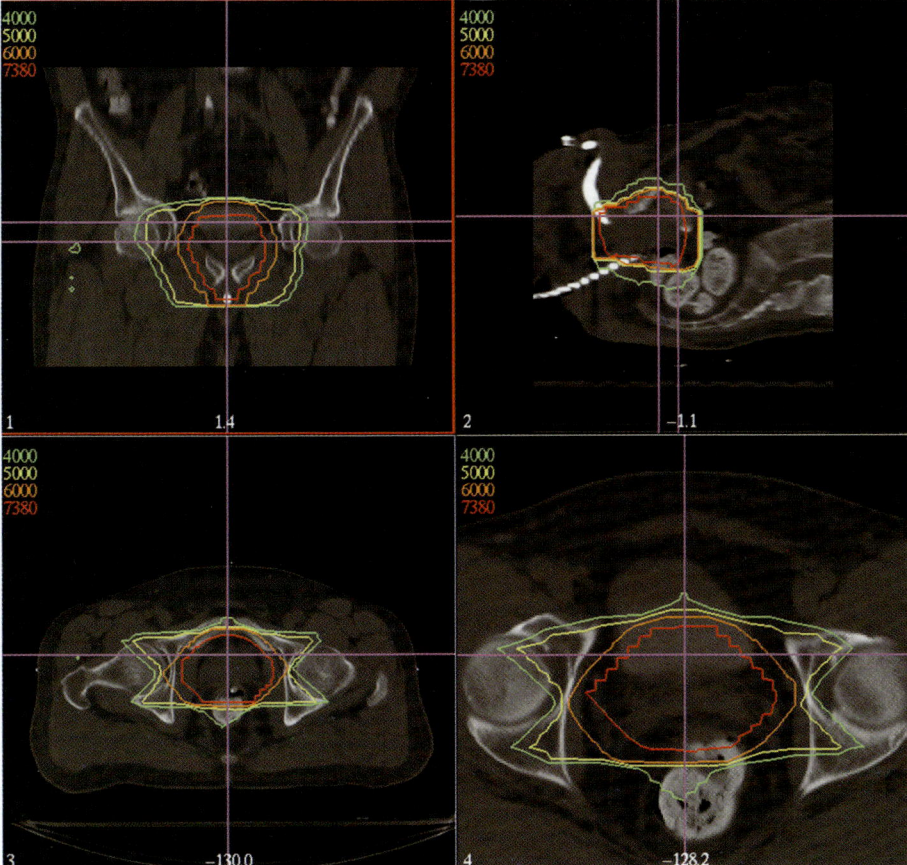

Fig. 9.12 Dose distribution displays for a patient with prostate cancer shows coronal, sagittal, and two axial CT sections with superimposed color-coded isodose lines (73.8, 60, 50, and 40 Gy). *Vertical and horizontal lines* displayed on each CT section indicate the positions of each section. Evaluating volumetric 3D dose distributions using this type-2D display only is difficult and time-consuming.

and REV displays. The arrangement is then modified based on the evaluation of the dose distribution using multi-level 2D displays showing isodose lines superimposed on CT images (Fig. 9.12), or as a spectrum of colors superimposed on the anatomic information represented by modulation of intensity (color wash), and REV 3D dose clouds, shown previously. However, because of the large amount of dosimetric data that must be analyzed when a 3D CRT plan is evaluated, methods for condensing and presenting the data in easily understandable formats have been developed. The most useful data reduction tool for 3D planning is the dose-volume histogram (DVH), which is discussed in detail in a later section (DRZYMALA et al. 1991). The planned dose distribution approved by the radiation oncologist is most often one in which a uniform dose is delivered to the target volume (e.g., +7% and –5% of prescribed dose) with doses to critical structures held below some tolerance level (e.g., as given by EMAMI et al.) that has been specified by the radiation oncologist (EMAMI et al. 1991).

9.2.6
Plan Implementation and Treatment Verification

Once the treatment plan has been designed, evaluated, and approved, documentation for plan implementation must be generated. Documentation includes beam parameter settings transferred to the treatment machine record and verify system, hardcopy block templates for the block fabrication room, or MLC parameters communicated over a network to the computer system that controls the MLC system of the treatment machine, DRR generation and printing, or transfer to an image database.

Quality-assurance checks used to confirm the validity and accuracy of the 3D CRT plan include an independent check of the plan and monitor unit calculation by a physicist, isocenter placement check on the treatment machine using orthogonal radiographs, field-apertures check using portal films or electronic portal images, and diode or MOSFET in vivo dosimetry check. A record and verify (R&V) system is now considered essential to help manage 3D CRT treatments; however, careful scrutiny must be given to insure that the input data into the R&V system are correct.

If a clinic is in the initial phases of implementing 3D CRT techniques (or if implementing non-conventional beam arrangements), a verification simulation procedure is recommended to confirm the geometric validity and accuracy of the 3D treatment plan. The DRRs generated by the 3D TPS are used for comparison with the verification simulation radiographs to confirm the correctness of the beam orientations in the physical implementation. When a beam orientation cannot be simulated, orthogonal radiographs may be taken and compared with similar DRRs to ensure correct isocenter positioning. The optical distance indicator is also useful in assessing the correctness of the setup of a particular beam. Documentation provides a depth of isocenter below the skin surface on the central ray of the beam, which can then be compared with the isocenter depth measured on the simulator or treatment machine after the beam is set up using the couch and gantry positions specified by the treatment plan.

9.3
Volume and Dose Specification for 3D CRT

The International Commission on Radiation Units and Measurements (ICRU) reports 50 and 62 gave the radiation oncology community a consistent language and a methodology for image-based volumetric treatment planning (ICRU 1993a, 1999). For 3D CRT planning, the physician must specify the volumes of known tumor, i.e., gross tumor volume (GTV), the volumes of suspected microscopic spread, i.e., clinical target volume (CTV), and the additional margin around the CTV/GTV necessary to account for setup variations and organ and patient motion, i.e., planning target volume (PTV).

9.3.1
Definition of Target Volumes

To achieve accurate radiation therapy, it is necessary to accurately relate the positions of target volumes and critical structures in the patient to the positions and orientation of beams used for planning imaging studies and for treatment. This requires the use of multiple coordinate systems, one directly related to the patient and those related to the imaging and treatment machines. The positions of target volumes

and critical structures are related to anatomical reference points or alignment marks in the coordinate system within the patient. The position and orientation of the imaging and treatment machines are defined in a coordinate system related to these machines. Reference points serve to link the patient and machine coordinate systems, since they can be defined in both patient and machine coordinates, and thus allow the coordinates of the target volumes and critical structures to be defined for treatment planning.

The ICRU report 50 and report 62 definitions of target volume are illustrated in Figure 9.13. The ICRU report 62 refines the GTV, CTV, and PTV concept by introducing the definition of an internal margin (IM) to take into account variations in size, shape, and position of the CTV, and the definition of a setup margin (SM) to take into account all uncertainties in patient-beam positioning. The IM is referenced to the patient's coordinate system using anatomical reference points and the SM is referenced to the treatment machine coordinate system. Report 62 argues that identification of these two types of margins is needed since they compensate for different types of uncer-

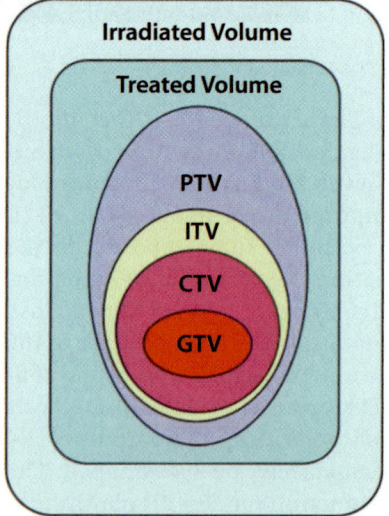

Fig. 9.13 International Commission on Radiation Units and Measurements (ICRU) reports 50 and 62 volumes used in 3D treatment planning. Gross tumor volume (*GTV*) is the volume(s) of known tumor. Clinical target volume (*CTV*) is the volume(s) of suspected microscopic tumor infiltration. Planning target volume (*PTV*) is the volume containing the CTV/GTV with enough margin necessary to account for setup variations and organ and patient motion. Internal target volume (*ITV*) represents the movements of the CTV referenced to the patient coordinate system by internal and external reference points.

tainties and refer to different coordinate systems. The IM uncertainties are due to physiological variations (e.g., filling of rectum, movements due to respiration, etc.) and are difficult or almost impossible to control from a practical viewpoint. The SM uncertainties are related largely to technical factors that can be dealt with by more accurate set up and immobilization of the patient and improved mechanical stability of the machine. The volume formed by the CTV and the IM is defined as the internal target volume (ITV) and represents the movements of the CTV referenced to the patient coordinate system by internal and external reference points, which preferably should be rigidly related to each other through bony structures. The volume formed by the CTV and the IM and SM combined is the PTV, as previously defined; however, exactly how mathematically these margins should be combined is not rigidly defined. Simple linear addition of the two margins will generally lead to an excessively large PTV that does not reflect the actual clinical consequences. This point will be discussed further, but for now it should be understood that the selection of an overall margin and delineation of the border of the PTV involves a compromise that requires the experience and the judgment of the radiation oncologist and the physicist.

Using the ICRU definitions, one sees that the GTV is the gross extent of the malignant growth as determined by palpation or imaging studies. The terms GTV$_{primary}$ and GTV$_{nodal}$ (or GTV-T and GTV-N) are typically used to distinguish between primary disease and other areas of macroscopic tumor involvement, such as involved lymph nodes, that are visible on imaging studies. The GTV together with this surrounding volume of local subclinical involvement constitutes the primary CTV and can be denoted as CTV-T. It is noteworthy that even if the GTV has been removed by radical surgery, and radiation therapy to the tumor bed is considered necessary for the tissues close to the site of the removed GTV, the volume should be designated as CTV-T. Additional volumes with presumed subclinical spread may also be considered for therapy, e.g., regional lymph nodes. These may be designated CTV-N (and if necessary CTV-N1, CTV-N2, etc.). Adding the letter T, N, or M to identify the volumes, may help better clarify their clinical significance. In specifying the CTV, the physician must not only consider microextensions of the disease near the GTV, but also the natural avenues of spread for the particular disease and site including lymph node, perivascular, and perineural extensions. The GTV and CTV are anatomical–clinical concepts that should be defined before a choice of treatment modality and technique is made.

Once GTV/CTV(s) are contoured, margins around the CTV must be specified to create the PTV, in order to account for geometric uncertainties. These margins do not form the block aperture. The PTV is a static, geometrical concept used for treatment planning, including dose prescription. Its size and shape depend primarily on the GTV, the CTV, and the effects caused by internal motion of organs and the tumor as well as the treatment technique (beam orientation and patient fixation) used. The PTV can be considered a 3D envelope, fixed in space, in which the tumor and any microscopic extensions reside. The GTV/CTV can move within this envelope, but not through it. Also, the penumbra of the beam(s) is not considered when delineating the PTV; however, when designing the beam apertures, the width of the penumbra is taken into account and the beam size must be enlarged accordingly to insure dosimetric coverage of the PTV.

In addition to the GTV, CTV, ITV, and PTV definitions, the ICRU defines two other volumes that are not anatomic, but instead are based on the dose distribution: (a) the treated volume, which is the volume enclosed by an isodose surface that is selected and specified by the radiation oncologist as being appropriate to achieve the purpose of treatment (e.g., 95% isodose surface); and (3) the irradiated volume, which is the volume that receives a dose considered significant in relation to normal tissue tolerance (e.g., 50% isodose surface). These volumes are used mainly for plan optimization and plan evaluation.

9.3.2
Definition of Organs at Risk

The ICRU reports 50 and 62 define organs at risk (ORs) as those normal critical structures (e.g., spinal cord) whose radiation sensitivity may significantly influence treatment planning and/or prescribed dose. Report 62 introduces the concept of the planning organ at risk volume (PRV), in which a margin is added around the OR to compensate for that organ's spatial uncertainties. The PRV margin around the OR is analogous to the PTV margin around the CTV. The use of the PRV concept is even more important for those cases involving IMRT because of the increased sensitivity of this type of treatment to geometric uncertainties. The PTV and the PRV may overlap, and often do so, which implies searching for a compromise in weighting the importance of each in the treatment-planning process.

9.3.3
Dose Reporting and Dose Prescription

The ICRU reports 50 and 62 define a series of doses, including the minimum, maximum, mean dose, and ICRU reference dose (defined at the ICRU reference point) for reporting dose. The ICRU reference point for a particular treatment plan should be chosen based on the following criteria: (a) be clinically relevant and can be defined in an unambiguous way; (b) be located where the dose can be accurately determined; and (c) be located in a region where there are no steep dose gradients. In general, this point should be in the central part of the PTV. In cases where the treatment beams intersect at a given point, it is recommended that the intersection point be chosen as the ICRU reference point.

It is noteworthy that ICRU reports 50 and 62 do not make strict recommendations regarding dose prescription; instead, ICRU states that "...the radiation oncologist should have the freedom to prescribe the parameters in his/her own way, mainly using what is current practice to produce an expected clinical outcome of the treatment" (ICRU 1993b).

With regard to dose homogeneity, ICRU report 50 does recommend that the dose coverage of the PTV be kept within specific limits, namely +7% and –5% of the prescribed dose (ICRU 1993b).This level of dose homogeneity may not be achieved in all cases (particularly for current IMRT techniques), and the reader is reminded that ICRU report 50 explicitly states that if this degree of homogeneity cannot be achieved, it is the responsibility of the radiation oncologist to decide whether the dose heterogeneity can be accepted or not, pointing out that in those parts of the PTV where the highest malignant cell concentration may be expected, i.e., GTV, a higher dose may even be an advantage.

9.3.4
Using the GTV, CTV, and PTV Concepts

Some limitations and practical issues must be clearly understood when the ICRU report 50/62 methodology is adopted (PURDY 1996b,c, 2000). Firstly, the physical treatment-planning process is dependent on the delineation of the three volumes (GTV, CTV, and PTV) and the prescription of the target dose. The GTV, CTV, and PTV must be specified by the radiation oncologist independent of the dose distribution: the GTV in terms of the patient's anatomy, the CTV in terms of the patient's anatomy or as a quantitative margin to be added to the GTV, and the PTV in terms of a quantitative margin to be added to the CTV to account for positional uncertainties.

When the GTV is delineated, it is important to use the appropriate CT window and level settings

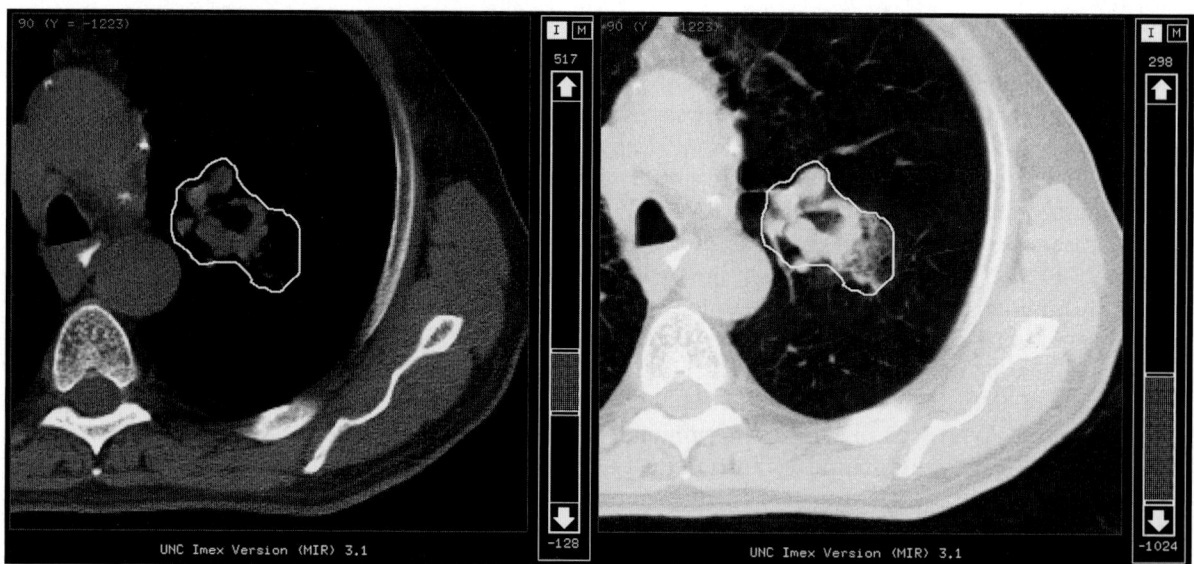

Fig. 9.14 A CT slice for a lung cancer patient shows that the appropriate CT window and level settings (*right frame*) must be used to determine the maximum dimensions of the gross tumor volume (GTV). Note that a much smaller GTV would have been contoured with the settings used in the *left frame*. (From PURDY 1997)

to determine the maximum dimension of what is considered to be potential gross disease (Fig. 9.14). Defining the CTV is even more difficult and must be performed by the radiation oncologist based on clinical experience because current imaging techniques cannot be used to directly detect subclinical tumor involvement. Radiation oncologists who, when first using a 3D TPS, are unfamiliar with defining target volumes and normal tissue on axial CT slices, should seek assistance from a diagnostic radiologist. Image-based cross-sectional anatomy training should be an essential component in radiation oncology residency training programs as the radiation oncologist needs to become much more expert in cross-sectional tissue anatomy and gross tumor changes to accurately define GTVs and CTVs.

The PTV margin is specified by the radiation oncologist, in consultation with the radiation oncol-ogy physicist, and is also based on clinical experience. Unfortunately, data for internal organ motion and setup error are lacking for most sites, although uncertainty studies addressing these issues for some sites (e.g., prostate) are increasingly being reported (LANGEN and JONES 2001). The radiation oncologist specifies the PTV margin as an estimate based on clinical experience, taking into account published literature and intramural uncertainty studies. When defining the PTV, the radiation oncologist should account for the asymmetric nature of positional uncertainties. For example, it is now recognized that prostate organ motion and daily setup errors may be anisotropic (side-to-side or rotational shifts of patients are likely to be different from movement in the anteroposterior direction); thus, the PTV margin around a CTV generally should not be uniform (Fig. 9.15).

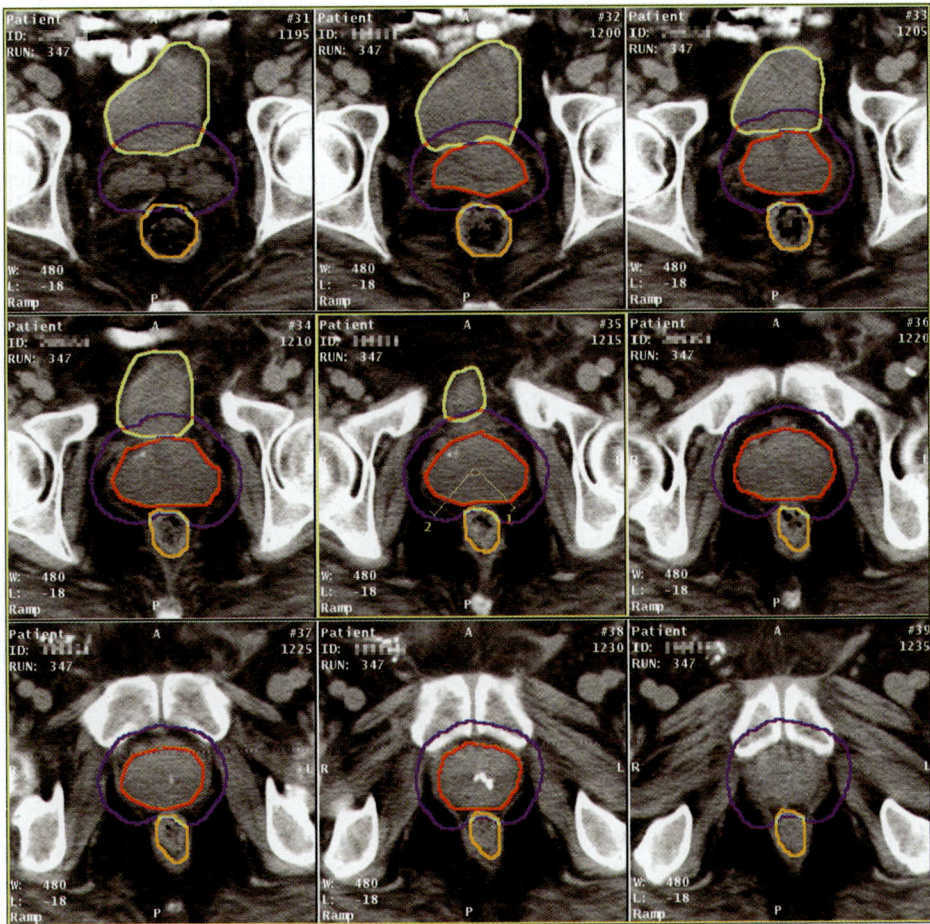

Fig. 9.15 The CT images of prostate cancer patient shows the contour outlines for the GTV, PTV, bladder, and rectum. The physician made the decision that no additional margin around the prostate for the CTV was required, i.e., CTV=GTV. Note that a non-uniform margin around the GTV/CTV was used to define the PTV in the region of the rectum (see *middle frame*). Also note the additional PTV contours needed to cap the GTV/CTV (*upper left* and *lower right frame*). (From PURDY 1998)

When the beam portal is defined, additional margin beyond the PTV is still needed to obtain dose coverage because of beam penumbra and treatment technique. Some physicians (and physicists) have confused this margin in the past by considering the PTV boundary as the required beam edge. Treatment portal margins needed around the PTV must be set according to the dosimetric characteristics of the beams being used. Typically, a 7- to 9-mm margin (port edge to PTV) is a good starting point, but one must determine the appropriate characteristics of the actual beams used to make this starting point determination. An additional point to understand is that in the case of coplanar treatment techniques, the margins required across the plane of treatment and the margins orthogonal to this plane will be different. To clarify this point, consider a four-field axial technique as an example. Portions of the lateral aspects of the PTV which are in the low-dose regions (near the penumbra) of the anteroposterior and posteroanterior fields will be in the high-dose regions (well away from the beam penumbra) of the lateral fields; however, the superior and inferior aspects of the PTV will always be in the same low-dose regions of all four fields, so there will be no dose filling from any other fields. This results in the need to have a larger portal margin in the inferior–superior dimension to ensure that the 95% isodose from all beams contains the PTV, while the lateral and anterior–posterior portal margins for each field may be reduced due to the other beams filling in the dose. The size of the margins will also be affected by the relative beam weighting; hence, making hard rules about margin sizes is impossible and requires some planning iteration to find the right mix of superior–inferior and lateral margins.

Another point of concern is the fact that some 3D TPS still do not possess accurate methods for providing a true 3D margin around the GTV/CTV (BEDFORD and SHENTALL 1998). The problem is illustrated in Figure 9.16. Typically, the margin expansion is drawn or specified in 2D around the GTV/CTV contour to get the PTV. For large contour differences in neighboring slices this will yield margins that are too small in the cranial–caudal direction (BEDFORD and SHENTALL 1998).

When a PTV overlaps with a contoured normal structure, a quandary is created as to which volume the overlapping voxels should be assigned for DVH calculations. Planning systems should have the feature that allows the planner to assign the overlapping voxels to both volumes. This ensures that the clinician is made aware of the potential for the high-dose region to include the normal structure as well

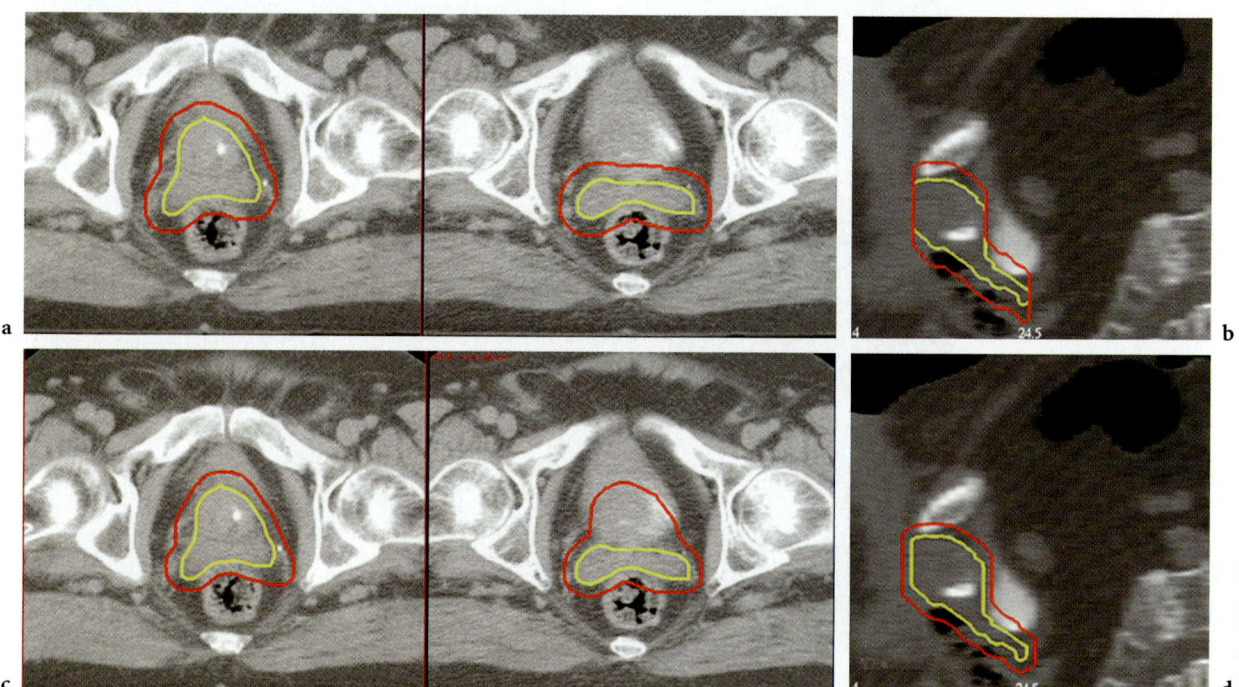

Fig. 9.16 a Multiple 2D margins around a prostate GTV in a transverse CT slice may yield margins that are too small in the cranio-caudal directions, as shown in a sagittal reconstruction (**b**). The 3D margins may appear too large in a transverse CT slice (**c**) but are actually correct as shown in a sagittal reconstruction (**d**). (From PURDY 1999)

as the PTV when reviewing the DVHs. In addition, most 3D TPS cannot account for a PTV contour that extends outside the skin surface. In those cases, the PTV should coincide with the surface, but the treating physician should be acutely aware of this approximation when setting or approving field margins.

Finally, one must understand that the PTV concept treats all points within the PTV as equally likely for the CTV to occupy all of the time, and this obviously does not occur in practice. A probabilistic approach, in which the positional uncertainties are convolved mathematically with the dose calculation, will likely evolve; however, until that time, the ICRU reports 50/62 PTV methodology provides the most practical way to account for positional uncertainties to ensure coverage of the CTV.

9.4
Integration of Multimodality Image Data for 3D Planning

While CT is the principal source of image data for 3D planning, there is a growing demand to incorporate the complementary information available from magnetic resonance imaging (MRI). The sharply demonstrated tumor–soft tissue interface often seen on an MRI scan, in tumors such as those in the brain, can be used to better define the GTV. Several groups have also demonstrated the value of MRI in distinguishing the prostate gland from surrounding normal structures (LAU et al. 1996; ROACH et al. 1996). Also, functional imaging modalities, such as single photon emission computed tomography (SPECT) and positron emission tomography (PET), are proving to be important in both the target definition phase of treatment planning and also in the follow-up studies needed to assess efficacy (AUSTIN-SEYMOUR et al. 1995). For example, MARKS et al. (1995, 1997) have reported on the use of SPECT lung perfusion scans to determine functioning regions of the lung. The functional lung volume data are used in calculating DVHs rather than the CT-defined anatomy, and are referred to as functional DVHs. Radiation beams are planned that minimize irradiation of these functioning areas. Similarly, PET imaging is increasingly being used to aid in defining the patient's lung cancer GTV/CTV (ERDI et al. 2002). MAH et al. (2002) have recently reported on the impact of ^{18}F-fluoro-deoxy-2glucose hybrid (FDG)-PET on target and critical organs in CT-based treatment planning of patients with poorly defined

non-small-cell lung carcinoma. They found that co-registration of planning CT and FDG-PET images made significant alterations to patient management and to the PTV. A recent review summarizes the current literature of PET-based treatment planning in lung cancer (BRADLEY et al. 2004).

The accurate co-registration of the MR, SPECT, or PET imaging studies with the planning CT is a crucial step in the use of multimodality imaging in 3D planning. This requires calculation of a 3D transformation that relates the coordinates of a particular imaging study to the planning CT coordinates. A variety of quantitative methods have been developed to determine transformation parameters, including point matching, line or curve matching, surface matching, and volume matching. Qualitative methods that rely on manual alignment of the data sets are also commonly used. Details on these methodologies are given by KESSLER and LI (2001), ROSENMAN et al. (1998), and HILL et al. (2001). Once determined, the 3D transformation is used to integrate or "fuse" information, such as anatomic structure contours, from the imaging study with the planning CT.

9.5
Three-Dimensional Dose Calculation Algorithms

In 2D TPS, the dose calculations are typically done assuming that all tissue densities are those of water. The change in dose due to the presence of tissue inhomogeneities, such as the lungs, bony structures, air cavities, and metal prostheses, is related to the perturbation of the transport of primary and scattered photons and that of the secondary electrons set in motion from photon interactions. Depending on the energy of the photon beam, the shape, the size, and the constituents of the inhomogeneity, the resultant change in dose can be substantial.

Perturbation of photon transport is more noticeable for the lower-energy beams, as appreciated by their larger mass attenuation coefficients. Usually, an increase in transmission, and therefore dose, occurs when the beam traverses a low-density inhomogeneity. The reverse applies when the inhomogeneity has a density higher than that of water. The change in dose usually is lessened, however, because of the concomitant decrease or increase in the scatter dose. For a modest lung thickness of 10 cm, the increase in the dose to the lung for 6-mV X-rays would be about 15%, and be reduced to about 5% for 18-mV X-rays (MACKIE et al. 1985a).

When there is a net imbalance of electrons leaving and entering the region near an inhomogeneity, a condition of electron disequilibrium is created. The effects are similar to those in the build-up region, near a beam edge, or in a small beam. Because electrons have finite travel, the resultant change in dose usually is local to the vicinity of the inhomogeneity and may be large. The effects are more noticeable for the higher photon energy beams because of their increased energy and range. Near the edge of the lungs and air cavities, the reduction in dose can be >15% (YOUNG and KORNELSEN 1983). For inhomogeneities with density greater than water, an increase in dose occurs locally simply because of the generation of more electrons. Most dense inhomogeneities, however, have atomic numbers higher than that of water such that the resultant dose perturbation is compounded further by the perturbation of the multiple coulomb scattering of the electrons. Near the interface between a bony structure and water-like tissue, large hot and cold dose spots can be present.

In the past, dose calculation methods have traditionally been based on parameterizing dose distributions measured in water phantoms under standard conditions and applying correction factors to the beam representations for the nonuniform surface contour of the patient or the obliquity of the beam, tissue heterogeneities, and beam modifiers such as blocks, wedges, and compensator; however, in the past decade, several more advanced models have been developed for 3D TPS that compute the dose more from first principles and only use a limited set of measurements to obtain a better fit of the model. Examples of the more advanced type algorithms include the superposition/convolution method developed by MACKIE et al. (1985b) and the differential pencil-beam method developed by MOHAN et al. (1986). The reader is referred to reviews for the rigorous mathematical formalism of dose calculation algorithms, as only a brief review is presented here (AHNESJÖ and ASPRADAKIS 1999).

These 3D TPS methods utilize convolution energy deposition kernels that describe the distribution of dose about a single primary photon interaction site. The convolution kernels are most often obtained by using the Monte Carlo method to interact monoenergetic primary photons at the origin in a phantom and to transport the charged particles and scattered and secondary photons that are set in motion. The energy that gets deposited about the primary photon interaction site is tabulated and stored for use in

the convolution method. In addition to describing how scattered photons contribute to dose absorbed at some distance away from the interaction site of primary photons, the convolution kernels take into account charged particle transport. This information can be used to compute dose in electronic disequilibrium situations such as occurs in the build-up region and in the beam penumbra.

Eventually, the direct use of Monte Carlo simulations will likely be the preferred method for 3D TPS dose computation. It is the only method capable of computing the dose accurately near interfaces of materials with very dissimilar atomic numbers, as, for example, near metal prostheses. Researchers are actively working on reducing the computation time to make this approach practical (CYGLER et al. 2004; FRAASS et al. 2003; LIU 2001; HARTMANN SIANTAR et al. 2001).

9.6
Dose-Volume Histograms

The large amount of dosimetric data that must be analyzed when a 3D CRT plan is evaluated has prompted the development of new methods of condensing and presenting the data in more easily understandable formats. One such data-reduction tool is the DVH (DRZYMALA et al. 1991). Two types of DVHs, differential and cumulative, are used in 3D CRT planning.

9.6.1
Differential Dose-Volume Histograms

Figures 9.17 and 9.18 illustrate how a differential DVH is generated for a defined volume that is subjected to an inhomogeneous dose distribution. Firstly, the volume under consideration is divided into a 3D grid of volume elements (voxels), the size of which is small enough so that the dose can be assumed to be constant within one voxel. The volume's dose distribution is then divided into dose bins and the voxels grouped according to dose bin without regard to anatomic location. A plot of the number of voxels in each bin (x-axis) vs the bin dose range (y-axis) is, by definition, a differential DVH. The size of the dose bins determines the height of each bin of the differential DVH. For example, if the bin widths were increased, the heights of the histogram bins generally would increase because more voxels would fall into any given bin; thus, it should be clearly

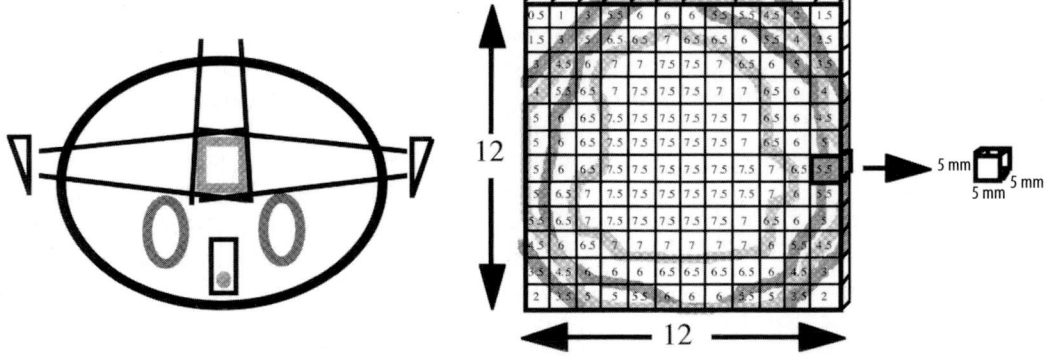

Fig. 9.17 Dose grid for a hypothetical plan. In this plan, an irradiated organ has been divided into 100 5-cm³ voxels, each of which receives 0–7.5 Gy. The number of voxels receiving a given dose range is indicated. For example, 22 voxels received ≥1 Gy but <2 Gy. (From Lawrence et al. 1996)

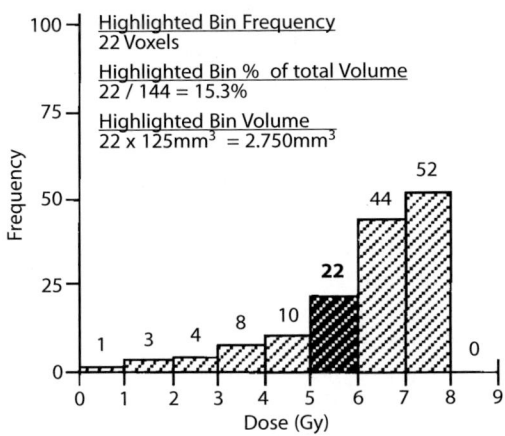

Fig. 9.18 Differential dose-volume histogram display of the voxels shown in Fig. 9.17. The abscissa shows the 1-Gy bin sizes. The ordinate is expressed in a variety of equally valid units: number of voxels (directly from Fig. 9.11–9.5), volume (cm³; equal to the voxel number ×5 cm³ per voxel), and volume (%; equal to the fraction of the total volume in that bin). For instance, 11 voxels (or 60 cm³ or 11% of the organ) received 2 Gy or more but <3 Gy. (From Lawrence et al. 1996)

understood that the detailed shape of a differential DVH depends on the bin choice, even though the underlying dose-volume data are not different.

9.6.2
Cumulative Dose-Volume Histograms

A cumulative DVH is a plot in which each bin represents the volume, or percentage of volume (y-axis), that receives a dose equal to or greater than an indicated dose (x-axis). An example of a cumulative DVH is shown in Figure 9.19, in which the value at any dose bin is computed by summing the number of voxels of the corresponding differential DVH to the

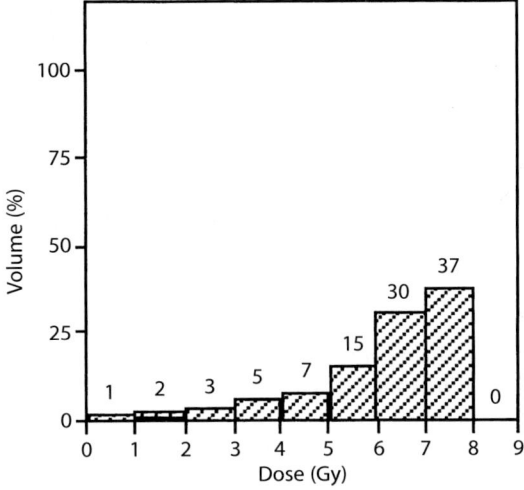

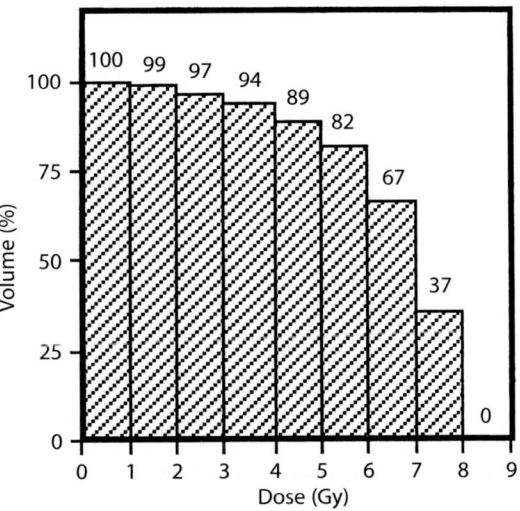

Fig. 9.19 Cumulative dose-volume histogram (DVH) display of the voxels shown in Fig. 9.11–17. This figure contains the same data as shown in Fig. 9.11–18, but now displayed as a cumulative DVH. For instance, 71 voxels (or 350 cm³ or 71% of the organ) received Gy or more. (From Lawrence et al. 1996)

right of that dose bin. The volume value for the first bin (dose origin) is the full volume of the structure because the total volume receives at least zero dose, and the volume for the last bin is that which receives the maximum dose bin.

9.6.3
Dose-Volume Statistics

Explicit values of dose-volume parameters can be extracted from the DVH data and are called dose-volume statistics, or simply, dose statistics. Examples include maximum point dose, minimum point dose, mean dose, percentage volume receiving greater than or equal to the prescription dose for target volumes and maximum point dose, mean dose, and percentage volume receiving greater than or equal to an established tolerance dose for organs at risk. There is some question as to whether point doses are meaningful clinically, and perhaps maximum dose should be reported for the dose averaged over a small but clinically significant volume.

9.6.4
Plan Evaluation Using Dose-Volume Histograms and Dose-Volume Statistics

The DVH is an essential tool used for 3D CRT plan evaluation. The planner can review the DVHs for the PTV and ORs for the dose distribution under review, or superimpose DVHs for a specific PTV or OR from several competing plans on one plot and compare them directly. The DVH display effectively points out the ORs that are overdosed as well as any target volume(s) that may be underdosed. It is useful to generate a DVH for what is called "unspecified tissues" (those voxels within the skin contour that are not contained within contours for which DVHs have been generated). The unspecified tissue DVH helps prevent the radiation oncologist from overlooking high-dose information that may be clinically significant.

While a set of DVHs provides a complete summary of the entire 3D dose matrix showing the amount of target volume or critical structure receiving more or less than a specified dose level, it does not provide any spatial information; thus, the DVH can only complement, and not replace, spatial dose-distribution displays.

Sometimes the differences between the DVHs of all of the volumes of interest of two compared plans are clear (Fig. 9.20a), and one can easily determine which is the better plan; however, this is not the case between DVHs for a normal tissue that crosses over in mid-range (Fig. 9.20b), with one being higher than the other at low doses and lower at high doses. This difficulty has prompted the development of the biological indices for plan evaluation, which are discussed in the next section.

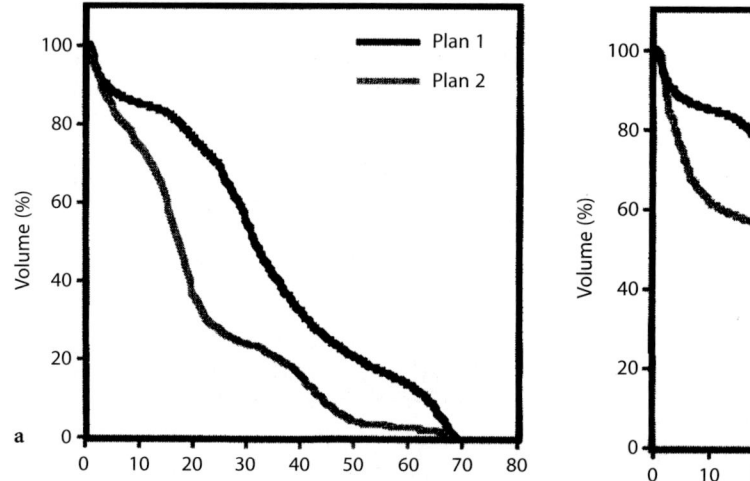
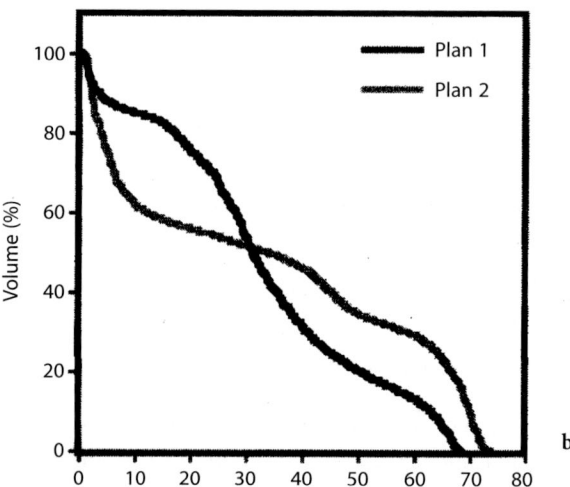

Fig. 9.20 a Cumulative DVHs of a normal tissue produced by two different plans in which one is completely to the left of the other. If these two plans give the same tumor coverage, then plan 2 will surely cause less toxicity. **b** Cumulative DVHs of normal tissue produced by two different plans in which the DVHs cross. In this example, when compared with plan 2, plan 1 treats less normal tissue with a low dose, but more normal tissue with a high dose. Plan 1 is typical of two opposed fields, whereas plan 2 represents multiple noncoplanar beams. The less toxic plan is not obvious. (From Lawrence et al. 1996)

9.7
Biological Models

Because 3D CRT plans provide both dose and volume information, the traditional practice of determining the best plan is proving extremely difficult. For example, it is not clear what degree of dose uniformity in the PTV can be tolerated as dose levels are escalated using 3D CRT or how high of a dose can be tolerated by a small portion of a normal structure. In the past, class solutions based on clinical experience were used because quantitative 3D dose-volume data were not available. Researchers are now developing biophysical models that attempt to translate the dose-volume information into estimates of biologic impact, i.e., tumor control probability (TCP) and normal tissue complication probability (NTCP) models. Presently, most agree that the TCP and NTCP models developed thus far are not accurate to the extent that the absolute values can be used to predict response, but they can be used to compare rival plans. In any case, such biological indices should be used only under protocol conditions until their utility has been firmly established for routine clinical use.

9.7.1
Normal Tissue Complication Probability

Currently, there are two different approaches in modeling normal tissue complication probability (NTCP): the empiric model introduced by LYMAN and WOLBARST (1987, 1989), and the functional models based on the functional subunit (FSU) concept (KÄLLMAN et al. 1992; OLSEN et al. 1994; WITHERS et al. 1988).

The Lyman model can be expressed by an error function of dose and volume:

$$NTCP = \frac{1}{\sqrt{2\pi}} \int_{-\infty}^{1} \exp(-t^2/2)dt \, ,$$

where

$$V = V/V_{ref} \, ,$$

$$t = (D - TD_{50}(v))/(m \cdot TD_{50}(v)) \, ,$$

and

$$TD_{50}(1) = TD_{50}(v) \cdot V^{-n}$$

$TD_{50}(1)$ is the tolerance dose for 50% complications for uniform whole-organ irradiation, whereas $TD_{50}(v)$ is the 50% tolerance dose for uniform partial-organ irradiation to the partial volume V. The arbitrary variables m and n are found by fitting tolerance doses for uniform whole and uniform partial-organ irradiation, where m characterizes the gradient (slope) of the dose-response function at TD_{50} and n characterizes the effect of volume. When n is near unity, the volume effect is large; conversely, when n is near zero, the volume effect is small. When NTCP is plotted against dose, the NTCP equation demonstrates a sigmoid shape.

Two methods are currently used to extend this method to nonuniform organ irradiation. The interpolation method, proposed by LYMAN and WOLBARST (1989), modifies the DVH to one in which the organ receives an effective dose, D_{eff}, which is less than or equal to the maximum organ dose. The effective volume method, proposed by KUTCHER and BERMAN (1989), modifies the DVH to one in which a fraction of the organ, v_{eff}, receives the maximum organ dose. With this method, a uniformly irradiated dose equivalent is calculated for each tissue that contains dose heterogeneities (Fig. 9.21). For example, each step in the histogram of height

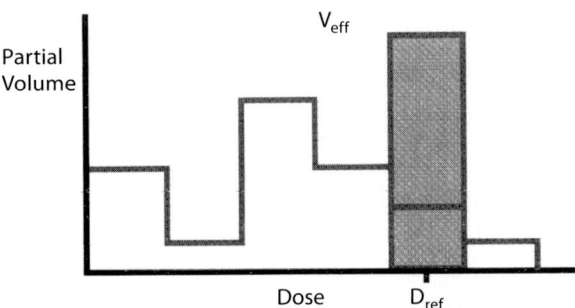

Fig. 9.21 The effective-volume DVH reduction scheme. For a differential DVH (**a**), volume elements at one dose level are transposed to an effective volume at the reference dose level through use of the power-law relationship. This results in the single-step DVH (**b**), predicted to yield NTCPs identical to the original nonuniform irradiation DVHs. (Modified from KUTCHER and BERMAN 1989)

ΔV_i and extension D_i is assumed to satisfy a power-law relationship so that it adjusts to one of smaller-volume V_{eff} and extension D_{max} using

$$V_{eff} = \Delta V_{max} + \Delta V_i (D_i / D_{max}) \frac{1}{n} + \Delta V_2 (d_2 / D_{max}) \frac{1}{n} + \dots$$

where n is a size parameter.

Other NTCP models include the critical volume model proposed by NIEMIERKO and GOITEIN (1991), which they applied to the appearance of nephritis. Its form is similar to that of LYMAN and WOLBARST (1987) but includes additional terms to account for the radiosensitivity of FSUs in the kidney. The group at the University of Michigan has proposed a simple phenomenological model for normal tissue complication, based on the sigmoid relationship derived by GOITEIN (1987). In their model, however, they have nested the sigmoid cell-killing function into another sigmoid relationship describing complication or functional damage of an organ.

9.7.2
Tumor Control Probability

Tumor control probabilities have been modeled by BRAHME (1984) and (GOITEIN 1987), who proposed that, for a uniform dose within a tumor volume, TCP may be computed using the following equation:

$$TCP = \frac{1}{\left(1 + \left(\dfrac{Dose}{D_{50}}\right)^{-K}\right)},$$

where

$$K = 4/\Gamma,$$

and Γ=slope of the dose response curve at 50% cell death. For a nonuniform dose distribution, the total volume is reduced to smaller volumes having "uniform" doses within. The TCPs are computed for each volume element and are then weighted according to their volume fractions and summed according to the following equation:

$$\ln TCP = \sum \left(\frac{\Delta V}{V_0}\right) \ln TCP(V_0, D(x,y,z)),$$

where V_0=total tumor volume, and ΔV=volume element of dose D.

When TCP is plotted against dose, the TCP equation demonstrates a sigmoid shape.

9.7.3
Equivalent Uniform Dose

Equivalent uniform dose (EUD) is a concept introduced by NIEMIERKO (1997a) for use in evaluating and reporting inhomogeneous dose distributions. This concept assumes that any two dose distributions are equivalent if they cause the same radiobiological effect and appears well suited for use in evaluating 3D CRT plans; however, McGARY et al. (1997, 2000) has pointed out that there are conditions in which EUD is not adequate as a single parameter to report or analyze inhomogeneous dose distributions, e.g., when the minimum dose is significantly lower than the mean dose (NIEMIERKO 1997b).

In summary, and as stated previously, the biological models developed thus far for 3D TPS are not accurate to the extent that the absolute values can be used to predict response; however, they can be used to compare rival plans.

9.8
Management of Three-Dimensional Treatment Planning Data

To perform the steps involved in 3D CRT, several forms of patient imaging and other data must be acquired, displayed, manipulated, and stored. Typically, patient image data acquired from several imaging subsystems must be communicated to a 3D TPS to permit these images to be used for treatment planning. Several software components also must be integrated so that the output of one processing step can be made available for use as input to the next step. The issues in data management in 3D CRT are complex and continue to be somewhat problematic.

It is important to understand that nontrivial differences exist in the way various 3D TPS describe the details of radiation treatment, including units of measure and coordinate systems for specifying the geometric relationship between the patient and treatment beams. Efforts are underway to define a consistent set of data objects and a representation for these objects that can be used to design, evaluate, execute, and verify a 3D CRT plan; these include the Radiation Therapy Oncology Group Tape/Network Format for Exchange of Treatment Planning Information (HARMS et al. 1997), which is based on work of BAXTER et al. (1982). This data exchange specification defines seven data objects: CT scans

(3D images); structures (volume contours); beam geometries; digital film images (simulation images, portal images, and DRRs); dose distributions (3D array); DVHs; and comments (free text).

Another significant approach uses the framework of the American College of Radiology (ACR)/NEMA DICOM 3.0 standard for representing and communicating digital medical image (BENNETT and MCINTYRE 1993). Image and treatment planning data are modeled as DICOM information objects, and services that act on these information objects (e.g., storage or printing) are identified. DICOM information objects are used to represent CT scans, structures, treatment plan specifications, dose distributions and DVHs, 2D radiographs, and treatment verification information.

Patient CT images are the basic input data for 3D TPS and dominate the storage requirements for data that support the treatment-planning process. The image data must be transferred from the imaging systems where they are acquired into the 3D TPS before any planning can take place. The image transfer process generally entails data transfer and format conversion. Transferring these data to a 3D TPS can be accomplished by several means, including physical media (disk or tape) exchange, point-to-point connection, or network communication. Because of their efficiency, flexibility, and low cost, computer networks have become the method of choice for communicating digital data into the 3D TPS.

Several hardware technologies exist for connecting computers together in a local area network (LAN). Ethernet, originally developed by Xerox Corp., is without question the most popular LAN technology used to interconnect 3D TPS workstations.

The rules governing the transmission of data over a network are referred to as a protocol. Several protocols have been developed to support the transfer of data files between computers. File Transfer Protocol (FTP) was developed for data file transport between computers on the Internet and is part of the Transmission Control Protocol/Internet Protocol suite of network protocols. The FTP client and server software has been implemented for a wide range of computer hardware and operating systems. Because of its flexibility, FTP has found application both locally (e.g., to transfer data files between a CT scanner computer and the main 3D TPS file server) and between computers at widely separated sites on the Internet (e.g., to exchange 3D TPS data for multi-institutional clinical studies; PURDY et al. 1996).

Once image data have been acquired and transferred from the imaging system to the 3D TPS workstation, two important steps still remain. Firstly, the image data must be converted into a format that can be used by the 3D TPS software, and secondly, the image data must be associated with the related, non-image information that is needed to interpret it.

If the format of image files is known, format conversion is a fairly straightforward task. Particular attention must be directed to correct interpretation of image header information and translation of image data values. Numeric representations of image pixel values vary from machine to machine. In addition, some manufacturers use compression schemes to reduce the space required to store images. In such cases, images must first be decompressed before they can be reformatted for use with the 3D TPS.

Non-image information that must be associated with the image comes from the image acquisition process as well as from the clinical environment. An example of the former is calibration information that permits interpretation of image pixel values, such as the CT values representing air and water and the size of image pixels. Clinical information includes patient identification and demographics, as well as the date and time of acquisition and identification of the image acquisition system.

Data compression can be used to reduce the size of files because managing the large volumes of data that must be stored is a major issue in 3D CRT. The amount of storage space saved by compressing files is strongly data dependent; however, readily available file compression utilities yield compression ratios of 2:1 (savings of 50%) or greater for many 3D TPS data files. The disadvantage of compressing data is that the cost of decompressing the file must be addressed when access is needed.

Another useful technique is the archiving of data on removable media, such as magneto-optical disks. Jukebox systems, which combine storage slots for several magneto-optical disk platters, a robotic changer, and a disk drive mechanism can be used to gain automated access to any of the disks. Experience with such a system has indicated that, because the overhead of swapping disks is costly, concurrent access to multiple disk platters in the jukebox should be avoided; thus, these systems are probably best regarded as off-line storage: data should be copied to faster magnetic disks before they are processed by 3D TPS software.

The protection of data security and the maintenance of data consistency in the presence of concurrent access are two of the issues driving a trend toward the use of database management systems (DBMS) rather than simple file systems to store 3D CRT data. DBMS, long used for hospital information management, are now finding application in managing treatment planning and treatment verification data.

Access to database contents can be controlled with more precision than is permitted by operating-system file access permissions, and database systems generally provide superior control of concurrent data accesses. Formal data models supported by many DBMS allow explicit representation of data entities and the relationships among them. Additionally, the ability to support queries that link 3D CRT image and treatment planning data to other clinical information has important benefits both for managing the care of patients and for investigating the relationship between treatment and outcome.

9.9
Summary and Conclusion

The use of 3D CRT treatment planning and treatment has had a major impact on the practice of radiation therapy. There are very few tumor types whose treatment has not been radically impacted by its use. For the most part this impact has been of major importance and benefit. But as with any major technical advance, the requirements for its use have had a major impact on the requirement for enhanced quality assurance from all members of the treatment team. If one carefully reads the previous sections related to the quality-assurance demands made on the entire treatment team, it is obvious that a great deal more time is required from all involved in the quest to achieve precise and increased dose to the tumor PTV, and to minimize the dose to the normal organs at risk PRV.

References

Ahnesjö A, Aspradakis MM (1999) Dose calculations for external photon beams in radiotherapy. Phys Med Biol 44: R99–R155

Austin-Seymour M et al. (1995) Tumor and target delineation: current research and future challenges. Int J Radiat Oncol Biol Phys 33(5):1041–1052

Baxter BS, Hitchner LE, Maguire J (1982) GQ AAPM report no. 10: A standard format for digital image exchange. American Institute of Physics, New York

Bedford JL, Shentall GS (1998) A digital method for computing target margins in radiotherapy. Med Phys 25(2):224–231

Bennett B, McIntyre J (1993) Understanding DICOM 3.0 version 1.0. Kodak Health Imaging Systems

Bradley J et al. (2004) Implementing biological target volumes in radiation treatment planning for non-small cell lung cancer. J Nucl Med 45 (Suppl 1):96S–101S

Brahme A (1984) Dosimetric precision requirements in radiation therapy. Acta Radiol Oncol 23:379–391

Cygler JE et al. (2004) Evaluation of the first commercial Monte Carlo dose calculation engine for electron beam treatment planning. Med Phys 31(1):142–153

Drzymala RE et al. (1991) Dose-volume histograms. Int J Radiat Oncol Biol Phys 21(1):71–78

Emami B et al. (1991) Tolerance of normal tissue to therapeutic irradiation. Int J Radiat Oncol Biol Phys 21(1):109–122

Erdi YE et al. (2002) Radiotherapy treatment planning for patients with non-small cell lung cancer using positron emission tomography (PET). Radiother Oncol 62(1):51–60

Fraass BA, Smathers J, Deye JA (2003) Summary and recommendations of a National Cancer Intitute workshop on issues limiting the clinical use of Monte Carlo dose calculation algorithms for megavoltage external beam radiation therapy. Med Phys 30(12):3206–3216

Garcia-Ramirez JL et al. (2002) Performance evaluation of an 85-cm-bore X-ray computed tomography scanner designed for radiation oncology and comparison with current diagnostic CT scanners. Int J Radiat Oncol Biol Phys 52(4):1123–1131

Goitein M (1987) The probability of controlling an inhomogeneously irradiated tumor.

Goitein M et al. (1983) Multi-dimensional treatment planning: II Beam's eye view, back projection, and projection through CT sections. Int J Radiat Oncol Biol Phys 9:789–797

Harms WB Sr, Bosch WR, Purdy JA (1997) An interim digital data exchange standard for multi-institutional 3D conformal radiation therapy trials. In: Twelfth International Conference on the Use of Computers in Radiation Therapy, Salt Lake City Utah. Medical Physics Publishing, Madison, Wisconsin,

Hartmann Siantar CL et al. (2001) Description and dosimetric verification of the Peregrine Monte Carlo dose calculation system for photon beams incident on a water phantom. Med Phys 28(7):1322–1337

Hill DLG et al. (2001) Medical image registration. Phys Med Biol 46:R1–R45

ICRU (1993a) ICRU, report 50. Prescribing, recording, and reporting photon beam therapy I.C.o.R.U.a. Measurements editor. International Commission on Radiation Units and Measurements, Bethesda, Maryland

ICRU (1993b) Prescribing, recording, and reporting photon beam therapy. International Commission on Radiation Units and Measurements, Bethesda, Maryland

ICRU (1999) ICRU report 62. Prescribing, recording, and reporting photon beam therapy (supplement to ICRU report 50). International Commission on Radiation Units and Measurements, Bethesda, Maryland

IMRT (2001) Collaborative Working Group. Intensity modulated radiation therapy: current status and issues of interest. Int J Radiat Oncol Biol Phys 51(4):880–914

Källman P, Lind BK, Brahme A (1992) An algorithm for maximizing the probability of complication free tumor control in radiation therapy. Int J Radiat Oncol Biol Phys 37:871–890

Kessler ML, Li K (2001) Image fusion for conformal radiation therapy. In: Purdy JA et al. (eds) 3-D conformal and modulated radiation therapy: physics and clinical applications. Advanced Medical Publishing, Madison, Wisconsin, pp 71–82

Kutcher G, Berman C (1989) Calculation of complication probability factors for non-uniform tissue irradiation: the effective volume method. Int J Radiat Oncol Biol Phys 16:1623–1630

Langen KM, Jones DTL (2001) Organ motion and its management. Int J Radiat Oncol Biol Phys 50(1):265–278

Lau HY et al. (19960 Short communication: CT–MRI image fusion for 3D conformal prostate radiotherapy: use in patients with altered pelvic anatomy. Br J Radiol 69(825):1165–1170

Lawrence TS, Kessler ML, Ten Haken RK (1996) Clinical interpretation of dose-volume histograms: the basis for normal tissue preservation and tumor dose escalation. In: Meyer JL, Purdy JA (eds) Frontiers of radiation therapy oncology, vol 29. Karger, Basel, pp 57–66

Liu HH (2001) Status of Monte Carlo dose calculation algorithms for three-dimensional treatment planning. In: Purdy JA et al. (eds) 3-D conformal and modulated radiation therapy: physics and clinical applications. Advanced Medical Publishing, Madison, Wisconsin, pp 201–220

Lyman JT, Wolbarst AB (1987) Optimization of radiation therapy III: a method of assessing complication probabilities from dose-volume histograms. Int J Radiat Oncol Biol Phys 13:103–109

Lyman JT, Wolbarst AB (1989) Optimization of radiation therapy IV: a dose-volume histogram reduction algorithm. Int J Radiat Oncol Biol Phys 17(2):433–436

Mackie TR et al. (1985a) Lung dose corrections for 6- and 15-MV X rays. Med Phys 12:327–332

Mackie TR, Scrimger JW, Battista JJ (1985b) A convolution method of calculating dose for 15-MV X-rays. Med Phys 12:188–196

Mackie TR et al. (1996) Photon beam dose computations in teletherapy: present and future. In: Palta J, Mackie TR (eds) Advanced Medical Publishing, College Park, Maryland, pp 103–136

Mah K et al. (2002) The impact of 18 FDG-PET on target and critical organs in CT-based treatment planning of patients with poorly defined non-small-cell lung carcinoma: a prospective study. Int J Radiat Oncol Biol Phys 52(2):339–350

Marks LB et al. (1995) The role of three dimensional functional lung imaging in radiation treatment planning: the functional dose-volume histogram. Int J Radiat Oncol Biol Phys 33(1):65–75

Marks LB et al. (1997) Quantification of radiation-induced regional lung injury with perfusion imaging. Int J Radiat Oncol Biol Phys 38(2):399–409

McGary JE et al. (1997) Comment on "Reporting and analyzing dose distributions: a concept of equivalent uniform dose [Med Phys 24:103–109 (1997)]. Med Phys 24(8):1323–1324

McGary JE, Grant W, Woo SY (2000) Applying the equivalent uniform dose formulation based on the linear-quadratic model to inhomogeneous tumor dose distributions: caution for analyzing and reporting. J Appl Clin Med Phys 1(4):126–137

McShan DL et al. (1979) A computerized three-dimensional treatment planning system utilizing interactive color graphics. Br J Radiol 52:478–481

Meyer JL, JA Purdy (eds) (1996) 3-D Conformal radiotherapy: a new era in the irradiation of cancer. Frontiers of radiation therapy and oncology, vol 29. Karger, Basel

Mohan R, Chui C, Lidofsky L (1986) Differential pencil beam dose computation model for photons. Med Phys 13:64–73

Mutic S et al. (2003) Quality assurance for computed-tomography simulators and the computed-tomography-simulation process: report of the AAPM Radiation Therapy Committee Task Group no. 68. Med Phys 30(10):2762–2792

Niemierko A (1997a) Reporting and analyzing dose distributions: a concept of equivalent uniform dose. Med Phys 24(1):103–110

Niemierko A (1997b) Response to "Comment on 'Reporting and analyzing dose distributions: a concept of equivalent uniform dose'" [Med Phys 24:1323–1324 (1997)]. Med Phys 24(8):1325–1327

Niemierko A, Goitein M (1991) Calculation of normal tissue complication probability and dose-volume histogram reduction schemes for tissues with a critical element architecture. Radiother Oncol 20:166–176

Olsen DR, Kambestad BK, Kristoffersen DT (1994) Calculation of radiation induced complication probabilities for brain, liver and kidney, and the use of a reliability model to estimate critical volume fractions. Br J Radiol 67:1218–1225

Perez CA et al. (1994) Design of a fully integrated three-dimensional computed tomography simulator and preliminary clinical evaluation. Int J Radiat Oncol Biol Phys 30(4):887–897

Purdy JA (1992) Photon dose calculations for three-dimensional radiation treatment planning. Semin Radiat Oncol 2(4):235–245

Purdy JA (1996a) 3-D radiation treatment planning: a new era. In: Meyer JL, Purdy JA (eds) Frontiers of radiation therapy and oncology 3-D conformal radiotherapy: a new era in the irradiation of cancer. Karger, Basel, pp 1–16

Purdy JA (1996b) Defining our goals: volume and dose specification for 3-D conformal radiation therapy. In: Meyer JL, Purdy JA (eds) Frontiers of radiation therapy and oncology 3-D conformal radiotherapy: a new era in the irradiation of cancer. Karger, Basel, pp 24–30

Purdy JA (1996c) Volume and dose specification, treatment evaluation, and reporting for 3D conformal radiation therapy. In: Palta J, Mackie TR (eds) Teletherapy: present and future. Advanced Medical Publishing, College Park, Maryland, pp 235–251

Purdy JA (1997) Advances in three-dimensional treatment planning and conformal dose delivery. Semin Oncol 24:655–672

Purdy JA (1998) Three-dimensional treatment planning and conformal dose delivery: a physicist's perspective. In: Mittal BB, Purdy JA, Ang KK (eds) Advances in radiation therapy. Kluwer Academic Publishers, Boston, pp 1–33

Purdy JA (1999) 3D treatment planning and intensity-modulated radiation therapy. Oncology 13:155–168

Purdy JA (2000) Dose-volume specification and reporting. In: Shiu AS, Mellenberg DE (eds) General practice of radiation oncology physics in the 21st century. Medical Physics Publishing, Madison, Wisconsin, pp 3–15

Purdy JA, G Starkschall (1999) A practical guide to 3-D planning and conformal radiation therapy. Advanced Medical Publishing, Madison, Wisconsin, p 369

Purdy JA et al. (1987) Three dimensional radiation treatment planning system. In: Proc 9th International Conference on the Use of Computers in Radiation Therapy. Elsevier, Scheveningen, The Netherlands

Purdy JA et al. (1993) Advances in 3-dimensional radiation treatment planning systems: room-view display with real time interactivity. Int J Radiat Oncol Biol Phys 27(4):933–944

Purdy JA et al. (1996) Multi-institutional clinical trials: 3-D conformal radiotherapy quality assurance. In: Meyer JL, Purdy JA (eds) Frontiers of radiation therapy and oncology 3-D conformal radiotherapy: a new era in the irradiation of cancer. Karger, Basel, pp 255–263

Reinstein LE et al. (1978) A computer-assisted three-dimensional treatment planning system. Radiology 127:259–264

Roach M et al. (1996) Prostate volumes defined by magnetic resonance imaging and computerized tomographic scans for three-dimensional conformal radiotherapy. Int J Radiat Oncol Biol Phys 35(5):1011–1018

Rosenman JG et al. (1998) Image registration: an essential part of radiation therapy treatment planning. Int J Radiat Oncol Biol Phys 40(1):197–205

Sherouse GW, Novins K, Chaney EL (1990) Computation of digitally reconstructed radiographs for use in radiotherapy treatment design. Int J Radiat Oncol Biol Phys 18(3):651–658

Siddon RL (1985) Fast calculation of the exact radiological path for a three-dimensional CT array. Med Phys 12:252–255

Withers HR, Taylor JMG, Maciejewski B (1988) Treatment volume and tissue tolerance. Int J Radiat Oncol Biol Phys 14:751–759

Young MEJ, Kornelsen RO (1983) Dose corrections for low-density tissue inhomogeneities and air channels for 10-MV X-rays. Med Phys 10(4):450–455

10 Intensity-Modulated Radiation Therapy

Daniel A. Low, Wei Lu, James A. Purdy, Carlos A. Perez, and Seymour H. Levitt

CONTENTS

10.1 Introduction 203
10.2 Basic Physical Principles of IMRT 207
10.3 Inverse Treatment Optimization
(Cost Function and Search Algorithm) 208
10.4 IMRT Treatment Delivery Systems 210
10.4.1 Historical Review 210
10.4.2 Arc-Based Fan-Beam Dynamic MLC IMRT 211
10.4.3 Fixed Portal Cone-Beam Dynamic MLC 211
10.4.4 Fixed-Portal X-ray Compensating Filter IMRT 211
10.5 IMRT Quality Assurance
(Commissioning and Clinical) 212
10.5.1 Dosimetry System 213
10.5.2 Commissioning 214
10.5.3 System Quality Assurance 214
10.5.4 Monitor Unit Verification 215
10.5.5 Clinical Quality Assurance 215
10.5.6 Tolerance Levels 216
10.5.7 Doses of Irradiation Outside
the Treatment Volume 216
10.6 Clinical Studies with IMRT 217
10.6.1 Head and Neck 218
10.6.2 Prostate 220
10.7 Discussion and Summary 224
10.8 Additional Challenges for
IMRT Implementation 224
References 225

10.1 Introduction

The 1990s saw dramatic changes in radiation therapy treatment planning and treatment delivery, driven largely by advances in computer hardware and software. These advances inspired the development of sophisticated three-dimensional (3D) radiation therapy treatment planning (3D RTTP) and delivery systems, making practical the implementation of 3D conformal radiation therapy (3D CRT). The purpose of 3D CRT was to conform the prescribed dose distribution to the 3D target volume (cancerous cells) shape while simultaneously minimizing the dose to neighboring normal patient structures (Meyer and Purdy 1996). Conventional 3D CRT delivery was implemented by conforming the incident-beam portal outlines to the target volume projections for a user-specified set of beam directions or during rotational beam delivery. The radiation beams had uniform intensity, or where appropriate, the intensity was varied by beam modifiers such as compensating filters or wedges. This treatment method was referred to as conventional 3D CRT.

3D CRT provides excellent conformation of the dose distribution with the tumor targets. By matching the radiation portal shape to the projected tumor shape, the radiation beams produces a dose distribution that encompasses the tumor shape. Because the radiation dose distribution is being delivered by relatively homogeneous fluences, the common practice is to assign the dose delivered by each single beam as the dose delivered by that beam at the linear accelerator isocenter. The isocenter point of each beam is placed at a common location within the patient's tumor. This allows a straightforward mechanism for categorizing the dose delivered to the patient, namely the sum of the individual radiation beam isocenter doses. This process helps to make the prescription and monitor unit calculation protocols relatively straightforward.

Even though conventional radiation beams are labeled as homogeneous, the dose distribution has slight variations throughout the portal, with a rapid drop-off near the projected portal boundary. When coupled with variations in the patient's surface contours and internal heterogeneities (defined as variations in the patient's electron density distribution relative to water), the total dose delivered to the target is heterogeneous. Prescription protocols,

D. A. Low, PhD, W. Lu, PhD, C. A. Perez, MD
Department of Radiation Oncology, Mallinckrodt Institute of Radiology, Washington University School of Medicine, 4921 Parkview Place, St. Louis, MO 63110, USA
J. A. Purdy, PhD
Department of Radiation Oncology, University of California Davis Medical Center, Sacramento, CA 95817, USA
S. H. Levitt, MD
Department of Therapeutic Radiation Oncology, University of Minnesota, Minneapolis, MN 55455, USA

therefore, often identify dose distribution variation limits, including the maximum acceptable dose hot spot and the minimum acceptable dose. Another criterion includes the percentage of the target encompassed covered by the prescription (isocenter) dose. This prescription method is often so rigid that the selection of radiation beam locations and orientations is constrained to keep the beam isocenters in a common location.

The 3D-CRT treatment planner has a few degrees-of-freedom with respect to developing the treatment plan. They can typically select from two or more beam energies, affecting the overall penetration of the beam, the depth of the build-up region, and the width of the beam-edge penumbra. They can select from a continuous spectrum of gantry angles and treatment couch angles to preferentially align the radiation beam within the patient's anatomy. They can select the number of beams that are used and their weights (typically expressed as the dose to isocenter delivered by each beam), and can use off-the-shelf beam-fluence modification tools such as wedges or simple custom-designed surface-compensation filters (ELLIS and MILLER 1944; ELLIS et al. 1959). Wedges are designed to tilt the radiation beam dose distribution relative to the central axis to compensate for abnormal incidence on the patient's surface or for the dose distribution caused by intersecting radiation beams. In many tumor sites, e.g., in the treatment of prostate cancer, class solutions are developed by clinics to improve efficiency in the treatment planning process.

In 3D CRT, while the radiation targets are able to be covered, there is no mechanism for selectively avoiding critical structures. These 3D-CRT dose distributions have convex surfaces and cannot "wrap around" critical structures. Simply shielding the projection of a critical structure by a radiation block or multileaf collimator produces heterogeneities in the resulting dose distribution that are impossible to compensate for by adjusting the other radiation portals. The relatively strict limitations on target dose distribution homogeneity precludes the partial blocking of the radiation portals. Solutions to this problem include limiting the target dose and mixing radiation beam modalities, such as photon and high-energy electron treatments. For head and neck treatments, lateral photon beams can adequately irradiate the target tissues but would over-irradiate the spinal cord. The treatment is typically divided into two courses: firstly, the spinal cord is irradiated along with the target tissues until the cord tolerance is reached. Then, the photon beam is blocked

to avoid the cord and electron beams are patched to the photon beams. The electron beam energy is selected so that the cord lays beyond the practical range of the electron beam (Fig. 10.1). Even with this level of sophistication, these treatments irradiate the salivary glands causing significant morbidity and a degradation in the quality of life.

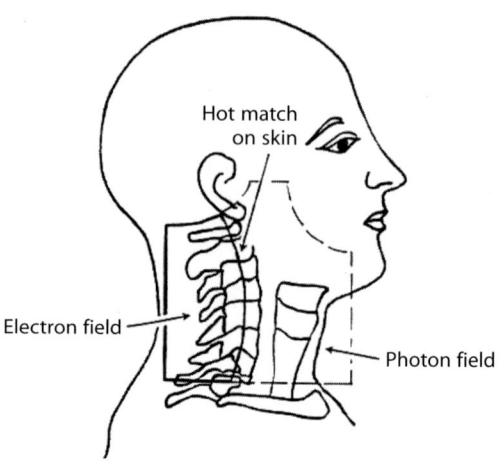

Fig. 10.1 Setup diagram shows a hot match of photon field and electron field on skin

Almost immediately after the introduction of 3D CRT, the radiation therapy physics community began to understand that the radiation dose distribution shape could be significantly altered to conform to target shapes while selectively avoiding critical structures. The underlying technology required was to remove the restriction of homogeneous incident fluences and allow the fluences to vary throughout the radiation portal, with selection of the fluence intensities typically made using sophisticated algorithms. This process is sometimes referred to as the "inverse problem" or "inverse method" of treatment planning (Fig. 10.2). While the use of fluence modulation does not remove radiation from critical structures, it does limit the dose to the structures directly within the beam's path. The resulting cold regions within the tumor are compensated by increasing the fluence from the other radiation beams. The addition of fluence modulation leads to the term intensity-modulated radiation therapy (IMRT) as the name for this modality.

The use of IMRT requires that the fluences be adjustable within the constraints of the dose-delivery technology. There are two general delivery technologies that can deliver IMRT dose distributions, both of which have hundreds to thousands of individual fluence elements (bixels) that require adjust-

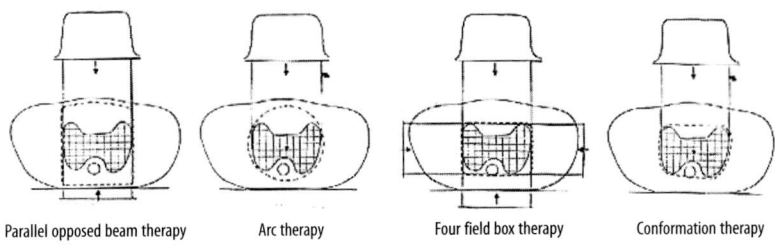

Parallel opposed beam therapy   Arc therapy   Four field box therapy   Conformation therapy

Non uniform beam radiotherapy

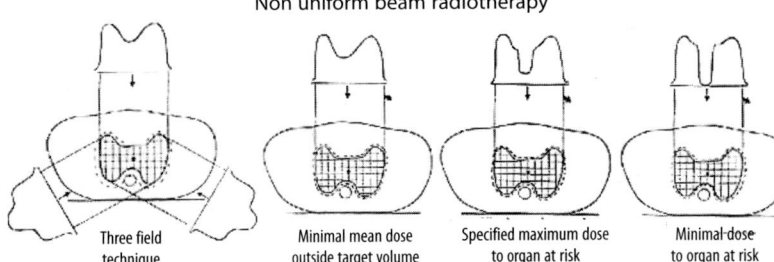

Three field technique   Minimal mean dose outside target volume   Specified maximum dose to organ at risk   Minimal dose to organ at risk

Fig. 10.2 Conventional and intensity-modulated radiation therapy (IMRT) dose distribution to a complex target volume (*hatched*) and critical structure geometry. The *upper four figures* indicate the dose distribution using conventional planning and delivery. The *lower four figures* indicate fixed-beam and arc-based IMRT with different optimization criteria

ment. The bixel intensities require adjustments that are too complex for individual treatment planners to conduct. The modification of single bixel intensity is coupled with the bixel intensities from other beams that intersect its path. With this level of complexity, the optimization of bixel intensities can only be conducted by computers, and the field of inverse planning is developed.

The change in radiation therapy treatments has been profound. For the first time, radiation doses could be developed that wrapped around the target volumes, selectively, avoiding critical structures. The radiation oncologists, physicists, and dosimetrists could describe their intent based on dose and the geometry of the target and organs, rather than by the delivery geometry. Of course, the radiation beam energy and incident angles have an effect on the resulting dose distribution, but the tone of the communication between these professionals changed from a technology- to a dose-based one. This revolution in treatment planning did not come without a price. The compensation of fluence reduction across critical structures was not perfect. This led to larger dose heterogeneities within the target than was customarily seen with 3D CRT. While treatment planners were comfortable with hot spots of a few percent, they now had to consider allowing hot spots of 10–15%, depending on the complexity of the dose distribution, the beam geometry, and the treatment-planning system. Cold spots were also often more severe and treatment planners often had to compromise their previous experiences. Tools to selectively modify and adjust these

dose heterogeneities are now available (ActiveRx, Corvus, Cranberry Township, Pa.), but their ultimate utility has not been proven.

Two methods for delivering the optimized fluences have been developed (see later). For the purposes of this discussion, their principal feature is that the radiation beam is delivered using a sequence of relatively small and complex portals. Before IMRT, there was significant scientific discussion regarding the accuracy of dose distribution calculation protocols for such fields. For example, the calibration and quality assurance of small fields used in stereotactic radiation therapy required specialized equipment and techniques to make sure that the measurements did not suffer from volume averaging due to the rapid variations of dose as a function of position. For conventional therapy, approximations were made in modeling the sources of the radiation fluences exiting the linear accelerator. For example, while most of the X-ray radiation comes from the electron beam target, the flattening filter, as well as the primary and secondary collimators, also provide radiation fluence. For larger fields, the influence of these different radiation sources is straightforward to model. For smaller fields, the relative contribution of these sources is more complex and requires more sophisticated methods for modeling the dose delivered from the linear accelerator head. In IMRT, the radiation fields can extend from relatively large to extremely small, all within the same treatment. This stretches dose modeling and verification measurement techniques to the limit.

The two delivery mechanisms differ by their method for temporally delivering the computer-optimized radiation fluences. The first, termed tomotherapy (CAROL et al. 1996; GRANT and CAIN 1998; Low 1998), uses a geometry similar to that of a computed tomography scanner. A narrow fan beam is generated using a linear accelerator mounted to the gantry such that its central axis is aligned with radii extending from the gantry rotation axis. The fan beam is oriented such that the wide extent is parallel to the plane of rotation. The fluence is modulated along the wide direction using a collimator, which is pneumatically operated. The collimation leaves move into and out of the beam in the direction parallel to the gantry axis of rotation. The delivery system moves the leaves into and out of the beam to provide the fluence modulation as the gantry is rotated. It is important to note that for any single treatment, the gantry moves at a constant rotation rate with the linear accelerator operating at a constant fluence rate. This causes a sequence that provides a constant fluence rate per angle bin. The modulation is conducted by subdividing the gantry angles into small bins (e.g., 5° bins) and leaving a leaf open a fraction of that angle bin corresponding to the requested relative fluence from that gantry angle. The first commercial application of this technique was produced by the Nomos Corporation and was called the Peacock system. It had the unique characteristic of being an add-on system to conventional linear accelerators. The system included the collimator and a treatment-planning system. By rotating the collimator around the patient using the linear accelerator gantry, the system delivered an intensity-modulated dose distribution to two nearly 1-cm-thick regions (often termed "slices"). When the tumor extended beyond the limit of the collimator, it was treated using multiple passes, moving the couch precisely to abut the dose distributions within the patient.

While the Peacock system was an add-on system to conventional linear accelerators, Tomotherapy, Inc., has produced a stand-alone tomotherapy linear accelerator (MACKIE et al. 1999, 2003). The accelerator (Fig. 10.3) looks like a conventional CT scanner. The radiation beam is produced using a 6-mV linear accelerator mounted to a CT gantry. The accelerator is aimed at the rotation axis of the gantry. A pneumatic MLC is used to modulate the radiation fluence in a method similar to the Peacock system. Unlike the Peacock system, the Tomotherapy system moves the patient couch at the same time the gantry is rotated, providing a delivery geometry that is helical in nature, rather than the serial delivery provided by the Peacock system. An added feature of the Tomotherapy system is an imaging system that is placed opposite the radiation source. This allows the Tomotherapy system to be used as a megavoltage CT system for on-board patient imaging. Because the radiation source is high energy (they detune the linear accelerator to approximately 4 mV), the radiation interaction is primarily through Compton scattering, so high-Z materials (contrast agents, dental fillings, and metal implants) do not cause image artifacts.

The second delivery process took advantage of the hardware on existing linear accelerators. For the "2D" and 3D conformal therapy processes, the method of conforming radiation to the tumor had been to fabricate custom-designed high-density metal collimators called blocks. Linear accelerator manufacturers marketed multileaf collimators (MLCs) to replace these custom-fabricated blocks. The use of MLCs provided a more efficient treatment process because the therapist did not have to enter the room between each portal to change the block. The MLC leaves (typically projecting to 1 cm

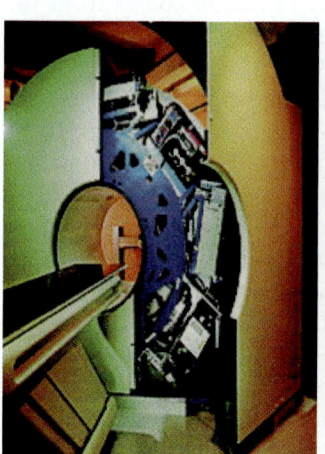 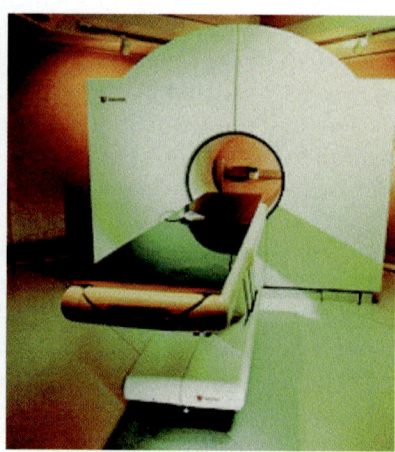

Fig. 10.3 The helical tomotherapy unit installed at the University of Wisconsin. The *left panel* is with a cover open. The *right panel* is looking along the CT table into the ring gantry bore.

at the linear accelerator isocenter plane) were computer controlled, in that they were each separately controlled and operated by the linear accelerator computer system. Implementation of IMRT required the delivery of the nonuniform optimized fluence distributions. With conventional linear accelerators, the most flexible modulation technique was to fix the gantry and collimator and deliver the fluence using a programmed set of portals such that the total fluence was equal to the desired fluence. The linear accelerators had not been designed to deliver dozens to hundreds of portals for each treatment, so during the early implementation, delivery of the fluence distribution took as long as 40 min. Current linear accelerators are capable of delivering these complex fluence distributions in 1–2 min per portal.

The main consequence of the conventional linear accelerator implementation of IMRT is that the radiation field geometry is similar to that in conventional conformal therapy in that relatively few beam directions are employed for each treatment. The selection of those directions has changed, however. Treatment planners often decide on portal directions more to spread the fluence than to avoid specific critical structures. The assumption is that the fluence modulation will take care of the normal structure dose limits.

The introduction of IMRT has changed the radiation therapy treatment-planning process. For 2D, the physician localizes therapy using bony and cavity anatomy, implanted markers, and contrast agents. The portals are selected based on the ability to define such anatomical references. For example, anteroposterior and lateral beams are preferred because the anatomical references are straightforward to identify. Wedges and compensating filters are used to homogenize the radiation dose of intersecting beams. Confirmation of the portal positioning and shape is conducted using kilovoltage X-ray simulator systems or using images of the treatment portals taken with the linear accelerator beam. With 3D, the beams are defined by the projections of targets in the 3D model of the tumor and normal anatomy provided, typically using CT. The negative dose distribution consequences of the interactions of these portals are still overcome with wedges and compensating filters. The connection of the portal orientation and the ability to verify the portal position began to be broken as clinics realized that the use of non-conventional beam angles provides significant improvements in some treatment sites. This led to the disconnection between portal shape verification and placement

verification. As long as a common isocenter was used, portal placement verification could be conducted using only the portals that provided useful anatomical references. Some clinics completely separated the portal and patient positioning verification, defining portals that were used only for positioning verification.

The dose prescription also changed with the introduction of the 3D treatment process. With a quantitative model of the tumor shape and size, and an accurate model of the linear accelerator, the clinician could be provided with a comprehensive review of the planned dose distribution and could determine if the selected beam orientations, energies, portal shapes, and relative intensities provided an adequate delivered dose distribution. The process of prescribing dose became more sophisticated, with maximum and minimum dose constraints along with the nominal prescription dose.

These trends continued with the introduction of IMRT. No longer could the portal dosimetry even be qualitatively determined by examining the portal orientation. Portals often overlapped the critical structures they were intended to protect. The physician's prescription was now an input to the planning process, rather than simply a manual feedback for the plan quality. The use of IMRT reduced the influence that portal orientations and energies had on the resulting treatment plan. The prescription criteria, however, had a profound influence on the treatment-plan quality, and each clinic had to develop appropriate dose-prescription criteria to enter into their treatment-planning system. In fact, the prescription criteria were often adjusted not to match the ultimate desired dose, but in order to force the planning system's optimization software to develop an adequate dose distribution.

10.2
Basic Physical Principles of IMRT

IMRT generates dose distributions that conform to complex target volume geometries in all three dimensions. The dose distributions can be considered to be produced by a superposition of individual beamlets, each consisting of a narrow incident photon beam (Fig. 10.4). Ideally, all beam entry angles and positions would be available for use, but limitations of the dose delivery devices constrain the incident radiation beam-fluence distributions to two types.

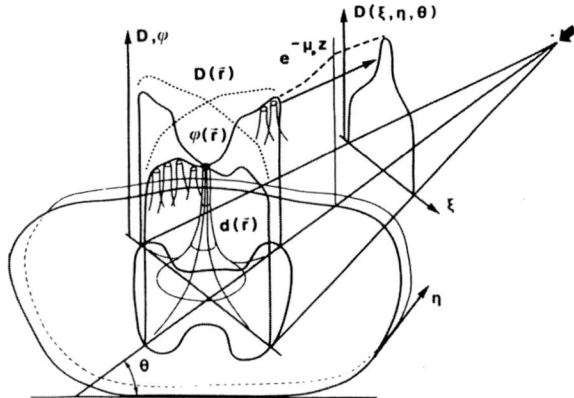

Fig. 10.4 Sketch of beamlet-generated dose distribution

For IMRT delivery, the fluence is modulated as a function of entry angle for each point within the target volume. The dose at any point within the patient is generated by a series of beamlets incident on that point, each with a unique entry angle. Currently, treatment accelerators are capable of simultaneously delivering fluence across a broad surface area of the patient in the form of a cone beam, with a unique patient entry angle for each position within the cone beam. The radiation fluence can be modulated within the cone beam using a physical modulator, a scanning dynamic MLC, a scanning "bremsstrahlung" photon beam, or a combination of these techniques (Convery and Rosenbloom 1992; Mackie et al. 1993; Chui et al. 1994; Grant and Bellezza 1994; Spirou and Chui 1994; Stein et al. 1994; Svensson et al. 1994; Carol 1995; Yu 1995; Yu et al. 1995; Geis et al. 1996; van Santvoort and Heijmen 1996; Hounsell and Wilkinson 1997; Richter et al. 1997; Stein et al. 1997; Webb et al. 1997); for each of these, significant time is required to deliver the incident fluence distribution. To limit the treatment to practical delivery times, the number of incident cone beams is limited at most to a dozen. The beamlet distribution at each point within the patient when cone-beam IMRT is used is therefore limited principally in number, rather than in direction. Alternatively, the accelerator can be operated using dynamic motion of one or more of the angular degrees of freedom (couch, collimator, and gantry); of these, only dynamic gantry motion is currently being investigated (Mackie et al. 1993; Yu 1995; Carol et al. 1996). When dynamic motion of the gantry is used, there is insufficient time to arbitrarily adjust the fluence at each gantry angle using standard MLC;

however, the fluence can be arbitrarily modulated along a one-dimensional array of beamlets using a rapidly actuating MLC (Woo et al. 1994; Carol et al. 1996; Webb and Oldham 1996). The collimator modulates the fluence by alternately opening and closing each leaf as the gantry rotates. This technique is termed fan-beam IMRT delivery. The fluence is modulated in time by opening and closing each of the multileaves over the course of a narrow gantry angle range. In this manner, modulated beams of 50 or more gantry angles are delivered to a roughly cylindrical volume with the cylinder thickness corresponding to the leaf opening size. A commercial MLC is designed using two adjacent independently actuated leaf banks, so two adjacent cylindrical volumes are simultaneously irradiated (Carol 1995; Wu et al. 1996; Low 1997). After the dose is delivered to the first volumes, the patient is precisely moved and the next two cylindrical volumes treated. The couch angle can also be adjusted to allow a broad range of entry angles. The beamlet distribution for each point within the patient is limited principally by incident angular distribution (the beamlets are coplanar) rather than by number.

Standard MLCs can also be employed when using dynamic gantry motion (Yu 1995). The MLC is moved during arc rotation such that an open field of predetermined shape is delivered for each gantry angle. To modulate the beam intensity for each beamlet, multiple arcs are delivered, each with a different portal shape such that the modulated final fluence is delivered. The beamlet distribution is similar to that of the fan-beam geometry.

In traditional radiation therapy, the photon beam energy selection is used to modify shallow doses and to limit dose heterogeneities along the beam path. In megavoltage therapy, shallow and deep tumors are treated with low and high beam energies, respectively; however, when beam-intensity modulation is used, much of the dose heterogeneity due to the depth dose can be removed by modulating nearly orthogonal beams (similar to the concept of a wedged pair); therefore, the need for higher energy beams (>10 mV), with their associated neutron contamination, may be limited. In addition, the conformation of dose within the patient relies on each beamlet independently depositing dose within the patient, which requires that the secondary electron range be small with respect to the beamlet size. Studies have so far concluded that the ideal beam energy for intensity modulation is between 4 and 10 mV.

10.3
Inverse Treatment Optimization (Cost Function and Search Algorithm)

As indicated previously, IMRT requires a computer optimization method of determining the nonuniform beam-fluence profiles. Early attempts at automated plan optimization methods were conducted by several groups, including HOPE and ORR (1965), REDPATH et al. (1976), and McDONALD and RUBIN (1977), but with limited acceptance. With the development of 3D CRT and the corresponding increase in image and graphic data, interest in computer optimization and automation has been renewed. Investigation of computer optimization has intensified because of the requirement for determining optimal nonuniform beam fluences for IMRT (BRAHME 1988; LIND and KALLMAN 1990; GUSTAFSSON et al. 1995).

In 1982, BRAHME et al. developed the concept of determining the modulated radiation field fluence distribution necessary to produce a desired dose distribution. An analytic solution of the beam-fluence profile required to treat a circularly symmetric target with a critical structure at its center was developed. In 1987, CORMACK (1987; CORMACK and CORMACK 1987) extended this approach by extending it to targets with an axis of symmetry. In 1990, BARTH further developed this approach by describing a mathematical solution for a target of arbitrary shape by circular component target decomposition. For general mathematical solutions, the optimized incident fluence profiles contained negative fluence beamlets, an unphysically realizable result (CORMACK and QUINTO 1989). Application of the non-negativity constraint has been included in recent efforts to develop numerical solutions to calculate the incident fluence profiles. Reported methods include: (a) exhaustive search (BARTH 1990); (b) gradient techniques (XING and CHEN 1996; SPIROU and CHUI 1998; BORTFELD 1999; SECO et al. 2002; ZHANG et al. 2004); (c) image reconstruction approaches (BRAHME 1988; BRAHME et al. 1988; Bortfeld et al. 1990; Holmes et al. 1991); (d) quadratic programming (DJORDJEVICH et al. 1990); (e) simulated annealing (WEBB 1989; MOHAN et al. 1992; DJAJAPUTRA et al. 2003); (f) superposition algorithm (SCHOLZ et al. 2003); (g) matrix inversion (GOLDMAN et al. 2005); (h) genetic algorithm (COTRUTZ and XING 2003; HOU et al. 2003); and (i) adaptive control algorithm (LOF et al. 1998).

The task of generating an optimal plan can be separated into two parts: (a) specification of an optimization criterion, and (b) the optimization algorithm used. The optimization criterion is expressed as a mathematical entity in the form of an objective or cost function. The objective function defines a plan's quality and is to be maximized or minimized, as appropriate, to satisfy a set of mathematical constraints. The objective function yields a single numerical value (sometimes referred to as a score) for the plan that is used for evaluating a set of competing treatment plans. Often an iterative computer search is made of the fluence distributions, with guidance by the objective function. Several types of objective functions have been investigated and have been shown to be useful in some, but not all, clinical situations (STARKSCHALL 1984; SCHULTHEISS and ORTON 1985; LANGER and LEONG 1987; LANGER et al. 1990; MOHAN et al. 1992; NIEMIERKO 1992; NIEMIERKO et al. 1992; ROSEN et al. 1992; LANGER et al. 1993; SODERTROM and BRAHME 1993; JAIN et al. 1994; DJAJAPUTRA et al. 2003; LIAN et al. 2003; YAN et al. 2003; BEAULIEUA et al. 2004; YANG and XING 2004; ARRANS et al. 2005; LEE et al. 2005). Historically, most objective functions use dose-based or dose-volume-based criteria, but recent efforts are examining the use of biologically based indices (e.g., tumor control probability and normal tissue complication probability; NIEMIERKO 1992, 1997, 1998; BRAHME 1995; GRAHAM et al. 1996; DEASY et al. 2002; WU et al. 2002; STAVREV et al. 2003).

Once the optimization criteria have been selected, an algorithm is used to automatically determine a set of plan parameters that optimizes the chosen objective function. There are a number of IMRT treatment-planning optimization algorithms under investigation. If the objective function and constraints can be described by linear or quadratic functions of the plan parameters, efficient mathematical optimization techniques can be used that are guaranteed to find the best score (LANGER and LEONG 1987; DJORDJEVICH et al. 1990; LANGER et al. 1990; ROSEN et al. 1992). If the objective function is not linear or quadratic as a function of the optimization parameters, non-linear optimization techniques must be used (COOPER 1978; XIAO et al. 2004; YANG and XING 2004; OLAFSSON et al. 2005). Other optimization approaches being used for IMRT arc adopted from computed tomography (CT) image reconstruction techniques, with filtered back projection being the principal technique (HOLMES et al. 1991; GUSTAFSSON et al. 1994; HOLMES and MACKIE 1994). If these techniques are used in an unmodified form, the optimized fluence distribution will have negative fluences. Non-negativity constraints impact on the quality of the resulting distribution, so iterative techniques are often also required. One iteration method is a stochastic technique termed "simulated annealing," originally developed from statistical mechanics for finding the global minima of non-linear objective functions (METROPOLIS et al. 1953; KIRKPATRIC 1985). WEBB (1989) applied this approach to radiation therapy plan optimization. Since then, several other researchers, as well as a commercial system, have used this method (MORRILL et al. 1990, 1991; WEBB 1991a,b, 1992; MOHAN et al. 1992; MAGERAS and MOHAN 1993; MOHAN et al. 1996; SAUER et al. 1997). The gradient techniques (XING and CHEN 1996; SPIROU and CHUI 1998; BORTFELD 1999; SECO et al. 2002; ZHANG et al. 2004), which require less computation time than simulated annealing, are also used in several commercial systems.

Optimization criteria and mathematical search algorithms are being investigated vigorously. As the capabilities of computers continue to improve, the time required to conduct an optimization search may become shorter. The biggest challenge may be the development of optimization criteria that will be appropriate for a wide variety of cancer sites and histologies and be able to account for individual physician preferences.

10.4
IMRT Treatment Delivery Systems

10.4.1
Historical Review

Early IMRT concepts were pioneered several decades ago. Seminal contributions were made by TAKAHASHI and colleagues (1965). Their work illustrated some of the important concepts in both 3D CRT and IMRT treatment delivery. Dynamic treatments were planned and delivered by TAKAHASHI et al.'s group using what may have been the first MLC system. The MLC used a mechanical control system to conform the beam aperture to the projected target shape as the machine was rotated around the patient.

In the late 1950s, another pioneering effort in IMRT was conducted by the group at the Massachusetts Institute of Technology Lahy Clinic, who independently developed an asynchronous portal-defining device similar to that of TAKAHASHI et al. (WRIGHT et al. 1959; PROIMOS 1960, 1966; TRUMP et al. 1961). The Royal Northern Hospital in England

also pioneered a 3D CRT effort (GREEN 1965). The group developed a series of cobalt-60 teletherapy machines in which the patient was automatically positioned during rotational therapy by moving the treatment couch and gantry during the radiation delivery using electromechanical systems. This was called the "Tracking Cobalt Project" because the planning and delivery system attempted to track around the path of disease spread and subsequently to conform the dose distribution. The work was extended in the 1970s and 1980s by Davy and Brace at the Royal Free Hospital in London (DAVY et al. 1975; DAVY and BRACE 1979; BRACE et al. 1981a,b; BRACE 1982, 1985; DAVY 1985), The 1970s saw important advances in IMRT from the Harvard Medical School by Bjarngard, Kijewski, and others (BJARNGARD and KIJEWSKI 1976; BJARNGARD et al. 1977; KIJEWSKI et al. 1978; CHIN et al. 1981, 1983). Unfortunately, computer technology had not yet advanced to the capacity required for practical implementation of traditional 3D CRT and IMRT. This work and that of the previous researchers led the way for modern IMRT.

The recent introduction of commercially available MLCs (BOYER et al. 1992; WEBB 1993; KLEIN 1994; MOHAN et al. 1996) and the simultaneous development of medical accelerator computer control systems (BRAHME 1987; FRAASS 1994; BOYER 1995; FRAASS et al. 1995; MOHAN et al. 1996) assisted the rapid and widespread development of practical IMRT. Until recently, medical accelerators were produced with relatively primitive beam collimation systems consisting of opposed jaw pairs that were constrained to move symmetrically to provide square or rectangular beam apertures. A low melting point lead alloy was used to fabricate a customized beam block for each radiation portal to provide static conformal field shaping (POWERS et al. 1973). This system was relatively inefficient, in that the radiation therapist was required to enter the treatment room and manually change the aperture block for each treatment portal. A limited improvement came when the collimator jaws were allowed to move independently, allowing the fabrication of lighter blocks. A rudimentary form of dynamic intensity modulation, the dynamic wedge, was developed using independent jaws (KIJEWSKI et al. 1978; LEAVITT et al. 1990, 2000; KLEIN et al. 1995, 1998; LEAVITT et al. 1997; EDLUND et al. 1999; KUBO and WANG 2000). The MLC replaced (or supplemented, depending on the accelerator manufacturer) the rectangular jaw system with a set of independently adjustable, relatively narrow, tungsten leaves (typically projecting to 1 or 1.25 cm wide at isocenter). The leaves were placed under computer control and were therefore able to create custom-shaped beam apertures and allow the change of portal shapes without the requirement of room entry by the radiation therapist. The computer control system was often used to automatically rotate the gantry (and collimator), position the treatment table, and control dose rate.

The IMRT delivery is made practical by using a MLC in a dynamic mode similar to dynamic wedge (i.e., moving each pair of MLC leaves with the beam on while at a fixed gantry angle or during a gantry rotation). This method, and three others, for delivering IMRT are discussed in the following sections.

10.4.2
Arc-Based Fan-Beam Dynamic MLC IMRT

A commercial implementation of a fan-beam approach to IMRT (tomotherapy) uses a mini-MLC system that is mounted to an unmodified linear accelerator, and treatment is delivered to a narrow slice of the patient during arc rotation (Carol 1992, 1995a,b; Carol et al. 1996; Low 1997; Verellen et al. 1997; Low 1998a-c; Al-Ghazi et al. 2001; Saw et al. 2001a,b). The beam is collimated to a narrow slit and beamlets are turned on and off by driving the mini-MLC leaves out and in the beam path, respectively, as the gantry rotates around the patient. A complete treatment is accomplished by sequential delivery to adjoining slices. As previously indicated, this type of IMRT system (Peacock, Nomos Corp., Pittsburgh, Pa.) was the first broadly implemented IMRT planning and delivery system (Woo et al. 1994).

The helical version of tomotherapy was developed by Mackie et al. and is commercially available from Tomotherapy, Inc. (Mackie et al. 1993a,b, 1997, 1999; Welsh et al. 2002; Jeraj et al. 2004). The dose is delivered using a narrow MLC and small linear accelerator mounted in a modified CT scanner gantry. The radiation is delivered while the patient is moved through the gantry in the same way that a helical CT study is conducted (Fang et al. 1997; Mackie et al. 1997).

10.4.3
Fixed Portal Cone-Beam Dynamic MLC

Other IMRT researchers have concentrated on using the MLC to provide a full-field or cone-beam modulation technique (Kallman et al. 1988; Convery and Rosenbloom 1992; Galvin et al. 1993; Bortfeld et

al. 1994; Svensson et al. 1994; van Santvoort and Heijmen 1996; Webb et al. 1997). For IMRT, the MLC is operated in a dynamic mode where the gap formed by each opposing leaf sweeps under computer control across the target volume to produce the desired fluence profile. The technique for setting the gap opening and its speed for each MLC leaf pair was first determined by a technique introduced by Convery and Rosenbloom (1992) and extended by Bortfeld et al. (1994) and Spirou and Chui (1994). The Memorial Sloan Kettering Cancer Center was one of the first centers to adopt a variation of this technique (Ling et al. 1996). Because of its implementation on conventional linear accelerators, this technique currently treats the greatest number of IMRT patients worldwide (Burman et al. 1997; de Neve et al. 1999; Ma et al. 1999; Mott et al. 1999; Wu et al. 2000; Budgell et al. 2001; Chui et al. 2001; Low et al. 2001).

10.4.4
Fixed-Portal X-ray Compensating Filter IMRT

The original and common use of compensating filters is to account for missing tissue deficits while maintaining the skin-sparing benefit of megavoltage photon beams. Filters designed using 3D RTTP are capable of compensating for not only the missing tissue deficit, but also for internal tissue heterogeneities. Filters can be designed by calculating a thickness along a ray line using an effective attenuation coefficient for the filter material and dose-ratio parameters for effective depths may consider scattered radiation from the filter. The filter construction process can be automated using numerically controlled milling machines, and can be extended for use when inverse treatment planning is employed, fabricating the compensation filter that generates the desired IMRT fluence profile when the filter is placed in the radiation beam. A carousel or stack device could be supplied to hold multiple filters that would be automatically moved into the beam under computer control. Being relatively simple, this method of delivering IMRT may prove to be the easiest to clinically implement. Comparisons of IMRT dose distributions delivered using physical modulators and MLC-based approaches are under investigation, with indications that the dose distributions provided by these modalities are similar (Stein et al. 1997). Stein et al. (1997) have developed a method for producing physical modulators by stacking layered filters up to 1.6 cm thick.

The filters are fabricated using low-melting-point alloy poured into foam molds, which are cut using a computer-controlled cutter. Other authors have also investigated the use of physical filters for IMRT (Low et al. 1996a,b).

10.5
IMRT Quality Assurance
(Commissioning and Clinical)

As with all treatment modalities, verification and characterization measurements are required before clinically implementing of IMRT. Dosimetric verification traditionally includes a comparison of measured and calculated dose distributions for selected test treatment plans. Acquisition of benchmark quality dose distributions for static-beam therapy relies on ionization chambers, which measure doses at individual points. Acquisition of isodose distributions relies on the assumption that the dose distribution being delivered during the measurements is constant in time (at least to within an overall constant dose rate, monitored by a reference ionization chamber), and that the chamber can be scanned throughout the dose distribution during the measurement. Dynamic IMRT dose distributions are delivered using a temporal sequence of incident fluences, so this assumption does not hold. As a consequence, the measurement of dose at a single point using an ionization chamber requires that the entire fluence delivery sequence be delivered, requiring measurement times as long as 20 min. Similar difficulties are seen with all electronic point dosimeters (e.g., diodes). Stand-alone integrating point dosimeters, such as thermoluminescent dosimeters (TLDs), allow the simultaneous measurement of dose at numerous locations, but with the additional workload required for off-line read-out. Practical application of real-time point dosimeters provide, at most, dozens of measurement points within the treatment volume.

Characteristics of IMRT dose distributions include complex 3D treatment volume geometries and steep spatial dose gradients. The precise characterization of these distributions using only point dosimeters is not practical. To obtain the large quantity of dose measurements necessary for the description of the dose distributions, at least planar dosimeters are necessary. Currently, the only practical planar dosimeter is film, either radiographic or radiochromic. Benchmark-quality dose distributions for megavoltage beams cannot be acquired using radio-

graphic film due to its energy response and relative dosimetric characteristics (WILLIAMSON et al. 1981; HALE et al. 1994; VAN BATTUM and HEIJMEN 1995; MAYER et al. 1997; DOGAN et al. 2002; ZHU et al. 2002). Radiochromic film uses a tissue-equivalent dye that changes color when irradiated (STEVENS et al. 1996; ZHU et al. 1997, 2002; NIROOMAND-RAD et al. 1998; DOGAN et al. 2002; ESTHAPPAN et al. 2002). Early implementations of radiochromic film had significant difficulties with response homogeneity and poor dose sensitivity. Current radiochromic film has homogeneous response (better than 5%) and sufficient sensitivity to use for single-fraction IMRT quality assurance. The cost of the film was also significantly reduced such that the cost of the film may be less than the cost of radiographic film when the processor purchase and maintenance is considered. The new radiochromic film is sufficiently novel that its full impact on IMRT quality assurance has not been determined.

Even considering the dosimetric limitations of radiographic film, it is at least capable of accurately localizing regions of steep spatial dose gradients. For example, in a region of a 5% mm^{-1} dose gradient, an error of 5% in the relative dose measurement results in a spatial error in the localization of that gradient region of only 1 mm. Radiographic film therefore provides an opportunity for accurately *localizing* the edges of a complex dose distribution within the 2D measured area.

To determine the spatial accuracy of the treatment planning and delivery system, the spatial location of the measured and calculated doses must be accurately and independently determined. Most treatment-planning systems use a Cartesian coordinate system to localize the calculated dose distribution. For example, many planning systems tie the coordinate system directly to treatment delivery. Before acquisition of the volumetric CT scan of noninvasively immobilized patients, radiopaque markers are placed on the immobilization system (e.g., thermoplastic masks), corresponding to the lateral and anterior (for supine patients) projection of a selected point, typically the treatment isocenter. During treatment planning, the isocenter location is identified on the CT-scan data set. Verification measurements need to be conducted having a quantitative alignment with the treatment plan isocenter.

The treatment-planning system may provide instructions to offset the patient position in a specified direction and distance to place the accelerator isocenter at the optimal location for treatment. Verification measurements should be made in the same way. Localization wires are placed on the dosimetry phantom identifying the projected origin. The wires are visualized on the CT-scan data set to determine the isocenter location within the phantom. The dosimeter position locations are determined relative to the origin using physical measurements or machine drawings of the phantom.

10.5.1
Dosimetry System

There is no one dosimetry system that conveniently measures all of the dose information necessary for quality assurance. Systems that balance thorough dosimetric measurements with labor and complexity have been developed in many academic centers (KLEIN et al. 1998; CHAO 2002; OLCH 2002; HIGGINS et al. 2003; AGAZARYAN et al. 2004; LETOURNEAU et al. 2004; MORAN et al. 2005; WIEZOREK et al. 2005; WINKLER et al. 2005; YAN et al. 2005); these exploit the optimal characteristics of point dosimeters (e.g., ionization chambers) and planar dosimeters. Radiographic film is used to measure the spatial position of the dose distribution (by aligning the steep-dose gradient regions) and ionization chambers are used to measure the absolute dose in relatively shallow dose-gradient regions.

Careful dosimetry phantom design is important to commissioning and quality assurance procedures. Anthropomorphic phantoms offer the advantage that they are shaped and sized similarly to patients; however, if the treatment-planning system does not accurately consider internal heterogeneities, determination of the cause of discrepancies between measured and calculated doses may be difficult, especially within and near bony anatomy. While TLDs provide convenient single-point measurements, the spacing between locations for TLD chips may be larger than desired. Film preparation is made more difficult with anthropomorphic phantoms due to irregular external contours. Radiogrphic film preparation may require careful cutting in the darkroom to conform the film shape to the phantom outline. Homogeneous cuboid dosimetry phantoms do not share these limitations.

Multipoint planar dosimeters, such as electronic portal imaging devices (EPIDs) and diode system (Sun Nuclear), have been developed that can greatly assist in the quality assurance process (McCURDY et al. 2001; VAN ESCH et al. 2001, 2004; CHANG et al. 2003, 2004; LETOURNEAU et al. 2004; BAKER et al. 2005; WIEZOREK et al. 2005). Rather than measure dose

within a phantom, the dosimeter system is placed at isocenter and irradiated using the delivered beams. A prediction of the incident dose from each beam is made by the treatment planning system and compared against the measurement. Typically, the comparisons are made port by port, but some planning systems are also capable of computing the summed dose, i.e., the dose delivered from all beams using a common gantry angle. While this does not reflect the delivered dose, gross delivery or planning errors would be detected by such a system. One of the greatest challenges of using these dosimeters is the selection of acceptance criteria. Because the dose delivered to these planar dosimeters does not reflect the dose to the patient, developing pass–fail criteria based on the total delivered dose is not possible. The user can only compare the measured planar dose with the corresponding calculated planning dose distribution, for the same beam and depth. For this purpose, two main analysis tools are used: overlay comparison of calculated and measured dose profiles; and 2D plot of measurement points that fail both measures of a two component user-selected criteria consisting of distance to agreement (DTA) and dose difference (percentage of dose difference; Low 1998; Low and DEMPSEY 2003). Commonly, the passing criteria are set at 3 mm for DTA and 3% for dose difference. The DTA criteria would then fail any measured dose point for which a corresponding point could not be found in the calculated distribution, with the same dose, within a 3-mm radius. A point measurement would fail the percentage of dose-difference criteria if the difference in dose between the measured point and the equivalent point in the calculated distribution was greater than 3%.

An ideal dosimeter would be capable of providing dose measurements in a 3D volume, with an energy response similar to that of water or muscle. BANG (bis, acrylamide, nitrogen, and gelatin) and other polyacrylamide gels have been investigated for use in dynamic IMRT dose-distribution measurements (MARYANSKI et al. 1994, 1996a,b, 1997; GORE et al. 1996; LOW et al. 2000; ISLAM et al. 2003). The dosimeters operate on the principal that monomers can be made that cross-link when irradiated by ionizing radiation. There is a subsequent increase in the solvent proton relaxation in the presence of the polymer. The increased proton relaxation rate ($R_2=1/T_2$) can be imaged using magnetic resonance imaging (MRI). The gel is irradiated using IMRT delivery and subsequently imaged using a clinical MRI unit. T1- and T2-weighted scans are obtained for each gel and the volumetric distribution of R_2 is determined using these scans. A monotonic relationship exists between R_2 and absorbed dose, but because the gel radiation sensitivity is batch dependent, some gels are irradiated to known doses and scanned to obtain a dose calibration curve. A measurement of the full 3D dose distribution is provided when this medium is used, with measurement voxels as small as $1\times1\times2$ mm^3. The requirement of MRI for readout has limited the use of this detector medium to a few dosimetry studies. In addition to its MRI properties, the gel is normally transparent and becomes opaque on irradiation. Optical absorption imaging has also been investigated as a method for extracting the dose (MARYANSKI et al. 1996a,b; KELLY et al. 1998; WOLODZKO et al. 1999), but gels have not enjoyed widespread clinical implementation.

10.5.2
Commissioning

The core of a treatment-planning system commissioning procedure is the comparison of calculated and measured dose distributions. Traditional commissioning procedures investigate beams incident on a water phantom for a variety of field sizes, blocking geometries, beam modulators (wedges), and incident angles. However, IMRT dose distributions are generated using a superposition of complex fluence distributions; therefore, commissioning of these systems relies on developing dose distributions and fluence-delivery sequences for a series of selected target and critical structure volume geometries. The volumes are selected based on expected sizes and shapes of clinical target volumes. Neighboring critical structures can also be defined to provide complex target geometries. Treatment plans used for commissioning should include a variety of target sizes, shapes, and locations within the test phantom, as well as at least one test using phantoms of different sizes. The test target volume can be cylindrical to simplify preparation and subsequent description of the target geometry. Both diameter and length of the targets should span the range of sizes used for clinical cases. Experiments should also include a broad range of optimization parameters.

Each commissioning experiment will include both film and absolute dose measurements (ionization chamber and/or TLDs, and possibly the use of radiochromic film). The results should be tabulated to show the dosimetric and spatial accuracy of the system.

10.5.3
System Quality Assurance

The accurate delivery of IMRT dose distributions depends on thorough accelerator- and delivery-system QA programs. A description of all accelerator QA procedures is beyond the scope of this chapter. A thorough review was published by the American Association of Physicists in Medicine (KUTCHER et al. 1994). Some of the specialized procedures specific to IMRT are mentioned in this chapter.

The accurate localization of the accelerator isocenter relative to the patient alignment fiducial markers is important for noninvasively immobilized patients. The isocenter position within the patient is aligned to the accelerator using the positioning lasers. As in all external beam therapy, if the lasers are not correctly aligned, the localization of the dose distribution within the patient will suffer. For IMRT treatment planning and delivery, dose distributions are generated using fluence distributions incident on the patient from a series of directions. The superposition of the fluence distributions generates the planned dose distribution, and the planning system assumes that the orientation and position of each beam angle is accurate. Dose-delivery errors can occur due to excessive gantry sag and gantry and collimator angle misalignments. Quality assurance tests have been developed that check beam and isocenter alignment.

For indexed sequential arc therapy, accurate patient positioning between arcs is extremely important. CAROL et al. (1996) determined that an incorrect placement of the patient between successive arc treatments will cause a 10% mm^{-1} dose heterogeneity in the abutment region. Consequently, the accuracy of the patient immobilization and placement system is critical to accurate dose delivery. Periodic testing of patient indexing system accuracy should be conducted. One method is to place a film at the plane of isocenter and irradiate it using the open collimated portal.

The film is then moved using the same apparatus as used in patient treatments, and the process is repeated six to ten times. Ideally, the film will have a homogeneous rectangular irradiated region, with no evidence of high- or low-dose regions. Overlaps or underlaps can serve to identify problems with either the positioning system or the treatment couch support structure or bearings. Tests are also being developed for the QA of dynamic MLC delivery (CHUI et al. 1996) and helical tomotherapy (YAN et al. 2005).

The implications of incorrect treatment setup are being investigated. Studies (CONVERY and ROSENBLOOM 1995; LOW 1997; AHMAD et al. 2005; HONG et al. 2005; SIEBERS et al. 2005) have shown that variations in the delivered-dose distribution can arise when the gantry, collimator, or couch angle are incorrectly set.

10.5.4
Monitor Unit Verification

A manual or simple computer calculation has been found useful to double check the monitor unit of a treatment in 3D CRT. In IMRT the treatment-planning system provides the MUs for each field. The MUs are verified either by direct measurement (LOW 1998; XING et al. 1999; LI et al. 2001) or by an independent calculation system (BOYER et al. 1999; CHEN et al. 2000; KUNG et al. 2000; MA et al. 2000, 2003; XING et al. 2000; PAWLICKI and MA 2001; CHEN et al. 2002; YANG et al. 2005). Independent computation is more efficient and less manpower intensive than direct measurement. The most accurate computation methods proposed for IMRT verification are based on Monte Carlo simulation (MA et al. 2000, 2003; PAWLICKI and MA 2001; YANG et al. 2005). Other computation methods include convolution of pencil beams (BOYER et al. 1999), self-consistent monitor unit and isocenter point-dose calculation (CHEN et al. 2002), scatter-summation algorithm (XING et al. 2000), ray tracing (CHEN et al. 2000), and modified Clarkson integration (KUNG et al. 2000).

10.5.5
Clinical Quality Assurance

The delivery of IMRT is similar in most respects to the delivery of standard conformal radiation therapy. The principal difference lies in the requirement to keep the patient stationary during the entire course of treatment. For traditional conformal therapy, the motion of the patient during treatment will result in minor dose variations within the portal outlines and a major dose variation near the portal boundaries; however, margins are typically applied to the tumor volumes to account for an expected amount of motion so that the targets will generally remain within the portal boundary. The exception to this rule is when compensating filters or wedges are used. In these cases, patient motion during treatment will alter the doses within the beams an amount propor-

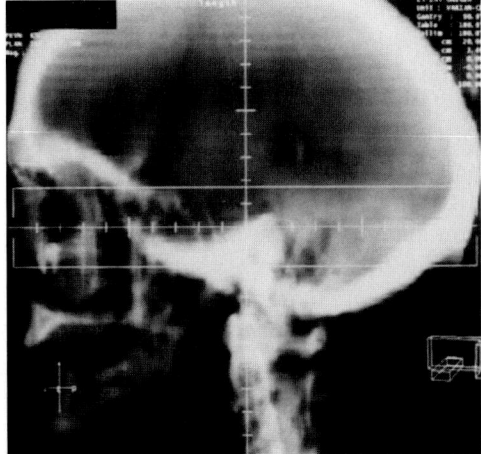

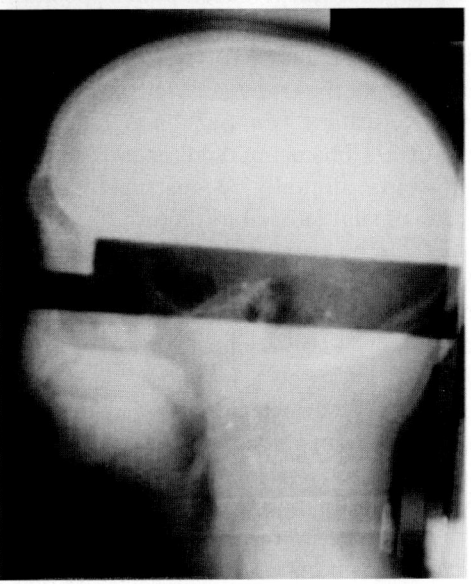

Fig. 10.5 a A digitally reconstructed radiograph of lateral projection depicts a representative arc through nasopharynx and sphenoid sinus. **b** A double-exposure portal film taken at the same geometry from the linear accelerator corresponds to the digitally reconstructed radiograph.

tional to the lateral fluence gradient. For example, the dose gradient for a 45° wedge is typically 4–5% cm^{-1}, so movement within the portal boundary of 5 mm will result in a dose-delivery error of only 2–3%, respectively.

In IMRT, the dose distribution is due to a delivery sequence of incident fluences, each with a potentially large lateral fluence gradient. YU et al. (1997) showed that the fluence delivery error can be as high as a factor of two when dynamic-MLC delivery is coupled with breathing motion; however, BORTFELD et al. (2002, 2004) have shown that, under certain assumptions, the dose-delivery errors in realistic treatments is significantly less and averages out over multiple fractions such that the overall effect of tumor motion is to blur the resulting dose distribution. Clearly these two results appear contradictory, and the truth in actual clinical cases probably lies between the two. More research is necessary using more sophisticated simulations before the actual effects are understood using four-dimensional techniques. Until then, clinicians are cautioned to avoid using IMRT for tumors that move due to breathing, and to provide good immobilization techniques for all patients (Yu et al. 1997; Bortfeld et al. 2002, 2004; Engelsman et al. 2005).

One consequence of IMRT is the lack of a convenient open portal for image-based patient position and treatment verification. In IMRT, the two concepts must be separated: the patient position is verified separately from the treatment delivery. Patient positioning verification can be conducted using orthogonal portal films using fixed open fields and comparing against digitally reconstructed radiographs (DRRs) or simulator films. Figure 10.5 shows examples of the DRRs and portal films for a head and neck patient (Chao et al. 2000). Determination of patient positioning accuracy is made by manually marking bony landmarks and overlaying the portal films and DRRs on a light box. Treatment verification can be conducted using measurement phantoms. The use of in vivo dosimetry may be limited to intracavitary dosimetry. Engler et al. (1997) used in vivo TLD dosimetry placed beneath bolus on the patient's skin to verify the delivered dose to arc-based fan-beam treatments; however, the acceptable dose tolerances were by necessity large due to the high spatial dose gradient near the patient surface.

10.5.6
Tolerance Levels

The consensus in the radiation therapy community is that the dose delivered to the tumor volume should be within 5% of the prescribed dose (Dische et al. 1993). The more complex IMRT systems have difficulty in meeting the 2% in relative dose accuracy in shallow-dose gradients or 2-mm spatial accuracy in regions with steep-dose gradients recommended by the International Commission of Radiation Units and Measurements (van Dyk et al. 1993). Palta et al. (2003) proposed a set of more appropriate criteria for IMRT plan validation. The over-uncertainty in delivered dose should be less than 5%. The proposed confidence limits for relative dose difference

in treatment planning are 3% for regions with high dose and shallow-dose gradient, 10% or 2 mm DTA (distance to agreement) for regions with high dose and steep-dose gradient, 4% for regions with low dose and shallow-dose gradient, and 2 mm DTA for dose fall-off regions. The proposed tolerance limits for delivery systems with MLC are: 1 for SMLC and 0.5 mm for DMLC for leaf position accuracy; 0.2 mm for leaf position reproducibility and for gap width reproducibility; 0.75 mm for isocenter; 2% for SMLC and 3% for DMLC for low MU (<2 MU); and 2% for low MU symmetry.

10.5.7
Doses of Irradiation Outside the Treatment Volume

Followill et al. (1997), based on the observation that with IMRT some form of X-ray attenuation to modulate intensity is required, suggested that the number of MUs used to deliver a given treatment is increased over that used for conventional radiation therapy, resulting in the likelihood of increased whole-body dose to the patient by leakage and scattering of X-rays and, for higher energies, leakage of neutrons through the collimation and treatment head assemblies of the linear accelerator. These authors made estimates of the leakage neutron dose equivalent or photon dose at isocenter in the patient plane at a point 5 cm away from the central axis of a pelvic field, and estimated the whole-body X-ray dose equivalent contribution for 50 cm from the center of a 20×20-cm treatment field based on previously measured data (Stovall et al. 1995). They also calculated the MUs required to deliver a treatment with conventional X-rays (wedged or unwedged) and with beam-intensity modulated irradiation from two different systems (Varian MLC or Nomos Peacock MLC). Average total MUs derived from treatment plans for six patients are shown in Table 10.1. The unwedged 6-mV beam required 20% more MUs than the 18-mV beam. With wedges, the 6-mV-beam MUs increased by 60% over that for 18 mV. With the Nomos Peacock system the total number of MUs per fraction increased with the number of arcs used. The total whole-body dose equivalent to a point 50 cm from the center of the pelvic field, when 70 Gy was delivered, was significantly higher with the intensity-modulated plans (Table 10.2). The authors also calculated the estimated risk of any fatal secondary cancer associated with scattered dose from the prescribed treatment as a percentage increase in likeli-

Table 10.1 The MU/cGy and total MU to deliver conventional and modulated beam-intensity radiotherapy. *MLC* multileaf collimator

	Beam energy (mV)	Conventional		Beam-intensity modulated	
		Unwedged	Wedged	Varian MLC modulated	Nomos MLC tomotherapy
		(MU/cGy)	(MU/cGy)	(MU/cGy)	(MU/cGy)
MU per dose[a]	6	1.2	2.4	3.4	9.7
	18	1.0	1.5	2.8	8.1
	25	1.0	1.5	2.8	8.1
		(MU)	(MU)	(MU)	(MU)
Total MU[b]	6	8,400	16,800	23,800	67,900
	18	7,000	10,500	19,600	56,700
	2.5	7,000	10,500	19,600	56,700

[a]The number of MU needed for the indicated technique to deliver 1 cGy to isocenter
[b]The total MU needed, using the indicated technique and energy, to deliver 70 Gy to isocenter

Table 10.2 The estimated total whole-body dose equivalent (mSv) from a total delivered dose of 70 Gy at isocenter

	6 mV		18 mV		25 mV	
	No wedges	Wedges	No wedges	Wedges	No wedges	Wedges
Conventional	67	134	326	488	602	903
MLC modulated	190	--	911	--	1686	--
Tomotherapy	543	--	2637	--	4876	--

hood in non-irradiated populations. The smallest increased risk was 0.4% for the 6 mV unwedged conventional technique, and the greatest risk was 24.4% with the tomotherapy technique using 25 MV X-rays. The authors caution against the use of 25 MV X-rays with intensity-modulated techniques, which may carry a seriously high risk of secondary cancers, and recommend that the use of X-ray energies greater than 10 MV for beam-intensity modulated conformal therapy should not be attempted until the X-ray and neutron leakage dose and risks associated with the treatment have been carefully evaluated in long-term studies. At our institution MUTIC and LOW (1997) also measured whole-body irradiation doses from arc-based IMRT to the head and neck region using a water-equivalent plastic block whole-body phantom and a polystyrene phantom, placing TLDs in multiple locations arranged into orthogonal linear arrays. To assess the leakage dose component, the MLC leaves remained closed during treatment delivery. The total midplane whole-body dose from internal scatter and leakage was approximately 2.5% of the total target dose, decreasing to 0.5% at 30 cm

from the target. The whole-body dose was entirely due to head irradiation leakage. The internal scatter dose was significant near the target but became negligible beyond 15 cm from the target in relation to the leakage dose; therefore, the total-body dose was proportional to the total MUs used to deliver a given treatment.

10.6
Clinical Studies with IMRT

Recently, GUERRERO URBANO and NUTTING (2004) presented two reviews on clinical use of IMRT in tumors of the head and neck region, central nervous system, lung, and in prostate, gynecological, breast, and gastrointestinal malignancies, as well as in other issues related to the clinical use of this new technique. The main indications at every site have been dose escalation and acceptable or reduced normal tissue toxicity. In general, a better dose distribution by IMRT has been well documented,

218

D. A. Low et al.

whereas its correlations with better clinical outcome or improved sparing and/or improvements in quality of life have not been evaluated on a large scale. Most available reports are small phase-I or phase-II trials. Although the data are promising, randomized clinical trial data are important to support its use. The authors concluded that IMRT delivery should remain in the context of clinical trials until such time as these improved dose distributions have proved clinical benefits for patients. Some clinical studies of IMRT for tumors in head and neck and in prostate are summarized below.

10.6.1
Head and Neck

KAM et al. (2003) compared IMRT with 2D RT and 3D CRT treatment plans in three patients with different stages of nasopharyngeal carcinoma (NPC). The NPC has been a challenge to radiation oncologists because of its unique histological features, stra-

tegic location, and high radiosensitivity. A split-field technique with seven coplanar beams separated at 50° apart was used for IMRT planning with Helios (Varian Medical Systems). A dynamic multileaf collimation (DMLC) technique was used for treatment delivery (Varian 600 CD or 2300 CD). Figure 10.6 shows that IMRT produced the best target coverage, target conformity, and dose homogeneity. The IMRT target dose conformed most accurately to the concave target, while a rapid dose fall-off was seen around the ear canals, temporomandibular joints, and parotid glands. Figure 10.7 shows the dose-volume histogram (DVH) curves for the gross tumor volume (GTV) and OARs. The mean parotid dose and D_{max} for the temporomandibular joints were significantly lower in the IMRT plan. The D_{max} for brain stem and temporal lobes was similar for the three plans; however, the volume receiving low-dose RT was greater with IMRT. Similar dose advantages were seen in the other two patients. The authors concluded that for early-stage disease IMRT provides better parotid gland sparing, whereas in

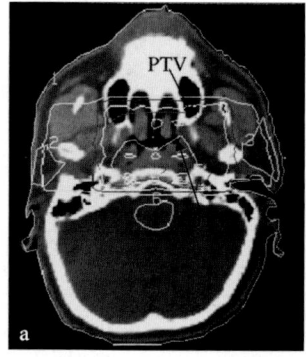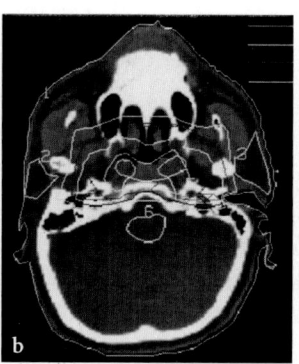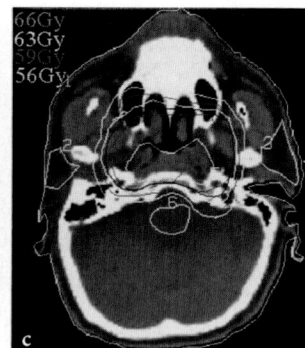

Fig. 10.6a-c. Comparison of isodose distribution among **a** 2D RT, **b** 3D CRT, and **c** IMRT in T1N0M0 NPC

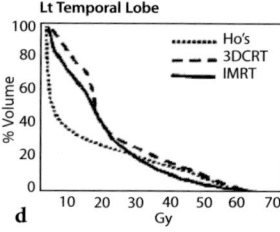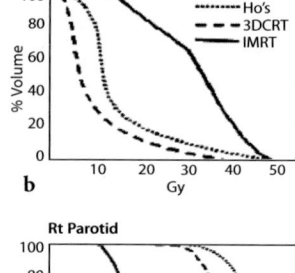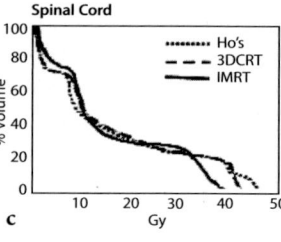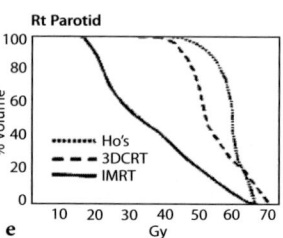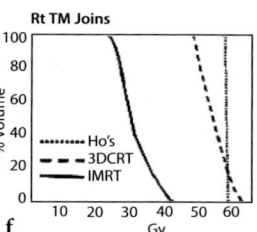

Fig. 10.7a-f. Dose-volume histogram curves of **a** gross tumor volume (*GTV*), **b** brain stem, **c** spinal cord, **d** left temporal lobe, **e** right parotid, and **f** right temporomandibular (*TM*) joints with three different plans in T1N0M0 NPC

Fig. 10.8 A composite magnified view of the resulting isodose curves in the junctional region indicates that the best plan (*Plan C*) with respect to parotid sparing yields the worst target coverage. *Plan F* seems to be the best compromise.

locally advanced disease, IMRT offers better tumor coverage, normal organ sparing, and room for dose escalation.

The same IMRT technique was used to treat 63 NPC patients in a subsequent study (KAM et al. 2004). The results showed a very high rate of locoregional control and favorable toxicity profile. Another important observation was that dose escalation above 66 Gy of IMRT-based therapy was a significant determinant of progression-free survival and distant metastasis-free survival for advanced T-stage tumors.

A system for patient immobilization, setup verification, and dose optimization of parotid sparing was implemented for tomotherapy-based IMRT at our institution (CHAO et al. 2000). The IMRT was delivered with a serial tomotherapy device on a 6 MV linear accelerator. To optimize the two competing goals of minimum dose to targets and maximum allowable dose to normal structures, several parameters, such as structure weights (from 0–2) and target priority, were specified for the treatment optimization. Figure 10.8 compares six plans where plans A–C (no target priority with increasing order of parotid weight) had significantly poorer coverage of the target volume than plans D–F (target priority with increasing order of parotid weight); however, the trade-off for better target coverage was a bigger volume of the parotid gland receiving a higher dose. For example, plan D provided the best target coverage (<2% of target receiving <95% of prescription dose); however, nearly 50% of parotid

glands received more than 30 Gy. In contrast, plan C spared the parotid gland most effectively (only 14% of parotid glands received more than 30 Gy), but 7% of the target volume received less than 95% of the prescribed dose. The area of undercoverage was usually located in the junction of the target and parotid volume where the dose transition took place. This dose gradient at the edge of the target volume and the adjacent critical normal tissue is a common feature of IMRT dosimetry. The best compromise seemed to be achieved by plan F, which yielded one of the lowest target underdose volumes (3.3±0.6%), with a relatively low fractional volume of parotid gland being overdosed (27±8%).

The same IMRT technique was used to treat postoperatively 14 patients and definitively without surgery 12 patients with carcinoma of the oropharynx (CHAO et al. 2001). The conventional radiation therapy (CRT) was used to treat preoperatively 109 patients, postoperatively 142 patients, and definitively 153 patients. The 2-year locoregional control values for the three CRT groups and the two IMRT groups were 78, 76, 68, 100, and 88%, respectively. The 2-year disease-free survival values were 68, 74, 58, 92, and 80%, respectively. The authors concluded that the dosimetric advantage of IMRT did translate into a significant reduction of late salivary toxicity in patients with oropharyngeal carcinoma.

NUTTING et al. (2001) performed a planning study for six patients with thyroid carcinoma with Corvus system (Corvus v3.0, Nomos Corporation, Pittsburgh, Pa.). Figure 10.9 shows typical dose dis-

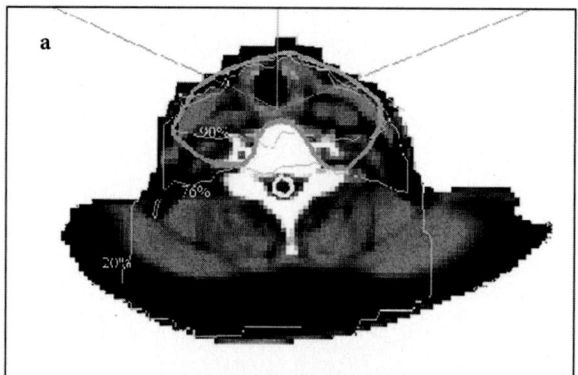

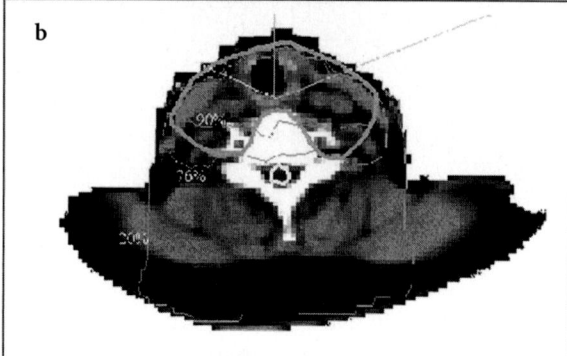

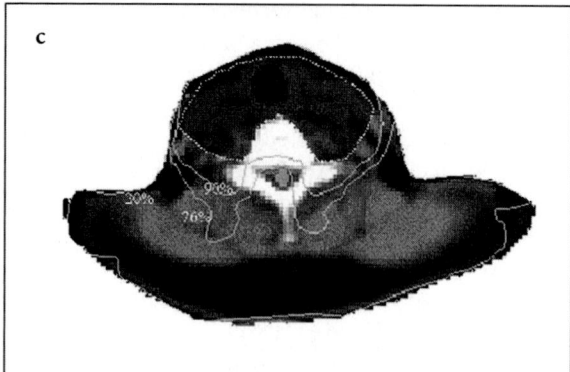

Fig. 10.9a-c. Typical dose distributions at the level of the iso-center produced by **a** conventional RT, **b** 3D CRT, and **c** nine-field IMRT techniques are shown for the single-phase thyroid-bed technique. The prescribed dose was 60 Gy to the isocenter: 90, 76, and 20% isodoses are shown. The spinal cord dose was constrained at 46 Gy (76%).

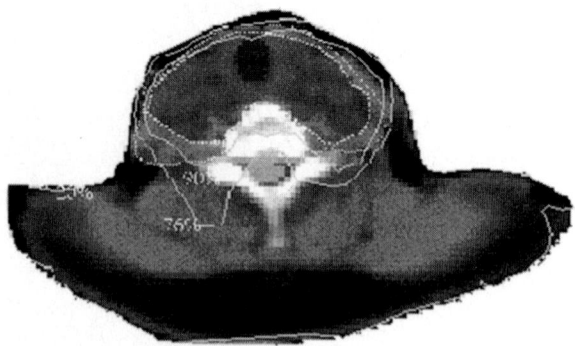

Fig. 10.10 The SIB using IMRT. The prescribed dose was 60 Gy to the boost region and 46 Gy to the loco-regional nodes: 90, 76, and 20% isodoses are shown. The spinal cord dose was constrained at 46 Gy (76%).

10.6.2
Prostate

Many researchers have reported improved tumor control through dose escalation using 3D CRT or IMRT for localized prostate cancers. Doses of 75 Gy and higher were considered favorable to a lower dose for local disease control. On the other hand, a higher dose is susceptible to high incidence of treatment-related toxicity; therefore, a high radiation dose may be used only when the amount of proximal critical structures (bladder and rectum) exposed to specific marker dose remains low. This requires more precise conformity and tighter normal dose constraints, both characteristics for which IMRT has proven useful.

ZELEFSKY et al. (2000) compared high-dose radiation (81 Gy) for prostate cancer delivered by 3D CRT for 61 patients and IMRT for 171 patients. Figure 10.11 shows the beam arrangements and dose distributions for a typical patient. For 3D CRT, the patients were treated with six-field coplanar beam arrangement to 72 Gy followed by a 9-Gy boost using a separate multi-field coplanar plan in which the rectum was blocked in each field. For IMRT, an isocentric five-field technique was used and the parameters included dose uniformity (100%) to the PTV, and limits of 40 and 58% of the prescription dose to the rectal wall and bladder wall, respectively. The isodose distributions indicated that the IMRT plan provided improved tumor coverage with 81 Gy. The DVH for the target (Fig. 10.12) in this patient showed that while 99% of the CTV received 81 Gy with IMRT, the same dose was delivered to only 94% of the CTV with the conventional plan. Figure 10.12

tributions produced by CRT, 3D CRT, and nine-field IMRT. In this study, 3D CRT reduced normal tissue irradiation compared with conventional techniques, but it did not improve planning tumor volume (PTV) or spinal cord doses. The IMRT improved the PTV coverage and reduced the spinal cord dose. In addition, the authors showed that simultaneous integrated boost technique with IMRT improved dose distribution (Fig. 10.10), although its clinical effect was uncertain.

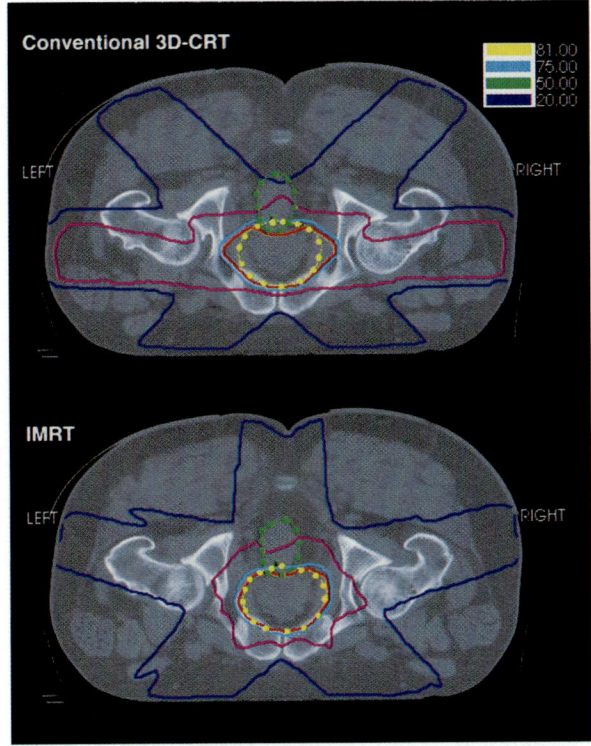

Fig. 10.11 Dose distributions of treatment plans designed for a prostate cancer patient. The figure shows the composite dose distribution. Planning was carried out as described in the materials and methods. Note the improved conformality of the planning tumor volume coverage by the 75- and 81-Gy isodose lines in the IMRT plan. Also note that the 50-Gy isodose line avoids the femoral heads in the IMRT plan.

also demonstrates that IMRT reduced the volume of the rectal wall carried to doses between 50 and 77 Gy, the bladder wall between 55 and 85 Gy, and of the femoral heads between 25 and 60 Gy. The authors concluded, based on comparison of 20 patients, that significantly larger volumes received the prescribed dose of 81 Gy with IMRT relative to the conventional 3D-CRT plan ($p < 0.01$). Furthermore, whereas 100% of the CTV received 75 Gy in both plans, 98±2% of the CTV received 81 Gy in the IMRT plan compared with 95±2% in the conventional 3D CRT plan ($p < 0.01$). For the normal tissues, the percentages of the rectal wall and bladder wall volumes carried to 75 Gy were significantly decreased with IMRT ($p < 0.01$). For all patients, acute and late urinary toxicities were not significantly different for the two methods; however, the combined rates of acute grade-1 and grade-2 rectal toxicities, and the risk of late grade-2 rectal bleeding, were significantly lower in the IMRT patients. The 2-year actuarial risk of grade-2 bleeding was 2% for IMRT and 10% for conventional 3D CRT ($p < 0.001$).

Zelefsky et al. (2002) further reported the toxicity and biochemical outcome in 772 patients with clinically localized prostate cancer treated with high-dose (81.0 Gy for 698 patients, 86.4 Gy for 74 patients) IMRT. The median follow-up time was 24 months. The results are summarized in Tables 10.3 and 10.4. High-dose IMRT was well tolerated acutely. As shown in Table 10.3, 35 patients

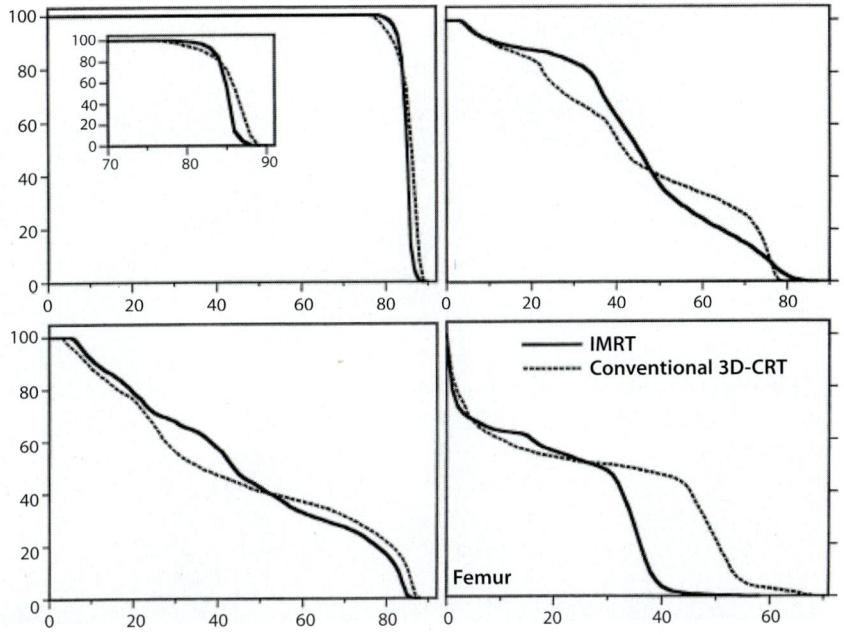

Fig. 10.12 Dose-volume histograms of the CTV, rectal wall, bladder wall, and femoral heads displayed for the treatment plans shown in Fig. 10.11. *Inset* for the DVH of the CTV demonstrates the high-dose region.

(4.5%) developed acute grade-2 rectal toxicity, and no patient experienced acute grade-3 or higher rectal symptoms. Two hundred seventeen patients (28%) developed acute grade-2 urinary symptoms, and one experienced urinary retention (grade 3) toward the end of his treatment course. The rates of late complications are shown in Table 10.4. Eleven patients (1.5%) developed grade-2 rectal bleeding at a median of 9 months after completion of IMRT. Four patients (0.5%) developed grade-3 rectal toxicity requiring one or more transfusions or a laser cauterization procedure. No grade-4 rectal complications were observed. The 3-year actuarial PSA relapse-free survival outcomes for favorable-, intermediate-, and unfavorable-risk-group patients were 92, 86, and 81%, respectively. These data suggest that the PSA outcome after high-dose IMRT is not inferior to what can be achieved with conventional 3D-CRT techniques.

As precise targeting is essential in IMRT, two approaches have been adopted to address the prostate motion issue. The first approach was the develop-

Table 10.3 Acute toxicity IMRT (*n*=772)

Toxicity grade	Gastrointestinal	Genitourinary
0	568 (74%)	258 (33%)
1	169 (22%)	296 (38%)
2	35 (4%)	217 (28%)
3	0	1

Table 10.4 Late-toxicity IMRT (*n*=772)

Toxicity grade	Gastrointestinal	Genitourinary
0	688 (89%)	570 (74%)
1	69 (9%)	121 (16%)
2	11 (1.5%)	76 (9.5%)
3	4 (0.5%)	5 (0.5%)

ment of an accurate daily localization method using ultrasound, CT, or implanted marker imaging; the other was immobilization of the prostate with the use of an inflatable rectal balloon. TEH et al. (2001) evaluated the relationship between dose-volume effects and acute toxicity in 100 patients immobilized with the latter approach. The Peacock system (Nomos Corp., Sewickley, Pa.) with the multivane intensity-modulating collimator (MIMiC) was used for treatment planning and delivery. A mean dose of 76 Gy was prescribed to the 85% isodose line for full coverage of the prostate. Figure 10.13 shows the dose distributions illustrating coverage of the prostate gland and seminal vesicles as well as dose delivered to surrounding normal structures. Figure 10.14 shows a representative DVH from a patient depicting two treatment targets (prostate and seminal vesicles) and two dose-limiting normal structures (bladder and rectum). This study showed that IMRT leads to decrease in treatment-related toxicity for prostate cancer. There was reduction in the volume of the critical avoidance structures receiving high-

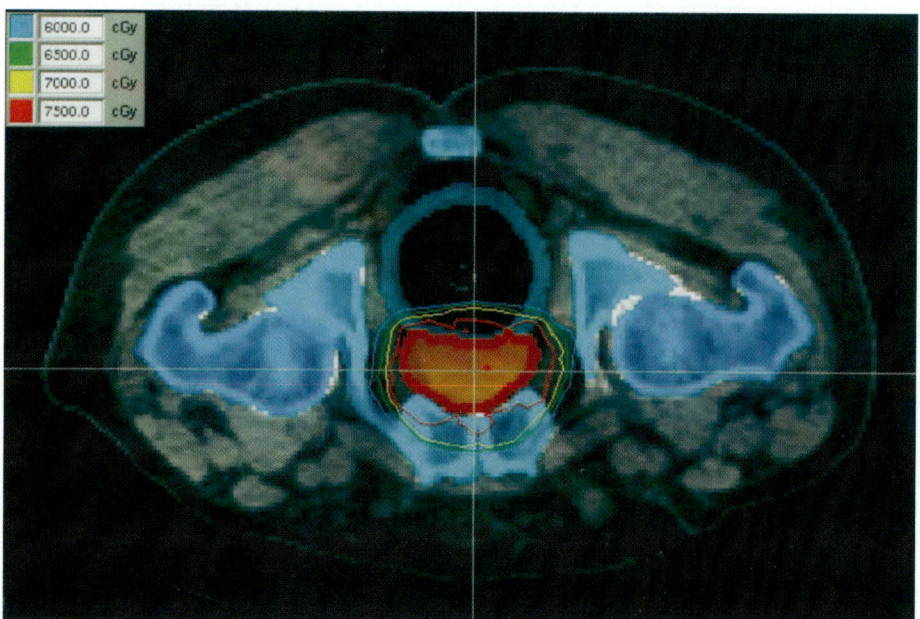

Fig. 10.13 Axial CT image (with the patient in the prone position) shows the rectal balloon (*green*) immobilizing prostate (*brown* with *red border*) against pubis (*light blue*). Also note the isodose coverage of the prostate.

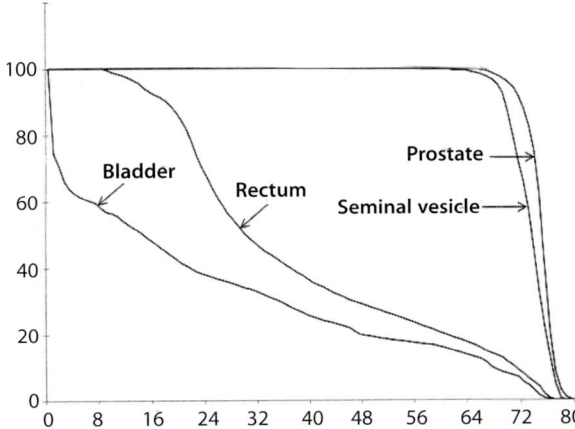

Fig. 10.14 Dose-volume histograms of prostate, seminal vesicle, bladder, and rectum

dose radiation and lower radiation dose delivered to the normal structures; however, the relationship between acute toxicity and mean dose to rectum and bladder or DVH was not statistically significant.

In another study, TEH et al. (2001) reported their experience on postprostatectomy (PPI) IMRT, and compared the acute genitourinary (GU) toxicity to primary (PI) IMRT for 40 patients with prostate cancer. Nine patients had adjuvant radiotherapy, while 31 patients had salvage radiotherapy. One difficulty in postoperative radiotherapy is that the shape of the target (prostate fossa) is irregular for which IMRT is an ideal solution. The PPI was initiated when the patient's urinary continence reached a plateau, usually within 3 months after radical prostatectomy. A dose of 60–66 Gy (mean dose 64–72 Gy) was prescribed using a 2-Gy daily fraction. The dose was prescribed to the 86% (84–89%) isodose line for coverage of the target volume. Figure 10.15 shows a typical dose distribution for PPI. The acute GU toxicity profile is more favorable in the PPI group with 82.5% of grade 0–1 and 17.5% of grade-2 toxicity compared with 59.2 and 40.8%, respectively, in the PI group ($p<0.001$). There was no grade-3 or higher toxicity in either group. The author concluded that PPI can be delivered with acceptable acute GU toxicity, and the favorable acute GU toxicity in PPI might be related to a combination of lower mean and maximum doses and smaller bladder volumes receiving >65 Gy.

Fig. 10.15a,b. The IMRT dose distribution in **a** axial plane (superior level). Note that the patient was treated in the prone position with an air-filled rectal balloon in place. Also note the "butterfly"-shaped target volume covering surgical clips. **b** Sagittal plane. Note rectal balloon posteriorly. *Brown:* prostatic fossa; *green:* rectum; *light blue:* pubic bone; *dark blue:* femoral head; *pink:* bladder

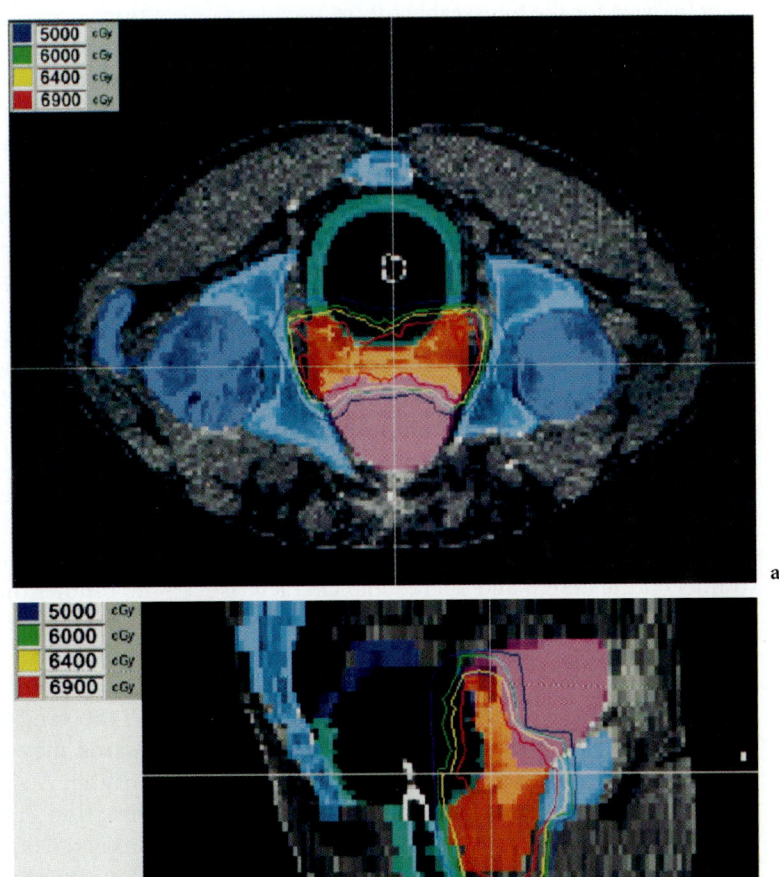

10.7
Discussion and Summary

The process of 3D-CRT planning is significantly different from traditional 2D treatment planning and requires the retraining of the radiation oncologist and treatment planner. Most of the difference lies in the additional tasks of identifying target volumes and critical structures in three dimensions. The introduction of IMRT significantly changes the roles of the radiation oncologist and treatment planner, principally because of the inverse-planning method. With inverse planning, the physician prescribes the optimal dose distribution using tools provided by the software manufacturer; these may be simple text descriptions of minimum target doses and maximum critical structure doses, or they may include the specification of ideal DVHs. The algorithms may have additional input parameters that specify the relative importance of sparing critical structures vs providing the requested target doses. While the process of inverse planning ideally removes the need for multiple plan development and evaluations, the lack of robust objective functions will continue to necessitate the execution of multiple plans, at least for the foreseeable future.

The IMRT has the potential for delivering more conformal dose distributions, treating complex 3D target volumes while sparing critical structures, and therefore improving the quality of radiation therapy; however, there are numerous clinical issues that remain unanswered in IMRT. Most important are the radiobiological consequences of non-standard dose fractionation caused by having multiple dose prescription levels simultaneously delivered. With IMRT, there is significantly greater dose heterogeneity within the target than for conventional conformal therapy. The premise that the ideal dose distribution includes a homogeneous dose within the target will require reevaluation.

The dosimetry instrumentation used for both treatment delivery and QA needs to undergo reevaluation for IMRT. The IMRT delivery places significant constraints on the design and operational characteristics of dosimetry equipment. The dynamic dose delivery and deficiencies of traditional QA techniques will leave many potential users cynical of this exciting development. Significant evidence of the clinical advantages and safeguards will be required before IMRT becomes routine clinical practice. Although significant efforts remain, the introduction of IMRT has been successful. The number of patients treated using commercial and developmental implementations of this modality will continue to increase.

The introduction of multi-institutional trials using IMRT is essential to document any clinical advantages. In the future, the further development of CT-based simulation, inverse-planning systems, and linear accelerator systems designed to deliver IMRT-based dose distributions will likely reduce the overall cost of radiation therapy.

10.8
Additional Challenges for IMRT Implementation

The use of IMRT presents some unique challenges for radiation therapy. Commercial treatment-planning systems assume that the patient geometry and position relative to the linear accelerator is the same during treatment as in the computer model. Variations of the internal position and shape of the tumor can cause misalignments between the tumor and the high dose gradient, although accurately defined target margins should limit underdoses caused by the variations. Similarly, the complex dose distributions provided by IMRT can avoid critical structures using steep dose gradients, but the efficacy of this avoidance can be compromised by misalignment of the patient or by internal shifting of the critical structures. Again, robust margin definitions should alleviate these problems; however, if tight conformance is required, e.g., due to proximity between critical structures and the tumor, accurate patient positioning becomes critical. While this has always been true to some extent, the complex 3D structure of the dose distribution makes alignment more difficult than for traditional therapies.

Another challenge is for the irradiation of tumors that lie near the skin. When there is a combination of en-face and tangential beams, the dose build-up of the en-face beams will be compensated for by additional fluence from the tangential beams; however, when the patient is misaligned, the dose build-up follows the patient, while the fluence modulation is fixed to the room coordinates. This can lead to unintended hot spots if the deeper portions of the tumor overlap the compensating increased fluence of the tangential beams. Similarly, the tangential beams may not extend beyond the projected skin outline. A shifting of the patient towards the flash direction may cause some of the tumor to miss the projected tangential beam, causing a cold spot. Most planning systems are incapable of compensating for this uncertainty.

References

Agazaryan N, Ullrich W, Lee SP, Solberg TD (2004) A methodology for verification of radiotherapy dose calculation. J Neurosurg 101 (Suppl 3):356–361

Ahmad S, Vlachaki MT, Teslow TN, Amosson CM, McGary J, Teh BS, Woo SY, Butler EB, Grant WH III (2005) Impact of setup uncertainty in the dosimetry of prostate and surrounding tissues in prostate cancer patients treated with Peacock/IMRT. Med Dosim 30:1–7

Al-Ghazi M, Kwon R, Kuo J, Ramsinghani N, Yakoob R (2001) The University of California, Irvine, experience with tomotherapy using the Peacock system. Med Dosim 26:17–27

Arrans R, Gallardo MI, Rosello J, Sanchez-Doblado F (2005) Additional dose constraints for analytical beam weighting optimization in IMRT. Radiother Oncol 75:224–226

Baker SJ, Budgell GJ, MacKay RI (2005) Use of an amorphous silicon electronic portal imaging device for multileaf collimator quality control and calibration. Phys Med Biol 50:1377–1392

Barth NH (1990) An inverse problem in radiation therapy. Int J Radiat Oncol Biol Phys 18:425–431

Beaulieua F, Beaulieu L, Tremblay D, Roy R (2004) Simultaneous optimization of beam orientations, wedge filters and field weights for inverse planning with anatomy-based MLC fields. Med Phys 31:1546–1557

Bjarngard B, Kijewski P (1976) The potential of computer control to improve dose distributions in radiation therapy. In: Sternick E (ed) Computer applications in radiation oncology. University Press, Hanover

Bjarngard B, Kijewski P, Pashby C (1977) Description of a computer-controlled machine. Int J Radiat Oncol Biol Phys 2:142

Bortfeld T (1999) Optimized planning using physical objectives and constraints. Semin Radiat Oncol 9:20–34

Bortfeld T, Burkelbach J, Boesecke R, Schlegel W (1990) Methods of image reconstruction from projections applied to conformation radiotherapy. Phys Med Biol 35:1423–1434

Bortfeld T, Kahler DL, Waldron TJ, Boyer AL (1994) X-ray field compensation with multileaf collimators. Int J Radiat Oncol Biol Phys 28:723–730

Bortfeld T, Jokivarsi K, Goitein M, Kung J, Jiang SB (2002) Effects of intra-fraction motion on IMRT dose delivery: statistical analysis and simulation. Phys Med Biol 47:2203–2220

Bortfeld T, Jiang SB, Rietzel E (2004) Effects of motion on the total dose distribution. Semin Radiat Oncol 14:41–51

Boyer A (1995) Present and future developments in radiotherapy treatment units. Semin Radiat Oncol 5:146–155

Boyer A, Ochran TG, Nyerick CE, Waldron TJ, Huntzinger CJ (1992) Clinical dosimetry for implementation of a multileaf collimator. Med Phys 19:1255–1261

Boyer A, Xing L, Ma CM, Curran B, Hill R, Kania A, Bleier A (1999) Theoretical considerations of monitor unit calculations for intensity modulated beam treatment planning. Med Phys 26:187–195

Brace JA (1982) A computer controlled tele-cobalt unit. Int J Radiat Oncol Biol Phys 8:2011–2013

Brace JA (1985) Computer systems for the control of teletherapy units. In: Orton CG (ed) Progress in medical radiation physics. Plenum, New York

Brace JA, Davy TJ, Skeggs DB (1981a) Computer-controlled cobalt unit for radiotherapy. Med Biol Eng Comput 19:612–616

Brace JA, Davy TJ, Skeggs DB, Williams HS. (1981b) Conformation therapy at the Royal Free Hospital. A progress report on the tracking cobalt project. Br J Radiol 54:1068–1074

Brahme A (1987) Design principles and clinical possibilities with a new generation of radiation therapy equipment. A review. Acta Oncol 26:403–412

Brahme A (1988) Optimization of stationary and moving beam radiation therapy techniques. Radiother Oncol 12:129–140

Brahme A (1995) Treatment optimization using physical and radiobiological objective functions. In: Smith A (ed) Medical radiology, radiation therapy physics. Springer, Berlin Heidelberg New York, pp 469–474

Brahme A, Roos JE, Lax I (1982) Solution of an integral equation encountered in rotation therapy. Phys Med Biol 27:1221–1229

Budgell GJ, Mott JH, Logue JP, Hounsell AR (2001) Clinical implementation of dynamic multileaf collimation for compensated bladder treatments. Radiother Oncol 59:31–38

Burman C, Chui CS, Kutcher G, Leibel S, Zelefsky M, LoSasso T, Spirou S, Wu Q, Yang J, Stein J, Mohan R, Fuks Z, Ling CC (1997) Planning, delivery, and quality assurance of intensity-modulated radiotherapy using dynamic multileaf collimator: a strategy for large-scale implementation for the treatment of carcinoma of the prostate. Int J Radiat Oncol Biol Phys 39:863–873

Carol M (1992) 3-D planning and delivery system for optimized conformal therapy (abstract). Int J Radiat Oncol Biol Phys 24:150

Carol M (1995a) Integrated 3D conformal planning/multivane intensity modulating delivery system for radiotherapy. In: Purdy J, Emami B (eds) 3D radiation treatment planning and conformal therapy. Medical Physics Publishing, Madison, Wisconsin, pp 435–445

Carol M (1995b) A system for planning and rotational delivery of intensity-modulated fields. Int J Imaging Syst Technol 6:56–61

Carol M, Grant WH III, Bleier AR, Kania AA, Targovnik HS, Butler EB, Woo SW (1996a) The field-matching problem as it applies to the peacock three dimensional conformal system for intensity modulation. Int J Radiat Oncol Biol Phys 34:183–187

Carol M, Grant WH III, Pavord D, Eddy P, Targovnik HS, Butler B, Woo S, Figura J, Onufrey V, Grossman R, Selkar R (1996b) Initial clinical experience with the Peacock intensity modulation of a 3-D conformal radiation therapy system. Stereotact Funct Neurosurg 66:30–34

Chang J, Mageras GS, Ling CC (2003) Evaluation of rapid dose map acquisition of a scanning liquid-filled ionization chamber electronic portal imaging device. Int J Radiat Oncol Biol Phys 55:1432–1445

Chang J, Obcemea CH, Sillanpaa J, Mechalakos J, Burman C (2004) Use of EPID for leaf position accuracy QA of dynamic multi-leaf collimator (DMLC) treatment. Med Phys 31:2091–2096

Chao KS (2002) Protection of salivary function by intensity-modulated radiation therapy in patients with head and neck cancer. Semin Radiat Oncol 12 (Suppl 1):20–25

Chao KS, Low DA, Perez CA, Purdy JA (2000) Intensity-modulated radiation therapy in head and neck cancers: the Mallinckrodt experience. Int J Cancer 90:92–103

Chao KS, Majhail N, Huang CJ, Simpson JR, Perez CA, Haughey B, Spector G (2001) Intensity-modulated radiation therapy reduces late salivary toxicity without compromising tumor control in patients with oropharyngeal carcinoma: a comparison with conventional techniques. Radiother Oncol 61:275–280

Chen Y, Boyer AL, Ma CM (2000) Calculation of X-ray transmission through a multileaf collimator. Med Phys 27:1717–1726

Chen Z, Xing L, Nath R (2002) Independent monitor unit calculation for intensity modulated radiotherapy using the MIMiC multileaf collimator. Med Phys 29:2041–2051

Chin LM, Kijewski P, Svensson GK, Chaffey JT, Levene MB, Bjarngard BE (1981) A computer-controlled radiation therapy machine for pelvic and para-aortic nodal areas. Int J Radiat Oncol Biol Phys 7:61–70

Chin LM, Kijewski PK, Svensson GK, Bjarngard BE (1983) Dose optimization with computer-controlled gantry rotation, collimator motion and dose-rate variation. Int J Radiat Oncol Biol Phys 9:723–729

Chui CS, LoSasso T, Spirou S (1994) Dose calculation for photon beams with intensity modulation generated by dynamic jaw or multileaf collimations. Med Phys 21:1237–1244

Chui CS, Spirou S, LoSasso T (1996) Testing of dynamic multileaf collimation. Med Phys 23:635–641

Chui CS, Chan MF, Yorke E, Spirou S, Ling CC (2001) Delivery of intensity-modulated radiation therapy with a conventional multileaf collimator: comparison of dynamic and segmental methods. Med Phys 28:2441–2449

Convery D, Rosenbloom M (1992) The generation of intensity-modulated fields for conformal radiotherapy by dynamic collimation. Phys Med Biol 37:1359–1374

Convery DJ, Rosenbloom ME (1995) Treatment delivery accuracy in intensity-modulated conformal radiotherapy. Phys Med Biol 40:979–999

Cooper RE (1978) A gradient method of optimizing external-beam radiotherapy treatment plans. Radiology 128:235–243

Cormack AM (1987) A problem in rotation therapy with X rays. Int J Radiat Oncol Biol Phys 13:623–630

Cormack AM, Cormack RA (1987) A problem in rotation therapy with X-rays: dose distributions with an axis of symmetry. Int J Radiat Oncol Biol Phys 13:1921–1925

Cormack AM, Quinto E (1989) On a problem in radiotherapy: questions on non-negativity. Int J Imaging Syst Technol 1:120–124

Cotrutz C, Xing L (2003) Segment-based dose optimization using a genetic algorithm. Phys Med Biol 48:2987–2998

Davy TJ (1985) Physical aspects of conformation therapy using computer-controlled tracking units. In: Orton CG (ed) Progress in medical radiation physics. Plenum, New York

Davy TJ, Brace J (1979) Dynamic 3-D treatment using a computer-controlled cobalt unit. Br J Radiol 53:612–616

Davy TJ, Johnson PH, Redford R, Williams JR (1975) Conformation therapy using the tracking cobalt unit. Br J Radiol 48:122–130

De Neve W, de Gersem W, Derycke S, de Meerleer G, Moerman M, Bate MT, van Duyse B, Vakaet L, de Deene Y, Mersseman B, de Wagter C (1999) Clinical delivery of intensity modulated conformal radiotherapy for relapsed or second-primary head and neck cancer using a multileaf collimator with dynamic control. Radiother Oncol 50:301–314

Deasy JO, Niemierko A, Herbert D, Yan D, Jackson A, Ten Haken RK, Langer M, Sapareto S (2002) Methodological issues in radiation dose-volume outcome analyses: summary of a joint AAPM/NIH workshop. Med Phys 29:2109–2127

Dische S, Saunders MI, Williams C, Hopkins A, Aird E (1993) Precision in reporting the dose given in a course of radiotherapy. Radiother Oncol 29:287–293

Djajaputra D, Wu Q, Wu Y, Mohan R (2003) Algorithm and performance of a clinical IMRT beam-angle optimization system. Phys Med Biol 48:3191–3212

Djordjevich A, Bonham DJ, Hussein EM, Andrew JW, Hale ME (1990) Optimal design of radiation compensators. Med Phys 17:397–404

Dogan N, Leybovich LB, Sethi A (2002) Comparative evaluation of Kodak EDR2 and XV2 films for verification of intensity modulated radiation therapy. Phys Med Biol 47:4121–4130

Edlund T, Leavitt DD, Gibbs FA Jr (1999) Dosimetric advantages of enhanced dynamic wedge in small field irradiation for the treatment of macular degeneration. Med Dosim 24:21–26

Ellis F, Miller H (1944) The use of wedge filters in deep X-ray therapy. Br J Radiol

Ellis F, Hall E, Oliver R (1959) A compensator for variations in tissue thickness for high energy beams. Br J Radiol 32:421–422

Engelsman M, Sharp GC, Bortfeld T, Onimaru R, Shirato H (2005) How much margin reduction is possible through gating or breath hold? Phys Med Biol 50:477–490

Engler M, Tsai J-S, Ling M, Wu J, Palano J, Koistinen M, Kramer B, Fagundes M, Dipertrillo T, Wazer D (1997) Physical and clinical aspects of the dynamic intensity modulated radiotherapy of 44 patients. The 38th annual meeting of the American Society for Therapeutic Radiology and Oncology, Los Angeles, California

Esthappan J, Mutic S, Harms WB, Dempsey JF, Low DA (2002) Dosimetry of therapeutic photon beams using an extended dose range film. Med Phys 29:2438–2445

Fang G, Geiser B, Mackie TR (1997) Software system for the UW/GH tomotherapy prototype. In: Leavitt DD, Starkschall G (eds) XIIth International Conference on the Use of Computers in Radiation Therapy. Medical Physics Publishing, Salt Lake City, pp 332–334

Followill D, Geis P, Boyer A (1997) Estimates of whole-body dose equivalent produced by beam intensity modulated conformal therapy. Int J Radiat Oncol Biol Phys 38:667–672

Fraass BA (1994) Computer-controlled three-dimensional conformal therapy delivery systems. In: Purdy J, Fraass BA (eds) Syllabus: a categorical course in physics, three-dimensional radiation therapy treatment planning. Radiological Society of North America, Oak Brook, Illinois, pp 93–100

Fraass BA, McShan DL, Kessler ML (1995) Computer-controlled treatment delivery. Semin Radiat Oncol 5:77–85

Galvin JM, Chen XG, Smith RM (1993) Combining multileaf fields to modulate fluence distributions. Int J Radiat Oncol Biol Phys 27:697–705

Geis P, Boyer AL, Wells NH (1996) Use of a multileaf collimator as a dynamic missing-tissue compensator. Med Phys 23:1199–1205

Goldman SP, Chen JZ, Battista JJ (2005) Feasibility of a fast inverse dose optimization algorithm for IMRT via matrix inversion without negative beamlet intensities. Med Phys 32:3007–3016

Gore JC, Ranade M, Maryanski MJ, Schulz RJ (1996) Radiation dose distributions in three dimensions from tomographic optical density scanning of polymer gels. I. Development of an optical scanner. Phys Med Biol 41:2695–2704

Graham MV, Jain NL, Kahn MG, Drzymala RE, Purdy JA (1996) Evaluation of an objective plan-evaluation model in the three dimensional treatment of nonsmall cell lung cancer. Int J Radiat Oncol Biol Phys 34:469–474

Grant W, Bellezza D (1994) Leakage considerations with a multi-leaf collimator designed for intensity-modulated conformal radiotherapy (abstract). Med Phys 21:921

Grant W III, Cain RB (1998) Intensity modulated conformal therapy for intracranial lesions. Med Dosim 23:237–241

Green A (1965) Tracking cobalt project. Nature 207:1311

Guerrero Urbano MT, Nutting CM (2004a) Clinical use of intensity-modulated radiotherapy, part I. Br J Radiol 77:88–96

Guerrero Urbano MT, Nutting CM (2004b) Clinical use of intensity-modulated radiotherapy, part II. Br J Radiol 77:177–182

Gustafsson A, Lind BK, Brahme A (1994) A generalized pencil beam algorithm for optimization of radiation therapy. Med Phys 21:343–356

Gustafsson A, Lind BK, Svensson R, Brahme A (1995) Simultaneous optimization of dynamic multileaf collimation and scanning patterns or compensation filters using a generalized pencil beam algorithm. Med Phys 22:1141–1156

Hale JI, Kerr AT, Shragge PC (1994) Calibration of film for accurate megavoltage photon dosimetry. Med Dosim 19:43–46

Higgins PD, Alaei P, Gerbi BJ, Dusenbery KE (2003) In vivo diode dosimetry for routine quality assurance in IMRT. Med Phys 30:3118–3123

Holmes T, Mackie TR (1994) A filtered backprojection dose calculation method for inverse treatment planning. Med Phys 21:303–313

Holmes T, Mackie TR, Simpkin D, Reckwerdt P (1991) A unified approach to the optimization of brachytherapy and external beam dosimetry. Int J Radiat Oncol Biol Phys 20:859–873

Hong TS, Tome WA, Chappell RJ, Chinnaiyan P, Mehta MP, Harari PM (2005) The impact of daily setup variations on head-and-neck intensity-modulated radiation therapy. Int J Radiat Oncol Biol Phys 61:779–788

Hope C, Orr J (1965) Computer optimization of 4 MeV treatment planning. Phys Med Biol 10:365–370

Hou Q, Wang J, Chen Y, Galvin JM (2003) Beam orientation optimization for IMRT by a hybrid method of the genetic algorithm and the simulated dynamics. Med Phys 30:2360–2367

Hounsell AR, Wilkinson JM (1997) Head scatter modelling for irregular field shaping and beam intensity modulation. Phys Med Biol 42:1737–1749

Islam KT, Dempsey JF, Ranade MK, Maryanski MJ, Low DA (2003) Initial evaluation of commercial optical CT-based 3D gel dosimeter. Med Phys 30:2159–2168

Jain N, Kahn M, Grahm M, Purdy JA (1994) 3D conformal radiation therapy. V. In: Hounsell AR, Wilkinson JM, Williams PC (eds) Decision-theoretic evaluation on radiation treatment plans. Proc XIth International Conference on the Use of Computers in Radiation Therapy. Manchester, UK, pp 8–9

Jeraj R, Mackie TR, Balog J, Olivera G, Pearson D, Kapatoes J, Ruchala K, Reckwerdt P (2004) Radiation characteristics of helical tomotherapy. Med Phys 31:396–404

Kallman P, Lind B, Eklof A, Brahme A (1988) Shaping of arbitrary dose distributions by dynamic multileaf collimation. Phys Med Biol 33:1291–1300

Kam MK, Chau RM, Suen J, Choi PH, Teo PM (2003) Intensity-modulated radiotherapy in nasopharyngeal carcinoma: dosimetric advantage over conventional plans and feasibility of dose escalation. Int J Radiat Oncol Biol Phys 56:145–157

Kam MK, Teo PM, Chau RM, Cheung KY, Choi PH, Kwan WH, Leung SF, Zee B, Chan AT (2004) Treatment of nasopharyngeal carcinoma with intensity-modulated radiotherapy: the Hong Kong experience. Int J Radiat Oncol Biol Phys 60:1440–1450

Kelly RG, Jordan KJ, Battista JJ (1998) Optical CT reconstruction of 3D dose distributions using the ferrous-benzoic-xylenol (FBX) gel dosimeter. Med Phys 25:1741–1750

Kijewski PK, Chin LM, Bjarngard BE (1978) Wedge-shaped dose distributions by computer-controlled collimator motion. Med Phys 5:426–429

Kirkpatric S (1985) Optimization by simulated annealing. Science 220:671–680

Klein E (1994) Implementation and clinical use of multileaf collimation. In: Purdy JA, Fraass BA (eds) Syllabus: a categorical course in physics, three-dimensional radiation therapy treatment planning. Radiological Society of North America, Oak Brook, Illinois

Klein EE, Low DA, Meigooni AS, Purdy JA (1995) Dosimetry and clinical implementation of dynamic wedge. Int J Radiat Oncol Biol Phys 31:583–592

Klein EE, Gerber R, Zhu XR, Oehmke F, Purdy JA (1998) Multiple machine implementation of enhanced dynamic wedge. Int J Radiat Oncol Biol Phys 40:977–985

Kubo HD, Wang L (2000) Compatibility of Varian 2100C gated operations with enhanced dynamic wedge and IMRT dose delivery. Med Phys 27:1732–1738

Kung JH, Chen GT, Kuchnir FK (2000) A monitor unit verification calculation in intensity modulated radiotherapy as a dosimetry quality assurance. Med Phys 27:2226–2230

Kutcher GJ, Coia L, Gillin M, Hanson WF, Leibel S, Morton RJ, Palta JR, Purdy JA, Reinstein LE, Svensson GK et al (1994) Comprehensive QA for radiation oncology: report of AAPM Radiation Therapy Committee Task Group 40. Med Phys 21:581–618

Langer M, Leong J (1987) Optimization of beam weights under dose-volume restrictions. Int J Radiat Oncol Biol Phys 13:1255–1260

Langer M, Brown R, Urie M, Leong J, Stracher M, Shapiro J (1990) Large scale optimization of beam weights under dose-volume restrictions. Int J Radiat Oncol Biol Phys 18:887–893

Langer M, Brown R, Kijewski P, Ha C (1993) The reliability of optimization under dose-volume limits. Int J Radiat Oncol Biol Phys 26:529–538

Leavitt DD, Martin M, Moeller JH, Lee WL (1990) Dynamic wedge field techniques through computer-controlled collimator motion and dose delivery. Med Phys 17:87–91

Leavitt DD, Huntzinger C, Etmektzoglou T (1997) Dynamic collimator and dose rate control: enabling technology for enhanced dynamic wedge. Med Dosim 22:167–170

Leavitt DD, Williams G, Tobler M, Moeller JH, Gibbs FA Jr, Gaffney DK (2000) Application of enhanced dynamic wedge to stereotactic radiotherapy. Med Dosim 25:61–69

Lee EK, Fox T, Crocker I (2005) Simultaneous beam geometry and intensity map optimization in intensity-modulated radiation therapy. Int J Radiat Oncol Biol Phys

Letourneau D, Gulam M, Yan D, Oldham M, Wong JW (2004)

Evaluation of a 2D diode array for IMRT quality assurance. Radiother Oncol 70:199–206

Li JS, Boyer AL, Ma CM (2001) Verification of IMRT dose distributions using a water beam imaging system. Med Phys 28:2466–2474

Lian J, Cotrutz C, Xing L (2003) Therapeutic treatment plan optimization with probability density-based dose prescription. Med Phys 30:655–666

Lind BK, Kallman P (1990) Experimental verification of an algorithm for inverse radiation therapy planning. Radiother Oncol 17:359–368

Ling CC, Burman C, Chui CS, Kutcher GJ, Leibel SA, LoSasso T, Mohan R, Bortfeld T, Reinstein L, Spirou S, Wang XH, Wu Q, Zelefsky M, Fuks Z (1996) Conformal radiation treatment of prostate cancer using inversely-planned intensity-modulated photon beams produced with dynamic multileaf collimation. Int J Radiat Oncol Biol Phys 35:721–730

Lof J, Lind BK, Brahme A (1998) An adaptive control algorithm for optimization of intensity modulated radiotherapy considering uncertainties in beam profiles, patient set-up and internal organ motion. Phys Med Biol 43:1605–1628

Low D, Dempsey JF (2003) Evaluation of the gamma dose distribution comparison method. Med Phys 30:2455–2464

Low D, Mutic S (1997) Abutment region dosimetry for sequential arc IMRT delivery. Phys Med Biol 42:1465–1470

Low D, Mutic S (1998) A commercial IMRT treatment-planning dose-calculation algorithm. Int J Radiat Oncol Biol Phys 41:933–937

Low D, Zhu X, Harms W, Purdy JA (1996a) Beam-intensity modulation using physical modulators (abstract). Med Phys 23:1001

Low D, Li Z, Klein EE (1996b) Verification of milled two-dimensional photon compensating filters using an electronic portal imaging device. Med Phys 23:929–938

Low D, Zhu XR, Purdy JA, Soderstrom S (1997) The influence of angular misalignment on fixed-portal intensity modulated radiation therapy. Med Phys 24:1123–1139

Low D, Gerber R, Mutic S, Purdy JA (1998a) Phantoms for IMRT dose distribution measurement and treatment verification. Int J Radiat Oncol Biol Phys 40:1231–1235

Low D, Chao KS, Mutic S, Gerber RL, Perez CA, Purdy JA (1998b) Quality assurance of serial tomotherapy for head and neck patient treatments. Int J Radiat Oncol Biol Phys 42:681–692

Low D, Harms WB, Mutic S, Purdy JA (1998c) A technique for the quantitative evaluation of dose distributions. Med Phys 25:656–661

Low D, Mutic S, Dempsey JF, Gerber RL, Bosch WR, Perez CA, Purdy JA (1998d) Quantitative dosimetric verification of an IMRT planning and delivery system. Radiother Oncol 49:305–316

Low D, Markman J, Dempsey JF, Mutic S, Oldham M, Venkatesan R, Haacke EM, Purdy JA (2000) Noise in polymer gel measurements using MRI. Med Phys 27:1814–1817

Low D, Sohn JW, Klein EE, Markman J, Mutic S, Dempsey JF (2001) Characterization of a commercial multileaf collimator used for intensity modulated radiation therapy. Med Phys 28:752–756

Ma CM, Pawlicki T, Jiang SB, Li JS, Deng J, Mok E, Kapur A, Xing L, Ma L, Boyer AL (2000) Monte Carlo verification of IMRT dose distributions from a commercial treatment planning optimization system. Phys Med Biol 45:2483–2495

Ma CM, Jiang SB, Pawlicki T, Chen Y, Li JS, Deng J, Boyer AL

(2003) A quality assurance phantom for IMRT dose verification. Phys Med Biol 48:561–572

Ma L, Boyer AL, Ma CM, Xing L (1999) Synchronizing dynamic multileaf collimators for producing two-dimensional intensity-modulated fields with minimum beam delivery time. Int J Radiat Oncol Biol Phys 44:1147–1154

Mackie TR, Holmes T, Swerdloff S, Reckwerdt P, Deasy JO, Yang J, Paliwal B, Kinsella T (1993) Tomotherapy: a new concept for the delivery of dynamic conformal radiotherapy. Med Phys 20:1709–1719

Mackie TR, Aldridge S, Angelos L, Balog J, Coon S, Fang G, Fitchard E, Geiser B, Glass M, Iosevich S, Kapatoes J, McNutt TR (1997) Tomotherapy: rethinking the process of radiotherapy. In: Leavitt DD, Starkschall G (eds) XIIth International Conference on the Use of Computers in Radiation Therapy. Medical Physics Publishing, Salt Lake City, pp 329–331

Mackie TR, Balog J, Ruchala K, Shepard D, Aldridge S, Fitchard E, Reckwerdt P, Olivera G, McNutt T, Mehta M (1999) Tomotherapy. Semin Radiat Oncol 9:108–117

Mackie TR, Kapatoes J, Ruchala K, Lu W, Wu C, Olivera G, Forrest L, Tome W, Welsh J, Jeraj R, Harari P, Reckwerdt P, Paliwal B, Ritter M, Keller H, Fowler J, Mehta M (2003) Image guidance for precise conformal radiotherapy. Int J Radiat Oncol Biol Phys 56:89–105

Mageras GS, Mohan R (1993) Application of fast simulated annealing to optimization of conformal radiation treatments. Med Phys 20:639–647

Maryanski MJ, Schulz RJ, Ibbott GS, Gatenby JC, Xie J, Horton D, Gore JC (1994) Magnetic resonance imaging of radiation dose distributions using a polymer-gel dosimeter. Phys Med Biol 39:1437–1455

Maryanski MJ, Ibbott GS, Eastman P, Schulz RJ, Gore JC (1996a) Radiation therapy dosimetry using magnetic resonance imaging of polymer gels. Med Phys 23:699–705

Maryanski MJ, Zastavker YZ, Gore JC (1996b) Radiation dose distributions in three dimensions from tomographic optical density scanning of polymer gels. II. Optical properties of the BANG polymer gel. Phys Med Biol 41:2705–2717

Maryanski MJ, Audet C, Gore JC (1997) Effects of crosslinking and temperature on the dose response of a BANG polymer gel dosimeter. Phys Med Biol 42:303–311

Mayer R, Williams A, Frankel T, Cong Y, Simons S, Yang N, Timmerman R (1997) Two-dimensional film dosimetry application in heterogeneous materials exposed to megavoltage photon beams. Med Phys 24:455–460

McCurdy BM, Luchka K, Pistorius S (2001) Dosimetric investigation and portal dose image prediction using an amorphous silicon electronic portal imaging device. Med Phys 28:911–924

McDonald SC, Rubin P (1977) Optimization of external beam radiation therapy. Int J Radiat Oncol Biol Phys 2:307–317

Metropolis N, Rosenbluth A, Rosenbluth M, Teller A, Teller E (1953) Equation of state calculations by fast computing machines. J Chem Phys 21:1087–1092

Meyer J, Purdy J (eds) (1996) 3-D conformal radiotherapy. Frontiers in radiation therapy and oncology. Karger, Basel, Switzerland

Mohan R, Mageras GS, Baldwin B, Brewster LJ, Kutcher GJ, Leibel S, Burman CM, Ling CC, Fuks Z (1992) Clinically relevant optimization of 3-D conformal treatments. Med Phys 19:933–944

Mohan R, Lovelock M, Mageras G, LoSasso T, Chui CS (1996a)

Computer controlled radiation therapy and multileaf collimation. In: Meyer J, Purdy JA (eds) A new era in the irradiation of cancer. Karger, New York, pp 123–138

Mohan R, Wang XH, Jackson A (1996b) Optimization of 3-D conformal radiation treatment plans. Front Radiat Ther Oncol 29:86–103

Moran JM, Roberts DA, Nurushev TS, Antonuk LE, El-Mohri Y, Fraass BA (2005) An Active Matrix Flat Panel Dosimeter (AMFPD) for in-phantom dosimetric measurements. Med Phys 32:466–472

Morrill S, Lane RG, Rosen I (1990) Constrained simulated annealing for optimized radiation therapy treatment planning. Comput Methods Progr Biomed 33:135–144

Morrill SM, Lane RG, Jacobson G, Rosen I (1991) Treatment planning optimization using constrained simulated annealing. Phys Med Biol 36:1341–1361

Mott JH, Hounsell AR, Budgell GJ, Wilkinson JM, Williams PC (1999) Customised compensation using intensity modulated beams delivered by dynamic multileaf collimation. Radiother Oncol 53:59–65

Mutic S, Low D (1997) Whole body dose from arc-based IMRT treatment (abstract). Int J Radiat Oncol Biol Phys 24:1368

Niemierko A (1992) Random search algorithm (RONSC) for optimization of radiation therapy with both physical and biological end points and constraints. Int J Radiat Oncol Biol Phys 23:89–98

Niemierko A (1997) Reporting and analyzing dose distributions: a concept of equivalent uniform dose. Med Phys 24:103–110

Niemierko A (1998) Radiobiological models of tissue response to radiation in treatment planning systems. Tumori 84:140–143

Niemierko A, Urie M, Goitein M (1992) Optimization of 3D radiation therapy with both physical and biological end points and constraints. Int J Radiat Oncol Biol Phys 23:99–108

Niroomand-Rad A, Blackwell CR, Coursey BM, Gall KP, Galvin JM, McLaughlin WL, Meigooni AS, Nath R, Rodgers JE, Soares CG (1998) Radiochromic film dosimetry: recommendations of AAPM Radiation Therapy Committee Task Group 55. American Association of Physicists in Medicine. Med Phys 25:2093–2115

Nutting CM, Convery DJ, Cosgrove VP, Rowbottom C, Vini L, Harmer C, Dearnaley DP, Webb S (2001) Improvements in target coverage and reduced spinal cord irradiation using intensity-modulated radiotherapy (IMRT) in patients with carcinoma of the thyroid gland. Radiother Oncol 60:173–180

Olafsson A, Jeraj R, Wright SJ (2005) Optimization of intensity-modulated radiation therapy with biological objectives. Phys Med Biol 50:5357–5379

Olch AJ (2002) Dosimetric performance of an enhanced dose range radiographic film for intensity-modulated radiation therapy quality assurance. Med Phys 29:2159–2168

Palta JR, Kim S, Li JG, Liu C (2003) Tolerance limits and action levels for planning and delivery of IMRT. In: Palta JR, Mackie TR (eds) Intensity-modulated radiation therapy: the state of the art. Medical Physics, Madison, Wisconsin

Pawlicki T, Ma CM (2001) Monte Carlo simulation for MLC-based intensity-modulated radiotherapy. Med Dosim 26:157–168

Powers WE, Kinzie JJ, Demidecki AJ, Bradfield JS, Feldman A (1973) A new system of field shaping for external-beam radiation therapy. Radiology 108:407–411

Proimos BS (1960) Synchronous field shaping in rotational megavolt therapy. Radiology 74:753–757

Proimos BS (1966) Beam-shapers oriented by gravity in rotational therapy. Radiology 87:928–932

Redpath AT, Vickery BL, Wright DH (1976) A new technique for radiotherapy planning using quadratic programming. Phys Med Biol 21:781–791

Richter J, Neumann M, Bortfeld T (1997) Dynamic multileaf collimator rotation techniques versus intensity modulated fixed fields. In: Leavitt D, Starkschall G (eds) XIIth International Conference on the Use of Computers in Radiation Therapy. Medical Physics Publishing, Salt Lake City, pp 335–337

Rosen II, Morrill SM, Lane RG (1992) Optimized dynamic rotation with wedges. Med Phys 19:971–977

Sauer O, Shepard D, Angelos L, Mackie TR (1997) A comparison of objective funtions for use in radiotherapy optimization. In: Leavitt D, Starkschall G (eds) XIIth International Conference on the Use of Computers in Radiation Therapy. Medical Physics Publishing, Salt Lake City

Saw CB, Ayyangar KM, Thompson RB, Zhen W, Enke CA (2001a) Commissioning of peacock system for intensity-modulated radiation therapy. Med Dosim 26:55–64

Saw CB, Ayyangar KM, Zhen W, Thompson RB, Enke CA (2001b) Quality assurance procedures for the Peacock system. Med Dosim 26:83–90

Scholz C, Nill S, Oelfke U (2003) Comparison of IMRT optimization based on a pencil beam and a superposition algorithm. Med Phys 30:1909–1913

Schultheiss TE, Orton CG (1985) Models in radiotherapy: definition of decision criteria. Med Phys 12:183–187

Seco J, Evans PM, Webb S (2002) An optimization algorithm that incorporates IMRT delivery constraints. Phys Med Biol 47:899–915

Siebers JV, Keall PJ, Wu Q, Williamson JF, Schmidt-Ullrich RK (2005) Effect of patient setup errors on simultaneously integrated boost head and neck IMRT treatment plans. Int J Radiat Oncol Biol Phys 63:422–433

Sodertrom S, Brahme A (1993) Optimization of the dose delivery in a few field techniques using radiobiological objective functions. Med Phys 20:1201–1210

Spirou SV, Chui CS (1994) Generation of arbitrary intensity profiles by dynamic jaws or multileaf collimators. Med Phys 21:1031–1041

Spirou SV, Chui CS (1998) A gradient inverse planning algorithm with dose-volume constraints. Med Phys 25:321–333

Starkschall G (1984) A constrained least-squares optimization method for external beam radiation therapy treatment planning. Med Phys 11:659–665

Stavrev P, Hristov D, Warkentin B, Sham E, Stavreva N, Fallone BG (2003) Inverse treatment planning by physically constrained minimization of a biological objective function. Med Phys 30:2948–2958

Stein J, Bortfeld T, Dorschel B, Schlegel W (1994) Dynamic X-ray compensation for conformal radiotherapy by means of multi-leaf collimation. Radiother Oncol 32:163–173

Stein J, Hartwig K, Levergrun S, Zhang G, Preiser K, Rhein B, Debus J, Bortfeld T (1997) Intensity-modulated treatments: compensators vs. multileaf modulation. In: Leavitt D, Starkschall G (eds) XIIth International Conference on the Use of Computers in Radiation Therapy. Medical Physics Publishing, Salt Lake City, pp 338–341

Stevens MA, Turner JR, Hugtenburg RP, Butler PH (1996) High-resolution dosimetry using radiochromic film and a document scanner. Phys Med Biol 41:2357–2365

Stovall M, Blackwell CR, Cundiff J, Novack DH, Palta JR, Wagner LK, Webster EW, Shalek RJ (1995) Fetal dose from radiotherapy with photon beams: report of AAPM Radiation Therapy Committee Task Group no 36. Med Phys 22:63–82

Svensson R, Kallman P, Brahme A (1994) An analytical solution for the dynamic control of multileaf collimators. Phys Med Biol 39:37–61

Takahashi S (1965) Conformation radiotherapy. Rotation techniques as applied to radiography and radiotherapy of cancer. Acta Radiol Diagn (Stockh) (Suppl) 242:241

Teh BS, Mai WY, Augspurger ME, Uhl BM, McGary J, Dong L, Grant WH III, Lu HH, Woo SY, Carpenter LS, Chiu JK, Butler EB (2001a) Intensity modulated radiation therapy (IMRT) following prostatectomy: more favorable acute genitourinary toxicity profile compared to primary IMRT for prostate cancer. Int J Radiat Oncol Biol Phys 49:465–472

Teh BS, Mai WY, Uhl BM, Augspurger ME, Grant WH, III, Lu HH, Woo SY, Carpenter LS, Chiu JK, Butler EB (2001b) Intensity-modulated radiation therapy (IMRT) for prostate cancer with the use of a rectal balloon for prostate immobilization: acute toxicity and dose-volume analysis. Int J Radiat Oncol Biol Phys 49:705–712

Trump JG, Wright KA, Smedal MI, Salzman FA (1961) Synchronous field shaping and protection in 2-million-volt rotational therapy. Radiology 76:275

Van Battum LJ, Heijmen BJ (1995) Film dosimetry in water in a 23-mV therapeutic photon beam. Radiother Oncol 34:152–159

Van Dyk J, Barnett RB, Cygler JE, Shragge PC (1993) Commissioning and quality assurance of treatment planning computers. Int J Radiat Oncol Biol Phys 26:261–273

Van Esch A, Vanstraelen B, Verstraete J, Kutcher G, Huyskens D (2001) Pre-treatment dosimetric verification by means of a liquid-filled electronic portal imaging device during dynamic delivery of intensity modulated treatment fields. Radiother Oncol 60:181–190

Van Esch A, Depuydt T, Huyskens DP (2004) The use of an aSi-based EPID for routine absolute dosimetric pre-treatment verification of dynamic IMRT fields. Radiother Oncol 71:223–234

Van Santvoort JP, Heijmen BJ (1996) Dynamic multileaf collimation without "tongue-and-groove" underdosage effects. Phys Med Biol 41:2091–2105

Verellen D, Linthout N, van den Berge D, Bel A, Storme G (1997) Initial experience with intensity-modulated conformal radiation therapy for treatment of the head and neck region. Int J Radiat Oncol Biol Phys 39:99–114

Webb S (1989) Optimisation of conformal radiotherapy dose distributions by simulated annealing. Phys Med Biol 34:1349–1370

Webb S (1991a) Optimization by simulated annealing of three-dimensional conformal treatment planning for radiation fields defined by a multileaf collimator. Phys Med Biol 36:1201–1226

Webb S (1991b) Optimization of conformal radiotherapy dose distributions by simulated annealing. 2. Inclusion of scatter in the 2D technique. Phys Med Biol 36:1227–1237

Webb S (1992) Optimization by simulated annealing of three-dimensional, conformal treatment planning for radiation fields defined by a multileaf collimator. II. Inclusion of two-dimensional modulation of the X-ray intensity. Phys Med Biol 37:1689–1704

Webb S (1993) The physics of three-dimensional radiation therapy. Institute of Physics Publishing, Bristol, p 373

Webb S, Oldham M (1996) A method to study the characteristics of 3D dose distributions created by superposition of many intensity-modulated beams delivered via a slit aperture with multiple absorbing vanes. Phys Med Biol 41:2135–2153

Webb S, Bortfeld T, Stein J, Convery D (1997) The effect of stair-step leaf transmission on the "tongue-and-groove problem" in dynamic radiotherapy with a multileaf collimator. Phys Med Biol 42:595–602

Welsh JS, Patel RR, Ritter MA, Harari PM, Mackie TR, Mehta MP (2002) Helical tomotherapy: an innovative technology and approach to radiation therapy. Technol Cancer Res Treat 1:311–316

Wiezorek T, Banz N, Schwedas M, Scheithauer M, Salz H, Georg D, Wendt TG (2005) Dosimetric quality assurance for intensity-modulated radiotherapy feasibility study for a filmless approach. Strahlenther Onkol 181:468–474

Williamson JF, Khan FM, Sharma SC (1981) Film dosimetry of megavoltage photon beams: a practical method of isodensity-to-isodose curve conversion. Med Phys 8:94–98

Winkler P, Zurl B, Guss H, Kindl P, Stuecklschweiger G (2005) Performance analysis of a film dosimetric quality assurance procedure for IMRT with regard to the employment of quantitative evaluation methods. Phys Med Biol 50:643–654

Wolodzko JG, Marsden C, Appleby A (1999) CCD imaging for optical tomography of gel radiation dosimeters. Med Phys 26:2508–2513

Woo SY, Sanders M, Grant W, Butler EB (1994) Does the "peacock" have anything to do with radiotherapy? Int J Radiat Oncol Biol Phys 29:213–214

Wright KA, Proimos BS, Trump JG, Smedal MI, Johnson D, Salzman FA (1959) Field shaping and protection in 2-million-volt rotational therapy. Radiology 72:101

Wu A, Johnson M, Chen AS, Kalnicki S (1996) Evaluation of dose calculation algorithm of the peacock system for multileaf intensity modulation collimator. Int J Radiat Oncol Biol Phys 36:1225–1231

Wu Q, Manning M, Schmidt-Ullrich R, Mohan R (2000) The potential for sparing of parotids and escalation of biologically effective dose with intensity-modulated radiation treatments of head and neck cancers: a treatment design study. Int J Radiat Oncol Biol Phys 46:195–205

Wu Q, Mohan R, Niemierko A, Schmidt-Ullrich R (2002) Optimization of intensity-modulated radiotherapy plans based on the equivalent uniform dose. Int J Radiat Oncol Biol Phys 52:224–235

Xiao Y, Michalski D, Censor Y, Galvin JM (2004) Inherent smoothness of intensity patterns for intensity modulated radiation therapy generated by simultaneous projection algorithms. Phys Med Biol 49:3227–3245

Xing L, Chen GT (1996) Iterative methods for inverse treatment planning. Phys Med Biol 41:2107–2123

Xing L, Curran B, Hill R, Holmes T, Ma L, Forster KM, Boyer AL (1999) Dosimetric verification of a commercial inverse treatment planning system. Phys Med Biol 44:463–478

Xing L, Chen Y, Luxton G, Li JG, Boyer AL (2000) Monitor unit calculation for an intensity modulated photon field by a simple scatter-summation algorithm. Phys Med Biol 45: N1–N7

Yan H, Yin FF, Guan HQ, Kim JH (2003) AI-guided parameter optimization in inverse treatment planning. Phys Med Biol 48:3565–3580

Yan Y, Papanikolaou N, Weng X, Penagaricano J, Ratanatharathorn V (2005) Fast radiographic film calibration procedure for helical tomotherapy intensity modulated radiation therapy dose verification. Med Phys 32:1566–1570

Yang J, Li J, Chen L, Price R, McNeeley S, Qin L, Wang L, Xiong W, Ma CM (2005) Dosimetric verification of IMRT treatment planning using Monte Carlo simulations for prostate cancer. Phys Med Biol 50:869–878

Yang Y, Xing L (2004) Clinical knowledge-based inverse treatment planning. Phys Med Biol 49:5101–5117

Yu CX (1995) Intensity-modulated arc therapy with dynamic multileaf collimation: an alternative to tomotherapy. Phys Med Biol 40:1435–1449

Yu CX, Symons MJ, Du MN, Martinez AA, Wong JW (1995) A method for implementing dynamic photon beam intensity modulation using independent jaws and a multileaf collimator. Phys Med Biol 40:769–787

Yu CX, Jaffray DA, Wong JW (1997) Calculating the effects of intra-treatment organ motion on dynamic intensity modu-

lation. In: Leavitt DD, Starkschall G (eds) XIIth International Conference on the Use of Computers in Radiation Therapy. Medical Physics Publishing, Salt Lake City, pp 231–233

Zelefsky MJ, Fuks Z, Happersett L, Lee HJ, Ling CC, Burman CM, Hunt M, Wolfe T, Venkatraman ES, Jackson A, Skwarchuk M, Leibel SA (2000) Clinical experience with intensity modulated radiation therapy (IMRT) in prostate cancer. Radiother Oncol 55:241–249

Zelefsky MJ, Fuks Z, Hunt M, Yamada Y, Marion C, Ling CC, Amols H, Venkatraman ES, Leibel SA (2002) High-dose intensity modulated radiation therapy for prostate cancer: early toxicity and biochemical outcome in 772 patients. Int J Radiat Oncol Biol Phys 53:1111–1116

Zhang X, Liu H, Wang X, Dong L, Wu Q, Mohan R (2004) Speed and convergence properties of gradient algorithms for optimization of IMRT. Med Phys 31:1141–1152

Zhu XR, Jursinic PA, Grimm DF, Lopez F, Rownd JJ, Gillin MT (2002) Evaluation of Kodak EDR2 film for dose verification of intensity modulated radiation therapy delivered by a static multileaf collimator. Med Phys 29:1687–1692

Zhu Y, Kirov AS, Mishra V, Meigooni AS, Williamson JF (1997) Quantitative evaluation of radiochromic film response for two-dimensional dosimetry. Med Phys 24:223–231

11 Stereotactic Radiosurgery and Radiotherapy

Joseph R. Simpson, Robert E. Drzymala, and Keith M. Rich

CONTENTS

11.1   Introduction   233
11.2   Origins of Radiosurgery   233
11.3   Descriptive Overview   234
11.4   Equipment Required for Stereotactic Radiosurgery
        and Stereotactic Radiotherapy   236
11.4.1 Stereotactic Devices   236
11.4.2 Medical Imaging   236
11.4.3 Computerized Treatment-Planning Systems   236
11.5   Specific Treatment Techniques   238
11.5.1 Converging Arc Technique   238
11.5.2 Montreal (Dynamic Rotational Arc) Technique   238
11.5.3 Gamma Knife   240
11.5.4 Linear Accelerator   241
11.6   Intracranial Stereotactic Radiosurgery Devices   242
11.6.1 Conventional Linac   242
11.6.2 Other Dedicated Stereotactic Treatment Units   243
11.6.2.1 CyberKnife   244
11.6.2.2 Varian Trilogy System   245
11.6.2.3 Gamma Knife Technique   245
11.7   Extracranial Stereotactic Radiotherapy Devices   248
11.8   Quality Assurance   250
11.8.1 Periodic checks   250
11.8.2 Mechanical Checks   250
11.8.3 Safety Procedures   250
11.9   Conclusion   251
        References   251

J. R. Simpson, MD, PhD
Associate Professor in Radiation Oncology, Department of Radiation Oncology, Washington University School of Medicine, 4921 Parkview Place, Campus Box 8224, St. Louis, MO 63110, USA
R. E. Drzymala, PhD
Associate Professor in Radiation Oncology Physics, Department of Radiation Oncology, Washington University School of Medicine, 4921 Parkview Place, Campus Box 8224, St. Louis, MO 63110, USA
K. M. Rich, MD
Associate Professor, Neurological Surgery, Department of Neurological Surgery, Washington University School of Medicine, 660 South Euclid Avenue, Campus Box 8057, St. Louis, MO 63110, USA

11.1 Introduction

Radiation therapy has evolved in a number of ways over the past 50 years to be able to deliver specific doses of ionizing radiation with great conformality in single or multiple fractions to a wide variety of body sites. Advances in medical imaging, computer technology, and software tools have enabled two of these techniques, stereotactic radiosurgery and stereotactic radiotherapy, to now be widely used around the world to deliver highly focal radiation treatments to both intracranial and extracranial sites. Through anatomically accurate imaging and restriction of target motion, or radiation-beam sequencing to correct for target and normal organ motion inside the body, such highly conformal treatment strategies are being incorporated into the multimodality management of cancer. The procedures are more complex than standard radiotherapy providing a high level of precision and accuracy with greater sparing of normal tissues. This chapter provides a clinical overview of the main techniques and devices available currently for stereotactic radiosurgery and radiotherapy.

11.2 Origins of Radiosurgery

Lars Leksell, a Swedish neurosurgeon, developed the first commercially available dedicated radiosurgical device called the "Gamma Knife" in 1968. This machine made it possible to precisely deliver a single, large dose of highly conformal radiation to any number of intracranial sites using 201 fixed cobalt sources aimed at a center point. This provided an alternative treatment to certain neurosurgical procedures which were then associated with significant morbidity (Leksell 1951, 1971). Conditions thought to be appropriate for radiosurgery included acoustic schwannomas, intracranial arteriovenous malformations, pituitary adenomas, metastatic tumors, and skull base meningiomas, in addition to certain functional disorders such as intractable pain, trigeminal neuralgia, and essential tremor (LEKSELL 1968). Once the efficacy of this approach was demonstrated, sources of radiations other than fixed ^{60}Co sources were incorporated into stereotactic systems, not only for single but also for fractionated intracranial treatments (DUNBAR et al. 1994). Additional methods for localization and treatment of a number of extracranial sites have since been developed. The principles of and several practical methods for stereotactic irradiation are presented.

11.3
Descriptive Overview

Both stereotactic radiosurgery and stereotactic radiotherapy require patient immobilization and repositioning with either an invasive frame, a non-invasive frame, or a frameless system to direct precise radiation beam targeting. Whereas stereotactic radiosurgery requires positioning accuracy of $\leq$ 1mm, stereotactic radiotherapy is usually accurate to about 2 mm. The Gill-Thomas-Cosman (GTC) frame has been frequently used for linac-based radiosurgery as well as stereotactic radiotherapy (KOOY et al. 1994). It has been modified by some to allow treatment of extracranial head and neck tumors involving the skull base, nasopharynx, and paranasal sinuses (KASSAEE et al. 2003). It can also be used with an eye fixation device to permit stereotactic radiotherapy of choroidal melanomas, e.g., giving 70 Gy in five fractions over 10 days (EMARA et al. 2004). A standard thermoplastic mask system (BrainLAB) has also been used for this purpose (DIECKMANN et al. 2003). This mask system was used to treat larger cavernous sinus meningiomas employing CT–MRI image fusion and conventional-size radiation doses (SELCH et al. 2004). Hypofractionated stereotactic radiotherapy has also been applied, using noninvasive skull fixation and a dose schedule of 35 Gy in four fractions to the isocenter over 4–6 days (AOYAMA et al. 2003), or 40 Gy in five fractions on consecutive days (LINDVALL et al. 2005) as an alternative approach for treating larger metastases and to avoid the discomfort of repeat invasive fixation for metachronous metastases. The possibility of less toxicity was suggested as the advantage of this technique over single-fraction stereotactic radiosurgery. A similar technique, using the Laitinen

stereoadapter and a conventional linac, has been used for treatment of arteriovenous malformations and (AVMs; LINDVALL et al. 2003), using a synthetic thermoplastic mask custom fitted and attached to a metal base frame and the Brown-Roberts-Wells (BRW) localizing ring, for fractionated treatment of acoustic schwannomas (WILLIAMS 2002, 2003). Similarly, KIM et al. (2003) reported a mean isocenter accuracy within 0.53 mm for frameless stereotactic radiotherapy in 43 patients using three 2-mm gold markers implanted in the cranium.

Extracranial stereotactic radiotherapy poses several different challenges for patient immobilization and tumor localization. For example, large body areas need to be accommodated and internal organ and target motion limited or tracked accurately. Both framed and frameless systems have been employed for this purpose, using both internal and external fiducials. A body cast and head mask system with a stereotactic body frame was developed at the German Cancer Research Center and used for stereotactic radiotherapy of paramedullary tumors in the thoracic and lumbar spine. An analysis of setup accuracy showed that this system allowed an overall mean three-dimensional vectorial patient movement, and therefore mean overall accuracy of $\leq$3.6 mm (LOHR et al. 1999). This group has also treated liver tumors with single-dose stereotactic irradiation using a vacuum pillow and an abdominal compression plate to reduce liver movement. They reported that deviations of the body in the cranio-caudal direction were always less than the CT slice thickness (<5 mm) but advocated that a control CT scan be performed immediately before therapy to confirm setup accuracy and prompt necessary corrections, which were required on 16 of 26 occasions (HERFARTH et al. 2000).

For lung cancer, a frameless approach to stereotactic radiotherapy using the FOCAL unit, a combination of linac, CT scanner, X-ray simulator, and carbon table has been reported. Shallow breathing with an oxygen mask and an abdominal compression belt, if necessary, allowed intrafractional tumor motion to generally be <5 mm (UEMATSU et al. 2001).

Finally, prostate cancers have been stereotactically treated using implanted fiducials, ultrasound localization, and computed tomography imaging of the pelvis for daily targeting. This stereotactic radiotherapy technique, using three fiducial markers placed in the apex, base, and mid-gland, has been reported to limit average prostate deviation from its planned position to 2 mm or less (MADSEN et al. 2003).

Stereotactic procedures in modern radiation therapy thus require a stereotactic device, medical imaging, computer treatment planning, and a radiation source. The radiation modalities in use include high-activity ^{60}Co sources, photons from low to medium energy linear accelerators, and proton beams. As indicated previously, stereotactic procedures in radiation therapy are divided into stereotactic radiosurgery (SRS) and stereotactic radiotherapy (SRT). This division is based on the radiobiological intent of the therapy. The intent of SRS is radioablation, an attempt to inactivate the growth potential of cells within a target volume using a single, high-dose fraction of irradiation. Stereotactic radiotherapy attempts to preserve the function of normal cells within the target volume and surrounding normal tissues by the use of multiple, smaller dose fractions, while using very tight margins around the intended target. In both cases, highly accurate positioning of radiation beams and rapid dose fall-off outside the target volume are of prime importance; the former is achieved through the use of a stereotactic device; the latter is accomplished either with secondary collimation close to the patient (Fig. 11.1) or with a micro-multileaf collimator (BENEDICT et al. 2001; COSGROVE et al. 1999).

Although the techniques of SRS and SRT depend greatly on the equipment used, the specific steps of the procedures have much in common (see flow chart in Fig. 11.2). Firstly, the stereotactic device is attached to the patient before images are obtained. The images are verified with respect to their spatial accuracy and transferred to a computerized treatment-planning system. The location of the stereotactic device is identified in the images to establish the spatial relationship of the image set with

Example Stereotactic Timelines

SRS
- Stereotactic frame placement (30 min)
- Pre-Tx imaging (30–90 min)
- Tx planning (15–120 min)
- Patient treatment (15–150 min, 1 Fx)

SRT
- Imaging and Tx planning (Day 1: 60 min and 360 min, respectively)
- Pre-Tx imaging (Days 4, 7, 9: 45 min per Fx)
- Tx isocenter adjustment (5–15 min per Fx)
- Patient treatment (30–45 min per Fx)

Fig. 11.2 Example of stereotactic time lines. **a** Stereotactic radiosurgery (*SRS*). **b** Stereotactic radiotherapy (*SRT*)

respect to the stereotactic coordinate system. The planning system is used to determine the desired amount, location, and shape of the dose distribution, in real time, intended to maximize dose in the target volume and minimize dose to surrounding tissues, through specifying the arrangement, size, shape, and weight of the beams utilized. The computer specifies the plan in the stereotactic coordinate system and calculates the irradiation time or monitor units needed for its execution. Evaluation of the treatment plan is performed by collaboration between the radiation oncologist, neurosurgeon, and medical physicist, who most often constitute the stereotactic team. Consideration is given to the type of target treated, conformity of dose to the target, and dose to nearby tissue using both dose volume histograms and isodose contours. Typically the goal is to cover >95% of the volume with the prescription dose. The "conformity index" is the volume of tissue in the matrix containing the target which receives the prescription dose divided by the target volume (MONK et al. 2003). Generally, it should be less than 2:1 according to the model for variables contributing to acute or chronic grade 3,4, or 5 toxicity in RTOG protocol 90-05 corresponding to the PIV/TV (a measure of dose conformity of the treatment relative to the target where PIV is the prescription isodose volume (in mm^3) and TV the tumor volume (in mm^3; SHAW et al. 1996). In addition to periodic maintenance quality assurance (QA), appropriate tests on treatment day are required to verify the alignment and proper function of the treatment device prior to treatment. Finally, the patient is treated according to the plan on a chair or treatment couch under the guidance of the stereotactic device, monitored by the stereotactic team. In the United States, the Nuclear Regulatory Commission requires the presence of an authorized medical physicist and authorized user (radiation oncologist) throughout treatment (CFR Part 35) with a Gamma Knife unit.

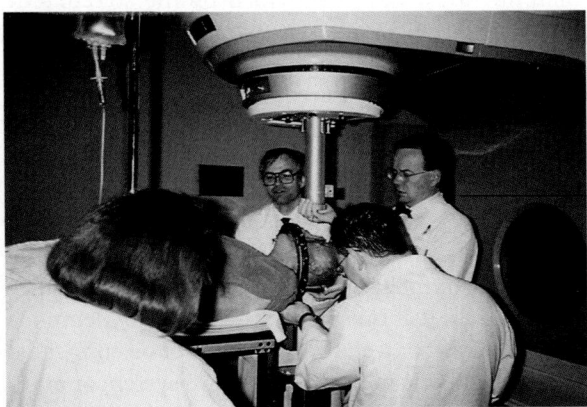

Fig. 11.1 Extended collimator used with linac radiosurgery

11.4
Equipment Required for Stereotactic Radio-surgery and Stereotactic Radiotherapy

11.4.1
Stereotactic Devices

The term, "stereotaxy," and its adjectival forms, stereotaxic and stereotactic, have a Greek and Latin etymology, in which *stereo* refers to three dimensions in space and *taxis* which means "to arrange." In this context, the word implies the use of a device that can precisely direct the radiation intervention to a specific site within the body. In the usual approach to stereotactic radiosurgery, a mechanical device, called the stereotactic frame (LEKSELL 1951), is attached to the patient's body. When fixed to the skull, the frame can immobilize the patient as well. Less invasive mechanical, optical, and radiographic methods, developed more recently, can be reliably reapplied for fractionated treatment (ASHAMALLA et al. 2003). Such optical and radiographic methods are designated "frameless" stereotactic techniques. The stereotactic devices in these cases may include a bite block that can be located in space with a stereo-optical camera (BOVA et al. 1998; KAI et al. 1998), a stereo-pair of kilo voltage fluoroscopy units (CHANG and ADLER 2001), or a combination of various imaging and tracking methods (YIN et al. 2002). In each case, the basic purpose of the stereotactic device is to provide an accurate coordinate system within which to direct a radiation beam to the target. The targeting precision for a stereotactic procedure is assumed to be on the order of 1 mm (LUTZ et al. 1988); however, sub-millimeter precision may be achieved with the use of many current stereotactic devices (BOVA et al. 1998).

11.4.2
Medical Imaging

Stereotactic procedures are image based, employing computed tomography (CT), magnetic resonance (MR) imaging, bi-plane angiography, positron emission tomography (PET), or other medical imaging modalities such as MR spectroscopy. Selection of the appropriate imaging modalities is of paramount importance to visualize patient anatomy. The clinicians use these images to define the anatomic volume to treat, differentiating it from surrounding normal tissues, and to determine the stereotactic coordinates at which the radiation beams will focus.

In many cases, a combination of imaging modalities is useful for computerized treatment planning and quality assurance (QA) checks (LEFKOPOULOS et al. 2001); the latter are covered in another section.

The imaging modality employed limits the accuracy of target volume and normal structure localization with respect to the stereotactic coordinate system. This may be a result of poor resolution, such as MR spectroscopy or spatial distortions, as may be found with MRI due to internal artifacts or poor shim of the imager's magnet. Furthermore, a major consideration for extracranial applications is the ability of the imaging modality to adequately represent the internal motion of the target and normal structures, especially for treatments in the thorax, abdomen, and pelvis. With the advent of 4D CT imaging (PAN 2005), much progress has been made in recording the movement of internal anatomy as a function of breathing. This is feasible for motion that occurs over a few seconds. Similarly, changes resulting from repositioning inaccuracies or from bladder and rectum filling can be adequately measured by imaging the patient on each treatment day.

11.4.3
Computerized Treatment-Planning Systems

A computerized treatment-planning system with software customized for the radiation source and its stereotactic device is usually sold as part of the SRS or SRT equipment package. There are common features of SRS and SRT treatment-planning systems that should be noted. In addition to the common peripherals found on radiotherapy treatment-planning systems, the stereotactic treatment-planning system must have the appropriate software and hardware needed to fully take advantage of the superior anatomic localization provided by the stereotactic device and multimodality medical imaging, while accomplishing the planning tasks in the shortest amount of time. As a stereotactic radiosurgery treatment is given in one session, usually on the same day as the imaging, there must be sufficient computing power and sufficient graphics capability to calculate isodose distributions and display them within a few seconds. Disk space must allow storage of data for a reasonable number of patients who typically require 80 MB each. Furthermore, convenience demands that the treatment-planning computer be connected to the diagnostic imagers via a PACS system or other network for rapid transfer of images. Most present radiotherapy-planning systems can accept transfer of images that are formatted using the ACR-NEMA DICOM standard (Low et al. 1995; NEUMANN 2002; WARRINGTON et al. 1994).

Commissioning the treatment-planning system requires entry of beam data that is specific to the radiotherapy device being used. The vendor may provide the beam data for such a device that is produced under tight quality control, especially if its radiation beam has invariant characteristics, such as the Gamma Knife. Only further spot checks of the data may be necessary. The vendor may perform this analysis as part of the installation included in the purchase price. Data for a linac-based device can be more variable and typically requires the customer to measure the appropriate data, especially if the system is assembled with components from multiple vendors.

The type of dosimetric data required for the treatment-planning system depends on the algorithm used. An algorithm that is frequently used for intracranial SRS treatment planning may be very simple. Since the head appears quite homogeneous to a megavoltage beam, tissue density correction is not necessary. The algorithm utilizing the ratio of tissue phantom ratios and off-axis ratio corrections (RTPROAR) to each point in the dose calculation matrix is simple and fast. The RTPROAR is therefore best used for SRS and SRT in the brain. Extracranial stereotactic treatments, however, require more complex algorithms that apply tissue density corrections and account for target motion, especially when treating tumors in the lung. A discourse on heterogeneity correction algorithms for computer-ized treatment planning is beyond the scope of this chapter. We therefore confine our discussion to the RTPROAR algorithm.

The RTPROAR algorithm calculates dose to each point in a volume of interest according to the following formalism:

$$Dose_p = Dose\ Rate_{(fs_{cal}, SPD_{cal}\ d_{cal})} \times \left(\frac{TPR_{(fs_{cal}, d_{cal})}}{TPR_{(fs_p, d_p)}} \right) \times OAR \times \left(\frac{SPD_{cal}}{SPD_p} \right)^2,$$

where, *Dose Rate* is the output at the reference point under calibration conditions, *fs* is the field size, *SPD* is the radiation source to point distance, *d* is depth from the surface, *TPR* is the tissue-phantom ratio, and *OAR* is the dose rate at an off-axis point relative to the central axis.

The subscripts, *cal* and *p*, refer to the calibration reference point and the arbitrary point under treatment conditions, respectively. An inspection of the formalism shows that the beam data needed for commissioning the RTPROAR algorithm is threefold for each collimator used: tissue-phantom ratios over the range of clinical depths, profiles taken orthogonally with respect to the central axis of the beam over the range of clinical depths, and the dose rate at the beam specification point. One beam energy (Co-60 γ-rays or 6-MeV X-rays) is usually commissioned for SRS and SRT; therefore, the amount of data required may not be too extensive. Examples of TPRs and OARs for the 6-MeV X-ray beam of a linac is shown in Figure 11.3. The TPR data contains within it the radiation scatter and attenuation properties of the beam, with water density typically assumed, whereas

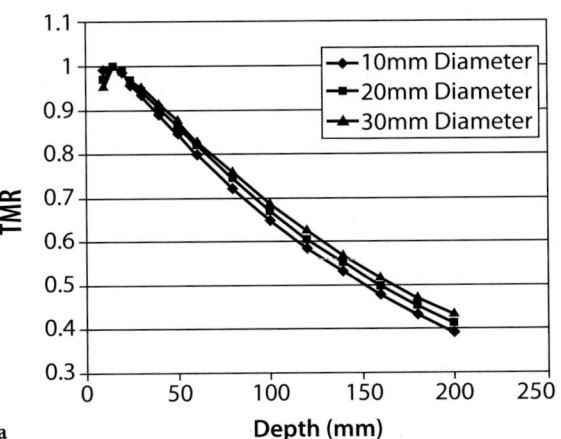

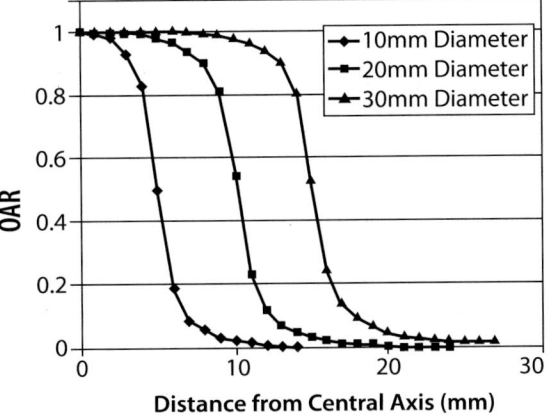

a   **b**

Fig. 11.3 Graphical representations of **a** tissue phantom ratios (*TPR*) and **b** dose rate at an off-axis point relative to the central axis (*OAR*)

the OAR data results from the flatness and symmetry of the radiation beam as well as the penumbra defined by the secondary collimator.

Since SRS and SRT field sizes can be as small as 0.4 cm, measurement of the calibrated dose rate can be a challenge. The use of finite size ionization chambers and the lack of lateral equilibrium have an impact on accurate calibration. Dosimeters have been used including radiochromic film (KELLERMANN et al. 1998; MACK et al. 2003; RAMANI et al. 1994), radiographic film (ROBAR and CLARK 1999), microchambers (DUGGAN and COFFEY 1996; LI et al., 2004), diodes (FIDANZIO et al. 2000; MCKERRACHER and THWAITES 1999; SOMIGLIANA et al. 1999), BANG gel (ERTL et al. 2000; FORONI et al. 2000; OLDHAM et al. 2001; SCHEIB and GIANOLINI 2002), and 1-mm^3 thermoluminescent dosimeters (ERTL et al. 1996). We recommend that the output of each collimated field size be calibrated with a sensor of the highest spatial resolution and sensitivity available and by more than one method, if possible. Figure 11.4

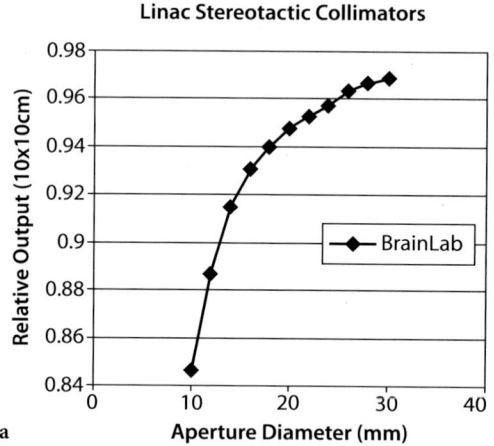

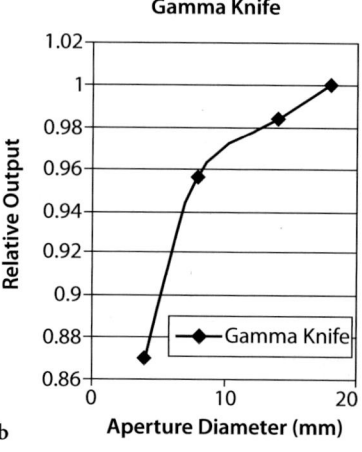

Fig. 11.4 Relative output vs field size. **a** Linac. **b** Gamma Knife

shows typical dose rates obtained for the conventional linac with a stereotactic collimation system and the Gamma Knife.

11.5
Specific Treatment Techniques

11.5.1
Converging Arc Technique

One of the first methods developed for linac or particle beam radiosurgery was the converging arc technique. The method utilizes a combination of couch and gantry arc angles about one or more isocenters placed within the target volume. For equally weighted and uniform arc lengths in combination with equally spaced couch angles, the result is an ellipsoidal isodose distribution centered at the isocenter. Figure 11.5a shows the combination of five arcs, 100° each and at the following couch angles: ±0°; ±20°; and ±40°. Since most tumors are not spherical, varying the couch angles and arc weights allows one to better conform dose to the tumor shape (see Fig. 11.5b: some arcs of the previous example weighted to zero). The use of multiple isocenters provides additional control over the dose distribution, albeit with an increase in dose heterogeneity in the target volume. Figure 11.5c shows a multi-isocentric plan for the same target volume as in Fig. 11.5a and b.

11.5.2
Montreal (Dynamic Rotational Arc) Technique

At McGill (PODGORSAK et al. 1988), a linac was also used for stereotactic treatment, while synchronizing the rotations of the couch and gantry. This resulted in a looping beam trajectory about the isocenter, similar in form to a baseball seam, which created a spheroidal dose distribution. Advantages of this method include: (a) no parallel-opposed beams which sharpens the dose fall-off outside the treated volume; (b) fewer entries into the treatment room by the therapist, since fewer arcs are required to achieve the desired dose distribution; and (c) shorter treatment times because of more efficient dose delivery. Various combinations of isocenter positions, collimator sizes, and treatment times can readily conform the dose distribution to the shape of a non-spherical target.

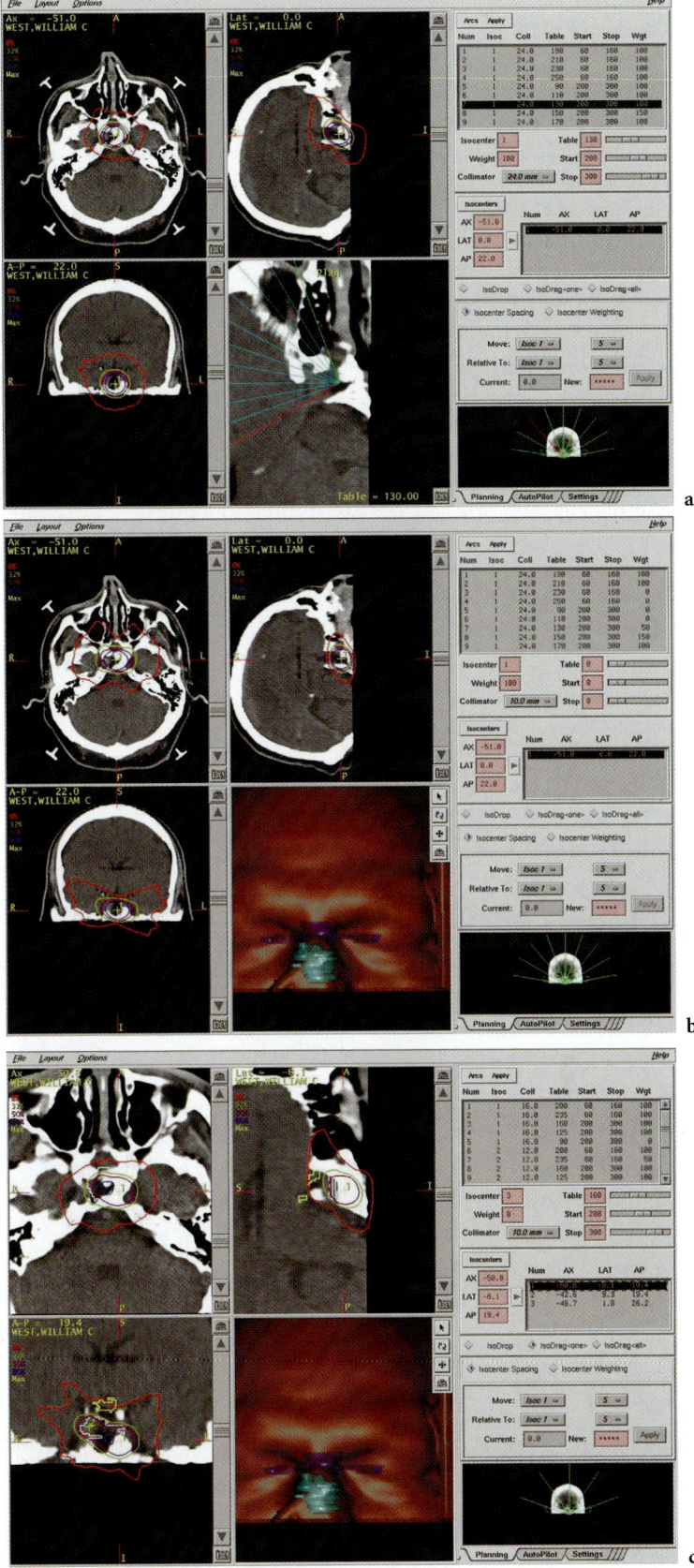

Fig. 11.5 a Isodose distribution from nine arc linac plans. **b** Modified linac arc plan. **c** Multi-isocentric plan for the same target volumes as in **a** and **b**

11.5.3
Gamma Knife

The Gamma Knife was designed to provide highly accurate radiation treatment of intracranial targets (LEKSELL 1968). The manufacturer specifies an overall treatment accuracy of 0.3 mm. The system has three basic components: a spherical source housing; four collimator helmets; and a couch with electronic controls (see image of the model C in Fig. 11.6). Different models of the Gamma Knife developed over the years vary mainly in the pattern of the source distribution within the housing, the couch path, hydraulic or electric motor driven couch movement, and whether the treatment is computer controlled with automatic patient positioning.

The source housing contains the 201 Co-60 gamma-ray emitting sources distributed in a quasi-hemispherical arrangement. The emitted photons have an average energy of 1.25 MeV and decay with a half-life of 5.26 years; therefore, the sources are usually exchanged every 5 years; otherwise, treatment times become overly long. Each source is constructed so that the Co-60 pellet is encapsulated within a welded stainless steel tube, which is enclosed in a stainless steel jacket and bushing (see image of disassembled source in Fig. 11.7). The radiation beam from each source converges to the "unit center point (UCP)" which is 40 cm away from each source. The UCP is analogous to the isocenter of a linac and is the location where the target volume must reside during treatment. This is accomplished by the three-axis coordinate system on the Leksell frame. Each source has an activity of approximately 30 Ci, when newly installed, and the 201 sources combined provide a dose rate of about 300 cGy/min at the UCP.

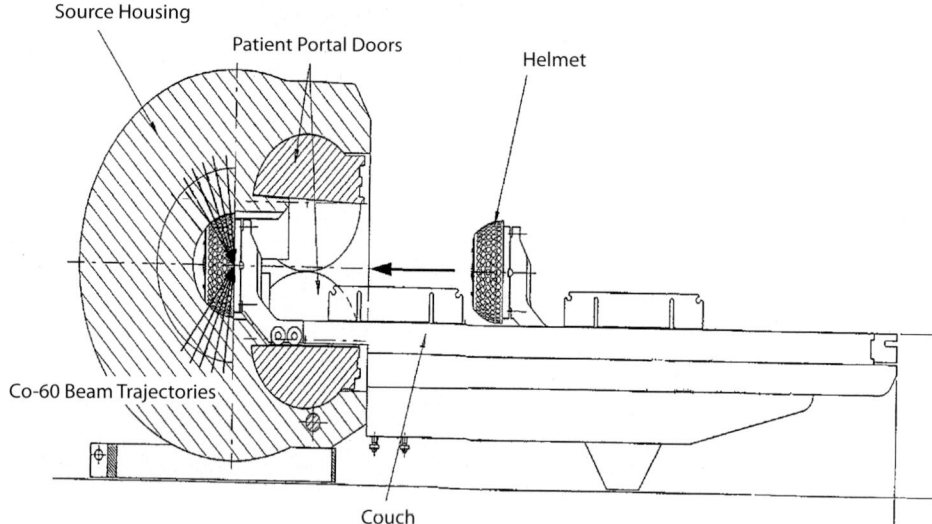

Fig. 11.6 Gamma Knife, Model C, cutaway schematics

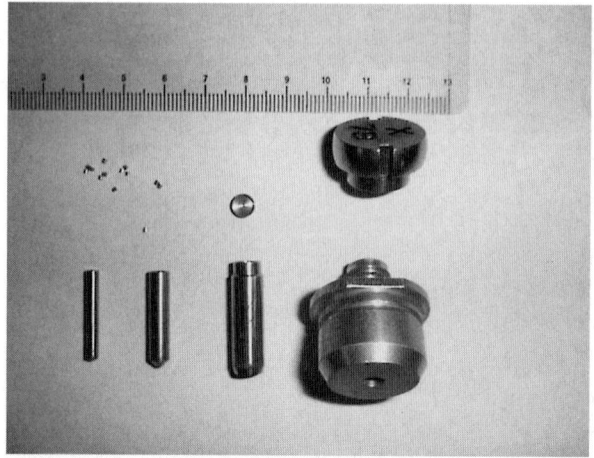

Fig. 11.7 a Disassembled and **b** assembled Gamma Knife source in bushing

Along the path to the UCP, the radiation beam from each source is collimated twice, once by a primary collimator and then by one of four secondary collimator helmets. For each helmet, 201 tungsten collimators define circular apertures that project a specific beam diameter of either 4, 8, 14, or 18 mm at the UCP. Not until the primary and secondary collimators align, when the couch docks the helmet in the source housing, does a therapeutic radiation dose reach the patient. In order to conform the radiation dose to the shape of the target in the patient, various combinations of aperture diameters, aperture blocking (plugging), irradiation times, and head positions, are utilized. Head positions must allow placement of the target at the UCP and include specification of the three axial coordinates (x, y, and z) and neck flexion or extension that is given by the gamma angle. A specific combination of these four parameters defines a "shot" in Gamma Knife terminology.

Although the Gamma Knife is well suited for targets in the cranium, the device does not have sufficient room within the helmet to provide extracranial treatments, with the exception of a few superiorly located head and neck sites. Furthermore, the Gamma Knife requires the use of the Leksell stereotactic frame, which is invasively fixed to the patient's skull. It is therefore not designed to provide fractionated treatments with the device (WALTON et al. 2000).

A major advance in the Gamma Knife design has been made recently with the introduction of computer control of the treatment steps. The automatic positioning system (APS) is a pair of computer-driven motors with feedback monitoring that moves the patient's head into the proper location prior to irradiation (see Fig. 11.8 for a close up view of the APS motors). Since target volumes for SRS and SRT are relatively small, the patient can be positioned without human intervention in many cases; however, when multiple target volumes are widely separated, which can occur when treating multiple metastases, the controlling software requires the therapist to observe the patient within the room during long traverses. Computerization of the Gamma Knife also permits integration of a record-and-verify system. For example, the Gamma Knife model C, with its current software version, monitors various treatment parameters: helmet aperture, patient position including gamma angle, and treatment time. Currently there is only manual checking of helmet plugging.

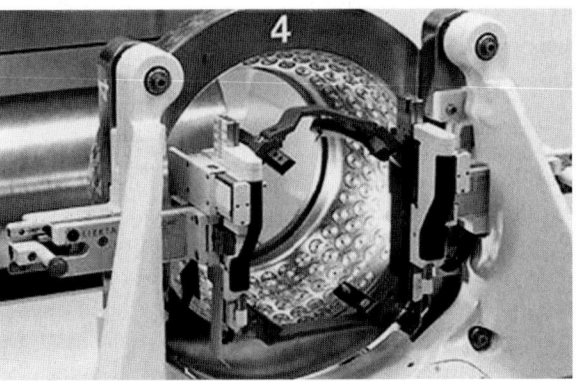

Fig. 11.8 Close-up of the Gamma Knife model C APS motors

11.5.4
Linear Accelerator

As indicated previously, an SRS or SRT treatment system may be based on a linear accelerator (linac; DAS et al. 1996; DELANNES et al. 1990), a specialized device employing multiple fixed ^{60}Co sources (WALTON et al. 2000; YAMAMOTO 1999), or a particle-beam source. The equipment usually produces low-energy megavoltage radiation beams in the range 1–6 MeV converging via multiple trajectories to a common point in space. An exception is the CyberKnife (CHANG et al. 1998; GERSZTEN et al. 2002; GERSZTEN and WELCH 2004; ISHIHARA et al. 2004; KUO et al. 2003; MURPHY 2004; ROCK et al. 2004), which employs a compact linac on a robotic arm and does not require a defined isocenter by design. A discussion of the other device used for stereotactic irradiation, the charged-particle accelerator, is beyond the scope of this chapter, and the reader is referred to other resources on this topic (LEVY et al. 1999).

In order to provide sub-millimeter beam-pointing accuracy, the mechanical tolerance of the stereotactic linear accelerator's isocenter must be tighter than that specified for a conventional radiation therapy linac. The radiation beam's central ray and the gantry rotation axis must remain aligned with the couch rotation axis to within a fraction of a millimeter over all rotational angles that are used clinically. In order to achieve this level of mechanical accuracy for the conventional linac, realignment and modifications may be necessary (LUTZ et al. 1988).

11.6
Intracranial Stereotactic Radiosurgery Devices

11.6.1
Conventional Linac

The conventional linac is the most commonly available source for SRS and SRT by virtue of its ubiquitous presence in the radiation therapy department; however, special stereotactic accessories must be obtained and fitted to the conventional linac for SRS or SRT; these include a secondary collimator system, a patient positioning and immobilization system, and a stereotactic device.

Figure 11.9a shows a conventional linear accelerator with stereotactic accessories attached. Observe the secondary collimator housing that hangs from the wedge slot and the BRW stand pinned to the couch bearing. Figure 11.9b shows a detailed view of the commercially available collimation system fabricated by BrainLAB, Inc. Note that the collimator inserts have circular apertures. The aperture diam-

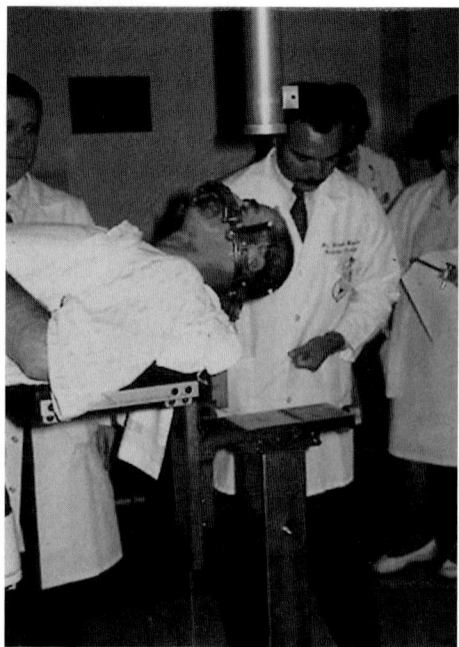

Fig. 11.10 Independent stereotactic frame support on the linac radiosurgery

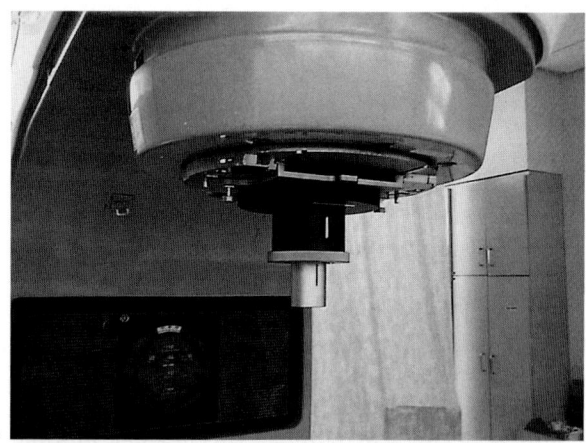

a

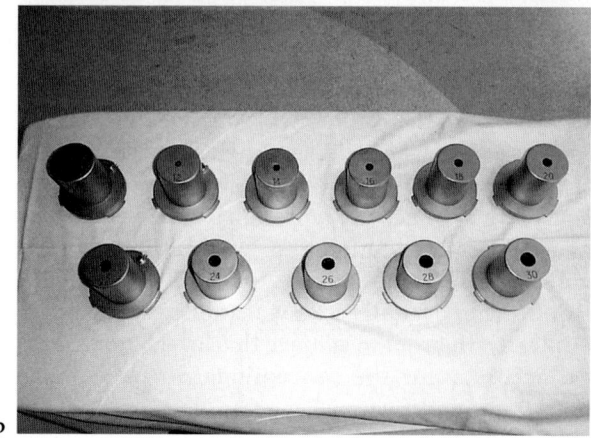

b

Fig. 11.9a,b. Linac collimators **a** in place and **b** various sizes

eters range from 10 to 30 mm in 2-mm increments for this set of inserts. During treatment, an insert resides in the collimator housing that is approximately 25 cm from the linac isocenter. This arrangement can provide a sharper penumbra than would be otherwise given by a block placed at a larger distance, while maintaining sufficient clearance, so that the housing will not collide with the patient.

With the system shown in Figure 11.10, the patient lies on the couch with the head independently supported by a stand attached to the couch bearing via the stereotactic frame; thus, the stand rotates with the couch. Since the patient's head is supported independently from the body, one must ensure that the couch cannot collapse under the patient. Additional couch supports have been used as a preventive measure (Fig. 11.11; DRZYMALA et al. 1994). Collision of the gantry with the stand is also a concern and limit switches prevent such movements (Fig. 11.12: switch panel on Clinac 600C).

In order to obtain a more precise isocenter when using a conventional linac, the University of Florida team developed an additional gantry to house the stereotactic collimators (FRIEDMAN and BOVA 1989). The gantry rests on the floor with the collimator housing attached by a gimbal bearing to the linac gantry head, which also drives its motion. Figure 11.13 shows the gantry designed and fabricated at the University of Florida. Its inventors claim

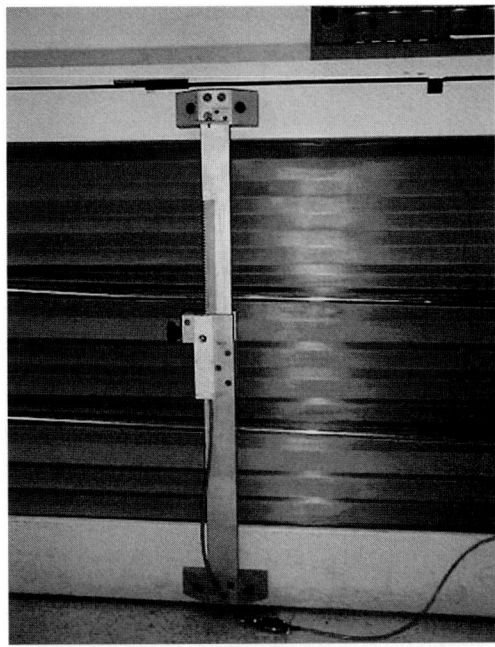

Fig. 11.11 Linac couch support legs to prevent accidental collapse during treatment

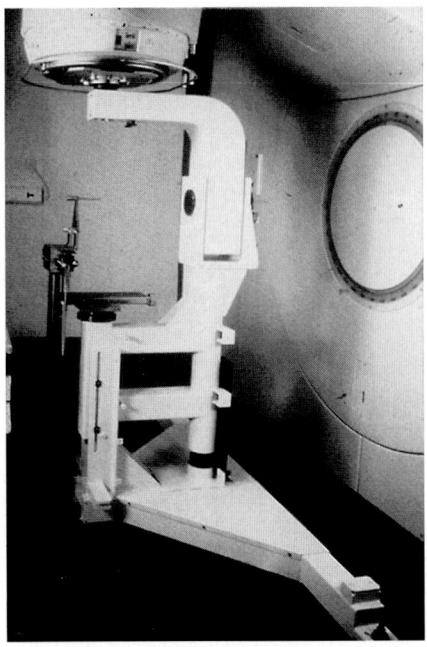

Fig. 11.13 The independent gantry support developed at the University of Florida

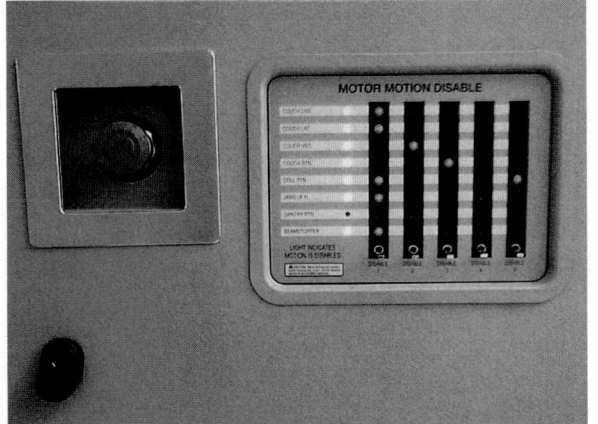
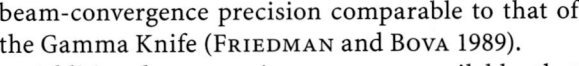

Fig. 11.12 Linac safety control panel for restricting motion couch gantry and collimator motion

Fig. 11.14 The BrainLab m3 collimator for Varian linac systems

beam-convergence precision comparable to that of the Gamma Knife (FRIEDMAN and BOVA 1989).

Additional accessories are now available that allow for stereotactic treatments on the conventional linac. The BrainLAB m3 high-resolution multileaf collimator (MLC) attachment (BENEDICT et al. 2001) replaces the circular aperture collimation system and allows for field shaping with a combination of 3-mm-thick inner leaves for tumors smaller than 3 cm, and 5-mm outer-leaf widths for field sizes up to 10 cm^2 (Fig. 11.14). The m3 is designed to attach to a Varian Medical Systems C-series linac.

11.6.2
Other Dedicated Stereotactic Treatment Units

The BrainLAB Novalis system (Novalis, Heimstetten, Germany) was a cooperative development project between Varian Medical Systems, Inc. and BrainLAB, AG. Novalis integrates the BrainLAB m3 MLC collimator into the head of a Varian 600C linac (see Fig. 11.15). As with the stand-alone m3, Novalis is limited to 10 cm^2 radiation field, which must be patched together in order to obtain larger field sizes. Because of the integrated m3, Novalis can provide

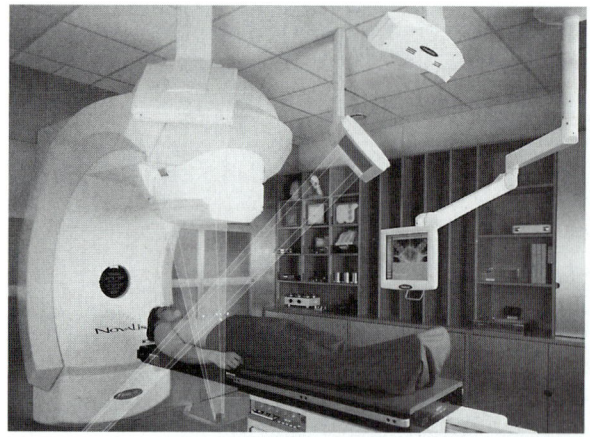

Fig. 11.15 The Novalis treatment system

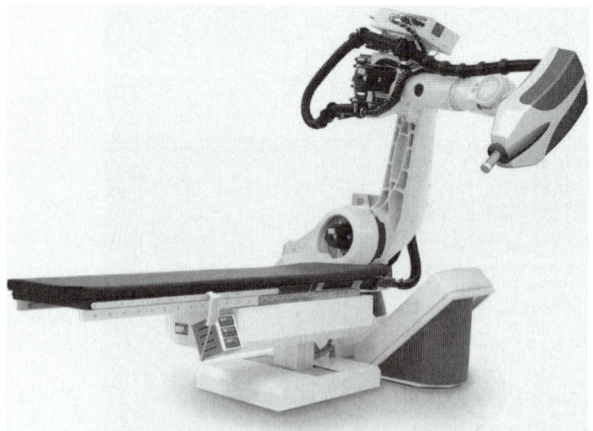

Fig. 11.16 The CyberKnife treatment system

dynamic intensity-modulated radiotherapy as well. The Novalis system is more than a radiation device, however. Included with the system are various positioning, tracking, and immobilization devices for the patient, such as on-board stereoscopic kilovoltage X-ray imaging with additional infrared and video motion-tracking systems (VERELLEN and SOETE 2003; YAN 2003). The manufacturer claims an overall treatment accuracy of 1–2 mm. A specialized treatment-planning system is also included (GREBE 2001; GROSU 2003; HAMM 2004; SOLBERG 2001).

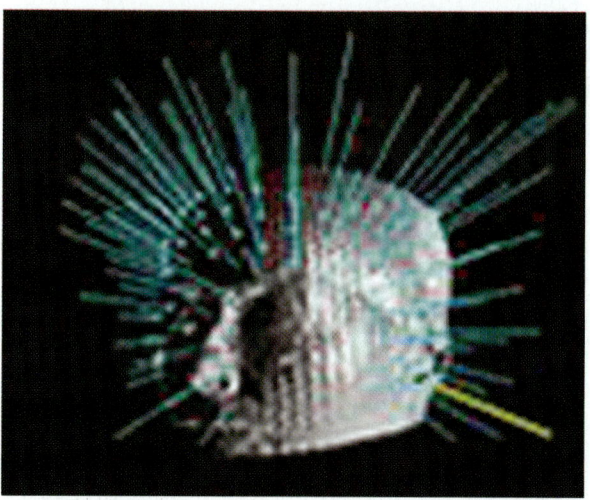

Fig. 11.17 Multiple-beam trajectories using the CyberKnife

11.6.2.1
CyberKnife

The CyberKnife, manufactured by Accuray, Inc. (Sunnyvale, Calif.), is a significant departure from the typical linac SRS/SRT system. The design couples an X-band, 6-mV linear accelerator to a robotic arm (Fig. 11.16). The robotic arm is capable of movement with 6° of rotational freedom, thereby allowing a radiation beam to approach up to 100 locations in the patient from up to 12 trajectories for each location without the need of a mechanical isocenter (Fig. 11.17). The manufacturer states an overall spatial treatment accuracy of 0.95 mm. The CyberKnife system uses frameless image guidance to locate the patient relative to the radiation beam. A patient lies on the couch with the anatomical region of interest positioned between two orthogonally placed fluoroscopic detectors (see Fig. 11.18). Two kilovoltage X-ray tubes reside in the ceiling and each X-ray beam projects upon flat-panel detector on the opposite side of the patient. With this arrangement the patient's position can

be monitored "live" any time before or during the treatment. Internal fiducials are sometimes placed to facilitate tracking. The "live" images are compared with pre-computed digitally reconstructed radiographs to detect patient movement. The vendor claims that extracranial targets, such as lung tumors, can be monitored using Dynamic Tracking Software during treatment to compensate for breathing motion, thereby potentially allowing for smaller radiation margins around the target volume.

A description of the installation and initial evaluation of the use and functionality of the CyberKnife at the University of Southern California has been published. A price tag of $3.2 million USD for purchase plus $0.5–0.74 million USD for site setup has been quoted (KUO et al. 2003).

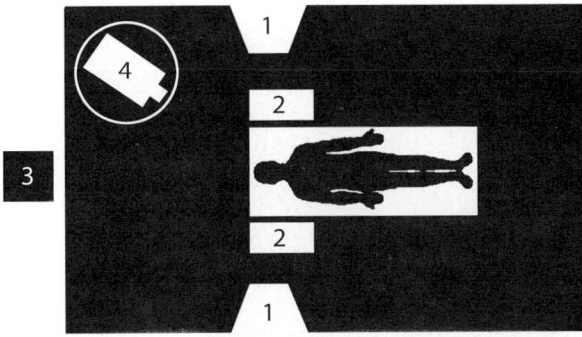

Fig. 11.18 CyberKnife component layout: *1* X-ray tubes; *2* flat panel imagers; *3* modulator cabinet; and *4)*linear accelerator

11.6.2.2
Varian Trilogy System

Another system that has stereotactic potential is the Varian Trilogy system (see Fig. 11.19). Components of Trilogy include on-board kilovoltage imaging orthogonal to its megavoltage imaging system and the Millennium 120 multileaf collimator. Imaging during rotation of the gantry can provide a cone-beam reconstruction of the treatment volume for final positioning of the patient on the treatment couch. The manufacturer claims stereotactic accuracy with Trilogy.

11.6.2.3
Gamma Knife Technique

A single shot by the Gamma Knife results in an elliptical isodose distribution similar to an oblate spheroid. This is because the sources are not evenly distributed over a hemisphere. Figure 11.20 shows orthogonal views of the isodoses resulting from a single 14-mm shot. Adjusting the gamma angle rotates the axial plane of the isodose distribution relative to anatomical structures. This is useful for avoiding the optic chiasm during a pituitary treatment (Fig. 11.21) or for directing the major axis of the ellipsoid along the trigeminal nerve (Fig. 11.22). Adjusting the gamma angle is also helpful to avoid collisions between the skull or stereotactic frame and the inside of the collimator helmet. The computerized treatment planning system specific to the Gamma Knife, Gamma Plan, allows the selection of the collimator aperture sizes, shot positions, treatment times, and plugging patterns (see below) appropriate for each patient.

Aperture plugging is a unique feature of the Gamma Knife, which allows some of the 201 collimators in a helmet to be replaced with solid tungsten plugs to block the radiation in appropriate patterns. With this technique, one can distort the ellipsoidal isodose distribution of the unplugged helmet in a variety of directions. This can increase isodose distribution conformality to the target shape or avoid organs at risk adjacent to the target. A comparison of an unplugged helmet (left) with a plugging pattern that shapes the isodose distribution into a butterfly shape (right) in the coronal plane is shown in (Fig. 11.23). This plugging pattern reduces dose to the optic chiasm and nerves.

Conforming the dose to target shape is best accomplished through the use of multiple shots with the Gamma Knife. Each shot may have its own position, collimator aperture size, treatment time, gamma angle, and plugging pattern. Figure 11.24 shows a multiple shot plan for a meningioma. The

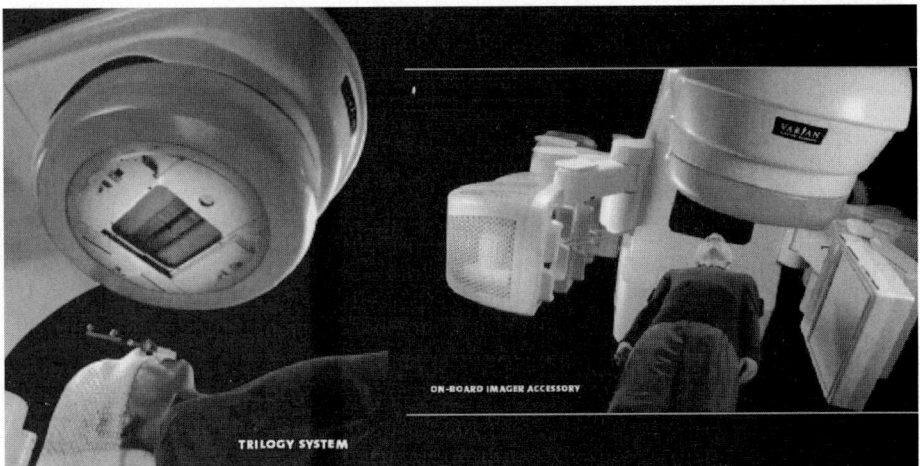

Fig. 11.19 The Trilogy system with on-board imager

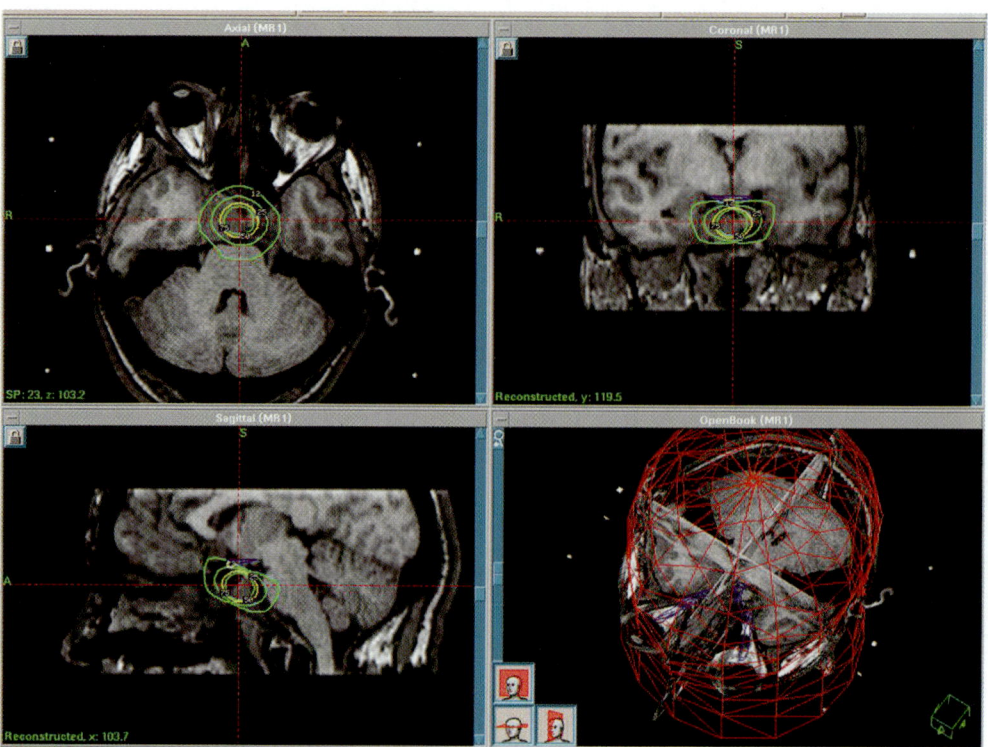

Fig. 11.20 Gamma Knife isodose distribution from a 14-mm pituitary shot with the gamma angle set to 110°

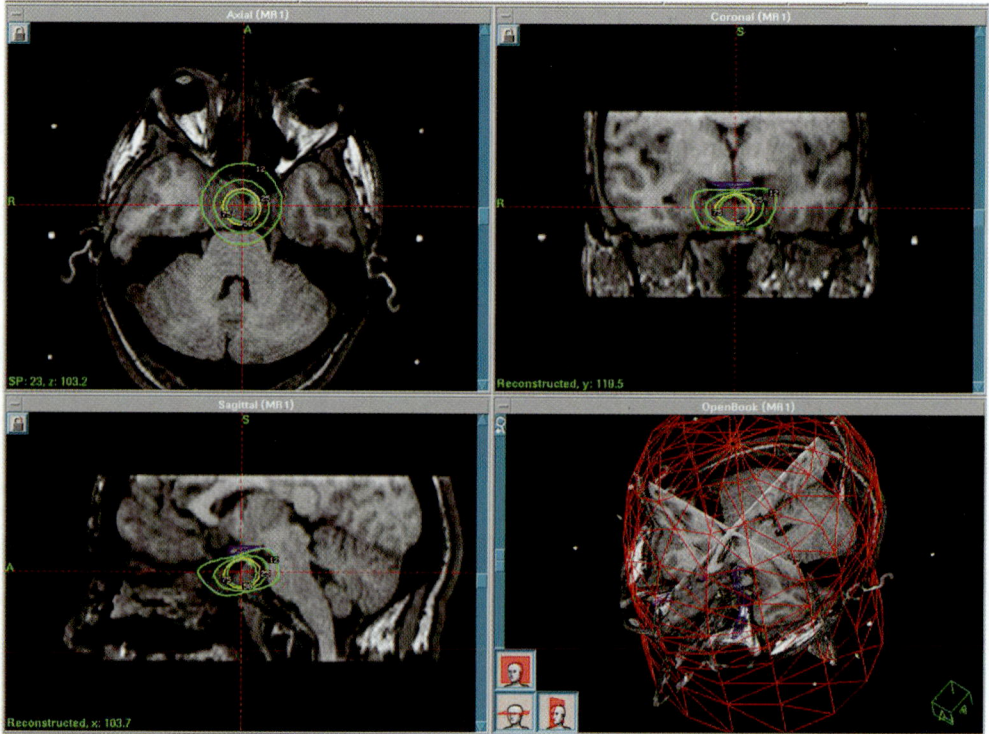

Fig. 11.21 Gamma Knife isodose distribution using the 14-mm helmet on a pituitary target with the gamma angle set to avoid the optic structures

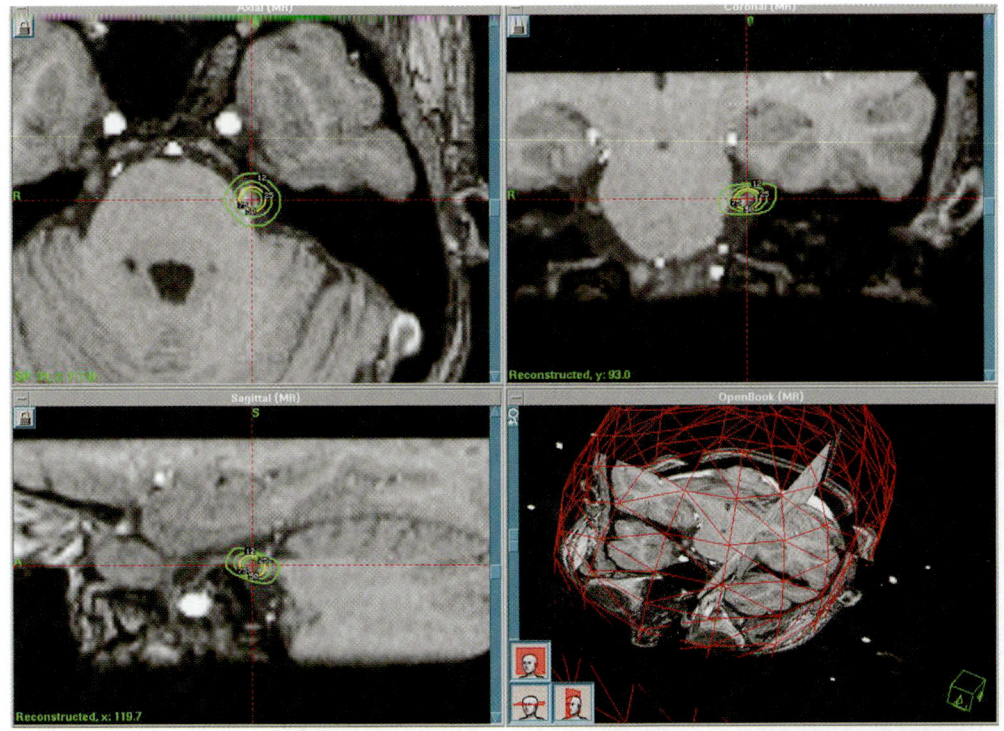

Fig. 11.22 Isodoses from the Gamma Knife using a 4-mm shot for trigeminal neuralgia

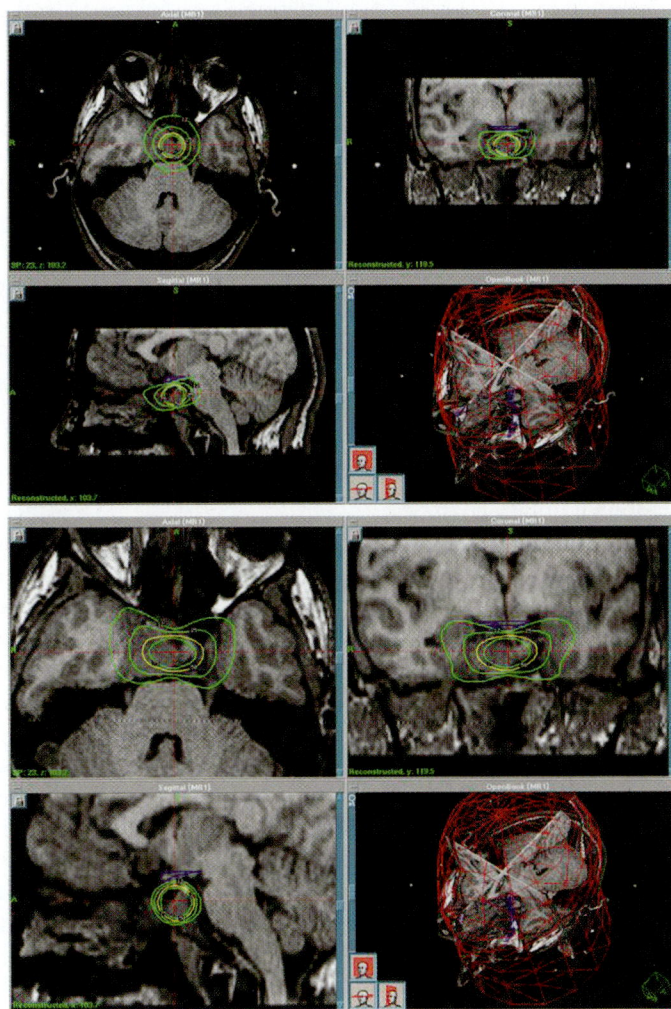

Fig. 11.23 A Gamma Knife plan for a pituitary treatment when using plugging

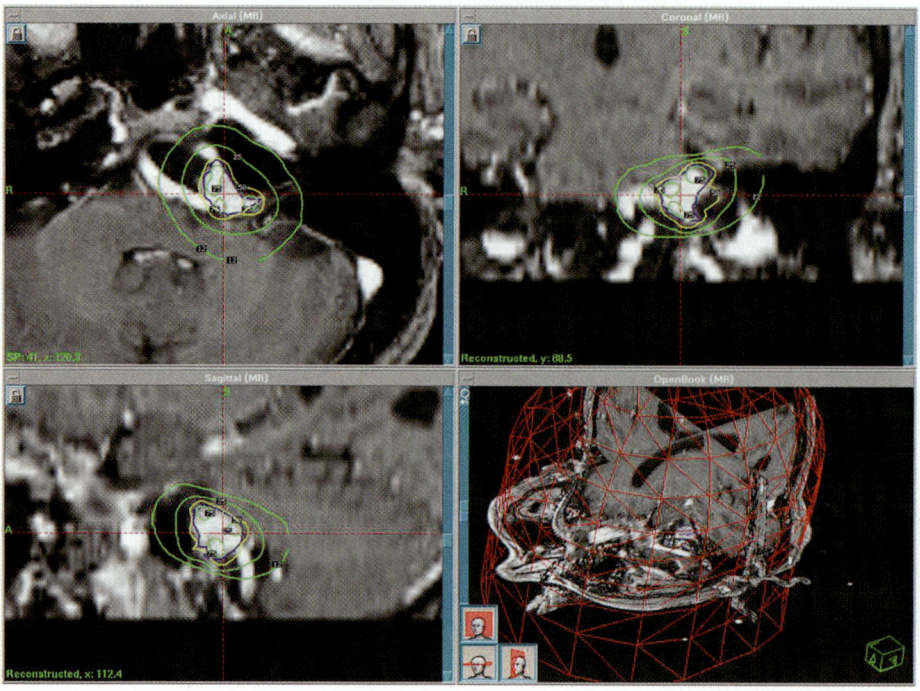

Fig. 11.24 A multi-shot Gamma Knife plan treating a meningioma

results are similar when using a conventional linac with a circular or oval stereotactic collimator. The CyberKnife uses its robotic arm to pour the dose into the target volume with multiple-beam trajectories. The approach can result in a high degree of conformity of the dose to the target through "inverse planning." Figure 11.25 shows the dose distribution from a CyberKnife treatment. Because of its mini-multileaf collimator, the Novalis system can combine beamlet-based IMRT (see Chap. 10 for a description of IMRT) with stereotaxy. This results in high-dose conformity with sharp dose gradients as shown in Fig. 11.26.

11.7
Extracranial Stereotactic Radiotherapy Devices

The conventional linac, the Novalis system, the CyberKnife, and the Varian Trilogy can also provide stereotactic treatment of extracranial targets. Tumors in any region of the body can be treated, such as the lung, paraspinal regions, liver, and pelvis. Typically, a special stereotactic frame or an imaging device localizes the target in the patient

during planning and treatment. A mold immobilizes the patient's body in the region of interest. The Novalis and CyberKnife systems use their onboard imaging systems in conjunction with CT or MR images.

A special device available for the conventional linac that facilitates extracranial stereotactic treatments is the Elekta stereotactic body frame (SBF; Elekta, Stockholm). Figure 11.27 shows a picture of the SBF. The SBF consists of a wooden box with honeycomb walls. Within the SBF a vacuum mold bag helps to immobilize the patient. A laser system attached to the SBF allows for repositioning of the patient during a fractionated treatment regimen. Panels with copper strips attached to the sidewalls of the SBF appear as fiducial markers when the patient is imaged using CT. The fiducial marker pattern is spatially related to rulers attached to the exterior of the SBF. These rulers serve as the coordinates to align with treatment room lasers, thereby positioning the SBF relative to the radiation beam of the linac. To restrict the effects of breathing motion, a diaphragm-compression device may be attached to the SBF. Alternatively, a breathing-control device can also be used (CHEUNG et al. 2003; DENISSOVA et al. 2005; ONISHI et al. 2004; REMOUCHAMPS et al. 2003).

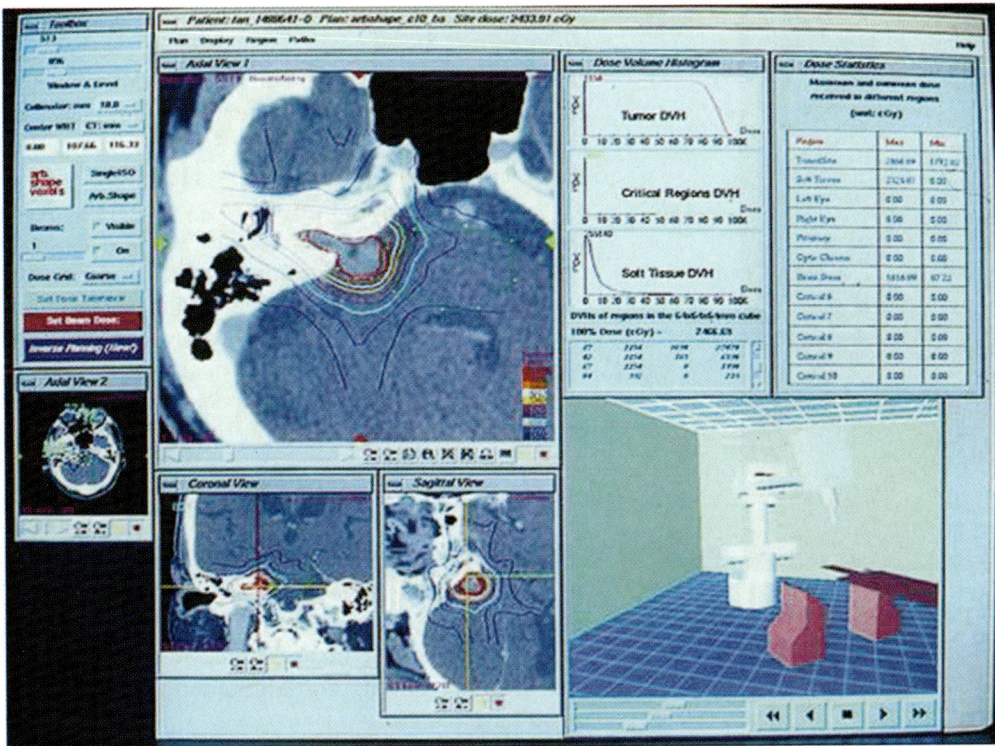

Fig. 11.25 An acoustic schwannoma treatment plan using the CyberKnife

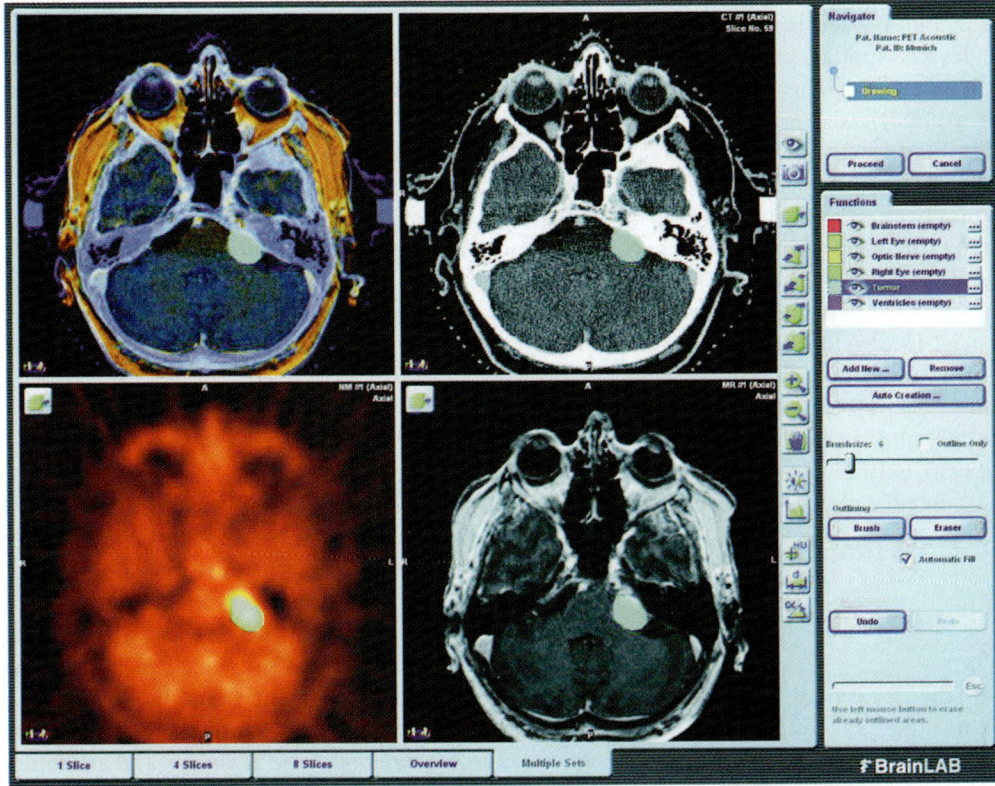

Fig. 11.26 Multimodality image fusion using BrainLab treatment planning software

Fig. 11.27 The Elekta Stereotactic Body Frame

11.8
Quality Assurance

Maintaining the highest accuracy for stereotactic radiosurgery and stereotactic radiotherapy requires careful assessment of error at each step in the treatment process. This includes assessing the integrity of the stereotactic device, spatial accuracy of the imaging modality, accuracy of the computerized treatment-planning system, radiation beam characteristics and alignment accuracy of the treatment machine, and the positional accuracy of the patient in the radiation beam. Final accuracy of the treatment is ultimately the combination of all these potential sources of error. Each stereotactic system is different and requires the expertise of a board-certified medical physicist to develop the appropriate quality-assurance program. Guidelines have been developed by the American Association of Physicists in Medicine and are published in Task Group Report no. 42 (Schell et al. 1995). Although written in 1995, the report remains applicable at present.

11.8.1
Periodic checks

Periodic checks of the treatment machine, its accessories, and associated procedures, should be performed daily, weekly, monthly, or annually, depending on the criticality of the issue and the time between expected failure.

For example, in order to assure that each treatment have sub-millimeter imaging accuracy, the images used to localize the target should be checked for spatial accuracy on a per-treatment-day basis. As the CT or MR scanner used for SRS and SRT target localization may not be under the supervision of the physicist in the radiation oncology department, if the radiology department does not specify the tight spatial tolerances needed for stereotactic imaging, good quality control may be hard to maintain. One simple solution to this problem is to compare the consistency of fiducial marker positions in patient treatment images with that expected from a model of the stereotactic localization frame; another is to fuse a suspect image set of the patient on MRI with one that is highly accurate, such as CT, and to check their co-registration. A more detailed analysis can be performed using an anthropomorphic phantom on an annual basis (Drzymala and Mutic 1999). Frameless stereotactic techniques are dependent on imaging during treatment and may require sophisticated image quality and alignment tests on a frequent basis.

11.8.2
Mechanical Checks

If the radiation device is not dedicated to stereotactic procedures and the accessories are reassembled prior to treatment, as is the case for a conventional linac, alignment of the radiation beam with the patient-positioning system or lasers will need confirmation prior to treatment. A radiographic alignment-verification procedure has been published for the conventional linac (Low et al. 1995; Warrington et al. 1994).

11.8.3
Safety Procedures

An important tool for quality assurance of patient treatment is a quality checklist. A list of critical items and the sequence in which these are to be performed is invaluable for maintaining efficiency and rigorous adherence during treatment (Drzymala et al. 1994). An understanding of the role of each member of the stereotactic treatment team in the overall procedure is of paramount importance. In-depth knowledge requires specific training and periodic drills for team members. This is especially true for emergency procedures where speed is essential.

11.9
Conclusion

Stereotactic radiosurgery and radiotherapy are two valuable and increasingly applied techniques combining sophisticated brain or body imaging with stereotactic guidance and computer treatment planning for precise radiation treatment programs given either as a single large, small number of moderately large, or multiple conventional size doses with maximal sparing of adjacent normal tissues. Several devices and treatment schedules have been tested and reported on in the literature, substantiating the validity of these approaches. They have substantially increased our knowledge of tumor response and normal tissue tolerance in a very quantitative sense, and have provided useful alternative and complimentary treatments for a number of conditions. Further progress is eagerly anticipated in combining these techniques with other targeted therapies to enhance our increasingly sophisticated anti-cancer armamentarium.

References

Aoyama H, Shirato H, Onimaru R et al. (2003) Hypofractionated stereotactic radiotherapy alone without whole-brain irradiation for patients with solitary and oligo brain metastasis using noninvasive fixation of the skull. Int J Radiat Oncol Biol Phys 56:793–800

Ashamalla H, Addeo D, Ikoro NC et al. (2003) Commissioning and clinical results utilizing the Gildenberg-Laitinen adapter device for X-ray in fractionated stereotactic radiotherapy. Int J Radiat Oncol Biol Phys 56:592–598

Benedict SH, Cardinale RM, Wu Q et al. (2001) Intensity-modulated stereotactic radiosurgery using dynamic micro-multileaf collimation. Int J Radiat Oncol Biol Phys 50:751–758

Bova FJ, Meeks SL, Friedman WA et al. (1998) Optic-guided stereotactic radiotherapy. Med Dosim 23:221–228

Chang SD, Adler JR (2001) Robotics and radiosurgery: the CyberKnife. Stereotact Funct Neurosurg 76:204–208

Chang SD, Murphy M, Geis P et al. (1998) Clinical experience with image-guided robotic radiosurgery (the CyberKnife) in the treatment of brain and spinal cord tumors. Neurol Med Chir (Tokyo) 38:780–783

Cheung PC, Sixel KE, Tirona R et al. (2003) Reproducibility of lung tumor position and reduction of lung mass within the planning target volume using active breathing control (ABC). Int J Radiat Oncol Biol Phys 57:1437–1442

Cosgrove VP, Jahn U, Pfaender M et al. (1999) Commissioning of a micro multi-leaf collimator and planning system for stereotactic radiosurgery. Radiother Oncol 50:325–336

Das IJ, Downes MB, Corn BW et al. (1996) Characteristics of a dedicated linear accelerator-based stereotactic radiosurgery–radiotherapy unit. Radiother Oncol 38:61–68

Delannes M, Daly N, Bonnet J et al. (1990) Laitinen's stereoadapter: application to the fractionated cerebral irradiation under stereotaxic conditions. Neurochirurgie 36:167–175

Denissova SI, Yewondwossen MH, Andrew JW et al. (2005) A gated deep inspiration breath-hold radiation therapy technique using a linear position transducer. J Appl Clin Med Phys 6:61–70

Dieckmann K, George D, Zehetmayer M et al. (2003) LINAC based stereotactic radiotherapy of uveal melanoma: 4 years clinical experience. Radiother Oncol 67:199–206

Drzymala RE, Mutic S (1999) Stereotactic imaging quality assurance using an anthropomorphic phantom. Comput Aided Surg 4:248–255

Drzymala RE, Klein EE, Simpson JR et al. (1994) Assurance of high quality linac-based stereotactic radiosurgery. Int J Radiat Oncol Biol Phys 30:459–472

Duggan DM, Coffey CW (1996) Use of a micro-ionization chamber and an anthropomorphic head phantom in a quality assurance program for stereotactic radiosurgery. Med Phys 23:513–516

Dunbar SF, Tarbell NJ, Kooy HM (1994) Stereotactic radiotherapy for pediatric and adult brain tumors: preliminary report. Int J Radiat Oncol Biol Phys 30:531–539

Emara K, Weisbrod DJ, Sahgal A et al. (2004) Stereotactic radiotherapy in the treatment of juxtapapillary choroidal melanoma: preliminary results. Int J Radiat Oncol Biol Phys 59:94–100

Ertl A, Hartl RF, Zehetmayer M et al. (1996) TLD array for precise dose measurements in stereotactic radiation techniques. Phys Med Biol 41:2679–2686

Ertl A, Berg A, Zehetmayer M et al. (2000) High-resolution dose profile studies based on MR imaging with polymer BANG (TM) gels in stereotactic radiation techniques. Magn Reson Imaging 18:343–349

Fidanzio A, Azario L, Miceli R et al. (2000) PTW-diamond detector: dose rate and particle type dependence. Med Phys 27:2589–2593

Foroni R, Gambraini G, Danesi U et al. (2000) New dosimetric approach for multidimensional dose evaluation in gamma knife radiosurgery. Technical note. J Neurosurg 93:239–242

Friedman WA, Bova FJ (1989) The University of Florida radiosurgery system. Surg Neurol 32:334–342

Gerszten PC, Ozhasoglu C, Burton SA et al. (2002) Feasibility of frameless single-fraction stereotactic radiosurgery for spinal lesions. Neurosurg Focus 13:e2

Gerszten PC, Welch WC (2004) CyberKnife radiosurgery for metastatic spine tumors. Neurosurg Clin N Am 15:491–501

Grebe G (2001) Dynamic arc radiosurgery and radiotherapy: commissioning and verification of dose distributions. Int J Radiat Oncol Biol Phys 49:1451–1460

Grosu A-L et al. (2003) Validation of a method for automatic image fusion (BrainLAB System) of CT data and 11C-methionine-PET data for stereotactic radiotherapy using a linac: first clinical experience. Int J Radiat Oncol Biol Phys 56:1450–1463

Hamm KD (2004) Stereotactic radiation treatment planning and follow-up studies involving fused multimodality imaging. J Neurosurg 101:326–333

Herfarth KK, Debus J, Lohr F et al. (2000) Extracranial stereotactic radiation therapy: set-up accuracy of patients treated for liver metastases. Int J Radiat Oncol Biol Phys 46:329–335

Ishihara H, Saito K, Nishizaki T et al. (2004) CyberKnife radiosurgery for vestibular schwannoma. Minim Invasive Neurosurg 47:290–293

Kai J, Shiomi H, Sasama T et al. (1998) Optical high-precision three-dimensional position measurement system suitable for head motion tracking in frameless stereotactic radiosurgery. Comput Aided Surg 3:257–263

Kassaee A, Das IJ, Tochner Z et al. (2003) Modification of Gill-Thomas-Cosman frame for extracranial head-and-neck stereotactic radiotherapy. Int J Radiat Oncol Biol Phys 57:1192–1195

Kellermann PO, Ertl A, Gornik E (1998) A new method of readout in radiochromic film dosimetry. Phys Med Biol 43:2251–2263

Kim KH, Cho M-J, Kim J-S et al. (2003) Isocenter accuracy in frameless stereotactic radiotherapy using implanted fiducials. Int J Radiat Oncol Biol Phys 56:266–273

Kooy HM, Dunbar SF, Tarbell NJ (1994) Adaptation and verification of the relocatable Gill-Thomas-Cosman frame in stereotactic radiotherapy. Int J Radiat Oncol Biol Phys 30:685–691

Kuo JS, Yu C, Petrovich Z et al. (2003) The CyberKnife stereotactic radiosurgery system: description, installation, and an initial evaluation of use and functionality. Neurosurgery 53:1235–1239

Lefkopoulos D, Foulquier JN, Petegnief Y et al. (2001) Physical and methodological aspects of multimodality imaging and principles of treatment planning in 3D conformal radiotherapy. Cancer Radiother 5:496–514

Leksell L (1951) The stereotactic method and radiosurgery of the brain. Acta Chir Scand 102:316–319

Leksell L (1968) Cerebral radiosurgery: I. Gammathalamotomy in two cases of intractable pain. Acta Chir Scand 134:585–595

Leksell L (1983) Stereotactic radiosurgery. J Neurol Neurosurg Psychiatry 46:797–803

Levy RP, Schulte RW, Slater JD et al. (1999) Stereotactic radiosurgery: the role of charged particles. Acta Oncol 38:165–169

Li S, Rashid A, He S et al. (2004) An new approach in dose measurement and error analysis for narrow photon beams (beamlets) shaped by different multileaf collimators using a small detector. Med Phys 31:2020w2032

Lindvall P, Bergstrom P, Lofroth P-O et al. (2003) Hypofractionated conformal stereotactic radiotherapy for arteriovenous malformations. Neurosurgery 53:1036–1043

Lindvall P, Bergstrom P, Lofroth P-O et al. (2005) Hypofractionated conformal stereotactic radiotherapy alone or in combination with whole-brain radiotherapy in patients with cerebral metastases. Int J Radiat Oncol Biol Phys 61:1460–1466

Lohr F, Debus J, Frank C et al. (1999) Noninvasive patient fixation for extracranial stereotactic radiotherapy. Int J Radiat Oncol Biol Phys 45:521–527

Low DA, Li Z, Drzymala RE (1995) Minimization of target positioning error in accelerator-based radiosurgery. Med Phys 22:443–448

Lutz W, Winston KR, Maleki N (1988) A system for stereotactic radiosurgery with a linear accelerator. Int J Radiat Oncol Biol Phys 14:373–381

Mack A, Mack G, Weltz D et al. (2003) High precision film dosimetry with GAFCHROMIC films for quality assurance especially when using small fields. Med Phys 30:2399–2409

Madsen BL, Hsi RA, Pham HT et al. (2003) Intrafractional stability of the prostate using a stereotactic radiotherapy technique. Int J Radiat Oncol Biol Phys 57:1285–1291

McKerracher C, Thwaites DW (1999) Assessment of new small-field detectors against standard-field detectors for practical stereotactic beam data acquisition. Phys Med Biol 44:2143–2160

Monk JE, Perks JR, Doughty D et al. (2003) Comparison of a micro-multileaf collimator with a 5-mm-leaf-width collimator for intracranial stereotactic radiotherapy. Int J Radiat Oncol Biol Phys 57:1443–1449

Murphy MJ (2004) Tracking moving organs in real time. Semin Radiat Oncol 14:91–100

Neumann M (2002) DICOM-current status and future developments for radiotherapy. Z Med Phys 12:171–176

Oldham M, Siewerdsen JH, Shetty A et al. (2001) High resolution gel-dosimetry by optical-CT and MR scanning. Med Phys 28:1436–1445

Onishi H, Kuriyama K, Komiyama T et al. (2004) Clinical outcomes of stereotactic radiotherapy for stage I non-small cell lung cancer using a novel irradiation technique: patient self-controlled breath-hold and beam switching using a combination of linear accelerator and CT scanner. Lung Cancer 45:45–55

Pan T (2005) Comparison of helical and cine acquisitions for 4D-CT imaging with multislice CT. Med Phys 32:627–634

Podgorsak EB, Olivier A, Pla M et al. (1988) Dynamic stereotactic radiosurgery. Int J Radiat Oncol Biol Phys 14:115–126

Ramani R, Lightstone AW, Mason DL et al. (1994) The use of radiochromic film in treatment verification of dynamic stereotactic radiosurgery. Med Phys 21:389–392

Remouchamps VM, Letts, Yan D et al. (2003) Three-dimensional evaluation of intra- and interfraction immobilization of lung and chest wall using active breathing control. A reproducibility study with breast cancer patients. Int J Radiat Oncol Biol Phys 57:968–978

Robar JL, Clark BG (1999) The use of radiographic film for linear accelerator stereotactic radiosurgical dosimetry. Med Phys 26:2144–2150

Rock JP, Ryu S, Yin FF et al. (2004) The evolving role of stereotactic radiosurgery and stereotactic radiation therapy for patients with spine tumors. J Neurooncol 69:319–334

Scheib SG, Gianolini S (2002) Three-dimensional dose verification using BANG gel: a clinical example. J Neurosurg 97:582–587

Schell MC, Bova FJ, Larson DA et al. (1995) AAPM report no. 42, Stereotactic radiosurgery, Report of Task Group 42, Radiation Therapy Committee. American Institute of Physics, New York

Selch MT, Ahn E, Laskari A et al. (2004) Stereotactic radiotherapy for treatment of cavernous sinus meningiomas. Int J Radiat Oncol Biol Phys 59:101–111

Shaw E, Scott C, Souhami L et al. (1996) Radiosurgery for the treatment of previously irradiated recurrent primary brain tumors and brain metastases: initial report of Radiation Therapy Oncology Group Protocol 90–05. Int J Radiat Oncol Biol Phys 34:647–654

Solberg TD (2001) Dynamic arc radiosurgery field shaping: a comparison with static field conformal and on-coplanar circular arcs. Int J Radiat Oncol Biol Phys 49:1451–1460

Somigliana A, Cattaneo GM, Fiorino C et al. (1999) Dosimetry of Gamma Knife and linac-based radiosurgery using radiochromic and diode detectors. Phys Med Biol 44:887–897

Uematsu M, Shioda A, Suda A et al. (2001) Computed tomography-guided frameless stereotactic radiotherapy for stage I non-small-cell lung cancer: a 5-year experience. Int J Radiat Oncol Biol Phys 51:666–670

Verellen D, Soete G (2003) Quality assurance of a system for improved target localization and patient set-up that combined real-time infrared tracking and stereoscopic X-ray imaging. Radiother Oncol 67:129–141

Walton L, Hampshire A, Roper A et al. (2000) Development of a relocatable frame technique for gamma knife radiosurgery. Technical note. J Neurosurg 93:198–202

Warrington AP, Laing RW, Brada M (1994) Quality assurance in fractionated stereotactic radiotherapy. Radiother Oncol 30:239–246

Williams JA (2002) Fractionated stereotactic radiotherapy for acoustic neuromas. Acta Neurochir 144:1249–1254

Williams JA (2003) Fractionated stereotactic radiotherapy for acoustic neuromas: preservation of function versus size. J Clin Neurosci 10:48–52

Yamamoto M (1999) Gamma Knife radiosurgery: technology, applications and future directions. Neurosurg Clin N Am 10:181–202

Yan H (2003) A phantom study on the positioning accuracy of the Novalis system. Med Phys 30:3052–3060

Yin FF, Zhu J, Yan H et al. (2002) Dosimetric characteristics of Novalis shaped beam surgery unit. Med Phys 29:1729–1738

12 Physics and Clinical Aspects of Brachytherapy

CONTENTS

12.1 Introduction 255
12.2 Classifications of Brachytherapy 256
12.2.1 Permanent Versus Temporary Implants 256
12.2.2 Interstitial, Intracavitary/Intraluminal, Topical/Mold 256
12.2.3 Hot Loading, Manual Afterloading, and Remote Afterloading 256
12.2.4 Low Dose Rate, High Dose Rate, Medium Dose Rate, and Pulse Dose Rate 257
12.3 Physical Characteristics of Brachytherapy Sources 258
12.3.1 Half-Life 258
12.3.2 Specific Activity 258
12.3.3 Average Energy 259
12.4 Sources Used in Brachytherapy 259
12.4.1 High-Energy Photon Emitters 259
12.4.2 Low-Energy Photon Emitters 260
12.4.3 Emerging Sources 261
12.5 Dose Calculations in Brachytherapy 261
12.5.1 The Superposition Principle 261
12.5.2 Source Strength Units 261
12.5.3 Single Source Dosimetry 262
12.5.3.1 Point Source Dosimetry 262
12.5.3.2 Line Source Dosimetry 263
12.5.3.3 Total Delivered Dose Calculations 265
12.6 Gynecological Intracavitary Implant 266
12.6.1 Applicators 266
12.6.2 Dose Specification for Cervical Cancer Treatments 266
12.6.3 ICRU Report 38 Recommendations 266
12.6.4 Volumetric Image-Based GYN Brachytherapy 266
12.7 Interstitial Implant Dosimetry Systems 268
12.7.1 Paterson–Parker (Manchester) System 269
12.7.2 Quimby System 269
12.7.3 Paris System 269
12.7.4 ICRU Report 58 Recommendations 270
12.8 Process of Brachytherapy Treatment Planning and Delivery 270
12.8.1 Preplanning 270
12.8.2 Source and Applicator Preparation 271
12.8.3 Applicator and Catheter Insertion 271
12.8.4 Source/Applicator Localization 272
12.8.5 Treatment Planning and Quality Assurance Review of Brachytherapy Treatment Plan 273
12.8.6 Source Loading and Treatment Delivery 274
12.9 Transrectal Ultrasound-Guided Permanent Prostate Brachytherapy 274
12.9.1 Dosimetric Goals of a Permanent Prostate Brachytherapy Treatment 274
12.9.2 Equipment 275
12.9.3 Volume Study 275
12.9.4 Treatment Planning 276
12.9.5 Treatment Plan Review and Pretreatment Quality Assurance Tests 277
12.9.6 Implant Procedure 278
12.9.7 Post-Implant Dosimetry 278
12.10 High Dose Rate Remote-Afterloading Brachytherapy 279
12.10.1 Equipment and Operating Principles 280
12.10.2 Clinical Application of HDR Brachytherapy: Interstitial Accelerated Partial Breast Irradiation 281
12.11 Radiation Protection and Regulatory Compliance in Brachytherapy 285
12.11.1 Licensing, Authorization, and Report of Medical Events 286
12.11.2 Radiation Safety Concerns in Permanent Implants 287
12.11.3 Safety and Regulatory Issues in HDR Brachytherapy 287
References 288

Z. Li, DSc
Department of Radiation Oncology, Washington University, St. Louis, MO 63110, USA

12.1
Introduction

Brachytherapy is the use of sealed radioactive sources placed in close proximity to the treatment target volume, either by directly inserting them into the tumor, or by loading them into instruments (applicators) which were previously inserted into cavities inside the body at close distance to the tumor. Brachytherapy may be used as a sole radiation treatment of the tumor, such as in the case of early-stage prostate and breast cancers. It is also often used in combination with external beam radiation therapy to deliver a boost radiation dose to the tumor, as in the case of gynecological tumors, later-stage prostate cancer, and many head and neck cancers. Following the surgical removal of the gross tumor, brachytherapy may be used to delivery a tumoricidal radiation dose to the tumor bed, where microscopic diseases remain. Due to the rapid falloff of dose away from the sources, brachytherapy allows the delivery of greater tumor doses than external beam radiation therapy, while retaining excellent sparing of neighboring critical organs. Compared with surgery, brachytherapy does not create a tissue deficit, thereby allowing potentially better cosmetic results.

12.2
Classifications of Brachytherapy

Brachytherapy modalities can be classified according to various criteria, including implant duration, the approach used to insert the sources into the patient, the technique used to load the sources, and the rate at which radiation dose is delivered to the target. These classifications hold significance not only as medical terms, but also in the selection of radioactive sources for a given brachytherapy treatment.

12.2.1
Permanent Versus Temporary Implants

Permanent brachytherapy implants are those where the sources are inserted into the patient, remaining permanently in the patient. Common permanent brachytherapy procedures include treatment of the prostate, head and neck cancers, lung, and sarcomas. I-125 and Pd-103 seeds are commonly used for permanent implants (NAG et al. 1999, 2000), although Au-198 seeds have also been used occasionally (CRUSINBERRY et al. 1985; HOCHSTETLER et al. 1995). As is discussed later, sources used for permanent implants need to have low energy, short half-lives, or a combination of both, so that the radiation exposure received by people that have either frequent or close contact with the patient is limited.

Temporary brachytherapy implants are those where the sources are implanted in the patient for a pre-determined length of time and then removed. Treatment times of temporary implants range from a few minutes, when the high dose rate afterloading technique is used, to a few days, for common low dose rate treatments. Patients may need to be admitted into the hospital for the duration of the treatment. Radiation exposure to hospital workers is therefore a significant concern when temporary implants are employed.

12.2.2
Interstitial, Intracavitary/Intraluminal, Topical/Mold

Depending on the approach used to insert the brachytherapy sources into the patient, brachytherapy can be classified into interstitial, intracavitary/intraluminal or topical/mold treatments.

In interstitial brachytherapy treatments, brachytherapy sources are introduced into the tissue, often with the use of needles and catheters of small diameters, used to minimize trauma to the normal tissue. Correspondingly, brachytherapy sources used for interstitial treatment need to have small dimensions to fit into the needles and catheters. Interstitial treatments are used for tumors such as prostate cancer, breast cancer, and sarcomas.

In intracavitary brachytherapy treatments, sources are loaded into applicators, which are positioned into cavities within the human anatomy adjacent to the target tissue. Treatment site-specific applicators are designed to fit into cavities and place the sources near the target tissues. Examples include the tandem and ovoid applicators for treatment of cervical cancer, the cylinder applicator for treatment of vaginal cancer, and the nasopharyngeal applicator for treatment of cancer of the nasopharynx. The applicators remain in the patient during the treatment and are removed at the completion of the treatment, so intracavitary brachytherapy treatments are usually temporary treatments.

12.2.3
Hot Loading, Manual Afterloading, and Remote Afterloading

Depending on the timing of source insertion relative to the surgical procedure to insert the applicators and/or needles, brachytherapy can be divided into hot loading, in which the sources are inserted in the operation room (OR) immediately after the applicators are inserted; manual afterloading, where the applicators are inserted into the patient in the OR, and sources are loaded after the patient's return from the recovery room to the patient's room; and remote afterloading, in which a computer-controlled device is used to load the sources automatically, thus eliminating manual handling of radioactive sources altogether. Hot loading is rarely used currently due to the high radiation exposure to OR, recovery room, and transportation personnel when compared against afterloading. Permanent implant treatments are typically hot-loaded, as are pre-fab-ricated eye plaque applicators containing ^{125}I or ^{103}Pd seeds for treatment of ocular melanoma. Use of afterloading techniques, either manual or remote (computer controlled), minimizes radiation exposure to hospital personnel, in addition to providing an opportunity for the treatment planner to optimize the source strength and loading distribution based on a retrospective review of the applicator position-ing relative to the target tissue.

12.2.4
Low Dose Rate, High Dose Rate, Medium Dose Rate, and Pulse Dose Rate

Brachytherapy treatments can also be classified according to the dose rate at which brachytherapy treatments are delivered:

1. Low dose rate (LDR): $\dot{D} < 120$ cGy/h

2. Medium dose rate (MDR):
   120 cGy/h $\leq \dot{D} < 1200$ cGy/h
3. High dose rate (HDR): $\dot{D} \geq 1200$ cGy/h

Much of the existing clinical brachytherapy expe-rience was for treatments delivered using the classic LDR regimen, at $\dot{D} \approx 45$ cGy/h. The biological effec-tiveness of brachytherapy treatment depends sig-nificantly on the dose rate at which the treatment is delivered. Much effort has been spent, therefore, on the biological effect of brachytherapy delivery at higher dose rates, such that a dose biologically equivalent to previous LDR treatments can be deliv-ered. Remote afterloading HDR units, equipped with a high-activity ^{192}Ir source, can deliver an entire treatment fraction in minutes. At such high dose rates, the advantage of normal tissue repair associated with LDR is lost, so HDR treatments must be fractionated, delivering a smaller total dose rela-tive to their LDR counterparts. This has led to the

Table 12.1 Common brachytherapy sources and their physical characteristics. (From WILLIAMSON 1998b). *LDR* low dose rate, *HDR* high dose rate

Element	Isotope	Energy (MeV)	Half-life	HVL-lead (mm)	Exposure rate con-stant $(\Gamma_\delta)^a$	Source form	Clinical application
Obsolete sealed sources of historic significance							
Radium	^{226}Ra	0.83 (avg)	1626 years	16	8.25[b]	Tubes and needles	LDR intracavitary and interstitial
Radon	^{222}Rn	0.83 (avg)	3.83 days	16	8.25[b]	Gas encapsulated in gold tubing	Permanent interstitial; temporary molds
Currently used sealed sources							
Cesium	^{137}Cs	0.662	30 years	6.5	3.28	Tubes and needles	LDR intracavitary and interstitial
Iridium	^{192}Ir	0.397 (avg)	73.8 days	6	4.69	Seeds in nylon ribbon; metal wires; encapsu-lated source on cable	LDR temporary interstitial; HDR interstitial and intracavitary
Cobalt	^{60}Co	1.25	5.25 years	11	13.07	Encapsulated spheres	HDR intracavitary
Iodine	^{125}I	0.028	59.6 days	0.025	1.45	Seeds	Permanent interstitial
Palladium	^{103}Pd	0.020	17 days	0.013	1.48	Seeds	Permanent interstitial
Gold	^{198}Au	0.412	2.7 days	6	2.35	Seeds	Permanent interstitial
Strontium	^{90}Sr–^{90}Y	2.24 β_{max}	28.9 years	–	–	Plaque	Treatment of superficial ocular lesions
Developmental sealed sources							
Americium	^{241}Am	0.060	432 years	0.12	0.12	Tubes	LDR intracavitary
Ytterbium	^{169}Yb	0.093	32 days	0.48	1.80	Seeds	LDR temporary interstitial
Californium	^{252}Cf	2.4 (avg) neutron	2.65 years	–	–	Tubes	High-LET LDR intracavitary
Cesium	^{131}Cs	0.030	9.69 days	0.030	0.64	Seeds	LDR Permanent implants
Samarium	^{145}Sm	0.043	340 days	0.060	0.885	Seeds	LDR temporary interstitial
Unsealed radioisotopes used for radiopharmaceutical therapy							
Strontium	^{89}Sr	1.4 β_{max}	51 days	–	–	SrCl^2, IV solution	Diffuse bone metastases
Iodine	^{131}I	0.61 β_{max} 0.364 MeV γ	8.06 days	–	–	Capsule NaI oral solu-tion	Thyroid cancer
Phosphorus	^{32}P	1.71 β_{max}	14.3 days	–	–	Chromic phosphate colloid instillation; Na$_2$PO$_2$ solution	Ovarian cancer seeding; peri-toneal surface; PVC, chronic leukemia

[a] No filtration in units of R $\times$ cm$^2 \times$ mCi$^{-1} \times$ hr^{-1} [b] 0.5-mm platinum filtration; units of R/cm^2/mCi^{-1}/hr

development of the pulsed dose rate delivery, in which the overall treatment time is equivalent to a traditional low dose rate treatment at 40–80 h. The sources, however, are only inserted into the patient for minutes during each hour of treatment, resulting in higher instantaneous dose rate, through the same dose delivered within each hour as a traditional LDR treatments. Several authors (BRENNER et al. 1996, 1997; CHEN et al. 1997; VISSER et al. 1996) have demonstrated biological equivalence of PDR relative to LDR treatments.

12.3
Physical Characteristics of Brachytherapy Sources

A brachytherapy source is characterized by the rate at which its strength decays (half-life), by how much radioactivity can be obtained for a given mass of the radioactive source (specific activity), and by the energies and types of the radiation particles that are emitted from the source (energy spectrum). These physical brachytherapy source characteristics will guide the clinical utilization. Table 12.1 lists the common radioactive sources used in brachytherapy, together with their physical characteristics.

12.3.1
Half-Life

The strength of a radiation source decays exponentially. Let the strength of the source at time 0 be A_0. The strength of the source $A(t)$ at time t is then given by the equation

$$A(t) = A_0 \times e^{-\mu t}, \tag{1}$$

where μ is the decay constant. μ describes the rate at which the source strength decays. Of particular use in brachytherapy is the time it takes for the source strength to decay to half of its initial value, i.e., $A(T_{1/2}) = A_0/2$, where $T_{1/2}$ is the half-life of the source. Substituting into Eq. (1), we obtain

$$A(T_{1/2}) = A_0/2 = A_0 \times e^{-\mu T_{1/2}}; \tag{2}$$

or

$$T_{1/2} = \frac{\ln(2)}{\mu}. \tag{2a}$$

A source's half-life is a fundamental quantity of the radioactive nuclide of the source. Common brachytherapy sources have half-lives ranging from days to years. The length of a given brachytherapy source's half-life determines its shelf life, namely, whether a source can be stored and used repeatedly over a long period of time. Sources with shorter half-lives, such as ^{125}I and ^{103}Pd sources, need to be purchased and received with an accurate knowledge of the source strength relative to the intended implant procedure date, so that the source strength on the day of implant is as prescribed, and that the desired initial dose rate (in terms of cGy/h or cGy/day) is achieved. Sources with longer half-lives, such as ^{137}Cs and ^{192}Ir sources, can be used for treatment of multiple patients before replacement, thereby reducing the cost of each treatment.

A source's half-life, together with its average energy, determines its suitability for use in permanent or temporary implants. When brachytherapy sources are permanently inserted into a patient and the patient is released from the hospital, the radiation exposure around the patient can pose a risk to people that are present within short distances from the patient. Sources with shorter half-lives can reduce these risks because the radiation exposure around the patient decreases rapidly with time. If necessary, the patient can be hospitalized in a private room for a short period of time.

The half-life of a brachytherapy source also impacts the implant dose calculation. The decay of the source may not need to be explicitly accounted for if the source has a sufficiently long halflife. For example, ^{137}Cs sources, with a half-life of 30 years, may be assumed to hold a constant source strength during the treatment period of a few days, whereas the dose calculation of an implant using ^{125}I sources, with a half-life of 59.8 days, needs to consider the decay of the sources during the implant.

12.3.2
Specific Activity

The strength of a brachytherapy source for practical applications is limited by its specific activity. The specific activity is the ratio of activity contained within a unit mass of the source. When a parent nuclide is activated within a neutron flux field, the number of radioactive nuclides per unit mass that may be obtained is limited by the neutron flux field strength, the parent nuclide's neutron cross section, and the source half-life. This is important for HDR interstitial brachytherapy applications, which require small source dimensions as well as high source strengths. The popularity of the ^{192}Ir source in modern brachytherapy is partly due to its high specific activity and high neutron cross section, thereby making it suitable as an HDR remote afterloading source. The small size of the source makes it useful for both interstitial and intracavitary brachytherapy treatments.

12.3.3
Average Energy

The average energy of a brachytherapy source determines the penetrability of the photon particles emitted from the source. The high-energy photon sources allow higher radiation dose to tissues at larger distances to the sources, such as the pelvis nodes in the treatment of cervical cancer. On the other hand, the high-energy photons require thicker shields for protection of hospital personnel. Permanent brachytherapy treatments often use low-energy photon emitting sources, such as ^{125}I and ^{103}Pd, as the photons from these sources are mostly attenuated by the patient tissue, resulting in very low radiation exposure rates around the patient. Patients treated with these sources can be released from the hospital without violating federal regulations on radiation exposure to members of the public from the implanted sources. When high-energy sources, such as ^{198}Au, are used for permanent implants, the patient needs to be confined in the hospital until the source strength decays to a suitable value, such that the radiation exposure from the sources outside the patient satisfies the limits of these regulations. For these reasons, ^{222}Rn and ^{198}Au are the only sources useful for permanent brachytherapy implant because their short half-lives of approximately 3 days allow adequate source decay during the patient's hospital stay. ^{125}I and ^{103}Pd sources, however, can be easily shielded by a thin lead foil, making them useful for treatments of shallowly located or superficial tumors such as ocular melanoma.

12.4
Sources Used in Brachytherapy

The use of radioactive sources for treatment of malignancies started shortly after the discovery of radium in 1898 by Madame Curie. ^{226}Ra, sealed in platinum tubes or needles, was used for interstitial and intra-cavitary temporary treatments. ^{222}Rn, the daughter product of ^{226}Ra, in a gas form sealed within a gold seed, was later used for permanent implants, due to its short half-life. While neither sources are currently used clinically, much of the current brachytherapy treatments derive the dose specification and prescription parameters from the earlier clinical experiences using ^{226}Ra and ^{222}Rn sources. Their historical importance therefore cannot be ignored.

12.4.1
High-Energy Photon Emitters

The following information is given about high-energy photon emitters:

1. Radium-226: ^{226}Ra was the first radionuclide isolated, and the first used in brachytherapy treatments. Radium-226 has a half-life of 1620 years. The γ-rays from radium and its decay products range in energy from 0.05 to 2.4 MeV, with an average energy of about 0.8 MeV. The active ^{226}Ra sources consist of a radium salt (sulfate) mixed with filler (usually barium sulfate), which is encapsulated in platinum cylinders to form radium tubes or needles. Radium tubes have a platinum wall thickness of 0.5 mm, are typically 22 mm long, and contain from 5 to 25 mg of radium in 15-mm active lengths. Radium tubes have a wall thickness of 1.0 mm and are often classified their strengths per centimeters of length. Full-intensity needles typically have 0.66 mg of radium per centimeter of length; half-intensity needles have 0.33 mg of radium per centimeter, and quarter-intensity needles have 0.165 mg of radium per centimeter.

2. Radon-222: ^{222}Rn, with a half-life of 3.83 days and average energy of 1.2 MeV, is a gas produced when radium decays. The radon gas was extracted and encapsulated in gold seeds, which were used for permanent brachytherapy. Both ^{226}Ra and ^{222}Rn have been replaced by newly developed isotopes for clinical brachytherapy, as discussed below.

3. Cesium-137: ^{137}Cs, a fission by-product, is a popular radium substitute because of its 30-year half-life. Its single γ-ray (0.66 MeV) is less penetrating (HVL$_{Pb}$=0.65 cm) than the γ-rays from radium (HVL$_{Pb}$=1.4 cm) or ^{60}Co (HVL$_{Pb}$=1.1 cm). Modern ^{137}Cs intracavitary tubes have been the mainstay for intracavitary treatment of gynecological malignancies. The radioactive material is distributed in insoluble glass microspheres, which produce far less hazard from ruptured sources than does the radon gas in a radium

tube. The active source material is then sealed in stainless steel encapsulation cylinders. Modern ^{137}Cs tubes usually have about a 2.65 mm external diameter with lengths of about 20 mm and active lengths between 14 and 20 mm, depending on the vendor's design. In addition, miniaturized sources equivalent to 10 mg Ra with external diameters of about 1.25 mm attached to the end of long metal stems (Heyman-Simon sources) are used to treat endometrial cancer. Cesium-137 needles were used as replacements for ^{226}Ra needles in interstitial implants; however, their use has been diminishing in favor of more popular remote afterloading systems.

4. Cobalt-60: ^{60}Co: is produced from thermal neutrons captured by ^{59}Co. The subsequent decay to ^{60}Ni releases two highly energetic γ-rays (1.17 and 1.33 MeV), but ^{60}Co has a relatively short half-life (5.26 years). Cobalt-60 tubes and needles were used for brachytherapy during the 1960s and 1970s. Because of its high specific activity, ^{60}Co spherical pellets are used for HDR intracavitary therapy in some centers.

5. Iridium-192: ^{192}Ir, which has a 74-day half-life and lower-energy γ-rays (average γ-ray energy, 0.4 MeV), is the most widely used source for temporary interstitial implants. In Europe, ^{192}Ir is used in the form of a wire containing an iridium–platinum radioactive core encased in a sheath of platinum. In the United States, ^{192}Ir is available as seeds (0.5-mm diameter by 3 mm long) with an active ^{192}Ir core cylinder contained in stainless steel or platinum encapsulation. The seeds are encapsulated in a 0.8-mm-diameter nylon ribbon and are usually spaced at 0.5 cm or 1 cm center-to-center intervals. ^{192}Ir ribbons and wires can be trimmed to the appropriate active length for each catheter. Finally, high-intensity ^{192}Ir sources are used in the latest-generation single stepping source HDR remote afterloading devices.

6. Gold-198: Insoluble ^{198}Au seeds, with a 2.7-day half-life and a 0.412 MeV γ-ray, are available for use as a radon seed substitute to perform permanent implants. Gold-198 seeds are 2.5 mm long and 0.8 mm in outer diameter, and have 0.15-mm-thick platinum encapsulation.

12.4.2
Low-Energy Photon Emitters

The following information is given about low-energy photon emitters:

1. Iodine-125: ^{125}I seeds emit γ-rays and X-rays with energies below 0.0355 MeV, has a half-life of 59.7 days, and are readily shielded by a few tenths of a millimeter of lead (HVL$_{Pb}$=0.002 cm). Many designs of ^{125}I seeds are currently available, all having external dimensions similar to the Oncura model 6711 seed, with an outer cylindrical encapsulation of titanium shell of 4.5 mm in length and 0.8 mm in diameter, as shown in Fig. 12.1. Iodine-125 seeds are used mostly for permanent implant treatments of cancers of the prostate, lung, sarcomas, as well as the temporary implant treatment of ocular melanoma when affixed in an eye plaque.

2. Palladium-103: ^{103}Pd, produced from thermal neutron capture in ^{102}Pd, is an alternative to ^{125}I for permanent implants. ^{103}Pd emits 20- to 23-keV characteristic X-rays and has a shorter half-life (16.9 vs 59.7 days). Because of the much higher dose rates at which ^{103}Pd doses are delivered, ^{103}Pd is thought to have a greater biological effect than ^{125}I (LING 1992; LING et al. 1995; NATH et al. 2005; ANTIPAS et al. 2001; WUU et al. 1996; WUU and ZAIDER 1998). On the other hand, implants that use 103Pd may be more sensitive to errors in source positioning due to the reduced penetration of the low-energy x rays. These errors primarily affect the target tissue, and in cases such as prostate cancer implants where seed insertion errors are a few millimeters, the dose errors can be substantial (DAWSON et al. 1994; NATH et al. 2000). Figure 12.2 shows a diagram of the Theragenics model 200 ^{103}Pd seed.

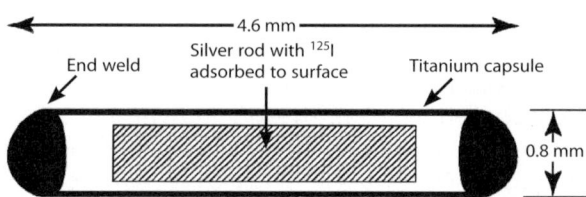

Fig. 12.1 Nycomed Amersham model 6711 ^{125}I seed. (From RIVARD et al. 2004)

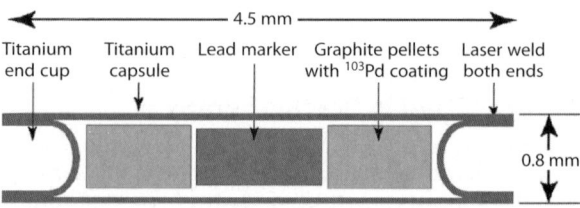

Fig 12.2 Theragenics model 200 ^{103}Pd seed. (From RIVARD et al. 2004)

3. Cesium-131: ^{131}Cs has an average energy of 31 keV and a half-life of 9 days, and is thought to combine the advantages of higher energy of ^{125}I and the high dose rate of ^{103}Pd for permanent implants (MURPHY et al. 2004; YUE et al. 2005). Cesium-131 seed sources, with external dimensions similar to ^{125}I and ^{103}Pd seed sources, have recently become available for permanent implant treatments.

12.4.3
Emerging Sources

Interest in use of ytterbium-169 as a brachytherapy sources dates back to the early 1990s (MASON et al. 1992; PERERA et al.1994; DAS et al. 1995). Ytterbium-169 has an average energy of approximately 90 keV, with a half-life of 31 days. The average energy of ^{169}Yb falls within a region of Compton scattering interaction in tissue, where the ratio of scattered photon energy to primary photon energy is nearly at a maximum. The ^{169}Yb sources can therefore deliver a higher dose to points distant from the source in comparison with traditional brachytherapy sources, such as ^{137}Cs and ^{192}Ir, which is considered an advantage for gynecological cancer treatments. At the same time, radiation shielding for ^{169}Yb requirements is much easier than for ^{137}Cs and ^{192}Ir, due to its lower energy and smaller half-value-layer value in lead (see Table 12.1).

12.5
Dose Calculations in Brachytherapy

12.5.1
The Superposition Principle

The clinical calculation of dose distribution from brachytherapy sources, as is currently practiced, is based on the superposition principle, i.e., the total dose distribution, at a given point of interest, from a group of brachytherapy sources is equal to the sum of the dose to that point by each of the brachytherapy sources in the group, or

$$\dot{D}(x,y,z) = \sum_{i=1,\dots n} \dot{D}_i(x,y,z),$$

where $\dot{D}_i(x, y, z)$ is the dose contribution from the ith source to point of interest (x,y,z).

The superposition principle assumes that the dose distribution to a point of interest is not affected by the presence of other sources. In reality, this assumption is only an approximation. The accuracy of this assumption, or the so-called interseed effect, depends on the average energy of the sources, as well as the distances of the points of interest to the sources. For low-energy sources, such as ^{125}I and ^{103}Pd seeds used in permanent prostate implants, this assumption has been shown to underestimate dose by several percent (MEIGOONI et al. 1992; DEMARCO et al. 1995; ZHANG et al. 2005; CHIBANI et al. 2005). Similar effects have also been demonstrated for intravascular brachytherapy applications, where high-energy beta emitting sources are used to deliver a prescription dose to points located within 2 mm from the center of the sources, an extremely short distance in brachytherapy applications (PATEL et al. 2002). For high-energy photon emitting sources, such as ^{137}Cs and ^{192}Ir, the interseed effect is negligible.

Assuming that the superposition principle holds for a clinical application, the brachytherapy dose calculation problem reduces to the calculation of single sources. i.e., calculation of the radiation dose distribution around a single brachytherapy source. Once such dose distribution parameters are obtained, they can be tabulated for a manual calculation or for computerized isodose distribution calculation for an implant using a group of sources.

12.5.2
Source Strength Units

Brachytherapy source specification protocols have evolved since its inception. The earliest unit for brachytherapy source strength was based on the *mass of radium*, which was used to define the unit of Curie (Ci) for activity:

1 g radium = 1 Ci = 3.7×10^{10} disintegrations/s.

While the unit of Ci, as defined in terms of elemental disintegration rate, is a measurable physical quantity, it cannot be easily applied to brachytherapy source strength specifications because the dose distribution around an encapsulated brachytherapy source depends on the attenuation and scattering of the photons by the encapsulation material. Specification of source strength based on the elemental disintegration rate is therefore usually referred to as *contained activity* in brachytherapy literature, and holds little interest to brachytherapy physicists except in the case of regulatory compliances, where the federal and state governments in the United States require accounting of radioactive material possession in terms of this quantity.

Brachytherapy source strength specifications, therefore, are usually based on what can be measured outside of the encapsulated source. The following units are often encountered in brachytherapy literature:

1. Milligram-radium-equivalent (mgRaEq): High-energy brachytherapy sources with average energy higher than 300 keV have dose distribution characteristics similar to that of radium. They are usually referred to as radium substitute sources; 1 mgRaEq of the radium substitute source is defined to be the amount of the radium substitute source that gives the same output as a 1 mg radium source encapsulated in 0.5 mm platinum in the same output measurement geometry. The measurement geometry specification includes a large distance between the source and the dosimeter, such that the radiation distribution is equivalent to a point source; that the dosimeter should be placed on the transverse axis of the source; and air attenuation and scattering should be corrected. The output quantity of a brachytherapy source used in the determination of mgRaEq is exposure, with units in Roentgen (R). An amount of 1 mgRaEq of a radium substitute source therefore will have the same exposure as 1 mg of radium with 0.5-mm platinum encapsulation, or 8.25 R cm^{-2} h.

   The quantity mgRaEq has a long use history in clinical brachytherapy. The product mgRaEq and the implant time, mgRaEq h, has been used as a prescription quantity for many temporary implants, such as in tandem and ovoids implant for the treatment of cervical cancer.

2. Apparent activity (A): Apparent activity is defined similarly to mgRaEq, with the exception that the encapsulated radium source is replaced by an unshielded source of the specified isotope and has the unit of Ci. A 1-Ci apparent activity of an encapsulated radioisotope source is defined to be the amount of encapsulated source that gives rise to the same output, or exposure in air, as an unencapsulated source of the same isotope of 1 Ci [contained] activity. Apparent activity, due to its not being based on radium sources, is applicable to non-radium-substitute sources such as ^{125}I and [103]Pd sources.

3. Air-kerma strength (S_k): Both milligram-radium-equivalent and apparent activity in mCi have served the brachytherapy community for a long

time and hold historical significance by their association with the clinical experiences accumulated over the years. They are both limited, however, in their applications, and are associated with historical variations in their conversion to exposure in air through the use of the exposure rate constant. In addition, for the calculation of dose in water, as is required for brachytherapy applications, an additional conversion factor between exposure in air and dose in water is required. The AAPM therefore recommended the use of air-kerma strength (S_k), defined as the dose in free air along the transverse axis of an encapsulated source, measured at a large distance from the source such that the source can be approximated by a point source. Air-kerma strength has the unit of cGy cm^2 h^{-1}, and is represented by the symbol U.

12.5.3
Single Source Dosimetry

12.5.3.1
Point Source Dosimetry

The dose distribution surrounding a point brachytherapy source will decrease with the square of the distance r from the source such that the dose rate $\dot{D} \propto 1/r^2$. Because the source strength specifications are defined for an output in air (exposure or air-kerma), a conversion factor from the quantity in air to dose in water is required. This is represented by the f_{med} factor for use with exposure or *dose-rate constant* for use with S_k. The dose falloff is also affected by the attenuation and scattering of photons in media. When these factors are combined, the dose rate at a distance of r centimeters away from a point brachytherapy source is

$$\dot{D}_{med}(r) = \frac{A \cdot (\Gamma_\delta)_x \cdot f_{med} \cdot T(r)}{r^2}.$$

This dose rate equation is appropriate for source strengths specified in apparent activity, where A is the source activity, $(\Gamma_\delta)_x$ is the exposure rate constant (converting the source strength to exposure in air), with δ specifying the lower limit of photon energy included in the determination of the exposure rate constant (photons with energy lower than δ are absorbed near source surface and do not contribute to doses at clinically significant target locations), and x specifying the isotope. The factor f_{med} has units of cGy/R and is specific to photon energy. The tissue attenuation and scatter factor, $T(r)$, accounts for the attenuation and scattering of photons from the source as they traverse the medium.

12.5.3.2
Line Source Dosimetry

Clinical brachytherapy sources have finite physical dimensions, typically in the shape of a cylinder, and are encapsulated in a metal shell of stainless steel, platinum, or titanium. Dose calculations around such sources therefore must include considerations of the geometric distribution of the source within the encapsulated source, as well as the attenuation and scattering of the encapsulation materials.

Sievert Integral. In its simplest form, the Sievert integral (WILLIAMSON et al. 1983; WILLIAMSON 1996; KARAISKOS et al. 2000) only considers the effect of active source geometric distribution within the encapsulated source on the dose distribution around the source by integrating over the active source particles. For a source that can be approximated by a line of length L and without encapsulation, the Sievert integral takes the form of

$$\dot{D}_{med}(P) = A \cdot (\Gamma_\delta)_x \cdot f_{med} \cdot \int_{l \in L} \frac{T(\vec{r})}{|\vec{r}|^2} dl \ ,$$

where $\vec{r}$ is the vector between a segment dl on the line source and the point of interest P, as shown in Fig. 12.3.

When the attenuation and scattering of the photons by the active source and the encapsulation materials are considered, as shown in Fig. 12.4, the dose at point P becomes

$$\dot{D}_{med}(P) = A \cdot (\Gamma_\delta)_x \cdot f_{med} \int_{l \in L} \frac{T(\vec{r})}{|\vec{r}|^2} \cdot e^{-\mu_1 \cdot t_1} \cdot e^{-\mu_2 \cdot t_2} \, dl \ ,$$

where t_1 and t_2 are the thicknesses of active source and encapsulation materials along the vector $\vec{r}$ to point P, and μ_1 and μ_2 represent the average linear attenuation coefficients for the average energy of the source photon spectrum in the active source and encapsulation materials, respectively.

The use of attenuation coefficients for the average energy of the source photon spectrum in Sievert integral has been shown to be highly accurate with some high-energy sources such as Cs-137 tubes (WILLIAMSON 1996), allowing its implementation in commercial brachytherapy treatment-planning systems; however, for sources with complex photon energy spectra, such as Ir-192, and low-energy

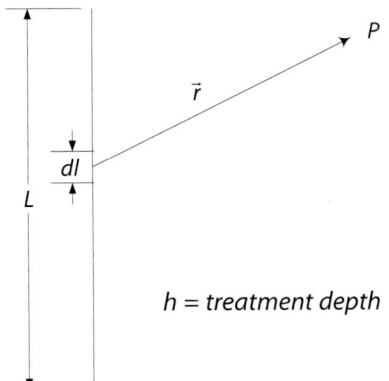

Fig. 12.3 Unfiltered Sievert line source integral. The contributions to dose at point P by each source segment dl are integrated over the entire active source length L without considering the attenuation and scattering of active source materials and source encapsulation.

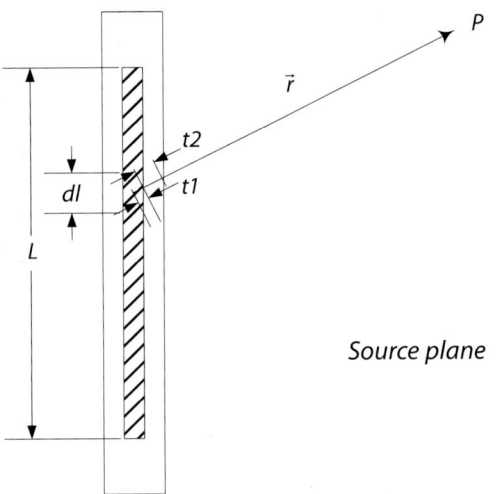

Fig. 12.4 Integration over source segments dl through the entire active source length, taking into account of active source and source encapsulation attenuation by segments t_1 and t_2 in a Sievert Integral

sources, such as I-125 and Pd-103, Sievert integral results in significant dose calculation errors. Modified forms of Sievert integral have been proposed to improve the accuracy of Sievert integral for Ir-192 sources (WILLIAMSON 1996; KARAISKOS et al. 2000), although those are at the present time not available in commercial treatment-planning systems.

For high-energy sources with minimal active source and encapsulation thicknesses, the attenuation and scatter of photons in the source material and in the medium can be assumed to minimally affect the dose distribution. The terms of $T(\vec{r})$, $e^{-\mu_1 \times t_1}$, and $e^{-\mu_2 \times t_2}$ can then be removed from the integral. For a point of interest located on the transverse axis of the source, this yields the following unfiltered line source approximation of Sievert integral:

$$\dot{D}(x,L) = \frac{A \cdot (\Gamma_\delta)_x f_{med}}{x \cdot L} \cdot 2 \cdot \tan^{-1}\left(\frac{L}{2 \cdot x}\right),$$

where x is the distance of the point of interest to the center of the source along its transverse axis. The unfiltered line source integral is often adequately accurate for certain clinical applications, such as in the manual calculation of dose at a point on the transverse axis of the source, for the purpose of double-checking a computer-generated dose distribution for a group of sources arranged in a line.

Away-Along Tables. While Sievert integral provides adequate calculation accuracy for high energy sources, the integration cannot be quickly done manually. A look-up table summarizing the dose distribution surrounding a line source has been a powerful tool for quality assurance. A point in an away-along table is identified by its distance away from the source projected to the transverse axis of the source, and its distance along the longitudinal axis of the source and projected to the source longitudinal axis. Table 12.2 shows such a table for the 3M model 6500 Cs-137 source. Clinically, the away and along distances of a point of interest relative to

a source can be measured off a radiograph of the implant and the dose contribution of the source to this point looked up on the table. This process is then repeated for all sources in the implant to obtain the total dose to the point of interest.

TG43 Formalism. The line source dose calculation formalisms discussed thus far, the Sievert integral and away-along tables, have served traditional brachytherapy dose calculation needs adequately for ^{226}Ra, ^{137}Cs, and ^{198}Au sources. Attempts have been made to apply these formalisms to newer brachytherapy sources such as ^{192}Ir, ^{125}I, and ^{103}Pd seeds. The use of Sievert integral for dose calculation of these sources has proven to be challenging, due to this source's complex photon emission spectrum (WILLIAMSON 1996; KARAISKOS et al. 2000), while computational errors increase rapidly due to interpolation for points near the sources and between the away-along table entries. In 1995, the American Association of Physicists in Medicine (AAPM) Radiation Therapy Committee Task Group 43 published its report (NATH et al. 1995) entitled "Dosimetry of Interstitial Brachytherapy Sources," subsequently revised in 2004 in the updated TG43 report (RIVARD et al. 2004). This report introduced a dose calculation formalism, commonly referred as the TG43 formalism. The TG43 formalism is based on using

Table 12.2 Away-along dose distribution table for 3M model 6500 Cs-137 source. Unit of the entries is cGy × h⁻¹/(μGy × m² × h⁻¹) source strength. (From WILLIAMSON 1998a)

Distance along (cm)	Distance away (cm)													
	0.00	0.25	0.50	0.75	1.00	1.50	2.00	2.50	3.00	3.50	4.00	5.00	6.00	7.00
7.00	0.0193	0.0189	0.0184	0.0180	0.0179	0.0178	0.0176	0.0170	0.0164	0.0156	0.0147	0.0129	0.0111	0.0094
6.00	0.0269	0.0263	0.0254	0.0249	0.0248	0.0247	0.0241	0.0231	0.0218	0.0204	0.0189	0.0160	0.0134	0.0112
5.00	0.0397	0.0386	0.0370	0.0365	0.0365	0.0359	0.0344	0.0322	0.0297	0.0272	0.0247	0.0201	0.0162	0.0131
4.00	0.0638	0.0614	0.0586	0.0584	0.0582	0.0559	0.0517	0.0468	0.0416	0.0367	0.0323	0.0249	0.0193	0.0151
3.50	0.0848	0.0811	0.0774	0.0773	0.0766	0.0719	0.0648	0.0570	0.0495	0.0428	0.0369	0.0276	0.0209	0.0162
3.00	0.118	0.112	0.107	0.107	0.105	0.0949	0.0824	0.0700	0.0591	0.0497	0.0420	0.0304	0.0226	0.0172
2.50	0.176	0.164	0.159	0.156	0.149	0.128	0.106	0.0863	0.0702	0.0575	0.0475	0.0332	0.0241	0.0181
2.00	0.290	0.265	0.257	0.246	0.225	0.178	0.137	0.106	0.0827	0.0657	0.0530	0.0359	0.0257	0.0189
1.50	0.580	0.516	0.489	0.427	0.360	0.249	0.176	0.128	0.0957	0.0737	0.0581	0.0383	0.0268	0.0196
1.00	–	1.580	1.135	0.799	0.582	0.34	0.217	0.149	0.107	0.0807	0.0625	0.0403	0.0278	0.0202
0.50	–	6.569	2.468	1.345	0.852	0.426	0.252	0.165	0.116	0.0855	0.0654	0.0415	0.0284	0.0205
0.00	–	7.806	3.039	1.594	0.973	0.462	0.266	0.171	0.119	0.0872	0.0664	0.042	0.0286	0.0206
−0.50	–	6.566	2.466	1.343	0.851	0.425	0.252	0.165	0.116	0.0855	0.0654	0.0416	0.0285	0.0205
−1.00	–	1.590	1.136	0.803	0.584	0.340	0.217	0.149	0.108	0.0807	0.0625	0.0403	0.0278	0.0202
−1.50	0.547	0.498	0.489	0.428	0.360	0.249	0.176	0.128	0.0958	0.0738	0.0582	0.0384	0.0269	0.0196
−2.00	0.273	0.251	0.256	0.247	0.226	0.178	0.137	0.106	0.0828	0.0657	0.0530	0.0360	0.0256	0.0189
−2.50	0.166	0.154	0.155	0.156	0.149	0.129	0.106	0.0863	0.0702	0.0575	0.0475	0.0333	0.0241	0.0181
−3.00	0.112	0.106	0.104	0.106	0.104	0.0949	0.0824	0.0701	0.0591	0.0497	0.0420	0.030393	0.0226	0.0172
−3.50	0.0802	0.0767	0.0745	0.0759	0.0760	0.0719	0.0648	0.0571	0.0495	0.0428	0.0369	0.027559	0.0209	0.0162
−4.00	0.0604	0.0582	0.0561	0.0570	0.0575	0.0557	0.0517	0.0468	0.0416	0.0367	0.0322	0.024838	0.0193	0.0151
−5.00	0.0376	0.0366	0.0352	0.0353	0.0358	0.0357	0.0344	0.0323	0.0298	0.0271	0.0246	0.019982	0.0161	0.0131
−6.00	0.0255	0.0250	0.0242	0.0239	0.0242	0.0244	0.0240	0.0231	0.0218	0.0204	0.0189	0.016003	0.0134	0.0111
−7.00	0.0183	0.018	0.0175	0.0172	0.0173	0.0175	0.0174	0.0170	0.0164	0.0156	0.0147	0.012847	0.0111	0.0094

air-kerma strength for source strength specification and is described by the following equation:

$$\dot{D}(r,\theta) = S_k \cdot \Lambda \cdot \frac{G_L(r,\theta)}{G_L(r_0,\theta_0)} \cdot g_L(r) \cdot F(r,\theta),$$

using the coordinate system shown in Fig. 12.5, where S_k is the source strength specified in air-kerma strength, in units of $U = cGy\ cm^2\ h^{-1}$. Clinically used brachytherapy sources should have their strength directly or secondarily traceable to a calibration standard established by the National Institute of Standards and Technology (NIST).

Λ is the dose rate constant of the source in water, in unit of $cGy\ h^{-1}\ U^{-1}$. It is defined to the dose rate at 1 cm from a source of 1 U strength, along the source transverse axis, give by $\Lambda = D(r_0,\theta_0)/S_k$. The dose rate constant for a given source must be evaluated carefully, using either well-validated calculation methods, such as Monte Carlo calculations, or measured using appropriate dosimeters, such as thermoluminescent dosimeters (TLD).

$G_L(r,\theta)$ is the geometry function that describes the effect of the active source material distribution within the source on the dose distribution outside the source, and is by the inverse square law $1/r^2$. The geometry function therefore can be calculated by integrating the inverse-square law over all active source particles within an encapsulated brachytherapy source. In practice, it is common to approximate the active source material distribution within a brachytherapy source by an idealized geometry such as a line. The values of the geometry function can then be analytically calculated, thereby avoiding interpolation errors at short distances to the source as may occur with the use of away-along tables. It is, however, crucial that the assumptions made in calculating the geometry function, such as the length of the idealized source distribution be consistent between the source dosimetry parameter derivation and the clinical applications of these parameters. Disagreement in the values of these assumptions may lead to significant dose calculation errors at points close to the source. The updated TG43 report emphasizes this point by using a subscript L in the symbol for the geometry function, indicating the use of a line source assumption for the calculation of geometry function values.

$g(r)$, the radial dose function, accounts for the effect of photon absorption and scattering on the dose distribution along the source transverse axis.

$F(r,\theta)$ is the anisotropy function, which describes the effect of anisotropic photon attenuation, either

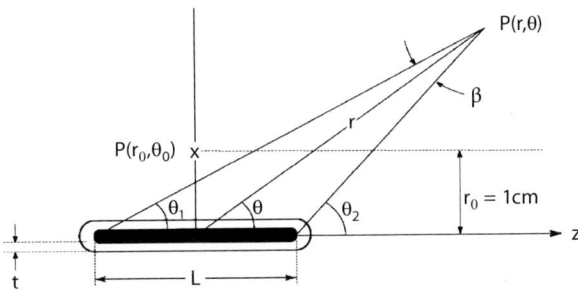

Fig. 12.5 Coordinate system of TG43 dose calculation formalism. (From Rivard et al. 2004)

by the source materials (active source core and source encapsulation) or at locations away from the source transverse axis.

12.5.3.3
Total Delivered Dose Calculations

Given the half-life value of a radioactive isotope and the initial dose rate $\dot{D}_0(r)$ at point r from the source, the total dose delivered to point r in the time interval $[0, t_1]$ can be calculated to be

$$D(r) = \int_0^{t_1} \dot{D}_0(r) \cdot e^{-\frac{\ln(2) \cdot t}{T_{1/2}}} dt = \dot{D}_0(r) \cdot T_{1/2}(1 - e^{-\frac{0.693 \cdot t_1}{T_{1/2}}})/\ln(2)$$

$$= \dot{D}_0(r) \cdot 1.443 \cdot T_{1/2}(1 - e^{-\frac{0.693 \cdot t_1}{T_{1/2}}}),$$

using the relation given in Eq. (2a).

For short treatments using sources with large half-lives, the term $e^{-\frac{0.693 \cdot t_1}{T_{1/2}}}$ can be adequately approximated by $(1 - \ln(2) \times t_1 / T_{1/2})$, resulting in $D(r) = \dot{D}_0(r) \times t_1$. This assumption, however, does not apply to some brachytherapy treatments using short half-lived isotopes, such as ^{125}I or ^{103}Pd seeds. Of particular interest is the use of these sources for permanent implants, where the total doses delivered to a point of interest is given by

$$D(r) = \dot{D}_0(r) \times 1.433 \times T_{1/2},$$

Because of the special importance of the term $1.433 \times T_{1/2}$, it is defined to be the average life of an isotope, i.e., $T_{avg} = 1.433 \times T_{1/2}$.

12.6
Gynecological Intracavitary Implant

12.6.1
Applicators

Brachytherapy implants are an integral part of the treatment of many gynecological cancers, including the cervix, uterine body, and vagina. Applicators are used to hold the brachytherapy sources in clinically defined configurations, or loading patterns. The applicators used for cervical and uterine cancer treatments typically include a tandem, to be inserted into the uterus, and two ovoids, to be positioned in the vaginal vault abutting the cervix. The Fletcher-Suit-Delclos applicator is one of such applicator sets commonly in clinical use in the United States (see Fig. 12.6). This applicator set has tandems of several curvatures to conform to the patient's anatomy, as well as ovoids of diameters of 2, 2.5, and 3 cm, with the larger diameters achieved by fitting plastic caps outside the 2-cm-diameter ovoids. Internally, the ovoids have tungsten shields in the anterior and posterior aspects of the ovoids, as shown in Fig. 12.7, to provide dose attenuation and reduced doses to the bladder and rectum.

12.6.2
Dose Specification for Cervical Cancer Treatments

In the United States, cervix cancer treatments using brachytherapy implants are typically prescribed by one of two methods: the total exposure method, as represented by the product of the total strength of sources implanted and the total source dwell time; and the point-A prescription method. The total exposure method is used at Washington University at St. Louis, where a typical prescription for a course of low-dose rate cervical cancer treatment includes two insertions to deliver nominally 8000 mgRaEq h exposure, with the actual delivered exposure modified based on the length of the tandem and the diameters of the ovoids (WILLIAMSON 1998b). It is noteworthy that, when prescription by total exposure is chosen for a treatment, the pattern of source loading, or the distribution of source strengths in the tandem and ovoids applicators, should adhere to institutional rules. WILLIAMSON (1998b) explained these rules in detail.

The traditional Manchester system for cervical cancer brachytherapy specifies the prescription dose at point A, defined to be the paracervical points at

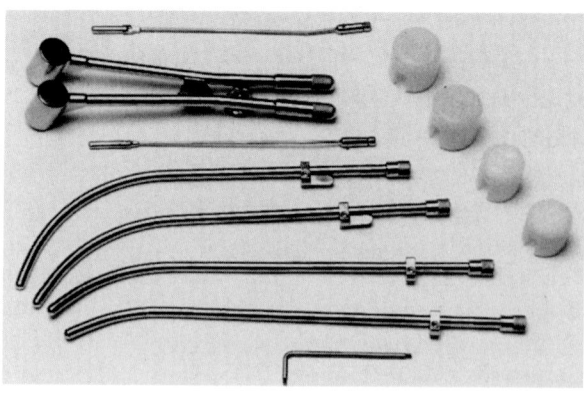

Fig. 12.6 Fletcher-Suit low-dose-rate cervix applicator set (Best Medical, Springfield, Virginia)

2 cm superior to the vaginal fornix, and 2 cm lateral from the uterine canal. The system also specifies point B, located at the same superior–inferior level as point A, but at 5 cm lateral from the patient's midline, intended to represent dose to the parametria. It is noted that point A is related to the orientation of the cervical canal, as localized by the tandem in a radiograph. Lateral distension of the cervical canal results in the corresponding shift of point A.

12.6.3
ICRU Report 38 Recommendations

The International Commission on Radiation Units and Measurements (ICRU) made several recommendations (ICRU 1985) on the reporting of cervix cancer brachytherapy treatment dosimetry, including the volume included by the 60 Gy isodose line, estimated by the product of the length, width, and height of this isodose line. In addition, ICRU report 38 clarified the reporting of bladder and rectum doses. The bladder dose is measured at the posterior-most aspect of a 7-cc Foley balloon in the bladder, pulled back such that the balloon is located on the bladder trigone. The rectum point is defined by the point bisecting the ovoid sources supero-inferiorly, and at 5 mm posterior to the posterior vaginal wall. Figure 12.8 shows the ICRU report 38 definitions of the bladder and rectal points, together with the Manchester system point-A and point-B definitions.

12.6.4
Volumetric Image-Based GYN Brachytherapy

The treatment planning of GYN tandem and ovoids, to date, are still often based on planar images,

Shielding Details

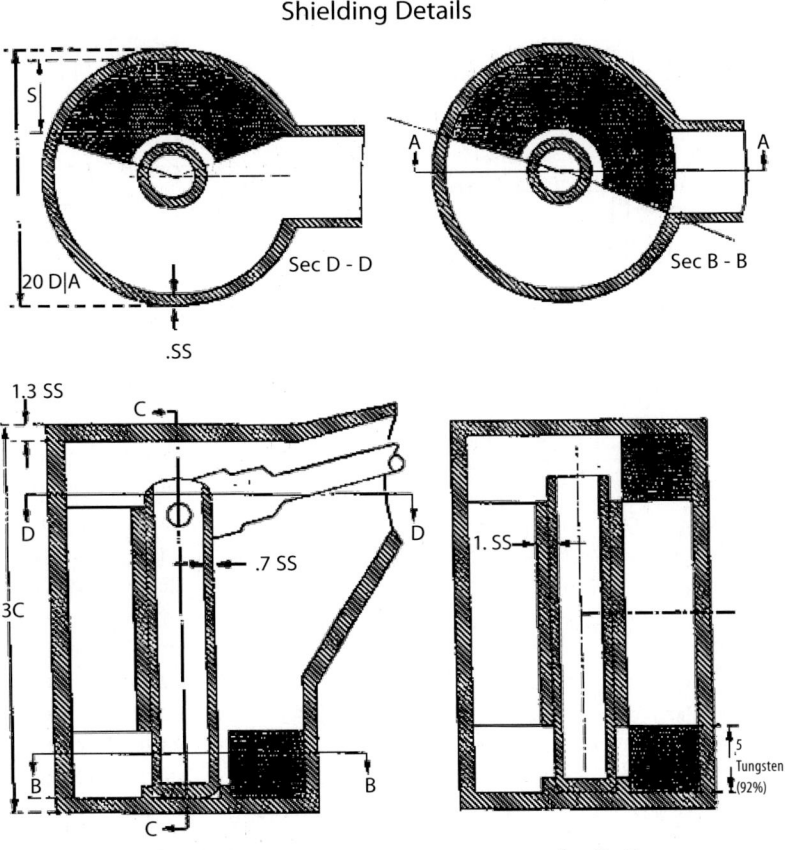

Fig. 12.7 Internal structures of Fletcher-Suit-Delclos ovoid applicator (Best Medical, Springfield, Virginia)

unable to reap the benefits of volumetric images in target and critical organ definitions. This lack of progress has largely been due to the lack of shielded ovoid applicators that are free of CT and MR imaging artifacts. Conventional ovoid applicators, with their tungsten rectal and bladder shields, create so many imaging artifacts that accurate critical organ and target segmentation are impossible. In addition, the progress of volumetric image-based GYN brachytherapy has been limited by CT's low specificity for delineating abdominal tumors.

WEEKS and MONTANA (1997) developed the first CT-compatible ovoid applicators with afterloadable shields. The rest of the applicators are made of aluminum, producing little CT imaging artifacts. The CT images of the implants using these applicators may be acquired first, before the shields and the sources are inserted. The applicators have large-diameter handles for the shields to pass through. MARTEL and NARAYANA (1998) used these applicators to perform 3D treatment planning of GYN tandem and ovoids implants, and obtained the first set of 3D dose distribution data, especially for the rectum and bladder, to show that the maxi-

mum doses that these critical organs receive in a tandem and ovoid implant are significantly higher than estimated by the ICRU report 38 rectal and bladder points. Commercial CT and MR compatible HDR tandem and ovoids, made of carbon fiber or titanium, are now available for use in volumetric image-based GYB tandem and ovoid implants, although those still suffer from the absence of high-density rectal and bladder shields, making it difficult to translate the large amount of clinical experience in GYN tandem and ovoid implants, established using applicators with rectal and bladder shields, into the implementation of this new technology.

Magnetic resonance imaging has been used recently for the target and critical organ delineation in tandem and ovoid implants. Compared with CT, MR imaging provides significantly higher specificity for tumor delineation, while preserving the ability to allow critical organ segmentation. For these reasons, the ABS has recommended its use for image-based GYN brachytherapy (NAG et al. 2004). Modifying the International Commission on Radiation Units and Measurements (ICRU) report 50 nomenclature

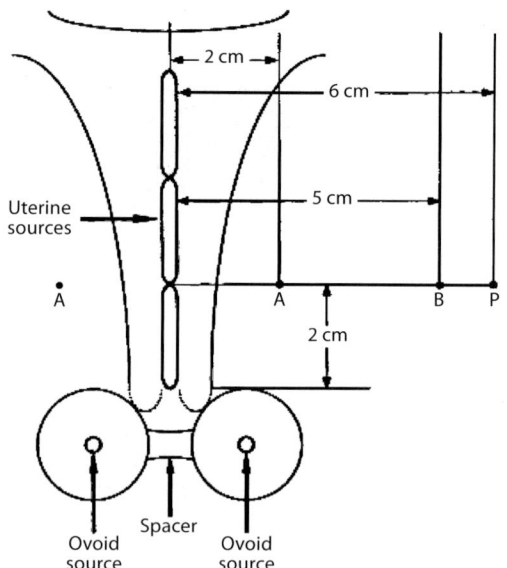

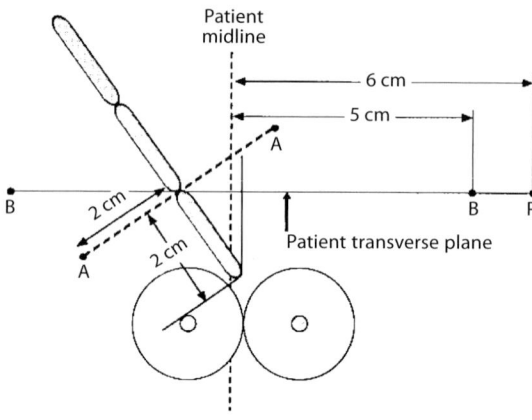

Fig. 12.8 ICRU report 38 definition of bladder and rectal points and Manchester system point A and point B (ICRU 1985)

(ICRU 1993), the ABS defines the following target volumes for image-based GYN brachytherapy:

1. $GTV_{(I)}$: tumor visible in imaging
2. GTV: $GTV_{(I)}$ + clinically visible or palpable tumor (including parametria)
3. GTV+cx: GTV plus the entire cervix
4. pCTV (primary CTV)
   a. For external beam, GTV+cx; entire uterus; parametria; and upper 2 cm of vagina
   b. For brachytherapy, GTV+cx plus 1-cm margin
5. rCTV (regional CTV) for external beam: pCTV plus regional lymph nodes
6. CTV for external beam: combination of pCTV for external beam and rCTV

Note that, in the above, separate target volumes have been defined for treatment planning using external beam techniques and brachytherapy treatments, respectively. The ABS further recommends that individual MR imaging sessions be performed prior to the initiation of external therapy treatments, and before each fraction of brachytherapy treatments following the insertion of brachytherapy applicators, as the tumor regresses through therapy, and as its shape and location are affected by the insertion.

Positron emission tomography (PET) has recently received increased attention for use in GYN brachytherapy (MUTIC et al. 2002; MALYAPA et al. 2002; WAHAB et al. 2004; LIN et al. 2005). The PET imaging has the potential to identify biologically active tumor regions, and, when co-registered with CT images, allows for biologically determined tumor delineation and anatomically segmented critical organs. The PET images, acquired prior to each brachytherapy fraction with the applicators inserted, demonstrate clearly tumor regression through the course of radiotherapy treatment, and allow intricate sculpting of the tumor, as represented by high-uptake regions on the PET images for planning of optimized HDR brachytherapy treatments.

12.7
Interstitial Implant Dosimetry Systems

Brachytherapy treatments using interstitial technique insert brachytherapy sources within the target volume, in order to deliver a prescribed target dose with acceptable dose distribution homogeneity. Prior to the development of computerized treatment-planning techniques, several classical implant systems were developed to calculate, for a given target volume, the total activity of the sources, number of sources, and the source distribution within the target volume, for a given prescription dose. The relation between the target dimensions and the total activity were given in tabular form for a nominal prescription dose, whereas the rules of source distribution were specified separately. While the importance of these classical systems has been reduced with the use of computerized treatment planning, they remain fundamental in the planning of interstitial brachytherapy treatments, both to help guide the pattern of source distribution within the target volume for improved dose distribution homogeneity, and to ensure the technical consistency of treatment delivery for all patients. In addition, the classical implant systems are often used as tools of independent quality assurance checks of the computer treatment plans.

12.7.1
Paterson–Parker (Manchester) System

The Paterson–Parker system, developed by Paterson and Parker in 1934, aims to deliver a uniform dose (±10% from the prescribed or stated dose) on the plane or surface of treated volume. The sources are distributed non-uniformly following specific rules, based on the size of the target volume, with more source strength concentrated in the periphery of the target volume. Such non-uniform distribution of source activities may be achieved either by use of sources of non-uniform strengths, or by varying the spacing of sources of uniform strengths. The Patterson–Parker dose tables give the cumulative source strength required to deliver 860 cGy, using current factors and dose units, as a function of the treated area (planar implants) or volume (WILLIAMSON 1998b).

Regarding single-plane implants, source catheters, arranged in a single plane at 1-cm spacing, can be used to deliver a prescribed dose to a target plane at 0.5 cm away and parallel to the source plane, as shown in Figure 12.9. Cross-end needles may be used, in which case the lengths of the needles in the plane may be reduced by 10% for each cross-end needle. The fractions of source strengths in the periphery of the implant depends on the total treated area: for areas less than 25 cm², two-thirds of the total activities are implanted in the periphery; changing to one-half for total areas between 25 and 100 cm² and one-third for total areas larger than 100 cm², respectively. The total activity is further increased for non-rectangular target areas. For thicker slabs of target, up to 2.5 cm thick, the needles may be arranged in two parallel planes. The total activity needed for double-plane implants are looked up using a single plane implant table, followed by application of a correction factor depending on the thickness of the target tissue. The total activities are then distributed between the two source planes in proportion to their relative areas.

Regarding volume implants, the Patterson–Parker system for volume implant is similar to the planar implant, in that the total activity is non-uniformly distributed between the periphery and the core (center) of the target volume. Typically, with all sides of a target volume implanted, the total activity

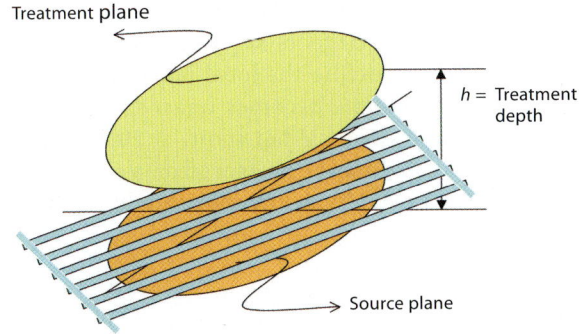

Fig. 12.9 Planar implant of Paterson-Parker system

is divided into eight parts, with only two parts of the total eight parts in the core volume of the target. The lengths of the implant needles may be reduced if cross-end needles are used, by 7.5% for each end of the volume implanted with cross-end needles.

12.7.2
Quimby System

This system, developed by QUIMBY in 1932, is based on a uniform distribution of source strength, allowing a higher dose in the center of the treatment volume than near the periphery. Typically, for equal dose delivery to similar size planar or volume implants, the total source strength required when using the Quimby system will be much greater than what is required by the Patterson–Parker system.

12.7.3
Paris System

The Paris system is used primarily for single- and double-plane implants using parallel, equidistant needles arranged in triangular or rectangular shapes when viewed on the needles' ends. All sources used in a Paris system-based implant are to have the same linear strength, although it is possible that sources of different lengths may be used. A central implant plane is defined that approximately bisects all implanted parallel needles. The prescribed dose is a percentage, typically 85%, of the average doses at minimum dose points within the central plane, or the so-called basal dose points. Geometrically, these minimum dose points are approximated by points equidistant to neighboring needles. The lengths and spacing of the needles are therefore dependent on the target thickness and width. Due to the use of no more than two-plane-implants, the Paris system may treat a target of typically no more than 2.2 cm thickness, before the local hot spots within the implanted volume becomes unacceptable. The dimensions of these local hot spots, at 200% of the prescription dose, may be measured on a computerized treatment plan. Typically they should not have diameters >1 cm.

12.7.4
ICRU Report 58 Recommendations

In 1998 the ICRU published its report no. 58 on the specification and reporting of interstitial brachytherapy (ICRU 1998). Continuing the conceptual path set forth by the ICRU report no. 50 (ICRU 1993), the report defines the gross target volume (GTV) as the tumor volume visible on imaging or clinically palpable target volume, and the clinical target volume (CTV) as the volumes of potential tumor spread or microscopic diseases. For brachytherapy treatments, the planning target volume (PTV) is considered to be identical to the CTV, as no source positioning uncertainties within the target volume is included. The report defines dosimetric concepts similar to those used in the Paris system, renaming the basal dose points as central dose points. It expands on the Paris system by allowing for use of multiple central planes when the implant is such that one central plane may not interest all implanted needles. Furthermore, it recommends the reporting of the low-dose (LD) volume, defined as the volume of the CTV receiving <90% of the prescription dose, and the high-dose (HD) region, defined to be the tissue receiving greater than 150% of the mean central dose (MCD), or the average of the doses to the central dose points. The ICRU report 58 also recommends reporting of the dose homogeneity index (DHI) of the implant, defined to be the ratio of the minimum target dose (MTD) to the MCD, echoing the Paris system concept of implanting a target such that 85% of the average basal dose should cover the target and therefore be used as the prescription dose.

12.8
Process of Brachytherapy Treatment Planning and Delivery

The successful delivery of a brachytherapy treatment is a coordinated effort among hospital person-nel across a wide spectrum of expertise, including the radiation oncologist, surgical oncologist, radiation physicist, medical dosimetrist, radiation therapist, the nursing staff of radiation oncology, surgery, recovery, and hospital floor units, and other support personnel. Careful planning, scheduling, and communication for a brachytherapy treatment are the keys to its successful completion. Various pre-surgery patient tests and exams must be performed before the patient is considered eligible for the surgical insertion of brachytherapy applicators. The results of such tests must be communicated to the radiation oncologist and radiation physicist in a timely manner, such that adequate time remains for the ordering, shipping, and receipt of the sources to be used for the patient. In an institution that provides a large number of brachytherapy treatments, such as permanent prostate implant or HDR partial breast treatment, it is often helpful to have a radiation oncology staff member designated for the overall scheduling of the patient. Such a person must coordinate with the surgical oncologist and the operating room (OR) staff for the coordination of the surgery aspects of the treatment, in addition to coordinating the activities germane to the brachytherapy aspects of the treatment as listed below. A checklist can be helpful to document the timely completion of each of the landmark events in the overall brachytherapy treatment, identifying the information required and its recipients at the completion of these landmark events.

12.8.1
Preplanning

The goal of preplanning for a brachytherapy treatment is to determine the applicators and sources required for the treatment. Given a specific treatment site, preplanning allows the identification of the appropriate applicators, applicator insertion technique, sources required, and their distribution patterns for the brachytherapy treatment. For transrectal ultrasound (TRUS)-guided permanent prostate implant treatments, the preplanning typically includes the acquisition of treatment-planning ultrasound images, or volume study, and an isodose plan based on these images specifying the number of needles and sources, source strength, and source and needle distribution pattern. Adequate time must be left between preplanning and surgery to allow for the ordering, shipping, and receipt of the sources.

While preplanning does not always result in a physical isodose plan for the intended treatment, the underlying principles of an isodose plan, i.e., establishment of the relation between sources and the dose distribution, is always a part of the preplanning. For example, the preplanning for an HDR interstitial treatment would estimate the number of catheters required for the treatment, as well as the spatial distribution and arrangement of the catheters. Such specifications of catheter number and distribution are typically based on one of the classical implant systems as previously discussed (Manchester, Quimby, and Paris), which establishes predictable relations between the catheter distributions and the resultant dose distributions.

12.8.2
Source and Applicator Preparation

The applicators and sources specified in preplanning are subsequently prepared. Such preparation may be as simple as making sure that the tandem and ovoids applicators, the associated surgical instruments such as dilation sets and sounding rules (a pliable ruler used to measure uterine depth and angle), and the sources are available. Custom applicators and templates may need to be fabricated or acquired for the more complex treatments. In all cases, the applicators and instruments are collected and sterilized so as to be ready for the procedure. A member of the nursing or radiation therapist staff should be designated for this task.

Many interstitial brachytherapy treatments require acquisition of sources as specified in the preplanning for the treatment. The federal government requires that an institution's radioactive material license be on file with a source vendor before the vendor may ship the sources to the institution. The radiation physicist or radiation safety officer should ensure that this has been done prior to source ordering.

A secure room must be available for temporary storage of the sources, equipped with a shielded source safe, an L-block shield, source handling tools, and source assay instruments, as source preparation is typically performed in this room. During source receipt and preparation, the radiation physicist performs source packaging radiation contamination tests to ensure its integrity. These tests include an inspection of the physical condition of the packaging, removable radiation contamination wipe tests, and measurement of radiation exposure rates at package surface and 1 m away from the package. Records of the results of these tests must be maintained.

The custody of the sources, upon receipt in the institution and through to final source disposal, should be ensured. Regulatory bodies at both the federal and state levels require that the chain of custody for radioactive materials be maintained, and the location of sources is known at all times. A source inventory is performed at the receipt of the sources in comparison with the source order. Any discrepancies must be investigated through communication with the vendor, and, in the event that a source is deemed to have been lost, reported to regulatory authorities immediately. Source accounting is further performed through the entire process of source preparation, with the locations of the sources and the times when the sources are moved between locations documented.

The strengths of the sources must be confirmed to be in agreement with the values ordered and the values used for treatment planning. The AAPM recommends that the institution perform an independent source strength assay, using a dosimetry system with secondary traceability to the NIST/ADCL standard (KUTCHER et al. 1994; NATH et al. 1997). Source strength assay is typically performed using a well-type ionization chamber and electrometer system. This system should be calibrated at the ADCL every 2 years, and its constancy verified using a long-half-life source, such as a ^{137}Cs source, at the beginning of each day when the system is to be used for source strength assay.

12.8.3
Applicator and Catheter Insertion

The insertion of applicators is usually performed in an OR or a procedure room under sterile or clean conditions. Accuracy of applicator positioning is crucial in realizing the preplanned dose distribution, and every effort should be made to ensure that the applicators are inserted into the preplanned locations, especially for complex treatments such as interstitial implants using multiple catheters. The radiation oncologist and physicist involved in the brachytherapy treatment should discuss in detail the catheter insertion procedure prior to performing the procedure, and ensure that each step of the procedure is well understood by all persons in the OR to minimize miscommunications and to achieve maximum efficiency for the successful insertion of the catheters. The aspects of catheter insertion procedure in the OR usually include the following:

1. Having the preplan available in the OR where applicable. The physicist should ensure that a preplan for a multiple catheter implant be available in an easily interpretable format in the OR, so that the position and depth of each catheter can be looked up quickly and without ambiguity. In addition, the physicist should be prepared to evaluate the positioning accuracy of catheters as they are inserted, such that their dosimetric effect may be estimated, and additional catheters may be inserted to achieve desired dose distribution.

2. Confirmation of target and critical organ localization. Surgical clips, markers, or other target localization media, such as contrast solution in a lumpectomy cavity for breast implant, should be placed in the target volume, such that they can be used to assess the adequacy of applicator insertion by imaging later. Similarly, markers may be placed near critical organs, such as blood vessels and nerves, such that radiation doses to these critical organs may be minimized.

3. Determination of catheter insertion approach on skin. For sarcoma and other interstitial implants, the catheter insertion approach may depend on the surgical approach for gross tumor removal. The entrance and exit points of the catheters should be evaluated to allow accurate catheter insertion, to minimize the number of catheters used, and to facilitate ease of source loading. If the treatment is to be delivered on an outpatient basis in multiple fractions, the termination of catheters should allow for easy securing of the catheters to avoid their movement within the tissue, and to minimize the possibility of their damage between treatment fractions. After the catheter insertion approach has been selected, a ruler should be used to measure the catheter entrance and exit points on skin to facilitate the accurate and efficient catheter insertion.

4. Assessment of catheter insertion accuracy. As the procedure proceeds, the accuracy of catheter insertion should be periodically reviewed, using imaging techniques appropriate for the procedure, including fluoroscopic X-rays, ultrasound, and other imaging techniques, such as CT or MRI, for CT/MRI-guided procedures. Catheters deviating significantly from preplanned positions should be re-inserted, or additional catheters inserted, so that the desired dose distribution as preplanned may be achieved. At the completion of applicator insertion, a set of images should be obtained and reviewed, to confirm that the all catheters are correctly inserted, and to serve as guidance for later dosimetry calculations. For treatments delivered in multiple fractions, the initial images of applicator and catheter positioning are often useful to evaluate the positioning of applicator and catheter insertion of subsequent treatment fractions. Deviations of subsequent applicator and catheter insertion from the initial images should be evaluated.

12.8.4
Source/Applicator Localization

Except for cases where the implant images obtained in the OR are used directly for dose distribution calculations, such as in intra-operative high dose rate prostate treatments (KINI et al. 1999), the dose distributions of the implant are usually calculated based on either planar X-ray images or volumetric CT or MR images acquired after the completion of applicator insertion. For multiple catheter implants, the individual catheters are labeled by identification numbers prior to imaging. As appropriate for the imaging modality used, radio-opaque markers are inserted into the catheters. Coded radio-opaque markers are commonly used for planar X-ray imaging. The correspondence between the radiographic markers and the catheter numbering should be recorded for reference during treatment planning.

Patient images, either volumetric or planar, need to be acquired to permit accurate applicator and catheter localization and dose calculations. For each type of brachytherapy implant, an imaging protocol should be developed and adhered to, with the parameters of these protocols chosen to minimize imaging artifacts, and to allow accurate applicator, target, and critical organ localization. Breathing artifacts in CT scans or orthogonal radiographs will significantly increase the uncertainties in applicator and catheter localization accuracy, as well as in the delineation of treatment target and critical organs. Figure 12.10 shows an orthogonal pair of radiographs for the HDR treatment of bile duct, where the patient breathing motion artifact caused up to 1-cm differences in the y-coordinates of the X-ray markers. The physicist needs to evaluate the dosimetric consequences of such imaging artifacts, and communicate with the treating physician to arrive at the optimal actions to be taken in the patient's treatment.

When volumetric imaging is used for brachytherapy treatment planning, the field of view (FOV),

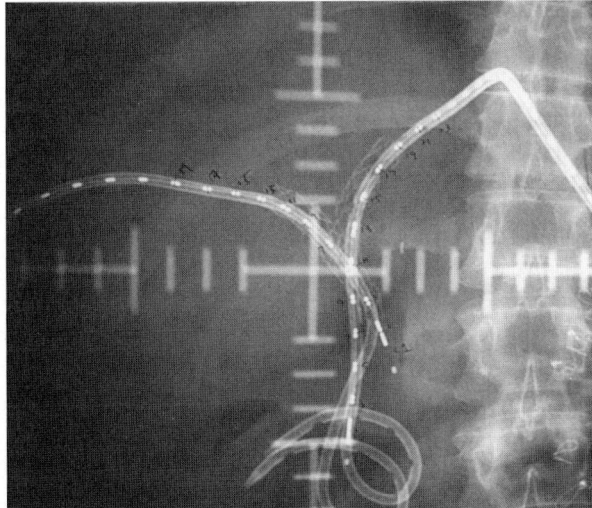
a

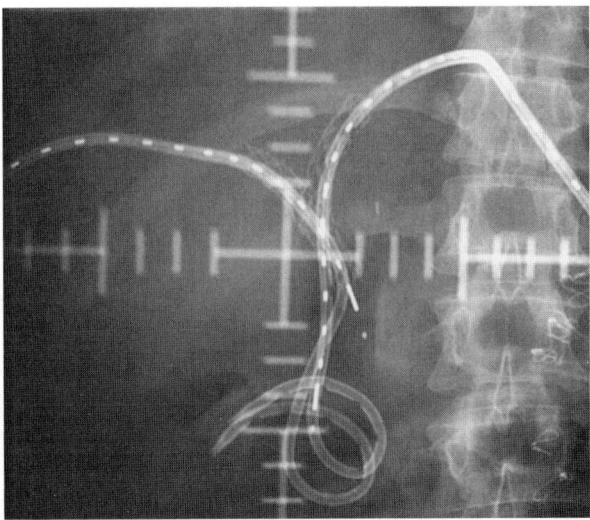
b

Fig. 12.10a,b. Breathing motion artifacts in the X-ray images of a bile-duct brachytherapy treatment. The catheter positions differ by up to 1 cm between the two anteroposterior/posteroanterior images supero-inferiorly. **a** Anteroposterior image of bile duct HDR treatment, patient free breathing. **b** Same image as **a**, except with patient holding breath. (From Li 2005)

slice thickness, table pitch, and gantry angle of the scans need to be reviewed to assure that they conform to the scanning protocol established for the particular brachytherapy treatment. Deviations from the established scanning parameters may reduce the accuracy of applicator, target and critical organ localization, and could result in significantly increased dose calculation uncertainties.

When orthogonal or other types of planar radiographs are used, the isocenter of the radiographs should be placed near the center of the target volume. The magnification factors of the radiographs and gantry angles need to be confirmed. The coordinate system used on the simulator, such as a Varian Ximatron unit, may be different from the IEC coordinate system used in some treatment-planning systems such as the Nucletron Plato TPS. The conversion of gantry angles between the two coordinate systems must be done correctly to avoid significant source localization errors. The filming technique used should allow clear and unambiguous recognition of applicators, surgical clips and radiographic markers on the film.

The quality of simulation images directly impacts the accuracy of subsequent dose distribution calculations. It is therefore important that treatment-site-specific imaging protocols be developed, and that a physicist or dosimetrist familiar with the treatment be present during the simulation process for complex, multiple-catheter implants.

12.8.5
Treatment Planning and Quality Assurance Review of Brachytherapy Treatment Plan

The simulation images of the brachytherapy implant, once approved by the treating physician, are used for calculation of dose distribution and treatment planning of the implant. The source applicators and source positions are localized on the images, and dose distributions are calculated based on the superposition principle as previously discussed. It is critical that the numbering of catheters on the images and the sources to be inserted into the catheters be accurately identified in the treatment plan, as miscorrelations in these parameters results in significant dose-calculation errors. The source strengths and their spatial distributions may be selected such that the calculated dose distributions are optimized to conform the prescribed isodose to the target, and to minimize critical organ doses. While this selection process is done manually through an iterative process for most of the low dose rate brachytherapy treatments, treatment planning for high dose rate remote afterloading treatments are typically performed on a treatment-planning system with automatic and semi-automatic optimization capabilities, thereby significantly improving the efficiency of treatment planning and quality of the resulting treatment plans.

Following the completion of dose distribution calculation for a brachytherapy implant, the treatment plan should be reviewed by a medical physicist for its appropriateness for the treatment, accuracy, and dose distribution quality (Li 2005). During this review, the treatment prescription is reviewed for appropriateness for the treatment by comparison with institutional treatment protocols or national treatment guidelines. The treatment-plan input parameters, including imaging parameters, source localization accuracy, and source dosimetric parameters, are evaluated for appropriateness and accuracy. The dose distribution is reviewed for homogeneity, adequate target dose coverage, and acceptable critical organ doses. Automatic optimization algorithms and parameters, when used, should be reviewed for the appropriateness of their selections.

12.8.6
Source Loading and Treatment Delivery

Depending on the complexity of the implant design, source loading may occur either before or after treatment planning is completed. In either case, the loading of sources into applicators and catheters starts delivery of the brachytherapy treatment. In the United States, federal and state regulations require that a written directive for the treatment must be completed by a physician authorized by the relevant regulatory bodies prior to start of the treatment. Throughout the treatment delivery process, the number of sources, source loading pattern, source strengths, and source dwell times must be verified to be in agreement with the prescription. A record of the sources removed from brachytherapy source storage room, and the time of source removal, must be kept as required by regulations and to ensure that all brachytherapy sources be accounted for at all times. Appropriate radiation surveys need to be performed, following the loading of brachytherapy sources, to ensure that all areas in close proximity to the patient room be free of radiation risk to both patient care personnel and to visitors to the hospital. The results of these radiation surveys, together with appropriate warning signs and instructions, should be posted. A shielded radioactive source container, together with source-handling tools, need to be kept in the patient room for handling of sources that may have dislodged from patient.

When the treatment is completed, the sources are removed from the applicators and catheters, making sure that all implanted sources are accounted for. The applicators and catheters are subsequently removed. A radiation survey of the treatment room is performed to ensure that all radiation sources have been removed. The sources are returned to the brachytherapy source room, and the appropriate source inventory documentation is completed for regulatory compliance.

12.9
Transrectal Ultrasound-Guided Permanent Prostate Brachytherapy

Transperineal permanent brachytherapy implant, using low-energy sources such as ^{125}I and ^{103}Pd seed sources and guided by real-time visualization of the prostate and its surrounding organs using transrectal ultrasound imaging (TRUS), has become a widespread option in the management of early-stage prostate cancer. It offers the advantages of a relatively minor surgical procedure compared with prostatectomy, and the convenience of a single outpatient procedure compared with external-beam irradiation, which typically requires up to 8 weeks of daily treatments. First introduced by Holm (1997), it has become one of the most-often performed brachytherapy procedures in the United States. The American Association of Physicists in Medicine (AAPM) published a report (Yu et al. 1999) that provided detailed discussions on the physical and clinical aspects of TRUS-guided permanent prostate brachytherapy.

12.9.1
Dosimetric Goals of a Permanent Prostate Brachytherapy Treatment

The American Brachytherapy Society (ABS) recommended that the prescription doses for permanent prostate implants be 144 and 115–120 Gy for seeds-alone treatment using ^{125}I and ^{103}Pd seeds, respectively (Nag et al. 2000). For boost treatments, where the brachytherapy implant is delivered after 40–50 Gy of external-beam radiotherapy, the prescription doses should be reduced to 100–110 Gy and 80–90 Gy for ^{125}I and ^{103}Pd treatments, respectively. Additional dosimetric parameters to be reviewed in permanent prostate brachytherapy dose distributions include the volume of prostate receiving larger than 150% of the prescription dose (V_{150} of the prostate), and critical organ doses such as doses to the urethra and rectum.

12.9.2
Equipment

A portable ultrasound unit, together with a transrectal probe capable of visualizing organs at up to 7- to 8-cm depths from probe surface, is an integral part of TRUS-guided permanent prostate brachytherapy. The probe is rigidly fixed to a stepper device mounted on a stabilizer such that during the brachytherapy procedure the probe may be moved longitudinally at desired step sizes, and no unintended motion of the probe is allowed. A template, shown in Fig. 12.11, with needle holes at 0.5-cm spacing, arranged in a rectangular fashion and labeled both vertically and horizontally, is mounted securely to the stabilizer. The TRUS unit displays an array of markers on the ultrasound image identical to the template holes, so that a corresponding relation is established between the internal anatomical geometry with the template. During volume study as well as the implant procedure, the patient is placed in an extended lithotomy position, requiring the use of a set of stirrups. Needle loading boxes are required for use with the preloaded needle technique, in which the sources are inserted into stainless steel implant needles at pre-calculated spacing. Alternatively, a Mick applicator may be used to insert the radioactive seeds into the planned positions one by one. A brachytherapy treatment-planning system capable of both pre-treatment planning and post-implant

dosimetry calculations is necessary. Newer versions of such systems are capable of interfacing directly with the TRUS probe position sensors on the stepper device for real-time treatment planning in the OR (NAG et al. 2001; LEE and ZAIDER 2003; MATZKIN et al. 2003; POTTERS et al. 2003; RABEN et al. 2004).

For the quality assurance testing throughout a TRUS-guided permanent prostate implant procedure, a well-type source calibration chamber and an electrometer, both with traceability to the National Institute of Standards and Technology (NIST) or the Accredited Dosimetry Calibration Laboratories (ADCL) of the AAPM, are used to verify the strengths of the seeds to be implanted. Additionally, ion-chamber survey meters and solid-state scintillation survey meters are used to measure the radiation exposure rates around the patient to ensure the safety of releasing the patient from the hospital without exceeding federal or state regulatory limits, and are used to locate a misplaced low-energy photon-emitting source.

12.9.3
Volume Study

Volume studies are to permanent prostate implants what simulations are to other radiation therapy modalities, i.e., a patient anatomy model is constructed, and the dose distribution calculated from this model, to be delivered to the patient. A set of TRUS images of patient's prostate and surrounding anatomy is obtained in the volume study, and the locations of sources to be inserted into the prostate are selected from all possible combinations of rows, columns, and insertion depths available, such that an optimized dose distribution satisfying the dosimetric goals of the treatment is achieved. It is therefore important that the volume study be performed under identical conditions as those to be used for treatment delivery. In addition, the patient setup and TRUS image acquisition should be such that they are easily reproducible in the OR The TRUS probe should be positioned such that the entire prostate, with adequate margins as desired, is encompassed by the grid pattern on the TRUS image, as no needle holes exist to guide the insertion of needles outside the grid pattern. The urethra should be placed in the center of the TRUS image whenever possible, and should not travel across the lateral direction of the image, as such travels prohibit the use of needle holes lateral to the urethra for source implantation and increase the possibility of insert-

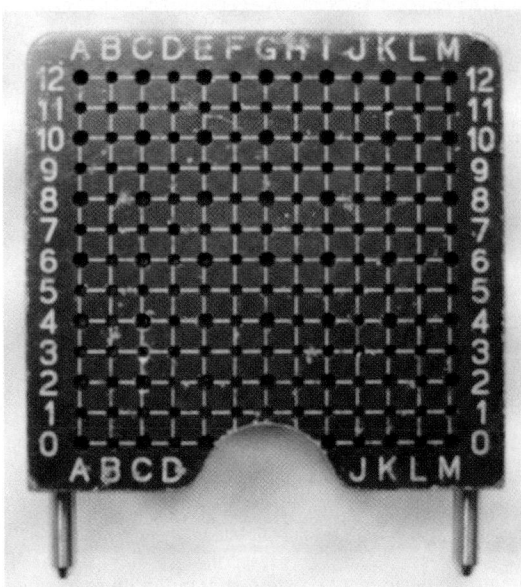

Fig. 12.11 Transperineal TRUS-guided permanent prostate implant template

ing sources into the urethra. The posterior surface of the prostate should be at adequate distance away from the posterior-most row of needle holes, so as to allow implantation of sources in the posterior aspect of the prostate while maintaining safe distances of the seeds from the rectum.

12.9.4
Treatment Planning

During treatment planning, the prostate capsule as visualized on the TRUS images is segmented as the GTV. A margin of 3–5 mm around the prostate, except for the posterior aspect, is then added to the GTV to obtain a planning target volume (PTV). The goals of treatment planning are therefore to calculate a dose distribution to provide adequate prescription dose coverage of the PTV (e.g., 95% or higher), while maintaining adequately low doses to the urethra and the rectum. In addition, the source arrangement of the treatment plan should be such that it is easily implementable in the OR and helps to minimize source implantation errors, such as having needles laterally symmetric and requiring minimal needle/TRUS probe retractions during the procedure. It is also often preferred for an institution to establish policies regarding the seed strength and average number of seeds per needle, such that

the chances of using sources of wrong strengths are minimized, and that treatment plans do not require excessively high numbers of implant needles.

The distribution patterns of sources in a permanent prostate implant can be divided into uniform loading, in which all sources are arranged at 1-cm spacing without conforming to the outline of the prostate capsule; modified uniform loading, in which the peripheral sources are arranged to achieve conformality with the prostate capsule, while maintaining 1-cm spacing for sources in the center region of the prostate; and peripheral loading, which improves upon the modified loading by reducing the number of sources in the center region of the prostate, thereby providing a degree of sparing of the urethra. A comparison of the uniform source-loading scheme and the peripheral source-loading scheme is shown in Figure 12.12. Modern permanent prostate brachytherapy typically uses a form of the peripheral-loading scheme, optimizing on PTV dose coverage while reducing urethra and rectum doses as well as dose distribution inhomogeneity, either by manually iterative optimization or by use of automatic optimization algorithms (EDMUNDSON et al. 1995; POULIOT et al. 1996; YU and SCHELL 1996; GALLAGHER and LEE 1997; YANG et al. 1998; CHEN et al. 2000; D'SOUZA et al. 2001; LEE and ZAIDER 2003; SUMIDA et al. 2004). Optimized dose distributions for permanent prostate implant will therefore have

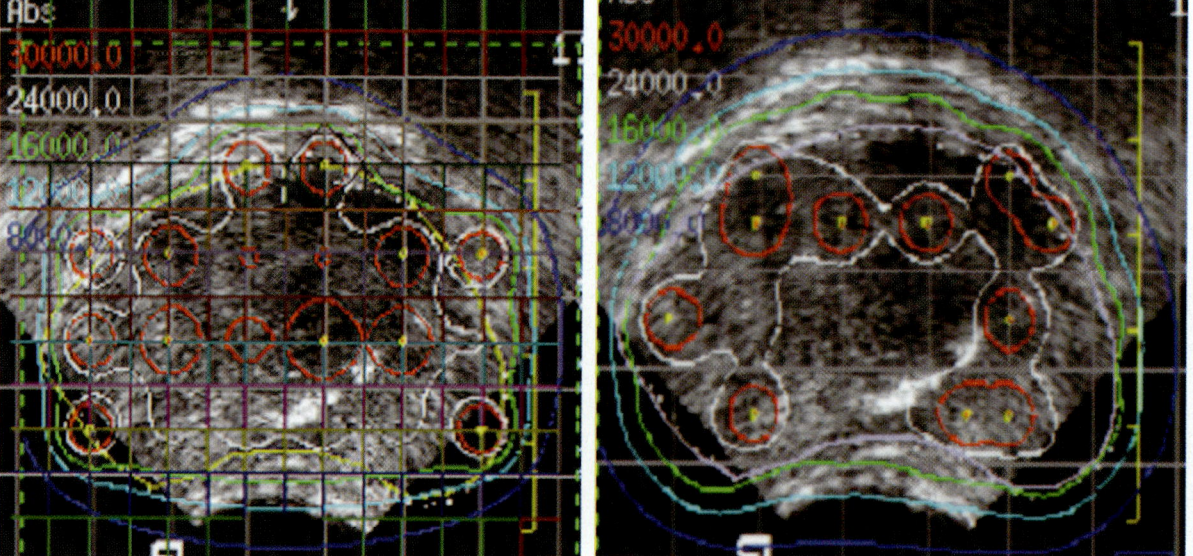

Fig. 12.12a,b. Comparison of uniform loading and peripheral loading schemes in permanent prostate implants, with prescribed dose of 16,000 cGy. **a** Uniform loading scheme, with high-dose region in the central part of the prostate, and less than conformal prescription isodose coverage (*green isodose line*) of the prostate. **b** Peripheral loading scheme, with reduced central region doses and improved conformality of prostate dose coverage

lower doses near the urethra and the region anterior to the urethra, and higher doses in the peripheral zones of the prostate, particularly if such regions are positive for tumor presence according to biopsy results.

Conventional dose calculation for permanent prostate implants use only the point-source approximation model of the TG43 formalism, or its traditional equivalents. In addition, the calculation does not consider the inter-seed attenuation effect on the overall dose distribution. While this has been adequate for clinical treatments so far, it has been shown that such approximation, compared with calculations done using a version of the line source models of the TG43 report, results in erroneous estimates of urethra dose and volume of hotspots of the treatment plan (LINDSAY et al. 2001; CHIBANI et al. 2005). In addition, the interseed attenuation effect (DAWSON et al. 1994; DeMARCO et al. 1999; CHIBANI et al. 2005) has been shown to decrease the overall prostate dose coverage by several percentage points as compared with the conventional calculations. Recent studies have shown that the Monte Carlo calculation method (CHIBANI et al. 2005; ZHANG et al. 2005), where all photon particles emitted from all the sources are tracked and their energy deposition within the treated volume accumulated, is able to provide more accurate dose calculations. It is possible that such more accurate calculation algorithms will become available clinically in the near future.

12.9.5
Treatment Plan Review and Pretreatment Quality Assurance Tests

The treatment plan of a permanent prostate implant should be reviewed, prior to implantation, for its accuracy and suitability for treatment. It is important to verify that the prescribed dose is appropriate for the treatment, as discussed previously, based on the brachytherapy source isotope used and whether the patient has received previous external beam radiation therapy treatments. The dose calculation accuracy and total activity used for the implant can be checked by performing an independent calculation using a nomogram, an equation relating the total activity required for a permanent prostate brachytherapy implant relative to the volume or average linear dimensions of the target, the isotope used, and the prescribed dose. It is of particular importance that an institution's permanent prostate brachytherapy program maintains consistency

across all patient treatments, even if an individual institution's treatment plans differ from published nomograms by a significant margin, as long as the dose calculation accuracy has been confirmed, and that acceptable clinical outcome has been achieved. Cohen et al. reported the use of nomograms for the independent calculation check of prostate-seed implants (COHEN et al. 2002). With d_{avg} as the average distances between pre-planned needles/seeds in the lateral, anterior–posterior and superior–inferior directions, the authors reported the following nomograms:

$$\frac{S_k}{U} = 1.524 \left(1.09 \frac{d_{avg} + 0.8}{cm} \right)^{2.2}$$

for ^{125}I seed implants with a prescription dose of 144 Gy, treated to the prostate volume with a 5-mm margin in all directions except posterior, and

$$\frac{S_k}{U} = 5.395 \left(1.09 \frac{d_{avg} + 0.8}{cm} \right)^{2.56}$$

for ^{103}Pd seed implants with a prescription of 140 Gy. Compared with computerized treatment plans generated using a genetic algorithm, the authors reported agreement of better than 10% in the total activity required.

Once the treatment plan for a permanent prostate implant has been reviewed and approved by a medical physicist, the seeds may be ordered. Many institutions order 10% more seeds than prescribed by the treatment plan, to be implanted to perceived cold spots, or areas with larger than planned seed spacing, estimated intraoperatively using ultrasound and X-ray imaging.

Upon receipt of the seeds, their strengths are assayed using a NIST or ADCL traceable well chamber and electrometer system as described previously. The seeds are then sterilized, loaded into sterile needles for the preloaded needle technique, or kept in sterile cartridges for the Mick-applicator technique. The glass vial holding the seeds should be plugged with gauze, so that the seeds do not escape from the vial during sterilization.

For treatments using preloaded needles, the accuracy of needle loading should be verified. Needle-loading devices of various designs are commercially available to facilitate such verifications as the needles are loaded. Alternatively, an autoradiograph of the loaded needles may be obtained, paying attention to maintain the sterility of the loaded needles. The number of seeds loaded into a needle, as well as their spacing, can be readily verified on such an autoradiograph, as shown in Fig. 12.13.

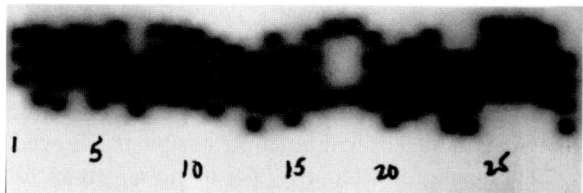

Fig. 12.13 Autoradiograph film of preloaded permanent prostate implant needles. The number of seeds in a given needle, as well as the seed spacing, can be readily visualized on the film

12.9.6
Implant Procedure

During a permanent prostate seed implant procedure, the patient is anesthetized and set up in a position reproducing that used during the volume study. All other aspects of the volume study are reproduced as well, including the use of Foley catheters. Radio-opaque contrast is injected into the bladder or the Foley balloon, whichever applicable, to help visualize the bladder neck. The positions of the prostate, urethra, and rectum as seen on the volume study ultrasound images are reproduced by proper adjustment of the ultrasound probe. Stabilization needles may be inserted to hold the prostate in position during the implant (FEYGELMAN et al. 1996; TASCHERAU et al. 2000). The seed implantation may then start, in the order that needles with the deeper insertion depths are implanted first. The ultrasound probes are therefore set to be at the base of the prostate first, retracted to the correct depth as the needle insertion depth decreases.

The transrectal ultrasound unit is used during the needle insertion to visualize the tip of the needle. For a needle inserted through a given hole on the template, its tip should appear at the corresponding grid position on the ultrasound image. Deviations larger than 2 mm typically require reinsertion of the needle. Many preloaded needles are designed with a beveled tip, with the opening of the tip marked on the hub of the needle. The bevel may be used to steer the needle, such that the needle travels toward the bevel opening direction, as can be identified by the mark on the needle hub. The tip of the needle is also visualized simultaneously on the fluoroscopy image, such that its position relative to the bladder neck is maintained. The person inserting the implant needles needs to take into consideration residue prostate motion, caused by the friction of needles with the prostate tissue, as well as the accuracy of seed deposition, typically at locations inferior to the intended ones. Needles that

are intended to deposit seeds at the base of the prostate are therefore often inserted to a distance past the intended position, then retracted a corresponding distance, before the seeds are deposited, either needle by needle in the preloaded needle technique, or one by one in the Mick-applicator technique. The sagittal view mode of the TRUS unit is often used to confirm deposition of seeds at the base of the prostate and to verify the spacing of seed positioning. The fluoroscopy unit should also be used to visualize the deposition of sources as it occurs, such that any seeds misplaced may be immediately identified, and additional seeds may be placed along the same needle track, if necessary.

When all planned seeds are implanted, a review of the seed positions, using both ultrasound and fluoroscopy X-ray images, is performed. Combined with intraoperative notes taken during the insertion of each implant needle, the regions of the prostate requiring additional seeds may be identified. While such evaluation of the needs for additional seeds is at best qualitative, it has proved to be useful in achieving acceptable dose distribution, as seen in post-implant dosimetry evaluations.

At the conclusion of the procedure, a cystoscopy is performed to identify and remove any seeds implanted into the patient's urethra or bladder. Patient radiation exposure and room radiation surveys are subsequently performed. The patient radiation survey, measuring the radiation exposure level at 1 m from the patient, confirms regulatory compliance for releasing the patient from the hospital without the need to hospitalize the patient due to radiation levels higher than 1 mrem/h at 1 m from the patient. The room radiation survey confirms absence of misplaced brachytherapy sources in the OR, so that the room may be released for housekeeping and preparation for next case. Results of these radiation surveys are recorded for regulatory compliance as well. Any seeds left unimplanted are brought back to the brachytherapy source room, and a final source count is performed to assure that all sources are accounted for.

12.9.7
Post-Implant Dosimetry

The dose distribution of a permanent prostate implant is evaluated by performing the post-implant dosimetry calculations, 2–4 weeks following the procedure. While planar films were the major modality of post-implant dosimetry imaging, they have been replaced by volumetric CT or MR imaging modalities, as it is possible to visualize the prostate and calculate its dose coverage with these latter modalities. Following the acquisition of the CT or MR images, the prostate is segmented, and the sources are identified. A set of CT or MR images of the patient's pelvic region is obtained, on which the prostate is segmented. The ^{125}I or ^{103}Pd sources are identified, using either manual or automatic methods (Brinkman and Kline 1998; Li et al. 2001; Tubic et al 2001; Liu et al. 2003; Tubic and Beaulieu 2005). Dose distribution of the implant is then calculated using the superposition principle as previously described.

The post-implant dosimetry of a permanent prostate implant provides an indication of the quality of the treatment, although its interpretation remains somewhat unclear. Outcome studies have shown that various dosimetric criteria, including D_{100}, D_{90}, and D_{80}, i.e., the dose covering 100, 90 and 80% of the prostate, respectively, may be useful in evaluation of the quality of a permanent prostate implant based on post-implant dosimetry studies. The ABS (Nag et al. 2000) recommends that the reporting of post-implant dosimetry for permanent prostate implant include the following:

1. The values of D_{100}, D_{90}, and D_{80}, (the dose that covers 100, 90, and 80% of the prostate, respectively)
2. The values of V_{200}, V_{150}, V_{100}, V_{90}, and V_{80}, (the fractional volume of the prostate that receives 200, 150, 100, 90, and 80% of the prescribed dose, respectively)
3. The total volume of the prostate (in cc) obtained from post-implant dosimetry
4. The number of days between implantation and the date of the imaging study used for dosimetric reconstruction
5. The urethral and rectal doses

It is likely that significant underdosage of the prostate, as identified from post-implant dosimetry, especially combined with pathological findings on the disease, may be used to guide remedial therapy, such as re-implantation of the underdosed region, or the addition of external-beam therapy.

The timing of performing post-implant dosimetry scans depends on the isotope used, as well as the speed at which the post-implant edema of the prostate resolves. The CT scans acquired for post-implant dosimetry, as well as the dose calculations based on these images, provide only a snapshot of a time-varying dose distribution, due to source decay,

in a spatially varying organ, due to the edema and its resolution. After a review of existing data on the effects of edema on post-implant dosimetry, the ABS concluded that the most reproducible dosimetric results would be obtained if post-implant CT scans for dosimetry were obtained at 1 month post-implant (Nag et al. 2000), although it acknowledged that this recommendation might not be practical for all patients, and that the clinical significance of having reproducible post-implant dosimetric results was not well understood.

The volume of the prostate segmented on CT scans should be compared with the prostate volume prior to treatment calculated from ultrasound images. It has been reported that prostate volumes from CT scans are often between 20 and 40% larger than those calculated from ultrasound images (Nag et al. 2000). Factors contributing to this large discrepancy may include the effect of prostate swelling post-implant, and the accidental inclusion of paraprostatic tissue into the prostate volume on CT scans; the latter can be alleviated by outlining the prostate volume to a few millimeters inside the higher-density visible prostate volume on CT scans.

12.10
High Dose Rate Remote-Afterloading Brachytherapy

High dose rate (HDR) remote-afterloading technology, using high activity ^{192}Ir sources and a computer-controlled mechanism to drive the source into a set of implanted applicators and catheters sequentially, has become well accepted for delivery of brachytherapy treatment for cancers of the cervix, breast, prostate, and other sites. The HDR treatments, compared with manually loaded brachytherapy treatments, offer the advantages of significantly lowered radiation risk to patient care personnel, because the radiation source is housed in a shielded source safely built into the remote afterloader unit located in a shielded treatment room, and is only driven out of the safe for patient treatment and machine quality assurance tests, while all patient care personnel are safely outside the treatment room. The radiation sources for HDR units are typically of small dimensions, an example of which is shown in Figure 12.14. During a treatment, the source is driven out of the HDR unit, and steps through pre-determined dwell positions within each treatment catheter, stopping at each dwell position for a pre-calculated length

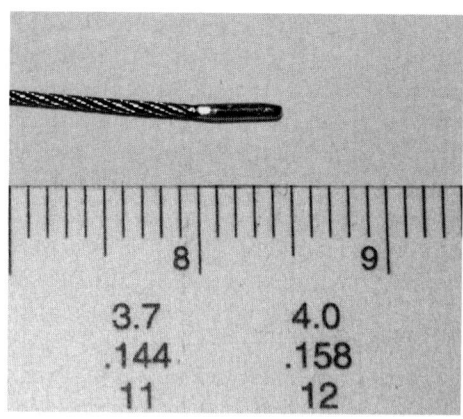

Fig. 12.14. An example of the ^{192}Ir source. The source capsule is welded to a flexible steel ribbon

of time, called dwell time, to deliver the planned treatment dose distribution. The HDR technology therefore offers significant flexibility in customizing the source dwell times at all dwell positions, equivalent to using a large number of sources of various strengths, to achieve optimized dose distribution for the treatment. In addition, various automatic optimization algorithms are available in all HDR treatment-planning systems that greatly facilitate the custom sculpting of dose distribution for a treatment (EDMUNDSON 1994; VAN DER LAARSE 1994; LESSARD and POULIOT 2001; MILICKOVIC et al. 2002; LAHANAS et al. 2003). It is noted, however, that while the flexibility of source dwell time and position optimization in HDR treatments, especially with the help of automatic optimization algorithms, provide unprecedented capabilities in sculpting dose distributions, they are fundamentally limited by the physical limits of applicator and catheter locations, and cannot be relied upon to correct applicator and catheter positioning errors. The use of optimization techniques to achieve acceptable dose distributions in an HDR treatment should be done with the utmost care, and the resulting dose distributions should be reviewed in detail for dose distribution homogeneity, especially for the presence of large-volume hot spots. Repositioning of the applicators and insertion of additional catheters are sometimes the only remedy for an implant to achieve acceptable dose distributions.

The widespread use of HDR technology in brachytherapy treatments has also been limited by the relative lack of clinical experiences with the increased biological effects of radiation dose delivery at higher than conventional dose rates (BRENNER and HALL 1991; ORTON 2001). It is therefore important that

institutional and/or national treatment guidelines be followed for the implementation of a new HDR treatment technique so as to assure its safety and efficacy.

12.10.1
Equipment and Operating Principles

A typically HDR remote afterloading unit comprises a high-activity (up to 12 Ci) ^{192}Ir source, a shielded source safe, a stepping motor to drive the source out of the safe to positions in applicators at sub-millimeter accuracy for pre-calculated dwell times, an emergency motor, and a backup battery for retraction of the source into the shielded safe in emergencies, even in the presence of a power failure. A check cable, driven by a separate motor, is usually included for testing the integrity of the applicator/catheters and their secure connection to the machine, prior to driving the active source out of the source safe. Multiple applicators and catheters, depending on the machine design, may be connected to the indexer of an HDR unit, which automatically selects the current channel of source travel based on the treatment plan. Transfer tubes of various designs specific to the applicator/catheter in use connect to the indexer and serve as source conduits between the HDR unit and the applicators. Multiple safety interlocks are integrated into the system to perform additional tests on source travel speed and position accuracy, and retract the source into the source safe in the event of unexpected source travel speed and positions, to prevent the source from being stuck within damaged applicators and catheters.

An HDR system usually is equipped with its own treatment-planning system, such that customized source dwell-time calculation algorithms can be used, and to allow the calculated source dwell positions and times to be digitally transferred to the treatment delivery unit. While a conventional brachytherapy treatment planning system may be modified to calculate the dose distributions from given HDR source dwell positions and dwell times, they usually lack the capabilities of automatic source dwell position and dwell-time optimization and digital interface with the HDR delivery unit, and are therefore unable to fully take advantage of the flexibilities afforded by a modern HDR-specific treatment-planning systems.

Because of the high-activity source used, HDR units must be housed within a shielded treatment room, such that the radiation exposure outside the treatment room during source-out sessions are below regulatory limits. The HDR treatment rooms may be either dedicated for HDR treatments or shared with an external-beam radiation treatment room. In the former case, the HDR treatment room may be designed to allow outpatient surgical procedures to further improve the efficiency of applicator insertion, simulation, and treatment delivery, by installation of appropriate surgical and imaging equipment.

12.10.2
Clinical Application of HDR Brachytherapy: Interstitial Accelerated Partial Breast Irradiation

Accelerated partial breast irradiation (APBI) has recently gained significant acceptance for the treatment of early-stage localized breast cancer (VICINI and ARTHUR 2005). This treatment modality delivers a total dose of 32–40 Gy to the treatment target in a breast cancer patient following resection of the tumor by lumpectomy, in no more than ten fractions delivered over 4–5 days, significantly reducing the number of days in a treatment course as required by a conventional external-beam whole-breast irradiation, which requires typically 6 weeks. In addition, APBI delivers a prescription dose to a target within the breast that may be significantly reduced from the conventional target of the entire breast, thereby offering the advantage of reduced normal tissue and critical organ doses. The techniques used to deliver APBI include conformal external-beam irradiation (VICINI et al. 2003), intracavitary brachytherapy irradiation using an implanted inflatable balloon (KEISCH 2005), and multi-catheter interstitial brachytherapy (KUSKE et al. 1994). An overview of interstitial APBI procedures practiced at Washington University is presented as follows:

1. Dosimetric goals of interstitial APBI: At the present time, there has been a lack of nationally agreed-upon dosimetric end points of interstitial APBI. Such goals may include parameters on percent target (clinical target volume) coverage by prescription dose; skin dose tolerance; dose distribution homogeneity as represented by the dose homogeneity index (DHI), or the ratio of prescription dose to the mean central dose, as defined by ICRU report 58; dose conformality index, defined as the ratio of total volume of tissue receiving prescription dose to that of the CTV; and volumes of high-dose regions of the entire implant, such as the volume receiving 150 and 200% of the prescription dose (V_{150} and V_{200} of the treated volume). A commonly used set of dosimetric goals, as described by an ongoing clinical trial by the NSABP and the RTOG, include the following (NSABP/RTOG, 2005): (a) D_{100} of the CTV, or percent of CTV volume receiving prescription dose, larger than 90%; (b) V_{150} of treated tissue less than 70 cc; (c) V_{200} of treated tissue less than 20 cc; and (d) a volume dose inhomogeneity index (DHI), defined by $(1 - V_{150}/V_{100})$, larger than 0.75.

   In addition, the DHI of the radio of prescription dose and the mean central dose, as defined by the ICRU report 58, should be reviewed and maintained to be larger than 0.70. The diameters of volumes receiving 200% of prescription dose should be evaluated and maintained to less than 1 cm in diameter, following the Paris Implant System.

2. Equipment. Single-leader flexible catheters of 1.9-mm outer diameter are used in interstitial APBI. Many such catheters are available from different vendors. The catheters need to be up to 30 cm long to allow insertion of catheters to the deep aspect of the breast near the chest wall. As CT imaging is used for interstitial APBI at Washington University, CT-compatible buttons that are non-radio-opaque are used. One button is fixed to the non-leader end of the catheter, while one additional button per catheter is available for fixation of the catheter at completion of the procedure. The buttons are designed to fit the catheters snugly, and do not require suturing to skin for fixation. Stainless steel needles of 17-G diameter are used for introduction of the catheters. The catheters, buttons, and needles are sterilized before implantation. A set of CT-compatible radiographic markers are used in CT imaging. Compared with conventional high-density X-ray markers, the CT-compatible markers are made of aluminum so as to minimize the imaging artifacts and improve source-localization accuracy. Figure 12.15 shows a set of such catheters, buttons, CT-compatible radiographic markers, and needles.

3. Catheter implantation. The implantation of catheters for interstitial APBI is guided by X-ray images or ultrasound images. At the University of Wisconsin, a template custom manufactured for APBI is for use together with a stereotactic mammography unit for X-ray-guided catheter implantation (DAS et al. 2004). Alternatively, a free-ultrasound-guided technique is used for

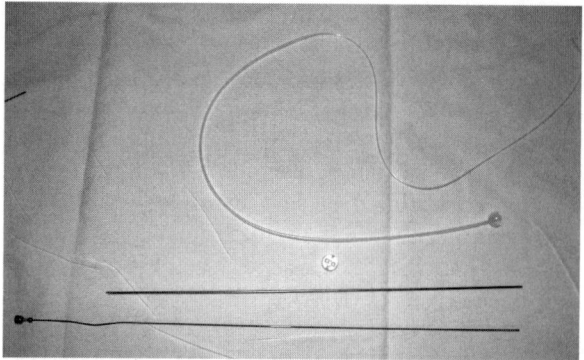

Fig. 12.15 Catheter, buttons, needle, and CT-compatible radiographic marker used for HDR interstitial accelerated breast irradiation

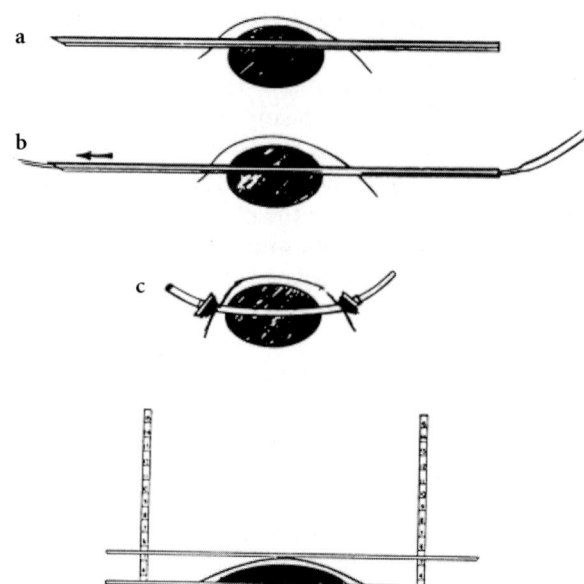

Fig. 12.16 a–d Technique used to insert flexible catheters for interstitial accelerated breast irradiation. (From HILARIS et al.1988)

catheter implantation. The lumpectomy cavity is identified via the ultrasound imaging, with its boundaries outlined on patient skin. Catheters arranged in multiple planes, at 1.0- to 1.5-cm catheter spacing and up to 2.0-cm plane spacing, are planned for the implant, depending on the dimensions of the CTV, defined to be the lumpectomy cavity plus a 2.0-cm margin in all directions except when the CTV approaches the skin and the pectoral muscle. In the latter cases, the CTV is drawn to leave a 5-mm margin from skin surface, and to not include the pectoral muscle. The planned catheter locations are thus selected to cover the entire CTV. The orientations of the catheters are selected such that the implanted catheters will be minimally interfering with patient's normal motions, and to allow easy catheter wrapping, between treatment fractions.

Under ultrasound guidance, the stainless steel needles are inserted following the entrance and exit points marked on skin. The leader of the single-leader catheters are then inserted through the hollow needles, and pulled through the breast tissue, bringing the needles out of the tissue with them, as shown in Figure 12.16. The catheters are thus implanted plane by plane, as visualized on ultrasound imaging, until all planned catheters are implanted. An additional snug-fit CT-compatible button is subsequently threaded onto each catheter and pushed against skin to prevent their motion. At the end of the procedure, the catheters are wrapped in gauze for their protection.

4. CT Simulation. The CT imaging simulation for interstitial APBI is typically performed on the day following catheter implantation so as to allow resolution of tissue swelling due to the implantation. Prior to the imaging session, the catheters

are cleaned and labeled with numbered tags. They are cut open using wire strippers so as to withdraw the plastic core stiffeners inside the catheters and to allow insertion of the CT-compatible radiographic markers. When cutting the catheters, it is important to leave adequate lengths of the catheters outside of skin, both to allow connection of HDR transfer tubes and for easier wrapping of the catheters between treatment fractions. The total lengths of each catheter-transfer tube combination are measured, for example, using a source simulator as shown in Figure 12.17. The measured lengths are recorded for use during treatment planning. Digital photographs are taken of the catheters, preferably in both anteroposterior (AP) and lateral directions, to be used for guiding catheter reconstruction during treatment planning, as shown in Figure 12.18. In addition, drawings of the AP and lateral views of the catheters are obtained as backups to the digital photographs. The snugness of catheter buttons and their proximity to skin surface are confirmed, and radiographic contrast solution are injected into the lumpectomy cavity as necessary to help visualization of the cavity on CT images. CT-compatible radiographic markers are inserted into the catheters, making sure that the markers are inserted to the ends of the catheters.

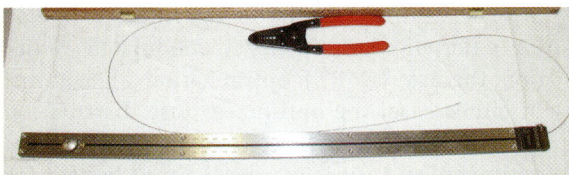

Fig. 12.17 Source simulator used to measure total lengths of catheter and transfer tube combinations for interstitial HDR brachytherapy, together with a wire-stripper used to cut flexible catheters open without cutting their inner stiffening cores

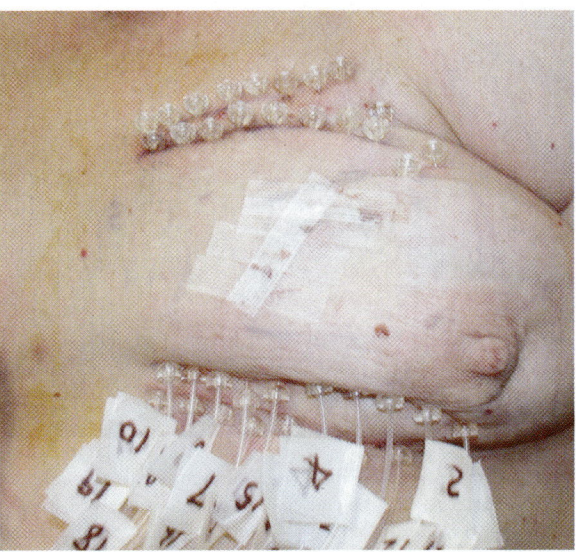

Fig. 12.18 Digital photograph of interstitial accelerated partial breast irradiation implants (APBI) for use during catheter localization in treatment planning

During CT imaging for simulation of the interstitial APBI treatment, the patient is placed on the CT table, sometimes at slightly tilted orientation relative to the table motion direction, so as to minimize the occurrences of catheters falling parallel with the CT scan plane for ease of catheter localization and reduce the potential positional errors caused by catheters falling in between the CT scan slices. A CT slice thickness of 2 mm is chosen to further reduce such potential errors. The CT scan is planned with 2- to 3-cm margins in the superior and inferior directions of the catheters and are acquired with a whole-body FOV. The patient is instructed to breathe shallowly to minimize motion artifacts, and the images are reviewed as they are acquired to confirm the absence of significant breathing motion artifacts. The CT images are reviewed at the end of simulation before the patient is taken off the table, and the images transferred to the treatment-planning system for contouring and planning. The inner stiffening cores of the catheters are re-inserted into the catheters for their protection before treatment starts. Occasionally, larger than desired catheter separations or potentially inadequate catheter coverage of the CTV are observed on the CT images. In such cases, additional catheters are inserted and CT images are re-acquired for final treatment planning.

5. Treatment planning for interstitial APBI. Once the CT images are available, the gross target volume (GTV) can be segmented. For APBI, the GTV includes the entire lumpectomy cavity, as represented by the seroma or contrast solution if available. The clinical target volume (CTV) is then defined based on the GTV, by expanding the GTV with a 2-cm margin. The automatically expanded CTV is revised to maintain a 5-mm distance between the CTV and the skin surface, if possible, except where the GTV is closer to the skin. In addition, the initial CTV is revised so as to not include the pectoral muscle.

The catheters are identified and reconstructed on the CT images, by tracing the CT-compatible radiographic markers, either starting from the tip or the connector ends of the catheters, although it is crucial that a consistent tracing pattern be maintained across all catheters. The numbering of the catheters in the treatment-planning system must be identical to those on the digital photographs and the hand-prepared drawings. As catheter reconstruction proceeds, the reconstructed catheters should be reviewed periodically to confirm that they appear smooth, without sudden direction changes between slices, as these indicate significant breathing motion artifacts, and may require re-acquisition of CT scans. A three-dimensional view of the catheters is often helpful in this confirmation, as shown in Figure 12.19. The indexer lengths, as measured using the source simulator device, are entered for each catheter.

The dosimetry planning of interstitial APBI at Washington University starts with selection of active dwell positions and placement of dose normalization points. The active dwell positions are chosen such that, together, they enclose the CTV with a 1- to 2-cm margin, either via manual or automatic activation. Similar to the Paris Implant System and following ICRU report 58 recommendations, central dose points, located in a central plane perpendicular to the catheters and geomet-

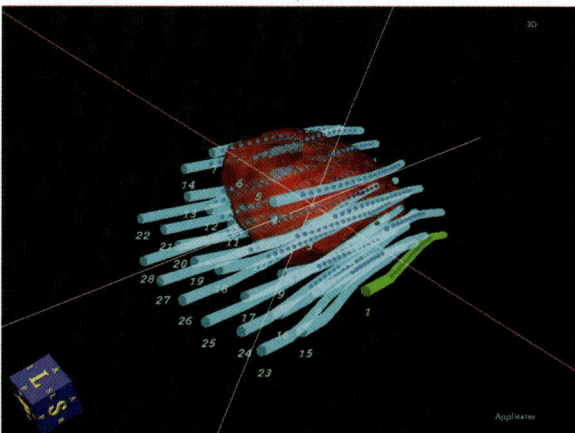

Fig. 12.19 Three-dimensional view of reconstructed catheters in interstitial APBI. The catheters should be reviewed for accuracy in numbering and for absence of sudden directional changes between CT slices to rule out breathing motion artifacts

rically equidistant to neighboring catheters, are used for normalization of the dose distribution, with the prescribed dose as a percentage of the mean central dose, i.e., the average dose values at the central dose points. The catheters are rotated on the treatment-planning system to allow identification of a central catheter, and the central dose points are selected manually, as shown in Fig. 12.20.

The dose optimization for interstitial APBI is done by first calculating a geometrically optimized dose distribution to optimize over the entire implant (EDMUNDSON 1994). This optimization algorithm assigns dwell times for all source dwell positions based on their respective distances to each other, such that the a dwell position at larger distances from other dwell positions receive larger dwell times, to compensate for the reduced dose contribution from the other dwell positions. While use of this automatic optimization algorithm does not assure adequate target coverage and minimal skin dose, it has served as an excellent first guess as to whether adequate dose distribution homogeneity is possible. The dose distribution homogeneity parameters, such as volume of high-dose (150 and 200% of prescription dose) regions, should be calculated for later comparison. Further manual and automatic optimization of the dose distributions may then proceed, using the tools available in a given treatment-planning system, such as graphic optimization, which allows manual dragging and dropping of isodose lines to desired locations, and

inverse planning, similar to algorithms available for external-beam intensity modulated radiation therapy (IMRT) optimization algorithms. In either case, the optimized dose distribution must be carefully reviewed, especially if manual optimization is used, as dwell time changes for optimized dose distribution locally may create unexpected dose distribution inhomogeneity in other parts of the implant. This optimization process is repeated until the dosimetric goals of the implant are met. The completed plan is then reviewed by an authorized medical physicist and the authorized treating physician, using tools including natural, differential, and cumulative dose volume histograms (LI 2005), as shown in Figure 12.21, before the plan parameters are transferred to the HDR treatment unit.

6. Pretreatment quality assurance. The goals of pretreatment quality assurance tests for interstitial APBI should include both a dosimetric evaluation of the accuracy of computerized treatment plan, and its geometric accuracy. An independent manual calculation of the relation between the target volume and the total exposure, expressed either in milligram–hours or Ci–seconds, i.e., the product of the source strength and the total treatment time, may be calculated and compared with the values obtained from traditional implant systems such as the Paterson-Parker system (DAS et al. 2004). Significant deviations between the TPS calculated total exposure and that predicted by the traditional implant system should be evaluated before the plan is deemed acceptable for treatment.

The geometric accuracy of the treatment plan is crucial in the delivery of interstitial APBI treatments. Aspects affecting a treatment plan's geometric accuracy include the accuracy of the indexer length settings, the dwell position step sizes (distance between dwell positions), the total active length of each catheter, and the correct correspondence of numbering between planned and implanted catheters. A good tool for a comprehensive evaluation of the geometric accuracy of an interstitial HDR treatment plan is the autoradiograph, in which all the treated catheters are connected to a film holder, and the film is irradiated using the treatment plan geometric parameters, leaving a film with high-density spots indicating the source dwell positions. The film is indexed such that the indexer lengths of all catheters can be measured and compared with the treatment plan to ensure that they are error-

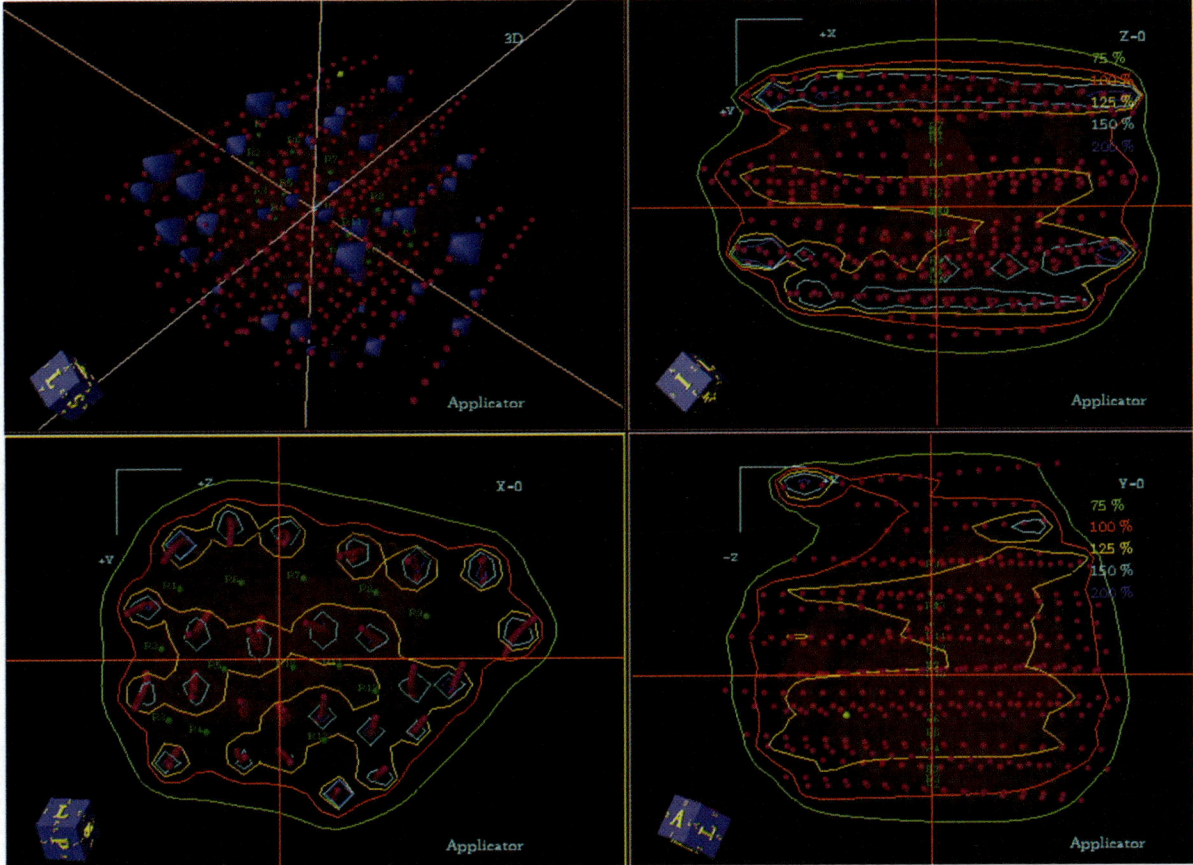

Fig. 12.20 Orthogonal views of the implanted catheters. The transverse view (*lower left*) allows selection of central dose points (*green dots*) for plan normalization. A 3D display of high-dose volumes (200% prescription dose) allows quick identification of unacceptably large hot spots

free. In addition, the total active lengths, as well as dwell position step sizes, may be measured from the autoradiographs. As geometric errors are among the most prevalent type, and perhaps the most devastating, errors in HDR brachytherapy, every effort should be made to ensure that a treatment plan be free of such errors before treatment starts.

7. Treatment delivery. The delivery of HDR interstitial ABPI treatments are typically done in a twice-a-day fractionation scheme. Prior to the initiation of each treatment, the implant should reviewed to ensure that the buttons be pushed against skin, as loose buttons may allow catheter longitudinal movements and result in geometric errors of treatment delivery. The connection of catheters to the treatment unit should be done in a systematic manner and under peer review, to prevent catheter misconnections. The numbering of the catheters should be reviewed, and replaced if necessary, using the digital photographs as guides. For treatments using a large number of catheters that exceeds the number of channels available on the treatment unit, the treatment may be divided into multiple parts to be delivered sequentially. A procedure should be devised to ensure that the correct parts are recalled from computer memory for the treatments.

12.11
Radiation Protection and Regulatory Compliance in Brachytherapy

Significant radiation protection issues exist with the clinical practice of brachytherapy due to its use of radioactive materials, which, unlike electrically activated radiation machines, cannot be turned off at will. Radioactive materials implanted in patients produce radiation exposures around the patient that must be limited, so as to avoid high radiation

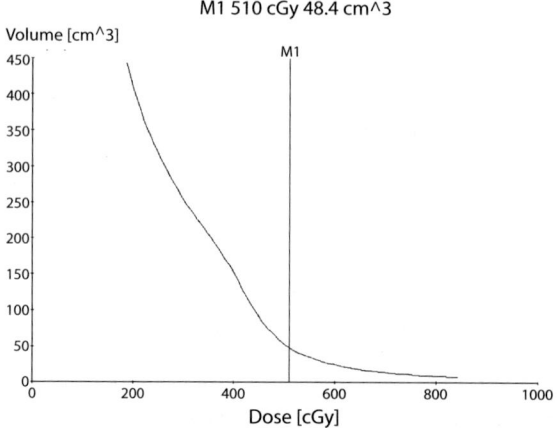

Prescription dose: 340 cGy   Natural prescr. dose: 345 cGy
Peak dose: 410 cGy   Quality index: 2.08
Low dose: 387 cGy   Uniformity index: 1.70
High dose: 517 cGy   Natural dose ratio: 1.01

dV/du [*10^6 cm^3/cGy^-1.5]

$u = \bar{D}^{-1.5}$

a   DVH_2: Natural DVH on implant. State: Consistent.

Volume gradient ratio: 1.27   M1: 500 cGy 554 mm^3/cGy

dV/dD [mm^3/cGy]

Dose [cGy]

b   DVH_2: Differential DVH on implant. State: Consistent.

M1 510 cGy 48.4 cm^3

Volume [cm^3]

Dose [cGy]

c   DVH_2: Cumulative DVH on implant. State: Consistent.

Fig. 12.21a–c Natural, differential, and cumulative dose-volume histograms (*DVH*) of an interstitial APBI HDR treatment plan. **a** Natural DVH. **b** Differential DVH. **c** Cumulative DVH. (From LI 2005)

exposures to the public and to patient care personnel. The sources need to be under strict control to prevent their misplacement and subsequent accident radiation exposure. Significant errors in dose calculation and deviations in treatment delivery from prescription and treatment plan may cause irreparable harm to the patient. For these reasons, the use of brachytherapy sources is under strict government regulatory control.

In the United States, the use of nuclear reactor by-product radioactive materials falls under the jurisdiction of the Nuclear Regulatory Commission (NRC), created by the Congress in 1954 under the Atomic Energy Act. Regulation of non-reactor-produced radioactive materials is governed by state agencies, although the brachytherapy application of such materials is limited to, in most cases, accelerator-produced isotopes such as [103]Pd seeds. The NRC has entered into agreements with state agencies to delegate its authority to the individual states, such states called the "agreement states." The agreement states exercise their authorities in compliance with NRC regulations, although such states often create regulations that apply uniformly to all therapeutic radioactive materials. Title 10, part 35 of the Code of Federal Regulations (10CFR35), available from the NRC, details the federal regulation requirements on the medical use of by-product radioactive materials, including brachytherapy sources.

12.11.1
Licensing, Authorization, and Report of Medical Events

A radioactive materials license is required for the possession of brachytherapy sources. This license is granted by the NRC or the agreement states after the applicant institution has established radiation protection procedures that satisfy requirements of the NRC or the agreement states. Depending on the type of license being applied for, i.e., general scope or limited scope, the regulatory requirements include identification of a radiation safety officer or the establishment of a radiation safety committee, with members including representatives from institution administration, authorized users, and other interested parties in the institution's radiation protection program. The radioactive isotopes, the maximum amounts, and their clinical uses are specified in the license.

The NRC and the agreement states require that all brachytherapy treatments be performed under the supervision of an authorized user for the specific modality of brachytherapy treatment, including general manually loaded brachytherapy, as well as low dose rate and high dose rate remotely afterloading brachytherapy. Furthermore, the technical aspects, such as source strength calibration and machine quality assurance review, must be done under the supervision of authorized medical physicists. The NRC specifies training requirements for authorized users and authorized medical physicists, depending on the modality of authorization being applied for, including didactic classroom training and practical clinical training. Delivery of brachytherapy treatments without the supervision of an authorized user or an authorized medical physicist, whichever applicable, constitutes violation of federal or state regulations, and may result in investigations by these agencies and subsequent civil penalties, up to revocation of the institution's license to provide brachytherapy treatment.

The regulations consider it a significant error in dose calculation or delivery in a brachytherapy treatment when, for example, there is a deviation of delivered dose by larger than 20% from the prescribed dose. This is termed a medical event. Detailed definitions of medical events are listed in 10CFR35. When a medical event occurs, the regulations require reporting to the appropriate governing bodies, the NRC or the appropriate agreement state agency, as well as the patient or the patient's referring physician, via telephone, within 24 h of the identification of the event. A written report is required within 15 days of the identification of such events. The occurrence of a medical event may trigger an investigational visit by the regulatory agency, during which the root cause of, as well as the proposed corrective actions for, the medical event are reviewed. Such an investigational visit often reviews the entire relevant radiation protection program at the institution, including, for example, all modalities of brachytherapy treatments, in order to identify other potential weaknesses of the program.

NRC permits release of patients with permanently implanted brachytherapy sources, as long as the radiation exposure levels from the sources near the patients are sufficiently low. For prostate implants, radiation exposure levels at less than 1 mrem/h at 1 m from the patient is sufficient for release of the patient, with written instructions given to the patient on how to minimize radiation exposures to his family members. Such written discharge instructions typically explain the radiation exposure risk associated with the sources implanted in the patient, and ask the patient to take precautions to further minimize such risks, such as sleeping in a separate bed from his spouse for a given period of time; keeping a safe distance (>1 m) from pregnant women and from children, except for short periods (e.g., 30 min) per day of close contact; and handling of discharged sources, i.e., seeds that may have migrated out of the prostate and into the bladder, and subsequently discharged in urine.

All brachytherapy sources for permanent prostate implants are encapsulated in thin titanium shells, making them easily damaged. Damaged ^{125}I seeds are of particular radiation protection concern, as they may release radio-iodine into the environment. Such seed sources must be handled with care, using tools such as reverse-action tweezers instead of sharp objects. For seeds embedded in absorbent sutures, such as the RapidStrand, extreme care must be exercised to avoid damaging the seeds when cutting the seed sutures.

With the heightened security measures in recent years, radiation monitors have been installed in transportation checkpoints and public access areas. The radioactive materials implanted in patients receiving permanent prostate implant have the potential to activate such monitors. It is therefore advisable that the patient be given a card, listing the procedure performed on the patient, the radioisotope, the total activity administered, its half-life, and contact telephone number of the radiation therapy center.

12.11.2
Radiation Safety Concerns in Permanent Implants

For permanent brachytherapy treatments, a primary concern is the radiation exposure produced by the radioactive sources that remain in the patient. As it is often impractical to keep such patients hospitalized under strict radiation-protection control, the

12.11.3
Safety and Regulatory Issues in HDR Brachytherapy

The HDR brachytherapy treatments, due to its use of high-activity radiation sources, are under even greater regulatory oversight by the NRC and the agreement states. The regulations require detailed daily and periodic quality assurance tests of the treatment unit performance, as well as associated radiation safety control interlocks and equipment. Authorized users, authorized medical physicists, as well as treatment unit operators are required to undergo annual radiation safety in-service training, which includes aspects of the HDR source characteristics, machine operating principles, and emergency procedures in cases where the source is stuck inside a treatment applicator or catheter. The records of such annual training must be maintained. The treatment-planning system must be tested for its functioning and its dose-calculation accuracy, and an independent calculation check of a treatment plan must be performed for each patient planned on the treatment-planning system. At the initiation of an HDR treatment, the physical presence of an authorized user, as well as an authorized medical physicist, is required. Once the treatment has started, the authorized user may be replaced by a physician trained in the emergency removal of the treatment applicators.

References

Antipas V, Dale RG, Coles IP (2001) A theoretical investigation into the role of tumour radiosensitivity, clonogen repopulation, tumour shrinkage and radionuclide RBE in permanent brachytherapy implants of 125I and 103Pd. Phys Med Biol 46:2557–2569

Brenner DJ, Hall EJ (1991) Fractionated high dose rate versus low dose rate regimens for intracavitary brachytherapy of the cervix. I. General considerations based on radiobiology. Br J Radiol 64:133–141

Brenner DJ, Hall EJ, Randers-Pehrson G, Huang Y, Johnson GW, Miller RW, Wu B, Vazquez ME, Medvedovsky C, Worgul BV (1996) Quantitative comparisons of continuous and pulsed low dose rate regimens in a model late-effect system. Int J Radiat Oncol Biol Phys 34:905–910

Brenner DJ, Schiff PB, Huang Y, Hall EJ (1997) Pulsed-dose-rate brachytherapy: design of convenient (daytime-only) schedules. Int J Radiat Oncol Biol Phys 39:809–815

Brinkmann DH, Kline RW (1998) Automated seed localization from CT datasets of the prostate. Med Phys 25:1667–1672

Chen CZ, Huang Y, Hall EJ, Brenner DJ (1997) Pulsed brachytherapy as a substitute for continuous low dose rate: an in vitro study with human carcinoma cells. Int J Radiat Oncol Biol Phys 37:137–143

Chen Y, Boyer AL, Xing L (2000) A dose-volume histogram based optimization algorithm for ultrasound guided prostate implants. Med Phys 27:2286–2292

Chibani O, Williamson JF, Todor D (2005) Dosimetric effects of seed anisotropy and interseed attenuation for 103Pd and 125I prostate implants. Med Phys 32:2557–2566

Cohen GN, Amols HI, Zelefsky MJ, Zaider M (2002) The Anderson nomograms for permanent interstitial prostate implants: a briefing for practitioners. Int J Radiat Oncol Biol Phys 53:504–511

Crusinberry RA, Kramolowsky EV, Loening SA (1985) Percutaneous transperineal placement of gold 198 seeds for treatment of carcinoma of the prostate. Prostate 11:59–67

Das RK, Mishra V, Perera H, Meigooni AS, Williamson JF (1995) A secondary air kerma strength standard for Yb-169 interstitial brachytherapy sources. Phys Med Biol 40:741–756

Das RK, Patel R, Shah H, Odau H, Kuske RR (2004) 3D CT-based high-dose-rate breast brachytherapy implants: treatment planning and quality assurance. Int J Radiat Oncol Biol Phys 59:1224–1228

Dawson JE, Wu T, Roy T, Gu JY, Kim H (1994) Dose effects of seeds placement deviations from pre-planned positions in ultrasound guided prostate implants. Radiother Oncol 32:268–270

DeMarco JJ, Smathers JB, Burnison CM, Ncube QK, Solberg TD (1999) CT-based dosimetry calculations for 125I prostate implants. Int J Radiat Oncol Biol Phys 45:1347–1353

D'Souza WD, Meyer RR, Thomadsen BR, Ferris MC (2001) An iterative sequential mixed-integer approach to automated prostate brachytherapy treatment plan optimization. Phys Med Biol 46:297–322

Edmundson GK (1994) Geometry-based optimization: an American viewpoint. In: Mould RJ, Batterman J (eds) Brachytherapy: from radium to optimization. Nucletron, Columbia, Maryland, pp 314–318

Edmundson GK, Yan D, Martinez AA (1995) Intraoperative optimization of needle placement and dwell times for conformal prostate brachytherapy. Int J Radiat Oncol Biol Phys 33:1257–1263

Feygelman V, Friedland JL, Sanders RM, Noriega BK, Pow-Sang JM (1996) Improvement in dosimetry of ultrasound-guided prostate implants with the use of multiple stabilization needles. Med Dosim 21:109–112

Gallagher RJ, Lee EK (1997) Mixed integer programming optimization models for brachytherapy treatment planning. Proc AMIA Annu Fall Symp, pp 278–282

Hilaris BS, Nori D, Anderson LL (eds) (1988) An atlas of brachytherapy. Macmillan, New York

Hochstetler JA, Kreder KJ, Brown CK, Loening SA (1995) Survival of patients with localized prostate cancer treated with percutaneous transperineal placement of radioactive gold seeds: stages A2, B, and C. Prostate 26:316–324

Holm HH (1997) The history of interstitial brachytherapy of prostatic cancer. Semin Surg Oncol 13:431–437

International Commission on Radiation Units and Measurements (ICRU) (1985) Report 38: Dose and volume specification for reporting intracavitary therapy in ynecology.

International Commission on Radiation Units and Measurements (ICRU) (1993) Report 50: Prescribing, recording, and reporting photon beam therapy

International Commission on Radiation Units and Measurements (ICRU) (1998) Report 58: Dose and volume specification for reporting interstitial therapy

Karaiskos P, Angelopoulos A, Baras P, Rozaki-Mavrouli H, Sandilos P, Vlachos L, Sakelliou L (2000) Dose rate calculations around 192Ir brachytherapy sources using a Sievert integration model. Phys Med Biol 45:383–398

Keisch M (2005) MammoSite. Expert Rev Med Devices 2:387–389

Kini VR, Edmundson GK, Vicini FA, Jaffray DA, Gustafson G, Martinez AA (1999) Use of three-dimensional radiation therapy planning tools and intraoperative ultrasound to evaluate high dose rate prostate brachytherapy implants. Int J Radiat Oncol Biol Phys 43:571–578

Kuske R, Bolton J, Wilenzick R et al. (1994) Brachytherapy as the sole method of breast irradiation in Tis, T1, T2, N0,1 breast cancer [Abstract]. Int J Radiat Oncol Biol Phys 30 (Suppl):245

Kutcher GJ, Coia L, Gillin M, Hanson WF, Leibel S, Morton RJ, Palta JR, Purdy JA, Reinstein LE, Svensson GK et al. (1994) Comprehensive QA for radiation oncology: report of AAPM Radiation Therapy Committee Task Group 40. Med Phys 21:581–618

Lahanas M, Baltas D, Giannouli S (2003) Global convergence analysis of fast multiobjective gradient-based dose optimization algorithms for high-dose-rate brachytherapy. Phys Med Biol 48:599–617

Lee EK, Zaider M (2003) Intraoperative dynamic dose optimization in permanent prostate implants. Int J Radiat Oncol Biol Phys 56:854–861

Lessard E, Pouliot J (2001) Inverse planning anatomy-based dose optimization for HDR-brachytherapy of the prostate using fast simulated annealing algorithm and dedicated objective function. Med Phys 28:773–779

Li Z (2005) QA review of brachytherapy treatment plans. In: Thomadsen BR, Rivard MJ, Butler WM (eds) Brachytherapy physics, 2nd edn. Medical Physics Publishing, Madison, Wisconsin

Li Z, Nalcacioglu IA, Ranka S, Sahni SK, Palta JR, Tome W, Kim S (2001) An algorithm for automatic, computed-tomography-based source localization after prostate implant. Med Phys 28:1410–1415

Lin LL, Mutic S, Malyapa RS, Low DA, Miller TR, Vicic M, Laforest R, Zoberi I, Grigsby PW (2005) Sequential FDG-PET brachytherapy treatment planning in carcinoma of the cervix. Int J Radiat Oncol Biol Phys 63:1494–1501

Ling CC (1992) Permanent implants using Au-198, Pd-103 and I-125: radiobiological considerations based on the linear quadratic model. Int J Radiat Oncol Biol Phys 23:81–87

Ling CC, Li WX, Anderson LL (1995) The relative biological effectiveness of I-125 and Pd-103. Int J Radiat Oncol Biol Phys 32:373–378

Lindsay P, Battista J, Van Dyk J (2001) The effect of seed anisotrophy on brachytherapy dose distributions using 125I and 103Pd. Med Phys 28:336–345

Liu H, Cheng G, Yu Y, Brasacchio R, Rubens D, Strang J, Liao L, Messing E (2003) Automatic localization of implanted seeds from post-implant CT images. Phys Med Biol 48:1191–1203

Martel MK, Narayana V (1998) Brachytherapy for the next century: use of image-based treatment planning. Radiat Res 150 (5 Suppl):S178–S188

Mason DL, Battista JJ, Barnett RB, Porter AT (1992) Ytterbium-169: calculated physical properties of a new radiation source for brachytherapy. Med Phys 19:695–703

Matzkin H, Kaver I, Bramante-Schreiber L, Agai R, Merimsky O, Inbar M (2003) Comparison between two iodine-125 brachytherapy implant techniques: pre-planning and intra-operative by various dosimetry quality indicators. Radiother Oncol 68:289–294

Meigooni AS, Meli JA, Nath R (1992) Interseed effects on dose for 125I brachytherapy implants. Med Phys 19:385–390

Milickovic N, Lahanas M, Papagiannopoulo M, Zamboglou N, Baltas D (2002) Multiobjective anatomy-based dose optimization for HDR-brachytherapy with constraint free deterministic algorithms. Phys Med Biol 47:2263–2280

Murphy MK, Piper RK, Greenwood LR, Mitch MG, Lamperti PJ, Seltzer SM, Bales MJ, Phillips MH (2004) Evaluation of the new cesium-131 seed for use in low-energy X-ray brachytherapy. Med Phys 31:1529–1538

Mutic S, Grigsby PW, Low DA, Dempsey JF, Harms WB, Laforest R, Bosch WR, Miller TR (2002) PET-guided three-dimensional treatment planning of intracavitary gynecologic implants. Int J Radiat Oncol Biol Phys 52:1104–1110

Nag S, Beyer D, Friedland J, Grimm P, Nath R (1999) American Brachytherapy Society (ABS) recommendations for transperineal permanent brachytherapy of prostate cancer. Int J Radiat Oncol Biol Phys 44:789–799

Nag S, Bice W, DeWyngaert K, Prestidge B, Stock R, Yu Y (2000) The American Brachytherapy Society recommendations for permanent prostate brachytherapy postimplant dosimetric analysis. Int J Radiat Oncol Biol Phys 46:221–230

Nag S, Ciezki JP, Cormack R, Doggett S, DeWyngaert K, Edmundson GK, Stock RG, Stone NN, Yu Y, Zelefsky MJ (2001) Intraoperative planning and evaluation of permanent prostate brachytherapy: report of the American Brachytherapy Society. Int J Radiat Oncol Biol Phys 51:1422–1430

Nag S, Cardenes H, Chang S, Das IJ, Erickson B, Ibbott GS, Lowenstein J, Roll J, Thomadsen B, Varia M (2004) Proposed guidelines for image-based intracavitary brachytherapy for cervical carcinoma: report from Image-Guided Brachytherapy Working Group. Int J Radiat Oncol Biol Phys 60:1160–1172

Nath R, Anderson LL, Luxton G, Weaver KA, Williamson JF, Meigooni AS (1995) Dosimetry of interstitial brachytherapy sources: recommendations of the AAPM Radiation Therapy Committee Task Group no. 43. American Association of Physicists in Medicine. Med Phys 22:209–234

Nath R, Anderson LL, Meli JA, Olch AJ, Stitt JA, Williamson JF (1997) Code of practice for brachytherapy physics: report of the AAPM Radiation Therapy Committee Task Group no. 56. American Association of Physicists in Medicine. Med Phys 24:1557–1598

Nath S, Chen Z, Yue N, Trumpore S, Peschel R (2000) Dosimetric effects of needle divergence in prostate seed implant using 125I and 103Pd radioactive seeds. Med Phys 27:1058–1066

Nath R, Bongiorni P, Chen Z, Gragnano J, Rockwell S (2005) Relative biological effectiveness of 103Pd and 125I photons for continuous low-dose-rate irradiation of Chinese hamster cells. Radiat Res 163:501–509

National Surgical Adjuvant Breast and Bowel Project (NSABP) B-39/Radiation Therapy Oncology Group (RTOG) 0413 (2005) A randomized phase III study of conventional whole breast irradiation (WBI) versus partial breast irradiation (PBI) for women with stage 0, I, or II breast cancer, web page http://atc.wustl.edu/protocols/nsabp/b-39/0413.html, last accessed 2 December 2005

Orton CG (2001) High-dose-rate brachytherapy may be radiobiologically superior to low-dose rate due to slow repair of late-responding normal tissue cells. Int J Radiat Oncol Biol Phys 49:183–189

Patel NS, Chiu-Tsao ST, Ho Y, Duckworth T, Shih JA, Tsao HS, Quon H, Harrison LB (2002) High beta and electron dose from 192Ir: implications for gamma intravascular brachytherapy. Int J Radiat Oncol Biol Phys 54:972–980

Perera H, Williamson JF, Li Z, Mishra V, Meigooni AS (1994) Dosimetric characteristics, air-kerma-strength calibration and verification of Monte Carlo simulation for a new Ytterbium-169 brachytherapy source. Int J Radiat Oncol Biol Phys 28:953–970

Potters L, Calguaru E, Thornton KB, Jackson T, Huang D (2003) Toward a dynamic real-time intraoperative permanent prostate brachytherapy methodology. Brachytherapy 2:172–180

Pouliot J, Tremblay D, Roy J, Filice S (1996) Optimization of permanent 125I prostate implants using fast simulated annealing. Int J Radiat Oncol Biol Phys 36:711–712

Raben A, Chen H, Grebler A, Geltzeiler J, Geltzeiler M, Keselman I, Litvin S, Sim S, Hanlon A, Yang J (2004) Prostate seed implantation using 3D-computer assisted intraoperative planning vs a standard look-up nomogram: improved target conformality with reduction in urethral and rectal wall dose. Int J Radiat Oncol Biol Phys 60:1631–1638

Rivard MJ, Coursey BM, DeWerd LA, Hanson WF, Huq MS, Ibbott GS, Mitch MG, Nath R, Williamson JF (2004) Update of AAPM Task Group no. 43 report: a revised AAPM protocol for brachytherapy dose calculations. Med Phys 31:633–674

Sumida I, Shiomi H, Oh RJ, Tanaka E, Isohashi H, Inoue T, Inoue T (2004) An optimization algorithm of dose distribution using attraction-repulsion model (application to low-dose-rate interstitial brachytherapy). Int J Radiat Oncol Biol Phys 59:1217–1223

Taschereau R, Pouliot J, Roy J, Tremblay D (2000) Seed misplacement and stabilizing needles in transperineal permanent prostate implants. Radiother Oncol 55:59–63

Tubic D, Beaulieu L (2005) Sliding slice: a novel approach for high accuracy and automatic 3D localization of seeds from CT scans. Med Phys 32:163–174

Tubic D, Zaccarin A, Pouliot J, Beaulieu L (2001) Automated seed detection and three-dimensional reconstruction. I. Seed localization from fluoroscopic images or radiographs Med Phys 28:2265–2267

Van der Laarse R (1994) The stepping source dosimetry system as an extension of the Paris system. In: Mould RJ, Batterman J (eds) Brachytherapy: from radium to optimization. Nucletron, Columbia, Maryland, pp 352–372

Vicini FA, Arthur DW (2005) Breast brachytherapy: North American experience. Semin Radiat Oncol 15:108–115

Vicini FA, Remouchamps V, Wallace M, Sharpe M, Fayad J, Tyburski L, Letts N, Kestin L, Edmundson G, Pettinga J, Goldstein NS, Wong J (2003) Ongoing clinical experience utilizing 3D conformal external beam radiotherapy to deliver partial-breast irradiation in patients with early-stage breast cancer treated with breast-conserving therapy. Int J Radiat Oncol Biol Phys 57:1247–1253

Visser AG, van den Aardweg GJ, Levendag PC (1996) Pulsed dose rate and fractionated high dose rate brachytherapy: choice of brachytherapy schedules to replace low dose rate treatments. Int J Radiat Oncol Biol Phys 34:497–505

Wahab SH, Malyapa RS, Mutic S, Grigsby PW, Deasy JO, Miller TR, Zoberi I, Low DA (2004) A treatment planning study comparing HDR and AGIMRT for cervical cancer Med Phys 31:734–743

Weeks KJ, Montana GS (1997) Three-dimensional applicator system for carcinoma of the uterine cervix. Int J Radiat Oncol Biol Phys 37455–463

Williamson JF (1996) The Sievert integral revisited: evaluation and extension to 125I, 169Yb, and 192Ir brachytherapy sources. Int J Radiat Oncol Biol Phys 36:1239–1250

Williamsom JF (1998a) Monte Carlo-based dose-rate tables for the Amersham CDCS.J and 3M model 6500 137Cs tubes. Int J Radiat Oncol Biol Phys 41:959–970

Williamson JF (1998b) Clinical brachytherapy physics. In: Perez CA, Brady LW (eds) Principles and practice of radiation oncology, 3rd edn. Lippincott Williams and Wilkins, Philadelphia, pp 405–467

Williamson JF, Morin RL, Khan FM (1983) Monte Carlo evaluation of the Sievert integral for brachytherapy dosimetry. Phys Med Biol 28:1021–1032

Wuu CS, Zaider M (1998) A calculation of the relative biological effectiveness of 125I and 103Pd brachytherapy sources using the concept of proximity function. Med Phys 25:2186–2189

Wuu CS, Kliauga P, Zaider M, Amols HI (1996) Microdosimetric evaluation of relative biological effectiveness for 103Pd, 125I, 241Am, and 192Ir brachytherapy sources. Int J Radiat Oncol Biol Phys 36:689–697

Yang G, Reinstein LE, Pai S, Xu Z, Carroll DL (1998) A new genetic algorithm technique in optimization of permanent 125I prostate implants. Med Phys 25:2308–2315

Yu Y, Schell MC (1996) A genetic algorithm for the optimization of prostate implants. Med Phys 23:2085–2091

Yu Y, Anderson LL, Li Z, Mellenberg DE, Nath R, Schell MC, Waterman FM, Wu A, Blasko JC (1999) Permanent prostate seed implant brachytherapy: report of the American Association of Physicists in Medicine Task Group no. 64. Med Phys 26:2054–2076

Yue N, Heron DE, Komanduri K, Huq MS (2005) Prescription dose in permanent 131Cs seed prostate implants. Med Phys 32:2496–2502

Zhang H, Baker C, McKinsey R, Meigooni A (2005) Dose verification with Monte Carlo technique for prostate brachytherapy implants with (125)I sources. Med Dosim 30:85–89

13 Radiobiology of Low- and High-Dose-Rate Brachytherapy

Eric J. Hall and David J. Brenner

CONTENTS

13.1   Absorption of Radiation   *291*
13.2   DNA Damage and Strand Breaks   *291*
13.3   Chromosomal Aberrations   *292*
13.4   Cell Survival Curves   *293*
13.5   The Dose-Rate Effect   *295*
13.6   The Inverse Dose-Rate Effect   *297*
13.7   The Dose-Rate Effect Summarized   *298*
13.8   Early- and Late-Responding Tissues   *298*
13.9   The Dose-Rate Effect and Clinical Data   *299*
13.10   Dose-Rate Effects from a Retrospective Analysis of
        the Iridium Implant Data   *300*
13.11   The Bias of Tumor Size and Dose Rate   *301*
13.12   Rationale for LDR Brachytherapy   *301*
13.13   Pulsed Brachytherapy   *302*
13.14   Experimental Validation of the PDR Concept   *302*
13.15   Practical Clinical Schedules for PDR   *304*
13.16   HDR Versus LDR   *305*
        References   *307*

13.1
Absorption of Radiation

When X-rays or γ-rays are absorbed in biological material, the first step is that part or all of the photon energy is converted into the kinetic energy of fast-moving electrons.

A fast-moving electron may interact directly with DNA causing an excitation or ionization; this is called the direct action of radiation (Fig. 13.1). It is the dominant process for high linear energy transfer (LET) radiations, such as α-particles. Alternatively, the electron may interact with other atoms or molecules in the cell, particularly water, to produce free radicals that can diffuse far enough to reach and damage DNA. This is referred to as the indirect action of radiation. A free radical is a free (not combined) atom or molecule that carries no electrical charge, but has an unpaired electron in the outer shell. This state is associated with a high degree of chemical reactivity. The most important radical produced from the interaction of radiation with water is the hydroxyl radical (OH·). There is evidence to support the notion that any OH· radicals produced within a cylinder of diameter of about 4 nm around the DNA double helix is able to diffuse to the DNA and cause damage (Fig. 13.1). About two-thirds of the biological damage caused by X-rays is mediated via free radicals. This has been described in more detail by Hall (2000).

13.2
DNA Damage and Strand Breaks

DNA consists of two strands that form a double helix. Each strand is composed of a series of deoxynucleotides, the sequence of which contains the genetic code. Sugar moieties and phosphate groups form the backbone of the double helix. The bases on opposite strands must be complementary; adenine pairs with thymine, while guanine pairs with cytosine. When cells are irradiated with X-rays, many breaks of a single strand occur. These can be observed and scored as a function of dose if the DNA is denatured and the supporting structure stripped away. In intact DNA, however, single-strand breaks are of little biological consequence as far as cell killing is concerned because they are readily repaired using the opposite strand as a template (Fig. 13.2). If the repair is incorrect (misrepair), it may result in a mutation. If both strands of the DNA are broken, and the breaks are well separated (Fig. 13.2), repair again occurs readily since the two breaks are handled separately.

E. J. Hall, D. Phil., DSc, FACR, FRCR
Higgins Professor of Radiation Biophysics, Center for Radiological Research, Columbia University, 630, West 168th St., New York, NY 10032, USA
D. J. Brenner, PhD, DSc
Professor of Radiation Oncology and Public Health, Center for Radiological Research, Columbia University Medical Center, 630, West 168th St., New York, NY 10032, USA

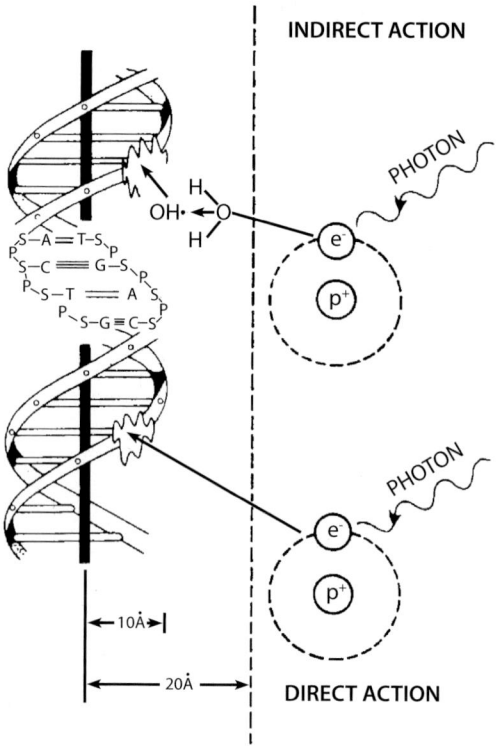

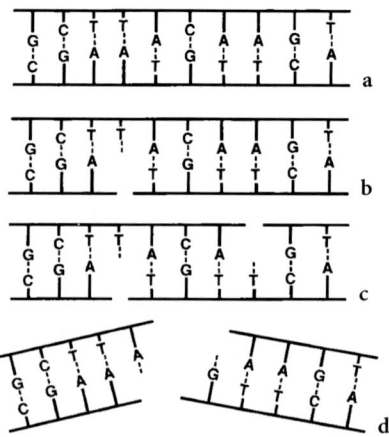

Fig. 13.2a–d. Diagrams of single- and double-strand DNA breaks caused by radiation. **a** Two-dimensional representation of the normal DNA double helix. The base pairs carrying the genetic code are complementary (i.e., adenine pairs with thymine, guanine pairs with cytosine). **b** A break in one strand is of little significance because it is readily repaired, using the opposite strand as a template. **c** Breaks in both strands, if well separated, are repaired as independent breaks. **d** If breaks occur in both strands and are directly opposite or separated by only a few base pairs, this may lead to a double-strand break (DSB) where the chromatin snaps into two pieces. [Drawn to illustrate concepts described by Dr. John Ward (1981, 1988)]

Fig. 13.1. Direct and indirect actions of radiation. The structure of DNA is shown schematically; the letters *S, P, A, T, G,* and *C* represent sugar, phosphorus, adenine, thymine, guanine, and cytosine, respectively. In direct action, a secondary electron resulting from absorption of an X-ray photon interacts with the DNA to produce an effect. In indirect action, the secondary electron interacts with a water molecule to produce a hydroxyl radical (*OH·*), which in turn diffused to the DNA to produce damage. The DNA helix has a diameter of about 2 nm. It is estimated that free radicals produced in a cylinder with a diameter double that of the DNA helix can affect the DNA. Indirect action is dominant for sparsely ionizing radiation, such as X-rays. [Redrawn from Radiobiology for the Radiologist (Hall 1994)]

13.3
Chromosomal Aberrations

When cells are irradiated with X-rays, DSBs occur as described above. The broken ends appear to be "sticky" and can rejoin with any other sticky end. It would appear, however, that a broken end cannot join with a normal, unbroken chromosome end (Evans 1962). Once breaks are produced, different fragments may behave in a variety of ways:

1. The breaks may restitute, that is, rejoin in their original configuration. In this case, of course, nothing amiss will be visible at the next mitosis.
2. The breaks may fail to rejoin and give rise to an aberration, which will be scored as a deletion at the next mitosis.
3. Broken ends may re-assort and rejoin other broken ends to give rise to chromosomes that appear to be grossly distorted when viewed at the following mitosis.

The aberrations seen at metaphase are of two classes: chromosome aberrations and chromatid aberrations. Chromosome aberrations result if a cell is irradiated early in interphase, before the chromosome material has been duplicated. In this case, the

By contrast, if the breaks in the two strands are opposite one another, or separated by only a few base pairs (Fig. 13.2), this may lead to a double-strand break (DSB); that is, the piece of chromatin snaps into two pieces. This picture of DNA damage has been described eloquently by Ward (1981, 1988). A DSB is believed to be the most biologically important lesion produced in chromosomes by radiation; the interaction of two DSBs may result in cell killing, mutation, or carcinogenesis.

In practice, the situation is probably much more complicated than this, since both free radicals and direct ionizations may be involved. In particular, other base damage may be involved with the DSB to form a multiply damaged site (Ward 1988).

radiation-induced break will be in a single strand of chromatin; during the DNA synthetic phase that follows, this strand of chromatin will lay down an identical strand next to itself and will replicate the break that had been produced by the radiation. If, however, the dose of radiation is given later in interphase, after the DNA material has doubled and the chromosomes consist of two strands of chromatin, then the aberrations produced are called chromatid aberrations.

Three types of aberrations are lethal to the cell, namely the dicentric and the ring, which are chromosome aberrations, and an anaphase bridge, which is a chromatid aberration. For simplicity, we will describe in detail the formation of a dicentric; similar considerations apply to rings and anaphase bridges.

The formation of a dicentric is illustrated (Fig. 13.3). This aberration involves an interchange between two separate chromosomes. If a break is produced in each one early in interphase and the sticky ends are close to one another, they may rejoin in an illegitimate way, as shown. This bizarre interchange will replicate during the DNA synthetic phase, and the result will be a grossly distorted chromosome with two centromeres (hence, dicentric). There will also be a fragment that has no centromere (acentric fragment). The dose – response relationship for exchange-type chromosomal aberrations, such as dicentrics, is a linear quadratic function of dose since they result from an interaction between breaks in two different chromosomes (Lea 1956). The linear component is a consequence of the two breaks resulting from a single charged particle. If the two breaks result from different charged particles, the probability of an interaction will be a quadratic function of dose. The dose – response relationship for the formation of dicentrics is shown in Figure 13.4.

13.4
Cell Survival Curves

A cell survival curve describes the relationship between the radiation dose and the proportion of

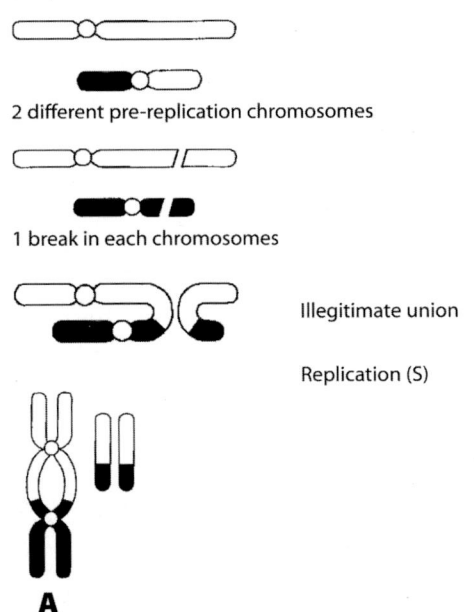

A

Dicentric chromosome plus acentric fragment

Fig. 13.3. The steps in the formation of a dicentric by irradiation of prereplication (i.e., G_1) chromosomes. A break is produced in each of two separate chromosomes. The "sticky" ends may join incorrectly to form an interchange between the two chromosomes. Replication then occurs in the DNA synthetic period. One chromosome has two centromeres – a dicentric. The other is an acentric fragment, which will be lost at a subsequent mitosis, since, lacking a centromere, it will not go to either pole at anaphase. [Redrawn from Hall (2000), Radiobiology for the Radiologist]

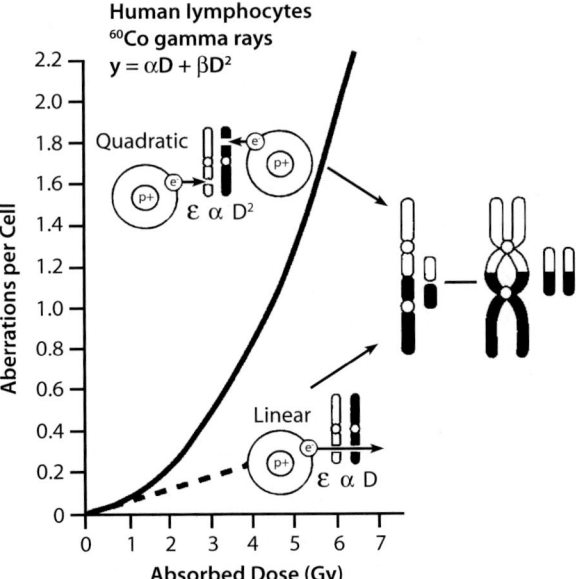

Fig. 13.4. The frequency of interchange-type chromosomal aberrations (dicentrics and rings) is a linear-quadratic function of dose because the aberrations are the consequence of the interaction of two separate breaks. At low doses, both breaks may be caused by the same electron; the probability of an exchange aberration is proportional to dose (D). At higher doses, the two breaks are more likely to be caused by separate electrons. The probability of an exchange aberration is then proportional to the square of the dose (D^2). [Redrawn from Radiobiology for the Radiologist, Hall 2000 (1)]

cells that survive. What is meant by "survival"? Survival is the opposite of death! For differentiated cells that do not proliferate – such as nerve, muscle, or secretory cells – death can be defined as the loss of a specific function. For proliferating cells, such as hematopoietic stem cells or cells grown in culture, loss of the capacity for sustained proliferation – that is, loss of reproductive integrity – is an appropriate definition. This is sometimes called reproductive death, and a cell that survives by this definition is said to be clonogenic because it can form a clone or colony.

This definition is generally relevant to the radiobiology of whole animals and plants and their tissues. It has particular relevance to the radiotherapy of tumors. For a tumor to be eradicated, it is only necessary that cells be "killed" in the sense that they are rendered unable to divide and cause further growth and spread of the malignancy.

The classic mode of cell death following exposure to radiation is "mitotic death". Cells die in attempting to divide because of chromosomal damage, such as the formation of a ring or a dicentric that causes loss of genetic material and prevents the clean segregation of DNA into the two progeny. Death does not necessarily occur at the first mitosis following irradiation; cells can often manage to complete several divisions, but death is inevitable if chromosomal damage is severe.

The other form of cell death is programmed cell death or apoptosis, first described by KERR and colleagues (1972). This is an important form of cell death during the development of the embryo and is implicated, for example, in the regression of the tadpole tail during metamorphosis. Is it also important in many facets of biology including cell renewal systems and hormone-related atrophy. Apoptosis is characterized by a sequence of morphological events; cells condense and the DNA breaks up into pieces of discrete size before the cells are phagocytosed and removed. The DNA "laddering", which is so characteristic of a cell undergoing apoptosis, is shown in Figure 13.5. In radiation biology, apoptosis is a dominant mode of radiation-induced cell death in cells of lymphoid origin, but of variable importance in other cell types.

Most cell survival curves have been obtained by growing cells in vitro in petri dishes. Many cell lines have been established from malignant tumors and from normal tissues taken from humans or laboratory animals. If cells are seeded as single cells, allowed to attach to the surface of a petri dish, and provided with culture medium and appropriate con-

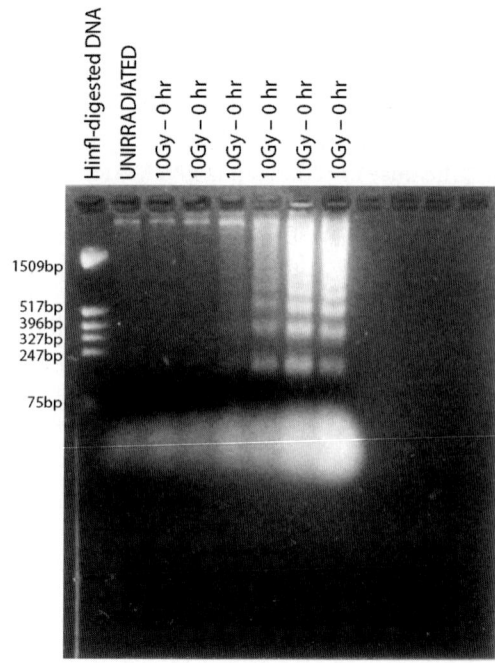

The expression of Nucleosome Ladders in deraded DNA from DU-145 cells at various times after irradiation with 10Gy Cs-137 γ-rays

Fig. 13.5. Illustrating the "laddering" of DNA as it breaks up into pieces of discrete sizes during the process of cell death by apoptosis. (Courtesy of Dr. Eileen Rakovitch)

ditions, each cell will grow into a macroscopic colony that is visible by eye in a period of a few weeks. If, however, the same number of cells are placed into a parallel dish and exposed to a dose of radiation just after they have attached to the surface of the dish, some cells will grow into colonies indistinguishable from those in the unirradiated dish, but others will form only tiny abortive colonies because the cells die after a few divisions. The surviving fraction is the number of macroscopic colonies counted in the irradiated dish divided by the number on the unirradiated dish. This process is repeated so that estimates of survival are obtained for a range of doses; surviving fraction is plotted on a logarithmic scale against dose on a linear scale. The result is the solid line shown in Figure 13.6.

Qualitatively, the shape of the survival curve can be described in relatively simple terms. At "low doses", the survival curve starts out straight on the log-linear plot with a finite initial slope; that is, the surviving fraction is an exponential function of dose. At higher doses, the curve bends. This bending or curving region extends over a dose range of a few Gy. At very high doses, the survival curve often

Linear-Quadratic Relation

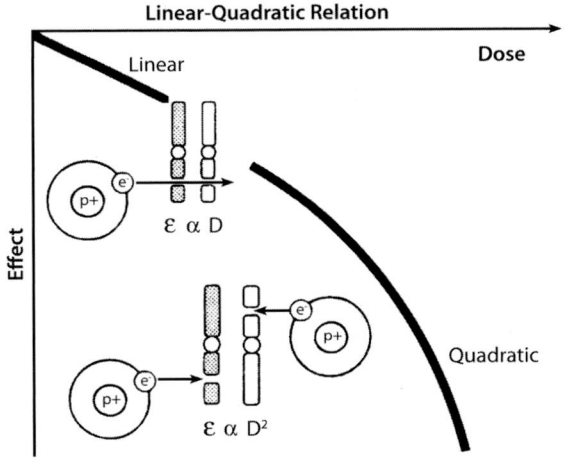

Fig. 13.6. Relationship between chromosome aberrations and cell survival. Cells that suffer exchange-type chromosome aberrations (such as a dicentric) are unable to survive and continue to divide indefinitely. At low doses, the two chromosome breaks are the consequence of a single electron set in motion by the absorption of X-rays or γ-rays. The probability of an interaction between the breaks is proportional to dose; this is the linear portion of the survival curve. At higher doses, the two chromosome breaks may result also from two separate electrons. The probability of an interaction is then proportional to $(dose)^2$. The survival curve bends when the quadratic component dominates

tends to straighten again; in general, this does not occur until doses are in excess of 10 Gy.

The linear quadratic model is nowadays the model of choice for cell survival curves. This model assumes that there are two components to cell killing by radiation, one that is proportional to dose and one that is proportional to the square of the dose. The notion of a component of cell inactivation that varies with the square of the dose reflects the fact that cell death is due largely to complex chromosomal exchange-type aberrations that are the result of breaks in two separate chromosomes as previously described (Fig. 13.6).

By this model, the expression for the cell survival curve is:

$$s = e^{-\alpha D - \beta D^2}$$

S is the fraction of cells surviving a dose D, and α and β are constants. The components of cell killing that are proportional to dose and to the square of the dose are equal when:

$$\alpha D = BD^2,$$
or
$$D = \alpha/\beta$$

This is an important point that bears repeating: the linear and quadratic contributions to cell killing are equal at a dose that is equal to the ratio of α/β (Fig. 13.7). The ratio is a measure of the "curviness" of the survival curve; if α/β is small, the survival curve is very curvy; if α/β is large, the survival curve tends to be straighter and less curvy.

13.5
The Dose-Rate Effect

For X-rays or γ-rays, dose rate is one of the principal factors that determines the biological consequences of a given absorbed dose. As the dose rate is lowered and the exposure time extended, the biological effect of a given dose is generally reduced. Continuous low dose-rate (CLDR) irradiation may be considered to be an infinite number of infinitely small dose fractions; consequently, the survival curve under these conditions would also be expected to have no shoulder and to be shallower than for single acute exposures at high dose rate (HDR), as illustrated by the

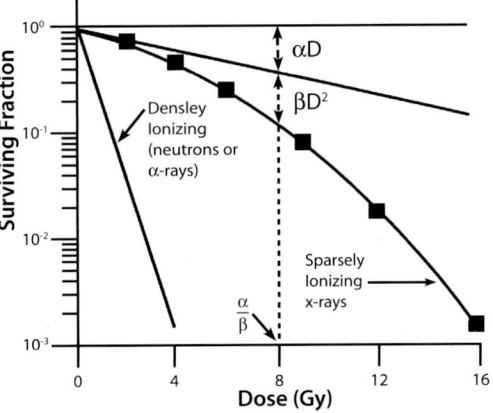

Fig. 13.7. Shape of survival curve for mammalian cells exposed to radiation. The fraction of cells surviving is plotted on a logarithmic scale against dose on a linear scale. For α-particles or low-energy neutrons (said to be densely ionizing) the dose-response curve is a straight line from the origin (i.e., survival is an exponential function of dose). The survival curve can be described by just one parameter, the slope. For X-rays or γ-rays (said to be sparsely ionizing), the dose–response curve has an initial linear slope, followed by a region that curves. The experimental data are fitted to a linear-quadratic function. There are two components of cell killing: one is proportional to dose (αD), while the other is proportional to the square of the dose (βD^2). The dose at which the linear and quadratic components are equal is the ratio α/β. The linear-quadratic curve bends continuously but is a good fit to experimental data for the first few decades of survival

dashed line in Figure 13.8. This is easy to understand in terms of the repair of chromosome damage. The linear component of cell damage will be unaffected by dose rate since the two chromosome breaks that interact to form a lethal lesion are caused by a single electron track. The quadratic component, however, is caused by two separate electron tracks; if there is a long time interval between the passage of the two electron tracks, then the damage caused by the first may be repaired before the second arrives. In this case an exchange-type aberration would not be caused and the cell would survive. Under these circumstances, the high dose component of the dose-response relationship is simply an extrapolation of the initial linear slope of the HDR survival curve as illustrated in Figure 13.8. Sub-lethal damage repair, in this model, equates to repair of one double-strand chromosome break. This idea has its origins in a hypothesis of GRAY, published in 1944, who wrote:

"It is postulated that in order to produce the biological effect under consideration, there must coexist in a considerable proportion of the irradiated cells two separate injuries, each produced by a separate ionizing particle and each capable of restitution. In advancing this postulate we have of course type C chromosome aberrations particularly in mind" (GRAY 1944).

The magnitude of the dose-rate effect from the repair of sublethal damage varies enormously among different types of cells. Cells characterized by a survival curve for acute exposures that has a small initial shoulder exhibit a modest dose-rate effect. This is to be expected, since both are expressions of the cell's capacity to accumulate and repair sublethal radiation damage. Cell lines characterized by a survival curve for acute exposures which has a broad initial shoulder exhibit a dramatic dose-rate effect. Survival curves for HeLa cells cultured in vitro over a wide range of dose rates – from 7.3 Gy/min to 0.535 cGy/min – are summarized in Figure 13.9, taken from the early work of HALL and BEDFORD (1964). As the dose rate is reduced, the survival curve becomes shallower and the shoulder tends to disappear (i.e., the survival curve becomes an exponential function of dose). The dose-rate effect caused by repair of sublethal damage is most dramatic between 0.01 Gy/min and 1 Gy/min. Above and below this dose-rate range, the survival curve changes little, if at all, with dose rate.

Continuing to view the huge variation among different types of cells, HeLa cells are characterized by a survival curve for acute exposures that has a small initial shoulder, which goes hand in hand with a modest dose-rate effect. Again, this is to be expected, since

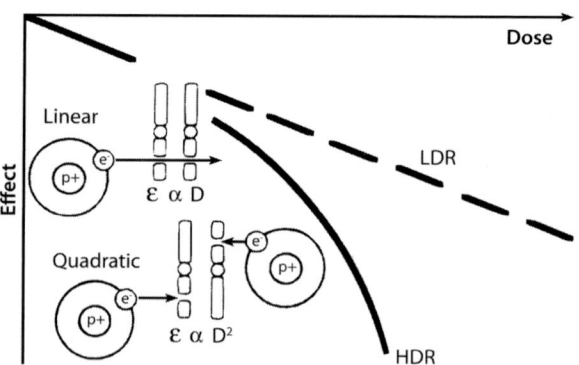

Fig. 13.8. Cell killing by radiation is due largely to aberrations caused by breaks in two chromosomes. The dose–response curve for high-dose rate irradiation (*HDR*) is linear-quadratic; the two breaks may be caused by the same electron (dominant at low doses) or by two different electrons (dominant at higher doses). For low-dose rate irradiation (*LDR*), where radiation is delivered over a protracted period, the principal mechanism of cell killing is by the single electron. Consequently, the LDR survival curve is an extension of the low-dose region of the HDR survival curve

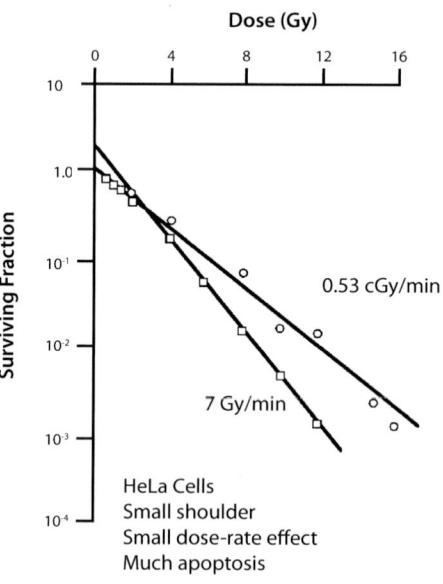

Fig. 13.9. Survival curves for HeLa cells exposed to X-rays at high dose-rate and low dose-rate. [Redrawn from the data of HALL and BEDFORD (1964)]

both are expressions of the cell's capacity to accumulate and repair sublethal radiation damage. By contrast, Chinese hamster (CHO) cells have a broad shoulder to their acute X-ray survival curve and show a corresponding large dose-rate effect. This is evident in Figure 13.10 (redrawn from the work of BEDORD and MITCHELL 1973), where a clear-cut difference in biological effect is demonstrated, at least at high doses, among dose rates of 1.07, 0.3 and 0.16 Gy/min.

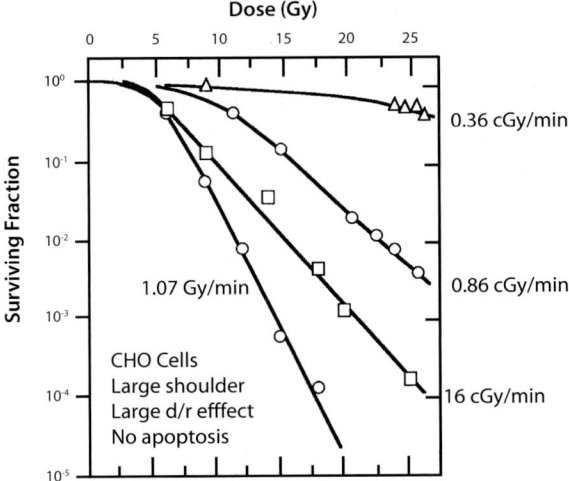

Fig. 13.10. Survival curves for Chinese hamster (*CHO*) cells grown in vitro and exposed to X-rays at various dose-rates. [Redrawn from the data of BEDFORD and MITCHELL (1973)]

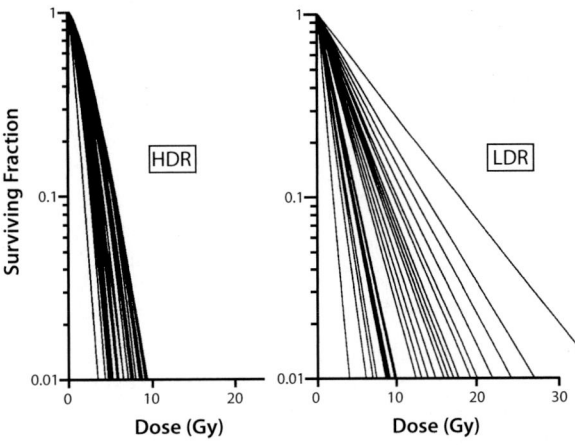

Fig. 13.11. Dose survival curves at high dose rate (*HDR*) and low dose rate (*LDR*) for a large number of cells of human origin cultured in vitro. Note that the survival curves fan out at low dose rate because, in addition to a range of inherent radiosensitivities (evident at *HDR*), there is also a range of repair times of sublethal damage. [Redrawn from HALL and BRENNER (1992)]

This difference in shoulder size and corresponding magnitude of the dose-rate effect correlates with the dominant mechanism of cell death. CHO cells (Fig.13.5), like many cell lines selected to grow in culture, have an abrogated p53 status and do not die an apoptotic death (LOWE et al. 1993). However, apoptosis is an important (but not the only) form of radiation-induced lethality in HeLa cells. This accounts for the smaller shoulder to the survival curve and the less dramatic dose-rate effect.

Figure 13.11 shows survival curves for 40 different cell lines of human origin, cultured in vitro and irradiated at HDR and LDR taken from HALL and BRENNER (1992). At LDR, the survival curves "fan out" and show a greater variation in slope because, in addition to the variation of inherent radiosensitivity evident at HDR, there is a range of repair times of sublethal damage. Some cell lines repair sublethal damage rapidly, some more slowly, and this is reflected in a fanning out of the survival curves at LDR.

13.6
The Inverse Dose-Rate Effect

In cells of human origin, an inverse dose-rate effect is often seen over a narrow range of dose rates, whereby decreasing the dose rate actually increases the efficacy of cell killing. This was first reported by MITCHELL, BEDFORD and BAILEY (1979) for HeLa cells. This is illustrated in Fig. 13.12. Decreasing the dose rate for this cell line from 1.53 Gy/h to 0.37 Gy/h

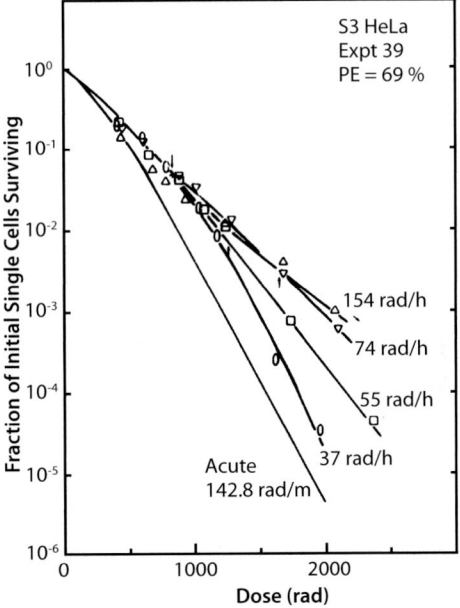

Fig. 13.12. The inverse dose-rate effect. A range of dose rates can be found for HeLa cells such that lowering the dose rate leads to more cell killing. At 1.54 Gy/h, cells are "frozen" in the various phases of the cycle and do not progress. As the dose rate is dropped to 0.37 Gy/h, cells progress to a block in G_2, a radiosensitive phase of the cycle. [Redrawn from the data of MITCHELL et al. (1979)]

increases the efficiency of cell killing, so that this LDR is almost as effective as an acute exposure. The explanation of this phenomenon is illustrated in Fig. 13.13, taken from HALL (1985; Henschke Memorial Lecture 1984). At about 0.3 Gy/h, cells tend to

Mechanism of the Inverse d/r Effect

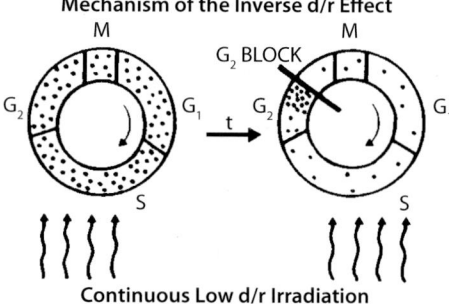

Continuous Low d/r Irradiation

Fig. 13.13. Explanation of the inverse dose rate effect. A range of dose rates can be found, at least for many cells of human origin, that allows cells to progress through the mitotic cycle to a block in late G_2. Under continuous low dose-rate irradiation, an asynchronous population becomes a population of radio-sensitive G_2 cells. [Redrawn from HALL (1985)]

progress through the cycle and become arrested in G_2, a radiosensitive phase of the cycle. At higher dose rates they are "frozen" in the phase of the cycle they are in at the start of the irradiation; at lower dose rates they continue to cycle during irradiation.

13.7
The Dose-Rate Effect Summarized

Figure 13.14 summarizes the entire dose-rate effect. For acute exposures at HDR, the survival curve has a significant initial shoulder. As the dose rate is lowered and the treatment time protracted, more and more sublethal damage can be repaired during the exposure. Consequently, the survival curve becomes progressively shallower, and the shoulder tends to

disappear. A point is reached at which all sublethal damage is repaired, resulting in a limiting slope. In at least some cell lines, a further lowering of the dose rate allows cells to progress through the cycle and accumulate in G_2. This is a radiosensitive phase, and so the survival curve becomes steeper again. This is the inverse dose-rate effect. A further reduction in dose rate will allow cells to pass through the G_2 block and divide. Proliferation may then occur during the radiation exposure if the dose rate is low enough and the exposure time is long compared with the length of the mitotic cycle. This may lead to a further reduction in biological effect as the dose rate is progressively lowered, because cell birth will tend to balance cell death.

13.8
Early- and Late-Responding Tissues

Clinical and laboratory data suggest that there is a consistent difference between early- and late-responding tissues in their response to changes in the time course over which radiation is delivered.

The dose–response relationship for late-responding tissues is more curved than that for early-responding tissues as first described by WITHERS et al. (1982). In terms of the linear-quadratic relationship between effect and dose, this translates into a larger α/β-ratio for early than late effects. The difference in the shapes of the dose–response relationships is illustrated in Figure 13.15. The α/β ratio is the dose at which cell killing by the linear (α) and quadratic (β) components are equal. This difference

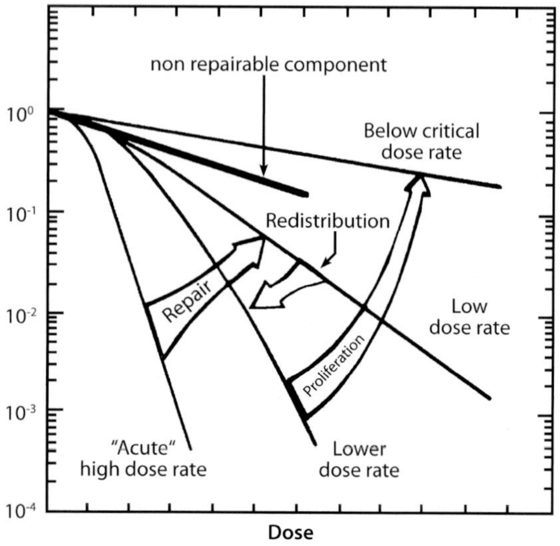

Fig. 13.14. The dose-rate effect due to repair of sublethal damage, redistribution in the cycle, and cell proliferation. The dose–response curve for acute exposures is characterized by a broad initial shoulder. As the dose rate is reduced, the survival curve becomes progressively shallower as more and more sublethal damage is repaired but cells are "frozen" in their positions in the cycle and do not progress. As the dose rate is lowered further and for a limited range of dose rates, the survival curve steepens again because cells can progress through the cycle to pile up at a block in G_2, a radiosensitive phase, but still cannot divide. A further lowering of dose rate allows cells to escape the G_2 block and divide; cell proliferation may then occur during the protracted exposure, and survival curves become shallower as cell birth from mitosis offsets cell killing from the irradiation. (Based on the ideas of Dr. Joel Bedford)

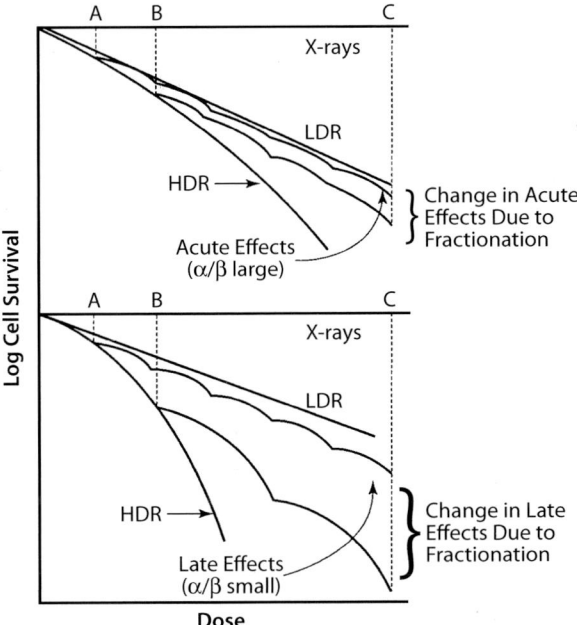

Fig. 13.15. For X-rays, late-responding tissues have a much more curved dose–response relationship and consequently show a much greater sparing in a multifraction regimen than early-responding tissues. Low dose-rate (*LDR*) is, effectively, an infinite number of infinitely small doses. LDR therefore gives the maximum differential in sparing late-responding normal tissues compared with early-responding tissues, which includes tumors

in shape translates into a difference in response to changes in fractionation or dose rate. Late-effect tissues, characterized by a very curvy dose–response relationship, show a much greater sparing in a multifraction or CLDR regime than do early-responding tissues, which include tumors

Consequently, it can be concluded that LDR results in the maximal differential in response between late-responding normal tissues and tumors. This is a significant biological advantage for brachytherapy, to be added to the physical advantage of good dose distribution resulting from the direct implantation of the tumor.

13.9
The Dose-Rate Effect and Clinical Data

Interstitial implants and intracavitary treatments typically involve the range of dose rates where laboratory data would suggest that the biological effect should vary critically with dose rate. However, reports in the literature have been controversial.

It was pointed out by Paterson in the 1960s, that the dose-limiting factor in the case of interstitial implants is the tolerance of normal tissues (PATERSON 1963). His philosophy was to push to the maximum dose tolerated by the normal tissues in order to maximize tumor control. Paterson published a curve relating total dose to overall time, with limiting normal tissue tolerance as the endpoint. Regarding 60 Gy in 7 days as the standard, he proposed that an implant of shorter duration should have a lower dose, and an implant of longer duration an augmented dose. The published curve represented his considerable clinical experience accumulated over many years and was unequivocally based on equalizing late effects in normal tissues. The experience was based on treatment with radium needles implanted according to the Manchester system. Ellis proposed an essentially identical scheme for use in clinical practice (ELLIS 1968); the curve of Paterson and Ellis is reproduced in Figure 13.16, together with theoretical curves relating equivalent dose to overall time based on radiobiological data for early- and late-responding tissues, normalized to 60 Gy in 7 days. It is interesting to note that the calculated curve based on the radiobiological parameters for late-responding tissues is virtually indistinguishable from the curves of Paterson and of Ellis who unequivocally based their judgment on obtaining equal late effects. The curve relating equivalent dose and overall time is steeper for late than for early-responding tissues because of the smaller value of α/β.

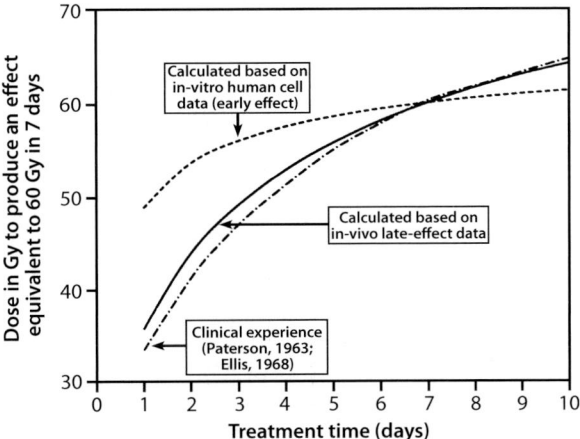

Fig. 13.16. Dose equivalent to 60 Gy (6000 rads) in 7 days as proposed by PATERSON (1963) and ELLIS (1968) based on clinical observation of normal tissue tolerance, or calculated from radiobiological principles based on α/β ratios and $t_{1/2}$ values characteristic of early- and late-responding tissues. [Redrawn from Radiobiology for the Radiologist (HALL 1994)]

The introduction of iridium-192 as a substitute for radium needles in interstitial brachytherapy allowed greater flexibility and patient comfort, but also resulted in a much larger variation of dose rate between individual implants. Two factors contribute to this:

1. As a consequence of the relatively short half-life of iridium-192 (74 days), the linear activity will vary significantly during the period of several months that the wires may be used or re-used.
2. The "Paris" system of dosimetry (PIERQUIN 1971) developed for iridium-192 implants, where all sources have the same linear activity with varying separation between wires for different lengths (i.e., greater separation for larger wires) results in a wider range of dose rates than was characteristic of radium implants using the Parker-Paterson dosimetry system, where internal needles had half or two-thirds of the linear activity of outer needles. Because all wires in an iridium-192 implant have the same linear activity, there is a correlation between implanted volume and dose rate, with larger volumes being associated with higher dose rates. The combination of those two factors results, in practice, in a three-fold variation in the overall irradiation time for the delivery of a given tumor dose. Nevertheless, PIERQUIN and his colleagues (1973) came to the conclusion that, in iridium-192 implants, the time factor – and therefore the dose rate – was unimportant.

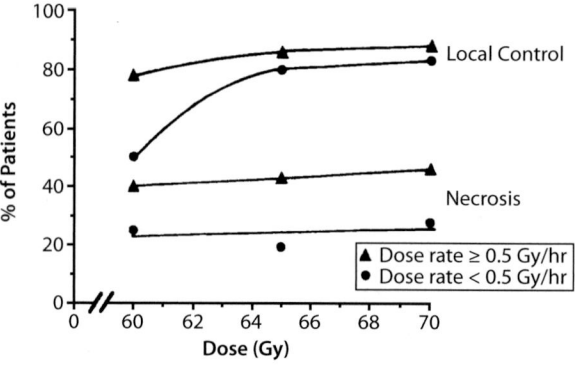

Fig. 13.17. Local tumor control and necrosis rate at 5 years as a function of dose in patients treated for T1-2 squamous cell carcinomas of the mobile tongue and the floor of the mouth with interstitial iridium-192 implants. The patients were grouped according to whether the implant was characterized by a high dose rate (above 0.5 Gy/h) or low dose rate (below 0.5 Gy/h). The necrosis rate is higher for the higher dose rate at all dose levels. Local tumor control did not depend on dose rate provided the total dose was sufficiently large, but did vary with dose rate for lower total doses. [Redrawn from the data of MAZERON et al. (1991b)]

Consequently, the Paris school recommended the same prescribed dose irrespective of overall time within the range 3–8 days. They were careful to point out that their conclusions were preliminary, but nevertheless concluded:
"We can, however, say with certainty that the variation in overall treatment time for the same tumor dose from 3 days to 8 days does not appear to influence the frequency of recurrence of necrosis."

Based on this conviction, many hundreds of patients were treated with iridium-192 implants using standard doses uncorrected for treatment time or dose rate, despite the fact that this conflicts with the previously published clinical experience of PATERSON and of ELLIS, and does not agree with the experimental radiobiological data that would predict a substantial dose-rate effect over the dose-rate range in question.

13.10
Dose-Rate Effects from a Retrospective Analysis of the Iridium Implant Data

The large series of patients treated in Paris with iridium wire implants have been followed carefully over the years, and two important papers have appeared describing a retrospective analysis of these data. In the first, MAZERON and colleagues (1991b) studied the incidence of local tumor control and necrosis in T_1 and T_2 squamous cell carcinoma of the mobile tongue and floor of mouth treated with interstitial iridium-192. The data are shown in Figure 13.17 and compare tumor control and necrosis in patients treated at dose rates above or below 0.5 Gy/h. Two principal conclusions can be drawn from this analysis.

1. There is little or no difference in local control between the two dose-rate ranges provided a sufficiently high total dose is used (65–70 Gy), but there is a clear separation at lower doses (around 60 Gy) with the lower dose rate being significantly less effective.
2. Over the entire range of doses used, there was a higher incidence of necrosis associated with the higher dose rate range.

The clinical data are in line with the predictions that would be expected based on radiobiological considerations, particularly the more critical dependence on dose rate of late-responding tissues.

In a second study, MAZERON and colleagues (1991a) analyzed data from a large group of patients with carcinoma of the breast who received an iridium-192 implant as a boost to external beam radiotherapy. A fixed standard total dose was used, regardless of the dose rate, and there is a clear correlation between the proportion of recurrent tumors and the dose rate, as illustrated in Figure 13.18. For a given total dose, a clear difference in tumor control could be seen between 0.3 Gy/h and 0.9 Gy/h, as predicted from radiobiological experiments with cells in vitro.

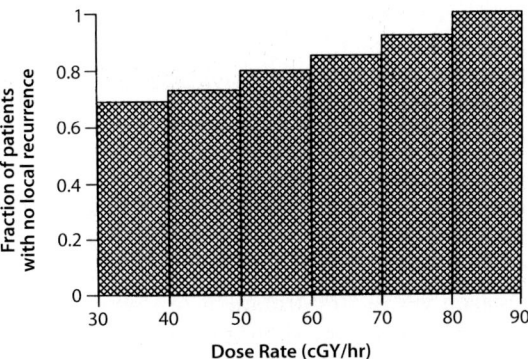

Fig. 13.18. Percentage of patients who showed no local recurrence as a function of dose rate in patients treated for breast carcinoma by a combination of external-beam irradiation plus iridium-192 interstitial implant. The implant was used to deliver a constant dose of 37 Gy; the dose rate varied by a factor of three owing to different linear activities of the iridium-192 wire and to different size volumes implanted. [Drawn from the data of MAZERON et al. (1991a)]

13.11
The Bias of Tumor Size and Dose Rate

A complication and confounding variable in the interpretation of these clinical data relating dose to produce an equivalent effect to implant time (and therefore to dose rate) is the fact that, for interstitial implants, the dose rate tends to increase as the size of the implant increases. This correlation is particularly true for implants using iridium-192 wires, as used in the Paris system, which are all of the same linear activity, but less so when there is a variation in linear activity, as in the Parker and Paterson system (MEREDITH 1967). The bias of larger tumors and larger volumes being associated with higher dose rates, while smaller tumors and smaller treatment volumes are associated with lower dose rates, was pointed out by PIERQUIN and his colleagues (1973). Larger tumors of course require a larger dose for a given level of local control, while the maximum dose that can be tolerated by normal tissues decreases as the volume implanted increases. This will tend to flatten the isoeffect curve for tumor control and steepen the isoeffect curve for normal tissue tolerance.

Based on these considerations, then, it is clear why the Paris school and the Paterson/Ellis school differed so radically in their prescriptions for dealing with dose-rate changes. First, the Paterson/Ellis recommendations were based on equalizing only late effects, where there is a clear change of equi-effect dose with dose rate (solid curves in Fig. 13.16). However, the Paris recommendations were based on an attempt to equalize late and early effects (with hindsight, it is clear that when the dose rate changes it is not possible to match both late and early effects). Second, the Paterson/Ellis recommendations were made based on data from the era of radium needles

when there was less correlation between volume and dose rate. However, the Paris recommendations were based on iridium wire implants where there is strong correlation between tumor volume and dose rate, which would tend to make the equi-effect curve for tumor control vary even less with dose rate.

13.12
Rationale for LDR Brachytherapy

There are three factors that contribute to the efficacy of LDR brachytherapy.

1. The dose distribution is favorable because the radioactive sources are either implanted directly into the tumor or contained in a body cavity close to the tumor. This allows a larger dose to be delivered to the tumor for a given dose to limiting normal tissues. It is interesting to note that this advantage for an implant was suggested by Alexander Graham BELL as early as 1903 in a letter to American Medicine. He wrote; Correspondence, American Medicine, Aug. 15th, 1903.

"Dear Dr. Sowers:
I understand from you that the Röntgen rays and the rays emitted by radium, have been found to have a marked curative effect upon external cancers, but that the effects upon deep-seated cancers have not thus far proved satisfactory.

It has occurred to me that one reason for the unsatisfactory nature of these latter experiments arises from the fact that the rays have been applied externally, thus having to pass through healthy tissues of various depths in order to reach the cancerous matter.

The Crookes' tube from which the Röntgen rays are emitted is of course too bulky to be admitted into the middle of a mass of cancer, but there is no reason why a tiny fragment of radium sealed up in a fine glass tube should not be inserted into the very heart of the cancer, thus acting directly upon the diseased, material. Would it not be worth while making experiments along this line?"

(signed) Alexander Graham Bell

2. LDR is effectively an infinite number of infinitely small dose fractions; this exploits to the full the difference between tumors and late-responding normal tissues that are a consequence of their different α/β ratios.
3. Brachytherapy is usually delivered over a shorter overall time period than conventional external beam radiotherapy; it constitutes accelerated treatment par excellence.

13.13
Pulsed Brachytherapy

A major innovation in brachytherapy during the past decade has been the introduction of pulsed dose-rate brachytherapy (PDR), first described by BRENNER and HALL (1991b). The principle of PDR is to replace the many individual wires, ribbons or sources in a conventional implant or intracavitary treatment by a single iridium-192 source of about 37 GBq. This source, under computer control from a remote afterloading device, steps through the implanted catheters with dwell times tailored to produce the dose distribution required. The principle is illustrated in Figure 13.19. Based on an analysis of a large body of data from cell lines of human origin, BRENNER and HALL (1991b) came to the conclusion that a 10-min pulse delivering 40–60 cGy and repeated every hour would adequately mimic CLDR irradiation at 40–60 cGy/h. They concluded that, as long as the dose/pulse is kept low, a reasonable equivalence would be achieved in terms of both early and late effects. Between individual pulses, the source is returned to the safe. This simple strategy leads to several important advantages.

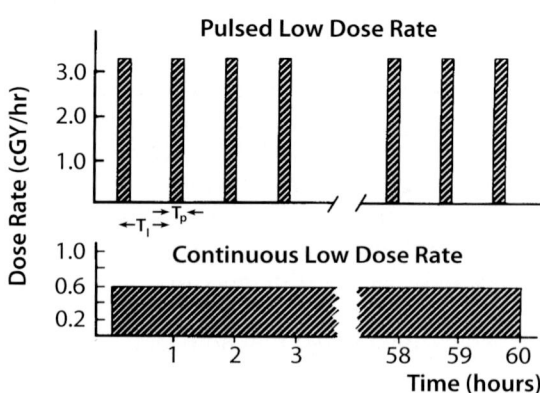

Fig. 13.19. Illustrating the concept of replacing continuous low dose-rate irradiation by a series of short pulses, so-called PDR

1. Improved radiation safety – since there is no individual source preparation beforehand, and during an implant the source can be returned to the safe while the patient is nursed, examined or visited.
2. A substantial cost saving – since only one source needs to be replaced instead of a whole inventory of sources.
3. Improved optimization of the dose distribution – due to a stepping source under computer control, with variable dwell time in each position.
4. The average dose rate can be kept constant for implants of different sizes and the iridium-192 source decays by the simple expedient of varying the pulse length. This is illustrated in Figure 13.20.

The pulsing schedule recommended by BRENNER and HALL (1991b) was very conservative and primarily designed to be "safe" for almost any conceivable set of biological response parameters characteristic of the relevant target tissues – whether early or late responding. For this reason, the proposed schedule maintained the same overall dose in the same overall time as the CLDR that the PDR replaced, and suggested frequent pulses with small doses per pulse. In this way, the technical advantages offered by the new generation of computer-controlled remote afterloaders could be combined with the true and trusted advantages of LDR.

13.14
Experimental Validation of the PDR Concept

Conditions for the equivalence of PDR and CLDR irradiation have been investigated by a number of groups, always with a view to discover the limits

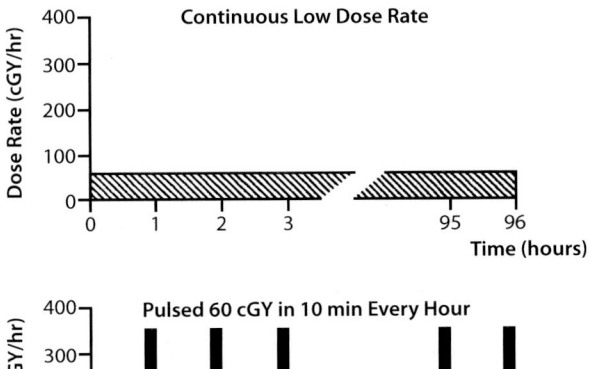

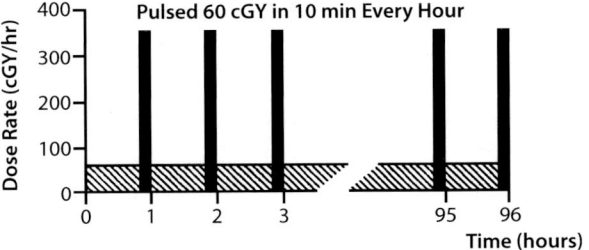

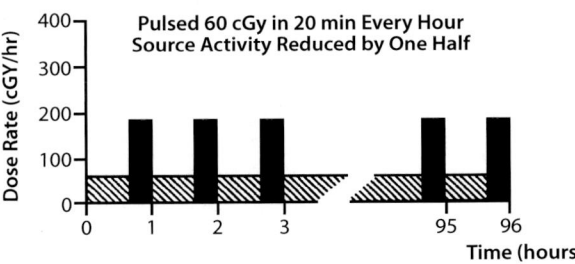

Fig. 13.20. Illustrating the principle of pulsed brachytherapy. Continuous low-dose-rate irradiation at (for example) 60 cGy/h is replaced by a relatively high dose rate pulse of 60 cGy delivered once per hour. The pulse, during which the single iridium-192 source steps through the implant, would take about 12 min, depending on the activity of the source and the size of the implant. Over a period of months as the activity of the iridium-192 source decays, the dose per pulse and, therefore, the average dose rate per hour, can be maintained by simply increasing the pulse length. After one half-life of the radionuclide, which is 70 days, the pulse length would be doubled to 24 min in each hour. [Redrawn from HALL and BRENNER (1992)]

of fraction number and dose per fraction where the equivalence breaks down, bearing in mind that fewer, more widely spaced fractions, would be logistically more convenient.

The first published experimental test of PDR was by Armour and colleagues (ARMOUR et al. 1992) using 9L rat gliosarcoma cells cultured in vitro. In this cell line, no difference in cell survival could be detected between CLDR at 0.5 Gy/h and pulsed schedules up to 3 Gy every 6 h. The equivalence broke down for a pulse schedule of 6 Gy every 12 h.

A later in vitro study by Chen and colleagues compared PDR and CLDR using three cell lines of human origin, one derived from a cervical carci-

noma and two from breast carcinomas (CHEN et al. 1997). There was no significant difference in cell survival between PDR and CLDR for any of the cell lines, when a pulse interval of 1 h (and a dose per pulse of about 0.6 Gy) was used, supporting the initial conservative recommendation of BRENNER and HALL (1991b). As the pulse interval was increased, PDR became progressively more effective than CLDR, for a given dose, and there were significant differences for pulse intervals of 6 h and 12 h, corresponding to large doses/pulse.

A comparison of CLDR and PDR was also published by Mason and colleagues from the MD Anderson Hospital (MASON et al. 1994) scoring regenerating crypts in the mouse jejunum – an in vivo early-responding end point. They found an hourly PDR schedule to be indistinguishable from CLDR at 0.7 Gy/h, whether the pulse was delivered in 10 min or 1 min. This is an important demonstration in vivo that agrees with the more extensive in vitro studies.

BRENNER and colleagues (1996) used a model late-effect system, namely cataract formation in the ocular lens of the rat, to compare PDR and CLDR. This is a true late effect, though not one that is generally dose limiting in radiation therapy. They found that, for the same total dose in the same overall time there was no difference in cataractogenic potential between CLDR and hourly pulses (0.62 Gy/pulse) or pulses repeated every 4 h (2.48 Gy/pulse). The likely explanation of the fact that fewer larger pulses are still equivalent to CLDR in this system (whereas they were not in in vitro experiments) lies in a relatively slow rate of repair of sub-lethal damage in this late-responding tissue. There is good evidence from the clinic that sublethal damage responsible for late effects in the human also repairs relatively slowly. This has previously been discussed (BRENNER et al. 1994).

ARMOUR and colleagues (1997) later used late rectal stenosis in the rat as an endpoint to evaluate PDR. This is probably a "consequential" late effect, but highly relevant to intracavitary brachytherapy where late rectal damage can be dose limiting. They found that CLDR (0.75 Gy/h) was indistinguishable from pulsed regimes consisting of 0.375 Gy every 0.5 h, 0.75 Gy every 1 h, or 1.5 Gy every 2 h. However, a 3-Gy pulse every 4 h was slightly more damaging while a 6-Gy pulse every 8 h was much more damaging than CLDR to the same total dose in the same overall time.

In summary, these experimental studies confirm the conservative recommendations of BRENNER and

HALL (1991b) that small pulses repeated every 1 h (or possibly 2 h) is indistinguishable from CLDR, but that bigger doses per pulse, with a longer separation between pulses, produce more severe biological damage. This is equally true for cells cultured in vitro and for tissues in vivo.

13.15
Practical Clinical Schedules for PDR

Whilst PDR prospered in Europe and elsewhere, in the US it foundered for some time on the Nuclear Regulatory Commission requirement that a physicist and/or radiotherapist (or some other suitably qualified person) be present throughout the treatment (that is, day and night) to deal with the possible, if unlikely, eventuality that the source becomes lodged inside the patient. This restriction has now been removed. A popular option is to restrict treatment pulses to "office hours", when the need for the presence of a physicist and/or radiation oncologist is not a problem. This can only be done by dropping the constraint that was considered prudent in the original paper on this topic (that both the total dose and overall treatment time must be the same as the conventional CLDR brachytherapy treatment) – and instead allowing somewhat longer overall treatment times.

In addition, in order for the PDR to be equivalent to the CLDR, in terms of both early- and late-responding tissues, it must be assumed that the rate of repair of sub-lethal damage is slower in late than in early-responding tissues. There is some evidence for this from animal experiments as well as from clinical data (MOULDER and FISH 1992; THAMES et al. 1984; TURESSON and THAMES 1989; VAN RONGEN et al. 1993). The existing data summarizing repair half times in normal tissues are summarized in Tables 13.1 and 13.2. If these various assumptions are made, it is possible to design PDR schedules in which pulses are given only during "office hours". An extreme example of this is the protocol of VISSER and colleagues (1996) who proposed pulses at 3-h intervals only during the working day. This is illustrated in Figure 13.21, which also shows how the overall pulsed schedule must be of longer duration than the CLDR protocol it replaces in order to achieve equivalence. Their early reports indicate no worse late effects in the patients treated with PDR, but since the implant was a "boost", with the majority of the total dose being delivered in highly fractionated

Table 13.1. Early- and late-responding skin damage (TURESSON and THAMES 1989)

	$t_{1/2}$ slow (min)	$t_{1/2}$ fast (min)
Early	75	25
Late	250	25

Table 13.2. Values of α,β, α/β and $t_{1/2}$ for late-responding normal tissues evaluated in vivo

Endpoint	α Gy^{-1}	β Gy^{-2}	α/β Gy	$t_{1/2}$ min
Skin telangiectasia				
Fast repair	0.1	0.024	4.1	24
Slow repair	0.1	0.024	4.1	210
Mouse lung (late damage)	0.31	0.072	4.3	39
Rat spinal cord (paralysis)	0.066	0.019	3.4	93

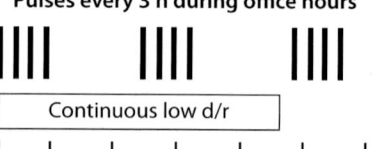

Fig. 13.21. To illustrate the PDR scheme proposed by VISSER and colleagues (1996)

external beam radiotherapy, this is not really a critical test of the idea. HARMS et al. (2005) described a day-time PDR regime, consisting of 0.5 Gy per pulse per hour, continued for 10 h per day, delivered on an out-patient basis for previously irradiated patients with recurrent esophageal cancer.

The total clinical experience with PDR to date is still limited. The first patient was treated in 1992 at the University of California at San Francisco. A survey conducted by MAZERON et al. (1997) concluded that, by the end of 1995, over 1100 patients had been treated with PDR in 20 centers in North America, Europe, and Australia. More recent applications of PDR have involved re-treatment of areas that have already received a substantial dose of radiation. RESCH et al. (2002) described the use of post-operative PDR for patients who had small recurrent breast carcinomas after initially undergoing breast-conserving therapy, which included surgery and post-operative radiation therapy. Additional PDR systems have become available in recent years as well as improved dosimetry (KARAISKOS et al. 2003; PEREZ-CALATAYUD et al. 2001). Advantages of the new technology, cited by the most frequent users, included:

- Better quality of treatment
- Technical verification
- Patient care during treatment
- Radiation protection

13.16
HDR Versus LDR

The move towards HDR brachytherapy has been fueled by a combination of many factors, including patient convenience, cost, radiation protection, and dose optimization made possible by the new generation of computer-controlled afterloading devices.

Since tumors are characterized in general by large α/β ratios (from 10 Gy to ∞), while late-responding normal tissues have smaller α/β ratios (closer to 2 Gy), it is simply impossible to replace a CLDR schedule with a HDR schedule consisting of a few fractions while preserving both local tumor control and normal tissue late effects. If doses are matched to produce the same tumor control, then late effects will inevitably be worse; if, on the other hand, doses are matched to result in equivalent normal tissue late effects, then tumor control will be jeopardized. This is an inevitable consequence of the biology involved. If the HDR is divided into a sufficient number of fractions (20 or more) to allow an approximate equivalence of both early and late effects, then all the advantages of patient convenience and cost would be lost. FOWLER (1989) summed it up well saying:

"There is, therefore, a certain loss of therapeutic ratio when regimes of logistic convenience (that is, HDR) are used".

What then is the place for HDR? There are two quite different situations where the convenience and cost savings associated with HDR can be exploited:

a) *The implants as a boost.* When the brachytherapy implant constitutes only one-third to one-half of the total radiation treatment, with the remainder of the treatment being delivered as highly fractionated external beam radiotherapy, then a few HDR fractions are tolerated. This has been the approach, for example, of Levendag and his colleagues in the Netherlands (VISSER et al. 1996). The HDR dose should be calculated to be equivalent to the CLDR regime it replaces for tumor control; the normal tissues are sufficiently "forgiving" that the slightly worse late effects will not be a problem because of all the other advantages of

brachytherapy – limited volume, good dose distribution, etc.

b) *Intracavitary brachytherapy for carcinoma of the uterine cervix.* What makes this situation different from almost all others is that the radiation dose that produces unwanted late sequelae is significantly less than the treatment dose prescribed to the tumor. This is because the dose-limiting organs at risk (rectum and bladder) are some distance away from the brachytherapy sources, in contrast to the more usual situation in which the dose-limiting normal tissue is adjacent to the treatment volume. A further factor must be considered, namely that the short treatment time characteristic of HDR allows packing and retraction of the sensitive normal tissues estimated by ORTON (1989; ORTON et al. 1992) to result in a further 20% reduction in dose to bladder/rectum compared with that associated with conventional LDR treatments.

ORTON and colleagues (1992) analyzed clinical data from a survey of 56 centers treating more than 17,000 cervical cancer patients with HDR. A wide range of doses and number of fractions were used. The average fractionation regime consisted of about five fractions of about 7.5 Gy each to point A, regardless of the stage of the disease. Fractionation of the HDR treatments significantly influenced toxicity. Morbidity rates were significantly lower for point A dose/fraction less than or equal to 7 Gy compared with greater than 7 Gy. This was true for both moderate and severe complications. The effect of dose/fraction on cure rates was equivocal. Some findings of the Orton survey are summarized in Table 13.3. Finally, the data showed that for conversion from LDR to HDR, the total dose to point A was reduced on average by a factor 0.54+/−0.06. The overall conclusion of the survey was that HDR resulted in a 5-year survival figure and complication rates that were at least as good as historical controls from the same centers treated with CLDR; however, none of these studies was a prospective randomized trial.

Table 13.3. Survey of high dose-rate schedules (ORTON et al. 1991)

	>7 Gy per fraction	<7 Gy per fraction
Mean no. of fractions	3.8	5.6
Mean total dose (Gy)	35	30
Mean 5-year survival	66.2-2.2	56.8-2.1
Complication rate (%)	9.8	8.6
Severe complications (%)	3.4	1.4

Several critiques of the Orton study have been published, as well as a summary of the benefits of conventional LDR by EIFEL (1992). In summary, EIFEL pointed out that HDR therapy can be expected to produce comparable results to LDR therapy only in situations where the dose to organ at risk for late complications is lower than the prescribed tumor dose. This leads to one of the clear guidelines as to when HDR is indicated as a good alternative to LDR.

The optimal number of fractions and the dose per fraction for an HDR treatment are still a matter of debate. Using the linear quadratic formalism and biological data from a battery of almost 40 cell lines of human origin, BRENNER and HALL (1991a) calculated HDR schemes that were designed to be equivalent to many of the LDR protocols in common use. Distributions of HDR doses predicted to yield comparable acute effects to 60 Gy in two conventional LDR treatments are shown in Figure 13.22. The arithmetic mean of these dose distributions as a function of the number of fractions is plotted in Figure 13.23. The radiobiological predictions would suggest a dose per fraction of about 7.5 Gy if five fractions are used, which is in remarkable agreement with the average of the Orton survey.

A system has been described by STITT and colleagues (1992)and THOMADSEN and colleagues (1992) which provides a set of dose schedules and dose specification points for treatment of carcinoma of the cervix with HDR brachytherapy plus external beam radiotherapy. These various protocols, based on sound radiobiological principles have been used to treat hundreds of patients. Local tumor control and complication rates are reported to be similar to LDR treatments, but these conclusions are based on historical controls.

The radiobiological principles summarized in this chapter lead to clear guidelines for the use of HDR brachytherapy for the uterine cervix.

1. When the dose to the dose limiting normal tissues is less than 75% of the prescribed tumor dose, for equal tumor control, HDR results in late effects that are comparable to and no worse than for LDR.
2. For patients in whom the dose to the bladder/ rectum is comparable to the prescribed dose, HDR is contra-indicated.
3. HDR protocols, comparable to LDR, should be designed based on matching early rather than late effects.

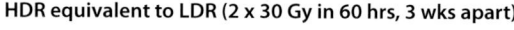

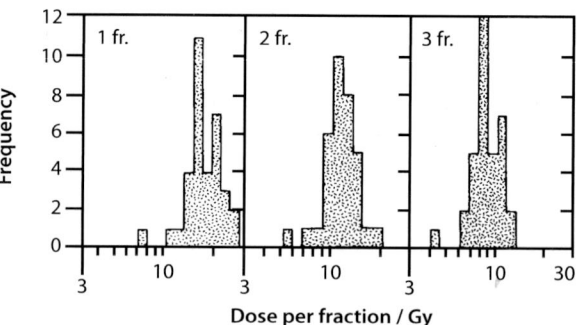

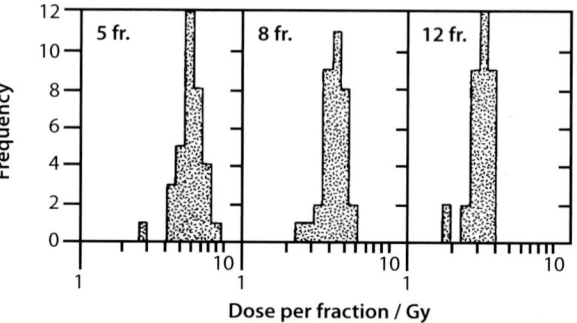

Fig. 13.22. Distribution of doses delivered in 1–12 HDR fractions, equivalent to two LDR treatments of 30 Gy in 60 h, based on data from cell lines of human origin as calculated by BRENNER and HALL (1991a)

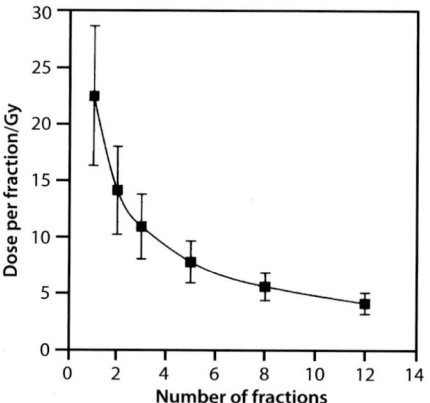

Fig. 13.23. Dose/fraction as a function of the number of HDR treatments to achieve equal biological effect. The *points* plotted are the arithmetic means of the distributions shown in Fig. 13.22 from the calculations of BRENNER and HALL (1991a)

References

Armour E, Wang Z, Corry P, Martinez A (1992) Equivalence of continuous and pulse simulated low dose rate irradiation in 9L gliosarcoma cells at 37° and 41°C. Int J Radiat Oncol Biol Phys 22:109–114

Armour E, White A Jr, Armin A, Lorry P, Coffey M, Dewitt C, Martinez A (1997) Pulsed LDR brachytherapy in a rat model; dependence of late rectal injury on radiation pulse size. Int J Radiat Oncol Biol Phys 38:825–834

Bedford JS, Mitchell B (1973) Dose-rate effects in synchronous mammalian cells in culture. Radiat Res 54:316–327

Bell AG (1903) Correspondence. American Medicine

Brenner DJ, Hall EJ (1991a) Fractionated high dose-rate versus low dose-rate brachytherapy of the cervix. I. General considerations based on radiobiology. Br J Radiat 64:133–141

Brenner DJ, Hall EJ (1991b) Conditions for the equivalence of continuous to pulsed low dose rate brachytherapy. Int J Radiat Oncol Biol Phys 20:181–190

Brenner DJ, Hall EJ, Huang Y, Sachs RK (1994) Optimizing the time course of brachytherapy and other accelerated radiotherapeutic regimes. Int J Radiat Oncol Biol Phys 29:893–901

Brenner DJ, Hall EJ, Randers-Pehrson G, Huang YP, Johnson GW, Miller RW, Wu B, Vazquez ME, Medvedovsky C, Worgul BV (1996) Quantitative comparisons of continuous and pulsed low dose-rate regimens in a model late-effect system. Int J Radiat Oncol Biol Phys 34:905–910

Chen CZ, Huang Y, Hall EJ, Brenner DJ (1997) Pulsed brachytherapy as a substitute for continuous low dose rate: an *in vitro* study with human carcinoma cells. Int J Radiat Oncol Biol Phys 37:137–143

Eifel PH (1992) High-dose-rate brachytherapy for carcinoma of the cervix: High tech or risk risk? Int J Radiat Oncol Biol Phys 24:383

Ellis F (1968) Dose time and fractionation in radiotherapy. In: Howard EM (ed) Current topics in radiation research, vol 4. Amsterdam, North Holland, pp 359–397

Evans HJ (1962) Chromosome aberrations induced by ionizing radiation. Int Rev Cytol 13:221–321

Fowler JF (1989) Dose rate effects in normal tissues. In: Mould RF (ed) Brachytherapy 2. Nucletron, The Netherlands, p 26

Gray LH (1944) Dose-rate in radiotherapy. Br J Radiol 17:327–335

Hall EJ (1985) The biological basis of endocurie therapy. The Henschke Memorial Lecture 1984. Endocurie Hypertherm Oncol 1:141–151

Hall EJ (2000) Radiobiology for the radiologist, 5th edn. Lippincott, Williams and Wilkins, Philadelphia

Hall EJ, Bedford JS (1964) Dose-rate: its effect on the survival of HeLa cells irradiated with gamma-rays. Radiat Res 22:305–315

Hall EJ, Brenner DJ (1992) The 1991 George Edelstyn Lecture: needles, wires and chips - advances in brachytherapy. Clin Oncol 4:249–256

Harms W, Krempien R, Grehn C, Berns C, Hensley FW, Debus J (2005) Daytime pulsed dose rate brachytherapy as a new treatment option for previously irradiated patients with recurrent oesophageal cancer. Br J Radiol 78:236–241

Karaiskos P, Angelopoulos A, Pantelis E, Papagiannis P, Sakelliou L, Kouwenhoven E, Baltas D (2003) Monte Carlo dosimetry of a new 192Ir pulsed dose rate brachytherapy source. Med Phys 30:9–16

Kerr JFR, Wyllie AH, Currie AR (1972) Apoptosis: a basic biological phenomenon with wide ranging implications in tissue kinetics. Br J Cancer 26:239–257

Lea DEA (1956) Actions of radiations on living cells, 2nd edn. Cambridge University Press, Cambridge

Lowe SW, Schmitt EM, Smith SW, Osborne BA, Jacks T (1993) P53 is required for radiation-induced apoptosis in mouse thymocytes. Nature 362:847–849

Mason KA, Thames HD, Ochran TG, Ruifrok AC, Janjan N (1994) Comparison of continuous and pulsed low dose rate brachytherapy: biological equivalence *in vivo*. Int J Radiat Oncol Biol Phys 28:667–671

Mazeron JJ, Simon JM, Crook J et al (1991a) Influence of dose-rate on local control of breast carcinoma treated by external beam irradiation plus iridium-192 implant. Int J Radiat Oncol Biol Phys 21:1173–1177

Mazeron JJ, Simon JM, Le Pechoux C, et al (1991b) Effect of dose-rate on local control and complications in definitive irradiation of T_{1-2} squamous cell carcinomas of mobile tongue and floor of mouth with interstitial iridium-192. Radiother Oncol 21:39–47

Mazeron JJ, Boisserie G, Baltas D (1997) Pulsed dose rate brachytherapy: a survey. Curr Oncol 4 [Suppl 1]:S4–S6

Meredith WJ, Editor (1967) Radium dosage: the Manchester system, 2nd edn. Livingstone, Edinburgh

Mitchell B, Bedford JS, Bailey SM (1979) Dose-rate effects in plateau-phase cultures of S3 HeLa and V79 cells. Radiat Res 79:520–536

Moulder JE, Fish BL (1992) Repair of sublethal damage in the rat kidney (abstract). In: Chapman, Dewey JD, Dewey WC, Whitmore GF (eds) Radiation research: a twentieth-century perspective, vol 1. Academic Press, San Diego CA

Orton CG (1989) Remote afterloading for cervix cancer: the physicist's point of view. In: Martinez AA, Orton, DG, Mould RF (eds) Brachytherapy HDR and LDR. Proceedings of brachytherapy meeting on remote afterloading: state of the art, Dearborn, Michigan May 1989. Nucletron, The Netherlands

Orton CG, Seyedsadr M, Somnay A (1991) Comparison of high and low dose rate remote afterloading for cervix cancer and the importance of fractionation. Int J Radiat Oncol Biol Phys 21:1425–1434

Paterson R (1963) Treatment of malignant disease by radiotherapy. Williams and Wilkins, Baltimore

Perez-Calatayud J, Ballester F, Serrano-Andres MA, Puchades V, Lluch JL, Limami Y, Casal F (2001) Dosimetry characteristics of the Plus and 12i Gammamed PDR 192Ir sources. Med Phys 12:2576–2585

Pierquin B (1971) Dosimetry: the relational system. Proceedings of a conference on afterloading in radiotherapy. US Department of Health, Education and Welfare. Publication number (FDA) 72–8024, Rockville, New York, pp 204–227

Pierquin B, Chassagne D, Baillet F et al (1973) Clinical observations on the time factor in interstitial radiotherapy using iridium-192. Clin Radiol 24:506–509

Resch A, Fellner C, Mock U, Handl-Zeller L, Biber E, Seitz W, Potter R (2002) Locally recurrent breast cancer: pulse dose rate brachytherapy for repeat irradiation following lumpectomy - a second chance to preserve the breast. Radiology 225:713–718

Stitt JA, Fowler JF, Thomadsen BR et al (1992) High dose rate intracavitary brachytherapy for carcinoma of the cervix: the Madison system. I. Clinical and radiological considerations. Int J Radiat Oncol Biol Phys 24:335–348

Thames Jr HD, Withers HR, Peters LJ (1984) Tissue repair capacity and repair kinetics deduced from multifractionated or continuous irradiation regimens with incomplete repair. Br J Cancer 49:263–269

Thomadsen BR, Shahabi S, Stitt JA et al (1992) High dose rate intracavitary brachytherapy for carcinoma of the cervix: the Madison system. II. Procedural and physical considerations. Int J Radiat Oncol Biol Phys 24:349–357

Turesson I, Thames HD (1989) Repair capacity and kinetics of human skin during fractionated radiotherapy: erythenam desquamation, and telangiectasia after 3 and 5 year's follow-up. Radiother Oncol 15:169–188

Van Rongen E, Thames HD, Travis EL (1993) Recovery from radiation damage in mouse lung: interpretations in terms of two rates of repair. Radiat Res 133:225–233

Visser AG, van den Aardweg JM, Levendag PC (1996) Pulsed dose rate and fractionated high dose rate brachytherapy: choice of brachytherapy schedules to replace low dose rate treatments. Int J Radiat Oncol Biol Phys 34:497–505

Ward JF (1981) Some biochemical consequences of the spatial distribution of ionizing radiation produced free radicals. Radiat Res 86:185–195

Ward JF (1988) DNA damage produced by ionizing radiation in mammalian cells: identities, mechanisms of formation and repairability. Prog Nucleic Acids Mol Biol 35:95–125

Withers HR, Thames HD, Peters LJ (1982) Differences in the fractionation response of acutely and late-responding tissues. In: Karcher KH, Kolgelnik HD, Reinartz G (eds) Progress in radio-oncology, vol II. Raven, New York, pp 287–296

14 Clinical Applications of Low Dose Rate and Medium Dose Rate Brachytherapy

CARLOS A. PEREZ, ROBERT D. ZWICKER, and ZUOFENG LI

CONTENTS

14.1 Introduction *310*
14.2 Radionuclides *310*
14.3 Afterloading Interstitial Brachytherapy *311*
14.3.1 Afterloading Iridium-192 Wires or Ribbons *311*
14.3.2 Through-and-Through Plastic Tubing Technique *311*
14.3.3 Suturing of Needles or Guides or Plastic Buttons *312*
14.3.4 Removable Iridium-192 Hairpin Technique *313*
14.3.5 Removable Iodine-125 Plastic Tube Implants *313*
14.3.6 Permanent Interstitial Iodine-125 Implants *314*
14.4 Templates *314*
14.4.1 Syed/Neblett Templates *314*
14.4.2 Martinez Universal Perineal Interstitial Template *315*
14.5 Molds *316*
14.6 Remote Control Afterloading *316*
14.6.1 Pulsed Dose Rate *317*
14.6.2 Financial Considerations *319*
14.7 Implantation Techniques *319*
14.7.1 Anesthesia *319*
14.7.2 Preoperative and Postoperative Orders *320*
14.7.3 Radiation Safety in the Operating Room *320*
14.7.4 Removal of Implants *320*
14.7.5 Feeding the Patient with a Head and Neck Implant *321*
14.8 Low Dose Rate Brachytherapy Techniques for Specific Sites *321*
14.8.1 Interstitial Brain Implants *321*
14.8.2 Implants with Multiple Iridium-192 Sources *322*
14.8.3 Eye *324*
14.8.3.1 Episcleral Plaque *324*
14.8.3.2 Pterygium *326*
14.8.4 Head and Neck *327*
14.8.4.1 Maxillary Sinus *327*
14.8.4.2 Nasal Vestibule *328*
14.8.4.3 Skin and Lip *329*
14.8.4.4 Nasopharynx *329*
14.8.4.5 Oral Cavity *331*
14.8.4.6 Tongue and Floor of Mouth *331*
14.8.4.7 Base of Tongue *333*
14.8.4.8 Tonsillar Region Including Faucial Arch (Oropharynx) *334*
14.8.5 Breast *335*
14.8.6 Lung and Mediastinum *339*
14.8.7 Esophagus *339*
14.8.8 Pancreas *340*
14.8.9 Biliary Tree *341*
14.8.10 Soft Tissue Sarcomas *343*
14.8.11 Uterine Cervix *345*
14.8.11.1 Applicators for Carcinoma of the Cervix *347*
14.8.12 Endometrium *352*
14.8.13 Iridium-192 Interstitial Brachytherapy for Locally Advanced or Recurrent Gynecological Malignancies *353*
14.8.14 Vagina, Vulva, and Female Urethra *353*
14.8.14.1 Vaginal Cylinders *353*
14.8.14.2 Carcinoma In Situ *355*
14.8.14.3 Stage I *355*
14.8.14.4 Stage II *356*
14.8.14.5 Stages III and IV *356*
14.8.14.6 Tumors of the Rectovaginal Septum *356*
14.8.14.7 Lesions of the Bladder or Proximal Female Urethra *357*
14.8.14.8 Tumors of the Vulva or Distal Urethra *358*
14.8.15 Anal Canal and Rectum *358*
14.8.16 Penis and Male Urethra *360*
14.8.17 Prostate *361*
14.8.17.1 Transperineal I-125 Implants *362*
14.8.17.2 Palladium 103 Implants *366*
14.8.17.3 Removable Interstitial Implants with Iridium-192 *366*
14.8.17.4 Interstitial Irradiation and Hyperthermia *367*
14.10 Quality Assurance and Radiation Safety in Brachytherapy *367*
14.10.1 Safety Regulations in the United States *367*
References *370*

C. A. PEREZ, MD
Professor, Department of Radiation Oncology, Washington University Medical Center, 4511 Forest Park Boulevard, Suite 200, St. Louis, MO 63108, USA
R. D. ZWICKER, PhD
Department of Radiation Oncology, University of Kentucky Medical Center, 800 Rose Street, Lexington, KY 40536, USA
Z. LI, DSc
Associate Professor, Division of Radiation Physics, Washington University Medical Center, 4921 Parkview Place, Campus Box 8224, Saint Louis, MO 63110, USA

This chapter is an update of a similar chapter published in *Principles and Practice of Radiation Oncology*, 4th edn., Lippincott Williams Wilkins (PEREZ et al. 2004).

14.1
Introduction

After decades of use of radium sources and radon seeds, reactor-produced isotopes became available for brachytherapy, including ^{198}Au, ^{60}Co, ^{137}Cs and ^{192}Ir, as well as ^{125}I and ^{252}Cf a few years later, and, more recently, ^{241}Am, ^{103}Pd, ^{169}Yb, ^{75}Se, and ^{145}Sm (IYER and SHANTA 1994). The combined administration of interstitial thermoirradiation in a variety of lesions has been reported (EMAMI and PEREZ 1991).

Medical imaging and computers have enhanced the ability to generate precise dosimetric calculations and to improve dose distribution in patients treated with brachytherapy. HALL and BRENNER (1992) published excellent reviews on the biological basis of brachytherapy and clinical implications of dose rate, including pulse dose rate (PDR).

The widespread use of remote afterloading devices has enhanced the clinical applications of brachytherapy and has practically eliminated radiation exposure to the operators. Furthermore, in many parts of the world, high dose rate (HDR) brachytherapy has supplanted low dose rate (LDR), with equivalent clinical results.

According to the INTERNATIONAL COMMISSION ON RADIATION UNITS (ICRU) Report No. 38 (1985), dose rates of 0.4–2 Gy/h are referred to as LDR, those in the range of 2–12 Gy/h as medium dose rate (MDR), and those greater than 12 Gy/h as HDR. PDR will be defined later in the chapter.

The distribution of dose around radioactive sources depends on the physical properties of the isotopes, including the encapsulation and activity of the sources, and the inverse-square law. At distances greater than three times the physical length of a source, the inverse-square law applies within practical approximation; at closer distances, the dosimetry is more complex.

GIAP and MASSULO (1999) used the linear-quadratic model based on Dale's formalism to derive the brachytherapy dose rate at which biological effectiveness is equivalent to that of external beam irradiation based on therapy relative effectiveness, dose rate, alpha/beta ratios, and implant duration. The isoeffect dose rate depends on the dose per fraction, sublethal damage repair constant, and the implant duration, but it does not depend on alpha/beta ratio. For sufficiently long implant duration greater than 10–15 h, the value for isoeffect dose rate approaches a constant value at approximately 40–50 cGy/h. However, DALE and JONES (2000) caution that such studies should be based on an averaged biologically effective dose (BED) representative of the entire treated volume rather than the lower prescribed dose at the surface.

14.2
Radionuclides

To meet all clinical situations, a variety of radionuclides must be available. The Amersham ^{137}Cs stainless-steel encapsulated sources have been widely used in LDR brachytherapy with manual or remote afterloading. CASAL et al. (2000) presented Monte Carlo calculations of absolute dose rate in water around this source using a Monte Carlo code in the form of along-away tables and in the TG43 formalism, which can be used as benchmark data to verify treatment-planning-system calculations or directly as input data for treatment planning.

ZHANG et al. (2004) recently reviewed the use of the American Association of Physicists in Medicine (AAPM) TG-13 dose formalism applied to a ^{137}Cs source modeled after the Nuclear Associates 67-809 series stainless-steel jacketed tube sources used for gynecological implants. The dose rate distribution through the center of the source using the AAPM TG-43 dose formalism was compared with the calculations obtained using the Sievert summation and Monte Carlo simulation. The three methods resulted in an agreement within less than 5%, or an isodose rate line agreement within 2 mm.

Afterloaded ^{192}Ir wires or seeds (in nylon strands) have been used in many sites. Iridium-192 has a relatively short half-life (72 days); thus, the wires must be calibrated often.

ANDERSON et al. (1985) developed a nomographic planning guide to be used for planar implants of ^{192}Ir seeds in ribbons. Planar interstitial dosimetry systems have also been described by ZWICKER et al. (1985, 1994) and KWAN et al. (1983). A simplified dosimetry system for ^{192}Ir volume implants was designed by OLCH et al. (1987). Dose parameters of ^{192}Ir and ^{125}I seeds have been reported by WEAVER et al. (1989). KARAISKOS et al. (2001) evaluated the AAPM Task Group 43 dosimetric formalism for ^{192}Ir wires used as interstitial sources in LDR brachytherapy applications and presented them in look-up tables that allow interpolation for dose rate calculations around all practically used wire lengths, with accuracy acceptable for clinical applications.

As longer iridium wires have become available, the necessity for crossing one or both ends of the implant has almost disappeared. Sources of 0.4–0.5 mCi Ra eq/cm are used for single-plane arrangements, and sources of 0.25–0.35 mgRaEq/cm linear intensity or seeds of equivalent activity are typically used for multiple-plane or volume implants. The two intensities have been combined for complex implants (DELCLOS 1984).

^{125}I seeds are widely used for permanent implants in the prostate and in less accessible areas and for tumors that require surgical exposure at laparotomy or thoracotomy, such as the pancreas or lung. Other isotopes, such as palladium-103 (^{103}Pd), americium-241 (^{241}Am), and californium-152 (^{152}Cf), were introduced in clinical practice.

14.3
Afterloading Interstitial Brachytherapy

The flexible carrier method was first used with radon seeds by HAMES (1937) in 1937. Afterloading was systematized by HENSCHKE et al. (1963) and SUIT and FLETCHER (DELCLOS 1982a).

In the early 1960s, PIERQUIN et al. (1971) popularized the Henschke techniques with modifications and contributed the use of "hairpins" for afterloading with thicker iridium wires (0.5 mm diameter), mainly for lesions of the oral cavity and oropharynx.

KOLOTAS and ZAMBOGLOU (2001) recently reviewed the current status of interstitial brachytherapy. Modern techniques involve the use of ^{192}Ir in computer-controlled remote afterloading machines, which can deliver HDR, PDR or LDR brachytherapy. Treatment planning is undertaken with computers and anatomical cross-section images using computed tomography (CT), ultrasound, or magnetic resonance imaging (MRI). Several reports in the literature describe techniques and instrumentation for the use of afterloading interstitial therapy with other isotopes, such as tantalum wires (MEERWALDT et al. 1989), ^{125}I seeds, and ^{103}Pd (ALLT and HUNT 1963; HENSCHKE 1956; MORPHIS 1960).

14.3.1
Afterloading Iridium-192 Wires or Ribbons

Removable implants are performed with either stainless-steel needles or semi-flexible Teflon or nylon catheters with metallic guides.

Stainless-steel or Teflon 16-gauge tubing (1.6 mm in outer diameter) is cut into the desired length. The distal end of the tubing is beveled at a 30- or 45-degree angle and is crimped but not closed to hold the afterloaded iridium insert in place, still allowing for repositioning should it be required. A nylon or Teflon ball or a metallic button is fitted snugly at the proximal end (DELCLOS 1980). In this Teflon ball, a further development of the glass ball used by HAMES (1937), or metallic button, a hole is inserted to thread the suture, and a lead bead is added for X-ray localization. Figure 14.1 shows the procedures followed for insertion and afterloading of the rigid guides in tumors approachable from one side only (e.g., tumors of the columella and nasal septum, base of the tongue, female urethra, and anal margin). This technique can also be used to treat the parotid gland, involved lymph nodes in the neck, or for primary breast cancer.

14.3.2
Through-and-Through Plastic Tubing Technique

This technique is used when a tumor can be transfixed from both sides (e.g., lower and upper lip, buccal mucosa, breast, or neck masses). In locations in which the guide can be placed through the tumor or normal tissues, the 16-gauge metallic guides are inserted at the appropriate distances to achieve the desired distribution (Fig. 14.2). With the guides in place, the narrow leads of the nylon tubes into which the ^{192}Ir ribbons will be loaded are inserted into the guides. The metallic guides and leads are then pulled through and out of the tissue, leaving the nylon tubes in their place. The tubes are then secured at the lead end by crimped metal buttons, Teflon ball washers, or similar means. When all of the nylon tubing has been implanted, the desired length of the active wire or nylon ribbon with seeds is measured by using a "dummy" wire (to 0.5 cm below the skin at the opposite end) and cut a few centimeters longer so that it will protrude beyond the skin and will be easier to manipulate. After localization images are taken with inactive wires (or seeds) used to determine the length and position of the tubes, the ^{192}Ir active sources are prepared and inserted, and the proximal end of the tubing is crimped with a metallic button. We identify each dummy and corresponding active source or wire with different color threads and buttons and specific radiopaque patterns to determine the position of each tube or loading on the patient or the implant radiographs.

Afterloading of the active sources with either stainless-steel needles or flexible guides can be done

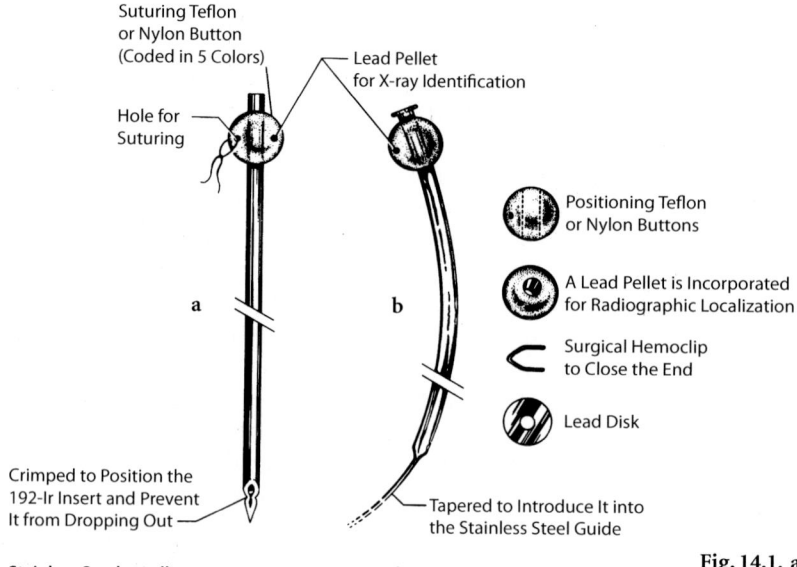

Fig. 14.1. a Stainless-steel afterloadable needle with plastic, Teflon, or nylon buttons. The needles are made to any desired length. **b** Teflon or nylon tube, closed-end variety, used for the through-and-through technique. A stainless-steel guide of the same outside diameter is inserted first (DELCLOS 1982b)

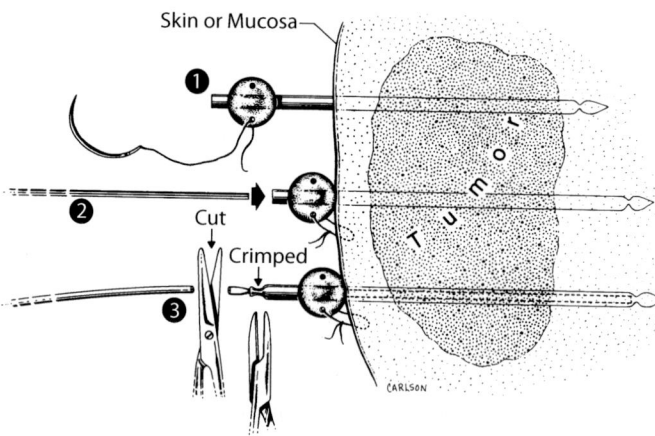

Fig. 14.2. One-end implant technique for tumors approached from one side. (*1*) Insertion of empty stainless-steel needle with nylon button. The needle is sutured to the skin or mucosa through a hole in the nylon button. (*2*) The iridium wire mounted in plastic tube carrier is introduced into the stainless-steel needle. (*3*) The open end is crimped to close it. The plastic tube carrier is cut, leaving approximately 0.5 cm protruding to facilitate removal of the iridium wire when indicated. To remove it, one can either cut the suture and remove the needle with the iridium inside, then deposit it in the leaded carrier for transportation and further manipulation in the laboratory, or uncrimp the end of the stainless-steel needle with a specially designed uncrimper and pull out the plastic tube carrier with the iridium insert inside (DELCLOS 1980)

after the patient is back in the hospital room. Radiation exposure within the operating and recovery rooms is thereby avoided (PEREZ et al. 1998).

14.3.3
Suturing of Needles or Guides or Plastic Buttons

Needles or guides are sutured to the implanted tissues in various ways (FLETCHER and MACCOMB 1962). Separate 2-0 silk or cotton sutures permanently attached to a half-circle taperpoint needle, which is threaded through the loop of the color-coded silk before insertion, are preferred.

Color-coded silk threads are used to identify the different lengths and strengths of the radioactive sources. This facilitates both the selection of sources at the time of the implant and the orderly removal of the implant.

Needles or buttons holding the catheters should be sutured systematically. For a double-plane implant, all suturing is done outside the needle rows to simplify removal.

14.3.4
Removable Iridium-192 Hairpin Technique

The physical characteristics of the Paris technique have been described. Metallic gutter guides have been constructed to facilitate insertion of the iridium wires (Fig. 14.3) (PIERQUIN et al. 1987a). The usual separation of the legs is 1.2 cm, although 0.9-cm or 1.5-cm separation can be used. The standard gutter length is 2.5, 3, 4, or 5 cm. Iridium wire ends are inserted along the gutters and held in place with a fine-tip clamp while the gutter guide is removed (Fig. 14.4). Gutter guides should allow for a predictable insertion of the hairpin, which will result in an acceptable geometry and homogeneous dose distribution of the implant (Fig. 14.5). The gutter guide technique is used primarily in smaller tumors of the oral cavity and in the anal region.

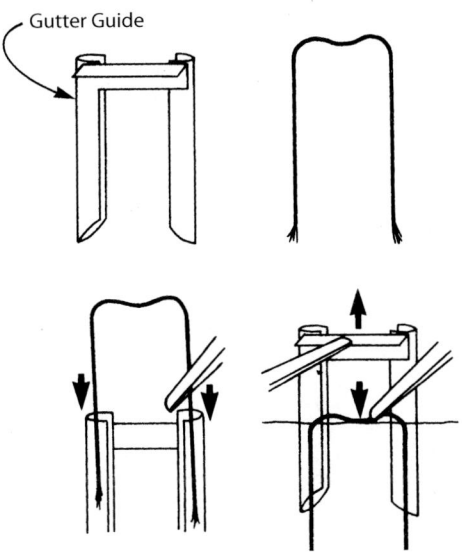

Fig. 14.4. Diagram showing basic design of gutter guide and technique for insertion of the iridium wire and subsequent removal of the guide while holding the wire with clamp

14.3.5
Removable Iodine-125 Plastic Tube Implants

CLARKE et al. (1989) described a temporary removable ^{125}I plastic tube implant technique. ^{125}I seeds were 4.5 mm in length, and the interseed spacing within the ribbons (from seed center to seed center) ranged from 4.5 mm (seeds back to back) to 12.5 mm (8-mm spacers). The operative technique, using hollow stainless-steel 18-gauge trocars, is identical to the ^{192}Ir implant procedure. ^{125}I dosimetry is somewhat more complex, since iodine-seed dose distributions are more anisotropic, fall off more rapidly with distance, and are more sensitive to tissue heterogeneities than ^{192}Ir sources. However, the ^{125}I tubes must have a greater diameter to house the ^{125}I seed ribbons, which are larger than the ^{192}Ir ribbons.

The seed ribbons are prepared by loading loose seeds into the hollow ribbons; the seeds are separated by spacers and held in position by a "pusher." The open end of the seed ribbon is heated for sealing. The seed separation varies depending on the activity, the geometry of the implant, and the desired dose rate, which is individualized for each patient and determined after the procedure in the operating room is completed. The most common clinical applications of temporary ^{125}I seeds are episcleral plaque therapy for ocular melanoma, volume implants in the brain, and occasionally for breast implants.

Compared with the ^{192}Ir implants, use of the ^{125}I seed ribbons requires additional physicist or

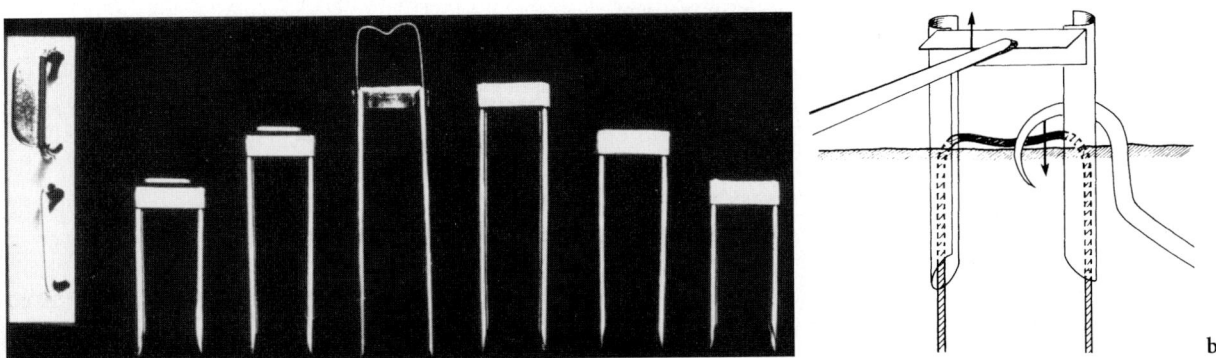

Fig. 14.3. a Hairpins of different sizes and iridium wire (*center*). **b** Diagram of gutter guide used by Pierquin and small hook to hold the iridium wire in place while the guide is being removed with a clamp (PIERQUIN et al. 1987b)

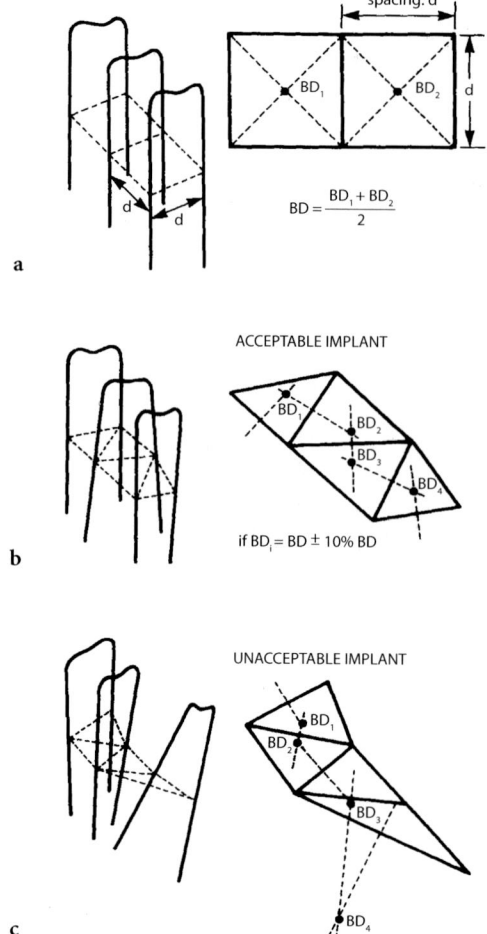

Fig. 14.5a–c. Diagrams illustrating dosimetry principles of Paris system. **a** For implants containing more than one plane, equidistant radioactive lines imply that the intersections of the lines with the central plane will be arranged as the apices of equilateral triangles or as the corners of squares. Calculation of basal dose rate (*BD*) is made at various points. Dose is specified along an isodose surface defined as a given proportion of the basal dose rate calculated inside the implant volume (reference isodose, which should encompass target volume as closely as possible). In practice, the value of the reference isodose is fixed at 85% of the basal dose rate. **b** Geometry of acceptable implant. **c** Geometry of unacceptable implant

dosimetrist time to assemble and disassemble the ribbons. However, this is offset by a compensatory decrease in other tasks that are required for the preparation of the ^{192}Ir seeds or wires. Because of the lower energy of ^{125}I, shielding is easily accomplished, which increases safety during the operation and decreases exposure to nurses caring for the patient.

14.3.6
Permanent Interstitial Iodine-125 Implants

With the widespread popularity of permanent ^{125}I prostate implants in recent years, there has been a remarkable increase in the number of providers for seeds and peripheral equipment for this modality. The Radiological Physics Center at M.D. Anderson, in conjunction with the AAPM, maintains a list of ^{125}I seeds for which they consider the dosimetric characterization to be complete and reliable.

In circumstances in which the supplied surgical needle is unsuitable, it can be replaced by a tie-on needle (e.g., a French spring-eye needle). The placement of the strands and spacing of the seeds should follow appropriate dosimetric considerations. The absorbable carrier material and ^{125}I seeds are implanted in the tumor tissues by successive advancing of the needle and gentle pulling of the carrier.

The carrier material is absorbed by body tissue; the rate depends on the nature of the implanted tissue. Intramuscular implantation studies in rats showed that the absorption of the carrier is minimal until the 40th postoperative day. Absorption is essentially complete between 60 days and 90 days.

GOFFINET et al. (1985) reported on 64 intraoperative ^{125}I implants with absorbable Vicryl suture carriers performed in 53 patients with head and neck cancers, many of them recurrent after initial definitive radiation therapy.

A variation of this technique was described by GREENBLATT et al. (1987), who sewed the ^{125}I suture material through Gelfoam, which in turn was secured to the tumor bed with special clips.

14.4
Templates

A variety of templates have been designed in an attempt to more easily place interstitial sources and to obtain more homogeneous doses with implants.

14.4.1
Syed/Neblett Templates

Several Syed/Neblett templates are primarily used for gynecological tumors. They consist of two Lucite plates joined by six screws, which tighten to grasp as many as 38 afterloading, hollow, stainless-steel needles. An additional six needles fit into grooves of a 2-cm diameter plastic vaginal cylinder, which is placed inside an opening in the middle of the template. These needles are arranged in concentric circles or arcs with a spacing of 1 cm between adjacent needles (ARISTIZABAL et al. 1985). A 4×10-cm area can be implanted in a butterfly distribution. The 18-gauge needles supplied with the templates are 20-cm long, but they can be shortened to treat more shallow areas. The vaginal cylinder has a central opening for placement of a tandem if desired.

A rectal template is similar to the one just described, but the two plates contain three concentric circular rings with a total of 36 needles with 1-cm spacing. Cylindrical volumes with diameters of 2, 4, or 6 cm can be implanted. A rectal tube can be placed in the central hole if necessary, but this hole can be left open if the template is placed in an area not covering the anus, such as the vulva.

SYED et al. (1983a) described a prostate template used to guide the insertion of metallic source guides transperineally. The template consists of two concentric rings with radii of 1 cm and 2 cm, containing 6 and 12 guide holes, respectively (Fig. 14.6) (PUTHAWALA et al. 1985; SYED et al. 1983a). Up to 18 metallic source guides (18-gauge, 20-cm-long needles) are inserted transperineally through the prostate and seminal vesicles as indicated. The tips of the guides are usually 1 cm above the level of the bladder neck. The template is fixed to the perineum by "00" silk sutures, and the space between the perineum and the template is filled with gauze soaked in antibiotic cream.

The urethral template has two concentric rings with a total of 18 needles with the same 1-cm spacing as the rectal template. A cylindrical volume with either a 2- or 4-cm diameter is implanted with this template. This is a single plate with no machine screws to other plates. A Foley urethral catheter is inserted through the central opening to drain the urinary bladder.

HÖCKEL and MULLER (1994) described a modified Syed/Neblett-type perineal template for HDR interstitial brachytherapy of gynecological malignancies. The template can easily be disassembled after insertion of the central needles into the pelvis, allowing cystoscopic and rectoscopic control of the needle positions. Needles penetrating the bladder or the rectum can be repositioned before reassembling the template, eliminating a high-irradiation zone in tumor-free bladder and rectum walls.

14.4.2
Martinez Universal Perineal Interstitial Template

The Martinez universal perineal interstitial template (MUPIT) was designed to treat locally advanced or recurrent tumors in the prostatic, anorectal, perineal, or gynecological areas. The device consists of two acrylic cylinders, one that can be placed in the vagina and the other in the rectum, an acrylic template with an array of holes that allows placement of the metallic guides in the tissues to be implanted, and a cover plate (Fig. 14.7) (MARTINEZ et al. 1985b). The cylinders are placed in the vagina, rectum, or both and fastened to the template so that a fixed geometric relationship among the tumor volume, normal structures, and source placement is preserved throughout the course of the implantation. When the MUPIT interstitial template is used, no central intracavitary sources are inserted, except in some patients requiring an intrauterine tandem (beyond the volume treated with the interstitial sources).

 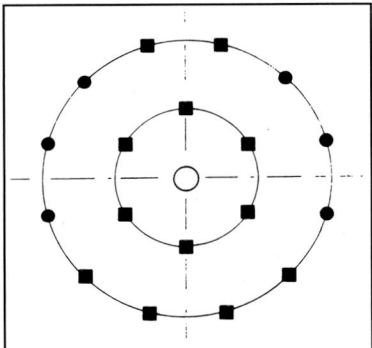 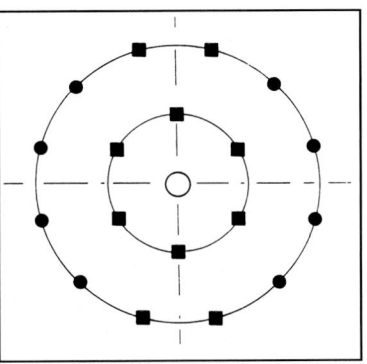
b

Fig. 14.6. a Syed/Neblett prostate template (SYED et al. 1983a). **b** Example of different intensity sources used with Syed template to decrease doses to urethra, bladder, and rectum (PUTHAWALA et al. 1985)

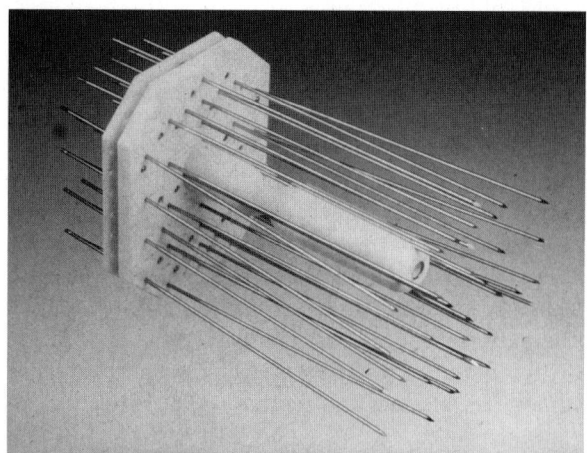

Fig. 14.7. a Martinez Universal Perineal Interstitial Template (MUPIT). (Courtesy of Dr. Alvaro Martinez, William Beaumont Hospital, Detroit, Michigan.) **b** Diagrammatic representation in coronal and sagittal planes of same template (MARTINEZ et al. 1985b)

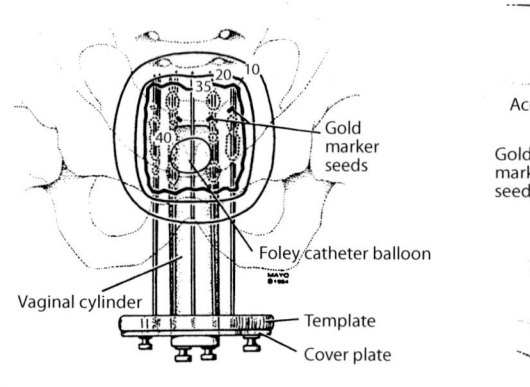

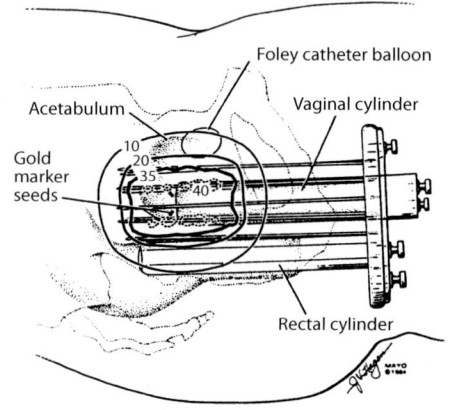

14.5
Molds

Molds have been used for the treatment of patients with skin cancer of the face or hands or other anatomic locations and for lesions of the lip and oral cavity (PATTERSON 1963). The mold can be constructed from plastic or acrylic, after initially obtaining the configuration of the anatomic area to be molded with a liquid plaster cast to form a negative plaster mold. Computations for the dose desired are carried out, and optimal placement of the sources is determined. Small holes are drilled in the mold to contain the nylon ribbons or catheters with the radioactive sources or the rigid cesium needles. These techniques were extensively used by PATTERSON (1963) and FLETCHER (1980).

MARCHESE et al. (1984) described a technique using ^{125}I sources embedded in Gelfoam to permanently implant small residual tumors or tumor margins in anatomic locations where standard implant techniques may not be feasible, for instance, at sites involving tissues adjacent to major blood vessels, the vertebral column, or the brain. The technique con-

sists of preparing an adequate size and thickness of Gelfoam and fixing it to the implant site. Catheters with ^{125}I seeds inserted into the Gelfoam are placed at 1-cm intervals. The absorbable Gelfoam mesh is sutured with catgut absorbable material, and both are absorbed over 6 weeks.

Acrylic molds have been used for the treatment of vaginal or uterine cervix lesions. LICHTER et al. (1978) described the use of thermoplastic vaginal molds for this purpose. The locations of the channels for insertion of the sources and for the central tandem (if desired) are determined by the topography of the tumor. The central tandem can be placed in the uterus or the vagina, through the vaginal mold, and locked into position.

14.6
Remote Control Afterloading

Remote afterloading brachytherapy for interstitial and intracavitary applications is used for both LDR and HDR implants. GLASGOW (1996) reviewed the developmental aspects of remote afterloading and described the characteristics of several commercially available systems.

LDR units use ^{137}Cs or ^{192}Ir, while HDR systems are built for either ^{60}Co or ^{192}Ir sources. Shielded rooms equivalent to ^{60}Co are necessary for HDR procedures.

The advantages of LDR remote control afterloading include:

- Radiation exposure to hospital personnel virtually eliminated
- Improved control of isodose distributions
- Low probability of misplacing or losing sources
- Less source preparation work for the source curator
- Medical and nursing staff not rushed; no fear of exposure while caring for the patient
- Source loading, unloading, and recording performed automatically

After the unloaded applicators are placed in the patient, the sources are loaded under pneumatic or mechanical control through hollow tubes connected to the applicators by a remotely activated system. A sorting and selection device and transport train for the sources are available. Safety mechanisms for checking correct connection of the applicator and the position of the sources are integral components of the system (WILLIAMSON 1991; WILLIAMSON et al. 1995). These units produce a hardcopy of the treatment technical parameters at completion of the procedure. Equipment for remote control afterloading brachytherapy is available for multiple anatomic sites and applications.

A special problem with remote afterloading equipment for gynecological use is reproduction of the isodose distributions obtained with standard 2-cm cesium tubes and the Fletcher-Suit-Delclos tandem, ovoids, and vaginal cylinders. A fixed source train decreases flexibility unless several source trains are in inventory. A system of active and inactive pellets (Nucletron Selectron) require the compilation of a dose-distribution atlas to duplicate the standard 2-cm cesium tube dose distribution of the Fletcher-Suit applicators (GRIGSBY et al. 1992).

ORTON et al. (1991), SCALLIET et al. (1993), FU and PHILLIPS (1990), and PETEREIT et al. (1999) compared HDR and LDR in gynecological brachytherapy, particularly regarding the conversion of LDR total dose into equivalent HDR dose per fraction and total dose. The reported clinical experience with HDR is equivalent to that of classic LDR. Treatment with LDR has proven to be quite tolerant to a lack of absolute precision, something that would be disastrous with HDR techniques.

14.6.1
Pulsed Dose Rate

PDR was proposed (BRINDLE et al. 1989) to exploit the advantages of HDR computer-controlled remote afterloading technology. By varying the dwell times of the stepping source, dose optimization could be achieved, maintaining the potential biological benefits of LDR with improved radiation protection. The inactive source times, when the sources are in the safe between pulses, should allow for better nursing care and visiting of the patients. A stepping source of 1 curie carries a sphere of "HDR" of a radius 20 mm within its track through tissue. The pulse initially delivers approximately 0.6 Gy per 10-min exposure every hour. As the dose rate gradually decreases because of radioactive decay of the source, somewhat longer periods of pulsed times are required (Fig. 14.8) (HALL and BRENNER 1992).

High ratios of PDR/LDR effect can be avoided by keeping dose per pulse below 1 Gy. Approximately

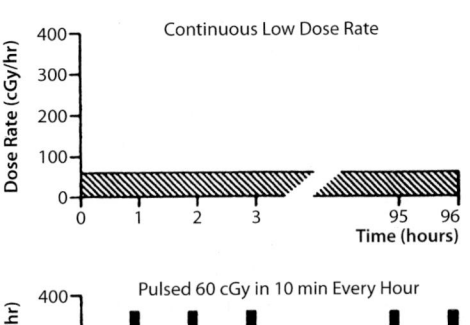

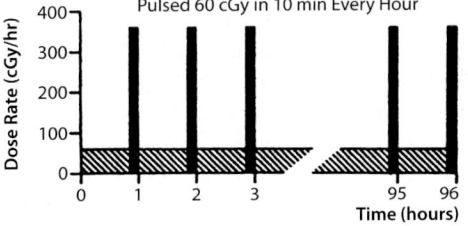

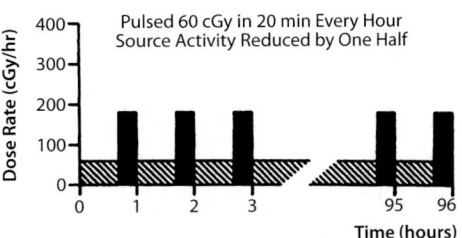

Fig. 14.8. Principles of pulsed brachytherapy. A continuous low-dose rate of 0.6 Gy/h, for example, is replaced by a pulse of 0.6 Gy delivered in 10 min. As the single ^{192}Ir source decays with a half-life of 74 days, the pulse length is adjusted to maintain the dose per pulse to precisely 0.6 Gy. Thus, the average dose rate is maintained, and the overall treatment time for a given total dose remains fixed (HALL and BRENNER 1992)

75% of the total dose is delivered at HDR in a PDR implant of moderate volume, reducing to 40% as a source decays from 1 curie to 0.3 curies. Even so, restricting the dose per pulse to 0.5 Gy or 0.6 Gy should avoid ratios of increased effect larger than approximately 10%. It appears that PDR delivered by stepping source might behave more like HDR than LDR, particularly for tissues with a substantial component of repair of very short half-life ($t_{1/2}$)

Linear quadratic formalism, in which the late normal tissue damage and tumor control were assumed by BRENNER et al. (1997) to be determined primarily by the level of cellular survival, was used. PDR schedules were designed in which pulses are delivered during "extended office hours" (0800 hours to 2000 hours) with no irradiation overnight. Generally, the proposed PDR regimes last the same number of treatment days as the corresponding LDR regimen, but the PDR treatment lasts longer on the final day. The protocols could allow the patient to go home overnight or to stay overnight in an adjacent medical inn or hospital-associated hotel, rather than in a hospital bed, which has major economic benefits. In such an economic situation, an extra treatment day for the daytime PDR could well be considered, which would virtually guarantee an improved clinical advantage relative to LDR.

Using the linear-quadratic formula, BRENNER and HALL (1991) determined the pulse lengths and frequencies based on radiobiological data that were equivalent to conventional continuous LDR irradiation. They noted that for a regimen of 30 Gy in 60 h, a 1-h period between 10-min pulses might produce up to a 2% increase in late effects probability.

VISSER et al. (1996) described a radiobiological model and equations to determine the HDR or PDR schedules equivalent to certain LDR schedules, similar to that proposed by BRENNER and HALL (1991), by applying probable ranges for the values for ∂/β ratio and repair time. They concluded that eight fractions of 1–1.5 Gy per 24 h, up to 3 h apart, would be equivalent to commonly used LDR treatment schedules.

ERICKSON and SHADLEY (1996), using in vitro irradiation experiments on rodent tumor cell lines, showed that there was a slight increase in cell killing with PDR relative to continuous LDR irradiation of hourly 5-, 10-, or 20-min pulses, or a 20-min pulse every 2 h. In no case were the increases statistically significant, and they did appear to be clinically indistinguishable as determined by the BRENNER and HALL criteria.

NARAYANA and ORTON (1999) developed a generalized extrapolated response dose (ERD) equation based on the linear quadratic model to account for the variation in the dose rate to maximize the therapeutic advantage (TA). They noted that with a careful choice of pulse length and frequency and using the ERD bioeffect dose model, TA values greater than 1 might be possible, depending upon the repair rate constants assumed for the tissues involved. Furthermore, for PDR treatments, the dose rate at a point of interest during each pulse is not uniform, since the treatment involves a single stepping source. NARAYANA and ORTON (1999) calculations indicated that PDR performed with 40 pulses in 120 h with an irradiation time of 30 min per pulse with a delay time of 2.5 h is the best replacement for a LDR treatment that delivers 60 Gy in 120 h.

FOWLER and VAN LIMBERGEN (1997) explored the possible increase of radiation effect in tissues irradiated by pulsed brachytherapy for local tissue dose rates between those "averaged over the whole pulse" and the instantaneous high dose rates close to the dwell position. Increased effect is more likely for tissues with short half-times of repair, of the order of a few minutes, similar to pulse durations. Calculations were done assuming the linear quadratic formula for radiation damage, in which only the dose-squared term is subject to exponential repair. A constant overall time of 140 h and a constant total dose of 70 Gy were assumed throughout, the continuous LDR of 0.5 Gy/h providing the unitary standard effects for each PDR condition. Effects of dose rates ranging from 4 Gy/h to 120 Gy/h (HDR at 1 Gy/min) were studied, covering the gap in an earlier publication. Four schedules were examined: doses per pulse of 0.5, 1, 1.5, and 2 Gy given at repetition frequencies of 1, 2, 3, and 4 h, respectively, each with a range of assumed half-times of repair of 4 min to 1.5 h. Ratios as high as 1.5 can be found for large doses per pulse (2 Gy) if the half-time of repair in tissues is as short as a few minutes. The major influences on biological effect are doses per pulse, half-time of repair in tissues, and – when $t_{1/2}$ is short – the instantaneous dose rate. Maximum ratios of PDR/LDR occur when the dose rate is such that pulse duration is approximately equal to $t_{1/2}$ As dose rate in the pulse is increased, a plateau of effect is reached for most $t_{1/2}$s, above 10–20 Gy/h, which is radiobiologically equivalent to the highest HDR.

In an editorial, HALL and BRENNER (1996) noted that although the linear-quadratic model has been widely used and accepted, it has not been tested in extreme cases and that the biological data needed for the model calculations are not very well known. They also pointed out that Visser et al. (1996) showed that the more different the proposed regimen is from continuous LDR the longer the overall treatment time needs to be extended to preserve the therapeutic ratio.

While PDR has prospered in Europe and Asia, unfortunately in the United States, it has floundered because the Nuclear Regulatory Commission (NRC) requires that a physicist and/or radiation oncologist (or other suitably qualified person) be present throughout the treatment, which is almost impossible to accomplish in a long treatment schedule in a hospital setting. Williamson et al. (1995) described the procedures and quality assurance regulations for PDR brachytherapy. The NRC has recently revised its position on the use of PDR units, requiring only that an authorized medical physicist and a trained physician be available on-call during PDR treatments. It is possible that the use of PDR techniques will become more widespread.

Swift et al. (1997) at University of California, San Francisco, CA, described results in 65 patients who underwent 77 PDR brachytherapy procedures as part of their treatment for pelvic malignancies. PDR brachytherapy showed no significant increased toxicity above that seen with the standard continuous low dose rate approach. Further trials will need to be carried out to determine if larger doses per pulse and shorter total treatment times have comparable therapeutic ratios.

De Pree et al. (1999) reported on 43 patients treated with PDR interstitial brachytherapy (24 with pelvic, 18 with head and neck, and 2 with breast cancer). Of 14,499 source and 14,399 dummy source transfer procedures, 3 technical machine failure events were observed (0.02%). Grade 3–4 late complications were observed in 4 of 41 (9.8%) patients.

Peiffert et al. (2001) prospectively evaluated PDR brachytherapy in 30 patients and concluded that PDR is feasible in patients with head and neck tumors but necessitates improvement of the quality of the plastic tubes.

14.6.2
Financial Considerations

Jones et al. (1994) compared the costs of HDR and LDR treatment (capital, maintenance, source, and operating costs) for Nucletron intracavitary equipment under alternative assumptions (three HDR fractions compared with one LDR fraction). The LDR-3 (Nucletron) was the most cost effective, practical machine for up to 40 patients per year; however, HDR would be recommended for more than 40 patients a year for practical reasons. Similarly, for 5 HDR compared with two LDR fractionations, LDR-3 was recommended for up to 20 patients per year and HDR for a greater number of patients. Recommendation was based on no cost sharing with other sites.

Bastin et al. (1993) compared HDR treatment cost with LDR intracavitary brachytherapy for gynecological malignancy. General anesthesia was used in 95% of applications with tandem and ovoid and in 31% for ovoid-only placement. Differences among private and academic practice respondents were minimal. At their institution, a 244% higher overall charge for LDR treatment was noted, primarily due to hospitalization and operating room expenses. In addition to its ability to save thousands of dollars per patient, HDR therapy generated a "cost-shift," increasing radiation therapy departmental billings by 438%. Capital investment, maintenance requirements, and depreciation costs for HDR brachytherapy are lower, since it is an outpatient procedure.

Grigsby and Baker (1995) reviewed the socioeconomic aspects of remote afterloading for both LDR and HDR, including required resources, reimbursement, cost-effectiveness, and a pro-forma analysis of a new facility and conversion of an existing facility.

14.7
Implantation Techniques

14.7.1
Anesthesia

Small implants can be done under local anesthesia (with or without monitored sedation); general anesthesia is sometimes preferable for good visualization and palpation of the tumor and for the patient's comfort.

General anesthetic is administered by nasotracheal intubation for implants of the oral cavity and lips. An elective tracheostomy is usually performed in patients with extensive oral cavity lesions requiring large implants and for all tumors of the glossopalatine sulcus, base of the tongue, or vallecula, because the associated edema may cause serious breathing difficulties (Delclos 1984).

Breast implants are done with local or general anesthesia.

For brachytherapy procedures in the pelvis, general or spinal anesthesia is administered. Occasionally a pudendal nerve block may be used.

WALLNER (2002) used local anesthesia for prostate brachytherapy without anesthesia personnel in attendance in 600 patients. The patient was brought into the simulator suite in the radiation oncology department, IV lines were started, a cardiac monitor was attached, and a urinary catheter was inserted. The patient was then placed in the lithotomy position, using stirrups. A 6- to 8-cm patch of perineal skin and subcutaneous tissue was anesthetized by local insertion of 1% lidocaine. The transrectal ultrasound (TRUS) probe was inserted and positioned to reproduce the planning images. A 3.0-inch 22-gauge spinal needle was used to inject the lidocaine up to the prostatic apex, in a pattern around the periphery of the prostate. Once the pelvic floor and prostatic apex were anesthetized, a 7.0-inch 22-gauge spinal needle was inserted through an 18-gauge 3-inch spinal needle in the peripheral planned needle tracts, monitored by TRUS. As the needles were advanced to the prostatic base, approximately 1.0 cc of lidocaine solution was injected in the intraprostatic track. A total of 200–500 mg of lidocaine was used. Patients tolerated brachytherapy under local anesthesia surprisingly well.

14.7.2
Preoperative and Postoperative Orders

The radiation oncologist must assess the condition of the patient before the brachytherapy procedure is performed and log in the chart detailed instructions for nursing personnel, including tests results to be obtained, medications to be administered, preparation procedures for the operating room, and radiation safety measures. After the procedure is completed, a description of it should be recorded in the chart, including a diagram illustrating the exact location and pattern of placement as well as characteristics of the sources (length, strength, etc.). Clear postoperative orders are necessary, including time of removal of radioactive sources, appropriate medications, and special precautions.

14.7.3
Radiation Safety in the Operating Room

If high-energy radioactive sources such as ^{192}Ir or ^{137}Cs are prepared in the operating room (rarely done in

the United States today), a workbench with shielding should be placed in one corner of the room so no one except the brachytherapy technician preparing them is exposed to radiation. The workbench is designed with a frontal working area with an L-shaped lead screen to protect the trunk, lower extremities, and medial aspect of the arms. In addition, a leaded-glass screen reduces exposure to the eyes.

Behind the barrier, there should be a lead well to store the remaining radioactive material while the individual needles, wire, grains, or seeds are being prepared for insertion into the patient. The bench is covered with sterile drapes.

Sterilization of the radioactive sources is done by soaking the cesium needles in a germicidal solution such as Cydex. Gold-grain magazines and iridium wires are sterilized by gas.

When using radioactive sources, the operating physician, assistants, and anesthesiologist should work behind individual lead barriers. Exposure to the eyes and hands can be reduced only by distance and by dexterity gained through experience. All radioactive sources should be handled with long instruments. Because most procedures are performed with afterloading techniques, exposure to the fingers during manipulation is minimal. After insertion of the sources and removal of the patient, the remaining sources should be carefully inventoried and the operating room surveyed using a Geiger–Müller detector.

The details of protective procedures used during the preparation and transportation of radioactive materials and the regulations governing them have been described. Compliance with NRC procedures and regulations is mandatory in the United States.

14.7.4
Removal of Implants

Interstitial needle sources can generally be removed in the patient's room. For patients with standard needles directly implanted in the posterior tongue and for less-than-cooperative patients, it is preferable, and sometimes essential, to remove the implant in the operating room, at times with the patient under general anesthesia; thus, adequate lighting, suction, and assistants are available. Bleeding at the time of needle removal is infrequent, but when it occurs, it may cause the patient or the assisting staff to panic. Firm and steady pressure with a finger on a compress over the bleeding point for several minutes usually is adequate treatment; occasionally, suturing of the blood vessel area with absorbable catgut may be necessary. It is not uncommon for the needle thread to be accidentally cut instead of the suture, and finding the needle requires an optimal surgical environment, because the task is complex and time-consuming. Radiographic localization of the needle may be required before the needle base can be surgically exposed.

The afterloading nylon tubing is more easily removed. For the sake of radiation protection, it is advisable, initially, to uncrimp the metallic buttons and carefully remove the radioactive sources, which are accounted for and immediately placed in a portable safe or shielded cart. After this is done, each individual tube is removed by freeing one end. For oral cavity or oropharynx implants, we prefer to cut the two ends of the tubing at the skin and pull them out through the oral cavity. A previously tied silk thread inside the cavity on the nylon tube loop is very helpful in this maneuver.

After all needles or tubes are removed, the implanted site may be gently palpated to verify that all implant materials have been removed. The patient and, after the radioactive sources are taken out of the room, the room should be surveyed with a Geiger–Müller counter or other radiation detector to make sure that there is no residual radioactivity.

Appropriate notes in the patient's chart, isotope form, and radiation survey form should be completed to record all procedures performed.

14.7.5
Feeding the Patient with a Head and Neck Implant

Although some patients undergoing head and neck implants can be allowed to sip a liquid formula through a straw, most are fed through a nasogastric tube. This is a strict necessity when the lips have been sutured together for implants involving the buccal commissure.

14.8
Low Dose Rate Brachytherapy Techniques for Specific Sites

14.8.1
Interstitial Brain Implants

Brachytherapy may allow delivery of interstitial radiation "boosts" to primary brain tumors after conventional external radiation therapy or may be used to treat recurrent brain tumors. At some institutions, permanent implants have been used; however, removable implants are more popular. In our opinion, the advantages of removable implants include: (1) greater control of the irradiation dose, since the source placement can be rearranged to improve dose distribution, and the time is controlled by the operator; (2) decreased possibility of migration of the radioactive sources by necrosis or fibrosis; (3) easy removal of the sources if emergency decompressive surgery is required; (4) less exposure to the patient's family and others coming in close proximity after hospital discharge; and (5) provision of dose rates greater than 0.3 Gy per h, which are necessary to treat fast-growing malignant brain tumors as suggested by some data (GUTIN et al. 1981).

Several techniques have been used for interstitial irradiation of the brain, some using multiple planar implants and ^{192}Ir wires or seeds and others with higher-intensity ^{125}I sources.

SNEED et al. (1997) noted that LDR brachytherapy (60–100 Gy given at 0.05–0.10 Gy/h) has been used for low-grade gliomas, resulting in 5- and 10-year survival of 85% and 83% for pilocytic astrocytomas and 61% and 51% for grade-II astrocytomas. Only 2.6% of patients had symptomatic radiation necrosis. For faster-growing high-grade gliomas, temporary implants delivering about 60 Gy at 0.40–0.60 Gy/h are generally used. Reported median survival times after brachytherapy are 12–13 months for recurrent malignant gliomas and 18–19 months for primary glioblastomas treated with external beam radiation therapy and brachytherapy boost. It is well known that higher irradiation doses may significantly increase risk of brain necrosis; over 50% of patients who undergo brachytherapy for malignant gliomas require reoperation for tumor progression and/or radiation necrosis. Strategies are under development to improve local tumor control without increasing radiation toxicity (SNEED et al. 1997).

LEIBEL et al. (1989) updated basic concepts on these techniques. They selected supratentorial, well-circumscribed lesions up to 6 cm in diameter for implants. They originally used the Leksell stereotaxic system but later changed to the Brown–Roberts–Wells frame. The procedure was performed with local anesthesia in adults and general anesthesia in children. The outer catheters were implanted in the tumor with stereotactic and CT-scan guidance through a small skin incision and burr holes or transcutaneous twist drill perforations. Implants were performed with ^{192}Ir- or ^{125}I-encapsulated sources encased in afterloading silicone catheters. Doses of 50–200 Gy were delivered with permanently implanted sources.

PRADOS et al. (1992) reported results in 56 patients with glioblastoma multiforme and 32 patients with anaplastic glioma treated with temporary ^{125}I interstitial implants and external irradiation (median, 59.4 Gy); most received concomitant hydroxyurea. Of the patients, 8 (14%) survived 3 years or longer, and 16 (29%) survived 2 years or longer. A second operation was necessary in 50% of patients to remove symptomatic necrosis produced by the implant. Prolonged steroid administration was used in many patients.

LARSON et al. (2004) updated reports on tumor resection and permanent, low-activity ^{124}I brachytherapy in 38 patients with progressive or recurrent glioblastoma multiforme and compared results with those of similar patients treated previously with temporary brachytherapy without tumor resection. Selection criteria were Karnofsky performance score greater than or equal to 60, unifocal, contrast-enhancing, well-circumscribed progressive or recurrent glioblastoma multiforme that was judged to be completely resectable, and no evidence of leptomeningeal or subependymal spread. The median brachytherapy dose 5 mm exterior to the resection cavity was 300 Gy (range, 150–500 Gy). Median survival was 52 weeks from the date of brachytherapy. Age, Karnofsky performance score, and preimplant tumor volume were all statistically significant on univariate analyses. Multivariate analysis for survival showed only age to be significant. Both univariate and multivariate analysis of freedom from progression showed only preoperative tumor volume to be significant. Comparison to temporary brachytherapy patients showed no apparent difference in survival time. Chronic steroid requirements were low in patients with minimal postoperative residual tumor.

SAW et al. (1989) evaluated the difference in dose distribution with various interstitial implant stereotactic techniques used at four institutions on an idealized tumor phantom 5 cm in diameter and 5 cm in length. Either 4 or 6 sources of ^{125}I or 9 or 24 ^{192}Ir catheters were used with different numbers of sources (Table 14.1) (SAW and SUNTHARALINGAM 1988). Quantitative evaluation of dose homogeneity using three volumetric irradiation indexes indicated that the dose homogeneity improved as the number of catheters and number of sources increased (Fig. 14.9). Dose volume histograms (area

under the histogram is target volume) demonstrated inhomogeneous irradiation. Institution A technique (4 catheters and 20 ^{125}I seeds) showed a highly inhomogeneous dose distribution compared with more homogeneous irradiation obtained with the technique from institution D (24 catheters and 140 ^{192}Ir seeds). The dose homogeneity of the implants from institutions B and C were between the others. The dose gradient outside the target volume was believed to be more dependent on the geometry of the implant than on the type of radionuclide. At distances beyond the first centimeter, the dose rate falls off at a lower rate for implants using ^{192}Ir sources than with ^{125}I.

VIOLA et al. (2004) described a method for verification of the position of implanted catheters with ^{125}I seeds after brachytherapy of brain tumors with fusion of the CT images used at planning and after the implantation of the catheters. The tumor volume covered by the prescribed dose and the normal tissue volume covered by the prescribed dose were compared between the plan and the actual result. The image fusion was performed by the Brain-Lab-Target 1.19 software on an Alfa 430 (Digital) workstation. The position of 16 of the 116 catheters (13.8%) required adjustment after the fusion of control images in the 70 cases studied.

14.8.2
Implants with Multiple Iridium-192 Sources

The technique described here was developed at Washington University using radioactive sources placed in Teflon catheters inserted into the brain under direct CT monitoring.

A radiolucent ring frame immobilizes the patient's head on the CT table top (Fig. 14.10a). Multiple burr holes are made in the brain at 1-cm intervals (patient under local anesthesia). The locations of the burr holes are determined by a template,

Table 14.1. Characteristics of brain implants from four institutions

Institution	A	B	C	D
Source type	^{125}I	^{125}I	^{192}Ir	^{192}Ir
No. of catheters used	4	6	9	24
No. of seeds used	20	42	36	140/28*
Seed spacing (cm)	0.95	1.0	1.5	1.0
Activity (mCi/seed)	30	12	2	0.79/0.54*

*The implant consisted of 140 and 28 seeds with seed activity of 0.79 mCi and 0.54 mCi, respectively (SAW et al. 1989)

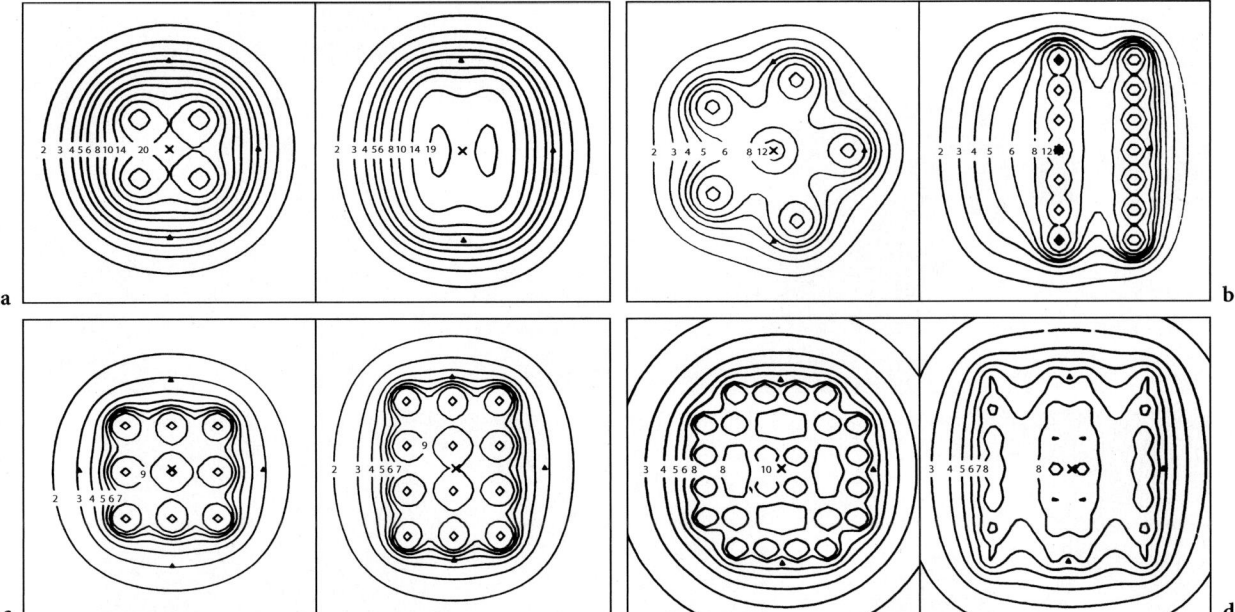

Fig. 14.9. a Dose distributions of brain implant at Institution A in two orthogonal planes. (*a*) the transverse plane and (*b*) the longitudinal plane, both bisecting the implant. Isodose rates in 0.1 Gy/h are labeled. **b** Dose distributions of the implant at Institution B in two orthogonal planes: (*a*) the transverse plane, taken 0.5 cm above the central plane, and (*b*) the longitudinal plane. Isodose rates in 0.1 Gy/h are labeled. **c** Dose distribution of the implant at Institution C in two orthogonal planes: (*a*) the transverse plane, bisecting the implant, and (*b*) the longitudinal plane. Isodose rates in 0.1 Gy/h are labeled. **d** Dose distributions of the implant at Institution D in two orthogonal planes: transverse plane and the longitudinal plane, both bisecting the implant. Isodose rates in 0.1 Gy/h are labeled (Saw et al. 1989)

which is attached to a stereotactic frame and to the patient's head. The template used is a thick acrylic block containing a 7×7-cm array of 49 holes spaced at 1-cm intervals. The holes along the diagonal axis of the template have slightly larger diameters to provide a method of orientation for each CT slice. The tumor is outlined on the CT screen with the aid of intravenously administered contrast material. The template is placed against the scalp at the site allowing best access to the tumor, usually a lateral surface. Intravenous contrast is administered and scanning performed with the scan plane parallel to the rows of the template. The target volume for the implant is the contrast-enhancing ring seen on CT scans, with a 1.0-cm margin. The number of catheters required to encompass the target at each level is determined at the CT console and 5/64-inch drill holes are made through the scalp, skull, and dura according to the matrix determined at the CT scan console. Next, 17-gauge catheters, 15 cm long, and calibrated at 1-cm intervals are then placed through the template into the brain to the desired depths with CT monitoring. Following the grid pattern, under CT observation, the Teflon Angiocath catheters with a metallic stylet are inserted through the burr holes into the brain

substance to ensure straight and parallel insertion (Fig. 14.10b).

After the tumor volume is implanted, the length of the radioactive sources is determined, and films, with the distribution of the catheters, are obtained for dosimetry calculations. Dummy seeds and ribbons are loaded in each of the catheters. Once the catheters are secured, the patient is transferred to the intensive care unit, where the dummy sources are replaced by ribbons of active ^{192}Ir seeds with a specific activity of about 0.6 mCi per seed. Metal buttons are attached to the catheters to fasten them to the scalp (Fig. 14.10c). Careful records are maintained of the position and length of all the catheters.

Computer-generated isodose calculations are used to determine the dose and distribution in the implant volume (Fig. 14.10d). The dose rate ranges from 0.5Gy/h to 0.8 Gy/h at 0.5 cm to 1 cm. In general, the implant duration is 70–100 hours, to deliver 60–70 Gy total dose to the entire tumor. Verification dosimetry with thermoluminescent dosimeters placed in catheters disclosed an agreement of ±5–10% between the computer calculations and the actual doses at any point within the irradiated

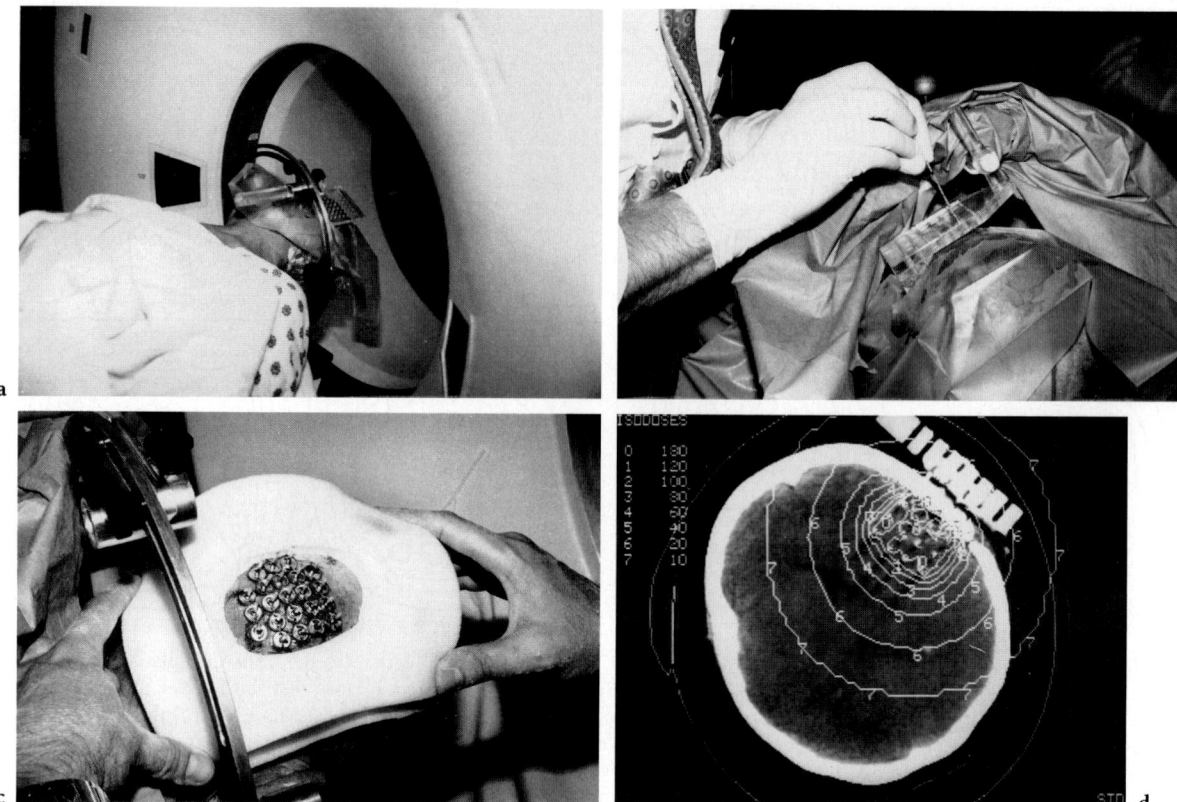

Fig. 14.10. a Patient on computed tomography (CT) scanner in position for ^{192}Ir brain implant. Stereotactic ring and plastic template to direct placement of catheters are shown. **b** Insertion of afterloading plastic catheter with metallic guide into the brain through small burr holes in the skull. A plastic template is used to determine exact positioning of catheters. **c** Patient after implant is finished, demonstrating metallic buttons sewn to scalp to secure catheters in place. **d** CT scan of skull with ^{192}Ir sources in place and isodose curve

volume (ABRATH et al. 1986). This method has been used in over 70 patients at Washington University, most of them with glioblastoma multiforme, sometimes recurrent after external irradiation, and in a few patients with solitary brain metastasis. Fatal intracranial bleeding has been rare (<5%), and edema is not severe enough to represent a significant management problem. Brain necrosis has been observed in approximately 25% of patients.

14.8.3
Eye

14.8.3.1
Episcleral Plaque

Episcleral plaque therapy is a cost-effective approach to treat localized intraocular malignancies such as retinoblastoma and choroidal melanoma. The technique consists of fabricating a small, spheri-

cally curved plaque containing radioactive sources, immobilizing the patient's eye, and suturing the plaque onto the sclera opposite the tumor, where it remains for 3–10 days. Because of the close proximity of the radioactive sources to the tumor, a highly localized and intense dose of irradiation is delivered to the tumor, which spares more normal tissue than is possible by conventional external-beam techniques and is competitive with the precision of heavy particle therapy. An interinstitutional randomized clinical trial through the Collaborative Ocular Melanoma Study (COMS) compares eye plaque therapy to enucleation with survival and preservation of vision as endpoints.

Historically, plaque therapy was delivered using the ^{60}Co plaque system originally developed by STALLARD (1961) for treatment of retinoblastoma. These plaques were available in a limited range of sizes (8–12 mm diameter), with both circular and semicircular notched configuration, for treatment of posterior lesions abutting the optic nerve.

PACKER et al. (1984) used ^{125}I seeds as a substitute for ^{60}Co plaques in the treatment of ocular melanoma. In the COMS clinical trial, ^{125}I seeds are being used in conjunction with standardized gold-alloy plaques ranging from 12 mm to 20 mm in diameter (CHIU-TSAO 1995). Each plaque is accompanied by a Silastic insert with precut channels for reproducibly positioning the seeds in concentric circles (Fig. 14.11) (HILARIS et al. 1988b). After the seeds are positioned in the insert, it is securely glued to the plaque so that the seeds are "sandwiched" between a 1-mm-thick layer of plastic and the gold backing of the plaque. A COMS plaque can be assembled within 30 min, almost entirely eliminates the possibility of seed loss during treatment, fixes the seeds in a rigid geometry, and retains a high degree of individualization. Notched or noncircular plaques can be fabricated using dental casting techniques.

LUXTON et al. (1988) and CHIU-TSAO et al. (1988) demonstrated that ^{125}I plaques give dose distributions very similar to those of ^{60}Co plaques. [125]I plaque therapy delivers retinal surface doses of 270–400 Gy for a prescribed dose of 100 Gy to the tumor apex (Fig. 14.12) (LUXTON et al. 1988). [125]I plaques offer several dosimetric advantages over ^{60}Co plaques. The 0.5-mm-thick gold plaque almost completely attenuates ^{125}I primary X-rays, providing a high degree of protection (95%) to tissue posterior to the eye. The 2.5- to 3.3-mm-high lip of the COMS plaque produces limited collimation of the [125]I X-rays, which reduces the area of the retina treated to a high dose. Moreover, a thin lead foil (0.2-mm thick) placed over the patient's eye affords substantial radiation protection, making it possible to treat with plaques on an outpatient basis. Intraoperative ultrasound localization has been used to help improve the position-

ing accuracy of eye plaques, potentially minimizing treatment failures (TABANDEH et al. 2000; HARBOUR et al. 1996; FINGER et al. 1998).

When using ^{125}I plaques, physicists and clinicians should be aware that WILLIAMSON (1988), WEAVER et al. (1989), and NATH et al. (1990) showed that conventional ^{125}I data overestimate the dose rate in water at 1 cm from a model 6711 seed by 13–20%. The AMERICAN ASSOCIATION OF PHYSICISTS IN MEDICINE (1995) has incorporated these differences into a new interstitial brachytherapy dosimetry protocol applicable to all tumor sites. WEAVER (1986) demonstrated that the gold backing of the plaque, which significantly reduces the volume of tissue contributing scatter dose to tissue anterior to the plaque, may reduce doses to points on the plaque axis by an additional 5–8%. CHIU-TSAO et al. (1986) have shown that the 1-mm-thick Silastic insert, which has an effective atomic number (11.2) higher than that of tissue, may reduce doses on the central axis of the plaque by 10%. Because currently used dosimetry algorithms and data take none of these effects into account, minimum tumor doses delivered by COMS plaques are probably no greater than 75% of the normally prescribed values (CHIU-TSAO 1995).

Prior to plaque fabrication, all relevant imaging studies should be examined to define the basal dimensions and location of the tumor. A-mode ultrasound studies are used to define the maximum height of the tumor. Fluorescein angiograms are often helpful in determining the posterior boundary of the tumor. The fundus view diagram, used by the ophthalmologist to record clinical impressions, represents a polar plot of the surface anatomy of the retina with its origin at the macula. When the ante-

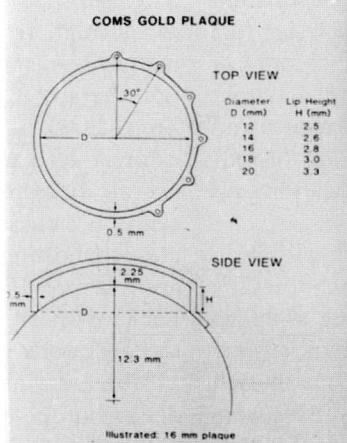 

Fig. 14.11. Drawings and photograph of the gold-alloy plaques used in the Collaborative Ocular Melanoma Study. A Silastic plastic insert, containing seed receptacles on its outer surface, is glued inside the plaque, which positions the ^{125}I seeds against the gold backing and maintains a treatment distance of 1.4 mm from seed to center to outer surface of the sclera (HILARIS et al. 1988b)

rior margin of the tumor is anterior to the equator, every attempt should be made to localize this margin relative to the ora serrata using transillumination. After the basal diameters, height, and location of the tumor are defined, a plaque is fabricated such that its diameter is 4–8 mm larger than the assumed diameter of the tumor. A dummy plaque of identical size is used to define the plaque position in the operating room using transillumination as the definitive guide to tumor localization and size. A small caliper should be available for measuring the orthogonal dimensions and location of the tumor relative to the ora serrata. These data should be used as the basis for the final treatment plan. Both fundus view isodose curves, which give the dose distribution on the retinal surface, and conventional transverse views are useful.

DAMATO and LECUONA (2004) reported on conservation of eyes in 1632 patients with choroidal melanoma treated with a multimodality approach of primary enucleation (35%), brachytherapy (31.3%), proton beam radiotherapy (16.7%), transscleral local resection (11%), endoresection (3.7%), transpupillary thermotherapy (2.5%), and photocoagulation (0.1%). Logistic regression showed the main predictive factors for primary enucleation to be: age more than 60 years [odds ratio (OR), 2.4], reduced visual acuity (OR, 2.5), posterior extension close to or involving the optic disc and fovea (OR, 3.5), circumferential spread around the ciliary body, iris, or angle (OR, 3.1), basal tumor diameter (OR 3.5), or tumor height (OR, 6.3).

The COMS group (2004) reported on 994 patients of 1003 enrolled in the COMS trial of preenucleation radiation and 1296 patients of 1317 enrolled in the COMS trial of ^{125}I brachytherapy. At 5 years after enrollment, 1307 of 2290 fellow eyes were examined; 358 eyes were examined 10 years after enrollment. Mean changes in visual acuity of fellow eyes from baseline to each examination was one letter (0.2 lines) or less. Cumulative 5-year incidence rates of cataract surgery and visually significant cataract in initially phakic eyes with good visual acuity and no lenticular opacity were 8% in both trials; 10-year rates were 18% in the trial of pre-enucleation and 15% in the trial of ^{125}I brachytherapy.

PUUSAARI et al. (2004) calculated radiation doses to intraocular tissues in ^{125}I brachytherapy for uveal melanoma in 96 patients, using a plaque simulator and a collimating plaque design, replacing the actual plaque with the modified one in each model. Median doses to tumor apex and base were 81 (range, 40–158 Gy) and 384 (range, 188–1143) Gy,

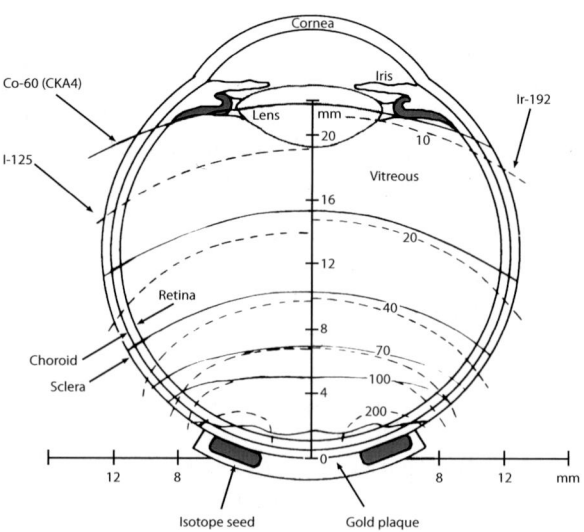

Fig. 14.12. Isodose curves in the transverse plane of the eye for a 12-mm diameter plaque. The solid lines indicate isodoses arising from the CKA-4 ^{60}Co plaque; whereas, the dashed lines on the right and left denote isodose curves arising from ^{192}Ir and ^{125}I seeds, respectively. All isodoses are normalized to 100% on the central axis 5 mm from the plaque surface (LUXTON et al. 1988)

respectively, and median dose rates at these points were 53 cGy/h and 289 cGy/h, respectively. Median doses to the lens, macula, and optic disc were 69, 79, and 83 Gy, respectively. Dose to the lens was associated with cataract [hazard ratio (HR) 1.15 for each 10-Gy increase, $P=0.002$], and dose to the optic disc with optic neuropathy (HR 1.08, $P=0.001$). Dose to the macula predicted a low vision (HR 1.06, $P=0.025$) and blindness (HR 1.10, $P=0.001$).

14.8.3.2
Pterygium

After surgical resection of the pterygium, because of the high recurrence rate (20–69%), it was, for many years, common practice to administer radiation therapy (COOPER 1978; VAN DEN BRENK 1968; ZOLLI 1979). In most institutions, a [90]Sr β-ray applicator is used for treatment of these patients. The overall diameter of the applicator is 12.7 mm; the center is a circular radioactive disk 5 mm in diameter, containing the isotope. The dose rate is generally approximately 5 Gy/min. In some models, a Lucite disk on the shaft of the applicator shields the operator's hands (Fig. 14.13a) (PIERQUIN et al. 1987b). A stainless-steel cover is available for some designs of the eye applicator and must be removed before the applicator is to be used for treatment.

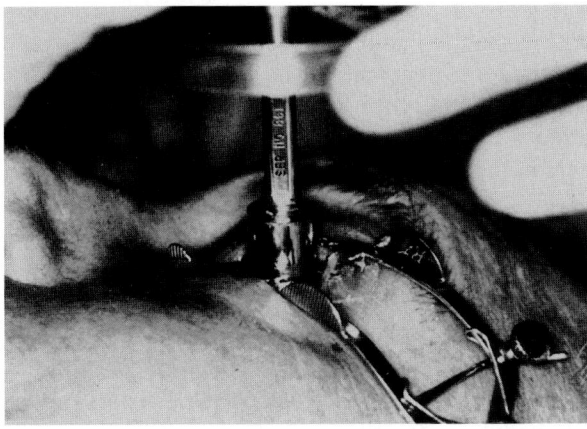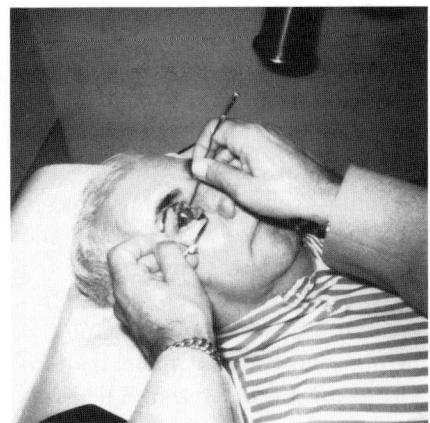

a   b

Fig. 14.13a,b. Patient undergoing ^{90}Sr application with eyelids retracted. Applicator has plastic shielding to protect operator's hands (PIERQUIN et al. 1987b)

Failure to remove this cover will result in severely diminished radiation dose to the pterygium. A set of steel collimators may also be available to shape the radiation field to conform to the irregular shape of the lesion. In the United States, the NRC has required that a Sr-90 eye applicator must be calibrated at an accredited calibration laboratory prior to its clinical use and that an authorized medical physicist must perform all decay calculations. Irradiation is begun within 24 h after resection, since failure increases with greater time delays (VAN DEN BRENK 1968).

The patient is placed in a comfortable supine position, with the head slightly tilted for optimal positioning of the medial portion of the eye. A lid retractor is inserted to hold the eye open. The cornea and conjunctiva are anesthetized with a few drops of 0.5–1% lidocaine. After 30 s to 1 min, to allow for the anesthetic to take effect, the applicator is carefully applied on the surface of the resected sclera (Fig. 14.13b). Doses of approximately 10 Gy are delivered. The application is repeated in three consecutive weekly fractions for a total of 30 Gy. If a larger area of resection is to be irradiated, it may require application to two contiguous areas, each receiving the same dose. The lens, which is located at the depth of 3.5–5 mm from the surface, receives less than 5% of the dose (GREENBERG 1987).

MONTEIRO-GRILLO et al. (2000) reported on 94 patients (100 eyes) treated with ^{90}Sr beta radiation; 37 eyes for primary and 63 for recurrent pterygium. Radiation doses were 30 Gy per 3 fractions for 5 days in 17 patients, 60 Gy per 6 fractions for 6 weeks in 80, and 20 Gy per 1 fraction in 3 patients. Of the 100 eyes treated, 14% developed a recurrence of the pterygium. The 5-year local control rates were 94% for

patients with primary and 76.9% for patients with recurrence. No late sequelae have been observed.

CONILL et al. (2004) evaluated interstitial ^{192}Ir brachytherapy in 24 carcinomas involving the eyelid tarsal structure in 23 patients (in the lower eyelid in 22 cases and in the upper eyelid in 2). The mean tumor size was 1.3 cm. Of the 24 tumors, 79.2% were basal cell carcinoma, 16.7% were squamous cell carcinoma, and 4.2% were adenocarcinoma. The total radiation dose was 40 Gy delivered to 20-mm depth (mean dose rate, 73 cGy/h). With a follow-up of 43 months, local control was obtained in 22 (91.6%) tumors. Good functional results were achieved in all patients.

14.8.4
Head and Neck

Brachytherapy may provide a useful method for the primary treatment or retreatment of patients with recurrent, persistent, or second primary head and neck malignant tumors in a previously irradiated region. FONTANESI et al. (1989) recommend a dose rate of 0.42 Gy/h or less to deliver total doses of 50–60 Gy to these lesions.

14.8.4.1
Maxillary Sinus

ROSENBLATT et al. (1996a) described the use of a surgical obturator made of vinyl polysiloxane as a carrier for afterloading ^{192}Ir seed ribbons to treat patients with maxillary antrum tumors after partial or total maxillectomy. An impression was made of the maxillary cavity 2 weeks after the surgical procedure, and the obturator mold was built. Multiple nylon catheters were inserted, depending on the geometry and dosimetry of the implant. The inner aspect of the impression was coated with a sheet of VLC denture material, the obturator was placed in an oven for 5 min, and the device was trimmed and polished. After the obturator was inserted in the patient, isodose distributions were obtained. Prescribed doses were 45–70 Gy (modal dose of 60 Gy) at 0.5 cm from the outermost source plane. Iridium seed activity was 0.7–1 mg Ra eq per seed. The obturator mold previously loaded with ^{192}Ir was carefully coated with acryl-methacrylate resin to secure it in place and prevent disturbance of the dosimetry once inserted in the surgical cavity. The approximate dose rate per day was 10–15 Gy.

14.8.4.2
Nasal Vestibule

Small lesions of the nasal vestibule can be adequately treated with either external or interstitial irradiation; whereas, more advanced lesions require a combination of both modalities. Irradiation is an excellent alternative to surgery in the treatment of these tumors, since tumor control can be very good and cosmetic results are better than with surgery (MENDENHALL et al. 1984, 1987). These tumors are implanted with single- or double-plane techniques using cesium needles or ^{192}Ir nylon tubing techniques. According to Mendenhall and associates (MENDENHALL et al. 1991), the distal vertical needles (perpendicular to the dorsum of the nose) in each plane may be mounted in a nylon bar to stabilize the distal needles and adequately cover the tumor involving the opening of the nasal vestibule (Fig. 14.14) (PARSONS et al. 1994).

LANGENDIJK et al. (2004) described the results of primary radiation therapy for squamous cell carcinoma of the nasal vestibule in 56 patients with stage-T1 and -T2 tumors (Wang classification) treated with external beam radiation therapy with or without a boost using endocavitary brachytherapy (32 treated with radiation therapy and an additional boost with intermediate-dose-rate brachytherapy and 9 with external-beam radiation therapy alone). The local tumor control at 2 years was 80%. Most failures could be successfully salvaged with surgery, with an ultimate local control rate of 95%. No statistically significant differences were noted among the different treatment approaches. Of the 56 patients, 12% developed lymph-node metastases.

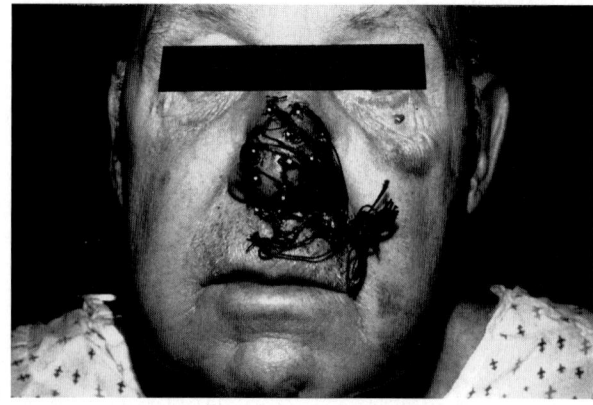

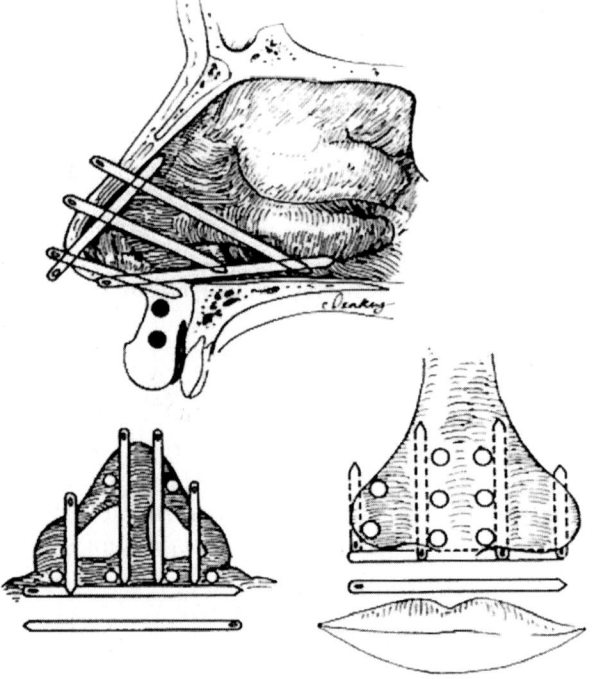

Fig. 14.14. a The basic interstitial treatment plan for nasal septum tumors consists of multiple planes of needles (usually two to four) inserted through the skin and cartilage of the nose perpendicular to the nasal bridge. Crossing needles parallel to the floor of the vestibule or bridge of the nose are necessary to ensure homogeneous irradiation of the tumor volume. One or two needles are also implanted in the upper lip even when it is clinically uninvolved (MILLION et al. 1989) **b** Treatment consisted of a radium needle implant with three planes of needles perpendicular to the nasal bridge plus two crossing needles in the upper lip and one needle in the floor of the nose; the dose was calculated to be 65 Gy in 130 h (MILLION et al. 1984)

14.8.4.3
Skin and Lip

Brachytherapy for treatment of skin and lip tumors was popular before the advent of external irradiation techniques. Interstitial single- or double-plane implants could be performed to encompass the tumor with a safe margin, following the basic principles of brachytherapy (Paterson–Parker, Quimby, or Paris technique). Doses of 50–70 Gy are delivered in 5–7 days. Carcinoma of the skin has been treated with surface molds or interstitial brachytherapy (MARCHESE et al. 1984). JORGENSEN et al. (1973) reported on 869 patients with squamous cell carcinoma of the lip for whom irradiation was the initial form of treatment in all but 25. Radium implants were used in 766 patients, with local tumor control rates of 93% in T1, 87% in T2, and 75% in T3 tumors. Similar results were described by PIGNEUX et al. (1979).

TOMBOLINI et al. (1998) described the technique and results in 57 patients with squamous cell carcinoma of the lower lip treated with LDR interstitial brachytherapy. The median tumor dose was 62 Gy (range, 44–96 Gy). The clinical N+ cases were irradiated to total doses of 65–70 Gy on the involved station. Actuarial disease-free survival at 10 years was 81%. Actuarial local tumor control was 90% at 3 years and 5 years, rising to 94% with salvage surgery.

14.8.4.4
Nasopharynx

ERICKSON and WILSON (1993) summarized the techniques for management of patients with carcinoma of the nasopharynx (Table 14.2). Some authors have used interstitial techniques, which are more laborious to carry out because of difficulty in positioning the applicator in the tumor area, dosimetric problems related to the irregular mucosal surface of the nasopharynx, and limitation of effective depth dose versus surface dose (ERICKSON and WILSON 1993). Palatal fenestration may be required in patients with lesions in the superior and high posterior nasopharyngeal walls, which are more difficult to reach through the nasal or oral cavities (ERICKSON and WILSON 1993; HARTER et al. 1975). SCOTT (1975) described temporary interstitial techniques traversing the lesion with sutures containing iridium seeds accessed through the oropharynx and subsequently removed. The use of ^{103}Pd seeds for permanent implant of nasopharyngeal tumors has been described by PORRAZZO et al. (1992).

WANG (1987) described use of intracavitary brachytherapy alone or combined with external irradiation to boost the dose to the nasopharynx, in conjunction with external-beam irradiation. Two pediatric endotracheal tubes with inner and outer diameters of 5 mm and 6.9 mm, respectively, each loaded with two 20 mg Ra eq ^{137}Cs sources were used. Local anesthesia of the nasal cavity was achieved with cocaine. The endotracheal tubes were introduced through the nares into the nasopharynx with the head hyperextended. Under fluoroscopic control on the simulator, the tips of the cesium sources were placed at the free edge of the soft palate posteriorly and behind the posterior wall of the maxillary sinus anteriorly. A 5-cc balloon attached to the distal end of the endotracheal tube was inflated for anchoring purposes and to improve the dose to the nasopharynx (by increasing the distance from the source). The dose reference point was 0.5 cm below the mucosa of the nasopharyngeal vault; the dose rate is approximately 1.2 Gy/h.

DENHAM et al. (1988) used intracavitary irradiation of the nasopharynx with afterloading catheters of different curvatures, which were introduced into the nasopharynx via the nasal cavity, with appropriate anesthesia. The major difficulty with this technique was the successful rigid anchoring of the catheters to prevent movements that could be potentially injurious to the nasal cavity or nasopharynx. For this purpose, a special plastic face mask was constructed with adjustable universal joint fittings for rigid attachment of the catheters with minimal discomfort to the patient. Because of asymmetry of the nasopharynx, different angle catheters can be used (22.5, 40, or 50 degrees).

LEVENDAG et al. (1997) designed an inexpensive, re-usable and flexible silicone applicator, tailored to the shape of the soft tissues of the nasopharynx, which can be used with either LDR brachytherapy or HDR (pulsed) controlled afterloaders. The applicator proved to be easy to introduce, patient friendly, and can remain in situ for the duration of the treatment (2–6 days).

LEE et al. (2002) used intracavitary brachytherapy as a boost in both primary and recurrent nasopharyngeal carcinoma in 55 patients treated with megavoltage external beam radiation (43 patients treated for initial disease and 12 for recurrence). Brachytherapy was routinely used for early cases of T1 and T2 lesions and selected cases of more advanced lesions, as well as recurrent lesions. Of the patients, 18 had concurrent chemotherapy. The brachytherapy applicators used were Rotterdam

Table 14.2. Intracavitary low dose rate/medium dose rate techniques in carcinoma of nasopharynx. *D* definitive, *E* elective boost, *P* persistent, *R* recurrent, *Fx* fraction. Modified from ERICKSON and WILSON (1993)

Series	Technique	Dose/prescription point	Timing	External beam dose	D/E/R
WANG and SCHULZ (1966)	^{226}Ra pack	20–40 Gy		±22–58 Gy	R
PATERSON (1948)	^{226}Ra in cork (15 mg)	80 Gy/7 days at 0.5 cm		±	D, P
MARTIN and MACCOMB (1937)	^{226}Ra needles (10 mg) or ^{226}Ra seeds in lead capsule ball	2500–3000 mgh (with external), 3500 mgh (definitive) over 20 days with 48-h break every 5 days	Before external	±30 Gy	D, E, R
HENSCHKE et al. (1963)	Plastic sphere (^{192}Ir, ^{60}Co)	1000 mgh/two applications		±	E, R
PRYZANT et al. (1992)	Teflon balls (^{137}Cs)	20–50 Gy/2–5 days at applicator surface	After external	+	R
SUIT et al. (1960)	Acrylic resin mold (^{226}Ra tubes)	30 Gy/2 Fx every 10 days; at 1 cm	After external	50 Gy	P
YAN et al. (1983); QIN et al. (1988)	Mold (^{226}Ra needles)	40 Gy/2 Fx at surface, 7–10 days apart	After external	± (R); 70 Gy (P)	E, P, R
MCNEESE and FLETCHER (1981)	Mold (^{226}Ra)	30–35 Gy at surface	After external	40 Gy	R
CHASSAGNE et al. (1963); PIERQUIN et al. (1968); GERBAULET et al. (1992)	Mold (^{192}Ir)	30 Gy (E); 40–98.5 Gy (P, R)	After external	45 Gy (E), ± (P, R)	E, P, R
DEUTSCH et al. (1973)	Mold (^{60}Co), palatal fenestration	60 Gy at 1.5 cm at 0.5Gy/h/14 h/day x 8 days		-	R
ASHAYERI et al. (1987)	Foley catheter (^{192}Ir)	21 Gy		-	E, R
WANG (1980)	Endotracheal tube (^{137}Cs)	20 Gy at 0.5 cm at 1.2 Gy/h, 1–2 Fx 1 week apart	10–14 days after external	40 Gy	R
WANG (1980)	Endotracheal tube (^{137}Cs)	7 Gy at 0.5 cm in 5 h	2–3 weeks after external	65 Gy	E
FLORES (1988)	Endotracheal tube (^{226}Ra, ^{137}Cs)	60 Gy/58 h at 1.5 cm		40 Gy	R
FLORES (1988)	Endotracheal tube (^{137}Cs)	26 Gy/1.77 h at 1.5 cm		±40 Gy	R
SHANKAR et al. (1991)	Endotracheal tube + mold (^{137}Cs + ^{192}Ir)	21 Gy/30 h at surface			P
HARRISON et al. (1980)	Endotracheal tube (^{192}Ir); plastic tube (^{192}Ir)	15–20 Gy		45 Gy	R
SHAM et al. (1989)	Latex tube (^{137}Cs)	20–30 Gy/2–3 Fx/every week 4–5 h/Fx		- (P); 40–50 Gy (R)	P, R
YAMASHITA et al. (1986)	Double lumen-plastic tube (^{60}Co)	8 Gy/2 Fx at 0.5 cm 1 week apart	During external (after 2 weeks)	66–70 Gy	E
DENHAM et al. (1988)	Two uterine tandems (^{192}Ir)	7.2 Gy/16 h		64.5 Gy	E
HAGHBIN et al. (1985)	Intracavitary (^{226}Ra, ^{137}Cs)	40–60 Gy			R

(n=24), balloon (n=16), ovoid (n=14), and ribbon (n=1). The dose rate was high (n=24), low (n=29), or pulsed (n=2). External beam doses ranged from 54 Gy to 72 Gy for primary disease and to 42 Gy for recurrent disease. Brachytherapy doses ranged from 5 Gy to 7 Gy for HDR and 10 Gy to 54 Gy for LDR. With a median follow-up of 36 months in those who were treated for primary carcinoma, the year estimate of local tumor control was 89%, and overall survival estimate was 86%. Recurrent patients had a median follow-up of 50 months; the 5-year estimate of local tumor control was 64%, and the overall survival estimate was 91%.

14.8.4.5
Oral Cavity

The oral cavity should be kept dry with adequate preanesthesia medication, including scopolamine, and suction. It is desirable to outline the tumor with gentian violet, Castellani's paint, or a surgical marker before starting the implantation of sources. A metric ruler should always be on the implant tray. When rigid needles are being implanted in the oral cavity, one assistant retracts the patient's lips and another either pulls or depresses the tongue while the operating radiation oncologist performs the implant.

The anterolateral needles of an implant of the oral cavity should be kept away from the thin mucous membrane that covers the bone in the upper and lower gum, as well as from the periosteum, teeth, and bone. To increase and maintain the distance, a regular fluoride carrier is thickened on the inside by one to four layers (one layer=2 mm) to increase the distance so that the unavoidable "hot spot" around each needle is kept away from the adjacent normal mucosa.

MIURA et al. (1998) reported on 103 patients with T1 or T2 tongue carcinoma treated by a single-plane implantation of iridium (^{192}Ir) pins (60 treated by brachytherapy alone, and the rest were combined with external irradiation and/or chemotherapy); 48 and 55 patients were given brachytherapy with and without a spacer, respectively. Spacers were individually made of acrylic resin according to a prosthetic technique to obtain the thickness of 7–10 mm at the lingual part of the implanted side. The spacer reduced approximately 50% of the absorbed dose at the lingual side surface of the lower gingival to that in the absence of a spacer. Absolute incidence of mandibular osteonecrosis was 2.1% (1 of 48) and 40.0% (22 of 55), with and without a spacer, respectively (P=0.0004).

14.8.4.6
Tongue and Floor of Mouth

Lesions beneath the tongue, or in the floor of the mouth, are usually implanted through the dorsum of the tongue, if standard needles (or substitutes) are used. The anterolateral needles emerge from the undersurface of the tongue and are reinserted into the floor of the mouth. The implants should extend beyond the visible or palpable tumor by at least 1 cm in all directions. A popular technique of interstitial implants with nylon tubing and ^{192}Ir sources for lesions of the oral tongue or floor of the mouth uses a submental or submaxillary approach for the insertion of metallic guides into the oral cavity with one hand. The exit points of the guides in the oral cavity are carefully verified with the index finger of the other hand (through-and-through technique).

The major nylon tubing is threaded through the metallic guides and looped around the dorsum of the tongue and exits through a parallel metallic guide. The metallic guides are pulled out externally, and the nylon thread is secured by crimping with a metallic button at one end. The procedure continues as described previously, leaving the other end open momentarily for insertion of the radioactive sources. To facilitate removal, we prefer to tie a silk thread on the loop of each nylon tube inside the oral cavity.

After position of the sources is verified on X-ray films using radiopaque inactive dummy sources, the appropriate ^{192}Ir wires (or seeds in nylon tubing) are inserted, and the other end of the larger nylon tube is crimped.

The sequence of needle implantation for lesions involving the oral tongue and the anterior floor of the mouth is illustrated in Figure 14.15.

MENDENHALL et al. (1991) described a template for floor of the mouth implants made of aluminum, stainless steel, or nylon that is individually customized to fit the lesion of each patient (MARCUS et al. 1980). The device is inserted into the floor of the mouth under general anesthesia and is secured by one suture through the submental area, which is tied to a cotton cigarette roll. The active ends of the cesium needles may be positioned above the level of the mucosa to ensure an adequate surface dose. Crossing is accomplished by placing a needle parallel to the mucosal surface on the implant device. The system is not afterloaded, but the procedure can be performed rapidly with predictable geometry, so that irradiation exposure to the operating staff is lower than with the hairpin technique. According to

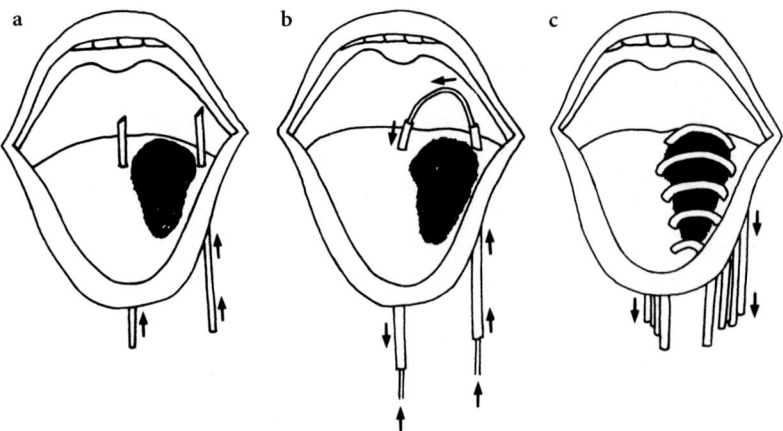

Fig. 14.15a–c. Diagram illustrating submental or submaxillary approach for insertion of Teflon catheters with metallic guides (for ^{192}Ir) into oral cavity for lesions for the floor of the mouth or lateral border of the tongue. **a** Metallic guides are introduced, with one hand guiding the position of the guide inside the oral cavity. **b** Introduction of nylon strand for placement of the ^{192}Ir wire or seeds, looped over dorsum of the tongue. **c** Various nylon tubes in position on dorsum of the tongue. At this point, the metallic guides have been withdrawn from the submaxillary region. After position of the radioactive sources is radiographically determined using dummy sources, active sources are inserted, and the ends of the plastic tubes are crimped with metallic buttons

the authors, the advantage of this technique over use of iridium hairpins is that all needles with the template are rigidly fixed in relationship to one another, and the isodose distributions can be calculated before the procedure or can be modified if necessary by adjusting the arrangement of the needles.

Implantation with rigid needles of the posterolateral border of the tongue via the oral cavity requires pulling the tongue forward to start the implant at the base of the tongue. The first needle is inserted pointing posteriorly and inferiorly at approximately 45 degrees; a lesser angle is used for successive needles. At the end of the implant, when the tongue returns to its normal position, the implant needles adopt a vertical position.

At the University of Florida, a technique has been used with radium or cesium needles mounted on rigid bars made of stainless steel or nylon; the number of needles depends on the diameter of the lesion (MENDENHALL et al. 1991). A crossing needle is usually added to one or both bars to ensure adequate irradiation dose to the dorsal mucosal surface of the tongue.

Superficial spacers may be used to ensure sufficient dose to the dorsal tongue (SCHMIDT-ULLRICH et al. 1991).

Another technique described by BAILLET et al. (1987) uses the ^{192}Ir hairpin technique. Inactive gutter guides are placed into the tongue, and under fluoroscopic control it is verified that the gutter guides are parallel. The iridium hairpins are afterloaded into the guides, which are removed at that

time. A suture is used to secure each hairpin to the tongue. A cotton roll sutured between the tongue and the mandible with either technique displaces the tongue medially and decreases the irradiation given to the mandible.

The advantage of the iridium hairpin technique over the radium or cesium rigid needles is that the overall source length is shorter for the same active length because of the 6-mm inactive tips at either end of the rigid needles. Furthermore, there are only two vertical sources per hairpin as opposed to three or four radium or cesium needles on each bar, so that it is easier to position the hairpins in the tongue (Fig. 14.16) (PARSONS et al. 1992). This is particularly helpful in patients with small mouths, trismus, or full dentition, where it is very difficult to adequately position the rigid needles.

BOURGIER et al. (2004) treated 279 patients with T1,2-N0 carcinoma of the mobile tongue using LDR ^{192}Ir brachytherapy alone or with neck dissection. A guide gutter or plastic tube technique was used to deliver a median dose of 60 Gy (median dose rate 0.5 Gy/h ^{192}Ir activity 0.59–1.6 mCi/cm). At 2 years, local tumor control was 79.1%, and complication rate was 16.5% (grade III, 2.9%).

An effort should be made to reduce treatment morbidity. LOZZA et al. (1997) noted mandibular necrosis in 10 of 100 patients with oral cancer treated with LDR brachytherapy; median follow-up was 38 months. No significant incidence of this complication was observed when tumor site (mobile tongue versus floor of mouth), dental status, or total

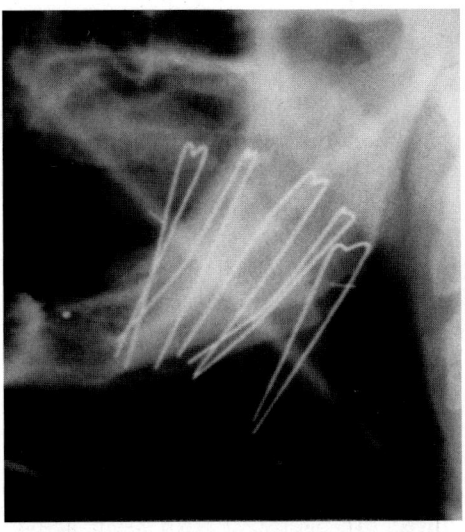
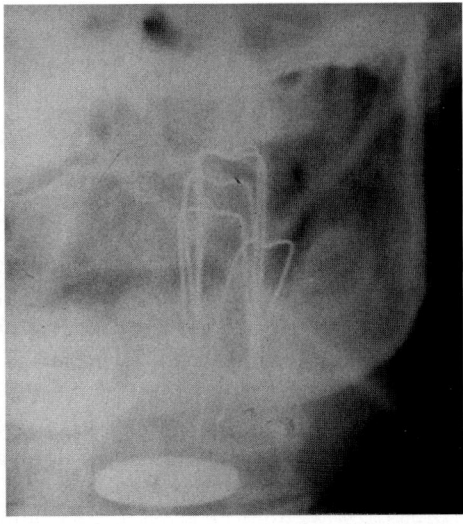

a   b

Fig. 14.16a,b. Roentgenograms of ^{192}Ir hairpin implant for carcinoma of the left side of the oral tongue, stage T2N0, measuring 3.5×2.0×2.0 cm, with submucosal extension to within 0.5 cm of the midline of the tongue. Treatment consisted of 30 Gy in ten fractions, followed by an ^{192}Ir implant using the gutter-guide technique with the patient in a sitting position. A gauze pack was secured onto the lateral floor of the mouth to displace the tongue medially away from the mandible. The implant delivered a 40-Gy tumor dose to the area of gross disease (0.55 Gy/h). The patient remained free of disease at 36 months (Parsons et al. 1992)

physical dose was considered. A significant correlation between the incidence of bone necrosis and two main parameters was found, i.e., dose-rate ($P<0.02$) and reference volume ($P<0.05$). A threshold value may be suggested both for dose-rate (50 cGy/h) and reference volume (25,000 mm^3).

14.8.4.7
Base of Tongue

Because of the possibility of airway obstruction, it is advisable to perform an elective temporary tracheostomy before the implant procedure is initiated.

Implantation of the base of the tongue (and sometimes the posterolateral border of the oral tongue) is best accomplished by using long metallic or Teflon catheters with guides inserted through the submaxillary/subdigastric region, with the index finger of the other hand in the oropharynx to verify the position of the guide at the exit point, the base of the tongue. As described earlier, the nylon thread is inserted through the tubing into the oropharynx, looped around, and brought out through the opposite guide, thus providing the equivalent of a "crossing needle" in the cephalad end of the implant (Fig. 14.17a). The metallic guides are withdrawn from the submental region (Fig. 14.17b–d), and the nylon tubes are secured externally with metallic buttons as described earlier. Occasionally, it is not

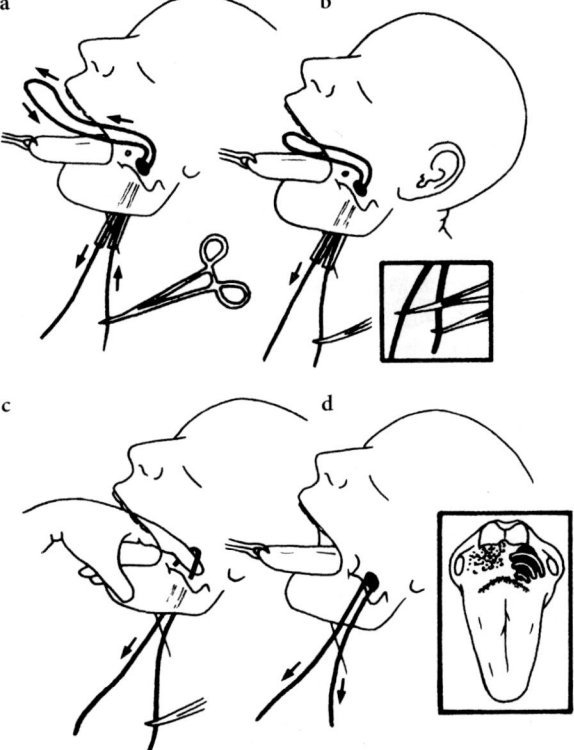

Fig. 14.17a–d. Diagram illustrating the use of stainless-steel guides inserted through the submaxillary region for patients with carcinoma of the left base of the tongue. **a, b** The nylon tubing is looped around the dorsum of the tongue, the stainless-steel guides are removed (**c**), and dummy and later radioactive sources are inserted and secured with metallic buttons (**d**)

possible to open the oral cavity adequately, and the only recourse is to perform a submandibular implant with metallic guides and afterloading [192]Ir (Fig. 14.18). Double-plane or volume implants can be easily performed. After implant localization, 3-D imaging with dummy sources are taken, afterloading ^{192}Ir wire or seeds in nylon threads are inserted into the nylon tubing or metallic guides, and isodose distributions are obtained using 3-D dosimetry (Fig. 14.19).

LEBORGNE et al. (2002) compared local tumor control and complications of external beam radiation therapy with or without interstitial ^{137}Cs needle brachytherapy boost doses in 201 patients with locally advanced (T3-T4) cancer of the tongue and floor of the mouth treated with definitive external beam irradiation with (n=78) or without (n=123) LDR interstitial brachytherapy. External beam irradiation was administered with conventional fractionation in 105 patients and administered with accelerated hyperfractionation in 96 patients. Grade 3–5 late effects were 25% and 35% for patients with and without brachytherapy, respectively (not statistically significant), although the incidence of mandibular radiation osteonecrosis was 10% and 1.6%, respectively (P=0.001).

14.8.4.8
Tonsillar Region Including Faucial Arch (Oropharynx)

FLETCHER and MacCOMB (1962) described a double-plane pterygomaxillary implant with radium needles to boost the dose in patients with carcinoma of the faucial arch or tonsillar region with extension into the tongue. [192]Ir hairpin or plastic tube techniques have been used by PIERQUIN et al. (1987c) and MAZERON et al. (1986). The nylon tube technique may also be used to implant the soft palate (ESCHE et al. 1988). The iridium hairpin technique is used with one gutter guide placed in the soft palate in the transverse plane and additional gutter guides placed vertically into the anterior tonsillar pillars, depending on the extent of the lesion. Iridium hairpins are afterloaded into the gutter guides, which are removed as described earlier. If the uvula is involved by tumor, it should be amputated before implantation (ESCHE et al. 1988).

MENDENHALL et al. (1991) reviewed the techniques for implantation of the anterior tonsillar pillars, soft palate, or tonsillar region using two nylon bars, each containing three full-intensity, 2- to 3-cm active-length radium or cesium needles implanted into the anterior tonsillar pillar and the other 1 cm medial to the tonsillar pillar bar, in the base of the tongue. A crossing needle was sometimes included in the anterior pillar bar to ensure adequate mucosal dose.

LEVENDAG et al. (2004) used brachytherapy in 104 patients with carcinoma of the tonsil and/or soft palate combined with external irradiation. At 5 years, local tumor control was 88%, and disease-free survival was 57%. Mucosal ulceration was observed in 39% of the patients and trismus in 1%.

CANO et al. (2004) treated 18 patients with locally advanced base of tongue carcinoma using external beam radiation therapy (54 Gy to primary tumor and 59.4 Gy to neck nodes) followed by a LDR brachytherapy ^{192}Ir implant in 52% of patients (24 Gy to primary tumor and 17.5 Gy to neck nodes); 68% of all patients received induction or concurrent chemotherapy. Local tumor control at 5 years was 89%;

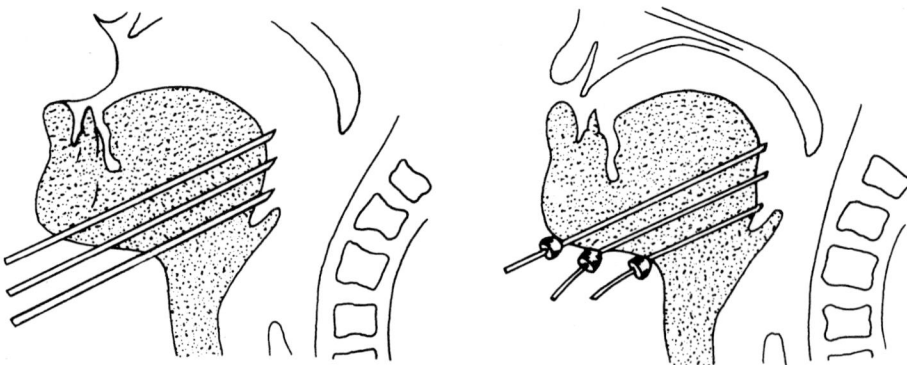

Fig. 14.18. Submandibular implant with metallic guides in which it was not possible to "loop around" nylon strands over the base of the tongue (one-end technique). The nylon tubing is cut to fit desired tumor volume to be implanted, and metallic buttons are used to secure nylon tubes and guides in position. The buttons are sutured to the skin to ensure the placement of the stainless-steel guides

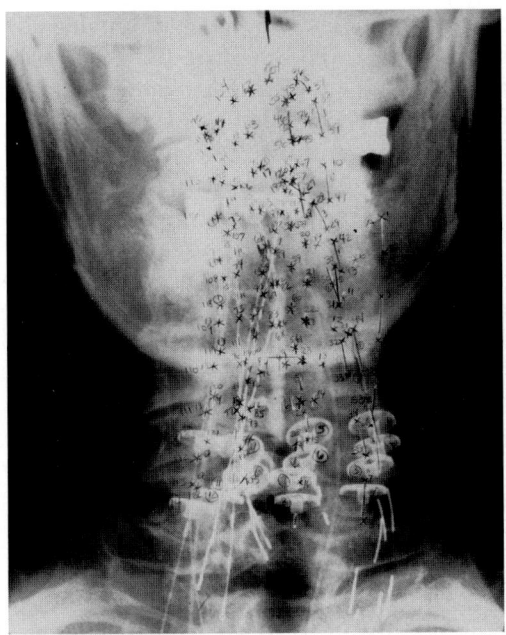
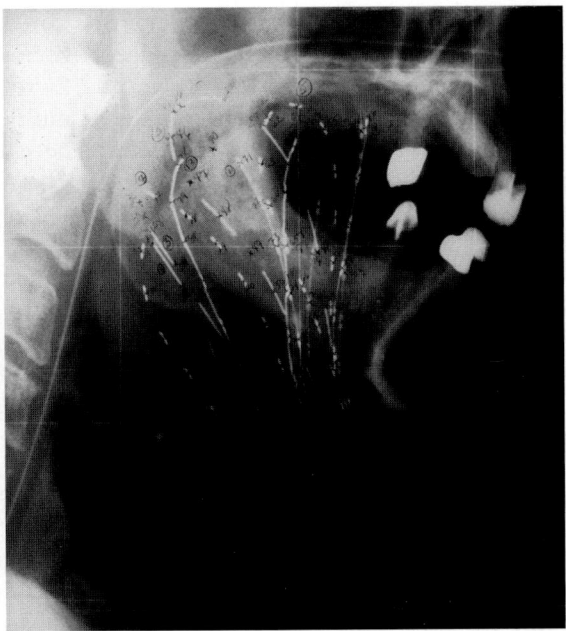

a

b

Fig. 14.19a,b. Position of stainless-steel guides for implantation of large T3 epidermoid carcinoma of the base of the tongue to deliver 30 Gy in approximately 65 h. Patient received 45 Gy with external-beam irradiation before the ^{192}Ir implant. AP (**a**) and lateral (**b**) radiographs illustrate positions of the dummy sources in base of tongue

there was a 4% incidence of soft tissue necrosis and 5% osteonecrosis. The 5-year disease-free survival rate was 67%, and overall survival was 52%.

14.8.5
Breast

Interstitial brachytherapy has long been in use as boost therapy after the whole breast has been given external beam irradiation, usually at 45–50 Gy. More recently, brachytherapy has been investigated as sole radiation therapy in conjunction with conservation surgery (KUSKE et al. 1994; VICINI et al. 1997), although this approach remains controversial. A phase-I/II Radiation Therapy Oncology Group (RTOG) trial evaluating the use of brachytherapy alone following lumpectomy in stage-I and -II breast cancer patients was completed (KUSKE et al. 1998), and several preliminary results of partial breast irradiation have been reported (KUSKE et al. 1998; VICINI et al. 1997; ARTHUR et al. 2003). The American Brachytherapy Society (ABS) has published detailed recommendations for both boost and sole modality partial breast brachytherapy (NAG et al. 2001b).

Selection of patients for breast brachytherapy is limited to those with adequate breast volume and lesions less than 4 cm in diameter. The ABS recommendations limit the use of sole modality brachytherapy to patients with T1 or T2, N0 disease with a lesion less than 3 cm in size, treated with lumpectomy and axillary dissection. In this chapter, only interstitial LDR implants using ^{192}Ir seeds or wires will be described, although ^{125}I seeds have also been used for this application.

Preplanning of the implant provides important information on the target geometry and required placement of the source guides, information critical in achieving acceptable target coverage, critical structure avoidance, and dose uniformity throughout the irradiated volume. A variety of modalities have been used for identification of the target volume in preparation for the implant. These include the use of surgical clips, ultrasound, CT, MRI, and intraoperative visualization of the lumpectomy cavity. DEBIOSE et al. (1997) reviewed 60 patients with early stage breast cancer, a portion of whom underwent ultrasonography (US)-assisted interstitial brachytherapy needles. The lumpectomy cavity was outlined in all dimensions, and corresponding skin marks were placed for reference at time of implantation. These dimensions were compared with the physician's clinical estimate of the location of the lumpectomy cavity, the patient's presurgical mammograms, and the position of the scar. In the intraoperative setting, the dimensions of the lumpectomy cavity were obtained, and the placement of the deep plane of interstitial needle was verified using US. The full extent of the lumpectomy cavity was underestimated by clinical examination (physical exam, operative report, mammographic information, and location of the scar) in 33 of 38 patients (87%). The depth of the chest wall was also incorrectly estimated in 34 patients (90%) when compared with US examination. Intraoperatively, US was performed on 9 patients and was useful in verifying the accurate placement of the deepest plane of interstitial brachytherapy needles. In 7 of the 9, the posterior extent of the lumpectomy cavity was visualized using the intraoperative US.

RTOG 95-18 required that at least six radiopaque clips be left in the cavity at the time of surgery to define the maximal extent of the cavity in three dimensions. Visualization of the clips under fluoroscopy allows the determination of an advantageous direction of approach for the source guides, which may achieve the most conformal coverage of the target volume with the minimum number of source ribbons or wires. Skin marks are then placed to guide the needle insertions. CT, MRI, or US can be used for the same purpose, with even better visualization of the excision cavity geometry. The volume used for planning treatment after the cavity has been delineated is somewhat controversial. In the RTOG study (KUSKE et al. 1998), a 2-cm margin was added to the cavity in directions parallel to the implanted source planes, and a 1-cm margin was added perpendicular to the planes. The ABS recommends a 2-cm margin all around the excision cavity but restricts the superficial margin to a depth of at least 0.5 cm under the skin and the deep margin to the surface of the chest wall.

After the clinical target volume has been decided and the optimal source guide approach determined, the number and placement of sources should be planned. Coverage of the target can usually be accomplished with a two-plane implant, although single- and multi-plane configurations are not uncommon. The spacing between source-bearing catheters in each plane should be between 1.0 cm and 1.5 cm. The closer spacing helps to avoid cold spots at the periphery of the target volume and can contribute to improved dose uniformity. The implants are typically carried out with seeds of uniform activity, but mixed activities can be used where deemed advantageous. For two-plane implants and uniform seed activities, the optimal interplanar spacing for conformal irradiation can be determined as a func-

tion of the target thickness from tables provided by ZWICKER (ZWICKER and SCHMIDT-ULLRICH 1994; ZWICKER et al. 1985) or by computer simulation. For multiplane implants, computer preplanning should be carried out to optimize the source geometry. The ABS recommends that sources should extend 1–2 cm outside the target volume in the direction of the source planes, but this is often not possible, as sources must never be placed closer than 0.5 cm to the skin. A minimum depth of 1–2 cm under the skin is recommended by the ABS to minimize the risk of hyperpigmentation and telangiectasia (NAG et al. 2001b).

The implant procedure may be performed with the patient under either general or local anesthesia. The implant can be carried out in conjunction with resection of the primary tumor (and when indicated, the axillary dissection) or as a separate operating room procedure. The former approach has the advantage of reducing the cost of treatment and allowing the surgeon and radiation oncologist to interact closely in determining the extent and location of the tumor but reduces the time available for planning the placement of the sources and requires the use of available source activities (WATERMAN et al. 1997a). Following the skin marks placed at the time of preplanning, (Fig. 14.20a) the guide needles (or Teflon catheters with rigid stylets) are inserted into the breast, passing through the tumor excision site until the end of the guide reaches the planned marks on the opposite portion of the breast (Fig. 14.20b). In general, if a double-plane implant is carried out, the guides for the deeper plane are inserted first, followed by those for the superficial plane. Next, the leads on the nylon afterloading catheters are inserted through the hollow guide needles, and the needles and leads are pulled though the breast, with the catheters following and remaining in the breast (Fig. 14.20b). The distal end of the each catheter may contain a button, which will secure the catheter against the breast, or the end may be secured by crimping metallic buttons against the skin (Fig. 14.20c). The proximal ends of the catheters are left open for the insertion of the sources but may be partially secured by means of tight-fitting plastic washers, which slide over the catheters to the level of the skin.

After the patient recovers from the anesthesia, radiopaque dummy sources are inserted in the catheters (Fig. 14.20d), and, if film planning is intended, orthogonal or stereo films of the breast are obtained. The ideal planning geometry is obtained with the film planes at 45 degrees and 135 degrees to the source planes, as this will help avoid extensive over-

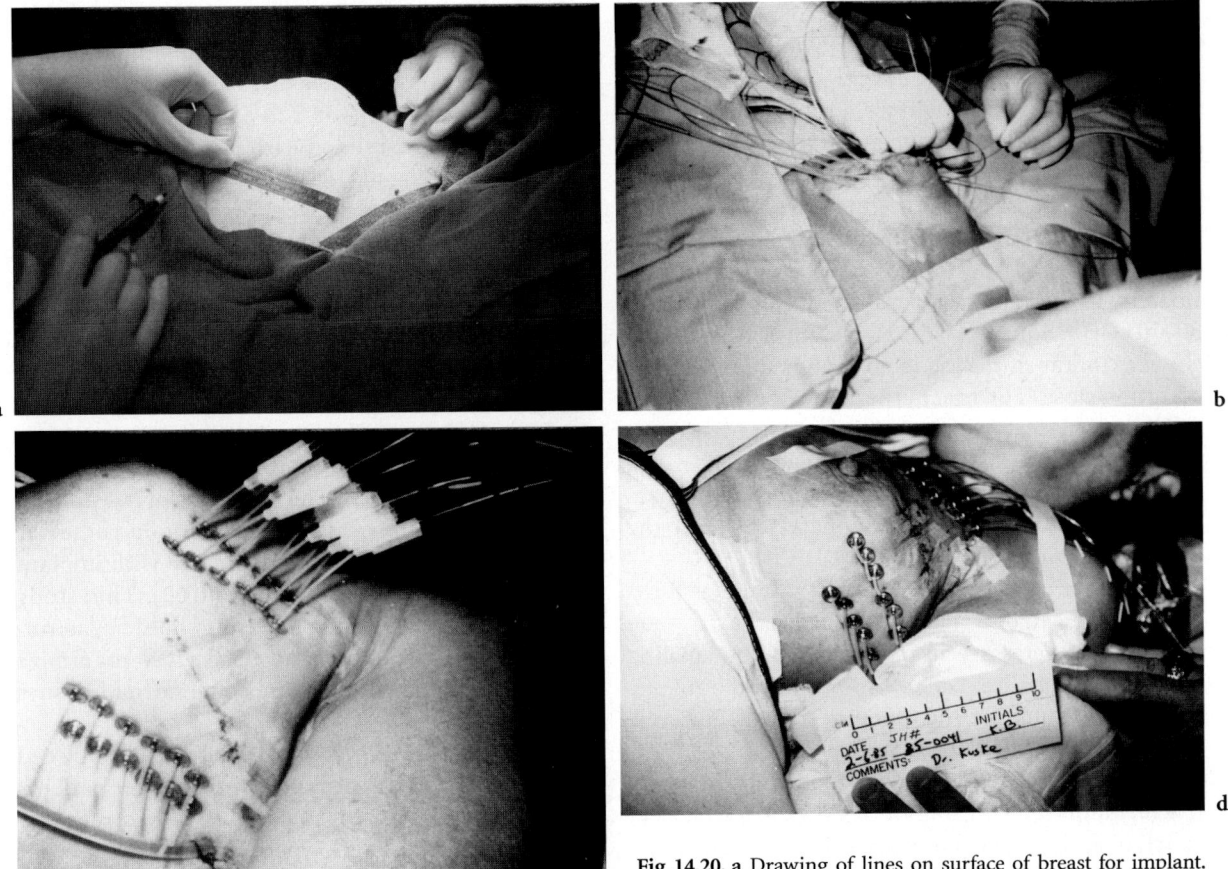

Fig. 14.20. a Drawing of lines on surface of breast for implant. **b** Placement of metallic or plastic guides for breast implant. **c** Plastic tubes are secured at the skin surface using metallic buttons. **d** Sources are inserted in proximal end of tubing

lap of sources on either of the two films. However, the range of useful film angles may be limited by the presence of underlying bony structures. Films parallel and perpendicular to the implant plane may also be helpful in visualizing the target volume and the locations of the chest wall and the skin overlying the sources. Radiopaque markers should be placed on the skin, especially in surgical defects or other regions that appear close to the superficial source plane. The lengths of the catheter sections inside the breast are found by measuring the lengths of the dummy seed ribbon sections inside the catheters and subtracting from this the length of catheter projecting outside the breast. From the measurement, the maximum length of ^{192}Ir seed ribbon or wire for each catheter can be determined, keeping in mind that no source should approach closer than 0.5 cm to the skin, and greater margins are strongly recommended.

For film-based treatment planning, isodose curves are typically generated for the central plane transverse to the direction of catheter insertion, for other planes parallel to the central transverse plane, for the central plane parallel to the implant source planes, and for a plane perpendicular to this and to the central transverse plane. Coverage of the target volume can be estimated to some extent from the positions of the surgical clips. Some means of quantifying dose uniformity should also be included in the calculations. In the RTOG 95-18 protocol (KUSKE et al. 1998), the average of local dose minima in the central transverse plane (i.e., the basal dose of the Paris system) was calculated, and the dose uniformity was judged to be acceptable if the prescription dose was at least 75% of this value.

Image-based treatment planning has substantial advantages in visualizing the target volume and surrounding structures and should be used whenever possible. For this application, the slice thickness and interval should be no greater than 0.5 cm, small enough to localize individual seeds. Correlation of the images with films can be useful in identifying individual dummy seeds. The clinical target volume should be outlined and isodose curves generated on each slice to assess target coverage. A dose–volume histogram of the target will yield information on the fractional target volume receiving the full prescription dose (V_{100}) and other dose levels of interest.

For LDR breast brachytherapy, the ABS has recommended a total dose of 45–50 Gy, recognizing that the dose has traditionally been given at approximately 10 Gy/day, with a range of approximately 30–70 cGy/h. Source positioning should be such that the maximum skin dose is no higher than the prescription dose. For brachytherapy used as a boost following 45–50 Gy of external beam radiation, a total dose of 10–20 Gy is recommended, also at a rate of 30–70 cGy/h. Typical maximum skin doses for boost implants are approximately 50% of the prescription dose.

A number of variants of the breast implant technique have been described in the literature. In some centers, templates are used to control the spacing of the guide needles. DELCLOS (1984) used only one point of entrance for the guide needles to improve the cosmetic results.

MANSFIELD et al. (1994) described an intraoperative technique placing four or five plastic tubes, 2 cm apart, in each of two planes separated by 2 cm at the time of the breast tumor excision. [192]Ir seeds spaced 0.5 cm and with activity of 0.5 mCi per seed per 1 cm were inserted after the wound was closed, and the position of the dummy sources was determined on localization radiographs. The plastic tubes were loaded with the active sources within 6 h of surgery. The dose rate was 0.3–0.5 Gy/h; usual dose was 20 Gy delivered in 50–60 h. Breast irradiation was begun 10 days later, with tangential fields, 6-MV photons, to deliver 45 Gy at 1.8 Gy per day. The 10-year local tumor control rates for stage T1 and T2 were 93% and 87%, respectively, and the 10-year disease-free survival rates were 82% and 75%, respectively.

MAZERON et al. (1991) used interstitial brachytherapy in the breast with rigid metallic needles inserted through a template in single or double planes after breast irradiation (45 Gy in 25 fractions). A boost to the primary tumor was prescribed at the 85% basal dose rate (Paris system). Intersource spacing varied from 1.5 cm to 2 cm. Implanted volume was adapted to tumor extent by varying the number of sources and active length according to the Paris system rules. Linear activity ranged from 1.3 mCi/cm to 1.8 mCi/cm. Mean dose rates were 0.53 Gy/h for patients with local recurrence and 0.56 Gy/h for recurrence-free patients

($P<0.01$). Of the patients, 58 were treated with single-plane and 340 with two-plane implants. The local tumor control rates at 15 years were 76% for T1 and T2a and 70% for T2b and T3 lesions. Local tumor control correlated with dose rate and tumor size (MAZERON et al. 1991).

Similar observations were reported by DEORE et al. (1993) in 118 T1 and 181 T2 lesions of the breast treated with radiation therapy after conservative surgery. External irradiation (43 Gy) was delivered with either 2.5 Gy or 1.8 Gy per fraction. A boost dose of 15–30 Gy was given to the primary tumor, using interstitial implants; dose rate varied between 0.2 Gy/h and 1.6 Gy/h. The local failure rate was significantly increased, with implant dose rate less than 0.3 Gy/h ($P<0.05$). The incidence of late normal tissue complications and poor cosmetic outcome was significantly higher in patients treated with implant dose rate greater than 1 Gy/h ($P<0.05$). This study indicates that the implant dose rate should be maintained between 0.3 Gy/h and 0.7 Gy/h to maximize local tumor control and reduce late normal tissues injury.

VICINI et al. (1997) used permanent ^{125}I seed implants after local tumor excision in 60 patients with breast cancer. In 18 patients, an intraoperative implant was performed, and the post-excision cavity was outlined with multiple radiopaque clips. In 42 patients, the implant was performed post-operatively with a closed cavity. Ultrasound was used to delineate the cavity, and the boundaries were marked on the skin. A preimplant virtual simulation was performed using a CT scan. Using a template, afterloading stainless-steel needles or later plastic catheters were introduced in the breast, and ^{125}I seeds were implanted. Seeds were kept at 5–7 mm from the skin surface. A dose of 50 Gy at 0.52 Gy/h was prescribed.

SMINIA et al. (2001) proposed PDR brachytherapy every hour for breast cancer patients; an office hours scheme was designed using radiobiological parameters, which included an alpha/beta value of 3 Gy for normal tissue late effects and 10 Gy for early normal tissue or tumor effects. Tissue repair halftime ranged from 0.1 h to 6 h. The reference LDR dose rate of 0.80 Gy/h was obtained from analysis of patients' data. The patient brachytherapy protocol (SMINIA et al. 2002) consisted of two treatment blocks separated by a night break; dose delivery was 20 Gy in two 10-Gy blocks and, for application of the 15-Gy boost, one 10-Gy block plus one 5-Gy block. The dose per pulse was 1.67 Gy applied within approximately 1.5 h.

14.8.6
Lung and Mediastinum

The group at Memorial Sloan–Kettering Cancer Center has published several reports (HILARIS and MARTINI 1988) on the use of ^{125}I seeds and ^{198}Au grains for permanent perioperative brachytherapy in patients with persistent or recurrent bronchogenic carcinoma after external irradiation or for residual disease after surgical resection. The radioactive seeds or grains are directly implanted in the tumor at the time of thoracotomy under general anesthesia.

Temporary removable implants of the mediastinum with or without resection followed by a moderate dose of postoperative external irradiation (35–40 Gy) have been used alone or combined with ^{125}I implantation of the known primary tumor (HILARIS et al. 1985).

CHOBE et al. (1996) treated 76 patients with LDR ^{192}Ir implants. In some patients, laser tumor debulking was performed before the brachytherapy insertion. Active sources were endoscopically placed in the bronchi with a manual technique. Active length of the sources ranged from 4 cm to 12 cm with a 5-mm space between sources. After radiographs were obtained and the placement of the dummy sources was believed to be satisfactory, the active ^{192}Ir sources were inserted through an afterloading catheter, which was secured as earlier described. A dose of 25–30 Gy was prescribed at an average depth of 1 cm. They compared results in 43 patients treated with HDR and 6 treated with both HDR and LDR therapy. Of 119 patients, 74 received external-beam irradiation. Tumor response was 74% in the LDR and 86% in the HDR groups. There was no difference in survival between the two groups. Morbidity of therapy was not reported.

Lo et al. (1995) described results in 110 patients (group 1) treated with LDR brachytherapy (133 procedures) and 59 patients (group 2) treated with HDR brachytherapy (161 procedures). In group 1, patients were treated with one or two sessions of 30–60 Gy each calculated at a 1-cm radius. In patients in group 2, three weekly sessions of 7 Gy each calculated at a 1-cm radius were used. External-beam radiation therapy had previously been given to 88% of patients in group 1 and to 85% of patients in group 2. Laser bronchoscopy was performed in 36% of patients in group 1 and in 24% of patients in group 2 before brachytherapy. Clinical or bronchoscopic improvement was noted in 72% of patients in group 1 and in 85% of patients in group 2 ($P>0.05$). Survival and complication rates, which were low, were equivalent in both groups.

RABEN and MYCHALCZAK (1997) reviewed the indications, techniques, and results of brachytherapy in the treatment of non-small cell lung cancer and selected chest neoplasms. Various isotopes and techniques are used to place radioactive sources directly into a tumor, tumor bed, or the chest with permanent interstitial volume or planar implants (radioactive sources permanently imbedded into the tumor or tumor bed) or temporary interstitial or endoluminal implants (where radioactive sources irradiate a tumor bed over a certain length of time and then are removed).

14.8.7
Esophagus

Irradiation, both external beam and intracavitary, has been used in the curative and palliative treatment of patients with esophageal cancer, either alone or combined with surgery. LDR intracavitary insertions have been performed using ^{226}Ra, ^{60}Co, ^{137}Cs, or ^{192}Ir sources. FLORES et al. (1989) outlined the advantages of intracavitary brachytherapy: radiation sources can be easily placed and removed at the desired tumor site; normal anatomy is preserved; radiation dose to the tumor is higher than to the adjacent tissues; and, with remote afterloading, radiation exposure to the staff can be eliminated. The insertion technique can be performed as an outpatient procedure under local anesthesia, usually xylocaine spray (1–2%), or mild sedation. A soft rubber bougie or French catheter (No. 24–26) is inserted, preferably through the nose. The rubber tube is removed, and a 260-cm Teflon-coated guidewire in a 60-cm FA-f10 cut-end feeding tube is inserted to the stomach. The cut-end feeding tube is removed, and the esophageal stricture is dilated to f32 by a balloon dilator (2 minutes required). The balloon dilator is removed, and the esophageal bougie containing dummy markers for intracavitary treatment is placed and secured in the desired position using fluoroscopy. After the position of the dummy sources is verified on radiograph, the patient is taken to the treatment room, where the remote-controlled afterloading device is connected for treatment. If LDR sources are used, the usual dose rate is 0.4 Gy/h at 0.5–1 cm. Depending on the external-beam dose given, the total intracavitary dose is prescribed to complete 65–70 Gy to the tumor volume. With higher dose rates, corresponding lower treatment times and total doses are used. Isoeffect curves are illustrated in Figure 14.21 (FLORES et al. 1989).

SYED et al. (1987) described a comparable technique used in 47 patients with carcinoma of the esophagus (37 with primary and 10 with recurrent lesions). After completion of external irradiation, patients received intraluminal brachytherapy to deliver 30–40 Gy at 0.5 cm from the surface of the applicator in two applications, 2 weeks apart. In patients with minimum residual tumor, only one application was used to deliver 20–25 Gy minimal tumor dose. Most patients also received concomitant 5-fluorouracil infusion. Intraluminal application was performed with a Syed–Puthawala–Hedger esophageal applicator consisting of a special tube made of silicon rubber (outer diameter 1 cm) with a conical smooth tip for easy insertion. The central nasogastric tube has six longitudinally placed afterloading catheters to accommodate various radionuclide sources. The total length of the applicator is 65 cm. Marked rings are present at 10-cm intervals from the tip of the applicator for identification on localization films. The central nasogastric tube can be used for both feeding and suction.

The procedure was performed under either general anesthesia or deep sedation and local anesthesia. Determination was made of the proximal and distal end of the tumor from the level of the incisor teeth, on endoscopy, and on review of the initial barium swallow X-ray films. A Robinson catheter (size 14 or 16) was inserted through one of the nostrils, and its tip was brought out through the mouth. The tip of the esophageal applicator was sutured to the proximal end of the Robinson catheter using "00" silk sutures (Fig. 14.22a). The Robinson catheter was pulled through the mouth until the tip of the esophageal applicator entered the oral cavity. The suture was cut, and the Robinson catheter was discarded, while the

tip of the esophageal applicator was inserted into the oropharynx and guided along the hypopharynx and esophagus into the stomach (Fig. 14.22b). The esophageal applicator was secured in position by "00" silk sutures through the nasal septum and around the applicator or adhered by adhesive tape. Orthogonal anteroposterior and lateral X-ray films of the chest were obtained after inactive dummy sources had been inserted into the afterloading catheters in the applicator. The location of the tumor was marked on the X-ray films, and appropriate margins were determined to carry out the dose calculations. Radioactive sources were spaced 0.5–1 cm, and margins of 3 cm above and below the tumor were allowed. When the dose calculations were completed, the treatment with the active sources was initiated (Fig. 14.22c). A total of 74 procedures were performed. Average survival was 13 months in patients with primary tumors. There were two esophageal strictures, and one patient died of tracheoesophageal fistula (tumor invaded trachea before treatment).

GASPAR et al. (2000) reported on an RTOG study delivering 50 Gy external beam irradiation (25 fractions in 5 weeks) followed 2 weeks later with esophageal brachytherapy, either HDR (6 Gy during weeks 8, 9, and 10, for a total of 18 Gy) or LDR (20 Gy during week 8) in patients with esophageal cancer; 45 (92%) had squamous histology and 4 (6%) had adenocarcinoma. Chemotherapy was given during weeks 1, 5, 8, and 11, with Cisplatin (75 mg/m^2 and 5-fluorouracil (1000 mg/m^2 per 24 h in a 96-h infusion); 47 patients (96%) completed external beam irradiation plus at least two courses of chemotherapy, whereas 34 patients (69%) were able to complete external beam radiation and at least two courses of chemotherapy. The estimated survival rate at 12 months was 49%. Life-threatening toxicity or treatment-related death occurred in 12 (24%) and 5 (10%) patients, respectively. Treatment-related esophageal fistulas occurred in 6 cases (12% overall, 14% of patients starting esophageal brachytherapy) at 0.5–6.2 months from the first day of brachytherapy, leading to death in 3 cases.

The ABS has published a set of consensus guidelines on the use of brachytherapy for esophageal cancer (GASPAR et al. 1997).

14.8.8
Pancreas

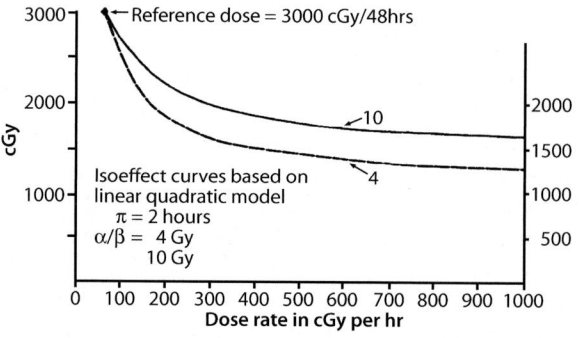

Fig. 14.21. Isoeffect curves correlated with dose rates (FLORES et al. 1989)

Interstitial irradiation, most frequently using ^{125}I permanent implants, has been used in patients with

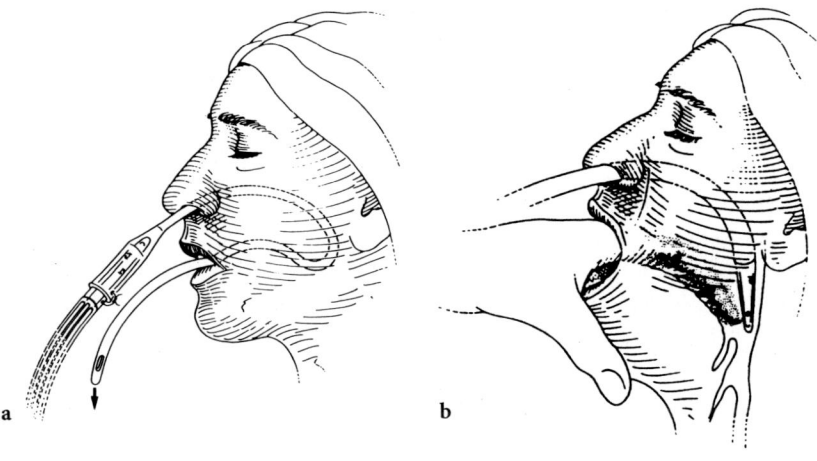

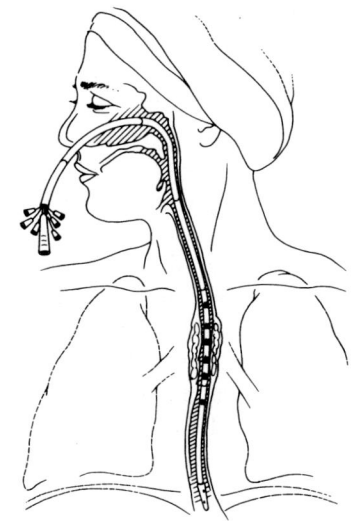

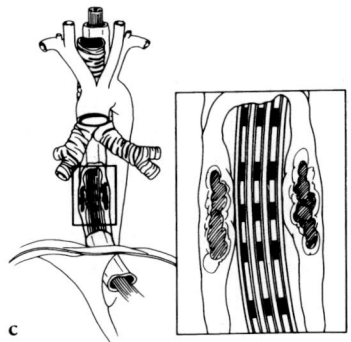

Fig. 14.22a–c. a The tip of the applicator is sutured to the Robinson catheter. b Esophageal applicator is being guided into the hypopharynx and the esophagus. c Radioactive sources shown against the tumor (SYED et al. 1987)

locally advanced unresectable carcinoma of the pancreas (DOBELBOWER et al. 1986; SHIPLEY et al. 1980; SYED et al. 1983a). With the patient under general anesthesia, after the tumor is exposed by the surgeon, and biopsies are performed, tumor volume is evaluated, and biliary and/or gastric bypasses are performed as required. Multiple seeds are implanted in the pancreas with ^{125}I implantation techniques (with a device such as the Mick applicator), usually at 0.5- or 1-cm intervals, depending on the volume to be implanted and intensity of the sources. After localization X-ray films are obtained by the stereoshift or orthogonal technique, computer dose calculations to determine the minimal peripheral dose are obtained.

In 98 patients described by PERETZ et al. (1989), the mean peripheral dose was 136.6 Gy, the mean activity of the implant was 35 mCi, and the mean volume was 53 cm^3. Of the patients, 10 (10%) survived more than 18 months, and 3 patients were long-term survivors (18, 19, and 45 months). Significant pain relief was observed in 37 of 57 patients (65%). Of the patients, 19 (20%) experienced postoperative complications: 1 patient died with a pancreatic fistula and generalized sepsis, and 8 patients (8%) experienced major complications that included fistula formation, gastrointestinal bleeding, gastrointestinal obstruction, and intraabdominal abscess. Similar survival results in groups of 12–18 patients have been reported by MOHIUDDIN et al. (1988), SHIPLEY et al. (1980), and SYED et al. (1983a). Because of putative potential biological disadvantages of ^{125}I (long half-life and low dose rate), PERETZ et al. (1989) introduced Pal-

ladium 103 (half-life of 18 days and 20–23 KeV) as a new isotope for pancreatic implants.

14.8.9
Biliary Tree

ERICKSON and NAG (1998) reviewed the use of intraluminal brachytherapy as definitive treatment for unresectable bile duct tumors or as adjuvant therapy after resection. External beam irradiation (45–50 Gy) is generally given. Brachytherapy can be given using LDR or HDR via an in-dwelling biliary drainage catheter to boost external beam doses. Brachytherapy alone is reserved for palliative therapy.

An increasingly popular technique is the insertion of radioactive sources in Teflon catheters in the biliary tree under fluoroscopic conditions. The main objective is to drain bile and palliate obstructive jaundice. Because these tumors have a tendency to spread to the periductal tissues and regional lymph nodes, intracatheter irradiation is considered a "boost," administered as a supplement to external irradiation to a larger volume.

A transhepatic cholangiogram is initially performed; in patients who have undergone a surgical procedure, the cholangiogram can be performed through the T-tube. In patients not treated surgically, a percutaneous cholangiogram is carried out under fluoroscopic control.

After the site of obstruction is identified, flexible catheters are inserted into the biliary tree to appropriate depths, under fluoroscopic control. A dual-lumen catheter or two separate catheters should be inserted, one for lodging the radioactive sources and the other for bile drainage. The patency of the biliary tree is monitored with injection of radioactive material under fluoroscopic control. Special care must be taken to maintain biliary drainage. Otherwise, the patient will develop pain and fever as a result of obstructive cholangitis. The catheter is sutured to the skin. Radiographs are obtained to determine the length of active radioactive sources to be inserted and the exact position of the catheter for dosimetric purposes (Fig. 14.23). Doses of 20–30 Gy are delivered at 1 cm from the catheter. This is combined with external irradiation (45–50 Gy) to encompass the periductal tissues and regional lymph nodes. If only intracavitary irradiation alone is used, the doses with this modality are 60–65 Gy at 1 cm (FIELDS and EMAMI 1987).

MEERWALDT et al. (1989) reported on 42 patients with bile duct tumors treated with one or two brachytherapy sessions and external irradiation. A dose of 15 Gy was delivered at each of two sessions or 25 Gy in one session, calculated at 1 cm from the wire, combined with external irradiation (40 Gy in 16 fractions). Of patients, 14% survived for 2 years or more. Fever occurred shortly after the insertion of the ^{192}Ir wire in 6 of 38 brachytherapy sessions; it was usually controlled with antibiotics.

ALDEN and MOHIUDDIN (1994) evaluated intra-luminal ^{192}Ir brachytherapy in 48 patients with cancer of the extrahepatic bile duct. Of the patients, 24 received a combined-modality approach, using external beam irradiation (46 Gy), brachytherapy implant, and chemotherapy, and 24 did not receive irradiation in the course of treatment. The implant was performed with ^{192}Ir ribbon sources (average activity, 29 mCi; active source length, 6 cm) to deliver a mean dose of 25 Gy at 1 cm. Chemotherapy consisted of 5-FU alone or combined with doxorubicin or mitomycin C. Patients treated with external irradiation had a 2-year survival of 30% (median, 12 months) versus 18% in the no-irradiation group (median, 5.5 months) (P=0.01). Those treated to more than 55 Gy experienced an extended 2-year survival of 48% (median, 24 months) versus 0% for those receiving less than 55 Gy (median, 6 months) (P=0.0003).

ISHII et al. (2004) reported on 25 patients with unresectable hilar or distal cholangiocarcinoma treated by external beam irradiation (30 Gy or 50 Gy) followed by intraluminal LDR brachytherapy (24–40 Gy). The biliary drainage tubes were removed from the patients who responded, and stenting was not performed in these patients. Full patency was achieved at the treated lesion in 19 (76%) patients,

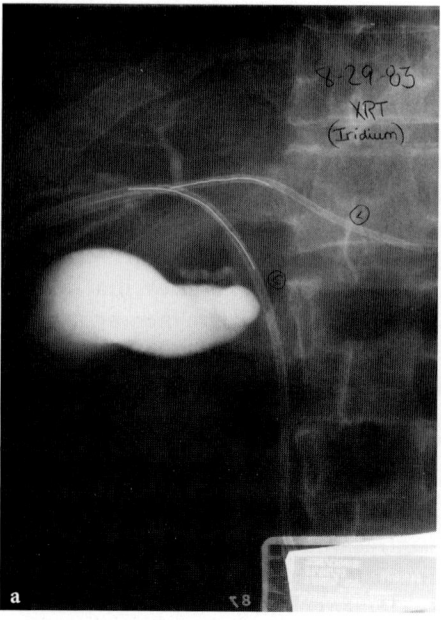
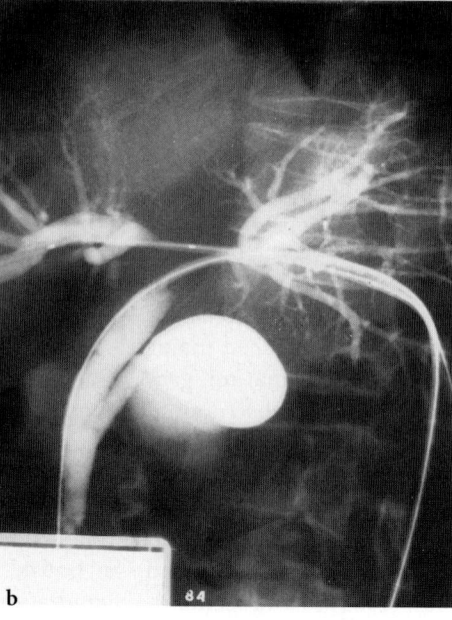

Fig. 14.23. a Radiograph of ^{192}Ir afterloading implant in common bile duct. **b** Biliary tree with contrast material

and they were tube free thereafter. The tube-free survival time in these patients ranged from 7 days to 468 days (median, 76 days). Cholangitis and gastroduodenal ulcer developed in 10 (40%) and 2 (8%) patients, respectively, as adverse events after the combined radiotherapy.

14.8.10
Soft Tissue Sarcomas

External radiation therapy in soft tissue sarcomas combined with limb-preservative surgery has been successful in achieving the same results as obtained with radical surgical resection. Interstitial brachytherapy also eradicates microscopic or minimal macroscopic residual sarcomas. The theoretic advantages of such a combination include: (1) less extensive surgery; (2) synchronous brachytherapy, which allows aggressive treatment of residual malignant cells at a time when these cells are still oxygenated and before they are embedded in scar tissue; (3) placing of the implant plane(s) on the residual tumor (bed), which ensures that this target will receive the highest irradiation dose; (4) short treatment (4–5 days) completed before the discharge of the patient from the hospital, which presents a considerable medical, psychological, and economic advantage; and (5) feasibility even when surgery and external-beam irradiation have previously failed (HILARIS et al. 1988a).

The ABS has published guidelines for the use of brachytherapy for patients with soft tissue sarcoma (NAG et al. 2001c), including patient selection criteria for both adjuvant monotherapy and boost brachytherapy. The ABS recommends brachytherapy as sole radiation therapy only for patients with completely resected intermediate- or high-grade sarcomas of the extremity or superficial trunk, with negative margins. Contraindications for sole therapy include inability to cover the entire target volume, normal tissue tolerance concerns, positive margins, or skin ulceration indicating extensive cutaneous spread via lymphatics. Brachytherapy as a boost to wide-field external beam therapy is recommended for patients with intermediate- or high-grade sarcoma with either negative or positive margins, and can be considered for patients with low-grade sarcoma and post-operatively for patients with small lesions having positive or uncertain margins, possible surgical field contamination, or deep lesions (NAG et al. 2001c).

[192]Ir seeds or wire are usually used for soft tissue sarcoma implants, but ^{125}I has also been used and may be advantageous, as doses to surrounding normal tissues are critical. Basic or sealed-end temporary implant technique can be used to implant these lesions, depending on the location of the tumor. After surgical removal of the tumor, the overlying skin and soft tissues collapse onto the underlying structures. This composite slab of tissues forms the clinical target volume (CTV). Radiopaque markers such as surgical clips should be placed at the time of surgery to identify the extent of the CTV radiographically. The dimensions of the area to be implanted should be measured and recorded. Usually a single-plane implant is satisfactory to cover the CTV.

To ensure a proper implant, the points of needle insertion are marked on the skin with a sterile pen. The guide needles are inserted through the normal skin (at least 1 cm from the incision) after surgical resection, but before completion of any reconstruction and wound closure. The parallel needles are spaced uniformly at 1.0–1.5 cm apart and embedded in the depth of the operative field. If pre-ordered seeds or wires of known activity are to be used, the needle spacing required to deliver a dose of approximately 10 Gy per day to the CTV can be estimated using the Anderson planar implant nomogram or comparable dosimetry system (THOMADSEN et al. 1997). There is currently no universal agreement on the extent of margins to be added to the CTV to ensure adequate coverage of suspected disease, but the ABS recommends that catheter placement should extend at least 1–2 cm outside the CTV in the direction lateral to the catheters and 2–5 cm along the catheter length (NAG et al. 2001c).

After placement of the guide needles, the closed end of each afterloading catheter (in the sealed end technique) is threaded through the needle until it emerges from the opposite end of the needle. The needle is withdrawn while holding the catheter in place until the needle is out of the skin. The process is repeated for the total planned number of afterloading tubes. Each catheter is secured in proper position in the tumor bed with no. 2 or no. 3 absorbable suture material.

Radiopaque clips are placed near the blind end of each afterloading catheter for later identification of this end on localization radiographs. The catheters are individually secured to the skin by means of a stainless-steel button that is threaded over the catheter, fixed to it by crimping, and anchored to the underlying skin by silk sutures. A plastic hemispheric bead cushions the button on the skin, protecting it from undue pressure.

Because of the anticipated effects of the radiation, wound closure requires extra planning and care to avoid undue tension predisposing to wound breakdown. To further diminish wound complications, the ABS recommends that the loading of the radioactive sources should be delayed at least 5 days after surgery for implants used as sole radiotherapy but that loading may take place within 2–3 days of surgery if a brachytherapy dose less than 20 Gy is to be given as a supplement to external beam therapy.

After the surgical procedure is completed, orthogonal radiographs of the CTV with dummy seeds inserted in the catheters are obtained for treatment-planning purposes. If available, CT or MRI images can provide more detailed information for treatment planning. Isodose curves should be generated in planes approximately perpendicular to the source ribbons at intervals of 1.0 cm or less. The CTV should be drawn on the isodose planes, and the isodose line giving satisfactory coverage of the CTV on all planes should be selected as the prescription. Sources can be specially ordered with activities selected to scale the prescription dose rate to deliver 10 Gy per day (or any other desired dose rate), or, if pre-ordered sources are used, the calculated prescription dose rate must be used to determine the required treatment time. For LDR brachytherapy as sole adjuvant radiotherapy, the ABS recommends a dose of 45–50 Gy delivered over 4–6 days. For implants providing a boost to an external beam dose of 45–50 Gy at 1.8–2.0 Gy per day, a boost dose of 15–25 Gy should be delivered in 2–3 days. For correlation of implant quality with clinical outcome, the ABS encourages the calculation of dose volume histograms for the CTV and suggests that the dose covering 90% and 100% of the CTV should be recorded along with the percentage of the CTV receiving 100%, 150%, and 200% of the prescribed dose (NAG et al. 2001c). For additional technical details, the reader is referred to the textbook by HILARIS et al. (1988a).

PISTERS et al. (1995) reported on 164 patients with soft tissue sarcomas randomized to receive or not receive brachytherapy after complete wide local tumor resection (78 and 86 patients in either group, respectively). A target region in the tumor bed was identified by adding 2 cm to the superior and inferior dimensions and 1.5–2 cm in the medial and lateral directions. Afterloading catheters were placed approximately 1 cm apart and were fixed in treatment position with absorbable sutures secured to the skin at the catheter exit site with buttons and nonabsorbable sutures. Implant dose was 42–45 Gy over 4–6 days using ^{192}Ir. Sources were loaded on the fifth or sixth postoperative day to decrease interference with wound healing. There were 13 local recurrences in 78 patients (16%) receiving brachytherapy and 25 in 86 patients (29%) treated with surgery only. Actuarial estimates of local recurrence at 60 months were 18% in the brachytherapy and 31% in the no irradiation group. It is highly likely that the prescribed dose of irradiation was not adequate to eliminate microscopic disease, and higher doses (55–60 Gy) would have been more effective.

LDR intraoperative brachytherapy has been used as a boost in primary tumors. The mean brachytherapy dose was 20 Gy and external beam irradiation dose 45 Gy. DELANNES et al. (2000) reported on a group of 58 patients with primary sarcomas treated by a combination of conservative surgery, intraoperative brachytherapy, and external irradiation. Most of the tumors were located in the lower limbs (46 of 58, 79%). Median size of the tumor was 10 cm, most of the lesions being stages T2–T3. With a median follow-up of 54 months, the 5-year actuarial survival and actuarial local control rates were 64.9% and 89%, respectively. Wound healing problems occurred in 20 of 58 patients, late side effects in 16 of 58 patients (7 neuropathies G2 to G4). No amputation was required.

POTTER et al. (1995) reported on 12 patients with soft-tissue sarcomas treated with HDR or PDR brachytherapy. Brachytherapy was part of the recurrence treatment in 8 patients and part of the primary treatment alone or combined with external-beam irradiation in 4 patients. With HDR, a dose of 15–43 Gy was delivered in 3–16 fractions, and, with PDR, 13–36 Gy in fractions of 1 Gy/h were used. In 6 patients with Ewing's sarcoma, brachytherapy was performed intraoperatively as a boost treatment after external-beam therapy (50–55 Gy), if no wide resection could be achieved. A dose of 10–12 Gy was applied in one fraction to a limited volume (20–50 cm^3) at the time of surgery. With a median follow-up of 21 months, all patients are disease free, and perioperative and subacute morbidity were not increased.

Brachytherapy (15–20 Gy with LDR) was combined with external irradiation (45–50 Gy) either pre- or postoperatively by SCHRAY et al. (1990) in patients with soft tissue sarcomas; 3-D reconstruction of the tumor or tumor bed was accomplished using CT scan or MRI. Margins beyond the tumor were 5–10 cm axially or along tissue planes and 2–4 cm radially or perpendicular to tissue planes. Brachytherapy was performed with standard nylon afterloading tubes positioned to encompass the boost volume with a 2- to 4-cm margin. The boost volume was considered to be the tumor bed after preoperative irradiation and the surgical bed and incision if no previous irradiation had been given. With rare exceptions (distal extremities, groin), needles were placed transversely to the incision (and axis of the extremity) under direct visualization of the tumor bed and with entry and exit points outside the tumor and surgical bed. Needles were commonly placed in contact with bone and neurovascular structures and were maintained in place by skin fascia and muscle; sutures were rarely used. After the implant was in place, skin was closed with sutures. Implants were loaded postoperatively with ^{192}Ir 72–96 h later; the sources were placed 1-cm deep to the skin surface along the tube axis. Average implant activity was 107 mCi, and a portion of the implants used two or more planes. The average dose rate was 48.6 cGy/h, average time was 44.2 h, and dose was 20.52 Gy. In 63 patients, 65 brachytherapy procedures were performed. With a median follow-up of 20 months, there were two local failures in 56 patients (4%) initially treated and in 3 of 9 patients treated for recurrent tumors. Of 40 implants, 2 (5%) performed at initial resection followed by postoperative irradiation led to wound complications, in contrast to 4 of 16 implants (25%) performed at resection after preoperative external irradiation.

14.8.11
Uterine Cervix

FLETCHER (1953) illustrated the importance of selecting the appropriate diameter for cylinders or colpostats and the length for intrauterine tandems. Use of a colpostat or vaginal cylinder with the largest clinically indicated diameter will yield the highest tumor dose at the depth, for a given mucosal dose. It is extremely important to keep in mind the surface dose, because excessive irradiation to the vaginal mucosa (maximum 150 Gy total dose to the proximal and 90 Gy total dose to the distal vagina) may result in severe mucosal atrophy, fibrosis, and vaginal stenosis or necrosis (HINTZ et al. 1980). Similarly, longer tandems will result in improved doses delivered to the lateral parametrium and pelvic lymph nodes.

Intracavitary insertions in carcinoma of the cervix are performed under general, spinal, or local (block) anesthesia. The patient is placed in the lithotomy position, and a complete bimanual pelvic and rectal examination is performed. After

adequate preparation, sterile fields are draped, the cervix is grasped with a tenaculum, and the uterus is sounded carefully to prevent a perforation. If the cervical os/canal is not identified, a small metallic probe may be used. Bimanual pelvic examination is extremely helpful in determining the position of the uterus and the probe or sound. In most patients, dilatation and curettage is performed at the time of the first intracavitary insertion (if not performed at initial workup). MAYR et al. (1998) described the use of osmotic dilators (laminarias) for gradual nontraumatic dilation of the cervical canal for brachytherapy in gynecological cancer patients without the use of general/regional anesthesia. Discomfort is minimal in all cases. Radiopaque markers (lead shots or metallic clips) are placed in the anterior and posterior lips of the cervix. The tandem is inserted in the uterus to the appropriate depth (as determined by a stopper), and, subsequently, each ovoid is gently inserted to prevent injury to the vaginal mucosa.

If ideally inserted in the patient, the tandem should be in the midline or as nearly as possible equidistant from the lateral pelvic wall, and the vaginal colpostats should be symmetrically positioned against the cervix in relation to the tandem (Fig. 14.24a). The tandem should be kept along the sagittal axis of the pelvis, equidistant from the pubis, sacral promontory, and lateral pelvic wall (Fig. 14.24b) as allowed by the geometry of the patient and the tumor to avoid overdosage to the bladder, rectosigmoid, or either ureter. CORN et al. (1994), in a retrospective evaluation of the technical quality of brachytherapy procedures with respect to ovoid and tandem placement, demonstrated a significantly worse outcome for patients whose implants were judged to be unacceptable.

After the tandem and colpostats positions are judged to be correct, careful packing of the vagina with iodoform gauze should be performed. A small mount of packing in front and behind the colpostats (making sure overpacking will not separate the cervix from the colpostats) will decrease the dose to the bladder base and the anterior rectal wall.

An indwelling Foley catheter should be inserted in the bladder; 7 ml of radiopaque contrast material in the Foley balloon will aid in determining a point dose to the bladder neck (ICRU 1985). After the patient recovers from anesthesia, anteroposterior and lateral X-ray films of the pelvis are obtained to document the position of the applicator, and isodose curves are generated.

CORN et al. (1993), in a prospective study of 15 patients with cervix cancer treated with external

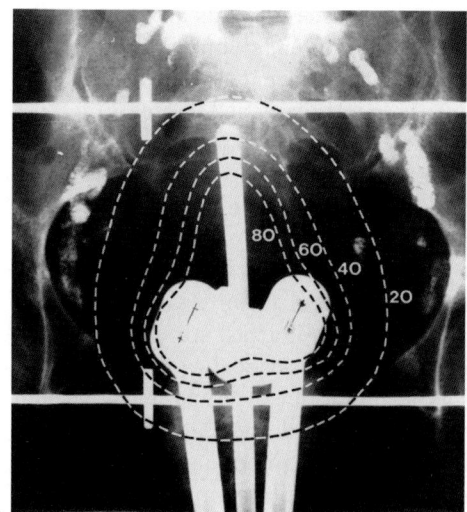
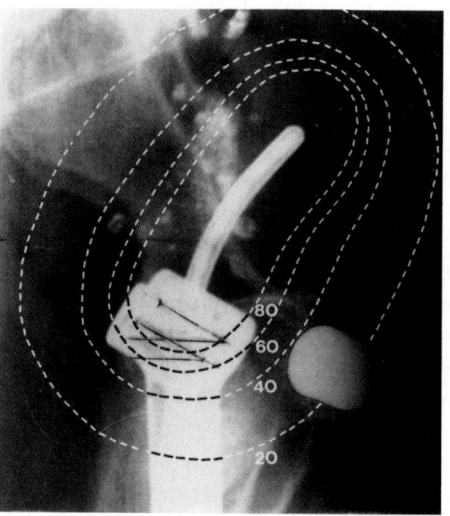

a

b

Fig. 14.24. a Anteroposterior view of intracavitary insertion for carcinoma of the uterine cervix. b Lateral view of same implant. Isodose curves (cGy/h) are superimposed

irradiation and brachytherapy on whom pelvic radiographs were obtained before afterloading and after removal of the ^{137}Cs sources (median duration of insertion, 56.5 h), documented an average 3-mm shift of the applicator. The changes in median dose resulting from source movement were 1.4% to point A, 1.7% to point B, 0.9% to pelvic lymph nodes, 1.9% to the bladder, and 2.6% to the rectum. Thus, applicator movement during LDR brachytherapy does not result in significant dose changes that could have an impact on tumor control or complication rate.

Brachytherapy Systems for Carcinoma of the Cervix
Initially, three systems for LDR brachytherapy in carcinoma of uterine cervix were developed: the Paris, the Swedish, and the Manchester systems (Fig. 14.25) (MEREDITH 1967; MOSS et al. 1979). The systems differ in the type of applicator used, strength of the source, and time of administration (Moss et al. 1979). In the United States, most systems used are derivations of the Manchester technique.

The Manchester intracavitary system, introduced by TOD and MEREDITH (1938) in 1938, used a dosimetric field quantity, total exposure at point A, to prescribe treatment rather than milligram hours. Point A was defined as being 2 cm above the mucous membrane of the lateral vaginal fornix and 2 cm lateral to the center of the uterine canal. Allegedly, this area corresponded to the paracervical triangle, in the medial edge of the broad ligament, where the uterine vessels cross the ureter. A subsequent arbitrary convention defined point A as being 2 cm above the external cervical os and 2 cm lateral to the midline. Yet another definition located point A 2 cm above the distal end of the lowest source in the cer-

vical canal and 2 cm lateral to the tandem. BATLEY and CONSTABLE (1967) illustrated how these modifications to the basic conventions affected the dose to point A because of the different definitions.

The two most vulnerable points in the pelvis were thought to be the vaginal mucosa and the rectovaginal septum, opposite the cervix. No more than 40% of total dose at point A could be delivered safely through the vaginal mucosa. The rectal dose should be 80% or less of the dose at point A; this rectal dose can usually be achieved by careful packing.

Point B was established at the same level as point A, 5 cm from the midline; this point was near the obturator lymph nodes and gave an indication of the lateral throw-off dose.

NAG et al. (2002) published the ABS recommendations for LDR brachytherapy for carcinoma of the cervix. The ABS strongly recommends that radiation treatment for cervical carcinoma (with or without chemotherapy) should include brachytherapy as a component. Precise applicator placement is essential. Doses given by external beam irradiation and brachytherapy depend upon the initial volume of disease, the ability to displace the bladder and rectum, the degree of tumor regression during pelvic irradiation, and institutional practice. Intracavitary brachytherapy is the standard technique for cervical carcinoma; interstitial brachytherapy should be considered for patients with disease that cannot be optimally encompassed by intracavitary brachytherapy. The ABS recommends completion of treatment within 8 weeks, since prolonging total treatment duration can adversely affect local tumor control and survival. Suggested dose and fractionation schemes for combining the external irradiation with LDR brachytherapy for each stage of

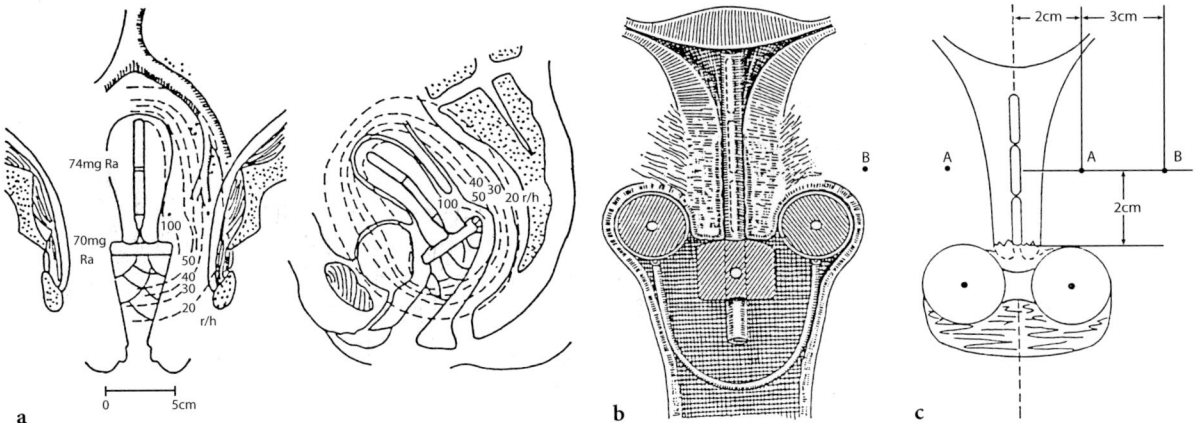

Fig. 14.25. a The Stockholm system. The intrauterine rod-shaped applicator is loaded with 53–88 mg of radium (74 mg in the example shown). The vaginal applicator usually consists of a flat box containing 60–80 mg of radium (70 mg in the example shown), but in special cases other forms of vaginal applicators may be used. Classically, the two applicators are not fixed to each other, but fixed or semifixed combinations have been developed. The vaginal applicator is held against the cervix and lateral fornices by careful and systematic gauze packing. Typically, two or three applications are given with 3-week intervals, each application lasting for 27–30 h (WALSTAM 1954). **b** The Paris system. Typical radium application for a treatment of cervix carcinoma consisting of three individualized vaginal sources (one in each lateral fornix and one central in front of the cervical os), one intrauterine source made of three radium tubes (in so-called tandem position). The active length of the sources is usually 16 mm, their linear activity is between 6 mg/cm and 10 mg/cm, and their strength is 10–15 mg of radium. The total activity is one of the lowest in use for such treatments and implies a typical duration of the application of 6–8 days. Typically, the ratio of the total activity of the vaginal sources to the total activity of the uterine sources should be 1 (with variations between 0.66 and 1.5) (PIERQUIN 1964). **c** The Manchester system. Definitions of points A and B in the classic Manchester system are found in the text. In a typical application, the loading of intrauterine applicators varies between 20 mg and 35 mg of radium and between 15 mg and 25 mg of radium for each vaginal ovoid. The resultant treatment time to get 8000 R at point A is 140 h (MEREDITH 1967)

disease are presented. Dose rates of 0.50–0.65 Gy/h are suggested for intracavitary brachytherapy. Dose rates of 0.50–0.70 Gy/h to the periphery of the implant are suggested for interstitial implant. Use of differential source activity minimizes excessive central dose rates. The dose prescription point (point A) is defined for intracavitary insertions. The ABS recommends reporting the following parameters.

For intracavitary insertions:
- The prescription, including the prescribed dose to point A
- Dose rate
- Implant duration
- Radionuclide used
- Sources' strengths and loading pattern
- Type of applicator used
- Doses to vaginal dose points Vs & Vd
- Doses to rectal and bladder points
- Dose to the pelvic wall, Point PW

For interstitial implants:
- The prescription, including the prescribed dose, dose rate, implant duration, radionuclide used, sources' strengths, and loading patterns

- Type of applicator used
- Volume encompassed by the prescribed isodose surface
- Maximum significant dose
- Rectal and bladder doses if assessed

Considerations for future image-based dosimetry are also noted. They emphasize that the responsibility for medical decisions ultimately rests with the treating radiation oncologists.

14.8.11.1
Applicators for Carcinoma of the Cervix

Applicators used to insert intracavitary sources in the uterus and vagina included rubber catheters and ovoids developed by French researchers, metallic tandems and plaques designed in Sweden, and plastic tandems and ovoids of the Manchester system. FLETCHER (1953) designed a preloadable colpostat, which SUIT et al. (1963) modified and made afterloading.

Intracavitary vaginal colpostats typically incorporate internal shielding to reduce dose to the bladder and rectum. MARKMAN et al. (2001) evaluated the dosimetric effects of inhomogeneities in brachytherapy using Monte Carlo calculations to model dose distributions about both a Fletcher-Suit-Delclos (FSD) LDR system and the microSelectron HDR remote afterloading system. Errors were largely dominated by the primary photon attenuation and were largest behind the shields and tandem. For the FSD applicators, applicator superposition showed differences ranging from a mean of 2.6% at high doses (greater than Manchester point A dose) to 4.3% at low doses (less than Manchester point A dose) compared with the full geometry simulation. Source-only superposition yielded errors higher than 10% throughout the dose range. For the HDR applicator system, applicator superposition induced errors ranging from 3.6% to 6.3% at high and low doses, respectively. Source superposition caused errors of 5–11%. These results indicated that precalculated applicator-based dose distributions can provide an excellent approximation of a full geometry Monte Carlo dose calculation for gynecological implants.

Plastic caps placed posteriorly over the 2.0-cm ovoids increase the diameter to 2.5 cm or 3 cm; typically the 2-cm diameter ovoids have a surface dose of 6.3 cGy/mgh and are loaded with 20 mg sources. If plastic caps are used with the regular ovoids, the surface dose with 2.5-cm ovoids is 4.2 cGy/mgh and 3.0 cGy/mgh with 3-cm ovoids. Therefore, 25-mg or 30-mg sources, respectively, are inserted.

ROSENBLATT et al. (1996b) modified the Fletcher–Suite applicator with two small inflatable balloons attached to the posterior end of each colpostat. The balloons are connected to catheters that emerge from the vagina attached to the colpostat's handles. The balloons were affixed to the colpostats with a plastic adaptor and are inserted empty. The balloons are filled with radiological contrast material and on lateral film typically shows a significant posterior displacement of the anterior rectal wall away from the vaginal sources. In 90 brachytherapy applications using this device for cervical cancer and vaginal applications for endometrial carcinoma following total abdominal hysterectomy, on average, the ICRU rectal point was displaced 14 mm posterior from the colpostats, reducing the dose rate by 60% and resulting in an average dose sparing of approximately 10 Gy to the anterior rectal wall.

The Fletcher tandems, about 6 mm in diameter, are available in three curvatures. A flange or stopper is used to keep the uterine tandem in the selected position; a keeled flange can be used to avoid rota-

tion of the tandem. A special yoke was designed to maintain the position between the intrauterine tandem and the colpostats (DELCLOS 1984). In general, the loading in the tandem is with 20-10-10 mg Ra eq ^{137}Cs sources.

It is extremely important when applicators are purchased to examine the design, to obtain radiographs to identify the position of the shielding (DELCLOS et al. 1978), and to take dosimetric measurements after determining the diameter and thickness of the walls of the applicator to exactly determine the dose distribution around the applicators (HAAS et al. 1985).

The total number of milligram hours prescribed depends on total dose (in Gy) desired at point A (according to tumor stage or volume), number and strength of sources inserted in the tandem and vaginal colpostats, number of insertions (one or two) performed, and whole-pelvis dose delivered with external irradiation.

14.8.11.1.1
Minicolpostats

Minicolpostats have a diameter of 1.6 cm and a flat inner surface to allow their insertion in patients on whom the only alternative would be a protruding vaginal source in the tandem (DELCLOS et al. 1978). Some miniovoids have no shielding. Due to the smaller diameter, its surface dose is significantly higher than with the regular ovoids (at Washington University, with 3M cesium sources, the surface dose is 9.8 cGy/mgh with the miniovoids in contrast to 6.3 cGy with the 2-cm diameter ovoids), and they are usually loaded with 10-mg sources. The 3M miniovoids have internal shielding. However, phantom measurements did not demonstrate a significant decrease in dose for the newer minicolpostats with rectal shielding for a source separation of 3 cm, which potentially could allow undue user confidence in the doses delivered.

KUSKE et al. (1988), in dosimetry studies with thermoluminescent dosimeters in phantom, showed that the measured dose to point A, bladder, and rectum with the minicolpostats is approximately 10% higher than with the regular ovoids. Because of the decreased capacity of the vaginal vault, packing may be more difficult, which results in the bladder and rectum being in closer proximity to the cesium sources. With 10-mg Ra eq sources in the miniovoids, the tandem in the minicolpostat system contributes 6–8% higher dose to point A and the surrounding structures than with regular colpostats. Evaluating the results of therapy in 99 patients with carcinoma of the cervix treated with miniovoids, KUSKE et al. (1988) noted a 15% incidence of grade-3 complications compared with 8% in a group of 194 patients treated during the same period with regular (2 cm) colpostats ($P=0.08$).

14.8.11.1.2
Henschke Applicator

The Henschke and other applicators are commercially available (DELCLOS et al. 1978). The basic configuration of the ovoids is hemispheres that are inserted parallel to the lateral wall of the vaginal vault and the intrauterine tandem. Three ovoid diameters and various tandems are available. Although this applicator's configuration conforms better to a narrow vaginal vault, the radioactive sources are placed parallel to the long axis of the bladder and the rectum and do not have any shielding, thus potentially delivering a higher dose to these organs. Users should familiarize themselves with the dosimetric aspects of these devices. DELCLOS et al. (1978) emphasized that the dosimetry with the Fletcher colpostats is unique and that treatment techniques and tables derived for this applicator should not be used with other applicators, because this might result in significantly higher doses to the vagina, bladder, or rectum. DELCLOS et al. (1978) illustrated the differences in doses delivered to the bladder or rectum with the Fletcher or the Henschke applicator for a normalized dose of 70 Gy to point A. Appropriate source loading and dose prescription will produce satisfactory clinical results.

14.8.11.1.3
Interstitial Implants for Cervical Carcinoma

Metallic needles containing ^{137}Cs or, more recently, afterloading metallic guides or Teflon catheters for insertion of ^{192}Ir wires or seeds have been implanted in the parametrium or cervix, using a transvaginal or transperineal approach (sometimes in lieu of intracavitary insertions when the cervical canal cannot be identified), frequently with the aid of templates (PREMPREE 1983).

The procedure is similar to that followed for intracavitary insertions. The operator should keep in mind the expected anatomic location of the major pelvic vessels, especially veins (since arteries are more difficult to pierce). The cervix should always be held firmly with a tenaculum. For implants in the cervix itself, the needles or nylon catheters with metallic guides (5- to 6-cm long) are inserted straight,

approximately 1.2 cm apart, following the position of the uterus (which can be verified with a finger in the rectum) in a single- or double-circle arrangement. If a single circle is used, full-intensity sources are required. If a double circle is implanted, the central one should have half-intensity sources (usually four), and the periphery should have full-intensity sources. At Washington University, the parametrial Teflon catheters (with metallic guides), usually 12–15 cm long, are inserted through the vaginal fornices. A double-plane or volume implant usually can be placed in each parametrium. The catheters are implanted starting at 1 o'clock on the patient's left side and at 11 o'clock on the right, directed parallel to the coronal plane of the patient and 5–10 degrees lateral toward the pelvic wall. The peripheral planes should be placed 1.2–1.5 cm lateral to the more medial planes, and the catheters should be inserted in the same fashion, approximately 10 degrees divergent in the cephalad direction from the midline.

Insertion of the needles into the bladder should be avoided, unless it is necessary to cover the tumor volume.

When the uterosacral ligament area is to be implanted, the catheters are directed 5–10 degrees posteriorly. In general, 6–10 catheters can easily be implanted in each parametrium. We prefer to implant the interstitial catheters alone, without vaginal colpostats or cylinders, to prevent displacement or enhanced penetration of the needles (Fig. 14.26). Gentle packing with iodoform gauze will keep the needles in place. Cystoscopy and a careful rectal examination at the completion of the procedure will help identify any misplaced needles, which should be withdrawn or replaced immediately. A digital rectal examination is performed (with a second glove, to be discarded) to ensure that there are no catheters where radioactive sources would be placed in the rectum.

ARISTIZABAL et al. (1985), MARTINEZ et al. (1985b), and SYED et al. (1983b) have popularized the use of interstitial implants, using perineal templates for guidance of spacing and alignment, with introduction of long metallic guides through the perineum into the parametrial tissues (Fig. 14.27). ^{192}Ir seeds or ^{137}Cs microspheres in nylon tubes are inserted in an afterloading fashion after X-ray films are obtained with dummy sources for dosimetry computations. ARISTIZABAL et al. (1985) modified their technique by deleting three anteriorly and three posteriorly placed needles in the central row; the central tandem was also omitted in an effort to decrease an initial high incidence of vesicovaginal or rectovaginal fistula. The authors reported approxi-

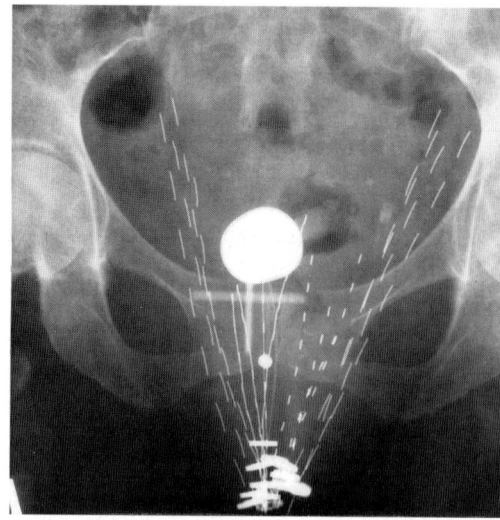
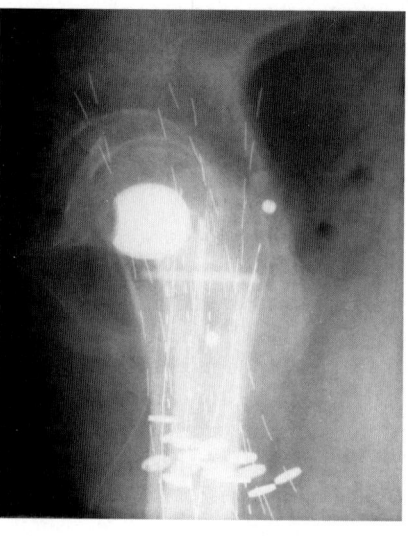

Fig. 14.26a,b. Anteroposterior (**a**) and lateral (**b**) radiographs of the pelvis illustrating bilateral parametrial implant (with sources extending into vaginal walls) for extensive carcinoma of the uterine cervix. The upper radiopaque marker indicates the position of the cervix. The lower radiopaque marker denotes the distal margin of vaginal tumor extension

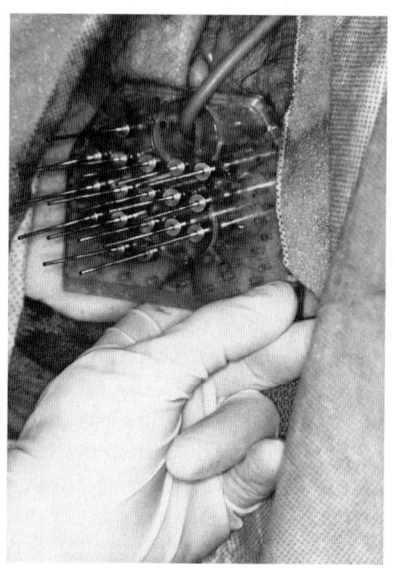
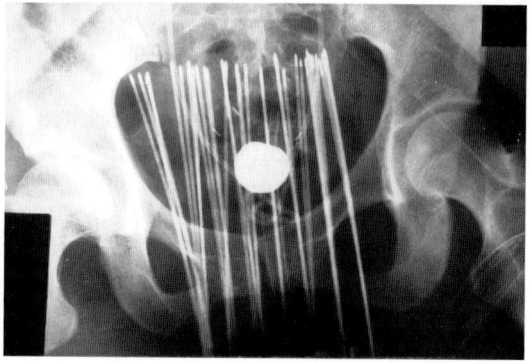

Fig. 14.27. a Martinez Universal Perineal Interstitial Template applicator and metallic guides in place. **b** Anteroposterior radiograph of the pelvis shows the position of the guides in the parametrium

mately 75% pelvic tumor control in 118 patients with stage-IIB and -III carcinoma of the uterine cervix. The major complication rate was 6% with less than 4500 mgh, 16% with 4500–4999 mgh, 28% with 5500 mgh, and 87% with higher intracavitary doses (combined with 45–50 Gy to the whole pelvis).

MARTINEZ et al. (1985b) described results in 104 patients with locally advanced or recurrent pelvic tumor using a universal perineal template combined with external irradiation (36 Gy to the whole pelvis and 14 Gy to the pelvic sidewall with midline

block using four-field techniques, 4- or 10-MV photons). Local tumor control was obtained in 82% of 63 patients with gynecological lesions. The major complication rate was 3.2%.

HUGHES-DAVIES et al. (1995) reported on 139 patients treated with transperineal template interstitial brachytherapy for locally advanced or recurrent pelvic cancer. Most patients received external pelvic irradiation (median dose, 42 Gy) followed by an implant (median dose, 30 Gy, 48 h). The dose rate was 0.4–1 Gy/h. Implant geometry was based on CT scan or MRI studies. An acrylic template was sutured in place, and a bladder catheter was inserted. Blind-ended hollow plastic afterloading catheters were inserted in the pelvic tissues. With median follow-up of 57 months in the survivors, the 5-year local tumor control rate was 25%, and the disease-free survival rate was 22%. Of the patients, 18 (13%) developed fistula. Late bladder complications were observed in 18 patients (12%), and bowel complications were seen in 28 patients (20%). In addition, 2 patients developed pathological fracture of the pubic ramus.

NAG et al. (1998a) used fluoroscopy to guide the needles for interstitial brachytherapy with [192]I using a Syed template to treat various gynecological malignancies. The brachytherapy dose (prescribed to the periphery of the implant) was 40–55 Gy when used alone (15 patients) and 22–40 Gy when used as a boost, combined with 34.2–59.4 Gy of pelvic external-beam radiotherapy. NAG et al. (1998b) also reported on 31 patients with carcinoma of the cervix and 8 patients with vaginal carcinoma treated with external beam radiation therapy and fluoroscopic-guided interstitial brachytherapy. Clinical indications for interstitial brachytherapy were extensive parametrial involvement in 22 patients, extensive vaginal involvement in 10, and poor vaginal anatomy in 7. With a median follow-up of 36 months, 16 patients (51%) with cervical carcinomas and 5 patients (62.5%) with vaginal carcinomas had local tumor control. Only 1 patient experienced grade-3 complications (2.5%).

Results reported by several authors using templates in locally advanced uterine carcinoma are shown in Table 14.3.

14.8.11.1.4
Other Brachytherapy Techniques in Carcinoma of the Cervix

A report by SHERRAH-DAVIES (1985) illustrates the importance of variations in techniques when new devices are introduced into clinical use. At the Christie Hospital, it was customary to deliver 75 Gy to point A using [226]Ra sources in two insertions of 70 h each, with a week-interval between insertions, without supplemental external irradiation in patients with stage-I and -IIA disease. In 1979, the sources were changed to ^{137}Cs, and patients were given one brachytherapy fraction of 37.5 Gy to point A with [226]Ra and a second insertion of 35 Gy with [137]Cs (to account for 10% higher dose rate for cesium sources). Subsequently, 12 patients were treated with

two ^{137}Cs insertions for a dose of 70 Gy to point A. After the short pilot study, patients were randomly allocated to be treated with 75 Gy with [226]Ra or 75 Gy to point A with ^{137}Cs sources. At the same time, patients received external irradiation using two techniques, one with wedges and the other with four hexagonal fields, with different doses of irradiation. The reader is referred to the original article for more technical details. The incidence of bowel damage in patients treated with ^{137}Cs alone was 27% (7 of 26 patients) compared with 3% (1 of 33) with [226]Ra alone or in the group combined with ^{137}Cs intracavitary insertion. It was concluded that dose rate may have contributed to the increased morbidity with the cesium sources, but to a smaller extent than radiobiology predicted. The author ascertained that bowel damage seemed to be associated with use of long (6 cm) intrauterine tubes in 98% of patients treated with ^{137}Cs to 75 Gy to point A. With new 40-degree angle tubes, less use of the long tubes, and a decreased dose to 65–70 Gy to point A, the incidence of bowel damage was reduced to 0.5%.

LEBORGNE et al. (1999) reported their experience with dose fractionation schedules using MDR brachytherapy (1–12 Gy/h) in 42 patients with stage-IB, -IIA, and -IIB carcinoma of the cervix. External irradiation with a central block was given to the pelvis (40 Gy at 2 Gy per fraction), and patients with stage-IIB disease received an additional 20 Gy to the whole pelvis without central shielding. The MDR group was treated at 1.6–1.7 Gy/h to point A; treatment factors are summarized in Table 14.4 (LEBORGNE et al. 1999). A control group of 102 patients was treated with LDR brachytherapy (average dose rate was 0.44 Gy/h, two 32.5-Gy fractions to point A in 74 h each, 2 weeks apart). Grade-2 and -3 sequelae at 2 years were noted in 1% of patients treated with LDR brachytherapy and in 2.4% treated with MDR. The average nominal BED for the various groups ranged from 78 Gy to 124 Gy. The incidence of late rectal complications was zero for patients receiving rectal BED of less than 50 Gy, 24–36% (53 of 184) for 50 Gy to 199 Gy, and 67% (4 of 6) for doses of 200 Gy BED or greater. The authors

Table 14.3. Results with external beam irradiation and template for locally advanced (stage IIb and IIIb) cervix cancer

Author	No. of patients	Local recurrence	Complications
ARISTIZABAL et al. (1985)	118	30 (25%)	25 (21%)
MARTINEZ et al. (1985b)	37	6 (16%)	2 (5.4%)
GADDIS et al. (1983)	51	18 (33%)	8 (16%)
AMPUERO et al. (1983)	24	9 (38%)	7 (29%)
Total	265	76 (29%)	45 (18%)

Table 14.4. Treatment factors in low dose rate or medium dose rate brachytherapy. LEBORGNE et al. (1996)

	Low dose rate	Medium dose rate 1	Medium dose rate 2	Medium dose rate 3	Medium dose rate 4
Dose rate (median Gy/h)	0.44	1.68	1.65	1.64	1.61
Brachytherapy fractions	2	2	2	3	6
Mean dose/fraction	32.6	31.3	24.1	15.3	7.7
Brachytherapy total dose	65.1	62.5	48.2	46.0	46.2
External dose to point A (two fractions)	15.2	18.9	10.0	12.4	9.3
Total dose to point A, coefficient of variation (=SD/mean)%	80.3, 13%	80.4, 10%	58.2, 18%	58.4, 22%	55.5, 21%

concluded that the safest schedule was to deliver 18 Gy to the whole pelvis with external irradiation, with brachytherapy delivering a dose rate to point A of 1.6 Gy/h, in six fractions of 8 Gy, two in each treatment day, 10 days apart. Two fractions are given on a single day, 6 h apart, to reduce the number of insertions to three. This study emphasizes the importance of conducting prospective dose-fractionation studies based on sound biological data.

14.8.12
Endometrium

Carcinoma of the endometrium may grow irregularly into the uterine cavity and produce deformity of the lumen from exophytic tumor, thickening of the uterine wall caused by myometrial infiltration, or uterine enlargement. It is important to determine the size and shape of the uterus; this can be accomplished by rotating the uterine sound and measuring the width and depth of the uterine cavity as well as by bimanual palpation or hysterogram. Special care should be taken to avoid a perforation because, if this occurs, packing with Heyman capsules should not be performed at that time. However, a carefully inserted tandem may be used, avoiding the site of perforation. Ultrasound may help in ascertaining the exact position of the tandem. RUTLEDGE and DELCLOS (1980) also cautioned against rupture (splitting) of the cervix, which may be caused by excessive careless dilatation.

Uterine packing with capsules was originally described by HEYMAN et al. (1941). The practice of introducing as many capsules as possible to stretch the wall of the uterus has several advantages, as outlined by RUTLEDGE and DELCLOS (1980): a bulky tumor can be flattened out, allowing the base of the lesion to be more effectively irradiated; stretching of the uterine wall to make it thinner permits higher doses to be delivered to the serosa of the organ; and a more uniform distribution of the radiation is delivered to the entire myometrium.

Afterloading Heyman–Simon capsules have been available in 6-, 8-, and 10-mm diameters and 2–3 cm length. Inactive metallic guides and later ^{137}Cs sources are inserted.

When capsules are used, it is convenient to insert an afterloading tandem to cover the lower uterine segment, because this permits more flexibility in the loading to obtain improved coverage of this portion of the uterus and the cervical canal. Afterloading colpostats should be routinely used to irradiate the vaginal cuff. A technical problem with the afterloading Heyman–Simon capsules is the relatively large thickness of the stems, which requires continued dilatation of the cervical canal (Hegar dilators) after a few capsules have been inserted.

It is critical to record the order of insertion of the capsules (by numbers that are printed on each capsule), so that removal is done in the reverse order of insertion. Otherwise, the capsules may become jammed, making removal more difficult. Ideally, a minimum of four capsules should be inserted. If fewer are allowed by the size of the endometrial cavity, it may be better to insert an afterloading tandem.

The dose of irradiation delivered with this system is somewhat empirically derived. In general, in preoperative insertions (currently rarely used), we have used 3500 mgh in the uterine cavity; however, cavities larger than 8 cm received doses of approximately 4000 mgh. Doses of approximately 65 Gy to the mucosal surface of the vagina are delivered (1900–2000 mgh) with 2-cm diameter vaginal ovoids. GRIGSBY et al. (1991) reported higher survival and fewer pelvic recurrences and distant metastases in patients with stage-I poorly differentiated endometrial carcinoma when doses higher than 3500 mgh were delivered in the uterus. A lesser beneficial impact was noted in moderately differentiated tumors.

In patients treated with radiation therapy alone (because of medical condition), higher intracavitary doses (in the range of 8000 mgh) are given in two or three insertions. This is combined with external irradiation (20 Gy whole pelvis and an additional 30 Gy to the parametria with midline shielding).

For postoperative irradiation in endometrial carcinoma, if no preoperative irradiation was delivered, in the past we used afterloading colpostats or cylinders to deliver 60–70 Gy to the vaginal mucosa (1900–2000 mgh) with LDR brachytherapy in patients with poorly differentiated tumors even in the absence of deep myometrial invasion. When there is deep myometrial invasion (>50%), regardless of the histological features, the intracavitary therapy is combined with external irradiation (20 Gy whole pelvis and additional 30 Gy to parametria with midline shielding). If a preoperative implant was performed, only external irradiation is administered postoperatively, as outlined. Well-differentiated or moderately differentiated tumors with less than 50% myometrial invasion do not benefit from postoperative irradiation. In many institutions, LDR has been replaced by HDR brachytherapy, with appropriate dose adjustments.

NAG et al. (1997b, 1998a) treated 15 patients with locally recurrent endometrial adenocarcinoma with perineal template interstitial irradiation with LDR brachytherapy (^{192}Ir or ^{137}Cs). Of the 7 previously unirradiated patients, 5 received pelvic external beam radiation therapy of 45–50 Gy, with standard fractionation followed by an interstitial brachytherapy boost dose of 30 Gy (range 25–35 Gy); the other 2 patients received only palliative brachytherapy (40–50 Gy). Eight previously irradiated patients received only brachytherapy of 50–55 Gy. After a median follow-up of 47 months, the actuarial local tumor control rate was 66.6% (with interstitial irradiation only was 64.3% with interstitial irradiation and external irradiation, 100%). Toxicity has been minimal, with 6 patients complaining of vaginal/rectal (RTOG) grade 1–3 complications (5 patients grade 1–2, 1 patient grade 3).

14.8.13
Iridium-192 Interstitial Brachytherapy for Locally Advanced or Recurrent Gynecological Malignancies

GUPTA et al. (1999) assessed treatment outcome in 69 patients with either locally advanced or recurrent malignancies of the cervix, endometrium, vagina, or female urethra treated using the MUPIT with (24 patients) or without (45 patients) interstitial hyperthermia. Of the patients, 54 had no prior treatment with radiation and received a combination of external beam irradiation and an interstitial implant. The combined median dose was 71 Gy (range 5–99 Gy), median external beam irradiation dose was

39 Gy (range 30–74 Gy), and the median implant dose was 32 Gy (range 18–40 Gy). In addition, 15 patients with prior radiation treatment received an implant alone. The total median dose, including previous external beam irradiation, was 91 Gy (range 70–130 Gy), and the median implant dose was 35 Gy (range 25–55 Gy). With a median follow-up of 4.7 years in survivors, the 3-year actuarial local control, disease-specific survival, and overall survival for all patients was 60%, 55%, and 41%, respectively. The clinical complete response rate was 78%, and, in these patients, the 3-year actuarial local control, disease-specific survival, and overall survival was 78%, 79%, and 63%, respectively. On univariate analysis for local tumor control, tumor volume and hemoglobin levels were found to be statistically significant. On multivariate analysis, however, only tumor volume remained significant ($P=0.011$). The grade-4 complication rate (small bowel obstruction requiring surgery, fistulas, soft tissue necrosis) for all patients was 14%. With a dose rate less than 70 cGy/h, the grade-4 complication rate was 3% versus 24% with dose rate greater or equal to 70 cGy/h ($P=0.013$).

14.8.14
Vagina, Vulva, and Female Urethra

Indications for and techniques of interstitial therapy for carcinoma of the vagina, vulva, and urethra have been described (PEREZ et al. 1988, 2004). Use of interstitial implants ideally should be limited to a volume encompassing 75% or less of the circumference of the vagina, particularly when the lesion involves the posterior wall and rectovaginal septum. The remaining normal tissues should be kept away from the implanted area as much as possible, with the judicious use of gauze packing, cylinders, or templates. Two rolls of gauze are placed on top of and between the thighs, so that when the legs are brought down from the lithotomy position (in which the implant is done), the inside surfaces of the thighs are separated as much as possible from the radioactive sources.

14.8.14.1
Vaginal Cylinders

Afterloading vaginal cylinders have a central, hollow metallic cylinder, in which the sources are placed, and plastic rings of varying lengths and diameters are inserted over the cylinder. Domed cylinders are used to irradiate the vaginal cuff homogeneously, when indicated. DELCLOS et al. (1980) recommended that a short cesium source be used at the top to obtain a uniform dose around the dome because a lower dose is noted at the end of the linear cesium sources. In some instances, the cylinders have lead shielding to protect selected portions of the vagina. A flange with a keel is placed over the tandem after the last plastic cylinder has been inserted to secure the system in place and avoid rotation.

The Bloedorn applicator consisted of a device incorporating the configuration of vaginal colpostats or a single midline ovoid and vaginal cylinders. Although extensively used, it was never described in detail; the Bloedorn applicator was later adapted for afterloading (DELCLOS et al. 1980).

PEREZ et al. (1990) designed a vaginal applicator that incorporated two ovoid sources and a central tandem that can be used to treat the entire vagina (alone or in combination with the uterine cervix). The applicator has vaginal apex caps and additional cylinder sleeves that allow for increased dimensions. The dosimetry and dose specifications for this applicator have been published (SLESSINGER et al. 1992), showing that the applicator delivers 1.1–1.2 Gy/h to the vaginal apex and 0.95–1 Gy/h to the distal vaginal surface when loaded with 20 mgRaEq ^{137}Cs tubes in each ovoid and 10, 10, and 20 mgRaEq tubes in the vaginal cylinder (Fig. 14.28). The tandem in the uterus can be used when clinically indicated with standard loadings, depending on the depth of the uterus (20-10-10 or 20-10 mg Ra eq). When the tandem and vaginal cylinder are used, the strength of the sources in the ovoids should always be 15 mg Ra eq. The vaginal cylinder or uterine tandem never carries an active source at the level of the ovoids.

ARMADUR et al. (1997) described a simple, inexpensive custom-made applicator for irradiation of localized areas of the vagina with intracavitary brachytherapy, which allowed the higher dose to be limited to the portion of the vagina at risk for residual disease. The applicator was fabricated from a clear case acrylic (Lucite) rod, 3.5 cm diameter×5 cm long. The applicator contained 11 parallel grooves, each 1.8 mm deep×2.2 mm wide, machined along the surface of the cylinder parallel to its long axis at 1.0-cm increments. Plastic needles (15 gauge) were inserted into the grooves along the surface of the acrylic cylinder and held in place with heat shrink tubing. The applicator was easily inserted and positioned without anesthesia. Standard LDR ^{192}Ir ribbons were inserted into the plastic needles after positioning the applicator in the vagina. Fabrication of this applicator requires a few weeks' notice, is a routine task for a workshop with a milling machine, and costs approximately US$150.

RYAN et al. (1992) described a vaginal obturator used in combination with implanted catheters to deliver microwave hyperthermia and brachytherapy to the vulva and vaginal wall. The obturator was modified to provide grooves for the mount-

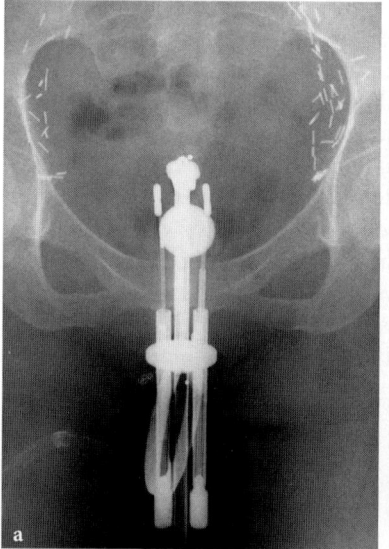 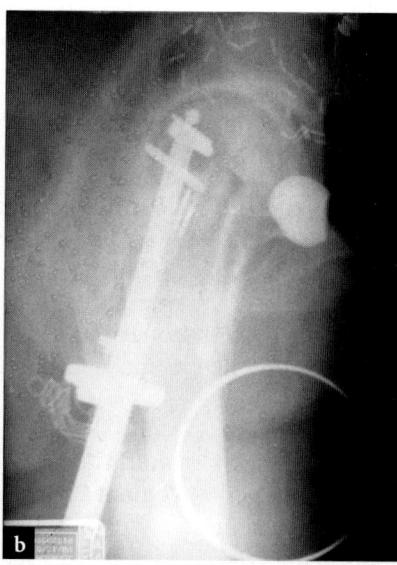 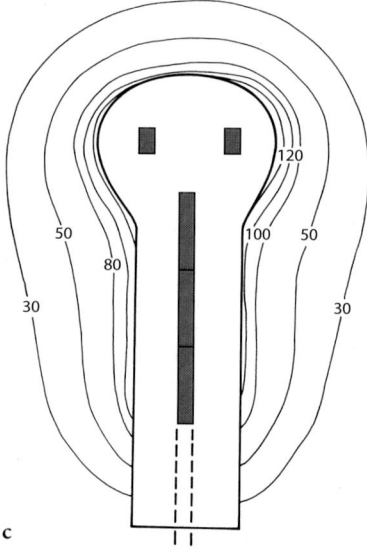

Fig. 14.28a–c. Anteroposterior (**a**) and lateral (**b**) radiographs depicting position of MIRALVA applicator for treatment of patient with vaginal recurrence of carcinoma of uterine cervix previously treated with a radical hysterectomy. **c** Isodose curves of the MIRALVA applicator (PEREZ et al. 1990)

ing of interstitial catheters into the outer wall and was matched with a template for circumferential implants. Two obturator catheters along with two freestanding catheters formed the obturator array. Four freestanding catheters formed the nonobturator array. Therapeutic heating was measured in the catheters on the obturator between antennas in contact with the vaginal mucosa.

KUCERA et al. (2001) compared the role of remote afterloaded HDR brachytherapy in 80 patients and 110 patients treated with intracavitary LDR brachytherapy (with or without external beam therapy). No significant differences were found between the two groups. Overall, actuarial 3-year survival and disease-specific survival rates for all patients in the HDR series were 51% and for LDR 66%, respectively. Complications were equivalent in the two groups.

Because there is substantial individualization in the management of patients with vaginal carcinoma, the treatment guidelines at Washington University are summarized for each stage.

14.8.14.2
Carcinoma In Situ

An intracavitary application with a vaginal cylinder or similar applicator (i.e., BURNETT, DELCLOS, MIRALVA) delivering approximately 75–80 Gy to the mucosa with LDR brachytherapy is adequate to control carcinoma in situ. Higher doses of irradiation may result in significant vaginal fibrosis and steno-

sis. Because of the multicentric nature of this tumor, the entire vaginal mucosa must be treated.

14.8.14.3
Stage I

The most superficial tumors are treated with an intracavitary insertion alone, usually with a cylinder 2.5–3 cm in diameter covering the entire vagina. If the lesion is thicker, a single-plane needle implant is used in addition to the intracavitary cylinder. This has the advantage of increasing the tumor depth dose without delivering excessive irradiation to the uninvolved vaginal mucosa, which receives 60–65 Gy. The gross tumor is treated with 65–70 Gy calculated 0.5 cm beyond the plane of the implant; the vaginal mucosa in this area receives an estimated 80–120 Gy, depending on size of lesion and tumor dose prescribed (Fig. 14.29).

CHOO et al. (2004), in a study of pathology biopsy slides, demonstrated that 95% of the lymphatics of the vagina are located within a 3-mm depth from the vaginal surface; thus, the dose prescription at 0.5 cm may be adequate for most patients with stage-I tumors.

At Washington University, more extensive tumors were treated with intracavitary and interstitial therapy supplemented with external-beam irradiation (whole pelvis dose of 10 Gy or 20 Gy, with additional parametrial dose with a midline block, to deliver a total of 45–50 Gy to the lateral pelvic wall). When

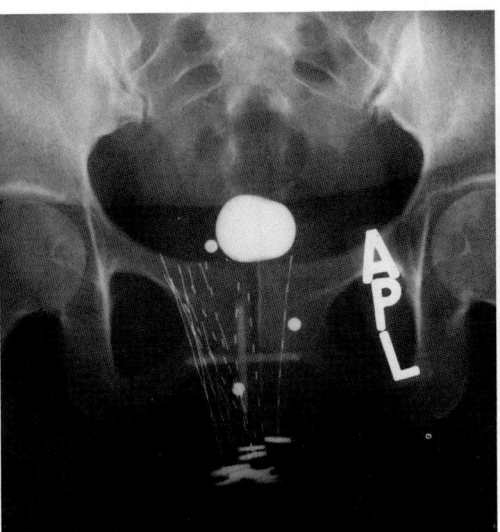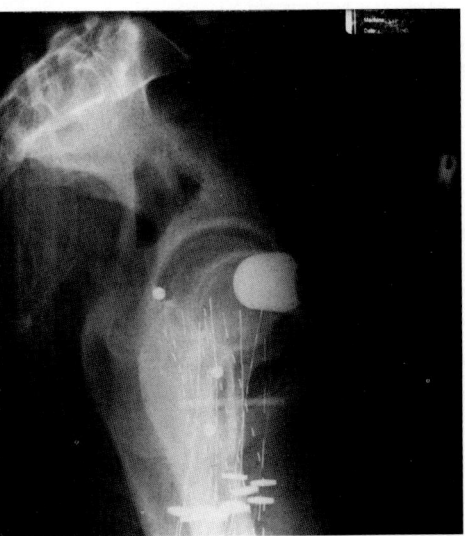

Fig. 14.29a,b. Anteroposterior (**a**) and lateral (**b**) radiographs of the pelvis illustrate interstitial implant performed in a patient with carcinoma of right vaginal wall (intracavitary cylinder omitted)

external irradiation is used, the brachytherapy dose is adjusted downward, adding it to the whole pelvis to achieve the prescribed total vaginal tumor dose.

14.8.14.4
Stage II

Patients with more advanced paravaginal tumors without extensive parametrial infiltration (stage-II lesions) were always treated with a greater external irradiation dose (20–40 Gy to the whole pelvis) and an additional parametrial dose with midline block, to deliver a total of 50–60 Gy to the lateral pelvic wall. In these patients, intracavitary LDR brachytherapy should also be used to deliver a total of 65 Gy to the entire vaginal mucosa and an interstitial implant to administer approximately 70 Gy to a volume 0.5–1 cm around the palpable tumor (dose includes whole pelvis external-beam contribution). Because of the more extensive tumor, double-plane or volume implants are frequently necessary.

14.8.14.5
Stages III and IV

For stage-III and -IV tumors, 40 Gy to the whole pelvis and a total of 55–60 Gy parametrial dose with a midline block are administered. As in stage-IIA tumors, a vaginal cylinder and an interstitial implant have been used with LDR sources to com-

plete total doses of 75–80 Gy to the tumor volume and 65–70 Gy to the uninvolved vaginal mucosa. If the tumor is located in the middle or lower third of the vagina, it is possible to combine the insertion of a cylinder with the ^{192}Ir implant along the vaginal walls in one procedure. However, if the tumor is in the upper third or involves the parametrium, we prefer to perform two procedures, because of concern that the cylinder will displace the interstitial catheters and distort the geometry of the implant. In patients with parametrial infiltration, in addition to the above doses, it is advisable to deliver an additional 15–20 Gy with an interstitial implant (Fig. 14.30).

14.8.14.6
Tumors of the Rectovaginal Septum

When the catheters and needles are inserted in the thin rectovaginal septum, one finger (covered with a second glove) should be inserted in the rectum to ensure that the catheters do not protrude beyond the rectal mucosa. If this occurs, the catheters should be withdrawn and reinserted in a satisfactory position.

When needles or stainless-steel guides are implanted for tumors of the posterior vaginal wall, the rectal ampulla is kept distended for the duration of the implant with a 30-ml Foley double-lumen catheter to minimize irradiation to the lateral and posterior rectal walls.

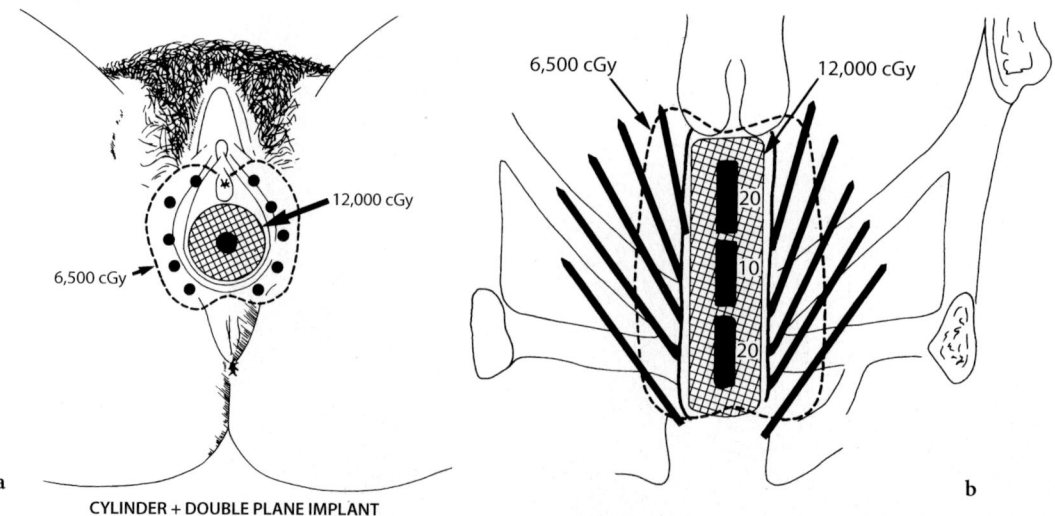

CYLINDER + DOUBLE PLANE IMPLANT

Fig. 14.30. a Cross-section (perineal) view of source arrangement for interstitial and intracavitary implant in a patient with involvement of the right and left lateral vaginal walls and paravaginal tissues. **b** Coronal illustration of interstitial and intracavitary implant for same patient

14.8.14.7
Lesions of the Bladder or Proximal Female Urethra

An open-bladder implant may be necessary for lesions of the bladder or proximal portion of the female urethra that extend into the bladder neck. This procedure also allows direct visualization of tumor extension into the bladder (DELCLOS 1982a). If the tumor has extended beyond the vesical wall, the implant procedure is stopped, and external irradiation is used.

In a series of 160 patients, the mean hospitalization was 36 days after the operation. In 10% of the patients, the abdomen had to be reopened to remove one or more needles (BATTERMAN and BOON 1988; BATTERMAN and SZABOL 1989).

VAN DER WERF-MESSING et al. (1983) used radium implants in 328 patients with T2 and 63 patients with T3 bladder tumors after preoperative irradiation (3.5 Gy for three fractions). The recurrence rates were 16% for the T2 and 28% for the T3 tumors. Disease-free survival rates were 75% and 62%, respectively. BATTERMANN and TIERIE (1986), using a similar technique, obtained local tumor control in 69 of 85 patients (81%) with T2 tumors and a 10-year disease-free survival rate of 56%. Subsequently, VAN DER WERF-MESSING and VAN PUTTEN (1989) used 40 Gy external irradiation followed by ^{137}Cs implants in 48 patients with T2 and 42 patients with T3 bladder cancer. The 5-year disease-free survival rate was 70%.

A different method using iridium wires was designed in France and modified by BATTERMANN and BOON (1989) to overcome most of the disadvantages of the rigid needle technique. After a lower abdominal incision, the bladder is opened to visualize the tumor area. Plastic carriers consisting of a hollow part and a thinner leading end are inserted 1.5 cm apart. The tubes penetrate the abdominal wall, are tunneled in the bladder muscle through the tumor and out of the bladder, and penetrate the abdominal wall again. The catheters should be placed in such a way that removal is feasible without a second laparotomy, although in more complex cases, this may be necessary. Dummy sources are introduced in the carriers to visualize the length of the source to be used while the bladder is still open. After the position of the sources is checked, the bladder is closed, and subsequently the abdomen is closed. A Foley catheter is placed for drainage. After film localization, the dose distribution is determined. The carriers are connected to the MicroSelectron, and the radioactive phase of the procedure is started. The tubes are well tolerated and, after completion of irradiation, can be removed easily. All patients receive preoperative external irradiation to prevent tumor seeding during operation (30 Gy). A dose of 40 Gy is given by brachytherapy at a dose rate of 0.3–0.5 Gy/h.

GERARD et al. (1989) reported a technique involving a combination of external irradiation (10.5 Gy in three fractions in 3 days), external iliac lymph-node dissection, and partial cystectomy to remove the tumor, in addition to an ^{192}Ir implant using a nylon thread technique or a specially designed curved needle to implant the nylon thread. Radiopaque markers help to accurately position the ^{192}Ir wires, which are 5–9 cm in length, with a linear activity of 1.2–2 mCi/cm. The thickness of the treated volume depends on the spacing of the wires (6–10 mm). The dose is calculated using the Paris system (40–50 Gy specified on the 85% isodose of the basal dose). The brachytherapy application lasts 2–5 days, depending on the dose desired, and removal is simply accomplished by pulling the plastic tubes. Somewhat comparable brachytherapy techniques for treatment of carcinoma of the bladder were described by MOONEN (1989), MAAT and VENSELAAR (1989), and WIJNMAALEN et al. (1989).

STRAUS et al. (1988) also used preoperative external irradiation (10–15 Gy for tumors less than 3 cm or 36–50 Gy for tumors 3–5 cm) and ^{192}Ir afterloading implants. MAZERON et al. (1988) used preoperative irradiation (8.5 Gy in a single fraction), partial cystectomy, and ^{192}Ir implants in the adjacent resection margins in 32 patients with T2 and 5 with T3 tumors. The 5-year disease-free survival rate in 20 patients with T2 tumors was 55%, and in 4 patients with T3 cancer, only 1 was alive and free of disease.

An interstitial implant was used for 46 patients with stage-T1 or -T2 and 1 with stage-T3 cancer of the urinary bladder (LYBEERT et al. 1994). Before implantation, 1 patient received no external radiation therapy; the other 46 patients were treated with either a low dose (40 patients, 12 Gy median) or an intermediate dose (6 patients, 38–40 Gy) of external irradiation. Locoregional relapse was observed in 14 of 47 patients (30%). The sites of locoregional relapse were the bladder in 11 patients and the immediate vicinity of the bladder in 3 patients. Only 4 patients died of uncontrolled locoregional disease. A salvage cystectomy was performed in five patients. Ulceration of the bladder mucosa was observed in 9 of 46

(19.6%), and bladder stone formation occurred in 3 of 46 patients (6.5%).

Wijnmaalen et al. (1997) evaluated the results of transurethral resection, external beam radiotherapy, and interstitial radiation with iridium-192, using afterloading technique in 66 patients with primary, solitary muscle invasive bladder cancer, aiming at bladder preservation. Immediately prior to interstitial radiation, 42 patients underwent a lymph-node dissection, and, in 16 cases, a partial cystectomy was performed. For interstitial radiation, two of five catheters were used, and interstitial radiation was started within 24 h after surgery. The majority of patients received 30 Gy with interstitial radiation (mean dose rate of 0.58 Gy/h). In 3 patients, additional external irradiation was applied following interstitial radiation. Follow-up consisted of cystoscopies, mostly done jointly by urologist and radiation oncologists, with urine cytology routinely performed. With a median follow-up of 26 months, the probability of remaining bladder-relapse-free at 5 years was 88%, and the bladder was preserved in 98% of the surviving patients. The 5-year overall and distant relapse-free survival rates were 48% and 69%, respectively. Surgical correction of a persisting vesicocutaneous fistula was necessary in two patients. Serious late toxicity (bladder, RTOG Grade 3) was experienced by only one patient.

Pos et al. (2004) treated 40 patients with T1G3 and T2 bladder carcinoma with 30 Gy external beam irradiation followed by interstitial HDR brachytherapy (32 Gy in ten sessions of 3.2-Gy fractions in two fractions daily with a 6-hour interfraction interval). The HDR schedule was designed to be biologically equivalent to the previously used LDR schedule with the linear-quadratic model. The local tumor control rate and toxicity were compared with a historical group of 108 patients treated with 30 Gy external beam radiation therapy followed by 40-Gy interstitial LDR brachytherapy. The local tumor control rate at 2 years was 72% for HDR versus 88% for LDR brachytherapy ($P=0.04$). In the HDR group, 5 of 30 evaluable patients had serious late toxicity; 4 developed a contracted bladder (<100 ml), and 1 patient required cystectomy because of a painful ulcer at the implant site. In the LDR group, only 2 of 84 patients developed serious late toxicity (one vesicocutaneous fistula and the other a urethral stricture). The difference in late toxicity for HDR and LDR was statistically significant ($P=0.005$). The increased late toxicity with the HDR schedule compared with the LDR schedule suggests a short repair half-time of 0.5–1 h for late-responding normal bladder tissue.

14.8.14.8
Tumors of the Vulva or Distal Urethra

Vulvar or distal urethral tumors can be treated with similar brachytherapy techniques. Erickson (1996) published a historical review of interstitial implants for vulvar carcinoma.

The patient is placed in a lithotomy position, and single, double-plane, or volume implants can be designed around the urethra or in the vulvar labia. We prefer to carefully place a no. 8 or no. 10 Hegar dilator in the urethra during the procedure for orientation of the planes of the implant. If the proximal urethra is involved, the radioactive sources must be inserted reaching the bladder. When the procedure is completed, the Hegar dilator is withdrawn; cystoscopy is performed to ascertain the position of the catheters in the bladder, and an indwelling three-way catheter is inserted. If there is intravesical bleeding, periodic irrigation of the bladder is necessary while the implant is in place, and it is preferable to leave a three-way catheter in place for a few days (up to 1 week) to facilitate bladder irrigation and avoid clot formation with bladder neck obstruction. When the vulva is involved, the sources must protrude into the perineum. If the tumor extends into the vagina, an intravaginal cylinder with some sources may be necessary to increase the dose to the vaginal mucosa (Fig. 14.31).

The design of the implant, placement of the radioactive sources, and tumor doses are similar to those for comparable lesions in the vagina.

14.8.15
Anal Canal and Rectum

Interstitial and intracavitary techniques have been used for many years for the treatment of anorectal carcinoma. Ideally, implants should be restricted to lesions that require implantation of no more than half the circumference of the anal canal for preservation of sphincter function. Single, double-plane, or volume implants may be necessary, depending on the extent of the tumor.

The catheters are inserted through the perianal area in the central plane 0.5 cm away from the anal or rectal mucosa with one finger (double gloved) in the rectum to verify appropriate placement. Peripheral planes are placed at 1- to 1.5-cm spacing. The anal canal is kept distended with a custom-designed rectal plug, which reduces the dose to the opposite side of the canal to less than

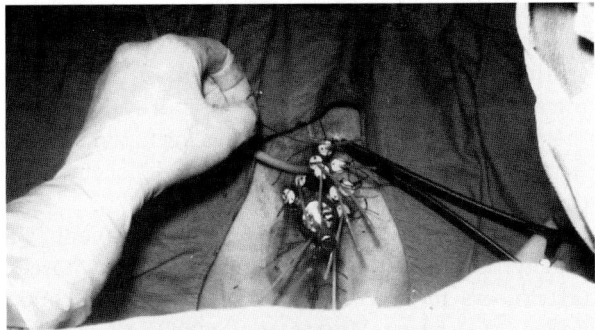

a

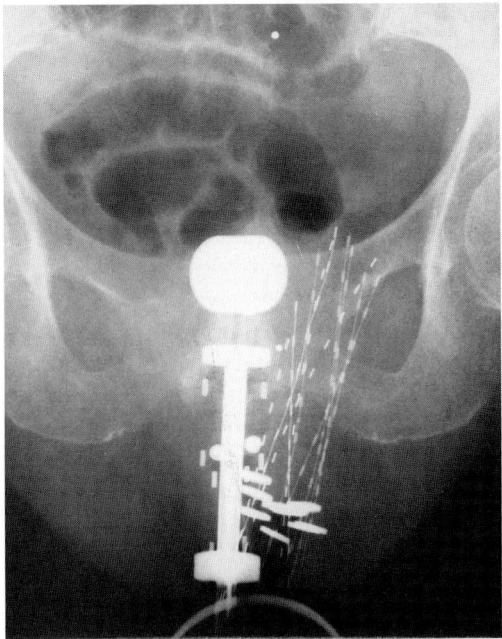

b

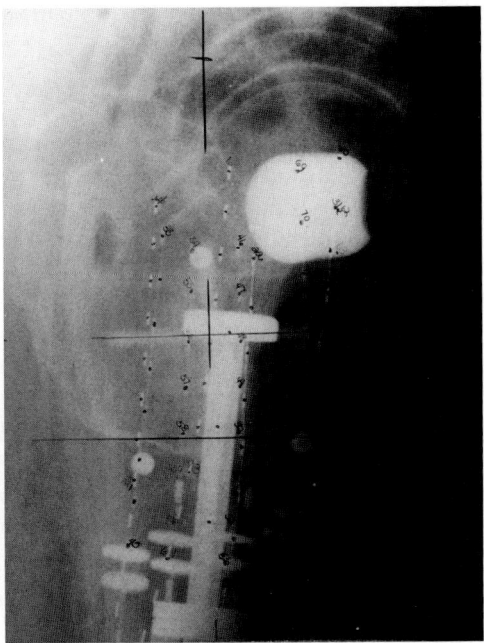

c

Fig. 14.31. a Patient at completion of interstitial implant and intracavitary insertion with stainless-steel guides for ^{192}Ir tubing and Delclos vaginal cylinder. A bladder catheter is in place. The metallic buttons on the plastic catheters are being sutured to the skin to secure the position of the implant. Anteroposterior (**b**) and lateral (**c**) radiographs of implant for urethral tumor with left paraurethral extension

15% of the minimum tumor dose at the implanted area (DELCLOS 1982a).

Although a colostomy may be avoided with diligent care of the implanted area, this is not always practical. It may be necessary to precede the implant with a temporary diverting colostomy, regardless of tumor size or lack of bowel constriction.

It is important to decrease irradiation of the adjacent buttock and thighs, as described in the section on vagina, vulva, and urethra. Ivalon or gauze is placed in the intergluteal space.

The MUPIT applicator has been used in the treatment of anorectal tumors with satisfactory results (MARTINEZ et al. 1985b).

KIN et al. (1988) used a template for insertion of hollow steel needles to place the ^{192}Ir and a rubber drain for treatment of patients with carcinoma of the anal canal, in some patients combined with external-beam irradiation. A ring-shaped template with the appropriate number and lengths of radioactive hollow steel needles was placed over the anus, and the needles were successfully implanted through the corresponding holes (about 1-cm apart) into the anal or rectal wall. The template and rubber drain were carefully withdrawn while the needles were maintained in place. The needles were secured, and the whole applicator was held in place by suturing the drain and the template to the perianal skin. Later orthogonal X-ray films were obtained, and dose calculations were performed. The total tumor dose from the external beam and interstitial irradiation ranged from 54 Gy to 80 Gy (mean 64.2 Gy). Of 32 patients treated, 24 (74%) had local tumor control. Of the patients, 4 had severe complications (two radionecrosis, one atony of the sphincter, and one severe rectitis); 1 patient required colostomy and another abdominoperineal resection. Of the other patients, 14 had less severe complications. The probability of preserving good or acceptable anal function was 69% (22 of 32).

PAPILLON et al. (1989) reported on 221 patients with epidermoid carcinoma of the anal canal treated with a combination of external irradiation (35 Gy) combined with 5-fluorouracil and mitomycin-C, followed by an ^{192}Ir implant 2 months later. The implants were performed with either a plastic template or a steel fork, using four to eight wires, 5–7 cm long, adapted to the tumor extent covering the quadrants of the anal circumference involved by the tumor. A minimum dose of 15–20 Gy was delivered in 15–28 h. Of 189 patients followed for 5 years, 118 (65.9%) were alive and well, and 110 (61.4%) had anal preservation. In patients with tumors less than 4 cm, 50 of 66 (75.7%) were alive with anal preservation at the time of the report, and only 5 (7.5%) died of cancer.

PAPILLON et al. (1989) also reported on 90 patients with T1 or T2 rectal carcinoma treated with contact X-ray endocavitary therapy followed by ^{192}Ir implant with an iridium fork. Doses were similar to those administered to the patients with anal carcinoma. The 5-year survival rate was 77.8%; 67 (74%) were alive with anal preservation, and only 10 (11.1%) died of cancer. They also reported on a third group of patients with more advanced, moderately infiltrating low-lying T2 or T3 tumors, who would have been treated by abdominoperineal resection but, because of age or poor operative risk, were treated with radiation therapy, including interstitial implants. At 4 years, 37 of 62 patients (59.6%) were alive, and 36 (58%) had anal preservation. Only 9 patients (14.5%) died of cancer; 3 had unresectable lesions, and 1 died after major surgery.

PRICE et al. (1988) described 44 patients with inoperable anorectal carcinoma treated with interstitial implants using ^{226}Ra or ^{137}Cs needles to doses of 50–60 Gy (in 5 patients preceded by external irradiation). They recommended a dose of 60 Gy at 0.5 cm when external irradiation is not used. Local tumor control was achieved in 16 of 31 patients (52%) assessed for response. Late morbidity was observed in 12 patients: 5 experienced occasional bleeding or diarrhea; 1 had mucoid discharge; 3 developed stricture requiring surgery; and 3 had necrosis requiring surgery. Most patients who developed complications received total tumor doses above 80 Gy.

Interstitial implants with 10- to 15-cm nylon catheters for ^{192}Ir ribbons are used to treat patients with recurrent carcinoma of the rectum in the perineal and presacral fossa after abdominoperineal resection. Care should be exercised to direct the metallic guides initially inserted or catheters with a posterior orientation (5–10 degrees from the horizontal plane).

In many instances, the needles find resistance from the sacrum; occasionally, the sources are inadvertently placed in the bladder. We have performed intraoperative implants at the time of resection of the recurrent tumor, which allows for better identification of the volume to be treated and placement of the catheters (Fig. 14.32).

A technique with intraoperative permanent ^{125}I brachytherapy was used in 29 patients for colorectal cancers recurrent in the pelvis and paraortic nodes (MARTINEZ-MONGE et al. 1998). All patients had undergone prior surgery; 72% had prior external beam radiation therapy. The implanted residual tumor volume was microscopic in 38% and gross in 62%. The implanted area (median 25 cc) received a median minimal peripheral dose of 140 Gy. An omental pedicle was used to minimize irradiation of the bowel. Of the patients, 5 received additional postimplant external beam radiation therapy (20–50 Gy; median 30 Gy). The 4-year actuarial local-regional control rate was 18%. Overall survival was better for patients with smaller volume implants ($P=0.007$), with a lower total activity implanted ($P=0.0003$), with a smaller number of implanted sites ($P=0.004$), and with microscopic residual disease ($P=0.01$). Patients receiving additional external beam radiation therapy also had a better prognosis ($P=0.005$). Of the patients, 13 (45%) experienced 15 toxic events, including 3 patients (10%) with enteric fistula.

14.8.16
Penis and Male Urethra

Carcinoma of the penis is rare in the United States; therefore, experience in its treatment is scarce. Interstitial therapy with single- or double-plane implants has been used for small lesions of the glans or distal penis. Doses of 60–70 Gy are delivered in 6–7 days with a dose rate of approximately 0.45–0.5 Gy/h. Molds have been used, particularly in earlier years in Europe, but tumor control and functional results were not as satisfactory as with other irradiation techniques (JACKSON 1966).

Small tumors of the proximal urethra can be treated with an intraurethral catheter containing radioactive sources, whereas distal urethral lesions are irradiated with techniques similar to those for penile tumors. Tumors larger than 2 cm should be treated with a combination of external irradiation and intracavitary or interstitial techniques or with a surgical procedure.

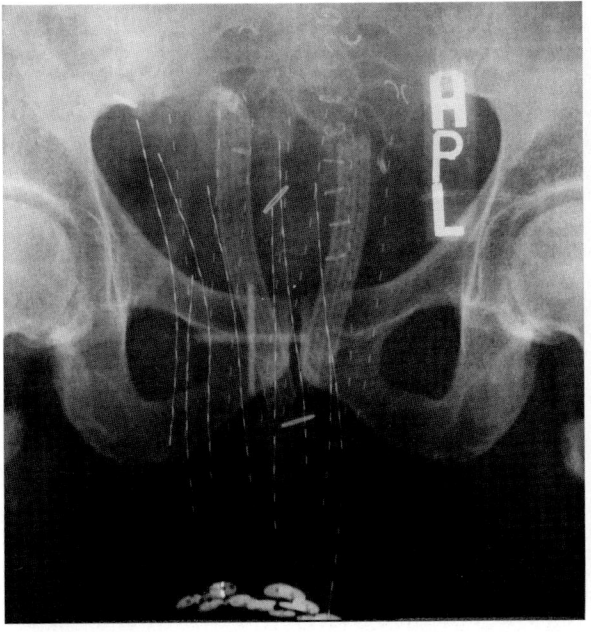
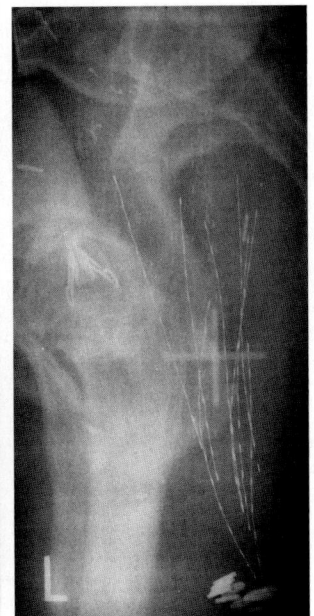

Isodoses (Rads/hr)

0	80
1	60
2	50
3	40
4	30
5	20

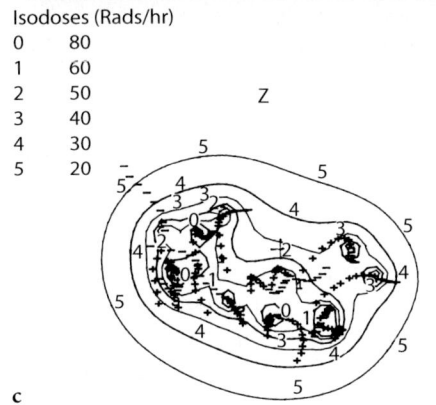

Fig. 14.32a–c. Anteroposterior (**a**) and lateral (**b**) radiographs of interstitial implant performed intraoperatively with plastic catheters and ^{192}Ir in a patient with recurrent carcinoma of the rectum in the posterior pelvis. The patient had received 45 Gy preoperatively a year earlier. **c** Cross-section isodose curves of implant showing dose rate of approximately 40 cGy/h. Additional 50 Gy was administered with the interstitial implant

14.8.17
Prostate

The older retropubic techniques have been replaced by ultrasound or CT-guided transperineal techniques. Prostate brachytherapy may be temporary or permanent, and the planning techniques for either approach are similar. Nori and Moni (1997) briefly discussed the advantages and limitations of each. Temporary techniques may be used with LDR or HDR applications. The basic steps include assessing prostate volume by any diagnostic modality (CT or US), determining total activity needed to encompass the gland and deliver the appropriate minimum peripheral dose, and determining the pattern of placement of seeds within the gland. Preplanning may be done either by US or by CT. The operative technique requires the visualization of the prostate in three dimensions and is usually performed using a combination of US and fluoroscopy. Special circumstances that necessitate neoadjuvant hormonal therapy include interference from the pubic arch and large volume glands. Potency is preserved in greater than 80% of patients. Patient selection criteria include the pretreatment prostate-specific antigen (PSA) level, tumor grade (Gleason), stage of disease, and presence or absence of bilateral positive biopsies and/or perineural invasion. We have divided patients with prostate cancer into good-, intermediate-, and poor-risk groups. We recommend brachytherapy as the sole procedure for good-risk patients, and a combination of external beam radiation therapy and brachytherapy for the intermediate-risk group. Selection criteria for permanent implants are summarized in Table 14.5 (Nag et al. 1999). Currently,

^{125}I or ^{103}Pa are used for permanent implants (Table 14.6).

Since publication of the AAPM Task Group No. 43 Report in 1995 (TG-43), both the utilization of permanent source implantation and the number of low-energy interstitial brachytherapy sources commercially available have dramatically increased (RIVARD et al. 2004). The National Institute of Standards and Technology introduced a new primary standard of air-kerma strength, and the brachytherapy dosimetry literature has documented both improved dosimetry methodologies and dosimetric

Table 14.5. Selection criteria for permanent brachytherapy of the prostate. Modified from NAG et al. (1999). *PSA* prostate-specific antigen, *EBRT* external beam radiation therapy

Brachytherapy as monotherapy:
  Stage T1 to T2a and
  Grade: Gleason sum 2–6 and
  PSA ≤10 ng/ml
Brachytherapy as a boost to EBRT:
  Stage clinical T2b, T2c *or*
  Grade: Gleason sum 7–10 *or*
  PSA >10 ng/ml
Brachytherapy (including boosting EBRT) in conjunction with androgen deprivation
  Patients with initially large prostate (>60 cc) that have downsized sufficiently
Clinical exclusion criteria:
  Life expectancy <5 years
  Large or poorly healed transurethral resection of the prostate defect
  Unacceptable operative risks
  Distant metastases
Relative contraindications for brachytherapy
Patients not ideal candidates for brachytherapy, but nevertheless have been successfully implanted. Beginners should not implant these patients at increased risk of developing complications:
  Large median lobes
  High American Urological Association score
  History of multiple pelvic surgeries
  Severe diabetes with healing problems
Technical difficulties which may result in inadequate dose coverage:
  Gland size >60 cc at time of implantation
  Positive seminal vesicles

characterization of particular source models. In response to these advances, the AAPM Low-Energy Interstitial Brachytherapy Dosimetry Subcommittee presented an update of the TG-43 protocol for calculation of dose-rate distributions around photon-emitting brachytherapy sources, TG-43U1, which includes (a) revised definition of air-kerma strength, (b) elimination of apparent activity for specification or source strength, (c) elimination of the anisotropy constant in favor of the distance-dependent one-dimensional anisotropy function, (d) guidance on extrapolating tabulated TG-43 parameters to longer and shorter distances, and (e) correction for minor inconsistencies and omissions in the original protocol and its implementation. In addition, the report recommends a unified approach to comparing reference dose distributions derived from different investigators to develop a single critically evaluated consensus dataset as well as guidelines for performing and describing future theoretical and experimental single-source dosimetry studies.

LDR brachytherapy is being used extensively for the treatment of prostate cancer. As of September 2003, there are a total of 13 ^{125}I and 7 ^{103}Pd sources that had calibrations from the National Institute of Standards and Technology and the Accredited Dosimetry Calibration Laboratories of the AAPM (DEWERD et al. 2004).

HOLM et al. (1983) developed the transperineal implant technique for prostate cancer. BLASKO et al. (1991) popularized the technique in the USA. NAG et al. (1997a) and POTTERS (2000) have reviewed the current status of permanent prostate brachytherapy.

14.8.17.1
Transperineal I-125 Implants

Several authors, including BLASKO et al. (1987), HOLM et al. (1983), and WALLNER et al. (1991), described the technique for ^{125}I or ^{103}Pd implants of the prostate

Table 14.6. Radioisotopes used for prostate brachytherapy. Modified from PORTER et al. (1995). *EBRT* external beam radiation therapy

	Energy (keV)	Half-life (days)	Initial dose rate (cGy/h)	Mean activity per seed (mCi)	Monotherapy dose (Gy)	Dose (Gy) combined with EBRT
Permanent:						
Iodine 125	27	60	8	0.42	145	110
Palladium 103	21	17	20	1.3	125	100
Gold 198	412	2.7	64		60	
Temporary:						
Iridium-192	340	70	Highly variable		Variable (about 60)	20–25

using a transperineal approach under TRUS. Implants are recommended for patients with prostate volume less than 60 cm³, no severe pubic arch interference, no severe urinary obstructive symptoms, clinically intraprostatic disease, and prior transurethral resection of the prostate has been discouraged. However, MORAN et al. (2004) recently used brachytherapy in 171 patients with stage T1a–T2b who had prior transurethral resection of the prostate (TURP). Of the patients, 100 (60%) returned complete surveys. Time of TURP before implant ranged from 2 months to 300 months (median, 6.5 years). Mean patient age was 74±5.2 years, follow-up time after implant ranged from 6.1 months to 50.9 months. Multivariate analysis revealed higher pretreatment International Prostate Symptom Scores have significant negative impact (P=0.001) on urinary function and bother scores. With accurate ultrasound identification of the urethral defect and precise dosimetry, brachytherapy could be performed in selected patients who have had prior TURP with resultant low impact on urinary function and bother scores. The ABS has issued a detailed set of guidelines for transperineal prostate implants (NAG et al. 1997a, 1999).

The standard technique for TRUS-guided prostate implants involves a two-step approach. The initial step is a volume study of the prostate obtained for treatment-planning purposes. For this purpose, the patient is placed in the lithotomy position, and a rectal ultrasound probe is inserted. The probe is part of a larger positioning system, which includes a special needle-guidance template, which will be locked into a fixed position relative to the probe during the implant procedure (Fig. 14.33). The projected position of the template relative to the prostate images is shown on the US screen. After a near-central transverse image of the prostate is obtained, the probe can be repositioned with the aim of centralizing the prostate in the image and minimizing distortion by the probe. Further images can be obtained and repositioning carried out to ensure that through the course of images the urethra position does not move significantly, relative to the template guide holes, which could obstruct the placement of needles in critical locations. After positioning is completed, a set of transverse images is recorded at 5-mm increments from the base to the apex (Fig. 14.34a). The patient is released, and treatment planning is carried out. Using the projected guide-hole locations as constraints, a seed distribution is generated, which is optimized in terms of prostate coverage, while limiting the doses to the urethra and rectum. This will generally require that a large percentage of the

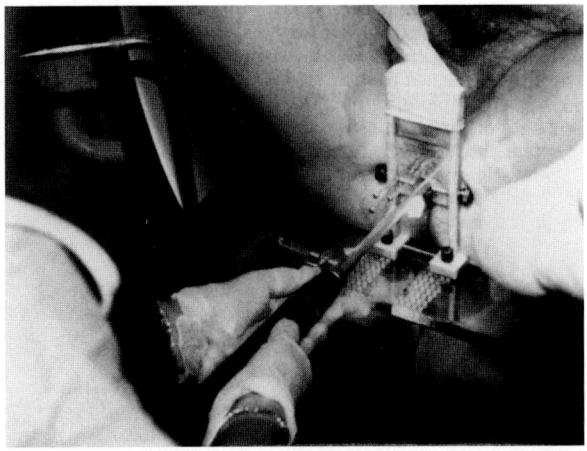

Fig. 14.33. Prostate implant system, including ultrasound unit with rectal tube, stepper unit controlling probe movement, and perineal template (PEREZ et al. 2004)

seeds be placed at the periphery of the prostate capsule (Fig. 14.34b). Recommended prescribed doses for ¹²⁵I prostate implants are 145 Gy for implants used as the sole radiotherapy and 100–110 Gy for implants used as a boost to 40–50 Gy of external beam radiation. Seeds of the required number and activity to deliver the prescribed dose are ordered and, on receipt, calibrated and loaded in sterile needles (typically 18 gauge) along with absorbable spacers, according to the treatment plan. Sterile bone wax or other biodegradable material is used to stopper the ends of the needles.

At the time of the implant, the patient is again placed in the lithotomy position, usually under spinal epidural anesthesia, and after appropriate sterilization of the skin and sterile draping, the rectal ultrasound probe is inserted and moved as nearly as possible to the same position used for the volume study. The template is locked into position against the perineum, and special fixation needles are inserted (through guide holes not planned for use in the implant) to hold the prostate in position during the implant. With the US probe imaging the cephalad portion of the target volume first, loaded needles that have been planned to reach that depth are inserted one at a time, their position verified under US, and then withdrawn while holding the stylet in place to leave the seeds and spacers behind in the prostate. The probe is then withdrawn to the next image plane (usually 5 mm), and needles reaching that depth are inserted. The procedure is repeated until the entire prostate is filled with seeds according to the treatment plan.

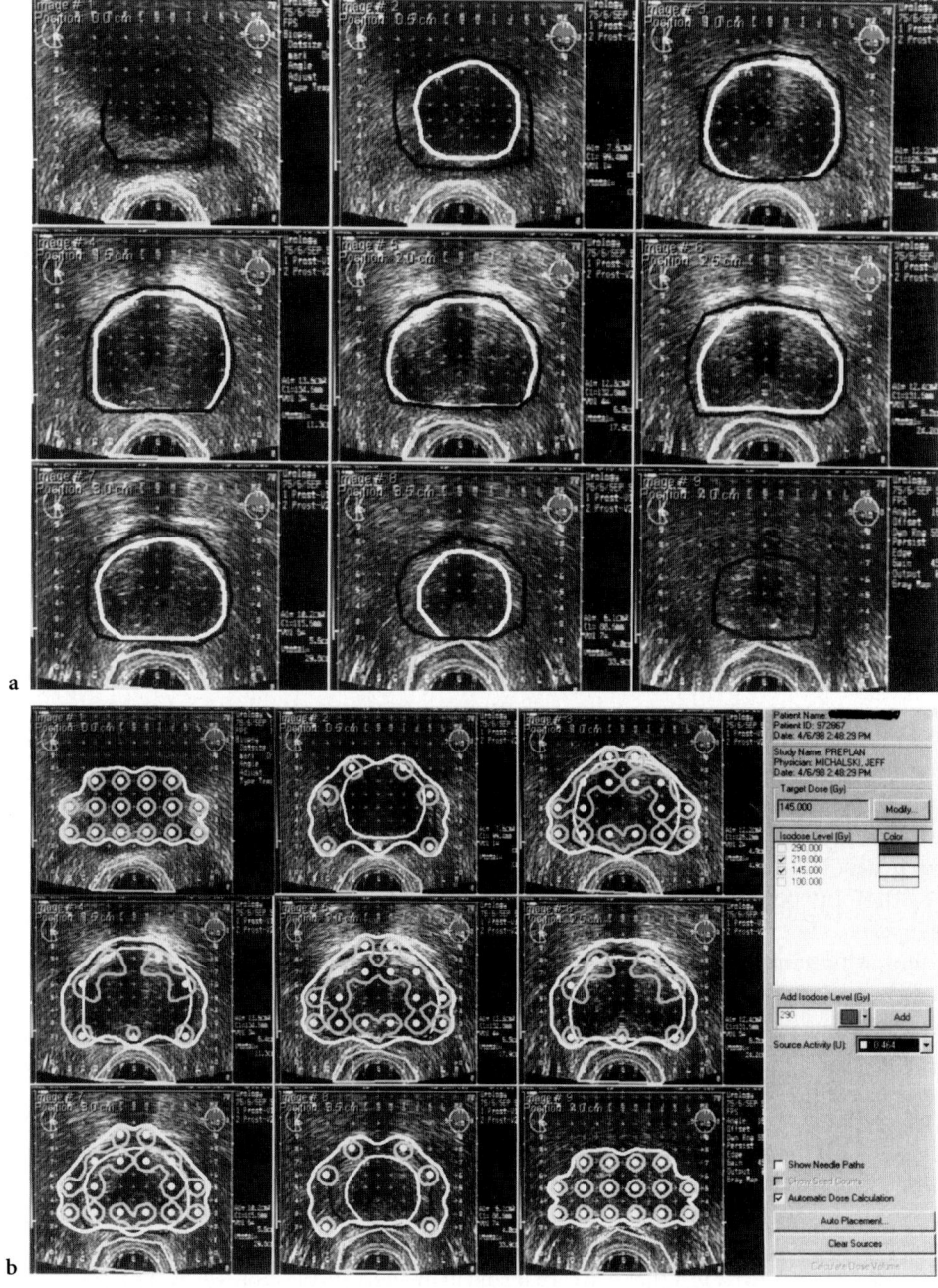

Fig. 14.34. a Ultrasound volume study for preimplant dosimetry. The prostate is contoured I white, and the planning target volume margin is in black. The margin larger at the prostate base and apex to account for three-dimensional changes in the gland contours. Posteriorly, there is no margin because the ultrasound probe contacts the rectal wall and prostate directly. **b** A preplant treatment plan using the modified uniform loading technique described by MERRICK et al. (2000). Isodose curves displayed are 150% and 100%. Source positions are represented by the points on the grid. Alternate slices are loaded heavily or peripherally. The periurethral sources have been removed to decrease the dose to that organ (PEREZ et al. 1999)

EAPEN et al. (2004) noted in 28 patients an association between the degree of prostate trauma during prostate ^{125}I brachytherapy and development of acute urinary toxicity. Median values were prostate volume 33 cc, number of needles per patient 32, and number of periurethral needle manipulations 42 (range, 17–65). The only significant association between urinary toxicity and these variables was the number of periurethral needle manipulations (P=0.025).

C-arm fluoroscopy is often used to check seed positions during and upon completion of the planned implant. Extra seeds are sometimes inserted at this point to avoid underdosing in regions that appear to be deficient in seeds, possibly due to needle bending or unintended movement of the prostate during the implant. A few seeds may be placed in the urethra, and they are usually expelled within 24 h, necessitating urine collection and monitoring after implantation. MERRICK et al. (2000) reported that approximately 10% of seeds lost from the pelvis were found to embolize to the lungs. The use of stranded seeds has minimized seed loss (REED et al. 2004). A cystoscopy is carried out after the implant is completed to remove misplaced seeds and detect other possible problems.

Many variants of the transperineal implant technique have been introduced, including the use of the Mick applicator for seed insertion instead of pre-loaded needles. WALLNER et al. (1991) described a CT method for transperineal implants that is integrated with transrectal ultrasound for verification of correct needle placement at implantation.

A more recent development features an intraoperative approach in which the volume study, treatment planning, applicator preparation, and implantation are carried out in a single operating room procedure. This technique offers a number of advantages over the two-step procedure. In addition to sparing the patient a second US probe insertion, it also eliminates the requirement for accurate repositioning of the patient and probe in the planned position prior to the implant. It further allows for possible modification of the treatment plan in response to needle divergence and other problems arising during the implant procedure. ZELEFSKY et al. (2000) noted significant improvement in target coverage using this technique. An ABS report on intraoperative planning and dosimetry for prostate implants has been published by NAG et al. (2001a). An intraoperative technique carried out under MRI has been described (CORMACK et al. 2000; D'AMICO et al. 1998). MRI offers the advantages of excellent soft tissue resolution with arbitrary imaging planes, although seed imaging with MRI is generally inferior to CT. A technique using magnetic resonance spectroscopy imaging has also been proposed for prostate implants (ZAIDER et al. 2000). This technique introduces the possibility of identifying and applying elevated doses to small, localized high-risk areas within the prostate capsule.

Postimplant-image-based dosimetry is usually performed on every patient. For this purpose, the patient returns a few weeks after the implant for an imaging procedure, usually CT and/or orthogonal films. The films may be used for planning but are more useful for seed count and position verification when CT or MRI planning is used. The CT images should include all relevant critical structures and seeds in proximity to the prostate. ABS guidelines (NAG et al. 2000) recommend that a margin of at least 2 cm should be added to the superior and inferior aspects of the prostate.

The optimal time interval between the implant procedure and postimplant dosimetry has not been resolved at this time. Postoperative edema appears to resolve with a half-life of approximately 10 days (NAG et al. 2000). It has been suggested that the most reproducible dosimetric results can be obtained with a postoperative interval of 1 month for implants using ^{125}I seeds (WATERMAN et al. 1997b; YUE et al. 1999). After the seeds have been identified on the cross-sectional images, isodose curves are generated on all slices imaging the prostate. The original ABS guidelines for transperineal prostate implants (NAG et al. 1999) recommended that the items to be recorded and correlated with outcome were the prescribed dose, the maximum permissible dose, the dose covering 90% of the prostate volume (D_{90}), and the percentage of the prostate volume receiving the prescription dose or greater (V_{100}). The later ABS report on postimplant dosimetry (NAG 2000) suggested a much more elaborate reporting scheme, based on detailed information from dose volume histograms, and included urethral and rectal doses.

Several technical details may have an impact on tumor control of prostate cancer treated with brachytherapy. Because of patient anatomy, the pubic bone can interfere with the direct placement of sources in the interior prostate, and this can potentially cause underdosing. ROY et al. (1991) recommend that needles be placed at oblique angles to cover the anterior prostate adequately. However, unless differential loading of source activity is used, it is possible that high doses may be delivered to the central and anterior portions of the gland (D'AMICO and COLEMAN 1996). However, rapid dose fall-off in the most peripheral portion of the gland may result in underdosing of posterior, peripheral prostate cancer or overdosing of the anterior rectal wall. ROY et al. (1993), in an analysis of 10 prostate implants with CT-planned and fluoroscopically guided radioactive ^{125}I seed placement, found that the 150 Gy prescription isodose line encompassed only 78–96% of the total prostate volume.

WALLNER et al. (1995), in review of 65 patients treated with transperineal ^{125}I implants for T1 and T2 prostatic carcinoma, noted that a greater incidence of urinary grade-2 and -3 morbidity was associated with maximum central urethral dose, length of urethra that received more than 400 Gy, and large prostate volume. Rectal ulceration was associated with irradiation of the rectal wall to doses of more than 100 Gy. Efforts should be made to keep the central urethral dose below 150% of the prescribed minimal peripheral dose and rectal surface dose below 80% to decrease toxicity.

14.8.17.2
Palladium 103 Implants

[103]Pd seeds are physically similar to ^{125}I seeds, so the equipment and implant techniques used with ^{125}I can also be used with [103]Pd. The main differences are that the [103]Pd seed has a slightly lower photon energy (21 keV versus 28 keV for ^{125}I) and a significantly shorter half-life at 18 days (compared with 60 days for ^{125}I). These differences in energy and half-life require corresponding changes in dose prescription and planning techniques, because tissue tolerance is a function of dose rate as well as total dose. Radiobiological modeling suggests that a prescription dose of 125 Gy with [103]Pd should be equivalent to the 145 Gy prescribed with ^{125}I for implants alone. For implants used as a boost to 40–50 Gy of external beam therapy, a [103]Pd prescription dose of 90–100 Gy is recommended (BLASKO et al. 2000; NAG et al. 1999; NATH et al. 1992). Because of its shorter half life, a higher initial activity of [103]Pd must be used, resulting in a typical initial dose rate of approximately 18–20 cGy/h for [103]Pd used as sole therapy, compared with approximately 7 cGy/h for [125]I. At 5 weeks after implantation, the [103]Pd will have delivered approximately 76% of the total dose, while a ^{125}I implant will have delivered only 33% of its intended total dose.

The lower photon energy of [103]Pd results in a greater degree of attenuation of the dose relative to [125]I. This difference in dose fall-off becomes more pronounced as the distance from the seed increases and must be taken into consideration in treatment planning for [103]Pd implants (BLASKO et al. 2000; NATH and MEIGOONI 1989; NATH et al. 1992). ABS guidelines recommend a seed spacing no greater than 1.7 cm when [103]Pd is used (NAG et al. 1999).

Because of the more rapid dose delivery with [103]Pd, this isotope has been more commonly used

for higher grade malignancy (Gleason score >6), while ^{125}I is preferred for lower grades (PRESTIDGE et al. 1998). While some studies have found a greater effectiveness of [103]Pd in poorly differentiated tumors (NAG et al. 1996), other reports (POTTERS et al. 2004) showed comparable outcome with either ^{125}I or [103]Pd.

14.8.17.3
Removable Interstitial Implants with Iridium-192

CHARYULU (1980), SYED et al. (1983a), and PUTHAWALA et al. (1985) described an interstitial implant technique using removable ^{192}Ir sources with a transperineal template for the treatment of carcinoma of the prostate. Activity of the iridium sources in the central guides was approximately 0.25–0.3 mg Ra eq and in the outer 12 guides, 0.4–0.5 mg Ra eq per seed. Dose rate per hour is 0.7–0.9 Gy, with the bladder neck and rectum receiving only 0.3–0.4 Gy/h. The implant was removed after 30–35 Gy was delivered (40–45 h).

MARTINEZ et al. (1985a) and BRINDLE et al. (1989), in an update of the initial publication, described implantation of the prostate with a perineal template (MUPIT). The implant dose was 33 Gy combined with external irradiation (5 Gy in one dose before the implant and 30 Gy in 18 fractions after the implant).

NICKERS et al. (2000) evaluated the feasibility of combining external beam radiation therapy and LDR brachytherapy in 71 patients with prostate cancer in a dose escalation trial from 74 Gy to 85 Gy performed in four groups. Shifting from intraoperative placement of source vectors (group I) to positioning under ultrasound controls (groups II–IV), improving the implantation shape and optimizing radiation delivery to urethral bed reduced the total dose to rectal wall under 65 Gy and to urethra under 100 Gy. Rectal/prostate dose ratio was lowered from 0.7 (groups I–II) to 0.58 (groups III–IV) while avoiding problems resulting from pelvic bone arch interference, prostate volume, or seminal vesicles location. The mean and median follow-up periods were 28 months and 18 months. In Groups III and IV, 85% of patients without hormonotherapy treated with 80–85 Gy had normalized PSA under 1 ng/ml within 6 months. No severe late effect was noted for patients implanted under echographic control. Longer follow-up, however, is needed, but the dose delivered up to 85 Gy was not expected to induce prohibitive side effects.

14.8.17.4
Interstitial Irradiation and Hyperthermia

BAGSHAW et al. (1993) reported a combination of ^{192}Ir and hyperthermia in 13 of 32 patients treated with brachytherapy. The technique is similar to that described by other authors, except for the introduction of the hyperthermia trocars, which are energized with 0.5 MHz microwave radiofrequency for 45 min to achieve 42–44.5°C throughout the prostate.

14.10
Quality Assurance and Radiation Safety in Brachytherapy

It is extremely important in the use of brachytherapy to formulate and strictly observe radiation safety procedures at each institution in compliance with U.S. NRC regulations.

The safety of personnel, patients, and visitors is based on three factors: time of radiation exposure as short as possible, distance as great as practically allowed between the radioactive sources and the operator, and shielding to diminish radiation exposure to all concerned.

Careful quality control procedures should be followed in the prescription and calculation of doses, preparation, calibration, and handling of radioactive sources, as well as verification of treatment parameters. If promptly discovered, an error in brachytherapy can be corrected, but it is more difficult to do than in external-beam irradiation.

At Washington University, formal procedures for brachytherapy have been established to minimize treatment errors. For temporary implants, source loadings are usually prescribed after the physician has reviewed the orthogonal dummy-source localization radiographs. The prescription is written on a form that is given to the brachytherapy-source curator, specifying the configuration of source strengths for intracavitary treatment or the array of active lengths and linear activity if iridium wires or seed ribbons are used for interstitial techniques. Treatment duration is generally determined after reviewing the computer planar isodose rate distributions and is double-checked with "hand calculations." The source curator documents the preparation of sources in a treatment logbook, on a source inventory sheet that is to be posted on the patient's door, and also on a magnetic source inventory board in the radioactive source room. If iridium is used, the vendor's lot identification code is also documented. A well-type re-entrant ion chamber is used to verify the source activity in accord with the recent AAPM recommendations (AMERICAN ASSOCIATION OF PHYSICISTS IN MEDICINE 1994; WILLIAMSON 1991). The U.S. NRC requires that all brachytherapy sources are calibrated using equipment traceable to an accredited national calibration laboratory, either at the user institution or by the vendors, and that all source calibration documents must be reviewed by an authorized medical physicist.

When manual intracavitary afterloading is used, for the sake of prompt patient loading, the various cesium tubes are color-coded. The attending physician or resident (after verifying the source loading) and the source curator load the applicator in the patient. The loading time is documented by the physician, and the curator or physicist measures the radiation exposure levels around the patient and arranges lead shields appropriately. The nursing division is also actively involved in checking every 3–4 h that applicators or sources do not become dislodged over the course of treatment. For further discussion of LDR brachytherapy quality assurance techniques and programs, the reader is referred to a review by WILLIAMSON (1991) and published AAPM recommendations (1993, 1994, 1995).

The physician's orders sheet contains the home telephone number and the beeper number of at least two physicians who can be contacted in an emergency if source removal is required. The attending physician or resident is responsible for the unloading of the implant. The physician counts the sources as they are removed and places them in a lead carrier. After removal of the sources, the patient is surveyed to ensure that no sources remain in the patient or in the patient's room. The time of unloading is documented, and all radiation warning signs are removed from the patient's door. The source curator checks that all sources have been recovered and returns the sources to their designated storage area. The magnetic inventory board is revised to show that the sources have been returned to their storage area. Additionally, source recovery is documented in the source logbook.

14.10.1
Safety Regulations in the United States

The U.S. NRC and states that have negotiated agreements with NRC regulate the use and safety of all reactor by-product materials (excluding naturally occurring radionuclides such as ^{226}Ra and electronically generated radiation). The basic NRC functions such as exposure control standards are outlined in Title 10 Code of Federal Regulations, Part 20 (10 CFR 20).

For the brachytherapy applications discussed in this chapter, the NRC requires that a written directive is prepared before initiating each treatment. The NRC further mandates that institutional written procedures be implemented to provide high confidence that: (1) the patient's identity is verified before brachytherapy treatment; (2) it is verified that the treatment is delivered according to the treatment plan and the written directive; and (3) both manual and computer-generated dose calculations are checked. The written directive is to be retained for 3 years. When a brachytherapy treatment is delivered with greater than 20% error, or 50% error in an HDR fraction, to the wrong patient, using the wrong isotope, to an unintended area resulting in greater than 50 cGy and greater than 50% of the dose that the area would have received in a correct treatment, or using a leaking source, the NRC considers it a "medical event". Medical events are to be reported to the NRC within 24 h of the detection of the event verbally, followed by a written report within 15 days. The patient and the referring physician should be informed verbally within 24 h and by written report within 15 days, or if informing the patient is medically harmful, a relative or friend of the patient must be selected to receive this information. For HDR procedures, the NRC requires the presence of an authorized radiation oncologist and an authorized medical physicist to be present when starting a treatment. An authorized medical physicist and a training physician must be present during the entire HDR treatment. For PDR procedures, the NRC requires that an authorized medical physicist and a training physician be physically present during the initiation of the treatment, and an authorized medical physicist and a trained physician must be available on-call through the treatment.

10CFR35.41:

Procedures for administrations requiring a written directive

a. For any administration requiring a written directive, the licensee shall develop, implement, and maintain written procedures to provide high confidence that:
   1. The patient's or human research subject's identity is verified before each administration; and
   2. Each administration is in accordance with the written directive.

b. At a minimum, the procedures required by paragraph (a) of this section must address the following items that are applicable to the licensee's use of byproduct material:
   1. Verifying the identity of the patient or human research subject;
   2. Verifying that the administration is in accordance with the treatment plan, if applicable, and the written directive;
   3. Checking both manual and computer-generated dose calculations; and
   4. Verifying that any computer-generated dose calculations are correctly transferred into the consoles of therapeutic medical units authorized by § 35.600.

c. A licensee shall retain a copy of the procedures required under paragraph (a) in accordance with § 35.2041

In addition, the NRC has eliminated the terms of misadministration and recordable events, replacing it with "medical events". The following defines a medical event and discusses the actions to be taken when a medical event occurs:

10CFR35.3045:

Report and notification of a medical event

a. A licensee shall report any event, except for an event that results from patient intervention, in which the administration of byproduct material or radiation from byproduct material results in:
   1. A dose that differs from the prescribed dose or dose that would have resulted from the prescribed dosage by more than 0.05 Sv (5 rem) effective dose equivalent, 0.5 Sv (50 rem) to an organ or tissue, or 0.5 Sv (50 rem) shallow dose equivalent to the skin; and
      I.   The total dose delivered differs from the prescribed dose by 20% or more;
      II. The total dosage delivered differs from the prescribed dosage by 20% or more or falls outside the prescribed dosage range; or
      III. The fractionated dose delivered differs from the prescribed dose, for a single fraction, by 50% or more.
   2. A dose that exceeds 0.05 Sv (5 rem) effective dose equivalent, 0.5 Sv (50 rem) to an organ or tissue, or 0.5 Sv (50 rem) shallow dose equivalent to the skin from any of the following:
      I.   An administration of a wrong radioactive drug containing byproduct material;
      II. An administration of a radioactive drug

containing byproduct material by the wrong route of administration;

III. An administration of a dose or dosage to the wrong individual or human research subject;

IV. An administration of a dose or dosage delivered by the wrong mode of treatment; or

V. A leaking sealed source.

3. A dose to the skin or an organ or tissue other than the treatment site that exceeds by 0.5 Sv (50 rem) to an organ or tissue and 50% or more of the dose expected from the administration defined in the written directive (excluding, for permanent implants, seeds that were implanted in the correct site but migrated outside the treatment site).

b. A licensee shall report any event resulting from intervention of a patient or human research subject in which the administration of byproduct material or radiation from byproduct material results or will result in unintended permanent functional damage to an organ or a physiological system, as determined by a physician.

c. The licensee shall notify by telephone the NRC Operations Center[3] no later than the next calendar day after discovery of the medical event.

d. By an appropriate method listed in § 30.6(a) of this chapter, the licensee shall submit a written report to the appropriate NRC Regional Office listed in § 30.6 of this chapter within 15 days after discovery of the medical event.

1. The written report must include:

I. The licensee's name;

II. The name of the prescribing physician;

III. A brief description of the event;

IV. Why the event occurred;

V. The effect, if any, on the individual(s) who received the administration;

VI. What actions, if any, have been taken or are planned to prevent recurrence; and

VII. Certification that the licensee notified the individual (or the individual's responsible relative or guardian), and if not, why not.

2. The report may not contain the individual's name or any other information that could lead to identification of the individual.

e. The licensee shall provide notification of the event to the referring physician and also notify the individual who is the subject of the medical event no later than 24 h after its discovery, unless the referring physician personally informs the licensee either that he or she will inform the indi-

vidual or that, based on medical judgment, telling the individual would be harmful. The licensee is not required to notify the individual without first consulting the referring physician. If the referring physician or the affected individual cannot be reached within 24 hours, the licensee shall notify the individual as soon as possible thereafter. The licensee may not delay any appropriate medical care for the individual, including any necessary remedial care as a result of the medical event, because of any delay in notification. To meet the requirements of this paragraph, the notification of the individual who is the subject of the medical event may be made instead to that individual's responsible relative or guardian. If a verbal notification is made, the licensee shall inform the individual, or appropriate responsible relative or guardian, that a written description of the event can be obtained from the licensee upon request. The licensee shall provide such a written description if requested.

f. Aside from the notification requirement, nothing in this section affects any rights or duties of licensees and physicians in relation to each other, to individuals affected by the medical event, or to that individual's responsible relatives or guardians.

g. A licensee shall:

1. Annotate a copy of the report provided to the NRC with the:

I. Name of the individual who is the subject of the event; and

II. Social security number or other identification number, if one has been assigned, of the individual who is the subject of the event; and

2. Provide a copy of the annotated report to the referring physician, if other than the licensee, no later than 15 days after the discovery of the event.

Review should include all recordable events and misadministrations and an audit of representative sample cases treated. For HDR or PDR procedures, the NRC requires the presence of an authorized radiation oncologist and physicist at all times when a procedure is being performed. Imaging or techniques must be in place to verify source position and accuracy before performing a procedure. A physicist must verify the accuracy of plan input data, dose calculation, and information transfer. Before treatment, the technologist verifies treatment site, isotope, total dose, dose per fraction and treatment modality, program sequence of source positions, and dwell times, which must agree with the treatment plan calculation; the technologist must also verify that HDR treatment channels are correctly connected to corresponding applicators. After treatment, the attending physician must review the record and sign forms as required.

For other (non-HDR) brachytherapy procedures, there are similar requirements. In addition, for manual afterloading intracavitary sources color coding or serial number must identify the sources; ideally a second person should verify the correct loading of the applicator. For ^{192}Ir, ^{103}Pd, ^{125}I, or other radionuclides, verification of batch number and a pot calibration check must be performed.

Misadministrations with remote afterloading devices occur, although fortunately they are rare. These devices and their operation are complex, and a mishap may have severe or fatal complications. To prevent these, strictly followed safety and quality assurance procedures are mandatory (GLASGOW 1996).

References

Abrath FG, Henderson SD, Simpson JR et al (1986) Dosimetry of CT-guided volumetric Ir-192 brain implant. Int J Radiat Oncol Biol Phys 12:539–363

Alden ME, Mohiuddin M (1994) The impact of radiation dose in combined external beam and intraluminal ^{192}Ir brachytherapy for bile duct cancer. Int J Radiat Oncol Biol Phys 28:945–951

Allt WEC, Hunt JW (1963) Experience with radioactive tantalum wire as a source for interstitial therapy. Radiology 80:581–587

American Association of Physicists in Medicine (1993) Remote Afterloading Technology, Report of the Radiation Therapy Task Group No 41 (G. Glasgow, Chairman). American Institute of Physics, New York, NY

American Association of Physicists in Medicine (1994) Comprehensive QA for Radiation Oncology: Report of the AAPM Radiation Therapy Committee Task Group 40 (J Kutcher, Chairman). Med Phys 21:581–618

American Association of Physicists in Medicine (1995) Dosimetry of interstitial brachytherapy sources: Recommendations of the AAPM Radiation Therapy Committee Task Group No 43. (R. Nath, Chairman). Med Phys 22:209–234

Ampuero F, Doss LL, Khan LM et al (1983) The Syed-Neblett interstitial template in locally advanced gynecological malignancies. Int J Radiat Oncol Biol Phys 9:1897–1903

Anderson LL, Hilaris BS, Wagner LK (1985) A nomograph for planar implant planning. Endocurie Hypertherm Oncol 1:9–15

Aristizabal SA, Valencia A, Ocampo G et al (1985) Interstitial parametrial irradiation in cancer of the cervix stage IIB-IIIB. Endocurie Hypertherm Oncol 1:41

Armadur RJ, Piontek R, Hadley VE et al (1997) A simple, inexpensive applicator for irradiation of localized areas of the vagina with intracavitary brachytherapy. Int J Radiat Oncol Biol Phys 37:965–969

Arthur DW, Koo D, Zwicker RD et al (2003) Partial breast brachytherapy after lumpectomy: low-dose-rate and high-dose-rate experience. Int J Radiat Oncol Biol Phys 56:681–689

Ashayeri E, Collier-Manning J, Nibhanupudy JR et al (1987) Localization technique and afterloading intracavitary irradiation in the treatment of nasopharyngeal carcinoma. Endocurie Hypertherm Oncol 3:115–119

Bagshaw MA, Kaplan ID, Cox RC (1993) Radiation therapy for localized disease. Cancer 71:939–952

Baillet F, Decroix Y, Mazeron JJ (1987) Oral tongue. In: Pierquin B, Wilson J-F, Chassagne D (eds) Modern brachytherapy. Masson, New York, pp 107–118

Bastin K, Buchler D, Stitt J et al (1993) Resource utilization: High dose rate versus low dose rate brachytherapy for gynecologic cancer. Am J Clin Oncol 16:256–263

Batley F, Constable WC (1967) The use of the "Manchester system" for treatment of cancer of the uterine cervix with modern after-loading radium applicators. J Can Assoc Radiol 18:396–400

Battermann JJ, Boon TA (1988) Interstitial therapy in the management of T2 bladder tumors. Endocurie Hypertherm Oncol 4:1–6

Battermann JJ, Boon TA (1989) Treatment of T2 bladder tumours with interstitial therapy: the role of lymph node dissection. In: Mould RF (ed) Brachytherapy 2. Nucletron International BV, Leersum, The Netherlands, pp 187–191

Battermann JJ, Szabol B (1989) Preliminary results of radiation therapy for carcinoma of the uterine cervix, using the Selectron afterloading machine. In: Mould RF (ed) Brachytherapy 2. Nucletron International BV, Leersum, The Netherlands, pp 229–234

Battermann JJ, Tierie AH (1986) Results of implantation for T1 and T2 bladder tumours. Radiother Oncol 5:85–90

Blasko JC, Radge H, Schumacher D (1987) Transperineal percutaneous Iodine-125 implantation for prostatic carcinoma using transrectal ultrasound and template guidance. Endocurie Hypertherm Oncol 3:131–139

Blasko JC, Ragde H, Grimm PD (1991) Transperineal ultrasound-guided implantation of the prostate: morbidity and complications. Scand J Urol Nephrol 137 [Suppl]:113–118

Blasko JC, Grimm PD, Sylvester JE et al (2000) Palladium-103 brachytherapy for prostate carcinoma. Int J Radiat Oncol Biol Phys 46:839–850

Bourgier C, Coche-Dequeant B, Fournier C et al (2004) LDR brachytherapy is the reference treatment for T2N0 tongue carcinoma: the Centre Oscar Lambret experience in 279 patients. Int J Radiat Oncol Biol Phys 60 [Suppl 1]:S493–S494

Brenner DJ, Hall EJ (1991) Conditions for the equivalence of continuous to pulsed low dose rate brachytherapy. Int J Radiat Oncol Biol Phys 20:181–190

Brenner DJ, Schiff PB, Huant Y et al (1997) Pulsed-dose-rate brachytherapy: design of convenient (daytime-only) schedules. Int J Radiat Oncol Biol Phys 39:809–815

Brindle JS, Martinez A, Schray M et al (1989) Pelvic lymphadenectomy and transperineal interstitial implantation of ^{192}Ir combined with external beam radiotherapy for bulky stage C prostatic carcinoma. Int J Radiat Oncol Biol Phys 18:1063–1066

Cano E, Johnson JT, Carrau R et al (2004) Brachytherapy in the treatment of stage IV carcinoma of the base of tongue. Brachytherapy 3:41–48

Casal L, Ballester F, Lluch JL et al (2000) Monte Carlo calculations of dose rate distributions around the Amersham CDCS-M-type ^{137}Cs source. Med Phys 27:132–140

Charyulu KKN (1980) Transperineal interstitial implantation of prostate cancer: a new method. Int J Radiat Oncol Biol Phys 6:1261–1266

Chassagne D, Janvier L, Pierquin B et al (1963) La plesiotherapie des cancers du cavum avec support-moule et iridium 192. Ann Radiol 6:719–726

Chiu-Tsao ST (1995) ^{125}I episcleral eye plaques for treatment of intra-ocular malignancies. In: Williamson JF, Thomadsen BR, Nath R (eds) Brachytherapy physics. Medical Physics Publishing Company, Madison, WI, pp 451–485

Chiu-Tsao ST, Tsao HS, Vialotti C et al (1986) Monte Carlo dosimetry for ^{125}I and ^{60}Co in eye plaque therapy. Med Phys 13:678

Chiu-Tsao ST, Anderson LL, Stabile L (1988) TLD dosimetry for ^{125}I eye plaque. Phys Med Biol 33:28

Chobe R, Paryani S, Scott W et al (1996) Bronchoscopic brachytherapy for endobronchial obstruction: high dose rate or low dose rate. Endocurie Hypertherm Oncol 12:18–24

Choo JJ, Scudiere J, Bitterman P et al (2004) Vaginal lymphatic duct location and implication on intracavitary brachytherapy radiation treatment (abstract). Int J Radiat Oncol Biol Phys 60 [Suppl 1]:S480

Clarke DH, Edmundson GK, Martinez A et al (1989) The clinical advantages of I-125 seeds as a substitute for Ir-192 seeds in temporary plastic tube implants. Int J Radiat Oncol Biol Phys 18:859–863

Collaborative Ocular Melanoma Study Group (2004) Ten-year follow-up of fellow eyes of patients enrolled in Collaborative Ocular Melanoma Study randomized trials: COMS report no 22. Ophthalmology 111:966–976

Conill C, Sanchez-Reyes A, Molla M et al (2004) Brachytherapy with ^{192}Ir as treatment of carcinoma of the tarsal structure of the eyelid. Int J Radiat Oncol Biol Phys 59:1326–1329

Cooper FS (1978) Postoperative irradiation of pterygia: ten more years of experience. Radiology 128:753–756

Cormack RA, Kooy H, Tempany CM et al (2000) A clinical method for real-time dosimetric guidance of transperineal I-125 prostate implants using interventional magnetic resonance imaging. Int J Radiat Oncol Biol Phys 46:407–214

Corn BW, Galvin JM, Soffen EM et al (1993) Positional stability of sources during low-dose-rate brachytherapy for cervical carcinoma. Int J Radiat Oncol Biol Phys 26:513–518

Corn BW, Hanlon AL, Pajak TF et al (1994) Technically accurate intracavitary insertions improve pelvic control and survival among patients with locally advanced carcinoma of the uterine cervix. Gynecol Oncol 53:294–300

Dale RG, Jones B (2000) Regarding Giap and Massullo. Int J Radiat Oncol Biol Phys 48:304–305

Damato B, Lecuona K (2004) Conservation of eyes with choroidal melanoma by a multimodal approach to treatment: An audit of 1632 patients. Ophthalmology 111:977–983

D'Amico AV, Coleman CN (1996) Role of interstitial radiotherapy in the management of clinically organ-confined prostate cancer: the jury is still out. J Clin Oncol 14:304–315

D'Amico AV, Cormack RA, Tempany CM et al (1998) The use of real time MR guided interstitial brachytherapy in select patients with localized prostate cancer. Int J Radiat Oncol Biol Phys 42:507–515

DeBiose DA, Horwitz EM, Martinez AA et al (1997) The use of Ultrasonography in the localization of the lumpectomy cavity for interstitial brachytherapy of the breast. Int J Radiat Oncol Biol Phys 28:755–759

Delannes M, Thomas L, Martel P et al (2000) Low-dose-rate intraoperative brachytherapy combined with external beam irradiation in the conservative treatment of soft tissue sarcoma. Int J Radiat Oncol Biol Phys 47:165–169

Delclos L (1980) Afterloading method for interstitial gamma-ray therapy. In: Fletcher GH (ed) Textbook of radiotherapy, 3rd edn. Lea and Febiger, Philadelphia

Delclos L (1982a) A second look at interstitial irradiation. In: Deeley TJ (ed) Topical reviews in radiotherapy and oncology 2. Wright, London

Delclos L (1982b) Interstitial irradiation of the penis. In: Johnson DE, Boileau MA (eds) Genitourinary tumors: fundamental principles and surgical techniques. Grune and Stratton, New York

Delclos L (1984) Interstitial irradiation techniques. In: Levitt SH, Tapley N duV (eds) Technological basis of radiation therapy: practical clinical applications. Lea and Febiger, Philadelphia, pp 55–84

Delclos L, Fletcher GH, Sampiere V et al (1978) Can the Fletcher gamma ray colpostat system be extrapolated to other systems? Cancer 41:970–979

Delclos L, Fletcher GH, Moore EB et al (1980) Minicolpostats, dome cylinders, other additions and improvements of the Fletcher-Suit after loadable system: indications and limitations of their use. Int J Radiat Oncol Biol Phys 6:1195–1206

Denham JW, Baldacchino AC, Gutte J et al (1988) Remote afterloading techniques for the treatment of nasopharyngeal and endometrial cancer. Int J Radiat Oncol Biol Phys 14:191–195

Deore SM, Sarin R, Dinshaw KA et al (1993) Influence of dose-rate and dose per fraction on clinical outcome of breast cancer treated by external beam irradiation plus iridium-192 implants: Analysis of 289 cases. Int J Radiat Oncol Biol Phys 26:601–606

De Pree C, Popowski Y, Weber D et al (1999) Feasibility and tolerance of pulsed dose rate interstitial brachytherapy. Int J Radiat Oncol Biol Phys 43:971–976

Deutsch M, Segall BW, Leen R et al (1973) Retreatment of recurrent nasopharyngeal malignancy using a radium mold. J Prosthet Dent 30:315–320

DeWerd LA, Huq MS, Das IJ et al (2004) Procedures for establishing and maintaining consistent air-kerma strength standards for low-energy, photon-emitting brachytherapy sources: recommendations of the Calibration Laboratory Accreditation Subcommittee of the American Association of Physicists in Medicine. Med Phys 31:675–581

Dobelbower RR, Merrick HW, Ahuja RK et al (1986) I-125 interstitial implant, precision high-dose external beam therapy, and 5-FU for unresectable adenocarcinoma of pancreas and extrahepatic biliary tree. Cancer 58:2185–2195

Eapen L, Kayser C, Deshaies Y et al (2004) Correlating the degree of needle trauma during prostate brachytherapy and the development of acute primary toxicity. Int J Radiat Oncol Biol Phys 59:1392–1394

Emami B, Perez CA (1991) Interstitial thermoradiotherapy in the treatment of malignant tumors. In: Sauer R (ed) Interventional radiation therapy techniques: brachytherapy. Springer, Berlin Heidelberg New York

Erickson BA (1996) Interstitial implantation of vulvar malignancies: an historical perspective. Endocurie Hypertherm Oncol 12:101–112

Erickson BA, Nag S (1998) Biliary tree malignancies. J Surg Oncol 67:203–210

Erickson BA, Shadley JD (1996) In vitro test of the cytotoxic equivalence between pulsed dose rate and continuous low dose rate. Radiat Oncol Invest 3:218–224

Erickson BA, Wilson JF (1993) Nasopharyngeal brachytherapy. Am J Clin Oncol 16:424–443

Esche BA, Haie CM, Gerbaulet AP et al (1988) Interstitial and external radiotherapy in carcinoma of the soft palate and uvula. Int J Radiat Oncol Biol Phys 15:619–625

Fields JN, Emami B (1987) Carcinoma of the extrahepatic biliary system: results of primary and adjuvant radiotherapy. Int J Radiat Oncol Biol Phys 13:331–338

Finger PT, Romero JM, Rosen RB et al (1998) 3-Dimensional ultrasonography of choroidal melanoma: localization of radioactive eye-plaques. Arch Ophthalmol 116:305–312

Fletcher GH (1953) Cervical radium applicators with screening in the direction of bladder and rectum. Radiology 60:77–84

Fletcher GH (1980) Oral cavity and oropharynx. In: Fletcher GH (ed) Textbook of radiotherapy, 3rd edn. Lea and Febiger, Philadelphia, pp 286–329

Fletcher GH, MacComb WS (1962) Radiation therapy in the management of cancers of the oral cavity and oropharynx. Thomas, Springfield, IL

Flores AD (1988) Remote afterloading intracavitary irradiation for carcinoma of the nasopharynx. In: Mould RF (ed) Brachytherapy 2. Nucleton, Leersum, The Netherlands, pp 49–66

Flores AD, Nelems B, Evans K et al (1989) Impact of new radiotherapy modalities on the surgical management of cancer of the esophagus and cardia. Int J Radiat Oncol Biol Phys 18:937–944

Fontanesi J, Hetzler D, Ross J (1989) Effect of dose rate on local control and complications in the reirradiation of head and neck tumors with interstitial Iridium-192. Int J Radiat Oncol Biol Phys 18:365–369

Fowler JF, van Limbergen EF (1997) Biological effect of pulsed dose rate brachytherapy with stepping sources if short half-time of repair are present in tissues. Int J Radiat Oncol Biol Phys 37:877–883

Fu KK, Phillips TL (1990) High-dose rate versus low-dose rate intracavitary brachytherapy for carcinoma of the cervix. Int J Radiat Oncol Biol Phys 19:791–796

Gaddis O Jr, Morrow CP, Klement V et al (1983) Treatment of cervical carcinoma employing a template for transperineal interstitial ^{192}Ir brachytherapy. Int J Radiat Oncol Biol Phys 9:819–827

Gaspar LF, Nag S, Herskovic A et al (1997) American Brachytherapy Sociey (ABS) consensus guidelines for brachytherapy of esophageal cancer. Int J Radiat Oncol Biol Phys 38:127–132

Gaspar LF, Winter K, Kocha WI et al (2000) A phase I/II study of external beam radiation, brachytherapy, and concurrent chemotherapy for patients with localized carcinoma of the esophagus. Cancer 88:988–995

Gerard JP, Rozan R, Mazeron JJ et al (1989) Iridium-192 brachytherapy in urinary bladder cancer: the French experience. In: Mould RF (ed) Brachytherapy 2. Nucletron International BV, Leerum, The Netherlands, pp 189–192

Gerbaulet A, Haie-Meder C, Marsiglia H et al (1992) The role of brachytherapy in treatment of head and neck cancer: Institut Gustave Roussy experience with 1140 patients. In: Mould RF (ed) International Brachytherapy. Nucletron, Leerum, The Netherlands, pp 49–66

Giap HB, Massulo V (1999) Derivation of isoeffect dose rate for low-dose-rate brachytherapy and external beam irradiation. Int J Radiat Oncol Biol Phys 45:1355–1358

Glasgow GP (1996) Radiation control, personnel training, and emergency procedures for remote afterloading units. Endocurie Hypertherm Oncol 12:67–79

Goffinet DR, Martinez A, Fee WE Jr (1985) ^{125}I Vicryl suture implants as a surgical adjuvant in cancer of the head and neck. Int J Radiat Oncol Biol Phys 11:399–402

Greenberg M (1987) Eye: Choroidal melanomas and pterygium. In: Pierquin B, Wilson J-F, Chassagne D (eds) Modern brachytherapy. Masson, New York, pp 301–314

Greenblatt DR, Nori D, Tankenbaum A et al (1987) New brachytherapy techniques using iodine-125 seeds for tumor bed implants. Endocurie Hypertherm Oncol 3:73–80

Grigsby PW, Baker S (1995) Socioeconomic aspects of remote afterloading. In: Williamson JF, Thomadsen BR, Nath R (eds) Brachytherapy physics. Medical Physics Publishing Company, Madison, WI, pp 699–708

Grigsby PW, Perez CA, Kuten A et al (1991) Clinical stage I endometrial cancer: results of adjuvant irradiation and patterns of failure. Int J Radiat Oncol Biol Phys 21:379–385

Grigsby PW, Williamson JF, Perez CA (1992) Source configuration and dose rates for the Selectron afterloading equipment for gynecologic applicators. Int J Radiat Oncol Biol Phys 24:423–430

Gupta AK, Vicini FA, Frazier AJ et al (1999) Iridium-192 transperineal interstitial brachytherapy for locally advanced or recurrent gynecological malignancies. Int J Radiat Oncol Biol Phys 43:1055–1060

Gutin PH, Phillips TL, Hosobuchi Y et al (1981) Permanent and removable implants for the brachytherapy of brain tumors. Int J Radiat Oncol Biol Phys 7:1371–1381

Haas JS, Dean RD, Mansfield CM (1985) Dosimetric comparison of the Fletcher family of gynecologic colpostats 1950–1980. Int J Radiat Oncol Biol Phys 11:1318–1321

Haghbin M, Kramer S, Patchefsky AS et al (1985) Carcinoma of the nasopharynx: A 25-year study. Am J Clin Oncol (CCT) 8:384–392

Hall EJ, Brenner DJ (1992) The dose-rate effect in interstitial brachytherapy: a controversy resolved. Br J Radiol 65:242–247

Hall EJ, Brenner DJ (1996) Pulsed dose rate brachytherapy: can we take advantage of new technology? Int J Radiat Oncol Biol Phys 34:511–512

Hames F (1937) A new method in the use of radon gold seeds. Am J Surg 38:235

Harbour JW, Murray TG, Byrne SF et al (1996) Intraoperative echographic localization of iodine 125 episcleral radioactive plaques for posterior uveal melanoma. Retina 16:129–134

Harrison LB, Nori D, Hilaris BS et al (1980) Nasopharynx. In: Interstitial Collaborative Working Group (eds) Interstitial brachytherapy. Raven, New York, NY, pp 95–109

Harter DJ, Delclos L, Johns MF (1975) Sealed sources in synthetic absorbable suture: a new method for permanent interstitial implantation. Radiology 116:721–723

Henschke UK (1956) Artificial radioisotopes in nylon ribbons for implantation in neoplasms. International conferences on the peaceful uses of atomic energy. United Nations, New York

Henschke UK, Hilaris BS, Mahan GD (1963) Afterloading in interstitial and intracavitary radiation therapy. Am J Roentgenol 90:386–395

Heyman J, Reuterwall O, Benner S (1941) The Radiumhemmet experience with radiotherapy in cancer of the corpus of the uterus: classification, method of treatment and results. Acta Radiol 11:11

Hilaris BS, Martini N (1988) The current state of intraoperative interstitial brachytherapy in lung cancer. Int J Radiat Oncol Biol Phys 15:1347–1354

Hilaris BS, Gomez J, Nori D et al (1985) Combined surgery, intraoperative brachytherapy, and postoperative external radiation in stage III non-small cell lung cancer. Cancer 55:1226–1231

Hilaris BS, Nori D, Anderson LL (1988a) Brachytherapy for soft tissue sarcomas. In: Hilaris BS, Nori D, Anderson LL (eds) Atlas of brachytherapy. McMillan, New York, pp 180–182

Hilaris BS, Nori D, Anderson LL (1988b) Brachytherapy of ocular melanoma. In: Hilaris BS, Nori D, Anderson LL (eds) Atlas of brachytherapy. McMillan, New York, pp 304–310

Hintz BL, Kagan AR, Chan P et al (1980) Radiation tolerance of the vaginal mucosa. Int J Radiat Oncol Biol Phys 6:711–716

Höckel M, Muller T (1994) A new perineal template assembly for high-dose-rate interstitial brachytherapy of gynecologic malignancies. Radiother Oncol 31:262–264

Holm HH, Juul N, Pedersen JF et al (1983) Transperineal ^{125}Iodine seed implantation in prostatic cancer guided by transrectal ultrasonography. J Urol 130:283–286

Hughes-Davies L, Silver B, Kapp KS (1995) Parametrial interstitial brachytherapy for advanced or recurrent pelvic malignancy: the Harvard/Stanford experience. Gynecol Oncol 58:24–27

International Commission on Radiation Units (1985) ICRU report no 38: dose and volume specification for reporting intracavitary therapy in gynecology. ICRU, Bethesda, MD, ICRU, 1–16

Ishii H, Furuse J, Nagase M et al (2004) Relief of jaundice by external beam radiotherapy and intraluminal brachytherapy in patients with extrahepatic cholangiocarcinoma: Results without stenting. Hepatogastroenterology 51:954–957

Iyer PS, Shanta A (1994) Update of radionuclides used in endocurietherapy. Endocurie Hypertherm Oncol 10:161–165

Jackson SM (1966) The treatment of carcinoma of the penis. Br J Surg 53:33–35

Jones C, Lukke H, O'Brien B (1994) High dose rate versus low dose rate brachytherapy for squamous cell carcinoma of the cervix: an economic analysis. Br J Radiol 67:1113–1120

Jorgensen K, Elbrond O, Andersen AP (1973) Carcinoma of the lip: a series of 869 cases. Acta Radiol Ther Phys Biol 12:187–190

Karaiskos P, Papagiannis P, Angelopoulas A et al (2001) Dosimetry of ^{192}Ir wires for LDR interstitial brachytherapy following the AAPM TG-43 dosimetric formalism. Med Phys 28:156–166

Kin NYK, Pigneux J, Auvray H et al (1988) Our experience of conservative treatment of anal canal carcinoma combining external irradiation and interstitial implant: 32 cases treated between 1973 and 1982. Int J Radiat Oncol Biol Phys 14:253–259

Kolotas C, Zamboglou N (2001) Role of interstitial brachytherapy in the treatment of malignant disease. Onkologie 24:222–228

Kucera H, Mock C, Knocke TH et al (2001) Radiotherapy alone for invasive vaginal cancer: outcome with intracavitary high dose rate brachytherapy versus conventional low dose rate brachytherapy. Acta Obstet Gynecol Scand 80:355–360

Kuske RR, Perez CA, Jacobs AJ et al (1988) Mini-colpostats in the treatment of carcinoma of the uterine cervix. Int J Radiat Oncol Biol Phys 14:899–906

Kuske RR, Bolton JS, Wilenzick RM et al (1994) Brachytherapy as the sole method os breast irradiation in T1S, T1, T2, N0-1 breast cancer. Int J Radiat Oncol Biol Phys 30 [Suppl 1]:245

Kuske RR, Bolton JS, Hanson W (1998) RTOG 95–18: a phase I/II trial to evaluate brachytherapy as the sole method of radiation therapy for stage I and II breast carcinoma. Radiation Therapy Oncology Group, Philadelphia, pp 1–34

Kwan DK, Kagan AR, Olch AJ et al (1983) Single and double plan iridium-192 interstitial implants: implantation guidelines and dosimetry. Med Phys 10:456–461

Langendijk JA, Poorter R, Leemans CR et al (2004) Radiotherapy of squamous cell carcinoma of the nasal vestibule. Int J Radiat Oncol Biol Phys 59:1319–1325

Larson DA, Suplica JM, Chang SM et al (2004) Permanent iodine 125 brachytherapy in patients with progressive or recurrent glioblastoma multiforme. Neuro Oncol 6:119–126

Leborgne F, Fowler JF, Leborgne JH et al (1996) Fractionation in medium dose rate brachytherapy of cancer of the cervix. Int J Radiat Oncol Biol Phys 35:907–914

Leborgne F, Fowler JF, Leborgne JH et al (1999) Medium-dose-rate brachytherapy of cancer of the cervix: preliminary results of a prospectively designed schedule based on the linear-quadratic model. Int J Radiat Oncol Biol Phys 43:1061–1064

Leborgne F, Leborgne JH, Zubizarreta E et al (2002) Cesium-137 needle brachytherapy boost after external beam irradiation for locally advanced carcinoma of the tongue and floor of the mouth. Brachytherapy 1:126–130

Lee N, Hoffman R, Phillips TL et al (2002) Managing nasopharyngeal carcinoma with intracavitary brachytherapy: one institution's 45-year experience. Brachytherapy 1:74–82

Leibel SA, Gutin PH, Sneed PK et al (1989) Interstitial irradiation for the treatment of primary and metastatic brain tumors. PPO Updates 3:1–11

Levendag FC, Peters R, Meelwis CA et al (1997) A new applicator design for endocavitary brachytherapy of cancer in the nasopharynx. Radiother Oncol 45:95–98

Levendag P, Nijdam W, Noever I et al (2004) Brachytherapy bersus surgery in carcinoma of the tonsillar fossa and/or palate: late adverse sequelae and performance status: can we be more selective and obtain better tissue sparing? Int J Radiat Oncol Biol Phys 59:713–724

Lichter AS, Dillon MB, Rosenshein NB et al (1978) The use of custom molds for intracavitary treatment of carcinoma of the cervix. Int J Radiat Oncol Biol Phys 4:873–879

Lo TC, Girshovich L, Healey GA et al (1995) Low-dose rate versus high-dose rate intraluminal brachytherapy for malignant endobronchial tumors. Radiother Oncol 35:193–197

Lozza L, Cerrota A, Gardani G et al (1997) Analysis of risk factors for mandibular bone radionecrosis after exclusive low dose-rate brachytherapy for oral cancer. Radiother Oncol 44:143–147

Luxton G, Astrahan MA, Liggett PE et al (1988) Dosimetric calculations and measurements of gold plaque ophthalmic irradiators using ^{192}Ir and ^{125}I seeds. Int J Radiat Oncol Biol Phys 15:167–176

Lybeert ML, Ribot JG, de Neve W et al (1994) Carcinoma of the urinary bladder: long-term results of interstitial radiotherapy. Bull Cancer Radiother 81:33–40

Maat B, Venselaar JLM (1989) Improved afterloading technique for interstitial brachytherapy of bladder cancer. In: Mould RF (ed) Brachytherapy 2, Nucletron International BV, Leerum, The Netherlands, pp 183–186

Mansfield CM, Komarnicky LT, Schwartz GF et al (1994) Perioperative implantation of iridium-192 as the boost technique for stage I and II breast cancer: Results of a 10-year study of 655 patients. Radiology 192:33–36

Marchese MJ, Nori D, Anderson LL et al (1984) A versatile permanent planar implant technique utilizing Iodine-125 seeds imbedded in Gelfoam. Int J Radiat Oncol Biol Phys 10:747–751

Marcus RB Jr, Million RR, Mitchell TP (1980) A preloaded, custom-designed implantation device for stage T1-T2 carcinoma of the floor of mouth. Int J Radiat Oncol Biol Phys 6:111–113

Markman J, Williamson JF, Dempsey JF et al (2001) On the validity of the superposition principle in dose calculations for intracavitary implants with shielded vaginal colpostats. Med Phys 28:147–155

Martin HE, MacComb WS (1937) Protracted irradiation by radium. Am J Roentgenol Radium Ther 37:224–233

Martinez A, Benson RC, Edmundson GK et al (1985a) Pelvic lymphadenectomy combined with transperineal interstitial implantation of Iridium-192 and external beam radiation for locally advanced prostatic carcinoma: technical description. Int J Radiat Oncol Biol Phys 11:841–847

Martinez A, Edmundson GK, Cox RS et al (1985b) Combination of external beam irradiation and multiple-site perineal applicator (MUPIT) for treatment of locally advanced or recurrent prostatic, anorectal, and gynecologic malignancies. Int J Radiat Oncol Biol Phys 11:391–398

Martinez-Monge R, Nag S, Martin EW (1998) ^{125}Iodine brachytherapy for colorectal adenocarcinoma recurrent in the pelvis and paraortics. Int J Radiat Oncol Biol Phys 42:545–550

Mayr NA, Sorosky JI, Zhen W et al (1998) The use of laminarias for osmotic dilation of the cervix in gynecological brachytherapy applications. Int J Radiat Oncol Biol Phys 42:1049–1053

Mazeron JJ, Lusichini A, Marinello G et al (1986) Interstitial radiation therapy for squamous cell carcinoma of the tonsillar region: the Creteil experience (1971–1981). Int J Radiat Oncol Biol Phys 12:895–900

Mazeron JJ, Crook J, Chopin D et al (1988) Conservative treatment of bladder carcinoma by partial cystectomy and interstitial iridium 192. Int J Radiat Oncol Biol Phys 15:1323–1330

Mazeron JJ, Simon J-M, Crook J et al (1991) Influence of dose rate on local control of breast carcinoma treated by external beam irradiation plus iridium 192 implant. Int J Radiat Oncol Biol Phys 21:1183–1187

McNeese MD, Fletcher GH (1981) Retreatment of recurrent nasopharyngeal carcinoma. Radiology 138:191–193

Meerwaldt JH, Veeze-Kuijpers B, Visser AG et al (1989) Combined modality radiotherapy in the treatment of bile duct carcinoma. In: Mould RF (ed) Brachytherapy 2. Nucletron International BV, Leersum, The Netherlands, pp 577–583

Mendenhall NP, Parsons JT, Cassisi NJ et al (1984) Carcinoma of the nasal vestibule. Int J Radiat Oncol Biol Phys 10:627–637

Mendenhall NP, Parsons JT, Cassisi NJ et al (1987) Carcinoma of the nasal vestibule treated with radiation therapy. Laryngoscope 97:626–632

Mendenhall WM, Parsons JT, Mendenhall JP et al (1991) Brachytherapy in head and neck cancer. Oncology 5:44–54

Meredith WJ (1967) Radium dosage: the Manchester system. Livingstone, Edinburgh, Scotland

Merrick GS, Butler WM, Dorsey AT et al (2000) Seed fixity in the prostate/periprostatic region following brachytherapy. Int J Radiat Oncol Biol Phys 462:215–220

Million RR, Cassisi NJ, Hamlin DJ (1984) Nasal vestibule, nasal cavity, and paranasal sinuses. In: Million RR, Cassisi NJ (eds) Management of head and neck cancer: a multidisciplinary approach. Lippincott, Philadelphia, p 432

Million RR, Cassisi NJ, Clark JR (1989) Cancer of the head and neck. In: DeVita VT Jr, Hellman S, Rosenberg SA (eds) Cancer: principles and practice of oncology, 3rd edn. Lippincott, Philadelphia, p 565

Miura M, Takeda M, Sasaki T et al (1998) Factors affecting mandibular complications in low dose rate brachytherapy for oral tongue carcinoma with special reference to spacer. Int J Radiat Oncol Biol Phys 41:763–770

Mohiuddin M, Canton RJ, Bierman W et al (1988) Combined modality treatment of localized unresectable adenocarcinoma of the pancreas. Int J Radiat Oncol Biol Phys 14:79–84

Monteiro-Grillo I, Gaspar L, Monteiro-Grillo M et al (2000) Postoperative irradiation of primary or recurrent pterygium: results and sequelae. Int J Radiat Oncol Biol Phys 48:865–869

Moonen L (1989) Brachytherapy in the management of bladder cancer. In: Mould RF (ed) Brachytherapy 2. Nucletron International BV, Leersum, Netherlands, pp 184–188

Moran BJ, Stutz MA, Gurel MH (2004) Prostate brachytherapy can be performed in selected patients after transurethral resection of the prostate. Int J Radiat Oncol Biol Phys 59:392–396

Morphis OL (1960) Teflon tube method of radium implantation. Am J Roentgenol 83:455–461

Moss WT, Brand WN, Battifora H (eds) (1979) Radiation oncology: rationale, technique, results, 5th edn. Mosby, St Louis, MO

Nag S (2000) Brachytherapy for prostate cancer: summary of American Brachytherapy Society Recommendations. Semin Urol Oncol 18:133–136

Nag S, Ribovich M, Cai JZ et al (1996) Palladium-103 vs Iodine-125 brachytherapy in the Dunning-PAP rat prostate tumor. Endocurie Hypertherm Oncol 12:119–124

Nag S, Baird M, Blasko J et al (1997a) American Brachytherapy Society (ABS) survey of current clinical practice for permanent brachytherapy of prostate cancer. J Brachy Int 13:243–251

Nag S, Martinez-Monge R, Copeland LJ et al (1997b) Perineal template interstitial brachytherapy salvage for recurrent endometrial adenocarcinoma metastatic to the vagina. Gynecol Oncol 66:16–19

Nag S, Martinez-Monge R, Ellis R et al (1998a) The use of fluoroscopy to guide needle placement in interstitial gynecological brachytherapy. Int J Radiat Oncol Biol Phys 40:415–420

Nag S, Martinez-Monge R, Selman AE et al (1998b) Interstitial brachytherapy in the management of primary carcinoma of the cervix and vagina. Gynecol Oncol 70:27–32

Nag S, Beyer D, Friedland J et al (1999) American Brachytherapy Society (ABS) recommendations for transperineal permanent brachytherapy of prostate cancer. Int J Radiat Oncol Biol Phys 44:789–799

Nag S, Bice W, DeWyngaert K et al (2000) American Brachytherapy Society recommendations for permanent prostate brachytherapy post-implant dosimetric analysis. Int J Radiat Oncol Biol Phys 46:221–223

Nag S, Ciezki JP, Cormack R et al (2001a) Intraoperative planning and evaluation of Permanent Prostate Brachytherapy: report of the American Brachytherapy Society. Int J Radiat Oncol Biol Phys 51:1422–1430

Nag S, Kuske RR, Vicini FA et al (2001b) Brachytherapy in the treatment of breast cancer: Recommendations from the American Brachytherapy Society. Oncology 15:195–205

Nag S, Shasha D, Janjan N et al (2001c) The American Brachytherapy Society: The American Brachytherapy Society recommendations for brachytherapy of soft tissue sarcomas. Int J Radiat Oncol Biol Phys 49:1033–1043

Nag S, Chao C, Erickson B, Fowler J et al (2002) The American Brachytherapy Society recommendations for low dose rate brachytherapy for carcinoma of the cervix. Int J Radiat Oncol Biol Phys 52:33–48

Narayana Y, Orton CG (1999) Pulsed brachytherapy: a formalism to account for the variation in dose rate of the stepping source. Med Phys 26:161–165

Nath R, Meigooni AS (1989) Some treatment planning considerations for Palladium-103 and Iodine-125 permanent interstitial implants (abstract). Endocurie Hypertherm Oncol 5:244

Nath R, Meigooni AS, Meli JA (1990) Dosimetry on the transverse axes of ^{125}I and ^{192}Ir interstitial brachytherapy sources. Med Phys 18:1032–1040

Nath R, Meigooni AS, Melillo A (1992) Some treatment planning considerations for Pd-103 and I-125 permanent interstitial implants. Int J Radiat Oncol Biol Phys 22:1131–1138

Nickers P, Coppers L, Beauduin M et al (2000) Feasibility study combining low dose rate ^{192}Ir brachytherapy and external beam radiotherapy aiming at delivering 80–85 Gy to prostatic adenocarcinoma. Radiother Oncol 55:41–47

Nori D, Moni J (1997) Current issues in techniques of prostate brachytherapy. Semin Surg Oncol 13:444–453

Olch AJ, Kagan AR, Wollin M et al (1987) A simple volume iridium implant dosimetry system. Endocurie Hypertherm Oncol 3:183–191

Orton CG, Seyedsadr M, Somnay A (1991) Comparison of high and low dose rate remote afterloading for cervix cancer and the importance of fractionation. Int J Radiat Oncol Biol Phys 21:1425–1434

Packer S, Rotman M, Salanitro P (1984) Iodine-125 irradiation of choroidal melanoma: clinical experience. Ophthalmology 91:1700–1708

Papillon J, Montbarbon JR, Gerard JP et al (1989) Interstitial curietherapy in the conservative treatment of anal and rectal cancers. Int J Radiat Oncol Biol Phys 18:1161–1169

Parsons JT, Mendenhall WM, Bova FJ et al (1992) Head and neck cancer. In: Levitt SL, Khan FM, Potish RA (eds) Levitt and Tapley's technological basis of radiation therapy: practical clinical applications, 2nd edn. Lea and Febiger, Philadelphia, PA, pp 203–231

Parsons JT, Stringer SP, Mancuso AA et al (1994) Nasal vestibule, nasal cavity and paranasal sinuses. In: Million RR, Cassisi NJ (eds) Management of head and neck cancer: a multidisciplinary approach, 2nd edn. Lippincott, Philadelphia, pp 551–598

Patterson R (1948) The treatment of malignant disease by radium and X-rays being a practice of radiotherapy. Williams and Wilkins, Baltimore, MD, pp 254–255

Patterson R (1963) The treatment of malignant disease by radiotherapy, 2nd edn. Williams and Wilkins, Baltimore

Peiffert D, Castelain B, Thomas L et al (2001) Pulsed dose rate brachytherapy in head and neck cancers. Feasibility study of a French cooperative group. Radiother Oncol 58:71–75

Peretz T, Nori D, Hilaris B et al (1989) Treatment of primary unresectable carcinoma of the pancreas with I-125 implantation. Int J Radiat Biol Phys 18:931–935

Perez CA, Camel HM, Galakatos AE et al (1988) Definitive irradiation in carcinoma of the vagina: long-term evaluation of results. Int J Radiat Oncol Biol Phys 15:1283–1290

Perez CA, Slessinger E, Grigsby PW (1990) Design of an afterloading vaginal applicator (MIRALVA). Int J Radiat Oncol Biol Phys 18:1503–1508

Perez CA, Grigsby PW, Williamson JF (1998) Clinical applications of brachytherapy I: low dose rate. In: Perez CA, Brady LW (eds) Principles and practice of radiation oncology, 3rd edn. Lippincott, Philadelphia, pp 487–559

Perez CA, Michalski J, Martinez AA (1999) Prostate. In: Levitt SH, Kahn FM, Potish RA, Perez CA (eds) Levitt and Tapley's technological basis of radiation therapy: clinical applications, 3rd edn. Lippincott Williams and Wilkins, Philadelphia, pp 435–465

Perez CA, Zwicker R, Williamson J (2004) Clinical applications of brachytherapy I. LDR and PDR. In Perez CA, Brady LW, Halperin EC, Schmidt-Ullrich RK (eds) Principles and practice of radiation oncology, 4th edn. Lippincott Williams and Wilkins, Philadelphia, pp 538–603

Petereit DG, Sarkaria JN, Potter DM, Schink JC (1999) High-dose rate versus low-dose rate brachytherapy in the treatment of cervical cancer: analysis of tumor recurrence - the University of Wisconsin experience. Int J Radiat Oncol Biol Phys 45:1267–1274

Pierquin B (1964) Precis de curietherapie, endocurietherapie et plesiocurietherapie. Masson, Paris

Pierquin B, Cachin Y, Chassagne D et al (1968) Etude de 49 cas de carcinomes epidermides du cavum traites a l'institut Gustav-Roussy de 1960 a 1965. Presse Med 76:1565–1566

Pierquin B, Chassagne D, Baillet F et al (1971) The place of implantation in tongue and floor of mouth cancer. JAMA 215:961–963

Pierquin B, Wilson J-F, Chassagne D (1987a) Radiation protection and the organizational plan of a brachytherapy department. In: Pierquin B, Wilson J-F, Chassgne D (eds) Modern brachytherapy. Masson, New York, NY, pp 43–59

Pierquin B, Wilson J-F, Chassagne D (eds) (1987b) Modern brachytherapy. Masson, New York

Pierquin B, Pernot M, Baillet F (1987c) Tonsillar region. In:

Pierquin B, Wilson J-F, Chassagne D (eds) Modern brachytherapy. Masson, New York, NY, pp 141–145

Pigneux J, Richaud PM, Largade C (1979) The place of interstitial therapy using ^{192}Ir in the management of carcinoma of the lip. Cancer 43:1073–1077

Pisters PWT, Harrison LB, Leung DHY et al (1995) Long-term results of a prospective randomized trial of adjuvant brachytherapy in soft tissue sarcoma. J Clin Oncol 14:859–868

Porrazzo MS, Hilaris BS, Moorthy CR et al (1992) Permanent interstitial implantation using palladium-103: the New York Medical College preliminary experience. Int J Radiat Oncol Biol Phys 23:1033–1036

Porter AT, Blasko JC, Grimm PD et al (1995) Brachytherapy for prostate cancer. CA Cancer J Clin 45:165–178

Pos FJ, Horenblas S, Lebesque J et al (2004) Low-dose rate brachytherapy is superior to high-dose rate brachytherapy for bladder cancer. Int J Radiat Oncol Biol Phys 59:696–705

Potter R, Knocke TH, Kovacs G et al (1995) Brachytherapy in the combined modality treatment of pediatric malignancies: principles and preliminary experience with treatment of soft tissue sarcoma (recurrence) and Ewing's sarcoma. Klin Padiatr 207:164–183

Potters L (2000) Permanent prostate brachytherapy: lessons learned, lessons to learn. Oncology 14:981–991

Potters L, Morgenstern C, Mullen EE et al (2004) Twelve year outcomes following permanent brachytherapy in patients with clinically localized prostate cancer. Int J Radiat Oncol Biol Phys 60 [Suppl 1]:S183

Prados MD, Gutin PH, Phillips TL et al (1992) Interstitial brachytherapy for newly diagnosed patients with malignant gliomas: the UCSF experience. Int J Radiat Oncol Biol Phys 24:593–597

Prempree T (1983) Parametrial implant in stage IIIB cancer of the cervix. III. A five-year study. Cancer 52:748–750

Prestidge BR, Prete JJ, Buchholz TA (1998) A survey of current clinical practice of permanent prostate brachytherapy in the United States. Int J Radiat Oncol Biol Phys 40:461–465

Price A, Kerr GR, Arnott SJ (1988) Radioactive needle implants in the treatment of anorectal cancer. Clin Radiol 39:186–189

Pryzant RM, Wendt CD, Delclos L et al (1992) Re-treatment of nasopharyngeal carcinoma in 53 patients. Int J Radiat Oncol Biol Phys 22:941–947

Puthawala AA, Syed AM, Tansey LA et al (1985) Temporary iridium-192 implant in the management of carcinoma of the prostate. Endocurie Hypertherm Oncol 1:25

Puusaari I, Heikkonen J, Kivela T (2004) Effect of radiation dose on ocular complications after iodine brachytherapy for large oveal melanoma: Empirical data and simularion of collimating plaques. Invest Ophthalmol Vis Sci 45:3425–2434

Qin D, Hu Y, Yan J et al (1988) Analysis of 1379 patients with nasopharyngeal carcinoma treated by radiation. Cancer 61:1118–1124

Raben A, Mychalczak B (1997) Brachytherapy for non-small cell lung cancer and selected neoplasms of the chest. Chest 112:276S–286S

Reed Dr, Mueller A, Wallner K et al (2004) RAPID Strand™ vs. loose (Iodine-125 seeds): a prospective randomized trial. Int J Radiat Oncol Biol Phys 60 [Suppl 1]:S461

Rivard MJ, Coursey BM, DeWerd LA et al (2004) Update of AAPM Task Group no 43 report: a revised AAPM protocol for brachytherapy dose calculations. Med Phys 31:633–674

Rosenblatt E, Rachmiel A, Blumenfeld I et al (1996a) Intracavitary mould brachytherapy in malignant tumors of the maxilla. Endocurie Hypertherm Oncol 12:25–34

Rosenblatt E, Cederbaum M, Yereslav N et al (1996b) Reduction of the rectal dose in gynecological brachytherapy: modification of the Fletcher-Suit applicator. Med Dosim 21:139–143

Roy JN, Wallner KE, Chiu-Tsao S et al (1991) CT-based optimized planning for transperineal prostate implant with customized template. Int J Radiat Oncol Biol Phys 21:483–489

Roy JN, Wallner KE, Harrington PJ et al (1993) A CT-based evaluation method for permanent implants: application to prostate. Int J Radiat Oncol Biol Phys 26:163–169

Rutledge FN, Delclos L (1980) Adenocarcinoma of the uterus. In: Fletcher GH (ed) Textbook of radiotherapy, 3rd edn. Lea and Fibeger, Philadelphia, PA, pp 798–808

Ryan TP, Taylor JH, Coughlin CT (1992) Interstitial microwave hyperthermia and brachytherapy for malignancies of the vulva and vagina. I. Design and testing of a modified intracavitary obturator. Int J Radiat Oncol Biol Phys 23:189–199

Saw CB, Suntharalingam S (1988) Reference dose rates for single- and double-plane ^{192}Ir implants. Med Phys 15:391–396

Saw CB, Suntharalingham N, Ayyangar K et al (1989) Dosimetric considerations of stereotactic brain implants. Int J Radiat Oncol Biol Phys 18:887–891

Scalliet P, Gerbaulet A, Dubray B (1993) HDR versus LDR in gynecological brachytherapy revisited. Radiother Oncol 28:118–126

Schmitt-Ullrich R, Zwicker RD, Kelly WA, Kelly K (1991) Interstitial ^{192}Ir implants of the oral cavity: planning and construction of volume implants. Int J Radiat Oncol Biol Phys 20:1079–1085

Schray MF, Gunderson LL, Sim FH et al (1990) Soft tissue sarcoma: integration of brachytherapy, resection, and external irradiation. Cancer 66:451–456

Scott WP (1975) Interstitial therapy using non-absorbable (Ir[192] nylon ribbon) and absorbable (I[125] "vicryl") suturing techniques. Am J Roentgenol 124:560–564

Sham JST, Wei WI, Choy D et al (1989) Treatment of persistent and recurrent nasopharyngeal carcinoma by brachytherapy. Br J Radiol 62:355–361

Shankar PG, Wolf D, Cytacki EP (1991) Brachytherapy as boost for advanced tumors involving the oropharynx. Endocurie Hypertherm Oncol 7:53–56

Sherrah-Davies E (1985) Morbidity following low-dose rate Selectron therapy for cervical cancer. Clin Radiol 36:131–139

Shipley WU, Nardi GL, Cohen AM et al (1980) Iodine-125 implant and external beam irradiation in patients with localized pancreatic carcinoma. Cancer 45:709–714

Slessinger ED, Perez CA, Grigsby PW et al (1992) Dosimetry and dose specification for a new gynecological brachytherapy applicator. Int J Radiat Oncol Biol Phys 22:1118–1124

Sminia P, Schneider CJ, van Tienhoven G et al (2001) Office hours pulsed brachytherapy boost in breast cancer. Radiother Oncol 59:273–280

Sneed PK, McDermott MW, Gutin PH (1997) Interstitial brachytherapy procedures for brain tumors. Semin Surg Oncol 13:157–166

Stallard HB (1961) Malignant melanoma of the choroid treated with radioactive applicators. Ann R Coll Surg Engl 29:180

Straus KL, Littman P, Wein AJ et al (1988) Treatment of bladder cancer with interstitial iridium-192 implantation and external beam irradiation. Int J Radiat Oncol Biol Phys 14:265–271

Suit HD, Lloyd RS, Andrews JR et al (1960) Technique for intracavitary irradiation of the nasopharynx. Am J Roentgenol 84:629–631

Suit HD, Moore EB, Fletcher GH et al (1963) Modifications of Fletcher ovoid system for afterloading using standard sized radium tubes (milligram and microgram). Radiology 81:126–131

Swift, PS, Purser P, Roberts LW et al (1997) Pulsed low dose rate brachytherapy for pelvic malignancies. J Radiat Oncol Biol Phys 37:811–818

Syed AM, Puthawala AA, Neblett DL (1983a) Interstitial iodine-125 implant in the management of unresectable pancreatic carcinoma. Cancer 52:808–813

Syed AMN, Puthawala AA, Tansey LA et al (1983b) Temporary iridium-192 implantation in the management of carcinoma of the prostate. In: Hilaris BS, Batata MA (eds) Brachytherapy oncology - 1983. Memorial Sloan-Kettering Cancer Center, New York, pp 83–91

Syed AMN, Puthawala AA, Severance SR et al (1987) Intraluminal irradiation in the treatment of esophageal cancer. Endocurie Hypertherm Oncol 3:105–113

Tabandeh Chaundhry NA, Murray TG et al (2000) Intraoperative echographic localization of iodine-125 episcleral plaque for brachytherapy of choroidal melanoma. Am J Ophthalmol 129:199–204

Thomadsen B, Ayyangar K, Anderson L et al (1997) Brachytherapy treatment planning In: Nag S (ed) Principles and practice of brachytherapy. Futura, Armonk, NY, pp 127–199

Tod MC, Meredith WJ (1938) A dosage system for use in the treatment of cancer of the uterine cervix. Br J Radiol 11:809

Tombolini V, Bonanni A, Valeriani M et al (1998) Brachytherapy for squamous cell carcinoma of the lip. The experience of the Institute of Radiology of the University of Rome "La Sapienza". Tumori 84:478–482

Van den Brenk HAS (1968) Results of prophylactic postoperative irradiation in 1300 cases of pterygium. Am J Roentgenol 103:723–733

Van der Werf-Messing B, Menon RS, Ho WCJ (1983) Cancer of the urinary bladder category T2, T3, (NxM0) treated by interstitial radium implant: Second report. Int J Radiat Oncol Biol Phys 9:481–485

Van der Werf-Messing B, van Putten WLJ (1989) Carcinoma of the urinary bladder category T2,3NxM0 treated by 40 Gy external irradiation followed by cesium 137 implant at reduced dose (50%). Int J Radiat Oncol Biol Phys 16:369–371

Vicini FA, Chen PY, Fraile M et al (1997) Low dose rate brachytherapy as the sole radiation modality in the management of patients with early stage breast cancer treated with breast conserving therapy: preliminary results of a pilot trial. Int J Radiat Oncol Biol Phys 38:301–310

Viola A, Major T, Julow J (2004) The importance of postoperative stereotactic CT image fusion verification of stereotactic interstitial irradiation for brain tumors. Int J Radiat Oncol Biol Phys 60:322–328

Visser AG, van den Aardweg GJMJ, Levendag PC (1996) Pulsed dose rate and fractionated high dose rate brachytherapy: choice of brachytherapy schedules to replace low dose rate treatments. Int J Radiat Oncol Biol Phys 34:497–505

Wallner K (2002) Prostate brachytherapy under local anesthesia; lessons from the first 600 patients. Brachytherapy 1:145–148

Wallner K, Chiu-Tsao S-T, Roy J et al (1991) An improved method for computerized tomography-planned transperineal ^{125}Iodine prostate implants. J Urol 146:90–95

Wallner K, Roy J, Harrison L (1995) Dosimetry guidelines to minimize urethral and rectal morbidity following transperineal I-125 prostate brachytherapy. Int J Radiat Oncol Biol Phys 32:465–471

Walstrom R (1954) The dosage distribution in the pelvis in radium treatment of carcinoma of the cervix. Acta Radiol 42:237

Wang CC (1980) Treatment of malignant tumors of the nasopharynx. Otolaryngol Clin North Am 13:477–481

Wang CC (1987) Re-irradiation of recurrent nasopharyngeal carcinoma: treatment techniques and results. Int J Radiat Oncol Biol Phys 13:953–956

Wang CC, Schultz MD (1966) Management of locally recurrent carcinoma of the nasopharynx. Radiology 86:900–903

Waterman FM, Mansfield CM, Komarnicky L et al (1997a) A dosimetry system for Ir-192 interstitial breast implants performed at the time of lumpectomy. Int J Radiat Oncol Biol Phys 37:229–235

Waterman FM, Yue N, Reisinger S et al (1997b) Effect of edema on the post-implant dosimetry of an I-125 prostate implant: a case study. Int J Radiat Oncol Biol Phys 38:335–339

Weaver K (1986) The dosimetry of ^{125}I seed eye plaques. Med Phys 13:78–83

Weaver K, Smith V, Huang D et al (1989) Dose parameters of ^{125}I and ^{192}Ir seed sources. Med Phys 16:636–643

Wijnmaalen AJ, Helle PA, Koper PCM et al (1989) Combined external beam and interstitial radiation for bladder cancer. In: Mould RF (ed) Brachytherapy 2. Nucletron International BV, Leerum, The Netherlands, pp 192–195

Wijnmaalen AJ, Helle PA, Koper PC et al (1997) Muscle invasive bladder cancer treated by transurethral resection, followed by external beam radiation and interstitial iridium-192. Int J Radiat Oncol Biol Phys 39:1043–1052

Williamson JF (1988) Monte Carlo evaluation of specific dose constants in water for ^{125}I seeds. Med Phys 15:686–694

Williamson JF (1991) Practical quality assurance in low-dose rate brachytherapy. Proceedings of American College of Medical Physics-Sponsored Symposium on Quality Assurance in Radiotherapy Physics. Medical Physics Publishing Company, Madison, WI, pp 139–182

Williamson JF, Grigsby PW, Meigooni AS et al (1995) Clinical physics of pulsed dose-rate remotely-afterloaded brachytherapy. In: Williamson JF, Thomadsen BT, Nath R (eds) Brachytherapy physics. Medical Physics Publishing Company, Madison, WI 1995, pp 577–616

Yamashita S, Kondo M, Inuyama Y et al (1986) Improved survival of patients with nasopharyngeal squamous cell carcinoma. Int J Radiat Oncol Biol Phys 12:307–312

Yan JH, Hu YH, Gu XZ (1983) Radiation therapy of recurrent nasopharyngeal carcinoma: Report on 219 patients. Acta Radiol Oncol 22:23–28

Yue N, Chen Z, Peschel R et al (1999) Optimum timing for image-based dose evaluation of ^{125}I and ^{103}Pd prostate seed implants. Int J Radiat Oncol Biol Phys 45:1063–1072

Zaider M, Zelefsky MJ, Lee EK et al (2000) Treatment planning for prostate implants using magnetic resonance spectroscopy imaging. Int J Radiat Oncol Biol Phys 47:1085–1096

Zelefsky JM, Yamada Y, Cohen et al (2000) Postimplantation dosimetric analysis of permanent transperineal prostate implantation: improved dose distributions with an intra-operative computer-optimized conformal planning technique. Int J Radiat Oncol Biol Phys 48:602–608

Zhang P, Beddar AS, Sibata CH (2004) AAPM TG-43 formalism for brachytherapy dose calculation of ^{137}Cs tube source. Med Phys 31:755–759

Zolli CI (1979) Experience with the avulsion technique in pterygium surgery. Ann Ophthalmol 11:1569–1576

Zwicker RD, Schmidt-Ullrich R (1994) Dose uniformity in a planar interstitial implant system. Int J Radiat Oncol Biol Phys 31:149–155

Zwicker RD, Schmidt-Ullrich R, Schiller B (1985) Planning of Ir-192 seed implants for boost irradiation to the breast. Int J Radiat Oncol Biol Phys 11:2163–2180

15 Clinical Applications of High-Dose-Rate Brachytherapy

Subir Nag

CONTENTS

15.1 Abstract 379
15.2 Introduction 379
15.3 How to Design a New HDR Protocol 381
15.4 Common Uses of HDR Brachytherapy 382
15.4.1 Carcinoma of the Cervix 382
15.4.2 Carcinoma of the Endometrium 383
15.4.3 Endobronchial Radiation 385
15.4.4 Cancer of the Esophagus 386
15.4.5 Carcinoma of the Prostate 386
15.4.6 Head and Neck Cancers 387
15.4.7 Soft Tissue Sarcomas 388
15.4.8 Pediatric Tumors 388
15.4.9 Breast Cancer 389
15.4.10 Skin Cancer 390
15.5 Discussion 390
References 391

It is expected that the use of HDR brachytherapy will greatly expand over the next decade and that refinements will occur primarily in the integration of imaging (computed tomography, magnetic resonance imaging, intraoperative ultrasonography) and optimization of dose distribution. It is anticipated that better tumor localization and normal tissue definition will help to optimize dose distribution to the tumor and reduce normal tissue exposure. The development of well-controlled randomized trials addressing issues of efficacy, toxicity, quality of life, and costs versus benefits will ultimately define the role of HDR brachytherapy in the therapeutic armamentarium.

15.1 Abstract

Brachytherapy has the advantage of delivering a high dose to the tumor while sparing the surrounding normal tissues. With proper case selection and delivery technique, high-dose-rate (HDR) brachytherapy has great promise, because it eliminates radiation exposure, allows short treatment times, and can be performed on an outpatient basis. Additionally, use of a single-stepping source allows optimization of dose distribution by varying the dwell time at each dwell position. However, when HDR brachytherapy is used, the treatments must be executed carefully, because the short treatment times do not allow any time for correction of errors, and mistakes can result in harm to patients. Hence, it is very important that all personnel involved in HDR brachytherapy be well trained and constantly alert.

S. Nag, MD, FACR, FACRO
Chief of Brachytherapy, Department of Radiation Medicine, Arthur G. James Cancer Hospital and Solove Research Institute, 300 West Tenth Avenue, The Ohio State University, Columbus, OH 43210, USA

15.2 Introduction

Brachytherapy has the advantage of delivering a high dose of radiation to the tumor while sparing the surrounding normal tissues. Brachytherapy procedures were previously performed by inserting the radioactive material directly into the tumor ("hot" loading), thereby giving high radiation exposure to the physicians performing the procedure. Manually afterloaded techniques were introduced to increase accuracy and reduce the radiation hazards. In afterloaded techniques, hollow needles, catheters, or applicators are first inserted into the tumor then loaded with radioactive materials. The introduction of remote-controlled insertion of sources eliminated radiation exposure to visitors and medical personnel. In this technique, the patient is housed in a shielded room, and the radiation therapist controls the treatment from outside the room. Hollow applicators, needles, or catheters are inserted into the tumor and connected by transfer tubes to the radioactive material, which is stored in a shielded safe within the HDR afterloader. The radiation source is driven through the transfer tubes and into the tumor by remote control.

Remote controlled brachytherapy can be performed using low-dose-rate (LDR), medium-dose-rate (MDR), or HDR techniques. The usual dose rate employed in current HDR brachytherapy units is about 100–300 Gy/h. The use of remote-controlled brachytherapy (whether it is LDR, MDR, or HDR brachytherapy) eliminates the hazards of radiation exposure. The use of HDR has the added advantage of the treatments taking only a few minutes, allowing them to be given on an outpatient basis with minimal risk of applicator movement and minimal patient discomfort. Additionally, use of a single-stepping source, as used in most modern HDR afterloaders, allows optimization of dose distribution by varying the dwell time at each dwell position. However, it should be emphasized that while optimization can improve the dose distribution, it should not be used to substitute for a poorly placed implant. NAG and SAMSAMI (2000) have provided examples of inappropriate optimization strategies, which can lead to suboptimal dosimetry plans and clinical problems. The advantages and disadvantages of HDR in comparison with LDR are enumerated in Table 15.1.

Table 15.1. Advantages and disadvantages of high-dose-rate (HDR) compared with low-dose-rate (LDR) brachytherapy

Advantages	Disadvantages
1. Radiation protection	1. Radiobiological
– HDR eliminates radiation exposure hazard for caregivers and visitors. Caregivers are able to provide optimal patient care without fear of radiation exposure – HDR eliminates source preparation and transportation – Since there is only one source, there is minimal risk of losing a radioactive source	– The short treatment times do not allow for the repair of sublethal damage in normal tissue, the redistribution of cells within the cell cycle, or reoxygenation of the tumor cells; hence, multiple treatments are required
2. Allows shorter treatment times	2. Limited experience
– There is less patient discomfort, since prolonged bed rest is eliminated – It is possible to treat patients who may not tolerate long periods of isolation and those who are at high risk of pulmonary embolism due to prolonged bed rest – There is less risk of applicator movement during therapy – There are reduced hospitalization costs, since outpatient therapy is possible – HDR may allow greater displacement of nearby normal tissues (by packing or retraction), which could potentially reduce morbidity – It is possible to treat a larger number of patients in institutions that have a high volume of brachytherapy patients but insufficient in-patient facilities (e.g., in some developing countries) – Allows intraoperative treatments, which are completed while patient is still in the operating room	– Few centers in the United States have long-term (greater than 20 years) experience – Until recently, standardized treatment guidelines were not available; however, the American Brachytherapy Society has recently provided guidelines for HDR at various sites (ARTHUR et al. 2002; NAG et al. 2000a; NAG et al. 2000b; NAG et al. 2000e; NAG et al. 2001a; NAG et al. 2001b; NAG et al. 2001c; RODRIGUEZ et al. 2001)
3. HDR sources are of smaller diameter than the Cesium sources that are used for intracavitary LDR	3. The economic disadvantage
– This reduces the need for dilatation of the cervix and therefore reduces the need for heavy sedation or general anesthesia – High-risk patients who are unable to tolerate general anesthesia can be more safely treated – HDR allows for interstitial, intraluminal and percutaneous insertions	– The use of HDR brachytherapy, compared with manual afterloading techniques, requires a large initial capital expenditure, since the remote afterloaders cost approximately $300,000 – There are additional costs for a shielded room, and personnel costs are higher, as the procedures are more labor intensive
4. HDR makes treatment dose distribution optimization possible	4. Greater potential risks
– Variations of the dwell times of a single stepping source allow an almost infinite variation of the effective source strengths, and the source positions allow for greater control of the dose distribution and potentially less morbidity	– Since a high activity source is used, there is greater potential harm if the machine malfunctions or if there is a calculation error. The short treatment times, compared with LDR, allow much less time to detect and correct errors

The advantages listed above have led to increased use of HDR worldwide; however, training and expertise is required for proper administration of these treatments. Scientific Societies, including the American Brachytherapy Society (ABS), have recently published guidelines and recommendations for the use of HDR at various sites (GASPAR et al. 1997; NAG et al. 2000a,b,e, 2001a,b,c, 2002, 2004a, RODRIGUEZ et al. 2001; ARTHUR et al. 2002). While this chapter incorporates many of these recommendations, the reader should refer to the original publications for details.

15.3
How to Design a New HDR Protocol

Most radiation oncologists are familiar with LDR brachytherapy. LDR (at 30–50 cGy/h) can be added to external beam radiation therapy (EBRT) doses (at 2 Gy/day) to obtain equivalent total doses. HDR brachytherapy is a relatively new modality that is distinct from LDR brachytherapy, and radiation oncologists who are accustomed to LDR techniques must realize that experience in LDR cannot be automatically translated into expertise in HDR. It is important to review the current literature and survey the experiences of centers that have been performing HDR. When converting from LDR to HDR, one must keep the other parameters (chemotherapy, EBRT field/dose, dose-specification point, applicators, patient population, etc.) the same, changing only the LDR to HDR.

Fractionation schemes for HDR are widely variable, and many radiation oncologists are not very familiar with the resultant biological effects. Empirical methods such as the nominal standard dose, time–dose factor, or a dose reduction factor of 0.6 have been used in the past to convert HDR doses to LDR equivalent doses. The linear-quadratic (LQ) equation (BARENDSEN 1982) can be used to guide development of HDR doses and fractionation schedules. However, the LQ mathematical calculations are tedious and may not be practical on a day-to-day basis. Hence, a simplified computer program was developed by NAG and GUPTA (2000) to obtain the biologically equivalent doses for HDR. The clinician needs only to enter the EBRT total dose and dose/fraction, HDR dose, and the number of HDR fractions. The computer program will automatically calculate the equivalent doses for tumor and normal tissue effects. Equivalent doses are expressed in clinically familiar terms (as if given at 2 Gy per fraction)

rather than as biologically equivalent doses, which are unfamiliar to clinicians. Furthermore, a dose-modifying factor (DMF) is applied to the normal tissues to account for the fact that doses to normal tissues are different from the doses to the tumor, thus providing a more realistic equivalent normal tissue effect. This program can be used to determine HDR doses that are equivalent to LDR brachytherapy doses used to treat various cancers. Alternatively, the program may be used to express the equivalent dose of different HDR dose-fractionation regimes, as shown for cervical cancer in Table 15.2. It is remarkable that the equivalent doses for tumor effects for the various fractionation regimes used for early-stage cervical cancers are so similar, ranging from 82 Gy to 85 Gy, while those used for advanced cancers are about 90 Gy (Table 15.2). The equivalent dose for normal tissue late effects depends on the assumed DMF (0.6, 0.7, or 0.9). For the fractionation scheme shown in Table 15.2, row 1, the equivalent late effect on normal tissue (bladder or rectum) would be 59.5, 71, or 98 Gy, respectively, if the doses to normal tissues were 60, 70, or 90% of the prescribed dose to point A.

Although the LQ biomathematical model can be helpful in determining equivalent doses, it has many limitations that must be kept in mind when using the program. The L-Q model accounts for the repair of sublethal damage and does not account for reoxygenation of hypoxic cells, reassortment within the cell cycle, or repopulation of tumor cells. These factors are generally small under normal circumstances. However, large doses per fraction do not allow reoxygenation of hypoxic tumor cells or reassortment of tumors from radioresistant S-phase. Hence, a large radiation dose will preferentially kill radiosensitive cells, leaving a high number of hypoxic, radioresistant cells. Therefore, the computer program will overestimate the tumor effect of a single large dose per fraction (unless a resensitization factor is introduced).

The LQ equation does not take into account the proliferation of tumor cells. This factor is small if the treatments are performed over a short duration. However, if the treatments are highly protracted (e.g., there is a long time interval between EBRT and HDR), or in cases of tumors with high proliferation rates, the LQ model will overestimate the actual tumor effect. It also must be noted that individual α/β values are very variable. The α/β values for early reactions vary from 6 to 13 (the default in the program is set at 10); the α/β values for late reactions vary from 1 to 7 (default being set at 3), while α/β

Table 15.2. American Brachytherapy Society suggested doses of external beam radiation therapy (EBRT) and high-dose-rate (HDR) brachytherapy to be used in treating early and advanced cervical cancer. The α/β ratio assumed for tumor equals ten. The α/β ratio assumed for normal tissue late effects equals three. *DMF* = dose modifying factor

Total EBRT dose (Gy) @ 1.8 Gy/fraction	No. of HDR fractions	HDR dose/fraction (Gy)	Equivalent dose (Gy) for tumor effects	Equivalent dose (Gy) for late effects with DMF=0.6	Equivalent dose (Gy) for late effects with DMF=0.7	Equivalent dose (Gy) for late effects with DMF=0.9
Early cervical cancer						
19.8	6	7.5	85.1	59.5	71.0	98.0
19.8	7	6.5	82.0	56.7	67.1	91.5
19.8	8	6.0	83.5	57.0	67.4	91.6
45	5	6.0	84.3	67.0	73.4	88.6
45	6	5.3	84.8	66.8	73.1	87.7
Advanced cervical cancer						
45	5	6.5	88.9	70.1	77.6	95.0
45	6	5.8	90.1	70.3	77.6	94.7
50.4	4	7.0	89.2	72.6	79.4	95.3
50.4	5	6.0	89.6	72.1	78.6	93.7
50.4	6	5.3	90.1	72.0	78.3	92.9

values for tumors vary from 0.4 to 13 (the default being set at 10). However, α/β values for a particular patient are not known and may vary even within the same tissue. The equivalent doses obtained will, therefore, depend on the α/β values used for that particular calculation. The LQ model assumes complete repair between fractions. If the time interval between fractions is too short (<6 h) or the half-time of repair is very long, the repair of normal tissues will be incomplete, and the LQ formula will underestimate the biological effect. Hence, it is important to have a sufficient time interval (at least 6 h) between treatment fractions.

The infinite variation of the dwell times that is possible with HDR (or pulsed dose rate) allows better optimization of the doses than can be achieved with LDR. Better packing or retraction of normal tissues is possible with HDR, due to the short treatment duration. This factor is not usually taken into account in the LQ model (unless the DMF is altered). Another difference not accounted for in the LQ model is that the dose stated in brachytherapy is generally the minimum tumor dose. The doses within the tumor are much higher. Hence, the effective dose (for tumor control probability) is much higher for brachytherapy than for EBRT.

In view of the many limitations of the LQ model, it must be stressed that, as with any mathematical model, the LQ model should be used judiciously as a guide only and should always be correlated with clinical judgment and outcome results. Caution is especially warranted whenever large fraction sizes are used, since their clinical results have not been well studied.

15.4
Common Uses of HDR Brachytherapy

Although HDR brachytherapy has been used in almost every site in the body, it is most commonly used to treat cancers of the cervix, endometrium, lung, and esophagus. Less common treated sites for HDR include the prostate, bile duct, breast, brain, rectum, head and neck, skin, soft tissues, and blood vessels (coronary and peripheral arteries). HDR is generally used as a component of a multi-modality treatment that includes EBRT and/or chemotherapy and surgery. A summary of the clinical uses of HDR at various sites is included this article.

15.4.1
Carcinoma of the Cervix

Brachytherapy is a necessary component in the curative treatment of cervical cancers (NAG et al. 2000b). HDR has gained popularity in the U.S. over the last decade due to the advantages alluded to earlier, specifically the possibility of therapy on an outpatient basis, avoidance of long-term bed rest, and avoidance of cervical dilation. Additionally, greater sparing of the rectum and bladder by temporary retraction, dose optimization, and integration with EBRT to the pelvis are possible. These advantages must be counterbalanced with the greater number of treatments required (typically five or six treatments lasting approximately 10–15 min each).

The ABS recommends keeping the total duration of treatment (EBRT and HDR) to less than 8 weeks, since prolongation adversely affects local control and survival (NAG et al. 2000b). Because the overall treatment duration would be unduly prolonged if HDR treatments (generally five or six fractions given once a week) were begun after completion of EBRT, the HDR is commonly given concurrently during the course of EBRT (but note that EBRT is not given on the day of HDR brachytherapy). The combined EBRT and HDR dose to point A (or point H) is an LDR equivalent of 80–85 Gy for early stage disease and 85–90 Gy for advanced stage (NAG et al. 2000b). Early disease is defined as non-bulky stage I/II less than 4 cm in diameter; advanced diseased is defined as stage I/II greater than 4 cm in diameter or stage IIIB. The pelvic side-wall dose recommendations are 50–55 Gy for smaller lesions and 55–60 Gy for larger ones. The HDR dose is dependent on the stage of the disease and the dose of EBRT. Most centers use a schedule of approximately 6–8 Gy per fraction in four to six fractions, although two or three fractions of 8–10 Gy have been used (a smaller number of fractions is used by those using larger doses per fraction) (FU and PHILLIPS 1990; ORTON et al. 1991; NAG et al. 2000b; SOOD et al. 2002; NAKANO et al. 2005; PATEL et al. 2005).

As mentioned earlier, HDR doses can be obtained from the LDR equivalent using the linear-quadratic equation. While recognizing that many efficacious HDR fractionation schedules exist, the ABS suggestions and the equivalent doses are given in Table 15.2 as a guide. The recommended HDR dose per fraction may vary by ±0.25 Gy. It is emphasized that extra care must be taken to ensure adequate bladder and rectal packing if high dose (>7 Gy) per fraction is used.

In certain difficult clinical situations (e.g., a narrow fibrotic vagina, bulky tumors, the inability to enter the cervical os, extension to the lateral parametria or pelvic side wall, lower vaginal extension and suboptimal applicator placement), the normal tissue tolerance may be exceeded if the above doses are used. In these situations, the HDR fraction size can be decreased (which requires an increase in the fraction number) or the EBRT dose increased while decreasing the HDR total dose. Alternatively, an interstitial implant (either LDR or HDR) may be used.

LDR brachytherapy has been used with good results in carcinoma of the cervix for almost 100 years. Hence, it is important to critically analyze whether the results obtained with HDR brachytherapy, which has a much shorter history, compare with those obtained with LDR. Unfortunately, most of the published reports have been non-randomized studies. Although large, multi-institutional,

randomized trials are not available, the available data from single-institution randomized trials, retrospective analyses and meta-analyses suggest that survival, local control, and morbidity of HDR treatments are equivalent to that of LDR (SHIGEMATSU et al. 1983; FU and PHILLIPS 1990; ORTON et al. 1991; ARAI et al. 1992; PATEL et al. 1993; PETEREIT et al. 1999; LORVIDHAYA et al. 2000; HAREYAMA et al. 2002; NAKANO et al. 2005; PATEL et al. 2005).

15.4.2
Carcinoma of the Endometrium

HDR brachytherapy is commonly used for adjuvant treatment of the vaginal cuff after hysterectomy in patients with intermediate and high risk for vaginal recurrence (high-grade, deep myometrial invasion, or advanced stage). Additionally, brachytherapy may be used for primary treatment in inoperable endometrial carcinoma and for treatment of recurrences after hysterectomy.

Patients at high risk for vaginal recurrences after hysterectomy (deep myometrial invasion, high histological grade and stage, cervical or extra-uterine spread, squamous cell or papillary histology) should receive adjuvant radiation therapy. There is controversy regarding the best method. Pelvic EBRT, vaginal vault brachytherapy or a combination can be used depending on the extent of pelvic lymph node dissection and whether chemotherapy will be added. EBRT has the advantage of irradiating the pelvic lymph nodes but takes approximately 5 weeks to perform and has some morbidity. Brachytherapy is more convenient and has low morbidity, but does not treat the lymph nodes. Hence, some centers combine both EBRT and brachytherapy, although it has not been proven that the combination yields any superior results. Others prefer "watchful waiting" and use salvage irradiation if there is a recurrence, since the final survival rate is not compromised. However, in cases of recurrence, a combination of pelvic EBRT and brachytherapy is required.

If pelvic EBRT is used, the dose is usually 40–45 Gy in 20–25 treatments (NAG et al. 2000a). A vaginal cylinder is commonly used to deliver HDR brachytherapy. The largest diameter cylinder that comfortably fits the vagina should be used to increase the depth dose. The length of vaginal vault treated varies. Some treat the superior 3 cm or 5 cm, while others treat the superior half or two-thirds of the vagina (MANDELL et al. 1985; LYBEERT et al. 1989; SORBE and SMEDS 1990; NORI et al. 1994; NAG et al. 2000a; ALEKTIAR et al. 2005). For serous and clear cell histologies, treatment of the entire vaginal canal should be considered. The dose distribution should be optimized to follow the curvature of the dome of the cylinder to deliver the prescribed dose either at the vaginal surface or at 0.5 cm depth, depending on the institutional policy. Regardless of the prescription point, doses to both of these points should be reported (NAG et al. 2000a). The ABS dose suggestions (NAG et al. 2000a) for HDR alone or in combination with 45 Gy EBRT are given in Table 15.3. Since some institutions specify the dose to the vaginal surface and others specify the dose at 0.5 cm depth, suggested HDR doses have been given for both specification methods.

The 5-year survival rates of HDR therapy vary from 72% to 97%, depending on the stage, grade, and depth of myometrial invasion (MANDELL et al. 1985; LYBEERT et al. 1989; SORBE and SMEDS 1990; NORI et al. 1994; ALEKTIAR et al. 2005). The severe (grades III or IV) late complication rate is usually less than 1% and depends on the dose per fraction (MANDELL et al. 1985; LYBEERT et al. 1989; SORBE and SMEDS 1990; NORI et al. 1994; ALEKTIAR et al. 2005). The incidence of vaginal shortening is also very much dose dependent, reportedly as high as 70% when 9 Gy per fraction was prescribed at 1 cm depth to 31% when the dose was reduced to 4.5 Gy per fraction (SORBE and SMEDS 1990). Other factors that increase morbidity include the use of a small (2 cm) diameter vaginal cylinder, the addition of pelvic EBRT, and dose specification point beyond 0.5 cm (MANDELL et al. 1985).

A combination of pelvic EBRT and brachytherapy is generally used to treat recurrences at the vaginal cuff. With distal vaginal recurrences, the entire vagina and medial inguinal nodes are included in the EBRT field. Intracavitary vaginal brachytherapy should be used only for non-bulky recurrences (thickness less than 5 mm after the completion of EBRT) (NAG et al. 2000a). Interstitial brachytherapy is to be used for bulky recurrences (thickness >5 mm after the completion of EBRT) and for previously irradiated patients. The ABS suggested doses for HDR brachytherapy (in combination with 45 Gy EBRT) are provided in Table 15.4 (NAG et al. 2000a).

Patients with adenocarcinoma of the endometrium who are not candidates for surgery because of severe medical problems are treated with radiation therapy. A combination of pelvic EBRT beam and brachytherapy is preferred whenever possible. However, many of the conditions that do not allow surgery in these cases are also relative contraindica-

Table 15.3. American Brachytherapy Society suggested doses of high-dose-rate (HDR) brachytherapy alone or in combination with external beam radiation therapy (EBRT) to be used for adjuvant treatment of post-operative endometrial cancer

EBRT (Gy) @ 1.8 Gy/fraction	No. of HDR fractions	HDR dose per fraction (Gy)	Dose specification point
0	3	7.0	0.5 cm depth
0	4	5.5	0.5 cm depth
0	5	4.7	0.5 cm depth
0	3	10.5	Vaginal surface
0	4	8.8	Vaginal surface
0	5	7.5	Vaginal surface
45	2	5.5	0.5 cm depth
45	3	4.0	0.5 cm depth
45	2	8.0	Vaginal surface
45	3	6.0	Vaginal surface

Table 15.4. American Brachytherapy Society suggested doses of high-dose-rate (HDR) brachytherapy to be used in combination with pelvic external beam radiation therapy (EBRT) for treating vaginal cuff recurrences from endometrial cancer

EBRT (Gy) @ 1.8 Gy/fraction	No. of HDR fractions	HDR dose per fraction (Gy)	Dose specification point
45	3	7.0	0.5 cm depth
45	4	6.0	0.5 cm depth
45	5	6.0	Vaginal surface
45	4	7.0	Vaginal surface

tions for EBRT and for LDR brachytherapy. These patients may be treated with HDR alone.

Numerous applicators can be used for treatment of primary endometrial cancer. A tandem and colpostat is often used; however, this applicator will not irradiate the uterine surface homogeneously. Others have used a curved tandem, turning it to the left and right in alternate insertions. A "Y"-shaped applicator irradiates the fundus more evenly. Other possibilities include modified Heyman capsules or multiple tandems. The dose is commonly specified at 2.0 cm from the source, although computed-tomography-based treatment planning to ensure a more homogeneous dose to the entire myometrium is preferred. The dose per fraction has ranged from 5 Gy to 12 Gy, and four to six fractions are commonly employed (SORBE and FRANKENDAL 1989; ROUANET et al. 1993; NAG 1996; NAG et al. 2000a). The dose and/or the dose per fraction is reduced if EBRT is added. The ABS-suggested doses for HDR brachytherapy alone or in combination with 45 Gy EBRT are given in Table 15.5 (NAG et al. 2000a). The survival at 5 years for stage I is approximately 70–80%, which is slightly lower than that obtained by surgery. The toxicity is higher (about 7%) when patients are treated with high doses per fraction (SORBE and FRANKENDAL 1989).

Table 15.5. American Brachytherapy Society suggested doses of high-dose-rate (HDR) brachytherapy alone or in combination with external beam radiation therapy (EBRT) for treatment of inoperable primary endometrial cancer

EBRT (Gy) @ 1.8 Gy/fraction	No. of HDR fractions	HDR dose per fraction (Gy)*
45	2	8.5 Gy
45	3	6.3 Gy
45	4	5.2 Gy
0	4	8.5 Gy
0	5	7.3 Gy
0	6	6.4 Gy
0	7	5.7 Gy

*HDR doses are specified at 2 cm from the midpoint of the intrauterine sources

15.4.3
Endobronchial Radiation

The use of HDR brachytherapy is well established for palliation of cough, dyspnea, pain, and hemoptysis in patients with advanced or metastatic lung cancer. The use of brachytherapy as a boost to EBRT in curative cases should be restricted to a select group of patients who have predominantly endobronchial disease, are medically inoperable, or have small/occult carcinomas of the lung.

An initial bronchoscopy is performed to evaluate the airway and locate the site of obstruction. While the proximal site of obstruction is usually easily visualized, the distal extent of the obstruction may have to be estimated. Either a 5- or 6-French catheter (inserted through the brush channel of the bronchoscope) can be used to deliver the brachytherapy. The catheter (with a radiopaque wire in the lumen) is passed through the obstructed segment and lodged into the distal bronchus. The bronchoscope is removed, leaving the catheter in position. Fluoroscopic guidance helps catheter positioning and minimizes inadvertent catheter dislodgment during bronchoscopic removal. The length to be irradiated usually includes the endobronchial tumor and 1.0- to 2.0-cm proximal and distal margins. If a single catheter is used, and there is minimal curvature of the catheter in the area to be irradiated, it is possible to minimize the treatment planning time using pre-planned dosimetry. For example, Ohio State University has pre-calculated treatment plans for 3-, 5-, 7-, and 10-cm lengths to be irradiated to 5 Gy or 7.5 Gy at 1 cm from the source using equal dwell times. This allows the treatment to be performed without any delay if standard lengths and doses are used. Individualized image-based treatment planning must be performed if multiple catheters are used.

Candidates for palliative endobronchial brachytherapy include (MEHTA et al. 1992; MEHTA et al. 1997; GASPAR 1998; SPEISER 1999; NAG et al. 2001a):
1. Patients with a significant endobronchial tumor component that causes symptoms such as shortness of breath, hemoptysis, persistent cough, and other signs of post-obstructive pneumonitis. Tumors with a predominantly endobronchial component are considered suitable, as opposed to extrinsic tumors that compress the bronchus or the trachea. Endobronchial brachytherapy can generally give quicker palliation of obstruction than EBRT. Furthermore, brachytherapy can be more convenient than 2–3 weeks of daily EBRT.
2. Patients who are unable to tolerate any EBRT because of poor lung function.
3. Patients with previous EBRT of sufficient total dose to preclude further EBRT.

A variety of doses have been successfully used by various centers. Total doses ranging from 15 Gy to 47 Gy HDR in 1–5 fractions calculated at 1.0 cm have been reported (MEHTA et al. 1997; GASPAR 1998; NAG et al. 2001a). The ABS suggests using three weekly fractions of 7.5 Gy each or 2 fractions of 10 Gy each or 4 fractions of 6 Gy each prescribed at 1.0 cm when HDR is used as the sole modality for palliation (NAG 2001a). These fractionation regimes have similar radiobiological equivalence using the linear quadratic model, and there is no evidence of superiority for one regime over the other. The benefits of fewer bronchoscopic applications should be weighed against the risks of a higher dose per fraction. Additional treatments or doses higher than those suggested can be considered for unirradiated patients or those who have received limited radiation. When HDR is used as a planned boost to supplement palliative EBRT of 30 Gy in 10–12 fractions, the ABS suggests using 2 fractions of 7.5 Gy each or 3 fractions of 5 Gy each or 4 fractions of 4 Gy each (prescribed at 1.0 cm) in patients with no previous history of thoracic irradiation (NAG 2001a). The interval between fractions is generally 1–2 weeks. The brachytherapy dose should be reduced when aggressive chemotherapy is given. Concomitant chemotherapy should be avoided during brachytherapy, unless it is in the context of a clinical trial.

The results from various centers show clinical improvement from 50% to 100% and bronchoscopy response from 59% to 100% (MEHTA et al. 1997; GASPAR 1998; NAG 2001a). Comparison of these results is difficult because of the differences in patient populations and the variability in dose and fractionation employed. Radiation bronchitis and stenosis may occur after endobronchial brachytherapy (SPEISER and SPARTLING 1993a), necessitating close follow-up.

Another more serious complication is fatal hemoptysis. The hemoptysis could be a radiation therapy complication resulting from the high dose delivered to the area of the pulmonary artery, or it could represent the failure of treatment due to the progression of disease (SPEISER and SPARTLING 1993b). Multiple courses of brachytherapy, a high previous EBRT dose, a left upper lobe location, or long-irradiated segments increase the rate of hemoptysis (KHANAVKAR et al. 1991; BEDWINEK et al. 1992). Incidence of fatal hemoptysis varies from 0% to 50%, with a median value of 8% (MEHTA et al. 1997).

The standard, definitive therapy for unresectable lung cancer is a combination of chemotherapy and EBRT. Select patients with predominantly endobronchial tumor may benefit from endobronchial brachytherapy as a boost to EBRT (AYGUN et al. 1992; MEHTA et al. 1992; HUBER et al. 1997). In cases of post-obstructive pneumonia or lung collapse, brachytherapy can be used to open up the bronchus and aerate the lung, which allows some sparing of normal lung from the EBRT field. Endobronchial brachytherapy alone with curative intent is indicated in patients with occult carcinomas of the lung confined to the bronchus or trachea who are medically inoperable because of decreased pulmonary function, advanced age, or refusal of surgery (PEROL et al. 1997; HUBER et al. 1997; FURUTA et al. 1999; MARSIGLIA et al. 2000; SAITO et al. 2000). Marsiglia et al. reported a survival rate of 78% with a median follow-up of 2 years in 34 patients treated with a HDR dose of 30 Gy in six fractions (5 Gy fractions given once a week) (MARSIGLIA et al. 2000).

Endobronchial brachytherapy can be used as adjuvant treatment in cases with minimal residual disease after surgical resection. MACHA et al. (1995) reported tumor-free survival up to 4 years in 19 patients with doses of 20.0 Gy delivered in four fractions at 1 cm from the source axis.

The ABS suggests a HDR dose of three 5-Gy fractions or two 7.5-Gy fractions as a boost to EBRT (either 60 Gy in 30 fractions or 45 Gy in 15 fractions) (NAG et al. 2001a). The HDR dose should be prescribed at a distance of 1.0 cm from the central axis of the catheter and given weekly. If endobronchial brachytherapy is used alone (in previously unirradiated patients), HDR doses of five 5-Gy fractions or three 7.5-Gy fractions prescribed to 1 cm may be used (NAG et al. 2001a).

15.4.4
Cancer of the Esophagus

The results of treatment for advanced cancer of the esophagus are dismal (5-year survival=6%); hence, treatment is essentially palliative. HDR brachytherapy can be used either alone or in combination with EBRT in the treatment of esophageal cancer (HISHIKAWA et al. 1987; GASPAR et al. 1997; SUR et al. 1998; SUR et al. 2002; SUR et al. 2004). Brachytherapy is relatively simple to perform, since a single catheter is used for the treatment. A nasogastric tube or a specially designed esophageal applicator is used to deliver the treatments. The largest diameter applicator that can be inserted easily (either intraorally or intranasally) should be used to minimize the mucosal dose relative to the dose at depth. The site to be irradiated, which includes the tumor and a margin of 2–5 cm, can be confirmed by fluoroscopy or endoscopy. The ABS recommends a HDR dose of 10 Gy in two fractions, prescribed at 1 cm from the source, to boost 50-Gy EBRT (GASPAR et al. 1997). HDR brachytherapy can be given before, concurrent with, or after EBRT. The advantage of giving brachytherapy after EBRT is that a more uniform dose can be delivered to the residual tumor after it has been reduced using EBRT. Brachytherapy given initially provides rapid relief of dysphagia. HDR brachytherapy at doses of 16 Gy in two treatments have been used without additional EBRT to palliate esophageal cancers (SUR et al. 1998; SUR et al. 2002).

Retrospective studies as well as prospective, randomized clinical trials show that there is improved local control and survival when HDR brachytherapy is added to EBRT and that HDR brachytherapy alone can be used for palliation of advanced esophageal cancers (HISHIKAWA et al. 1987; GASPAR et al. 1997; SUR et al. 1998; SUR et al. 2002; SUR et al. 2004). Since a high dose is delivered to the esophageal mucosa, side effects may include ulcerations, fistulae, and esophageal strictures.

15.4.5
Carcinoma of the Prostate

Currently, permanent implantation of iodine-125 or palladium-103 seeds is the most common type of prostate brachytherapy. However, several centers have used HDR brachytherapy, usually as a boost to EBRT for the treatment of prostate cancer, with encouraging results (BORGHEDE et al. 1997; DINGES et al. 1998; MARTINEZ et al. 2000; MARTINEZ et al. 2001; GALALAE et al. 2004; ASTROM et al. 2005; DEMANES et al. 2005). One of the major advantages of HDR is that the dose distribution can be intra-operatively optimized by varying the dwell times at various dwell positions (EDMUNDSON et al. 1995), potentially allowing reliable and reproducible delivery of the prescribed dose to the target volume while keeping the doses to normal structures, i.e., rectum, bladder, and urethra, within acceptable limits. Another potential advantage of HDR brachytherapy in prostate cancer is the theoretical consideration that prostate cancer cells behave more like late-reacting tissue with a low alpha–beta ratio and they should, therefore, respond more favorably to higher dose fractions rather than to the lower dose rate delivered in LDR brachytherapy (DUCHESNE and PETERS 1999; FOWLER et al. 2001).

Standard fractionation EBRT (39.6–50.4 Gy) is given concurrent with or within 2 weeks before or after HDR brachytherapy. The minimum volume treated should include the entire prostate and seminal vesicles with margin, with or without pelvic lymph nodes. The HDR dose is given in multiple fractions in one or two implant procedures. A variety of dose and fractionation schemes have been used for same-stage disease as shown in Table 15.6 (BORGHEDE et al. 1997; DINGES et al. 1998; MARTINEZ et al. 2000; MARTINEZ et al. 2001; GALALAE et al. 2004; ASTROM et al. 2005; DEMANES et al. 2005;). The HDR fractions are generally given twice a day with a minimum of 6 h between fractions. The most commonly encountered acute genito-urinary morbidities include urinary irritative symptoms, hematuria, hematospermia, and/or urinary retention, similar to LDR permanent implants. HDR brachytherapy has also been used as monotherapy in a few centers, but long-term results are awaited. HDR doses of 38 Gy delivered in four fractions, two times daily for 2 days or 54 Gy in nine fractions given twice a day over 5 days have been reported (GRILLS et al.; YOSHIOKA et al. 2003).

15.4.6
Head and Neck Cancers

Brachytherapy, especially using manually afterloaded iridium-192, has been widely used to treat head and neck cancers. HDR brachytherapy has been used in selected cases to reduce radiation exposure and permit optimization as summarized in Table 15.7 (DONATH et al. 1995; INOUE et al. 1996; LAU et al. 1996; YU et al. 1996; DIXIT et al. 1997; LEUNG et al. 1998; NAG et al. 2001). However, these advantages are offset by the need for multiple fractionation, since the head and neck area does not tolerate high doses per fraction. The nasopharynx is a site within the head and neck area that is easily accessed by an intracavitary HDR applicator (GAO et al. 1992; LEVENDAG et al. 1994). Doses of 18 Gy in six fractions are delivered by a special nasopharynx applicator to boost 46–60 Gy of EBRT (LEVENDAG et al. 1994).

The use of HDR brachytherapy catheters incorporated in removable dental molds allows repeated, highly reproducible, fractionated outpatient brachytherapy of superficial (less than 0.5-cm thick) tumors without requiring repeated catheter insertion into the tumor (JOLLY and NAG 1992). Suitable sites for mold therapy include the scalp, face, pinna, lip, buccal mucosa, maxillary antrum, hard palate, oral cavity, external auditory canal, and the orbital cavity after exenteration. HDR can be used as the sole modality or in conjunction with EBRT. A total HDR dose equivalent to approximately 60 Gy LDR (prescribed at 0.5-cm depth) is recommended when used as the sole modality (NAG et al. 2000e). The HDR can also be used as a boost to 45–50 Gy EBRT, in which case, the HDR doses are appropriately reduced to LDR equivalent doses of 15–30 Gy. The actual HDR dose per fraction and number of fractions can be varied to suit individual situations (including site and treatment volume). Biomathematical (linear-quadratic) modeling can be used to assist in the conversion of LDR to HDR (NAG and GUPTA 2000).

Table 15.6. Dose fractionation and equivalent doses (as if given at 2 Gy per fraction) of common combined external beam radiation therapy (EBRT) and high-dose-rate (HDR) doses used for treating prostate cancer

EBRT dose (Gy)	Total HDR dose (Gy)	HDR dose/ fraction (Gy)	No. of HDR fractions	Equivalent dose (Gy)* (α/β=1.5)	Equivalent dose (Gy)* (α/β=5)	Equivalent dose (Gy)* (α/β=10)
39.6	22	5.5	4	81	72	67
45	18	6	3	81	72	68
39.6	26	6.5	4	97	81	75
50.4	19.5	6.5	3	92	81	76
46	19	9.5	2	106	85	77

Table 15.7. High-dose-rate (HDR) brachytherapy for head and neck cancers. *LC* local control, *EBRT* external beam radiation therapy

Author	EBRT dose (Gy)	Fraction size (Gy)	No. fractions	Equivalent dose* (Gy)	No. of patients	LC
DIXIT et al. 1997	0	3	20	65	3	–
LAU et al. 1996	0	6.5	7	63	27	53%
INOUE et al. 1996	0	6	10	80	14	100%
DONATH et al. 1995	0	4.5–5	10	54–63	13	90%
LEUNG et al. 1998	0	5.5–6	10	71–80	13	100%
YU et al. 1996	50	2.7	6	67	12	79%
DIXIT et al 1997	40–48	3	7	63–71	18	80%

*Equivalent dose for tumor effects as if given at 2 Gy/day using the linear quadratic model with an α/β ratio of 10 (NAG and GUPTA 2000)

Another innovative approach is the use of intraoperative HDR brachytherapy, which permits normal tissues to be retracted or shielded during brachytherapy. Intraoperative HDR brachytherapy can reach many sites in the head and neck area that are difficult to treat or are inaccessible by either LDR brachytherapy or intraoperative electron beam radiation. The catheters are removed immediately after the single dose of radiation, hence, minimizing inconvenience and permitting the use of brachytherapy in areas such as the base of skull (NAG et al. 2004b; NAG et al. 2005). Doses of 7.5–15.0 Gy are given when EBRT of 45–50 Gy can be added. In recurrent tumors where no further EBRT can be given, a single intraoperative dose of 15–20 Gy can be given.

15.4.7
Soft Tissue Sarcomas

Excellent results are obtained with a combination of wide excision of the tumor and adjuvant EBRT. However, irradiation of large volumes after surgery gives rise to morbidity, especially normal tissue fibrosis. A few centers have used LDR brachytherapy to minimize morbidity and improve local control (HARRISON et a. 1992; ALEKTIAR et al. 2002). The major problem with LDR brachytherapy of large volumes is the radiation exposure involved. Hence, a few centers are investigating the use of HDR brachytherapy for soft tissue sarcomas (ALEKHTEYAR et al. 1994; CROWNOVER et al. 1997; KOIZUMI et al. 1999; ALEKTIAR et al. 2000). HDR brachytherapy catheters are implanted approximately 1 cm apart along the tumor bed, and radio-opaque clips indicate the margins. A 2- to 5-cm margin proximally and distally is used after gross excision of tumor. Optimized treatment planning can be used to deliver a more homogeneous dose. Doses of 40–50 Gy are given in 12–15 fractions if the HDR is given alone. If EBRT (45–50 Gy) is added, the brachytherapy dose

is limited to 18–25 Gy in 4–7 fractions. An alternative technique not widely available is intraoperative HDR brachytherapy (HDR-IORT) (ALEKTIAR et al. 2000). A HDR-IORT dose of 12–15 Gy is given to the tumor bed in a single fraction intraoperatively to boost EBRT doses of 45–50 Gy. Nerve tolerance to high dose per fraction is poor, and HDR should be used with caution when catheters have to be placed in contact with neurovascular structures. The ABS suggests the following interventions to minimize morbidity in soft tissue sarcomas (NAG et al. 2001):

1. When brachytherapy is used as adjuvant monotherapy, the source loading should start no sooner than 5–6 days after wound closure. However, the radioactive sources may be loaded earlier (as soon as 2–3 days after surgery) if doses of less than 20 Gy are given with brachytherapy as a supplement to EBRT.

2. Minimize dose to normal tissues (e.g., gonads, breasts, thyroid, skin) whenever possible, especially in children and patients of childbearing age.

3. Limit the allowable skin dose – the 40-Gy isodose line (LDR) to less than 25 cm^2 and the 25-Gy isodose line to less than 100 cm^2.

15.4.8
Pediatric Tumors

LDR brachytherapy has been used in children to reduce the deleterious effects of EBRT (FLAMANT et al. 1990). However, LDR brachytherapy is difficult to perform in young children and infants because they require prolonged sedation and immobilization with close monitoring, which increases the risk of radiation exposure to nursing staff and parents. HDR is, therefore, very appealing for infants and younger children and is currently undergoing trials at Ohio State University (NAG et al. 1995, 2003; NAG and TIPPIN 2003). The recommended dose for HDR as monotherapy is 36 Gy in 12 fractions given at 3 Gy (prescribed at 0.5 cm) twice a day (NAG et al. 2001c; NAG and TIPPIN 2003). The interval between fractions is at least 6 h. There are no published dose recommendations for HDR as a boost to EBRT. The linear-quadratic model (NAG and GUPTA 2000) can be used to calculate a fractionation scheme equivalent to that of a LDR implant boost-dose of 15–25 Gy (prescribed at 0.5 cm). The recommended dose for intraoperative HDR brachytherapy is 10–15 Gy (prescribed at 0.5 cm), as a boost to 30–40 Gy EBRT, depending on the extent of residual disease (NAG et al. 1998; MERCHANT et al. 1998; NAG et al. 2001c,d). Although the long-term morbidity of HDR brachytherapy in young children is not fully known, one may expect preservation of organ function similar to that seen with LDR brachytherapy (FLAMANT et al. 1990). Due to the complexities involved in pediatric HDR brachytherapy, it is recommended that the use of HDR brachytherapy in pediatric tumors be limited to centers that have experience with pediatric implants (NAG et al. 2001c).

15.4.9
Breast Cancer

EBRT is the standard radiation modality used after lumpectomy in the conservative management of breast cancer. Recently, there has been interest in using brachytherapy as the sole modality of treatment (KUSKE et al. 1998; POLGAR et al. 2004; SHAH et al. 2004; STRAND et al. 2004) to decrease the 6-week treatment duration required for a course of EBRT to approximately 5 days. Table 15.8 lists the patients in whom an accelerated (4–5 days) brachytherapy treatment course can be an attractive alternative to 6 weeks of EBRT (NAG et al. 2001b). The ABS recommends a total dose of 34 Gy in ten fractions to the clinical target volume when HDR brachytherapy is used as the sole modality (NAG et al. 2001b). The HDR treatments of 3.4 Gy are generally given at two fractions per day separated by at least 6 h. This was also the dose used in a Phase-II Radiation Therapy Oncology Group trial (KUSKE et al. 1998). Depending on the selection criteria, final pathological assessment is necessary to completely evaluate a patient for partial breast brachytherapy, and, therefore, the ABS does not advocate intraoperative treatment delivery at this time (ARTHUR et al. 2002). The use of a single-channel Mammosite applicator has simplified the brachytherapy procedure (SHAH et al. 2004). Further clinical studies are required to further define the most appropriate candidates for breast brachytherapy as a sole modality treatment.

Brachytherapy has been used to boost the EBRT dose in select high-risk patients (MANNING et al. 2000; POORTMANS et al. 2004; ROMESTAING et al 1997). Because brachytherapy is an invasive procedure, it should be used selectively as a boosting technique. Situations in which brachytherapy may be advantageous as a boost are listed in Table 15.8. The brachytherapy boost can be given before or after EBRT, usually with a 1- to 2-week gap between EBRT and brachytherapy. The ABS recommends a dose fractionation scheme that yields early and late effects approximately equivalent to those of 10–20 Gy LDR following 45–50 Gy EBRT (NAG et al. 2001b). Biomathematical models are often used to estimate equivalent HDR regimens (BARENDSEN

Table 15.8. Brachytherapy in the conservative management of breast cancer. *EBRT* external beam radiation therapy, *CTV* clinical target volume

Indications for brachytherapy as the sole modality	Selection criteria for brachytherapy as the sole modality	Indications for brachytherapy as a boost to EBRT
1. The patient lives a long distance from radiation oncology treatment facilities	1. All patients should be appropriate candidates for standard breast conservation therapy	1. For patients with close, positive, or unknown margins
2. The patient lacks transportation	2. Unifocal, invasive ductal carcinoma	2. For patients with extensive intraductal component
3. The patient is a professional whose schedule will not accommodate a 6-week course of therapy	3. ≤3 cm in size	3. For younger patients
4. The patient is elderly, frail, or in poor health and therefore unable to travel for a prolonged course of daily treatment	4. Negative microscopic surgical margins of excision	4. For deep tumor location in a large breast
5. The patient's breasts are sufficiently large that they may have unacceptable toxicity with EBRT	5. Axillary node negative by level I/II axillary dissection or sentinel node evaluation	5. For CTV of irregular thickness

et al. 1982; NAG and GUPTA 2000). For example, a HDR regimen of five fractions of 310 cGy per fraction should approximate the early and late effects of 20 Gy LDR delivered at 0.5 Gy/h. Although biomathematical models can be used to estimate the appropriate dose, there is no standardized HDR fractionation schedule that can be recommended for the use of HDR as a boost. Controlled clinical studies are required to further define the most appropriate doses to be used for boost treatment.

15.4.10
Skin Cancer

The widespread availability of HDR remote afterloading brachytherapy units allows the use of surface molds as an alternative to electron beam and, for cases where surface irregularity, proximity to bone, or poor intrinsic tolerance of tissues do not allow for satisfactory treatment using electron beam. For most cases, a satisfactory mold can be made from 5-mm-thick sheets of wax, with the HDR catheters spaced 1-cm apart. A simpler alternative is to use commercially available surface template applicators (e.g., Freiburg flab from Nucletron Corporation, Columbia, MD and HAM applicator from Mick Radionuclear Instruments Inc, Bronx, NY), which are used for intraoperative HDR brachytherapy (SVOBODA et al. 1995; NAG et al. 1999; GUIX et al. 2000).

There is a wide range of recommended doses and fractionation schemes for treating skin cancer. Doses in the range of 3500 cGy in five fractions to 5000 cGy in ten fractions have been used with success in HDR molds. Standard, more prolonged fractionation schemes with 180–200 cGy daily or twice daily fractions can also be used. The linear quadratic radiobiological model can be used to determine the total dose for a given fractionation scheme (FU and PHILLIPS 1990; NAG and GUPTA 2000).

15.5
Discussion

Brachytherapy has gradually evolved over the past century. Initially, it was performed by inserting the radioactive sources directly into the tumor ("hot" loading), which exposed personnel to high doses of radiation, and hence, brachytherapy did not gain much popularity. Manual afterloaded procedures,

in which hollow catheters are initially inserted into the tumor and then loaded with radioactive materials after proper positioning, significantly reduced radiation exposure to personnel. In remote afterloading, an operator outside the room loaded the radioactive material into the catheters using remote control. In HDR brachytherapy, the elimination of the radiation exposure and the short treatment times allow for outpatient brachytherapy. Intraoperative HDR brachytherapy allows a single dose of radiation to be delivered during surgery. Doses of 10–20 Gy are usually given as a single fraction over 10–60 min. The advantages of intraoperative HDR brachytherapy over perioperative brachytherapy or electron beam IORT are listed in Table 15.9. Unfortunately, the relative scarcity of shielded operating rooms has currently limited its availability to just a few centers (MERCHANT et al. 1998; NAG et al. 1999, 2004b; NAG and HU 2003).

Although brachytherapy is a very effective modality, case selection and proper patient evaluation are essential. If the tumor is very large or widely metastatic, one is doomed to fail due to the physics of dose distribution in the former case and due to the biology of the tumor in the latter case. There are some differences between the various brachytherapy modalities (Table 15.9). These differences should be kept in mind when selecting the brachytherapy modality in a particular situation.

HDR has special relevance for developing countries, where resources may be scarce. In this regard, the International Atomic Energy Agency has recently issued recommendations for the use of HDR brachytherapy in developing countries (NAG et al. 2002). A brief summary is given here; however, readers interested in the details are referred to the original article. A HDR treatment system should be purchased as a complete unit that includes the ^{192}Ir radioactive source, source loading unit, applicators, treatment planning system, and control console. Infrastructure support may require additional or improved buildings and procurement of or access to new imaging facilities. A supportive budget is needed for quarterly source replacement and the annual maintenance necessary to keep the system operational. The radiation oncologist, medical physicist, and technologist should be specially trained before HDR can be introduced. Training for the oncologist and medical physicist is an ongoing process, as new techniques or sites of treatment are introduced. Procedures for quality assurance of patient treatment and the planning system must be introduced. Emergency procedures with adequate training of all associated

Table 15.9. Comparison of different brachytherapy techniques. *LDR* low-dose-rate brachytherapy. *MDR* medium-dose-rate brachytherapy, *PDR* pulsed dose rate, *HDR* high-dose-rate brachytherapy, *IOHDR* intraoperative high dose rate, *EBRT* external beam radiation therapy

	LDR Ir-192	LDR remote	MDR	PDR	HDR	IOHDR
Dose rate	Low	Low	Medium	High	High	High
Duration of each treatment	2–6 days	2–4 days	1 days	Minutes	Minutes	Minutes
Overall duration of treatment	2–6 days	2–4 days	1 days	2–4 days	3–5 weeks	Minutes
Radiation hazards	High	Small	Small	Small	Small	Small
Availability (worldwide)	++	–	–	–	+	–
Ease of optimization	–	–	–	+	+	+
Dose as sole modality (Gy)	60	60	40	60	30–40	15–20
Dose as boost to EBRT (Gy)	20–40	20–40	20–30	20–40	20–30	10–15

personnel must be in place. The decision to select HDR in preference to alternate methods of brachytherapy is influenced by the ability of the machine to treat a wide variety of clinical sites. In departments with personnel and budgetary resources to support this equipment appropriately, economic advantage becomes evident only if large numbers of patients are treated. With HDR, it is possible to treat a large number of patients in institutions that have a high volume of brachytherapy patients but insufficient in-patient facilities for LDR brachytherapy or insufficient finances for the purchase of iodine-125 or palladium seeds for permanent implants. Intangible benefits of source safety, personnel safety, and easy adaptation to fluctuating demand for treatments also require consideration when evaluating the need to introduce this treatment system.

When HDR brachytherapy is used, the treatments must be executed carefully, because the short treatment times do not allow any time for correction of errors, and mistakes can result in harm to patients. Hence, it is very important that all personnel involved in HDR brachytherapy be well trained and be constantly alert. However, with proper case selection and delivery technique, HDR brachytherapy has great promise and convenience, because it eliminates radiation exposure, allows short treatment times, and can be performed on an outpatient basis.

It is expected that the use of HDR brachytherapy will greatly expand over the next decade and that refinements will occur primarily in the integration of imaging (computed tomography, magnetic resonance imaging, intraoperative ultrasonography) and optimization of dose distribution (Li et al. 2003; Nag et al. 2004a). It is anticipated that better tumor localization and normal tissue definition will help to optimize dose distribution to the tumor and reduce normal tissue exposure (Nag et al. 2004a). The development of well-controlled randomized trials addressing issues of efficacy, toxicity, quality of life, and costs versus benefits will ultimately define the role of HDR brachytherapy in the therapeutic armamentarium.

References

Alekhteyar KM, Porter AT, Herskovic AM (1994) Preliminary results of hyperfractionated high dose rate brachytherapy in soft tissue sarcoma. Endocuriether Hypertherm Oncol 10:179–184

Alektiar KM, Hu K, Anderson L et al (2000) High dose rate intraoperative radiation therapy (HDR-IORT) for retroperitoneal sarcomas. Int J Radiat Oncol Biol Phys 47:157–163

Alektiar KM, Leung D, Zelefsky MJ et al (2002) Adjuvant brachytherapy for primary high-grade soft tissue sarcoma of the extremity. Ann Surg Oncol 9:48–56

Alektiar KM, Ventakatraman E, Chi DS, Barakat RR (2005) Intravaginal brachytherapy alone for intermediate-risk endometrial cancer. Int J Radiat Oncol Biol Phys 62:111–117

Arai T, Nakano T, Morita S, Sakashita K, Nakamura YK, Fukuhisa K (1992) High-dose-rate remote afterloading intracavitary radiation therapy for cancer of the uterine cervix: a 20-year experience. Cancer 69:175–180

Arthur DW, Vicini F, Kuske RR, Wazer DE, Nag S (2002) Accelerated partial breast irradiation: an updated report from the American Brachytherapy Society. Brachytherapy 1:184–190

Astrom L, Pederson D, Mercke C et al (2005) Long-term outcome of high dose rate brachytherapy in radiotherapy of localised prostate cancer. Radiother Oncol 74:157–161

Aygun C, Weiner S, Scariato A et al (1992) Treatment of non-small cell lung cancer with external beam radiotherapy and high dose rate brachytherapy. Int J Radiat Oncol Biol Phys 23:127–132

Barendsen GW (1982) Dose fractionation, dose rate, and isoeffect relationships for normal tissue responses. Int J Radiat Oncol Biol Phys 8:1981–1997

Bedwinek J, Petty A, Bruton C et al (1992) The use of high dose rate endobronchial brachytherapy to palliate symptomatic endobronchial recurrence of previously irradiated bronchogenic carcinoma. Int J Radiat Oncol Biol Phys 22:23–30

Borghede G, Hedelin H, Holmang S, Johansson KA (1997) Combined treatment with temporary short-term high dose rate iridium-192 brachytherapy and external beam radiotherapy for irradiation of localized prostatic carcinoma. Radiother Oncol 44:237–244

Chang LFL, Horvath J, Peyton W (1994) High dose rate afterloading intraluminal brachytherapy in malignant airway obstruction of lung cancer. Int J Radiat Oncol Biol Phys 28:589–596

Crownover RL, Marks KE, Zehr RJ (1997) Initial results with high dose rate brachytherapy for soft-tissue sarcomas. Sarcoma 1:196–205

Demanes JD, Rodriguez RR, Schour L et al (2005) High dose rate intensity-modulated brachytherapy with external beam radiotherapy for prostate cancer: California Endocurietherapy's 10 year results. Int J Radiat Oncol Biol Phys 61:1306–1316

Dinges S, Deger S, Koswig S, Boehmer D (1998) High-dose rate interstitial with external beam irradiation for localized prostate cancer–results of a prospective trial. Radiother Oncol 48:197–202

Dixit S, Baboo HA, Rakesh V (1997) Interstitial high dose rate brachytherapy in head and neck cancers: preliminary results. J Brachyther Int 13:363–370

Donath D, Vuong T, Shnouda G et al (1995) The potential uses of high-dose-rate brachytherapy in patients with head and neck cancer. Eur Arch Otorhinolaryngol 252:321–354

Duchesne GM, Peters LJ (1999) What is the α/β ratio for prostate cancer? Rationale for hypofractionated high-dose-rate brachytherapy. Int J Radiat Oncol Biol Phys 44:747–748

Edmundson GK, Yan D, Martinez A (1995) Intraoperative optimization of needle placement and dwell times for conformal prostate brachytherapy. Int J Radiat Oncol Biol Phys 33:1257–1264

Flamant F, Gerbaulet A, Nihoul-Fekete C et al (1990) Long-term sequelae of conservative treatment by surgery brachytherapy and chemotherapy for vulval and vaginal rhabdomyosarcoma in children. J Clin Oncol 8:1847–1853

Fowler J, Chappell R, Ritter M (2001) Is α/β for prostate tumors really low? Int J Radiat Oncol Biol Phys 50:1021–1031

Fu K, Phillips T (1990) High-dose-rate versus low-dose-rate intracavitary brachytherapy for carcinoma of the cervix. Int J Radiat Oncol Biol Phys 19:791–796

Furuta M, Tsukiyama I, Ohno T et al (1999) Radiation therapy for roentgenographically occult lung cancer by external beam irradiation and endobronchial high dose rate brachytherapy. Lung Cancer 25:183–189

Galalae RM, Martinez A, Mate T et al (2004) Long-term outcome by risk factors using conformal high dose rate brachytherapy (HDR-BT) boost with or without neoadjuvant androgen suppression for localized prostate cancer. Int J Radiat Oncol Biol Phys 58:1048–1055

Gao Li, Xu Guo-zhen, Yin Wei-bo et al (1992) Preliminary experience in HDR brachytherapy for 72 nasopharyngeal carcinoma patients. In: Mould RF (ed) Brachytherapy in the Peoples Republic of China. Kowloon, Nucletron Far East, pp E76–E81

Gaspar LE (1998) Brachytherapy in lung cancer. J Surg Oncol 67:60–70

Gaspar LE, Nag S, Herskovic A, Mantravadi P, Speiser B (1997) American Brachytherapy Society (ABS) consensus guidelines for brachytherapy of esophageal cancer. Int J Radiat Oncol Biol Phys 38:127–132

Grills IS, Martinez AA, Hollander M et al (2004) High-dose rate brachytherapy as prostate cancer monotherapy reduces toxicity compared to low dose rate palladium seeds. J Urology 171:1098–1104

Guix B, Finestres F, Tello J, Palma C (2000) Treatment of skin carcinomas of the face by high-dose-rate brachytherapy and custom-made surface molds. Int J Radiat Oncol Biol Phys 47:95–102

Hareyama M, Sakata K, Oouchi A et al (2002) High-dose-rate versus low-dose-rate intracavitary therapy for carcinoma of the uterine cervix. A randomized trial. Cancer 94:117–124

Harrison LB, Franzese F, Gaynor JJ, Brennan MF (1992) Long-term results of a prospective randomized trial of adjuvant brachytherapy in the management of completely resected soft tissue sarcomas of the extremity and superficial trunk. Int J Radiat Oncol Biol Phys 27:259–265

Hishikawa Y, Kamikonya N, Tanaka S, Miura T (1987) Radiotherapy of esophageal carcinoma: role of high dose rate intracavitary irradiation. Radiother Oncol 9:13–20

Huber RM, Fischer R, Haútmann H et al (1997) Does additional brachytherapy improve the effect of external irradiation? A prospective, randomized study in central lung tumors. Int J Radiat Oncol Biol Phys 38:533–540

Inoue T, Inoue T, Teshima T, Murayama S (1996) Phase III trial of high and low dose rate interstitial radiotherapy for early oral tongue cancer. Int J Radiat Oncol Biol Phys 36:1201–1204

Jolly DE, Nag S (1992) Technique for construction of dental molds for high-dose-rate remote brachytherapy. Spec Care Dentist 12:219–224

Khanavkar B, Stern P, Alberti W, Nakhosteen JA (1991) Complications associated with brachytherapy alone or with laser in lung cancer. Chest 99:1062–1065

Koizumi M, Inoue T, Yamazaki H et al (1999) Perioperative fractionated high-dose rate brachytherapy for malignant bone and soft tissue tumors. Int J Radiat Oncol Biol Phys 43:989–993

Kuske RR, Bolton JS, Harrison W (1998) RTOG 95-17. A phase I/II trial to evaluate brachytherapy as the sole method of radiation therapy for stage I and II breast carcinoma. Radiation Therapy Oncology Group. Philadelphia, PA

Lau HY, Hay JH, Flores AD et al (1996) Seven fractions of twice daily high dose-rate brachytherapy for node-negative carcinoma of the mobile tongue results in loss of therapeutic ratio. Radiother Oncol 39:15–18

Leung TW, Wong VYW, Wong CM et al (1998) Technical hints for high dose rate interstitial tongue brachytherapy. Clin Oncol 10:231–236

Levendag PC, Vikram B, Flores AD et al (1994) High dose rate brachytherapy for cancer of the head and neck. In: Nag S (ed). High dose rate brachytherapy: a textbook. Futura Publishing Co., Armonk, NY, pp 237–273

Li S, Frassica D, DeWeese T (2003) A real-time image-guided intraoperative high-dose-rate brachytherapy system. Brachytherapy 2:5–16

Lorvidhaya V, Tonusin A, Changwiwit W, Chitapanarux I (2000) High-dose-rate afterloading brachytherapy in carcinoma of the cervix: and experience of 1992 patients. Int J Radiat Oncol Biol Phys 46:1185–1191

Lybeert MLM, van Putten WLJ, Ribot JG, Crommelin MA (1989) Endometrial carcinoma: high dose rate brachytherapy in combination with external irradiation – a multivariate analysis of relapses. Radiother Oncol 16:245–252

Macha HN, Wahlers B, Reichle C, von Zwehl (1995) Endobronchial radiation therapy for obstructing malignancies: ten years' experience with iridium-192 high-dose radiation brachytherapy afterloading technique in 365 patients. Lung 173:271–280

Mandell LM, Nori D, Anderson LL, Hilaris BS (1985) Post-operative vaginal radiation in endometrial cancer using a remote afterloading technique. Int J Radiat Oncol Biol Phys 11:473-478

Manning MA, Arthur DW, Schmidt-Ullrich RK et al (2000) Interstitial high dose rate brachytherapy boost: The feasibility and cosmetic outcome of a fractionated outpatient delivery scheme. Int J Radiat Oncol Biol Phys 48:1301–1306

Marsiglia H, Baldeyrou P, Lartigau E et al (2000) High-dose-rate brachytherapy as a sole modality for early-stage endobronchial carcinoma. Int J Radiat Oncol Biol Phys 47:665–672

Martinez A, Kestin L, Stromberg J (2000) Interim report of image guided conformal high dose rate brachytherapy for patients with unfavorable prostate cancer: the William Beaumont phase II dose escalating trial. Int J Radiat Oncol Biol Phys 47:343–352

Martinez A, Pataki I, Edmundson G et al (2001) Phase II prospective study of the use of conformal high-dose rate brachytherapy as monotherapy for the treatment of favorable stage prostate cancer: a feasibility report. Int J Radiat Oncol Biol Phys 49:61–69

Mehta MP, Petereit DG, Chosy L et al (1992) Sequential comparison of low dose rate and hyperfractionated high dose rate endobronchial radiation for malignant airway occlusion. Int J Radiat Oncol Biol Phys 23:133–139

Mehta MP, Lamond JP, Nori D, Speiser BL (1997) Brachytherapy of lung cancer. In: Nag S (ed). Principles and practice of brachytherapy. Futura Publishing Co., Armonk, NY, pp 323–349

Merchant TE, Zelefsky MJ, Sheldon JM et al (1998) High-dose rate intraoperative radiation therapy for pediatric solid tumors. Med Ped Oncol 30:34–39

Nag S (1996) Modern techniques of radiation therapy for endometrial cancer. Clin Obstet Gynecol 39:728–744

Nag S, Samsami N (2000) Pitfalls of inappropriate optimization. J Brachyther Int 16:187–198

Nag S, Gupta N (2000) A simple method of obtaining equivalent doses for use in HDR brachytherapy. Int J Radiat Oncol Biol Phys 46:507–513

Nag S, Hu KS (2003) Intraoperative high-dose-rate brachytherapy. Surg Oncol Clin N Am 12:1079–1097

Nag S, Tippin D (2003) Brachytherapy for pediatric tumors. Brachytherapy 2:131–138

Nag S, Olson T, Ruymann F et al (1995) High dose rate brachytherapy in childhood sarcomas: a local control strategy preserving bone growth and function. Med Ped Oncol 25:463–469

Nag S, Martínez-Monge R, Ruymann FB, Bauer C (1998) Feasibility of intraoperative high-dose rate brachytherapy to boost low dose external beam radiation therapy to treat pediatric soft tissue sarcomas. Med Ped Oncol 31:79–85

Nag S, Gunderson L, Harrison L (1999) Techniques of intraoperative radiation therapy vs. intraoperative high dose rate brachytherapy. In: Intraoperative irradiation: techniques and results. Gunderson LL, Willet CG, Harrison LV, Calvo FA (eds) Humana Press, Totowa, NJ, pp 111–130

Nag S, Erickson B, Parikh S, Gupta N, Varia M, Glasgow G (2000a) The American Brachytherapy Society recommendations for HDR brachytherapy for carcinoma of the endometrium. Int J Radiat Oncol Biol Phys 48:779–790

Nag S, Erickson B, Thomadsen B, Orton C, Demanes JD, Petereit D (2000b) The American Brachytherapy Society recommendations for HDR brachytherapy of the cervix. Int J Radiat Oncol Biol Phys 48:201–211

Nag S, Kelly JF, Horton JL, Komaki R, Nori D (2001a) The American Brachytherapy Society recommendations for HDR brachytherapy for carcinoma of the lung. Oncology 15:371–381

Nag S, Kuske RR, Vicini F, Arthur DW, Zwicker RD (2001b) The American Brachytherapy Society recommendations for brachytherapy for carcinoma of the breast. Oncology 15:195–207

Nag S, Shasha D, Janjan N, Petersen I, Zaider M (2001c) The American Brachytherapy Society recommendations for brachytherapy of soft tissue sarcomas. Int J Radiat Oncol Biol Phys 49:1033–1043

Nag S, Tippin D, Ruymann FB (2001d) Intraoperative high-dose-rate brachytherapy for the treatment of pediatric soft tissue sarcomas. Int J Radiat Oncol Biol Phys 51:729–735

Nag S, Vikram B, Demanes JD, Cano E, Puthawala AA (2001e) The American Brachytherapy Society recommendations for HDR brachytherapy for head and neck carcinoma. Int J Radiat Oncol Biol Phys 50:1190–1198

Nag S, Dally M, De la Torre M et al (2002) Recommendations for implementation of high dose rate ^{192}Ir brachytherapy in developing countries by the advisory group of international atomic energy agency. Radiother Oncol 64:297–308

Nag S, Tippin D, Ruymann FB (2003) Long term morbidity in children treated with fractionated high dose rate brachytherapy for soft tissue sarcomas. J Pediatr Hematol Oncol 25:448–452

Nag S, Cardenes H, Chang S et al (2004a) Proposed guidelines for image-based intracavitary brachytherapy for cervical carcinoma: a report from the image-guided brachytherapy working group. Int J Radiat Oncol Biol Phys 60:1160–1172

Nag S, Tippin D, Grecula J, Schuller D (2004b) Intraoperative high dose rate brachytherapy for paranasal sinus tumors. Int J Radiat Oncol Biol Phys 58:155–160

Nag S, Koc M, Schuller D, Tippin D, Grecula JC (2005) Intraoperative single fraction high dose rate brachytherapy for head and neck cancers. Brachytherapy 4:217–223

Nakano T, Kato S, Ohno T et al (2005) Long-term results of high-dose-rate intracavitary brachytherapy for squamous cell carcinoma of the uterine cervix. Cancer 103:92–101

Nori D, Merimsky O, Batata M et al (1994) Postoperative high-dose-rate intravaginal brachytherapy combined with external irradiation for early stage endometrial cancer: A long term follow-up. Int J Radiat Oncol Biol Phys 30:831–837

Orton CG, Seyedsadr M, Somnay A (1991) Comparison of high and low dose rate remote afterloading for cervix cancer and the importance of fractionation. Int J Radiat Oncol Biol Phys 21:1425–1434

Patel FD, Sharma SC, Negi PS et al (1993) Low dose rate versus high dose rate brachytherapy in the treatment of carcinoma of the uterine cervix: A clinical trial. Int J Radiat Oncol Biol Phys 28:335–341

Patel FD, Rai B, Mallick I, Sharma SC (2005) High dose rate brachytherapy in uterine cervical carcinoma. Int J Radiat Oncol Biol Phys 62:125–130

Perol M, Caliandro R, Pommier P, Malet C (1997) Curative irradiation of limited endobronchial carcinomas with high-dose-rate brachytherapy. Results of a pilot study. Chest 111:1417–1423

Petereit DG, Sarkaria JN, Potter DM, Schink JC (1999) High-dose-rate versus low-dose-rate brachytherapy in the treatment of cervical cancer: analysis of tumor recurrence – the University of Wisconsin experience. Int J Radiat Oncol Biol Phys 45:1267–1274

Polgar C, Major T, Somogyi A, Fodor J et al (2004) High-dose-rate brachytherapy alone versus whole breast radiotherapy with or without tumor bed boost after breast-conserving surgery: seven-year results of a comparative study. Int J Radiat Oncol Biol Phys 60:1173–1181

Poortmans P, Bartelink H, Horiot JC et al (2004) The influence of the boost technique on local control in breast conserving treatment in the EORTC 'boost versus no boost' randomised trial. Radiother Oncol 72:25–33

Rodriguez R, Nag S, Mate T, Martinez, A, Shasha D, Orton C, Tripuraneni P, Linares L, Kelly D, Gustafson G, Devlin P, Alteiri G, Gribble M, Rao J, Perez C, Syed N (2001) High Dose rate brachytherapy for prostate cancer: assessment of current clinical practice and recommendations of the American Brachytherapy Society. J Brachyther Int 17:265–282

Romestaing P, Lehingue Y, Carrie C, Coquard R, Montbarbon X, Ardiet JM, Mamelle N, Gerard JP (1997) Role of a 10 Gy boost in the conservative treatment of early breast cancer: results of a randomized clinical trial in Lyon, France. J Clin Oncol 15:963–968

Rouanet P, Dubois JB, Gely S et al (1993) Exclusive radiation therapy in endometrial carcinoma. Int J Radiat Oncol Biol Phys 26:223–228

Saito M, Yokoyama A, Kurita Y, Uematsu T, Tsukada H, Yamanoi T (2000) Treatment of roentgenographically occult endobronchial with external beam radiotherapy and intraluminal low dose rate brachytherapy: second report. Int J Radiat Oncol Biol Phys 47:673–680

Shah NM, Tenenholz T, Arthur D et al (2004) MammoSite and interstitial brachytherapy for accelerated partial breast irradiation: factors that affect toxicity and cosmesis. Cancer 101:727–734

Shigematsu Y, Nishiyama K, Masaki N, Inoue T (1983) Treatment of carcinoma of the uterine cervix by remotely controlled afterloading intracavitary radiotherapy with high-does rate: a comparative study with a low-dose rate system. Int J Radiat Oncol Biol Phys 9:351–356

Sood BM, Gorla G, Gupta S et al (2002) Two fractions of high dose rate brachytherapy in the management of cervix cancer: clinical experience with and without chemotherapy. Int J Radiat Oncol Biol Phys 53:702–706

Sorbe B, Frankendal B (1989) Intracavitary irradiation of endometrial carcinoma stage I by a high dose rate afterloading technique. Gynecol Oncol 33:135–145

Sorbe BG, Smeds AC (1990) Postoperative vaginal irradiation with high dose rate afterloading technique in endometrial carcinoma stage I. Int J Radiat Oncol Biol Phys 18:305–314

Sorbe B, Kjelligren O, Stenson S (1990) Prognosis of endometrial carcinoma stage I in two Swedish regions. A study with special regard to the effects of intracavitary irradiation with high dose rate afterloading technique. Acta Oncol 29:29–37

Speiser BL (1999) Brachytherapy in the treatment of thoracic tumors. Lung and esophageal. Hematol Oncol Clin North Am 13:609–634

Speiser B, Spratling L (1993a) Fatal hemoptysis: complication or failure of treatment. Int J Radiat Oncol Biol Phys 25:925

Speiser BL, Spartling L (1993b) Radiation bronchitis and stenosis secondary to high dose rate endobronchial irradiation. Int J Radiat Oncol Biol Phys 25:589–597

Strnad V, Ott O, Potter R et al (2004) Interstitial brachytherapy alone after breast conserving surgery: interim results of a German–Austrian multicenter phase II trial. Brachytherapy 3:115–119

Sur R, Donde B, Falkson C et al (2004) Randomized prospective study comparing high dose rate intraluminal brachytherapy (HDRILBT) alone to HDRILBT and external beam radiotherapy in the palliation of advanced esophageal cancer. Brachytherapy 3:191–195

Sur RK, Levin CV, Donde B et al (2002) Prospective randomized trial of HDR brachytherapy as a sole modality in palliation of advanced esophageal carcinoma – an international atomic energy agency study. Int J Radiat Oncol Biol Phys 53:127–133

Svoboda VH, Kovarik J, Morris F (1995) High dose-rate microselectron molds in the treatment of skin tumors. Int J Radiat Oncol Biol Phys 31:967–972

Yoshioka Y, Nose T, Yoshida K et al (2003) High-dose rate brachytherapy as monotherapy for localized prostate cancer: a retrospective analysis with special focus on tolerance and chronic toxicity. Int J Radiat Oncol Biol Phys 56:213–220

Yu L, Vikram B, Chadha M (1996) High dose rate interstitial brachytherapy in patients with cancers of the head and neck. Endocuriether Hyperther Oncol 12:1–6

16 Quality Assurance in Radiation Oncology

James A. Purdy, Eric Klein, Srinivasan Vijayakumar, Carlos A. Perez, and Seymour H. Levitt

CONTENTS

16.1 Introduction 395
16.2 Goals and Structure of a QA Program 396
16.2.1 Physics Staffing 396
16.2.2 Training 397
16.2.3 Structure of a QA Program 398
16.3 Dosimetry Instrumentation QA 399
16.4 Medical Linear Accelerator QA 400
16.4.1 Linac Annual QA Tests 405
16.5 Linac Advanced-Technology QA 405
16.5.1 Linac Computer Control System 406
16.5.2 Asymmetric Jaws (Independent Collimation) 407
16.5.3 Dynamic Wedge 408
16.5.4 Multileaf Collimation 408
16.5.5 Online Electronic Portal Imaging 409
16.5.6 Intensity-Modulation Radiation Therapy 411
16.6 Quality Assurance of Cobalt Teletherapy Units 412
16.7 Treatment-Machine Maintenance 413
16.8 Treatment-Planning Computer System QA 414
16.9 Treatment-Planning QA 415
16.9.1 Patient Immobilization and Data Acquisition 415
16.9.2 Critical Structure, Tumor, Target Volume
 Delineation 416
16.9.3 Designing Beams and Field Shaping 416
16.9.4 Dose Calculation 416
16.9.5 Computation of Monitor Units (Time) 416
16.9.6 Plan Evaluation 417
16.9.7 Treatment-Plan Review 417
16.9.8 Beam Modifiers 417
16.9.9 Plan Implementation and Verification 417
16.10 Clinical Aspects of QA 418
16.10.1 Planning Conference 418
16.10.2 Chart Checking 418
16.10.3 Port Film/Image Verification Review 419
16.11 Conclusion 420
 References 421

J.A. Purdy, PhD, S. Vijayakumar, MD
Department of Radiation Oncology, University of California
Davis Medical Center, Sacramento, CA 95817, USA
E. Klein, MS, C.A. Perez, MD
Department of Radiation Oncology, Mallinckrodt Institute
of Radiology, Washington University School of Medicine, St.
Louis, Missouri 63110, USA
S.H. Levitt, MD
Department of Therapeutic Radiation Oncology, University of
Minnesota, Minneapolis, MN 55455, USA

16.1 Introduction

Dose precision in radiation therapy is expected to be on the order of ±5%, based on the fact that certain tumors and normal tissues exhibit steep dose response curves (Herring and Compton 1971). Delivery of radiation with this criterion places great demands on the entire process, although such a level is believed to be achievable (ICRU 1976).

Uncertainties in treatment are due to many factors including: (a) dose calibration at a point in phantom; (b) patient-specific data used for treatment planning; (c) dose calculation in the patient; (d) transfer of the treatment plan to the radiation therapy machine; and (e) day-to-day variations in patient positioning and internal motion of tumor volume and organs at risk. These uncertainties may be categorized as systematic and random. *Random uncertainties* vary in magnitude and sign and cannot be totally controlled (e.g., the position of the radiation field on the patient may vary from day to day by a few millimeters). Moreover, the degree with which treatments can be reproduced differs among clinical sites and between institutions. *Systematic uncertainties* maintain their magnitude and direction over a period of time. For example, the use of an incorrect factor in the calibration of a treatment unit would have the same effect on the dose delivered to all patients. Systematic errors, in principle, should be controllable: for example, the degree of misregistration of field defining apertures can be reduced with periodic review of beam localization films; however, many systematic errors remain, e.g., the approximations used in dose calculation algorithms.

We first need to establish some definitions. *Quality assurance* (QA) is defined as the set of policies and procedures instituted to ensure the proper and safe delivery of the prescription dose to the patient. *Quality control* constitutes the actual tests taken to maintain and improve the quality of the treatment. We must also understand that a QA program is an interdisciplinary effort involving radiation oncologists, radiation physicists, dosimetrists, and radiation therapists. Although medical physicists and radiation therapists are more involved in the technical aspects of QA, and radiation oncologists in the medical aspects, the efforts of each group substantially overlap.

Material presented in this chapter relies heavily on recommendations given in various AAPM Task Groups, including TG 35 Report on Medical Accelerator Safety Considerations (PURDY et al. 1993), TG 40 Report on Comprehensive QA for Radiation Oncology (KUTCHER et al. 1994), World Health Organization (WHO 1988), American College of Radiology (ACR 2004), American College of Medical Physics (ACMP) (9), government regulations (NRC 2003), and the authors' previous QA reports (PARRINO and PURDY 1983; PURDY 1983, 1991a,b; PURDY et al. 1986, 1995; VAN DYK and PURDY 1999).

16.2
Goals and Structure of a QA Program

A series of publications known as the "Blue Book" have provided a strong rationale for the development, purpose, and need for QA in radiation oncology. Five such reports were published over the period 1968 to 1991 including: *A Prospect for Radiation Therapy in the United States* (1968); *A Proposal for Integrated Cancer Management in the United States: The Role of Radiation Oncology* (1972); *Criteria for Radiation Oncology in Multidisciplinary Cancer Management* (1981); *Radiation Oncology in Integrated Cancer Management* (1986); and *Radiation Oncology in Integrated Cancer Management* (1991). The 1991 version published by The Inter-Society Council for Radiation Oncology (ISCRO) provides the following statement regarding the purpose of a QA program (ISCRO 1991):

"The purpose of a Quality Assurance Program is the objective, systematic monitoring of the quality and appropriateness of patient care. Such a program is essential for all activities in Radiation Oncology. The Quality Assurance Program should be related to structure, process and outcome, all of which can be measured. Structure includes the staff, equipment and facility. Process covers the pre- and post-treatment evaluations and the actual treatment application. Outcome is documented by the frequency of accomplishing stated objectives, usually tumor control, and by the frequency and seriousness of treatment-induced sequelae."

The report emphasizes that the complexity of radiation therapy requires a teamwork approach among radiation oncologists, medical physicists, dosimetrists, nurses, and therapists as no one individual has all the skills necessary. This series of publications has not been updated in over a decade, and now more than ever, there is a definite need to do so. Most important is for administrators to understand the need for a robust radiation oncology QA program, and to work with the radiation oncology team and ensure that adequate funding is available to support such a program. For a QA program to be effective, all of the faculty and staff involved with providing radiation therapy to patients must be committed to the QA program.

16.2.1
Physics Staffing

Appropriate physics staffing is an essential component of the radiation oncology QA program. In the past, staffing guidelines were promulgated via the Blue Book and were based on patient load and treatment equipment. The 1991 Blue Book recommended at least one clinical physicist per center for up to 400 patients treated annually (Table 16.1a; ISCRO 1991). Additional clinical physicists are recommended in the ratio of one per 400 patients treated annually. This report makes clear that these staffing levels are for clinical duties only, and additional full-time equivalent (FTEs) medical physicists will be required for translational research, teaching, and administration duties; however, the present physics staffing levels must also take into account the complexity of treatments being performed in the clinic such as IMRT, brachytherapy, and stereotactic radiosurgery, as such procedures are physics intensive.

The most detailed information currently available regarding medical physics work effort is in the reports by ABT ASSOCIATES (1995, 2003). These reports were the result of the American College of Medical Physics (ACMP) and the American Association of Physicists in Medicine (AAPM) engaging Abt Associates to conduct a study that measured what was termed Qualified Medical Physicist (QMP) work for medical physics services, and to develop a relative work value scale depicting the relative amount of QMP work required for each medical physics service. The results of that survey were published in 1995 (ABT ASSOCIATES 1995). This report was updated in 2003 due to the recognition that the many changes in medical physics practice and technology

Table 16.1a. Minimum personnel requirements for clinical radiation therapy. (From 1991 Blue Book: Radiation Oncology in Integrated Cancer Management: Report of the Inter-Society Council for Radiation Oncology)

Category	Staffing
Radiation oncologist-in-chief	One per program
Staff radiation oncologist	One additional for each 200–250 patients treated annually. No more than 25–30 patients under treatment by a single physician
Radiation physicist	One per center for up to 400 patients annually; additional in ratio of 1 per 400 patients treated annually
Treatment planning staff	
Dosimetrist or physics assistant	One per 300 patients treated annually
Physics technologist (mold room)	One per 600 patients treated annually
Radiation therapy technologist supervisor	One per center
Staff (treatment)	Two per megavoltage unit up to 25 patients treated daily per unit, 4 per megavoltage unit up to 50 patients treated daily per unit
Staff (simulation)	Two for every 500 patients simulated annually
Staff (brachytherapy)	As needed
Treatment aid	As needed, usually one per 300-400 patients treated annually
Nurse[a]	One per center for up to 300 patients treated annually and an additional one per 300 patients treated annually
Social worker	As needed to provide service
Dietitian	As needed to provide service
Physical therapist	As needed to provide service
Maintenance engineer/electronics technician	One per 2-mV units or 1-mV unit and a simulator if equipment serviced "in house"

Additional personnel will be required for research, education, and administration. For example, if 800 patients are treated annually with three accelerators, one 60Co teletherapy unit, a superficial X-ray machine, one treatment-planning computer, the clinical allotment for physicists would be two to three. A training program with eight residents, two technology students, and a graduate student would require another 1–1.5 FTEs. Administration of this group would require 0.5 FTE. If the faculty had 20% time for research, a total of five to six physicists would be required

[a] For direct patient care. Other activities supported by LVNs and nurse aides

that had occurred since the original report may have affected QMP work-related values (ABT ASSOCIATES 2003). HERMAN et al. used the data from the first Abt report along with a manpower study conducted by the American College of Medical Physics (ACMP) and AAPM, accounting for IMRT and other special procedures, to show that current reimbursement models do not adequately support the needed physics QA effort, particularly for those clinics involved with IMRT (HERMAN et al. 2003). HERMAN et al. at the 2005 AAPM Annual Meeting presented an algorithm making use of the Abt-II study and the previously referenced work survey data to determine medical physics FTE recommendations depending on number of procedures and types of procedures. An example of this type of FTE estimation is given in Table 16.1b (HERMAN et al. 2005).

Blue Book recommendation on dosimetrist staffing is 1 per 300 patients treated annually (ISCRO 1991). This number appears to still be valid; however, it should be noted that in some institutions, dosimetrists not only provide treatment-planning services, but also participate in QA tests, and in some cases, perform the simulations. In those cases, staffing levels must be increased accordingly.

16.2.2
Training

The education and training of the radiation oncology team and their continuing education are of critical importance to a QA program. In the past, clinical physics training and dosimetrists' training

Table 16.1b. Example of full-time equivalent (FTE) needs for a modern radiation therapy clinic having three treatment machines (including advanced technology procedures such as IMRT), and including physician residency training and one medical physics resident. A total of 9.4 FTE medical physicists were required. (From HERMAN et al. 2005)

Procedure	Quantity
Total new patients	800
IMRT	250
HDR	50
Radiosurgery	60
Prostate implants	40
TBI	30

Task	FTE
Patient procedures	6.67
Commissioning and QA	1.04
Education	0.11
Research	0.78
Administration	0.78
Total FTE	9.38

have been the weakest link, as the training programs lacked organized clinical training beyond individual apprenticeships or self-training on the job. This was probably adequate in the early days of physics involvement in radiation oncology; however, as radiation oncology has become increasingly more sophisticated and complex, this strategy is no longer acceptable. The practice of hiring inadequately trained medical physicists, who are allowed to perform patient-related tasks, must be discontinued. The lack of proper clinical training of medical physicists reached a serious level in the late 1980s. There was, and continues to be, an acute shortage of qualified clinical physicists, e.g., physicists with adequate clinical training and board certification. There was, and continues to be, a growing abundance of physics graduates (including medical physics graduates) with inadequate clinical training applying for hospital positions. The AAPM recognized this problem, and in 1988–1989, developed a comprehensive document entitled AAPM report no. 36, "Essentials and Guidelines for Hospital-Based Medical Physics Residency Training Programs," which sets down the educational and administrative requirements for a hospital-based residency training program (AAPM 1990). The AAPM report recommends 2 years of clinical physics training beyond an M.S. or Ph.D. degree in physics or a closely related field. The organization of the recommended program was patterned after

physician residency programs. In the words of the committee that developed the recommendations, "This document will hopefully encourage the development of a high-quality clinical medical physics instructional environment on a nationwide basis and make an important contribution to the protection of the public health, safety, and welfare."

In October 1992, the Barnes-Jewish Hospital/Washington University Radiation Oncology Center established the first Radiation Oncology Physics Residency Program accredited by the Commission on Accreditation of Medical Physics Education Program (CAMPEP) in the United States. More physics residency programs are now being instituted, although at a much slower pace than needed.

Certification boards for physicists exist, but the entry requirements for the examination still do not mandate residency-type clinical training as is required of the physicians. As a result, there is an unchecked influx of inadequately trained physicists into the field.

16.2.3
Structure of a QA Program

At the heart of any modern QA program is a QA/continuing quality improvement (QA/CQI) committee. The need for such a committee is contained in reports by several United States radiation oncology organizations including the ACR (ACR 2004) and the AAPM (KUTCHER et al. 1994), and international organizations such as the World Health Organization (WHO 1988). To properly function, the QA/CQI committee should be created by the radiation oncology departmental chairman and should report directly to the chairman and the hospital administration. Its function is to design, implement, and maintain a multidisciplinary QA/CQI program whose goal is to improve the quality of patient care. The QA/CQI committee should meet regularly, preferably on a monthly (at least quarterly) basis to review the ongoing QA/CQI program, and these deliberations should be reported to the department chairman in writing. It is important that the committee have the full support of the chairman of radiation oncology and all of the faculty and staff; otherwise, maintaining quality and implementing improvements in care will prove extremely difficult.

The structure of a typical QA/CQI committee is shown schematically in Figure 16.1. The committee's work encompasses numerous areas, and needs the participation of many individuals, including the

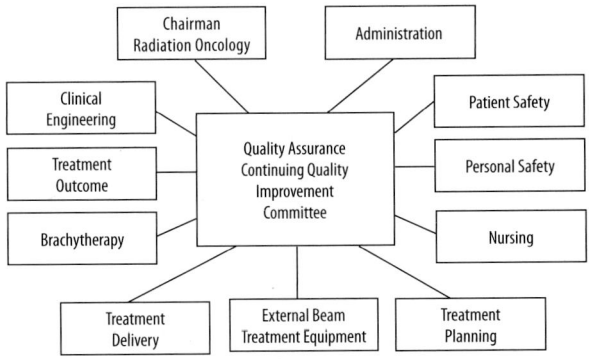

Fig. 16.1 Structure of a quality assurance (QA) committee illustrating typical areas addressed

committee chairman (preferably a radiation oncologist or medical physicist), radiation oncologists, radiation therapists, dosimetrists, nurses, and physicists involved in simulation and treatment machine quality assurance, treatment planning, brachytherapy, nursing, and radiation safety. The number of committee members varies, and in smaller institutions, the committee might be comprised of only three individuals: a radiation oncologist; a medical physicist; and a radiation therapist.

Depending on the institution, the responsibilities of the QA/CQI committee members will differ. Moreover, there are many gray zones in responsibilities in the radiation oncology practice. In fact, one of the roles of the committee is to define the lines of demarcation between tasks so that some important functions are not overlooked. Typical QA/CQI functions and lines of demarcation are listed in Table 16.2.

It is also suggested by the ACR that there should be an annual review of outcome (ACR 2004). Such a review is highly complex, since it includes an understanding and acceptance of standards of care in the sites under review.

16.3
Dosimetry Instrumentation QA

A list of the type of equipment typically considered most useful in a QA program for treatment machines is given in Table 16.3. Special QA devices for treatment machines are now commercially available for checking beam alignment, field symmetry, and the output of the machine. Generally, the radiation QA devices consist of an array of ionization chambers or diodes positioned in a plastic phantom (see

Fig. 16.2). Also, a plastic constancy phantom that can be attached quickly to the treatment machine, allowing measurements to be made at a fixed and reproducible distance, is a useful QA device as it allows measurements to be performed on a daily basis with a minimum of setup time. (see Fig. 16.3).

Accurate data acquisition with automated beam data scanning systems and scanning film densitometers requires that the systems be subjected to a systematic performance test prior to use and also undergo periodic QA tests thereafter. Details of acceptance testing and QA of such devices have been reported in the literature (MELLENBERG et al. 1990).

Film is satisfactory for assessing beam symmetry and flatness, but one must be cautious if it is used to measure dose (WILLIAMSON et al. 1981). For exam-

Table 16.2 Typical QA responsibilities and functions

Review by a medical physicist of ongoing QA of external beam equipment including treatment units, calibration equipment, and imaging modalities (CT simulators, conventional simulators, port films or electronic portal imaging devices, cone-beam CT, megavoltage CT, ultrasound, etc.)
Review by a medical physicist of ongoing QA in brachytherapy, including instrumentation, handling of sources, treatment planning, and remote afterloaders/applicators operation
Review by a medical physicist and a medical dosimetrist of QA in treatment planning which includes the treatment planning system and peripherals, graphical planning, in vivo dosimetry, and plan review
Review by a radiation therapist of procedures for dose delivery to the patient, which includes verification of patient treatment setup parameters, chart checking, portal films/images, and patient safety
Review by a physician of patient QA procedures which includes weekly chart review, port film/images review, and mortality and morbidity assessment (typically reviewed in a separate committee)
Review of any cases in which there have been deviations outside the action levels set by the QA/CQI committee, department, hospital, and regulatory bodies
Regularly scheduled audits of charts, films, and QA procedures. The results should be presented to the QA/CQI committee
Recommend actions to be taken as a result of the problems encountered in the reports and audits. Approve or suggest modifications to corrective actions
Assurance that the recommended actions have been taken
Report the QA/CQI committee's results, actions, and recommendations to the chairman of radiation oncology and to the hospital QA committee
Supervision of staff continuing education programs

Table 16.3 Typical QA instrumentation/equipment needed for a modern radiation oncology clinic

Secondary standard dosimetry system
Field use dosimetry system
Parallel plate ionization chamber
Standard and field use barometers and thermometers
Plastic phantom for output calibration constancy checks
Small water phantom with movable fixed ion chamber holder
Array detector device to monitor beam symmetry
Polystyrene, solid water, substitute tissue heterogeneity stack phantoms
Specialized phantoms and instrumentation for treatment-planning systems
Specialized phantoms and instrumentation for simulator, CT simulator, and linac onboard imaging systems
Anthropomorphic phantom
Film densitometer system
Beam data scanning system
In vivo dosimetry system (Diodes, MOSFETs, and/or TLD)

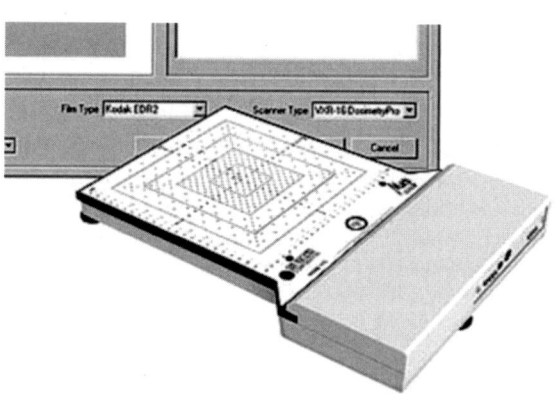

Fig. 16.2 Example of an array detector (MapCHECK) used for radiation therapy QA measurements. (Courtesy Sun Nuclear Corporation)

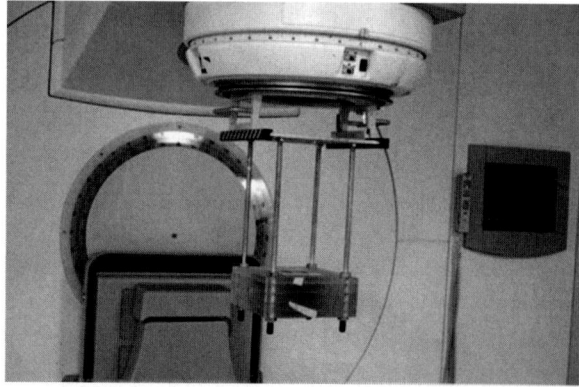

Fig. 16.3 Plastic photon output calibration constancy phantom. (Courtesy Valiant Instruments)

ple, for some types of radiographic film the dose is not a linear function of optical density so that it is necessary to calibrate the film and then make corrections to the densitometer readings. It is also important to make sure that processing conditions are maintained so that the measurements are reproducible. Another problem with the film technique is that the results of measurements are not immediately known, i.e., after exposing the film, it must be developed and then read on a densitometer.

Accurate data acquisition with automated beam data scanning systems and scanning film densitometers requires that the systems be subjected to a systematic performance test prior to use. Details of acceptance testing of such devices have been reported in the literature (HOLMES and MCCULLOUGH 1983; MCCULLOUGH and HOLMES 1985). These devices should also undergo periodic QA tests thereafter.

16.4
Medical Linear Accelerator QA

The decision to purchase and implement a megavoltage radiation therapy linear accelerator for clinical use carries with it a commitment to provide adequate staff, test equipment and instrumentation, and allow the necessary machine time in order to assure that the unit is performing according to specifications (NATH et al. 1994).

An effective QA program for treatment machines establishes criteria for optimum machine performance, monitors adherence to established criteria, ensures the accuracy of the dose delivered, minimizes treatment machine downtime, and enhances communication between radiation oncologists, physicists, dosimetrists, therapists, and maintenance technicians.

The prescribed dose planned and the dose delivered by a linear accelerator is dependent on several parameters including the dose calibration, percentage depth dose, and other dose ratios used in calculating the dose distribution and machine monitor unit settings, off-axis beam characteristics, wedge and block factors, multileaf collimator (MLC), calibration, etc. All parameters must be carefully determined when the therapy machine is installed during machine commissioning. Quality assurance procedures must then be implemented to ensure the accuracy and reproducibility of the dose delivered by the linear accelerator.

A QA program for linear accelerators requires certain key ingredients if it is to be successful; these include: (a) a commitment by the staff to QA; (b) participation of radiation, electronic, and mechanical specialists; (c) active radiation therapists participation; (d) regularly scheduled QA and preventive maintenance inspections of the linac; (e) agreed on QA machine performance tests and acceptance criteria; (f) adequate test instrumentation; (g) accurate and complete documentation of the linac, and (h) commitment to good maintenance and radiation calibration record keeping (bound archival records or computer database). It should be recognized that a QA program for treatment machines is very much a team effort and will at some stage call on the expertise of the physicist, dosimetrist, therapist, maintenance technician, and radiation oncologist.

A typical linac QA program is designed to provide testing at different levels and frequencies; these typically include: (a) daily checks, performed each morning by the radiation therapist who normally operates the machine; (b) weekly checks performed by a physicist, dosimetrist, therapist, or a physics resident; (c) monthly checks performed by a dosimetrist, physics resident, or physicist, and preventive maintenance inspections performed by an accelerator maintenance technician (radiation oncology clinical engineer); and (d) annual full calibration performed by a qualified medical physicist.

The responsibility of performing the various tasks is divided among physicists, physics residents, dosimetrists, maintenance technicians, and therapists, but the exact distribution is not critical. What is essential is that each individual competently perform and record the results of their tests on a regular basis, and that the overall responsibility for a machine QA program be assigned to one individual, generally the medical physicist.

All measurements should be recorded chronologically in bound notebooks or in a computer database. Such records and the data contained therein are a valuable resource in maintaining the treatment machine. All parties involved should receive periodic reports on the QA measurement results.

The QA acceptance criterion should be established for each of the constancy checks performed.

Table 16.4 Quality assurance of simulators. (From AAPM TG 40)

Frequency	Procedure	Tolerance
Daily	Localizing lasers	2 mm
	Distance indicator (ODI)	2 mm
Monthly	Field-size indicator	2 mm
	Gantry/collimator angle indicators	1°
	Cross-hair centering	2-mm diameter
	Focal spot-axis indicator	2 mm
	Fluoroscopic image quality	Baseline
	Emergency/collision avoidance	Functional
	Light/radiation field coincidence	2 mm or 1%
	Film processor sensitometry	Baseline
Annual	Mechanical checks	
	Collimator rotation isocenter	2-mm diameter
	Gantry rotation isocenter	2-mm diameter
	Couch rotation isocenter	2-mm diameter
	Coincidence of collimator, gantry	2-mm diameter
	Couch axes and isocenter	2 mm
	Table-top sag	2 mm
	Vertical travel of couch	
	Radiographic checks	
	Exposure rate	Baseline
	Table-top exposure with fluoroscopy	Baseline
	kVp and mAs calibration	Baseline
	High and low contrast resolution	Baseline

The tolerances mean that the parameter exceeds the tabulated value (e.g., the measured isocenter under gantry rotation exceeds 2-mm diameter)

Table 16.5. Quality assurance of medical linear accelerators (from AAPM TG 40)

Frequency	Procedure	Tolerance
Daily	Dosimetry	
	X-ray output constancy	3%
	Electron output constancy	3%
	Mechanical checks	
	Localizing lasers	2 mm
	Distance indicator (ODI)	2 mm
	Safety Interlocks	
	Door interlock	Functional
	Audiovisual monitor	Functional
Monthly	Dosimetry	
	X-ray output constancy	2%
	Electron output constancy	2%
	Backup monitor constancy	2%
	X-ray central axis dosimetry parameter (PDD, TAR) constancy	2%
	Electron central axis dosimetry parameter constancy (FDD)	2 mm at therapeutic depth
	X-ray beam flatness constancy	2%
	Electron beam flatness constancy	3%
	X-ray and electron symmetry	3%
	Safety interlocks	Functional
	Emergency off switches	Functional
	Wedge, electron cone interlocks	Functional
	Mechanical checks	
	Light/radiation field coincidence	2 mm or 1% on a side
	Gantry/collimator angle indicators	1°
	Wedge position	2 mm (or 2% change in transmission factor)
	Tray position	2 mm
	Applicator position	2 mm
	Field-size indicators	2 mm
	Cross-hair centering	2-mm diameter
	Treatment couch position indicators	2 mm/1°
	Latching of wedges, blocking tray	Functional
	Jaw symmetry	2 mm
	Field light intensity	Functional

Frequency	Procedure	Tolerance
Annual	Dosimetry	
	X-ray/electron output calibration constancy	2%
	Field-size dependence of X-ray output constancy	2%
	Output factor constancy for electron applicators	2%
	Central axis parameter constancy (PDD, TAR)	2%
	Off-axis factor constancy	2%
	Transmission factor constancy for all treatment accessories	2%
	Wedge transmission factor constancy	2%
	X-ray output constancy vs gantry angle	2%
	Electron output constancy vs gantry angle	2%
	Off-axis factor constancy vs gantry angle	2%
	Arc mode	Manufacturer's specifications
	Safety interlocks	
	Follow manufacturers test procedures	Functional
	Mechanical checks	
	Collimator rotation isocenter	2-mm diameter
	Couch rotation isocenter	2-mm diameter
	Coincidence of collimetry, gantry, couch axes with isocenter	2-mm diameter
	Coincidence of radiation and mechanical isocenter	2-mm diameter
	Table-top sag	2 mm
	Vertical travel of table	2 mm

The tolerances listed should be interpreted to mean that if a parameter either (a) exceeds the tabulated value (e.g., the measured isocenter under gantry rotation exceeds 2 mm diameter), or (b) the change in the parameter exceeds the nominal value (e.g., the output changes by >2%), then an action is required. The distinction is emphasized by the use of the term constancy for the latter case. Moreover, constancy, percent values are + the deviation of the parameter with respect its nominal value; distances are referenced to the isocenter or nominal SSD. All electron energies need not be checked daily, but all electron energies are to be checked at least twice weekly. Whichever is greater should also be checked after change in light-field source. Jaw symmetry is defined as difference in distance of each jaw from the isocenter. Most wedges' transmission factors are field size and depth dependent

The frequency of each QA procedure to be performed depends primarily on the stability of the parameter tested, based on one's own experience. Tables 16.4 and 16.5, adapted from the AAPM's TG 40, lists the recommended QA tests with frequency and tolerance values for simulators and medical linacs, respectively (KUTCHER et al. 1994).

For some tests, only a very quick "observation-only" type test is required. For example, a weekly "light-radiation field congruence" test result can be analyzed by simply looking at a radiograph showing the light field and radiation field edges and observing whether or not the agreement appears reasonable. Other tests require a more careful quantitative analysis. For example, a quantitative "light-radiation field congruence" test would require that the film be analyzed using a film densitometer and that the results be carefully plotted on graph paper in order to determine the edge agreement precisely.

The QA tests should be designed to be quick and reproducible checks on key parameters, if they are to be accepted and performed faithfully. The discussion provided below is for general guidance and the actual QA tests required at a particular institution must be developed by that institution.

Recommended daily checks are listed in Table 16.5. The manufacturer's instructions for start-up and operation of the accelerator should be followed and readings of the various meters, dials, and gauges recommended for monitoring should be recorded. The daily readings should be maintained in a logbook or computer database. These data provide performance trends of a particular component which are helpful in isolating faults and may even alert one to a developing problem before full component failure occurs.

Daily treatment room checks include testing the functionality of the treatment room door interlock, intercom and closed circuit monitor system, and radiation warning lights. Treatment checks include the accuracy of the optical distance indicator (ODI), the alignment of the laser localization lights, and the radiation output calibration constancy for all photon treatment modes.

The photon-beam radiation output calibration constancy for each of the photon energies used should be checked daily for the reference geometry (e.g., 10×10 cm, 100 cm SSD or SAD). This can be accomplished efficiently using an ion chamber in a simple plastic phantom which contains a fitted hole at a standard depth from the top surface for the ion chamber, and which attaches to the accelerator at the standard SSD. A cylindrical ion chamber (e.g.,

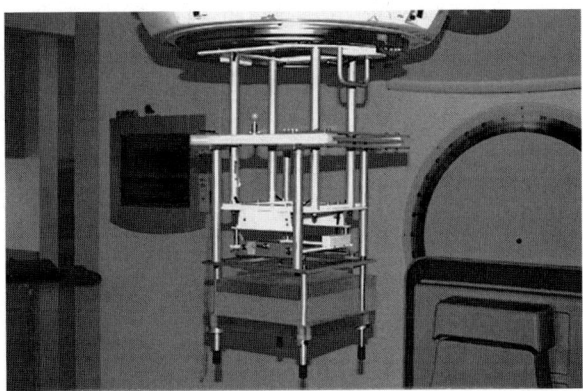

Fig. 16.4 Plastic electron output calibration constancy phantom. (Courtesy Valiant Instruments)

Farmer type) or other type of ion chamber can be used for the test. The ion chamber reading should be corrected for temperature and pressure and converted to dose using predetermined factors and the output value should be compared with the value established at the time of the last full calibration.

The electron beam radiation output calibration constancy for each of the electron energies used for the reference geometry applicator (e.g., 14×14 cm) should be checked once or twice weekly. This can be accomplished efficiently using a plastic stack phantom and ion chamber dosimetry system (Fig. 16.4).

The plastic constancy phantom used for the photon beam output check is also convenient for checking the ODI and the laser localization lights. A visual inspection of where the ODI indicator image strikes the output constancy phantom surface generally suffices as a daily check. Tests using a mechanical front pointer can be performed when a more quantitative test is needed. The vertical and sagittal lines of the laser localization lights should pass through the central axis of the beam. This may also be checked using marks on the top surface of the plastic constancy phantom and observing the intersection of the laser lines with the image of the cross hair.

Light field radiation congruence and radiation field symmetry is typically checked monthly but, in some cases, may need to be checked more often. This test can be performed by exposing a film placed perpendicular to the central axis of the beam. The film is aligned with metal markers placed on the edges of the light field (or alternatively pressure marks from a sharp device or pin prick) to mark the position of the light field. A plastic sheet of adequate thickness to provide electronic build-up is placed over the film. The developed film can be analyzed visually for the weekly test and with a film densitometer for a monthly quantitative check.

The alignment of the intersection of the cross hairs with the center of the light and radiation field should be checked regularly. This is easily accomplished by placing a film at the isocenter perpendicular to the beam and marking (e.g., pin prick) the intersection of the cross-hairs on the film. The relative motion of the cross-hairs around the pin-prick should be observed as the collimator is rotated ±90°. The center of the radiation field can be determined from the exposed film and compared with the position of the pin prick.

Symmetry can be checked using a device containing an array of detectors (usually four or five) and compared with the values obtained at the last full calibration. Symmetry can also be checked with a film in a plastic phantom placed perpendicular to the beam central axis for a large field. It should be noted that for "bent-beam" linear accelerators, where the electron scattering foil and the photon flattening filter are moved in and out of position, more frequent checks of beam symmetry may need to be performed.

The QA checks discussed above require about 15–20 min per treatment machine to complete.

Additional monthly checks for the linac are listed in Table 16.6. For example, the monthly preventative maintenance program for each machine should include regular safety checks for all electrical and mechanical interlocks. All "emergency off" switches on the machine should be checked periodically.

Collimator and gantry angles, field size indicators, and the mechanical distance indicator should be checked at least monthly. Using a simple level and rotating the gantry, the gantry angle indicators can be checked at the vertical and horizontal positions. The collimator angle indicators and the field size indicators at selected field sizes can be checked using graph paper placed on the treatment couch at the reference distance. Collimator rotation can be quickly checked by observing the movement of the image of the cross hair as the collimator is rotated ± 90°.

Photon beam energy can be checked by measuring depth dose at two specified depths, although a more sensitive test is to measure the beam profile in a plastic phantom at a specified depth. For electrons, the relative ionization measured at two depths is usually a sufficient check.

In addition, the radiation output calibration should be checked using a different dosimetry system than that used for the daily output constancy

Table 16.6 Recommended monthly QA checks

General conditions of treatment-unit checks: (key switch, monitors, machine movements, pendant, accessories, treatment aids, audio-visual/communication, room condition)
Review of daily/weekly check logs (daily machine operation check log, daily photon/electron calibration constancy check log, weekly machine maintenance inspection log)
Safety-features checks [operating instructions at machine, emergency instructions displayed, radiation warning sign, treatment-room door interlock, treatment-room operability, beam condition indicator lights (door), beam condition indicator lights (monitor), emergency offs]
Mechanical checks (gantry rotation angle, mechanical and digital; rotation arc check of MU/degree; collimator rotation angle indicator check; mechanical and digital; cross-wire deviation)
Radiation/light-field check (visual inspection for ±2-mm tolerance for light/radiation field congruence and central plane overlap/gap)
Photon beam energy/off-axis factor check
Radiation output calibration constancy check (treatment unit monitor chamber, assigned dosimetry system)
Electron beam energy check (ionization depth ratio)
Electron beam output calibration constancy check
MLC QA for IMRT
EPID mechanical and imaging checks
Onboard imaging mechanical and imaging checks

check. This serves as a redundancy check on both the treatment machine monitor chamber calibration (cGy/MU) and the daily system used to check the output calibration constancy.

Accessories such as wedge filters, electron beam applicators, and blocking tray assemblies --including the mounting slots and micro-switches -- should be examined for any cracks and potential malfunctions.

If onboard imaging is performed, additional checks related to mechanical (collision, interlocks, readout accuracy, centering, etc.), imaging quality (contrast, sensitivity, constancy, etc.), and related software, such as magnification accountability, should be checked on a monthly basis. These tests are discussed in more detail in a later section.

Also, one should ensure that an up-to-date machine operator's manual is located at the treatment machine console. In addition, complex treatment techniques require that detailed, unambiguous written procedural instructions also be available at the control console. One should also ensure that the posting of radiation warning signs and emergency instructions have not been removed.

The monthly checks may take from 1 to 2 h depending on the number of tests performed and the number of modes and energies available. These times clearly show that a QA program for treatment machines designed to insure the delivery of dose can be implemented in any size radiation therapy clinic at an acceptable cost in terms of time, staff, and equipment. When such a program is neglected and problems are not fixed as they arise, the treatment machines will inevitably deteriorate and the quality of patient treatments will be compromised.

16.4.1
Linac Annual QA Tests

A full calibration of the treatment machine should be performed annually. Suggested tests are listed in Tables 16.5 and 16.7. The basic calibration should be performed in a water phantom using an ion chamber according to an appropriate protocol, e.g., AAPM (AAPM 1983; ALMOND et al. 1999). The stability of the dose per monitor unit and the beam symmetry should be checked at different gantry angles. Verification of the output factors and central axis depth dose should be done for several different field sizes. In addition, current values for off-axis factors,

Table 16.7 Recommended annual QA checks

Emergency off switches and interlocks
Mechanical and digital indicators (gantry, collimator, field size, couch)
Inspection of mechanical parts of accelerator including blocking tray and treatment aids
Machine alignment (isocenter check)
Light-radiation field congruence
Radiation beam symmetry for all treatment modalities
Monitor chamber linearity and end effect
Dose calibration (cGy/monitor unit) for all treatment modalities
Output field size dependence for all treatment modalities
Percent depth doses for several field sizes for all treatment modalities
Wedge factors for all treatment modalities
Tray factors for all treatment modalities
Off-axis factors for all treatment modalities
MLC checks
Special procedure modes (Arc therapy, TBI, TSEI, IMRT)
Onboard mechanical and imaging checks

monitor linearity, monitor end effect, all wedge and tray factors, and bolus and comp filter attenuation factors should be verified.

In addition, various mechanical alignments should be checked annually. For example, the mechanical isocenter can be checked by observing the position of the front pointer tip in relation to a 2-mm-diameter rod as the gantry is rotated through 360°. A "star pattern" is sometimes produced to check radiation isocenter, i.e., a film is placed parallel to the radiation beam and one set of collimator jaws is closed to a narrow slit and exposures are made at different gantry angles. All couch movements and table-top sag underload should also be evaluated.

Continuing education lectures on the machine operation, safety, and QA should be presented to the staff on an annual basis. It is important that emergency procedures be reviewed periodically with the staff to ensure proper interpretation and understanding. A thorough hands-on training period for all therapists is essential following instruction about the operation of the equipment and prior to assuming treatment responsibilities. Written instructions should be provided to guide therapists as to a safe response when equipment malfunctions or exhibits unexpected behavior, or after any component has been changed or readjusted.

With a good QA and preventive maintenance program in which the parameters are measured and adjusted on a regular basis as outlined herein, treatment machines can be kept running in good operating condition. These QA checks and adjustments are generally simple to learn and easy to implement.

16.5
Linac Advanced-Technology QA

Exciting technical developments for improving dose delivery have occurred over the past two decades as a result of the development of 3D conformal radiation therapy (3D CRT) and intensity-modulated radiation therapy (IMRT; PURDY et al. 2001). These developments are based on advanced computer hardware and software technology and include 3D treatment planning systems (3D TPS), computer-controlled treatment machines, asymmetric collimators, dynamic wedge, multileaf collimators (MLC), beam-intensity modulation, imaging devices (MV electronic portal imaging devices (EPIDs), and mV and kV cone-beam CT) for treatment verification. Peripheral to the clinic are other daily localization systems. These advanced technologies provide for radiation therapy techniques that will likely improve therapeutic ratios through the use of conformal physical dose distributions that cannot be achieved using 2D planning, delivery, and verification methods; however, ensuring that the radiation therapy process is safe and accurate when these advanced technologies are used is much more difficult than ensuring the 2D process that uses a treatment machine with simple electromechanical controls, 2D planning systems, standard treatment-aid devices (alloy blocks, physical wedges, etc.), and weekly port-film treatment verification. In many cases, the complexity of interactions between hardware and software in a near real-time environment makes it virtually impossible to demonstrate with certainty that the operation of the advanced technology systems is correct and that all possible failure modes have been eliminated. Exhaustive testing of all possible combinations of inputs, in all possible sequences, and from all possible sources, cannot be realistically accomplished; therefore, it is essential that a well-planned and rigorous approach to QA tests and safety procedures be practiced when these advanced technologies are implemented in the clinic (KLEIN et al. 1996).

16.5.1
Linac Computer Control System

Modern medical linear accelerators utilize computer control systems. Incidents in the past have shown that such accelerators have the potential for massive overdoses to the patient as a result of software flaws (JOYCE 1986; KARZMARK 1987). This poses a major problem for the radiation therapy community since standard QA tests on accelerators are not designed to catch software flaws. We caution the physicist to scrutinize carefully the computer operation of the linac during the acceptance testing period, paying particular attention to verifying what happens when beam setup parameters are edited.

An AAPM task group report discuss the safety considerations stemming from the increased use of computer logic and microprocessors in the control systems of treatment units (PURDY et al. 1993). It suggests how procedures and operator responses can be improved to reduce or obviate risks associated with hardware and software failures in radiation therapy equipment. Two other publications address testing and QA of computer-controlled accelerators

(WEINHOUS et al. 1990; ROSEN and PURDY 1992). Recommendations from these three publications are summarized in this section.

Acceptance testing procedures for new software updates and/or new computer-control features should be designed specifically to test the software and control aspects of the system. Safety interlocks and new functionality should be tested rigorously after review of all vendor documentation and testing information provided by the vendor. The reader should also note that while it is much easier to test safety interlocks in service mode, such tests do not necessarily properly predict the accelerator's behavior in clinical modes; therefore, to ensure safe operation, it is important that interlocks be tested in the clinical modes used to treat patients.

Routine updates of software for a computer-controlled machine should be treated as if it includes the possibility of major changes in system operation. All vendor information supplied with the update should be studied carefully, and a detailed software/control system test plan should be created. All safety interlocks and dosimetry features should be carefully tested, regardless of the scope of the changes implied by the update documentation. All tests suggested by the manufacturer to confirm correct operation of the new software should be performed. Treatment beam parameters which may be affected by the software changes should be verified. Near full-acceptance testing may be necessary depending on the nature and extent of the software changes.

Safety interlocks also may have to be tested following non-trivial repairs. Because software and hardware are intimately linked in a computer-controlled machine, even minor changes in hardware can produce aberrations in the operation of the machine if there is a flaw in the software design or implementation. Integrity of software and data should be verified using appropriate tools supplied by the manufacturer. If repairs are extensive or involve critical components, near full-acceptance testing may again be necessary to ensure proper operation.

Computer-assisted setup features should be verified. If possible, return of the machine to a safe condition in the event of a computer or computer-related hardware failure should be verified. If power conditioning and isolation for the computer is not used, the computer and machine operation should be carefully monitored for any adverse effects of occasional power transients.

Routine scheduled maintenance and testing should be performed to minimize hardware malfunctions that can occur over time due to normal wear of components and environmental stresses such as radiation damage. In computer-controlled machines, hardware changes may also affect correct software operation by corrupting essential data or software; therefore, even though the software may have passed acceptance testing without demonstrable errors, latent bugs may appear as the hardware changes with age. Even minor changes in hardware can produce aberrations in the operation of the machine if there is a flaw in the software design or implementation; therefore, constant vigilance is necessary. In particular, safety interlocks may have to be tested following non-trivial repairs. If repairs are extensive or involve critical components, full acceptance testing may again be necessary to ensure proper operation.

In the case of software updates, the integrity of all safety interlocks and the software and database should be verified following installation. Treatment beam parameters that may be affected by software changes should be verified. Full-acceptance testing may be necessary depending on the nature and extent of the software changes. To assist users in properly verifying new versions of software, the documentation for software updates should include: (a) reasons for all changes, including bug fixes; (b) details of modifications made; (c) details of planned or expected operational changes following installation of the update; (d) effects on site-dependent and user-accessible data and/or software; (e) suggested procedures for testing operations affected by the update; (f) revised design specifications, support documentation, and/or operations manuals; and (g) results of beta tests.

Despite extensive in-house and field testing by the manufacturer, new problems are occasionally discovered by users in the field. Manufacturers should provide procedures for reporting such problems. These procedures should clearly describe the information to be submitted with the report. Manufacturers should respond to problem reports with a written acknowledgment, followed by a timely response evaluating the severity of the problem, a recommended temporary solution or a recommendation to suspend treatments, and a proposed permanent solution with time schedule for implementation. Dissemination of significant problem reports to all users should be done in a timely manner.

16.5.2
Asymmetric Jaws (Independent Collimation)

All modern medical accelerators have collimator jaws that move independently. For example, the Varian linear accelerators, the Y-Jaws (upper), can be move independently 10 cm beyond isocenter, whereas the X-jaws (lower) can move independently 2 cm past isocenter. Independent jaw capability allows the isocenter to be positioned at locations other than the treatment field center. This flexibility allows simplified patient positioning and improved safety by avoiding overlapping field abutments without the necessity of using heavy beam-splitting blocks. For example, breast irradiation techniques commonly use the independent jaw feature (KLEIN et al. 1994).

Monitor unit calculations are only slightly more complex for independent jaws than for symmetric jaws (SLESSINGER et al. 1993). An off-axis correction factor can be used that depends only on the distance from the machine's central axis to the center of the independently collimated open field. The influence of backscattered photons into the monitor chamber may influence the output for very elongated symmetric fields (PALTA et al. 1988). The dose distributions of asymmetric fields defined by jaws are quite similar to those defined by alloy blocks. Only slight differences are seen at the field edges due to the distance between the patient and the jaws.

Treatment record forms that denote each jaw setting (Y_1, Y_2, X_1, X_2) should be used. We also recommend that the four jaws be identified by labels placed on the treatment machine, simulator collimator, and the block trays. This is especially useful when treatment techniques call for collimator rotation or involve the use of MLC or dynamic wedge, which are oriented in a particular direction along an independent jaw set (i.e., MLC - X_1 or X_2, dynamic wedge Y_1 or Y_2).

The QA checks include a monthly check of each independent jaw by comparing jaw setting vs light field position vs 50% radiation value (edge) for fields designed as quadrants (two non-divergent edges). We recommend specifying the jaw position accuracy to an accuracy of 1 mm for setting vs light field at all positions (note that this exceeds TG 40 recommendations) and a specification of 0.5 mm for light vs radiation field. An effective QA test is to irradiate a film superimposing each quadrant of the field separately. Ideally, the composite film should exhibit no distinct regions of overlapped regions or gaps. On a monthly basis, the simulator's asymmetric jaw (wire) settings accuracy should be checked for the 0.0- and 10.0-cm settings. An accuracy of 1 mm is expected, which is the same criteria used for symmetric fields.

Annually, the same quadrant test should be performed at the four cardinal gantry angles. An isodensity scan of the superimposed film will show dose homogeneity across all intersections. A dose inhomogeneity of $\leq \pm 5\%$ (when compared with a point away from the junction) over a distance of ≤ 2 mm is acceptable. Annually, OAFs (used for asymmetric jaw monitor unit calculations) should be spot checked to ensure agreement within 0.5% of the tabulated OAFs. The output at d_{max} off-axis when the Y-jaw is 10.0 cm beyond isocenter should be checked annually to ensure that backscattering to the monitor chamber that can cause an eventual decrease in output is not occurring. This measurement (at the respective d_{max} depths) is compared with calculations using the appropriate OAF. Annually, corresponding quadrant films are also taken on the simulator.

16.5.3
Dynamic Wedge

Dynamic wedge technology takes advantage of asymmetric jaw technology in conjunction with control of the dose rate over the course of one treatment (LEAVITT et al. 1990). An initial field is set along with a desired isodose angle (wedge angle) with a particular wedge direction (heel to toe). After a specific number of monitor units have been delivered, the designated collimator jaw begins to move with a varying speed while the dose rate is varied simultaneously. This type technology as generally replaced physical wedges. The authors' experience is based primarily on the Varian system and will be used for discussion purposes, but the recommended testing should be applicable to other manufacturers dynamic wedge systems. The variations in jaw position and dose rate are driven by computer files called segmented treatment tables (STTs), which are unique for each energy, wedge angle, and field size. This customization of each dynamic wedge angle for each field size yields excellent-wedged isodose distributions when compared with physical wedges. The obvious practical advantages of dynamic wedging include no lifting of heavy wedges over patients, no blocking of light field during setup, and larger wedge field sizes (up to 30 cm field for 60° wedges). Dosimetric advantages include no beam harden-

ing, less scatter outside of the field, slightly shorter treatment times, lower-intensity hot spots (in most cases), no wedge tray "play," and improved wedged isodoses for all field sizes.

The QA program appropriate for dynamic wedge technology have been reported in the literature and are summarized here (KLEIN et al. 1995, 1998). Periodic checks on the accelerators should include spot checking the dynamic wedge dosimetry. These checks include verification of wedge factors for selected field sizes and wedge angles. The wedge factors should be checked for each wedge orientation (Y1-in and Y2-out). Spot checks of the isodoses (or profiles) should be performed to ensure that the dose distributions have not changed.

The ability of the dynamic wedge to be completed during interruptions should be tested by terminating and restarting the beam during a dynamic wedge run. Patient-specific checks include a diode check on the patient during the first fraction. A dual-diode system works best, particularly for wedge treatments, with one placed at the central axis on the patient's skin surface and the other at an off-axis point toward an anatomically noted direction that corresponds to the "heel or toe" of the wedge. The diode electrometer reading should be corrected by factors that depend on SSD, field size, and diode response for that day. The corrected reading corresponds to the dose at d_{max}. Deviations of 5% or less are considered acceptable due to the diode's systematic limitations in terms of spatial resolution and placement by the therapists. A second reading and investigation should be performed for larger deviations. Therapists should also be instructed to illuminate the light field at the end of each fraction to confirm that the remaining light field strip corresponds to the "toe" of the wedge. An R&V system provides a direct check of the dynamic wedge angle and orientation and a visual check need only be confirmed during the first fraction; however, there should be an independent check confirming the R&V entry of the wedge angle and direction before the first fraction. On the first fraction, the therapist still should check the indicated position and light field strip after treatment. The R&V system provides ongoing checks for the remaining fractions.

16.5.4
Multileaf Collimation

Multileaf collimators have now become the state-of-the-art method for generating irregularly shaped fields for photon-beam radiation therapy (BOYER et al. 1992, 2001; KLEIN et al. 1995). The authors' QA experience is based on use of a Varian MLC system (KLEIN et al. 1995, 1996; KLEIN and Low 2001). In that system, the leaf settings for each field are sent to a dedicated MLC computer, interfaced to the treatment machine, which drives the leaves via a controller system. The leaf settings may be obtained by two different methods. Firstly, computer software and hardware (called the "Shaper") is provided with which the user can digitize a portal shape drawn on a simulation film or using a 3D treatment-planning system. In either case, the patient's MLC configuration files are sent over a local area network to the MLC computer.

Rigorous QA is essential when MLC is clinically implemented. When first implemented in a clinic, all MLC fields should be checked visually (light field vs skin marks) and imaged using film or an electronic portal imaging device (EPID) on a daily basis.

Specific QA tests performed include the daily running of a sampling of actual clinical fields during the morning checkout of the accelerator. In addition, the following periodic (quarterly) checks are recommended: (a) testing of MLC settings vs light field vs radiation field for selected gantry and collimator angles; (b) network testing; (c) check of active patient files; and (d) interlock checks (carriage under jaw, leaf spread, leaf movement during electron and/or port film modes, etc.). Particular attention must be paid to the testing of MLC files sent over a computer network. When first implemented, it is prudent to test the network by illuminating every MLC field configuration onto the original simulation film or digitally reconstructed radiograph. The AAPM TG-50 report on basic applications of MLCs describes the QA program of recommended patient and quarterly and annual checks. The annual check of accelerators equipped with MLC include the above-listed monthly MLC tests and also film scans to review interleaf leakage, abutted leaf transmission, penumbra dependence on leaf position, and a review of procedures with clinical staff. A summary of the QA checks recommended is given in Table 16.8.

16.5.5
Online Electronic Portal Imaging

Online portal imaging systems consist of a suitable radiation detector, usually attached through a manual or semi-robotic arm to the linac and capable of transferring the detector information to a computer that will process it and convert it to an image. These systems use a variety of detectors, all producing computer-based images of varying degrees of quality. Currently these systems include: (a) fluoroscopic detectors; (b) ionization chamber detectors; and (c) amorphous silicon detectors. The QA issues for this technology can be separated into five categories: (a) physical operation and safety; (b) spatial and contrast resolution; (c) image storage, analysis,

Table 16.8 Multileaf collimation QA. (From KLEIN et al. 1996 and AAPM TG-50 report). *DRR* digitally reconstructed radiographs

Frequency	Test	Tolerance
Patient specific	Check of MLC-generated fields vs simulator film/DRR before each field is treated	2 mm
	Double check of MLC field by therapists for each fraction	Expected field
	Online imaging verification for patient on each fraction	Physician's discretion
	Port-film approval before second fraction	Physician's discretion
Quarterly	Setting vs light field vs radiation field for two designated patterns	1 mm
	Testing of network system	Expected fields over network
	Check of interlocks	All must be operational
Annual	Setting vs light field vs radiation field for patterns over range of gantry and collimator angles	1 mm
	Water scan of set patterns	50% radiation edge within 1 mm
	Film scans to evaluate interleaf leakage and interleaf, abutted leaf transmission	Leakage <3%, abutted leakage <25%
	Review of procedures and in-service with radiation therapists	

and handling; (d) reference image acquisition, and (e) clinical applications.

Physical operation and safety checks consider the physical motions of the imager, as well as patient and operator safety considerations. The imager may be detachable or permanently attached to the gantry. Interlocks should be installed so that motion of the gantry is restricted during conditions when gantry motion could cause the imager to disengage or slip during gantry rotation. Additional interlocks should be installed to prevent a collision between the patient support assembly (PSA) and the EPID or between a PSA-supported object (e.g., the patient) and the EPID. These interlocks will disable PSA motion but may allow limited motion to disengage the collided objects. Additionally, the manufacturer may provide override buttons or other hardware to bypass these interlocks. The interlocks and bypasses should be tested for functionality both during commissioning and periodically (e.g., weekly or monthly) for correct operation.

The EPID support may provide the user with flexibility of the location of the EPID sensitive surface relative to the accelerator beam. The identification of the EPID location may be through a manual readout system, or may be transmitted to the EPID acquisition system for storage and display with other relevant image information. This information is critical to quantitative utility of the portal images. For example, if the user determines that the patient position is five pixels from the intended position by examining the image, the scale factor of the image must be known before a suitable patient shift can be determined. Similarly, lateral adjustments may be available to place the EPID in an optimal location relative to the irradiated beam. Some alignment software algorithms may require the magnitude of the EPID offset. The accuracy of the readout systems must be determined and monitored periodically, as needed.

The stability of the EPID location is also important. If the EPID is not held by a stable support, the pixel location of isocenter will be a function of the gantry angle. Certain alignment algorithms may depend on the stability of isocenter to determine the accuracy of portal placement. Similarly, if a component of a stable system slips, the detection of portal misalignment may suffer. A simple test to assess the stability of isocenter is to image a graticule tray at each of four principal gantry angles (0, 90, 180, and 270°). Finally, a periodic inspection of the cassette (with protective covers removed) will reveal any hidden damage caused, for example, by an unreported collision.

While the above tests are essential for safety, the heart of an EPID QA program must center on maintaining its ability to acquire a useful portal image within the physical constraints of radiation treatments. Standard methods for determining imager response include measurement of the modulation transfer function (MTF); however, this requires knowledge of the imager sensitivity over the range of fluences and beam energies used for the measurement. The linearity of the EPID response (pixel value vs incident radiation fluence rate) can be assessed for EPIDs with an adjustable target-to-imager-distance. A measurement of the relative photon fluence at the chamber surface can be made by placing an ionization chamber at the appropriate distance with the phantom in place. The pixel value (averaged over a region of interest to reduce the effects of random noise) is then correlated with the measured relative photon fluence.

Techniques have been developed to measure the spatial and contrast resolutions under low-contrast conditions and with phantoms in place to simulate clinical conditions (Low et al. 1996). There are two advantages for using a low-contrast phantom. Firstly, primary photon fluence changes are linear with respect to changes in phantom thickness, and secondly, the EPID response is linear within a small range of photon fluences. The spatial and contrast resolution measurement techniques use geometries that irradiate the EPID with a small range of fluences.

One method for providing a split-field measurement in a low-contrast geometry is to use a single-step phantom, with the step intersecting the central axis. The step should yield a fluence change across the phantom that is within the linearity limits of the EPID. While scattered photon radiation from the step will perturb the overall profile, the sharp-gradient region will not be significantly affected. The spatial resolution is obtained by taking the spatial derivative of the profile.

Contrast and spatial resolutions can be determined qualitatively by imaging a Las Vegas phantom or equivalent. The Las Vegas phantom consists of a block of aluminum drilled with a series of holes of varying diameter and depth. An image of the phantom is used to indicate the ability of the EPID to resolve small and low-contrast features.

Images must be stored with correct patient and acquisition data and be retrieved with the data intact. While a test of the entire image handling software is impractical, a few critical tests should be performed before clinical use of the EPID. Tests should be conducted to ensure that images are being stored and retrieved correctly. This is especially important when the acquisition software compresses the file before storage and decompresses the file on retrieval. The EPID software may also provide a quantitative evaluation of pixel values and averages. These features should be tested for reproducibility and accuracy and then used to analyze an image prior to storage and after retrieval.

The EPID systems may provide a means of digitizing radiographic film for comparison with online images. Tests should be conducted to ensure that the scale factor and aspect ratio of the digitized films are correct. One method of doing this is to digitize a transparency with a printed rectangular grid. The pixel values of the grid intersections should all be the same distance apart throughout the image.

Clinical application refers to a number of special techniques developed for an EPID. The open-field calibration image is carried out with 5-cm Solid Water placed at isocenter to provide some photon scatter to simulate a patient. The minimum number of monitor units required to obtain an image as a function of accelerator repetition rate and beam energy should be measured and tabulated to aid physicians in designing imaging studies. For EPIDs that contain sensitive electronics, the maximum imageable field size should be displayed at the accelera-

tor console and on the imager body. Recommended QA checks are listed in Table 16.9 and more details regarding EPID QA can be found in the AAPM TG-58 report including recommendations for QA tests of EPID systems along with test details (HERMAN et al. 2001).

16.5.6
Intensity-Modulation Radiation Therapy

The latest development in external beam radiation therapy treatment implementation exploits the use of fields in which intensity is varied across the beam. (The reader is referred to Chap. 10 in this textbook for more details on this subject.) Also, there have been several consensus documents published (2001; EZZELL et al. 2003; GALVIN et al. 2004).

The QA for IMRT techniques is still being developed and as yet there are no specific national QA recommendations or tolerance criteria published. Increased vigilance is required and the users of this technology must accept an even greater responsibility with regard to patient safety. The IMRT techniques require both geometric and dosimetric verification to a phantom before a patient's treatment is delivered. New technologies for these QA tasks that increase efficiency are needed.

Table 16.9 Electronic portal image quality assurance. (Modified from AAPM TG-50 report). *SNR* signal-to-noise ratio

Frequency	Test	Responsible individual
Daily	Inspect imager housing	Therapist
	Test collision interlock	Therapist
	Acquire day's first image during machine warm-up	Therapist
	Procedure to verify operation and image quality	Therapist
	Verify sufficient data capacity for day's images	Physicist
Monthly	Acquire image and inspect for artifacts	Physicist
	Perform constancy check of SNR, resolution, localization	Physicist
	Review image quality	Physicist
	Perform image/disk maintenance	Physicist
	Mechanical inspection (latches, collision sensors)	Physicist
	Optical components	Clinical engineer
	Electrical connections	Clinical engineer
	Test collision interlock	Physicist
	Hardcopy output	Physicist
Annual	Perform full check of geometric localization accuracy	Physicist

The user should also be aware that the dose to the patient outside the treated volume may be increased due to increase leakage radiation levels. This is increased with beam modulation systems because the number of monitor units required to deliver a specific dose to the patient is substantially increased (MUTIC and LOW 1998).

Table 16.10 (modified from AAPM TG-35 report) lists a model QA and safety program for a medical linear accelerator facility (PURDY et al. 1993).

16.6
Quality Assurance of Cobalt Teletherapy Units

The United States Nuclear Regulatory Commission (USNRC) has established, in Title 10, Part 35 of the Code of Federal Regulations, Subpart I, special requirements for the full calibrations and monthly spot check measurements of these licensed ^{60}Co teletherapy units (NRC 2003). The full cali-

Table 16.10 Model quality assurance and safety program for medical accelerator facilities. *ALARA* as low as reasonably acceptable. (Modified from AAPM TG-35 report)

Use of the medical accelerator
- The unit should be operated only by authorized personnel who are trained in the safe operation of the unit. This typically includes radiation therapists, radiation oncology physicists, dosimetrists, physics residents, and machine maintenance personnel
- Instructions on how the unit is to be operated should be maintained at the console

Safety devices
- The console and room radiation warning lights and door interlock should be checked daily by the radiation therapist. Their status should be recorded in the unit's daily log
- All ancillary equipment, including but not limited to that for patient aural and visual communication, should be in good working order and regularly tested at appropriate intervals as part of a continuing QA program. Treatment should not proceed if specific ancillary equipment essential to treatment is inoperative
- All computer-controlled treatment machines with the associated ancillary high technology devices (MLC, dynamic wedge, etc.) should be equipped with a record-and-verify system

Personnel dosimetry
- Appropriate personnel monitors (e.g., film badges) should be provided by the institution's Radiation Safety Office
- The personnel monitors should be supplied with a specified frequency, e.g., monthly
- The personnel dose reports should be reviewed by Radiation Safety Office staff and reported values exceeding the investigative levels of the institution's ALARA program should be referred to the ALARA investigator for timely review. Monthly personnel reports should be posted conveniently for ready access by involved personnel

Procedures for securing the medical accelerator
- The treatment room should be secured during non-working hours and when left unattended

Instrument calibration and checks
- Radiation survey meters should be calibrated annually. A description of the sources calibration frequency and equipment procedures should be documented

- The dosimetry system used for full calibration should be calibrated every 2 years by an AAPM Accredited Dosimetry Calibration Laboratory (ADCL)
- The dosimetry system(s) used for periodic QA checks should be calibrated on a yearly basis by a qualified radiation oncology physicist by intercomparison with a dosimetry system calibrated by an ADCL

Acceptance testing and full calibration of medical accelerator
- Testing and full calibration should be performed by a qualified radiation oncology physicist following the procedures given in the AAPM Code of Practice for Radiotherapy Accelerators (TG 45 report) and the AAPM TG 21 or 51 Protocol

Software QA and testing
- Acceptance testing procedures for new software and/or new computer-control features should be designed specifically to test the software and control aspects of the system. All safety interlocks and new functionality should be tested rigorously after review of all vendor documentation and testing information which is available. See AAPM TG-35 report
- Routine updates of software for a computer-controlled machine should be treated as if it includes the possibility of major changes in system operation. All vendor information supplied with the update should be studied carefully, and then a detailed software/control system test plan created. All safety interlocks and dosimetry features should be carefully tested, regardless of the scope of the changes implied by the update documentation. See AAPM TG-35 report

Periodic QA measurements of medical accelerator
- The QA measurements should be performed following procedures and frequencies recommended by the AAPM TG 40 report
- The results of the spot-check measurements should be reviewed (and signed/initialed) by the radiation oncology physicist

Servicing and inspection of the medical accelerator
- Only persons or firms specifically authorized by the physicist in charge at the institution should perform any maintenance or repair of the unit

- Appropriate dosimetry measurements should be performed after any maintenance or service is performed. The person responsible for releasing the accelerator to clinical service after maintenance is the radiation oncology physicist
- A full inspection of the medical accelerator should be made by the manufacturer at intervals not to exceed 3 years

Radiation survey
- A radiation survey should be performed by a qualified physicist before initial use and whenever any changes are made in the shielding, location, or use of the unit that could affect radiation levels in surrounding areas

Emergency procedures
- Emergency procedures should be posted at the medical accelerator control console
- All new treatment personnel should be trained in emergency procedures as soon as they report to duty. Practice drills in emergency procedures should be conducted by the radiation oncology physicist or his designee with all appropriate personnel at least once a year

Procedure for notifying the proper person in the event of an accident or an unusual occurrence
- In the case of an accelerator malfunction, individuals listed on facility's "Emergency Procedures" should be notified
- In the case of a suspected treatment misadministration, follow guidelines listed in section 4 of the AAPM TG-35 report

Radiation therapist training
- Before a therapist starts working with a treatment machine, he/she should be given extensive training on (a) the normal operation of the machine, (b) the meanings of the various interlocks and fault lights, as well as the appropriate response to their occurrence, (c) any unusual aspects of the machine that could be important during routine treatments, and (d) QA tests
- Continuing education should be provided at regular intervals on subjects pertinent to machine operation, QA, and safety to supplement previous training

Record keeping
- Copies of the following records should be maintained by the institution: (a) results of safety device checks; (b) records of personnel dose monitoring; (c) survey instrument calibration reports; (d) calibration report for dosimetry system used for full calibration measurements; (e) calibration or intercomparison report for dosimetry system(s) used for periodic QA measurements; (f) acceptance test and full calibration reports; (g) periodic QA test results; (h) training and experience credentials of the qualified radiation oncology physicist; (i) training of new personnel and annual refresher training of personnel; (j) full inspection and all maintenance work performed; (k) radiation survey(s) reports; (l) copies of applicable regulatory statutes

bration measurements must be performed by a qualified teletherapy physicist in accordance with the procedures recommended by the AAPM using a secondary standard dosimetry system calibrated within the previous 2 years by the National Institute of Standards and Technology (NIST) or by an AAPM-Accredited Dosimetry Calibration Laboratory (ADCL). The USNRC does allow for the use of a secondary standard dosimetry system calibrated only every 4 years provided certain authorized intercomparisons are made. The reader is referred to the reference for details required for this intercomparison (NRC 2003).

The mechanical integrity of the unit should be checked during the annual full calibration. Typical tests and procedures are given in the Hospital Physicists Association report no. 3, "A Suggested Procedure for the Mechanical Alignment of Telegamma and Megavoltage X-ray Beam Units" (HPA 1970).

Additionally, the USNRC requires that the dose rate determined at the full calibration be corrected mathematically for physical decay at intervals not exceeding 1 month. These corrections must also be made by a qualified teletherapy physicist.

The USNRC also requires that a semi-annual leak test be performed for teletherapy units (NRC 2003).

The USNRC also requires that monthly spot-check measurements be performed for teletherapy units (NRC 2003). While a qualified teletherapy physicist is not required to perform these measurements, the results of the spot-check measurements must be reviewed by the qualified teletherapy physicist within 15 days. These measurements may be made either with the secondary standard dosimetry system calibrated as previously described, or alternately, with a field instrument intercompared with a secondary standard system within the previous year.

There are also certain daily checks that should be performed; these include a check of the treatment-room radiation monitor, beam condition indicator lights, and door interlock. A log book should be kept documenting that these items were checked daily prior to treating the first patient.

16.7
Treatment-Machine Maintenance

The division of labor between the maintenance and physics groups can create significant difficulties. An effective QA program must coordinate maintenance of the treatment machines and it must be clearly understood that the release of the treatment machine following repair is the responsibility of the medical physicist.

Maintenance should absolutely avoid by-passing interlocks and document clearly any changes made to the unit. Often unnecessary tweaking (fine tuning) of the accelerator is done which can put the system into marginally unstable or erratic operation. This can cause unnecessary service calls, confusion, and delay in restoring a system to proper operation.

It is important that maintenance be kept updated to new developments and system changes. There is also a clear need for good documentation of testing and repair, operator and maintenance staff training, and vendor support in equipment maintenance.

16.8
Treatment-Planning Computer System QA

It is very important that the user check that the planning system is accurately functioning before it is used clinically. The treatment-planning system, in this respect, is no different than other medical devices. Procedures for testing treatment-planning systems have been described in the literature (JACKY and WHITE 1990; VAN DYK et al. 1993; FRAASS et al. 1998). One class of tests consists of a comparison of treatment plans calculated for standard phantoms to dose measurements for the same phantoms. Such tests are recommended because they check the dose delivered under treatment conditions, but they are time-consuming because of the large number of parameters. These comparisons will not only test the system, but will also deepen the user's understanding of the computation models. Furthermore, it may reveal errors in coding of the algorithms that cannot be obtained directly from dosimetric measurements; however, it should be appreciated that all possibilities can never be tested, nor can the manufacturer assure the user of a bug-free system. Instead, the system should be tested over a range of parameters which would be typical of those used in the clinic, and the following items should be checked: (a) consistency of I/O data; (b) monitor units (time); (c) relative dose distributions; (d) graphical data including BEV display and field apertures; and (e) output data including isodose curves, DVHs, and digitally reconstructed radiographs (DRRs). The test procedures and results should be fully documented. They will provide the basic test results to which further tests can be referred.

Tests should always be performed on the system after program modification. The tests should include a set of typical treatment-planning examples which can be compared to the initial acceptance testing results. The user should test the operation of the treatment-planning system as a whole, even if only one module is modified since changes in one part of the code can lead to unexpected results elsewhere.

Periodic QA tests should be performed on a regular basis. For example, daily tests of the consistency of I/O should be confirmed. This procedure will often reveal nonlinearities and other errors in digitizers and optical systems. In addition, an annual test should be performed in standard geometries with beam arrangements used during acceptance tests. These procedures may reveal changes in the way the system is used and inadvertent modifications in the treatment-planning programs or beam library.

The manufacturer should provide full documentation needed to operate and understand the hardware and software components of the system. This documentation includes, but is not limited to, the following: (a) beam data library; (b) description of the dose calculation models; (c) procedures for entering patient data and machine parameters into the system; and (d) system test procedures.

The user must have the ability to produce and have access to a beam data library. The format of this library, and how to produce and edit it, should be clearly documented. If beam data is provided by the treatment-planning system manufacturer, the source of the data should be fully documented. It is preferable for the user to acquire their own basic data set. If the manufacturer's data is used, then it should be carefully checked.

It is important that the manufacturer provide a complete description of the dose calculation models used in the planning system. The documentation should include a description of the expected accuracy of the dosimetric calculations for various treatment-planning conditions and a discussion of the limitations of the dose calculation models. Furthermore, it should include a description of geometry calculations used in the system.

There should be complete documentation on procedures for entering patient data and machine parameters into the system. Equally important is a description of the definition of the output parameters such as field size, gantry angle, etc. The manufacturer should also provide examples illustrating the use of the system. For example, it should be clear to the user how the different weighting procedures operate and whether wedge transmission factors are internally used during the dose computation process or externally applied by the user.

The manufacturer should make available documentation on their software testing of the treatment-planning system. The manufacturer should also supply information on the error rate discovered in field operation of the system. The schedule for software upgrading should also be provided. Furthermore, the manufacturers should specify their procedures for documenting and correcting software bugs discovered in the field.

16.9
Treatment-Planning QA

When considering QA procedures in treatment planning one must consider three distinct methods of planning:

1. *Non-graphical planning*, often used for single or parallel opposed fields; in this approach, the monitor units (time) for the prescribed dose to a point on the central axis is usually calculated using central axis depth dose or TMR's and calibration tables. Furthermore, the apertures for the treatment, which define the treatment volume, are usually obtained during simulation.

2. *Traditional graphical planning* is used for many patients. A target volume is defined from CT or orthogonal simulation films, field arrangements are designed, and dose distributions are calculated on a limited number of axial cross sections. The patient's contour is either obtained mechanically or with CT. The portals are usually defined from simulation films and the dose is prescribed to a point or an isodose curve.

3. *Three-dimensional planning*, where the physician defines target volumes, critical structure volumes, and surface contours obtained directly from CT or MR scans. The field apertures are defined using beams-eye-view (BEV) techniques (virtual simulation). Moreover, 3D systems produce DRRs reconstructed from the CT data set. Doses are evaluated volumetrically using 2D isodose planes, 3D isodose surfaces, and dose volume histograms (DVHs).

Treatment planning should be considered as a process that begins with patient data acquisition and continues through graphical planning, plan implementation, and treatment verification.

It entails interactions between the treatment planners, radiation oncologists, residents, and radiation therapists (subsequently referred to as the treatment-planning team), and uses a large number of software programs and hardware devices. Each step of the complex treatment-planning process consists of a number of issues relevant to QA.

16.9.1
Patient Immobilization and Data Acquisition

It is important to immobilize the patient in some fashion to ensure position reproducibility for daily treatment and to ensure maintaining the patient in a fixed position during the course of imaging and treatment. A number of techniques may be used for immobilization; these vary from taping the patient's head to the couch to applying complex body moulds.

Diagnostic units, including simulators, CT, MRI, and ultrasound, are used for acquiring patient contours and target and normal organ volumes. There are a number of additional demands placed on imaging units which are specific to QA in treatment planning (MUTIC et al. 2003). Since the patient needs to be repositioned reproducibly, special couch attachments simulating the treatment machines and imaging devices are useful. Immobilizing devices should be constructed so that they can be attached to the diagnostic and treatment couches. In some instances the composition of these devices is important, for example, low atomic number materials are necessary on CT scanners. In addition, patient motion can distort MRI and CT images and can cause changes in the linear attenuation coefficients derived from CT. Furthermore, the position of the patient in the CT scanning ring can lead to errors in the CT numbers which are used to derive the linear attenuation coefficients of the patient. Moreover, nonlinearities in the video chain can cause magnification and distortion errors in hard-copy CT output. The size of contours obtained from CT may be affected by the contrast setting.

It is good quality-control practice to test the geometrical consistency of patient data obtained from CT, MRI, other diagnostic devices, simulators, and treatment units. This can be accomplished by imaging suitably designed phantoms on the various machines and comparing the results. Special care should be given to MRI units which may suffer from appreciable spatial distortions. In addition, for CT it is necessary to obtain or confirm the relationship between CT number and electron density.

One older method of obtaining body outlines is to place solder wire on the body of the patient, to transfer the contouring device to a sheet of paper, and to trace the patient's contour. With this method, it is important to measure the distance between at least three points which have been marked on the contour. In this respect, the calipers used to measure distance should be checked regularly since offset errors are not uncommon. Specialized mechanical devices can also be used to obtain patient contours and may have greater accuracy and reproducibility. The CT simulation has now become the standard methodology to obtain patient contour data (Mutic et al. 2003). Geometrical accuracy should be checked for CT contouring, as previously discussed.

Patient data can be entered into the treatment-planning computer from plane film or hard-copy CT with a digitizer. Data-transfer errors can occur because of digitizer nonlinearities and malfunctions. The accuracy of this form of data entry should be checked regularly, perhaps on a daily basis. Alternatively, the transfer can occur directly via tape, CDs, DVDs, or computer network. Data-transfer routines should be designed to check the integrity of the transfer. Systematic errors can also occur: for example, an error in the specification of a scan diameter can lead to geometric distortions of the image. Computer data transfer systems should be confirmed on a regular basis.

16.9.2
Critical Structure, Tumor, Target Volume Delineation

Uncertainties in the target volume are related to uncertainties in the size of the tumor mass and the extent of the microscopic spread of the disease; therefore, high-quality imaging on treatment simulators and other imaging devices is important. With CT, for example, the contrast setting can affect the size of the target volume. The CT-defined target volumes created online at a video monitor are preferable to those defined on hard-copy images with a grease pencil. In designing clinical target volumes, an additional margin should be included to account for internal motion, patient motion, and setup uncertainty (ICRU 1993). The size of this margin is usually based on local experience.

Normal organs are sometimes difficult to localize on plane films. Procedures should be in place to assure that contrast agents are used to localize critical organs. For example, if the small bowel is imaged, pelvic fields can be designed to minimize small bowel toxicity. Although most internal organs are imaged on CT with high contrast, it is possible to improperly define normal organs due to faulty or incomplete CT procedures. For example, the base of the brain may be better defined if sagittal reconstructions are also available.

16.9.3
Designing Beams and Field Shaping

In 2D treatment planning, block apertures are defined from simulation films; therefore, accurate specification of the magnification factors is important. Three-dimensional treatment-planning systems are more complex in that apertures can be defined interactively using BEV computer displays. In this approach all volumes of interest are projected onto a plane (usually passing through the isocenter) along ray lines that originate the source. Block apertures or MLC settings can be entered to encompass the projected target. Geometrical distortions due to programming bugs can cause serious errors in the irradiated volume due to mis-registration of the treatment fields and the target volume. The BEV accuracy as a function of gantry angle, collimator angle, field size, and isocenter distance should be confirmed prior to use and checked at a regular intervals and after any software modification.

16.9.4
Dose Calculation

The accuracy of dose distribution calculations depends on a number of factors, including: machine input data; approximations made in the dose calculation algorithm used; patient data including inhomogeneities; and the accuracy with which treatment machine parameters, such as flatness and symmetry, are maintained.

16.9.5
Computation of Monitor Units (Time)

The number of monitor units (or time) to realize the prescription is obtained either directly from the dose calculation system, from tables and graphs and "hand calculations," or from a combination of both. The accuracy of these calculations is affected by each of the parameters used in the dose calculation model

The most important elements are (a) contours, (b) collimator settings, (c) calibration as a function of collimator settings, (d) calculation depth, (e) target surface and target isocenter distances, (f) relative dose factors (depth dose, TMRs, etc.), (g) field size and SSD, (h) wedge and compensator transmission, and (i) blocking-tray transmission. Each of the components in the calculation should be part of the QA chain.

16.9.6
Plan Evaluation

Treatment-plan evaluation include review of dose distributions on a video monitor or hard copy and, for 3D systems, DVHs. The fidelity of the output, which is not only used for plan evaluation but also for dose prescription, depends upon additional factors in addition to the accuracy of the dose calculation algorithms. For example, nonlinearities in plotting can lead to distortions in the dose distributions and patient anatomy: it is good practice to have fixed length scales printed out in order to check the geometrical accuracy of the output. Furthermore, the dose distribution calculations can be sensitive to the grid size, and DVHs can additionally be sensitive to the dose bin size. All output data, including those presented in graphical format, should be regularly checked.

16.9.7
Treatment-Plan Review

All treatment plans should be reviewed, signed, and dated by the treatment planner. In addition, all plans should be independently checked by an individual who did not produce the plan, preferably within 48 h. The independent plan check should assure that setup instructions have been properly recorded, e.g., field size, gantry angle, etc. In addition, planning errors may be revealed if the dose at one point in the plan (preferably at the isocenter or at the center of the tumor) is independently calculated. The dose at the specified point should be calculated for each field using the prescribed monitor units (minutes) and tabulated TMRs, inverse square, and other relevant factors. Care must be taken if dose weight points are placed too close to a beam of blockage, or if located in the presence of heterogeneity if heterogeneity calculation corrections are used. An action level should be established based upon the accuracy

of the computer algorithm and independent dose calculation procedure.

16.9.8
Beam Modifiers

Beam modification is most often produced with alloy blocks. Errors can occur because of: incorrect specification of the magnification factor; inaccuracies in the block cutting system; and human error. Grid plates, when inserted in the radiation field produce regularly spaced marks on port films, are useful in distinguishing between patient positioning problems and block cutting errors.

The block-cutting system should be checked regularly by cutting a standard block outline and evaluating the projected aperture. The geometrical scales used in the system should also be checked on a regular basis.

Systems for fabricating compensators can be quite complex. It is important to check on a regular basis that compensating filters are accurately fabricated. In addition, the system for registering the position of the compensating filter on the tray in relation to the patient should be carefully designed and reviewed regularly, as it can be a serious source of error.

16.9.9
Plan Implementation and Verification

It is important to check that all parameters in the treatment plan are properly implemented. This is best accomplished by having the treatment-planning team available at the treatment machine during first day setup so that any detected ambiguities or problems are corrected immediately. Double-exposure localization (portal) films should be reviewed by the radiation oncologist and when necessary, in conjunction with the radiation therapist and treatment planner. Special care should be taken to assure that all beam-modifying devices are correctly positioned. Although errors in block fabrication and mounting should be observed when reviewing the port films, wedge or compensator misalignment is much more insidious, and may only be revealed by careful observation during patient setup.

A record-and-verify system should be used to assure that the same parameters (within tolerance limits) are used each day. Such systems are valuable for the recording and verification of at least the following parameters: (a) monitor units; (b) energy;

(c) mode; (d) collimator settings (including independent jaws and in the future MLC); (e) collimator angle; (f) gantry angle; (g) wedge number and orientation; and (h) blocking-tray number; however, record-and-verify systems should be used with care since they can give the user a false sense of security. For example, if a setup error is made on the first day, the system will faithfully verify the setting from day to day. To reduce the chance of this occurring, it is preferable to set up the patient according to the treatment plan, rather than the record-and-verify system, and then to verify the parameters with the record-and-verify system after the setup is complete. One report by PATTON et al. (2003) showed propagation of errors where pretreatment checks of the record-and-verify data entry were not performed. Another study by KLEIN et al. (2005) examined the frequency, longevity, and dosimetric impact of errors in a process where record-and-verify entries were checked pre-treatment; however, as vendor implementation of DICOM data exchange matures, direct transfer of data from treatment planning to the record-and-verify system will undoubtedly reduce errors, although QA reviews are still recommended.

Thermoluminescent dosimeters (TLD) are often used for in vivo dosimetry, because they are small and relatively easy to calibrate. Diodes may also be used and, unlike TLD, have instantaneous readout (AAPM 2005). This is useful, since if an error is observed, the treatment can be immediately discontinued and the source of the error found; however, diodes are more difficult to calibrate than TLDs, since their readings depend on dose rate, beam direction, beam energy, and temperature. In vivo dosimetry is usually used in the United States for checking the dose for unusual treatment conditions or for critical structures in or near the treatment volume; however, LEUNENS et al. (1990) has used in vivo dosimetry on a more systematic basis and has demonstrated that systematic errors in treatment techniques can be identified with a carefully controlled in vivo diode dosimetry system.

16.10
Clinical Aspects of QA

The clinical aspects of QA refer here to those areas which link the work of radiation oncologists, medical physicists, radiation therapists, and dosimetrists. This interaction is fulfilled in three specific tasks, namely new patient planning, chart review, and port-film review.

16.10.1
Planning Conference

One of the most important components of a QA program is the establishment of a planning conference that is attended by radiation oncologists, medical physicists, radiation therapists, and dosimetrists. During patient planning conference, the pertinent medical history, physical and diagnostic findings along with the tumor staging and the plan of treatment, and the prescription, should be presented by the attending radiation oncologists or residents. Ideally, all patients seen in consultation should be discussed, although this may not always be practical. The background information presented in planning conference is important for the treatment planner, since significant medical problems in the past or intended surgery or chemotherapy will all have an influence on the design of the treatment plan. In addition, this information is important for the radiation therapist, since it may influence scheduling, and the requirements for setup and treatment, and may alert the therapist to be aware of changes in the patient during the course of treatment. For each patient, the possible treatment plans should be discussed including a description of the prescribed dose, critical organ doses, and suggested source/applicator arrangements.

16.10.2
Chart Checking

The items recorded in the radiation oncology chart (see Table 16.11) are reviewed by a number of individuals at different times during the patient's treatment. For example, the radiation therapist usually refers to the chart on a daily basis, and errors are not infrequently discovered by the therapist during this daily routine. Furthermore, radiation oncologists refer to the chart on a frequent basis, especially during medical examinations. Given the complexity of modern-day radiotherapy, and the great differences in functioning of various departments, it is not possible to define when and where each item in a chart should be reviewed; however, it is necessary for each department's QAC to recommend and implement a program which clearly and unambiguously defines the following: (a) which items are to

Table 16.11 Suggested components of radiotherapy patient treatment record/electron medical record

Patient name, ID, photograph
Consent form
Medical background (diagnosis, stage, etc.)
Ongoing medical tests during treatment
Nursing reports
Prescription (dose, time, and fractionation)
Simulation form (patient setup parameters)
Treatment plan
Treatment field documentation (field photographs, portal films)
Monitor unit (time) calculations
Total cumulative dose (tumor and normal tissue)
Medical progress
Quality assurance forms
Treatment verification films/images
In vivo dosimetry reports

Table 16.12 Suggested treatment-plan review

Data consistency (e.g., prescribed dose is same as dose used to calculate monitor units)
Graphical plan
Monitor unit (time) calculation
Treatment record (e.g., monitor units and graphical plan parameters are properly transferred to chart and record and verify system)
Critical doses (e.g., the cord dose is correctly calculated and recorded)
Matching fields (e.g., field gaps are correctly calculated and implemented)
Potential overdoses (e.g., prior treatment of the areas currently irradiated)
Cumulative tumor and other specified doses

be reviewed; (b) who is to review them; (c) when are they to be reviewed; and (d) what actions are to be taken in the event of an error.

The frequency of these checks varies between institutions, but charts should probably be reviewed on a weekly basis by the medical physicist and the radiation oncologist. In some departments a review of this type occurs during weekly port-film conference, in others during special chart-review sessions. The extent of a weekly check depends, in part, on the functioning of the department and the rate and extent of other reviews and checks of the treatment chart. If information resides in both the paper and electronic charts, they both should be reviewed simultaneously for consistency and accuracy. If a clinic relies primarily or exclusively on an electronic record, it is imperative that a backup plan be in place in the event that the server or the network puts the electronic environment in a nonfunctioning mode. During this weekly review, the treatment plan should also be reviewed as listed in Table 16.12.

In addition to the weekly checks, all monitor unit (time) calculations, whether derived from charts or graphical plans, should be checked by an individual who did not perform the original calculation within 48 h.

16.10.3
Port Film/Image Verification Review

In general, two types of imaging techniques are used to assess radiation field position and target volume coverage, namely, portal single- or double-exposure verification images taken prior to treatment. Portal imaging assesses the position of the patient over a time interval much shorter than the treatment, and therefore does not assess the effect of patient motion during treatment. Another difficulty is that, although the dose per portal field is low (<5 cGy), it represents an additional dose to the patient, particularly if a significant number of images are needed. Single-exposure verification images, since they image the entire radiation delivery for each field, are a record of what occurred during treatment, including motion of the patient and the presence of radiation field modifiers; however, it is generally more difficult to analyze these films due to less inherent contrast, patient motion, field modifiers, and the limited view of the anatomy.

Online electronic portal imaging devices (EPIDs) are rapidly being implemented and provide improved patient imaging before and during treatment (HERMAN et al. 2001). Results are now available on a video monitor within seconds, making review of the radiation fields more expeditious. This makes it possible to obtain images of the patient on a more frequent basis. Furthermore, these devices are capable of enhancing images and improving bony landmark contrast.

With the continuing implementation of even more advanced onboard imaging and other data localization systems (i.e., ultrasound, video surfacing, static kV imaging, kV cone-beam CT, and mV cone beam CT), clinical work-flow processes must be clearly established with emphasis on a quality

Table 16.13 Operational recommendations for medical accelerator facilities. (Modified from AAPM TG-35 report)

Remove the patient from the treatment room as a first step when uncertainty in normal treatment unit operation occurs. Err on the side of safety rather than staying on schedule

Establish and maintain good communication among the radiation therapists who operate the treatment units in a given facility and between them and the radiation oncologists, physicists, engineers, and service personnel. Maintain a continuing informal dialogue on the running characteristics of the treatment units. Therapists, who are most directly involved with machine operation, provide the most important first line of defense against accidents

Identify a primary technical expert, preferably involved in the dialogue noted above, to whom therapist have recourse in the event of unexpected behavior of the treatment unit. This would normally be a maintenance engineer or qualified radiation oncology physicist. Involve the expert promptly and restrict the therapist's ability to easily resume treatment when significant malfunctions occur

Prepare written procedures with specific steps to be followed by the therapist in case of specific malfunctions. Review this with the therapist. The technical expert would be called only when these steps fail to solve the problem

Ensure that all ancillary equipment, including but not limited to that for patient's aural and visual communication, is in good working order and regularly tested at appropriate intervals as part of a continuing QA program. Treatment must not proceed if specific ancillary equipment essential to treatment is inoperative

Conduct periodic reviews of all relevant safety procedures with the radiation therapists. Reiterate the location and function of emergency off buttons. Maintain an archival log and/or database describing routine operating conditions and QA tests for each treatment unit together with a description of technical problems and their solutions. Be alert to changes in performance, both gradual and sudden

Have a full-time, qualified radiation oncology physicist available at all facilities which employ dual or multimodality megavoltage treatment equipment. A part-time consulting physicist does not provide an adequate safeguard against the hazards addressed here

Incorporate redundancy with no common failure modes where safety is involved. For example, confirm computer actions with manual methods

Equip all computer-controlled treatment machines equipped with the associated ancillary high-technology devices, such as asymmetric collimators, dynamic wedge, MLC, and IMRT, with a record-and-verify system

The technical, medical, and administrative heads of the organization responsible for radiation treatment should endorse the QA program in writing

streamline review process. All such systems require check procedures that provide clear instructions as to (a) localization procedure, (b) therapists instructions and tolerance criteria to move (or not move) a patient, (c) whether post-imaging is required when a move is made, (d) subsequent reviews by physicians regarding timing and instructions after such reviews, and (e) a peer-review process. Item (b) is especially important, as some necessary movements of a patient may be too small (i.e., <2 cm) or too large (i.e., >2 cm), which perhaps could be indicative of a poor setup.

16.11
Conclusion

Radiation oncology requires a strong commitment to QA including the active participation of physicians, medical physicists, dosimetrists, radiation therapists, and electronic and mechanical maintenance specialists. It is essential that regularly scheduled QA and preventive maintenance inspections of the treatment units be performed. These QA machine-performance tests and acceptance criteria must be well defined and understood by the appropriate personnel. The staff must have adequate test instrumentation and machine documentation. Also, there must be a commitment to good maintenance and radiation calibration record keeping. The institution must have a strong commitment to adequate staffing, continuing education, and the implementation of any needed complementary safety and QA devices such as record-and-verify systems. Table 16.13 (modified from AAPM report 35) gives a list of operational recommendations for medical linear accelerator facilities (PURDY et al. 1993).

Finally, it must recognized that radiation oncology is facing a new set of challenges as computer-controlled machines and related high-tech ancillary devices are now the standard of practice, not only at major medical centers, but even at small clinics. The QA tests and procedures for these high-tech devices will continue to evolve. There is no doubt that the use of computer-controlled medical linear accelerators and the associated high-tech ancillary devices are advancing the state of the art in radiation therapy by providing the opportunity for both more sophisticated treatments and more complete and thorough safety systems; however, it must be stressed that the skills of the radiation oncologist, radiation oncology physicist, dosimetrist, and radiation therapist can never be entirely replaced by technological advances. The radiation oncology team must be constantly vigilant because no computer system can compensate for a team member's error in judgment, misunderstanding of physical concepts or technological limitations, or unsatisfactory planning and delivery of radiation therapy.

References

AAPM (1983) Task Group 21, Radiation Therapy Committee, American Association of Physicists in Medicine: A protocol for the determination of absorbed dose from high-energy photon and electron beams. Med Phys 10:741–771

AAPM (1990) Report number 36: essentials and guidelines for hospital-based medical physics residency training programs

AAPM (2005) Report 87: diode in vivo dosimetry for patients receiving external beam radiation therapy: report of Task Group 62 of the Radiation Therapy Committee

Abt Associates (1995) Abt study of medical physicist work effort. Prepared for the American College of Medical Physics and American Association of Physicists in Medicine by Abt Associates Inc., 1110 Vermont Avenue NW, Suite 610, Washington DC, 20005

Abt Associates (2003) Abt study of medical physicist work values for radiation oncology physics services: round II (final report). Prepared for American College of Medical Physics and American Association of Physicists in Medicine by Abt Associates Inc., 1110 Vermont Avenue NW, Suite 610, Washington DC, 20005

ACR (2004) American College of Radiology: practice guidelines and technical standards. American College of Radiology, Reston, Virginia

Almond PR, Biggs PJ, Coursey BM, Hanson WF, Huq MS, Nath R, Rogers DWO (1999) AAPM's TG-51 protocol for clinical reference dosimetry of high-energy photon and electron beams. Med Phys 26:1847–1870

Boyer A, Ochran TG, Nyerick CE, Waldron TJ, Huntzinger CJ (1992) Clinical dosimetry for implementation of a multileaf collimator. Med Phys 19:1255–1261

Boyer A, Biggs P, Galvin J, Klein E, LoSasso T, Low D, Mah K, Yu C (2001) AAPM report 72: basic applications of multileaf collimators, report of Task Group 50. American Association of Physicists in Medicine, Medical Physics Publishing, Madison, Wisconsin

Ezzell GA, Galvin JM, Low D, Palta JR, Rosen I, Sharpe MB, Xia P, Xiao Y, Xing L, Yu CX (2003) Guidance document on delivery, treatment planning, and clinical implementation of IMRT: report of the IMRT Subcommittee of the AAPM Radiation Therapy Committee. Med Phys 30:2089–2115

Fraass B, Doppke K, Hunt M, Kutcher G, Starkschall G, Stern R, van Dyk J (1998) AAPM Radiation Therapy Committee Task Group 53: quality assurance for clinical radiotherapy treatment planning. Med Phys 25:1773–1829

Galvin JM, Ezzell G, Eisbrauch A, Yu C, Butler B, Xiao Y, Rosen I, Rosenman J, Sharpe M, Xing L, Xia P, Lomax T, Low DA, Palta J (2004) Implementing IMRT in clinical practice: a joint document of the American Society for Therapeutic Radiology and Oncology and the American Association of Physicists in Medicine. Int J Radiat Oncol Biol Phys 58:1616–1634

Herman MG, Balter JM, Jaffray DA, McGee KP, Munro P, Shalev S, van Herk M, Wong JW (2001) Clinical use of electronic portal imaging: report of AAPM RadiationTherapy Committee Task Group 58. Med Phys 28:712–737

Herman MG, Mills MD, Gillin MT (2003) Reimbursement versus effort in medical physics practice in radiation oncology. J Appl Clin Med Phys 4:179–187

Herman MG, Klein EE, Mills MD, Boyer A(2005) Estimating medical physicist FTE using the 2003 Abt Survey and procedure volumes in radiation therapy (Abstract). Med Phys 32:2052

Herring DF, Compton DMJ (1971) The degree of precision required in radiation dose delivered in cancer radiotherapy. In: Glicksman AJ, Cohen M, Cunningham JR (eds) Computers in radiotherapy. Br J Radiol Special Report Series No. 5

Holmes T, McCullough EC (1983) Acceptance testing and quality assurance of automated scanning film densitometers used in dosimetry of electron and photon therapy beams. Med Phys 10:698–700

HPA (1970) Hospital Physics Association Report: a suggested procedure for the mechanical alignment of telegamma and megavoltage X-ray beam units

ICRU (1976) Report no. 24: Determination of absorbed dose in a patient irradiated by beams of X or gamma rays in radiotherapy procedures. International Commission on Radiation Units and Measurements, Bethesda, Maryland

ICRU (1993) Report no. 50: Prescribing, recording, and reporting photon beam therapy. International Commission on Radiation Units and Measurements, Bethesda, Maryland

IMRTCWG (2001) NCI IMRT Collaborative Working Group: intensity modulated radiation therapy: current status and issues of interest. Int J Radiat Oncol Biol Phys 51:880–914

ISCRO (1991) Radiation oncology in integrated cancer management: report of the Inter-Society Council for Radiation Oncology. Philadelphia, Pennsylvania

Jacky J, White CP (1990) Testing a 3-D radiation therapy planning program. Int J Radiat Oncol Biol Phys 18:253–261

Joyce E (1986) "Malfunction 54": unraveling deadly medical mystery of computerized accelerator gone awry. Am Med News

Karzmark CJ (1987) Procedural and operator error aspects of radiation accidents in radiotherapy. Int J Radiat Oncol Biol Phys 13:1599–1602

Klein EE, Low DA (2001) Interleaf leakage for 5- and 10-mm dynamic multileaf collimation systems incorporating patient motion. Med Phys 28:1703–1710

Klein EE, Taylor M, Michaletz-Lorenz M, Zoeller D, Umfleet W (1994) A mono-isocentric technique for breast and regional nodal therapy using dual asymmetric jaws. Int J Radiat Oncol Biol Phys 28:753–760

Klein EE, Harms WB, Low DA, Willcut V, Purdy JA (1995) Clinical implementation of a commercial multileaf collimator: dosimetry, networking, simulation, and quality assurance. Int J Radiat Oncol Biol Phys 33:1195–1208

Klein EE, Low DA, Maag D, Purdy JA (1996) A quality assurance program for ancillary high technology devices on a dual-energy accelerator. Radiother Oncol 38:51–60

Klein EE, Gerber RG, Zhu XR, Oehmke F, Purdy JA (1998) Mutliple machine implementation of enhanced dynamic wedge. Int J Radiat Oncol Biol Phys

Klein EE, Drzymala RE, Purdy JA, Michalski J (2005) Errors in radiation oncology: a study in pathways and dosimetric impact. J Appl Clin Med Phys 6:81–94

Kutcher GJ, Coia L, Gillin M, Hanson WF, Leibel S, Morton RJ, Palta JR, Purdy JA, Reinstein LE, Svensson GK, Weller M, Wingfield L(1994) Comprehensive QA for radiation oncology: report of AAPM Radiation Therapy Committee Task Group 40. Med Phys 21:581–618

Leavitt DD, Martin M, Moeller JH, Lee WL (1990) Dynamic wedge field techniques through computer-controlled collimator motion and dose delivery. Med Phys 17:87–91

Leunens G, van Dam J, Dutreix A, van der Schueren A (1990) Quality assurance in radiotherapy by in vivo dosimetry. I. Entrance dose measurements, a reliable procedure. Radiother Oncol 17:141

Low DA, Klein EE, Maag DK, Umfleet WE, Purdy JA (1996) Commissioning and periodic quality assurance of a clinical electronic portal imaging device. Int J Radiat Oncol Biol Phys 34:117–123

McCullough EC, Holmes TW (1985) Acceptance testing computerized radiation therapy treatment planning systems: direct utilization of CT scan data. Med Phys 12:237–242

Mellenberg DE, Dahl RA, Blackwell CR (1990) Acceptance testing of an automated scanning water phantom. Med Phys 17:311–314

Mutic S, Low DA (1998) Whole-body dose from tomotherapy delivery. Int J Radiat Oncol Biol Phys 42:229–232

Mutic S, Palta JR, Butker EK, Das IJ, Huq MS, Loo L-H D, Salter BJ, McCollough CH, van Dyk J (2003) Quality assurance for computed-tomography simulators and the computed-tomography-simulation process: report of the AAPM Radiation Therapy Committee Task Group no 68. Med Phys 30:2762–2792

Nath R, Biggs PJ, Bova FJ, Ling CC, Purdy JA, van de Geijn J, Weinhous MS (1994) AAPM code of practice for radiotherapy accelerators: report of AAPM Radiation Therapy Task Group no 45. Med Phys 21:1093–1121

NRC (2003) Nuclear Regulatory Commission 10 CFR part 35, medical use of byproduct material

Palta JR, Ayyangar KM, Suntharalingam N (1988) Dosimetric characteristics of a 6-mV photon beam from a linear accelerator with asymmetric collimator jaws. Int J Radiat Oncol Biol Phys 14:383–387

Parrino PA, Purdy JA (1983) The role of maintenance in quality assurance. In: Starkschall G (ed) Proc Symposium on Quality Assurance of Radiotherapy Equipment. American Institute of Physics, New York, pp 157–168

Patton GA, Gaffney DK, Moeller JH (2003) Facilitation of radiotherapeutic error by computerized record and verify systems. Int J Radiat Oncol Biol Phys 56:50–57

Purdy JA (1983) Quality assurance of electron linear accelerators. In: Starkschall G (ed) Proc Symposium on Quality Assurance of Radiotherapy Equipment. American Institute of Physics, New York, pp 86–106

Purdy JA (1991a) Organization of a quality assurance program: a large radiotherapy facility with affiliated clinics. In: Starkschall G, Horton J (eds) Quality assurance in radiotherapy physics. Proc American College of Medical Physics Symposium. Medical Physics Publishing, Madison, Wisconsin, pp 299–317

Purdy JA (1991b) QA of external beam megavoltage radiotherapy equipment. In: Starkschall G, Horton J (eds) Quality assurance in radiotherapy physics. Proc American College of Medical Physics Symposium. Medical Physics Publishing, Madison, Wisconsin, pp 23–43

Purdy JA, Harms WB, Gerber RG (1986) Report on a long-term quality assurance program. In: Kereiakes JG, Elson HR, Born CG (eds) Radiation oncology physics 1986. American Institute of Physics, New York, NY, pp 91–109

Purdy JA, Biggs PJ, Bowers C, Dally E, Downs W, Fraass BA, Karzmark CJ, Khan F, Morgan P, Morton R, Palta J, Rosen II, Thorson T, Svensson G, Ting J (1993) Medical accelerator safety considerations: report of AAPM Radiation Therapy Committee Task Group no 35. Med Phys 20:1261–1275

Purdy JA, Klein EE, Low DA (1995) Quality assurance and safety of new technologies for radiation oncology. Semin Radiat Oncol 5:156–165

Purdy JA, Grant W, Palta JR, Butler EB, Perez CA (2001) 3-D conformal and intensity modulated radiation therapy: physics and clinical applications. Advanced Medical Publishing, Madison, Wisconsin

Rosen II, Purdy JA (1992) Computer controlled medical accelerators. In: Purdy JA (ed) Advances in radiation oncology physics: dosimetry, treatment planning, and brachytherapy. American Institute of Physics, New York, pp 1–18

Slessinger ED, Gerber RG, Harms WB, Klein EE, Purdy JA (1993) Independent collimator dosimetry for a dual photon energy linear accelerator. Int J Radiat Oncol Biol Phys 27:681–687

Van Dyk J, Purdy JA (1999) Clinical implementation of technology and the quality assurance process. In: Dyk JV (ed) The modern technology of radiation oncology: a compendium for medical physicists and radiation oncologists. Medical Physics Publishing, Madison, Wisconsin, pp 19–51

Van Dyk J, Barrett RB, Cygler JE, Shraggo PC (1993) Commissioning and quality assurance of treatment planning computers. Int J Radiat Oncol Biol Phys 26:261–273

Weinhous MS, Purdy JA, Granda CO (1990) Testing of a medical linear accelerator's computer-control system. Med Phys 17:95–101

World Health Organization (1988) Quality assurance in radiotherapy. World Health Organization, Geneva

Williamson JF, Khan FM, Sharma SC (1981) Film dosimetry of megavoltage photon beams: a practical method of isodensity-to-isodose curve conversion. Med Phys 8:94–98

Part II:
Practical Clinical Applications

17 Central Nervous System Tumors

Volker W. Stieber, Kevin P. McMullen, Michael T. Munley, and Edward G. Shaw

CONTENTS

17.1   Introduction 425
17.2   Natural History 425
17.2.1 Anatomy of the Brain 425
17.2.2 Epidemiology 427
17.2.2.1 Primary CNS Tumors 427
17.2.2.2 Tumors Metastatic to the CNS 428
17.3   Workup and Staging 430
17.4   General Management 431
17.4.1 Medical Management 431
17.4.2 Surgical Management 431
17.4.3 Radiation Therapy 432
17.4.3.1 Definitive Radiation Therapy 432
17.4.3.2 Palliative Radiation Therapy 432
17.5   General Concepts of Modern Radiation
        Therapy Technique 434
17.5.1 Principles of Imaging-Based
        Treatment Planning 434
17.5.2 Target Volumes and Organs at Risk Specifications 434
17.5.3 Dose Reporting 436
17.5.4 Intensity-Modulated Radiation Therapy 436
17.6   Simulation Procedures 437
17.6.1 General Concepts of Positioning,
        Immobilization and Simulation 437
17.6.2 Specific Examples of Treatment Techniques 439
17.6.2.1 Pituitary Region 439
17.6.2.2 Meningiomas 440
17.6.2.3 Temporal Lobe Lesions 441
17.6.2.4 Frontal or Parietal Lobe Lesions 441
17.6.2.5 Central and Thalamic Lesions 442
17.6.2.6 Posterior Fossa 442
17.6.2.7 Whole Brain Irradiation 443
17.6.2.8 Craniospinal Irradiation 444
17.6.2.9 Spinal Tumors 446
17.7   Simulation Films and Portal Films 447
17.8   Dose Prescriptions 447
17.8.1 Primary CNS Tumors 447
17.8.2 Metastatic CNS Tumors 448
17.9   Future Directions 448
        References 449

V. W. Stieber, MD, Assistant Professor;
K. P. McMullen, MD, Assistant Professor;
M. T. Munley, PhD, Associate Professor;
E. G. Shaw, MD, Professor and Chairman
Department of Radiation Oncology, Wake Forest University
School of Medicine, Medical Center Boulevard, Winston-Salem, NC 27157-1030, USA

17.1 Introduction

The central nervous system (CNS) is comprised of the brain and spinal cord and their coverings and may be affected by both primary and metastatic tumors. The outcomes of treatment are highly variable, dependent on diagnosis. Patients with benign diseases are often able to live out their natural life span, while those with malignant tumors frequently have survival measured in weeks or months. It is incumbent upon the clinician to use an appropriate treatment technique for all patients with CNS disease in order to minimize the chance of significant acute toxicities (or late toxicities in survivors) and maximize the therapeutic benefit.

17.2 Natural History

17.2.1 Anatomy of the Brain

Knowledge of the basic topographical and functional anatomy of the brain is important for accurate communication of tumor location within the CNS as well as defining areas of functional eloquence that need to be considered when planning therapy. Generally, the brain can be considered to have three major divisions: the cerebrum, cerebellum and brain stem. When considering tumor location it is also common to distinguish between supratentorial (cerebral hemispheres and midline structures) and infratentorial (cerebellum, lower brain stem).

The longitudinal cerebral fissure divides the cerebrum into two hemispheres, Each hemisphere is the separated by the major sulci into six lobes: frontal, parietal, occipital and temporal, and the midline central and limbic lobes (Fig. 17.1).

The prominent central sulcus (of Rolando) separates the frontal lobe from the parietal lobe. The

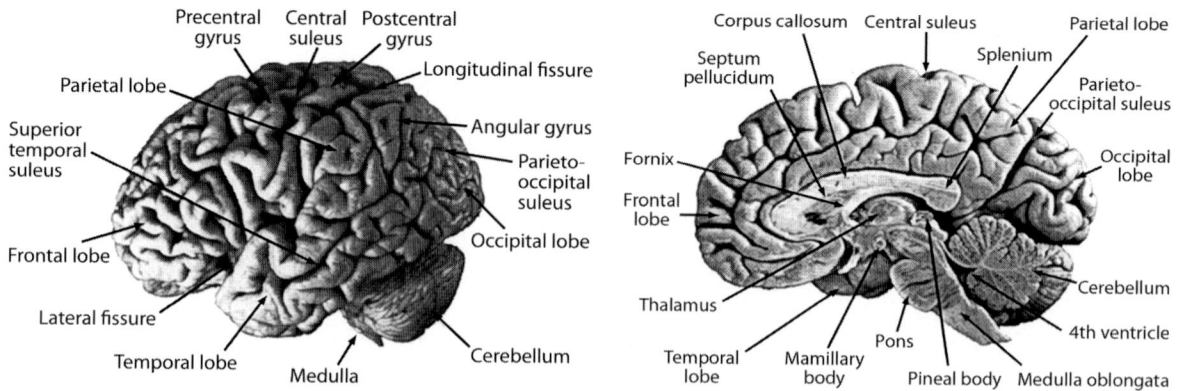

Fig. 17.1. a Lateral surface of the brain including cerebral hemisphere, cerebellum, and brainstem. The major sulci divide the cerebral cortex into four lateral lobes: frontal, parietal, occipital, and temporal. **b** Medial surface of the left brain demonstrating midline structures of the central and limbic lobes

parietal–occipital fissure separates parietal lobe from occipital lobe. The lateral fissure (of Sylvius) defines the temporal lobe boundaries. The cerebral hemispheres are connected by the corpus callosum, beneath which are the midline structures (third ventricle, pineal body, and midbrain) and the deep paramedian structures (lateral ventricles, caudate nucleus, lentiform nucleus, thalamus, and hypothalamus).

A basic understanding of the functional anatomy of the cerebral hemispheres can be approached in three ways. The first is a regional or "lobe-by-lobe" consideration of function. The occipital lobe is primarily involved with vision and its dependent functions. The temporal lobe processes sound, vestibular sensations, sights, smells, and other perceptions into complex "experiences" important for memory. Wernicke's area is located on the posterior portion of the superior temporal gyrus and plays a critical role in receptive speech. The parietal lobe, specifically the postcentral gyrus, is critically involved in somatosensory function. Sensory integration (body image) and Gnostic (perceptive) functions also reside within the parietal lobe. The frontal lobe is associated with higher level cognitive functions such as reasoning and judgment. The frontal lobe also contains the primary motor cortex (precentral gyrus) and Broca's area (inferior third of the frontal gyrus), important in expressive speech. The limbic lobe mediates memories, drives, and stimuli. It affects visceral functions central to emotional expression, such as sexual drive. Finally, the central lobe (insula) is important in visceral sensation and motility.

The second approach to functional neuroanatomy is the schema of BRODMANN, which numbers areas

of structural specialization (BRODMANN 1908a,b). These numbered areas, in some cases, correspond to the functional location of important primary sensory and motor areas. The 52 numbered areas provide both an anatomical and functional "road map" of the brain by which tumor location can be described (Fig. 17.2).

The final and most eloquent method to describe functional neuroanatomy is through various techniques of functional mapping (Fig. 17.3).

While classic mapping utilizes microelectrode stimulation of the cortical surface directly, new non-invasive techniques such as functional magnetic resonance imaging (fMRI), positron emission tomography (PET), and magnetoencephalography (MEG) are increasingly being integrated into clinical practice (BABILONI et al. 2004; BARNES et al. 1997; CHOI et al. 2005). These procedures allow for precise mapping of function in an individual and can accurately predict deficits related to injury of a given area by tumor or therapy.

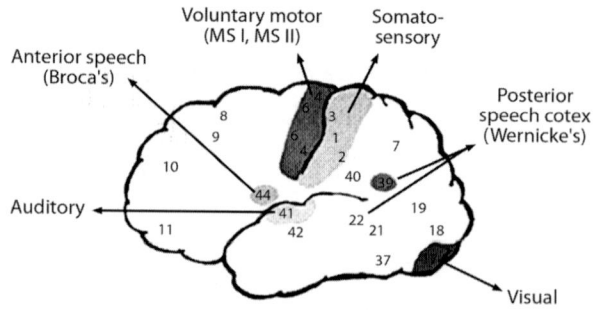

Fig. 17.2. Location of the major motor, sensory, and speech areas of the cerebral cortex with reference to Brodmann's area numbers

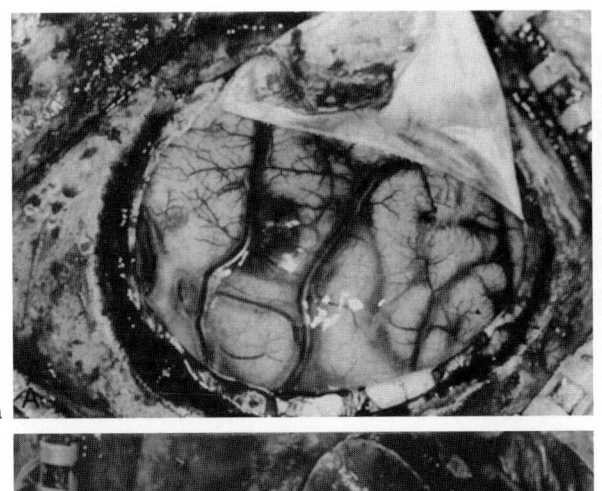

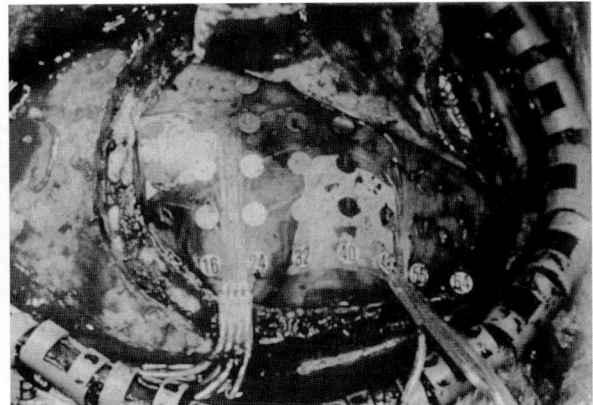

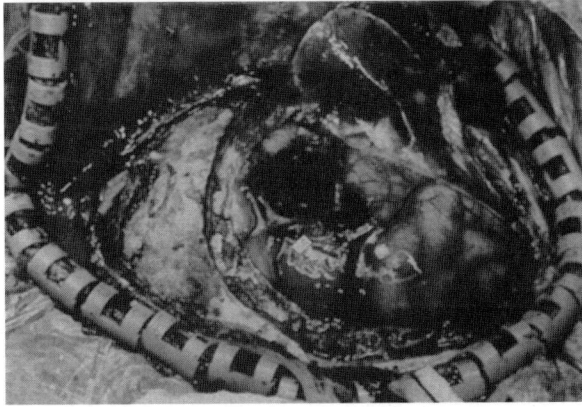

Fig. 17.3a–c. Cortical surface electrode mapping of the speech area in a patient undergoing resection of recurrent glioma. **a** Exposed cortical surface. **b** Mapping array in place. **c** Resection cavity with electrode number 5 demarcating speech area. Patient had normal speech postoperatively

17.2.2
Epidemiology

17.2.2.1
Primary CNS Tumors

The incidence rate of all primary non-malignant and malignant (including pilocytic astrocytomas) brain and CNS tumors is 14.1 cases per 100,000 person-years (6.8 per 100,000 person-years for benign and borderline tumors and 7.3 per 100,000 person-years for malignant tumors). The overall incidence rates increased from 13.5 per 100,000 person-years in 1997 to 14.7 per 100,000 person-years in 2001 (CBTRUS 2004). The rate is higher in females (14.3 per 100,000 person-years) than males (13.9 per 100,000 person-years). An estimated 41,130 new cases of primary non-malignant and malignant brain and CNS tumors were expected to be diagnosed in 2004 (CBTRUS 2004). The worldwide incidence rate of primary malignant brain and CNS tumors, age-adjusted using the world standard population, is 3.6 per 100,000 person-years in males and 2.5 per 100,000 person-years in females. The incidence rates are higher in more developed countries (males 5.9 per 100,000 person-years; females 4.1 per 100,000 person-years) than in less developed countries (males 2.8 per 100,000 person-years; females 2.0 per 100,000 person-years) (FERLAY et al. 2001). The incidence rate of childhood primary non-malignant and malignant brain and CNS tumors is 4.0 cases per 100,000 person-years. The rate is higher in males (4.2 per 100,000 person-years) than females (3.8 per 100,000 person-years) (CBTRUS 2004). An estimated 18,500 deaths in 2005 will be attributed to primary malignant brain and CNS tumors (AMERICAN CANCER SOCIETY 2005).

The majority of primary CNS tumors (31%) are located within the frontal, temporal, parietal, and occipital lobes of the brain. Tumors in other locations in the cerebrum, ventricle, cerebellum, and brainstem account for 3, 2, 4, and 2% of all tumors, respectively. Other tumors of the brain account for 16% of all tumors. Tumors of the meninges represent 24% of all tumors. The cranial nerves and the spinal cord/cauda equina account for 6% and 4% of all tumors, respectively. The pituitary and pineal glands account for about 7% of tumors. Olfactory tumors of the nasal cavity and other CNS tumors, NOS (not otherwise specified), each account for less than 1% of tumors (CBTRUS 2004).

The overall incidence of primary spinal cord tumors is approximately 10–19% that of all primary brain tumors (CONNOLY 1982). The incidence ratio of intracranial to intraspinal tumors is up to four times higher in pediatric patients than in adults, the frequency of specific spinal cord tumors being quite different from that of their counterpart brain tumors. The incidence ratios of intracranial to intraspinal astrocytomas, ependymomas and meningiomas are approximately 10:1, 3:1, and 18:1, respectively (SASANELLI et al. 1983). Schwannoma and meningioma account for approximately 60% of primary spinal tumors, with schwannoma being slightly more frequent; both types occur primarily in adults. Regional differences are also noted. Gliomas constitute 46% of primary intracranial tumors but only 23% of spinal tumors. Most primary spinal gliomas are ependymomas with a predilection for the cauda equina.

In 2000, the World Health Organization (WHO) updated their comprehensive classification of primary CNS neoplasms (INTERNATIONAL AGENCY FOR RESEARCH ON CANCER 2000). Table 17.1 shows the WHO's pathological classification system of common primary CNS tumors. The most frequently reported histology is a predominately benign tumor, meningioma, which accounts for over 29% of all tumors, followed closely by glioblastomas and astrocytomas. The predominately benign nerve sheath tumors and pituitary tumors account for 8% and 6% of all tumors, respectively. Acoustic neuromas account for 54% of all nerve sheath tumors. Gliomas are tumors that arise from glial cells, and include astrocytomas, glioblastomas, oligodendrocytomas, ependymomas, mixed gliomas, malignant gliomas NOS, and neuroepithelial tumors. The broad category glioma represents 42% of all tumors (CBTRUS 2004). Of gliomas, 61% occur in the frontal, temporal, parietal, and occipital lobes of the brain. Glioblastomas account for the majority of gliomas, while astrocytomas and glioblastomas account for three-quarters of gliomas (CBTRUS 2004). Table 17.2 shows a comparison of different grading systems that have been used to classify the malignant gliomas. The most common spinal cord intramedullary tumors are those that are derived from glial precursors (astrocytes, ependymocytes, and oligodendrocytes) (PRESTON-MARTIN 1990).

17.2.2.2
Tumors Metastatic to the CNS

Metastatic brain tumors are the most common intracranial neoplasms in adults and are a significant cause of morbidity and mortality. The estimate of the incidence rate of metastatic brain tumors varies from 8.3–11 per 100,000 (PERCY et al. 1972; WALKER et al. 1985). In two cohorts of patients who were diagnosed with colorectal, lung, breast, or kidney carcinoma or melanoma, brain metastases were diagnosed in 8.5–9.6% of patients (BARNHOLTZ-SLOAN et al. 2004; SCHOUTEN et al. 2002). The cumulative incidence was estimated at 16.3–19.9% in patients with lung carcinoma, 6.5–9.8% in patients with renal carcinoma, 6.9–7.4% in patients with melanoma, 5.0–5.1% in patients with breast carcinoma, and 1.2–1.8% in patients with colorectal carcinoma.

The spine is the most common site of bony metastases overall, with a reported incidence in cancer patients of 40% (BYRNE 1992). Malignant spinal cord compression (MSCC) from epidural metastases occurs in 5–10% of cancer patients and in up to 40% of patients with preexisting nonspinal bone metastases (BILSKY et al. 1999; BYRNE 1992; HEALEY and BROWN 2000; WONG et al. 1990). Of those with bony spinal disease, 10–20% develop symptomatic spinal cord compression, resulting in between 14,100 and 28,200 cases per year (GERSZTEN and WELCH 2000; LOBLAW et al. 2003; SCHABERG and GAINOR 1985). The overall incidence of MSCC within 5 years of death from cancer is 2.5% (LOBLAW et al. 2003). Symptoms depend on location of the compression and can involve the spinal cord at any level. The incidence of MSCC by vertebral level is 10–16% cervical, 35–40% in T1-6, 44–55% in T7-12 and 20% lumbar (GILBERT et al. 1978; PATCHELL et al. 2003; PIGOTT et al. 1994). Metastatic lesions present initially at multiple, noncontiguous levels in 10–38% of cases (GILBERT et al. 1978; O'ROURKE et al. 1986; RUFF and LANSKA 1989).

The histology of MSCC follows the incidence patterns of malignant disease, with the most common histological diagnoses (breast, lung and prostate) accounting for approximately half of all cases (AMERICAN CANCER SOCIETY 2005; BYRNE 1992). Approximately 25% of all patients with MSCC have breast cancer, 15% lung cancer, and 10% prostate carcinomas. Overall, 5.5% of breast cancer patients, 2.6% of lung cancer patients, 7.2% of prostate cancer patients, and 0.8% of colorectal cancer patients experience a MSCC (LOBLAW et al. 2003). Other commonly reported histological diagnoses in adults

Table 17.1. World Health Organization Classification of Tumors of the Nervous System (INTERNATIONAL AGENCY FOR RESEARCH ON CANCER 2000)

Tumors of neuroepithelial tissue
Astrocytic tumors
Diffuse astrocytoma
• Fibrillary astrocytoma
• Protoplasmatic astrocytoma
• Gemistocytic astrocytoma
Anaplastic astrocytoma
Glioblastoma
• Giant cell glioblastoma
• Gliosarcoma
Pilocytic astrocytoma
Pleomorphic xanthoastrocytoma
Subependymal giant cell astrocytoma

Oligodendroglial tumors
Oligodendroglioma
Anaplastic oligodendroglioma

Mixed gliomas
Oligoastrocytoma
Anaplastic oligoastrocytoma

Ependymal tumors
Ependymoma
• Cellular
• Papillary
• Clear cell
• Tanycytic
Anaplastic ependymoma
Myxopapillary ependymoma
Subependymoma

Choroid plexus tumors
Choroid plexus papilloma
Choroid plexus carcinoma

Glial tumors of uncertain origin
Astroblastoma
Gliomatosis cerebri
Chordoid glioma of the third ventricle

Neuronal and mixed neuronal–glial tumors
Gangliocytoma
Dysplastic gangliocytoma of cerebellum (Lhermitte–Duclos)
Desmoplastic infantile astrocytoma/ganglioglioma
Dysembryoplastic neuroepithelial tumor
Ganglioglioma
Anaplastic ganglioglioma
Central neurocytoma
Cerebellar liponeurocytoma
Paraganglioma of the filum terminale

Neuroblastic tumors
Olfactory neuroblastoma (Esthesioneuroblastoma)
Olfactory neuroepithelioma
Neuroblastomas of the adrenal gland and sympathetic nervous system

Pineal parenchymal tumors
Pineocytoma
Pineoblastoma

Pineal parenchymal tumor of intermediate differentiation

Embryonal tumors
Medulloepithelioma
Ependymoblastoma
Medulloblastoma
• Desmoplastic medulloblastoma
• Large cell medulloblastoma
• Medullomyoblastoma
• Melanotic medulloblastoma
Supratentorial primitive neuroectodermal tumor (PNET)
• Neuroblastoma
• Ganglioneuroblastoma
Atypical teratoid/rhabdoid tumor

Tumors of peripheral nerves
Schwannoma (neurilemmoma, neurinoma)
Cellular
Plexiform
Melanotic

Neurofibroma
Plexiform

Perineurioma
Intraneural perineurioma
Soft tissue perineurioma

Malignant peripheral nerve sheath tumor (MPNST)
Epithelioid
MPNST with divergent mesenchymal and/or epithelial differentiation
Melanotic
Melanotic psammomatous

Tumors of the meninges
Tumors of meningothelial cells
Meningioma
• Meningothelial
• Fibrous (fibroblastic)
• Transitional (mixed)
• Psammomatous
• Angiomatous
• Microcystic
• Secretory
• Lymphoplasmacyte-rich
• Metaplastic
• Clear cell
• Chordoid
• Atypical
• Papillary
• Rhabdoid
• Anaplastic meningioma

Mesenchymal, non-meningothelial tumors
Lipoma
Angiolipoma
Hibernoma
Liposarcoma (intracranial)
Solitary fibrous tumor

Fibrosarcoma
Malignant fibrous histiocytoma
Leiomyoma
Leiomyosarcoma
Rhabdomyoma
Rhabdomyosarcoma
Chondroma
Chondrosarcoma
Osteoma
Osteosarcoma
Osteochondroma
Hemangioma
Epithelioid hemangioendothelioma
Hemangiopericytoma
Angiosarcoma
Kaposi sarcoma

Primary melanocytic lesions
Diffuse melanocytosis
Melanocytoma
Malignant melanoma
Meningeal melanomatosis

Tumors of uncertain histogenesis
Hemangioblastoma

Lymphomas and hemopoietic neoplasms
Malignant lymphomas
Plasmacytoma
Granulocytic sarcoma

Germ cell tumors
Germinoma
Embryonal carcinoma
Yolk sac tumor
Choriocarcinoma
Teratoma
• Mature
• Immature
• Teratoma with malignant transformation
Mixed germ cell tumors

Tumors of the sellar region
Craniopharyngioma
• Adamantinomatous
• Papillary
Granular cell tumor

Table 17.2. Grading of astrocytic tumors. A given tumor may not fall under the same designation in all three systems (DAUMAS-DUPORT et al. 1988; HUNTINGTON et al. 1965; INTERNATIONAL AGENCY FOR RESEARCH ON CANCER 2000)

WHO designation	WHO grade	Kernohan grade	St. Anne/ Mayo grade	St. Anne/Mayo grade criteria
Pilocytic astrocytoma	I	I	(excluded)	(n/a)
Astrocytoma	II	I, II	1	0 criteria present*
			2	1 criterion present: usually nuclear atypia
Anaplastic astrocytoma	III	II, III	3	2 criteria present: usually nuclear atypia and mitosis
Glioblastoma multiforme	IV	III, IV	4	3–4 criteria present: usually the above and/or endothelial proliferation and/or necrosis

*Criteria include nuclear atypia, mitosis, endothelial proliferation and necrosis

include, by order of cumulative incidence, multiple myeloma, nasopharynx, renal cell, melanoma, small-cell lung, lymphoma and cervix (BYRNE 1992; LOBLAW et al. 2003; SCHIFF et al. 1998).

17.3
Workup and Staging

For CNS tumors, both benign and malignant, MRI with and without contrast remains the gold standard for imaging (RICCI and DUNGAN 2001). The preferred slice thickness of MRI is 5 mm or less with a 2.5-mm or lower slice sampling. T1-weighted images with contrast allow excellent visualization of contrast-enhancing tumors such as meningiomas, glioblastoma multiforme and brain metastases. T2-weighted images demonstrate areas of edema that reflect involvement by infiltrative low- or high-grade gliomas. T1-weighted FLAIR images are also useful in this regard. MRI fusion or registration with the treatment planning computed tomography (CT) scan should be utilized for target definition; this is described in the section Principles of Imaging-Based Treatment Planning. Other imaging studies such as MRI spectroscopy, fMRI, PET scans, and single photon emission tomography (SPECT) scans better reflect biological characteristics of CNS tumors, such as tumor metabolism, proliferation, oxygenation and blood flow, and function of surrounding normal brain. Functional and biological imaging data are currently incorporated into decisions as to whether ablative procedures such as surgery, radiosurgery, or brachytherapy can be safely considered. Also, PET scans and MRI spectroscopy may allow for differentiating active tumor versus radionecrosis after radiation therapy. The integration of MR spectroscopy and PET imaging into radiation therapy treatment planning is a current

topic of research (MUNLEY et al. 2002; NUUTINEN et al. 2000; PIRZKALL et al. 2000).

CNS tumors rarely metastasize outside the CNS but may spread within it. For example, medulloblastomas, primitive neuroectodermal tumors, anaplastic ependymomas, choroids plexus carcinomas, pineoblastomas, germ cell tumors, and lymphomas may involve the cerebrospinal fluid (CSF), leptomeninges (i.e., coverings of the brain), or spinal cord. Studies that stage the extent of these tumors include MRI of the entire neural axis and CSF cytology.

In patients who present emergently, CT can be obtained rapidly, and provide information on ventricular obstruction, hemorrhage or edema, although imaging of the parenchyma is inferior to MRI. Lumbar puncture should be avoided if at all possible until intracranial pressure has normalized, due to the risk of herniation and death. The most important modality in the workup of suspected MSCC is gadolinium-enhanced MRI of the entire spinal axis. With the exception of a primary paraspinal or neuraxis tumor, MSCC occurs most often in the setting of disseminated disease from a distant primary tumor site. A potential pitfall in the initial evaluation of a patient with suspected spinal cord compression is imaging only the symptomatic area of the spine. Frequently, patients with lower body or extremity symptoms, and/or radicular pain in the lumbar distribution present for evaluation and management with only partial spine imaging. However, 25% of patients have spinal cord compression verified at multiple levels by MRI, and approximately two-thirds of these have involvement of different regions of the spine (HUSBAND et al. 2001). When a sensory level is present on patient evaluation, it may be two or more segments different from the actual lesion on MRI in 28% of patients, and four or more levels distant from the lesion in 21% of patients (HUSBAND et al. 2001).

17.4
General Management

Multimodality therapy for CNS tumors routinely takes the form of medical treatment followed by surgical resection for durable decompression and to obtain a tissue diagnosis. Patients who are medically inoperable, refuse surgery, or have multiple and/or unresectable lesions often receive whole brain radiation therapy for palliation of their symptoms.

17.4.1
Medical Management

Medical treatment generally consists of steroids with or without mannitol (SARIN and MURTHY 2003). Both drugs have previously been shown to decrease peritumoral brain edema by different mechanisms of action (BELL et al. 1987). Mannitol is used in steroid refractory patients. A common regimen of mannitol is 20–25% solution given intravenously (i.v.) over approximately 30 min dosed at 0.5–2.0 g/kg (QUINN and DEANGELIS 2000). Patients who present with emergent symptoms from intracranial malignancy are typically treated with dexamethasone. Response to therapy is usually noted within 12–18 h of administration, and over 80% of patients show dramatic improvement by 3–4 days after initiation of therapy (FRENCH 1966; LONG et al. 1966). A common regimen in patients receiving radiation therapy is high-dose dexamethasone (10–25 mg i.v.) followed by maintenance on oral steroids (4–6 mg p.o. every 6 h), with tapering initiated at the clinician's discretion, usually over 1–2 months following the completion of radiation therapy (SARIN and MURTHY 2003; VECHT et al. 1994; WOLFSON et al. 1994). In the setting of MSCC from solid tumors, dexamethasone has been shown to improve rates of surviving with intact gait function (KALKANIS et al. 2003; SORENSEN et al. 1994).

Given the concerns over the incidence of steroid-induced toxicity with steroid dosing longer than 21 days in duration, higher doses and longer tapering schedules should be based on the physician's assessment of symptom severity and response (HEIMDAL et al. 1992; SORENSEN et al. 1994; WEISSMAN et al. 1991). Side effects of intermediate- to long-term steroid use include: hyperglycemia; insomnia; emotional lability; thrush; gastric irritation, ulceration and possibly perforation; proximal muscle wasting; weight gain and adiposity (moon facies, buffalo hump, centripetal obesity), osteoporotic com-

pression fractures; and aseptic necrosis of the hip joints (BILSKY and POSNER 1993). Select patients with MSCC may not require steroids during treatment (MARANZANO et al. 1996). This management strategy may be considered reasonable if a patient is at high risk of complication from steroids due to underlying medical comorbidities such as peptic ulcer disease, uncontrolled diabetes, or other medical problems that may cause severe or life-threatening problems if steroids are initiated.

Stabilization of the patient in status epilepticus in order to perform imaging and make management decisions is critical (WORKING GROUP ON STATUS EPILEPTICUS AND EPILEPSY FOUNDATION OF AMERICA 1993). After securing the airway and stabilizing the patient, seizure activity must be terminated as rapidly as possible. Phenytoin and rapid onset/short-acting benzodiazepines are commonly used to quickly control seizure activity. Recommended initial regimens include 0.1 mg/kg at 2 mg/min lorazepam or 0.2 mg/kg diazepam at 5 mg/min. Phenytoin infusion of 15–20 mg/kg at 50 mg/min or less in adults is indicated for seizure activity refractory to benzodiazepines or after truncation of seizures with diazepam (WORKING GROUP ON STATUS EPILEPTICUS AND EPILEPSY FOUNDATION OF AMERICA 1993). Failure to control seizures can potentially lead to physical injuries, airway compromise and secondary brain hypoxia/injury, or coma (EPILEPSY FOUNDATION OF AMERICA 1993; QUINN and DEANGELIS 2000). However, there is no evidence to support routine use of prophylactic anticonvulsants for patients diagnosed with a brain tumor (FORSYTH et al. 2003; GLANTZ et al. 1996).

17.4.2
Surgical Management

Surgical resection and/or placement of a shunt is often required for emergent management of brain tumors causing life threatening hydrocephalus, mass effect, or profound neurological impairment and may relieve symptoms enough that other treatment modalities may be considered as alternate or adjunctive management strategies. Symptoms are usually related to mass effect, so resection or debulking are often the only logical choices if medical therapy fails to provide improvement in neurological symptoms. Rapid surgical decompression is the treatment of choice for such problems when surgery can be safely performed based on patient performance status or tumor location. If no neurosurgical team is available, transfer of the patient should be initiated while medical measures are undertaken to stabilize the patient.

Laminectomy has historically been the standard surgery for MSCC, with most series in the literature showing no benefit to laminectomy-treated patients over patients managed with radiation therapy (Byrne 1992; Gilbert et al. 1978; Loblaw and Laperriere 1998; Young et al. 1980). In reality, many patients with spinal cord compression are not surgical candidates and are best treated with steroids and radiation therapy as their primary modality. Even patients with very poor initial performance or mobility/continence status can be helped by receiving emergent radiation therapy. A randomized trial evaluating the benefit of adding surgical decompression to the radiotherapeutic management of symptomatic metastatic spinal cord compression showed that patients who underwent decompressive surgery had a significantly improved median time of gait retention and ability to regain gait function (Patchell et al. 2003). Overall survival was not significantly different. These data indicate that all patients presenting with MSCC of short duration should be evaluated by a neurosurgeon for emergent decompression prior to initiating radiation therapy.

17.4.3
Radiation Therapy

17.4.3.1
Definitive Radiation Therapy

Radiation therapy is a mainstay of the management of most malignant and a significant number of benign primary CNS tumors. Table 17.3 provides an overview of the most common primary CNS tumors and literature-based guidelines for treatment. For tumors treated with a "shrinking-field" technique, ICRU (International Commission on Radiation Unit) definitions of treatment volumes for both the initial and boost fields are given, together with general dosing guidelines. Selected outcome endpoints are provided along with the appropriate references. A more detailed description of treatment technique is provided for selected tumors later in the chapter.

17.4.3.2
Palliative Radiation Therapy

The fact that 60–70% of patients who present with brain metastases have multiple lesions makes radiotherapy the primary modality for palliation in the majority of cases (Hazuka et al. 1993; Patchell 2003). Many patients will respond dramatically to medical therapy and radiation in the emergent setting with an improvement in their performance status, particularly if their symptoms are largely caused by edema. With this in mind, radiation therapy dosing schedules for the treatment of emergent patients can be tailored to patient parameters such as initial response to steroids, extent of extracranial disease, primary site and purported response of primary to systemic therapy. Two randomized trials comparing radiotherapy with or without surgical resection in the management of a solitary brain metastasis have documented a survival advantage with the addition of surgery over radiation alone (Patchell et al. 1990; Vecht et al. 1993). A third randomized trial was negative (Mintz et al. 1996). There is no level-I evidence demonstrating any survival benefit from operating on patients with multiple metastases. However, patients with multiple lesions and severe neurological symptoms from a dominant metastasis that is unresponsive to medical therapy may benefit from a craniotomy for reasons described above. An improvement in the patient's performance status may then allow further aggressive therapy with external beam radiation therapy and/or radiosurgery. Leukemic brain infiltration causing acute mental status changes and/or impending herniation is a rare entity treated with whole-brain radiation therapy.

Patients with malignant glioma who require emergent treatment are typically treated with regimens similar to those used for brain metastases. Surgical debulking is the mainstay of emergent treatment, as is the initiation of steroid therapy. Patients who are unable to undergo surgical debulking may be treated with a short course of whole-brain radiation similar to that used for brain metastases.

Although no randomized trials exist comparing radiotherapy over best supportive care or medical therapy alone, every published series for MSCC has shown efficacy of radiation therapy in helping patients relieve pain and retain/regain lost function. Morbidity is generally low and well tolerated even by patients with a poor performance status. Approximately 89% of patients who are ambulatory prior to radiation therapy can expect to retain gait function, while an average of 39% of paretic patients and only 10% of paraplegic patients will remain ambulatory (Loblaw et al. 2003).

Table 17.3. Suggested definitions of ICRU volumes by diagnosis, based on magnetic resonance imaging and dose ranges (in Gray; Gy) delivered to these volumes. *GTV* gross tumor volume, *CTV* clinical target volume, *GTV* gross tumor volume, *+C* with contrast, *LC* local control, *MS* median survival, *DFS* disease-free survival, *OS* overall survival, *LC* local control. The numbers in subscript denote years

Diagnosis	Definition of initial treatment field (CTV)	Usual dose to CTV	Definition of final field (to GTV)	Usual dose to GTV	Dose to craniospinal axis (if indicated)	MS	DFS$_5$	OS$_1$	OS$_5$	LC$_5$	Selected references
WHO grade-I glioma	(n/a)	(n/a)	Enhancing tumor (T1 + C); 1 cm margin	45–50.4 Gy	(n/a)		95%	95%			Brown et al. (2004)
WHO grade-II glioma	Enhancing tumor (T1 + C); Edema (T2/FLAIR) 2 cm margin	45 Gy	Enhancing tumor (T1+C); 2 cm margin	50.4–54 Gy	(n/a)		37–50%		58–73%		Karim et al. (1996, 2002); Shaw et al. (2002)
WHO grade-III glioma	Enhancing tumor (T1+C); edema (T2/FLAIR); 2-cm margin	45–50.4 Gy	Enhancing tumor (T1+C); 2-cm margin	59.4 Gy	Leptomeningeal spread on MRI: 30–39.6 Gy. Bulky disease: 55.8–59.4 Gy	17.5–58.6 months			38%		Prados et al. (2004); Scott et al. (1998); Tortosa et al. (2003)
WHO grade-IV glioma	Enhancing tumor (T1+C); edema (T2/FLAIR); 2-cm margin	45–50.4 Gy	Enhancing tumor (T1+C); 2-cm margin	59.4–64.8 Gy	Leptomeningeal spread on MRI: 30–39.6 Gy. Bulky disease: 55.8–59.4 Gy	17.5–17.1 months		28–70%	0–14%		Shaw et al. (2003)
Meningioma, benign/atypical	(n/a)	(n/a)	Enhancing tumor (T1+C); 1-cm margin	52.2–64.8 Gy	(n/a)		48–89%		58–85%		Condra et al. (1997); Goldsmith et al. (1994); Stafford et al. (1998)
Meningioma, malignant	Enhancing tumor (T1+C); edema (T2/FLAIR); 2-cm margin	45–50.4 Gy	Enhancing tumor (T1+C); 2-cm margin	55.8–59.4 Gy	(n/a)	1.5 years					Perry et al. (1999)
Pituitary adenoma	(n/a)	(n/a)	Enhancing tumor (T1+C); 0.7–1 cm margin	45–50.4 Gy (non-functioning); 45–54 Gy (functioning)	(n/a)					90–95% (33–95% biochemical control)	Stieber and deGuzman (2003)
Ependymoma	Enhancing tumor (T1+C); edema (T2/FLAIR); 2-cm margin	45 Gy	Enhancing tumor (T1+C); 2-cm margin	50.4–55.8 Gy	Negative CSF: 30 Gy. Positive CSF: 36 Gy. Gross leptomeningeal spread on MRI: 39.6 Gy. Bulky disease: 54 Gy			67–100%		95–100%	Stieber et al. (2005)
Chordoma, chondrosarcoma	Enhancing tumor (T1 + C); tumor bed; 2-cm margin	50.4 Gy	Enhancing tumor (T1+C); 2-cm margin	59.4–70.2 Gy	(n/a)		36–72%		75–80%	40–75%	Stieber et al. (2005)
Central neurocytoma	(n/a)	(n/a)	FLAIR changes; 1-cm margin	50.4–55.8 Gy	(n/a)				98%	98%	Rades et al. (2003b, 2005)
Vertebral hemangioma	Enhancing vascular lesion Entire involved vertebral body ≥1 cm margin	(n/a)	(n/a)	36–45 Gy	(n/a)					82%	Rades et al. (2003a)
Brain metastases	Whole brain	30–37.5 Gy (adjuvant 40–50.4 Gy)	(n/a)	(n/a)	(n/a)	3.8–7.1 months		12–32%			Gaspar et al. (1997, 2000)
Spinal cord metastasis	Enhancing tumor (T1+C); tumor bed; surgical track, including scar; One vertebral body above and below	30–37.5 Gy	(n/a)	(n/a)	(n/a)	3.3–4.2 months					Patchell et al. (2003)
Leukemic brain infiltration	Whole brain	24–30 Gy in 1.8–3.0 Gy fractions	(n/a)	(n/a)	18 Gy	9 months					Sanders et al. (2004)

17.5
General Concepts of Modern Radiation Therapy Technique

17.5.1
Principles of Imaging-Based Treatment Planning

The current standard of care for radiation therapy of primary CNS tumors is to use CT- and MRI-based 3-D treatment planning. CT simulation is still necessary, since most commercial treatment planning systems perform dose calculations from CT data sets. Fusion of a MRI scan data set (see below) with the treatment planning CT scan allows for optimum definition of the extension of the tumor and accurate delineation of surrounding structures of interest including the optic apparatus. MRI scans are essential to define treatment-planning volumes based on the reports of the ICRU 50 (INTERNATIONAL COMMISSION ON RADIATION UNITS AND MEASUREMENTS I 1993) and ICRU 62 (INTERNATIONAL COMMISSION ON RADIATION UNITS AND MEASUREMENTS I 1999), which are described in detail in the section on target volume definition, below (Table 17.3). Imaging of tumor biology or physiology may provide additional information for radiation therapy treatment planning. Functional MRI scans that image cerebral blood flow can show regions of normal brain function, e.g., motor strip, expressive and receptive language areas (RICCI and DUNGAN 2001). With three-dimensional MR spectroscopy, the choline-to-N-acetylaspartate ratio or index (CNI) appears to be both sensitive and specific at differentiating tumor from normal tissue when the proper threshold is selected (MCKNIGHT et al. 2002). [^{11}C]-Methionine PET imaging also shows promise in improving the anatomical delineation of low-grade gliomas (NUUTINEN et al. 2000). Currently, the main utility of these functional and metabolic imaging studies is to better define the anatomical

extent of the tumor for radiation therapy treatment planning. Biological or physiological data are not yet imported and incorporated into the planning process, but ultimately will be when so-called "bioanatomic" radiation therapy treatment planning becomes widely available (CARSON et al. 2003; MORRIS et al. 2001).

All diagnostic information, but particularly MRI scans (including T2 and FLAIR images) and CT scans, as well as clinical and surgical findings, should be combined to define the tumor volume and critical structures. Depending on diagnosis, some patients may enjoy survival measured in years rather than months, so it is especially critical to respect dose-tolerances of normal structures in order to limit late toxicities. Normal tissues to be contoured include, at a minimum, the eyes (including lacrimal glands and lenses), optic nerves, optic chiasm, pituitary gland, brainstem, and temporal lobes. Using dose–volume histograms (DVHs), dose to the tumor can be maximized and normal tissue dose minimized by analyzing competing treatment plans.

Table 17.4 shows normal tissue tolerance of critical normal intracranial structures to radiation therapy. Comparison of dose tolerances to the doses required to control disease demonstrated clearly that at the "standard" treatment doses of 54–60 Gy the probability of causing serious toxicity is quite low as long as care is taken with planning. Beam energies of no less than 6 MV should be used in order to spare surrounding structures, most notably the temporal lobes. A very good balance between depth dose and penumbra width is provided by 10-MV photons.

17.5.2
Target Volumes and Organs at Risk Specifications

Treatment planning is based on the three-dimensional volumes of interest described by the INTER-

Table 17.4. Normal tissue tolerance of intracranial organs at risk. All doses are in cGy at conventional fractionation (EMAMI et al. 1991)

Organ	TD5/5 (Volume)			TD50/5 (Volume)			Endpoint
	1/3	2/3	3/3	1/3	2/3	3/3	
Brain	6000	5000	4500	7500	6500	6000	Necrosis, infarction
Brainstem	6000	5300	5000			6500	Necrosis, infarction
Optic nerve			6000				Blindness
Optic chiasm			5400				Blindness
Eye (lens)			1000				Cataracts
Eye (retina)			4500				Blindness
Lacrimal gland			3000				Dry eye syndrome
Ear (mid/external)	3000	3000	5500	4000	4000	4000	Acute serous otitis
	5500	5500	5500	6500	6500	6500	Chronic serous otitis
Pituitary			<4500			6000	Panhypopituitarism

NATIONAL COMMISSION ON RADIATION UNITS AND MEASUREMENTS (ICRU) (INTERNATIONAL COMMISSION ON RADIATION UNITS AND MEASUREMENTS I 1993, 1999). Figure 17.4 shows a graphical representation of these volumes as described below.

Based on the ICRU 50 and 62 reports, gross tumor volume (GTV) represents the grossly visible disease burden (INTERNATIONAL COMMISSION ON RADIATION UNITS AND MEASUREMENTS I 1993, 1999). For grade-IV astrocytoma (glioblastoma multiforme) this is a T1-enhancing abnormality on MRI. If there is no residual abnormality after a surgical resection, the tumor resection cavity is defined to be the GTV. Surrounding edema is not considered part of the GTV. The clinical target volume (CTV) is subclinical microscopic malignant disease, often seen as T2 or FLAIR abnormality (which does include edema) on MRI (INTERNATIONAL COMMISSION ON RADIATION UNITS AND MEASUREMENTS I 1993). Suggested definitions of and doses to be delivered to GTV and CTV are given in Table 17.3 by histological diagnosis.

The planning target volume (PTV) is also referred to as "dosimetric margin". The dosimetric margin of the PTV takes two additional margins into consideration (INTERNATIONAL COMMISSION ON RADIATION UNITS AND MEASUREMENTS I 1999). The PTV1 is the CTV plus a dosimetric margin; the smaller PTV2 is the GTV plus dosimetric margin. The internal margin is defined so as to take into account variations in size, shape, and position of the CTV in relation to anatomical reference points. The set-up margin is added to take into account uncertainties in patient-beam positioning. Segregating the internal margin and the set-up margin reflects the differences in the source of uncertainties. The internal margin is due mainly to physiological variations that are difficult or impossible to control, such as (potential) fluctuations in the mass effect from cerebral edema which may occur over the course of treatment. In contrast, the set-up margin is added because of uncertainties related mainly to technical factors that can be reduced by more accurate positioning and immobilization of the patient (such as stereotactic positioning), as well as improved mechanical stability of the machine. The addition of uniform margins that take into account all types of uncertainties would generally lead to an excessively large PTV, which could result in exceeding normal tissue tolerances. Thus, the balance between disease control and risk of complications may require an evaluation based on the experience and the judgment of the clinician in order to avoid serious treatment-related complications. The PTV may therefore be reduced in areas near critical structures. Dosimetric margins as low as 3–5 mm may be acceptable with appropriate immobilization devices. The target is usually considered to be appropriately treated if the PTV is enclosed within the 95–105% isodose line. For plans emphasizing homogeneous dose delivery, typically no more than 20% of the PTV should exceed 110% of the prescription dose.

Organs at risk (OARs) are critical normal structures that are at risk for significant toxicity in the judgment of the treating physician. Such OARs are normal tissues, of which the radiation sensitivity and proximity to the CTV may significantly influ-

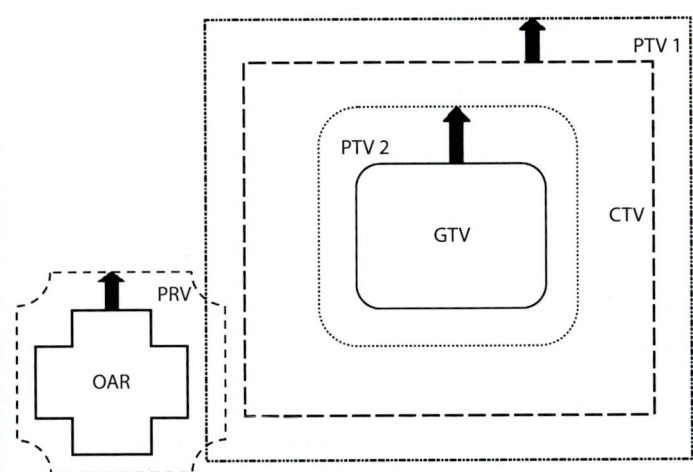
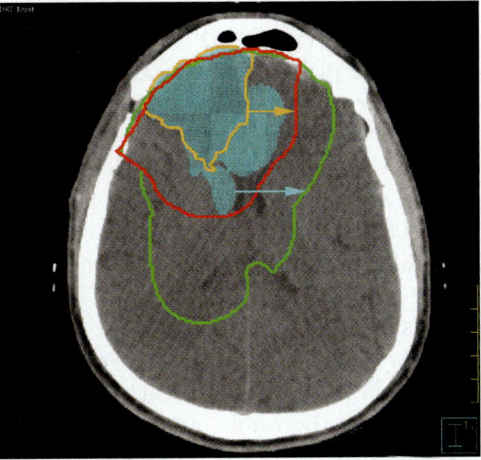

Fig. 17.4. Diagrammatic representation of ICRU 50 and 60 volumes. Abbreviations are defined in the text. *Solid arrows* expansion of one volume to define another. The gross tumor volume (GTV; *orange*) expands to the planning target volume (PTV)2 (*red*). The clinical target volume (CTV; *blue*) expands to the PTV1 (*green*)

ence the prescribed dose and the treatment planning strategy. When possible to do so without compromising the dose to the GTV, attempts should be made to limit the dose to any part of the following normal structures as follows: optic chiasm (54 Gy); optic nerves (60 Gy); optic globes including retina (50 Gy); brainstem to include mid-brain, pons, and medulla (54 Gy); pituitary gland (50 Gy); and spinal cord (50 Gy). In pediatric patients, doses as low as 18 Gy have been implicated in neurocognitive deficits (SILBER 1992; JANKOVIC 1994). A more detailed guideline to dose tolerances is provided in Table 17.4. It is estimated that, with conventionally fractionated irradiation (1.8–2.0 Gy per fraction, five fractions per week), at 5 years the incidence of myelopathy is 5% for doses in the range of 57–61 Gy (tolerance dose $TD_{5/5}$) and 50% for doses of 68–73 Gy ($TD_{50/5}$) (SCHULTHEISS et al. 1995). There is no convincing evidence that the cervical and thoracic cords differ in their radiosensitivity, and there appears to be little change in tolerance with variations in the length of cord irradiated (SCHULTHEISS et al. 1995). A dose of 45–50.4 Gy in 25–28 fractions over 5–5.5 weeks is usually considered to be safe, the risk of myelopathy being less than 1%, well below the steep portion of the dose–response curve (MARCUS and MILLION 1990; SCHULTHEISS et al. 1995).

Figure 17.5 plots a range of isomorbid fractionation schemes, all of which carry a 5% risk of radiation myelopathy. The tolerance of the lumbosacral nerve roots appears to be somewhat higher than that of the spinal cord. Most series report a 0% complication rate if patients are treated with doses of 70 Gy (or equivalent) as long as fraction sizes are kept at or below 2 Gy (FULLER and BLOOM 1988; PIETERS et al. 1996; SCHOENTHALER et al. 1993).

The ICRU Report 62 also describes the concept of the planning organ at risk volume (PRV) (INTERNATIONAL COMMISSION ON RADIATION UNITS AND MEASUREMENTS I 1999). The relationship between a PRV and the OAR for a given structure is analogous to the PTV and the CTV for a given target. For reporting, the description of the PRV (like that of the PTV) should include the extent of the margins in all directions. The PTV and the PRV may overlap, and often do so, which requires a compromise as discussed above when determining the allowable dose. For each OAR, when part of the organ or the whole organ are irradiated above the accepted tolerance level, the maximum dose should be reported (INTERNATIONAL COMMISSION ON RADIATION UNITS AND MEASUREMENTS I 1993). The volume receiving more than the maximum

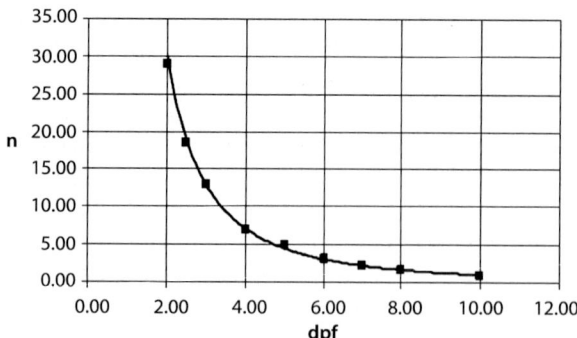

Fig. 17.5. Fractionation schemes with a 5% risk of radiation-induced spinal cord myelopathy. *n* number of fractions, *Dpf* dose per fraction (COHEN and CREDITOR 1981; JEREMIC et al. 1991; MACBETH et al. 1996; MARCUS and MILLION 1990; McCUNNIFF and LIANG 1989; NIEDER et al. 2000; NIEWALD et al. 1998; SCHULTHEISS 1999; WARA et al. 1975)

allowable dose should be evaluated using the corresponding DVH.

17.5.3
Dose Reporting

Recommendations contained in ICRU Report 50 for dose specification reporting are maintained in ICRU Report 62 (INTERNATIONAL COMMISSION ON RADIATION UNITS AND MEASUREMENTS I 1999). First, the absorbed dose at the ICRU reference point should be reported. Second, the maximum and the minimum doses to the PTV should be reported. Furthermore, any additional relevant information should be given when available, for example, DVHs. The absorbed doses to the OARs should also be given. When reporting doses in a series of patients, the treatment prescription or protocol should be described in detail, including the volumes, absorbed dose levels, and fractionation. The treatments should be reported following the above recommendations, and the deviations from the prescription should be stated. In particular, the proportion of patients in whom the dose variation is less than ±5%, ±5–10%, and more than ±10% of the prescribed dose at the ICRU reference point should be reported.

17.5.4
Intensity-Modulated Radiation Therapy

Intensity-modulated radiation therapy (IMRT) is a treatment delivery method that may be used to further optimize the dose distribution. The NCI defines IMRT as "a dose plan and treatment delivery that is optimized using inverse or forward planning techniques for modulated beam delivery, using either a binary collimator, or with a conventional MLC system using either "sliding window" (DMLC) or "step and shoot" (SMLC) modes; note, this definition also includes techniques with compensators designed by inverse planning techniques, to create a highly conformal dose distribution. The IMRT plan includes dose planning objectives and constraints, criteria for target and critical structure expansions, 3-D dose distributions, DVH analysis for targets and critical structures, and plan verification." (DEYE et al. 2005).

There are two aspects of treatment that theoretically may benefit from the use of IMRT. First, since multiple critical, sensitive organs are located near the CNS, one may reason that improved dose distribution should allow the dose to these structures to be minimized. Second, since most primary CNS tumors typically recur locally, IMRT should allow the exploration of anatomical/biological treatment planning and delivery in order to optimize different doses to different cell populations within heterogeneous volumes (STIEBER and MUNLEY 2004). The benefits of IMRT are currently under investigation at multiple institutions.

In selected cases, IMRT should be considered to improve dose delivery to target volumes. If the biology of the tumor suggests a homogeneous cell population, such as is the case with a WHO grade-I meningioma, IMRT may allow for more homogeneous dose delivery with decreased dose to normal surrounding tissues, especially with irregularly shaped lesions. Conversely, WHO grade-IV gliomas are typically comprised of heterogeneous cell populations, and dose escalation with increased dose per fraction to the GTV compared with the CTV is quite feasible using IMRT techniques. In general, the goal is to use IMRT to more conformally treat irregularly shaped tumors. Furthermore, when treating patients with craniospinal axis irradiation, IMRT may improve homogeneous dose delivery to the contents of the spinal canal and allow improved conformality of any boost volumes.

IMRT should be ideally suited to sparing of normal organs. Dose-limiting structures of interest within the cranium include the optic chiasm, right and left optic nerves, both globes, the brain stem, the right and left inner ear, the area postrema, and uninvolved normal brain, especially optic cortex and right and left temporal lobes (MILLER and LESLIE 1994). Reducing the dose to the area postrema may reduce the incidence of treatment-related nausea (MILLER and LESLIE 1994). An indication to use IMRT to treat patients should involve an assessment of which dose-limiting structures are uninvolved by tumor and therefore do not need to receive a clinically significant dose. If conventional treatment planning suggests unacceptable dose delivery to any of these structures (dose per fraction; total dose; or both), IMRT should be considered in order to limit dose to these structures. This decision should include an honest assessment as to the patient's expected life span, as many late toxicities will not manifest until after 6 months from the time of treatment. Patients with poor performance status, for example, are unlikely to survive long enough to benefit from such organ-sparing approaches (GASPAR et al. 2000; SHAW et al. 2003).

Based on anatomical location of the treatment volumes, one may imagine several examples where IMRT could be of marked benefit. Patients with a concave or otherwise irregularly shaped target in a frontal lobe may require IMRT in order to spare the adjacent globe and any uninvolved optic apparatus. In patients with well-lateralized tumors involving the brain parenchyma, maximal sparing of the contralateral hemisphere is a desirable goal. Patients with infiltrative gliomas traditionally have large margins placed around the treatment volumes, and these may often encompass uninvolved critical normal structures (usually the optic chiasm and the brain stem); in these cases, IMRT allows non-uniform reduction of the treatment volume around these structures. Patients with large, well-circumscribed lesions near the base of the skull (e.g., meningiomas, acoustic neuromas, chordomas, and chondrosarcomas) should be also considered for IMRT in order to minimize the dose to the brain stem, inner ear, and posterior fossa. In the setting of retreatment (e.g., previous whole-brain radiation therapy for metastases) an IMRT approach may be considered if the clinical situation suggests that these patients may be long-term survivors with aggressive treatment.

17.6
Simulation Procedures

17.6.1
General Concepts of Positioning, Immobilization and Simulation

The first step in the treatment process is the decision of how to position the patient for simulation and treatment. There are two positioning issues for CNS tumor patients. The first is whether to have the patient lie supine or prone; the second is whether to the have the head and neck in a neutral, flexed or extended position. In general, the supine position with a neutral head position is the most stable and most frequently used when treating the majority of brain tumors. Exceptions include patients with posteriorly located tumors, such as those in the occipital lobes or posterior fossa, patients requiring craniospinal axis irradiation, and most spinal cord tumor patients. Historically, with two-dimensional treatment planning, a neutral position was often not optimal for brain tumors in certain locations; for example, anteroposterior (AP) or posteroanterior (PA) beams would enter or exit through the eyes when treating a centrally located target such as a pituitary tumor. Flexion of the head would rotate the eyes inferiorly relative to the target, thus allowing utilization of AP and PA beams. When treating brain tumors located in the posterior fossa, such as pontine gliomas, the supine position with neck extension allowed a PA beam to be utilized, since the eyes were rotated superiorly and out of the exit of the beam. Now, three-dimensional radiation treatment planning (3DRTP) and IMRT allow for neutral head and neck position in most situations as non-coplanar beams can be used to avoid entry and exit dose to critical structures. Lastly, positioning of the head, neck, and body should be such that the anterior and lateral setup marks are in locatable and reproducible positions. Marks on steeply sloping surfaces, ears, nose, lips, and chin should be avoided whenever possible.

Once positioned, the patient must be immobilized. There are a variety of commercially available custom-made head immobilization devices, most of which utilize thermoplastic or other materials such as expandable foam or plastic beads in a vacuum bag. They are adaptable for flexion or extension when the patient is in the supine position. Similar devices can be used for brain tumor patients requiring the prone position. Variability of setup should be no more than 2 mm or 3 mm with contemporary immobilization such as a thermoplastic mask. When more accurate and/or rigid head positioning and immobilization is desired, such as for fractionated stereotactic radiation therapy or radiosurgery, a modified stereotactic head frame with non-invasive multiple-point head fixation or a bite-block and tilting-head baseplate system, which is referenced to stereotactic space by means of infrared positioning may be used. Once the patient is placed in the positioning device, they

are scanned, typically with radio-opaque reference markers placed at the setup isocenter. The isocenter may be placed at a standard location on every patient with a planned isocenter shift taking place after simulation; alternatively, the clinician may decide *a priori* at the time of simulation where the isocenter should be placed. After completion of treatment planning in virtual reality (see below), verification films should be taken on the conventional simulator before treatment; these should include orthogonal radiographs to verify the isocenter and films of any cerrobend blocks to be used. If multi-leaf collimation will be employed, isocenter verification films still need be taken on the simulator. Portal films must be periodically obtained to verify accuracy of the treatment setup.

Generally, special immobilization devices are not used for patients with spinal cord tumors, although custom devices made of thermoplastic, expandable foam or of plastic beads in vacuum bags are available. The exception is in the arena of spinal radiosurgery, where very complex devices such as the Elekta stereotactic body frame may be utilized. Regardless of the immobilization device chosen, it must be designed to fit the physical dimensions of the CT or MRI scanner and constructed of materials that are compatible with the imaging modalities to prevent image artifact or distortion.

For CT simulation involving the brain, the patient is placed in a positioning device and scanned with three radio-opaque reference markers placed on the thermoplastic mask. Patients are usually treated in the supine position with a neutral head position. All plans in this section assume that patients are simulated in the supine position using three-dimensional CT-based planning; exceptions will be clearly indicated. Prone setup may be considered for posterior fossa tumors. For intracranial disease, single-field or opposed-beam, two-field arrangements are usually not considered acceptable, as they deliver excessive dose to normal tissues in the beam paths. The exception is the short course of palliative radiation therapy typically using an opposed-lateral beam arrangement. An optimum beam arrangement usually consists of multiple (usually from 3–7 in number) non-opposed shaped beams. The contralateral uninvolved hemisphere should be spared completely or as much as possible. A true vertex beam should be avoided if possible due to exit dose into the body, especially in female patients of child-bearing age; a 5–10° gantry rotation should be considered instead. Typically, a homogeneous dose distribution within the target volume is desirable, with no more than a 5–10% inhomogeneity in the irradiated volume. When IMRT is used, a practical approach is to utilize the beam arrangement from an optimum non-coplanar 3-D plan and then invoke the treatment planning system's IMRT module. This can significantly shorten the treatment planning effort and allows for greater freedom of optimization than a coplanar IMRT plan, which is conceptually two-dimensional.

Beam arrangements suggested in the text are described by a commonly used 3D naming system resulting in a unique name for each field (Jasper et al. 2004). The primary body planes are transverse, sagittal, and coronal, with body axes anterior, posterior, right, left, superior, and inferior. Isocentric treatment machines rotate 360° around a transverse plane of the patient's body, and treating in a coronal or sagittal plane requires the treatment couch to be rotated. Whether a patient lies in a supine or prone position, beams entering the patient's body are always named relative to the patient's anatomy. In this 3D naming system, a beam is named based on the plane and the axis closest to its entry and angle of deviation, which are rules and steps 1 and 2. Rule 3 states that the angle of deviation used to describe the beam will never exceed 45°. A beam lying in a primary plane will only be described by three descriptive parameters. To name a non-axial beam, distinguished by the entry of the central axis between three axes in two planes, one additional step is necessary, with a total of two additional rules. Thus, each field name defines the beam's direction and relation to the patient, literally spelling out the correct gantry and couch positions to be used for treatment. These rules are summarized in Table 17.5, and specific examples are provided in the section on Specific Examples of Treatment Techniques. Differential weighting of beams and the use of wedges or compensators is assumed and will not be described in detail, due to considerable

Table 17.5. Naming rules for 3-dimensional beam arrangements (Jasper et al. 2004)

Rule	Step	Description
1	1	Identify the closest primary axis relative to the beam's entry
2	2	Describe the angle of deviation toward the next closest axis
3		The angle of deviation shall be less than or equal to 45°
4	3	Describe the angle of deviation in the 3rd axis
5		The beam's name should be listed with gantry angle first, followed by any couch angle

variability from patient to patient. Most lesions can be treated well with 6-MV photons; for some deeper seated targets, 10-MV photons may provide slightly better dose distribution with respect to normal tissues, although with small targets penumbra issues may dominate. For very lateralized tumors, 10-MV photons may be used for contralateral beams.

17.6.2
Specific Examples of Treatment Techniques

17.6.2.1
Pituitary Region

17.6.2.1.1
Simulation

Since these tumors tend to be well-demarcated and are located near organs at risk (visual apparatus), stereotactic positioning may be useful to minimize the PTV. Fusion of a contrast-enhanced MRI scan with the contrasted treatment CT scan allows for optimum definition of the suprasellar optic apparatus and extension of the tumor. The treatment volume should be slightly larger to include a margin for error in estimating tumor volume and for variation in day-to-day setup. With invasive tumors, such as those involving the sphenoid or cavernous sinuses or other intracranial structures, there is greater uncertainty, which must be considered in determining the volume to be included in the CTV and PTV.

The GTV is the pituitary adenoma, including any of its extension into adjacent anatomical regions. Generally, the entire contents of the sella and, if appropriate, the entire cavernous sinus are included in the CTV. Since pituitary adenomas are non-infiltrating tumors, a CTV expansion of 0.5 cm, i.e., 1.0- to 1.5-cm margin to block edge may be adequate to determine the PTV. With fractionated stereotactic treatment setups, block edge margins in the order of 7 mm can provide excellent dose distribution with minimal dose to surrounding tissues. OARs to be contoured include the eyes (lenses), optic nerves, optic chiasm, brainstem, and temporal lobes.

17.6.2.1.2
Beam Arrangement

Historically, two-dimensional plans with three static beams have often been used, including wedged opposed laterals and an anterior or vertex beam superior to the eyes. Due to unfavorably high temporal lobe dose, this approach can be modified by rotating the couch so that the beams enter from a superior–lateral angle. Two 110° arcs rotating in the coronal plane at the level of the ears can be utilized. Typically, a 30° wedge is used for each arc. Care must be taken to flip the wedge between arcs. A more complex setup includes five non-coplanar shaped static beams: two laterally oriented and angled superiorly so that they are non-opposed and three oriented along the sagittal plane; each of the two lateral beams can be split into two non-coplanar beams to further optimize the dose distribution. IMRT may be useful to further improve dose distribution, especially for irregularly shaped lesions involving a cavernous sinus. A useful arrangement is LG15S, RG15S, SG30A, SG60P, P (Fig. 17.6).

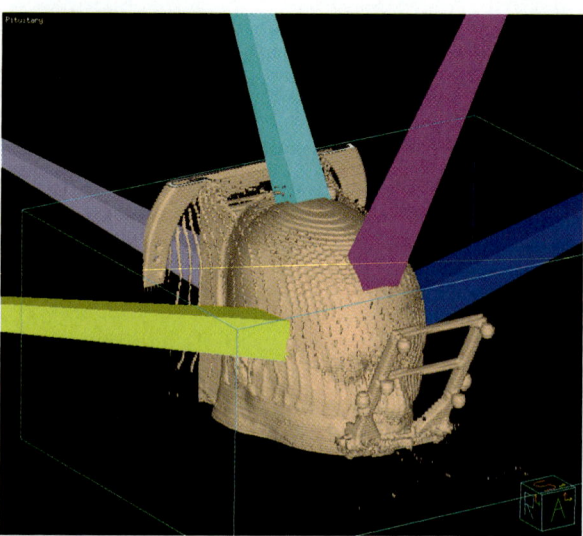

Fig. 17.6. Suggested beam arrangement for a tumor in the pituitary region

17.6.2.1.3
Selection of Beam Energies

Beam energies of no less than 6 MV should be used in order to spare surrounding structures, most notably the temporal lobes. Photons of 10 MV provide a good balance between depth dose and penumbra width; although, for stereotactic plans with small margins, 6-MV photons may be better.

17.6.2.2
Meningiomas

17.6.2.2.1
Simulation

Meningiomas often present a special challenge to treatment planning. The majority of these tumors are benign and well circumscribed but they often have very irregular shapes. Cavernous sinus, cerebello-pontine angle, and convexity meningiomas require careful 3D treatment planning in order to minimize dose to what otherwise would be unnecessarily large normal tissue volumes, especially given that patients otherwise enjoy a normal life span. For well-demarcated lesions, the GTV and CTV are identical, and stereotactic positioning may be useful to minimize the PTV. Fusion of a contrast-enhanced MRI scan with the non-contrast-enhanced treatment CT scan allows for optimum definition of the meningioma and adjacent OARs. The treatment volume should be slightly larger to include a margin for error in estimating tumor volume, errors inherent to image fusion, and for variation in day-to-day

setup. With invasive tumors, such as those involving the sphenoid or cavernous sinuses or other intracranial structures, there is greater uncertainty, which must be considered in determining the boundaries of the CTV. OARs that must be delineated include any of the structures listed in Table 17.4, depending on the clinical situation.

17.6.2.2.2
Beam Arrangement

Beam arrangements vary depending on the region of the brain in which the index lesion is found; we refer the reader to the individual sections by region below. For large convexity lesions, it may be useful to start with three coplanar beams angled at 120° from one another, or a cruciform arrangement of four coplanar beams. The couch and gantry are then rotated so that each beam is shifted approximately 10° inferiorly with respect to the patient (Fig. 17.7). This minimizes the profile of the lesion in the beam's-eye view. IMRT is recommended for any complicated (i.e. non-spheroid) shape.

17.6.2.2.3
Selection of Beam Energies

Meningiomas may be relatively superficially located, as in the case of convexity tumors, or more deeply situated, such as cavernous sinus lesions. Photon beam energies of 6 MV and 10 MV are typically used.

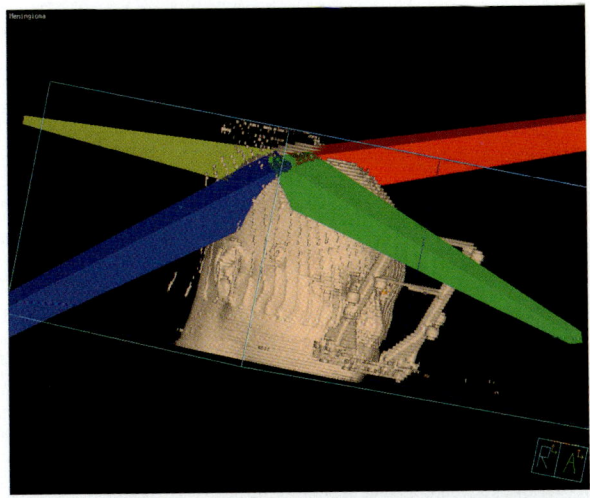

Fig. 17.7. Suggested beam arrangement for a convexity meningioma

17.6.2.3
Temporal Lobe Lesions

17.6.2.3.1
Simulation

For well-lateralized lesions, the goal of treatment planning is to minimize the dose to the contralateral hemisphere. Tumors in this location present a particular challenge due to the proximity of OARs, most significantly the ipsilateral eye, optic nerve, optic chiasm, pituitary gland, and brainstem. Often, meningiomas are encountered in this region (i.e., cavernous sinus). Since these are typically well-demarcated, non-infiltrative tumors, the CTV is equal to the GTV. Hence, stereotactic positioning may be useful, to keep the expansion of margin to form the PTV to as low as 7 mm. For infiltrative tumors, where larger volumes are used, one may reduce the PTV along true anatomical boundaries (e.g., bone) to spare critical normal structures. OARs to be contoured include the optic globes including lenses, optic nerves, optic chiasm, pituitary gland, brainstem, and contralateral temporal lobe.

17.6.2.3.2
Beam Arrangement

Coplanar arrangements are generally to be avoided because of dose to the contralateral temporal lobe. A 3-field arrangement is often utilized, consisting of lateral angled fields with a vertex added. An additional posterior beam angled to avoid exit through the eye may provide additional benefit. A useful arrangement is R10S, L10S, S5A, P101 (Fig. 17.8).

17.6.2.3.3
Selection of Beam Energies

Photons of 6 MV are typically used for the ipsilateral beams, with 10-MV or 18-MV photons used for vertex or superior beams and for beams entering from the side opposite the target.

17.6.2.4
Frontal or Parietal Lobe Lesions

17.6.2.4.1
Simulation

Again, as with temporal lobe lesions, when these tumors are well-lateralized, minimizing dose to the contralateral brain, especially the frontal lobe, is highly desirable. This can be quite challenging with infiltrative gliomas, especially if there is extension across the corpus callosum. When designing the PTV for an infiltrative glioma, one should remember that the falx is a true anatomical boundary between lobes and hence it may be used to limit expansion of the PTV. OARs to be contoured include the optic globe including lenses, optic nerves, optic chiasm, pituitary gland, brainstem, contralateral frontal or parietal lobe, and both temporal lobes. In addition,

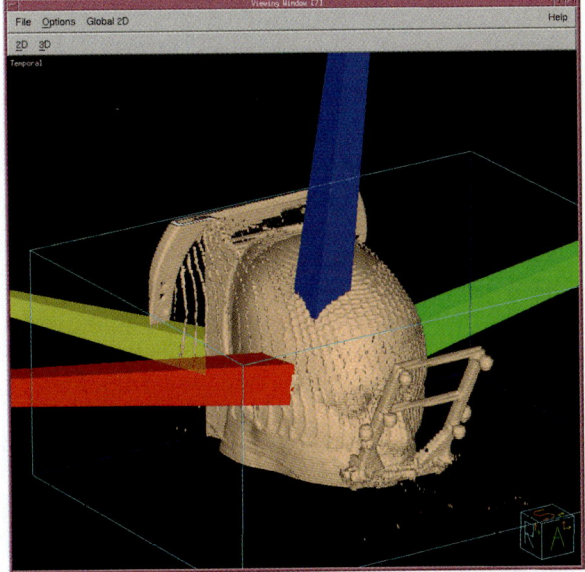

Fig. 17.8. Suggested beam arrangement for a tumor in the temporal region

minimizing dose to the cribriform plate and the area postrema may minimize treatment-related acute toxicities. Defining the motor strip as well as Broca's and Wernicke's areas (possibly using functional MR if available) may be of benefit if these structures are clearly uninvolved by tumor and could be spared unnecessary dose using an IMRT-based approach.

17.6.2.4.2
Beam Arrangement

A starting approach is to utilize an anterior and a lateral beam from the ipsilateral side. Having the anterior beam enter from a superior–anterior approach may be useful for frontal lobe lesions so as to spare the eye and lacrimal gland. An angled vertex beam and/or a non-coplanar posterior beam may be added to further improve the dose distribution; again, exit through the eye should be avoided. Sometimes a beam from the contralateral side may be unavoidable in order to improve dose homogeneity, in which case we again recommend not directly opposing it with the ipsilateral lateral beam. A useful arrangement is A15I, S15A, L15S, R15S (Fig. 17.9).

17.6.2.4.3
Selection of Beam Energies

Photons of 6 MV are typically used for the ipsilateral beams, with 10-MV or 18-MV photons used for vertex or superior beams and for beams entering from the side opposite the target.

17.6.2.5
Central and Thalamic Lesions

17.6.2.5.1
Simulation

Thalamic tumors are usually infiltrative low- or high-grade gliomas. Due to their paracentral location, they often spread into the surrounding lobes and into the brainstem. Thus, PTV expansion is typically circumferential. Craniopharyngiomas or other well-demarcated centrally located lesions may benefit from the stereotactic approach. OARs to be contoured include the optic globes including lenses, optic nerves, optic chiasm, pituitary gland, brainstem, and temporal lobes. In addition, minimizing dose to the cribriform plate and the area postrema hypothetically may minimize treatment-related acute toxicities.

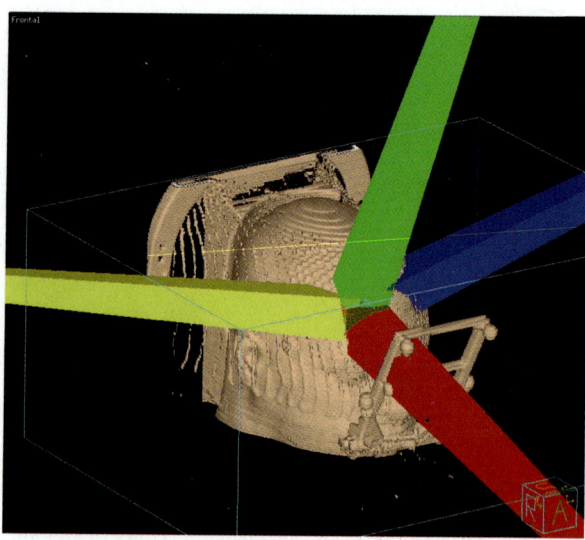

Fig. 17.9. Suggested beam arrangement for a tumor in the frontal region

17.6.2.5.2
Beam Arrangement

For thalamic lesions, we recommend a four-field noncoplanar arrangement: A40R, P40R, S45A, S45A25R, P30R20I (Fig. 17.10). Well-demarcated centrally located lesions may be conceptually approached in a fashion similar to pituitary tumors.

17.6.2.5.3
Selection of Beam Energies

These lesions are usually treated with 6-MV photons. Photons of 10 MV may provide a slightly better dose distribution with respect to normal tissues, although with small targets penumbra issues may dominate.

17.6.2.6
Posterior Fossa

17.6.2.6.1
Simulation

Clinical simulation based on bony anatomy has long been known to be anatomically inaccurate and all posterior fossa simulations should utilize 3D imaging (CT with or without MRI) (Solit and Goldwein 1995). Frequently performed in conjunction with craniospinal axis irradiation, patients are often best simulated in the prone position. This requires care-

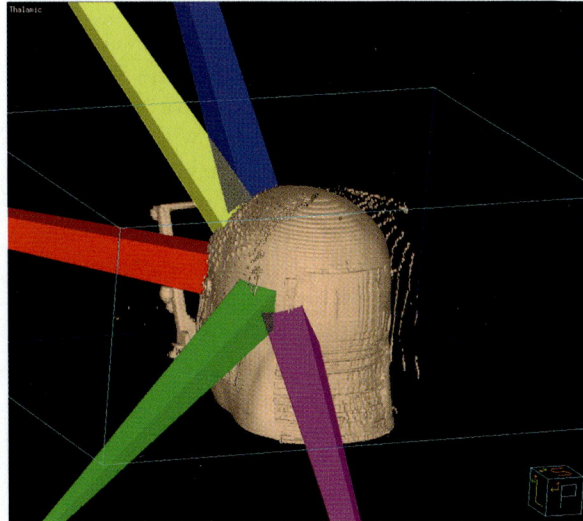

Fig. 17.10. Suggested beam arrangement for a tumor in the thalamic region

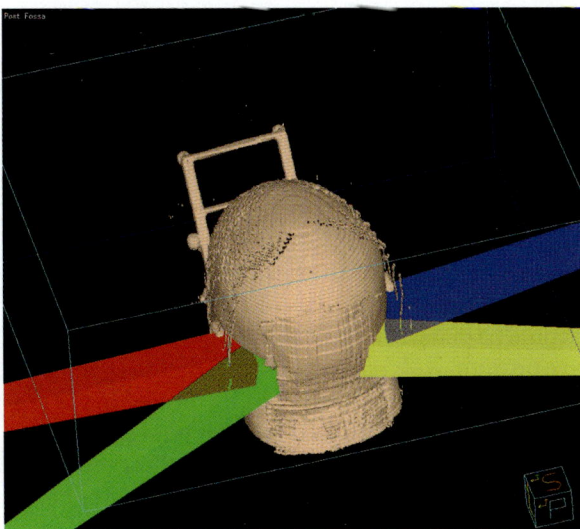

Fig. 17.11. Suggested beam arrangement for a tumor in the posterior fossa

ful attention to immobilization technique and reproducibility. However, for instances where only the posterior fossa will be treated, the standard supine position is usually used, sometimes with extension of the neck to allow for posterior beams that do not exit through the eyes, though this is not always necessary with 3D treatment planning delivery. OARs to be contoured include the brainstem, temporal lobes, and middle and inner ear regions.

17.6.2.6.2
Beam Arrangement

Due to the risk of toxicity to the ear structures, true opposed lateral beams should not be used. Instead, two opposed beams should be angled anteriorly; when four beams are used, a "scissor" arrangement is useful: R20A, R20P, L20A, L20P.

The use of a posterior beam may be useful when treating a volume smaller than the entire posterior fossa, although exit through the eyes should be avoided and exit dose through the pharynx minimized (Fig. 17.11). The use of a vertex beam, posterior oblique wedged fields or IMRT has been shown to significantly reduce the risk of hearing loss by decreasing the radiation dose delivered to the cochlea and eighth cranial nerve (auditory apparatus) in pediatric patients treated for medulloblastomas. Minimizing cochlear doses, particularly in patients receiving potentially ototoxic chemotherapy is strongly recommended for all patients receiving posterior fossa irradiation for anything except

palliative intent (HUANG et al. 2002). Field matching to spinal irradiation fields is described in the section on craniospinal irradiation, below.

17.6.2.6.3
Selection of Beam Energies

The posterior fossa is typically best treated with 6-MV photons.

17.6.2.7
Whole Brain Irradiation

There are several techniques that can be utilized for whole-brain irradiation. Typically, patients are treated with two opposed lateral fields. A clinical setup places a rectangular treatment field over the cranial vault with the collimator angled so as to place the inferior border 1–2 cm below a line drawn from the medial canthus to the mastoid tips. The superior, anterior, and posterior borders are simply set up with a light border flashing over the skull. A small block is carefully placed to shield the globes while still allowing adequate coverage of the anterior–inferior extent of the cranial contents.

Alternatively, a custom setup can be designed with a shaped block that follows the contours of the base of the brain with a 1- to 2-cm margin. When drawing the custom block, one should take care to correctly identify the floor of the anterior cranial fossa including the cribriform plate, middle cranial fossa, posterior fossa, and skull base including the foramina through which the cranial nerves exit. The cribriform plate is often misidentified and hence underdosed (GRIPP et al. 2004; WEISS et al. 2001), which is especially critical when treating the craniospinal axis, for tumors that seed the subarachnoid space and for whole-brain treatment of leukemia or lymphoma. The patient should be simulated in the supine position with the head immobilized, usually with a thermoplastic cast. Radio-opaque markers should be placed on each lateral canthus to document eye position with respect to the radiation field. If the isocenter is chosen to be centrally located in the cranium, the canthus markers will not be aligned due to beam divergence. The opposite eye will receive dose from each lateral beam unless this divergence is corrected. One may calculate the number of degrees to compensate for beam divergence using the formula:

$$\theta = \tan^{-1}\left(\frac{0.5 \times L}{SAD}\right) \quad (1)$$

where L=the length of the field (assuming a symmetrical field), so (0.5*L) equals the distance from the isocenter to the canthus and SAD the source-axis distance. One should verify this calculation by rotating the gantry under fluoroscopic guidance until the canthus markers are superimposed and verifying the correct angle. Alternatively, one may place the beam axis along the canthi: if the patient has been set up straight, the canthi should then be superimposed and there will be no beam divergence into the contralateral eye at the central axis (Fig. 17.12).

The selection of beam energy must be taken into consideration for whole brain irradiation. With photon energies above 10 MV, the peripheral tissues of the brain (e.g., the temporal lobes) and the meninges may be underdosed, depending on the thickness of the scalp and skull. Conversely, due to the curvature of the skull, there will be a relatively high dose region toward the "Mohawk" of the skull. This inhomogeneity is increased with lower energies. When treating whole brain fields, the clinically relevant issue is to provide a set minimum dose to the entirety of the cranial contents. Therefore, a photon energy of 6 MV is usually recommended. If desired, a compensator can be manufactured or wedges employed to improve the dose distribution.

17.6.2.8
Craniospinal Irradiation

Certain neoplasms, such as medulloblastomas and other primitive neuroectodermal tumors, high-grade ependymomas, some germ cell tumors, pineoblastomas, and some CNS lymphomas require treatment to the entire craniospinal axis. This technique can also be used for patients with craniospinal leptomeningeal carcinomatosis or gliomatosis. Several modifications of this approach are used in clinical practice (SHIU et al. 2003). Patients may be treated either in the supine or prone position, often in an immobilization cast to ensure daily positional reproducibility. The neck should be slightly flexed so as to avoid unnecessary exit dose to the mandible, maxilla, and oral cavity. The fleshy canthi are visualized by radio-opaque skin markers. The intracranial contents, including the upper one or two segments of the cervical cord, are treated through opposed lateral fields, usually positioned so that the isocenter is at midline with the beam axes passing through the lateral canthi to minimize divergence into the con-

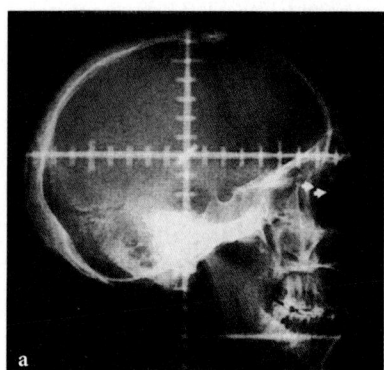 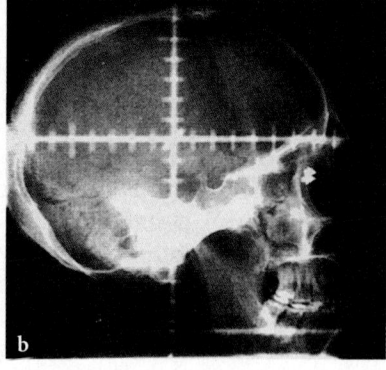 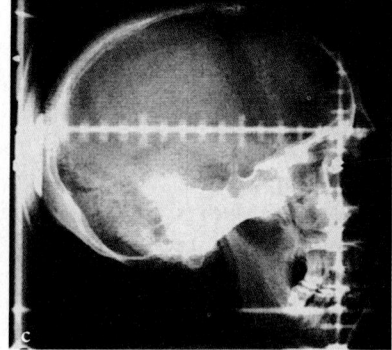

Fig. 17.12a–c. Simulation films of a skull with arrows on the bony canthus and a small radio-opaque marker (commonly referred to as a "BB") representing the lens. **a** Centrally located isocenter showing beam divergence creating misalignment of the canthi. **b** Centrally located isocenter with 4° of gantry rotation creating non-divergent beam geometry through the eyes. **c** Asymmetric jaws with beam axes aligned on canthus markers

tralateral eye. Customized blocks protect the normal head and neck tissues from the primary radiation beam; as mentioned above, care must be taken not to underdose the cribriform plate. The inferior border of the short field is placed around C2–3, leaving adequate room for subsequent shifts in the match with the upper spine field (see below). This is commonly referred to as "feathering the gap".

The spine is treated through one or two posterior fields, depending on the length of the spine. It is customary to maximize the field length of the upper spinal field (40 cm at 100 cm source-to-skin distance) and minimize the length of the lower spinal field, as this simplifies planning for junction shifts (see below). If 40 cm or less of length covers the spine, inclusive of the end of the thecal sac (as defined by MRI, typically near the level of S3), a lower spine field is not necessary. All fields' central axes remain fixed; it is only the fields' lengths which are changed. Therefore, the caudal border of the lower PA spine field should be set inferior to S3 by a length equal to the two field shifts, then blocked back to S3 using asymmetric collimators or custom blocking.

Matching the upper border of the spine field to the lower border of the cranial field requires strict attention to accuracy, since overlap (i.e., overdosing) in the upper cervical cord may have catastrophic outcomes for the patient. In one method, the collimator for the lateral cranial fields is angled to match the divergence of the upper border of the adjacent spinal field, and the treatment couch is angled so that the inferior border of the cranial field is perpendicular to the superior edge of the spinal field ("exact-match" technique). Both the rotation of the collimator and degree of couch rotation are calculated from Eq. 1, and typically range from 9° to 11°. The drawback to this technique is that the couch rotation displaces the contralateral eye cephalad so that it cannot be blocked without blocking frontal brain tissue (Fig. 17.13).

This technique may also result in underdosing of the temporal lobes and cribriform plate.

Alternatively, one may dispense with couch rotation by calculating appropriate skin gaps (the collimator should still be rotated for accurate alignment). The gap is calculated so that the 50% isodose lines meet at the level of the anterior spinal cord ("Gap-Match" technique).

$$G = \frac{\frac{1}{2}(CL_1 + CL_2) \times d}{SAD} \qquad (2)$$

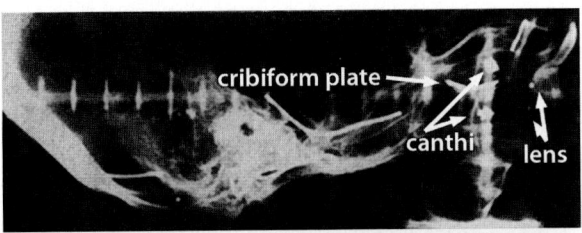

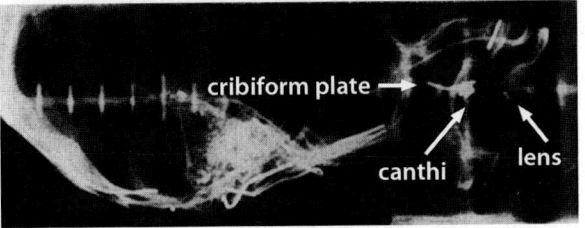

Fig. 17.13a,b. Lateral radiographs of a skull with arrows placed on the bony canthi. BBs are placed on the lens. A piece of solder wire was placed in the cribriform plate and is visible at the intersection of the beam axes. **a** Neutral table angle showing beam divergence and misalignment of the canthi and lens. **b** As above with 8° of couch rotation. Note alignment of canthi and lens. Also, note closer proximity of lens to cribriform plate, making adequate coverage of the cribriform plate, or shielding of the lens difficult

where CL is the field length from isocenter to the field edge to be matched, d is the depth at the match point, and SAD is the source-axis distance (Fig. 17.14).

Ideally, the gap width and collimator rotation should be recalculated for each junction shift and one must remember that d may vary significantly along the extent of the spinal cord. Typically, it is safest to match at the anterior edge of the spinal cord, although this does create a small "cold" triangle posteriorly. Clinical verification of correct calculation of the gap may be achieved by lowering the treatment couch so that the beam edges for the upper and lower spine fields project onto the skin, where they should match exactly. The upper spine field edge is clinically verified in a similar fashion except that the matching lower edge of the cranial field is verified by moving the patient laterally so that the match line projects onto the skin.

All junction lines are moved 0.5–1.0 cm every 8–12 Gy (typically twice during the craniospinal portion of neuraxis irradiation) to avoid overdosing or underdosing segments of the cord. This is accomplished by shortening the inferior margin of the lateral cranial fields, symmetrically lengthening the superior and inferior margins of the posterior spine field, and shortening the cranial margin of the caudal spinal field; as mentioned previously, the

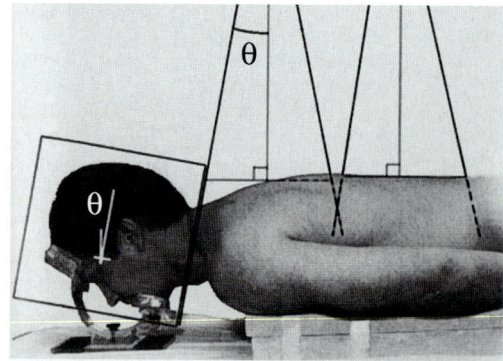

a

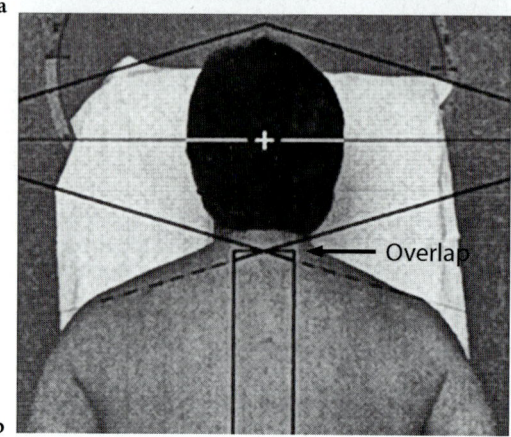

b

Fig. 17.14. a Lateral view of the treatment fields used for craniospinal axis irradiation. The collimator of the lateral brain fields must be rotated by the angles shown to match the divergence of the upper spine field. **b** Posteroanterior view of the match between the lateral brain fields and the upper spine field. Note the areas of overlap and gap between the inferior border of the brain fields and the superior border of the upper spine field

inferior margin of the caudal spinal field remains unchanged at the same location.

Another difficulty is the varying depth of the spinal cord along the length of the patient and the difference in separation distance between head and neck. If this is not taken into consideration, considerable hot or cold spots may occur along the length of the treatment field. A true 3D CT simulation allows for more accurate dosimetric calculation than a sagittal contour or calculation from lateral simulation films (Fig. 17.15).

Typically, in order to achieve homogeneous dose distribution, some form of compensation must be used.

17.6.2.9
Spinal Tumors

Common treatment approaches include a single posterior field (PA), opposed lateral fields, a PA field with opposed laterals, opposed anterior–posterior (AP/PA) fields and oblique wedge-pair fields (MICHALSKI 1998; MINEHAN et al. 1995). Normal tissue constraints and potential toxicities must be considered when defining field arrangements. For tumors in the cervical region, an opposed lateral beam approach can be employed to minimize dose to the anterior neck. For tumors in the cervicothoracic region, a split beam approach is often used in anticipation of matching another palliative treatment field in the future, with the central axis placed just above the shoulders. Opposed lateral beams are used to treat the upper spine, while a PA field is used for the area of the spine below the central axis (Fig. 17.16).

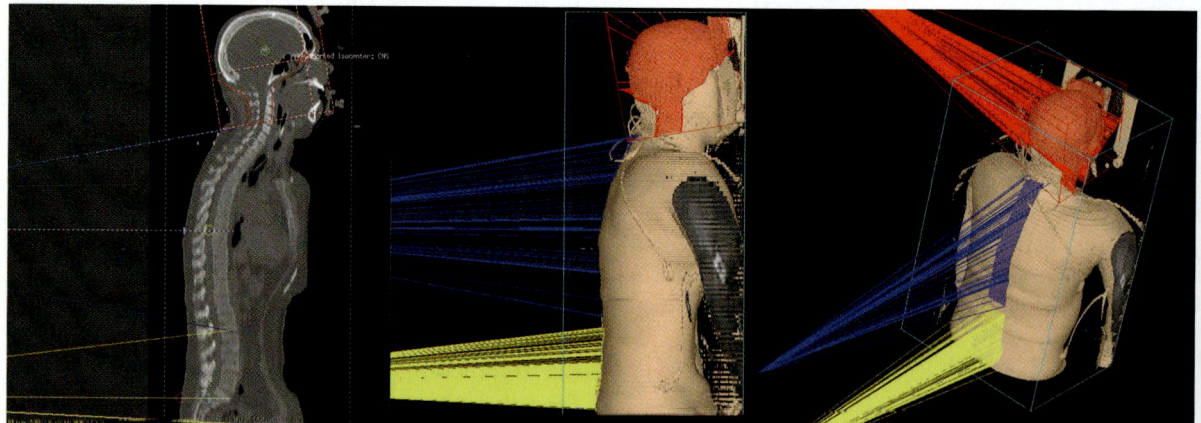

Fig. 17.15. 3-D planning of cranio-spinal irradiation (CSI)

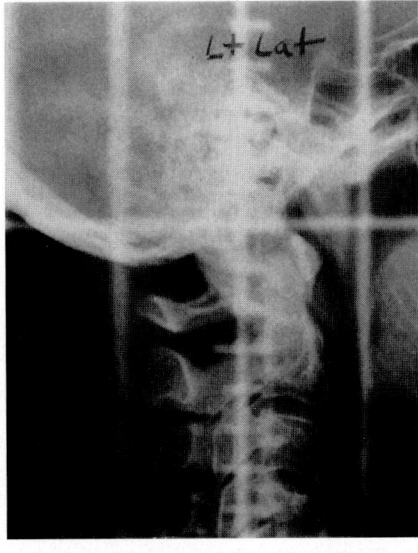
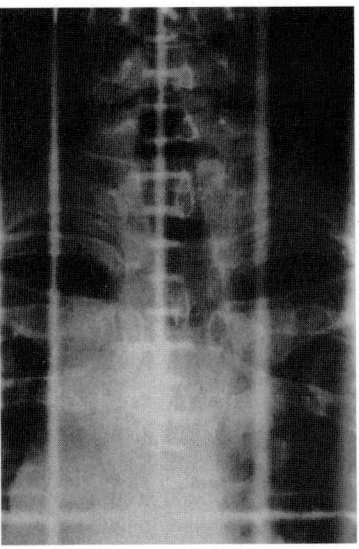

a
b

Fig. 17.16a,b. Simulation films of the beam split technique for cervicothoracic irradiation. The central axis is placed just above the shoulders (C4). **a** Opposed lateral fields for the mid and upper spine. **b** Single posteroanterior field for the low cervical and upper thoracic spine. Potential overlap of the abutting fields must be addressed

Tumors in the thoracic region can be treated with opposed lateral beams, a 3-field approach using a PA field and opposed lateral beams, a 2-field approach using AP-PA beams, or a posterior beam prescribed to an appropriate depth. The tolerance of dose-limiting organs, most commonly the spinal cord and esophagus, may need to be taken into account by the radiation oncologist, depending on the clinical scenario. In the lumbar region, care should be taken to minimize the dose to the kidneys; AP/PA or PA fields are often used here, but a four-field approach using AP/PA and opposed lateral beams with the AP/PA beams preferentially weighted may be useful. Comparison of differing treatment setups by means of DVHs is strongly recommended.

17.7
Simulation Films and Portal Films

After completion of treatment planning in virtual reality, verification films should be taken on a conventional simulator before treatment; these should include orthogonal radiographs to verify the isocenter and films of any cerrobend blocks to be used. If multi-leaf collimation will be employed, portal images should be taken before beginning treatment to verify beam shape (if static) and orientation of the field with respect to the patient as well as orthogonal isocenter verification films. Portal films are

typically obtained weekly to verify accuracy of the treatment setup.

17.8
Dose Prescriptions

The technical principles of prescribing dose to ICRU volumes are summarized earlier in this chapter in the section titled "General Concepts of Modern Radiation Therapy Technique", as has the concept of comparing DVHs. The following section will briefly discuss dosing guidelines by diagnosis.

17.8.1
Primary CNS Tumors

When treated definitively, most patients with a diagnosis of a primary CNS tumor (benign or malignant) will typically require between 5 weeks and 7 weeks of conventionally fractionated radiation therapy. Table 17.3 gives an overview of the typical doses delivered to the ICRU volumes by diagnosis. Details and references are given in the sections for each individual diagnosis.

For patients diagnosed with primary malignant gliomas who are treated with palliative intent, varying fractionation schedules ranging from 30 Gy in 10 fractions to 50 Gy in 20 fractions have been described (Chang et al. 2003; McAleese et al. 2003; Phillips et al. 2003; Roa et al. 2004). Typically, the prognosis for these patients is quite poor; if they survive long enough to complete their initial course of treatment, their survival time typically ranges from 3 months to 9 months under the most favorable circumstances.

17.8.2
Metastatic CNS Tumors

The Radiation Therapy Oncology Group (RTOG) has reported on several large trials with differing fractionation schedules of whole brain radiotherapy for brain metastases (Gaspar et al. 1997, 2000). In conventional fractionation, 50.4 Gy has been used for post-operative treatment (Patchell 2003) but, in the typical palliative setting, larger daily fraction sizes are typically used to achieve a more rapid response. Doses of 30–37.5 Gy using fraction sizes in the range of 2.5–3 Gy per day are commonly used. One should keep in mind that larger fraction sizes do seem to predict a higher incidence of radiation-induced late effects if the patient survives for an extended period of time (Patchell 2003; Quinn and DeAngelis 2000). Patients presenting with leukemic brain involvement requiring emergent palliative therapy are typically treated with total doses to the brain ranging from 24 Gy to 30 Gy in 1.8- to 3.0-Gy fractions, with a median dose of 18 Gy given to the spine (Sanders et al. 2004).

In the setting of MSCC, treatment outcomes reported in the literature vary only a small amount from series to series regardless of fractionation schedule. An overview is given in Table 17.6. Multiple fractionation schedules ranging from 8 Gy x 1 to 2 Gy x 20 have been proposed and evaluated both prospectively and retrospectively (Greenberg et al. 1980; Helweg-Larsen et al. 2000; Hoskin et al. 2003; Maranzano and Latini 1995; Rades et al. 2002; Rades and Karstens 2002). Regimens such as 30 Gy in 10 fractions and 37.50 Gy in 15 fractions are commonly used, but no compelling data are available to point to poorer outcomes with hypofractionated regimens (Rades et al. 2002). Preliminary data of a randomized trial comparing 16 Gy in 2 fractions to 30 Gy split-course (15 Gy in 3 fractions followed by 15 Gy in 5 fractions for responders) show no difference in efficacy or toxicity (Maranzano et al. 2002). Reirradiation of a spinal metastasis may be necessary in some long-term survivors with recurrent disease. In one series with a median initial dose of 30 Gy in 10 fractions and a median reirradiation dose of 22 Gy in 11 fractions, and an average time of 9.1 months elapsing between treatment courses, 88% of patients ambulatory at the end of reirradiation remain ambulatory at the last documented post-reirradiation follow-up (Schiff et al. 1995).

17.9
Future Directions

Integration of biological data (e.g., MR spectroscopy and proliferation or hypoxia-specific PET data) into the treatment planning process may allow further optimization of dose based on biological parameters (Munley et al. 2002; Pirzkall et al. 2000). Dose–function histogram (DFH) analysis using this technology can display the relative function of a structure versus dose and may provide additional

Table 17.6. Typical fractionation schedules for treatment of spinal cord compression and subsequent functional outcomes. *Gy* Gray

Total dose	Fraction size	Overall percentage of patients ambulatory after irradiation	Reference
15 Gy, followed by 15 Gy for responders	5 Gy, followed by 3 Gy for responders	57–75%	Greenberg et al. (1980); Maranzano et al. (1992)
28 Gy	4 Gy	61–72%	Helweg-Larsen et al. (2000); Sorensen et al. (1994)
8 Gy	8 Gy	71%	Hoskin et al. (2003)
20 Gy	4 Gy		
16 Gy (1 week split)	8 Gy	63%	Maranzano et al. (1997)
15 Gy, then 15 Gy	5 Gy, then 3 Gy	76%	Maranzano and Latini (1995)
30 Gy	3 Gy		
30 Gy	3 Gy	56–60%	Rades et al. (2004)
40 Gy	2 Gy		

data needed to obtain and/or evaluate desired heterogeneous dose distributions. The dose could be conformed either directly or inversely proportional to the biological properties of a target and/or normal tissues, depending on what information the corresponding functional imaging set represented for a particular structure. Integrating magnetoencephalography data into treatment planning together with IMRT implementation may allow further sparing of normal functional regions of the brain (BABILONI et al. 2004). Finally, late effects of IMRT, which theoretically may expose more normal tissue to lower doses, are not yet known and work is underway to biologically model these responses (NIEMIERKO 1997).

Acknowledgement:
Dr. Stieber would like to respectfully dedicate this chapter to the late Jimmy O. Fenn, PhD, teacher, mentor and friend.

References

American Cancer Society (2005) Cancer facts and figures 2005. American Cancer Society, Atlanta, Georgia

Babiloni F, Mattia D, Babiloni C et al (2004) Multimodal integration of EEG, MEG and fMRI data for the solution of the neuroimage puzzle. Magn Reson Imaging 22:1471–1476

Barnes D, Egan G, O'Keefe G et al (1997) Characterization of dynamic 3-D PET imaging for functional brain mapping. IEEE Trans Med Imaging 16:261–269

Barnholtz-Sloan JS, Sloan AE, Davis FG et al (2004) Incidence proportions of brain metastases in patients diagnosed (1973 to 2001) in the Metropolitan Detroit Cancer Surveillance System. J Clin Oncol 22:2865–2872

Bell BA, Smith MA, Kean DM et al (1987) Brain water measured by magnetic resonance imaging. Correlation with direct estimation and changes after mannitol and dexamethasone. Lancet 1:66–69

Bilsky M, Posner JB (1993) Intensive and postoperative care of intracranial tumors. In: Ropper AH (ed) Neurological and neurosurgical intensive care. Raven Press, New York, pp 309–329

Bilsky MH, Lis E, Raizer J et al (1999) The diagnosis and treatment of metastatic spinal tumor. Oncologist 4:459–469

Brodmann K (1908a) Beiträge zur histologischen Lokalisation der Grosshirnrinde. VI. Die Cortexgliederung des Menschen. J Psychol Neurol 10:287–334

Brodmann K (1908b) Beiträge zur histologischen Lokalisation der Grosshirnrinde. VI. Die Cortexgliederung des Menschen. J Psychol Neurol 10:231–246

Brown PD, Buckner JC, O'Fallon JR et al (2004) Adult patients with supratentorial pilocytic astrocytomas: a prospective multicenter clinical trial. Int J Radiat Oncol Biol Phys 58:1153–1160

Byrne TN (1992) Spinal cord compression from epidural metastases. N Engl J Med 327:614–619

Carson PL, Giger M, Welch MJ et al (2003) Biomedical imaging research opportunities workshop: report and recommendations. Radiology 229:328–339

CBTRUS (2004) Statistical Report: Primary Brain Tumors in the United States, 1997–2001 Years Data Collected. Central Brain Tumor Registry United States

Chang EL, Yi W, Allen PK et al (2003) Hypofractionated radiotherapy for elderly or younger low-performance status glioblastoma patients: outcome and prognostic factors. Int J Radiat Oncol Biol Phys 56:519–528

Choi SJ, Kim JS, Kim JH et al (2005) [(18)F]3'-deoxy-3'-fluorothymidine PET for the diagnosis and grading of brain tumors. Eur J Nucl Med Mol Imaging (in press)

Cohen L, Creditor M (1981) An iso-effect table for radiation tolerance of the human spinal cord. Int J Radiat Oncol Biol Phys 7:961–966

Condra KS, Buatti JM, Mendenhall WM et al (1997) Benign meningiomas: primary treatment selection affects survival. Int J Radiat Oncol Biol Phys 39:427–436

Connoly ES (1982) Spinal cord tumors in adults. In: Winn RH (ed) Youmans' neurological surgery. Saunders, Philadelphia, p 3196

Daumas-Duport C, Scheithauer B, O'Fallon J et al (1988) Grading of astrocytomas. A simple and reproducible method. Cancer 62:2152–2165

Deye J, Abrams J, Coleman N (2005) The National Cancer Institute Guidelines for the use of intensity-modulated radiation therapy in clinical trials. DCTD, NCI. 1-14-2005. http://www3.cancer.gov/rrp/imrt2004.pdf

Emami B, Lyman J, Brown A et al (1991) Tolerance of normal tissue to therapeutic irradiation. Int J Radiat Oncol Biol Phys 21:109–122

Epilepsy Foundation of America (1993) Treatment of convulsive status epilepticus. Recommendations of the Epilepsy Foundation of America's Working Group on Status Epilepticus. JAMA 270:854–859

Ferlay J, Bray F, Pisani P, Parkin DM (2001) GLOBOCAN 2002: cancer incidence, mortality and prevalence worldwide, version 1.0. IARC CancerBase 5. IARC, Lyon

Forsyth PA, Weaver S, Fulton D et al (2003) Prophylactic anticonvulsants in patients with brain tumour. Can J Neurol Sci 30:106–112

French L (1966) The use of steroids in the treatment of cerebral edema. Bull NY Acad Med 42:301–311

Fuller DB, Bloom JG (1988) Radiotherapy for chordoma. Int J Radiat Oncol Biol Phys 15:331–339

Gaspar L, Scott C, Rotman M et al (1997) Recursive partitioning analysis (RPA) of prognostic factors in three Radiation Therapy Oncology Group (RTOG) brain metastases trials. Int J Radiat Oncol Biol Phys 37:745–751

Gaspar L, Scott C, Murray K et al (2000) Validation of the RTOG recursive partitioning analysis (RPA) classification for brain metastases. Int J Radiat Oncol Biol Phys 47:1001–1006

Gerszten PC, Welch WC (2000) Current surgical management of metastatic spinal disease. Oncology (Huntingt) 14:1013–1024

Gilbert RW, Kim JH, Posner JB (1978) Epidural spinal cord compression from metastatic tumor: diagnosis and treatment. Ann Neurol 3:40–51

Glantz MJ, Cole BF, Friedberg MH et al (1996) A randomized, blinded, placebo-controlled trial of divalproex sodium prophylaxis in adults with newly diagnosed brain tumors. Neurology 46:985–991

Goldsmith BJ, Wara WM, Wilson CB et al (1994) Postoperative irradiation for subtotally resected meningiomas. A retrospective analysis of 140 patients treated from 1967 to 1990. J Neurosurg 80:195–201

Greenberg HS, Kim JH, Posner JB (1980) Epidural spinal cord compression from metastatic tumor: results with a new treatment protocol. Ann Neurol 8:361–366

Gripp S, Kambergs J, Wittkamp M et al (2004) Coverage of anterior fossa in whole-brain irradiation. Int J Radiat Oncol Biol Phys 59:515–520

Hazuka MB, Burleson WD, Stroud DN et al (1993) Multiple brain metastases are associated with poor survival in patients treated with surgery and radiotherapy. J Clin Oncol 11:369–373

Healey JH, Brown HK (2000) Complications of bone metastases: surgical management. Cancer 88:2940–2951

Heimdal K, Hirschberg H, Slettebo H et al (1992) High incidence of serious side effects of high-dose dexamethasone treatment in patients with epidural spinal cord compression. J Neurooncol 12:141–144

Helweg-Larsen S, Sorensen PS, Kreiner S (2000) Prognostic factors in metastatic spinal cord compression: a prospective study using multivariate analysis of variables influencing survival and gait function in 153 patients. Int J Radiat Oncol Biol Phys 46:1163–1169

Hoskin PJ, Grover A, Bhana R (2003) Metastatic spinal cord compression: radiotherapy outcome and dose fractionation. Radiother Oncol 68:175–180

Huang E, Teh BS, Strother DR et al (2002) Intensity-modulated radiation therapy for pediatric medulloblastoma: early report on the reduction of ototoxicity. Int J Radiat Oncol Biol Phys 52:599–605

Huntington HW, Kernohan JW, Stapley LA (1965) Tumors of the central nervous system. Ariz Med 22:683–686

Husband DJ, Grant KA, Romaniuk CS (2001) MRI in the diagnosis and treatment of suspected malignant spinal cord compression. Br J Radiol 74:15–23

International Agency for Research on Cancer (2000) World Health Organization classification of tumours: pathology and genetics. Tumours of the nervous system, 2nd edn. Oxford Univ Press, Lyon

International Commission on Radiation Units and Measurements I (1993) ICRU report 50, prescribing, recording, and reporting photon beam therapy. Nuclear Technology Publishing, Bethesda, MD, USA

International Commission on Radiation Units and Measurements I (1999) ICRU report 62: prescribing, recording and reporting photon beam therapy (supplement to ICRU report 50). Nuclear Technology Publishing, Bethesda, MD, USA

Jankovic M, Brouwers P, Valsecchi MG, Van Veldhuizen A, Huisman J, Kamphuis R et al. (1994) Association of 1800 cGy cranial irradiation with intellectual function in children with acute lymphoblastic leukaemia. ISPACC. International Study Group on Psychosocial Aspects of Childhood Cancer. Lancet 344:224–227

Jasper KR, Hummel SM, Laramore GE (2004) 3D naming system. Med Dosim 29:97–103

Jeremic B, Djuric L, Mijatovic L (1991) Incidence of radiation myelitis of the cervical spinal cord at doses of 5500 cGy or greater. Cancer 68:2138–2141

Kalkanis SN, Eskandar EN, Carter BS et al (2003) Microvascular decompression surgery in the United States, 1996 to 2000: mortality rates, morbidity rates, and the effects of hospital and surgeon volumes. Neurosurgery 52:1251–1261

Karim AB, Maat B, Hatlevoll R et al (1996) A randomized trial on dose-response in radiation therapy of low-grade cerebral glioma: European Organization for Research and Treatment of Cancer (EORTC) Study 22844. Int J Radiat Oncol Biol Phys 36:549–556

Karim AB, Afra D, Cornu P et al (2002) Randomized trial on the efficacy of radiotherapy for cerebral low-grade glioma in the adult: European Organization for Research and Treatment of Cancer Study 22845 with the Medical Research Council study BR04: an interim analysis. Int J Radiat Oncol Biol Phys 52:316–324

Loblaw DA, Laperriere NJ (1998) Emergency treatment of malignant extradural spinal cord compression: an evidence-based guideline. J Clin Oncol 16:1613–1624

Loblaw DA, Laperriere NJ, Mackillop WJ (2003) A population-based study of malignant spinal cord compression in Ontario. Clin Oncol (R Coll Radiol) 15:211–217

Long DM, Hartmann JF, French LA (1966) The response of experimental cerebral edema to glucosteroid administration. J Neurosurg 24:843–854

Macbeth FR, Wheldon TE, Girling DJ et al (1996) Radiation myelopathy: estimates of risk in 1048 patients in three randomized trials of palliative radiotherapy for non-small cell lung cancer. The Medical Research Council Lung Cancer Working Party. Clin Oncol (R Coll Radiol) 8:176–181

Maranzano E, Latini P (1995) Effectiveness of radiation therapy without surgery in metastatic spinal cord compression: final results from a prospective trial. Int J Radiat Oncol Biol Phys 32:959–967

Maranzano E, Latini P, Checcaglini F et al (1992) Radiation therapy of spinal cord compression caused by breast cancer: report of a prospective trial. Int J Radiat Oncol Biol Phys 24:301–306

Maranzano E, Latini P, Beneventi S et al (1996) Radiotherapy without steroids in selected metastatic spinal cord compression patients. A phase II trial. Am J Clin Oncol 19:179–183

Maranzano E, Latini P, Perrucci E et al (1997) Short-course radiotherapy (8 Gy x 2) in metastatic spinal cord compression: an effective and feasible treatment. Int J Radiat Oncol Biol Phys 38:1037–1044

Maranzano E, Frattegiani A, Rossi R, Bagnoli R, Mignogna M, Bellavista R, Perrucci E, Lupattelli M, Ponticelli P, Latini P (2002) Randomized trial of two different hypofractionated radiotherapy (RT) schedules (8 Gyx2 vs 5 Gyx3; 3 Gyx5) in metastatic spinal cord compression (MSCC) [abstract]. Radiother Oncol 64:82–83

Marcus RB Jr, Million RR (1990) The incidence of myelitis after irradiation of the cervical spinal cord. Int J Radiat Oncol Biol Phys 19:3–8

McAleese JJ, Stenning SP, Ashley S et al (2003) Hypofractionated radiotherapy for poor prognosis malignant glioma: matched pair survival analysis with MRC controls. Radiother Oncol 67:177–182

McCunniff AJ, Liang MJ (1989) Radiation tolerance of the cervical spinal cord. Int J Radiat Oncol Biol Phys 16:675–678

McKnight TR, dem Bussche MH, Vigneron DB et al (2002) Histopathological validation of a three-dimensional magnetic resonance spectroscopy index as a predictor of tumor presence. J Neurosurg 97:794–802

Michalski JM (1998) Spinal canal. In: Leibel SA, Phillips TL (eds) Textbook of radiation oncology. Saunders, Philadelphia, pp 860–875

Miller AD, Leslie RA (1994) The area postrema and vomiting. Frontiers Neuroendocrinol 15:301–320

Minehan KJ, Shaw EG, Scheithauer BW et al (1995) Spinal cord astrocytoma: pathological and treatment considerations. J Neurosurg 83:590–595

Mintz AH, Kestle J, Rathbone MP et al (1996) A randomized trial to assess the efficacy of surgery in addition to radiotherapy in patients with a single cerebral metastasis. Cancer 78:1470–1476

Morris D, Bourland J, Rosenman J et al (2001) Three-dimensional conformal radiation treatment planning and delivery for low- and intermediate-grade gliomas. Semin Radiat Oncol 11:124–137

Munley MT, Kearns WT, Hinson WH et al (2002) Bioanatomic IMRT treatment planning with dose function histograms. Int J Radiat Oncol Biol Phys 54:126–126

Nieder C, Milas L, Ang KK (2000) Tissue tolerance to reirradiation. Semin Radiat Oncol 10:200–209

Niemierko A (1997) Reporting and analyzing dose distributions: a concept of equivalent uniform dose. Med Phys 24:103–110

Niewald M, Feldmann U, Feiden W et al (1998) Multivariate logistic analysis of dose-effect relationship and latency of radiomyelopathy after hyperfractionated and conventionally fractionated radiotherapy in animal experiments. Int J Radiat Oncol Biol Phys 41:681–688

Nuutinen J, Sonninen P, Lehikoinen P et al (2000) Radiotherapy treatment planning and long-term follow-up with [C-11]methionine pet in patients with low-grade astrocytoma. Int J Radiat Oncol Biol Phys 48:43–52

O'Rourke T, George CB, Redmond J III et al (1986) Spinal computed tomography and computed tomographic metrizamide myelography in the early diagnosis of metastatic disease. J Clin Oncol 4:576–583

Patchell RA (2003) The management of brain metastases. Cancer Treat Rev 29:533–540

Patchell RA, Tibbs PA, Walsh JW et al (1990) A randomized trial of surgery in the treatment of single metastases to the brain. N Engl J Med 322:494–500

Patchell RA, Tibbs PA, Regine WF et al (2005) Direct decompressive surgical resection in the treatment of spinal cord compression caused by metastatic cancer: a randomised trial. Lancet 366:643–648

Percy AK, Elveback LR, Okazaki H et al (1972) Neoplasms of the central nervous system. Epidemiologic considerations. Neurology 22:40–48

Perry A, Scheithauer BW, Stafford SL et al (1999) "Malignancy" in meningiomas: a clinicopathologic study of 116 patients, with grading implications. Cancer 85:2046–2056

Phillips C, Guiney M, Smith J et al (2003) A randomized trial comparing 35 Gy in ten fractions with 60 Gy in 30 fractions of cerebral irradiation for glioblastoma multiforme and older patients with anaplastic astrocytoma. Radiother Oncol 68:23–26

Pieters RS, O'Farrell D, Fullerton B (1996) Cauda Equina tolerance to radiation therapy [abstract]. Int J Radiat Oncol Biol Phys 36:359

Pigott KH, Baddeley H, Maher EJ (1994) Pattern of disease in spinal cord compression on MRI scan and implications for treatment. Clin Oncol (R Coll Radiol) 6:7–10

Pirzkall A, Larson DA, McKnight TR et al (2000) MR-Spectros-copy results in improved target delineation for high-grade gliomas. Int J Radiat Oncol Biol Phys 48:115–115

Prados MD, Seiferheld W, Sandler HM et al (2004) Phase III randomized study of radiotherapy plus procarbazine, lomustine, and vincristine with or without BUdR for treatment of anaplastic astrocytoma: final report of RTOG 9404. Int J Radiat Oncol Biol Phys 58:1147–1152

Preston-Martin S (1990) Descriptive epidemiology of primary tumors of the spinal cord and spinal meninges in Los Angeles County, 1972–1985. Neuroepidemiology 9:106–111

Quinn JA and DeAngelis LM (2000) Neurologic emergencies in the cancer patient. Semin Oncol 27:311–321

Rades D, Karstens JH (2002) A comparison of two different radiation schedules for metastatic spinal cord compression considering a new prognostic factor. Strahlenther Onkol 178:556–561

Rades D, Karstens JH, Alberti W (2002) Role of radiotherapy in the treatment of motor dysfunction due to metastatic spinal cord compression: comparison of three different fractionation schedules. Int J Radiat Oncol Biol Phys 54:1160–1164

Rades D, Bajrovic A, Alberti W et al (2003a) Is there a dose-effect relationship for the treatment of symptomatic vertebral hemangioma? Int J Radiat Oncol Biol Phys 55:178–181

Rades D, Fehlauer F, Schild S et al (2003b) Treatment for central neurocytoma: a meta-analysis based on the data of 358 patients. Strahlenther Onkol 179:213–218

Rades D, Fehlauer F, Hartmann A et al (2004) Reducing the overall treatment time for radiotherapy of metastatic spinal cord compression (MSCC): 3-year results of a prospective observational multi-center study. J Neurooncol 70:77–82

Rades D, Fehlauer F, Lamszus K et al (2005) Well-differentiated neurocytoma: what is the best available treatment? Neuro Oncol 7:77–83

Ricci PE, Dungan DH (2001) Imaging of low- and intermediate-grade gliomas. Semin Radiat Oncol 11:103–112

Roa W, Brasher PM, Bauman G et al (2004) Abbreviated course of radiation therapy in older patients with glioblastoma multiforme: a prospective randomized clinical trial. J Clin Oncol 22:1583–1588

Ruff RL, Lanska DJ (1989) Epidural metastases in prospectively evaluated veterans with cancer and back pain. Cancer 63:2234–2241

Sanders KE, Ha CS, Cortes-Franco JE et al (2004) The role of craniospinal irradiation in adults with a central nervous system recurrence of leukemia. Cancer 100:2176–2180

Sarin R, Murthy V (2003) Medical decompressive therapy for primary and metastatic intracranial tumours. Lancet Neurol 2:357–365

Sasanelli F, Beghi E, Kurland LT (1983) Primary intraspinal neoplasms in Rochester, Minnesota, 1935–1981. Neuroepidemiology 2:156–163

Schaberg J, Gainor BJ (1985) A profile of metastatic carcinoma of the spine. Spine 10:19–20

Schiff D, Shaw EG, Cascino TL (1995) Outcome after spinal reirradiation for malignant epidural spinal cord compression. Ann Neurol 37:583–589

Schiff D, Batchelor T, Wen PY (1998) Neurologic emergencies in cancer patients. Neurol Clin 16:449–483

Schoenthaler R, Castro JR, Petti PL et al (1993) Charged particle irradiation of sacral chordomas. Int J Radiat Oncol Biol Phys 26:291–298

Schouten LJ, Rutten J, Huveneers HA et al (2002) Incidence of brain metastases in a cohort of patients with carcinoma of the breast, colon, kidney, and lung and melanoma. Cancer 94:2698–2705

Schultheiss TE (1999) The radiation dose response of the human cervical spinal cord [abstract]. Int J Radiat Oncol Biol Phys 45:174

Schultheiss TE, Kun LE, Ang KK et al (1995) Radiation response of the central nervous system. Int J Radiat Oncol Biol Phys 31:1093–1112

Scott CB, Scarantino C, Urtasun R et al (1998) Validation and predictive power of Radiation Therapy Oncology Group (RTOG) recursive partitioning analysis classes for malignant glioma patients: a report using RTOG 90-06. Int J Radiat Oncol Biol Phys 40:51–55

Shaw E, Arusell R, Scheithauer B et al (2002) Prospective randomized trial of low - versus high-dose radiation therapy in adults with supratentorial low-grade glioma: initial report of a North Central Cancer Treatment Group/Radiation Therapy Oncology Group/Eastern Cooperative Oncology Group Study. J Clin Oncol 20:2267–2276

Shaw EG, Seiferheld W, Scott C et al (2003) Reexamining the radiation therapy oncology group (RTOG) recursive partitioning analysis (RPA) for glioblastoma multiforme (GBM) patients. Int J Radiat Oncol Biol Phys 57:S135–S136

Shiu AS, Chang EL, Ye JS et al (2003) Near simultaneous computed tomography image-guided stereotactic spinal radiotherapy: an emerging paradigm for achieving true stereotaxy. Int J Radiat Oncol Biol Phys 57:605–613

Silber JH, Radcliffe J, Peckham V, Perilongo G, Kishnani P, Fridman M et al. (1992) Whole-brain irradiation and decline in intelligence: the influence of dose and age on IQ score. J Clin Oncol 10:1390–1396

Solit DB, Goldwein JW (1995) Posterior fossa: analysis of a popular technique for estimating the location in children with medulloblastoma. Radiology 195:697–698

Sorensen S, Helweg-Larsen S, Mouridsen H et al (1994) Effect of high-dose dexamethasone in carcinomatous metastatic spinal cord compression treated with radiotherapy: a randomised trial. Eur J Cancer 30A:22–27

Stafford SL, Perry A, Suman VJ et al (1998) Primarily resected meningiomas: outcome and prognostic factors in 581 Mayo Clinic patients, 1978 through 1988. Mayo Clin Proc 73:936–942

Stieber VW, deGuzman AF (2003) Pituitary. In: Perez C, Brady L, Schmidt-Ullrich R, Halperin E (eds) Principles and practice of radiation oncology. Lippincott Williams and Wilkins, New York

Stieber VW, Munley M (2004) Central nervous system tumors. In: Mundt AJ, Roeske JC (eds) Intensity-modulated radiation therapy: a clinical perspective. Decker, Ontario

Stieber V, Tatter S, Shaw EG (2005) Primary spinal tumors. In: Schiff D (ed) Cancer of the nervous system: principles and practice of neuro-oncology. McGraw-Hill, Columbus

Tortosa A, Vinolas N, Villa S et al (2003) Prognostic implication of clinical, radiologic, and pathologic features in patients with anaplastic gliomas. Cancer 97:1063–1071

Vecht CJ, Haaxma-Reiche H, Noordijk EM et al (1993) Treatment of single brain metastasis: radiotherapy alone or combined with neurosurgery? Ann Neurol 33:583–590

Vecht CJ, Hovestadt A, Verbiest HB et al (1994) Dose-effect relationship of dexamethasone on Karnofsky performance in metastatic brain tumors: a randomized study of doses of 4, 8, and 16 mg per day. Neurology 44:675–680

Walker AE, Robins M, Weinfeld FD (1985) Epidemiology of brain tumors: the national survey of intracranial neoplasms. Neurology 35:219–226

Wara WM, Phillips TL, Sheline GE et al (1975) Radiation tolerance of the spinal cord. Cancer 35:1558–1562

Weiss E, Krebeck M, Kohler B et al (2001) Does the standardized helmet technique lead to adequate coverage of the cribriform plate? An analysis of current practice with respect to the ICRU 50 report. Int J Radiat Oncol Biol Phys 49:1475–1480

Weissman DE, Janjan NA, Erickson B et al (1991) Twice-daily tapering dexamethasone treatment during cranial radiation for newly diagnosed brain metastases. J Neurooncol 11:235–239

Wolfson AH, Snodgrass SM, Schwade JG et al (1994) The role of steroids in the management of metastatic carcinoma to the brain. A pilot prospective trial. Am J Clin Oncol 17:234–238

Wong DA, Fornasier VL, MacNab I (1990) Spinal metastases: the obvious, the occult, and the impostors. Spine 15:1–4

Working Group on Status Epilepticus and Epilepsy Foundation of America (1993) Treatment of convulsive status epilepticus. Recommendations of the Epilepsy Foundation of America's Working Group on Status Epilepticus. JAMA 270:854–859

Young RF, Post EM, King GA (1980) Treatment of spinal epidural metastases. Randomized prospective comparison of laminectomy and radiotherapy. J Neurosurg 53:741–748

18 Head and Neck Cancer

WILLIAM M. MENDENHALL, ROBERT J. AMDUR, and JATINDER R. PALTA

CONTENTS

18.1   Larynx 454
18.1.1   Glottic Larynx 454
18.1.1.1 Stage T1–T2 454
18.1.1.2 Stage T3–Favorable T4 454
18.1.1.3 Supraglottic Larynx 456
18.2   Hypopharynx 456
18.2.1   Pyriform Sinus 456
18.2.2   Pharyngeal Wall 456
18.2.2.1 Postoperative Irradiation of Laryngeal and
          Hypopharyngeal Tumors 456
18.3   Oropharynx 457
18.4   Base of Tongue 459
18.5   Tonsillar Area 459
18.6   Soft Palate 460
18.7   Nasopharynx 460
18.8   Oral Cavity 462
18.8.1   Lip 462
18.9   Floor of Mouth 463
18.9.1   External-Beam Irradiation 463
18.9.2   Interstitial Irradiation 464
18.9.2.1 Intraoral Cone Irradiation 464
18.10   Oral Tongue 465
18.10.1   External-Beam Irradiation 465
18.10.1.1 Interstitial Irradiation 467
18.10.1.2 Postoperative Irradiation for Tumors of Oral
          Cavity and Oropharynx 468
18.11   Nasal Cavity, Paranasal Sinuses, and
          Nasal Vestibule 468
18.12   Major Salivary Gland 472
18.12.1   Parotid Gland 472
18.12.1.1 External-Beam RT 472
18.12.1.2 Interstitial RT 472
18.12.2   Submandibular Gland 474
18.13   Unknown Primary 474
18.14   Design of Low Neck Portal 475
18.14.1   Oropharynx, Nasopharynx, and
          Oral Cavity Cancers 475
18.14.1.1 Larynx and Hypopharynx Cancers 476
18.15   IMRT 478
18.15.1.1 Guidelines for Contouring Target Structures 479
18.15.1.2 Prioritization Rules When Structures Overlap 480

18.15.1.3 Guidelines for Dose Prescription and
          Normal Tissue Dose Analysis 480
18.15.1.4 Dose Constraints for Target PTVs 480
18.15.1.5 Guidelines for Treatment of
          Regional Lymphatics 480
          References 483

This chapter deals with the design of radiation portals for patients with head and neck cancer who have been selected to receive radiotherapy (RT). For a discussion of the rationale for selecting RT and the appropriate sequencing of the various treatment methods available for treating head and neck cancer, the reader is referred to more comprehensive sources (FLETCHER 1980; MILLION et al. 1994; MENDENHALL et al. 2004).

Following is a discussion of the treatment design used at the University of Florida for the most common head and neck primary sites. The staging system is based on the 2002 manual of the American Joint Committee on Cancer (AJCC 2002).

RT techniques can be broadly stratified into: (1) 3-dimensional conformal radiotherapy (3DCRT) and (2) intensity-modulated radiotherapy (IMRT). Definition of the target volume and the normal tissues that must be protected is based on data obtained via treatment planning computed tomography (CT) for both techniques. 3DCRT employs forward treatment planning and is much like conventional 2-dimensional RT except that the treatment plans are based on CT-defined 3-dimensional anatomy rather than a 2-dimensional radiograph and surface anatomy. Parenthetically, it is still essential to check the relationship of the portals with the surface anatomy on the treatment table. In contrast with 3DCRT, inverse treatment planning is used with IMRT, which may yield a more conformal treatment plan, thus reducing the dose to normal tissues and thereby decreasing the likelihood of acute and late toxicity. However, RT is much like cooking and considerable variability exists so that whether IMRT results in a plan that

W. M. MENDENHALL, MD; R. J. AMDUR, MD; J. R. PALTA, PhD
Department of Radiation Oncology, University of Florida
College of Medicine, Health Science Center, P.O. Box 100385,
Gainesville, FL 32610-0385, USA

is superior to that which could be achieved with 3DCRT depends on the site and extent of the tumor and the experience and expertise of the treatment team (radiation oncologist, physicists, and radiation therapists). One of the major disadvantages of IMRT is that, because the dose distribution is more conformal and the dose gradient much sharper, there may be an increased risk of developing a marginal recurrence. Additional disadvantages may include: (1) less homogeneous dose distribution, (2) more beam "on-time" leading to increased total body dose due to scatter, (3) increased treatment planning time for the physicists and radiation oncologists, and (4) increased cost. Therefore, there must be a clear potential advantage for proceeding with IMRT in a particular patient. The most common reason is to reduce the dose to the salivary gland(s) in patients receiving RT to both sides of the neck thereby minimizing xerostomia. Another indication is to reduce the dose to the temporal lobes in patients treated for nasopharyngeal cancer. If a clear indication for IMRT does not exist, the patient is likely better treated with 3DCRT. Treatment planning with 3DCRT for head and neck cancer is relatively complex and performed suboptimally by many radiation oncologists. Accordingly, this chapter will focus on 3DCRT techniques; a general discussion of IMRT for head and neck cancer will follow.

Primary fields are the portals used to deliver radiation treatment to the primary site of the cancer; neck fields are additional portals used to treat cervical lymph nodes not included in the primary fields. Clinically positive neck nodes (N+) are cervical lymph nodes believed to harbor metastatic cancer on the basis of physical and/or radiographic findings; a clinically negative neck (N0) has no such findings and may contain no metastatic tumor or only subclinical (clinically undetectable) deposits of cancer.

18.1
Larynx

18.1.1
Glottic Larynx

18.1.1.1
Stage T1–T2

Because the risk of subclinical disease in the cervical lymphatics is remote, the portals are limited to the primary lesion (Fig. 18.1) (MILLION et al. 1994c; MENDENHALL et al. 2001). Although a common practice is to treat early vocal cord cancer with a standard field size (e.g., 6×6 cm), our preference is to design the portal to fit the specific lesion. The patient is treated in the supine position with the neck extended and the head immobilized in an aquaplast mask. The physician at the treatment machine checks the field each day according to palpable anatomic landmarks. This practice allows the treatment volume to be kept at a minimum, while virtually eliminating the risk of geographic miss. Overall treatment time, as well as total dose, is critical in obtaining maximum control rates (PARSONS 1984a; MENDENHALL et al. 1988b, 2001); failure to use the smallest field size consistent with adequate coverage of the tumor usually means that either the total dose or dose per fraction must be compromised to limit acute reactions and/or late effects. The patient is treated with parallel opposed 6-MV X-ray fields weighted 3:2 to the side of the tumor if it is lateralized. An anterior boost field is usually employed to deliver approximately 5–10% of the total dose to reduce the high dose distribution laterally. ^{60}Co or 4-MV X-ray beams are ideal but are not available in most radiation oncology facilities. The typical borders for a T1N0 cancer are the middle of the thyroid notch, the bottom of the cricoid cartilage, 1 cm posterior to the thyroid ala, and 1.5 cm anterior to the skin of the anterior neck. The portals may be modified depending on the precise extent of the tumor.

CT helps determine tumor extent, and therefore portal design, in patients with large T2 cancers. It is useful in detecting subglottic spread, which may be submucosal and difficult to detect by direct laryngoscopy. T1 and early T2 cancers are often superficial and inapparent on the planning CT. The dose fractionation schedule is 63 Gy in 28 once-daily fractions for T1–T2a cancers and 65.25 Gy in 29 fractions for T2b tumors (MENDENHALL et al. 2001). The prescribed dose is the minimum target dose (MTD). The maximum dose in the irradiated volume is typically less than 103%.

18.1.1.2
Stage T3–Favorable T4

The initial portals for T3–T4N0 true vocal cord cancer are shown in Figure 18.2 (PARSONS et al. 1989). Patients selected for definitive RT have

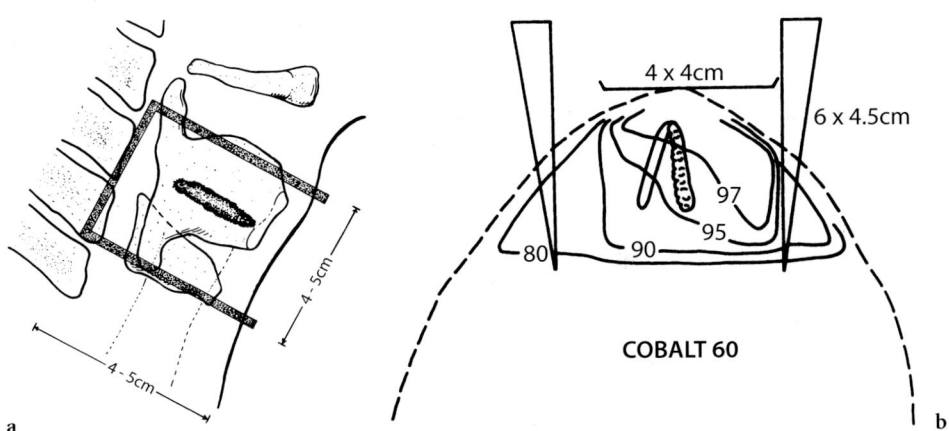

Fig. 18.1a,b. Radiation treatment technique for carcinoma of glottic larynx, stage T1–T2. **a** For T1 cancer, the superior border of field usually is at mid-thyroid notch (height of notch typically is about 1.0 cm or slightly more in male adults). If ventricle or false vocal cords are minimally involved, top of notch (which corresponds to cephalad portion of thyroid lamina as palpated just off midline) is often selected; more advanced lesions call for greater superior coverage. If only anterior half of vocal cord is involved, posterior border is placed at back of midportion of thyroid lamina. If posterior portion of cord is involved, border is 1.0 cm behind lamina. If anterior face of arytenoid is also involved, posterior border is placed 1.5 cm behind cartilage. If no subglottic extension is detected, inferior border of irradiation portal is at bottom of cricoid arch as palpated at midline. If computed tomography demonstrates subglottic extension, portal is adjusted accordingly. Anteriorly, beam falls off (by 1.5 cm) over patient's skin (from MILLION et al. 1999, Fig. 21, p 464). **b** Three-field technique (two lateral wedge fields and an anterior open field). Lateral fields are differentially weighted to the involved side. Anterior field, which usually measures 4×4 cm, is centered approximately 0.5 cm lateral to midline in patients with one cord involved and typically delivers about 5% of total tumor dose (usually on last two treatment days) after treatment from lateral portals is completed. Anterior portal is essentially reduced portal that centers high dose to the tumor. Isodose line at which dose is specified is that which covers gross disease. By appropriate field weightings, encompassing the tumor within 95–97% of maximum isodose line is virtually always possible (from LEVITT et al. 2nd edn., 1999)

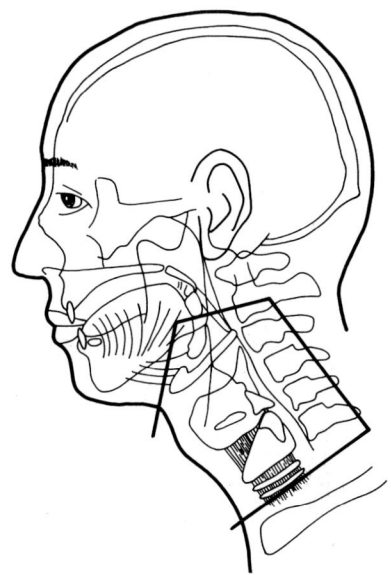

Fig. 18.2. Radiation treatment technique for carcinoma of glottic larynx, stage T3–T4N0. Patient is treated in the supine position, and field is shaped with Lipowitz's metal. Anteriorly, field is allowed to fall off. The entire pre-epiglottic space is included by encompassing the hyoid bone and epiglottis. The superior border (just above angle of mandible) includes jugulodigastric (level II) lymph nodes. Posteriorly, portion of spinal cord must be included within field to ensure adequate coverage of midjugular (level III) lymph nodes; spinal accessory (level V) lymph nodes themselves are at low risk of involvement. Lower border is slanted (1) to facilitate matching with low-neck field and (2) to reduce length of spinal cord in high-dose field. Inferior border is placed at bottom of cricoid cartilage if patient has no subglottic spread; in presence of subglottic extension, inferior border must be lowered according to disease extent (from PARSONS et al. 1989, Fig. 1, p 124; LEVITT et al. 2nd edn., 1999, Fig. 15-2)

favorable low-volume cancers without significant cartilage destruction (MENDENHALL et al. 2003a). Because of a 20–25% risk of subclinical involvement of the jugulodigastric (level II) or midjugular (level III) lymph nodes, these areas are electively treated with 45.6–50 Gy MTD. A small low-neck portal treats the low jugular (level IV) lymph nodes with a 50-Gy given dose (at Dmax) over a duration of 5 weeks (see subsequent section, "Design of the Low Neck Portal"). Primary fields are then reduced, and the treatment is continued to the final tumor dose with fields that are usually slightly larger than those described for early vocal cord cancer. Most patients currently receive 74.4 Gy MTD at 1.2 Gy twice a day with a minimum 6-h interfraction interval. Care must be taken not to underdose tumor that extends anteriorly through the thyrocricoid membrane for patients with favorable T4 tumors who are treated with 6-MV X-rays. Patients with high-volume unfavorable tumors are usually not treated with RT alone or combined with concomitant chemotherapy because of a relatively low probability of cure with a functional larynx (MENDENHALL et al. 2003b).

18.1.1.3
Supraglottic Larynx

RT alone produces high control rates for T1, T2, and low-volume T3 supraglottic cancers (MENDENHALL et al. 2003a). Unfavorable T3 and T4 tumors are often treated with laryngectomy and adjuvant RT; those who receive definitive RT also receive concomitant chemotherapy. The treatment volume is similar to that shown in Figure 18.2, with the exception that the beam generally is not allowed to "fall off" over the anterior skin surface, except in thin patients, those with very bulky lymphadenopathy that extends anteriorly, or those who have lesions involving the infrahyoid epiglottis near the anterior commissure. Shielding even a few millimeters of the anterior skin, subcutaneous tissues, and lymphatic vessels reduces the likelihood of desquamation (particularly in patients who receive concomitant chemotherapy) and may lessen the risk of serious laryngeal edema.

The inferior border of the portal is adjusted according to disease extent. For a false cord or infrahyoid epiglottic cancer, the bottom of the cricoid cartilage is usually chosen. For an epiglottic tip cancer, the lower border may be placed at or above the level of the true cords, depending on the extent and growth pattern (infiltrative versus exophytic) of disease.

If the neck is clinically negative and tumor does not extend beyond the larynx, only the level-II and level-III nodes are treated. If the base of the tongue or pyriform sinus is involved or if neck disease is extensive, the primary portal includes the entire jugular chain, spinal accessory chain (level V) and retropharyngeal nodes. In all situations, the low neck is treated with an anterior en face portal, the size and shape of which vary according to the N stage and laterality of disease (see subsequent section, "Design of the Low Neck Portal").

18.2
Hypopharynx

18.2.1
Pyriform Sinus

The portals used for the initial treatment volume of early and moderately advanced pyriform sinus cancer are shown in Figure 18.3. In addition to the entire jugular lymphatic chain, the lateral retropharyngeal lymph nodes (medial to the carotid arteries, usually located just in front of the C1–C2 vertebral bodies) and level-V nodes are also at risk and are treated even if the neck is clinically negative. The pyriform sinus lies posteriorly in the pharynx, extending from its upper limit on the pharyngoepiglottic fold to its apex located between the superior and inferior borders of the cricoid cartilage. It is rarely necessary to allow anterior "falloff" over the anterior skin of the midline. If the primary lesion is so extensive as to require "falloff" anteriorly to achieve adequate coverage, total laryngopharyngectomy is usually the treatment of choice.

18.2.2
Pharyngeal Wall

Most pharyngeal wall lesions involve only the posterior wall. With CT, some lesions are seen to extend posterolaterally, in effect wrapping around the anterior vertebral body (Fig. 18.4) (MENDENHALL et al. 1988a). If the posterior edge of the reduced portal splits the middle of the vertebral body, geographic miss may result. The initial treatment volume is much like that shown in Figure 18.3 for pyriform sinus cancer. The upper margin of the port should be at the base of the skull, to include the retropharyngeal nodes. The initial lower portal margin should include the entire pharyngeal wall because of these tumors' propensity to have "skip lesions." The reduced portals are shown in Figure 18.4. The posterior field edge is placed at the posterior vertebral body.

18.2.2.1
Postoperative Irradiation of Laryngeal and Hypopharyngeal Tumors

In this situation, the larynx has been removed. The primary fields are treated through lateral parallel-opposed portals (Fig. 18.5a) (AMDUR et al. 1989) to include the anterior and posterior neck from the

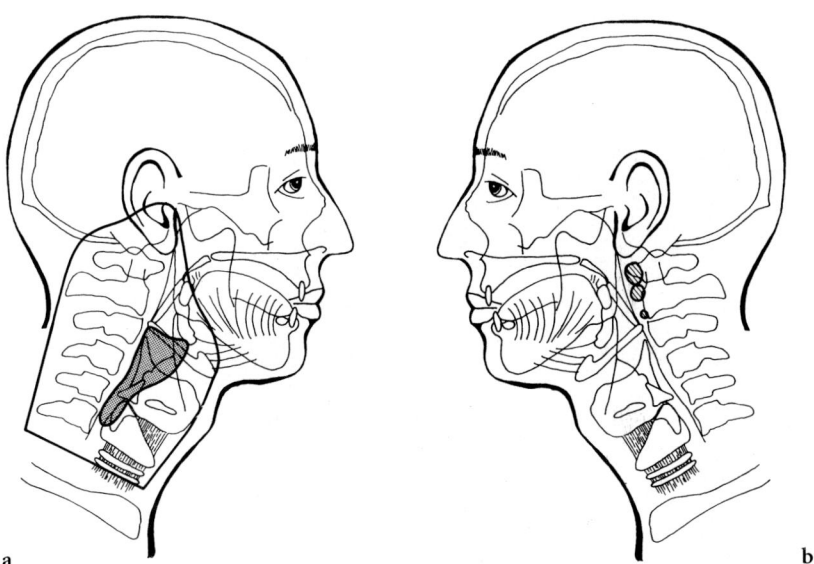

Fig. 18.3. a Portals used for initial treatment volume in a patient with carcinoma (stippled area) of pyriform sinus. Superiorly, portal covers lymph nodes at base of skull, then sweeps anteroinferiorly to cover posterior tongue base and level-II lymph nodes. Anteriorly, at least 1 cm of skin and subcutaneous tissues (as viewed from lateral projection) is usually spared. Inferior border is 2–3 cm below bottom of cricoid cartilage and is slanted to facilitate matching with low-neck portal and to avoid irradiating shoulders. Posterior field edge usually encompasses spinous process of C2 vertebral body. As treatment progresses, several field reductions are made (to shield spinal cord and to limit volume of mucosa that receives high dose irradiation). **b** Location of lateral retropharyngeal lymph nodes in relation to C1–C2 vertebral bodies (from LEVITT et al. 2nd edn, 1999, Fig. 16.3a,b)

base of skull to the top of the tracheal stoma. Techniques with either anterior or anterior and posterior portals have the disadvantages of underdosage of lymph nodes at the base of the skull in the region of the mastoid and unnecessary irradiation of a large volume of brain tissue in the posterior cranial fossa. The field is reduced after approximately 45 Gy MTD so that the spinal cord is no longer in the treatment field; the dose to high-risk areas behind the plane of the spinal cord may be boosted with 8- to 10-MV electrons.

The low-neck portal, which usually includes the stoma, is treated as shown in Figure 18.5b (AMDUR et al. 1989). The dose at Dmax (the dose at maximum buildup) in most patients is 50 Gy in 25 fractions. In patients at high risk for recurrence in the low neck (i.e., who have positive level-IV nodes), a boost dose is occasionally given through a reduced field. In patients with subglottic extension, the dose to the stoma and peristomal tissues is boosted with electrons, usually 12 MV. The energy selected should be high enough to deliver an adequate dose to the tracheoesophageal groove (level VI) lymph nodes.

Cobalt 60 is the beam of choice because of its buildup characteristics. Alternatively, 4-MV or 6-MV X-rays may be employed. A petrolatum gauze

bolus is placed on all scars and over drain sites. All scars, suture holes, and drain sites are treated with generous (2–3 cm) margins.

18.3
Oropharynx

There are two important points in the design of portals for the oropharynx. One, the risk of lymph-node metastases in both the upper and lower neck is significant, and both areas should be treated even when the neck is clinically disease free. Two, the use of long lateral primary fields that include the larynx and a longer length of spinal cord than necessary is inappropriate. Dividing the treatment volume into upper (primary) and anterior low-neck fields is essential so that the larynx can be shielded from irradiation and the spinal cord dose is reduced; this is true even in the presence of a large lymph node that is bisected by the lower border of the primary field. Although many radiation oncologists explain that treating the larynx with 50 Gy is acceptable because "it can tolerate it," there is little excuse for exposing a vital structure

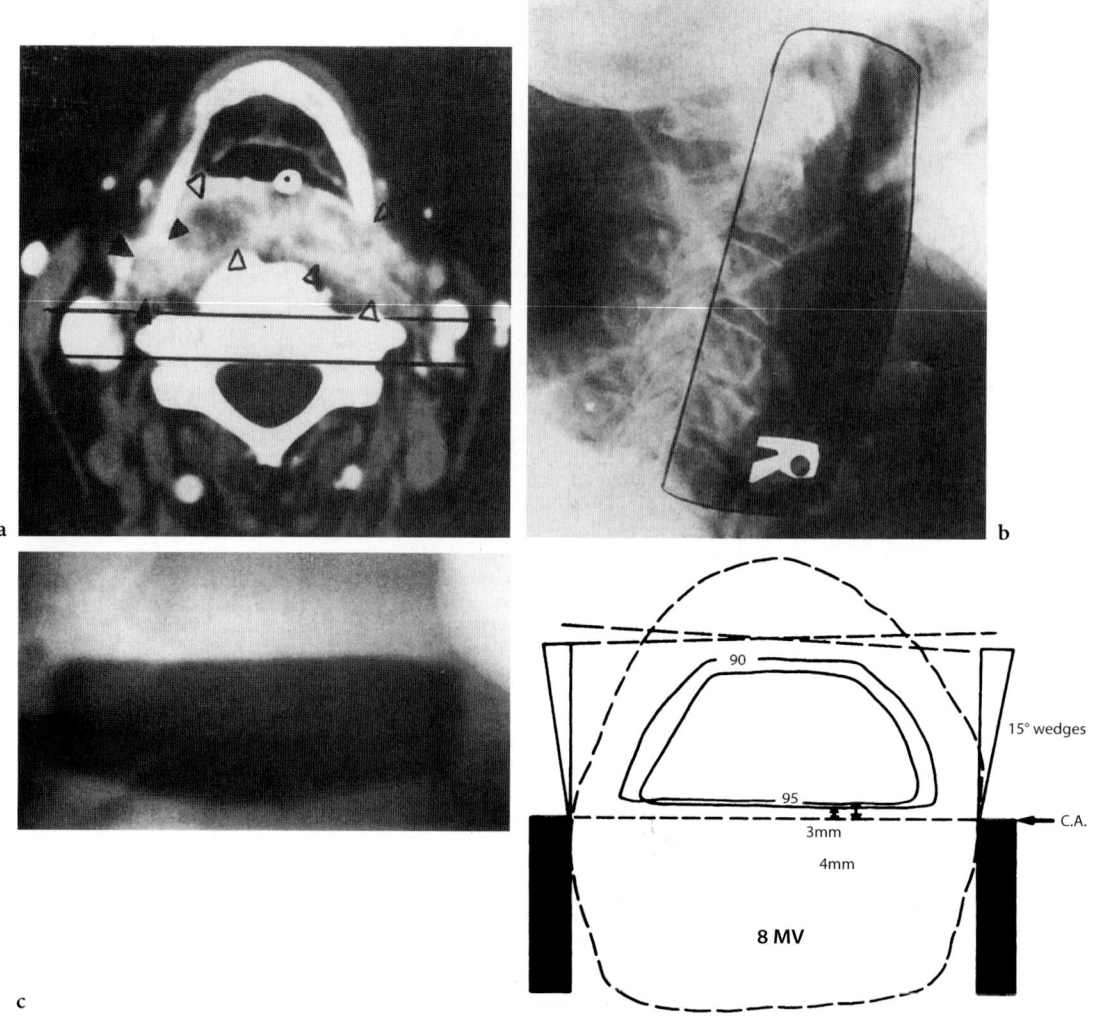

Fig. 18.4a–c. Radiation treatment techniques for carcinoma of posterior pharyngeal wall. **a** Computed tomogram of T3 posterior pharyngeal wall cancer (*arrowheads*). Horizontal lines represent two possible placements of posterior field edge. If field edge bisects vertebral body when spinal cord is shielded, part of cancer will be in penumbra or altogether outside irradiated volume. Entire width of vertebral body is always treated in patients with posterior pharyngeal wall tumors. **b** Simulation film shows first field reduction (to shield spinal cord) for a patient with T3N0 cancer. Note that with reduced portals, little of larynx remains within treatment volume. Only epiglottis and part of arytenoids and aryepiglottic folds cannot be excluded. **c** Isodose plots for reduced portals. In our practice, 6-MV X-rays are ideal energy. Because of characteristics of high-energy (e.g., 20 MV) X-rays near field edge, isodose distributions are constricted compared with low energy beams (2). Result is reduced dosage to cancer near posterior field edge, which is undesirable in treatment of posterior pharyngeal wall cancer. Central axis (CA) is placed at posterior field edge to provide nondivergent posterior field edge. Wedges are used to reduce dose anteriorly and to pull isodose distribution slightly posteriorly. SSD source-to-surface distance (a,b from MENDENHALL et al. 1988a, Fig. 3b,c p 210; c from LEVITT et al. 2nd edn., 1999, Fig. 16.4)

to a moderately high dose of irradiation when it is avoidable. Because the neck is thinner at the level of the larynx than of the oropharynx, the involved larynx often receives a higher total dose at a higher dose per fraction than the primary tumor (AMDUR et al. 2004). Inclusion of the larynx within the primary portals causes more severe mucositis, which often leads to unplanned treatment interruptions or requires treatment with low total daily doses, either of which results in poor tumor control (MENDENHALL et al. 2003b). Treatment of the larynx also produces some edema and dries the mucous membranes resulting in unnecessary chronic morbidity. Such treatment also complicates the management of the patient who later develops a second primary cancer in the larynx.

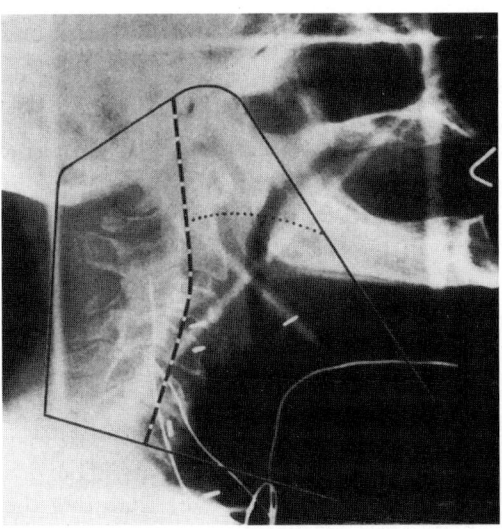 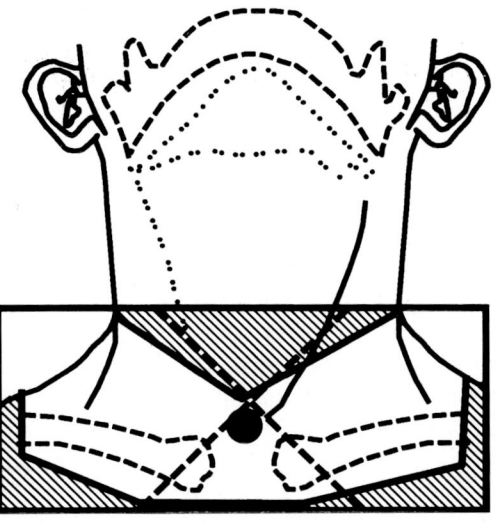

a   b

Fig. 18.5. a Typical postoperative simulation film of a patient with advanced-stage cancer of the laryngopharynx. Dashed line initial field reduction (after 50 Gy, to shield spinal cord), dotted line final reduction (after 60 Gy). Wires mark surgical scars and stoma. Slanting line used on lower border reduces length of spinal cord treated by primary field, allows better caudal coverage of mucosal surfaces while simultaneously bypassing shoulders, and facilitates matching with low-neck field. **b** Low-neck field. Beam is vertical (0°). Rectangle (solid line) light field, dashed line central axis, shaded areas blocked portions of field (stacked lead blocks). Superior border of neck field is inferior border of primary field. Actual line is treated only with primary field. Upper border of low-neck field assumes V shape. In midline of patient, apex of V generally is at or close to central axis, so portion of beam that irradiates the spinal cord is nondivergent. At junction of the three fields, short (2–3 cm) segment of spinal cord remains untreated through any of the fields (from AMDUR et al. 1989, Fig. 1, p 27; LEVITT et al. 2nd edn, 1999, Fig. 16-5)

18.4
Base of Tongue

Typical initial and reduced portals for treating a cancer of the base of the tongue are shown in Figure 18.6 (PARSONS et al. 1999b). A variety of ways are used to administer the final "boost" dose. All lesions currently are managed by external-beam irradiation alone at our institution. Concomitant chemotherapy is employed for unfavorable T3–T4 and/or advanced neck disease. Iridium-192 implantation is a popular boost technique in some treatment centers. However, no convincing evidence is available at this time that supports the notion that the implant produces superior local control rates or reduced complications when compared with the results of external-beam irradiation alone (MENDENHALL et al. 2000).

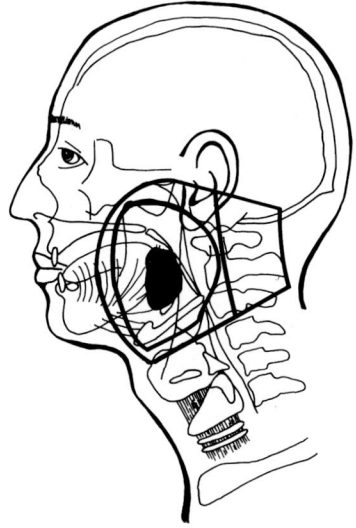

Fig. 18.6. Radiation treatment technique for carcinoma of base of tongue. Superiorly, portal treats jugular and spinal accessory (level V) lymph nodes to base of skull. Posterior border is behind spinous process of C2. Inferior border is at or just below thyroid notch, depending on disease extent. Antero-inferiorly, skin and subcutaneous tissues of submentum are shielded, except in case of advanced disease. The portals are usually reduced off of the spinal cord at approximately 45 Gy and a second reduction occurs at 60 Gy. The portals are usually equally weighted 3:2 towards the side of the lesion, if it is lateralized (from PARSONS et al. 1998, Fig. 42.1)

18.5
Tonsillar Area

The minimum initial treatment volume for early cancers of the tonsillar region includes the retro-molar trigone, tonsillar pillars, soft palate, base of the tongue, and entire tonsillar fossa. The anterior margin of the portal varies according to the anterior extent of disease; in patients with cancer at an early stage, it is at the level of the second molar tooth. If the buccal mucosa is invaded, anterior coverage should be generous for the first 50 Gy MTD to avoid geographic miss. Anterior extension into the oral tongue is appreciated by digital palpation. For T1–T2, well-lateralized lesions of the tonsillar region with no tongue invasion, ipsilateral treatment with a wedge pair using 4- to 6-MV X-ray beams is used to preserve salivary flow on the contralateral side. It has become easier to plan the appropriate treatment volume, and avoid geographic miss, with the availability of 3DCRT.

The RT technique for more advanced tonsillar cancers is by parallel-opposed portals, similar to those shown in Figure 18.6 which are used for lesions of the base of the tongue. Portals are usually weighted 3:2 or 1:1, depending on the anatomic distribution of disease. The pterygoid plates up to the base of the skull should be covered in those patients with advanced disease; if trismus is present, the pterygoids should remain within the treatment volume for the entire course of irradiation. Nasopharyngeal or hypopharyngeal extension must be recognized to avoid geographic miss with the reduced fields.

The failure rate after RT alone for T1–T2 lesions of the anterior tonsillar pillar is higher than for tonsillar fossa lesions. External beam (45 Gy) plus iridium implantation (30 Gy) has produced a high rate of success (MAZERON 1986). Alternatively, the dose may be boosted with an intraoral cone using either 250-Kvp X-rays or electrons.

18.6
Soft Palate

The RT technique by parallel opposed fields for soft palate cancer is demonstrated in Fig. 18.7 (PARSONS et al. 1999b). Figure 18.8 (PARSONS et al. 1999b) depicts the treatment of early, discrete lesions that can be encompassed by an intraoral cone. Intraoral cone RT has the advantage of delivering a high dose to a limited tissue volume in a short overall treatment time. Intraoral cone RT is administered before the start of external-beam RT, while the lesion is still clearly visible and before the onset of mucositis; it is not recommended as the sole treatment because of the risk of lymph node metastases.

18.7
Nasopharynx

A typical portal for a patient with an advanced-stage cancer of the nasopharynx is outlined in Figure 18.9. The basic plan must be individualized according to disease extent. The retropharyngeal lymph nodes are usually involved and are best seen by magnetic resonance imaging (MRI). They incidentally fall within the treatment volume. Because of the high density of capillary lymphatics in the nasopharynx, the spinal accessory and jugular lymph node chains are irradiated in their entirety, even in the N0 setting.

After shielding the spinal cord, the final dose may be administered via parallel-opposed lateral portals or oblique portals (PARSONS 1984b). Planning the appropriate boost technique for the nasopharyngeal cancer is one of the most challenging tasks that the radiation oncologist faces in the treatment of head and neck cancer. The boost must be designed to encompass all areas of primary disease and retropharyngeal lymphadenopathy, while sparing as much normal tissue as possible. Treatment is highly individualized; high-quality CT and MRI scans are essential. Use of CT for simulation and treatment

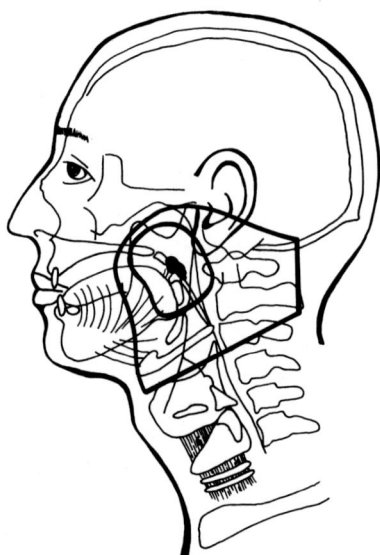

Fig. 18.7. Initial and reduced lateral fields for treatment of carcinoma of the soft palate. Usual technique involves parallel-opposed portals. Minimum treatment volume for early-stage disease includes entire soft palate and adjacent pillars. Timing and extent of field reductions after 50 Gy depend on status of neck as well as extent of primary lesion. If primary lesion extends to midline or if clinically positive lymph nodes are present, both sides of lower neck are irradiated. (from PARSONS et al. 1998, Fig. 42-3)

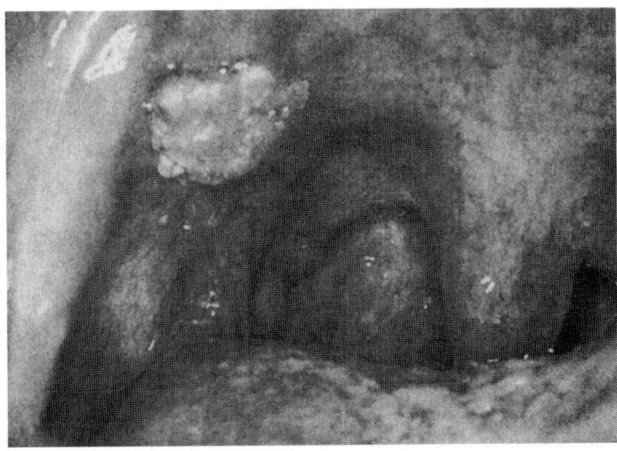
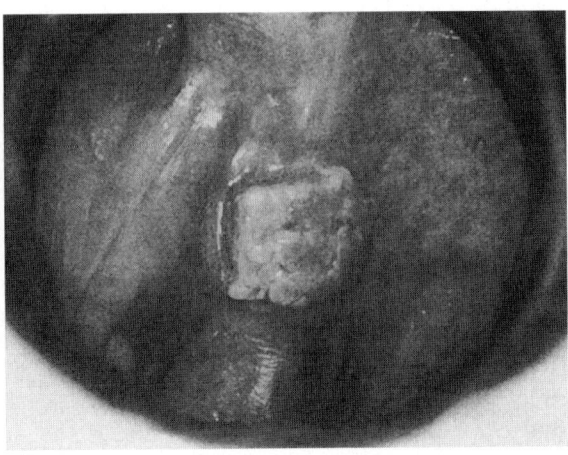

Fig. 18.8a,b. Intraoral cone treatment technique. **a** Exophytic invasive squamous cell carcinoma, 1.5 cm, of right side of soft palate. Discrete lesions in cooperative patients receive 15–20 Gy (2.5–3 Gy per fraction) with intraoral cone before external beam irradiation. **b** View through intraoral cone shows adequate coverage of lesion. If lesion does not extend deeply into the tongue and neck shows no evidence of involvement, treatment is completed by ipsilateral wedge pair field arrangement using 4-MV or 6-MV X-rays that encompass primary lesion and upper neck nodes, to a dose of approximately 50 Gy. The ipsilateral low neck is treated with a separate en face portal (from PARSONS et al. 1998, Fig. 42-4)

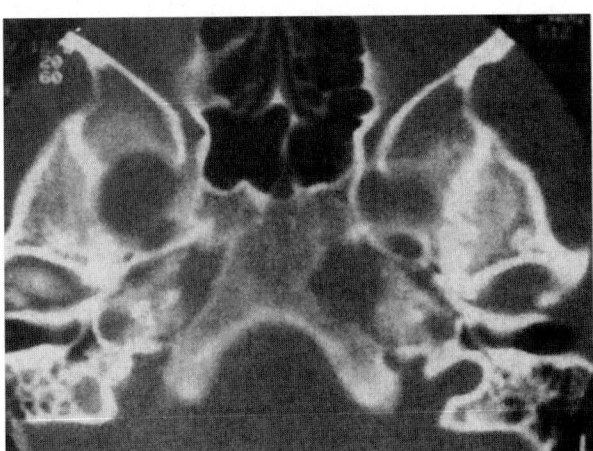
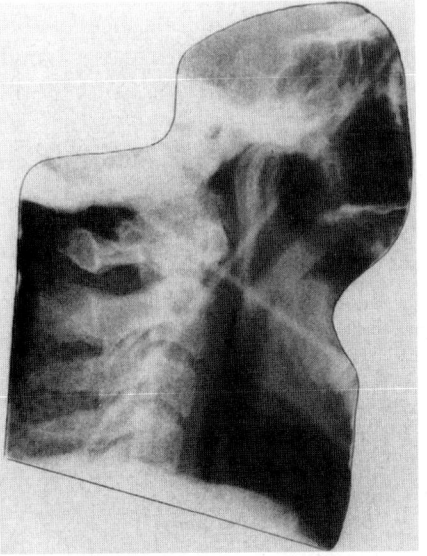
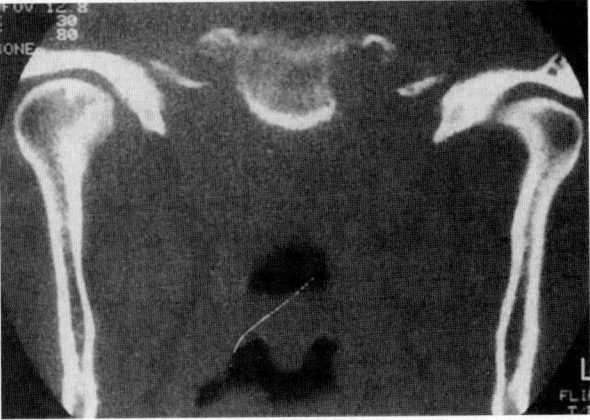

Fig. 18.9a–c. Radiation treatment technique for carcinoma of the nasopharynx. Axial (**a**) and coronal (**b**) computed tomograms of patient with T4N2 squamous cell carcinoma of nasopharynx. Note bone destruction at petroclival junction. **c** Beam angled 5° posteriorly to avoid exit irradiation through posterior pole of contralateral eye. Because of destruction of base of skull, superior border of treatment volume is above pituitary fossa; in less advanced presentations, superior border often passes through anterior and posterior clinoid processes, thereby placing optic nerve in penumbra or out of radiation beam. Lymph nodes are included up to jugular foramen or about 2 cm above tip of mastoid. Posteriorly, the level-V chain is irradiated; in presence of large or multiple lymph nodes, posterior coverage can be more generous. Inferior border excludes larynx, except in rare circumstance of tumor extension down lateral pharyngeal wall into hypopharynx. Submandibular (level I) lymph nodes are at risk in patients with extensive lymph node metastases. Anterior border is designed to shield segment of posterior mandible while still encompassing tonsillar area. The portals include the posterior 2 cm of the nasal cavity. If necessary, the portal may be bowed anteriorly to cover more of the nasal cavity (from LEVITT et al. 2nd edn., 1999, Fig. 16-12a-c)

planning facilitates the planning of oblique-entry portals. IMRT has significantly improved the ability to boost the tumor volume, particularly in the retrostyloid parapharyngeal space and posterolateral clivus, while minimizing the dose to the normal tissues such as the temporal lobes and brain stem.

18.8
Oral Cavity

18.8.1
Lip

Lip cancer may be treated by external beam, interstitial implant, or both. External-beam techniques use orthovoltage X-rays or electrons; the former is preferred because there is less beam constriction and the maximum dose is at the surface. A lead shield placed behind the lip limits RT to the mandible and oral cavity. The shield consists of two sheets of lead (each 1/8 inch thick), overlaid with one sheet of aluminum (1/64 inch) and is coated with wax or vinyl to prevent excessive exposure from low-energy-

scattered electrons to tissue adjacent to the shield. Dose schemes are similar to those for skin cancer. Protracted treatment schedules (4–6 weeks) are preferred over short regimens because short courses are more likely to cause progressive radiation changes with passing years.

Bulky cancers are often treated first by external beam (30–50 Gy MTD), followed by interstitial cesium or ^{192}Ir implant once the lesion has flattened [Fig. 18.10 (MILLION et al. 1994b)]. We prefer preloaded implant devices that allow rapid, accurate positioning if cesium needles are employed (ELLINGWOOD et al. 1976). Iridium using the plastic tube technique has an advantage because a larger volume can be easily implanted, if necessary, and it can be afterloaded.

The regional lymphatics are not electively treated in patients with cancer in its early stages because the risk of metastasis is low. Patients with advanced, poorly differentiated, or recurrent cancers should receive elective neck treatment because the risk of lymphatic involvement is substantial. The risk of involvement also increases in patients with tumors that extend onto the wet mucosa of the lip or the buccal mucosa. Lymphatic spread is to the level-I and level-II lymph nodes and rarely to a facial node.

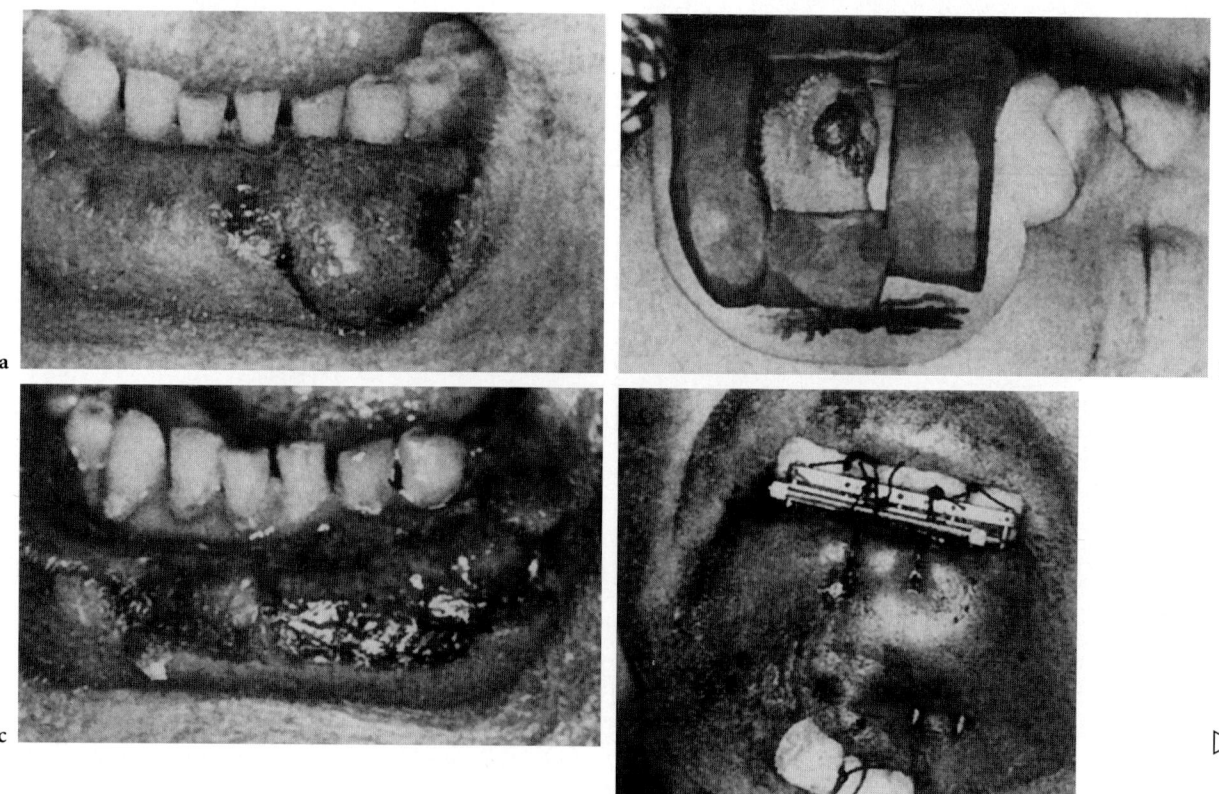

a

c

b

d

▷

18.9
Floor of Mouth

At the University of Florida, most floor of the mouth cancers are treated with resection. If RT is used, availability of intraoral cone or interstitial therapy is essential to obtain maximum local control rates. Intraoral cone therapy is preferred when the lesion is suitable. Megavoltage external-beam RT alone gives inferior control results, even for T1 lesions.

18.9.1
External-Beam Irradiation

External-beam portals for cancer of the anterior floor of the mouth usually are opposed lateral portals. If the lesion is small and confined to the floor of the mouth, the tip of the tongue is elevated out of the portal with a cork [Fig. 18.11 (PARSONS et al. 1999a)]. If the lesion has grown into the tongue, the tongue is flattened to reduce the superior border of the portal

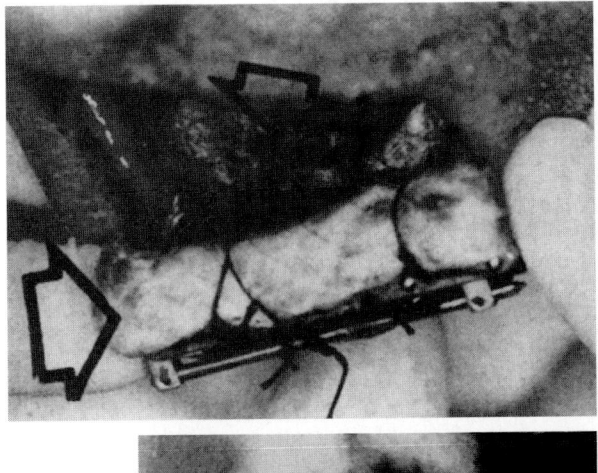

e

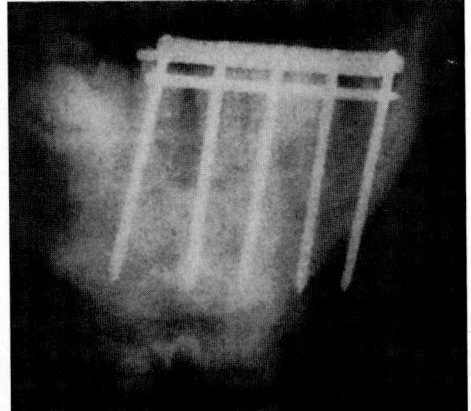

f

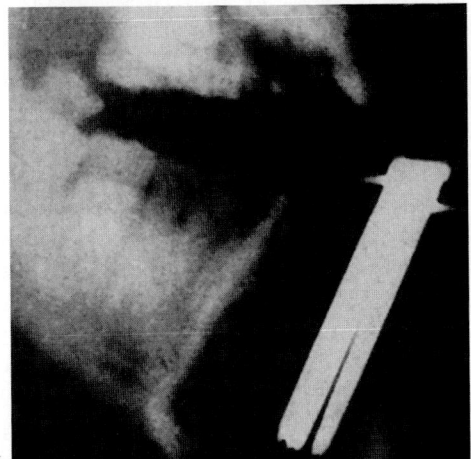

g

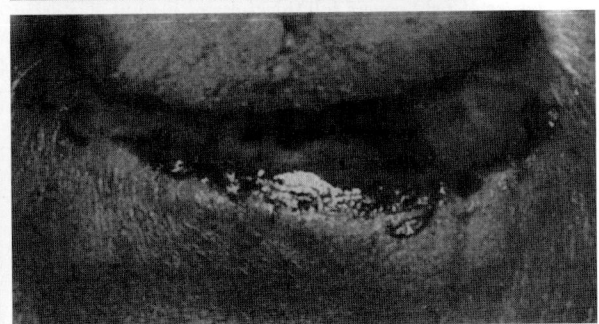

h

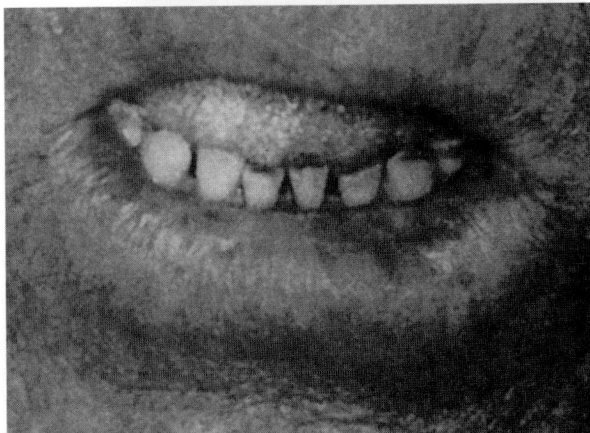

i

Fig. 18.10a–i. A 67-year-old man had T2N0 squamous cell carcinoma of the lower lip. **a** The lesion measured 3.0×2.0×1.5 cm. Radiation therapy was elected because of functional deficit likely to result from excision of large lesion. **b** Lead mask, 2 mm thick, designed to outline portal. Lead putty was added to shield to reduce transit irradiation to less than 1%. Separate lead shield covered with beeswax was inserted behind lower lip. Patient received 30 Gy in 2 weeks, 3 Gy per fraction, 250 kV (0.5 mm Cu). **c** By completion of 30 Gy, he had brisk mucositis of lip and approximately 60–70% regression of obvious tumor. **d** Single-plane radium needle implant with double crossing. Pack was tied to top of bar to displace upper lip away from radiation, and chin pack anchored gingivolabial pack in place (see **e**). **e** Gauze pack (*arrows*) sewn into gingivolabial gutter to displace radium from mandible, teeth, and gums. **f** and **g** Anteroposterior and lateral views of implant. Implant added 35 Gy at 0.5 cm. **h** 2.5 weeks after implantation. Note superficial ulceration. **i** 22 months after treatment. No evidence of disease, and lip was completely healed. Nine-year follow-up revealed no evidence of disease (from MILLION et al. 1994b, Fig. 16-7)

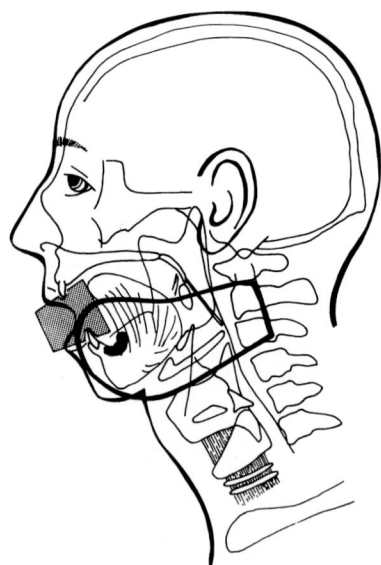

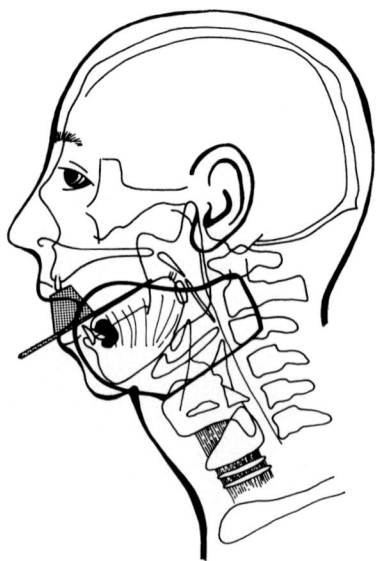

Fig. 18.11. Portal for irradiation of limited anterior floor of mouth carcinoma (no tongue invasion; N0 or N1 neck disease) by parallel-opposed ^{60}Co, 4-MV X-ray, or 6-MV X-ray fields. Two notches on a cork ensure it is held in same position between upper and lower incisors during every treatment session; tip of tongue is displaced from treatment field. Anterior border of field covers full thickness of mandibular arch. Lower field edge is at thyroid cartilage, ensuring adequate coverage of submandibular (level I) lymph nodes. Subdigastric lymph (level II) nodes are covered adequately by including entire width of vertebral bodies posteriorly. Superior border is shaped so oropharynx, much of oral cavity, and parotid glands are out of portal. Minimum tumor dose is specified at primary site (i.e., not along central axis of portal) (from Parsons et al. 1998, Fig. 35-8; Levitt et al. 2nd edn, 1999, Fig. 16-16)

Fig. 18.12. The treatment portal for carcinoma of the floor of mouth with tongue invasion. Tongue is depressed into floor of mouth with tongue blade and cork (from Parsons et al. 1998 Fig. 35-9)

[Fig. 18.12 (Parsons et al. 1999a)]. A small stainless-steel pin is inserted into the posterior border of the tumor and serves as a marker on the treatment planning (simulation) film for external-beam therapy and for confirmation of coverage at the time of interstitial implantation. About 1 cm of skin and subcutaneous tissue can usually be shielded in the submental area; the submental lymph nodes are at little risk of involvement. In far-advanced cases, direct invasion of the skin and subcutaneous tissues of the mental and submental areas requires inclusion of these areas.

(Marcus et al. 1980) and 18.14 (Million et al. 1994b)] (Ellingwood et al. 1976). It is used for T1 and T2 lesions and holds the cesium needles in a fixed position, thus ensuring near-perfect geometry. The location of the needles relative to the gingiva is tailored according to the distribution of tumor. The arrangement of needles for early stage lesions is usually a modified, curved, teardrop-shaped, two-plane implant with a single crossing needle that lies close to the mucosal surface. Homogeneity of dose is better than can generally be achieved by free-hand implantation with either active sources or afterloading techniques because the computer implant dosimetry is available before implantation so the arrangement of sources can be modified as necessary. Use of the implant device avoids piercing and unnecessarily irradiating the tongue, as is often necessary when free-hand techniques are used. Implantation is completed in less than 1 min with minimal exposure to personnel and minimal implant trauma. The operating time is substantially reduced compared with that required for freehand techniques.

18.9.2
Interstitial Irradiation

A preloaded, custom-designed, metal or nylon implant device for cesium needles has been in use at the University of Florida since 1976 [Figs. 18.13

18.9.2.1
Intraoral Cone Irradiation

An orthovoltage intraoral cone can be used instead of an interstitial implant for small, anterior, super-

Fig. 18.13. Custom-made implant device for stage T1–T2 cancers of floor of mouth. Note single crossing needle (*arrow*) through center. Devices machined from nylon are also available. Cesium needles usually are used (2.0 cm active length, 3.2 cm actual length). Intensity of needles is adjusted so dose rate is approximately 0.4 Gy/h to area of gross disease. To ensure adequate surface dose, height of implant device (9 mm) is such that the active ends of the needles extend above the mucosal surface. Crossing needle is also 3 mm above mucosal surface (i.e., at active ends of needles) (from Marcus et al. 1980, Fig. 2, p 112; Levitt et al. 2nd edn, 1999, Fig. 16-18)

ficial lesions in the edentulous patient with a low alveolar ridge. Alternatively, electron beam intraoral cone may be employed. Tiny lesions can be treated solely by cone (e.g., 50 Gy over 3 weeks or 60 Gy over 4 weeks; given dose). Larger lesions receive 20–25 Gy (2.5–3 Gy per treatment) via intraoral cone, followed by external-beam RT (45–50 Gy MTD) to the primary tumor and first-echelon lymph nodes. The technique has the advantages of delivering a smaller dose to the mandible than an implant and avoiding hospitalization. Lesions more than 1 cm thick may be underdosed because of the rapid falloff in depth dose from short treatment–distance orthovoltage cones. The cones used at the University of Florida are poured from lead and can be trimmed individually to adapt the cone to the anatomy. A variety of cone sizes, with straight or beveled edges, must be available. Intraoral cone therapy requires daily, meticulous positioning by the physician, because the margin for error is less than with other techniques of irradiation. The patient must be cooperative. After the treatment cone is centered on the tumor, its position is verified with the use of a periscopic localizer

to ensure adequacy of coverage [Fig. 18.15 (Million and Cassisi 1984)]. The end of the cone is in contact with the oral mucosa during treatment. A dental appliance may be used to ensure reproducibility of the treatment setup.

18.10
Oral Tongue

The ability to control the primary tumor is enhanced by giving all or part of the treatment by either interstitial implant or intraoral cone. Megavoltage external-beam RT alone produces poor results, even for T1 lesions. Patients referred after excisional biopsy of small lesions (TX) or with T1 lesions measuring less than 1.0 cm often are treated by interstitial or intraoral cone therapy alone. More advanced lesions receive a component of the treatment from external beam.

18.10.1
External-Beam Irradiation

Before external-beam RT, the cancer is photographed and diagrammed to document its extent at the time of the implant. Sometimes, the anterior and posterior borders of the lesion are tattooed with two tiny (1–2 mm) marks. Under no circumstances should the ink that is used to tattoo the patient be injected under pressure (e.g., with a syringe), because the ink may diffuse over a large area.

Portals are usually shaped to exclude part of the parotid gland. If teeth with metal fillings lie against the tongue or buccal mucosa, a thin layer of gauze (a few millimeters thick) is inserted between the teeth and tongue or buccal mucosa to prevent a high-dose effect secondary to scattered low-energy electrons. Common dose schedules for T1 or T2 lesions consist of 32 Gy MTD in 20 fractions (1.6 Gy twice a day) followed (after 3–5 days) by an interstitial implant (35–40 Gy). If the intraoral cone is used, generally the dose is 25 Gy given dose in 10 fractions with the cone, followed by 32–38.4 Gy MTD at 1.6 Gy twice a day, depending on the lesion size and presence or absence of infiltrating characteristics.

N0 Situation. If the lesion is well lateralized, it is treated with a single ipsilateral portal with an intraoral lead block to reduce the dose to contralateral minor salivary glands and mucosa. The subdigastric (level II) and submandibular (level I) lymph

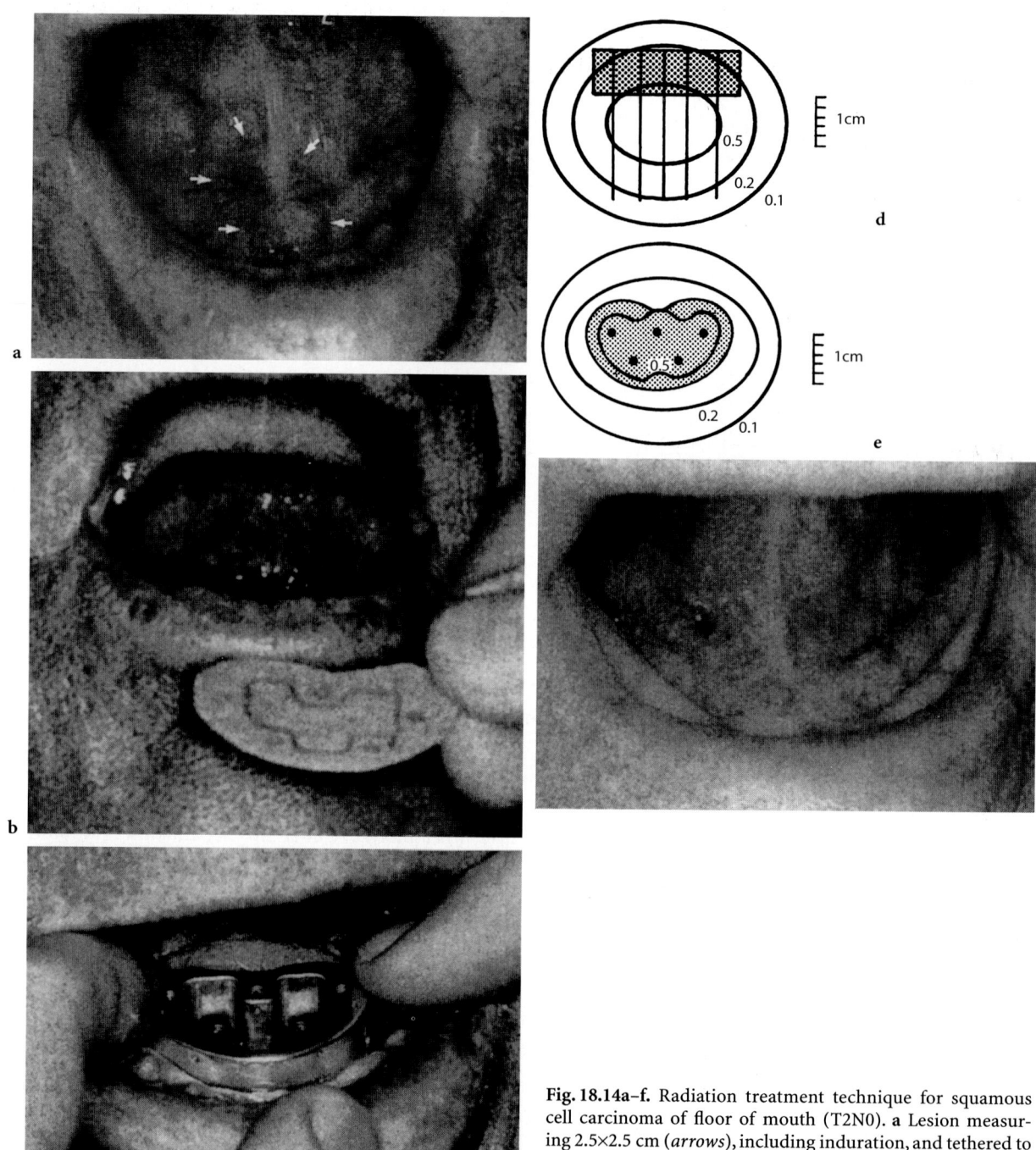

Fig. 18.14a–f. Radiation treatment technique for squamous cell carcinoma of floor of mouth (T2N0). **a** Lesion measuring 2.5×2.5 cm (*arrows*), including induration, and tethered to periosteum at midline. Treatment plan is 50 Gy over 5 weeks with parallel-opposed portals that include submandibular (level I) and subdigastric (level II) lymph nodes. Midjugular (level III) lymph nodes are treated with anterior portal. Implant is planned to add 15–20 Gy. **b** Cardboard template for design of cesium needle holder. **c** One day preoperatively, before securing cesium needles to implant device, implant holder is placed into floor of mouth to ensure adequate fit and to check adequacy of tumor coverage. At surgery, device is sutured with two 1-0 silk sutures passed on long curved needle through submentum into floor of mouth. Five, 2.0-cm active length, full-intensity cesium needles without crossing are used. **d** Coronal isodose distribution. The 0.5-Gy/h line is selected for specification of dose; implant remains in place for 30 h. Stippled area implant device. **e** Transverse isodose distribution through middle of needles. The 0.5-Gy/h isodose line is approximately 2 mm outside needles. Highest dose rate to anterior lingual gingiva would be about 0.3–0.35 Gy/h, or at least 4.5 Gy lower than minimum tumor dose. **f** Patient is free of disease at 4 years 8 months with no complications (from MILLION et al. 1994b, Fig 16-25, pp 352–353; LEVITT et al. 2nd edn, 1999, Fig. 16-19 a-f)

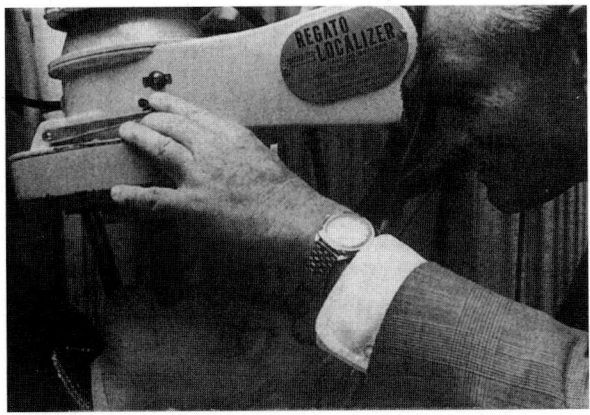

Fig. 18.15. Positioning of lead cone used for orthovoltage intra-oral therapy is checked each day by physician. A good localizer is essential for final positioning (from MILLION et al. 1984, Fig. 6-7; LEVITT et al. 2nd edn, 1999, Fig. 16-20)

nodes are included in the primary portal, and the ipsilateral low neck nodes receive 35 Gy (at Dmax) in 10 fractions via a separate en face field (Fig. 18.16) (PARSONS et al. 1999a).

If tumor extends near the midline of the tongue, the upper neck is irradiated on both sides through parallel-opposed fields, and an en face field encompasses the low neck, on both sides.

N+ Situation. In general, the submandibular (level I) and subdigastric (level II) lymph nodes are irradiated bilaterally, and both sides of the low neck are treated; a neck dissection is added 4–6 weeks after irradiation.

18.10.1.1
Interstitial Irradiation

Small-volume implants are readily performed with cesium needles on rigid implant devices [Fig. 18.17 (ELLINGWOOD et al. 1976)]. The implants are virtually always double plane; because of frequent subclinical infiltrative extensions of tumor that occur even in apparently superficial lesions, single-plane implants are not recommended. The implant is performed after the administration of general anesthesia with a short operating time and minimal exposure to operating personnel. The dose to the mandible and gingiva can be reduced significantly by inserting a pack into the floor of the mouth to displace the tongue medially [Fig. 18.18 (MILLION et al. 1994b)].

For larger volume implants, rigid cesium implant devices are difficult to manipulate (particularly the medial plane, if it contains more than three needles). Iridium implants may be performed under general

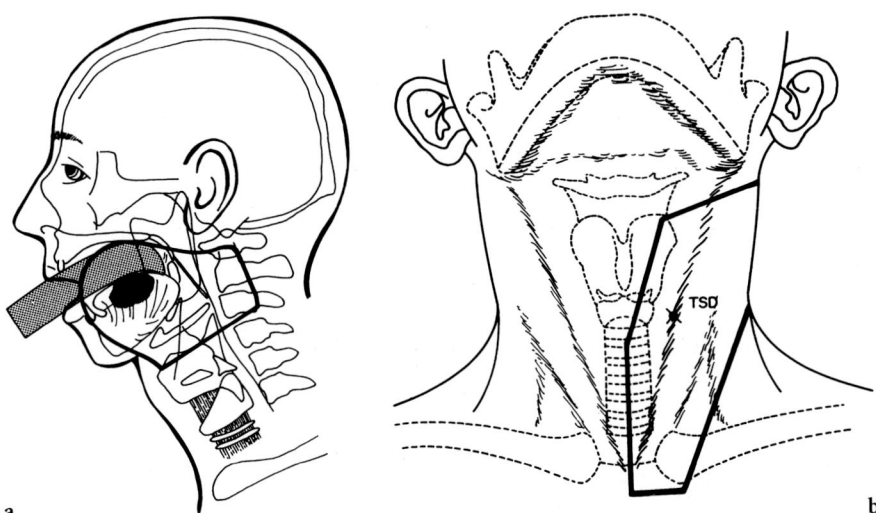

a   b

Fig. 18.16a,b. Superficial lateralized squamous cell carcinoma of the oral tongue; N0 neck. **a** Single ipsilateral field encompasses submandibular (level I) and subdigastric (level II) lymph nodes; entire width of vertebral body is included to ensure adequate posterior coverage of level-II lymph nodes. Stainless-steel pins inserted into most anterior and posterior aspects of lesion aid in localizing cancer on treatment planning (simulation) radiograph and confirm coverage by interstitial implant. Larynx is excluded from radiation field. Anterior submental skin and subcutaneous tissues are shielded when possible. Upper border is shaped to exclude most of parotid gland. Intraoral lead block (stippled area) shields contralateral mucosa and is coated with beeswax to prevent high dose effect on adjacent mucosa from scattered low-energy electrons from metal surface. The usual preinterstitial tumor dose is 30 Gy over 10 fractions with ^{60}Co. For larger lesions that extend near midline, treatment is by parallel opposed portals without intraoral lead block. **b** For patients with clinically negative necks, only ipsilateral low neck field is irradiated. TSD tumor-to-source distance (from PARSONS et al. 1998, Fig. 35-17a, b)

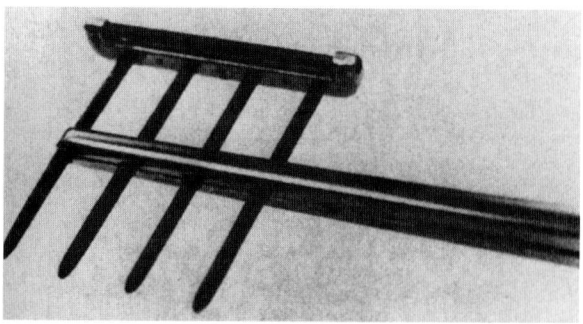

Fig. 18.17. Cesium needles mounted in rigid device for single-plane implantation of oral tongue cancer. Note single crossing needle. Holders originally were of stainless steel or aluminum; nylon has proved more satisfactory. Needles are secured to bar with half-hard stainless-steel wire passed through eyelets. Allen forceps have been drilled at 1-cm intervals to grasp needles during surgery (from Ellingwood et al. 1976, Fig. 6a; LEVITT et al. 2nd edn, 1999, Fig. 16-22)

anesthesia or a combination of regional and local anesthesia, with the patient sedated and in a sitting position. Iridium hairpins (Fig. 18.19) or the plastic tube technique are used.

18.10.1.1.1
Intraoral Cone Irradiation

Intraoral cone treatment is useful in well-selected lesions and may be used instead of an implant. When intraoral cone therapy is used, our preference is to administer this treatment before the external-beam treatment, while the lesion is clearly visible and before the mouth becomes sore.

18.10.1.2
Postoperative Irradiation for Tumors of Oral Cavity and Oropharynx

In this situation, the larynx is intact, and the portal arrangements are similar to those used when RT is the primary treatment. The primary fields are parallel opposed portals [Fig. 18.20a (AMDUR et al. 1989)], and the low neck is treated by means of a single anterior en face portal [Fig. 18.20b (AMDUR et al. 1989)]. The junction of the primary and low-neck portals facilitates shielding of the larynx by a midline block.

The choice of beam, the bolus technique, and details of treatment volume are similar to those described for the larynx and hypopharynx (see "Postoperative Irradiation for Tumors of the Larynx and Hypopharynx").

18.11
Nasal Cavity, Paranasal Sinuses, and Nasal Vestibule

The external-beam techniques for nasal cavity, ethmoid sinus, and maxillary sinus cancers are similar. Treatment emphasizes an anterior portal with one or two lateral portals that are angled 5° posteriorly (frequently with the use of wedges). Even when the lesion is considered localized, treating a large initial volume is preferable to relying too greatly on the findings of radiography and physical examination.

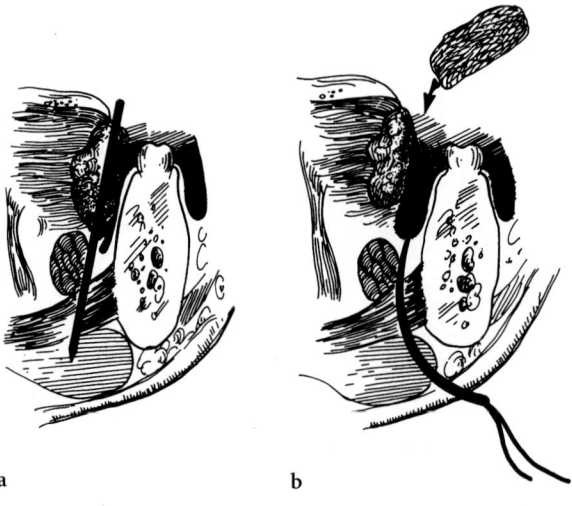

a   b   c

Fig. 18.18a–c. Cesium needle implant for cancer of lateral border of oral tongue. Gauze packing displaces tongue from mandible and thus reduces dose to bone. **a** Implant without packing. **b** Large curved needle inserted through skin into lateral floor of mouth. **c** Gauze pack tied to suture and secured between mandible and tongue after implant is completed (from MILLION et al. 1994b, Fig. 16-40; LEVITT et al. 2nd edn, 1999, Fig. 16-23 a-c)

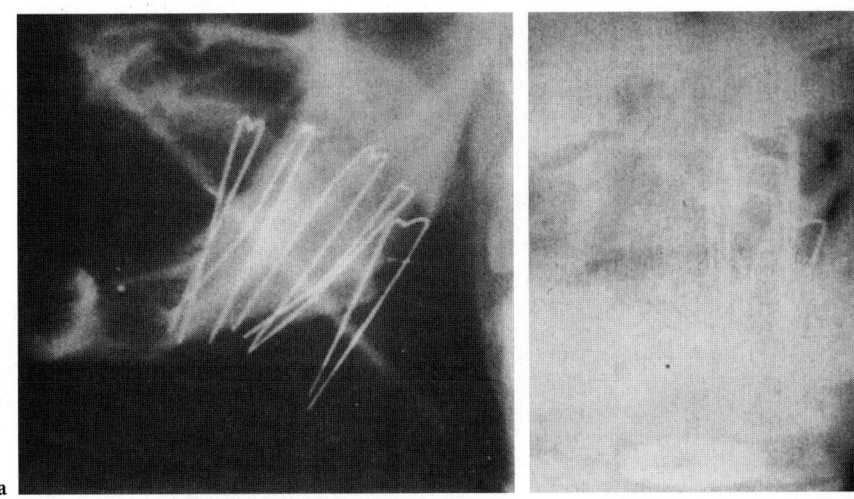

Fig. 18.19a,b. Lateral (**a**) and anteroposterior (**b**) views of ^{192}Ir implant for carcinoma of left side of oral tongue, stage T2N0, measuring 3.5×2.0×2.0 cm, with submucosal extension to within 0.5 cm of midline of tongue. Treatment consists of 30 Gy in 10 fractions, followed by ^{192}Ir implant the next week. Gutter-guide technique is used with patient sitting; local anesthesia and regional nerve block are administered. Fluoroscopy in the operating room verifies accurate source spacing and alignment. Implant sources are 4 cm in length. Gauze pack secured into lateral floor of mouth displaces tongue medially away from mandible. Implant remains in place for 73 h and delivers 40 Gy tumor dose to area of gross disease (0.55 Gy/h) (from LEVITT et al. 2nd edn, 1999, Fig. 16-24)

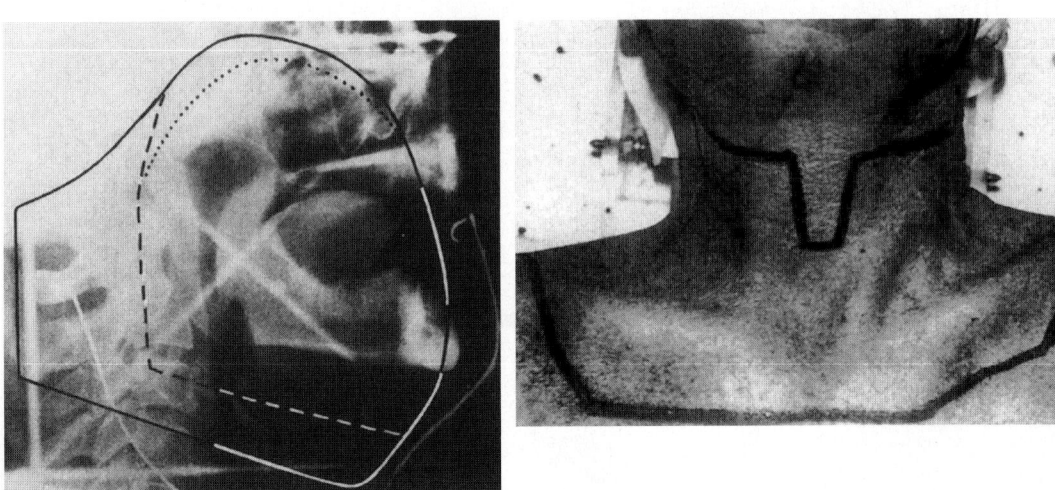

Fig. 18.20a,b. Typical portal for irradiation after hemimandibulectomy, partial maxillectomy, and radical neck dissection for pathological stage T4N0 retromolar trigone lesion. **a** Field reductions made at 45 Gy (*dashed line*) and 60 Gy (*dotted line*). **b** Low neck receives 50 Gy given dose (at D_{max}) in 25 fractions. Larynx and a segment of spinal cord are shielded by tapered midline block (from AMDUR et al. 1989, Fig. 1)

Fields may be reduced to the area of initial gross disease, with a margin, after 50 Gy MTD.

For limited cancers of the nasal cavity, the initial treatment volume includes the medial maxillary sinus, ethmoid sinus, medial portion of the orbit, nasopharynx, sphenoid sinus, and base of skull (PARSONS et al. 1988).

Ethmoid sinus and advanced nasal cavity cancers are similarly managed [Fig. 18.21 (ELLINGWOOD and MILLION 1979; PARSONS et al. 1992, 1994d)]. Treatment is heavily weighted toward the anterior field. The weighting of tumor doses administered to the anterior versus the lateral fields is usually 8:1 or 10:1 in favor of the anterior portal. Wedges are used to achieve a satisfactory dose distribution. A reduced anterior open field often is incorporated into the treatment plan to concentrate the dose to the major bulk of disease. Orbital invasion is common when

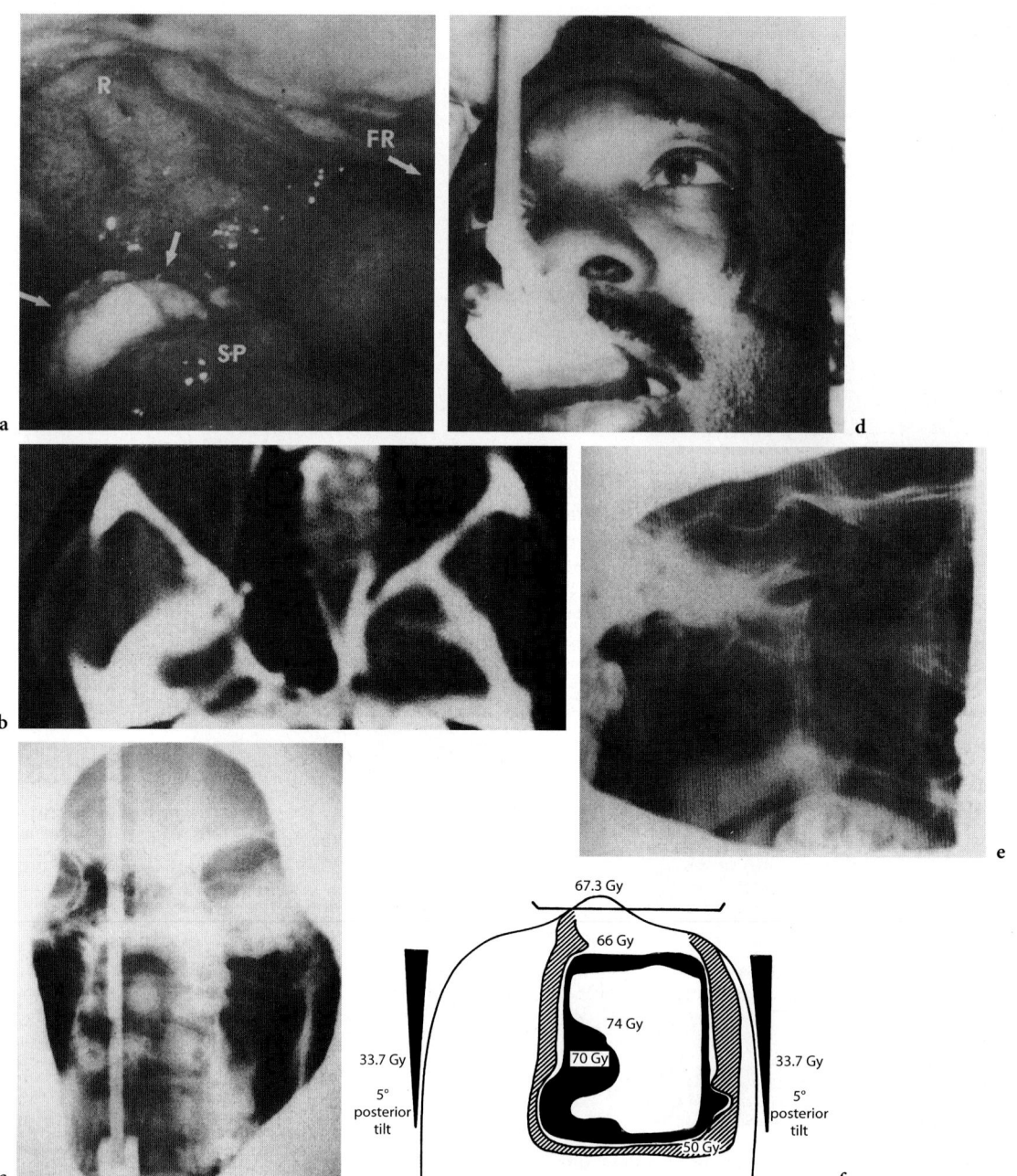

Fig. 18.21a–f. Radiation treatment technique for squamous cell carcinoma filling entire nasal cavity and extending into naso-pharynx. **a** Tumor mass extends into nasopharynx (*arrows*). *R,* roof; *SP,* soft palate; *FR,* right fossa of Rosenmuller. **b** Computed tomogram at level of orbit and ethmoids. Mass bulges into medial aspect of left orbit. Additional views showed opacification of left sphenoid sinus and left maxillary sinus, erosion of left pterygoid plates, and possible erosion of cribriform plate. Site of origin could have been nasal cavity or maxillary antrum. Tumor is considered unresectable because of involvement of nasopharynx and possible sphenoid sinus invasion. **c** Anteroposterior view (simulation) of anterior portal. Straight white line is aluminum support for bite block. Patients can be immobilized with customized Aquaplast masks from which windows for portals are cut to accomplish skin sparing (WFR/Aquaplast Corp, PO Box 635, Wyckoff, NJ 07481). **d** Radiation is delivered through anterior and left and right lateral portals. Left upper lateral eyelid and lacrimal gland are shielded because only medial orbit is involved by tumor. **e** Simulation film of lateral portal (5° posterior tilt). Treatment volume encompasses base of skull, posterior ethmoid and maxillary sinuses, posterior nasal cavity, sphenoid sinus, nasopharynx, posterior one-third of both orbits, pterygoid plates, infratemporal fossa, and parapharyngeal lymph nodes. Posterior border of portal is just anterior to external auditory canal, thereby excluding cervical spinal cord and brain stem. **f** Treatment plan is 70 Gy minimum (77 Gy maximum) tumor dose over 7 weeks. Weighting of given doses is 2 to 1 in favor of anterior portal. Right and left upper neck receives 40.5 Gy over 3 weeks through anterior portal with midline shielding. Visible tumor disappeared during therapy in this patient. Patient returned to full-time work as a truck driver at 4 months. A cataract developed in left eye at 36 months. After cataract extraction, visual acuity was only "counts fingers at 2 feet" because of radiation retinopathy. Patient was free of disease at 8.5 years (a, b, and d from PARSONS et al. 1994d, Fig. 22-34 and Fig. 22-35b; c and e from ELLINGWOOD and MILLION 1979, Fig. 3b, c; f from PARSONS et al. 1992, Fig. 29-5d)

tumor involves the ethmoid sinus. When such invasion is minimal, the major lacrimal gland and lateral upper eyelid are shielded on the anterior portal; more advanced orbital invasion requires irradiation of all of the orbital contents. An example of a two-field technique for advanced nasal cavity or ethmoid sinus cancers with invasion of the orbit is shown in Figure 18.22 (PARSONS et al. 1994a,b,c; MENDENHALL et al. 2004). In recent years, the patient's head has usually been immobilized with slight neck extension, so that the orbital floor par-

allels the angle of entry of the anterior portal, thus allowing greater sparing of the intraorbital contents (PARSONS et al. 1992). The patient is treated with the eyes open. An eyelid retractor is sometimes useful to displace some of the upper lateral lid from the treatment field. Too narrow a margin around the eye may result in geographic miss. The anterior portal extends 1.5–2.0 cm across the midline to encompass the entire nasal cavity and ethmoid–sphenoid complex and the medial aspect of the contralateral orbit. The superior margin encompasses the roof of the ethmoid sinuses, the cribriform plate, and all or part of the frontal sinus. The inferior margin is low enough to cover the floor of the nose, the maxillary antrum, and the upper gum; the inferior border generally extends to the lip commissure. The tongue is displaced out of the treatment field by a tongue blade and cork [Fig. 18.23 (PARSONS et al. 1992)].

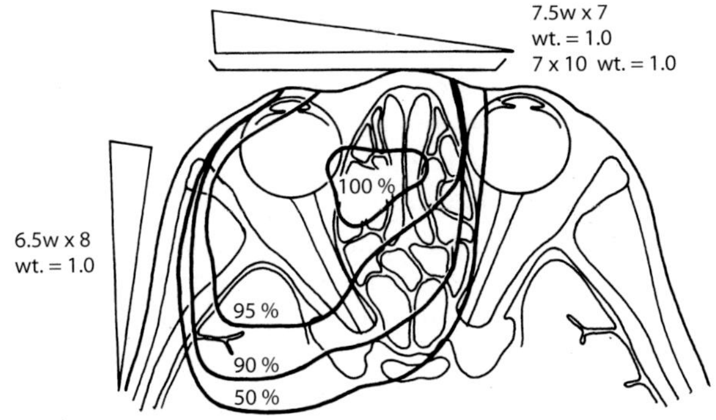

Fig. 18.22. Isodose distribution for carcinoma of ethmoid sinus with invasion of orbit. Lateral portal is angled 5° posteriorly (from MILLION et al. 1989, Fig. 21-26)

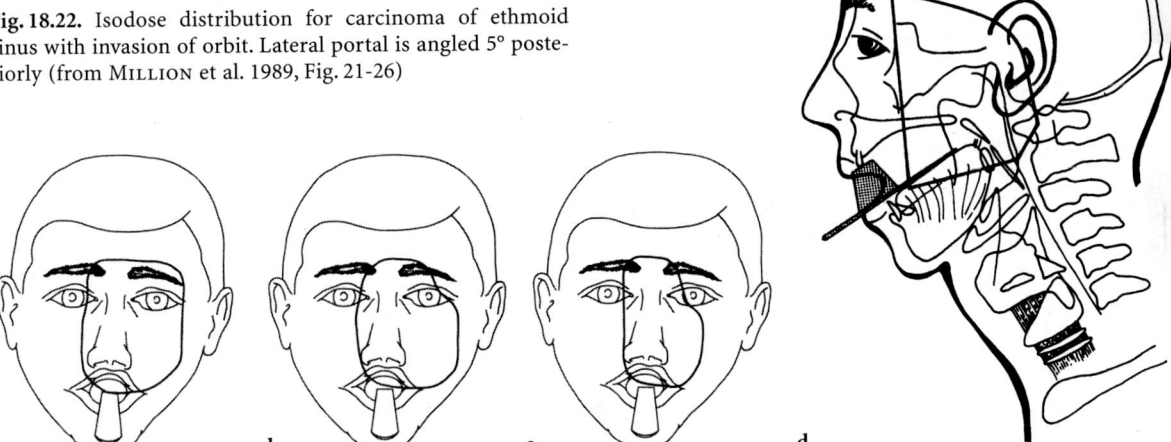

Fig. 18.23a–d. Portals used to treat patients with tumors of nasal cavity and paranasal sinuses. **a** In patients with extensive orbital invasion (palpable orbital mass, proptosis, or blindness), all orbital contents are irradiated. **b** In patients with limited orbital invasion, major lacrimal gland is shielded. **c** Portal primarily used for limited lesions of nasal cavity or as reduced field for ethmoid sinus primary lesion. In patients with lesions of nasal cavity without orbital invasion, field edge is placed at medial limbus. With evidence of gross disease in ethmoid sinuses, field is reduced with great caution because of both high incidence of subclinical tumor extension through lamina papyracea and anatomic configuration of sinus relative to orbit. Although upper lateral walls of ethmoid sinuses are parallel, inferiorly and posteriorly they diverge to conform to cone-shaped orbit. If eyeball is totally shielded from anterior portal, some posteroinferior ethmoid air cells also are shielded. The same principle applies to roof of maxillary sinus, which slopes upward as it proceeds from anterior to posterior. **d** Typical lateral portal for treatment of paranasal sinus and nasal cavity tumors. Beam is angled 5° posteriorly to avoid exit irradiation to contralateral eye. Anterior border is at lateral bony canthus; thus some of posterior pole of ipsilateral eyeball is included within treatment volume. Superior border is adjusted according to extent of disease, generally 1.0 cm above roof of ethmoid sinuses, but may be raised to cover known or suspected intracranial extension. Inferior border is usually at level of lip commissure, covering floor of antrum, which is below floor of nasal cavity. Cork and tongue blade are used to depress tongue out of field. Posterior and posterosuperior borders are shaped to exclude spinal cord and brain stem, respectively. Usually, posterior border is at or near tragus and bisects vertebral bodies. Posterosuperior border is usually 2–3 mm posterior to clivus. If spinal cord and brain stem are encompassed by lateral portal(s) for initial 50 Gy, total dose to these structures exceeds 50 Gy at completion of "typical" course of irradiation (e.g., 65–70 Gy); shielding brain stem from reduced anterior field after 50 Gy is not possible. These structures should be encompassed within lateral portals only if tumor extension involves area posterior to plane of cord. Patient must then be apprised of increased risk of neurological sequelae. (a–c from LEVITT et al. 2nd edn., 1999, Fig. 16-28 a–c; d from Parsons et al. 1992, Fig. 29-6c)

The anterior portal for maxillary sinus cancers resembles that used for nasal cavity and ethmoid sinus lesions. The inferior border must be shaped to cover the lowest extent of disease (e.g., tumor tracking down the buccal mucosa from the gingivobuccal sulcus or tumor in the low parapharyngeal or tonsillar regions must be recognized). If the temporal fossa is grossly invaded, the lateral border (of the anterior portal) is usually allowed to fall off.

The lateral fields for nasal cavity, ethmoid sinus, and maxillary sinus lesions are similar (Fig. 18.21). The superior border is at least 1.0 cm above the base of the skull. If intracranial extension is demonstrated or suspected, the superior border is raised 2–3 cm.

RT to the nasal vestibule may be delivered by external-beam therapy [Figs. 18.24 (MENDENHALL et al. 2004) and 18.25 (MILLION et al. 1984)], interstitial therapy (cesium or iridium) [Fig. 18.26 (MILLION et al. 1985)], or a combination of both. Currently, the most common external-beam technique at the University of Florida uses an anterior portal with both high-energy electrons and photons. Virtually all patients at the University of Florida receive external-beam RT (e.g., 50 Gy MTD) followed by an implant (20–25 Gy).

18.12
Major Salivary Gland

18.12.1
Parotid Gland

RT plays its major role as an adjunct to surgery and is usually administered postoperatively, although preoperative treatment may be considered in special situations. The minimum treatment volume includes the parotid bed and upper neck nodes. The entire ipsilateral neck is electively irradiated for high-grade lesions or when tumor is found in lymph nodes in the neck dissection specimen.

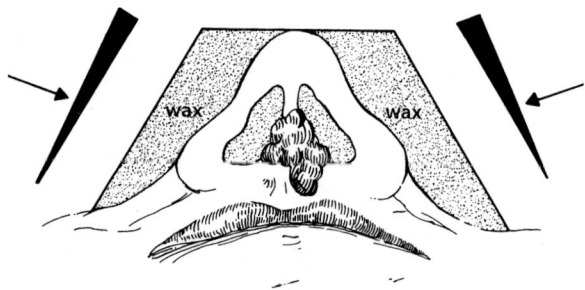

Fig. 18.24. Treatment plan for external-beam irradiation of nasal vestibule carcinoma (from Million et al. 1985, Fig. 18.23)

18.12.1.1
External-Beam RT

Treatment is administered by one of three external-beam techniques. One technique involves a wedge pair, with the portals aimed either superiorly and inferiorly (to direct the exit dose away from the orbits and oral cavity) or anteriorly and posteriorly (with the portals angled so that the beams pass below the level of the eyes). The latter technique facilitates matching the low-neck portal to the primary fields and is preferred. The wedge pair technique can treat a generous portion of the base of skull in a homogeneous manner and is particularly useful when perineural spread is present or suspected, as in adenoid cystic carcinoma. Fields are best designed with the aid of three-dimensional virtual simulation and treatment planning.

A second basic technique uses ipsilateral portals shaped to fit the anatomy (Fig. 18.27). A treatment scheme using a combination of photons and high-energy electrons produces a homogeneous dose distribution and delivers 30 Gy or less to the opposite salivary glands. The advantages of the technique are the ability to shape and reduce the fields easily and the ease with which an ipsilateral low-neck field may be adjoined to the primary portal. A disadvantage, especially in patients with adenoid cystic carcinoma, is underdosage of possible perineural tumor extensions deep in the temporal bone because of inadequate penetration of electrons in dense bone. Because electrons are so subject to perturbations from tissue inhomogeneity, the risk of deep geographic miss must always be kept in mind.

When tumor involves the deep lobe or otherwise extends near the midline, a third technique, parallel opposed photon portals weighted to the side of the lesion, may be necessary.

Calculation of the brain stem–spinal cord dose must be precise in all three techniques. Field reductions are highly individualized.

18.12.1.2
Interstitial RT

Interstitial implants may be added if the primary tumor is located in the preauricular portion of the superficial lobe, or when there are positive margins or unresectable or locally recurrent disease. A modified two-plane cesium needle implant (the deep plane extends into the retromandibular area) will cover the tumor bed, but may give inadequate coverage

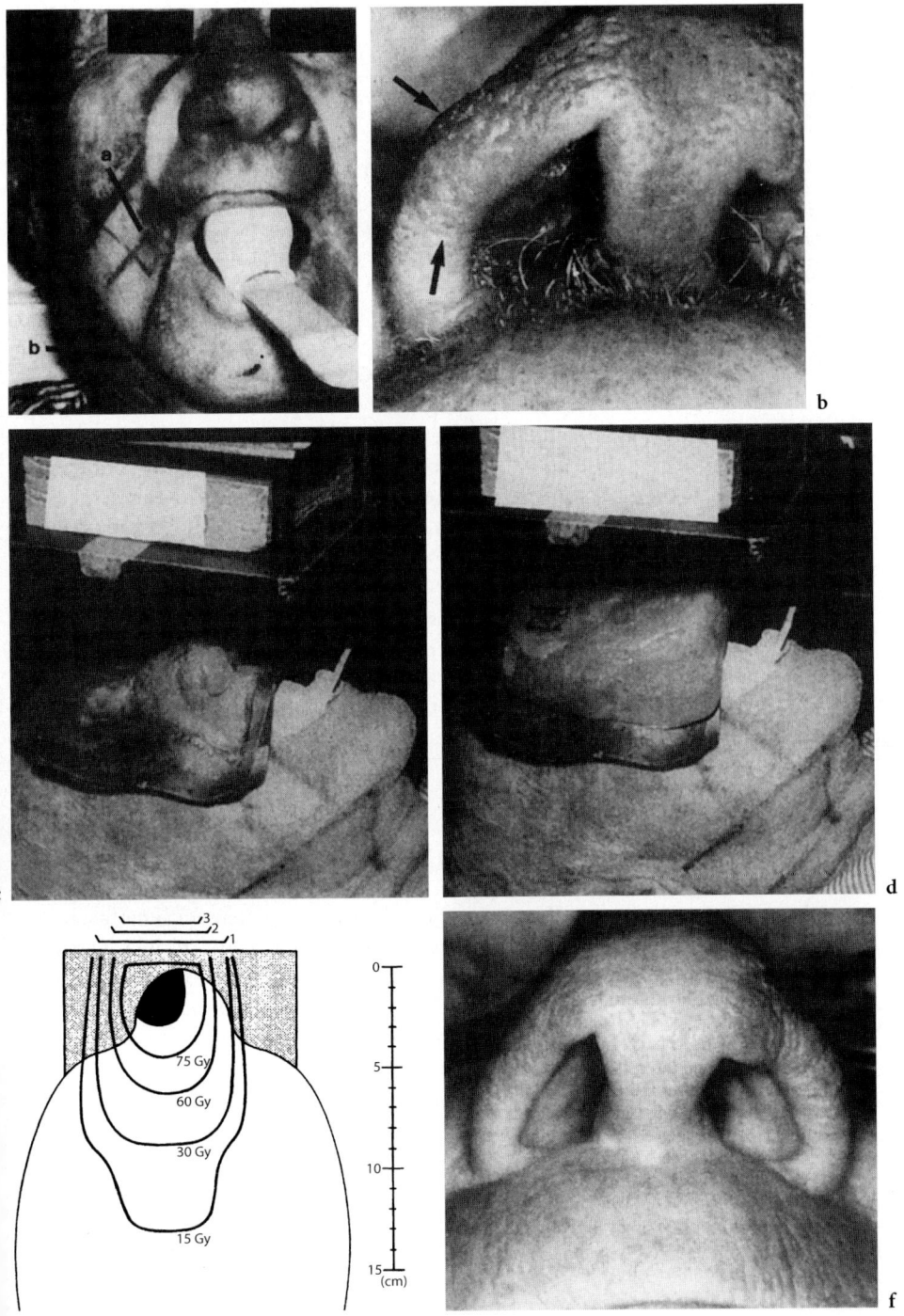

Fig. 18.25a-f. An 84-year-old man with 3-month history of nosebleeds had a 1.5-cm tumor on the right lateral nasal vestibule with erythema and induration extending to overlying skin of tip and ala of nose and just into lip. Invasion of lateral alar carti-lage was likely. He also had squamous cell carcinoma of vocal cord (T1N0). **a** Squamous cell carcinoma of right lateral wall of nasal vestibule (*arrows* indicate skin invasion). **b** Outline of treatment portals. Transit lymphatics and facial lymph nodes were treated with electrons (*a*), and submandibular lymph nodes are treated electively (*b*) because of significant dermal extension and undesirability of neck dissection in 84-year-old patient. **c** Treatment setup with lead shield, wax plugs in nose, and tongue depressor. **d** Wax bolus in place. Electron beam is used, collimated by Lipowitz's metal block on tray. Treatment plan is 75 Gy over 8 weeks using both photons and electrons. Usually, no bolus is applied over tip of nose unless skin at this site is infiltrated by tumor. **e** Isodose distribution. Lightly stippled area beeswax bolus/compensator, darker stippled area extent of gross tumor. **f** At 2 years, no evidence of disease. Patient remained free of disease at 8 years 4 months (from MILLION et al. 1984, Fig. 23-24)

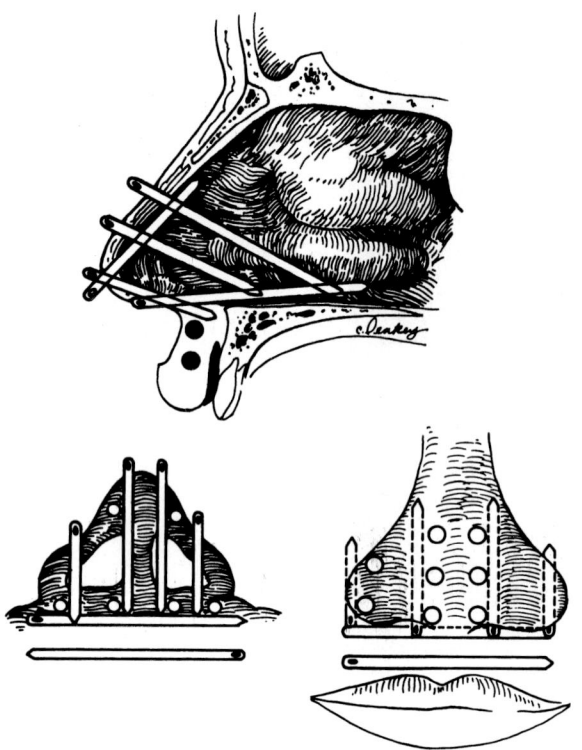

Fig. 18.26. Interstitial implant for carcinoma of nasal vestibule (from MILLION et al. 1989, Fig. 21.25)

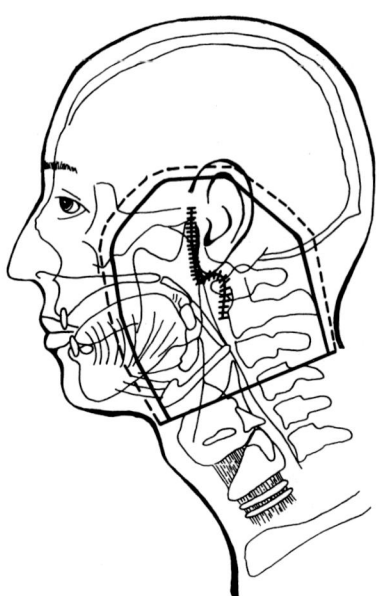

Fig. 18.27. Portal for postoperative irradiation of parotid gland cancer. Anterior border is usually at anterior border of masseter muscle; inferior border is at top of thyroid cartilage. Superiorly and posteriorly, entire parotid and surgical bed are included. Electron portal (dashed lines) is 1.0 cm larger than photon portal, because of constriction of electron isodose lines at depth (from LEVITT et al. 2nd edn, 1999, Fig. 16.32)

of the retromandibular portion of the parotid bed. To ensure an adequate treatment margin along the external ear canal, the needles usually penetrate the tragal cartilage. The implantation of the retromandibular deep lobe area is a "blind" procedure, but can be done safely and with relative ease [Fig. 18.28 (MENDENHALL et al. 1994)]; a preoperative CT scan can be used to help plan the implant and avoid major blood vessel injury.

18.12.2
Submandibular Gland

Ipsilateral external-beam portals are tailored to the extent of disease found in the surgical specimen. The possible sites of local recurrence include the submandibular triangle, adjacent oral cavity, pterygomaxillary fossa, base of the skull, parotid gland, and neck. The entire ipsilateral neck is always included; the opposite side of the neck is usually not treated. The energy used depends on the depth at risk. An electron beam, photon beam, or a combination of both is selected, depending on the situation.

18.13
Unknown Primary

Treatment planning for the patient with metastases to the nodes in the neck from a primary site that cannot be located after multiple physical examinations, CT, and direct laryngoscopy with biopsies depends on the location of the lymph nodes; occasionally histology also plays a role in determining the treatment volume. Involvement of the level-II lymph nodes indicates elective RT of the nasopharynx and oropharynx, via parallel opposed portals (Fig. 18.29), and low neck RT to the level of the clavicles via an anterior field. The primary site is almost always located in the tonsillar fossa or tongue base. The retropharyngeal nodes are included and the portals are enlarged to a modest degree to include the nasopharynx in the unlikely event that the primary site is located there. The portals are extended to include the supraglottis and hypopharynx if the presentation is primarily in the level-III nodes. Sparing a midline strip of skin on the neck is important to avoid lymphedema. When a solitary lymph node without extracapsular extension is involved, treatment with neck dissection and observation is preferred to avoid RT-related morbidity and because the chance of cure is not compro-

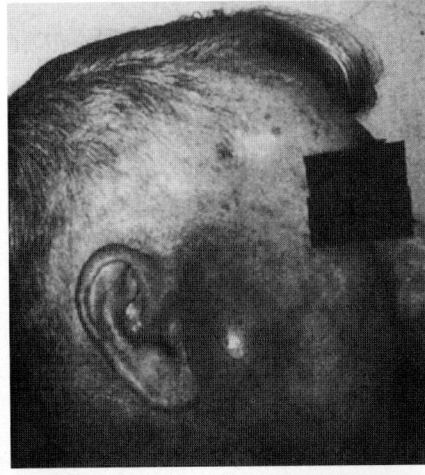
a

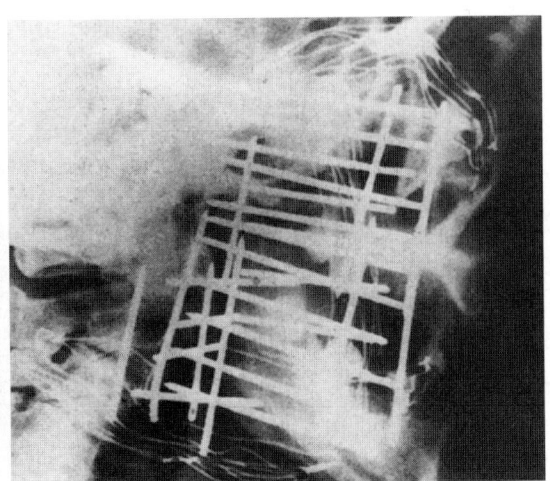
b

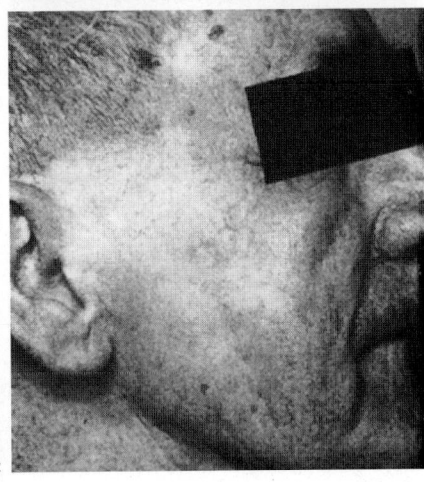

Fig. 18.28. a Squamous cell carcinoma metastatic to parotid lymph node (squamous cell carcinoma of skin of temple had been treated 2 years previously). Note tragal involvement. **b** Modified three-plane radium needle implant. One plane (four needles) is in retromandibular deep lobe. **c** Complete healing at 1 year (no evidence of disease at 9 years) (courtesy of R.L. Lindberg, MD, Department of Radiation Therapy, MD Anderson Hospital and Tumor Institute, Houston, TX) (from MENDENHALL et al. 1994, Fig. 25.22)

mised if the patient is closely followed. A preauricular lymph node(s) containing squamous cell carcinoma typically represents metastasis from a skin cancer and is treated by a combination of parotidectomy and RT or RT alone if surgery is not feasible. Supraclavicular nodes are irradiated through a generous regional portal, which should include the adjacent apex of the axilla. Low-neck presentation almost always arises from a primary site below the clavicles and the treatment is palliative.

18.14
Design of Low Neck Portal

18.14.1
Oropharynx, Nasopharynx, and Oral Cavity Cancers

The low-neck portal is designed according to external anatomic landmarks and findings from the clinical and radiographic examination of the neck. No simulation films or portal verification films are taken. The lines are drawn on the patient's skin, and lead blocks are stacked freehand on a tray positioned above the patient to shape the beam to the desired volume. Alternatively, a monoisocentric technique may be employed.

For all base of tongue, midline soft palate, advanced tonsil, all nasopharyngeal, and oral cavity cancers that require parallel opposed portals, both sides of the neck are irradiated, even in the N0 situation. Failure to irradiate the low neck in patients with the lesions just listed will result in at least a 10% rate of failure in the level-III and level-IV nodes even when the upper neck is clinically negative. The low neck is irradiated through an anterior field only. The basic portal design in all of these situations is similar [Fig. 18.30 (MILLION et al. 1994d; PARSONS et al. 1999b)].

For patients with early-stage cancer of the tonsillar region or lateralized tumors of the oral tongue or retromolar trigone, only the ipsilateral upper and lower neck require RT in the N0 situation.

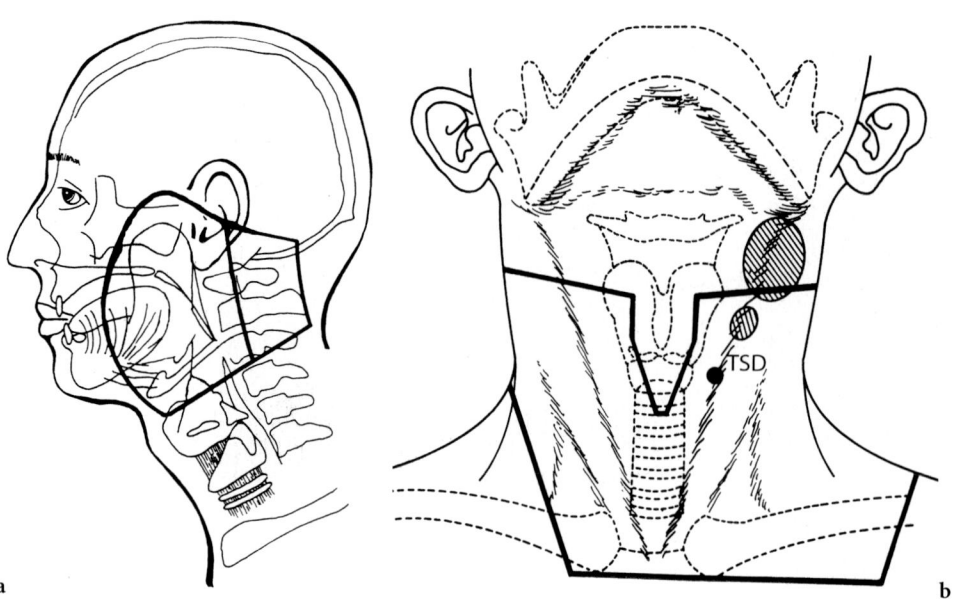

Fig. 18.29. a Radiation treatment technique for carcinoma from an unknown primary site. Superiorly, the portal treats the nasopharynx and the jugular and spinal accessory lymph nodes to the base of skull. The posterior border is behind the spinous process of C2. The inferior border is at the thyroid notch. Anteroinferiorly, the skin and subcutaneous tissues of the submentum are shielded, except in the case of advanced neck disease. The anterior tongue margin is set so as to obtain a 2-cm margin on the base of the tongue and tonsillar fossa, as well as the nasopharynx. One portal reduction is shown. **b** Fields for bilateral lower neck radiotherapy. The larynx shield should be carefully designed. Because the internal jugular vein lymph nodes lie adjacent to the posterolateral margin of the thyroid cartilage, the shield cannot cover the entire thyroid cartilage without producing a low-dose area in these nodes. A common error in the treatment of the lower neck is to extend the low neck portal laterally out to the shoulders, encompassing lateral supraclavicular lymph nodes that are at negligible risk while partially shielding the high-risk level-III and -IV lymph nodes with a large, rectangular laryngeal block. The inferior extent of the shield is at the bottom of the cricoid cartilage or first or second tracheal ring. The inferior extent of the shield is at the midline as the lower neck is approached. Lateral borders of the low neck portals are set to cover only the lymph nodes in the root of the neck when the risk of low-neck disease on that side is small (i.e., stage N0 or N1 disease). If there are clinically positive lymph nodes in the lower neck, or if major disease is present in the upper neck, the lateral border of the low-neck field is widened on that side to cover the entire supraclavicular region out to the junction of the trapezius muscle with the clavicle (a from MENDENHALL et al. 2001, Fig. 18.1; b from MILLION et al. 1994d)

18.14.1.1
Larynx and Hypopharynx Cancers

The principles of portal design are the same as those described in the previous section (Fig. 18.31).

18.14.1.1.1
Boost Technique for Large or Fixed Lymph Nodes

Some patients have small or unknown primary lesions that require only 60–65 Gy, but have a large, fixed nodal mass (e.g., 7–8 cm or more in size) that requires a higher dose, e.g., 70–80 Gy even when neck dissection is planned. In many treatment centers, the common practice is to treat the neck node with elec-

trons after the primary lesion has been irradiated. We prefer to use anterior and posterior parallel-opposed wedged portals to spare the mucosal surfaces and to avoid the excessive skin reaction and fibrosis produced by high-energy electrons [Fig. 18.32 (PARSONS and MILLION 1987; MILLION et al. 1994a)]. Only the large mass, and not the entire neck, receives the high-dose boost, even if there are other involved lymph nodes in the neck. Neck dissection is usually performed 6 weeks after RT, unless the nodal mass completely regresses based on both physical examination and CT performed 3–4 weeks after the completion of RT. We believe that wound healing after neck dissection is less likely to be a problem when the patient has been irradiated with photons rather than electrons.

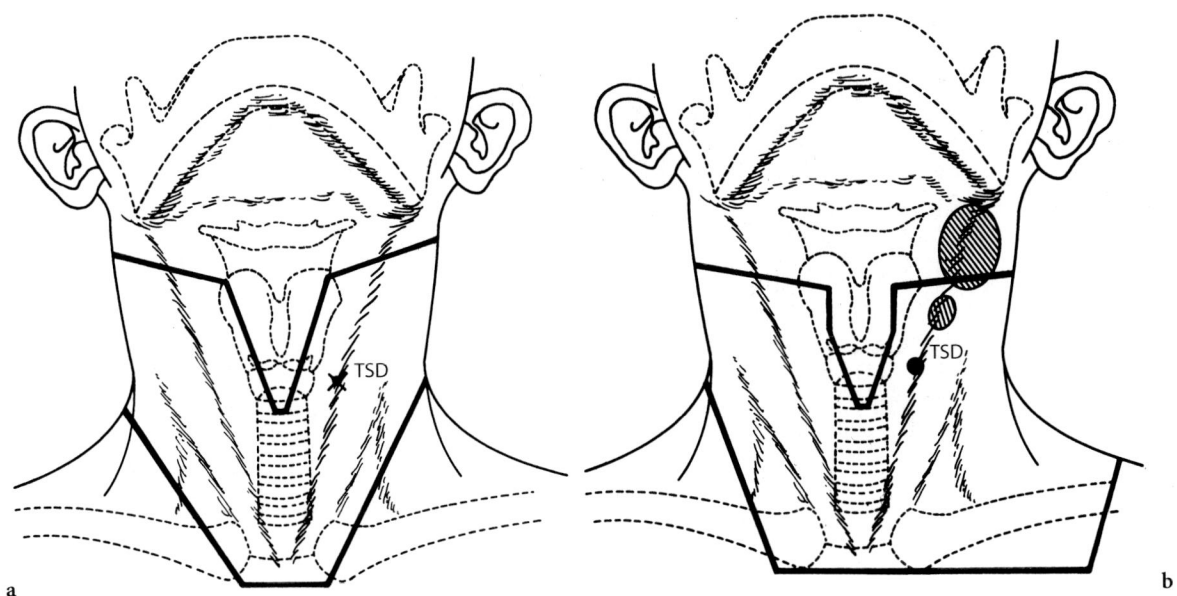

Fig. 18.30a,b. Fields for bilateral lower neck irradiation in patients with base of tongue, midline soft palate, advanced tonsil, nasopharynx, or advanced oral cavity cancers. **a** N0 neck. Larynx shield design is important. Because midjugular lymph nodes lie adjacent to posterolateral margin of thyroid cartilage, attempts to shield entire thyroid cartilage with 4- to 5-cm-wide block produce low dose area in these nodes. Inferior extent of larynx shield is usually at cricoid cartilage or first or second tracheal ring; shield is tapered; nodes in low neck may lie close to midline. If larynx block is extended for entire length of low-neck portal, it should probably cast a shadow no wider than 1.0–1.5 cm in suprasternal notch region. TSD tumor-to-source distance. In N0 setting, lateral supraclavicular lymph nodes are at little risk of involvement, except possibly for patients with cancer of nasopharynx. Usually, only root of neck is included. The most common error observed in design of low neck portal is actually a combination of mistakes that results in underdosage of high-risk areas and unnecessary treatment of low risk zones by (1) shielding larynx with large, square or rectangular block, (2) blocking midline with wide (3–4 cm) block down to level of suprasternal notch, and (3) irradiating entire supraclavicular fossa bilaterally. **b** Neck with clinical evidence of disease. Treatment to each side of neck is individualized. If extensive neck disease is limited to one side, entire neck, including all supraclavicular lymph nodes (out to junction of trapezius muscle and clavicle), is irradiated on that side. If both sides are involved, treatment on each side is modified according to disease extent (a from PARSONS et al. 1998, Fig. 42-7a; b from MILLION et al. 1994d, Fig. 6.55a)

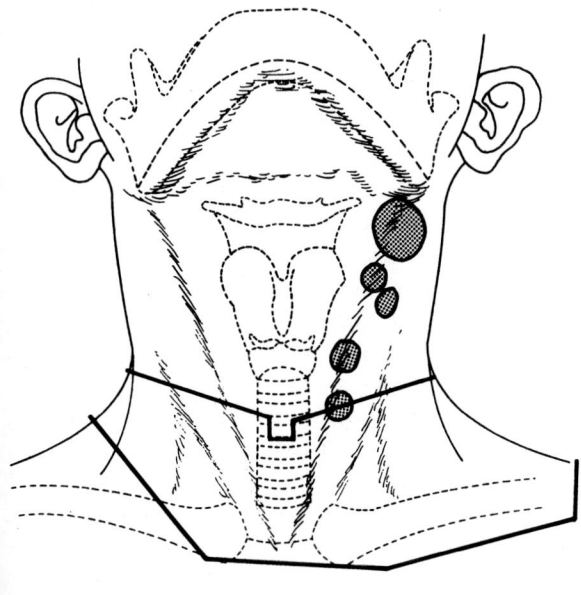

Fig. 18.31. Anterior portal for treatment of low neck in patients with hypopharyngeal or laryngeal cancer. Low neck is irradiated bilaterally. Level of superior border of portal varies according to primary lesion treated and may be as high as cricothyroid membrane (e.g., lesion of suprahyoid epiglottis) or as low as 2–3 cm below inferior border of cricoid cartilage (e.g., for advanced pyriform sinus cancer). Matchline is treated in primary portals but excluded from low-neck field. Usually, a 1×1–cm midline block is introduced at upper edge of field, except in postoperative patients in whom tracheal stoma is at risk (AMDUR et al. 1989). Each side of low-neck portal is individualized according to risk and/or presence of lymph node metastases on that side (from LEVITT et al. 2nd edn, 1999, Fig. 16-36)

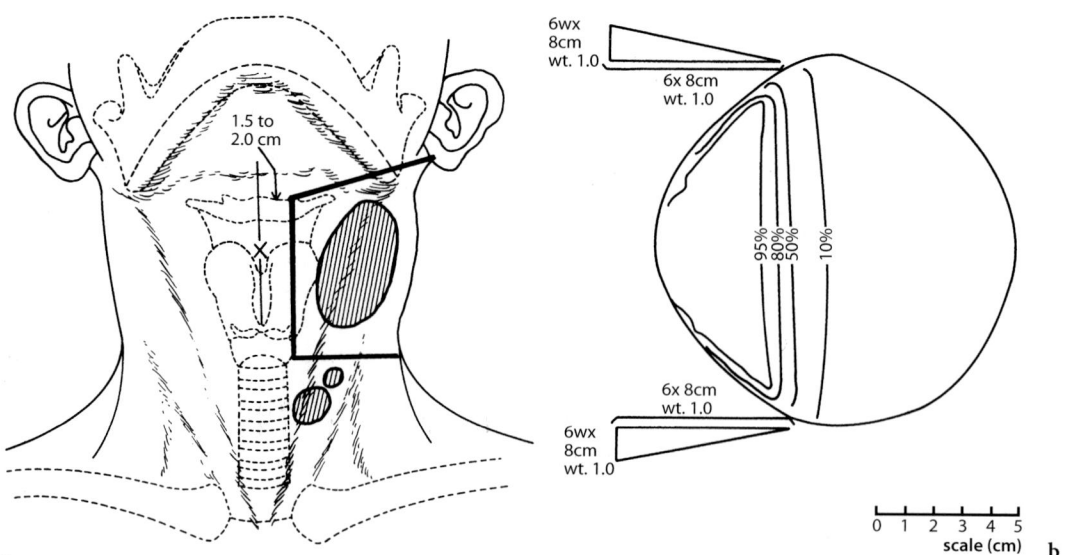

Fig. 18.32a,b. Photon portals for boosting dose to neck node after completion of treatment to primary lesion by lateral portals. **a** Parallel opposed anterior and posterior neck portals with wedges. Medial border is usually 1.5–2.0 cm from midline. Technique spares much normal mucosa and cervical spinal cord from high-dose irradiation and is particularly useful in a patient with a small primary cancer (e.g., T1) that requires a dose of 60 Gy and a large neck node (e.g., N3) that requires a high dose (e.g., 70–75 Gy). **b** Dose distribution produced by anterior and posterior portals with equally weighted ^{60}Co wedge pair. Portals may be differentially weighted or high-energy X-rays may be used to produce a variety of dose distributions (a from Parsons and Million 1987, Fig. 15-37; b from Million et al. 1994a, Fig. 15-4d, Levitt et al. 2nd edn, 1999, Fig. 16-37)

18.15
IMRT

IMRT may be useful to preserve salivary function when both sides of the neck are thought to be at more than a 10% risk for harboring subclinical disease. IMRT is not used to preserve salivary function when clinically positive lymph nodes are present in both sides of the neck because it is difficult to adequately treat the nodes and spare at least one parotid gland. IMRT is used infrequently in the postoperative setting because of uncertainty in target definition and the potential for unusual patterns of lymphatic drainage due to surgical disruption of the lymphatics. IMRT may also be particularly useful to treat the primary site and minimize the dose to surrounding normal tissues compared with 3DCRT in situations where preservation of salivary function is not a goal. One example is nasopharyngeal cancer where IMRT may be used to reduce the dose to the temporal lobes and brainstem. Another situation where IMRT is useful is for cancers located near the thoracic inlet (i.e., thyroid carcinoma, cervical esophageal cancer) where the contour of the shoulders make treatment with 3DCRT difficult.

There are two options for treating the low neck: (1) use a separate conventional anterior portal and (2) include the low neck in the IMRT fields used to treat the primary tumor and upper neck. The former is preferred and the low-neck portal is junctioned with the IMRT portals where the junction would usually be placed if 3DCRT was employed. The match line is moved once at 24 Gy if the low neck done is limited to 50 Gy. The match line is moved twice at 20 Gy and 40 Gy if part or all of the low neck receives more than 50 Gy. The match line is moved in 3-mm increments by moving the inferior jaw superiorly so that the jaw moves superiorly across the central axis into the IMRT field. A separate low-neck portal is not employed for tumors arising near the sternal notch where IMRT is used to treat the primary site and all of the at risk regional lymphatics.

Treatment planning is done with a CT scan done with 3-mm contiguous slices from the vertex of the skull to the clavicles. Non-ionic intravenous (IV) contrast is used to visualize vascular structures unless there is a medical contraindication. In cases where MRI is useful for defining the target volume, such as nasopharyngeal cancers, a MRI is also obtained using a Gd-DPTA-enhanced T1-weighted imaging protocol.

Patients are immobilized with a minimum of an aquaplast mask; a customized dental impression/bite plate device combined with a commercial infrared camera realignment system is used for all dentulous patients.

The following normal tissues are outlined for planning purposes in all patients: skin surface, mandible, spinal cord, and brain stem. Additional structures are outlined if relevant to the treatment plan: oral cavity, glottis, parotid glands, submandibular glands, lacrimal glands, globe/retinas, lenses, optic nerves, and optic chiasm.

A modified M.D. Anderson concomitant boost plan is employed. The subclinical and gross disease receives 54 Gy in 30 once-daily fractions. During the last 12 days of treatment, the gross disease is boosted an additional 18 Gy in 12 twice-daily fractions with a minimum 6-h interfraction interval. Thus, the gross tumor receives 72 Gy in 42 fractions over 30 treatment days with RT administered 5 days per week in a continuous course. The initial plan covering the subclinical and gross disease aims for approximately 49.5 Gy at 1.65 Gy per fraction in 30 fractions to areas of low risk subclinical disease, and 54 Gy or more in 30 fractions for high-risk subclinical and gross disease. The high-risk subclinical and gross disease receives the additional 18 Gy in 12 daily fractions for a total dose of 72 Gy or more.

The following are guidelines for contouring target structures and for prioritization when structures overlap.

18.15.1.1
Guidelines for Contouring Target Structures

- Clinical target volume (CTV) boost = CTV 7200: the physician outlines the volume containing the gross (primary and nodal) and high-risk subclinical disease based on imaging and physical examination. This is the volume that will receive 72 Gy or more.

- CTV subclinical disease = CTV 4950: this volume encompasses all tissues that are considered to have a 10% or higher risk of tumor spread.
- CTV 5400: there are situations in which we want an area to get more than 49.50 Gy but it is not necessary to give 72 Gy. The most common example of this situation is a suspicious contralateral level-II node. When an intermediate dose is desired, a third CTV is delivered and is called CTV 5400. CTV 7200, CTV 5400, and CTV 4950 are treated simultaneously with the same IMRT plan for 30 consecutive days. CTVs 7200 and 5400 each receive 1.8 Gy per fraction while CTV 4950 receives 1.65 Gy per fraction. A second IMRT plan is then used to "boost" CTV 7200 with a second afternoon fraction of 1.50 Gy for 12 consecutive days.
- The CTV-planning target volume (PTV) expansion value is currently set at 3 mm in all directions for all structures, both target and normal tissue.
- Normal tissue volumes: the International Commission on Radiation Units and Measurement, Inc. (ICRU) report 62 defines the terms "organ at risk volume" (ORV) and "planning organ at risk volume" (PRV). The ORV is defined by the edge of the organ as visualized on the planning scan. The PRV includes margin for setup uncertainty and organ motion.

The ORV is outlined as part of the planning process. Currently, a uniform margin of 3 mm is added to all ORVs to obtain the PRV. This is probably not accurate for all situations, especially those that involve organ motion. In the future, customized PRV expansions may be developed. The dose limits for normal tissues are outlined in Tables 18.1 and 18.2.

Table 18.1. University of Florida head and neck intensity-modulated radiation therapy dose limits for normal tissues

	No concomitant chemotherapy	Concomitant chemotherapy
Unspecified tissue	≤110% of prescribed dose	Same
Larynx**	10 Gy max dose*	Same
Mandible	70 Gy max dose*	Same
Brainstem	54 Gy max dose*	50 Gy max dose*
Spinal cord	45 Gy max dose*	40 Gy max dose*
Optic nerve	55 Gy max dose*	50 Gy max dose*
Optic chiasm	50 Gy max dose*	50 Gy max dose*
Lens (anterior chamber)	12 Gy max dose*	Same
Retina (posterior globe)	40 Gy max dose*	40 Gy max dose*
Lacrimal gland***	26 Gy max dose*	Same

*In all discussions related to IMRT normal tissue dose parameters, the terms "maximum dose", "max dose", or "Dmax" refer to the maximum dose received by 0.1 cc of the tissue in question. Maximum doses based on point dose calculations are not used to make decisions regarding IMRT delivery

**Larynx: when IMRT is used to treat what would be in the low-neck field such that the larynx would be shielded with a midline block if we were not using IMRT

***Lacrimal gland dose limit is different from parotid and submandibular gland because the lacrimal gland is too small and too ill defined to specify a mean or volume

Table 18.2. Parotid and submandibular gland: same with or without concomitant chemotherapy

Both neck N–0	Ipsilateral and contralateral glands: 50% volume ≤30 Gy or mean ≤26 Gy
Ipsilateral neck N+; contralateral neck N–0	No constraint on ipsilateral glands. Contralateral gland: 50% volume ≤30 Gy or mean ≤26 Gy
Both necks N+	Individualized

An important aspect of the University of Florida IMRT program is that all dose volume histogram (DVH) displays and plan evaluations are based on both PTVs and PRVs. This makes it more difficult to achieve dosimetric goals but represents a more rigorous approach to the planning process.

18.15.1.2
Prioritization Rules When Structures Overlap

It is not unusual for there to be areas of overlap of structures that have been defined for treatment planning purposes. There may be overlap between a PTV and PRV when a target area is adjacent to a normal tissue structure that is also being contoured (e.g., level-II nodes and the parotid). In many cases, overlap is the result of the PTV and PRV expansions, meaning the CTV and ORVs are adjacent to each other but do not overlap. When structures overlap, a prioritization hierarchy must be defined for the treatment planning process. Prioritization guidelines in regions of overlap are as follows:

- Target-spinal cord or brainstem overlap: PTV wins over PRV but ORV wins over PTV. In other words, the plan will spare the cord or brainstem in an area where the 3-mm PTV expansion extends into the spinal cord.
- Target-parotid, submandibular gland, lacrimal gland, mandible, oral cavity, or skin overlap: PTV wins over PRV and ORV.
- Target-optic nerve, optic chiasm, globe, retina, or lens overlap: Individualized based on the situation. The physician must make a decision about the assignment of overlap areas prior to treatment planning.

18.15.1.3
Guidelines for Dose Prescription and Normal Tissue Dose Analysis

The dose is prescribed to PTV targets. DVH displays and corresponding analyses for targets and normal tissues record the dose to the PTV and PRV, respectively. We do not display or analyze dose–volume data related to the ORV, CTV, or GTV. All IMRT plans are heterogeneity corrected.

18.15.1.4
Dose Constraints for Target PTVs

We do not define a "hot spot" limit – meaning we do not limit the volume that may receive greater than a certain percentage of the prescription dose. However, the plan optimization goal is to limit the "hot spot" to less than 120%.

We define two parameters to insure that the target is adequately treated: (1) 95% or more of the PTV receives the prescription dose and (2) 99% or more of the PTV receives 93% or more of the prescription dose.

18.15.1.5
Guidelines for Treatment of Regional Lymphatics

The extent to which the regional lymphatics are irradiated depends on the site and extent of the primary tumor and the location and extent of clinically positive lymph nodes. Guidelines for treatment of the regional lymphatics are outlined in Figures 18.33–18.35. The imaging-based nodal classification is depicted in Table 18.3 (SOM et al. 1999).

Fig. 18.33. N0 Bilaterally

Node Levels and Laterality														
Primary Site	IA	IB		II		III		IV		V		VI	Retropharyngeal	
	-	Ip	C	Ip	C	Ip	C	Ip	C	Ip	C	-	Ip	C
				R	R	R	R	R	R	R	R		R	R
Nasopharynx				R	R	R	R	R	R				R	R
Soft palate				R	R	R	R	R	R					
Anterior tonsillar pillar				R	Only if to midline	R	Only if to midline	R	Only if to midline					
Tonsil				R	Only if to midline	R	Only if to midline	R	Only if to midline					
Base of tongue				R	R	R	R	R	R					
Pharyngeal wall				R	R	R	R	R	R			Only if esophagus or apex PS + but spare larynx	R	R
Pyriform sinus				R	R	R	R	R	R			Only if apex+ but spare larynx	R	Only if to midline
Postcricoid				R	R	R	R	R	R			Only if esophagus or apex PS+	R	R
Larynx				R	R	R	R	R	R			Only if subglottic +		

IP = ipsilateral neck; C = contralateral neck; apex PS = apex of pyriform sinus; R = Radiotherapy to nodes.
Image borrowed with permission from the University of Florida Department of Radiation Oncology treatment planning guidelines.

Fig. 18.34. N–2b or Unilateral N3

Node Levels and Laterality														
Primary Site	IA	IB		II*		III		IV		V		VI	Retropharyngeal	
	-	Ip	C	Ip	C	Ip	C	Ip	C	Ip	C	-	Ip	C
Nasopharynx		½*		R	R	R	R	R	R	R	R		R	R
Soft palate		½*		R	R	R	R	R	R	R			R	R
Anterior tonsillar pillar		½*		R	R	R	R	R	R	R			R	
Tonsil		½*		R	R	R	R	R	R	R			R	
Base of tongue		½*		R	R	R	R	R	R	R			R	
Pharyngeal wall		½*		R	R	R	R	R	R	R		Only if esophagus or apex PS + but spare larynx	R	R
Pyriform sinus		½*		R	R	R	R	R	R	R		Only if apex+ but spare larynx	R	Only if to midline
Postcricoid		½*		R	R	R	R	R	R	R		R	R	R
Larynx		½*		R	R	R	R	R	R	R		R	R	Only if to midline

II* = cover jugular foramen if positive nodes in level II, V, or retropharyngeal.
½* = in the absence of high volume level II disease we usually cover only the posterior ½ of level Ib: 1cm anterior to the submandibular gland.
IP = ipsilateral neck; C = contralateral neck; apex PS = apex of pyriform sinus; R = Radiotherapy to nodes.
Image borrowed with permission from the University of Florida Department of Radiation Oncology treatment planning guidelines.

Fig. 18.35. N–2c (Bilateral Neck Nodes)

Node Levels and Laterality
Note: Intensity-modulated radiation therapy is rarely used with bilateral adenopathy. We have used IMRT with N2c disease in a nasopharynx primary and low volume adenopathy confined to level II-III and the retropharyngeal nodes.

Primary Site	IA	IB		II*		III		IV		V		VI	Retropharyngeal	
	-	Ip	C	Ip	C	Ip	C	Ip	C	Ip	C	-	Ip	C
Nasopharynx		½*		R	R	R	R	R	R	R	R		R	R
Soft palate		½*		R	R	R	R	R	R	R	R		R	R
Anterior tonsillar pillar		½*		R	R	R	R	R	R	R	R		R	R
Tonsil		½*		R	R	R	R	R	R	R	R		R	R
Base of tongue		½*		R	R	R	R	R	R	R	R		R	R
Pharyngeal wall		½*		R	R	R	R	R	R	R	R	Yes but spare larynx	R	R
Pyriform sinus		½*		R	R	R	R	R	R	R	R	Yes but spare larynx	R	R
Postcricoid		½*		R	R	R	R	R	R	R	R	R	R	R
Larynx		½*		R	R	R	R	R	R	R	R	R	R	R

II* = cover jugular foramen if positive nodes in level II, V, or retropharyngeal. ½* = in the absence of high volume level II disease we usually cover only the posterior ½ of level Ib: 1cm anterior to the submandibular gland. IP = ipsilateral neck; C = contralateral neck; apex PS = apex of pyriform sinus; R = Radiotherapy to nodes. Image borrowed with permission from the University of Florida Department of Radiation Oncology treatment planning guidelines.

Table 18.3. Imaging-based nodal classification (From: Som et al. 1999)

Level I	The submental and submandibular nodes. They lie above the hyoid bone, below the mylohyoid muscle, and anterior to the back of the submandibular gland
Level IA	The submental nodes. They lay between the medial margins of the anterior bellies of the digastric muscles
Level IB	The submandibular nodes. On each side, they lie lateral to the level-IA nodes and anterior to the back of each submandibular gland
Level II	The upper internal jugular nodes. They extend from the skull base to the level of the bottom of the body of the hyoid bone. They are posterior to the back of the submandibular gland and anterior to the back of the sternocleidomastoid muscle
Level IIA	A level-II node that lies either anterior, medial, lateral, or posterior to the internal jugular vein. If posterior to the vein, the node is inseparable from the vein
Level IIB	A level-II node that lies posterior to the internal vein and has a fat plane separating it and the vein
Level III	The middle jugular nodes. They extend from the level of the bottom of the body of the hyoid bone to the level of the bottom of the cricoid arch. They lie anterior to the back of the sternocleidomastoid muscle
Level IV	The low jugular nodes. They extend from the level of the bottom of the cricoid arch to the level of the clavicle. They lie anterior to a line connecting the back of the sternocleidomastoid muscle and the posterolateral margin of the anterior scalene muscle. They are also lateral to the carotid arteries
Level V	The nodes in the posterior triangles. They lie posterior to the back of the sternocleidomastoid muscle from the skull base to the level of the bottom of the cricoid arch and posterior to a line connecting the back of the sternocleidomastoid muscle and the posterolateral margin of the anterior scalene muscle from the level of the bottom of the cricoid arch to the level of the clavicle. They also lie anterior to the anterior edge of the trapezius muscle
Level VA	Upper level-V nodes extend the skull base to the level of the bottom of the cricoid arch
Level VB	Lower level-V nodes extend from the level of the bottom of the cricoid arch to the level of the clavicle, as seen on each axial scan
Level VI	The upper visceral nodes. They lie between the carotid arteries from the level of the bottom of the body of the hyoid bone to the level of the top of the manubrium
Level VII	The superior mediastinal nodes. They lie between the carotid arteries below the level of the top of the manubrium and above the level of the innominate vein
Supraclavicular nodes	They lie at, or caudal to, the level of the clavicle and lateral to the carotid artery on each side of the neck, as seen on each axial scan
Retropharyngeal nodes	Within 2 cm of the skull base, they lie medial to the internal carotid arteries

* The parotid nodes and other superficial nodes are referred to by their anatomical names

References

Amdur RJ, Parsons JT, Mendenhall WM et al. (1989) Postoperative irradiation for squamous cell carcinoma of the head and neck: an analysis of treatment results and complications. Int J Radiat Oncol Biol Phys 16:25–36

Amdur RJ, Li JG, Liu C et al. (2004) Unnecessary laryngeal irradiation in the IMRT era. Head Neck 26:257–264

American Joint Committee on Cancer (2002) Head and neck sites. AJCC Cancer Staging Manual, 6th edn. Springer, Berlin Heidelberg New York, pp 17–22

Ellingwood KE, Million RR (1979) Cancer of the nasal cavity and ethmoid/sphenoid sinuses. Cancer 43:1517–1526

Ellingwood KE, Million RR, Mitchell TP (1976) A preloaded radium needle implant device for maintenance of needle spacing. Cancer 37:2858–2860

Fletcher GH (1980) Textbook of radiotherapy, 3rd edn. Lea and Febiger, Philadelphia

Levitt S, Potish RA, Khan FM, Perez CA (eds) (1999) Levitt and Tapley's technological basis of radiation therapy: clinical applications, 3rd edn. Lippincott, Williams & Wilkins, Philadelphia

Marcus RB, Jr., Million RR, Mitchell TP (1980) A preloaded, custom-designed implantation device for stage T1-T2 carcinoma of the floor of mouth. Int J Radiat Oncol Biol Phys 6:111–113

Mazeron JJ (1986) Interstitial radiation therapy for squamous cell carcinoma of the tonsillar region: the Creteil experience (1971–1981). Int J Radiat Oncol Biol Phys 12:895–900

Mendenhall WM, Parsons JT, Mancuso AA et al. (1988a) Squamous cell carcinoma of the pharyngeal wall treated with irradiation. Radiother Oncol 11:205–212

Mendenhall WM, Parsons JT, Million RR et al. (1988b) T1-T2 squamous cell carcinoma of the glottic larynx treated with radiation therapy: relationship of dose-fractionation factors to local control and complications. Int J Radiat Oncol Biol Phys 15:1267–1273

Mendenhall WM, Million RR, Mancuso AA, Cassisi NJ, Flowers FP (1994) Carcinoma of the skin. In: Million RR, Cassisi NJ (eds) Management of head and neck cancer: a multidisciplinary approach, 2nd edn. J.B. Lippincott Company, Philadelphia, pp 643–691

Mendenhall WM, Stringer SP, Amdur RJ et al. (2000) Is radiation therapy a preferred alternative to surgery for squamous cell carcinoma of the base of tongue? J Clin Oncol 18:35–42

Mendenhall WM, Amdur RJ, Morris CG et al. (2001) T1-T2N0 squamous cell carcinoma of the glottic larynx treated with radiation therapy. J Clin Oncol 19:4029–4036

Mendenhall WM, Morris CG, Amdur RJ et al. (2003a) Parameters that predict local control following definitive radiotherapy for squamous cell carcinoma of the head and neck. Head Neck 25:535–542

Mendenhall WM, Riggs CE, Amdur RJ et al. (2003b) Altered fractionation and/or adjuvant chemotherapy in definitive irradiation of squamous cell carcinoma of the head and neck. Laryngoscope 113:546–551

Mendenhall WM, Riggs CE, Cassisi NJ (2004) Treatment of head and neck cancers. In: DeVita VT, Jr., Hellman S, Rosenberg SA (eds) Cancer: principles and practice of oncology, 7th edn. Lippincott Williams & Wilkins, Philadelphia, pp 662–732

Million RR, Cassisi NJ (1984) General principles for treatment of cancers in the head and neck: radiation therapy. In: Million RR, Cassisi NJ (eds) Management of head and neck cancer: a multidisciplinary approach, 1st edn. Lippincott, Philadelphia, pp 77–90

Million RR, Cassisi NJ (1994) Management of head and neck cancer: a multidisciplinary approach, 2nd edn. J.B. Lippincott Company, Philadelphia

Million RR, Cassisi NJ, Hamlin DJ (1984) Nasal vestibule, nasal cavity, and paranasal sinuses. In: Million RR, Cassisi NJ (eds) Management of head and neck cancer: a multidisciplinary approach, 1st edn. J.B. Lippincott, Philadelphia, pp 407–444

Million RR, Cassisi NJ, Wittes RE (1985) Cancer of the head and neck. In: DeVita VT, Jr., Hellman S, Rosenberg SA (eds) Cancer: principles & practice of oncology, 2nd edn. J B Lippincott, Philadelphia, pp 407–506

Million RR, Cassisi NJ, Mancuso AA (1994a) The unknown primary. In: Million RR, Cassisi NJ (eds) Management of head and cancer: a multidisciplinary approach, 2nd edn. J.B. Lippincott Company, Philadelphia, pp 311–320

Million RR, Cassisi NJ, Mancuso AA (1994b) Oral cavity. In: Million RR, Cassisi NJ (eds) Management of head and neck cancer: a multidisciplinary approach, 2nd edn. J.B. Lippincott Company, Philadelphia, pp 321–400

Million RR, Cassisi NJ, Mancuso AA. Larynx (1994c) Larynx. In: Million RR, Cassisi NJ, editors. Management of head and neck cancer: a multidisciplinary approach, 2nd edn. J. B. Lippincott Company, Philadelphia, pp 431–497

Million RR, Cassisi NJ, Mancuso AA, Stringer SP, Mendenhall WM, Parsons JT (1994d) Management of the neck for squamous cell carcinoma. In: Million RR, Cassisi NJ (eds) Management of head and neck cancer: a multidisciplinary approach, 2nd edn. J.B. Lippincott Company, Philadelphia, pp 75–142

Parsons JT (1984a) Time-dose-volume relationships in radiation therapy. In: Million RR, Cassisi NJ (eds) Management of head and neck cancer: a multidisciplinary approach, 1st edn. J. B. Lippincott Company, Philadelphia, pp 137–172

Parsons JT (1984b) The effect of radiation on normal tissues of the head and neck. In: Million RR, Cassisi NJ (eds) Management of head and neck cancer: a multidisciplinary approach, 1st edn. J.B. Lippincott Company, Philadelphia, pp 173–207

Parsons JT, Million RR (1987) Treatment of tumors of the oropharynx: radiation therapy. In: Thawley SE, Panje WR (eds) Comprehensive management of head and neck tumors, 1st edn. W.B. Saunders, Philadelphia, pp 684–699

Parsons JT, Mendenhall WM, Mancuso AA et al. (1988) Malignant tumors of the nasal cavity and ethmoid and sphenoid sinuses. Int J Radiat Oncol Biol Phys 14:11–22

Parsons JT, Mendenhall WM, Mancuso AA et al. (1989) Twice-a-day radiotherapy for T3 squamous cell carcinoma of the glottic larynx. Head Neck 11:123–128

Parsons JT, Mendenhall WM, Stringer SP, Cassisi NJ, Million RR (1992) Nasal cavity and paranasal sinuses. In: Perez CA, Brady LW (eds) Principles and practice of radiation oncology, 2nd edn. J.B. Lippincott Company, Philadelphia, pp 644–656

Parsons JT, Bova FJ, Fitzgerald CR et al. (1994a) Radiation optic neuropathy after megavoltage external-beam irradiation: analysis of time-dose factors. Int J Radiat Oncol Biol Phys 30:755–763

Parsons JT, Bova FJ, Fitzgerald CR et al. (1994b) Radiation retinopathy after external-beam irradiation: analysis of time-dose factors. Int J Radiat Oncol Biol Phys 30:765–773

Parsons JT, Bova FJ, Fitzgerald CR et al. (1994c) Severe dry-eye syndrome following external beam irradiation. Int J Radiat Oncol Biol Phys 30:775–780

Parsons JT, Stringer SP, Mancuso AA, Million RR (1994d) Nasal vestibule, nasal cavity, and paranasal sinuses. In: Million RR, Cassisi NJ (eds) Management of head and neck cancer: a multidisciplinary approach, 2nd edn. J.B. Lippincott Company, Philadelphia, pp 551–598

Parsons JT, Mendenhall WM, Moore GJ, Million RR (1999a) Radiotherapy of tumors of the oral cavity. In: Thawley SE, Panje WR, Batsakis JG, Lindberg RD (eds) Comprehensive management of head and neck tumors, 2nd edn. W.B. Saunders Company, Philadelphia, pp 695–719

Parsons JT, Mendenhall WM, Moore GJ, Million RR (1999b) Radiotherapy of tumors of the oropharynx. In: Thawley SE, Panje WR, Batsakis JG, Lindberg RD (eds) Comprehensive management of head and neck tumors, 2nd edn. W.B. Saunders Company, Philadelphia, pp 861–875

Som PM, Curtin HD, Mancuso AA (1999) An imaging-based classification for the cervical nodes designed as an adjunct to recent clinically based nodal classifications. Arch Otolaryngol Head Neck Surg 125:388–396

19 Breast Cancer

Sabin B. Motwani, Eric A. Strom, Marsha D. McNeese, and Thomas A. Buchholz

CONTENTS

19.1 Introduction 485
19.2 Epidemiology 486
19.3 Anatomy 486
19.3.1 Lymphatics 486
19.3.2 Lymphatic Drainage 486
19.4 Patterns of Spread 488
19.4.1 Locoregional 488
19.4.2 Systemically 488
19.5 Work-Up 489
19.5.1 History and Physical Examination 489
19.5.2 Mammography 489
19.5.3 Biopsy 489
19.6 Staging 490
19.6.1 Axillary Lymph Node Dissection and Sentinel Biopsy 491
19.6.2 Other Studies 492
19.6.3 Surgery 492
19.7 Radiation Therapy – General Concepts 492
19.7.1 Boost Treatment – General Concepts 493
19.8 Breast Conservation Radiation Therapy 493
19.8.1 Indications 493
19.8.2 Treatment to Breast Only 493
19.8.3 Simulation: The University of Texas M. D. Anderson Cancer Center Method, Supine Breast 494
19.8.4 Virtual Simulation 495
19.8.5 Dosimetry 495
19.9 Regional Nodes 495
19.10 Boost to Primary Site 495
19.11 Supraclavicular Nodal Irradiation 497
19.12 Prone Position Breast Conservation Radiation Therapy 499
19.12.1 General Concepts 499
19.12.2 Dosimetry 499
19.12.3 Set-Up 499
19.13 Partial Breast Irradiation 500
19.13.1 General Concepts 500
19.13.2 Standard Brachytherapy 500
19.13.3 Multiplane, Multineedle Implant 500
19.13.4 MammoSite™ 501
19.13.5 Conformal External Beam 501
19.14 Postmastectomy Irradiation 501
19.14.1 Indications 501
19.15 Regional Nodal Failure 501
19.15.1 Internal Mammary Chain Radiation – General Concepts 502
19.16 Postoperative Chest Wall Tangential Photon Fields 503
19.16.1 Field Borders 503
19.16.2 Dosimetry 503
19.17 Electron Beam Chest Wall Fields 504
19.17.1 Field Borders 504
19.17.1.1 Dosimetry 505
19.17.1.2 Energy 505
19.18 Partly-Deep Tangential Pair Fields 505
19.18.1 General Concepts 505
19.18.1.1 Treatment Planning 505
19.18.1.2 Dosimetry 505
19.18.1.3 Energy 505
19.19 Anterior Supraclavicular–Axillary Photon Field 505
19.19.1 Field Borders 505
19.19.2 Dosimetry 506
19.20 Supraclavicular and Axillary Apex Field (Electron Beam Technique) 507
19.21 Internal Mammary Chain Radiation 507
19.21.1 Field Borders 507
19.21.2 IM Dosimetry 508
19.22 Axillary Irradiation 508
19.22.1 Indications 508
19.23 Posterior Axillary Field 508
19.24 Postmastectomy Boost 509
19.24.1 General Concepts 509
19.24.2 Dosimetry 509
References 509

S. B. Motwani, MD; E. A. Strom, MD; M. D. McNeese, MD;
T. A. Buchholz, MD
Department of Radiation Oncology, The University of Texas,
M.D. Anderson Cancer Center, 1515 Holcombe Boulevard, Unit
1202, Houston, TX 77030, USA

19.1 Introduction

Radiation oncologists may be called on to treat all or some of the following in breast cancer patients: the breast or postmastectomy chest wall, axilla, ipsilateral internal mammary (IM) lymph nodes, and the supraclavicular lymph nodes depending on the clinicopathological features of an individual patient's disease. This chapter will begin with a brief review of epidemiology, breast anatomy and lymphatic drainage patterns, patterns of spread, work-up and staging of breast cancer, and will focus on the technical aspects of irradiation to the intact breast and in the postmastectomy setting.

19.2
Epidemiology

Breast cancer is the most common cancer of women, with 211,240 estimated new cases reported (2005) annually in the U.S. and 40,410 estimated deaths (JEMAL et al. 2005). It is estimated that a woman has a one in eight chance of developing breast cancer sometime in their lifetime (FEUER et al. 1993). The risk of breast cancer is associated with increasing age, particularly in women 55 years or older (DEVESA et al. 1995). Additionally, genetic, endocrine and reproductive factors, and family histories all contribute to assessing an individual's risk for developing breast cancer. It is interesting that there is up to a tenfold difference in breast cancer incidence depending on geography and socioeconomic background (PARKIN and MUIR 1992).

19.3
Anatomy

A thin layer of mammary or breast tissue extends from the edge of the sternum medially to the latissimus dorsi muscle laterally, up to the inferior edge of the clavicle superiorly. The breast mound usually extends from the anterior 2nd rib to anterior 6th or 7th rib, depending on the size. This protuberance is from near midline to the anterior axillary line, although the largest volume of tissue is located in the upper-outer quadrant. Often times the breast has a thick portion of tissue extending into the axilla, known as the axillary tail of Spence. A common site for primary tumors, this area can become so prominent that it can form an apparent axillary mass or give the appearance of a secondary breast mound.

19.3.1
Lymphatics

There is a well-developed lymphatic system within the breast with primary drainage to the axilla, supraclavicular, infraclavicular, and IM lymph nodes as shown in Fig. 19.1. Generally, this drainage moves superiorly and laterally toward the axillary lymph nodes. These then drain into the confluence of the internal jugular and subclavian veins in the base of the neck under the insertion of the sternocleidomastoid muscle to the head of the clavicle. The lymphatic trunks are therefore located in the medial

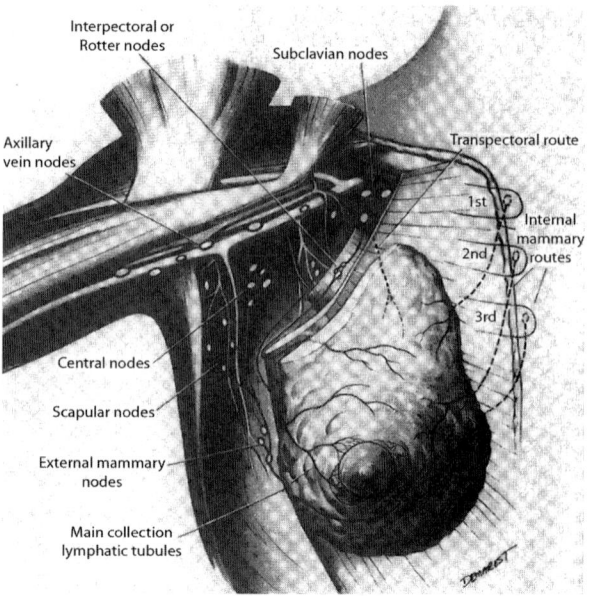

Fig. 19.1. Distribution of lymphatic trunks draining breast and chest wall musculature. Intercostal lymphatics are probably also directing route of spread. (Reprinted with permission from HAAGENSEN 1986)

portion of the supraclavicular area. There are three trunks of primary importance: the subclavian trunk from the axilla, the jugular trunk from the neck and supraclavicular area, and the bronchomediastinal trunks. All the trunks combine in a variety of ways to empty on the right side into a short common trunk, the right lymphatic trunk, and on the left side into the internal jugular vein, subclavian vein, or the thoracic duct.

The axillary lymph nodes are separated into three anatomically continuous levels using the pectoralis minor muscle as a point of reference as shown in Figure 19.2. This muscle has its origin from the third, fourth, and fifth ribs inserting onto the coracoid process of the scapula. The three levels are:
- Level I (proximal) – nodes located inferior (lateral) to the pectoralis minor muscle
- Level II (middle) – nodes directly beneath the muscle
- Level III (distal) – nodes superior (medial) to the muscle

19.3.2
Lymphatic Drainage

The primary drainage of the breast tissue is to the ipsilateral lower axilla. Analysis of the degree of axillary lymphatic involvement by the tumor pro-

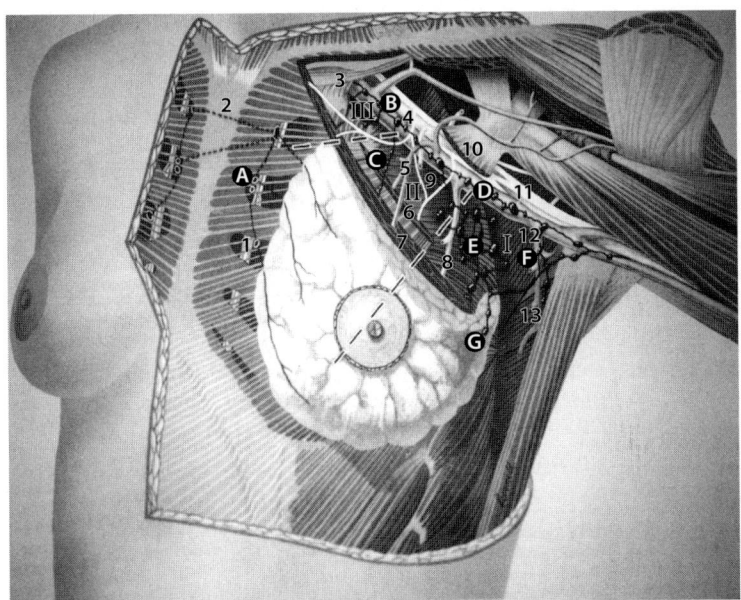

Fig. 19.2. Anatomy of axilla (pectoralis major and minor muscles partially removed to demonstrate anatomic levels of lymph nodes). Internal mammary artery and vein (*1*), substernal cross drainage to contralateral internal mammary lymphatic chain (*2*); subclavian muscle and Halsted's ligament (*3*); lateral pectoral nerve (from lateral cord) (*4*); pectoral branch from thoracoacromial vein (*5*); pectoralis minor muscle (*6*); pectoralis major muscle (*7*); lateral thoracic vein (*8*); medial pectoral nerve (from medial cord) (*9*); pectoralis minor muscle (*10*); median nerve (*11*); subcapsular vein (*12*); thoracodorsal vein (*13*); internal mammary lymph nodes (*A*); apical lymph nodes (*B*); interpectoral (Rotter's) lymph nodes (*C*); axillary vein lymph nodes (*D*); central lymph nodes (*E*); scapular lymph nodes (*F*); external mammary lymph nodes (*G*). *Level I lymph nodes* lateral to lateral border of pectoralis minor muscle; *level II lymph nodes* behind pectoralis minor muscle; *level III lymph nodes* medial to medial border of pectoralis minor muscle. (Reprinted with permission from OSBORNE 1987)

vides the most important single prognostic factor for breast carcinoma. In general, axillary lymphatic involvement occurs in an orderly fashion. It is more common to see level-I and -II involvement than I and III or III alone. Generally, only level-I and -II nodes are routinely dissected for staging purposes. In a series of 1446 patients, VERONESI et al. (1990) showed that the incidence of skip metastases is quite rare, with only 1.2% showing invasion in level II without involvement at level I. Conversely, when the nodes at level I were positive, approximately 40% of higher level nodes were also involved. The distribution of nodal metastases by level is shown in Table 19.1.

The other major route of lymphatic spread is via the ipsilateral internal mammary chain (IMC). These nodes are small, 2–5 mm in diameter, and can be structural nodes or poorly organized collections of lymphocytes in areolar tissue. These nodes are grouped around and travel with the IM artery and vein and generally within a few centimeters of the lateral edge of the sternum. They usually lie on the thoracic fascia in the interspaces between the costal cartilages with rare exceptions underneath the edge

Table 19.1. Distribution of 839 cases according to metastatic involvement by level

Levels involved	Number of cases (%)
I	455 (54.2)
I+II	187 (22.3)
I+II+III and I+III	186 (22.2)
Total number of cases with regular distribution	828 (98.7)
II	10 (1.2)
III	1 (0.1)
Total number of cases with 'skip' distribution	11 (1.3)
Total	839 (100.0)

of the costal cartilage. They are primarily found in the first three intercostal spaces with fewer lymph nodes identified in the fourth and fifth interspaces. Ultrasound and lymphoscintigraphy can be used to determine the location of these structures, and ultrasound can detect small volume metastases based on architectural changes. Retromanubrial lymphatic connecting at the level of the first interspace occurs between the right and left lymphatic trunk, 20% of the time (HAAGENSEN 1971).

Supraclavicular nodal involvement generally represents stages of advanced regional disease and carries a poorer prognosis. The major route of cancer spread to the supraclavicular nodes is via the axillary lymph nodes. When lymph becomes obstructed in the lymphatic trunks or the internal jugular–subclavian venous confluence, retrograde spread may occur into the lateral supraclavicular nodes and spread more posteriorly or in a cephalad direction. The supraclavicular nodes at greatest risk are difficult to examine, located behind the sternocleidomastoid muscle, while the supraclavicular nodes more laterally are easier to detect on physical examination.

The frequent use of sentinel lymph node surgery has re-established preoperative lymphoscintigraphy as an important diagnostic tool. SHAHAR et al. (2005) looked at 297 breast carcinoma patients with at least one positive sentinel lymph node. Of these patients, 279 had drainage to the regional lymph node basins seen on lymphoscintigraphy. Dual drainage to both the IMC and axilla was seen in 59 patients, and 4 patients had drainage to the IMC only. IMC drainage is correlated with tumor location. IMC drainage rates differed significantly between upper and lower tumors [lower 36.4% versus central 28.4% versus upper 14.6%, but not between medial and lateral tumors (SHAHAR et al. 2005)]. IMC drainage patterns can be found in Tables 19.2 and 19.3.

19.4
Patterns of Spread

19.4.1
Locoregional

Locoregional tumor spread occurs both by direct extension and lymphatic transit. Locally, the cancer can be found in the overlying skin, the pectoralis muscles, or even the chest wall. Therefore, it is important to assess the size and mobility of the mass in relation to the skin and chest wall on physical examination. True, independent multicentric tumor development is probably uncommon. With inflammatory carcinoma, assessment of the full extent of cutaneous changes, particularly erythema and skin edema, is necessary prior to initiating therapy since radiographic studies typically underestimate the extent of disease. Regionally, the two most common areas of spread are the IMC and the axillary nodes. IMCs are usually involved much less than the axillary nodes. Resection of the IMCs does not affect overall

Table 19.2. The frequency of internal mammary chain (IMC) drainage as a function of primary tumor location ($P=0.017$) (SHAHAR ET AL. 2005)

Tumor quadrant	Number of patients (%)	Rate of drainage to IMC (%)
Upper outer	128 (46)	14.1
Upper inner	30 (11)	16.7
Lower outer	19 (7)	31.6
Lower inner	14 (5)	42.9
Central	88 (31)	28.4

Table 19.3. Relationship between tumor location and drainage to internal mammary chain (IMC) lymph nodes. Subgrouping 1, $P=0.003$; subgrouping 2, $P=0.077$ (SHAHAR ET AL. 2005)

Variable	Number of patients (%)	Rate of drainage to IMC (%)
Analysis: upper versus lower	-	-
Lower quadrants	33 (12)	36.4
Central breast	88 (31)	28.4
Upper quadrants	158 (57)	14.6
Analysis: medial versus lateral	-	-
Medial quadrants	44 (16)	25.0
Central breast	88 (31)	28.4
Lateral quadrants	147 (53)	16.3

survival though. In early stages of breast cancer, when lymphatic mapping has been shown to include IM drainage, these nodes probably also need to be targeted. Since nodal status (and hence stage) is the primary prognostic determinant in breast cancer, accurate clinical staging is critical for patients who will not undergo initial surgery. Of breast cancer cases, 40% have some axillary node involvement. As the number of involved lymph nodes increases, survival decreases substantially. The risk of lymph node involvement is also dependent on the size of the mass and whether or not it is clinically palpable (SILVERSTEIN et al. 1995).

19.4.2
Systemically

Systemically, breast cancer can spread to any area of the body, but well-recognized patterns are seen. Two-thirds of the first recurrences after a mastectomy are systemic, and the remaining third is locoregional (CROWE et al. 1991). Since the mechanism of these two events are distinct, it is recommended that systemic events be labeled as metastasis and the label "recurrence" be reserved for locoregional events. The most common areas of systemic metastases are bone, lung, liver, and the central nervous system – in decreasing order of frequency.

The timing of breast cancer recurrences deserves special attention. Recurrence is likely to occur earlier in women with large, node positive disease than in those with small, node negative disease. First-time recurrences typically occur between the second and fifth years after treatment (HELLMAN and HARRIS 2000). Overall, up to 80% of the patients will have had their recurrence within 5 years and 95% of the patients will have had it by 10 years, but the recurrence rate is not absolutely zero even after 15 years (HELLMAN and HARRIS 2000).

The radiation oncologist needs to be aware of these patterns of spread so they can educate their patients and their families. They need to maintain routine follow-up for women with non-breast malignancies. Finally, they need to be aware of any prior radiation, especially if it occurred at a young age, as this prior exposure poses a major risk for radiation-induced breast cancer. The latent period for induction is at least 10–15 years (VAN LEEUWEN et al. 2000). For example, patients who had Hodgkin's disease at a very young age require thoughtful follow-up as a population at increased risk to develop new cancers.

19.5
Work-Up

19.5.1
History and Physical Examination

The routine use of screening mammography has resulted in the routine presentation of breast cancer as an asymptomatic finding. The first and most important step in the work-up of breast cancer is to obtain a clear history and appropriate physical examination. Important elements of the history and physical exam can be found on Tables 19.4 and 19.5 (MORROW et al. 2002). Any symptoms suggesting systemic disease, such as bone pain (especially in the spine, pelvis, hips, and ribs), unusual cough or shortness of breath, headache, dizziness, or change in vision or mental status should also be documented.

19.5.2
Mammography

The initial work-up of a breast cancer begins with the bilateral diagnostic mammogram. All areas of

Table 19.4. Elements of a breast cancer specific history

- Family history (any first-degree relatives or extended family), age at diagnosis
- Any history of ovarian carcinoma
- History of any prior irradiation
- Presence of implants (type and location)
- Last menstrual period
- Systemic symptoms
- Nipple discharge; spontaneous versus induced, color of discharge

Table 19.5. Elements of a breast physical exam

- Tumor location and size, if palpable
- Fixation to skin
- Ratio of breast size to tumor size
- Axillary node status: size and mobility
- Supraclavicular nodal status
- Evidence of locally advanced disease:
  - Skin ulceration
  - Peau d'orange
  - Skin erythema, edema
  - Fixed axillary nodes
  - Lymphedema of the ipsilateral arm
- Nipple changes
- Appearance of the opposite breast and axilla

the breast must be imaged, with palpable areas of concern marked. Magnification or spot compression views of any suspicious areas should be performed during the examination. Breast lesions are measured in two dimensions, and skin thickening should be noted. If the mass is associated with microcalcifications, there should be documentation of the extent of microcalcification and its relationship to any mass lesions. Clustering and type of microcalcification should also be noted. It is also important to document multicentricity and multifocality since it may influence the direction of treatment.

19.5.3
Biopsy

It is highly desirable to sample any suspicious or ambiguous lesions including secondary lesions in patients with documented cancers. Short interval re-imaging (i.e., 6 months) should only be used to follow findings that are most probably benign (BIRADS 2–3) (SICKLES 1995). Stereotactic core biopsy and fine needle aspiration cytology are the two most common methods to obtain tissue. During a core biopsy, mammographic guidance is used to obtain tissue samples with minimal disturbance of the breast.

Detailed pathological features are described on tissue specimens collected from biopsy. Histological type and grade; estrogen receptor, progesterone receptor, and Her2/neu receptor status; presence or absence of tumor necrosis; vascular and lymphatic invasion; an inflammatory infiltrate; the presence of DCIS in association with an invasive ductal carcinoma surgical margin; and pathological nodal status should all be noted. Tumor necrosis, lymphatic and vascular invasion, and an inflammatory infiltrate have been known to increase breast cancer recurrence rates, up to 10–15% at 5 years (KURTZ et al. 1990). Noting an extensive intraductal component (EIC) associated with the primary invasive cancer on mammography is also useful. Patients with microcalcifications greater than 3 cm are at higher risk of EIC, and this alerts the surgeon to anticipate wider margins (KURTZ et al. 1990). The presence of EIC with pathologically clear margins does not lead to higher risk of locoregional recurrence (HURD et al. 1997). Mammography should be obtained after excision, especially if there was a presence of microcalcifications before excision, in order to ensure all the suspicious lesions are removed.

19.6
Staging

Once a diagnosis has been pathologically confirmed, the tumor needs to be appropriately staged. As previously mentioned, clinical staging is used for patients who have not undergone initial surgery. Pathological staging is used for patients who have already undergone surgery. Good clinical staging is assessed by a thorough, focused breast physical exam (Table 19.5). Appropriate staging combined with precise location of tumor facilitates proper radiation treatment field planning.

The current staging system standard for breast cancer is a collaborative work of the American Joint Committee on Cancer (AJCC) and the Union Internationale Centre Cancer (UICC). This system is a TNM system and patients may be staged pre-operatively (clinically) or postoperatively (pathologically). Presented in Tables 19.6, 19.7, and 19.8 are the 6th edition (2002) TNM staging classification schemes and stage groupings (SINGLETARY et al. 2002). The two major changes from the prior 5th edition (1997) are: (1) clinical evidence of metastases to the infraclavicular or supraclavicular lymph nodes are classified as N3 instead of M1 and (2) pathological lymph node status is now based on the number of axillary nodes involved.

Table 19.6. Sixth Edition AJCC TNM clinical staging system for breast cancer

Primary Tumor (T)

T_x: Primary tumor cannot be assessed
T_0: No evidence of primary tumor
T_{is}: Carcinoma in situ (DCIS, LCIS, Paget's disease with no invasive tumor)
T_{is} (DCIS): ductal carcinoma in situ
T_{is} (LCIS): lobular carcinoma in situ
T_{is} (Paget's): Paget's disease of the nipple with no tumor
Note: Paget's disease associated with a tumor is classified according to the size of the tumor

T_1: ≤2 cm in greatest dimension
T_{1mic}: microinvasive ≤0.1 cm
T_{1a}: >0.1–0.5 cm
T_{1b}: >0.5–1 cm
T_{1c}: >1–2 cm
T_2: >2–5 cm
T_3: >5 cm

T_4: Tumor of any size with direct extension to (a) chest wall or (b) skin, only as described below:
T_{4a}: Extension to chest wall, not including pectoralis muscle
T_{4b}: Edema/peau d'orange or ulceration of the skin of the breast, or satellite skin nodules confined to the same breast
T_{4c}: Both T4a and T4b
T_{4d}: Inflammatory carcinoma

Regional (Ipsilateral) Lymph Nodes (N)

Nx: cannot be assessed (e.g., previously removed)
N_0: No lymph node metastasis
N_1: Metastasis to movable ipsilateral axillary lymph nodes
N_2: Metastasis in ipsilateral axillary lymph nodes fixed of matted, or in clinically apparent* ipsilateral internal mammary nodes in the *absence* of clinically evident axillary lymph node metastasis
N_{2a}: Metastasis in ipsilateral axillary lymph nodes fixed to one another (matted) or to other structures
N_{2b}: Metastasis only in clinically apparent* ipsilateral internal mammary nodes in the *absence* of clinically evident axillary lymph node metastasis
N_3: Metastasis in ipsilateral infraclavicular lymph nodes, with or without axillary lymph node involvement, or in clinically apparent* ipsilateral internal mammary lymph nodes in the *presence* of clinically evident axillary lymph node metastasis; or metastasis in ipsilateral supraclavicular lymph nodes, with or without axillary or internal mammary lymph node involvement
N_{3a}: Metastasis in ipsilateral infraclavicular lymph nodes
N_{3b}: Metastasis in ipsilateral internal mammary lymph nodes and axillary lymph nodes
N_{3c}: Metastasis in ipsilateral supraclavicular lymph nodes

Distant Metastasis (M)

M_x: Distant metastases cannot be assessed
M_0: No distant metastases
M_1: Distant metastases present

* *Clinically apparent* is defined as detected by imaging studies (excluding lymphoscintigraphy) or by clinical examination or grossly visible pathologically

Table 19.7. Sixth Edition AJCC TNM pathological staging system for breast cancer

Primary Tumor (T)

T_x: Primary tumor cannot be assessed

T_0: No evidence of primary tumor

T_{is}: Carcinoma in situ (DCIS, LCIS, Paget's disease with no invasive tumor)

T_{is} (DCIS): Ductal carcinoma in situ

T_{is} (LCIS): Lobular carcinoma in situ

T_{is} (Paget's): Paget's disease of the nipple with no tumor

Note: Paget's disease associated with a tumor is classified according to the size of the tumor

T_1: ≤2 cm in greatest dimension

T_{1mic}: Microinvasive ≤0.1 cm

T_{1a}: >0.1–0.5 cm

T_{1b}: >0.5–1 cm

T_{1c}: >1–2 cm

T_2: >2–5 cm

T_3: >5 cm

T_4: Tumor of any size with direct extension to (a) chest wall or (b) skin, only as described below:

T_{4a}: Extension to chest wall, not including pectoralis muscle

T_{4b}: Edema/peau d'orange or ulceration of the skin of the breast, or satellite skin nodules confined to the same breast

T_{4c}: Both T4a and T4b

T_{4d}: Inflammatory carcinoma

Regional (Ipsilateral) Lymph Nodes (N)

pNx: Regional lymph nodes cannot be assessed (e.g., previously removed, or not removed for pathological study)

pN0: No regional lymph node metastasis histologically, no additional examination for isolated tumor cells (ITC)

pN0(i-): No regional lymph node metastasis histologically, negative immunohistochemistry (IHC)

pN0(i+): No regional lymph node metastasis histologically, positive IHC, no IHC cluster >0.2 mm

pN0(mol-): No regional lymph node metastasis histologically, negative molecular findings (reverse transcriptase/polymerase chain reaction)

pN0(mol+): No regional lymph node metastasis histoologically, positive molecular findings (reverse transcriptase/polymerase chain reaction)

pN_1: Metastasis in 1 to 3 axillary lymph nodes, or in internal mammary nodes with microscopic disease detected by sentinel lymph node dissection, or both, but not clinically apparent[a]

pN_{1mi}: Micrometastasis (>0.2 mm, none >2.0 mm)

pN_{1a}: Metastasis in 1 to 3 axillary lymph nodes

pN_{1b}: Metastasis in internal mammary nodes with microscopic disease detected by sentinel lymph node dissection but not clinically apparent[a]

pN_{1c}: Metastasis in 1 to 3 axillary lymph nodes and in internal mammary nodes, with microscopic disease detected by sentinel lymph node dissection, but not clinically apparent[a] (if associated with more than 3 positive axillary lymph nodes, the internal mammary nodes are classified as pN3b to reflect increased tumor burden)

pN_2: Metastasis in 4 to 9 axillary lymph nodes, or in clinically apparent[b] internal mammary lymph nodes in the *absence* of axillary lymph nodes metastasis

pN_{2a}: Metastasis in 4 to 9 axillary lymph nodes (at least 1 tumor deposit >2.0 mm)

pN_{2b}: Metastasis in clinically apparent[b] internal mammary lymph nodes in the absence of axillary lymph nodes metastasis

pN_3: Metastasis in 10 or more axillary lymph nodes, or in infraclavicular lymph nodes, or in clinically apparent[b] ipsilateral internal mammary lymph nodes in the *presence of* 1 or more positive axillary lymph nodes; of in more than 23 axillary lymph nodes with clinically negative microscopic metastasis in internal mammary lymph nodes; or in ipsilateral supraclavicular lymph nodes

pN_{3a}: Metastasis in 10 more axillary lymph nodes (at least 1 tumor deposit >2.0 mm), or metastasis to the infraclavicular lymph nodes

pN_{3b}: Metastasis in clinically apparent[b] ipsilateral internal mammary lymph nodes in the *presence of* 1 or more positive axillary lymph nodes; or in more than 3 axillary lymph nodes and in internal mammary lymph nodes with microscopic disease detected by sentinel lymph node dissection, but not clinically apparent[a]

pN_{3c}: Metastasis in ipsilateral supraclavicular lymph nodes

Distant Metastasis (M)

M_x: Distant metastases cannot be assessed

M_0: No distant metastases

M_1: Distant metastases present

[a] *Not clinically apparent* is defined as detected by imaging studies (excluding lymphoscintigraphy) or by clinical examination or grossly visible pathologically. [b] *Clinically apparent* is defined as detected by imaging studies (excluding lymphoscintigraphy) or by clinical examination

19.6.1
Axillary Lymph Node Dissection and Sentinel Biopsy

Traditionally, the standard approach to the axilla was a level-I and -II axillary lymph node dissection (ALND). This procedure provided useful prognostic information, treated involved lymph nodes, and determined whether systemic treatment was necessary or not. Even though there are many advantages from an ALND, a patient's quality of life may be diminished. Patients often have pain and numbness in the axilla accompanied by decreased range of motion and risk of chronic lymphedema.

The sentinel node biopsy has become the de facto standard to determine nodal status in patients with

Table 19.8. Stage grouping for breast cancer

Stage	Tumor	Nodes	Metastasis
0	T_{is}	N_0	M_0
I	T_1	N_0	M_0
IIA	T_0	N_1	M_0
-	T_1	N_1	M_0
-	T_2	N_0	M_0
IIB	T_2	N_1	M_0
-	T_3	N_0	M_0
IIIA	T_0	N_2	M_0
-	T_1	N_2	M_0
-	T_2	N_2	M_0
-	T_3	N_1	M_0
-	T_3	N_2	M_0
IIIB	T_4	Any N	M_0
IIIC	Any T	N_3	M_0
IV	Any T	Any N	M_1

a clinically negative axilla. Sentinel nodes are the first nodes that the mass primarily drains to. If these nodes are negative, the probability of involvement to other axillary lymph nodes is small. Sentinel node dissection is indicated for small primary tumors with clinically negative axillary lymph nodes and no prior axillary surgery. Blue dye and radioactive colloid are typically used to identify sentinel lymph nodes. Sentinel nodes have been identified more than 90% of the time independently of the technique used (KRAG et al. 1998). Usually only one node was identified 67–93% of the time and the false negative rate was 0–11.9% (WOLMARK and RISHER 1981). If the sentinel node is positive, the current standard of care remains a formal level-I and -II ALND. In the absence of sentinel lymph node biopsy, accurate lymph node staging requires a minimum of six lymph nodes identified in an axillary dissection, and the mean number of identified nodes at M.D. Anderson Cancer Center is 17 (KATZ et al. 2000). In the NSABP trial B-4, not one patient with more than six nodes removed experienced an axillary failure (WOLMARK and RISHER 1981). The more lymph nodes removed, the more reliable the number of involved nodes becomes.

19.6.2
Other Studies

Other studies every breast cancer patient should have are a chest X-ray, a complete blood count, and a biochemical profile with particular attention to liver function tests and lactate dehydrogenase (LDH) levels. Other limited staging studies may be warranted depending on the patient's symptoms and stage of disease. Bone scan and com-

puted tomography (CT) of the chest and abdomen may be useful image modalities if the patients are asymptomatic or have locoregionally advanced breast cancer.

19.6.3
Surgery

The historic treatment for invasive breast cancer was mastectomy accompanied by a level-I and -II ALND. Multiple clinical trials have shown an equal disease-free and overall survival between mastectomy and breast conservation therapy. Thus, today, most women with early breast cancer elect to have lumpectomies followed by radiation therapy (Fig. 19.3) (FISHER et al. 2002).

19.7
Radiation Therapy – General Concepts

Adjuvant radiation therapy should begin after the integration of mammographic, pathological, and surgical information from the patient's disease. It has a good therapeutic index by decreasing locoregional recurrence and avoiding late complications. In addition, radiotherapy planning must take into account the extent and location of the tumor, the size of the breast, and the patient's concerns about recurrence, cosmesis, and side effects. If radiation is to be given concurrently with chemotherapy, there needs to be constant communication and open dialogue between the radiation oncologist and the medical oncologist so that both treatments are integrated smoothly.

As soon as the patient has healed from surgery, it is appropriate for radiation therapy to begin. Typically, in an uncomplicated breast-conserving surgery, this time period is 2–4 weeks after surgery. It is important that the radiation oncologist demonstrates reproducibility with regard to patient set-up, treatment planning, and equipment to assure dose homogeneity. Typically 6-MV photons are used on the intact breast, but higher energy photons may be used for women with larger breasts to assure better dose homogeneity. Each field is treated Monday through Friday, typically once a day. Radiation pneumonitis can be reduced as long as no more than 3–3.5 cm of lung is irradiated, and often only 1–1.5 cm is required (LINGOS et al. 1991). At our institution, M.D. Anderson Cancer Center, we

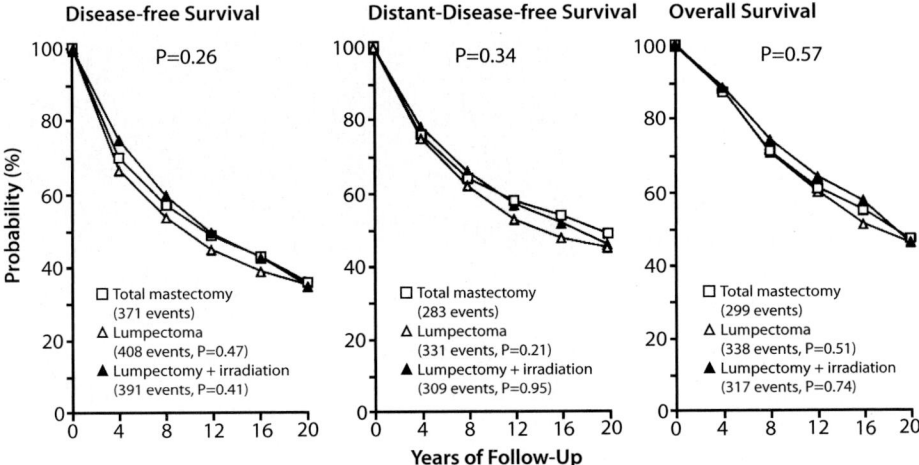

Fig. 19.3. NSABP B-06 data showing equivalent disease-free survival, distant-disease-free survival, and overall survival between patients who had lumpectomy followed by external beam radiation therapy compared with women who had a total mastectomy. (Reprinted with permission from FISHER et al. 2002)

believe the requirement of lung irradiation is zero. The lung and all other critical structures should be avoided whenever possible. Ideally, a thoughtful assessment of the volume at risk, focusing particularly on the tumor bed, permits delineating target volume according to the patient's anatomy. For left-sided lesions, effort is necessary to minimize the amount of heart irradiation in the tangential fields, which should ideally be nil. Typically, in an intact breast, radiation therapy is delivered using opposed tangential fields to a dose of 45–50 Gy (1.8–2.0 Gy per fraction).

19.7.1
Boost Treatment – General Concepts

Although there has been controversy regarding whether or not a boost is required for treatment, recent clinical trials have substantiated its usefulness (BARTELINK et al. 2001). Among these, the EORTC has reported the favorable impact of boosts on local recurrent rates (VRIELING et al. 1999). Boost treatment is clearly indicated for patients with focally positive or close margins of resection. The total dose to the primary tumor site is increased to approximately 60–66 Gy. Selection of the boost dose and volume is based on the patient's surgical and pathological information. A boost may not be required for those patients with more favorable tumors whose margins are clearly negative as long as the whole breast is treated to at least 50 Gy (KURTZ 2001).

19.8
Breast Conservation Radiation Therapy

19.8.1
Indications

The primary target of radiation therapy and breast conservation therapy is to eradicate microscopic residual disease adjacent to the original site of tumor as well as to eliminate any evidence of multicentric disease. Since 80% of the early failures in this patient group are in the same quadrant as the original primary, achieving adequate coverage in that area is imperative. The risk of distal failure in the intact breast due to multicentric disease is in part a function of the volume of the breast and the techniques used. Distant failures are typically low when small volumes of breast tissue are left unirradiated, particularly if that volume is well away from the primary site.

CT permits the easy evaluation of the operative bed, identification of the majority of the remaining breast tissue as well as adjacent avoidance structures such as heart and lung. CT-based planning is highly preferred over fluoroscopic planning and is rapidly becoming a standard of care.

19.8.2
Treatment to Breast Only

Irradiation is administered to the breast alone in all patients who have undergone excision ("lumpectomy" or segmental resection) of early-stage invasive carcinomas. Current standards also recommend the use of post-excision radiation for most patients with ductal carcinoma in situ (DCIS), although selective avoidance is considered in patients with "favorable" DCIS – those with very small low-grade lesions with excessive wide to clear margins. Treatment of regional lymphatics is not indicated in lymph node-negative patients with an adequate axillary assessment and is generally not useful in patients with early-stage disease. If regional nodal irradiation is contemplated, its intent must be integrated into the original planning sessions. For patients with incompletely assessed axillae or those with microscopic involvement of sentinel nodes, it is technically possible to include the level-I and level-II axilla in opposing pairs using "high" tangents (SCHLEMBACH et al. 2001). Tumors at the extreme edges of the breast are particularly at risk for a geographic miss of the peritumoral tissue. Full delineation of the tumor bed is most easily accomplished with patients with recent lumpectomies and the placement of surgical clips remains useful for the planning of breast conservation therapy. If long time intervals pass from surgery to radiation planning, as is common for adjuvant chemotherapy, there may be almost no visible operative bed to guide radiotherapy planning.

19.8.3
Simulation: The University of Texas M. D. Anderson Cancer Center Method, Supine Breast

The basic technique for supine breast tangential treatment is surprisingly variable. Subtle differences in target delineation, immobilization techniques, dose specification, and missing tissue compensation suggest that there is no such thing as a single "standard" breast treatment. At UT M.D. Anderson Cancer Center, patients are immobilized in an L-shaped Vac-Lok™ cradle with the arm highly abducted. The cradle is fabricated on a slant board with cut-outs that permit only a single placement of the Vac-Lok™ cradle on the slant board (Fig. 19.4). Subsequently, supports that result in 5, 10, 15 or 20° inclination are inserted. Lastly, all of these devices are fixed to the table using an Exact™ Lok-Bar (with a cam-lock locking mechanism). Thin-cut spiral CT images, typically 3-mm slices, are obtained from the mid-neck to the upper abdomen (Fig. 19.5). It may be useful to place delineating wires on incision sites and to delineate the clinically determined breast

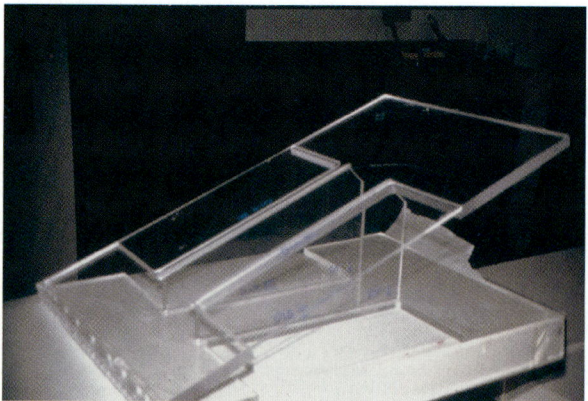

Fig. 19.4. Slant board. The slant board is used along with a vacuum cradle to ensure proper immobilization of the body and arm as well as daily reproducibility of the patient's treatment position

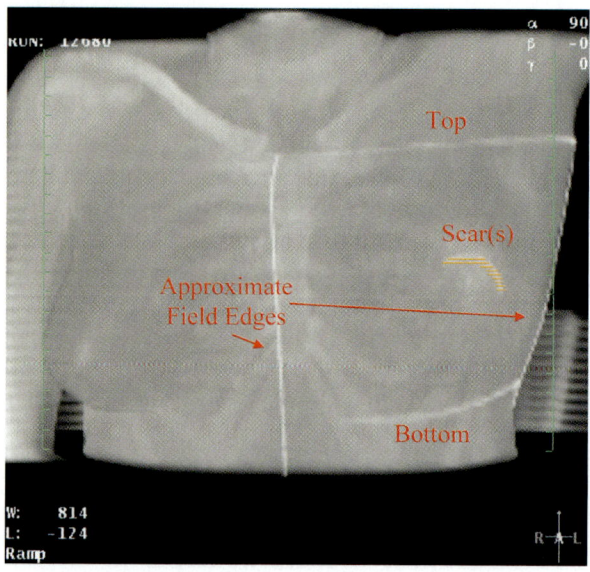

Fig. 19.5. Pre-computed tomography (CT) simulation skin marks including approximate field borders, incision sites or drains can be useful during 3-D treatment planning

mound. After reconstruction of the CT images, two separate fiducials are placed on the patient's skin. The first represents laser lines coincident with the central axis and the second is placed at the CT-determined likely field edges (Fig. 19.6). The relationship between these two fiducials is a known quantity and permits subsequent verification of positioning at the initial treatment set-up (Fig. 19.7). Subsequently, virtual simulation of the tangential fields is performed after the patient has been released. Careful documentation of all the parameters and recording of set-up instructions is the key to success.

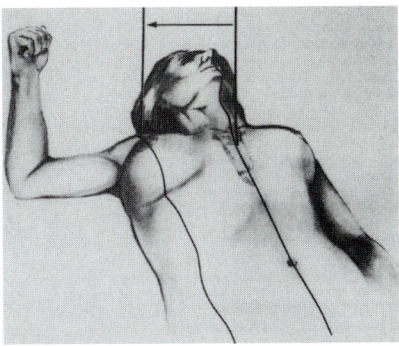

Fig. 19.6. Sagittal laser beam used for setting up patient relative to anatomic landmarks. Isocenter point first positioned in sagittal plane and then shifted laterally through pre-calculated distance to position within breast to define treatment isocenter

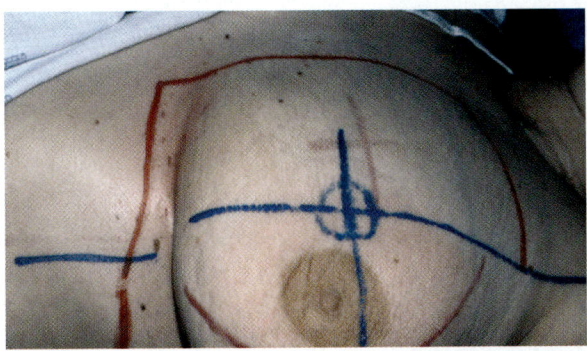

Fig. 19.7. Treatment fields drawn on intact breast. *Red lines* represent the primary field and central axis. *Blue lines* represent the set-up lasers

19.8.4
Virtual Simulation

Prior to beginning the final development of the tangential field set, contouring of the incision site and operative bed is performed, any adjacent avoidance structures may also be contoured (Fig. 19.8). The development of a tangential pair with a non-coplanar posterior edge is then determined by the physics and dosimetry staff along with the clinical radiation oncologist to obtain optimum coverage of the tumor bed adjacent structures and the majority of the breast mound, at the same time minimizing adjacent structures not at risk (Fig. 19.9). Generally, no cardiac structure is included within the tangential fields and it is usually possible to minimize the amount of adjacent lung to 1.5–2 cm (Fig. 19.10).

There is ample evidence to suggest that cardiac irradiation is detrimental, although cardiac consequences of breast irradiation have long latencies estimated to only begin to be seen at 15 years after

delivery of radiation therapy because patients with early breast cancer have a high probability of long-term survival. This is a critical aspect of treatment planning. The Oxford overview suggests that in the absence of careful treatment planning, a small survival benefit for radiation therapy, even in the context of early breast cancer, becomes offset by increased cardiovascular deaths (Fig. 19.11).

19.8.5
Dosimetry

The dose is specified at the pectoral surface and with appropriate compensation for missing tissue or lung heterogeneity. This results in the specified dose becoming the breast minimum dose. Typically 50 Gy in 25 fractions is prescribed at the dose specification isodose. We use intensity modulation with a field-in-field "step and shoot" technique, which results in excellent homogeneity of dose within the target volume (Fig. 19.12). Additional advantages include the need for fewer monitor units than comparable wedged plans and hence enhanced treatment times as well as decreased dose to distant normal tissue, such as the opposite breast (HONG et al. 1999).

19.9
Regional Nodes

When regional nodal irradiation is combined with breast tangential fields, additional complexities are involved. Junctional variation, always an area of dosimetric uncertainty, should be minimized through standardized techniques and superior immobilization. Breast field planning with non-divergent field edges are required at field abutments. Although it has some drawbacks, the rod and chain technique provides a non-divergent cephalad border with great flexibility in application and verification (Fig. 19.13).

The techniques used to treat the supraclavicular, axillary and IMCs are identical to those described below for the postmastectomy setting.

19.10
Boost to Primary Site

If margins are close or unknown, re-excision should be considered, provided the amount of breast tissue

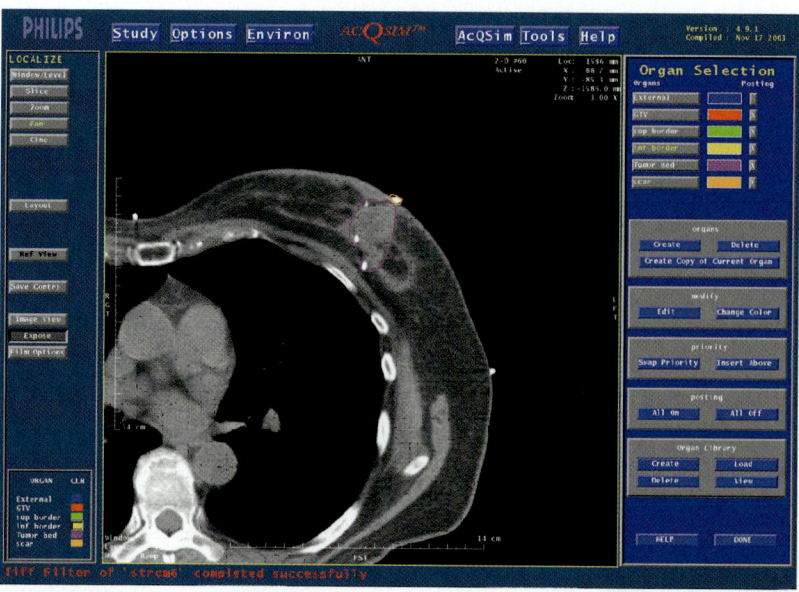

Fig. 19.8. Contouring of a tumor bed and incision site on 3-D treatment planning

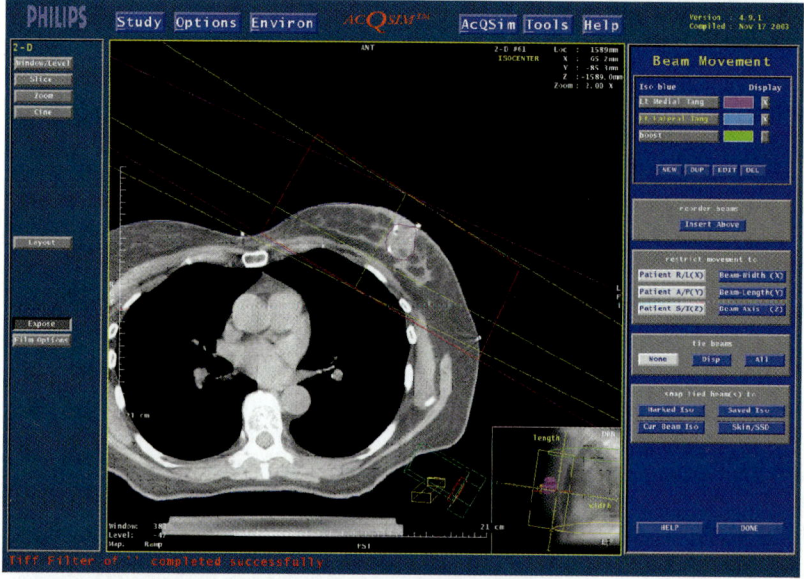

Fig. 19.9. Intermediate step in development of tangent pair. After opposing medial field, the new lateral field has a diverging deep edge. Additional gantry rotation will result in the desired coplanar deep field edge

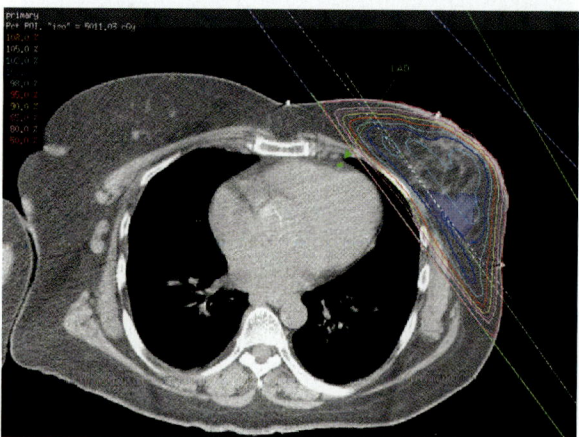

Fig. 19.10. Left-sided tumor in lateral breast with contoured left anterior descending (*LAD*) artery, showing minimal irradiation of the heart and lung in the tangential fields

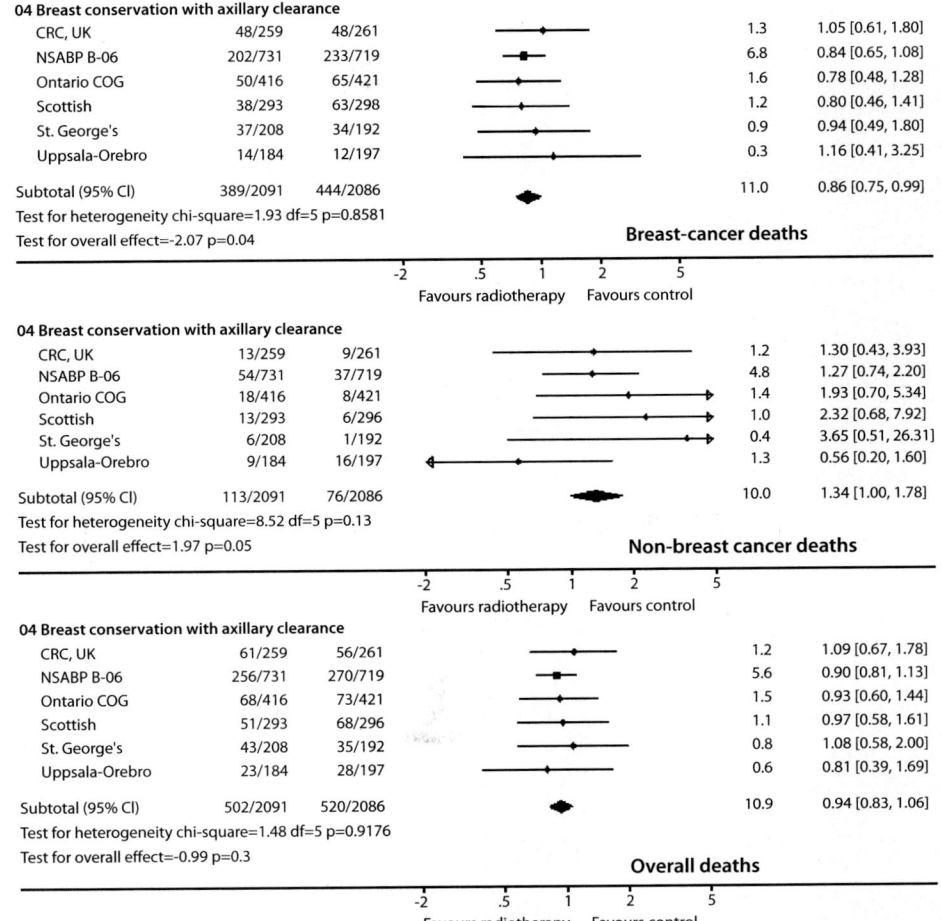

04 Breast conservation with axillary clearance

CRC, UK	48/259	48/261		1.3	1.05 [0.61, 1.80]
NSABP B-06	202/731	233/719		6.8	0.84 [0.65, 1.08]
Ontario COG	50/416	65/421		1.6	0.78 [0.48, 1.28]
Scottish	38/293	63/298		1.2	0.80 [0.46, 1.41]
St. George's	37/208	34/192		0.9	0.94 [0.49, 1.80]
Uppsala-Orebro	14/184	12/197		0.3	1.16 [0.41, 3.25]
Subtotal (95% CI)	389/2091	444/2086		11.0	0.86 [0.75, 0.99]

Test for heterogeneity chi-square=1.93 df=5 p=0.8581
Test for overall effect=-2.07 p=0.04

Breast-cancer deaths

-2   .5   1   2   5
Favours radiotherapy   Favours control

04 Breast conservation with axillary clearance

CRC, UK	13/259	9/261		1.2	1.30 [0.43, 3.93]
NSABP B-06	54/731	37/719		4.8	1.27 [0.74, 2.20]
Ontario COG	18/416	8/421		1.4	1.93 [0.70, 5.34]
Scottish	13/293	6/296		1.0	2.32 [0.68, 7.92]
St. George's	6/208	1/192		0.4	3.65 [0.51, 26.31]
Uppsala-Orebro	9/184	16/197		1.3	0.56 [0.20, 1.60]
Subtotal (95% CI)	113/2091	76/2086		10.0	1.34 [1.00, 1.78]

Test for heterogeneity chi-square=8.52 df=5 p=0.13
Test for overall effect=1.97 p=0.05

Non-breast cancer deaths

-2   .5   1   2   5
Favours radiotherapy   Favours control

04 Breast conservation with axillary clearance

CRC, UK	61/259	56/261		1.2	1.09 [0.67, 1.78]
NSABP B-06	256/731	270/719		5.6	0.90 [0.81, 1.13]
Ontario COG	68/416	73/421		1.5	0.93 [0.60, 1.44]
Scottish	51/293	68/296		1.1	0.97 [0.58, 1.61]
St. George's	43/208	35/192		0.8	1.08 [0.58, 2.00]
Uppsala-Orebro	23/184	28/197		0.6	0.81 [0.39, 1.69]
Subtotal (95% CI)	502/2091	520/2086		10.9	0.94 [0.83, 1.06]

Test for heterogeneity chi-square=1.48 df=5 p=0.9176
Test for overall effect=-0.99 p=0.3

Overall deaths

-2   .5   1   2   5
Favours radiotherapy   Favours control

Fig. 19.11. Oxford review showing data of overall deaths, breast cancer deaths, and non-breast cancer deaths. (Reprinted with permission from EARLY BREAST CANCER TRIALISTS COOPERATIVE GROUP 2003)

is sufficient, to insure the lowest failure rate and avoid the consequences of higher doses of irradiation (SCHNITT et al. 1994). With most invasive carcinomas – especially with any extensive or multifocal intraductal component, lymphatic invasion, or anaplastic nuclear grade, or in any patient who has irradiation deferred until after chemotherapy – the field is reduced after 50 Gy. An additional 10 Gy in five fractions with electrons is delivered to include the site and scar of the excision biopsy. The choice of energy depends on the thickness of the breast in treatment position. Figure 19.14 shows how a boost is drawn on a patient with a T2 inner quadrant primary. Appropriate electron energy is selected to allow the 90% isodose line to encompass the target volume, and we currently use a 2-cm radial expansion of the combined operative bed and incision site to set the field borders (Fig. 19.15). If the tumor bed is more than 4–5 cm deep within the breast, an interstitial implant may be preferable because of decreased dose to the skin with the implant compared with the dose that would be delivered by electrons. Use of a staggered double plane technique is usually best because of the possibility of a geographic miss with a single-plane implant. For patients with tumors deep in the breast, additional external irradiation may be delivered by means of a compression technique or by turning the patient into the lateral decubitus position for lateral tumors. Higher doses (typically 14–16 Gy) may be used if the margins are close or focally positive, or if the breast tangent dose is reduced to less than 50 Gy.

19.11
Supraclavicular Nodal Irradiation

Close attention must be given to the medial border of the supraclavicular field because involved nodes may be located under the junction of the sternocleidomastoid muscle and the head of the clavicle.

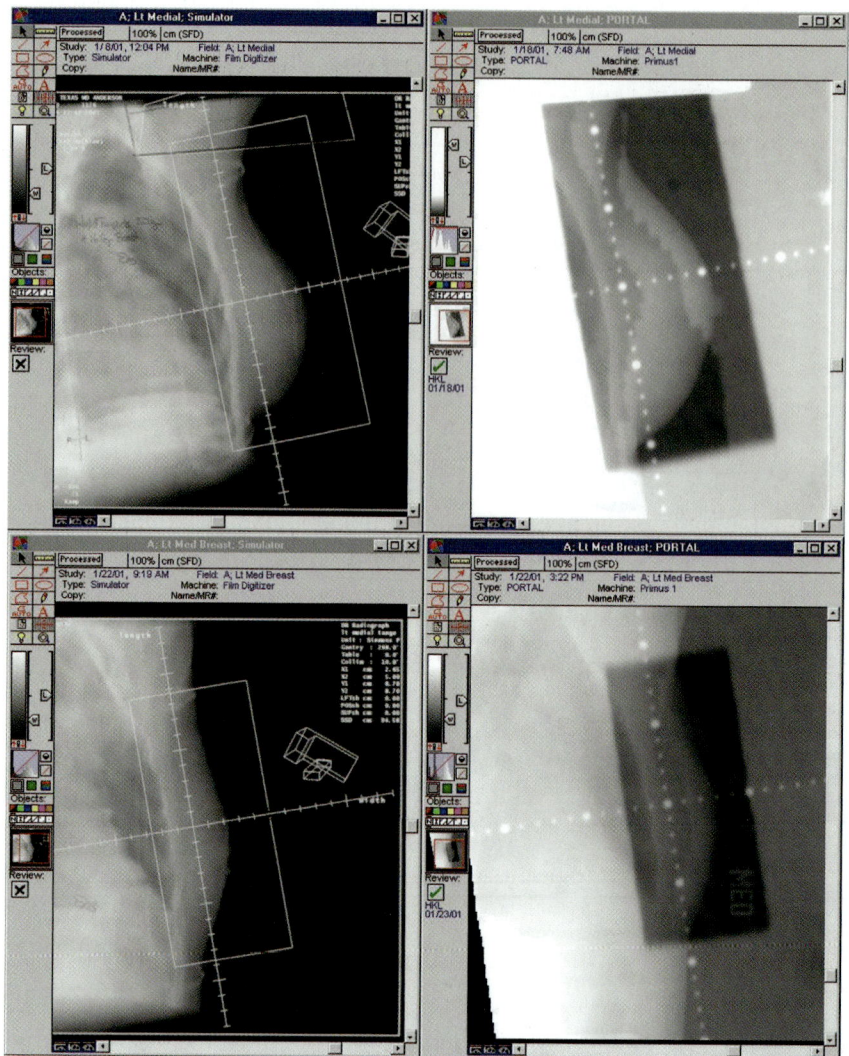

Fig. 19.12. Left medial tangent digital reconstructed radiograph (DRR) and left medial tangent multilayer IMRT film port

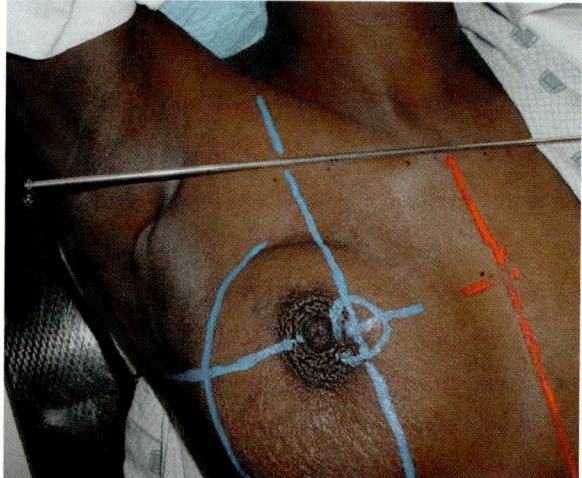

Fig. 19.13. Rod and chain to provide non-divergent cephalad border of tangent pair

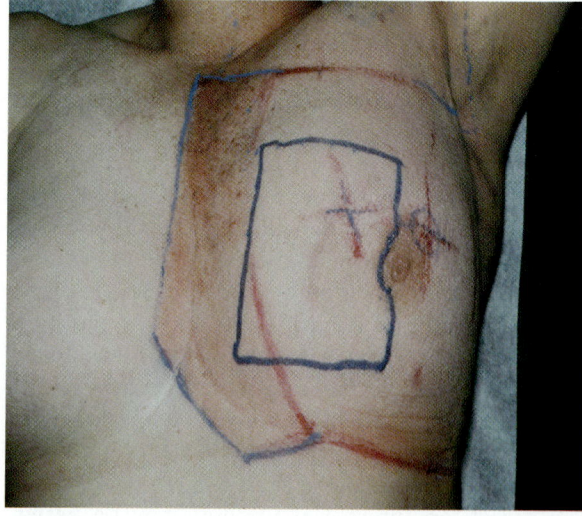

Fig. 19.14. Electron boost field drawn on intact breast

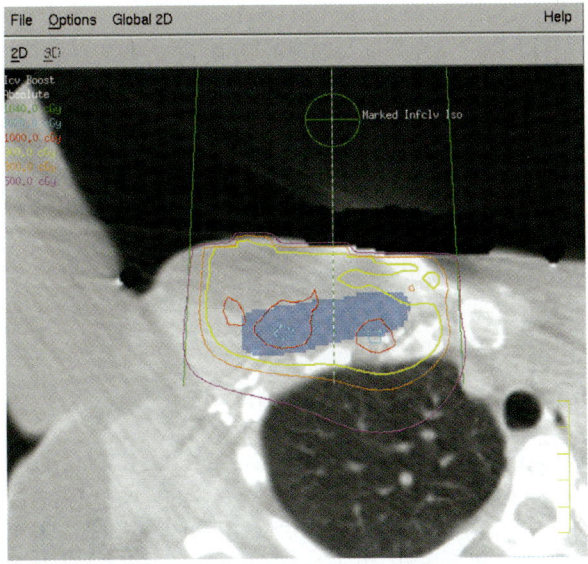

Fig. 19.15. Electron planning of an infraclavicular boost. The target is encompassed within the distal 90% isodose line. Electrons are typically prescribed at the 100% isodose line

The upper medial border of the field should be at or slightly beyond midline. The field covers supraclavicular, subclavicular apical, and low jugular nodes. For electron beam radiation, separate fields are used for the supraclavicular and IMCs because the differing depths of the target volumes determine the energy needed (Fig. 19.16). The patient is placed on a slant board on the treatment table so that the plane of the sternum is relatively parallel to the treatment table. The angle of the tilt is documented for daily duplication of treatment.

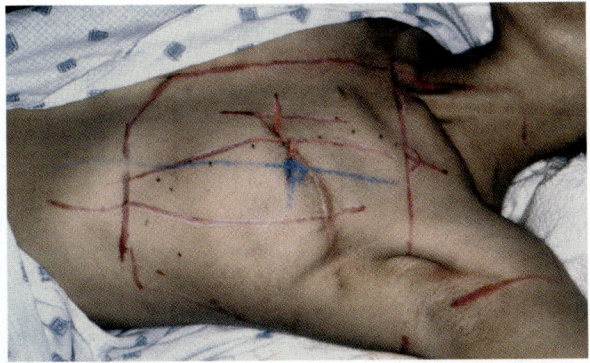

Fig. 19.16. Internal mammary chain (*IMC*) and lateral tangent field arrangement for post-mastectomy irradiation

19.12
Prone Position Breast Conservation Radiation Therapy

19.12.1
General Concepts

Historically, large pendulous breasts have been a contraindication for breast conservation therapy. When women with such breasts have been treated in the standard supine position, it has led to an excessive amount of lung and heart irradiation, and the consequences of tissue folds and large separations resulted in excessive skin reaction and late fibrosis. Treating these patients in the prone position minimizes breast separation and dose inhomogeneity within the treatment field and removes most skin folds. In addition, it is frequently possible to reduce irradiation to the heart, lung, and contralateral breast when prone, since the tumor bed moves away from these structures (GRANN et al. 2000).

19.12.2
Dosimetry

Radiotherapy is given with tangential fields to the entire breast. Depending on the shape of the breast, most patients require field compensation to achieve optimal dose homogeneity. The dose and fractionation scheme, of 45–50 Gy in 1.8- to 2.0-Gy fractions, 5 days a week, is the same as that for a patient treated in the supine position. Treatment is usually delivered using 6-MV photons. Higher energy photons may be used to improve dose inhomogeneity.

19.12.3
Set-Up

The patient lies in a prone position on a platform board at the time of simulation (Fig. 19.17). An aperture in the board allows the breast to fall naturally away from the chest wall. This position allows excess skin folds to be minimized. The contralateral breast is abducted, and the patient is rotated slightly to allow the ipsilateral chest wall to extend to the board aperture. A custom cradle is used to assure a reproducible position. Almost all patients have had clips placed in their lumpectomy cavity at the time of surgery, and we use these clips and any seroma as landmarks to map the primary tumor bed. Marking wires are placed at the breast edges both laterally

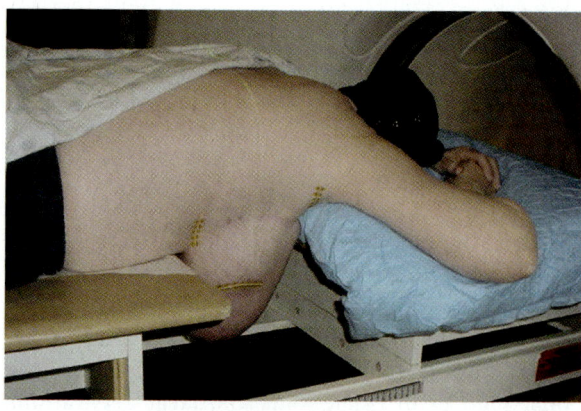

Fig. 19.17. Set-up for prone breast irradiation

and medially. Final marks include central axis lasers (medial and lateral), set-up lasers on torso, and lateral field projections. The ribs are used as a reliable landmark to ensure how much heart, lung, and chest wall are within the treatment fields.

19.13
Partial Breast Irradiation

19.13.1
General Concepts

Partial breast irradiation can be achieved using brachytherapy or hypofractionated conformal radiation therapy, or intraoperatively using either electrons or a linear accelerator. Each technique has its advantages and limitations. There are some reports of phase-I and -II studies treated with accelerated partial breast irradiation after breast-conserving surgery; however, there is limited long-term follow-up and selection bias with these studies despite potential comparable outcomes with regard to toxicity, cosmesis, and local control with those patients who had breast-conserving surgery followed by standard whole-breast irradiation (Keurer et al. 2004). Partial breast irradiation is under active clinical investigation.

19.13.2
Standard Brachytherapy

Brachytherapy can be delivered using low-dose-rate (LDR) or high-dose-rate (HDR) radiation sources. A dose of 45–50 Gy is usually delivered to a target volume at a rate of 30–70 cGy/h for LDR implants.

The patient is admitted to the hospital for 3–5 days. HDRs have become increasingly popular and are delivered on an outpatient basis. A dose of 34 Gy is delivered as twice-daily fractions of 3.4 Gy over a total of 5 days. At the present time, brachytherapy as the sole type of radiation therapy after breast conservation therapy remains largely investigational and has not been compared with standard whole breast irradiation.

19.13.3
Multiplane, Multineedle Implant

When contemplating definitive partial breast irradiation using a needle- or catheter-based approach, a larger volume implant is required than usually used for boost purposes. Definitive catheter-based brachytherapy, while the most validated of the partial breast techniques, is also technically very challenging. Typically, a minimum of two planes are required – more if the surgical volume is extensive. The deep plane is placed at the level of and parallel to the bottom of the excision cavity. The superficial plane is at the level of the superficial border of the cavity. Within a given plane, the separation between catheters is 1.0–1.5 cm and catheters extend at least 2 cm beyond the edge of the clips or seroma. The number of catheters is chosen so that a least one catheter is placed 1 cm or more beyond the edge of the target. As with all planar implants, the catheters are parallel and as straight as possible (Fig. 19.18).

To achieve adequate target volume coverage, there should be at least two LDR seeds or 1.5–2.0 cm of HDR dwell positions beyond the edge of the target volume. This implies that three to four LDR seeds

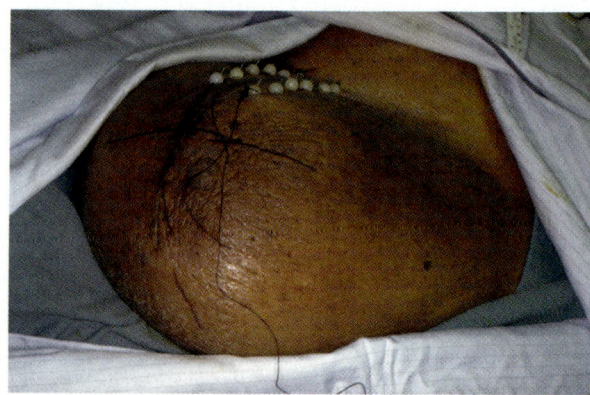

Fig. 19.18. Multiplane, multineedle implant

or 3.5–4.0 cm of HDR dwell positions are required beyond the surgical clips. Sources should not protrude or be located near the skin entry points to prevent late consequences. An acceptable implant results in a dose homogeneity index (ratio of peripheral dose to mean central dose) of at least 0.85. For LDH implants, 45 Gy is specified with an acceptable dose rate of 30–60 cGy/h. For HDR implants, 34 Gy is prescribed to the target volume and delivered in two fractions per day, separated by at least six hours, for a total of ten fractions of 3.4 Gy each.

19.13.4
MammoSite™

The MammoSite™ applicator may be placed during the local excision or a few weeks after the surgical procedure. The applicator should be selected to conform to the seroma cavity without large air pockets or residual seroma fluid. Ideally, the applicator is 1 cm or more from the skin surface, and focal skin necrosis has been noted with less than 7 mm separation from the surface of the skin to the surface of the implant. After verifying that no significant distortion of the applicator has occurred and that the source travels to the geometric center of the applicator, 34 Gy is prescribed at 1 cm from the balloon surface and delivered in two fractions per day, separated by at least six hours, for a total of ten fractions of 3.4 Gy each.

19.13.5
Conformal External Beam

Three to five non-coplanar photon beams have also been used to achieve partial breast irradiation. The clinical target volume (CTV) is defined by uniformly expanding the excision cavity volume by 10–15 mm. The CTV is limited to at least 5 mm from the skin surface and lung–chest wall interface. The CTV is further expanded an additional 10 mm (less if breathing motion studies have been performed) to generate a planning target volume (PTV). A total of 38.5 Gy is prescribed to the isocenter, in ten fractions of 3.85 Gy delivered BID with at least a six-hour interfraction interval. Ideally, less than 25% of the whole breast receives the prescribed dose and less than 50% of the whole breast receives 50% or more of the prescribed dose. The heart, contralateral lung, and contralateral breast should receive less than 5% of the dose (Fig. 19.19).

19.14
Postmastectomy Irradiation

19.14.1
Indications

Postmastectomy radiation continues to be an important therapy for patients with more advanced cancers. It substantially reduces locoregional recurrence rates and contributes to improved disease-specific survival. It is indicated in the following subset of patients (RECHT et al. 2001):
1. Four or more axillary lymph nodes
2. T3/T4 tumors with positive axillary nodes and patients with operable stage-III tumors
3. Positive margins or gross (>2mm) extranodal extension

In addition, postmastectomy radiation therapy may be indicated for selected patients with some of the following features:

1. Tumor is located in the central or inner quadrant and associated with positive axillary nodes
2. Lymphatic vascular space invasion (WHITE et al. 2004)
3. 1–3 Axillary lymph nodes (MARKS and PROSNITZ 1997)

19.15
Regional Nodal Failure

Since most of the recurrences in patients with high-risk breast cancer involve the supraclavicular fossa/axillary apex and/or chest wall, these are the two obligate targets for postmastectomy radiation therapy. Factors predictive of locoregional recurrence in the supraclavicular fossa/axillary apex parallel those for the chest wall and include the presence of four or more axillary lymph nodes, or more than 20% of the axillary lymph nodes involved with tumor. These patients have a 15–20% risk of failure in the supraclavicular fossa/axillary apex and should therefore be offered adjuvant radiation therapy to this region as well as to the chest wall (STROM et al. 2005).

The indications for regional nodal irradiation to the low–mid axilla and supraclavicular fossa/axillary apex remain controversial. At M.D. Anderson Cancer Center, 1031 patients treated with mastectomy, including an ALND and doxorubicin-based systemic therapy without radiation, were studied for

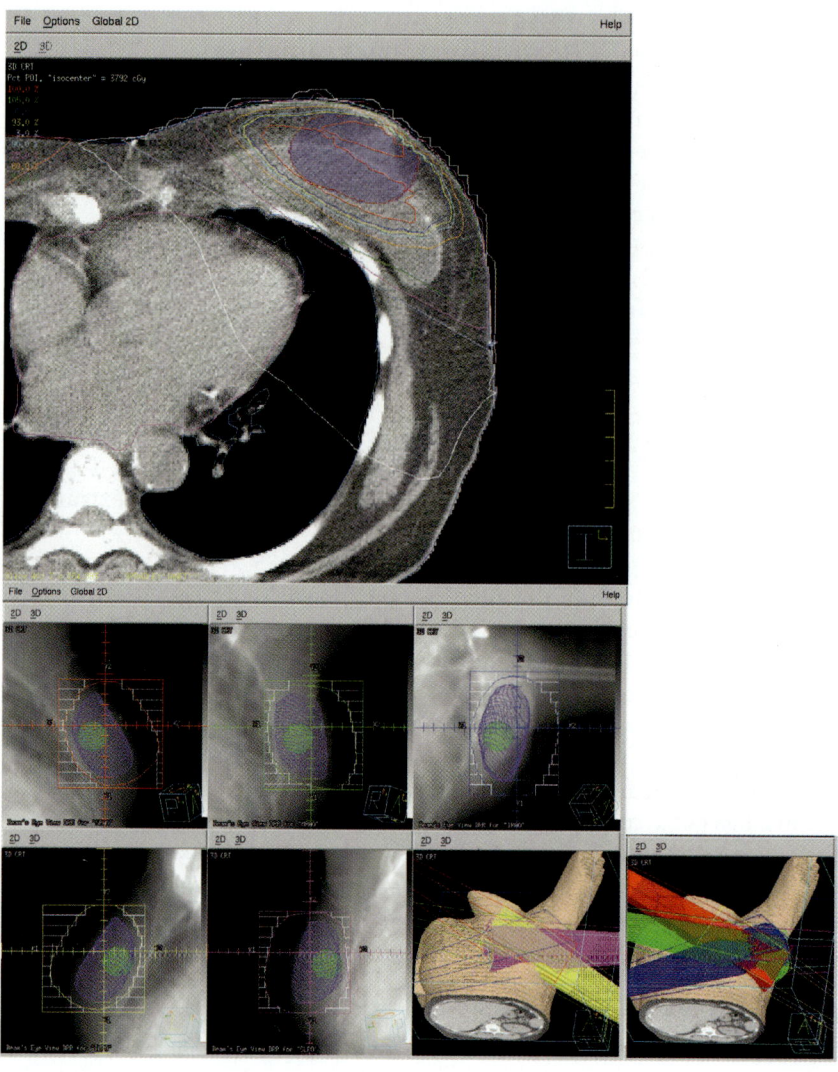

Fig. 19.19. 3-D conformal treatment plan for partial breast irradiation

regional nodal failure patterns. Failure in the low–mid axilla was an uncommon occurrence (3% at 10 years) and supplemental radiotherapy is probably not warranted for most patients (BUCHHOLZ 2000). The risk of failure was not higher for patients with increased number of axillary lymph nodes, increasing percentage of involved axillary lymph nodes, larger nodal size or gross extranodal extension than for patients who had these features.

19.15.1
Internal Mammary Chain Radiation – General Concepts

Treatment of the IMC, in the context of postmastectomy radiation therapy, continues to be a con-

troversial topic (FREEDMAN et al. 2000; BUCHHOLZ 2000). Although IMC recurrences rarely happen and are even more rarely detected, women with locally advanced breast cancer have rates of IMC lymph node involvement up to 50% (URBAN and MARJANI 1971). While it may not be useful to treat the IMC in the majority of early breast cancer patients, it should be considered for patients with advanced presentations, inner or central tumors, axillary node-positive disease, or early stage disease with primary drainage to the IMC on lymphoscintigraphy. Historically, two-dimensional (2D) fluoroscopic simulations did not allow direct visualization of the IMC. Conversely, with the advent of 3D CT simulation, accurate localization of these nodal regions is possible while minimizing cardiac toxicity.

19.16
Postoperative Chest Wall Tangential Photon Fields

Tangential fields are commonly used and are required when chest wall flaps are too thick or too irregular for electron beam therapy (Fig. 19.20). These fields usually abut an IM electron field, the use of which not only treats nodes but also decreases the amount of lung and/or heart in the tangents.

19.16.1
Field Borders

The medial border coincides with the lateral border of the IM field. Because of the chest wall curvature and tangential angle of the beam, the medial border rarely is a straight line. On most patients, the lateral border of the IM field is shaped to match this curvature. An overlap of 0.5 cm is acceptable to avoid a cold spot, and the IM electron field can be angled if necessary. The lateral border is at the midaxillary line. If the surgical scar extends to or beyond the midaxillary line, the lateral border is moved posteriorly to obtain adequate margins; and if moving the lateral border would require excessive radiation to lung or heart, an additional appositional electron beam field may be used for the portion of the scar beyond the midaxillary line. The superior border coincides with the lower border of the supraclavicular field. Since postmastectomy patients have been selected for irradiation because of high risk of locoregional recurrence, the contiguous line is treated by both chest wall and supraclavicular fields, which may result in a slight overdose. The inferior border is a horizontal line at the level of the tip of xiphoid, including scar extension with a 1- to 2-cm margin. For scars that extend further inferiorly than the level of the xiphoid, a separate electron beam field is used to treat that portion of the scar. The isocenter is placed in a location that permits easy validation of the set-up, usually offset medially from mid-separation. Otherwise, the technique is essentially that described for simulation of the intact breast.

19.16.2
Dosimetry

The dose prescribed is 50 Gy over 5 weeks at 2.0 Gy/fraction usually with 6-MV photons. The energy is

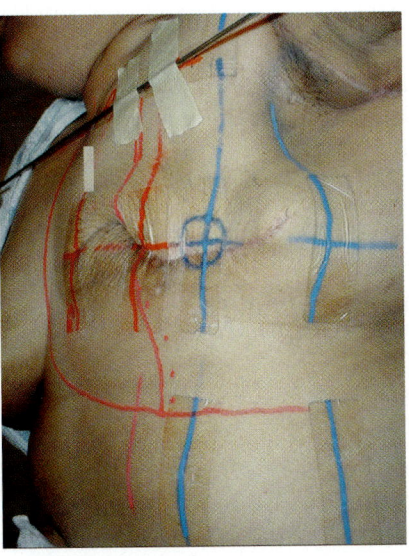

Fig. 19.20. Post-mastectomy patient with thick, irregular flaps which are best treated by photons. *Lines* on abdomen are for laser set-up. Note curve of *match line* of the internal mammary chain and chest wall fields and the *dotted lines* indicating the supraclavicular field

increased if medial lateral field separation for tangential fields exceeds 24 cm. Both medial and lateral fields are treated each day. The tumor depth dose percentage is determined from either contour distributions of the chest wall or by CT-based dosimetry to a point specified as two-thirds the distance from the skin to the lung/chest wall interface. Treatments with bolus are required to increase skin dose usually every other day for the first 2 weeks, then re-evaluated beginning after the next week. The chest wall adjacent to the scar is boosted an additional 10 Gy with electrons to include at least 3 cm on either side of the scar, again using bolus as necessary to achieve a brisk erythema.

In contrast to the standard postmastectomy setting, patients with locoregional recurrences have higher locoregional failure rates when treated to 45–50 Gy plus a boost (BALLO et al. 1999). Since the late 1990s, these patients have received a 10% dose escalation. All areas treated prophylactically are treated to 54 Gy in 27 fractions, and all areas boosted because of microscopic disease receive an additional 12 Gy in 6 fractions, totaling at least 66 Gy in the boost volume.

Patients with inflammatory breast cancer receive twice daily postmastectomy radiation. A dose of 51 Gy in 34 fractions is delivered at 1.5 Gy per fraction BID over 17 treatment days to the chest wall and regional lymph nodes, followed

by a 15-Gy chest wall boost in 10 fractions over a
period of 5 treatment days, totaling 66 Gy. Accelerated fractionation is used to treat most patients
unless the patient cannot comply with the twice-
daily schedule.

19.17
Electron Beam Chest Wall Fields

19.17.1
Field Borders

The entire chest wall may sometimes be treated
with electron beam fields. Although this technique may permit improved conformation of the
treatment volume to the target volume, there is an
increased risk of geographic miss of the tumor or
an excessive amount of lung irradiation (Fig. 19.21).
Patients with irregular chest surface contours make
treatment planning with this technique extremely
difficult. Mastectomy scar extension onto the
opposite breast, upper abdomen, or arm can also
be treated with electrons. Drain sites are included
if possible; however, recurrences are rarely noted
at these sites.

The medial border is contiguous with the lateral border of the IM field. The superior border is
contiguous with the inferior border of the supraclavicular field, at the level of the second costal
cartilage. The inferior border is a horizontal line

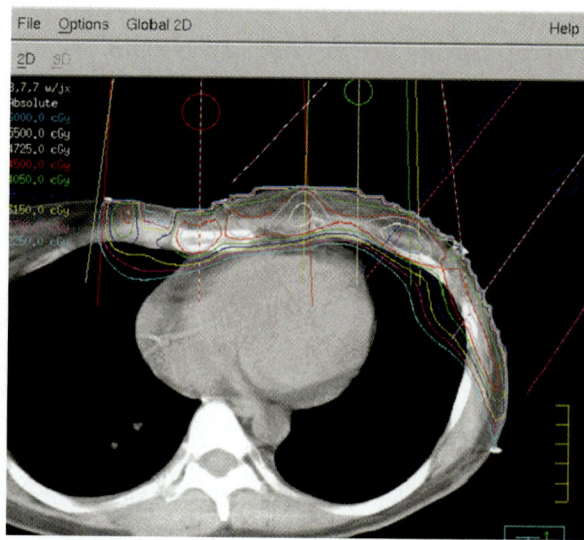

Fig. 19.21. Electron chest wall treatment plan

at the level of the xiphoid. The lateral border follows the midaxillary line. Because of the slope
of the chest wall, it is usually separated into two
fields; medial and lateral, with the lateral angled at
about 35°. The medial field is not angled in order
to avoid too large a "hot spot" from overlap into
the higher energy IMC field. To avoid dose buildup
at the junction of the medial and lateral chest wall
fields, the junction is moved 0.5 cm laterally each
week (Fig. 19.22).

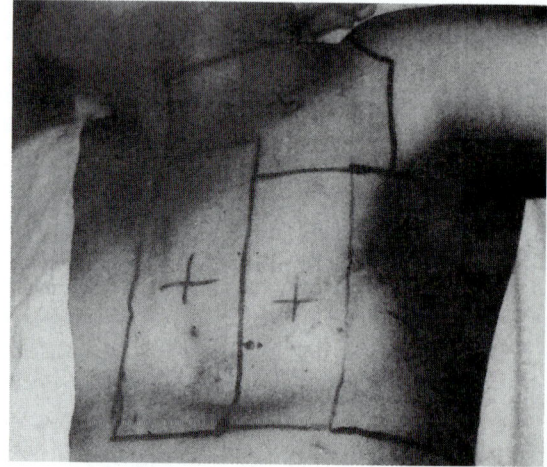

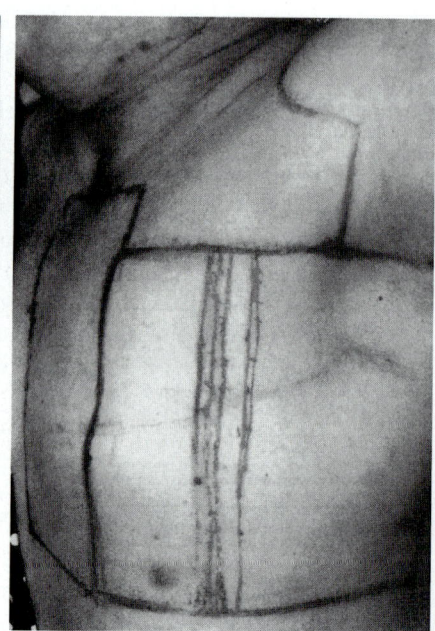

Fig. 19.22a,b. Chest wall and lymphatic fields in a patient with fairly curved
chest wall. Two separate low energy fields are used with the anterior field treated
straight on and the lateral field at a 35° lateral tilt, with junctions moved weekly.
a The appearance of the two fields at the beginning of treatment. **b** A patient
after 5 weeks of treatment, with junctions moved weekly

19.17.1.1
Dosimetry

The dose is 50 Gy to D_{max} over 5 weeks followed by a 10-Gy boost to a reduced field that encompasses the mastectomy scar and the original tumor bed.

19.17.1.2
Energy

Usually, 5–8 MeV is sufficient for most chest wall tissue after a modified radical mastectomy. CT planning is used to determine chest wall thickness to avoid harm to excess lung tissue. With the advent of the skin-sparing ability in newer accelerators, the use of bolus, usually 0.5 in thickness, for some of the fractions (for example, every other day, for the first 2 weeks of treatment) increases the surface dose for those few days. Re-evaluation of the skin reaction is carried out in the fourth week of treatment to consider the use of additional days of bolus. The objective is to obtain a degree of skin reaction commensurate with the stage of the original tumor yet not precipitate an unplanned treatment break. Bolus is also used during the boost treatment, unless the patient has already developed brisk erythema in the reduced field.

19.18
Partly-Deep Tangential Pair Fields

19.18.1
General Concepts

The partly-deep tangential approach targets the IMC along with all the chest wall, breast reconstruction or intact breast in a single pair of fields. This technique is used primarily in the context of intermediate to advanced breast cancers when a separate IMC field cannot be employed. Because more of the contralateral breast, ipsilateral lung and occasionally cardiac silhouette are included in the high dose volume than would be the case for a three-field approach, this technique is primarily useful for patients with unusual anatomical considerations. For example, this approach is appropriate for postmastectomy irradiation when an immediate reconstruction has been performed.

19.18.1.1
Treatment Planning

After acquisition of the planning CT in the usual manner, the IM vessels and the space medial to them are contoured in the first three intercostal spaces. If appropriate, other targets such as lumpectomy cavity or partly dissected axilla are also contoured. The upper portion of the tangential pair is designed to encompass the contoured IMC target with additional margin for set-up and breathing variation. Ideally, no more than 3 cm of lung is present in this portion of the tangential fields. The lower portion of the tangential pair is more shallow and is designed in the usual manner, except that the deep, nondivergent edge is delineated with a block or multileaf collimator (Fig. 19.23).

19.18.1.2
Dosimetry

The dose is 45–50 Gy at 1.8–2.0 Gy per fraction. The dose is specified at the pectoral surface with heterogeneity correction employed. Attention to the dose distribution in the intercostal spaces is required to prevent an underdose in this area. If desired, a dose–volume histogram (DVH) of the ipsilateral lung may be useful to assess for clinically relevant lung consequences.

19.18.1.3
Energy

A minimum of 6-MV photons are used, and higher energy photons should be considered for baseline separations greater than 23–25 cm. Field compensation using wedges or 3-D compensators are almost always required.

19.19
Anterior Supraclavicular–Axillary Photon Field

19.19.1
Field Borders

The anterior supraclavicular–axillary field covers the low and central axilla, subclavicular/axillary apex nodes, and the supraclavicular fossa, includ-

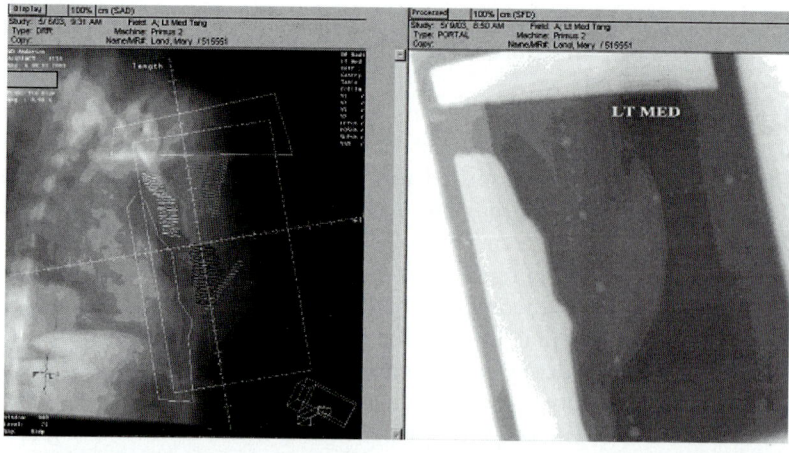

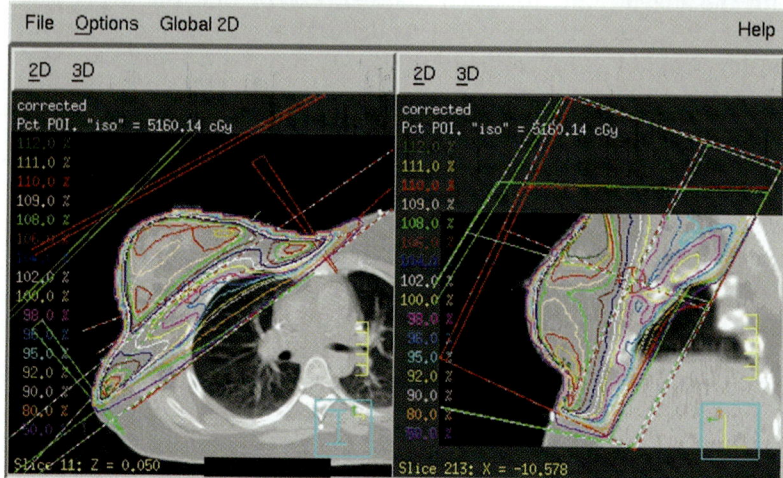

Fig. 19.23a,b. Partly deep tangent pair fields

ing the deep inferior cervical nodes and low jugular nodes. The medial border is a vertical line across the midline extending from the second costal cartilage to the thyrocricoid groove, following the inner border of the sternocleidomastoid muscle. The superior border extends laterally across the neck and the trapezius, just avoiding fall-off to the acromial process. The lateral border is at the acromioclavicular joint and is drawn across the shoulder to exclude the shoulder joint. The line then follows the pectoral fold, just avoiding fall-off. The inferior border is a horizontal line at the level of the second costal cartilage, abutting the medial edge of the IM field, which extends upward to over the first intercostal space. The junction of the inferior supraclavicular–anterior axially field and the chest wall is set up using a rod and chain and a half-beam block to avoid divergence (CHU et al. 1990). It is important to abduct the ipsilateral arm minimizing tissue folds in both the supraclavicular fossa and axilla in order to reduce erythema to the skin and other acute skin effects. For some patients, a supplemental dose using a posterior field may be necessary to the region of the mid-axilla since the level-III axilla is often deeper than the supraclavicular fossa.

The beam typically is tilted 15° laterally to avoid irradiating part of the trachea, esophagus, and spinal cord and to irradiate the deep inferior cervical nodes medial to the sternocleidomastoid muscle. The angulation additionally ensures that nodes close to the margin of the pectoral muscle are included without requiring fall-off of the beam, which may produce moist desquamation in the axilla.

19.19.2
Dosimetry

The dose is 50 Gy to D_{max} over 5 weeks in 25 fractions usually with 6-MV photons.

19.20
Supraclavicular and Axillary Apex Field (Electron Beam Technique)

The medial border extends superiorly from the superomedial corner of the IM field 1 cm across the midline to the level of the thyrocricoid groove. The superior border extends laterally across the neck and trapezius to the acromial process. The lateral border crosses the acromioclavicular joint and extends to meet the inferior border. The inferior border is a horizontal line extending laterally at the level of the second intercostal space. Electron beam energy is sufficient for 90% depth dose at 2- to 2.5-cm depth (usually 9–10 MeV). The dose is 50 Gy to D_{max} over 5 weeks, 5 fractions/week. A combination of electrons and photons usually (4:1) may be substituted to achieve less of a skin reaction. Patients with thicker body habitus requiring higher electron energies are probably better treated with photons alone to avoid apical nodes and to allow more skin sparing.

19.21
Internal Mammary Chain Radiation

With CT-based planning, the ipsilateral IMC can be contoured on the CT data set by visualizing the patient's anatomy in axial, coronal, and sagittal views (Fig. 19.24). Coverage can be accomplished by either extending the deep tangential field border to cover the IMC or creating a separate appositional IMC electron field that is matched to the edge of the medial tangential field. IMC nodes are treated with electron beam irradiation to avoid affecting underlying cardiac and mediastinal tissues. Our policy at the University of Texas M.D. Anderson Cancer Center (MDACC) does not include bilateral IMC irradiation for the following reasons: (1) experience with IMC metastases has been gained with ipsilateral surgical dissection, and contralateral parasternal nodules are extremely rare; (2) adequate irradiation to nodes of both IMCs would entail a 12-cm wide portal, which substantially increases the volume of mediastinal irradiation even with electron beam; and (3) contralateral IMC irradiation would interfere with future irradiation in those patients who develop a contralateral breast cancer.

19.21.1
Field Borders

The portal consists of a single rectangular electron beam field, usually 7 cm in width, covering the ipsilateral IMCs and the confluence of venous drainage under the insertion of the sternocleidomastoid muscle into the head of clavicle. The medial border is 1 cm across midline because of constriction of electron isodose curves at depth (Fig. 19.25). Thus, the upper border includes the head of the clavicle and the lower border is usually at the same level as the lower border of the chest wall. The usual IMC target includes the first three interspaces; however, the fields may be extended inferiorly depending on the drainage pattern seen on lymphoscintigraphy.

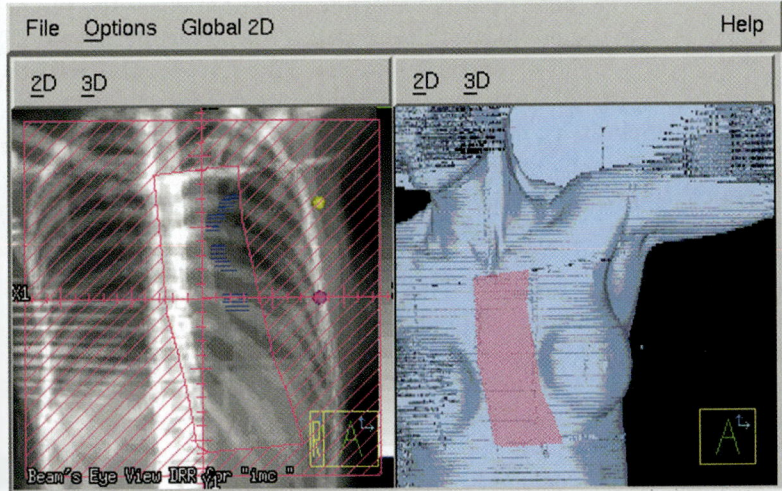

Fig. 19.24. Contouring of the internal mammary chain (*IMC*) and 3D treatment plan of an IMC treatment field

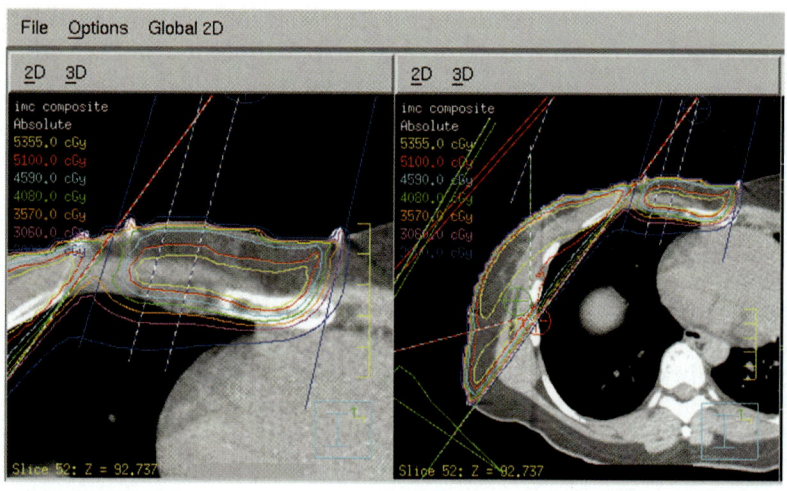

Fig. 19.25. Dosimetry of a non-overlapping internal mammary chain (*IMC*) and tangent pair

19.21.2
IM Dosimetry

The dose is 50 Gy to D_{max} over 5 weeks, 5 fractions/week, 2 Gy/fraction, when using electrons. It may not be possible to give the entire dose with electrons since skin doses vary according to the accelerator being used. As an alternative, a combination of electrons and photons may be necessary to achieve a lower dose to the skin. At least 50% of the total dose delivered should be given with electrons. Energy is sufficient for approximately 90% depth dose at 3–4 cm (usually 10- to 13-MeV electrons). CT planning or ultrasound is highly desirable to measure the depth to the pleural interface.

19.22
Axillary Irradiation

19.22.1
Indications

Specific radiation treatment of the axilla is rarely required since failure in this area is very rare after modified radical mastectomy and chemotherapy. Therefore, supplemental radiotherapy to the dissected axilla is not warranted in most patients. However, this technique is used: (1) anytime the axilla requires irradiation after simple or segmental mastectomy or excisional biopsy without axillary dissection, (2) for pre-operative irradiation [reserved for inoperable (N2) patients], or (3) in patients who require axillary irradiation because of inadequate axillary dissection (only a few nodes are recovered

and are positive). It may be considered for gross (>2 mm) extranodal axillary disease, or matted or fixed axillary nodes (N2 disease) at presentation.

19.23
Posterior Axillary Field

The machine is rotated 180° without moving the patient, and the thickness of the immobilization device is taken into account. The superior border splits the clavicle extending medially to inside the spine of the scapula. The verification film should show a small amount of lung in the upper inner portion of the field. The lateral superior border goes to the humoral head, as in the anterior field. The lateral and inferior borders match the anterior field. The beam is vertical.

The posterior axillary portal supplements the dose from the anterior portal to deliver the tumor dose to the thickest part of the axilla. Because axillary recurrences are rare after level-I and -I axillary dissection, and increases in the dose to the axilla can increase the risk of arm edema, the dose contribution from AP and PA fields should not exceed 2.0 Gy/fraction at midplane to minimize risk of arm edema and other late events.

For pre-operative treatment (designed to reduce matted or fixed nodes to facilitate axillary dissection), a midplane dose of 50 Gy is necessary. For postoperative patients in whom extranodal extension in the axilla is found, but no gross residual disease remains, the midaxillary dose is reduced to 40–45 Gy in an attempt to decrease risk of arm edema.

19.24
Postmastectomy Boost

19.24.1
General Concepts

The target for boost in the postmastectomy setting are the central operative flaps. Rigid conformation to the incision site – the effect of sometimes naming this a "scar" boost – is not required. Boosts are typically done in the postmastectomy setting using electron energies so that the isodose line encompasses the target volume.

19.24.2
Dosimetry

The boost dose is used to bring the total tumor dose to a minimum of 60 Gy.

Acknowledgements
This work was supported by a generous grant from the Stanford and Joan Alexander foundation.

References

Ballo MT, Strom EA, Prost H et al (1999) Local-regional control of recurrent breast carcinoma after mastectomy: does hyperfractionated accelerated radiotherapy improve local control? Int J Radiat Oncol Biol Phys 44:105–112

Bartelink H, Horiot JC, Poortmans P et al (2001) Recurrence rates after treatment of breast cancer with standard radiotherapy with or without additional radiation. N Engl J Med 345:1378–1387

Buchholz TA (2000) Internal mammary lymph nodes: to treat or not to treat. Int J Radiat Oncol Biol Phys 46:801–803

Chu JC, Solin LJ, Hwang CC et al (1990) A nondivergent three field matching technique for breast irradiation. Int J Radiat Oncol Biol Phys 19:1037–1040

Crowe J, Gordon N, Antunez A et al (1991) Local-regional breast cancer recurrence following mastectomy. Arch Surg 126:429–432

Devesa SS, Blot WJ, Stone BJ et al (1995) Recent cancer trends in the United States. J Natl Cancer Inst 87:175–182

Early Breast Cancer Trialists Cooperative Group (2003) Radiotherapy for early breast cancer. Cochrane Library, Issue 4

Feuer EJ, Wun LM, Boring CC et al (1993) The lifetime risk of developing breast cancer. J Natl Cancer Inst 85:892–897

Fisher B, Anderson S, Bryant J et al (2002) Twenty-year follow-up of a randomized trial comparing total mastectomy, lumpectomy, and lumpectomy plus irradiation for the treatment of invasive breast cancer. N Engl J Med 347:1233–1241

Freedman GM, Fowble BL, Nicolaou N et al (2000) Should internal mammary lymph nodes in breast cancer be a target for the radiation oncologist? Int J Radiat Oncol Biol Phys 46:805–814

Grann A, McCormick B, Chabner ES et al (2000) Prone breast radiotherapy in early-stage breast cancer: a preliminary analysis. Int J Radiat Oncol Biol Phys 47:319–325

Haagensen C (1971) Disease of the breast, 2nd edn. Saunders, Philadelphia, PA, pp 576–584

Haagensen CD (1986) Diseases of the breast, 3rd edn. Saunders, Philadelphia

Hellman S, Harris JR (2000) Natural history of breast cancer. In: Harris J, Lippman ME, Morrow M, Osborne CK (eds) Diseases of the breast. Lippincott Williams and Wilkins, Philadelphia, p 407

Hong L, Hunt M, Chui C et al (1999) Intensity-modulated tangential beam irradiation of the intact breast. Int J Radiat Oncol Biol Phys 44:1155–1164

Hurd TC, Sneige N, Allen PK et al (1997) Impact of extensive intraductal component on recurrence and survival in patients with stage I or II breast cancer treated with reast conservation therapy. Ann Surg Oncol 4:119–124

Jemal A, Murray T, Ward E et al (2005) Cancer statistics, 2005. CA Cancer J Clin 55:10–30

Katz A, Strom EA, Buchholz TA et al (2000) Locoregional recurrence patterns after mastectomy and doxorubicin-based chemotherapy: implications for postoperative irradiation. J Clin Oncol 18:2817–2827

Keurer HM, Julian TB, Strom EA et al (2004) Accelerated partial breast irradiation after conservative surgery for breast cancer. Ann Surg 239:338–351

Krag D, Weaver D, Ashikaga T et al (1998) The sentinel node in breast cancer a multicenter validation study. N Engl J Med 339:941–946

Kurtz JM, Jacquemier J, Amalric R et al (1990) Risk factors of breast recurrence in premenopausal and postmenopausal patient with ductal cancers treated by conservation therapy. Cancer 65:1867–1878

Kurtz JM (2001) Which patients don't need a tumor bed boost after whole-breast radiotherapy? Strahlenther Onkol 177:33–36

Lingos TI, Recht A, Vicini F et al (1991) Radiation pneumonitis in breast cancer patients treated with conservative surgery and radiation therapy. Int J Radiat Oncol Biol Phys 21:355–360

Marks LB, Prosnitz LR (1997) "One to three" or "four or more"? Selecting patients for postmastectomy radiation therapy. Cancer 79:668–670

Morrow M, Strom EA, Basset LW et al (2002) Standard for breast conservation therapy in the management of invasive breast carcinoma. CA Cancer J Clin 52:277–300

Osborne MP (1987) Breast development and anatomy. In: Harris JR, Hellman S, Henderson IC, Kinne DW (eds) Breast disease. Lippincott, Philadelphia

Parkin DM, Muir CS (1992) Cancer incidence in five continents. Comparability and quality of data. IARC Sci Publ 120:45

Recht A, Edge SB, Solin LJ et al (2001) Postmastectomy radiotherapy: clinical practice guidelines of the American Society of Clinical Oncology. J Clin Oncol 19:1539–1569

Schlembach PJ, Buchholz TA, Ross MI et al (2001) Relationship of sentinel and axillary level I-II lymph nodes to tangential fields used in breast irradiation. Int J Radiat Oncol Biol Phys 51:671–678

Schnitt SJ, Abner A, Gelman R et al (1994) The relationship between microscopic margins of resection and the risk of local recurrence in patients with breast cancer treated with breast-conserving surgery and radiation therapy. Cancer 74:1746–1751

Shahar KH, Buchholz TA, Delpassand E et al (2005) Lower and central tumor location correlates with lymphoscintigraphy drainage to the internal mammary lymph nodes in breast carcinoma. Cancer 103:1323–1329

Sickles EA. (1995) Management of probably benign breast lesions. Radiol Clin North Am 33:1123–1123

Silverstein MJ, Gierson ED, Waisman JR et al (1995) Predicting axillary node positivity in patients with invasive carcinoma of the breast by using a combination of T category and palpability. J Am Coll Surg 180:700–704

Singletary SE, Allred C, Ashley P et al (2002) Revision of the American Joint Committee on Cancer Staging System for Breast Cancer. J Clin Oncol 20:3628–3636

Strom EA, Woodward WA, Katz A et al (2005) Regional nodal failure patterns in breast cancer patients treated with mastectomy without radiotherapy. Int J Radiat Oncol Biol Phys 63:1508–1513

Urban JA, Marjani MA (1971). Significance of internal mammary lymph node metastases in breast cancer. Am J Roentgenol Radium Ther Nucl Med 111:130–136

Van Leeuwen F, Klokman W, Aleman B (2000) Long-term risk of second malignancy in survivors of Hodgkin's disease treated during adolescence or young adulthood. J Clin Oncol 18:487–497

Veronesi U, Luini A, Galimberti V (1990) Extent of metastatic axillary involvement in 1446 cases of breast cancer. Eur J Surg Oncol 16:127–133

Vrieling C, Collette L, Fourquet A et al (1999) The influence of the boost in breast-conserving therapy on cosmetic outcome in the EORTC "boost versus no boost" trial. EORTC Radiotherapy and Breast Cancer Cooperative Groups. European Organization for Research and Treatment of Cancer. Int J Radiat Oncol Biol Phys 45:677–685

White J, Moughan J, Pierce LJ et al (2004) Status of postmastectomy radiotherapy in the United States: a patterns of care study. Int J Radiat Oncol Biol Phys 60:77–85

Wolmark N, Risher B (1981) Surgery in the primary treatment of breast cancer. Breast Cancer Res Treat 1:339–348

20 Carcinoma of the Esophagus

JEFFREY D. BRADLEY and SASA MUTIC

CONTENTS

20.1   Natural History of the Disease   511
20.1.1 Anatomical Classification and Staging   511
20.1.2 Prognosis and Predictive Factors   512
20.1.3 General Management of Carcinoma of
       the Esophagus   513
20.1.4 Palliation   514
20.1.5 Postoperative Adjuvant Radiation Therapy   515
20.1.6 Preoperative Chemoradiation Therapy   515
20.1.7 Definitive Chemoradiation Therapy   516
20.1.8 Preoperative Chemoradiation Versus
       Definitive Chemoradiation Therapy   517
20.1.9 Superficial Esophageal Cancer   518
20.2   Radiation Therapy Techniques   518
20.3   Simulation Procedures   518
20.4   Target Volume Definition   518
20.5   Dose Prescription   519
20.6   Beam Selection and Design   519
20.7   Plan Evaluation   520
20.8   Plan Implementation and Quality Assurance   521
20.9   Simulation Films and Portal Films   522
20.10 Future Directions   522
       References   522

20.1
Natural History of the Disease

Esophageal cancer constitutes 3% of all carcinomas; approximately 14,000 patients are diagnosed per year in the United States (CANCER STATISTICS 2003). Unfortunately, most die of their disease. Maximum incidence occurs in males between the ages of 60 years and 70 years. The ratio of males to females is approximately 2–4:1 for squamous cell carcinomas and 7:1 for adenocarcinomas (CREW and NEUGUT 2004).

J. D. BRADLEY, MD
Department of Radiation Oncology, Washington University School of Medicine, 4921 Parkview Place – Lower Level, Mail Stop #90-38-635, Saint Louis, MO 63110, USA
S. MUTIC, MS
Department of Radiation Oncology, Washington University School of Medicine, Saint Louis, MO 63110, USA

There has been a dramatic shift in the distribution and histology of esophagus and gastric cancers in Western countries over the past 25 years. Whereas, squamous cell carcinoma of the upper or mid esophagus and adenocarcinoma of the gastric body were previously more common, adenocarcinomas of the distal esophagus, gastroesophageal (GE) junction, and gastric cardia now predominate. In Caucasian males of the United States and other Western countries, the incidence of adenocarcinoma of the esophagus has risen faster than that of any other malignancy (BLOT et al. 1991). The reasons for this epidemiological shift are not completely clear.

Radiation therapy is commonly employed for either primary or palliative management and is an important means to obtain local control. Patients typically present with obstructive symptoms and weight loss. Local tumor progression leads to complete obstruction, inability to swallow liquids or saliva, and invasion of adjacent structures (i.e., trachea or bronchus). Thus, local control efforts, including radiation therapy, contribute significantly to a patient's quality of life.

In this chapter, we will present the techniques of delivering radiation therapy to the esophagus.

20.1.1
Anatomical Classification and Staging

The esophagus is a muscular tube measuring approximately 25 cm in length. During endoscopic examinations, lengths are typically measured from the incisors. Distances vary with patient size, with the carina at approximately 25 cm and the GE junction at 40 cm. Segments of the esophagus are traditionally surgically defined. The cervical portion extends from the bottom of the cricoid to the suprasternal notch. The upper and middle thoracic esophagus extends from the suprasternal notch to the inferior pulmonary vein. The lower thoracic esophagus extends from the inferior pulmonary vein to the diaphragm, and the abdominal esophagus extends from the diaphragm to the GE junction.

Pathological staging classification is given in Table 20.1. Carcinoma of the esophagus commonly spreads submucosally along the muscular layers and can extend for lengths of 10 cm or more. The esophagus has no serosal layer, so extension outside of the esophagus and to adjacent structures is fairly common. The esophagus also has a rich lymphatic supply. Lymph nodes with metastatic carcinoma can be involved throughout the length of the structure. Figure 20.1 shows the patterns of lymph-node involvement according to the location of the primary tumor in the surgical series by AKIYAMA et al. (1981), employing an extensive lymph-node dissection. Most commonly, cancers of the distal esophagus or GE junction involve lymph nodes in the periesophageal, gastrohepatic, or celiac nodal regions. Thoracic esophagus lesions typically involve mediastinal nodal stations and may spread to involve the supraclavicular and/or abdominal lymph nodes.

The staging workup includes a history and physical examination, routine laboratories, barium swallow, upper endoscopy, and a computed tomography (CT) scan of the chest. Bronchoscopy is indicated for patients with upper and mid esophageal cancers adjacent to the trachea or bronchus. More recently, the staging workup has expanded to include endoscopic ultrasound (EUS) and positron emission tomography (PET). The primary role of EUS is to determine the depth of tumor invasion into or through the wall of the esophagus, particularly for separating T1–2 from T3–4 lesions (Fig. 20.2). EUS has been shown to have accuracy, sensitivity, and specificity values of 87, 91, and 82%, respectively, for detecting T3–4 disease (RICE et al. 2003a). EUS also has the advantage of providing access to regional lymph nodes for biopsy. EUS with fine-needle aspiration (FNA) of lymph nodes has accuracy, sensitivity, and specificity values of 93, 100, and 93%, respectively, compared with 63, 81, and 70%, respectively, for EUS alone (VASQUEZ-SEQUEIROS et al. 2001). However, EUS is not possible in patients with complete obstruction. EUS–FNA of lymph nodes is not possible if the nodes are distant from the esophageal lumen.

PET is helpful for assessing the primary cancer and distant nodal or other metastases. In nearly all studies, PET with 2-[18F]fluoro-2-deoxy-D-glucose (FDG-PET) has been shown to detect primary esophagus cancer with a higher sensitivity than that of CT (95–100% versus 81–92%) (FLANAGAN et al. 1997; LUKETICH et al. 1997; FLAMEN et al. 2000; KOLE et al. 1998; MCATEER et al. 1999; YEUNG et al. 1999). However, it may not be very helpful for cancers that are limited to the mucosa (HIMENO et al. 2002).

With regard to distant metastases, FDG-PET may detect an additional 3–37% of metastases compared with conventional staging methods (FLANAGAN et al. 1997; LUKETICH et al. 1997; KOLE et al. 1998).

20.1.2
Prognosis and Predictive Factors

Treatment decisions regarding surgical versus nonsurgical approaches for esophagus cancer have not depended on the location of the primary cancer within the esophagus. The primary determinants of outcome are the degree of invasion (T stage), nodal status (N stage), number of regional nodes involved (IGAKI et al. 2003), presence of distant metastases (M status) (CHAU et al. 2004), histology, patient-performance status (CHAU et al. 2004), weight loss (THOMAS et al. 2004), and medical operability.

Table 20.1. Tumor node metastasis staging for esophageal cancer

Stage	Definition
Primary tumor (T)	
Tx	Primary tumor cannot be assessed
T0	No demonstrable tumor
TIS	Carcinoma in situ
T1	Tumor invades the lamina propria or submucosa
T2	Tumor invades the muscularis propria
T3	Tumor invades the adventitia
T4	Tumor invades adjacent structures
Regional lymph nodes (N)	
Nx	Regional lymph nodes cannot be assessed
N0	No regional lymph node involvement
N1	Regional lymph node metastasis
Distant metastasis (M)	
MX	Distant metastases cannot be assessed
M0	No distant metastases
M1	Distant metastases
Tumors of the lower thoracic esophagus	
M1a	Metastasis in celiac lymph nodes
M1b	Other distant metastasis
Tumors of the mid-thoracic esophagus	
M1a	Not applicable
M1b	Non-regional lymph nodes and/or other distant metastasis
Tumors of the upper thoracic esophagus:	
M1a	Metastasis in the cervical lymph nodes
M1b	Other distant metastasis
Clinical-diagnostic classification for cervical esophagus	
Stage 0	TIS, N0, M0
Stage I	T1, N0, M0
Stage IIA	T2, N0, M0, T3, N0, M0
Stage IIB	T1, N1, M0, T2, N1, M0
Stage III	T3, N1, M0, T4, Any N, M0
Stage IVA	Any T, any N, M1a
Stage IVB	Any T, any N, M1b

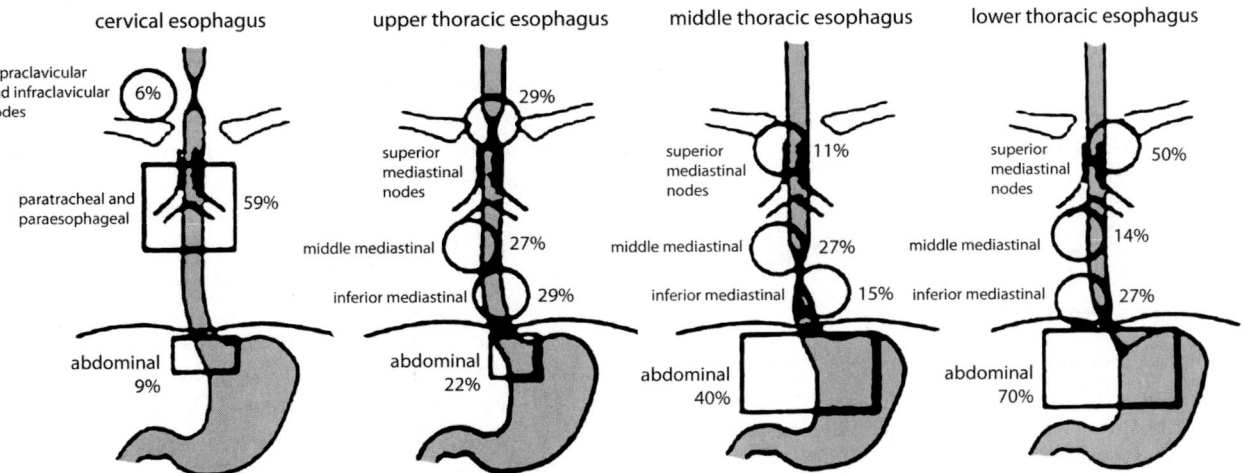

Fig. 20.1. Positive lymph-node distribution according to the location of the primary tumor. (Modified from AKIYAMA et al. 1981 and DORMANS 1939, with permission)

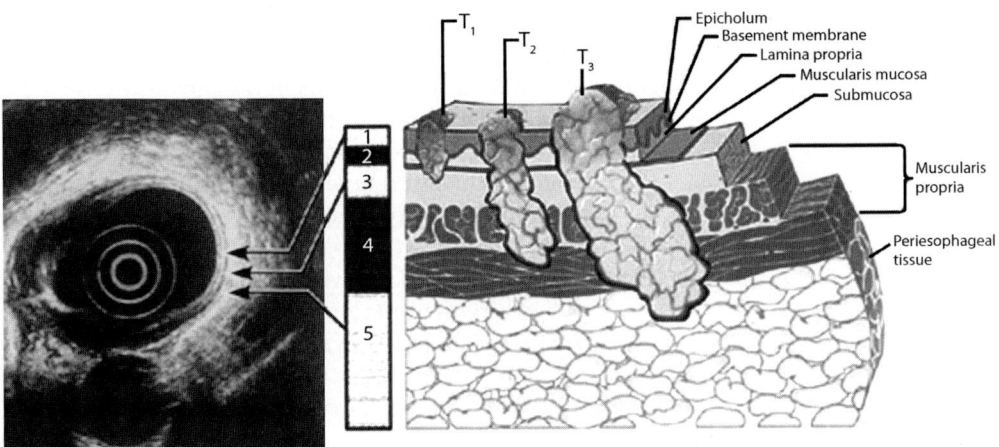

Fig. 20.2. The esophageal wall is visualized as five alternating layers of differing echogenicity by means of endoscopic ultrasound. The fourth ultrasonographic layer is critical to differentiating T1, T2, and T3 carcinomas. *Left and middle* The first (inner) layer is hyperechoic (white) and represents the superficial mucosa (epithelium and lamina propria). The second layer is hypoechoic (black) and represents the deep mucosa (muscularis mucosae). The third layer is hyperechoic and represents the submucosa. The fourth layer is hypoechoic and represents the muscularis propria. The fifth layer is hyperechoic and represents the periesophageal tissue. *Right* T1 carcinoma: no invasion beyond the submucosa. The tumor is confined to the first three ultrasonographic layers. It does not involve the fourth ultrasonographic layer. T2 carcinoma: invasion into but not beyond the muscularis propria. The tumor is confined to the fourth ultrasonographic layer. T3 carcinoma: invasion into the periesophageal tissue. The tumor breaches the fourth ultrasonographic layer (RICE et al. 2003a)

20.1.3
General Management of Carcinoma of the Esophagus

In a very broad sense, the treatment aim can be divided into curative and palliative. In both categories, all three modalities of radiotherapy, surgery, and chemotherapy can be used in combination with one another. In general, patients with T1–4, N0–1 carcinomas are considered curable. There is some discrepancy as to whether patients with M1a disease are potentially curable. The paucity of literature on this group suggests that patients with distal esophagus cancers involving the abdominal nodes are not curable. RICE et al. (2003b) reported institutional esophagectomy results and showed no difference in survival between patients with M1a and M1b disease (Fig. 20.3). Likewise, the two nonoperative trials defining the role of chemoradiation for esophageal cancer, Radiation Therapy Oncology Group (RTOG) 8501 and 9205, allowed but did not enroll any patients with M1a disease (personal com-

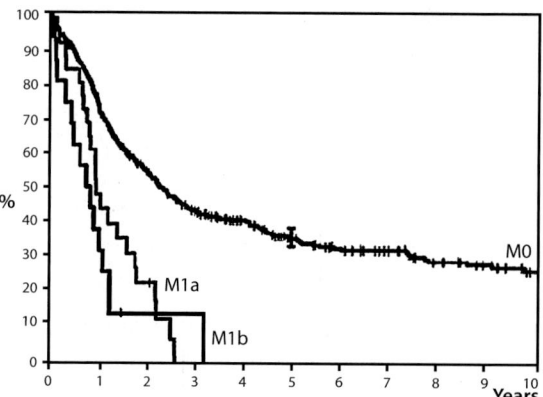

Fig. 20.3. Survival after esophagectomy stratified by M1a (*n*=26, none alive at 5 years) and M1b (*n*=16, none alive at 5 years) (RICE et al. 2003b)

munication with RTOG statistical group). While the results of these two RTOG trials are discussed later in this chapter, the relatively poor overall survival in patients with M0 disease indicates that patients with M1a disease have negligible 5-year survival rates. More investigation is needed as to whether this patient population is potentially curable.

20.1.4
Palliation

Radiation therapy is often used for palliation of obstructive symptoms in patients with carcinoma of the esophagus. Most patients with esophageal cancer present with locally advanced or metastatic disease. Of those treated with curative intent, approximately 80% will eventually require palliation due to symptoms from locally recurrent disease (HANCOCK and GLATSTEIN 1984). The options for treatment include external beam radiation therapy, chemotherapy and radiation therapy, intracavitary brachytherapy, and esophageal stent placement. The goal of palliation is to relieve symptoms and to avoid causing moderate to severe side effects in the process. The decision of which palliative option to choose can be complicated, especially as merging treatment modalities increase the rate of expected side effects.

Metallic esophageal stents and endoluminal brachytherapy are both valid options for palliation of esophageal obstruction. Stents are placed during endoscopy and may be coated. They come in varying diameters and lengths. Stents have the advantage of producing rapid results. A potential disadvantage is that patients may experience GE reflux symptoms if the stent traverses the GE junction. Stents with anti-

reflux valves are available, though they may not prevent acid reflux (HOMS et al. 2004).

Endoluminal brachytherapy is a good option for patients with an esophageal lesion that can be traversed during endoscopy. In the United States, high dose rate (HDR) catheters are available in diameters of 6 mm and 12 mm. These should be placed under endoscopy to minimize the risk of perforation or bleeding from the procedure. Metallic clips can be placed by the endoscopist at the proximal and distal ends of the lesion. These are visualized during fluoroscopy to facilitate prescribing the length of the esophagus to be irradiated. These clips often remain in place and are used for subsequent dose fractions. Brachytherapy doses range from single fractions of 12 Gy to three fractions of 5–6 Gy each.

HOMS et al. (2004) recently published a randomized comparison of single-dose brachytherapy versus metal stent placement for palliation of dysphagia from nine centers in the Netherlands. Patients on the HDR brachytherapy arm received a single dose of 12 Gy prescribed to 1 cm depth using Iridium[192]. Dysphagia improved more rapidly after stent placement than after brachytherapy, but longterm relief of dysphagia was better after brachytherapy. There was no difference in overall survival between groups. However, the dysphagia-adjusted survival was superior for brachytherapy for time points beyond 120 days from treatment. Major complications occurred in 25% of patients with stents and 13% of patients receiving brachytherapy. SUR et al. (2002) published a prospective randomized trial from the International Atomic Energy Agency, comparing brachytherapy schemes of 16 Gy in two fractions versus 18 Gy in three fractions. Doses were prescribed to a 1-cm depth and were given 1 week apart. Both fractionation schemes resulted in similar dysphagia-free survival and complication rates. Figure 20.4 shows the improvement in dysphagia by treatment scheme and shows whether or not the patient was experiencing symptoms at the time of the procedure. Based on this trial, SUR el al. conducted a second randomized prospective study of brachytherapy alone (8 Gyx2 fractions) versus brachytherapy followed by external beam radiation therapy (3 Gyx10 fractions) (SUR et al. 2004). There was no benefit to the addition of external beam radiation therapy following brachytherapy.

External beam radiation therapy can effectively palliate pain and dysphagia in 60–80% of patients (WARA et al. 1976; ROSENBERG et al. 1981). Doses range from 30 Gy in 10 fractions over 2 weeks to 50 Gy in 25 fractions over 5 weeks. The fraction-

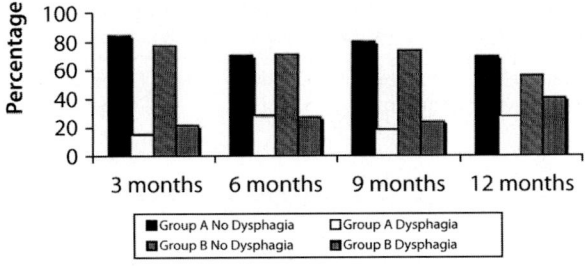

Fig. 20.4. Improvement in dysphagia after brachytherapy. *Black bars* group A without dysphagia; *white bars* group A with dysphagia; *cross-hatched bars* group B without dysphagia, group B with dysphagia (Sur et al. 2002)

ation schema employed should consider whether or not chemotherapy is planned. Fraction sizes larger than 2 Gy given with concurrent chemotherapy may cause unwanted side effects in this debilitated patient population. The presence of a tracheoesophageal fistula is a relative contraindication for radiation therapy, though limited data suggest that such therapy may be safe (Muto et al. 1999). At the Siteman Cancer Center, we routinely require bronchoscopy and request that a tracheal or endobronchial stent be placed if the patient is suspected of having a tracheobronchial fistula.

20.1.5
Postoperative Adjuvant Radiation Therapy

Adjuvant radiation therapy following surgical resection for tumors of the upper, mid, or lower esophagus is generally used for patients with positive longitudinal margins or gross residual disease. A tumor at the circumferential resection margin is a controversial indication with limited data available. For cancers involving or limited to the esophagogastric junction, postoperative therapy is given for patients with T2 or node-positive disease.

There have been two randomized trials addressing the value of postoperative adjuvant radiation therapy for squamous cell carcinoma (Teniere et al. 1991; Fok et al. 1993). Chemotherapy was not used in either trial. In the French trial, 221 patients having a complete resection were randomized to adjuvant radiation therapy or not (Teniere et al. 1991). The presence of regional lymph-node involvement was similar in both groups. Adjuvant radiation therapy did not improve survival in either the N0 or N+ subgroups. Fok et al. randomized 130 patients following either curative or palliative esophagectomy to adjuvant radiation therapy or not (Fok et al. 1993). Patients

with residual tumor in the mediastinum following surgery benefited from radiation therapy through reduced local recurrence rates. However, there was a survival benefit in favor of surgery alone due to excess complications seen in the radiation therapy group. Taken together, these two trials indicate that the role of radiation therapy for squamous cell carcinoma of the esophagus following surgery is limited to patients with residual microscopic or gross disease.

There have been no randomized trials addressing adjuvant radiation therapy for adenocarcinomas of the esophagus. However, a prospective randomized intergroup study testing adjuvant chemotherapy and radiation therapy for gastric and esophagogastric junction adenocarcinomas showed a survival benefit for patients receiving adjuvant chemoradiation compared with surgery alone (Macdonald et al. 2001). With respect to esophagus cancer, this trial is applicable for tumors involving the esophagogastric junction. This trial randomized 556 patients to postoperative chemotherapy and radiation therapy or not. Inclusion criteria included patients with stages IB through IV (M0) disease, a performance status of two or lower, and adequate daily caloric intake. Approximately 20% of patients had GE-junction cancers. Analysis of the whole group demonstrated 3-year survival rates of 50% and 41% (P=0.005) for the adjuvant therapy and control groups, respectively. Outcomes specific to the subgroup of patients with GE-junction tumors have not been reported.

20.1.6
Preoperative Chemoradiation Therapy

Preoperative chemoradiation therapy has become routine for patients with either T3, N1, or M1a carcinomas that are otherwise operable. Several randomized, controlled studies have been published comparing preoperative chemoradiation therapy followed by esophagectomy to esophagectomy alone (Nygaard et al. 1992; Apinop et al. 1994; LePrise et al. 1994; Walsh et al 1996; Busset et al. 1997, Urba et al. 2001). In general, these trials show that preoperative therapy increases the resectability rate. However, collectively these trials have failed to show a convincing survival advantage to preoperative therapy. One reason for this is because of a lack of statistical power due to the small size of the trials. Fiorica et al. (2004) recently published a meta-analysis of these six trials and showed a survival advantage in favor of preoperative therapy (Tables 20.2 and 20.3).

Table 20.2. Therapeutic regimen of all trials included in the meta-analysis of chemoradiation followed by surgery versus surgery alone. *BED* biological equivalent dose, *CDDP* cisplatin, *BLM* bleomycin, *5-FU* 5-fluorouracil, *VNB* vinblastine

Study (reference)[a]	Total dose (Gy)	Fractions (no/days)	Daily dose (Gy)	BED (Gy)	BED corrected by time (Gy)	Drugs	Dosage (mg/m^2)	Schedules (day)	Interval between end of irradiation and surgery (weeks)
1) NYGAARD et al. 1992	35	20/28	1.75	41.12	26.7[b]	CDDP, BLM	20/5	1–5; 15–19	3
2) LE PRISE et al. 1994	20	10/12	2	24	17.8[b]	CDDP, 5FU	100/600	1; 21, 2–5; 22–25	2.5
3) APINOP et al. 1994	40	20/28	2	48	33.6[b]	CDDP, 5FU	100/1000	1; 29, 1–4; 29–32	4
4) WALSH et al. 1996	40	15/21	2.67	50.7	42.6[c]	CDDP, 5FU	75/15 mg/kg	7; 42, 1–5; 36–40	2
5) BOSSET et al. 1997	37	10/24	3.7	50.7	38.4[b]	CDDP, CDDP	80/20	0–2; 19–21, 1–5; 17–21	2–4, 3
6) URBA et al. 2001	45	30/21	1.5x2	51.7	43.6[c] adeno-carcinoma 40.9[b] squamous carcinoma	CDDP, 5FU, VNB	20/300/1	1–5; 17–21, 1–21, 1–4; 17–20	3

[a] For expansion of the study names, see corresponding reference
[b] $Tpot_{squamo}$ = 4.5 days   [c] $Tpot_{adeno}$ = 6 days, α = 0.3

Table 20.3. Results of meta-analysis of chemoradiation followed by surgery versus surgery alone. *OR* odds ratio, *CI* confidence interval

Study	Treatment (n/N)	Control (n/N)	OR (95% CI random)	Weight (%)	OR (95% CI random)
NYGAARD et al. 1992	44/53	40/50		13.4	0.54 [0.17, 1.75]
APINOP et al. 1994	26/53	27.34		14.1	0.75 [0.24, 2.31]
LE PRISE et al. 1996	33/41	39/45		13.6	0.63 [0.20, 2.02]
WALSH et al. 1996	39/58	52/55		11.7	0.12 [0.03, 0.43]
BOSSET et al. 1997	94/151	97/146		30.2	0.83 [0.52, 1.34]
URBA et al. 2001	34/50	42/50		17.1	0.40 [0.15, 1.06]
Total (95% CI)	270/388	302/380		100.0	0.53 [0.31, 0.89]
Test for heterogeneity χ^2=8.84, d f=5 P=0.12					
Test for overall effect, z=-2.41, P=0.02					

0.01   0.1   1   10   100
Better CRT + surgery   Better surgery

20.1.7
Definitive Chemoradiation Therapy

Medically inoperable patients with localized or locally advanced regional disease are best managed with definitive chemotherapy and radiation therapy. Patients with locally advanced esophageal cancers who are ineligible for chemotherapy are eligible for palliative radiation therapy alone. Three randomized trials have compared radiation therapy alone with concurrent chemotherapy and radiation therapy (ARAUJO et al. 1991; ROUSSELL et al. 1988; AL-SARRAF et al. 1996) (Table 20.4). Two of these trials suffered from design flaws. They were underpowered and employed suboptimal doses of chemotherapy. In RTOG 8501, patients were randomized to 64 Gy of radiation alone versus 50 Gy of radiation concurrent with chemotherapy. The

chemotherapy consisted of four cycles of infusional 5-FU (1,000 mg/m^2) given over 4 days and cisplatinum (75 mg/m^2) given on day 1 of weeks 1, 5, 8, and 11. The results demonstrated a survival advantage for concurrent chemotherapy and radiation therapy despite the reduced radiation dose. Because of a large difference in survival between the two arms, the trial was stopped early, and an additional number of patients received chemoradiation on a third confirmatory arm. At 5 years, there were no survivors on the radiation therapy alone arm versus 30% on the chemoradiation arm. The results from the non-randomized confirmatory arm were similar to the randomized chemoradiation arm, though the 5-year survival in this arm was slightly lower as more patients on this arm had T3 disease.

The question of increasing the radiation dose beyond 50 Gy with concurrent chemotherapy was

Table 20.4. Studies of chemoradiation alone or chemoradiation versus radiation alone

Study	Patients	Local failure (%)	Distant failure (%)	Median survival (months)	Survival (%)
ARAUJO et al. 1991					
Radiation alone	31	84	23	N/A	6 (5 years)
Chemoradiation	28	61	32	N/A	16 (5 years)
ROUSSELL et al. 1998					
Radiation alone	69	N/A	N/A	N/A	6 (3 years)
Chemoradiation	75	N/A	N/A	N/A	12 (3 years)
AL-SARRAF et al. 1996					
Radiation alone	60	69	40	8.9	10 (2 years), 0 (5 years)
Chemoradiation					
Randomized	61	49	23	17	38 (2 years), 30 (5 years)
Non-randomized	69	45	12		

addressed by RTOG 9405/Intergroup Trial 0123 (MINSKY et al. 2002). This trial randomized patients with locally advanced, medically inoperable carcinoma of the esophagus to either 50.4 Gy or 64.8 Gy, both arms using the same concurrent 5-FU and cisplatinum chemotherapy. This trial was stopped after an early interim analysis showed a higher toxicity and treatment-related mortality on the 64.8 Gy arm. Unfortunately, 7 of 11 treatment-related deaths experienced on this arm occurred at or below the 50.4-Gy dose level. Thus, it was not the higher radiation dose that resulted in these 7 deaths. Nevertheless, the standard radiation dose prescribed using concurrent chemotherapy is 50.4 Gy, based on RTOG 8501 and 9405. Table 20.5 shows the local failure rates for three cooperative group randomized trials of chemoradiation therapy. Note that the crude rates of local failure are similar for total radiation doses ranging from 50 Gy to 65 Gy. Actuarial 2-year local failure rates are slightly higher for higher doses. Further investigations to increase radiation dose are warranted.

20.1.8
Preoperative Chemoradiation Versus Definitive Chemoradiation Therapy

Recently, a randomized comparison of preoperative chemoradiation followed by surgery versus chemoradiation alone for squamous cell carcinoma of the esophagus was published (STAHL et al. 2005). Patients randomized to induction therapy received two cycles of bolus 5-FU, leucovorin, etoposide, a cisplatin (FLEP) followed by cisplatin and etoposide concurrently with 40 Gy. Radiation on this arm was delivered in 2-Gy fractions, 5 days per week. Patients randomized to the chemoradiation arm received the same induction chemotherapy followed by chemoradiation to a higher radiation dose. Radiation therapy consisted of 50 Gy in 2-Gy fractions over 5 weeks followed by a boost. The boost could be delivered with external beam therapy consisting of 1.5 Gy twice daily for 5 days to bring the total dose to 65 Gy. The alternative was to deliver an additional 10 Gy of external beam treatment (60 Gy) followed by a brachytherapy boost.

There were 172 patients allocated to treatment. Overall survival was equivalent between the two treatment arms. Two-year overall survival rates were 39.9% for the surgery arm versus 35.4% for the non-surgery arm. Median survival rates were also similar between arms (16.4 months versus 14.9 months). However, the 2-year freedom from local progression was superior in the surgery arm (63% versus 40.7%, $P=0.003$). Based on these data, the authors concluded that patients should be selected for surgical salvage based on response.

Table 20.5. Phase-III trials with definitive radiochemotherapy in esophageal cancer. *RTOG* Radiation Therapy Oncology Group, *INT* Intergroup, *German ECSG* German Esophageal Cancer Study Group

Trial	No. of patients	Proportion of patients with T3–4 tumors (%)	Radiation dose (Gy)	Crude rate of local failure (%)	Local failure at 2 years (%)
RTOG 85-01 (AL SARRAF et al. 1996)	61	8	50	45	47
INT 0123 (MINSKY et al. 2002)	109	43	50	55	52
INT 0123 (MINSKY et al. 2002)	109	48	64	50	56
German ECSG (STAHL et al. 2005)	86	199	>65	51	58

20.1.9
Superficial Esophageal Cancer

Patients with endoscopic ultrasound-staged Ta or T1 cancers are typically treated surgically. However, a limited number of patients with superficial esophageal cancers are not candidates for surgery because of medical comorbidities. Recent literature suggests that brachytherapy alone or combined with external beam is an option for patients in this category (NEMOTO et al. 2001; MAINGON et al. 2000). For brachytherapy alone, MAINGON et al. employed weekly doses of 5–7 Gy prescribed to a 5-mm depth using a 13-mm diameter applicator.

20.2
Radiation Therapy Techniques

In general, the techniques of radiation therapy are similar for patients with an intact esophagus, where pre-operative or definitive radiation therapy is planned. The techniques of radiation therapy are slightly different in the setting of post-operative radiation therapy, where the target is focused on residual microscopic or gross disease with or without inclusion of adjacent nodal regions. Both situations will be addressed in the following sections.

20.3
Simulation Procedures

CT simulation is used for treatment planning. Patients are positioned supine in a body cast with chin slightly extended to assure it is removed from the treatment field. Arms are folded above the head and may rest on the forehead or a 5-cm or 7-cm sponge. When positioning the patient, it should be kept in mind that the isocenter will be placed midplane at the highest anterior–posterior (AP) separation in the thorax. Most patients are given approximately 100 ml of oral contrast immediately prior to the scan to help delineate the esophagus mass on axial images. Scan limits are from the chin to the top of kidneys. Common scan parameters are 120–130 kVp, 230–300 mAs, and 3-mm-thick images with 3-mm image spacing. Thicker images and spacing can be used away from the region of interest.

In absence of a CT simulator, conventional simulation can be used with orthogonal radiographs obtained in the treatment position and with oral barium. It should be kept in mind that the barium only represents the lumen and does not indicate the size and shape of the tumor.

20.4
Target Volume Definition

For patients with an intact esophagus, three separate targets are identified: gross tumor volume (GTV), clinical target volume (CTV), and planning target volume (PTV). The GTV consists of the primary tumor and involved regional lymph nodes identified using the information from the endoscopy report, endoscopic ultrasound, CT, and PET imaging. The oral contrast aids in identification of the upper and lower edges of the tumor, particularly if the tumor extends into the stomach. If the endoscopy report does not describe the lowermost aspect of the tumor, then we generally contact the endoscopist to obtain this information. Nodal GTV is generally contoured separately, using information from the CT and PET. Any lymph nodes with short axis diameters of 1 cm or greater are included within the GTV.

The CTV consists of the GTV plus a 1-cm axial margin and 3-cm longitudinal margins. The posterior edge of the CTV is modified to exclude the anterior vertebral body from this volume. For lesions of the lower esophagus and GE junction, the CTV includes the gastrohepatic and celiac nodal regions, due to the high risk of disease in these nodes (AKIYAMA et al. 1981). For lesions of the upper esophagus (above the carina), the supraclavicular nodes are included in the CTV. The PTV consists of a 1-cm margin around the CTV in all dimensions. Beyond 45 Gy, the PTV may be reduced to a 2-cm margin around the GTV.

Contoured normal tissue structures include the remaining esophagus, right and left lung, heart, spinal cord, and carina. The heart is contoured, beginning at the base where the aortic root emerges from the left ventricle. The carina is identified to check the craniocaudal isocenter position on the digitally reconstructed radiograph compared with verification or portal films. Treatment plans are optimized to limit dose to normal lung and heart. We aim to keep the volume of normal lung receiving 20 Gy (V20) to less than 20% or a mean lung dose less than 18 Gy. Heart doses are limited so that the

1/3, 2/3, and 3/3 heart volumes are less than 45, 40, and 30 Gy, respectively. The spinal cord is limited to a maximum dose less than 48 Gy. For cancers of the stomach, attention must be paid to the liver and both kidneys. These organs are not generally of concern for cancers of the esophagus.

In the postoperative setting, a CTV is contoured to identify the positive margin of concern. If gross residual disease exists, then a GTV is contoured and is expanded by 1 cm to generate a CTV. The PTV is a 1-cm volumetric expansion of the CTV. In situations where the patient has had an esophagectomy and gastric conduit with a positive upper margin at the anastomosis and residual abdominal nodal disease, a large volume of radiation would be necessary to treat the PTV within one field. A viable alternative option is to identify separate PTVs, using different fields for the anastomosis and nodal disease.

20.5
Dose Prescription

In the preoperative setting, we prescribe 45 Gy in 1.8-Gy daily fractions to cover the PTV, using heterogeneous dose prescriptions. Energies of 6–18 MV are used. For definitive therapy, the prescribed dose is 50.4 Gy to cover the PTV. In the postoperative setting, 45–50.4 Gy is used to cover the PTV (or PTVs). The following represents our practice at the Siteman Cancer Center/Washington University:

I. Definitive preoperative chemotherapy plus radiation therapy – with intent to proceed with radical esophagectomy
   Chemotherapy:
      Cisplatin – 75 mg/m^2, day 1, week 1
      5-FU – 1000 mg/m^2×4 days, week 1
   Radiation therapy:
      45 (25 fractions/5–5.5 weeks)
      Surgery to be performed week 4–6 if patient has recovered hematologically

II. Postoperative chemoradiation for those with positive surgical margins or residual gross disease
   Chemotherapy:
      Cisplatin – 75 mg/m^2, day 1, weeks 1, 5, 8, 11
      5-FU – 1000 mg/m^2×4 days, weeks 1, 5, 8, 11
   Radiation therapy:
      50–54 Gy/25 fx/5 weeks for negative margins
      60–66 Gy/30 fx/6 weeks for positive margins

III. Definitive chemoradiotherapy for patients who are not surgical candidates but still have localized disease:
   Chemotherapy:
      Cisplatin – 75 mg/m^2 – first day of weeks 1, 5, 8, 11
      5-FU – 1000 mg/m^2×4 days – weeks 1, 5, 8, 11
   Radiation therapy – 45 Gy (9–10 Gy/wk) to the primary tumor and regional lymphatics plus a 5 Gy boost to the primary and nodal GTV with oblique fields

IV. For patients who will not receive chemotherapy: Radiation therapy dose is as above with the difference being that the total tumor dose is 64.8 Gy (after 45 Gy to volume I and 19.8 Gy to volumes II and III with gradually reduced fields)

20.6
Beam Selection and Design

Most of the tumors require relatively large fields. Treatment field arrangements are designed to meet the spinal cord dose tolerance and spare the lungs and the heart. Due to large depths of these tumors, high-energy beams should be used to reduce the integral lung dose. Technical difficulties for this treatment site include irregular patient topology along the thorax, curving spinal cord canal, and angle of the esophagus. Typical beam arrangements are designed to limit the dose to the spinal cord to 4500–4700 cGy maximum dose. In our clinic, AP/posteroanterior (PA) beams are used to deliver an initial dose of 3600 cGy to the PTV. AP field usually requires a wedge to reduce the spinal cord dose in the superior portion of the treatment field. An AP and posterior oblique are then used to deliver an additional 1440 cGy. The longitudinal margins of the AP and off-cord oblique fields are reduced to GTV+2 cm after 4500 cGy. This field arrangement usually brings the spinal cord to a tolerance level of 4500–4700 cGy. We typically choose a left posterior oblique field to minimize exit dose to the heart (Fig. 20.5a). Opposed oblique fields (Fig. 20.5b) are routinely used to deliver a final dose of 5040 cGy to minimize radiation dose to the anterior right lung. If the planning target volume is situated to the left of midline, coverage of the PTV is sometimes better with a three-field arrangement consisting of an anterior and bilateral posterior oblique beams (Fig. 20.5c). Wedge angle for the AP field is usually best determined by evaluating dose distributions in

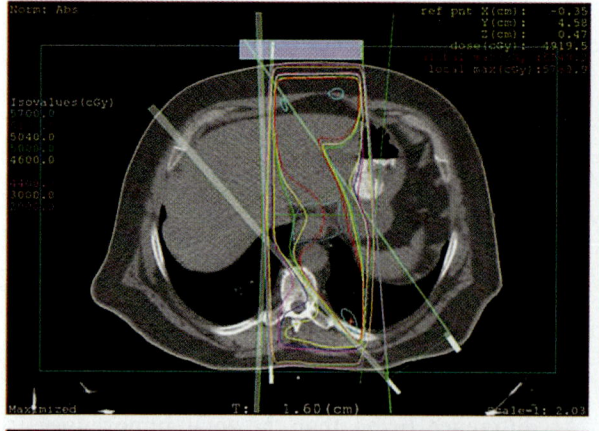

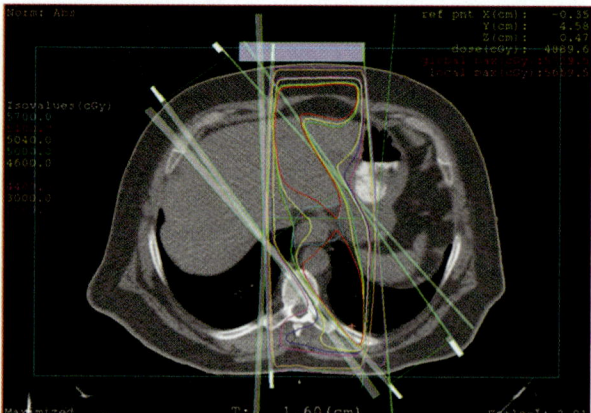

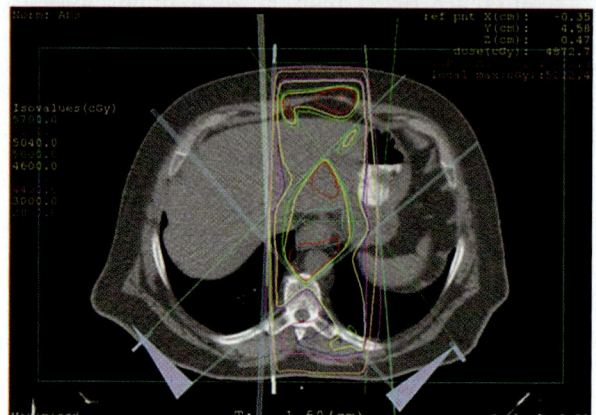

Fig. 20.5a-c. a Axial computed tomography (CT) image showing the isodose lines generated using an AP/PA/RPO field arrangement. **b** Axial CT image showing the isodose lines generated using an AP/PA/RPO/LAO field arrangement. We generally limit the use of opposed oblique fields in order to minimize the normal lung tissue dose. **c** Axial CT image showing the isodose lines generated using an anterior–posterior/right posterior oblique/left posterior oblique field arrangement. This technique is sometimes used for cancers to the left of midline

the sagittal plain bisecting the spinal canal. The cord tolerance isodose in this view should follow the curvature of the spinal cord and should be maintained at a safe distance away from the cord (Fig. 20.6).

20.7
Plan Evaluation

In our clinic, treatment plans are evaluated with the treating physician and dosimetrist at the treatment-planning computer. This enables a thorough review of slice-by-slice isodose curves and dose volume histogram (DVH) analysis of the target and critical normal structures. Our checklist is as follows: PTV coverage, maximum spinal cord dose, normal lung-tissue dose, and heart dose. If heterogeneity calculation algorithms are used, dose is prescribed to the PTV. Optimally, 100% of the prescribed dose will cover 100% of the PTV. Treatment plans are acceptable if 95% of the dose covers 100% of the PTV. Our maximum acceptable spinal cord dose is 47 Gy. Mean lung dose or V20 (the volume of lung at or exceeding 20 Gy) is used to evaluate normal lung

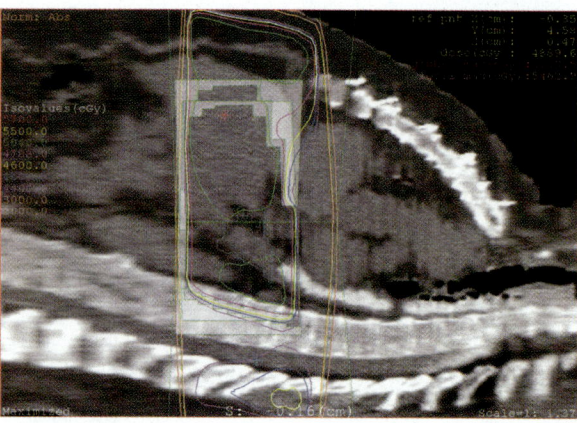

Fig. 20.6. Sagittal computed tomography image showing iso-dose lines generated by the treatment plan in Fig. 20.4a. In this example, the maximum cord dose is 4788 cGy

dose. Ideally, mean lung dose is kept below 18 Gy and V20 below 30%. For heart dose, we try to keep two-thirds of the heart below 45 Gy. Figure 20.7 shows a comparison of dose volume histograms for the three plans shown in Figure 20.5. For all practical purposes, the two plans (Figs. 20.5a and 20.5c) result in equivalent DVHs (Fig. 20.7). Analysis of the isodose lines shows that the main difference between the three plans is the location of the hot spots (maximum doses) in the plan. Then, the best plan is determined by the location of the tumor with respect to other normal structures. The hot spot should be kept away from the heart and the spinal cord.

20.8
Plan Implementation and Quality Assurance

Technically, these treatments are relatively straightforward to implement on treatment machine. In our clinic, treatment settings are exported from the treatment-planning computer directly to the record and verify system. One of the main concerns with verification of these treatments is that the wedge on the AP beam is used and that it is oriented properly with the heel toward the patient's head. Absence of the wedge or incorrect orientation can lead to serious overdose of the spinal cord especially in the superior portion of

the treatment field. To ensure that the wedge is present and oriented correctly, we perform in vivo dosimetry for the AP field on the first day of treatment, using a diode dosimeter. The dosimeter is placed 2–4 cm away from the central axis, toward the superior portion of the field. This off-axis reading allows us to determine the orientation of the wedge with respect to the treatment field (measuring across the dose gradient in the wedged direction). In addition to the AP field, diode dosimetry is performed for all other fields as well. Our tolerance for agreement between diode measured and calculated doses is ±5%. Once the satisfactory measurements are obtained with diode dosimetry, accuracy of patient treatments are verified with port films and weekly review of patient's electronic treatment record. In addition to this, source to surface distances on central axis are checked weekly for all treatment fields. In our clinic, it is a policy to independently calculate maximum spinal cord dose to verify treatment plan calculations. This verification can be performed using manual calculations or using a monitor unit calculator.

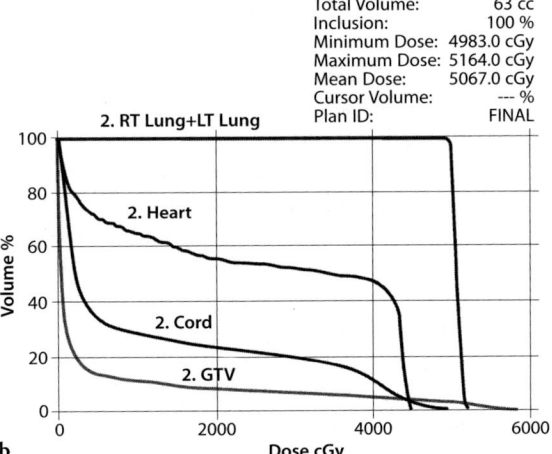

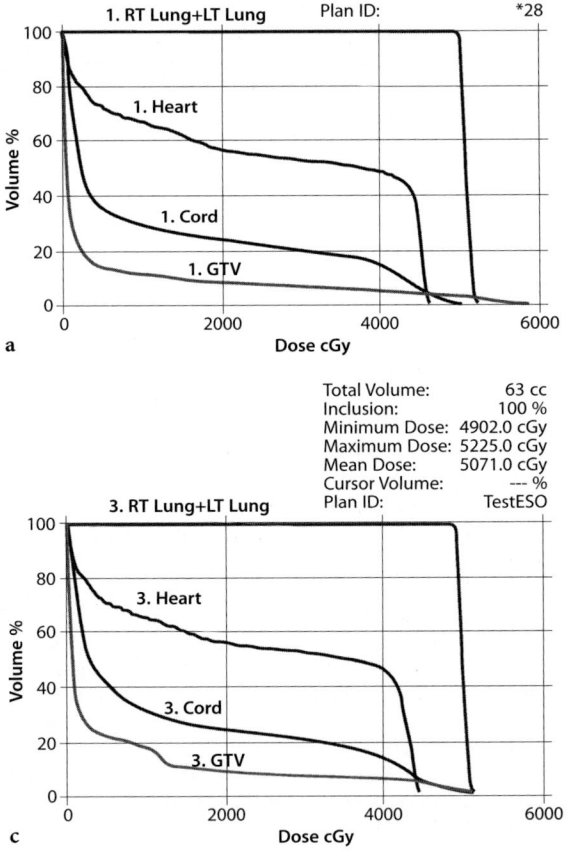

Fig. 20.7a-c. Dose volume histogram comparison of plans 20.4. In this particular patient, the plans generated using anterior–posterior/posterior–anterior/left posterior oblique or anterior–posterior/right posterior oblique fields have similar dose volume histograms. The treatment plan used is based on planning target volume coverage and the location of D_{max} (maximum dose) (away from critical structures)

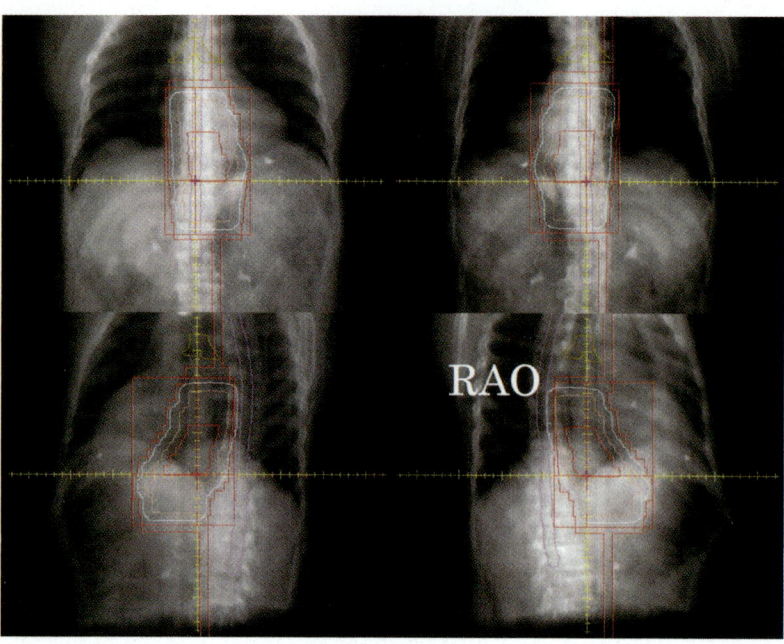

Fig. 20.8. Typical treatment portals (anterior–posterior, posterior–anterior and left posterior oblique) for a patient with a carcinoma of the distal esophagus. The clinical stage for this patient was T3N0M0

20.9
Simulation Films and Portal Films

For treatment simulation films, digitally reconstructed radiographs (DRRs) or conventional simulator films can be used. Figure 20.8 shows a set of DRRs for a typical patient treatment. For verification of beam shapes and patient positioning, these films are compared with port films from treatment machines. For AP/PA field arrangement, port films for both treatment fields are acquired on the first day of treatment and, subsequently, every five fractions. For AP and oblique field arrangements, all fields are also filmed on the first day of treatment and, subsequently, every five fractions. Lateral setup films can also be acquired, if needed, to verify patient positioning and spinal cord depth.

20.10
Future Directions

The prognosis for patients with locally advanced (T3 or N1) or inoperable esophagus cancer remains poor. Efforts to improve survival under investigation include the testing of laparoscopic surgical techniques, newer (i.e., taxanes) or triple chemotherapy combinations, and developmental targeted biological agents. Specific to radiation therapy, the application of PET-based radiation-treatment planning is likely to have an influence on the planning target volume. Other investigators are testing the addition of a brachytherapy boost to external beam therapy using applicators with a centering mechanism and having diameters of 10 mm or more. Applicators with larger diameters may avoid the fistula complications seen previously using applicators with smaller diameters and, therefore, delivering an exceedingly high dose to the esophageal wall (RTOG 9207) (GASPAR et al. 1995).

References

Akiyama H, Tsurumaru M, Kawamura T et al (1981) Principles of surgical treatment for carcinoma of the esophagus: an analysis of lymph node involvement. Ann Surg 194:438-446

Al-Sarraf M, Martz K, Herskovic A et al (1996) Superiority of chemo-radiotherapy (CT-RT) vs. radiotherapy (RT) in patients with esophageal cancer. Final report of an intergroup randomized and confirmed study (abstract). Proc Am Soc Clin Oncol 15:206

Apinop C, Puttisak P and Preecha N (1994) A prospective study of combined therapy in esophageal cancer. Hepatogastroenterology 41:391-393

Araujo CM, Souham L, Gil RA et al (1991) A randomized trial comparing radiation therapy versus concomitant radiation therapy and chemotherapy in carcinoma of the thoracic esophagus. Cancer 67:2258-2261

Blot WJ, Devesa SS, Kneller RW et al (1991) Rising incidence of adenocarcinoma of the esophagus and gastric cardia. JAMA 265:1287-1289

Bosset J-F, Gignoux M, Triboulet J-P et al (1997) Chemoradiotherapy followed by surgery compared with surgery alone in squamous-cell cancer of the esophagus. N Engl J Med 337:161-167

Cancer Statistics (2003) Cancer statistics, 2003. CA Cancer J Clin 53:5-64

Chau I, Norman AR, Cunningham D et al (2004) Multivariate prognostic factor analysis in locally advanced and metastatic esophago-gastric cancer-pooled analysis from three multicenter, randomized, controlled trials using individual patient data. J Clin Oncol 22:2395-2403

Crew KD, Neugut AI (2004) Epidemiology of upper gastrointestinal malignancies. Semin Oncol 31:450-464

Dormans E (1939) Das Oesophaguscarcinoma: Ergebnisse der unter Mitarbeit von 39 pathologischen Instituten Deutschlands durchgeführten Erhebung über das Oesophaguscarcinoma (1925-1933). Z Krebforsch 49:86

Fiorica F, Di Bona D, Schepis F et al (2004) Preoperative chemoradiotherapy for oesophageal cancer: a systematic review and meta-analysis. Gut 53:925-930

Flamen P, Lerut A, van Cutsem E et al (2000) Utility of positron emission tomography for the staging of patients with potentially operable esophageal carcinoma. J Clin Oncol 18:3202-3210

Flanagan FL, Dehdashti F, Siegel BA et al (1997) Staging of esophageal cancer with 18F-fluorodeoxyglucose positron emission tomography. AJR Am J Roentgenol 168:417-424

Fok M, Sham JS, Choy D et al (1993) Postoperative radiotherapy for carcinoma of the esophagus: a prospective, randomized controlled study. Surgery 113:138-147

Gaspar LE, Quian C, Koeha WI et al (1995) A phase I/II study of external beam radiation, brachytherapy and concurrent chemotherapy in localized cancer of the esophagus (RTOG 9207): preliminary toxicity report. Int J Radiat Oncol Biol Phys 32 [Suppl]:160

Hancock SL, Glatstein E (1984) Radiation therapy of esophageal cancer. Semin Oncol 11:144

Himeno S, Yasuda S, Shimada H et al (2002) Evaluation of esophageal cancer by positron emission tomography. Jpn J Clin Oncol 32:340-346

Homs MY, Steyerberg EW, Eijkenboom WM et al (2004) Single-dose brachytherapy versus metal stent placement for the palliation of dysphagia from oesophageal cancer: multicentre randomised trial. Lancet 364:1497-1504

Homs MY, Wahab PJ, Kuipers EJ et al (2004) Esophageal stents with antireflux valve for tumors of the distal esophagus and gastric cardia: a randomized trial. Gastrointest Endosc 60:695-702

Igaki H, Kato H, Tachimori Y et al (2003) Prognostic evaluation of patients with clinical T1 and T2 squamous cell carcinomas of the thoracic esophagus after 3-field lymph node dissection. Surgery 133:368-374

Kole AC, Plukker JT, Nieweg OE et al (1998) Positron emission tomography for staging oesophageal and gastroesophageal malignancy. Br J Cancer 78:521-527

LePrise E, Etienne PL, Meunier B et al (1994) A randomized study of chemotherapy, radiation therapy, and surgery versus surgery for localized squamous cell carcinoma of the esophagus. Cancer 73:1779-1784

Luketich JD, Schauer PR, Meltzer CC et al (1997) Role of positron emission tomography in staging esophageal cancer. J Thorac Cardiovasc Surg 64:765-769

Macdonald JS, Smalley SR, Benedetti J et al (2001) Chemoradiotherapy after surgery compared with surgery alone for adenocarcinoma of the stomach or gastroesophageal junction. N Engl J Med 345:725-730

Maingon P, d'Hombres A, Truc G et al (2000) High dose rate brachytherapy for superficial cancer of the esophagus. Int J Radiat Oncol Biol Phys 46:71-76

McAteer D, Wallis F, Couper G et al (1999) Evaluation of 18F-FDG positron emission tomography in gastric and oesophageal carcinoma. Br J Radiol 72:525-529

Minsky BD, Pajak TF, Ginsberg RJ et al (2002) INT 0123 (Radiation Therapy Oncology Group 94-05) phase III trial of combined modality therapy for esophageal cancer: high dose versus standard-dose radiation therapy. J Clin Oncol 20:1167-1174

Muto M, Ohtsu A, Miyamoto S et al (1999) Concurrent chemoradiotherapy for esophageal carcinoma patients with malignant fistulae. Cancer 86:1406-1413

Nemoto K, Yamada S, Hareyama M et al (2001) Radiation therapy for superficial esophageal cancer: a comparison of radiotherapy methods. Int J Radiat Oncol Biol Phys 50:639-644

Nygaard K, Hagen S, Hansen HS et al (1992) Pre-operative radiotherapy prolongs survival in operable esophageal carcinoma: a randomized, multicenter study of pre-operative radiotherapy and chemotherapy. The second Scandinavian trial in esophageal cancer. World J Surg 16:1104-1109

Rice TW, Blackstone EH, Adelstein DJ et al (2003a) Role of clinically determined depth of tumor invasion in the treatment of esophageal carcinoma. J Thorac Cardiovasc Surg 125:1091-1102

Rice TW, Blackstone EH, Rybicki LA et al (2003b) Refining esophageal cancer staging. J Thorac Cardiovasc Surg 125:1103-1113

Rosenberg JC, Franklin R, Steiger Z (1981) Squamous cell carcinoma of the thoracic esophagus: an interdisciplinary approach. Curr Probl Cancer 6:5

Roussell A, Jacob JH, Haegele P (1988) Controlled clinical trial for treatment of patients with inoperable esophageal carcinoma: a study of EORTC Gastrointestinal Tract Cancer Cooperative Group. Rec Res Cancer Res 110:21-29

Stahl M, Stuschke M, Lehmann N, Meyer HJ, Walz MK, Seeber S, Klump B, Budach W, Teichmann R, Schmitt M, Schmitt G, Franke C, Wilke H (2005) Chemoradiation with and without surgery in patients with locally advanced squamous cell carcinoma of the esophagus. J Clin Oncology 23:2310-2317

Sur R, Donde B, Falkson C et al (2004) Randomized prospective study comparing high-dose-rate intraluminal brachytherapy (HDRILBT) alone with HDRILBT and external beam radiotherapy in the palliation of advanced esophageal cancer. Brachytherapy 3:191-195

Sur RK, Levin CV, Donde B et al (2002) Prospective randomized trial of HDR brachytherapy as a sole modality in palliation of advanced esophageal carcinoma-an International Atomic Energy Agency study. Int J Radiat Oncol Biol Phys 53:127-133

Teniere P, Hay JM, Fingerhut A et al (1991) Postoperative radiation therapy does not increase survival after curative resection for squamous cell carcinoma of the middle and lower esophagus as shown by a multicenter controlled trial. Surg Gynecol Obstet 173:123-130

Thomas CR Jr, Berkey BA, Minsky BD et al (2004) Recursive partitioning analysis of pretreatment variables of 416 patients with locoregional esophageal cancer treated with definitive concomitant chemoradiotherapy on Intergroup and Radiation Therapy Oncology Group trials. Int J Radiat Oncol Biol Phys 58:1405-1410

Urba SG, Orringer MB, Turrisi A et al (2001) Randomized trial of preoperative chemoradiation versus surgery alone in patients with locoregional esophageal carcinoma. J Clin Oncol 19:305-313

Vaszquez-Sequeiros E, Norton ID, Clain JE et al (2001) Impact of EUS-guided fine-needle aspiration on lymph node stag-

ing in patients with esophageal carcinoma. Gastrointest Endosc 53:751-757

Walsh TN, Noonan N, Hollywood D et al (1996) A comparison of multimodal therapy and surgery for esophageal adeno-carcinoma. N Engl J Med 335:462-467

Wara WM, Mauch PM, Thomas AN (1976) Palliation for carci-noma of the esophagus. Radiology 121:717

Yeung HWD, Macapinlac HA, Mazumdar M et al (1999) FDG-PET in esophageal cancer*1: incremental value over com-puted tomography. Clin Positron Imaging 2:255-260

21 Carcinoma of the Lung

Samir Narayan and Srinivasan Vijayakumar

CONTENTS

21.1   Introduction   525
21.2   Epidemiology   525
21.3   Anatomy and Pathologic Features   525
21.4   Natural History and Symptoms   526
21.5   Staging   527
21.6   Prognosis   528
21.7   Diagnostic Work-up   528
21.8   General Management of
        Non-Small Cell Lung Cancer   528
21.8.1 Surgery   528
21.8.2 Chemotherapy   529
21.9   Radiation Therapy Management   529
21.9.1 Stage I   529
21.9.2 Stage II/III   529
21.9.3 Postoperative Care   530
21.10   Dose, Fractionation, Volume   531
21.10.1 Tumor Dose   531
21.10.2 Fractionation   532
21.10.3 Elective Nodal Irradiation   532
21.11   Radiation Therapy Technique   533
21.11.1 Introduction   533
21.11.2 Simulation   533
21.11.3 Target Delineation   534
21.11.4 Standard and Conformal Treatment Planning   534
21.11.5 Energy and Lung Heterogeneity   535
21.11.6 Normal Tissue Considerations   536
21.12   Patterns of Failure   536
21.13   Management of Small Cell Carcinoma   537
21.13.1 Thoracic Radiotherapy   537
21.13.2 Dose   537
21.13.3 Fractionation   537
21.13.4 Timing with Systemic Therapy and
        Treatment Volume   538
21.13.5 Prophylactic Cranial Irradiation   538
        References   539

S. Narayan, MD; S. Vijayakumar, MD
Department of Radiation Oncology, University of California, Davis Medical Center, Sacramento, CA 95817, USA

21.1 Introduction

Lung cancer continues to be the leading cause of cancer-related mortality in both men and women in the United States (Jemal et al. 2005). Despite technologic advances over the past three decades that have resulted in improved diagnosis and therapy, the overall cure rate of patients with lung cancer has not significantly changed and is <15% at 5 years (Jemal et al. 2005). Recent clinical trials employing a combination of therapeutic modalities, however, have improved outcomes in certain subsets of patients (Farray et al. 2005).

21.2 Epidemiology

Lung cancer was diagnosed in approximately 171,000 people in the United States in 2005 (Jemal et al. 2005), third highest among all cancers. Although the incidence of lung cancer has been higher in men than women, the magnitude of this difference has been declining. Cigarette smoking has been reported to be a factor in causing 80–90% of lung cancer (Wingo et al.1999). Exposure to second-hand smoke, asbestos, polycyclic aromatic hydrocarbons, nickel, arsenic, and radon may also be related to the development of lung cancer.

21.3 Anatomy and Pathologic Features

The right lung is composed of the upper, middle, and lower lobes, which are separated by the major and minor fissures. The left lung is composed of two lobes separated by a single fissure. The trachea enters the superior mediastinum and bifurcates approximately at the level of the fifth thoracic vertebra. The hila contain the bronchi, pulmonary arteries, and veins, various branches from the pulmonary plexus, bronchial arteries and veins, and lymphatics.

The lung has a rich network of lymphatic vessels throughout its loose interstitial connective tissue, ultimately draining into various lymph node stations. Fourteen regional lymph node stations are defined for non-small cell lung carcinoma (NSCLC) and classified into N1, distal to the mediastinal pleural reflection; N2, within the ipsilateral mediastinal envelope; and N3, mediastinal or hilar nodes which are contralateral to the primary tumor or any scalene or supraclavicular nodes (Fig. 21.1). N1 nodes (stations 10–14) include the intrapulmonary, along the secondary bronchi or in the bifurcation of branches of the pulmonary artery; and the bronchopulmonary lymph nodes, situated either alongside the lower portions of the main bronchi (hilar lymph nodes) or at the bifurcations of the main bronchi into lobar bronchi (interlobar nodes). The N2 nodes are the mediastinal lymph nodes, including the upper paratracheal, prevascular, lower paratracheal nodes (azygos nodes), para-aortic and a group of nodes located in the aortic window. Inferior to the carina are the subcarinal, paraesophageal, and pulmonary ligament nodes. Although the nodal stations have

been adapted based on anatomic definitions based on surgery, a cross-sectional depiction on CT has recently been developed (Ko et al. 2000).

Lymph from the right upper lobe flows to the hilar and tracheobronchial lymph nodes. Lymph from the left upper lobe flows to the venous angle of the same side and to the right superior mediastinum. The right and left lower lobe lymphatics drain into the inferior mediastinal and subcarinal nodes and from there to the right superior mediastinum. The left lower lobe also may drain into left superior mediastinum.

Lung carcinomas are broadly divided into NSCLC and SCLC. The NSCLC compromises over 80% of lung cancer and is further subdivided into squamous cell carcinoma, adenocarcinoma, and large cell carcinoma. Bronchoalveolar cell carcinoma is classified as a subtype of adenocarcinoma.

21.4
Natural History and Symptoms

Bronchogenic carcinoma usually originates in secondary to tertiary bronchial divisions. Even before the tumor has reached a clinically detectable size, invasion of the regional lymphatics and the blood vessels may occur, resulting in widespread lymphatic and hematogenous dissemination (LINE and DEELEY 1971). This information is important in designing radiation therapy portals. Ipsilateral hilar node metastasis occurs in between 50 and 60% of patients (CARTER and EGGLESTON 1980; GOLDBERG et al. 1974; BAIRD 1965). Mediastinal adenopathy is noted in 40–50% of operative specimens (DILLMAN et al. 1996). Metastasis to the scalene (supraclavicular) nodes is predominantly from primary sites in ipsilateral upper lobes or from superior mediastinal lesions.

Symptoms resulting from lung cancer can be related to the primary lung lesion or to intrathoracic spread, distant metastasis, or paraneoplastic syndromes. The most common symptom of lung cancer is cough. Dyspnea may occur from airway obstruction, postobstructive pneumonia or atelectasis, lymphangitic spread, and pleural or pericardial effusion. Hemoptysis has been reported in 27–57% of patients with lung cancer (MIDTHUN and JETT 1996). Chest pain can occur from invasion of the chest wall, vertebrae, or mediastinal structures. Pancoast syndrome results from a superior sulcus

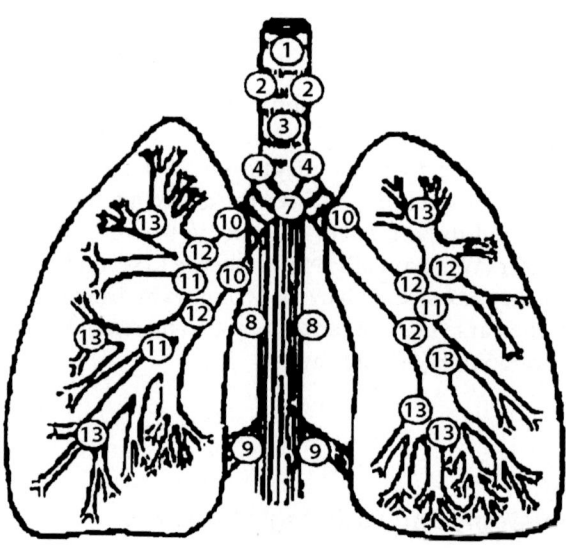

Fig. 21.1 Sites of lymph nodes in lungs and mediastinum. *1* Highest mediastinal, *2* upper paratracheal, *3* prevascular and retrotracheal, *4* lower paratracheal, *5* subaortic (aorto-pulmonary window); *6* para-aortic (ascending aorta or phrenic), *7* subcarinal, *8* paraesophageal (below carina), *9* pulmonary ligament, *10* hilar, *11* interlobar, *12* lobar, *13* segmental, *14* subsegmental (not depicted). (From EMAMI et al. 1998)

tumor invading into the apical chest wall, causing brachial plexopathy and ipsilateral Horner's syndrome.

Enlarged mediastinal adenopathy may cause left vocal cord paralysis and hoarseness due to injury to the left recurrent laryngeal nerve as it passes under the aortic arch in the aorto-pulmonary window. Although more common with small cell carcinoma,

superior vena cava syndrome may occur from primary tumor or nodal compression of the superior vena cava.

Metastatic disease is often symptomatic in the brain and bones, but may also spread to the liver, adrenals, and to the other lobes of the lung. The most frequent paraneoplastic syndromes of NSCLC are hypercalcemia, clubbing, and hypertrophic pulmonary osteoarthropathy.

Weight loss has been reported in 8–68% of patients with lung cancer. It may occur with localized disease, metastatic disease, or as a consequence of a paraneoplastic syndrome. Weight loss represents a negative prognostic indicator in lung cancer.

Table 21.1 The TNM classification of the lung

Primary tumor (T)

TX   Primary tumor cannot be assessed, or tumor proven by presence of malignant cells in sputum or bronchial washings but not visualized by imaging or bronchoscopy

T0   No evidence of primary tumor

Tis   Carcinoma in situ

T1   Tumor 3 cm or less in greatest dimension, surrounded by lung or visceral pleura, without bronchoscopic evidence of invasion more proximal than the lobar bronchus

T2   Tumor with any of the following features of size or extent: More than 3 cm in greatest dimension
Involves main bronchus, 2 cm or more distal to the carina
Invades the visceral pleura
Associated with atelectasis or obstructive pneumonitis that extends to the hilar region but does not involve the entire lung

T3   Tumor of any size that directly invades any of the following: Chest wall (including superior sulcus tumors), diaphragm, mediastinal pleura, parietal pericardium; or tumor in the main bronchus <2 cm distal to the carina but without involvement of the carina; or associated atelectasis or obstructive pneumonitis if the entire lung

T4   Tumor of any size that invades any of the following: mediastinum; heart; great vessels; trachea; esophagus; vertebral body; carina; separate tumor nodules in the same lobe; or tumor with a malignant pleura

Lymph node (N)

NX   Regional lymph nodes cannot be assessed

N0   No regional lymph node metastasis

N1   Metastasis to ipsilateral peribronchial and/or ipsilateral hilar lymph nodes, and intrapulmonary nodes involved by direct extension of the primary tumor

N2   Metastasis in ipsilateral mediastinal and/or subcarinal lymph node(s)

N3   Metastasis in contralateral mediastinal, contralateral hilar, ipsilateral or contralateral scalene or supraclavicular lymph nodes

Distant metastases (M)

MX   Presence of distant metastasis cannot be assessed

M0   No distant metastasis

M1   Distant metastasis

21.5
Staging

The sixth edition of the AJCC Staging System for NSCLC remains unchanged from 1997. Staging grouping by T, N, and M with 5-year survivals are shown in Tables 21.1 and 21.2. It is noted that the survival results of the new staging system are based primarily upon surgical staging and treatment. Patient outcome and survival may be significantly affected by surgical procedure and additional combined modality therapy (GINSBERG and RUBINSTEIN 1995; KRIS et al. 1995; MARTINI et al. 1997).

Table 21.2 Sixth edition, AJCC staging and 5-year survival

Overall stage	T	N	M	Estimated 5-year survival (%)
0	Tis	N0	M0	–
IA	T1	N0	M0	60–70
IB	T2	N0	M0	~40
IIA	T1	N1	M0	~35
IIB	T2	N1	M0	25–35
	T3	N0		
IIIA	T1	N2	M0	~10
	T2	N2	M0	
	T3	N1–N2	M0	
IIB	T1–T4	N3	M0	~5
	T4	N0-3	M0	
IV	Any T	Any N	M1	~2

21.6
Prognosis

Several tumor and clinical characteristics have been found to be significant predictors for survival. The impact of stage on survival has been well documented by MOUNTAIN et al. (1997). The two most important clinical factors are weight loss and performance status (FEINSTEIN 1964; PATER and LOEB 1982; BLACKSTOCK et al. 2002; STANLEY 1980; HOANG et al. 2005).

21.7
Diagnostic Work-up

The goal of work-up is pathological confirmation of malignancy and accurate staging. Staging is not only prognostic, but also helps determine the most appropriate treatment. In addition to a detailed history and thorough physical examination, complete blood count, liver function tests, and an electrolyte panel are necessary. Chest X-ray is often the initial radiographic imaging used in the evaluation of suspected bronchogenic carcinoma. Suspicious abnormalities should then be evaluated with a CT chest to reveal the size and local invasion of the primary tumor, effusions, extent of hilar and mediastinal adenopathy, and metastatic disease in the bones and other lobes of the lung. Staging should also include the liver and adrenals, which are the common sites of metastatic disease.

Determining the status of the N2 nodes has both important prognostic and treatment decision-making implications. Surgery has not been shown to improve survival compared with definitive radiation and concurrent chemotherapy (ALBAIN et al. 2003). Based on CT, a 1-cm short axis is generally regarded as the upper limit of normal for mediastinal lymph node diameter, and nodes exceeding this size are presumed to be pathological (GLAZER et al. 1985); however, using such size criteria, the sensitivity and specificity of mediastinal nodes on CT is 61 and 79%, respectively (GOULD et al. 2003).

Positron emission tomography (PET) is a functional imaging tool which, when combined with CT, improves the sensitivity and specificity of mediastinal node detection to 85 and 90%, respectively (GOULD et al. 2003).The most commonly used radioisotope is (18)F-2-fluoro-2-deoxy-D-glucose (FDG), a glucose analog tagged with a positron-emitting isotope of fluorine, (18)F. Metabolically active cells, such as malignant cells, uptake more glucose than other tissues (NOLOP et al. 1987). Despite the improved mediastinal staging accuracy of PET, mediastinoscopy remains the gold standard for evaluation of these nodes (GDEEDO et al. 1997).

Magnetic resonance imaging (MRI) may provide additional value in certain situations when the superior sulcus structures or vertebral invasion needs to be assessed. Computed tomography or MRI brain, as well as bone scan, may be ordered if suspicious symptoms are present. For SCLC, both bone scan and either CT or MRI of brain are included in the staging. Compared with conventional staging, PET has been shown to detect occult distant metastases in 19% of patients otherwise felt to have curable disease (MACMANUS et al. 2001). Pulmonary function tests should include spirometry and diffusion capacity.

Tissue diagnosis can be obtained through sputum cytology or several other methods of varying invasiveness. Since the presence of a malignant pleural effusion dramatically alters the prognosis and treatment, pleuracentesis should be performed when appropriate. Bronchoscopy provides valuable information for staging and treatment planning. Washings and brushings of the central airways and transbronchial needle aspiration using a Wang needle biopsy of subcarinal and hilar nodes can be used to obtain a diagnosis, especially for centrally located tumors. Percutaneous transthoracic needle aspiration is useful for obtaining tissue diagnosis, but offers limited staging information.

Cervical mediastinoscopy offers the ability to pathologically stage the paratracheal and subcarinal nodes. A left anterior mediastinotomy (Chamberlain procedure) allows access to the station-5 and station-6 nodes. Occasionally, video-assisted thoracoscopic surgery (VATS), which allows exploration of hemithorax, is generally used to obtain an excisional biopsy when less invasive means are not successful.

21.8
General Management of Non-Small Cell Lung Cancer

21.8.1
Surgery

The increasing use of combined modality therapy (radiation therapy, chemotherapy, surgery) for lung cancer underlies the need for accurate staging and for interdisciplinary evaluation of the newly diagnosed lung cancer patient. A determination of resectability should be made by an experienced thoracic surgeon. Surgery should involve lobectomy or pneumonectomy, if a lung sparing sleeve resection is not feasible, as well as an evaluation of the N1 and N2 nodes. It is estimated that only about 20% of all newly diagnosed patients are appropriate for surgical resection (SHIELDS 1993). Stage, extent of resection, performance status, comorbidity, and cardiopulmonary evaluation determine resectability. Pulmonary function tests, including spirometry and diffusion capacity, should be performed as part of the preoperative evaluation.

21.8.2
Chemotherapy

Chemotherapy has been shown to improve survival in the multi-modality treatment of NSCLC. For patients with stage-II or stage-III unresectable disease, chemotherapy combined with radiation therapy has been shown to be superior than radiation therapy alone (BRADLEY et al. 2005a; SOCINSKI 2005; TURRISI et al. 2005). Furthermore, a recent trial has demonstrated that adjuvant chemotherapy following resection of stages I–III NSCLC has also been shown to improve survival over resection alone (ARRIAGADA et al. 2004). For patients with metastatic or recurrent cancer, chemotherapy may improve survival and quality of life compared with best supportive care (NSCLC Collaborative Group 1995). In general, patients with poor performance status and advanced weight loss may not receive a benefit from chemotherapy. Combination chemotherapy regimens using a platinum-based agent have been shown to be most effective; however, no particular combination has been shown to be optimal (SCHILLER et al. 2002).

21.9
Radiation Therapy Management

21.9.1
Stage I

Surgery alone for stage I results in a 5-year OS of 60–70%. The numbers of patients with stage-I disease can be expected to increase with the increasing use of spiral screening CT. Adjuvant chemotherapy has been found to improve survival in patients with resected stages I–III disease (ARRIAGADA et al. 2004).

Definitive radiation therapy is the appropriate alternative to surgery for early stage patients who have medical contraindications to surgery. In this population of patients with poor pulmonary function and severe medical comorbidities, radiation therapy alone has produced 5-year overall survivals of 8–15% (DOSORETZ et al. 1992; HAFFTY et al. 1988; KASKOWITZ et al. 1993; NOORDIJK et al. 1988; SANDLER et al. 1990; SIBLEY et al. 1998; KUPELIAN et al. 1996); however, due to competing risks of death, the 5-year disease-specific survival of these patients is in the range of 30–55%.

Recent studies support the concept of treating only the primary tumor and not the elective nodal regions to reduce the volume of irradiated lung tissue to doses of ≥65 Gy, conventionally fractionated (EMAMI 1994; KROL et al. 1996; SLOTMAN et al. 1996). Also, some studies appear to demonstrate a dose–response relationship for tumor control (SIBLEY et al. 1998).

Early results with stereotactic hypofractionated radiotherapy have shown improved local control rates compared with conventionally fractionated radiation (BLOMGREN et al. 1995). WULF et al. (2001) reported a 76% local control rate for lung tumors treated with three fractions of 10 Gy each. TIMMERMAN et al. (2003) conducted a phase-I dose-per-fraction escalation of a three-fraction course of radiation using an extracranial stereotactic radioablation. A dose of 6000 cGy in three fractions was tolerable and is currently being evaluated in a phase-I/II Radiation Therapy Oncology Group (RTOG) study of stereotactic body radiation therapy (SBRT) for medically inoperable stage-I/II NSCLC. Further validation of the early results of SBRT are eagerly awaited.

21.9.2
Stage II/III

The standard of care for patients with stage-III disease is combined-modality treatment with chemotherapy and radiation therapy (DILLMAN et al. 1996; LE CHEVALIER et al. 1991; SAUSE et al. 2000). Some subsets of stage-III disease or superior sulcus tumors may be considered for neoadjuvant chemoradiation followed by resection (ALBAIN et al. 2003). Medically inoperable patients with stage-II disease should be managed similarly to stage-III patients, i.e., with chemoradiation.

The different, mostly non-overlapping toxicities of radiation therapy and chemotherapy allow them to both be given with minimal compromise of either treatment modality. Toxicities increase when the two modalities are combined, but not excessively. The main toxicity that has often become dose limiting is esophagitis (KELLY et al. 1998).

Conceptually, radiation therapy attempts to eradicate locoregional disease while chemotherapy sensitizes cancer cells to radiation injury as well as attempts to eradicate distant micro-metastatic disease. A variety of approaches to combining chemotherapy and radiation therapy have been utilized in unresectable stage-III disease, including induction chemotherapy followed by radiation (DILLMAN et al. 1996; LE CHEVALIER et al. 1991; SAUSE et al. 2000), concurrent chemoradiotherapy (FURUSE et al. 1999), induction chemotherapy followed by concurrent chemoradiotherapy, and concurrent chemoradiotherapy followed by consolidative chemotherapy (Fig. 21.2; GANDARA et al. 2003).

A platinum-based regimen with concurrent radiation therapy is recommended for patients with good performance status and minimal weight loss. Cisplatin/etoposide or carboplatin/paclitaxel are the most commonly used regimens during thoracic radiation therapy. Consolidation chemotherapy has been associated with 3-year survival rates of 37% (GANDARA et al. 2003). The optimal method of integrating novel therapeutic agents, such as inhibitors of the epidermal growth factor receptor and angiogenesis pathways, remains to be studied (GANDARA et al. 2005).

As new chemotherapeutic agents are incorporated into multimodality regimens, the traditional parameters of radiation therapy dose, fractionation, and volume will need to be addressed. Caution needs to be used as increased and unexpected toxicities may arise (CHOY et al. 1997; GRAHAM et al. 1996; GREGOR et al. 1997; LANGER et al. 1997; ROSENTHAL et al. 1997).

21.9.3
Postoperative Care

The role of postoperative radiation therapy has been diminishing because randomized trials have failed to show a survival advantage; however, none of the trials reported to date have had the statistical power to prove or disprove a survival effect in N2 patients (DAUTZENBERG et al. 1995; THE LUNG CANCER STUDY GROUP 1986); many have shown a significant local tumor and disease-free survival improvement (DAUTZENBERG et al. 1995; THE LUNG CANCER STUDY GROUP 1986). On the other hand, a meta-analysis of 2128 patients who received postoperative radiation therapy in nine trials showed an adverse effect on survival for stages I and II. The results with the few patients with stage-III disease showed no significant benefit or detriment to radiation therapy. This meta-analysis has several major criticisms, including antiquated radiation techniques and inadequate treatment volumes. In addition, the role of adjuvant chemotherapy is rapidly developing, further complicating the role of adjuvant radiation therapy. Patients with positive surgical margins or N2 disease are usually offered adjuvant radiation therapy. Both the dose, which varies from 50 to 66 Gy, and the treatment volume, depend on the pathological findings.

The RTOG recently reported a phase-II trial of postoperative radiation therapy with concurrent carboplatin and paclitaxel for resected stage-II and stage-IIIA patients. The treatment toxicity was considered acceptable. The promising 3-year survival of 61% with this regimen merits further evaluation in clinical trials (BRADLEY et al. 2005b).

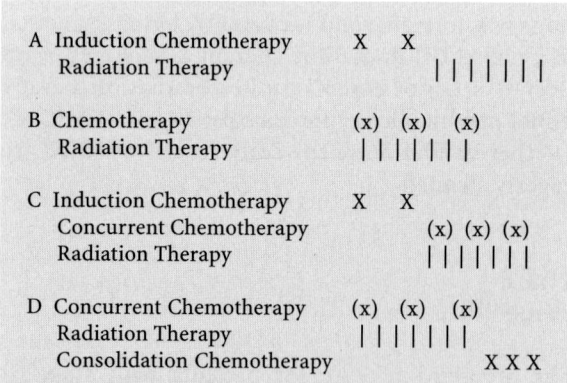

Fig. 21.2 Common combined modality regimens used to treat unresectable non-small cell lung cancers: induction chemotherapy followed by radiation therapy (A), concurrent chemoradiation (B), induction chemotherapy followed concurrent chemoradiation (C), concurrent chemoradiation followed by consolidative chemotherapy (D). Reduced dose chemotherapy (x), full-dose chemotherapy (X), radiation therapy (|)

21.10
Dose, Fractionation, Volume

21.10.1
Tumor Dose

Prior to recent studies showing the benefit of combining chemotherapy and radiation, radiation therapy alone had been the standard treatment for unresectable stage-III NSCLC. Based on a randomized trial by the RTOG 7301 (PEREZ et al. 1980), a dose of 60 Gy became accepted as the standard minimal tumor dose. In this study, patients with stage-III NSCLC were randomized to four radiation therapy regimens: 40-Gy split course; and 40-, 50-, and 60-Gy continuous course (2 Gy/day, 5 days per week). The survival rates were 20% for the 50- and 60-Gy arms and 10% for the 40-Gy arms. The complete and partial response rates were 56, 53, and 48%, respectively, for the 60-, 50-, and 40-Gy continuous schedules. The intrathoracic failure rates trended towards improvement with increasing dose from 40 to 60 Gy on the continuous schedules. Tumors less than 3 cm in diameter had a tumor control of 60% as opposed to only 40% for larger lesions (PEREZ et al. 1987). These observations support the need for higher doses of radiation in order to control larger tumors, although this increase must be tempered by the effect of large doses of radiation on surrounding normal tissues and the possibility of serious complications.

Dose escalation may be a useful means to improve local control and survival for NSCLC, if a critical relationship exists between the dose of radiation administered to the tumor and the probability of controlling the lesion (MEHTA et al. 2001). On the basis of basic principles advocated by FLETCHER (1973), doses in the range of 80–100 cGy might be required to sterilize tumors of the size frequently treated in bronchogenic carcinoma. Some authors, such as SALAZAR et al. (1976), CHOI and DOUCETTE (1981), MANTRAVADI et al. (1989), and PEREZ et al. (1982) have shown improved survival rates in patients receiving higher doses of radiation up to 3 years after radiation. Two retrospective studies have shown improved 5-year survival with 70 Gy/7 weeks vs lower doses (BALL et al. 1993; WURSCHMIDT et al. 1994). Definitive doses of radiation therapy currently are approximately 61–70 Gy with conventional fractionation; however, the optimal dose of radiation for treatment of non-oat cell bronchogenic carcinoma has not been well defined.

A major goal of 3D CRT is to escalate dose to the tumor to improve local control while minimizing toxicity. The development of 3D conformal radiation therapy (3D CRT) in the late 1980s using CT-based planning techniques is an important treatment advance for the treatment of lung cancer. CT-based planning offers more accurate definition of the tumor volume and normal tissues than radiographs. A beam's-eye-view (BEV) capability allows optimization of portal design and beam arrangements with the aim of more precise dose delivery to the target while minimizing irradiation of normal tissue (GRAHAM et al. 1994; VIJAYAKUMAR et al. 1991). Delineation of normal tissue has led to the rapid development of dosimetric models to predict normal tissue complication probabilities.

Several treatment-planning studies have reported the feasibility of increasing doses to NSCLC without increasing toxicity (GRAHAM et al. 1994; ARMSTRONG et al. 1993; HA et al. 1993). GRAHAM (1996, 1997) found that pneumonitis could be best predicted by dose volume histogram (DVH) analysis of the total lung volume. The percent volume of the total lung exceeding 20 Gy (V20) has been highly predictive in assessing risk of pneumonitis.

The primary objective of the University of Michigan phase-I dose escalation study was to establish the maximum tolerated dose using conventional fractionation (HAYMAN et al. 2001). A secondary objective was to better define the relationship between pulmonary toxicity, dose, and volume. From 1992 to 1999, 104 patients were entered and stratified based on their effective volume (V_{eff}), a measure of the amount of irradiation to normal lung tissue and risk of pneumonitis (KUTCHER and BURMAN 1989). In patients with the smallest V_{eff}, doses of 102.9 Gy were safely tolerated without significant pulmonary toxicity (NARAYAN et al. 2004).

RTOG 9311 was a multi-institutional phase-I/II trial of radiation therapy dose escalation using 3D CRT for inoperable stage-I to stage-III NSCLC (BRADLEY et al. 2005a). Although a few patients received induction chemotherapy, concurrent chemotherapy was not allowed. Patients were stratified to three groups with individual dose escalation levels based on the V20. For V20 <25%, dose levels were 70.9 Gy/33 fractions, 77.4 Gy/36 fractions, 83.8 Gy/39 fractions, and 90.3 Gy/42 fractions. For V20 between 25 and 37%, dose levels were 70.9 and 77.4 Gy. Dose escalation for patients with V20 >37% closed early due to poor accrual. Dose was safely escalated to 83.8 and 77.4 Gy for V20 of <25% and 25–37%, respectively.

The use of conformal dose escalation combined with chemotherapy has also been investigated. The Carolina Conformal Therapy Consortium has found approximately 80 Gy to be the maximum tolerated dose when used an accelerated hyperfractionated regimen following induction chemotherapy. Using once daily fractions, investigators at the University of North Carolina have safely increased dose to 90 Gy in 2 Gy/day with concurrent weekly carboplatin and paclitaxel following full-dose induction chemotherapy. The phase-II component of RTOG 0117 is a multi-institutional study of the efficacy of concurrent weekly carboplatin and paclitaxel plus high-dose conformal radiation therapy to 74 Gy given in 2 Gy/day. Dose escalation beyond 70 Gy with concurrent chemotherapy for NSCLC remains an active area of investigation.

21.10.2
Fractionation

Multiple daily fractions have been advocated for the delivery of higher doses of radiation to the tumor without enhancing morbidity in the normal tissue (Cox and BAUER 1988). Theoretically, repair of sublethal damage occurs between the fractions, when separated by 4–6 h. Because the total dose of radiation is given in a shorter period of time, and higher doses can be delivered, a greater biological effect is anticipated.

The RTOG conducted a multi-institutional phase-I/II prospective dose-escalation study evaluating the effect of hyperfractionation (HFX) in tumor control and the survival of patients with NSCLC (Cox et al. 1990). Fractions of 1.2 Gy were administered twice daily (at 4- to 6-h intervals). Patients were randomized to escalating doses between 60 and 79.2 Gy. Among 519 patients, 248 had a favorable prognosis (performance status 70–100 and weight loss of <5%) and 271 had an unfavorable prognosis (performance status 50–69 or weight loss >5%). Response rates were not reported in this study. No significant difference in disease-free survival was found among the five arms in the group of patients with an unfavorable prognosis. Patients with a favorable prognosis showed significant benefit (P=0.04) in survival and disease-free survival with 69.6 Gy compared with the lower total doses (median 14.8 months and 2-year survival of 33%). Grade 3 or greater pulmonary toxicity was reported in 5–10% of patients receiving ≥69.6 Gy.

Subsequently, however, the hyperfractionated regimen of 69.6 Gy has not been shown to improve survival compared with conventionally fractionated regimens of radiation (SAUSE et al. 2000). Furthermore, the addition of concurrent chemotherapy to HFX has not been shown to be superior to chemoradiation (RTOG 9410). One possible explanation for the lack of benefit of hyperfractionated regimens is the treatment duration. Evidence suggests that epithelial tumors have a doubling time of approximately 5 days. Such rapidly proliferating tumors may repopulate over a prolonged treatment duration of 6–7 weeks (WITHERS 1982); therefore, accelerated fractionation (decreasing the treatment duration) regimens have been developed for NSCLC.

SAUNDERS et al. (1999) reported results of a large prospective trial of continuous hyperfractionated accelerated radiotherapy (CHART) vs conventional 60 Gy/30 fx for inoperable NSCLC. The CHART regimen delivery scheme was 1.5 Gy three times daily for 36 fractions (total dose 54 Gy). This CHART regimen resulted in an improved overall 2-year survival, 20 vs 29% (p=0.008) and a 21% reduction in the relative risk of local progression.

The Eastern Cooperative Oncology Group compared a sequential approach of full-dose induction chemotherapy followed by conventional fractionated radiation therapy (64 Gy/32 fractions) vs 57.6 Gy of hyperfractionated accelerated radiation therapy (HART; 1.5 Gy three times daily for 2.5 weeks). Unlike CHART, HART did not include weekend treatment. This trial closed early due to slower-than-expected accrual and increased esophagitis. Although not statistically significant, the 2- and 3-year overall survival rates favored the HART regimen (44 and 24% vs 34 and 18%, respectively). The trend towards improved survival of HART merits additional investigation in rapidly proliferating tumor cell types.

21.10.3
Elective Nodal Irradiation

Based on knowledge of patterns of spread and pathological information, standard radiation therapy portals (Fig. 21.3) for NSCLC have included not only sites of gross disease, but also ipsilateral hilar, bilateral, mediastinal and, occasionally, ipsilateral supraclavicular lymph node regions. These fields are treated to 45–50 Gy followed by an additional 15–20 Gy to sites of gross disease. The initial fields are treated using an anteroposterior/posteroanterior beam arrangement.

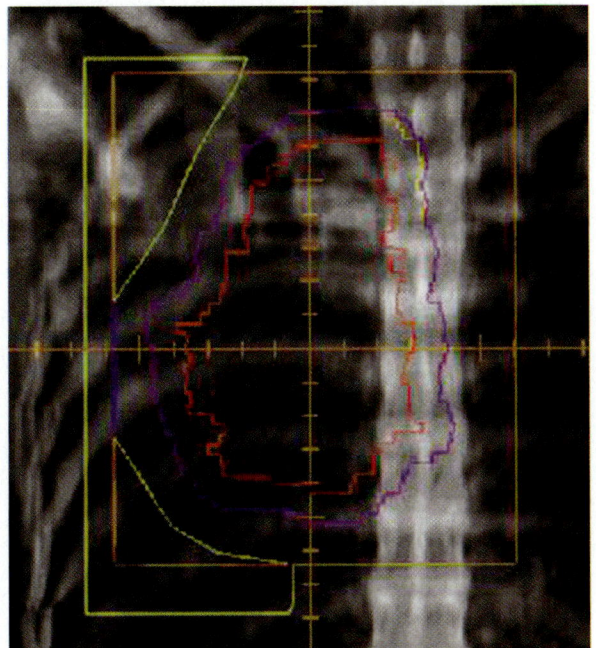

Fig. 21.3 Example of anteroposterior treatment portal for non-small cell lung cancer. A 66-year-old woman with a T4N2M0 (IIIB) adenocarcinoma of the right upper lobe invading into the pulmonary artery with large ipsilateral hilar and mediastinal nodes was treated with radiation and concurrent chemotherapy. Her initial fields were anteroposterior/posteroanterior to include the primary right upper lobe mass, right hilum, and mediastinum. Careful attention is paid to delineate the gross tumor volume (GTV) at the primary site and nodes. Blocks are drawn to include a margin of 1.5–2.0 cm on the GTV and to include the remaining hilar and mediastinal nodal regions

The ability to accurately define the tumor volume is of paramount importance for 3D CRT; however, the sensitivity and specificity of CT scanning for lymph node metastasis is only in the range of 60 and 79%, respectively (GOULD et al. 2003).

The use of PET with CT for staging the mediastinal nodes has improved sensitivity and specificity to 80–90% (DWAMENA et al. 1999). The use of both PET and CT for treatment planning should help the radiation oncologist more accurately contour the involved hilar and mediastinal nodes. The use of PET for radiation treatment planning has been shown to alter the stage of disease and radiation field design (BRADLEY et al. 2004b; ERDI et al. 2004; MAH et al. 2002); however, until the optimal imaging parameters to define the tumor volume on PET are further elucidated, the incorporation of PET data into radiation treatment planning should be individualized (ASHAMALLA et al. 2005).

21.11
Radiation Therapy Technique

21.11.1
Introduction

Treatment planning for thoracic tumors is complex. In order to ensure safe and effective treatment, several issues must be considered: (a) accurate target volume delineation; (b) proximity of sensitive structures (lung, esophagus, spinal cord, and heart) to the tumor; (c) sloping surface of the chest; (d) presence of tissues with nonuniform densities, such as lung and bone, in the treatment volume; (e) frequent requirement for irregular field dose calculations; and (f) respiratory motion of the target and normal structures such as heart and lung.

21.11.2
Simulation

Simulation is performed with the patient supine and arms placed above the head. A thorax board or other immobilization device helps reproduce the treatment position. A CT scan of the entire thoracic cavity, including lung apices to base, allows for 3D delineation of tumor and normal tissues which can be used for treatment planning and dose volume analysis. Fluoroscopic observance of tumor motion

The remaining treatments are given using off-cord oblique and cone-down fields to limit the dose to the spinal cord to 45 Gy.

Recent studies have challenged the traditional ideas of electively treating the noninvolved, node-bearing areas in the hilum, mediastinum, and supraclavicular fossa (EMAMI et al. 2003). The local control of gross disease is disappointingly low at 60 Gy (BRADLEY 2005). Most recurrences are within the sites of gross disease (CHEUNG et al. 2000). Isolated failures in the elective nodal areas are uncommon occurrences (e.g., RTOG 93-11; BRADLEY et al. 2004a; ROSENZWEIG et al.2001; SENAN et al. 2002). Furthermore, elective nodal irradiation does increase the volume of normal lung and esophagus irradiated; therefore, methods to intensify the dose to the gross tumor volume (GTV) to improve local control while sparing normal tissue irradiation are warranted. One method to accomplish treatment intensification is radiation dose escalation.

or diaphragm excursion during normal breathing helps determine an appropriate margin of expansion to account for respiratory motion.

21.11.3
Target Delineation

Newer imaging modalities are increasingly being used in the diagnosis and staging of lung cancer. Radiation oncologists are called upon to have an increased ability to interpret CT data and define tumor targets vs normal anatomy (VIJAYAKUMAR et al. 1991). Modern treatment-planning systems allow the fusion of complementary imaging modalities, such as PET and single photon emission CT, for improved target volume delineation (PURDY et al. 2000). The International Commission on Radiation Units report no. 50 guidelines for defining targets can be applied to lung cancers (ICRU 1993). The GTV includes the extent of primary lung tumor and any enlarged hilar or mediastinal nodes as defined by imaging studies such as CT or PET, bronchoscopy, or mediastinoscopy (Fig. 21.4). The clinical target volume (CTV) includes potential sites of microscopic disease either surrounding the GTV or also including hilar and mediastinal nodes. Although this margin is typically 5 mm, a study reported by GIRAUD et al. (2000) found that 5 mm would cover the microscopic extension in only 80% of the adenocarcinomas and 91% of the squamous cell carcinomas evaluated. The planning target volume (PTV) accounts for intrafraction organ motion and daily setup uncertainty.

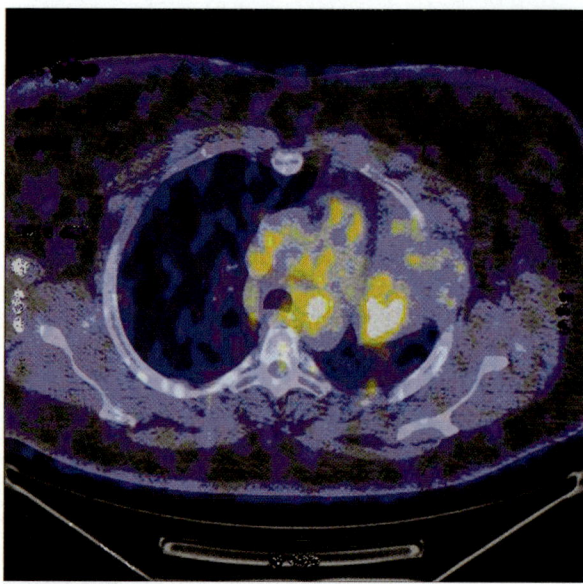

Fig. 21.4 Example of how combined PET/CT may help to define the tumor volume for non-small cell lung cancer. A 48-year-old woman presented with shortness of breath. A chest X-ray revealed a left hilar opacity and left upper lobe collapse. A CT scan of the chest demonstrated an endobronchial lesion near the origin of the left upper lobe, left upper lobe collapse, and multiple nodes measuring 1–1.5 cm in the left hilar, left paratracheal regions as well as lateral to the aortic arch. On bronchoscopy, a left mainstem endobronchial mass within 1 cm of the carina was encountered with biopsy revealing squamous cell carcinoma. A PET/CT demonstrated increased FDG uptake in the left hilar mass, and paratracheal and peri-aortic nodes. An anterior mediastinoscopy revealed six positive nodes in station 6. She was treated with radiation and concurrent chemotherapy. Her PET/CT images were used to help delineate the primary tumor from regions of atelectasis as well as the extent of involvement of the mediastinal nodes.

21.11.4
Standard and Conformal Treatment Planning

Traditional treatment planning based on anterior and posterior radiographs (anteroposterior/posteroanterior) calls for a 2-cm margin around visible gross disease and a 1-cm margin around electively treated nodal regions in the hilum and mediastinum (BRADLEY et al. 2003), using irregularly shaped blocks (Fig. 21.3). After 45–50 Gy to the elective nodes, the treatment portals are reduced and the remainder of the dose is given to the gross disease at the primary and nodal sites (Table 21.3).

To maintain the spinal cord dose at or below 45 Gy, off-cord field arrangements, such as obliques, are used to deliver the remainder of the dose. Depending on the energy and patient separation, the dose deliv-

ered to the spinal cord may vary and it may often be necessary to switch to an oblique field sooner than 45 Gy. Posterior spinal cord blocks should no longer be used.

The sloping surface of the chest, especially at the thoracic inlet, results in varying source-tumor distances over the treatment field and must be accounted for in treatment planning. The sloping-surface effect may be corrected by the use of wedges or compensating filters. Several of these devices have been described, and some are commercially available.

The goal of 3D conformal treatment planning is to maximize the dose to the tumor while minimizing the dose to adjacent normal tissues. The major advantages of conformal planning systems (Fig. 21.5) are the abilities to accurately define the tumor volume and normal tissue, use multiple –

Table 21.3 Standard radiotherapy guidelines for stage-III A/B non-small cell lung cancer

Volume	Dose (cGy)	Fractionation (cGy)
I. Subclinical disease: "Large Volume Target" (uninvolved mediastinum and hilar nodes; supraclavicular nodes in selected cases)	4500–5000	180–200
II. Primary and involved nodes with 1- to 2-cm margin: "Involved Target Volume"	6000–6600	180–200

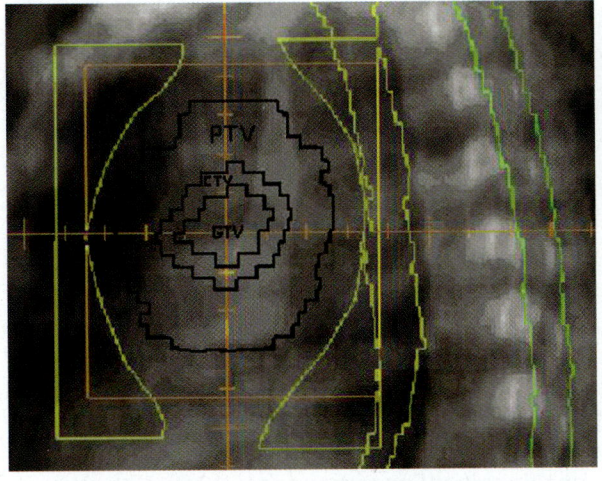

a

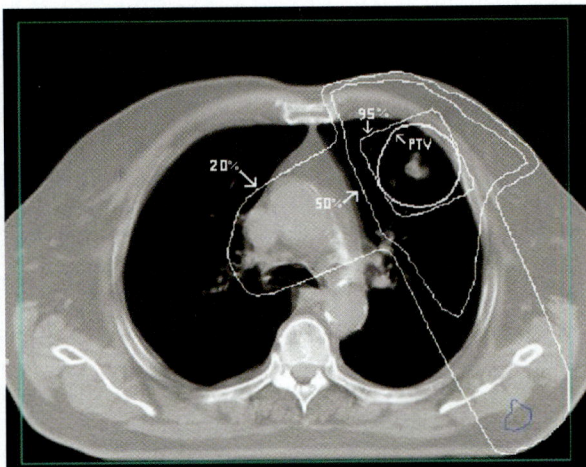

b

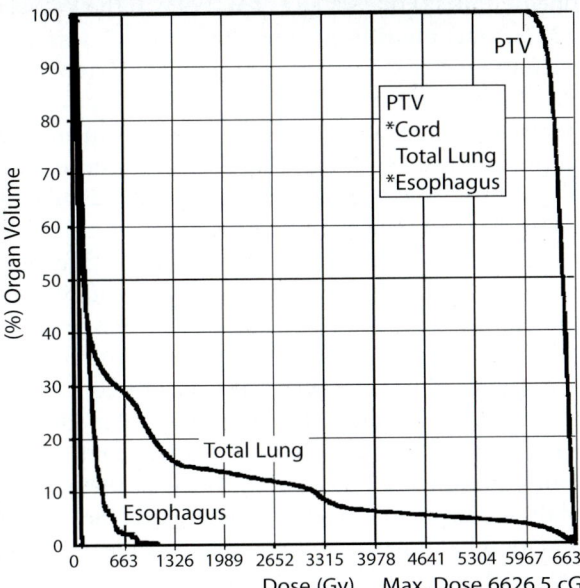

c

Fig. 21.5a-c. Example of a three-dimensional treatment plan depicting the beam's-eye-view display of a treatment portal (**a**), conformal dose distribution (**b**), and dose-volume histogram (**c**). A 77-year-old man with significant comorbid conditions, including ischemic cardiomyopathy, congestive heart failure, and emphysema, was diagnosed with a T1N0M0 (IA) non-small cell lung cancer of the left upper lobe, measuring 1.9 cm in largest dimension. Due to his comorbidities and FEV1 <0.9 l/min, he was considered medically inoperable. The GTV was defined using both a recent diagnostic CT as well as lung windows on planning CT. The GTV was expanded by 5 mm for CTV. To account for the respiratory motion and daily setup variation of the patient, an additional 1.5 cm in the cranial and caudal directions, and 1 cm radially, were added for PTV (**a**). A right anterior oblique, left anterior oblique, and left posterior oblique beam arrangement was designed to encompass the PTV by the 95% isodose surface (**b**). A total dose of 6500 cGy in 26 fractions was given with minimal acute toxicity. In order to minimize the dose to the surrounding lung in this patient, elective nodal regions were not treated. The dose volume histogram (**c**) reveals a V20 of 14%, V30 of 10%, and mean lung dose of 834 cGy

including noncoplanar – beams based on the shape of the target, and iteratively adjust beam weights and wedges for optimized target coverage, normal tissue sparing, and more accurate dose calculation. Dose volume histograms for the lungs, heart, esophagus, and spinal cord assist the treatment planner and physician in choosing an appropriate plan and reducing the chance of complications.

21.11.5
Energy and Lung Heterogeneity

The appropriate beam energy and use of tissue density correction for intrathoracic tumor is controversial (YORKE et al. 1996; EKSTRAND and BARNES 1990; MCDONALD et al. 1976; MACKIE et al. 1985; RICE et al. 1988; WHITE et al. 1996). Because of the large path length of the thorax, many health care practitioners prefer the use of high-energy photons (≥10 mV). Higher-energy photons allow reduced hot spots in the chest wall and surrounding critical organs such as the spinal cord and heart; however, there are several disadvantages to high energies. High-energy photons may have problems with equilibrium re-establishment of small tumors which are surrounded by lung tissue (MACKIE et al. 1985; KLEIN et al. 1997). YORKE et al. (1996) have demonstrated the effects of small fields for 10 mV where tumors have a 10% lower dose at the periphery vs the center. This problem is a result of the loss of electronic equilibrium in tumors embedded within the periphery of the lung, but is most significant when small field sizes are used. Furthermore, high-energy photons also result in a widened penumbra within low-density tissue (EKSTRAND and BARNES 1990; KLEIN et al. 1997). This phenomenon may increase the risk of pneumonitis and may not be reflected in some current treatment-planning isodose displays (KAN et al. 1995).

The routine implementation of lung-density corrections has remained controversial. The main reasons are that most clinical experience with lung cancer has assumed a homogeneous density, most cooperative group trials from the RTOG and SWOG have continued to use homogeneous treatment planning, and no standard method for prescribing dose with heterogeneity correction has been accepted. KLEIN et al. (1997) have recommended caution when applying corrections for lung cancer radiotherapy planned with 3D RTP. The authors reported overestimation of correction factors, especially at tumor borders. Furthermore, incorporating heterogeneity correction into dose calculations resulted in underdosing due to loss of electronic equilibrium with 15-mV photons (MACKIE et al. 1985).

Nevertheless, the merits of using lung-density correction have been debated for years. Modern radiation treatment-planning systems and newer dose-calculation algorithms take into account tissue heterogeneity (PAPANIKOLAOU and KLEIN 2000). One study from MD Anderson Cancer Center demonstrated superior PTV coverage with the use of lung-density corrections (FRANK et al. 2003); however, its routine implementation has remained controversial.

21.11.6
Normal Tissue Considerations

Treatment planning must provide adequate dose coverage of the target while not exceeding the tolerance doses of the sensitive normal organs in the thorax, such as the spinal cord, lung, esophagus, and heart. EMAMI (1994) published a consensus opinion on radiation tolerance doses and volumes based on clinical experience. Most clinicians agree that a maximum dose of 45 Gy to the spinal cord is associated with a very low incidence of injury. A report from the Princess Margaret Hospital showed no incidence of myelopathy in patients treated with once daily fractionated radiotherapy to an equivalent dose of 50 Gy in 25 fractions (WONG et al. 1994). With traditional treatment planning, planners are cautioned to keep portals to ≤1.0–2.0 cm around the primary and nodal disease to keep lung toxicity within acceptable limits (BYHARDT et al. 1993); however, the dose and volume relationship regarding partial organ irradiation of the lung and esophagus is less well understood.

Recently, several authors have developed dosimetric parameters to predict the development of pneumonitis (GRAHAM 1997; BOERSMA et al. 1992; FORASTIERE et al. 1993; KWA et al. 1998; MARKS et al. 1997; MARTEL et al. 1994). GRAHAM (1997) reported the percent volume of the total lung exceeding 20 Gy (V_{20}) to be the best predictor of the development of pneumonitis (FORASTIERE et al. 1993). If the V_{20} was <32%, pneumonitis greater than grade 3 was not encountered. Similarly, MARKS et al. (1997) reported the V_{30} to be most predictive of the development of pneumonitis. KWA et al. (1998) related the incidence of pneumonitis with the total lung mean dose. The V_{20}, V_{30}, and mean lung dose can be easily calculated using 3D treatment-planning systems.

MARTEL et al. (1994) described the stratification of patients for their development of pneumonitis on the effective volume (V_{eff}) and normal tissue complication probability (NTCP) from lung-dose volume histograms. Their data were subsequently confirmed by OETZEL et al. (1995). The uses of NTCP models as well as predictors of esophageal toxicity continue to evolve (SINGH et al. 2003; MAGUIRE et al.1999). Each of these parameters gives clinicians guidelines to use and develop improved radiation treatment plans, especially when using conformal techniques.

21.12
Patterns of Failure

PEREZ et al. (1984) reported results of an early RTOG trial that the tumor failure rate within the irradiated volume was 48% with 40 Gy delivered continuously, 38% with 40 split course or 50 Gy continuously, and

27% with 60 Gy continuously. The failure rate in the nonirradiated lung ranged from 25 to 30% in various groups (TUCKER et al. 1997). Distant metastases, most commonly involving the brain, bone, liver, and adrenals, occur in 50–80% of patients. Even in early stage disease, distant recurrence is the most common failure pattern (FELD et al. 1984; IMMERMAN 1981).

Of particular concern to the radiation oncologist, however, were the results of trials where biopsy by bronchoscopy was required after combined chemotherapy and radiation therapy (65 Gy; LE CHEVALIER et al. 1994). Local control was only 15–17%. Furthermore, in patients achieving complete or partial response, the incidence of brain metastasis was 16% for squamous cell carcinoma and 30–40% for adenocarcinoma and large-cell undifferentiated carcinoma. Whether prophylactic cranial irradiation for NSCLC provides benefit is currently being investigated in a large, cooperative group study.

21.13
Management of Small Cell Carcinoma

The incidence of small cell lung cancer has decreased to approximately 15% of lung cancer cases in the United States. Untreated small cell lung cancer is rapidly metastatic and fatal; indeed, two-thirds of patients present with metastatic disease. The remaining one-third usually present with bulky, centrally located tumors confined to the chest. Currently, the staging of small cell lung cancer is either limited or extensive, depending on whether the disease extent can be safely encompassed in a radiation portal. In extensive-stage disease the role is limited to palliative treatment of symptomatic disease failing to respond to chemotherapy; however, thoracic radiation therapy has an important role in the curative treatment of limited stage small cell lung cancer (LOOPER and HORNBACK 1984; TURRISI 1997). In limited disease the central issues include (a) dose, (b) fractionation, (c) treatment volume and timing with systemic therapy, and (d) prophylactic cranial irradiation (PCI).

21.13.1
Thoracic Radiotherapy

In the 1970s and 1980s, several small phase-III trials compared chemotherapy alone vs thoracic radiotherapy (TRT) plus chemotherapy (CHUTE et al. 1997; KIES et al. 1987; PERRY et al. 1987; WARDE and PAYNE 1992; MIRA et al. 1982). Individually, the small size of the trials resulted in unclear benefit and small differences in outcome; however, two meta-analyses both concluded that the addition of TRT to systemic therapy resulted in improved local control and overall survival compared with systemic chemotherapy alone (PIGNON et al. 1992; WORK et al. 1997). PIGNON et al. (1992) reported an improved 3-year survival for 14.3% with CT plus TRT vs 8.9% with CT alone. Local control is also improved with TRT plus chemotherapy. PERRY et al. (1987) reported a 50% 3-year actuarial local control with 50-Gy concurrent TRT vs 10% without TRT.

21.13.2
Dose

CHOI and CAREY (1989), in a retrospective review, reported only 50% local control with 30–35 Gy, and 70% local control with doses of 40–50 Gy. These data suffer from their retrospective nature and inherent biases of patient selection, era of treatment, etc. Prospective studies using 40–50 Gy showed local control rates of only 30–50% (PERRY et al. 1987; WORK et al. 1997; COY et al. 1988). Common practice of the era used posterior spinal cord blocks to protect the spinal cord, although it is clear that this resulted in underdosing to the mediastinum and probably tumor. This practice is no longer considered acceptable. Even with the use of modern radiation therapy techniques and concurrent chemotherapy, local control has been only 48% with 45 Gy (TURRISI et al. 1999). Recent dose escalation studies have been completed indicating that doses of 61.2–70 Gy can be delivered safely (KOMAKI et al. 2005; CHOI et al. 1998). KOMAKI et al. (2005) reported a 68% complete response rate with doses ranging from 50.4 to 64.8 Gy using an accelerated hyperfractionated concomitant boost regimen.

21.13.3
Fractionation

Besides dose escalation, altered fractionation regimens have been shown to improve local control. Laboratory studies have suggested that typical SCLC cell lines have radiation survival curves with little to no shoulders and have suggested that accelerated fractionation schemas would, therefore, be advantageous.

Indeed, accelerated fractionation regimens have generated promising results with median survivals of 18–27 months and 2-year survivals of 19–60% (JOHNSON et al. 1993, 1996; TURRISI et al. 1988). The favored regimen from several phase-II trials combining BID TRT with cisplatin and etoposide is 45 Gy/30 fractions over 3 weeks. Each fraction is 1.5 Gy on a twice-daily schedule. A multi-institutional trial compared cisplatin/etoposide (four cycles) with once-daily TRT (45 Gy/35 fx every 5 weeks) vs twice-daily TRT 45 Gy (30 fx/3 weeks) in 471 patients with limited stage SCLC (TURRISI et al. 1999). Although the incidence of grade-3 esophagitis was worse, 27 vs 11%, both the local control and survival rates significantly favored the accelerated treatment arm. The 5-year overall survival was improved in the twice-daily arm, 26 vs 16% (p=0.04). The local failure rates were improved from 52 to 36% in the twice-daily arm. Despite the improved results with accelerated twice-daily radiation compared with a once-daily regimen to 45 Gy, the reluctance to use it widely as well as the optimal dose to use for once-daily treatment have been discussed elsewhere (TURRISI 2004).

21.13.4
Timing with Systemic Therapy and Treatment Volume

The issue of timing of radiation therapy (early vs late) has been studied in several randomized trials (SAUSE et al. 2000; PERRY et al. 1987; WORK et al. 1997; JEREMIC et al. 1997; LEBEAU et al. 1993; MURRAY et al. 1993; TAKADA et al. 2002). Three trials reported a survival benefit to early thoracic radiation therapy (JEREMIC et al. 1997; MURRAY et al. 1993; TAKADA et al. 2002). In the Japanese Clinical Oncology group trial, 231 patients were randomized to concurrent cisplatin, etoposide for four cycles, and radiation therapy beginning on day 2 vs the same chemotherapy with TRT beginning with the fourth cycle of chemotherapy. The complete response rates were 40 vs 27% in favor of the early TRT arm. The 5-year overall survival was 23.7 vs 18.3% in favor of the early TRT (p=0.097).

If TRT is delayed until after chemotherapy, the issue arises as to whether to treat the prechemotherapy volume of disease vs the smaller postchemotherapy volume. Some authors advocate generous portals that would include the primary with generous margins as well as both hilar regions, the entire mediastinum, and both supraclavicular areas. Other authors, however, argue that only limited portals

encompassing the prechemotherapy primary tumor with a 1-cm margin and high-risk nodal areas are adequate, because effective chemotherapy takes care of subclinical or microscopic disease and thus eliminates the need for generous portals. Moreover, the use of limited portals may reduce complications resulting from a combined therapeutic approach with radiation and chemotherapy. The Southwest Oncology Group (SWOG) addressed this question in the only randomized trial of its nature. Patients achieving a partial response without CT were randomized to initial vs postchemotherapy volume (KIES et al. 1987). No benefit was seen for the larger target volume. In a retrospective review, LIENGSWANGWONG et al. (1994) found no benefit to larger-volume TRT and no negative impact on survival to post-chemotherapy TRT volumes. A typical AP port for early treatment of SCLC is shown in Figure 21.6.

21.13.5
Prophylactic Cranial Irradiation

Prophylactic cranial irradiation has been a controversial component of treatment for SCLC. Much of the controversy centers around previous reports of late neurocognitive impairment and a lack of definitive evidence of survival improvement. Certainly, the brain is a common site of failure and may be as high as 80% for 2-year survivors (KOMAKI et al. 1981). Furthermore, brain metastasis impairs quality of life and shortens survival (FELLETTI et al. 1985).

Low-dose PCI has been shown to reduce the incidence of brain metastasis; however, demonstration of a benefit in overall survival has been lacking (KOMAKI et al. 1981; ROSENSTEIN et al. 1992), until recently. In a meta-analysis of seven trials comparing prophylactic cranial radiation (PCI) to no PCI for complete responders from chemotherapy with or without radiation therapy (AUPERIN et al. 1999), a statistically significant survival improvement for PCI was found. At 3 years, those receiving PCI had a 20.7% survival compared with 15.4% for those not receiving PCI. Disease-free survival was also improved from 13.3 to 22.1% (p<0.0001) as was the incidence of brain metastasis.

Several reports have reported on the neuropsychological sequelae after PCI in long-term survivors (LISHNER et al. 1990; FLECK et al. 1990). The frequency of mental deterioration, including memory deficits, cognitive decline, and language difficulties, was reported in as many as 86% of survivors

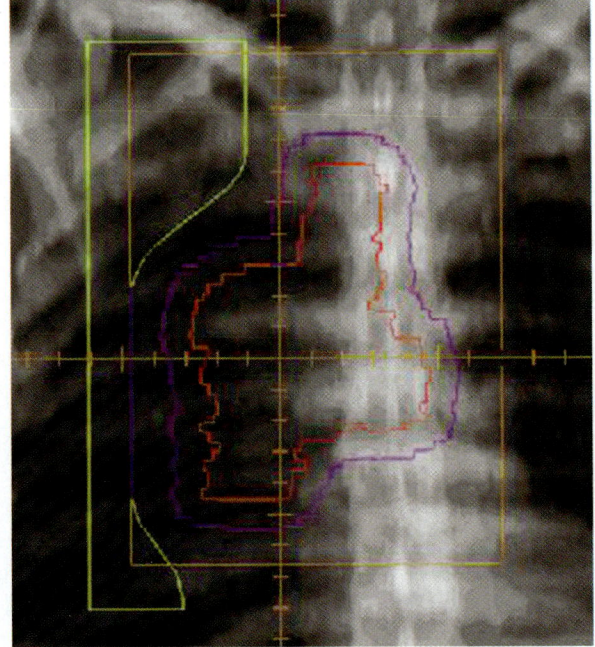

Fig. 21.6 Example of anteroposterior treatment portal for small cell lung cancer (SCLC). A 58-year-old man with limited-stage SCLC involving the right hilum and right paratracheal regions was treated with once-daily thoracic radiation to 6100 cGy and concurrent chemotherapy according to an institutional protocol. The GTV, which included the right hilar mass and enlarged paratracheal nodes, was expanded to create a PTV. His initial fields were anteroposterior/posteroanterior using 6 mV to a dose of 3960 cGy. A 540 cGy off-cord oblique fields were used to bring the mediastinal dose to 4500 cGy. The same oblique fields were coned down to treat the PTV for 1600 cGy in eight fractions

living 6–13 years after treatment (JOHNSON et al. 1990); however, KOMAKI et al. (1995) reported that over 90% of patients with SCLC demonstrate evidence of cognitive dysfunction even prior to PCI. These impairments involve memory, frontal lobe functions, and fine motor coordination. Furthermore, the authors report no significant neurological deterioration in 30 patients receiving PCI with follow-up of 6–20 months. Certainly, the neurocognitive effects seen in these patients is multifactorial in nature and could be partly related to treatment factors such as radiation fraction size (JOHNSON et al. 1990), use of concurrent chemotherapy with PCI, and type of chemotherapy (LEE et al. 1986; KANARD et al. 2004).

Two recent large randomized prospective trials provide data to show that PCI does not impair cognitive function or quality of life (GREGOR et al. 1997; ARRIAGADA et al. 1995). Both trials randomized patients with SCLC in complete remission to PCI or no PCI and both employed formal prospective neuropsychological testing. ARRIAGADA et al. (1995) found that PCI (24 Gy/8 fractions) significantly reduced the cumulative incidence of brain metastasis with a trend toward survival improvement. There were no differences in terms of neuropsychological function or radiographic abnormalities between the two groups. Likewise, GREGOR et al. (1997) found that PCI reduced the incidence of brain metastases with a trend toward improved survival.

The optimal dose and fractionation for PCI remains uncertain. Several regimens are currently used in practice in North America, including 25 Gy in 10 fractions, 24 Gy in 8 fractions, 30 Gy in 10 fractions, and 36 Gy in 18 fractions. There currently seems to be no consensus on what is the best dose, or if there is any clear relationship between hypofractionated regimens or total dose and neurological sequelae. An ongoing phase-II/III study coordinated by RTOG will compare low-dose vs higher-dose and daily vs hyperfractionated PCI regimens.

References

Albain KS et al (2003) Phase III comparison of concurrent chemotherapy plus radiotherapy (CT/RT) and CT/RT followed by surgical resection for stage IIIA(pN2) non-small cell lung cancer (NSCLC): initial results from intergroup trial 0139 (RTOG 93-09; abstr 2497). Proc Am Soc Clin Oncol 22:621

Armstrong JG et al (1993) Three-dimensional conformal radiation therapy may improve the therapeutic ratio of high dose radiation therapy for lung cancer. Int J Radiat Oncol Biol Phys 26:685–689

Arriagada R et al (1995) Prophylactic cranial irradiation for patients with small-cell lung cancer in complete remission. J Natl Cancer Inst 87:183–190

Arriagada R et al (2004) Cisplatin-based adjuvant chemotherapy in patients with completely resected non-small-cell lung cancer. N Engl J Med 350:351–360

Ashamalla H et al (2005) The contribution of integrated PET/CT to the evolving definition of treatment volumes in radiation treatment planning in lung cancer. Int J Radiat Oncol Biol Phys Jun 23 [Epub ahead of print]

Auperin A et al (1999) Prophylactic cranial irradiation for patients with small-cell lung cancer in complete remission. Prophylactic cranial irradiation overview Collaborative Group. N Engl J Med 341:476–484

Baird JA (1965) The pathways of lymphatic spread of carcinoma of the lung. Br J Surg 52:868–875

Ball D et al (1993) Longer survival with higher doses of thoracic radiotherapy in patients with limited non-small cell lung cancer. Int J Radiat Oncol Biol Phys 25:599–604

Blackstock AW et al (2002) Outcomes among African-American/non-African-American patients with advanced non-small-cell lung carcinoma: report from the Cancer and Leukemia Group B. J Natl Cancer Inst 94:284–290

Blomgren H et al (1995) Stereotactic high dose fraction radiation therapy of extracranial tumors using an accelerator. Clinical experience of the first thirty-one patients. Acta Oncol 34:861–870

Boersma LJ et al (1992) Three-dimensional superimposition of SPECT and CT data to quantify radiation-induced ventilation and perfusion changes of the lung; as a function of the locally delivered dose. In: Schmidt HA, Hofer R (eds) Proceedings of European Nuclear Medicine Congress, Vienna. New York, Schattauer

Bradley J (2005) A review of radiation dose escalation trials for non-small cell lung cancer within the Radiation Therapy Oncology Group. Semin Oncol 32:S111–S113

Bradley J, Govindan R, Komaki R (2003) Lung. In: Perez CA, Brady LW, Halperin EC, Schmidt-Ullrich RK (eds) Principles and practice of radiation oncology, 4th edn, Lippincott-Raven Publishers, New York, Chap. 44

Bradley J et al (2004a) Impact of FDG-PET on radiation therapy volume delineation in non-small-cell lung cancer. Int J Radiat Oncol Biol Phys 59:78–86

Bradley JD et al (2004b) Positron emission tomography in limited-stage small-cell lung cancer: a prospective study. J Clin Oncol 22:3248–3254

Bradley J et al (2005a) Toxicity and outcome results of RTOG 9311: a phase I–II dose-escalation study using three-dimensional conformal radiotherapy in patients with inoperable non-small-cell lung carcinoma. Int J Radiat Oncol Biol Phys 61:318–328

Bradley JD et al (2005b) Phase II trial of postoperative adjuvant paclitaxel/carboplatin and thoracic radiotherapy in resected stage II and IIIA non-small-cell lung cancer: promising long-term results of the Radiation Therapy Oncology Group–RTOG 9705. J Clin Oncol 23:3480–3487

Byhardt RW et al (1993) The influence of field size and other treatment factors on pulmonary toxicity following hyperfractionated irradiation for inoperable non-small cell lung cancer (NSCLC)-analysis of a Radiation Therapy Oncology Group (RTOG) protocol. Int J Radiat Oncol Biol Phys 27:537–544

Carter D, Eggleston JC (1980) Tumors of the lower respiratory tract. Armed Forces Institute of Pathology, Washington DC

Cheung PC et al (2000) Involved-field radiotherapy alone for early-stage non-small cell lung cancer. Int J Radiat Oncol Biol Phys 48:703–710

Choi NC, Doucette JA (1981) Improved survival of patients with unresectable non-small-cell bronchogenic carcinoma by an innovated high-dose en-bloc radiotherapeutic approach. Cancer 48:101–109

Choi NC, Carey RW (1989) Importance of radiation dose in achieving improved loco-regional tumor control in limited stage small-cell lung carcinoma: an update. Int J Radiat Oncol Biol Phys 17:307–310

Choi NC et al (1998) Phase I study to determine the maximum-tolerated dose of radiation in standard daily and hyperfractionated-accelerated twice-daily radiation schedules with concurrent chemotherapy for limited-stage small-cell lung cancer. J Clin Oncol 11:3528–3536

Choy H et al (1997) Preliminary analysis of a phase II study

of paclitaxel, carboplatin and hyperfractionated radiation therapy for locally advanced inoperable nonsmall cell lung cancer. Semin Oncol 24:S12-21–S12-26

Chute JP et al (1997) Outcome of patients with small-cell lung cancer during 20 years of clinical research at the U.S. National Cancer Institute. Mayo Clin Proc 72:901–912

Cox JD, Bauer M (1988) Therapeutic ratio and fractionation in cancer of the lung. Front Radiat Ther Oncol 22:121–126

Cox JD et al (1990) A randomized phase I/II trial of hyperfractionated radiation therapy with total doses of 60.0 Gy to 79.2 Gy: possible survival benefit with greater than or equal to 69.6 Gy in favorable patients with Radiation Therapy Oncology Group stage III non-small-cell lung carcinoma: report of Radiation Therapy Oncology Group 83–11. J Clin Oncol 8:1543–1555

Coy P et al (1988) The effect of dose of thoracic irradiation on recurrence in patients with limited stage small cell lung cancer. Initial results of a Canadian multicenter randomized trial. Int J Radiat Oncol Biol Phys 14:219–226

Dautzenberg B et al (1995) Adjuvant radiotherapy versus combined sequential chemotherapy followed by radiotherapy in the treatment of resected nonsmall cell lung carcinoma. A randomized trial of 267 patients. GETCB (Groupe d'Etude et de Traitement des Cancers Bronchiques). Cancer 76:779–786

Dillman RO et al (1996) Improved survival in stage III non-small-cell lung cancer: seven-year follow-up of cancer and leukemia group B (CALGB) 8433 trial. J Natl Cancer Inst 88:1210–1215

Dosoretz D et al (1992) Radiation therapy in the management of medically inoperable carcinoma of the lung: results and implications for future treatment strategies. Int J Radiat Oncol Biol Phys 24:3–9

Dwamena BA et al (1999) Metastases from non-small cell lung cancer: mediastinal staging in the 1990s meta-analytic comparison of PET and CT. Radiology 213:530–536

Ekstrand KE, Barnes WH (1990) Pitfalls in the use of high energy X rays to treat tumors in the lung. Int J Radiat Oncol Biol Phys 18:249–252

Emami B (1994) Management of hilar and mediastinal lymph nodes with radiation therapy in the treatment of lung cancer. In: Meyer JL (ed) The lymphatic system and cancer. Karger, Basel

Emami B, Graham MV (1998) Lung. In: Perez CA, Brady LW (eds) Principles and practice of radiation oncology, 3rd edn. Lippincott-Raven, Philadelphia, pp 1181–1220

Emami B et al (2003) The impact of regional nodal radiotherapy (dose/volume) on regional progression and survival in unresectable non-small cell lung cancer: an analysis of RTOG data. Lung Cancer 41:207–214

Erdi YE et al (2004) The CT motion quantitation of lung lesions and its impact on PET-measured SUVs. J Nucl Med 45:1287–1292

Farray D, N Mirkovic, Albain KS (2005) Multimodality therapy for stage III non-small-cell lung cancer. J Clin Oncol 23:3257–3269

Feinstein AR (1964) Symptomatic patterns, biologic behavior, and prognosis in cancer of the lung. Practical application of boolean algebra and clinical taxonomy. Ann Intern Med 61:27–43

Feld R, Rubinstein LV, Weisenberger TH (1984) Sites of recurrence in resected stage I non-small-cell lung cancer: a guide for future studies. J Clin Oncol 2:1352–1358

Felletti R et al (1985) Social consequences of brain or liver relapse in small cell carcinoma of the bronchus. Radiother Oncol 4:335–339

Fleck JF et al (1990) Is prophylactic cranial irradiation indicated in small-cell lung cancer? J Clin Oncol 8:209–214

Fletcher GH (1973) Clinical dose-response curves of human malignant epithelial tumours. Br J Radiol 46:1–12

Forastiere AA et al (1993) Preoperative chemoradiation followed by transhiatal esophagectomy for carcinoma of the esophagus: final report. J Clin Oncol 11:1118–1123

Frank SJ et al (2003) Treatment planning for lung cancer: traditional homogeneous point-dose prescription compared with heterogeneity-corrected dose-volume prescription. Int J Radiat Oncol Biol Phys 56:1308–1318

Furuse K et al (1999) Phase III study of concurrent versus sequential thoracic radiotherapy in combination with mitomycin, vindesine, and cisplatin in unresectable stage III non-small-cell lung cancer. J Clin Oncol 17:2692–2699

Gandara DR et al (2003) Consolidation docetaxel after concurrent chemoradiotherapy in stage IIIB non-small-cell lung cancer: phase II Southwest Oncology Group Study S9504. J Clin Oncol 21:2004–2010

Gandara D et al (2005) Integration of novel therapeutics into combined modality therapy of locally advanced non-small cell lung cancer. Clin Cancer Res 11:5057–5062

Gdeedo A et al (1997) Prospective evaluation of computed tomography and mediastinoscopy in mediastinal lymph node staging. Eur Respir J 10:1547–1551

Ginsberg RJ, Rubinstein LV (1995) Randomized trial of lobectomy versus limited resection for T1 N0 non-small cell lung cancer. Lung Cancer Study Group. Ann Thorac Surg 60:615–622

Giraud P et al (2000) Evaluation of microscopic tumor extension in non-small-cell lung cancer for three-dimensional conformal radiotherapy planning. Int J Radiat Oncol Biol Phys 48:1015–1024

Glazer GM et al (1985) Normal mediastinal lymph nodes: number and size according to American Thoracic Society mapping. AJR Am J Roentgenol 144:261–265

Goldberg EM, Shapiro CM, Glicksman AS (1974) Mediastinoscopy for assessing mediastinal spread in clinical staging of lung carcinoma. Semin Oncol 1:205–215

Gould MK et al (2003) Test performance of positron emission tomography and computed tomography for mediastinal staging in patients with non-small-cell lung cancer: a meta-analysis. Ann Intern Med 139:879–892

Graham MV (1996) Three-dimensional conformal radiotherapy for lung cancer: the Washington University Experience. In: Meyer JL, Purdy JA (eds) Frontiers of radiation therapy and oncology. Karger, Basel

Graham MV (1997) Predicting radiation response. Int J Radiat Oncol Biol Phys 39:561–562

Graham MV et al (1994) Three-dimensional radiation treatment planning study for patients with carcinoma of the lung. Int J Radiat Oncol Biol Phys 29:1105–1117

Graham MV et al (1996) Results of a trial with topotecan dose escalation and concurrent thoracic radiation therapy for locally advanced, inoperable nonsmall cell lung cancer. Int J Radiat Oncol Biol Phys 36:1215–1220

Gregor A et al (1997) Prophylactic cranial irradiation is indicated following complete response to induction therapy in small cell lung cancer: results of a multicentre randomised trial. United Kingdom Coordinating Committee for Cancer Research (UKCCRC) and the European Organization for Research and Treatment of Cancer (EORTC). Eur J Cancer 33:1752–1758

Ha CS, Kijewski PK, Langer MP (1993) Gain in target dose from using computer controlled radiation therapy (CCRT) in the treatment of non-small cell lung cancer. Int J Radiat Oncol Biol Phys 26:335–339

Haffty BG et al (1988) Results of radical radiation therapy in clinical stage I, technically operable non-small cell lung cancer. Int J Radiat Oncol Biol Phys 15:69–73

Hayman JA et al (2001) Dose escalation in non-small-cell lung cancer using three-dimensional conformal radiation therapy: update of a phase I trial. J Clin Oncol 19:127–136

Hoang T et al (2005) Clinical model to predict survival in chemonaive patients with advanced non-small-cell lung cancer treated with third-generation chemotherapy regimens based on eastern cooperative oncology group data. J Clin Oncol 23:175–183

ICRU (International Commission on Radiation Units and Measurements) (1993) Report 51, Quantities and units in radiation protection dosimetry, Oxford University Press, London

Immerman SC et al (1981) Site of recurrence in patients with stages I and II carcinoma of the lung resected for cure. Ann Thorac Surg 32:23–27

Jemal A et al (2005) Cancer statistics, 2005. CA Cancer J Clin 55:10–30

Jeremic B et al (1997) Initial versus delayed accelerated hyperfractionated radiation therapy and concurrent chemotherapy in limited small-cell lung cancer: a randomized study. J Clin Oncol 15:893–900

Johnson BE et al (1990) Neurologic, computed cranial tomographic, and magnetic resonance imaging abnormalities in patients with small-cell lung cancer: further follow-up of 6- to 13-year survivors. J Clin Oncol 8:48–56

Johnson DH et al (1993) Alternating chemotherapy and twice-daily thoracic radiotherapy in limited-stage small-cell lung cancer: a pilot study of the Eastern Cooperative Oncology Group. J Clin Oncol 11:879–884

Johnson BE et al (1996) Patients with limited-stage small-cell lung cancer treated with concurrent twice-daily chest radiotherapy and etoposide/cisplatin followed by cyclophosphamide, doxorubicin, and vincristine. J Clin Oncol 14:806–813

Kan MW et al (1995) A comparison of two photon planning algorithms for 8 mV and 25 mV X-ray beams in lung. Australas Phys Eng Sci Med 18:95–103

Kanard A, Frytak S, Jatoi A (2004) Cognitive dysfunction in patients with small-cell lung cancer: incidence, causes, and suggestions on management. J Support Oncol 2:127–132

Kaskowitz L et al (1993) Radiation therapy alone for stage I non-small cell lung cancer. Int J Radiat Oncol Biol Phys 27:517–523.

Kelly K et al (1998) Phase I study of daily carboplatin and simultaneous accelerated, hyperfractionated chest irradiation in patients with regionally inoperable non-small cell lung cancer. Int J Radiat Oncol Biol Phys 40:559–567

Kies MS et al (1987) Multimodal therapy for limited small-cell lung cancer: a randomized study of induction combination chemotherapy with or without thoracic radiation in complete responders; and with wide-field versus reduced-field radiation in partial responders: a Southwest Oncology Group Study. J Clin Oncol 5:592–600

Klein EE et al (1997) A volumetric study of measurements and calculations of lung density corrections for 6 and 18 mV photons. Int J Radiat Oncol Biol Phys 37:1163–1170

Ko JP et al (2000) CT depiction of regional nodal stations for lung cancer staging. Am. J Roentgenol 174:775–782

Komaki R, Cox JD, Whitson W (1981) Risk of brain metastasis from small cell carcinoma of the lung related to length of survival and prophylactic irradiation. Cancer Treat Rep 65:811–814

Komaki R et al (1995) Evaluation of cognitive function in patients with limited small cell lung cancer prior to and shortly following prophylactic cranial irradiation. Int J Radiat Oncol Biol Phys 33:179–182

Komaki R et al (2005) Phase I study of thoracic radiation dose escalation with concurrent chemotherapy for patients with limited small-cell lung cancer: report of Radiation Therapy Oncology Group (RTOG) protocol 97–12. Int J Radiat Oncol Biol Phys 62:342–350

Kris MG et al (1995) Effectiveness and toxicity of preoperative therapy in stage IIIA non-small cell lung cancer including the Memorial Sloan-Kettering experience with induction MVP in patients with bulky mediastinal lymph node metastasis (clinical N2). Lung Cancer 12:S47–S57

Krol AD et al (1996) Local irradiation alone for peripheral stage I lung cancer: Could we omit the elective regional nodal irradiation? Int J Radiat Oncol Biol Phys 34:297–302

Kupelian PA, Komaki R, Allen P (1996) Prognostic factors in the treatment of node-negative nonsmall cell lung carcinoma with radiotherapy alone. Int J Radiat Oncol Biol Phys 36:607–613

Kutcher GJ, Burman C (1989) Calculation of complication probability factors for non-uniform normal tissue irradiation: the effective volume method. Int J Radiat Oncol Biol Phys 16:1623–1630

Kwa SL et al (1998) Radiation pneumonitis as a function of mean lung dose: an analysis of pooled data of 540 patients. Int J Radiat Oncol Biol Phys 42:1–9

Langer CJ et al (1997) Induction paclitaxel and carboplatin followed by concurrent chemoradiotherapy in patients with unresectable, locally advanced non-small cell lung carcinoma: report of Fox Chase Cancer Center study 94-001. Semin Oncol 24:S12-89–S12-95

Lebeau B et al (1993) A randomized trial of delayed thoracic radiotherapy in complete responder patients with small-cell lung cancer. Petites Cellules Group. Chest 104:726–733

Le Chevalier T et al (1991) Radiotherapy alone versus combined chemotherapy and radiotherapy in nonresectable non-small-cell lung cancer: first analysis of a randomized trial in 353 patients. J Natl Cancer Inst 83:417–423

Le Chevalier T et al (1994) Radiotherapy alone versus combined chemotherapy and radiotherapy in unresectable non-small cell lung carcinoma. Lung Cancer 10 (Suppl 1): S239–S244

Lee JS et al (1986) Neurotoxicity in long-term survivors of small cell lung cancer. Int J Radiat Oncol Biol Phys 12:313–321

Liengswangwong V et al (1994) Limited-stage small-cell lung cancer: patterns of intrathoracic recurrence and the implications for thoracic radiotherapy. J Clin Oncol 12:496–502

Lishner M et al (1990) Late neurological complications after prophylactic cranial irradiation in patients with small-cell lung cancer: the Toronto experience. J Clin Oncol 8:215–221

Looper JD, Hornback NB (1984) The role of chest irradiation in limited small cell carcinoma of the lung treated with combination chemotherapy. Int J Radiat Oncol Biol Phys 10:1855–1860

Mackie TR et al (1985) Lung dose corrections for 6- and 15-mV X rays. Med Phys 12:327–332

MacManus MP et al (2001) High rate of detection of unsuspected distant metastases by pet in apparent stage III non-small-cell lung cancer: implications for radical radiation therapy. Int J Radiat Oncol Biol Phys 50:287–293

Maguire PD et al (1999) Clinical and dosimetric predictors of radiation-induced esophageal toxicity. Int J Radiat Oncol Biol Phys 45:97–103

Mah K et al (2002) The impact of (18)FDG-PET on target and critical organs in CT-based treatment planning of patients with poorly defined non-small-cell lung carcinoma: a prospective study. Int J Radiat Oncol Biol Phys 52:339–350

Mantravadi RV et al (1989) Unresectable non-oat cell carcinoma of the lung: definitive radiation therapy. Radiology 172:851–855

Marks LB et al (1997) Physical and biological predictors of changes in whole-lung function following thoracic irradiation. Int J Radiat Oncol Biol Phys 39:563–570

Martel MK et al (1994) Dose-volume histogram and 3-D treatment planning evaluation of patients with pneumonitis. Int J Radiat Oncol Biol Phys 28:575–581

Martini N, Kris MG, Ginsberg RJ (1997) The role of multimodality therapy in locoregional non-small cell lung cancer. Surg Oncol Clin N Am 6:769–791

McDonald SC, Keller BE, Rubin P (1976) Method for calculating dose when lung tissue lies in the treatment field. Med Phys 3:210–216

Mehta M et al (2001) A new approach to dose escalation in non-small-cell lung cancer. Int J Radiat Oncol Biol Phys 49:23–33

Midthun DE, Jett JR (1996) Clinical presentation of lung cancer. In: Pass HI et al (eds) Lung cancer: principles and practice. Lippincott-Raven, Philadelphia, p. 421

Mira JG et al (1982) Influence of chest radiotherapy in frequency and patterns of chest relapse in disseminated small cell lung carcinoma. A Southwest Oncology Group Study. Cancer 50:1266–1272

Mountain CF (1997) Revisions in the international system for staging lung cancer. Chest 111:1710–1717

Murray N et al (1993) Importance of timing for thoracic irradiation in the combined modality treatment of limited-stage small-cell lung cancer. The National Cancer Institute of Canada Clinical Trials Group. J Clin Oncol 11:336–344

Narayan S et al (2004) Results following treatment to doses of 92.4 or 102.9 Gy on a phase I dose escalation study for non-small cell lung cancer. Lung Cancer 44:79–88

Nolop KB et al (1987) Glucose utilization in vivo by human pulmonary neoplasms. Cancer 60:2682–2689

Noordijk EM et al (1988) Radiotherapy as an alternative to surgery in elderly patients with resectable lung cancer. Radiother Oncol 13:83–89

Oetzel D et al (1995) Estimation of pneumonitis risk in three-dimensional treatment planning using dose-volume histogram analysis. Int J Radiat Oncol Biol Phys 33:455–460

Papanikolaou N, Klein EE (2000) Heterogeneity corrections should be used in treatment planning for lung cancer. Med Phys 27:1702–1704

Pater JL, Loeb M (1982) Nonanatomic prognostic factors in carcinoma of the lung: a multivariate analysis. Cancer 50:326–331

Perez CA et al (1980) A prospective randomized study of various irradiation doses and fractionation schedules in the treatment of inoperable non-oat-cell carcinoma of the lung. Preliminary report by the Radiation Therapy Oncology Group. Cancer 45:2744–2753

Perez CA et al (1982) Impact of irradiation technique and tumor extent in tumor control and survival of patients with unresectable non-oat cell carcinoma of the lung. Report by the Radiation Therapy Oncology Group. Cancer 50:1091–1099

Perez CA et al (1984) Randomized trial of radiotherapy to the thorax in limited small-cell carcinoma of the lung treated with multiagent chemotherapy and elective brain irradiation: a preliminary report. J Clin Oncol 2:1200–1208

Perez CA et al (1987) Long-term observations of the patterns of failure in patients with unresectable non-oat cell carcinoma of the lung treated with definitive radiotherapy. Report by the Radiation Therapy Oncology Group. Cancer 59:1874–1881

Perry MC et al (1987) Chemotherapy with or without radiation therapy in limited small-cell carcinoma of the lung. N Engl J Med 316:912–918

Pignon JP et al (1992) A meta-analysis of thoracic radiotherapy for small-cell lung cancer. N Engl J Med 327:1618–1624

Purdy JA et al (2000) Three-dimensional conformal radiation therapy and intensity modulated radiation therapy: practical potential benefits and pitfalls. In: Perez C, Brady L (eds) Updates to principles and practice of radiation oncology. Lippincott Williams and Wilkins, New York, pp 1–13

Rice RK, Mijnheer BJ, Chin LM (1988) Benchmark measurements for lung dose corrections for X-ray beams. Int J Radiat Oncol Biol Phys 15:399–409

Rosenstein M et al (1992) A reappraisal of the role of prophylactic cranial irradiation in limited small cell lung cancer. Int J Radiat Oncol Biol Phys 24:43–48

Rosenthal DI et al (1997) Seven-week continuous-infusion paclitaxel plus concurrent radiation therapy for locally advanced non-small cell lung cancer: a phase I study. Semin Oncol 24:S12-96–S12-100

Rosenzweig KE et al (2001) Elective nodal irradiation in the treatment of non-small-cell lung cancer with three-dimensional conformal radiation therapy. Int J Radiat Oncol Biol Phys 50:681–685

Salazar OM et al (1976) The assessment of tumor response to irradiation of lung cancer: continuous versus split-course regimes. Int J Radiat Oncol Biol Phys 12:1107–1118

Sandler HM, Curran WJ Jr, Turrisi AT III (1990) The influence of tumor size and pre-treatment staging on outcome following radiation therapy alone for stage I non-small cell lung cancer. Int J Radiat Oncol Biol Phys 19:9–13

Saunders M et al (1999) Continuous, hyperfractionated, accelerated radiotherapy (CHART) versus conventional radiotherapy in non-small cell lung cancer: mature data from the randomised multicentre trial. CHART Steering committee. Radiother Oncol 52:137–148

Sause W et al (2000) Final results of phase III trial in regionally advanced unresectable non-small cell lung cancer. Radiation Therapy Oncology Group, Eastern Cooperative Oncology Group, and Southwest Oncology Group. Chest 117:358–364

Schiller JH et al (2002) Comparison of four chemotherapy regimens for advanced non-small-cell lung cancer. N Engl J Med 346:92–98

Senan S et al (2002) Can elective nodal irradiation be omitted in stage III non-small-cell lung cancer? Analysis of recurrences in a phase II study of induction chemotherapy and involved-field radiotherapy. Int J Radiat Oncol Biol Phys 54:999–1006

Shields TW (1993) Surgical therapy for carcinoma of the lung. Clin Chest Med 14:121–147

Sibley GS et al (1998) Radiotherapy alone for medically inoperable stage I non-small-cell lung cancer: the Duke experience. Int J Radiat Oncol Biol Phys 40:149–154

Singh AK, Lockett MA, Bradley JD (2003) Predictors of radiation-induced esophageal toxicity in patients with non-small-cell lung cancer treated with three-dimensional conformal radiotherapy. Int J Radiat Oncol Biol Phys 55:337–341

Slotman BJ, Antonisse IE, Njo KH (1996) Limited field irradiation in early stage (T1-2 N0) non-small cell lung cancer. Radiother Oncol 41:41–44

Socinski MA (2005) Combined modality trials in unresectable stage III non-small cell lung cancer: the Cancer and Leukemia Group B experience. Semin Oncol 32:S114–S118

Stanley KE (1980) Prognostic factors for survival in patients with inoperable lung cancer. J Natl Cancer Inst 65:25–32

Takada M et al (2002) Phase III study of concurrent versus sequential thoracic radiotherapy in combination with cisplatin and etoposide for limited-stage small-cell lung cancer: results of the Japan Clinical Oncology Group Study 9104. J Clin Oncol 20:3054–3060

The Lung Cancer Study Group (1986) Effects of postoperative mediastinal radiation on competely resected stage II and stage III epidermoid cancer of the lung. N Engl J Med 315:1377–1381

Timmerman R et al (2003) Extracranial stereotactic radioablation: results of a phase I study in medically inoperable stage I non-small cell lung cancer. Chest 124:1946–1955

Tucker MA et al (1997) Second primary cancers related to smoking and treatment of small-cell lung cancer. Lung Cancer Working Cadre. J Natl Cancer Inst 89:1782–1788

Turrisi AT III (1997) Concurrent chemoradiotherapy for limited small-cell lung cancer. Oncology 11:31–37

Turrisi AT III (2004) Limited-disease small-cell lung cancer research: sense and nonsense. Int J Radiat Oncol Biol Phys 59:925–927

Turrisi AT III, Glover DJ, Mason BA (1988) A preliminary report: concurrent twice-daily radiotherapy plus platinum-etoposide chemotherapy for limited small cell lung cancer. Int J Radiat Oncol Biol Phys 15:183–187

Turrisi AT III, Kim K, Blum R (1999) Twice-daily compared with once-daily thoracic radiotherapy in limited small-cell lung cancer treated concurrently with cisplatin and etoposide. N Engl J Med 340:265–271

Turrisi AT et al (2005) Southwest Oncology Group: two decades of experience in non-small cell lung cancer. Semin Oncol 32:S119–S121

Vijayakumar S et al (1991) Optimization of radical radiotherapy with beam's eye view techniques for non-small cell lung cancer. Int J Radiat Oncol Biol Phys 21:779–788

Warde P, Payne D (1992) Does thoracic irradiation improve survival and local control in limited-stage small-cell carcinoma of the lung? A meta-analysis. J Clin Oncol 10:890–895

White PJ, Zwicker RD, Huang DT (1996) Comparison of dose homogeneity effects due to electron equilibrium loss in lung for 6 mV and 18 mV photons. Int J Radiat Oncol Biol Phys 34:1141–1146

Wingo PA et al (1999) Annual report to the nation on the status of cancer, 1973–1996, with a special section on lung cancer and tobacco smoking. J Natl Cancer Inst 91:675–690

Withers HR et al (1982) Hyperfractionation. Int J Radiat Oncol Biol Phys 8:1807–1809

Wong CS et al (1994) Radiation myelopathy following single courses of radiotherapy and retreatment. Int J Radiat Oncol Biol Phys 30:575–581

Work E et al (1997) Randomized study of initial versus late chest irradiation combined with chemotherapy in limited-stage small-cell lung cancer. Aarhus Lung Cancer Group. J Clin Oncol 15:3030–3037

Wulf J et al (2001) Stereotactic radiotherapy of targets in the lung and liver. Strahlenther Onkol 177:645–655

Wurschmidt F et al (1994) Inoperable non-small cell lung cancer: a retrospective analysis of 427 patients treated with high-dose radiotherapy. Int J Radiat Oncol Biol Phys 28:583–588

Yorke E et al (1996) Dosimetric considerations in radiation therapy of coin lesions of the lung. Int J Radiat Oncol Biol Phys 34:481–487

22 Cancers of the Colon, Rectum, and Anus

James A. Martenson, Jr., Michael G. Haddock, and Leonard L. Gunderson

CONTENTS

22.1   Diagnostic Evaluation   545
22.1.1 Colon and Rectal Cancer   545
22.1.2 Anal Cancer   546
22.2   Anatomy   546
22.2.1 Rectum   546
22.2.2 Colon   547
22.3   Pathways of Spread   547
22.3.1 Colorectal Cancer   547
22.3.2 Anal Cancer   547
22.4   Patterns of Recurrence after
       Potentially Curative Resection   548
22.4.1 Rectal Cancer   548
22.4.2 Colon Cancer   548
22.4.3 Anal Cancer   549
22.5   Adjuvant Irradiation for Rectal Cancer   549
22.6   Locally Advanced and Locally Recurrent
       Colorectal Cancer   552
22.7   Primary Treatment of Anal Cancer   553
22.8   Therapeutic Ratio   555
       References   557

J. A. Martenson, Jr., MD
Consultant, Division of Radiation Oncology, Mayo Clinic, Professor of Oncology, Mayo Clinic College of Medicine, Department of Radiation Oncology, 200 2nd St. SW DeskR, Rochester, MN 55905, USA
M. G. Haddock, MD
Consultant, Division of Radiation Oncology, Mayo Clinic, Scottsdale, Arizona, Associate Professor of Oncology, Mayo Clinic College of Medicine, Department of Radiation Oncology, 200 2nd St. SW DeskR, Rochester, MN 55905, USA
L. L. Gunderson, MD
Chair, Department of Radiation Oncology, Mayo Clinic, Scottsdale, Arizona, Professor of Oncology, Mayo Clinic College of Medicine, Department of Radiation Oncology, 200 2nd St. SW DeskR, Rochester, MN 55905, USA

Combined radiotherapy and chemotherapy is often used as an adjuvant to surgical resection in selected patients with rectal cancer. These treatments are also used as the definitive procedures in patients with locally advanced rectal and colon cancer. Combined radiotherapy and chemotherapy has replaced abdominoperineal resection as the principal form of treatment for anal cancer. Appropriate radiotherapeutic management of the patient with lower gastrointestinal cancer includes proper patient selection and diagnostic evaluation, close cooperation by all physicians participating in the patient's care, and the use of proper radiotherapeutic techniques.

22.1
Diagnostic Evaluation

22.1.1
Colon and Rectal Cancer

The diagnostic evaluation of a patient with colorectal cancer begins with a history and physical examination. In taking the history of a patient with large-bowel cancer, particular attention should be given to rectal bleeding, change in bowel habits, and abdominal pain (Postlethwait 1949). Other presenting features of large-bowel cancer include nausea, vomiting, weakness, and abdominal mass. Loss in performance status, fatigue, weight loss, anorexia, and sweats may indicate distant metastatic disease. Patients found incidentally to have microcytic anemia should be considered to have large-bowel cancer until another cause can be proven.

Laboratory studies should include liver function tests and a complete blood cell count. The preoperative concentration of carcinoembryonic antigen (CEA) is an independent prognostic factor in that patients with high concentrations have a worse prognosis. Some physicians also obtain a baseline CEA concentration preoperatively so that serial measurements can be used postoperatively to identify disease progression in asymptomatic patients (MARTIN et al. 1977; WANEBO et al. 1978). The usefulness of CEA in this context is limited because most patients with recurrence have symptoms before the CEA concentration increases (BEART and O'CONNELL 1983), and most patients with recurrence do not have an abnormally high CEA concentration (MOERTEL et al. 1978). Moreover, patients found to have recurrent disease on the basis of serial CEA measurements are unlikely to be cured (MARTIN et al. 1977; PATTERSON and ALPERT 1983; MINTON et al. 1985; FLETCHER 1986, 1993; MOERTEL et al. 1993; NORTHOVER et al. 1994). Despite these limitations, CEA is considered the most cost-effective test for detecting potentially curable recurrent disease, and periodic monitoring is recommended by the American Society of Clinical Oncology in appropriately selected patients (BENSON et al. 2000).

The radiation oncologist is sometimes consulted after a patient's malignant tumor has been resected. In this situation, it is necessary to use information from preoperative studies as well as operative and pathological findings in the design of radiotherapy fields. Preoperative studies that are helpful in the evaluation of local disease include digital examination, proctoscopy or colonoscopy (or both), computed tomography (CT), and a barium enema study, including cross-table lateral views. Although endoscopic procedures are of value in diagnosis, a barium enema study is more helpful to the radiation oncologist in determining the preoperative tumor volume. When endoscopy is performed, a description of the lesion should be provided, including its position on the bowel wall, its distance from the anal verge, its size, the degree of circumference involved, and whether the lesion is exophytic or ulcerative.

If the lesion is palpable, the physician should note the inferior extent relative to the anal verge, the location of the tumor on the bowel wall, the degree of circumference involved, and whether the lesion is clinically mobile or fixed to extrarectal structures. If lesions are immobile or fixed, CT or magnetic resonance imaging (MRI) of the pelvis may be helpful in assessing resectability. If CT or MRI findings demonstrate that the tumor is unresectable for cure, consultation with a radiation oncologist is appropriate for consideration of moderate-dose preoperative irradiation (i.e., approximately 5,040 cGy in 28 fractions) delivered simultaneously with 5-fluorouracil (5-FU)-based chemotherapy, with the goal of shrinking the tumor so that it becomes resectable.

22.1.2
Anal Cancer

The evaluation of a patient with anal cancer begins with a thorough history. The patient should be questioned about the common presenting symptoms, such as mass sensation in the anal region, pain, or bleeding. Anal cancer occurs most commonly in elderly women. Most patients do not have a recent history of multiple sex partners or a history of intravenous drug abuse or other factors that place them at risk for human immunodeficiency virus (HIV) infection or sexually transmitted disease. However, a sexual and drug-abuse history should be obtained from all patients; anal cancer may develop in some patients because of these risk factors.

The physical examination should give particular attention to evaluation of the abdomen, inguinal lymph nodes, anus, and rectum. In addition, a pelvic examination should be performed in females. During examination of the anus and rectum, the size and location of the tumor and whether any perirectal lymph nodes are palpable should be noted.

Laboratory studies should include liver function tests and a complete blood cell count. Routine testing for HIV is unnecessary in most cases, but all patients with a history that places them at risk for this infection should be tested. Identification of patients with overt acquired immunodeficiency syndrome (AIDS) is important because they may not tolerate conventional combined modality therapy for anal cancer.

Imaging studies should include chest radiography and CT of the abdomen and pelvis. Except for very large lesions, CT is inferior to physical examination for assessing the primary lesion. Nevertheless, CT is useful for assessing regional and para-aortic lymph nodes and for evaluating the liver.

22.2
Anatomy

22.2.1
Rectum

The rectum begins in the upper to middle presacrum as a continuation of the sigmoid colon. Whereas the sigmoid colon has a complete peritoneal covering and mesentery, the upper rectum is covered by peritoneum only anteriorly and laterally. The lower one-half to two-thirds of the rectum is not covered by peritoneum and is surrounded by fibro-fatty tissue, organs, and structures, including the bladder, prostate, ureters, vagina, sacrum, nerves, and vessels, that can be involved by direct extension of the tumor.

Lymphatic and venous drainage of lesions limited to the rectum depends on the level of the lesion. The lymphatic system of the upper rectum follows the inferior mesenteric system via the superior hemorrhoidal veins. The middle and lower rectum can, in addition, drain directly to internal iliac and presacral nodes. Lesions that extend to the anal canal only rarely spread to inguinal nodes (TAYLOR et al. 2001), and lesions that extend beyond the rectal wall theoretically may spread through the lymphatic system of the invaded tissue or organ.

22.2.2
Colon

The ascending and descending colon and the splenic and hepatic flexures share some anatomical features with the rectum. They are relatively immobile structures that lack a mesentery and usually do not have a peritoneal covering on the posterior and lateral surfaces. Lesions that extend through the entire bowel wall in these locations may have narrow radial operative margins, especially tumors that invade through the posterior or lateral colonic wall. Lesions on the anterior wall or medial wall of the retroperitoneal colon have closer access to a free peritoneal surface.

The sigmoid and transverse colons are intraperitoneal organs with a complete mesentery and serosa. Each is freely mobile except for its proximal and distal segments, where extracolonic extension may result in narrow surgical margins. If the lesion is in the midtransverse or midsigmoid colon, excellent surgical margins can usually be obtained unless the tumor is adherent to or invades adjacent organs or structures.

The cecum has a variable mesentery and some mobility. When cecal lesions extend posteriorly, it may be difficult to obtain tumor-free surgical margins in the region of the iliac wing with its associated musculature and blood vessels.

The lymphatic and venous drainage of the colon is by the inferior mesenteric system for the left colon and superior mesenteric system for the right colon. If organs or structures adjacent to the primary lesion are involved, the lymphatic drainage of these areas may also be at risk for development of regional metastatic lesions. For example, if lesions in the sig-

moid colon invade the bladder, the iliac nodes may be at risk. Extrapelvic lesions that involve the posterior abdominal wall can spread directly to para-aortic lymph nodes. If the anterior abdominal wall is involved at or below the level of the umbilicus, inguinal nodes may be at risk for metastatic involvement.

22.3
Pathways of Spread

22.3.1
Colorectal Cancer

Colorectal cancers can metastasize hematogenously, by surgical implantation, to the peritoneum or to regional lymph nodes. Peritoneal spread is relatively rare with rectal lesions because most of the rectum is below the peritoneal reflection. With colonic lesions, direct extension to a free peritoneal surface may occur more easily.

Within the bowel, extension of tumor beyond the gross lesion is unusual. In one analysis, for example, only 4 of 103 patients had microscopic intramural spread more than 0.5 cm from the gross lesion and the maximal extent of longitudinal spread was 1.2 cm (BLACK and WAUGH 1948). Because primary venous and lymphatic channels originate in submucosal layers of the bowel, there is little risk for either venous or lymphatic dissemination in patients with lesions limited to the mucosa.

Lymph-node involvement is found in about 50% of patients and is usually orderly and predictable. Skip metastasis or retrograde spread is associated with an ominous prognosis. It occurs in only 1–3% of these patients and is usually related to lymphatic blockage (GRINNELL 1966). The major spread through lymphatic channels is cephalad, except for lesions 8 cm or less above the anal verge, where both lateral and distal flow can occur. In females, this latter pattern of flow places the posterior vaginal wall at risk for involvement by tumor (ENQUIST and BLOCK 1966).

22.3.2
Anal Cancer

The anal canal extends from the dentate line to the anal verge. It has an average length of 2.1 cm (NIVATVONGS et al. 1981). The dentate line is located at the lower border of the anal valves (MORSON 1960;

Stearns et al. 1980). Lymphatic drainage from the anal region is to the inguinal nodes, the lateral pelvic side wall nodes along the path of the middle hemorrhoidal vessels, and the inferior mesenteric nodes (Stearns et al. 1980). The most distal portion of the rectum and the proximal anal canal share a plexus that drains to lymphatics that accompany the inferior rectal and internal pudendal blood vessels and ultimately drain to iliac nodes. Carcinomas of the lower rectum or those that extend into the anal canal may metastasize to superficial inguinal nodes through connections to efferent lymphatics that drain the lower anus. Tumors below the anal verge drain primarily to the inguinal system.

22.4
Patterns of Recurrence after Potentially Curative Resection

22.4.1
Rectal Cancer

The risk of local recurrence after complete surgical resection is related to the degree of disease extension beyond the rectal wall and the extent of nodal involvement. The incidence of local recurrence for lesions with involved nodes but with tumor confined to the wall varies in most studies from 20% to 40%. This is similar to the local recurrence risk for patients without nodal involvement who have extension beyond the wall. Lesions that have both tumor extension beyond the rectal wall and lymph-node involvement have nearly an additive risk of local recurrence - 40–65% in clinical studies (Gilbert 1978; Walz et al. 1981; Gunderson et al. 1983c; Mendenhall et al. 1983) and 70% in a reoperative series (Gunderson and Sosin 1974). A recent publication of combined results from 3,791 patients treated on cooperative group trials over the last 25 years has confirmed the independent importance of tumor penetration through the wall and nodal status (Gunderson et al. 2004).

22.4.2
Colon Cancer

The study of patterns of recurrence in patients with colon cancer is principally of interest because of the role of radiotherapy in patients with locally advanced disease.

Data from clinical studies of patterns of failure suggest that one-third of patients in whom tumor relapse develops after curative resection have recurrences solely in the liver (Welch and Donaldson 1978). Patterns-of-recurrence studies that do not routinely use reoperation or autopsy, however, may underestimate the incidence of local-regional recurrence in patients with colon cancer. Autopsy and reoperative series suggest that liver-only relapse occurs in fewer than 10% of cases (Welch and Donaldson 1979; Gunderson et al. 1985b). Data from autopsy and reoperative patterns-of-recurrence studies must also be interpreted with caution because only a subset of patients who have a potentially curative operation subsequently undergo reoperation or autopsy.

Patterns of recurrence in colon cancer have been analyzed in autopsy, clinical, and reoperation series (Cass et al. 1976; Welch and Donaldson 1978; Russell et al. 1984; Willett et al. 1984a,b; Gunderson et al. 1985b; Minsky et al. 1988). Data from these series suggested that local failure is an important problem after resection of colon cancer in selected patients. Local recurrence is highest among patients with tumors that adhere to surrounding structures and among patients who have both tumor extension beyond the bowel wall and metastatic involvement of lymph nodes. In a retrospective study, the local recurrence rate among patients with these pathological characteristics was 42%.

In a colorectal reoperative series from the University of Minnesota, failures in the tumor bed or lymph nodes were most common with rectal lesions but did occur with primary lesions at other bowel sites (Gunderson et al. 1985b). Peritoneal seeding was least common with primary lesions of the rectum, probably because these tumors are less accessible to the peritoneal cavity than most colon cancers. The incidence of hematogenous spread was similar for all sites, although the distribution differed. With primary rectal lesions, hematogenous failures were fairly evenly divided between the liver and lung. This distribution is explained by the pattern of venous drainage through both the inferior mesenteric system, which drains to the liver through the portal vein, and the internal iliac system, which ultimately drains to the lungs through the inferior vena cava. With primary tumors of the colon, initial hematogenous failures were usually in the liver. This is consistent with the colon's pattern of venous drainage, which is initially to the liver through the portal system.

22.4.3
Anal Cancer

Although primary surgical management of anal cancer has been replaced largely by sphincter-sparing therapy, analysis of patterns of failure from surgically treated patients is informative for treatment planning. Of the patients who have abdominoperineal resection, approximately 35% have metastatic involvement of pelvic lymph nodes, and approximately 13% have recurrence in the inguinal lymph nodes (BOMAN et al. 1984). According to an analysis of 118 patients who had abdominoperineal resection with curative intent, 46 (39%) had recurrence after resection (BOMAN et al. 1984). Local-regional recurrence occurred in 23% of the patients. Furthermore, 5% of the patients experienced both local and distant recurrence, and 6% had distant metastasis without local recurrence. In 7% of patients, the site of tumor recurrence could not be determined.

22.5
Adjuvant Irradiation for Rectal Cancer

The foundation of treatment for patients with resectable rectal cancer is surgery. When radiotherapy, with or without chemotherapy, is offered as an adjuvant to patients who are candidates for surgical resection or who have undergone a potentially curative surgical resection, it is, by definition, being administered to a person who may already be cured by surgery alone. Therefore, a high standard of scientific evidence for the efficacy of adjuvant treatment, which is potentially toxic, expensive, and inconvenient, is needed before its use in routine clinical practice can be justified.

Randomized clinical trials that have compared preoperative radiotherapy with surgery alone have demonstrated improved local control, and one study demonstrated a survival advantage for this approach (SWEDISH RECTAL CANCER TRIAL 1997). Several randomized trials of postoperative adjuvant radiotherapy and chemotherapy have demonstrated improved local control and survival compared with surgery alone or surgery followed by adjuvant radiotherapy without chemotherapy (GASTROINTESTINAL TUMOR STUDY GROUP 1985; DOUGLASS et al. 1986; KROOK et al. 1991; TVEIT et al. 1997). Continuous-infusion 5-FU during postoperative radiotherapy has been found to be more effective than bolus 5-FU during radiotherapy (O'CONNELL et al. 1994).

Considerable debate has focused on the issue of whether adjuvant radiotherapy and chemotherapy should be given preoperatively or postoperatively to patients with rectal cancer. The presentation by SAUER (2003) at the 2003 annual meeting of the American Society for Therapeutic Radiology and Oncology of the landmark German phase-III study has largely ended this debate. In this study, 823 patients with T3 or T4 or node-positive rectal cancer received 5,040 cGy in 28 fractions, given with two courses of 5-FU (1 g/m^2 per day for 120 h) at the beginning and end of radiotherapy. Patients were randomly allocated to receive this treatment either preoperatively or postoperatively. Additional chemotherapy was administered in both arms of the study. Patients in the preoperative arm experienced fewer local recurrences than those in the postoperative arm (7% versus 11%, $P=0.02$). Also, better sphincter preservation was reported in preoperatively irradiated patients with low rectal cancers (39% versus 19%, $P=0.004$). Sauer also reported that patients in the preoperative arm had less acute and long-term toxicity than in the postoperative arm. No survival difference was demonstrated.

On the basis of these results, most patients who require adjuvant therapy for rectal cancer should receive preoperative radiotherapy and chemotherapy. However, a limited number of patients will still require postoperative adjuvant therapy. For example, if preoperative endorectal ultrasonographic staging indicates that a rectal cancer is T2 N0, the patient is at low risk for recurrence and should be treated with "up front" surgery without preoperative radiotherapy and chemotherapy. However, if the pathological specimen shows a more advanced tumor than indicated by preoperative staging, the patient should be considered for postoperative adjuvant therapy. Because preoperative staging is not entirely accurate, it is important for radiation oncologists to be able to give adjuvant therapy both preoperatively and postoperatively.

Radiotherapy fields used in the adjuvant treatment of rectal cancer should include the primary tumor or tumor bed, with 3-cm to 5-cm margins, and the regional lymph nodes. In most institutions, internal iliac and presacral nodes are not routinely dissected during surgery for rectal cancer. These lymph nodes should be included in the initial irradiation volume. External iliac nodes are not a primary nodal drainage site and should not be included in the radiation fields. The exception to this is when pelvic organs with major external iliac drainage, such as the prostate, bladder, vagina, cervix, and uterus, are involved by direct extension from a rectal cancer. Treatment of external iliac lymph nodes has generally been recommended in such cases. A recent study, however, calls into question the importance of treating lymph-node groups that are at risk because of contiguous spread of rectal cancer to other pelvic organs (TAYLOR et al. 2001). TAYLOR and colleagues found that patients with contiguous spread of rectal cancer to the anus (which drains to the inguinal lymph nodes) who did not receive elective inguinal lymph-node treatment had only a 4% risk of inguinal lymph-node recurrence at 5 years. Accordingly, although treatment of lymph-node groups beyond the internal iliac nodes should be considered in patients with spread of rectal cancer to other pelvic organs, the ultimate decision is a matter of individual physician judgment.

Most tumor bed recurrences are in the posterior one-half to two-thirds of the true pelvis (GILBERTSEN 1960). The internal iliac and presacral nodes are located posteriorly in the pelvis (GUNDERSON et al. 1985a). Therefore, lateral fields can be used for a portion of the treatment to reduce the dose of radiation to anterior normal tissues such as the small bowel (Fig. 22.1). Bladder distention and prone position are useful techniques for providing additional displacement of the small bowel out of the high-dose radiation field. The use of a three-field technique (a posterior field and opposed lateral fields) can also spare anterior structures, particularly the small bowel and external genitalia in males. When a three-field technique is used, wedges, with the heels posterior, should be used on the lateral fields (Fig. 22.2). The field size is progressively reduced, with initial radiotherapy fields designed to treat the primary tumor volume and regional lymph nodes to a dose of 4,500 cGy in 25 fractions. Smaller fields can then be used to treat the primary tumor bed to an additional 540–900 cGy in three to five fractions, as clinically indicated. Isodose curves for anterior, posterior, and opposed lateral fields are shown in Figure 22.3. Simulation films obtained after the use of oral contrast medium can be used to demonstrate the amount of small bowel in the radiation field. These films are particularly helpful in the design of radiation boost fields. Imaging with contrast in the small bowel is often helpful in assessing the usefulness of bladder distention (GUNDERSON et al. 1985a) or other measures (SHANAHAN et al. 1989) to decrease the volume of intestine in radiation fields.

Posterior and anterior radiotherapy fields (Fig. 22.1) should cover the pelvic inlet with a 2-cm margin. The superior margin is usually 1.5 cm above

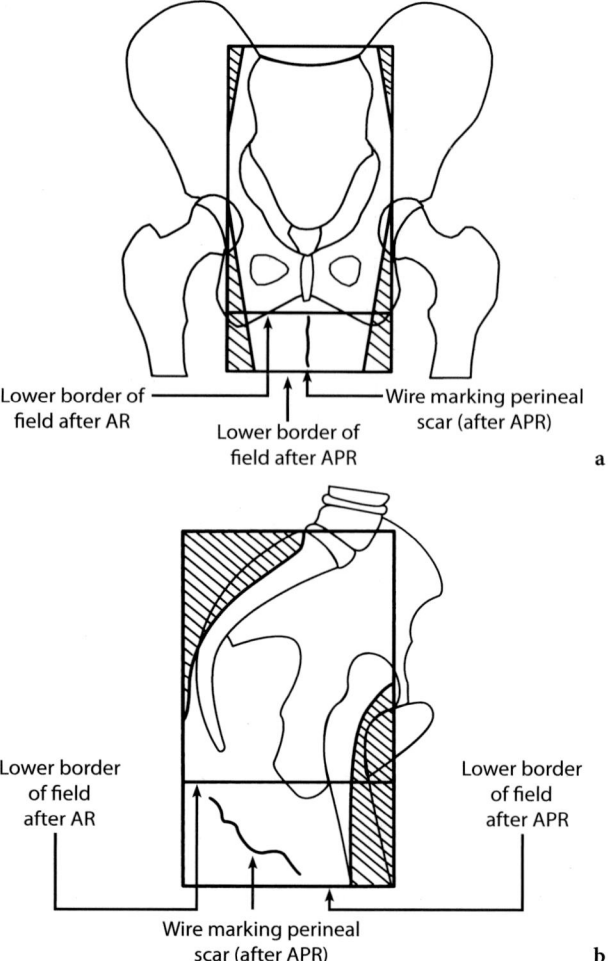

Fig. 22.1a,b. Posterior (**a**) and lateral (**b**) pelvic radiotherapy fields used in adjuvant radiotherapy for rectal cancer. In patients with tumor adherence to organs drained by external iliac lymph nodes, the anterior border of the lateral field is modified to place it anterior to the symphysis pubis. *AR* anterior resection, *APR* abdominoperineal resection (from MARTENSON et al. 1998; by permission of the publisher)

the level of the sacral promontory. In patients who have had an anterior resection, the usual inferior margin is below the obturator foramina or approximately 3 cm below the most inferior portion of the tumor bed.

The posterior field margin for lateral fields is critical because the rectum and perirectal tissues lie just anterior to the sacrum and coccyx. Accordingly, the posterior field margin should be at least 1.5–2 cm behind the anterior bony sacral margin (Fig. 22.1, 22.4, and 22.5). The entire sacral canal with a 1.5-cm margin should be included in patients with locally advanced disease to avoid sacral recurrence from tumor spread along nerve roots. The anterior

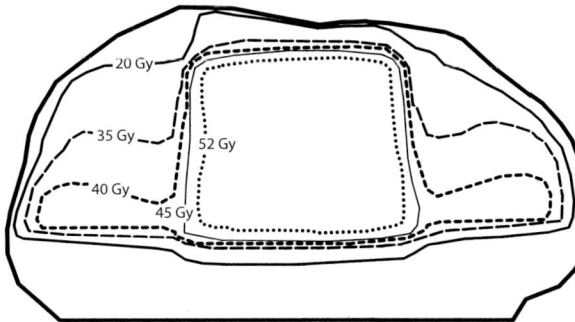

Fig. 22.2. Isodose curves for adjuvant pelvic radiotherapy for rectal cancer with use of posterior and opposed lateral fields. The total dose at isocenter is 5,220 cGy in 29 fractions. Wedges, with heels posterior, are used on the lateral fields to increase dose homogeneity (from MARTENSON et al. 1998; by permission of the publisher)

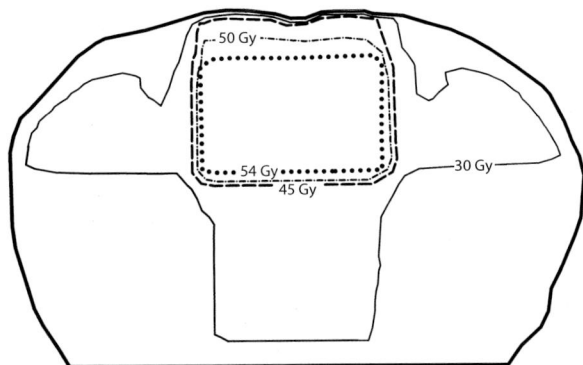

Fig. 22.3. Isodose curves for adjuvant pelvic radiotherapy for rectal cancer with use of anterior, posterior, and opposed lateral fields. The total dose at isocenter is 5,400 cGy in 30 fractions (from MARTENSON et al. 1998; by permission of the publisher)

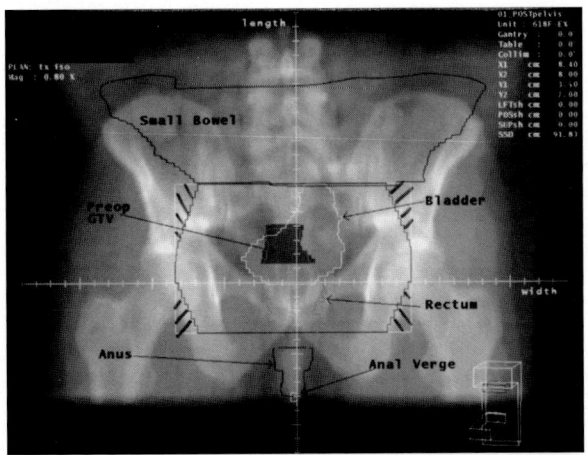

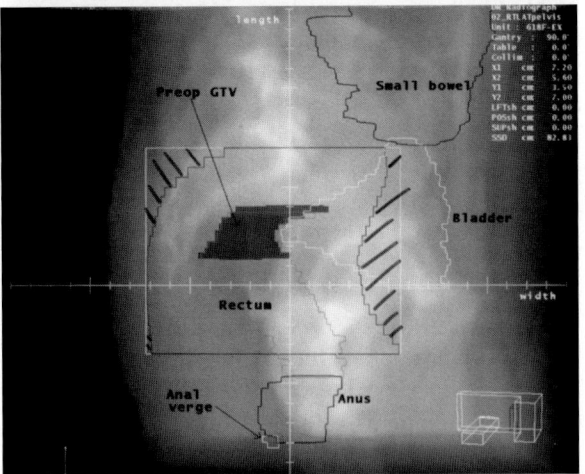

Fig. 22.4a,b. Posterior (**a**) and lateral (**b**) fields used for adjuvant treatment of rectal cancer, designed using computed tomographic simulation. *GTV* gross tumor volume, *Preop* preoperative

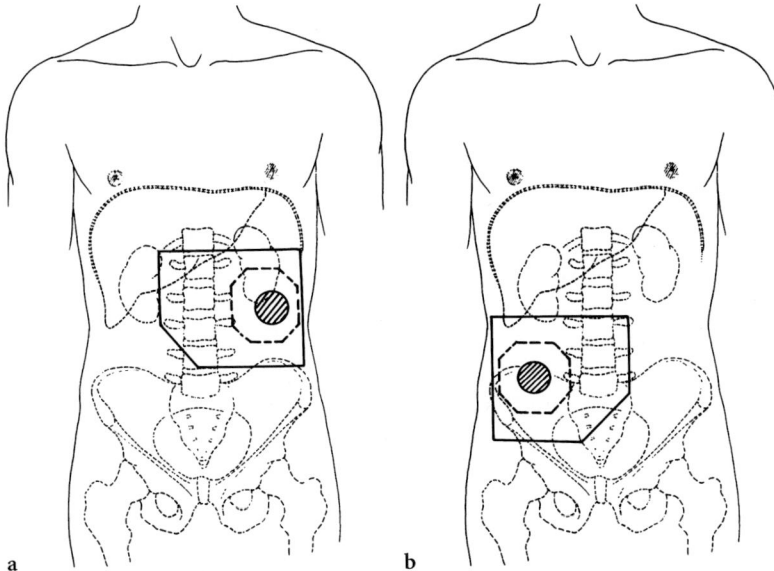

Fig. 22.5a,b. Idealized external radiotherapy fields used in the treatment of locally advanced or locally recurrent colon cancer. After 4,500 cGy in 25 fractions to the large field, a boost field (*broken lines*) may be used to deliver an additional 540-900 cGy in 3 to 5 fractions. **a** A field designed to treat a lesion in the distal descending colon includes the ipsilateral iliac and para-aortic nodes. **b** A field designed to treat a lesion in the mid-ascending colon includes the immediately adjacent regional nodes and para-aortic nodes

margin can sometimes be shaped to decrease the amount of radiation to the head of the femur and bladder inferiorly and the small bowel superiorly. Anteriorly, the lower one-third of the rectum abuts the posterior vaginal wall and prostate, and the posterior portion of these structures should be included in the radiotherapy field. In females, inclusion of the vagina can be verified at simulation using a contrast-soaked tampon.

After abdominoperineal resection, the perineum should be included with the tumor bed and nodal volumes to prevent marginal recurrences from surgical implantation of tumor (GUNDERSON and SOSIN 1974; RICH et al. 1983; HOSKINS et al. 1985; SCHILD et al. 1989). In a study from Massachusetts General Hospital, the incidence of perineal recurrence with surgery alone was 8.5% (RICH et al. 1983). The incidence of perineal recurrence was only 1.7% for patients who received postoperative irradiation (HOSKINS et al. 1985). In a Mayo Clinic analysis of patients with rectal cancer who received postoperative irradiation, the incidence of a perineal component of recurrence was 2% after abdominoperineal resection followed by 4,000 cGy or more to the perineum, but was 23% when the perineum was not adequately irradiated after abdominoperineal resection ($P<0.05$) (SCHILD et al. 1989). A radiopaque marker should be used to delineate the entire extent of the perineal scar (Fig. 22.1). The inferior and posterior field edges should include a margin extending 1.5 cm beyond the perineal scar. Inferolaterally, the margin should be the lateral aspect of the ischial tuberosities. For treatment of the posterior field, bolus material should be placed over the perineal incision (thickness depends on beam energy) to allow delivery of an adequate dose to the scar surface. If pelvic drains exited through the buttocks instead of the perineal wound, bolus material should also be placed over these drain sites. All patients in whom the perineum is included within the radiation fields experience perineal discomfort during treatment. This can be mitigated by a three-field technique with posterior and lateral fields.

The perineum can usually be treated to a dose level of 4,500 cGy in 25 fractions over a 5-week period with acceptable short- and long-term tolerance. Because of skin reactions, patients occasionally require a 7- to 10-day rest during treatment, sitz baths, topical anesthetic (such as topical lidocaine), and protective agents such as petrolatum ointment (e.g., Aquaphor). Most patients finish on schedule, and limited skin reactions generally improve markedly within 1–2 weeks after completion.

22.6
Locally Advanced and Locally Recurrent Colorectal Cancer

Generally, 4,500–5,040 cGy in 25 to 28 fractions is delivered to radiation treatment fields designed to include the tumor and the regional lymph nodes. This treatment can be followed by a boost of 540–900 cGy in 3 to 5 fractions in selected patients. Doses greater than 5,040 cGy are rarely administered when using external radiotherapy unless the small bowel can be completely excluded from the radiotherapy field after 5,040 cGy. Boost pelvic fields are usually treated with opposed lateral fields or three fields (posteroanterior and lateral fields, with wedges on the lateral fields, heels posterior). Field shaping of the lateral boost fields can often be used to reduce or eliminate the volume of small intestine in the radiotherapy field anteriorly and superiorly. Bladder distention may be extremely useful in displacing small-bowel loops superiorly and anteriorly out of both large and boost fields (GUNDERSON et al. 1980, 1983a,b, 1985a; GALLAGHER et al. 1986). Imaging with small-bowel contrast can help to identify patients in whom immobile loops remain in an area at high risk (GREEN et al. 1975; GREEN 1983; GUNDERSON et al. 1980, 1983a,b, 1985a; GALLAGHER et al. 1986). In such instances, the radiation oncologist must limit the dose to conform to small-bowel tolerance.

For patients with residual, recurrent, or fixed pelvic lesions in posterior or lateral locations, it is important to include the sacral canal in the target volume for the initial 4,500–5,000 cGy (GUNDERSON et al. 1985a). Including this area is indicated because of the increased risk of tumor spread along nerve roots. Failure to do so may result in a marginal recurrence in the sacral canal (GUNDERSON et al. 1983a).

For patients with locally advanced or recurrent colon cancer, the initial external beam radiotherapy fields should include the primary tumor, immediately adjacent lymph nodes, and adjacent para-aortic nodes (Fig. 22.5). These fields should receive 4,500 cGy in 25 fractions. Smaller boost fields can then be considered for an additional 540–900 cGy in 3 to 5 fractions. In general, total cumulative doses more than 5,040 cGy are not recommended unless all the small bowel can be excluded from fields considered for such doses. A small-bowel study obtained on the simulator with the patient in treatment position is helpful for determining the position of the small bowel for this purpose. These films sometimes demonstrate that a lateral decubitus position for a portion of the treatment may be useful to decrease exposure or to exclude small bowel from boost fields.

After administration of 4,500–5,040 cGy of external beam radiation, an intraoperative electron beam can be used to give a boost of 1,000–2,000 cGy to areas of residual tumor, with the goal of improving disease control and survival. For patients who completed a course of external beam radiotherapy, surgical debulking, and an intraoperative electron boost, 5-year survival rates of approximately 20% for locally recurrent disease and 45% for primary locally advanced disease have been reported from Massachusetts General Hospital and Mayo Clinic (Suzuki et al. 1995; Gunderson et al. 1996a,b; Schild et al. 1997). A less favorable outcome has been reported for patients who have a previous history of radiation to the site of recurrent disease. In a series of patients from Mayo Clinic who had radiotherapy before their recurrence, treatment with intraoperative radiotherapy resulted in a 5-year survival rate of only 12% (Haddock et al. 2001).

Better outcomes may be possible for patients who have intraoperative radiotherapy after gross or complete resection of the tumor (Suzuki et al. 1995; Schild et al. 1997; Mannaerts et al. 1999; Lindel et al. 2001; Wiig et al. 2002; Haddock et al. 2003). Treatment of patients with advanced nodal disease from colon and rectal cancer also provides an opportunity for long-term survival in a substantial minority of patients. At Mayo Clinic, 48 patients with advanced nodal metastases from rectal and colon cancer received intraoperative radiotherapy as a component of treatment. Advanced nodal disease presented as recurrent disease in 79% of these patients. The 5-year survival rate for this group was 34% (Haddock et al. 2003).

22.7
Primary Treatment of Anal Cancer

The results of several recently published clinical trials have added substantially to our understanding of appropriate treatment planning for anal cancer. The Radiotherapy Oncology Group (RTOG) and Eastern Cooperative Oncology Group (ECOG) conducted a randomized, inter-group clinical trial comparing radiotherapy plus 5-FU with radiotherapy, 5-FU, and mitomycin C (Flam et al. 1996). A continuous 5-FU intravenous infusion, 1,000 mg/m^2

per day, was given on days 1–4 of radiotherapy and repeated on days 29–32 of radiotherapy. Mitomycin C, 10 mg/m^2 intravenously, was given on day 1 and day 29 of radiotherapy to patients who were randomly allocated to receive this drug. All patients received 3,600 cGy in 20 fractions to the primary tumor, pelvic lymph nodes, and inguinal lymph nodes, followed by a field reduction to include the primary tumor with a 10×10-cm field, which was then treated to an additional 900 cGy in 5 fractions for a total dose of 4,500 cGy in 25 fractions. For patients thought to have residual tumor after 4,500 cGy, the final boost field was continued to a total cumulative dose of 5,040 cGy in 28 fractions.

The combination of radiotherapy, 5-FU, and mitomycin C resulted in a lower colostomy rate than radiotherapy and 5-FU without mitomycin C. At 4 years, the colostomy rate was 9% for patients who received mitomycin C and 23% for those who did not ($P=0.002$). Persistent or recurrent tumor was by far the most common cause of colostomy: residual tumor was found in the surgical specimen from 97% of colostomy patients who did not receive mitomycin C and in 85% of colostomy patients who received mitomycin C. A statistically significant difference in survival between patients who received mitomycin C and those who did not was not observed. The RTOG-ECOG study demonstrated that mitomycin C is an important component of combined modality therapy for anal cancer.

Although the RTOG-ECOG trial provided critical information about combined modality therapy for anal cancer, it did not definitively address whether this form of treatment is superior to high-dose radiotherapy alone. Two randomized trials that compared radiation alone with radiotherapy, 5-FU, and mitomycin C for cancer of the anal canal recently provided important data on this point. A phase-III trial, reported by the United Kingdom Coordinating Committee on Cancer Research (UKCCCR), compared radiotherapy alone with radiotherapy combined with a regimen of 5-FU and mitomycin C (UKCCCR Anal Cancer Trial Working Party 1996). All patients in this study received 4,500 cGy over a 4- to 5-week period, and most received a boost of 1,500–2,500 cGy after a 6-week break. Patients randomly allocated to combined modality therapy generally received 5-FU, 1,000 mg/m^2 per day, on the first 4 days of radiotherapy and for 4 days during the fourth or fifth week of radiotherapy. A single dose of mitomycin C, 12 mg/m^2, was usually given on the first day of radiotherapy in the combined modality therapy group. Some variation on these standard doses was allowed for selected patients (UKCCCR ANAL CANCER TRIAL WORKING PARTY 1996). The local recurrence rate at 3 years was 61% for patients receiving radiotherapy alone and 39% for those receiving combined modality therapy. There was no difference in overall survival between patients in the radiotherapy group and those in the combined modality therapy group.

The European Organization for the Research and Treatment of Cancer (EORTC) randomized trial compared radiotherapy alone with radiotherapy, 5-FU, and mitomycin C in patients with T3, T4, or lymph-node-positive anal cancer. The treatment program was similar to that of the UKCCCR study. Local control and colostomy-free survival rates were superior for patients randomly allocated to receive combined modality therapy (BARTELINK et al. 1997).

These randomized studies provide strong evidence that combined modality therapy should be administered to patients with anal cancer, with the goal of sphincter preservation and cure. Radiotherapy without chemotherapy (MARTENSON and GUNDERSON 1993) should be reserved for patients who are unable to tolerate combined modality therapy, such as those with serious co-morbid illnesses. Because most patients with anal cancer are treated with mitomycin C and an initial 4-day infusion of 5-FU concurrent with the initiation of radiotherapy, close coordination with the patient's medical oncologist is needed before treatment begins. To maximize the interaction of radiotherapy and chemotherapy, patients should receive treatment with radiation on each day that 5-FU is infused. Accordingly, it is preferable to begin treatment on a Monday or Tuesday. Alternatively, special arrangements can be made for patients to receive radiation treatments on the first weekend after the initiation of radiotherapy if treatment is started later in the week.

Radiotherapy fields should be designed to include the primary lesion and regional pelvic and inguinal lymph nodes. A portion of the inguinal lymph-node chain is superficial to the head and neck of the femur. Radiotherapy fields should be designed to avoid giving a full dose to these structures. Anteroposterior and posteroanterior fields or four-field box techniques that include the inguinal lymph nodes with the head and neck of the femur may place patients at risk for subsequent treatment-induced fracture (MARTENSON and GUNDERSON 1993). Factors that increase the chance of this complication are of particular concern in a population of patients that includes a large number of elderly women, many of whom are already at risk for fracture because of osteoporosis. Radiation techniques should be used that minimize the dose to the head and neck of the femur by treating lateral superficial inguinal nodes through anterior fields only. This can be accomplished by treating the primary tumor, pelvic nodes, and inguinal nodes with an anterior photon field that encompasses all these structures (Fig. 22.6 a). The posterior photon field includes only the primary tumor and pelvic lymph nodes (Fig. 22.6 b). Electron fields are used to supplement the dose to the portion of the lateral superficial inguinal nodes not included in the posterior photon field (Fig. 22.6 c). The medial borders of the lateral electron fields are the same as the lateral border of the posterior photon field at its exit point on the patient's anterior abdominal wall. This border is determined with the aid of radiopaque markers placed on the anterior abdominal wall under fluoroscopic guidance while the posteroanterior photon field is being simulated (Fig. 22.6 b, c). Isodose curves for this technique are shown in Figure 22.7. An alternative to this technique is to use CT-based planning to determine precisely the inguinal node-bearing areas along the femoral vessels in an effort to limit the volume of the head and neck of the femur within the irradiated fields (Fig. 22.8). Intensity-modulated radiotherapy may also be useful in limiting the femoral head and neck dose.

The radiation treatment regimen used in the most effective arm of the RTOG-ECOG study was different from that used in the EORTC and UKCCCR studies because a lower total radiation dose without a treatment break together with a somewhat more intensive chemotherapy regimen was used, whereas the EORTC and UKCCCR studies used a higher total dose with a 6-week treatment break after 4,500 cGy. Definitive recommendations are not possible for the preferred treatment regimen, because these regimens have not been directly compared scientifically. Data from a preliminary RTOG study of high-dose radiotherapy, however, suggest that a treatment regimen including a planned treatment break may result in an inferior outcome (JOHN et al. 1995). Accordingly, a treatment regimen based on the one used in the RTOG-ECOG randomized trial, with total radiation doses of approximately 4,500–5,040 cGy in 25 to 28 fractions together with two courses of 5-FU and mitomycin C, is generally accepted as the standard of care in the United States. Treatment programs that use substantially higher radiation doses or different chemotherapy combinations (MARTENSON et al. 1996b) should be confined to peer-reviewed clinical trials.

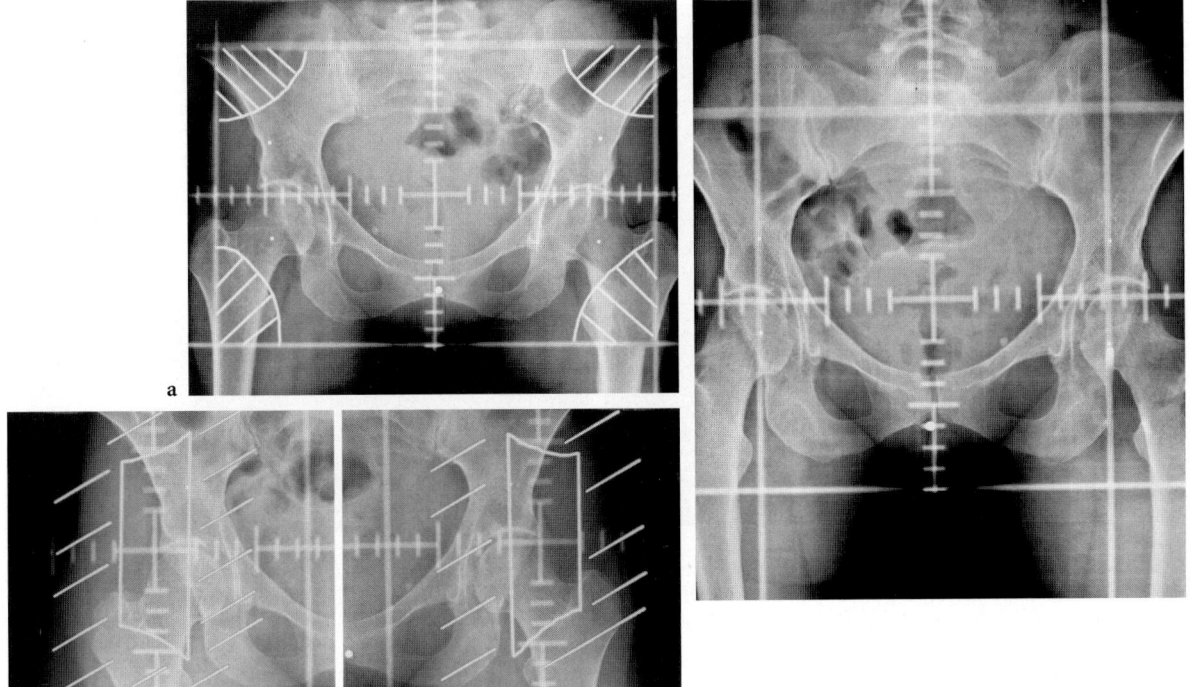

Fig. 22.6a–c. Anterior (**a**) and posterior (**b**) radiotherapy photon fields used for treatment of anal cancer. **c** Electron fields are used to supplement lateral inguinal nodes not included in the posterior photon field. The medial borders of the electron fields are determined by placing radiopaque markers on the anterior abdominal wall at the exit point of the lateral border of the posterior photon field (from MARTENSON et al. 1998; by permission of the publisher)

22.8
Therapeutic Ratio

The potential for optimizing the therapeutic ratio is enhanced by close cooperation among the surgeon, medical oncologist, and radiation oncologist (GUNDERSON et al. 1980; COHEN et al. 1981). The use of radiopaque clips to mark the tumor or tumor bed areas is particularly helpful in the design of high-dose boost volumes. Reconstruction techniques that exclude or minimize the volume of small bowel in the irradiated field are also helpful.

Several techniques can be used by radiation oncologists to potentially improve the therapeutic ratio. For both rectal and colon cancers, shrinking-field techniques should be used after a dose of 4,500 cGy. With rectal cancers and proximal sigmoid cancers, lateral fields should be used for a portion of the treatment to avoid as much small bowel as possible. Treatment with the bladder distended is appropriate unless the distention displaces the tumor outside the radiation field. In patients with colon cancer, it may be possible to reduce the volume of small bowel within the field, often by placing the patient in the lateral decubitus position for a portion of the treatment.

In studies of patients with rectal cancer who are given postoperative radiation as adjuvant therapy, the risk of small-bowel obstruction requiring reoperation seems to be affected by treatment technique. When pelvic and para-aortic fields were treated with an anterior and posterior opposed technique at M.D. Anderson Hospital, the incidence of small-

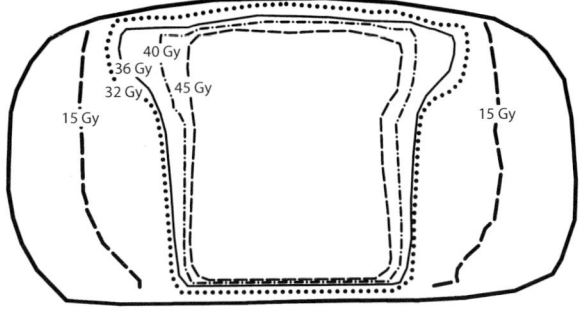

Fig. 22.7. Isodose curves for pelvic radiotherapy fields used in the primary treatment of anal cancer. The total dose at the isocenter for the photon fields is 4,500 cGy. Lateral inguinal nodes receive 3,600 cGy through a combination of the anterior photon field and supplementary electron fields (Fig. 22.6); (from MARTENSON et al. 1998; by permission of the publisher)

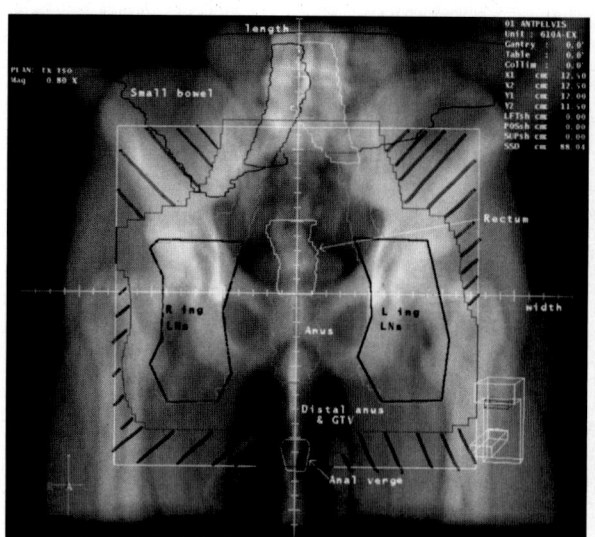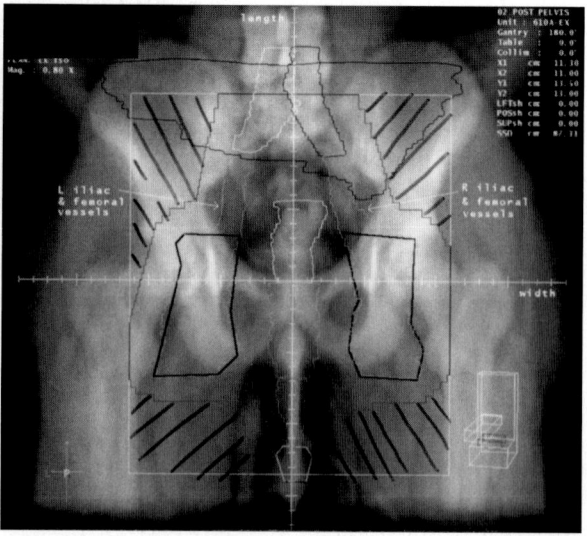

Fig. 22.8a,b. Anterior (**a**) and posterior (**b**) fields for treatment of anal cancer, designed using computed tomographic simulation. Precise definition of the inguinal lymph nodes allows some sparing of the femur, particularly with the posterior field. *GTV* gross tumor volume, *ing* inguinal, *LNs* lymph nodes

bowel obstruction requiring reoperation was 17.5%, compared with 5% with surgery alone (ROMSDAHL and WITHERS 1978; WITHERS et al. 1981). When the superior extent of the field was shifted down to L5, the incidence of complications requiring operative intervention decreased to about 12%. At Massachusetts General Hospital, multi-field techniques and bladder distention were used. The incidence of small-bowel obstruction requiring surgical intervention in patients receiving postoperative radiotherapy was 6%, which was nearly equal to that of patients treated with surgery alone.

When multi-field irradiation techniques are used in combination with chemotherapy in the adjuvant treatment of rectal cancer, no apparent increase occurs in the risk of severe small-bowel complications (GUNDERSON et al. 1986). In an analysis of the North Central Cancer Treatment Group randomized trial, with minimum 3-year follow-up, the incidence of severe small-bowel complications was less than 5% with either irradiation alone or irradiation plus chemotherapy.

A large retrospective analysis of patients who received radiotherapy for high-risk, completely resected colon cancer or for incompletely resected colon cancer found that acute enteritis resulting in hospitalization or a break from treatment occurred in 16 of 203 patients (8%). Long-term toxicity requiring surgery was observed in 9 patients (4.4%). Nonsurgical complications such as chronic abdominal pain were not assessed.

Some reassurance about the risk of surgical complications resulting from adjuvant radiotherapy for rectal cancer is provided by the above data from Mayo Clinic, M.D. Anderson Hospital, and Massachusetts General Hospital. However, the risk of functionally important long-term toxicity not requiring surgical correction after pelvic radiotherapy and chemotherapy is high. In the retrospective study of KOLLMORGEN and colleagues (1994), for example, bowel function was assessed in patients who either had or had not received postoperative adjuvant radiotherapy and chemotherapy after anterior resection for rectal cancer. Consistently worse bowel function was found in the patients who had received radiotherapy and chemotherapy. For example, 56% of the patients who had received adjuvant treatment reported at least occasional fecal incontinence, in contrast to only 7% of those who did not receive adjuvant treatment ($P<0.001$). These results have been corroborated in studies of patients with rectal cancer treated with either preoperative or postoperative radiotherapy in the context of phase-III clinical trials. LUNDBY and colleagues (1997) found statistically significant worse rectal function in patients who received postoperative radiotherapy than in those who received no further treatment postoperatively. Of the patients receiving postoperative pelvic radiotherapy, 49% experienced fecal incontinence, compared with only 5% of those who did not receive this therapy ($P<0.001$). Similar findings have been reported by DAHLBERG and colleagues (1998) in patients with rectal cancer who were randomly allocated to receive preoperative radiotherapy or immediate surgery.

The risk of complications is high following treatment with intraoperative radiotherapy (TEPPER et al. 1984; NOYES et al. 1992; MANNAERTS et al. 2002). For example, in one study of functional outcome in patients treated with intraoperative radiotherapy, 44% experienced fatigue, 42% experienced perineal pain, 36% experienced difficulty walking, and 42% experienced voiding problems (MANNAERTS et al. 2002). In another study, peripheral neuropathy was observed in 32% of patients following intraoperative radiotherapy (SHAW et al. 1990). Although these risks are sobering, it is important to recognize that morbidity is often high in patients with locally advanced colorectal cancer, regardless of treatment. Retrospective studies suggest that morbidity may be similar for patients who received intraoperative radiotherapy and those who did not (TEPPER et al. 1984; NOYES et al. 1992).

No treatment has been demonstrated clearly to be effective in the management of complications of radiotherapy. Therefore, decreasing the risk and severity of complications by minimizing the volume of normal tissue within the radiotherapy field is very important. Clinical trials to assess the value of olsalazine and cholestyramine in mitigating radiation-related side effects have demonstrated that these agents have unacceptable toxicity (CHARY and THOMSON 1984; MARTENSON et al. 1996a). Sucralfate appeared to be a more promising agent, and a European study suggested that it may reduce both acute and long-term adverse effects of pelvic radiotherapy (HENRIKSSON et al. 1992). In a confirmatory randomized trial undertaken by the North Central Cancer Treatment Group, no beneficial effect was observed with sucralfate administered to patients who had received pelvic radiotherapy, and several measures of gastrointestinal function were made worse by the use of this agent (MARTENSON et al. 2000). The oncology community will have the best chance of improving the therapeutic ratio for patients with lower gastrointestinal cancer if radiation oncologists and other oncologists are committed to entering patients into well-designed prospective studies to assess promising new ways of improving treatment.

References

Bartelink H, Roelofsen F, Eschwege F et al (1997) Concomitant radiotherapy and chemotherapy is superior to radiotherapy alone in the treatment of locally advanced anal cancer: results of a phase III randomized trial of the European Organization for Research and Treatment of Cancer Radiotherapy and Gastrointestinal Cooperative Groups. J Clin Oncol 15:2040–2049

Beart RW Jr, O'Connell MJ (1983) Postoperative follow-up of patients with carcinoma of the colon. Mayo Clin Proc 58:361–363

Benson AB III, Desch CE, Flynn PJ et al (2000) 2000 Update of American Society of Clinical Oncology colorectal cancer surveillance guidelines. J Clin Oncol 18:3586–3588

Black WA, Waugh JM (1948) Intramural extension of carcinoma of descending colon, sigmoid, and rectosigmoid: pathologic study. Surg Gynecol Obstet 84:457–464

Boman BM, Moertel CG, O'Connell MJ et al (1984) Carcinoma of the anal canal: a clinical and pathologic study of 188 cases. Cancer 54:114–125

Cass AW, Million RR, Pfaff WW (1976) Patterns of recurrence following surgery alone for adenocarcinoma of the colon and rectum. Cancer 37:2861–2865

Chary S, Thomson DH (1984) A clinical trial evaluating cholestyramine to prevent diarrhea in patients maintained on low-fat diets during pelvic radiotherapy. Int J Radiat Oncol Biol Phys 10:1885–1890

Cohen AM, Gunderson LL, Welch CE (1981) Selective use of adjuvant radiotherapy in resectable colorectal adenocarcinoma. Dis Colon Rectum 24:247–251

Dahlberg M, Glimelius B, Graft W et al (1998) Preoperative irradiation affects functional results after surgery for rectal cancer: results from a randomized study. Dis Colon Rectum 41:543–549

Douglass HO Jr, Moertel CG, Mayer RJ et al (1986) Survival after postoperative combination treatment of rectal cancer (letter). N Engl J Med 315:1294–1295

Enquist IF, Block IR (1966) Rectal cancer in the female: selection of proper operation based upon anatomic studies of rectal lymphatics. Prog Clin Cancer 2:73–85

Flam M, John M, Pajak TF et al (1996) Role of mitomycin in combination with fluorouracil and radiotherapy, and of salvage chemoradiation in the definitive nonsurgical treatment of epidermoid carcinoma of the anal canal: results of a phase III randomized intergroup study. J Clin Oncol 14:2527–2539

Fletcher RH (1986) Carcinoembryonic antigen. Ann Intern Med 104:66–73

Fletcher RH (1993) CEA monitoring after surgery for colorectal cancer. When is the evidence sufficient? (Editorial.) JAMA 270:987–988

Gallagher MJ, Brereton HD, Rostock RA et al (1986) A prospective study of treatment techniques to minimize the volume of pelvic small bowel with reduction of acute and late effects associated with pelvic irradiation. Int J Radiat Oncol Biol Phys 12:1565–1573

Gastrointestinal Tumor Study Group (1985) Prolongation of the disease-free interval in surgically treated rectal carcinoma. N Engl J Med 312:1465–1472

Gilbert SG (1978) Symptomatic local tumor failure following abdomino-perineal resection. Int J Radiat Oncol Biol Phys 4:801–807

Gilbertsen VA (1960) Adenocarcinoma of the rectum: incidence and locations of recurrent tumor following present-
day operations performed for cure. Ann Surg 151:340–348

Green N (1983) The avoidance of small intestine injury in
gynecologic cancer. Int J Radiat Oncol Biol Phys 9:1385–
1390

Green N, Iba G, Smith WR (1975) Measures to minimize small
intestine injury in the irradiated pelvis. Cancer 35:1633–
1640

Grinnell RS (1966) Lymphatic block with atypical and retrograde lymphatic metastasis and spread in carcinoma of the
colon and rectum. Ann Surg 163:272–280

Gunderson LL, Sosin H (1974) Areas of failure found at reoperation (second or symptomatic look) following "curative
surgery" for adenocarcinoma of the rectum: clinicopathologic correlation and implications for adjuvant therapy.
Cancer 34:1278–1292

Gunderson LL, Cohen AM, Welch CE (1980) Residual, inoperable or recurrent colorectal cancer: interaction of surgery
and radiotherapy. Am J Surg 139:518–525

Gunderson LL, Cohen AC, Dosoretz DD et al (1983a) Residual,
unresectable, or recurrent colorectal cancer: external beam
irradiation and intraoperative electron beam boost +/−
resection. Int J Radiat Oncol Biol Phys 9:1597–1606

Gunderson LL, Meyer JE, Sheedy PF et al (1983b) Radiation
oncology. In: Margulis AR, Burhenne HJ (eds) Alimentary tract radiology, vol 2, 3rd edn. CV Mosby, St Louis,
pp 2409–2446

Gunderson LL, Tepper JE, Dosoretz DE et al (1983c) Patterns
of failure after treatment of gastrointestinal cancer. Cancer
Treat Symp 2:181–197

Gunderson LL, Russell AH, Llewellyn HJ et al (1985a) Treatment planning for colorectal cancer: radiation and surgical techniques and value of small-bowel films. Int J Radiat
Oncol Biol Phys 11:1379–1393

Gunderson LL, Sosin H, Levitt S (1985b) Extrapelvic colon:
areas of failure in a reoperation series; implications for
adjuvant therapy. Int J Radiat Oncol Biol Phys 11:731–741

Gunderson LL, Collins R, Earle JD, et al (1986) Adjuvant treatment of rectal cancer: randomized prospective study of
irradiation +/− chemotherapy: a NCCTG, Mayo Clinic
study (abstract). Int J Radiat Oncol Biol Phys 12 [Suppl
1]:169

Gunderson LL, Martenson JA, Haddock MG (1996a) Indications for and results of irradiation +/− chemotherapy for
rectal cancer. Ann Acad Med Singapore 25:448–459

Gunderson LL, Nelson H, Martenson JA et al (1996b) Intraoperative electron and external beam irradiation with or
without 5-fluorouracil and maximum surgical resection for
previously unirradiated, locally recurrent colorectal cancer.
Dis Colon Rectum 39:1379–1395

Gunderson LL, Sargent DJ, Tepper JE et al (2004) Impact of
T and N stage and treatment on survival and relapse in
adjuvant rectal cancer: a pooled analysis. J Clin Oncol
22:1785–1796

Haddock MG, Gunderson LL, Nelson H et al (2001) Intraoperative irradiation for locally recurrent colorectal cancer
in previously irradiated patients. Int J Radiat Oncol Biol
Phys 49:1267–1274

Haddock MG, Nelson H, Donohue JH et al (2003) Intraoperative electron radiotherapy as a component of salvage therapy for patients with colorectal cancer and advanced nodal
metastases. Int J Radiat Oncol Biol Phys 56:966–973

Henriksson R, Franzen L, Littbrand B (1992) Effects of sucralfate on acute and late bowel discomfort following radiotherapy of pelvic cancer. J Clin Oncol 10:969–975

Hoskins RB, Gunderson LL, Dosoretz DE et al (1985) Adjuvant
postoperative radiotherapy in carcinoma of the rectum
and rectosigmoid. Cancer 55:61–71

John MJ, Pajak TJ, Flam MS et al (1995) Dose acceleration in
chemoradiation (CRX) for anal cancer: preliminary results
of RTOG 92-08 (abstract). Int J Radiat Oncol Biol Phys 32
[Suppl 1]:157

Kollmorgen CF, Meagher AP, Wolff BG et al (1994) The long-
term effect of adjuvant postoperative chemoradiotherapy
for rectal carcinoma on bowel function. Ann Surg 220:676–
682

Krook JE, Moertel CG, Gunderson LL et al (1991) Effective
surgical adjuvant therapy for high-risk rectal carcinoma.
N Engl J Med 324:709–715

Lindel K, Willett CG, Shellito PC et al (2001) Intraoperative
radiotherapy for locally advanced recurrent rectal or rectosigmoid cancer. Radiother Oncol 58:83–87

Lundby L, Jensen VJ, Overgaard J et al (1997) Long-term
colorectal function after postoperative radiotherapy for
colorectal cancer. Lancet 350:564

Mannaerts GH, Martijn H, Crommelin MA et al (1999) Intraoperative electron beam radiotherapy for locally recurrent
rectal carcinoma. Int J Radiat Oncol Biol Phys 45:297–308

Mannaerts GH, Rutten HJ, Martijn H et al (2002) Effects on
functional outcome after IORT-containing multimodality
treatment for locally advanced primary and locally recurrent rectal cancer. Int J Radiat Oncol Biol Phys 54:1082–
1088

Martenson JA Jr, Gunderson LL (1993) External radiotherapy
without chemotherapy in the management of anal cancer.
Cancer 71:1736–1740

Martenson JA Jr, Hyland G, Moertel CG et al (1996a) Olsalazine
is contraindicated during pelvic radiotherapy: results of a
double-blind, randomized clinical trial. Int J Radiat Oncol
Biol Phys 35:299–303

Martenson JA, Lipsitz SR, Wagner H Jr et al (1996b) Initial
results of a phase II trial of high dose radiotherapy, 5-
fluorouracil, and cisplatin for patients with anal cancer
(E4292): an Eastern Cooperative Oncology Group Study.
Int J Radiat Oncol Biol Phys 35:745–749

Martenson JA, Schild SE, Haddock MG (1998) Cancers of
the gastrointestinal tract. In: Khan FM, Potish RA (eds)
Treatment planning in radiation oncology. Williams and
Wilkins, Baltimore, pp 319–342

Martenson JA, Bollinger JW, Sloan JA et al (2000) Sucralfate in
the prevention of treatment-induced diarrhea in patients
receiving pelvic radiotherapy: a North Central Cancer
Treatment Group phase III double-blind placebo-controlled trial. J Clin Oncol 18:1239–1245

Martin EW Jr, James KK, Hurtubise PE et al (1977) The use of
CEA as an early indicator for gastrointestinal tumor recurrence and second-look procedures. Cancer 39:440–446

Mendenhall WM, Million RR, Pfaff WW (1983) Patterns of
recurrence in adenocarcinoma of the rectum and rectosigmoid treated with surgery alone: implications in treatment planning with adjuvant radiotherapy. Int J Radiat
Oncol Biol Phys 9:977–985

Minsky BD, Mies C, Rich TA et al (1988) Potentially curative
surgery of colon cancer: patterns of failure and survival. J
Clin Oncol 6:106–118

Minton JP, Hoehn JL, Gerber DM et al (1985) Results of a 400-patient carcinoembryonic antigen second-look colorectal cancer study. Cancer 55:1284–1290

Moertel CG, Schutt AJ, Go VL (1978) Carcinoembryonic antigen test for recurrent colorectal carcinoma: inadequacy for early detection. JAMA 239:1065–1066

Moertel CG, Fleming TR, Macdonald JS et al (1993) An evaluation of the carcinoembryonic antigen (CEA) test for monitoring patients with resected colon cancer. JAMA 270:943–947

Morson BC (1960) The pathology and results of treatment of squamous cell carcinoma of the anal canal and anal margin. Proc R Soc Med 53:416–420

Nivatvongs S, Stern HS, Fryd DS (1981) The length of the anal canal. Dis Colon Rectum 24:600–601

Northover J, Houghton J, Lennon T (1994) CEA to detect recurrence of colon cancer (letter). JAMA 272:31

Noyes RD, Weiss SM, Krall JM et al (1992) Surgical complications of intraoperative radiotherapy: the Radiotherapy Oncology Group experience. J Surg Oncol 50:209–215

O'Connell MJ, Martenson JA, Wieand HS et al (1994) Improving adjuvant therapy for rectal cancer by combining protracted-infusion fluorouracil with radiotherapy after curative surgery. N Engl J Med 331:502–507

Patterson DJ, Alpert E (1983) Tumour markers of the gastrointestinal tract. In: Hodgson HJF, Bloom SR (eds) Gastrointestinal and hepatobiliary cancer. Chapman and Hall, London, pp 189–205

Postlethwait RW (1949) Malignant tumors of the colon and rectum. Ann Surg 129:34–46

Rich T, Gunderson LL, Lew R et al (1983) Patterns of recurrence of rectal cancer after potentially curative surgery. Cancer 52:1317–1329

Romsdahl MM, Withers HR (1978) Radiotherapy combined with curative surgery: its use as therapy for carcinoma of the sigmoid colon and rectum. Arch Surg 113:446–453

Russell AH, Tong D, Dawson LE et al (1984) Adenocarcinoma of the proximal colon: sites of initial dissemination and patterns of recurrence following surgery alone. Cancer 53:360–367

Sauer R (2003) Adjuvant versus neoadjuvant combined modality treatment for locally advanced rectal cancer: first results of the German Rectal Cancer Study (CAO/ARO/AIO-94) (abstract). Int J Radiat Oncol Biol Phys 57 [Suppl]:S124–S125

Schild SE, Martenson JA Jr, Gunderson LL et al (1989) Postoperative adjuvant therapy of rectal cancer: an analysis of disease control, suvival, and prognostic factors. Int J Radiat Oncol Biol Phys 17:55–62

Schild SE, Gunderson LL, Haddock MG et al (1997) The treatment of locally advanced colon cancer. Int J Radiat Oncol Biol Phys 37:51–58

Shanahan TG, Mehta MP, Gehring MA et al (1989) Minimization of small bowel volume utilizing customized "belly board" mold (abstract). Int J Radiat Oncol Biol Phys 17 [Suppl 1]:187–188

Shaw EG, Gunderson LL, Martin JK et al (1990) Peripheral nerve and ureteral tolerance to intraoperative radiation therapy: clinical and dose-reponse analysis. Radiother Oncol 18:247–255

Stearns MW Jr, Urmacher C, Sternberg SS et al (1980) Cancer of the anal canal. Curr Probl Cancer 4:1–44

Suzuki K, Gunderson LL, Devine RM et al (1995) Intraoperative irradiation after palliative surgery for locally recurrent rectal cancer. Cancer 75:939–952

Swedish Rectal Cancer Trial (1997) Improved survival with preoperative radiotherapy in resectable rectal cancer. N Engl J Med 336:980–987. Erratum in: N Engl J Med 1997, 336:1539

Taylor N, Crane C, Skibber J et al (2001) Elective groin irradiation is not indicated for patients with adenocarcinoma of the rectum extending to the anal canal. Int J Radiat Oncol Biol Phys 51:741–747

Tepper JE, Gunderson LL, Orlow E et al (1984) Complications of intraoperative radiotherapy. Int J Radiat Oncol Biol Phys 10:1831–1839

Tveit KM, Guldvog I, Hagen S, et al (1997) Randomized controlled trial of postoperative radiotherapy and short-term time-scheduled 5-fluorouracil against surgery alone in the treatment of Dukes B and C rectal cancer. Br J Surg 84:1130–1135

UKCCCR Anal Cancer Trial Working Party (1996) Epidermoid anal cancer: results from the UKCCCR randomised trial of radiotherapy alone versus radiotherapy, 5-fluorouracil, and mitomycin. Lancet 348:1049–1054

Walz BJ, Green MR, Lindstrom ER et al (1981) Anatomical prognostic factors after abdominal perineal resection. Int J Radiat Oncol Biol Phys 7:477–484

Wanebo HJ, Rao B, Pinsky CM et al (1978) Preoperative carcinoembryonic antigen level as a prognostic indicator in colorectal cancer. N Engl J Med 299:448–451

Welch JP, Donaldson GA (1978) Detection and treatment of recurrent cancer of the colon and rectum. Am J Surg 135:505–511

Welch JP, Donaldson GA (1979) The clinical correlation of an autopsy study of recurrent colorectal cancer. Ann Surg 189:496–502

Wiig JN, Tveit KM, Poulsen JP et al (2002) Preoperative irradiation and surgery for recurrent rectal cancer: will intraoperative radiotherapy (IORT) be of additional benefit? A prospective study. Radiother Oncol 62:207–213

Willett C, Tepper JE, Cohen A et al (1984a) Local failure following curative resection of colonic adenocarcinoma. Int J Radiat Oncol Biol Phys 10:645–651

Willett CG, Tepper JE, Cohen AM et al (1984b) Failure patterns following curative resection of colonic carcinoma. Ann Surg 200:685–690

Withers HR, Cuasay L, Mason KA et al (1981) Elective radiotherapy in the curative treatment of cancer of the rectum and retrosigmoid colon. In: Stroehlein JR, Romsdahl MM (eds) Gastrointestinal cancer. Raven, New York, pp 351–362

23 Bladder Cancer – Technical Basis of Radiation Therapy

Alan R. Schulsinger, Ron R. Allison, Walter H. Choi, and Marvin Rotman

CONTENTS

23.1 Natural History of the Disease
(Patterns of Spread) 561
23.2 Work-Up and Staging 562
23.3 Prognostic and Predictive Factors 563
23.4 General Management 564
23.5 Radiation Therapy Techniques
(General Description) 568
23.6 Simulation 569
23.7 Target Volume and Organs at Risk (Critical Structures)
−Specifications (Including Tolerance Doses) 569
23.8 Dose Prescription Beam Selection/Design Isodoses
Plan Evaluation/Implementation 570
23.8.1 Simulation/CT Simulation Procedures 570
23.9 Future Directions 573
References 575

23.1 Natural History of the Disease (Patterns of Spread)

Bladder cancer is the sixth most common cancer in the United States (Cancer Statistics 2004). It accounts for approximately 4% of all cancers, which translates into 60,000 new cases a year. The 12,000 deaths per year attributable to bladder malignancies are comparable to the yearly mortality rates of brain, stomach, and esophageal cancers.

A. R. Schulsinger, MD, Associate Professor of Radiation Oncology, Department of Radiation Oncology, State University of New York, Health Science Center at Brooklyn, 450 Clarkson Avenue, Box 1211, Brooklyn, NY 11203-2098, USA and Director of Radiation Oncology at the Long Island College Hospital
M. Rotman, MD, Distinguished Service Professor and Chairman, Department of Radiation Oncology, State University of New York, Health Science Center at Brooklyn, 450 Clarkson Avenue, Box 1211, Brooklyn, NY 11203-2098, USA
R. R. Allison, MD, Professor and Chairman, Department of Radiation Oncology, Leo Jenkins Cancer Center, Brody School of Medicine, Greenville, NC 27834, USA
W. H. Choi, MD, Department of Radiation Oncology, SUNY Downstate at Brooklyn, 450 Clarkson Avenue, Box 1211, Brooklyn, NY 11203, USA

Over the past 25 years, we have witnessed a nearly twofold increase in the incidence of bladder cancer, but overall survival rates have remained essentially unchanged.

Risk factors for transitional cell cancer of the bladder include exposure to chemical carcinogens (i.e., aniline dyes), tobacco (estimated to account for half of all cases that occur in men in the United States and one-third of all cases that occur in women), coffee, artificial sweeteners, and phenacetin-containing analgesics (Whitmore et al. 1977). Chronic irritation by foreign bodies (i.e., indwelling Foley catheters, calculi, and *Schistosoma hematobium* in endemic areas) are risk factors for squamous cell cancers. Exstrophy of the bladder is the main risk factor for adenocarcinoma (Morison and Cole 1976).

Some of the more common clinical presentations of bladder cancer include (1) painless hematuria, which occurs in up to 80% of patients; (2) bladder irritability, such as urinary frequency, urgency, and dysuria (all of which are suggestive of muscle-invasive disease); and (3) recurrent urinary tract infections, particularly in men. Some of the less common clinical presentations include (1) flank pain or anemia associated with a pelvic mass; (2) pelvic mass associated with lower extremity weakness, weight loss, abdominal pain, or bone pain; and (3) suprapubic pain.

Anatomically, the bladder is a hollow muscular organ that lies in the anterior half of the pelvis. When full, it contains about 0.5 l of urine. It occupies a triangular space bound anteriorly and laterally by the symphysis pubis and the diverging walls of the pelvis, respectively. The posterior border is the rectum/rectovaginal septum. The lateral and inferior portions of the bladder are supported by the obturator internus and levator ani muscles, respectively. In males, the prostate lies between the levator ani and the bladder. The superior surface of the bladder is covered by peritoneum.

The interior inferior surface of the bladder is lined by a loosely attached mucus membrane, except in the trigone region where the mucus membrane is firmly attached. The urinary epithelium lining the bladder is thrown into many folds in the relaxed state. This allows the bladder to expand with the filling of urine. Deep to the epithelium, the wall of the bladder consists of three loosely arranged smooth muscle and elastic fiber layers which contract during micturition. These are the inner longitudinal, outer spiral, and outer longitudinal layers. The outer longitudinal layer is surrounded by the outer adventitial coat. This coat contains arteries veins and lymphatic channels. Because these lymphovascular channels reside most abundantly in the outer layer of the bladder, depth of penetration of tumor cells is correlated with the incidence of locoregional lymph node metastasis. The trigone region that leads into the bladder neck is defined by the ureteral orifices posterolaterally and by the urethral aperture at the inferior/anterior angle. Transitional cell epithelium lines the bladder and is contiguous into the ureters (urothelium).

The bladder contains submucosal plexus of lymphatics that are most abundant in the region of the trigone. These lymphatics usually drain into channels that pierce the muscular layers and then organize from the superior and inferolateral surfaces of the bladder to ultimately drain into the external iliac lymph nodes. From the posterior surface of the bladder, lymphatic channels drain to both the external iliac and internal iliac lymph node chains. Lymphatic vessels from the bladder neck may combine with some prostatic lymphatic vessels in males, which can ultimately drain to the presacral and common iliac lymph nodes.

The most common histology in the United States is transitional cell carcinoma, which makes up approximately 90% of cases, followed by squamous cell (7%), and adenocarcinoma (less than 1%) (PEARSE 1994). Sarcomas, lymphomas, carcinoid, and small cell tumors are rarely seen. About 30% of bladder cancers present as multiple lesions. Adjacent carcinoma in-situ (CIS) is also common.

Tumors of the bladder may be papillary in appearance – which are generally not deeply invasive – or solid in appearance – which are generally deeply invasive. Most transitional cell tumors are found at the trigone, followed in frequency by the lateral and posterior walls and then the bladder neck (MOSTOFI et al. 1988). Adenocarcinoma also most frequently arises at the trigone (JOHNSON et al. 1972).

Tumors progress by further muscle invasion and by lymphatic involvement to the external iliac lymph nodes. About 40% of patients with muscle-invasive cancers have involved lymph nodes at presentation (SKINNER et al. 1982). Almost all of these patients will ultimately die of distant metastasis (SKINNER et al. 1982). Of note is that metastases is rarely seen in squamous cell histology.

The most common sites of spread are lung, bone, and liver.

23.2
Work-Up and Staging

The basis of the work-up in bladder cancer is to determine whether the disease is a superficial non-invasive cancer, a locally invasive lesion, or metastatic disease. In addition to evaluating the bladder, the ureters and kidneys are also examined for lesions, since multiple tumors are not uncommon. Perhaps this is related to common carcinogen exposure or embryology.

Cystoscopy and urethroscopy allow for excellent visualization and biopsy of lesions. A bladder diagram should be completed at the time of cystoscopy to record pertinent findings (Fig. 23.1). This information can be quite valuable to the radiation oncologist for treatment planning, as a precise knowledge of the tumor location is critical. In addition, computed tomography (CT) and/or magnetic resonance imaging (MRI) can be used to evaluate for bladder wall thickening and invasion as well as for lymphadenopathy. It should be kept in mind that edema and hemorrhage seen on MRI and CT scan obtained shortly after transurethral bladder resection (TURBT) may easily be confused with tumor (BARENTSZ et al. 1996, 2000). For this reason, we advise imaging studies be performed prior to TURBT. All patients should also have blood taken for a complete blood count and serum chemistries, including liver functions tests, as this may offer clues to systemic spread (Table 23.1). Chest and abdominal CTs are also recommended for invasive disease. Positron emission tomography (PET) scanning may also help delineate disease spread.

Bimanual examination under anesthesia should be performed both before and after transurethral bladder resection of the visualized lesion to get a better appreciation of the size, consistency, and location of the tumor. This allows for estimation of the extent of local infiltration into the surrounding tissue by assessing whether the mass is freely mobile, tethered, or fixed, before and after maximal transurethral bladder resection.

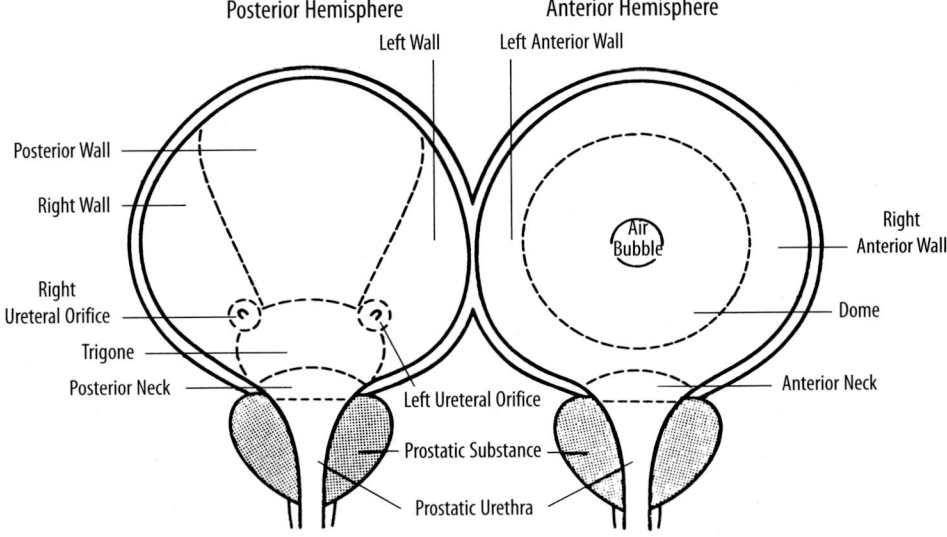

Fig. 23.1. A bladder diagram is completed at the time of cystoscopy and used to record the cystoscopy findings, biopsy, information, and tumor characteristics

Table 23.1. Diagnostic Work-Up for Bladder Cancer

Routine
  History and physical examination
  Pelvic/rectal examination
Laboratory studies
  Complete blood cell count
  Live function tests and chemistries
  Urinalysis and Urine Cytology
Imaging
  Computed tomography or magnetic resonance scan of pelvis and abdomen
  Intravenous pyelography
  Chest radiograph
  Radioisotope bone scan
Cystourethroscopy
Bimanual pelvic/rectal examination under anesthesia
Biopsies of bladder and urethra
Transurethral resection, if indicated

Determination of muscular wall invasion is the most important aspect of staging. It is often not possible for the pathologist examining TURBT specimens to determine whether the tumor is confined to the superficial muscle layers of the muscularis propria, which is the first muscle layer reached after tumor cells have already invaded the connective tissue of the lamina propria, or whether the tumor has penetrated further to involve the deeper muscles layers. This inability to distinguish deep from superficial muscle invasion leads to both understaging

and, less frequently, overstaging. These limitations to accurate staging have undoubtedly compounded the difficulty of showing the benefit of effective bladder-sparing treatments.

A further difficulty to accurate staging is the use of both clinical and pathological staging systems in clinical trials. Caution is therefore required in the interpretation and comparisons of trial results. The two most widely used staging systems are presented in Table 23.2. In the American Joint Committee on Cancer (AJCC) staging system, if staging is based on evaluation of the cystectomy specimen, the stages are preceded by the lower case letter p.

23.3
Prognostic and Predictive Factors

Approximately 70% of patients have Tis, Ta, or T1 disease at presentation. Approximately 20% have stage T2–T4 disease and another 10% present with metastatic disease. Although the majority of tumors are superficial, they can behave aggressively and locally they recur repeatedly, often with further and deeper invasion. Eventually, penetration of the bladder muscular layers occur. Once muscle is involved, lymphatic and blood vessel invasion is common. It is generally reported that pathological T2 disease has a 30% risk of nodal involvement, as does early T3 disease. Patients with advanced T3 or T4 tumors have a 50–80% risk of nodal involvement (SHIPLEY et al.

Table 23.2. Comparsion of Marshall and AJC Staying System for Bladder Cancer

	Marshall modification of Jewett-Strong classification	AJCC
Tumor Extent		
Confined to Mucosa	0	
• nonpapillary, noninvasive		TIS
• papillary, noninvasive		Ta
Not Beyone Lamina Propria (no mass palpable after complete TUR)	A	T1
Invasion of superficial muscle (inner half) (no duration after complete TUR)	B1	T2a
Invasion of deep muscle (outer half) (induration after complete TUR)	B2	T2b
Invasion into perivesical fat (mobile mass after TUR)–microscopic	C	T3a
–macroscopic		T3b
Invasion of neighboring structures: Muscle invasion present		
Substance of prostate, vagina, uterus	D1[a]	T4a
Pelvic sidewall fixation or invading abdominal wall	D1[a]	T4b
Nodal involvement (N)		
Minimum requirements to assess the regional nodes cannot be met		Nx
No involvement of regional lymph nodes		No
Involvement of a single lymph node, 2 cm or less in size		N1
Involvement of a single lymph node >2 cm or less but <5 cm or multiple lymph nodes measuring <5 cm		N2
Lymph node mestastasis >5 cm in diameter		N3
Distant metastasis (M)		
Minimum requirements to assess the presence of distant metastasis cannot be met		Mx
No distant metastasis		Mo
Distant metastasis		M1

[a] In the Marshall modification of the Jewett-Strong staging system, D1 disease may involve lymph nodes below the sacral promontory (bifurcation of the common iliac artery). D2 implies distant metastases or more extensive lymph node metastases.
TUR, transurethral resection

1985). As nodal metastasis is in part an indicator of potential for systemic spread, a similar frequency of distant metastasis is eventually noted for these patients also.

Poor prognostic factors at presentation include deeply invasive tumors (there is an increasing probability of perivesicular and pelvic nodal metastasis as the depth of invasion increases), associated CIS, vascular invasion, positive lymph nodes, tumors greater than 6 cm in size, urethral obstruction/obstructive uropathy, solid tumor morphology, a palpable mass present on bimanual examination, visible tumor following TURBT, solid/flat surface tumor histology as opposed to papillary histology which is a more favorable characteristic, high-grade (poorly differentiated) tumors, hemoglobin of less than 12 gm/dl, stage T3b or T4 tumors, and multiple tumors (SHIPLEY et al. 1985).

23.4
General Management

Optimal therapeutic options for bladder cancer depend on histology and stage of disease. Patients with squamous cell cancer tend to experience failure locally. Management of these individuals should be with a course of preoperative radiation therapy to the pelvis followed by radical cystectomy. This approach yields an approximate 50% 5-year survival rate and is generally considered to offer the best chance for cure (AWWAD et al. 1979; GHONEIM et al. 1985). The management of transitional cell cancers should be based on whether the patient has non-muscle- or muscle-invading disease. For non-muscle-invading disease [stage 0 (Tis), and stage A (T1)], acceptable local control rates and 5-year survival rates have been obtained with a variety of interventions including transurethral resection and fulguration of the bladder, partial cystectomy, interstitial implants, intraoperative irradiation, intravesicular chemotherapy, and Bacillus Calmette Guerin (BCG) following TUR. Early-stage patients (Tis, Ta, T1) with non-muscle-invasive disease are generally managed by maximal TURBT followed by intravesicular BCG instillation. For muscle-invasive bladder cancers, survival results and morbidity remain poor with cystectomy. This surgery usually includes permanent ileal conduit with loss of sexual potency and is considered "standard treatment" in the United States for muscle-invasive disease. Although preoperative external beam irradiation may improve outcome (SILVERMAN et al. 1992), and the combination of chemotherapy with irradiation may allow for organ preservation, only a fraction of patients are offered these options.

Transurethral resection in the management of non-muscle-invasive disease is often an outpatient procedure in which transurethral visualization of the lesion(s) in question is obtained and a biopsy, if not a resection, is accomplished. Tumor removal may be by scalpel, heat, or laser source. The goal is to remove tumor down to uninvolved tissue. TURBT is a well-tolerated procedure; however, after multiple TURBTs have been performed, the bladder is typically fibrotic and contracted. Perforations are rare, although at times they do occur and may require surgical repair. Most of these patients undergo one or more TURBT, generally followed by intravesicular BCG. Local control rates of 60–80% are obtained (HERR et al. 1995). A significant minority of patients will go on to develop invasive bladder cancer or a second urethral malignancy. For this reason, close follow-up with cystoscopy performed at 3- to 4-month intervals is recommended.

Interstitial and intraoperative radiation treatment of non-muscle-invasive bladder cancer has excellent outcome for selected patients, but is rarely employed in the United States. Matsumoto, using intraoperative radiation therapy delivered by an electron beam, achieved an impressive 95% local control rate for early-stage solitary lesions (MATSUMOTO et al. 1981). VAN DER WERF-MESSING et al. (1981) implanted the bladder by brachytherapy and reported 80–85% local control rates. Given that understaging is such a significant problem after TURBT, it is possible that intravesicular therapy may be undertreating some of these patients with "non-muscle-invasive" disease. Patients receiving some form of radiation are adequately treated, as the not-infrequent muscle-invasive component is responsive to radiation but

not chemotherapy or immunotherapy, neither of which penetrates into the full thickness of the bladder as completely as radiation does. To date there are no randomized trials comparing intravesicular therapies to radiation. There is, however, prospective, nonrandomized data from the Dutch South Eastern Bladder cancer study "that suggests that if radiotherapy is used routinely and not restricted to unfavorable subgroups, the results are probably better than with adjuvant intravesical therapy" (RODEL et al. 2005).

Perhaps due to the inherent inadequacies of clinical staging, the optimal management for muscle-invasive disease remains unclear. For these patients, however, TUR alone is usually unacceptable due to high local failure rates. The exception to this is in selected patients with single, small, superficially muscle-invasive tumors (T2a) not associated with CIS (HERR 1987; SOLSONA et al. 1992). Treatment options for the remaining patients include cystectomy (partial in selected cases) and combined modality therapy with a view to bladder preservation (i.e., maximal TUR followed by irradiation with chemotherapy). In patients with T2 disease, most commonly, cystectomy alone is employed and offers a 60% survival in pathologically staged patients (RESNICK and O'CONNOR 1972; BRANNAN et al. 1978). The role of radiation in this stage of disease in not yet well defined; however, VAN DER WERF-MESSING reported that selected T2 patients implanted with radium needles achieved an 80% disease-free survival at 5 years (VAN DER WERF-MESSING et al. 1983). These results were replicated by BATTERMAN and DENUE (1986). External beam radiation series for T2 disease is composed mainly of clinically staged patients. Clinical staging is inherently inaccurate and often includes individuals with pathologically more advanced tumors. Further, these outcomes are generally based on medically inoperable or elderly and frail patients. Despite these shortcomings, good local control and survival rates are possible as summarized in Table 23.3.

Table 23.3. Outcomes of trials of radiation alone to treat bladder cancer

Series (ref)	No. of Patients	Complete T2 (%)	Response T3 (%)
Blandy et al. (47)	704	48	42
Duncan et al. (48)	889	49	41
Smaaland (49)	146	69	36
Greven et al. (50)	116	36	18
Vale et al. (51)	60	79	46

For most patients with T2 disease, "surgical option" means radical cystectomy combined with pelvic lymphadenectomy. Radical cystectomy includes resection of the bladder, distal ureters, perivesicular fat, and the regional peritoneum. In men, the prostate, seminal vesicles, vas deferens, and proximal urethra are also removed. Up to 40% of men undergoing radical cystectomy for bladder cancer have been found to have prostate cancer (NIXON et al. 2002; WOOD et al. 1989). It has been suggested that men not at high risk of either bladder cancer involvement of the prostatic urethra or a second primary prostate cancer should be considered for prostate sparing cystectomy with a view toward improved urinary control and sexual potency (COOKSON 2005). In women, the uterus, fallopian tubes, ovaries, anterior vaginal wall, and urethra are resected. Urinary diversion may be by ileal conduit with external appliance or an internal stoma reservoir that may even maintain continence. Regarding the extent and completeness of the lymphadenectomy, there has been a renewed interest in extending the resection above the "traditional" level of the bifurcation of the iliac arteries, as this may impact on the disease specific survival even in patients without apparent lymph node involvement, up to the level of the inferior mesenteric artery (COOKSON 2005). It must be remembered, however, that there is an increased risk of lymphedema in extending the level of resection. Results of developing multi-institutional randomized studies comparing "traditional" with extended lymphadenectomy will hopefully resolve this issue.

Operative mortality is still about 1–2%, mainly due to pulmonary emboli, myocardial infarction and stroke. There is also significant blood loss associated with the procedure, and overall 30% of patients require transfusion with a median requirement of two units of packed red blood cells (COOKSON 2005). Additionally, there are major lifestyle changes brought about by this procedure, including vaginal dryness, incontinence (depending on the method of reconstruction), and the loss of sexual function.

A small minority of patients may be eligible for partial cystectomy. These are individuals with solitary well-defined tumors that allow at least 2 cm of margin all around the resection plane. Preoperative pelvic irradiation should be considered when there is a significant likelihood of microscopically involved pelvic lymph nodes.

Radical cystectomy is also the most widely selected therapy for patients in the United States with stage T3 disease. Cystectomy alone offers a 20–40% 5-year survival and similar local control rates (GREVEN et al. 1992; MONTIE et al. 1984; MORABITO et al. 1979; DRAGO and ROHNER 1983; MARSHALL and McCARRON 1977). In an attempt to improve results, several randomized studies involving preoperative external beam radiation therapy have been employed. In general, 4500 centigray (cGy) are delivered (BLOOM et al. 1982; BATATA et al. 1981; TIMMER et al. 1985; WOEHERE et al. 1993). Despite the fact that these studies have shown improved local control and survival rates, in the United States, most patients undergoing radical cystectomy do not receive preoperative radiation therapy.

Radiation therapy alone is usually unsuccessful in patients with T3 disease, with 5-year survivals reported in the 20% range (BLOOM et al. 1982; GOFFINET et al. 1975; QUILTY and DUNCAN 1986; POLLACK et al. 1994; EDSMYR et al. 1985; DeWEERD and COLBY 1973). This may be due, in part, to the fact that many of the patients who are chosen for "definitive radiation therapy" are often patients who initially failed multiple TURBTs with BCG and either refused salvage cystectomies or were deemed medically inoperable. These patients typically have scarred, contracted bladders to begin with, and perhaps have biologically more aggressive tumors as evidenced by their history of recurrences. It should also be kept in mind that a significant number of patients undergoing cystectomy for clinical stage T3 lesions are found to have T4 lesions on pathological analysis (MARSHALL 1952; RICHIE et al. 1975; WHITMORE et al. 1977). A significant minority are also downstaged (MARSHALL 1952; RICHIE et al. 1975; WHITMORE et al. 1977). For this reason, comparing results from clinically staged series to pathologically staged series is difficult. Some radiation series do reveal fair pelvic local control rates (Table 23.3) but often it is based on salvage cystectomy in patients able to undergo this procedure.

Various clinical trials suggest that bladder cancer is a chemoresponsive tumor. Recent reports of chemotherapy integrated with radiation therapy suggest that results can be improved over radiation therapy alone and may obviate the need for radical cystectomy with its resultant compromised quality of life.

Chemotherapy has been successful in improving outcome for both early and advanced patients. Compared with TURBT alone, single agents instilled into the bladder – or BCG employment for early-stage patients following maximal transurethral bladder resection – clearly increase local control. For patients with muscle-invasive disease, chemotherapy plays an important role in enhancing the effects of radiation therapy. Numerous trials demonstrate that the addition of radiosensitizing doses of chemotherapy [i.e., 5-fluorouracil (5-FU), cisplatin, etc.] improve complete response rates by more that 50% when compared with radiation therapy alone (ROTMAN et al. 1987; REIBISCHUNG et al. 1992; TESTER et al. 1993; CERVAK et al. 1993; DUNST et al. 1994).

Incorporation of chemotherapy into the management paradigm of muscle-invasive bladder cancer offers the theoretical advantage of "spatial cooperation", where the primary role of chemotherapy is to control micrometastasis at distant sites, while either surgery, irradiation, or irradiation with radiosensitizing doses of chemotherapy are used to address the localized primary tumor.

The major side effect of chemotherapy is generally hematological toxicity. Also, many patients who were referred for radiation therapy with chemotherapy are often sent because they are medically inoperable. Frequently, these patients have coronary artery disease, and certain chemotherapeutic agents may cause coronary artery spasm (i.e., 5-FU) (DEVITA et al. 1989). Although it is unclear whether or not the addition of chemotherapy will ultimately increase the rate of long-term complications, this has not yet been reported in any prospective randomized studies. There may be increased frequency of diarrhea in patients who are treated with concomitant 5-FU, but these side effects are frequently prevented by prophylactically placing patients on a combination of Metamucil and Pepto-Bismol prior to initiation of treatment.

As distant metastases remain the most common cause of treatment failures for patients with muscle-invasive bladder carcinoma, it seems reasonable to try to incorporate a systemic component into the treatment regimen in an attempt to control micrometastases at distant sites.

Results of the South West Oncology Group trial 8710 (INT-00800) – a phase-III trial of neoadjuvant methotrexate, vinblastine, doxorubicin, cisplatin (MVAC) plus cystectomy versus cystectomy alone in patients with locally advanced bladder cancer – have shown improved 5-year survival figures for patients treated with the neoadjuvant chemotherapy (NATALE et al. 2001). Two other recently published randomized trials including the Medical Research Council/European Organization for Research and Treatment of Cancer (MRC/EORTC) and US Intergroup studies have also suggested both improvement in locoregional control and metastatic relapse

(HUSSAIN and JAMES 2005). These studies suggest that neoadjuvant chemotherapy does in fact have an impact on the control of micrometastases.

Despite numerous trials by groups utilizing cisplatin, it is becoming increasingly clear that cisplatin may not be the ideal drug of choice for chemoradiotherapy in the treatment of bladder cancer. As pointed out by JAMES and HUSSAIN (2005; HUSSAIN and JAMES 2005), "... a significant proportion of patients (with bladder cancer) have impaired renal function, and administration would require inpatient stay and hydration. Only about 50% of patients were fit to receive cisplatin at their institution at the doses used in the Canadian study." In studies reported by Rotman et al., patients with clinically staged bladder cancer underwent high doses of external beam radiation therapy in combination with sensitizing doses of 5-FU chemotherapy (ROTMAN et al. 1990). The majority of patients retained functioning bladders, and minimal toxicities were noted. More importantly, survival rates were excellent. This and other studies provide evidence that chemotherapy also improves local tumor complete response rates.

Table 23.3 summarizes the complete response rates for various studies employing the use of radiation therapy alone (BLANDY et al. 1980; DUNCAN and QUILTY 1986; SMAALAND et al. 1991; GREVEN et al. 1990; VALE et al. 1993). Table 23.4 summarizes the results of studies combining aggressive TURBT followed by combining concomitant chemotherapy and radiation therapy (SAVER et al. 1990; ROTMAN et al. 1987; JAKSE et al. 1985; TESTER et al. 1993; RICHARDS et al. 1983; CERVAK et al. 1993). Comparison of these results strongly suggests that the addition of chemotherapy improves results over standard radiation therapy alone.

In patients treated with nonsurgical bladder-sparing approaches, maximal TURBT is usually the initial step and is followed by radiation therapy combined with chemotherapy. When such strategies have been employed, response rates of 70% or greater have been obtained (DEVITA et al. 1989; DUNST et al. 1994; EAPEN et al. 1989; HOUSSET et al. 1993).

Hyperfractionated trials draw from the encouraging results from the Royal Marsden Hospital where local control for muscle-invading bladder cancer was enhanced by accelerated multiple daily treatments (COLE D DURANT et al. 1992; HORWICH et al. 1995). Incorporating this benefit into combined modality therapy with maximal TURBT and chemotherapy has resulted in impressive results in both French and Italian trials (HOUSSET et al. 1993; DANESI et al. 2004).

Table 23.4 Outcomes of trials of chemotherapy and radiation to treat bladder cancer. *CR* complete response, *conc* concomitant, *neo* neoadjuvant

Reference	No.	Stage	Chemotherapy	Radiation dose	Sequence	Median F/U	CR(%)	Survival (%)
ROTMAN et al. (1990)	19	T2–4	5-FU±MMC	60–65 Gy	Conc	38 months	74–89	53.6
RUSSELL et al. (1990)	34		5-FUx2	40 Gy	Conc	18 months	81	64
HOUSSET et al. (1993)	54	T2–4	CDDP+5-FUx2	24 Gy/8Fx/ 4 days+20 Gy boost	Conc	27 months	74	59 (3 years)
TESTER et al. (1996)	91	T2–4A	MCV⇒CDDPx2	39.6 Gy	Neo+Conc		75	62 (4 years)
SAUER et al. (1998)	115	T1–4	CDDP or Carbo x2	45 Gy	Conc	7.5 years	70–85	57-69 (5 years)
BIRKENHAKE et al. (1999)	25	T3–4	CDDP+5-FUx2	59.4 Gy	Conc	38 months	88	80
RADOSEVIC-JELIC et al. (1999)	67	T3–4	Carbo weekly	65 Gy	Conc		92.5	55 (5 years)
HUSSAIN et al. (2004)	41	T3–4	5-FU+MMCx2	55 Gy in 20Fx	Conc	51 months	71	36 (5 years)
Rodel et al. (2002)	45	T1–4	5-FUx2	54–59.4 Gy	Conc	31 months	87	67
SAUER et al. (1998)	67	T1–4	CDDPx2	50.4 Gy	Conc		75	66 (3 years)
DANESI et al. (2004)	77	T2–4A	MCV⇒CDDP+5-FU PVI	69 Gy 3Fx/day	Neo+Conc	82 months	90.3	58.5 (5 years)

Danesi and Arcangeli recently reported long-term results on a phase I/II trial in which a series of invasive bladder carcinomas were treated with or without an initial two cycles of methotrexate, cisplatin, vinblastine (MCV) followed by concomitant continuous infusion of 5-FU and continuous infusion of cisplatin and irradiation (DANESI et al. 2004). Treatment consisted of three 100-cGy fractions per day of radiation, 5 days per week, for a total dose of 5000 cGy in 3.5 weeks. If required, due to residual disease, a consolidative dose of 2000 cGy was given in 1 week. A complete response rate of 90% (65 patients) was obtained in a series of 77 patients. High-grade toxicity was uncommon. With a median follow-up of 82.2 months, 44 of 65 patients who had an initial complete response were still alive, and 33 (57.1%) of these patients remain with a tumor-free bladder (61.5). The 5-year overall, bladder-intact, tumor-specific, disease-free, and cystectomy-free survival rates for all 77 patients were 58.5%, 46.6%, 75%, 53.5%, and 76.1% respectively.

Patients with T4 disease have poor survival no matter what treatment is employed. Radical surgery has few 5-year survivals, and preoperative irradiation delivered prior to radical cystectomy has not improved on this. Some investigators have examined multi-agent chemotherapy for these patients. However, long-term results are not yet available. Possibly, organ preservation chemoradiation protocols may be an option (Table 23.4)

23.5
Radiation Therapy Techniques (General Description)

A wide variety of planning, dosing, and actual therapy techniques are available in the treatment of bladder cancer. The ultimate goal for each of these is to optimally define the tumor volume, while at the same time minimizing dose to normal tissues. This requires accurate definition of the critical normal structures in relation to the tumor volume so that uninvolved organs can be maximally shielded during treatment to minimize morbidity.

Precise knowledge of tumor location on a daily basis is critical information for the radiation oncologist treating patients with bladder tumors. Cystogram, CT, and MRI have all been relied on for obtaining this information. PET/CT can also be very helpful in this regard. ROTHWELL et al. (1983) have demonstrated that CT localization is superior to cystogram localization and additionally points out that when cystogram is used alone, up to 85% geographic miss occurred. GRAHAM et al. (2003) compared MRI with CT planning and concluded that CT was sufficient if the whole bladder is to be treated. However, they advocate the use of MRI information in treating partial bladder volumes. It should again be reiterated that edema and hemorrhage seen on MRI obtained shortly after TURBT may easily be confused with tumor. For this reason, we advise on relying on MRI obtained prior to TURBT in addition to utilizing bladder-mapping information obtained at the time of the initial cystoscopy.

Treatment volume used during planning must be the same during daily treatment set-ups. There are advantages and disadvantages to treating with an empty as opposed to full bladder. The advantage is that patients are more comfortable with an empty bladder and the bladder location is more certain. The disadvantage is that less of the small bowel is pushed out of the field of treatment (potentially resulting in increased treatment morbidity), and it is also possible to spare more normal bladder mucosa when the bladder is full and the tumor volume maximally displaced from uninvolved bladder mucosa.

23.6
Simulation

One must always remember that patient comfort is a priority when setting up a treatment (i.e., simulation). One half-hour prior to simulation, the patient may be given an oral contrast to drink so that the small bowel can be adequately visualized during the simulation process. Patients may be treated supine or prone. When the regional lymph nodes are to be covered for the initial 4500 cGy of treatment, we recommend that the patient be treated prone on a belly board, with the bladder fully distended. If this pushes the small bowel out of the lateral treatment portals, small bowel toxicity may be minimized; otherwise, supine treatment may be more comfortable to the patient. During simulation, an alpha cradle is fashioned, or landmarks are identified, for each patient so that the individual can be optimally and accurately repositioned for the precise daily treatment setup. Once positioned, a Foley catheter is inserted into the bladder with a sterile technique,

and 7 cc of Hypaque is used to inflate the Foley catheter balloon.

The Foley catheter is pulled down to ensure that the balloon is at the base of the bladder. This critical step is required to ensure identification of the location of the bladder base. A solution of Hypaque mixed with saline in a one to two ratio is then instilled into the bladder. Generally, 25 cc of this mixture is instilled. Subsequent to this, approximately 25 cc of air is also injected into the bladder and the Foley catheter is clamped. Patients should be informed that a small quantity of air will be injected into the bladder so that they will not become alarmed when, after the procedure is over, they note that air is being passed from their bladder. The information obtained from this air contrast cystogram is combined with information previously obtained from examination, bladder mapping obtained at the time of initial cystoscopy, CT scan, and possibly cystogram, if previously performed, to optimally define the bladder location and the tumor and target volume for treatment. At some institutions, a rectal tube is placed at the distal end of the anal canal to identify this anatomy (a rectal balloon may also limit rectal movement but if it is used for this purpose it must be used on a daily basis during the course of treatment). The rectal tube is then connected to a Twomey syringe that has been previously filled with 25 cc of barium paste mixed with 25 cc of water. Please note that the rectal tube should be inserted into the rectum empty; No barium should be inserted into the rectum until later during the procedure when the lateral fields are simulated. The barium may obscure the outline of the bladder on anterior posterior simulation films. An anal canal marker should be placed at the distal end of the anal canal.

At this point, patient positioning should be re-verified by fluoroscopy as these manipulations may have induced misalignment.

23.7
Target Volume and Organs at Risk (Critical Structures)–Specifications (Including Tolerance Doses)

Morbidity should be analyzed by envisioning the tissue that will be traversed by the treatment beams. Consider the consequences of the beam as it passes through skin, small bowel, rectum, bone, and bladder.
The target volume in the treatment of bladder cancer includes the bladder and regional lymphat-

ics up to the level of the common iliac lymph nodes. It should be mentioned that there are institutions that treat the bladder only to 4500 cGy and then either continue on to 6500 cGy or cone down to the tumor only. The primary critical structures of major concern include the femoral heads (tolerance dose 4500 cGy), small bowel (tolerance dose of 4500 cGy), and rectum (tolerance dose of 6000 cGy) (EMAMI et al. 1991). These tolerances are for patients treated using standard fractionation schemes in which the bladder is treated with a 2-cm margin. Institutions utilizing hypofractionation schemes typically use smaller margins of 1 cm to compensate for the larger fraction sizes in terms of side effects (MURREN et al. 2004).

The radiation oncologist, when considering morbidity related to treatment, should think in terms of both acute and chronic morbidity. With optimal treatment planning, acute morbidity can be minimized to less that 10%, and long-term morbidity to the normal surrounding structures can be brought down to less than 5%.

23.8
Dose Prescription Beam Selection/Design Isodoses Plan Evaluation/Implementation

23.8.1
Simulation/CT Simulation Procedures

External beam therapy is most commonly delivered by means of a four-field box technique (Figs. 23.2–23.4); however, multiple conformal fields outlining the bladder can also be employed. In a four-field treatment plan, matched anterior/posterior and lateral portals are employed. The anteroposterior/posteroanterior (AP/PA) field encompasses the bladder as outlined by information obtained from both diagnostic studies and during simulation, and may be expanded to cover the regional lymph nodes if needed. When the regional lymph nodes are covered for the initial 4500 cGy of treatment, the patient should be treated prone on the belly board with the bladder full to push the small bowel out of the field, minimizing small-bowel toxicity. In general, these fields are defined superiorly by the S1/S2 interspace (midsacroiliac joint) to cover pelvic nodes up to the level of the common iliac lymph nodes.

If this volume should encompass a significant amount of small bowel despite the patient being prone and on a belly board, then the upper border

should be lowered accordingly to minimize the volume of small bowel in the treatment field. This generally requires the upper border to be placed at the lower sacroiliac joints. However, it must be kept in mind that to adequately cover the bladder, the upper border should extend approximately 2 cm above the dome of the bladder as visualized by the air contrast cystogram. The inferior border of the AP/PA field is placed at the lower border of the obturator foramen, which allows for good nodal and bladder coverage. The lower border of the field should be placed at the lower border of the obturator foramen only when there is no clinical suspicion or cystoscopic evidence of involvement at the base of the bladder or proximal urethra.

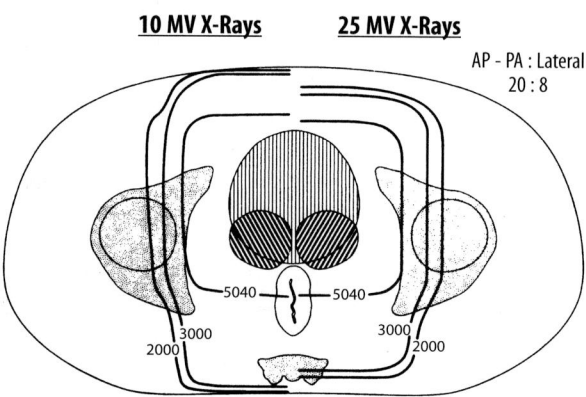

Fig. 23.2. Composite isodose curves for whole pelvic irradiation (isocentric four field) to 50.4 Gy. Compares 10-MV with 25-MV photons

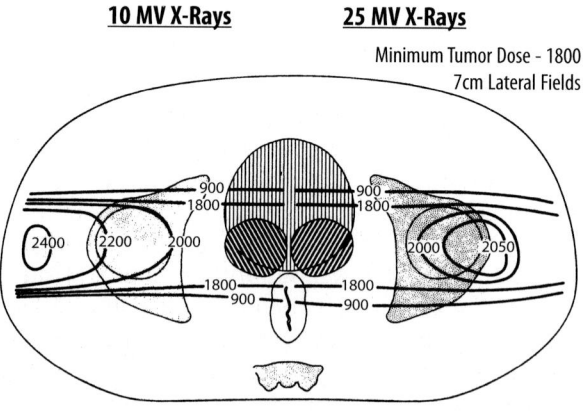

Fig. 23.3. Isodose distributions for boost portion of treatment delivered through opposed lateral fields. These do not reflect effects of beam width improving device. Compares 10-MV with 25-MV photons

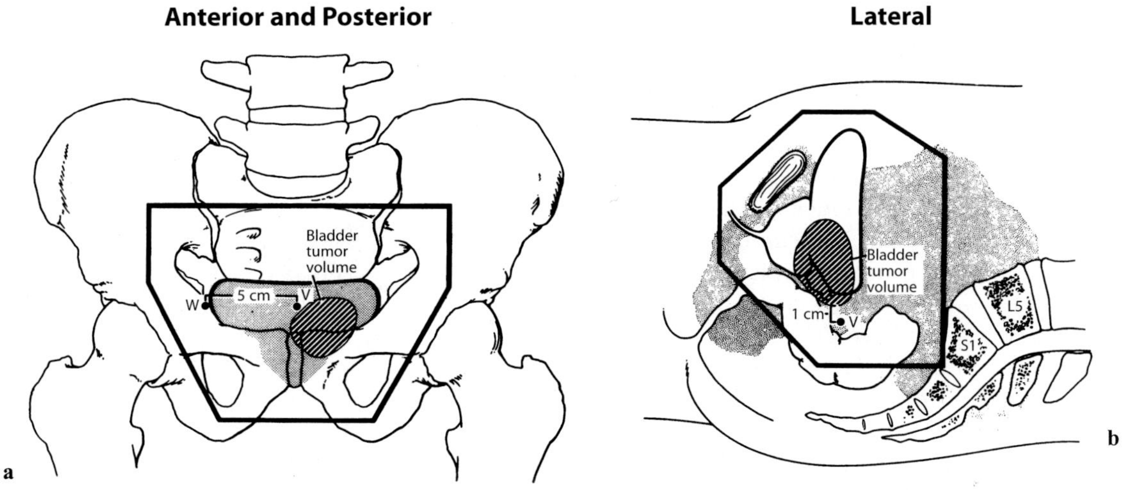

Anterior and Posterior   **Lateral**

Fig. 23.4a,b. Radiation fields for initial whole pelvic treatment. **a** Anterior/posterior; **b** laterals

In cases where the tumor is at the bladder neck or disease is noted near or involving the proximal urethra, the border should be extended inferiorly, generally to the level of the ischial tuberosity, to adequately cover disease in this region. If there is any suspicion of urethral involvement, the entire length of the proximal urethra should be covered. Frequently in this situation, the lower border will be at the bottom of the ischial tuberosities.

Laterally, the anterior field borders of the AP/PA fields are placed 1.5–2 cm lateral to the bony pelvis to allow coverage of the iliac lymph nodes. Custom blocking is employed to shield the femoral heads and prepubertal soft tissues. If the lymphatics are not to be included, these fields should be diminished to outline the bladder with a 2-cm margin. In this clinical situation, the patient should be treated with an empty bladder to minimize the treatment volume.

The lateral fields, superior and inferior borders are set at the same anatomical levels of the AP/PA fields, hence the term four-field box technique. The anterior border on the lateral field should be placed 2 cm above the bladder as outlined on the air contrast cystogram and also include the external iliac lymph nodes on the lateral fields. The only way to accurately locate the external iliac lymph nodes is to perform a bipedal lymphangiogram prior to simulation. This technique is not frequently employed, but we have found that the external iliac lymph nodes are adequately covered if the anterior border on the lateral field is defined by a line extending from the tip of the pubic symphysis to a point 2.5 cm anterior to the bony sacral promontory. After 4500 cGy

of radiation has been delivered, an attempt to shield a portion of the pubic symphysis on the lateral films should be made to prevent osteoradionecrosis or fracture from developing. The posterior border of the lateral field is also determined by the air contrast cystogram and places 2 cm beyond where the bladder is outlined. During lateral simulation, the barium paste mixed with saline should be injected into the rectum so that it can be accurately defined on the lateral simulation field and optimally blocked by custom blocking. Also note that the small bowel, which is opacified by the oral contrast given to the patient prior to simulation, should be shielded as outlined on the lateral field. The entire anal canal as well as the soft tissue anterior/inferior to the pubic symphysis should also be shielded. During design of these custom blockings to shield both the rectum and the small bowel, care should be taken so that the tumor is not inadvertently blocked. After the AP/PA and lateral radiographs have been obtained and reviewed to the satisfaction of the attending radiation oncologist, the field borders and field centers of the AP and lateral fields are marked on the patient.

If available, we recommend obtaining a computer treatment plan to optimize dose homogeneity. To accomplish this, generally a treatment position CT scan cut is obtained at the isocenters of the field. A physicist will use the CT scan to optimally select appropriate wedges and field weighting to minimize dose inhomogeneity to less than 5–10% around the target volume while also minimizing the dose to the normal surrounding critical organs. A typical four-field box technique isodose is shown in Fig. 23.2. In general, these large fields are treated to 4500–5040 cGy at 180 cGy per day. If the patient is being treated by chemoradiation and any small bowel is present in the treatment fields, we routinely reduce the fields by employing custom blocking to shield the small bowel after 4000 cGy had been delivered. This minimizes the chance of small-bowel toxicity even if it means potentially blocking the external iliac lymph nodes. Further, if chemotherapy is used during treatment, we generally limit our treatment to the bladder and lymphatics to 4500 cGy in 5 weeks. After this dose has been delivered, a boost field is constructed to encompass regions at risk such as residual disease or the premaximal TURBT tumor volume.

Information obtained from pre-TURBT CT scan, examination under anesthesia, and cystoscopy findings are used to define the boost volume for treatment planning optimally (Fig. 23.5). This volume is taken to be the tumor bed with margin. In some institutions, boost is delivered with bilateral 120° arc rotations as seen in Figure 23.6. Several composite examples of radiation treatment technique are provided in Figure 23.7 and Figure 23.8. The boost fields are generally treated to a total dose of 6500 cGy.

It is of utmost importance that the normal tissue tolerances of critical organs such as the rectum, small bowel, and femoral heads are respected. We regard the tolerance of the entire rectum to be 6000 cGy (1/2 of 6.5 weeks) and try to limit our dose to the rectum to no more than 5500 cGy, especially when chemotherapy is employed. We also keep the dose to the femoral heads and small bowel below 4500 cGy. As previously mentioned, we limit the

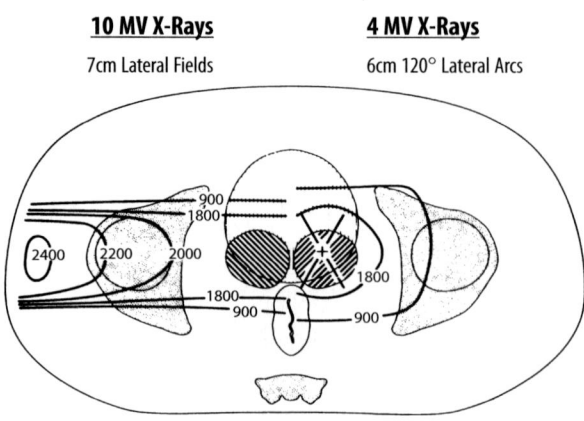

Fig. 23.6. Isodose distribution for boost portion of treatment compares 120° arc rotation using 4-MV photons with opposed lateral fields using 10-MV photons

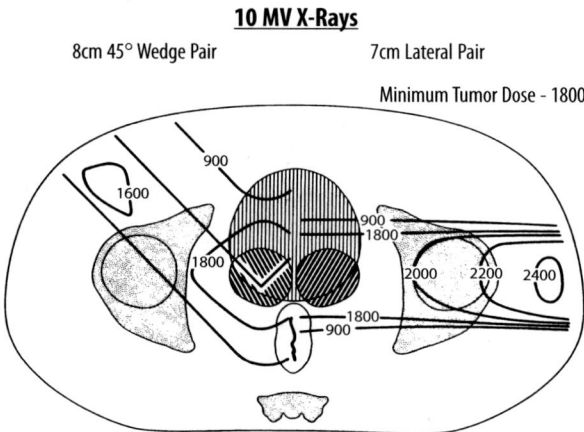

Fig. 23.7. Isodose distribution for boost portion of treatment using 10-MV photons delivered through opposed laterals or an anterior wedged pair

dose to the small bowel to 4000 cGy when chemotherapy is employed.

As part of the entire radiation treatment or as part of the boost field, conformal three-dimensional (3D) radiation therapy is an option if it is available. Figure 23.9 is an example of a 3D Beam's eye view treatment plan for fields conformally outlining the bladder. Normal tissues such as bowel, rectum, prostate and femoral bones are outlined, as is the bladder. With the resultant anatomical information and localization, a treatment plan can be created that offers high precision and dosing to the bladder while constructing appropriate and precise blocks to shield as much normal anatomy as possible. Since this anatomical information is readily available in all dimensions, not just AP/PA

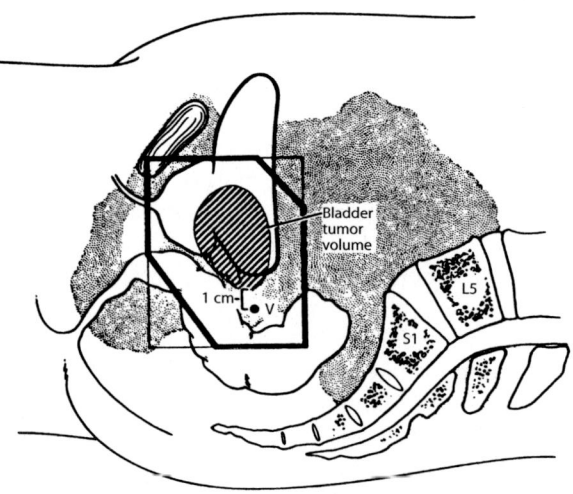

Fig. 23.5. Lateral boost field

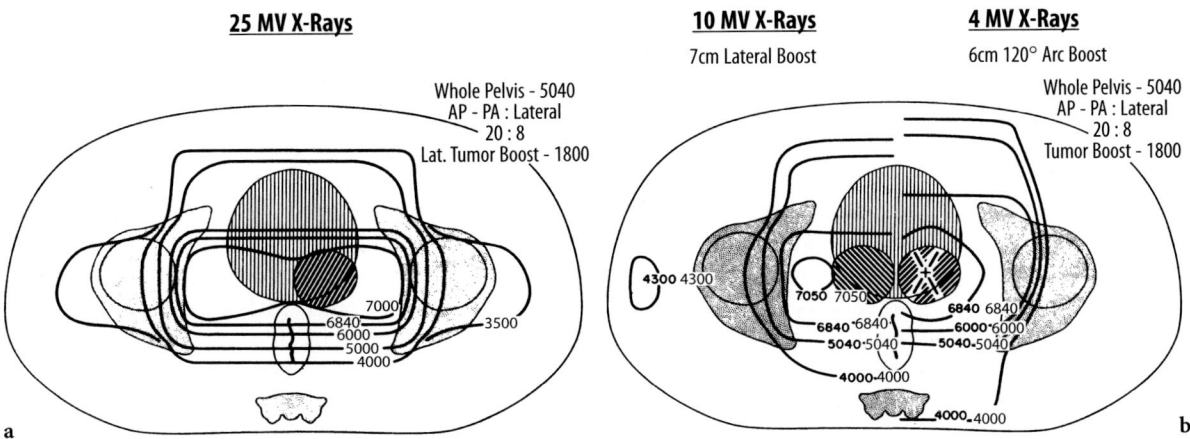

Fig. 23.8.a,b Composite isodose distributions for whole pelvic irradiation through four-field technique (50.4 Gy) plus various boost methods (18 Gy) in definitive plan for T2-T3 bladder carcinoma

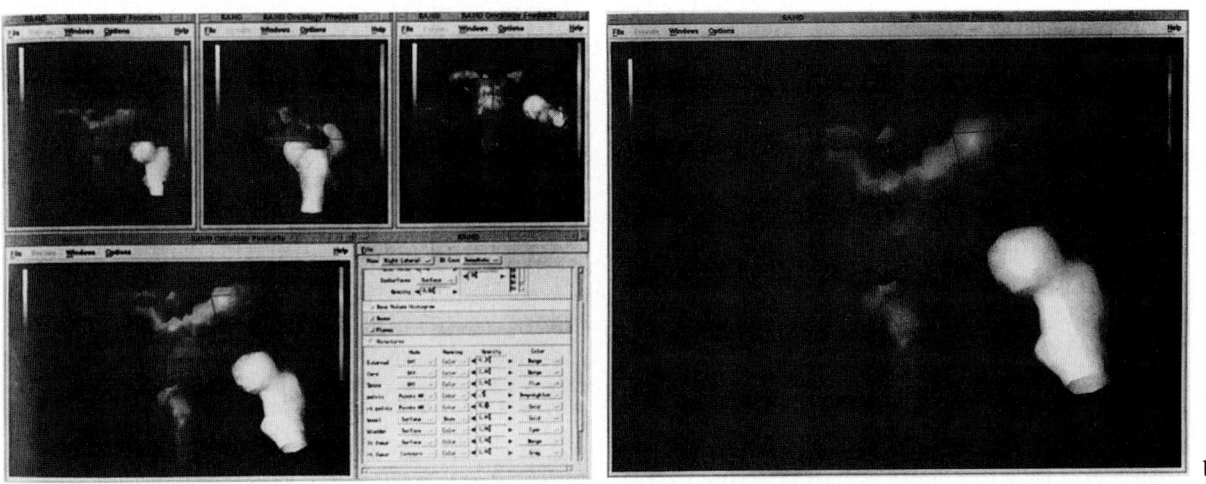

Fig. 23.9a,b. a Conformal 3D treatment plan contours bladder and surrounding anatomy in sagittal, coronal, and transverse views. **b** The single perspective view reveals superb isodose lines contouring the bladder and sparing normal anatomy. (Courtesy of RAHD oncology products, St. Louis, MO)

and lateral as is the usual case for a box field simulation, the flexibility to create an improved course of therapy using multiple noncoplanar portals is at hand.

In summary, multiple CT scan slices are obtained from the pelvis in the treatment position. The bladder is outlined on each slice and digitized into an appropriate 3D conformal treatment planning computer. At that point, custom blocks are created based on the reconstructed digitized bladder anatomy so that four, six and possibly eight or more fields can be employed to treat the bladder and spare a maximal amount of normal surrounding tissues. Many linear accelerators are directly linked to the 3D treatment planning device and create the blocking necessary using multi-leaf collimators.

23.9
Future Directions

The goal of the radiation oncologist in any treatment is to deliver an adequate dose of radiation to destroy tumor cells within a given target volume while avoiding injury to the surrounding normal tissue. A high enough dose of radiation can eradicate just about any cancerous mass. The cost, however, is injury to the surrounding normal tissue which limits the delivery of high doses of radiation. Intensity modulated radiation therapy (IMRT) can potentially allow for dose escalation in the treatment of bladder cancer without an increase in morbidity, while resulting in improvement in local control.

It potentially offers the possibility of further improvements in bladder preservation. IMRT is a computer-generated plan that employs the use of multiple beams of varying intensity coming from different angles to deliver maximal radiation to the tumor target while minimizing the irradiation of healthy tissue and organs at risk. In many cases, IMRT is an improvement over conformal CT planning. This is particularly true in cases in which the tumors are concave in physical nature and not well separated from the surrounding normal organs at risk (i.e., tumor wraps itself around an organ). In these cases IMRT delivers a greater dose to the tumor while limiting the radiation dose to the adjacent healthy tissue. This is accomplished by the delivery of hundreds of tiny radiation beams of varying intensity delivered through many angles rather than delivery of radiation by larger beams of uniform intensity through more limited angles.

IMRT utilizes more numerous beams with "inverse" planning. Inverse planning is a method that begins with the required dose distribution and works "backward" through a computer algorithm to produce the necessary beam profiles to accomplish the desired dose prescriptions and constraints.

For IMRT to be worthwhile, it is critical for the tumor/target to be defined as precisely as possible and for the dose of radiation to be delivered with greater accuracy to a more precisely defined target volume. Patient and organ motion can easily negate the benefits of IMRT. Therefore, rigid immobilization and real time target verification with on line portal imaging are highly desirable when treating organs, such as the bladder, which move considerably. In addition to the greater sensitivity of IMRT over other radiation techniques to patient and organ motion, other potential disadvantages include higher integral dose of normal tissues (higher aggregate volume of normal tissue exposure); prolonged treatment delivery time; possibility that close delineation of the radiation field to the tumor leaves parts of the tumor untreated; greater vulnerability to the physical uncertainties in defining the tumor volume; and greater risk of error due to the increased complexity of planning, delivery, quality assurance, and portal verification. Patients selected as candidates for IMRT should therefore clearly benefit more from this modality than from conventional techniques. The best way to be certain of this is to compare IMRT

plans with conventional plans on a given patient prior to treatment.

Special markers such as small ball bearings, rods, cross hatches, etc., known as "fiducial markers," are placed on the patient and/or the patient's immobilization device. This enables the planning computer to localize any point within the patient on a 3D grid. Target volumes and normal tissue are then outlined on CT images for treatment planning. For gross tumor, the clinician outlines this gross tumor volume (GTV). When warranted, microscopic targets can be added to the GTV (i.e., to include draining lymphatics at high risk for harboring micrometastatic disease). This combination of GTV with microscopic targets is referred to as the clinical target volume (CTV). If there is no gross tumor present (e.g., after

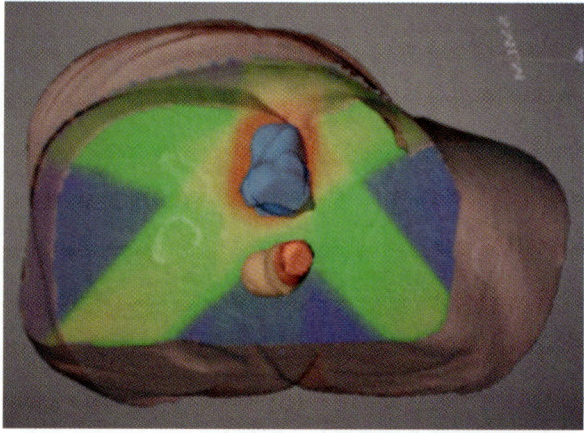

Fig. 23.10. Five-field 3D conformal treatment plan. Excellent bladder coverage with minimal rectal dosing. (Courtesy of East Carolina University Department of Radiation Oncology)

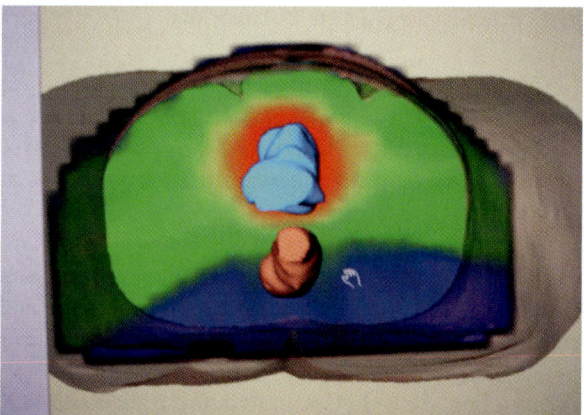

Fig. 23.11. Multi-segment intensity modulated (IMRT) treatment. Note dramatic dose fall off of dose (red to green) to surrounding structures and homogeneous dose distribution (red volume) to the bladder. (Courtesy of East Carolina University Department of Radiation Oncology)

complete TURBT), previously obtained tumor mapping obtained prior to TURBT should be consulted. In cases with no gross tumor present, only a CTV may be applicable. An advantage of IMRT is that multiple GTVs and subclinical target volumes can be outlined. This allows the clinician to treat each volume with a different daily dose. For the actual treatment, an additional margin may be added to the CTV to allow for variations in organ motion [the internal target volume (ITV)] and in patient set-up [the planning target volume (PTV)]. The next step is to specify the desired dose to the targets and dose limiting normal tissues and organs. There are several methods to accomplish this. One method is to specify a maximal dose limit to the tumor and normal critical structures. Each structure is then assigned a relative weighting that identifies to the planning system the relative importance of the structure. Typically, the tumor volume receives the highest weighting, but a critical structure may supersede this if the toxicity to that organ would be inadmissible, such as bowel perforation or myelopathy. At the State University of New York, Health Science at Brooklyn (SUNY-HSCB), a typical weighting would be rectum/colon 6000 cGy maximal dose with a weighting of 0.8–1.0, small bowel 4500 cGy maximal dose with a weighting of 1–1.2, and GTV 6500 cGy minimal dose with a weighting of 2.

Another method of defining dose criteria is to incorporate dose–volume constraints to each structure. The idea here is that structure injury in most organs is a function of dose received by a certain volume of the structure. These numbers are being defined empirically based on experience of observed side effects and tumor complete response rates in treated patients.

Planned treatment volumes should take into account organ motion in the ITV. This can be accomplished by ultrasound-guided target volume identification or portal image guided set-up with implanted fiducial insertions. Newer linacs may combine kilo and/or megavoltage CT scanners to allow for image-guided radiation therapy (IGRT). This will allow for treatment modification in real time with improved targeting accuracy as well as conformal avoidance of normal tissue.

Given the above considerations, particularly those of organ motion and the need for image-guided therapy during actual treatment, it will likely be several more years before IMRT is used as the standard in the treatment of bladder cancer. At the time of this writing, there is only one report in the literature on IMRT being used in the treatment of bladder cancer (BUDGELL et al. 2001). In this series, a standard field arrangement and set-up were employed to achieve a somewhat homogeneous dose within the same treatment volume, with some additional sparing of normal tissue compared with that expected with 3-D conformal CT planning.

We also expect advances to be made with the successful incorporation of newer chemotherapeutic radiosensitizers into bladder preservation paradigms. As pointed out by RODEL et al. (2005) gemcitabine and paclitaxel are both potent radiosensitizers that have shown significant activity against urothelial tumors. The Radiation Therapy Oncology Group (RTOG 99-06) is conducting a trial using twice-daily radiation therapy with concomitant paclitaxel and cisplatin followed by either selective bladder preservation or radical cystectomy and adjuvant chemotherapy with gemcitabine and cisplatin. Mention is also made that the "future aspects of radiosensitization" may "relate to the potential inhibition of oncogene products frequently activated and overexpressed in bladder cancer, such as H-ras and c-erbB-1." "Inhibition of EGF receptor activity with small molecule tyrosine kinase inhibitors or antibodies against receptors may also increase tumor radiosensitization in bladder cancer" (RODEL et al. 2005).

Finally, proper use of molecular markers are likely to be of more help in the future as bladder cancer is one malignancy with "extensive information regarding molecular pathogenesis and genetic predictors of natural history as well as response to various modalities of treatment based on molecular profiles" (from RODEL et al. 2005; HUSSAIN and JAMES 2005). In these reviews the authors stress the need to investigate the use of these molecular prognostic markers, such as p53, bcl-2, bax, Rb, and p21, to predict sensitivity to chemotherapy, radiotherapy, and overall survival in randomized trials, pointing out that it may ultimately become possible to perform chemo/radiosensitivity testing similar to that used for antibiotics (RODEL et al. 2005).

References

Awwad H, Abd El, Baki HA El et al (1979) Preoperative irradiation of T3 carcinoma in bilharzial bladder. A comparison between hyperfractionation and conventional fractionation. Int J Radiat Oncol Biol Phys 5:787

Barentsz JO, Jager GJ, van Vierzen PB et al (1996) Staging urinary bladder cancer after transurethral biopsy: value of fast dynamic contrast-enhanced MR imaging. Radiology 201:185-193 (abstract)

Barentsz JO, Jager GJ, Witjes JA (2000) MR imaging of the

urinary bladder. Magn Reson Imaging Clinic North Am 8:853-867

Batata MA, Chu FCH, Hilaris BS et al (1981) Pre-operative whole pelvis versus true pelvis irradiation and/or cystectomy for bladder cancer. Int. J Radiat Oncol Biol Phys 7:1349-1355

Battermann JJ,Tierie AH (1986) Results of implantation for T1 and 2 bladder tumors. Radiother Oncol 5:85-90

Birkenhake S, Leykamm S, Martus P, Sauer R (1999) Concomitant radiochemotherapy with 5-FU and cisplatin for invasive bladder cancer. Acute toxicity and first results. Strahlenther Onkol 175:97-101

Blandy JP, England HR, Evans SJW et al (1980) T3 bladder cancer, the case for salvage cystectomy. Br J Urol 52:502-506

Bloom JJG, Hendry WF, Wallace DM et al (1982) Treatment of T3 bladder cancer: controlled trial of pre-operative radiotherapy and radical cystectomy versus radical radiotherapy. Second report and review (for the Clinical Trials Group, Institute of Urology). Br J Urol 11:136-151

Brannan W, Ochsner MG, Fuselier HA Jr et al (1978) Partial cysectomy in the treatment of transitional cell carcinoma of the bladder. J Urol 119:213-215

Budgell GJ, Mott JHG, Logue JP, Hounsell AR (2001) Clinical implementation of dynamic multileaf collimation for compensated bladder treatments. Radiother Oncol 59:31-38

Cancer Statistics (2004) 54:9

Cervak J, Cufer T, Kragelj B (1993) Sequential transurethral surgery, multiple drugs, and radiation therapy for invasive bladder carcinoma. Int J Radiat Oncol Biol Phys 25777-782

Cole D Durrant K, Robert J et al (1992) A pilot study of accelerated fractionation in the radiotherapy of invasive carcinoma of the bladder. Br J Radiol 65:792-798

Cookson MS (2005) The surgical management of muscle invasive bladder cancer: a contemporary review. Semin Radiat Oncol 15:10-18

Coppin CM, Gospodarowicz MK, James K, Tannock IF, Zee B, Carson J, Pater J, Sullivan LD (1996) Improved local control of invasive bladder cancer by concurrent cisplatin and preoperative or definitive radiation. The National Cancer Institute of Canada Clinical Trials Group. J Clin Oncol Nov 14(11):2901-2907

Danesi DT, Arcangeli G, Cruciani E et al (2004) Conservative treatment of invasive bladder carcinoma by transurethral resection, protracted intravenous infusion chemotherapy, and hyperfractionated raiotherapy: long term results. Cancer 101:2540-2548

Devita J. Hellman S, Rosenberg S (1989) Cancer principles and practice of oncology, 3rd edn. Lippincott, Philadelphia, pp 361-362

DeWeerd JH, Colby MY Jr (1973) Bladder carcinoma treated by irradiation and surgery: Interval report. J Urol 109:409-413

Drago JR, Rohner TJ Jr (1983) Bladder cancer: results of radical cystectomy for invasive and recurrent superficial tumors. J Urol 130:460-462

Duncan W, Arnott SJ, Jack WJL et al (1985) A report of a randomized trial of d (13) +Be neutrons compared with megavoltage x-ray therapy of bladder cancer. Int J Radiat Oncol Biol Phys 11(12):1985-2049

Duncan W, Quilty PM (1986) The results of a series of 963 patients with transitional cell carcinoma of the bladder primarily treated by radical megavoltage x-ray therapy. Radiother Oncol 7:299-310

Dunst J, Saur R, Schrott KM et al (1994) An organ-sparing treatment of advanced bladder cancer: a 10 year experience. Int J Radiat Oncol Biol Phys 30:261-266

Eapen L, Stewart D, Danjoux C et al 1989;Intraarterial cisplatin and concurrent radiation for locally advanced bladder cancer. J Clin Oncol 7:230-235

Edsmyr F, Anderson L, Eposti PL et al (1985) Irradiation therapy with multiple small fractions per day in urinary bladder cancer. Radiother Oncol 4:197-203

Emami B, Lyman J, Brown A et al (1991) Tolerance of normal tissue to therapeutic irradiation. Int J Radiat Oncol Biol Phys 21:109-122

Ghoneim MA, Ashaella AC, Awaad HK et al (1985) Randomized trial of cystectomy with or without pre-operative radiotherapy for carcinoma of the bilharzial bladder. J Urol 134:266

Goffinet DR, Schneider MJ, Glatstein EJ et al (1975) Bladder cancer:results of radiation therapy in 384 patients. Radiology 117:149-153

Graham J Gee A, Hilton S et al (2003) Geometric uncertainties in radiotherapy of the prostate and bladder. In: Geometric uncertainties in radiotherapy. British Institute of Radiology, London

Greven KM, Solin Hanks GE (1990) Prognostic factors in patients with bladder carcinoma treated with definitive irradiation. Cancer 65:908-912

Greven KM, Spera JA, Solin LJ et al (1992) Local recurrence after cystectomy alone for bladder carcinoma. Cancer 69:2767-2770

Herr HW, Schwalb DM, Zhaang ZF et al (1995) Intravesical bacillus Calmette-Guerin therapy prevents tumor progression and death form superficial bladder cancer. Ten aear follow-up of a prospective randomized trial. J Clin Oncol 13:1404-1408

Herr HW (1987) Conservative treatment of muscle-infiltrating bladder cancer: prospective experience. J Urol 138:1162

Horwich A, Pendlebury S, Dearnaley DP (1995) Organ conservation in bladder cancer. Eur J Cancer 31 [Suppl 5]:208

Housset M, Maulard C, Chretien Y et al (1993) Combined radiation and chemotherapy for invasive transitional-cell carcinoma of the bladder: a prospective study. J Clin Oncol 11:2150-2157

Hussain SA, James NS (2005) Molecular markers in bladder cancer. Semin Radiat Oncol 15:3-9

Hussain SA, Stocken DD, Peake DR, Glaholm JG, Zarkar A, Wallace DM, James ND (2004) Long-term results of a phase II study of synchronous chemoradiotherapy in advanced muscle invasive bladder cancer. Br J Cancer 90:2106-2111

Jakse G, Fromhold H, Zurmedden D (1985) Combined radiation and chemotherapy for locally advanced transitional cell carcinoma of the urinary bladder. Cancer 55:1659-1664

James N, Hussain SA (2005) Management of muscle invasive bladder cancer-British approaches to organ conservation. Semin Radiat Oncol 15:19-27

Johnson DE, Hogan JM, Ayala AG (1972) Primary adenocarcinoma of the urinary bladder. South Med J 65:527-530

Marshall VF, McCarron JP Jr (1977) The curability of vesical cancer: greater no or then? Cancer Res 37:2753-2755

Marshall VF (1952) The relation of the preoperative estimate to the pathologic demonstration of the extent of vesicle neoplasms. J Urol 68:714

Matsumoto K, Kakizoe T, Mikuriya S et al (1981) Clinical evaluation of intraoperative radiotherapy for carcinoma of the urinary bladder. Cancer 47:509-513

Montie JE, Straffon RA, Stewart BH (1984) Radical cystectomy without radiation therapy for carcinoma of the bladder. J Urol 131:477-482

Morabito RA, Kandzari SJ, Milam DF (1979) Invasive bladder carcinoma treated by radical cystecomy: survival of patients. Urology 14:478-481

Morison AS, Cole P (1976) Epidemiology of bladder cancer. Urol Clin North Am 3:13-29

Mostofi FK, Davis CJ Jr, Sesterhenn IA (1988) Pathology of tumors of the urinary tract. In: Skinner DG, Lieskovskky G (eds) Diagnosis and management of genitourinary cancer. Saunders, Philadelphia, pp 83-117

Murren LP, Redpath AT, Mclaren DB et al (2004) Treatment margins and treatment fractionation in conformal radiotherapy of muscle-invading urinary bladder cancer. Radiother Oncol 71:65-71

Natale RB, Grossman HB, Blumenstein BB et al (2001) SWOG 8710 (INT-00800): Randomized Phase III trial of neoadjuvant MVAC plus cystectomy versus cystectomy alone in patients with locally advanced bladder cancer. Proc Am Soc Clin Oncol 20:2a (abstract 3)

Nixon RG, Chang SS, Lafleur BJ et al (2002) Carcinoma in situ and tumor multifocality predict the risk of prostatic urethral involvement at radical cystectomy in men with transitional cell carcinoma of the bladder. J Urol 167:502-505

Pearse HD (1994) The urinary bladder. In: Cox JD (ed) Moss' radiation oncology: rationale, technique results. Mosby, St Louis, p 522

Pollack A, Zagars GK, Swanson DA (1994) Muscle-invasive bladder cancer treated with external beam radiotherapy: prognostic factors. Int J Radiat Oncol Biol Phys 30:267-277

Quilty PM, Duncan W (1986) Primary radical radiotherapy for T3 transitional cell cancer of the bladder: an analysis of survival and control. Int J Radiat Oncol Biol Phys 12:853-860

Radosevic-Jelic L, Pekmezovic T, Pavlovic-Cvetkovic L, Radulovic S, Petronic V (1999) Concomitant radiotherapy and carboplatin in locally advanced bladder cancer. Eur Urol 36:401-405

Reibischung JL, Vannetjel ZM, Fournier F (1992) Cyclic concomitant chemoradiotherapy for invasive bladder cancer. Phase I study with organ preservation. Proc Am Soc Clin Oncol 11:208

Resnick MI, O'Conor VJ Jr (1972) Segmental resection for carcinoma of the bladder: review of 102 patients. J Urol 109:1007-1010

Richards B, Bastabla JR, Freedman L et al (1983) Adjuvant chemotherapy in T3NxMo bladder cancer treated with radiotherapy. Br J Urol 55:386-391

Richie JP, Skinner DG, Kaufman JJ (1975) Radical cystectomy for carcinoma of the bladder: 16 years of experience. J Urol 113:186

Rodel C, Grabenbauer GG, Kuhn R, Zorcher T, Papadopoulos T, Dunst J, Schrott KM, Sauer R (2002) Organ preservation in patients with invasive bladder cancer: initial results of an intensified protocol of transurethral surgery and radiation therapy plus concurrent cisplatin and 5-fluorouracil. Int J Radiat Oncol Biol Phys 52:1303-1309

Rodel C, Weiss C, Sauer R (2005) Organ preservation by combined modality treatment in bladder cancer: the European perspective. Semin Radiat Oncol 15:28-35

Rothwell RI, Ash DV, Jones WG (1983) Radiation treatment planning for bladder cancer: a comparison of cystogram localization with computed tomography. Clin Radiol 34:103-111

Rotman M, Macchia R, Silverstein R (1987) Treatment of advanced bladder cancer with irradiation and 5 fluorouracil infusion. Cancer 59:710-714

Rotman M, Aziz H, Porrazo M et al (1990) Treatment of advanced transitional cell carcinoma of the bladder with irradiation and concomitant 5-FU infusion. Int J Radiat Oncol Phys 18:1131-1137

Russell KJ, Boileau MA, Higano C, Collins C, Russell AH, Koh W, Cole SB, Chapman WH, Griffin TW (1990) Combined 5-fluorouracil and irradiation for transitional cell carcinoma of the urinary bladder. Int J Radiat Oncol Biol Phys 19:693-699

Sauer R, Birkenhake S, Kuhn R, Wittekind C, Schrott KM, Martus P (1998) Efficacy of radiochemotherapy with platin derivatives compared to radiotherapy alone in organ-sparing treatment of bladder cancer. Int J Radiat Oncol Biol Phys 40:121-127

Saver R, Dunst J, Alternddor F et al (1990) Radiotherapy with and without Cisplatinum in bladder cancer. Int J Radiat Oncol Biol Phys 19:687-691

Shipley WU, Rose MA, Perrone TL et al (1985) Full dose irradiation for patients with invasive bladder carcinoma. Clinical and histologic factors prognostic of improved survival. J Urol 134:679

Silverman OT, Hartage P, Morrison S et al (1992) Epidemiology of bladder cancer. Hematol Oncol Clin North Am 10:1-30

Skinner DG, Tift JP, Kaufman JJJ (1982) High dose, short course preoperative radiation therapy and immediate single stage radical cystectomy with pelvic node dissection in the management of bladder cancer. J Urol 127:671

Smaaland R, Aksle A, Tonder B, Mehus A, Lote K, Albreksten G (1991) Radical radiation treatment of invasive and locally advanced bladder carcinoma in elderly patients. Br J Urol 67:61-69

Solsona E, Iborra I, Ricos JV et al (1992) Feasibility of transurethral resection for muscle-infiltrating carcinoma of the bladder: prospective study. J Urol 147:1513-1515

Tester W, Porter A, Asbell S (1993) Combined modality program with possible organ preservation for invasive bladder cancer: results of RTOG protocol 85-12. Int J Radiat Oncol Biol Phys 25:783-790

Tester W, Caplan R, Heaney J, Venner P, Whittington R, Byhardt R, True L, Shipley W (1996) Neoadjuvant combined modality program with selective organ preservation for invasive bladder cancer: results of Radiation Therapy Oncology Group phase II trial 8802. J Clin Oncol 14:119-126

Timmer PR, Hartlief HA, Hooijkass JAP (1985) Bladder cancer: pattern of recurrence in 142 patients. Int J Radiat Oncol Biol Phys 11:899-905

Vale JA, A'hern Liu K et al (1993) Predicting the outcome of radical radiotherapy for invasive bladder cancer. Eur Urol 24:24:48-51

Van der Werf-Messing B, Hop WCJ (1981) Carcinoma of the urinary bladder (category T1NxMO) treated either by radium implant or by transurethral resection only. Int J Radiat Oncol Biol Phys 7:199-303

Van der Werf-Messing B, Menon RS, Hop WCJ (1983) Cancer of the urinary bladder category T2, T3 (NxMo) treated by interstitial radium implant: second report. Int J Radiat Oncol Biol Phys 9:481-485

Whitmore WF Jr, Batata MA, Ghoneim MA et al (1977) Radical cystectomy with or without prior radiation in the treatment of bladder cancer. Gen Urol 188:184-187

Woehere H, Ous S, Klevmark B et al (1993) A bladder cancer multi-institutional experience with total cystectomy for muscle-invasive bladder cancer. Cancer 72:3044-3051

Wood DP Jr, Montie JE, Pontes JE et al (1989) Transitional cell carcinoma of the prostate in cystoprostatectomy specimens removed for bladder cancer. J Urol 141:346-349

24 Radiation Therapy for Cervical Cancer

KATHRYN E. DUSENBERY and BRUCE J. GERBI

CONTENTS

24.1   Natural History of Cervical Cancer and Patterns of Spread 579
24.2   Workup and Staging 580
24.3   General Management 581
24.4   Radiotherapy Techniques 581
24.4.1 General Description 581
24.5   Simulation/CT Simulation Procedures 584
24.5.1 Target Volume and Organs at Risk 584
24.5.2 Dose Prescription 585
24.6   LDR Brachytherapy 586
24.6.1 Equipment 586
24.6.2 LDR Procedure 587
24.6.3 Milligram Hour 587
24.6.4 The Manchester System 588
24.6.5 The International Commission on Radiographic Units and Measurements 590
24.6.6 University of Minnesota 592
24.6.7 Postoperative Radiotherapy 593
24.6.8 Dose to Bladder and Rectum 593
24.7   Sequelae 594
24.8   Results 595
24.9   Quality Management Program 596
24.10 Conclusion 596
      References 596

K. E. DUSENBERY, MD
University of Minnesota Medical School, Department of Radiation Oncology, MMC436, 420 Delaware St. S.E., Minneapolis, MN 55455, USA
B. J. GERBI, PhD
Associate Professor, Therapeutic Radiology – Radiation Oncology, University of Minnesota, Mayo Mail Code 494, 420 Delaware St SE, Minneapolis, MN 55455, USA

24.1 Natural History of Cervical Cancer and Patterns of Spread

Cervical cancer is the third most common gynecological malignancy after ovarian and uterine corpus cancers. In the United States in 2002 there were 13,000 new cases and about 4,000 deaths (COMMITTEE ON PRACTICE 2002). Worldwide, it is an enormous health problem, with 500,000 women dying each year. In the United States, largely because of the widespread use of the Papanicolaou (Pap) smear, often patients are diagnosed with early and pre-invasive lesions and are reliably cured with surgery alone (HERZOG 2003; WAGGONER 2003). The peak age at diagnosis is around 47 years; however, 50% of patients are under age 35 years when diagnosed. Only 10% of patients are older than 60 years (HERZOG 2003; WAGGONER 2003). Cervical cancer is associated with known risk factors including a higher number of sexual partners (>4), first intercourse at an early age (<16 years), history of genital warts and lower social economic class with poor access to health care (MENDENHALL et al. 1984; CHEN et al. 1999; KAROLEWSKI et al. 1999; GASINSKA et al. 2002; DUNST et al. 2003).

More than 90% of cervical squamous cell cancers contain DNA evidence of human papilloma (HPV) virus, acquired through sexual activity. Certain types of HPV are particularly oncogenic these; include types 16, 18, 31, 33 and 35, 68, 52, 59, 45, 53, 66, 73, mm4, mm8 and mm7 and are often associated with moderate to severe dysplasia or carcinoma in situ (AU et al. 2003; HERNANDEZ-HERNANDEZ et al. 2003; DUNLEAVEY 2004; FEY and BEAL 2004). High-risk subtypes of HPV produce viral proteins E6 and E7 thought to be important for malignant transformation by binding and inactivating products of the retinoblastoma gene which ultimately allows unchecked cell-cycle progression in cells infected with HPV16 or 18 (HARIMA et al. 2002). Tobacco smoking as well as possibly second hand exposure to smoke are independent risk factors for cervical dysplasia and cancer. Immunosuppression from any cause [organ transplantation, acquired immunode-ficiency syndrome (AIDS), chronic steroid usage] increases the risk of cervical cancer (WAGGONER 2003).

24.2
Workup and Staging

The Pap smear is recommended for all sexually active women or women after the age of 18 years. Abnormal Pap smears are generally reported as either normal, ascus (atypical squamous cells of uncertain significance), ascus-h (atypical cells of uncertain significance: cannot rule out high grade), low-grade squamous intraepithelial lesion (lgsil), high-grade squamous intraepithelial lesion (hgsil) or carcinoma (LIEU 1996; MASHBURN and SCHARBO-DeHAAN 1997; PERLMAN 1999). Any Pap smear that is not normal requires further evalua-tion, which can range from simply repeating it, to testing for the presence of high-risk HPV subtypes or to colposcopy and directed biopsies, depending on the clinical situation

If a visible lesion is identified by means of pelvic exam, a Pap smear is not recommended and these patients should have a biopsy. The most common invasive cancer by far will be a squamous cell cancer, diagnosed in at least 80% of patients. Adenocarcino-mas make up about 20% of cervical cancers, and less than 1% of patients have rare types such as a true small cell (neuroendocrine), clear cell, melanoma or lymphoma.

Once a biopsy reveals the diagnosis of inva-sive cervical cancer, the staging workup follows (Table 24.1). In general, cervical cancers are clini-cally staged. The FIGO staging system (Table 24.2) is the most widely used, and the cornerstone of the system is a thorough careful pelvic examination, often done under general anesthesia. Adjuncts to the pelvic examination include either an intravenous pyelogram (IVP) or computed tomography (CT) scan with intravenous (IV) contrast to determine whether there is ureteral obstruction and hydrone-phrosis. Additionally, a chest X-ray is usually part of the initial workup.

Recently, the American College of Radiology Imaging Network completed a multicentre trial to assess the diagnostic performance of magnetic reso-nance imaging (MRI) and CT compared with that of clinical staging. Thus far, these scans have not proved superior to clinical evaluations. If, however,

pelvic or para-aortic lymph nodes appear enlarged, they should be investigated by means of fine-needle aspiration or retroperitoneal node dissection. If pos-itive nodes are found, treatment can be individually designed to encompass known sites of disease.

In the future, positron emission tomography (PET) scans with fluorodeoxyglucose may have a role in the staging and follow-up of cervical cancer (GREVEN et al. 1999; MUTIC et al. 2003; TSAI et al. 2004). Recently, SINGH et al. (2003) demonstrated PET imaging in the periaortic nodal chains in patients with FIGO IIIB disease to correlate with survival. Additionally, persistence of PET positiv-ity after completion of therapy may portend a worse prognosis and may be useful in designing follow-up treatment for high-risk patients (GRIGSBY et al. 2004). However, at the current time, PET must be considered investigational.

There are many known prognostic factors in cer-vical cancer. The FIGO stage and volume of tumor are the most obvious. Other factors that affect the prog-nosis include the depth of stromal invasion which is associated with a higher incidence of metastasis to pelvic lymph, higher grade of tumor and whether there is unilateral or bilateral hydronephrosis. At the University of Minnesota we have routinely used extraperitoneal lymph node sampling to determine the extent and spread of the cervical cancer. In our experience, the presence of grossly positive pelvic nodes that are unresectable is a very poor prognostic factor; whereas, patients who have only microscopic spread to their pelvic nodes do as well as patients who have grossly positive nodes that can be resected

Table 24.1. Staging workup for cervical cancer

Physical examination
Lymph node assessment (supraclavicular, inguinal)
Pelvic examination
Radiographic studies
    Chest X-ray
    Intravenous pyelogram or CT scan of abdomen and pelvis
    Barium enema
Laboratory evaluation
    Creatinine
    Hemoglobin
Procedures
    Cervical biopsy
    Cystoscopy
    Proctoscopy
Other studies [not allowed for clinical (FIGO) staging]
    CAT scan or MRI
    PET scan
    Ultrasound
    Bone scan
    Extraperitoneal lymph node sampling

Table 24.2. FIGO staging

Stage 0 carcinoma in situ

Stage I confined to the cervix
IA microscopic tumor
 IA1 no more than ≤3 mm of invasion, no wider than 7 mm
 IA2 more than >3 mm but ≤5 mm of invasion, or wider than 7 mm
IB visible tumor
 IB1 tumor ≤4 cm
 IB2 tumor greater than 4 cm

Stage II tumor beyond cervix, but not to pelvic side wall
IIA vaginal involvement but not to lower two-thirds of vagina
IIB parametrial involvement, but not to pelvic side wall

Stage III
IIIA vaginal involvement to lower third of vagina
IIIB extension to the pelvic side wall or hydronephrosis

Stage IV
IVA spread to bladder or rectum
IVB spread to distant organs

(COSIN et al. 1998). Whether this reflects tumor biology or is actually therapeutic is unknown. Other prognostic factors include anemia at diagnosis, number of positive nodes and patient age.

24.3
General Management

Once the diagnosis and staging workup are complete, a management plan can be outlined. For asymptomatic patients with normal appearing cervix, the diagnosis is usually made after conization. If there is less than 3 mm of invasion below the basement membrane (stage IA1), the risk of pelvic nodal spread is less than 1%. The treatment options include a simple hysterectomy or, if preservation of fertility is desired, cervical conization and careful follow-up.

If the focus of invasion extends 3 mm or more or if there is lymph vascular space involvement (stage IA2), the risk of nodal spread increases to 2–8% and most oncologists recommend a (modified) radical hysterectomy with pelvic lymphadenectomy or definitive radiotherapy. The advantage to a surgical approach is the possible preservation of ovarian function, the fact that the entire uterus is removed for analysis and the lack of long-term radiation side effects. Alternatively, patients with FIGO stage-I disease can be managed with definitive radiotherapy. Advantages to this approach include the lack of a need for prolonged general anesthesia, especially if

high-dose-rate (HDR) brachytherapy is to be used. The only randomized trial to address this issue was reported by LANDONI et al. (1997). In their study, patients with FIGO stage IB or II were randomized to either radical hysterectomy followed by tailored radiotherapy or definitive radiotherapy. In that trial of over 300 patients, the disease-free survival and overall survival were equivalent. However, in the patients in the surgery arm, there was an excessive complication rate of 20%. Critics of this study have pointed to the high rate of needing postoperative radiotherapy in the surgical arm and the lack of concomitant chemotherapy in the radiation therapy arm as weaknesses of this study. It is however, the only recent randomized study to look at this important question.

For patients beyond stage I, the management is generally definitive radiotherapy with both external beam and brachytherapy. Low-dose-rate (LDR) brachytherapy has the longest experience record but, recently, HDR brachytherapy has become widely available. There are advantages and disadvantages to each (Table 24.3). Although there are no good randomized trials comparing these two modalities, it appears that in trained hands the acute and long-term side effects are roughly equivalent (Tables 24.4, 24.5).

The addition of cis-platinum-based chemotherapy during the course of definitive radiotherapy for patients with IB2-IVA disease has been shown to improve outcomes (LUKKA et al. 2002; NAG et al. 2002). Five randomized trails have recently demonstrated significant improvement in local control and survival when concurrent chemotherapy was added to radiation therapy in patients with early-stage disease and high risk for recurrence, as well as in patients with advanced-stage disease (KEYS et al. 1999; ROSE et al. 1999; WHITNEY et al. 1999). Four of these studies evaluated concurrent chemotherapy combined with definitive radiotherapy (Table 24.6).

24.4
Radiotherapy Techniques

24.4.1
General Description

Design of the radiation treatment program depends on the extent and volume of the tumor. Most patients receive a combination of external beam treatments and brachytherapy, although very early lesions may be treated with brachytherapy alone.

Table 24.3. Advantages of low- and high-dose-rate brachytherapy

Low-Dose-Rate Advantages
Long history of use
Ability to predict rate of late complications
Improves chances of catching tumors in sensitive phase of cell cycle
Longer treatment times allow for leisurely review of and potential modifications to
the treatment plan prior to the delivery of a significant portion of treatment
Favorable dose-rate effect on repair of normal tissues
Infrequent replacement and calibration of sources because of long isotope half-life

High-Dose-Rate Advantages
No short- or long-term confinement to bed
No indwelling bladder catheters and risk of bladder infection
Not labeled „radiation risk zone" to relatives, visitors and staff
Avoidance of several anesthesias (possibly)
Maintain position of the sources during the brief treatment
Ability to retract sensitive structures (rectum) during short treatment time
Patient preparation simpler
Ability to treat greater patient loads (high output of patients on each machine)
Ease of purchasing HDR iridium sources compared with cesium sources
Short treatment times and minimal radiation protection problems
Possibility of optimizing dose distribution by altering the well times of the source
at different locations

Pulsed Dose Rate Advantages
Complication rate profile more similar to that of LDR
Between fractions, patient is not radioactive, allowing for near continuous nursing care
More predictable time for removal of the applicator than remote after loading
LDR where the sources are retracted whenever someone enters the patient's room

Table 24.4. Survival data determined in the meta-analysis of the carcinoma of the
cervix brachytherapy results from 56 institutions. Modified from Orton CG. High-
and low-dose-rate brachytherapy for cervical carcinoma. *Acta Oncol* 1998;37:117-125
(ORTON 1998)

Stage	HDR patients	LDR patients	5 year survival		P value
			HDR %	LDR %	
I	1327	630	82.7	82.4	>0.05
II	2891	1271	66.6	66.8	>0.05
III	2721	1464	47.2	42.6	0.005
IV	221	56	20.4	14.3	>0.05
Overall	7468	4738	60.8	59.0	0.045

Table 24.5. Late complication rates in the meta-analysis of the carcinoma of the cervix
brachytherapy from 56 institutions. Modified from Orton CG. High and low dose-rate
brachytherapy for cervical carcinoma. *Acta Oncol* 1998;37:117-125. (ORTON 1998)

Severity	Number of patients		Complication rate		P value
	HDR	LDR	HDR	LDR	
Grade IV	10331	5274	2.2%	5.3%	<0.001
Grades III and IV	10887	4709	9.1%	20.7%	<0.001

The number of external beam treatments relative
to brachytherapy can be determined by general
guidelines, although experience provides optimal
treatment determinations. In general, advanced
tumors require more external beam therapy. This
is in part because the periphery of large tumors are
inadequately treated with brachytherapy due to the
rapid decrease in dose incurring at a distance from
the implant. To treat advanced tumors, the majority of the external beam therapy is given prior to
initiating brachytherapy to shrink the tumor. This
leads to a technically superior brachytherapy application and may result in radiobiological advantages, including the possibility of better tumor
oxygenation and, therefore, more radio sensitivity
as the tumor involutes.

Table 24.6. Summary of results of chemoradiation therapy for definitive therapy for cervical cancers. *XRT* definitive pelvic radiotherapy, *H* hydroxyurea, *5FU* 5-flourouracil, *CDDP* cisplatin

Study	FIGO stage	Overall survival	Progression-free survival	Relative risk of progression
KEYS et al; GOG 123**	IB2			
XRT (control)		74%	63%	
XRT plus weekly CDDP		83%	79%	0.51
WHITNEY et al.; GOG 85	IIB-IVA*			
XRT + H (control)		43%	47%	
XRT plus CDDP/5FU		55%	57%	0.79
ROSE et al.; GOG 120	IIB-IVA*			
XRT +H (control)		50%	41%	
XRT + weekly CDDP		66%	62%	0.57
XRT+H/5FU/CDDP		67%	61%	0.55

*Negative para-aortic nodes by extraperitoneal laparotomy
**Also had adjuvant hysterectomy

Design of the external beam fields takes into account the fact that cancer of the uterine cervix spreads in a very predictable manner, first spreading laterally to the para cervical nodes, then to the internal common iliac and finally to the para-aortic nodes. At the University of Minnesota, we prefer extraperitoneal lymph node staging to delineate the borders of the external beam fields (POTISH et al. 1984, 1989). Patients with negative pelvic nodes receive treatment to a small pelvic field with a superior border approximately at the mid sacroiliac joint level. Patients with involved pelvic nodes, but negative common iliac or periaortic nodes, receive radiotherapy that encompasses what we call the "low periaortic" area. The superior border of this field is approximately at the L2–L3 interspace. For patients with positive common iliac or periaortic nodes the fields are extended to include the "high periaortic" areas and the field extends to the T9–T10 interspace (Fig. 24.1).

PET imaging may replace extraperitoneal lymph node sampling, and recently we have begun designing fields based on PET or PET/CT results if extraperitoneal nodal sampling is not performed. Alternatively, if extraperitoneal node sampling or PET scans are not available, the extent of external beam fields can be determined by combining CT or MRI results with risk rates of pelvic nodal spread. Approximately 15% of patients with FIGO stage-I disease will be found to have positive pelvic nodes, 30% of those with stage II and up to 45% of those with stage III. The risk of positive para-aortic nodes is roughly half that of the pelvic node rate (6% in stage I, 12% in stage II and 24% in stage III). This information can be used to plan the external beam fields. For a small tumor, which is stage I, the pelvis alone is usually adequate external beam volume. For patients with

more advanced disease, one could consider treating extended fields to include either the common iliac or paraaortic nodes. The RTOG reported their 10-year results of prophylactic extended field radiotherapy in patients with stages-IIB and bulky -1B and -IIA

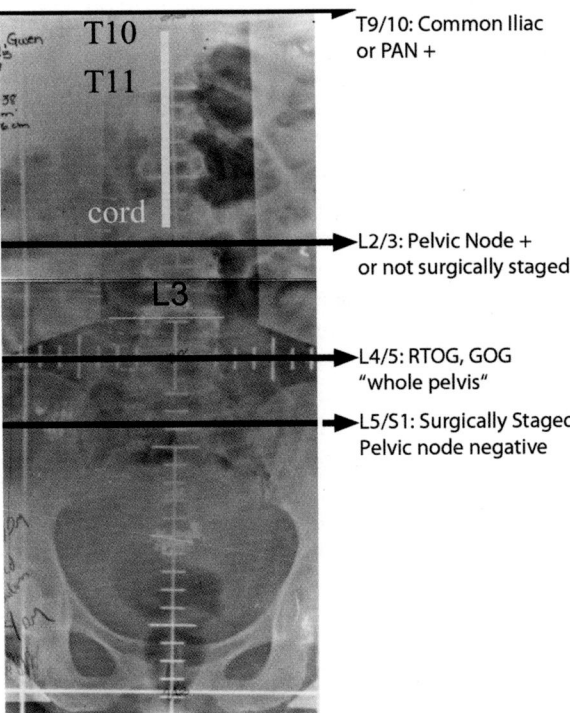

Fig. 24.1. Anteroposterior (AP) radiograph showing the possible locations for the superior border of the external beam field. Small pelvis field used if pelvic and periaortic nodes are free of metastasis at the time of surgical staging. Low periaortic field is used if pelvic nodes are involved with metastasis but periaortic nodes are negative. Note that it does not include the inferior-most extent of the spinal cord. Entire periaortic nodes are treated when shown to be involved by metastasis either by surgical staging or needle biopsy

cervical cancers (GRIGSBY et al. 2001). In that series, the extended radiotherapy patients had an approximately 10% improvement in 10-year survival (44% versus 55%).

The inferior border of the external beam field must be tailored to the extent of disease. A cervical gold seed implanted at the time of the exam under anesthesia is an excellent time to place gold seeds to mark either the cervix or the inferior extent of vaginal spread. If there is vaginal extension of the tumor, the inferior border should give at least a 2-cm margin on the most inferior extent of the vaginal extension. Additionally, for tumors that involve the lower third of the vagina, the inguinal nodes are at risk and should be included in the external beam fields. Appropriate measures must be taken to ensure that they receive adequate dose, such as using mixed energy beams and ensuring that the field is wide enough to include them.

In the past, bony landmarks were often used to delineate the width of the pelvic field. On an AP radiograph, if between 1.5 cm and 2 cm exist between the widest point of the bony pelvis and the field edge, it was thought that the pelvic nodes would easily be included. However, now with the advent of CT simulations it is known that often these margins are not adequate and it is superior to perform a treatment planning CT with both IV and oral contrast agents. The actual location of the vessels and surrounding nodes can then be delineated and the field more accurately rendered.

The pelvis is usually treated with a four-field external beam arrangement and care must be taken in designing the lateral fields so that the entire uterus is compassed and the utero-sacral ligaments, which attach at S1 and S2, are included. A common mistake is to try to block large portions of the rectum and, in doing so, shield the tumor extent posteriorly. Additionally, the uterus is often anteverted and a tight anterior margin can block some of the uterus. For this reason, also treatment planning, CAT scans are quite useful and more accurate than just relying on radiographs (Fig. 24.2).

24.5
Simulation/CT Simulation Procedures

At the time of the simulation or CT simulation, the patient is usually placed supine on the simulator or CT couch. Attempts have been made to try to spare more small bowel with a prone position on some type of false table type device, but with the

location of the uterus, cervix and nodal spread, it is unclear whether the advantages of a prone position are outweighed by the instability of this position (GHOSH et al. 2001). The cervix is marked either with a gold seed that has previously been placed or with a vaginal marker (Fig. 24.2). Oral contrast with the CT is highly useful, as is IV contrast, to delineate the vessel location. The approximate field borders are set and recorded, and the CT scan transferred to the treatment planning computer.

24.5.1
Target Volume and Organs at Risk

The target volume is the cervix, uterus, uterosacral ligaments and nodes deemed at risk or known to harbor metastatic disease. The uterus is easily seen by means of CT scan or MRI. More difficult to visualize are ligaments which need to be included, especially in more advanced disease states. The bladder and rectum are outlined, as is the small bowel and kidneys (Fig. 24.3). Usually a four-field arrangement

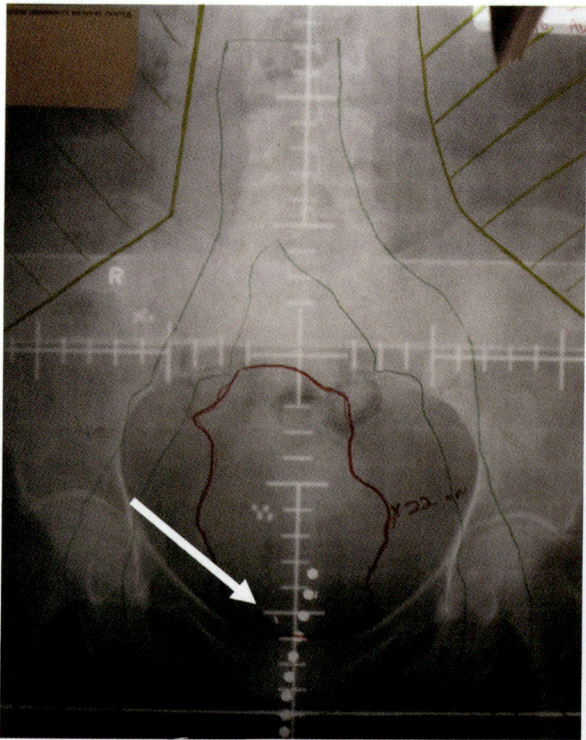

Fig. 24.2. Gold seeds are placed at the most inferior extent of vaginal involvement, if present, or in the cervix to act as a marker for field placement. Lateral field posterior border set at S2 so as to include the uterosacral ligament and presacral nodes

CT simulation/Digitally Reconstructed Radiograph (DDR)

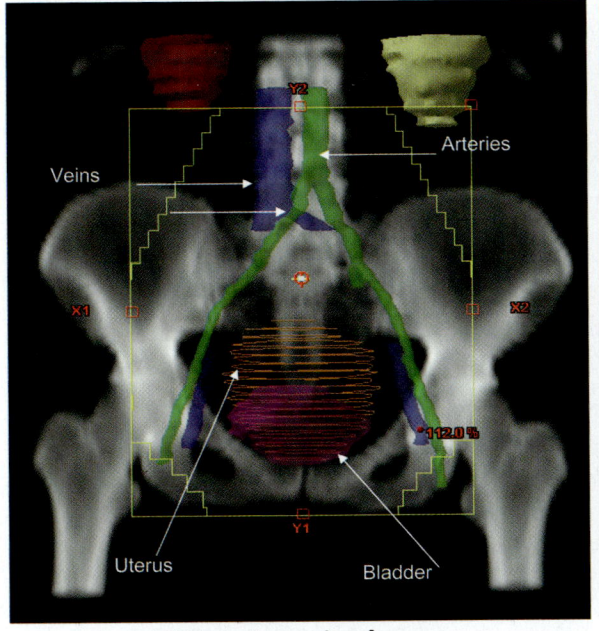

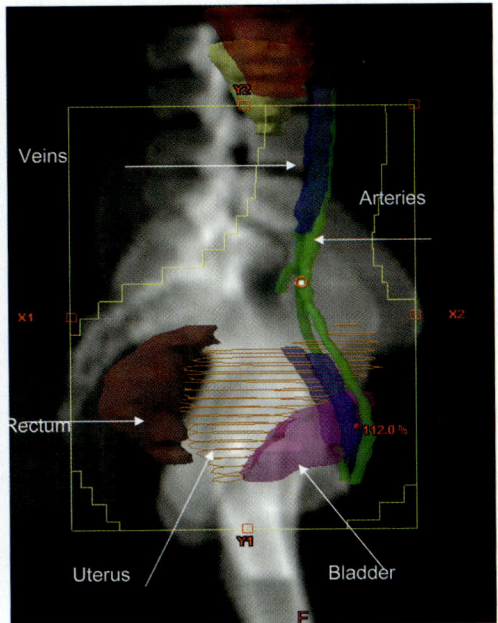

DDR – Anterior beam

DDR – Right Lateral beam

a

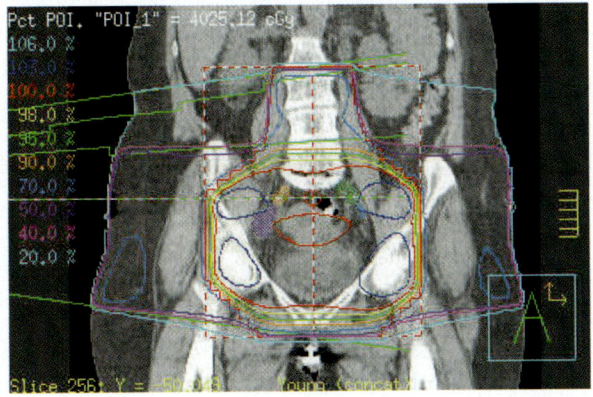

b

Fig. 24.3a,b. Digitally reconstructed radiograph computed tomography (CT) scan with contours of external and internal iliac vessels outlined as well as the small bowel, bladder and rectum. In general, a 2-cm margin on the vessels will encompass the associated lymph nodes and can be considered the clinical target volume (CTV). A margin placed around this volume for day to day set-up variation can be considered the planning target volume (PTV)

gives excellent dose distributions and does allow for some sparing of small bowel and bladder and possibly some of the rectum. As stated earlier, care must be taken to not spare bowel at the extent of blocking the uterosacral ligament. If extended fields are used, the kidney location must be identified and avoided. If a four-field arrangement is chosen, the lateral fields may treat a large proportion of the kidneys unless there is judicious use of blocking.

24.5.2
Dose Prescription

Patients with very early lesions can be treated with brachytherapy alone, although often these patients undergo radical hysterectomy and are referred for definitive radiotherapy only if there are intermediate- or high-risk pathological features. Everyone else receives a combination of EBT and brachytherapy.

Table 24.7 gives the recommendations for treatment regimens used at the University of Minnesota. For a patient treated with brachytherapy alone, 10,000 mg h^{-1} in two applications is usually prescribed. As the FIGO stage and bulk of disease increase, more external beam therapy is delivered and less brachytherapy is delivered. At the University of Minnesota, we generally deliver 175 cGy/day, but at most institutions between 180 cGy and 200 cGy is standard. The initial phase is given with concurrent chemotherapy, usually cis-platinum (40 mg/m^2 per week.) The platinum is usually delivered early in the week so that

Table 24.7. ICRU reporting data

Description of technique
○ Source used (radionuclide, reference air kerma rate, shape and size of source, and filtration)
○ Simulation of linear source for point or moving sources
○ Applicator geometry (rigidity, tandem curvature, vaginal uterine connection, source geometry, shielding material

Total reference air kerma

Time dose pattern (application duration)

Description of reference volume
○ Reference isodose level
○ Isodose width, height, and thickness

Dose at reference points
○ Bladder
○ Rectum
○ Lymphatic trapezoid
  (lower periaortic, common iliac, external iliac)
○ Pelvic wall

there is maximum potential for radiosensitization. After 4–5 weeks, the first of two LDR brachytherapy applications is delivered. This is followed 2 weeks later by the second LDR brachytherapy application. In the interval of time between the two applications, external beam treatments continue with blocking of the central area where the high-dose brachytherapy application was concentrated. This parametrial boost can be delivered to one or both sides. At the University of Minnesota, our split pelvis block is generally 4.5 cm wide. At other institutions, partial transmission blocks are designed that correspond to the isodose lines of the brachytherapy application. The height of the block is determined by the height of the tandem and roughly corresponds to the height of the 30 cGy/h isodose line from the brachytherapy application. It is important to keep the treatment course to less than 8 weeks, as protraction has been associated with a worse pelvic control rate (ERRIDGE et al. 2002; NAG et al. 2002).

At the completion of therapy, the periaortic nodes, if involved, will have received between 4500 cGy and 5000 cGy if the periaortic nodes were negative; but pelvic nodes involved the "low periaortic" nodes will have received between 4500 cGy and 5000 cGy. The whole pelvis and dose split pelvis dose will total between 4500 cGy and 5500 cGy, depending on the bulk of disease, response to treatment and contribution to pelvic dose of the brachytherapy.

24.6
LDR Brachytherapy

24.6.1
Equipment

Manual after-loading Fletcher-Suit-Delclos applicators are used for the majority of cervical cancer patients at the University of Minnesota (Fig. 24.4). Over 100 other applicators are available worldwide. LDR and pulse rate remote after-loading systems, which minimize the exposure to medical personnel, are available from manufacturers in several countries. For both manual and remote after-loading, the clinical, radiobiological, and physics principles of LDR are the same. Only the size of the applicators, radiation protection features and possibly the ability to limit applicator movement during treatment differs. Integral to the applicator itself is an implantation system that has evolved around the use of those applicators.

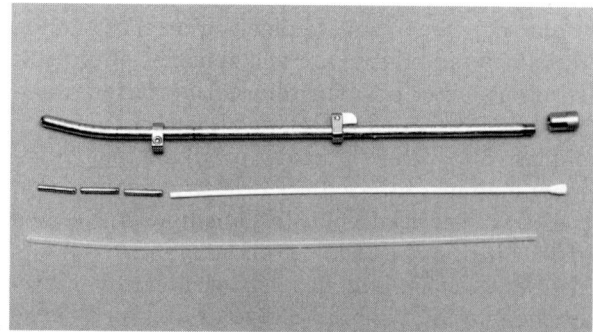 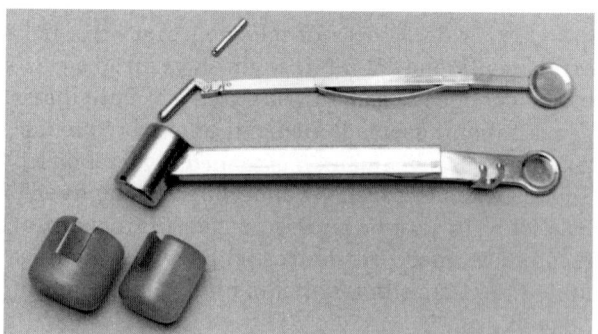

a

Fig. 24.4. a Fletcher-Suit-Delclos tandems no. 1 showing flange that can be sutured to the cervix. The cesium sources are placed in tandem in a straw which is placed in the metal tandem. A "pusher" ensures that the sources are snug in the tip of the tandem. **b** A colpostat (also called ovoid) is shown. The diameter of the bare metal colpostat is 2 cm. Caps can be placed over the metal colpostat increasing the diameter to either 2.5 cm or 3 cm. Caps effectively push the vaginal mucosa away from the source and improve the ratio of dose at the vaginal mucosal surface to the dose deeper below the vaginal surface. The cesium source is placed into the metal device which slides into the colpostat and locks in place

24.6.2
LDR Procedure

The LDR applicators are generally placed with the patient in the operating room under general or spinal anesthesia. HDR applicators can be placed under general or regional anesthesia, or just IV sedation. Once the patient is anesthetized and an examination performed, the cervical os is identified and the intrauterine canal serially dilated (often not required for HDR). The appropriate length tandem is inserted into the uterine canal and affixed to the cervix by suturing it to the tandem through the collar. In order to ascertain that the tandem is in the intrauterine canal, an intraoperative transabdominal ultrasound is performed. (Fig. 24.5). The ovoids are then placed in each fornix and affixed to the tandem (we use a rubber band). Packing with gauze soaked in an antiseptic, a radio-opaque wire is used to push the bladder and rectum away from the applicators, taking care not to let the packing extend superior to the ovoids. Foley catheters with balloons are inflated

with diluted contrast material and placed in both the bladder and rectum for later determination of IRCU bladder and rectal point doses. Intraoperative radiographs are performed. The desired geometry of the tandem relative to the ovoids and relative to the pelvis is shown in Figure 24.6. Examples of poor implant geometry are shown in Figure 24.7. If a poor application is achieved, the packing is removed and packing is repeated and re-filmed until satisfactory. Figure 24.8 shows a radiograph of the desired orientation of the tandem to ovoids and pelvis. Once this is achieved, the anesthesia is reversed and the patient recovers in the recovery room and is then transported to the radiotherapy department for simulation.

24.6.3
Milligram Hour

In the pre-computer era, when only radium sources were available, gynecological brachytherapy applications were specified by the simple mathematical product of the number of milligrams of radium times the duration (number of hours) of the implant. Thus, an implant of five 10-mg radium sources left in place for 48 h would yield a "dose" of 2400 mg h^{-1} (10×5×48). Since its initial use in the early 1900s in Europe, the dose prescription system evolved and was refined at M.D. Anderson Hospital by Dr. Gilbert Fletcher and his colleagues. This system is still used today. A 7% (8.25/7.71) correction factor is used to convert the "Fletcher milligram" hours to ICRU milligram hours, since the milligram hours recommended by Dr. Fletcher are for implants with radium and encapsulated with 1 mm of platinum (exposure

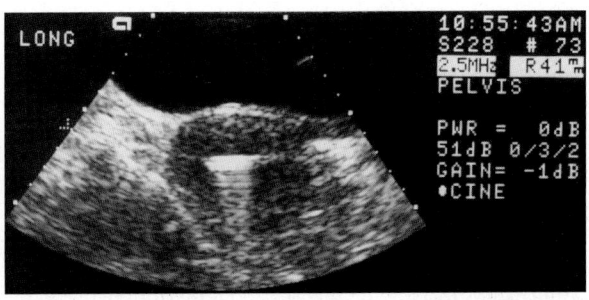

Fig. 24.5. Longitudinal ultrasound image through the uterus demonstrating bladder (B), myometrium (M) and correct position of tandem within the uterine cavity (T)

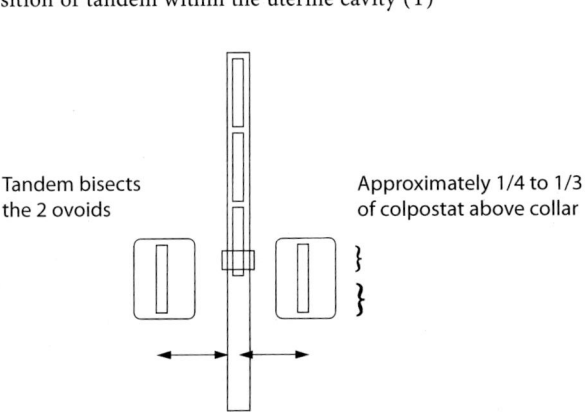

Tandem bisects the 2 ovoids

Approximately 1/4 to 1/3 of colpostat above collar

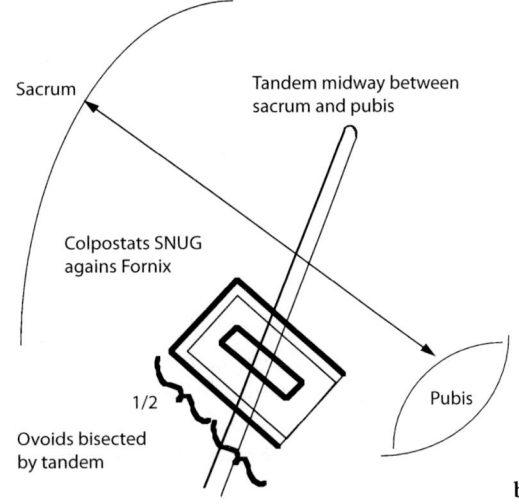

Sacrum

Tandem midway between sacrum and pubis

Colpostats SNUG agains Fornix

1/2

Ovoids bisected by tandem

Pubis

b

Fig. 24.6. a Diagram illustrating the desired orientation of the tandem relative to the ovoids as seen on an anterior view. b Diagram illustrating the desired orientation of the tandem relative to the ovoids as seen on a lateral view

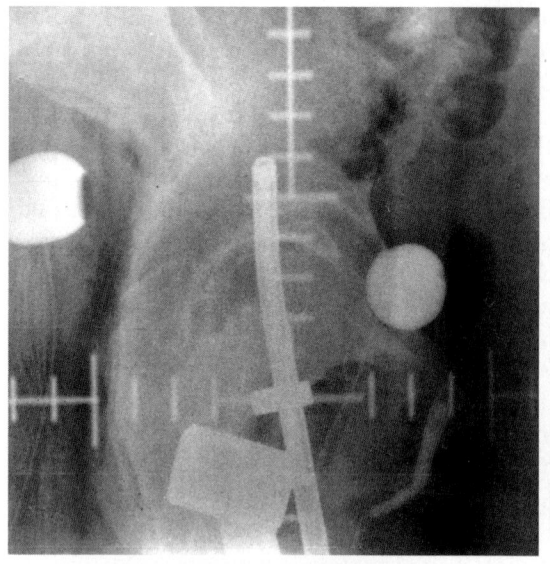
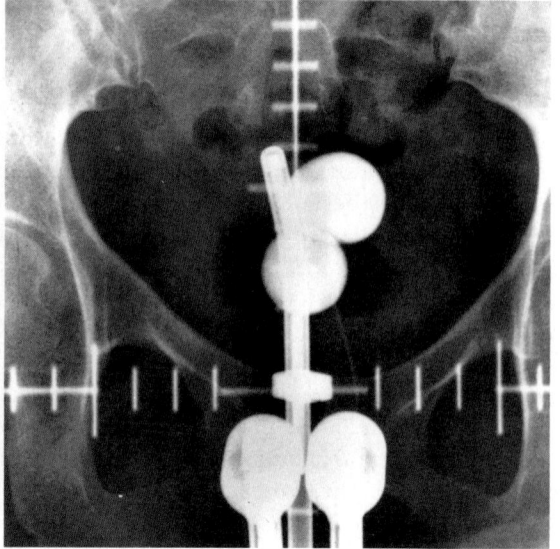
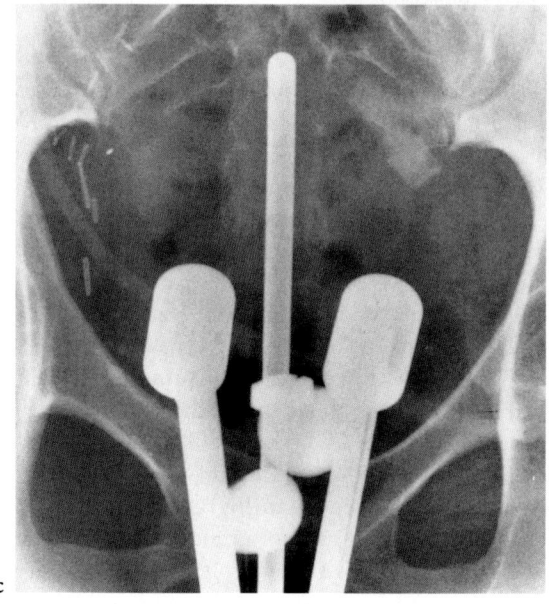

Fig. 24.7a–c. Radiographic examples of poor geometry. **a** X-ray demonstrating ovoids that are placed too inferior and posterior relative to the tandem. Re-packing is indicated. **b** Radiograph demonstrating ovoids that are too inferior and posterior relative to the tandem. If caps are on the ovoids, removing them might improve the geometry. **c** The ovoids are superior relative to the tandem. Note that the ovoids are widely spaced and placing caps on the ovoids may help

rate constant 7.71 R/cm^2 h^{-1} mg^{-1}). The ICRU specifies that radium source with 0.5 mm of platinum filtration be used as the standard (exposure rate constant 8.25 R/cm^2 h^{-1} mg^{-1}).

Although the milligram hour system is easy to use, it does not give any information about the dose distribution around the application, and the reason the system "works" is because it specifies a particular geometry between the tandem and ovoid, and sources are loaded in a rigidly prescribed manner. It is only applicable when both the tandem and ovoid are implanted. For a tandem loaded with a protruding vaginal source (no ovoids), the milligram hour dose specification is not applicable.

24.6.4
The Manchester System

The second dose prescription system that evolves specifies the dose to four specific points and space around the applicator – point A, point B, bladder and rectum. Originally developed in Manchester, England, this system is widely used. Point A was initially defined as the point 2 cm superior to the vaginal fornix and 2 cm lateral to the cervical canal. Point B was 3 cm lateral to point A. Points A and B were said to represent critical anatomical structures, with point A representing the site where the uterine vessels cross the ureter and point B representing the location of the more lateral pelvic nodes. Although most descrip-

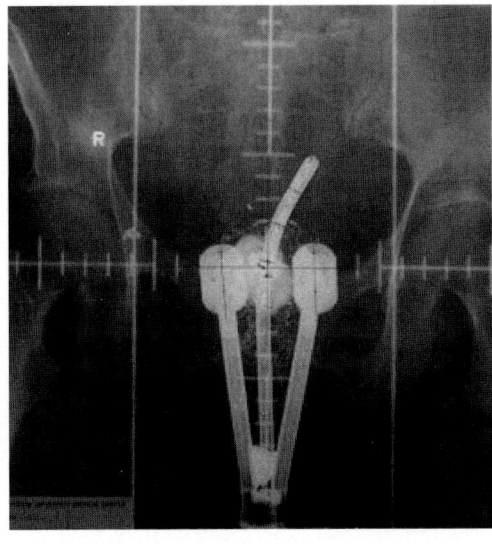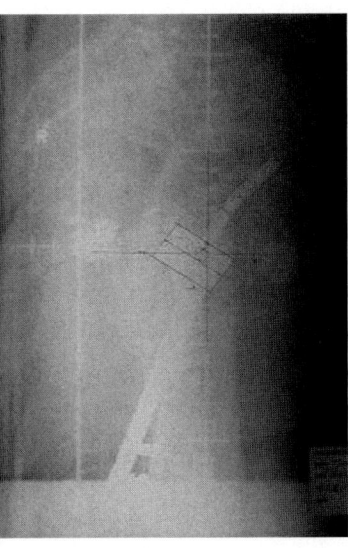

Fig. 24.8. Radiographs of desired orientation of the tandem and ovoids

tions of point A are 2 cm superior and 2 cm lateral to a specified origin, the definition of the origin is not standardized and there are at least six different origins in the literature (Fig. 24.9). Depending on which origin is chosen, the point A dose can vary widely, especially if the ovoids are displaced superiorly or inferiorly (POTISH and GERBI 1986). A relatively small displacement of the ovoid results in a large change in the dose rate to point A. Since dose rate falls off rapidly with distance from the brachytherapy sources, any point surrounding the applicator will be expected to be in a rapid fall off area, another disadvantage of using the Manchester system. For small tumors, point A may even lay outside the tumor volume whereas for larger tumors the tumor may extend significantly more lateral than point A. Since some definitions of point A are radiographic, point A may end up outside of the cervix altogether. Figure 24.10 is an isodose distribution of a standard Fletcher suit tandem and ovoid application. Note that point "A" is in an area of sharp dose fall off.

An association between point A, point B and milligram hours does exist. In a perfect Fletcher suit tandem loaded as described above, point A is almost always between 50 cGy and 60 cGy per hour. Point B usually lies between 10 cGy and 20 cGy per hour and is largely dependent on the source strengths in the ovoids. Because these dose rates for points A and B are usually confined to a narrow range, as are implant durations, it is not surprising that there is an association between these total point doses and milligram hours. In an evaluation of almost 100 brachytherapy applications performed at the University of Minnesota, a fairly high correlation between milligram hours of radium and doses at point A and B was reported, with correlation coefficients of 0.73 and 0.89, respectively. However, it was pointed out that the point-A dose was markedly affected by the position of the colpostats relative to the tandem collar and that there was considerable inter-patient variability making the routine translation between Fletcher and Manchester systems too unpredictable for clinical use (POTISH et al. 1982). Despite these limitations, it is useful to translate milligram hours into point A or point B doses. BHATNAGAR et al. (1980) has suggested a formula that roughly calculates the point A dose based on the milligrams of radium loaded:

Point A (cGy/h)=1.1 (mg in tandem) + (mg in ovoids/2) cGy/h

Likewise, the following formula roughly converts milligram hours into the total point B dose:

Point B (total dose in cGy)=(total milligram hours in two implants)/4

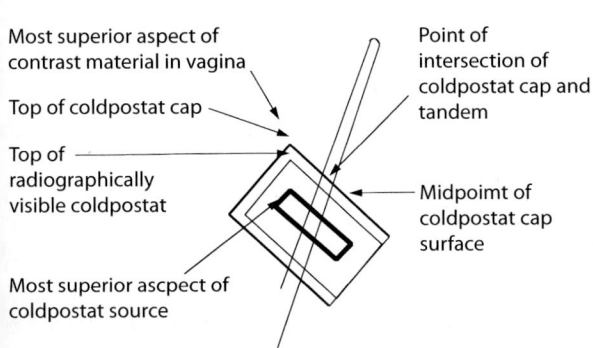

Fig. 24.9a,b. a Possible origins of point A on anterior view. **b** Possible origins of point A on lateral view

Isodose Curves for Tandem and Ovoids:
Standard Fletcher Loading (in cGy/hour dose rate)

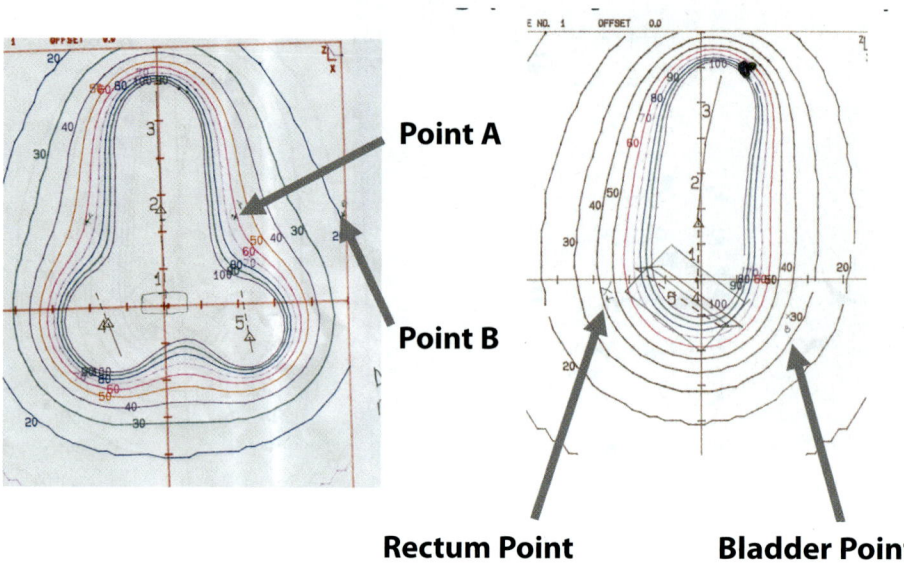

Point A

Point B

Rectum Point

Bladder Point

Fig. 24.10. Brachytherapy isodose plot. Note that point "A" is in an area where the isodoses change rapidly and a small variation in the location of point "A" can make a big difference in the resultant dose rate

24.6.5
The International Commission on Radiographic Units and Measurements

In 1985, the ICRU recommended the system of LDR brachytherapy dose and volume specifications aimed at redefining and standardizing brachytherapy reporting terminology (Table 24.7). This has not been widely adopted in the US. Three reporting approaches were proposed, the reference air kerma rate, the absorbed dose at certain reference points (Figs. 24.11–24.14) and isodose reference volume.

The reference air kerma rate was proposed to introduce international units into the brachytherapy reporting. The reference air kerma rate is expressed in mGy/h at 1 m. The total reference air kerma is, therefore, the sum of the products of the reference air kerma rate and the duration of the application. The Fletcher milligram hours can easily be converted into reference air kerma using the following formula:

1 mg h^{-1}=6.5 mGy total reference **air kerma** (for filtration of 1 mm of platinum). The total reference **air kerma,** therefore, has the same limitations as milligram hour dose specification. The ICRU recommended calculating the absorbed dose at certain reference points (rectal, bladder, pelvic wall, trapezoid, node points but fell short of trying to standardize the definitions for point A or B.

The most radical change proposed by the ICRU was to specify reporting parameters for the pear-

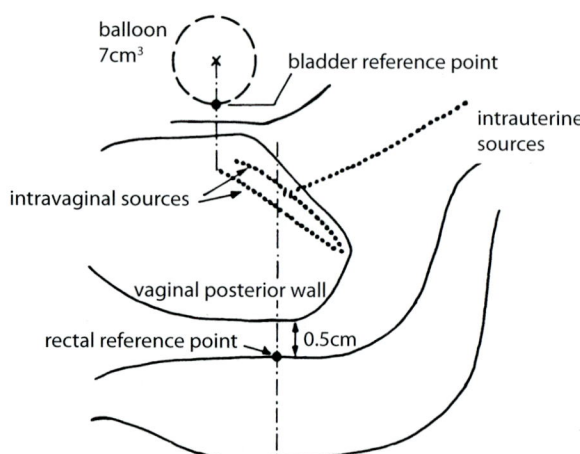

Fig. 24.11. Location of bladder and rectum points as specified by the ICRU (ICRU 1985). From ICRU report no. 38. International Commission on Radiation Units and Measurements, Bethesda, MD, 1985. Used with permission

shape reference volume. The report states that "an absorbed dose level of 60 Gy is widely accepted as the appropriate reference level for classical low-dose-rate brachytherapy" and therefore the 60-Gy isodose reference volume was suggested. This reference volume is determined by measuring the width, height and thickness of the tissue encompassed by the 60-Gy isodose curve. The ICRU chose to subtract any external beam therapy from this 60-Gy reference volume to choose the relevant isodose. To choose

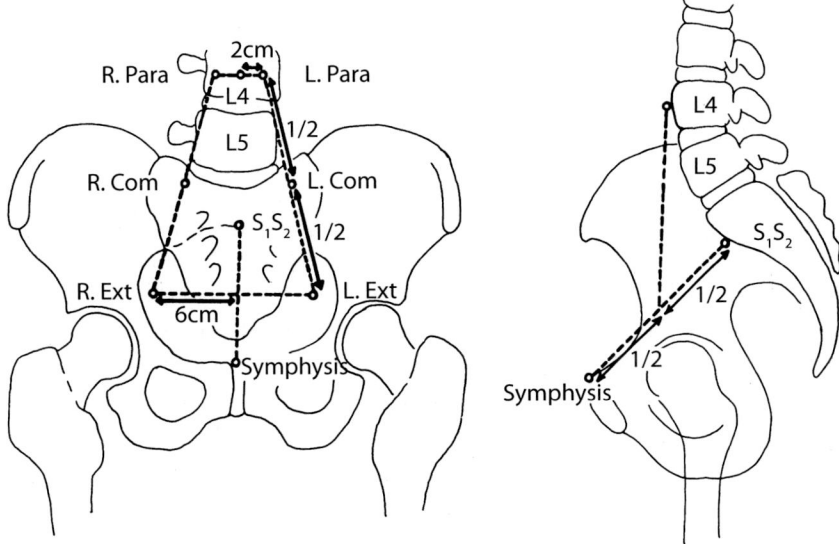

Fig. 24.12. Determination of reference points corresponding to the lymphatic trapezoid of Fletcher. From ICRU report no. 38. International Commission on Radiation Units and Measurements, Bethesda, MD, 1985. Used with permission

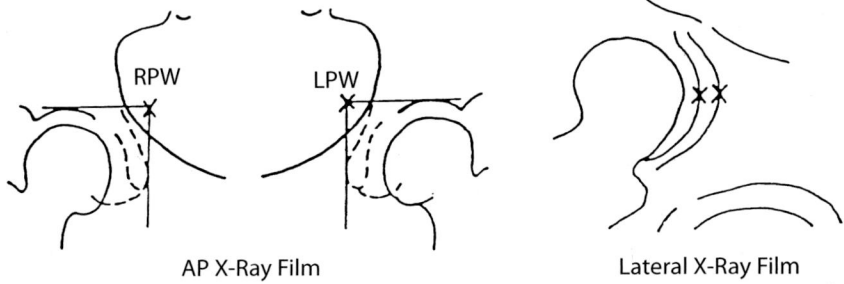

Fig. 24.13. Definition of pelvic wall points. Position of right pelvic wall (RPW) and left pelvic wall (LPW) are diagrammed. From ICRU report no. 38. International Commission on Radiation Units and Measurements, Bethesda, MD, 1985. Used with permission

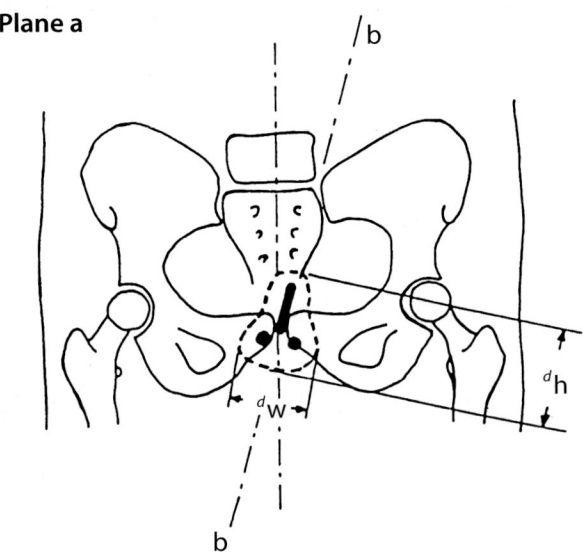

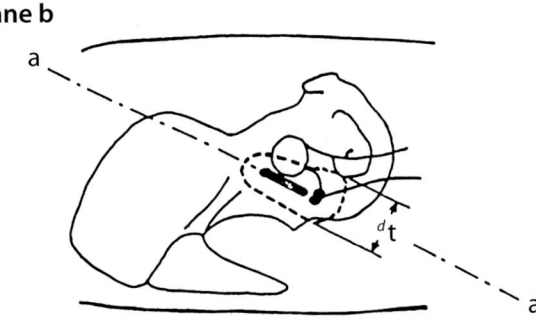

Fig. 24.14. ICRU reference volume is determined by finding the reference isodose surface dimensions, dw (width) dh (height) and dt (thickness). From ICRU dose volume specification for reporting intracavitary therapy in gynecology. From ICRU report no. 38. International Commission of Radiation Units and Measurements, Bethesda, MD, 1985. Used with permission

the relevant isodose surface, the EBRT is subtracted from the 60-Gy volume, and the remainder is divided by the implant duration to obtain the isodose surface by which to calculate the reference volume, for instance, if a patient has received 40 Gy of external beam therapy and 80 h of brachytherapy:

60 Gy–40 Gy=20 Gy
20 Gy/80 h=0.25 Gy/h

The 25-cGy/h isodose surface is chosen.

The attempt by the ICRU recommendation was that this "reference volume" would further describe the brachytherapy application and shed light on the volume of tissue receiving a particular level of dose. Unfortunately, instead of being helpful, the concept has only caused confusion and criticism, being helpful neither for dose specification nor for dose reporting.

In recent years, image-guided radiation therapy for cervical cancer has had some success. If the tumor volume can be delineated with adequate imaging, the ICRU concept of reference volume, especially if it is the tumor reference volume, might be very helpful. Further work is sorely needed in this area.

24.6.6
University of Minnesota

At the University of Minnesota, we rely heavily on the Fletcher milligram hour dose specification system (FLETCHER 1973). When this is inappropriate (protruding vaginal source, vaginal cylinder), point calculations (A_v, or vaginal surface dose) are used

(Table 24.8) (POTISH and GERBI 1986). The Manchester points are calculated for all patients but are not routinely used to specify dose. We find them useful as a double check to assure ourselves that both an adequate tumor dose has been achieved (remembering that point A may have little relationship to the tumor) and that bladder or rectal tolerances have not been exceeded (remembering that a point dose to the rectum or bladder provides no information about the dose to the rest of the rectum or bladder received).

Microinvasive (FIGO IA) tumors are unlikely to have spread to the pelvic lymph nodes. Patients with these tumors may be treated with intracavitary therapy alone. Doses of up to 10,000 mg h^{-1} in two applications are used. For small (1–3 cm) IB tumors, up to 3000-cGy pelvis external beam whole pelvic external beam therapy is given depending on the extent of disease with up to 2000 cGy of split pelvis radiation. This is followed by LDR brachytherapy applications of up to 7000 mg h^{-1}. For more advanced tumors, 4000-cGy whole pelvis EBRT is used, and up to 6500 mg h^{-1} of cesium may be used in two implants. The parametrial areas may be boosted with an additional 1000-cGy EBRT using a split pelvis field, either with a straight split pelvis block (usually 4.5 cm in diameter) or with a step wedge block. For even larger tumors, more whole pelvis (±periaortic nodal) EBRT is given (4500 cGy), and the two-brachytherapy applications are shorter (5000–6000 mg h^{-1}). Once whole pelvis EBRT doses of over 4500 EBRT are exceeded, however, it is difficult to add significant amounts of brachytherapy. Therefore, an examination under anesthesia is performed at the 4000-cGy dose level. If there has been good tumor shrinkage, the first implant is performed at that time.

Table 24.8. General guidelines used at University of MN for treatment of carcinoma of the intact cervix in surgically staged patients

FIGO stage	Parametrial extension*	Tumor size (cm)	Whole pelvis EBRT dose (cGy)**	Maximum mg h^{-1}‡	Point A♦ (cGy)	Point B† (cGy)
IB, IIA	No	2–5	3500–4000	6000–6500	7500+	5500–6000
		5+	3500–4500	5000–6000	8000–9000	6000
IIB	Yes	2–5	3500–4000	6000–6500	8000+	6000
						6000
		5+	3500–4500	6000–6500	8000–9000	6000
III	Yes		4000–5000	5500–6000	8500–9500	6000
IV A	Yes		Individualize			

** Dose to the whole pelvis (± periaortic nodes) given with EBRT given depends on bulk of tumor, FIGO stage, and amount of regression at the time of the first brachytherapy application. If the periaortic nodes are included, periaortic nodal EBRT doses range from 4000-5000 cGy.

† Point „B" is boosted to 6000 cGy for all patients with pathologically proved involvement of pelvic nodes or tumors larger than 2 cm in size. A split pelvis block is 4.5 cm wide with height determined by height of the tandem is used.

‡ Given in 2 applications, with tandem and ovoids 2 weeks apart. Does not take into account 1.07-filtration factor correction for cesium in .5 mm platinum.

♦ This maximum is when nominal strengths of cesium of 10, 15, 20 and 25 mgs are used. If cesium of lower activity is used, the hours may be increased accordingly.

EBRT = external beam therapy given at 175 cGy/day

If the tumor is still large after an initial 4000-cGy whole pelvis EBRT and there is concern that the implant will not have good geometry, additional EBRT (with doses up to 1000 cGy) may be given or an interstitial implant can be considered. For tumors with significant vaginal involvement, the implantation volume should include the extent of disease in the vagina. Usually the EBRT leads to shrinkage of the vaginal involvement so that at the time of the first brachytherapy application there is little palpable tumor left. Often the previous tumor is replaced with vague induration. If there is more tumor than 0.5 cm of induration remaining, implanting the vaginal component of the tumor with iridium should be considered. Tolerance of the vaginal wall to high doses of radiation probably depends on the length of vagina treated as well as on the total dose and dose rate. In general, the vaginal apex tolerates more than the distal vagina, with the apex tolerating doses of 10,000 cGy or more. More distal portions of the vagina should probably not receive more than 7000–8000 cGy.

The American Brachytherapy Society has given recommendations on the combination of external beam irradiation and LDR brachytherapy (NAG et al. 2002). For completeness, the summary table has been reproduced in Table 24.9.

24.6.7
Postoperative Radiotherapy

In cases of radical hysterectomy, recent randomized trials have delineated patients who might be at greater risk for recurrence and who benefit from postoperative radiation therapy either alone (SEDLIS et al. 1999) or with concurrent chemotherapy (PETERS et al. 2000). In the first of these trials, patients with "intermediate" risk factors were randomized to either observation or pelvic radiotherapy. The DFS was better for the pelvic radiotherapy group (Table 24.10). In the second trial, "high-risk" patients were randomized to either pelvic radiation or pelvic radiation and chemotherapy. The DFS was better for the combined modality group. The risk of bowel or bladder complications is high in this setting and care must be taken to minimize the risk of a late radiation complication as much as possible. The American Brachytherapy Society has published guidelines for postoperative radiation therapy for cervical cancer (NAG et al. 2002), which is summarized in Table 24.11.

24.6.8
Dose to Bladder and Rectum

The ICRU suggests a method to choose which point to calculate the rectal and bladder dose. These point doses may have little relationship to the dose received by other parts of the bladder or rectum and may not be the highest dose area (CROOK et al. 1987; ESCHE et al. 1987; KATZ and EIFEL 2000). The localization of the bladder and rectum can be achieved by either placing Hypaque in a Foley catheter, which has been placed in the bladder or rectum, or by the

Table 24.9. Carcinoma of the uterine cervix – the ABS suggested doses of external beam irradiation and low-dose-rate (LDR) intracavitary brachytherapy. The panel making these recommendations acknowledge that a range of doses can be suitable depending on individual patient circumstances. Modified from reference NAG et al. (2002)

Tumor stage	Tumor extent	External irradiation (Gy)		Parametrial boost (Gy)	LDR brachytherapy (Gy)	
		Whole pelvis	Pelvic wall		Dose to point A	Total dose to point A (Gy)
IA1		0	0	0	50–60	50–60
1A2	Superficial ulceration; less than 1cm in diameter or involving fewer than two quadrants	0	0	0	60–70	60–70
Selected IB1						
IB1		19.8 or 45	50.4 or 45	0	55 or 30–35	75 or 75–80
IB2, IIA,*		45	45	0	40	85
IIB*		45	45	9–15	40	85
III*		45 50	45–50	9–15	40	85–90
IIB, IIIB, IV	Poor pelvic anatomy; patient not readily treated with intracavitary insertions (barrel-shaped cervix not regressing; inability to locate external os)	50	50	9–15	40	90
	Or interstitial	39.6–45	39.6–45	0–15	35–40**	75–85**

*The alternative approach is to increase brachytherapy contribution to point A by giving whole pelvic EBRT of 19.8–30.6 Gy. This is followed by whole pelvic EBRT with a step wedge midline shield for an additional 19.8–30.6 Gy and intracavitary brachytherapy to bring point A dose to the recommended level described in the table

Table 24.10. Randomized trials of postoperative irradiation after radical hysterectomy. *PFS* progression free survival

Author	Inclusion criteria	Tx ARMS	Outcome
PETERS et al. (2000) "High risk"			
	(+) LN	RT	64% 4-year PFS
	(+) Margin parametrial extension	RT & CT	80% 4-year PFS
SEDLIS et al. (1999) "Intermediate risk"			
	Lymph space invasion	Observation	79% PFS at 2 years
	Deep stromal invasion, large size	RT	88% PFS at 2 years

Table 24.11. The ABS general guidelines for postoperative radiation therapy for cervical cancer**

Tumor status	Whole pelvis (Gy)	Pelvic wall (Gy)	Tumor boost (Gy)	Vaginal brachytherapy dose (Gy)	Total vaginal mucosa dose (Gy)
Radiation therapy after simple hysterectomy					
>3-mm invasion, margin clear, nodal status unknown	45–50.4	45–50.4	–	0–15	45–60
Microscopically positive vaginal margins, or LVSI	45–50.4	45–50.4	–	20–30	70–75
Gross residual tumor or recurrent disease	45–50.4	45–50.4	–	30–35	80
Or with interstitial	45	45	–	30–35*	75–80*
Radiation therapy after radical hysterectomy					
Positive pelvic lymph nodes	45–50.4	45–50.4	–		45–50.4
Deep stromal invasion (≥10 mm or ≥70%, and ≥4 cm tumor) microscopically positive vaginal margins or positive LVSI	50	50	–	20	70
Microscopically positive parametrial or paravesical margins	45–50.4	50.4	9–15	0	45–60
Gross residual tumor or recurrent disease	45–50.4	45–50.4	–	30–35	80
Or with interstitial	45	45	–	30–35*	75–80*

*The interstitial brachytherapy dose is the ICRU #58 reference dose and not the dose to vaginal mucosa
**Modified from ref (NAG et al. 1999)

use of rectally inserted ionization chambers. This method, however, is probably no more reliable than the point calculation method.

Generally, the bladder can tolerate more radiation than the rectum. No absolute point dose limit can be set, but it is preferable to keep the maximum bladder dose below 8000 cGy and the maximum rectal dose below 7500 cGy. Therefore, it is optimal if the rectal dose rate is lower than the bladder dose rate for the brachytherapy application. Additionally, it is also desirable for the dose rate to the bladder point to be 0.8 or less than that of the dose rate to point A.

24.7
Sequelae

Acute sequelae including diarrhea, cystitis, fatigue and lowered peripheral blood counts are common, but usually resolve within weeks after treatment. Ovarian failure occurs in nearly all patients, unless ovarian transposition outside the pelvis has been performed (BELINSON et al. 1984; HUSSEINZADEH et al. 1984). Shrinkage of the vagina can be minimized

by daily use of a dilator during and after a course of radiation therapy.

The risk of developing a major complication depends on multiple factors. Patient-related factors include the stage and extent of the disease, weight, age (KUCERA et al. 1986), smoking history (KUCERA et al. 1987) and number of previous abdominal surgical procedures (POTISH et al. 1989; COIA et al. 1990). Treatment-related factors include the volume of EBRT field treated, fraction size, the dose of EBRT, the brachytherapy used and the technique of implantation used (UNAL et al. 1981; HANKS et al. 1983; PEREZ et al. 1984; CROOK et al. 1987; POURQUIER et al. 1987; MONTANA and FOWLER 1989; DEORE et al. 1991).

The Patterns of Care study reported that from 8% to 15% of patients treated for cervical cancer with definitive irradiation required hospitalization for a complication, half of which required a surgical intervention. Others have reported similar percentages. In a review of 1784 patients treated at M.D. Anderson Hospital with FIGO stage-IB cervical cancer, the risk of a major complication was 9.3% at 5 years and 14.4% at 20 years (EIFEL et al. 1995). The most common gastrointestinal complications include proctitis, rectal ulceration, sigmoid stricture or small-bowel obstruction. Urinary complications include cystitis or ureteral stricture. Rectovaginal or vesicovaginal fistulas are uncommon.

In a recent review of M.D. Anderson Hospital data by Jhingran et al, among the 4043 patients who had been treated with LDR intracavitary brachytherapy, only 11 (0.3%) had documented or suspected cases of thromboembolism resulting in four deaths (JHINGRAN and EIFEL 2000). Other life-threatening perioperative complications included myocardial infarction (one death in 5 patients), cerebrovascular accident (2 patients), congestive heart failure or atrial fibrillation (3 patients) and halothane liver toxicity (two deaths in 2 patients). Intraoperative complications included uterine perforation (2.8%), and vaginal laceration (0.3%), which occurred more frequently in patients older than 60 years.

PEREZ et al. (1999) reported the morbidity results of 1456 patients treated with external radiotherapy plus two LDR intracavitary insertions to deliver 70–90 Gy to point A. In stage IB, the frequency of patients developing grade-2 morbidity was 9% and grade-3 morbidity 5%; in stages IIA, IIB, III and IVA, grade-2 morbidity was 10–12% and grade 3 was 10%. The most frequent grade-2 sequelae were cystitis and proctitis (0.7–3%). The most common grade-3 sequelae were vesicovaginal fistula (0.6–2% in patients with stage I–III tumors), rectovaginal fistula (0.8–3%) and intestinal obstruction (0.8–4%). In this study, multivariate analysis showed that dose to the rectal point was the only factor influencing rectosigmoid sequelae, and dose to the bladder point affected bladder morbidity.

24.8
Results

The Patterns of Care Study reported 4-year disease-free survivals of 87%, 66% and 28% for FIGO stages I, II and III, respectively (COIA et al. 1990). Results from other individual institutions are shown in Table 24.12. Although results are often reported by FIGO stage, other prognostic factors such as tumor volume or extent of nodal spread may be more important prognosticators of outcome, not currently reflected in the clinical staging system of FIGO. Patients with FIGO stage-IA or small -IB disease have an excellent prognosis with 5-year disease-free survival estimates from 80% to 100% (MENDENHALL et al. 1984; PEREZ et al. 1984; WILLEN et al. 1985; MONTANA et al. 1987; COIA et al. 1990; GERBAULET et al. 1992). For larger IB lesions, 5-year disease-free survival estimates range from 75% to 90%. Stage-IIB disease-free survival results range from 60% to 75% (MONTANA et al. 1985; HORIOT et al. 1988; KIM et al. 1989) and for IIIB the 5-year disease-free survival rates drop to 30–50% (LEIBEL et al. 1987; KIM et al. 1989).

The in-field failure rate increases with increasing initial FIGO stage (Table 24.13). For stage-I patients, an in-field recurrence rate of 5–9% has been reported (ADCOCK et al. 1984; KIM et al. 1989; POTISH et al. 1989; SOMMERS et al. 1989; EIFEL et al. 1990; MONTANA et al. 1991). For stage-II patients, the in-field recurrences range between 10% and 23%, and for stage-III patients up to a 61% in-field recurrence rate has been reported.

Table 24.12. Five-year disease-free survival rates reported for carcinoma of the cervix

Author	Total number of patients	Stage				
		IB	IIA	IIB	III	IVA
PEREZ et al. (1998)[c]	1499	82	65	65	40	4
COIA et al. (1990)	565	74	–	56	33	–
HORIOT et al. (1988)	1383	9	85	75	50	20
MONTANA et al. (1985, 1986, 1987)	533	83	76	62	33	–
POTISH et al. (1989)	153	67	71	70	–	–
LEIBEL et al. (1987)	119	–	–	–	–	–
GERBAULET et al. (1992)	441	89	78	–	–	–
KRAMER et al. (1989)	48	–	–	–	–	18
MENDENHALL et al. (1984)	264	70[a]	71	70[a]	–	–
		68[b]		43[b]	–	–
KIM et al. (1989)	569	82	78	65	48	27
THOMS et al. (1992)	371	56[b]	49[b]	53[b]	–	–
WILLEN et al. (1985)	168	88	77	68	–	–

[a]<6 cm; [b]>6 cm; [c]10-year survival

Table 24.13. Percentage of in-field failure rates (expressed as a percentage)

Author	FIGO stage			
	I	II	III	IV
Coia et al. (1990)	12	27	51	–
Kim et al. (1989)	11.2	8.2 (IIA), 30 (IIB)	52	69
Horiot et al. (1988)	8	12 (IIA), 20 (IIB)	37 (IIIA), 43 (IIIB)	82
Montana et al. (1983, 1985, 1986)	11	16 (IIA), 34 (IIB)	61 (IIIA), 47 (IIIB)	–
Sasaoka et al. (2001)	5	9	17	–

24.9
Quality Management Program

A quality management program is crucial to the safe functioning of a brachytherapy program. This cannot be accomplished without dedicated physics support. The various components of a quality management program are covered in another chapter. However, several issues relative to intracavitary cesium applications deserve emphasis. To assure accurate preparation of the sources that are to be loaded into the patient, a series of checks should be in place. We use color-coded cesium sources and check each source in a well-ionization chamber. Two people must prepare the sources to cross check each other. Sources must be logged in and out of the department (along with a visual count of the remaining sources left in the cesium safe) and a cesium transport card always accompanies all sources leaving the isotope room. Regulations mandated by the Nuclear Regulatory Commission are followed.

Quality assurance tests for the brachytherapy applicators need to be performed at specified intervals (Table 24.14). A rattling noise when the ovoids are gently shook suggests that the rectal or bladder shields have dislodged and the applicator should be radiographed immediately to verify this. Reports of tandem tips coming off while implanted and remaining within the uterine cavity make it prudent to check the integrity of the tandem tip with each insertion. The AAPM Task Group 40 recommended quality assurance procedures for brachytherapy applicators (Kutscher 1994). In addition to these tests, we radiograph the applicators yearly to confirm correct position of the shields.

24.10
Conclusion

In the early stages, cancer of the uterine cervix is highly curable with either surgery or radiation

Table 24.14. Quality assurance tests for intracavitary brachytherapy applicators: frequency of performance and acceptable tolerance limits. *I* initial use or following malfunction and repairs, *Y* yearly, *D* documented and correction applied or noted in report of measurement when appropriate

Test	Frequency	Tolerance
Source location	I, Y	D
Coincidence of dummy and active sources	I	1 mm
Location of shields	I, Y*	D

* Before each use the applicator may be shaken to listen for loose parts (Kutscher et al. 1994)

therapy. As the tumor advances, however, the best chance of cure is with aggressive, definitive radiation therapy consisting of a combination of external beam treatments and brachytherapy applications. Both of the components are important; however, it is probably the skilled use of intracavitary brachytherapy that is the most crucial to a successful outcome. This skill is achieved through experience as well as meticulous attention to the details of the treatment outlined here.

References

Adcock LL et al. (1984) Carcinoma of the cervix, FIGO Stage IB: treatment failures. Gynecol Oncol 18:218–225

Au WW, Sierra-Torres CH, Tyring SK (2003) Acquired and genetic susceptibility to cervical cancer. Mutation Res 544:361–364

Belinson JL, Doherty M, McDay JB (1984) A new technique for ovarian transposition. Surg Gynecol Obstet 159:157–160

Bhatnagar JP, Kaplan C, Specht R (1980) A rule of thumb for dose rate at point A for the Fletcher-Suit applicator. Br J Radiol 53:511–512

Chen RJ et al. (1999) Influence of histologic type and age on survival rates for invasive cervical carcinoma in Taiwan. Gynecol Oncol 73:184–190

Coia L et al. (1990) The patterns of care outcome study for cancer of the uterine cervix. Results of the Second National Practice Survey. Cancer 66:2451–2456

Committee on Practice (2002) B.-G., ACOG practice bulletin. Diagnosis and treatment of cervical carcinomas, number 35, May 2002. Obstet Gynecol 99:855–867

Cosin JA et al. (1998) Pretreatment surgical staging of patients with cervical carcinoma: the case for lymph node debulking. Cancer 82:2241–2248

Crook JM et al. (1987) Dose–volume analysis and the prevention of radiation sequelae in cervical cancer. Radiother Oncol 8:321–332

Deore SM et al. (1991) The severity of late rectal and rectosigmoid complications related to fraction size in irradiation treatment of carcinoma cervix stage III B. Strahlenther Onkol 167:638–642

Dunleavey R (2004) Incidence, pathophysiology and treatment of cervical cancer. Nursing Times 100:38–41

Dunst J et al. (2003) Anemia in cervical cancers: impact on survival, patterns of relapse, and association with hypoxia and angiogenesis. Int J Radiation Oncol Biol Phys 56:778–787

Eifel PJ et al. (1990) Adenocarcinoma of the uterine cervix. Prognosis and patterns of failure in 367 cases. Cancer 65:2507–2514

Eifel PJ et al. (1995) Time course and incidence of late complications in patients treated with radiation therapy for FIGO stage IB carcinoma of the uterine cervix. Int J Radiat Oncol Biol Phys 32:1289–1300

Erridge SC et al. (2002) The effect of overall treatment time on the survival and toxicity of radical radiotherapy for cervical carcinoma. Radiother Oncol 63:59–66

Esche BA et al. (1987) Reference volume, milligram-hours and external irradiation for the Fletcher applicator. Radiother Oncol. 9:255–261

Fey MC and Beal MW (2004) Role of human papilloma virus testing in cervical cancer prevention. J Midwifery Women's Health 49:4–13

Fletcher GH (1973) Female pelvis: squamous cell carcinoma of the uterine cervix. In: Fletcher GH (ed) Textbook of radiotherapy. Saunders, London, p 620

Gasinska A et al. (2002) Prognostic significance of intratumour microvessel density and haemoglobin level in carcinoma of the uterine cervix. Acta Oncologica 41:437–443

Gerbaulet AL et al. (1992) Combined radiotherapy and surgery: local control and complications in early carcinoma of the uterine cervix–the Villejuif experience, 1975–1984. Radiother Oncol 23:66–73

Ghosh K et al. (2001) Using a belly board device to reduce the small bowel volume within pelvic radiation fields in women with postoperatively treated cervical carcinoma. Gynecol Oncol 83:271–275

Greven K et al. (1999) Current developments in the treatment of newly diagnosed cervical cancer. Hematol Oncol Clin North Am 13:275–303

Grigsby PW et al. (2001) Long-term follow-up of RTOG 92-10: cervical cancer with positive para-aortic lymph nodes. Int J Radiat Oncol Biol Phys 51: 982–987

Grigsby PW et al. (2004) Posttherapy [18F] fluorodeoxyglucose positron emission tomography in carcinoma of the cervix: response and outcome. J Clin Oncol 22:2167–2171

Hanks GE, Herring DF, Kramer S (1983) Patterns of care outcome studies. Results of the national practice in cancer of the cervix. Cancer 51:959–967

Harima Y et al. (2002) Human papilloma virus (HPV) DNA associated with prognosis of cervical cancer after radiotherapy. Int J Radiat Oncol Biol Phys 52:1345–1351

Hernandez-Hernandez DM et al. (2003) Association between high-risk human papillomavirus DNA load and precursor lesions of cervical cancer in Mexican women. Gynecol Oncol 90:310–317

Herzog TJ (2003) New approaches for the management of cervical cancer. Gynecol Oncol 90:S22–S27

Horiot JC et al. (1988) Radiotherapy alone in carcinoma of the intact uterine cervix according to G.H. Fletcher guidelines: a French cooperative study of 1383 cases. Int J Radiat Oncol Biol Phys 14:605–611

Husseinzadeh N et al. (1984) The preservation of ovarian function in young women undergoing pelvic radiation therapy. Gynecol Oncol 18:373–379

ICRU (1985) Dose and volume specification for reporting intracavitary therapy in gynecology. ICRU Report 38:1

Jhingran A, Eifel PJ (2000) Perioperative and postoperative complications of intracavitary radiation for FIGO stage I–III carcinoma of the cervix. Int J Radiat Oncol Biol Phys 46:1177–1183

Karolewski K et al. (1999) Prognostic significance of pretherapeutic and therapeutic factors in patients with advanced cancer of the uterine cervix treated with radical radiotherapy alone. Acta Oncologica 38:461–468

Katz A, Eifel PJ (2000) Quantification of intracavitary brachytherapy parameters and correlation with outcome in patients with carcinoma of the cervix (see comment). Int J Radiat Oncol Biol Phys 48:1417–1425

Keys HM et al. (1999) Cisplatin, radiation, and adjuvant hysterectomy compared with radiation and adjuvant hysterectomy for bulky stage IB cervical carcinoma. N Engl J Med 340:1154–1161

Kim RY et al. (1989) Radiation alone in the treatment of cancer of the uterine cervix: analysis of pelvic failure and dose response relationship. Int J Radiat Oncol Biol Phys 17:973–978

Kramer C et al. (1989) Radiation treatment of FIGO stage IVA carcinoma of the cervix. Gynecol Oncol 32:323–326

Kucera H et al. (1986) Prognosis of primary radiotherapy of cervix cancer in younger females (in German). Geburtshilfe Frauenheilkd 46:800–803

Kucera H et al. (1987) The influence of nicotine abuse and diabetes mellitus on the results of primary irradiation in the treatment of carcinoma of the cervix. Cancer 60:1–4

Kutcher GJ et al. (1994) Comprehensive QA for radiation oncology: report of AAPM Radiation Therapy Committee Task Group 40. Med Phys 21:581–618

Landoni F et al. (1997) Randomised study of radical surgery versus radiotherapy for stage Ib-IIa cervical cancer (see comment). Lancet 350:535–540

Leibel S et al. (1987) Radiotherapy with or without misonidazole for patients with stage IIIB or stage IVA squamous cell carcinoma of the uterine cervix: preliminary report of a Radiation Therapy Oncology Group randomized trial. Int J Radiat Oncol Biol Phys 13:541–549

Lieu D (1996) The Papanicolaou smear: its value and limitations (see comment). J Fam Pract 42:391–399

Lukka H et al. (2002) Concurrent cisplatin-based chemotherapy plus radiotherapy for cervical cancer – a meta-analysis (see comment). Clin Oncol (Royal College of Radiologists) 14:203–212

Mashburn J, Scharbo-DeHaan M (1997) A clinician's guide to Pap smear interpretation (see comment). Nurse Practitioner 22:115–118

Mendenhall WM et al. (1984) Prognostic and treatment factors affecting pelvic control of Stage IB and IIA-B carcinoma of the intact uterine cervix treated with radiation therapy alone. Cancer 53:2649–2654

Montana GS, Fowler WC (1989) Carcinoma of the cervix: analysis of bladder and rectal radiation dose and complications. Int J Radiat Oncol Biol Phys 16:95–100

Montana GS et al. (1983) Carcinoma of the cervix stage IB: results of treatment with radiation therapy. Int J Radiat Oncol Biol Phys 9:45–49

Montana GS et al. (1985) Analysis of results of radiation therapy for stage II carcinoma of the cervix. Cancer 55:956–962

Montana GS et al. (1986) Carcinoma of the cervix, stage III. Results of radiation therapy. Cancer 57:148–154

Montana GS et al. (1987) Analysis of results of radiation therapy for stage IB carcinoma of the cervix. Cancer 60:2195–2200

Montana GS, Martz KL, Hanks GE (1991) Patterns and sites of failure in cervix cancer treated in the U.S.A. in 1978. Int J Radiat Oncol Biol Phys 20:87–93

Mutic S et al. (2003) PET-guided IMRT for cervical carcinoma with positive para-aortic lymph nodes – a dose-escalation treatment planning study. Int J Radiat Oncol Biol Phys 55:28–35

Nag S et al. (1999) The American brachytherapy society survey of brachytherapy practice for carcinoma of the cervix in the United States. Gynecol Oncol 73:111–118

Nag S et al. (2002) The American Brachytherapy Society recommendations for low-dose-rate brachytherapy for carcinoma of the cervix (see comment). Int J Radiat Oncol Biol Phys 52:33–48 (erratum 52:1157)

Orton CG (1998) High and low dose-rate brachytherapy for cervical carcinoma. Acta Oncol 37:117–125

Perez CA et al. (1984) Radiation therapy alone in the treatment of carcinoma of the uterine cervix. II. Analysis of complications. Cancer 54:235–246

Perez CA et al. (1998) Tumor size, irradiation dose, and long-term outcome of carcinoma of uterine cervix. Int J Radiat Oncol Biol Phys 41:307–317

Perez CA et al. (1999) Radiation therapy morbidity in carcinoma of the uterine cervix: dosimetric and clinical correlation. Int J Radiat Oncol Biol Phys 44:855–866

Perlman SE (1999) Pap smears: screening, interpretation, treatment. Adolescent Medicine State of the Art Reviews 10:243–254

Peters WA, 3rd, et al. (2000) Concurrent chemotherapy and pelvic radiation therapy compared with pelvic radiation therapy alone as adjuvant therapy after radical surgery in high-risk early-stage cancer of the cervix. J Clin Oncol 18:1606–1613

Potish RA (1992) Cervix cancer. In: Lewitt SH (ed) Technological basis of radiation therapy. Lea & Febiger, Philadelphia, p 289

Potish RA, Gerbi BJ (1986) Role of point A in the era of computerized dosimetry. Radiology 158:827–831

Potish RA, Twiggs LB (1989) An analysis of adjuvant treatment strategies for carcinoma of the cervix. Am J Clin Oncol 12:430–433

Potish RA, Deibel FC Jr., Khan FM (1982) The relationship between milligram-hours and dose to point A in carcinoma of the cervix. Radiology 145:479–483

Potish RA et al. (1984) The utility and limitations of decision theory in the utilization of surgical staging and extended field radiotherapy in cervical cancer. Obstet Gynecol Surv 39:555–562

Potish RA et al. (1989) The role of surgical debulking in cancer of the uterine cervix. Int J Radiat Oncol Biol Phys 17:979–984

Pourquier H et al. (1987) A quantified approach to the analysis and prevention of urinary complications in radiotherapeutic treatment of cancer of the cervix. Int J Radiat Oncol Biol Phys 13:1025–1033

Rose PG et al. (1999) Concurrent cisplatin-based radiotherapy and chemotherapy for locally advanced cervical cancer. N Engl J Med 340:1144–1153

Sasaoka M et al. (2001) Patterns of failure in carcinoma of the uterine cervix treated with definitive radiotherapy alone. Am J Clin Oncol 24:586–590

Sedlis A et al. (1999) A randomized trial of pelvic radiation therapy versus no further therapy in selected patients with stage IB carcinoma of the cervix after radical hysterectomy and pelvic lymphadenectomy: A Gynecologic Oncology Group Study (see comment). Gynecol Oncol 73:177–183

Singh AK et al. (2003) FDG-PET lymph node staging and survival of patients with FIGO stage IIIb cervical carcinoma. Int J Radiat Oncol Biol Phys 56:489–493

Sommers GM et al. (1989) Outcome of recurrent cervical carcinoma following definitive irradiation. Gynecol Oncol 35:150–155

Thoms WW Jr., et al. (1992) Bulky endocervical carcinoma: a 23-year experience. Int J Radiat Oncol Biol Phys 23:491–499

Tsai CS et al. (2004) Preliminary report of using FDG-PET to detect extrapelvic lesions in cervical cancer patients with enlarged pelvic lymph nodes on MRI/CT. Int J Radiat Oncol Biol Phys 58:1506–1512

Unal A et al. (1981) An analysis of the severe complications of irradiation of carcinoma of the uterine cervix: treatment with intracavitary radium and parametrial irradiation. Int J Radiat Oncol Biol Phys 7:999–1004

Waggoner SE (2003) Cervical cancer. Lancet 361:2217–2225

Whitney CW et al. (1999) Randomized comparison of fluorouracil plus cisplatin versus hydroxyurea as an adjunct to radiation therapy in stage IIB-IVA carcinoma of the cervix with negative para-aortic lymph nodes: a gynecologic Oncology Group and Southwest Oncology Group study. J Clin Oncol 17:1339–1348

Willen H et al. (1985) Invasive squamous cell carcinoma of the uterine cervix. VIII. Survival and malignancy grading in patients treated by irradiation in Lund 1969-1970. Acta Radiol Oncol 24:41–50

25 Technical Aspects of Radiation Therapy in Endometrial Carcinoma

Higinia R. Cardenes and Brent Tinnel

CONTENTS

25.1   Introduction 599
25.2   Anatomy 599
25.3   Epidemiology and Risk Factors 600
25.4   Clinical Presentation and Diagnostic Evaluation 600
25.5   Histopathological Classification 601
25.6   Clinical and Surgical Staging – FIGO
        Pathological Staging System 601
25.6.1 Clinico-Pathological Prognostic Factors 603
25.7   Adjuvant Therapy – Early Stage (FIGO I–II)
        Endometrial Adenocarcinoma 604
25.7.1 Randomized Trials in Early Stage (I and Occult II)
        Endometrial Cancer 604
25.7.2 Risk Groups in Early Stage Endometrial Cancer
        – Table 25.7 605
25.8   Adjuvant Therapy – Advanced Stage (FIGO III–IV)
        Endometrial Adenocarcinoma 607
25.9   RT Techniques 607
25.9.1 Preoperative Irradiation 607
25.9.1.1 Preoperative ICB 607
25.9.1.2 Preoperative EBRT 609
25.9.2 Postoperative Irradiation 610
25.9.2.1 Postoperative ICB 610
25.9.2.2 Postoperative EBRT Techniques 613
25.10   Medically Inoperable Early Stage
        Endometrial Cancer 615
25.11   Recurrent Endometrial Cancer 620
25.11.1 External Beam RT 620
25.11.2 Brachytherapy in
        Recurrent Endometrial Cancer 621
25.11.3 Salvage Surgery for
        Recurrent Endometrial Cancer 622
25.12   Palliative Therapy in Endometrial Cancer 622
25.12.1 Palliative Surgery 622
25.12.2 Palliative RT 624
25.12.3 Palliative Systemic Chemotherapy 624
25.13   Uterine Papillary Serous and
        Clear Cell Carcinoma 624
25.13.1 Intraperitoneal Radioactive Chromic Phosphate
        Suspension (^{32}P) Administration 626
        References 628

H. R. Cardenes, MD, PhD
Clinical Associate Professor, Indiana University School of Medicine, Department of Radiation Oncology, 535 Barnhill Drive, RT 041, Indianapolis, IN 46202, USA
B. Tinnel, MD
Indiana University School of Medicine, Department of Radiation Oncology, 535 Barnhill Drive, RT 041, Indianapolis, IN 46202, USA

25.1 Introduction

Endometrial cancer (EC) is the most common gynecological malignancy in the United States and is expected to remain so for some time given current demographic trends of an aging population and an increased incidence of obesity. The National Cancer Institute Surveillance, Epidemiology and End Results (SEER) Program estimated there would be 40,880 new cases of the uterine corpus diagnosed in 2005, with an estimated 7,310 deaths (American Cancer Society 2005). Currently, EC is the fourth most common cancer in females, ranking behind breast, bowel and lung cancers. In about 75% of the cases, the disease is confined to the uterus and cervix at the time of diagnosis, and uncorrected survival rates of 75% or greater are expected (FIGO 1989).

In the last 20 years, the treatment of EC has evolved from almost routine use of preoperative radiotherapy (RT) – generally intracavitary brachytherapy (ICB) – followed by hysterectomy to upfront surgical staging followed by tailored adjuvant therapy based on histopathological findings, as recommended by the International Federation of Gynecology and Obstetrics (FIGO) (FIGO 1989).

The purpose of this chapter is to discuss the potential role and technical aspects of external beam RT (EBRT) and brachytherapy in the management of EC. We will briefly review the rationale for patient selection for treatment with these techniques in the adjuvant setting, as well as definitive therapy for non-surgical candidate patients and the role of RT for recurrent disease.

25.2 Anatomy

The uterus, located in the pelvis between the rectum and the bladder, is divided into the body (corpus) and the cervix, separated by the isthmus. The uterus is attached to the pelvis primarily by the broad (lateral) and round (antero-lateral) ligaments; in addition, the utero-sacral ligaments at the lower uterine segment and the cardinal ligaments at the upper-lateral margin of the cervix contribute to supporting the uterus.

The main artery supplying the uterus is the uterine artery, a branch of the hypogastric artery. The uterus has a rich lymphatic network; the lower and mid-third of which drain laterally along the parametrium into the paracervical lymph nodes and from here to the external iliac nodes (obturator nodes are the innermost component) and hypogastric nodes; subsequently, the pelvic lymphatics drain into the common iliac and peri-aortic lymph nodes. However, the lymphatics from the upper corpus and fundus pass laterally across the broad ligaments continuous with those of the ovary directly into the peri-aortic and upper abdominal lymph nodes. Finally, there are lymphatic channels that drain along the round ligaments to the femoral nodes. The anatomic distribution of the lymphatics represents the basis for the RT delivery.

25.3
Epidemiology and Risk Factors

EC is primarily a disease of postmenopausal women, with a peak incidence at 50–70 years of age. Approximately, 25% of cases occur in premenopausal patients; however, only 5% occur in patients younger than 40 years of age. In general, the incidence is highest in Western countries, lower in Eastern Europe, and lowest in South Asia and India.

More specifically, there appear to be two different forms of EC showing distinct biological and clinical behaviors, while stemming from different etiologies. The most common form, type I, is estrogen related and likely accounts for regional, anthropometric and menstrual associated differences. This type is associated with endometrial hyperplasia and typically presents as a lower grade endometrioid tumor (BOKHMAN 1983). These changes result from excess estrogen and can be from exogenous exposure, such as estrogen replacement therapy or tamoxifen, or endogenous, such as obesity, anovulatory cycles, early menarche or estrogen-secreting tumors.

Type-II EC appears to be unrelated to estrogen or endometrial hyperplasia and tends to present with higher grade tumors or poor prognostic cell types, such as uterine papillary serous carcinomas (UPSCs)

or clear cell carcinomas (CCCs). These patients are more often multiparous, older and less likely to have large body habitus (SLOMOVITZ et al. 2003). The risk factors for these patients are not well defined.

Familial or genetic predisposition has been associated with increased risk of EC in the setting of the Lynch syndrome II, also known as hereditary non-polyposis colorectal cancer syndrome (HNPCC), in which afflicted females may have up to 43% risk of EC by age 70 years (AARNIO et al. 1995).

25.4
Clinical Presentation and Diagnostic Evaluation

The most common and classically defined presentation of EC is abnormal uterine bleeding, and the potential of carcinoma deserves exclusion in any women with postmenopausal bleeding unrelated to hormonal therapy. Patients may also present with vaginal discharge, pelvic and lower back pain. In more advanced disease, there may be urinary or rectal bleeding, frequency, or urgency. Peritoneal spread may present as abdominal distention and/or ascites. Metastatic spread to the lung can cause cough or hemoptysis.

In order to facilitate early diagnosis, clinical suspicion must be high in postmenopausal bleeding, or pre- and perimenopausal women with menometrorrhagia, particularly if they have other risk factors of anovulation. Although the Papanicolaou (Pap) smear is not a reliable screening test for EC, the presence of endometrial cells or atypical glandular cell changes can signify endometrial disease. In the presence of abnormal bleeding or suspicious changes on Pap smear, the diagnosis is most easily made by office sampling using an endometrial, or Pipelle, biopsy, as this does not require anesthesia and is generally, well tolerated. Hysteroscopy with fractional dilation and curettage (D&C) remains the gold standard when a Pipelle examination is non-diagnostic and suspicion remains high. However, the potential for under grading by either of these methods can be significant, with up to 30% of cases having higher grade at the time of hysterectomy (LARSON et al. 1995).

There is no recognized screening program for the general population. Endometrial biopsy is relatively uncomfortable and equivocal tests may lead to additional unnecessary evaluation. However, this approach has been recommended by the American Cancer Society for women carrying the HNPCC mutation, or those related to carriers (SMITH et al. 2001). Transvaginal ultrasonography is useful in women who experience bleeding, as an abnormal endometrial thickness is associated with gradually increasing risk of neoplasia, but it is relatively expensive and the positive predictive value is too low for a screening tool.

25.5
Histopathological Classification

Amongst the histopathological subtypes of EC, the most prevalent types can be grouped together as endometrioid adenocarcinomas (EACs), comprising 75–80%. They can be further subdivided and described as shown in Table 25.1 (SILVERBERG and KURMAN 1992). The differentiation of a carcinoma

Table 25.1. International Society of Gynecologic Pathologists (ISGP) and the World Health Organization (WHO) classification of uterine tumors (SILVERBERG and KURMAN 1992)

Pathological subtypes	Incidence
Endometrioid adenocarcinoma	75–80%
Papillary villoglandular	
Secretory adenocarcinoma	
Ciliated carcinoma	
Adenocarcinoma with squamous:	
Differentiation	
Adenoacanthoma	
Adenosquamous carcinoma	
Uterine papillary serous	10%
Clear cell carcinoma	4%
Mucinous carcinoma	1%
Squamous cell carcinoma	<1%
Mixed cell type	~10%
Undifferentiated carcinoma	<1%
Glassy-Cell Carcinoma	
Metastatic Carcinoma to the Endometrium	~1%

is expressed as its grade. The FIGO grading system is shown in Table 25.2. Well-differentiated EAC is characterized microscopically by a proliferation of back-to-back endometrial glands without intervening stroma, but there is no more than 5% solid growth. Moderately differentiated EAC has between 6% and 50% solid tumor without glands. Poorly differentiated EAC contains more than 50% solid component. In addition, nuclear atypia increases the grade of the lesion by 1.

Many non-EAC histopathological subtypes behave more aggressively and are shown in Table 25.1. The most common of these poorer prognosis types are UPSC and CCC. UPSC accounts for 5–10% of EC and is hallmarked by a papillary architecture rich with high-grade, pleomorphic cells containing hyperchromatic nuclei with prominent nucleoli and supported by a fibroblastic stroma, not different from that seen in the ovarian counterpart. The aggressiveness of this tumor is manifested by its propensity for early spread to the abdominal peritoneum. CCC occurs in 1–5% of ECs. The appearance can be described as cells with clear cytoplasm that demonstrate a glandular, papillary or solid pattern of growth, and like the ovarian counterpart can be composed of cysts lined by flattered epithelium. Like UPSC, this type of EC shows a propensity for distant spread and failure, with series showing 75% of failures outside of the pelvis, most commonly to the upper abdomen, lungs and liver.

25.6
Clinical and Surgical Staging – FIGO Pathological Staging System

Once the diagnosis of EC has been established by endometrial biopsy or curettage, the patient should

Table 25.2. Histopathology–degree of differentiation. FIGO and ISGP-WHO definitions

Histological grade	Definition
G1: Well differentiated	5% or less of a nonsquamous or nonmorular solid growth pattern
G2: Moderately differentiated	6–50% of a nonsquamous or nonmorular solid growth pattern
G3: Poorly differentiated	More than 50% of a nonsquamous or nonmorular solid growth pattern

Notes on pathological grading: Notable nuclear atypia, inappropriate for the architectural grade, raises the grade of grade 1 or grade 2 tumor by 1. In serous adenocarcinomas, clear-cell adenocarcinomas and squamous cell carcinomas, nuclear grading takes precedence. Adenocarcinomas with benign squamous differentiation are graded according to the nuclear grade of the glandular component.
Rules related to staging: Because corpus cancer is now staged surgically, procedures previously used for determination of stages are no longer applicable (e.g., findings from fractional D&C) to differentiate between stage I and stage II). It is appreciated that there may be a small number of patients with corpus cancer who will be treated primarily with radiation therapy. If that is the case, the clinical staging adopted by FIGO in 1971 still would apply, but designation of that staging system would be noted; ideally width of the myometrium should be measured along with the width of tumor invasion

undergo a preoperative evaluation that should include a thorough history and physical examination, in particular in the presence of medical comorbidities, primarily intended to evaluate the surgical fitness. Routine laboratory studies and a chest X-rays [posteroanterior (PA) and lateral] are recommended. The benefit of more extensive work-up in patients with uterine-limited disease is questionable since most patients will be offered surgical therapy and, therefore, computerized tomography (CT) of the abdomen and pelvis is not indicated in the absence of symptoms or physical findings suspicious of extrauterine disease. Although trans-vaginal ultrasonography and magnetic resonance imaging (MRI) have been used to assess the depth of myometrial invasion, this can only be accurately determined by histological examination of the hysterectomy specimen. Symptoms or exam findings suspicious of advanced disease or local organ invasion can be further evaluated with colonoscopy or cystoscopy. Serum levels of CA-125 are elevated in

most patients with advanced or metastatic EC. Pre-operative levels greater than 40 U/ml can be considered an indication for full pelvic and peri-aortic lymphadenectomy at the time of surgical staging (HSIEH et al. 2002).

Assuming the patient is a surgical candidate and there is no vaginal or bulky cervical involvement, initial surgical management/staging is the recommended approach. The surgical staging as per FIGO criteria is highlighted in Table 25.3. The result of surgical staging is the categorization of the stage of the disease as per the FIGO 1988 update (Table 25.4). The prognostic utility of surgical–pathological stage has been confirmed by multiple studies, being the single strongest predictor of outcome in patients with EC. For those patients who are not surgical candidates, the clinical staging system adopted by FIGO in 1971 (Table 25.5) would still apply. Retrospective evaluation of both staging systems have demonstrated the superiority of the surgical staging system in predicting outcome and allowing more accurate assessment of extent of disease and individualization of adjuvant therapy. Approximately 15–20% of the patients will be upstaged because of a complete surgical staging.

Subsequent to the FIGO-endorsed change which led to wider, but not universal, adoption of surgical staging, MORROW et al. (1991) reported that, of 895 surgically staged patients, only 48 were found to have positive para-aortic nodes, and 47 of these 48 were shown to have either para-aortic nodes suspicious of disease, grossly positive pelvic nodes, involved adnexa or deep myometrial invasion. It was concluded that "it is logical to limit the surgical evaluation of aortic nodes to those patients with sus-

Table 25.3. FIGO surgical staging

Adequate abdominal incision (usually vertical)
Peritoneal/pelvic washings for cytology
Surgical exploration of all peritoneal surfaces with biopsies and/or excision of any suspicious lesion
Total extrafascial abdominal hysterectomy and bilateral salpingo-oophorectomy
The uterus should be bivalved in the operating room for gross assessment of myometrial infiltration
Omental biopsies (omentectomy if papillary serous carcinoma or clear cell carcinoma)
Pelvic and peri-aortic lymph node selective sampling versus dissection in all patients except those with stage IA–B grade-1 disease

Table 25.4. Endometrial cancer surgical staging system, FIGO (1988)

Stages/grades	Definition
Stage I	Tumor limited to the uterus
IA G1,2,3	Tumor limited to the endometrium
IB G1,2,3	Invasion to <50% of the myometrium
IC G1,2,3	Invasion to ≥50% of the myometrium
Stage II	Extension to the cervix but not beyond the uterus
IIA G1,2,3	Endocervical glandular involvement only
IIB G1,2,3	Cervical stromal invasion
Stage III	Extension outside of the uterus/cervix +/- regional metastasis
IIIA G1,2,3	Tumor invades serosa or adnexae or positive peritoneal cytology
IIIB G1,2,3	Vaginal metastasis
IIIC G1,2,3	Metastasis to pelvic and/or peri-aortic lymph nodes
Stage IV	
IVA G1,2,3	Tumor invades bladder and/or bowel mucosa
IVB G1,2,3	Distant metastasis including intra-abdominal and/or inguinal lymph nodes

Table 25.5. Endometrial cancer clinical staging system, FIGO (1971)

Stages /grades	Definition
Stage I	Tumor confined to the uterus
IA	Length of the uterine cavity ≤8 cm
IB	Length of the uterine cavity >8 cm
Histological subtypes of adenocarcinoma	
G1	Highly differentiated adenomatous carcinoma
G2	Differentiated adenomatous carcinoma with partly solid areas
G3	Predominantly solid or entirely undifferentiated carcinoma
Stage II	Extension to the cervix but not beyond the uterus
Stage III	Extension outside of the uterus/cervix but not outside of the true pelvis
Stage IV	Extension outside of the true pelvis or involvement of the bladder and/or rectum. Tumor invades bladder and/or bowel mucosa. Distant metastasis including intra-abdominal and/or inguinal lymph nodes

pect aortic nodes on palpation or high risk factors such as grossly positive pelvic nodes, gross adnexal masses or outer-third myometrial invasion". Such high risk factors were seen in only 25% of patients but accounted for 98% of those with positive para-aortic nodes (MORROW et al. 1991). In the previous GOG 33 study (CREASMAN et al. 1987), of 621 clinical stage FIGO I and occult stage-II patients, 11% presented with metastasis to pelvic and peri-aortic nodes. The main prognostic factor related to the presence of positive nodes was the depth of myometrial invasion. Regardless of grade, only 1% of patients with endometrial involvement was found to have metastasis to either pelvic or peri-aortic nodes; however, the relative frequency of pelvic or peri-aortic nodal involvement increased to 23% and 17%, respectively, for deep myometrial invasion (CREASMAN et al. 1987).

The Society of Gynecologic Oncology (SGO) Practice Guidelines recommend extensive surgical staging (ESS) for those with high-risk histologies (e.g., clear cell, papillary serous), high-grade endometrioid lesions (grades 2–3), deep myometrial invasion (> 50%), clinical evident extrauterine disease, suspect nodes or cervical involvement (CHEN et al. 1999).

There is currently significant controversy regarding the appropriate extent of the surgical staging as well as its potential therapeutic benefit. A more extensive discussion of this topic is out of the scope of this chapter. It is also important to realize that a significant proportion of patients with EC will not be candidates for ESS because of large body habitus and/or medical co-morbidities. Alternatively, they may be operated on by non-gynecology oncologists in community-hospital-type settings, where expertise in lymph-node dissections is potentially unavailable.

According to Naumann's survey of the SGO membership, wherein 65% of the responders stated they believed lymphadenectomy to be therapeutic, only 45% of surgeons carry out complete lymph node dissection routinely, 19% do not sample obturator nodes and 30% do not routinely sample para-aortic nodes (NAUMANN et al. 1999).

25.6.1
Clinico-Pathological Prognostic Factors

Prognostic factors identified in EC include (1) patient-related factors such as age and medical co-morbidities and (2) tumor-related factors, in particular, pathological stage, depth of myometrial invasion, histological grade and ploidy, cell type (endometrioid versus non-endometrioid tumors, i.e., UPSC and CCC) and presence of lymphovascular space invasion (LVSI). The pathological stage is the most significant predictor of outcome.

A significant minority, up to 20–25%, of clinical stage-I EC patients have extrauterine disease (CREASMAN et al. 1987). These patients are considered at high risk for local recurrence and distant spread and should be evaluated for both adjuvant local and systemic therapy. This was demonstrated by the GOG, by showing a relationship between histological grade, depth of myometrial invasion and presence of positive nodes (CREASMAN et al. 1987). In addition, histological grade is an independent prognostic factor for tumor recurrence, whereas the presence of positive nodes, adnexal metastasis and deep myometrial invasion are correlated with the presence of positive peri-aortic nodes (MORROW et al. 1991). ZAINO et al. (1996) analyzing more than 1000 patients entered on a GOG protocol developed two models of survival for patients with clinical and pathological stage I–II EC. For clinical stage I–II tumors, the cell type, histological grade, depth of myometrial invasion, peritoneal cytology, age and

LVSI were all found to be independent prognostic factors for survival. However, in patients with pathological stages I–II, the only two significant independent prognostic factors found for survival were age and depth of myometrial invasion (ZAINO et al. 1996).

25.7
Adjuvant Therapy – Early Stage (FIGO I–II) Endometrial Adenocarcinoma

Given its early presentation and perceived excellent results for patients with disease confined to the uterine corpus/cervix, the therapy rendered to patients has been highly variable, dependent on the expertise and preferred approaches of the local providers. The two current therapeutic paradigms are: (1) total abdominal hysterectomy and bilateral salpingo-oophorectomy (TAH-BSO), without ESS, followed by a more liberal use of postoperative RT based on histopathological factors noted in the uterine specimen and (2) TAH-BSO with ESS followed by a more restricted use of postoperative RT. A more detailed discussion regarding the current controversy is beyond the scope of this chapter.

25.7.1
Randomized Trials in Early Stage (I and Occult II) Endometrial Cancer

AALDERS et al. randomized 540 patients with EC of stages IB–IC undergoing TAH-BSO and complete surgical staging to receive adjuvant vaginal brachytherapy, after which the patients were randomized to observation versus EBRT. A significant reduction on local recurrence rates was observed with the addition of pelvic RT (1.9% versus 6.9%, respectively). There was no difference in survival between the two groups, although in the group of patients with IC grade-3 disease there was a cause-specific survival advantage with the addition of pelvic RT (AALDERS et al. 1980).

The Post-Operative Therapy in Endometrial Carcinoma (PORTEC) trial (which did not require lymph node sampling) randomized 715 patients with grade 1, 50% or more myometrial invasion; grade 2, any invasion; and grade 3, less than 50% myometrial invasion to receive adjuvant pelvic RT or no further treatment (NFT) (CREUTZBERG et al. 2000). PORTEC investigators felt that patients with more than 50%

invasion and grade-3 disease represent a "higher risk" subgroup presumed to benefit from pelvic RT, and these patients were thus excluded from the trial.

The Gynecologic Oncology Group (GOG-99) conducted a similar trial in 448 patients with "intermediate-risk" EC, defined as stages IB–IC and occult II disease, after complete surgical staging. As per protocol design, this was considered satisfactory, even if only one lymph was contained in the surgical specimen. The investigators subsequently defined a "high-intermediate-risk" group (HIRG), 132 patients (about one-third of the study population), based on the presence of grade 2–3 disease, presence of LVSI and/or deep myometrial involvement. Patients younger than 50 years old with the three risk factors, 51–69 years old with two factors or older than 70 years old with any of those risk factors were included in the HIRG and analyzed separately. Age was found an independent prognostic factor for survival.

Both of these trials demonstrated the ability of adjuvant pelvic RT to decrease pelvic and vaginal recurrences. Eight-year actuarial local–regional relapse (LRR) was 15% in the NFT versus 4% in the RT patients in the PORTEC trial (CREUTZBERG et al. 2003). In the GOG-99, the 5-year LRR values were 8% (NFT) and 2% (RT), with the larger difference being in the HIRG with a LRR of 5% in the RT arm and 13% in the observation arm (KEYS et al. 2004). In both trials, the majority of the recurrences in the observation arm (up to 70% in the PORTEC trial and around 72% in the GOG-99 trial) were in the vagina. The success of salvage therapy for vaginal relapses in the PORTEC study was reported by Creutzberg in 2003 with a 61% durable local control and an overall 49% salvage rate (some patients failed distantly despite local control) (CREUTZBERG et al. 2003). Overall survival, however, was not statistically improved with the addition of pelvic RT. In the PORTEC trial, 8-year actuarial survivals were 71% for the RT and 77% for the NFT arms. Similarly, in the GOG-99 trial, the estimated 4-year survival was 92% in the RT arm and 86% in the observation arm. The high-intermediate risk group accounted for nearly two-thirds of the recurrences and cancer-related deaths. However, disease-free survival (DFS) favored the RT arm in the GOG-99 trial. A comparison of the 5-year cumulative incidence of recurrence/death for the PORTEC and GOG-99 trials is shown in Table 25.6.

Adjuvant pelvic RT is associated with toxicity, which must be balanced against the benefit derived from its use. Most of the toxicity will be primarily

Table 25.6. GOG-99 and PORTEC trials. 5-year cumulative incidence of recurrence and death. *HR* high–intermediate risk group, *LR* low risk group, *RT* radiotherapy, *DM* distant metastasis

	PORTEC	GOG 99, ALL	GOG 99, HR	GOG 99, LR
LRR – RT	4.2%	2%	5%	0%
LRR – no RT	13.7%	8%	13%	5%
DM – RT	7.9%	6%	10%	4%
DM – no RT	7%	8%	19%	9%
Any – RT	9.4%	7%	13%	4%
Any – no RT	17.2%	14%	27%	7%
Death – RT	19.3%	9%	13%	7%
Death – no RT	14.9%	15%	27%	8%

gastrointestinal, genitourinary and hematological. The majority of acute toxicity is self-limited, and treatment interruptions are relatively rare. The PORTEC investigators found that pelvic RT was associated with a higher risk of grade 1–2 late gastrointestinal (17% versus 1%) and genitourinary (8% versus 4%) toxicity. Grade 3–4 toxicity was rare in both arms, but it all occurred in the RT arm (3% of patients) (CREUTZBERG et al. 2000). The GOG study (KEYS et al. 2004) reported results similar to the PORTEC, with statistically significant differences in gastrointestinal, genitourinary, hematological and cutaneous toxicities between the treatment arms. As in the PORTEC study, the majority of recorded toxicities were grades 1–2 in the RT group. Surgical staging did change the toxicity profile in the GOG study; patients in both the surgery and RT arms were noted to have lymphatic complications (primarily chronic lymphedema, occurring in 2.5% of the control patients and 5% of the RT patients). This complication was not noted in patients from the PORTEC study, where lymph node dissection was not utilized.

In view of the results from the randomized trials, it seems that the use of postoperative pelvic RT should be limited to those patients with sufficiently high risk of LRR (≥15%). Therefore, it is imperative to define risk groups based on the known histopathological risk factors (stage, histological grade, depth of myometrial infiltration, presence of LVSI, cervical stroma involvement) in addition to the extent of the surgical staging (were the lymph nodes sampled or was a lymph node dissection performed?) and adequate knowledge regarding patterns of failure (vagina only, regional nodal or pelvic recurrence or systemic failures) in order to be able to tailor adjuvant therapy and potentially minimize morbidity (MCCORMICK et al. 2002).

25.7.2
Risk Groups in Early Stage Endometrial Cancer – Table 25.7

1. *Low-Risk Group*
   a. Stage-IA disease, grades 1–2, without evidence of LVSI, has a recurrence rate of less than 10% (MORROW et al. 1991). The risk of positive pelvic nodes for this group is also very low, less than 5% (CREASMAN et al. 1987). These patients can be managed with surgery alone, even without ESS (MARIANI et al. 2000) without adjuvant therapy.
2. *Intermediate-Low Risk Group*
   a. Stage IA, grade 3, represents a small subgroup (only 8 patients in GOG 33) for which potential options in the adjuvant setting, with or without surgical staging, may include observation versus vaginal brachytherapy alone.
   b. Stage IB and probably stage IIA (with less than 50% myometrial invasion), grades 1–2, have excellent outcome with recurrence rates between 2% and 4% after ESS (STRAUGHN et al. 2002) or with adjuvant vaginal brachytherapy without routine lymph node staging (ALEKTIAR et al. 2002).
      I. In the absence of LVSI, patients with stage-IB, grade-1 disease could be observed, whereas patients with IB grade-2 and patients with IIA disease, grades 1–2, could be offered vaginal brachytherapy alone (mainly in older women) versus observation (after ESS).

Table 25.7. Prognostic groups in early stage endometrial cancer

Low risk	Intermediate–low risk	Intermediate–high risk	High risk
Stage IA, G1	Stage IA, G3 Negative LVSI	Stage IB, G3 Stage IC, any grade Stage IIA, grade 2–3, >50% MI +LVSI	Stage IIB, any grade
Stage IA, G2	Stage IB, G1 Stage IB, G2, negative LVSI	1/3 above w/age ≥70 2/3 above w/age ≥50 3/3 above w/age <50	Papillary or clear cell carcinoma
	Stage IIA, grade 1, <50% MI, negative LVSI		Stage III–IV

3. *Intermediate-High Risk Group*
   a. Stage-IB, grade-3 patients represent a subgroup for which there is limited data. In the PORTEC study (non-surgically staged patients) there was a 14% 5-year loco-regional failure in 37 patients with these characteristics, included in the non-RT arm, with all the local failures occurring in the vagina. Potential options after surgery include:
      I. In non-surgically staged patients:
         1. Absence of LVSI, vaginal brachytherapy alone
         2. Presence of LVSI, EBRT (patients older than 50 years) or vaginal brachytherapy (younger patients)
      II. In surgically staged patients, vaginal brachytherapy alone since these patients seem to be at similar risk of local and distant failures (STRAUGHN et al. 2002). Potential trials involving the use of vaginal brachytherapy and chemotherapy are needed.
   b. Stage IC have an estimated incidence of positive pelvic nodes of 26% and peri-aortic nodes of 17% (CREASMAN et al. 1987), stressing the importance of adequate ESS in this patient population. The recurrence rate for this group is around 15–20% (MORROW et al. 1991).
      I. In non-surgically staged patients, EBRT was shown to improve the loco-regional control rate without a significant impact on overall survival in the PORTEC trial (CREUTZBERG et al. 2000, 2003). However, only patients with stage IC grades 1–2 were included, and the LRR rate was 13% (77% vaginal only), being less than 5% in patients younger than 60 years old. Therefore:
         1. Patients with stage IC, grades 1–2, older than 50–60 years old (cut out for age as a risk factor variable between studies) and without LVSI should be considered for at least vaginal brachytherapy.
         2. Ongoing PORTEC study 2 randomizes patients with stage IB grade 3, stage IC grades 1–2, and stage IIA grades 1–2 or grade 3 (less than 50% myometrial invasion), of any age, to receive pelvic EBRT versus vaginal brachytherapy alone.
         3. Ongoing National Cancer Institute of Canada (NCIC) randomizes patients with grade-3, any depth, myometrial invasion and grade-2 lesions with more than 50% myometrial invasion to RT (EBRT +/- brachytherapy) versus observation.

Similarly, the Medical Research Council Trial, ASTEC trial, is evaluating the role of ESS and adjuvant RT in patients with stages IC–IIA or grade-3 disease, including UPSC and CCC histologies.

   II. In surgically staged patients, GOG-99 (KEYS et al. 2004) defined a high-intermediate risk group as those patients with stages IB, IC and occult II (a) younger than 50 years old with grades 2–3, presence of LVSI and deep myometrial infiltration, (b) older than 50 years old with two prognostic factors or (c) older than 70 years old with one prognostic factor. The 2-year recurrence rate for this high-risk group was 6% for those patients receiving adjuvant EBRT and 26% for those in the NFT arm, respectively. There was no difference in survival between each group. Again, the majority of recurrences took place in the vagina. The incidence of distant metastasis was 10% for the RT group and 19% for the NFT group (KEYS et al. 2004). Therefore, potential options may include:
      1. Vaginal brachytherapy alone since the majority of recurrences are in the vagina, if the patient has been adequately staged. Retrospective data seem to indicate that these patients do well with vaginal treatment only (ALEKTIAR et al. 2005), although most of the series include a limited number of patients with stage-IC and grade-3 disease.
      2. EBRT based on the results of GOG. Although up to date, EBRT versus vaginal brachytherapy alone have not been compared in a prospective randomized study.
   III. Similar recommendations could probably be made for patients with stage IIA (over 50% myometrial infiltration) grades 1–2.
4. *High-Risk Group*
   a. Stage-IIA (more than 50% myometrial invasion) grade-3 disease and stage-IIB disease, any grade, should be offered EBRT in addition to vaginal brachytherapy. Unfortunately, the grade-3, stage-IIB groups of any grade represented less than 10% of patients included in GOG 99. These patients may benefit from EBRT+/-vaginal brachytherapy, primarily in younger patients with evidence of LVSI.
   b. Patients with unfavorable histologies, UPSC and CCC, should also be considered high-risk patients requiring adjuvant therapy. These will be discussed later in the chapter.

25.8
Adjuvant Therapy – Advanced Stage (FIGO III–IV) Endometrial Adenocarcinoma

Patients with advanced EC represent a very heterogeneous group, with survival rates ranging from about 10% for patients with stage IVB to over 90% for those patients with stage IIIA, with positive cytology only (MARIANI et al. 2002a). In addition, within stage-III disease, it is possible to distinguish a "favorable" group that includes those patients with isolated adnexal involvement (CONNELL et al. 1999) or pelvic nodes positive with negative retroperitoneal sampling/dissection (NELSON et al. 1999). These patients generally do well with adjuvant pelvic RT or extended field RT (EFRT), with survival rates between 65% and 85%. However, patients with uterine serosal involvement do have a worse prognosis with a high rate of distant relapse (around 50%) and a 5-year DFS of less than 50% in some series (ASHMAN et al. 2001).

Patients with stage-IIIC disease also represent a heterogeneous group with very different outcome depending on the presence of isolated pelvic nodes without any other significant risk factors, positive periaortic nodes or presence of multiple sites of extrauterine disease. Patients with positive nodes only without any other site of extrauterine disease have 5-year DFS values between 55% and 70%, whereas in the presence of other extrauterine disease the survival drops dramatically to around 30% (MARIANI et al. 2002b).

Patients with advanced EC have been treated using different approaches in the adjuvant setting including pelvic RT, EFRT (pelvic + peri-aortic RT), whole abdominal irradiation (WAI) and chemotherapy. A more extensive discussion of the outcome with these modalities is beyond the scope of this chapter. It is important to mention that, currently, treatment selection for patients with advanced disease is strongly influenced by the recent reporting of the GOG-122 clinical trial, the only randomized trial conducted in this patient population (RANDALL et al. 2003). This trial randomized patients with stage-III to -IV disease (without evidence of distant metastasis) after surgical staging and optimal surgical debulking to receive WAI versus adjuvant chemotherapy (doxorubicin and cisplatin), demonstrating the superiority of chemotherapy over WAI in terms of progression-free survival and overall survival. Unfortunately, the recurrence rates were quite high, indicating the need for a better approach. The clinical trial GOG-184 investigating the role of "tailored radiotherapy" – this is pelvic RT only or EFRT depending upon the extent of nodal involvement – followed by chemotherapy (randomization between two different chemotherapy regimens, doxorubicin + cisplatin versus cisplatin + doxorubicin + taxol) has been completed, and data analysis is pending.

25.9
RT Techniques

There are two types of RT that can be used either alone or in combination in the management of EC. These are ICB and EBRT. Over the last four decades, there has been a shift from preoperative RT (EBRT+/- ICB) to postoperative EBRT+/-intravaginal brachytherapy (AALDERS et al. 1980), to EBRT alone in the early 1990s (PORTEC, GOG). More recently, since the establishment by FIGO in 1988 of surgical staging, there has been an increased interest in the use of postoperative ICB alone as well as a tendency toward "more tailored" adjuvant therapy based on the histopathological features of the surgical specimen.

25.9.1
Preoperative Irradiation

Although preoperative irradiation is less commonly used nowadays, it is still advocated for some patients such as those with gross involvement of the cervix (FIGO clinical stage IIB) or vagina (FIGO clinical stage III). Generally it involves the use of ICB alone or in combination with pelvic EBRT. Patients undergoing ICB only in the preoperative setting could undergo extrafascial hysterectomy as soon as 1–3 days from completion of the implant. If EBRT is required, surgery should be done no sooner than 4–6 weeks after completion of therapy when most of the radiation-associated inflammation has subsided.

Although uncommon, patients may present with gross pelvic or retroperitoneal nodal disease, in the absence of distant metastasis. These patients could be considered for preoperative EFRT and brachytherapy to be followed by surgical staging.

25.9.1.1
Preoperative ICB

25.9.1.1.1
Low-Dose-Rate Brachytherapy

For preoperative intracavitary insertions, in addition to afterloading tandem and vaginal ovoids, it is the practice at Mallinckrodt Institute of Radiology (MIR) to pack the uterine cavity with afterloading Heyman-Simon capsules. The implant is performed in the operating room with the patient under general or spinal anesthesia. Following the exam under anesthesia (EUA), a Foley catheter is placed in the bladder and the bulb is inflated with 7 cc of contrast material. Subsequently, a D&C is performed with or without cervical biopsies if they had not previously been obtained. Radio-opaque seed markers are placed on the cervix at 12 o'clock and 6 o'clock positions. The Simon-Heyman capsules have diameters of 6, 8 and 10 mm. The capsules are numbered on their distal end and the order of capsule placement is recorded at the time of placement since they are to be removed in reverse order of their placement. Simon-Heyman capsules are placed through the cervical os into the uterine fundus to fill the body of the uterus. Depending on the uterine cavity sounding, small capsules (i.e., 6 mm) are to be used for patients with hysterometry between 6 cm and 8 cm and large capsules (i.e., 8–10 mm) with that between 8 cm and 12 cm. The number of capsules placed in the uterus, generally 5–7, would depend on its size, although large uterine cavities could potentially accommodate as many as 16. The intrauterine tandem is placed into the uterus such that the tip of the tandem abuts the lowest Simon-Heyman capsule, generally at the lower uterine segment with the flange at 4–6 cm. Vaginal ovoids are then placed in the vaginal fornices. The size of the ovoids should always be the largest diameter that fills the lateral fornices (usually 2 cm or 2.5 cm). Once all the applicators are in adequate position, the vagina is packed with Ray-tack vaginal packing soaked in AVC cream. Vulvar labial sutures are sometimes placed to prevent the implant from sliding down.

Localization radiographs [anteroposterior (AP) and lateral orthogonal films] are obtained in the regular simulator after the procedure is completed, for dosimetry purposes (Fig. 25.1).

In the absence of Simon-Heyman capsules, one or two intrauterine tandems can be used in combination with vaginal ovoids (Fig. 25.2). If there is tumor extension to the vagina, the entire length of the vagina should be treated because of the propensity of advanced EC to metastasize to the vagina through the submucosal venous and lymphatic plexuses. If tumor thickness is less than 0.5 cm, vaginal Delclos cylinders can be used. However, for more extensive disease, the use of Syed-Nebblett interstitial implant is recommended to ensure adequate tumor coverage.

When using low-dose-rate (LDR) brachytherapy, the implant remains in place for 48–72 h. During that time, the patients are kept in the hospital on radiation precautions. While an inpatient, it is important that adequate pain control is obtained; for this, we use a patient-controlled anesthesia (PCA) machine with (morphine or fentanyl). In addition, careful attention should be paid to deep venous thrombosis prophylaxis using subcutaneous heparin and sequential compression devices (SCDs), as well as by encouraging patients to do leg exercises. Finally, patients are kept on a low residue diet and Imodium while in bed and they are encouraged to do incentive spirometry to minimize the risk of atelectasis and subsequent respiratory infections. Prophylactic use of antibiotics is not recommended unless, at the time of the D&C, the patient is found to have a pyometra or if uterine perforation is suspected. Once the implant time is completed, the radioactive sources will be removed in the patient's room and returned to storage. Subsequently, the vaginal packing, ovoids tandem and Simon-Heyman capsules are

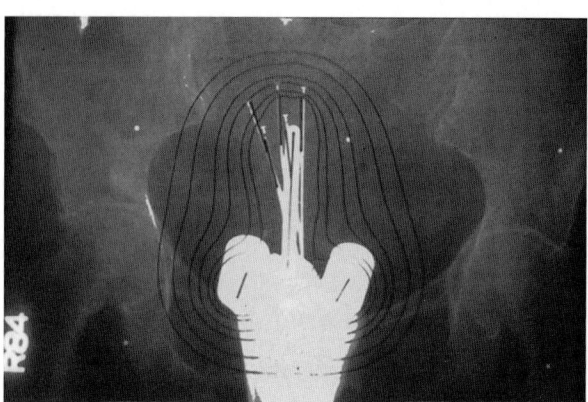
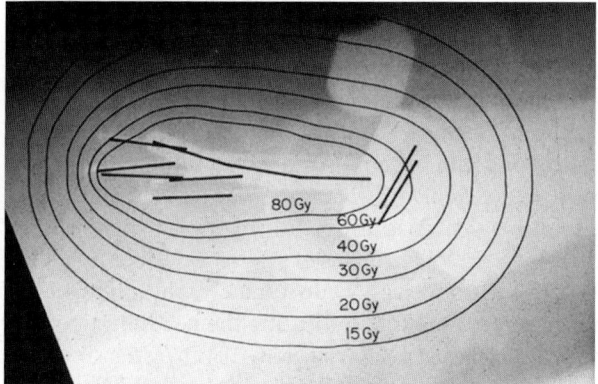

Fig. 25.1. a Anteroposterior radiograph: tandem+ovoids+Heyman capsules. **b** Lateral radiograph: tandem+ovoids+Heyman capsules

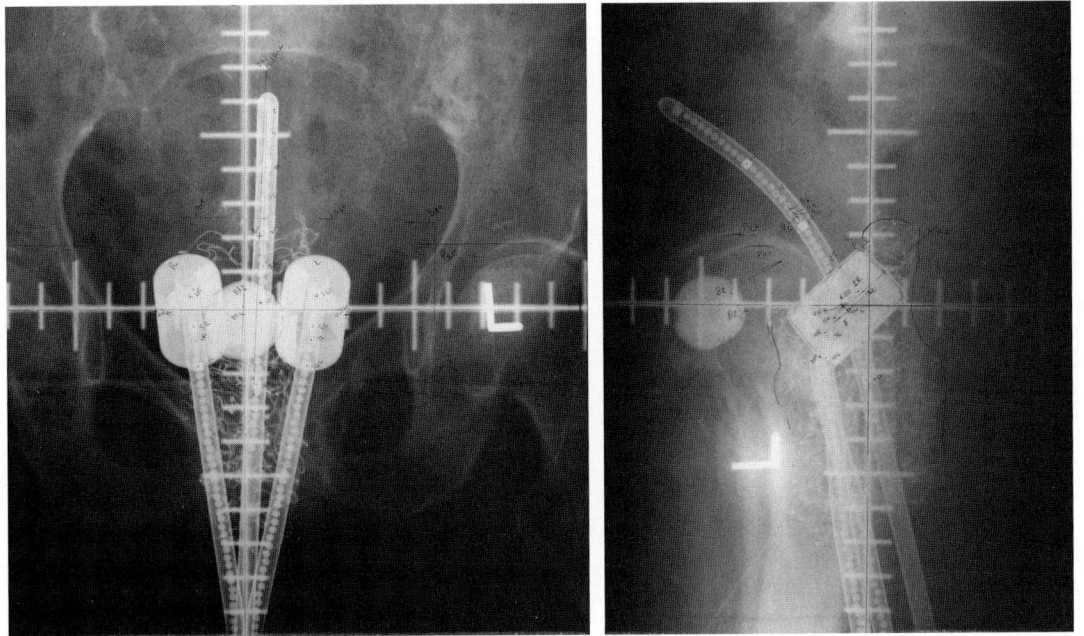

Fig. 25.2. a Anteroposterior radiograph: preoperative tandem and ovoids. **b** Lateral radiograph: preoperative tandem and ovoids

removed (the latter in reverse order of their placement). Generally, patients are given a "bolus" of pain medication while on the PCA, approximately 30 min before implant removal. Careful attention should be paid to potential vaginal bleeding after removal of the implant. The patients will be discharged once they are able to void without difficulties, have had a bowel movement and are able to ambulate without assistance.

The typical prescription at the MIR, empirically established, is of 3500–4000 mg h RaEq to be delivered to the uterus and 6500 cGy surface dose to the vagina. The Simon-Heyman capsules are loaded with 10 mgRaEq each; the tandem is loaded generally with three sources (for 6-cm-long tandem), 10-20-20 mgRaEq (adequate dose to the lower uterine segment); each one of the ovoids with be loaded with 20 mgRaEq (2 cm ovoids) or 25 mgRaEq (2.5 cm ovoids). Classical point-A dose rates will be calculated as for cervical cancer brachytherapy implants, as well as points B, P, rectal and bladder doses.

25.9.1.1.2
High-Dose-Rate Brachytherapy

The procedure for high-dose-rate (HDR) preoperative implant is the same as for a LDR implant and, preferably, it should take place in the operating room. The prescription follows as well the same rules in terms of dose distribution as with LDR. Generally, implants are performed once a week, for three consecutive weeks for a total of 3300 mg h RaEq.

25.9.1.2
Preoperative EBRT

Patients with gross cervical involvement or more advanced disease could potentially benefit from EBRT to the pelvis in addition to ICB. Generally, patients will receive a dose between 20 Gy and 40 Gy (180 cGy/fraction), using photon energies greater than 10 MV with a four-field technique (AP/PA and lateral fields). CT scan simulation is encouraged. Oral contrast should be used to opacify the small bowel. A vaginal marker should be placed in addition to seed markers at the most distal tumor extent in the vagina, to be able to adequately encompass the gross tumor volume (GTV). This also includes the entire uterus and cervix as well as any gross regional lymphadenopathy. The clinical target volume (CTV) includes the GTV as well as the pelvic lymph node areas potentially harboring microscopic disease; this is the obturator, external and internal iliacs as well as the lower common iliacs. Careful attention should be paid to include the external iliac anteriorly as well as the presacral nodes posteriorly (S2–S3 level), in the lateral fields, since these nodal areas are at high risk in the presence of gross cervical involvement.

Any attempt should be made to minimize the volume of small bowel included in the radiation fields by using bladder distension or by placing the patient in the prone position. Similarly, as much as possible of bladder and rectum volume should be spared without compromising the coverage of the CTV.

At the MIR, after a single ICB as described above, the patients would receive 20 Gy to the whole pelvis and 30 Gy to the split fields with a midline step wedge covering the high dose of the ICB.

At our institution for those selected patients with clinical stage-IIB uterine cancer with gross cervical infiltration, we have used a combination of EBRT, 3960 cGy in 22 fractions, followed by a single ICB implant using intrauterine tandem and vaginal ovoids to deliver a dose of 3500 cGy to classical point A. After completion of the implant, the split fields (using a central customized block) receive a boost of 540 cGy to deliver a total pelvic side wall dose of 4500 cGy.

Grigsby et al. (1992) reported on 858 patients with clinical stage-I EC treated with preoperative RT followed by TAH-BSO. The patients received 2500–4000 mg h RaEq to the uterus with Heyman capsules and intrauterine tandem, and 6500 cGy surface dose to the upper vagina. This was followed by TAH-BSO within 3 days to 6 weeks. When deep myometrial invasion was present, patients received postoperative EBRT (2000 cGy to the whole pelvis and an additional 3000 cGy to the parametrium with a midline step wedge), in order to deliver a total dose of 5000 cGy to the pelvic side-wall and lymph node regions. The 5-year survival rate for all patients was 84%, with a 5-year progression-free survival rate of 92% for FIGO clinical stage IA and 86% for stage IB. Survival was clearly related to tumor grade and degree of myometrial invasion.

25.9.2
Postoperative Irradiation

Adjuvant EBRT in early stage EC, either after TAH-BSO alone or after ESS has been shown in three prospective randomized trials to improve significantly pelvic control rates and DFS, when compared with observation, without significant impact in overall survival (Aalders et al. 1980; Creutzberg et al. 2000, 2003; Keys et al. 2004), In addition, EBRT to the pelvis is associated with acute and long-term toxicity, primarily gastrointestinal and urinary, which must be balanced against the absolute ben-

efit derived from its use. Furthermore, the fact that most of the loco-regional failures, 60–70%, are in the vagina, which can be successfully salvaged in a high proportion of cases (Creutzberg et al. 2003; Jhingran et al. 2003), have limited the use of EBRT in the adjuvant setting in patients with early stage disease. In the last decade, there has been a tendency toward increasing use of vaginal brachytherapy alone or even observation for the majority of patients with intermediate risk EC. The use of EBRT is still recommended in those patients with deeply invasive tumors (stage IC), poorly differentiated histologies – including unfavorable types such as IPSC and CCC – and in patients with pathological stage IIB without ESS, mainly if there is evidence of LVSI in the surgical specimen. Whether some of these high-risk patients could be managed with vaginal brachytherapy alone is unknown.

Patients with advanced disease, pathological stages III–IVA, have been treated with a variety of approaches including: pelvic EBRT +/- EFRT to cover the peri-aortic nodes in those with positive nodes, Whole Adominal Irradiation (WAI) and intraperitoneal (IP) radioisotopes (^{32}P).

25.9.2.1
Postoperative ICB

Intracavitary vaginal brachytherapy (IVB) is an integral component in the adjuvant management of selected patients with early stage EC. Although traditionally delivered using LDR techniques, the use of HDR techniques has become increasingly more common worldwide. However, there are no standardized treatment recommendations. Potential advantages of IVB when compared with EBRT include lower costs, lower morbidity and patient convenience; the main disadvantage is that it does not address the pelvis and, therefore, should be limited to patients in whom the pelvic failure rate is estimated to be small and the vagina represents the organ at risk for recurrence. There have been numerous series published in the literature using IVB alone in the adjuvant setting for patients in the low and intermediate risk groups, whether or not ESS has been performed. Although an exhaustive review of the literature is beyond the scope of this chapter, it is important to point out that the recurrence rate after IVB alone is 3–10% (Kucera et al. 1990; Eltabbakh et al. 1997; Weiss et al. 1998; MacLeod et al. 1998; Petereit et al. 1999; Chadha et al. 1999; Anderson et al. 2000; Ng et al. 2000; Fanning 2001; Alektiar et al. 2005).

25.9.2.1.1
LDR Intracavitary Brachytherapy

LDR-IVB is generally performed in the operating room with the patient under anesthesia. After the EUA, a Foley catheter is placed in the bladder and the bulb is inflated with 7 cc of radiopaque contrast material. A radiopaque seed marker is placed in the center of the colpectomy scar. Fletcher-Suit afterloading colpostats are placed in contact with the suture line; we should always use the larger diameter ovoids, which can be accommodated comfortably to the patient's anatomy. Subsequently, the applicators are kept in place by carefully packing in front and behind the ovoids, to improve the separation from the recto-vaginal and vesico-vaginal septum. The packing of the entire vagina is completed, and often two labial sutures are placed in the vulva to further avoid displacement of the applicators. After implant is completed in the operating room, a set of orthogonal radiographs (AP and lateral) are obtained to document applicator placement and obtain dosimetry (Fig. 25.3). Generally, a dose of 6500 cGy is prescribed to the vaginal surface. Point doses to the rectum and bladder should also be calculated and documented.

25.9.2.1.2
HDR Intracavitary Brachytherapy

Being an outpatient procedure with much shorter treatment times, HDR-IVB has become more popular primarily because of the convenience to the patients. Generally, two to three procedures are performed at 1-week intervals. Vaginal ovoids or more commonly vaginal cylinders are used. These are commercially available and similar in design to the LDR applicators. The majority of patients in the adjuvant setting can nicely accommodate vaginal cylinders of different diameters, depending on patients' geometry. For patients with more retracted vaginal cuff ("dog-ear" shaped), vaginal ovoids are preferred since they allow a better contact with the vaginal cuff surface. Vaginal cylinders are available in diameter ranging from 2 cm to 4 cm to treat a variety of patients (Fig. 25.4). It is imperative for the vaginal mucosa to be in contact with the applicator surface in order to obtain an effective dose distribution and, therefore, the largest diameter cylinder or ovoids that can comfortably fit in the vagina should be used. It is important to confirm that the applicator, when using vaginal cylinders in the outpatient setting, is in close contact with the vaginal apex by gently pushing it cephalad while assessing patient

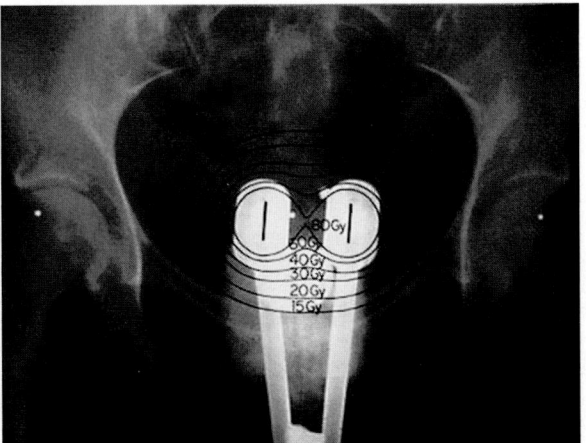
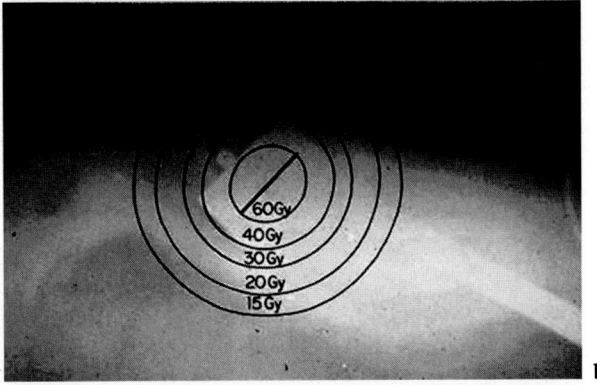

Fig. 25.3. a Anteroposterior radiograph of low-dose-rate vaginal ovoids. b Lateral radiograph of low-dose-rate vaginal ovoids

Fig. 25.4. High-dose-rate vaginal cylinders

discomfort. Subsequently, the applicator should be secured in position, in the midline of the patient. The position of the applicator should be rechecked before each treatment and adjusted as necessary. It is important to avoid backward placement of the applicators since this would contribute to higher doses to the rectum.

Localization radiographs should be obtained in the treatment position for dosimetry purposes (Fig. 25.5a,b). Although this is not a generalized

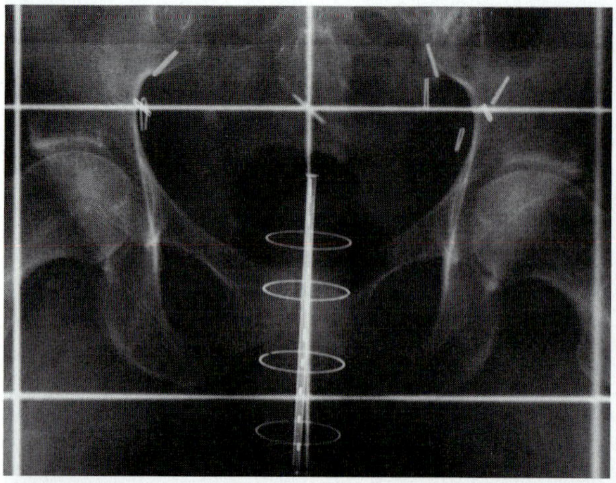

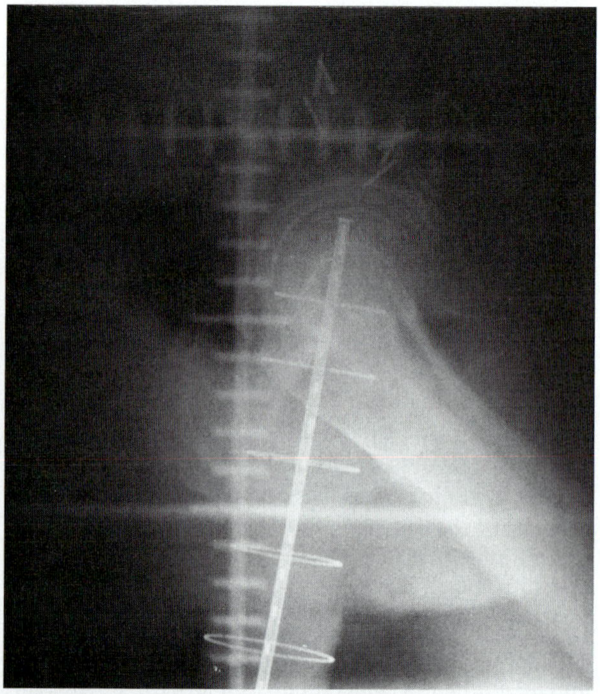

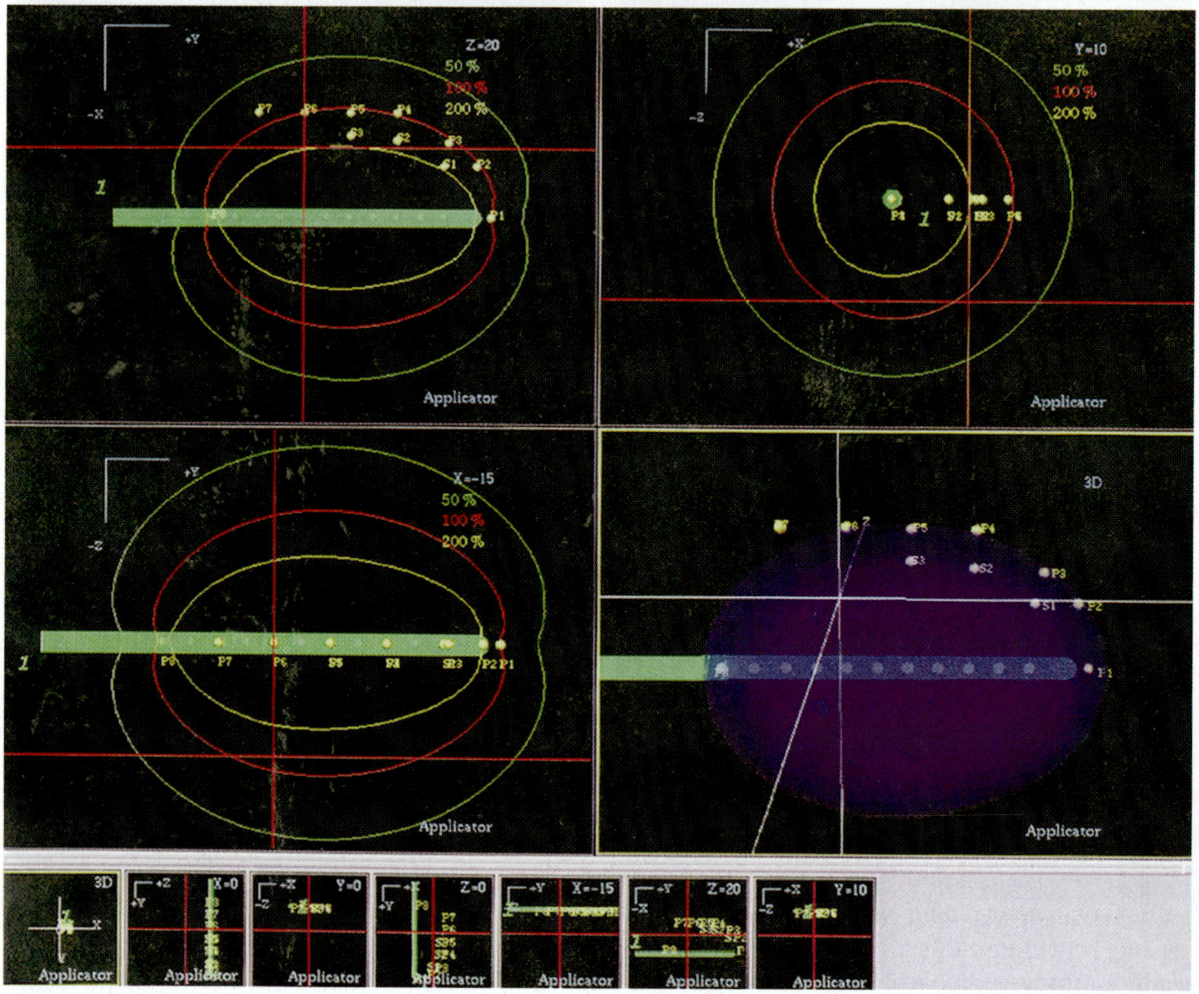

Fig. 25.5. a Anteroposterior radiograph. Vaginal cylinders, 3 cm diameter. **b** Lateral radiograph. Vaginal cylinders, 3 cm diameter. **c** Optimization and dose distribution of 3 cm vaginal cylinders. 700 cGy prescribed to 0.5 cm from the vaginal surface

practice when using vaginal cylinders, we strongly recommend doing it for documentation. In addition, although it would be ideal to obtain a customized plan prior to each fraction, the films and dosimetry are generally obtained only prior to the first application, since they are time consuming and when using an applicator with a fixed geometry, repeating the plan prior to each fraction seems unnecessary (assuming the geometry of the implant remains the same for each fraction).

Generally, the proximal 3–5 cm of the vagina is treated. At our institution, we treat the entire length of the vagina only in those patients with unfavorable histologies, such as UPSC and CCC. In those situations, special attention should be paid to avoid protruding a source at the introitus. The last active source position should be placed at least 1 cm from the introitus. Placement of a radiopaque marker at the introitus as well as at the vaginal cuff is highly recommended.

The HDR dose depends on the dose specification point and whether EBRT is also used. We strongly recommend reviewing the "American Brachytherapy Society (ABS) Recommendations for HDR Brachytherapy for Carcinoma of the Endometrium" published by Dr. Nag in 2000 (NAG et al. 2000). The most commonly used regimen when using HDR-IVB alone is 700 cGy per fraction, per week, prescribed to a depth of 0.5 cm from the vaginal surface. The dose distribution should be optimized to deliver the prescribed dose to the points of interest. Optimization points should be placed not only along the lateral vaginal surface but also at the apex (or 0.5 cm depth, depending upon prescription point), along the curved portion of the vaginal cylinder to assure adequate distribution (Fig. 25.5c).

Table 25.8. ABS recommendations for adjuvant high-dose-rate (HDR) vaginal brachytherapy in endometrial cancer. Modified from NAG et al. (2000). *EBRT* external beam radiotherapy

No. of fractions	HDR dose/Fx (Gy)	Dose-specific point	Equivalent tumor dose (Gy)	Equivalent late effect dose
Adjuvant HDR vaginal brachytherapy alone				
3	7.0	0.5 cm depth	29.8	23.2
4	5.5	0.5 cm depth	28.4	21.1
3	10.5	Vaginal surface	53.8	45.6
Adjuvant EBRT – 4500 cGy – and HDR vaginal brachytherapy				
2	5.5	0.5 cm depth	58.5	53.7
3	6	Vaginal surface	68.3	61.3
Definitive EBRT – 4500 cGy – and HDR vaginal brachytherapy for recurrent disease				
3	7	0.5 cm depth	74.0	66.4
4	6	0.5 cm depth	76.3	67.4

When patients have received external beam, the most common fractionation used at our institution is two fractions of 500–550 cGy prescribed at a depth of 0.5 cm or three fractions of 600 cGy prescribed to the surface of the vagina. We generally use this second regimen in higher risk patients, such as those in stage IIB or with evidence of LVSI. Table 25.8 shows some of the ABS recommended schedules for HDR-ICB when used as the only adjuvant therapy or after whole pelvic RT. For additional information regarding other potential schedules please refer to Dr. Nag's publication (NAG et al. 2000).

Occasionally, patients with stage-III disease are given HDR-IVB after EBRT. At our institution, we limit the use of IVB in advanced disease for those patients with evidence of lower uterine segment or cervical involvement, positive vaginal margins or LVSI. Patients with stage-III disease optimally debulked, candidates for the GOG-184 clinical trial, or treated outside of the study, as per protocol design, will receive a single fraction of HDR-IVB delivering a dose of 700 cGy prescribed to a depth of 0.5 cm from the vaginal surface after an EBRT dose of 5040 cGy.

25.9.2.2
Postoperative EBRT Techniques

CT simulation is encouraged for better delineation of target volumes and sparing normal tissues, primarily small bowel, rectum and bladder. The four-field technique is the most commonly used, with treatment delivered using Linear Accelerator with a peak photon energy of 10 MV or greater.

Whole pelvis EBRT fields in patients with negative pelvic and peri-aortic nodes or positive pelvic nodes with negative sampled peri-aortic nodes should encompass all the potential nodal areas of microscopic involvement, with adequate margins, as well as the proximal half to two-thirds of the vagina. At the time of simulation, a vaginal marker should be used as well as oral contrast to opacify the small bowel. If possible, maneuvers such as full bladder or the use of "belly boards" may be explored, to exclude as much as possible of the small bowel in the radiation fields.

The superior border of the *AP/PA pelvic fields* will be through the L4–5 interspace unless the target volume (e.g., need for inclusion of common iliac nodes in patients with positive pelvic nodes) would not be encompassed adequately in a cephalad direction. In the latter case, a 2-cm margin should be added to the highest level of pathological abnormality, but should not be cephalad to the L-3/L-4 interspace. The lateral border will be 2 cm beyond the lateral margins of the bony pelvis. The inferior border will be inferior to the obturator foramen or the lowest extension of vaginal disease with at least a 3-cm margin. The inferior aspect of any potential vaginal extension should be marked so that the inferior border of disease can be documented and treated with at least 3-cm margins.

The anterior border of the *lateral pelvis fields* should be determined by the location of the external iliac vessels as drawn at the time of the CT simulation with a minimum margin of 1.5 cm (generally anterior to the symphysis pubis). The posterior border should be placed at least 1.5 cm behind the anterior surface of the sacrum to adequately include the sacral nodes, at risk in those patients with lower uterine and cervical tumor extension. Attention should be paid to exclude as much as possible of the sacral plexus. The posterior wall of the rectum should be excluded providing that the vagina is adequately covered (as delineated by vaginal marker) with a 2- to 3-cm margin. Superior and inferior borders will be the same as for the anterior and posterior fields. If clips are present from the lymph node dissection to document the position of the lymph nodes, then these should be used as a guide when anterior blocks are designed to shield the small bowel. At least 3 cm should not be blocked anterior to the L-5 vertebral body. In addition, the anterior two-thirds of the L-5 vertebral body should not be blocked in order to

adequately cover the common iliac nodes with 1.5- to 2-cm margins (Fig. 25.6).

More recently, intensity modulated radiation therapy (IMRT) has been incorporated in the adjuvant treatment of EC for those patients requiring EBRT. Recent series have shown significantly less small bowel, rectum and bladder being irradiated when using IMRT than with more conventional two- or four-field techniques. However, it has also been recognized that this technique requires accurate target delineation, highly reproducible patient immobilization and a clear understanding of internal-organ motion in order to achieve optimal advantage in the use of IMRT over conventional methods of post-hysterectomy pelvic RT (AHAMAD et al. 2005).

Extended pelvic and peri-aortic fields should be treated in patients found to have positive peri-aortic nodes, when pelvic nodes are positive and no peri-aortic lymph node sampling has been performed or in patients without ESS with positive pelvic nodes assessed by CT scan. The use of CT simulation is even more crucial when treating extended fields, since this allows an accurate delineation of the kidneys, small bowel and liver, as well as the CTV, which should encompass the peri-caval, inter aorto-cava and para-aortic areas with a minimum margin of 1.5 cm (microscopic disease) or 2–2.5 cm for gross residual disease.

The superior border of the *AP/PA extended fields* will be at the T11–T12 or L1–L2 interspace, depending on the extent of the nodal disease; the inferior

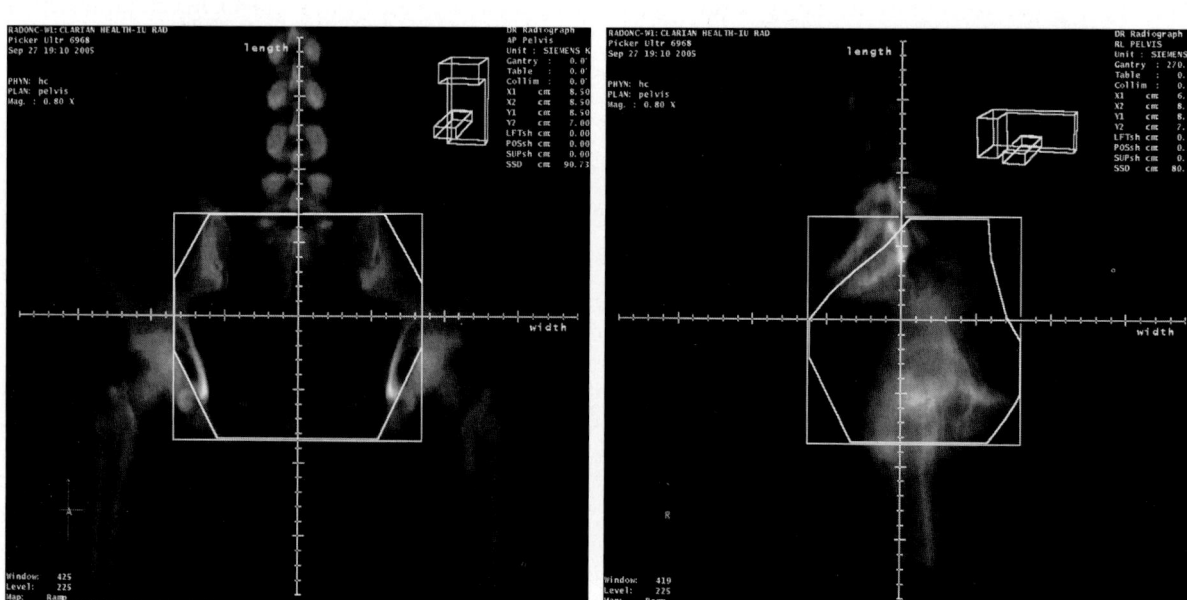

Fig. 25.6. a Anteroposterior whole pelvis field. **b** Lateral whole pelvis field

border will be as the inferior border of the pelvic fields described above. Lateral borders should include the entire peri-aortic region as delineated by CT scan simulation, including any potential gross lymphadenopathy with a margin between 1.5–2 cm (in uninvolved areas) and 2–2.5 cm if gross disease is present. A minimum width of 5 cm is required. Attention should be paid to exclude as much as possible of the kidney volume. At the level of the pelvis, the lateral borders should be as described above, with a margin between 2 cm and 2.5 cm beyond any grossly involved pelvic node or 1.5 cm and 2 cm for microscopic disease only.

The superior and inferior borders of the *lateral extended fields* are the same as for AP/PA fields. The anterior border should cover the entire peri-aortic regions, as delineated by CT scan with a 1.5-cm margin in uninvolved areas and 2- to 2.5-cm margins around any grossly involved pelvic or peri-aortic nodes. The posterior border should be anterior to the spinal canal in order to protect the spinal cord and lumbo-sacral plexus. At least half to two-thirds anterior of the vertebral body should be included in the field in order to adequately encompass the peri-aortic region (Fig. 25.7).

Patients treated with extended fields will receive 4500 cGy delivered to the pelvic and peri-aortic fields at 150 cGy per fraction, using a four-field technique, in 29 fractions. The pelvis will receive a total dose of 5040 cGy in 28 fractions at 180 cGy/fraction. Therefore, a daily boost to the pelvis of 30 cGy, via AP/PA fields (superior edge of the pelvic field at L5–S1 or L4–L5) will be required in order to deliver 180 cGy/ fraction to the pelvis while the peri-aortic region would receive 150 cGy/fraction. It is important to deliver the boost dose with AP/PA fields rather than lateral fields in order to preserve some of the bone marrow in the iliac bones. After 28 fractions, the peri-aortic region treatment should receive 1 additional fraction to complete a total dose of 4500 cGy. In the case of positive resected peri-aortic nodes, an additional boost of 500 cGy could be delivered to the high-risk area, generally using four-field techniques. The placement of intraoperative vascular clips in the areas of gross disease could be very helpful in delineating the boost areas.

WAI has been used for the adjuvant treatment of advanced EC, pathological stages III–IVA, as well as unfavorable histologies such as UPSC and CCC known to have a pattern of spread similar to ovarian cancer with a greater incidence of intra-abdominal failures (MARTINEZ et al. 2003). However, since the presentation of the results of GOG 122 (RANDALL et al. 2003), demonstrating the superiority of systemic chemotherapy (Doxorubicin and Cisplatin) over WAI, this modality has fallen out of favor.

The ability of WAI to alter failure patterns by decreasing upper abdominal relapse is determined by the adequacy of the technique, emphasizing the necessity of covering the diaphragm with adequate margin during all phases of normal respiration. This requires that liver shielding be limited or absent. Appropriate kidney localization and blocking should be undertaken to keep total doses within tolerance. CT simulation and possibly IMRT may more precisely shield the liver and kidneys (Fig. 25.8).

The WAI dose is generally 2500–3000 cGy (at 150 cGy/fraction). The peri-aortic region generally receives a boost to 4200–4500 cGy, and the pelvis is treated to 5040–5100 cGy. The kidney dose should be limited to 2000 cGy or less via a 100% posterior transmission block. MARTINEZ et al. (2003) advocate the use of a liver transmission block to keep the total liver dose to 2250 cGy. This is not universally accepted and it was not required in the GOG-122 trial.

Most of the toxicity encountered with WAI is gastrointestinal, up to 10–15% grades 3–4, in some series (MARTINEZ et al. 2003; RANDALL et al. 2003). In addition, grade-3 renal and/or liver toxicity has been reported in about 2% of patients (MARTINEZ et al. 2003; SUTTON et al. 2005). Hematological toxicity was reported in 4% of patients undergoing WAI in the GOG 122 trial (RANDALL et al. 2003).

25.10
Medically Inoperable Early Stage Endometrial Cancer

Although surgery, including TAH-BSO, is the definitive treatment of choice for most patients with EC, in approximately 10–20% of cases, patients are morbidly obese, are very advanced in age and have severe co-morbid cardiovascular or other diseases that contraindicate extirpative abdominal surgery (FISHMAN et al. 1996). In these non-operable cases, primary RT has been used with success varying with the stage of disease and the individual experience of the treating institution, using a combination of brachytherapy with or without pelvic irradiation. The ABS (NAG et al. 2000) recommends determining the thickening of the myometrial wall and depth of infiltration using CT or preferably MRI or ultrasound whenever possible.

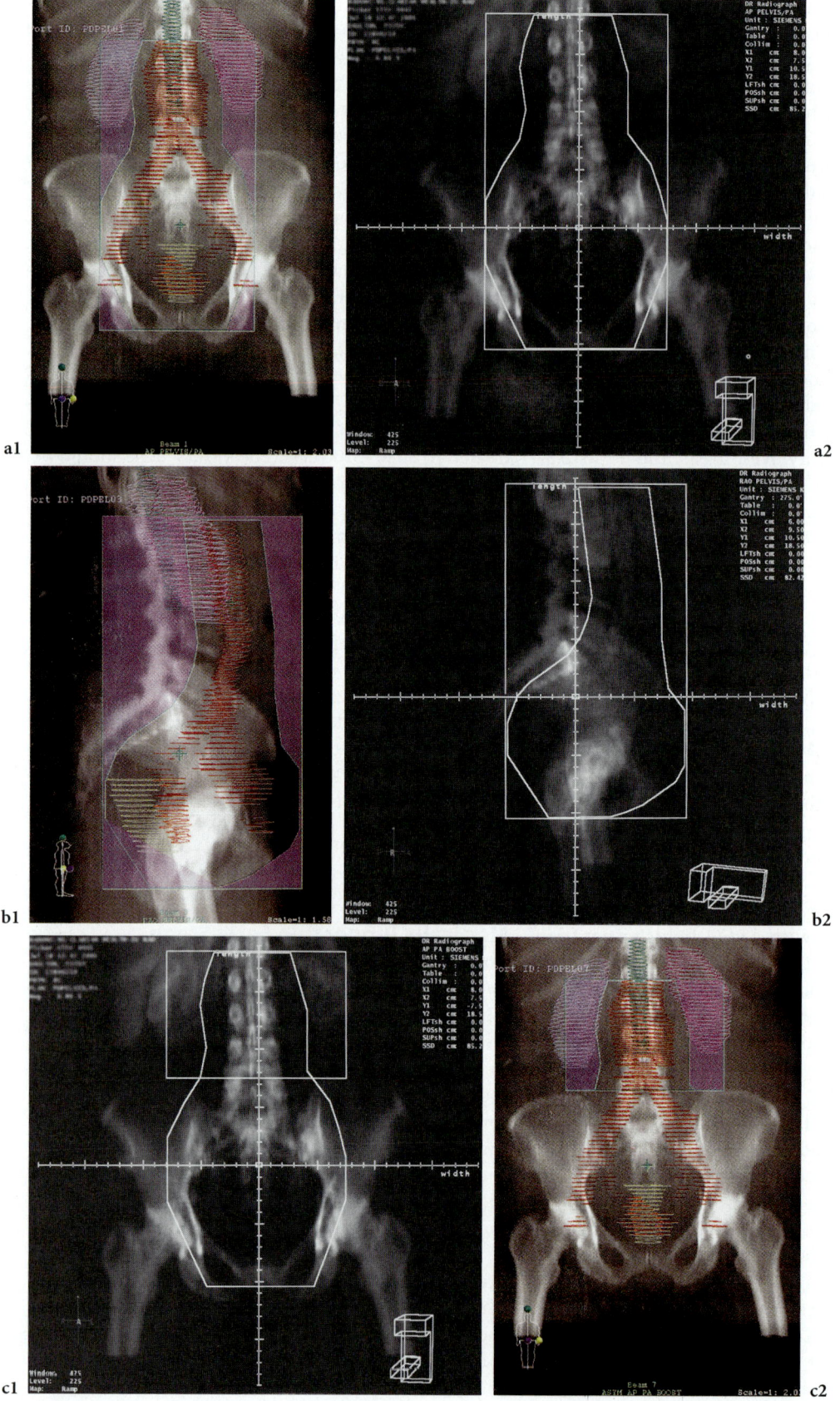

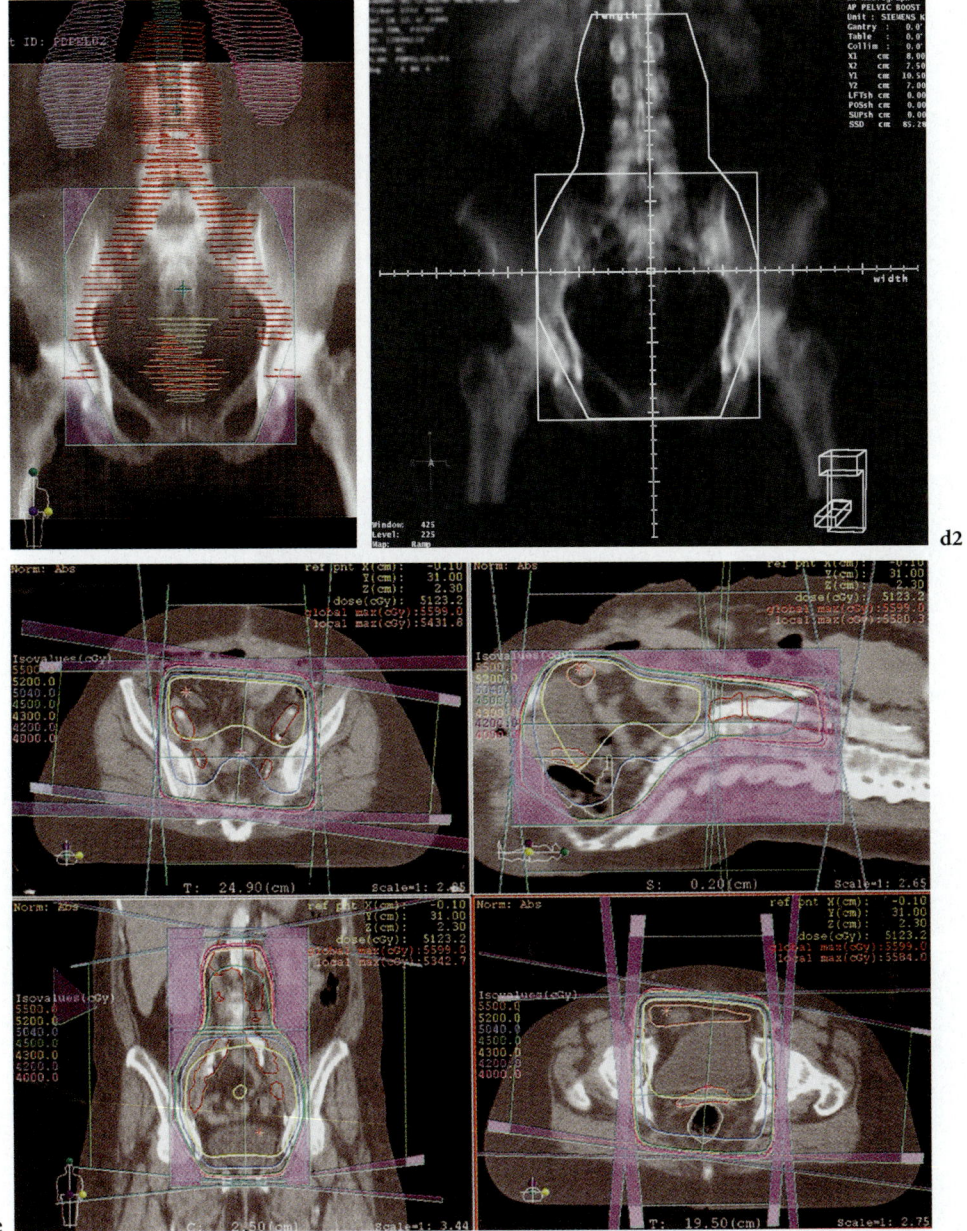

Fig. 25.7a–e. a1 Anteroposterior (AP) pelvic+peri+aortic irradiation. a2 AP pelvis and peri-aortic. b1 Lateral pelvic and peri-aortic irradiation. b2 Lateral pelvic+peri-aortic. c1 Peri-aortic boost digital reconstructed radiograph (DRR). c2 Peri-aortic boost. d1 Pelvic boost. d2 Pelvic boost DRR. e Isodose distribution of pelvic and peri-aortic irradiation

Patients with stage-IA well- or moderately differentiated tumors, without evidence of myometrial infiltration or lymph node metastasis by CT scan, can be treated with ICB alone. This is generally performed with Fletcher-Suit applicators, with one or two tandems depending on the uterus size, and/or Simon-Heyman capsules in combination with vaginal ovoids, using a technique similar to the described preoperative ICB technique. There is no consensus regarding optimal dose or prescription points. In some institutions, one or two applications are performed to deliver a dose of 7000–7500 cGy to point A, when using LDR-ICB. The loading of the tandem(s) is different from that in cervical cancer, in order to provide adequate dose distribution laterally and superiorly. Image-based brachytherapy, although not routinely employed, would potentially provide a more accurate understanding of the dose

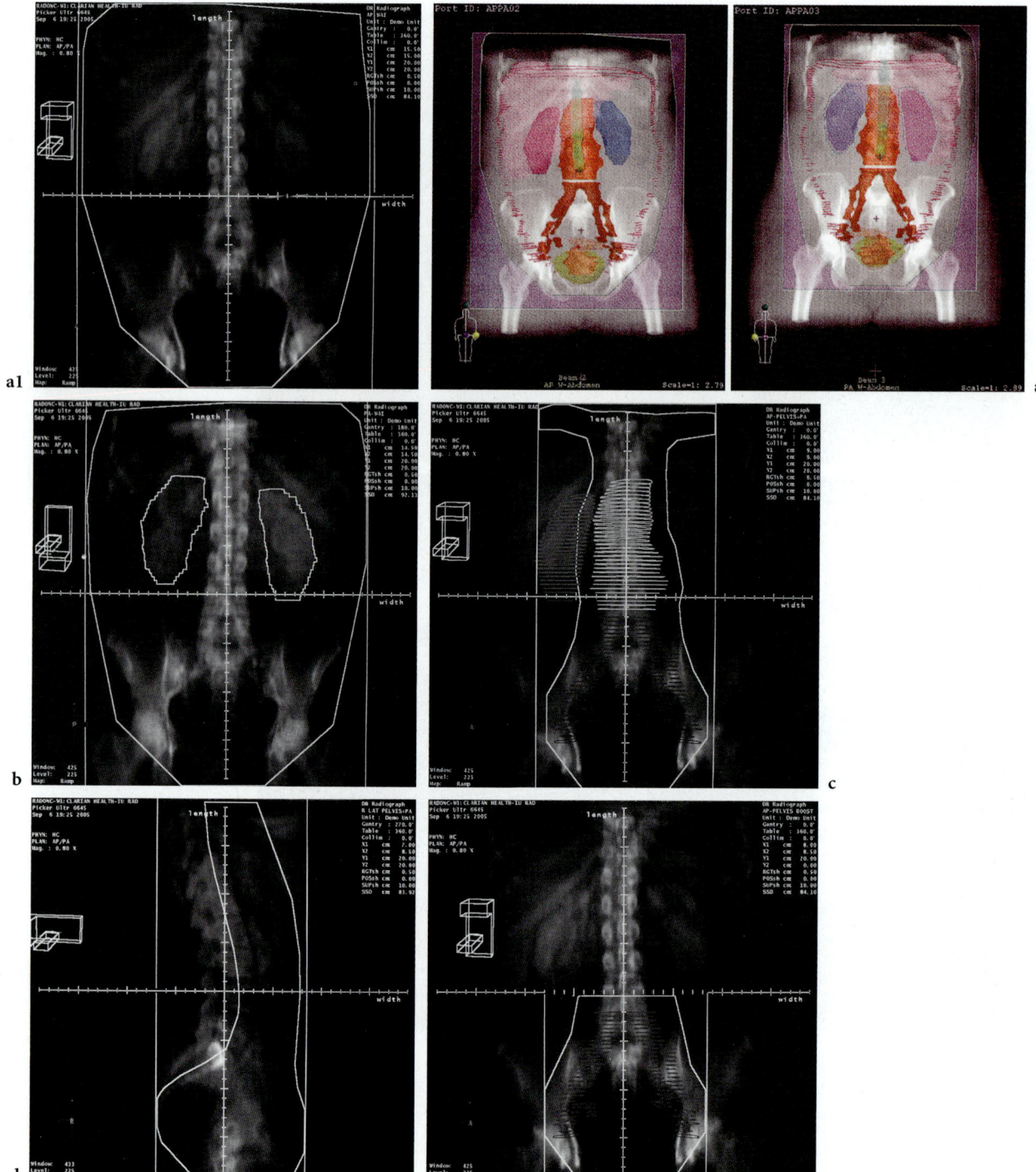

Fig. 25.8a–f. a1 Anteroposterior (AP)–whole abdominal irradiation (WAI) digital reconstructed radiograph (DRR). **a2** AP–PA WAI. **b** PA–WAI DRR. **c** AP peri-aortic+diaphragmatic DRR. **d** Lateral peri-aortic+diaphragmatic. **e** AP lateral pelvic boost DRR. **f** Isodose distribution WAI

received by the GTV (as delineated by MRI) and CTV (defined as the entire uterus, cervix and proximal 2–3 cm of the vagina). The ABS recommends prescribing the dose at a point 2 cm from the central axis at the midpoint along the intrauterine sources (Nag et al. 2000). However, in the Mallinckrodt system, the dose is prescribed such that the total activity/exposure from the intrauterine sources (generally Simon-

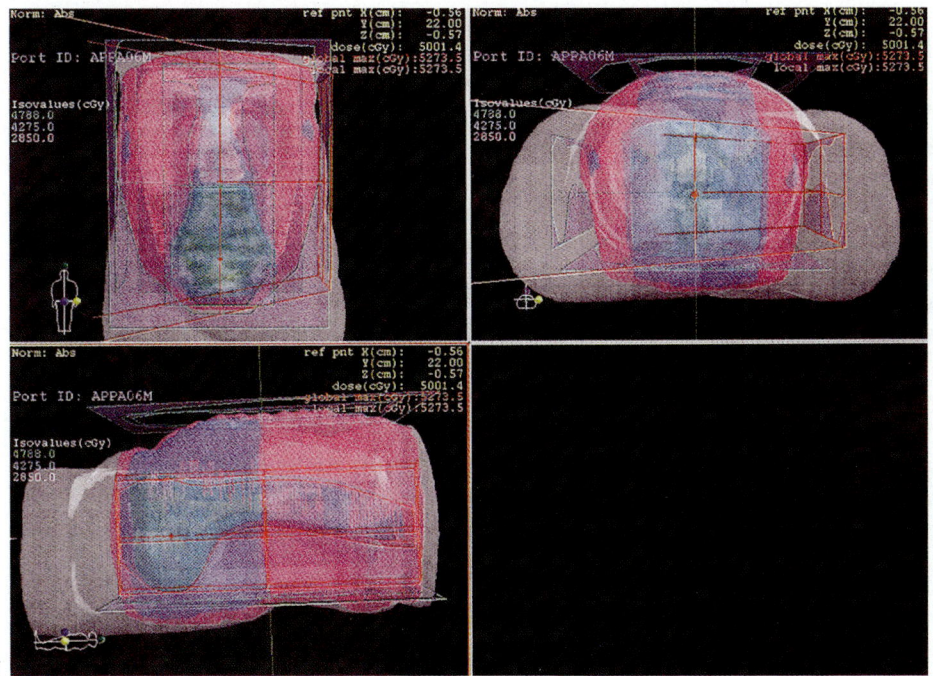

f

Heyman capsules and intrauterine tandem) would be around 5000 mg h RaEq, and from the vaginal ovoids approximately 3000 mg h RaEq with excellent control and 5-year progression-free survival rates, with pelvic control rates of 100% for patients with stage IA with ICB alone, and 88% for patients with stage-IB disease, generally with a combination of EBRT and ICB (GRIGSBY et al. 1986; CHAO et al. 1996). Chao et al. demonstrated a reasonable mortality of 2.1% and life-threatening complication rate of 4.2% (CHAO et al. 1996) using this approach.

Patients with stage-I disease, poorly differentiated or deeply invasive tumors, as well as patients with inoperable stage-II disease, will be treated with a combination of EBRT (4500–5000 cGy) and ICB (3000–3500 cGy to point A), to bring the total point A dose to 8000–8500 cGy.

When there is clinical involvement of the uterine cervix, FIGO clinical stage II, most series show DFS outcomes with definitive RT to be 10–20% worse than those for clinical stage-I disease (TAGHIAN et al. 1988). As the risk of lymphatic and extra-uterine spread increases with cervical involvement, the use of combination EBRT and ICB is appropriate in those patients who have fairly good overall health and performance status. The combination and technique for EBRT and ICB are similar to those described for higher risk clinical stage-I cancer outlined above.

Advanced stage EC is more difficult to definitively control with RT alone, as the risk for systemic spread dramatically increases and the local bulk of disease is usually significant. However multiple series have shown long-term freedom from EC relapse in a significant proportion of patients, with disease-specific survivals in up to 50% of patients at 5 years (KUPELIAN et al. 1993). Indeed, many clinical stage-III patients are so-staged on the bases of small vaginal nodules, which can be adequately treated with the ICB. In these patients, the entire vaginal surface should be treated with Delclos vaginal cylinders. Potentially, patients with inoperable stage-III disease could receive treatment to the pelvis only, followed by ICB in the absence of nodal pelvic or retroperitoneal metastasis, or by EFRT and ICB in those patients with nodal disease, followed by chemotherapy or enrollment in a clinical trial if the patient has adequate performance status. This situation is very uncommon and, generally, patients with advanced pelvic and retroperitoneal disease present multiple co-morbidities that preclude them from such an aggressive approach, being, often times, considered for palliative therapy only.

More recently, HDR-ICB has also been employed in the management of clinically inoperable EC. The ABS recommendations regarding HDR-ICB alone or in combination with EBRT in terms of prescription points, number of fractions, dose per fraction and optimization have been extensively reviewed by NAG et al. (2000). A careful review of their guidelines is strongly recommended prior to implementation of such an approach.

25.11
Recurrent Endometrial Cancer

The rates of pelvic recurrence in patients with early stage disease after surgery alone range between 5% and 15%, most of which are isolated vaginal recurrences (CREUTZBERG et al. 2000, 2003; KEYS et al. 2004). The treatment of recurrent EC after surgery requires a combination of EBRT and brachytherapy. The reported salvage rate with radical irradiation is 65–80% in patients with vaginal relapses. The results are less favorable in patients with pelvic and/or regional recurrences (<50% salvage rate). This is probably a reflection of being able to deliver higher doses of radiation with a combination of EBRT and brachytherapy to isolated vaginal lesions than central recurrences, which often times are not amenable to brachytherapy boost. ACKERMAN et al. (1996) performed a retrospective review of 54 patients with recurrent EC in order to identify patterns of relapse, determine the outcome of salvage treatment and examine the factors predictive of effective salvage. Of 28 patients with pelvic relapses, only 16 experienced recurrence in the vagina. With a minimum follow-up of 5 years, 67% had pelvic control until death (79% of patients with vaginal recurrences and 43% of patients with pelvic recurrences), after definitive RT.

Updated results of the PORTEC trial were published by CREUTZBERG et al. in 2003 with a median follow-up of 73 months, showing 8-year actuarial LRR rates of 4% and 15% in the adjuvant RT and control groups, respectively (P<0.0001). Again, there was no difference in the actuarial overall survival rates (71% and 77%, respectively) or distant metastasis rates (10% and 6%, respectively) between the two groups. The majority of the LRR in the control group were in the vagina (32 of 46 patients). Of the 46 (13%) LRRs in the control group, only 22 were salvaged and remained without evidence of disease at the time of the last follow-up (48%). In addition, CREUTZBERG et al. have shown significantly better 3-year survival rates after vaginal recurrence alone than after pelvic or distant relapse (73%, 8%, and 14%, respectively, P<0.01). Similar results have been reported by the M.D. Anderson group (JHINGRAN et al. 2003) with 5-year local control and survival of 75% and 43%, respectively, in 91 patients treated with definitive doses of RT for isolated vaginal recurrences. Unfortunately, much higher doses of RT are required for the successful treatment of LRR in EC than those delivered in the adjuvant setting, mainly when only IVB is used. This translates into

a higher toxicity profile – 9% grade-4 complications in the M.D. Anderson series. Given the more limited use of adjuvant RT in the management of early stage EC, we may become more involved than in the past in the treatment of patients with recurrent EC.

25.11.1
External Beam RT

The volume irradiated will include the totality of the gross disease locally and regionally, as visualized by CT scan, in addition to the lymphatic regions including the obturator, hypogastric, external and internal iliac lymph nodes. A margin of 2–3 cm should be given around the gross disease and 1.5–2 cm around uninvolved lymph nodes. Generally, the pelvis is treated to a dose of 4500–5000 cGy (Fig. 25.9). CT simulation is encouraged in order to more adequately delineate the GTV and CTV. Prior to the simulation, placement of radiopaque markers delineating the GTV or at least the most inferior extent of the disease in the vagina is very helpful to assure adequate coverage. In general, it is recommended to treat the entire length of the vagina. Those lesions involving the distal third of the vagina are to be treated with at least a 2- to 3-cm distal margin, which often times requires the inclusion of vulva or perineum in the external beam fields. In that particular situation, inclusion of the inguino-femoral lymph node regions is highly recommended (please refer to Chap 27, Carcinoma of the Vagina, for details regarding the technique for treating inguino-femoral nodes).

Patients with pelvic/regional recurrences nonamenable to brachytherapy implant should receive additional boost using conformal techniques (we prefer the use of multiple coplanar or non-coplanar beam arrangement) in order to deliver a total dose to the gross disease around 6500 cGy, providing the small bowel can be spared doses over 4500 cGy.

Patients with extensive pelvic lymphadenopathy extending to the common iliacs and/or peri-aortic nodal disease – in the absence of systemic failure as documented by CT scan of the chest, abdomen and pelvis – should be considered for EFRT with or without brachytherapy if associated with vaginal recurrence. These patients are at very high risk of distant failures. Although there are no data to support the use of adjuvant chemotherapy in this setting, after completion of RT, patients with residual disease after completion of EFRT should be considered to be enrolled in clinical trials for advanced/recurrent

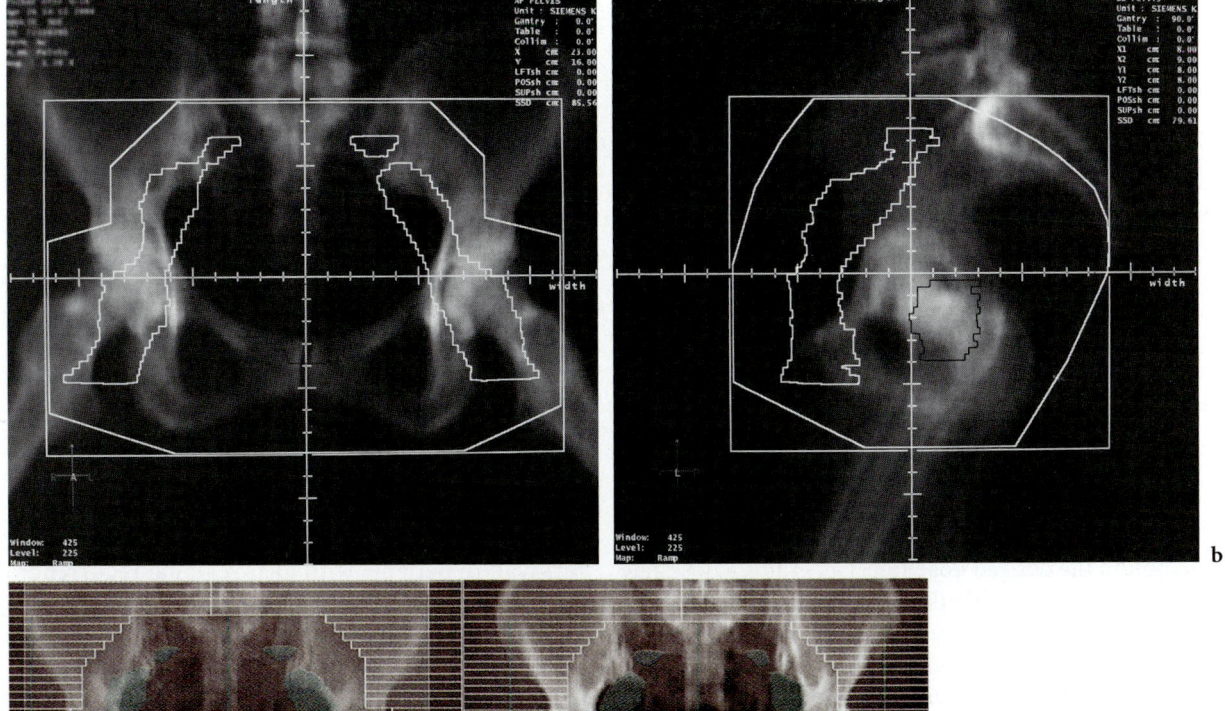

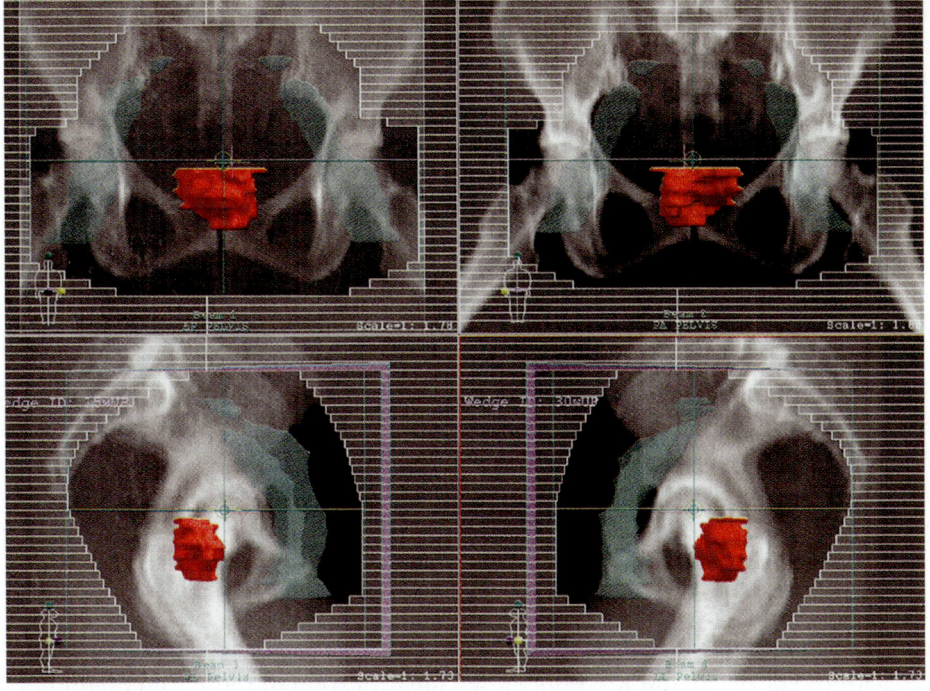

Fig. 25.9a–c. a Anteroposterior digital reconstructed radiograph (DRR) pelvic field in recurrent endometrial cancer extending to involve the mid-third of the vagina. **b** Lateral DRR pelvic field in recurrent endometrial cancer. **c** Multi-leaf collimator (MLC) fields; pelvic irradiation in recurrent endometrial cancer

disease, rather than waiting until progression. An alternative approach will be a combination of "tailored EBRT" followed by systemic therapy similar to that investigated in the GOG 184 clinical trial.

25.11.2
Brachytherapy in Recurrent Endometrial Cancer

Following the completion of EBRT, the patient will receive one or two implants, using a combination of interstitial +/– intracavitary techniques. At our institution, the entire surface of the vagina receives a dose of 6000 cGy. The GTV, as defined by physical exam and/or CT scan, will receive a dose between 7500 Gy and 8000 Gy, including the contribution of the EBRT and the intracavitary and interstitial implants, depending on the size and depth of infiltration of the recurrence.

ICB alone will be allowed in patients with vaginal lesions of less than 5 mm thickness. If so, dose should be prescribed to a depth of 0.5 cm unless only superficial mucosal infiltration present. We mostly use LDR-ICB in this setting at Indiana University. However, HDR-ICB has been used in combination with EBRT for patients with small vaginal cuff recurrences amenable to ICB alone; see the publication by NAG et al. (2000) regarding the American Brachytherapy Society recommendation (Table 25.8).

Interstitial brachytherapy should be used for lesions of 5 mm thickness of greater. The interstitial techniques used in the management of isolated vaginal recurrences from EC are similar to those described in Chap. 27, Carcinoma of the Vagina. Essentially, the first step in designing an interstitial implant is adequate target volume delineation primarily by performing a detailed EUA in addition to the CT imaging data. Once the target volume has been defined, a preoperative plan is obtained to determine the position, number and strength of the radioactive sources to obtain the desired dose rate. Generally, a dose rate of 40–50 cGy/h is calculated to the periphery of the implant. When performing an interstitial procedure, the flexiguides of funnel needles are placed into the tumor, in a 1×1-cm matrix. It is important to pass the catheters beyond the most superior extension of the tumor, which often times is possible by careful rectal examination at the time of the procedure, to assure an adequate cephalad margin of minimum 1 cm. Generally, the catheters are secured in place with the aid of a template, which also helps to assure adequate separation between the catheters. We often use the modified Syed-Neblett (DISAIA et al. 1990) applicator, which consists of a perineal template, vaginal obturator and 17-gauge hollow guides of various lengths. The vaginal obturator is 2 cm in diameter and 12 cm or 15 cm in length. The vaginal obturator has seven grooves on its surface for the placement of guide needles and is centrally drilled so it can allow the placement of a tandem to be loaded with ^{137}Cs sources, if indicated. This allows simultaneous combination of an interstitial and intracavitary application (Fig. 25.10). Generally, one implant is performed, delivering a dose of around 2000–2500 cGy after EBRT, in order to achieve a total tumor dose of 7000–7500 cGy. Those patients with bulky residual disease after external beam (larger than 2 cm) would require doses in the range of 3000–3500 cGy, for which two implants may be necessary, 2 weeks apart (Fig. 25.11).

Given the fact that the prognosis of patients with regional recurrences from EC is poor, with low salvage rate using RT alone, and high risk for distant failures, there is a need for new therapies in order to improve outcome. The GOG will hopefully soon activate a phase-II trial designed to determine the

Fig. 25.10. Modified templates for interstitial brachytherapy

response rate, duration of remission and patterns of relapse, in addition to survival and progression-free survival, in patients with loco-regional recurrences of EC limited to the pelvis treated with RT in combination with weekly cisplatin.

25.11.3
Salvage Surgery for Recurrent Endometrial Cancer

Isolated pelvic recurrences after definitive RT are rare and generally associated with distant failures. Patients with no evidence of metastatic disease, with recurrences limited to the central pelvis with no extent to the pelvic side wall, may be considered for exenterative surgical salvage – although the benefit is unclear, and the toxicity could be quite significant – however, the median survival time, is generally in the order of months (BARAKAT et al. 1999). Early recurrences are often associated with a very poor prognosis, and the role of additional surgery is generally limited. Although successful palliation can be achieved surgically in a selected subset of women, the identification of those who might benefit remains a challenge.

25.12
Palliative Therapy in Endometrial Cancer

25.12.1
Palliative Surgery

The role of surgery in the palliative setting in EC involves primarily: (1) debulking large masses in an

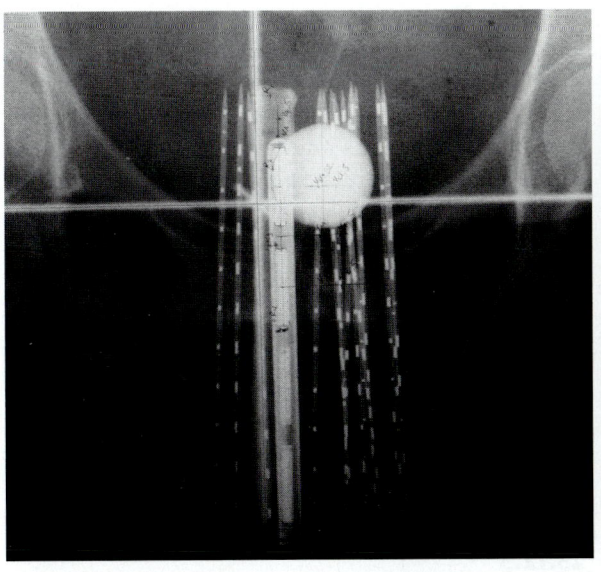

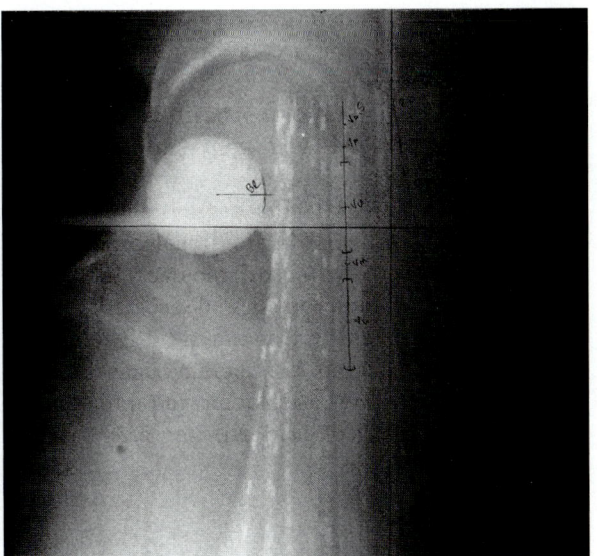

a

b

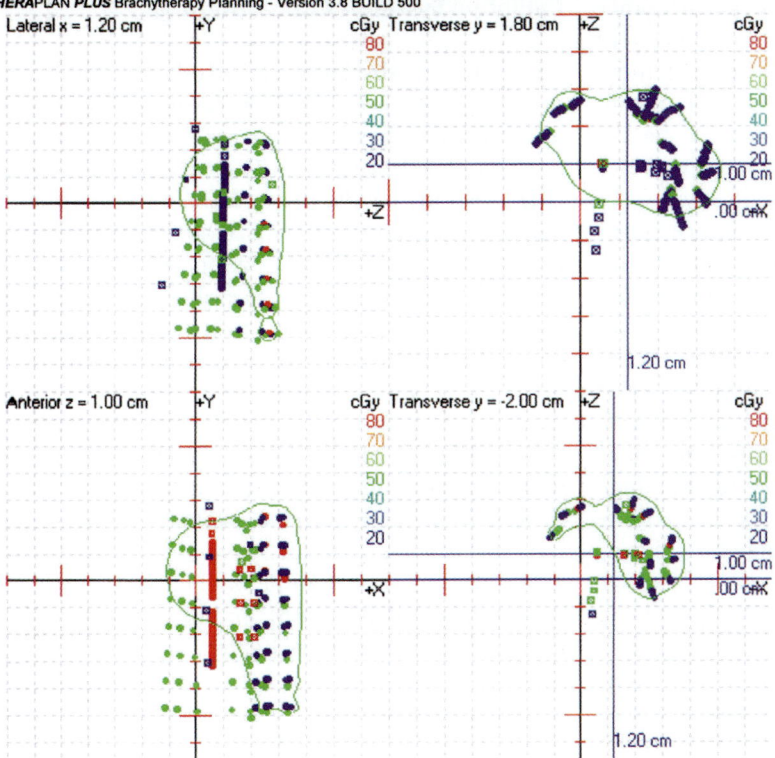

c

THERAPLAN PLUS Brachytherapy Planning - Version 3.8 BUILD 500

Fig. 25.11a–c. a Anteroposterior radiograph interstitial implant in recurrent endometrial cancer. **b** Lateral radiograph interstitial implant in recurrent endometrial cancer. **c** Isodose distribution for the interstitial implant

attempt to improve patients' symptoms and quality of life and (2) an attempt to relieve intestinal obstruction generally related to a mechanical obstruction from recurrent masses. Careful preoperative evaluation is indicated before any surgical intervention, including assessment of patient's performance and nutritional status and history of prior RT. Imaging studies may help to delineate the number and location of obstructions as well as the extent of intra-abdominal disease, in order to guide the surgical decision making process. Patients with recurrent disease and short life expectancy are unlikely to benefit from exploration. For patients in whom surgery may be indicated, the procedure to be performed needs to be carefully evaluated. It is important to exercise good judgment and expertise to maximize outcome for these patients with short life expectancy in order to improve their quality of life.

25.12.2
Palliative RT

Patients with grossly recurrent and metastatic EC following surgery and RT often have significant symptoms unresponsive to further systemic therapy. Symptoms may be due to recurrent disease in the pelvis, causing pain and/or bleeding. Distant recurrence in the brain, chest, groin and other areas may also require palliation.

There are no curative options for patients who present with stage-IVB disease. Many of these patients suffer from severe pelvic pain or bleeding. The selection of treatment modality and doses of RT in the palliative setting depend primarily on the extent of the disease in the pelvis, prior RT, patient's performance status and estimated length of survival. Patients with large volume disease in the pelvis with limited metastatic disease and good performance status, with an estimated survival of around 1 year or longer, could potentially benefit from a more protracted course of EBRT (4500–5000 cGy) in combination with or without brachytherapy, to be followed by systemic therapy. The techniques of EBRT to the pelvis have been described previously.

However, patients with poor performance status and/or extensive distant disease, who present with local symptoms such as bleeding or pain, should be treated using shorter RT regimens. If vaginal bleeding is the main concern, brachytherapy using endocavitary and/or interstitial techniques, when feasible, often offers good symptom control with relatively low morbidity. For patients who have received prior RT, intracavitary doses in the range of 35–40 Gy tumor dose may suffice to palliate symptoms. For those patients who may not be candidates for brachytherapy, a short course of EBRT using high dose fractionation schedules have been used. SPANOS et al. (1989) reported on a phase-II study (RTOG 85-02) of daily multifraction split-course EBRT in patients with recurrent or metastatic disease. The regimen consisted of 370 cGy per fraction given twice daily for two consecutive days and repeated at 2- to 6-week intervals, for a total of three courses (tumor dose, 4440 cGy). Occasionally, this regimen was combined with an ICB (4500 mg h), with a midline block in the last 1440 cGy. In patients completing three courses of irradiation (59%), the rate of complete or partial response was 45%. Of the patients, 27 survived longer than 1 year. Late complications were significantly less than expected, with a projected actuarial rate of 5% at 12 months (SPANOS et al. 1989). The same investigators (SPANOS

et al. 1993) have shown that using a 2-week interval between fractions improved the response rates mainly because more patients would complete all three courses of therapy. This schedule offers significant logistic benefits and has been shown to result in good tumor regression and excellent palliation of symptoms. SPANOS et al. (1994) reported a trend toward increased acute toxicity in patients with shorter rest periods, but late toxicity was not significantly different in the two groups. When using this regimen it is important to use CT simulation, since this would allow us a better target delineation and the use of more limited fields, very conformal, in an attempt to reduce not only acute but also long-term toxicity (Fig. 25.12).

25.12.3
Palliative Systemic Chemotherapy

Phase-II chemotherapy trials in women with advanced or recurrent EC have identified doxorubicin, cisplatin and carboplatin as active agents with response rates as single agents of 30–35%. The most commonly used chemotherapy combinations have been doxorubicin and cisplatin (THIGPEN et al. 2004), which was shown recently in a phase-III trial conducted by the GOG to be similar in terms of response rate, progression-free survival or overall survival to the triplet paclitaxel, doxorubicin and cisplatin (FLEMING et al. 2004). A preferred regimen, because of more limited toxicity, is the combination of carboplatin and paclitaxel (HOSKINS et al. 2001), which seems to be associated with response rates in the order of 60–80%. A phase-III trial would be required to evaluate a progression-free survival and overall survival benefit. A more extensive review of the role of chemotherapy in the management of advanced and/or recurrent EC is beyond the scope of this chapter.

25.13
Uterine Papillary Serous and Clear Cell Carcinoma

As previously mentioned, UPSC and CCC are aggressive histological subtypes that differ from typical EC by their poor prognosis, even in early stage disease, and a propensity for peritoneal as well as lymphatic and systemic spread. Therefore, it is imperative to perform a complete surgical staging, because of the expected high rate of surgical upstaging.

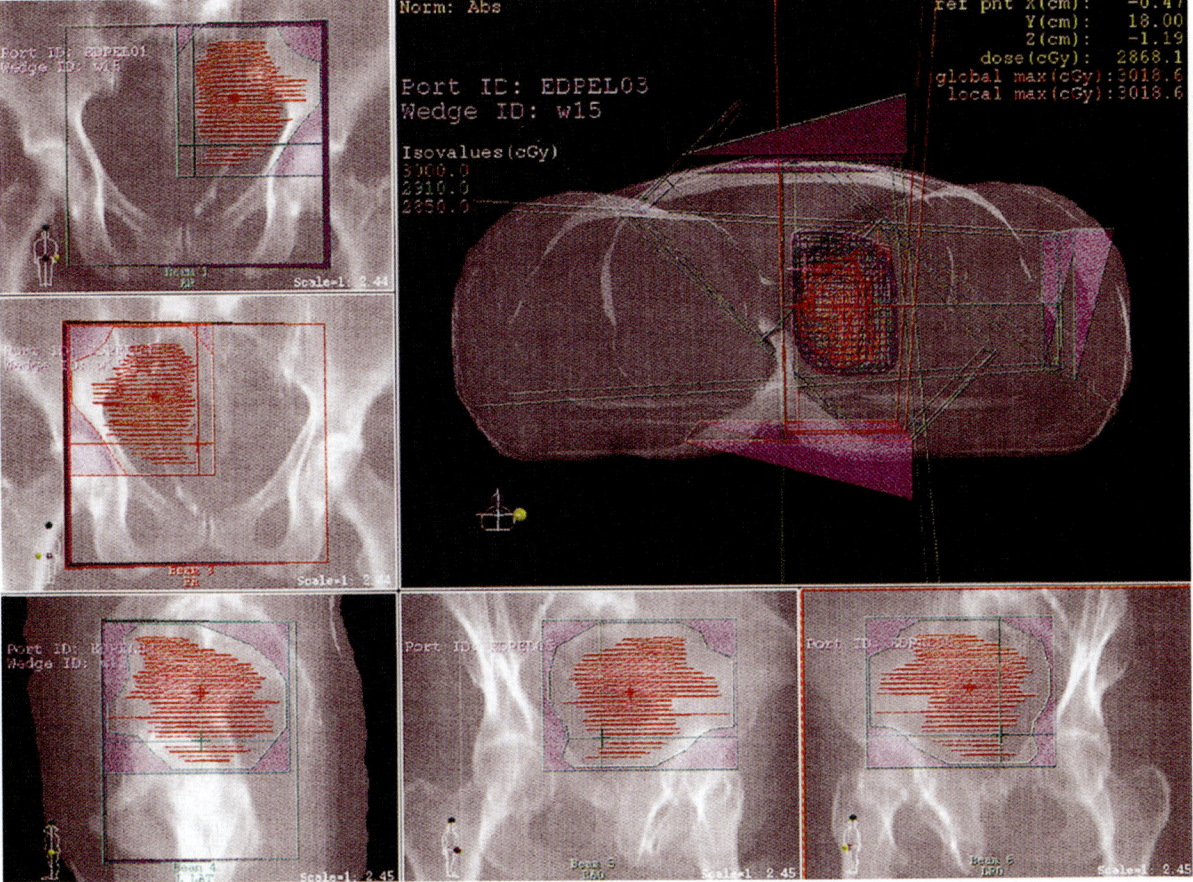

Fig. 25.12. Five-field coplanar beam arrangement for palliative external beam radiotherapy

The rarity of these subtypes makes prospective studies, and thus definitive treatment recommendations, difficult. Published adjuvant therapeutic approaches in patients with stage-I and -II disease, after complete surgical staging, include observation (HUH et al. 2003; CRAIGHEAD et al. 2000), systemic chemotherapy and vaginal cuff brachytherapy (TURNER et al. 1998; KELLY et al. 2004), and limited fields EBRT (MEHTA et al. 2003). However, the incidence of abdominal recurrences has led some investigators to study the impact of adjuvant WAI on these subtypes, with mixed results in single institution series. In a relatively large retrospective series of patients treated in the British Columbia Cancer Agency, 78 patients underwent hysterectomy and were found to have uterine confined disease or positive washings only (stages I–IIIA). There were 58 patients treated with adjuvant WAI similar to the technique described above, and they had a 5-year disease-specific survival of 74.9%, compared with 41.3% in 20 patients receiving less than WAI (LIM et al. 2001).

MARTINEZ et al. (2003) have also reported similar results with "high dose WAI" in patients with UPSC and CCC, with 5-year cause-specific survival of 80%; whereas, the 5-year DFS was only 49%, given the high-recurrence rates intra-abdominal as well as distant failures. The GOG did conduct a prospective trial (GOG 94), in which all stages of UPSC and CCC underwent surgical staging with optimal cytoreduction to less than 2 cm residual disease. Subsequently, they received adjuvant WAI (30 Gy) followed by a peri-aortic boost to 45 Gy, and a 49.8-Gy boost to the pelvis. The overall survival at 5 years, reported in abstract form, was 65% for stage I and stage II, and 33% for stage III and stage IV (AXELROD et al. 1995). SUTTON et al. (2005) have recently published the results of GOG 94 using WAI as adjuvant therapy for patients with EC stages III–IV, including 103 patients with UPSC and CCC histologies. The 3-year recurrence-free survival and overall survival for this group was 27% and 35%, respectively.

However, WAI is well recognized to have serious potential toxicities and, as discussed above,

GOG 122 prospectively compared WAI to AP chemotherapy, finding WAI to be inferior to AP chemotherapy, including stage-III and stage-IV UPSC or CCC (Randall et al. 2003). The inclusion of the poor prognosis subtypes in GOG 122 will likely also decrease interest in WAI in these patients.

The role of adjuvant RT, then, is probably best viewed using the principles of volume-directed adjuvant RT. In patients with advanced stages of UPSC and CCC that have spread to the lymph nodes (stage-IIIC disease), the potential for metastatic spread, combined with the results of GOG 122, suggest a prominent role for systemic therapy. However, overall outcomes with chemotherapy alone are still undesirable and volume-directed EBRT with or without vaginal cuff brachytherapy boost, followed by systemic chemotherapy, an approach similar to the recently closed GOG 184 protocol, may still provide a local control benefit.

In patients with negative nodes, comprehensively staged, without residual disease, the potential risk for peritoneal spread can be addressed with less toxicity using IP treatment with radioactive colloids. IP radioactive phosphorus (^{32}P) has been used for similar adjuvant therapy in the ovarian counterpart of UPSC with low morbidity (Young et al. 1990). Likewise, the vaginal cuff recurrence pattern can be addressed with IVB with acceptably low toxicity (Alektier et al. 2005).

Such a combined approach has been prospectively studied at Indiana University School of Medicine by the Hoosier Oncology Group (Fakiris et al. 2005). In this study, 22 patients were treated with up-front TAH-BSO, including surgical staging/lymph node dissections and omentectomy or omental biopsies in addition to peritoneal/pelvic washings. Maximal surgical debulking was required with patients having no residual disease of more than 3 mm, no positive lymph nodes and no evidence of disease outside of the abdomen. Their surgical stages ranged from IA to IVB, with the majority (17) having stage-I or -II disease. The treatment consisted of IP administration of 15 mCi of ^{32}P followed by three HDR-vaginal brachytherapy procedures. The results from this experience were encouraging, as acute toxicity was limited to grade 1, and no late toxicity was observed. In addition, with a median follow-up of 39.6 months, the disease-free survival data compares favorably with only five patients (20.3%) suffering recurrence (Fakiris et al. 2005). Of note, two recurrences were in the distal vagina early in the experience when we routinely treated the proximal vagina only. After these recurrences, the protocol was altered to treat

the entire vaginal length, and thereafter no vaginal recurrences have been seen. Therefore, we recommend treating the entire vaginal length for these patients when considering adjuvant IVB.

25.13.1
Intraperitoneal Radioactive Chromic Phosphate Suspension (^{32}P) Administration

Radioactive phosphorous-32 (^{32}P) seems to be the most attractive radiocolloid for IP administration because it is a pure β emitter that avoids the hazard of γ radiation. ^{32}P has a half life of 14.3 days, average tissue penetration of 1.4–3.0 mm, and maximum and average β energies of 1.71 MeV and 0.69 MeV, respectively. Chromic ^{32}P is a blue–green, chemically inert colloidal form of ^{32}P, used for intracavitary instillation. Unfortunately, the precise distribution and dose delivered by ^{32}P to the peritoneal surface is unknown and often unpredictable. The IP isotope distribution should be tested prior to instillation with radioactive technetium sulfur colloid (^{99m}Tc) or after radiocolloid administration with scintigraphic imaging of Bremsstrahlung photons.

The best available data supporting that 10–15 mCi of ^{32}P delivers superficial but therapeutic dosages to the peritoneal surfaces is from Currie et al. (1981). There appears to be predominantly an abdominal distribution with much smaller systemic absorption. The majority of the ^{32}P is absorbed either by the peritoneal surface or the macrophages lining the peritoneal cavity, or is phagocytized by free-floating macrophages. The rest of the ^{32}P is carried by the abdominal current to the right hemidiaphragm where it passes through the diaphragmatic lymphatics and enters the mediastinal lymphatics. It then passes to the right subclavian vein via the right thoracic trunk and enters the general circulation where it is rapidly cleared by the liver and deposited to a lesser extent in other tissues (spleen and bone marrow). Pelvic and para-aortic lymph nodes receive relatively low nontherapeutic doses. Several studies using imaging techniques confirm that the distribution of chromic ^{32}P is dynamic for the first 6–24 h but thereafter is fixed. Boye et al. (1984) reported that the estimated peritoneal surface dose from 10 mCi of ^{32}P is approximately 30 Gy, although the uptake and distribution of ^{32}P in the peritoneal cavity often shows significant inhomogeneity. They noted an increase in measurable levels of ^{32}P in the blood for 7 days, following which it declined; the estimated maximum dose to the peripheral blood even at its peak was 1.2 cGy.

The dose to the bone marrow was higher by a factor of two to five, but the maximum dose was still very low, in the order of 6 cGy.

The protocol of instillation and distribution of ^{32}P at Indiana University is outlined in Table 25.9. Generally, the administration takes place within 3 weeks of surgery, although occasionally we have treated patients up to 8 weeks from the surgical staging. Ideally, the catheter for IP administration should be placed at the time of the surgery, before adhesions are formed. However, the practical matter is that a decision regarding therapy is not completely made until a full pathological evaluation of the specimen is submitted.

At our institution, following placement of an IP catheter, generally within 3 weeks from surgery, a distribution study is performed using Tc^{99m} to ensure that there are no adhesions causing loculations, which could result in an uneven distribution of the radioisotope throughout the peritoneal surfaces (Fig. 25.13). In our experience, only 1 of 22 patients had a loculation preventing peritoneal therapy. Subsequently, the IP administration of 15 mCi of ^{32}P dissolved in 1500 cc of normal saline, is performed, as indicted in Table 25.9.

At 1 week after the IP administration, we proceed with IVB, using either HDR, 2100 cGy in three fractions prescribed to 0.5 cm depth, or LDR, 6500 cGy in one to two fractions prescribed to the vaginal surface. At the time the patient returns for her first HDR-IVB the suture from the catheter placement is removed and a CBC with differential and platelets is obtained to evaluate any potential hematological toxicity from the IP-32P administration.

Table 25.9. Procedure for administration of intraperitoneal ^{32}P at Indiana University

Patient admitted to the Hospital the day prior to or the same day of the administration. Patient needs to be NPO after midnight.

One multi-perforated peritoneal dialysis catheter (Tenckhoff's catheter) to be placed in the right or left lower quadrants.

Transportation to the Nuclear Medicine Department where approximately 2 mCi of ^{99m}Tc is inserted into the catheter, followed by approximately 100 cc of normal saline.

Patient instructed to roll from side to side and, preferably, to lie on the abdomen in order to distribute the radioisotope. A small detectable external marker should be placed on the patient's skin to identify the xiphoid process and the pubic symphysis.

The abdomen is then scanned, AP and lateral, following the right side injection.

If the distribution is poor, the procedure is terminated and no ^{32}P is administered.

15 mCi of ^{32}P will be dissolved in 250 cc of normal saline (NS) and injected into the abdominal cavity using an intravenous line (running full flow) through the catheter.

Approximately 1000–1500 cc of NS is then injected into the peritoneal cavity following the radioisotope administration.

Intraperitoneal catheter is removed at the completion of the procedure. The catheter insertion site should be closed with a purse-string suture.

The patient is transported to her room and instructed to turn to her left side, onto her back in Trendelenburg and reverse Trendelenburg positions, onto her right side and her abdomen every 10 min for about 2 h.

Intramuscular compazine may be given routinely for the first 24 h to prevent the occurrence of nausea and vomiting.

The patient can be discharged home with adequate radiation precautions once the rotational schedule is completed.

Radiation precautions are recommended within the area of administration and patient's room, with regard to linens saved and room decontamination procedures.

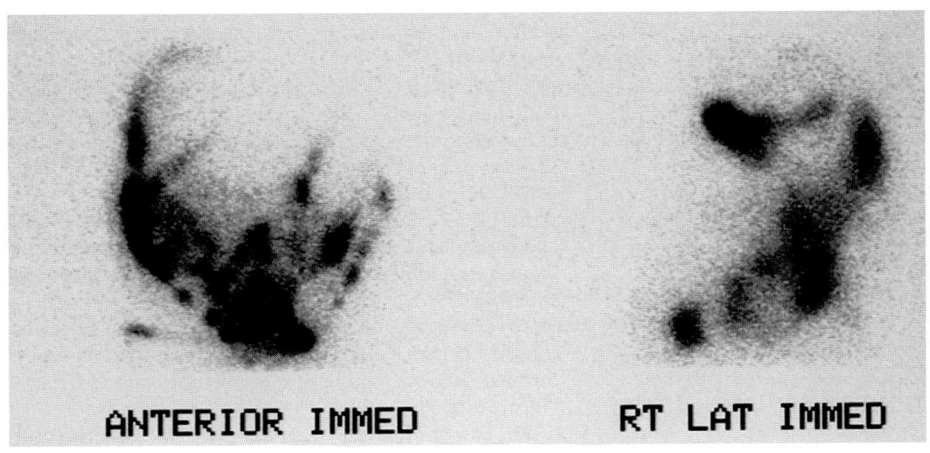

ANTERIOR IMMED   RT LAT IMMED

Fig. 25.13. Distribution study prior to administration of intraperitoneal ^{32}P

References

Aalders J, Abeler V, Kolstad P et al (1980) Postoperative external irradiation and prognostic parameters in stage I endometrial carcinoma: clinical and histopathologic study of 540 patients. Obstet Gynecol 56:419-427

Aarnio M, Mecklin JP, Aaltonen LA et al (1995) Life-time risk of different cancers in hereditary non-polyposis colorectal cancer (HNPCC) syndrome. Int J Cancer 64:430-433

Ackerman I, Malone S, Thomas G et al (1996) Endometrial carcinoma-relative effectiveness of adjuvant irradiation vs therapy reserved for relapse (see comment). Gynecol Oncol 60:177-183

Ahamad A, D'Souza W, Salehpour M et al (2005) Intensity-modulated radiation therapy after hysterectomy: comparison with conventional treatment and sensitivity of the normal-tissue-sparing effect to margin size. Int J Radiat Oncol Biol Phys 62:1117-1124

Alektiar KM, McKee A, Venkatraman E et al (2002) Intravaginal high-dose-rate brachytherapy for Stage IB (FIGO Grade 1, 2) endometrial cancer. Int J Radiat Oncol Biol Phys 53:707-713

Alektiar KM, Venkatraman E, Chi DS et al (2005) Intravaginal brachytherapy alone for intermediate-risk endometrial cancer. Int J Radiat Oncol Biol Phys 62:111-117

American Cancer Society (2005) Cancer facts and figures, p 21

Anderson JM, Stea B, Hallum AV et al (2000) High-dose-rate postoperative vaginal cuff irradiation alone for stage IB and IC endometrial cancer. Int J Radiat Oncol Biol Phys 46:417-425

Ashman JB, Connell PP, Yamada D et al (2001) Outcome of endometrial carcinoma patients with involvement of the uterine serosa. Gynecol Oncol 82:338-343

Axelrod J, Bundy J, Roy T et al (1995) Advanced endometrial carcinoma treated with whole abdominal irradiation: a GOG Study (abstract). Gynecol Oncol 56:135-136

Barakat RR, Goldman NA, Patel DA et al (1999) Pelvic exenteration for recurrent endometrial cancer (abstract). Gynecol Oncol 250:163

Bokhman JV (1983) Two pathogenetic types of endometrial carcinoma. Gynecol Oncol 15:10-17

Boye E, Lindegaard MW, Paus E et al (1984) Whole-body distribution of radioactivity after intraperitoneal administration of 32P colloids. Br J Radiol 57:395-402

Chadha M, Nanavati PJ, Liu P et al (1999) Patterns of failure in endometrial carcinoma stage IB grade 3 and IC patients treated with postoperative vaginal vault brachytherapy (see comment). Gynecol Oncol 75:103-107

Chao CK, Grigsby PW, Perez CA et al (1996) Medically inoperable stage I endometrial carcinoma: a few dilemmas in radiotherapeutic management. Int J Radiat Oncol Biol Phys 34:27-31

Chen LM, McGonigle KF, Berek JS (1999) Endometrial cancer: recent developments in evaluation and treatment. Oncology 13:1665-1675

Connell PP, Rotmensch J, Waggoner S et al (1999) The significance of adnexal involvement in endometrial carcinoma. Gynecol Oncol 74:74-79

Craighead PS, Sait K, Stuart GC et al (2000) Management of aggressive histologic variants of endometrial carcinoma at the Tom Baker Cancer Centre between 1984 and 1994. Gynecol Oncol 77:248-253

Creasman WT, Morrow CP, Bundy BN et al (1987) Surgical pathologic spread patterns of endometrial cancer. A Gynecologic Oncology Group Study. Cancer 60:2035-2041

Creutzberg CL, van Putten WL, Koper PC et al (2000) Surgery and postoperative radiotherapy versus surgery alone for patients with stage 1 endometrial carcinoma: multicentre randomised trial PORTEC Study Group Post Operative Radiation Therapy in Endometrial Carcinoma (see comment). Lancet 355:1404-1411

Creutzberg CL, van Putten WL, Koper PC et al (2003) Survival after relapse in patients with endometrial cancer: results from a randomized trial (see comment). Gynecol Oncol 89:201-209

Currie JL, Bagne F, Harris C et al (1981) Radioactive chromic phosphate suspension: studies on distribution, dose absorption, and effective therapeutic radiation in phantoms, dogs, and patients. Gynecol Oncol 12:193-218

Disaia PJ, Syed N, Puthwala AA (1990) Malignant neoplasia of the upper vagina. Endocurietherapy / Hyperthermia Oncol 6:251-256

Eltabbakh GH, Piver MS, Hempling RE et al (1997) Excellent long-term survival and absence of vaginal recurrences in 332 patients with low-risk stage I endometrial adenocarcinoma treated with hysterectomy and vaginal brachytherapy without formal staging lymph node sampling: report of a prospective trial. Int J Radiat Oncol Biol Phys 38:373-380

Fakiris AJ, Moore DH, Reddy SR et al (2005) Intraperitoneal radioactive phosphorus (32P) and vaginal brachytherapy as adjuvant treatment for uterine papillary serous carcinoma and clear cell carcinoma: a phase II Hoosier Oncology Group (HOG 97-01) study. Gynecol Oncol 96:818-823

Fanning J (2001) Long-term survival of intermediate risk endometrial cancer (stage IG3, IC, II) treated with full lymphadenectomy and brachytherapy without teletherapy. Gynecol Oncol 82:371-374

Fishman DA, Roberts KB, Chambers JT et al (1996) Radiation therapy as exclusive treatment for medically inoperable patients with stage I and II endometrioid carcinoma with endometrium. Gynecol Oncol 61:189-196

Fleming GF, Brunetto VL, Cella D et al (2004) Phase III trial of doxorubicin plus cisplatin with or without paclitaxel plus filgrastim in advanced endometrial carcinoma: a Gynecologic Oncology Group Study. J Clin Oncol 22:2159-2166

Grigsby PW, Kuske RR, Perez CA et al (1986) Medically inoperable stage I adenocarcinoma of the endometrium treated with radiotherapy alone. Int J Radiat Oncol Biol Phys 13:483–488

Grigsby PW, Perez CA, Kuten A et al (1992) Clinical stage I endometrial cancer: prognostic factors for local control and distant metastasis and implications of the new FIGO surgical staging system. Int J Radiat Oncol Biol Phys 22:905-911

Hoskins PJ, Swenerton KD, Pike JA et al (2001) Paclitaxel and carboplatin, alone or with irradiation, in advanced or recurrent endometrial cancer: a phase II study. J Clin Oncol 19:4048-4053

Hsieh CH, ChangChien CC, Lin H et al (2002) Can a preoperative CA 125 level be a criterion for full pelvic lymphadenectomy in surgical staging of endometrial cancer? Gynecol Oncol 86:28-33

Huh WK, Powell M, Leath CA 3rd et al (2003) Uterine papillary serous carcinoma: comparisons of outcomes in surgical Stage I patients with and without adjuvant therapy (see comment). Gynecol Oncol 91:470-475

International Federation of Gynecology and Obstetrics (FIGO) (1971) Classification and staging of malignant tumors in the female pelvis. Int J Gynaecol Obstet 9:172

International Federation of Gynecology and Obstetrics (FIGO) (1989) Corpus cancer staging. Int J Gynecol Obstet 28:189-190

Jhingran A, Burke TW, Eifel PJ (2003) Definitive radiotherapy for patients with isolated vaginal recurrence of endometrial carcinoma after hysterectomy. Int J Radiat Oncol Biol Phys 56:1366-1372

Kelly MG, O'Malley D, Hui P et al (2004) Patients with uterine papillary serous cancers may benefit from adjuvant platinum-based chemoradiation. Gynecol Oncol 95:469-473

Keys HM, Roberts JA, Brunetto VL et al (2004) A phase III trial of surgery with or without adjunctive external pelvic radiation therapy in intermediate risk endometrial adenocarcinoma: a Gynecologic Oncology Group study (see comment; erratum appears in Gynecol Oncol 2004, 94:241-242). Gynecol Oncol 92:744-751

Kucera H, Vavra N, Weghaupt K (1990) Benefit of external irradiation in pathologic stage I endometrial carcinoma: a prospective clinical trial of 605 patients who received postoperative vaginal irradiation and additional pelvic irradiation in the presence of unfavorable prognostic factors. Gynecol Oncol 38:99-104

Kupelian PA, Eifel PJ, Tornos C et al (1993) Treatment of endometrial carcinoma with radiation therapy alone. Int J Radiat Oncol Biol Phys 27:817-824

Larson DM, Johnson KK, Broste SK et al (1995) Comparison of D&C and office endometrial biopsy in predicting final histopathologic grade in endometrial cancer. Obstet Gynecol 86:38-42

Lim P, Al Kushi A, Gilks B et al (2001) Early stage uterine papillary serous carcinoma of the endometrium: effect of adjuvant whole abdominal radiotherapy and pathologic parameters on outcome. Cancer 91:752-757

MacLeod C, Fowler A, Duval P et al (1998) High-dose-rate brachytherapy alone post-hysterectomy for endometrial cancer. Int J Radiat Oncol Biol Phys 42:1033-1039

Mariani A, Webb MJ, Keeney GL et al (2000) Low-risk corpus cancer: is lymphadenectomy or radiotherapy necessary? Am J Obstet Gynecol 182:1506-1519

Mariani A, Webb MJ, Keeney GL et al (2002a) Assessment of prognostic factors in stage IIIA endometrial cancer. Gynecol Oncol 86:38-44

Mariani A, Webb MJ, Keeney GL et al (2002b) Stage IIIC endometrioid corpus cancer includes distinct subgroups. Gynecol Oncol 87:112-117

Martinez AA, Weiner S, Podratz K et al (2003) Improved outcome at 10 years for serous-papillary/clear cell or high-risk endometrial cancer patients treated by adjuvant high-dose whole abdomino-pelvic irradiation. Gynecol Oncol 90:537-546

McCormick TC, Cardenes H, Randall ME (2002) Early-stage endometrial cancer: is intravaginal radiation therapy alone sufficient therapy? Brachytherapy 1:61-65

Mehta N, Yamada SD, Rotmensch J et al (2003) Outcome and pattern of failure in pathologic stage I-II papillary serous carcinoma of the endometrium: implications for adjuvant radiation therapy. Int J Radiat Oncol Biol Phys 57:1004-1009

Morrow CP, Bundy BN, Kurman RJ et al (1991) Relationship between surgical-pathological risk factors and outcome in clinical stage I and II carcinoma of the endometrium: a Gynecologic Oncology Group study. Gynecol Oncol 40:55-65

Nag S, Erickson B, Parikh S et al (2000) The American Brachytherapy Society recommendations for high-dose-rate brachytherapy for carcinoma of the endometrium. Int J Radiat Oncol Biol Phys 48:779-790

Naumann RW, Higgins RV, Hall JB (1999) The use of adjuvant radiation therapy by members of the Society of Gynecologic Oncologists (see comment). Gynecol Oncol 75:4-9

Nelson G, Randall M, Sutton G et al (1999) FIGO stage IIIC endometrial carcinoma with metastases confined to pelvic lymph nodes: analysis of treatment outcomes, prognostic variables, and failure patterns following adjuvant radiation therapy. Gynecol Oncol 75:211-214

Ng TY, Perrin LC, Nicklin JL et al (2000) Local recurrence in high-risk node-negative stage I endometrial carcinoma treated with postoperative vaginal vault brachytherapy (see comment). Gynecol Oncol 79:490-494

Ng TY, Nicklin JL, Perrin LC et al (2001) Postoperative vaginal vault brachytherapy for node-negative Stage II (occult) endometrial carcinoma. Gynecol Oncol 81:193-195

Petereit DG, Tannehill SP, Grosen EA et al (1999) Outpatient vaginal cuff brachytherapy for endometrial cancer. Int J Gynecol Cancer 9:456-462

Randall ME, Brunetto VL, Muss H et al. (2003) Whole abdominal radiotherapy versus combination doxorubicin-cisplatin chemotherapy in advanced endometrial carcinoma: a randomized phase III trial of the Gynecologic Oncology Group. Proc Am Soc Clin Oncol 22:2. Abstract no. 3

Silverberg S, Kurman R (1992) Tumors of the uterine corpus and gestational trophoblastic disease, vol 3. Armed Forces Institute of Pathology, Washington, DC

Slomovitz BM, Burke TW, Eifel PJ et al (2003) Uterine papillary serous carcinoma (UPSC): a single institution review of 129 cases (see comment). Gynecol Oncol 91:463-469

Smith RA, von Eschenbach AC, Wender R et al (2001) American Cancer Society guidelines for the early detection of cancer: update of early detection guidelines for prostate, colorectal, and endometrial cancers Also: update 2001-testing for early lung cancer detection (erratum appears in CA Cancer J Clin 2001, 51:150). Ca Cancer J Clin 51:38-75; quiz 77-80

Spanos W Jr, Guse C, Pere C et al (1989) Phase II study of multiple daily fractionations in the palliation of advanced pelvic malignancies: preliminary report of RTOG 8502. Int J Radiat Oncol Biol Phys 17:659-661

Spanos WJ Jr, Perez CA, Marcus S et al (1993) Effect of rest interval on tumor and normal tissue response-a report of phase III study of accelerated split course palliative radiation for advanced pelvic malignancies (RTOG-8502) (see comment). Int J Radiat Oncol Biol Phys 25:399-403

Spanos WJ Jr, Clery M, Perez CA et al (1994) Late effect of multiple daily fraction palliation schedule for advanced pelvic malignancies (RTOG 8502). Int J Radiat Oncol Biol Phys 29:961-967

Straughn JM Jr, Huh WK, Kelly FJ et al (2002) Conservative management of stage I endometrial carcinoma after surgical staging (see comment). Gynecol Oncol 84:194-200

Sutton G, Axelrod JH, Bundy BN et al (2005) Whole abdominal radiotherapy in the adjuvant treatment of patients with stage III and IV endometrial cancer: a Gynecologic Oncology Group Study. Gynecol Oncol 97:755-763

Taghian A, Pernot M, Hoffstetter S et al (1988) Radiation therapy alone for medically inoperable patients with adenocarcinoma of the endometrium. Int J Radiat Oncol Biol Phys 15:1135-1140

Thigpen JT, Brady MF, Homesley HD et al (2004) Phase III trial of doxorubicin with or without cisplatin in advanced endometrial carcinoma: a gynecologic oncology group study. J Clin Oncol 22:3902-3908

Turner BC, Knisely JP, Kacinski BM et al (1998) Effective treatment of stage I uterine papillary serous carcinoma with high dose-rate vaginal apex radiation (192Ir) and chemotherapy. Int J Radiat Oncol Biol Phys 40:77-84

Weiss E, Hirnle P, Arnold-Bofinger H et al (1998) Adjuvant vaginal high-dose-rate afterloading alone in endometrial carcinoma: patterns of relapse and side effects following low-dose therapy. Gynecol Oncol 71:72-76

Young RC, Walton LA, Ellenberg SS et al (1990) Adjuvant therapy in stage I and stage II epithelial ovarian cancer Results of two prospective randomized trials (see comment). N Engl J Med 322:1021-1027

Zaino RJ, Kurman RJ, Diana KL et al (1996) Pathologic models to predict outcome for women with endometrial adenocarcinoma: the importance of the distinction between surgical stage and clinical stage - a Gynecologic Oncology Group study (erratum appears in Cancer 1997, 79:422). Cancer 77:1115-1121

26 Vulva

Carlos A. Perez and Imran Zoberi

CONTENTS

26.1   Anatomy 631
26.2   Natural History 632
26.3   Diagnostic Workup 633
26.3.1 Lymph Node Evaluation 633
26.4   Staging 634
26.5   Pathological Classification 634
26.6   Prognostic Factors 636
26.6.1 Capillary Lymphatic Space Involvement and
        Lymph Node Metastasis 637
26.6.2 Extracapsular Extension of
        Metastatic Lymph Nodes 637
26.7   General Management 637
26.7.1 Stage III and IV Tumors 638
26.7.2 Management of Patients with Positive Nodes 639
26.8   Radiation Therapy Techniques 640
26.8.1 Irradiation Alone 640
26.8.2 Regional Lymphatics 641
26.8.3 Preoperative Radiation Therapy 641
26.8.4 Postoperative Radiation Therapy 642
26.8.5 Treatment for Recurrent Lesions 642
26.8.6 Patient Positioning and Simulation 642
26.8.7 Beams and Energies 642
26.8.8 Lymphatic Irradiation 644
26.8.9 Brachytherapy 646
26.8.10 Three-Dimensional Conformal and Intensity
        Modulated Radiation Therapy 646
26.8.11 Patterns of Failure after Treatment 646
26.9   Sequelae of Treatment 648
26.10   Carcinoma of the Vulva and Pregnancy 649
26.11   Clinical Trials 649
26.12   Treatment of Other Vulvar Malignant Neoplasias 649
26.12.1 Adenocarcinoma 649
26.12.2 Vulvar Sarcoma 649
26.12.3 Malignant Melanoma 650
26.13   Chemotherapy 650
26.13.1 Combination of Chemotherapy and Irradiation 650
        References 652

C. A. Perez, MD
Professor, Department of Radiation Oncology, Washington University Medical Center, 4511 Forest Park Boulevard, Suite 200, St. Louis, MO 63108, USA
I. Zoberi, MD
Department of Radiation Oncology, Campus Box 8224, 4921 Parkview Place, St. Louis, MO 63110, USA

26.1 Anatomy

The vulva consists of the mons pubis, clitoris, labia majora and minora, vaginal vestibule, and their supporting subcutaneous tissues and blends with the urinary meatus anteriorly and with the perineum and anus posteriorly. The mons pubis is a prominent mound of subcutaneous connective and adipose tissue located anterior to the pubic symphysis; after puberty it is covered with pubic hair. The labia majora are two elongated skin folds that course posteriorly from the mons pubis and blend into the perineal body. The skin of the labia majora is pigmented and contains both hair follicles and sebaceous glands. The labia minora are a smaller pair of skin folds located between the labia majora; they extend posteriorly to form the margin of the vaginal vestibule. Anteriorly, the labia minora separate into two components that course above and below the clitoris, fusing with those of the opposite side to form the prepuce and frenulum, respectively. The skin of the minora contains numerous sebaceous glands but no hair follicles and has no underlying adipose tissue. The clitoris is supported by the fusion of the labia minora and is about 2–3 cm anterior to the urethral meatus. It is composed of erectile tissue organized into the glans, body, and two crura, which course laterally, covered by the ischiocavernosus muscles, and attach to the ischial rami.

The vaginal vestibule is in the center of the vulva and is demarcated laterally by the labia minora and posteriorly by the perineal body. Anteriorly, numerous small vestibular glands are located beneath the mucosa and open onto its surface adjacent to the urethral meatus. The Bartholin's glands, two small mucous-secreting glands situated within the subcutaneous tissue of the posterior labia majora, have ducts that open onto the posterolateral portion of the vestibule. The perineal body is a 3- to 4-cm band of skin, located between the posterior extension of the labia majora, which separates the vaginal vestibule from the anus and forms the posterior margin of the vulva (Burke et al. 1996).

Lymphatics of the labia drain into the superficial inguinal and femoral lymph nodes, located anterior to the cribriform plate and fascia lata. The lymphatics subsequently penetrate the cribriform fascia and reach the deep femoral nodes. Lymphatics of the fourchette, perineum, and prepuce follow the lymphatics of the labia. The lymph from the glans clitoris, on the other hand, can drain not only to the inguinal nodes but also to deep femoral nodes and pelvic lymph nodes. Some lymphatics originating in the clitoris may enter the pelvis directly, connecting with the obturator and external iliac lymph nodes, bypassing the femoral area (Fig. 26.1a). About ten superficial inguinal lymph nodes lie along the saphenous vein and its branches between Camper's fascia and the cribriform fascia overlying the femoral vessels. The superficial nodes are located within the triangle formed by the inguinal ligament superiorly, the border of the sartorius muscle laterally, and the border of the adductor longus muscle medially (Fig. 26.1b). There are usually three to five deep nodes, the most superior of which, located under the inguinal ligament, is known as Cloquet's node (BURKE et al. 1996). From these, the lymph drains into the pelvic lymphatics (external and common iliac lymph nodes).

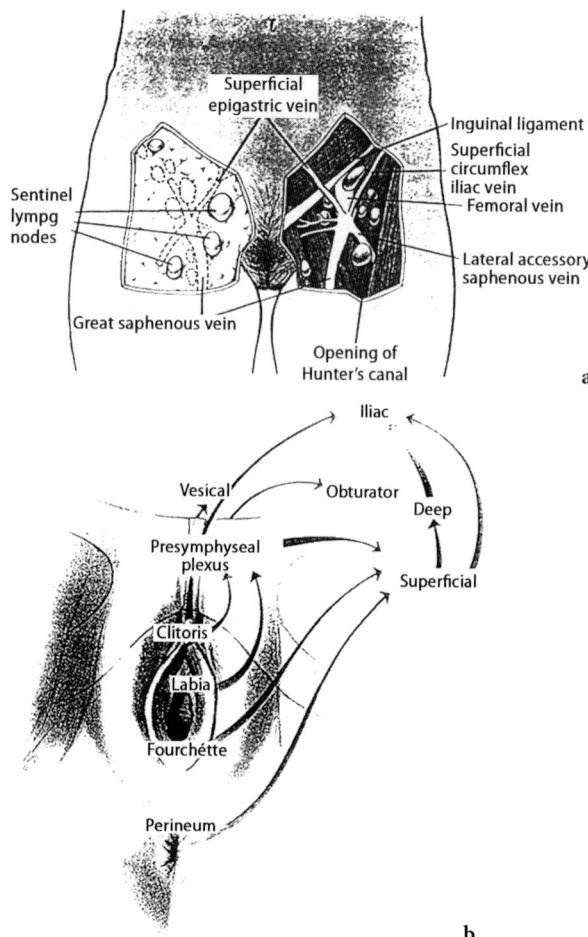

Fig. 26.1. a The lymphatic drainage of the vulva initially flows to the superficial inguinal nodes, then to the deep femoral and iliac groups. Drainage from midline structures may flow directly beneath the symphysis to the pelvic nodes (PLENTL and FRIEDMAN 1971). b The superficial inguinal lymph nodes comprise eight to ten subcutaneous nodes located between Camper's fascia and the cribriform fascia. These nodes are immediately adjacent to the saphenous vein and its branches (DISAIA et al. 1979)

26.2
Natural History

Over 70% of vulvar malignancies arise in the labia majora and minora, 10–15% in the clitoris, and 4–5% in the perineum and fourchette. The vestibule, Bartholin's gland, and the clitoral prepuce are unusual primary sites, each accounting for less than 1% of vulvar cancers (PLENTL and FRIEDMAN 1971).

Carcinomas arising in the vulvar area ordinarily follow a predictable pattern of spread to the regional lymph nodes. Superficial inguinofemoral lymph nodes are involved first, followed by the deep inguinofemoral nodes. Metastasis to the contralateral inguinal or pelvic lymph nodes is very unusual in the absence of ipsilateral inguinofemoral node metastasis. Although lesions arising in or involving the glans clitoris or urethra theoretically can spread to pelvic lymph nodes through the channels that bypass the inguinal areas, such metastases without inguinal node involvement occur infrequently (FRANKLIN and RUTLEDGE 1971, KRUPP and BOHM 1978).

The incidence of inguinal lymph node metastasis in surgically staged patients varies from 6% to 50%, depending on tumor invasion (Table 26.1) (RUTLEDGE et al. 1970; BOUTSELIS 1972; PARKER et al. 1975; DONALDSON et al. 1981; HACKER et al. 1981; HOFFMAN et al. 1988; WANG et al. 1996). PLENTL and FRIEDMAN (1971) reported a 62% incidence of lymph node metastases in patients with clinically palpable adenopathy and 35% without clinically palpable lymph nodes. In a review of clinical staging, FRANKLIN (1972) noted that approximately 75% of patients with clinically suspicious lymph nodes proved to have nodal metastasis, and nodes that were clinically negative were found to be positive for metastasis in 11–43% of cases. In women with T1 or T2 carcinoma of the vulva who underwent radical vulvectomy and bilateral inguinofemoral lymphadenectomy, 23 of 104 (22.1%) with unilateral tumor had lymph node metastases; 21 (91%) had unilateral, one contralateral, and one bilateral metastases (Table 26.2); the contralateral node may have been from a previously treated endometrial carcinoma (BURGER et al. 1996). Approximately 20–30% of patients with histologically proven involvement of femoral nodes show deep pelvic lymph node involvement if pelvic lymphadenectomy is performed (BOUTSELIS 1972). Hematogenous dissemination is unusual and is a manifestation of late disease (STERN and KAPLAN 1969). The most common metastatic sites are lung, liver, and bone.

26.3
Diagnostic Workup

Clinical history and a complete physical examination are essential. In addition to the vulvar and anal area and perineum, the physical examination should include the vagina and cervix, which should be thoroughly inspected. Careful bimanual pelvic examination is mandatory. Besides careful determination of the extent and depth of the primary lesion (size, fixation, etc.), essential in physical examination is assessment of the regional lymph nodes; although, because of inflammatory lymphadenopathy in the inguinal area, lymph node assessment in vulvar tumors has a substantial rate of error. A Papanicolaou smear of the cervix and vagina should be performed.

Chest radiographs should be routinely obtained. Other studies include cystoscopy, proctosigmoidoscopy, barium enema, and intravenous pyelogram when indicated. Computed tomography (CT) or magnetic resonance imaging (MRI) aid in the definition of tumor extent and in evaluating the inguinal and pelvic/periaortic lymph nodes. Radiographic evaluation of regional lymphatics in carcinoma of the vulva is of limited value and is rarely used. Preoperative lymphography was evaluated at the Mallinckrodt Institute of Radiology in 32 patients with vulvar carcinoma; correlation with the surgical specimens showed an overall accuracy of 54.5%, with a sensitivity of 15.7% and a specificity of 66.1% (WEINER et al. 1986). The standard workup for these patients is described in Table 26.3.

26.3.1
Lymph Node Evaluation

Recently, positron emission tomography (PET) was used to detect inguinal node metastasis is 15 patients prior to exploration of 29 groins (COHN et al. 2002). Of these patients, 6 had positive scans, suggesting nodal metastasis in 8 groins. On pathological exam, 5 patients had metastasis in 9 groins; PET demonstrated the metastasis in 4 of the patients and in 6 groins. The positive predictive value of PET was 80% and negative predictive value 90%. The authors concluded that PET was relatively insensitive in predicting groin lymph node metastases, but the high specificity (90%) made it useful in planning radiation therapy.

Table 26.1. Incidence of lymph node involvement correlated with primary tumor size and extent. Data from RUTLEDGE et al. 1970; BOUTSELIS 1972; PARKER et al. 1975; KRUPP and BOHM 1978; DONALDSON et al. 1981; HACKER et al. 1981; HOFFMAN et al. 1988; and PEREZ et al. 1997

Primary tumor size or depth of invasion	No. of patients	Number of patients with positive lymph nodes (%)
Depth		
ä1 mm	120	0 (0)
1.1–2 mm	121	8 (6.6)
2.1–3 mm	97	8 (8.2)
3.1–4 mm	50	11 (22)
4.1–5 mm	40	10 (25)
Size		
>5 mm	32	12 (37.5)
>2 cm	168	77 (45.8)
Any size primary tumor extending beyond vulva	70	38 (54.2)

Table 26.2. Relationship between laterality of primary tumor and laterality of inguinofemoral lymph node metastases (n=180) (BURGER et al. 1996)

	Laterality of primary tumor		
Laterality of inguinofemoral metastases	Unilateral left	Unilateral right	Central or bilateral
No metastases	44	37	43
Unilateral left	10	1	12
Unilateral right		11	9
Bilateral		1	12

Table 26.3. Diagnostic workup for vulva tumors (PEREZ et al. 1997)

General
– History
– Physical examination, including careful bimanual pelvic examination
Special studies
– Exfoliative cytology of cervix and vagina
– Colposcopy and directed biopsies (including Schiller's test)
– Biopsies and examination under anesthesia to determine tumor extent
– Cytoscopy
– Proctosigmoidoscopy (as indicated)
Radiographic studies
– Standard
  – Chest radiographs
  – Intravenous pyelogram
– Complementary
  – Barium enema
  – Lymphangiogram
  – Computed tomography or magnetic resonance imaging of pelvis and abdomen
Laboratory studies
– Complete blood count
– Blood chemistry
– Urinalysis

Surgical assessment of superficial femoral (inguinal) lymph nodes continues to be a widely accepted procedure for determination of therapeutic strategy, since survival is closely correlated with the pathological status of the inguinal nodes (at 5 years, 80–85% for patients with negative nodes and 40–50% with positive nodes). It is generally agreed that pelvic lymph nodes do not need to be treated in patients without femoral lymph node involvement, yet the identification of patients with involved inguinal nodes for therapeutic management directed at the pelvis is of obvious importance. Many surgeons routinely dissect the pelvis in patients with positive inguinal areas; however, CURRY et al. (1980) noted that none of their patients with three or fewer unilaterally positive groin nodes had positive deep pelvic nodes. Similar findings have been observed by HACKER et al. (1983). Although deep pelvic node involvement is an ominous sign, one-quarter to one-third of the patients are still salvageable, particularly if only a few nodes are involved (FRANKLIN and RUTLEDGE 1971; BOUTSELIS 1972; CURRY et al. 1980). Multiple prognostic factors were identified by RUTLEDGE et al. (1991).

26.4
Staging

The International Federation of Gynecology and Obstetrics (FIGO) adopted a modified surgical staging system for vulvar cancer in 1989 (SHEPHERD 1989; CREASMAN 1990). Staging errors led to acceptance of the current surgical evaluation of the inguinal lymph nodes. Tumor assessment is based on physical examination with endoscopy in cases of bulky disease. Nodal status is determined by the surgical evaluation of the groins.

A microinvasive substage (IA) was defined by FIGO for tumors less than 2 cm in diameter with depth of invasion less than 1 mm (PARKER et al. 1975).

The stages correlate well with treatment results. The use of 2.0 cm as the size to determine assignment of the primary lesion to T1 or T2 has been criticized, however. KRUPP et al. (1975) and DONALDSON et al. (1981) suggested that a 3.0-cm size be the determining factor for assignment of T category. The American Joint Committee (GREEN et al. 2002) and FIGO staging systems are shown in Table 26.4 and Figure 26.2 (DuBESHTER et al. 1993).

26.5
Pathological Classification

Preinvasive forms of vulvar malignancy include carcinoma in situ (Bowen's disease or erythroplasia of Queyrat and Paget's disease). Paget's disease is equivalent to the same entity in the breast and is associated with invasive apocrine carcinoma in approximately 20–30% of cases (HELWIG and GRAHAM 1963; KAUFMAN and GARDNER 1965; ULBRIGHT et al. 1983; BARNHILL et al. 1988; HOSKINS and PEREZ 1989). Unsuspected invasion in patients with intraepithelial vulva neoplasia was described in 13 of 69 patients (18.8%) and superficial invasion in 8 patients (CHAFE et al. 1988).

Squamous cell carcinoma comprises over 90% of invasive lesions of the vulva. Histologically, most of these tumors are well differentiated with keratin formation; 5–10% are anaplastic (GOSLING et al. 1961).

Two variants of squamous cell carcinoma infrequently described are adenosquamous (LASSER et al. 1974) and basaloid carcinoma (LUCAS et al. 1974).

Verrucous carcinoma of the vulva is extremely rare. Until 1988, the number of cases involving the female genital tract was 89 (ANDERSON and SORENSON 1988). Verrucous carcinoma is a slowly growing, nonaggressive tumor with pushing margins and a rather benign histological appearance. The incidence of lymph node metastasis is very low. The preferred treatment is wide surgical excision.

Table 26.4. TNM and staging classifications for carcinoma of the vulva. TNM tumor-node-metastasis; FIGO International Federation of Gynecology and Obstetrics (GREENE FL et al. 2002)

TNM (AJCC) 2002		Staging (FIGO) 1988		
T primary tumor				
		Stage 0	Tis	Carcinoma in situ; intraepithelial carcinoma
Tis	Preinvasive carcinoma (carcinoma in situ)	Stage I	T1 N0 M0	Tumor confined to the vulva and/or *perineum* – 2 cm or less in greatest dimension. *No nodal metastases*
T1	Tumor confined to the vulva and/or *perineum* – 2 cm or less in diameter			
T2	Tumor confined to the vulva and/or *perineum* – more than 2 cm in diameter	Stage II	T2 N0 M0	Tumor confined to the vulva and/or *perineum* – more than 2 cm in greatest dimension. *No nodal metastases*
T3	Tumor of any size with adjacent spread to the urethra, vagina, anus, or all of these			
T4	Tumor of any size infiltrating the bladder mucosa or the rectal mucosa or both, including the upper part of the urethral mucosa or fixed to the anus	Stage III	T3 N0 M0, T3 N1 M0, T1 N1 M0	Tumor of any size with the following: (1) adjacent spread to the lower urethra, the vagina, the anus, and/or (2) *unilateral regional lymph node metastases*
N regional lymph nodes		Stage IVA	T1 N2 M0	Tumor invades any of the following:
N0	No nodes palpable		T2 N2 M0	Upper urethra, bladder mucosa, rectal
N1	Unilateral regional lymph node metastases		T3 N2 M0	mucosa, pelvic bone, and/or *bilateral*
N2	Bilateral regional lymph node metastases		T4 any N M0	*regional node metastases*
M distant metastases		Stage IVB	Any T, any N, M1	Any distant metastases, including pelvic lymph nodes
M0	No clinical metastases			
M1	Distant metastases *(including pelvic lymph node metastases)*			

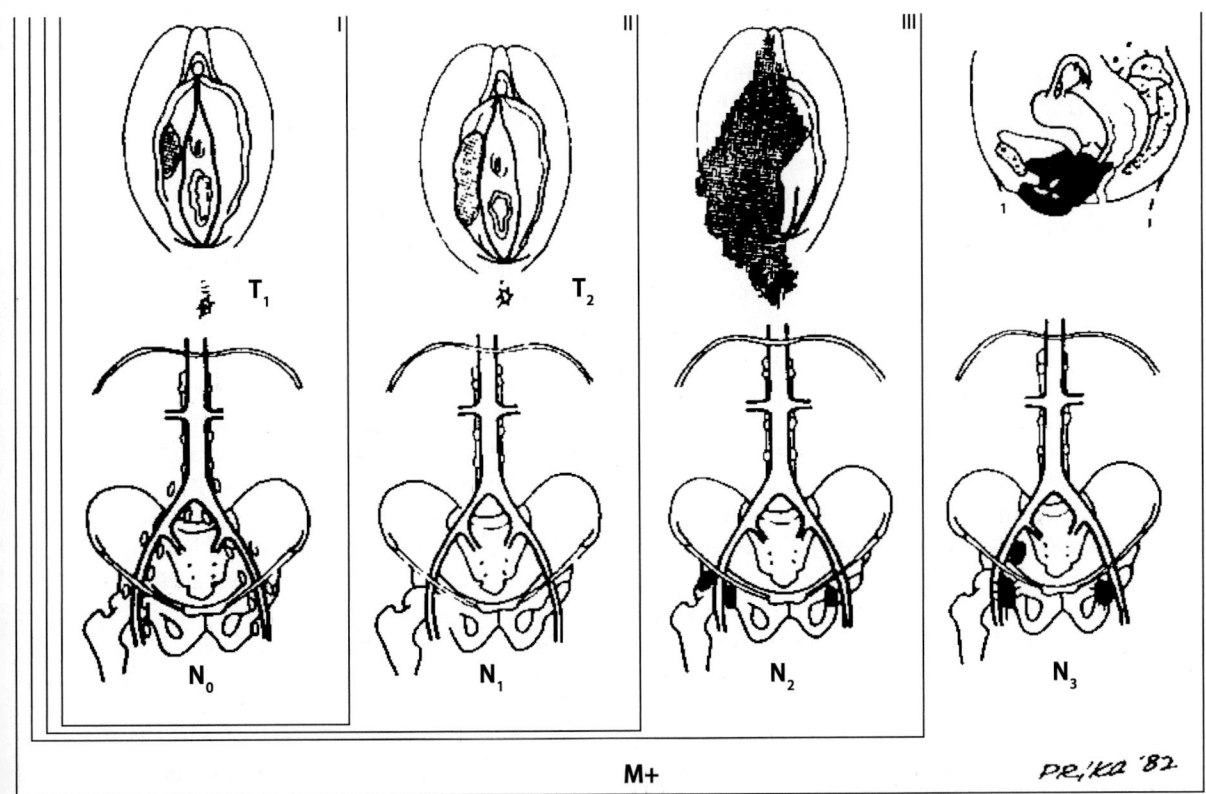

Fig. 26.2. Anatomic staging for vulva cancer (DuBESHTER B et al. 1993)

Basal cell carcinoma of the vulva is occasionally reported (BREEN et al. 1975; PERRONE et al. 1987; HOFFMAN et al. 1988).

Adenoid cystic carcinoma of the Bartholin's gland constitutes about 10% of all carcinomas of this gland and approximately 0.1% of all vulvar malignancies. Small numbers of patients with this tumor have been treated with a combination of surgery (wide local excision or hemivulvectomy) combined with postoperative irradiation with doses ranging from 40 Gy to 54 Gy. About 75% of the patients are alive and well, with follow-up ranging from 2 years to over 13 years. Late distant metastases may occur (ROSENBERG et al. 1989; LELLE et al. 1994).

Adenocarcinoma usually originates either in Bartholin's gland or from bulboadnexal structures, although rarely arise from the periurethral Skene's glands (LEUCHTER et al. 1982). Occasionally Bartholin's gland carcinoma may be squamous cell when it originates near the orifice of the duct, papillary if it arises from the transitional epithelium of the duct, or adenocarcinoma when it arises from the gland itself.

CARDOSI et al. (2001) treated 12 women with Bartholin's gland carcinoma; 11 patients were reported. Squamous cell carcinoma was most common. Of the patients, 10 were treated with primary surgery, followed by adjuvant radiation in 7 for inadequate resection margins or lymphatic metastases; 1 patient was treated with primary chemoirradiation. Recurrence was seen in 54.5% during a mean follow-up time of 73.5 months. Overall survival was 58.3%.

In contrast to this, BALAT et al. (2001) analyzed 18 patients with advanced primary carcinoma of the Bartholin's gland (median follow-up was 9 years): 7 were treated with wide local excision followed by radiation therapy, 9 had radical vulvectomy followed by radiation therapy to the vulvar and inguinofemoral and pelvic node areas, and 2 were treated with radiation therapy alone after biopsy of the tumor. The 5-year disease-free survival rates were 86%, 78%, and 50%, respectively, for the three groups. Of 2 patients treated with radiation therapy alone, 1 lived for 6 years with no evidence of disease, and the other lived for 20 months. The rate of local tumor control was 100% for all three treatment groups. There were no significant differences among the treatment groups in rate of primary tumor control or 5-year disease-free survival rate ($P=0.1300$).

Melanoma represents 2–9% of vulva malignancies; two varieties, nodular and superficial spreading melanoma, are described. As in other locations, the depth of invasion correlates with the patterns of spread and prognosis (CHUNG et al. 1975; BALTZER et al. 1986; BENDA et al. 1986; WOOLCOTT et al. 1988; BRAND et al. 1989; HOSKINS and PEREZ 1989).

Sarcomas of the vulva are extremely rare; leiomyosarcoma is the most common. Neurofibrosarcoma, rhabdomyosarcoma, fibrosarcoma, and angiosarcoma have been reported (TAVASSOLI and NORRIS 1979; ULBRIGHT et al. 1983; BARNHILL et al. 1988).

Metastatic carcinoma to the vulva from the uterine cervix (most common), the endometrium, or the ovary and extension or metastases from the urethra or the vagina have been described (DEHNER 1973).

26.6
Prognostic Factors

Local recurrence is related to the adequacy of the surgical resection margins. HEAPS et al. (1990), in an analysis of formalin-fixed tissue specimens, demonstrated a sharp rise in the incidence of local recurrence for tumors with microscopic margins less than 8 mm. They suggested that this would correspond to a minimum margin of 1 cm in fresh, unfixed tissue. Inadequate resection margins are an important consideration when planning treatment for large tumors and those involving midline structures such as the urethra, vagina, and anus.

ROSÉN and MALSTROM (1997) reported a retrospective analysis of 328 patients with primary invasive vulvar cancer, 277 of whom were treated primarily with surgery and 189 with the addition of irradiation. Of the 328, 18 were treated primarily with irradiation and 5 of these in combination with chemotherapy. The most important prognostic factors were tumor stage, histological differentiation, and patient age (69 years).

Extension of the primary tumor to the urethra, vagina, and anal area is associated with an increased incidence of nodal involvement and worsening of prognosis. Treatment usually involves either exenterative surgery or a combination of surgery and irradiation.

The size, depth of invasion, and histological subtype of the primary tumor, as well as degree of lymphatic and vascular invasion, correlate closely with the incidence of regional lymph node involvement and prognosis (RUTLEDGE et al. 1970; DONALDSON et al. 1981; MALFETANO et al. 1985; SHIMM et al. 1986; SEDLIS et al. 1987). KURZL and MESSERER (1989) carried out a multivariate analysis of 124 patients with various stages of vulvar carcinoma treated with simple vulvectomy and local/inguinal irradiation

(40 Gy). No inguinal lymphadenectomy was performed. They found that age, disassociated growth, lymphatic spread, tumor thickness, and ulceration were relevant prognostic factors.

The incidence of lymph node involvement correlates well with the FIGO clinical stage. In the experience of DONALDSON et al. (1981), 15% of patients with clinical stage I disease, 40% with stage II, 80% with stage III, and 100% of patients with stage IV disease had confirmed regional lymph node involvement. Similar involvement was noted by SEDLIS et al. (1987), who found regional node involvement with stages I, II, III, and IV to be 8.9%, 25.3%, 31.1%, and 62.5%, respectively.

Lymph node metastasis is the single most important prognostic factor in women with vulvar cancer. The presence of inguinal node metastases routinely results in a 50% reduction in long-term survival (FIGGE et al. 1985; FARIAS-EISNER et al. 1994).

26.6.1
Capillary Lymphatic Space Involvement and Lymph Node Metastasis

In the detailed analysis of Gynecologic Oncology Group (GOG) protocol no. 36, two significant risk factors were identified for recurrence in the vulva: (1) tumor size greater than 4 cm and (2) capillary lymphatic space involvement. If either of these factors was present, the risk of vulva recurrence after radical vulvectomy was 20.7% (30 of 184); but, if neither factor was present, the risk was only 9.2% (37 of 404). The depth of invasion did not add significantly to the prediction of vulvar failure.

However, in another study, depth of invasion of 1, 2, and 3 mm corresponded to 4.3%, 7.8%, and 17% incidence of nodal involvement, respectively. Perineural invasion was strongly associated with lymph node metastasis (ROWLEY et al. 1988).

WHARTON et al. (1974) proposed eliminating groin dissection for patients with small tumors that invaded less than 5 mm. Later reports showed that 10–20% of these patients had occult groin metastases, making the elimination of some form of inguinal lymphadenectomy or irradiation undesirable (PARKER et al. 1975; DONALDSON et al. 1981; HACKER et al. 1984b; BINDER et al. 1990). The current consensus is that only tumors with less than 1 mm invasion fulfill this criterion (SEDLIS et al. 1987; BERMAN et al. 1989; KELLEY et al. 1991); this is reflected in FIGO's recent decision to classify tumors invading less than 1 mm into substage IA.

GOG protocol no. 37 demonstrated that two major poor prognostic factors were clinically suspicious or fixed ulcerated groin nodes and more than one positive groin node. Nodal metastases did not influence the probability of vulvar relapse, but did increase the risk of relapse in the groin, pelvis, or other sites. The irradiated group had statistically fewer groin recurrences (3 of 59, 51%) than the patients who received no postoperative irradiation (13 of 55, 23.6%). The difference in survival for 114 evaluable patients significantly favored the adjuvant radiation therapy group (P=0.03).

26.6.2
Extracapsular Extension of Metastatic Lymph Nodes

For patients with vulvar carcinoma, ORIGONI et al. (1992) noted that extracapsular tumor extension was of prognostic value, even in those with a single positive lymph node. This observation was confirmed by the study of VAN DER VELDEN et al. (1995), in which this variable was the only independent predictive factor for survival. It suggested that extranodal extension is an indication for adjuvant radiation therapy even in patients with single-node metastatic disease.

26.7
General Management

In the current treatment of women with vulvar cancer, greater emphasis is placed on prognostic factors and organ preservation. The preinvasive forms of vulvar malignancies (carcinoma in situ and Paget's disease) and microinvasive tumors can be treated with topical chemotherapy, cryosurgery, or surgical resection. The preferred method of treatment for invasive carcinoma is surgery, which varies from wide local excision to partial vulvectomy (HACKER et al. 1984a; FIORICA et al. 1988; HUSSEINZADEH et al. 1989), depending on extent and multiplicity of intraepithelial lesions and the patient's wish to preserve the vulva. BURKE et al. (1996) summarized this subject.

The traditional management of patients with invasive stage I and II carcinomas of the vulva consisted of radical vulvectomy with inguinofemoral lymphadenectomy (KURZL and MESSERER 1989), which involved removal of the entire vulva from the perineum to the upper margin of the mons pubis and bilateral excision of the tissues in the femoral triangle and those overlying the inguinal ligament. The urethral meatus is generally left in situ.

Many gynecologists have proposed limited resections for smaller invasive vulvar tumors considered to represent early or low-risk disease (DiSAIA et al. 1979; BERMAN et al. 1989; BURKE et al. 1990). DiSAIA et al. (1979) described conservative resection in 18 of 20 cancers measuring 1 cm or less with invasion of less than 5 mm. Others include patients with larger lesions and more significant invasion (WHARTON et al. 1974; BERMAN et al. 1989; BURKE et al. 1990; STEHMAN et al. 1992a).

While many surgeons have performed pelvic lymphadenectomy, the current policies at some institutions reserve this procedure only for patients with clinically positive inguinofemoral lymph nodes. Attempts are being made to refine patient selection, omitting inguinal lymphadenectomy in patients with small, low-grade primary lesions and omitting pelvic lymphadenectomy in patients with three or fewer involved inguinal nodes, provided the primary lesion does not invade the clitoris, urethra, vagina, or anal region.

Because of the morbidity associated with radical vulvectomy and to enhance psychosexual rehabilitation and based on the biological and therapeutic principles already validated in the head and neck, breast, and soft-tissue sarcomas, there is increasing use of wide local excision or partial vulvectomy to remove the primary tumor (usually T1) and, if necessary, an inguinofemoral lymph node dissection in patients with clinically positive nodes combined with moderate doses of radiation therapy to the remaining vulva and regional lymph node-bearing areas (50 Gy for subclinical disease with a boost of 10–15 Gy through reduced portals or brachytherapy for microscopically involved areas).

Resection of the primary lesion with a 1- to 2-cm margin of normal tissue and carrying the dissection to the deep perineal fascia are recommended. This is combined with a more conservative surgical approach to the groin, the ipsilateral superficial groin nodes being used as the sentinel group for lymphatic metastases (MORRIS 1977; DiSAIA et al. 1979; STEHMAN et al. 1992a). Bilateral superficial dissections are performed in patients whose tumors involve midline structures (clitoris or perineal body) (BURKE et al. 1995). Recently, sentinel lymph node biopsies have been introduced to decrease treatment morbidity (DE HULLU et al. 2004). In patients with pathologically negative inguinal nodes, no further

dissection or postoperative therapy is used. Patients with positive nodes can undergo additional nodal dissection of the deep nodes and the contralateral groin or be treated with postoperative irradiation, or both. In a review of 67 patients with clinically suspicious and 160 with clinically negative inguinal lymph nodes, KATZ et al. (2003) noted that 119 were treated with lymph node dissection alone, 57 with lymph node dissection and radiation therapy, and 51 with radiation therapy alone. Treatment guidelines for lateralized or central operable tumors are summarized in Fig. 26.3.

BUSCH et al. (2000) described results in 92 patients treated for vulvar cancer with simple vulvectomy, electrocoagulation, or local excision and radiation therapy to the vulva (0–90 Gy). All patients received irradiation to the inguinal lymph nodes, ranging from 30 Gy to 60 Gy. In T1 patients, 5-year survival rates were 71% and in T2 patients 43%. Doses of 45 Gy or more to the vulva were sufficient to increase the 5-year cause-specific survival rate from 55% to 88%.

26.7.1
Stage III and IV Tumors

Some of these more extensive tumors can be completely resected by radical vulvectomy or some variation of pelvic exenteration and vulvectomy. Radical surgery is frequently ineffective in curing patients with bulky tumors or positive groin nodes. However, most recent therapeutic efforts have focused on preoperative multimodality treatment that combines radiation therapy or chemoirradiation with less-radical surgery (BURKE et al. 1996). Management options are illustrated in Figure 26.4. Vulvar cancers are radioresponsive, and function-sparing operations are feasible in some patients with advanced disease (Fig. 26.5).

The role of radiation therapy alone in the primary management of carcinoma of the vulva remains controversial, primarily because of lack of long-term data on the results of treatment with modern techniques and because of the traditional belief that vulvar tissues could not tolerate high doses of irradiation (over 60 Gy). However, this misconception has been corrected in recent reports (PEREZ et al. 1993) and, with reduced fields, doses of 65–70 Gy in 7–8 weeks are delivered to gross tumor volumes (HELGASON et al. 1972; KUIPERS 1975). Radiation therapy is often used for palliation or for treatment of patients who are not amenable to radical surgical

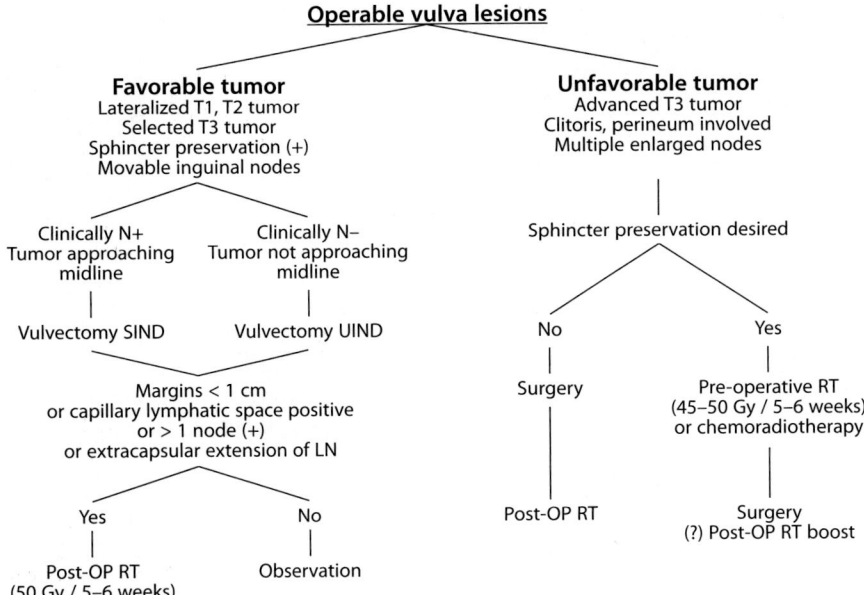

Fig. 26.3. Algorithm illustrating the various therapeutic options for patients with favorable or unfavorable operable carcinoma of the vulva (PEREZ et al. 1997)

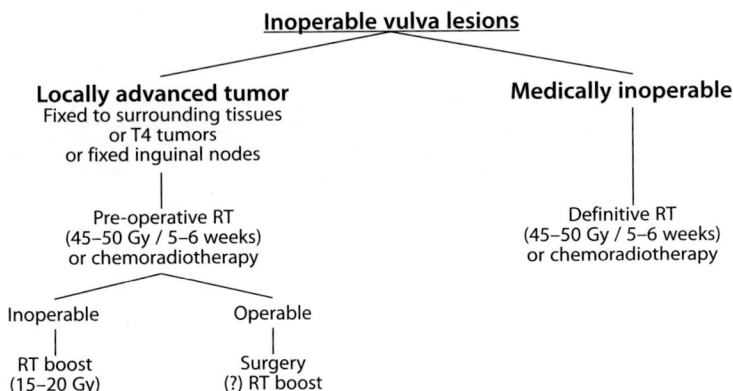

Fig. 26.4. Algorithm illustrating therapeutic options for patients with locally advanced or inoperable carcinoma of the vulva (PEREZ et al. 1997)

resection. Table 26.5 summarizes treatment guidelines at Washington University.

Misonidazole was used as a radiosensitizer combined with high-dose radiation therapy (usually 48–58 Gy) in ten patients with advanced vulvovaginal carcinoma. Nine patients had tumor control and five were tumor free 9–30 months after diagnosis. Toxicity was minimal, with only one patient developing peripheral neuropathy (NORI et al. 1983).

26.7.2
Management of Patients with Positive Nodes

Patients with clinically positive inguinal nodes may benefit from a course of preoperative irradiation (45–50 Gy) as suggested by BORONOW (BORONOW 1982; BORONOW et al. 1987) or, in a clinical trial, combined with chemotherapy. Patients who undergo bilateral inguinofemoral lymphadenectomy as ini-

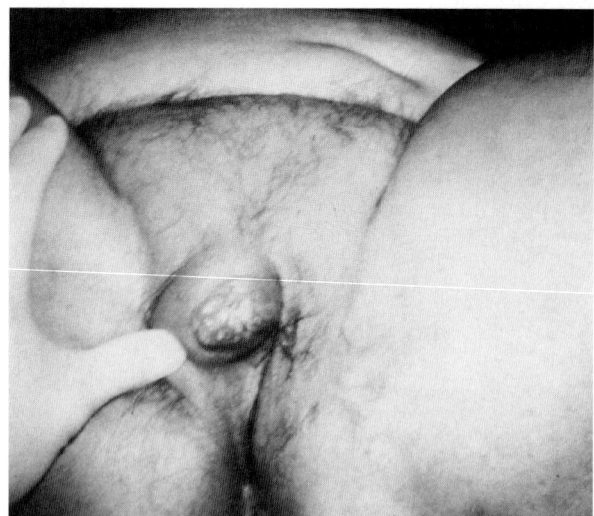

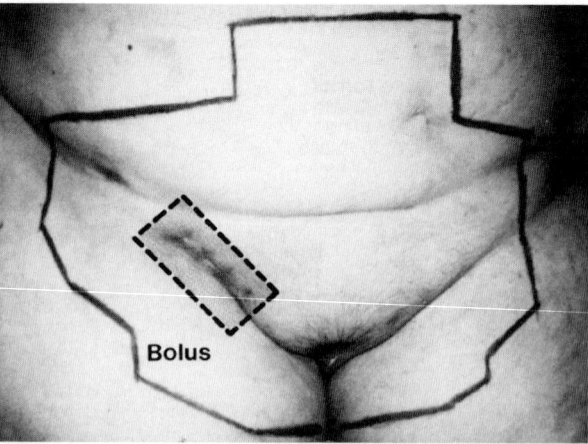

Fig. 26.5. a Patient with a 4-cm epidermoid carcinoma in the right labia and clitoris and a 4×4×3-cm right inguinal lymph node. Wide local excision of the primary tumor was carried out. **b** Portal used to deliver external irradiation to treat pelvic and vulvar areas to 50 Gy. Bolus (2 cm thick) was used over right inguinal areas. Additional 15 Gy was delivered with 12-MeV electrons to right tumor volume. **c** Posttreatment photograph 3 years later showing excellent cosmetic results. Patient is tumor free (PEREZ et al. 1997)

tial therapy and are found to have positive nodes – particularly more than one positive node – will benefit from postoperative irradiation to the groins and pelvis (HOMESLEY et al. 1986). Treatment algorithms are shown in Figures 26.3 and 26.4. Radiation therapy is superior to surgery in the management of patients with positive pelvic nodes as documented by HOMESLEY et al. (1986). The morbidity (primarily lymphedema) of combining superficial and deep inguinal lymphadenectomy with irradiation is significant.

26.8
Radiation Therapy Techniques

In general, irradiation will encompass the vulvar area and the inguinofemoral and, in some patients,

the pelvic lymph nodes, while minimizing the dose to the femoral heads.

26.8.1
Irradiation Alone

In the occasional medically inoperable patient, small superficial lesions may be controlled with 60–65 Gy. For larger tumors, the primary lesion needs to be irradiated with reduced fields to a dose of approximately 70 Gy. It is important to use daily fractionation of 1.6–1.8 Gy in five weekly fractions to enhance tolerance to treatment. Usually parallel opposed anterior and posterior portals are used, preferentially loaded anteriorly (or a high-energy photon single anterior beam with bolus), that cover the vulva and the regional lymphatics to deliver 45–50 Gy to an appropriate depth (PEREZ et al. 1998).

Table 26.5. Carcinoma of the vulva: recommended treatment guidelines

Surgical therapy has been the standard in vulvar carcinoma.
Carcinoma in situ or microinvasion (≤5 mm): wide local excision
Invasive carcinoma
Stage I (superficial, <2 cm in diameter): wide local excision or hemi/simple vulvectomy
Other stage I or stage II: radical vulvectomy with inguinal lymph node dissection
If clinically negative nodes, a reasonable alternative is no lymph node dissection, elective node irradiation
Radiation therapy doses:
Negative lymph nodes, simple vulvectomy: 50 Gy with 6- to 18-MV photons and appropriate bolus
Wide local excision, pathologically negative margins: as above. Perineal electron beam boost to bring vulva excision site dose to 60 Gy
Pathologically positive margins: after 60 Gy, additional boost to positive margins or positive lymph nodes (5 Gy) with electrons or interstitial implant
Stage III: after radical vulvectomy and lymphadenectomy, indications for postoperative irradiation
a. Primary tumor ≥4 cm
b. Positive surgical margins
c. Three or more positive lymph nodes
Dose: 50 Gy to vulva and inguinal areas; boost to positive margins (10–15 Gy) via perineal portal or interstitial implant; boost to inguinal region via anteroposterior (AP) field (10–15 Gy).
Stage IV: pelvic exenteration
Preoperative irradiation: 45 Gy to pelvis and inguinal areas with radical vulvectomy and complete inguinal lymph node dissection. *Postoperative boost to primary (10–15 Gy) via interstitial and/or intracavitary and/or appositional electrons, when indicated

*In patients with palpable inguinal nodes, superficial inguinal node dissection and inguinal/pelvic irradiation may be an acceptable, less-mutilating alternative (PEREZ et al. 1997)

Bolus material over the areas of the skin (vulva) at risk for tumor involvement is essential (PAO et al. 1988). Interruption of the irradiation course is frequently necessary in the third or fourth week of treatment to prevent severe moist desquamation and maceration of the perineal skin. After a dose of 45–50 Gy is delivered to the vulvar area, a 6- to 9-MeV electron beam or low-energy photon beam (4–6 MV) aimed directly at the vulva is used to deliver an additional 10–20 Gy to gross or microscopic tumor volumes. An interstitial implant may also be considered to deliver the boost dose to the primary tumor (MIYAZAWA et al. 1983; CARLINO et al. 1984).

Palpable metastatic inguinofemoral lymph nodes receive an additional dose (15–20 Gy), preferably with electrons to decrease the dose to the underlying bones. The energy (12–20 MeV) is determined with CT scans to define the depth of the lymph nodes.

26.8.2
Regional Lymphatics

In patients with primary lesions less than 2 cm in diameter, the probability of nodal involvement is low, and irradiation of the pelvic lymph nodes may be omitted if only the inguinofemoral nodes are being treated (Fig. 26.5a).

In patients with primary tumors larger than 2 cm and no clinical evidence of regional lymphatic involvement, the inguinal and pelvic lymph nodes may be treated electively to a dose of 45–50 Gy in 1.8- to 2-Gy fractions per day in lieu of lymph node dissection (Fig. 26.5b).

If palpable inguinal lymph nodes are present, the dose to the inguinofemoral lymph nodes needs to be 65–70 Gy (with reduced fields after 50 Gy), depending on the size of the involved nodes. When there is evidence of spread to the pelvic nodes, the dose must be increased to 60 Gy. Because some patients with involved pelvic lymph nodes are potentially curable, irradiation of the lower periaortic chain in the presence of pelvic lymph node involvement might be appropriate.

In patients in whom the pelvic nodes must be treated, anterior and posterior portals covering the vulvar and regional lymphatic volumes are required (Fig. 26.5c).

26.8.3
Preoperative Radiation Therapy

Patients with advanced primary lesions involving surrounding structures, either of questionable resectability or clearly unresectable, are sometimes treated with preoperative irradiation (BORONOW et al.

1987). Moderate doses of 45–50 Gy in 5–6 weeks may increase the resectability rate and also avoid mutilating procedures such as exenteration (Boronow 1982; Boronow et al. 1987).

26.8.4
Postoperative Radiation Therapy

Increasingly, radiation therapy is used in combination with surgery. Postoperative irradiation consisting of 50 Gy (1.75–1.8 Gy daily) is indicated in patients who have undergone radical resection of the primary lesion and inguinofemoral lymph nodes and are considered at high risk of recurrence because of inadequate resection margins; in patients with positive inguinal nodes (in lieu of pelvic lymph node dissection); or in patients treated with wide local tumor excision. If the resection margins are microscopically involved or if there is gross residual tumor, an additional dose of 15–20 Gy needs to be administered with reduced portals or an interstitial implant.

When an inguinal lymph node dissection is performed and only superficial node involvement (three or more) is detected, postoperative irradiation is given only to the inguinofemoral lymph nodes (50 Gy at 4–6 cm); a boost of 5–10 Gy may be administered with electron beam, depending on the number of nodes, size, or extracapsular extension of the tumor (Fig. 26.6a).

Patients with metastatic deep inguinofemoral nodes found at node dissection may benefit from postoperative irradiation, including the pelvic lymphatics, consisting of 50 Gy at the midplane of the pelvis in 5–6 weeks (Fig. 26.6b, c).

26.8.5
Treatment for Recurrent Lesions

Some recurrent lesions may be treated surgically (Burke et al. 1996). Recurrences after surgical resection remain potentially curable and must be treated aggressively in the manner described earlier to deliver tumor doses of 65–70 Gy with reducing fields.

26.8.6
Patient Positioning and Simulation

The patient is usually treated in the supine "frog-leg" position, with the knees apart and the feet together

(Fig. 26.7). A cast or alpha cradle in the treatment position facilitates everyday repositioning.

When the patient is simulated, wires should be used to identify surgical scars or palpable or visible lesions. If the vagina or perineum is involved, a radiopaque marker should be placed to identify the tumor margins on the radiographs. For tumors involving the urethra, a urethral catheter, if tolerated by the patient, may aid in tumor localization.

During simulation, consideration should be given to designing and constructing compensating filters to achieve a more homogeneous dose in the entire target volume to compensate for the sloping surface and decreased diameter of the perineum.

26.8.7
Beams and Energies

Depending on the available equipment, either an anteroposterior (AP) beam or differentially loaded parallel opposed AP–posteroanterior (PA) beams or electron beam (for part of the treatment) can be used. Helpful in the design of portals covering the inguinal regions is a mapping of distribution of gross inguinal lymph nodes marked with radiopaque lead wire in patients with various tumors of the anus, low rectum, and gynecologic organs (Fig. 26.8). If only the vulva and inguinofemoral lymph nodes are treated, ^{60}Co or 4- to 6-MV X-rays through an anterior portal and high-energy (>10 MV) X-rays through a posterior portal may be adequate. Care should be taken to deliver adequate doses not only to the primary tumor area and superficial inguinal nodes but also to the femoral and deep pelvic nodes (Fig. 26.9).

Higher-energy photon beams (15–18 MV) with unequal loading (AP3:PA2) and bolus on the anterior portal can be used. An alternative is to treat the anterior portal with ^{60}Co or 4- to 6-MV X-rays and the posterior portal with a high-energy photon beam (15–18 MV).

Kalnicki et al. (1987) and King et al. (1993) described a technique using a partial transmission block that reduces the dose to the femoral heads. CT is very helpful both for evaluation of size and depth of tumor and regional lymph nodes and for treatment planning.

Gross tumor in the vulva or nodal areas may be boosted with en face electron fields; the energy for the vulva is 6–9 MeV using a 1- to 1.5-cm bolus. The electron-beam energy used to irradiate lymph nodes, which are usually 4–6 cm or even 8 cm deep

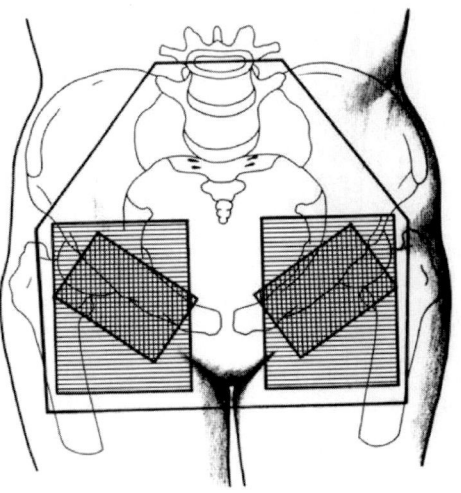

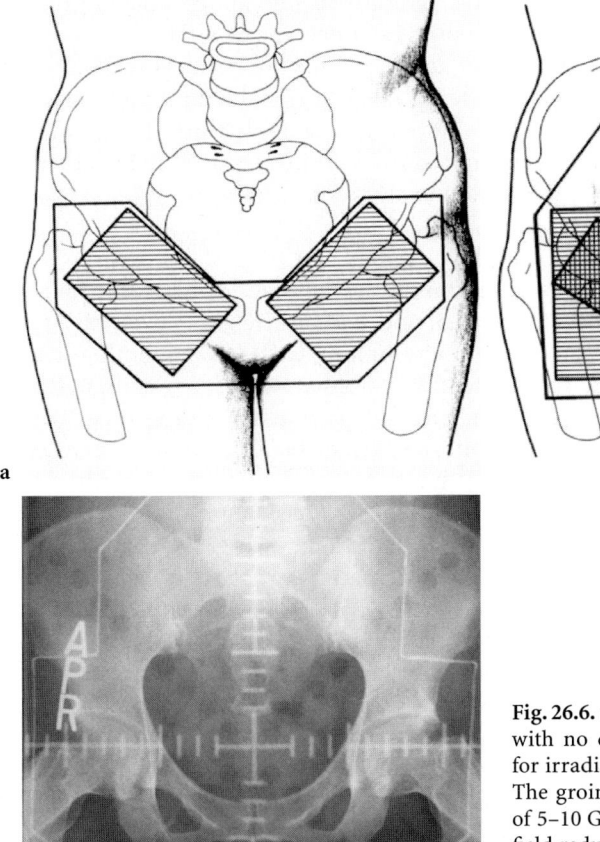

Fig. 26.6. a Portal for elective irradiation of regional lymphatics in patients with no clinical evidence of inguinal lymph node involvement. b Portal for irradiation of pelvic and inguinofemoral lymph nodes and vulvar area. The groins are subsequently boosted to a total of 60–65 Gy. A final boost of 5–10 Gy to the positive inguinal lymph nodes may be given with further field reduction, bringing the total dose to that area to approximately 70 Gy. c Simulation film of portal covering pelvic and inguinofemoral lymph nodes and vulva (PEREZ et al. 1997)

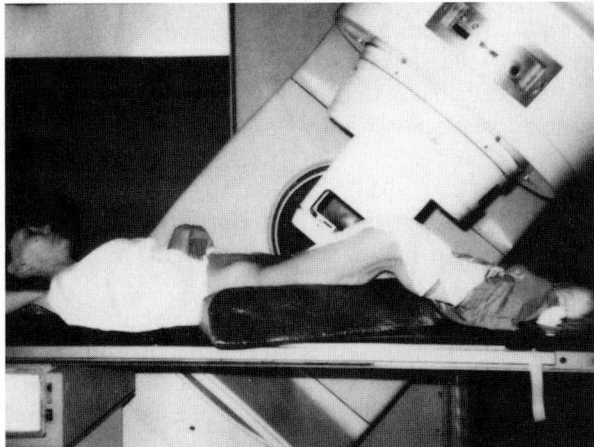

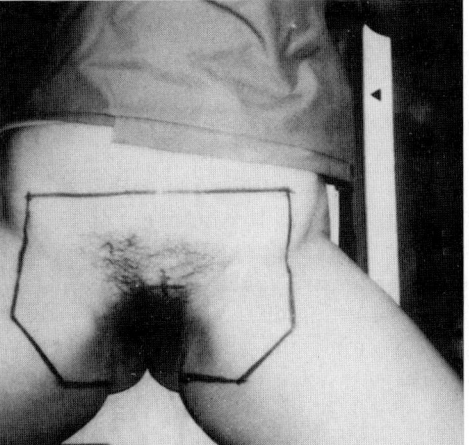

Fig. 26.7. a Supine frog-leg position for irradiation of vulva and inguinal pelvic lymph nodes. The patient's thighs are abducted and outwardly rotated so that the inguinal folds are stretched flat. b Example of portal used for irradiation of vulva and inguino-femoral lymph nodes with patient in treatment position (PAO et al. 1988)

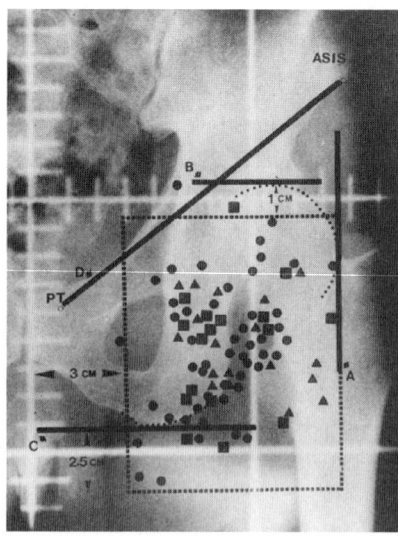

Fig. 26.8. Topographic distribution of inguinal lymph node metastases in patients with carcinoma of the vulva–vagina–cervix (*triangles*), urethra (*squares*), or anus–low rectum (*circles*) (WANG et al. 1996)

(KOH et al. 1993a; McCALL et al. 1995), depends on the depth of the inguinal nodes, which must be determined by CT scanning. In some cases, interstitial implants or en face electron fields may be used to boost the dose to the primary site (CARLINO et al. 1984; MIYAZAWA et al. 1983). Because of the sloping surface of the perineum, higher doses may be delivered to this area when the tumor dose is calculated at the central axis of the portal, and no compensators are used (Fig. 26.10).

Although dry and moist desquamation are frequently observed, we still favor the use of bolus in areas potentially involved with tumor.

26.8.8
Lymphatic Irradiation

The value of elective irradiation of the pelvic lymph nodes was demonstrated by HOMESLEY et al. (1986), who reported on a randomized study involving 114 patients with invasive squamous carcinoma of the vulva and positive inguinal lymph nodes treated with radical vulvectomy and bilateral groin lymphadenectomy. Patients were randomized to receive irradiation of 45–50 Gy to the pelvic and inguinal regions over a 5- to –6-week period or pelvic lymph node dissection. No irradiation was given to the central vulvar area. Of 59 patients randomized to irradiation, the 53 who were treated had a 68% 2-year

survival rate. In contrast, the nonirradiated group had a 54% survival rate. The benefit of radiation therapy was noted only in the patients with N2 or N3 lymphadenectomy. Regional lymph node recurrences were noted in 7 of 59 patients (11.9%) allocated to irradiation in contrast to 14 of 55 patients (25.4%) treated with surgery only.

STEHMAN et al. (1992b) published results of a GOG protocol that randomized 58 patients with squamous carcinoma of the vulva and nonsuspicious (N0 or N1) inguinal nodes to receive either groin dissection or groin irradiation, each in conjunction with radical vulvectomy. Radiation therapy consisted of 50 Gy given in daily 2-Gy fractions to a depth of 3 cm below the anterior skin surface. The study was closed prematurely because of an excessive number of groin relapses in the irradiation group. In the groin dissection group, 5 of 25 patients (20%) had positive groin nodes; these patients received postoperative irradiation. There were five groin relapses (18.5%) among the 27 patients on the groin-irradiation regimen and none on the groin-dissection regimen. The groin-dissection regimen had a significantly better progression-free interval (P=0.03) and survival (P=0.04). This study has been vehemently criticized because the protocol prescription required that 50% of the tumor dose be delivered with 9-MeV electrons, which are grossly inadequate to treat the deep inguinal lymph nodes (90% depth dose at 3.1 cm).

Noteworthy, McCALL et al. (1995) evaluated the depth of inguinal lymph nodes with CT scans in 100 women without palpable inguinal adenopathy or prior inguinal surgery. The tumor doses the patients would have received were determined using isodose curves constructed according to the guidelines in GOG protocol no. 88. Only 18% of women had all inguinal lymph nodes measured at a depth of 3 cm or less. More than half of all women would have received less than 60% of the prescribed irradiation dose because their inguinal lymph nodes were deeper than 5 cm.

KOH et al. (1993a) also quantified inguinofemoral node depths using pretreatment CT scans in 50 patients with gynecological cancer. The distance of each femoral vessel beneath the overlying skin surface was determined as an indicator of groin nodal depth. Correlative data regarding height and weight were used to calculate the Quetelet index, defined as (weight in kg)/(height in m)^2. Individual femoral vessel depths ranged from 2.0 cm to 18.5 cm. When the depths of all four femoral vessels were averaged in each patient, the mean "four-vessel average" nodal depth was 6.1 cm. Recalculation of doses provided

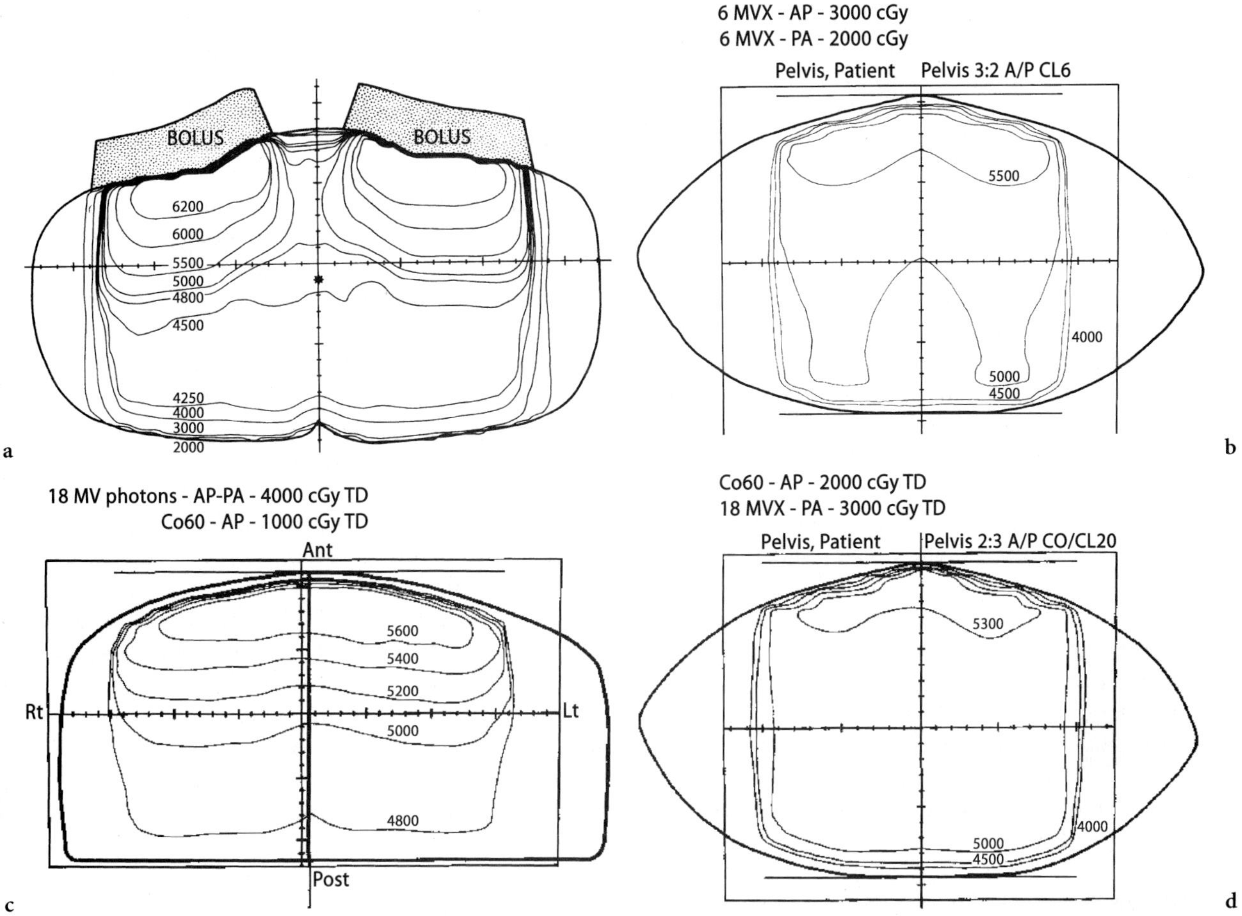

6 MVX - AP - 3000 cGy
6 MVX - PA - 2000 cGy

Pelvis, Patient   Pelvis 3:2 A/P CL6

BOLUS   BOLUS

6200
6000
5500
5000
4800
4500

4250
4000
3000
2000

a

5500

4000

5000
4500

b

18 MV photons - AP-PA - 4000 cGy TD
Co60 - AP - 1000 cGy TD

Ant

5600
5400
5200
5000

Rt   Lt

4800

Post

c

Co60 - AP - 2000 cGy TD
18 MVX - PA - 3000 cGy TD

Pelvis, Patient   Pelvis 2:3 A/P CO/CL20

5300

5000
4500

4000

d

Fig. 26.9a–d. Representative treatment plans for irradiation of vulvar region and regional lymphatics. **a** Parallel opposed 18-MV photon beams, preferentially loaded anteriorly (27 Gy anteriorly, 18 Gy posteriorly); bolus is added over the inguinal areas to improve dose distribution in subcutaneous tissues in that area. A 15-Gy boost using 16-MeV electrons (without bolus) is added to the groin. **b–d** Alternative setups with different beam energies and loadings (Perez et al. 1997)

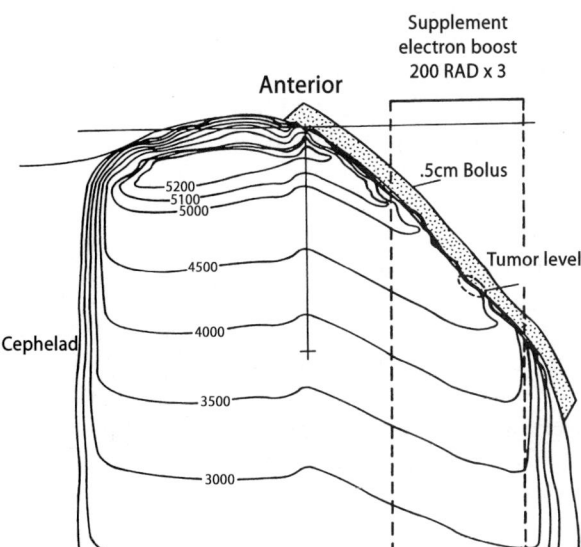

Supplement
electron boost
200 RAD x 3

Anterior

.5cm Bolus

5200
5100
5000

Tumor level

4500

Cephelad

4000

3500

3000

Fig. 26.10. Sagittal isodose contours on a patient. Note the relative underdosage of the posterior vulvar region, thereby requiring a supplemental en face perineal electron beam field (Pao et al. 1988)

to 5 patients failing prophylactic groin irradiation in the GOG study showed that all had received nodal tumor doses of less than 47 Gy, with 3 patients being underdosed by more than 30%.

PETEREIT et al. (1993) updated a study that compared inguinofemoral irradiation to lymphadenectomy for N0 and N1 vulvar carcinomas. Because of the skin reaction and possible underdosing of deep femoral nodes with electrons, as used in the GOG study, opposed AP–PA photon fields to 50 Gy were used; 48 patients underwent radical vulvectomy followed by either lymphadenectomy (25 patients) or inguinofemoral irradiation (23 patients). Actuarial nodal tumor control was 100% in the first group and 91% in the latter ($P=0.14$). In addition, there was no difference in cause-specific survival at 3 years (96% and 90%, respectively) ($P=0.47$). The morbidity of lymphadenectomy included lymphedema (16%), seroma (16%), infection (44%), and wound separation (68%). In the irradiated patients, 16% developed lymphedema, and only 9% had a significant skin reaction. Thus, irradiation of the N0 or N1 inguinofemoral nodes is an alternative to lymphadenectomy for squamous cell carcinoma of the vulva if the proper irradiation technique and dose are used. The acute and late morbidity are less than with lymphadenectomy.

VAN DER VELDEN and ANSINK (2001) carried out a literature search using the criteria set by the Cochrane Gynaecological Cancer Group. Of nine reviewed papers, only three met the selection criteria. From these studies, it became clear that the incidence of groin recurrences after primary radiation therapy was higher than after surgery, and survival was also worse in the radiation therapy group. Morbidity after primary irradiation was lower than with surgery. The conclusion was criticized, on the grounds that the depth of 3 cm used in the radiation therapy is too shallow to administer an optimal dose to the deeper groin nodes. This means that surgery ought still to be considered the cornerstone of therapy for the groin nodes in women with vulvar cancer. Individual patients not fit enough to withstand surgery can be treated with primary radiation therapy.

26.8.9
Brachytherapy

A few reports have described the use of interstitial irradiation in vulvar cancer. POHAR et al. (1995) reported on 34 patients treated with ^{192}Ir brachy-

therapy for vulvar cancer (21 at first presentation when surgery was contraindicated or declined); 12 patients had FIGO stage III or IV disease, 8 had stage II, 1 had stage I, and 1 had stage 0. Of these patients, 13 were treated for recurrent disease. Paris system brachytherapy rules were followed; the median reference dose was 60 Gy (range, 53–88 Gy). With a median follow-up of 31 months, 10 of 34 patients (29%) were alive. Actuarial 5-year local tumor control was 47%, and locoregional tumor control was 45%; 5-year disease-specific survival was 56% and overall survival 29%. Subset analysis disclosed higher actuarial 5-year locoregional tumor control in patients treated at first presentation (80% versus 16%) ($P=0.01$).

At Washington University, 20 patients had an interstitial implant as part of therapy, usually as a boost at the primary tumor site, occasionally for positive lymph nodes. The locoregional tumor control was 100% for T1 or T2 (4 patients), 80% for T3 (10 patients), and 50% for 6 patients with T4 or recurrent tumors. None of 5 patients receiving a total dose of 60 Gy or less had significant sequelae, in contrast to 8 of 15 (53%) treated with higher doses. Patients treated with external irradiation only had 11–26% moderate or severe sequelae ($P=0.16$).

26.8.10
Three-Dimensional Conformal and Intensity Modulated Radiation Therapy

Recently, a few patients have been treated with 3-D conformal radiation therapy to minimize irradiation to normal tissues, including the bone marrow in the pelvic bones (HARPER et al. 2004).

An example of target delineation, including iliac and inguinal lymphatics, portals defined by MLC, and dose distributions are illustrated in Figure 26.11 for patient with primary vulvar cancer.

26.8.11
Patterns of Failure after Treatment

The experience of Princess Margaret Hospital demonstrated that 80–95% of relapses after radical vulvectomy and bilateral groin node dissection were locoregional and could be encompassed in the locoregional irradiation fields (BRYON et al. 1991).

PEREZ et al. (1993) updated the results on 50 patients with primary invasive and 17 with recurrent histologically confirmed vulvar carcinoma treated with radiation therapy for locoregional disease. Results in 68 patients with primary and 18 with recurrent tumors (unpublished data) are further updated in this chapter. Of the patients with primary tumors, 18 were treated with biopsy and irradiation alone, 13 with wide excision plus radiation therapy, 11 with partial or simple vulvectomy, and 24 with radical vulvectomy followed by irradiation to the operative fields and inguinofemoral/pelvic lymph nodes. In patients with primary tumors treated with biopsy or local excision, local tumor control was 92–100% for T1-3N0 disease, 40% for similar stages with N1-3, and 27% for recurrent tumors. Among patients treated with partial or radical vulvectomy and radiation therapy, primary tumor control was 90% in those with T1-3 tumors and any nodal stage, 33% in those with any T stage and N3 lymph nodes, and 66% in patients with recurrent tumors. The actuarial 5-year disease-free survival rates were 87% for patients with T1N0 disease, 62% for those with T2-3N0 disease, 30% for those with T1-3N1 disease, and 11% for patients with recurrent tumors. There was no significant impact of type of vulvectomy on outcome. There were no long-term survivors with T4 or N2-3 disease. Of 17 patients treated for post-

vulvectomy recurrent disease, 4 remain disease-free after local tumor excision and radiation therapy. In patients with T1 or T2 tumors treated with biopsy or wide tumor excision and irradiation with doses less than 50 Gy, the local tumor control rate was 75% (3 of 4 patients), in contrast to 100% (13 patients) with 50.01–65 Gy. With T3 or T4 tumors treated with local excision and radiation therapy, tumor control occurred in none of 3 patients with doses less than 50 Gy and in 66% (6 of 9) with 50.01–65 Gy. In patients with T1 or T2 tumors treated with partial or radical vulvectomy and irradiation, local tumor control was 75% (6 of 8), regardless of dose level; in T3 and T4 tumors, it was 67% (4 of 6 patients) with 50–60 Gy and 86% (6 of 7) with 65–70 Gy. Differences were not statistically significant. There was no significant dose response for tumor control in the inguinofemoral lymph nodes with doses of 50–60 Gy for elective treatment of nonpalpable lymph nodes, yielding 91.6% tumor control (33 of 36), and 60–70 Gy controlling tumor growth in 75–80% of patients with positive nodes when administered postoperatively after partial or radical lymph node dissection.

DUSENBERY et al. (1994) reported on 27 patients irradiated postoperatively after radical vulvectomy

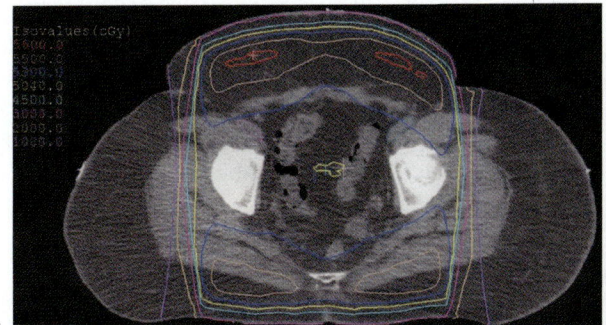

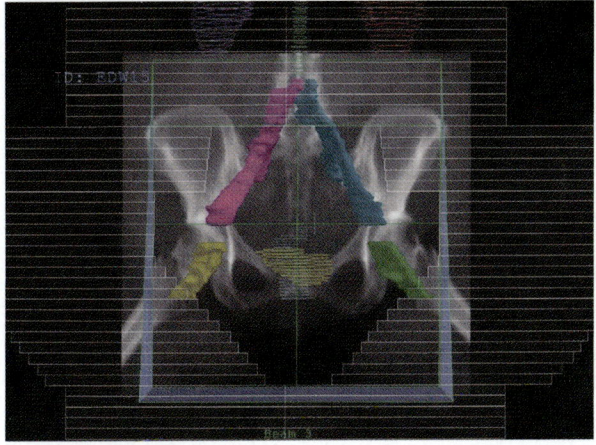

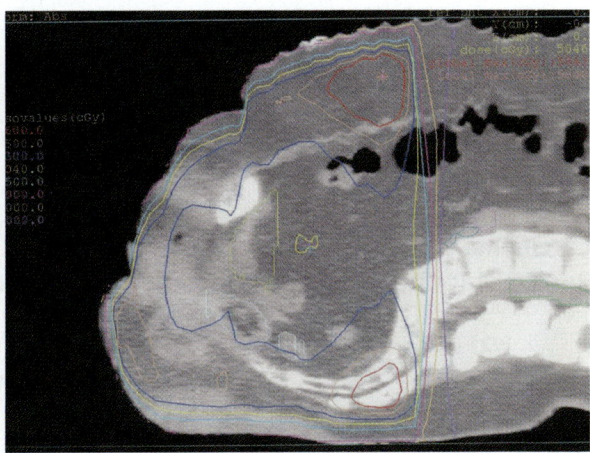

Fig. 26.11a–c. Example of 3-D conformal treatment planning for a patient with carcinoma of the vulva treated with anteroposterior–posteroanterior (AP–PA) pelvic portals including regional lymphatics. A total of 50.40 cGy was delivered to the planning target volume (PTV) in 1.8-Gy daily fractions using 18-MV photons and 15° dynamic wedges. a Inguinofemoral and pelvic common iliac lymphatic volumes are outlined. b Low pelvis cross section dose distribution. c Sagittal dose distribution

and bilateral lymphadenectomy (25 patients), radical vulvectomy and unilateral lymphadenectomy (1), or hemivulvectomy and bilateral lymphadenectomy (1). There were 14 FIGO stage III, 8 stage IVA, and 5 stage IVB patients. Inguinal lymph nodes were involved with tumor in all patients. Postoperative irradiation was directed at the bilateral groin and pelvic nodes (19 patients), unilateral groin and pelvic nodes (6 patients), or unilateral groin only (1 patient); the midline area was blocked in all patients. One patient received irradiation to the entire pelvis and perineum. Doses ranged from 10.8 Gy to 50.7 Gy (median, 45.5 Gy). Actuarial 5-year overall and disease-free survival rates were 40% and 35%, respectively. Recurrences developed in 63% (17 of 27 patients) (median of 9 months from surgery), and 15 of these died; two patients with recurrences were surviving at 24 months and 96 months after further surgery and radiation therapy. Central recurrences (under the midline block) were present in 13 of these 17 patients (76%), either as central only (8 patients), central and regional (4 patients), or central and distant (1 patient) (Fig. 26.12). Additionally, 3 patients developed regional recurrences, and 1 patient developed a concurrent regional and distant relapse. Use of a midline block resulted in a 48% central recurrence rate (13 of 27), much higher than the 8.5% rate previously reported using this technique. Routine use of the midline block should be abandoned, and postoperative irradiation volumes should be tailored to the tumor in each patient.

In 86 patients treated at Washington University, with either irradiation alone or combined with local excision or a partial or radical vulvectomy (always irradiating the vulvar area), the incidence of recurrence was about 38% at the primary site, 10% in the inguinofemoral lymph node region, and 23% distant metastases (PEREZ et al. 1993).

26.9
Sequelae of Treatment

Common sequelae associated with radical surgery are those related to wound problems, primarily infection and necrosis. The reported incidence of wound infection varies greatly. IVERSEN et al. (1980) observed it in 5.7% of patients and BOUTSELIS (1972) in 50% of patients. The incidence of wound dehiscence and necrosis varies in most reports from 30% to 50% (RUTLEDGE et al. 1970; BOUTSELIS 1972). Leg edema is a serious complication of nodal dissection. Transient edema occurred in approximately 14% of patients reported on by BOUTSELIS (1972) and RUTLEDGE et al. (1970). Chronic (persistent) edema was reported in 71% of patients reported on by BOUTSELIS (1972) and in 20% of those reported on by RUTLEDGE et al. (1970). Operative mortality in most series varies from 3% to 6%.

With a tumor at the skin or mucosal surface, which requires that the peak dose be at the surface, it is to be expected that literally all patients will have a significant acute cutaneous and mucosal reaction. Of more concern, however, is the incidence of late (chronic) sequelae, some of which can be attributed to the fractionation scheme used. SCHULZ et

Carcinoma of vulva
Radical vulvectomy and postoperative irradiation

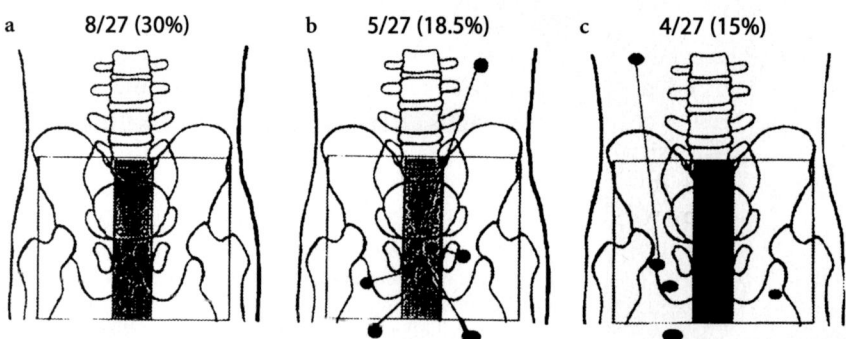

Fig. 26.12a–c. Patterns of failure in 27 patients irradiated after radical vulvectomy in whom the vulva was shielded with a central block. *Dots* represent a single patient. If relapsing in two or more sites, *dots* are connected by *lines*. **a** Isolated central (*under the midline block*) recurrences. **b** Combined central and regional (*inguinal or thigh*) recurrences. **c** Remainder of first recurrences consisted of either isolated regional recurrences or regional and distant recurrences (DUSENBERY et al. 1994)

al. (1980), for example, reported a very high incidence of complications with 5-Gy fractions. The rate has been consistently low in patients treated with 2 Gy per day or similar fractionation schemes (BACKSTROM et al. 1972; NOBLER 1972). Significant treatment morbidity in 86 patients treated at Washington University included one rectovaginal fistula, one case of proctitis, one rectal stricture, four bone or skin necroses, five vaginal necroses, and one groin abscess. Occasionally, necrosis and fracture of the femoral head/neck may be observed; GRIGSBY et al. (1995) reported a 5% actuarial 5-year incidence of fractures in patients receiving doses of 50 Gy or greater. Cosmetic results with conservation surgery and irradiation may be very rewarding if appropriate surgical and irradiation techniques are applied.

In a review of 73 patients, BALAT et al. (2000) compared complications with wide local excision plus postoperative radiation therapy or radical vulvectomy and bilateral lymphadenectomy plus pre- or postoperative radiation therapy. There were no significant differences among these treatments in terms of primary tumor control, 5-year disease-year survival, or overall survival. The best treatment for advanced vulvar cancer was wide local excision plus radiation therapy, as this method retained a high survival with less morbidity.

26.10
Carcinoma of the Vulva and Pregnancy

Extremely infrequently encountered is the association of pregnancy and vulvar carcinoma. GITSCH et al. (1995) reported on two women with stage II and III disease; the first patient, who had a 5x2-cm ulcerated lesion under the clitoris, was treated by radical anterior vulvectomy and bilateral groin dissection when 22 weeks pregnant. She had an uneventful pregnancy; a Cesarean section was performed, and she was free of disease 29 months after initial surgery. The second patient, who had multifocal lesions with a maximal depth of 1.9 mm of tumor invasion, underwent modified radical vulvectomy with bilateral groin dissection when 17 weeks pregnant. Because of a grossly positive groin lymph node, she underwent radiation therapy after Cesarean section. She was alive without invasive cancer 28 months after diagnosis. Early histological diagnosis and treatment are mandatory in the management of suspicious vulvar lesions during pregnancy, delaying irradiation until the pregnancy is terminated.

26.11
Clinical Trials

Only a small number of clinical trials are available for the treatment of vulvar carcinoma. A phase II trial to document the rates and patterns of recurrence in early stage I patients, all treated with ipsilateral superficial inguinal lymphadenectomy and modified radical hemivulvectomy, is in progress by the Gynecologic Oncology Group (GOG-74). These patients must have pathologically negative ipsilateral superficial inguinal lymph nodes and resection margins at the dermis of greater than 5 mm.

Patients with advanced or recurrent vulvar cancer that is refractory to curative therapy or established treatments are eligible for GOG no. 26. This phase II protocol screens new agents and drug combinations for activity in patients with advanced or recurrent pelvic malignancies.

26.12
Treatment of Other Vulvar Malignant Neoplasias

26.12.1
Adenocarcinoma

Most primary adenocarcinomas of the vulva arise in the Bartholin's gland (LEUCHTER et al. 1982; COPELAND et al. 1986; BURKE et al. 1996). Other potential tissues of origin include skin appendages or ectopic breast tissue.

Radical wide excision, hemivulvectomy, or radical vulvectomy has been used to treat these patients (COPELAND et al. 1986). The incidence of groin node metastases is about 30% (LEUCHTER et al. 1982), thus inguinal lymphadenectomy should be included in primary surgical resection. Radiation therapy may have a role in enhancing local tumor control for women with large primary tumors or inguinal node metastases (COPELAND et al. 1986).

26.12.2
Vulvar Sarcoma

Primary sarcomas are rare tumors that can arise from any mesenchymal tissues of the vulva; leiomyosarcoma and rhabdomyosarcoma predominate (HAYS et al. 1988; TAVASSOLI and NORRIS, 1979). Usual treatment consists of aggressive resection of either primary or locally recurrent disease; postoperative irradiation may decrease the incidence of local recurrence. Rhabdomyosarcoma may be responsive to both chemotherapy and irradiation; the current treatment of choice is to combine chemoirradiation (using vincristine, dactinomycin, cyclophosphamide, and doxorubicin and 50–55 Gy in 1.6–1.8-Gy fractions) with limited surgical resection of residual tumor (BORONOW 1973; KRUPP et al. 1975).

26.12.3
Malignant Melanoma

Superficial malignant melanoma of the vulva can be treated with wide local excision. The current accepted management for more invasive vulvar melanoma is radical vulvectomy and bilateral inguinal lymph node dissection for levels of invasion greater than Clark's level 2. Although some gynecology oncologists recommend ipsilateral inguinal lymph node dissection, CHUNG et al. (1975) found no lymph node metastases in 8 patients with level-2 malignant melanoma, and none of these patients died of the disease. Survival was closely related to depth of invasion; only 2 patients of 25 with Clark levels 3 through 5 melanoma survived 5 years. The most common sites of recurrence were the groin (9 patients), perineum, vagina, and urethra (two cases in each location), and cervix and rectum (one each). All patients who died of melanoma had widespread metastases to lungs, adrenals, brain, liver, kidneys, or retroperitoneal lymph nodes.

WOOLCOTT et al. (1988) reported on 50 patients with primary melanoma of the vulva, 42 of whom were treated with definitive therapy: 16 with wide local excision, 2 with unilateral inguinofemoral and 2 with bilateral inguinofemoral lymphadenectomy, 2 with hemivulvectomy and inguinofemoral lymph node dissection, and 1 with simple vulvectomy alone. Twenty-three patients were treated with radical vulvectomy, combined with bilateral inguinofemoral lymphadenectomy in 22 patients. Two patients also had pelvic lymphadenectomy. Of 42 patients treated with curative intent, 3 received adjuvant radiation therapy. Of 50 patients, 17 (34%) were alive and free of disease at the last follow-up. The 5-year survival rate for 32 eligible patients was 43.8%. Depth of melanocytic penetration was the main prognostic factor. Local recurrence developed in 7 of the 23 patients treated with radical vulvectomy; 4 had inguinofemoral nodal recurrences,

and 8 developed distant metastases. Of 16 patients treated with wide local excision, 3 developed local recurrence, 4 had inguinofemoral recurrences, and 2 had distant metastases.

26.13
Chemotherapy

Topical chemotherapy has been used for treatment of selected patients with vulvar or vaginal intraepithelial neoplasia; 5-fluorouracil (5-FU) has been the most commonly used agent (SILLMAN et al. 1985). All 25 patients treated in this manner were free of disease for 3 months to more than 40 months after treatment, although 3 required retreatment at 3, 9, and 11 months (CALGAR et al. 1981).

Several drug regimens have been used in squamous cell vulvar cancer, most frequently incorporating bleomycin, vincristine, cisplatin, mitomycin-C, or methotrexate in various three- or four-agent combinations with limited activity in phase II studies (BURKE et al. 1996).

26.13.1
Combination of Chemotherapy and Irradiation

THOMAS et al. (1989) described results in 33 patients with vulvar cancer treated with a combination of 5-FU and/or mitomycin-C; 32 had squamous cell carcinoma. Of 18 patients with primary disease, 9 were treated in an adjuvant setting (5 after radical vulvectomy and inguinal node dissection; 4 had local excision only). Several patients had positive surgical margins; 6 of 9 received vulvar irradiation (40–64 Gy in daily fractions of 1.6–1.8 Gy), and 5 of 6, in addition, received inguinal lymph node irradiation. Of 9 patients, 3 were treated to inguinal nodes only. At 5–45 months, 6 of 9 patients remained disease-free. Of the other patients with primary disease, 9 received chemotherapy and irradiation as definitive management; 5 had clinically suspicious inguinal lymph nodes. All 9 patients received irradiation to the vulva, and 4 of 5 with positive nodes also underwent treatment to the inguinal areas. After a follow-up of 5–43 months, 6 of the 9 patients were disease-free. Of 6 patients with residual or recurrent vulvar carcinoma after initial chemoirradiation, 5 had salvage surgery (local excision in 4 and radical vulvectomy in 1). Of 9 patients, 7 had vulvar primary tumor control after salvage therapy.

No patient developed nodal or distant metastases. After initial surgery, 15 patients were treated for recurrence; 10 had a previous radical vulvectomy and bilateral node dissection. At 5–45 months, 7 of these patients remain alive without disease. A complete response was never demonstrated in 3 patients treated for inguinal node recurrence, and all died of disease, 2 also having developed lung metastases. Of 11 patients treated for recurrence in the vulva only, 6 were alive without disease. Four patients who developed pulmonary metastases died.

Combined therapy was well tolerated, except for the expected oropharyngeal mucositis and hematological toxicity of 5-FU. One patient developed severe proctitis after receiving 55 Gy in 35 fractions with electrons to the vulva, and one patient developed a vascular hip necrosis after a dose of 47 Gy in 27 fractions.

The experience from Princess Margaret Hospital (Thomas et al. 1989) using concurrent chemoirradiation for vulvar cancer showed that late vulvar fibrosis, atrophy, telangiectasia, and necrosis can be almost totally avoided if the irradiation fraction size is kept below 1.6–1.8 Gy and excessive total dose is not used; disease recurred in two of six patients receiving less than 50 Gy, and in two of seven patients with doses greater than 50 Gy. Dose exceeding 60 Gy may substantially increase treatment morbidity, particularly if surgical salvage is required. If residual disease persists after 60 Gy, an attempt should be made to remove it surgically. The irradiation volume is tailored to the extent of disease.

Of patients with advanced squamous cell carcinoma of the vulva, 42 were treated with a combination regimen of bleomycin (180 mg) and external irradiation (30–45 Gy) (Scheistroen and Trope 1993); 22 patients had primary and 22 had recurrent disease. Of 15 patients with primary disease, 5 showed complete and 10 partial response; 4 underwent surgery. Of these, 1 was alive after 60 months with no evidence of disease, 2 died of unrelated causes without signs of recurrence, and 17 relapsed and died of carcinoma of the vulva. Of 22 patients treated for recurrence, 2 had complete and 11 had partial responses; none underwent surgery. All of these patients died of carcinoma of the vulva; median survival was 6.5 months. Toxicity was acceptable, and there were no treatment-related deaths. Even when the fact that patients had very advanced disease is taken into account, the results are disappointing. Increased irradiation dose beyond 45 Gy and more aggressive surgery might have improved the results.

Russell et al. (1992) observed 16 complete responses in 25 women with locally advanced or recurrent squamous cell carcinoma of the vulva treated with 5-FU continuous infusion for 96 h (in 11 patients combined with three doses of cisplatin at 100 mg/m^2) and pelvic irradiation (median dose 54 Gy). Of 18 previously untreated patients, 12 were cancer-free at 2–52 months; 3 patients developed intermittent urinary incontinence and 4 developed leg edema.

Wahlen et al. (1995) described 19 patients with locally advanced vulvar cancer (4 stage II and 15 stage III); all had clinically negative inguinal lymph nodes except for 2 patients who had ipsilateral inguinal nodes removed before treatment. The patients received 45–50 Gy to the pelvis and inguinal nodes with concurrent chemotherapy (5-FU in a 96-h continuous infusion, 1000 mg/m^2 per day, during week 1 and week 5 of radiation therapy). Boosts with implants or electrons were received by 10 patients, and 6 others underwent local excision. With median follow-up of 34 months, combined therapy resulted in a local tumor control rate of 75% (14 of 19); all five failures occurred within 6 months of treatment. Of these patients, 4 were rendered disease free by radical vulvectomy and/or exenteration, for an overall local control rate of 95% (18 of 19).

Han et al. (2000) published results in 54 patients with locally advanced vulvar cancer treated with radiation therapy, among which 20 received chemotherapy and radiation therapy, while 34 patients received radiation therapy alone. Of the 20 patients, 14 were treated for primary or recurrent disease, and after vulvectomy for high-risk disease. Of the 34 patients, 12 were treated primarily and 22 received adjuvant treatment. Chemotherapy consisted of two courses of 5-FU and mitomycin-C administered during radiation therapy. Six patients received cisplatin in place of mitomycin-C. In chemoirradiation groups, radiation was administered to the vulva, pelvic, and inguinal lymph nodes to a median dose of 45 Gy, with an additional 6–17 Gy to gross disease. In radiation therapy groups, the median dose to the microscopic disease was 45 Gy. Nine patients received external beam boost and 16 patients received supplementary brachytherapy. Overall survival was superior in the patients treated with chemoirradiation versus irradiation alone (P=0.04). There was also a statistically significant improvement in disease-specific (P=0.03) and relapse-free (P=0.01) survival, favoring chemoirradiation. No statistically significant trends of improved survival rates favoring chemoirradiation over adjuvant radiation therapy were observed.

Table 26.6. Results of chemotherapy and irradiation in locally advanced or recurrent vulvar cancer. 5-FU 5-fluorouracil, Bleo bleomycin, C cisplatin, M mitomycin-C, RV radical vulvectomy, DFS disease-free survival. Modified from PEREZ et al. (1997)

Author	No. of patients	Chemo-therapy	Irradiation (Gy)	Complete response (%)	Partial response (%)	Locoregional tumor control	Disease-free survival	DFS time (months)
EIFEL et al. 1995	12	5-FU, C		90	90		6/12	17–30
IVERSEN 1982	15	Bleo	36–40	15		4/15	15%	48
KOH et al. 1993b	20	5-FU, C, M	40–54	50	40	10/20	59%	60
MONTANA et al. 2000	46	C, 5-FU	47.6	76	–	29/38	50%	
RUSSELL et al. 1992	25	5-FU, C	36–72	64	–	16/18 primary	67%	
						4/7 recurrent	29%	2–52
SCHEISTROEN and TROPE 1993	42	Bleo	30–45	33	40	5/22	5%	60
THOMAS et al. 1989	33	5-FU, M	40–64			7/9 primary, 7/10 post-RV		5–45

MONTANA et al. (2000) reported on a GOG study of 46 patients with vulvar cancer and N2-3 lymph nodes. Patients underwent a split course of radiation therapy – 4760 cGy to the primary and lymph nodes with concurrent chemotherapy, cisplatin/5-FU followed by surgery. The chemoirradiation was not completed by 4 patients because 3 died and 1 refused to complete the treatment. The 4 patients who completed chemoirradiation did not undergo surgery because 2 died of non-cancer-related causes and, in the other 2, the lymph nodes remained unresectable. Following chemoirradiation, the disease in the lymph nodes became resectable in 38 of 40 patients. Because of pulmonary metastasis, 2 patients who completed chemoirradiation did not undergo surgery as per protocol. One underwent radical vulvectomy and unilateral node dissection and the other radical vulvectomy only. The specimen of the lymph nodes was histologically negative in 15 of 37 patients (40%). Recurrent and/or metastatic disease developed in 19 patients. The sites of failure were as follows: primary area only, 9; lymph node area only, 1; primary area and distant metastasis, 1; distant metastasis only, 8. Local control of the disease in the lymph nodes was achieved in 36 of 37 and in the primary area in 29 of 38 patients. Of the patients, 20 are alive and disease-free, and 5 have died without evidence of recurrence or metastasis; 2 patients died of treatment-related complications.

Preliminary results of chemotherapy and irradiation, summarized in Table 26.6, are promising and suggest that patients in high-risk groups can be treated effectively with combinations of surgery, irradiation, and systemic chemotherapy. Further assessment of long-term results is in progress, and prospective clinical trials are strongly encouraged.

References

Anderson ES, Sorenson IM (1988) Verrucous carcinoma of the female genital tract: Report of a case and review of the literature. Gynecol Oncol 30:427-434

Backstrom A, Edsmyr F, Wicklund H (1972) Radiotherapy of carcinoma of the vulva. Acta Obstet Gynecol Scand 51:109-115

Baltzer JE, Kurzl RY, Lohe KJ, et al. (1986) Melanoma of the vulva. J Reprod Med 31:825-827

Barnhill DR, Boling R, Nobles W, et al. (1988) Vulvar dermatofibrosarcoma protuberans. Gynecol Oncol 30:149-152

Balat O, Edwards CL, Delclos L (2000) Complications following combined surgery (radical vulvectomy versus wide local excision) and radiotherapy for the treatment of carcinoma of the vulva: Report of 73 patients. Eur J Gynaecol Oncol 21:501-503

Balat O, Edwards CL, Delclos L (2001) Advanced primary carcinoma of the Barthlin gland: Report of patients. Eur J Gynaecol Oncol 22:46-49

Benda JA, Platz CE, Anderson B (1986) Malignant melanoma of the vulva: a clinical-pathologic review of 16 cases. Int J Gynecol Pathol 5:202-216.

Berman ML, Soper JT, Creasman WT, et al. (1989) Conservative surgical management of superficially invasive stage I vulvar carcinoma. Gynecol Oncol 35:352-357

Binder SW, Huang I, Fu YS, et al. (1990) Risk factors for the development of lymph node metastasis in vulvar squamous carcinoma. Gynecol Oncol 37:9-16

Boronow RC (1973) Therapeutic alternative to primary exenteration for advanced vulvovaginal cancer. Gynecol Oncol 1:233

Boronow RC (1982) Combined therapy as an alternative to exenteration for locally advanced vulvo-vaginal cancer: rationale and results. Cancer 49:1085-1091

Boronow RC, Hickman BT, Regan MT (1987) Combined therapy as an alternative to exenteration for locally advanced vulvovaginal cancer. II. Results, complications, and dosimetric and surgical considerations. Am J Clin Oncol (CCT) 10:171-181

Boutselis JG (1972) Radical vulvectomy for invasive squamous cell carcinoma of the vulva. Obstet Gynecol 39:827-836

Brand E, Fu YS, Lagasse LD, et al. (1989) Vulvovaginal melanoma: Report of seven cases and literature review. Gynecol Oncol 33:54-60

Breen JL, Neubecker RD, Greenwald E (1975) Basal cell carcinomas of the vulva. Obstet Gynecol 46:122-129

Bryon SC, Dembo AJ, Colgan TJ, et al. (1991) Invasive squamous cell carcinoma of the vulva: Defining low and high risk groups for recurrence. Int J Gynecol Can 1:25-31

Burger MPM, Holleman H, Bouma J (1996) The side of groin node metastases in unilateral vulvar carcinoma. Int J Gynecol Cancer 6:318-322

Burke TW, Eifel P, McGuire W (1996) Vulva. In Hoskins WJ, Perez CA, Young RC (eds) Principles and practice of gynecologic oncology, 2nd edn. Lippincott-Raven, Philadelphia

Burke TW, Levenback C, Coleman RC, et al. (1995) Surgical therapy of T1 and T2 vulvar carcinoma: Further experience with radical wide excision and selective inguinal lymphadenectomy. Gynecol Oncol 57:215-220

Burke TW, Stringer CA, Gershenson DM (1990) Radical wide excision and selective inguinal node dissection for squamous cell carcinoma of the vulva. Gynecol Oncol 38:328-332

Busch M, Wagener B, Schaffer M, et al. (2000) Long-term impact of postoperative radiotherapy in carcinoma of the vulva FIGO I/II. Int J Radiat Oncol Biol Phys 48:213-218

Calgar H, Hertzog AW, Hreschyshyn MM (1981) Topical 5-fluorouracil treatment of vaginal intraepithelial neoplasia. Obstet Gynecol 5:580

Cardosi RJ, Sprights A, Fiorica JW, et al. (2001) Bartholin's gland carcinoma: A 15-year experience. 82:247-251

Carlino G, Parisi S, Montemaggi P, et al. (1984) Interstitial radiotherapy with ^{192}Ir in vulvar cancer. Eur J Gynaecol Oncol 3:183-185

Chafe W, Richards A, Morgan L, et al. (1988) Unrecognized invasive carcinoma in vulvar intraepithelial neoplasia (VIN). Gynecol Oncol 31:154-162

Chung AF, Woodruff JM, Lewis JL Jr (1975) Malignant melanoma of the vulva. Obstet Gynecol 45:638-646

Cohn DE, Dehdashti F, Gibb RK, et al. (2002) Prospective evaluation of prositron emission tomography for the detection of groin node metastases from vulvar cancer. Gynecol Oncol 85:179-184

Copeland LJ, Sneige N, Gershenson DM, et al. (1986) Bartholin gland carcinoma. Obstet Gynecol 67:794-801

Creasman WT (1990) New gynecologic cancer staging. Obstet Gynecol 75:287-288

Curry SL, Wharton JT, Rutledge F (1980) Positive lymph nodes in vulvar squamous carcinoma. Gynecol Oncol 9:63-67

Dehner LP (1973) Metastatic and secondary tumors of the vulva. Obstet Gynecol 52:47-57

de Hullu JA, Oonk MH, van der Zee AG (2004) Modern management of vulvar cancer. Curr Opin Obstet Gynecol 16:65-72

DiSaia PJ, Creasman WT, Rich WM (1979) An alternative approach to early cancer of the vulva. Am J Obstet Gynecol 133:825-832

Donaldson ES, Powell DE, Hanson MB, et al. (1981) Prognostic parameters in invasive vulvar cancer. Gynecol Oncol 11:184-190

DuBeshter B, Lin J, Angel C, et al. (1993) Gynecologic tumors, 7th edn. In: Rubin P (ed) Clinical oncology for physicians and medical students: a multidisciplinary approach. WB Saunders, Philadelphia

Dusenbery KE, Carlson JW, LaPorte RM, et al. (1994) Radical vulvectomy with postoperative irradiation for vulvar cancer: Therapeutic implications of a central block. Int J Radiat Oncol Biol Phys 29:989-998

Eifel PH, Morris M, Burke TW, et al. (1995) Prolonged continuous infusion cisplatin and 5-fluorouracil with radiation for locally advanced carcinoma of the vulva. Gynecol Oncol 59:51-56

Farias-Eisner R, Cirisano FD, Grouse D, et al. (1994) Conservative and individualized surgery for early squamous carcinoma of the vulva: The treatment of choice for stage I and II (T1-2 N0-1M0) disease. Gynecol Oncol 53:55-58

Figge DC, Tamimi HK, Greer BE (1985) Lymphatic spread in carcinoma of the vulva. Am J Obstet Gynecol 152:387-394

Fiorica JV, Cavanagh D, Marsden DE, et al. (1988) Carcinoma in situ of the vulva: 24 years' experience in Southwest Florida. South Med J 81:589-593

Franklin EW (1972) Clinical staging of carcinoma of the vulva. Obstet Gynecol 40:277

Franklin EW, Rutledge FD (1971) Prognostic factors in epidermoid carcinoma of the vulva. Obstet Gynecol 39:892-901

Gitsch G, van Eijkeren M, Hacker NF (1995) Surgical therapy of vulvar cancer in pregnancy. Gynecol Oncol 56:213-215

Gosling JRG, Abell MR, Drolette BM, et al. (1961) Infiltrative squamous carcinoma of the vulva. Cancer 14:330-343

Greene FL, Page DL, Fleming ID, et al. (eds) (2002) AJCC Cancer Staging Manual, 6th edn. Springer-Verlag, New York

Grigsby PW, Roberts HL, Perez CA (1995) Femoral neck fracture following groin irradiation. Int J Radiat Oncol Biol Phys 32:63-67

Hacker NF, Berek JS, Juillard GJF, et al. (1984a) Preoperative radiation therapy for locally advanced vulvar cancer. Cancer 54:2056-2061

Hacker NF, Berek JS, Lagasse L, et al. (1983) Management of regional lymph nodes and their prognostic influence in vulvar cancer. Obstet Gynecol 61:408-412

Hacker NF, Berek JS, Lagasse LD, et al. (1984b) Individualization of treatment for stage I squamous cell vulvar carcinoma. Obstet Gynecol 63:155-162

Hacker NF, Leuchter RS, Berek JS, et al. (1981) Radical vulvectomy and bilateral inguinal lymphadenectomy through separate groin incisions. Obstet Gynecol 58:574-579

Han SC, Kim DH, Higgins SA, et al. (2000) Chemoradiation as primary or adjuvant treatment for locally advanced carcinoma of the vulva. Int J Radiat Oncol Biol Phys 47:1235-1244

Harper JL, Jenrette JM, Goddu SM, et al. (2004) Vulvar cancer in a patient with Fanconi's anemia, treated with 3-D conformal radiotherapy. Am J Hematol 776:148-151

Hays DM, Shimada H, Raney RB, et al. (1988) Clinical staging and treatment results in rhabdomyosarcoma of the female genital tract among children and adolescents. Cancer 61:1893-1903

Heaps JM, Fu YS, Montz FJ, et al. (1990) Surgical-pathologic variables predictive of local recurrence in squamous cell carcinoma of the vulva. Gynecol Oncol 38:309-314

Helgason NM, Hass AC, Latourette HB (1972) Radiation therapy in carcinoma of the vulva. Cancer 30:997-1000

Helwig EB, Graham JH (1963) Anogenital (extramammary) Paget's disease. Cancer 16:387-403

Hoffman MS, Roberts WS, Ruffolo EH (1988) Basal cell carcinoma of the vulva with inguinal lymph node metastases. Gynecol Oncol 29:113-119

Homesley HD, Bundy BN, Sedlis A, et al. (1986) Radiation therapy versus pelvic node resection for carcinoma of the vulva with positive groin nodes. Obstet Gynecol 68:733-740

Hoskins W, Perez CA (1989) Gynecologic tumors. In: DeVita

VT, Hellman S, Rosenberg SA (eds) Cancer: Principles and practice of oncology, 3rd edn. JB Lipppncott, Philadelphia

Husseinzadeh N, Newman NJ, Wesseler TA (1989) Vulvar intraepithelial neoplasia: A clinicopathological study of carcinoma in situ of the vulva. Gynecol Oncol 33:157-163

Iversen T (1982) Irradiation and bleomycin in the treatment of inoperable vulval carcinoma. Acta Obstet Gynecol Scand 61:195-197

Iversen T, Aalders JG, Christensen A, et al. (1980) Squamous cell carcinoma of the vulva: A review of 424 patients, 1956-1974. Gynecol Oncol 9:271-279

Kalnicki S, Zide A, Maleki N, et al. (1987) Transmission block to simplify combined pelvic and inguinal radiation therapy. Radiology 164:578-580

Katz A, Eifel PJ, Jhingran A, et al. (2003) The role of radiation therapy in preventing regional recurrences of invasive squamous cell carcinoma of the vulva. Int J Radiat Oncol Biol Phys 57:409-418

Kaufman RH, Gardner HL (1965) Intraepithelial carcinoma of the vulva. Clin Obstet Gynecol 8:1035-1050

Kelley JL III, Burke TW, Tornos C, et al. (1991) Minimally invasive vulvar carcinoma: An indication for conservative surgical therapy. Gynecol Oncol 144:240-244

King GC, Sonnik DA, Dalend AM, et al. (1993) Transmission block technique for the treatment of the pelvis and perineum including the inguinal lymph nodes: Dosimetric considerations. Med Dosim 18:7-12

Koh WJ, Chiu M, Stelzer KJ, et al. (1993a) Femoral vessel depth and the implications for groin node radiation. Int J Radiat Oncol Biol Phys 27(4):969-974

Koh WJ, Wallace JH III, Greer BE, et al. (1993b) Combined radiotherapy and chemotherapy in the management of local-regionally advanced vulvar cancer. Int J Radiat Oncol Biol Phys 26:809-816

Krupp PJ, Bohm JW (1978) Lymph gland metastases in invasive squamous cell cancer of the vulva. Am J Obstet Gynecol 130:943-952

Krupp PJ, Lee FYL, Bohm JW, et al. (1975) Prognostic parameters and clinical staging criteria in epidermoid carcinoma of the vulva. Obstet Gynecol 46:84-88

Kuipers T (1975) Carcinoma of the vulva. Radiol Clin 44:475-483

Kurzl R, Messerer D (1989) Prognostic factors in squamous cell carcinoma of the vulva: A multivariate analysis. Gynecol Oncol 32:143-150

Lasser A, Cornorg IJT, Morris JM (1974) Adenoid squamous carcinoma of the vulva. Cancer 33:224-227

Lelle RJ, Davis KP, Roberts JA (1994) Adenoid cystic carcinoma of the Bartholin's gland: The University of Michigan experience. Int J Gynecol Cancer 4:145-149

Leuchter RS, Hacker NF, Vopet RL, et al. (1982) Primary carcinoma of the Bartholin gland: A report of 14 cases and review of the literature. Obstet Gynecol 60:361-368

Lucas WE, Bernischke K, Lebherz TB (1974) Verrucous carcinoma of the female genital tract. Am J Obstet Gynecol 119:435-440

Malfetano JH, Piver S, Tsukada Y, et al. (1985) Univariate and multivariate analyses of 5-year survival, recurrence, and inguinal node metastases in stage I and II vulvar carcinoma. J Surg Oncol 30:124-131

McCall AR, Olson MC, Potkul RK (1995) The variation of inguinal lymph node depth in adult women and its importance in planning elective irradiation for vulvar cancer. Cancer 75(9):2286-2288

Miyazawa K, Nori D, Hilaris BS, et al. (1983) Role of radiation therapy in the treatment of advanced vulvar carcinoma. J Reprod Med 28:539-541

Montana GS, Thomas GM, Moore DH, et al. (2000) Preoperative chemo-radiation for carcinoma of the vulva with N2/N3 nodes: A Gynecologic Oncology Group study. Int J Radiat Oncol Biol Phys 48:1007-1013

Morris JM (1977) A formula for selective lymphadenectomy. Obstet Gynecol 50:152-158

Nobler MP (1972) Efficacy of a perineal teletherapy portal in the management of vulvar and vaginal cancer. Radiology 103:393-397

Nori D, Cain JM, Hilaris BS, et al. (1983) Metronidazole as a radiosensitizer and high-dose radiation in advanced vulvovaginal malignancies: A pilot study. Gynecol Oncol 16:117-128

Origoni M, Sideri M, Garsia S, et al. (1992) Prognostic value of pathological patterns of lymph node positivity in squamous cell carcinoma of the vulva stage III and IVA FIGO. Gynecol Oncol 45:313-316

Pao WM, Perez CA, Kuske RR, et al. (1988) Radiation therapy and conservation surgery for primary and recurrent carcinoma of the vulva: Report of 40 patients and a review of the literature. Int J Radiat Oncol Biol Phys 14:1123-1132

Parker RT, Duncan I, Rampone J, et al. (1975) Operative management of early invasive epidermoid carcinoma of the vulva. Am J Obstet Gynecol 123:349-355

Perez CA, Grigsby PW, Chao KSC, et al. (1997) Vulva. In: Perez CA, Brady LW (eds) Principles and practice of radiation oncology, 3rd edn. Lippincott-Raven, Philadelphia, pp 1915-1942

Perez CA, Grigsby PW, Galakatos A, et al. (1993) Radiation therapy in management of carcinoma of the vulva with emphasis on conservation therapy. Cancer 71(11):3707-3716

Perrone T, Twiggs LB, Adcock LL, et al. (1987) Vulvar basal cell carcinoma: An infrequently metastasizing neoplasm. Int J Gynecol Pathol 6:152-165

Petereit DG, Mehta MP, Buchler DA, et al. (1993) Inguinofemoral radiation of N0, N1 vulvar cancer may be equivalent to lymphadenectomy if proper radiation technique is used. Int J Radiat Oncol Biol Phys 27:963-967

Plentl AA, Friedman EA (1971) Lymphatic system of the female genitalia. WB Saunders, Philadelphia

Pohar S, Hoffstetter S, Peiffert D, et al. (1995) Effectiveness of brachytherapy in treating carcinoma of the vulva. Int J Radiat Oncol Biol Phys 32:1455-1460

Rosén C, Malmström H (1997) Invasive cancer of the vulva. Gynecol Oncol 656:213-217

Rosenberg P, Simonses E, Risberg D (1989) Adenoid cystic carcinoma of Bartholin's gland: A report of five new cases treated with surgery and radiotherapy. Gynecol Oncol 34:145-147

Rowley KC, Gallion HH, Donaldson SE, et al. (1988) Prognostic factors in early vulvar cancer. Gynecol Oncol 31:43-49

Russell AH, Mesic JB, Scudder SA, et al. (1992) Synchronous radiation and cytotoxic chemotherapy for locally advanced or recurrent squamous cancer of the vulva. Gynecol Oncol 47:14-20

Rutledge F, Smith JP, Franklin EW (1970) Carcinoma of the vulva. Am J Obstet Gynecol 106:1117-1130

Rutledge, FN, Mitchell MF, Munsell MF, et al. (1991) Prognostic indicators for invasive carcinoma of the vulva. Gynecol Oncol 42:239-244

Scheistroen M, Trope C (1993) Combined bleomycin and irradiation in preoperative treatment of advanced squamous cell carcinoma of the vulva. Acta Oncol 32:657-661

Schulz U, Callies R, Kruger KG (1980) Effizienz der postoperativen elektronentherapie des lokalisierten vulvakarzinoms. Strahlentherapie 156:326-330

Sedlis A, Homesley H, Bundy BN, et al. (1987) Positive groin lymph nodes in superficial squamous cell vulvar cancer. Am J Obstet Gynecol 156:1159-1164

Shepherd JH (1989) Revised FIGO staging for gynecological cancer. Br J Obstet Gynaecol 96:889-892

Shimm DS, Fuller AF, Orlow EL, et al. (1986) Prognostic variables in the treatment of squamous cell carcinoma of the vulva. Gynecol Oncol 24:343-358

Sillman FH, Sedlis A, Boyce JG (1985) A review of lower genital intraepithelial neoplasia and the use of topical 5-fluorouracil. Obstet Gynecol Surg 40:190-220

Stehman FB, Bundy BN, Dvoretsky PM, et al. (1992a) Early stage I carcinoma of the vulva treated with ipsilateral superficial inguinal lymphadenectomy and modified radical hemivulvectomy: A prospective study of the Gynecologic Oncology Group. Obstet Gynecol 79:490-497

Stehman FB, Bundy BN, Thomas C, et al. (1992b) Groin dissection versus groin radiation in carcinoma of the vulva: A Gynecologic Oncology Group study. Int J Radiat Oncol Biol Phys 24(2):389-396

Stern BD, Kaplan L (1969) Multicentric foci of carcinoma arising in structures of cloacal origin. Am J Obstet Gynecol 104:255-266

Tavassoli FA, Norris HJ (1979) Smooth muscle tumors of the vulva. Obstet Gynecol 53:213-217

Thomas G, Dembo A, DePetrillo A, et al. (1989) Concurrent radiation and chemotherapy in vulvar carcinoma. Gynecol Oncol 34:263-267

Tod MC (1949) Radium implantation treatment of carcinoma vulva. Br J Radiol 22:508

Ulbright TM, Brokaw SA, Stehman FB, et al. (1983) Epithelioid sarcoma of the vulva. Cancer 52:1462-1469

van der Velden J, van Lindert AC, Lammes FB, et al. (1995) Extracapsular growth of lymph node metastases in squamous cell carcinoma of the vulva. Cancer 75:2885-2890

van der Velden K, Ansink A (2001) Primary groin irradiation vs primary groin surgery for early vulvar cancer. Cochrane Database Syst Rev 4:CD002224

Wahlen SA, Slater JD, Wagmer RJ, et al. (1995) Concurrent radiation therapy and chemotherapy in the treatment of primary squamous cell carcinoma of the vulva. Cancer 75:2289-2294

Wang CJ, Chin YY, Leung SW, et al. (1996) Topographic distribution of inguinal lymph nodes metastasis: Significance in determination of treatment margin for elective inguinal lymph nodes irradiation of low pelvic tumors. Int J Radiat Oncol Biol Phys 35:133-136

Weiner SA, Lee JKT, Kao MS, et al. (1986) The role of lymphangiography in vulvar carcinoma. Am J Obstet Gynecol 5:1073-1075

Wharton JT, Gallagher S, Rutledge FN (1974) Microinvasive carcinoma of the vulva. Am J Obstet Gynecol 118:159-162

Woolcott RJ, Henry RJW, Houghton CRS (1988) Malignant melanoma of the vulva. J Reprod Med 33:699-702

27 Carcinoma of the Vagina

Higinia R. Cardenes

CONTENTS

27.1   Anatomy 657
27.2   Pathology 658
27.3   Natural History 658
27.4   Clinical Presentation 659
27.4.1   Diagnostic Work-up 659
27.4.2   Staging 659
27.5   Prognostic Factors Influencing Choice of
         Treatment 660
27.6   General Management: Treatment Options 662
27.6.1   RT Techniques 663
27.6.1.1 External Beam Radiotherapy 663
27.6.1.2 Low-Dose-Rate Intracavitary Brachytherapy 665
27.6.1.3 High-Dose-Rate Intracavitary Brachytherapy 666
27.6.1.4 Interstitial Brachytherapy 667
27.6.2   SCC: Treatment Options and Outcome by
         FIGO Stages 669
27.6.2.1 FIGO Stage 0: VAIN – CIS 669
27.6.2.2 Invasive SCC 670
27.6.3   Chemotherapy and Radiation 670
27.6.4   Patterns of Failure in SCC 674
27.6.4.1 Potential RT-Related Factors Influencing
         Outcome 674
27.7   Clear Cell Carcinoma of the Vagina 675
27.8   Salvage Therapy 675
27.8.1   Surgical Considerations 676
27.8.2   RT Considerations 676
27.9   Treatment Complications and their
         Management 678
27.10   Palliative RT 680
27.11   Chemotherapy in
         Advanced-Recurrent Vaginal Cancer 682
         References 682

Primary vaginal cancer is a rare entity and accounts for only 1–2% of all female genital neoplasias (Herbst et al. 1970), ranking next to last in frequency among gynecological malignancies. Most carcinomas found in the vagina represent direct extension or metastasis from other primary gynecological (cervix or vulva) and non-gynecological sites, most commonly breast and gastrointestinal tract. There are a number of controversies in terms of epidemiology, staging and diagnostic evaluation as well as management of vaginal cancer. Given the lack of prospective, randomized studies in patients with vaginal cancer, given its rarity, it is difficult to establish strong, evidence-based recommendations. Therefore, the decisions regarding therapeutic options should be based on the best retrospective data and individual assessment using general principles derived from clinical experience in the management of cancer at other sites. Most of the available data refer to the treatment of primary invasive squamous cell carcinoma (SCC) of the vagina, since this represents the most common histology.

27.1
Anatomy

The vagina is a dilatable tubular structure averaging 7.5 cm in length, lined by non-keratinizing, stratified, squamous epithelium extending from the vestibule to the cervix of the uterus. It lies dorsal to the base of the bladder and urethra, and ventral to the rectum. The upper fourth of the posterior wall is separated from the rectum by a reflection of peritoneum, the pouch of Douglas. The vaginal wall is composed of three layers: the mucosa, muscularis and adventitia. Beneath the mucosa lies a submucosal layer of elastin and a double muscularis layer, highly vascularized with a rich innervation and lymphatic drainage. The muscularis layer is composed of smooth muscle fibers, arranged circularly on the inner portion and longitudinally in the outer por-

H. R. Cardenes, MD, PhD
Clinical Associate Professor, Department of Radiation Oncology, RT 041, Indiana University School of Medicine, 535 Barnhill Drive, Indianapolis, IN 46202, USA

tion. The adventitia is a thin, outer connective tissue layer that merges with that of adjacent organs.

The proximal vagina is supplied by the vaginal artery branch from the uterine or cervical branch of the uterine artery. It runs along the lateral wall of the vagina and anastomoses with the inferior vesical and middle rectal arteries from the surrounding viscera (SEDLIS and ROBBOY 1987). The accompanying venous plexus, running parallel to the arteries, ultimately drains to the internal iliac vein. The lumbar plexus and pudendal nerve, with branches from the sacral roots 2 to 4, provide innervation to the vaginal vault.

The lymphatic drainage of the vagina is complex, consisting of an extensive inter-communicating network. Fine lymphatic vessels coursing through the submucosa and muscularis coalesce into small trunks running laterally along the walls of the vagina. The upper anterior vagina drains along cervical channels to the interiliac chain; the posterior vagina drains into the inferior gluteal, presacral, and anorectal nodes. The distal vagina lymphatics drain into the inguinal and femoral nodes and from there to the pelvic nodes. Lymphatic flow from lesions in the mid-vagina may drain either way (PLENTL and FRIEDMAN 1971). However, because of the presence of inter-communicating lymphatics along the terminal branches of the vaginal artery and near the vaginal wall, the external iliac nodes are at high risk, even in lesions of the lower third of the vagina. Such a complex lymphatic drainage pattern has significant implications for therapeutic planning. Therefore, bilateral pelvic nodes should be considered at risk in any invasive vaginal carcinoma, and bilateral groin nodes considered at risk in those lesions involving the distal third of the vagina.

27.2
Pathology

SCC represents about 80–90% of primary vaginal cancers (ZAINO et al. 2002). These tumors occur in older women and are most often located in the upper, posterior wall of the vagina. Histologically, keratinizing, non-keratinizing, basaloid, warty and verrucous variants have been described. Tumors may also be graded, moderately or poorly differentiated, most cases being moderately differentiated and non-keratinizing. Vaginal intraepithelial neoplasia (VAIN) is a precursor of SCC and is graded from I to III, depending on the degree of nuclear atypia

and crowding, and the proportion of the epithelium involved. The true incidence of VAIN and its rate of progression to invasive carcinoma are unknown, ranging in several series from 9% to 28% (AHO et al. 1991; BRINTON et al. 1990; HOFFMAN et al. 1991).

Clear-cell adenocarcinoma (CCA) is associated with intrauterine diethylstilbestrol (DES) exposure (ANTONIOLI and BURKE 1975; KAUFFMAN et al. 1982; MAASSEN et al. 1993; ROBBOY et al. 1977, 1982, 1984). These tumors have a predilection for the upper third of the vagina and the exocervix. Most are exophytic and superficially invasive (HERBST et al. 1974). Other adenocarcinomas that could involve the vagina include: mucinous type (EBRAHIM et al. 2001; HIROI et al. 2001; MUNKARAH et al. 1994), endometrioid adenocarcinoma – usually arising with endometriosis (HASKEL et al. 1989), mesonephric (HINCHEY et al. 1983) and papillary serous adenocarcinoma (RIVA et al. 1997).

CREASMAN et al. (1998) published in 1998 the National Cancer Data Base (NCDB) report on 4885 patients with primary diagnosis of vaginal cancer registered from 1985 to 1994. Approximately 92% of the patients were diagnosed with in-situ or invasive SCC or adenocarcinomas; 4% with melanomas; 3% with sarcomas; and 1% with other or unspecified types of cancer. In the NCDB report, invasive carcinomas accounted for 72% of the carcinoma cases, or 66% of all vaginal cancers. In-situ carcinomas accounted for 28%; SCC represented 79% of invasive vaginal carcinomas; and adenocarcinomas represented 14% (CREASMAN et al. 1998).

27.3
Natural History

The majority (57–83%) of vaginal primaries occur in the upper third or at the apex of the vault, most commonly in the posterior wall; the lower third may be involved in as many as 31% of patients (GALLUP et al. 1987; RUBIN et al. 1985; STOCK et al. 1995). Lesions confined to the middle third are uncommon. Vaginal tumors may spread along the vaginal walls to involve the cervix or the vulva. A lesion on the anterior wall may infiltrate the vesico–vaginal septum and/or the urethra; those on the posterior wall may eventually involve the recto–vaginal septum and, subsequently, infiltrate the rectal mucosa. Vaginal cancer may invade the parametrium and paracolpal tissues, extending into the obturator fossa, cardinal ligaments, lateral pelvic walls and the uterosacral ligaments.

The issue of regional nodal metastasis, both the incidence of occult nodal disease and the anatomic pathways of lymphatic spread, are somewhat controversial. The incidence of positive pelvic nodes at diagnosis varies with the stage and location of the lesion. There does seem to be a significant risk of nodal metastasis for patients with disease beyond stage I. Although data on staging lymphadenectomy are sparse, two studies reported a significant incidence of nodal disease in early stage vaginal carcinoma. In Al-Kurdi's series (AL-KURDI and MONAGHAN 1981), the incidence of pelvic nodal metastasis was 14% and 32% for stages I and II, respectively, whereas in Davis's series (DAVIS et al. 1991) the incidence was 6% and 26% for stages I and II, respectively. The incidence is expected to be higher for stage III, although no substantial data are available. Involvement of inguinal nodes is most common when the lesion is located in the lower third of the vagina.

Distant metastasis may occur, primarily in patients with advanced disease at presentation, or in those who experienced tumor recurrence after primary therapy. In Perez series (PEREZ et al. 1988), the incidence of distant metastasis was 16% in stage I, 31% in stage IIA, 46% in stage IIB, 62% in stage III and 50% in stage IV.

27.4
Clinical Presentation

VAIN is most often asymptomatic (LENEHAN et al. 1986) and is usually detected by means of cytology. In patients with invasive disease, irregular vaginal bleeding and/or discharge (often post-coital) is the most common presenting symptom, followed by vaginal discharge and dysuria. Pelvic pain is a relatively late symptom, generally related to tumor extent beyond the vagina (GALLUP et al. 1987; RUBIN et al. 1985; TJALMA et al. 2001).

27.4.1
Diagnostic Work-up

In general, in patients with suspected vaginal malignancy, thorough physical exam with detailed speculum inspection, digital palpation, colposcopic and cytological evaluation and biopsy constitute the most effective procedure for diagnosing primary, metastatic or recurrent carcinoma of the vagina. In

symptomatic patients, biopsy of any abnormal exophytic or endophytic lesion noted at the time of the exam is indicated. Examination under anesthesia is recommended for the thoroughness of evaluation of all of the vaginal walls and local extent of the disease, primarily if the patient is in great discomfort because of advanced disease, in order to obtain a biopsy. Biopsies of the cervix, if present, are recommended to rule out primary cervical tumor. It is important that the speculum is slowly withdrawn from the vaginal fornix so that the total vaginal mucosa may be visualized.

The patient with a history of pre-invasive or invasive carcinoma of the cervix found to have abnormal cytology following prior hysterectomy or radiotherapy (RT) should be offered colposcopy with application of acetic acid to the entire vault, followed by biopsies as indicated by areas of white epithelium, mosaicism, punctation or atypical vascularity. It can be very helpful for the menopausal patient or the patient previously irradiated to use a short course of topical estrogen (Premarin) into the vault once or twice a week for 1 month prior to the colposcopy, in order to foster epithelial maturation. Another method of identifying the area(s) most in need of biopsy would be, after application of the acetic acid, to place half-strength Schiller's iodine to determine whether the Schiller-positive (non-staining) areas correspond with the involved areas identified after acetic acid application.

27.4.2
Staging

At present, primary malignancies of the vagina are all staged clinically. In addition to a complete history and physical examination, routine laboratory evaluations including complete blood cell count (CBC) with differential and platelets and assessment of renal and hepatic function should be undertaken. In order to determine the extent of disease, the following tests are allowed according to the International Federation of Gynecology and Obstetrics (FIGO) criteria: chest X-ray, a thorough bimanual and recto–vaginal exam, cystoscopy, proctoscopy and intravenous pyelogram. If the patient is in significant discomfort, the exam should be conducted under anesthesia, preferably by a radiation oncologist and gynecological oncologist. However, it can be difficult even for the experienced examiner to differentiate between disease confined to the mucosa (stage I) and disease spread to the submucosa (stage II) (BALL and BERMAN 1982; RUBIN et al. 1985). Cystoscopy and proctoscopy are generally performed on patients with symptoms or clinical findings indicative of bladder or rectal infiltration, respectively.

Pelvic computed tomography (CT) scan is generally performed to evaluate inguino-femoral and/or pelvic lymph nodes, as well as extent of local disease. In patients with vaginal melanoma or sarcoma, chest, abdomen, and pelvis CT scans are often part of the work-up. Magnetic resonance imaging (MRI) has emerged as a potentially important imaging modality in the evaluation of vaginal cancers, predominantly the T1-weighted with contrast and T2-weighted images. An additional role of MRI is differentiation of tumor from fibrotic tissue in patients with suspected recurrent vaginal carcinoma (CHANG et al. 1988).

The two commonly used staging systems for carcinoma of the vagina are the FIGO (Table 27.1a) (PECORELLI et al. 2000) and the American Joint Commission on Cancer (TNM) (Table 27.1b) (AJCC 2002) classifications. According to the FIGO guidelines, cases should be classified as vaginal carcinomas only when "the primary site of the growth is in the vagina". A tumor of the vagina that involves the cervix or vulva should be classified as a primary cervical or vulvar cancer, respectively. Additionally, in the setting of a prior gynecological malignancy, a neoplasm would be classified as primary carcinoma of the vagina if the current vaginal tumor occurred 5 years or more after the initial cancer diagnosis and if there is no other clinical evidence of recurrence of the initial gynecological lesion (ZAINO et al. 2002). It may be difficult or impossible histologically to distinguish a primary vaginal SCC from recurrent cervical or vulvar disease. In this setting, it is unclear whether the vaginal lesion represents a new carcinoma of the vagina, recurrent cervical cancer, or a human papilloma-virus (HPV)-related field effect in these patients.

PEREZ et al. (1973) proposed in 1973 that FIGO stage-II vaginal cancer should be subdivided into stage IIa (tumor infiltrating the subvaginal tissues but not extending into the parametrium) and stage IIb (tumor infiltrating the parametrium but not extending to pelvic side walls). However, most authors do not use this classification, and there is limited published data to support the prognostic significance of this subclassification (PEREZ et al. 1988; PREMPREE and AMOMMAN 1985). In addition, FIGO does not assign a specific stage for those patients with inguino-femoral lymphadenopathy. Some

authors assign these patients to stage III, whereas others consider them stage IVB. In the AJCC staging system, patients with T1–3 and positive nodes (pelvic or inguinal) are assigned to stage III (AJCC 2002).

27.5
Prognostic Factors Influencing Choice of Treatment

As with most primaries, stage of disease is the dominant prognostic factor in terms of ultimate outcome (CHU and BEECHINOR 1984; CHYLE et al. 1996; DANCUART et al. 1988; DELCLOS 1984; DIXIT et al. 1993; EDDY et al. 1991; KIRKBRIDE et al. 1995; LEE et al. 1994; LEUNG and SEXTON 1993; MACNAUGHT et al. 1986; PEREZ et al. 1988, 1999; PETERS et al. 1985). In Perez series, including 165 patients with primary vaginal carcinomas treated with definitive RT, the 10-year actuarial disease-free survival (DFS) was 94% for stage 0, 75% for stage I, 55% for stage IIA, 43% for stage IIB, 32% for stage III and 0% for those with stage IV (PEREZ et al. 1988).

The impact of lesion location has been controversial. Several authors (ALI et al. 1996; CHYLE et al. 1996; KUCERA and VAVRA 1991; TARRAZA et al. 1991; URBANSKI et al. 1996) have shown better survival and decreased recurrence rates for patients with cancers involving the proximal half of the vagina when compared with those in the distal half, or those involving the entire length of the vagina. CHYLE et al. (1996) observed a 17% pelvic relapse in patients with upper vagina lesions, 36% with mid- or lower vagina tumors and 42% with whole vagina involvement. In addition, lesions of the posterior wall have worse prognosis than those involving other vaginal walls (10-year recurrence rates of 32% and 19%, respectively), which probably reflects the greater difficulty of performing adequate brachytherapy procedures in this location.

The prognostic importance of lesion size has been controversial, with an adverse impact noted by TJALMA et al. (2001) and CHYLE et al. (1996) contrary to the findings of PEREZ et al. (1983). In the CHYLE et al. (1996) series, lesions measuring less than 5 cm in maximum diameter had a 20% 10-year local recurrence rate, compared with 40% for those lesions larger than 5 cm. Similarly, in the Princess Margaret Hospital (PMH) experience, tumors larger than 4 cm in diameter fared significantly worse than smaller lesions (KIRKBRIDE et al. 1995). In the Perez

Table 27.1a. FIGO staging system for carcinoma of the vagina

Stage	Description
Stage 0	Carcinoma in situ, intraepithelial neoplasia grade III
Stage I	Limited to the vaginal wall
Stage II	Involvement of the subvaginal tissue but without extension to the pelvic side wall
Stage III	Extension to the pelvic side wall
Stage IV	Extension beyond the true pelvis or involvement of the bladder or rectal mucosa. Bullous edema as such does not permit a case to be allotted to stage IV
Stage IVa	Spread to adjacent organs and/or direct extension beyond the true pelvis
Stage IVb	Spread to distant organs

From Pecorelli et al. (2000)

Table 27.1b. American Joint Commission on Cancer (AJCC) staging of vaginal cancer

Primary tumor(T)

Tx	Primary tumor cannot be assessed
T0	No evidence of primary tumor
Tis/0	Carcinoma in situ
T1/I	Tumor confined to the vagina
T2/II	Tumor invades paravaginal tissues but not to the pelvic wall
T3/III	Tumor extends to the pelvic wall
T4/IVA	Tumor invades mucosa of the bladder or rectum and/or extends beyond the pelvis (bullous edema is not sufficient to classify a tumor as T4)

Regional lymph nodes (N)

Nx	Regional lymph nodes cannot be assessed
N0	No regional lymph nodes
N1/IVB	Pelvic or inguinal lymph node metastasis

Distant metastasis (M)

Mx	Distant metastasis cannot be assessed
M0	No distant metastasis
M1/IVB	Distant metastasis

AJCC stage groupings

Stage 0	Tis N0 M0
Stage I	T1 N0 M0
Stage II	T2 N0 M0
Stage III	T1–3 N1 M0, T3 N0 M0
Stage IVa	T4, any N, M0
Stage IVb	Any T, any N, M1

From American Joint Committee on Cancer (2002)

series (Perez et al. 1999), stage was an important predictor of pelvic tumor control and 5-year DFS, but the size of the tumor in stage-I patients was not a significant prognostic factor. However, in stage-IIA disease, lower pelvic tumor control and survival were noted with tumors larger than 4 cm. In stages IIB–III, tumor size was not a significant prognostic factor, probably related to the difficulty in assessing size, and the fact that higher doses of RT were delivered for larger tumors. Stock et al. (1995) reported that disease volume, a likely surrogate for stage or lesion size, adversely impacted survival, as well as local control.

Age was a significant prognostic factor in Urbanski's series (Urbanski et al. 1996), with 5-year survival of 63.2% for patients below the age of 60 years, compared with 25% for those over 60 years of age (*P*<0.001). Similar findings were reported by Eddy et al. (1991) and in the NCDB of the American College of Surgeons (Creasman et al. 1998) with better survival in younger patients (90% versus 30%, respectively). However, most of these series do not correct for death secondary to intercurrent disease in the elderly population. No statistical significance of age to survival was found in the series of Dixit et al. (1993) and Perez et al. (1999).

With regard to the histological type and grade, several series (KIRKBRIDE et al. 1995; KUCERA and VAVRA 1991; URBANSKI et al. 1996) have shown the histological grade to be an independent, significant predictor of survival. However, the histology of the tumor (SCC versus other) has not been found to be a prognostic factor for NED survival among the patients with invasive tumors. CHYLE et al. (1996) noted a higher incidence of local recurrences in adenocarcinomas when compared with SCC (52% and 20%, respectively, at 10 years), higher distant metastasis rate (48% and 10%, respectively) and lower 10-year survival (20% versus 50%, respectively). An increased propensity for distant metastases to the lung and supraclavicular nodes has been reported in patients with CCA (ROBBOY et al. 1974). Patients with vaginal melanoma (REID et al. 1989) and malignant mesenchymal tumors (TAVASSOLI and NORRIS 1979) have a worse prognosis than patients with SCC due to a higher propensity for development of local recurrence and distant metastases.

Lymph node metastasis at diagnosis portends a poor prognosis. However, this has not been extensively evaluated in vaginal cancer. The only report of outcome based on lymph node status is by PINGLEY et al. (2000). They reported a 56% DFS for patients without lymph node involvement and 33% for those with metastatic lymphadenopathy at presentation.

27.6
General Management: Treatment Options

Due to its rarity, data concerning the natural history, prognostic factors and treatment of vaginal carcinoma derive from small, retrospective studies. Most of the currently available literature in terms of radiotherapeutic and surgical techniques refer to primary SCC of the vagina. In general, SCC of the vagina has been treated by means of RT. However, several surgical series have reported acceptable to excellent outcomes in well-selected patients, with survival rates after radical surgery for stage-I disease ranging from 75% to 100% (BALL and BERMAN 1982; CREASMAN et al. 1998; DAVIS et al. 1991; RUBIN et al. 1985; TJALMA et al. 2001), although few studies directly compared the two treatment modalities. Cases in which surgery may be the preferred treatment include selected stage I–II patients with lesions at the apex and upper-third of the posterior or lateral vagina that could be approached with radical hysterectomy,

upper vaginectomy, and pelvic lymphadenectomy providing adequate margins (BALL and BERMAN 1982; DAVIS et al. 1991; GALLUP et al. 1987; RUBIN et al. 1985; STOCK et al. 1995) and very superficial lesions that may be removed with wide local excision. If the margins are found to be close or positive after resection, adjuvant RT is recommended. However, for lesions at other sites and in those cases requiring more extensive resection, definitive RT is the treatment of choice in order to maximize cure and improve quality of life; generally, those patients with isolated central failures are offered exenteration (RUBIN et al. 1985).

Furthermore, most patients are elderly, and a radical surgical approach is often not feasible. Local excision and partial and complete vaginectomy have given way to a more individualized approach that takes into consideration the patient's age, the extent of the lesion, and whether it is localized or multicentric. In younger patients with early stage disease, treatment can also depend on the desire to preserve a functional vagina as well as ovarian function. Radical surgery in the past precluded vaginal function, but this situation has been improved significantly by the use of split-thickness grafts, intestinal segments and myocutaneous flap reconstruction (MAGRINA and BASTERSON 1981). CREASMAN noted superior survival in those undergoing surgery (CREASMAN et al. 1998). However, he and Tjalma (TJALMA et al. 2001) recognized that there may be bias in surgical series. Younger, healthier patients with better performance status are more likely to be offered radical surgery, whereas older patients with multiple co-morbid medical conditions are offered RT.

In most patients, the primary treatment modality is RT, as noted by the Society of Gynecologist Oncologists in practice guidelines published in 1998 (CREASMAN et al. 1998). RT provides excellent tumor control in early and superficial lesions, with satisfactory functional results. This makes it imperative that RT techniques yielding optimal tumor control and functional results are utilized. Despite the acceptance of RT as the treatment of choice for this disease, in particular for patients with lesions involving the mid-third of the vagina or stage II and greater, the optimal therapy for each stage is not well-defined in the literature. Intracavitary and interstitial irradiation is used in small superficial stage-I disease. A combination of external beam RT (EBRT), intracavitary brachytherapy (ICB) and/or interstitial brachytherapy (ITB) with or without chemotherapy is used in more extensive stage-I and stages II–IV disease.

27.6.1
RT Techniques

Radiation prescription and technique varies according to stage, disease bulk and anatomic location. In recent years, several series have reported improved outcomes with techniques emphasizing the higher radiation doses that can be achieved when all or part of the therapy is accomplished with interstitial technique (LEUNG and SEXTON 1993; PEREZ et al. 1999; PINGLEY et al. 2000; PUTHAWALA et al. 1983).

27.6.1.1
External Beam Radiotherapy

EBRT is advisable in patients with deeply infiltrating or poorly differentiated stage-I lesions, and in all patients with stage II–IVA disease. The treatment is generally delivered using opposed anterior and posterior fields (AP/PA). The pelvis receives between 20 Gy and 45 Gy, depending on the stage of the disease. High-energy photons (>10 MV) are usually preferred. Treatment portals cover at least the true pelvis with a 1.5- to 2-cm margin beyond the pelvic rim. Superiorly, the field extends to either L4–L5 or L5–S1 to cover the pelvic lymph nodes up to the common iliacs, and extends distally to the introitus to include the entire vagina. Lateral fields, if used, should extend anteriorly to adequately include the external iliac nodes, anterior to the pubic symphysis, and at least to the junction of S2–S3 posteriorly (Fig. 27.1).

In patients with tumors involving the middle and lower vagina with clinically negative groins, the bilateral inguino-femoral lymph node regions

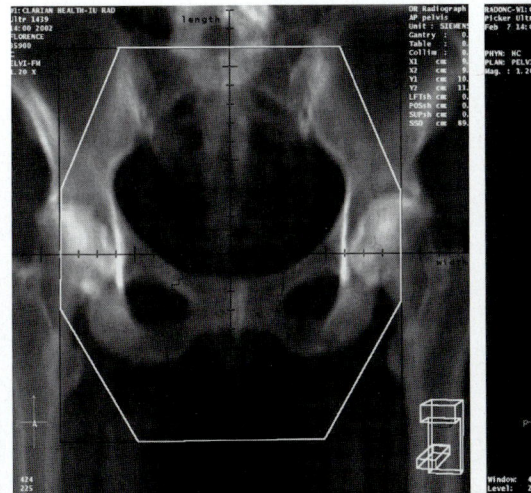
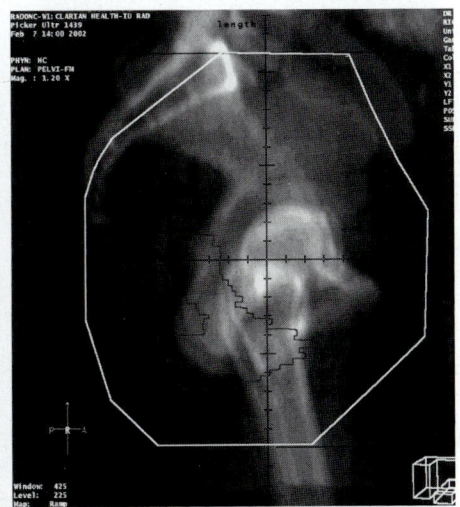
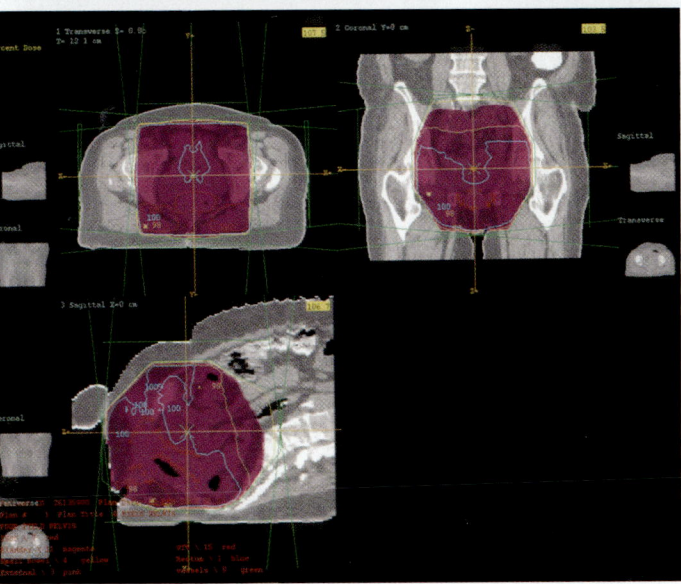

Fig. 27.1a–c. Invasive Squamous cell carcinoma of the proximal vagina with Squamous-cell carcinoma in-situ in the mid-third of the vaginal squamous cell carcinoma in situ, without involvement of inguino-femoral nodes. Digital reconstructed radiographs (*DRRs*): **a** AP/PA whole pelvis fields. **b** Right/left lateral whole pelvis fields, intended to treat the entire length of the vagina. **c** Axial, sagittal and coronal isodose distributions

should be treated electively to 45–50 Gy. Planning CT is recommended to adequately determine the depth of the inguino-femoral nodes. A number of techniques have been used to treat the areas at risk without over-treating the femoral necks. Some of the most commonly used techniques include the use of unequal loading (2:1, AP/PA), a combination of low- and high-energy photons (4–6 MV, AP, and 15–18 MV, PA), or equally weighted beams with a transmission block in the central AP field, utilizing small AP photon or electron beams to deliver a daily boost to the inguino-femoral nodes. A technique has been developed and implemented at Indiana

University which uses a narrow PA field to treat the pelvis, and a wider AP field encompassing the pelvis and inguino-femoral nodes, with daily AP photon boost to the inguinal nodes being delivered using the asymmetric collimator jaws (DITTMER and RANDALL 2001). Advantages of this technique include simplicity of set-up and treatment (single isocenter, no need for transmission block), dose homogeneity, reduced dose to the femoral necks, low potential risk of nodal underdose, and elimination of dosimetric difficulties inherent in electron boosts (Figs. 27.2). In patients with clinically palpable inguinal nodes, additional doses of 15–20 Gy (calculated at a depth

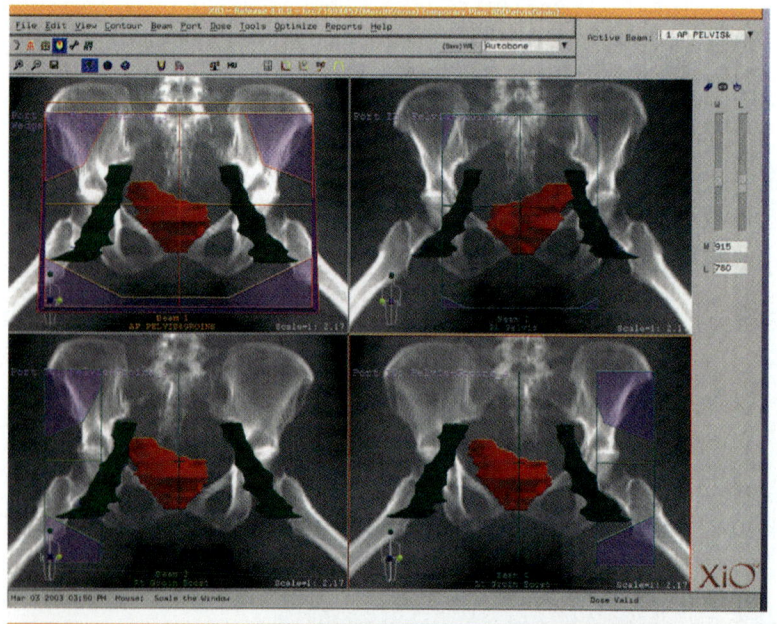

a

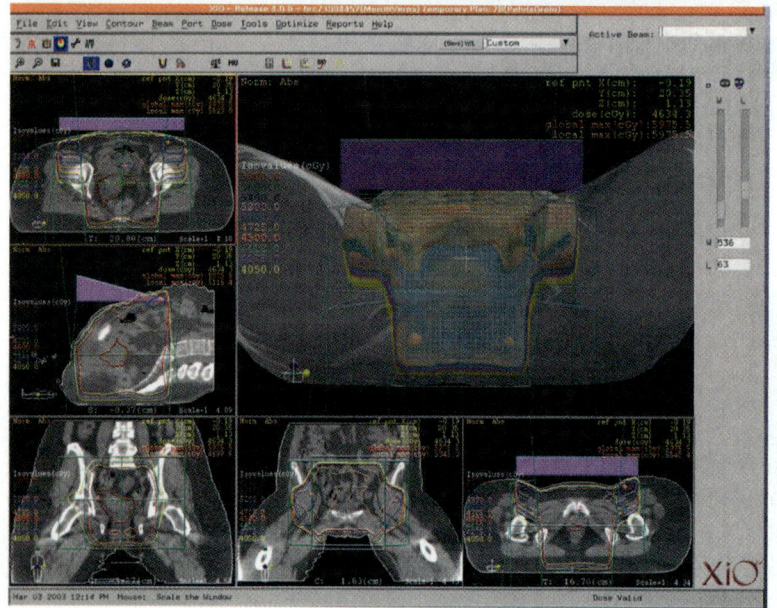

b

Fig. 27.2a,b. Vaginal cancer with distal third vaginal involvement, squamous cell carcinoma. Technique for pelvic and inguino-femoral nodal irradiation. **a** Digital reconstructed radiographs (*DRRs*). AP field intended to treat pelvis and groins; PA field intended to treat the pelvis only, in order to decrease the dose to the femoral heads; small AP photon field for additional daily boost to the inguino-femoral nodes. **b** Axial, sagittal and coronal isodose distributions

determined by CT scan) are necessary with reduced portals. This is generally achieved using low-energy photons or electron beam (12–18 MeV).

For patients with positive pelvic nodes, or those patients with advanced disease not amenable to interstitial implant, additional boost to the areas of gross disease, as defined by CT scan, should be given using conformal therapy to deliver a total dose between 65 Gy and 70 Gy with high-energy photons (Fig. 27.3).

27.6.1.2
Low-Dose-Rate Intracavitary Brachytherapy

VAIN and small T1 lesions with less than 0.5 cm depth can be adequately treated with ICB alone. Low-dose-rate ICB (LDR-ICB) is performed using vaginal cylinders such as Burnett, Bloedorn, Delclos (Delclos et al. 1980) or MIRALVA (Perez et al. 1990) loaded with cesium-137 (^{137}Cs) radioactive sources. Delclos afterloading vaginal cylinders have a central hollow metallic cylinder in which the sources are placed, and plastic rings of varying diameter (2.5–4 cm), 2.5 cm in length, are inserted over the cylinder. Domed cylinders are used to irradiate the vaginal cuff homogeneously, when indicated. Delclos (Delclos et al. 1980) recommended a short ^{137}Cs source to be used at the top to obtain a uniform dose around the dome, because a lower dose occurs at the end of the linear cesium source. Some cylinders have a lead shielding to protect selected portions of the vagina, the bladder and/or the rectum. The largest possible diameter that can be comfortably accommodated by the patient should be used to improve the ratio of mucosa to tumor dose and eliminate vaginal rogations (Fig. 27.4). The cylinders can be mounted in the vaginal component of an intrauterine tandem or along the stem of a dome cylinder. Before the cylinders are mounted over the vaginal component of an intrauterine tandem, this is inserted into the uterus (when present) and the cylinders are fitted along the protruding tandem. To minimize the rotation of the tandem, a flat, round flange with keel is placed below the last cylinder and is kept in position with some packing if required. In general, the vulva is sutured-closed with proline or silk for the duration of the implant in order to secure the position of the applicators (Fig. 27.4c, d).

In patients with upper vagina lesions with less than 0.5 cm depth of invasion, vaginal colpostats alone (after hysterectomy) or in combination with intrauterine tandem, loaded with ^{137}Cs sources similar to that used in treatment of cervical cancer, can be used to treat the proximal vagina to a minimum

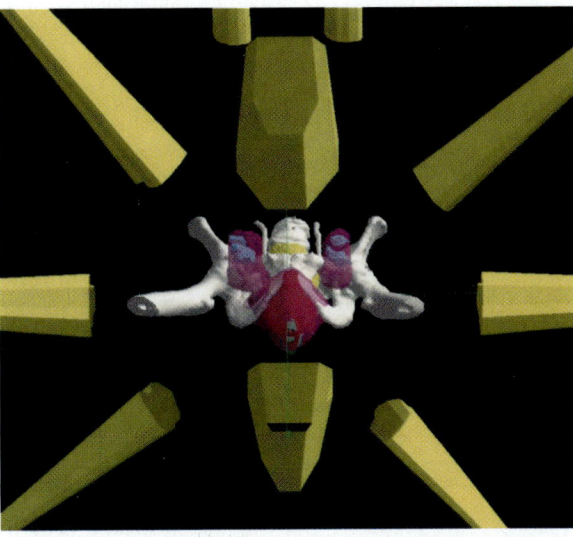

Fig. 27.3. Locoregionally advanced vaginal cancer. External irradiation – beam arrangement including initial whole pelvis and inguino-femoral fields and three-dimensional conformal radiotherapy boost. In *magenta*, the isodose distribution corresponds to the 65 Gy isodose line

dose of 65–70 Gy, estimated to 0.5 cm depth, including the contribution of EBRT if given. When indicated, the remainder of the vagina can be treated by performing a subsequent implant using vaginal cylinders generally 60 Gy total dose to the vaginal surface will be delivered including the contribution of the EBRT and the intracavitary implant. It is important to avoid the placement of a protruding source over the vulva, with the subsequent increased risk of complications. The use of LDR remote control afterloading technology allows the reduction of radiation exposure to hospital personnel and optimization of the isodose distribution. When appropriate dose specification points are chosen, a very uniform dose distribution over the entire length of the vagina can be obtained.

Perez et al. (1990; Slessinger et al. 1992) designed and constructed a vaginal applicator, MIRALVA, which incorporates two ovoid sources and a central tandem that can be used to treat the entire vagina (alone, or in combination with the uterine cervix). The applicator has vaginal apex caps and additional cylinder sleeves to allow for increased dimensions (Fig. 27.5a). A tandem in the uterus can be used if clinically indicated. The dosimetry and dose specifications for this applicator have been published (Slessinger et al. 1992), showing that the applicator delivers 1.1–1.2 Gy/h to the vaginal apex and 0.95–1 Gy/h to the distal vaginal surface when loaded with 20 mgRaEq ^{137}Cs tubes in each ovoid

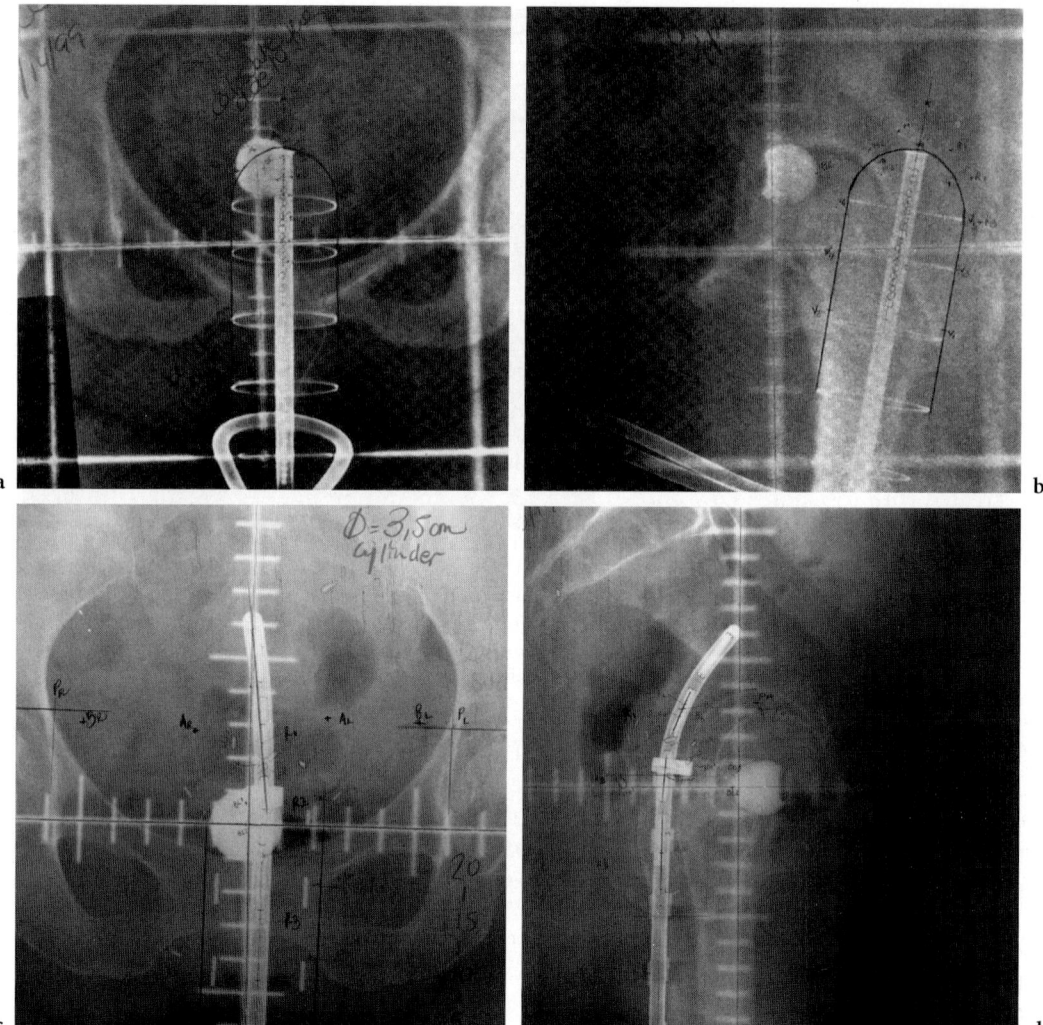

Fig. 27.4a–d. AP (**a**) and lateral (**b**) views of vaginal cylinders only. AP (**c**) and lateral (**d**) views of intrauterine tandem and vaginal cylinders

and 10–10–20 mgRaEq tubes in the vaginal cylinder (Fig. 27.5b). When the tandem and vaginal cylinder are used, the strength of the sources in the ovoids should always be 15 mgRaEq. The vaginal cylinder or uterine tandem never carries an active source at the level of the ovoids to prevent excessive doses to the bladder or rectum.

27.6.1.3
High-Dose-Rate Intracavitary Brachytherapy

High-dose-rate ICB (HDR-ICB) is typically performed with a 10 Ci single iridium-192 (^{192}Ir) source (Micro-Selectron HDR, Nucletron). The applicators are similar to those described for LDR, consisting of vaginal cylinders of 2.5–4 cm in diameter.

Little information regarding HDR-ICB in the treatment of primary carcinoma of the vagina is available (NANAVATI et al. 1993; STOCK et al. 1992). Few patients have been treated, follow-up is short, publication bias is likely and there is no agreement on treatment regimen. With HDR, there is a need to adjust the total dose, and additional fractionation is necessary, compared with LDR brachytherapy, because of the biologically equivalent dose considerations. Generally, the number of insertions ranges from 1 to 6 (median 3), with the dose per fraction ranging from 300 cGy to 800 cGy (median 700 cGy). NANAVATI et al. (1993) reported 13 patients with primary vaginal cancer (5 St I, 4 St IIA and 4 St IIB) treated with external beam RT (45 Gy) and HDR-ICB (20–28 Gy in three to four fractions, calculated at 0.5 cm from the surface of the applicator). All 13

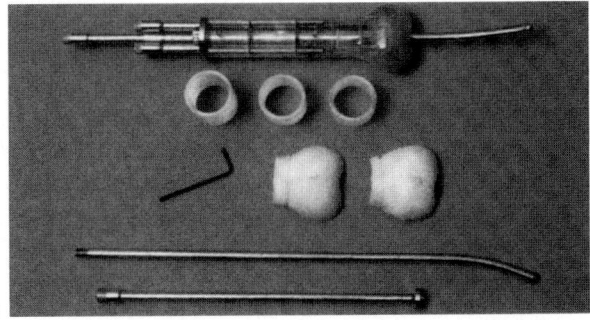

Fig. 27.5. a MIRALVA structure. b Coronal, sagittal and axial isodose distributions for the applicator loaded with Cesium-137 radioactive sources dose distribution, A' B' without and C' D' with intrauterine tandem. Reprinted with permission from PEREZ et al. (1990) and SLESSINGER et al. (1992)

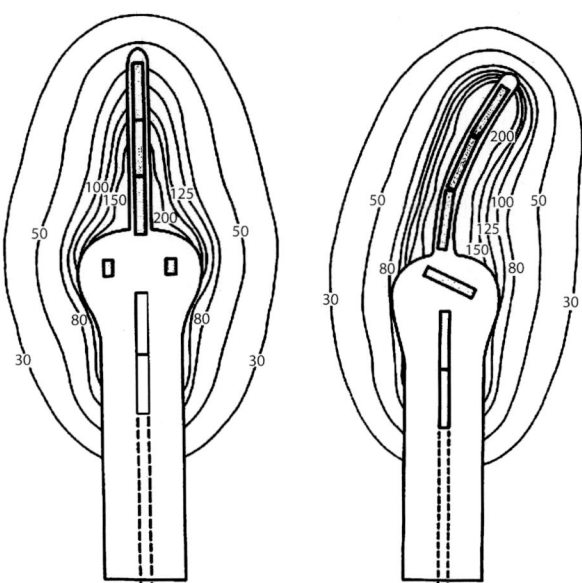

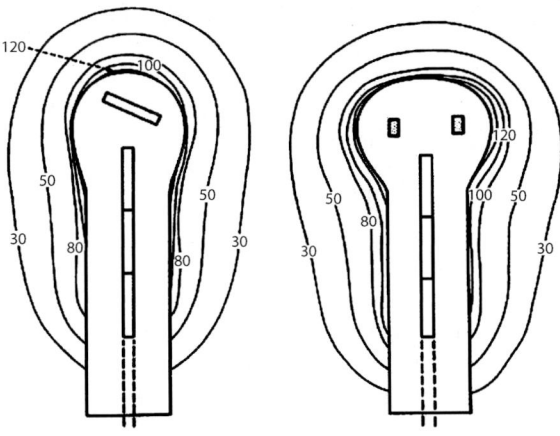

patients had a complete response, and local control was achieved in 92% of the patients with a median follow-up of only 2.6 years (range 0.7–5.2 years). Originally, the planned dose was 2100 cGy in three fractions, 700 cGy/fraction, but because of reports of decreased complications with increased fractionation, the planned dose was changed to 2000 cGy in four equal fractions. They did not observe any acute or chronic intestinal or bladder grade 3 or 4 toxicity. However, moderate to severe vaginal stenosis occurred in 46% of the patients. The authors recognize that "late-occurring toxicity could be missed at a medium follow-up of 2.6 years." In a recent report, MOCK et al. (2003) showed similar outcome and toxicity rate, stage by stage, in 86 patients with primary vaginal cancer treated with a variety of external beam and HDR brachytherapy.

Many aspects remain unknown or not well-understood in the use of HDR-ICB. These include the radiobiological equivalency of HDR to LDR, fractionation schedule, total dose, specification of dose prescription, and how to combine HDR with EBRT and/or LDR brachytherapy. In addition, optimization approaches and methods of dose calculation, such as the inclusion of anisotropic corrections, are not well-described in the sparse literature available to date (GORE et al. 1995; LI et al. 1998). LI et al. (1998)

have shown that when optimized dose distribution at a distance from the cylinder surface is calculated using an accurate dose calculation model, the vaginal mucosa dose becomes significantly higher than calculated, and therefore should be carefully monitored. These factors could result in an increased incidence of severe complications, such as vaginal necrosis, and recto–vaginal or vesico-vaginal fistulas (RUTKOWSKI et al. 2002; TYREE et al. 2002).

In the opinion of the authors, until further data are available with longer follow-up, as well as a better understanding of the physical and radiobiological principles involved in the HDR-ICB, this should not be routinely used in the management of primary vaginal carcinoma. We strongly encourage the continued use of LDR-ICB, given its excellent results and widely documented long-term outcome and complications (CHYLE et al. 1996; KIRKBRIDE et al. 1995; PEREZ et al. 1988, 1999).

27.6.1.4
Interstitial Brachytherapy

ITB is an important component in the treatment of more advanced primary vaginal carcinomas, typically in combination with EBRT and/or ICB. In the first place, a careful definition of the "target volume", this is gross tumor volume (based on clinical, radiological and operative findings) and a margin of adjoining normal tissue, is required. Other considerations include whether a permanent ([198]Gold or [125]Iodine) or temporary implant (^{192}Ir) is optimal, the geometry of the implant (e.g., single or double plane or volume implant), source distribution, dose rate and total dose, based on tumor size, location, local extent and proximity of normal structures (HILARIS et al. 1987). The principal advantages of temporary implants are readily controlled distribution of the radioactive sources and easier modification of the dose distribution. The main advantages of a permanent seed implant include relative safety/simplicity, easy applicability, cost-effectiveness and ability, in most cases, to be performed under local anesthesia. As a general rule, temporary implants are more commonly used in the curative treatment of larger gynecological malignancies, whereas permanent implants are usually performed for smaller volume disease.

The number and strength of the radioactive sources and their intended distribution within the target volume are determined pre-operatively, making use of available guidelines such as nomograms, tables and computer-assisted optimization techniques. Following this, it is necessary to specify an approximate dose rate to the target volume, which requires careful localization of the sources and computer calculation of the three-dimensional radiation dose distribution. Finally, a dose prescription, based on the treatment volume, tumor sensitivity, dose rate, prior treatments and tolerance of normal surrounding tissues, is required (HILARIS et al. 1987).

When performing an interstitial procedure, free-hand implants or template systems designed to assist in pre-planning and to guide and secure the position of the needles in the target volume can be employed. Popular commercially available templates include the Syed-Neblett device (SNIT) (Alpha Omega Services, Bellflower, CA) (SYED et al. 1986), the modified Syed-Neblett (DISAIA et al. 1990) and the "MUPIT" (Martinez Universal Perineal Interstitial Template) (MARTINEZ et al. 1984). All rely on pelvic examination to help guide the location and depth of needle placement.

The Syed-Neblett device consists of two identical superimposed Lucite plates, each about 1 cm thick, held together by six screws. Both plates are drilled in an identical pattern of predetermined needle positions that can be afterloaded with iridium-192

(^{192}Ir) sources (SYED et al. 1986). The modified Syed-Neblett (DISAIA et al. 1990) applicator consists of a perineal template, vaginal obturator, and 17-gauge hollow guides of various lengths. The vaginal obturator is 2 cm in diameter, and 12 cm or 15 cm in length. The vaginal obturator has seven grooves on its surface for the placement of guide needles and is centrally drilled so it can allow the placement of a tandem to be loaded with ^{137}Cs sources. This makes possible to combine an interstitial and intracavitary application simultaneously (Fig. 27.6).

A similar afterloading applicator, referred to as "MUPIT" (Fig. 27.7) was developed by MARTINEZ et al. (1984). The device basically consists of two acrylic cylinders, an acrylic template with predrilled holes that serve as guides for trocars and a cover plate. Some of the guide holes on the template are angled outward to permit a wide lateral coverage without danger of striking the ischium. The cylinders have an axial hole large enough to pass a central tandem or suction tube for drainage of secretions. Thus, the device allows for the interstitial placement of ^{192}Ir ribbons as well as the intracavitary placement of either ^{137}Cs tubes or ^{192}Ir ribbons. In use, the cylinders are placed in the vagina and rectum and then fastened to the template, so that a fixed geometric relationship among the tumor volume, normal structures and source placement is preserved throughout the course of implantation.

The major advantage of these systems is greater control of the placement of the sources relative to tumor volume and critical structures due to the fixed geometry provided by the template and cylinders. In addition, improved dose-rate distributions are obtained by means of computer-assisted optimization of the source placement and strength during the planning and loading phases. Due to the inac-

Fig. 27.6. Syed-Neblett and modified templates

curacies of pelvic examination and close proximity of the rectum and bladder to the target volume, there exists a serious risk of either underdosing the target volume or causing bladder and rectal morbidity. In order to improve the accuracy of target localization and needle placement, several investigators have explored performing ITI under transrectal ultrasound (TRUS) (STOCK et al. 1997), CT, MRI-planned implants with endorectal coil (CORN et al. 1995), laparotomy and laparoscopic guidance (CHILDERS and SURWIT 1993; CORN et al. 1995). While laparotomy facilitates the displacement of bowel during the procedure using slings or tissue expanders and/or lysis of adherent bowel, there is some degree of associated morbidity, such as ileus, bleeding and increased operative time. Laparoscopy is a shorter and less invasive procedure. Real time TRUS-guided Syed-Neblett template implantation technique was reported by STOCK et al. (1997). TRUS allows the ultrasound (US) probe to be in closer proximity to the pelvic structures (cervix, parametria, vagina) than trans-abdominal US and can more accurately guide needle placement into tumor and avoid needle insertion into critical surrounding normal tissues. Transverse US imaging is used to assure that the needles cover the target area and do not enter the bladder, rectum or small bowel. The longitudinal mode of the US probe is equally important in the implant procedure due to its ability to guide the optimum depth of needle insertion. With this technique, invasive laparotomy and/or laparoscopy can often be avoided, providing an interactive, non-invasive technique allowing for highly accurate needle placement (STOCK et al. 1997).

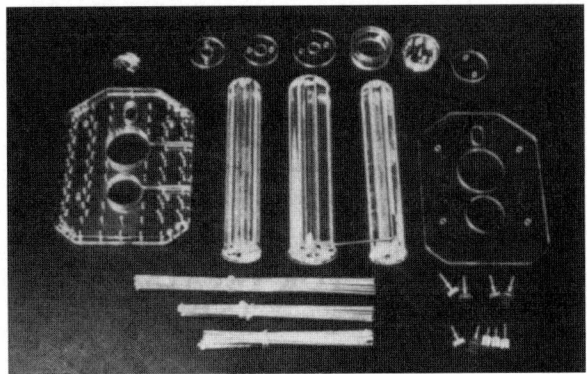

Fig. 27.7. A. MUPIT template structure. Reprinted with permission from MARTINEZ et al. (1984)

27.6.2
SCC: Treatment Options and Outcome by FIGO Stages

27.6.2.1
FIGO Stage 0: VAIN – CIS

VAIN has been approached both surgically and medically by multiple investigators. Treatment options include local excision, partial or complete vaginectomy, laser vaporization, topical 5% fluorouracil (5-FU) administration, or ICB alone. Overall, the reported control rates are very similar among the different approaches, ranging from 48% to 100% for laser (DIAKOMANOLIS et al. 2002; HOFFMAN et al. 1991; JOBSON and HOMESLEY 1983; JULIAN et al. 1992; STAFL et al. 1977; TOWNSEND et

al. 1982), 52% to 100% for colpectomy (CREASMAN et al. 1998; FANNING et al. 1999; HOFFMAN et al. 1992; ROBINSON et al. 2000), 75% to 100% for topical 5-FU (KREBS 1989; PETRILLI et al. 1980; PIVER et al. 1979; WOODRUFF et al. 1975) and 83% to 100% for RT (CHYLE et al. 1996; KIRKBRIDE et al. 1995; PEREZ et al. 1977). The degree of VAIN and the age and general health of the patient are important treatment considerations. RT has a long history of documented efficacy and has a significantly better therapeutic ratio than other modalities (CHYLE et al. 1996; KIRKBRIDE et al. 1995; PEREZ et al. 1977; PREMPREE and AMOMMAN 1985). Using conventional LDR-ICB techniques, the entire vaginal mucosa should receive between 60 Gy and 70 Gy in one or two implants (PEREZ et al. 1977). PEREZ et al. (1977) reported only one distal local failure in the 20 patients treated for CIS. This recurrence developed distally to the vaginal vault in a patient inadequately treated with vaginal ovoids only.

There have been some reports in the literature regarding the use of HDR-ICB for patients with VAIN-3. OGINO et al. (1998) reported 6 patients treated with HDR to a mean dose of 23.3 Gy (range 15–30 Gy), none of whom developed recurrent disease. Limited rectal bleeding and moderate to severe vaginal mucosa reactions were noted in patients treated to the entire length of the vagina. MACLEOD et al. (1997) used HDR-ICB to treat 14 patients with VAIN 3 with a dose of 34–45 Gy in 4.5-Gy to 8.5-Gy fractions, with a local control of 78.5%. With a median duration of follow-up of 46 months, two patients developed grade-3 vaginal toxicity. At the present time, no definite conclusions can be drawn from the limited data published in the literature regarding the use of HDR-ICB. Based on the excellent local control and functional results obtained with LDR-ICB, this remains, in the authors' opinion, the treatment of choice when definitive RT is used.

27.6.2.2
Invasive SCC

Most authors emphasize that brachytherapy alone is adequate for superficial FIGO stage-I patients, for whom 95–100% local control has been achieved with intracavitary and interstitial techniques (CHU and BEECHINOR 1984; KUCERA and VAVRA 1991; LEUNG and SEXTON 1993; PEREZ et al. 1988; PETERS et al. 1985; REDDY et al. 1991; STOCK et al. 1995; URBANSKI et al. 1996). Superficial lesions can be adequately treated with ICB alone using afterloading vaginal cylinders. Mucosal doses of 80–120 Gy are typically delivered, depending on the diameter of the cylinders, when prescribing 65–70 Gy at 0.5 cm depth beyond the vaginal surface (PEREZ et al. 1977). For lesions thicker than 0.5 cm at the time of implantation, it is advisable to combine ICB and ITB in order to deliver tumor dose in the range of 65–70 Gy, calculated to the base of the lesion, limiting the proximal and distal vaginal mucosal doses to 140 Gy and 100 Gy, respectively (Figs. 27.8 and 27.9).

There are no well-established criteria regarding the use of EBRT in patients with stage-I disease. PEREZ et al. (1988, 1999) did not find a significant correlation between the technique of irradiation used and the probability of local or pelvic recurrence, probably since the treatment technique varied based on tumor-related factors. There is general consensus that EBRT (20–50 Gy) is advisable for larger, more infiltrating or poorly differentiated tumors that may have a higher risk of lymph node metastasis. CHYLE et al. (1996) recommended EBRT in addition to brachytherapy for stage-I disease to cover at least the paravaginal nodes and, in larger lesions, to cover the external and internal iliac nodes. The 5-year survival for patients with stage-I disease treated with RT alone ranges from 70% to 95%.

Patients with FIGO stage-II disease are uniformly treated with EBRT, followed either by ICB and/or ITB (Fig. 27.8a–c). PEREZ et al. (1999) showed that in stage IIA, the local tumor control was 70% (37/53) in patients receiving brachytherapy combined with EBRT, compared with 40% (4/10) in patients treated with either brachytherapy or EBRT alone. In stage IIB, the local-regional control was also superior with combined EBRT and brachytherapy (61% versus 50%, respectively). Generally, 40–50 Gy is delivered to the whole pelvis, followed by an additional boost

of 30–35 Gy given with brachytherapy. Patients with lesions limited to the upper third of the vagina can be treated with an intrauterine tandem and vaginal ovoids or cylinders. In patients with parametrial infiltration, a "boost" with EBRT and/or an interstitial implant is advisable to deliver a minimum tumor dose of 70–75 Gy and 55–60 Gy to the pelvic side wall (Fig. 27.8a–c). The 5-year survival for patients with stage-II disease treated with RT alone ranges between 35% and 70% for stage IIA and 35% and 60% for stage IIB. The results of several series published in the literature using RT with or without limited surgical resection for the treatment of stage-I and -II vaginal cancer are shown in Table 27.2 (CHYLE et al. 1996; CREASMAN et al. 1998; DAVIS et al. 1991; KIRKBRIDE et al. 1995; KUCERA and VAVRA 1991; PEREZ et al. 1999; STOCK et al. 1995; URBANSKI et al. 1996).

Generally, patients with FIGO stages III–IV disease will receive 45–50 Gy EBRT to the pelvis and, in some cases, additional parametrial dose with midline shielding to deliver up to 60 Gy to the pelvic side walls. Ideally, ITB brachytherapy boost is performed, if technically feasible, to deliver a minimum tumor dose of 75–80 Gy (Fig. 27.8a–c). If brachytherapy is not feasible, a shrinking-field technique can be used, with fields defined using the three-dimensional treatment planning capabilities to deliver a tumor dose around 65–70 Gy (Fig. 27.3). The overall cure rate for patients with stage-III disease is 30–50%. Stage IVA includes patients with rectal or bladder mucosa involvement or, in most series, positive inguinal nodes. Although some patients with stage-IVA disease are curable, many patients are treated palliatively with EBRT only. Pelvic exenteration can also be curative in highly selected stage-IV patients with small volume central disease. Table 27.3 (CHYLE et al. 1996; CREASMAN et al. 1998; KIRKBRIDE et al. 1995; KUCERA and VAVRA 1991; PEREZ et al. 1999; STOCK et al. 1995; URBANSKI et al. 1996) shows the treatment results in patients with advanced disease. However, each of these series reported a greater number of patients with similar stage disease treated with RT, which represents the preferred approach in contemporary practice (PEREZ et al. 1988; PREMPREE and AMOMMAM 1985).

27.6.3
Chemotherapy and Radiation

The control rate in the pelvis for stages-III to -IV patients is relatively low, and about 70–80% of the

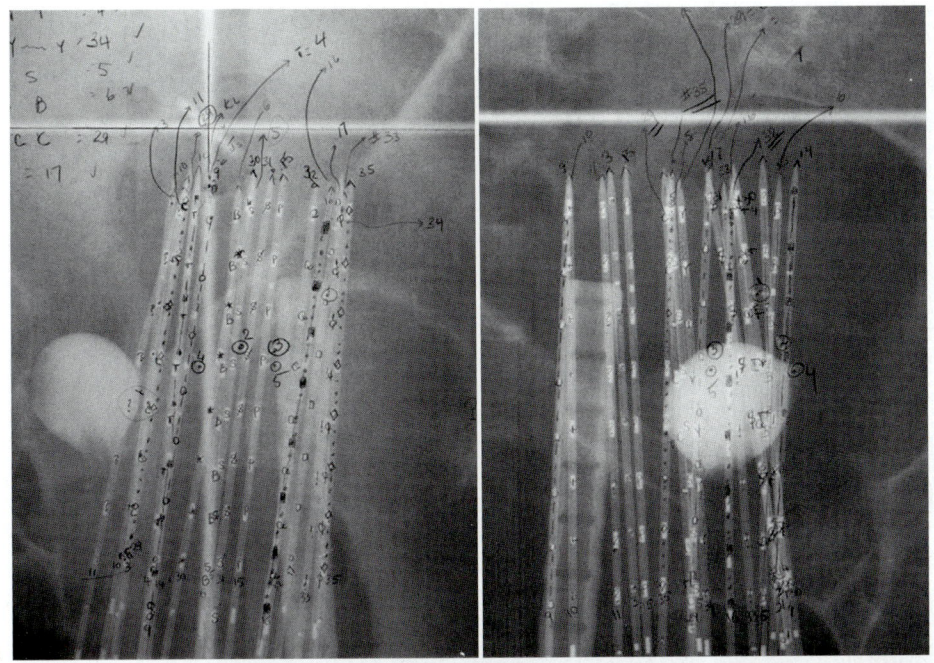

a1 a2

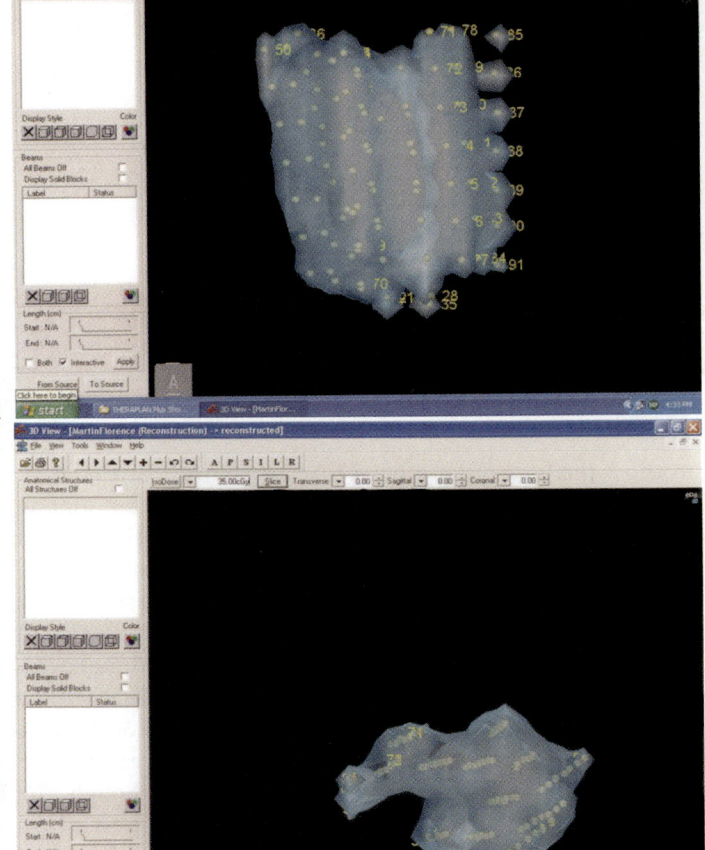

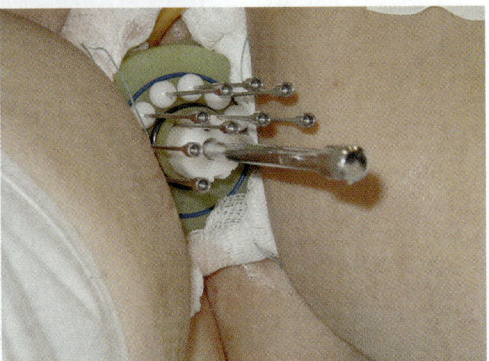

c

Fig. 27.8a–c. Interstitial brachytherapy boost in a patient with locally advanced vaginal cancer. Left anterior oblique (**a1**) and right anterior oblique (**a2**) radiographs of the implant. Coronal (**b1**) and sagittal (**b2**) isodose distributions. **c** Placement of funnel needles using the modified Syed-Neblett template

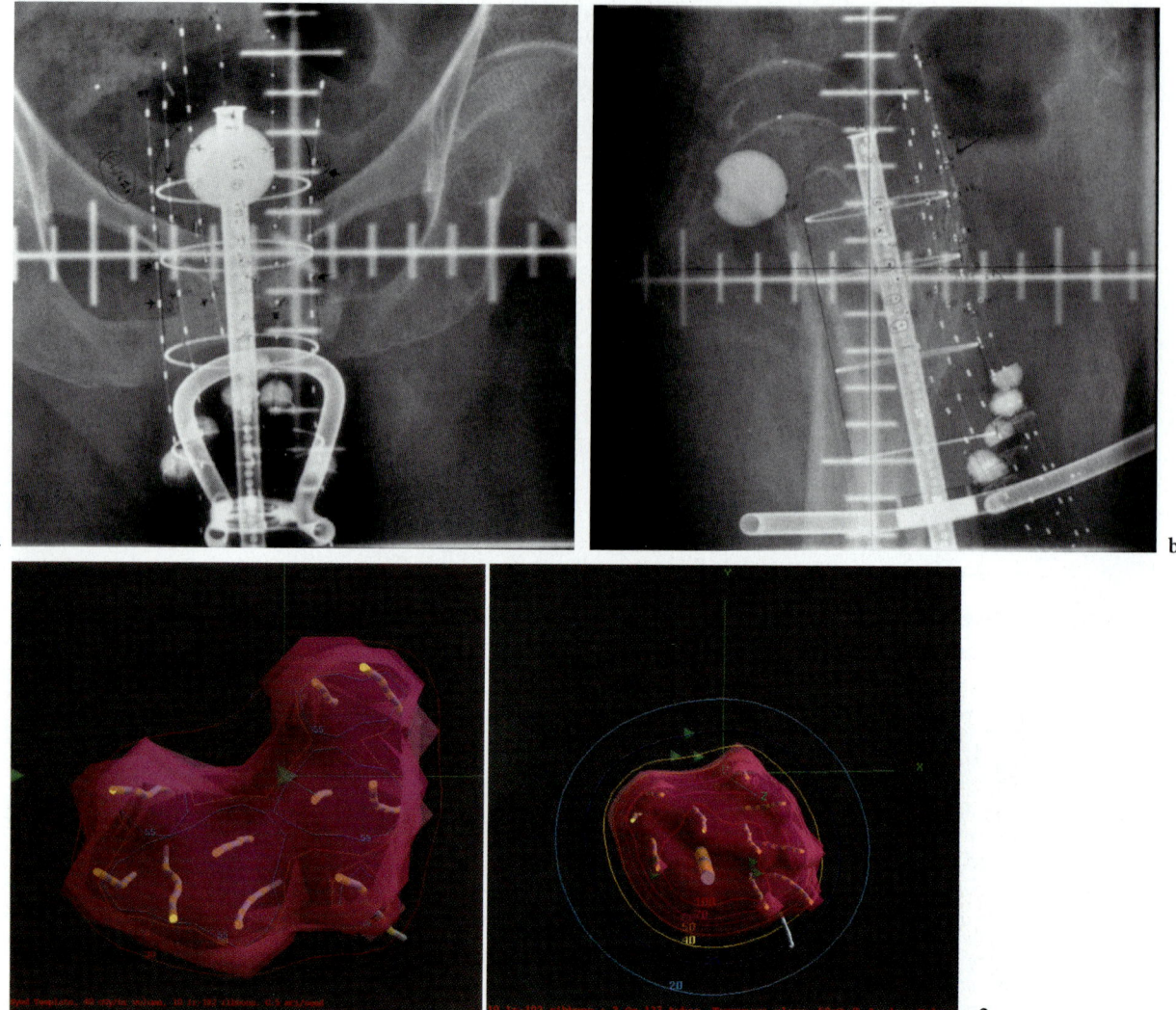

Fig. 27.9a–c. AP (**a**) and lateral (**b**) radiographs of an interstitial and intracavitary implant in carcinoma of the vagina using the modified Syed-Neblett template. **c** Three-dimensional isodose distribution: **c1** Ir^{192} only; **c2** Ir^{192} plus Cs^{137}

patients have persistent or recurrent disease in the pelvis, despite high doses of external beam RT and brachytherapy. Failure in distant sites does occur in about 25–30% of the patients with locally advanced tumors, much less than pelvic recurrences. Therefore, there is a need for better approaches to the management of advanced disease, such as the use of concomitant chemo-radiotherapy. Agents such as 5-FU, mitomycin and cisplatin have shown promise when combined with RT, with complete response rate as high as 60–85% (EVANS et al. 1988; ROBERTS et al. 1991) but long-term results of such therapy have been variable. In these small studies, many of the patients had advanced (stage III) disease at the initiation of combined modality therapy, perhaps explaining

the lack of long-term disease control. EVANS et al. (1988) found no local recurrences, however, among patients achieving a complete response with RT and 5-FU plus mitomycin-C (12 of 25 patients), with a median follow-up period of 28 months, suggesting that local control may be improved with combined modality therapy since local failure is common with radiation alone in large volume pelvic disease. The survival for the entire population was 56% (66% for patients with primary vaginal cancer). Only 2 patients had severe complications, although the authors recognize that longer follow-up is probably required to assess the true incidence of late effects. More sobering are the data from ROBERTS et al. (1991) who reported 67 patients with advanced

Table 27.2 FIGO stages I–II vaginal cancer. Treatment outcome with radiation therapy with/without surgery

Author	No. of patients	Outcome (survival)
CHYLE et al. (1996)	59 stage I	10 years 76%
	104 stage II	10 years 69%
CREASMAN et al. (1998)	169 stage I	5 years 73%; 79% S+RT (47), 63% RT (122)
	175 stage II	5 years: 58%; 58% S+RT (39), 57% RT (136)
DAVIS et al. (1991)	19 stage I	5 years 100% S+RT (5), 65% RT (14)
	18 stage II	5 years 69% S+RT (9), 50% RT (9)
KIRKBRIDE et al. (1995)	40 stage I	5 years 72%
	38 stage II	5 years 70%
KUCERA and VAVRA (1991)	16 stage I	5 years 81%
	23 stage II	5 years 43.5%
PEREZ et al. (1999)	59 stage I	10 years 80%
	63 stage IIA	10 years 55%
	34 stage IIB	10 years 35%
STOCK et al. (1995)	8 stage I	5 years 100% S+RT, 80% RT
	35 stage II	5 years 69% S+RT, 31% RT
URBANSKI et al. (1996)	33 stage I	5 years 73%
	37 stage II	5 years 54%

Table 27.3 FIGO stages III–IV vaginal cancer. Treatment outcome with radiation therapy with/without surgery

Author	No. of patients	Outcome (survival)
CHYLE et al. (1996)	55 stage III	10 years 47%
	16 stage IV	10 years 27%
CREASMAN et al. (1998)	180 stage III–IV	5 years 36%; 60%-S+RT (36), 35%-RT (144)
KIRKBRIDE et al. (1995)	42 stage III–IV	5 years 53%
KUCERA and VAVRA (1991)	46 stage III	5 years 35%
	19 stage IVA	5 years 32%
PEREZ et al. (1999)	20 stage III	10 years 38%
	15 stage IV	0%
STOCK et al. (1995)	9 stage III	5 years 0%
	8 stage IV	0%
URBANSKI et al. (1996)	40 stage III	5 years 22.5%
	15 stage IVA	0%

cancers of the vagina, cervix and vulva treated with concurrent 5-FU, cisplatin and RT. Although 85% experienced a complete response, 61% of them experienced a tumor recurrence, with a median time to recurrence of only 6 months, and an overall survival at 5 years of 22%. Further, of 67 patients, severe late complications developed in 9 (13%), 8 of whom required surgeries. KERSCH et al. (1990) reported that 5 of 8 vaginal cancer patients achieved local control with combined modality therapy. Studies of primary chemo-radiation in primary vaginal cancer are small or heterogeneous populations including cervical and vulvar cancers, making it difficult to truly assess the role of combined modality therapy in the management of locally advanced disease. No randomized trials comparing radiation with or without chemotherapy have been reported.

DALRYMPLE et al. (2004) published recently a small study including 14 patients, primarily stages I and II SCC of the vagina treated with reduced doses of RT (median 63 Gy) concurrently with different 5-FU-based chemotherapeutic regimens. They report a 93% control rate, probably reflecting a more favorable stage distribution. Interestingly, none of the patients required interstitial implants and no patients developed fistulas. The authors indicated that this approach, similar to that used in the management of anal and vulvar cancer, would allow reducing the RT dose with the subsequent improvement in organ function and late toxicity.

Further investigation is needed to determine the therapeutic efficacy of the concurrent chemoradiotherapy and the optimal chemotherapy regimen. Recently published data on locally advanced cervical cancer have demonstrated an advantage in loco-regional control, overall survival and DFS for patients receiving cisplatin-based chemotherapy concurrently with RT (Keys et al. 1999; Morris et al. 1999; Rose et al. 1999; Whitney et al. 1999). The only drug common to all the studies was cisplatin, suggesting it may be the only agent needed to improve radiation sensitivity. Based on these data, as well as data on loco-regionally advanced vulvar cancer (Moore et al. 1988), consideration should be given to a similar approach in patients with advanced vaginal cancer. Randomized trials comparing RT alone to chemo-radiation therapy, however, are unlikely due to small patient numbers.

27.6.4
Patterns of Failure in SCC

Of those patients who experienced a tumor recurrence, at least 85% will have loco-regional failure, and the vast majority of these recurrences will be confined to the pelvis and vagina (Chyle et al. 1996; Davis et al. 1991; Kirkbride et al. 1995; Kucera and Vavra 1991; Perez et al. 1999; Stock et al. 1995; Tabata et al. 2002; Urbanski et al. 1996). The rate of loco-regional recurrence in stage I is approximately 10–20% versus 30–40% in stage II (Davis et al. 1991; Perez et al. 1999; Tabata et al. 2002). The pelvic control rate for patients with stage III and stage IV is relatively low, and about 50–70% of the patients have recurrences or persistence, despite well-designed RT (Perez et al. 1999; Tabata et al. 2002). The median time to recurrence is 6–12 months. Tumor recurrence is associated with a dismal prognosis, with only a few long-term survivors after salvage therapy. Failure in distant sites alone or associated with loco-regional failure does occur in about 25–40% of patients with locally advanced tumors (Chyle et al. 1996; Davis et al. 1991; Kirkbride et al. 1995; Kucera and Vavra 1991; Perez et al. 1999; Tabata et al. 2002; Urbanski et al. 1996).

27.6.4.1
Potential RT-Related Factors Influencing Outcome

It is important to recognize that analysis of RT doses and techniques and their impact on local/pelvic tumor control is fraught with difficulty, since the available data is retrospective, and not the result of prospective randomized or dose-escalation studies. Given the fact that more than 75% of the recurrences are local, the necessity for optimizing local therapy is clear. Clinical experience dictates that higher doses of RT, greater than 70–75 Gy, when feasible, are generally prescribed for more advanced stages of the disease (Andersen 1989; Chyle et al. 1996; Perez et al. 1999; Spirtos et al. 1989). Perez et al. (1988, 1999) reported increased tumor control in patients with stages IIA–IVA with EBRT and brachytherapy, compared with patients receiving brachytherapy alone. In patients with stage-I disease, no correlation was found between the technique of RT used and the incidence of local or pelvic recurrences. In addition, they suggested that doses in the range of 70–75 Gy to the primary tumor volume and 55–65 Gy to the medial parametria for patients with more advanced disease are necessary to optimize tumor and pelvic control. Furthermore, of 100 patients with primary tumors involving the upper and middle third of the vagina who received no elective irradiation to the groin, none developed metastatic inguino-femoral lymph nodes, in contrast to 3 of 29 (10%) with lower third primaries, and 1 of 20 with tumors involving the entire length of the vagina. Of 7 patients with initially palpable inguinal lymph nodes treated with doses in the range of 60 Gy, only 1 developed a nodal recurrence. The authors recommended that elective RT of the inguinal lymph nodes should be carried out only in patients with primary tumors involving the lower third of the vagina.

Stock et al. (1995) found a significant increase in local control and 5-year survival for patients receiving EBRT and brachytherapy, compared with those treated with EBRT alone. The 5-year actuarial local control and survival in the EBRT and brachytherapy group were 44% and 50%, respectively, compared with 12% and 9%, respectively, in the EBRT alone group. However, these two groups were not evenly matched with a large percentage of stage-IV lesions in the EBRT alone group compared with the brachytherapy group.

Lee et al. (1994) identified overall treatment time as the most significant treatment factor predicting pelvic tumor control in 65 patients with carcinomas of the vagina treated with definitive RT. If the entire course of RT, including EBRT and brachytherapy, was completed within 9 weeks, pelvic tumor control was 97%, in contrast with only 57% when treatment time extended beyond 9 weeks ($P<0.01$). Similarly, Pingley et al. (2000) reported that DFS rate was reduced from 60% to 30% if the overall treatment time was prolonged. Conversely, PEREZ et al. (1999) did not find a significant impact of prolongation of treatment time on pelvic tumor control. Nevertheless, these authors advocate completion of treatment within 7–9 weeks.

27.7
Clear Cell Carcinoma of the Vagina

Since Herbst's first report (HERBST and SCULLY 1970) of 7 adenocarcinomas arising in the vagina of adolescent females after in-utero exposure to DES, there have been several reports limited to DES-related vaginal CCA (HERBST and ANDERSON 1990; HERBST et al. 1971, 1972, 1979). In 1979, HERBST et al. (1979) reported 142 cases of stage-I CCA of the vagina. An 8% risk of recurrence was seen after radical surgery (n=117), and an 87% survival was achieved. There was a 36% risk of recurrence after RT for stage-I lesions; however, the authors acknowledged that, in general, RT was reserved for large stage-I lesions which involved more of the vault and were less amenable to surgical resection. As the majority of CCAs occur in the upper third of the vault, the largest series (HERBST et al. 1979; SENEKJIAN et al. 1987, 1988) addressing the surgical approach to these lesions have advocated radical hysterectomy, pelvic and para-aortic lymphadenectomy, and sufficient colpectomy to achieve negative margins. Senekjian has also reported a series of exenterations done for CCA (SENEKJIAN et al. 1989). However, there have been efforts to also attempt fertility-sparing radical resections (HUDSON et al. 1983, 1988) or more limited wide local excisions followed by some form of RT (SENEKJIAN et al. 1987).

SENEKJIAN et al. (1987) reported a series of 219 stage-I CCA cases with 92% overall 5-year and 88% 10-year survival rates, respectively, in 176 patients receiving conventional therapy (identical to 43 who had undergone local therapy). Of the 176 treated conventionally, 128 underwent radical hysterectomy and vaginectomy; 16 had the same operation followed by adjuvant RT; and 32 were treated with RT alone. Because of the risk of node metastases, 14 of 43 patients treated with local therapy underwent extraperitoneal pelvic lymphadenectomy. Of the 43 patients treated with local therapy, 9 had vaginectomy, 17 had local excision alone, 6 had brachytherapy alone and 11 had combined local excision and brachytherapy. The 10-year actuarial recur-

rence rate was an unsatisfactory 45% in those who underwent local excision alone, versus only 16% if they had received conventional therapy and 27% if they had received local excision followed by RT. SENEKJIAN et al. (1987) advocated a combination of wide local excision and extraperitoneal node dissection followed by brachytherapy for patients desirous of fertility preservation.

In a subsequent report, SENEKJIAN et al. (1988) reviewed the experience with 76 cases with stage-II CCA from the Registry for Research on Hormonal Transplacental Carcinogenesis. The overall 5- and 10-year survival rates were 83% and 65%, respectively. Of the 76 patients, 22 received surgery exclusively (either radical hysterectomy with vaginectomy, 13 patients, or exenterative type procedure, 9 patients), 38 received RT alone, 12 received combination therapy and 4 underwent other approaches. Patients treated with primary RT achieved an 87% 5-year survival rate versus 80% for those treated with surgery and 85% for those receiving both treatments. The authors concluded that most patients with stage-II vaginal CCA should be treated with combination EBRT and brachytherapy; however, small, easily resectable lesions in the upper fornix might undergo resection, allowing better preservation of coital and ovarian function (SENEKJIAN et al. 1988). In 1989, SENEKJIAN et al. (1989) reported their experience of 20 pelvic exenterations for CCA of the vagina, including 13 for primary lesions and 7 for recurrent disease. They reported a 72% success rate if the exenterations were done as part of primary therapy. The authors advocated reserving exenterative approaches for those who have failed RT in order to maximize quality of life for the greatest number of patients.

There are few published reports regarding the use of systemic therapy for this tumor. FOWLER et al. (1979) reported one complete and one partial response after treatment with melphalan (1mg/kg qd×5 days). ROBBOY et al. (1974) reported responses in recurrent disease to both 5-FU and vinblastine.

27.8
Salvage Therapy

In general, the patient with recurrent cancer of the lower female genital tract presents a difficult clinical dilemma. Optimal therapy for patients with recurrent gynecological cancer after potentially curative therapy has not been completely defined, partly due to the difficulty of conducting prospective randomized trials in this heterogeneous population. It must be determined whether the disease is amenable to curative salvage therapy, implying some reasonable chance of cure, or whether palliation is the primary goal. Treatment selection factors include primary therapy, extent of the disease at presentation, site of recurrence, extent of the recurrence, disease-free interval, evidence of metastatic disease, patient age, performance status and co-existing medical conditions. The presence of distant metastasis portends a poor prognosis and, although chemotherapy may result in objective responses and improvements in short-term survival, the current lack of curative systemic treatments focuses therapeutic attempts on symptom palliation and quality of life.

In most cases, only patients with small volume local recurrences and no metastatic disease are curable. Therefore, careful work-up to establish extent of disease is crucial. When salvage therapies are contemplated, local recurrences should be confirmed by biopsy, and, when possible, parametrial recurrences should be documented pathologically. Pelvic side wall involvement can almost always be diagnosed in the presence of a symptom triad of sciatic pain, leg edema and hydronephrosis. It is important to evaluate for regional and/or distant metastasis by means of physical examination and imaging studies such as CT or MRI scans. More recently, positron emission tomography (PET) scan has been used to document the extent of recurrent disease (Sun et al. 2001) but both false-positive and false-negative results have been reported.

Generally, patients with isolated pelvic or regional recurrences after definitive surgery who have not received prior RT are managed with EBRT, often in conjunction with brachytherapy (Davis et al. 1991; Kirkbride et al. 1995; Stock et al. 1995). Concurrent cisplatin-based chemotherapy may also be recommended (Urbanski et al. 1996). Salvage options for patients with central recurrence after definitive or adjuvant RT are limited to radical surgery, usually exenterative (Davis et al. 1991; Kirkbride et al. 1995; Stock et al. 1995; Urbanski et al. 1996). In selected patients with small-volume disease, re-irradiation using interstitial radiation implants or highly conformal three-dimensional EBRT could be considered. Response rates with chemotherapy are low, and the impact on survival is limited. Further, response to chemotherapy in central pelvic recurrences following RT tends to be less common than response at distant sites. Additionally, prior high-

dose RT compromises bone marrow tolerance of many agents that are active in this tumor (e.g., ifosfamide and doxorubicin). However, chemotherapy-responsive patients can obtain meaningful palliation in many cases.

27.8.1
Surgical Considerations

Despite thorough clinical evaluation of patients considered excellent candidates for salvage surgery, this will be aborted in over 25% of the cases because of advanced disease found at the time of the exploratory laparotomy (Miller et al. 1993). Pelvic exenteration results in long-term functional and psychological changes that have not been adequately studied (Ratliff et al. 1996). Surgical refinements have done much to improve body image changes associated with pelvic exenteration. The purposes of vaginal and perineal reconstruction following radical pelvic surgery for recurrent gynecological cancer are primarily twofold: to restore or create vulvo–vaginal function, thereby minimizing effects of surgical treatment on body image and normal sexual activity; and to minimize postoperative complications by transferring to the pelvic defect healthy tissue with good blood supply (Burke et al. 1995; Magrina and Basterson 1981). Detailed review of urinary diversion and pelvic reconstruction techniques are beyond the scope of this chapter.

27.8.2
RT Considerations

Those patients who have not received prior RT should receive whole-pelvis EBRT followed, when feasible, by brachytherapy. Generally, the whole pelvis receives a dose of 40–50 Gy. Inguino-femoral lymph node regions should be included in patients with involvement of the distal third of the vagina or with vulvar recurrences. The gross tumor volume in the vagina, paravaginal tissues and/or parametrium should receive an additional boost, preferably with an interstitial implant, to bring the total tumor dose to 75–80 Gy. The role of combined chemo-radiotherapy in the management of patients with recurrent disease is unknown. Given the rarity of vaginal carcinoma and the heterogeneity within the population with recurrent disease, large randomized studies intended to answer this question will probably never be conducted. However, by extrapolation from the available data for locally advanced cervical and vulvar cancer (KEYS et al. 1999; MOORE et al. 1988; MORRIS et al. 1999; ROSE et al. 1999; WHITNEY et al. 1999) it seems that combined modality approach may improve the loco-regional control and survival in patients with isolated pelvic recurrences.

Re-irradiation in previously irradiated patients must be undertaken with extreme caution. However, selected patients who are medically inoperable, technically unresectable or refuse to undergo exenterative surgery are appropriately considered for re-irradiation to limited volumes. A variety of techniques are available, and the choice is based on patient and tumor-related factors, as well as the experience of the radiation oncologist. When using EBRT, multiple beam arrangements utilizing three-dimensional treatment planning is favored. Only limited doses are possible, and the physician might consider a hyper-fractionated regimen in an attempt to decrease the incidence of late toxicity.

In patients with small, well-defined vulvo–vaginal or pelvic recurrences, re-irradiation using primarily interstitial techniques has been attempted with control rates between 50% and 75%, and grade 3 or higher complication rates between 7% and 15% (CHARRA et al. 1998; GUPTA et al. 1999; RANDALL et al. 1993; RUSSELL et al. 1987; WANG et al. 1998). The rationale, logistics and selection of implant technique when performing an ITI were reviewed earlier in this chapter. Permanent radioactive seed implants (e.g., ^{198}Gold) in patients with small vaginal recurrences often provides long-lasting tumor control in elderly or medically debilitated patients previously treated with definitive doses of RT. Advantages and disadvantages of surgery and re-irradiation as sal-

vage therapies are shown in Table 27.4 (RANDALL et al. 1993).

Other potential treatment options include the use of surgery and intraoperative RT (IORT), which allows direct visualization of the target volume, displacement and/or shielding of the surrounding normal tissues. Several approaches have been used. Intraoperative electron beam and HDR have been used for treatment of isolated central and nodal recurrences (HADDOCK et al. 1999). However, the published series are generally small, including a wide spectrum of patients with different gynecological malignancies, varying amounts of residual disease and disparate initial therapies. The loco-regional recurrence and distant metastasis rates after IORT vary between 20% and 60% and 20% and 58%, respectively. The 3- to 5-year actuarial survival is poor, ranging from 8% to 25%. Grade 3 or higher toxicity has been reported in about 35% of patients (GARTON et al. 1997; HADDOCK et al. 1999). In the Memorial Sloan-Kettering Cancer Center experience using radical surgical resection and HDR-IORT, patients with complete gross resection had a 3-year local control rate of 83%, compared with 25% in patients with gross residual disease. Interestingly, most of the failures in the microscopic group were distant, perhaps indicating a potential role for adjuvant chemotherapy (GEMIGNANI et al. 2001).

HOCKEL et al. (1996) described a combined operative and radiotherapeutic treatment (CORT) for the treatment of recurrent gynecological malignancies infiltrating the pelvic side wall. The procedure involves gross complete re-section of the tumor and a single plane interstitial implant. In order to improve the therapeutic index, well-vascularized tissue is

Table 27.4 Advantages and disadvantages of salvage surgery and interstitial re-irradiation

Salvage therapy	Advantages	Disadvantages
Surgery	• Ability to assess the extent of disease and act accordingly • Applicable to larger volume recurrences	• Perioperative morbidity and mortality, particularly after previous RT • Prolonged hospitalization • High rate of re-operation • Expense • Applicable only to selected patients with good performance status/good general condition Detrimental to patient self-image
Re-Irradiation	• Little perioperative morbidity or mortality • Little or no hospitalization for permanent implants unless laparotomy is required • Relatively inexpensive • Preserves structure and function in most patients • Applicable to patients who are medically infirm or aged	• Extent of disease difficult to assess in some cases • Risk of late radiation injury • Applicable only to small-volume recurrences if excessive complication rate is to be avoided

Reprinted with permission from RANDALL et al. (1993)

transposed to the pelvis to protect the hollow organs and reduce the late effects of RT. Reconstruction of pelvic organs is performed as with exenteration. The tumor bed is irradiated postoperatively, days 10–14, using HDR brachytherapy. In a total of 48 patients treated using this technique, the overall severe complication rate was 33% at 5 years. The 5-year survival rate was 44%, and the absolute local control rate was 60% for the first 20 patients and 85% for the last 28 treated patients.

Stereotactic body radiotherapy (SBRT), also known as extracranial stereotactic radioablation (ESR), is a novel treatment paradigm that delivers a small number of high-dose fractions to extracranial targets using a linear accelerator with highly precise, accurate and reproducible target localization, based on the same principles as that of the gamma-knife therapy. By means of better target localization and patient immobilization, smaller margins or normal tissue surrounding the gross tumor volume are required, which allow treatment complications to be minimized. BLOMGREN et al. (1998) reported on 15 patients with 19 extrahepatic abdominal tumors that had a mean survival of 17.7 months. The toxicity was more often self-limited, except for 4 patients with gastrointestinal bleeding. The authors concluded that this treatment, which is non-invasive, painless, rapid and does not require hospitalization, does not impair the quality of life of the patients when used properly.

27.9
Treatment Complications and their Management

The anatomic location of the vagina places the lower gastrointestinal and genitourinary tracts at greatest risk for complications after surgery or RT. Although in most of the retrospective series the authors comment on the nature of the complications encountered, little information is typically given regarding their prevention or management (BALL and BERMAN 1982; GALLUP et al. 1987; HOFFMAN et al. 1992; PETERS et al. 1985; RUBIN et al. 1985; STOCK et al. 1995). In modern oncology, survival rate is the primary end-point in treatment evaluation, but the analysis of treatment complications and quality of life are of crucial importance. Clearly, the knowledge of common acute and late complications with standard RT and consideration of risk factors may improve the therapeutic ratio of RT for gynecologi-

cal malignancies in general and for vaginal cancer in particular (CARDENES et al. 2001).

The acute and chronic pathophysiology of vaginal RT has been well described by GRIGSBY et al. (1995). As an immediate response to high-dose RT, there is loss of most or all of the vaginal epithelium, especially in areas in proximity to brachytherapy sources. Clinically, the severity of the acute effects (edema, erythema, moist desquamation, and confluent mucositis with or without ulceration) vary in intensity and duration depending on patient age, hormonal status, tumor size, stage, RT dose and personal hygiene. These effects usually resolve within 2–3 months after completion of therapy. In some patients, there is progressive vascular damage with the subsequent ulcer formation and mucosal necrosis, which may require up to 8 months for healing. Chemotherapy concurrently with RT enhances the acute mucosal response to both EBRT and brachytherapy. The effects of chemotherapy on the incidence of late complications, if any, are unclear. Over time, most patients will develop some degree of vaginal atrophy, fibrosis and stenosis. Telangiectasis is commonly seen in the vagina. Vaginal narrowing, shortening, paravaginal fibrosis, loss of elasticity and reduced lubrication often result in dyspareunia. More severe complications include necrosis, with ulceration that can progress to fistula formation (recto–vaginal, vesico–vaginal, urethro–vaginal).

The RT tolerance limits of the entire vagina are ill-defined, given the variety of techniques employed for the treatment of vaginal cancers. An irradiation tolerance level of the proximal vagina was suggested by HINTZ et al. (1980) based on a study of 16 patients who received a maximum surface dose of 140 Gy, none of whom developed severe complications or necrosis of the upper vagina. Based on their previous observation of a patient who developed a vesico-vaginal fistula after receiving a dose of 150 Gy mucosal dose to the anterior vaginal wall, they recommended a tolerance dose level of 150 Gy (direct summation of EBRT dose and ICB) to the anterior upper vaginal mucosa. They also recommended keeping the total dose to the distal vagina less than 98 Gy. In addition, it was also observed that the posterior wall of the vagina is more prone to radiation injury than the anterior or lateral walls, and that the dose should be kept below 80 Gy in order to minimize the risk of recto–vaginal fistula. RUBIN and CASARET (1986) suggested that the tolerance of the vaginal mucosa (TD 5/5: 5% necrosis within 5 years) is around 90 Gy for ulceration, and more than 100 Gy for fistula formation. This tolerance limit has been specified as a direct summation of dosage given by LDR-ICB and EBRT in the treatment of cervical cancer. Within the low-dose-rate range, whether a correction for brachytherapy dose rate is necessary remains controversial. In a more recent series from Washington University, the traditional LDR tolerance dose of 150 Gy to the mucosa of the proximal vagina was shown to yield a nominal 11% and 4% grades 1–2 and 3 sequelae, respectively (Au and Grigsby 2003).

The incidence of grade 2 or higher complications has been reported to be 15–25%, (Chyle et al. 1996; Kirkbride et al. 1995; Kucera and Vavra 1991; Perez et al. 1999; Peters et al. 1985; Rubin et al. 1985; Stock et al. 1995; Urbanski et al. 1996) with the average of severe complications (those requiring surgery for correction or necessitating hospitalization) around 8–10%.

Host factors that may increase the risk of complications include prior pelvic surgery, pelvis inflammatory disease, immunosuppression status, collagen vascular disease, low body weight, patient age, significant smoking history and co-morbid illness (e.g., diabetes, hypertension and cardiovascular disease) (Perez et al. 1988, 1999).

Lee et al. (1994) showed that the total dose to the primary site was the most significant factor predicting the development of a severe complication (9% in patients receiving 80 Gy or less compared with 25% in those receiving higher doses). Perez et al. (1988) reported an increase in the rate of severe complications with higher clinical stage, probably reflecting the higher doses delivered with EBRT and brachytherapy.

Ball and Berman (1982) reported 58 patients with carcinoma of the vagina, including 30 who underwent surgery. There were four recto–vaginal fistulae (one following RT, and three after exenterative surgery) and two vesico–vaginal fistulae (one following radical vaginectomy and the other following a recurrence, being managed with cystectomy and diversion). The single uretero–vaginal fistula occurred after radical vaginectomy and partial cystectomy, and was managed with ureteroneocystostomy.

In Peters' report (Peters et al. 1985) of 86 vaginal primaries, there were two fistulae in the 57 patients who received primary RT. However, there was a 44% rate of fistulae formation in the nine patients who underwent re-irradiation after having previously received RT for a previous cancer. Rubin et al. (1985) reported a 23% incidence of complications after RT, including a 13% rate of fistula formation and a 10% rate of cystitis/proctitis. Although 2 patients developed fistulae following combination therapy, the authors did not think that the rate of complications following combination therapy was greater than that seen following RT alone.

In Stock's series (Stock et al. 1995) of 100 patients with vaginal carcinoma, there was a 16% actuarial complication rate at 10 years. All patients undergoing vaginectomies or exenterations lost vaginal function. None of the patients was offered vaginal reconstruction in this series. Stock et al. (1995) emphasized that therapeutic options needed to be individualized such that surgery is offered only to those most likely to benefit and least likely to suffer complications.

Treatment options for acute radiation vaginitis include daily vaginal douching with a diluted hydrogen peroxide/water mixture. This should continue for 2–3 months, or until the mucosal reactions have subsided. Patients are then advised to continue douching once or twice per week for several months. Regular vaginal dilation is recommended as a way for patients to maintain vaginal health and good sexual function, although the compliance rate is low. The lack of resolution of vaginal ulceration or necrosis after several months of adequate therapy must be appropriately evaluated, considering the possibility of recurrent tumor. The use of topical estrogens following completion of RT appears to stimulate epithelial regeneration more than systemic estrogens.

Some patients with severe radiation sequelae, such as fistula formation, will respond to conservative treatment with antibiotics and periodic limited debridement of necrotic tissue. More recently, Delanian et al. (2003) published a randomized trial demonstrating the effectiveness of the combination of pentoxifylline and vitamin E in the regression of radiation-induced fibrosis.

Patients with more severe gastrointestinal or urinary late effects will require urinary or fecal diversion with possible delayed re-anastomosis. Occasionally, repair of the fistula may be attempted by employing a myocutaneous graft, in which the skin, subcutaneous fat and muscle are mobilized using a vascular pedicle to maintain the blood supply to the pedicled graft (Martius flap), or by excision of the necrotic tissue with reestablishment of organ continuity (such as in the treatment of high recto–vaginal fistula). A detailed review of the pathogenesis and management of potential late effects of treatment is beyond the scope of this chapter and may be found elsewhere in this textbook.

It is likely that improvements in modern practice such as advancements in surgical techniques (such as more generous use of myocutaneous flaps) (Burke et al. 1995; Magrina and Basterson 1981), improved supportive care during the immediate post-operative stay, use of more sophisticated RT field setting (three-dimensional conformal therapy) and treatment delivery, more accurate brachytherapy techniques and dose calculations have potential to lessen complication rates post-therapy, regardless of which modality is used.

27.10
Palliative RT

At the present time, there is no curative option for patients who present with stage-IVB disease. Many of these patients suffer from severe pelvic pain or bleeding. If vaginal bleeding is the main concern, ICB, if feasible, often offers a good symptom control with relatively low morbidity. For patients who have received prior RT, intracavitary doses in the range of 35–40 Gy to point A should be prescribed.

A short course of EBRT using high-dose fractionation schedules have been used, including single doses of 10 Gy per fraction, times three, with an interval of 4–6 weeks between courses, combined with misonidazole, RTOG clinical trial 79–05, resulting in significant palliation in selected patients with advanced gynecological malignancies. The overall response rate was 41% for patients completing the three courses; however, the actuarial 45% incidence of grade 3–4 late gastrointestinal toxicity was unacceptable (Spanos et al. 1987).

Spanos et al. (1989) reported on a phase-II study (RTOG 85-02) of daily multifraction split-course EBRT in patients with recurrent or metastatic disease. The regimen consisted of 3.7 Gy per fraction given twice daily for two consecutive days, and repeated at 3- to 6-week intervals, for a total of three courses (tumor dose, 44.4 Gy). Occasionally, this regimen was combined with an ICI (4500 mgh), with a midline block in the last 14.4 Gy. Complete tumor response was noted in 15 patients (10.5%) and a partial response in 32 (22.5%) (Fig. 27.10a–e2). In patients completing three courses of irradiation

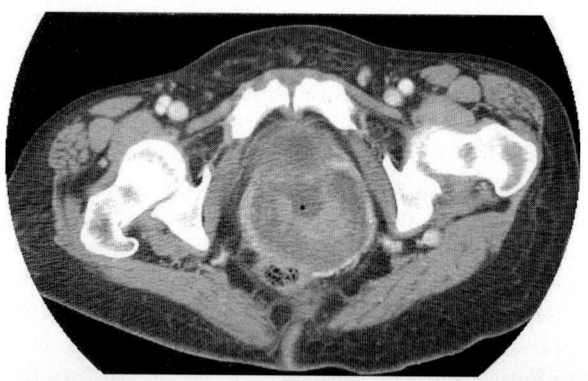

Fig. 27.10a–c. Palliative course of external beam irradiation using the RTOG 85-02 regimen. **a** Computed tomography (CT) scan prior to initiation of therapy. **b** AP and lateral digital reconstructed radiographs (*DRRs*) of the initial fields. **c** CT scan after two split courses of hypofractionated irradiation (2950 cGy) in 1 week, prior to a brachytherapy boost. **c** AP (**c1**) and lateral (**c2**) radiographs of the brachytherapy implant

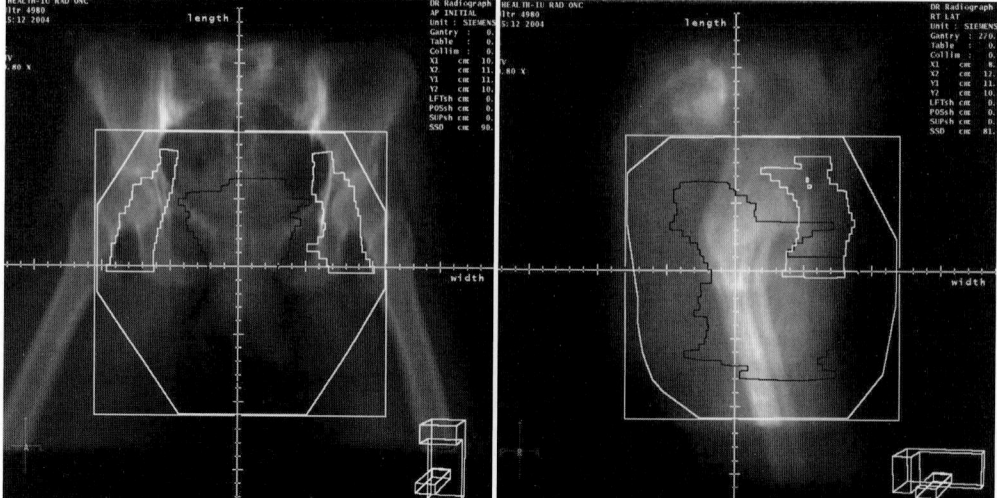

a

b1   b2

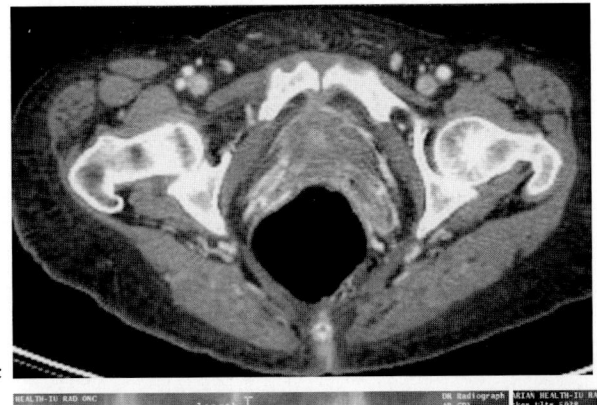

c

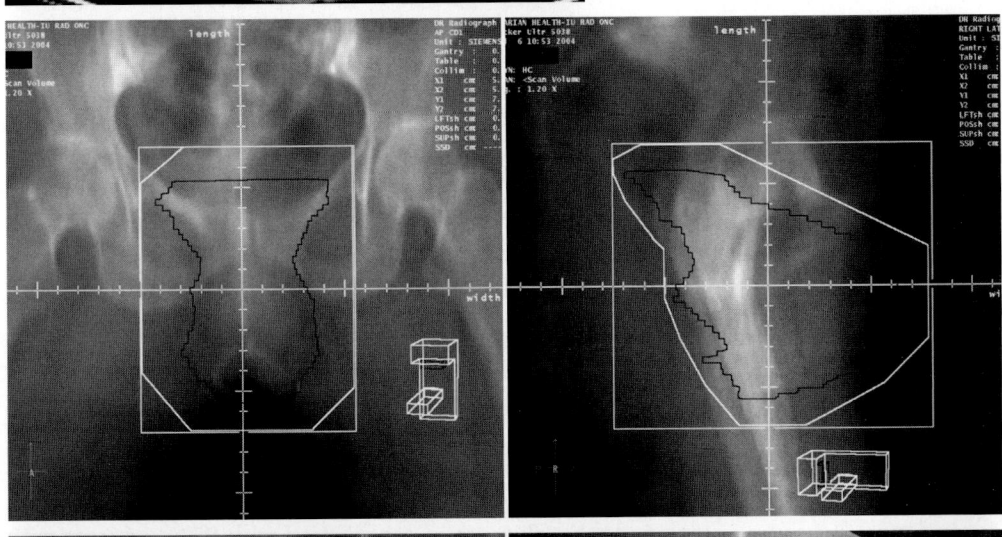

d1

d2

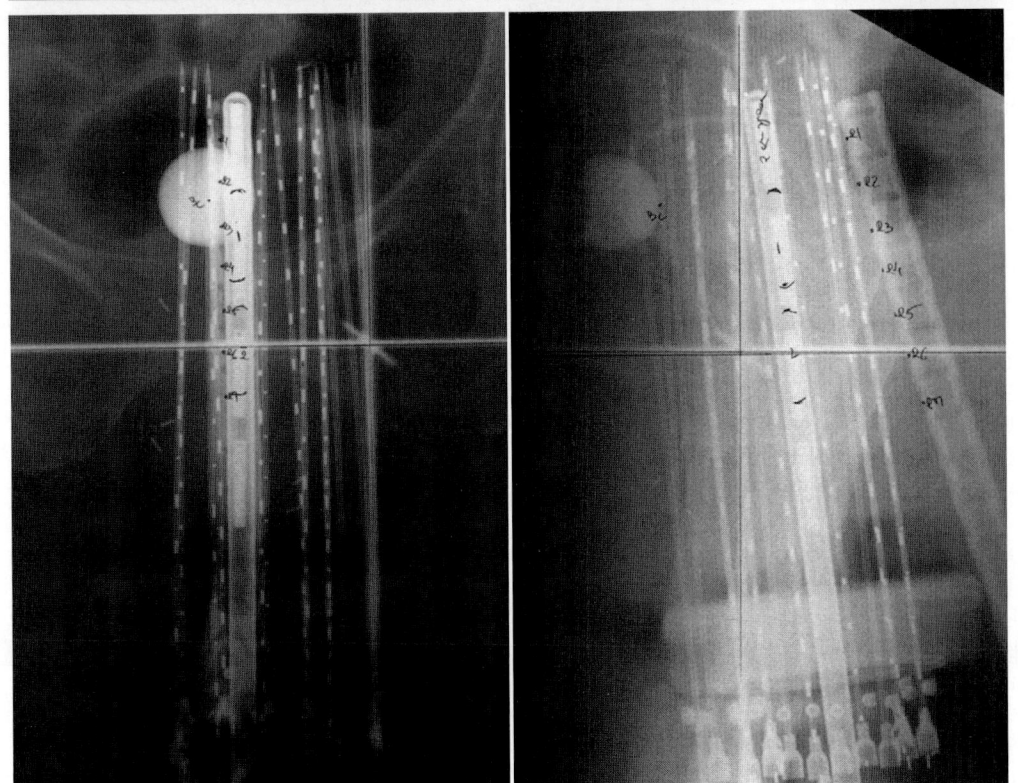

e1

e2

(59%), the rate of complete or partial response was 45%. Of the patients, 27 survived longer than 1 year. Late complications were significantly less, with a projected actuarial rate of 5% at 12 months.

In a subsequent phase-III study (SPANOS et al. 1993), 136 patients were randomized between rest intervals of 2 weeks versus 4 weeks between the split courses of RT. Decreasing the interval between courses did not result in a significant improvement in tumor response (34% versus 26%). More patients in the 2-week rest group completed the three courses of therapy and, not surprisingly, patients completing all three courses had a higher overall response rate than patients completing fewer than three courses (42% versus 5%), and a higher complete response rate (17% versus 1%). This schedule offers significant logistic benefits and has been shown to result in good tumor regression and excellent palliation of symptoms. SPANOS et al. (1994) reported a trend toward increased acute toxicity in patients with shorter rest periods, but late toxicity was not significantly different in the two groups.

27.11
Chemotherapy in Advanced-Recurrent Vaginal Cancer

Given its rarity, most chemotherapy reports for treatment of metastatic disease in vaginal cancer are anecdotal, or combined with reports of treatment of advanced or recurrent cervical cancer. Concurrent chemo-radiation is frequently employed in clinical practice in the treatment of unresectable loco-regionally advanced disease. Various chemotherapeutic agents have been used with limited success (EVANS et al. 1988; ROBERTS et al. 1991). EVANS et al. (1988) reported 7 patients with vaginal cancers who were treated with a combination of 5-FU 1000 mg/m^2/d×4 days, mitomycin C 10 mg/m^2 day 1, and primary irradiation, receiving 2000–6500 cGy. All of the vaginal cancer patients responded, and 66% were alive with a median follow-up of 28 months.

Treatment of recurrent or metastatic disease is confined to a handful of phase-II clinical trials and anecdotal reports. In general, regimens that are active in cervical cancer are usually active in vaginal cancer. THIGPEN et al. (1986) reported the results of a phase-II trial of cisplatin 50 mg/m^2 every 3 weeks in 26 patients with advanced or recurrent vaginal cancer. There were 22 evaluable patients, 16 with SCC, 2 adenosquamous carcinoma, 1 clear cell car-

cinoma, 1 leiomyosarcoma and 2 unspecified. Of the 16 SCC patients, there was 1 with complete response (6.2%). It should be noted that these patients, for the most part, had received prior surgery and RT. MUSS et al. (1989) reported no responses in 19 evaluable patients who were treated with mitoxantrone 12 mg/m^2 every 3 weeks. Median survival of patients with vaginal cancer was 2.7 months. Other anecdotal reports of responses in trials that included advanced cervical cancer include a report by LONG et al. (1995) in which 3 patients with advanced vaginal SCC received treatment with methotrexate, vinblastine, doxorubicin and cisplatin (MVAC). All 3 patients achieved a complete response of short duration. PATTON et al. (1991) reported results of intra-arterial chemotherapy with mitomycin C, bleomycin, and cisplatin and vincristine, including 6 patients with primary vaginal cancer and 40 patients with cervical cancer; 76% responded to intra-arterial chemotherapy and subsequently received primary RT. The report did not give details as to site of relapse impact on DFS or overall survival.

At the present time, systemic treatment of advanced vaginal cancer outside of a clinical trial is purely anecdotal, although it might be reasonable to extrapolate from the experience reported with SCC of the cervix and vulva. Although published response rates are low, standard therapy should include cisplatin alone or in conjunction with RT in patients with loco-regionally advanced vaginal cancer.

References

Aho MK, Vesterinen E, Meyer B et al (1991) Natural history of vaginal intraepithelial neoplasia. Cancer 68:195–197

Ali MM, Huang DT, Goplerud DR et al (1996) Radiation alone for carcinoma of the vagina. Variation in response related to the location of the primary tumor. Cancer 77:1934–1939

Al-Kurdi M, Monaghan JM (1981) Thirty-two years experience in management of primary tumors of the vagina. Br J Obstet Gynaecol 88:1145–1150

American Joint Committee on Cancer (AJCC) (2002) Vagina. In: Greene FL, Page DL, Fleming ID, Fritz AG, Balch CM, Haller DG, Morrow M (eds) AJCC cancer staging manual, 6th edn. Springer, Berlin Heidelberg New York, pp 251–257

Andersen ES (1989) Primary carcinoma of the vagina. Gynecol Oncol 33:317–320

Antonioli DA, Burke L (1975) Vaginal adenosis: analysis of 325 biopsy specimens from 100 patients. Am J Clin Pathol 64:625

Au Samuel P, Grigsby PW (2003) The irradiation tolerance dose of the proximal vagina. Radiother Oncol 67:77–85

Ball HG, Berman ML (1982) Management of primary vaginal carcinoma. Gynecol Oncol 14:154–163

Blomgren H, Lax I, Goranson H et al (1998) Radiosurgery of tumors in the body: clinical experience using a new method. J. Radiosurg 1:63–74

Brinton LA, Nasca PC, Mallin K et al (1990) Case-control study of in situ and invasive carcinoma of the vagina. Gynecol Oncol 38:49–54

Burke TW, Morris M, Roh MS et al (1995) Perineal reconstruction using single gracilis myocutaneous flaps. Gynecol Oncol 57:221–225

Cardenes HG, Song M, Randall (2001) Late sequelae of radiation therapy in the management of gynecological malignancies. Current medical literature. Gynecol Oncol 2:1–10

Chang YCF, Hricak H, Thurnher S et al (1988) Vagina: evaluation with MR imaging. Radiology 169:175–179

Charra C Roy P Coquard R et al (1998) Outcome of treatment of upper third vaginal recurrences of cervical and endometrial carcinomas with interstitial brachytherapy. Int J Radiat Oncol Biol Phys 40:421–426

Childers JM, Surwit EA (1993) Current status of operative laparoscopy in gynecologic malignances. Oncology 7:47–57

Chu AM, Beechinor R (1984) Survival and recurrence patterns in the radiation treatment of carcinoma of the vagina. Gynecol Oncol 19:298–307

Chyle V, Zagars GK, Wheeler JA et al (1996) Definitive radiotherapy for carcinoma of the vagina. Int J Radiat Oncol Biol Phys 35:891–905

Corn BW, Lanciano RM, Rosenblum N et al (1995) Improved treatment planning for the Syed-Neblett template using endorectal-coil magnetic resonance and intraoperative (laparotomy/laparoscopy) guidance: A new integrated technique for hysterectomized women with vaginal tumors. Gynecol Oncol 56:255–261

Creasman WT, Phillips JL, Menck HR (1998) The National Cancer Data Base report on cancer of the vagina. Cancer 83:1033–1040

Dalrymple JL, Russell AH, Lee SW et al (2004) Chemoradiation for primary invasive squamous carcinoma of the vagina. Int J Gynecol Cancer 14:110–117

Dancuart F, Delclos L, Wharton JT et al (1988) Primary squamous cell carcinoma of the vagina treated by radiotherapy: a failures analysis - the MD Anderson hospital experience 1955–1982. Int J Radiat Oncol Biol Phys 14:745–749

Davis KP, Stanhope CR, Garton GR et al (1991) Invasive vaginal carcinoma: analysis of early stage disease. Gynecol Oncol 42:131–136

Delanian S, Porcher R, Balla-Mekias et al (2003) Randomized, placebo-controlled trial of combined Pentoxifylline and Tocopherol for regression of superficial radiation-induced fibrosis. J Clin Oncol 21:2545–2550

Delclos L (1984) Gynecological cancers: pelvic examination and treatment planning. In: Levitt SH, Tapley N (eds) Technological basis of radiation therapy practical clinical applications. Lea and Febiger, Philadelphia, pp 193–227

Delclos L, Fletcher GH, Moore EB et al (1980) Minicolpostats, dome cylinders, other additions and improvements of the Fletsher-Suit afterloadable system: indications and limitations of their use. Int J Radiat Oncol Biol Phys 6:1195–1206

Diakomanolis E, Rodolakis A, Boulgaris Z et al (2002) Treatment of vaginal intraepithelial neoplasia with laser ablation and upper vaginectomy. Gynecol Obstet Invest 54:17–20

Disaia PJ, Syed N, Puthwala AA (1990) Malignant neoplasia of the upper vagina. Endocurietherapy/Hyperthermia Oncol 6:251–256

Dittmer PH, Randall ME (2001) A technique for inguinal node boosts using photon fields defined by asymmetric collimators jaws. Radiother Oncol 59:61–64

Dixit S, Singhal S, Baboo HA (1993) Squamous cell carcinoma of the vagina. A review of 70 cases. Gynecol Oncol 48:80–87

Ebrahim S, Daponte A, Smith TH et al (2001) Primary mucinous adenocarcinoma of the vagina. Gynecol Oncol 80:89

Eddy GL, Marks RD, Miller MC et al (1991) Primary invasive vaginal carcinoma. Am J Obstet Gynecol 165:292–298

Evans LS, Kersh CR, Constable WC et al (1988) Concomitant 5-fluorouracil, mitomycin-C and radiotherapy for advanced gynecological malignancies. Int J Radiat Oncol Biol Phys 15:901–906

Fanning J, Manahan KJ, McLean SA (1999) Loop electrosurgical excision procedure for partial upper vaginectomy. Am J Obstet Gynecol 181:1382–1385

Fowler WC, Brantley JC, Edelman DA (1979) Clear cell adenocarcinoma of the genital tract. South Med J 72:15–17

Gallup DG, Talledo OE, Shah KJ et al (1987) Invasive squamous cell carcinoma of the vagina. A 14-year study. Obstet Gynecol 69:782–785

Garton GR, Gunderson LL, Webb MJ et al (1997) Intraoperative radiation therapy in gynecologic cancer: update of the experience at a single institution. Int J Radiat Oncol Biol Phys 37:839–843

Gemignani ML, Alektiar KM, Leitao M et al (2001) Radical surgical resection and high-dose intraoperative radiation therapy (HDR-IORT) in patients with gynecologic cancers. Int J Radiat Oncol Biol Phys 50:687–694

Gore E, Gillin MT, Albano K et al (1995) Comparison of high dose-rate and low dose-rate dose distributions for vaginal cylinders. Int J Radiat Oncol Biol Phys 31:165–170

Grigsby PW, Russell A, Bruner D et al (1995) Late injury of cancer therapy on the female reproductive tract. Int J Radiat Oncol Biol Phys 31:1281–1299

Gupta AK, Vicini FA, Frazier AJ et al (1999) Iridium-192 transperineal interstitial brachytherapy for locally advanced or recurrent gynecological malignances. Int J radiat Oncol Biol Phys 43:1055–1060

Haddock MG, Martinez-Monge R, Petersen IA et al (1999) Locally advanced primary and recurrent gynecologic malignancies. EBRT with or without IORT or HDR-IORT. In: Gunderson LL, Calvo F, Harrison LB et al (eds) Current clinical oncology: intraoperative irradiation: techniques and results. Humana Press, New Totowa, NJ, pp 397–419

Haskel S, Chen SS, Spiegel G (1989) Vaginal endometrioid adenocarcinoma arising in vaginal endometriosis: a case report and literature review. Gynecol Oncol 34:232

Herbst AL, Anderson D (1990) Clear cell adenocarcinoma of the vagina and cervix secondary to intrauterine exposure to diethylstilbestrol. Semin Surg Oncol 6:343–346

Herbst AL, Scully RD (1970) Adenocarcinoma of the vagina in adolescence: a report of seven cases including six clear-cell carcinomas (so-called mesonephromas). Cancer 25:745–757

Herbst AL, Green TH Jr, Ulfelder H (1970) Primary carcinoma of the vagina. Am J Obstet Gynecol 106:210

Herbst AL, Ulfelder H, Poskanzer DC (1971) Adenocarcinoma of the vagina: association of maternal stilbestrol therapy with tumor appearance in young women. N Engl J Med 284:878–881

Herbst AL, Kurman RJ, Scully RE et al (1972) Clear cell adeno-carcinoma of the genital tract in young females. Registry report. N Engl J Med 287:1259

Herbst AL, Robboy SJ, Scully RE et al (1974) Clear-cell adeno-carcinoma of the vagina and cervix in girls: analysis of 170 registry cases. Am J Obstet Gynecol 119:713–724

Herbst AL, Norusis MJ, Rosenow PJ et al (1979) An analysis of 346 cases of clear cell adenocarcinoma of the vagina and cervix with emphasis on recurrence and survival. Gynecol Oncol 7:111–122

Hilaris BS, Nori D, Anderson LL (1987) Brachytherapy treat-ment planning. Front Radiat Ther Oncol 21:94–106

Hinchey WW, Silva EG, Guarda LA et al (1983) Paravaginal wolffian duct (mesonephros) adenocarcinoma: a light and electron microscopic study. Am J Clin Pathol 80:539

Hintz BL, Kagan AR, Gilbert HA et al (1980) Radiation toler-ance of the vaginal mucosa. Int J Radiat Oncol Biol Phys 6:711–716

Hiroi H, Yasugi T, Matsumoto K et al (2001) Mucinous adeno-carcinoma arising in a neovagina using the sigmoid colon thirty years after operation: a case report. J Surg Oncol 77:61

Hockel M, Schlenger K, Hamm H et al (1996) Five-year expe-rience with combined operative and radiotherapeutic treatment of recurrent gynecologic tumors infiltrating the pelvic wall. Cancer 77:1918–1933

Hoffman MS, Roberts WS, LaPolla JP et al (1991) Laser vapor-ization of grade 3 vaginal intraepithelial neoplasia. Am J Obstet Gynecol 165:1342–1344

Hoffman MS, DeCesare SL, Roberts WS et al (1992) Upper vaginectomy for in situ and occult, superficially invasive carcinoma of the vagina. Am J Obstet Gynecol 166:30–33

Hudson CN, Crandon AJ, Baird PJ et al (1983) Preservation of reproductive potential in diethylstilbestrol-related vaginal adenocarcinoma. Am J Obstet Gynecol 145:375–377

Hudson CN, Findlay WS, Roberts H (1988) Successful preg-nancy after radical surgery for diethyl-stilboestrol (DES)-related vaginal adenocarcinoma. Case report. Br J Obstet Gynecol 95:818–819

Jobson V, Homesley HD (1983) Treatment of vaginal intraepi-thelial neoplasia with the carbon dioxide laser. Obstet Gynecol 62:90–93

Julian TM, O'Connell BJ, Gosewehr JA (1992) Indications, tech-niques, and advantages of partial laser vaginectomy. Obstet Gynecol 80:140–143

Kaufman RH, Korhonen MO, Strama T et al (1982) Develop-ment of clear cell adenocarcinoma in DES-exposed off-spring under observation. Obstet Gynecol 59:68S

Kersch CR, Constable W, Spaulding C et al (1990) A phase I–II trial of multimodality management of bulky gyneco-logic malignancy. Combined chemoradiosensitization and radiotherapy. Cancer 66:30–34

Keys HM, Bundy BN, Stehman FB et al (1999) Cisplatin, radia-tion and adjuvant hysterectomy compared with radiation and adjuvant hysterectomy for bulky stage IB cervical car-cinoma. N Engl J Med 340:1154–1161

Kirkbride P, Fyles A, Rawlings GA et al (1995) Carcinoma of the vagina - experience at the Princess Margaret Hospital (1974–1989). Gynecol Oncol 56:435–443

Krebs HB (1989) Treatment of vaginal intraepithelial neopla-sia with laser and topical 5-fluorouracil. Obstet Gynecol 73:657–660

Kucera H, Vavra N (1991) Radiation management of pri-mary carcinoma of the vagina: clinical and histopatho-logical variables associated with survival. Gynecol Oncol 40:12–16

Lee WR, Marcus RB Jr, Sombeck MD et al (1994) Radiotherapy alone for carcinoma of the vagina: the importance of overall treatment time. Int J Radiat Oncol Biol Phys 29:983–988

Lenehan PM, Meffe F, Lickrish GM (1986) Vaginal intraepi-thelial neoplasia: biologic aspects and management. Obstet Gynecol 68:333–337

Leung S, Sexton M (1993) Radical radiation therapy for car-cinoma of the vagina – impact of treatment modalities on outcome: Peter MacCallum Cancer Institute experience 1970–1990. Int J Radiat Oncol Biol Phys 25:413–418

Li Z, Liu C, Palta JR (1998) Optimized dose distribution of a high dose rate vaginal cylinder. Int J Radiat Oncol Biol Phys 41:239–244

Long HJ 3rd, Cross WG, Wieand HS et al (1995) Phase II trial of methotrexate, vinblastine, doxorubicin, and cisplatin in advanced/recurrent carcinoma of the uterine cervix and vagina. Gynecol Oncol 57:235–239

Maassen V, Lampe B, Untch M et al (1993) Adenocarcinoma and adenosis of the vagina. On the histogenesis, diagno-sis and therapy of a rare genital neoplasm. Geburtshilfe Fraunenheilkd 53:308

MacLeod C, Fowler A, Dalrymple C et al (1997) High-dose-rate brachytherapy in the management of high-grade intraepi-thelial neoplasia of the vagina. Gynecol Oncol 65:74–77

MacNaught R, Symonds RP, Hole D et al (1986) Improved con-trol of primary vaginal tumors by combined external beam and interstitial brachytherapy. Clin Radiol 37:29–32

Magrina JF, Basterson BJ (1981) Vaginal reconstruction in gynecologic oncology. A review of techniques. Obstet Gynecol Surv 36:1–10

Martinez A, Cox RS, Edmudson GK (1984) A multiple-site perineal applicator (MUPIT) for treatment of prostatic, anorectal and gynecologic malignancies. Int J Radiat Oncol Biol Phys 10:297–305

Miller B, Morris M, Rutledge E et al (1993) Aborted exentera-tive procedures in recurrent cervical cancer. Gynecol Oncol 50:94–99

Mock U, Kucera H, Fellner C et al (2003) High Dose Rate (HDR) brachytherapy with or without external beam radiother-apy in the treatment of primary vaginal carcinoma: long term results and side effects. Int J Radiat Oncol Biol Phys 56:950–957

Moore DH, Thomas GM, Montana GS et al (1988) Preopera-tive chemoradiation for advanced vulvar cancer. A phase II study of the Gynecologic Oncology Group. Int J Radiat Oncol Biol Phys 42:79–85

Morris M, Eifel PJ, Lu J et al (1999) Pelvic irradiation with concurrent chemotherapy compared with pelvic and para-aortic radiation for the high-risk cervical cancer. N Engl J Med 340:1137–1143

Munkarah A, Malone JM Jr, Budey HD et al (1994) Mucinous adenocarcinoma arising in a neovagina. Gynecol Oncol 52:272

Muss HB, Bundy BN, Christopherson WA (1989) Mitoxan-trone in the treatment of advanced vulvar and vaginal car-cinoma: a Gynecologic Oncology Group study. Am J Clin Oncol 12:142–144

Nanavati PJ, Fanning J, Hilgers RD et al (1993) High-dose-brachytherapy in primary stage I and II vaginal cancer. Gynecol Oncol 51:67–71

Ogino I, Kitamura T, Okajima H et al (1998) High-dose-rate intracavitary brachytherapy in the management of cervical and vaginal intraepithelial neoplasia. Int J Radiat Oncol Biol Phys 40:881–887

Patton TJ Jr, Kavanagh JJ, Delclos L et al (1991) Five-year survival in patients given intra-arterial chemotherapy prior to radiotherapy for advanced squamous carcinoma of the cervix and vagina. Gynecol Oncol 42:54–59

Pecorelli S, Beller U, Heintz AP et al (2000) FIGO annual report on the results of treatment in gynecological cancer. J Epidemiol Biostat 24:56

Perez CA, Arneson AN, Galakatos A (1973) Radiation therapy in carcinoma of the vagina. Cancer 31:36–44

Perez CA, Korba A, Sharma S (1977) Dosimetric considerations in irradiation of carcinoma of the vagina. Int J Radiat Oncol Biol Phys 2:639–649

Perez CA, Bedwinek JM, Breaux SR (1983) Patterns of failure after treatment of gynecologic tumors. Cancer Treat Rep 2:217

Perez CA, Camel HM, Galakatos AE et al (1988) Definitive irradiation in carcinoma of the vagina: long-term evaluation and results. Int J Radiat Oncol Biol Phys 15:1283–1290

Perez CA, Slessinger ED, Grigsby PW (1990) Design of an afterloading vaginal applicator (MIRALVA). Int J Radiat Oncol Biol Phys 18:1503–1508

Perez CA, Grigsby PW, Garipagaoglu M et al (1999) Factors affecting long-term outcome of irradiation in carcinoma of the vagina. Int J Radiat Oncol Biol Phys 44:37–45

Peters WA, Kumar NB, Morley GW (1985) Carcinoma of the vagina. Factors influencing treatment outcome. Cancer 55:892–897

Petrilli ES, Townsend DE, Morrow CP et al (1980) Vaginal intraepithelial neoplasia: biologic aspects and treatment with topical 5-fluorouracil and the carbon dioxide laser. Am J Obstet Gynecol 138:321–328

Pingley S, Shrivastava SK, Sarin R et al (2000) Primary carcinoma of the vagina: Tata Memorial Hospital experience. Int J Radiat Oncol Biol Phys 46:101–108

Piver MS, Barlow JJ, Tsukada Y et al (1979) Postirradiation squamous cell carcinoma in situ of the vagina: treatment by topical 20% 5-fluorouracil cream. Am J Obstet Gynecol 135:377–389

Plentl AA, Friedman EA (1971) Lymphatic system of the female genitalia. In: Plentl AA, Friedman EA (eds) The morphologic basis of oncologic diagnosis and therapy. Saunders, Philadelphia, pp 51–74

Prempree T, Amommam R (1985) Radiation therapy of primary carcinoma of the vagina. Acta Radiol Oncol 24:51–56

Puthawala A, Syed AM, Nalick R et al (1983) Integrated external and interstitial radiation therapy for primary carcinoma of the vagina. Obstet Gynecol 62:367–372

Randall ME, Evans L, Greven KM et al (1993) Interstitial re-irradiation for recurrent gynecological malignancies: results and analysis of prognostic factors. Gynecol Oncol 48:23–31

Ratliff CR, Gershenson DM, Morris M et al (1996) Sexual adjustment of patients undergoing gracilis myocutaneous flap vaginal reconstruction in conjunction with pelvic exenteration. Cancer 78:2229–2235

Reddy S, Saxena VS, Reddy S et al (1991) Results of radiotherapeutic management of primary carcinoma of the vagina. Int J Radiat Oncol Biol Phys 21:1041–1044

Reid GC, Schmidt RW, Roberts JA et al (1989) Primary melanoma of the vagina. A clinicopathologic analysis. Obstet Gynecol 74:190–199

Riva C Fabbri A Facco C et al (1997) Primary serous papillary adenocarcinoma of the vagina: a case report. Int J Gynecol Pathol 6:286

Robboy SJ, Herbst AL, Scully RE (1974) Clear cell adenocarcinoma of the vagina and cervix in young females: analysis of 37 tumors that persisted or recurred after primary therapy. Cancer 34:606–614

Robboy SJ, Scully RE, Welch WR et al (1977) Intrauterine diethylstilbestrol exposure and its consequences: pathologic characteristics of vaginal adenosis, clear cell adenocarcinoma and related lesions. Arch Pathol Lab Med 101:1

Robboy SJ Welch WR Young RH et al (1982) Topographic relation of cervical ectropion and vaginal adenosis to clear cell adenocarcinoma. Obstet Gynecol 60:546–551

Robboy SJ, Young RH, Welch WR et al (1984) Atypical vaginal adenosis and cervical ectropion: association with clear cell adenocarcinoma in diethylstilbestrol-exposed offspring. Cancer 54:869–875

Roberts WS, Hoffman MS, Kavanagh JJ et al (1991) Further experience with radiation therapy and concomitant intravenous chemotherapy in advanced carcinoma of the lower female genital tract. Gynecol Oncol 43:233–236

Robinson JB, Sun CC, Bodurka-Bevers D et al (2000) Cavitational ultrasonic surgical aspiration for the treatment of vaginal intraepithelial neoplasia. Gynecol Oncol 78:235–242

Rose PG, Bundy BN, Watkins EB et al (1999) Concurrent cisplatin-based radiotherapy and chemotherapy for locally advanced cervical cancer. N Engl J Med 340:1144–1153

Rubin P, Casarett GW (1986) The female tract genital. In: Rubin P, Casarett GW, (eds) Clinical radiation pathology. Saunders, Philadelphia, pp 396–342

Rubin SC, Young J, Mikuta JJ (1985) Squamous carcinoma of the vagina: treatment, complications and long-term follow-up. Gynecol Oncol 20:346–353

Russell AH, Koh WJ, Markette K et al (1987) Radical reirradiation for recurrent or second primary carcinoma of the female reproductive tract. Gynecol Oncol 27:226–232

Rutkowski T, Bialas B, Rembielak A et al (2002) Efficacy and toxicity of MDR versus HDR brachytherapy for primary vaginal cancer. Neoplasma 49:197–200

Sedlis A, Robboy SJ (1987) Diseases of the vagina. In: Kurman RJ (ed) Blaustein's pathology of the female genital tract, 3rd edn. Springer, Berlin Heidelber New York, pp 98–140

Senekjian EK, Frey KW, Anderson D et al (1987) Local therapy in stage I clear cell adenocarcinoma of the vagina. Cancer 60:1319–1324

Senekjian EK, Frey KW, Stone C et al (1988) An evaluation of stage II vaginal clear cell adenocarcinoma according to substages. Gynecol Oncol 31:56–64

Senekjian EK, Frey K, Herbst AL (1989) Pelvic exenteration in clear cell adenocarcinoma of the vagina and cervix. Gynecol Oncol 34:413–416

Slessinger ED, Perez CA, Grigsby PW et al (1992) Dosimetry and dose specification for a new gynecological brachytherapy applicator. Int J Radiat Oncol Biol Phys 22:1117–1124

Spanos WJ, Wasserman T, Meoz R et al (1987) Palliation of advanced pelvic malignant disease with large fraction pelvic radiation and misonidazole: Final report of RTOG phase I/II study. Int J Radiat Oncol Biol Phys 13:1479–1482

Spanos WJ, Guse C, Perez CA et al (1989) Phase II study of multiple daily fractionations in the palliation of advanced pelvic malignancies. Preliminary report of the RTOG 85-02. Int J Radiat Oncol Biol Phys 17:659–662

Spanos WJ, Perez CA, Marcus S et al (1993) Effect of rest interval on tumor and normal tissue response. A report of phase III study of accelerated split-course palliative radiation for advanced pelvic malignancies (RTOG 85-02). Int J Radiat Oncol Biol Phys 25:399–403

Spanos WJ, Clery M, Perez CA et al (1994) Late effect of multiple daily fraction palliation schedule for advanced pelvic malignancies (RTOG 85-02). Int J Radiat Oncol Biol Phys 29:961–967

Spirtos NM, Doshi BP, Kapp DS et al (1989) Radiation therapy for primary squamous cell carcinoma of the vagina: Stanford University experience. Gynecol Oncol 35:20–26

Stafl A, Wilkinson EJ, Mattingly R (1977) Laser treatment of cervical and vaginal neoplasia. Am J Obstet Gynecol 128:128–134

Stock RG, Mychalczak B, Asmstrong JG et al (1992) The importance of the brachytherapy technique in the management of primary carcinoma of the vagina. Int J Radiat Oncol Biol Phys 24:747–753

Stock RG, Chen ASJ, Seski J (1995) A 30-year experience in the management of primary carcinoma of the vagina: analysis of prognostic factors and treatment modalities. Gynecol Oncol 56:45–52

Stock RG, Chen K, Terk M (1997) A new technique for performing Syed-Neblett template interstitial implants for gynecological malignancies using transrectal-ultrasound guidance. Int J Radiat Oncol Biol Phys 37:819–825

Sun SS, Chen TC, Yen RF et al (2001) Value of whole body 18F-fluoro-2-deoxyglucose positron emission tomography in the evaluation of recurrent cervical cancer. Anticancer Res 21:2957–2961

Syed AMN, Puthawala AA, Neblett D et al (1986) Transperineal interstitial-intracavitary "Syed-Neblett" applicator in the treatment of carcinoma of the uterine cervix. Endocuriether Hypertherm Oncol 2:1–13

Tabata T, Takeshima N, Nishida H et al (2002) Treatment failure in vaginal cancer. Gynecol Oncol 84:309–314

Tarraza MH Jr, Muntz H, Decain M et al (1991) Patterns of recurrence of primary carcinoma of the vagina. Eur J Gynecol Oncol 12:89–92

Tavassoli FA, Norris HJ (1979) Smooth muscle tumors of the vagina. Obstet Gynecol 53:689–693

Thigpen JT, Blessing JA, Homesley HD et al (1986) Phase II trial of cisplatin in advanced or recurrent cancer of the vagina: a Gynecologic Oncology Group study. Gynecol Oncol 23:101–104

Tjalma W, Monaghan JM, de Barros lopes A et al (2001) The role of surgery in invasive carcinoma of the vagina. Gynecol Oncol 81:360–365

Townsend DE, Levine RU, Crum CP et al (1982) Laser therapy of vaginal intra-epithelial neoplasia with the carbon dioxide laser. Am J Obstet Gynecol 143:565–568

Tyree WC, Cardenes H, Randall M et al (2002) High-dose rate brachytherapy for vaginal cancer: learning from treatment complications. Int J Gynecol Cancer 12:27–31

Urbanski K, Kojs Z, Reinfuss M et al (1996) Primary invasive vaginal carcinoma treated with radiotherapy: analysis of prognostic factors. Gynecol Oncol 60:16–21

Wang X, Cai S, Ding Y et al (1998) Treatment of late recurrent vaginal malignancy after initial radiotherapy for carcinoma of the cervix: an analysis of 73 cases. Gynecol Oncol 69:125–129

Whitney CW, Sause W, Bundy BN et al (1999) Randomized comparison of fluorouracil plus cisplatin versus hydroxyurea as an adjunct to radiation therapy in stage IIB-IVA carcinoma of the cervix with negative para-aortic lymph nodes: a Gynecologic Oncology Group and Southwest Oncology Group Study. J Clin Oncol 17:1339–1348

Woodruff JD, Parmley THE, Julian CG (1975) Topical 5-fluorouracil in the treatment of vaginal carcinoma in situ. Gynecol Oncol 3:124–132

Zaino RJ, Robboy SJ, Kurman RJ (2002) Diseases of the vagina. In: Kurman RJ (ed) Blaustein's pathology of the female genital tract, 5th edn. Springer, Berlin Heidelberg NewYork, pp 151–206

28 Prostate

Jeff M. Michalski, Gregory S. Merrick, and Sten Nilsson

CONTENTS

28.1   Anatomy 687
28.2   Natural History 688
28.2.1   Local Growth Patterns 688
28.2.2   Regional Lymph Node Involvement 690
28.2.3   Distant Metastases 694
28.3   Prognostic Factors 694
28.3.1   Tumor Stage, Pretreatment PSA, and
          Histological Features 694
28.3.2   Tumor Volume 695
28.4   Radiation Therapy Techniques 695
28.4.1   General Treatment Guidelines 695
28.4.2   External Irradiation 696
28.4.3   Conventional Radiation Therapy Techniques 696
28.4.3.1 Volume Treated 696
28.4.3.2 Beam Energy and Dose Distribution 698
28.4.3.3 Standard Tumor Doses
          (Non-Conformal Treatment Planning) 699
28.4.4   Conformal Radiation Therapy Simulation and
          Treatment Planning 699
28.4.4.1 Patient Immobilization and Positioning 699
28.4.4.2 Clinical Target Volume Definition 700
28.4.4.3 Planning Target Volume Definition 702
28.4.4.4 Organs-at-Risk Definition 706
28.4.4.5 Beam Selection and Shaping 706
28.4.5   Intensity-Modulated Radiation Therapy 709
28.4.6   Dose Prescription for 3D CRT or IMRT 711
28.4.7   Considerations for Postoperative Irradiation 713
28.4.8   Postoperative Radiation Therapy Techniques 714
28.5   Prostate Brachytherapy 715
28.5.1   Introduction 715
28.5.1.1 Patient Selection 715
28.5.1.2 Prostate Size 715
28.5.1.3 Transition Zone 716
28.5.1.4 Median Lobe Hyperplasia 716
28.5.1.5 International Prostate Symptom Score 717
28.5.1.6 Prostatitis 717
28.5.1.7 Pubic Arch Interference 717

28.5.1.8 Transurethral Resection of the Prostate 717
28.5.1.9 Tobacco 718
28.5.1.10 Inflammatory Bowel Disease 718
28.5.1.11 Adverse Pathological Features 718
28.5.1.12 Prostatic Acid Phosphatase 718
28.5.2   Brachytherapy Planning 719
28.5.3   Post-Implant Evaluation 719
28.5.3.1 Isotope 720
28.5.4   Supplemental Therapies 721
28.5.4.1 External-Beam Radiation Therapy 721
28.5.4.2 Androgen Deprivation Therapy 721
28.5.5   PSA Spikes 721
28.5.6   Approaches to Minimize Morbidity 721
28.5.6.1 Urinary 721
28.5.5.2 Rectal Morbidity 722
28.5.5.3 Erectile Dysfunction 723
28.6   High-Dose-Rate Brachytherapy 724
28.6.1   Introduction 724
28.6.2   Technological Aspects 724
28.6.3   Radiobiological and Dose Fractionation Aspects 725
28.6.4   Clinical Outcome Data from HDR-BT Plus EBRT 725
28.6.5   Health-Related Quality-of-Life Data After
          HDR-BT Plus EBRT 727
28.6.6   HDR-BT Monotherapy: Initial Experience 727
28.6.7   The Role of Neoadjuvant and Concomitant
          Endocrine Therapy 728
28.6.8   Discussion/Future Aspects 729
          References 729

J. M. Michalski, MD, MBA
Associate Professor in Radiation Oncology, Department of Radiation Oncology, Washington University School of Medicine, 4921 Parkview Place, St. Louis, MO 63110, USA
G. S. Merrick, MD
Schiffler Cancer Center and Wheeling Jesuit University, Wheeling, WV 26003, USA
S. Nilsson, MD, PhD
Profesor, Institution for Oncology Pathology, Karolinska University Hospital, 17176 Stockholm, Sweden

28.1
Anatomy

The prostate gland is a walnut-shaped solid organ that surrounds the male urethra between the base of the bladder and the urogenital diaphragm, and weighs about 20 g. The prostate is attached anteriorly to the pubic symphysis by the puboprostatic ligament. It is separated from the rectum posteriorly by Denonvilliers' fascia (retrovesical septum), which attaches above to the peritoneum and below to the urogenital diaphragm. The seminal vesicles and the vas deferens pierce the posterosuperior aspect of the gland and enter the urethra at the verumontanum (Fig. 28.1). The lateral margins of the prostate, usu-

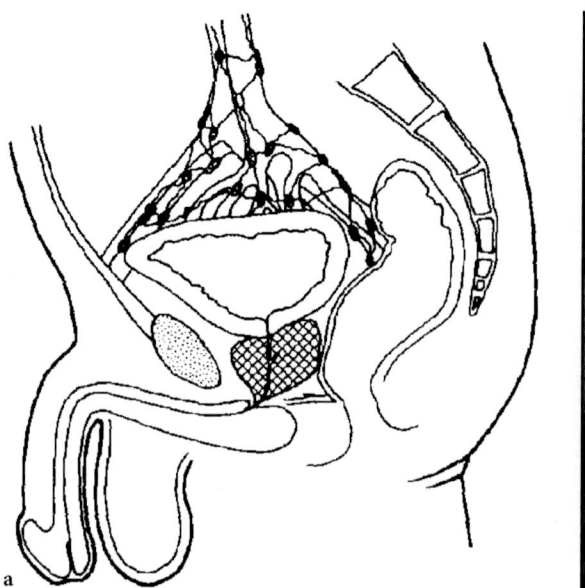
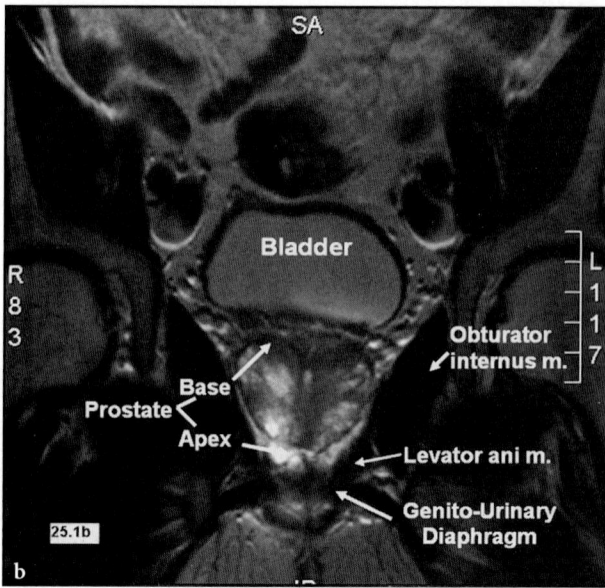

Fig. 28.1. a Sagittal diagram of pelvis illustrates anatomic relationships of the prostate. **b** Coronal MRI of prostate illustrates close relationship of prostate and bladder. (From PEREZ 1998)

ally delineated against the levator ani muscles, form the lateral prostatic sulci.

In young men the prostate gland can be divided into four distinct zones (MCLAUGHLIN et al. 2005). The central zone (CZ) surrounds the ejaculatory ducts. The anterior fibromuscular stroma (AFS) is an anterior band of fibromuscular tissue contiguous with the bladder muscle and external sphincter. The transition zone (TZ) is the central component of the prostate that tends to hypertrophy with age. The hypertrophied TZ will compress the CZ and periurethral glandular tissue, making it nearly impossible to identify these zones on ultrasound, CT, or MRI. The hypertrophied TZ is often called the central gland and is readily recognized on MRI (VILLEIRS et al. 2005). While in young men the peripheral zone (PZ) may make up 70% of the prostate tissue, it becomes condensed by an enlarged TZ in men with benign prostatic hypertrophy.

MYERS et al. (1987), in a study of 64 gross prostatectomy specimens, noted variations in the shape and exact location of the prostatic apex and pointed out that the configuration of the external striated urethral sphincter was related to the shape of the prostatic apex. Two basic prostatic shapes were recognized distinguished by the presence or absence of an anterior apical notch, depending on the degree of lateral lobe development and the position of its anterior commissure. Observations by these authors that the urethral sphincter is a striated muscle in contact with the urethra from the base of the bladder

to the perineal membrane corroborates the previous description by OELRICH (1980), who pointed out that there is no distinct superior fascia of the so-called urogenital diaphragm separating the sphincter muscle from the prostate. The anatomy of the male pelvis at right angles to the perineal membrane, through the membranous urethra, is illustrated in Fig. 28.2.

28.2
Natural History

28.2.1
Local Growth Patterns

Almost all clinically significant prostatic carcinomas develop in the peripheral zone of the prostate, whereas benign prostatic hyperplasia arises predominantly from the central (periurethral) portions (MCNEAL 1969). In recent years, more attention is being paid to tumor arising in the transitional zone. Transition zone cancers can grow relatively large before extending beyond the confines of the fibromuscular stroma. These tumors can produce substantial elevations of prostate-specific antigen (PSA). In a series of 148 cases of TZ prostate cancer, 70% were clinical stage T1c with a preoperative PSA of greater than 10 ng/ml in nearly two-thirds of the patients. On pathology review of the radical pros-

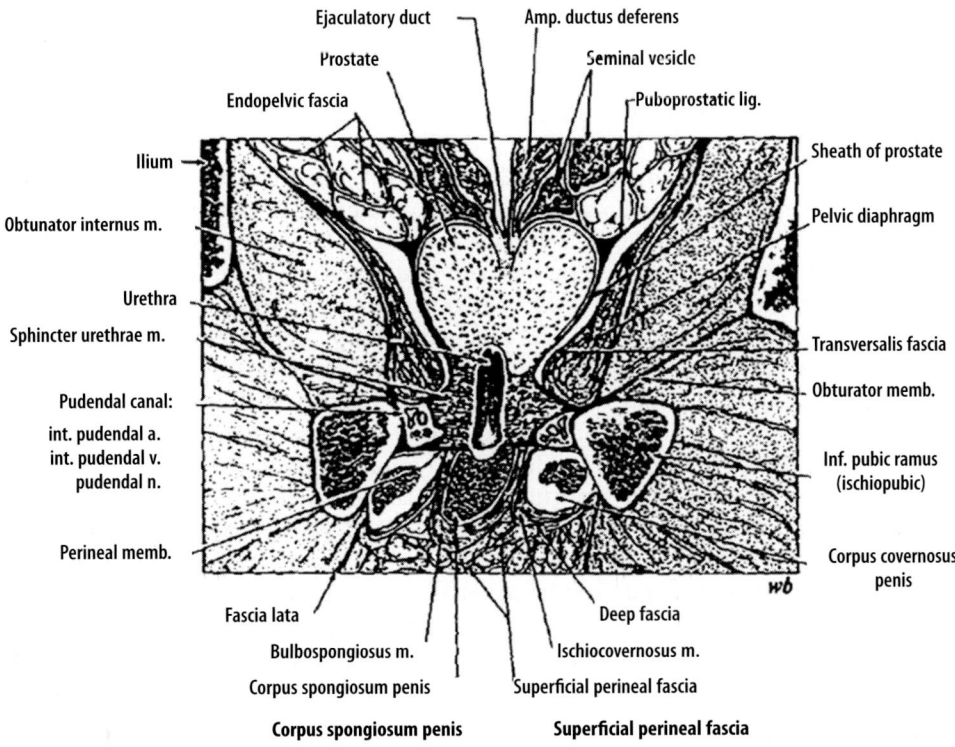

Ejaculatory duct — Amp. ductus deferens
Prostate Seminal vesicle
Endopelvic fascia Puboprostatic lig.
Ilium → Sheath of prostate
Obtunator internus m. Pelvic diaphragm
Urethra Transversalis fascia
Sphincter urethrae m. Obturator memb.
Pudendal canal:
int. pudendal a. Inf. pubic ramus
int. pudendal v. (ischiopubic)
pudendal n.
Perineal memb. Corpus covernosus penis
Fascia lata Deep fascia
Bulbospongiosus m. Ischiocovernosus m.
Corpus spongiosum penis Superficial perineal fascia
Corpus spongiosum penis **Superficial perineal fascia**

Fig. 28.2. Frontal section of male pelvis at right angles to the perineal membrane. (From OELRICH 1980)

tatectomy specimens 80% of cancers originating in the TZ had organ-confined disease (NOGUCHI et al. 2000).

BRESLOW et al. (1977) found that 64% of 350 carcinoma were present in a slice taken 5 mm from the distal end of the prostate; therefore, the urethra must be transected distal to the prostate to avoid leaving prostatic cancer behind (BLENNERHASSETT and VICKERY 1966; McNEAL 1969). This is an important detail in the design of external irradiation portals or in brachytherapy of prostate cancer.

JEWETT (1980) reported that multiple foci of tumor were found throughout the prostate in 77% of prostatectomy specimens. ANDRIOLE et al. (1992) described bilateral lobe pathological involvement in 13 of 15 patients (87%) with clinical stage-A1 (T1a) cancer. In a series of 486 patients, WISE et al. (2002) reported that 83% of the cancers were multifocal; therefore, the entire gland (with a margin) had to be treated.

As the tumor grows, it may extend into and through the capsule of the gland, invade seminal vesicles and periprostatic tissues, and later involve the bladder neck or the rectum. Tumor may invade the perineural spaces, the lymphatics, and the blood vessels producing lymphatic or distant metasta-

ses. The incidence of microscopic tumor extension beyond the capsule of the gland (at time of radical prostatectomy) in patients with clinical stages A2 or B ranges from 18 to 57% (CATALONA and SMITH 1994; VILLERS et al. 1991).

OESTERLING et al. (1994), in an analysis of patients with stage-T1c disease treated with radical prostatectomy, noted that 53% had pathologically organ-confined tumors; 35% had extracapsular extension, and 9% had seminal vesicle invasion. In the latter group, 66% of patients had positive surgical margins, an incidence comparable to that of clinical stage-T2 tumors. In a similar group of patients with stage-T1c tumors EPSTEIN et al. (1994) found that 34% had established extracapsular extension, 6% seminal vesicle invasion, and 17% positive surgical margins.

STONE et al. (1995) reported none of 13 patients with prostate-specific antigen (PSA) <4 ng/ml, 11 of 99 patients (11%) with PSA of 4.1–20 ng/ml, and 12 of 45 patients (27%) with PSA >20 ng/ml having positive seminal vesicle biopsy. None of the patients with stages T1a–T1c had positive seminal vesicle biopsies, compared with 2 of 33 (6%) with stage T2a, 14 of 80 (17.5%) with stage T2b, and 7 of 23 (30%) with stage T2c tumors. Seminal vesicle involvement has been observed in 10% of patients with stage-A2 tumors to 30% of patients with stage-B2 lesions (CATALONA and BIGG 1990).

STOCK et al. (1995), in 120 patients with clinical stage-T1b to stage-T2c carcinoma of the prostate, in whom transrectal ultrasound-guided needle biopsies of the seminal vesicles were performed, reported on 99 who also underwent laparoscopic lymph node dissection. The incidence of seminal vesicle involvement was correlated with PSA level, Gleason score, and clinical stage (Table 28.1). When PSA level and Gleason score were correlated, none of the patients with Gleason scores of 4 or lower showed seminal vesicle involvement, regardless of PSA level. Patients with Gleason scores of 5 and 6 and PSA of 4–20 ng/ml had 10–11%, and those with PSA >20 ng/ml had 14% positive seminal vesicle biopsy. With Gleason scores of 7 or higher, 25% of patients with PSA of 4–20 ng/ml and 53% of those with PSA >20 ng/ml had seminal vesicle involvement.

D'AMICO et al. (1995), in a pathological evaluation of 347 radical prostatectomy specimens, reported none of 38 patients with PSA <4 ng/ml or less having seminal vesicle involvement, in contrast to 6% of 144 patients with PSA of 4–10 ng/ml, 11% of 101 with PSA of 10–20 ng/ml, 36% of 45 with PSA of 20–40 ng/ml, and 42% of 19 with PSA >40 ng/ml. The incidence of positive surgical margins was 11, 20, 33, 56, and 33%, respectively.

ROACH (1993) proposed the following formula based on analysis of radical prostatectomy specimens to estimate the probability of seminal vesicle involvement:

$$SVI = PSA + (Gleason\ score\ minus\ 6) \times 10$$

Table 28.1. Correlation of prostate-specific antigen (*PSA*) levels, Gleason score, and clinical stage with positive seminal vesicle biopsy specimen. (From STOCK et al. 1995)

Parameter	No. of positive seminal vesicle biopsy specimens	p value
PSA ≤10	3 (6)	
PSA >10	15 (21)	0.02
PSA ≤20	8 (9)	
PSA >20	10 (37.5)	0.005
Grade <7	6 (7)	
Grade ≥7	12 (37.5)	<0.0001
T1a–T2a	2 (5)	
T2b–T2c	16 (20)	0.03

Values in parentheses are percentages

28.2.2
Regional Lymph Node Involvement

Tumor size and degree of differentiation affect the tendency of prostatic carcinoma to metastasize to regional lymphatics. The incidence of lymph node metastases was as high as 12–28% in clinical stage-T2 disease in the pre-PSA era (MIDDLETON 1988). FOWLER and WHITMORE (1981) reported that 40% of 300 patients with apparently localized prostate cancer had pelvic lymph node metastases upon surgery. In the era of PSA screening, there has been a decrease in the incidence of pelvic lymph node metastases. Modern series report less than a 10% incidence of lymph node involvement at radical prostatectomy (DANELLA et al. 1993).

OHORI et al. (1995), in 478 patients treated with radical prostatectomy, reported no pelvic lymph node metastases in 70 patients with stages T1a,b, 1 of 43 (2%) in patients with stage T1c, 5 of 96 (5%) with stage T2a, and 19 of 269 (7%) with stages T2b,c. The incidence of seminal vesicle invasion was 6, 11, 5, and 17%, respectively.

In a review of 2439 patients treated with radical prostatectomy, PISANSKY et al. (1996b) reported positive pelvic nodes in 12 of 457 (2.6%) with stages T1a–c, 15 of 456 (3.3%) with stage T2a, 130 of 1206 (10.8%) with stage T2b,c, and 81 of 320 (25%) with stage-T3 tumors.

STOCK et al. (1995), in 99 patients who underwent laparoscopic lymph node dissection (Table 28.2),

Table 28.2. Correlation of PSA levels, Gleason score, clinical stage, and status of positive seminal vesicle biopsy specimen with incidence of positive pelvic lymph nodes. (From STOCK et al. 1995)

Parameter	No. of positive seminal vesicle biopsy specimens	p value
PSA ≤10	2 (6)	
PSA >10	7 (15)	0.3
PSA ≤20	2 (3)	
PSA >20	7 (24)	0.003
Grade <7	1 (2)	
Grade ≥7	8 (35)	<0.0001
T1a–T2a	0 (0)	
T2b–T2c	9 (18)	0.03
Positive seminal vesicle biopsy specimen	9 (50)	
Negative seminal vesicle biopsy specimen	0 (0)	<0.0001

Values in parentheses are percentages

correlated incidence of positive nodes with PSA, Gleason score, stage, and involvement of seminal vesicles. None of the patients with a Gleason score of 4 or lower, even with >20 ng/ml, had positive pelvic lymph nodes, and 8% in the group with Gleason scores of 5 or 6 had PSA levels of 4–10 ng/ml and had positive nodes; however, the incidence of positive lymph nodes increased significantly (24%) in patients with PSA >20 ng/ml.

BLUESTEIN et al. (1994), NARAYAN et al. (1994), PARTIN et al. (1993, 1997, 2001), and SPEVACK et al. (1996) have offered comparable models based on pathological data that may predict risk for lymph node metastases, to decide whether the patient should be subjected to a staging lymphadenectomy (including laparoscopic technique) or considered for irradiation of the pelvic lymph nodes.

STONE et al. (1995) reported that none of 11 patients with PSA <4 ng/ml, 4 of 77 (9%) with PSA of 4–20 ng/ml, and 10 of 42 (24%) with PSA >20 ng/ml had positive nodes. When correlated with clinical stage, none of the patients with stage T1b,c or stage T2a had positive nodes, compared with 10 of 69 (15%) with T2b and 4 of 17 (24%)

with T2c tumors. Eleven of 23 patients (48%) with positive seminal vesicles also had positive nodes, compared with 3 of 107 (3%) with negative seminal vesicle biopsy.

PARTIN et al. (1997, 2001) analyzed surgical data from three academic institutions to develop validated nomograms that predict lymph node involvement based on clinical stage, preoperative PSA and Gleason score. These nomograms are useful in predicting the risk of lymph node involvement and aid in the decision to use elective pelvic lymph node radiation (Table 28.3).

ROACH (1993) suggested a revised formula based on pathological findings in prostatectomy specimens incorporating clinical stage, to estimate the incidence of metastatic pelvic lymph nodes:

Risk of node positive =
2/3 PSA + [(GS-6) + TG-1.5] × 10,

where GS is Gleason score and TG is clinical tumor group; TG is as follows: TG 1 (stages T1c and T2a) is assigned a value of 1, TG 2 (T1b and T2b) is given a value of 2, and TG 3 (T2c and T3) is assigned a value of 3.

Table 28.3a. Clinical stage-T1c disease (nonpalpable, PSA elevated). (From PARTIN et al. 2001)

PSA range (ng/ml)	Pathological stage	Gleason score				
		2–4	5–6	3+4 = 7	4+3 = 7	8–10
0–2.5	Organ confined	95 (89–99)	90 (88–93)	79 (74–85)	71 (62–79)	66 (54–76)
	Extraprostatic extension	5 (1–11)	9 (7–12)	17 (13–23)	25 (18–34)	28 (20–38)
	Seminal vesicle (+)	–	0 (0–1)	2 (1–5)	2 (1–5)	4 (1–10)
	Lymph node (+)	–	–	1 (0–2)	1 (0–4)	1 (0–4)
2.6–4.0	Organ confined	92 (82–98)	84 (81–86)	68 (62–74)	58 (48–67)	52 (41–63)
	Extraprostatic extension	8 (2–18)	15 (13–18)	27 (22–33)	37 (29–46)	40 (31–50)
	Seminal vesicle (+)	–	1 (0–1)	4 (2–7)	4 (1–7)	6 (3–12)
	Lymph node (+)	–	–	1 (0–2)	1 (0–3)	1 (0–4)
4.1–6.0	Organ confined	90 (78–98)	80 (78–83)	63 (58–68)	52 (43–60)	46 (36–56)
	Extraprostatic extension	10 (2–22)	19 (16–21)	32 (27–36)	42 (35–50)	45 (36–54)
	Seminal vesicle (+)	–	1 (0–1)	3 (2–5)	3 (1–6)	5 (3–9)
	Lymph node (+)	–	0 (0–1)	2 (1–3)	3 (1–5)	3 (1–6)
6.1–10.0	Organ confined	87 (73–97)	75 (72–77)	54 (49–59)	43 (35–51)	37 (28–46)
	Extraprostatic extension	13 (3–27)	23 (21–25)	36 (32–40)	47 (40–54)	48 (39–57)
	Seminal vesicle (+)	–	2 (2–3)	8 (6–11)	8 (4–12)	13 (8–19)
	Lymph node (+)	–	0 (0–1)	2 (1–3)	2 (1–4)	3 (1–5)
>10.0	Organ confined	80 (61–95)	62 (58–64)	37 (32–42)	27 (21–34)	22 (16–30)
	Extraprostatic extension	20 (5–39)	33 (30–36)	43 (38–48)	51 (44–59)	50 (42–59)
	Seminal vesicle (+)	–	4 (3–5)	12 (9–17)	11 (6–17)	17 (10–25)
	Lymph node (+)	–	2 (1–3)	8 (5–11)	10 (5–17)	11 (5–18)

Table 28.3b. Clinical stage-T2a disease (palpable, <50% of one lobe). *PSA* prostate-specific antigen. (From Partin et al. 2001)

PSA range (ng/ml)	Pathological stage	Gleason score				
		2–4	5–6	3+4 = 7	4+3 = 7	8–10
0–2.5	Organ confined	91 (79–98)	81 (77–85)	64 (56–71)	53 (43–63)	47 (35–59)
	Extraprostatic extension	9 (2–21)	17 (13–21)	29 (23–36)	40 (30–49)	42 (32–53)
	Seminal vesicle (+)	–	1 (0–2)	5 (1–9)	4 (1–9)	7 (2–16)
	Lymph node (+)	–	0 (0–1)	2 (0–5)	3 (0–8)	3 (0–9)
2.6–4.0	Organ confined	85 (69–96)	71 (66–75)	50 (43–57)	39 (30–48)	33 (24–44)
	Extraprostatic extension	15 (4–31)	27 (23–31)	41 (35–48)	52 (43–61)	53 (44–63)
	Seminal vesicle (+)	–	2 (1–3)	7 (3–12)	6 (2–12)	10 (4–18)
	Lymph node (+)	–	0 (0–1)	2 (0–4)	2 (0–6)	3 (0–8)
4.1–6.0	Organ confined	81 (63–95)	66 (62–70)	44 (39–50)	33 (25–41)	28 (20–37)
	Extraprostatic extension	19 (5–37)	32 (28–36)	46 (40–52)	56 (48–64)	58 (49–66)
	Seminal vesicle (+)	–	1 (1–2)	5 (3–8)	5 (2–8)	8 (4–13)
	Lymph node (+)	–	1 (0–2)	4 (2–7)	6 (3–11)	6 (2–12)
6.1–10.0	Organ confined	76 (56–94)	58 (54–61)	35 (30–40)	25 (19–32)	21 (15–28)
	Extraprostatic extension	24 (6–44)	37 (34–41)	49 (43–54)	58 (51–66)	57 (48–65)
	Seminal vesicle (+)	–	4 (3–5)	13 (9–18)	11 (6–17)	17 (11–26)
	Lymph node (+)	–	1 (0–2)	3 (2–6)	5 (2–8)	5 (2–10)
>10.0	Organ confined	65 (43–89)	42 (38–46)	20 (17–24)	14 (10–18)	11 (7–15)
	Extraprostatic extension	35 (11–57)	47 (43–52)	49 (43–55)	55 (46–64)	52 (41–62)
	Seminal vesicle (+)	–	6 (4–8)	16 (11–22)	13 (7–20)	19 (12–29)
	Lymph node (+)	–	4 (3–7)	14 (9–21)	18 (10–27)	17 (9–29)

Table 28.3c. Clinical stage-T2b disease (palpable, >50 of one lobe, not on both lobes). (From Partin et al. 2001)

PSA range (ng/ml)	Pathological stage	Gleason score				
		2–4	5–6	3+4 = 7	4+3 = 7	8–10
0–2.5	Organ confined	88 (73–97)	75 (69–81)	54 (46–63)	43 (33–54)	37 (26–49)
	Extraprostatic extension	12 (3–27)	22 (17–28)	35 (28–43)	45 (35–56)	46 (35–58)
	Seminal vesicle (+)	–	2 (0–3)	6 (2–12)	5 (1–11)	9 (2–20)
	Lymph node (+)	–	1 (0–2)	4 (0–10)	6 (0–14)	6 (0–16)
2.6–4.0	Organ confined	80 (61–95)	63 (57–69)	41 (33–48)	30 (22–39)	25 (17–34)
	Extraprostatic extension	20 (5–39)	34 (28–40)	47 (40–55)	57 (47–67)	57 (46–68)
	Seminal vesicle (+)	–	2 (1–4)	9 (4–15)	7 (3–14)	12 (5–22)
	Lymph node (+)	–	1 (0–2)	3 (0–8)	4 (0–12)	5 (0–14)
4.1–6.0	Organ confined	75 (55–93)	57 (52–63)	35 (29–40)	25 (18–32)	21 (14–29)
	Extraprostatic extension	25 (7–45)	39 (33–44)	51 (44–57)	60 (50–68)	59 (49–69)
	Seminal vesicle (+)	–	2 (1–3)	7 (4–11)	5 (3–9)	9 (4–16)
	Lymph node (+)	–	2 (1–3)	7 (4–13)	10 (5–18)	10 (4–20)
6.1–10.0	Organ confined	69 (47–91)	49 (43–54)	26 (22–31)	19 (14–25)	15 (10–21)
	Extraprostatic extension	31 (9–53)	44 (39–49)	52 (46–58)	60 (52–68)	57 (48–67)
	Seminal vesicle (+)	–	5 (3–8)	16 (10–22)	13 (7–20)	19 (11–29)
	Lymph node (+)	–	2 (1–3)	6 (4–10)	8 (5–14)	8 (4–16)
>10.0	Organ confined	57 (35–86)	33 (28–38)	14 (11–17)	9 (6–13)	7 (4–10)
	Extraprostatic extension	43 (14–65)	52 (46–56)	47 (40–53)	50 (40–60)	46 (36–59)
	Seminal vesicle (+)	–	8 (5–11)	17 (12–24)	13 (8–21)	19 (12–29)
	Lymph node (+)	–	8 (5–12)	22 (15–30)	27 (16–39)	27 (14–40)

Table 28.3d. Clinical stage-T2c disease (palpable on both lobes). (From Partin et al. 2001)

PSA range (ng/ml)	Pathological stage	Gleason score				
		2–4	5–6	3+4 = 7	4+3 = 7	8–10
0–2.5	Organ confined	86 (71–97)	73 (63–81)	51 (38–63)	39 (26–54)	34 (21–48)
	Extraprostatic extension	14 (3–29)	24 (17–33)	36 (26–48)	45 (32–59)	47 (33–61)
	Seminal vesicle (+)	–	1 (0–4)	5 (1–13)	5 (1–12)	8 (2–19)
	Lymph node (+)	–	1 (0–4)	6 (0–18)	9 (0–26)	10 (0–27)
2.6–4.0	Organ confined	78 (58–94)	61 (50–70)	38 (27–50)	27 (18–40)	23 (14–34)
	Extraprostatic extension	22 (6–42)	36 (27–45)	48 (37–59)	57 (44–70)	57 (44–70)
	Seminal vesicle (+)	–	2 (1–5)	8 (2–17)	6 (2–16)	10 (3–22)
	Lymph node (+)	–	1 (0–4)	5 (0–15)	7 (0–21)	8 (0–22)
4.1–6.0	Organ confined	73 (52–93)	55 (44–64)	31 (23–41)	21 (14–31)	18 (11–28)
	Extraprostatic extension	27 (7–48)	40 (32–50)	50 (40–60)	57 (43–68)	57 (43–70)
	Seminal vesicle (+)	–	2 (1–4)	6 (2–11)	4 (1–10)	7 (2–15)
	Lymph node (+)	–	3 (1–7)	12 (5–23)	16 (6–32)	16 (6–33)
6.1–10.0	Organ confined	67 (45–91)	46 (36–56)	24 (17–32)	16 (10–24)	13 (8–20)
	Extraprostatic extension	33 (9–55)	46 (37–55)	52 (42–61)	58 (46–69)	56 (43–69)
	Seminal vesicle (+)	–	5 (2–9)	13 (6–23)	11 (4–21)	16 (6–29)
	Lymph node (+)	–	3 (1–6)	10 (5–18)	13 (6–25)	13 (5–26)
>10.0	Organ confined	54 (32–85)	30 (21–38)	11 (7–17)	7 (4–12)	6 (3–10)
	Extraprostatic extension	46 (15–68)	51 (42–60)	42 (30–55)	43 (29–59)	41 (27–57)
	Seminal vesicle (+)	–	6 (2–12)	13 (6–24)	10 (3–20)	15 (5–28)
	Lymph node (+)	–	13 (6–22)	33 (18–49)	38 (20–58)	38 (20–59)

At least two modern European series of extended lymph node dissection did discover a significantly increased number of men with lymph node metastases, compared with a similar group who had undergone standard lymph node dissection (Heidenreich et al. 2002; Wawroschek et al. 2003). In both these series, PSA screening was not prevalent in the patient population. The incidence of lymph node metastases discovered by the extended pelvic lymph node dissection was 26.2–26.8%. This finding raises the question as to whether or not the incidence of lymph node metastases is underestimated in modern series.

Clark et al. (2003) performed a randomized trial of standard lymph node dissection vs an extended lymph node dissection in a series of 123 patients. In that trial each patient served as his own control, with one side having a standard dissection and the other side an extended dissection. There was no difference in the rate of lymph node metastases discovered with the extended dissection. The patient population was generally low risk with 72% clinical stage T1c, 84.6% having a PSA of <10 ng/ml and 67% with cancers of Gleason score 6 or less (Clark et al. 2003).

The pattern of lymph node metastases has been described by well-known series. Periprostatic and obturator nodes are involved first, followed by external iliac, hypogastric, common iliac, and periaortic nodes (Pistenma et al. 1979). While the incidence of metastases is less than in the past, the pattern of involvement is unchanged. Wawroschek reported a series of 194 patients who underwent an extended pelvic lymph node dissection. The overall rate of lymph node involvement was 26.8%. Table 28.4 demonstrates the number of node-positive patients who would have been detected, provided only that lymph nodes from different regions had been histologically investigated. These data suggest that 98% of the involved lymph nodes would be covered by fields that encompassed the obturator, external, and internal iliac chains. Only 2% of lymph nodes would be missed if the presacral, pararectal, and paravesical nodes were omitted (Wawroschek et al. 2003).

Prognosis is closely related to the presence of regional lymph node metastases. Patients with positive pelvic lymph node metastasis have a significantly greater probability (>85% at 10 years) of developing distant metastasis than those with nega-

Table 28.4. Percentage of node-positive patients who would have been detected if only lymph nodes from different regions had been histologically investigated (based on the results of sentinel lymphadenectomy in 194 patients with or without additional pelvic lymphadenectomy and with serial sections and immunohistochemistry of all sentinel lymph nodes.) (From WAWROSCHEK et al. 2003)

Regions of lymphadenectomy	Node-positive patients (%)
Obturator fossa, external and internal iliac region, presacral, pararectal, paravesical	100 (93.2–100)
Obturator fossa, external and internal iliac region	98 (89.7–100)
Obturator fossa, internal iliac region	82.7 (69.7–91.8)
Obturator fossa, external iliac region	65.4 (50.9–78)
Obturator fossa	44.2 (30.5–58.7)

tive nodes (<20%; GERVASI et al. 1989). GERVASI et al. (1989) reported that the risk of metastatic disease at 10 years was 31% for patients with negative lymph nodes compared with 83% for those with positive nodes. The risk of dying of prostate cancer was 17 and 57%, respectively.

RUKSTALIS et al. (1996) compiled data from the literature that documented the strong prognostic implications of the number of involved lymph nodes. In general, patients with a single metastatic lymph node have a 5-year survival rate ranging from 60 to 80%, whereas in those with more involved lymph nodes, the 5-year survival is 20–54%.

Metastases to the peri-aortic nodes are seen in 5–25% of patients, depending on tumor stage and histological differentiation (PISTENMA et al. 1979). They are associated with a higher incidence of distant metastases and lower survival (LAWTON et al. 1997).

28.2.3
Distant Metastases

Prostatic carcinoma metastasizes to the skeleton, liver, lungs, and occasionally to the brain or other sites, PEREZ et al. (1988) reported an overall incidence of distant metastases of 20% in stage B, 40% in stage C, and 65% in stage D1.

28.3
Prognostic Factors

28.3.1
Tumor Stage, Pretreatment PSA, and Histological Features

The strongest prognostic indicators in carcinoma of the prostate are clinical stage, pretreatment PSA level, and pathological tumor differentiation (PEREZ 1998). This is a consequence of the more aggressive behavior and greater incidence of lymphatic and distant metastases in the larger and less-differentiated tumors. BASTACKY et al. (1993) noted that perineural invasion on prostate biopsies is correlated with a higher probability of capsular penetration, and BONIN et al. (1997) reported a lower 5-year biochemical failure-free survival rate in the presence of perineural invasion (39 vs 65%; $p = 0.0009$) in patients with PSA <20 ng/ml treated with three-dimensional conformal radiation therapy (3D CRT).

D'AMICO described a three-tiered prognostic risk-group categorization that utilizes pretreatment PSA, biopsy Gleason score, and clinical stage (D'AMICO et al. 1998). Low-risk patients are those with a PSA ≤10 ng/ml and a Gleason Score of 2–6 and stages T1–T2c. Intermediate-risk patients have PSA of 10 ng/ml to ≤20 ng/ml and/or Gleason score 7, and no high-risk features. High-risk patients have PSA >20 ng/ml and/or Gleason score 8–10 and/or stage ≥T3. These risk categories were significantly associated with an increasing rate of biochemical failure using the ASTRO failure definition (1997). In a pooled data set of 4839 patients, KUBAN (2003) reported that pretreatment PSA, Gleason score, radiation dose, tumor stage, and year of treatment were all significant prognostic factors in a multivariate analysis. The risk groups were defined as follows: low risk (group 1), stages T1b, T1c, or T2a, Gleason score ≤6, and PSA ≤10 ng/ml; intermediate risk (group 2), stage T1b, T1c, or T2a, Gleason score ≤7, and PSA >10 ng/ml but ≤20 ng/ml or stage T2b or T2c, Gleason score ≤7, and PSA ≤20 ng/ml; and high risk (group 3), Gleason score 8–10 or PSA >20 ng/ml. The 5-year clinical DFS rate was 78, 66, and 49% for low-, intermediate-, and high-risk patients, respectively.

KATTAN (2000) has developed a nomogram from patients treated at Memorial Sloan-Kettering Cancer Center to predict outcome after 3D conformal radiation therapy that employs clinical parameters including stage, biopsy Gleason score, pretreatment PSA level, and whether neoadjuvant hormone therapy was administered and the radiation dose administered. The nomogram has been validated using clinical data from the Cleveland Clinic.

28.3.2
Tumor Volume

Tumor volume is a powerful predictor for patient outcome. In a series of 151 patients undergoing radical prostatectomy, investigators found that the number positive biopsy sites, tumor bilaterally, and the percentage of biopsy sites positive for disease were all useful predictors of tumor volume in surgical specimens (POULOS et al. 2004). SELEK reported that the percentage positive prostate biopsy (PPPB) was a predictor of post-external beam radiotherapy PSA outcome in clinically localized prostate cancer. The 5-year PSA failure-free survival rate was 79% vs 69% ($p = 0.02$) and the clinical disease-free survival

rate was 97% vs 86% ($p = 0.0004$) for patients with <50% vs ≥50% PPPB (SELEK et al. 2003).

28.4
Radiation Therapy Techniques

28.4.1
General Treatment Guidelines

The treatment volume guidelines used at Washington University are outlined in Table 28.5. Patients are treated according to their risk category. Some patients may be a candidate for either external-beam radiation therapy or brachytherapy. Unless there are contraindications for one type of treatment, the choice of primary treatment modality often depends on patient choice. Elective pelvic lymph node irradiation is used along with neoadjuvant hormone therapy when patients have a risk of lymph node

Table 28.5. Washington University 3D and IMRT Radiation Treatment Volume Guidelines. *EBRT* external-beam radiation therapy, *HDR* high dose rate, *LN* lymph node *SV* seminal vesicle, *CTV* clinical target volume, *PTV* planning target volume.

Risk group	Prognostic factors[a]	Treatment	T stage	LN risk (%)[b]	Pelvic node field size (cm)[c]	CTV boost	PTV margin (mm)[d]	Total PTV dose (EBRT; Gy)
Low risk	Stage T1b, T1c, or T2a, Gleason score ≤6, *and* PSA ≤10 ng/ml	EBRT to prostate or permanent seed brachytherapy or HDR brachytherapy	T1–T2a	–	–	Prostate	5–10	70.2–75.6
Intermediate risk	Stage T1b, T1c, or T2a, Gleason score ≤7 *and* PSA >10–20 ng/ml; or stage T2b or T2c, Gleason score ≤7 *and* PSA ≤20 ng/ml	EBRT to prostate and seminal vesicles; external beam RT to prostate and seminal vesicles (45 Gy), and brachytherapy boost (permanent or HDR); elective pelvic irradiation if LN risk >15% Neoadjuvant hormone therapy to be considered if LN risk >15%	T1–T2a	<15	–	Prostate and proximal SV	5–10	70.2–75.6
			T2b–c	<15	–	Prostate and SV	5–10	73.8–75.6
			T1–T2a	≥15	16.5×16.5	Prostate and proximal SV	5–10	70.2–75.6
			T2b–c	≥15	16.5×16.5	Prostate and SV	5–10	73.8–75.6
High risk	Stage T3, Gleason score 8–10, PSA >20 ng/ml	Neoadjuvant hormone therapy; adjuvant hormone therapy for Gleason score 8–10; elective pelvic irradiation; prostate and seminal vesicle boost (EBRT or brachytherapy)	T1–T2a	–	16.5×16.5	Prostate and proximal SV	5–10	70.2–75.6
			T2b–T3	–	16.5×20.0	Prostate and SV and EPE	5–10	73.8–75.6

[a] From KUBAN et al. 2003
[b] LN risk = 2/3 PSA + [(GS-6) + TG-1.5] × 10
[c] Pelvic fields receive 45 Gy/25fractions. All treatments are in 1.8 Gy per fraction
[d] PTV margin varies by localization method and patient stability

metastases that exceeds 15% as determined by the Roach equation (ROACH et al. 2003).

28.4.2
External Irradiation

With the advent of megavoltage equipment, an increase in the use of external irradiation rapidly emerged for the treatment of patients with carcinoma of the prostate (DEL REGATO et al. 1993). Various techniques have been used, ranging from parallel anteroposterior (AP) portals with a perineal appositional field to lateral portals (box technique) or rotational fields to supplement the dose to the prostate (BAGSHAW et al. 1988; DEL REGATO et al. 1993; McGOWAN 1981), In recent years, 3D conformal and intensity-modulated radiation therapy (IMRT) techniques have been increasingly used (HANKS et al. 1996; LEIBEL et al. 1994; PEREZ 1998).

In patients in whom transurethral resection of prostate (TURP) has been carried out for relief of obstructive lower urinary tract symptoms, 4 weeks should elapse before radiation therapy begins in order to decrease sequelae (e.g., urinary incontinence, urethral strictures; PEREZ 1998).

28.4.3
Conventional Radiation Therapy Techniques

28.4.3.1
Volume Treated

When the pelvic lymph nodes are treated, the anterior and posterior field size is 16.5×16.5 cm at isocenter. Patients younger than 71 years of age with clinically localized disease and a risk of lymph node metastases that exceeds 15%, as well as all patients with stage-C (T3) lesions, are treated to the whole pelvis with four fields (45 Gy) and additional dose to complete 70 Gy or higher to the prostate, with a six-field 3D conformal or IMRT technique. For node-positive disease, the pelvic field size is increased to 16.5×20.5 cm at isocenter to cover the common iliac lymph nodes. The inferior margin of the field can be determined using an urethrogram with 25% radiopaque iodinated contrast material. The inferior field edge usually is 1.5–2 cm distal to the junction of the prostatic and membranous urethra (usually at or caudal to the bottom of the ischial tuberosities). The lateral margins should be about 1–2 cm from the lateral bony pelvis (Fig. 28.3a).

With lateral portals, which are used with the box technique (including the lymph nodes) or to irradiate the prostate with two-dimensional (2D) stationary fields or rotational techniques, it is important to delineate anatomic structures of the pelvis and the location of the prostate in relation to the bladder, rectum, and bony structures with computed tomography (CT) scan or magnetic resonance imaging (MRI).

The initial lateral fields encompass a volume similar to that treated with AP-posteroanterior (PA) portals. The anterior margins should be 1.5 cm posterior to the projection of the anterior cortex of the pubic symphysis (Fig. 28.3b). Some of the small bowel may be spared anteriorly, keeping in mind the anatomic location of the external iliac lymph nodes. Posteriorly, the portals include the internal iliac nodes anterior to the S1–S2 interspace, which allows for some sparing of the posterior rectal wall distal to this level.

Figure 28.4 shows examples of digitally reconstructed radiograph (DRR) simulation films outlining the AP and lateral portals used for the box technique with coverage of the iliac nodes. For the boost with 2D treatment planning, the upper margin is 3–5 cm above the pubic bone or acetabulum, depending on extent of disease and volume to be covered (i.e., prostate with or without seminal vesicles). The anterior margin is 1.5 cm posterior to the anterior cortex of the pubic bone. The inferior margin is 1.5 cm inferior to the genitourinary diaphragm as demonstrated by urethrogram. The posterior margin is 2 cm behind the marker rod in the rectum.

The reduced fields for treatment of the prostatic volume can be about 8×10 cm at isocenter for stages T1a–T2b to 10×12 or 12×14 cm for stages T2c–T3 or T4 (Fig. 28.3c) or ideally are anatomically shaped fields, using CT scans or MRI volume reconstructions of the prostate and seminal vesicles. The seminal vesicles are located high in the pelvis and posterior to the bladder, which is particularly critical when reduced fields are designed in patients with clinical or surgical stage-T3b tumors. PEREZ et al. (1993) demonstrated a correlation between size of the reduced portal and probability of pelvic tumor control.

The boost portal configuration and size should be individually determined for each patient, depending on clinical and radiographic assessment of tumor extent. After the appropriate portals have been determined, the central axis and some corners of the reduced portals for both portals are tattooed on the patient with India ink.

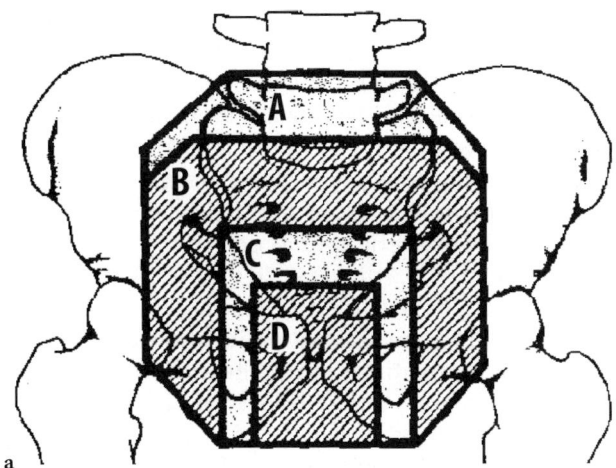

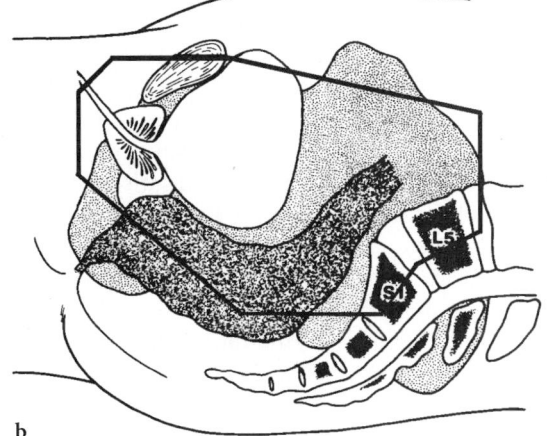

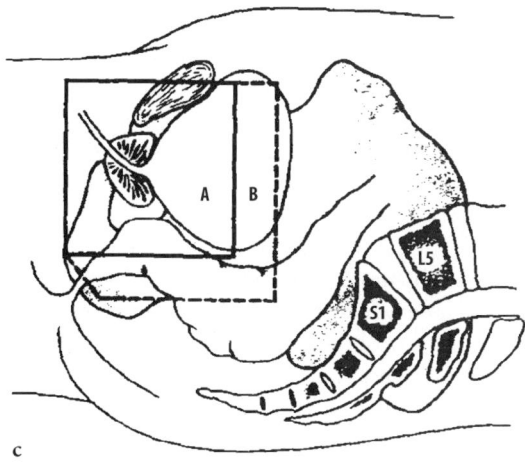

Fig. 28.3. a Diagrams of the pelvis show volumes used to conventionally irradiate the pelvic lymph nodes, when indicated, and the prostate. Lower margin is at or even 1 cm below ischial tuberosities. At Washington University, 15×15-cm portals at source-to-skin distance are used for selected stage-T1b, stage-T2, and all stage-T3 diseases, and for high-risk postoperative patients, whereas for stage-N1 disease, 18×15-cm portals are used, when necessary, to cover all lymph nodes up to the bifurcation of the common iliac vessels. Sizes of reduced fields are larger (up to 12×14 cm) when the seminal vesicles and/or periprostatic tumor is irradiated compared with prostate boost only (up to 8×10 cm) or larger for patients with stage-T3 tumors. **b** Lateral portals used in conventional box technique to irradiate pelvic tissues and prostate. The anterior margin is 1 cm posterior to projected cortex of pubic symphysis. Presacral lymph nodes are included down to S2; inferiorly the posterior wall of rectum is spared. **c** Boost fields, lateral projection, are used to irradiate the prostate with conventional radiation therapy.

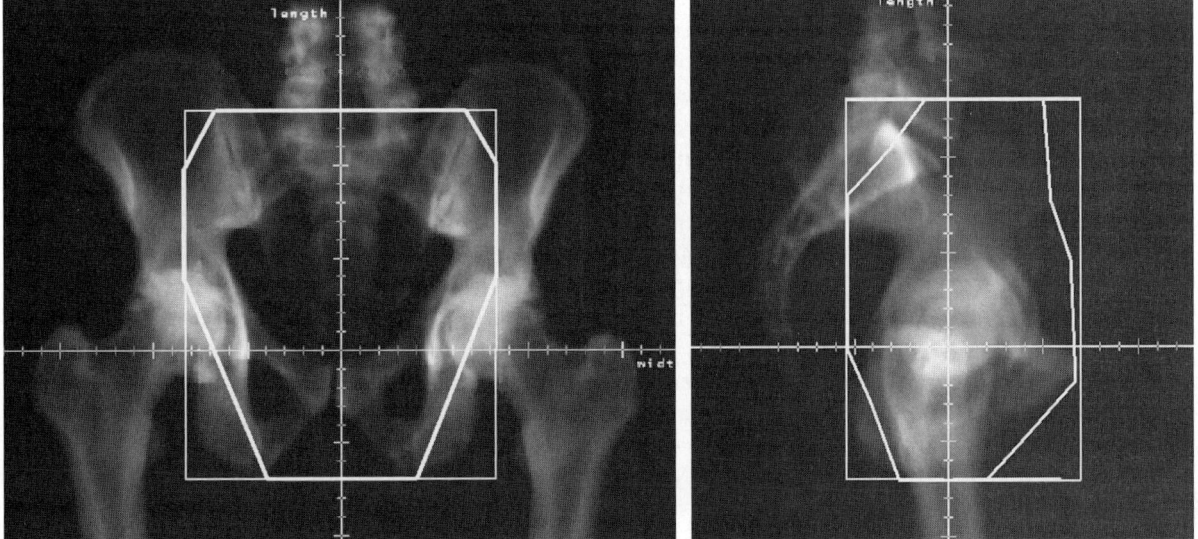

Fig. 28.4a,b. Anteroposterior (**a**) and lateral (**b**) virtual simulation digitally reconstructed radiographs (DRRs) for carcinoma of the prostate treating the pelvic lymph nodes. Note the relationship of the portals to the roof of the acetabulum, the pubic symphysis anteriorly, and the ischial tuberosities posteriorly.

To simulate these portals with the patient in the supine position, a small plastic tube is inserted in the rectum to localize the anterior rectal wall. After thorough cleansing of the penis and surrounding areas with Betadine, using sterile technique, 28–40% iodinated contrast material is injected in the urethra until the patient complains of mild discomfort, and AP and lateral radiographs are taken after the position of the small portals is determined under fluoroscopic examination. For 3D CRT a topogram and a CT scan of the pelvis are performed. The urethrogram documents the junction of the prostatic and bulbous urethra and accurately localizes (within 1 cm) the apex of the prostate, which may be difficult to identify on CT scans without contrast. Care should be taken to avoid overdistension of the bulbous urethra with contrast. MALONE et al. (2000) reported that the prostate could be displaced an average of 6.1 mm due to the urethrogram. Others have described minimal movement due to the urethrogram, suggesting that it may be related to technique of urethrography and not to the contrast itself (LIU et al. 2004).

A great deal of controversy has developed in reference to the most accurate anatomic location of the prostate apex. In a study of 115 patients, none of the urethrograms showed the urethral sphincter to be caudal to the ischial tuberosities; 10% were located <1 cm cephalad to a line joining the ischial tuberosities. If 2 cm or more are arbitrarily considered, 42.5% of patients would have received unnecessary irradiation to small volumes of normal tissues (SADEGHI et al. 1996). Cox et al. (1994) evaluated urethrogram and CT scans of the pelvis for treatment planning in prostate cancer. Interobserver identification of the prostatic apex varied in 70% of cases. This variability resulted in an inadequate margin (<1 cm) beneath the urogenital diaphragm in 5% of patients. In contrast, placing the inferior border of the portal at the ischial tuberosities or the base of the penis, as seen on CT scans, ensured an adequate margin for all patients. They concluded that urethrography is more accurate than CT scanning in determining the inferior extent of the urogenital diaphragm.

WILSON et al. (1994) determined the anatomic location of the apex of the prostate in 153 patients undergoing 128-I implants by direct surgical exposure (133 patients) or transrectal ultrasound (TRUS; 20 patients). There was excellent agreement in the estimate of location of the prostatic apex between the two methods. It was located 1.5 cm or more above the ischial tuberosities in approximately 95% of patients and within 1 cm in 98% (150 of 153).

ALGAN et al. (1995) reviewed the location of the prostatic apex in 17 patients in whom an MRI scan was obtained in addition to retrograde urethrogram and CT scan of pelvis for 3D treatment planning. The location of the prostatic apex as determined by the urethrogram alone was, on average, 5.8 mm caudad to the location on the MRI, whereas the location of the prostatic apex as determined by CT/urethrogram was 3.1 mm caudad to that on MRI. If the prostatic apex is defined as 12 mm instead of 10 mm above the urethrogram tip (junction of membranous and prostatic urethra), the difference between the urethrogram and MRI locations of the prostatic apex is no longer present.

CROOK et al. (1995), in 55 patients with localized carcinoma of the prostate, placed one gold seed under TRUS at the base of the prostate near the seminal vesicles, at the posterior aspect, and the apex of the prostate. At the time of first simulation an urethrogram was performed, and the rectum was opacified with 10–15 cc of barium. The tip of the urethrogram cone varied in position from 0 to 2.8 cm above the most inferior aspect of the ischial tuberosities. At initial simulation the apex of the prostate was <2 cm above the ischial tuberosities in 42% of patients, <1.5 cm in 19%, and <1 cm in 8% of patients. Because of variability in the thickness of the urogenital diaphragm, only 12 of 22 (55%) of these low-lying prostates would have been detected by urethrogram.

28.4.3.2
Beam Energy and Dose Distribution

Ideally, high-energy photon beams (>10 mV) should be used to treat these patients, which simplify techniques and decrease morbidity.

With photon-beam energies below 18 mV, lateral portals are always necessary to deliver part of the dose in addition to the AP–PA portals (box technique). In our experience, an advantage of using the box technique is a decrease in erythema and skin desquamation in the intergluteal fold, which occurs more frequently with exclusively AP–PA portals. The additional prostate dose is administered with anatomically shaped lateral and oblique or rotational portals.

For the reduced volume, a reasonable dose distribution is obtained with bilateral 120° arc rotation, skipping the midline anteriorly and posteriorly (60° vectors). Figure 28.5 illustrates the dose distribution for 8×10-cm bilateral 120° arcs using the 3D treat-

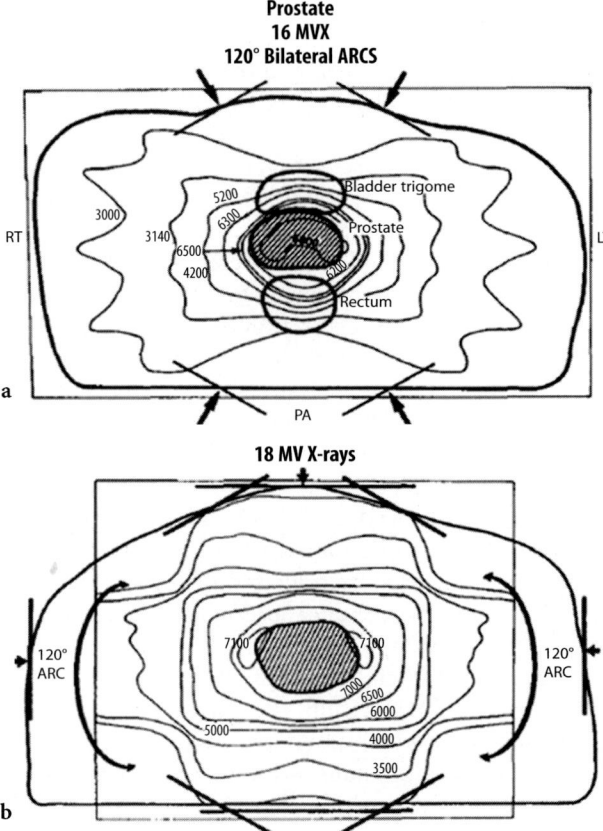

**Prostate
16 MVX
120° Bilateral ARCS**

RT   LT

a

PA

18 MV X-rays

120°
ARC

120°
ARC

b

Fig. 28.5. a Isodose curves to deliver 68 Gy to the prostate with 18-mV photons, bilateral 120° arcs, and skipping 60° anterior and posterior vectors. **b** Isodose curves for irradiation of pelvic lymph nodes and prostatic volume using anteroposterior/posteroanterior and lateral fields to deliver 45 Gy to the pelvis and 26 Gy to prostate with reduced fields with bilateral 120° arcs. (From PEREZ 1998)

ment-planning software. It is readily apparent that this technique irradiates significantly more volume of bladder and rectum than 3D CRT.

28.4.3.3
Standard Tumor Doses (Non-Conformal Treatment Planning)

A frequently used minimum tumor dose to the prostate is 66–70 Gy for stage T1a when these patients are irradiated, 70–72 Gy for T1b, c-T2b tumors, and approximately 74 Gy for stage C. For stage T4 or node-positive lesions, treatment is usually palliative, and the minimum tumor dose can be held at 60–65 Gy to decrease morbidity. Most institutions treat with daily fractions of 1.8–2 Gy, five fractions per week (PEREZ 1983; TAYLOR et al. 1979). Occa-

sionally, four weekly fractions of 2.25 Gy have been used (BAGSHAW et al. 1985). At least two portals should be treated daily to improve tolerance to irradiation.

BIGGS and RUSSELL (1988) described an average dose decrease of approximately 2% for patients with metallic hip prostheses who were treated with lateral portals, and an average increase of 2% for 10-mV X-rays and 5% for [60]Co.

The usual dose for the pelvic lymph nodes (when the latter are to be irradiated) is 45 Gy, with a boost (24–26 Gy) to the prostate or enlarged lymph nodes (5 Gy) through reduced fields (PEREZ 1998).

28.4.4
Conformal Radiation Therapy Simulation and Treatment Planning

The scientific basis and process of 3D CRT are described in Chapter 9. The process of 3D CRT planning entails patient positioning and immobilization followed by acquisition of a treatment-planning CT data set. Defining target volumes and organs at risk is accomplished by contouring anatomy on a slice-by-slice basis. Radiation beams are created with virtual simulation software tools that are analogous to the operation of a conventional fluoroscopic isocentric radiation therapy simulator. The radiation beams or apertures are shaped using a beam's eye view (BEV) display and the contributing dose from each beam is entered into the treatment-planning software. Finally, the plan is reviewed using a variety of dose display and analysis tools.

28.4.4.1
Patient Immobilization and Positioning

Patient immobilization devices are more frequently used in 3D CRT compared with traditional treatment planning. Some investigators have reported improved treatment setup accuracy with these devices, whereas others have reported no significant advantage with their use (NUTTING et al. 2000; ROSENTHAL et al. 1993). In a randomized study, KNEEBONE demonstrated that the average simulation-to-treatment deviation of the isocenter position was 8.5 mm in a control group and 6.2 mm in an immobilized group ($p<0.001$). The use of immobilization devices reduced isocenter deviations exceeding 10 mm from 30.9 to 10.6% in the immobilized arm ($p<0.001$). The average deviations in the anteroposterior, right–left, and superior–inferior directions were reduced to 2.9, 2.1, and 3.9 mm for the immobilized group, respectively (KNEEBONE et al. 2003). It is recommended that each radiation therapy center study and review the treatment setup variations with either method and choose the one that gives the best reproducibility.

There is significant debate regarding the appropriate positioning for patients treated with localized carcinoma prostate. Some groups advocate a prone position which they claim minimizes prostate positional uncertainty and decreases volume of rectum irradiated with conformal radiation therapy when the seminal vesicles are irradiated. Furthermore, in the prone position the seminal vesicles fall forward and increase the separation from the rectum (ZELEFSKY et al. 1997b). More recent data suggest that a prone position is associated with greater prostate motion accompanying normal ventilation. The increased intra-abdominal pressure associated with breathing in a prone position results in significant movement of the prostate and seminal vesicles. DAWSON et al. (2000) evaluated the impact of breathing on the position of the prostate gland in four patients treated in four different positions in whom radiopaque markers were implanted in the periphery of the prostate using transrectal ultrasound guidance. Fluoroscopy was performed in four different positions: prone in foam cast cradle; prone in thermoplastic mold; supine on a flat table; and supine with a false table under the buttocks. During normal breathing maximum movement of prostate markers seen in the prone position (cranial–caudal) ranged from 0.9 to 5.1 mm and anterior–posterior ranged up to 3.5 mm. In the supine position prostate movements during normal breathing was <1 mm in all directions; however, deep breathing resulted in movements of 3.8–10.5 mm in the cranial–caudal direction in the prone position (with and without thermoplastic mold). This range was reduced to 2.7 mm in the supine position and to 0.5–2.1 mm with the use of the false tabletop. Deep breathing resulted in anterior–posterior skeletal movements of 2.7–13.1 mm in the prone position, whereas in the supine position these variations were negligible.

MALONE et al. (2000) also characterized inaccuracies in prostatic gland location due to respiration observed fluoroscopically in 28 patients in whom three gold fiducial markers were implanted under ultrasound guidance at the apex, posterior wall, and base of prostate. Patients were immobilized on a customized thermoplastic shell placed on a rigid pelvic board. A second group of 20 patients were evaluated both prone (with or without thermoplastic shell) and supine (without immobilization shell). When the patients were immobilized prone in the thermoplastic shell, the prostate moved synchronously with respiration a mean distance of 3.3±1.8 mm (range 1–10 mm). In 9 of 40 observations (23%) the displacements were 4 mm or greater. The respiratory-associated prostate movement decreased significantly when the thermoplastic shells were removed. Other investigators have confirmed this finding of prostate movement with respiration being significantly less in patients placed in a supine position (DAWSON et al. 2000; KITAMURA et al. 2002; LITZENBERG et al. 2002; MALONE et al. 2000; McLAUGHLIN et al. 1999; WEBER et al. 2000). BAYLEY (2004) conducted a randomized trial of the supine vs prone position in patients undergoing conformal radiation therapy. Twenty-eight patients were randomized to commence radiation therapy in the prone or supine position and then change to the alternate position midway through their treatment course. After placement of fiducial markers in the prostate for daily prostate localization, the patients underwent CT simulation and treatment planning in both positions. Observed prostate motion was significantly less in the supine position than the prone position. Pretreatment positioning corrections were required more often for patients in the prone position. A DVH (dose-volume histogram) analysis revealed more bladder wall, rectal wall, and small bowel in the high-dose volumes when patients were in the prone position than in the supine position. Finally, patients were more comfortable in the supine position than the prone position. Seven patients that started in the supine position refused to be treated in the prone position due to discomfort.

28.4.4.2
Clinical Target Volume Definition

Because prostate cancer is often found to be multifocal at the time of radical prostatectomy, the entire gland is commonly considered the gross tumor volume for radiation treatment-planning purposes. The clinical target volume (CTV) may expand the gross tumor volume to account for direct extension or the CTV can be extended to encompass adjacent organs or regions of spread. In prostate cancer, the CTV may encompass the seminal vesicles and possibly the regional pelvic lymph nodes.

Additional CTV margin for possible extraprostatic extension (EPE) has been recommended by several authors. In a study of radical prostatectomy specimens, DAVIS et al. (1999) demonstrated that EPE was present in 28% of 376 patients. The average radial EPE distance from the prostate capsule was 0.8 mm with a range of 0.04–4.4 mm. SOHAYDA and colleagues (2000) reported EPE in 35% of 265 cases. When EPE was present, it extended posterolateral in 53%, lateral in 24%, and posterior in 13%. The 90th percentile of EPE distance was 3.3 mm for patients with clinical T1–T2, pretreatment PSA of ≤10 ng/ml, and biopsy Gleason score of 6 or less. For patients with more unfavorable tumors, the 90th percentile of EPE extended to 3.9 mm. Both these series did not correct for possible specimen shrinkage related to formalin fixation. In another series of 712 prostatectomy specimens, TEH (2003) reported that the mean radial distance from the capsule was 2.93 mm (accounting for formalin fixation). The decision to add additional CTV margin may be a function of both the total dose prescription and the degree of conformality expected to be achieved. With 3D CRT and escalated radiation doses, the tissues immediately adjacent to the prostate receive a dose of radiation that would be expected to control microscopic disease. On the other hand, with highly conformal radiation plans with IMRT or proton therapy, and small uncertainty margins for the PTV, marginal misses may be anticipated unless an adequate CTV margin is added to account for microscopic EPE.

In selected patients it is necessary to outline the seminal vesicles, which in most patients are well demonstrated on the cross-section CT scans of the pelvis, superior, lateral, and posterior to the base of the prostate. The nomograms developed by PARTIN et al. (1993), PISANSKY et al. (1996a) and ROACH et al. (1997) may be used to determine the probability of seminal vesicle or pelvic lymph node involvement using clinical stage, pretreatment PSA, and Gleason score. KESTIN et al. (2002) published an analysis of 344 radical prostatectomy specimens in which they measured the length of seminal vesicles, length of involvement by carcinoma, and percentage of seminal vesicle involved. They found an excellent correlation between the various prognostic parameters and the probability of seminal vesicle involvement. Also, in 81 patients with positive seminal vesicle involvement, the median length of tumor presence was 1 cm. In the entire population only 7% of patients had seminal vesicle involvement beyond 1 cm. They concluded that in selected patients seminal vesicles need to be treated and only 2.5 cm (approximately 60% of the seminal vesicle) should be included within the CTV, unless there is radiographic evidence of involvement.

When there is evidence of extraprostatic extension on physical examination or imaging modalities, such as MRI (clinical stage T3), the seminal vesicles should be included for the total radiation dose prescription. In cases where the disease is confined to the gland (clinical stages T1–T2) and the risk of seminal vesicle invasion exceeds 15%, we define two clinical target volumes; the first encompasses the prostate and the seminal vesicles, and with the second, the boost target volume is the prostate alone. In these cases a radiation dose that controls subclinical disease is prescribed to the first target volume and a higher dose is intended for the prostate itself.

As described in the section on conventional radiation therapy treatment planning, identification of the prostatic apex can be facilitated with a retrograde urethrogram, which can be performed using 25% iodinated contrast material. The prostatic apex is 3–13 mm above the most proximal aspect of the urogenital diaphragm as defined by the urethrogram (RASCH et al. 1999; WILDER et al. 1997). Care should be taken not to overinflate the urethra with iodinated contrast, as this may distend or move the prostate from its relaxed position.

Some authors feel that MRI may be more accurate in delineating the prostate and seminal vesicles (PERROTTI et al. 1996). Several publications have shown some discrepancy (0.5–1 cm) in defining the location of the apex of the prostate using CT scanning or MRI (ALGAN et al. 1995; COX et al. 1994; CROOK et al. 1995). PARKER et al. (2003) studied co-registration of CT and MR images in the radiation treatment planning of six patients with localized prostate cancer to assist with GTV delineation and identification of prostate position during radiation therapy. The overall magnitude of contoured GTV was similar for MR and CT; however, there were spatial discrepancies in contouring between the two modalities. The greatest systematic discrepancy was at the posterior apical prostate border, which was 3.6 mm more posterior on MR- than CT-defined contouring.

At Washington University we treat the entire prostate as our CTV for patients with low-risk disease. For patients with intermediate-risk disease we treat the prostate and seminal vesicles to 55.8 Gy, followed by a boost to the prostate alone (with appropriate PTV margin, *vide infra*). When using IMRT we utilize a single CTV for these patients to avoid the need for two IMRT plans. In this case we treat the prostate and the first 1–2 cm of seminal vesicle tissue in our CTV (Fig. 28.6). For patients with stage-T3 disease, we treat the prostate along with the seminal vesicles to the total prescription dose.

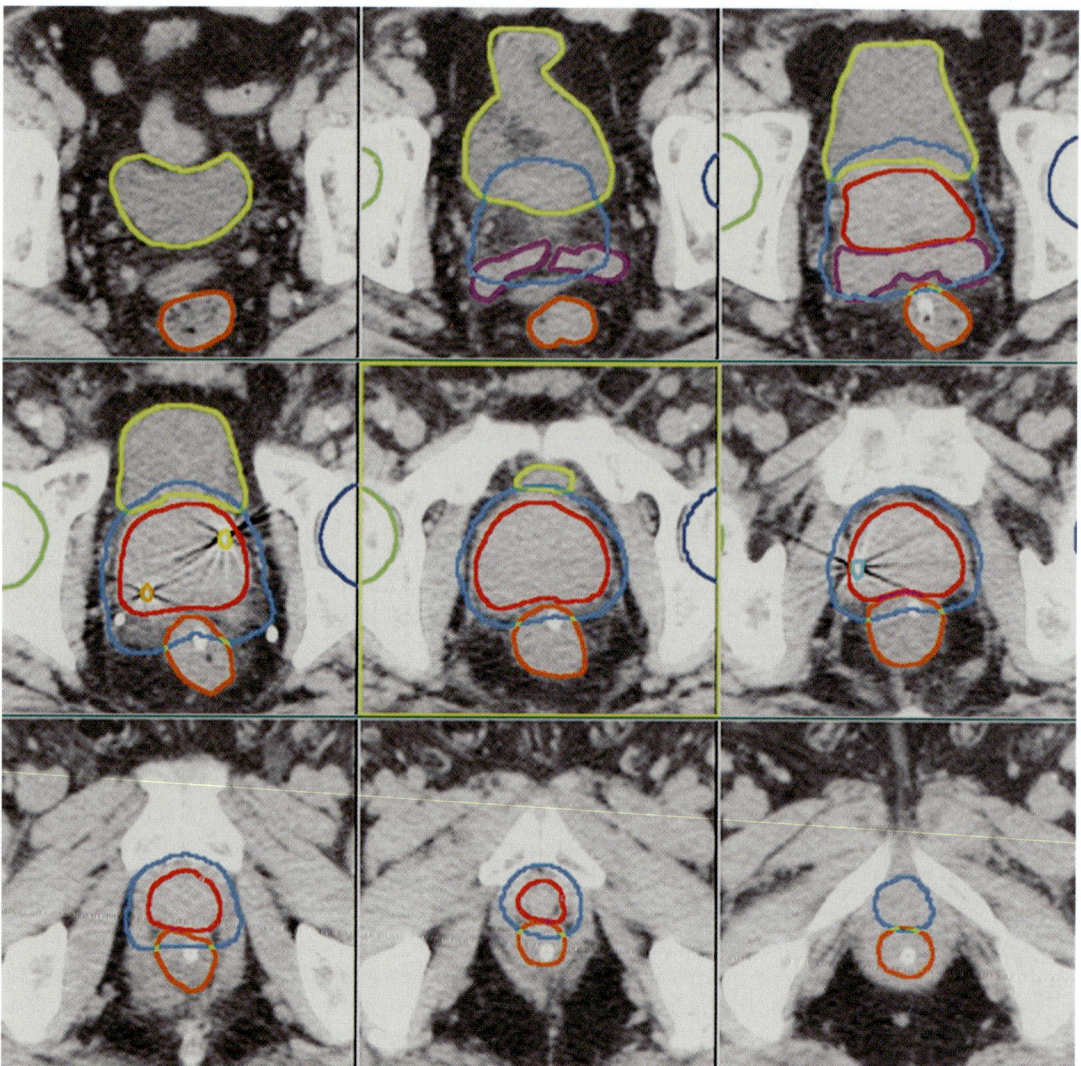

Fig. 28.6. Target volumes and organs at risk are defined on serial images of a treatment-planning CT scan. For illustration purposes only every fourth CT image is displayed. The bladder (*yellow*) and rectum (*orange*) are contoured as solid organs. The prostate (*red*) is contoured from its base superiorly to the apex at the genitourinary diaphragm. In this patient the first 1 cm of seminal vesicle tissue (*violet*) is included with the CTV and no attempt is made to treat them above this level. The PTV (*cyan*) margin is 7 mm from the prostate border. In this example the fiducial markers are seen inside the prostate gland

28.4.4.3
Planning Target Volume Definition

The magnitude of the PTV margin is dependent upon several treatment-related factors. A technologist's inability to precisely reproduce a patient's position on a daily basis is one contributing factor. Some patients may move while on the treatment table because of fatigue or discomfort. Internal organs, including the prostate gland, can shift because of variable filling of the bladder and rectum. The shifts can be anisotropic with most movement occurring

in the anterior and posterior directions. Langen (2001) summarized institutional experiences studying setup errors and internal-organ motion. These studies investigated both internal organ motions and setup errors to determine an appropriate margin for the planning target volume. Increasing the size of the PTV margin increases the probability of encompassing the CTV by the prescribed isodose from a complete course of radiation therapy. In order to assure that an adequate radiation dose is encompassing all areas at risk of harboring disease, the radiation oncologist needs to include an appropriate

PTV margin. There is a trade-off between assuring nearly 100% coverage during each treatment and the volume of adjacent organs irradiated unnecessarily.

ANTOLAK et al. (1998), in a study of 17 patients with prostate cancer who underwent CT scanning for treatment planning and three subsequent CT scans obtained at approximately 2-week intervals during external-beam fractionated irradiation, observed CTV motion of 0.09 cm (left to right), 0.36 cm (cranial–caudal), and 0.41 cm (anterior–posterior) directions. Prostate mobility was not significantly correlated with bladder volume, but both prostate and seminal vesicle positions were significantly influenced by rectal volume. From this study, the authors concluded that margins between the CTV and PTV necessary to enclose 95% of the PTV were 7 mm in the lateral and cranial–caudal directions and 11 mm in the anterior–posterior direction.

In a study of 11 patients with repeated CT scans during a course of radiation therapy for prostate cancer, VAN HERK et al. (1995) showed that the largest deviation in prostate position was 2.7 mm in the anterior–posterior direction with a rotation about the left–right axis of 4°. They found that the apex of the prostate does not move, and the majority of any motion is rotation around the apex.

ZELEFSKY et al. (1999) evaluated prostate and seminal vesicle motion in 50 patients treated in the prone position using CT scans for initial treatment planning and three scans obtained throughout the course of radiation therapy. Before the initial CT scans, patients had an enema and were given 250 ml of bowel contrast by mouth. Patients had an empty bladder and 10 cc of air was inserted into the rectum via rectal catheter. Prior to all CT scans, patients voided, and no additional procedures were performed. Relative to the initial planning CT, mean displacements of the prostate were –1.2±2.9 mm in the AP, –0.5±3.3 mm in the superior/inferior, and –0.6±0.8 mm in the lateral direction. The seminal vesicle displacements were –1.4±4.9 mm, 1.3±5.5 mm, and –0.8±3.1 mm in the AP, superior–inferior (SI), and lateral directions, respectively. (Negative values indicated displacements to the posterior, inferior, and left directions.) A combination of rectal volume >60 cc or a bladder volume >40 cc was found to be predictive for systematic deviations of the prostate and seminal vesicles of more than 3 mm. Based on the data and the prescription of irradiation dose to achieve at least 93% coverage of the CTV, ZELEFSKY et al. (1999) calculated the margins be added to the CTV for defining the PTV. Based on these reports,

it is apparent that beyond the CTV, additional margins of 5–8 mm are necessary to provide adequate coverage of the prostate and 6–11 mm for adequate coverage of the seminal vesicles when there is no organ distension that would result in a systematic error.

Strategies to reduce the uncertainty in daily treatment delivery, and therefore to reduce the magnitude of the PTV margin, have been successfully introduced. Each of these methods employs daily imaging of the prostate with the patient in the treatment position on the linear accelerator treatment table. Some institutions have employed radiopaque implanted fiducial markers that are subsequently imaged with electronic portal imaging devices. CROOK et al. (1995) evaluated prostate motion in 55 patients in whom gold seeds were implanted at the base of the gland. Initial simulation was obtained in a supine position with a full bladder and was repeated after patients received 40 Gy. Prostate motion was observed in the posterior direction (5.6±4.1 mm) and in the inferior direction (5.9±4.5 mm). In 30% of patients the base of the prostate was displaced posteriorly and in 11% in the inferior direction by >10 mm. VIGNEAULT et al. (1997) investigated inter- and intrafraction daily motion of the prostatic apex relative to pelvic bony structures in 11 patients using radiopaque markers implanted under ultrasound guidance near the prostatic apex. After the completion of treatment, a transrectal ultrasound performed in 8 of 11 patients verified that the position of the radiopaque markers had not shifted. Marker displacements up to 1.6 cm were measured between two consecutive days of treatment on portal images. The marker displacement relative to the center of the irradiation field was sensitive to both setup variations and internal prostate motion. The authors created treatment films (six consecutive electronic portal images during the same treatment fraction) and observed no visible intratreatment displacement of the markers. The position of the fiducial markers visualized on daily port films or electronic portal images can be used to align isocenter on a daily basis to assure precise and accurate treatment of the PTV. Paired images need to be acquired in order to calculate the 3D position of the center of three fiducial markers. We have begun to employ anterior oblique portal images for daily target localization because the density of the hips makes viewing the markers on lateral radiographs difficult (Fig. 28.7).

Transabdominal ultrasound has been used to localize the prostate for treatment planning and

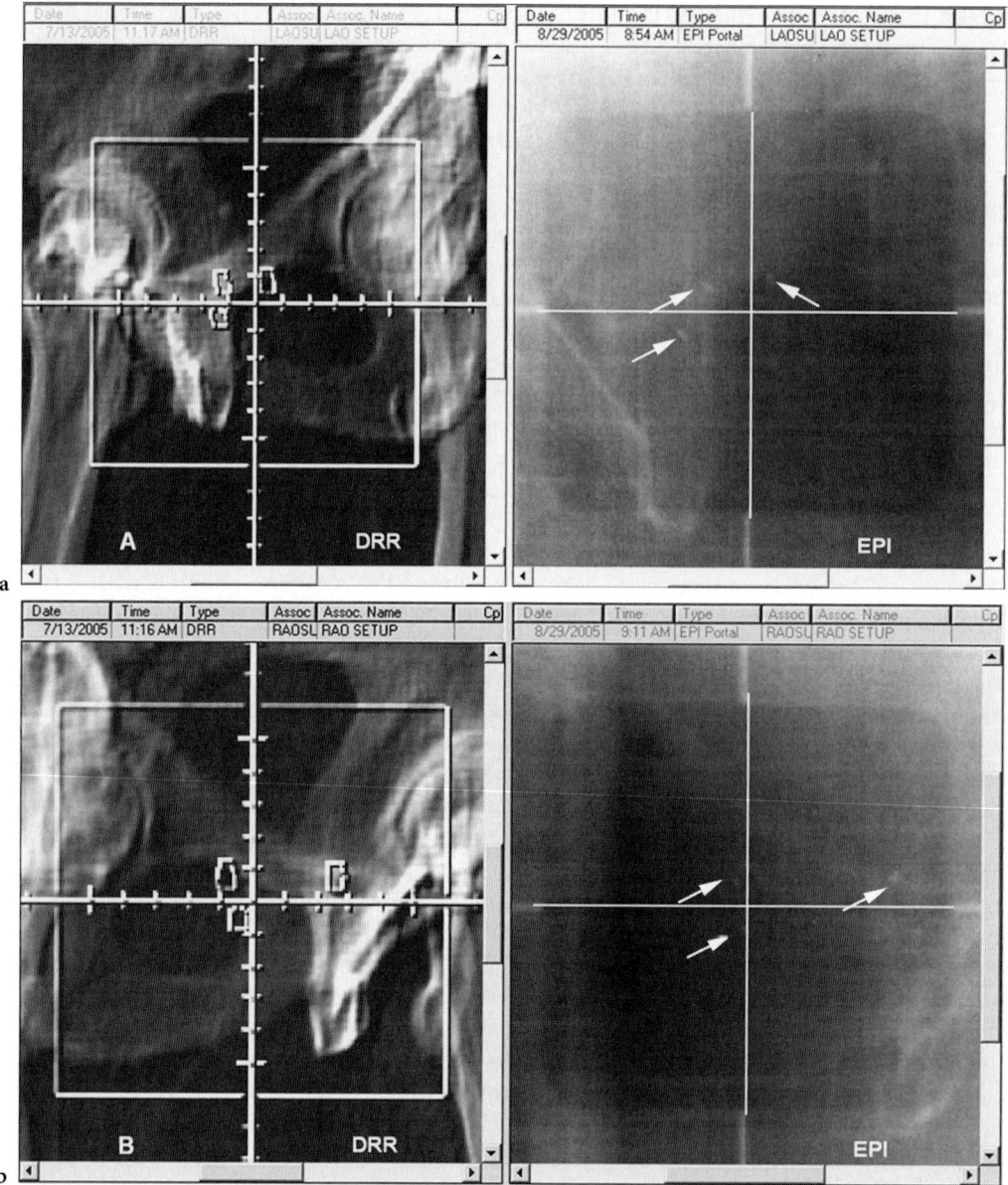

Fig. 28.7a,b. A pair of **a** left and **b** right anterior oblique digitally reconstructed radiographs (*DRR*) with corresponding electronic portal images demonstrating positions of intraprostatic fiducial markers for daily setup localization. *White arrows* indicate the seed positions which have been previously defined on the treatment-planning CT, making their positions visible on the DRR. If necessary, offsets of the isocenter are calculated and the patient table position is shifted to bring the epicenter of the markers in alignment with the beam isocenter

during daily radiation therapy delivery with an accuracy parallel to that of CT scanning of the pelvis (Fig. 28.8). Lattanzi et al. (1999, 2000) studied 23 patients with CT simulations in whom prostate-only fields based on CT scans were created with no PTV margin. Ten of the patients also had prostate localization with a transabdominal ultrasound system. The absolute magnitude difference in CT and ultrasound was small (AP mean 3±1.8 mm,

lateral mean 2.4±1.8 mm, superior/inferior mean 4.6±2.8 mm). The authors felt that transabdominal ultrasound was simple and expeditious, and they improved their ability to localize the position of the prostate with the patient at the treatment machine for daily irradiation.

Yan et al. (2000) have developed a strategy of adaptive radiation therapy. Patients undergo daily CT imaging during the first week of radiation ther-

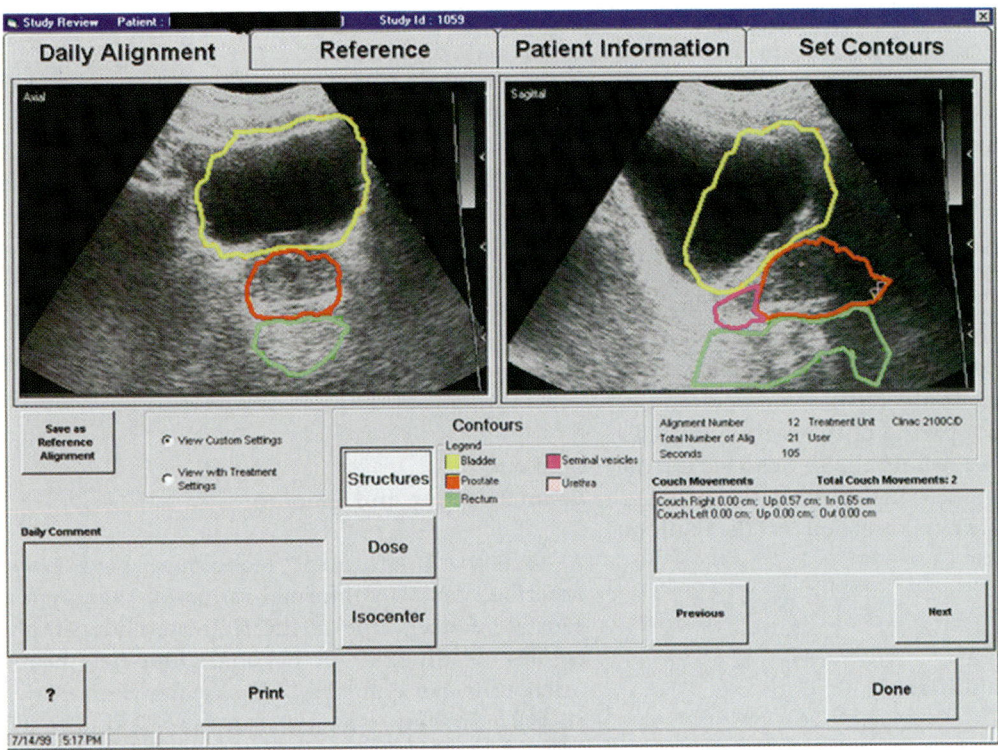

Fig. 28.8. Transabdominal ultrasound images of the prostate with transverse and sagittal projections. The CT-defined contours of the bladder (*yellow*), rectum (*green*), prostate (*red*), and seminal vesicles (*pink*) are superimposed on the ultrasound images to determine whether a table shift is necessary to make the treatment isocenter identical to the planned isocenter position

apy. A patient-specific PTV is then calculated and the patients are reassessed. A patient-specific margin allows reduction of the PTV, which is otherwise calculated based on population averages. This strategy does mean that some patients will be treated to larger PTVs if their CTV error is larger than average. Those patients may require other strategies to account for organ motion if dose escalation is necessary.

DE CREVOISIER (2005) reported that a large rectal volume defined at the time of treatment planning was associated with a decrease in biochemical tumor control. It has been suggested that a distended rectum at the simulation results in a systematic anterior displacement of the posterior prostate. Subsequent days of treatment with a less-distended or empty rectum results in a geographical miss, unless the PTV is enlarged posteriorly to accommodate the expected prostate location. An enema prior to the simulation, along with a rubber catheter or hollow tube to deflate flatus in the rectum at the treatment-planning CT, helps avoid the introduction of systematic errors in organ and target definition due to rectal distension (TINGER et al. 1998). The enema prior to simulation will empty the rectum and allow the prostate to move to its most posterior position.

This allows use of a tighter uncertainty margin posteriorly, as it eliminates the random error associated with a variably filled rectum.

Several authors have reported the use of an endorectal balloon during each daily treatment to stabilize the position of the gland. The balloon also moves the prostate anteriorly, allowing more complete shielding of the posterior rectal wall. Unlike a rectum distended by stool and flatus, the endorectal balloon is a controlled intervention that can be reproduced during the course of radiation therapy. Air in the balloon decreases the rectal surface dose by decreasing the electronic build-up and equilibrium at the air–soft tissue interface. TEH (2002) and associates reported results of 100 consecutive patients treated with IMRT and an endorectal rectal balloon. Ten of those patients also participated in a prostate motion study following gold-seed implantation. Each of these ten patients underwent ten CT scans during the course of their radiation therapy. The mean and standard deviation of superior–inferior (SI) target displacements were 0.92 and 1.78 mm, respectively. Of the 100 patients treated with a rectal balloon, 80% had no rectal complaints and 11 and 6% had grade-1 or grade-2 acute toxicity, respectively. They measured the radiation dose at a balloon–tissue interface using a phantom. The dose at the air–tissue interface is approximately 15% lower than the dose at the same point without an air cavity. The dose builds up rapidly so that at 1 and 2 mm away from the interface, the dose is only approximately 8 and 5% lower, respectively. WACHTER et al. (2002) demonstrated in 10 patients that the dose to the posterior wall of the rectum could be significantly reduced with the use of an endorectal balloon when the prostate only was boosted. The advantage of the balloon was lost when the seminal vesicles were treated. PATEL et al. (2003) demonstrated significant dosimetric sparing of the rectum with 3D CRT or IMRT when a rectal balloon was used during an entire course of radiation therapy in five patients. Patients tolerated daily insertion of the balloon exceptionally well.

28.4.4.4
Organs-at-Risk Definition

Organs at risk need to be defined for the treatment-planning process. In the treatment of prostate cancer, the organs at risk include the bladder, rectum, femoral heads, panile bulb, and occasionally the small bowel. There can be considerable variation in the definition of these organs from physician to physician. When comparing dosimetric constraints and clinical outcomes from various series, a consistent definition of these structures is important. The Radiation Therapy Oncology Group (RTOG) has described their method for normal organ definition (MICHALSKI et al. 2000). The femoral heads are contoured from the level of the ischial tuberosities to their proximal joint at the pelvis. The rectum and bladder are both defined as solid organs. Inferiorly, the rectum is defined from the level of the ischial tuberosities. It extends superiorly until the colon moves anteriorly toward the sigmoid. The bladder is contoured inferiorly from the prostate base to the dome. Radiation to the penile bulb tissue of the corpora spongiosa inferior to the prostatic apex and urogenital diaphragm appears to play a role in the development of radiation induced erectile dysfunction (FISCH et al. 2001). Magnetic resonance imaging allows more accurate definition of the prostate apex, urogenital diaphragm, and penile bulb (STEENBAKKERS et al. 2003).

It has been argued that the bladder and rectum, being hollow organs, should have the inner contents subtracted from volumetric dose information. The remaining volume then represents the volume of the organ wall. Unfortunately, the calculation of wall volume is not a trivial task and requires the added work of contouring both the inner and rectal or additional software that is not available on many 3D CRT computer planning systems. As a result, most institutions and cooperative groups report dosimetry data to the whole organ volume. Early clinical outcomes suggest that wall volume may indeed be a more important parameter with respect to 3D dosimetry in the whole organ volume (KOPER et al. 2004).

28.4.4.5
Beam Selection and Shaping

A variety of treatment techniques have been described for 3D conformal radiation therapy for prostate cancer. One of the first methods simply applies traditional orthogonal four-field-beam orientation but employs BEV shaping (FIORINO et al. 1997; SOFFEN et al. 1991). A four-field 3D CRT technique confers a significant advantage over a conventional four-field "box" or bilateral 120° arc technique (MAGRINI et al. 1999; TEN HAKEN et al. 1989). A four-field 3D CRT technique shields significant portions of the bladder and rectum from the primary beam in the lateral projections. This leads to significant sparing of these organs from a high radiation dose.

A six-field technique has been favored by many institutions (Fig. 28.9). This technique employs parallel opposed anterior and posterior oblique fields in addition to the traditional lateral fields. Typically the oblique fields are angled 45° from the lateral fields, although they can be at shallower angles to minimize treatment to the organs at risk. Compared with a four-field conformal technique employing prescription doses of 70–74 Gy, the six-field technique does not appear to confer a clear dosimetric advantage; however, institutions that have delivered radiation doses in excess of 74 Gy have preferred a six-field technique, as it may reduce dose bladder, rectum, and the bilateral femoral heads (MAGRINI et al. 1999; PEREZ et al. 1997; TEN HAKEN et al. 1989). At Washington University unequally weighted beams are used in the six-field technique with 40% of the dose contribution coming from the lateral fields and the other 60% of the dose divided evenly between the anterior and posterior oblique fields. The lateral beams are weighted more heavily in our six-field technique because this direc-

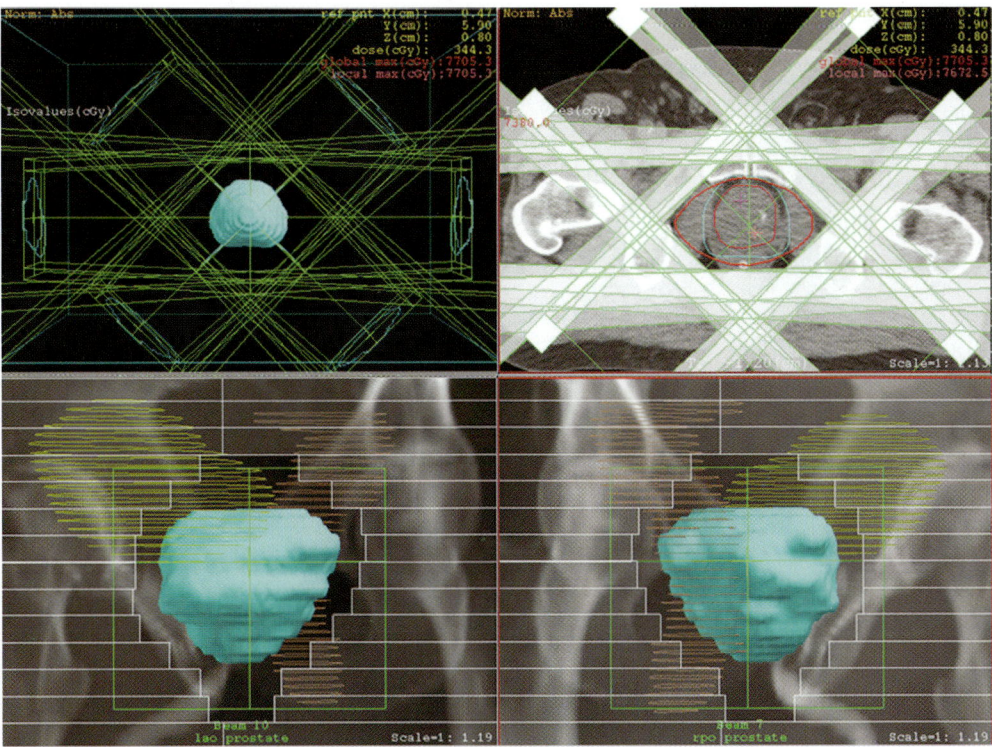

Fig. 28.9. Beam's eye view displays on left anterior oblique and right posterior oblique (*lower panels*) digitally reconstructed radiographs. Graphic representations of multileaf collimator leaves demonstrate beam shaping relative to the contour of the PTV (*solid light blue*) and shielding of the rectum (*brown wire cage*) and bladder (*yellow wire cage*). The *upper left panel* displays a six-field 3D conformal beam arrangement viewed along the superior–inferior axis. The upper right panel shows the 7380-cGy isodose line covering the PTV and prostate in the isocenter axial plane

tion allows maximum shielding of the bladder and rectum. Beam's eye view treatment planning with a six-field technique significantly improves PTV coverage and provides better bladder and rectum sparing compared with traditional bilateral arcs (Figs. 28.10, 28.11).

PEREZ compared 174 patients planned with either 3D conformal radiation therapy or standard radiation therapy and demonstrated that when field sizes were adjusted to adequately encompass the planning target volumes there was a significant reduction in the volumes of bladder and rectum irradiated with conformal technique (Table 28.6; PEREZ et al. 1997).

A noncoplanar technique has been described and used at the University of Michigan since 1992. This technique employs two lateral fields paired with noncoplanar anterior inferior oblique fields. The direction of the anterior inferior oblique fields is selected on a patient-by-patient basis using angles that optimally exclude the maximum amount of bladder and rectum. This technique has been demonstrated to effectively reduce the volume of the bladder and

rectal walls that receive significant radiation dose (MARSH et al. 1992). A technique employing blocked arc fields has been used at the University of California San Francisco. This technique has been used in patients with prostate treatment without the seminal vesicles. Bilateral 120° arc fields are blocked in the corners with a 2-cm margin on the PTV. This technique results in a lower dose to the rectum and a tighter dosimetric margin around the prostate compared with either a six-field or four-field fixed-beam 3D CRT technique (AKAZAWA et al. 1996; ROACH et al. 1994).

The choice of technique for any given institution may depend as much on the availability of technical resources as on any dosimetric advantage that the various techniques may or may not have. For example, the non-coplanar method may not be possible on all linear accelerators due to the presence of a beam stop or limited clearance of the gantry and collimator to the patient. In some busy clinics, the longer treatment times with six fields, as compared with four, may not be justified if doses <74 Gy are used.

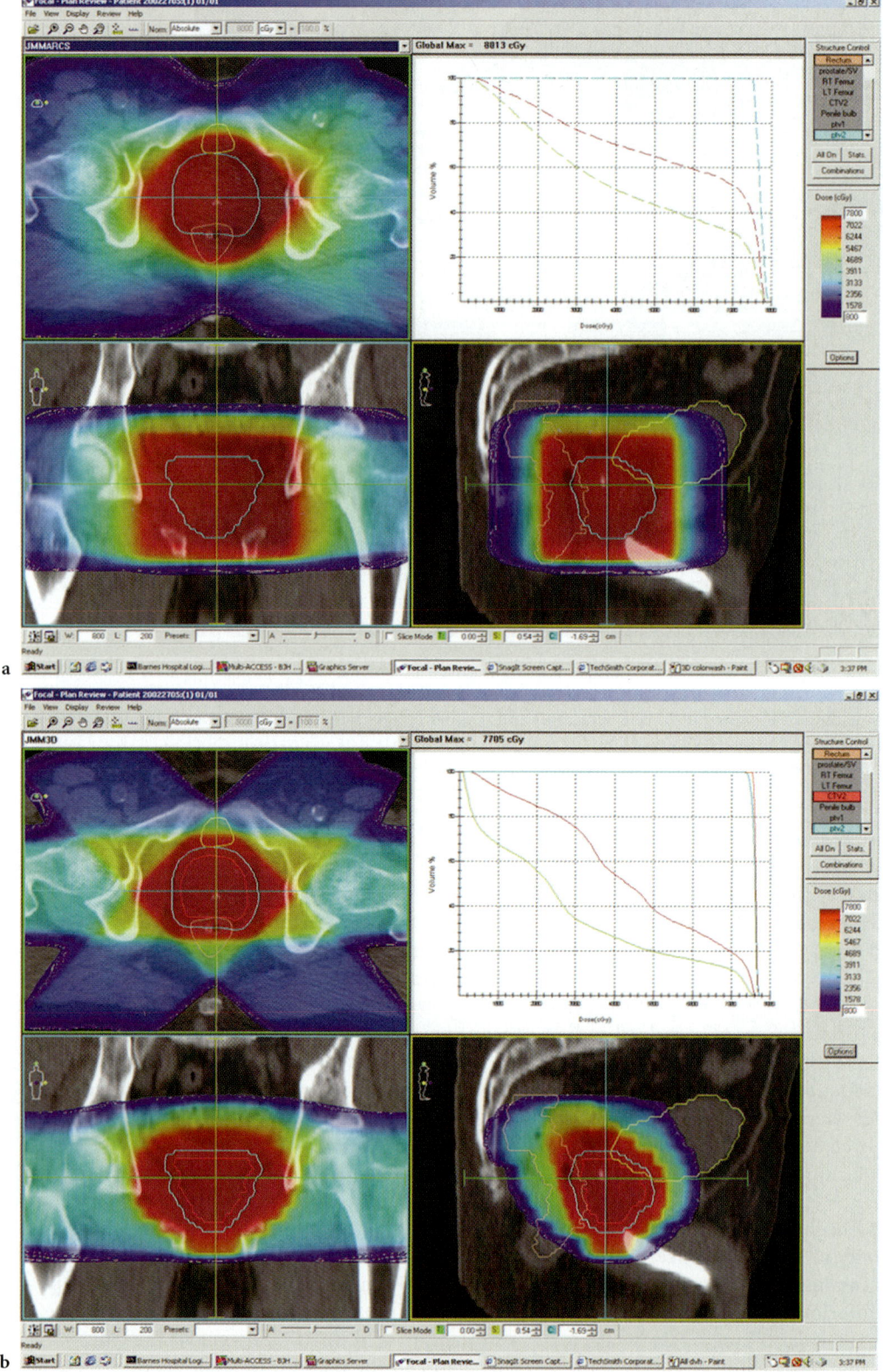

Fig. 28.10a,b. Radiation dose distribution is represented by a color-wash display on axial, sagittal, and coronal CT reconstructions for **a** 2D "arc" plan and **b** 3D six-field conformal plan. The *upper right panel* is a dose-volume histogram of the bladder, rectum, and PTV. (From SANDLER and MICHALSKI 2005)

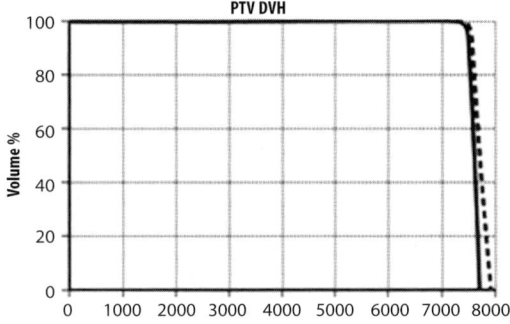

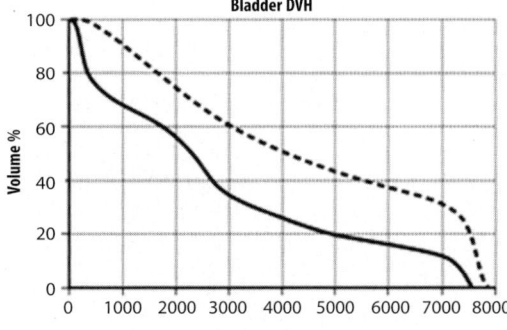

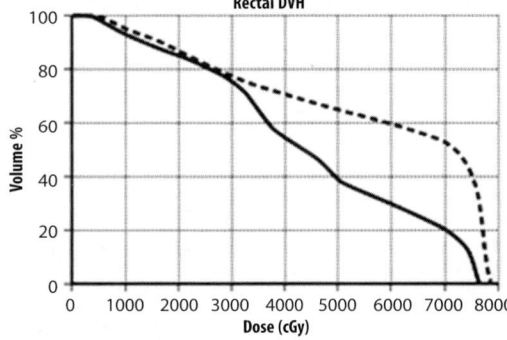

Fig. 28.11. Cumulative dose-volume histograms (*DVH*) demonstrate coverage of the PTV and a reduction in dose to the bladder and rectum using 3D CRT (*solid curve*) compared with a 2D plan with bilateral 120° arcs (*dashed curve*). (From SANDLER and MICHALSKI 2005)

28.4.5
Intensity-Modulated Radiation Therapy

Intensity-modulated radiation therapy planning begins in nearly an identical manner to that of forward-planned 3D CRT; however, patient positioning and reproducibility is far more critical due to the sharp dose gradients that are seen with this modality. We feel that a daily target localization method is critical in patients receiving IMRT for prostate cancer. Suitable methods include transabdominal ultrasound, intraprostatic fiducial markers with daily megavoltage portal or X-ray imaging, endorec-

Table 28.6. Comparison of mean dosimetric parameters for 3D conformal or standard bilateral arc rotation in carcinoma of prostate. *PTV* planning target volume, *ICRU* International Commission on Radiation Units and Measurements. (From PEREZ et al. 1997)

Parameter	Prostate irradiation only	
	3D conformal therapy	Standard therapy
No. of observations	87	87
Percentage of PTV receiving the prescribed dose or more	92.9±13.9	92.9±10.8
ICRU dose (Gy)	69.1±2.6	69.2±2.6
Minimum tumor dose (Gy	66.3±5.3	63.5±8.6
Mean tumor dose (Gy)	69.8±2.6	69.7±2.8
Maximum dose (Gy)	71.7±2.4	71.3±2.8
Percentage of volume rectum ≥65 Gy	33.7±15	62.7±21
Percentage of volume rectum ≥70 Gy	8.5±11.8	28.8±28.9
Percentage of volume bladder ≥65 Gy	22.3±12.5	50.5±22.8
Percentage of volume bladder ≥70 Gy	6.3±8.4	19.4±24.4

tal balloon immobilization, or daily in-room CT imaging. Figure 28.12 demonstrates the close proximity of the prescribed isodose to the edge of the PTV and emphasizes the importance of precise daily localization.

The target definition for IMRT proceeds in a manner similar to that of 3D CRT. Due to the significant time and effort required to execute quality assurance of the dose distribution for IMRT, cone down boosts requiring two separate IMRT plans are commonly avoided. For this reason at Washington University for our intermediate-risk patients we include the first 1 cm of seminal-vesicle tissue in a single CTV. This allows treatment for the seminal vesicle volume, which is most at risk in these patients without a separate boost plan for the prostate only.

The definition of organs at risk should include all radiosensitive structures in the pelvis including any small bowel near the PTV and the penile bulb inferior to the prostate and genitourinary diaphragm.

The IMRT treatment planning requires defining dose constraints for the target and each critical structure. The IMRT creates more heterogeneity of dose than 3D CRT, and the planning prescription needs to define a minimum dose to cover a predetermined volume of the PTV as well as a maximum dose to a small volume inside the PTV. Dose limits to organs at risk need to take into account both upper dose limits as well as the volume of those organs that are allowed to exceed those limits. The dose guidelines utilized in the RTOG randomized trial are listed in Table 28.7. (This trial is described in more detail in the next section.) The minimum dose prescription is the dose that encompasses at least 98% of the PTV. To minimize the heterogeneity within the PTV, we attempt to have no more than 2% of the PTV exceed 10% of the prescription dose.

Treatment techniques depend highly on the planning system and treatment delivery system. (See Chapter 10 for details about the various methods used for delivering IMRT including slice-based tomotherapy, dynamic MLC, step-and-shoot static MLC, and compensator-based systems.)

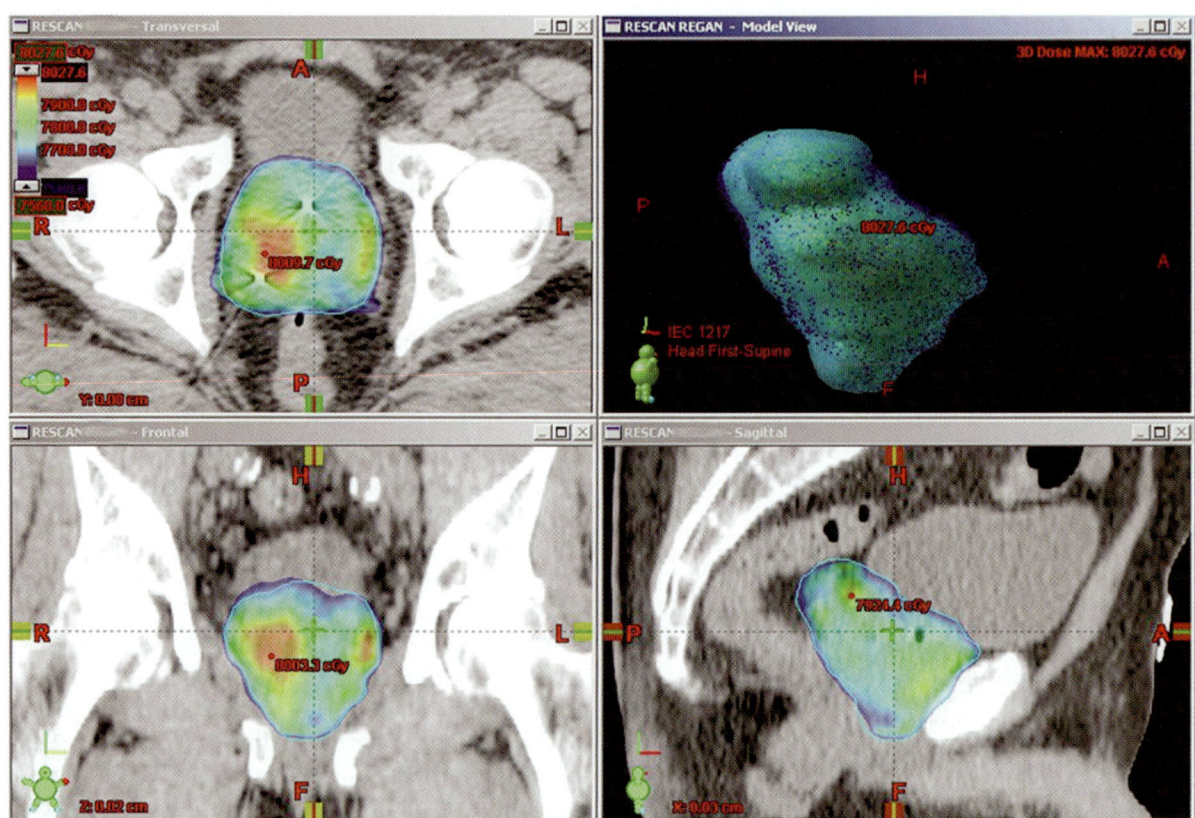

Fig. 28.12. An intensity-modulated radiation therapy dose distribution demonstrates the 7560-cGy isodose in a color-wash display in the axial (*upper left*), coronal (*lower left*), and sagittal (*lower right*) planes. The dose distribution is tighter against the PTV than the similar six-field 3D CRT plan demonstrated in Figs. 28.10 and 28.11. More heterogeneity is also demonstrated with "hot spots" in this target exceeding 8000 cGy for the prescribed dose of 7560 cGy

Table 28.7. Normal organ-dose guidelines utilized in the randomized dose trial, RTOG-0126

Normal organ limit[a]	No more than 15% volume receives dose that exceeds	No more than 25% volume receives dose that exceeds	No more than 35% volume receives dose that exceeds	No more than 50% volume receives dose that exceeds
Bladder constraint	80 Gy	75 Gy	70 Gy	65 Gy
Rectum constraint	75 Gy	70 Gy	65 Gy	60 Gy
Penile bulb (mean dose ≤52.5 Gy)				

[a]The volume of the organ that should not exceed the dose limit

28.4.6
Dose Prescription for 3D CRT or IMRT

Several institutions have reported a radiation dose response for localized prostate cancer based on retrospective data analyses. The dose response outcomes for selected institutions are listed in Table 28.8. In some cases there has not been a benefit for doses in excess of 70 Gy for low-risk prostate cancer (HURWITZ et al. 2002; POLLACK et al. 2004; SYMON et al. 2003). POLLACK et al. (2004) reported no benefit to doses above 77 Gy compared with 67–77 Gy for low-risk patients. The Fox Chase Cancer Center experience also did not suggest a dose response for doses higher than 72 Gy. HURWITZ et al. (2002) did not see a difference in biochemical disease-free survival for patients receiving more than 68 Gy compared with those patients treated to lower doses. This lack of benefit to higher doses may be due to the small local tumor burden that is readily controlled with conventional doses. On the other hand, investigators from the Cleveland Clinic and Memorial Sloan-Kettering have shown a biochemical disease-free survival benefit for patients with low-risk disease who receive escalated doses with 3D CRT or IMRT (KHUNTIA et al. 2004; POLLACK et al. 2000; ZELEFSKY et al. 2001). In their dose escalation trial, ZELEFSKY (2001) reported 5-year biochemical disease-free survival rate of only 77% for patients receiving 64.8–70.2 Gy compared with 90% for low-risk patients receiving 75.6–86.4 Gy. KHUNTIA et al. (2004) from the Cleveland Clinic reported a 5-year biochemical disease-free survival of 52, 82, and 93% for low-risk patients receiving ≤68, 68–72, and >72 Gy, respectively.

Intermediate-risk disease has been shown to benefit from escalated radiation doses in most retrospective analyses (HURWITZ et al. 2002; KHUNTIA et al. 2004; POLLACK et al. 2000, 2004; ZELEFSKY et al. 2001). POLLACK et al. (2004) reported a 5-year biochemical disease-free survival of 24, 65, and 79% for patients receiving isocenter doses of <72, 72–75.9, and

Table 28.8 Institutional experience with dose escalation

Institution	Dose response	Dose levels, 5-year biochemical disease-free survival			Comment
Low-risk patients					
FCCC (POLLACK et al. 2000)	No	<72 Gy 81%	72–75.9 Gy 78%	≥76 Gy 82%	3D CRT for all patients; dose prescribed to isocenter for all patients
CCF (KHUNTIA et al. 2004)	Yes	≤68 Gy 52%	68–72 Gy 82%	>72 Gy 93%	Conventional RT for lowest dose, 3D CRT or IMRT for high doses, 3D CRT prescribed to isocenter, IMRT isodose prescribed to cover PTV
MSKCC (ZELEFSKY et al. 2001)	Yes	64–70.2 Gy 77%		75.6–86.4 Gy 90%	3D CRT for low dose, 3D CRT or IMRT for high doses, minimum dose prescribed to PTV (exclusive of normal organ overlap)
Intermediate-risk patients					
FCCC (POLLACK et al. 2000)	Yes	<72 Gy	72–75.9 Gy	≥76 Gy	Dose prescribed as above, FCCC
PSA <10, GS 7–10		–	50%	83%	
PSA 10–19.9		24%	65%	79%	
CCF (KHUNTIA et al. 2004)	Yes	≤68 Gy 27%	68–72 Gy 51%	>72 Gy 83%	Dose prescribed as above, CCF
MSKCC (ZELEFSKY et al. 2001)	Yes	64–70.2 Gy 50%		75.6–86.4 Gy 70%	Dose prescribed as above, MSKCC
High-risk patients					
FCCC (POLLACK et al. 2000)	No	<72 Gy –	72–75.9 Gy 27%	≥76 Gy 34%	Dose prescribed as above, FCCC
CCF (KHUNTIA et al. 2004)	Yes	≤68 Gy 21%	68–72 Gy 29%	>72 Gy 71%	Dose prescribed as above, CCF
MSKCC (ZELEFSKY et al. 2001)	Yes	64–70.2 Gy 21%		75.6–86.4 Gy 47%	Dose prescribed as above, MSKCC

≥76 Gy, respectively. KHUNTIA et al. (2004) reported 5-year biochemical disease-free survival rates of 27, 51, and 83% for radiation doses of ≤68, 68–72, and >72 Gy, respectively. In the Cleveland Clinic series the low-dose patients were treated with either standard techniques or 3D CRT with doses prescribed to the isocenter. The intermediate-dose and high-dose patients were treated with either 3D CRT or IMRT. Radiation dose was prescribed to an isodose line that covered the PTV when IMRT was utilized at the Cleveland Clinic. ZELEFSKY et al. (2002) reported a 5-year biochemical disease-free survival in intermediate-risk patients of 50 and 70% in patients receiving 64–70.2 or 75.6–86.4 Gy, respectively.

Patients with high-risk disease do not uniformly demonstrate a benefit to escalated radiation doses (POLLACK et al. 2004; SYMON et al. 2003). This may be due to the greater burden of subclinical metastases in patients with high pretreatment PSA or high Gleason scores. KHUNTIA et al. (2004) reported improved 5-year biochemical disease-free survival rates for high-risk patients of 21, 29, and 71% for radiation doses of ≤68, 68–72, and >72 Gy, respectively. ZELEFSKY reported a 5-year biochemical disease-free survival in intermediate-risk patients of 21 and 47% in patients receiving 64–70.2 or 75.6–86.4 Gy, respectively (LEVEGRUN et al. 2002; ZELEFSKY et al. 2002). CHEUNG et al. (2003) from M.D. Anderson suggested a dose response for biochemical disease-free survival for high-risk patients. They suggested that a 5-Gy dose increase beyond 78 Gy may improve PSA control for these patients.

Several randomized trials have been undertaken to demonstrate whether there is a benefit to high-dose 3D CRT. The first completed and published study from M.D. Anderson demonstrated a significant in biochemical disease-free survival in patients randomized to receive 78 Gy compared with 70 Gy (POLLACK et al. 2002). In this trial all patients began with radiation to a limited pelvic field with a standard four-field arrangement. Patients were then randomized to receive a conventional field boost to a total isocenter dose of 70 Gy or a 3D CRT boost to a total isocenter dose of 78 Gy. The largest gain from this 8-Gy dose increase was seen in the patients with pretreatment PSA >10 ng/ml. In those intermediate-risk patients the 5-year biochemical disease-free survival was 72 vs 44% for 78 or 70 Gy, respectively. The RTOG is conducting a randomized trial of 70.2 vs 79.2 Gy in intermediate-risk patients with pretreatment PSA of 10–20 and Gleason score 6 or pretreatment PSA of 0–15 with Gleason score 7. The radiation doses are prescribed as a minimum to the PTV with either 3D CRT or IMRT. The study is powered to detect an overall survival difference between the two arms. Quality-of-life end points will also be evaluated.

Normal tissue dose limits play an important role in 3D CRT and are critical in IMRT treatment planning. Reports of morbidity following 3D CRT provide radiation oncologists with dose guidelines for acceptable conformal radiation treatment plans that minimize the risk of complications. In the RTOG prospective dose escalation trial, MICHALSKI et al. (2000, 2004, 2005) and RYU et al. (2001) reported the administration of minimum PTV doses from 64.8 to 79.2 Gy in 1.8-Gy/day fractions and 74–78 Gy in 2-Gy/day fractions with lower than expected incidence of grade-3 or worse intestinal or urinary toxicity based on comparisons with historical controls.

BOERSMA et al. found a significantly higher actuarial incidence of severe rectal bleeding in patients where more than 40 and 50% of the rectal wall volume received at least 65 Gy ($p < 0.02$) than in patients in whom these volumes were smaller. Other significant cut-off levels were a rectal wall volume of 30% receiving at least 70 Gy ($p < 0.008$), and a rectal wall volume of 5% receiving at least 75 Gy ($p < 0.02$; BOERSMA et al. 1998). Based on their review of rectal toxicity in patients who received 3D CRT in the Memorial Sloan-Kettering dose escalation study, JACKSON et al. (2001) recommended keeping ≤60% of the rectal wall to receive 40 Gy and ≤30% of the rectal wall to receive ≥75.6 Gy to minimize the risk of grade-2 or greater rectal toxicity. In a recent update on their IMRT experience, ZELEFSKY et al. (2002) reported that the rate of late grade-2 rectal bleeding was 1.5%, and only 0.5% experienced grade-3 toxicity requiring one or more transfusions or laser cauterization procedures. Late grade-2 urethritis occurred in 10% of patients and another 0.5% experienced grade-3 urethral stricture. Dose constraints for this study were 100% of the prescription to the PTV (excluding overlap with normal organs) and limits of 40 and 58% of the prescription dose to the rectal wall and bladder wall, respectively. In the overlap region between the PTV and these critical organs, the constraint was set at 88% of the prescription dose for the rectum and 98% for the bladder. The prescription doses to the PTV ranged from 81.0 to 86.4 Gy in 1.8-Gy fractions (ZELEFSKY et al. 2002).

Data from M.D. Anderson indicate that volumes receiving relative high radiation dose are of major importance. The incidence of grade-II/III toxicity at 3 years decreased from 28 to 12%, if <25% of rectum volume was 70 Gy or more. HUANG et al. (2002)

recommended additional volume cut-off points at other dose levels: 41% not to exceed 60 Gy, 16% not to exceed 75.6 Gy, and 5% not to exceed 78 Gy (POLLACK et al. 2002; STOREY et al. 2000).

In a dose escalation trial employing an adaptive radiotherapy process, BRABBINS et al. (2005) escalated dose to the PTV as long as normal organ dose constraints could be met. Minimum dose prescriptions to the PTV ranged between 70.2 and 79.2 Gy based upon the following rectal and bladder constraints: <5% of the rectal wall has a dose >82 Gy, <30% of the rectal wall has a dose >75.6 Gy, <50% of the bladder volume has a dose >75.6 Gy, and the maximum bladder dose is 85 Gy. Employing these dose constraints and a patient specific "confidence-limited" PTV margin determined by daily CT scans acquired during the first week of radiation, they were able to administer a dose of 70.2–72 Gy in 49 of 277 patients (18%), >72–75.6 Gy in 131 patients (47%), and >75.6–79.2 Gy in 100 patients (36%). Chronic toxicity of patients receiving >75.6 Gy was not significantly worse than it was in patients receiving doses in the lower dose groups.

KUPELIAN (2002) reported a dose–volume relationship between rectal bleeding and rectal dosimetry in 128 patients treated with either 3D CRT or hypofractionated IMRT. The actuarial rate of rectal bleeding at 24 months was only 8% with a prescribed dose of 78 Gy in 2-Gy fractions with 3D CRT or 70 Gy in 2.5-Gy fractions with IMRT. Patients receiving hypofractionated IMRT were localized daily using a transabdominal ultrasound. The IMRT PTV margins were 4 mm posterior, 8 mm laterally, and 5 mm in all other directions. The absolute volume of rectum receiving the prescription dose was significantly associated with a risk of rectal bleeding. The relative volume of rectum exceeding the prescription dose was not associated with toxicity, but the length of rectum was variably contoured and dependent on the height of the PTV. The rectum was only contoured 1 cm above and below the PTV and not for a uniform length or anatomic definition. KUPELIAN (2002) recommends keeping the absolute volume of rectum receiving the prescription dose <15 cm^3.

Many institutions report dose constraints by a volume threshold exceeding a specified dose. The V_N refers to the volume of an organ that exceeds that dose. For example, a V_{25} of 70 Gy means only 25% of the organ receives more than 70 Gy. FIORINO (2003) reported acceptable rectal toxicity rates in 245 patients treated with 3D CRT and doses ranging from 70 to 78 Gy to the ICRU reference point. With a median follow-up of 2 years they had 23 patients with late grade-2 or worse rectal bleeding. To keep the rate of severe bleeding below 5–10% they recommend keeping the rectal V50 to 60–65%, the V60 to 45–50%, and the V70 to 25–30%.

28.4.7
Considerations for Postoperative Irradiation

Postoperative radiation therapy has an important role in maintaining tumor control in patients who have high-risk disease on pathological examination of the surgical specimen, or in patients who have biochemical failure following radical prostatectomy. While some authors have argued that adjuvant or salvage radiation therapy offers no benefit (CADEDDU et al. 1998), there is substantial evidence from both retrospective series and prospective clinical trials that radiation may improve both biochemical and clinical disease-free survival.

LEIBOVICH (2000b) reported a matched control study of 76 patients who had a single positive surgical margin with pathological T2N0 prostate cancer. Patients who received radiation therapy in the adjuvant setting (within 3 months of surgery with an undetectable postoperative PSA) had a significant improvement in their biochemical, local, or systemic rates of failure. VALICENTI (1999) reported a matched-pair analysis of 149 patients with pathological stage-T3 disease. Fifty-two of these patients received postoperative radiation therapy to a median dose of 64.8 Gy. Patients who received radiation therapy had a significant reduction in the rate of biochemical failure compared with the matched controls.

The EORTC has conducted a randomized trial of adjuvant irradiation (60 Gy in 30 fractions over 6 weeks) vs observation in 1005 men with pathologically confirmed T3 prostate cancer and at least one of the following risk factors: capsule perforation; positive resection margin; or invasion of the seminal vesicles. With a median follow-up of 5 years, there was a significant benefit of radiotherapy in biochemical disease-free survival and clinical locoregional failure rate: 52.6–74.0% ($p \leq 0.0001$) and 15.4–5.4% ($p = 0.0009$), respectively (BOLLA et al. 2005). A similar randomized trial was conducted in Germany and it demonstrated an advantage over adjuvant radiation therapy (WIEGEL et al. 2005). Data from a similar North American trial conducted by the Southwest Oncology Group is expected to be reported soon.

Salvage radiation should be considered for men who have had prior prostatectomy and now have a detectable or rising PSA. Patients that have a persistent PSA or a rising PSA within the first year after surgery do not have as good a response rate to salvage radiation therapy as men who have an initially undetectable PSA that later becomes detectable (McCarthy et al. 1994). Those patients may have occult distant metastases and therefore radiation therapy is less effective. The level of PSA at the time of salvage radiation therapy is an important predictor for response. Nudell et al. (1999) reported the results of 68 patients that received salvage radiation therapy with a 60% 3-year relapse-free survival if the PSA was <1.0 compared with only 25% if the preradiation therapy PSA was >1.0. Zelefsky et al. (1997a) reported 2-year relapse-free survival of 74% in men with a preradiation therapy PSA of <1.0 compared with 17% with a higher PSA.

Stephenson et al. (2004) reported the results of salvage therapy in 501 patients from five institutions who had failed prior radical prostatectomy. With salvage radiation therapy, the 4-year progression-free survival was 45%. They identified several prognostic factors that were associated with a poor response to radiation therapy, including a surgical Gleason score of 8–10, preradiation therapy PSA of >2.0, negative surgical margins, a postoperative PSA doubling time of <10 months, and seminal vesicle invasion. Patients with none of these adverse features had a 4-year progression-free survival of 77% following salvage radiation therapy. Some subsets of patients with Gleason score of 8–10 benefited from salvage radiation therapy if they had a pretreatment PSA of <2.0, positive surgical margins, and a PSA doubling time of >10 months.

28.4.8
Postoperative Radiation Therapy Techniques

When indicated, as in patients with Gleason score 7 or higher, seminal vesicle involvement, or positive pelvic nodes, the pelvis may be irradiated to the bifurcation of the common iliac vessels with AP–PA and right/left lateral fields to a dose of 45–50 Gy (1.8-Gy per fraction). In the past the prostate bed and periprostatic tissues were irradiated with reduced fields (8–10 cm wide and 10–12 cm cephalad to caudal dimensions) using four stationary fields (box technique; Fig. 28.13) or bilateral 120° arc rotation. In recent years 3D CRT/IMRT techniques have been used because of the low treatment morbidity and potential for dose escalation. After pelvic irradiation (45–50 Gy), additional dose is delivered to the

prostatic bed to complete 65 Gy minimal tumor dose in 1.8- to 2-Gy fractions. When only the prostatic bed is treated, the minimal tumor dose is 64–66 Gy in 1.8- to 2-Gy fractions. Figure 28.14 demonstrates the boost postoperative clinical target volume (CTB) on a treatment-planning CT scan.

Valicenti (1998) reported on 86 patients with pathological stage-T3N0 prostate cancer who received post-radical prostatectomy radiation therapy (55.8–70.2 Gy, median dose 64.8 Gy) to the prostate and seminal vesicle bed. For 52 patients with an undetectable preirradiation PSA level, the 3-year chemical disease-free survival rates were 91% for patients irra-

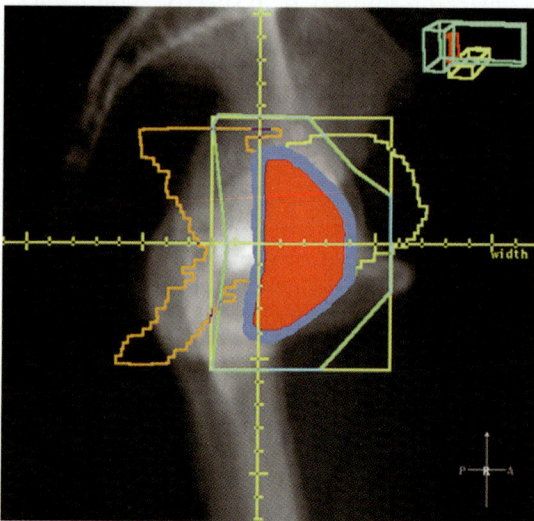

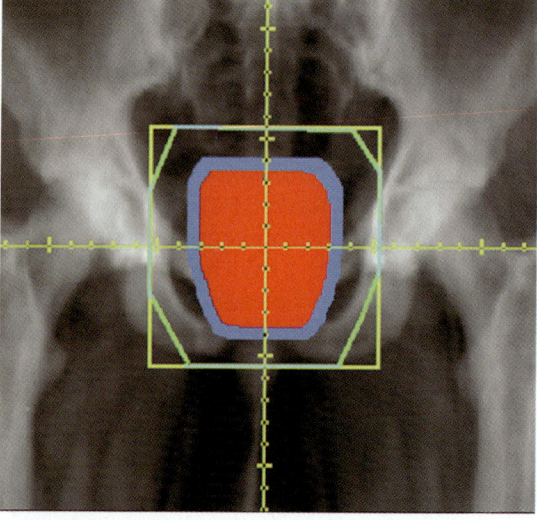

Fig. 28.13. Digitally reconstructed radiographs (AP and right lateral) demonstrate boost fields for patients receiving postoperative radiation therapy utilizing a four-field box technique. The CTV (*red*) is encompassed by a PTV margin (*blue*). In the lateral field the contours of the bladder and rectum are shielded by the block edges

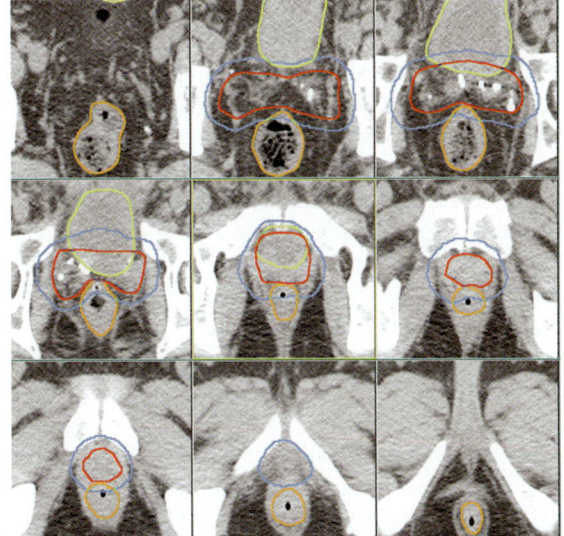

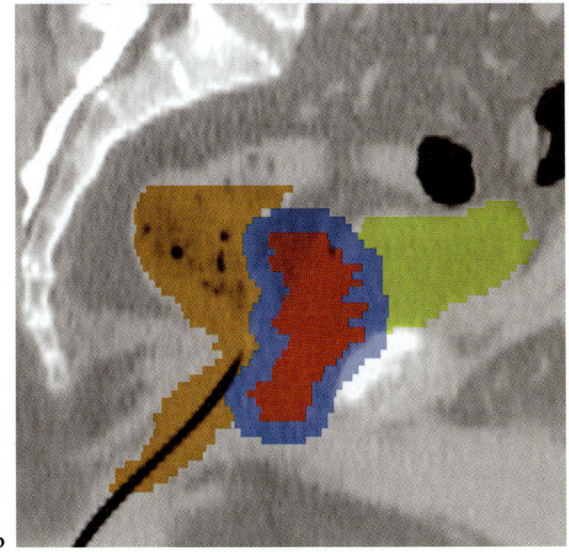

Fig. 28.14a,b. Target volumes and critical structures displayed on CT (**a**) and sagittal reconstruction (**b**) for 3D or IMRT treatment planning. For illustration purposes, only every fourth CT image is displayed. CTV (*red*) encompasses the region of the seminal vesicles and the bladder (*yellow*) neck inferiorly. The PTV (*blue*) encircles the CTV. The rectum (*orange*) is posterior

diated to 61.5 Gy or higher doses and 57% for those irradiated to lower doses ($p = 0.01$). For the 21 patients with preirradiation PSA levels >0.2 ng/ml but <2 ng/ml, the 3-year biochemical disease-free survival rates were 79% for patients irradiated to 64.8 Gy or greater doses and 33% for those irradiated to a lower dose ($p = 0.02$). They concluded that higher postoperative irradiation doses were more effective in patients with postoperative PSA levels <2 ng/ml. Data from Duke University Medical Center demonstrated improved

biochemical disease-free survival for patients receiving >65 Gy to the prostate fossa (ANSCHER et al. 2000). Because of the risk of occult metastatic disease, patients with PSA levels >2.0 ng/ml are less likely to benefit from higher radiation doses. Strong consideration should be given to treating them with hormone therapy in addition to radiation therapy.

28.5
Prostate Brachytherapy

28.5.1
Introduction

Prostate brachytherapy represents the ultimate 3D conformal therapy, permitting dose escalation far exceeding other radiation modalities with cancericidal treatment margins substantially larger than those obtainable with radical prostatectomy (RP). The resurgence of interest in brachytherapy was due to the availability of transrectal ultrasonography, the development of a closed transperineal approach and sophisticated treatment-planning software. These imaging and planning advances dramatically improved the accuracy of seed placement. In addition, computed tomography (CT)-based postoperative dosimetry provided the ability to evaluate implant quality and proactively influence outcome (MERRICK et al. 2003i).

28.5.1.1
Patient Selection

Although not all patients are acceptable candidates for brachytherapy, a reliable set of pretreatment criteria for predicting outcome has not been formulated (BLASKO et al. 2002; CROOK et al. 2001; D'AMICO et al. 1998; HAKENBERG et al. 2003; MERRICK et al. 2003h,i; NAG et al. 1999). The continued elucidation and adoption of evidence-supported planning philosophies, intraoperative techniques, and medical management should further improve biochemical results and quality-of-life (QOL) domains.

28.5.1.2
Prostate Size

Despite the absence of a clear relationship between prostate size and a greater incidence of urinary morbidity (MERRICK et al. 2000d, 2003d; SHERERTZ et al. 2001; STONE and STOCK 2000), large prostate size remains a relative contraindication to brachytherapy due to technical concerns and the perception that large prostate size increases the risk for acute and prolonged urinary morbidity (ASH et al. 2000; PEDLEY 2002). Multiple investigators, however, have successfully implanted large prostate glands with acceptable morbidity (MERRICK et al. 2000d, 2003d; SHERERTZ et al. 2001; STONE and STOCK 2000). In a study using a validated patient-administered instrument (Expanded Prostate Cancer Index Composite, EPIC), long-term urinary function did not correlate with prostate size (MERRICK et al. 2003d). On the other hand, favorable dosimetry with minimal urinary morbidity has been reported for patients with small glands (<20 cm^3; MERRICK et al. 2001a, 2003d).

28.5.1.3
Transition Zone

The transition zone (TZ) has consistently correlated with brachytherapy-related urinary morbidity (Fig. 28.15; HINERMAN-MULROY et al. 2004; MERRICK et al. 2001c; THOMAS et al. 2000). Harvard University investigators reported TZ volume to be the most important predictor of acute urinary retention following brachytherapy (THOMAS et al. 2000). In patients receiving neoadjuvant androgen deprivation therapy (ADT) for cytoreduction, International Prostate Symptom Score (IPSS) normalization, and prolonged catheter dependency, and the

need for a postbrachytherapy transurethral resection (TURP) were best predicted by the percentage of decrease in TZ volume (HINERMAN-MULROY et al. 2004). In particular, prolonged urinary morbidity was extremely unlikely in patients with a >30% decrease in TZ volume (HINERMAN-MULROY et al. 2004). The currently available data suggest that TZ may have a greater predictive power for urinary dysfunction than any other single parameter. TZ volume should always be assessed at the time of preimplant transrectal ultrasound.

28.5.1.4
Median Lobe Hyperplasia

Median lobe hyperplasia (MLH; the protrusion of hypertrophied prostate tissue into the bladder) is a relative contraindication to brachytherapy because of concerns of an increased risk of postimplant urinary morbidity and/or technical difficulties associated with the implantation of intravesical tissue (PEDLEY 2002). In a small series, complete dosimetric coverage of the median lobe was reported; however, 38% of the patients developed prolonged postimplant urinary retention, whereas others experienced persistent IPSS elevation (WALLNER et al. 2000). Although MLH should not be considered an absolute contraindication to brachytherapy, such patients should be approached with caution. It is conceivable that preimplant resection of the intravesical component may reduce the incidence of brachytherapy-related urinary morbidity.

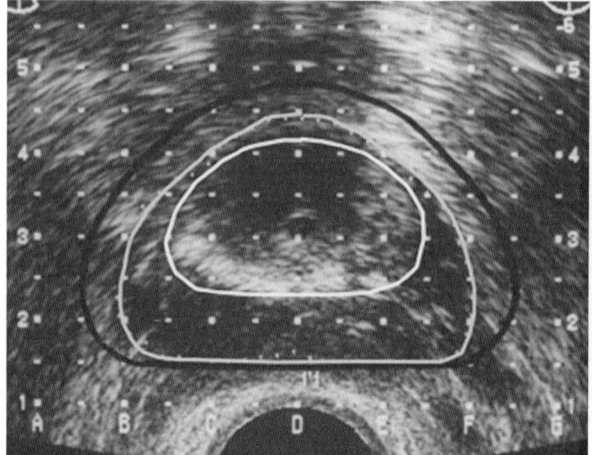
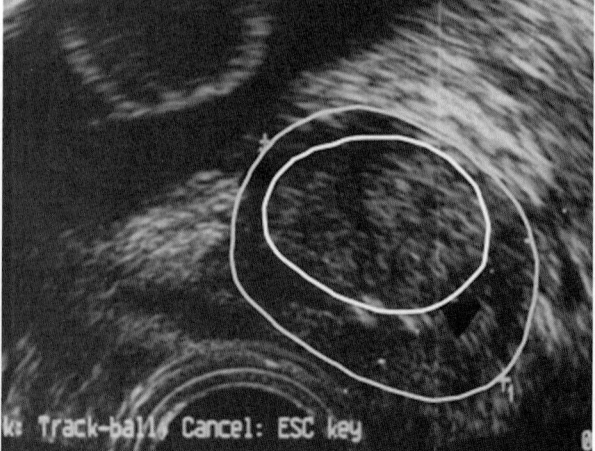

a

Fig. 28.15a,b. Transverse (**a**) and sagittal (**b**) ultrasound images. The prostate is outlined in *gray* and the transition zone in *white*. The transverse image also illustrates the planning target volume in *black*

28.5.1.5
International Prostate Symptom Score

The role of IPSS in predicting urinary morbidity (including urinary retention) has been studied extensively with conflicting conclusions (Ash et al. 2000; Bucci et al. 2002; Gelblum et al. 1999; Hakenberg et al. 2003; Landis et al. 2002; Merrick et al. 2000d, 2003d; Pedley 2002; Terk et al. 1998). Although almost all brachytherapy patients develop urinary irritative/obstructive symptoms with up to 34% developing acute urinary retention (Gelblum et al. 1999; Han et al. 2001a; Landis et al. 2002; Merrick et al. 2000d; Terk et al. 1998; Thomas et al. 2000), <5% require a urinary catheter for >1 week (Merrick et al. 2000d, 2002e). Preimplant IPSS predicted the duration of postimplant obstructive symptomatology (Bucci et al. 2002; Merrick et al. 2000d; Terk et al. 1998) but did not correlate with long-term urinary QOL (Merrick et al. 2003d). In contrast to three recently published patient-selection guidelines (Ash et al. 2000; Hakenberg et al. 2003; Pedley 2002), prospective studies have demonstrated little correlation between preimplant IPSS, urodynamic studies, postvoid residual urine or preimplant cystourethroscopy, and acute or long-term urinary function (Gray et al. 2000; Landis et al. 2002).

28.5.1.6
Prostatitis

A recent study reported no relationship between the presence or severity of prostatitis and the incidence of urinary retention or prolonged IPSS elevation following implantation (Hughes et al. 2001).

28.5.1.7
Pubic Arch Interference

Pubic arch interference (the obstruction of anterior needle placement insertion by a narrow pubic arch) is a relative contraindication to brachytherapy. Surprisingly, prostate volume has proved to be a poor predictor of pubic arch interference (Bellon et al. 1999). With the use of the extended lithotomy position and/or veering needles around the arch, almost all patients can be successfully implanted with favorable postimplant dosimetry regardless of the degree of pubic arch interference (Bellon et al. 1999).

28.5.1.8
Transurethral Resection of the Prostate

In contemporary brachytherapy series with the use of peripheral source loading and limitation of the radiation dose to the transurethral resection of the prostate (TURP) defect to approximately 110% of prescription dose, the risk of incontinence in patients with a preimplant TURP has been reported to be ≤6% (Stone et al. 2000; Wallner et al. 1997). Using EPIC, patients with a preimplant TURP have been found to have urinary QOL approaching that of non-TURP brachytherapy patients (Merrick et al. 2004a).

Following brachytherapy, approximately 2% of patients develop prolonged urinary retention with the vast majority eventually spontaneously urinating (Merrick et al. 2004a). If a postimplant TURP or transurethral incision of the prostate (TUIP) is necessary, it should be delayed for as long as possible (ideally a minimum of 12 months). Significant urinary morbidity has been demonstrated in approximately half of patients undergoing a postimplant TURP. Patients requiring a pre- and postimplant TURP have an especially high risk of urinary incontinence (Merrick et al. 2004a).

To minimize postbrachytherapy TURP-related incontinence, the preservation of the bladder neck at the 5 and 7 o'clock positions with minimal use of cautery is essential to maintain sufficient prostatic–urethral blood supply (Stone and Stock 2002).

Age

Although there remains a surgical bias to recommend RP for younger patients, patient age alone should not influence treatment decisions. In fact, patient age may be a stronger predictor of prostate cancer curability than differences in preimplant PSA (Carter et al. 1999). Older patients have been reported to be at increased risk for extracapsular extension, higher Gleason scores, and a greater propensity for distant metastases when compared with younger patients (Carter et al. 1999; Herold et al. 1998; Smith et al. 2000). Following brachytherapy, outstanding biochemical outcomes (median PSA <0.1 ng/ml) have been reported for hormone-naïve patients ≤62 years of age (Merrick et al. 2004b). On the other hand, older patients tolerate brachytherapy as well as younger men (Merrick et al. 2000d, 2003d).

Obesity

Obesity presents substantial procedural difficulties for RP and external-beam radiation therapy (XRT), but only relatively minor problems for brachytherapy (Merrick et al. 2005a). Favorable biochemical and QOL outcomes have been demonstrated for brachytherapy patients regardless of body mass index (BMI; Merrick et al. 2002c, 2005a).

28.5.1.9
Tobacco

Consistent with a correlation between cigarette smoking and aggressive prostate cancer, tobacco consumption has correlated with a trend for poorer biochemical progression-free survival following brachytherapy (Merrick et al. 2004d). In addition, tobacco consumption has been demonstrated to delay wound healing, compromise small blood vessel function, and exacerbate brachytherapy-related morbidity. Although the absolute differences were small, tobacco was a robust predictor for adverse late urinary and rectal QOL following brachytherapy (Merrick et al. 2003d,f). Following RP and XRT, the role of tobacco in treatment-related morbidity has not been explored but is likely to be consistent with tobacco-related brachytherapy outcomes.

28.5.1.10
Inflammatory Bowel Disease

Inflammatory bowel disease, ulcerative colitis, and regional enteritis (Crohn's disease) have been considered relative contraindications to radiation therapy; however, in a small brachytherapy series, no increased risk of gastrointestinal morbidity was reported (Grann and Wallner 1998).

28.5.1.11
Adverse Pathological Features

High Gleason score, perineural invasion, and percentage of positive biopsies correlate with a greater likelihood of extracapsular extension leading some to conclude that patients with high-risk features may not be adequately treated with brachytherapy (Crook et al. 2001; D'Amico et al. 1998; Hakenberg et al. 2003). In contrast, multiple brachytherapy studies have demonstrated favorable biochemical outcomes for hormone-naïve patients at high risk of extraprostatic extension (Blasko et al. 2000; Dattoli et al. 2003; Merrick et al. 2001b, 2003a, 2004e). Pathological evaluation of RP specimens demonstrate that extraprostatic extension is almost always limited to within 5 mm of the prostate capsule (Davis et al. 1999). As such, high-quality brachytherapy with or without supplemental XRT should sterilize extraprostatic extension (Fig. 28.16; Merrick et al. 2003b). The relative resiliency of brachytherapy to adverse pathological features is likely a result of intraprostatic dose escalation with therapeutic radiation dose delivery to the periprostatic region (Merrick et al. 2001a, 2003b, 2004e).

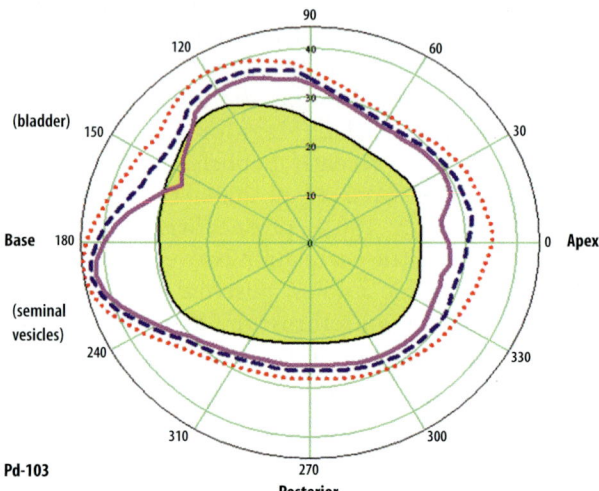

Fig. 28.16. Mean postimplant dosimetric margins >6 mm at the 100% isodose line

28.5.1.12
Prostatic Acid Phosphatase

Pretreatment enzymatic prostatic acid phosphatase (PAP) is predictive of biochemical outcome and/or disease-specific survival following potentially curative local treatment (Dattoli et al. 2003; Han et al. 2001b). In a contemporary RP series, PAP did not predict for organ-confined disease and/or lymph node status but was an independent predictor of biochemical recurrence (Han et al. 2001b). Similarly, PAP was the strongest predictor of freedom from biochemical progression for intermediate and high-risk brachytherapy patients (Dattoli et al. 2003). Although a substantial minority of brachytherapy patients with an elevated pretreatment PAP remain curable, it is conceivable that these patients may benefit from adjuvant systemic therapy.

28.5.2
Brachytherapy Planning

Favorable biochemical results have been obtained with a variety of planning and intraoperative techniques, of which no method has been proven superior (BUTLER et al. 2000b; MERRICK and BUTLER 2000). In general, four seed loading philosophies (uniform loading, peripheral loading, modified uniform loading, and modified peripheral loading) have been utilized with peripherally based planning philosophies most commonly used (PRETE et al. 1998). Although quality is easy to conceptualize, it is more difficult to quantitate. It is universally accepted that an adequate implant should encompass the target volume, but no consensus exists as to what represents the target. In addition, urethral and rectal tolerances have not been well defined, and the significance and degree of dose homogeneity throughout the implant region remains unclear. A multi-institutional analysis revealed that although prostate brachytherapy prescription doses are uniform, substantial variability exists regarding target volume, seed strength, dose homogeneity, treatment margins, and extracapsular seed placement (MERRICK et al. 2005c).

Brachytherapy planning entails either preplanning (in which a transrectal ultrasound volumetric study of the prostate gland is obtained before implantation) or intraoperative planning using nomograms or real-time techniques (MERRICK and BUTLER 2000; STOCK et al. 1998). In our preplanned approach, the planning target volume (the prostate gland with a periprostatic margin) is determined by a 3- to 8-mm enlargement of each ultrasound slice with a resultant planning target volume approximately 1.75 times the ultrasound-determined volume (MERRICK and BUTLER 2000). The rationale for treatment margin is based on pathological measures of the probability of microscopic extracapsular disease and estimates that seed placement uncertainty is approximately 5 and 3 mm in the longitudinal and transverse directions, respectively (ROBERSON et al. 1997). In addition, the utilization of treatment margins significantly decreases the effect of edema on postoperative dosimetry (BUTLER et al. 2000a; WATERMAN et al. 1998). At the periphery of the implant target volume, the radiation dose decreases by as much as 20 Gy/mm (DAWSON et al. 1994). Our planning technique mandates that the radiation dose be prescribed to the planning target volume with margin (BUTLER et al. 2000b; MERRICK and BUTLER 2000). Table 28.9 outlines criteria by

Table 28.9. Preimplant dosimetric evaluation criteria

Evaluated quantity	Parameter[a]	Value	
		I-125	Pd-103
Patient-specific needs	Tx volume, TURP[b], etc.	Primary importance	
Coverage of the planning volume	V_{100} D_{90}	>99.8% volume 125–140% mPD	
Dose homogeneity	V_{150}	40–55% volume	55–70% volume
High dose volume	V_{200}	<15% volume	15–25% volume
Urethra volume coverage	UV_{125}	80–100% volume	50–100% volume
	UV_{150}	<15% volume	<25% volume
Urethra dose	UD_{50}	130–145% mPD	120–140% mPD
	UD_{10}	140–150% mPD	130–160% mPD

[a] V_{100}, V_{150}, and V_{200} are the percentages of the planning target volume (PTV) covered by 100, 150, and 200% of the prescribed dose (mPD), respectively. D_{90} is the minimum dose covering 90% of the PTV. UV_{125} and UV_{150} are percentage of volume of the urethra at the defined mPD. UD_{50} and UD_{10} are the minimum doses covering 50 and 10%, respectively, of the urethra volume
[b] TURP, transurethral resection of prostate

which all plans are evaluated based on DVH of the planning target volume and urethra.

28.5.3
Post-Implant Evaluation

Postoperative CT-based dosimetric analysis provides detailed information regarding the coverage and uniformity of an implant, affords the ability to compare various intraoperative techniques, and provides a sound basis for future improvement. Although CT determination of prostate volume is widely accepted for external beam planning, the use of CT for brachytherapy purposes remains controversial (CROOK et al. 2004). Accurate delineation of prostate contours on postimplant CT may be difficult because of postoperative edema, degradation of the image due to implanted metallic seeds, and a possible tendency to overestimate prostate volume from CT compared with transrectal ultrasound; however, a close correlation has been demonstrated for CT and ultrasound-determined prostate volumes, provided that the levator ani muscles and the anterior venous complex are not included in the CT volume (Badiozamani et al. 1999; Merrick et al. 1999a). When provided with instructions for prostate volume determination, postbrachytherapy prostate volumes were comparable among multiple brachytherapists (Han et al. 2003). Fortunately, if implants are designed and executed with generous periprostatic margins, the determination of post-implant CT prostate volume does not significantly influence the dosimetric outcome (Merrick et al. 1999a).

The timing of CT after implantation remains controversial. Some groups recommend CT evaluation be obtained 30 days following implantation to allow for resolution of edema, whereas others propose day-0 dosimetry to provide information about edema when it is close to its maximum extent and for prompt closure of the learning curve (Merrick et al. 2001a). For intraoperative dosimetric evaluation, a detailed knowledge of day-0 threshold dosimetric parameters is essential to evaluate the advisability of corrective seed placement.

Biochemical progression-free survival is directly related to implant quality. Stock et al. (1998) reported a dose-response curve for patients undergoing monotherapeutic I-125 with superior biochemical results with a day-30 D90 ≥140 Gy. Detailed day-0 dosimetry demonstrated that prescribed radiation doses are routinely obtainable with day-0 evaluation using brachytherapy target volumes that include

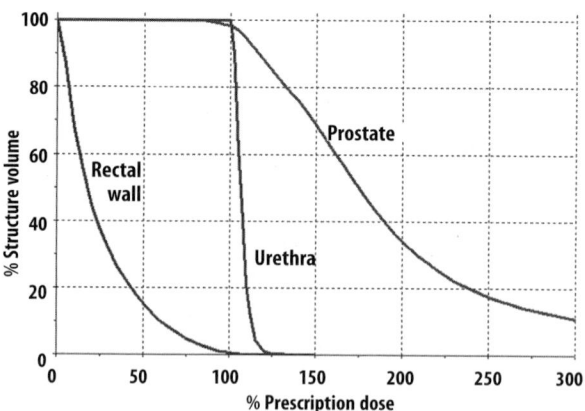

Fig. 28.18. Cumulative dose volume histograms of the prostate, urethra, and the volume encompassed by the rectal wall. Prostate dosimetric summary parameters of interest are V100 = 98.3% of the 38.7 cm^3 day-0 CT volume. The prostate: V150 = 68.8% and V200 = 34.0% of the volume, and D90 = 116.7% of the 125-Gy monotherapy Pd-103 dose prescribed; the urethra: V100 = 100%, V120 = 1.0%, and V150 = 0%. The volume of the rectal wall receiving 100% of the prescribed dose was 0.18 cm^3

generous periprostatic margins and urethral sparing techniques (Figs. 28.17, 28.18; Merrick et al. 2001a, 2003b). Extracapsular treatment margins of 6.5 and 9.6 mm at the 100 and 75% isodose lines, respectively, have been reported (Fig. 28.16; Merrick et al. 2003b). Except for the bladder neck and posterior prostate border, the 100% isodose margin was 5 mm or greater. Approximately 35% of the seeds are placed in extracapsular locations with a seed fixity rate >98% (Merrick et al. 2000b). There is a strong suggestion that day-0 brachytherapy treatment margins may be as important as the traditional dosimetric parameters (i.e., V100, D90) for predicting PSA outcomes after prostate brachytherapy (Choi et al. 2004; Merrick et al. 1999b). Generous periprostatic treatment margins are useful in patients with any risk of extracapsular extension and a low risk of pelvic lymph node involvement/distant metastases.

28.5.3.1
Isotope

Although traditionally I-125 has been used for Gleason scores 2–6 and Pd-103 for Gleason scores 7–10, no definitive data support the potential curative superiority of one isotope over another for any clinical stage, Gleason score, or pretreatment PSA (Cha et al. 1999; Wallner et al. 2002a, 2003).

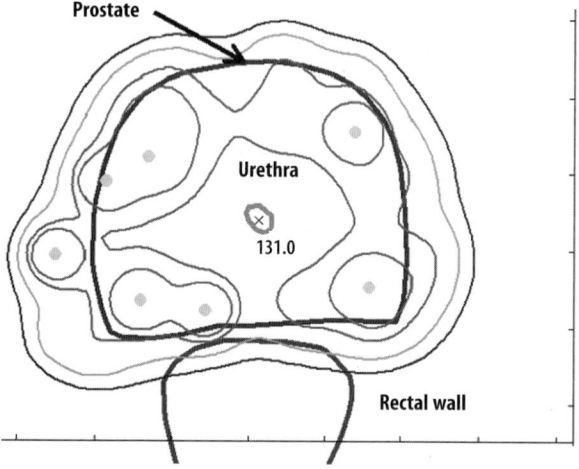

Fig. 28.17. Transverse postimplant isodose coverage of the prostate, urethra, and rectal wall for a 125-Gy monotherapy implant of a prostate which had a preimplant ultrasound volume of 25.9 cm^3 and a day-0 CT volume of 38.7 cm^3. The isodoses displayed are 65, 100, 50, and 200% of the prescribed dose. The dose point shown is at the center of the urethra. *Tick marks* on the border are at 1-cm increments

28.5.4
Supplemental Therapies

Both XRT and adjuvant ADT are commonly used as supplements to brachytherapy in patients with adverse pathological and/or biochemical features. Contemporary series, however, strongly suggest that with high-quality brachytherapy dose distributions and generous periprostatic treatment margins (BLASKO et al. 2000; MERRICK et al. 2003b), the vast majority of brachytherapy patients can be managed with implant alone.

28.5.4.1
External-Beam Radiation Therapy

The rationale for supplemental XRT combined with brachytherapy is to enhance the coverage of periprostatic tissue, escalate the dose to the intraprostatic tumor, supplement inadequate radiation dose distributions, and/or treat locoregional disease. The ABS has recommended monotherapy for patients with clinical stage T1–T2a, PSA ≤10 ng/ml, and Gleason score ≤6 with the addition of supplemental XRT for all other patients (NAG et al. 1999). Recently, however, the utility of supplemental XRT has been questioned by favorable biochemical outcomes following monotherapeutic brachytherapy in patients with higher pretreatment PSAs and/or Gleason scores (BLASKO et al. 2000). Supplemental XRT has also been demonstrated to exacerbate brachytherapy-related morbidity (MERRICK et al. 2003d,f).

28.5.4.2
Androgen Deprivation Therapy

Despite recent reports detailing favorable biochemical outcomes for hormone-naïve brachytherapy patients with higher-risk features (DATTOLI et al. 2003; DAVIS et al. 1999; MERRICK et al. 2003b), intermediate- and high-risk brachytherapy patients often receive adjuvant ADT as an extrapolation from the conventional XRT dose (65–70 Gy) in the literature. The favorable interaction between conventional dose XRT and ADT likely results from the inability of conventional XRT doses to sterilize large bulky prostate cancers and as such may not be applicable to brachytherapy. In a large retrospective matched-pair analysis, no benefit for ADT with brachytherapy was discerned for any risk group, Gleason score, pretreatment PSA, or clinical stage (POTTERS et al.

2000). Although subgroups of high-risk brachytherapy patients may benefit from adjuvant ADT, that patient population has not been definitively identified. In addition, ADT has been implicated in brachytherapy-related morbidity (MERRICK et al. 2003d,h, 2004f; TERK et al. 1998).

28.5.5
PSA Spikes

Following brachytherapy, up to one-third of all hormone-naïve patients develop a temporary PSA spike without adverse impact on long-term biochemical control (MERRICK et al. 2002d, 2003g). It has been proposed that spikes are a result of any mechanism compromising membrane integrity in PSA-producing epithelium (MERRICK et al. 2002d, 2003g). The PSA spikes have been correlated with younger patient age, clinical stage, benign prostatic hyperplasia, baseline postimplant PSA, and postoperative dosimetric parameters (MERRICK et al. 2002d, 2003g). The PSA spikes are least common in patients with a posttreatment PSA ≤0.2 ng/ml (MERRICK et al. 2002d). For patients receiving neoadjuvant ADT, postbrachytherapy "PSA progression" is often observed; however, the absolute increase is usually minimal (<0.5 ng/ml; MERRICK et al. 2004c).

Posttreatment prostate biopsies to differentiate viable cancer from a PSA spike can be misleading. Previous studies of prostate biopsies following XRT have demonstrated a slow resolution in the histological evidence of prostate cancer which may require 3 years of follow-up for accurate pathological assessment. Despite an increasing PSA and a biopsy positive for recurrent cancer, prostate brachytherapy patients have been reported to normalize serum PSA without additional therapeutic intervention (REED et al. 2003).

28.5.6
Approaches to Minimize Morbidity

28.5.6.1
Urinary

An enlarging body of data demonstrates that brachytherapy-related urinary morbidity can be lessened with refinements in patient selection, medical intervention, and intraoperative techniques.

Urinary morbidity in the immediate postimplant setting has been well documented with preimplant

IPSS correlating with the duration of postimplant obstructive symptoms (Desai et al. 1998; Merrick et al. 2000d, 2002e; Terk et al. 1998). Alpha blockers are widely used to ameliorate brachytherapy-related urinary morbidity, and the timing of their initiation may substantially influence their effect (Merrick et al. 2000d, 2002e). Following a policy of prophylactic and prolonged medication use, IPSS values returned to baseline values 3 months following brachytherapy (Merrick et al. 2000d, 2002e). In contrast, without the use of prophylactic alpha blockers, a timeline of approximately 12 months was reported for IPSS resolution (Desai et al. 1998); however, the use of alpha blockers did not correlate with the incidence of prolonged urinary catheter dependency or the need for postimplant surgical intervention (Merrick et al. 2002e). The initiation of alpha blockers 2–3 weeks before implantation, with continuation until IPSS normalization, maximizes their beneficial effect. In terms of isotope, a prospective randomized trial comparing Pd-103 with I-125 reported a statistically faster rate of IPSS resolution in the Pd-103 arm (Wallner et al. 2002a).

Although dysuria is a relatively common event during the first few years following brachytherapy, only rarely is it severe in frequency or intensity (Merrick et al. 2003e). To date, no significant factors for the prediction of dysuria have been reported. In addition, the use of alpha blockers has not significantly diminished the duration of dysuria; however, anti-inflammatory agents (low-dose prednisone, 5–10 mg daily) and the avoidance of bladder irritants (i.e., caffeine, alcohol) may alleviate the intensity and frequency of treatment-related dysuria.

Wallner and colleagues (1995) reported an association between urinary morbidity and urethral doses >250% minimum peripheral dose (mPD). In contemporary series, urethral doses have not been correlated with urinary morbidity because of sophisticated treatment planning with urethral sparing (100–140% mPD; Allen et al. 2005). While urethral doses >150% mPD should be minimized, underdosage (<100% mPD) should be avoided. Leibovich et al. (2000a) reported that the mean distance from the urethra to the nearest foci of cancer was 3 mm with 17% of all prostate cancers abutting the urethra.

It is conceivable that certain segments of the urethra may be more sensitive to radiation-induced morbidity; however, detailed urethral dosimetry did not substantially improve the ability to predict urinary morbidity (Allen et al. 2005). Radiation doses of 100–140% mPD were well tolerated by all segments of the prostatic urethra with resultant

tumoricidal doses to foci of periurethral cancer (Allen et al. 2005).

Urethral strictures occur in 5–12% of patients and are directly related to overimplantation of the periapical region (Merrick et al. 2002b). Strictures typically involve the bulbomembranous urethra and are easily managed with dilatation. Day-0 CT-based dosimetry demonstrated that radiation doses to the bulbomembranous urethra were significantly greater in patients with strictures than in those without (Merrick et al. 2002b). With careful attention to implant technique, including extensive use of the sagittal plane for deposition of the seeds, it is possible to implant the apex with a 5-mm margin without "dragging" seeds into the region of the bulbomembranous urethra. An ADT of longer than 6-month duration also increased the likelihood of urethral strictures (Merrick et al. 2002b).

28.5.5.2
Rectal Morbidity

Rectal complications primarily consist of mild self-limiting proctitis with an incidence of 4–12% and usually resolve spontaneously (Gelblum and Potters 2000; Merrick et al. 2003c,f; Snyder et al. 2001). Snyder et al. (2001) reported grade-II proctitis to be volume dependent for a given dose with no case developing more than 36 months following implantation. With day-30 dosimetry, an 8% rate of grade-II proctitis was reported when <1.8 cm^3 of rectum was exposed to 160 Gy following I-125 monotherapy, whereas the risk increased to 25% when >1.8 cm^3 of rectum was exposed. In a prospective randomized trial, the minimum dose received by 5% of the rectum best correlated with brachytherapy-related rectal dysfunction (Merrick et al. 2003c). A limited number of errant perirectal sources did not increase the risk of rectal bleeding, provided that the overall rectal wall doses were within acceptable values (Mueller et al. 2004). To compare rectal doses between series, the timing of postimplant CT must be documented. Waterman and Dicker (1999) reported that the minimum dose that encompassed 10% of the surface area of the rectum increased on average by 68% from day 0 through day 30.

Rectal ulceration and fistula formation have occasionally been reported (Howard et al. 2001). Although dose is associated with rectal bleeding, no correlation between rectal dose and the development of a fistula were identified with the subsequent conclusion that severe complications may occur in an unpredictable manner typically unrelated to known clinical, treatment, or dosimetric parameters (HOWARD et al. 2001). Although no studies have correlated constipation with rectal toxicity, constipation significantly increases the radiation dose to the rectum (MERRICK et al. 2000a). Postimplant attention to bowel habits for two half-lives of the implanted isotope will minimize rectal distention and decrease the dose to the anterior rectal wall. Bowel function assessment by patient-administered questionnaires has documented that long-term bowel dysfunction following brachytherapy is relatively uncommon (MERRICK et al. 2003c,f; TALCOTT et al. 2001).

Intraoperatively, careful attention to implant technique and ultrasound anatomy will reduce the dose to the anterior rectal wall and minimize bowel dysfunction. Extensive use of both transverse and sagittal images to confirm appropriate needle placement and the use of multiple ultrasound frequencies helps ensure proper seed placement. Higher transducer frequencies results in clearer definition of anatomy closer to the probe. Posterior-row needle placements performed with the 7.5-MHz setting help ensure that needles are placed just within the posterior prostate capsule and not in the rectal wall.

28.5.5.3
Erectile Dysfunction

The penile erectile bodies (the paired corpora cavernosa and midline corpora spongiosum) represent potential targets for structure-specific radiation-associated erectile dysfunction (ED; Fig. 28.19). Detailed reports illustrating the image-based anatomy of the proximal penis have been published (WALLNER et al. 2002b). The penile bulb is best visualized on T2-weighted MR images and appears as an oval-shaped, hyperdense midline structure located on average 10–15 mm inferior to the apex of the prostate gland.

In XRT studies, dose to the proximal penis has been related to posttreatment ED with the conclusion that XRT either directly or indirectly damaged the vascular supply of the erectile tissue as well as the nerves that supply the cavernosa smooth muscles (CARRIER et al. 1995; FISCH et al. 2001; ROACH et al. 2004). Subsequent histological examination of proximal penile shaft specimens demonstrated that the number of nitric oxide synthase-containing nerve fibers per corpora cavernosa were significantly decreased in the irradiated groups (CARRIER et al. 1995). Following brachytherapy, MERRICK et al.

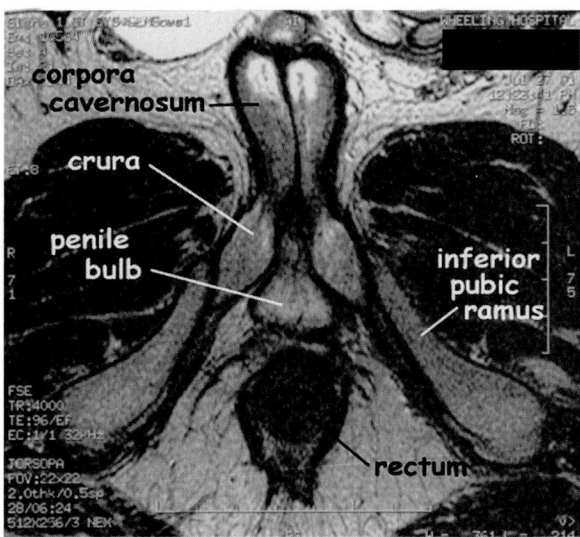

Fig. 28.19. Transverse magnetic resonance image of the proximal penis

(2001d, 2005b) reported that treatment-induced ED was highly dependent on the radiation dose to the proximal penis. With day-0 dosimetry, maximum potency preservation was dependent on limiting the penile bulb D50 to ≤30% of prescription dose and the proximal crura D50 to ≤50% of prescription dose (MERRICK et al. 2005b). With adherence to these cutpoints, >70% of patients maintained potency vs only 30% in patients who received higher doses.

Following RP, ED has been correlated with surgical trauma to the neurovascular bundles (NVB). It is conceivable that excessive NVB radiation could also represent a potential mechanism of brachytherapy-induced ED (DIBIASE et al. 2000); however, in both prospective and retrospective studies, no relationship between NVB radiation doses and the development of brachytherapy-induced ED was discerned (MERRICK et al. 2000c, 2001e). Although initial reports have been negative, it is possible that with additional follow-up, radiation dose to the NVB may contribute to brachytherapy-related erectile dysfunction.

From a clinical perspective, potency preservation after brachytherapy is most closely related to preimplant erectile function (MERRICK et al. 2002a, 2005b; STOCK et al. 2001). The role of neoadjuvant ADT has been mixed (MERRICK et al. 2002a, 2005b; POTTERS et al. 2001; STOCK et al. 2001), whereas the choice of isotope appears unrelated (MERRICK et al. 2002a, 2005b; STOCK et al. 2001). A prospective randomized trial reported that only preimplant erectile function and dose to the proximal penis were statistically significant predictors of brachytherapy-related ED

(MERRICK et al. 2005b). Fortunately, brachytherapy-related ED responds favorably to sildenafil citrate (MERRICK et al. 1999c, 2002a).

A lack of erectile activity may be deleterious to erectile function, and as such, patients should be encouraged to develop regular erections with or without sexual relations (MULHALL 2001). In the absence of routine penile erections, the corporal smooth muscle experiences chronic hypoxia with a resultant loss of elasticity and distensibility which may lead to a venous leak (MORELAND 1998). The early institution of phosphodiesterase (PDE)-5 inhibitors in patients with compromised or absent erections may help restore erectile function (MULHALL 2001).

Although brachytherapy-induced ED is likely multifactorial, the available data strongly support the proximal penis as an important site-specific structure. Suboptimal seed placement (either due to poor planning or poor implementation) of periapical radiation sources results in excessive radiation doses to the proximal penis. Refinements in implant technique to avoid overaggressive periapical implantation, along with the extensive use of sagittal ultrasonography during implantation, will decrease the radiation dose to the proximal penis and ultimately may improve potency preservation.

Although the majority of the brachytherapy literature has demonstrated biochemical results and morbidity profiles that compare favorably with competing local modalities, it has become increasingly apparent that efficacy and morbidity are highly dependent on implant quality. Sophisticated dosimetric analyses have demonstrated that cure rates, urinary and rectal complications, and potency preservation are related to specific source placement patterns and the subsequent dose gradients produced. Our upcoming challenges include the development of intraoperative planning and dosimetry to improve implant quality, improved intraoperative technique to include manual dexterity and imaging, and the development of evidence-based algorithms.

28.6
High-Dose-Rate Brachytherapy

28.6.1
Introduction

The recognition that the radiation dose is an independent determinant of biochemical outcome has been reported in several retrospective, sequential, prospective, and now randomized trials (POLLACK et al. 2002; VICINI et al. 2002). This is especially true for patients with intermediate- and high-risk prostate cancer (for overviews, see MANGAR et al. 2005; NILSSON et al. 2004). The development of sophisticated technologies has contributed to the resurgence of prostate cancer brachytherapy as a viable option for the treatment of localized prostate cancer. With high-dose-rate brachytherapy (HDR-BT) the delivery of highly conformal radiotherapy can be achieved to the target and with comparatively low radiation burden to the adjacent normal tissues (e.g., rectum and bladder). The apparent advantages with hypofractionation in prostate cancer and the total doses achievable with HDR-BT are discussed below.

28.6.2
Technological Aspects

The HDR-BT techniques using temporary iridium-192 implantation were introduced in the early 1980s (CHARYULU 1980; MARTINEZ et al. 1985; SYED et al. 1983), and encouraging long-term outcome data have been reported (BORGHEDE et al. 1997; KOVACS et al. 1999; MARTINEZ et al. 2003; SYED et al. 2001). The introduction of transrectal ultrasound (TRUS)-guided implantation technologies and 3D dose-planning systems have further revolutionized this treatment modality. A large variety of dose-planning systems are currently being used, both commercially available systems and systems developed at different treating centers. There are also different optimization programs available, such as, for example, dose-point optimization, geometrical optimization, and inverse optimization programs (EDMUNDSON 1994; HSU et al. 2004; KOLKMAN-DEURLOO et al. 1994; LESSARD and POULIOT 2001; YOSHIOKA et al. 2005). Many centers use intraoperative optimization dose-planning systems, whereas others perform pre-therapy dose planning. No prospective randomized studies have been performed between different dose planning systems and, thus, no firm recommendations can yet be given. Most dose-planning techniques are TRUS-based, but CT- and MRI-based dose-planning systems are being advocated at other centers. Briefly, when using the TRUS-based techniques, the patient is placed in an extended lithotomy position and a Foley catheter is inserted into the bladder to visualize the urethra. An ultrasound transducer is introduced to the rectum to evaluate the prostate, urethra, urinary bladder, and seminal vesicles. A matrix indicating possible positions for the needles (according to the template to be used for the transperineal insertion of needles) is overlaid in the TRUS image. During TRUS, axial cross sections are acquired in 5-mm steps using a stepping device and transferred to the treatment-planning software. The dose-planning system is used to optimize the dose in such a way that, in each 5-mm step, it enables coverage of the target without exceeding tolerance of rectum, urethra, and bladder. The planning target volume (PTV) is defined as the entire prostate without the central area around the urethra. One of the obvious advantages with HDR-BT is, in comparison with permanent seed implantation, the greater ability to optimize dose distribution by varying source dwell times along the needles/catheters.

Most patients have spinal anesthesia for pain management, and the treatment is performed with the patient placed in the extended lithotomy position with a Foley catheter inserted into the bladder. The treatment is usually delivered with the use of 17-G needles which are inserted transperineally into the prostate through the template mounted on the probe stand. (Some centers prefer the use of flexible catheters instead of needles.) The positioning (x–y axis) of each needle according to the template–matrix coordinates and the insertion length (z axis) are verified with the TRUS. The positioning of each needle should be within 0–2 mm from the position recommended by the dose plan. Typically, 9–20 needles are used depending on the size and anatomy of the prostate and adjacent organs at risk.

The treatment is performed using a computer-based remote afterloading technique in which the iridium-192 source is inserted into each of the needles utilizing stop positions and dwell times according to the dose plan. There is no consensus on dose constraints to normal tissues; however, the Radiation Therapy Oncology Group (RTOG) phase-II HDR study (p0321) has limited to 1 cc the volume of urethra receiving 125% of the prescription dose and to 1 cc of rectum and bladder receiving 75% of prescription dose. Usually it is possible to achieve these dose constraints while obtaining at least 90–95% of the prescribed dose to the clinical target volume.

28.6.3
Radiobiological and Dose Fractionation Aspects

There is ample clinical data supporting the assumption that prostate tumors have exceptionally low α/β values, i.e., reflecting a stronger enhancement of

tumor effect than late complications for larger and fewer radiation dose fractions (BRENNER and HALL 1999; BRENNER et al. 2002; FOWLER et al. 2001, 2003; FOWLER 2005). The careful dose escalation studies by MARTINEZ and co-workers (2001), and the remarkably good outcome data, have resulted in the first direct clinical evidence that α/β is as low as 1.2–1.5 Gy (BRENNER et al. 2002). In most studies, HDR-BT has been used as a radiation boost technique in combination therapy with external-beam radiation therapy (EBRT). A wide range of HDR fractionation schedules have been reported in these studies. A review of the literature shows that the median brachytherapy dose used in the larger trials is 20 Gy and the median EBRT dose 45 Gy in 5 weeks (MORTON 2005). Two different approaches to HDR fractionation have evolved: separate catheter insertions for each brachytherapy treatment or, alternatively, one single insertion followed by two to four fractions delivered over 2 days. The approach can be performed on an outpatient basis, whereas the latter, which is more commonly used in the U.S., requires hospital admission. The treatment regimens, technological aspects, and calculated total dose equivalents at 2 Gy per fraction from the largest treatment series are summarized in Table 28.10.

28.6.4
Clinical Outcome Data from HDR-BT Plus EBRT

The majority of patients treated with combined HDR-BT and EBRT have had relatively unfavorable prognostic factors (Table 28.11). The follow-up time reported differs markedly between different series (range from 30 months to 8 years), which makes outcome comparisons difficult. Despite a high proportion of patients with unfavorable prognostic factors, the reported disease-free survival probability is high, ranging between 68 and 93%. The local control rate, usually based on biopsy results, is >90% in most series. Outcome data from the larger single-institution treatment series are summarized in Table 28.11. Only one prospective randomized study has been performed thus far between HDR-BT and EBRT. The study was performed by SATHYA and co-workers (2005) at Hamilton Regional Cancer Center, Ontario. Patients with localized T2 and T3 disease were randomly assigned to EBRT of 66 Gy in 33 fractions during 6.5 weeks ($n = 53$) or to HDR-BT of 35 Gy delivered to the prostate during 48 h plus EBRT of 40 Gy in 20 fractions during 4 weeks ($n = 51$). The primary outcome consisted of biochemical or clinical failure (BCF). The BCF was

Table 28.10. Treatment regimens, technological aspects, and calculated total dose equivalents at 2 Gy per fraction using $\alpha/\beta = 1.5$ Gy and $\alpha/\beta = 3$ Gy. (Adapted and modified from Morton 2005)

Reference	Center	HDR per fraction (Gy) × no. of fractions	No. of catheter insertions	External-beam dose (Gy)	Dose equivalent ($\alpha/\beta = 1.5$ Gy)	Dose equivalent ($\alpha/\beta = 3$ Gy)
Kovacs and Galalae (2003)	Kiel	15×2	2	50 (40 Gy to prostate)	115–191	102–158
Martinez et al. (2002)	William Beaumont	5.5×3	3	46	79–131	74–113
Borghede et al. (1997)	Göteborg	10×2	2	50	115	102
Mate et al. (1998)	Seattle	3–4×4	1	50.4	63–73	63–71
Deger et al. (2002)	Berlin	10×2, 9×2	2	40–50.4	101–105	91–92
Syed et al. (2001)	Long Beach	5×3, 5.5×4, 6.5×4	1	39.6–45	70–79	67–87
Pellizzon et al. (2003, 2004)	Sao Paulo	4×4, 5×4	1	45	67–80	65–75
Martin et al. (2004b)	Offenbach	5–7×4	4	39.6–45	79–105	75–94
Curran et al. (2000)	Burlington	6×3	1	50	89	82
Hiratsuka et al. (2004)	Kawasaki	5.5×3, 5.5×4	1	45, 41.8	75–83	71–78
Chiang et al. (2004)	Kaohsiung	4.2×3	2	50.4–54	68–71	67–70

Table 28.11. Patient characteristics by stage, Gleason score, and initial PSA, and outcome data from single-institution series. (Adapted and modified from Morton 2005)

Reference	Center	Stage (%)				Grade (%)			Initial PSA (%)			Median follow-up	Disease-free survival (%)	Late toxicity (%; >grade 2)
		N	T1	T2	T3	1	2	3	<10	10-20	>20			
Kovacs and Galalae (2003)	Kiel	144	1	67	32	15	49	36	42	38[a]	16[a]	8 years	82.6	2 (GU), 4.1 (GI)
Martinez et al. (2002)	William Beaumont	207	17	73	10	39	42	19	59	31	10	53 months	68	8 (GU), 1 (GI)
Borghede et al. (1997)	Göteborg	50	6	68	26	28	60	12	60	24	16	45 months	84	4 (dysuria), 2 (proctitis), 8 (diarrhea)
Mate et al. (1998)	Seattle	104	30	60	10	79	16	5	52	28	20	46 months	84 (PSA ≤20), 50 (PSA >20)	8 (urethral stricture)
Deger et al. (2002)	Berlin	230	7	35	58	23	60	17	Median 12.8			40 months	100 (T1), 70 (T2), 65 (T3)	12.2 (GU), 3 (incontinence)
Syed et al. (2001)	Long Beach	200	14	65	21	14	75	11	'Average of 10'			30 months	93	2 (GU), 1.5 (GI)
Pellizzon et al. (2003)	Dao Paolo	119	52	35	13	14	57	29	48	34	18	41 months	75.3	No late grade 3 GU or GI
Martin et al. (2004a)	Offenbach	102	2	68	30	28	45	27	26	35	39	2.6 years	82	4.9 (GU), 1 (GI)
Hiratsuka et al. (2004)	Kawasaki	71	17	58	25	64	18	18	35	35	30	44 months	93	7 (urethral stricture)
Total		1227	16	57	27	31	50	19	47	33	20			

[a] Refers to PSA values 10–40 ng/ml and >40 ng/ml. Grade reported as either WHO grade or Gleason score usually divided as <7; and 8–10

defined by biochemical failure, clinical failure, or death as a result of prostate cancer. Secondary outcomes included 2-year postradiation biopsy positivity, toxicity, and survival. The median follow-up was 8.2 years. In the HDR-BT plus EBRT arm, 17 patients (29%) experienced BCF compared with 33 patients (61%) in the EBRT arm (hazard ratio 0.42; $p = 0.0024$). Eighty-seven patients (84%) underwent a postradiation biopsy; 10 (24%) of 42 in the HDR-BT plus EBRT arm had biopsy positivity compared with 23 (51%) of 45 in the EBRT arm (odds ratio 0.30; $p = 0.015$). It was concluded that the combination of HDR-BT plus EBRT was superior to EBRT alone for BCF and postradiation biopsy. The trial provides evidence that higher doses of radiation delivered in a shorter duration of time results in improved local as well as biochemical control in locally advanced prostate cancer (SATHYA et al. 2005).

28.6.5
Health-Related Quality-of-Life Data After HDR-BT Plus EBRT

Long-term outcome data in terms of health-related quality of life (HRQoL) have been presented by GALALAE and co-workers (2004) at Beaumont Hospital and by WAHLGREN and co-workers (2004) at the Karolinska Hospital. Both studies showed that the levels HRQoL were generally high. The Karolinska Hospital prospective study showed that the HRQoL data were comparable to those of normative and did not change over time (WAHLGREN et al. 2004). The negative contribution from late neoadjuvant androgen deprivation therapy on symptom development was substantial but mostly transitory. Jo and co-workers (2005) at the Kawasaki Medical School have compared HRQoL data from prostate cancer patients after combination therapy with HDR-BT and EBRT with prostate cancer patients after radical prostatectomy (RP). The study was performed between 1997 and 2002. A total of 182 men diagnosed with T1c to T3bN0M0 disease had RP ($n = 89$) or HDR-BT combined with EBRT ($n = 93$), and were followed for 6 months or more. A postal survey was sent in which HRQoL was assessed using the 36-item Short-Form Health Survey (SF-36) QoL questionnaire, and disease-specific QoL using the University of California Los Angeles Prostate Cancer Index (UCLA-PCI). Questionnaire responses were obtained from 151 of 182 patients. There was no significant difference in SF-36 scale scores between men treated with RP or HDR-BT plus EBRT. In

the UCLA-PCI, the HDR-BT plus EBRT group had better urinary function ($p < 0.001$) and sexual function scores ($p = 0.043$). Men treated with RP had better bowel bother scores ($p = 0.027$). In patients with 2 years or more of follow-up, urinary function ($p < 0.001$) and sexual function ($p = 0.029$) were better for men treated with HDR-BT plus EBRT than for men treated with RP. Men treated with HDR-BT plus EBRT had significantly better urinary function ($p = 0.009$) and sexual function ($p = 0.013$) than 30 men treated with unilateral nerve-sparing RP. The conclusions drawn by the investigators were that in terms of HRQoL, RP and HDR-BT plus EBRT did not differ, but HDR-BT plus EBRT resulted in better urinary and sexual function than RP (Jo et al. 2005).

28.6.6
HDR-BT Monotherapy: Initial Experience

Several HDR-BT schedules are currently being explored as monotherapy for patients with low-risk (organ-confined) prostate cancer. The schedule of MARTINEZ and co-workers (2001) is based on four HDR fractions of 10 Gy each given with a catheter technique (see above) in two successive days and the doses being at least 6 h apart. The NTD_{2Gy} dose equals 131 Gy to the prostate, assuming $\alpha/\beta = 1.5$ Gy. The schedule described by YOSHIOKA et al. (2000) uses nine HDR insertions of 6 Gy each, giving all nine fractions in 5 days at two fractions per day, equaling an NTD_{2Gy} dose of 116 Gy. The monotherapy schedule used at the Karolinska Institute and at the Sahlgrenska University Hospital in Gothenburg involves two fractions of 15 Gy with 2 weeks apart, equaling an NTD_{2Gy} dose of 141 Gy. Table 28.12 summarizes the radiobiological data from HDR monotherapy regimens used in clinical practice or in prospective trials at four American and European centers.

GRILLS and co-workers (2004) at the Beaumont Hospital have reported on a comparison between HDR monotherapy and LDR permanent seed (palladium-103) brachytherapy (120 Gy). A total of 149 patients with early-stage prostate cancer were consecutively treated with either HDR-IMBT (65 patients) or LDR BT (84 patients). The median follow-up time was 35 months. The majority of patients had clinical stage-2, stage-T1c, or stage-T2a disease, iPSA <10 ng/ml, and Gleason score ≤6. Neoadjuvant hormone therapy (NHT) was used in 36% of patients for prostate volume reduction. The two treatment groups were well balanced with respect

Table 28.12. Brachytherapy as monotherapy for early-stage prostate cancer. *BED* biologically effective dose, *NTD* normalized total dose. (Adapted from FOWLER 2005)

Reference/schedule	Tumor		Late rectal reactions		
	BED Gy$_{1,5}$	NTD$_{2Gy}$	Rectal dose as percentage of tumor dose	BED Gy$_3$	NTD$_{2Gy}$
	$\alpha/\beta = 1.5$ Gy			$\alpha/\beta = 3$ Gy	
Martinez et al. (2001)					
Monotherapy 4F×9.5 Gy	278	119	75%	128	77
Monotherapy 4F×10 Gy	306	131	75%	140	84
Gustafson et al. (2003)					
Monotherapy 6F×7 Gy	238	102	If 100%	140	84
			If 70%	80	49
Yoshioka et al. (2000)					
Monotherapy 9F×6 Gy	271	116	70%	96	61
Nilsson et al. (unpublished data)					
Monotherapy 2F×15 Gy	330	141	60%	72	43

to age, clinical stage, iPSA, Gleason score, use of NHT, pretreatment genitourinary (GU) symptoms, implanted gland volume, and length of follow-up. Biochemical control according to the ASTRO definition was 98 and 97% for HDR and LDR, respectively. The conclusions reached from this prospective, but non-randomized, study were that the use of HDR brachytherapy as monotherapy was associated with decreased rates of acute urinary frequency, urgency, dysuria, and rectal pain, as compared with LDR. Chronic urinary frequency, urgency, and grade-2 rectal toxicities were also decreased with HDR. A pronounced decrease, 66%, was noted in the rate of sexual impotency with HDR. The investigators state that HDR-IMBT monotherapy is an accepted, convenient, cost-effective method of prostate BT for prostate cancer patients with favorable-risk disease. Encouraging data have also been obtained by MARTIN and co-workers(2004a) from a pilot study on 52 patients with localized disease. The treatment-associated toxicity was low.

28.6.7
The Role of Neoadjuvant and Concomitant Endocrine Therapy

Several prospective randomized trials have demonstrated that hormonal manipulation, in conjunction with conventional doses (65–70 Gy) of EBRT, results in improvements in disease-free survival and overall survival in patients with locally advanced prostate cancer (for reviews see GOTTSCHALK and ROACH 2004; NILSSON et al. 2004). D'AMICO and co-workers

(2004) have also shown that a 6-month course of androgen suppression can prolong survival when given in combination with EBRT compared with EBRT alone in men with clinically localized disease; however, the possible role for hormone therapy in conjunction with combined HDR-BT and EBRT is less clear. MARTINEZ and co-workers (2005) have presented outcome data on this topic from a large series of consecutive patients ($n = 1260$) treated between 1986 and 2000 at William Beaumont Hospital, Kiel University Hospital, and California Endocurietherapy Cancer Center, respectively. The biologically equivalent EBRT dose used in this study ranged between 90 and 123 Gy (median 102 Gy) using an α / β of 1.2. Patient eligibility criteria included iPSA ≥ 10, Gleason score ≥ 7, or clinical stage $\geq$T2b. A total of 934 of 1260 patients met the inclusion criteria and were included in the study. These patients were divided up for analysis between the 406 who received up to 6 months of androgen deprivation therapy and the 528 patients who did not. Median follow-up time was 4.4 years for androgen-deprivation therapy patients and 4.9 for radiation alone. There was no difference at 5 and 8 years in overall survival, cause-specific survival, or biochemical control. The corresponding 8-year rates with and without androgen deprivation therapy were as follows: biochemical control 85 and 81%; overall survival 83 and 78%; cause-specific survival 89 and 94%; and metastatic rates of 16.6 and 7.3%. The authors concluded that at 8 years, the addition of a course of 6 months or more of neoadjuvant/concurrent androgen-deprivation therapy to a very high radiation dose did not confer a therapeutic advantage (MARTINEZ et al. 2005). The question that remains to be answered is whether the favorable interaction between hormone therapy and radiation seen with respect to disease-free and overall survival in the large prospective trials mentioned above may be a result of the inability of conventional doses of EBRT as a monotherapeutic approach to sterilize large prostate cancers, and as such it may not be applicable to brachytherapy. Carefully designed prospective studies stratifying between low-, intermediate-, and high-risk groups will hopefully give a conclusive answer as to whether or not short-term neo-adjuvant hormone therapy increases the cure rate also after combination therapy HDR-BT and EBRT or not.

28.6.8
Discussion/Future Aspects

High-dose-rate brachytherapy provides an accurate means of delivering very high doses to the prostate with great dose conformality. The rapid dose fall-off beyond the gland enables sparing of the adjacent normal tissues. The source positions and dwell times allow a wide range of doses that can be delivered from a single source position and the technique can thus, with the use of modern dose-planning systems, be characterized in terms of an intensity-modulated brachytherapy (IMBT) modality or in terms of 4D conformal radiotherapy (4D CRT), where the dwell time represents the fourth dimension. The outcome data both in locally advanced and in organ-confined disease are excellent with high local radicality. Future trials need to focus on improved predictive markers in order to tailor the treatment even better in each individual case, and to combine it with systemic endocrine and/or chemotherapeutic regimens in cases with high risk for micrometastatic disease.

References

Akazawa PF, Roach M III, Pickett B et al (1996) Three dimensional comparison of blocked arcs vs four and six field conformal treatment of the prostate. Radiother Oncol 41:83–88

Algan O, Hanks GE , Shaer AH (1995) Localization of the prostatic apex for radiation treatment planning. Int J Radiat Oncol Biol Phys 33:925–930

Allen ZA, Merrick GS, Butler WM et al (2005) Detailed urethral dosimetry in the evaluation of prostate brachyther-apy-related urinary morbidity. Int J Radiat Oncol Biol Phys 62:981–987

American Society for Therapeutic Radiology and Oncology Consensus Panel (1997) Consensus statement: guidelines for PSA following radiation therapy. Int J Radiat Oncol Biol Phys 37:1035–1041

Andriole GL, Ponas SH , Catalona WJ (1992) The implication of focal well differentiated prostate cancer (CaP) in men with elevated serum PSA and palpably normal prostates. J Urol 147:442A

Anscher MS, Clough R , Dodge R (2000) Radiotherapy for a rising prostate-specific antigen after radical prostatectomy: the first 10 years. Int J Radiat Oncol Biol Phys 48:369–375

Antolak JA, Rosen II, Childress CH et al (1998) Prostate target volume variations during a course of radiotherapy. Int J Radiat Oncol Biol Phys 42:661–672

Ash D, Flynn A, Battermann J et al (2000) ESTRO/EAU/EORTC recommendations on permanent seed implantation for localized prostate cancer. Radiother Oncol 57:315–321

Badiozamani KR, Wallner K, Cavanagh W et al (1999) Comparability of CT-based and TRUS-based prostate volumes. Int J Radiat Oncol Biol Phys 43:375–378

Bagshaw MA, Ray GR, Cox RS (1985) Radiotherapy of prostatic carcinoma: long- or short-term efficacy (Stanford University experience). Urology 25:17–23

Bagshaw MA, Cox RS, Ray GR (1988) Status of radiation treatment of prostate cancer at Stanford University. NCI Monogr 7:47–60

Bastacky SI, Walsh PC , Epstein JI (1993) Relationship between perineural tumor invasion on needle biopsy and radical prostatectomy capsular penetration in clinical stage B adenocarcinoma of the prostate. Am J Surg Pathol 17:336–341

Bayley AJ, Catton CN, Haycocks T et al (2004) A randomized trial of supine vs prone positioning in patients undergoing escalated dose conformal radiotherapy for prostate cancer. Radiother Oncol 70:37–44

Bellon J, Wallner K, Ellis W et al (1999) Use of pelvic CT scanning to evaluate pubic arch interference of transperineal prostate brachytherapy. Int J Radiat Oncol Biol Phys 43:579–581

Biggs PJ, Russell MD (1988) Effect of a femoral head prosthesis on megavoltage beam radiotherapy. Int J Radiat Oncol Biol Phys 14:581–586

Blasko JC, Grimm PD, Sylvester JE et al (2000) Palladium-103 brachytherapy for prostate carcinoma. Int J Radiat Oncol Biol Phys 46:839–850

Blasko JC, Mate T, Sylvester JE et al (2002) Brachytherapy for carcinoma of the prostate: techniques, patient selection, and clinical outcomes. Semin Radiat Oncol 12:81–94

Blennerhassett JB, Vickery AL Jr (1966) Carcinoma of the prostate gland. An anatomical study of tumor location. Cancer 19:980–984

Bluestein DL, Bostwick DG, Bergstralh EJ et al (1994) Eliminating the need for bilateral pelvic lymphadenectomy in select patients with prostate cancer. J Urol 151:1315–1320

Boersma LJ, van den Brink M, Bruce AM et al (1998) Estimation of the incidence of late bladder and rectum complications after high-dose (70–78 Gy) conformal radiotherapy for prostate cancer, using dose-volume histograms. Int J Radiat Oncol Biol Phys 41:83–92

Bolla M, van Poppel H, Collette L et al (2005) Postoperative radiotherapy after radical prostatectomy: a randomised controlled trial (EORTC trial 22911). Lancet 366:572–578

Bonin SR, Hanlon AL, Lee WR et al (1997) Evidence of increased failure in the treatment of prostate carcinoma patients who have perineural invasion treated with three-dimensional conformal radiation therapy. Cancer 79:75–80

Borghede G, Hedelin H, Holmang S et al (1997) Combined treatment with temporary short-term high dose rate iridium-192 brachytherapy and external beam radiotherapy for irradiation of localized prostatic carcinoma. Radiother Oncol 44:237–244

Brabbins D, Martinez A, Yan D et al (2005) A dose-escalation trial with the adaptive radiotherapy process as a delivery system in localized prostate cancer: analysis of chronic toxicity. Int J Radiat Oncol Biol Phys 61:400–408

Brenner DJ, Hall EJ (1999) Fractionation and protraction for radiotherapy of prostate carcinoma. Int J Radiat Oncol Biol Phys 43:1095–1101

Brenner DJ, Martinez AA, Edmundson GK et al (2002) Direct evidence that prostate tumors show high sensitivity to fractionation (low alpha/beta ratio), similar to late-responding normal tissue. Int J Radiat Oncol Biol Phys 52:6–13

Breslow N, Chan CW, Dhom G et al (1977) Latent carcinoma of prostate at autopsy in seven areas. The International Agency for Research on Cancer, Lyons, France. Int J Cancer 20:680–688

Bucci J, Morris WJ, Keyes M et al (2002) Predictive factors of urinary retention following prostate brachytherapy. Int J Radiat Oncol Biol Phys 53:91–98

Butler WM, Merrick GS, Dorsey AT et al (2000a) Isotope choice and the effect of edema on prostate brachytherapy dosimetry. Med Phys 27:1067–1075

Butler WM, Merrick GS, Lief JH et al (2000b) Comparison of seed loading approaches in prostate brachytherapy. Med Phys 27:381–392

Cadeddu JA, Partin AW, DeWeese TL et al (1998) Long-term results of radiation therapy for prostate cancer recurrence following radical prostatectomy. J Urol 159:173–178

Carrier S, Hricak H, Lee SS et al (1995) Radiation-induced decrease in nitric oxide synthase-containing nerves in the rat penis. Radiology 195:95–99

Carter HB, Epstein JI, Partin AW (1999) Influence of age and prostate-specific antigen on the chance of curable prostate cancer among men with nonpalpable disease. Urology 53:126–130

Catalona WJ, Bigg SW (1990) Nerve-sparing radical prostatectomy: evaluation of results after 250 patients. J Urol 143:538–544

Catalona WJ, Smith DS (1994) 5-year tumor recurrence rates after anatomical radical retropubic prostatectomy for prostate cancer. J Urol 152:1837–1842

Cha CM, Potters L, Ashley R et al (1999) Isotope selection for patients undergoing prostate brachytherapy. Int J Radiat Oncol Biol Phys 45:391–395

Charyulu KK (1980) Transperineal interstitial implantation of prostate cancer: a new method. Int J Radiat Oncol Biol Phys 6:1261–1266

Cheung R, Tucker SL, Dong L et al (2003) Dose-response for biochemical control among high-risk prostate cancer patients after external beam radiotherapy. Int J Radiat Oncol Biol Phys 56:1234–1240

Chiang PH, Fang FM, Jong WC et al (2004) High-dose rate iridium-192 brachytherapy and external beam radiation therapy for prostate cancer with or without androgen ablation. Int J Urol 11:152–158

Choi S, Wallner KE, Merrick GS et al (2004) Treatment margins predict biochemical outcomes after prostate brachytherapy. Cancer J 10:175–180

Clark T, Parekh DJ, Cookson MS et al (2003) Randomized prospective evaluation of extended versus limited lymph node dissection in patients with clinically localized prostate cancer. J Urol 169:145–148

Cox JA, Zagoria RJ, Raben M (1994) Prostate cancer: comparison of retrograde urethrography and computed tomography in radiotherapy planning. Int J Radiat Oncol Biol Phys 29:1119–1123

Crook J, Raymond Y, Salhani D et al (1995) Prostate motion during standard radiotherapy as assessed by fiducial markers. Radiother Oncol 37:35–42

Crook J, Lukka H, Klotz L et al (2001) Systematic overview of the evidence for brachytherapy in clinically localized prostate cancer. CMAJ 164:975–981

Crook J, McLean M, Yeung I et al (2004) MRI-CT fusion to assess postbrachytherapy prostate volume and the effects of prolonged edema on dosimetry following transperineal interstitial permanent prostate brachytherapy. Brachytherapy 3:55–60

Curran MJ, Healey GA, Bihrle W III et al (2000) Treatment of high-grade low-stage prostate cancer by high-dose-rate brachytherapy. J Endourol 14:351–356

D'Amico AV, Whittington R, Malkowicz SB et al (1995) A multivariate analysis of clinical and pathological factors that predict for prostate specific antigen failure after radical prostatectomy for prostate cancer. J Urol 154:131–138

D'Amico AV, Whittington R, Malkowicz SB et al (1998) Biochemical outcome after radical prostatectomy, external beam radiation therapy, or interstitial radiation therapy for clinically localized prostate cancer. J Am Med Assoc 280:969–974

D'Amico AV, Manola J, Loffredo M et al (2004) 6-month androgen suppression plus radiation therapy vs radiation therapy alone for patients with clinically localized prostate cancer: a randomized controlled trial. J Am Med Assoc 292:821–827

Danella JF, deKernion JB, Smith RB et al (1993) The contemporary incidence of lymph node metastases in prostate cancer: implications for laparoscopic lymph node dissection. J Urol 149:1488–1491

Dattoli M, Wallner K, True L et al (2003) Long-term outcomes after treatment with external beam radiation therapy and palladium 103 for patients with higher risk prostate carcinoma: influence of prostatic acid phosphatase. Cancer 97:979–983

Davis BJ, Pisansky TM, Wilson TM et al (1999) The radial distance of extraprostatic extension of prostate carcinoma: implications for prostate brachytherapy. Cancer 85:2630–2637

Dawson JE, Wu T, Roy T et al (1994) Dose effects of seeds placement deviations from pre-planned positions in ultrasound guided prostate implants. Radiother Oncol 32:268–270

Dawson LA, Litzenberg DW, Brock KK et al (2000) A comparison of ventilatory prostate movement in four treatment positions. Int J Radiat Oncol Biol Phys 48:319–323

De Crevoisier R, Tucker SL, Dong L et al (2005) Increased risk of biochemical and local failure in patients with distended rectum on the planning CT for prostate cancer radiotherapy. Int J Radiat Oncol Biol Phys 62:965–973

Deger S, Boehmer D, Turk I et al (2002) High dose rate brachytherapy of localized prostate cancer. Eur Urol 41:420–426

Del Regato JA, Trailins AH , Pittman DD (1993) Twenty years follow-up of patients with inoperable cancer of the prostate (stage C) treated by radiotherapy: report of a national cooperative study. Int J Radiat Oncol Biol Phys 26:197–201

Desai J, Stock RG, Stone NN et al (1998) Acute urinary morbidity following I-125 interstitial implantation of the prostate gland. Radiat Oncol Invest 6:135–141

DiBiase SJ, Wallner K, Tralins K et al (2000) Brachytherapy radiation doses to the neurovascular bundles. Int J Radiat Oncol Biol Phys 46:1301–1307

Edmundson GK (1994) Volume optimization: an American viewpoint. In: Mould RF et al (eds) Brachytherapy from radium to optimization. Nucletron International, Wageningen, The Netherlands, pp 314–318

Epstein JI, Walsh PC, Carmichael M et al (1994) Pathologic and clinical findings to predict tumor extent of nonpalpable (stage T1c) prostate cancer. J Am Med Assoc 271:368–374

Fiorino C, Reni M, Cattaneo GM et al (1997) Comparing 3-, 4- and 6-fields techniques for conformal irradiation of prostate and seminal vesicles using dose-volume histograms. Radiother Oncol 44:251–257

Fiorino C, Sanguineti G, Cozzarini C et al (2003) Rectal dose-volume constraints in high-dose radiotherapy of localized prostate cancer. Int J Radiat Oncol Biol Phys 57:953–962

Fisch BM, Pickett B, Weinberg V et al (2001) Dose of radiation received by the bulb of the penis correlates with risk of impotence after three-dimensional conformal radiotherapy for prostate cancer. Urology 57:955—959

Fowler JF (2005) The radiobiology of prostate cancer including new aspects of fractionated radiotherapy. Acta Oncol 44:265–276

Fowler J, Chappell R, Ritter M (2001) Is alpha/beta for prostate tumors really low? Int J Radiat Oncol Biol Phys 50:1021–1031

Fowler JE Jr, Whitmore WF Jr (1981) The incidence and extent of pelvic lymph node metastases in apparently localized prostatic cancer. Cancer 47:2941–2945

Fowler JF, Ritter MA, Chappell RJ et al (2003) What hypofractionated protocols should be tested for prostate cancer? Int J Radiat Oncol Biol Phys 56:1093–1104

Galalae RM, Loch T, Riemer B et al (2004) Health-related quality of life measurement in long-term survivors and outcome following radical radiotherapy for localized prostate cancer. Strahlenther Onkol 180:582–589

Gelblum DY, Potters L (2000) Rectal complications associated with transperineal interstitial brachytherapy for prostate cancer. Int J Radiat Oncol Biol Phys 48:119–124

Gelblum DY, Potters L, Ashley R et al (1999) Urinary morbidity following ultrasound-guided transperineal prostate seed implantation. Int J Radiat Oncol Biol Phys 45:59–67

Gervasi LA, Mata J, Easley JD et al (1989) Prognostic significance of lymph nodal metastases in prostate cancer. J Urol 142:332–336

Gottschalk AR, Roach M III (2004) The use of hormonal therapy with radiotherapy for prostate cancer: analysis of prospective randomised trials. Br J Cancer 90:950–954

Grann A, Wallner K (1998) Prostate brachytherapy in patients with inflammatory bowel disease. Int J Radiat Oncol Biol Phys 40:135–138

Gray G, Wallner K, Roof J et al (2000) Cystourethroscopic findings before and after prostate brachytherapy. Tech Urol 6:109–111

Grills IS, Martinez AA, Hollander M et al (2004) High dose rate brachytherapy as prostate cancer monotherapy reduces toxicity compared to low dose rate palladium seeds. J Urol 171:1098–1104

Gustafson G, Demanes D, Rodriguez R (2003) High dose rate (HDR) monotherapy for early stage prostate cancer: toxicity results utilizing the common toxicity criteria. Int J Radiat Oncol Biol Phys 57:S230–S231

Hakenberg OW, Wirth MP, Hermann T et al (2003) Recommendations for the treatment of localized prostate cancer by permanent interstitial brachytherapy. Urol Int 70:15–20

Han BH, Demel KC, Wallner K et al (2001a) Patient reported complications after prostate brachytherapy. J Urol 166:953–957

Han M, Piantadosi S, Zahurak ML et al (2001b) Serum acid phosphatase level and biochemical recurrence following radical prostatectomy for men with clinically localized prostate cancer. Urology 57:707–711

Han BH, Wallner K, Merrick G et al (2003) The effect of interobserver differences in post-implant prostate CT image interpretation on dosimetric parameters. Med Phys 30:1096–1102

Hanks GE, Lee WR, Hanlon AL et al (1996) Conformal technique dose escalation for prostate cancer: biochemical evidence of improved cancer control with higher doses in patients with pretreatment prostate-specific antigen ≥10 ng/ml. Int J Radiat Oncol Biol Phys 35:861–868

Heidenreich A, Varga Z, Knobloch R von (2002) Extended pelvic lymphadenectomy in patients undergoing radical prostatectomy: high incidence of lymph node metastasis. J Urol 167:1681–1686

Herold DM, Hanlon AL, Movsas B et al (1998) Age-related prostate cancer metastases. Urology 51:985–990

Hinerman-Mulroy A, Merrick GS, Butler WM et al (2004) Androgen deprivation-induced changes in prostate anatomy predict urinary morbidity after permanent interstitial brachytherapy. Int J Radiat Oncol Biol Phys 59:1367–1382

Hiratsuka J, Jo Y, Yoshida K et al (2004) Clinical results of combined treatment conformal high-dose-rate iridium-192 brachytherapy and external beam radiotherapy using staging lymphadenectomy for localized prostate cancer. Int J Radiat Oncol Biol Phys 59:684–690

Howard A, Wallner K, Han B (2001) Clinical course and dosimetry of rectal fistulas after prostate brachytherapy. J Brachyther Int 17:37–42

Hsu IC, Lessard E, Weinberg V et al (2004) Comparison of inverse planning simulated annealing and geometrical optimization for prostate high-dose-rate brachytherapy. Brachytherapy 3:147–152

Huang EH, Pollack A, Levy L et al (2002) Late rectal toxicity: dose-volume effects of conformal radiotherapy for prostate cancer. Int J Radiat Oncol Biol Phys 54:1314–1321

Hughes S, Wallner K, Merrick G et al (2001) Preexisting histologic evidence of prostatitis is unrelated to postimplant urinary morbidity. Int J Cancer 96:79–82

Hurwitz MD, Schnieder L, Manola J et al (2002) Lack of radiation dose response for patients with low-risk clinically localized prostate cancer: a retrospective analysis. Int J Radiat Oncol Biol Phys 53:1106–1110

Jackson A, Skwarchuk MW, Zelefsky MJ et al (2001) Late rectal bleeding after conformal radiotherapy of prostate cancer. II. Volume effects and dose-volume histograms. Int J Radiat Oncol Biol Phys 49:685–698

Jewett HJ (1980) Radical perineal prostatectomy for palpable, clinically localized, non-obstructive cancer: experience at the Johns Hopkins Hospital 1909–1963. J Urol 124:492–494

Jo Y, Junichi H, Tomohiro F et al (2005) Radical prostatectomy versus high-dose rate brachytherapy for prostate cancer: effects on health-related quality of life. Br J Urol Int 96:43–47

Kattan MW, Zelefsky MJ, Kupelian PA et al (2000) Pretreatment nomogram for predicting the outcome of three-dimensional conformal radiotherapy in prostate cancer. J Clin Oncol 18:3352–3359

Kestin L, Goldstein N, Vicini F et al (2002) Treatment of prostate cancer with radiotherapy: Should the entire seminal vesicles be included in the clinical target volume? Int J Radiat Oncol Biol Phys 54:686–697

Khuntia D, Reddy CA, Mahadevan A et al (2004) Recurrence-free survival rates after external-beam radiotherapy for patients with clinical T1–T3 prostate carcinoma in the prostate-specific antigen era: What should we expect? Cancer 100:1283–1292

Kitamura K, Shirato H, Seppenwoolde Y et al (2002) Three-dimensional intrafractional movement of prostate measured during real-time tumor-tracking radiotherapy in supine and prone treatment positions. Int J Radiat Oncol Biol Phys 53:1117–1123

Kneebone A, Gebski V, Hogendoorn N et al (2003) A randomized trial evaluating rigid immobilization for pelvic irradiation. Int J Radiat Oncol Biol Phys 56:1105–1111

Kolkman-Deurloo IK, Visser AG, Niel CG et al (1994) Optimization of interstitial volume implants. Radiother Oncol 31:229–239

Koper PC, Heemsbergen WD, Hoogeman MS et al (2004) Impact of volume and location of irradiated rectum wall on rectal blood loss after radiotherapy of prostate cancer. Int J Radiat Oncol Biol Phys 58:1072–1082

Kovacs G, Galalae R (2003) Fractionated perineal high-dose-rate temporary brachytherapy combined with external beam radiation in the treatment of localized prostate cancer: is lymph node sampling necessary? Cancer Radiother 7:100–106

Kovacs G, Galalae R, Loch T et al (1999) Prostate preservation by combined external beam and HDR brachytherapy in nodal negative prostate cancer. Strahlenther Onkol 175:87–88

Kuban DA, Thames HD, Levy LB et al (2003) Long-term multi-institutional analysis of stage T1–T2 prostate cancer treated with radiotherapy in the PSA era. Int J Radiat Oncol Biol Phys 57:915–928

Kupelian PA, Reddy CA, Carlson TP et al (2002) Dose/volume relationship of late rectal bleeding after external beam radiotherapy for localized prostate cancer: Absolute or relative rectal volume? Cancer J 8:62–66

Landis D, Wallner K, Locke J et al (2002) Late urinary function after prostate brachytherapy. Brachytherapy 1:21–26

Langen KM, Jones DT (2001) Organ motion and its management. Int J Radiat Oncol Biol Phys 50:265–278

Lattanzi J, McNeeley S, Pinover W et al (1999) A comparison of daily CT localization to a daily ultrasound-based system in prostate cancer. Int J Radiat Oncol Biol Phys 43:719–725

Lattanzi J, McNeeley S, Hanlon A et al (2000) Ultrasound-based stereotactic guidance of precision conformal external beam radiation therapy in clinically localized prostate cancer. Urology 55:73–78

Lawton CA, Winter K, Byhardt R et al (1997) Androgen suppression plus radiation versus radiation alone for patients with D1 (pN+) adenocarcinoma of the prostate (results based on a national prospective randomized trial, RTOG 85-31). Radiation Therapy Oncology Group. Int J Radiat Oncol Biol Phys 38:931–939

Leibel SA, Heimann R, Kutcher GJ et al (1994) Three-dimensional conformal radiation therapy in locally advanced carcinoma of the prostate: preliminary results of a phase I dose-escalation study. Int J Radiat Oncol Biol Phys 28:55–65

Leibovich BC, Blute ML, Bostwick DG et al (2000a) Proximity of prostate cancer to the urethra: implications for minimally invasive ablative therapies. Urology 56:726–729

Leibovich BC, Engen DE, Patterson DE et al (2000b) Benefit of adjuvant radiation therapy for localized prostate cancer with a positive surgical margin. J Urol 163:1178–1182

Lessard E, Pouliot J (2001) Inverse planning anatomy-based dose optimization for HDR-brachytherapy of the prostate using fast simulated annealing algorithm and dedicated objective function. Med Phys 28:773–779

Levegrun S, Jackson A, Zelefsky MJ et al (2002) Risk group dependence of dose-response for biopsy outcome after three-dimensional conformal radiation therapy of prostate cancer. Radiother Oncol 63:11–26

Litzenberg D, Dawson LA, Sandler H et al (2002) Daily prostate targeting using implanted radiopaque markers. Int J Radiat Oncol Biol Phys 52:699–703

Liu YM, Ling S, Langen KM et al (2004) Prostate movement during simulation resulting from retrograde urethrogram compared with "natural" prostate movement. Int J Radiat Oncol Biol Phys 60:470–475

Magrini SM, Cellai E, Rossi F et al (1999) Comparison of the conventional "box technique" with two different "conformal" beam arrangements for prostate cancer treatment. Cancer Radiother 3:215–220

Malone S, Crook JM, Kendal WS et al (2000) Respiratory-induced prostate motion: quantification and characterization. Int J Radiat Oncol Biol Phys 48:105–109

Mangar SA, Huddart RA, Parker CC et al (2005) Technological advances in radiotherapy for the treatment of localised prostate cancer. Eur J Cancer 41:908–921

Marsh LH, Ten Haken RK, Sandler HM (1992) A customized non-axial external beam technique for treatment of prostate carcinomas. Med Dosim 17:123–127

Martin T, Baltas D, Kurek R et al (2004a) 3-D conformal HDR brachytherapy as monotherapy for localized prostate cancer. A pilot study. Strahlenther Onkol 180:225–232

Martin T, Roddiger S, Kurek R et al (2004b) 3D conformal HDR brachytherapy and external beam irradiation combined with temporary androgen deprivation in the treatment of localized prostate cancer. Radiother Oncol 71:35–41

Martinez A, Benson RC, Edmundson GK et al (1985) Pelvic lymphadenectomy combined with transperineal interstitial implantation of iridium-192 and external beam radiotherapy for locally advanced prostatic carcinoma: technical description. Int J Radiat Oncol Biol Phys 11:841–847

Martinez A, Pataki I, Edmundson G et al (2001) Phase II prospective study of the use of conformal high-dose-rate brachytherapy as monotherapy for the treatment of favorable stage prostate cancer: a feasibility report. Int J Radiat Oncol Biol Phys 49:61–69

Martinez A, Gustafson G, Gonzalez J et al (2002) Dose escalation using conformal high-dose-rate brachytherapy improves outcome in unfavorable prostate cancer. Int J Radiat Oncol Biol Phys 53:316–327

Martinez A, Gonzalez J, Spencer W et al (2003) Conformal high dose rate brachytherapy improves biochemical control and cause specific survival in patients with prostate cancer and poor prognostic factors. J Urol 169:974–979

Martinez A, Demanes DJ, Galalae R et al (2005) Lack of benefit from a short course of androgen deprivation for unfavorable prostate cancer patients treated with an accelerated hypofractionated regime. Int J Radiat Oncol Biol Phys 62:1322–1331

Mate TP, Gottesman JE, Hatton J et al (1998) High dose-rate afterloading 192Iridium prostate brachytherapy: feasibility report. Int J Radiat Oncol Biol Phys 41:525–533

McCarthy JF, Catalona WJ, Hudson MA (1994) Effect of radiation therapy on detectable serum prostate specific antigen levels following radical prostatectomy: early versus delayed treatment. J Urol 151:1575–1578

McGowan DG (1981) The value of extended field radiation therapy in carcinoma of the prostate. Int J Radiat Oncol Biol Phys 7:1333–1339

McLaughlin PW, Wygoda A, Sahijdak W et al (1999) The effect of patient position and treatment technique in conformal treatment of prostate cancer. Int J Radiat Oncol Biol Phys 45:407–413

McLaughlin PW, Troyer S, Berri S et al (2005) Functional anatomy of the prostate: implications for treatment planning. Int J Radiat Oncol Biol Phys 63:479–491

McNeal JE (1969) Origin and development of carcinoma in the prostate. Cancer 23:24–34

Merrick GS, Butler WM (2000) Modified uniform seed loading for prostate brachytherapy: rationale, design, and evaluation. Tech Urol 6:78–84

Merrick GS, Butler WM, Dorsey AT et al (1999a) The dependence of prostate postimplant dosimetric quality on CT volume determination. Int J Radiat Oncol Biol Phys 44:1111–1117

Merrick GS, Butler WM, Dorsey AT et al (1999b) Potential role of various dosimetric quality indicators in prostate brachytherapy. Int J Radiat Oncol Biol Phys 44:717–724

Merrick GS, Butler WM, Lief JH et al (1999c) Efficacy of sildenafil citrate in prostate brachytherapy patients with erectile dysfunction. Urology 53:1112–1116

Merrick GS, Butler WM, Dorsey AT et al (2000a) The effect of constipation on rectal dosimetry following prostate brachytherapy. Med Dosim 25:237–241

Merrick GS, Butler WM, Dorsey AT et al (2000b) Seed fixity in the prostate/periprostatic region following brachytherapy. Int J Radiat Oncol Biol Phys 46:215–220

Merrick GS, Butler WM, Dorsey AT et al (2000c) A comparison of radiation dose to the neurovascular bundles in men with and without prostate brachytherapy-induced erectile dysfunction. Int J Radiat Oncol Biol Phys 48:1069–1074

Merrick GS, Butler WM, Lief JH et al (2000d) Temporal resolution of urinary morbidity following prostate brachytherapy. Int J Radiat Oncol Biol Phys 47:121–128

Merrick GS, Butler WM, Dorsey AT et al (2001a) Effect of prostate size and isotope selection on dosimetric quality following permanent seed implantation. Tech Urol 7:233–240

Merrick GS, Butler WM, Galbreath RW et al (2001b) Perineural invasion is not predictive of biochemical outcome following prostate brachytherapy. Cancer J 7:404–412

Merrick GS, Butler WM, Galbreath RW et al (2001c) Relationship between the transition zone index of the prostate gland and urinary morbidity after brachytherapy. Urology 57:524–529

Merrick GS, Wallner K, Butler WM et al (2001d) A comparison of radiation dose to the bulb of the penis in men with and without prostate brachytherapy-induced erectile dysfunction. Int J Radiat Oncol Biol Phys 50:597–604

Merrick GS, Wallner K, Butler WM et al (2001e) Short-term sexual function after prostate brachytherapy. Int J Cancer 96:313–319

Merrick GS, Butler WM, Galbreath RW et al (2002a) Erectile function after permanent prostate brachytherapy. Int J Radiat Oncol Biol Phys 52:893–902

Merrick GS, Butler WM, Tollenaar BG et al (2002b) The dosimetry of prostate brachytherapy-induced urethral strictures. Int J Radiat Oncol Biol Phys 52:461–468

Merrick GS, Butler WM, Wallner K et al (2002c) Permanent prostate brachytherapy-induced morbidity in patients with grade II and III obesity. Urology 60:104–108

Merrick GS, Butler WM, Wallner KE et al (2002d) Prostate-specific antigen spikes after permanent prostate brachytherapy. Int J Radiat Oncol Biol Phys 54:450–456

Merrick GS, Butler WM, Wallner KE et al (2002e) Prophylactic versus therapeutic alpha-blockers after permanent prostate brachytherapy. Urology 60:650–655

Merrick GS, Butler WM, Galbreath RW et al (2003a) Does hormonal manipulation in conjunction with permanent interstitial brachytherapy, with or without supplemental external beam irradiation, improve the biochemical outcome for men with intermediate or high-risk prostate cancer? Br J Urol Int 91:23–29

Merrick GS, Butler WM, Wallner KE et al (2003b) Extracapsular radiation dose distribution after permanent prostate brachytherapy. Am J Clin Oncol 26:e178–e189

Merrick GS, Butler WM, Wallner KE et al (2003c) Rectal function following brachytherapy with or without supplemental external beam radiation: results of two prospective randomized trials. Brachytherapy 2:147–157

Merrick GS, Butler WM, Wallner KE et al (2003d) Long-term urinary quality of life after permanent prostate brachytherapy. Int J Radiat Oncol Biol Phys 56:454–461

Merrick GS, Butler WM, Wallner KE et al (2003e) Dysuria after permanent prostate brachytherapy. Int J Radiat Oncol Biol Phys 55:979–985

Merrick GS, Butler WM, Wallner KE et al (2003f) Late rectal function after prostate brachytherapy. Int J Radiat Oncol Biol Phys 57:42–48

Merrick GS, Butler WM, Wallner KE et al (2003g) Prostate-specific antigen (PSA) velocity and benign prostate hypertrophy predict for PSA spikes following prostate brachytherapy. Brachytherapy 2:181–188

Merrick GS, Wallner KE, Butler WM (2003h) Minimizing prostate brachytherapy-related morbidity. Urology 62:786–792

Merrick GS, Wallner KE , Butler WM (2003i) Permanent interstitial brachytherapy for the management of carcinoma of the prostate gland. J Urol 169:1643–1652

Merrick GS, Butler WM, Wallner KE et al (2004a) Effect of transurethral resection on urinary quality of life after permanent prostate brachytherapy. Int J Radiat Oncol Biol Phys 58:81–88

Merrick GS, Butler WM, Wallner KE et al (2004b) Permanent interstitial brachytherapy in younger patients with clinically organ-confined prostate cancer. Urology 64:754–759

Merrick GS, Butler WM, Wallner KE et al (2004c) Temporal effect of neoadjuvant androgen deprivation therapy on PSA kinetics following permanent prostate brachytherapy with or without supplemental external beam radiation. Brachytherapy 3:141–146

Merrick GS, Butler WM, Wallner KE et al (2004d) Effect of cigarette smoking on biochemical outcome after permanent prostate brachytherapy. Int J Radiat Oncol Biol Phys 58:1056–1062

Merrick GS, Butler WM, Wallner KE et al (2004e) Prognostic significance of percent positive biopsies in clinically organ-confined prostate cancer treated with permanent prostate brachytherapy with or without supplemental external-beam radiation. Cancer J 10:54–60

Merrick GS, Wallner KE, Butler WM (2004f) Patient selection for prostate brachytherapy: more myth than fact. Oncology (Williston Park) 18:445–452, 455–457

Merrick GS, Butler WM, Wallner KE et al (2005a) Influence of body mass index on biochemical outcome after permanent prostate brachytherapy. Urology 65:95–100

Merrick GS, Butler WM, Wallner KE et al (2005b) Erectile function after prostate brachytherapy. Int J Radiat Oncol Biol Phys 62:437–447

Merrick GS, Butler WM, Wallner KE et al (2005c) Variability of prostate brachytherapy preimplant dosimetry. A multi institution analysis. Brachytherapy 4:241–251

Michalski JM, Purdy JA, Winter K et al (2000) Preliminary report of toxicity following 3D radiation therapy for prostate cancer on 3DOG/RTOG 9406. Int J Radiat Oncol Biol Phys 46:391–402

Michalski JM, Winter K, Purdy JA et al (2004) Toxicity after three-dimensional radiotherapy for prostate cancer with RTOG 9406 dose level IV. Int J Radiat Oncol Biol Phys 58:735–742

Michalski JM, Winter K, Purdy JA et al (2005) Toxicity after three-dimensional radiotherapy for prostate cancer on RTOG 9406 dose Level V. Int J Radiat Oncol Biol Phys 62:706–713

Middleton RG (1988) Value of and indications for pelvic lymph node dissection in the staging of prostate cancer. NCI Monogr 7:41–43

Moreland RB (1998) Is there a role of hypoxemia in penile fibrosis: a viewpoint presented to the Society for the Study of Impotence. Int J Impot Res 10:113–120

Morton GC (2005) The emerging role of high-dose-rate brachytherapy for prostate cancer. Clin Oncol (R Coll Radiol) 17:219–227

Mueller A, Wallner K, Merrick G et al (2004) Perirectal seeds as a risk factor for prostate brachytherapy-related rectal bleeding. Int J Radiat Oncol Biol Phys 59:1047–1052

Mulhall JP (2001) Minimizing radiation-induced erectile dysfunction. J Brachyther Int 17:221–227

Myers RP, Goellner JR, Cahill DR (1987) Prostate shape, external striated urethral sphincter and radical prostatectomy: the apical dissection. J Urol 138:543–550

Nag S, Beyer D, Friedland J et al (1999) American brachytherapy society (ABS) recommendations for transperineal permanent brachytherapy of prostate cancer. Int J Radiat Oncol Biol Phys 44:789–799

Narayan P, Fournier G, Gajendran V et al (1994) Utility of pre-operative serum prostate-specific antigen concentration and biopsy Gleason score in predicting risk of pelvic lymph node metastases in prostate cancer. Urology 44:519–524

Nilsson S, Norlen BJ, Widmark A (2004) A systematic overview of radiation therapy effects in prostate cancer. Acta Oncol 43:316–381

Noguchi M, Stamey TA, Neal JE et al (2000) An analysis of 148 consecutive transition zone cancers: clinical and histological characteristics. J Urol 163:1751–1755

Nudell DM, Grossfeld GD, Weinberg VK et al (1999) Radiotherapy after radical prostatectomy: treatment outcomes and failure patterns. Urology 54:1049–1057

Nutting CM, Khoo VS, Walker V et al (2000) A randomized study of the use of a customized immobilization system in the treatment of prostate cancer with conformal radiotherapy. Radiother Oncol 54:1–9

Oelrich TM (1980) The urethral sphincter muscle in the male. Am J Anat 158:229–246

Oesterling JE, Suman VJ, Zincke H (1994) PSA-detected 9clinical state T1c or B1 prostate cancer. Urol Clin North Am 20:293–302

Ohori M, Wheeler TM, Kattan MW et al (1995) Prognostic significance of positive surgical margins in radical prostatectomy specimens. J Urol 154:1818–1824

Parker CC, Damyanovich A, Haycocks T et al (2003) Magnetic resonance imaging in the radiation treatment planning of localized prostate cancer using intra-prostatic fiducial markers for computed tomography co-registration. Radiother Oncol 66:217–224

Partin AW, Yoo J, Carter HB et al (1993) The use of prostate specific antigen, clinical stage and Gleason score to predict pathological stage in men with localized prostate cancer. J Urol 150:110–114

Partin AW, Kattan MW, Subong EN et al (1997) Combination of prostate-specific antigen, clinical stage, and Gleason score to predict pathological stage of localized prostate cancer. A multi-institutional update. J Am Med Assoc 277:1445–1451

Partin AW, Mangold LA, Lamm DM et al (2001) Contemporary update of prostate cancer staging nomograms (Partin Tables) for the new millennium. Urology 58:843–848

Patel RR, Orton N, Tome WA et al (2003) Rectal dose sparing with a balloon catheter and ultrasound localization in conformal radiation therapy for prostate cancer. Radiother Oncol 67:285–294

Pedley ID (2002) Transperineal interstitial permanent prostate brachytherapy for carcinoma of the prostate. Surg Oncol 11:25–34

Pellizzon AC, Nadalin W, Salvajoli JV et al (2003) Results of high dose rate afterloading brachytherapy boost to conventional external beam radiation therapy for initial and locally advanced prostate cancer. Radiother Oncol 66:167–172

Pellizzon AC, Salvajoli JV, Maia MA et al (2004) Late urinary morbidity with high dose prostate brachytherapy as a boost to conventional external beam radiation therapy for local and locally advanced prostate cancer. J Urol 171:1105–1108

Perez CA (1983) Presidential address of the 24th annual meeting of the American Society of Therapeutic Radiologists:

carcinoma of the prostate, a vexing biological and clinical enigma. Int J Radiat Oncol Biol Phys 9:1427–1438

Perez CA (1998) Prostate. In: Brady LW (eds) Principles and practice of radiation oncology. Lippincott Raven, Philadelphia, pp 1583–1694

Perez CA, Pilepich MV, Garcia D et al (1988) Definitive radiation therapy in carcinoma of the prostate localized to the pelvis: experience at the Mallinckrodt Institute of Radiology. NCI Monogr 7:85–94

Perez CA, Lee HK, Georgiou A et al (1993) Technical and tumor-related factors affecting outcome of definitive irradiation for localized carcinoma of the prostate. Int J Radiat Oncol Biol Phys 26:581–591

Perez CA, Michalski J, Ballard S et al (1997) Cost benefit of emerging technology in localized carcinoma of the prostate. Int J Radiat Oncol Biol Phys 39:875–883

Perrotti M, Kaufman RP, Jr., Jennings TA et al (1996) Endorectal coil magnetic resonance imaging in clinically localized prostate cancer: Is it accurate? J Urol 156:106–109

Pisansky TM, Blute ML, Suman VJ et al (1996a) Correlation of pretherapy prostate cancer characteristics with seminal vesicle invasion in radical prostatectomy specimens. Int J Radiat Oncol Biol Phys 36:585–591

Pisansky TM, Zincke H, Suman VJ et al (1996b) Correlation of pretherapy prostate cancer characteristics with histologic findings from pelvic lymphadenectomy specimens. Int J Radiat Oncol Biol Phys 34:33–39

Pistenma DA, Bagshaw MA, Freiha FS (1979) Extended-field radiation therapy for prostatic adenocardinoma: status report of a limited prospective trial. In: Samuels ML (ed) Cancer of the genitourinary tract. Raven, New York, pp 229–247

Pollack A, Smith LG, von Eschenbach AC (2000) External beam radiotherapy dose response characteristics of 1127 men with prostate cancer treated in the PSA era. Int J Radiat Oncol Biol Phys 48:507–512

Pollack A, Zagars GK, Starkschall G et al (2002) Prostate cancer radiation dose response: results of the MD Anderson phase III randomized trial. Int J Radiat Oncol Biol Phys 53:1097–1105

Pollack A, Hanlon AL, Horwitz EM et al (2004) Prostate cancer radiotherapy dose response: an update of the fox chase experience. J Urol 171:1132–1136

Potters L, Torre T, Ashley R et al (2000) Examining the role of neoadjuvant androgen deprivation in patients undergoing prostate brachytherapy. J Clin Oncol 18:1187–1192

Potters L, Torre T, Fearn PA et al (2001) Potency after permanent prostate brachytherapy for localized prostate cancer. Int J Radiat Oncol Biol Phys 50:1235–1242

Poulos CK, Daggy JK, Cheng L (2004) Prostate needle biopsies: multiple variables are predictive of final tumor volume in radical prostatectomy specimens. Cancer 101:527–532

Prete JJ, Prestidge BR, Bice WS et al (1998) A survey of physics and dosimetry practice of permanent prostate brachytherapy in the United States. Int J Radiat Oncol Biol Phys 40:1001–1005

Rasch C, Barillot I, Remeijer P et al (1999) Definition of the prostate in CT and MRI: a multi-observer study. Int J Radiat Oncol Biol Phys 43:57–66

Reed D, Wallner K, Merrick G et al (2003) Clinical correlates to PSA spikes and positive repeat biopsies after prostate brachytherapy. Urology 62:683–688

Roach M III (1993) Re: The use of prostate specific antigen, clinical stage and Gleason score to predict pathological stage in men with localized prostate cancer. J Urol 150:1923–1924

Roach M III, Akazawa PF, Pickett B et al (1994) Bilateral arcs using "averaged beam's eye views": a simplified technique for delivering 3-D based conformal radiotherapy. Med Dosim 19:159–168

Roach M III, Faillace-Akazawa P, Malfatti C (1997) Prostate volumes and organ movement defined by serial computerized tomographic scans during three-dimensional conformal radiotherapy. Radiat Oncol Invest 5:187–194

Roach M III, DeSilvio M, Lawton C et al (2003) Phase III trial comparing whole-pelvic versus prostate-only radiotherapy and neoadjuvant versus adjuvant combined androgen suppression: Radiation Therapy Oncology Group 9413. J Clin Oncol 21:1904–1911

Roach M, Winter K, Michalski JM et al (2004) Penile bulb dose and impotence after three-dimensional conformal radiotherapy for prostate cancer on RTOG 9406: findings from a prospective, multi-institutional, phase I/II dose-escalation study. Int J Radiat Oncol Biol Phys 60:1351–1356

Roberson PL, Narayana V, McShan DL et al (1997) Source placement error for permanent implant of the prostate. Med Phys 24:251–257

Rosenthal SA, Roach M III, Goldsmith BJ et al (1993) Immobilization improves the reproducibility of patient positioning during six-field conformal radiation therapy for prostate carcinoma. Int J Radiat Oncol Biol Phys 27:921–926

Rukstalis DB, Lawton CA, Brendler CB (1996) Management options for patients with lymph node metastases from prostate cancer. In: Shipley WU (ed) Comprehensive textbook of genitourinary oncology. Williams and Wilkins, Baltimore, pp 838–853

Ryu JK, Winter K, Michalski JM (2001) Preliminary report of toxicity following 3D conformal radiation therapy (3DCRT) for prostate cancer on 3DOG/RTOG 9406, level III (79.2 Gy). Int J Radiat Oncol Biol Phys 51:136–137

Sadeghi A, Kuisk H, Tran L et al (1996) Urethrography and ischial intertuberosity line in radiation therapy planning for prostate carcinoma. Radiother Oncol 38:215–222

Sandler H, Michalski J (2005) Treatment of early stage prostate cancer: 3D conformal radiotherapy for localized prostate cancer. In: Vogelzang NJ, Scardino PT, Shipley WU, Debruyne FMJ, Linehan WM (eds) Comprehensive textbook of genitourinary oncology, 3rd edn. Lippincott, Williams and Wilkins, Philadelpphia, pp 203–217

Sathya JR, Davis IR, Julian JA et al (2005) Randomized trial comparing iridium implant plus external-beam radiation therapy with external-beam radiation therapy alone in node-negative locally advanced cancer of the prostate. J Clin Oncol 23:1192–1199

Selek U, Lee A, Levy L et al (2003) Utility of the percentage of positive prostate biopsies in predicting PSA outcome after radiotherapy for patients with clinically localized prostate cancer. Int J Radiat Oncol Biol Phys 57:963–967

Sherertz T, Wallner K, Wang H et al (2001) Long-term urinary function after transperineal brachytherapy for patients with large prostate glands. Int J Radiat Oncol Biol Phys 51:1241–1245

Smith CV, Bauer JJ, Connelly RR et al (2000) Prostate cancer in men age 50 years or younger: a review of the Department of Defense Center for Prostate Disease Research multicenter prostate cancer database. J Urol 164:1964–1967

736   J. M. Michalski et al.

Snyder KM, Stock RG, Hong SM et al (2001) Defining the risk of developing grade 2 proctitis following 125I prostate brachytherapy using a rectal dose-volume histogram analysis. Int J Radiat Oncol Biol Phys 50:335–341

Soffen EM, Hanks GE, Hwang CC et al (1991) Conformal static field therapy for low volume low grade prostate cancer with rigid immobilization. Int J Radiat Oncol Biol Phys 20:141–146

Sohayda C, Kupelian PA, Levin HS et al (2000) Extent of extracapsular extension in localized prostate cancer. Urology 55:382–386

Spevack L, Killion LT, West JC Jr et al (1996) Predicting the patient at low risk for lymph node metastasis with localized prostate cancer: an analysis of four statistical models. Int J Radiat Oncol Biol Phys 34:543–547

Steenbakkers RJ, Deurloo KE, Nowak PJ et al (2003) Reduction of dose delivered to the rectum and bulb of the penis using MRI delineation for radiotherapy of the prostate. Int J Radiat Oncol Biol Phys 57:1269–1279

Stephenson AJ, Shariat SF, Zelefsky MJ et al (2004) Salvage radiotherapy for recurrent prostate cancer after radical prostatectomy. J Am Med Assoc 291:1325–1332

Stock RG, Stone NN, Ianuzzi C et al (1995) Seminal vesicle biopsy and laparoscopic pelvic lymph node dissection: implications for patient selection in the radiotherapeutic management of prostate cancer. Int J Radiat Oncol Biol Phys 33:815–821

Stock RG, Stone NN, Tabert A et al (1998) A dose-response study for I-125 prostate implants. Int J Radiat Oncol Biol Phys 41:101–108

Stock RG, Kao J, Stone NN (2001) Penile erectile function after permanent radioactive seed implantation for treatment of prostate cancer. J Urol 165:436–439

Stone NN, Stock RG (2000) Prostate brachytherapy in patients with prostate volumes ≥50 cm(3): dosimetic analysis of implant quality. Int J Radiat Oncol Biol Phys 46:1199–1204

Stone NN, Stock RG (2002) Complications following permanent prostate brachytherapy. Eur Urol 41:427–433

Stone NN, Stock RG, Unger P (1995) Indications for seminal vesicle biopsy and laparoscopic pelvic lymph node dissection in men with localized carcinoma of the prostate. J Urol 154:1392–1396

Stone NN, Ratnow ER, Stock RG (2000) Prior transurethral resection does not increase morbidity following real-time ultrasound-guided prostate seed implantation. Tech Urol 6:123–127

Storey MR, Pollack A, Zagars G et al (2000) Complications from radiotherapy dose escalation in prostate cancer: preliminary results of a randomized trial. Int J Radiat Oncol Biol Phys 48:635–642

Syed AM, Puthawala A, Transey LA (1983) Temporary iridium-192 implantation in the management of carcinoma of the prostate. Memorial Sloan-Kettering Cancer Center, New York

Syed AM, Puthawala A, Sharma A et al (2001) High-dose-rate brachytherapy in the treatment of carcinoma of the prostate. Cancer Control 8:511–521

Symon Z, Griffith KA, McLaughlin PW et al (2003) Dose escalation for localized prostate cancer: substantial benefit observed with 3D conformal therapy. Int J Radiat Oncol Biol Phys 57:384–390

Talcott JA, Clark JA, Stark PC et al (2001) Long-term treatment related complications of brachytherapy for early prostate cancer: a survey of patients previously treated. J Urol 166:494–499

Taylor WJ, Richardson RG , Hafermann MD (1979) Radiation therapy for localized prostate cancer. Cancer 43:1123–1127

Teh BS, McGary JE, Dong L et al (2002) The use of rectal balloon during the delivery of intensity modulated radiotherapy (IMRT) for prostate cancer: More than just a prostate gland immobilization device? Cancer J 8:476–483

Teh BS, Bastasch MD, Mai WY et al (2003) Predictors of extracapsular extension and its radial distance in prostate cancer: implications for prostate IMRT, brachytherapy, and surgery. Cancer J 9:454–460

Ten Haken RK, Perez-Tamayo C, Tesser RJ et al (1989) Boost treatment of the prostate using shaped, fixed fields. Int J Radiat Oncol Biol Phys 16:193–200

Terk MD, Stock RG, Stone NN (1998) Identification of patients at increased risk for prolonged urinary retention following radioactive seed implantation of the prostate. J Urol 160:1379–1382

Thomas MD, Cormack R, Tempany CM et al (2000) Identifying the predictors of acute urinary retention following magnetic-resonance-guided prostate brachytherapy. Int J Radiat Oncol Biol Phys 47:905–908

Tinger A, Michalski JM, Cheng A et al (1998) A critical evaluation of the planning target volume for 3-D conformal radiotherapy of prostate cancer. Int J Radiat Oncol Biol Phys 42:213–221

Valicenti RK, Gomella LG, Ismail M et al (1998) Effect of higher radiation dose on biochemical control after radical prostatectomy for PT3N0 prostate cancer. Int J Radiat Oncol Biol Phys 42:501–506

Valicenti RK, Gomella LG, Ismail M et al (1999) The efficacy of early adjuvant radiation therapy for pT3N0 prostate cancer: a matched-pair analysis. Int J Radiat Oncol Biol Phys 45:53–58

Van Herk M, Bruce A, Kroes AP et al (1995) Quantification of organ motion during conformal radiotherapy of the prostate by three dimensional image registration. Int J Radiat Oncol Biol Phys 33:1311–1320

Vicini FA, Martinez A, Hanks G et al (2002) An interinstitutional and interspecialty comparison of treatment outcome data for patients with prostate carcinoma based on predefined prognostic categories and minimum follow-up. Cancer 95:2126–2135

Vigneault E, Pouliot J, Laverdiere J et al (1997) Electronic portal imaging device detection of radioopaque markers for the evaluation of prostate position during megavoltage irradiation: a clinical study. Int J Radiat Oncol Biol Phys 37:205–212

Villeirs GM, Verstraete KL, de Neve WJ et al (2005) Magnetic resonance imaging anatomy of the prostate and periprostatic area: a guide for radiotherapists. Radiother Oncol 76:99–106

Villers AA, McNeal JE, Freiha FS et al (1991) Development of prostatic carcinoma. Morphometric and pathologic features of early stages. Acta Oncol 30:145–151

Wachter S, Gerstner N, Dorner D et al (2002) The influence of a rectal balloon tube as internal immobilization device on variations of volumes and dose-volume histograms during treatment course of conformal radiotherapy for prostate cancer. Int J Radiat Oncol Biol Phys 52:91–100

Wahlgren T, Brandberg Y, Haggarth L et al (2004) Health-related quality of life in men after treatment of localized prostate cancer with external beam radiotherapy combined with (192)ir brachytherapy: a prospective study of 93 cases using the EORTC questionnaires QLQ-C30 and QLQ-PR25. Int J Radiat Oncol Biol Phys 60:51–59

Wallner K, Roy J, Harrison L (1995) Dosimetry guidelines to minimize urethral and rectal morbidity following transperineal I-125 prostate brachytherapy. Int J Radiat Oncol Biol Phys 32:465–471

Wallner K, Lee H, Wasserman S et al (1997) Low risk of urinary incontinence following prostate brachytherapy in patients with a prior transurethral prostate resection. Int J Radiat Oncol Biol Phys 37:565–569

Wallner K, Smathers S, Sutlief S et al (2000) Prostate brachytherapy in patients with median lobe hyperplasia. Int J Cancer 90:152–156

Wallner K, Merrick G, True L et al (2002a) I-125 versus Pd-103 for low-risk prostate cancer: morbidity outcomes from a prospective randomized multicenter trial. Cancer J 8:67–73

Wallner K, Merrick GS, Benson ML et al (2002b) Penile bulb imaging. Int J Radiat Oncol Biol Phys 53:928–933

Wallner K, Merrick G, True L et al (2003) 125I versus 103Pd for low-risk prostate cancer: preliminary PSA outcomes from a prospective randomized multicenter trial. Int J Radiat Oncol Biol Phys 57:1297–1303

Waterman FM, Dicker AP (1999) Effect of post-implant edema on the rectal dose in prostate brachytherapy. Int J Radiat Oncol Biol Phys 45:571–576

Waterman FM, Yue N, Corn BW et al (1998) Edema associated with I-125 or Pd-103 prostate brachytherapy and its impact on post-implant dosimetry: an analysis based on serial CT acquisition. Int J Radiat Oncol Biol Phys 41:1069–1077

Wawroschek F, Wagner T, Hamm M et al (2003) The influence of serial sections, immunohistochemistry, and extension of pelvic lymph node dissection on the lymph node status in clinically localized prostate cancer. Eur Urol 43:132–137

Weber DC, Nouet P, Rouzaud M et al (2000) Patient positioning in prostate radiotherapy: Is prone better than supine? Int J Radiat Oncol Biol Phys 47:365–371

Wiegel T, Bottke D, Willich N (2005) Phase III results of adjuvant radiotherapy (RT) versus "wait and see" (WS) in patients with pT3 prostate cancer following radical prostatectomy (RP). ASCO, Orlando, Florida

Wilder RB, Fone PD, Rademacher DE et al (1997) Localization of the prostatic apex for radiotherapy treatment planning using urethroscopy. Int J Radiat Oncol Biol Phys 38:737–741

Wilson LD, Ennis R, Percarpio B et al (1994) Location of the prostatic apex and its relationship to the ischial tuberosities. Int J Radiat Oncol Biol Phys 29:1133–1138

Wise AM, Stamey TA, McNeal JE et al (2002) Morphologic and clinical significance of multifocal prostate cancers in radical prostatectomy specimens. Urology 60:264–269

Yan D, Lockman D, Brabbins D et al (2000) An off-line strategy for constructing a patient-specific planning target volume in adaptive treatment process for prostate cancer. Int J Radiat Oncol Biol Phys 48:289–302

Yoshioka Y, Nose T, Yoshida K et al (2000) High-dose-rate interstitial brachytherapy as a monotherapy for localized prostate cancer: treatment description and preliminary results of a phase I/II clinical trial. Int J Radiat Oncol Biol Phys 48:675–681

Yoshioka Y, Nishimura T, Kamata M et al (2005) Evaluation of anatomy-based dwell position and inverse optimization in high-dose-rate brachytherapy of prostate cancer: a dosimetric comparison to a conventional cylindrical dwell position, geometric optimization, and dose-point optimization. Radiother Oncol 75:311–317

Zelefsky MJ, Aschkenasy E, Kelsen S et al (1997a) Tolerance and early outcome results of postprostatectomy three-dimensional conformal radiotherapy. Int J Radiat Oncol Biol Phys 39:327–333

Zelefsky MJ, Happersett L, Leibel SA et al (1997b) The effect of treatment positioning on normal tissue dose in patients with prostate cancer treated with three-dimensional conformal radiotherapy. Int J Radiat Oncol Biol Phys 37:13–19

Zelefsky MJ, Crean D, Mageras GS et al (1999) Quantification and predictors of prostate position variability in 50 patients evaluated with multiple CT scans during conformal radiotherapy. Radiother Oncol 50:225–234

Zelefsky MJ, Fuks Z, Hunt M et al (2001) High dose radiation delivered by intensity modulated conformal radiotherapy improves the outcome of localized prostate cancer. J Urol 166:876–881

Zelefsky MJ, Fuks Z, Hunt M et al (2002) High-dose intensity modulated radiation therapy for prostate cancer: early toxicity and biochemical outcome in 772 patients. Int J Radiat Oncol Biol Phys 53:1111–1116

29 Testicular Cancer

Maria Pearse and Gerard C. Morton

CONTENTS

29.1   Introduction   739
29.2   Anatomy and Natural History   740
29.2.1   Anatomy   740
29.2.2   Natural History   740
29.3   Seminoma: General Management   740
29.3.1   Staging   741
29.3.1.1 Stage I Seminoma   741
29.3.1.2 Stage II Seminoma   742
29.4   Radiotherapy Treatment   742
29.4.1   Target Volume and Field Borders   742
29.4.2   Simulation and Beam Arrangement   743
29.4.3   Dose and Fractionation   745
29.4.3.1 Stage I Seminoma   745
29.4.3.2 Stage II Seminoma   745
29.4.4   Organs at Risk   745
29.4.5   Treatment Delivery   746
29.4.5.1 Testicular Shielding   746
29.4.5.2 Treatment verification   746
29.5   Intratubular Germ Cell Neoplasia   746
29.6   Special Considerations   747
29.6.1   Scrotal Invasion/Scrotal Interference   747
29.6.2   Contralateral Germ Cell Tumors   747
29.6.3   Management of a Residual Mass Following
          Chemotherapy   748
29.6.4   Prophylactic Contralateral Pelvic Lymph Node
          Irradiation   748
29.7   Treatment Sequelae   748
29.7.1   Acute Side Effects   748
29.7.2   Late Side Effects   749
29.7.3   Impaired Spermatogenesis   749
29.7.4   Second Malignant Neoplasms   749
          References   750

M. Pearse, MD, MB, ChB, FRANZCR
Division of Radiation Oncology, Toronto-Sunnybrook Regional
Cancer Centre, University of Toronto, 2075 Bayview Avenue,
Toronto, Ontario M4N 3M5, Canada
G. C. Morton, MD, MRCPI, FRCPC
Assistant Professor, Division of Radiation Oncology, Toronto-
Sunnybrook Regional Cancer Centre, University of Toronto,
2075 Bayview Avenue, Toronto, Ontario M4N 3M5, Canada

29.1 Introduction

The majority of testicular tumors occur in young men with a peak incidence at 30 years of age. Although testicular tumors are uncommon, the incidence is increasing (Thompson et al. 1999; Stone et al. 1991; Power et al. 2001; Dos et al. 1999), with most cases occurring in white males (Daniels et al. 1981). While nonseminomatous germ cell tumors (NSGCTs) are primarily treated with surgery and systemic chemotherapy, radiation treatment continues to have a major role in the management of seminoma.

Over 95% of testicular malignancies are germ cell tumors. These are separated into two histological subgroups – seminoma and non-seminoma. NSGCTs include teratoma, embryonal carcinoma, endodermal (yolk sac) tumors, choriocarcinoma and mixed tumors.

There is increasing evidence that intratubular germ cell neoplasia (IGCN) is a precursor of all types of germ cell tumors except spermatocytic seminoma and infantile testicular cancer. In patients with invasive germ cell tumors, IGCN is identified adjacent to the invasive component in 90–99% of cases (Coffin et al. 1985; Dieckmann and Skakkebaek 1999; Jacobsen et al. 1981), and approximately 5% of patients with unilateral germ cell neoplasms have IGCN in the contralateral testicle (Dieckmann and Loy 1996).

With the advent of effective cisplatin chemotherapy, the role of radiotherapy in NSGCTs has dramatically diminished. Radiotherapy still plays an important role in the treatment of stage I and II seminoma, IGCN and residual disease following chemotherapy. Radiotherapy is also utilized in the palliation of distant metastases.

29.2
Anatomy and Natural History

29.2.1
Anatomy

Evaluation of lymphangiograms, surgical series of retroperitoneal lymph node dissections and anatomical studies in cadavers have provided valuable information on the lymphatic drainage of the testis (RAY et al. 1974). In the developing embryo, the testes originate from the genital ridge located near the second lumbar vertebra. Accompanied by their blood supply and lymphatics, they descend into the scrotum via the inguinal canal. As a result, the primary lymphatic drainage from the testis is to the retroperitoneal lymph nodes. The lymphatic vessels first drain into the collecting trunks at the hilum of the testicle. These lymphatic trunks accompany the testicular artery, vein and spermatic cord to the internal ring, and then continue proximally to the retroperitoneal lymph nodes. The retroperitoneal lymph nodes are situated anterior to the T11 to L4 vertebral bodies, although are concentrated at the L1–L3 level. On the left, the lymphatics drain primarily into the pre-aortic and para-aortic lymph nodes around the left renal hilum and thence to the inter-aortocaval nodes. On the right, the first echelon of nodes is in the inter-aortocaval region, followed by the pre-aortic and para-aortic lymph nodes. Early lymphographic studies demonstrated rapid crossover from right to left as well as from left to right. Clinically, however, contralateral spread is mainly seen with right-sided tumors and rarely with left-sided.

From the retroperitoneal nodes, the lymph drains into the cisterna chyli, thoracic duct, posterior mediastinum and the left supraclavicular fossa. The thoracic duct drains into the left subclavian vein in the left supraclavicular region. In 5–10% of patients, drainage into the right supraclavicular area can occur.

Aberrant lymphatic drainage may occur. Herniorrhaphy alters the drainage of the testicle. The testicular lymphatic vessels anastomose with the regional lymph vessels resulting in drainage into the ipsilateral inguinal and iliac lymph nodes (PEREZ et al. 2005). In addition, the testicular trunks may abandon the spermatic vessels at the internal inguinal ring and pass posteriorly and superiorly into the external iliac lymph nodes. The scrotum drains directly into the inguinal and external iliac lymph nodes.

29.2.2
Natural History

In the majority of patients, pathological examination of the radical orchidectomy specimen reveals the tumor confined to the testis. Occasionally, in advanced disease invasion of the epididymis, rete testis or spermatic cord can occur. Rarely, the tumor extends through the tunica albuginea to involve the scrotum.

Seminoma has an orderly and predictable pattern of spread. Locoregional lymphatics are the first site of metastatic disease. From the retroperitoneal lymph nodes, seminoma spreads proximally to involve the next echelon, the mediastinal lymph nodes, and then the supraclavicular lymph nodes. Very occasionally, metastases from retroperitoneal lymph nodes can drain directly via the thoracic duct to the supraclavicular fossa, resulting in supraclavicular metastases in the absence of mediastinal disease.

Hematogenous metastases are rare in pure seminoma, being much more common with NSGCT. Lung is the most common site of distant disease, although bone, liver and brain may also be involved.

29.3
Seminoma: General Management

If the clinical and radiological features are consistent with a testicular tumor, a radical orchidectomy is performed. Via an inguinal incision, the testicle and spermatic cord are removed en bloc with high ligation of the spermatic cord at the deep inguinal ring. An inguinal approach is used to minimize the risk of aberrant lymphatic spread and local contamination. A transscrotal approach risks development of alternative lymphatic drainage to the inguinal and pelvic lymph nodes. In addition, the spermatic cord remains in place from the external to the internal inguinal ring.

Laboratory investigations include a complete blood count, renal and liver function tests and serum tumor markers including alpha-fetoprotein (AFP), human chorionic gonadotropin (HCG) and lactate dehydrogenase (LDH). To assess the extent of metastatic disease, a chest X-ray and computed tomography (CT) scan of the abdomen and pelvis are obtained. Of patients diagnosed with seminoma, 80% have stage I disease at presentation.

29.3.1
Staging

A number of staging systems for testicular cancer are used. The commonly used AMERICAN JOINT COMMITTEE ON CANCER (AJCC) 2002 classification is outlined in Table 29.1. This staging system incorporates the features of the primary tumor (T), nodes (N), metastases (M) and level of serum tumor marker (S). The nodal staging is based on the greatest dimension of the largest involved lymph node and does not take into account the total bulk of lymphadenopathy, a well-recognized prognostic factor.

29.3.1.1
Stage I Seminoma

Following orchidectomy alone, relapse occurs in 12–20% of patients with stage I seminoma. The majority of these relapses occur in the para-aortic lymph nodes (HORWICH et al. 1992a; MAIER et al. 1968; MIKI et al. 1998; VON DER MASSE et al. 1993; WARDE et al. 1993). Management approaches include adjuvant radiotherapy and surveillance. Although adjuvant chemotherapy is not considered standard treatment for stage I seminoma, single agent carboplatin chemotherapy is presently under investigation.

Table 29.1. AMERICAN JOINT COMMITTEE ON CANCER (AJCC) (2002) Staging for Testicular Neoplasms

Primary Tumor (T) (pathological classification)

Tx	Primary tumor cannot be assessed
T0	No evidence of primary tumor (e.g., histological scar in testis)
Tis	Intratubular germ cell neoplasia (carcinoma *in situ*)
pT1	Tumor limited to testis and epididymis without vascular/lymphatic invasion; tumor may invade into the tunica albuginea but not the tunica vaginalis
pT2	Tumor limited to testis and epididymis with vascular/lymphatic invasion or tumor extending through the tunica albuginea with involvement of the tunica vaginalis
pT3	Tumor invades the spermatic cord with or without vascular/lymphatic invasion
pT4	Tumor invades scrotum with or without vascular/lymphatic invasion

Lymph Node (N)

N0	No regional node metastasis
N1	Metastasis within a lymph node mass 2 cm or less in greatest dimension; or multiple nodes no more than 2 cm in greatest dimension
N2	Metastasis within a lymph node mass that is >2 cm but not more than 5 cm in greatest dimension or multiple lymph nodes 2–5 cm, any one mass greater than 2 cm but not more than 5 cm in greatest dimension
N3	Metastasis within a lymph node mass that is more than 5 cm in maximum diameter

Distant Metastasis (M)

M0	No distant metastasis
M1a	Non-regional lymph node or pulmonary metastasis
M1b	Non-pulmonary visceral metastasis

Serum Tumor Markers (S)

Sx	Serum tumor markers not performed
S0	Serum tumor markers within normal limits
S1	LDH <1.5xN and HCG <5,000 and AFP <1,000
S2	LDH 1.5–10xN or HCG 5,000–50,000 or AFP 1,000–10,000
S3	LDH >10xN or HCG >50,000 or AFP >10,000

Staging Groupings
From the TNM system, patients are grouped into either stage 1, 2 or 3

IA	T1 N0 M0 S0
IB	T2–4 N0 M0 S0
IS	Any T N0 M0 S1–3
IIA	Any T N1 M0 S0/1
IIB	Any T N2 M0 S0/1
IIC	Any T N3 M0 S0/1
IIIA	Any T, any N M1a S0/1
IIIB	Any T, any N M1a S2
IIIC	Any T, any N M1b or S3

The standard post-operative management of patients with stage I seminoma has been adjuvant radiotherapy to the para-aortic and ipsilateral pelvic lymph nodes (the "dog-leg" or "hockey stick" radiation field). In view of the exquisite radiosensitivity of seminoma, this treatment results in an excellent recurrence-free survival of 97% and a disease-specific survival greater than 99% (PEREZ et al. 2005). Several retrospective reviews have reported similar disease-free and overall survival (OS) in patients with stage I seminoma treated with para-aortic radiotherapy alone (KIRICUTA et al. 1996; MELCHIOR et al. 2001; SANTONI et al. 2003; SULTANEM et al. 1998; TAYLOR et al. 2001; READ and JOHNSTON 1993). The Medical Research Council performed a randomized controlled trial comparing dog-leg and para-aortic irradiation in stage I seminoma (FOSSA et al. 1999). At a median follow-up of 4.5 years, there was no difference in the 3-year relapse-free survival (RFS) or OS. Less than 2% of patients in the para-aortic alone arm relapsed in the pelvis. Radiotherapy was better tolerated in the para-aortic alone arm with a reduction in the severity and frequency of acute gastrointestinal and hematological toxicity. The post-treatment peptic ulcer rate was similar in both arms, but sperm counts within the first 18 months were significantly higher in the para-aortic alone arm. Reducing the clinical target volume (CTV) to include the para-aortic lymph nodes alone is an option for many patients.

Prior to radiotherapy, sperm analysis and sperm banking may be carried out in patients who wish to preserve fertility.

29.3.1.2
Stage II Seminoma

Only 10–20% of patients are diagnosed with stage II disease, mostly of small volume.

Patients with stage IIA and IIB disease treated with radiotherapy alone have an excellent disease-free survival (88–93%) and OS (95–100%) (BAUMAN et al. 1998b; CHUNG et al. 2004; CLASSEN et al. 2003b; SCHMIDBERGER et al. 1997; WARDE et al. 1998). However, following radiotherapy alone, the relapse rate ranges between 20% and 50% in patients with stage IIC disease (WARDE et al. 1998) and between 33% and 100% in patients with retroperitoneal adenopathy greater than 10 cm (ANSCHER et al. 1992; BALL et al. 1982; SPEER et al. 1995; THOMAS 1997; WARDE et al. 1998).

Although radiotherapy provides adequate local control for patients with bulky stage II disease, the risk of distant relapse is high. Primary chemotherapy is therefore the treatment of choice providing excellent local control and treating possible distant micrometastases. In addition, radiotherapy in bulky stage II disease may be technically difficult, particularly if nodal disease covers a significant portion of the kidney.

29.4
Radiotherapy Treatment

29.4.1
Target Volume and Field Borders

For stage I disease, the CTV consists of the inter-aortocaval, pre-aortic and para-aortic nodes. The left renal hilar nodes are included for left-sided tumors. The ipsilateral external iliac and common iliac nodes may also be included, particularly if there is concern about aberrant drainage. Inclusion of the inguinal scar, inguinal lymph nodes or hemiscrotum is not warranted in the routine treatment of stage I disease. For stage II disease, a gross tumor volume is identified from diagnostic imaging, and the CTV also includes the ipsilateral pelvic nodes.

The planning target volume (PTV) includes the CTV plus a margin to account for positional and set-up uncertainties. To cover the known location of the retroperitoneal and iliac lymph nodes with an appropriate margin, standard anatomical field borders have been used. This is commonly referred to as the "dog-leg" or "hockey stick" field. The superior border is placed between the T9 and T10 vertebral bodies, with the inferior border at the top of the obturator foramen. The field is approximately 10–12 cm wide and usually covers the transverse processes (Fig. 29.1). On the left, the lateral border is extended to include the left renal hilum and customized shielding is positioned to reduce the amount of kidney irradiated (Fig. 29.2). At the mid L4 level the field is extended laterally to cover the ipsilateral external iliac nodes. Shielding is placed forming the "dog-leg" configuration. Multileaf collimators now largely replace lead blocks to define the field shape.

Once these borders are placed, the planning CT can be used to ensure adequate coverage of the target. Distance from the PTV to the field border is 8–15 mm depending on field size, energy, separa-

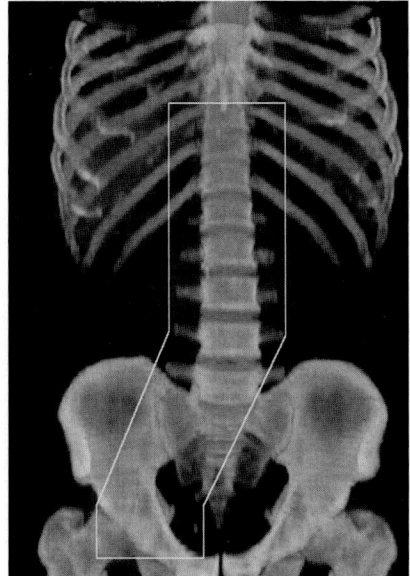
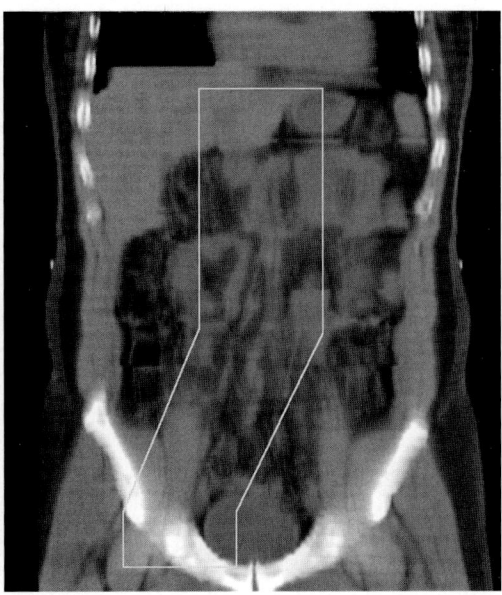

a   b

Fig. 29.1a,b. Radiotherapy field – stage I seminoma, right-sided tumor. The field includes the para-aortic and right pelvic lymph nodes ("dog-leg" field). **a** The superior border placed at the T9/T10 interspace, lateral borders at the edge of the transverse processes and inferior border above the obturator foramen. **b** A soft tissue image showing the location of the vessels within the radiotherapy field

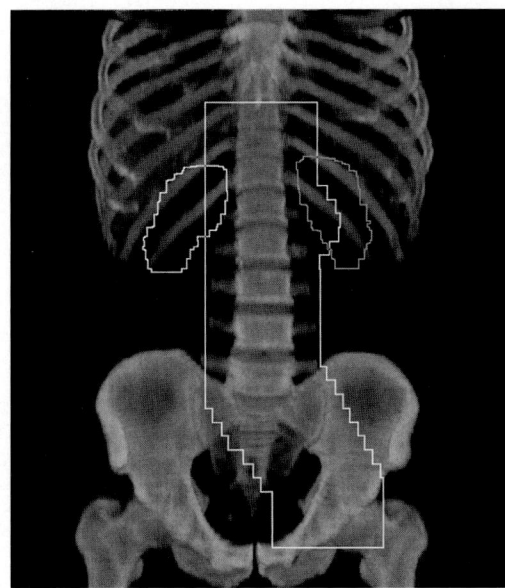

Fig. 29.2. Radiotherapy field – stage I seminoma, left-sided tumor. The field is extended to include the left renal hilar lymph nodes

tion and shielding. Originally, separate para-aortic and iliac fields were used to encompass this large volume and were matched at the L5/S1 junction. As the iliac nodes are anterior to the para-aortic nodes, this technique allowed differential weighting of the fields. A "dog-leg" field has the advantage of avoid-ing a field junction between the iliac and para-aortic nodes.

If retroperitoneal nodes alone are to be treated, the superior and lateral borders are as described above, although some use a lower superior border, e.g., T10/11. The inferior border is placed at the L5/S1 disc space (Fig. 29.3).

In stage II disease, the PTV includes the CTV plus an appropriate margin. A modified "dog-leg" field covers the macroscopic retroperitoneal nodal dis-ease and optionally the contralateral common iliac nodes if there is low-lying retroperitoneal adenopa-thy (Fig. 29.4). Organs at risk, including the kidney and liver, can be identified. Shielding is customized to allow adequate coverage of disease, while reduc-ing the dose to normal tissues. If the retroperitoneal lymphadenopathy is greater than 4 cm in size, a two-phase technique is used. The first phase encom-passes the PTV as described above. A second phase includes the gross disease with a tight margin.

29.4.2
Simulation and Beam Arrangement

The patient is simulated in the supine position with arms at his sides. This simple position is comfort-able. Testicular shielding (often in the form of a clam-shell device) is used. Foot stocks and knee

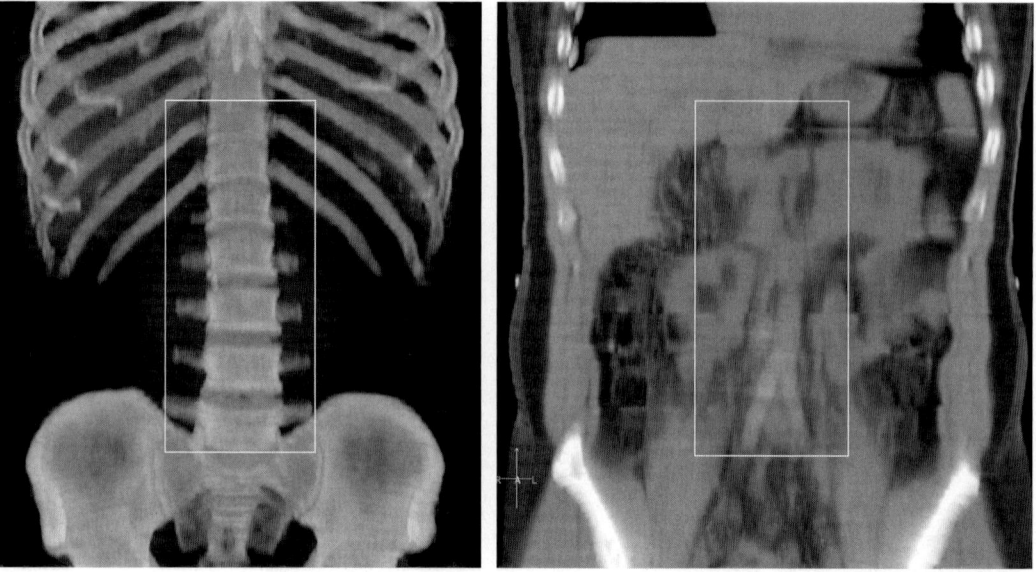

a   b

Fig. 29.3a,b. Para-aortic nodes alone. **a** The superior border is placed either at the T10/T11 interspace (although some use T9/T10), with the inferior border at the L5/S1 interspace. The lateral borders are at the edge of the transverse processes. The left renal hilum is included for left-sided tumors. **b** A soft tissue image showing the location of the vessels within the radiation field

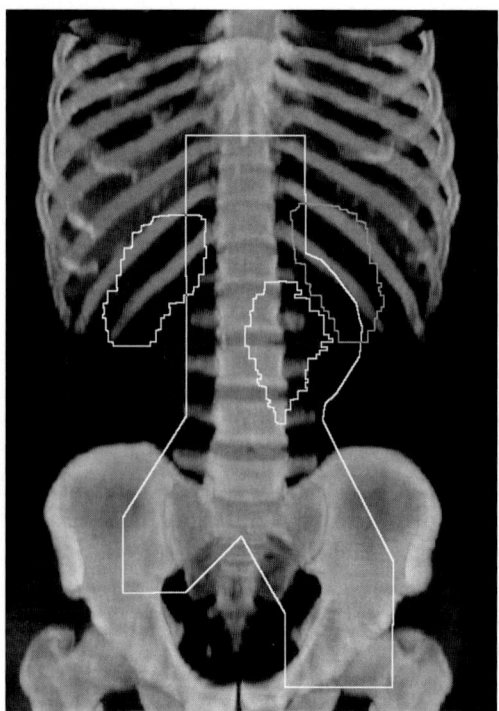

Fig. 29.4. Radiotherapy field – stage II seminoma. The field is extended to cover the para-aortic nodal disease and often the contralateral proximal external iliac lymph nodes

or ankle restraints may be used to immobilize the patient and decrease external movement. A volumetric planning CT scan is acquired with the patient in this position. Contiguous 5-mm CT slices are

taken from the seventh thoracic vertebra to 2 cm below the ischial tuberosities. Utilization of a CT-based planning technique enables visualization of the location of the lymph node regions, adjacent tissues and critical normal structures including the kidneys and liver. In addition, the beams eye view (BEV) allows evaluation of the coverage of the PTV, and shielding can be appropriately placed. If a planning CT is not available, an intravenous urogram is performed to identify the position of the kidneys.

Tattoos are placed at the center, superior and inferior borders of the field. A lateral tattoo is also placed at mid-plane. The interplanar distance is measured at the central tattoo using a sliding rod if a planning CT is not available.

Treatment is delivered with a linear accelerator using an anterior and posterior parallel opposed pair. The beams are equally weighted. Both fields are treated daily, 5 days per week. Depending on the separation, 6- to 18-MV photons are utilized. If the separation is greater than 24 cm, an energy of greater than 6 MV provides less dose inhomogeneity. The patient is treated with either an isocentric, source axis distance (SAD) technique or, alternatively, a standard source skin distance (SSD) technique. In most patients, standard SSD is adequate, although if the field is longer than 40 cm, the patient is treated with extended SSD. With extended SSD, the patient may need to change position between fields. If extended SSD is employed, the width of the penumbra is increased, and this must be taken into account when selecting the field borders.

Rarely dose inhomogeneity due to contour obliquity occurs and a compensator is placed in the beam. For a 10-MV beam at 100 cm SSD, a difference in separation of 7 cm results in a variation in dose of approximately 10%. A compensator or wedge is recommended if there is a variation in dose of 10% or greater.

For stage II disease, a parallel opposed anterior and posterior pair is used. However, depending on the location of the mass and the position of organs at risk relative to the mass, a CT planned technique with oblique fields may result in a reduced dose to normal structures.

29.4.3
Dose and Fractionation

29.4.3.1
Stage I Seminoma

Radiation doses between 25 Gy and 40 Gy at 1.25–2.0 Gy per fraction have been most commonly used in the past. A 1986 consensus statement recommended an adjuvant dose of 25 Gy in 20 fractions (THOMAS 1986), which still remains the standard.

More recently, several authors have reported similar RFS, OS and infield failure rates with lower doses (GIACCHETTI et al. 1993; GURKAYNAK et al. 2003; LOGUE et al. 2003; NIEWALD et al. 1995). The U.K. Medical Research Council (MRC TE18) completed a randomized controlled trial of 30 Gy in 15 fractions over 3 weeks or 20 Gy in 10 fractions over 2 weeks. RFS was similar in both groups. Acutely, there was significantly more moderate or severe lethargy and inability to carry out normal work in the group that received 30 Gy. However, by 12 weeks, there were no differences between the two groups (JONES et al. 2001).

Radiation dose is generally prescribed to a point in the mid-line, and ideally a homogeneous dose distribution is obtained with a variation in the coverage of the PTV of –5% and +7% (Fig. 29.5).

29.4.3.2
Stage II Seminoma

1. Retroperitoneal lymphadenopathy less than 4 cm. 25 Gy in 20 fractions is delivered over 4 weeks.
2. Retroperitoneal lymphadenopathy greater than 4 cm.
   - Phase I – 25 Gy in 20 fractions is delivered over 4 weeks.
   - Phase II – 10 Gy in 5–8 fractions is delivered to the residual mass.

29.4.4
Organs at Risk

Table 29.2 shows the organs in the treatment field and their tolerance doses (TDs) (EMAMI et al. 1991) for treatment delivered at 1.8–2 Gy per fraction. The dose resulting in a 5% risk of nephritis at 5 years is 23 Gy (TD 5/5=23 Gy, whole kidney) or 27 Gy at 1.25 Gy per fraction.

It is also necessary to minimize the radiation dose to the remaining testis. In these relatively young men, fertility and hormonal function are important. For the endpoint of infertility, the TD5/5 for the testis at standard fractionation is around 1 Gy.

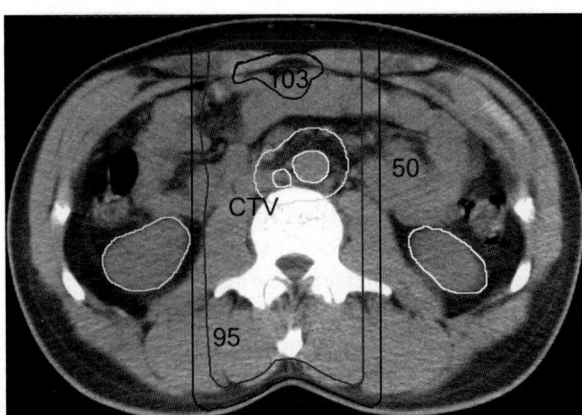

Fig. 29.5. Stage I seminoma. Axial computed tomography (CT) slice at the level of the kidneys showing the isodose distribution (parallel opposed fields). The 95% isodose curve covers the CTV

Table 29.2 Tolerance doses

Organ	TD 5/5 Volume			Endpoint
	1/3	2/3	3/3	
Kidney	50	30	23	Clinical nephritis
Liver	50	35	30	Liver failure
Small bowel	50		40	Obstruction/perforation/fistula
Stomach	60	55	50	Ulceration/perforation
Heart	60	45	40	Pericarditis
Bladder		80	65	Symptomatic bladder contracture and volume loss

From EMAMI et al. (1991)

29.4.5
Treatment Delivery

29.4.5.1
Testicular Shielding

During a fractionated course of radiotherapy to the retroperitoneal and ipsilateral iliac lymph nodes, the dose to the remaining testis ranges between 0.3 Gy and 1.5 Gy. A variety of factors contribute to this dose. A component is from leakage from the head. External scatter is generated from the collimator, field-shaping blocks and air. A significant proportion of the scattered dose is produced internally within the patient. A number of effective shielding devices have been described (BIERI et al. 1999; FRAASS et al. 1985; KUBO and SHIPLEY 1982). Most departments use simple forms of gonadal shielding such as the clam-shell device. The testis is placed in the 1-cm thick lead gonadal cups. This technique reduces the amount of internal scatter to the remaining testis to approximately 1% of the mid-plane dose. Shielding devices are available that shield both the penis and testis, although shielding the penis does not impact on the gonadal dose. In addition, these shielding devices are large, more cumbersome and less convenient when compared with the clam-shell device. BIERI et al. (1999) reported a greater than 50% reduction in testicular dose with gonadal shielding, and GORDON et al. (1997) noted the testicular dose was 5.1% of the target dose for patients with no or pipe cap type shields and 1.6% for patients with clam-shell type shields.

In addition to shielding, distance from the field is an important factor. JACOBSEN et al. (1997) report a significant correlation between the symphysis-to-testicle distance and gonadal dose. The authors suggest the ideal placement of the testicle is at 30 degrees rotation from the patient's long axis. This allows the testicle to lie in the scrotum as distally as possible from the field edge.

29.4.5.2
Treatment verification

An electronic portal image or port film is obtained on day 1 and optionally at weekly intervals to verify the field placement. This is compared with the simulation film or digitally reconstructed radiograph, which is important to verify the geometric set-up. The patient is reviewed weekly to assess toxicity.

29.5
Intratubular Germ Cell Neoplasia

Untreated, 50% of patients with IGCN will progress to invasive disease at 5 years (VON DER MASSE et al. 1986b) and 70% at 7 years (CLASSEN et al. 1998). In view of this, treatment is usually recommended. Both orchidectomy and radiotherapy yield excellent local control rates (DIECKMANN et al. 1993; DIECKMANN and LOY 1994; VON DER MASSE et al. 1986a), and it is often the clinical situation that dictates which approach is employed. Unilateral disease is treated with orchidectomy. In the situation of bilateral disease or in patients with IGCN in a solitary testis, an organ preservation approach using radiotherapy is the preferred option. Initial reports using chemotherapy were promising (VON DER MASSE et al. 1985). However, recently several authors have reported the persistence or recurrence of IGCN in patients treated with primary chemotherapy (CHRISTENSEN et al. 1998; DIECKMANN 1988; VON DER MASSE et al. 1988). Residual germ cell cancer is often found in orchidectomy specimens after cisplatin-based chemotherapy (GREIST et al. 1984).

Intratubular germ cell neoplasm is a radiosensitive tumor. The aim of treatment is to eradicate the IGCN while preserving hormone function. The CTV is the whole testis.

At simulation, the patient is supine, with the thighs abducted and soles together (this is commonly referred to as the "frog-leg" position). A lead shield is placed posteriorly to shield the perineum and immobilize the testis. The penis is taped out of the field, usually over the symphysis pubis.

The optimal radiotherapy technique to treat the testis is not defined. A number of different modalities have been employed. Due to the position of the testis and scrotum, a direct anterior field is preferred. Although parallel opposed beams provide a more homogeneous dose distribution, in this setting, the addition of a posterior beam would irradiate a large area of perineum resulting in extra toxicity.

Historically, a direct orthovoltage beam was commonly used to treat the testis. Although this provided a simple technique, there were several disadvantages. For a standard orthovoltage beam, the maximum dose is at or very close to the skin. The 90% depth dose for 270 KV is at approximately 2 cm. In view of this, a single direct field does not adequately treat to the depth required, and the posterior testis is underdosed. In addition, the high skin dose results in an increase in the skin reaction.

A direct electron beam has the advantage of a rapid fall off in dose with depth, reducing the dose to the perineum. A testicular ultrasound is acquired with the patient in the treatment position. This ultrasound provides a measurement of the thickness of the testis in the anterior–posterior plane. This measurement is used to determine the depth required for treatment. The electron energy is then selected. Commonly, the energy ranges between 9 and 15 megaelectron volts (MeV). A customized lead cut out is made to ensure that the testis with a 1-cm margin is treated and any adjacent tissue is shielded. Bolus is used to provide a homogeneous dose distribution, avoiding hot spots, and a rapid fall off in dose at the field edge. The bolus covers the entire testis, which also provides backscatter, thereby improving the dosimetry.

A direct cobalt beam has also been used as a treatment modality in this setting. As with orthovoltage, the depth dose characteristics of a cobalt beam results in under dosing of the posterior part of the target volume. For a 10×10-cm field, the D_{max} is at approximately 5 mm and 93% at approximately 2 cm. A further disadvantage is the relatively wide penumbra when compared with that of a linear accelerator. This must be taken into account when defining the lateral field borders.

In summary, the optimal treatment modality and technique for irradiation of the testis is unclear. However, electrons appear to have the advantage of delivering an adequate dose at depth. The curved contour of the testis can be adjusted for using bolus resulting in a homogeneous dose distribution.

Treatment with low dose radiotherapy (18–20 Gy) is able to eradicate IGCN and preserve hormone function avoiding life-long hormone supplementation (DIECKMANN et al. 1993). There is, however, evidence that with longer-term follow-up some impairment of androgen synthesis occurs (DIECKMANN and LOY 1994; DIECKMANN and SKAKKEBAEK 1999). In addition, patients with contralateral IGCN have a higher rate of baseline Leydig cell dysfunction than patients with a normal contralateral testis: 11/24 (45.8%) and 2/30 (6.6%), respectively (PETERSEN et al. 1999).

IGCN is a radiosensitive tumor and there is emerging evidence that doses less than 20 Gy in 10 fractions may be adequate to eradicate this tumor. SEDLMAYER et al. (2001) reported eradication of IGCN in nine patients treated with 13 Gy in 10 fractions. PETERSEN et al. (2002) evaluated radiation doses of 14 Gy, 16 Gy, 18 Gy and 20 Gy delivered at 2 Gy per fraction, 5 days per week. IGCN was eradi-

cated in all patients treated with doses of 16 Gy or more. However, CLASSEN et al. (2003a) compared a dose of 18 Gy in 9 fractions with 16 Gy in 8 fractions. One patient relapsed with IGCN after 16 Gy and two other patients had persistent spermatogonia following 16 Gy and 18 Gy.

Although patient numbers in the above-mentioned series are small and the follow-up period short, the data suggests that 16 Gy or more may eradicate a high proportion of IGCN. However, in a study by DIECKMANN et al. (2002), two patients subsequently developed a germ cell cancer (one patient a seminoma and the other patient a mixed seminoma and embryonal carcinoma) despite a total dose of 20 Gy. These tumors occurred at 5 years and 7 years after completion of radiotherapy. At present, 20 Gy in 10 fractions over 2 weeks is a widely accepted fractionation schedule for IGCN.

29.6
Special Considerations

29.6.1
Scrotal Invasion/Scrotal Interference

Although most patients present with T1 disease, testicular GCT can invade locally to involve the rete testis, epididymis and spermatic cord. As the tunica albuginea acts as a natural barrier, direct invasion of the scrotum is rare and occurs late. Historically, the hemiscrotum and inguinal lymph nodes were included in the treatment portal in patients with tunica albuginea invasion or scrotal interference. This practice has now been abandoned, as the risk of relapse is low and treatment would result in high dose to the remaining testis. However, if scrotal invasion occurs, adjuvant radiotherapy to the hemiscrotum and ipsilateral inguinal lymph nodes is recommended. The scrotal field is matched to the tattoo at the inferior border of the dog-leg field. Electrons are commonly used and a lead cut out is custom made to limit the dose to the remaining testis.

29.6.2
Contralateral Germ Cell Tumors

The prevalence of bilateral testicular cancer ranges between 1.5% and 5% (BOKEMEYER et al. 1993; FOSSA et al. 1989; HAMILTON et al. 1986; HAY et al. 1984; OHYAMA et al. 2002; VALLIS et al. 1995).

Although synchronous tumors occur, the majority are metachronous occurring at a median time interval of 5–8 years (BOKEMEYER et al. 1993; HAY et al. 1984; OHYAMA et al. 2002). A similar incidence of a second testicular tumor is observed in patients on surveillance (BOKEMEYER et al. 1993).

Standard treatment is radical orchidectomy. Following this procedure, the patient is sterile and requires life-long hormone replacement. Organ-sparing techniques have been reported. KAZEM and DANELLA (1999) described two patients that developed seminoma in the contralateral testis. Both patients were treated with an organ-sparing technique. Local excision of the tumor was followed by radiotherapy to the remaining testis. One patient received 19.8 Gy in 11 fractions and the other 20 Gy in 10 fractions. After more than 3 years, both patients have no evidence of disease. Androgen production is preserved but reduced and virility is retained. The authors conclude that this approach provides an alternative to radical orchidectomy with the advantage of avoiding long-term hormone replacement.

29.6.3
Management of a Residual Mass Following Chemotherapy

Following chemotherapy for metastatic seminoma, a residual mass may persist in up to 80% of patients (DUCHESNE et al. 1997; HORWICH et al. 1992b, 1997; MOTZER et al. 1988; PECKHAM et al. 1985; PUC et al. 1996; SCHULTZ et al. 1989). A variety of strategies have been proposed including surveillance, surgery and radiotherapy. Pathological examination of the resected specimen reveals viable tumor in only 10–20% of patients. Surgery is technically difficult following chemotherapy as a desmoplastic reaction occurs and the residual mass is densely adherent. In view of these factors, surveillance with careful monitoring of the residual mass has been suggested (CLEMM et al. 1986).

A number of authors have observed that the size of the residual mass (>3 cm) predicts the risk of viable tumor (FOSSA et al. 1989; HERR et al. 1997; MOTZER et al. 1987; PUC et al. 1996), although this is not supported by others (HORWICH et al. 1992b; SCHULTZ et al. 1989). The radiographic appearance of the residual mass (well defined versus poorly defined) has also been reported to predict residual tumor. In a series by RAVI et al. (1999), 6 of 11 (54.5%) patients with a well-defined residual mass of 3 cm or more

had positive histology versus only 1 of 14 (7.1%) patients with a poorly defined mass of similar size. None of the patients with a residual mass less than 3 cm had a viable tumor.

Routine radiotherapy is not recommended and does not improve OS (DUCHESNE et al. 1997). However, if the residual mass increases in size and surgery is not technically feasible, radiotherapy is the preferred option. If radiotherapy is employed, a CT planned technique allows accurate delineation of the gross tumor volume and organ at risk. In the post-chemotherapy setting, the CTV is generally limited to the gross disease identified on imaging. Technique is individualized. The total dose is 35–40 Gy in 1.6- to 2-Gy fractions, depending on bulk and location of disease.

29.6.4
Prophylactic Contralateral Pelvic Lymph Node Irradiation

The contralateral pelvic lymph nodes may be included in the CTV for patients with stage II disease if retrograde spread is a concern, although the risk of contralateral iliac lymph node involvement is extremely low (MASON and KEARSLEY 1988).

29.7
Treatment Sequelae

29.7.1
Acute Side Effects

Low dose infradiaphragmatic radiotherapy is well tolerated acutely. From the available literature, some nausea occurs in up to 100% of patients (AASS et al. 1992; BAUMAN et al. 1998a; KHOO et al. 1997; SOMMER et al. 1990; VALLIS et al. 1995), with vomiting and diarrhea in up to 80% of patients (AASS et al. 1992; BAUMAN et al. 1998a; KHOO et al. 1997; VALLIS et al. 1995). These side effects are generally mild in most patients, with the incidence of grade-3 and grade-4 acute toxicity ranging between 1.5% (SOMMER et al. 1990) and 2.5% (VALLIS et al. 1995). Most of the data reporting toxicity was collected in an era when higher total dose and dose per fraction were used. These rates of nausea, vomiting and diarrhea are not typically seen today with the current total dose and fractionation regime.

29.7.2
Late Side Effects

Late effects are uncommon. Analyzing the records of 1,026 patients treated with infradiaphragmatic radiotherapy, COIA and HANKS (1988) observed the 3-year actuarial complication rate was 4% for major complications and 14% for any complication. With increasing dose, there was a statistically significant increase in complications (P<0.01). The risk of major bowel complications increased from 1% for doses less than 35 Gy to 3% for doses of 35 Gy or more (P=0.03). Gastrointestinal injury including peptic ulceration, hemorrhage, chronic diarrhea and intestinal obstruction were the most frequent complications (COIA and HANKS 1988).

Table 29.3 shows the incidence of late effects following infradiaphragmatic radiotherapy in patients treated for testicular GCT. The incidence of second malignancies and radiation-induced impairment in spermatogenesis is discussed separately.

Peptic ulceration is relatively common, occurring in up to 16% of patients at a median follow-up of 12 months (AKIMOTO et al. 1997). It is more often seen with a higher dose and in patients with prior abdominal surgery, a prior history of dyspepsia or in those who had significant acute toxicity (HAMILTON et al. 1986; AASS et al. 1992). Others have reported no late toxicity (SOMMER et al. 1990; ZAGARS and BABAIAN 1987).

29.7.3
Impaired Spermatogenesis

Following radiotherapy or chemotherapy for testicular germ cell tumors, impairment of spermatogenesis can occur. A number of authors have reported that less than 50% of patients have a normal sperm count following radical orchidectomy even without radiotherapy (GORDON et al. 1997; HAHN et al. 1982; HANSEN et al. 1990; JACOBSEN et al. 1997; NIJMAN et al. 1987).

A radiotherapy-induced reduction in sperm count can take several weeks. The spermatogonia are more radiosensitive than the differentiated stages and, at low doses of radiation, aspermia occurs at approximately 10–12 weeks (HAHN et al. 1982).

The gonadal dose impacts on the recovery of sperm count (GORDON et al. 1997; HAHN et al. 1982; HANSEN et al. 1990). GORDON et al. (1997) reported recovery within 12 months if the gonadal dose was less than 0.79 Gy, but this was delayed for more than

Table 29.3. The incidence of late complications following infradiaphragmatic radiotherapy for testicular cancer

Complication	Incidence
Peptic ulcer disease	0–16%
Chronic diarrhea	0–2.6%
Dyspepsia	0–27.7%

AASS et al. (1992); AKIMOTO et al. (1997); HAMILTON et al. (1986); SOMMER et al. (1990); YEOH et al. (1993); ZAGARS and BABAIAN (1987)

2 years in patients who received a gonadal dose of more than 0.79 Gy. In a further study, HAHN et al. (1982) observed that recovery took between 21 weeks and 41 weeks with doses below 60 rad and 47–88 weeks with doses of 60–148 rad. Other factors that may prolong the recovery time include age over 25 years, low pretreatment sperm count and the addition of chemotherapy (HANSEN et al. 1990). Nevertheless, there are several reports in the literature describing survivors of testicular cancer fathering healthy infants (CENTOLA et al. 1994; GORDON et al. 1997; HAHN et al. 1982; AKIMOTO et al. 1997). In a series by AKIMOTO et al. (1997), 79% of patients who wanted to have children after postorchidectomy radiotherapy were successful.

29.7.4
Second Malignant Neoplasms

Following exposure to ionizing radiation, there is a latent period prior to the development of a second malignant neoplasm (FOSSA et al. 1990; HAY et al. 1984; MOLLER et al. 1993; SMITH and DOLL 1982). The incidence of solid tumors increases with time from exposure.

The relative risk of second cancer ranges between 1.5 and 7.5 (BOKEMEYER and SCHMOLL 1995). Patients diagnosed with testicular cancer may already be at an increased risk of developing a second malignancy (KLEINERMAN et al. 1985). Although some reports find no difference in the incidence of second malignant neoplasm within or outside the radiation field (HANKS et al. 1992; HAY et al. 1984), most second cancers occur within or at the margins of the radiation field (FOSSA et al. 1990; JACOBSEN et al. 1993).

The increased risk of solid malignancies following radiotherapy for testicular cancer appears to be higher at certain sites, notably the genitourinary tract (FOSSA 2004; HAY et al. 1984; MOLLER et al. 1993; WANDERAS et al. 1997), gastrointestinal tract (BOKEMEYER and SCHMOLL 1995; FOSSA 2004; MOLLER et al. 1993; TRAVIS et al. 1997;

Table 29.4. The relative risk of second malignancy in survivors of testicular cancer

Site of second malignancy	Author	Relative risk
All sites	Moller et al. (1993)	1.6
	Wanderas et al. (1997)	3.54
	Travis et al. (1997)	1.43
	Van Leeuwen et al. (1993)	1.6
	Hanks et al. (1992)	3.4
	Hay et al. (1984)	1.55
	Fossa et al. (1990)	1.58
Bladder	Moller et al. (1993)	2.1
	Wanderas et al. (1997)	2.1
	Travis et al. (1997)	2.02
Kidney	Moller et al. (1993)	2.3
Gastric cancer	Moller et al. (1993)	2.1
	Wanderas et al. (1997)	2.46
	Travis et al. (1997)	1.95
	Van Leeuwen et al. (1993)	3.7
Pancreatic cancer	Moller et al. (1993)	2.3
	Travis et al. (1997)	1.5
Gallbladder	Horwich and Bell (1994)	8.3
Colon	Moller et al. (1993)	1.5
	Travis et al. (1997)	1.27
Rectum	Travis et al. (1997)	1.41
Leukemia	Horwich and Bell (1994)	6.2
	Van Leeuwen et al. (1993)	5.1
	Moller et al. (1993)	2.4
	Travis et al. (1997)	2.13
Non-Hodgkin's lymphoma	Travis et al. (1997)	1.88
Connective tissue/sarcoma	Jacobsen et al. (1993)	4.0
	Travis et al. (1997)	3.16
	Wanderas et al. (1997)	9.2
Non-melanomatous skin cancer	Moller et al. (1993)	2.0
Melanoma	Travis et al. (1997)	1.69
	Fossa et al. (1990)	3.89
Unknown primary	Hay et al. (1984)	8.36
Lung	Wanderas et al. (1997)	2.19
Thyroid	Travis et al. (1997)	2.92t

van Leeuwen et al. 1993; Wanderas et al. 1997) and in connective tissue (Fossa 2004; Travis et al. 1997; Wanderas et al. 1997) (Table 29.4). The risk of leukemia is also increased following radiotherapy (Fossa 2004; Hay et al. 1984; Horwich and Bell 1994; van Leeuwen et al. 1993). Other rare second cancers may also occur. Amin et al. (2001) reported two cases of malignant peritoneal mesothelioma many years after previous abdominal radiation therapy for testicular carcinoma. Saiki et al. (1997) described a patient with metastatic testicular cancer who developed a glioblastoma multiforme after radiotherapy for a brain metastasis.

Cytotoxic treatment is associated with an increased risk of second malignant neoplasm. Not only do patients treated for testicular cancer have an increased incidence of second malignancy, most of these malignancies are significant neoplasms resulting in an increased mortality (Bokemeyer and Schmoll 1993; Hanks et al. 1992; Zagars et al. 2004). In view of this serious late complication of radiotherapy, long-term follow-up is necessary.

References

American Joint Committee on Cancer (2002) Cancer staging manual, 6th edn. Springer, Berlin Heidelberg New York

Aass N, Fossa SD, Host H (1992) Acute and subacute side effects due to infra-diaphragmatic radiotherapy for testicular cancer: a prospective study. Int J Radiat Oncol Biol Phys 22:1057–1064

Akimoto T, Takahashi I, Takahashi M et al (1997) Long-term outcome of postorchidectomy radiation therapy for stage I and II testicular seminoma. Anticancer Res 17:3781–3785

Amin AM, Mason C, Rowe P (2001) Diffuse malignant mesothelioma of the peritoneum following abdominal radiotherapy. Eur J Surg Oncol 27:214–215

Anscher MS, Marks LB, Shipley WU (1992) The role of radiotherapy in patients with advanced seminomatous germ cell tumors. Controversies in management, part 2. Oncology (Huntingt) 6:97–104

Ball D, Barrett A, Peckham MJ (1982) The management of metastatic seminoma testis. Cancer 50:2289–2294

Bauman GS, Venkatesan VM, Ago CT et al (1998a) Postoperative radiotherapy for stage I/II seminoma: results for 212 patients. Int J Radiat Oncol Biol Phys 42:313–317

Bauman GS, Venkatesan VM, Ago CT, Radwan JS, Dar AR, Winquist EW (1998b) Postoperative radiotherapy for stage I/II seminoma: results for 212 patients. Int J Radiat Oncol Biol Phys 42:313–317

Bieri S, Rouzaud M, Miralbell R (1999) Seminoma of the testis: is scrotal shielding necessary when radiotherapy is limited to the para-aortic nodes? Radiother Oncol 50:349–353

Bokemeyer C, Schmoll HJ (1993) Secondary neoplasms following treatment of malignant germ cell tumors. J Clin Oncol 11:1703–1709

Bokemeyer C, Schmoll HJ (1995) Treatment of testicular cancer and the development of secondary malignancies. J Clin Oncol 13:83–292

Bokemeyer C, Schmoll HJ, Schoffski P et al (1993) Bilateral testicular tumors: prevalence and clinical implications. Eur J Cancer 29A:874–876

Centola GM, Keller JW, Henzler M et al (1994) Effect of low-dose testicular irradiation on sperm count and fertility in patients with testicular seminoma. J Androl 15:608–613

Christensen TB, Daugaard G, Geertsen PF et al (1998) Effect of chemotherapy on carcinoma in situ of the testis. Ann Oncol 9:657–660

Chung PW, Gospodarowicz MK, Panzarella T et al (2004) Stage II testicular seminoma: patterns of recurrence and outcome of treatment. Eur Urol 45:754–759

Classen J, Dieckmann KP, Loy V et al (1998) Testicular intraepithelial neoplasms (TIN). An indication for radiotherapy? Strahlenther Onkol 174:173–177

Classen J, Dieckmann K, Bamberg M et al (2003a) Radiotherapy with 16 Gy may fail to eradicate testicular intraepithelial neoplasia: preliminary communication of a dose-reduction trial of the German Testicular Cancer Study Group. Br J Cancer 88:828–831

Classen J, Schmidberger H, Meisner C et al (2003b) Radiotherapy for stages IIA/B testicular seminoma: final report of a prospective multicenter clinical trial. J Clin Oncol 21:1101–1106

Clemm C, Hartenstein R, Willich N et al (1986) Vinblastine-ifosfamide-cisplatin treatment of bulky seminoma. Cancer 58:2203–2207

Coffin CM, Ewing S, Dehner LP (1985) Frequency of intratubular germ cell neoplasia with invasive testicular germ cell tumors. Histologic and immunocytochemical features. Arch Pathol Lab Med 109:555–559

Coia LR, Hanks GE (1988) Complications from large field intermediate dose infradiaphragmatic radiation: an analysis of the patterns of care outcome studies for Hodgkin's disease and seminoma. Int J Radiat Oncol Biol Phys 15:29–35

Daniels JL Jr, Stutzman RE, McLeod DG (1981) A comparison of testicular tumors in black and white patients. J Urol 125:341–342

Dieckmann KP (1988) Residual carcinoma-in-situ of contralateral testis after chemotherapy. Lancet 1:765

Dieckmann KP, Besserer A, Loy V (1993) Low-dose radiation therapy for testicular intraepithelial neoplasia. J Cancer Res Clin Oncol 119:355–359

Dieckmann KP, Lauke H, Michl U et al (2002) Testicular germ cell cancer despite previous local radiotherapy to the testis. Eur Urol 41:643–649

Dieckmann KP, Loy V (1994) Management of contralateral testicular intraepithelial neoplasia in patients with testicular germ-cell tumor. World J Urol 12:131–135

Dieckmann KP, Loy V (1996) Prevalence of contralateral testicular intraepithelial neoplasia in patients with testicular germ cell neoplasms. J Clin Oncol 14:3126–3132

Dieckmann KP, Skakkebaek NE (1999) Carcinoma in situ of the testis: review of biological and clinical features. Int J Cancer 83:815–822

Dos SS, I Swerdlow AJ, Stiller CA et al (1999) Incidence of testicular germ-cell malignancies in England and Wales: trends in children compared with adults. Int J Cancer 83:630–634

Duchesne GM, Stenning SP, Aass N et al (1997) Radiotherapy after chemotherapy for metastatic seminoma – a diminishing role. MRC Testicular Tumour Working Party. Eur J Cancer 33:829–835

Emami B, Lyman J, Brown A et al (1991) Tolerance of normal tissue to therapeutic irradiation. Int J Radiat Oncol Biol Phys 21:109–122

Fossa SD (2004) Long-term sequelae after cancer therapy – survivorship after treatment for testicular cancer. Acta Oncol 43:134–141

Fossa SD, Aass N, Kaalhus O (1989) Radiotherapy for testicular seminoma stage I: treatment results and long-term post-irradiation morbidity in 365 patients. Int J Radiat Oncol Biol Phys 16:383–388

Fossa SD, Langmark F, Aass N et al (1990) Second non-germ cell malignancies after radiotherapy of testicular cancer with or without chemotherapy. Br J Cancer 61:639–643

Fossa SD, Horwich A, Russell JM et al (1999) Optimal planning target volume for stage I testicular seminoma: a Medical Research Council randomized trial. Medical Research Council Testicular Tumor Working Group. J Clin Oncol 17:1146

Fraass BA, Kinsella TJ, Harrington FS et al (1985) Peripheral dose to the testes: the design and clinical use of a practical and effective gonadal shield. Int J Radiat Oncol Biol Phys 11:609–615

Giacchetti S, Raoul Y, Wibault P et al (1993) Treatment of stage I testis seminoma by radiotherapy: long-term results – a 30 year experience. Int J Radiat Oncol Biol Phys 27:3–9

Gordon W Jr, Siegmund K, Stanisic TH et al (1997) A study of reproductive function in patients with seminoma treated with radiotherapy and orchidectomy: (SWOG-8711). Southwest Oncology Group. Int J Radiat Oncol Biol Phys 38:83–94

Greist A, Einhorn LH, Williams SD et al (1984) Pathologic findings at orchiectomy following chemotherapy for disseminated testicular cancer. J Clin Oncol 2:1025–1027

Gurkaynak M, Akyol F, Zorlu F et al (2003) Stage I testicular seminoma: para-aortic and iliac irradiation with reduced dose after orchiectomy. Urol Int 71:385–388

Hahn EW, Feingold SM, Simpson L et al (1982) Recovery from aspermia induced by low-dose radiation in seminoma patients. Cancer 50:337–340

Hamilton C, Horwich A, Easton D et al (1986) Radiotherapy for stage I seminoma testis: results of treatment and complications. Radiother Oncol 6:115–120

Hanks GE, Peters T, Owen J (1992) Seminoma of the testis: long-term beneficial and deleterious results of radiation. Int J Radiat Oncol Biol Phys 24:913–919

Hansen PV, Trykker H, Svennekjaer IL et al (1990) Long-term recovery of spermatogenesis after radiotherapy in patients with testicular cancer. Radiother Oncol 18:117–125

Hay JH, Duncan W, Kerr GR (1984) Subsequent malignancies in patients irradiated for testicular tumors. Br J Radiol 57:597–602

Herr HW, Sheinfeld J, Puc HS et al (1997) Surgery for a post-chemotherapy residual mass in seminoma. J Urol 157:860–862

Horwich A, Bell J (1994) Mortality and cancer incidence following radiotherapy for seminoma of the testis. Radiother Oncol 30:193–198

Horwich A, Alsanjari N, A'Hern R et al (1992a) Surveillance following orchidectomy for stage I testicular seminoma. Br J Cancer 65:775–778

Horwich A, Dearnaley DP, A'Hern R et al (1992b) The activity of single-agent carboplatin in advanced seminoma. Eur J Cancer 28A:1307–1310

Horwich A, Paluchowska B, Norman A et al (1997) Residual mass following chemotherapy of seminoma. Ann Oncol 8:37–40

Jacobsen GK, Henriksen OB, von der Masse H (1981) Carcinoma in situ of testicular tissue adjacent to malignant germ-cell tumors: a study of 105 cases. Cancer 47:2660–2662

Jacobsen GK, Mellemgaard A, Engelholm SA et al (1993) Increased incidence of sarcoma in patients treated for testicular seminoma. Eur J Cancer 29A:664–668

Jacobsen KD, Olsen DR, Fossa K et al (1997) External beam abdominal radiotherapy in patients with seminoma stage I: field type testicular dose and spermatogenesis. Int J Radiat Oncol Biol Phys 38:95–102

Jones WG, Fossa SD, Mead GM (2001) A randomized trial of two radiotherapy schedules in the adjuvant treatment of stage I seminoma (MRC TE18). Eur J Cancer 37:5157

Kazem I, Danella JF (1999) Organ preservation for the treatment of contralateral testicular seminoma. Radiother Oncol 53:45–47

Khoo VS, Rainford K, Horwich A et al (1997) The effect of antiemetics and reduced radiation fields on acute gastrointestinal morbidity of adjuvant radiotherapy in stage I seminoma of the testis: a randomized pilot study. Clin Oncol (R Coll Radiol) 9:252–257

Kiricuta IC, Sauer J, Bohndorf W (1996) Omission of the pelvic irradiation in stage I testicular seminoma: a study of post-orchiectomy paraaortic radiotherapy. Int J Radiat Oncol Biol Phys 35:293–298

Kleinerman RA, Liebermann JV, Li FP (1985) Second cancer following cancer of the male genital system in Connecticut 1935–1982. Natl Cancer Inst Monogr 68:139–147

Kubo H, Shipley WU (1982) Reduction of the scatter dose to the testicle outside the radiation treatment fields. Int J Radiat Oncol Biol Phys 8:1741–1174

Logue JP, Harris MA, Livsey JE et al (2003) Short course paraaortic radiation for stage I seminoma of the testis. Int J Radiat Oncol Biol Phys 57:1304–1309

Maier JG, Sulak MH, Mittemeyer BT (1968) Seminoma of the testis: analysis of treatment success and failure. Am J Roentgenol Radium Ther Nucl Med 102:596–602

Mason BR, Kearsley JH (1988) Radiotherapy for stage II testicular seminoma: the prognostic influence of tumor bulk. J Clin Oncol 6:1856–1862

Melchior D, Hammer P, Fimmers R et al (2001) Long term results and morbidity of paraaortic compared with paraaortic and iliac adjuvant radiation in clinical stage I seminoma. Anticancer Res 21:2989–2993

Miki T, Nonomura N, Saiki S et al (1998) Long-term results of adjuvant irradiation or surveillance in stage I testicular seminoma. Int J Urol 5:357–360

Moller H, Mellemgaard A, Jacobsen GK et al (1993) Incidence of second primary cancer following testicular cancer. Eur J Cancer 29A:672–676

Motzer R, Bosl G, Heelan R et al (1987) Residual mass: an indication for further therapy in patients with advanced seminoma following systemic chemotherapy. J Clin Oncol 5:1064–1070

Motzer RJ, Bosl GJ, Geller NL et al (1988) Advanced seminoma: the role of chemotherapy and adjunctive surgery. Ann Intern Med 108:513–518

Niewald M, Waziri A, Walter K et al (1995) Low-dose radiotherapy for stage I seminoma: early results. Radiother Oncol 37:164–166

Nijman JM, Schraffordt KH, Kremer J et al (1987) Gonadal function after surgery and chemotherapy in men with stage II and III nonseminomatous testicular tumors. J Clin Oncol 5:651–656

Ohyama C, Kyan A, Satoh M et al (2002) Bilateral testicular tumors: a report of nine cases with long-term follow-up. Int J Urol 9:173–177

Peckham MJ, Horwich A, Hendry WF (1985) Advanced seminoma: treatment with cis-platinum-based combination chemotherapy or carboplatin (JM8). Br J Cancer 52:7–13

Perez CA, Brady LW, Halperin EC, Schmidt-Ullrich RK (2005) Principles and practice of radiation oncology, 4th edn. Lippincott Williams and Wilkins, Philadelphia

Petersen PM, Giwercman A, Hansen SW et al (1999) Impaired testicular function in patients with carcinoma-in-situ of the testis. J Clin Oncol 17:173–179

Petersen PM, Giwercman A, Daugaard G et al (2002) Effect of graded testicular doses of radiotherapy in patients treated for carcinoma-in-situ in the testis. J Clin Oncol 20:1537–1543

Power DA, Brown RS, Brock CS et al (2001) Trends in testicular carcinoma in England and Wales 1971–1999. BJU Int 87:361–365

Puc HS, Heelan R, Mazumdar M et al (1996) Management of residual mass in advanced seminoma: results and recommendations from the Memorial Sloan-Kettering Cancer Center. J Clin Oncol 14:454–460

Ravi R, Ong J, Oliver RT et al (1999) The management of residual masses after chemotherapy in metastatic seminoma. BJU Int 83:649–653

Ray B, Hajdu SI, Whitmore WF Jr (1974) Proceedings: distribution of retroperitoneal lymph node metastases in testicular germinal tumors. Cancer 33:340–348

Read G, Johnston RJ (1993) Short duration radiotherapy in stage I seminoma of the testis: preliminary results of a prospective study. Clin Oncol (R Coll Radiol) 5:364–366

Saiki S, Kinouchi T, Usami M et al (1997) Glioblastoma multiforme after radiotherapy for metastatic brain tumor of testicular cancer. Int J Urol 4:527–529

Santoni R, Barbera F, Bertoni F et al (2003) Stage I seminoma of the testis: a bi-institutional retrospective analysis of patients treated with radiation therapy only. BJU Int 92:47–52

Schmidberger H, Bamberg M, Meisner C et al (1997) Radiotherapy in stage IIA and IIB testicular seminoma with reduced portals: a prospective multicenter study. Int J Radiat Oncol Biol Phys 39:321–326

Schultz SM, Einhorn LH, Conces DJ Jr et al (1989) Management of postchemotherapy residual mass in patients with advanced seminoma: Indiana University experience. J Clin Oncol 7:1497–1503

Sedlmayer F, Holtl W, Kozak W et al (2001) Radiotherapy of testicular intraepithelial neoplasia (TIN): a novel treatment regimen for a rare disease. Int J Radiat Oncol Biol Phys 50:909–913

Smith PG, Doll R (1982) Mortality among patients with ankylosing spondylitis after a single treatment course with x rays. Br Med J (Clin Res Ed) 284:449–460

Sommer K, Brockmann WP, Hubener KH (1990) Treatment results and acute and late toxicity of radiation therapy for testicular seminoma. Cancer 66:259–263

Speer TW, Sombeck MD, Parsons JT et al (1995) Testicular seminoma: a failure analysis and literature review. Int J Radiat Oncol Biol Phys 33:89–97

Stone JM, Cruickshank DG, Sandeman TF et al (1991) Trebling of the incidence of testicular cancer in victoria Australia (1950–1985). Cancer 68:211–219

Sultanem K, Souhami L, Benk V et al (1998) Para-aortic irradiation only appears to be adequate treatment for patients with Stage I seminoma of the testis. Int J Radiat Oncol Biol Phys 40:455–459

Taylor MB, Carrington BM, Livsey JE et al (2001) The effect of radiotherapy treatment changes on sites of relapse in stage I testicular seminoma. Clin Radiol 56:116–119

Thomas GM (1986) 'The role of radiation therapy in all stages and extents of seminoma. Germ cell tumors II. Proceedings of the 2nd germ cell tumor conference, Leeds, 15–19 April 1985. Pergamon, Oxford

Thomas GM (1997) Over 20 years of progress in radiation oncology: seminoma. Semin Radiat Oncol 7:135–145

Thompson IM, Optenberg S, Byers R et al (1999) Increased incidence of testicular cancer in active duty members of the Department of Defense. Urology 53:806–807

Travis LB, Curtis RE, Storm H et al (1997) Risk of second malignant neoplasms among long-term survivors of testicular cancer. J Natl Cancer Inst 89:1429–1439

Vallis KA, Howard GC, Duncan W et al (1995) Radiotherapy for stages I and II testicular seminoma: results and morbidity in 238 patients. Br J Radiol 68:400–405

Van Leeuwen FE, Stiggelbout AM, van den Belt-Dusebout AW et al (1993) Second cancer risk following testicular cancer: a follow-up study of 1909 patients. J Clin Oncol 11:415–424

Von der Masse H, Berthelsen JG, Jacobsen GK et al (1985) Carcinoma-in-situ of testis eradicated by chemotherapy. Lancet 1:98

Von der Masse H, Giwercman A, Skakkebaek NE (1986a) Radiation treatment of carcinoma-in-situ of testis. Lancet 1:624–625

Von der Masse H, Rorth M, Walbom-Jorgensen S et al (1986b) Carcinoma in situ of contralateral testis in patients with testicular germ cell cancer: study of 27 cases in 500 patients. Br Med J (Clin Res Ed) 293:1398–1401

Von der Masse H, Meinecke B, Skakkebaek NE (1988) Residual carcinoma-in-situ of contralateral testis after chemotherapy. Lancet 1:477–478

Von der Masse H, Specht L, Jacobsen GK et al (1993) Surveillance following orchidectomy for stage I seminoma of the testis. Eur J Cancer 29A:1931–1934

Wanderas EH, Fossa SD, Tretli S (1997) Risk of subsequent non-germ cell cancer after treatment of germ cell cancer in 2006 Norwegian male patients. Eur J Cancer 33:253–262

Warde P, Gospodarowicz MK, Goodman PJ et al (1993) Results of a policy of surveillance in stage I testicular seminoma. Int J Radiat Oncol Biol Phys 27:11–15

Warde P, Gospodarowicz M, Panzarella T et al (1998) Management of stage II seminoma. J Clin Oncol 16:290–294

Yeoh E, Razali M, O'Brien PC (1993) Radiation therapy for early stage seminoma of the testis. Analysis of survival and gastrointestinal toxicity in patients treated with modern megavoltage techniques over 10 years. Australas Radiol 37:367–369

Zagars GK, Babaian RJ (1987) Stage I testicular seminoma: rationale for postorchiectomy radiation therapy. Int J Radiat Oncol Biol Phys 13:155–162

Zagars GK, Ballo MT, Lee AK et al (2004) Mortality after cure of testicular seminoma. J Clin Oncol 22:640–647

30 Extremity Soft Tissue Sarcoma in Adults

Thomas F. DeLaney, David C. Harmon, Andrew E. Rosenberg, and Francis J. Hornicek

CONTENTS

30.1   Abstract   755
30.2   Introduction   756
30.3   Clinical Evaluation   756
30.3.1   Clinical History   756
30.3.2   Anatomical Site, Sex, and Age   756
30.3.3   Physical Examination   757
30.3.4   Laboratory Investigations   757
30.3.5   Radiographic Evaluation   757
30.3.6   Biopsy   759
30.4   Pathology   759
30.4.1   Histological Classification   759
30.4.2   Grading   760
30.5   Staging   761
30.6   Management of the Primary Tumor   762
30.6.1   Overview   762
30.6.2   Surgical Considerations   763
30.6.3   Selection of Patients for Treatment with Conservative Surgery Alone   764
30.6.4   Combining Surgery with RT   764
30.6.5   Preoperative (Neoadjuvant) Versus Postoperative (Adjuvant) Radiotherapy   765
30.6.6   Brachytherapy   766
30.6.7   EBRT Planning   767
30.6.8   Radiation Treatment Volumes and Dose   768
30.6.9   Intensity-Modulated Photon RT   770
30.6.10   Proton Beam Radiotherapy   771
30.6.11   Neoadjuvant Doxorubicin-Based Chemotherapy Plus RT   772
30.6.12   STSs of the Hand and Foot   773
30.6.13   Wound Healing After Surgery and Radiation   773

30.6.14   Preoperative Radiation and Wound Complications   773
30.6.15   BRT and Wound Complications   774
30.6.16   Strategies to Reduce Wound Morbidity   774
30.6.17   Adjuvant Chemotherapy   775
30.6.17.1   Rhabdomyosarcoma   775
30.6.17.2   Adjuvant Chemotherapy for Other STSs   775
30.6.18   Functional Outcome   777
30.6.19   Treatment of Local Recurrence   777
30.6.20   Treatment of Unresectable or Locally Advanced Soft Tissue Sarcoma   777
30.7   Treatment of Metastatic Disease   778
30.7.1   Overview   778
30.7.2   Resection of Pulmonary Metastases   778
30.7.3   Chemotherapy for Metastatic Disease   778
30.8   Summary   779
   References   780

30.1 Abstract

When treating soft tissue sarcomas (STSs) of the extremities, the major therapeutic goals are survival, local tumor control, optimal function, and minimal morbidity. Surgical resection of the primary tumor is the essential component of treatment for virtually all patients. However, local control by surgery alone is poor for the majority of patients with extremity lesions, unless the procedure removes large volumes of grossly normal tissue, i.e., widely negative margins are attained because sarcomas tend to infiltrate normal tissue adjacent to the evident lesion. Thus, removal of the gross lesion by a simple excision alone (only a narrow margin) is followed by local recurrence in 60–90% of patients. Radical resections are associated with a reduction in the local recurrence rate of 10–30%, but they may compromise limb function. The combination of function-sparing surgery and radiation achieves better outcomes than either treatment alone for nearly all patients with STSs. Because both surgical and radiation technique are both critically important for optimizing local control of tumor and functional outcome, it is important to manage these patients in dedicated

T. F. DeLaney, MD
Medical Director, Northeast Proton Therapy Center, Massachusetts General Hospital, Harvard Medical School, 30 Fruit Street, Boston MA 02114, USA
D. C. Harmon, MD
Hematology Oncology, Department of Medicine, Massachusetts General Hospital, Harvard Medical School, Yawkey 7944, 30 Fruit Street, Boston MA 02114, USA
A. E. Rosenberg, MD
Surgical Pathology, Department of Pathology, Massachusetts General Hospital, Harvard Medical School, Warren 2, 30 Fruit Street, Boston MA 02114, USA
F. J. Hornicek, MD, PhD
Orthopedic Oncology, Department of Orthopedic Surgery, Massachusetts General Hospital, Harvard Medical School, Yawkey 3B, 30 Fruit Street, Boston MA 02114, USA

multispecialty clinics comprised of physicians with expertise in sarcomas, including orthopedic and general oncology surgeons, radiation oncologists, medical oncologists, sarcoma pathologists, and bone and soft tissue diagnostic radiologists. Radiation therapy (RT) can be given by external beam radiation (EBRT) or brachytherapy (BRT) or combination thereof. EBRT can be given either preoperatively or postoperatively. The clinical considerations and the outcome data that must be considered in choosing the most appropriate treatment technique for the individual patient will be discussed.

30.2
Introduction

Sarcomas are malignant tumors that arise from skeletal and extraskeletal connective tissues, including the peripheral nervous system. The term sarcomas of soft tissues embraces all of the malignant tumors that arise from the mesenchymal tissues, excluding bone, i.e., malignant fibrous histiocytoma, liposarcoma, leiomyosarcoma, synovial sarcoma, rhabdomyosarcoma, epithelioid sarcoma, angiosarcoma, fibrosarcoma, etc. In addition, malignant tumors of peripheral nerve sheaths are included despite being ectodermal in origin, as their clinical behavior is not measurably different from the other sarcomas.

STSs are rare, with an estimated incidence in the United States of approximately 8,680 diagnosed annually, representing less than 1% of all newly diagnosed malignant tumors (JEMAL et al. 2005). Approximately 37% of these patients are expected to die of this disease. Review of the statistics of recent years suggests an increase in the incidence of STSs, although it is not clear whether this represents a true increase or merely reflects more accurate diagnosis and increasing interest in these tumors (WEISS and GOLDBLUM 2001). Although the malignant tumors of soft tissue are rare, benign tumors are common. It is estimated that the frequency of benign tumors is 100 times that of the malignant lesions (WEISS and GOLDBLUM 2001). To appreciate the rarity of the sarcomas, note that during the same period, there are expected to be 211,240 newly diagnosed cases of breast cancer in women in the US, as compared with only 3,890 STSs in women. Thus, women are approximately 54 times more likely to develop a carcinoma of the breast than a sarcoma.

Sites of appearance of STSs, in order of frequency, are: lower extremity (46%), torso (19%), upper extremity (14%), retroperitoneum (13%), head/neck (8%)

(ABBAS et al. 1981; POTTER et al. 1985; LAWRENCE et al. 1987; TOROSIAN et al. 1988). The small number of cases seen and the great diversity in anatomic site, histopathology, and biology complicate understanding of the natural history of these tumors and their response to diverse therapies. This discussion will be focused on STSs arising in the extremities and treatment strategies designed to maximize cure rates while optimizing post-treatment function.

30.3
Clinical Evaluation

30.3.1
Clinical History

The most frequent initial complaint is that of a painless lump with a duration of a few weeks to several months. Occasionally, pain or tenderness precede the detection of a lump. With progressive growth of tumor, symptoms that are secondary to infiltration of or pressure on adjacent structures (e.g., tendons, muscles, nerves) or organs appear. Occasionally, symptoms secondary to the metabolic effects of the tumor products are seen, e.g., fever, anemia, lethargy, weight loss, histamine-like reactions. These are not rare in patients with malignant fibrous histiocytoma (WEISS and GOLDBLUM 2001). To accrue clinical genetic data in sarcoma patients, the history should include details of the cause of death and history of malignant disease in siblings, parents, grandparents, and progeny.

30.3.2
Anatomical Site, Sex, and Age

As previously mentioned, the sites of appearance of STSs, in order of frequency, are: lower extremity (46%), torso (19%), upper extremity (14%), retroperitoneum (13%), head/neck (8%) (ABBAS et al. 1981; POTTER et al. 1985; LAWRENCE et al. 1987; TOROSIAN et al. 1988). There is a very slight preponderance of STSs in males. They are more common in older people, with 40% in persons 55 years of age or older and 15% in patients 15 years of age or younger (WEISS and GOLDBLUM 2001). Rhabdomyosarcomas almost always arise in children, synovial sarcomas develop in late adolescence and young adulthood, and liposarcoma and malignant fibrous histiocytoma usually occur during mid and late adulthood.

30.3.3
Physical Examination

There must be a complete physical examination with particular attention paid to the region of the primary lesion: definition of size, site of origin (superficial or deep, attached to or fixed to deep structures), solitary or multinodular, involvement or discoloration of overlying skin, functional status of vessels and nerves, presence of distal edema, muscular strength, range of motion of affected part, etc. If the patient has had prior excision, the operated site should be examined for presence of ecchymosis, status of wound healing, palpable evidence of residual tumor, and location of drain site. The regional and distant lymph-node groups need to be examined with care in all patients, especially those with large grade-2 and -3 sarcomas. In an analysis of the experience at Massachusetts General Hospital, no patient with grade-1 sarcoma developed regional node involvement. The incidence in patients with grade-2 and -3 sarcomas were 2% and 12%, respectively. Of the patients with grade-3 sarcoma, the incidences were 3% and 15%, respectively, for lesions 5 cm or less and those more than 5 cm (MAZERON and SUIT 1987). FONG et al. (1993) found a slightly lower frequency of metastasis to regional nodes in the Memorial Hospital series. Involvement of regional nodes is relatively frequent in patients with rhabdomyosarcoma and epithelioid sarcoma but uncommon in patients with fibrosarcoma and malignant fibrous histiocytoma. Myxoid liposarcoma may also metastasize to soft tissue or bone.

30.3.4
Laboratory Investigations

Laboratory studies need not go beyond a complete blood count and blood urea nitrogen/creatinine [if IV contrast is to be administered for the chest computed tomography (CT) scan or for radiation-planning CT scan] unless the patient is to receive chemotherapy, in which case liver-function tests should be performed as well.

30.3.5
Radiographic Evaluation

For the primary site, the radiographic evaluation should include plain films and magnetic resonance imaging (MRI) scanning. The most useful radio-

logical study to evaluate the primary site is the MRI (BLAND et al. 1987); CT can be useful to evaluate the relatively uncommon situation where there is suspected bony involvement. Plain radiographs are helpful in evaluation of soft tissue tumors by demonstrating bone involvement and soft tissue masses arising from bone tumors. However, unlike bone tumors, imaging studies cannot be used to assess the biological behavior of soft tissue tumors (KRANSDORF et al. 1993). Indeed, specific diagnosis remains impossible for many soft tissue lesions, regardless of the choice of imaging (O'KEEFE et al. 1990). Imaging studies alone cannot definitively distinguish malignant from benign soft tissue lesions. MRI (Fig. 30.1) may prove of clinical value in planning the biopsy site, either using needle or incisional techniques. Furthermore, the demonstration of necrotic regions appears to be of diagnostic importance in discrimination between benign and malignant tumors (GUSTAFSON et al. 1992). Positron emission tomography (PET) scanning (Fig. 30.2) has been shown to be useful in discriminating between benign and high-grade lesions, although is unsuitable for distinguishing between benign and low- to intermediate-grade lesions (NIEWEG et al. 1996). In the future, PET may be of substantial value in defining response to preoperative therapy (PROSNITZ et

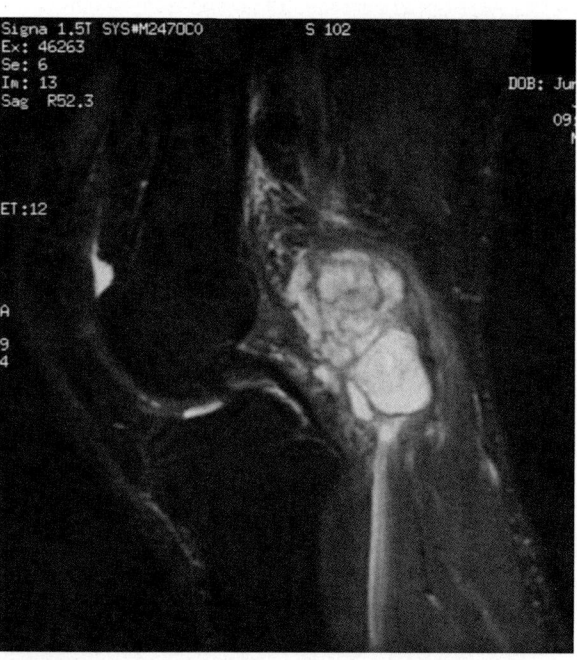

Fig. 30.1. T1-weighted, fat-suppressed magnetic resonance image of a high-grade, biphasic synovial sarcoma in the right popliteal fossa of a 27-year-old man following gadolinium administration

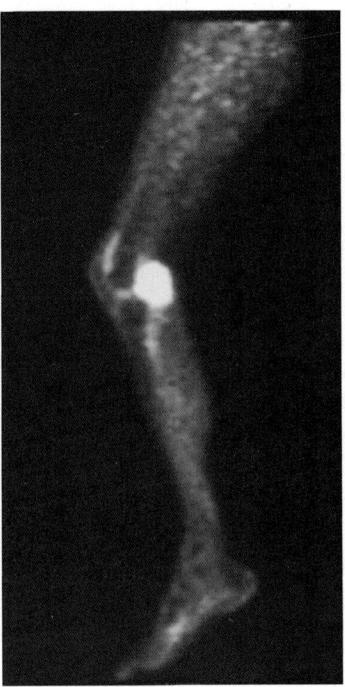

Fig. 30.2. FDG (F18-fluorodeoxyglucose)–positron emission tomography (PET) scan demonstrating markedly increased FDG avidity at the site of the high-grade, biphasic synovial sarcoma also seen on magnetic resonance imaging (MRI) scan in Fig. 30.1

al. 1999). MRI studies should always include T2-weighted sequences, as these provide the optimum contrast between lesion and muscle. Contrast-enhanced T1-weighted images (especially with fat suppression techniques) are also helpful. MRI often provides a clearer demonstration of the anatomical location of the lesion and the pattern of local extensions. The findings from these scans should be correlated with those of the physical examination to assess the details of the anatomical site of the tumor. It should be determined whether the lesion is in the subcutaneous tissue, transgressing the fascia, intermuscular or intramuscular, displacing or enveloping major vessels or nerves, abutting or invading bone, etc. Depending on the pattern of presentation and the nature of any planned surgery, an arteriogram or CT angiogram may be of value. For rhabdomyosarcoma, epithelioid sarcoma, high-grade synovial and unclassified sarcomas, PET or PET/CT evaluation of the regional nodes should be obtained. Bone scans need not be performed unless specifically indicated. We do not consider a positive bone scan of bone near or adjacent to a STS to be proof of invasion of periosteum or bone. For a diagnosis of

invasion of bone, there must be clear radiographic evidence of destruction of cortical bone. The single most important examination for distant metastasis is whole-lung CT; this should be obtained in all patients with intermediate or high-grade tumors. This has been extensively confirmed by the study of Peuchot and Libshitz (Peuchot and Libshitz 1987), who reported that, of the nodules detected using CT but not by chest X-ray and those biopsied, 94% were metastatic tumors.

Imaging of the response to treatment has been generally disappointing up to the present. Decrease in tumor size may occur (Sanchez et al. 1990) but does not correlate well with successful radiation or chemotherapy. Furthermore, although tumor volume can be approximated from two-dimensional images (CT or plain films), reliable algorithms to objectively distinguish tumor from surrounding tissues are not available, and, thus, three-dimensional (3-D) imaging techniques have not, as yet, found a role. Changes in MRI signal characteristics have been unreliable. Absence of high signal intensity on T2 images has been shown to indicate freedom from tumor. However, residual high signal may be due to tumor, edema, or fibrosis (Vanel et al. 1987). MRI spectroscopy has been used to detect high-energy phosphate metabolism in the lesions. This has helped in the distinction between malignant and benign tumors (Negendank et al. 1989). Several studies with phosphorus-31 magnetic resonance spectroscopy have shown changes in high-energy phosphate metabolism after effective chemotherapy (Dewhirst et al. 1990; Koutcher et al. 1990; Redmond et al. 1992), but the range of variation in sarcoma is large, and, because of limitations in spatial resolution, the procedure cannot be done reliably unless a large soft tissue mass is present. Schuetze and colleagues evaluated 46 patients with high-grade sarcoma who received neoadjuvant chemotherapy and reported that reductions in SUV_{max} correlated with the degree of necrosis in the resection specimen (Schuetze et al. 2005). They also noted that tumors with a baseline SUV_{max} of 6 or more had a higher risk of developing subsequent metastatic disease with disease recurrence. Patients with a 40% or greater decrease in tumor SUV_{max} had a significantly lower risk of disease recurrence and of metastasis. All 4 patients with local disease recurrence as the initial event had less than 40% reduction in tumor SUV_{max} after chemotherapy. A reduction in the pre-surgery sarcoma SUV_{max} of greater than or equal to 40% relative to the baseline SUV_{max} correlated with overall survival. A metabolic response determined using F-18 fluorodeoxygluose–PET scans correlated with a pathological response to preoperative doxorubicin-based chemotherapy but it was a stronger determinant of distant recurrence of disease than the pathological response.

30.3.6
Biopsy

An adequate biopsy must be obtained so that the best feasible histopathological diagnosis as to tumor type and grade can be made. The optimal treatment strategy is based on the correct diagnosis. Under optimal conditions, the biopsy procedure should be performed by an experienced surgeon, who is part of the multidisciplinary care-taking team and will be responsible for the definitive surgery. Prior to the biopsy, the imaging studies should be carefully studied to ascertain the most logical approach to the lesion, with explicit consideration of the regions to be traversed in subsequent surgical procedures (including marginal or wide resection or amputation), so that the biopsy track will not interfere with either surgery or radiation field.

The most commonly employed biopsy technique is currently the core needle biopsy. For palpable lesions with a superficial component that is away from the neurovascular bundle, a Tru-Cut needle biopsy can often be obtained in the clinic. For nonpalpable, deeper lesions and those near the neurovascular bundle, core biopsies can be obtained under CT or ultrasound guidance. There is experience in some centers with cytological diagnosis by 'skinny' needle aspiration. The amount of tissue obtained using these procedures is limited, and this should be considered before deciding which procedure to perform; incisional biopsy may be required in some cases to histologically classify a tumor or obtain sufficient tissue for immunohistochemical, electron microscopic, and/or gene arrays. Needle biopsy is particularly useful to confirm metastatic or recurrent tumor. Biopsy by the needle technique may also be used in those anatomical situations where incisional biopsy would require a major procedure.

When incisional biopsies are performed, **the incision for the biopsy of lesions on an extremity should be longitudinal (there is almost never a reason for a transverse incision on an extremity)**. The incision should be as short as possible, yet long enough to avoid excessive retraction of tissue or to make dissection and hemostasis difficult. The biopsy track should go through a muscle belly rather than along

fascial planes (the former tends to keep the tumor 'spill' within an anatomical compartment while the latter allows transgression of two or more compartments) and careful attention should be paid to achieve hemostasis in order to avoid ecchymosis or a hematoma. The wound should be closed in layers with a narrow skin closure; as a rule, drains should not be utilized (the tract of the drain is considered to be contaminated with tumor and may greatly extend the planes of subsequent surgery or the radiation treatment volume).

For the occasional small lesion in a readily accessible site (e.g., wrist or ankle), an excisional biopsy may be the approach of choice. The surgeon should adhere to the same principles as for the definitive surgery (see below) if complications are to be avoided.

The biopsy specimen needs to be of sufficient volume to be certain that it is representative. A pathological assessment of a frozen section is useful in assuring that the tissue obtained is from the lesion and is adequate for the diagnostic evaluation. Cultures should always be obtained. Specimens are processed for hematoxylin and eosin staining and various immunohistochemical stains considered necessary to aid in the diagnosis. A small portion of the tissue is set aside for electron microscopy, cytogenetics, and increasingly gene arrays.

30.4
Pathology

30.4.1
Histological Classification

The rationale for developing a well-defined, comprehensive, and flexible classification system of soft tissue tumors is to provide morphological guidelines that expand our understanding of neoplasia, predict biological behavior, and facilitate the development of more effective treatment. Originally, classification schemes were descriptive in nature and based on tumor cell configuration. Subsequently, they have evolved through the concept of histogenesis or 'cell of origin' to the current belief that a primitive or stem-like mesenchymal cell undergoes neoplastic transformation and, depending on the genetic code translated, differentiates along one or multiple cell lines.

Light microscopy in most instances is the modality of choice for determining whether a soft tissue tumor is benign or malignant and determining the subtype, grade, margins, and presence or absence of vascular invasion. However, electron microscopy, immunohistochemistry, DNA flow cytometry, cytogenetics, and molecular analysis can provide valuable information that substantiates the histological interpretation. Cytogenetic analysis is now being performed more frequently on sarcomas. It is a diagnostically useful technique, because some sarcomas have specific cytogenetic alterations that appear to be pathognomonic. As noted above, characteristic translocations are found in Ewing's sarcoma/primitive neuroectodermal tumors alveolar rhabdomyosarcomas (DeChiara et al. 1993; Delattre et al. 1994; Parham et al. 1994), myxoid liposarcomas (Aman et al. 1992), clear cell sarcomas, extraskeletal myxoid chondrosarcomas, and synovial sarcomas (Clark et al. 1994). Molecular biology will play a more important role in evaluating STSs in the near future, as the technology becomes more widely available. The identification of specific DNA and RNA gene sequences and oncogenes will help to diagnose and predict the biological behavior of sarcomas. For example, the expression of the gene product MYO D1 has already proved to be helpful in recognizing tumors showing skeletal muscle differentiation (Angervall and Kindblom 1993; Dias et al. 1994).

Currently, the most widely used classification system is the Enzinger and Weiss modification of the World Health Organization formulation (Weiss and Goldblum 2001). In this system, soft tissue tumors, including non-neoplastic tumor-like lesions, are categorized into three broad groups: (1) tumors that differentiate along cell or tissue lines and have normal counterparts, i.e., fibrous tissue, fat, vessels, smooth muscle, skeletal muscle, nerve, ganglia, synovium, bone, and cartilage; (2) tumors whose lines of differentiation have no normal counterpart but are consistent and recognized by a distinctive morphology, i.e., myxoma, epithelioid sarcoma, and alveolar soft part sarcoma; and (3) tumors that are so poorly differentiated and morphologically unique that they defy classification. The vast majority of tumors fall into the first two groups. Overall, there are approximately 200 different entities, of which 80 are malignant.

As intensive study of these tumors is rapidly expanding and new diagnostic procedures are increasingly employed, there is inevitably some flux in the diagnostic criteria for the diverse groups of soft tissue tumors. Examples include: (1) the reclassification of most adult pleomorphic rhabdomyosarcomas and many pleomorphic liposarcomas to malignant fibrous histiocytoma; (2) the recognition that a granular cell tumor is a Schwann cell neoplasm; (3) clear cell sarcoma is a malignant melanoma primary to the soft tissues; and (4) extraskeletal Ewing's sarcoma is a primitive neuroectodermal tumor.

The distribution of histological types of sarcoma of soft tissue from several large series is presented in Table 30.1.

30.4.2
Grading

The histological typing of soft tissue tumors does not provide sufficient information *per se* on which to base therapeutic decisions. Tumor grading is based on the concept that morphology reflects biological behavior. The specific microscopic characteristics of soft tissue tumors that best predict their aggressiveness, i.e., the potential for regional and distant

Table 30.1. Frequency of histopathological types of soft tissue sarcoma in a large series of patients with extremity and trunk soft tissue sarcomas. MPNST malignant peripheral nerve sheath tumor, NR not reported, MGH Massachusetts General Hospital, MSKCC Memorial Sloan-Kettering Cancer Center. MFH malignant fibrous histiocytoma (DeLaney et al. 2003a); Milan (Gronchi et al. 2005); Lund (Engellau 2004); MSKCC (Pisters et al. 1996b)

	MGH 1994	Milan 2005	Lund 2004	MSKCC 1996
Number of patients	738	911	298	1041
MFH	22%	8%	13%	25%
Liposarcoma	16%	31%	14%	29%
Fibrosarcoma	11%	NR	13%	10%
Leiomyosarcoma	10%	14%	33%	8%
Sarcoma, not otherwise specified	9%	NR	9%	NR
Synovial sarcoma	8%	15%	6%	12%
MPNST	10%	10%	7%	5%
Rhabdomyosarcoma	3%	NR	NR	NR
Vascular sarcomas	NR	5%	1%	NR
Other	11%	16%	4%	12%

metastasis, can be identified, integrated, and represented by grade.

The American Joint Committee on Cancer (AJCC) and the International Union Against Cancer staging systems for sarcoma of soft tissue are based on classification of the tumors into low- and high-grade tumors. The World Health Organization employs a four-tiered grading system: G1 well differentiated, G2 moderately differentiated, G3 poorly differentiated, and G4 undifferentiated; with the G1 and G2 lesions considered low grade and the G3 and G4 considered high grade. Some institutions employ a three-step grading, i.e., low-, intermediate-, and high-grade neoplasms, with the intermediate- and high-grade designations considered to be high grade for staging purposes. The designation of grade is based on a consideration and integration of each of these morphological features: degree of cellular differentiation, extent of necrosis, number of mitoses, cellularity, pleomorphism or anaplasia, quantity of matrix, vascularity, hemorrhage, vascular invasion, and encapsulation (Suit et al. 1975; Myhre-Jensen et al. 1975; Rydholm et al. 1984; Costa et al. 1984; Trojani 1984; Lack et al. 1989; Kulander et al. 1989). Among these variables, necrosis, mitoses,

and degree of differentiation appear to be the best predictors of outcome.

Despite some lack of agreement on the number of grades employed and the significance of individual morphological parameters (there is inevitably a subjective component in assigning grade, and only a part of the tumor is examined), grading, more than any clinical and pathological parameter available, is the most important prognosticator (Russell et al. 1977). High-power views of a STS, grades 1, 2, and 3 of 3 are shown in Figure 30.3a–c. The problems with current grading systems are that their criteria are not precisely defined, application and interpretation is subjective, and implementation is complex. Consistent grading requires adequate tissue and experienced pathologists.

30.5
Staging

The Task Force on Soft Tissue Sarcomas of the AJCC Staging and End Result Reporting has established a staging system for STSs, which is an extension of the tumor node metastasis (TNM) system to include G

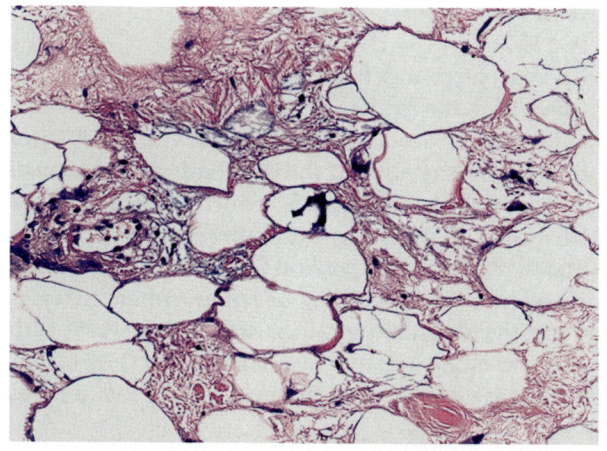
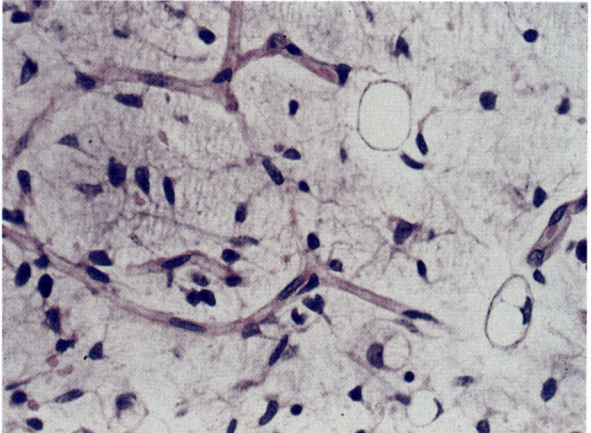
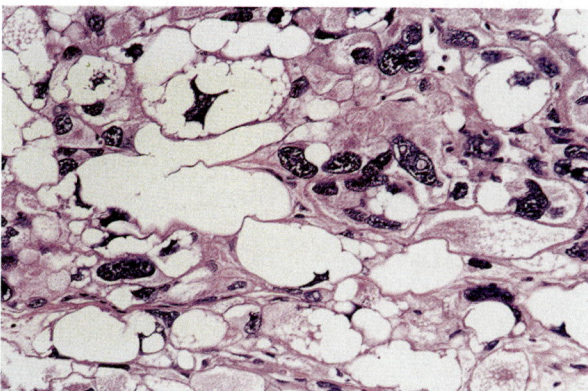

a

b

c

Fig. 30.3. Photomicrographs showing low-, intermediate-, and high-grade soft tissue sarcoma

for histological grade. Grade, size, depth, and presence of nodal or distant metastases are the determinants of stage. This staging system is applied to all sarcomas of soft tissue except rhabdomyosarcoma (for which there is a special staging system), Kaposi's sarcoma, dermatofibrosarcoma, desmoid, and sarcoma arising from the dura mater, brain, parenchymatous organs, or hollow viscera. The staging system was revised in 1998 with the addition of subgroupings of the T stage to designate superficial and deep lesions and the assignment of patients with nodal involvement to stage IV. Superficial lesions do not involve the superficial investing fascia in extremity lesions. The current system, last updated in 2002, is outlined in Table 30.2. Grade of sarcoma is determined on the basis of the histological features of the individual tumor. No tumor is assigned to a grade because of the histological type. The grading system cited in the AJCC staging manual is outlined in Table 30.2. T stage is determined on the basis of size and depth. As evidence for the importance of size as a determinant of frequency of distant metastasis, we present, in Table 30.3, an analysis of distant metastasis versus tumor size among patients who have achieved local control. For patients with grade-1 sarcomas, distant metastases are quite uncommon. The pooled data for G2 and G3 lesions show regular increases in distant metastases with tumor size. From these data, there clearly is importance in stratification of patients according to grade *and* size in attempts to compare efficacy of different modes of treatment, defining the natural history of various histological types or assessing the role of site, patient age, sex, etc.

30.6
Management of the Primary Tumor

The intent of treatment is to eradicate tumor while optimizing limb function.

30.6.1
Overview

Because sarcomas tend to infiltrate normal tissue adjacent to the evident lesion, simple excision alone is followed by local recurrence (LR) in 60–90% of patients (Markhede et al. 1981). Radical resection of a wider margin of apparently normal tissue around the tumor reduces the local failure rate to approximately 25–30% (Simon and Enneking 1976). More recently, with the advent of compartmental resections, the local failure rate has fallen to 10–20% with surgery alone (Simon and Enneking 1976). One study that reported a zero local failure rate derives

Table 30.2. American Joint Committee on the Staging of Cancer Staging System for Soft Tissue Sarcomas (2002)

Primary tumor (T)
TX: primary tumor cannot be assessed
T0:   no evidence of primary tumor
T1:   tumor 5 cm or less in greatest dimension
T1a: superficial tumor[a]
T1b: deep tumor[a]
T2:   tumor more than 5 cm in greatest dimension
T2a: superficial[a]
T2b: deep[a]

Regional lymph nodes (N)
N0:   no regional lymph node metastasis
N1:   regional lymph node metastasis

Distant metastases (M)
M0:   no distant metastasis
M1:   distant metastasis

Histological grade (G)
G1:   well differentiated
G2:   moderately differentiated
G3:   poorly differentiated
G4:   undifferentiated

Stage grouping[b]
I: low grade, small or large, superficial or deep: G1-2, T1a-2b, N0, M0
II: high grade, small, superficial or deep or large superficial: G3-4, T1a-b, T2a N0, M0
III: high grade, large, deep: G3-4, T2b, N0, M0
IV: any metastasis, any G or T, at least N1 or M1 or both

[a]Superficial tumor is located exclusively above the superficial fascia without invasion of the fascia; deep tumor is located either exclusively beneath the superficial fascia, superficial to the fascia with invasion of or through the fascia or both superficial and beneath the fascia
[b]This staging system is not to be used for Kaposi's sarcoma, dermatofibrosarcoma (protuberans), fibrosarcoma grade I (desmoid tumor) and sarcoma arising from the dura mater, brain, parenchymatous organs, or hollow viscera

Table 30.3. The 5-year actuarial distant metastasis (DM) probability in 501 consecutive local control patients as a function of tumor size for grades 2 and 3 in series from Massachusetts General Hospital (treatment by radiation and surgery) (DeLaney et al. 2003a)

Size (mm)	No. of patients	DM (%)
<25	58	3
26–50	128	22
51–100	177	34
101–150	68	43
151–200	49	58
200	21	57
Total	501	35

from the amputation arm of the National Cancer Institute trial comparing amputation with limb salvage treatment (ROSENBERG et al. 1982).

The combination of surgery and radiation achieves better outcomes than either treatment alone for nearly all STS (SUIT et al. 1975; LINDBERG et al. 1981). The rationale for combining radiation with surgery is to avoid the functional and cosmetic deformity associated with radical resection and the late consequences of high radiation doses to large volumes of normal tissue in patients treated with primary radiation alone. Radiation at moderate dose levels (60–65 Gy) is as effective as radical resection in eradicating the microscopic extensions beyond the gross lesion, resulting in similar high rates of local control. This has allowed maximization of functional and cancer-related outcome without the significant morbidity of radical surgery. Most centers report local control rates of approximately 90% for high-grade extremity STS and 90–100% for low-grade STS depending on the size (ROSENBERG et al. 1982; POTTER et al. 1985; KARAKOUSIS et al. 1986; BRANT et al. 1990; HARRISON et al. 1993; DELANEY et al. 2003a; EILBER et al. 2003).

In addition to its benefit in improving local control rates, adjunctive radiotherapy has also had a significant impact on limb salvage for extremity sarcomas. As an example, in the 1970s, 50% of patients with extremity sarcoma underwent amputation; those patients treated by wide excision alone with limb preservation experienced a 30% rate of LR. With the subsequent application of radiotherapy and advanced reconstructive techniques, the rate of amputation at major centers has been reduced to less than 10%, and the incidence of LR with limb preservation has been reduced to 10–15% without any measurable fall in overall survival (OS) (ROSENBERG et al. 1982; WILLIARD et al. 1992; LEVAY et al. 1993; KARAKOUSIS and DRISCOLL 1999). A single, prospective randomized trial showed similar rates of disease-free survival and OS for patients treated with amputation or the combination of limb-sparing surgery and radiotherapy for extremity STS (ROSENBERG et al. 1982).

The success of a conservative surgical approach has, as mentioned above, resulted in an amputation rate at major centers of only 5% in patients with extremity STS. The current indications for amputation include: massive disease such that a functional limb is not achievable, as well as severely compromised normal tissues due to age, peripheral vascular disease, and other comorbidities. The functional and cosmetic results of conservative procedures are dependent on the size and anatomic location of the tumor, the magnitude of the surgical procedure, extent to which muscles, tendons or nerves must be sacrificed, volume of tissues irradiated, and the radiation dose administered.

30.6.2
Surgical Considerations

The most important surgical variable that influences local control is the presence or absence of tumor cells at the surgical margins (ROSENBERG et al. 1982; WILLIARD et al. 1992; LEVAY et al. 1993; SADOSKI et al. 1993; TANABE et al. 1994; PISTERS et al. 1996b; TROVIK et al. 2000; ZAGARS et al. 2003b). In series that report radical resection with clear margins, such as the Scandinavian Sarcoma Group, the local failure rates are quite low (8%) (ALVEGARD et al. 1989). In contrast, in a second study of 559 patients who were treated with surgery alone from the same group, an inadequate surgical margin led to a 2.9-fold greater risk of LR than did clear surgical margins (TROVIK et al. 2000). Distant metastases are extremely uncommon for low-grade lesions but occur with high-grade lesions with a frequency that is influenced by the size of the lesion (POTTER et al. 1985) and whether local control is achieved. In the Scandinavian Sarcoma Group experience, local recurrence was identified as a risk factor for distant metastasis (4.4-fold higher) (ALVEGARD et al. 1989).

The status of the surgical margins also influences the local recurrence rate in patients treated with combined surgery and radiation (LEVAY et al. 1993; SADOSKI et al. 1993; ZAGARS et al. 2003b). In one review of 132 consecutive patients with STS of the extremities who were treated with preoperative RT followed by resection, the 5-year actuarial local control rates were 97% and 81%, respectively, for patients with negative and positive margins (SADOSKI et al. 1993). Local control was not a function of sarcoma size in patients with negative surgical margins. In a second series of 225 patients, all of whom received combined surgery and radiotherapy (either preoperative, postoperative, or both), local control rates at 5 years were 88, 76, and 64% for patients with negative, uncertain, and positive margins, respectively (ZAGARS et al. 2003b).

The exact size of the negative margin that is optimal for local control is not known. In one study, the local control rate did not differ in patients with a margin of 1 or less or more than 1 mm (local control 96% versus 97%) (SADOSKI et al. 1993). Most clinicians recommend that if surgery is used as the sole modality of treatment, the margin should be at least 1 cm in all directions (EILBER and ECKARDT 1997) or, if less, include a supervening fascial barrier. If surgery is combined with RT, the surgical margin can probably be safely reduced to 0.5 cm without compromising the rate of local control (SADOSKI et al. 1993).

The guiding principle of surgery is total en bloc excision of the primary tumor without cutting into tumor tissue. Tissues should be cut outside of the tumor pseudocapsule, if one exists, through normal uninvolved tissue. Violation of the tumor results in a higher local failure rate. In one report, for example, the local control rate in 95 patients with extremity STS was 47% if tumor violation occurred compared with 87% without violation (TANABE et al. 1994). The majority of STSs do not involve bone; as a result, it is seldom necessary to resect adjacent bone. It is also rarely necessary to resect a major nerve unless the tumor is a neurogenic sarcoma. Nonamputative surgery is now accomplished in more than 90% of patients.

In planning primary therapy for a patient who has had a suboptimal resection by a non-oncology surgeon and/or insufficient imaging with preoperative CT or MRI, it is important to consider re-resection. Approximately 37–68% of such patients will have residual tumor in a re-resection specimen (NORIA et al. 1996; KARAKOUSIS and DRISCOLL 1999; ZAGARS et al. 2003a). A partial excision of the tumor before referral to a tertiary center does not appear to compromise limb preservation, local control, or survival rates in such patients (KARAKOUSIS and DRISCOLL 1999), although the re-resection may entail a larger procedure than a de novo procedure and impact on the functional result. In one series of 295 patients who underwent re-resection at a single institution (final resection margins negative in 87%), local control rates at 5, 10, and 15 years were 85, 85, and 82%; the corresponding values for those who did not undergo re-resection were 78, 73, and 73%, respectively (ZAGARS et al. 2003a). A similar degree of benefit for re-resection was apparent for metastasis-free and disease-specific survival.

30.6.3
Selection of Patients for Treatment with Conservative Surgery Alone

Because of potential acute and late morbidity from RT, it is important to select patients who may be effectively treated with conservative surgery alone.

Several published series have evaluated wide-excision, limb-sparing surgery alone. In one report, 119 selected patients with extremity STS were grouped according to anatomic location as subcutaneous ($n = 40$), intramuscular ($n = 30$), or extramuscular ($n = 49$) (RYDHOLM et al. 1991). The 70 patients with subcutaneous and intramuscular tumors were all treated by local surgery, and a wide margin, requiring a cuff of fat tissue around the tumor and inclusion of the deep fascia beneath the tumor, was obtained in 56. These patients were followed without postoperative radiation. During a median follow-up of 5 years (range, 3.5–10 years), only 4 had a local recurrence, despite the fact that 84% had high-grade tumors. The authors concluded that postoperative radiation may not be necessary in this subgroup. In another study, 74 patients with localized STS of the extremity or trunk underwent function-sparing surgery without radiation (BALDINI et al. 1999). The overall 10-year actuarial local control rate was 93% and was dependent on the adequacy of surgical margins (87% versus 100% for patients with margins of less than 1 cm and 1 cm or more, respectively). The 10-year survival rate was 73%. This approach may be appropriate for carefully selected patients with small (<5 cm), superficial tumors that can be resected with all margins 1 cm or more.

30.6.4
Combining Surgery with RT

The recommended treatment for patients who are medically and technically operable is the combination of function-preserving surgery and radiation, with the exception of that minority of patients with small, superficial lesions that can be widely excised with secure margins and good functional results. In most instances, the probability of tumor control and the late functional and cosmetic result is clearly superior following this combined modality approach. Radiation is an effective treatment for STS, as the radiation sensitivity of cell lines derived from sarcomas is not less than that of epithelial cell lines (RUKA et al. 1996). For small sarcomas, good local control rates can be achieved by radiation alone. However, local control probabilities of more than 90% for tumors of estimated volumes of 15–65 ml (approximately a sphere of 3–5 cm in diameter) requires high radiation doses (>75 Gy) (TEPPER and SUIT 1985). For unresected sarcomas, there appears to be an advantage for doses above 63 Gy (KEPKA et al. 2005). As most treatment volumes are relatively large, the late normal tissue changes resulting from these dose levels are clinically important in nearly all patients. In animal models, a significantly lower radiation dose is required to achieve local control when radiation is combined with simple excision than with radiation alone (TODOROKI and SUIT 1985).

The impact of combined modality treatment that includes EBRT on both local control and survival has been evaluated in only one prospective randomized trial. In this study, 91 patients with high-grade lesions were randomly assigned to surgery plus postoperative chemotherapy with or without postoperative adjuvant EBRT, and 50 with low-grade lesions were randomized to surgery plus adjuvant EBRT or surgery alone (O'SULLIVAN et al. 2002). In the patients with high-grade lesions, there were no LRs in the patients randomized to EBRT, while the patients receiving only adjuvant chemotherapy had a 22% actuarial local failure rate at 10 years. In patients with low-grade sarcoma, the LR rates were 4% versus 33% in the postoperative EBRT and surgery alone groups, respectively. There was no influence of postoperative radiation on OS for either high- or low-grade tumors.

30.6.5
Preoperative (Neoadjuvant) Versus Postoperative (Adjuvant) Radiotherapy

There are potential advantages to both preoperative and postoperative administration of radiation. Preoperative RT might be expected to reduce tumor burden prior to resection, theoretically allowing more conservative surgical therapy. Radiation fields can be limited to the tumor and adjacent tissues at risk for microscopic infiltration, a volume that is considerably smaller than that which must be treated following surgery, where the entire surgical bed is included in the initial target volume irradiated to 50 Gy. Radiation doses are lower (50 Gy preoperative versus 60–66 Gy postoperative). Postoperative radiotherapy allows histological examination of the tumor specimen, especially the margins, aiding in further treatment planning; it is also associated with fewer acute wound complications.

There is one randomized, controlled study comparing preoperative versus postoperative radiotherapy. This study was designed to evaluate the incidence of acute wound healing complications in patients with potentially curable extremity STS (O'SULLIVAN et al. 1999). In this Canadian trial, 190

patients were randomly assigned to either preoperative (50 Gy preoperative for all 94 patients randomized to this arm with 16–20 Gy postoperative boost reserved for the 14 patients in this arm with a positive margin) or postoperative (50 Gy initial field + 16–20 Gy boost field for all patients) radiotherapy. Complications were defined as secondary wound surgery, hospital admission for wound care, or the need for deep packing or prolonged wound dressings within 120 days of tumor resection.

The study was terminated when a highly significant result was obtained at the time of a planned interim analysis. With a median follow-up of 3.3 years, a significantly higher percentage of preoperatively treated patients had acute wound complications (35% versus 17%). Other factors associated with wound complications were the volume of resected tissue and lower limb location of the tumor. Late morbidity was initially not reported. Because the RT fields for the postoperative RT were larger, and the dose delivered for most patients was higher, the authors indicated that more follow-up would be needed to assess whether these larger radiation volumes and higher radiation doses would lead to more late treatment effects in these patients.

In a later publication, the local recurrence rate, regional or distant failure rate, progression-free survival, and functional outcome did not differ between the groups (DAVIS et al. 2002). This data has now been updated with a median follow-up of 6.9 years. There remain no differences in local control between the patients in the two arms of the study with over 90% local control. The regional and distant failure rates as well as the progression-free and overall survival rates are also no different between the two arms of the study. The postoperative patients, however, have now developed more grade-2 to -4 late toxicity (86%) when compared with the preoperative patients (68%), $P = 0.02$. Notably, grade 3 (severe induration and loss of subcutaneous tissue or field contracture greater than 10% linear measurement) or grade 4 (necrosis) was significantly more common in the postoperative group, 36% versus 23%, $P = 0.02$ (O'SULLIVAN et al. 2004).

There is, thus, a difference in the morbidity profile between preoperative and postoperative radiotherapy, with a higher rate of generally reversible acute wound healing complications in the patients receiving preoperative treatment, offset by a higher rate of generally irreversible late complications, including grade 3–4 fibrosis, in those patients receiving postoperative RT. Because very few acute wound-healing complications occurred in either group when the tumor was in the upper extremity, it would seem prudent to treat these patients with preoperative RT. We have also favored preoperative radiation for the majority of lower extremity patients, because acute wound complications can usually be managed and will go on to heal, whereas the late treatment effects are generally irreversible. For patients with lower extremity lesions, this study, however, makes it clear that new strategies are needed to (1) reduce the risk of acute wound healing problems when patients receive preoperative RT and (2) reduce the risk of late-treatment-induced effects when higher dose, larger field postoperative radiation is given.

30.6.6
Brachytherapy

Compared with EBRT, BRT minimizes the radiation dose to surrounding normal tissues, maximizes the radiation dose delivered to the tumor, and shortens treatment times. In the usual dosage schedule, treatment is completed within 6 days and requires only one hospitalization. Afterloading catheters are

placed in a target area of the tumor operative bed, defined by the surgeon, and spaced at 1-cm intervals to cover the entire area of risk (Fig. 30.4). BRT can also be used for delivery of a boost to the tumor bed in conjunction with EBRT (SCHRAY et al. 1990).

A phase-III trial of postoperative BRT versus no BRT was conducted in 126 patients who had complete resection of either extremity or superficial trunk STS (HARRISON et al. 1993). The BRT dose was 45 Gy. The 5-year local control rates were 82% and 67% for the BRT and surgery alone groups, respectively. The advantage of BRT was seen only in the high-grade sarcomas (HARRISON et al. 1993; PISTERS et al. 1996a). It was limited to local control, since there was no difference between the groups in distant metastasis or disease-specific survival (HARRISON et al. 1993).

Although it is unclear whether BRT is associated with a higher risk of wound complications (see "Wound healing after surgery and radiation" below), the rate of wound reoperation may be higher (ALEKTIAR et al. 2000). BRT has been combined with free flap construction as a means of enhancing primary healing in difficult anatomic situations

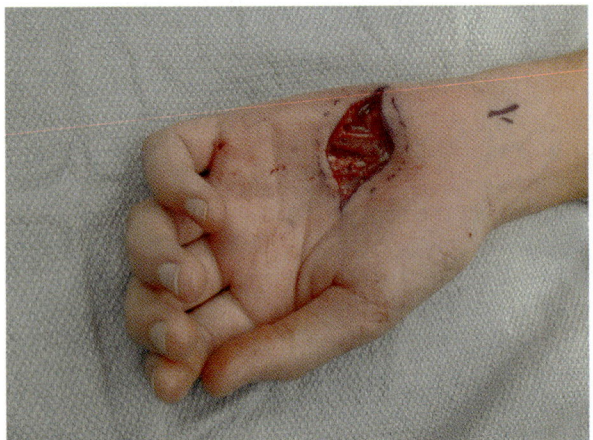

a

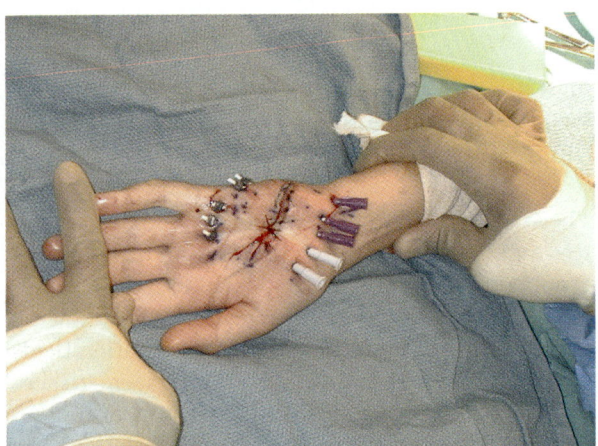

b

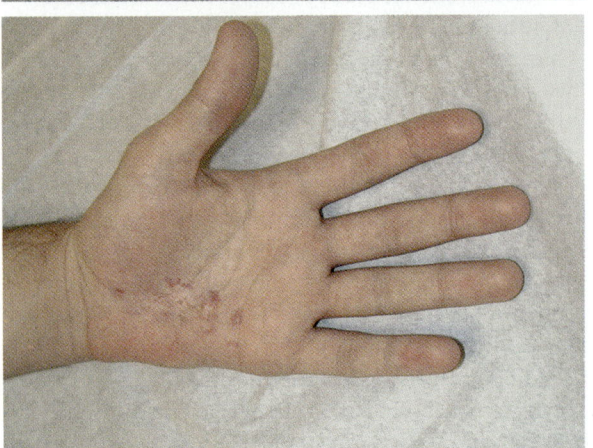

c

Fig. 30.4a–c. This boy with a recurrent epithelioid sarcoma on the hypothenar eminence of his left hand following an excisional biopsy 1 year earlier (lesion mistakenly called a benign fibrous histiocytoma at that time by an inexperienced pathologist) was treated at age 11 years with a combination of excision (a), 45 Gy of low dose rate iridium-192 brachytherapy (b), and 20 Gy delivered following the implant with 6-MeV electrons (c). At 5 years, he is free of any evidence of disease with normal hand function without an increase in the incidence of wound break-down (PANAGOPOULOS et al. 1996). There have been no randomized comparisons of the relative efficacy or morbidity of EBRT compared with BRT.

BRT for sarcomas has traditionally been given by low dose rate. There is some preliminary information on the use of fractionated high dose rate schedules (NAG et al. 2001; KRETZLER et al. 2004; PETERA et al. 2004). High dose rate BRT has been used in conjunction with EBRT for the tumor bed boost in doses of 15–24 Gy, often hyperfractionated at 2.3–4 Gy b.i.d. (NAG et al. 2001). One report using high dose rate BRT alone in doses of 40 Gy at 2.3–3 Gy b.i.d unfortunately reported poor local control of only 20%, in contrast with 100% when BRT was combined with EBRT (PETERA et al. 2004).

30.6.7
EBRT Planning

The radiation treatment technique should be carefully planned so that the tissues being irradiated are only those judged to be at risk. To utilize smaller treatment volumes, the part to be irradiated must be securely and reproducibly immobilized. We have special immobilization devices prepared for the individual patient. This may require casting, especially for hand, foot, or elbow sites (Fig. 30.5). For some sites, the part is placed in standard plastic supports and the extremity fastened tightly in place using a Velcro fastener. Others describe their experience with casts and polyurethane foam systems (NIEWALD et al. 1990).

The principal tasks involved in the development of a treatment plan are to:
- Design an immobilization device and a means to assure that the target is on the beam (Fig. 30.6).
- CT scan the immobilized, affected extremity for radiotherapy planning. This is facilitated by the availability of a large bore scanner that allows maximum flexibility in arranging the limb such that the contralateral extremity and the trunk will be out the beam.
- Define the target volume(s) (on each section of the CT/MRI of the affected region).
- Define non-target critical structures in the treatment volume and specify dose constraints for each such structure.
- Estimate the distribution of number of tumor clonogens/unit volume of tissue throughout the target volume.
- Define a series of target volumes to realize the

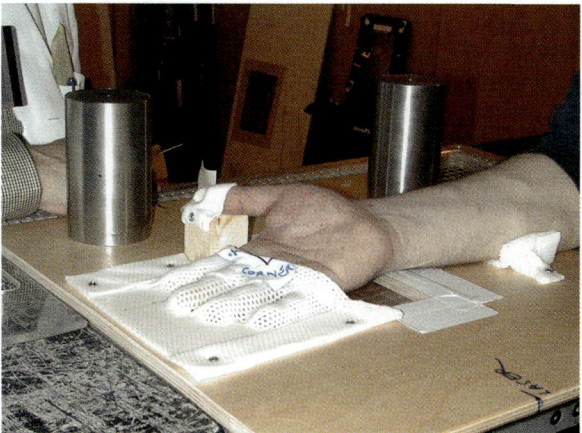

Fig. 30.5. Customized immobilization device for a 34-year-old physician with a G3/3T1bN0MO (stage II) monophasic synovial sarcoma of the thenar eminence of his right hand undergoing 50 Gy of external beam photon radiation therapy prior to tumor bed excision

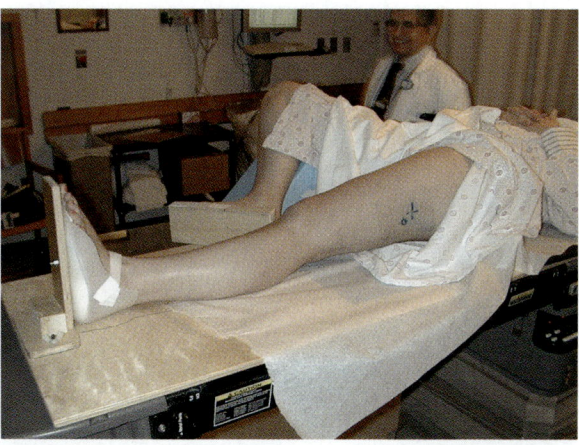

Fig. 30.6. Customized immobilization device for a 23-year-old female undergoing 60 Gy of postoperative radiation therapy for treatment of grade-1–2/3 myxoid liposarcoma of the medial, left thigh, low-to-intermediate grade, pT2BN0 M0 (stage III) with negative but close (0.2 cm) medial and deep resection margins. Involved lower extremity was rigidly immobilized for computed tomography (CT) simulation, while contralateral extremity was placed in a neutral position to allow it to get through a conventional-width CT simulator bore. For treatment, the involved extremity remains in same rigid immobilization, while additional immobilizer was added to frog-leg the contralateral extremity out of the treatment beam

appropriate dose distribution using "shrinking treatment volume methods".
- Design treatment techniques that achieve the closest feasible conformation of treatment to target volume. This may require complex field arrangements, treatment angles, gapped fields, wedge filters, tissue compensators, and/or bolus.

- Avoid inclusion of an entire joint space.
- Avoid full dose irradiation of adjacent bone to reduce the risk of pathological fracture.
- Utilize wedges and tissue compensators as needed to account for tissue heterogeneities and minimize dose inhomogeneity.
- Review the treatment plan at multiple levels along the extremity to assess dose homogeneity to the target and normal tissues (Fig. 30.7).

30.6.8
Radiation Treatment Volumes and Dose

The extent of normal tissue to be irradiated adjacent to the tumor bed in the case of preoperative RT and adjacent to the surgical bed in the case of postoperative RT is not definitively known. Few patterns of failure studies to relate the extent of the RT field to the site of local tumor recurrence have been reported. Because sarcomas are judged to infiltrate along rather than through tissue planes, longitudinal margins proximally and distally have traditionally been considerably more generous than radial margins. Historically, fields that extended from the muscle origin to insertion (TEPPER et al. 1982) or provided generous proximal/distal margins on the tumor were employed (SUIT et al. 1988). In some centers from the 1970s through the mid 1990s, 5- to 10-cm proximal and distal block margins were used for large grade-1 and small grade-2 lesions and more generous fields with 10- to 15-cm margins were encompassed for large grade-2 to -3 lesions (SUIT et al. 1988). The advent of improved MRI delineation of tumor extent and subsequent surgical experience with high rates of local control when surgical margins 1 cm or more could be obtained prompted radiation oncologists to employ proximal/distal margins of 5 cm or less for small grade-1 lesions and 5- to 7-cm proximal/distal margins for larger, higher grade lesions. Newer (3-D) treatment planning systems appear to allow smaller and more accurate treatment volumes in patients with extremity STS.

There are very few studies in the literature looking at the target volume used when planning radio-

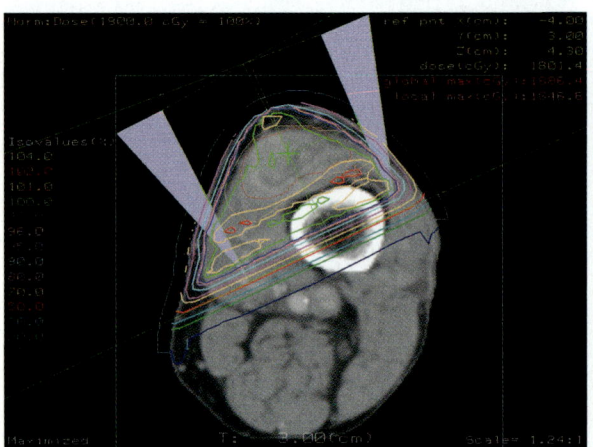

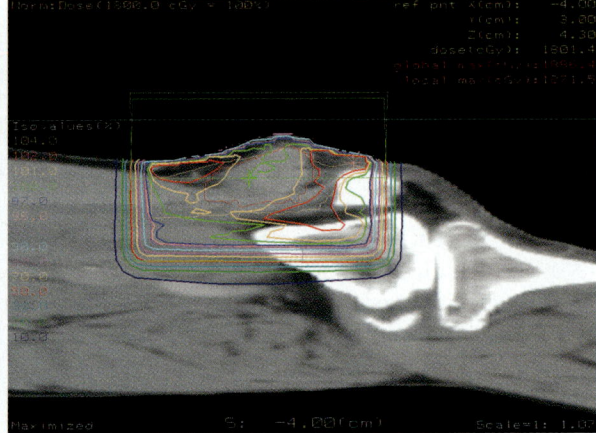

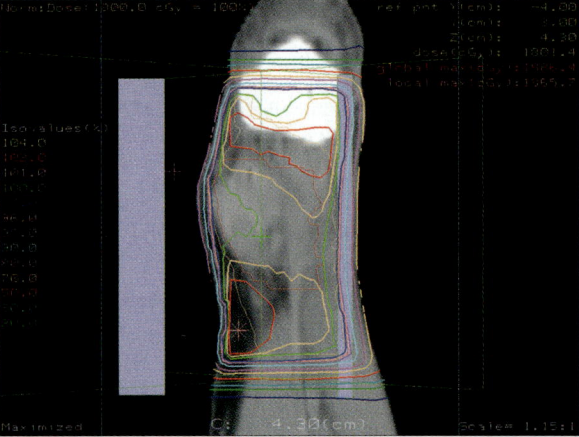

Fig. 30.7a-c. A 3-D conformal radiation therapy plan for a patient with a grade-3/3 T2bN0M0 (stage III) malignant fibrous histiocytoma along the anteromedial aspect of the left knee just above the patella and superficial to the quadriceps tendon undergoing preoperative radiation therapy of 50 Gy in 25 fractions

therapy. This tends to be poorly reported. One group found a remarkable difference in 5-year local control where the margin was less than 5 cm (30%) or at least 5 cm (93%) (MUNDT et al. 1995). This conflicts with the BRT data, where acceptable results are achieved using 4-cm margins longitudinally and 2 cm laterally. A recent publication from the Royal Marsden Hospital has suggested that, as in other tumor sites, the great majority of local recurrences occur within the high dose volume (CLEATOR et al. 2001). This raises the question as to whether the large-volume phase 1 is necessary or whether the boost is necessary where an adequate surgical margin has been achieved. This question is particularly relevant now because of the advent of techniques such as intensity-modulated radiotherapy (IMRT) and protons that allow us to selectively spare normal tissues. It will be important to determine the volumes that can safely be spared before implementing these techniques. Such studies are currently being considered by cooperative groups in the US and Europe. One recent paper presented some very provocative findings that will need to be very carefully considered in the discussion of RT volumes. This study, performed by investigators in Toronto, provided a histological assessment of peritumoral edema as demonstrated by increased T2-weighted signal intensity on MRI scans performed preoperatively on 15 patients with high grade extremity or truncal sarcomas ranging from 3.1 cm to 30.1 cm (mean, 13.8 cm) who did not undergo any neoadjuvant therapy prior to resection (WHITE et al. 2005). The extent of peritumoral T2-weighted signal intensity changes beyond the tumor ranged from 0 cm to 7.1 cm (mean, 2.5 cm); contrast enhancement ranged from 0 cm to 5.3 cm (mean 1.1 cm). Tumor cells were identified histologically in the tissues beyond the gross tumor in 10 of 15 cases. In 6 cases, the tumor cells were located within 1 cm of the tumor margin, and in 4 cases, malignant cells were found at a distance greater than 1 cm and up to 4 cm. The location of the tumor cells did not correlate with tumor size or extent of peritumoral changes on the MRI scans. In 9 of 10 cases, however, the tumor cells were identified histologically in areas with corresponding high T2-weighted signal changes on MRI. With ever-increasing ability of the available radiation oncology technology to conform the radiation dose to the target, this study has very significant implications for radiotherapy target design and must be considered in future studies of radiotherapy volumes in this disease.

The radial margins should be viewed with respect to the direction of most likely spread; the beam edges can be designed to deliver the prescription dose to a clinical target volume consisting of a margin of 1 cm radial to the gross tumor volume with approximately 0.5 cm for daily set-up variation. This 1-cm radial margin is rationally derived from the surgical experience with high rates of local control with greater than or equal to 1-cm margins. Because of the penumbra of the beam, this usually means that the radial block edge is approximately 2 cm from the edge of gross tumor. Where there is intervening bone, interosseous membrane, or major fascial planes, and these planes are intact in the imaging studies, the full prescription dose can be delivered to the surface of these structures, which approximates the tumor, again allowing for daily set-up variation. When a fascial plane has been violated, wider margins are appropriate to cover areas of potential contamination by tumor.

The use of 5-cm proximal and distal block margins for the first 50 Gy (on the tumor for preoperative radiotherapy and the surgical bed for postoperative radiotherapy) and 2-cm radial block margins provided very high rates of local control in the randomized National Cancer Institute Canada trial discussed above.

For patients receiving preoperative radiation, 50 Gy is administered over 5 weeks, followed 3–5 weeks later by a conservative resection. For patients with negative surgical margins and no other unfavorable prognostic features such as tumor cut-through or satellite lesions after prior surgical interventions, 50 Gy of preoperative RT appears sufficient to provide local control in a very high proportion of patients. SADOSKI et al. (1993) analyzed 132 consecutive patients with STS of the extremities treated with preoperative radiotherapy and resectional surgery and found that: (1) the 5-year actuarial local control rates were 97% and 81% for patients with negative margins and positive margins, respectively (this difference is highly significant); (2) there was no difference between the various sub-categories of negative margins [negative at less than 1 mm (96%), negative at greater than 1 mm (97%), not measured (94%), and no tumor in the specimen (100%)]; (3) there was no difference in local control for treatment of primary and locally recurrent lesions (after previous surgery alone) when the tumors were stratified for margin status; and (4) for the patients with negative margin, local control was not a function of sarcoma size.

For patients with positive margins following preoperative RT, it is recommended to use a "shrinking treatment volume technique" with delivery of either BRT or a postoperative EBRT boost dose of

16–18 Gy to the tumor bed once the surgical wound has healed. A boost dose to 66 Gy is given postoperatively or intraoperatively for microscopically positive margins and to 75 Gy if there is gross residual disease. In patients with frozen section evidence of close or positive margins, a boost dose can be administered intraoperatively using BRT or electron beam. We prefer BRT and use a low dose rate of 16 Gy for microscopically positive margins (or more recently, HDR of 14–16 Gy given as 3.5–4 Gy b.i.d.) and 25 Gy for gross residual tumor.

For patients undergoing postoperative RT, irradiation usually begins 14–20 days following surgery, once the wound is healed. Following resection of large tumors, it may be necessary to wait 3–4 weeks to allow resorption of the seroma. The initial volume must include all tissues handled during the surgical procedure, including the drain site, often encompassing the surgical bed with 5-cm proximal/distal block margins and 2-cm radial block margins. The dose to the initial volume is 50 Gy. Progressively shrinking treatment volumes are then employed to encompass the tumor bed and, if needed, areas of positive margins or gross residual disease; the final dose is 60 Gy for volumes with negative margins, 66–68 Gy for areas of positive margins or locally recurrent disease (Zagars and Ballo 2003), and 75 Gy for gross residual sarcoma.

The available information on a dose–response relationship for the local control of sarcomas treated with surgery and postoperative RT is somewhat conflicting. Mundt reported that local control was dose dependent. While postoperative patients receiving less than 60 Gy had a worse local control than those receiving 60 Gy or more, no difference was seen in local control between patients receiving 60–63.9 Gy (74.4%) and those receiving 64–66 Gy (87.0%) ($P = 0.5$). Severe late sequelae were more frequent in patients treated with doses 63 Gy or more than in patients treated with lower doses (23.1% versus 0%) (Mundt et al. 1995).

Fein noted that patients receiving less than 62.5 Gy had a 5-year local control of 78% versus 95% where the dose was greater than 62.5 Gy (Fein et al. 1995). In a multivariate analysis of patients undergoing postoperative radiotherapy, Zagars and Ballo (2003) identified dose as an independent variable for local control. Doses of 64 Gy or more correlated with improved local control. Recognizing that the effectiveness of a particular dose was also related to other factors influencing local control, such as margin status, anatomic site, and locally recurrent presentation, they recommended postoperative

doses of 60 Gy for patients with negative margins and otherwise favorable prognostic features, while suggesting increasing doses for less favorable presentations, up to doses of 68 Gy for positive margins. In contrast, other investigators have not been able to demonstrate a clear dose–response relationship in their reviews of their patients undergoing postoperative RT (Bell et al. 1989; Pao and Pilepich 1990; LeVay et al. 1993; Robinson et al. 1990). In practice, most centers give 60 Gy of postoperative radiotherapy for patients with negative margins and 66–68 Gy for positive margins, using shrinking fields as described above.

With regard to preoperative RT, Robinson et al. failed to demonstrate a variation in local control according to dose, although the response rate to preoperative radiotherapy was clearly dose dependent (Robinson et al. 1992). The accepted dose for preoperative radiotherapy is 50 Gy.

For treatment of an extremity lesion, a good functional result demands that only a portion of the cross section of the extremity be irradiated to any worthwhile dose level (Stinson et al. 1991). Thus, some tissue should not be irradiated to provide for lymphatic drainage. For very large tumors that are treated with wide resection, there may be persistent leg edema, requiring the use of a pressure-type stocking, even though the radiation treatment volume is less than circumferential. This can be a problem for patients with large (>10 cm) sarcomas of the medial thigh.

When postoperative radiation is combined with adjuvant chemotherapy, radiation daily dose has generally been reduced by 10% from 200 cGy to 180 cGy; radiation is not given concurrently with doxorubicin. Instead, 2–3 days are allowed between the doxorubicin and radiation. Some preoperative protocols have interdigitated chemotherapy and RT (see below); total preoperative radiation has been reduced (i.e., 44 Gy) in this setting.

30.6.9
Intensity-Modulated Photon RT

The purpose of RT is to maximize the dose delivered to the tumor while minimizing the exposure of dose-sensitive critical structures to high dose. This has been achieved traditionally by shaping the spatial distribution of the high radiation dose to conform to the target volume (hence, 3D conformal RT), thereby reducing the dose to the non-target structures. Although this approach is satisfactory in the treatment of targets that are roughly convex in shape, it is less than optimal for targets that contain complex concavities or that wrap around critical structures (VERHEY 1999). Growing experience suggests that intensity-modulated photon RT (IMRT) plans produce dose distribution to the patient superior to 3D conformal plans, both in terms of dose conformity in the tumor and dose reduction to the specified critical normal structures, albeit at the cost of increased integral dose to the normal tissues (Fig. 30.8). Recent dosimetric studies comparing IMRT and conformal radiotherapy for STS have been reported. When evaluating sarcomas arising in the extremities, pelvis, trunk, and paranasal sinuses, IMRT plans were more conformal. In the extremities, bone and subcutaneous doses were reduced by up to 20%. A conformal-IMRT comparative planning study has been reported for a large extraskeletal chondrosarcoma of the extremity (CHAN et al. 2001). Not surprisingly, IMRT produced excellent conformal treatment plans for this complex target volume, with a greater reduction of the maximum dose to the bone than the 3D-photon plan. HONG et al. (2004) performed treatment-planning comparisons of IMRT and 3-D conformal radiotherapy for ten patients with STS of the thigh. They were able to document a reduction in femur dose

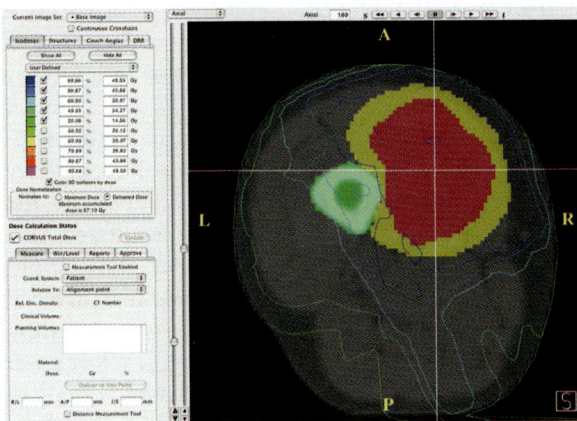

Fig. 30.8. Axial intensity-modulated radiotherapy (IMRT) treatment plan for an 82-year-old gentleman with a history of type-2 insulin-dependent diabetes mellitus treated for a 5.7-cm, grade-2–3/3 malignant fibrous histiocytoma arising deep in his left anterior thigh adjacent to the femur, clinical stage T2B N0 M0 (stage III). He underwent preoperative radiation therapy to a total dose of 48.6 Gy at 1.8 Gy, with reduction in fraction size and total dose because of his diabetes. An IMRT plan was used to contour the dose around the femur, to reduce the risk of late pathological fracture. The IMRT plan was able to accomplish this, but as would be anticipated, more low-dose radiation was delivered to the posterolateral thigh normal tissues than would have been the case for a three-dimensional conformal plan

without compromise in tumor coverage. In addition, IMRT reduced hot spots in the surrounding soft tissues and skin.

It is worth noting, however, that IMRT treatment plans often have localized areas within the high dose volumes, where dose inhomogeneities can be in the range of 10–15% above the prescription dose. Because there can also be dose inhomogeneities in the range of 5% below the target dose, treatment plans may be normalized to the 95% isodose line, meaning that selected areas of the treatment volume are receiving daily fractions and total doses of 15–20% above the target dose. Depending on the location of these "hot spots", there can be unanticipated acute normal tissue toxicity (LEE et al. 2002). Because of the multiple field angles employed with IMRT, integral doses to the extremity will likely also be higher with IMRT, meaning that more of the extremity will see some radiation dose, albeit relatively lower levels. Whether there are late effects attributable to these focal areas of high dose or the higher integral dose remains to be seen.

30.6.10
Proton Beam Radiotherapy

The rationale for the use of protons (or other charged particles) rather than photons (i.e., X-rays, which have traditionally been used for RT) is the superior dose distribution that can be achieved with protons. Protons and other charged particles deposit little energy in tissue until near the end of the proton range, where the residual energy is lost over a short distance, resulting in a steep rise in the absorbed dose, known as the Bragg peak (Fig. 30.9) (WILSON 1946; AUSTIN-SEYMOUR et al. 1989). The Bragg peak is too narrow for practical clinical applications, so for the irradiation of most tumors, the beam energy is modulated by superimposing several Bragg peaks of descending energies (ranges) and weights to create a region of uniform dose over the depth of the target; these extended regions of uniform dose are called "spread-out Bragg peaks" (Fig. 30. 9). Although the beam modulation to spread out the Bragg peaks does increase the entrance dose, the proton dose distribution is still characterized by a lower dose region in normal tissue proximal to the tumor, a uniform high dose region in the tumor, and **zero dose beyond the tumor** (Fig. 30.9).

Although protons have been extensively employed for sarcomas of the skull base and spine/paraspinal tissues, there are clearly opportunities to employ

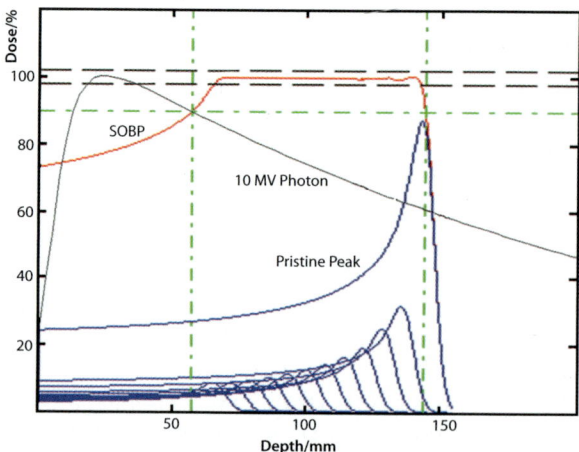

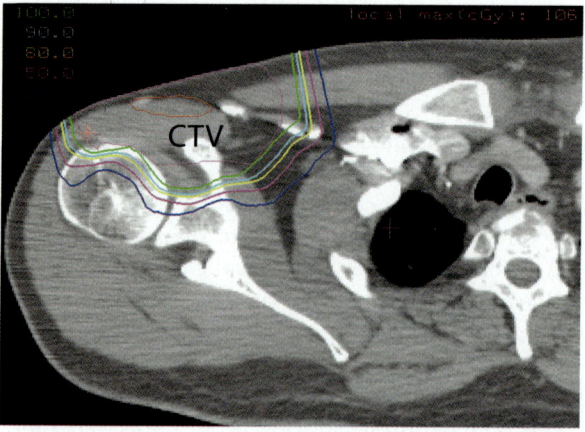

Fig. 30.9. Depth-dose distributions for a spread-out Bragg peak (SOBP, *red*), its constituent pristine Bragg peaks (*blue*), and a 10-MV photon beam (*black*). The SOBP dose distribution is created by adding the contributions of individually modulated pristine Bragg peaks. The penetration depth, or range, measured as the depth of the distal 90% of plateau dose, of the SOBP dose distribution is determined by the range of the most distal pristine peak (labeled "Pristine Peak"). The *dashed lines* (*black*) indicate the clinically acceptable variation in the plateau dose of +/−2%. The *dot-dashed lines* (*green*) indicate the 90% dose and spatial, range and modulation width, intervals. The SOBP dose distribution of even a single field can provide complete target volume coverage in depth and lateral dimensions, in sharp contrast to a single photon dose distribution; only a composite set of photon fields can deliver a clinical target dose distribution. Note the absence of dose beyond the distal fall-off edge of the SOBP. (Reprinted with permission from Levin WP, Kooy H, Loeffler JS, and DeLaney TF. Proton beam therapy. Br J Cancer 2005; 93:849–854)

Fig. 30.10. Proton dose distribution for a 38-year-old man with newly diagnosed alveolar sarcoma involving the soft tissues anterior to the right shoulder. The use of protons allowed sparing of the shoulder joint, which was not in the clinical target volume (CTV contoured in fine purple line). The posterior portion of the clinical target volume was too deep for the use of electrons, and use of intensity-modulated radiotherapy would have markedly increased the integral dose received by the patient

protons with very significant sparing of normal tissues in patients with extremity STSs. Large, medial proximal thigh lesions can be effectively treated with sparing of the femur, hip joint, genitalia, and anorectal tissue. Lesions around the shoulder can be treated without irradiating the lung apex and avoiding the shoulder (Fig. 30.10). With the recent or anticipated completion of proton beam facilities in major sarcoma centers in the United States (Massachusetts General Hospital, MD Anderson Cancer Center, and University of Florida), it is anticipated that a larger proportion of these patients will be treated with protons.

30.6.11
Neoadjuvant Doxorubicin-Based Chemotherapy Plus RT

The UCLA group popularized preoperative regional chemotherapy and RT followed by limb salvage sur-

gery in patients with high-grade sarcomas (EILBER et al. 2001). They were able to achieve a high rate of primary limb salvage, low rate of local recurrence (approximately 9%), and long-term survival in 65% of patients. The current regimen consists of doxorubicin (30 mg per day for 3 days) followed by radiation given at 28 Gy in 8 fractions.

A number of other groups have utilized this regimen, also obtaining low rates of local recurrence with varying degrees of toxicity. As an example, the Southeastern Cancer Study Group evaluated this protocol in 66 patients with nonmetastatic high-grade extremity sarcoma who received intraarterial doxorubicin infused directly into the vessel feeding the tumor (30 mg per 24 h for 3 days) (WANEBO et al. 1995). Concurrent RT was administered (30 Gy in 10 fractions, 35 Gy in 10 fractions, or 46 Gy in 23–25 fractions). Limb-sparing surgery was possible in 60 of 66 patients; an additional 2 patients required amputation due to wound healing complications. The 5-year survival and disease-free survival rates were 59% and 44%, respectively. The local failure rate was 1.5%.

It is not clear that intraarterial administration provides added benefit to intravenous doxorubicin. One study compared these two methods of administration (EILBER et al. 1995). The intraarterial route was thought to be associated with a higher incidence of complications and no improvement in survival or function. Another report evaluated two separate protocols utilizing preoperative treatment with intravenous doxorubicin and ifosfamide with or without intraarterial cisplatin; the histological response and local failure rate following surgery were better with the all-intravenous regimen (MERIMSKY et al. 1999).

Combination chemotherapy regimens such as MAID (mesna, doxorubicin, ifosfamide, and dacarbazine) may provide better anti-tumor activity than single-agent doxorubicin. Interesting early results have been noted with neoadjuvant MAID plus RT (KRAYBILL et al. 2001; DeLANEY et al. 2003b). The experience with preoperative MAID chemotherapy interdigitated with 44 Gy radiation and followed by surgery, postoperative MAID, and further radiation (16 Gy) for those with positive margins was reported in a series of 48 patients with high-grade extremity sarcomas greater than or equal to 8 cm (DeLANEY et al. 2003b). Despite the low objective response rate to preoperative therapy (partial response in five, stable disease in 36), all patients were able to undergo limb-sparing surgery initially, with 7 having positive margins. Median tumor necrosis was 95%, suggesting that conventional imaging in this setting may underestimate the degree of response to therapy. Of the patients, 25% required hospitalization for febrile neutropenia at some time during treatment. Wound healing complications occurred in 14 of 48 MAID patients (29%). One MAID patient developed late fatal myelodysplasia. The 5-year rates of local control (92% versus 86%), freedom from distant metastases (75% versus 44%), disease-free survival (70% versus 42%), and overall survival (87% versus 58%) all compared favorably with the outcomes of a cohort of historical control patients who were matched for tumor size, grade, patient age, and era of treatment.

Similar results were noted when this regimen was utilized in a multicenter United States Cooperative Group trial, in which 66 patients with primary high-grade STS 8 cm or larger in diameter received a modified MAID regimen plus granulocyte colony-stimulating factor and radiation, followed by resection and postoperative chemotherapy (KRAYBILL et al. 2001). In a preliminary report, although preoperative chemotherapy and radiation was successfully completed by 79% and 89% of patients, respectively, grade-4 hematological and non-hematological toxicity was experienced by 80% and 23% of patients. Delayed wound healing was noted in 31%. With a median follow-up of 2.75 years, the estimated 3-year survival, disease-free survival and local control rates

were 75, 55, and 79%, respectively. Of the patients, 2 developed late myelodysplasia. It remains to be confirmed in randomized studies whether these aggressive interdigitated approaches offer benefit to the subgroup of patients with large, high-grade sarcomas, who are at the highest risk of treatment failure.

30.6.12
STSs of the Hand and Foot

The 5-year survival rate for sarcoma of the hand and foot is approximately 80%, better than that usually given for extremity STSs (TALBERT et al. 1990; JOHNSTONE et al. 1994; KARAKOUSIS et al. 1998). This is likely related to the smaller size of these lesions at presentation. With surgical excision and the use of adjunctive radiotherapy when the minimum surgical margin is narrow (less than 2 mm), limb amputation can be avoided as primary therapy in most patients, and up to two-thirds of patients can retain a normal or fairly normal extremity (Fig. 30.11).

30.6.13
Wound Healing After Surgery and Radiation

In general, the use of adjunctive radiation is associated with a higher frequency of wound complications. Quantification of the impact of radiation on wound healing is difficult because of the significant complications seen with surgery alone. In addition, there is much heterogeneity among patients with STS with respect to anatomic site, histological type, lesion size, prior surgery, medical status, and age. The use of adjunctive radiotherapy can also be associated with joint stiffness, edema, and decreased range of motion (BUJKO et al. 1992; YANG et al. 1998). In one trial, extremity radiation resulted in significantly worse limb strength, edema, and range of motion than with surgery alone for extremity STS, but the symptoms were transient and did not affect global quality of life (YANG et al. 1998).

30.6.14
Preoperative Radiation and Wound Complications

Preoperative radiation is associated with a higher incidence of acute wound complications (O'SULLIVAN et al. 2002). In the randomized study of preopera-

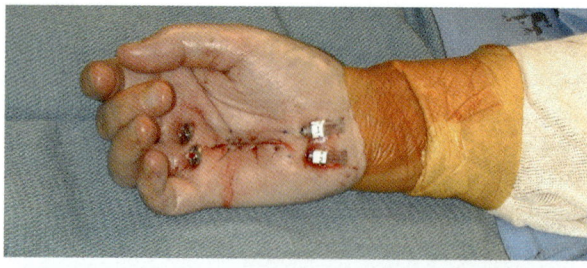

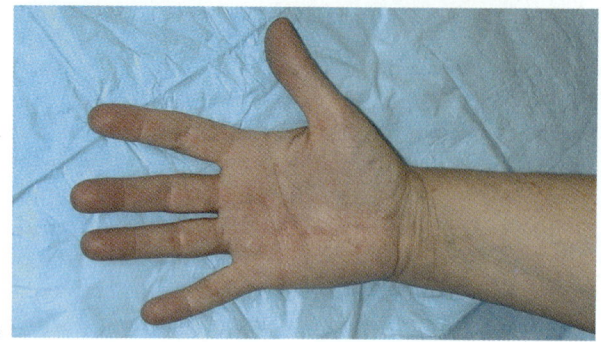

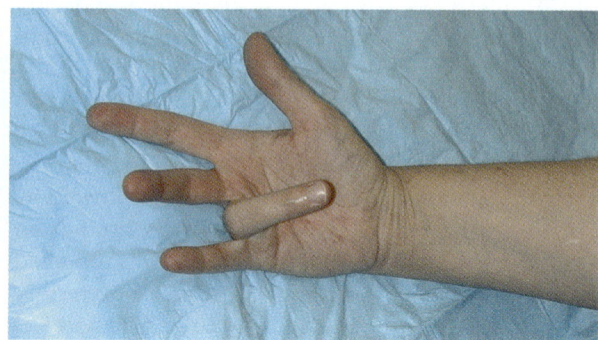

Fig. 30.11a–c. A 51-year-old right-handed composer and pianist with low-grade spindle cell neoplasm with features most suggestive of fibrosarcoma involving the flexor tendon sheath of the fourth finger, initially excised with a positive margin, subsequently treated with tumor bed re-excision and low-dose rate Ir-192 brachytherapy (**a**) of 45 Gy over 4.5 days. She returned to playing the piano approximately 2 months later. Photos taken 8 months after her brachytherapy document excellent functional and cosmetic results, as seen in **b** and **c**

tive versus postoperative RT, a significantly higher percentage of preoperatively treated patients had acute wound complications (35% versus 17%). In another series of 202 patients undergoing preoperative RT, the overall wound complication rate was 37% (Bujko et al. 1992). Of the patients, 1 died with necrotizing fasciitis, and 33 (17%) required secondary surgery, including 6 (3%) who required amputation. In another report, wound morbidity was 25% (4 of 16) in patients treated with preoperative EBRT plus BRT at the time of surgery but only 5% (2 of 40) in those treated postoperatively with EBRT plus BRT (Schray et al. 1990).

30.6.15
BRT and Wound Complications

The use of perioperative BRT may increase the incidence of wound complications. In one study of 105 patients with extremity and truncal sarcomas, major wound complications occurred in 9 of 41 (22%) cases treated with BRT compared with 2 of 64 (3%) non-BRT-treated patients (Arbeit et al. 1987). Patients treated with BRT also had a higher total number of complications as well as a higher combined frequency of major and moderate wound complications (44% versus 14%), and a longer time to wound healing (189 days versus 49 days).

However, the findings were different in a randomized trial of adjuvant BRT versus no BRT in 164 patients with resected extremity or truncal STS

(Arbeit et al. 1987). The incidence of serious wound complications was not significantly increased in the group receiving BRT (24% versus 14%, P=0.133), but the incidence of wound reoperation was increased (6% versus 0%) (Alektiar et al. 2000).

30.6.16
Strategies to Reduce Wound Morbidity

Based on published experience, the following strategies are suggested to reduce acute wound morbidity in patients being treated with preoperative radiation (Bujko et al. 1992) or perioperative BRT (Alektiar et al. 2000):

- Gentle handling of tissue during surgery
- Meticulous attention to achieving hemostasis before wound closure
- Avoidance of closure under tension
- Elimination of all wound dead space, using a rotated flap to fill the space, if necessary
- Wound drainage with tubes remaining in place until drainage is decreasing in a satisfactory manner
- Use of compression dressings
- Immobilization of the affected part for approximately 7 days
- Delineation of a subgroup of patients where postoperative boost dose can be omitted, which would include patients with negative margins and no tumor cut-through, complete tumor necrosis, or absence of tumor in the resection specimen

30.6.17
Adjuvant Chemotherapy

Adjuvant chemotherapy is standard treatment for rhabdomyosarcoma (CRIST et al. 2001), osteosarcoma (BRAMWELL 1997), and Ewing's sarcoma (ROSEN et al. 1981) but is not definitively established in other adult STSs (ANTMAN 1997; SCURR and JUDSON 2005).

30.6.17.1
Rhabdomyosarcoma

The vast majority of patients with rhabdomyosarcoma are children. In this population, the use of postoperative multiagent chemotherapy (typically vincristine, dactinomycin, and cyclophosphamide) contributes to cure rates in 65–75% of children with localized rhabdomyosarcoma (stages I–III) who are also treated with surgery and/or irradiation (CRIST et al. 2001). This is the standard of care for these patients. Ifosfamide has also found a role as part of the chemotherapy for patients with poorer prognosis disease, while the use of high-dose induction chemoradiotherapy followed by autologous bone marrow transplantation is being evaluated in younger patients with high-risk disease (BOULAD et al. 1998).

30.6.17.2
Adjuvant Chemotherapy for Other STSs

Many randomized trials and meta-analyses have addressed the potential benefit of adjuvant chemotherapy in resected extremity STS in adults. The vast majority of available studies have utilized doxorubicin, either alone or with dacarbazine. Among 14 previously published randomized trials, 2 reported a significant overall survival advantage for combination chemotherapy, while 3 noted higher survival in the observation arm (ANTMAN 1997). The remainder shows no significant difference when patients receiving adjuvant therapy are compared with controls.

Individual patient data from these studies, which involved 1,568 adults with localized resectable STS (only some of which were on the extremities), were analyzed by the Cochrane Collaborative and published in 1997 (SARCOMA META-ANALYSIS COLLABORATION 1997). All evaluated studies randomly assigned patients postoperatively to receive or not receive adjuvant chemotherapy containing doxorubicin. The following benefits were noted in the chemotherapy group:

- Longer local recurrence-free interval – hazard ratio 0.73 [95% confidence interval (CI) 0.56–0.94]
- Longer distant recurrence-free interval – hazard ratio 0.70 (95% CI 0.57–0.85)
- Higher overall recurrence-free survival – hazard ratio 0.75 (95% CI 0.64–0.87), corresponding to a significant absolute benefit of 6–10% at 10 years
- For overall survival, the hazard ratio of 0.89 was not significant but potentially represented an absolute benefit of 4% (95% CI -1–9) at 10 years

There was no consistent evidence of a difference in effect according to age, sex, stage, site, grade, histology, extent of resection, tumor size, or exposure to radiotherapy. However, the strongest evidence of a beneficial effect on survival was shown in patients with sarcoma of the extremities. Among these patients, the hazard ratio was 0.80 (P=0.029), equivalent to a 7% absolute benefit at 10 years.

In summary, doxorubicin-based adjuvant chemotherapy appears to significantly improve time to local and distant recurrence and overall recurrence-free survival in adults with localized resectable STS of the extremities. There is some evidence of a trend toward improved overall survival.

Since the publication of the meta-analysis, three additional randomized trials have explored the benefit of doxorubicin or ifosfamide-based chemotherapy in extremity STS (BRODOWICZ et al. 2000; FRUSTACI et al. 2001; PETRIOLI et al. 2002), two of which have shown a survival benefit for adjuvant chemotherapy (FRUSTACI et al. 2001; PETRIOLI et al. 2002):

- In a trial in which two-thirds of the enrolled patients had either synovial sarcoma or liposarcoma (two particularly chemosensitive STS histologies), 144 patients with high-grade large (5 cm or larger) or recurrent spindle cell sarcomas involving the extremities or girdles were randomly assigned to no postoperative therapy or to five cycles of a dose-intensive epirubicin/ifosfamide combination (epirubicin 60 mg/m^2 on day 1 and day 2 plus ifosfamide 1.8 g/m^2 on days 1–5) with mesna and granulocyte colony-stimulating factor support (FRUSTACI et al. 2001). Accrual was prematurely discontinued at 2 years, when a significant difference in the cumulative incidence of distant metastasis was found (45% versus 28%, favoring the chemotherapy group); the overall survival difference (85% versus 72%), also favoring the chemotherapy group, did not reach the level of statistical significance.

When the trial was reported, the 4-year overall survival rates were significantly higher in favor of chemotherapy (69% versus 50%), although the distant relapse rates were by then similar in both groups (44% and 45% for the chemotherapy and control groups, respectively). It is difficult to interpret these results, since the main benefit of adjuvant systemic chemotherapy is expected to be in reducing the rate of distant relapse. There was a trend toward improved local control in the adjuvant chemotherapy group, with a 17% local failure rate at 4 years in the control group compared with 6% in the adjuvantly treated group, $P = 0.07$.

- The second trial, also from Italy, randomly assigned 88 patients with high-risk STS to surgery with or without RT ($n = 43$) or to surgery plus chemotherapy ($n = 45$, 26 with epirubicin alone, and 19 to epirubicin plus ifosfamide) with or without RT (PETRIOLI et al. 2002). The 5-year survival rate of patients treated with chemotherapy was significantly higher than that of patients who did not receive chemotherapy (72% versus 47%). However, the large number of treatment variables and the small number of studied patients makes interpretation of this result impossible.

The importance of long-term follow-up in assessing benefit from chemotherapy was shown in a report of the combined experience of two major cancers (Memorial Sloan Kettering and MD Anderson) that included 674 consecutive adults undergoing resection of a stage-III extremity STS between 1984 and 1999 (CORMIER et al. 2004). Adjuvant doxorubicin-based chemotherapy was administered to 336 (50%), while the remainder received local therapy only. Although not a randomized trial, there were no significant differences between the chemotherapy and non-chemotherapy groups with respect to tumor size, anatomic site, histopathological subtype, or resection margin status. With a median follow-up of 6.1 years, the effect of chemotherapy appeared to vary over time. During the first year, the hazard ratio (HR) for disease-specific survival for chemotherapy versus no chemotherapy was 0.37 (95% CI 0.20–0.69); thereafter, the HR was 1.36 (95% CI 1.02–1.81).

Another recent study from Memorial Sloan Kettering Cancer Center retrospectively analyzed the relationship between neo-adjuvant chemotherapy (NAC) and outcome in patients with high-grade, deep, greater than 5-cm extremity sarcomas. Patients diagnosed between 1990 and 2001 were treated with surgery only ($n = 282$) or NAC containing doxorubicin/ifosfamide/mesna (AIM) ($n = 74$). NAC was associated with improved disease-specific survival for this cohort of patients ($P = 0.02$). This overall improvement appears to be driven by the benefit of NAC on disease-specific survival for patients with tumors greater than 10 cm. The 3-year disease-specific survival for tumors of this size was 0.62 (95% CI: 0.53–0.71) for patients not receiving NAC and 0.83 (95% CI: 0.72–0.95) for patients receiving NAC (GROBMYER et al. 2004). A study of concurrent, interdigitated neoadjuvant MAID chemotherapy and RT in patients with extremity sarcomas 8 cm or larger showed higher disease-free and overall survival than that of a historical matched patient cohort (DeLANEY et al. 2003b). These data emphasize the need for further prospective clinical studies of neo-adjuvant or adjuvant chemotherapy for patients with large high-grade extremity sarcomas.

Despite many randomized trials, the role of adjuvant chemotherapy remains uncertain and cannot be adopted as the standard of practice for all extremity sarcomas. The positive Italian trial is the only study that enrolled predominantly patients with chemosensitive types of STS (i.e., liposarcomas and synovial sarcomas) (FRUSTACI et al. 2001). However, the lack of a difference in the distant metastatic rate at 4 years (the time point at which the survival benefit was most pronounced) calls into question the validity of the conclusion that adjuvant chemotherapy improves survival in patients with high-grade or recurrent sarcomas. If there is a survival benefit for adjuvant chemotherapy, it is probably small (no more than 4–8% absolute increase in survival at 5–10 years). Although it has been proposed that patients be selected for adjuvant chemotherapy based on poor prognostic tumor characteristics such as histology [i.e., synovial sarcoma (FERRARI et al. 2004]), grade, or size, there is no evidence to date that this approach leads to improved outcome in any subset of patients (COINDRE et al. 1996). The absolute benefit, patient selection, and optimal regimen remain to be defined. A large, randomized study of postoperative adjuvant adriamycin and ifosfamide was recently completed by the European Oncology Research and Treatment Cooperative (EORTC) Group. The results of this study, which are pending, may help to better define the role for adjuvant chemotherapy in these patients.

30.6.18
Functional Outcome

There are increasing data available on the functional outcome of patients undergoing limb salvage procedures (ROBINSON et al. 1991; STINSON et al. 1991; LEVAY et al. 1993; DAVIS et al. 1999). The majority of patients have good or excellent functional outcome. In one series of 88 patients treated with surgery and either preoperative or postoperative RT, 68 had acceptable functional results, and 61 returned to work (LEVAY et al. 1993). Large tumors, neural sacrifice, proximal thigh tumors, and postoperative complications were associated with poor outcome. Subcutaneous tumors have a more favorable functional outcome (GERRAND et al. 2004). In a single institution series of 145 patients who underwent limb-sparing surgery plus RT, long-term treatment complications included bone fracture in 6%, contracture in 20%, significant edema in 19%, moderate to severe decrease in muscle strength in 20%, requirement for a cane or crutch in 7%, and tissue induration in 57% (STINSON et al. 1991). Of the patients, 3 (2%) required amputations for treatment-related complications. The percentage of patients ambulating without assistive devices and with mild or no pain was 84%. Higher doses of RT, a long radiation portal, and irradiation of more than 75% of the extremity diameter were associated with increased complications. Another study examined issues related to quality of life in patients with STS of the lower limb (ROBINSON et al. 1991). Although RT was associated with reduced muscle power and range of motion, compared with the use of surgery alone, most patients retained good to excellent limb function and quality of life.

The functional outcome is often not as good in patients requiring amputation. In a matched case-control study of patients with lower extremity sarcoma undergoing amputation ($n = 12$) or a limb-sparing approach ($n = 24$), there was a trend toward increased disability and handicap for those in the amputation group (DAVIS et al. 1999). Of the 12 amputees, 7 reported ongoing problems with the soft tissue overlying the stump. A few studies have assessed quality of life issues in amputees who had been treated for STS with amputation and chemotherapy compared with patients who underwent limb salvage with radiotherapy and chemotherapy (WEDDINGTON et al. 1985; SUGARBAKER et al. 1982). Contrary to expectations, there were no significant differences in measures of psychological outcome. Thus, a psychological advantage of limb-salvage surgery over amputation has yet to be demonstrated.

30.6.19
Treatment of Local Recurrence

Approximately 10–15% of patients with extremity STS who are treated with complete resection and adjuvant radiation will develop a local tumor failure, the majority within the first 2 years (LEVAY et al. 1993; SADOSKI et al. 1993; KARAKOUSIS and DRISCOLL 1999; ZAGARS et al. 2003b). The approach to the patient with an isolated local recurrence is similar to that for primary disease with some modifications. As with primary treatment, the goal is to provide limb salvage with conservative resection. However, approximately 10–25% of patients with locally recurrent disease will require amputation (CATTON et al. 1996a,b; KARAKOUSIS et al. 1996; UEDA et al. 1997). Surgery is an important component of successful salvage therapy (CATTON et al. 1996b). For patients whose primary treatment was surgery alone, re-excision combined with adjuvant radiation is the treatment of choice. If RT was used in primary treatment, further radiation may not be possible because the maximal tolerance for adjacent normal tissues would have to be exceeded, resulting in problems with wound healing and radiation fibrosis, although additional radiation given by BRT (mean dose 47.2 Gy) has been employed in these cases, with a 52% local control and a 33% disease-free survival rate in one series of 26 patients (PEARLSTONE et al. 1999a).

Optimal treatment for a local recurrence may require both surgery and radiation. This was illustrated in one report of salvage therapy using surgery alone or surgery plus re-irradiation for 25 patients with locally recurrent extremity STS (CATTON et al. 1996b). Of the patients, 18 underwent surgery alone; 11 were treated by a conservative procedure and 7 required amputation. Of these 18 patients, 7 relapsed. Of the 10 patients treated with surgery plus radiation, none experienced relapse, with a median follow-up of 24 months. Of these patients, 6 (60%) experienced significant wound healing complications, but 3 recovered completely.

30.6.20
Treatment of Unresectable or Locally Advanced Soft Tissue Sarcoma

In patients with advanced STS in whom the tumor has progressed beyond surgical resectability, treatment options depend on the site of tumor involvement. For patients with unresectable disease limited to the extremity, isolated limb perfusion protocols have been applied with striking early success. Selected patients can have their tumors, especially when small, controlled with RT with or without chemotherapy. For patients treated with a dose of 6,400 cGy or greater, Tepper and Suit noted control of unresected STSs in 87.5% of cases where tumors were less than 5 cm in diameter (TEPPER and SUIT 1985). Treatment was less effective for larger tumors, with local control falling to 53% for lesions 5–10 cm diameter (53%) and 30% for those greater than 10 cm. KEPKA and colleagues recently updated and expanded this experience, reporting on the efficacy of radiation on 112 patients with unresected STSs. For patients receiving 63 Gy or more, local control at 5 years was 72.4% in patients with lesions 5 cm or less, 42.4% for lesions 5–10 cm, and 25.4% for lesions greater than 10 cm (KEPKA et al. 2005).

Several centers have reported local control rates of approximately 50% with fast neutron irradiation of inoperable STS (PICKERING et al. 1987; SCHMITT et al. 1990; STEINGRABER et al. 1996). In addition, several radiation sensitizers have been used to treat patients with extensive sarcoma with promising early preliminary results (GOFFMAN et al. 1991; RHOMBERG et al. 1996).

30.7
Treatment of Metastatic Disease

30.7.1
Overview

Metastatic disease rarely occurs in patients with low-grade sarcomas but occurs at an appreciable frequency, which is related to grade and size, in patients with high-grade sarcomas (COINDRE et al. 1996). With intermediate or high-grade sarcomas, this risk may exceed 50% when the tumor is larger than 10 cm (PISTERS et al. 1996b; TROVIK et al. 2000; GRONCHI et al. 2005). For extremity sarcomas, the lung is the most common site of metastatic disease (POTTER et al. 1985; GADD et al. 1993). Some histologies – notably myxoid liposarcoma, which can metastasize to abdominal sites and bone (PEARLSTONE et al. 1999b; SPILLANE et al. 1999), and epithelioid sarcomas, which manifest regional nodal failure – are exceptions to the more general pattern. While most patients with metastatic sarcoma will ultimately die from their tumor, a modest proportion of patients will be long-term survivors after management with surgery and/or chemotherapy.

30.7.2
Resection of Pulmonary Metastases

The median survival of patients with pulmonary metastases is in the range of 15 months. Patients whose lung metastases can be resected fare better than those who are unresectable – the greatest impact on survival after lung metastasis. In one series, patients treated with resection had a median survival after complete resection of 33 months (BILLINGSLEY et al. 1999). Their 3-year actuarial survival rate was 46%, with a 5-year actuarial survival rate of 37%. The patients who did not undergo resection had a median survival of 11 months and a 3-year actuarial survival rate of 17%.

Patients to be considered for pulmonary resection are medically fit patients with controlled primary tumors without pleural effusion or hilar disease. The procedure generally involves wedge resections of the nodules. Patients with limited number of nodules fare better, but there is no consensus on the upper limit of nodules that should be considered for resection. The role for additional adjuvant chemotherapy after resection is not settled. A subset of patients may benefit from repeat thoracotomy for recurrent disease in the chest (POGREBNIAK et al. 1991). There are reports of resection of isolated metastatic disease in liver and also other extrapulmonary sites (LANG et al. 2000).

30.7.3
Chemotherapy for Metastatic Disease

For most patients with metastatic disease that is not respectable, treatment with chemotherapy is likely to be palliative in outcome. A small number of patients will be long-term survivors. An analysis of 1888 patients treated on studies organized by the Soft Tissue and Bone Sarcoma Group of the EORTC reported 66 5-year survivors, which translates to an 8% 5-year survival rate (BLAY et al. 2003). A brief discussion of chemotherapy for metastatic disease will be presented here; readers can find more complete discussion in recent reviews on this subject (JUDSON 2004; MAKI 2004).

Doxorubicin and ifosfamide have been demonstrated to be the most active chemotherapy agents in widely disseminated STS (CLARK et al. 2005). For doxorubicin, objective response rates between 20% and 40% for the single agent have been reported; few are complete responses, and response duration averages 8 months. A steep dose-response curve for objective responses was described by O'BRYAN et al. (1977). There is also a dose–response for ifosfamide (CHRISTMAN et al. 1993; VAN OOSTEROM et al. 2002). Dacarbazine (DTIC) by itself has a modest response rate of approximately 16% (GOTTLIEB et al. 1976). Cyclophosphamide appears less active in adults than in children and less active than the related compound ifosfamide (BRAMWELL et al. 1993). Gemcitabine with or without a taxane is active in a subset of patients with sarcomas (HENSLEY et al. 2002). Taxanes are active against angiosarcomas (FATA et al. 1999). Newer agents such as ecteinascidin-743 have at least some activity against sarcomas (YOVINE et al. 2004). Other less active agents include methotrexate (SUBRAMANIAN and WILTSHAW 1978; PINEDO and VERWEIJ 1986) and cisplatin (KARAKOUSIS et al. 1979; GRABOIS et al. 1994).

Many combination chemotherapy regimens for metastatic disease have been studied in phase-II trials. Responses are higher than with single agents. Most of these trials include doxorubicin (or epirubicin) and an alkylating agent. Adding DTIC to doxorubicin improved the response rate to 41%, as described by GOTTLIEB et al. (1972), but the response rate has decreased over time (GOTTLIEB et al. 1976). Comparisons have shown that the addition of less active drugs necessitate lower doses of doxorubicin and, accordingly, reduces overall effectiveness (CRUZ et al. 1979; SCHOENFELD et al. 1982). Adding ifosfamide, however, seems to be clearly beneficial as reported by BLUM et al. (1993) and SCHUTTE et al. (1993). Nevertheless, the improvement in response rate has not clearly translated into an improvement in overall survival. A prospective randomized trial was performed by EORTC, which compared single-agent doxorubicin with a combination of doxorubicin and ifosfamide and also the four-drug CYVADIC (cyclophosphamide/vincristine/doxorubicin/dacarbazine) combination. There was no improvement in progression-free or overall survival associated with combination therapies, although they were significantly more toxic. Similar findings were reported by Eastern Cooperative Group, which conducted a three-arm trial comparing doxorubicin alone, doxorubicin plus ifosfamide, and mitomycin plus doxorubicin plus cisplatin (SANTORO et al. 1995). Objective tumor regression occurred more frequently in the combination arms than in the single agent arm (20% with doxorubicin alone, 34% in doxorubicin plus ifosfamide, and 32% in the mitomycin plus doxorubicin plus cisplatin arm). However, the combination regimens resulted in significantly greater myelosuppression, e.g., 80% of the doxorubicin/ifosfamide group had grade-3 or greater myelosuppression. Most notably, no significant survival differences were observed among the three treatment regimens (EDMONSON et al. 1993). A popular regimen adds ifosfamide to adriamycin and DTIC (ELIAS and ANTMAN 1986). A combination of ifosfamide with mesna, doxorubicin and dacarbazine (MAID regimen) has resulted in response rates in measurable metastatic sarcomas as high as 47%, with complete response rates as high as 10% (ELIAS et al. 1989). In the absence of any clear benefit for the combination regimens, many clinicians favor initiation of chemotherapy with single-agent Adriamcyin (JUDSON 2004).

Attempts to intensify treatment by increasing the dose of doxorubicin in combination with ifosfamide, although promising in phase-II studies (STEWARD et al. 1993), were not confirmed in a subsequent randomized trial compared with the standard doses (LE CESNE et al. 2000). High-dose therapy with growth factor support has been evaluated in several investigational studies, but the data to date demonstrate increased toxicity with clear evidence of therapeutic gain; thus, this is still considered investigational treatment (DUMONTET et al. 1992; SCHWELLA et al. 1998).

30.8
Summary

Treatment for STS requires individual tailoring of the approach because of the wide variety of clinical situations that can arise from a tumor that involves a variety of anatomic sites with a range of histologies of variable grade and size. Nevertheless, the following suggestions can serve as useful guides. Surgery is always indicated, but the use of adjuvant therapy can vary according to the anatomic site, size, and histological grade.

- In general, patients with superficial low-grade tumors that are less than 5 cm in diameter can be treated with surgical excision alone when negative margins of ≥ 1 cm are achieved and can expect excellent local control and survival rates approximating 90%.
- In patients with intermediate grade lesions, surgical excision with negative margins in combination with radiotherapy has achieved excellent local control, with overall survival rates approximating 80%. For larger, deep-seated tumors, preoperative RT appears to be more effective than postoperative radiation to prevent local tumor recurrence.

Acute wound healing complications are higher with preoperative RT for lower extremity lesions, but generally irreversible late complications, including grade 3–4 fibroses, are more common in those patients receiving postoperative RT.

- In patients with high-grade STS greater than 5 cm, excellent local control can be achieved with surgery and radiotherapy, but at least 50% of these patients will develop metastatic disease. In this setting, the use of neoadjuvant chemotherapy may benefit some and should be considered in the context of a clinical trial, to be followed by definitive surgery combined with either pre- or postoperative radiotherapy or BRT.

- BRT can provide excellent local control and functional results in appropriately selected patients.

References

Abbas JS, Holyoke ED, Moore R et al (1981) The surgical treatment and outcome of soft-tissue sarcoma. Arch Surg 116:765-769

Alektiar KM, Zelefsky MJ, Brennan MF (2000) Morbidity of adjuvant brachytherapy in soft tissue sarcoma of the extremity and superficial trunk. Int J Radiat Oncol Biol Phys 47:1273-1279

Alvegard TA, Sigurdsson H, Mouridsen H et al (1989) Adjuvant chemotherapy with doxorubicin in high-grade soft tissue sarcoma: a randomized trial of the Scandinavian Sarcoma Group. J Clin Oncol 7:1504-1513

Aman P, Ron D, Fioretos T et al (1992) Rearrangement of the transcription factor gene CHOP in myxoid liposarcomas with t(12;16)(q13;p11). Genes Chromosomes Cancer 5:278

American Joint Committee on the Staging of Cancer Staging System for Soft Tissue Sarcomas (2002) AJCC cancer staging manual. Springer, Belrin Heidelberg New York

Angervall L, Kindblom LG (1993) Principles for pathologic-anatomic diagnosis and classification of soft-tissue sarcomas. Clin Orthop 289:9-18

Antman KH (1997) Adjuvant therapy of sarcomas of soft tissue. Semin Oncol 24:556-560

Arbeit JM, Hilaris B, Brennan MF (1987) Wound complications in the multimodality treatment of extremity and superficial truncal sarcomas. J Clin Oncol 5:480-488

Austin-Seymour M, Munzenrider J, Goitein M et al (1989) Fractionated proton radiation therapy of chordoma and low grade chondrosarcoma of the base of skull. J Neurosurg 70:13-17

Baldini EH, Goldberg J, Jenner C et al (1999) Long-term outcomes after function-sparing surgery without radiotherapy for soft tissue sarcoma of the extremities and trunk. J Clin Oncol 17:3252-3259

Bell RS, O'Sullivan B, Liu FF et al (1989) The surgical margin in soft tissue sarcoma. J Bone Joint Surg 71:370-375

Billingsley KG, Lewis JJ, Leung DHY et al (1999) Multifactorial analysis of the survival of patients with distant metastasis arising from primary extremity sarcoma. Cancer 85:389-395

Bland KI, McCoy DM, Kinnard RE et al (1987) Application of magnetic resonance imaging and computerized tomography as an adjunct to the surgical management of soft tissue sarcomas. Ann Surg 205:473-481

Blay J-Y, van Glabbeke M, Verweij J et al (2003) Advanced soft-tissue sarcoma: a disease that is potentially curable for a subset of patients treated with chemotherapy. Eur J Cancer 39:64-69

Blum RH, Edmonson J, Ryan L et al (1993) Efficacy of ifosfamide in combination with doxorubicin for the treatment of metastatic soft-tissue sarcoma. The Eastern Cooperative Oncology Group. Cancer Chemother Pharmacol 31 [Suppl 2]:S238-S240

Boulad F, Kernan NA, LaQuaglia MP et al (1998) High-dose induction chemoradiotherapy followed by autologous bone marrow transplantation as consolidation therapy in rhabdomyosarcoma, extraosseous Ewing's sarcoma, and undifferentiated sarcoma. J Clin Oncol 16:1697

Bramwell VH (1997) The role of chemotherapy in the management of non-metastatic operable extremity osteosarcoma. Semin Oncol 24:561

Bramwell VH, Mouridsen HT, Santoro A et al (1993) Cyclophosphamide versus ifosfamide: a randomized Phase 11 trial in adult soft-tissue sarcomas. The European Organization for Research and Treatment of Cancer [EORTC], Soft Tissue and Bone Sarcoma Group. Cancer Chemother Pharmacol 31 [Suppl 2]:S180-S184

Brant TA, Parsons JT, Marcus RB et al (1990) Preoperative irradiation for soft tissue sarcomas of the trunk and extremities in adults. Int J Radiat Oncol Biol Phys 19:899-906

Brodowicz T, Schwameis E, Widder J et al (2000) Intensified adjuvant IFADIC chemotherapy for adult soft tissue sarcoma. A prospective randomized feasibility trial. Sarcoma 4:151

Brown R, Marshall CJ, Pennie SG et al (1984) Mechanism of activation of an N-ras gene in the human fibrosarcoma cell line HT1080. EMBO J 3:1321-1326

Bujko K, Suit HD, Springfield S et al (1992) Wound healing after surgery and preoperative radiation for sarcoma of soft tissues. Surg Gynecol Obstet 176:124-134

Catton C, O'Sullivan B, Bell R et al (1996a) Chordoma: long-term follow-up after radical photon irradiation. Radiother Oncol 41:67-72

Catton C, Davis A, Bell R et al (1996b) Soft tissue sarcoma of the extremity. Limb salvage after failure of combined conservative therapy. Radiother Oncol 41:209-214

Chan MF, Chui CS, Shupak K et al (2001) The treatment of large extraskeletal chondrosarcoma of the leg: comparison of IMRT and conformal radiotherapy techniques. J Appl Clin Med Phys 2:3-8

Christman KL, Casper ES, Schwartz JK et al (1993) High-intensity scheduling of ifosfamide in adult patients with soft-tissue sarcoma. Proc Ann Meet Am Soc Clin Oncol 12: A1642

Clark J, Rocques PJ, Crew AJ et al (1994) Identification of novel genes, SYT and SSX, involved in the t(X;18)(p11.2;q11.2) translocation found in human synovial sarcoma. Nat Genet 7:502

Clark MA, FIsher C, Judson I, Thomas JM (2005) Soft-tissue sarcomas in adults. N Engl J Med 22:591-604

Cleator SJ, Cottril C, Harmer C et al (2001) Pattern of local recurrence after conservative surgery and radiotherapy for soft tissue sarcoma. Sarcoma 5:83-88

Coindre JM, Terrier P, Bui B et al (1996) Prognostic factors in adult patients with locally controlled soft tissue sarcoma. A study of 546 patients from the French Federation of Cancer Centers Sarcoma Group. J Clin Oncol 14:869-877

Cormier JN, Huang X, Xing Y et al (2004) Cohort analysis of patients with localized, high-risk, extremity soft tissue sarcoma treated at two cancer centers: chemotherapy-associated outcomes. J Clin Oncol 22:4567

Costa J, Wesley RA, Glatstein E et al (1984) The grading of soft tissue sarcomas. Results of a clinicohistopathologic correlation in a series of 163 cases. Cancer 53:530-541

Crist WM, Anderson JR, Meza JL et al (2001) Intergroup rhabdomyosarcoma study-IV: results for patients with nonmetastatic disease. J Clin Oncol 19:3091-3102

Cruz AB, Thames EA Jr, Aust JB et al (1979) Combination chemotherapy for soft tissue sarcomas: a phase III study. J Surg Oncol 11:313-323

Davis AM, Devlin M, Griffin AM et al (1999) Functional outcome in amputation versus limb sparing of patients with lower extremity sarcoma: a matched case-control study. Arch Phys Med Rehabil 80:615-618

Davis AM, O'Sullivan B, Bell RS et al (2002) Function and health status outcomes in a randomized trial comparing preoperative and postoperative radiotherapy in extremity soft tissue sarcoma. J Clin Oncol 20:4472-4477

DeChiara A, T'Ang A, Triche TJ (1993) Expression of the retinoblastoma susceptibility gene in childhood rhabdomyosarcomas. J Natl Cancer Inst 85:152-157

DeLaney TF, Rosenberg AE, Harmon DC et al (2003a) Soft tissue sarcomas. In: Price PM, Sikora K (eds) Treatment of cancer. Arnold, London, pp 869-907

DeLaney TF, Spiro IJ, Suit HD et al (2003b) Neoadjuvant chemotherapy and radiotherapy for large extremity soft-tissue sarcomas. Int J Radiat Oncol Biol Phys 56:1117-1127

Delattre O, Zucman J, Merlot T et al (1994) The Ewing family of tumors - a subgroup of small round cell tumors defined by specific chimeric transcripts. N Engl J Med 331:294-299

Dewhirst MW, Boatman HD, Leopold KA et al (1990) Soft-tissue sarcomas: MR imaging and MR spectrosopy for prognosis and therapy monitoring. Radiology 174:847-853

Dias P, Dealing M, Houghton P (1994) The molecular basis of skeletal muscle differentiation. Semin Diagn Pathol 11:3-14

Dumontet C, Biron P, Bouffet E et al (1992) High dose chemotherapy with ABMT in soft tissue sarcomas: a report of 22 cases. Bone Marrow Transplant 10:405

Edmonson JH, Ryan LM, Blum RH et al (1993) Randomized comparison of doxorubicin alone versus ifosfamide plus doxorubicin or mitomycin, doxorubicin, and Cisplatin against advanced soft tissue sarcomas. J Clin Oncol 11:1269-1275

Eilber FR, Eckardt J (1997) Surgical management of soft tissue sarcomas. Semin Oncol 24:526-533

Eilber F, Eckardt J, Rosen G et al (1995) Preoperative therapy for soft tissue sarcoma. Hematol Oncol Clin North Am 9:817-823

Eilber FC, Rosen G, Eckardt J et al (2001) Treatment-induced pathologic necrosis: a predictor of local recurrence and survival in patients receiving neoadjuvant therapy for high-grade extremity soft tissue sarcomas. J Clin Oncol 19:3203-3209

Eilber FC, Rosen G, Nelson SD et al (2003) High-grade extremity soft tissue sarcomas: factors predictive of local recurrence and its effect on morbidity and mortality. Ann Surg 237:218-226

Elias A, Ryan L, Sulkes A et al (1989) Response to mesna, doxorubicin, ifosfamide and dacarbazine in 108 patients with metastatic or unresectable sarcoma and no prior chemotherapy. J Clin Oncol 7:1208-1216

Elias AD, Antman KH (1986) Doxorubicin, ifosfamide, and dacarbazine (AID) with mesna uroprotection for advanced untreated sarcoma: a phase I study. Cancer Treatment Rep 70:827-833

Engellau J (2004) Prognostic factors in soft tissue sarcoma: tissue microarray for immunostaining, the importance of whole-tumor sections and time-dependence. Acta Orthop Scand 75 [Suppl 314]:1-52

Fata F, O'Reilly E, Ilson D et al (1999) Paclitaxel in the treatment of patients with angiosarcoma of the scalp or face. Cancer 86:2034-2037

Fein DA, Lee WR, Lanciano RM et al (1995) Management of extremity soft tissue sarcomas with limb-sparing surgery and postoperative irradiation: do total dose, overall treatment time, and the surgery-radiotherapy interval impact on local control? Int J Radiat Oncol Biol Phys 32:969-976

Ferrari A, Gronchi A, Casanova M et al (2004) Synovial sarcoma: a retrospective analysis of 271 patients of all ages treated at a single institution. Cancer 101:627-634

Fong Y, Coal DG, Woodruff JM et al (1993) Lymph node metastasis from soft tissue sarcoma in adults. Analysis of data from a prospective data base of 1772 sarcoma patients. Ann Surg 217:72-77

Frustaci S, Gherlinzoni F, DePaoli A et al (2001) Adjuvant chemotherapy for adult soft tissue sarcomas of the extremities and girdles: results of the Italian Cooperative Trial. J Clin Oncol 19:1238-1247

Gadd MA, Casper ES, Woodruff J et al (1993) Development and treatment of pulmonary metastases in adult patients with extremity soft tissue sarcoma. Ann Surg 218:705-712

Gerrand CH, Wunder JS, Kandel A et al (2004) The influence of anatomic location on functional outcome in lower-extremity soft-tissue sarcoma. Ann Surg Oncol 11:476-482. Epub 12.4.2004

Goffman T, Tochner Z, Glatstein E et al (1991) Primary treatment of large and massive adult sarcomas with iododeoxyuridine and aggressive hyperfractionated irradiation. Cancer 67:572-576

Gottlieb JA, Baker LH, Quagliana JM et al (1972) Chemotherapy of sarcomas with a combination of adriamycin and dimethyl triazeno imidazole carboxamide. Cancer 30:1632-1638

Gottlieb JA, Benjamin RS, Baker LH et al (1976) Role of DTIC (NSC45388) in the chemotherapy of sarcomas. Cancer Treat Rep 60:199-203

Grabois M, Frappaz D, Bouffet E et al (1994) High-dose VP-16 cisplatinum in soft tissue sarcomja of children. Cancer Chemother Pharmacol 33:355-357

Grobmyer SR, Maki RG, Demetri GD et al (2004) Neo-adjuvant chemotherapy for primary high-grade extremity soft tissue sarcoma. Ann Oncol 15:1667-1672

Gronchi A, Casali PG, Mariani L et al (2005) Status of surgical margins and prognosis in adult soft tissue sarcomas of the extremities: a series of patients treated at a single institution. J Clin Oncol 23:96-104

Gustafson P, Herrlin K, Biling L et al (1992) Necrosis observed on CT enhancement is of prognostic value in soft tissue sarcoma. Acta Radiol 33:474-476

Harrison LB, Franzese F, Gaynor JJ et al (1993) Long-term results of a prospective randomized trial of adjuvant brachytherapy in the management of completely resected soft tissue sarcomas of the extremity and superficial trunk. Int J Radiat Oncol Biol Phys 27:259-265

Hensley ML, Maki RG, Venkatraman E et al (2002) Gemcitabine and docetaxel in patients with unresectable leiomyosarcoma: results of a phase II trial. J Clin Oncol 20:2824-2831

Hong L, Alektiar KM, Hunt M et al (2004) Intensity-modulated radiotherapy for soft tissue sarcoma of the thigh. Int J Radiat Oncol Biol Phys 59:752-759

Jemal A, Murray T, Ward E et al (2005) Cancer statistics, 2005. CA Cancer J Clin 55:10-30

Johnstone PAS, Wexler LH, Venzon DJ et al (1994) Sarcomas of the hand and foot: analysis of local control and functional result with combined modality therapy in extremity preservation. Int J Radiat Oncol Biol Phys 29:735-745

Judson I (2004) Systemic therapy of soft tissue sarcoma: an improvement in outcome. Ann Oncol 15 [Suppl 4]:iv193-iv196

Karakousis CP, Holtermann OA, Holyoke ED (1979) Cis-dichlorodiammineplatinum (II) in metastatic soft tissue sarcomas. Cancer Treat Rep 63:2071-2072

Karakousis CP, Emrich LJ, Rao U et al (1986) Feasibility of limb salvage and survival in soft tissue sarcomas. Cancer 57:484-491

Karakousis CP, Proimakis C, Rao U et al (1996) Local recurrence and survival in soft-tissue sarcomas. Ann Surg Oncol 3:255-260

Karakousis CP, de Young C, Driscoll DL et al (1998) Soft tissue sarcomas of the hand and foot: management and survival. Ann Surg Oncol 5:238-240

Karakousis CP, Driscoll DL (1999) Treatment and local control of primary extremity soft tissue sarcomas. J Surg Oncol 71:155-161

Kepka L, DeLaney TF, Goldberg SI et al (2005) Results of radiation therapy for unresected soft tissue sarcomas. Int J Radiat Oncol Biol Phys 63:852–859

Koutcher JA, Ballon D, Graham M et al (1990) 31P NMR spectra of extremity sarcomas: diversity of metabolic profiles and changes in response to chemotherapy. Magn Reson Imaging Med 16:19-34

Kransdorf MJ, Jelinek JS, Moser RP et al (1993) Imaging of soft tissue tumors. Radiol Clin North Amer 31:359-371

Kraybill WG, Spiro I, Harris J et al (2001) Radiation Therapy Oncology Group (RTOG) 95-14: a phase II study of neoadjuvant chemotherapy (CT) and radiation therapy (RT) in high risk (HR), high grade, soft tissue sarcomas (STS) of the extremities and body wall: a preliminary report. Proc ASCO 20:348a

Kretzler A, Molls M, Gradinger R et al (2004) Intraoperative radiotherapy of soft tissue sarcoma of the extremity. Perioperative fractionated high-dose rate brachytherapy in the treatment of soft tissue sarcomas. Strahlenther Onkol 180:365-370

Kulander BG, Polissar L, Yang CY et al (1989) Grading of soft tissue sarcomas: necrosis as a determinate of survival. Modern Pathol 2:205-208

Lack EE, Steinberg SM, White DE et al (1989) Extremity soft tissue sarcomas: analysis of prognostic variables in 300 cases and evaluation of tumor necrosis as a factor in stratifying higher grade sarcomas. J Surg Oncol 263:73

Lang H, Nussbaum KT, Kaudel P et al (2000) Hepatic metastases from leiomyosarcoma: a single-center experience with 34 liver resections during a 15-year period. Ann Surg 231:500

Lawrence W Jr, Donegan WL, Natarajan N et al (1987) Adult soft tissue sarcomas. A pattern of care survey of the American College of Surgeons. Ann Surg 205:349-359

Le Cesne A, Judson I, Crowther D et al (2000) Randomized phase III study comparing conventional-dose doxorubicin plus ifosfamide versus high-dose doxorubicin plus ifosfamide plus recombinant human granulocyte-macrophage colony-stimulating factor in advanced soft tissue sarcomas: a trial of the European Organization for Research and Treatment of Cancer/Soft Tissue and Bone Sarcoma Group. J Clin Oncol 18:2676-2684

Lee N, Chuang C, Quivey JM et al (2002) Skin toxicity due to intensity-modulated radiotherapy for head-and-neck carcinoma. Int J Radiat Oncol Biol Phys 53:630-637

LeVay J, O'Sullivan B, Catton C et al (1993) Outcome and prognostic factors in soft tissue sarcoma in the adult. Int J Radiat Oncol Biol Phys 27:1091

Lindberg RD, Martin R. Romsdahl M et al (1981) Conservative surgery and postoperative radiotherapy in 300 adults with soft-tissue sarcomas. Cancer 47:2391-2397

Maki RG (2004) Role of chemotherapy in patients with soft tissue sarcomas. Exp Rev Anticancer Ther 4:229-236

Markhede G, Angervall L, Stener B et al (1981) A multivariate analysis of the prognosis after surgical treatment of malignant soft tissue tumors. Cancer 40:1721-1733

Mazeron JJ, Suit HD (1987) Lymph nodes as sites of metastasis from sarcomas of soft tissue. Cancer 60:1800-1808

Merimsky O, Meller I, Issakov J et al (1999) Adriamycin-ifosfamide induction chemotherapy for extremity soft tissue sarcoma: comparison of two non-randomized protocols. Oncol Rep 6:913-920

Mundt AJ, Awan A, Sibley S et al (1995) Conservative surgery and adjuvant radiation therapy in the management of adult soft tissue sarcoma of the extremities: clinical and radiobiological results. Int J Radiat Oncol Biol Phys 32:977-985

Myhre-Jensen, Kaae S, Madsen EH et al (1975) Histopathological grading in soft tissue tumors. Acta Pathol Microbiol Immunol Scand [A] 91:145-150

Nag S, Shasha D, Janjan N et al (2001) The American Brachytherapy Society recommendations for brachytherapy of soft tissue sarcomas. Int J Radiat Oncol Biol Phys 49:1033-1043

Negendank WG, Crowley MG, Ryan JR et al (1989) Bone and soft tissue lesions: diagnosis with combined H-1 MR imaging and P-31 spectroscopy. Radiology 173:181-187

Niewald M, Berberich W, Schnabel K et al (1990) A simple method for positioning and fixing the extremities during the radiotherapy of soft-tissue sarcomas. Strahlenther Onkol 166:295-296

Nieweg OE, Pruim J, van Ginkel J et al (1996) Fluorine-18-Flurodeoxyglucose PET imaging of soft-tissue sarcoma. J Nucl Med 37:257-261

Noria S, Davis A, Krandel R et al (1996) Residual disease following unplanned excision of soft-tissue sarcoma of an extremity. J Bone Joint Surg Am 78:650-655

O'Bryan RM, Baker LH, Gottlieb JE et al (1977) Dose response evaluation of adriamycin in human neoplasia. Cancer 39:1940-1948

O'Keefe F, Lorigan JG, Wallace S et al (1990) Radiological features of extraskeletal Ewing sarcoma. Br J Radiol 63:456-460

O'Sullivan B, Davis A, Bell R et al (1999) Phase III randomized trial of pre-operative versus post-operative radiotherapy in the curative management of extremity soft tissue sarcoma. A Canadian Sarcoma Group and NCI Canada Clinical Trials Group study. Proc ASCO 18:2066A

O'Sullivan B, Davis AM, Turcotte R et al (2002) Preoperative versus postoperative radiotherapy in soft-tissue sarcoma of the limbs: a randomized trial. Lancet 359:2235-2241

O'Sullivan B, Davis AM, Turcotte R et al (2004) Five-year results of a randomized phase III trial of pre-operative vs. post-operative radiotherapy in extremity soft tissue sarcoma. Proc Am Soc Clin Oncol 23:815

Panagopoulos I, Hoglund M, Mertens F et al (1996) Fusion of EWS and CHOP genes in myxoid liposarcoma. Oncogene 12:489-494

Pao WJ, Pilepich MV (1990) Postoperative radiotherapy in the treatment of extremity soft tissue sarcomas. Int J Radiat Oncol Biol Phys 19:907-911

Parham DM, Shapiro DN, Downing R et al (1994) Solid alveolar rhabdomyosarcomas with the t(2;13). Report of two cases with diagnostic implications. Am J Surg Pathol 18:474-478

Pearlstone DB, Janjan NA, Feig W et al (1999a) Re-resection with brachytherapy for locally recurrent soft tissue sarcoma arising in a previously radiated field (see comments). Cancer J Sci Am 5:26-33

Pearlstone DB, Pisters PW, Bold J et al (1999b) Patterns of recurrence in extremity liposarcoma: Implications for staging and follow-up. Cancer 85:85-92

Petera J, Neumanova R, Odrazka K et al (2004) Perioperative fractionated high-dose rate brachytherapy in the treatment of soft tissue sarcomas. Neoplasma 51:59-63

Petrioli R, Coratti A, Correale P et al (2002) Adjuvant epirubicin with or without Ifosfamide for adult soft-tissue sarcoma. Am J Clin Oncol 25:468-473

Peuchot M, Libshitz HI (1987) Pulmonary metastatic disease: radiologic-surgical correlation. Radiology 164:719-722

Pickering DG, Stewart JS, Rampling R et al (1987) Fast neutron therapy for soft tissue sarcoma. Int J Radiat Oncol Biol Phys 13:1489-1495

Pinedo HM, Verweij J (1986) The treatment of soft tissue sarcomas with a focus on chemotherapy: a review. Radiother Oncol 6:193-205

Pisters PW, Harrison LB, Leung DH et al (1996a) Long-term results of a prospective randomized trial of adjuvant brachytherapy in soft tissue sarcoma. J Clin Oncol 14:859-868

Pisters PW, Leung DH, Woodruff J et al (1996b) Analysis of prognostic factors in 1041 patients with localized soft tissue sarcomas of the extremities. J Clin Oncol 14:1679-1689

Pogrebniak HW, Roth JA, Steinberg M et al (1991) Reoperative pulmonary resection in patients with metastatic soft tissue sarcoma. Ann Thorac Surg 52:197-203

Potter DA, Glenn J, Kinsella T et al (1985) Patterns of recurrence in patients with high-grade soft tissue sarcomas. J Clin Oncol 3:353-366

Prosnitz LR, Maguire P, Anderson JM et al (1999) The treatment of high-grade soft tissue sarcomas with preoperative thermoradiotherapy. Int J Radiat Oncol Biol Phys 45:941-949

Redmond OM, Bell E, Stack JP et al (1992) Tissue characterization and assessment of preoperative chemotherapeutic response in musculoskeletal tumors by in vivo 31p magnetic resonance spectroscopy. Magn Reson Med 27:226-237

Rhomberg W, Hassenstein EO, Gefeller D et al (1996) Radiotherapy vs. radiotherapy and razoxane in the treatment of soft tissue sarcomas: final results of a randomized study. Int J Radiat Oncol Biol Phys 36:1077-1084

Robinson M, Barr L, Fisher C et al (1990) Treatment of extremity soft tissue sarcomas with surgery and radiotherapy. Radiother Oncol 18:221-233

Robinson MH, Ball AB, Schofield J et al (1992) Preoperative radiotherapy for initially inoperable extremity soft tissue sarcomas. Clin Oncol (R Coll Radiol) 4:36-43

Robinson MH, Spruce L, Eeles R et al (1991) Limb function following conservation treatment of adult soft tissue sarcoma. Eur J Cancer 27:1567-1574

Rosen G, Caparros B, Nirenberg A et al (1981) Ewing's sarcoma: ten-year experience with adjuvant chemotherapy. Cancer 47:2204-2213

Rosenberg SA, Tepper J, Glatstein E et al (1982) The treatment of soft-tissue sarcomas of the extremities: prospective randomized evaluations of (1) limb-sparing surgery plus radiation therapy compared with amputation and (2) the role of adjuvant chemotherapy. Ann Surg 196:305-315

Ruka W, Taghian A, Gioioso D et al (1996) Comparison between the in vitro intrinsic radiation sensitivity of human soft tissue sarcoma and breast cancer cell lines. J Surg Oncol 61:290-294

Russell WO, Cohen J, Enzinger F et al (1977) A clinical and pathological staging system for soft tissue sarcomas. Cancer 40:1562-1570

Rydholm A, Berg NO, Gullberg B et al (1984) Epidemiology of soft tissue sarcoma in the locomotor system. Acta Pathol Microbiol Immunol Scand [A] 92:363-374

Rydholm A, Gustafson P, Rooser B et al (1991) Limb-sparing surgery without radiotherapy based on anatomic location of soft tissue sarcoma. J Clin Oncol 9:1757-1765

Sadoski C, Suit HD, Rosenberg A et al (1993) Preoperative radiation, surgical margins, and local control of extremity sarcomas of soft tissues. J Surg Oncol 52:223-230

Sanchez RB, Quinn SF, Walling A et al (1990) Musculoskeletal neoplasms after intraarterial chemotherapy: correlation of MR images with pathologic specimens. Radiology 174:237-240

Santoro A, Tursz T, Mouridsen H et al (1995) Doxorubicin versus CYVADIC versus doxorubicin plus ifosfamide in first-line treatment of advanced soft tissue sarcomas: a randomized study of the European Organization for Research and Treatment of Cancer Soft Tissue and Bone Sarcoma Group. J Clin Oncol 13:1537-1545

Sarcoma Meta-Analysis Collaboration (1997) Adjuvant chemotherapy for localised resectable soft-tissue sarcoma of adults: meta-analysis of individual data. Lancet 350:1647-1654

Schmitt G, Pape H, Zamboglou M et al (1990) Long term results of neutron- and neutron-boost irradiation of soft tissue sarcomas. Strahlenther Onkol 166:61-62

Schoenfeld DA, Rosenbaum C, Horton J et al (1982) A comparison of adriamycin versus vincristine and adriamycin, and cyclophosphamide versus vincristine, actinomycin-D, and cyclophosphamide for advanced sarcoma. Cancer 50:2757-2762

Schray MF, Gunderson LL, Sim FH et al (1990) Soft tissue sarcomas. Integration of brachytherapy, resection, and external irradiation. Cancer 66:451-456

Schuetze SM, Rubin BP, Vernon C et al (2005) Use of positron emission tomoography in localized extremity soft tissue sarcoma treated with neoadjuvant chemotherapy. Cancer 103:339-348

Schutte J, Mouridsen HT, Steward W et al (1993) Ifosfamide plus doxorubicin in previously untreated patients with advanced soft-tissue sarcoma. Cancer Chemother Pharmacol 31 [Suppl 2]:S204-S209

Schwella N, Rick O, Meyer O et al (1998) Mobilization of peripheral blood progenitor cells by disease-specific chemotherapy in patients with soft tissue sarcoma. Bone Marrow Transplant 21:863-868

Scurr M, Juson I (2005) Neoadjuvant and adjuvant therapy for extremity soft tissue sarcoma. Hematol Oncol Clin North Am 19:489-500

Simon MA, Enneking WF (1976) The management of soft-tissue sarcomas of the extremities. J Bone Joint Surg Am 58:317-327

Spillane AJ, Fisher C, Thomas JM et al (1999) Myxoid liposarcoma - the frequency and the natural history of nonpulmonary soft tissue metastases. Ann Surg Oncol 6:389-394

Steingraber M, Lessel A, Jahn U et al (1996) Fast neutron therapy in treatment of soft tissue sarcoma - the Berlin-Buch study. Bull Cancer Radiother 83 [Suppl]:122s-124s

Steward WP, Verweij J, Somers R et al (1993) Granulocyte-macrophage colony-stimulating factor allows safe escalation of dose-intensity of chemotherapy in metastatic adult soft tissue sarcomas: a study of the European Organization for Research and Treatment of Cancer Soft Tissue and Bone Sarcoma Group. J Clin Oncol 11:15-21

Stinson SF, DeLaney TF, Greenberg J et al (1991) Acute and long term effects on limb function of combined modality limb sparing therapy for extremity soft tissue sarcomas. Int J Radiat Oncol Biol Phys 21:1493-1499

Subramanian S, Wiltshaw E (1978) Chemotherapy of sarcoma. Lancet 1:683-686

Sugarbaker PH, Barofsky I, Rosenberg SA et al (1982) Quality of life assessment of patients in extremity sarcoma clinical trials. Surgery 91:17-23

Suit HD, Russell WO, Martin RG et al (1975) Sarcoma of soft tissue: clinical and histopathologic parameters and response to treatment. Cancer 35:1478-1483

Suit HD, Mankin HJ, Wood WC et al (1988) Treatment of the patient with stage M0 sarcoma of soft tissue. J Clin Oncol 6:854-862

Talbert ML, Zagars GK, Sherman NE et al (1990) Conservative surgery and radiation therapy for soft tissue sarcoma of the wrist, hand, ankle, and foot. Cancer 66:2482-2491

Tanabe KK, Pollock RE, Ellis LM et al (1994) Influence of surgical margins on outcome in patients with preoperatively irradiated extremity soft tissue sarcomas. Cancer 73:1652-1659

Tepper JE, Suit HD (1985) Radiation therapy alone for sarcoma of soft tissue. Cancer 56:475-479

Tepper J, Rosenberg SA, Glatstein E et al (1982) Radiation therapy technique in soft tissue sarcomas of the extremity - policies of treatment at the National Cancer Institute. Int J Radiat Oncol Biol Phys 8:263-273

Todoroki T, Suit HD (1985) Therapeutic advantage in preoperative single-dose radiation combined with conservative and radical surgery in different-size murine fibrosarcomas. J Surg Oncol 29:207-215

Torosian MH, Friedrich C, Godbold J et al (1988) Soft-tissue sarcoma: initial characteristics and prognostic factors in patients with and without metastatic disease. Semin Surg Oncol 4:13-19

Trojani M (1984) Staging system for soft tissue and bone. Int J Cancer 33:37-42

Trovik CS, Bauer HC, Alvegard A et al (2000) Surgical margins, local recurrence and metastasis in soft tissue sarcomas: 559 surgically-treated patients from the Scandinavian Sarcoma Group Register. Eur J Cancer 36:710-716

Ueda T, Yoshikawa H, Mori S et al (1997) Influence of local recurrence on the prognosis of soft-tissue sarcomas. J Bone Joint Surg Br 79:553-557

Van Oosterom AT, Mouridsen HT, Nielsen S et al (2002) EORTC Soft Tissue and Bone Sarcoma Group. Results of randomised studies of the EORTC Soft Tissue and Bone Sarcoma Group (STBSG) with two different ifosfamide regimens in first- and second-line chemotherapy in advanced soft tissue sarcoma patients. Eur J Cancer 38:2397-2406

Vanel D, Lacombe MJ, Couanet D et al (1987) Musculoskeletal tumors follow-up with MR imaging after treatment with surgery and radiation therapy. Radiology 164:243-245

Verhey LJ (1999) Comparison of three-dimensional conformal radiation therapy and intensity-modulated radiation therapy systems. Semin Radiat Oncol 9:78-98

Wanebo HJ, Temple WJ, Popp MB et al (1995) Preoperative regional therapy for extremity sarcoma. A tricenter update. Cancer 75:2299-2306

Weddington WW Jr, Segraves KB, Simon MA et al (1985) Psychological outcome of extremity sarcoma survivors undergoing amputation or limb salvage. J Clin Oncol 3:1393-1399

Weiss SW, Goldblum JR (2001) Enzinger and Weiss's soft tissue tumors. Mosby, Philadelphia

White LM, Wunder JS, Bell RS et al (2005) Histologic assessment of peritumoral edema in soft tissue sarcoma. Int J Radiat Oncol Biol Phys 61:1439-1445

Williard WC, Hajdu SI, Casper ES et al (1992) Comparison of amputation with limb-sparing operations for adult soft tissue sarcoma of the extremity269. Ann Surg 215:389-396

Wilson RR (1946) Radiological uses of fast protons. Radiology 47:487-491

Yang JC, Chang AE, Baker AR et al (1998) Randomized prospective study of the benefit of adjuvant radiation therapy in the treatment of soft tissue sarcomas of the extremity. J Clin Oncol 16:197-203

Yovine A, Riofrio M, Blay JY et al (2004) Phase II study of ecteinascidin-743 in advanced pretreated soft tissue sarcoma patients. J Clin Oncol 22:890-899

Zagars GK, Ballo MT (2003) Significance of dose in postoperative radiotherapy for soft tissue sarcoma. Int J Radiat Oncol Biol Phys 56:473-481

Zagars GK, Ballo MT, Pisters PW et al (2003a) Surgical margins and reresection in the management of patients with soft tissue sarcoma using conservative surgery and radiation therapy. Cancer 97:2544-2553

Zagars GK, Ballo MT, Pisters PW et al (2003b) Prognostic factors for patients with localized soft-tissue sarcoma treated with conservation surgery and radiation therapy: an analysis of 225 patients. Cancer 97:2530-2543

31 Total Body Irradiation Conditioning Regimens in Stem Cell Transplantation

Kathryn E. Dusenbery and Bruce J. Gerbi

CONTENTS

31.1 Conditioning Regimen 786
31.2 Fractionation and Dose Rate 787
31.3 Sequence 788
31.4 Technical Aspects 788
31.5 Right and Left Lateral TBI 788
31.6 Simulation and Patient Measurements 788
31.6.1 Compensators for TBI 790
31.7 Compensator Design 791
31.7.1 Patient Treatment 793
31.7.2 Dose Verification 793
31.7.3 Dose Prescription 794
31.8 Anteroposterior TBI 794
31.8.1 Patient Treatment Technique 797
31.9 Normal Tissue Shielding 798
31.9.1 Lung Shielding 798
31.9.2 Kidney Shielding 799
31.10 Gonad Shielding 800
31.10.1 Thymus Shielding 800
31.10.2 TomoTherapy 800
31.10.2.1 Special Considerations for TBI in
          Young Children 801
31.10.2.2 Complications Following Preparation With TBI 802
          References 802

Since the first successful bone marrow transplantation was performed at the University of Minnesota in 1960 (Gatti et al. 1968), bone marrow and stem cell transplantation (SCT) has gained prominence as a therapy for a variety of diseases as outlined in Table 31.1. The majority of bone marrow transplants are carried out in an effort to eradicate malignant disease, but a growing number of patients with auto-

immune (Thomas 1997) or genetic disorders are being offered transplantation (Iannone et al. 2003; Peters et al. 2003). The rationale for SCT differs depending on the disease treated and the source of bone marrow cells. In both autologous and allogeneic SCT for malignant diseases, the rationale for SCT is to allow chemotherapeutic dose escalation. The SCT "rescues" the patient from what otherwise would be a lethal dose of chemotherapy.

In an allogeneic SCT, a healthy donor marrow regenerates in its place and the infused donor lymphocytes have a proven anti-tumor effect (graft versus leukemic effect) (Schleuning 2000; Remberger et al. 2002; Zecca et al. 2002). Donor lymphocytes can also serve to supply absent enzyme for patients with inborn errors of metabolism. In autologous SCT, the infused marrow may be contaminated with malignant cells. Various methods have been used to purge the marrow of residual malignant cells (Freedman et al. 1998; Colombat et al. 2000; Schouten et al. 2000; van Besien et al. 2003). As there is no anti-tumor (graft versus leukemia) effect, various cytokines are being tried in an effort to mimic the graft versus leukemia effect and improve the efficiency of autologous transplantation (Hawley et al. 1996; Imamura et al. 1996; Klingemann 1996; Leshem et al. 2000). In the future, the autologous cells may be manipulated with genes that confer relative chemotherapeutic resistance (Wood and Prior 2001), thus allowing for additional post transplantation chemotherapy with little damage to the new bone marrow (Heslop et al. 1995).

The development of HLA (human lymphocyte antigen) and MLC (mixed lymphocyte culture) assays allowed physicians to determine whether potential marrow donors were "histocompatible" (i.e., matched by HLA antigens and non-reactive to MLC cultures) and therefore less likely to develop graft-versus-host disease (GVHD) (Mahmoud et al. 1985; Parr et al. 1991; Petersdorf et al. 1998). Initially, transplants were only performed between HLA-matched related donors, but, with

K. E. Dusenbery, MD
University of Minnesota Medical School, Department of Radiation Oncology, MMC436, 420 Delaware St. S.E., Minneapolis, MN 55455, USA
B. J. Gerbi, PhD
Associate Professor, Therapeutic Radiology – Radiation Oncology, University of Minnesota, Mayo Mail Code 494, 420 Delaware St SE, Minneapolis, MN 55455, USA

Table 31.1. Diseases treated by stem cell transplantation

Acute myeloid leukemia
Acute lymphoblastic leukemia
Chronic myelogenous leukemia
Chronic lymphocytic leukemia
Myelodysplasia
Lymphoma
    Non-Hodgkin's
    Hodgkin's disease
Multiple myeloma
Aplastic anemia
    Idiopathic
    Fanconi
    Paroxysmal nocturnal hemoglobinuria
Congenital/immunodeficiency
    SCIDS
Autoimmune disease
    Rheumatoid arthritis
    Systemic lupus erythematosus

Osteopetrosis
Leukencephalopathies
Hurlers syndrome
Other inborn errors of metabolism
Sickle cell anemia/thalassemia

the advent of effective therapies directed at decreasing the probability and severity of GVHD, matched unrelated donor transplants have become more common. The National Marrow Donor Program (NMDP) types potential bone marrow and stem cell donors. More than 16,000 transplants have been performed using unrelated donors provided by the NMDP (McCullough et al. 1989; Karanes 2003; Cornetta et al. 2005). With more than four million donors listed in the registry, over 70% of patients can now find an HLA-A, -B, -DR phenotypic match at the initial search.

In recent years, in addition to related and unrelated bone marrow donor sources, additional sources of pluripotential stem cells have been investigated, including the use of umbilical cord blood (Cornetta et al. 2005), a rich source of stem cells. A bank of HLA typed umbilical cord blood harvests has also been established (Krishnamurti et al. 2003).

Another source of pluripotential stem cells are circulating blood stem cells. These cells can be harvested through leukapheresis, frozen for later use, then thawed and reinfused. In addition to sparing the donor the discomfort of a bone marrow harvest, these peripheral stem cell harvests usually result in a more prompt engraftment than occurs with bone marrow infusions, resulting in a shorter period of pancytopenia and thus less risk of infection (Korbling et al. 1991; Steingrimsdottir et al. 2000).

31.1
Conditioning Regimen

Presumed desired endpoints of the pre-transplantation conditioning regimen are to eradicate the recipient's native bone marrow, immune suppress the recipient sufficiently to avoid rejection of the donor transplant, and to do this with minimal toxicity to other tissues. In some situations, these three goals are attainable with a chemotherapy-only conditioning regimen; however, multiple variables need to be considered including the age of the patient, the underlying disease, the source of the donor marrow, and whether the donor marrow is manipulated (i.e., T depleted) (Fehr and Sykes 2004).

At the University of Minnesota, total body irradiation (TBI) is generally part of the conditioning regimens for the situations outlined in Table 31.2. These situations include unrelated marrow or cord blood donor transplants, certain underlying malignancies that are considered radiosensitive [acute lymphocytic leukemia (ALL), multiple myeloma], and for patients in whom a TBI-containing conditioning regimen has been shown to be superior to a chemotherapy-only conditioning regimen [acute myeloid leukemia (AML) in second remission] (Dusenbery et al. 1996).

There are theoretical advantages and disadvantages of a TBI-containing preparative regimen. Most often patients undergoing SCT have been exposed to multiple chemotherapeutic regimens; therefore, potential SCT recipients may be relatively chemotherapy "resistant." As most patients undergoing SCT have not been irradiated previously, their malignant cells may be more radiation sensitive than chemotherapy sensitive. Additionally, there are known sanctuary sites where chemotherapy does not penetrate well, such as the central nervous system (CNS) or testicles. There are no sanctuary sites for irradiation and, in certain situations such as relapsed ALL, a TBI-containing regimen may be especially beneficial. Lastly, chemotherapy which is usually given intravenously (busulfan may be given orally) needs to be metabolized and eliminated from the body. It is known that chemotherapy pharmacokinetics differ among patients resulting in some areas of the body exposed to higher or lower concentrations of drug. TBI requires no metabolism for clearance and all areas of the body receive the same dose of irradiation. The disadvantages of using a TBI regimen are the potential late side effects, such as sterility, cataracts, and growth retardation, as well as the potential neurological toxicity that may occur with irradiation.

Table 31.2. University of Minnesota protocols utilizing total body irradiation or total lymphoid irradiation in the conditioning regimen (CR)

1320 cGy in eight fractions over 4 days:
- Acute lymphoblastic leukemia in second or subsequent CR
- Acute lymphoblastic leukemia high risk in first CR
- Chronic myelogenous leukemia
- Acute myeloid leukemia (some children receive chemotherapy only conditioning)
- Myelodysplastic syndrome
- Non-Hodgkin's lymphoma (depending on dose of prior irradiation)
- Unrelated donor transplantation
- Cord blood transplantation

Other regimens
- Fanconi anemia: 450 cGy in single fraction, ±thymus block
- Non-myeloablative transplants: 200 cGy in a single fraction
- Sickle cell anemia or thalassemia: total lymphoid irradiation, block gonads
- Osteopetrosis: TLI to include spleen, liver and mesenteric lymph nodes

31.2
Fractionation and Dose Rate

Initial trials with TBI use a single fraction of up to 10 Gy. Subsequently, it was suggested that the therapeutic ratio (i.e., increased leukemic cell kill and decreased toxicity of late responding tissues, such as lung, heart, spinal cord, kidney, and CNS) of TBI might be improved by either going to a very low dose rate (not practical since treatment times of up to 24 h would be required) or using a fractionated schema. This was based on radiobiological data suggesting that leukemic cells (and their normal counterparts, the hematopoietic stem cells) were relatively radiosensitive, with a narrow "shoulder" on the survival curve and with little capacity of sublethal radiation damage repair capacity (HALL 1994). However, late responding tissues are better able to repair sublethal damage and have a relatively "broad shoulder" on the dose–response curve. Formulations based on dose survival models have been proposed to evaluate the biological equivalence of various doses and fractionation schedules. Assumptions are based on the linear-quadratic model that takes into account the α and β (non-reparable and reparable) components of cell kill. The values for the α and β components of cell kill can be derived experimentally, but are not available for many human tissues. Extrapolating from animal data and cell cultures, it has been found that the ratio of α/β is a useful indicator of the effect of fractionation on cell damage. Tissues with a high α/β include the gastrointestinal tract,

skin, and bone marrow cells ("early" responding); whereas, tissues with a low α/β include spinal cord, kidney, brain, and lung ("late" responding) (HALL 1994).

Although not completely applicable in the setting of TBI schemes, the biological effective dose (BED) of one TBI regimen can be compared with another regimen if several estimates are taken into consideration. The units are arbitrary but allow one to compare the theoretical effects of different TBI regimens on different tissues.

$$BED = n \times d \; (1+d/(\alpha/\beta))$$

where n=number of fractions, d=dose per fraction (Gy/fraction) and α/β=estimate (10 for early tissues, 3 for late tissues).

In engineering an optimal TBI regimen, the goal is to cause minimal damage to late responding tissues, while having a high probability of damaging the early responding bone marrow cells (and malignant cells). Regimens can be evaluated for their potential effect on late responding tissues or on early responding tissues. If evaluating for late effects of different TBI regimens:

Assume an α/β of 3	BED
750 cGy in 1 fraction	26
750 cGy in 3 fractions	14
1200 cGy in 6 fractions	20
1320 cGy in 8 fractions	20
1395 cGy in 12 fractions	19

One can see that fractionation would be expected to spare late effects. One could go to a higher total dose, without increasing the probability of late effects. If considering the early effects of different TBI regimens:

Assume an α/β of 10	BED
750 cGy in 1 fraction	13
750 cGy in 3 fractions	9
1200 cGy in 6 fractions	14
1320 cGy in 8 fractions	15
1375 cGy in 12 fractions	16

It becomes apparent that fractionation spares early effects. Since bone marrow ablation is desired, the total dose must be increased to 1200 cGy or more to get the same myeloablative effects as a single fraction of 750 cGy.

These theoretical equations are supported by data from reports of bone marrow transplantation regimens. For example, the risk of cataract formation (a late responding tissue) is substantially higher when the TBI is given in a single fraction of TBI than when it is fractionated (ARISTEI et al. 2002).

Although no randomized clinical trials exist, the majority of retrospective reviews looking at the rate of interstitial pneumonitis after single fraction TBI compared with multiple fraction TBI strongly suggest an advantage for a fractionated TBI schedule (SHANK et al. 1983; CARDOZO et al. 1985; KIM et al. 1985; MOLLS et al. 1986; STANDKE 1989; VALLS et al. 1989; CARLSON et al. 1994). One must keep in mind the fact that these trials are non-randomized and usually compare recently transplanted patients on fractionated schemas with former single fraction patients. Numerous variables are potentially implicated in the development of interstitial pneumonitis, including other TBI variables such as dose rate, use of lung shielding, and timing of the TBI (before or after chemotherapy) (MOLLS et al. 1986).

Initially single fraction TBI was delivered with cobalt units at extended distances. The dose rate achievable was only 5–10 cGy/min and it required several hours to deliver the dose. When using fractionated TBI, it is not clear that the dose rate needs to be this low, although the risk of pneumonitis may be higher with higher dose rate (KIM et al. 1985; CARRUTHERS and WALLINGTON 2004). However, most institutions have kept the dose rate low (under 10 cGy/min), since a fraction of 200 cGy can be delivered in a reasonable amount of time (20 min in this case).

31.3
Sequence

TBI can either precede or follow the chemotherapy portion of the conditioning regimen. An advantage to delivering the TBI first is that, with the appropriate use of antiemetics, it can be given as an outpatient treatment and thus reduce inpatient costs. Following completion of TBI, patients are then hospitalized for the chemotherapy portion of the conditioning regimen. Clinical data are lacking, however, on whether TBI is less toxic or more effective when given before chemotherapy, although it is theoretically possible that the variety of cytokines released during chemotherapy may influence the incidence of pneumonitis.

31.4
Technical Aspects

Numerous techniques for irradiation of the entire body are described in the literature. At the University of Minnesota, two general TBI techniques are currently in use, with modifications of the techniques for certain situations. The vast majority of our patients are treated using the first technique, which involves right and left lateral fields with the patient semi-recumbent at an extended distance on a specially designed couch. The second technique is an anterior–posterior treatment technique patterned after that developed at Memorial Sloan Kettering (SHANK et al. 1983). For the latter, adult patients are treated in a standing position with anterior and posterior beams, while younger (smaller) patients are treated in a reclined position if they can fit within the available field size at the floor of the treatment room. The goal of both of these techniques is to deliver a uniform dose to the entire body within ±10% of the dose at the prescription point. This second technique has the advantage that certain organs such as the thymus or testicles can be blocked.

31.5
Right and Left Lateral TBI

This technique uses lateral photon beams with the patient in a semi-recumbent position, as described by KHAN et al. (1980). The treatment is delivered at a source to patient midline distance of 410 cm, which produces a field approximately 120 cm wide at the 95% isodose line. Aluminum compensators are used to produce a uniform dose through all body regions to within ±10% of the dose specified at the umbilicus.

31.6
Simulation and Patient Measurements

Pretreatment measurements for TBI are performed in the simulator room to accurately reproduce the treatment position within the treatment room and to calculate the size and thickness of the compensating filters. The simulation procedure consists of three steps. During the first step, an anterior chest film is taken to determine the amount of lung traversed by the treatment fields. The radiograph is taken with the patient seated on the simulator couch with his or her back resting against the film cassette holder. The source–skin distance (SSD) to the patient's chest is measured and because the source–film distance (SFD) is known, the magnification factor for the size of the lungs can be determined. Using this information, the thickness of the lung inhomogeneity can be computed. This information is used later when the need for lung compensation is evaluated.

The second step is to establish the patient treatment position so that the entirety of the body fits within the 95% isodose line. An overhead projector is used to cast an isodose pattern that represents the uniformity of the actual treatment field (Fig. 31.1). The head and back of the patient are positioned within the 98% line, while the toes of the feet are within the 95% isodose line of the radiation field. Once this position has been established, setup measurements are recorded on the form, as illustrated in Figure 31.2.

To describe the position of the lower extremities, measurements are made from our reference point [the anterosuperior point of the iliac spine (ASIS)] to the knee and to the back of the heel. The length of the feet is also recorded. The distance from the sternal notch to the top of the knee or patella is documented to describe how compressed the patient is within the field. Finally, with the arms in the treatment position,

the distance from the middle of the shoulder to the top of the head is measured. The latter information is used to scale the size of the head compensator. The chin extension, measured from the sternal notch to the point of the chin, is also recorded. Additionally, the need for positional devices, such as pillows for the back of the head, foam sponges underneath the hips, or sandbags under the feet, is recorded on the form. For future reference, the names of the individuals who made and checked the measurements are recorded.

The third step is to measure the right–left lateral thickness of the patient at certain anatomical locations. These measurements are recorded on the form shown in Figure 31.3 and constitute the basic data needed to determine the thickness of the compensators at these locations. The key measurement is the width of the patient at the umbilicus since this is the location where the dose is prescribed. Values of lateral thickness are also measured at the head, neck, shoulders, mid-mediastinum, pelvis, knees, and ankles. The mid-mediastinal thickness is measured midway between the sternal notch and the xiphoid and includes the thickness of the arms.

LATERAL TOTAL BODY IRRADIATION (TBI) SET-UP DIAGRAM

DIAGNOSIS:
TREATMENT REGIME:
TREATMENT UNITS: Philips 10 MV,
    Varian 6/2500,
    or Varian 6/100

NAME: TBI Sample case
UM HOSPITAL #:
DATE:

	Philips 10 MV	Varian 6/2500	Varian 6/100
GANTRY ANGLE:	90°	270°	2700°
COLLIMATOR SETTING:	38 x 38 cm²	40 x 40 cm²	40 x 40 cm²
FIELD SIZE @ Rx DISTANCE	156 x 156 cm²	164 x 164 cm²	144 x 144 cm²
TREATMENT SOURCE-AXIS DISTANCE:	410 cm	410 cm	360 cm

OTHER MEASUREMENTS TO BE RECORDED ON THE DIAGRAM:
ANTERIOR CHEST FILM:
    SSD = 117 cm
    SFD = 140 cm
CHIN EXTENSION = 9 cm

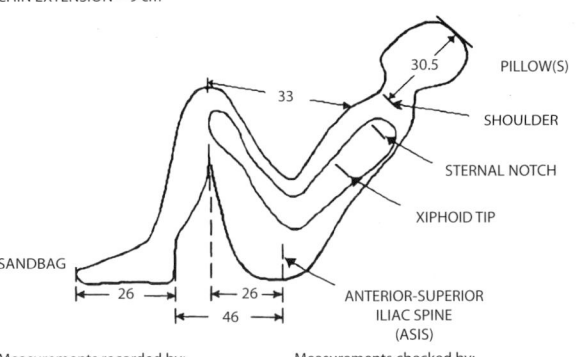

PILLOW(S)
SHOULDER
STERNAL NOTCH
XIPHOID TIP
ANTERIOR-SUPERIOR ILIAC SPINE (ASIS)
SANDBAG

30.5
33
26 26
46

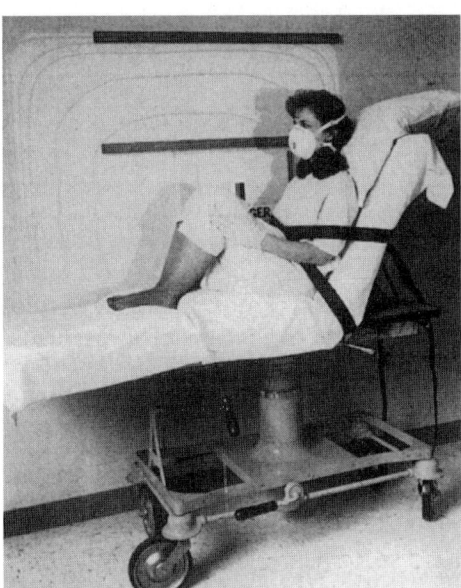

Fig. 31.1. For the bilateral technique, the patient is positioned within the homogeneous portion of the beam. The simulation is performed in the simulator room and an overhead projector is used to produce a representation of the treatment field

Measurements recorded by: _____   Measurements checked by: _____

Fig. 31.2. The form used to document the patient treatment position when the bilateral body technique is used

---

CALCULATIONS FOR TBI COMPENSATORS

NAME: TBI Sample case
UM Hospital #:
DATE

Umbilicus thickness (L_{ref}): 32.5

Lateral magnification $= \dfrac{\text{source-tray distance}}{\text{source-axis distance} - \dfrac{L_{ref}}{2}} = \dfrac{72}{410 - \dfrac{L_{ref}}{2}} = 0.183$

Compensator material: Aluminium

Thickness-ratio correction $= \dfrac{\tau}{\rho comp} = \dfrac{0.70}{2.70} = 0.259$

Site	Lateral thickness (L) (cm)	Lung thickness (cm)	Off-axis ratio (K)[4]	Tissue deficit[2] (g/cm²)	Compensator thickness (L_c)[3]	Percent Midline Dose
Head	16 cm	0 cm	1.00	16.5	2.1	100%
Neck	12.5 cm	0 cm	1.00	20	20.6	100%
Shoulders	45 cm	0 cm	1.00	-12.5	-1.6	85.6%
Mid-mediastinum	50 cm	27.6 cm[1]	1.00	3.2	0.4	100%
Umbilicus	32.5 cm	0 cm	1.00	0	0	100%
Pelvis	39 cm	0 cm	1.00	-6.5	-0.8	92.5%
Knees	25 cm	0 cm	1.00	7.5	1.0	100%
Ankles	14 cm	0 cm	1.00	18.5	2.4	100%

Start Compensator 15 cm below iliac crest

[1] from AP chest film

[2] TD (g/cm²) $= L_{ref} - L + (1 - \rho_{lung}) L_{lung}$

[3] $L_c = \dfrac{1}{2} \dfrac{\tau}{\rho comp} \times TD - \left| \dfrac{\ln K}{\mu_{eff}} \times \dfrac{\tau}{\rho comp} \right|$

$= 0.13 \times TD - |10.8 \times \ln K|$

$\rho_{lung} = 0.25$ g/cm³
$\mu_{eff} = 0.024$ cm²/g

Calc. by: _____

Checked by: _____

[4] normalized to off-axis ratio of 1.0

Fig. 31.3. The form used to record the values of right–left lateral thickness of the patient. This information is used to calculate the tissue deficits that exist at various body locations versus the umbilicus thickness. Values of compensator thickness are subsequently calculated from these data. The final column provides a location to document the percentage of the prescribed dose delivered to the midline of the indicated regions

Once the lateral separations are recorded, the point where the lower extremities compensator is to start must be determined. Since the dose is prescribed for the midline thickness at the umbilicus, and the pelvis is usually of greater thickness, the compensator must be started at some point below the pelvis. The location where the compensator is to begin is that point on the legs that has the same separation as the umbilicus. As final documentation, a photograph is taken with the patient in the treatment position with respect to the radiation field.

31.6.1
Compensators for TBI

Compensator thickness determination is calculated from measurements taken at the time of simulation. The form shown in Figure 31.3 also serves as the calculation sheet for the determination of compensator thickness at the different locations. The compensators are usually designed in three pieces: one for the lower extremities, one for the head and neck region, and one for the lungs. In most cases, a lung compensator is not required since the effective thickness at the mid-mediastinum is usually greater than that at the umbilicus. The arms are deliberately positioned in line with the lungs and act to increase the total thickness in this region.

The first step in designing tissue compensators is to determine the tissue deficit (TD), the difference in tissue-equivalent thickness between the prescription point (which in our case is the umbilicus) and the other locations. The following equation is used to calculate tissue deficit:

$$TD = L_{ref} - L + (1 - \rho_{lung}) L_{lung} \qquad (1)$$

where L_{ref} is the lateral separation at the umbilicus, L is the lateral separation at that particular anatomical location, L_{lung} is the separation of the lung determined from the anterior radiograph, and ρ_{lung} is the density of the lung. For ρ_{lung}, a value of 0.25 g/cm³ is used as the average lung density for healthy lung tissue (VAN DYK et al. 1982). Equation 1 is used only for the mid-mediastinal location where lung tissue is present. At all other locations, L_{lung} is zero and the tissue deficit can be obtained using Eq. 2:

$$TD = L_{ref} - L \qquad (2)$$

L_{lung} is determined using the anterior chest radiograph that was taken in the simulator. The lung thickness is determined at a point midway between the sternal notch and the most superior aspect of the domes of the diaphragms as seen on the radiograph. As shown in Figure 31.4, two lateral measurements are made for both the right and left lobes of the lung: the first measurement, represented by L_{Rt1} and L_{Lt1}, extends from the most lateral aspect to the most medial portion of the lung. The second measurement, represented by L_{Rt2} and L_{Lt2}, spans from the most lateral extent of the lung to the mediastinum. The lung thickness is calculated using the following equation:

$$L_{lung} = \frac{\left(L_{rt1} + L_{rt2} + L_{lt1} + l_{lt2}\right)}{2} \times \frac{\left(\dfrac{1}{2} \times (SSD + SFD)\right)}{SFD} \qquad (3)$$

where SSD is the source–skin distance and SFD is the source–film distance measured when the anterior chest radiograph was obtained during the first step of the simulation. In Eq. 3, the terms in the second parentheses serve to de-magnify the lung dimensions measured on the chest radiograph to life-size at the midplane of the patient.

The compensator thickness, L_c, is determined using the following equation:

$$L_c = \frac{1}{2} \times \frac{\tau}{\rho_{comp}} \times TD - \left(\frac{\ln K}{\mu_{eff}} \times \frac{\tau}{\rho_{comp}} \right) \quad (4)$$

In this expression, τ is the thickness ratio (KHAN et al. 1970; KIRBY et al. 1988). For beam energies from cobalt 60 (Co-60) to 10 MV, a value for τ of 0.70 is a good approximation for compensator distances greater than 20 cm from the surface (KHAN et al. 1980). The density of the compensating material is ρ_{comp} (aluminum in this case), K is the off axis correction factor that accounts for both the decreases in beam intensity away from the central axis and effective scattering field size for the various locations, and μ_{eff} is the broad-beam linear attenuation coefficient in tissue for this beam energy. The effective field size at various locations can be determined by Clarkson integration. The data in Table 31.3 were calculated in this manner for a Rando phantom and represents the equivalent scattering field at various locations of the body as a function of beam energy for an average adult (KIRBY et al. 1988). This data can be used to determine accurate values for K in Eq. 4.

An alternate method to determine compensator thickness for TBI has been suggested by the American Association of Physicists in Medicine (AAPM) (1986). Comparisons made between the system of

Table 31.3. The equivalent square field size for various anatomical locations. Determined at midline in a Rando phantom. The lateral dimensions of the phantom at these locations is also listed in addition to the equivalent lengths required to obtain the calculated field sizes by equivalent square calculation. Reproduced from KHAN et al. (1970)

Photon energy	Side of equivalent square (cm)				
	Head	Neck	Chest	Umbilicus	Hips
Co-60	17	22	31	30	28
4 MV	16	20	30	29	23
6 MV	18	23	29	27	26
10 MV	17	22	33	27	26
18 MV	>18	>18	>18	>18	>18
Mean	17	22	31	28	26
Lateral dimension	15	12–16	31	27	31
Equivalent length	19	30	31	29	20

compensator thickness determination described above and that suggested by the AAPM (1986) show that the two systems produce compensators the thickness of which varies by less than 1 mm of aluminum. Finally, the percentage difference in dose between the prescription point and other locations of the body are calculated and recorded in the last column of the form.

31.7
Compensator Design

The compensators are designed to be located at a distance of 72 cm from the virtual source of the accelerator so that appropriate compensation can be provided without the devices becoming excessively large and difficult to handle. Thus, the measurements recorded on the setup sheet in Figure 31.2 must be de-magnified from the treatment distance to the location of their use. Using the information supplied in Figures 31.2–31.4, the size of the compensators is determined in the following manner. For the lower extremities compensator, the base length for section A in Figure 31.5 is obtained by taking the ASIS to knee distance, subtracting the distance below the ASIS that the compensator is to start, and multiplying that dimension by the lateral magnification factor:

Lateral magnification factor =

$$\frac{\text{Source-Compensator tray distance}}{\left(\text{Source-axis distance-} \dfrac{L_{ref}}{2} \right)} \quad (5)$$

Section B of the lower extremities compensator is the de-magnified distance from the knee to the back of the heels, while section C is the de-magnified length of the feet plus an additional 2 cm to ensure adequate coverage of the feet. Section D is simply an additional 2 cm of aluminum that is needed to clamp the compensator to the compensator tray during the actual treatment. The thickness of the compensator is obtained from the compensator thickness column as shown on Figure 31.3 for the corresponding anatomical location.

The head and neck compensator is designed in much the same manner. The base length of section E of this compensator is the de-magnified distance from the middle of the shoulder to the top of the head. Section F is an additional 1 cm of material to ensure adequate coverage, while section G is provided for clamping. The thicknesses of the compen-

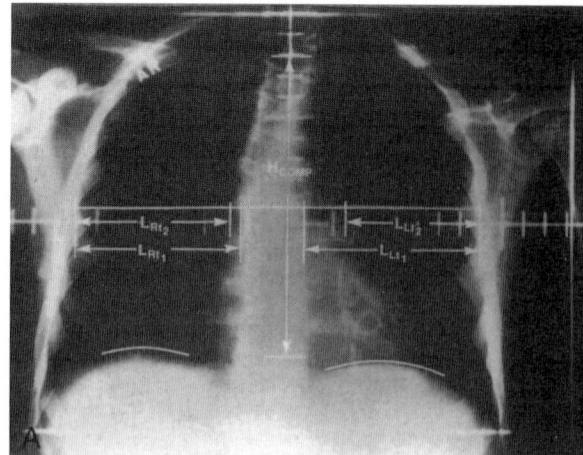

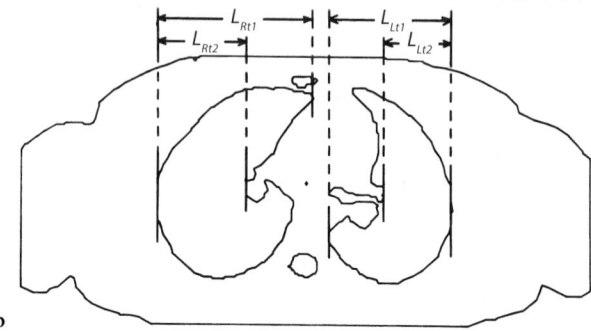

Fig. 31.4a,b. a The anterior chest radiograph obtained during simulation indicating the lateral measurements taken to determine lung thickness, L_{lung}. Also shown is the level where these measurements are taken, the midpoint between the sternal notch and the xiphoid. This dimension, H_{comp}, is also the one used if a lung compensator is required. **b** The inset, which is a diagrammatic representation of a transverse CT scan through the chest, illustrates the rationale behind these measurements

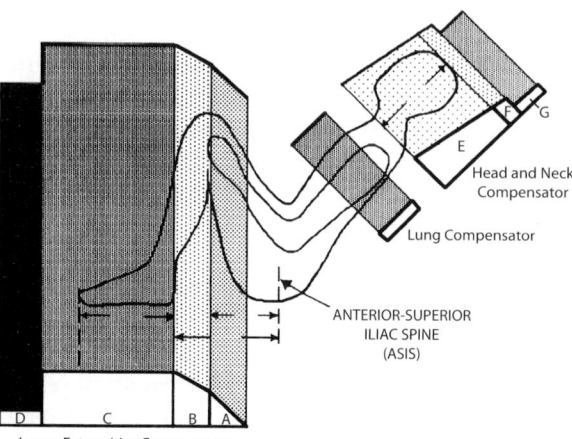

Fig. 31.5. A schematic diagram showing the relationship between the compensators used for the bilateral total-body technique and the patient treatment position

sator are again obtained from Figure 31.3 for the head and neck regions.

The length of the lung compensator is obtained by de-magnifying H_{comp} from the chest radiograph (Fig. 31.4) to life size using (1/2)(SSD–SFD)/SFD, then again de-magnifying these dimensions to the treatment position of the compensators using Eq. 5. The compensator thickness is obtained from Figure 31.3.

The width of the compensators (the dimension of the compensator that is not shown in Fig. 31.5) is typically 11 cm for the lower extremities compensator, 6.5 cm for the head and neck compensator, and 7.5 cm for the lung compensator.

Figure 31.6 shows the compensator design form that is sent to the machine shop for fabrication. Figure 31.7a shows the finished aluminum compensators and Figure 31.7b illustrates the compensators in use and how they are attached to the tray using specially designed clamps. The lung compensator, when required, is attached to the tray with double-sided tape.

Machine calibration and treatment calculation. The linear accelerator is calibrated according to the protocol outlined by the AAPM (ALMOND et al. 1999). To determine the total number of monitor units (MUs) for TBI, the calculation is done as an isocentric treatment at an extended distance. Equation 6 is used in the determination as follows:

$$MU = \left(\frac{TD \times STF}{k \times S_c(r_o) \times S_p(r_o) \times TMR(d,r_e) \times \left(\frac{f}{f'}\right)^2} \right) \quad (6)$$

In this equation, k is the machine calibration factor equal to 1 cGy per MU in tissue at d_{max} depth at the standard calibration distance, which is f, for the calibration field size of $10{\times}10$ cm^2; $S_c(r_o)$ represents the collimator scatter correction factor for r_o; the collimator field size, $S_p(r_e)$, is the phantom scatter factor for the effective scattering field, r_e; at the umbilicus, $(f/f')^2$ is the inverse square factor from the calibration distance, f, to the treatment distance, f', set to the midline of the patient; and TMR(d,r_e) is the tissue maximum ratio for the midline depth, d, for the effective field size. The accuracy of the TMR values taken at 100 cm source-axis distance has been verified at the extended treatment distance. Finally, a combined spoiler plus tray factor (STF) for both the 1-cm acrylic beam spoiler and the blocking trays that support the compensators is included.

The beam spoiler or degrader is necessary because of the large degree of skin sparing that is still present

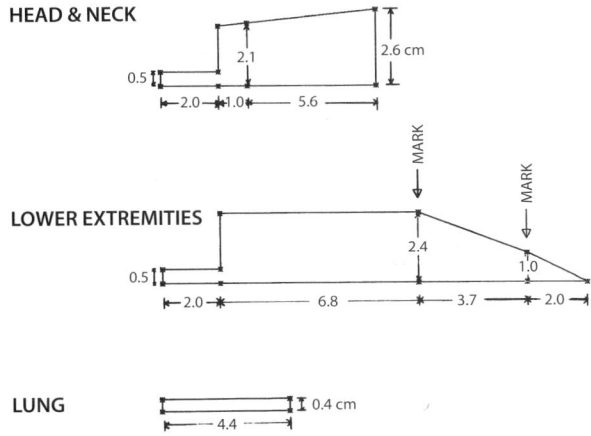

HEAD & NECK

2.1   2.6 cm

0.5

2.0 — 1.0 — 5.6

LOWER EXTREMITIES

MARK   MARK

2.4

1.0

0.5

2.0 — 6.8 — 3.7 — 2.0

LUNG

0.4 cm

4.4

Fig. 31.6. The compensator design from showing the information provided to the machine shop for fabrication of the aluminum compensators. For the position on the lower extremities compensator indicated by "Mark", a small peg is inserted into the side of the compensator. This peg aids in aligning the compensator with the knee and the ankle when the patient is in the treatment position

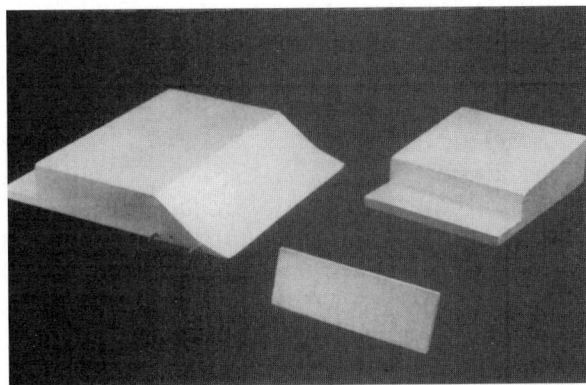

a

b

Fig. 31.7a,b. The completed aluminum compensators used for the bilateral total body technique. **a** The aluminum compensators for the lower extremities, lung, and head and neck (from left to right). **b** An illustration of the aluminum compensators attached to the block tray of the linear accelerator. The head and lower extremities compensators are attached using clamps while the lung compensator is mounted on the tray using double-sided tape

for the large field sizes and extended treatment distances employed in TBI. Figure 31.8a illustrates the buildup characteristics of 10-MV X-rays for a single incident beam with and without the beam spoiler in place. The measurements are normalized to d_{max} for the single field in this figure. Figure 31.8b shows the percentage surface dose for parallel-opposed 10-MV beams, for both open and degraded fields, normalized to the dose delivered to the midplane of a 25-cm thick patient. For both data sets, the beam spoiler was placed at a distance of 20 cm from the phantom surface. Without the beam spoiler, the dose delivered to the superficial regions of the patient could be inadequate.

31.7.1
Patient Treatment

To ensure that all the information is accurately transferred to the treatment room, the first setup of every patient is rigorously checked. The treatment position is checked for accuracy versus the data recorded on the setup sheet (Fig. 31.2). Next, it is verified that the patient is positioned within the uniformly flat portion of the radiation field. It is verified that the upper arms are properly positioned to provide shielding for the lungs so that they do not receive excessive dose, and that the forearms and hands are in line with the thighs. The fit, size, and positioning of the compensating filters are also checked to ensure that the proper amount of compensation is being applied to each anatomical region. For the lower extremities compensator, positioning is accomplished by aligning the pegs on the side of the compensator (indicated by "Mark" on the compensator design form, Fig. 31.6) with the knee and the back of the heel. The head and neck compensator is positioned so that compensation begins at the mid-shoulder and extends beyond the top of the head. The lung compensator, when required, is placed with the superior border of the compensator at the sternal notch and perpendicular to the back of the treatment couch. The final check before irradiation is to make sure that the beam spoiler is in place and that it is 20 cm or closer to the patient's most proximal surface.

31.7.2
Dose Verification

The dose delivered to the patient during TBI has been verified using both lithium fluoride (LiF) thermoluminescent dosimetry (TLD) chips and encapsulated powder. The dosimeters for the head and neck region were taped to the side of these regions and covered by 2.5 cm of wax bolus. The location of the TLDs for the mid-mediastinal readings was between the upper arm and the chest wall. The TLDs for the lower extremities were all placed between the legs at the indicated locations. The results of these measurements shown in Table 31.4 illustrates that there is fairly good dose uniformity throughout the entire treatment region when using parallel-opposed high-energy photon beams.

For routine treatment, TLD powder capsules embedded between 1-cm slabs of plastic are placed between the patient's legs as close to the groin for the first treatment. This is done to ensure that the proper dose is being delivered.

31.7.3
Dose Prescription

The usual dose prescribed using this technique is 165 cGy twice daily for 4 days for a total of eight fractions. This results in a cumulative dose of 1320 cGy. Each fraction is separated by at least 6 h. The dose rate is between 10 Gy/min and 19 cGy/min. A summary of the patient treatment schedule is shown in Table 31.5.

31.8
Anteroposterior TBI

An anteroposterior (AP) TBI technique used at the University of Minnesota is adapted from that developed at the Memorial Sloan Kettering Hospital in New York (SHANK et al. 1983). The patients are treated in either a standing or reclining position, alternating anterior and posterior surfaces for each fraction. The prescription dose is 1375 cGy to the

midplane of the pelvis delivered in 11 125-cGy fractions using 3 fractions per day at approximately 4.5-h intervals. 6-MV X-rays are used at a dose rate of 10–19 cGy/min at the midline of the pelvis, which is the prescription point. For each X-ray treatment, 2.1-cm thick cerrobend lung blocks are used to reduce the dose to the lungs by approximately 50%. The chest wall overlying the lungs, which were shielded by the lung blocks, are given an additional 600 cGy to d_{max} using electron beams of appropriate energy. The electron energy used for these chest wall boost fields is selected to place the 90% isodose line at the lung–chest wall interface. In addition, all male patients receive a testicular boost of 400 cGy to d_{max} on day 1. The electron energy for this boost is chosen to set the 90% isodose at the posterior surface of the scrotum. A summary of the patient treatment schedule is shown in Table 31.6.

The workup for the TBI patients consists of a simulation procedure to obtain lung block shape and position during treatment, measurements of patient thickness, and locating the CT scan region that will be used to determine the optimum electron energy for the chest wall boost. Details of the patient workup and treatment procedures are given below.

Simulation and patient measurements. Three steps are associated with the simulation of the patient: (1) fit the patient within the available field size, (2) take both an anterior and a posterior chest radiograph for the location of lung blocks, and (3) measure and record AP patient thickness at specific anatomical sites.

Patient positioning within the treatment field. The simulation of the patient is performed inside the treatment room with the gantry rotated to the lateral treatment position. The gantry is rotated to provide the best coverage of the patient within the visible field. However, the gantry angulation should not deviate by more than 2° from the lateral treatment position so that the proper treatment distance is maintained. The treatment room is preferred over the simulator because fitting the patient within

Table 31.4. Lithium fluoride thermoluminescent chip and disposable powder capsule measurement showing percentage of prescribed dose to various anatomical locations for the right and left lateral TBI technique. Aluminum compensators were used to account for differences in thickness. These results are for the 10-MV right–left lateral technique

	Anatomical location							
	Head	Neck	Chest wall	Pelvis	Thigh	Knee	Ankle	Oral cavity
Mean	95.5	99.8	97.8	102.1	97.3	97.2	99.9	109.0
Standard deviation	5.94	7.20	5.76	5.11	6.04	6.54	6.63	10.6
Maximum	110	122	116	111	118	113	116	143
Minimum	84	88	83	90	87	85	88	93
Number	36	35	35	36	36	36	35	20

Table 31.5. Summary of patient treatment schedule for right and left lateral total body irradiation treatment

Radiation therapy: BMT day			
Day 1: -4 TBI fractions 1, 2: 18-MV or 25-MV X-rays, 165 cGy/fraction	Day 2: -3 TBI fractions 3, 4: 18-MV or 25-MV X-rays, 165 cGy/fraction	Day 3: -2 TBI fractions 4, 6: 18-MV or 25-MV X-rays, 165 cGy/fraction	Day 4: -1 TBI fractions 7, 8: 18-MV or 25-MV X-rays, 165 cGy/fraction

Table 31.6. Summary of patient treatment schedule for standing total body technique showing timing of total body X-ray and electron treatments

BMT day			
Day 1: -7 TBI fractions 1, 2, 3: 6-MV X-rays, 125 cGy/fraction Fraction 1: testicular electron boost, 400 cGy/fraction	Day 2: -6 TBI fractions 4, 5, 6: 6-MV X-rays, 125 cGy/fraction	Day 3: -5 TBI fractions 7, 8, 9: 6-MV X-rays, 125 cGy/fraction Fraction 7: electron chest wall boost, 300 cGy/fraction	Day 4: -4 TBI fractions 10, 11: 6-MV X-rays, 125 cGy/fraction Fraction 10: electron chest wall boost, 300 cGy/fraction

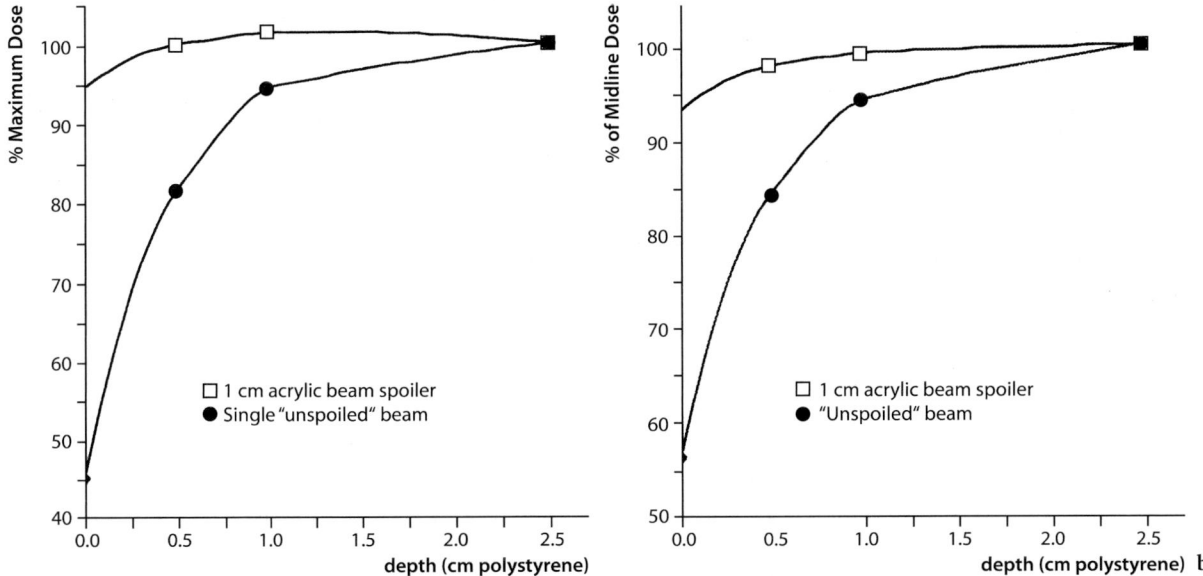

Fig. 31.8a,b. a The buildup characteristics of 10-MV X-rays for a single incident beam both with and without the beam spoiler in place. The measurements are normalized to d_{max} for the single field in this figure. **b** The percentage surface dose for parallel-opposed 10-MV beams for both open and degraded fields normalized to the dose delivered to the midplane of a 25-cm-thick patient

the available field size is a crucial step at our institution. For our treatment distance of 410 cm, the diagonal field size is 170 cm (5'8") inside the 90% isodose line. Patients shorter than 170 cm (5'8") can be easily treated in the standing position. However, it is necessary for taller patients to sit on the seat of the treatment stand in order to fit within the treatment field. Although this is not the optimum treatment position, acceptable dose uniformity can still be achieved.

Films for lung blocks. Once the position of the patient within the treatment field has been determined, chest radiographs are taken. The anterior film is taken with a small BB placed at the tops of the diaphragms. The distance of this BB below the sternal notch is measured and recorded on the form illustrated in Figure 31.9. Additionally, a posterior radiograph is taken with a BB placed at C7. The location of the BB is indicated on the form and marked with a tattoo. Also recorded are the gantry angle

STANDING TOTAL BODY IRRADIATION TECHNIQUE SETUP SHEET

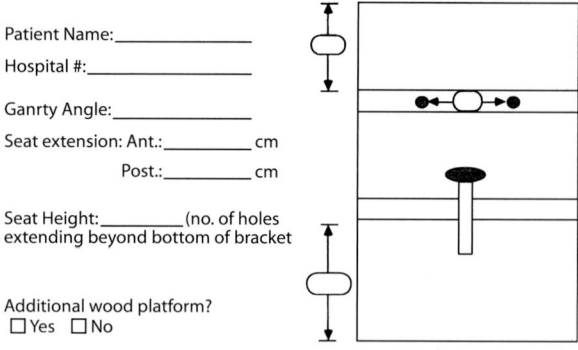

Patient Name:_____

Hospital #:_____

Ganrty Angle:_____

Seat extension: Ant.:_____ cm

Post.:_____ cm

Seat Height:_____ (no. of holes extending beyond bottom of bracket

Additional wood platform?
☐ Yes ☐ No

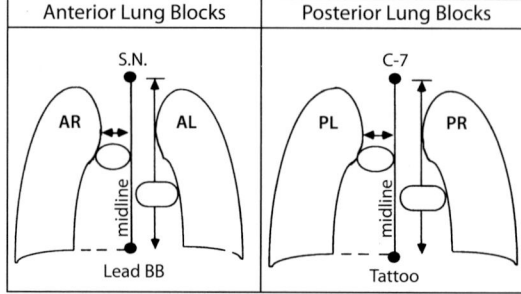

Anterior Lung Blocks	Posterior Lung Blocks
S.N.	C-7
AR AL	PL PR
midline	midline
Lead BB	Tattoo

Fig. 31.9. The form used to document the patient treatment position when using the standing total body technique

of the accelerator, the seat extension and height, the separation of the supports located under the arms, and the position of the hand rests. For shorter patients, an additional wooden platform is placed below their feet to position them more in the center

of the beam. The information on this form is later used to duplicate the patient treatment position.

Patient thickness measurements. Following the radiographs, AP separations are measured at the head, neck, sternal notch, mid-mediastinum, umbilicus, pelvis, knees, and ankles. The target dose is prescribed at the midplane thickness of the pelvis. The names of the individuals who made and checked the measurements are recorded.

Treatment planning computed tomography (CT) scans. The staff physician next outlines the lung blocks on both the anterior and posterior chest radiographs. For adult patients, the blocks are drawn so that there is a 2-cm margin between both the diaphragm and edge of the vertebrae and a 1.5-cm gap between the edge of the block and the rib cage. Once these lung blocks are indicated on the films, the patient is taken for treatment planning CT scans.

A CT scan is then taken through the region of the lung blocks. Treatment planning is performed on the CT scans to ensure proper dose coverage. Ultrasound scans are occasionally performed instead of CT scans, when the values of chest wall thickness of the patient lying supine compared with in the standing position are significantly different, for instance in women with pendulous breasts.

Once the scanning is completed, computerized treatment planning is performed to determine the appropriate electron beam energy for the electron chest wall boost. This is done by placing the 90% isodose line at the lung chest wall interface. A typical treatment plan is illustrated in Figure 31.10.

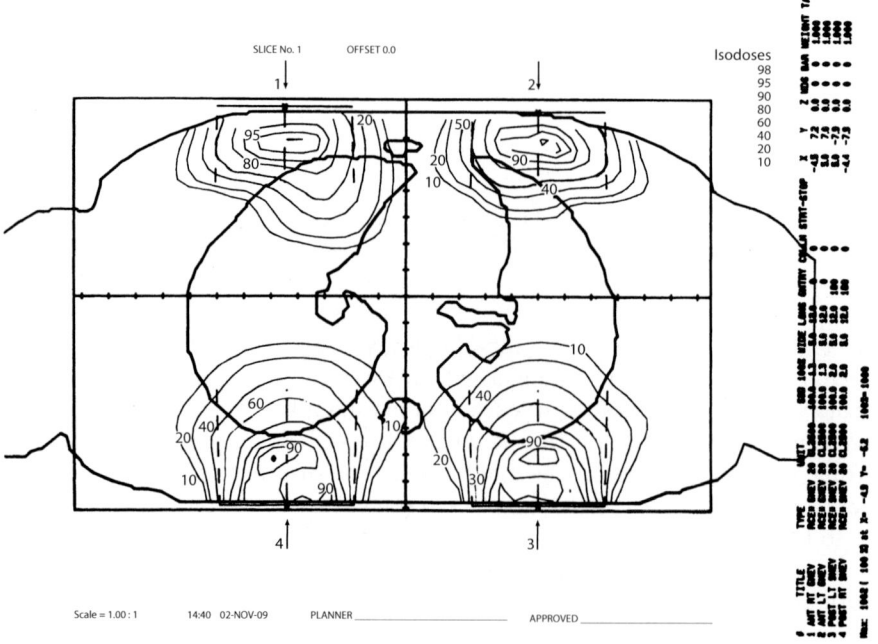

Fig. 31.10. A computerized treatment plan done to determine the proper electron energy for the chest wall boost fields. The objective is to place the 90% isodose line at the lung–chest wall interface

Construction of lung blocks and electron cut-outs. The lung blocks are constructed from the outlines drawn on the anterior and posterior chest films and are held together by a plastic plate that maintains the proper block separation (Fig. 31.11a). The cerrobend blocks are 2.1 cm thick, which is approximately the half-value thickness, including scatter, for 6-MV photons. The cerrobend cut-out used for the electron chest wall boost fields is an exact negative of the anterior and posterior lung blocks (Fig. 31.11b).

31.8.1
Patient Treatment Technique

Total body photon irradiation. The treatment is delivered using 6-MV X-rays with a dose rate between 10 cGy/min and 19 cGy/min at the midplane of the pelvis. The patient is treated in the standing position resting against the back plate of the total body treatment stand that was specifically designed for this treatment (Fig. 31.12). The treatment distance for this particular setup is 410 cm source–axis distance (SAD) to the midline of the pelvis.

The lungs are shielded with the appropriate lung blocks throughout the total body photon portion of the treatment. The lung blocks are hung by a Lexan hook from the anterior plate of the total body treatment stand. The location of the top of the lung blocks is positioned at the level of the skin tattoos and is verified with measurements from bony landmarks. Verification films confirm the positioning of the blocks during the photon treatments. A 1-cm-thick acrylic beam spoiler is placed between the patient and the beam to produce a high dose on the patient's skin surface. The screen is located 20 cm or less from the patient surface. The skin dose with this location of the beam spoiler is approximately 92% of the delivered midline dose for the 6-MV beam.

Electron chest wall boost. For fractions 7 and 10, an electron chest wall boost is given to that portion of the chest wall that was shielded by the lung blocks. A special couch extension has been designed so that both adult and pediatric patients are in the same upright treatment position for the chest wall boost fields as they were for the standing total body treatments (Fig. 31.13). The prescribed dose is 600 cGy to d_{max}, delivered in two 300-cGy fractions. The selection of electron energy and the need for bolus is based on the results of computerized treatment planning using the CT scan so that the 90% isodose line is placed at the lung–chest wall interface.

Fig. 31.11a,b. The apparatus used to shield the lungs during standing total body irradiation. **a** The proper separation of the two cerrobend lung blocks is maintained by the acrylic plate. The Lexan hook is used to suspend the blocks from the plastic plate that is placed in front of the patient when positioned in the total body treatment stand. This arrangement allows easy adjustment of both the height of the blocks and their right–left placement with respect to the patient. **b** The cerrobend insert that is used for the chest wall electron boost treatment exactly matches the shape of the cerrobend lung blocks

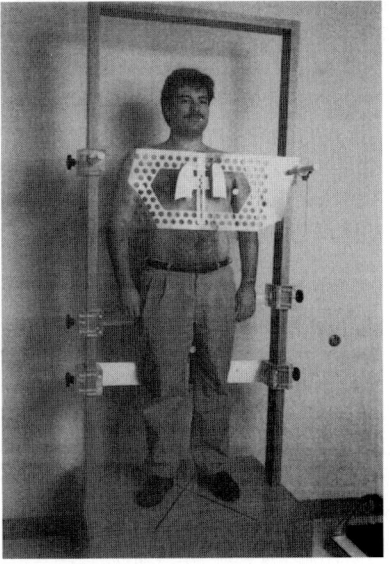

Fig. 31.12. The standing total body treatment position with the back of the patient resting against the back plate of the total body treatment stand. The lung blocks are shown, in position, hanging from the front acrylic plate

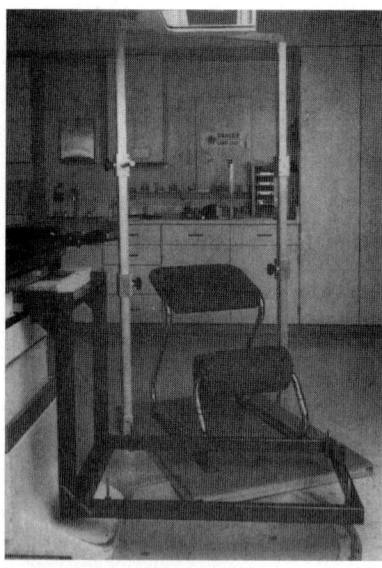

Fig. 31.13. A special couch extension designed to reproduce the standing TBI treatment position when treating the electron chest wall boost fields. The device is separated into two pieces for ease of handling and, since it is attached directly to the couch, it has the same range of motion as the couch. Patients up to approximately 5 feet tall can be treated in the standing position, while taller patients are treated in a seated position. This style of chair keeps the back of the patient in about the same orientation when seated as when they are standing

Testicular boost. Male patients are given a testicular electron boost on the first day of treatment. The prescribed dose is 400 cGy to d_{max} in one fraction. The patient is treated in the supine position with a sheet of lead placed under the testes to minimize the dose to the rectal area. A 6-mm-thick sheet of wax bolus is placed between the lead and the posterior surface of the scrotum to reduce the amount of backscatter from the lead. The electron energy is based on the thickness of the testes and is chosen so that the 90% isodose line is at the posterior surface of the scrotum.

Infant irradiation. Total body treatments for infants are done with the patient supine on a separate treatment couch positioned on the floor. We have found that the treatments are best performed with sedation or anesthesia. The gantry is directed vertically down for these cases and the collimator is rotated 45° to produce the largest available field size. A 1-cm acrylic beam spoiler is positioned approximately 20 cm above the torso of the patient both to provide a high surface dose and to support the lung blocks used for the anterior and posterior X-ray fields. The lower extremities are simply bolused to provide a high skin surface dose (Fig. 31.14). The chest wall

electron boost is also delivered with the patient in the supine position.

LiF TLD was performed on several patients to establish the homogeneity of dose throughout the treatment field. The TLD chips were covered by approximately 1 cm of bolus to indicate the dose at d_{max} at these locations and were placed at the same locations for both anterior and posterior treatments. The results of the measurements, shown in Table 31.7, indicate an acceptable level of dose homogeneity for this treatment technique.

31.9
Normal Tissue Shielding

Shielding of normal tissues must be carefully considered in TBI because shielding may potentially reduce the dose to the target volume (bone marrow cells, leukemic cells, and circulating stem cells). Despite this concern, there are situations in which partial shielding of critical tissues, including the lungs, kidneys, eyes (lens), and brain, is considered.

31.9.1
Lung Shielding

Because pneumonitis is a leading cause of death after SCT, with total dose of TBI implicated as a potential contributing cause, partial blocking of the lung has been advocated. The dose received by the

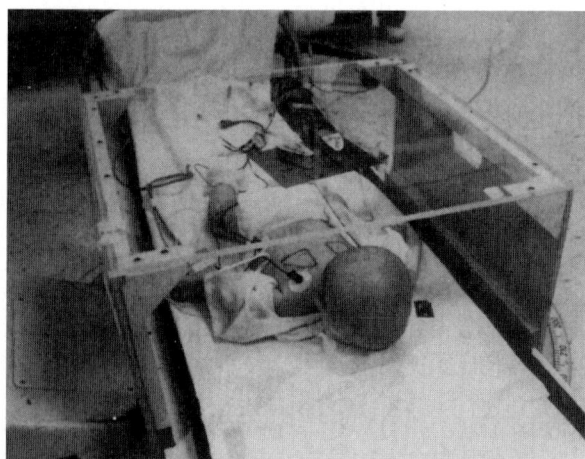

Fig. 31.14. An illustration showing the treatment position used for pediatric patients. The lung blocks are placed on top of the acrylic beam spoiler. The lower extremities are bolused to provide a high dose at the surface

Table 31.7. Lithium fluoride thermoluminescent dosimeters (TLD) establish homogeneity of dose for standing TBI technique. The TLD chips were covered by approximately 1 cm of bolus to indicate the dose at d_{max} at these locations. The chips were in place during both anterior and posterior treatments. *SD* standard deviation, *D* deviation

Anatomical Location	Prescription dose (%)				
	Mean	SD	Highest D	Lowest D	*n*
Umbilicus	97.4	4.21	104	90	27
Right palm – opposite knuckles	113.9	9.15	134	95	27
Right palm – heel of hand	104.7	8.76	117	90	6
Between breasts	101.0	5.97	118	92	27
Right hip	113.0	9.41	137	98	26
Left inner thigh	107.9	9.59	131	93	25
Perineal	105.2	6.38	120	93	23
Left outer ankle	112.3	11.6	130	103	7
Sternal notch	103.6	6.36	113	85	27
Forehead	97.0	7.44	109	83	26
Left lateral calf	111.7	8.77	128	92	27
Top of head	109.7	11.6	136	96	18
Under lung block	62.5	9.79	86	48	26
Neck-thyroid notch	100.8	6.37	110	92	6

lungs is influenced by both the irradiation geometry as well as lung density. At the University of Minnesota, when delivering TBI with right and left lateral fields, the arms are placed at the sides; the thickness of the arms is considered in determining whether additional compensation is needed to reduce the lung dose to within 10% of the dose received at the prescription point (level of the umbilicus at midplane). Often no additional compensation is needed to achieve this goal, but if needed, tissue compensators are placed to reduce the lung dose to within 10% of the prescription dose.

Using a similar right and left lateral technique, in addition to using the arms to decrease lung dose, some institutions use partial blocks to reduce the lung dose further, usually to an arbitrary amount (for instance 1000 cGy).

For AP–posteroanterior (PA) fields, partial attenuation blocks (80–90% transmission) or thicker blocks (usually one half-value layer; HVL) can be placed in front of the beam to decrease the lung dose to the desired amount. With one HVL block, the underlying ribs receive approximately half of the prescription dose and electron beams of the appropriate energy can be used to "boost" the underlying ribs. A CT scan through the lung can be used to determine the appropriate electron energy. This technique was initially reported at Memorial Sloan Kettering Cancer Center (SHANK et al. 1983) and was used at the University of Minnesota for about 10 years, although has now largely been abandoned as it is considerably more difficult to administer. Some institutions, in an effort to decrease the risk

of pneumonitis, omit the electron beam chest wall boost. No increased risk of leukemia relapse has been noted, but a prospective trial is lacking.

31.9.2
Kidney Shielding

The risk of renal injury after SCT is dependent on multiple factors, including previous chemotherapy use of nephrotoxic antibiotics, and therapies directed at the prevention and treatment of GVHD (MOULDER et al. 1987, 1988; LAWTON et al. 1989, 1992; MOULDER and FISH 1989, 1991; EMMINGER et al. 1991; COWEN et al. 1992; COLE et al. 1994; MIRALBELL et al. 1996). In a recent review of the incidence of acute renal failure in patients treated at the University of Minnesota SCT program, up to 30% of patients undergoing SCT in 1993 had acute renal failure, defined as a doubling of creatinine over the baseline creatinine (LANE et al. 1994). Of these patients, 10% required dialysis.

Late-onset renal failure occurs in up to 20% of survivors of SCT. On the beneficial effect of partial kidney blocking in the setting of T-depleted SCT, Lawton (LAWTON et al. 1992) found that the incidence of chronic renal failure was reduced from 26% to 6% when posterior 1-HVL renal blocks were placed, reducing the estimated kidney dose from 14 Gy to 12 Gy (given at 200 cGy twice a day). In another series of 79 patients transplanted with TBI-containing regimens, Miralbell et al. (MIRALBELL et al. 1996) reported that the 18-month probability of renal dysfunction-free survival decreased from 95, to 74, to 55% for patients conditioned with 10, 12, and 13.5 Gy, respectively. The other factor that predicted for renal dysfunction was the risk of developing GVHD. Renal dysfunction-free survivals were 93% for patients at lower risk of GVHD and 52% for patients with a high GVHD risk (e.g., unrelated allogeneic SCT, absence of T-cell depletion).

31.10
Gonad Shielding

A common late complication after SCT is sterility. In certain diseases, such as acute lymphoblastic leukemia, the gonads are considered sanctuary sites, and shielding would possibly increase the risk of relapse (QUARANTA et al. 2004). In other situations, especially SCT for non-malignant diseases, there is probably less risk. The challenge in shielding either of the testes is to use sufficient attenuating material and to do it in such as way as to minimally shield marrow sites. The challenges in shielding the ovaries are even more complicated because they lie in the pelvis near a rich supply of marrow; they are difficult to visualize, especially in young girls; and they are mobile and may move between the planning process and treatment days. Despite these obstacles, we have attempted to decrease the dose to the gonads on occasion, usually at the request of a parent or as part of a protocol. For patients transplanted for sickle cell anemia or thalassemia, the SCT goal is to provide a supply of normal red blood cells. Even partial engraftment ameliorates the symptoms of the disease. For this protocol, we localize the ovaries by ultrasound and use five HVL cerrobend blocks anteriorly and posteriorly to decrease the dose to the ovaries. The testicles are placed in a cerrobend "clam shell". Instead of TBI, total lymphoid irradiation to a dose of 500 cGy is used.

31.10.1
Thymus Shielding

At the University of Minnesota, we are conducting a trial in which the thymus is blocked in the hopes of speeding immune reconstitution in patients with Fanconi Anemia undergoing unrelated donor transplantation (STOREK et al. 2003). The thymus is difficult to visualize, especially in older children. For this protocol, a treatment planning CT scan is performed

with the patient in the TBI treatment position, if possible. Intravenous contrast is given and the thymus is delineated. Patients treated via this protocol are treated with AP and PA total body fields with 5-HVL cerrobend thymus blocks positioned over the thymus gland in both the anterior and posterior fields (Fig. 31.15). To design these blocks, a 1-cm margin is placed around the outline of the thymus on the AP and PA CT contours. Additionally, aluminum compensators are placed over the lungs with both the AP and PA fields to diminish the lung dose to be no more than the prescription point dose. Thus far, only a handful of patients have been treated on this regimen and whether there is speedier immune reconstitution remains to be seen.

31.10.2
TomoTherapy

It is theoretically desirable to deliver radiation only to the immune organs and bone marrow spaces while sparing sensitive structures such as the brain, lens, lung, and kidneys. Intensity modulated radiotherapy (IMRT) planning could accomplish this, but most systems are limited by field size issues. Additionally, accurate IMRT depends on a reproducible patient position, which is complicated when considering treating the entire marrow spaces. The

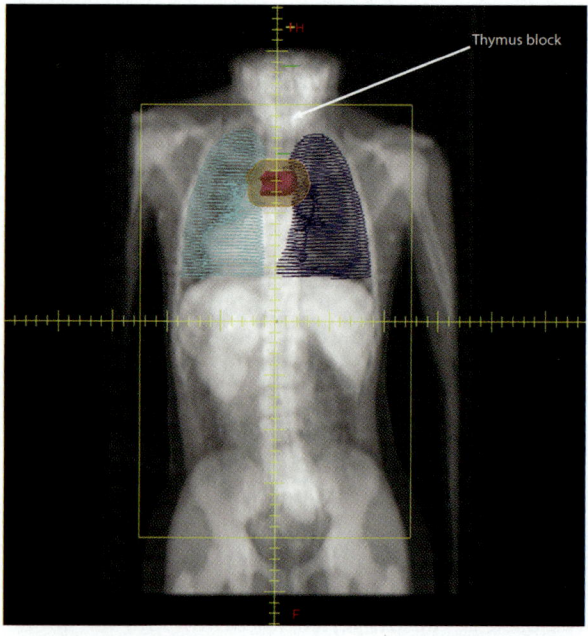

Fig. 31.15. Distally reconstructed radiograph showing location of thymus block (5 HVL) and lung block (1 HVL) used in Fanconi anemia

technology to accomplish this is now available with TomoTherapy (Mackie et al. 2003; Beavis 2004). The TomoTherapy radiation system is a linear accelerator mounted in the head of a spiral CT unit. IMRT can be delivered as beams spiral down the axis of a patient supine on the treatment couch. The beams can be planned to deliver dose to the bones and bone marrow, liver and spleen as well as major nodal groups and to relatively spare the lungs and kidneys (Fig. 31.16). Prior to treatment, a conformation CT is performed and the patients position on the treatment couch verified and adjusted, so as to match the position the patient was in for planning. There are technical hurdles to overcome in order to accomplish this, but it holds the promise of possibly being able to decrease the toxicity of TBI and increase the dose to the tumor and marrow sites thus decreasing the risk of non-engraftment or relapse.

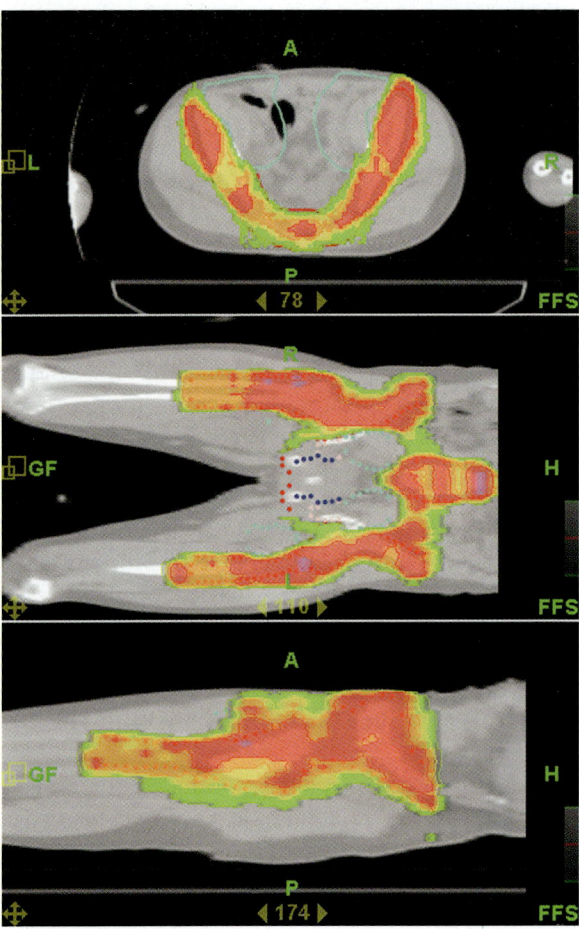

Fig. 31.16. Example of the isodose distribution achieved with TomoTherapy. Because this is delivered locally, an extremely conformal dose distribution can be achieved, thus delivering the dose only to the narrow spaces, while avoiding irradiation of sensitive structures (lung, kidney, liver)

31.10.2.1
Special Considerations for TBI in Young Children

A fundamental requirement for TBI is immobility during treatment. Whereas many children even as young as 3 years of age are able to cooperate and remain immobile with the encouragement of their parents, many children must be anesthetized. We usually try to determine at the time of simulation whether a patient will require anesthesia. Clues about whether a patient will be cooperative include whether the patient willingly leaves his or her parents for the measurements, whether they listen to the instructions the therapist gives, and how anxious they appear. Additionally, the parents usually have a good indication of how their child has done with other medical procedures. In an effort to avoid anesthesia, we sometimes arrange for a potential TBI patient come to the department for several consecutive days so he or she can become acquainted with our therapists. The therapist will spend 10–15 min with the patient in the treatment room practicing for the TBI treatment. As there are intercoms and video monitors, the therapist can place the patient in the treatment position, leave the room, and talk to the patient over the intercom. With practice, even young patients are often able to be spared anesthesia.

Obviously, there are situations where anesthesia is necessary. Anesthesia for TBI presents unique situations not ordinarily encountered by most anesthesiologists. Foremost is the fact that the anesthesiologist cannot be in the treatment room during the TBI. Additionally, for the AP/PA technique, the

patient is prone, and the airway is more difficult to keep patent.

If TBI under general anesthesia is scheduled, the patient fasts for 6 h before the scheduled procedure. For infants, an interval of 4 h from intake of formula is sufficient. On arrival to the radiation therapy room, atropine and propofol are administered intravenously. Dolasetron is effective at preventing nausea during and after the treatment.

As the patient loses consciousness, a blood pressure cuff is placed around the upper arm and the electrocardiogram (ECG) is monitored continuously. Pulse oximetry is used for continuous monitoring of oxygen saturation. Supplemental oxygen is administered with nasal prongs. A continuous drip of propofol is started. The patient is placed in the treatment position. The child's head is fixed firmly in position by a sponge rubber donut or Styrofoam. Adhesive tape is used to secure the head in the appropriate position so that airway patency and ventilation are secured.

Two closed-circuit television cameras are focused on the patient and on the physiological monitor console. When the patient is ready for irradiation, all attendant personnel withdraw from the treatment room. During the treatment, airway and respiratory adequacy are observed constantly by means of a zoom television monitor system, while blood pressure, ECG, and pulse oximetry are monitored on the second television monitor.

After the treatment is complete, the patient is transferred to the post-anesthetic recovery room, where surveillance is continued until full arousal occurs. Patients who receive several treatments on consecutive days show increased tolerance to propofol and the dose may need to be increased accordingly.

31.10.2.2
Complications Following Preparation With TBI

The major causes of morbidity and mortality following a SCT are infectious complications (AUNER et al. 2002). Additionally, interstitial pneumonitis develops in up to 20% of transplanted patients depending on the source of the marrow and previous therapies received (CARDOZO et al. 1985; ABRAHAM et al. 1999; EMMANOUILIDES et al. 2003; CARRUTHERS and WALLINGTON 2004). Acute side effects of TBI include nausea and vomiting, alopecia, diarrhea, low grade fever, mucositis, and pancytopenia. Intermediate side effects include interstitial pneumonitis, veno-occlusive disease, and nephrotoxicity. Late side effects include restrictive lung disease, possible decreased growth, endocrine abnormalities (especially hypothyroidism) sterility, cataracts, chronic renal failure, and neurological damage. Sanders reported that boys given single-fraction TBI were significantly shorter than boys given fractionated TBI ($P<0.03$). The same (non-significant) trend was demonstrated in girls.

Few studies of neuropsychiatric testing of patients treated after TBI exist. One might expect lower cumulative doses to be associated with less impairment, but data are lacking. Younger patients seem to be at a higher risk of neurological damage (FARACI et al. 2002; RUBIN et al. 2005).

The incidence of second tumors after SCT is low; Seattle reported 4 in 1800 patients. At the University of Minnesota, 53 second malignant neoplasms developed among 2150 patients for an estimated risk of 9.9% at 1–3 years after transplantation (BHATIA et al. 1996). Second neoplasms were more common in patients likely to have GVHD.

References

AAPM, American Association of Physicists in Medicine (task Group 2) (1986) The physical aspects of total and half body photon irradiation. AAPM Report no 17

Abraham R et al. (1999) Intensification of the stem cell transplant induction regimen results in increased treatment-related mortality without improved outcome in multiple myeloma. Bone Marrow Transplant 24:1291–1297

Almond PR et al. (1999) AAPM's TG-51 protocol for clinical reference dosimetry of high-energy photon and electron beams. Med Phys 26:1847

Aristei C et al. (2002) Cataracts in patients receiving stem cell transplantation after conditioning with total body irradiation. Bone Marrow Transplant 29:503–507

Auner HW et al. (2002) Infectious complications after autologous hematopoietic stem cell transplantation: comparison of patients with acute myeloid leukemia, malignant lymphoma, and multiple myeloma. Ann Hematol 81:374–377

Beavis AW (2004) Is tomotherapy the future of IMRT? Br J Radiol 77:285–295

Bhatia S et al. (1996) Malignant neoplasms following bone marrow transplantation. Blood 87:3633–3639

Cardozo BL et al. (1985) Lung damage following bone marrow transplantation: I. the contribution of irradiation. Int J Radiat Oncol Biol Phys 11:907–914

Carlson K et al. (1994) Pulmonary function and complications subsequent to autologous bone marrow transplantation. Bone Marrow Transplant 14:805–811

Carruthers SA, Wallington MM (2004) Total body irradiation and pneumonitis risk: a review of outcomes. Br J Cancer 90:2080–2084

Cole CH et al. (1994) Intensive conditioning regimen for bone marrow transplantation in children with high-risk haematological malignancies. Med Pediatr Oncol 23:464–469

Colombat P et al. (2000) Value of autologous stem cell transplantation with purged bone marrow as first-line therapy for follicular lymphoma with high tumor burden: a GOELAMS phase II study. Bone Marrow Transplant 26:971–977

Cornetta K et al. (2005) Umbilical cord blood transplantation in adults: results of the prospective Cord Blood Transplantation (COBLT). Biol Blood Marrow Transplant 11:149–160

Cowen D et al. (1992) Regimen-related toxicity in patients undergoing BMT with total body irradiation using a sweeping beam technique. Bone Marrow Transplant 10:515–519

Dusenbery KE et al. (1996) Autologous bone marrow transplantation in acute myeloid leukemia: the University of Minnesota experience. Int J Radiat Oncol Biol Phys 36:335–343

Emmanouilides C et al. (2003) Localized radiation increases morbidity and mortality after TBI-containing autologous stem cell transplantation in patients with lymphoma. Bone Marrow Transplant 32:863–867

Emminger W et al. (1991) Is treatment intensification by adding etoposide and carboplatin to fractionated total body irradiation and melphalan acceptable in children with solid tumors with respect to toxicity? Bone Marrow Transplant 8:119–123

Faraci M et al. (2002) Severe neurologic complications after hematopoietic stem cell transplantation in children. Neurology 59:1895–1904 [erratum (2003) 60:1055]

Fehr T, Sykes M (2004) Tolerance induction in clinical transplantation. Transplant Immunology 13(2): p. 117–30.

Freedman AS et al. (1998) High-dose chemoradiotherapy and anti-B-cell monoclonal antibody-purged autologous bone marrow transplantation in mantle-cell lymphoma: no evidence for long-term remission (comment). J Clin Oncol 16:13–18

Gatti RA et al. (1968) Immunological reconstitution of sex-linked lymphopenic immunological deficiency. Lancet 2:1366–1369

Hall EJ (1994) Radiobiology for the radiologist, 4th edn. J.B. Lippincott, Philadelphia. xii, p 478

Hawley RG et al. (1996) Retroviral vectors for production of interleukin-12 in the bone marrow to induce a graft-versus-leukemia effect. Ann NY Acad Sci 795:341–345

Heslop HE, Rooney CM, Brenner MK (1995) Gene-marking and haemopoietic stem-cell transplantation. Blood Rev 9:220–225

Iannone R et al. (2003) Results of minimally toxic nonmyeloablative transplantation in patients with sickle cell anemia and beta-thalassemia. Biol Blood Marrow Transplant 9:519–528

Imamura M, Hashino S, Tanaka J (1996) Graft-versus-leukemia effect and its clinical implications. Leuk Lymphoma 23:477–492

Karanes C (2003) Unrelated donor stem cell transplant: donor selection and search process. Pediatr Transplant 7[Suppl 3]:59–64

Khan FM, Moore VC, Burns DJ (1970) The construction of compensators for cobalt teletherapy. Radiology 96:187–192

Khan FM et al. (1980) Basic data for dosage calculation and compensation. Int J Radiat Oncol Biol Phys 6:745–751

Kim TH et al. (1985) Interstitial pneumonitis following total body irradiation for bone marrow transplantation using two different dose rates. Int J Radiat Oncol Biol Phys 11:1285–1291

Kirby TH, Hanson WF, Cates DA (1988) Verification of total body photon irradiation dosimetry techniques. Med Phys 15:364–369

Klingemann HG (1996) Role of postinduction immunotherapy in acute myeloid leukemia. Leukemia 10[Suppl 1]:S21–S22

Korbling M et al. (1991) Autologous blood stem cell (ABSCT) versus purged bone marrow transplantation (pABMT) in standard risk AML: influence of source and cell composition of the autograft on hemopoietic reconstitution and disease-free survival. Bone Marrow Transplant 7:343–349

Krishnamurti L et al. (2003) Availability of unrelated donors for hematopoietic stem cell transplantation for hemoglobinopathies. Bone Marrow Transplant 31:547–550

Lane PH et al. (1994) Outcome of dialysis for acute renal failure in pediatric bone marrow transplant patients. Bone Marrow Transplant 13:613–617

Lawton CA et al. (1989) Technical modifications in hyperfractionated total body irradiation for T-lymphocyte deplete bone marrow transplant. Int J Radiat Oncol Biol Phys 17:319–322

Lawton CA et al. (1992) Influence of renal shielding on the incidence of late renal dysfunction associated with T-lymphocyte deplete bone marrow transplantation in adult patients. Int J Radiat Oncol Biol Phys 23:681–686

Leshem B, Vourka-Karussis U, Slavin S (2000) Correlation between enhancement of graft-versus-leukemia effects following allogeneic bone marrow transplantation by rIL-2 and increased frequency of cytotoxic T-lymphocyte precursors in murine myeloid leukemia. Cytokines Cell Mol Ther 6:141–147

Mackie TR et al. (2003) Image guidance for precise conformal radiotherapy. Int J Radiat Oncol Biol Phys 56:89–105

Mahmoud HK et al. (1985) Bone marrow transplantation for chronic granulocytic leukaemia. Klin Wochenschr 63:560–564

McCullough J et al. (1989) The National Marrow Donor Program: how it works, accomplishments to date. Oncology (Huntingt) 3:63–68

Miralbell R et al. (1996) Renal toxicity after allogeneic bone marrow transplantation: the combined effects of total-body irradiation and graft-versus-host disease. J Clin Oncol 14:579–585

Molls M, Budach V, Bamberg M (1986) Total body irradiation: the lung as critical organ. Strahlentherapie und Onkologie 162:226–232

Moulder JE, Fish BL (1989) Late toxicity of total body irradiation with bone marrow transplantation in a rat model. Int J Radiat Oncol Biol Phys 16:1501–1509

Moulder JE, Fish BL (1991) Influence of nephrotoxic drugs on the late renal toxicity associated with bone marrow transplant conditioning regimens. Int J Radiat Oncol Biol Phys 20:333–337

Moulder JE, Fish BL, Abrams RA (1987) Renal toxicity following total-body irradiation and syngeneic bone marrow transplantation. Transplantation 43:589–592

Moulder JE, Fish BL, Holcenberg JS, Cheng M (1988) Effect of total-body irradiation with bone marrow transplantation on toxicity of cisplatin. NCI Monogr 6:29–33

Parr MD, Messino MJ, McIntyre W (1991) Allogeneic bone marrow transplantation: procedures and complications. Am J Hospital Pharmacy 48:127–137

Peters C et al. (2003) Hematopoietic cell transplantation for inherited metabolic diseases: an overview of outcomes and practice guidelines. Bone Marrow Transplant 31:229–239

Petersdorf E et al. (1998) Effect of HLA matching on outcome of related and unrelated donor transplantation therapy for chronic myelogenous leukemia. Hematol Oncol Clin North Am 12:107–121

Quaranta BP et al. (2004) The incidence of testicular recurrence in boys with acute leukemia treated with total body and testicular irradiation and stem cell transplantation. Cancer 101:845–850

Remberger M et al. (2002) The graft-versus-leukaemia effect in haematopoietic stem cell transplantation using unrelated donors. Bone Marrow Transplant 30:761–768

Rubin J et al. (2005) Acute neurological complications after hematopoietic stem cell transplantation in children. Pediatr Transplant 9:62–67

Schleuning M (2000) Adoptive allogeneic immunotherapy – history and future perspectives. Transfusion Sci 23:133–150

Schouten HC et al. (2000) The CUP trial: a randomized study analyzing the efficacy of high dose therapy and purging in low-grade non-Hodgkin's lymphoma (NHL). Ann Oncol 11[Suppl 1]:91–94

Shank B et al. (1983) Hyperfractionated total body irradiation for bone marrow transplantation. Results in seventy leukemia patients with allogeneic transplants. Int J Radiat Oncol Biol Phys 9:1607–1611

Standke E (1989) Fundamentals, trends and our experiences with total body irradiation (TBI) before bone marrow transplantation (BMT). Folia Haematologica – Internationales Magazin fur Klinische und Morphologische Blutforschung 116:481–485

Steingrimsdottir H et al. (2000) Immune reconstitution after autologous hematopoietic stem cell transplantation in relation to underlying disease, type of high-dose therapy and infectious complications. Haematologica 85:832–838

Storek J et al. (2003) Interleukin-7 improves CD4 T-cell reconstitution after autologous CD34 cell transplantation in monkeys. Blood 101:4209–4218

Thomas ED (1997) Pros and cons of stem cell transplantation for autoimmune disease. J Rheumatol 48[Suppl]:100–102

Valls A et al. (1989) Total body irradiation in bone marrow transplantation: fractionated vs single dose. Acute toxicity and preliminary results. Bulletin du Cancer 76:797–804

van Besien K et al. (2003) Comparison of autologous and allogeneic hematopoietic stem cell transplantation for follicular lymphoma. Blood 102:3521–3529

Van Dyk J, Keane TJ, Rider WD (1982) Lung density as measured by computerized tomography: implications for radiotherapy. Int J Radiat Oncol Biol Phys 8:1363

Wood KJ, Prior TG (2001) Gene therapy in transplantation. Curr Opin Mol Therapeut 3:390–398

Zecca M et al. (2002) Chronic graft-versus-host disease in children: incidence, risk factors, and impact on outcome. Blood 100:1192–1200

32 Radiotherapy for Hodgkin's Disease

Chung K. K. Lee

CONTENTS

32.1   Diagnostic Evaluation and Staging   805
32.2   Histopathological Classification   808
32.3   General Treatment Considerations   809
32.3.1   Treatment of Stage-I and -II Favorable Disease   809
32.3.2   Special Considerations for LMM   811
32.3.3   Delivery and Dose of CMT   811
32.3.4   Advanced Stage Hodgkin's Disease   812
32.4   Radiation Therapy Techniques   814
32.4.1   Mantle Fields   815
32.4.2   Simulation of Mantle Field   815
32.4.2.1 Anterior Fields   818
32.4.2.2 Posterior Fields   819
32.4.3   Design Shielding   820
32.4.4   Subdiaphragmatic Fields   823
32.4.5   Preauricular and Waldeyer's Ring Fields   823
32.4.6   Gap of Matching Fields   824
32.5   Limited Field (Involved Field/Regional Field)   824
32.6   Limited Field Above the Diaphragm   826
32.6.1   Cervical/Supraclavicular Region   826
32.6.1.1 Unilateral Cervical/Supraclavicular Region   826
32.6.1.2 Bilateral Cervical/Supraclavicular Region   827
32.6.2   Mediastinum/Hilar/Axillary Region   827
32.7   Limited/Regional Field Below the Diaphragm   828
32.7.1   Inguinal/Femoral/External Iliac
         Lymph Node Area   828
32.7.2   Paraaortic Lymph Nodes   828
32.7.3   Spleen   829
32.8   Radiation Dose and Fractionation   829
32.8.1   Radiation Therapy Alone   829
32.8.2   Radiation Dose in CMT   829
32.8.3   Radiation Dose in Salvage Treatment   830
32.9   Normal Tissue Tolerance and Complications   830
32.9.1   Lung   830
32.9.2   Heart   830
32.9.3   Central Nervous System   830
32.9.4   Thyroid   831
32.9.5   Liver   831
32.9.6   Gastrointestinal Tract   831
32.9.7   Head and Neck   831
32.9.8   Reproductive Organs   831
32.9.9   Secondary Neoplasms   831
         References   832

Chung K.K. Lee, MD
University of Minnesota Medical School, MMC494,
420 Delaware St.S.E., Minneapolis, MN 55455, USA

Improved understanding of the biological mechanisms of Hodgkin's disease, along with advances in staging and treatment, have made it one of the most successfully treated malignancies. This chapter focuses on the role of radiotherapy in the treatment of Hodgkin's disease, particularly on modern radiation therapy techniques used in conjunction with combination chemotherapy in the treatment of early and advanced disease. Of emphasis is the need to individualize treatment to achieve optimal outcomes while minimizing long-term complications.

32.1 Diagnostic Evaluation and Staging

Critical for the successful treatment of Hodgkin's disease is a careful evaluation of prognostic factors predictive of the ultimate outcome of the disease – factors such as disease stage, histopathology, performance status, bulk of disease, number and location of involved sites (including extra nodal sites), as well as age and gender of the patient.

A diagnostic work-up begins with a complete physical examination including documentation of any B symptoms (e.g., fever, night sweats, >10% weight loss during previous 6 months) and other symptoms such as pruritus intolerance, fatigue, respiratory problems, and alcohol intolerance. Evaluation of all nodal sites, including tonsillar and other lymphoid tissue-containing sites, is mandatory. For patients who may need radiation therapy that includes the oral cavity, it is essential to have a pre-radiation evaluation of the teeth and complete oral cavity by a dentist. After a physical examination, laboratory assessment is necessary and should include a complete differential blood count, erythrocyte sedimentation rate (ESR), serum electrolytes, liver and renal function tests, serum alkaline phosphatase, and beta2-lactate dehydrogenase. Other optional blood tests that may be useful include serum copper, microglobulin, and various cell surface cytogenetic analyses (FRIEDMAN et al. 1988; RAY et al. 1973; TUBIANA et al. 1984; AGNARSSON and KADIN 1989).

To identify the extent of the disease, radiological studies are essential and should include a chest X-ray and computed tomography (CT) scans of the chest, abdomen, and pelvis. Magnetic resonance imaging (MRI) and positron emission tomography (PET) should be done as needed. PET-CT is a new tool to identify the active disease.

Over 60% of Hodgkin's disease patients present with initial radiographic evidence of intrathoracic disease. Because of the ability of CT scans to detect small, unsuspected masses, determine the full extent of large masses, and detect involvement of the lung parenchyma, it is most often the test of choice (CASTELLINO et al. 1986; ROSTOCK et al. 1983). Evaluation of the extent of mediastinal disease is necessary to quantify the potential for intrathoracic relapse; a greater risk of relapse, particularly in the intrathoracic area and transnodal sites, is seen in patients with large mediastinal masses (LMMs). Several definitions are used to define a LMM. At the University of Minnesota, a large mass is defined as a mass with an MT ratio greater than 0.35. The MT ratio is defined as the largest transverse diameter of the mediastinal mass divided by the transverse diameter of the thorax at the level of T5-6 (LEE et al. 1980) (Fig. 32.1). Other institutions define a LMM as a mass whose greatest diameter is greater than one-third of the largest diameter of the thorax at the diaphragm on an upright posteroanterior chest radiograph (MAUCH et al. 1988) or a mass 5–10 cm in size using the transverse diameter of the mass.

Despite improvements in diagnosis with inclusion of CT scans, detection of occult abdominal and pelvic disease remains challenging. Imaging studies show a false-negative rate of 20–25% in detection of occult disease in these areas. This is largely due to the difficulty of detecting occult disease in the spleen (LEIBENHAUT et al. 1989; MAUCH et al. 1990). Available radiological studies that result in a much better yield in the detection of intra-abdominal disease are MRI, PET, gallium scans combined with single photon-emission computed tomography (SPECT), and bipedal lymphangiography (NG et al. 2002). The role of these ancillary tests is still under investigation. Some evidence suggests that gallium scans, particularly when combined with SPECT, are useful in assessing residual masses (FRONT and ISRAEL 1995). Other studies suggest a continued role for bipedal lymphangiography to assess lymph node size and internal architecture, despite the increas-

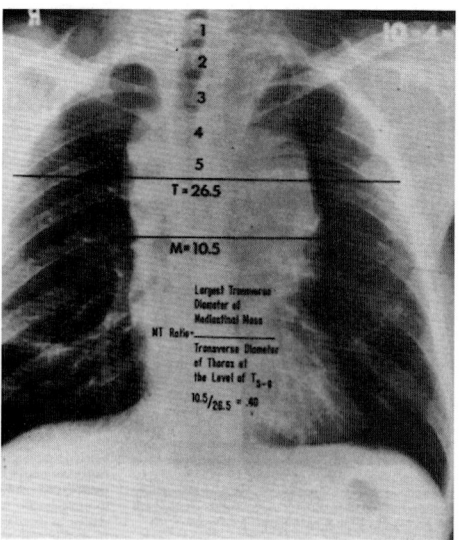

Fig. 32.1. The MT ratio (largest transverse diameter of the mediastinal mass divided by the transverse diameter of the thorax at the level of T5-6) is determined by measuring the mediastinal on the posteroanterior (PA) chest film (Definition used at the University of Minnesota). (From LEE et al. 1980. Copyright © (1980) American Cancer Society. Reprinted by permission of Wiley-Liss, Inc., a subsidiary of John Wiley & Sons, Inc.)

ingly less frequent use of this procedure due to the diminishing skill of physicians in performing and interpreting the results of this test (CASTELLINO et al. 1984). All of these imaging methods, however, remain limited in their ability to accurately identify occult abdominal disease. Whole-body PET using 18F-fluorodeoxy-glucose (FDG-PET) is a new imaging method currently under investigation and has shown some promise in improving overall diagnostic accuracy (NG et al. 2002; HUELTENSCHMIDT et al. 2001). It may also aid in evaluating response after systemic treatment in patients with a positive PET scan prior to treatment (DE WIT et al. 2001; SPAEPEN et al. 2001; SCHODER et al. 2001).

Although surgical staging with laparotomy and splenectomy was once relied on to provide the most precise way to determine abdominal involvement, this procedure is no longer used in most parts of the world because it does not greatly impact the eventual treatment strategy and involves an invasive procedure with possible morbidity in patients treated with combined modality. Surgical staging may still play a role in selecting patients for treatment if the patient is to be treated with radiation therapy alone (NG et al. 2002). Traditional historical surgical staging includes inspection, palpation, and biopsy of nodes in the abdomen and pelvis; wedge and needle biopsy of the liver; and the placement of splenic pedicle clips. Premenopausal women also usually undergo a bilateral midline oophoropexy in anticipation of pelvic irradiation.

The use of staging laparotomy in Hodgkin's disease has resulted in a better understanding of the natural evolution of the disease. The disease appears to spread contiguously to adjacent lymph nodes first. There is frequent extension to the spleen during the early course of the disease before it spreads to other visceral organs such as the liver. In 20–30% of clinical stage IA–IIA patients and 35% of IB–IIB patients, occult splenic or upper abdominal disease may be identified at staging laparotomy that is not detected on presurgical clinical staging studies (LEIBENHAUT et al. 1989; MAUCH et al. 1990; CASTELLINO et al. 1984; FRONT and ISRAEL 1995; BRADA et al. 1986; RUTHERFORD et al. 1980; ARAGON DE LA CRUZ et al. 1989). By removing the spleen during staging laparotomy, the volume of irradiation is reduced significantly and radiation to the left kidney can be avoided. In patients with negative laparotomy and other favorable prognostic factors, the radiation field can be confined to above the diaphragm (TUBIANA et al. 1989; HAYBITTLE et al. 1985; SUTCLIFF et al. 1985; MADELLI et al. 1986; MAUCH et al. 1995a). Staging laparotomy has allowed for the selection of early-stage patients who could be treated with radiation alone and has helped identify the selective criteria for determining the low incidence of abdominal disease. These criteria include clinical stage-IA and -IIA female patients, patients younger than 26 years of age, and clinical stage-IA male patients with lymphocyte predominance (LP) histology.

Staging laparotomy is associated with potential morbidity and mortality. Small-bowel obstruction, development of wound or subdiaphragmatic abscess, and postoperative bleeding are the major complications but are as low as 3% (TAYOR et al. 1985). Following splenectomy, patients are also at increased risk for infection with encapsulated bacteria (MOLRINE et al. 1995; SIBER et al. 1986). Vaccinations against pneumococcus and meningococcus or prophylactic antibiotics should be used to decrease risk. An approximate twofold increased risk of leukemia following splenectomy has been reported in some studies, especially in patients who received chemotherapy following splenectomy (VAN DER VELDEN et al. 1988). However, the mechanisms for this finding are poorly understood, and the increase is not recognized by all observers.

Table 32.1 summarizes the procedures recommended for proper work-up and staging of Hodgkin's disease.

Table 32.1. Diagnostic and staging procedure

Mandatory
– Biopsy of any mass or lymph nodes

History
– Age and gender
– Evaluation of systemic B symptoms
  • Unexplained fever
  • Night sweats
  • Weight loss >10% body weight in last 6 months
– Other symptoms
  • Alcohol intolerance
  • Pruritus
  • Respiratory problems
  • Easily fatigued

Physical examination
– Lymphadenopathy (note number, size, location, shape, consistency, and mobility of nodes)
– Palpable liver, spleen, and other masses

Laboratory studies
– Standard
  • Complete blood count including platelet counts
  • Liver and renal function
  • Blood chemistry
  • Erythrocyte sedimentation rate
– Optional
  • Serum copper
  • β_2 microglobulin

Radiographic
  • Standard
    Chest radiograph: posteroanterior and lateral
    Thoracic, abdominal and pelvic computerized tomography
  • Optional
    Bipedal lymphogram
    Gallium scan-67 (with high-dose SPECT)
    Technicium-99 bone scan
    Magnetic resonance imaging
    Positron emission tomography scan
    Echo-cardiography

Special tests
– Pulmonary function test
– Cardiac function test
– Immunophenotyping
– Molecular genetic analysis
– Morphological and immunophenotype features
– Cytological examination of effusions, if present
– Bone marrow, needle biopsy (especially subdiaphragmatic disease of B symptoms)
– Optional
  • Percutaneous or computed tomography-guided liver biopsy
  • Peritoneoscopy
  • Staging laparotomy with splenectomy, liver biopsy, selected lymph node biopsies, and open bone marrow biopsy

In 1971, the Ann Arbor classification for Hodgkin's disease was established (CARBONE et al. 1971) (Table 32.2). This staging system was used for over

Table 32.2. Hodgkin's disease staging classification

Stage	Definition
I	Involvement of a single lymph node region (I) or of a single extralymphatic organ (I_E)
II	Involvement of two or more lymphatic regions on the same side of the diaphragm (II), or localized extralymphatic involvement as well as involvement of one or more regional lymphatic sites on the same side of the diaphragm
III	Involvement of lymphatic regions on both sides of the diaphragm (III); such involvement may include splenic involvement (III_S), localized extra lymphatic disease (III_E), or both (III_{SE})
IV	Diffuse or disseminated involvement of one or more extralymphatic organs or tissues, with or without nodal involvement. The absence or presence of unexplained fever, night sweats, or loss of 10% or more of body weight in the 6 months preceding diagnosis are designated by the suffix letters A or B, respectively. Biopsy-proven involvement of extralymphatic sites is designated by letter suffixes: bone marrow M+; lung L+; liver H+; pleura P+; bone O+; skin and subcutaneous tissue D+

[a] Adopted at the workshop on the staging of Hodgkin's disease held at Ann Arbor, MI in April, 1971 (reprinted with permission from CARBONE et al. 1971)

two decades for both clinical and pathological staging of Hodgkin's and non-Hodgkin's lymphoma. However, inadequacies of the Ann Arbor staging system, including failure to account for bulk and extent of disease and to precisely define extralymphatic involvement, led to a modified classification system proposed at a 1989 international meeting at Cotswolds, England. The new classification relied on anatomic regions based on the knowledge that Hodgkin's disease spreads along the lymphatic channel in an orderly fashion (Fig. 32.2). The modifications incorporated some important prognostic factors, such as bulk of disease and a more precise definition of extralymphatic involvement (LISTER et al. 1989) (Table 32.3).

32.2
Histopathological Classification

Histopathological classification of Hodgkin's disease includes four histological subtypes as defined by the Rye modification of the Lukes and Butler system: (1) LP, (2) nodular sclerosis (NS), (3) mixed cellularity (MC), and (4) lymphocyte depletion (LD) Hodgkin's disease (LUKES and BUTLER 1966).

LP Hodgkin's disease comprises 5–10% of all Hodgkin's disease and is often localized to a single peripheral nodal region. Only 8% of these patients have mediastinal and abdominal involvement in early-stage disease (MAUCH et al. 1993). This subtype is found most often in male patients younger than 15 years of age or older than 40 years. It contains an abundance of benign-appearing cells and frequent variant lymphocytic and histiocytic cells with multilobulated nuclei (popcorn cells). Some

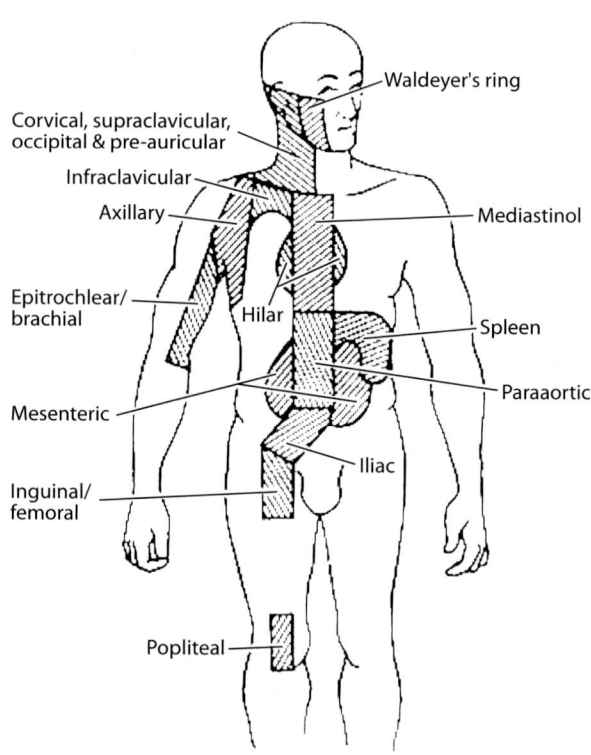

Fig. 32.2. Clinical lymphoid regions, as defined by the Ann Arbor Staging System

investigators have proposed classifying nodular LP histology as a separate clinically and histologically distinct entity (MASON et al. 1994).

Nodular sclerosis Hodgkin's disease is the most common histology and accounts for 40–60% of all cases. It usually affects patients between the age of 15 years and 40 years, and affects males and females equally. It presents with central nodal regional involvement in 80–90% of cases (LUKES and BUTLER 1966).

Table 32.3. Cotswold modifications to Ann Arbor Staging Classification

I	Suffix "X" to designate bulky disease as >1/3 widening of the mediastinum or >10 cm maximum dimension of nodal mass
II	The number of anatomical regions involved should be indicated by a subscript (e.g., II_3)
III	Stage III may be subdivided into: III_1: with or without splenic, hilar, celiac, or portal nodes III_2: with para-aortic, iliac, mesenteric nodes
IV	Staging should be identified as clinical stage (CS) or pathological stage (PS)
V	A new category of response to therapy, unconfirmed/uncertain complete response ($CR_{(U)}$) can be introduced because of the persistent radiological abnormalities of uncertain significance

(From CA-A Cancer Journal for Clinicians, ANONYMOUS 1993)

About 25% of Hodgkin's disease cases have MC histology. Patients with MC Hodgkin's disease are older and more likely to present with systemic symptoms and advanced stages. Both malignant Reed-Sternberg cells and pleomorphic variant cells in an inflammatory background are seen more commonly in this histology (LUKES 1971).

Only 5% of Hodgkin's disease patients have LD histology. Generally, these patients have advanced disease with systemic symptoms (NEIMAN et al. 1973).

In addition to the above four major subtypes of histopathology, interfollicular Hodgkin's disease presents with an uncommon pattern of focal involvement of a lymph node in the interfollicular zone. It may be easily confused with reactive hyperplasia (DOGGETT et al. 1983).

32.3
General Treatment Considerations

The goal of radiotherapy in treating Hodgkin's disease is to treat all of the involved and potentially involved lymphatic chains with an adequate irradiation dose to increase the potential for tumor eradication while minimizing long-term treatment-related morbidity. With increasing reliance on combined chemotherapy regimens to treat early as well as advanced disease, the role of radiation therapy has evolved from a single modality mainstay of treatment for early disease to its current role as combination therapy with chemotherapy regimens. Use of radiation alone is currently reserved largely for patients who wish to avoid chemotherapy and a special category of patient with low risk factors as described previously. (NG et al. 2002). The current controversies surrounding radiotherapy in the treatment of Hodgkin's disease are determination of optimal radiation field size (involved or extended field?) and radiation dose (NG et al. 2002).

32.3.1
Treatment of Stage-I and -II Favorable Disease

Radiation therapy alone. Although not commonly practiced, radiation alone, either by mantle field or subtotal nodal field plus splenic irradiation (STNI), can be used to treat stage-I and -II patients without the following features: LMM with or without hilar disease, bulky disease, systemic symptoms, four or more sites of involvement, advanced age (defined as older than 40 years), elevated ESR, male gender, and MC or LD histologies (FRIEDMAN et al. 1988; MAUCH et al. 1988; SOMERS et al. 1989; CRNKOVICH et al. 1987; SPECHT and NISSEN 1988; LEE et al. 1990; HENRY-AMAR et al. 1991; TUBIANA et al. 1982; HOPPE et al. 1982b).

Historical data suggest that patients with pathological stage-IA and -IIA disease who were treated with sequential mantle and para-aortic fields or the mantle alone had an expected 10-year free-from-failure rate of 75–80% (LEE et al. 1990; MAUCH et al. 1988; HOPPE et al. 1982a).

Most relapses occur within the first 3 years after radiation therapy, although up to 10% of patients relapse after 3 years. Prolonged late relapses beyond 5–10 years are uncommon. Following STNI, there is recurrence in the pelvis and inguinal femoral nodal region in 5–15% of patients.

In the past, the standard approach of most centers in the United States was to require pathological staging of Hodgkin's disease prior to recommending radiation therapy alone. However, Canadian and European studies have shown excellent overall survival for patients who were selected for radiation therapy alone based on clinical staging. Therefore, clinically staged patients with favorable prognostic factors may be treated with radiation therapy alone. The following subgroups have less than 10% risk of infradiaphragmatic involvement: clinical stage-IA females (6%), patients with involvement of the mediastinum alone (0)%, stage-I males with LP histology (4%), and young (<27 years of age) females with limited (fewer than four supradiaphragmatic sites) stage-II disease (9%) (SPECHT and NISSEN 1988). These patients may be treated effectively with a supradiaphragmatic field only. At the University of Minnesota Hospital, patients with clinical stage-I and -II disease with favorable features are selected for treatment with radiation therapy alone without staging laparotomy. Data from the European Organization for Research and Treatment of Cancer (EORTC) support the efficacy of this strategy. Treating extended or total nodal field has become an almost historical approach. It is important however to know the evolution of the treatment approach.

Influence of pathological and clinical staging on the outcome of stage-I and -II Hodgkin's patients following radiotherapy has been studied by the EORTC lymphoma cooperative group. Both groups of patients were treated with STNI with spleen included in clinical stage patients. At 10 years, there was no statistical difference in recurrence-free survival (clinical stage 68% versus pathological stage 73%); however, a higher number of patients with positive findings at laparotomy relapsed compared with those with negative laparotomy (56% versus 83%, respectively) (FRIEDMAN et al. 1988).

Randomized trials conducted to define the proper fields for early-stage Hodgkin's disease show better relapse-free survival for patients treated with extended field than for those treated with involved field irradiation (ROSENBERG and KAPLAN 1966; KAPLAN 1980). Results of mantle or limited field irradiation alone in early-stage Hodgkin's disease have been disappointing with an increasing risk of relapse in the abdomen, except in patients who present with very favorable features (SPECHT and NISSEN 1988).

Treatment for stage-I and -II infradiaphragmatic Hodgkin's disease is less well studied than for supradiaphragmatic disease. The prognosis of patients with clinical para-aortic lymph node involvement (stage II) is probably worse than those with a single site of peripheral nodal disease (stage I). The former patients are likely to have more disease in another site in the abdomen or have B-symptoms, or MC or LD histology. Staging laparotomy should be done if radiation therapy alone is carried out. Pathological stage-IA patients can be treated with radiation using an inverted-Y field only.

Radiation treatment alone using extended fields includes sequential mantle and para-aortic irradiation, including the spleen if it is not removed

(STNI). The mantle field includes the cervical, axillary, infraclavicular, mediastinal, and hilar lymph node regions. Most of the lungs and part of the heart (mainly the left ventricles) are shielded in the mantle field. The infradiaphragmatic field includes the abdominal nodes and spleen. A 4-week break is usually given between the mantle and infradiaphragmatic treatment. Since clinical stage-I and -II disease may be associated with a 25% risk of occult abdominal involvement, the infradiaphragmatic field is treated prophylactically except in the very favorable group who have a less than 10% incidence of occult infradiaphragmatic involvement.

Combined modality therapy. Currently combined modality therapy (CMT) is the main treatment approach for most Hodgkin's disease patients. Management of most patients with favorable stage-I and -II disease will be based on clinical staging, CMT, and chemotherapy regimens that are less toxic. Several randomized trials show improved free-from-failure survival rates with the use of CMT in favorable early-stage Hodgkin's patients (ZITTOUN et al. 1985; COSSET et al. 1992; BONADONNA 1994; FULLER et al. 1988; HOPPE et al. 1982a). However, a survival benefit has not been definitively shown because of the good salvage rates following radiation failures.

Previous data from the review of randomized trials of more versus less extensive radiotherapy, with or without chemotherapy, suggested the use of less extensive radiation fields and resulted in similar survival rates to those achieved with more intensive treatment (SPECHT et al. 1998). Recent evidence suggests that limited field irradiation in CMT is also feasible without compromising outcomes while reducing the risk of long-term toxicity. A study that examined the efficacy of adriamycin, bleomycin, vinblastine, and dacarbazine (ABVD) chemotherapy followed by limited field irradiation showed a high 5-year actuarial overall survival (100%) and progressive-free survival (97%) with no observation of secondary malignancies and only mild pulmonary toxicity (KARMIRIS et al. 2003).

For the group of favorable patients, the overall survival rates are good regardless of the treatment modality used. Of importance in the choice of treatment for these patients is the risk of long-term treatment-related toxicities and side effects. Radiation therapy has long been employed in the treatment of Hodgkin's disease, and its long-term toxicities are well described. On the contrary, the long-term side effects of chemotherapy and its combination with radiation therapy have not been as well evaluated or understood. If the long-term toxicities of CMT are demonstrated to be equal to or less than radiation therapy alone, CMT would be the treatment of choice even if the benefit in outcome is shown only in terms of free-from-failure rates. To reduce long-term toxicities, the field size and dose of the radiation may have to be reduced.

32.3.2
Special Considerations for LMM

For patients with stage-I and -II disease with unfavorable prognostic features, CMT is used. Altering treatment modalities may change the prognostic significance of some unfavorable factors.

Patients with a LMM require special therapeutic attention. Although the definition of large or bulky mediastinal disease varies, it is usually defined as a mass measuring greater than one-third of the largest transverse chest diameter (MAUCH 1994) or the transverse diameter of the mass at the T5-6 level divided by the largest transverse diameter of the chest (LEE et al. 1980) (Fig. 32.1).

In patients treated with mantle field irradiation, the majority of failures occur outside or at the edge of the radiation port in the intrathoracic region (MAUCH et al. 1988; LEE et al. 1990). This suggests that there are geometric difficulties in treatment volumes when trying to shield the lung parenchyma. This problem has led to modifications in treatment techniques, including low-dose lung irradiation to treat the microscopic disease and the use of a shrinking field technique as the size of the lung mass is reduced. The addition of low-dose (15–18 Gy) whole or hemi-lung irradiation has resulted in excellent clinical outcomes with acceptable side effects (MAUCH et al. 1983, 1988; MAI et al. 1991; MAUCH 1994; ROSENBERG and KAPLAN 1985; LEE et al. 1990).

The increased use of chemotherapy in the treatment of Hodgkin's disease coupled with the greater risk of relapse after standard mantle field irradiation has led to the wide acceptance of CMT in the treatment of patients with LMM. CMT has lowered the local recurrence rate but has not significantly impacted overall survival. Chemotherapy treats sub-clinical disease and decreases the bulk of disease. This allows for the use of smaller fields and lower doses of radiation.

Chemotherapy as the sole treatment for a poor-risk group of patients frequently fails to achieve complete response (CR) and results in greater relapse than when radiation therapy is used alone (ZITTOUN et al. 1985; BITI et al. 1992). Therefore, radiation is absolutely needed to maximize local regional control in this situation (PAVLOVSKY et al. 1988; BITI et al. 1992).

32.3.3
Delivery and Dose of CMT

Prior to the administration of chemotherapy, CT, gallium, and PET scans are recommended to permit evaluation of the post-treatment response and discern the need for additional treatment.

The optimal regimen for CMT remains unsettled. Multiple chemotherapeutic regimens with various numbers of cycles have been used. Largely based on the results of prospective trials in advanced Hodgkin's disease, ABVD has become the standard regimen for patients with stage-I or -II unfavorable prognosis. Other regimens used include MOPP (mechlorethamine, Oncovin, procarbazine, and prednisone), MOPP/ABV and BEACOPP (bleomycin, etoposide, adriamycin, cyclophosphamide, Oncovin, procarbazine, and prednisone). Most trials have incorporated four to six cycles of chemotherapy (SOMERS et al. 1989).

Results from the German Hodgkin's Study Group HD8 showed comparable outcomes among patients randomized to extended field irradiation versus involved field irradiation after two cycles of COPP/ABVD, with reduced toxicity in the involved field treatment arm (ENGERT et al. 2001).

The Milan group recently presented the result of their trial comparing 4 cycles of ABVD with STNI and the same chemotherapy with involved field radiation therapy for stage I–II unfavorable Hodgkin's disease (BONFANTE et al. 2001). No difference was found between the two groups with regard to overall survival and free-from-progression. Currently, the National Cancer Institute of Canada is conducting an ongoing trial comparing two cycles of ABVD followed by either extended mantle or mantle plus para-aortic irradiation versus four to six cycles of ABVD with the same radiation regimen for those with stage I–II Hodgkin's disease with unfavorable features.

Since the ABVD (doxorubicin, bleomycin, vinblastine, dacarbazine) era, combined modality (chemotherapy and radiotherapy) has become almost the standard approach in the treatment of disease for unfavorable as well as favorable early-stage Hodgkin's disease.

To find the optimal dose and better regimen of chemotherapy and radiation port and dose, numerous studies are being carried out throughout the world. The total radiation dose to the involved area of pre-chemotherapy region is 25–30 Gy with no detectable disease after chemotherapy, and 30–36 Gy with small residual disease. After chemotherapy, bulky residual disease receives a total radiation dose of 40 Gy. Establishing the extent of the disease prior to receiving chemotherapy is important for RT planning.

Multiple trials and studies have provided evidence that radiation fields may be safely limited to an involved or regional field in most combined modality programs if four to six or fewer cycles of chemotherapy are given. Table 32.4 aims to identify appropriate radiation volume for unfavorable Hodgkin's lymphoma. The treatment of stage-I and -II unfavorable Hodgkin's disease with CMT should result in a survival in the range of 90%. The long-term toxicities of CMT are not fully defined because of the relatively short follow-up.

At the University of Minnesota, unless the patient is being treated in a particular study, irradiation of the modified mantle or involved field with doses of 25–36 Gy is used following four to six cycles of systematic therapy. The site of initial bulky disease such as LMM is boosted up to 30–36 Gy if a CR is achieved, and to a dose of 40 Gy if a CR is not achieved. In patients with initial pericardial invasion, the initial field includes the entire heart with a cone-down to a mantle field after 15 Gy (Fig. 32.3). For patients with chest wall invasion, 25–30 Gy is given to the pre-chemotherapy volume in patients achieving CR. A total dose of 36–40 Gy is given if there is residual disease. Whenever possible, shrinking field techniques should be used (Fig. 32.3, 32.4)

Table 32.5 reveals some recommendations for primary treatment outside clinical trials.

32.3.4
Advanced Stage Hodgkin's Disease

Chemotherapy is the primary treatment modality for advanced stage Hodgkin's disease (ASHD). The role of radiation in these patients is a controversial

Table 32.4 Randomized clinical trials in unfavorable-prognosis stage-I and -II Hodgkin's lymphoma: trials to identify the appropriate radiation volume (used with permission from DE VITA et al. 2005). *ABV* doxorubicin (Adriamycin), bleomycin, vinblastine; *ABVD* doxorubicin, bleomycin, vinblastine, dacarbazine; *COPP* cyclophosphamide, vincristine (Oncovin), procarbazine, prednisone; *CS* clinical stage; *DFS* disease-free survival; *EFRT* extended-field radiotherapy; *EORTC* European Organization for Research and Treatment of Cancer; *ESR* erythrocyte sedimentation rate; *FFP* freedom from progression; *FFTF* freedom from treatment failure; *GELA* Groupe d'Etude des Lymphomes de l'Adulte; *GHSG* German Hodgkin Study Group; *IFRT* involved-field radiotherapy; *MOPP* mechlorethamine, vincristine (Oncovin), procarbazine, prednisone; *NS* not significant; *RFS* relapse-free survival; *STLI* subtotal nodal irradiation; *SV* survival

Trial	Eligibility	Treatment regimens	No. of patients	Outcome
French Cooperative, 1976–1981	CS I–II *without* age >45 years; ≥ 3 involved areas; bulky disease	A: 3 MOPP + IFRT (40 Gy) + 3 MOPP	82	DFS, 87%; SV (6 years), 92%
		B: 3 MOPP + EFRT (40 Gy) + 3 MOPP	91	DFS, 93%; SV (6 years), 91%
				(DFS: *P*=NS; SV: *P*=NS)
Istituto Nazionale Tumori, Milan, 1990–1997	All CS I–II	A: 4 ABVD + STLI	65	FFP, 96%; SV (5 years), 93%
		B: 4 ABVD + IFRT	68	FFP, 94%; SV (5 years), 94% (FFP: *P*=NS; SV: *P*=NS)
EORTC/GELA H8U, 1993–1998	CS IA–IIB *with* age ≥ 50 y; ESR ≥ 50 mm/h in A, ≥ 30 mm/h in B; ≥ 4 involved sites; large mediastinal disease	A: 6 MOPP/ABV + IFRT (36 Gy)	335	RFS, 94%; SV (4 years), 90%
		B: 4 MOPP/ABV + IFRT (36 Gy)	333	RFS, 95%; SV (4 years), 95%
		C: 4 MOPP/ABV + STLI	327	RFS, 96%; SV (4 years), 93% (RFS: *P*=NS; SV: *P*=NS)
GHSG HD8, 1993–1998	CS IA–IIB *with* ESR ≥ 50 mm/h in A, ≥ 30 mm/h in B; ≥ 3 involved sites; large mediastinal disease	A: 4 COPP/ABVD + EFRT	532	FFTF, 86%; SV (5 years), 91%
		B: 4 COPP/ABVD + IFRT	532	FFTF, 84%; SV (5 years), 92% (FFTF: *P*=NS; SV: *P*=NS)

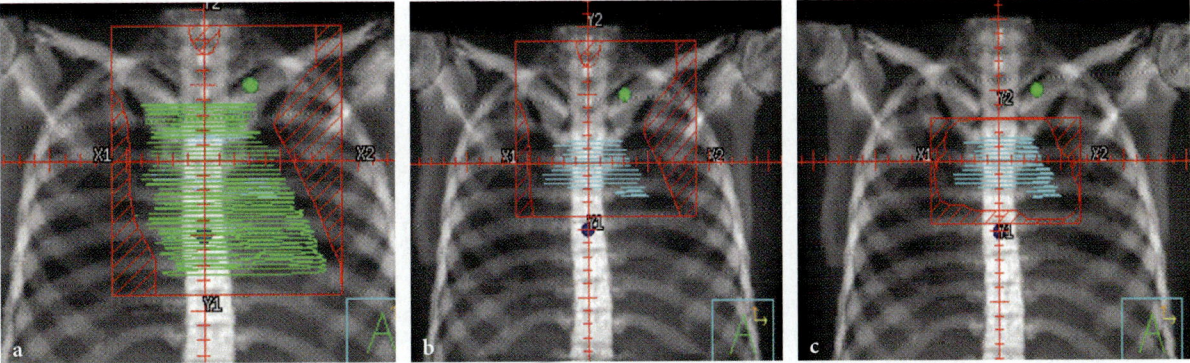

Fig. 32.3a–c. A 19-year-old female patient with clinical stage 2a, nodular sclerosis Hodgkin's disease with a large mediastinal mass extending left pericardial border. Positron emission tomography (PET) scans became negative after receiving six cycles of adriamycin, bleomycin, vinblastine, dacarbazine (ABVD) chemotherapy with remaining soft tissue density in the mediastinum. **a** Radiation port including left cardial border up to 1500 cGy with 100 cGy per day, treated anteroposterior–posteroanterior (AP–PA) fields. **b** Shielded left cardial border and lower mediastinum carried up to 2550 cGy. **c** Final boost field carried up to 3600 cGy

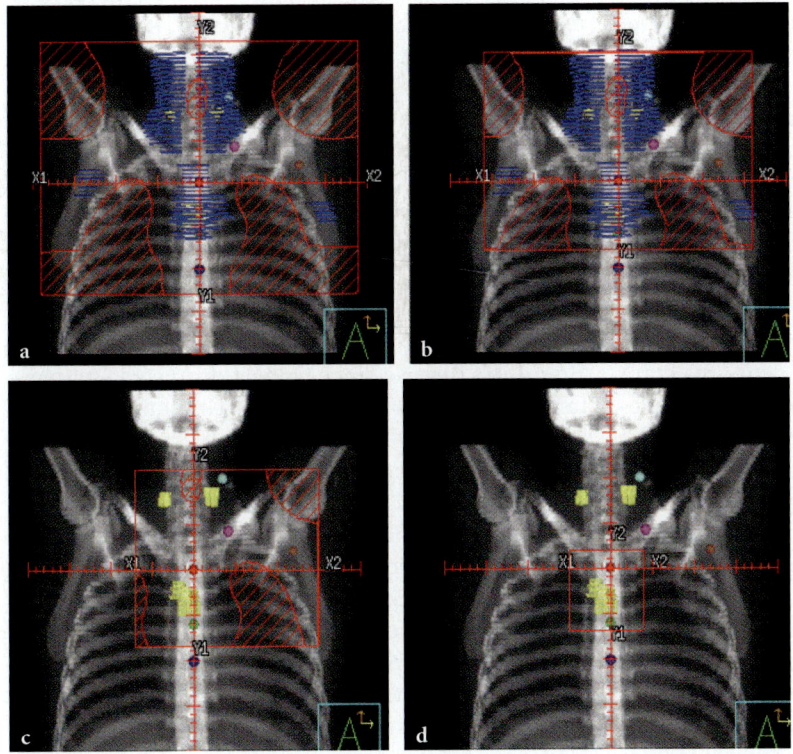

Fig. 32.4a–d. A 27-year old male with clinical stage 2a treated with six cycles of adriamycin, bleomycin, vinblastine, dacarbazine (ABVD) chemotherapy. **a** Field designed to include pre-chemotherapy sites and volume except mediastinum, which is designed using post-chemotherapy remaining soft tissue width. Treated up to 1650 cGy using 150-cGy daily dose. **b** Shielded low mediastinum and carried up to 2550 cGy. **c,d** Field design using post-chemotherapy volume carried up to 3600 cGy at the central axis point using 180 cGy daily dose. Neck, supraclavicular, and axillary regions were shielded as off-axis calculation points accumulated up to desired total dose by the Clarkson off-axis calculation point

issue. With combination chemotherapy, about 20% of patients fail to achieve CR and about one-third of patients who achieve a CR will eventually relapse (Longo et al. 1986; Raemaekers et al. 1997). The majority (80–90%) of failure occurs in previous disease sites, especially bulky and nodal disease areas (Fabian et al. 1994). For these reasons, radiation therapy has been added in ASHD patients, espe-

Table 32.5. Recommendations for primary treatment outside clinical trials (used with permission from DE VITA et al. 2005). *CS* clinical stage; *CT* chemotherapy; *EFRT* extended-field radiotherapy; *IFRT* involved-field radiotherapy; *RF* risk factors (see Table 41.5-7); *RT* radiotherapy

Group	Stage	Recommendation
Early stages (favorable)	CS I–II A/B no RF	EFRT (30–36 Gy) or 4–6 cycles CT[a] + IFRT (20–36 Gy)
Early stages (unfavorable)	CS I–II A/B + RF	4–6 cycles CT[b] + IFRT (20–36 Gy)
Advanced stages	CS IIB + RF; CS III A/B; CS IV A/B	6–8 cycles CT[c] + RT (20–36 Gy) to residual lymphoma and bulk

[a]ABVD [doxorubicin (Adriamycin), bleomycin, vinblastine, and dacarbazine], EBVP (epirubicin, bleomycin, vinblastine, and prednisone), or VBM (vinblastine, bleomycin, and methotrexate)

[b]ABVD, Stanford V (mechlorethamine, adriamycin, vinblastine, vincristine, etoposide, bleomycin, and prednisone), or MOPP/ABV [mechlorethamine, vincristine (Oncovin), procarbazine, and prednisone/Adriamycin, bleomycin, vinblastine]

[c]ABVD, MOPP/ABV, ChlVPP/EVA (chlorambucil, vinblastine, procarbazine, and prednisone/etoposide, vincristine, and Adriamycin), or BEACOPP (bleomycin, etoposide, adriamycin, cyclophosphamide, vincristine, procarbazine, and prednisone) escalated

cially those with bulky disease. Several phase-II and -III studies have explored whether adding radiation treatment improves disease-free or overall survival in these patients. Controversy still exists over proper total radiation dose and field size because of the questionable efficacy and the toxicity of consolidative RT. The type, intensity, duration, and dose of chemotherapy add difficulty to the assessment of the efficacy of RT in this setting.

Radiation therapy is employed in three different clinical settings in ASHD: (1) as consolidative treatment after CR post-chemotherapy; (2) as an integrated part of a CMT program, possibly with reduced dose chemotherapy; and (3) as a non-cross-resistant agent for treatment after partial response from chemotherapy. Most large prospective trials include radiation therapy in the treatment program.

Many retrospective studies suggest that, compared with chemotherapy alone, low-dose (20–30 Gy) consolidative RT increases survival benefits with the use of CMT (FABIAN et al. 1994; DIEHL et al. 1995). Since these studies are all retrospective they are subject to potential selection bias. Some of the studies used chemotherapy consolidation instead of radiotherapy. The potential contribution of radiotherapy is dependent on several factors such as patient characteristics, various prognostic factors, and response and duration of the chemotherapy program.

The guidelines for the dose and volume of radiation therapy in CMT are not well defined. Low-dose irradiation (15–30 Gy) has been employed based on the hypothesis that a lower dose of radiation may be all that is needed in the adjuvant setting. Table 32.6 summarizes randomized clinical trials in ASHD treatment combined modality radiation and chemotherapy. At the University of Minnesota institution,

patients with ASHD receive a dose of 25–30 Gy to the involved field if there is bulky disease (defined as mass >3–5 cm) but a CR is achieved. If there is less than a CR, a boost to 30–40 Gy will be given to the residual tumor mass.

32.4
Radiation Therapy Techniques

There has been a gradual evolution in concept and application in the use of radiation therapy to treat Hodgkin's disease since Gilbert proposed "segmental roentgen therapy" in 1939 (GILBERT 1939) and Peters proposed "radical radiation" in 1950 (PETERS 1950). Knowledge of the predictable patterns of relapse, the contiguous character of regional lymph node involvement, and the availability of megavoltage beam techniques have led to the development of current techniques and reasonably standardized radiation fields for the treatment of Hodgkin's disease (KAPLAN 1962). Rosenberg and Kaplan demonstrated in 1966 that in the vast majority of untreated patients with disease limited to lymph nodes only contiguous areas were involved, which proved the orderly progression in which Hodgkin's disease spreads (ROSENBERG et al. 1966).

Widely accepted terms to denote treatment fields – such as mantle, para-aortic, inverted Y field, pelvic field, Waldeyer's ring, preauricular field, spade-shape field, extended field (mantle and para-aortic fields), total nodal irradiation (mantle and inverted Y fields), and involved field – reflect the variation and growing standardization of these fields. Despite this apparent standardization, differences exist in the actual techniques used by different institutions.

An example of the techniques used at the University of Minnesota for extended field and total nodal field with liver irradiation are shown in Figures 32.3 and 32.4. Differences in treatment technique may account for different outcomes reported in the literature. A Pattern of Care survey of 163 treatment facilities found that recurrence significantly correlated with technique (involved field versus extended field), treatment machine (less than 80 cm cobalt-60 [^{60}Co] versus greater than 80 cm ^{60}Co, linear accelerator), simulation, and presence of splenic pedicle clips (HANKS et al. 1983).

One way to ensure quality control in the treatment of Hodgkin's disease is by routine field simulation and frequent film verification to verify that involved tissues are adequately treated and sensitive structures properly shielded. Portal films examined in the Patterns of Care Study demonstrated and increased overall recurrence rate (54% versus 14%, $P<0.001$) and an infield or marginal recurrence rate (33% versus 7%, $P<0.001$) between patients treated with inadequate margins between the protective lung and cardiac blocks and the tumor, and patients treated with adequate margins.

Careful follow-up of patients in large-scale clinical trials has permitted a rapid advance in our knowledge and optimal treatment of Hodgkin's disease.

Radiation therapy planning involves the radiation oncologist, radiation physicist, medical dosimetrist, and radiation therapist. The patient undergoes a simulation procedure where treatment parameters are set. Films are taken for the design of customized blocks and the points for dose measurements are marked. After the measurements are obtained, computerized planning determines the dose delivered to each reference point. A compensator is computer generated if necessary to achieve a uniform dose distribution to different reference points. If treatment to both the mantle and the para-aortic area is necessary, extra caution is exercised to match the two fields and avoid overdosing areas, especially spinal cord, as a result of the possible field overlap.

Radiation therapy will be delivered by standard megavoltage (4–10 MV) techniques, utilizing shaped fields with blocks individualized to the specific patient. In general, parallel opposed fields will be most appropriate. The minimum source-skin distance or source-axis distance should be 80 cm. Radiation treatments are administered in 150- to 180-cGy fractions, 5 days a week. There has been controversial data and opinion regarding total radiation dose in Hodgkin's disease. Dose has been strongly influenced by data from Stanford that initially used 40–44 Gy (1 Gy=100 cGy). This recommendation was derived from a retrospective analysis of in-field control in the early 1960s. Subsequent reports demonstrated excellent results with 30–36 Gy. Data from another comprehensive retrospective study on dose–response showed that a 98% in-field control rate could be achieved with 37.5 Gy. With megavoltage radiotherapy, the doses required for 98% in-field control for subclinical disease and disease of less than 6 cm and greater than 6 cm are 32.4 Gy, 36.9 Gy, and 37.4 Gy, respectively. Data from German Hodgkin's Disease Study Group showed that 30 Gy was adequate for the control of subclinical disease. In our institution, 30 Gy in 20 to 24 fractions is delivered to subclinical disease and 36–40 Gy to clinically detectable disease sites with special attention given to the placement of a subcarinal block to protect the heart. The para-aortic area and spleen receive a dose of 30 Gy in 150- to 180-cGy fractions.

32.4.1
Mantle Fields

Mantle fields were first used at Stanford in 1956 and since then modifications to mantle fields have been adopted by various institutions. Typical mantle fields include all the major lymph node bearing areas above the diaphragm, neck, axilla, mediastinal, and occipital. Preauricular lymph nodes and Waldeyer's ring area are included if needed. The mantle field is used to treat the major supradiaphragmatic nodal chains that are at high risk for the involvement of Hodgkin's disease, while maximally shielding the lungs. Preauricular lymph nodes are treated when there is high neck disease. Optimal design of the mantle field relies on imaging studies. A chest CT scan for patients with significant mediastinal adenopathy provides important information on disease extension to the lung, pericardium, chest wall, and internal mammary or pericardial lymph node. Incorporation of the CT scan into treatment planning decisions for patients treated with radiotherapy alone results in treatment field changes in about 15% of patients (SOMERS et al. 1989).

32.4.2
Simulation of Mantle Field

The following techniques are used at the University of Minnesota to simulate mantle fields.

Table 32.6. Randomized clinical trials in advanced-stage Hodgkin's lymphoma: major trials for which results have been recently published or not yet published (used with permission from DE VITA et al. 2005). *ABV* doxorubicin (Adriamycin), bleomycin, vinblastine; *ABVD* doxorubicin, bleomycin, vinblastine, dacarbazine; *ASCT* autologous stem cell transplantation; *BEACOPP* bleomycin, etoposide, doxorubicin, cyclophosphamide, vincristine (Oncovin), procarbazine, prednisone; *BEACOPP-14* BEACOPP regimen with 14-day interval between courses; *ChlVPP* chlorambucil, vinblastine, procarbazine, and prednisone; *CR* complete response; *CS* clinical stage; *CT* computed tomography; *ECOG* Eastern Cooperative Oncology Group; *EFS* event-free survival; *EORTC* European Organization for Research and Treatment of Cancer; *EPO* erythropoietin; *EBMT* European Bone Marrow Transplant Registry; *EVA* etoposide, vincristine, and doxorubicin; *FFS* failure-free survival; *FFTF* freedom from treatment failure; *GELA* Groupe d'Etude des Lymphomes de l'Adulte; *GHSG* German Hodgkin Study Group; *HDCT* high-dose chemotherapy; *IFRT* involved-field radiotherapy; *IPS* international prognostic score (International Prognostic Factors Project); *MOPP* mechlorethamine, vincristine (Oncovin), procarbazine, prednisone; *OS* overall survival; *PABlOE* prednisolone, doxorubicin, bleomycin, vincristine, etoposide; *PET* positron emission tomography; *PR* partial response; *PVACEBOP* prednisolone, vinblastine, doxorubicin, chlorambucil, etoposide, bleomycin, vincristine (Oncovin), procarbazine; *resid* residual; *RT* radiotherapy; *SNLG* Scotland and Newcastle Lymphoma Group; *Stanford V* mechlorethamine, doxorubicin, vinblastine, prednisone, vincristine, bleomycin, VP-16; *SV* survival; *SWOG* Southwest Oncology Group; *TTF* time to treatment failure; *UKLG (BNLI)* United Kingdom Lymphoma Group (British National Lymphoma Investigation); *VAPEC-B* doxorubicin, cyclophosphamide, etoposide, vincristine, bleomycin, prednisolone

Trial	Eligibility	Treatment regimens	No. of patients	Outcome
Manchester (Radford et al.)	—	A: 6 ChlVPP/EVA ± RT (bulk/resid)	144	FFTF, 82%; SV (5 years), 89%
		B: 11 VAPEC-B ± RT (bulk/resid)	138	FFTF, 62%; SV (5 years), 79%
				(FFTF: $P=0.006$; SV: $P=0.04$)
GHSG HD9 (Diehl)	CS IIB with large mediastinal involvement, massive splenic involvement, or E lesions; CS III, IV	A: 8 COPP/ABVD ± RT (bulk/resid)	260	FFTF, 69%; SV (5 years), 83%
		B: 8 BEACOPP baseline ± RT (bulk/resid)	469	FFTF, 76%; SV (5 years), 88%
		C: 8 BEACOPP escalated ± RT (bulk/resid)	466	FFTF, 87%; SV (5 years), 91%
				($P<0.0001$; A versus C: $P<0.002$)
GHSG HD12 (Diehl)	CS IIB with large mediastinal involvement or E lesions; CS III, IV	A: BEACOPP (escalated × 8) ± RT (bulk/resid)	Began 01/1998; planned: $n=1200$	Final analysis planned for 2006
		B: BEACOPP (escalated × 8)		
		C: BEACOPP (escalated × 4 + baseline × 4) ± RT (bulk/resid)		
		D: BEACOPP (escalated × 4 + baseline × 4)		
GHSG HD15 (Diehl)	CS IIB with large mediastinal involvement or E lesions; CS III, IV	A: 8 BEACOPP escalated + EPO/placebo + IFRT (30 Gy for PET-positive PR)	Opened 01/2003	Open
		B: 6 BEACOPP escalated + EPO/placebo + IFRT (30 Gy for PET-positive PR)		
		C: 8 BEACOPP-14 + EPO/placebo + IFRT (30 Gy for PET-positive PR)		

Study	Stage/risk	Treatment	n	Results
EORTC 20884 (09/1989–03/2000)	CS III and IV	6–8 MOPP/ABV + (if CR after 6 cycles): →A: IFRT (24–30 Gy) →B: No further treatment; Not randomized PR	Total: 736 / 172 / 161 / 85 / 250	EFS, 79%; SV (5 years), 91% / EFS, 84%; SV (5 years), 85% (EFS: $P=0.35$; SV: $P=0.10$) / EFS, 79%; SV, 87%
GELA (Fermé & Diviné)	IPS 0–2	ne	ne	ne
GELA/EBMT H96-1 (Fermé & Diviné)	IPS 3+	A: Brief intensified CT then HDCT + ASCT ± RT (bulk) / B: 8 ABVD ± RT (bulk)	83 / 80	FFS, 75%; SV (5 years), 86% / FFS, 82%; SV (5 years), 88% (FFS: $P=0.4$; SV: $P=0.6$)
UKLG (BNLI) LY09 (Hancock)	Disease requiring systemic therapy: CS I–II with bulk or >3 sites; CS III–IV	A: 6–8 ABVD ± RT (bulk/resid) / B: 6–8 ChlVPP/PABlOE ± RT (bulk/resid)or 6–8 ChlVPP/EVA ± RT (bulk/resid)	Recruitment: 04/1997– 09/2001; n=807	SV (3 years), 91% / SV (3 years), 88%
UKLG (BNLI) phase II study (Hancock)	CS IIE, III, IV with large mediastinal involvement or ≥2 extranodal sites	A: 6–8 ABVD ± RT (bulk) / B: Stanford V ± RT (bulk)	Began 03/1998; planned: n=80	End point: response; phase-III trial to follow
SNLG HDIII (Proctor et al. [193])	SNLG index <0.5 (high risk)	3 PVACEBOP ± RT (bulk) + (if response): →A: Melphalan,VP-16 then HDCT + ASCT →B: 2 PVACEBOP	n=107, randomized n=65	Median follow-up, 6 years / OS, 78% (all patients) TTF (5 years), arms not different / TTF, 79±11% / TTF, 85±7% (TTF: $P=0.35$)
SWOG/ECOG 2496 (Horning & Coltman)	CS II with bulk and III and IV; IPS 0–2	A: 6–8 ABVD ± RT (bulk) / B: Stanford V ± RT	Open	Open
Global study, EORTC 20012	CS III and IV	A: 8 ABVD / B: 4 BEACOPP escalated + 4 BEACOPP baseline No radiation	Opened 01/2003	Open
SWOG/ECOG	CS II with bulk and III and IV; IPS 3+	A: ABVD / B: ABVD then HDCT (BCNU, VP-16, cyclophosphamide) + ASCT	—	—

32.4.2.1
Anterior Fields

1. Patients are placed in a supine position with arms above the shoulder. The arm position is 90° from the axilla and should eliminate any skin folds of the axilla. Others have used various arm positions as shown in Figure 32.5. Different arm positions may be used but attention must be paid to how the lymph nodes move in relation to the arm position for a given situation. Upper extremity lymphangiograms show that the elevation of the arms results in changes to the location of the axillary lymph nodes (CRNKOVICH et al. 1987; GRANT and JACKSON 1973; WEISENBERGER and JUILLARD 1977). When the hands are over the head with a greater than 90° angle at the axillae, axillary lymph nodes are moved away from the lung area, which allows for the design of lung and shoulder shielding.

2. The head position should allow the mandible to be perpendicular to the table top to avoid the possibility of unnecessary radiation through the oral cavity and mandible Figure 32.6.

3. Palpable lymph nodes can be outlined with a thin wire at the time of simulation to help design the field.

4. Field boundaries are placed superiorly to the level of mastoid tip through the chin line and inferiorly at T9–10 or T10–11 vertebral interspace. Inferior border should be extended as needed to the level of T11–12, to include the mediastinum. In patients with an intact spleen, special consideration should be given to the inferior border of the mantle because the spleen is usually included in the para-aortic field. In this situation, lung and diaphragmatic movement by respiration should be watched. In patients who receive whole lung irradiation, lower margins of the mantle field are inferior to the diaphragm to include the lung parenchyma.

5. Laterally, the field includes the axillae as determined clinically.

6. The central axis is defined at the center of the treatment and lies usually close to or at the sternal notch. Simulation radiographs are taken including all borders. This will often require two radiographs to show all the field margins. If the field size to cover the treatment area is too large, an extended source-skin distance may be used.

7. Mark the field boundaries and central axis on the patient. A tattoo may be placed at the central axis and inferior border to help daily setup and for future reference when the infraradiodiaphragmatic field is setup. Tattoos may also be placed about 10 cm lateral to the right and left of the central axis line, which helps to confirm the arm position relative to the central axis line.

8. When all the field borders and the central axis positions have been determined, a mantle measurement sheet for irregular field calculations is prepared. Patient separation is taken at the central axis, axillae, midneck, supraclavicular, midmediastinum, and low mediastinum, about 3 cm above the bottom of the field (Fig. 32.7). An irregular field point dose calculation is performed at midseparation for each point using a computerized Clarkson method. These point dose calculations help to achieve the desired cumulative dose to all areas of interest (Fig. 32.8).

Lymph Node Position Relative to Akimbo Arm Position	Lymph Node Position Relative to Arm Extension Position	Lymph Node Position Relative to Regular Mangle Arm Position

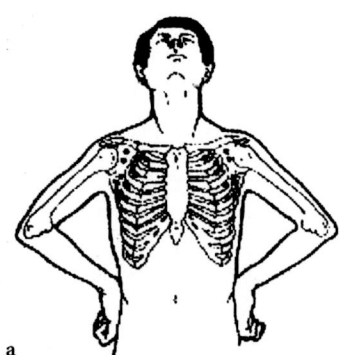
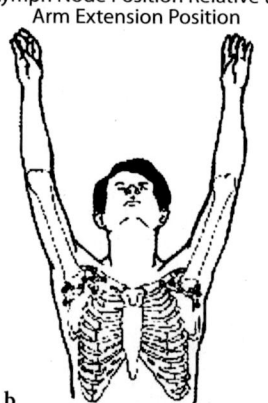
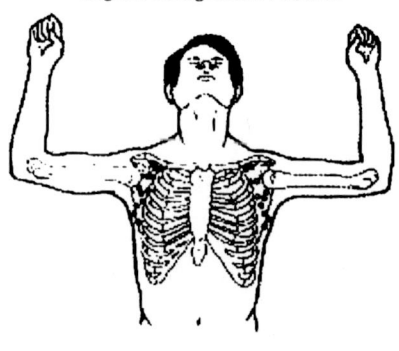

a

b

c

Fig. 32.5a–c. Arm positions for mantle fields. **a** Arm positions bring axillary lymph nodes medially close to the lateral part of the lung. Shielding the lung properly can be difficult. **b** Arm extension position makes the lymph nodes move over the shoulder area. Shielding the humeral head and shoulder joint can be difficult. **c** Right angle position gives reasonable room to shield the shoulder and place the lung block without including too much lateral strip of lung

32.4.2.2
Posterior Fields

If the treatment machine and the table permit, anterior and posterior treatments should be delivered in the same positions. The patient should remain in the supine position and the beam rotated 180° from the anterior field to set up the posterior field. In this case, a posteroanterior Beam's eye view will include more of the oral cavity, which requires special attention to shield the excess oral cavity in the field (Fig. 32.9). Treating the patient in the supine position, however, makes it difficult to evaluate pos-

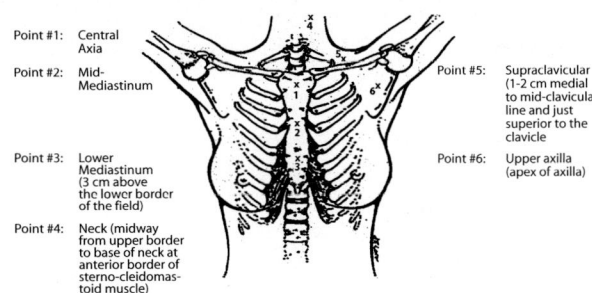

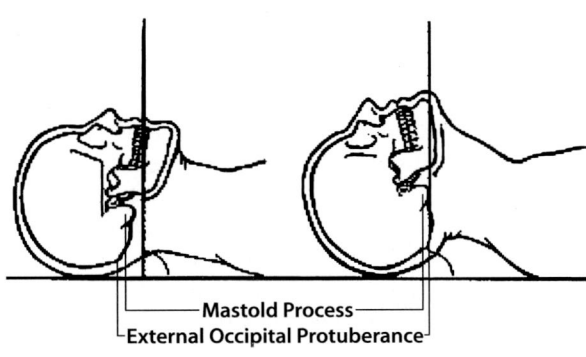

Mastold Process
External Occipital Protuberance

Fig. 32.6. Proper chin extension prevents excess dose to the oral cavity and mandible

Fig. 32.7. Measurement sheet for mantle fields for various calculations is completed at the time of simulation

Irregular Field Daily Dose, Output for: on: 3–Apr–90

Point #	1	2	3	4	5	6
	75.6%	72,5%	69.2%	87.3%	79.0%	75.1%
	CA	Mid Media	Low Media	Neck	S. Clav.	Axilla
RX #						
1	150	144	137	173	157	149
2	300	288	275	346	313	298
3	450	432	412	520	470	447
4	600	575	549	693	627	596
5	750	719	687	866	784	745
6	900	863	824	1039	940	894
7	1050	1007	961	1213	1097	1043
8	1200	1151	1098	1386	1254	1192
9	1350	1295	1236	1559	1411	1341
10	1500	1438	1373	1732	1567	1490
11	1650	1582	1510	1905	1724	1639
12	1800	1726	1648	2079	1881	1788
13	1950	1870	1785	2252	2038	1937
14	2100	2014	1922	2425	2194	2086
15	2250	2158	2060	2598	2351	2235
16	2400	2302	2197	2771	2508	2384
17	2550	2445	2334	2945	2665	2533
18	2700	2589	2471	3118	2821	2682
19	2850	2733	2609	3291	2978	2831
20	3000	2877	2746	3464 upper Neck	3135	2980
21	3150	3021	2883	3687	3292	3129
22	3300	3165	3021	3860	3448	3278
23	3450	3309	3158	3984	3605	3427
24	3600	3452	3295	4157	3762	3576
25	3750	3596	3433	4330	3919	3725
26	3900	3740	3570	4504	4075	3874
27	4050	3884	3707	4677	4232	4023
28	4200	4028	3844	4850	4389	4172
29	4350	4172	3982	5023	4546	4321
30	4500	4315	4119	5196	4702	4470
31	4650	4459	4256	5370	4859	4619

Fig. 32.8. Computer output sheet showing cumulative midthickness doses for mantle fields

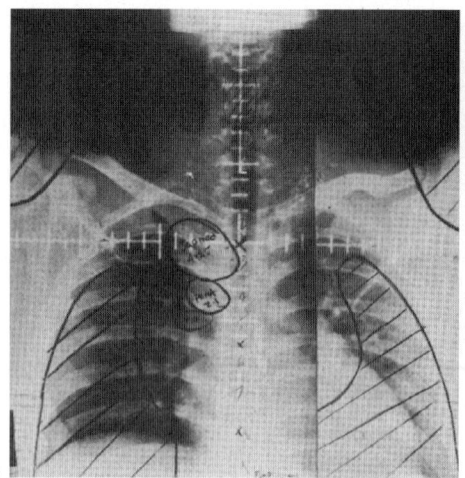
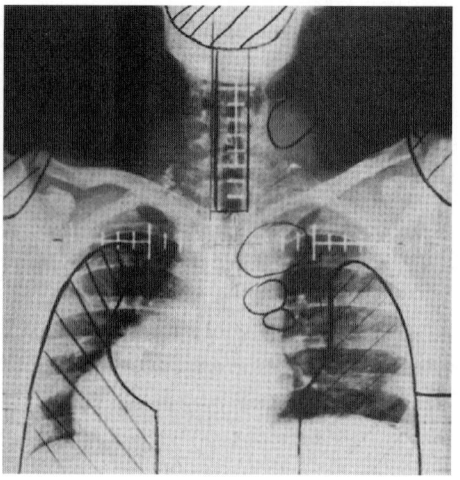

a   b

Fig. 32.9a,b. Techniques used at the University of Minnesota. **a** Anterior mantle field. **b** Posterior mantle field

terior neck nodes and posterior beam exit. Using the same superior border for the posterior field, as in the anterior radiograph, may miss the posterior neck node.

If simulation is done in the prone position for the posterior field, the same steps are followed as described above. Since the junction point for the anterior and posterior field is at the midthickness point, the posterior field border in the prone position will usually fall at one vertebrate higher than the anterior field, as seen in the simulation radiograph because the spine is located in the posterior part of the body relative to the middle of the antero-posterior thickness.

32.4.3
Design Shielding

The key to good mantle field setup is in the placement of lung blocks. Lung blocks must protect as much lung and heart as possible while irradiating macroscopic and microscopic disease to minimize the risk of treatment failure. The most important aspect of the mantle field design is the individualization that is needed in shaping lung blocks to conform to the specific contours of a given patient. Careful design and placement of shielding blocks protects the pulmonary parenchyma from the effects of excessive dose.

On simulation film, or port film, individualized diversion blocks should be designed for both anterior and posterior films that are not necessarily identical but that include identical nodal groups. The actual blocks are custom designed by following the outline

of the block as drawn on the simulation film as well as by the additional use of thoracic CT scans. For uncomplicated cases, the anterior field lung blocks should allow about 2 cm below the medial end of the clavicle to include the infraclavicular lymph node and leave the strip of the lateral part of the lung to include the axillary lymph node with adequate margins. To include the axillary lymph node in the strip, one should pay attention to the patient's arm position (GRANT and JACKSON 1973). Posteriorly, the lung block can be higher than the anterior field and leaves less strip of lung in the infraclavicular area since the lymph nodes in this area are anteriorly located. To include the axilla, there should be falloff outside of the lateral chest wall. It is also necessary to shield soft tissue of the lower lateral chest wall below the level of the fourth intercostal space, unless there is low axillary lymph node involvement.

In the design of lung blocks, medially at least a 1- to 1.5-cm margin should be allowed for any mediastinal shadow – except for the heart – or any mass shadow. The hilum is hard to define radiographically as its anatomical delineation. Therefore, the medial edge of the lung blocks around the hilum can be a source of variation among different institutions. Stanford advocates whole pericardial irradiation to the lower prophylactic doses (15 Gy) for mediastinal disease (CARMEL and KAPLAN 1976; PAGE et al. 1970). This practice is not universally followed.

If there is no gross midline neck disease, trapezoidal or ovoid anterior larynx and posterior cervical spine blocks in addition to lung blocks are recommended. Stanford also recommends an entire posterior thoracic spine block after the delivery of 2000–2500 cGy, as well as a subcarinal block. At the University of Minnesota, a 2-cm-wide posterior cervical block placed above the level of the thoracic inlet after delivery of 2000 cGy to the central axis is used (Fig. 32.10). If there is no presence of subcarinal disease, the subcarinal block can be used after the delivery of 3000 cGy. Radiation dose at each off axis point provides information on when to shield each necessary region. The subcarinal block is placed at the lower edge of the field extending cephalad to within 6 cm of the carina. This appears to significantly reduce the risk of radiation-induced heart disease. When the humeral head is shielded in patients with bulky axillary disease, extreme caution must be used.

After all of the blocks have been drawn, dosimetric calculations should be made at the multiple points of interest. Doses received in these areas can vary considerably because of varying patient thickness in the large field and in the different scatter contributions from the blocks. For example, nodes in the axilla and the high neck receive a greater dose than nodes in the lower mediastinum. Therefore, the dose differences at each site should be calculated (Fig. 32.7) and compensation should be provided by placing the shielding block earlier for the particular anatomic area or by using compensators (Fig. 32.11).

In the presence of inferior mediastinal disease or pericardial extension, the entire cardiac silhouette is irradiated to 15 Gy and then changed to shield the apex of the heart. After a dose of 30–35 Gy is delivered, a block is placed about 5 cm below the carina to provide more cardiac and pericardial protection. However, in no instance do we treat areas of clinical involvement of Hodgkin's disease with doses of less than 36 Gy if radiotherapy alone is planned. Figure. 32.12 represents examples of field modifications and shielding of the extended field which has been used in the CALGB and SWOG inter-group study.

Patients with massive mediastinal disease and/or hilar disease who are treated with radiation alone require further modifications to the lung blocks.

When hilar adenopathy is present, there is substantial risk of subclinical Hodgkin's disease in the lung. There is also an increased risk of subsequent pulmonary relapse; therefore, low-dose lung irradiation has been recommended using thin lung blocks that transmit 37% of the total 1650-cGy dose delivered to the central axis (PAGE et al. 1970; LEE et al. 1979).

In patients with LMM, locoregional recurrence is the major cause of failure (LEE et al. 1980). These

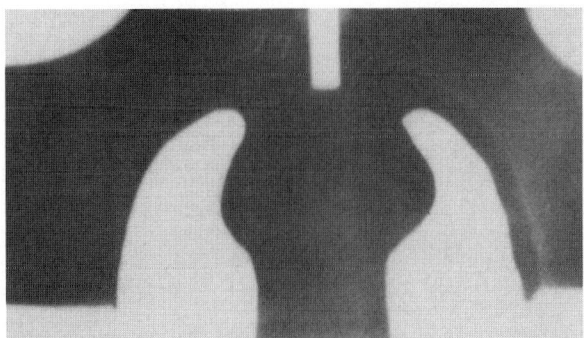

Fig. 32.10. Posterior mantle field port film with posterior cervical spine block

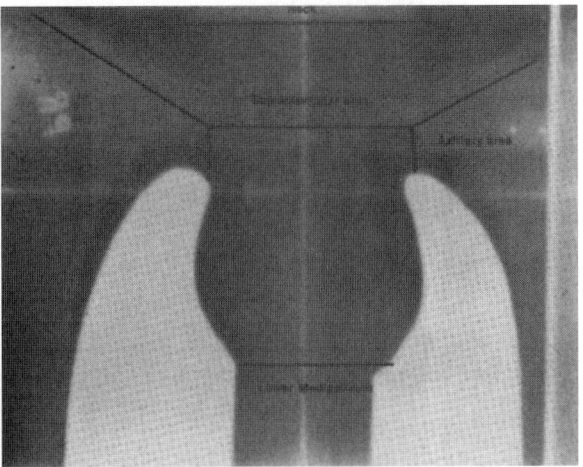

Fig. 32.11. Schematic figure for the shielding of the off-axis areas according to cumulative dose output

patients are treated with low-dose bilateral or unilateral lung irradiation to eliminate microscopic lung disease as well as to shrink tumor mass (Fig. 32.13). Total dose to the lung parenchyma of 1000–1600 cGy in daily doses of 100 cGy is given (LEE et al. 1979). The shrinking field technique should be used several times during the treatment course. One can consider a 10- to 14-day treatment break during the course of radiation treatment to allow for tumor shrinkage, as Stanford originally introduced (JOHNSON et al. 1976) (Fig. 32.14). The technique of reshaping and enlarging lung blocks as the mass responds to treatment reduces the risk of pulmonary toxicity.

Although historically whole lung irradiation was the best radiation treatment to provide optimal survival rates for patients with LMM, most patients currently receive CMT with modifications to the radiation fields and dose to deliver less aggressive treatment (LEOPOLD et al. 1989). An unresolved issue is what volume should be used in patients with LMM

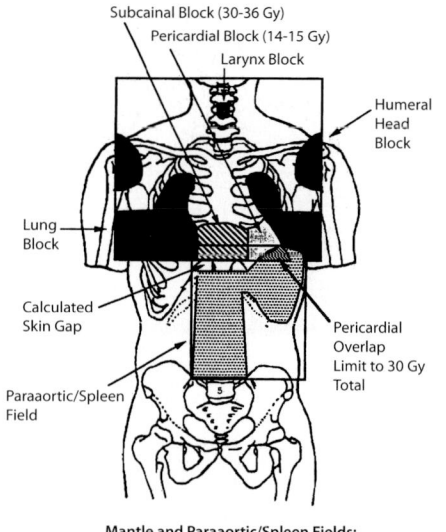

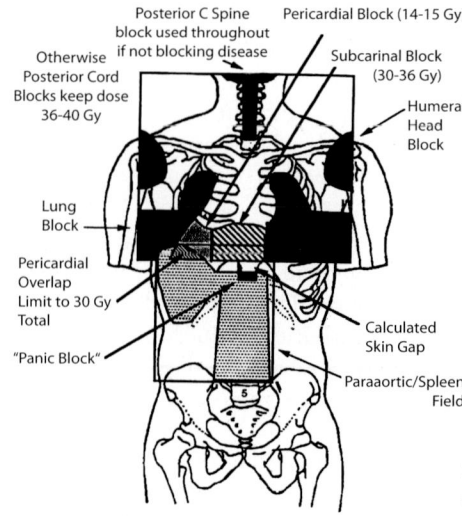

Fig. 32.12. Example of modified mantle and para-aortic spleen fields used in CALGB #9497 and SWOG #9133

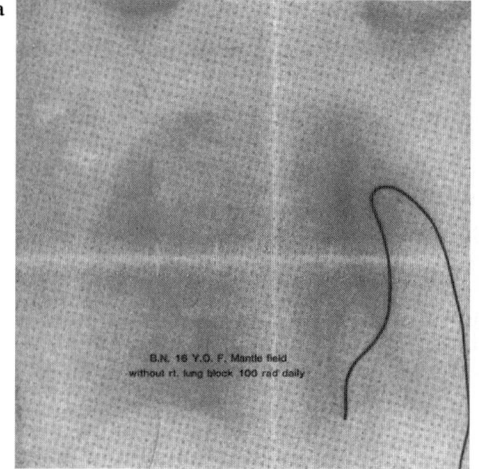

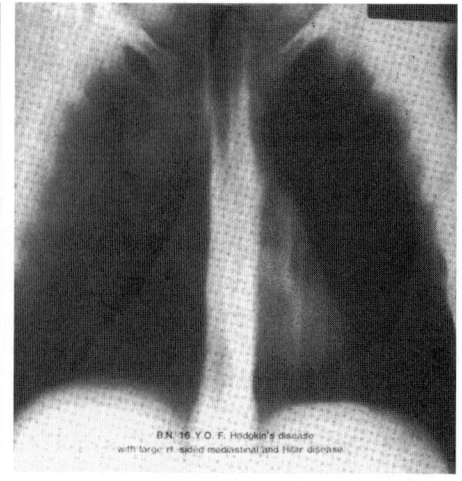

Fig. 32.13a,b. Unilateral lung irradiation for a patient with protruding mediastinal disease on the right side

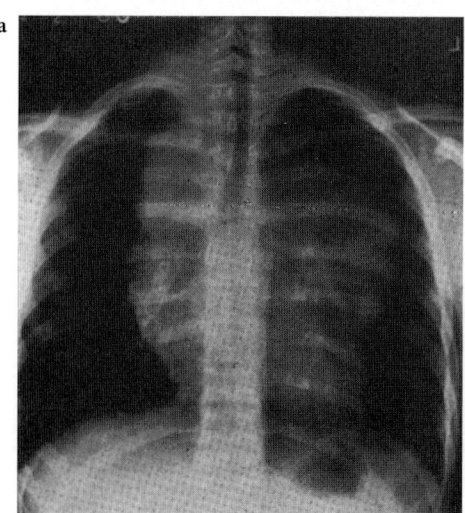

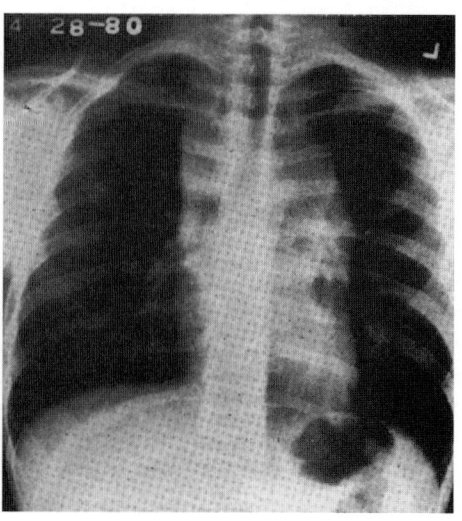

Fig. 32.14a,b. Reduction in size of large mediastinal mass after 1500 cGy to the whole lung using 100-cGy daily fractions

who have received CMT – the volume of post-chemotherapy disease or the original volume of disease. A possible solution is to treat the pre-chemotherapy volume with a low dose (15–18 cGy) followed by a boost to residual disease from 25 Gy to 36 cGy. To reduce lung complications, it is generally accepted to use the post-chemotherapy width of the mediastinal mass (transverse diameter) to reduce the lung volume to be treated with radiation. The longitudinal volume of the field is designed by the extension of the pre-chemotherapy disease volume.

32.4.4
Subdiaphragmatic Fields

The para-aortic field encompasses the para-aortic nodes with splenic pedicle as well as the spleen in clinically staged patients (Fig. 32.15). If the patient is to have a staging laparotomy with splenectomy, surgical clips should be placed to mark the splenic pedicle. The splenic pedicle field is designed with a 2.5- to 3-cm margin around the clips of the left side of the para-aortic field. In patients who did not have a splenectomy but who are receiving STNI, the spleen needs to be included in the field. An ultrasound or CT scan of the abdomen should be used to localize the spleen and to allow blocking of the left kidney as much as possible.

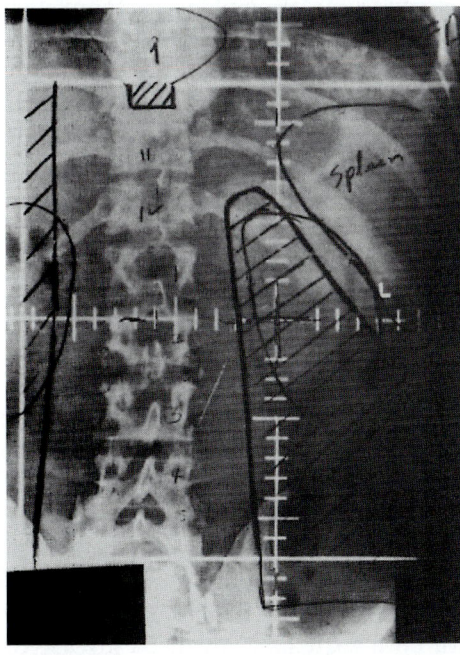

Fig. 32.15. View of a para-aortic field that includes the spleen. University of Minnesota technique

The lower margin of the para-aortic field usually extends down to the L4–L5 interface. In some cases, it can extend below the bifurcation of the aorta to include the common iliac nodes (spade-shape field).

The inverted-Y field is used when the patient has indications for total nodal irradiation. This field includes the para-aortic field and pelvic and inguinal femoral lymph nodes. With the development of CMT, the total nodal field is rarely used. In some situations, the pelvic nodal field is treated separately.

Both Stanford and the University of Minnesota have reported excellent results of low-dose liver irradiation in pathological stage IIIA patients (HOPPE et al. 1980; LEE et al. 1984) (Fig. 32.4). In recent practice, this is almost historic since most of these patients will receive CMT.

Careful blocking is required when the pelvic region is treated. The fields must be shaped carefully to minimize the amount of marrow treated. If a lymphogram has been performed, a more precise delineation of the pelvic field is possible and more bone marrow can be spared. Gonadal toxicity may also be an issue in women, since the ovaries normally overlie the iliac lymph nodes, and oophoropexy must be performed to avoid irradiation-induced amenorrhea (LEFLOCH et al. 1976). The surgeon marks the ovaries with radiopaque sutures or clips and places them medially and as low as possible behind the uterine body. A double thickness [10 half-value layer (HVL)] midline block is then recommended and its location is guided by the position of the opacified nodes and transposed ovaries. When the ovaries are at least 2 cm from the edge of this block, the dose is decreased to 8% of that delivered to the iliac nodes (LEFLOCH et al. 1976).

In males, the testicular dose may be as high as 10% of the dose delivered to the inguinal femoral nodes if no special blocking is provided for the testes. The dose is largely from internal scatter due to the proximity of the position of the testes in relation to the inferior margin of the inguinal femoral field (CARMEL et al. 1976). Use of a double thickness midline block and a specially constructed testicular shield can reduce this dose to 0.75–3.0%.

32.4.5
Preauricular and Waldeyer's Ring Fields

The preauricular lymph nodes are treated either when there is involvement of the preauricular nodes or of the high cervical lymph nodes above the level of the thyroid notch. The superior border is placed at the top of the zygomatic arch such that the sphenoid sinus is included in the field. The posterior border is at the external auditory canal and anteriorly the field extends up to the third molar. The inferior border is matched, on skin, to the divergence of the mantle field. Most commonly, this area is treated with about 9 MeV unilateral electron field to spare the parotid in the uninvolved side.

Waldeyer's Ring is treated in patients with involvement of the region or with bulky superior cervical lymph nodes. This volume is treated with opposed lateral photon beams. The treatment volume includes occipital, preauricular, and submandibular lymph nodes as well as the nasopharynx, tonsillar fossa, and base of the tongue. The superior border is placed at the top of the zygomatic arch, the anterior border includes the submandibular triangle, and the posterior border extends beyond the spinous processes. The inferior border is a direct match to the superior border of the mantle field. The match line is placed inferior to any palpable cervical disease to avoid underdosing the disease or overdosing the spinal cord (Fig. 32.16). With disease regression after 2500–3000 cGy, the Waldeyer's field is discontinued at some institutions. The treatment volume is changed to a mantle with a high superior border and matched to a unilateral preauricular field, thereby decreasing the risk of xerostomia.

32.4.6
Gap of Matching Fields

When two fields are matching on the skin, a potential overlap area will be created between the fields. With sequential treatment of separate mantle, para-aortic, or inverted Y fields, calculation of the gap between mantle and subdiaphragmatic fields is very important. Beam diversion from supra- and infradiaphragmatic fields creates the potential for field overlap and subsequent radiation doses that exceed spinal cord tolerance. A well-planned matching technique is essential and appropriate gap calculations should be applied. If the patient is being treated in the supine and prone positions, a four-field match can be created at the patient's midseparation.

A number of different matching techniques have been published (Lutz and Larsen 1983; Fraass et al. 1983). The gap can be calculated at the midline of the body or at the spine to create a junction at the same vertebrae. The depth of the matching point length of source-skin distance or source-axis distance and field sizes (both mantle and subdiaphragmatic fields) will dictate the divergence of the beam and will show the different levels of vertebrae between the anterior and posterior radiographic fields.

When mantle (AP/PA) and subdiaphragmatic fields are treated sequentially and field length of upper and lower fields is often different, longer fields diverge into the opposing smaller fields. When four fields meet, three fields can be overlapped and total dose of the overlapped area may exceed the central axis dose and it can be in the middle of the body thickness or at the vertebrae. This is a concern especially if the overlapped area is the spinal cord. Therefore, it is important to use small blocks to shield the spinal cord area for the second pair of the fields in addition to calculating the proper skin gap using the formula in Figure 32.17. This figure shows the geometry of adjacent beams that join at a given depth to calculate skin gap. Isodose distribution at the midline is almost uniform but cold spots are created anterior and posterior at the junction point (Fig. 32.18).

The tattoo at the inferior margin of the mantle field usually helps to show where the skin gap should be given when subdiaphragmatic fields are simulated with a gap. It is very important to verify the inferior margin of the mantle fields at the time of simulation.

Actual treatment of the patient is an important factor after precise simulation plans are completed. In a review of portal films from five major universities in the United States, the Patterns of Care study showed an 11% variation among the institutions (Kinzie et al. 1983). The most frequent inadequate margins were the mediastinum, the hilum, and the axillae. Infield or marginal recurrence was 8% in patients with adequate margins and 32% in patients with inadequate margins.

32.5
Limited Field
(Involved Field/Regional Field)

In recent years, limited or regional irradiation or treatment to the involved field has been used with combination modality treatment. Involved field has become the standard field for irradiation in patients with both early and advanced disease to reduce morbidity and achieve a compatible outcome to extended field radiation. Currently, there

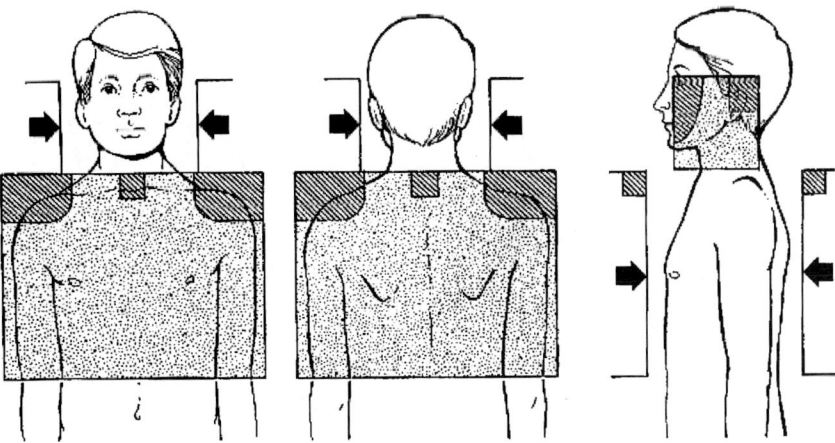

Fig. 32.16. Matching for head and neck and mini-mantle fields

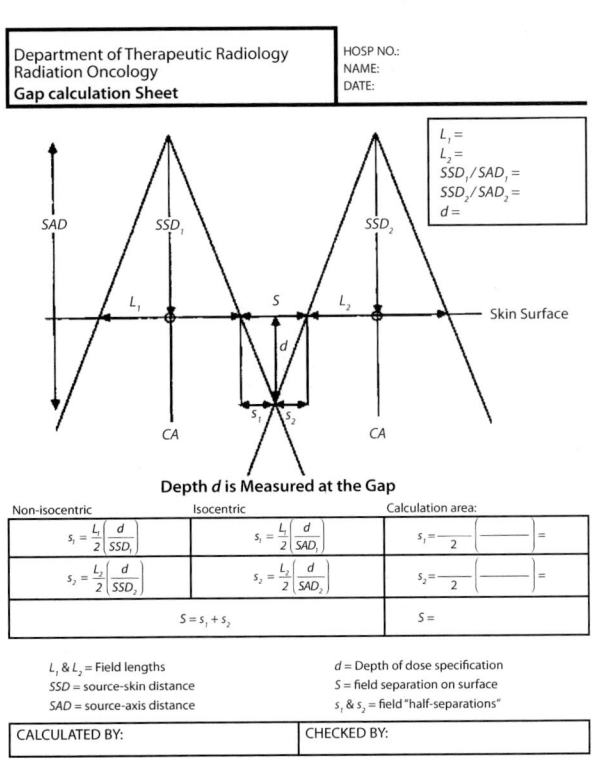

Fig. 32.17. Gap calculation measurement sheet

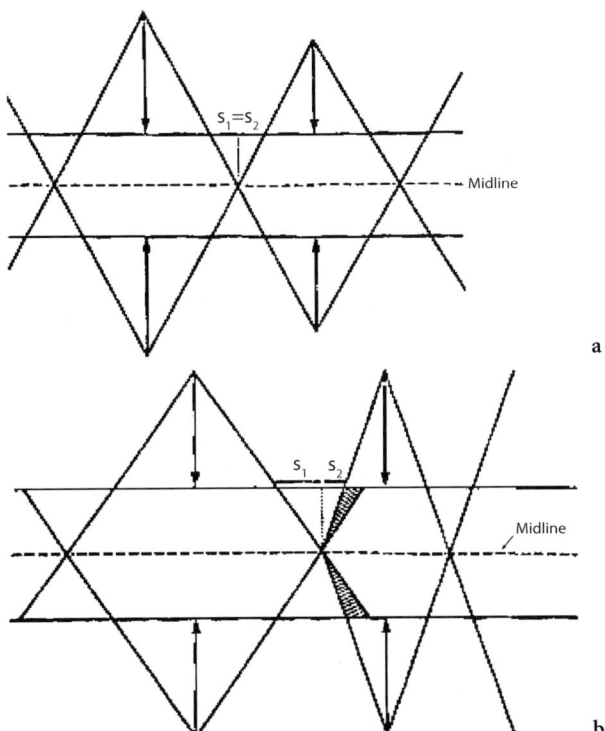

Fig. 32.18a,b. Two pairs of parallel opposed fields. Adjacent fields are separated on the surface so that they all join at the point on the midline. **a** Ideal geometry in which there is no three-field overlap. **b** Arrangement in which there are two regions (shaded) of three-field overlap. (From KHAN 1994, p 335)

is a trend toward decreasing the radiation field size in the form of consolidation therapy following chemotherapy in aggressive lymphoma. Results from a recent study that examined the role of consolidation radiotherapy following chemotherapy in patients who achieved CR after six cycles of ABVD showed that the addition of RT to CT significantly improved event-free survival and overall survival compared with chemotherapy alone (89% versus 76%, $P=0.01$; 100% versus 88%, $P=0.002$, respectively). Consolidation RT was particularly beneficial in younger patients (<15 years) and in those with B symptoms, advanced stage, and bulky disease (LASKAR et al. 2004).

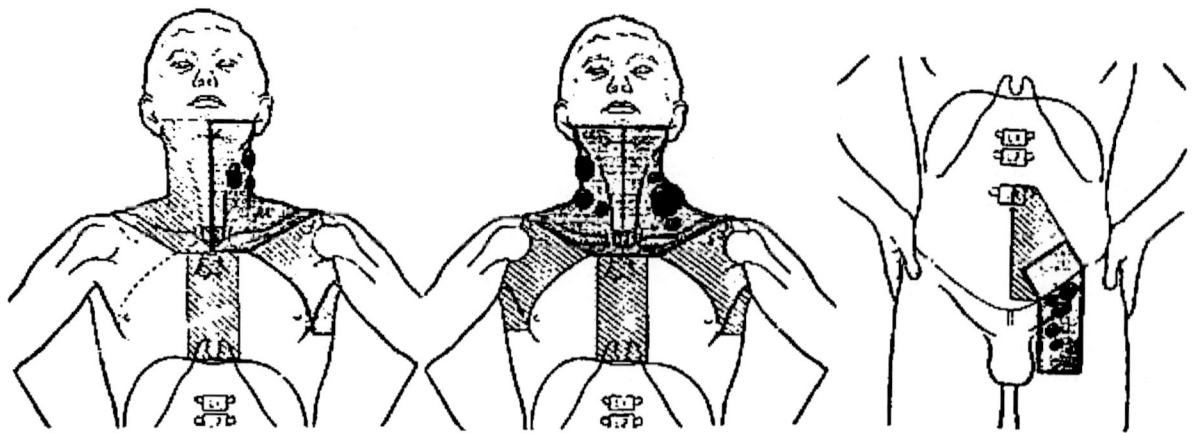

Fig. 32.19. Limited and regional fields used by the BNL1 study for early stage Hodgkin's disease. *Slash-marked area* indicates prophylactic treatment area. These fields have been used in combined modality treatment for primary or relapsed disease

Figure 32.19 shows an example of local and regional fields used by the British National Lymphoma Study (HAYBITTLE et al. 1985). Involved fields are usually designed with the concept of the lymph node regions diagram presented at the RYE Symposium in 1966 for staging purposes. The design is to treat the whole region of the lymph node chain, not just a single lymph node. With one positive cervical lymph node in the neck, the involved field is designed to include the entire ipsilateral neck and the superclavicular area.

The question arises whether to treat the pre-chemotherapy versus post-chemotherapy volume. The pre-chemotherapy volume and site is usually used as the initial port with the exception of the bulky mediastinal and para-aortic lymph nodes. Involved fields usually allow 2- to 5-cm margins around the tumor sites. Whenever possible, shrinking field techniques should be used (Figs. 32.3 and 32.4). It is important to review pre-chemotherapy and post-chemotherapy imaging studies at the time of radiation field design to identify the pathological disease site.

The definition of bulky disease can vary, ranging from 3 cm to 5 cm in diameter. Bulky mediastinal mass is usually defined as a mass of one-third the thoracic transverse diameter. It is also difficult to define what is residual disease. Some patients have residual soft tissue density on CT scan, yet PET scan does not show any activity. If there is any soft tissue densities that have been gradually reduced by chemotherapy, these areas should be irradiated.

There are no clear guidelines established on the size and volume of the radiation field for post-che-

motherapy field design. The Cancer and Leukemia Group B has made some guidelines for CALGB Protocol 59905 (ECOG 2496/SWOG2496). The University of Minnesota approach is similar to these guidelines.

32.6
Limited Field Above the Diaphragm

Schematic guidelines of limited fields above the diaphragm are shown in Fig. 32.20.

32.6.1
Cervical/Supraclavicular Region

The entire ipsilateral cervical/supraclavicular regions should be treated if bulky disease is present in any part of this region. If the disease extends to the midline and/or both sides of neck, the bilateral cervical/supraclavicular regions should be treated.

32.6.1.1
Unilateral Cervical/Supraclavicular Region

If there is involvement at any cervical lymph node with or without supraclavicular disease, arms can be in the extended position, akimbo, or at the sides. The upper border is placed 1–2 cm above the lower tip of the mastoid process and through the lower border of the mandible. If tumor is at the mid-

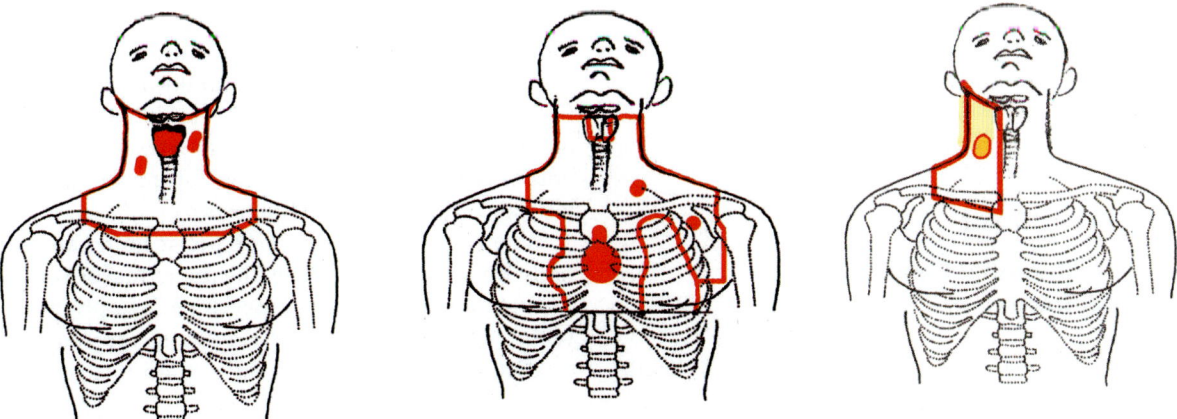

Fig. 32.20. Schematic guidelines of limited fields above the diaphragm

or lower neck area, the upper border should be below the mastoid tip. The lower border is 2 cm below the bottom of the clavicle. The lateral border includes the medial two-thirds of the clavicle. If the supraclavicular nodes are not involved, the medial border of the field is placed at the edge of ipsilateral vertebrae. When nodes close to the vertebral bodies or near midline of the neck are seen on the initial staging of the next CT scan, the entire vertebral body should be included and allow margins around the region. When the unilateral supraclavicular nodes are involved, the field border should be placed at the contralateral traverse processes of vertebrae. If the larynx is included in the field, it may be blocked at the start of the treatment or the dose of 15–20 Gy if it is at all possible. The humeral heads may be blocked anteriorly and posteriorly. A posterior cervical cord block is required if the cord dose exceeds 40 Gy. Mid-neck calculations should be performed to determine the maximum cord dose, especially when the central axis is in the mediastinum.

32.6.1.2
Bilateral Cervical/Supraclavicular Region

Treat both cervical and supraclavicular regions as described above, regardless of the extent of disease on each side. Posterior cervical cord and larynx blocks should be used as described above. Use a posterior mouth block if treating the patient supine (preferably with an extended travel couch at a source skin distance greater than 100 cm) to block the upper field divergence through the mouth. The chin should be marked anteriorly with a radio-opaque material to aid in drawing the block.

32.6.2
Mediastinum/Hilar/Axillary Region

The mediastinal field should encompass the potential residual disease present after the completion of 4–6 cycles of chemotherapy and should be a shaped field encompassing the mediastinum. The field should have a 1.5-cm margin laterally and inferiorly and can be extended at least 2–3 cm below the lower extent of disease at the time of initial presentation, or 5 cm of the post-chemotherapy volume. If there is disease in the mid-mediastinum or the carina area, the lower margin of the field should be placed 5 cm below the carina and include the bilateral hilar regions. The entire cardiac silhouette need not be treated, except when disease is extending along the cardiac border. The dose to the pericardium is limited to 1500 cGy with a 100-cGy daily dose.

When there is bulky upper mediastinal disease, the portal should also include the bilateral supraclavicular regions, even if uninvolved. The superior margin is at the level of the superior border of the larynx if supraclavicular adenopathy is present without other neck disease or at the inferior border of the larynx if there is no supraclavicular adenopathy. Intrathoracic sites of extralymphatic extension (e.g., lungs, pleura, chest wall, pericardium) should also be included if these extralymphatic sites of extension are considered to be part of the patient's initial bulky disease.

When there is mediastinum, cervical, and supraclavicular lymph node disease, the shape of the field is a modified mantle field without the axillary region. When there is additional axillary disease, the modified mantle field will be used. In case of axillary involvement only, infraclavicular and supraclavicular regions are treated with the arm position of regular mantle field. The lower border is 2 cm below the scapula or insertion of latissimus dorsi muscle.

32.7
Limited/Regional Field Below the Diaphragm

Schematic guidelines of limited fields below the diaphragm are shown in Figure 32.21.

32.7.1
Inguinal/Femoral/External Iliac Lymph Node Area

The most common field below the diaphragm is the inguinal/femoral/external iliac lymph node area. The superior border of this field is placed at the

superior border of the sacroiliac joint area down to 5 cm below the lesser trochanter to include the femoral triangle inferiorly. The lateral border extends to the lateral aspect of the greater trochanter or 2 cm outside the initial gross disease site. The medial border is placed at the obturator foramen or at 2 cm medial to the initial gross disease.

When common iliac or high pelvic nodes are initially involved with disease, the upper margin of the field should include lower periaortic lymph nodes. When both pelvic and groin areas are required to be treated, one should pay special attention to ovarian and testicular doses.

32.7.2
Paraaortic Lymph Nodes

The paraaortic field is designed to include lymph nodes at the level of the T10–T11 to L4–5 vertebral interspace. The width of the para-aortic field should conform to the volume necessary to treat residual

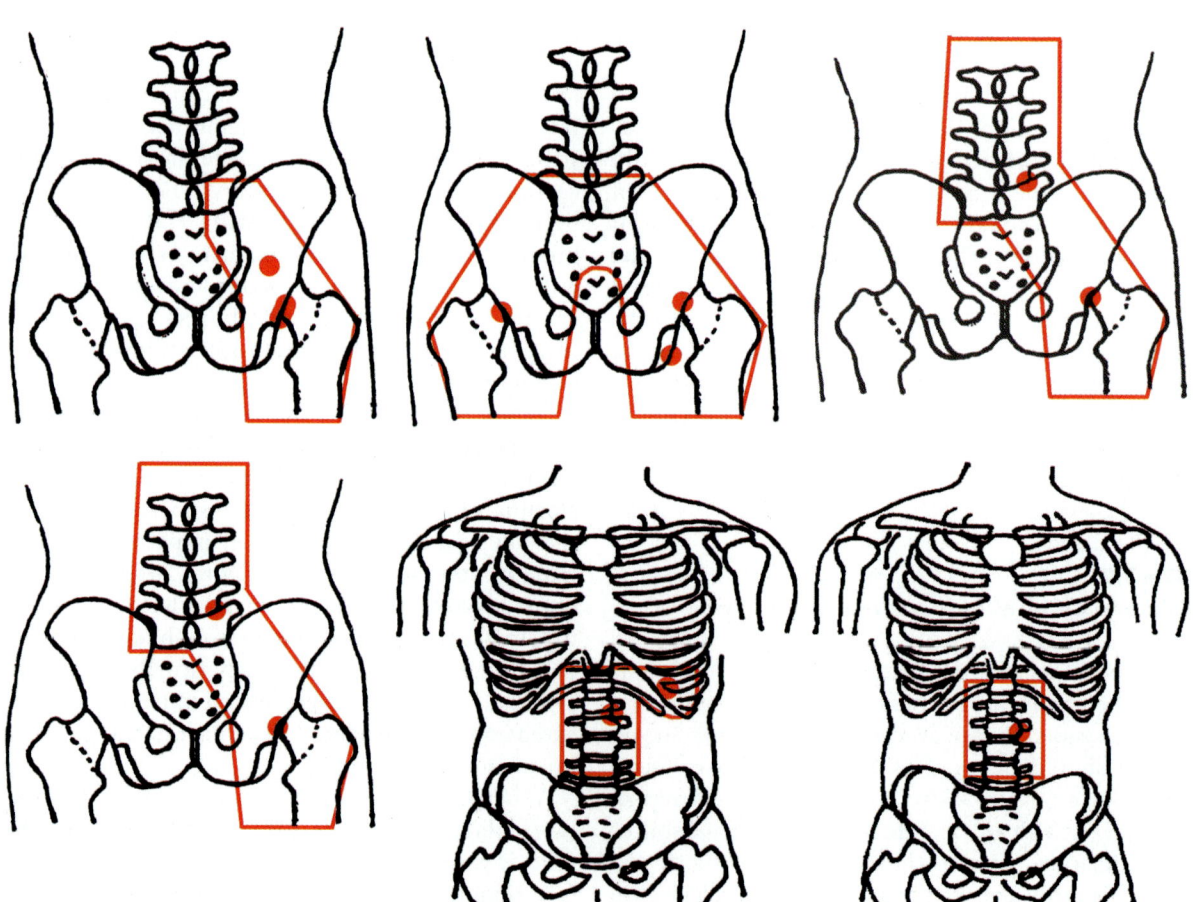

Fig. 32.21. Schematic guidelines of limited fields below the diaphragm

disease after chemotherapy with a 1.0- to 1.5-cm margin. The superior margin will be matched to the mediastinal field if treated.

32.7.3
Spleen

The spleen is one of the most common sites of disease in the abdomen. If the spleen needs to be treated, one should pay special attention to the left kidney. It should be outlined on the computer planning CT scan and every effort should be made to avoid including more than 40% of the left kidney. The spleen will move with respiration and diaphragmatic motion. This should be weighed as much as possible, especially if a small localized lesion in the spleen is being treated.

32.8
Radiation Dose and Fractionation

An optimal radiation dose results in a high rate of local disease control and acceptable side effects. Several factors are involved in determining radiation dose, including disease extent (macroscopic or microscopic), treatment modality (radiation alone or combined with chemotherapy), total dose delivered, and number of daily fractions.

32.8.1
Radiation Therapy Alone

There has been controversy regarding the total curative doses for the treatment of Hodgkin's disease (SCHEWE et al. 1989; VIJAYAKUMAR et al. 1989). Early studies by Peters showed that improvement in the 5-year survival in 319 stage-I patients treated with prophylactic irradiation to the subclinical area depended on the total dose delivered and the extent of treatment (PETERS 1966). Involved areas received more than 2500 cGy, and the adjacent uninvolved area received prophylactic irradiation of less than 2500 cGy. A review of the literature in 1966 by Kaplan reported that local control of Hodgkin's disease patients with radiation therapy was dose dependent (KAPLAN 1966, 1980). Doses of at least 4400 cGy over 4–5 weeks achieved a 98.6% infield control rate and doses of 3500–4000 cGy resulted in a recurrence rate of 4.4%. Subsequent analysis of

most of the same data by Fletcher and Shukovsky showed a 93% control rate with 3000 cGy or less and a 97% control rate for the doses 3100–4000 cGy. It was concluded that a dose of 3500 cGy in 6 weeks was needed to control Hodgkin's disease (FLETCHER and SHUKOVSKY 1975).

A recent compilation of dose–response data by Vijayakumar from a megavoltage series demonstrated a 90% control rate of subclinical disease with 22.1 Gy and 95% control for 27 Gy (VIJAYAKUMAR and MYRIANTHOPOULOS 1992). In patients with lymph nodes smaller than 6 cm, the corresponding doses for 50%, 90%, and 95% control were 26.7 Gy, 33.4 Gy, and 35.2 Gy. For nodes larger than 6 cm, the majority of patients received doses of more than 36 Gy. This retrospective analysis of patients treated at the University of Chicago was performed to assess the dose–response in patients treated in a more uniform manner in a single institution (MYRIANTHOPOULOS et al. 1995). Local control of subclinical disease was 100% with 25–35 Gy. For gross disease, there was also no dose dependence in the range of 35–55 Gy. Local control in gross disease was 98.3% with 35–45 Gy.

A trial from the German Hodgkin's Lymphoma Group randomized early-stage Hodgkin's disease patients to 40-Gy extended field irradiation versus 30-Gy extended field plus 10-Gy involved field irradiation (DUEHMKE et al. 1995). No infield relapses were noted in the extended field volumes, and there was no significant difference in overall survival or freedom from failure. Patients with bulky disease (mass of greater than 6 cm or LMM) commonly receive more than 36 Gy, although the need for a higher dose has not been well studied.

In summary, based on available data on the treatment of Hodgkin's disease with radiation, the American College of Radiology in their Patterns of Care Study recommended that optimal doses for local control were 36–44 cGy for treatment of involved portions of the field and 30–36 cGy for prophylactic portions of the fields (HOPPE 1990-1991).

32.8.2
Radiation Dose in CMT

The doses recommended for patients undergoing CMT are not as specific and well established. Most of the dose–response data with CMT are from data on pediatric and advanced Hodgkin's disease. Based on the experience of the German Austrian Pediatric Study Group, 18–20 Gy to subclinical sites after chemotherapy was effective as high dose radiation (SCHELLONG et al. 1988). Dose–response data from this cooperative group for childhood Hodgkin's disease show about a 95–97% control rate with 15–25 Gy to gross disease sites in patients treated with CMT (SCHELLONG et al. 1994; DONALDSON and LINK 1987; OBERLIN et al. 1992). When treated with CMT, patients who respond poorly to initial chemotherapy show a poor rate of control using 25 Gy. In adults, it is general practice to use 25–30 Gy for a completely responded area of previous bulky disease and 36–40 Gy to gross disease despite previous chemotherapy.

32.8.3
Radiation Dose in Salvage Treatment

For patients who relapse after primary treatment (either chemotherapy or radiation therapy) and who are treated with salvage chemotherapy, low doses of radiation can be used to sites previously irradiated if maximum tissue tolerance has not been exceeded. Higher doses of radiation should be considered if no previous radiation therapy was given to the involved sites. The radiation field size is usually confined to the loco-regional area. Total dose of radiation for these patients is similar to that for patients treated with primary chemotherapy (PROSNITZ 1990; NOORDIJK et al. 1994).

32.9
Normal Tissue Tolerance and Complications

32.9.1
Lung

The lung parenchyma is the most sensitive organ in the thorax to irradiation. Doses as low as 15 Gy can cause acute pulmonary reactions. Areas of lung treated with 35–44 cGy often show fibrosis during long-term follow-up, with varying degrees of severity depending on the volume treated, total dose used, and fraction size. Both lungs may be treated safely with a dose as high as 15–16 cGy, especially if partial transmission lung blocks are used or open fields at a 75- to 100-cGy daily dose rate (HENRY-AMAR et al. 1991; PALOS et al. 1971; LEE et al. 1979). When patients received 1650 cGy in 10 fractions, 35% developed pulmonary symptoms whereas only 15% developed these symptoms after the same dose in 20 fractions.

Radiation pneumonitis may develop 6–12 weeks after completion of mantle irradiation (CARMEL et al. 1976). The likelihood of radiation-related pulmonary complications may increase with the additional use of chemotherapy, such as bleomycin (COSSET et al. 1991).

32.9.2
Heart

Symptomatic pericarditis is rare with entire pericardium doses of 30 cGy or less. However, subclinical injury to the pericardium may occur at lower doses. Partial field blocking (apical portion of heart at 15 cGy, subcarinal portion of the heart at 30 cGy) is associated with a decreased risk of pericarditis. The myocardium is slightly more resistant than the pericardium, but doses of 35 cGy can cause a higher risk of coronary heart disease.

Heart irradiation modestly increases the relative risk of coronary artery disease and myocardial infarction. To modify the risk, attention should be paid to the total dose, the radiation port used, the size of the left ventricular block, and the fraction size. Long-term cardiovascular complications include coronary artery disease, pericarditis, pancarditis, and valvular disease (HANCOCK et al. 1993b; HANCOCK and HOPPE 1996). In an analysis of factors affecting the risk of cardiac-related deaths with long-term follow-up, patients who received a dose of 30 Gy to the mediastinum had a 3.5 times relative risk of cardiac-related mortality compared with a 2.6 relative risk in patients who received less than 30 Gy (HANCOCK and HOPPE 1993). The risk of coronary artery disease appears to be primarily related to mediastinal irradiation; however, chemotherapeutic agents, such as anthracyclines, are also potential cardiotoxic agents. Therefore, it is a concern when combined modality that includes cardiotoxic chemotherapy and radiation is considered.

32.9.3
Central Nervous System

The Lhermitte sign develops in about 10–20% of patients after mantle field irradiation. It may be related to transient demyelinization of the spinal cord occurring 1–2 months after completion of treatment and usually resolves completely 2–6 months after.

Spinal cord tolerance is reported to be reached at dose levels of 45–50 Gy and above, with the length and location of the area involved affecting total dose tolerance. Although spinal cord tolerance should not be an issue with standard lymphoma therapy, extreme attention to treatment details should be followed. These details, all of which demand conservative dose guidelines, include attention to the length of the spinal cord included in the field, the expected success of treatment, a long follow-up, and the use of contiguous fields that overlap the spinal cord. Lateral simulator radiographs or treatment planning CT scans may be used to determine the doses at key points along the length of the spinal cord. The cord dose should be limited to no more than 40 cGy in posterior spinal cord blocking.

32.9.4
Thyroid

Thyroid glands are usually included when the neck is treated. Subclinical hypothyroidism develops in about half of the patients who receive regular mantle fields (HANCOCK and HOPPE 1996). It is important to check the long-term thyroid function after mantle field treatment.

32.9.5
Liver

A portion of the left lobe of the liver is included in the para-aortic field. This may cause a transient elevation of the serum alkaline phosphatase, but it is not associated with long-term sequelae. If necessary, the entire liver may be treated safely with a dose up to 25 cGy using protracted fractionation; higher doses should be avoided.

32.9.6
Gastrointestinal Tract

The gastrointestinal tract (GI) generally tolerates doses up to 44 Gy. After doses of 35 Gy or more, there is some risk. Small-bowel obstruction could occur where intra-abdominal adhesions have formed after a staging laparotomy. The Patterns of Care review on complications after intradiaphragmatic field irradiation reported that 6% of major bowel complications occurred after 4000–4500 cGy (COIA and HANKS 1988). The incidence of GI complications was

especially high in patients with previous GI problems (19%).

32.9.7
Head and Neck

Xerostomia may develop after mantle or head and neck field irradiation, and prophylaxis including daily fluoride treatments is recommended. A dental examination before irradiation should be done for all patients who receive mantle field and Waldeyer's field irradiation to minimize long-term complications.

32.9.8
Reproductive Organs

The ovaries are the most sensitive organ in the pelvis, particularly in those women over age 30 years. The gonads are sensitive. Even with oophoropexy and shielding for the relocated ovaries and with testicular shielding, the scattered dose of irradiation may be sufficient to develop menopause and aspermia. Mantle field irradiation delivers 0.2–0.5% of the prescribed dose to the testis, 0.5–1.0% from the para-aortic field, and 5–10% from the pelvic field (MILLER et al. 1995). Alkylating chemotherapy agents also affect the gonads, especially in women older than 30 years (HORNING et al. 1988). Combined modality with chemotherapy and pelvic irradiation should be cautiously designed and used because of possible further damage to the gonads as well as to bone marrow reservation.

32.9.9
Secondary Neoplasms

One of the long-term sequelae of radiation is the development of secondary cancers, including solid tumors and leukemia-lymphoma. Overall, there is a 6.4 times relative risk of developing a secondary malignancy after treatment for Hodgkin's disease and an 84% absolute risk (HANCOCK and HOPPE 1996). The incidence of breast cancer in female patients has been increasing after mantle irradiation. Secondary solid tumors, such as lung, breast, and GI cancers, are most likely related primarily to radiation therapy and develop with a longer latent period of 7–10 years than soft cancers, such as lymphoma or leukemia (HANCOCK et al. 1993a;

BIRDWELL et al. 1995; VAN LEEUWEN et al. 1995; MAUCH et al. 1995b).

Long-term follow-up of treated patients provides important information on outcomes and complications of treatment, which allow for modification of irradiation techniques to improve results (HANCOCK and HOPPE 1996). Continual attempts should be made to reduce any life-threatening consequences of treatment and to balance complications with the benefits of treatment.

References

Agnarsson BA, Kadin ME (1989) The immunophenotype of Reed-Sternberg cells: a study of 50 cases of Hodgkin's disease using fixed frozen tissues. Cancer 63:2083-2087

Anonymous (1993) The Ann Arbor Staging Classifications. CA: A Cancer Journal for Clinicians 43:333

Aragon de la Cruz G, Cardenes H, Otero J et al (1989) Individual risk of abdominal disease in patients with stages I and II supradiaphragmatic Hodgkin's disease. Cancer 63:1799-1803

Birdwell SH, Hancock SL, Varghese A et al (1995) Gastrointestinal cancer after treatment of Hodgkin's disease. Int J Radiat Oncol Biol Phys 37:67-73

Biti Ga, Cimino G, Cartoni C et al (1992) Extended-field radiotherapy is superior to MOPP chemotherapy for the treatment of pathologic stage I-II A Hodgkin's disease: eight-year update of an Italian prospective randomized study. J Clin Oncol 10:378-382

Bonadonna G (1994) Modern treatment of malignant lymphomas: a multi-disciplinary approach? Ann Oncol 5 [Suppl 2]:5-16

Bonfante V, Viviani S, Devizz IL et al (2001) Ten years experience with ABVD plus radiotherapy: subtotal nodal (STNI) vs involved-field (IFRT) in early stage Hodgkin's disease (abstract). Proceedings of ASCO 281a

Brada M, Easton D, Horwich A et al (1986) Clinical presentation as a predictor of laparotomy findings in supradiaphragmatic stage I and II Hodgkin's disease. Radiother Oncol 5:15-22

Carbone PP, Kaplan HS, Musshoff K et al (1971) Report of the committee on Hodgkin's staging classification. Cancer Res 31:1860

Carmel RJ, Kaplan HS (1976) Mantle irradiation in Hodgkin's disease. An analysis of technique, tumor eradication and complications. Cancer 37:2813-2825

Carmel RJ, Palos BB, Duggan JP et al (1976) Testicular shielding of patients receiving inverted-Y irradiation. Int J Radiat Oncol Biol Phys 1 [Suppl 11]:61

Castellino RA, Hoppe RT, Blank N et al (1984) Computed tomography, lymphography, and staging laparotomy: correlations in initial staging of Hodgkin's disease. Am J Radiol 143:37-41

Castellino RA, Blank N, Hoppe RT et al (1986) Hodgkin's disease: contributions of chest CT in the initial staging evaluation. Radiology 160:603-605

Coia LR, Hanks GE (1988) Complications from large field intermediate dose infradiaphragmatic radiation: an analysis of the Patterns of Care Outcomes Studies for Hodgkin's disease and seminoma. Int J Radiat Oncol Biol Phys 15:29-35

Cosset JM, Henry-Amar M, Meerwaldt JH et al (1992) The E.O.R.T.C. trials for limited stage Hodgkin's disease. Eur J Cancer 11:1847-1850

Cosset JM, Henry-Amar M, Meerwaldt JH (1991) Long-term toxicity of early stages of Hodgkin's disease therapy: the EORTC experience. EORTC Lymphoma Cooperative Group. Ann Oncol 2:77-82

Crnkovich MJ, Leopold K, Hoppe RT et al (1987) Stage I to IIB Hodgkin's disease: the combined experience at Stanford and the Joint Center for Radiation Therapy. J Clin Oncol 5:1041-1049

De Vita VT, Hellman J, Rosenberg SA et al (2005) Cancer principles and practice of oncology, 7th edn. Williams and Wilkins, Lippincott

De Wit M, Bohuslavizki KH, Buchert R et al (2001) 18FDG-PET following treatment as valid predictor for disease-free survival in Hodgkin's lymphoma. Ann Oncol 12:29-37

Diehl V, Loeffler M, Pfreundschuh M et al (1995) Further chemotherapy versus low-dose involved-field radiotherapy as consolidation of complete remission after six cycle of alternating chemotherapy in patients with advanced Hodgkin's disease. Ann Oncol 6:901-910

Doggett Rs, Colby TV, Dorfman RF (1983) Interfollicular Hodgkin's disease. Am J Surg Pathol 7:145-149

Donaldson SS, Link MP (1987) Combined modality treatment with low-dose radiation and MOPP chemotherapy for children with Hodgkin's disease. J Clin Oncol 5:742-749

Duehmke E, Diehl V, Loeffler M et al (1995) Randomized trial with early stage Hodgkin's disease testing 30 Gy vs. 40 Gy extended field radiotherapy alone (abstract). Int J Radiat Oncol Biol Phys 32 [Suppl 1]:213

Engert A, Schiller P, Pfistner B et al (2001) Involved field (IF) radiotherapy is as effective as extended field (EF) radiotherapy after 2 cycles of COPP/ABVD in patients with intermediate-stage HD (abstract). Blood 98:768a

Fabian CJ, Mansfield CM, Dahlberg S et al (1994) Low-dose involved field radiation after chemotherapy in advanced Hodgkin Disease. Ann Intern Med 120:903-912

Fletcher GH, Shukovsky LJ (1975) The interplay of radio-curability and tolerance in the irradiation of human cancers. J Radiol Electro 56:383-400

Fraass B, Tepper J, Glatstein E, van de Geijn J (1983) Clinical use of matchline wedge for adjacent megavoltage radiation field matching. Int J Radiat Oncol Biol Phys 9:209

Friedman S, Henry-Amar M, Casset JM et al (1988) Evolution of erythrocyte sedimentation rate as predictor of early relapse in post-therapy early-stage Hodgkin's disease. J Clin Oncol 6:596

Front D, Israel O (1995) The role of Ga-67 scintigraphy in evaluating the results of therapy of lymphoma patients. Semin Nucl Med 25:60-71

Fuller L, Hagemeister F, North L et al (1988) The adjuvant role of two cycles of MOPP and low-dose lung irradiation in stage IA through IIB Hodgkin's disease: preliminary results. Int J Radiat Oncol Biol Phys 14:683-692

Gilbert R (1939) Radiotherapy in Hodgkin's disease (malignant granulomatosis). Am J Roentgenol 41:198-241

Grant L, Jackson W (1973) An investigation of the mantle technique. Clin Radiol 24:254-262

Hancock SL, Hoppe RT (1996) Long-term complications of treatment and causes of mortality after Hodgkin's disease. Semin Radiat Oncol 6:225-242

Hancock SL, Tucker MA, Hoppe RT (1993a) Breast cancer after treatment of Hodgkin's disease. J Natl Cancer Inst 85:25-31

Hancock SL, Tucker MA, Hoppe RT (1993b) Factors affecting late mortality from heart disease after treatment of Hodkin's disease. JAMA 270:1949-1955

Hanks GE, Kinzie JJ, White RL et al (1983) Patterns of care outcome studies: results of the national practice in Hodgkin's disease. Cancer 51:569-573

Haybittle J, Easterling M, Bennett M et al (1985) Review of British National Lymphoma Investigation studies of Hodgkin's disease and development of prognostic index. Lancet 1:967-972

Henry-Amar M, Friedman S, Hayat M et al (1991) Erythrocyte sedimentation rate predicts early survival in early-stage Hodgkin's disease. The EORTC Lymphoma Cooperative Group. Ann Intern Med 114:361-365

Hoppe RT (1990-1991) Hodgkin's disease. Patterns of care study newsletter no 3. American College of Radiology, Philadelphia

Hoppe RT, Rosenberg SA, Kaplan HS, Cox RS (1980) Prognostic factors in pathological stage IIIA Hodgkin's diseases. Cancer 46:1240-1246

Hoppe RT, Coleman CN, Cox RS et al (1982a) The management of stage I-II Hodgkin's disease with irradiation alone or combined modality therapy: The Stanford experience. Blood 59:455-465

Hoppe RT, Cox RS, Rosenberg SA et al (1982b) Prognostic factors in pathologic stage III Hodgkin's disease. Cancer Treat Rep 66:743-749

Horning SJ, Hoppe RT, Hancock SL et al (1988) Vinblastine, bleomycin, and methotrexate: an effective adjuvant in favorable Hodgkin's disease. J Clin Oncol 6:1822-1831

Hueltenschmidt B, Sautter-Bihl ML, Lang O et al (2001) Whole body positron emission tomography in the treat of Hodgkin disease. Cancer 91:302-310

Johnson RE, Ruhl U, Johnson SK, Glover M (1976) Split-course radiotherapy of Hodgkin's disease. Local tumor control and normal tissue reactions. Cancer 37:1713-1717

Kaplan HS (1962) The radical radiotherapy of regionally localized Hodgkin's disease. Radiology 89:553-561

Kaplan HS (1966) Evidence for a tumoricidal dose level in the radiotherapy of Hodgkin's disease. Cancer Res 26:1221-1224

Kaplan HS (1980) Hodgkin's disease, 2nd edn. Harvard University Press, Cambridge, MA

Karmiris TD, Grigoriou E, Tsantekidou M et al (2003) Treatment of early clinically staged Hodgkin's disease with a combination of ABVD chemotherapy plus limited field radiotherapy. Leuk Lymph 44:1523-1528

Khan FM (1994) The physics of radiation therapy, 2nd edn. William and Wilkins, Baltimore

Kinzie JJ, Hanks GE, Maclean GJ et al (1983) Patterns of care study: Hodgkin's disease relapse rates and adequacy of portals. Cancer 52:2223-2226

Laskar S, Gupta T, Vimal S et al (2004) Consolidation radiation after complete remission in Hodgkin's disease following six cycles of doxorubicin, bleomycin, vinblastine, and dacarbazine chemotherapy: is there a need? J Clin Oncol 22:62-68

Lee CKK, Bloomfield CD, Levitt SH (1979) Prophylactic whole lung irradiation for extensive mediastinal Hodgkin's disease. Proc Am Soc Clin Oncol 20:4411

Lee CK, Bloomfield CD, Goldman AJ et al (1980) Prognostic significance of mediastinal involvement in Hodgkin's disease treated with curative radiotherapy. Cancer 46:2403-2409

Lee CK, Bloomfield CD, Goldman A et al (1984) Liver irradiation in stage IIIA Hodgkin's disease patients with splenic involvement. Am J Clin Oncol 7:149-157

Lee CK, Aeppli DM, Bloomfield CD et al (1990) Curative radiotherapy for laparotomy-staged IA, IIA, IIIA Hodgkin's disease. An evaluation of the gains achieved with radical radiotherapy. Int J Radiat Oncol Biol Phys 19:547-549

LeFloch O, Donaldson SS, Kaplan HS (1976) Pregnancy following oophoropexy and total nodal irradiation in women with Hodgkin's disease. Cancer 38:2263-2268

Leibenhaut MH, Hoppe R, Efron B et al (1989) Prognostic indicators of laparotomy findings in clinical stage I-II supradiaphragmatic Hodgkin's disease. J Clin Oncol 7:81-91

Leopold KA, Canellos GP, Rosenthal D et al (1989) Stage IA-IIB Hodgkin's disease: Staging and treatment of patients with large mediastinal adenopathy. J Clin Oncol 7:1059-1065

Lister TA, Crowther D, Sutcliffe S et al (1989) Report of a committee convened to discuss the evaluation and staging of patients with Hodgkin's disease: Coswolds meeting. J Clin Oncol 7:1630-1636

Longo DL, Young RC, Wesley M et al (1986) Twenty years of MOPP therapy for Hodgkin's disease. J Clin Oncol 4:1295-1306

Lukes R (1971) Criteria for involvement of lymph node, bone marrow, spleen, and liver in Hodgkin's disease. Cacer Res 31:1755

Lukes RJ, Butler JJ (1966) The pathology and nomenclature of Hodgkin's disease. Cancer Res 26:1063-1081

Lutz W, Larsen R (1983) Technique to match mantle and para-aortic fields. Int J Radiat Oncol Biol Phys 9:1753

Madelli F, Anselmo A, Cartoni C et al (1986) Evaluation of therapeutic modalities in the control of Hodgkin's disease. Int J Radiat Oncol Biol Phys 12:1617-1620

Mai DHW, Peschel RE, Portlock D et al (1991) Stage I and II subdiaphragmatic Hodgkin's disease. Cancer 68:1467-1481

Mason D, Banks P, Chan J et al (1994) Nodular lymphocyte predominance Hodgkin's disease: a distinct clinicopathologic entity. Am J Surg Pathol 18:526-530

Mauch PM (1994) Controversies in the management of early stage Hodgkin's disease. Blood 2:318-329

Mauch P, Greenberg J, Lewin A et al (1983) Prognostic factors in patients with subdiaphragmatic Hodgkin's disease. Hematol Oncol 1:205-214

Mauch P, Tarbell N, Weinstein H et al (1988) Stage IA and IIA supradiaphragmatic Hodgkin's disease. Prognostic factors in surgically staged patients treated with mantle and para-aortic irradiation. J Clin Oncol 6:1576-1583

Mauch P, Larson D, Osteen R et al (1990) Prognostic factors positive surgical staging in patients with Hodgkin's disease. J Clin Oncol 8:257-265

Mauch P, Kalish L, Kadin M et al (1993) Patterns of presentation of Hodgkin's disease. Cancer 71:2062-2071

Mauch P, Canellos G, Shulman L et al (1995a) Mantle irradiation alone for selected patients with laparotomy-staged IA to IIA Hodgkin's disease: preliminary results of a prospective trial. J Clin Oncol 13:947-952

Mauch P, Kalish LA, Marchus KC et al (1995b) Long-term survival in Hodgkin's disease: relative impact of mortality, second tumors infection, and cardiovascular disease. Cancer J Sci Am 1:33-42

Miller RW, van de Geijn J, Raubitschek AA et al (1995) Dosimetric considerations in treating mediastinal disease with mantle fields: characterization of the dose under mantle blocks. Int J Radiat Oncol Biol Phys 32:1083-1095

Molrine D, George S, Tarbell N et al (1995) Antibody responses to polysaccharide-conjugate vaccines following treatment for Hodgkin's disease. Ann Intern Med 123:824-828

Myrianthopoulos L, Nautiyal J, Powers C et al (1995) A re-evaluation of dose response in Hodgkin's disease (abstract). Radiology 197:359

Nag AK, Li S, Neuberg D et al (2004) Factors influencing treatment recommendations in early-stage Hodgkin's disease: a survey of physicians. Ann Oncol 15:261-269

Neiman RR, Rosen PJ, Lukes RJ (1973) Lymphocyte-depletion Hodgkin's disease. N Engl J Med 288:751-755

Ng AK, Li S, Neuberg D et al (2004) Factors influencing treatment recommendations in early-stage Hodgkin's disease: a survey of physicians. Ann Oncol 15:261-269

Noordijk EM, Carde P, Mandard AM et al (1994) Preliminary results of the EORTC-GPMC controlled clinical trial H7 in early stage Hodgkin's disease. Ann Oncol 5 [Suppl 2]: S107-S112

Oberlin O, Leverger G, Pacquement H et al (1992) Low-dose radiation therapy and reduced chemotherapy in childhood Hodgkin's disease: the experience of the French Society of Pediatric Oncology. J Clin Oncol 10:1602-1608

Page V, Gardner A, Karzmark CJ (1970) Physical and dosimetric aspects of the radiotherapy of malignant lymphomas. I. The mantle technique. Radiology 96:609-626

Palos B, Kaplan HS, Karzmark CJ (1971) The use of thin lung shields to deliver limited whole-lung irradiation during mantle-field treatment of Hodgkin's disease. Radiology 101:441-442

Pavlovsky S, Maschio M, Santarelli MT et al (1988) Randomized trial of chemotherapy versus chemotherapy plus radiotherapy for stage I-II Hodgkin's disease. J Natl Cancer Inst 80:1466-1473

Peters M (1950) A study of survival in Hodgkin's disease treated radiologically. Am J Roentgenol 63:299-311

Peters MV (1966) Prophylactic treatment of adjacent areas in Hodgkin's disease. Cancer Res 26:1232-1243

Prosnitz LR (1990) Hodgkin's disease: the right dose. Int J Radiat Oncol Biol Phys 19:803-804

Raemakers J, Burgers M, Henry-Amar et al (1997) Patients with stage III/IV Hodkin's disease in partial remission after MOPP/ABV chemotherapy have excellent prognosis after additional involved-field radiotherapy: interim results from the ongoing EORTC-LCG and GPMC phase III trial. The EORTC Lymphoma Cooperative Group and Groupe Pierre-et-Marie-Curie. Ann Oncol 8[Suppl 1]:111-114

Ray GR, Wolf PH, Kaplan HS (1973) Value of laboratory indicators in Hodgkin's disease: Preliminary results. NCI Monogr 36:315

Rosenberg SA, Kaplan HS (1966) Evidence for an orderly progression in the spread of Hodgkin's disease. Cancer Res 26:1225-1231

Rosenberg SA, Kaplan HS (1985) The evolution and summary results of the Stanford randomized clinical trials of the management of Hodgkin's disease: 1962-1984. Int J Radiat Oncol Biol Phys 11:5-22

Rostock RA, Siegelman SS, Lenhard RE et al (1983) Thoracic CT scanning for mediastinal Hodgkin's disease. Results and therapeutic implications. Int J Radiat Oncol Biol Phys 9:1451-1457

Rutherford C, Desforges J, Davies B et al (1980) The decision to perform staging laparotomy in symptomatic Hodgkin's disease. Br J Haematol 44:347-358

Schellong G, Brämswig JH, Schwarze EW et al (1988) An approach to reduce treatment and invasive strategy in childhood Hodgkin's disease. The sequence of the German DAL multi-center studies. Bull Cancer 75:41-51

Schellong G, Brämswig JH, Hörnig-Franz I et al (1994) Hodgkin's disease in children: combined modality treatment for stages IA, IB, and IIA. Results in 356 patients of the German/Austrian Pediatric Study Group. Ann Oncol 5 [Suppl 2]:113-155

Schewe KL, Kun L, Cox J (1989) A step toward ending the controversies in Hodgkin's disease. Int J Radiat Oncol Biol Phys 17:1123-1124

Schoder H, Meta J, Yap C et al (2001) Effect of whole-body (18F)FDG-PET imaging on clinical staging and management of patients with malignant lymphoma. J Nucl Med 42:1139-1143

Siber G, Gorham C, Martin P et al (1986) Antibody response to pretreatment immunization and posttreatment boosting with bacterial polysaccharide vaccines in patients with Hodgkin's disease. Ann Intern Med 104:467-475

Somers R, Tubiana M, Henry-Amar M (1989) EORTC Lymphoma Cooperative Group Studies in clinical stage I-II Hodgkin's disease 1963-1987. Rec Res Cancer Res 117:175-181

Spaepen K, Stroobants S, Dupont P et al (2001) Prognostic value of positron emission tomography (PET) with fluorine-18 fluorodeoxyglucose ([18F]FDG) after first-line chemotherapy in non-Hodgkin's lymphom: is [18F]FDG-PET a valid alternative to conventional diagnostic methods? J Clin Oncol 19:414-419

Specht L, Nissen NI (1988) Hodgkin's disease stages I and II with infradiaphragmatic presentation: a rare and prognostically unfavorable combination. Eur J Haematol 40:396-402

Specht L, Gray RG, Clarke MJ et al (1998) Influence of more extensive radiotherapy and adjuvant chemotherapy on long-term outcome of early-stage Hodgkin's disease: a meta-analysis of 23 randomized trials involving 3,888 patients. J Clin Oncol 16:830-843

Sutcliff S, Gospodarowicz, Bergsagel D et al (1985) Prognostic groups for management of localized Hodgkin's disease. J Clin Oncol 3:393-401

Taylor M, Kaplan H, Nelson T (1985) Staging laparotomy with splenectomy for Hodgkin's disease: the Stanford experience. World J Surg 9:449-460

Tubiana M, Henry-Amar M, Hayat M et al (1982) Prognostic significance of the number of involved areas in the early stages of Hodgkin's disease. Cancer 54:95-104

Tubiana M, Henry-Amar M, Burgers MV et al (1984) Prognostic significance of erythrocyte sedimentation rate in clinical stages I-II of Hodgkin's disease. J Clin Oncol 2:194

Tubiana M, Henry-Amar M, Carde P et al (1989) Toward comprehensive management tailored to prognostic factors of patients with clinical stages I and II Hodgkin's disease. The

EORTC Lymphoma Group controlled clinical trials: 1964-1987. Blood 73:47-56

Van der Velden JW, van Putten WLJ, Guinee VF et al (1988) Subsequent development of acute non-lymphocytic leukemia in patients treated for Hodgkin's disease. Int J Cancer 42:252-255

Van Leeuwen FD, Klokman WJ, Stovall M et al (1995) Roles of radiotherapy and smoking in lung cancer following Hodgkin's disease. J Natl Cancer Inst 87:1530-1537

Vijayakumar S, Myrianthopolos LC (1992) An updated dose–response analysis of Hodgkin's disease. Radiother Oncol 24:1-13

Vijayakumar S, Rosenberg I, Brandt T et al (1989) The effect of posterior midline spinal corde block (PMSB) on Hodgkin's disease therapeutic dose estimates: a dosimetric study (abstract). Int J Radiat Oncol Biol Phys 17:170

Weisenberger TH, Juillard GJF (1977) Upper extremity lymphangiography in the radiation therapy of lymphomas and carcinoma of the breast. Radiology 122:227-230

Zittoun R, Audebert A, Hoerni B et al (1985) Extended field versus involved field irradiation combined with MOPP chemotherapy in early clinical stages of Hodgkin's disease. J Clin Oncol 3:207-214

33 Techniques of Intravascular Brachytherapy

Sri Gorty and Prabhakar Tripuraneni

CONTENTS

33.1   Introduction   *837*
33.2   Beta-Cath System for
        Coronary In-Stent Restenosis   *838*
        References   *841*

33.1
Introduction

Coronary artery disease (CAD) is an important factor in the morbidity and mortality of Americans. There are many ways to treat CAD, including medications, percutaneous transluminal coronary angioplasty (PTCA), stent placement, and coronary artery bypass surgery. Recent advances include intravascular brachytherapy (IVB) and drug-eluting stents (DESs).

The advent of PTCA has created the subspecialty of interventional cardiology. The deployment of stents has significantly decreased the acute complication of acute closure and dissection, resulting in the wider use of PTCA. However, stent placement has increased the risk of neointimal hyperplasia.

Restenosis of the coronary artery is a significant problem in the United States. In 2000, there were an estimated 882,000 coronary artery procedures in the United States. Of these, 6% involved a PTCA alone, 30–50% of which were at risk of restenosis. Of the procedures, 76% involved the use of a stent. The risk of restenosis in these individuals was 20–30%. Finally, 18% of the procedures were carried out on patients with an in-stent restenosis. In this subset, the risk of restenosis was very high at 40–80%. It is clear that, regardless of the intervention, restenosis remains a major problem. There are several reasons restenosis may occur: immediate recoil of the artery after PTCA, remodeling of the artery over time causing recoil of the artery, and scar formation in the PTCA area.

Radiotherapy has long been successfully used in treating benign conditions such as keloids, heterotopic bone formation, pterygia, and more recently macular degeneration. As radiation delays the healing process, it has been postulated that radiotherapy could be used to delay the scar formation within a coronary artery and therefore decrease the risk of restenosis.

In the late 1990s, brachytherapy proved to be effective in decreasing neointimal hyperplasia and subsequent restenosis after stenting in animal models. In 1995, the first human randomized trial was carried out at the Scripps Clinic in La Jolla, CA, confirming the efficacy of gamma irradiation in using iridium-192 seeds in decreasing in-stent restenosis after repeat angioplasty (Teirstein et al. 1997). Over the next 5 years, three major systems were developed: the Checkmate system using iridium-192 seeds, the Galileo system using phosphorus-32 wire, and the Beta-Cath system using strontium-90 (Sr-90) seeds (Fig. 33.1; Leon et al. 2001; Popma et al. 2002; Sieber et al. 2005; Urban et al. 2003). These three systems were approved by the Food and Drug Administration (FDA) for use in humans for coronary in-stent restenosis after repeat angioplasty (Bhargava et al 2004). A slight increase in sub-acute thrombosis was noted and was effectively treated with anti-platelet agents with no other significant complications (Giap et al.1999a,b).

Radioisotope-coated stents were studied in clinical trials and were found to be ineffective due to candy wrapper failure at the edges of the stents. A large trial was carried out for de novo stenosis randomizing between PTCA with or without stenting versus the same with IVB using the Beta-Cath system. This trial demonstrated no added benefit from IVB. There were many trials for peripheral ves-

S. Gorty, MD
Radiation Oncologist, Robert and Beverly Lewis Cancer Care Center, Pomona Valley Hospital Medical Center, 1910 Royalty Drive, Pomona, California, USA
P. Tripuraneni, MD, FACR
Scripps Clinic/Scripps Green Hospital, 10666 N. Torrey Pines Rd., MSB-1, La Jolla, CA 92037-1092, USA

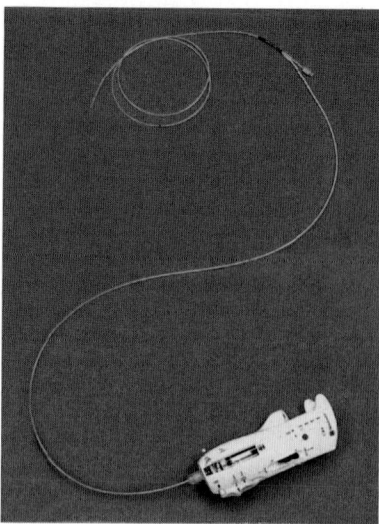

Fig. 33.1. Beta-Cath system

sels, mainly for femoral popliteal arteries, mostly in Vienna, Austria, other parts of Europe, and in the United States. The pivotal trial for the femoral popliteal artery, testing the efficacy of brachytherapy using a centering balloon "Paris" catheter with a high dose rate remote afterloader, did not show any superior efficacy with radiation over angioplasty. However, the single institution and multi-institution trials in Europe have shown the efficacy of the same technique in both de novo and in-stent restenosis of femoral popliteal arteries both with and without centering catheters. IVB has not been approved for use in peripheral vessels in the United States. A few prospective trials were done using external beam radiation both for coronary in-stent restenosis and femoral popliteal stenosis with mixed results. The prevailing opinion is that external beam is not the way to go (TRIPURANENI et al. 2001)!

It is estimated that, in 2002, there were approximately 50,000–80,000 cases of IVB done in the US alone and more around the world. At about that time, DESs using rapamycin and later taxol were coming into clinical trials. The early clinical trials and the subsequent randomized trials for coronary de novo stenosis confirmed the benefit of DESs in significantly reducing in-stent restenosis from about 15–18% to less than 2–4%. This is a good example of a new "destructive technology" that evolved into clinical use, obviating the need for an older technology that is quite useful. The newer DES technology significantly decreased the need for IVB by dramatically decreasing the in-stent restenosis occurrence. Since DESs have significantly decreased or eliminated the need for IVB for coronary in-stent reste-

nosis, the major manufacturers, Johnson & Johnson and Guidant, have discontinued the manufacturing and support of their respective systems, Checkmate and Galileo. Therefore, with the lack of availability of delivery catheters and radioisotopes for the two discontinued systems in 2005, the only system that is currently available for IVB is the Beta-Cath system and it will be reviewed in this chapter.

IVB is a truly multi-disciplinary procedure involving an interventional cardiologist, a radiation oncologist, and a medical physicist. The cardiologist is responsible for obtaining vascular access, determining the location and the extent of the in-stent restenotic lesion, performing the PTCA (preferably without further stent placement), and finally positioning the brachytherapy delivery catheter in the appropriate position. The radiation oncologist is responsible for obtaining informed consent for the radiation portion of the procedure, determining the details of the IVB delivery, and prescribing and delivering IVB. The medical physicist is responsible for the initiation of the Cath Lab Radiation Safety Quality Assurance Program, radiation delivery time calculations, and assistance with the delivery of IVB and safekeeping of the equipment (TRIPURANENI et al. 2001).

33.2
Beta-Cath System for Coronary In-Stent Restenosis

The Novoste Beta-Cath system uses Sr-90 as its radioactive source. The Sr-90 has a half-life of 28 years. It decays into yttrium-90(Y90) and a 0.54-MeV beta particle. The Y90 has a half-life of 64 h and decays into zirconium-90 and a 2.27-MeV beta particle. The zirconium-90 is a stable isotope and does not undergo any further decay.

The Sr-90 of the Beta-Cath system has several advantages. It has a long half-life, so the source does not have to be replaced frequently (Azeem et al. 2005). The dose rate is high enough that the dose is delivered in less than 5 min. As the penetration of beta particles is small, the physicians and catheter lab staff can stay with the patient during the procedure with manageable radiation protection precautions. The dose to the surrounding organs is minimal.

The Beta-Cath system consists of four main components: the source train, the transfer device, the Beta-Cath delivery catheter, and accessories (Fig. 33.2). It is a hydraulic delivery system. It comes

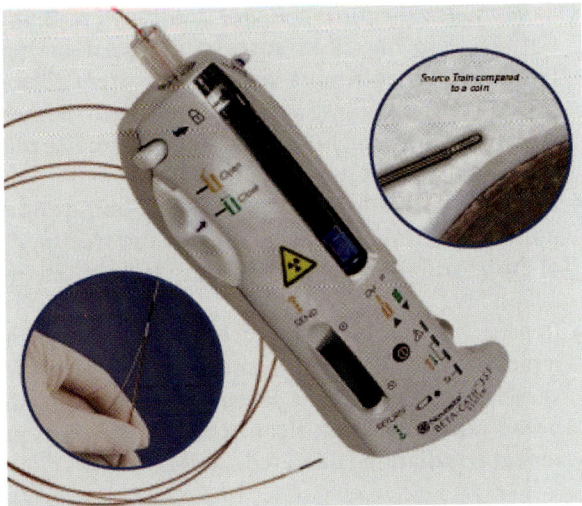

Fig. 33.2. Close up view of the transfer device and the delivery catheter

in two diameters, with an outer diameter of 5 French (slightly less than 2 mm in diameter) and 3.5 French (a bit more than 1 mm in diameter). The larger diameter of the 5-F system comes in three lengths of 30, 40, and 60-mm source trains. The smaller diameter of the 3.5-F system comes in only two lengths of 30 mm and 40 mm. The source train consists of Sr-90 seeds, each 2.5 mm in length. There are non-radioactive marker seeds at both the proximal and distal ends of the radioactive sources and are useful in fluoroscopic verification of the appropriate placement of the radioactive seeds and also the visual verification of the return of the all sources into the delivery device (Fig. 33.3). For practical purposes, we use a 40-mm long 3.5-F system and a 40-mm and 60-mm-long 5-F system. The smaller diameter catheter is used for smaller and tortuous vessels. This should cover the majority of the instances of IVB for both native coronary and saphenous vein graft in-stent restenosis.

The Beta-Cath delivery catheter is a triple lumen catheter that allows the use of a guidewire to place it into the coronary artery. The transfer device houses the radiation sources with the hydraulic delivery system. Once the Beta-Cath delivery catheter is placed into the coronary artery to be irradi-

ated, the transfer device is connected to the catheter making it a closed loop system. With the help of hydraulic pressure, the sources are sent into the delivery catheter end and radiographically verified. They are kept in place for the designated length of time to deliver the desired dose. At the end of the procedure, the switch is reversed and the seeds are brought back into the transfer device with hydraulic pressure. After visual verification of source return, the transfer device is disconnected from the delivery catheter. The delivery catheter is then removed from the coronary artery.

The first step involves placing the Beta-Cath system into a sterile bag. The bag is then closed. In the next step, we attach a syringe filled with saline through the opening in the bag and into the Beta-Cath system. We have found that attaching two syringes using a three-way stopcock seems to assure that we have enough saline for the procedure, but this makes the unit slightly harder to handle. The catheter is then attached to the system and the system is primed. Once the system is primed, the unit is ready.

During the angioplasty, the reference vessel diameter (RVD) is estimated to determine the dose to be delivered. After the angioplasty has been accomplished, the brachytherapy catheter is placed in the appropriate position. This catheter is placed over the existing guidewire and through the guide catheter. There are radio opaque markers on the brachytherapy catheter which can be seen on fluoroscopy and help with positioning of the catheter. The treatment portion of the catheter should be placed across the entire area of injury caused by the angioplasty, not just the area of restenosis. In addition, it is important to add margins at both the proximal and distal ends of at least 5 mm and more to minimize edge restenosis.

Once the catheter is in the appropriate position, the saline-filled syringe is used to deliver hydraulic pressure to move the source train into the appropriate position. Once the source positioning in the target is fluoroscopically confirmed, constant pressure on the syringe will keep the sources in the appropriate position. This can be verified by periodic fluoroscopy and by watching the pressure monitor on the

Distal Marker   **24 Active Sources**   **Proximal Marker**

Fig. 33.3. Strontium-90 seeds of 2.5 mm length (60-mm long source train) and radio opaque markers at both ends for radiographic visualization

Beta-Cath system. After the prescribed dose is delivered (usually in less than 5 min), the syringe is used to apply hydraulic pressure to return the source to the unit. The return of the sources needs to be verified visually before disconnecting the delivery catheter from the transfer device.

The IVB team must be ready to handle both cardiac and radiation emergencies in the catheter lab and be able to quickly remove the radioactive sources. A quality management program, with appropriate training, needs to be in place.

- *Dose*: 18.4 Gy is prescribed at a 2-mm radius from the center of the source axis for vessels with a reference vessel diameter between 2.7 mm and 3.3 mm. If the RVD is greater than 3.3 mm and less than or equal to 4 mm, then 23 Gy is delivered. With the source activity, the typical delivery times are in the range of 2–5 min. With the long half-life of strontium, the delivery times are adjusted once in 6 months (Tripuraneni et al. 2001).
- *Volume to be irradiated*: The gross target volume is the in-stent restenotic length of the vessel. The clinical target volume is the dilated portion of the vessel. The planning target volume includes margins of at least 5 mm, if not more, at both the proximal and distal ends of the dilated portion of the vessel (Fig. 33.4; Giap et al. 2001a; Giap et al. 2001b, Tripuraneni et al. 2002).
- *Longer lesions*: For lesions longer than 40 mm in the injured length of the vessel, the sequential positioning and pullback technique has been successfully used. Care should be taken to avoid any significant overlap or gap in order to minimize hot or cold spots. The branch vessels are useful in the positioning; so also is careful review of the cine angiograms in the same position of the table and the gantry (Crocker et al. 2001).

- *Saphenous vein graft in-stent restenosis*: The Beta-Cath system has been successfully used in the treatment of saphenous vein graft in-stent restenosis. Typically, these vessels have larger diameters in the range of 3–5 mm, so the dose may need to be adjusted (Schiele et al. 2003).
- *Bifurcations*: The sequential positioning and pullback technique may be used in the treatment of bifurcation in-stent restenosis (Costa et al. 2003).
- *Repeat irradiation*: The success rate of repeat irradiation of the same segment of the restenotic vessel after first IVB is lower. There appears to be no additional significant complications from repeat irradiation (Bae et al. 2004).

Enrollment is complete in two major ongoing trials that randomize patients having in-stent restenosis and will compare the use of drug-coated stents with the standard arm of IVB. The results are eagerly awaited and will determine the fate of any further continued use of IVB in the US (Tripuraneni 2003).

There are several small studies reviewing the efficacy of IVB in the management of instances of in-stent restenosis after the use of DESs. It appears that the efficacy of IVB is somewhat lower in this group of drug resistant in-stent restenosis. The results of studies with a larger number of patients and longer follow-up are awaited.

In summary, IVB has revolutionized the entry of brachytherapy in the management of cardiac disease and was briefly the most common brachytherapy technique used. For a year or two, IVB was also the most common radiotherapy technique used for non-malignant conditions. However, with the advent of DESs, the incidence of in-stent restenosis

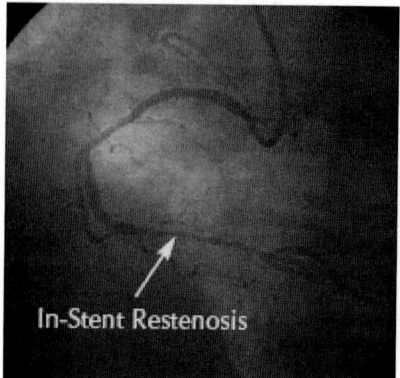
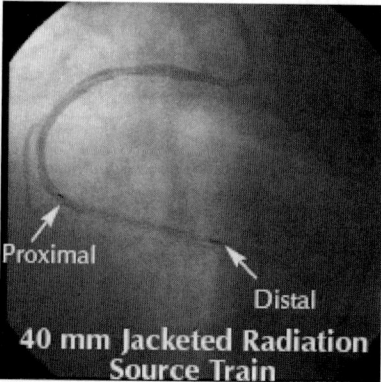
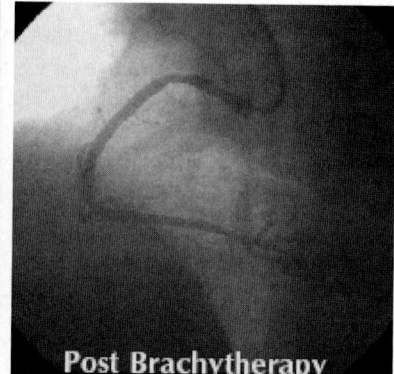

Fig. 33.4. Adequate margins at both ends of dilated segment of in-stent restenosis are included for irradiation

of coronary arteries has dramatically decreased, thereby obviating the need for IVB. The results of the randomized trials comparing the efficacy of IVB with drug-coated stents and the long-term results of a large group of patients with drug-resistant in-stent restenosis will determine the future of IVB, if any!

References

Azeem T, Adlam D, Gershlick A (2005) Evolution of vascular brachytherapy over time: data from the RENO-registry analysis. Int J Cardiol 100:225–228

Bae JW, Koo BK, Kim KI, et al (2004) Two-year outcomes of repeated brachytherapy in patients with restenosis after intracoronary radiation therapy. Am J Cardiol 94:1061–1063

Bhargava B, Karthikeyan G, Tripuraneni P (2004) Intravascular brachytherapy: indications and management of adverse events. Am J Cardiovasc Drugs 4:385–394

Costa R, Joyal M, Harel F, et al (2003) Treatment of bifurcation in-stent restenotic lesions with beta radiation using strontium 90 and sequential positioning pullback technique: procedural details and clinical outcomes. J Invasive Cardiol 15:469–474

Crocker I, Joyal M, Fox T, et al (2001) Treatment of long, diffuse, in-stent restenotic lesions with beta radiation using strontium 90 and sequential positioning "pullback" technique: procedural details and clinical outcomes. J Invasive Cardiol 13:782–787

Giap H, Massullo V, Teirstein P, et al (1999) Theoretical assessment of late cardiac complication from endovascular brachytherapy for restenosis prevention. Cardiovasc Radiat Med 1:233–238

Giap H, Tripuraneni P, Teirstein P, et al (1999) Theoretical assessment of dose-rate effect in endovascular brachytherapy. Cardiovasc Radiat Med 1:227–232

Giap HB, Bendre DD, Huppe GB, et al (2001) Source displacement during the cardiac cycle in coronary endovascular brachytherapy. Int J Radiat Oncol Biol Phys 49:273–277

Giap H, Teirstein P, Massullo V, et al (2001) Barotrauma due to stent deployment in endovascular brachytherapy for restenosis prevention. Int J Radiat Oncol Biol Phys 47:1021–1024

Leon MB, Teirstein PS, Moses JW, et al (2001) Localized intracoronary gamma-radiation therapy to inhibit the recurrence of restenosis after stenting. N Engl J Med 344:250–256

Popma JJ, Suntharalingam M, Lansky AJ, et al (2002) Randomized trial of 90Sr/90Y beta-radiation versus placebo control for treatment of in-stent restenosis. Circulation 106:1090–1096

Schiele TM, Regar E, Silber S, et al (2003) Clinical and angiographic acute and follow up results of intracoronary beta brachytherapy in saphenous vein bypass grafts: a subgroup analysis of the multicentre European registry of intraluminal coronary beta brachytherapy (RENO). Heart 89:640–644

Silber S, Popma JJ, Suntharalingam M, et al (2005) Two-year clinical follow-up of 90Sr/90 Y beta-radiation versus placebo control for the treatment of in-stent restenosis. Am Heart J 149:689–694

Teirstein PS, Massullo V, Jani S, et al (1997) Catheter-based radiotherapy to inhibit restenosis after coronary stenting. N Engl J Med 336:1697–1703

Tripuraneni P, Parikh S, Giap H, et al (2000) How long is enough? Defining the treatment length in endovascular brachytherapy. Catheter Cardiovasc Interv 51:147–153

Tripuraneni P, Jani S, Minar E, Leon M (2001) Intravascualr brachytherapy: from theory to practice. ReMedica Publishing, London

Tripuraneni P (2002) Coronary artery radiation therapy for the prevention of restenosis after percutaneous coronary angioplasty. II: Outcomes of clinical trials. Semin Radiat Oncol 12:17–30

Tripuraneni P (2003) The future of CART in the era of drug eluting stents: "It's not over until it's over". Counterpoint. Brachytherapy 2:74–76

Urban P, Serruys P, Baumgart D, et al (2003) A multicentre European registry of intraluminal coronary beta brachytherapy. Eur Heart J 24:604–612

Subject Index

A

Abt-II study 397
abutment geometry 154
accelerated partial breast irradiation (APBI) 281
- dose optimization 284
- interstitial treatment 283
accelerator isocenter 212
acentric fragment 293
acquired immunodeficiency syndrome (AIDS) 580
acrylic
- mold 316
- plate 151
acute
- genitourinary toxicity 223
- lymphocytic leukemia (ALL) 786
- mucosal response 23
- myeloid leukemia (AML) 786
adenoid cystic carcinoma 472
adjacent field 100
afterloading
- nylon tubing 321
- silicone catheter 322
air cavity 145
air-kerma strength 262
alloy block 91, 93, 172, 417
alpha-fetoprotein (AFP) 740
American Brachytherapy Society (ABS) 274
anal canal 358, 547
- epidermoid carcinoma 360
anal cancer 48, 546, 549, 553
Anderson planar implant nomogram 343
androgen deprivation therapy (ADT) 716, 721
anesthesia 319
angiogenesis 61
anisotropy 265
Ann Arbor staging 808
anorectal carcinoma 358
anterior
- chest radiograph 790, 792
- fibromuscular stroma (AFS) 688
anteroposterior (AP) beam 438
anterosuperior point of the iliac spine (ASIS) 789
anthropomorphic phantom 213
AP/PA extended field 614
apoptosis 65, 294, 297
apparent activity 262
applicator insertion 271
aquaplast mask 454, 478
arc length 238

ASTRO failure definition 694
astrocytic tumor 430
asymmetric
- collimator jaw 664
- jaw 78
- X-ray collimator 77
atom bomb survivor 37
attenuation correction factor (ACF) 117
automatic
- optimization algorithm 274, 276
- positioning system (APS) 241
away-along table 263
axillar anatomy 487
axillary
- irradiation 508
- lymph nodes 486
- - dissection (ALND) 491
azygos node 526

B

Bacillus Calmette Guerin (BCG) 565
Bartholin's glands 631, 636
basal dose point 269
base of tongue 459
beam
- angulation 163
- arrangement design 184
- geometry 205
- modification 417
- penumbra 192
- quality 77
beam-edge penumbra 203
beamlet distribution 208
beam's eye view (BEV) 108, 179, 531, 699
belly board 613
benign prostatic hyperplasia 688
Beta-Cath system 837, 838
bifurcation 840
bilateral inguinofemoral lymphadenectomy 639
bile duct tumor 341, 342
biliary tree 341
bioanatomic radiation therapy 434
biological target volume (BTV) 119
biologically effective dose (BED) 6, 310, 787
- acute mucosal 17, 25
- late complications 14, 25
- tolerance zone 14

biophysical model 196
bite block 168
– registration 97, 98
bixel 203
– intensity 205
bladder 706
– cancer 47, 357, 474, **561**
– – chemotherapy 566
– – composite isodose curve 570
– – IMRT 573
– – isodose distribution 570
– – staging system 564
– – target volume 569
– – therapeutic options 564
– diagram 562, 563
– distension 550
– exstrophy 561
– irritability 561
– mucosa 569
– preservation 574, 575
– tumor, see also bladder cancer 427
– – location 568
– wall 221
Bloedorn applicator 354
blood
– cell count (CBC) 659
– test 805
body immobilization system 169
bone
– density 144
– interface 85
– scan 758
bone marrow transplantation 785
bone–tissue interface 145
bony
– metastasis 428
– sacral promontory 571
boost
– brachytherapy 343
– treatment 493
bore-size limitation 182
Bowen's disease 634
brachytherapy 255
– applicators 596
– dose rate 257
– endoluminal 514
– high dose rate (HDR) 279, 305, 379
– interstitial technique 268
– line source dosimetry 263
– low dose rate (LDR) 306, 338
– of the breast 335
– permanent implant 256
– point source dosimetry 262
– preplanning 270
– quality control procedures 367
– radiation protection 285
– remote controlled 380
– source
– – half life 258
– – strength 261
– strength specification 262
– temporary implant 256
– transperineal permanent implant 274

– treatment planning 273
brain
– anatomy 425
– metastasis 432, 448
– metastatic tumor 428
BrainLab m3 collimator 243
breast anatomy 486
breast cancer 389, **485**
– anterior supraclavicular-axillary photon field 505
– biopsy 490
– brachytherapy 335
– electron beam technique 507
– field planning 495
– irradiation 495
– locoregional tumor 488
– partial irradiation 500
– partly-deep tangential pair fields 505
– staging 490
– standard brachytherapy 500
– supine tangential treatment 494
bremsstrahlung photon beam 208
Brodmann schema 426
bronchogenic carcinoma 339, 526
– imaging 528
– X-ray 528
bronchoscopy 385, 512
Brown-Roberts-Wells (BRW)
– frame 321
– localizing ring 234
buccal mucosa 150, 460
bulky
– cancer 462
– disease 826
– – upper mediastinum 827

C

camptomthecin 43
cancer
– of the laryngopharynx 459
– of the nasopharynx 460
– of the tonsillar region 459
carboplatin 43, 45
carcinoembryonic antigen (CEA) 545
carcinoma
– of nasal vestibule 474
– of posterior pharyngeal wall 458
– of the base of tongue 459
– of the cervix 345, 382, 595
– – applicators 347
– – survival rates 595
– of the endometrium 352, 383
– of the esophagus 340, 386
– – chemotherapy 516
– – clinical target volume (CTV) 516
– – dose volume histogram (DVH) 520
– – gross target volume (GTV) 516
– – metal stent placement 514
– – palliation 514
– – plan implementation 521
– – planning target volume (PTV) 516
– – quality assurance 521
– – radiochemotherapy 516

– of the glottic larynx 455
– of the lung
– – tumor dose 531
– of the nasopharynx 329, 461
– of the pancreas 341
– of the penis 360
– of the prostate
– – external irradiation 696
– of the tongue 331
– of the tonsillar region 475
– of the uterine cervix 35, 305, 346, 593
– of the vagina 657
– – high-dose-rate intracavitary brachytherapy 666
– – lymph node metastasis 662
– – staging 661
– – treatment options 662
– of the vulva 632, 625
– – treatment guidelines 641
cardiac irradiation 495
cataract 303
catheter
– insertion 271, 272, 281
– misconnection 285
– position 273
cecum 547
cell survival curve 9, 293
cell-cycle mechanism 41
central nervous system (CNS) 425
– menigioma 440
– metastatic tumors 448
– neoplasm 428
– primary tumors 447
– tumor 427
– – imaging 430
– – imaging-based treatment planning 434
– – medical treatment 431
– – radiation therapy 432
– – surgical management 431
cerrobend 91, 148
– block 172, 447
– calm shell 800
– insert 160
cervical cancer, see also carcinoma of the cervix, cervix
    cancer 266, 305, **579**
– beam fields 583
– complications 594
– FIGO staging 580
– LDR brachytherapy 586
– radiation treatment 581
cervical spine block 821
cervix
– biopsy 659
– cancer 47
cesium
– implant 467
– needle 332, 468
cesium-137 259, 260
chamber and electrometer system 277
chemoirradiation 39, 652
– non-small cell lung cancer 44
– small cell lung carcinoma 44
chest wall 153
– boost field 796

– electron beam field 504
– tangential photon field 503
Chinese hamster (CHO) cell 296
choline-to-N-acetylaspartate ratio 434
choroidal melanoma 324
chromatid aberration 292
chromatin 292
chromosome
– aberration 292
– damage 296
cigarette smoking 525
cisplatin 47, 519, 567, 624
Clarkson integration 791
clinical target volume (CTV) 147, 283, 343, 479
Cloquet's node 632
cobalt beam 747
cobalt-60 260, 457
– teletherapy 82
– – machine 210
Collaborative Ocular Melanoma Study (COMS) 324
– trial of preenucleation radiation 326
collimation system 112, 170
collimator
– angle indicators 404
– jaw 170, 407
– scatter factor 73
colon cancer 545, 547, 548
– therapeutic ration 555
color-coded silk thread 312
colorectal cancer 50, 547
colostomy 48, 359, 553
colpectomy scar 611
colpostat 345, 347, 384
commissioning 214
compensating filter 95, 211
compensation system 172, 173
computed tomography (CT) 58
– 16-slice scanner 112
– 4D imaging 58
– CT-compatible radiographic marker 281
– large-bore scanner 111
– multislice scanner 110
– scan parameters 124
– simulation 108, 109, 126, 282
– – instruction 121
– slice thickness 283
– topogram 181
concomitant boost 21
concurrent chemotherapy 43
cone
– beam 208
– – modulation technique 211
– system 158
conformity index 235
continuous
– hyperfractionated accelerated radiotherapy (CHART) 532
– low dose-rate (CLDR) 295
Continuous Accelerated Irradiation (CAIR) 18, 21
contrast 128
conventional
– linac 242
– radiation therapy (CRT) 219
– simulator 110, 124

convexity meningioma 441
convolution kernel 194
core needle biopsy 759
coronary artery disease 837
Corvus system 219
couch
– angles 238
– coordinate 129
Coulomb force interaction 136
craniospinal
– field arrangement 161
– irradiation (CSI) 102, 160, 444, 446
craniocaudal isocenter position 518
Crohn's disease 718
cross-sectional tissue anatomy 191
Cu-ATSM 119
custom compensating bolus 151, 174
CyberKnife 241
– treatment system 244
cystectomy 47, 565, 566, 568
cystogram 568, 569
cystoscopy 278, 349, 562, 572

D

dacarbazine 775
DAHANCA curve 20, 21
Dale's formalism 310
data
– acquisition 400, 415
– transfer error 416
database management system (DBM) 200
delayed feedback 13
Delclos cylinder 608, 619, 665
delta ray 137
Denonvillier's fascia 678
dental acrylic 150
18F-deoxyglucose (FDG) 64
dexamethasone 431
dicentric 293
DICOM 130, 199
diethylenetriamine pentaacetic acid (DTPA) 61
digitally
– composite radiograph (DCR) 185
– reconstructed radiograph (DRR) 108, 216, 522
distal urethral tumor 358
distance to agreement (DTA) 213
DNA double-strand breaks (DSB) 40
documentation 187
dog-leg configuration 742
Dolasetron 801
Domed cylinder 665
donor lymphocytes 785
dose
– calculation 416
– – algorithm 186, 194
– distribution homogeneity 284
– falloff 160
– homogeneity 190
– – index (DHI) 281
– per monitor unit 74
– pertubation factor (DPF) 85

– precision 395
– profile 73
– – horns 82
– specification 99
– volume histogram (DVH) 70, 194, 322, 434, 479
dose-modifying factor (DMF) 381
dose-rate effect 296
dose-volume statistics 196
dosimetric
– margin 435, 707
– parameter 70
– verification 212
dosimetrist
– staffing 397
– training 397
dosimetry
– check 187
– instrumentation 224
– system 213
double-strand break (DSB) 292
doxorubicin 624, 772, 775, 778
drug–radiation interaction 40
dummy
– source 323
– wire 311
dwell
– position 280
– time 280, 284
dynamic
– multileaf collimation (DMLC) technique 218
– tracking software 244
– wedge 173
– – technology 408
dysuria 722

E

early responding tissue 298
ecchymosis 757
effective pathlength (EPL) 79
EGFR 43
electrocardiogram (ECG) 801
electron
– arc therapy 159
– beam 135, 144
– – bolus 151
– – lung tissue 144
– – obliquity factor 143
– – treatment
– chest wall boost 797
– cone 158
– density 174
– depth–dose distribution 143
– disequilibrium 194
– dose profile 140
– field 152
– percentage depth dose 138
– shielding 149
electronic portal imaging
– device (EPID) 213
– – online 419
– quality assurance 411

electron-photon field matching 156
Elekta stereotactic body frame 438
endobronchial radiation 385
endoluminal brachytherapy 514
endometrial cancer (EC) 352, **599**
– clinical staging system 603
– palliation 624
– prognosis 605
endometrial hyperplasia 600
endometrioid adenocarcinoma 601
endoscopic ultrasound (EUS) 512
energy spectrum 258
en-face beam 224
enhanced dynamic wedge (EDW) 86, 87, 94
enteritis 556
EPID support 410
epidermal growth factor receptor (EGRF) 12
epidermoid carcinoma 640
– of the anal canal 360
epididymis 740
epiglottic tip cancer 456
episcleral plaque therapy 324
epithelioid sarcoma 778
equivalent dose 6
equivalent uniform dose (EUD) 198
erectile dysfunction 723
erythroplasia of Queyrat 634
erythropoietin 46
esophageal
– cancer, see also carcinoma of the esophagus 47, 339, 386,
     511, 512
– wall 513
esophagectomy 49, 514, 519
esophagitis 45, 530
esophagus
– adenocarcinoma 515
– cancer 48
estrogen 600
estrogen-secreting tumor 600
Ethernet 199
ethmoid sinus cancer 468, 469
exact-match technique 445
exit dose 83
exophytic invasive squamous cell carcinoma 461
external beam radiation therapy (EBRT) 344, 381, 725
extracranial
– stereotactic radioablation (ESR) 678
– target 248
extrapolated
– response dose (ERD) 11, 318
– tolerance dose (ETD) 10
extraprostatic extension (EPE) 700
extraskeletal chondrosarcoma of the extremities 771
eye shield 164
eyelid cancer 471

F

facemask 98
fan-beam IMRT delivery 208
fecal incontinence 556
fibrous histiocytoma 756

fiducial marker 574
field
– abutment 152
– size 76
– – indicator 404
File Transfer Protocol (FTP) 199
first-echelon lymph node 465
fistula
– recto-vaginal 678
– tracheobronchial 515
– tracheoesophageal 515
– uretero-vaginal 679
– vesico-vaginal 679
flatness 139
flattening filter 205
Fletcher milligram hours 587, 590, 592
Fletcher-Suit afterloading colpostat 611
Fletcher-Suit applicator 617
Fletcher-Suit-Declos (FSD)
– applicator 266
– LDR system 348
– tandem 317, 586
floor of the mouth cancer 463, 464
fluorinated pyrimidine 39
18F-fluorodeoxy-glucose 806
– PET scan 759
fluoroscopy 125, 278
foam device 169
FOCAL unit 234
Foley
– balloon 266
– catheter 278, 345, 569, 593
formalin fixation 701
French spring-eye needle 314
frog-leg position 642
5-FU 43, 49, 553, 567
functional mapping 426

G

gadolinium (Gd) 61
Gamma Knife 233, 237, 238, 240
– isodose distribution 246
gamma slope 15
gamma-37 15
gamma-50 15, 26
gantry 208, 242
Gap-Match technique 445
gastric cancer 511
Gelfoam 316
gemcitabine 46
genetic disorder 33
genitourinary diaphragm 709
germ cell tumor 739
Gill-Thomas-Cosman (GTC) frame 234
Gleason score 690, 694, 720
glioblastoma 321
– multiforme 51, 322
glioma 51, 428
glottic larynx cancer 454
GOG protocol 603, 605

gold
- fiducial marker 700
- seed 584
- - implantation 705
gold-198 260
gonadal shielding 746
Gortec II 22
graft-versus-host disease (GVHD) 785
groin node 646
gross target volume (GTV) 147, 283
gutter guide technique 313
GYN tandem 267
gynecological cancer 266
gypsum 174

H

half-value layers (HVL) 171
hand calculation 416
head and neck
- cancer 45, 387, **453**
- - RT techniques 453
- compensator 793
- intensity-modulated radiation therapy 480
- schedule 16, 18, 19
- tumor 327
heart irradiation 830
HeLa cell 296
hematuria 561
hemiscrotum 747
hemivulvectomy 649
hemoptysis 386
Henschke technique 311, 349
Heyman-Simon capsule 352, 608
High Dose Accelerated Dose-Per-Fraction Escalation
   (HARDE) 18, 21
high dose rate (HDR) system 280
- fractionation schemes 381
high-energy electron 135
high-Z material 206
hilar adenopathy 821
Hodgkin's disease 36, **805,** 807
- advanced stage 812
- Ann Arbor classification 807
- combined modality therapy 810
- infradiaphragmatic 810
- mantle fields 815
- spade-shape fiels 823
- subdiaphragmatic fields 823
hot loading 390
hot spot 481
human
- chorionic gonadotropin (HCG) 740
- lymphocyte antigen (HLA) 785
- papilloma virus (HPV) 579
hydronephrosis 580, 676
hydroxyl radical 291
Hypaque 593
hyperfractionation (HFX) 16, 17, 532
hyperthermia 367
hypofractionation 24
hypopharynx cancer 476, 477
hypothyroidism 831

hypoxia 41, 65
hysterectomy 383, 581, 585, 593, 608, 675
hysterogram 352

I
ICRU report 38 266
ifosfamide 778
ileal conduit 565
image
- transfer 129
- verification review 419
image-guided radiation therapy (IGRT) 167, 575
imaging chain 110
immobilization system 97, 120, 168
- of the breast 120
implant
- of the mediastinum 339
- removal 320
inelastic collision 137
infant irradiation 798
in-field sarcoma 35
inflammatory bowel disease 718
infrahyoid epiglottic cancer 456
inguinal lymph node 674
- dissection 642
intensity modulated radiation therapy (IMRT) 26, 27, 36, 57,
   167, 180, 203, 436, 478
interface dosimetry 83
interfraction 181
interlobar node 526
internal
- mammary chain (IMC) 499
- - radiation 507
- shielding 149, 175
- target volume (ITV) 189
International Commission on Radiation Units and Measurement (ICRU) 188
interseed attenuation 277
interstitial pneumonitis 788
intracavitary
- brachytherapy (ICB) 599
- cylinder 355
- radiation 157
- vaginal brachytherapy (IV) 610
intrafraction 181
intraoperative radiation therapy (IORT) 147, 157, 158
intraoral cone irradiation 464, 468
intraperitoneal radioisotope 610
intrathoracic disease 806
intratubular germ cell neoplasia (IGCN) 739, 746
intrauterine tandem 345, 666
introitus 613
inverse
- method 203
- planning 248
- square law 310
iodine-125 260
- permanent interstitial implants 314
- plastic tube implants 313
ion chamber 73, 403
iridium 329
- fork 360
- hairpin 332

– implant 467
– wire 300, 301, 310, 313
iridium-192 260, 300
– implants 366, 459
– interstitial brachytherapy 353
– – gynecological malignancy 353
irregular skin surface 142
irregular-field calculation 76
ischial tuberosities 552, 698, 706, 744
isocenter 168
isodose
– location 125
– charts 81
– curve 73, 140
– distribution 89, 153, 212
– line 139, 142
– shift method 79, 80
– surface 189
isotope 720

J

jaw
– setting 94
– system 171

K

Kaposi's sarcoma 176
Karnofsky performance score 322
kidney blocking 799
kilovoltage X-ray simulator system 207

L

labia majora 631
lactate dehydrogenase (LDH) 492, 740
laminectomy 432
laparatomy 357, 806
large mediastinal mass (LMM) 806
laryngectomy 46
larynx cancer 476, 477
Las Vegas phantom 410
late responding tissue 298
lateral
– fluence gradient 215
– pelvis field 614
L-block shield 271
lead
– cutout 148
– shield 175
lead time bias 57
leaf position 216
leaf-center insertion 93
Leksell frame 240
length time bias 57
leukapheresis 786
leukemia 444
Lhermitte sign 830
lidocaine 320, 552

lifestyle 33
LiF-Teflon dosimeter 84
light-radiation field congruence test 403
limb
– perfusion protocol 777
– salvage procedure 777
line source dosimetry 263
linear accelerator (linac) 37, 136, 207, 241
– computer control system 406
– radiosurgery 235
linear energy transfer (LET) radiation 291
linear quadratic (LQ)
– equation 383
– formalism 318
– formula 6
– – Seven Steps 8
– – simple 9
– model 295
lip cancer 462
liposarcoma 756, 775
Lipowitz metal 91, 172
lithium fluoride (LiF) 793
lithotomy position 320, 363
liver transmission block 615
lobectomy 529
local
– anesthesia 320
– area network (LAN) 199
log-linear plot 294
lumpectomy 272, 335, 493
– cavity 282
lung
– block 795, 797, 820, 821
– cancer 193, 234, 386, 525
– – fractionation 532
– – metastatic 385
– – TNM classification 527
– compensator 792, 793
– heterogeneity 535
– interface 84
– parenchyma 830
– toxicity 536
– tumor
– – target delineation 534
lung–chest wall interface 147, 797
lung-density corrections 536
Lyman model 197
lymph node
– chain 826
– in lung 526
– in mediastinum 526
lymphadenectomy 47, 633, 675
lymphangiogram 571, 740
lymphatic drainage 486
lymphedema 474, 566, 646
lymphoma 44
lymphoscintigraphy 487, 507

M

machine callibration 792
MACH-NC study 45

magnetic resonance imaging (MRI) 59
– diffusion-weighted (DWMRI) 64
– functional (fMRI) 64
– simulator 114
malignancy, treatment-related 34
malignant
– fibrous histocytoma 768
– melanoma 23
Mallinckrodt system 618
mammography 489
MammoSite applicator 389, 501
Manchester system 299, 347, 588
mantle field 815, 819
– arm positions 818
Martinez universal perineal interstitial template (MUPIT) 315
Martius flap 679
mass of radium 261
mastectomy scar 504
match-line wedge 101
mathematical models 7
maxillary sinus cancer 468
maxillectomy 328
median lobe hyperplasia (MLH) 716
mediastinoscopy 528
mediastinum 827
medical accelerator facilities
– quality assurance 412
medical event 287, 368
medullablastoma 101
meningococcus 807
menometrorrhagia 600
mesorectal excision 50
metallic esophageal stent 514
metastatic disease 778
Mick applicator 275, 277
milligram hour system 588
milligram-radium equivalent 262
minicolpostat 348
miniovoid 348
misadministration 370
mitomycin 553
mitotic death 294
mixed lymphocyte culture (MLC) 785
modulation
– of intensity 187
– transfer function (MTF) 410
mold 316
monitor unit 215
– calculation 75
mons pubis 631
Monte Carlo simulation 194, 310
mucosal tolerance 13
mucosis fungoides 176
mucositis 20, 458
multileaf collimator (MLC) 78, 92, 111, 171, 179, 203, 206, 408
multiple coplanar beam arrangement 620
multiple-beam arrangement 88
multislice CT scanner 110
MUPIT applicator 359, 668
myelopathy 436
myxoid liposarcoma 760

N

nasal
– cavity cancer 468, 469
– vestibule 328
– – carcinoma 472
nasopharyngeal carcinoma (NPC) 46, 218, 454
nasopharynx 387
National Marrow Donor Program (NMDP) 786
neck dissection 476
necrosis 649
needle
– insertion 278
– loading boxes 275
neoadjuvant hormone therapy (NHT) 727
nephritis 745
neuraxis irradiation 445
neutron flux field 258
nominal standard dose (NSD) 7
non-coplanar
– beam arrangement 440, 620
– method 707
– photon beam 501
non-Hodgkin's lymphoma 808
nonseminomatous germ cell tumor (NSGCT) 739
non-small cell lung cancer 44, 193, 530, 534
nonsquare beam 70
non-uniform
– beam fluences 209
– distribution 269
normal tissue complication 69
– probability (NTCP) 197
Novalis treatment system 244
Nuclear Regulatory Commission (NRC) 286
nucleosid analog 41
nylon catheter 311

O

ocular melanoma 325
off-axis factor 73
off-axis ratio (OAR) 78, 139
off-center ratio (OCR) 73, 78
oncological imaging 57
oophoropexy 823, 831
open-bladder implant 357
operating room (OR) 256
– requirements 158
optic chiasm 245, 437
optical distance indicator (ODI) 403
optimization
– algorithm 209, 284
– criterion 209
oral
– cavity 331
– tongue 465
orchidectomy 740, 748
organs at risk (OAR) 147, 435
oropharynx 333, 457
orthogonal field 103
orthovoltage
– beam 746

– X-ray unit 89
Orton study 306
overlapping voxel 192
ovoid 349, 587, 608
– implant 267
oxaliplatin 43

P

paclitaxel 45
PACS system 236
Paget's disease 634
palladium-103 260
– implants 366
palliative external beam radiotherapy 625
pancytopenia 786
Pap smear 580
paraaortic lymph node 828
paraffin wax 174
parametrium 349, 660
paravaginal tumor 356
Paris system 269, 300, 313, 338
Parker-Paterson dosimetry system 300
parotid gland 472
– cancer 474
parotidectomy 475
partial tumor volume (PTV) 220
particle beam radiosurgery 238
Paterson-Parker system 269
patient
– immobilization 119, 415
– positioning 97, 99, 114, 119, 125
– – reference point 168
– support assembly (PSA) 410
PDR procedure 368
Peacock system 206, 222
pediatric tumor 388
pelvic
– cancer 350
– field 216
– irradiation 346
– lymph node 641, 696
– – dissection 693
– – elective irradiation 644
– – metastases 690
– lymphadenectomy 566
– radiotherapy field 555
– recurrence 622, 674
– treatment
– – radiation field 571
– wall point 591
pelvis 587
pencil-beam method 194
penile bulb 723
penumbra 93, 101, 141, 189
peptic ulcer 742
percentage
– depth dose (PDD) 70, 86
– positive prostate biopsy (PPPB) 695
percutaneous transluminal coronary angioplasty (PTCA) 837
pericarditis 830
permanent prostate implant

– treatment planning 276
– volume study 275
PET/CT
– combination scanner 59
– combined 116
– simulator 116
pharyngeal wall lesion 456
phenytoin 431
phosphocholine 61
phosphoethanolamine 61
phospholipid anabolite 61
photon
– beam 193
– – target definition 146
– contamination 137
– field 152, 161
photon-beam
– dosimetry 81
– radiation output calibration constancy 403
physic staffing 396
physical wedges 173
physics training 397
pilot image 126
Pipelle examination 600
pituitary adenoma 439
planar dosimeter 213
planning
– conference 418
– organ at risk volume (PRV) 147, 189
– target volume (PTV) 147
plaster casting technique 98
plastic
– constancy phantom 403
– tandem 347
pneumococcus 807
pneumonectomy 529
pneumonitis 531, 536, 798
point source dosimetry 262
polyethylene foam 169
polymer 213
polystyrene 160
– phantom 217
polyurethane foam system 98, 767
port film verification review 419
PORTEC trial 620
positron emission tomography (PET) 64, 512
postbrachytherapy prostate volume 720
posterior fossa simulation 442
posteroanterior (PA) beam 438
post-implant dosimetry 279
postmastectomy 156
– irradiation 499. 501
Post-Operative Therapy in Endometrial Carcinoma (PORTEC) trial 604, 606
postprostatectomy (PPI) IMRT 223
post-radical prostatectomy radiation therapy 714
potency 361
potentially lethal damage repair (PLDR) 40
pouch of Douglas 657
preauricular field 823
Premarin 659
pretreatment quality assurance 284
proctitis 651

prone pillow 169
prostate
- brachytherapy 361, 386, 715
- - radioisotopes 362
- - selection criteria 362
- - TRUS-guided permanent 275
- cancer 22, 34, 59, 60, 220, 234, 386, **687**
- - brachytherapy 724
- - cigarette smoking 718
- - clinical target volume (CTV) 700
- - dose prescription 711
- - external beam radiation therapy 721
- - HDR-BT monotherapy 727
- - neoadjuvant hormone therapy (NHT) 727
- - obesity 718
- - organs at risk 706
- - patient positioning 699
- - postoperative radiation therapy 713
- - PSA spikes 721
- diaphragm 709
- gland 687
- implant 287
- - radiation exposure 287
- mobility 703
- size 715
- TRUS-guided implants 363
- volume 365
prostatectomy 689, 701, 727
prostate-specific antigen (PSA) 688, 690
prostatic
- acid phosphatase 718
- carcinoma 688
- - regional lymph node involvement 690
prostatitis 717
prostheses 85
proton 37
- beam radiotherapy 771
PSA spike 721
psychosexual rehabilitation 638
pterygium 326
pterygomaxillary implant 334
pubic arch interference 717
pulmonary metastases 778
pulse dose rate (PDR) 310, 317
- brachytherapy (PDR) 302
- - schedule 304
pyrifrom sinus cancer 456
pyrimidine 41

Q

QA/CQI committee 398
Qualified Medical Physicist (QMP) 396
quality
- checklist 250
- control 395
- management program 596
quality assurance 214, 235, 250, 395
- clinical aspects 418
- education lectures 405
- IMRT 411
- linac

- - advanced technology 405
- - annual test 405
- medical linear accelerator 400
- of simulators 401
- test 403
- treatment planning
- - computer system 414
- - non-graphical 415
- - three-dimensional 415
- - traditional graphical 415
- typical responsibilities 399
Quetelet iindex 644
Quimby system 269

R

radiation
- beam location 203
- block 203
- direct action 291
- exposure risk 287
- necrosis 321
- oncology chart 418
- pneumonitis 830
- safety 302, 320
- - control 288
- therapy simulator 107
Radiation Therapy Oncology Group (RTOG) 17
radical prostatectomy 727
radioactive
- materials license 286
- phosporus-32 626
radiochronic film 212, 238
radionuclide 310
radio-opaque
- contrast 278
- dummy source 336
- marker 272, 337, 438, 444, 552
radiotherapy technology 109
radium 259, 332
- needle 301
- source 587
radium-226 259
radon-222 259
random uncertainty 395
Randoman phantom 37
ratio of tissue phantom ratios and off-axis ratio corrections (RTPROAR) 237
- algorithm 237
Ray-tack vaginal packing 608
rectal
- balloon 222
- cancer 49, 50, **545**
- - chemotherapy 549
- - irradiation 549
- - surgery 549
- - therapeutic ration 555
- morbidity 722
- mucosa 356, 358
- ulceration 366
- ultrasound probe 363
- wall 221, 548

rectovaginal fistula 649, 678
rectum 706
recurrent endometrial cancer 621
Reed-Sternberg cell 809
reference
– air kerma rate 590
– mark 128
– vessel diameter (RVD) 839
regional lymphatics 481
registration
– image-based 130
– point-based 130
– surface-based 130
relative effectiveness 11
remote control afterloading 317
removable media 199
reporting dose 190
reproductive death 294
restenosis of the coronary artery 837
retinoblastoma 324
– gene 579
rhabdomyosarcoma 756, 775
Robinson catheter 340
room's eye view (REV) 184
– display 184
rotation axis 241
rotational therapy technique 89

S

sacral canal 550
safety interlock 406
sagittal laser 114
saphenous vein graft in-stent restenosis 840
sarcoma 272
– high-grade 779
– of soft tissues 756
– of the extremities 771
– of the hand and foot 773
– radiation-induced 35
scan
– field of view (SFOV) 112
– limit 128
– protocol 126, 128
scanner laser 114
scatter–air ratio (SAR) 72
scatter–maximum ratio (SMR) 72
Schiller's iodine 659
sciatic pain 676
scopolamine 331
scrotal interference 747
scrotum 740
second malignant neoplasm 749
secure room 271
SEER program 34, 35, 599
segmented treatment table (STT) 408
seminoma 739
sentinel lymph node surgery 488
setup margin (SM) 188
Seven Steps to LQ Heaven 8
sexual activity 579
shallow breathing 117

shielding 91
– block 184
shift method 129
Sievert
– integral 263
– summation 310
sigmoid colon 546
Silastic plastic insert 325
Simon-Heyman capsule 617
simulated annealing 210
simulation
– request form 127
– software 110
simulator
– film 216
– geometry 118
– tabletop 113
single-plane implant 269
single-strand break 291
Skene's glands 636
skin and lip tumor 329
skin cancer 316, 390
small cell lung carcinoma 44, 537
– cell lines 537
– prophylactic cranial irradiation 538
soft palate cancer 460
soft tissue sarcoma (STS) 343, 388, **755**
– biopsy 759
– brachytherapy 766
– comination chemotherapy regimen 773
– grading 760
– neoadjuvant chemotherapy 776
– of the extremities 755
– positive/negative margins 763
– proton beam radiotherapy 771
– radiation treatment technique 767
– radical resection 762
– rhabdomyosarcoma 775
– surgical margins 763
– wound complication 773
– wound healing 773
software update 406
source
– curator 367
– housing 240
– loading 274
source-axis distance 82, 444, 744
source-film distance (SFD) 789
source-skin distance (SSD) 136, 141, 744, 789
source–skin ratio 79
spatial cooperation 39
spermatogenesis impairment 749
spermatogonia 749
sphincter preservation 554
spinal
– cord
– – myelopathy 436
– – tolerance 831
– – tumor 428
– metastasis 448
– tumor 446
spleen 829
splenectomy 806, 823

spoiler plus tray factor (STF) 792
spread-out Bragg peak 771
squamous cell carcinoma 634
– of floor of mouth 466
– of right lateral wall of nasal vestibule 473
– of the lower lip 463
– of the oral tongue 467
– of the proximal vagina 663
– of the vagina 657
stage migration 57
stainless steel 311
stand-alone PET 116
standard fractionation 16
Stanford technique 162, 176
stem cell transplantation 785
stereotactic
– body frame (SBF) 248
– body radiotherapy (SBRT) 678
– collimation system 238
– frame 236
– radiosurgery 233
stereotaxy 236
sternocleidomastoid muscle 497
Stockholm system 347
stretcher 176
Styrofoam 172
– wegdes 120, 170
subcutaneous tumor 777
sublethal damage repair (SLDR) 40
superposition principle 261
supraclavicular nodal irradiation 497
Syed-Neblett
– applicator 622
– interstitial implant 608
– template 314, 668
Syed-Puthawala-Hedger esophageal applicator 340
synovial sarcoma 756, 775
systematic uncertainty 3995

T

tamoxifen 600
tandem 384
target
– delineation 113
– geometry 214
– specification 99
– volume 188
– – uncertainties 416
taxane 41, 43, 46, 48
Teflon
– 16-gauge tubing 311
– Angiocath catheter 323
– catheter 336, 349
teleangiectasis 678
template 314
temporal lobe lesion 441
testicular
– boost 798
– cancer 739
– – staging 741
– shielding 743. 746

tetraazacyclododecane tetraacetic acid (DOTA) 61
TG43 formalism 263, 277
thalamic tumor 442
thermoluminescent dosimeter 83, 160, 163, 212, 265, 418
– powder capsule 794
thermoplastic mask 168, 211, 234
thoracic radiotherapy (TRT) 537
– treatment planning 533
thoracotomy 339
thorax board 533
three dimensional
– conformal radiation therapy (3D CRT) 179
– treatment planning 180
thromboembolism 595
through-and-through technique 331
thymus block 800
thyroid
– carcinoma 219
– gland 831
tibial osteosarcoma 60
time-dose factor (TDF) 6, 7
tissue
– deficit (TD) 790
– inhomogeneities 79
– – correction 79
– toxicity 217
tissue–air ratio (TAR) 71, 79, 80
– power law method 80
tissue-equivalent bolus 96
tissue–maximum ratio (TMR) 71
tissue–phantom ratio (TPR) 71
titanium shell 287
tobacco 718
tolerance zone 18
– of early BED 26
tomotherapy 90, 206, 211, 800
tongue carcinoma 331
tonsillar region 334
topical chemotherapy 650
total
– abdominal hysterectomy and bilateral salpingo-oophorectomy (TAH-BSO) 604
– dose 6
– limb irradiation 155
– scalp treatment 157
– skin electron therapy 162, 176
– skin irradiation 148
total body irradiation (TBI) 167, 175, 786
– compensators 790
– – design 791
– compensators 790
– – thickness 791
– dose verification 793
– gonad shielding 800
– in young children 801
– kidney shielding 799
– lung shielding 798
– normal tissue shielding 798
– technique setup sheet 796
Tpot 13
tracheal stoma 457
tracheobronchial fistula 515
tracheoesophageal fistula 515

Tracking Cobalt Project 210
transabdominal ultrasound 703
- images of the prostate 705
transhepatic cholangiogram 342
transition zone cancer 688
transperineal implant technique 362
transrectal ultrasound (TRUS) 270, 274, 320, 669, 724
transurethral
- bladder resection (TURBT) 562, 565
- resection 358
- – of the prostate (TURP) 363, 696, 717
transvaginal ultrasonography 601
treatment
- machines maintenance 414
- planning 69, 96, 417
treatment-planning computer system
- quality assurance 414
true pelvis 550
truncal sarcoma 769
T-stage tumor 219
tumor
- control probability (TCP) 11, 197, 198
- failure rate 536
- hypoxia 65
- in the cervicothoracic region 446
- in the temporal region 441
- in the thoracic region 447
- node metastasis (TNM) system 761
- – soft tissue sarcoma 761
- of nasal cavity 471
- of oral cavity 468
- of oropharynx 468
- of paranasal sinuses 471
- of the head and neck 327
- of the meninges 427
- of the proximal urethra 360
- of the rectovaginal septum 356
- phenotype 118
- radioenhancement 43
- repopulation 12, 25
- response 20
- shape 203
- skin and lip 329
tumor-soft tissue interface 193
tungsten
- leaves 210
- shield 266
tunica albuginea 747

U

ultrasmall superparamagnetic iron oxide (USPIO) 62
uncertainty
- random 395
- systematic 395
unspecified tissue 196
until center point (UCP) 240

uretero-vaginal fistula 679
urethral stricture 358, 722
urethritis 712
urethrogram 698
urethroscopy 562
urinary
- epithelium 562
- morbidity 721
- toxicity 364
- tract infection 561
uterine
- bleeding 600
- tandem 354
uterus 599
uveal melanoma 326

V

Vac-Lok cradle 494
vacuum form body immobilization 170
vagina
- adenocarcinoma 675
- clear cell adenocarcinoma 658
- clear cell carcinoma 675
- lymphatic drainage 658
vaginal
- applicator 354
- cancer 657
- – chemotherapy 682
- – FIGO stages 673
- – inguinal-femoral nodal irradiation 664
- – interstitial brachytherapy 671
- carcinoma 351
- cuff 611
- cylinder 345, 353, 354, 611, 665, 666
- fistula 678
- intraepithelial neoplasia (VAIN) 650, 658, 659
- – FIGO staging 669
- – staging 659
- length 626
- mucosa 346
- surface 384, 613
- – dose 592
- vestibule 631
vaginectomy 662
α/β value 382
Varian Trilogy system 245
vasculogenesis 61
velocity effect 159
vesico-vaginal fistula 679
video-assistes thoracoscopic surgery (VATS) 528
vinyl polysiloxane obturator 327
virtual simulation software 110, 117
- capability 130
vocal cord cancer 454
volume study 275
voxel 194

vulva
– cancer 358, 635
– brachytherapy 646
– chemoirradiation 651
– irradiation 638
– lymph node metastasis 637
– organ preservation 637
  pregnancy 649
– wound infection 648
– intraepithelial neoplasia 650
– malignant melanoma 650
– melanoma 636
– recurrence 676
– sarcoma 636
– sarcoma 649
– tumor diagnosis 634
vulvectomy 633, 638, 647
vulvo-vaginal recurrence 677

W

Waldeyer's ring field 823
wall laser 114
Wang classification 328
wedge
– angle 74
– factor 74
– filter 74
– – dosimetry 85
– isodose curves 74
whole
– abdominal irradiation (WAI) 607, 618
– brain irradiation 443
– pelvis field 614

whole-body irradiation 76
whole-body phantom 217
World Health Organization formulation 760
wound
– healing problem 344
– morbidity 774

X

x-axis 194, 195
xerostomia 454, 824, 831
xiphoid 503
X-jaw 95, 171, 407
X-ray 33, 107137, 217, 291
– collimator 81, 160
– films 125
– simulator 180
– tube 120

Y

y-axis 195
Y-jaw 95, 171, 407
y-ray 37, 291
Y-shaped applicator 384
ytterbium-169 261
yttrium-90 838

Z

Z value 150
zirconium-90 838

List of Contributors

RON R. ALLISON, MD
Professor and Chairman
Department of Radiation Oncology
Leo Jenkins Cancer Center
Brody School of Medicine
Greenville, NC 27834
USA

ROBERT J. AMDUR, MD
Department of Radiation Oncology
University of Florida College of Medicine
Health Science Center
P.O. Box 100385
Gainesville, FL 32610-0385
USA

JEFFREY D. BRADLEY, MD
Department of Radiation Oncology
Washington University School of Medicine
4921 Parkview Place – Lower Level
Mail Stop #90-38-635
St. Louis, MO 63110
USA

DAVID J. BRENNER, PhD, DSc.
Professor of Radiation Oncology
and Public Health
Center for Radiological Research
Columbia University Medical Center
630, West 168th St.
New York, NY, 10032
USA

THOMAS A. BUCHHOLZ, MD
Department of Radiation Oncology
The University of Texas
M.D. Anderson Cancer Center
1515 Holcombe Boulevard, Unit 1202
Houston, TX 77030
USA

HIGINIA R. CARDENES, MD, PhD
Clinical Associate Professor
Department of Radiation Oncology, RT 041
Indiana University School of Medicine
535 Barnhill Dr
Indianapolis, IN 46202
USA

WALTER H. CHOI, MD
Department of Radiation Oncology
SUNY Downstate at Brooklyn
450 Clarkson Avenue, Box 1211
Brooklyn, NY 11203
USA

L. CHINSOO CHO, MD
Department of Radiation Oncology
The University of Texas Southwestern
Medical Center at Dallas
5801 Forest Park
Dallas, TX 75390-9183
USA

HAK CHOY, MD
Professor and Chair
Department of Radiation Oncology
Nancy B. and Jake L. Hamon Distinguished
Chair in Basic Cancer Research
The University of Texas Southwestern
Medical Center at Dallas
5801 Forest Park
Dallas, TX 75390-9183
USA

FILIP CLAUS, MD, PhD
Memorial Sloan-Kettering Cancer Center
1275 York Avenue
New York, NY 10021
USA

THOMAS F. DELANEY, MD
Medical Director
Northeast Proton Therapy Center
Massachusetts General Hospital
Harvard Medical School
30 Fruit Street
Boston, MA 02114
USA

KATHRYN E. DUSENBERY, MD
University of Minnesota Medical School
Department of Radiation Oncology
MMC436
420 Delaware St. S.E.
Minneapolis, MN55455
USA

ROBERT E. DRZYMALA, MD
Associate Professor in Radiation Oncology Physics
Department of Radiation Oncology
Wahington University School of Medicine
4921 Parkview Place
Campus Box 8224
St. Louis, MO 63110
USA

JACK F. FOWLER, DSc, PhD
Emeritus Professor
of Human Oncology and Medical Physics
Medical School of University of Wisconsin –
Madison, WI, USA
Former Director of the Gray Laboratory,
Nothwood, London, UK
Present address:
150 Lambeth Rd.
London SE1 7DF
UK

DAN P. GARWOOD, MD
Department of Radiation Oncology
UT Southwestern Medical Center at Dallas
5801 Forest Park
Dallas, Texas 75390-9183
USA

BRUCE J. GERBI, PhD
Associate Professor
Therapeutic Radioly – Radiation Oncology
University of Minnesota
Mayo Mail Code 494
420 Delaware St SE
Minneapolis, MN 55455
USA

SRI GORTY, MD
Radiation Oncologist
Robert and Beverly Lewis Cancer Care Center
Ponoma Valley Hospital Medical Center
1910 Royalty Drive
Pomona, CA
USA

LEONARD L. GUNDERSON, MD
Chair, Department of Radiation Oncology
Mayo Clinic, Scottsdale, Arizona
Professor of Oncology
Mayo Clinic College of Medicine
Department of Radiation Oncology
200 2nd St. SW DeskR
Rochester, MN 55905
USA

MICHAEL G. HADDOCK, MD
Consultant, Division of Radiation Oncology
Mayo Clinic, Scottsdale, Arizona
Associate Professor of Oncology
Mayo Clinic College of Medicine
Department of Radiation Oncology
200 2nd St. SW DeskR
Rochester, MN 55905
USA

ERIC J. HALL, D.Phil., D.Sc. FACR, FRCR
Higgins Professor of Radiation Biophysics
Center for Radiological Research
Columbia University
630, West 168th St.
New York, NY, 10032
USA

DAVID C. HARMON, MD
Hematology Oncology, Department of Medicine
Massachusetts General Hospital
Harvard Medical School
Yawkey 7944
30 Fruit Street
Boston, MA 02114
USA

Francis J. Hornicek, MD, PhD
Orthopedic Onocology
Department of Orthopaedic Surgery
Massachusetts General Hospital
Harvard Medical School
Yawkey 3B
30 Fruit Street
Boston, MA 02114
USA

HEDVIG HRICAK, MD, PhD
Chairman, Department of Radiology
Carroll and Milton Petrie Chair
Professor of Radiology, Weill Medical College
Cornell University, Memorial Sloan-Kettering
Cancer Center
1275 York Avenue
New York, NY 10021
USA

ERIC KLEIN, MS
Department of Radiation Oncology
Washington University School of Medicine
St. Louis, MO 63110
USA

CHUNG KYU KIM LEE, MD
University of Minnesota Medical School
MMC494
420 Delaware St.S.E.
Minneapolis, MN 55455
USA

SEYMOUR H. LEVITT, MD
Professor, Department of Therapeutic Radiation
Oncology MMC 436
University of Minnesota Hospital
420 Delaware St SE
Minneapolis, MN 55455
USA
and
Foreign Adjunct Professor
Karolinska Institute Stockholm, Sweden

ZUOFENG LI, DSc
Associate Professor
Division of Radiation Physics
Washington University Medical Center
4921 Parkview Place, Campus Box 8224
St. Louis, MO 63110
USA

DANIEL A. LOW, PhD
Associate Professor
Division of Radiation Physics
4921 Parkview Place, Campus Box 8224
St. Louis, MO 63110
USA

WEI LU, PhD
Department of Radiation Oncology
4921 Parkview Place, Campus Box 8224
St. Louis, MO 63110
USA

JAMES A. MARTENSON JR., MD
Consultant, Division of Radiation Oncology
Mayo Clinic
Professor of Oncology
Mayo Clinic College of Medicine
Department of Radiation Oncology
200 2nd St. SW DeskR
Rochester, MN 55905
USA

KEVIN P. MCMULLEN, MD,
Assistant Professor, Department of
Radiation Oncology
Wake Forest University School of Medicine
Medical Center Blvd.
Winston-Salem, NC 27157-1030
USA

MARSHA D. MCNEESE, MD
Department of Radiation Oncology
The University of Texas
M.D. Anderson Cancer Center
1515 Holcombe Boulevard, Unit 1202
Houston, TX 77030
USA

WILLIAM M. MENDENHALL, MD
Department of Radiation Oncology
University of Florida College of Medicine
Health Science Center
P.O. Box 100385
Gainesville, FL 32610-0385
USA

GREGORY S. MERRICK, MD
Schiffler Cancer Center
and Wheeling Jesuit University
Wheeling, WV 26003
USA

JEFF M. MICHALSKI, MD, MBA
Associate Professor in Radiation Oncology
Department of Radiation Oncology
Washington University School of Medicine
4921 Parkview Place, Campus Box 8224
Saint Louis, MO 63110
USA

GERARD C. MORTON, MD, MRCPI FRCPC
Assistant Professor, Division of Radiation Oncology
Toronto-Sunnybrook Regional Cancer Centre,
University of Toronto
2075 Bayview Avenue
Toronto, Ontario M4N 3M5
Canada

Sabin B. Motwani, MD
Department of Radiation Oncology
The University of Texas
M.D. Anderson Cancer Center
1515 Holcombe Boulevard, Unit 1202
Houston, TX 77030
USA

Michael T. Munley, PhD
Associate Professor, Department of Radiation
Oncology
Wake Forest University School of Medicine
Medical Center Blvd.
Winston-Salem, NC 27157-1030
USA

Sasa Mutic, MS
Department of Radiation Oncology
Washington University School of Medicine
4921 Parkview Place
St. Louis, MO 63110
USA

Subir Nag, MD, FACR, FACRO
Chief of Brachytherapy
Department of Radiation Medicine
Arthur G. James Cancer Hospital and Solove
Research Institut
The Ohio State University
300 West 10th Avenue, # 072
Columbus, Ohio 43210
USA

Samir Narayan, MD
Department of Radiation Oncology
University of California Davis Medical Center
4501 X Street, Suite G126
Sacramento, CA 95817
USA

Sten Nilsson, MD
Professor, Institution for Oncology Pathology
Karolinska University Hospital
17176 Stockholm
Sweden

Jatinder R. Palta, PhD
Department of Radiation Oncology
University of Florida College of Medicine
Health Science Center
P.O. Box 100385
Gainesville, FL 32610-0385
USA

Maria Pearse, MD, MB ChB FRANZCR
Division of Radiation Oncology
Toronto-Sunnybrook Regional Cancer Centre,
University of Toronto
2075 Bayview Avenue
Toronto, Ontario M4N 3M5
Canada

Carlos A. Perez, MD
Professor, Department of Radiation Oncology
Washington University Medical Center
4511 Forest Park Boulevard, Suite 200
St. Louis, MO 63108
USA

James A. Purdy, PhD
Professor and Vice Chairman
Department of Radiation Oncology
Chief, Physics Section
University of California Davis Medical Center
4501 X Street, Suite G140
Sacramento, CA 95817
USA

Keith M. Rich, MD
Associate Professor, Neurological Surgery
Department of Neurological Surgery
Wahington University School of Medicine
660 South Euclid Avenue
Campus Box 8057
St. Louis, MO 63110
USA

Andrew E. Rosenberg, MD
Surgical Pathology, Department of Pathology
Massachusetts General Hospital
Harvard Medical School
Warren 2
30 Fruit Street
Boston, MA 02114
USA

Marvin Rotman, MD
Distinguished Service Professor and Chairman
Department of Radiation Oncology
State University of New York
Health Science Center at Brooklyn
450 Clarkson Ave., Box 1211
Brooklyn, NY 11203-2098
USA

ALAN R. SCHULSINGER, MD
Clinical Associate Professor of Radiation Oncology
Department of Radiation Oncology
State University of New York
Health Science Center at Brooklyn
Box 1211
450 Clarkson Ave.
Brooklyn, NY 11203-2098
USA
and
Director of Radiation Oncology at
Long Island College Hospital

EDWARD G. SHAW, MD
Professor and Chairman, Department of
Radiation Oncology
Wake Forest University
School of Medicine
Medical Center Blvd.
Winston-Salem, NC 27157-1030
USA

JOSEPH R. SIMPSON, MD
Associate Professor in Radiation Oncology
Department of Radiation Oncology
Campus Box 8224
4921 Parkview Place
St. Louis, MO 63110
USA

VOLKER W. STIEBER, MD
Assistant Professor, Department of
Radiation Oncology
Wake Forest University
School of Medicine
Medical Center Blvd.
Winston-Salem, NC 27157-1030
USA

ERIC A. STROM, MD
Department of Radiation Oncology
The University of Texas
M.D. Anderson Cancer Center
1515 Holcombe Boulevard, Unit 1202
Houston,TX 77030
USA

BRENT TINNEL, MD
Indiana University School of Medicine
Department of Radiation Oncology
535 Barnhill Drive, RT 041
Indianapolis, IN 46202
USA

PRABHAKAR TRIPURANENI, MD
Scripps Clinic/Scripps Green Hospital
10666 N. Torrey Pines Rd., MSB-1
La Jolla, CA 92037-1092
USA

SRINIVASAN VIJAYAKUMAR, MD
Chair, Department of Radiation Oncology
University of California Davis Medical Center
4501 X Street, Suite G126
Sacramento, CA 95817
USA

IMRAN ZOBERI, MD
Department of Radiation Oncology
Campus Box 8224
4921 Parkview Place
St. Louis, MO 63110
USA

ROBERT D. ZWICKER, PhD
Department of Radiation Oncology
University of Kentucky Medical Center
800 Rose Street
Lexington, KY 40536
USA

RADIATION ONCOLOGY

Lung Cancer
Edited by C.W. Scarantino

Innovations in Radiation Oncology
Edited by H. R. Withers
and L. J. Peters

**Radiation Therapy
of Head and Neck Cancer**
Edited by G. E. Laramore

**Gastrointestinal Cancer –
Radiation Therapy**
Edited by R.R. Dobelbower, Jr.

**Radiation Exposure
and Occupational Risks**
Edited by E. Scherer, C. Streffer,
and K.-R. Trott

Radiation Therapy of Benign Diseases
A Clinical Guide
S. E. Order and S. S. Donaldson

**Interventional Radiation
Therapy Techniques – Brachytherapy**
Edited by R. Sauer

Radiopathology of Organs and Tissues
Edited by E. Scherer, C. Streffer,
and K.-R. Trott

**Concomitant Continuous Infusion
Chemotherapy and Radiation**
Edited by M. Rotman
and C. J. Rosenthal

**Intraoperative Radiotherapy –
Clinical Experiences and Results**
Edited by F. A. Calvo, M. Santos,
and L.W. Brady

**Radiotherapy of Intraocular
and Orbital Tumors**
Edited by W. E. Alberti and
R. H. Sagerman

**Interstitial and Intracavitary
Thermoradiotherapy**
Edited by M. H. Seegenschmiedt
and R. Sauer

Non-Disseminated Breast Cancer
Controversial Issues in Management
Edited by G. H. Fletcher and
S.H. Levitt

**Current Topics in
Clinical Radiobiology of Tumors**
Edited by H.-P. Beck-Bornholdt

**Practical Approaches to
Cancer Invasion and Metastases**
A Compendium of Radiation
Oncologists' Responses to   40 Histories
Edited by A. R. Kagan with the
Assistance of R. J. Steckel

Radiation Therapy in Pediatric Oncology
Edited by J. R. Cassady

Radiation Therapy Physics
Edited by A. R. Smith

Late Sequelae in Oncology
Edited by J. Dunst and R. Sauer

Mediastinal Tumors. Update   1995
Edited by D. E. Wood
and C. R. Thomas, Jr.

**Thermoradiotherapy
and Thermochemotherapy**
Volume   1:
Biology, Physiology, and Physics
Volume   2:
Clinical Applications
Edited by M.H. Seegenschmiedt,
P. Fessenden, and C.C. Vernon

Carcinoma of the Prostate
Innovations in Management
Edited by Z. Petrovich, L. Baert,
and L.W. Brady

**Radiation Oncology
of Gynecological Cancers**
Edited by H.W. Vahrson

Carcinoma of the Bladder
Innovations in Management
Edited by Z. Petrovich, L. Baert,
and L.W. Brady

**Blood Perfusion and
Microenvironment of Human Tumors**
Implications for
Clinical Radiooncology
Edited by M. Molls and P. Vaupel

Radiation Therapy of Benign Diseases
A Clinical Guide
2nd Revised Edition
S. E. Order and S. S. Donaldson

**Carcinoma of the Kidney and Testis,
and Rare Urologic Malignancies**
Innovations in Management
Edited by Z. Petrovich, L. Baert,
and L.W. Brady

**Progress and Perspectives in the
Treatment of Lung Cancer**
Edited by P. Van Houtte,
J. Klastersky, and P. Rocmans

**Combined Modality Therapy of
Central Nervous System Tumors**
Edited by Z. Petrovich, L. W. Brady,
M. L. Apuzzo, and M. Bamberg

Age-Related Macular Degeneration
Current Treatment Concepts
Edited by W. A. Alberti, G. Richard,
and R. H. Sagerman

**Radiotherapy of Intraocular
and Orbital Tumors**
2nd Revised Edition
Edited by R. H. Sagerman,
and W. E. Alberti

Modification of Radiation Response
Cytokines, Growth Factors,
and Other Biolgical Targets
Edited by C. Nieder, L. Milas,
and K. K. Ang

Radiation Oncology for Cure and Palliation
R. G. Parker, N. A. Janjan,
and M. T. Selch

**Clinical Target Volumes in Conformal and
Intensity Modulated Radiation Therapy**
A Clinical Guide to Cancer Treatment
Edited by V. Grégoire, P. Scalliet,
and K. K. Ang

**Advances in Radiation Oncology
in Lung Cancer**
Edited by Branislav Jeremić

New Technologies in Radiation Oncology
Edited by W. Schlegel, T. Bortfeld,
and A.-L. Grosu

Technical Basis of Radiation Therapy
Practical Clinical Applications
4th Revised Edition
Edited by S. H. Levitt, J. A. Purdy,
C. A. Perez, and S. Vijayakumar

**Multimodal Concepts for Integration of
Cytotoxic Drugs**
Edited by J. M. Brown, M. P. Mehta,
and C. Nieder

 Springer

DIAGNOSTIC IMAGING

Innovations in Diagnostic Imaging
Edited by J. H. Anderson

Radiology of the Upper Urinary Tract
Edited by E. K. Lang

The Thymus - Diagnostic Imaging, Functions, and Pathologic Anatomy
Edited by E. Walter, E. Willich, and W. R. Webb

Interventional Neuroradiology
Edited by A. Valavanis

Radiology of the Pancreas
Edited by A. L. Baert, co-edited by G. Delorme

Radiology of the Lower Urinary Tract
Edited by E. K. Lang

Magnetic Resonance Angiography
Edited by I. P. Arlart, G. M. Bongartz, and G. Marchal

Contrast-Enhanced MRI of the Breast
S. Heywang-Köbrunner and R. Beck

Spiral CT of the Chest
Edited by M. Rémy-Jardin and J. Rémy

Radiological Diagnosis of Breast Diseases
Edited by M. Friedrich and E.A. Sickles

Radiology of the Trauma
Edited by M. Heller and A. Fink

Biliary Tract Radiology
Edited by P. Rossi, co-edited by M. Brezi

Radiological Imaging of Sports Injuries
Edited by C. Masciocchi

Modern Imaging of the Alimentary Tube
Edited by A. R. Margulis

Diagnosis and Therapy of Spinal Tumors
Edited by P. R. Algra, J. Valk, and J. J. Heimans

Interventional Magnetic Resonance Imaging
Edited by J.F. Debatin and G. Adam

Abdominal and Pelvic MRI
Edited by A. Heuck and M. Reiser

Orthopedic Imaging Techniques and Applications
Edited by A. M. Davies and H. Pettersson

Radiology of the Female Pelvic Organs
Edited by E. K.Lang

Magnetic Resonance of the Heart and Great Vessels
Clinical Applications
Edited by J. Bogaert, A.J. Duerinckx, and F. E. Rademakers

Modern Head and Neck Imaging
Edited by S. K. Mukherji and J. A. Castelijns

Radiological Imaging of Endocrine Diseases
Edited by J. N. Bruneton in collaboration with B. Padovani and M.-Y. Mourou

Trends in Contrast Media
Edited by H. S. Thomsen, R. N. Muller, and R. F. Mattrey

Functional MRI
Edited by C. T. W. Moonen and P. A. Bandettini

Radiology of the Pancreas
2nd Revised Edition
Edited by A. L. Baert
Co-edited by G. Delorme and L. Van Hoe

Emergency Pediatric Radiology
Edited by H. Carty

Spiral CT of the Abdomen
Edited by F. Terrier, M. Grossholz, and C. D. Becker

Liver Malignancies
Diagnostic and Interventional Radiology
Edited by C. Bartolozzi and R. Lencioni

Medical Imaging of the Spleen
Edited by A. M. De Schepper and F. Vanhoenacker

Radiology of Peripheral Vascular Diseases
Edited by E. Zeitler

Diagnostic Nuclear Medicine
Edited by C. Schiepers

Radiology of Blunt Trauma of the Chest
P. Schnyder and M. Wintermark

Portal Hypertension
Diagnostic Imaging-Guided Therapy
Edited by P. Rossi
Co-edited by P. Ricci and L. Broglia

Recent Advances in Diagnostic Neuroradiology
Edited by Ph. Demaerel

Virtual Endoscopy and Related 3D Techniques
Edited by P. Rogalla, J. Terwisscha Van Scheltinga, and B. Hamm

Multislice CT
Edited by M. F. Reiser, M. Takahashi, M. Modic, and R. Bruening

Pediatric Uroradiology
Edited by R. Fotter

Transfontanellar Doppler Imaging in Neonates
A. Couture and C. Veyrac

Radiology of AIDS
A Practical Approach
Edited by J.W.A.J. Reeders and P.C. Goodman

CT of the Peritoneum
Armando Rossi and Giorgio Rossi

Magnetic Resonance Angiography
2nd Revised Edition
Edited by I. P. Arlart, G. M. Bongratz, and G. Marchal

Pediatric Chest Imaging
Edited by Javier Lucaya and Janet L. Strife

Applications of Sonography in Head and Neck Pathology
Edited by J. N. Bruneton in collaboration with C. Raffaelli and O. Dassonville

Imaging of the Larynx
Edited by R. Hermans

3D Image Processing
Techniques and Clinical Applications
Edited by D. Caramella and C. Bartolozzi

Imaging of Orbital and Visual Pathway Pathology
Edited by W. S. Müller-Forell

MEDICAL RADIOLOGY Diagnostic Imaging and Radiation Oncology
Titles in the series already published

Pediatric ENT Radiology
Edited by S. J. King
and A. E. Boothroyd

**Radiological Imaging
of the Small Intestine**
Edited by N. C. Gourtsoyiannis

Imaging of the Knee
Techniques and Applications
Edited by A. M. Davies
and V. N. Cassar-Pullicino

Perinatal Imaging
From Ultrasound to MR Imaging
Edited by Fred E. Avni

Radiological Imaging of the Neonatal Chest
Edited by V. Donoghue

**Diagnostic and Interventional
Radiology in Liver Transplantation**
Edited by E. Bücheler, V. Nicolas,
C. E. Broelsch, X. Rogiers,
and G. Krupski

Radiology of Osteoporosis
Edited by S. Grampp

Imaging Pelvic Floor Disorders
Edited by C. I. Bartram
and J. O. L. DeLancey
Associate Editors: S. Halligan,
F. M. Kelvin, and J. Stoker

Imaging of the Pancreas
Cystic and Rare Tumors
Edited by C. Procacci
and A. J. Megibow

**High Resolution Sonography
of the Peripheral Nervous System**
Edited by S. Peer and G. Bodner

Imaging of the Foot and Ankle
Techniques and Applications
Edited by A. M. Davies,
R. W. Whitehouse,
and J. P. R. Jenkins

Radiology Imaging of the Ureter
Edited by F. Joffre, Ph. Otal,
and M. Soulie

Imaging of the Shoulder
Techniques and Applications
Edited by A. M. Davies and J. Hodler

Radiology of the Petrous Bone
Edited by M. Lemmerling
and S. S. Kollias

Interventional Radiology in Cancer
Edited by A. Adam, R. F. Dondelinger,
and P. R. Mueller

**Duplex and Color Doppler Imaging
of the Venous System**
Edited by G. H. Mostbeck

Multidetector-Row CT of the Thorax
Edited by U. J. Schoepf

Functional Imaging of the Chest
Edited by H.-U. Kauczor

**Radiology of the Pharynx
and the Esophagus**
Edited by O. Ekberg

**Radiological Imaging
in Hematological Malignancies**
Edited by A. Guermazi

**Imaging and Intervention in
Abdominal Trauma**
Edited by R. F. Dondelinger

Multislice CT
2nd Revised Edition
Edited by M. F. Reiser, M. Takahashi,
M. Modic, and C. R. Becker

**Intracranial Vascular Malformations
and Aneurysms**
From Diagnostic Work-Up
to Endovascular Therapy
Edited by M. Forsting

Radiology and Imaing of the Colon
Edited by A. H. Chapman

Coronary Radiology
Edited by M. Oudkerk

**Dynamic Contrast-Enhanced Magnetic
Resonance Imaging in Oncology**
Edited by A. Jackson, D. L. Buckley,
and G. J. M. Parker

**Imaging in Treatment Planning
for Sinonasal Diseases**
Edited by R. Maroldi and P. Nicolai

Clinical Cardiac MRI
With Interactive CD-ROM
Edited by J. Bogaert,
S. Dymarkowski, and A. M. Taylor

Focal Liver Lesions
Detection, Characterization,
Ablation
Edited by R. Lencioni, D. Cioni,
and C. Bartolozzi

Multidetector-Row CT Angiography
Edited by C. Catalano
and R. Passariello

Paediatric Musculoskeletal Diseases
With an Emphasis on Ultrasound
Edited by D. Wilson

Contrast Media in Ultrasonography
Basic Principles and Clinical Applications
Edited by Emilio Quaia

**MR Imaging in White Matter Diseases of the
Brain and Spinal Cord**
Edited by M. Filippi, N. De Stefano,
V. Dousset, and J. C. McGowan

Diagnostic Nuclear Medicine
2nd Revised Edition
Edited by C. Schiepers

Imaging of the Kidney Cancer
Edited by A. Guermazi

**Magnetic Resonance Imaging in
Ischemic Stroke**
Edited by R. von Kummer and T. Back

Imaging of the Hip & Bony Pelvis
Techniques and Applications
Edited by A. M. Davies, K. J. Johnson,
and R. W. Whitehouse

**Imaging of Occupational and
Environmental Disorders of the Chest**
Edited by P. A. Gevenois and
P. De Vuyst

Contrast Media
Safety Issues and ESUR Guidelines
Edited by H. S. Thomsen

Virtual Colonoscopy
A Practical Guide
Edited by P. Lefere and S. Gryspeerdt

Vascular Embolotherapy
A Comprehensive Approach
Volume 1
Edited by J. Golzarian. Co-edited by
S. Sun and M. J. Sharafuddin

Vascular Embolotherapy
A Comprehensive Approach
Volume 2
Edited by J. Golzarian. Co-edited by
S. Sun and M. J. Sharafuddin

 Springer

Printing and Binding: Stürtz GmbH, Würzburg